中国区域地质志

海南志

海南省地质调查院　编著

地质出版社
·北　京·

内容简介

本书以海南省1:20万、1:25万、1:5万区域地质调查成果为基础，全面总结了海南省区域地质调查、矿产勘查和专题研究的最新成果，特别是国土资源大调查实施以来所取得的新资料、新成果，以活动论为指导，应用地质新技术、新方法，开展了综合研究。全面、系统地总结了海南省区域地层、侵入岩、火山岩及火山作用、变质岩及变质作用、区域地质构造、地球物理地球化学、第四纪地质与地貌、地质资源等方面的基本特征。

本书资料丰富，论证充分，反映了海南省的区域地质研究现状和地质科学理论水平，可作为省内地质调查研究的工具书之一，亦可供广大地质工作者和地质科研、教学及有关部门参考。

图书在版编目（CIP）数据

中国区域地质志·海南志/海南省地质调查院编著.
—北京：地质出版社，2017.7
ISBN 978-7-116-10508-9

Ⅰ.①中… Ⅱ.①海… Ⅲ.①区域地质-概况-海南
Ⅳ.①P562

中国版本图书馆CIP数据核字（2017）第184839号

Zhongguo Quyu Dizhizhi · Hainanzhi

责任编辑：田 野 祁向雷 郑长胜 郁秀荣 袁亚琴
责任校对：李 玫
出版发行：地质出版社
社址邮编：北京海淀区学院路31号，100083
电　　话：（010）66554528（邮购部）；（010）66554631（编辑室）
网　　址：http://www.gph.com.cn
传　　真：（010）66554686
印　　刷：中煤地西安地图制印有限公司
开　　本：889 mm×1194 mm 1/16
印　　张：59　插　页：5页　附　图：4幅
字　　数：1800千字
版　　次：2017年7月北京第1版
印　　次：2017年7月西安第1次印刷
定　　价：680.00元
审 图 号：琼S（2016）022号、琼S（2016）023号、琼S（2016）024号、琼S（2016）025号
书　　号：ISBN 978-7-116-10508-9

《中国区域地质志》编纂指导委员会
（第一届）

The first advisory committee of *the Regional Geology of China*

《中国区域地质志》编辑委员会

Editorial board of *the Regional Geology of China*

总　序

20世纪80~90年代，我国编纂出版了第一代省级区域地质志，系统总结了20世纪80年代以前各省（区、市）区域地质调查研究成果，为促进我国地质事业和服务经济社会的发展发挥了重要作用。

随着改革开放的不断深入，地质调查和各类科研项目的实施，我国地质事业又获得蓬勃发展，取得了丰硕的调查研究成果。在青藏高原、大兴安岭等地，完成了1∶25万比例尺区域地质调查和航空磁测，实现了全国陆域中比例尺区域地质调查和航空磁测的全覆盖。开展了各种比例尺的区域地球物理和区域地球化学的调查研究，实施了“深部探测技术与实验研究”专项，拓展了环境地质、农业地质、城市地质等新领域，获得了丰富的基础地质资料，解决了许多基础地质问题，找到了一大批矿产，大幅度地提高了我国的地质研究程度和地质科学技术水平。

《国务院关于加强地质工作的决定》指出，要提高基础地质调查（研究）程度，建立地质图文更新机制。《全国地质勘查规划》（2005~2020）提出，要开展全国基础地质资料综合研究和成果集成工作，编制新一代全国系列基础地质图件，努力创造有宏观影响的大成果。为此，专门组织有关专家，分层次编纂了各省（区、市）、跨省（区）成矿区带和重要经济区，以及全国层次的新一代系列地质志书，集中反映了我国区域地质调查和地质科学研究工作的重大进展和成果，创造一项有宏观影响的大成果，为地质工作和国家经济社会发展提供一套基础地质资料。

为国家经济建设和社会发展提供矿产资源和地质资料，是地质工作服务经济社会发展和人民生产生活的两项基本任务。通过编纂新一代区域地质志书，把全国各部门多年地质调查研究所积累的丰富地质、地球物理、地球化学等资料进行综合研究和集成，使之系统化、规律化和理论化，不仅为地质工作和经济社会发展提供了一套整装的地质资料，而且为发展地质科学理论奠定扎实的资料基础。

党的十八届三中全会指出，要紧紧围绕建设美丽中国深化生态文明体制改革，加快建立生态文明制度，健全国土空间开发、资源节约利用、生态环境保护的体制机制，推动形成人与自然和谐发展现代化建设新格局。这对地质工作提出了新的更高要求。举凡矿产资源的合理开发利用、国土空间的规划、地质环境的保护、地下水污染的治理、地质公园的建设、地球化学对人体健康的影响，以及土壤性质及农产品的优化等，都需要丰富的基础地质数据和地质信息服务。中国区域地质志既可以满足一个省（区、市）、一个成矿区（带）及经济区（带）和全国地质矿产勘查、地质工作规划、部署和管理的需求，又可以在城乡建设、交通、水利、能源和环境保护等重大工程规划、设计、建设和科学管理中发挥重要作用。

新一代《中国区域地质志》是对新时期我国基础地质调查工作的系统总结和高度浓缩，注重在结构和体例上尽量与第一代区域地质志保持有机衔接、汲取其优点和精粹的同时，着力坚持创新，突出反映近30年来取得的新成果和新资料，体现了新的水平和研究高度。

这套志书的编纂和陆续出版是一项浩繁的科学工程，参加单位和人员多、涉及专业领域广、持续时间长，浓缩了几代地质工作者数十年辛勤劳动的成果，是我国地质界的一件大事。参与编纂出版的地质工作者以对历史、对人民负责的高度责任感，孜孜以求，辛勤劳作，力争使志书做到资料翔实、内容精湛、质量上乘、富于创新，打造成高水平的精品著作。在这里，我要向为这项工作付出艰苦努力、做出贡献的地质工作者以及关心、支持这项工作的各有关方面致以崇高敬意和衷心感谢！

中华人民共和国国土资源部副部长

中国地质调查局局长

汪民

2014年2月

General Preface

Compiled in the 1980s and 1990s, Regional Geology of China of the first generation systematically summarized the research results of regional geology before the 1980s in provinces, autonomous regions or municipalities, and played vital roles in serving social and economic development, and geological advancement in China.

With the deepening of reforms and opening-up, the geological undertakings in China embrace another chance of prosperity. Abundant results of geological survey and research have been achieved along with the implementation of geological survey programs and various scientific research programs. In areas such as the Qinghai-Tibet Plateau and Great Khingan Mountains, regional geological survey and aeromagnetic survey at the scale of 1 : 250 000 have been completed, which means that geological and aeromagnetic survey on medium scales have achieved a full coverage of the whole land area in China. Regional geophysical and geochemical surveys on different scales have been conducted. Sino-Probe, a special program for deep exploration research in China, has been carried out as well. New research fields such as environmental geology, agricultural geology and urban geology have been expanded. All these efforts have brought us with substantial fundamental geological materials, and put a series of geological problems into settlement. Besides, a number of mineral deposits have been found. The intensity of geological survey and the level of geological science and technology have been improved significantly.

As is stated in *Decision of the State Council on Strengthening Geological Work*, the intensity of geological survey and research should be enhanced, and an update mechanism concerning geological materials, including geological maps and reports, should be established. *National Geological Exploration Plan (Program)* (2005~2020) proposes that comprehensive studies of nationwide geological data and integration of research results should be conducted, national series of fundamental geological maps of a new generation should be compiled, and we should strive to create outstanding achievements with macro impact. Under the above mentioned directions, experts in relative disciplines are organized to compile the new series of Regional Geology that successively cover provincial (municipal or autonomous regional) level, cross-provincial (or cross-regional) metallogenic belts or important economic zones level, and national level. The new series of Regional Geology principally represent the significant advancements and achievements of geological survey and research in China. The compilation of the series of Regional Geology constitutes an achievement with remarkable impact, and provides a set of geological data for economic and social development.

Providing mineral resources and supplying geological data for economic construction and social development of China, are two basic tasks of geological work which serves social and economic development and people's production and life. By means of compiling the new series of Regional Geology, abundant geological, geophysical and geochemical data that have been accumulated by different departments of the entire country for many years are integrated systematically, regularly and theoretically. The comprehensive study not only provides an entire set of geological data for both geology itself and economic and social development, but also lays concrete foundation for developing geological theory.

In the Third Plenary Session of the 18th CPC Central Committee, it is proposed that we must deepen ecological environmental management reform by centering on constructing a beautiful China, and that we should accelerate system building to promote ecological progress, improve institutions and mechanisms for developing land space, conserving resources and protecting the ecological environment, and promoting modernization featuring harmonious development between man and nature. This raises greater demand on geological work. The tasks such as rational development and utilization of mineral resources, land space planning, protection of geological environment, groundwater pollution control, geo-park construction, evaluation of geochemical impact on human health and optimization of soil properties and agricultural products, all depend on rich fundamental geological data and geological information services. *The Regional Geology of China* can not only fulfill the requirement for mineral resources exploration, and planning, deployment and management of geological work in provincial (municipal or autonomous regional) level, metallogenic zone (belt) and economic zone level and national level, but also hold an important role in planning, design, construction and scientific management of major projects in fields such as urban and rural construction, transportation, water conservancy, energy development and environmental protection.

The Regional Geology of China of the new generation systematically summarizes and highly condenses the fundamental geological survey work of the new era. To consistent with the first generation in structure and style as much as possible, the new series follow its advantages. Moreover, the series stick to innovation, and focus on presenting the new advancement and new materials acquired in past 30 years, which stands for the new height and achievement in research.

The compilation and successive publication of the Series is a complex scientific project, in which tremendous time and labor have been input. Scholars of a wide range of professional fields from different institutions were organized for the project. Results of hard work obtained by geologists of several generations in the past decades are condensed in the Series. It is a significant event for the geological community in China. Geologists participating in the project, with strong sense of responsibility for history and people, devoted themselves earnestly and unreservedly to creating excellent works with a high quality. They strive to make the books faithful, informative, sophisticated and innovative. Here, I would like to extend my heartfelt thanks to and sincere respect for all the geologists who worked hard and contributed to the project and all concerned who wholeheartedly support this project.

Wang Min,
Vice Minister of the Ministry of Land and Resources, PRC
Director of China Geological Survey

王民

February 2014

总 前 言

2008年，中国地质调查局下达了重编中国区域地质志的任务，计划项目名称为“中国地质构造区划综合研究与区域地质调查综合集成”。其总体目标任务是：通过对区域地质调查资料和地学研究成果的综合研究与集成，全面总结30多年来我国区域地质调查、矿产勘查和专题研究所取得的成果，特别是地质大调查实施以来的新资料、新进展、新成果。以活动论和先进的地质科学理论为指导，应用地质编图制图新技术、新方法，按省（市、自治区）、构造单元（成矿区带、经济区等）和全国3个层次编纂中国区域地质志，编制相关地质图件，建立空间数据库；针对综合研究过程中存在的全国性、区域性重大基础地质问题，有重点地开展野外调查和专题研究；进行区域地质调查和地质科学研究资料的系统总结，提高全国地质研究程度和水平，实现地质调查资源共享，提升地质调查和科学研究成果的应用效果和影响，为地质调查、矿产勘查、有关经济和工程建设、地质环境及地质灾害评估，以及地质科研和地质教育等，提供基础地质资料，使地质调查研究成果更好地服务于地质事业和经济社会的可持续发展。

2006年，《国务院关于加强地质工作的决定》指出：要“提高基础地质调查（研究）程度”，“系统开展区域地质、地球物理、地球化学和遥感地质等调查，建立地质图文更新机制，为社会提供有效快捷的地质信息服务”。经国务院同意颁布的《全国地质勘查规划》（2005～2020）指出：要“开展重大基础地质问题研究和重点地区基础地质综合研究，力争实现基础地质研究的重大突破。开展全国基础地质资料综合研究和成果集成工作。编制新一代全国系列基础地质图件，开展区域基础地质、矿产、地质环境等系列图件编制与各时代主要地层以及多学科的综合型立典研究”，“努力创造有宏观影响的大成果，不断提高地质勘查的精度、深度和广度”。

为经济建设和社会发展提供矿产资源和基础地质资料，一向是地质工作服务经济和社会发展的两项基本任务。通过编纂区域地质志，把各个部门、各种类型地质调查研究取得的丰富地质、地球物理、地球化学、遥感地质等资料进行综合研究和集成，使之系统化、规律化和理论化，可以为国家各有关部门和广大民众提供一套整装的地质资料，并为发展地球科学理论开辟通途。因此，国土资源部、中国地质调查局决定重编中国区域地质志是一个具有远见卓识的英明决定，是一项带有里程碑意义的战略性综合研究工作，必将为我国地质事业和国家经济社会发展做出重大贡献，在我国地质事业发展史上谱写出光辉的一页。

如果从1916年国民政府农商部改中央地质调查所为地质调查局算起，中国区域地质调查研究已经走过了将近100年的时间。在中国地质调查研究的历史上，中国地质界具有编纂地质矿产志的光荣传统，早在20世纪前半叶，就先后编辑出版了《苏浙皖矿志》、《中国矿产志》、《福建矿务志略》、《江苏地质志》、《北京西山地质志》、《四川西康地质志》等（陈宝国等，2011）。20世纪八九十年代，原地质矿产部主持编纂的中国区域地质志系统地总结了20世纪80年代末以前我国区域地质调查及专题研究成果，为我国地质事业和经济社会发展做出了重要贡献，建树了历史的功绩。

近30多年来，我国各有关部门进行了各种类型的地质矿产调查研究，积累了丰富的实际资料，获得大量地质、地球物理、地球化学、遥感地质，特别是高精度同位素年代及地球化学、岩石地球化学数据，解决了一系列长期悬而未决的地质问题，大幅度地提高了我国的地质研究程度，为中国区域地质志的编纂创造了良好条件。

为适应各方面的需求，提供更多的地质信息，项目任务要求开展3方面的调查研究，提交3方面的成果：①32个省（市、自治区）地质志（广东省地质志含香港、澳门特别行政区）；15个左右重要地区（构造单元、成矿区带、重要经济区、行政大区等）地质志；全国性地质志（地质、地球物理、地球化学系列专著）。②系列基础地质图件及空间数据库。地质图件包括地质图、地质构造图、岩浆岩地质图、第四纪地质与地貌图、航磁异常图、重力异常图（内部使用），各省（市、自治区）及各地区可根据地区特点与需要编制其他地质图件，如基岩地质图、岩相古地理图、变质地质图、断裂系统图等。③结合志书编写和系列图件编制开展必要的专题研究，解决一些重要的疑难地质问题，旨在统一地质划分对比和深化对这些疑难地质问题的认识，提高志书的质量。

编纂中国区域地质志涉及的范围大、学科领域广、参加的单位和人员多。这是一项对百万地质大军几十年乃至上百年地质调查研究成果的一次大综合、大集成。这样大规模的综合地质研究可以而且应该创造出有宏观影响的大成果。因此，我们提出以下5方面的要求，作为提交有宏观影响大成果的主要标志。

第一，在地质成果上要大综合和大集成。在充分搜集、分析各个部门、各种类型地质调查研究成果基础上，进行地质、地球物理、地球化学相结合的综合研究和成果集成，使志书内容丰富、资料翔实、数据准确、特色鲜明，如实反映当前地质调查研究的程度，反映志书地域内各种地质现象的认知水平。

第二，在地质研究程度上有大幅度的提高。不断提高全国和一个地区的地质研究程度，是地质工作永恒追求的目标，也是地质工作服务于经济社会发展的必然要求。地质研究程度的提高，是靠不断地进行地质调查研究和持续地开展综合研究和成果集成而实现的。因此，通过编纂地质志，从地质、地球物理、地球化学、遥感地质以及地表与深部地质结合上开展精细的综合研究，大幅度地提高编志地区的地质研究程度和水平，为地质工作和经济社会发展提供高质量的地质资料。

第三，在地质工作和经济社会发展中发挥大的作用。随着人类社会的进步，地质工作的社会功能和服务领域日益宽广，地质工作几乎是兴利于各个部门和行业，造福于人类的方方面面。作为一套完整而翔实的基础地质资料，既可以满足一个省（市、自治区）、一个成矿区（带）及经济区（带）和全国地质矿产勘查、地质工作规划、部署和管理的需求，又可以在城乡建设、交通、水利、能源和环境保护等重大工程规划、设计、建设和科学管理中发挥重要作用；既可以为地质科研和地质教育开辟捷径，又可以为丰富地质信息库和普及地质科学知识提供翔实地质资料，充分体现地质工作服务于经济社会发展的强度和广度。

第四，在地质规律的概括和地质理论创新上有大的提升。地质工作是一项科学技术密集的调查研究工作，地质科学基本上属于认识性科学。矿产资源勘查能否取得重大突破，地质环境及地质灾害的评估和预测是否准确，对诸多工程建设规划、设计和实施能否发挥更大作用，归根结底取决于对地质规律的认识水平和地质理论及技术方法的创新程度。地质调查研究的目的，在很大程度上是总结地质规律、认识地质规律和利用地质规律。抓住了地质规律，地质理论的创新也就应运而生了。地质志质量的好坏、水平的高低、贡献的大小，在很大程度上取决于对编志地区地质规律和地质理论概括、认识的深度和高度。因此，在地质志编纂过程中，在详细记录地质事实的基础上，通过实际资料的归纳、分析和总结，概括出地质规律，凝练出新发现、新认识，大幅度地提高对编志地区地质规律和地质总体特征的认知水平。

第五，在国内外地质界得到广泛应用和产生大的影响。一项科学成果，其在社会上应用的广度和影响的大小，主要取决于其实用价值和科学价值，其实用价值和科学价值越大，它的影响力就越大。中国地质构造十分复杂而又十分特殊，有许多地质现象为世界罕见，如高大的青藏高原、北方新生代的黄土堆积、南方岩溶（喀斯特）地质及地貌、中部延展数千千米的高压—超高压变质带、某些标准的地层剖面和罕见的古生物群落、某些特殊类型的矿床等，既是中国地质的特色和区位优势，又是诞生地质科学新理论的温床，引起国内外地质科学家的关注和竞相探索，以期地质科学理论上的突破。通过编纂区域地质志认真总结中国地质、特别是上述特色地质的总体特征及其演化规律，给予理论的概括和解释，不但会对地质勘查实践和经济社会发展发挥更大作用，而且会对发展地质科学理论奠定基础和做出重要贡献，也必将引起国内外地质界的广泛兴趣，产生重大影响。

编纂中国区域地质志是地质工作中一项最基础的综合研究工作，其成果将在国土资源部门及其他有关部门和外国各有关地质单位长期使用。因此，确保质量是对这套志书的必然要求。为了确保志书质量，要求做到“高起点、高要求、高水平”。

所谓“高起点”，就是在充分掌握编志地区地质研究程度、地质特点和地质规律的高点上着手进行地质志的编写。为此，首先要广泛收集编志地区地质资料和成果，包括报告、专著、图件和论文，特别是近些年来的地质调查研究的成果。其次，对收集到的资料进行地质、地球物理、地球化学相结合的综合研究，力求洞悉本区地质特点、地质规律和存在的主要地质问题。最后，利用先进的地质理论和技术方法进行志书编写。

所谓“高要求”，就是在编志过程中，对各项工作，包括资料的收集、整理、样品分析测试、志书编写和图件编制等，都要高标准、严要求，严格按“项目总体设计”、“工作指南”、“项目管理办法”和国土资源部、中国地质调查局的有关规程、规范执行，切实保证成果质量。

所谓“高水平”，就是力争把这套志书打造成高水平的精品著作，在科学内容上精益求精，忠实记录地质事实，科学地总结地质特征和地质规律，使志书做到资料翔实，内容精湛，质量上乘，富于创新。在表现形式上也要精益求精，文字描述要言简意赅，文图并茂；附图、附表要主题鲜明，清晰美观；度量单位和名词术语使用等要符合国标、行标和国际标准及惯例。

为了丰富志书内容，提高志书水平，为地矿行业、社会和公众提供更多的地质信息，充分发挥志书的功能，要求：一要实行地质、地球物理、地球化学、遥感等探测研究资料相结合，从宏观和微观尺度鉴别、认识地质现象，解决地质问题；二要实行地表地质与深部地质相结合，从三维角度揭示各种地质现象、成矿作用和地质灾害的深部根源；三要实行大陆地质与海洋地质相结合，把海陆作为一个整体加以综合研究；四要实行志书、图件和数据库建设相结合，更全面、形象地显示编志地区地质、地貌特点，更好地发挥志书功能；五是实行综合研究与专题研究相结合，在开展综合研究的同时，就一些重大的、有争议的地质问题进行专题研究，统一认识，统一划分对比，解决争论问题，提高志书质量。在指导思想上坚持继承和创新相结合，在继承基础上创新，在创新指导下继承，既要继承所有好的成果，又要大胆创新，敢于改变传统的认识和结论。同时，通过编志，以老带新，互教互学，培养一批学科性和综合性地质科技人才。

地质工作和国民经济发展规划、部署，地质资源勘查及地质环境与地质灾害评估、预测，城乡建设及重大工程规划、设计，普及地质科学知识，提高全民族科学文化素质和水平，开展国内外地质科学技术合作与交流，以及发展地球科学理论等，都需要丰富的地质信息。《中国区域地质志》可以为上述诸方面提供厚实的基础地质资料。

这次地质志的编纂规模大、要求高，参加的单位和人员多，持续的时间长，特别需要加强领导和组织协调，加强项目科学管理，全国、各省（市、自治区）及各地质区地质志，都要求建立领导机构和学术指导组织。

在项目编纂指导委员会和编辑委员会的领导下，设立了中国区域地质志综合研究项目（即“中国地质构造区划综合研究与区域地质调查综合集成”），由中国地质科学院地质研究所承担实施，负责《中国区域地质志》编纂的科学技术指导和组织协调工作，参加人员及分工如下。

一、项目负责人

项目负责人：李廷栋

项目副负责人：卢民杰　丁孝忠　姚建新　刘　勇

二、特聘专家

组长：肖庆辉

成员（按姓氏笔画排序）：王乃文　陆松年　陈克强　洪大卫　袁学诚　葛肖虹　潘桂棠

三、项目办公室

主任：庞健峰（2010~2013 年 8 月为耿元生，2013 年 8 月~2015 年 12 月为耿树方）

工作人员（按姓氏笔画排序）：王振洋　白　金　刘　平　庞健峰　郑　宁　寇彩化　韩坤英

四、项目专业组

区域地质组：李廷栋　丁孝忠　耿树方　陈炳蔚

区域地层组：王乃文　姚建新　高林志　郑　宁

区域岩浆岩组：王　涛　薛怀民　童　英　王晓霞　刘　勇　张　磊　寇彩化

区域构造组：刘　训　游国庆　王　军　张　恒

第四纪地质与地貌组：闵隆瑞　王　永　迟振卿　董　进

沉积与岩相古地理组：刘燕学　马丽芳　彭　阳

前寒武纪地质组：耿元生　任留东　杨崇辉　杜利林　王彦斌

同位素年代组：万渝生　董春燕　颉颃强

地球物理及深部地质组：管　烨　袁学诚　郑洪伟　张季生

地理底图、数据库与图件设计组：范本贤　韩坤英　剧远景　张庆合　王丽亚
王振洋　庞健峰　丁伟翠

六大区区域地质综合研究组：王洪亮　尹福光　杨祝良　牛志军　赵凤清　邢德和

中国区域地质志的编纂获得国土资源部、财政部、科技部及有关司局、中国地质调查局及有关部、处和中国地质科学院及有关处、室的指导和大力支持；获得中国地质调查局6个地质调查中心、中国地质科学院有关研究所和地质研究所及有关处、室、各省（市、自治区）国土资源厅、地质勘查局、地质调查院（所）和有关地质院校的积极参与和大力支持；地矿行业各部门、全国及各省（市、自治区）地质资料、地质图书等单位为志书编写提供了丰硕的地质资料和信息；参与志书编写的专家、学者和管理人员孜孜以求、精益求精，付出了艰辛的劳动和智慧。在这里，向所有关怀、指导志书编写的部门和领导、向所有关心、帮助志书编写的部门和人士、向所有参与志书编写的专家、学者和管理者，致以衷心的感谢和敬意。

地质工作是一项永恒的事业，只要有人类存在，就有地质工作；哪里有人类存在，哪里就有地质工作。这是因为地质工作服务的领域很广，而且越来越广。地质工作为经济社会发展服务的主要途径之一就是不断地为社会提供有效快捷的地质信息，这也是编纂中国区域地质志的主要宗旨之一。

我们希望这套志书能够集我国地质调查研究成果之大成，走向志书的深处、高处和远处，打造出志书的精品，为我国地质事业的发展做出更大贡献！为优化国土生态环境、建设美丽中国做出更大贡献！为我国经济社会的可持续发展和实现中华民族伟大复兴的中国梦做出更大贡献！为世界地球科学的繁荣、昌盛做出更大贡献！

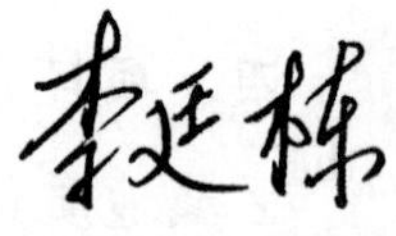

2014 年 10 月

General Foreword

In 2008, A project concerning recompilation of *the Regional Geology of China*, entitled "Comprehensive Study of Geological and Tectonic Division and Synthesized Integration of Regional Geological Survey in China", was assigned by China Geological Survey. The overall objective of the project is to fully summarize achievements acquired in regional geological survey, mineral resources exploration and monographic studies, especially the new materials, new advancement and new achievement since the implementation of the Great Geological Survey in China, by means of comprehensive study and integration of regional geological survey data and geoscience research results. Under the guidance of Mobilism and advanced geoscience theories, *the Regional Geology of China* and relative geological maps are compiled in three levels respectively: provincial (municipal or autonomous regional) level, tectonic unit (metallogenic belts or economic zones) level and national level, while spatial database are established, with new technologies and methods of geological mapping. Field surveys and monographic studies are conducted with priority to solve the national or regional fundamental geological problems encountered in comprehensive study. Systematical summaries of regional geological survey data and geoscience research materials are done so as to improve intensity and level of geological survey and research, to realize the sharing of geological survey resources and to raise the applied effectiveness and impact of geological survey and research achievements. *The Regional Geology of China* can provide fundamental geological materials for geological survey, mineral exploration, relative economic development and engineering construction, assessment of geological environment and hazard, as well as geological research and education. That enable achievements of geological survey to better serve geological cause and sustainable development of economy and society.

In 2006, *Decision of the State Council on Strengthening Geological Work* put forward that "the intensity of geological survey and research should be enhanced", and that "regional geological, geophysical, geochemical and geological remote sensing surveys should be conducted systematically and an update mechanism concerning geological maps and texts should be established, to provide effective and efficient geological information service". According to *National Geological Exploration Program* (2005 ~ 2020) promulgated with the State Council's consent, "researches on major geological problems and comprehensive study of fundamental geology in key areas should be carried out so as to achieve crucial breakthrough in fundamental geological researches. Comprehensive studies of nationwide geological data and integration of research results should be conducted. The new series of fundamental geological maps should be compiled; compilation of map series of regional fundamental geology, mineral resources, geological environment and multi-disciplinary comprehensive typical studies of various major strata should be conducted", and we should "strive to create outstanding achievements with great impact, and to improve accuracy, depth and breadth of geological exploration".

Providing mineral resources and supplying geological data for economic construction and social development of the country, have always been two basic tasks of geological work that serves for social and economic development. Through the compilation of *the Regional Geology of China*, abundant geological, geophysical, geochemical and remote sensing data collected in all kinds of geological surveys from various institutions are used for comprehensive research and integration. Geological materials that have been systematized, regularized and theorized, can serve relevant government departments and the general public with a whole set of data, and beat a path for developing theories of earth science. Therefore, to recompile *the Regional Geology of China* is a wise decision with visionary transcendence by China Geological Survey, Ministry of Land and Resources of China; it is a strategic synthetic study with monumental significance, which is bound to contribute greatly to geological cause and development of economy and society, and to write a glorious page in the history of geological development in China.

Dating back to 1916 when the Ministry of Commerce and Agriculture renamed Central Geological Institute as Geological Survey, geological survey has gone through a history of nearly 100 years in China. Compiling regional geology has always been a glorious tradition upheld by geological community in the history of geological survey of China. Early in the first half of 20th century, *Minerals Geology of Jiangsu*, *Zhejiang and Anhui*, *Minerals Geology of China*, *Mining Records of Fujian*, *Geology of Jiangsu*, *Xishan Geology of Beijing*, *Xikang Geology of Sichuan* were published successively (Chen Baoguo *et al.*, 2011). *The Regional Geology of China* compiled in the 1980s and 1990s, hosted by the former Ministry of Geology and Mineral Resources systematically summarized results of regional geological survey and monographic studies before the late 1980s in China, which made great contributions to geological cause and development of economy and society, and acquired historical achievements in development of China.

In the past three decades, geological mineral survey and research conducted by relevant departments have accumulated a wealth of practical materials, and have acquired generous data of geology, geophysics, geochemistry, remote sensing geology, and in particular data of high-precision isotopic chronology and rock geochemistry. Besides, a series of pending geological problems have been solved and the intensity of geological survey and research have been enhanced tremendously. All these have created favorable conditions for compiling the new *the Regional Geology of China*.

For the purpose of providing more geological data and meeting requirements of various aspects, surveys and researches in 3 aspects should be conducted and 3 kinds of achievements should be submitted in the project: ① regional geology of 32 provinces (municipalities and autonomous regions) (The Regional Geology of Guangdong Province covers HongKong Special Administrative Region and Macao Special Administrative Region), regional geology of 11 to 15 major regions (tectonic units, metallogenic belts, key economic zones, and major administrative regions), and national geology (series of geological, geophysical and geochemical monographs). ② the series of fundamental geological maps and spatial database. The geological maps include geological map, tectonic map, magmatic rock map, Quaternary geology and geomorphology map, aeromagnetic anomaly map and gravity anomaly map (restricted for internal use only). Each province (municipality or autonomous region) or region can adjust to compile other types of maps according to its features and demands, such as geological map of bed rocks, lithofacies and paleogeographic map, metamorphic map and map of fault systems. ③ to address major problems of geology, necessary monographic studies should be carried out in the course of compiling regional geology and maps. It is intended to unify recognition on geological division and correlation, to deepen recognition on difficult geological problems, and finally to improve the quality of the series of regional geology.

The compilation work covers a wide range of regions and gets scholars of broad discipline fields from many departments involved. The project synthetically integrate achievements of geological survey and researches obtained by a million geologists in decades of years or even a hundred years. It is believed that a comprehensive geological study of such a scale might and should create achievements with significant influence. Therefore, we propose the following 5 requirements, as major indicators of achievements with macro-influence.

Firstly, geological achievements should be broadly synthesized and integrated. Comprehensive study and integration of geological, geophysical and geochemical data should be conducted on the basis of fully collecting and analyzing various results of geological survey and research from different departments, to make the books informative, faithful, accurate and featured, and to present truthfully the level of geological survey as well as research and cognition of all geological phenomena in target areas.

Secondly, the level of geological survey and research should be highly raised. To continuously raise the level of geological survey and research of nationwide and an area is the eternal pursuit of geological work, and an inevitable requirement of economic and social development. The raise of level of geological research lies in continuous geological survey and comprehensive study as well as synthetic integration. By way of compiling the series of regional geology, sophisticated researches are conducted with integration of geological data, geophysical data, geochemical data, remote sensing data and combination of surface and deep geological data, which will significantly improve the intensity and level of geological researches and provide geological work and social and economic development with geological data in high quality.

Thirdly, *the Regional Geology of China* should play a vital role in geological development and economic and social development. With the progress of human society, the social functions and service sectors of geological work have been expanded. Geology almost facilitates to all various departments and industries and benefits all aspects of humankind. As a whole set of detailed and faithful fundamental geological materials, the *Series* can not only meet the requirements of mineral exploration, and planning, arrangement and management of geological work in provincial (municipal or autonomous regional) level, in level of metallogenic zone (belt) and economic zone (belt), or in national level, but also play important roles in planning, design, construction and scientific management of major projects in fields such as urban and rural construction, transportation, water conservancy, energy development and environmental protection. Meanwhile, the series will not only beat a shortcut for geological research and education, but also provide informative data for enrichment of geological repository and popularization of geological knowledge. All these embody the service intensity and breadth of geology to social and economic development.

Fourthly, there should be great improvement in summarization of geological regularity and innovation of geological theory. Geological work is an investigative work with intensive feature of science and technique. Geo-science essentially belongs to cognitive science. Whether a significant breakthrough in mineral resources exploration will be made or not, whether evaluation and prediction of geological environment and hazards will be accurate or not, and whether geology will play a greater role in planning, design and implementation of project construction, all these questions, in the last analysis, depend on the cognitive level of geological regularity and innovative extent of geological theory and technical method. The purpose of geological survey is to summarize, to recognize and to utilize geological regularity in a large degree. Once geological regularity is obtained, innovation in geological theory will come into being naturally. The quality of the series of *the Regional Geology of China*, and their achievements and contribution, largely depend on the depth and breadth of summarization and cognition of geological regularity and theory for the studied area. Therefore, in the course of compiling the series of *Regional Geology*, practical materials are concluded and analyzed on the basis of detailed record of geological facts so as to summarize the geological regularity and to extract new discoveries and new recognitions, resulting in significant enhancement of cognition on geological regularity and general geological features in the target area of the regional geology.

Fifthly, the achievements should be widely used by geological community both in China and abroad, and great impact should be achieved. The applied breadth and impact of a scientific achievement mainly depend on its practical value and scientific value. The higher its practical value and scientific value are, the greater its influence will be. Completely complex and quite special, geological structures in China hold lots of geological phenomena rare in the world, including the lofty Qinghai-Tibet Plateau, Cenozoic loess accumulation in the North, Karst geology and landform in the South, high or ultra-high pressure metamorphic belt that extend thousands of kilometers in the Middle, standard stratigraphic sections and rare paleo-biocenoses, and deposits of special types. These geological phenomena not only embody features and regional advantages of geology in China, but also serve as the cradle where new theories of geoscience are incubating. They attract attention of geoscientists both of China and abroad who race to quest for theoretical breakthroughs in geoscience. General features and evolution regularities of geology of China, especially the geological phenomena mentioned above, are concluded and explained theoretically by means of compiling the series. Such practices will not only play a greater role in geological exploration and economic and social development, but also lay concrete theoretical foundation for development of geoscience and have important contribution to it. It is believed that interests of geoscientists both at home and abroad will be attracted and enormous influence will be extended.

Compiling *the Regional Geology of China* is the most fundamental synthetic studies in geological work; the achievement will be used for a long time by Departments of Land and Resources and other relevant departments or even geological departments abroad. So, to ensure the quality of the *Series* is the inevitable requirements. For this purpose, requirements of high starting point, high requirement and high level are imposed.

The so called "high starting point" means that the compilation of the maps and books should be commenced from the starting point that the intensity of geological researches, geological features and regularities of the target area have been quite grasped. Hence geological materials and achievements of the target area, including reports, monographs, geological maps, papers and especially achievements of geological survey in recent years, are collected first of all. Secondly, in the compilation of the maps and books, synthetic studies integrated with geological, geophysical and geochemical data that have been gathered, are conducted and seek to penetrate geological features, regularities and major geological problems that existed in the target area. Finally, the books are written and the maps are compiled with advanced geological theories and technical methods.

The so called "high requirement" means that all procedures, including data collecting and sorting, sample analyzing and testing, book and map compiling, should be conducted strictly according to "General Design of the Project", "Work Guide", "Measures for Project Administration" and relevant regulation issued by MLR and CGS, to ensure the quality of the results.

The so called "high level" means that we should strive to create an excellent series of books with high level, to make sure that the books are excellent, informative in scientific content and faithful in record geological facts, and scientifically and creatively summarize geological features and regularities. We pursuit excellence in manifestation of the books as well. The descriptive text should be concise and illustrated with proper figures. The figures and tables should be outstanding in theme, and clear and beautiful in form. Unit of measure and terminology should conform to national, sectoral and international standards and convention.

To enrich contents and improve quality of the books, and to fully perform function of the books, the books should provide more geological information for geological and mineral industry as well as the public, demands are proposed as follows: firstly, integrated researches of geological, geophysical, geochemical and remote sensing data should be done so that geological phenomena can be distinguished and recognized on both macro and micro scales and geological problems can be solved; secondly, surface geology and deep geology should be combined with each other so that the deep roots of geological phenomena, mineralization and geo-hazards can be uncovered from three-dimensional perspective; thirdly, continental geology and marine geology should be associated as a whole for comprehensive study; fourthly, book and map compiling should be connected with construction of database, so that geological and geomorphological features of the target area are manifested completely and vividly and the functions of the series are better performed; fifthly, comprehensive studies and monographic studies should be integrated; while conducting comprehensive studies, we should carry out monographic studies concerning the major controversial geological problems as well, to unify cognition and to solve disputed problems, and to improve the quality of the books. Adhering to the ideology that inheritance combines with innovation, we seek to innovate on the basis of inheriting and to inherit with the guidance of innovation. We will not only inherit all the great achievements, but also innovate boldly to change traditional understanding and conclusions. Meanwhile, through the compilation of maps and books, a number of comprehensive geoscientists in various disciplines have been cultivated by means that the senior instruct the junior and they learn from each other.

Aspects such as planning and deployment of geological work and national economic development, geological mineral exploration, geological environment evaluation and hazards forecast, planning and design of urban and rural construction and major engineering projects, popularization of geoscience, improvement of scientific and cultural quality and level of the people, international cooperation and communication of geoscience and technology, and development of theories of earth science, all require abundant geological information. *The Regional Geology of China* can provide substantial fundamental geological data for the aspects mentioned above.

The project of compiling the series, because of its large scale and strict requirements, claims tremendous time and labor input. Hence it is necessary to strengthen leadership and coordination, and to reinforce scientific management of the project. It is required that leading organs and academic guiding organization should be established in national level, in provincial (municipal or autonomous regional) level and in level of geological areas.

Under the leadership of Advisory Committee and Editorial Board, a team responsible for Comprehensive Study Project of the Regional Geology of China (known as "Comprehensive Study of Tectonic Division and Synthesized Integration of Regional Geological Survey in China" as well) was set up and by the Institute of Geology, Chinese Academy of Geological Sciences to implement. Participants of the project and members of the team who are responsible for academic guiding, coordinating and organizing in the compilation of *the Regional Geology of China* are as follows:

a. Project Leaders

Chief Project Leader: Li Tingdong

Deputy Project Leaders: Lu Minjie, Ding Xiaozhong, Yao Jianxin and Liu Yong

b. Distinguished Expert Group

Group Leader: Xiao Qinghui

Members (by surname stroke order): Wang Naiwen, Lu Songnian, Chen Keqiang, Hong Dawei, Yuan Xuecheng, Ge Xiaohong and Pan Guitang

c. Project Office

Director: Pang Jianfeng (Geng Yuansheng from 2010 to August, 2013; Geng Shufang from August, 2013 to December, 2015)

Staff (by surname stroke order): Wang Zhenyang, Liu Ping, Bai Jin, Zheng Ning, Pang Jianfeng, Kou Caihua and Han Kunying

d. Project Professional Groups

Group of Regional Geology: Li Tingdong, Ding Xiaozhong, Geng Shufang and Chen Bingwei

Group of Regional Stratigraphy: Wang Naiwen, Yao Jianxin, Gao Linzhi and Zheng Ning

Group of Regional Magmatic Rocks: Wang Tao, Xue Huaimin, Tong Ying, Wang Xiaoxia, Liu Yong, Zhang Lei and Kou Caihua

Group of Regional Tectonics: Liu Xun, You Guoqing, Wang Jun and Zhang Heng

Group of Quaternary Geology and Geomorphology: Min Longrui, Wang Yong, Chi Zhenqing and Dong Jin

Group of Sedimentology and Lithofacies Palaeogeography: LiuYanxue, Ma Lifang and Peng Yang

Group of Precambrian Geology: Geng Yuansheng, Ren Liudong, Yang Chonghui, Du Lilin and Wang Yanbin

Group of Isotopic Chronology: Wan Yusheng, Dong Chunyan and Xie Hangqiang

Group of Geophysics and Deep Geology: Guan Ye, Yuan Xuecheng, Zheng Hongwei and Zhang Jisheng

Group of GeographicBasemap, Database and Map Design: Fan Benxian, Han Kunying, Ju Yuanjing, Zhang Qinghe, Wang Liya, Wang Zhenyang, Pang Jianfeng and Ding Weicui

Comprehensive Study Group of Regional Geology for the Six Major Regions: Wang Hongliang, Yin Fuguang, Yang Zhuliang, Niu Zhijun, Zhao Fengqing and Xing Dehe

The compilation work has gained full support and energetic assistance from MLR, MOF, MOST and its relevant departments and bureaus, from CGS and its relevant departments, and from Chinese Academy of Geological Sciences (CAGS) and its relevant departments. Six Centers of CGS, relevant institutes of CAGS, relevant departments of Institute of Geology, provincial (municipal or autonomous regional) Departments of Land and Resources, provincial (municipal or autonomous regional) Bureau of Geology and Mineral Exploration, provincial (municipal or autonomous regional) Bureau (Institute) of Geological Survey and relevant universities or colleges have actively participated in the job. Various departments of the geological and mining industry, National Geological Library of China, provincial (municipal or autonomous regional) Geological Libraries, National Geological Archives of China, and provincial (municipal or autonomous regional) Geological Archives have provided substantial geological materials and information. All experts, scholars and management personnel who participate in the compilation have devoted their wisdom and hard work in pursuit of excellence. Hereby, I would like to extend my sincere gratitude to and respect for departments and leaders who exercise care and guide the compilation of the series, to departments and persons who concern and help in the compilation, and to all experts, scholars and management personnel who involved in the compilation of the series.

Geological work is an eternal cause, and where there are human beings, there is geological work. The reason of this lies in that the fields for which geological work serve is wide and will be wider and wider. One of the main paths by which geological work serve for social and economic development is to provide effective and efficient geological information. It is one of the major reasons why we strive to compile the series of books.

It is our hope that the series of *the Regional Geology of China* can integrate all great achievements of geological survey in China and mark the culmination of years of hard work. It is our pursuit to create a series of books with excellent quality, to make greater contribution to the development of geological cause in China, to make a great contribution to the optimization of ecological environment and the construction of beautiful China, to make greater contribution to sustainable development of economy and society and realization of the Chinese dream of Chinese Nation's great rejuvenation, and to make greater contribution to the prosperity of Earth Science in the world.

Li Tingdong

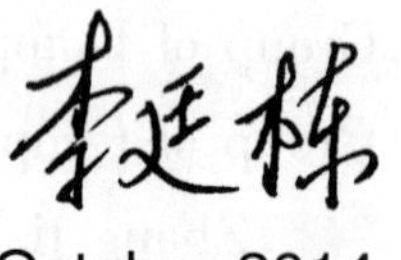

October 2014

《中国区域地质志·海南志》专家咨询委员会

主　任：黄香定

委　员：林起玉　陈哲培　何圣华　钟盛中

编制委员会

主　任：廖香俊

副主任：陈颖民　吴国爱　傅杨荣

委　员：陈沐龙　杨昌松　王大英　龙文国

编　写　组

主　编：李孙雄

副主编：云　平　林义华　陈哲培

编　委：周进波　朱耀河　魏昌欣　谢顺胜　杨　奕

龙文国　蔡水库　官　军　莫位任　袁勤敏

胡在龙　王　超　吴剑波　江宏伟　汪焰华

刘　兵　李正廷　吕嫦艳　曾雁玲　龚　丹

吴小洁　岑如南　陈方颖

Advisory Committee of *the Regional Geology of China*, *Hainan Province*

Director: Huang Xiangding

Members: Lin Qiyu Chen Zhepei He Shenghua Zhong Shengzhong

Organization Committee of *the Regional Geology of China*, *Hainan Province*

Director: Liao Xiangjun

Deputy Directors: Chen Yingmin Wu Guo'ai Fu Yangrong

Members: Chen Mulong Yang Changsong Wang Daying Long Wenguo

Compilers of *the Regional Geology of China*, *Hainan Province*

Chief Editor: Li Sunxiong

Deputy Editors: Yun Ping Lin Yihua Chen Zhepei

Members: Zhou Jinbo Zhu Yaohe Wei Changxin Xie Shunsheng
Yang Yi Long Wenguo Cai Shuiku Guan Jun
Mo Weiren Yuan Qinmin Hu Zailong Wang Chao
Wu Jianbo Jiang Hongwei Wang Yanhua Liu Bing
Li Zhengting Lü Changyan Zeng Yanling Gong Dan
Wu Xiaojie Cen Runan Chen Fangying

目　　录

第一篇　区域地层

第二篇　侵　入　岩

第三篇　火山岩及火山作用

第四篇　变质岩及变质作用

第五篇　区域地质构造

第六篇　地球物理地球化学特征

第七篇　地 质 资 源

CONTENTS

PART 2 Intrusive Rocks

PART 3 Volcanic Rocks and Volcanism

PART 4 Metamorphic Rocks and Metamorphism

PART 5 Regional Tectonics

PART 6 Geophysical and Geochemcial Characteristics

PART 7 Geologic Resources

绪　言

“海南省地质系列图件编制与综合研究”（《海南省区域地质志》）原项目名称为“海南省1:25万地质系列图编制与数据库建设”，项目编号为：1212010610713，工作起止年限2006年1月至2010年12月。所属计划项目为“中国地质构造区划综合研究与区域地质调查综合集成”，由中国地质科学院地质研究所实施、海南省地质调查院承担完成。项目总体目标任务是全面总结海南省已完成的1:25万区域地质调查及其他已有成果资料，针对海南岛重大基础地质问题和地质图编制过程中存在的问题，开展野外地质调查和综合研究工作；以活动论为指导，编制海南省1:25万数字地质图，建立地质图空间数据库，为海南省经济社会发展提供基础资料。

工作中严格遵照《〈中国区域地质志〉工作指南》和《1:25万区域地质调查技术要求》（暂行）以及其他有关规范、指南执行，紧密围绕项目目标任务开展工作，一是全面总结海南省已完成的1:25万区域地质调查及其他已有成果资料，并部分查阅了1:20万区调、1:5万区调地质资料及其他与海南省地质有关的科研文献和论文等，尤其是近年来一些科研人员的测试结果，为图件编制提供了翔实的资料依据；二是针对海南岛重大基础地质问题和地质图编制过程中存在的问题（海南岛结晶基底地层——长城系抱板群问题、海南岛南、北地体分野的界线构造问题、海南岛古生代变基性火山岩时代问题、海南岛北东向和北西向构造形迹形成时代及其相互作用关系问题、海南岛燕山期花岗岩体含矿性及其成矿时代问题、海南省乐东县丘糖岭古近纪—新近纪地层特征及岩石地层单位问题、海南省屯昌县南棍园地区含炭浅变质岩的时代及岩石地层单位问题、海南岛花岗岩统一清理、归并问题等等），开展了系统的野外地质调查和综合研究工作，深化海南岛基础地质的研究，提升项目的科研水平和地质成果。通过项目组几年来广泛的资料查阅和收集，从写实的角度，编制完成了《海南省1:50万地质图》、《海南省1:25万地质图》、《海南省1:50万地质构造图》、《海南省1:50万岩浆岩地质图》、《海南省1:50万地貌与第四纪地质图》、《海南省1:50万布格重力异常图》、《海南省1:50万航磁异常图》等系列地质图件，并编写了《海南省1:25万地质系列图件编制及综合研究》报告及各专题报告。本书即是以上工作成果的精练和升华。

海南省位于中国最南端，包括海南岛（含周缘小岛屿）、西沙群岛、中沙群岛、南沙群岛等岛礁及其领海，陆地（主要包括海南岛和西沙群岛、中沙群岛、南沙群岛）总面积约35400km^2，海域面积约200万km^2。海南岛是海南省最大的岛屿，面积约33900km^2，是全省政治、经济、文化和交通中心。岛内交通便利，有海榆东、中、西线三条国道、环岛高速公路及东环铁路等，全省公路网的密度高于全国平均水平；有海口、八所、三亚、洋浦四大港口和马村港以及数十个小港口，海上交通亦较方便；航空有海口美兰国际机场和三亚凤凰国际机场。岛内地形地貌呈中部隆起的穹隆形山体，中部高四周低，以五指山（海拔1867.1m）为隆起核心，向外围逐级下降，分别由山地、丘陵、台地、阶地和平原等构成环形层状地貌，梯形结构明显。海南岛主要河流有昌化江、南渡江、万泉河等。岛内地处热带、亚热带，属海洋气候和季风气候，高温多雨为其主要气候特征。全岛多年平均气温23.8℃，南部三亚、乐东沿海地区大于25℃，琼北地区小于24℃。1月份最冷，平均气温18.0℃左右，极端最低气温0.1℃；7月份最热，平均气温28.2℃，极端最高气温可达39℃。年平均降雨量为1639mm，每年的5~10月是多雨季，占全年总降雨量的70%~90%，每年11月至翌年4月为少雨季，干旱常常发生。截至2009年底，全省常住人口为864.07万人，其中，归侨、侨眷100多万人，是全国著名的侨乡。省内民族有汉族、黎族、苗族、壮族、回族等。语言种类多，主要使用的方言有10种，主要以讲海南话、黎话、临高话、儋州话、军话等为主。省内各类资源比较丰富，旅游资源极具特色，主要有海岸景观、山岳和热带原始丛林、珍奇异兽、大河、瀑布、水库风光、火山、溶洞、温泉、古迹名胜、民族风情、热带作物及田园风光等。粮食作物主要为水稻，次为旱稻、红薯、玉米等；经济作物有甘蔗、瓜菜、花生、芝麻、茶叶；热带作物有橡胶、油棕、咖啡、胡椒、剑麻、香茅；热带水果有菠萝、椰子、芒果、菠萝蜜等。海南省自1988年建省以来，经济飞速发展，形成了热带高效农业、工业、旅游业三足鼎立的局面。而随着海南国际旅游岛的建设，“一海、两市、三区”的发展构想（“一海”为中国南海世界明珠，“两市”为海口市、三亚市两个旅游城市，“三区”为东部中国椰树海岸区、中部海南岛国家公园区和西部海南岛西岸区），将为海南省的经济发展注入新的发展动力。

海南省开展过大量系统的地质调查，涉及区域地质、矿产、物探、化探、遥感、水工环和海洋地质领域，形成了大量的地质报告；因其所处的独特地质构造位置而被许多地质工作者所关注和涉足，发表了许许多多的学术论文，出版了很多学术专著。这些都为本次工作提供了大量翔实的地质资料。

区域地质调查包括1:100万、1:20万、1:5万、1:25万等系列不同比例尺。1:100万全岛区域地质调查工作由李承三等人于1929年完成，编制了海南岛1:100万地质图。1958~1964年，广东省地质局区测大队开展了海南岛1:20万区域地质调查，对海南岛地层、岩浆岩、构造进行了相对系统的研究，但由于种种原因，其报告及图件未能正式出版。直到1984年，海南地质大队在充分收集该区调资料的基础上，编制了海南岛1:20万地质图。1981年始，为解决一些重要的基础地质问题、重要经济区建设问题，为地质找矿提供重要地质资料，于“六五”至“九五”期间（1981~2000年），共完成志仲幅、海口市幅、千家幅等38幅1:5万区调图幅，总面积14264km^2，占海南岛总面积的42%；长流市、海口市、白莲市、灵山市4个1:5万区调图幅并进行了片区总结。2006年以来，又先后相继完成了8个1:5万图幅区调（面积3019km^2）。截至目前，完成的1:5万图幅数共为46幅，总面积17283km^2，约占海南岛面积的51%。“十五”至“十一五”期间，海南省启动了1:25万区域地质调查工作，截至2008年底，已全面完成海南岛1:25万图幅区调地质工作，为本次工作提供了重要的技术支撑。

海南省矿产调查起步较早，至少可追溯至清朝光绪年间。系统的调查工作是新中国成立以后，大体上可划分为5个阶段：

1950~1957年，从已知矿产地入手，就矿找矿，以扩大矿区远景，主要是石碌铁矿和东面沿海锆钛砂矿、蓬莱铝土矿的调查。

1958~1965年，主要是在上阶段新发现矿产地的基础上，深化矿产调查工作，开展了东海岸分布的以钛铁矿、锆英石、独居石矿为主的大型矿床和蓬莱铝土矿的调查工作，并做了全面的评价；还详查评价了新隆大型玻璃砂矿、烟塘石墨矿和勘探了长昌煤矿。

1966~1978年，受“文革”影响，矿产工作几乎陷于停顿状态，仅仅是对以往发现的矿产和矿化异常补充进行工作，矿种有铝土矿、铜矿、铅锌矿、煤矿、水晶矿、磷矿、钴土矿、菱铁矿等。

1979~1988年，根据国家建设的需要和成矿地质条件，采用多兵种作战模式，融区调、物化探和矿产普查为一体，加强了区域成矿地质规律的研究。工作矿种有铜钼矿、钼矿、硅藻土矿、铅锌矿、钨锡矿、沸石、膨润土矿、石墨矿、石灰岩矿、蓝宝石矿等。

1989年至今，主要是开展金矿、钼矿、铅锌矿和石英砂矿、花岗石材矿、石灰岩矿、萤石矿工作。

物探、化探、遥感工作相对比较薄弱。物探工作主要有1:10万、1:5万航磁和1:50万、1:20万重力。1:10万的航磁测量由地质部905队1957~1958年完成，覆盖全海南岛；1:5万航磁测量开展于1976~1978年，分别由广东省地质局物探队（岛西部和西北部，面积8594.5km^2）、二机部703队（岛东部，面积4000km^2）和冶金工业部物探公司航测队三分队（岛南、西、东部，面积约12866.4km^2）完成。海南岛1:50万重力测量由广州海洋地质队1970年完成，1996年海南省地质调查院完成了海南岛1:20万区域重力调查。此外，1992~1995年海南地质综合勘察院完成了全岛1:100万物化遥综合解释成果编图工作。化探主要是海南岛1:20万水系沉积物测量（江西省地矿局物化探大队1985~1988年完成）和海南岛1:25万多目标地球化学调查（2004年至今，由海南省地质调查院完成），部分地区在开展1∶5万区调（东方幅、玉道幅）时同步开展了1∶5万土壤测量。海南省的遥感工作开展起步较晚，1980~1986年先后由中国科学院遥感研究所和中国卫星地面站分别出版了《海南航空照片判读文集》、《海南岛1∶20万TM卫星影像集》。1990~1999年，北京大学遥感研究所、国家地震局、地质矿产部地质遥感中心、国家测绘局在海南岛开展了不同程度的工作。北京大学遥感研究所编制了1∶10万的地质、地貌、植被、土地利用、交通解译图；国家地震局在海南进行1∶10万机载侧视雷达成像，开展了地质构造解译和地震研究工作；地质矿产部地质遥感中心在海口、洋浦、八所等地区进行了1∶2万的彩色红外航空摄影，编制出版了海口市和洋浦开发区1∶1万航空遥感彩色红外影像图、海南岛1∶20万TM卫星影像集和海南岛遥感综合调查报告；国家测绘局在海南全岛范围进行了1∶3.5万的彩色红外航片摄影，同时编制了全岛的1∶1万的黑白正射影像图。2002年以来，海南省地质调查院先后开展了海南省国土资源遥感综合调查、海口市城市环境地质遥感调查和海南非金属矿集区遥感地质综合调查等遥感调查项目的工作。

水工环地质工作始于1961年，20世纪90年代以前，主要以区域性水文地质普查为主，水文地质工作研究程度较低。90年代以来，水工环地质工作随着环境地质问题的日益突出，逐步转向环境地质、地质灾害、生态环境地质调查等方面的工作。1961~1965年，开展了海南岛1∶20万水文地质普查和沿海部分地区的1∶5万水文地质普查，并重点对海口—演丰地区开展了1∶5万水文地质勘探。1966~1978年，水文地质工作继续完成沿海地区1∶5万水文地质普查；后期又进行了琼山龙桥供水水源地勘探，发现了大型玄武岩孔隙裂隙水水源地；为满足地方、军工、厂矿等部门供水需要，进行了大量的供水点勘探；对热泉进行了补充调查，发现了海口、崖城等地的低温热水。1979~1991年，完成了海南岛1∶20万水文地质普查、三江地区农田供水及海口城市供水勘探、东北部地区火山岩系及琼海东平红层盆地水文地质详查，完成了洋浦、三亚、八所等港口的供水勘探；其间编制了《海南岛1∶20万海岸工程地质调查报告》和1∶5万三亚地区水文工程地质综合勘查。1997~2000年，先后完成了海口市、琼山市、澄迈县、乐东县、东方市、临高县、儋州市等沿海县市1∶10万区域水文地质调查和1∶5万海南省乐东县莺歌海地区供水水文勘查和海南岛1∶50万环境地质调查。2000~2005年，相继完成了1∶10万海南省琼中县地质灾害调查与区划和1∶10万海南省屯昌县地质灾害调查与区划，以及1∶25万琼海县幅生态环境地质调查和1∶25万海南岛生态环境地质综合研究。2006年以来，开展了海南省主要城市环境地质调查评价。海南省自建省以来，工程地质调查转向于服务社会建设和经济发展，缺乏大范围系统的调查工作。

涉及海南省的科研著作、论文众多，本次共计收集科研著作、论文及相关地质资料达200余篇，资料时间从1964年至2010年（见参考文献）。专著《广东省区域地质志》（广东省地矿局，1986）首次对海南岛基础地质和矿产进行了系统总结，尽管完成时间相对早，工作方法和研究手段与当前大相径庭，但一些直观的认识主张尚值得提倡，如九所-陵水断裂带的北东走向认识、海南岛古生代中弧状构造带的认识等等；《海南岛前寒武纪地质》（张仁杰等，1989）首次对海南岛前寒武纪地质进行研究，深化了对海南岛前寒武纪地质的认识；《海南岛若干重要基础地质问题的研究》（宜昌地质矿产研究所、海南省地矿局，1991，1992）更是对海南岛基础地质的一次系统研究，取得了许多成果认识，解决了一些基础地质问题；《海南岛结晶基底——抱板群层序与时代》（马大铨等，1997）、《海南省岩石地层［全国地层多重划分对比研究（46）］》（陈哲培等，1997）、《海南乐东抱伦金矿地质及矿产预测》（丁式江等，2007）则采用专题研究形式深化对海南岛地质的认识，其中许多观点主张被引用到本书中；《南海地质》（刘昭蜀等，2002）和《中国海域构造地质学》（张训华等，2008）为本次建立相对完整的海南省地层系列和开展地层区划、南海构造单元划分提供了重要资料的来源。公开发表的论文中，最早是围绕矿区地质开展研究工作，如陈国达（1977）、袁奎荣（1977，1980）、王赐银（1979）、严正（1979）、夏邦栋（1979，1990，1991）、贾精一（1985）、石桂华（1988）、冯连顺（1988）、吕古贤（1988）、侯威（1992）、涂绍雄（1993）、梁新权（1995）、何龙清（1996）等；之后，区域性基础地质研究兴起，涉及不同的领域，如杨树锋（1989）、张仲英（1989）、陈哲培（1985，1991，1998）、付建明（1991，1997）、张业明（1997，1998，2005）、马大铨（1997）、符国祥（1999）、姚华舟（1999）、龙文国（1999，2001，2002，2006）、李志宏（2000）、李献华（2000）、丘元禧（2000）、许德如（2000，2001，2007）、陈柏林（2001，2004）、徐义刚（2002）、刘昭蜀（2002）、云平（2002，2003，2005）、葛小月（2003）、李孙雄（2003，2004，2005，2006，2007）、曾庆銮（2003，2004）、傅杨荣（2005）、雷裕红（2005）、谢才富（2005，2006，2007）、陈新跃（2006）、陈沐龙（2008）、廖香俊（2008）等。

上述成果是本项目地质背景研究过程中不可多得的主要资料，对于海南省构造岩浆期次划分、成矿作用机制等方面的研究将带来有益的启发。但不可否认的是，相同的地质问题由于作者的研究时间不同以及样品测试方法的不一，并受到当时认识程度的制约，在对同一个地区同一地质问题开展研究时，不同人得出的结论往往不完全一致，有的大相径庭。一些作者的主张与当前的认识相差甚远，而未采用。

根据项目性质、特点，把项目划分为设计编写阶段（2006 年 1~11 月）、野外调查阶段（2006 年 12 月至 2010 年 4 月）、地质系列图编制阶段（2010 年 5~7 月）、野外验收和野外补课阶段（2010 年 8~10 月）、报告编写和成果验收阶段（2010 年 11~12 月）5 个阶段实施。野外调查阶段，主要是针对存在的地质问题进行野外补充调查工作。其中，野外观测点 628 个，采集岩矿鉴定样品 390 件，稀土分析、微量元素分析、硅酸盐分析样品各 88 件，重砂分析样品 12 件，电子探针样品 29 件、年龄测定样品 18 件。完成的实物工作量能满足任务书和项目设计要求。

编写人员分工如下：前言：李孙雄，第一篇区域地层：李孙雄、陈哲培、朱耀河、龙文国，第二篇侵入岩：云平、魏昌欣，第三篇火山岩及火山作用：林义华、官军，第四篇变质岩及变质作用：林义华、袁勤敏、莫位任，第五篇区域地质构造：周进波、李孙雄、胡在龙，第六篇地球物理地球化学特征：谢顺胜、蔡水库、吴小洁、杨奕，第七篇地质资源：李孙雄、云平、林义华、周进波、朱耀河。插图编制：王超、吴剑波、江宏伟、汪焰华、刘兵、李正廷，数据库建设：莫位任、曾雁玲、龚丹、岑如南，岩矿鉴定：吕嫦艳。英文翻译：陈方颖。

在项目运作过程中，严格遵照 ISO9001：2008 国际质量标准体系和《〈中国区域地质志〉工作指南》、《1：25 万区域地质调查技术要求》（暂行），以及其他有关规范、指南执行。海南省地质调查院对项目实施开展院、项目组、各专业小组三级质量管理，并成立了《中国区域地质志·海南志》编制委员会及专家咨询委员会，对项目进行全程技术指导。

此外，为提高项目成果质量，2011 年 1 月 9~15 日，全国地质志项目专家组（肖庆辉、陆松年、刘训、洪大卫、王乃文、葛肖虹、陈克强、丁孝忠、高林志等）莅临海南省开展指导工作，通过野外考察和调研，专家组提出了许多独到中肯的意见，对提升项目的地质认识水平起到了非常重要的作用。

2011 年 9 月 7~9 日，中国地质调查局组织专家（李廷栋、肖庆辉、翟刚毅、卢民杰、廖香俊、耿元生、刘训、伍光英、刘建雄、丁孝忠、姚建新、牛志军、赵小明、祁向雷、迟振卿）在海口市对成果报告进行了评审。结论是：首次系统总结和集成了海南省区域地质调查与研究的最新成果和进展，在地层系统的厘定与完善、构造单元划分与演化、岩浆构造系统等方面的研究取得了重要进展和新认识，为海南省国土资源管理、资源勘查、环境保护、区域经济发展及地球科学研究提供了重要基础支撑。评审委员会一致同意通过验收，并按照中国地质调查局成果报告质量等级评分标准，项目成果报告综合质量评级为优秀级。

第一篇　区域地层

第一章　区域地层区划

第一节　区域地层概述

海南省地层（含火山岩地层）最早可下溯至中元古代。经历了多旋回的沉积和多阶段的构造运动改造，地层多呈孤岛状残存于花岗岩中或以隐伏状分布于南海诸岛及海南岛周缘盆地中。中元古代以来，除缺失泥盆系（?）、侏罗系外，其他时代地层均有分布。元古宇主要出露于海南岛，西沙群岛和南沙群岛也有分布；古生界、中生界分布遍及海南岛和南海陆缘盆地；古近系—新近系基本上隐伏于新生代盆地中。

海南省以往地质工作以海南岛地质研究程度较高，而南海地区地层由于大面积淹没于海水之下，其研究程度相对较低。海南岛已有的正式岩石地层单位有59个，本次新增1个（丘糖岭组）。地层总出露面积约为12240km^2（约占海南岛面积的36%），前中生界多发生不同程度的区域变质，而中生界以来未见区域变质作用的改造。为进一步完善全省岩石地层单位及其序列，通过系统的资料收集，采用中国地质科学院地矿所三室（1976）、中国地质科学院王崇友（1979）、李云通等（1984）、刘昭蜀（2002）等人资料，引入利米内康组、宝滩组（原东坡组）、阳明组、忠孝组、南湾组、永暑组、南沙组、南海组、西沙组、宣德组、永兴组、永发组、琛航组、石岛组、西沙洲组等15个三叠纪以来的岩石地层单位。

在充分分析、综合研究前人地质资料（区调资料、科研报告）的基础上，结合本项目工作要求，从断代地层角度进行全省地层区划，按多重地层划分原则，对全省地层开展了划分和区域对比，建立了全省地层格架，为全面认识其沉积作用史提供系统的基础性地质资料。

第二节　区域地层区划

一、现有地层区划

海南省现有的地层区划主要是以海南岛古生代地质演化为主线而进行，属于单一阶段的地层区划方法。其地层区划以东西向九所-陵水构造带为界，划分出两个一级地层区：北面的华南地层大区和南面的南海地层大区。华南地层大区在划分出东南地层区（二级）的基础上，以东西向王五-文教构造带为界，划分出两个三级地层分区：南面的五指山地层分区和北面的雷琼地层分区，雷琼地层分区大致以琼州海峡断裂（?）为界划分出南面的海口地层小区。南海地层大区包括三亚地层区（二级?）和莺歌海盆地（三级?）。

二、本次地层区划方案

《〈中国区域地质志〉工作指南》（2009）明确要求地层区划采用多阶段的断代地层法进行。共开展了6个断代地层区划，即太古宙—古元古代地层区划、中—新元古代地层区划、古生代地层区划、侏罗纪—古近纪地层区划、新近纪地层区划、第四纪地层区划。

海南省的地层具有多阶段、多旋回的地质演化特点。中元古代是海南岛（甚至是全省）结晶基底形成的重要地质时期，之后，先后经历了新元古代—三叠纪板内多阶段裂解与造山作用、侏罗纪—白垩纪陆内盆-山演化和新生代裂解等重要地质演化阶段。参考《〈中国区域地质志〉工作指南》（2009）中

表 1-1 海南省断代地层区划及其岩石地层单位序列表

Table 1-1 Stratigraphic division and list of lithostratigraphic units of Hainan Province

第四纪阶段

地质年代			岩石地层						
纪	世		华南地层大区						中央海地层大区
			中南-东南地层区						
			南沙群岛地层分区	曾母地层分区	西沙群岛地层分区	莺歌海盆地地层分区	秀英地层分区		
第四纪	全新世		南海组	北康组	西沙洲组	(未建组)	烟墩组		(未建组)
							琼山组		
							万宁组	石山组	
	更新世	晚	南沙组		石岛组		八所组	道堂组	
		中			琛航组		北海组	多文组	
		早			永发组		秀英组		

古近纪—新近纪阶段

地质年代			岩石地层									
纪	世		菲律宾-南沙地层大区	青藏-南海地层大区								中央海盆地层大区
			南沙地层区	桂粤-南海地层区								
			礼乐滩盆地地层分区	曾母盆地地层分区	西沙地层分区	莺歌海地层分区	长昌地层分区		雷琼地层分区			
									海口地层小区			
新近纪	上新世		永暑组	北康组	永兴组	莺歌海组	望楼港组	石门村沟组	海口组			(未建组)
	中新世	晚	南湾组	南康组		黄流组	佛罗组	石马村组	灯楼角组			
		中		海宁组	宣德组	梅山组			长坡组	角尾组		
		早		曾母组	西沙组	三亚组	?			下洋组		
古近纪	渐新世		忠孝组			陵水组	丘糖岭组		涠洲组			
	始新世		阳明组	拉让群		不详	瓦窑组		流沙港组			
							长昌组					
	古新世		宝滩组				昌头组		长流组			

侏罗纪—白垩纪阶段

地质年代		岩石地层			
纪	世	西北-华南地层大区		菲律宾-南沙群岛地层大区	
		东南地层区		南沙群岛地层区	
		海南地层分区		礼乐滩-北巴拉望地层分区	曾母-万安地层分区
白垩纪	晚白垩世	报万组		(未建组)	丹脑组
	早白垩世	岭壳村组	鹿母湾组		
		汤他大岭组			
		六罗村组			
侏罗纪	晚侏罗世				
	中侏罗世				
	早侏罗世			(未建组)	

新元古代—三叠纪阶段				
纪	世	印支-南海地层大区		羌塘-扬子-华南地层大区
		三亚地层区	南海地层区	华南地层区
				五指山地层分区
三叠纪	晚三叠世		(未建组)	
	中三叠世		利米内康组	
	早三叠世			岭文组
二叠纪	晚二叠世		(未建组)	南龙组
	中二叠世			鹅顶组
	早二叠世			峨查组
石炭纪	晚石炭世			青天峡组
	早石炭世			南好组
泥盆纪				
志留纪	普里道利世			
	罗德洛世			
	文洛克世			
	兰多维列世			足赛岭组 靠亲山组 大干村组 空列村组 陀烈组
奥陶纪	晚奥陶世	干沟村组 尖岭组 榆红组 沙塘组		南碧沟组
	中奥陶世	牙花组		
	早奥陶世	大葵组		
寒武纪	芙蓉世			?
	武陵世	大茅组		
	黔东世	孟月岭组		
	滇东世			美子林组
震旦纪				石灰顶组
南华纪				
青白口纪				石碌群

中 元 古 代 阶 段		
纪	世	闽-粤东地层大区
		海南地层区
待建纪		
蓟县纪		
长城纪		峨文岭组
		戈枕村组

断代地层划分方案，结合海南省具体地质情况，把海南省地层区划阶段划分为中元古代、新元古代—三叠纪、侏罗纪—白垩纪、古近纪—新近纪、第四纪等5个阶段。尽管海南省新元古代—三叠纪阶段存在着多个不整合面，作为一个断代欠妥，但考虑本次工作情况，主要参照《〈中国区域地质志〉工作指南》（2009）进行，不再作进一步的划分。此外，海南岛三叠纪仅有早三叠世陆相磨拉石、类磨拉石建造沉积，与二叠系海相碎屑岩为角度不整合接触，作为一个断代的原因是：海南岛华力西期—印支期造山作用是一个连续的过程，三叠纪陆相沉积代表了造山晚期地壳隆升阶段的产物，三叠纪地质无疑是古生代地质演化的延续，因此，把古生代和三叠纪作为一个地质发展阶段是合理的。不同阶段地层区划及其岩石地层单位见表1-1。

（一）中元古代地层区划

在《〈中国区域地质志〉工作指南》（2009）中，海南省中元古代地层区划分为闽-粤东地层大区（一级）。但包含在地层大区中的广东、广西基底地层与海南岛基底地层并不完全相同（马大铨等，1998），并缺乏14亿年同造山期花岗岩的记录。因此，本项目以遂溪断裂（？）为界，划分出海南地层区。其范围涉及海南岛和南海广大地区，区内岩石地层单位为长城系戈枕村组、长城系—蓟县系峨文岭组。由于中元古代地层研究程度低，资料有限，因此海南地层区难以再进行三、四级地层区划分。

（二）新元古代—三叠纪地层区划

按照《〈中国区域地质志〉工作指南》（2009）中的划分方案，新元古代—三叠纪地层区划以东西向九所-陵水构造带为界，北属羌塘-扬子-华南地层大区华南地层区，南为印支-南海地层大区（？）三亚地层区。

华南地层区中，广东、广西地区岩石地层单位与海南岩石地层单位并不相同（陈哲培等，1997），沉积作用、变质作用、构造活动等方面也有明显的差异性。本次以遂溪断裂（？）为界，划分出南面五指山地层分区。其编图单位为青白口系石碌群、南华系—震旦系石灰顶组、寒武系滇东统美子林组、奥陶系南碧沟组和志留系兰多弗里统陀烈组、空列村组、大干村组、靠亲山组、足赛岭组，以及下石炭统南好组、上石炭统青天峡组、下二叠统峨查组、中二叠统鹅顶组、中—上二叠统南龙组和下三叠统岭文组。

印支-南海地层大区包括中南半岛和南海地区，地层由中寒武统（武陵统?）—三叠纪海相碎屑岩组成。其最大的特点是岩石未经区域变质作用的改造。刘昭蜀等（2002）资料显示，中南半岛分布有中寒武统（武陵统?）含三叶虫化石的石英质砂岩和砂质泥板岩，往上为泥盆系—下石炭统砂岩、灰岩夹页岩和上石炭统一套海陆交互相沉积，缺乏二叠纪沉积。南沙群岛上古生界二叠系零星分布，与元古宇不整合接触，岩石为灰岩和燧石、硅质碎屑岩、长石杂砂岩，其磁场强度（-70～-100nT）、层速度（5.4~5.7km/s）与北部海区基本一致。北巴拉望盆地中生代早期为中三叠统利米内康组深海相含放射虫燧石岩，与下伏二叠系和上覆早三叠世—晚侏罗世灰岩不整合接触。礼乐滩见早三叠世—晚侏罗世三角洲相-开阔浅海相砂泥岩及硅质页岩、粉砂岩分布。海南岛三亚地区则出露寒武系黔东统、黔东统—武陵统和奥陶系等一套滨—浅海相碎屑岩，缺乏志留纪—三叠纪的沉积记录。西沙群岛永兴岛钻孔资料揭示该区缺乏新元古代—三叠纪的沉积。由此显示，印支-南海地层大区沉积作用比较复杂，各地沉积作用并不相同，初步认为属于不同地层区的沉积。结合刘昭蜀等（2002）资料，本次以北西向红河断裂、南北向越东滨外断裂、北东向陆坡北缘断裂和广雅滩西断裂为界，划分出印支地层区、三亚地层区和南海地层区（图1-1）。

（三）侏罗纪—白垩纪地层区划

本阶段，海南省为西北-华南地层大区东南地层区。但据刘昭蜀等（2002）资料，菲律宾群岛（菲律宾板块）出露最老的地层为石炭系—二叠系，中生代沉积了一套含放射虫碎屑岩、灰岩。南海地区曾母盆地出露有上白垩统丹脑组蛇绿岩建造；西巴拉望盆地上白垩统—渐新统下部为燧石、细碧岩和基性、超基性侵入岩及变质岩、沉积岩，与下伏晚侏罗世—早白垩世浅海相沉积为不整合接触；

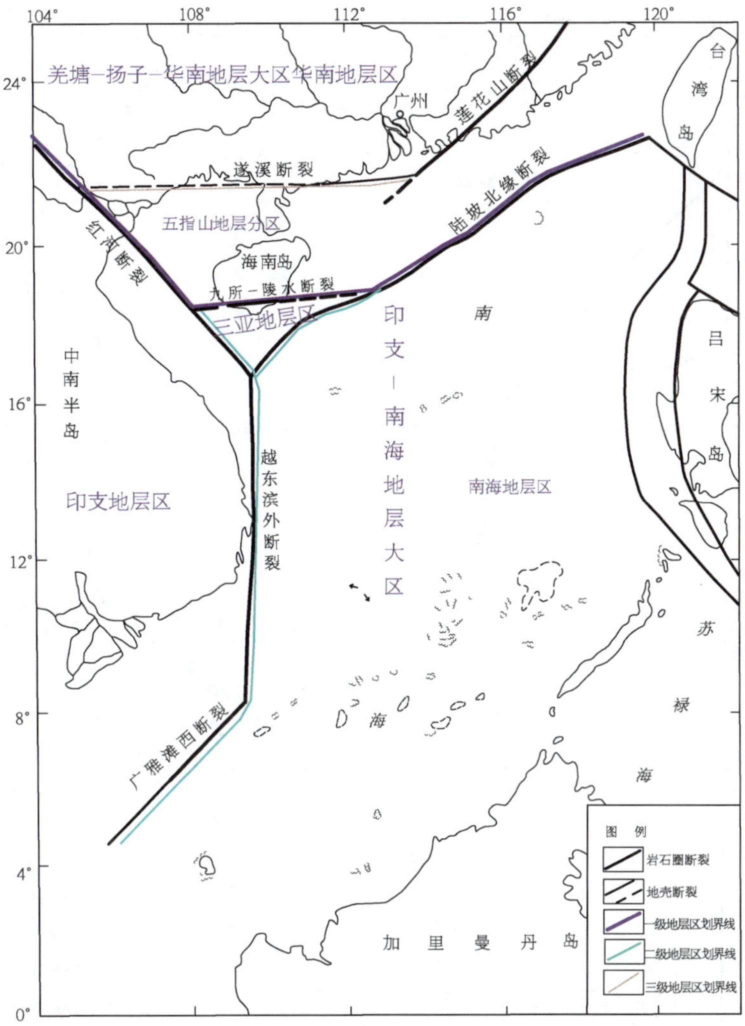

图 1-1　海南省新元古代—三叠纪地层区划图

Fig. 1-1　Neoproterozoic-Triassic stratigraphic division of Hainan Province

礼乐滩上三叠统至下侏罗统为三角洲相-开阔浅海相砂泥岩及硅质页岩、粉砂岩，与上覆早白垩世滨—浅海相夹蛇绿岩含煤砂页岩和砾岩呈不整合接触。而此时海南岛则为陆相盆地沉积。显示海南省南、北沉积明显不同。因此，应该属于两个不同的地层大区。本次以北东向南沙海槽北缘断裂和北西向万安滩西南断裂、广雅滩断裂为界，北为西北-华南地层大区东南地层区，南为菲律宾-南沙群岛地层大区（图 1-2）。

东南地层区中，海南与华南三叠纪的沉积作用明显不同，本次以遂溪断裂为界划分出海南地层分区。其岩石地层单位为下白垩统鹿母湾组、六罗村组、汤他大岭组、岭壳村组和上白垩统报万组。

菲律宾-南沙群岛地层大区以巴拉望南侧断裂、西布断裂为界划分出北面的南沙群岛地层区和南面的菲律宾地层区。南沙地层区进一步以广雅滩断裂和北西向万安滩西南断裂、广雅滩断裂为界，划分出礼乐滩-北巴拉望地层分区和曾母-万安地层分区。

（四）古近纪—新近纪地层区划

按照《〈中国区域地质志〉工作指南》（2009）划分方案，古近纪—新近纪地层为青藏-南海地层大区桂粤-南海地层区。

古近纪—新近纪时，随着南海扩张，中央海盆成型，整个南海被中央海盆分割，形成南北两个陆缘-陆坡相沉积。北面海区涉及海南省的盆地有雷琼断陷盆地、莺歌海断陷盆地、琼东南盆地、西沙海槽、中沙海槽等，以及分布于海南岛中的长昌盆地和长坡盆地等内陆盆地。南部海区以南沙裂陷最醒

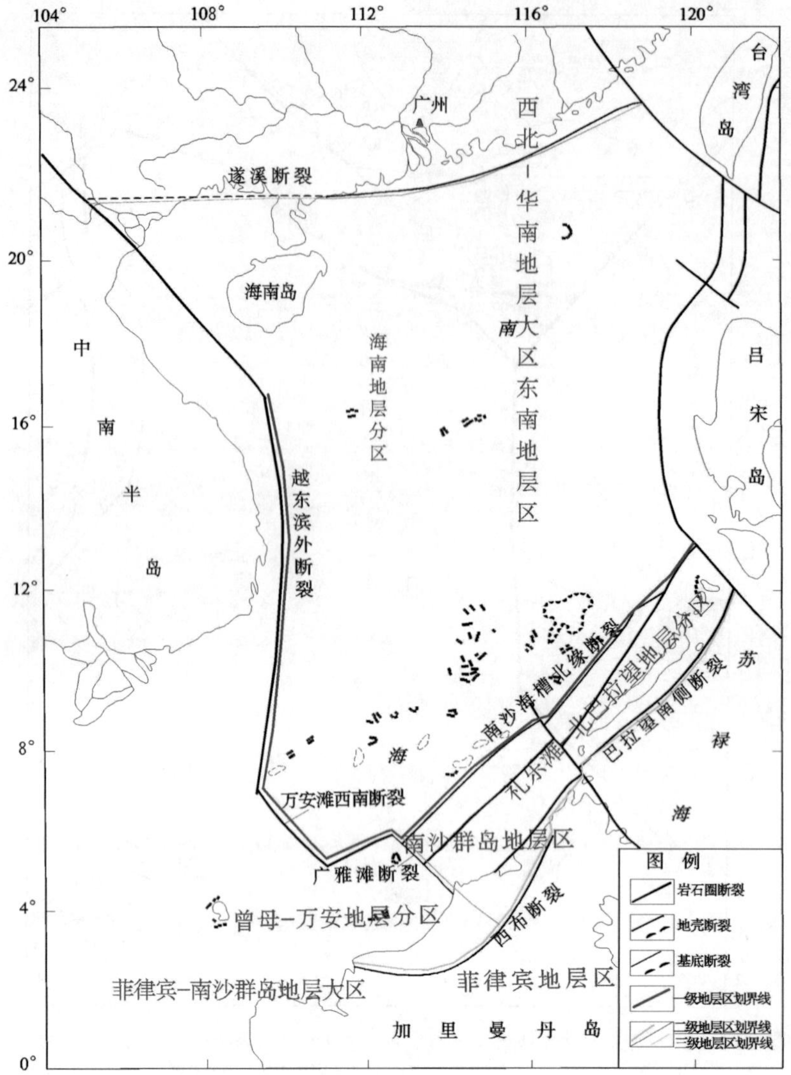

图 1-2 海南省中生代地层区划图

Fig. 1-2 Mesozoic stratigraphic division of Hainan Province

目，有礼乐滩盆地、西巴拉望盆地、北巴拉望盆地、曾母盆地、万安盆地等。由此，以中央海盆南、北缘断裂为界，划分出三个一级地层大区，即青藏-南海地层大区、中央海盆地层大区、菲律宾-南沙地层大区。

青藏-南海地层大区按照全国划分方案，海南岛和北部海区属于桂粤-南海地层区。进一步划分出雷琼地层分区、长昌地层分区、莺歌海地层分区、西沙地层分区。雷琼地层分区介于东西向遂溪断裂与王五-文教断裂之间；琼山地层分区位于东西向雷琼裂陷、北西向莺歌海裂陷和北东向琼东南裂陷间；莺歌海地层分区限于北西向莺歌海裂陷和北东向琼东南裂陷内；西沙地层分区则位于北西向莺歌海裂陷和北东向琼东南裂陷以南地区。

中央海盆地层大区由于资料有限，不再进行地层区划分。

菲律宾-南沙地层大区中，刘昭蜀等人（2002）资料显示，南沙群岛一带，地质作用变得更加复杂化。不同盆地沉积作用差异性明显，形成了不同的岩石组合或岩石地层单位；不同盆地构造运动作用时间也有所不同。但总体属于中生代盆地基础上的演化发展。依照中生代的地层区划，以巴拉望南侧断裂、西布断裂为界划分出北面的南沙地层区和南面的菲律宾地层区。南沙群岛地层区以北东向南沙海槽南缘断裂和北西向万安滩西南断裂、广雅滩断裂为界，划分出礼乐滩盆地地层分区和曾母盆地地层分区（图 1-3）。

（五）第四纪地层区划

第四纪时期，全省沉积作用格局与古近纪、新近纪基本一致。此时除了海南岛因沉积作用类似而不能进行南北地层分区外，其地层分区与古近纪、新近纪基本一致（图 1-4）。

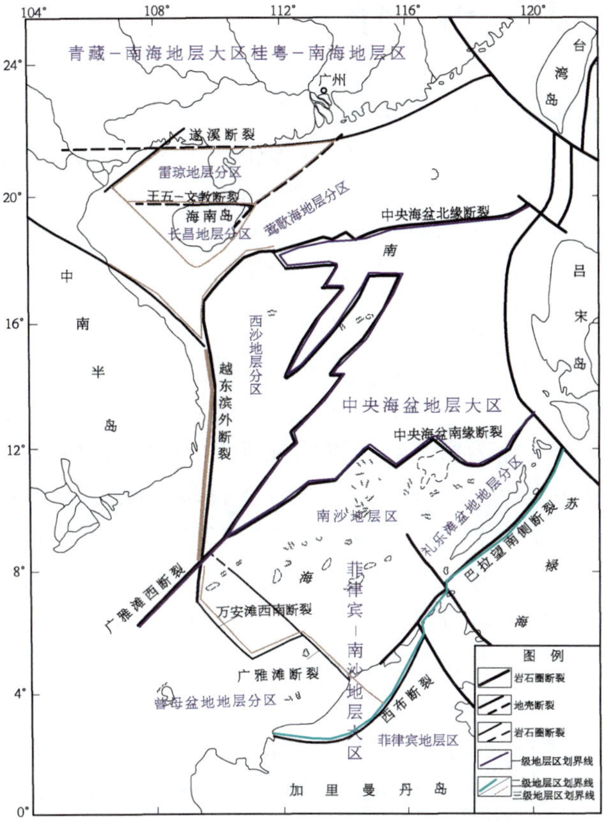

图 1-3　海南省古近纪—新近纪地层区划图

Fig. 1-3　Paleogene-Neogene stratigraphic division of Hainan Province

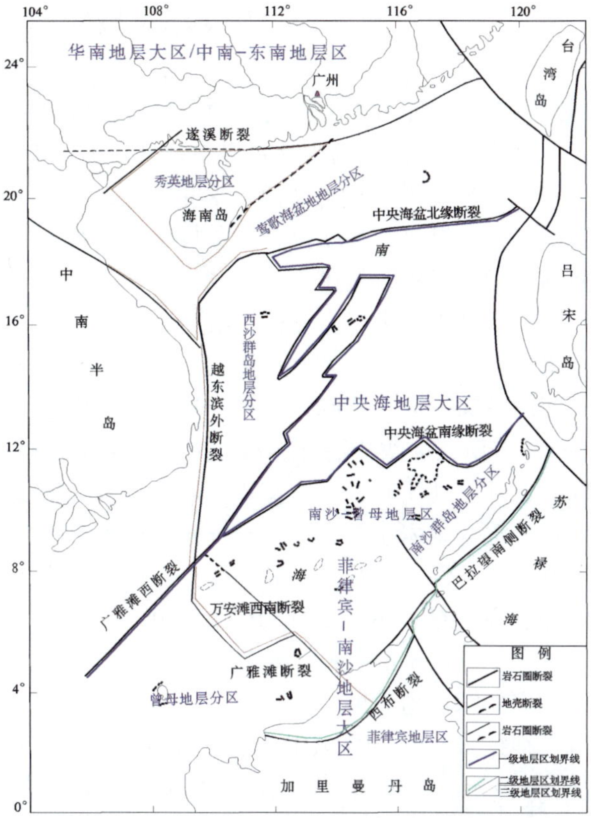

图 1-4　海南省第四纪地层区划图

Fig. 1-4　Quaternary stratigraphic division of Hainan Province

第二章 中元古界

海南省中元古界为深变质的表壳岩，目前主要见于海南岛和西沙群岛永兴岛中（南沙群岛也有中元古界分布）。海南岛中元古界由片麻岩类、混合岩类、石英岩类、片岩类岩石组成。西沙群岛永兴岛中元古界为花岗片麻岩（西永1井钻遇，下伏于新生界之下，矿物 Rb-Sr 等时线年龄为 1465Ma）。海南岛中元古界研究程度较高，资料比较丰富，而永兴岛中元古界资料少而零星。因此，本次仅能对海南岛中元古界进行叙述。

海南岛中元古界由戈枕村组和峨文岭组组成。戈枕村组和峨文岭组是从“抱板群”中划分出来的组级岩石地层单位。“抱板群”一词最早出现在赵鸿（1964）一篇短文中，把海南岛西部抱板地区一套变质较深的混合片麻岩命名为抱板群。广东区测大队（1964）将抱板、大蟹岭、冲卒岭、黄竹岭等4套深变质杂岩改称混合岩田，与陀烈群为渐变过渡，时代定为寒武纪。夏邦栋等（1979）在研究石碌铁矿外围地质时，将这套深变质岩系称为抱板群，沿用至今。1991年，马大铨等在琼中县上安乡又发现有混合杂岩出露，进一步查明了抱板群的空间分布。此后，马大铨等（1993，1997，1998）对抱板群进行了深入系统的研究，取得的成果较为丰富，进一步把抱板群划分为戈枕村组和峨文岭组，年龄分别为 1800～1600Ma 和 1600～1400Ma。这一成果认识被地学工作者广泛地引用。之后，随着1：5万和1：25万区调的进行，对分布于各地的抱板群进行了系统的剖面测制，为全面深入研究抱板群提供了翔实的基础地质资料。

根据《〈中国区域地质志〉编制项目“江南古陆”元古宙地层野外地质考察与专题讨论会会议纪要》，中元古代划分为待建纪（1400～1000Ma）、蓟县纪（1600～1400Ma）、长城纪（1800～1600Ma）。本次通过年代学资料测定，认为抱板群时代介于 1800Ma 至 1400Ma，结合地层的上、下叠覆关系，把戈枕村组和峨文岭组分别归入长城系和长城系—蓟县系（表2-1）。

表2-1 海南岛抱板群划分沿革表

Table 2-1 History of division of the Baoban Group，Hainan Island

<table>
<tr><th colspan="2">广东区测队（1964）</th><th colspan="2">夏邦栋等（1979）</th><th colspan="2">广东地矿局（1988）</th><th colspan="2">单惠珍（1990）</th><th colspan="2">张仁杰等（1991）</th><th colspan="3">马大铨、黄香定等（1993）</th><th colspan="3">陈哲培、钟盛中等（1997）</th><th colspan="3">1：25万乐东、陵水幅、琼海县幅、东方县幅（2002～2007）</th><th colspan="2">本书</th></tr>
<tr><td rowspan="2">寒武系</td><td rowspan="2">混合岩田</td><td rowspan="2">中奥陶统—寒武系</td><td rowspan="2">抱板群</td><td rowspan="2">震旦系</td><td rowspan="2">抱板群</td><td rowspan="2">中元古界</td><td rowspan="2">抱板群</td><td rowspan="2">蓟县系</td><td rowspan="2">抱板群</td><td rowspan="2">长城系</td><td rowspan="2">抱板群</td><td>峨文岭组</td><td rowspan="2">长城系</td><td rowspan="2">抱板群</td><td>峨文岭组</td><td rowspan="2">长城系</td><td rowspan="2">抱板群</td><td>峨文岭组</td><td>蓟县系—长城系</td><td>峨文岭组</td></tr>
<tr><td>戈枕村组</td><td>戈枕村组</td><td>戈枕村组</td><td>长城系</td><td>戈枕村组</td></tr>
</table>

第一节 岩石地层特征

海南省中元古界由戈枕村组和峨文岭组组成，属于海南岛的结晶基底地层。主要分布于琼西东方市抱板镇（广东区测队1964年所称的抱板混合岩田分布区）、乐东县峨文岭、琼中县上安乡、定安县黄竹岭及文昌市烟塘镇等地区，局部隐伏于琼北雷琼裂陷中（图2-1）。由于受到后期多期次的地质作用改造，导致抱板群呈“残留体状”产出，是控制海南岛金矿、石墨矿、伟晶岩型白云母矿产出的重要层位。

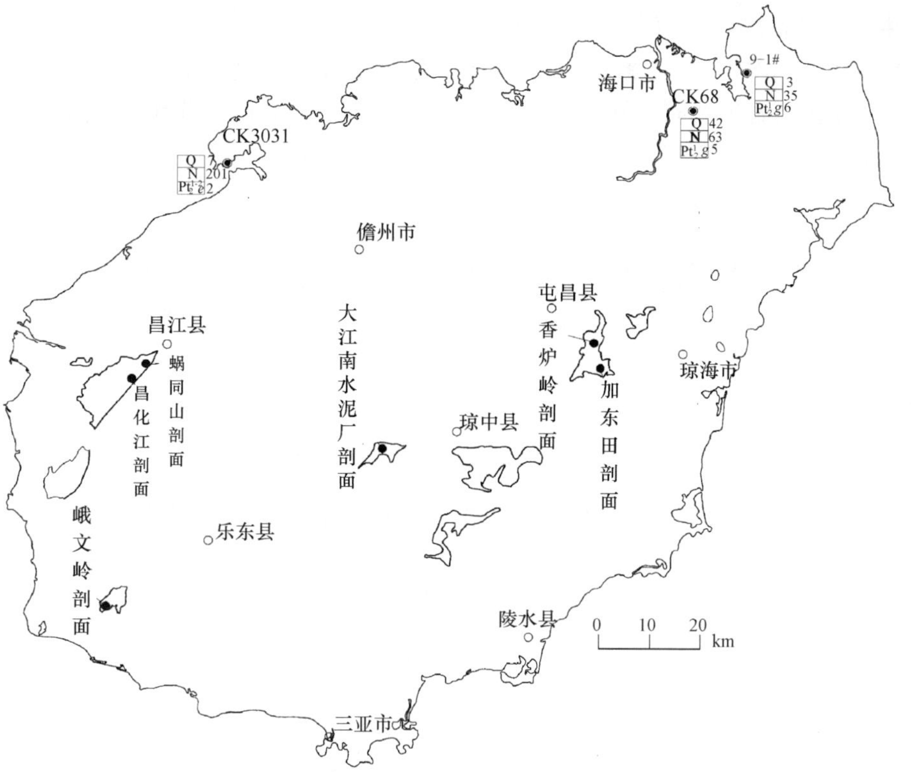

图2-1 海南岛中元古界分布及实测剖面位置简图

Fig. 2-1 Sketch map of Mesoproterozoic strata distribution and measured section positions of Hainan Island

一、剖面描述

（一）戈枕村组剖面

戈枕村组剖面主要有昌江县昌化江剖面、蜗同山剖面、屯昌县香炉岭剖面、乐东县峨文岭剖面等。主要剖面情况如下。

1. 昌江县昌化江剖面

该剖面为戈枕村组代表性剖面。系马大铨等人1991~1993年开展《海南岛结晶基底——抱板群层序与时代》（科研报告，1998）科研时测制。剖面位于昌江县叉河镇戈枕村东侧的昌化江畔，地理坐标：东经108°50′58″，北纬19°14′37″。1997年陈哲培等人在开展全国地层划分对比研究之《海南岛岩石地层》时，对该剖面进行了修订。据陈哲培等（1997），剖面自上而下为：

峨文岭组

7. 白云母石英片岩和石英白云片岩，偶见石榴子石变斑晶假象。夹有部分糜棱片岩、千糜岩 250m

6. 含石榴子石变斑晶的石英云母片岩夹云母石英片岩、变粒岩 170m

══════ 断　层 ══════

戈枕村组 总厚度2381m

5. 黑云母斜长片麻岩夹条纹状混合质黑云母斜长片麻岩 150m

4. 条纹状混合质黑云母斜长片麻岩 21m

3. 斑点状和部分条带状混合质角闪黑云斜长片麻岩夹混合质黑云母斜长片麻岩 230m

2. 条纹状混合质（角闪）黑云斜长片麻岩，偶见云母含量较高的二云母斜长片麻岩夹层 1480m

1. 混合花岗闪长岩及黑云母斜长质混合片麻岩 500m

未见底

2. 屯昌县中建农场香炉岭剖面

该剖面系海南省地调院 1997～2000 年开展 1：5 万《黎母岭、枫木市、琼中县、白马岭》幅区调测制，2001～2003 年海南省地调院在开展 1：25 万《琼海县》幅区调工作中补做了部分工作，主要是控制戈枕村组剖面。剖面起点位于香炉岭 68°方向 510m 处，地理坐标：东经 100°12′55″，北纬 19°1′55″；剖面长约 2294m；自上而下为（图 2-2）：

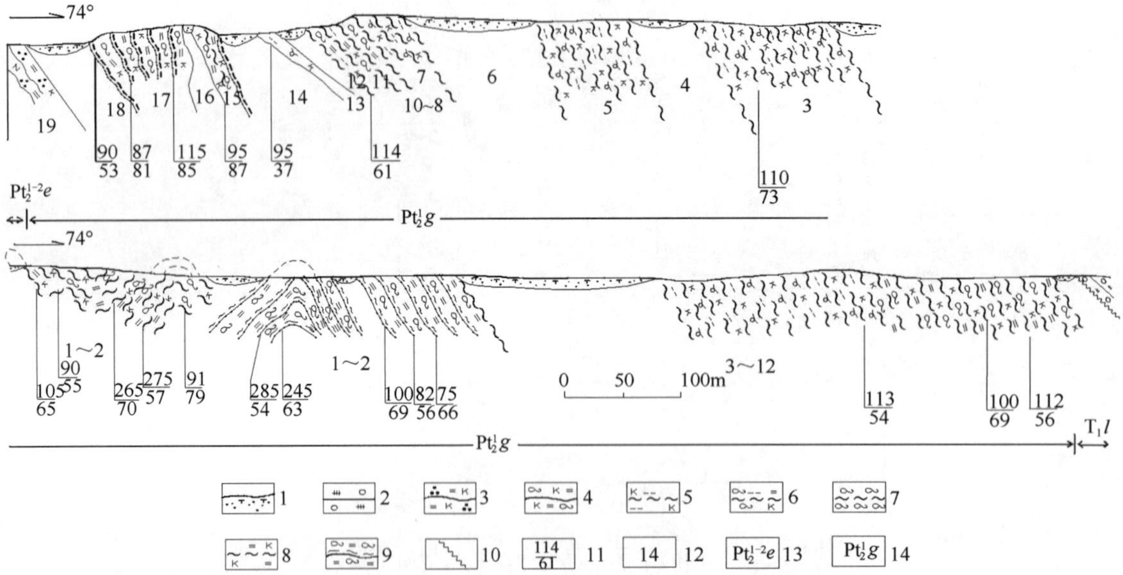

图 2-2 屯昌县中建农场香炉岭戈枕村组实测剖面图

Fig. 2-2 Measured section of the Gezhencun Formation, Xiangluling, China State Construction Farm, Tunchang County

1—浮土；2—复成分砾岩；3—长石白云母石英片岩；4—条带状白云母钾长片麻岩；5—长英质黑云母钾长（眼球状）混合岩；6—长英质二云母钾长条带状混合岩；7—条带状混合岩；8—长英质白云母钾长（条带状）混合岩；9—条带状白云母混合岩；10—不整合界线；11—产状；12—分层号；13—峨文岭组；14—戈枕村组

上覆地层 下三叠统岭文组复成分砾岩

～～～～～ 不整合 ～～～～～

峨文岭组（$Pt_2^{1-2}e$）

19. 浅灰白色长石白云母石英片岩 8.16m

18. 掩盖 51.42m

———— 推测为整合？————

戈枕村组（Pt_2^1g） 771.97m

17. 白云母钾长片麻岩 133.89m

16. 掩盖 46.72m

15. 白云母钾长片麻岩夹长英质黑云母钾长（眼球状）混合岩 27.59m

14. 掩盖 27.63m

13. 长英质黑云母钾长眼球状混合岩 5.87m

12. 掩盖 18.79m

11. 长英质二云母钾长条带状混合岩 14.86m

10. 掩盖 5.42m

9. 条带状混合岩 21.21m
8. 掩盖 5.17m
7. 长英质黑云母钾长眼球状混合岩 32.76m
6. 掩盖 106.70m
5. 长英质黑云钾长眼球状混合岩 85.73m
4. 掩盖 17.31m
3. 长英质黑云母钾长眼球状混合岩 50.15m
2. 糜棱岩化长英质白云母钾长条带状混合岩 55.48m
1. 白云母条带状混合岩 116.69m

未见底

（二）峨文岭组剖面

1. 乐东县冲坡镇峨文岭剖面

该剖面由海南地质大队1996～1998年在开展1∶5万《黄流市、千家》幅区调时测制。西端起于冲坡镇峨文岭290°方向1000m处，东端止于冲坡镇红墓林50°方向850m处；起点坐标：东经108°52′35″，北纬18°34′12″；导线总方向81°；剖面导线总长5558m；自上而下为（图2-3）：

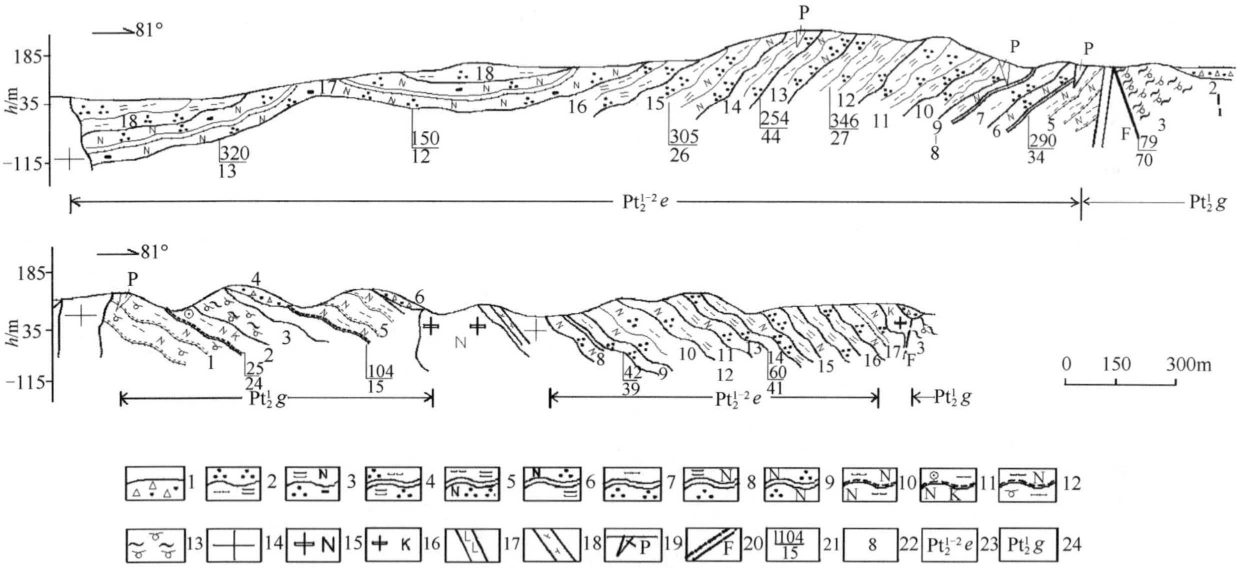

图2-3 乐东县峨文岭戈枕村组和峨文岭组实测剖面图

Fig. 2-3 Measured section of the Gezhencun and Ewenling Formations, Ewenling, Ledong County

1—浮土；2—二云母石英片岩；3—含石墨长石石英岩；4—石英二云母片岩；5—二云母长石石英片岩；6—长石石英二云母片岩；7—黑云母石英片岩；8—白云母长石石英岩；9—长石石英岩；10—黑云母斜长片麻岩；11—石榴黑云二长片麻岩；12—眼球状黑云母斜长片麻岩；13—眼球状混合岩；14—中粗粒钾长花岗岩；15—细粒斜长斑岩；16—细粒黑云母钾长花岗岩；17—煌斑岩脉；18—闪长玢岩脉；19—伟晶岩脉；20—推/实测断层；21—产状；22—分层号；23—峨文岭组；24—戈枕村组

未见顶

长城系抱板群峨文岭组（$Pt_2^{1-2}e$） 总厚度>681.26m

18. 二云母石英片岩 >52.14m
17. 含石墨长石石英岩 73.08m
16. 石英二云母片岩 68.30m
15. 二云母长石石英片岩 141.18m
14. 石英二云母片岩 53.69m
13. 二云母石英片岩 29.40m

12. 长石石英二云母片岩 107.09m

11. 二云母长石石英片岩 24.54m

10. 长石石英二云母片岩 57.35m

9. 黑云母长石石英片岩 22.67m

8. 长石石英岩 20.34m

7. 黑云母石英片岩 18.09m

6. 长石石英岩 13.38m

抱板群戈枕村组（Pt_2^1g） 总厚度>257.28m

5. 黑云母斜长片麻岩 112.00m

4. 掩盖 50.80m

3. 眼球状混合岩夹黑云母斜长片麻岩、浅粒岩 65.32m

2. 石榴子石黑云斜长片麻岩 7.19m

1. 眼球状黑云母斜长混合岩夹眼球状黑云母斜长片麻岩 >21.94m

~~~~~~~~ 侵入接触 ~~~~~~~~

下伏地质体 中粗粒斑状黑云母钾长花岗岩

2. 琼海市南俸农场加东田剖面

剖面系海南省地调院 1997~2000 年进行 1∶5 万《黎母岭、枫木市、琼中县、白马岭》幅区调测制，2001~2003 年海南省地调院进行 1∶25 万《琼海县》幅区调工作中补做了部分工作，主要是控制峨文岭组。剖面起点位于南俸育种站一区南西 216.8m 山包 270°方向 610m 处，地理坐标：东经 110°13′11″，北纬 19°11′35″；剖面长约 4053m；自上而下为（图 2-4）：

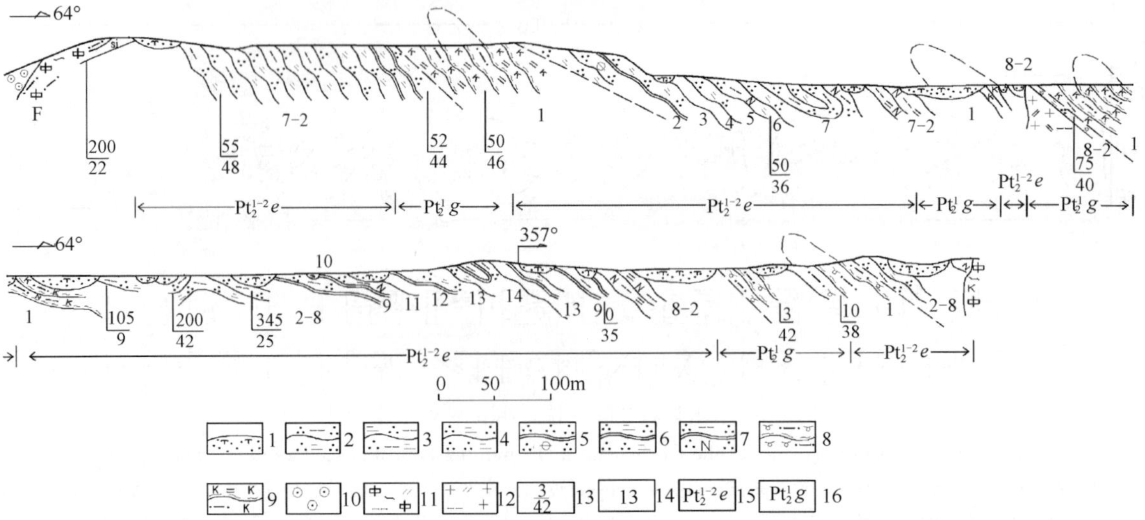

图 2-4 琼海市南俸农场加东田峨文岭组实测剖面图

Fig. 2-4 Measured section of the Ewenling Formation, Jiadongtian, Nanfeng Farm, Qionghai City

1—浮土；2—二云母石英片岩；3—石英二云母片岩；4—白云母石英片岩；5—细粒含石榴子石黝帘石石英岩；6—细粒绢云母石英岩；7—细粒黑云母长石石英岩；8—糜棱岩化混合岩化眼球状片麻岩；9—糜棱岩化白云母钾长片麻岩；10—碎裂岩；11—糜棱岩化片麻状斑状黑云母二长花岗岩；12—细粒黑云母二长花岗岩；13—产状；14—分层号；15—峨文岭组；16—戈枕村组

峨文岭组（$Pt_2^{1-2}e$） 309.1m

14. 浅灰白色细粒石英岩与浅灰白色石英二云母片岩互层 22.88m

13. 浅灰白色细粒（绢云母）石英岩 50.08m

12. 掩盖 26.71m

11. 浅灰白色细粒黑云母石英岩 5.04m
~~~~~~~~

10.	掩盖	5.38m
9.	浅灰白色细粒黑云母石英岩	7.38m
8.	白云母石英片岩	16.50m
7.	石榴子石二云母石英片岩，石榴子石含量10%，主要为铁铝榴石，大小1~3mm	42.46m
6.	浅灰白色含长石二云母石英片岩	13.47m
5.	浅灰白色白云母石英片岩	24.30m
4.	掩盖	15.51m
3.	浅白色白云母石英片岩	14.45m
2.	浅灰色长石白云母石英片岩，其中局部夹浅灰白色细粒含黝帘石石英岩	65.39m

——————— 推测为整合？———————

戈枕村组（Pt_2^1g）

1.	糜棱岩化、(眼球状）片麻状白云母钾长片麻岩、黑云母钾长片麻岩	71.38m

未见底

二、岩石地层特征

（一）戈枕村组

由马大铨等（1993）创名，黄香定等（Huang Xiangding et al.，1996）介绍。陈哲培等（1997）指定昌江县昌化江剖面为戈枕村组代表性剖面，乐东县峨文岭剖面为主要参考性剖面。剖面上部以黑云母斜长片麻岩或混合质黑云母斜长片麻岩为主，下部为混合花岗闪长岩及黑云母斜长混合片麻岩；未见底；厚度大于2381m；与上覆峨文岭组呈整合接触；主要分布于乐东县峨文岭、东方市抱板、大蟹岭及定安县黄竹岭、琼中县上安乡等地区。整体为一套经过混合岩化作用改造的深变质岩，以混合岩和片麻岩组合为特征，靠近边部并往往出现糜棱岩化现象。混合岩的脉体呈条带状、眼球状，眼球体大小均一、分布均匀、排列方向有序，与所夹的片麻岩呈渐变过渡关系，有时可见片麻岩与混合岩无规律叠置，片麻岩并见混合岩化现象（图版Ⅰ-1-A、图版Ⅰ-1-B、图版Ⅰ-1-C、图版Ⅰ-1-D）。

（二）峨文岭组

由马大铨等（1993）创名，黄香定等（Huang Xiangding et al.，1996）介绍。陈哲培等（1997）指定峨文岭剖面为峨文岭组代表性剖面，整合于戈枕村组之上，以云母石英片岩、石英云母片岩为主，夹数层长石石英岩及石墨矿层；厚度大于1181m；与上覆奥陶系南碧沟组呈断层接触；空间分布常与戈枕村组相随。主要见于乐东县峨文岭、东方县大蟹岭、抱板镇和琼中县上安乡、定安县黄竹岭等地区。

第二节　地层划分与对比

一、戈枕村组

琼西昌化江剖面为戈枕村组代表性剖面，戈枕村组下部为混合花岗闪长岩及黑云斜长质混合片麻岩，上部为条纹状黑云母斜长片麻岩或混合质黑云母斜长片麻岩，局部见含石榴子石二云母斜长片麻岩分布。未见底，与上覆峨文岭组呈整合接触。厚度大于2381m。

琼南乐东峨文岭剖面，戈枕村组上部为黑云斜长片麻岩夹混合岩化片麻岩，下部为眼球状混合岩夹混合岩化片麻岩，二者间呈渐变过渡关系，无明显界线。岩石有：黑云母斜长片麻岩、黑云母石英片麻岩、黑云母钾长片麻岩、含石榴子石黑云母二长片麻岩、混合岩化片麻岩、混合岩化黑云母二长片麻岩、眼球状混合岩、眼球状斜长混合岩、眼球状黑云母斜长混合岩以及少量浅粒岩。未见底，顶部与峨文岭组呈整合接触。厚度大于1807m。

琼中屯昌中建农场香炉岭剖面，戈枕村组下部为条带状混合岩，中部为长英质黑云母钾长眼球状混合岩，上部为条带状白云母钾长片麻岩夹长英质黑云母钾长（眼球状）混合岩；未见底，推测与上覆峨文岭组为整合接触。厚度大于771m。岩石为长英质白云母钾长（条带状）混合岩、长英质黑云母钾长眼球状混合岩、长英质二云母钾长条带状混合岩、条带状白云母钾长片麻岩、条带状白云母片麻岩。眼球状混合岩中眼球含量变化于1%～20%，多呈透镜状，定向排列，部分眼球向两端尖灭渐变为脉体。局部地带混合岩受糜棱岩化作用，形成糜棱岩化混合岩。

总体来看，海南岛各地出露的戈枕村组岩石组合特征基本可以对比，以片麻岩和混合岩组合为特征。但在岩石构造和成分上略有一定的差异性。琼南乐东峨文岭剖面，混合岩主要是眼球状构造，而琼中和琼西混合岩则多为条带状-条纹状构造；琼中屯昌中建农场香炉岭剖面片麻岩、混合岩钾长石含量高，为钾长混合岩或钾长片麻岩，而琼西和琼南则为斜长片麻岩-二长片麻岩，可能与混合岩化作用有关。

二、峨文岭组

琼南乐东县冲坡镇峨文岭剖面为峨文岭组代表性剖面，峨文岭组下部为长石石英岩，夹黑云母石英片岩；中部为长石云母片岩和长石石英片岩；上部为含石墨长石石英岩、云母片岩。未见顶，厚度大于1181m。整合于戈枕村组黑云母斜长片麻岩之上。

琼中琼海市南俸农场加东田剖面，峨文岭组下部为石英片岩，夹石英岩；中部为细粒石英岩；上部为浅灰白色细粒石英岩与浅灰白色石英二云母片岩互层。推测与下伏戈枕村组为整合接触，厚约391m。

整体来看，各地出露的峨文岭组岩石组合特征基本可以对比，主要以石英岩和片岩组合为特征，但其岩性空间横向变化明显。峨文岭剖面岩石普遍见长石矿物，并含多层石墨矿，而其余地方都没有石墨矿分布，基本上也没有见长石矿物，可能与沉积作用差异性及后期变质温度差异性有关。

第三节　年代地层探讨

自从夏邦栋（1979）首次公开定义抱板群以来，对其时代归属及区域对比始终争论不休。夏邦栋、任震鹏在创立抱板群时认为它是该区最古老的地层，时代定为寒武纪—中奥陶世，但缺乏古生物化石及同位素年龄依据。按夏邦栋、任震鹏创立抱板群的含义，其主要岩石组合类型为各种混合岩、片岩、角闪岩及片麻状混合岩、片麻岩、千枚岩、绢云母石英粉砂岩等（夏邦栋等，1979）。

重新定义后的戈枕村组和峨文岭组岩石变质较深，已达低—高角闪岩相程度。由于目前缺乏古生物化石资料，其时代确定主要（或唯一）是根据同位素年龄数据。但因采样岩性、地点及实验方法等的差异，导致前人于不同时期获得的年代学数据存在一定的差异。

吴学益等（1991）在戈枕村组片麻岩中获得Pb-Pb法等时线年龄为1751Ma。

俞受鋆等（1992）在昌江县叉河农场十七队和十八队花岗岩中采样，获得锆石U-Pb一致曲线年龄为1441Ma。

马大铨等（1993，1997，1998）在戈枕村组混合岩化黑云母斜长片麻岩中选用锆石作U-Pb一致曲线，获得上交点年龄为（1824±77）Ma，认为代表原岩锆石结晶年龄或火山岩形成年龄；在峨文岭地区戈枕村组混合岩脉体中获得Rb-Sr等时线年龄为（1637±31）Ma（相关系数0.9994），认为代表混合岩化年龄；在东方县二甲村东约300m水沟侵入峨文岭组的片麻状二长花岗岩中获锆石U-Pb一致曲线上交点年龄为1456.8Ma，下交点年龄262.8Ma，主张前一年龄即锆石的结晶年龄，代表二长花岗岩的形成年龄，后一年龄值则表示锆石在该年龄前经受了一次变质作用的铅丢失。用同一样品同时作蒸发法测定$^{207}Pb/^{206}Pb$年龄，结果为（1407±20）Ma，与一致曲线年龄误差范围一致。

王可伏等（1992）在土外山与云母片岩互层产出的斜长角闪（片）岩（火山岩）中获得Sm-Nd等时线年龄为1693.97Ma。侯威等（1992）、梁新权（1995）在斜长角闪（片）岩也分别获得Sm-Nd等时线年龄为（1699.4±3）Ma、（1700.9±20）Ma。而谭忠福等（1991）、马大铨等（1998）分别获得斜长角闪（片）岩的Sm-Nd等时线年龄为（975±8.6）Ma、（955±54）Ma。野外观测显示，该类岩石至少存在两种产出形式：顺层状和侵入状。顺层状产出的斜长角闪（片）岩与混合岩片麻理产状协调一致，并发生香肠化拉断变形；侵入状斜长角闪（片）岩产状与围岩（片麻岩、片岩）不协调（图版Ⅰ-1-E、图版Ⅰ-1-F）。因此，本项目研究认为上述年龄是不同时期形成的斜长角闪（片）岩的体现。

丁式江（1995）、龙文国等（2001）于片麻状花岗岩中获锆石蒸发法$^{207}Pb/^{206}Pb$年龄，获数据的峰值为（1407±20）Ma。

李正祥等（2002）于东方红泉农场片麻状花岗岩中采样，获锆石 SHRIMP 年龄 1430Ma，同时获得锆石核（core）年龄为 1770Ma，分别代表古侵入体的结晶年龄和花岗岩源岩的年龄（也许为戈枕村组片麻岩的形成年龄）。

综合上述，马大铨等（1993，1997，1998）以混合岩化事件和片麻状花岗岩侵入事件作为划分戈枕村组和峨文岭组的依据，认为戈枕村组形成年龄为（1800±100）~（1637±31）Ma。根据峨文岭组没有发生混合岩化作用，并被约 1400Ma 花岗岩侵入，其年龄应在（1420±36）~（1637±31）Ma 之间。这种认识一直沿用至今。但混合岩化事件的动力学背景、花岗岩侵位的变质变形作用效应以及戈枕村组与峨文岭组接触关系、海南省构造阶段划分等系列问题一直悬而未决。

本项目组在红林农场 19 队旁昌化江中戈枕村组混合岩化片麻岩中获得其岩浆锆石 SHRIMP 年龄为 1437Ma，于老烈村南部昌化江中片麻状花岗岩中获得岩浆锆石 SHRIMP 年龄为（1439±7）Ma（图 2-5），可分别代表混合岩化事件年龄和花岗岩的形成年龄。全国项目组高林志等人在乐东红五亚南甫沟眼球状混合岩中获得锆石 SHRIMP 年龄为（1438±5.3）Ma。此外，胡健民也曾在东方二甲一带斜长片麻岩中也获得锆石 SHRIMP 年龄为（1422±16）Ma（内部资料）。因此戈枕村组年龄应介于 1430Ma 至（1800±100）Ma 之间。峨文岭组自然叠覆于戈枕村组之上，而被片麻状花岗岩侵入，其形成时代应晚于戈枕村组的形成年龄，而大于片麻状花岗岩的形成年龄（1430Ma）。野外地质特征、岩石学、岩相学、岩石地球化学及同位素地球化学研究表明：①抱板杂岩中的长英质片麻岩类按其成因可划分为混合岩化片麻岩及片麻状花岗岩，二者的外貌具有某些相似性，但前者中普遍含有石榴子石，系表壳岩变质作用产物（原岩为中酸性火山岩），后者系深成作用而成的古侵入体；②混合岩化作用形成时代为（1437±7）Ma，花岗岩体侵位时代为（1439±7）Ma，花岗岩化与混合岩化作用的时间大致相同，表明二者属同一构造热事件的产物，两者片麻理协调一致（图 2-6）；③混合岩与花岗岩时间与空间上的紧密相关性表明混合岩形成于上地壳的下部，其形成（部分熔融）与花岗岩浆于下地壳形成后侵位于上地壳中而带来的热量与流体有关；④尽管片麻状花岗岩侵入峨文岭组，但片麻状花岗岩片麻理与围岩新生面理方向基本协调一致（图 2-7），也显示出区域变质与片麻状花岗岩侵位属于同一次构造-热事件的产物。

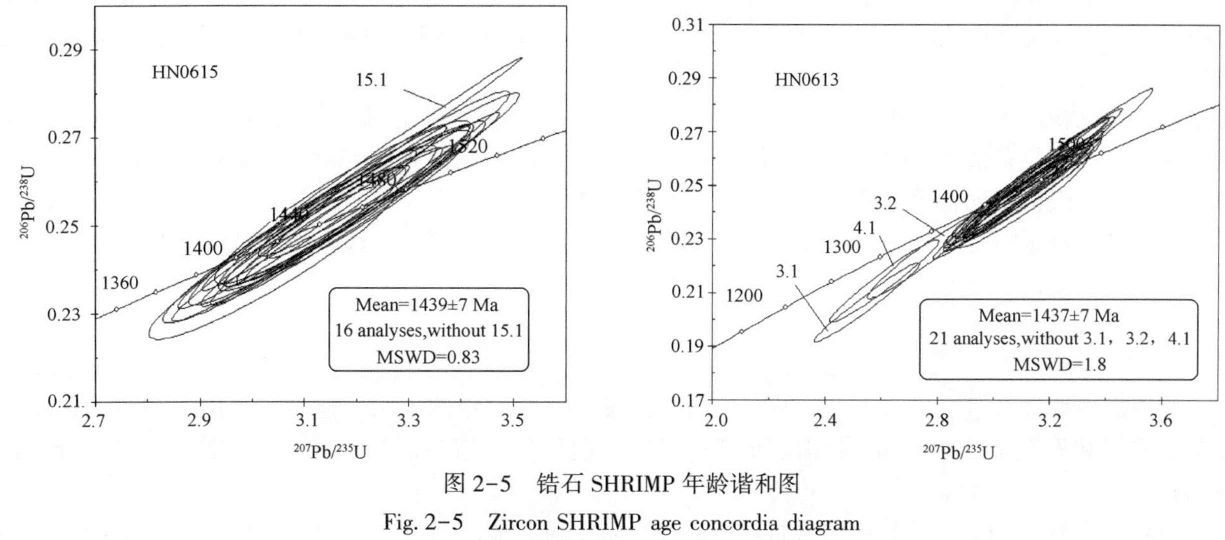

图 2-5 锆石 SHRIMP 年龄谐和图

Fig. 2-5 Zircon SHRIMP age concordia diagram

HN0615—片麻状花岗岩；HN0613—混合片麻岩

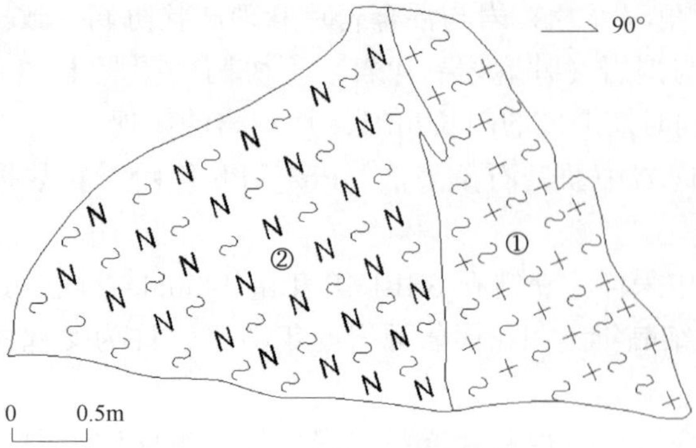

图 2-6 片麻状二长花岗岩侵入混合花岗闪长岩素描图

Fig. 2-6 Sketch of gneissic monzonitic granite and intrusive magmatized granodiorite

（据马大铨等，1998）

①混合花岗闪长岩；②片麻状二长花岗岩

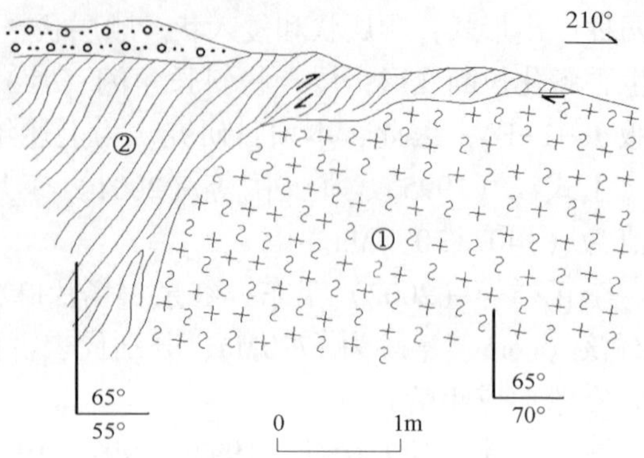

图 2-7 片麻状二长花岗岩侵入混合花岗闪长岩素描图

Fig. 2-7 Sketch of gneissic monzonitic granite and intrusive magmatized granodiorite

（据马大铨等，1998）

①片麻状二长花岗岩；②片岩

值得一提的是，叶伯丹等（1990）曾在东方二甲金矿区 ZK3601 孔 178.4m 采混合岩岩浆样品（岩石经马大铨等 1993 年检查，认为是片麻状二长花岗岩），获锆石 U-Pb 一致曲线年龄为（1364±26）Ma，与本次获得的混合岩化片麻岩岩浆锆石 SHRIMP 年龄（1437Ma）接近。

由此看来，混合岩化事件与花岗岩侵位事件应该密切有关。但由于样品数量有限，且采样位置（琼西戈枕）与马大铨等（1993，1998）混合岩化年龄采样位置（琼西南红五）并不一致，因此，不能由此而否定马大铨等（1993，1998）的认识（即混合岩化事件发生事件约为 1600Ma）。本次结合戈枕村组、峨文岭组空间上紧密相随和峨文岭组自然叠覆于戈枕村组之上的事实，暂时把戈枕村组、峨文岭组时代分别确定为长城纪（Pt_2^1）和长城纪—蓟县纪（Pt_2^{1-2}）。

第四节 岩相古地理特征

由于岩石经历了区域角闪岩相变质作用和和深层次变形作用的改造，原生岩石结构构造基本上已变得面目全非，尤其是缺乏古生物资料，从而给沉积环境恢复和岩相古地理恢复带来了极大的困难。

目前认为海南岛在 1400Ma 存在一次俯冲-碰撞事件（许德如等，2002），而中国境内其他地区，尤其是华南地区元古宙造山带形成时间集中于 1.0~0.8Ga（李江海等，1999；Li X.-H，1999），显示海南岛与中国其他地区造山作用时间明显不同。Li Z. X et al.（1995，1996，1999，2002）等认为海南岛是连接南极洲、澳大利亚东部、劳伦大陆西部的纽带，位于 R dinian 超级大陆的核心位置。但其与劳伦大陆在沉积作用、岩浆作用间依然具有一定的差异性（陆松年老师讲课资料）。因此，海南岛中元古代大地构造位置依然未决，直接制约了其沉积作用、火山作用及其地质演化的综合认识。

本次研究仅根据现有的资料，结合前人的科研成果，利用稀土元素、主量元素、微量元素和同位素特征对其进行粗略的浅析和简单的讨论，目的是为认识海南省中元古代大地构造属性以及中元古代块体演化特征提供有益的资料。

一、戈枕村组

琼西抱板戈枕村组原岩主体为钙碱性火山岩，属于一套中酸性火山岩建造。火山岩成分大体相当于英安质岩和石英安山岩（熔岩和凝灰岩），具有现代岛弧或陆缘弧火山岩系的特征（马大铨等，1998）。

琼东（中建农场）戈枕村组中岩石投影点在 Niggli 四面体图解中部分落入火成岩区，部分落入黏土质沉积岩区；在 Mg-K 图解中，除了一个样品落入富钾火山岩区外，其余 6 件样品均分布在黏土岩区或该岩区外侧；在（al-alk）-c 图解中，有 3 件样品落入英安质凝灰岩区，2 件样品落入黏土岩区，2 件样品落入钙质泥灰岩区；在 TiO_2-SiO_2 图解中，投影点都分布在沉积岩区；在 F A M图解中，投影点均落在钙碱性系列区。其不同岩石（不含斜长角闪片岩类岩石）稀土元素、过渡族元素及大离子亲石元素比值蛛网配分曲线变化趋势相当一致，说明各火山岩形成构造环境可能是相同的。所有样品的化学成分在里特曼-戈蒂里图解中的投影点均落入造山带火山岩区。在 FeO_t-MgO-Al_2O_3 图解中的投影点主要落入扩张中心岛屿与造山带之间。此外，利用某些微量元素所构成的构造环境判别图中，在 Hf-Th-Ta 图解和 Hf-Th-Nb 图解中的投影点主要落入岛弧的构造背景区，而在 Yb+Ta-Rb、Y+Nb-Rb、Ta-Yb 和 Y-Nb 等图解中的投影点主要落入板内、火山弧和同碰撞带的构造背景区，而且大离子亲石元素比值蛛网图总体上与造山带正常大陆弧花岗岩具有相似的标准化组成模式。由此反映戈枕村组的原岩性质较为复杂，但总体可能是以中酸性火山岩为主，局部夹有杂砂岩类以及泥岩等。形成环境为活动大陆边缘或岛弧构造环境（详见第四篇变质岩）。

琼中上安及琼南峨文岭、大蟹岭等地，戈枕村组中上部主要由混合岩化黑云（角闪）斜长片麻岩、黑云（角闪）斜长变粒岩、斜长角闪（片麻）岩组成；下部为紫苏（二辉）斜长片麻岩、麻粒岩。根据岩石组合关系、野外产出特征，结合岩石化学方法判别，片麻岩、变粒岩的原岩主要为中酸性火山岩，斜长角闪（片麻）岩和麻粒岩的原岩主要为中基性火山岩（其岩相和地球化学特征详见第四篇变质岩）。黑云（角闪）斜长片麻岩和黑云（角闪）斜长变粒岩的主要氧化物含量为：SiO_2 58.94%～75.39%，TiO_2 0.19%～1.30%，$Al_2O_3$12.5%～16.8%，MgO 0.36%～3.75%，FeO 0.57%～6.86%，CaO 0.79%～5.13%，Na_2O 2.04%～4.32%，K_2O 1.63%～5.03%；成分上大多类似于英安岩，少数相当于流纹岩，个别属于安山岩。斜长角闪（片麻）岩和麻粒岩的主要氧化物含量为：SiO_2 49.83%～55.88%，TiO_2 0.79%～1.87%，Al_2O_3 12.57%～17.18%（多大于 15%），FeO 6.50%～13.05%（多小于 9%），MgO 3.91%～9.21%，CaO 7.06%～9.88%，Na_2O 0.50%～3.76%，K_2O 0.68%～2.56%；成分上类似于玄武岩或玄武安山岩。多数样品属钙碱性系列，总体上构成单峰式的高钾钙碱性玄武岩-玄武安山岩-英安岩组合，初步判断其形成于岛弧环境。值得注意的是，产于琼南峨文岭、大蟹岭的斜长角闪（片麻）岩和少数片麻岩、变粒岩显示拉斑系列趋势，相对于琼中长征地区的样品来说，在 SiO_2 含量相近时，其 Na_2O+K_2O 含量较低，而 MgO、FeO 和 CaO 含量较高。由西往东的这种成分变化可能显示了自削减带向大陆内部的环境转变，即琼西位于大洋一侧，而琼东靠近大陆。许德如（2003）认为产于琼西抱板一带中元古界和中元古代花岗岩中的包体属于英云闪长岩-花岗闪长岩组合，具有类 TTG 岩石组合（英云闪长岩-奥长花岗岩-花岗闪长岩）特征。从另一个角度可能支持上述认识。

二、峨文岭组

全岛各地分布的峨文岭组原岩基本相似，主要为片岩和石英岩组合，局部夹有火山岩（斜长角闪片岩）或钙质岩等。岩石变质结晶较好，变余结构构造罕见，仅在个别地段还能见到由石英岩→石英片岩→云母片岩组成小型韵律层理。其副矿物磷灰石呈细微浑圆状，锆石呈浑圆粒状，电气石（含硼）呈微粒状。片岩投影点在 Niggli 四面体图解中落入泥质沉积岩区，原岩可能为泥岩、砂质泥岩、砂岩，片岩常分布有自形—半自形的含硼电气石，少量岩石并见石墨分布，表明原岩形成于海相还原环境。

许德如（2002）研究认为，峨文岭组具有大陆岛弧和安第斯型陆缘沉积特点。白云母石英片岩类岩石中稀土总含量低（85.71μg/g），Sm/Nd 比值高（0.25）；亲铁、亲硫元素和高场强元素含量低，亲石元素偏高，具有陆缘沉积的特点。石英二云母片岩类岩石稀土总量高（209.0μg/g），Sm/Nd 比值低（0.205），LREE/HREE 与 La/Yb 比值高，亲铁、亲硫元素和高场强元素含量高，亲石元素低，具有大洋岛弧沉积的特点。两者稀土分布模式基本相似，属于重稀土亏损，轻稀土富集，是在统一的扩张弧后（或弧间）盆地沉积的产物，属于陆相向海相过渡的古沉积环境。

另外，根据前人研究结果（马大铨等，1993，1998），呈夹层状产出的斜长角闪岩的原岩为基性火成岩，部分为火山岩。梁新权（1995）并获得斜长角闪岩 Sm-Nd 年龄为 1700Ma。该年龄小于戈枕村组年龄（1824±77）Ma，而大于混合事件年龄（1637±31）Ma。在乐东县红五公路边，可见斜长角闪片岩呈拉断状顺层产于戈枕村组眼球状混合岩中，与围岩发生同步变形。上述年龄与野外观测情况基本吻合。该类岩石在 Pl-Px-Ol（标准分子）辉长岩分类图解中，其投影点落在辉长苏长岩区；在基性超基性侵入岩的 $FeO^*-MgO-Fe_2O_3$ 图解中，部分样品落入洋中脊区，部分样品却落入大洋岛区，代表了裂解的产物。王可伏等（1994）认为与云母片岩互层产出的斜长角闪片岩 $\varepsilon_{Nd}(t)=+2.588$，来源于亏损地幔，属于岛弧低钾拉斑玄武岩系列。

综合上述资料分析，本书研究认为戈枕村组可能形成于陆缘弧裂解环境，是从劳伦大陆裂解出来的产物。峨文岭组属于裂解背景下的延续，沉积了一套具有韵律性层理的砂、泥质岩，随着裂解的加深，局部出现来源于亏损地幔的低钾拉斑玄武岩。

第三章　新元古界—三叠系

新元古界—古生界包括两个不同的地层大区。涉及海南省的地层区划主要是印支-南海地层大区三亚地层区、南海地层区和羌塘-扬子-华南地层大区华南地层区五指山地层分区（见表1-1）。其中，南海地层区由于资料少，本次未作描述。仅能对三亚地层区和五指山地层分区进行叙述。

第一节　三亚地层区

一、岩石地层特征

三亚地层区内仅出露寒武系和奥陶系，缺乏志留系及其上古生界沉积，主要分布在大东海榆林港至田独一线以西的鹿回头、狗岭、死马岭、三公夹一带，以及晴坡岭、尖岭、大茅峒、落笔洞等地。出露面积约67km^2，厚度可达7128.6m。寒武系和奥陶系呈假整合接触。其中，寒武系仅发育黔东统孟月岭组和黔东统—武陵统大茅组，奥陶系自下而上为下奥陶统大葵组、中奥陶统牙花组、中—上奥陶统沙塘组、上奥陶统榆红组、尖岭组和干沟村组。

（一）剖面描述

1. 孟月岭组剖面

（1）三亚市荔枝沟镇红花剖面

该剖面为孟月岭组的代表性剖面，由海南地质大队1962年开展《海南岛崖县大茅磷矿区地质详查》时测制，何阳芝等（1980）略加修订。剖面位于三亚市荔枝沟镇红花，起点坐标：东经109°31′58″，北纬18°21′41″；自上而下为：

上覆地层　大茅组石英砂岩

——————　整　　合　——————

孟月岭组	总厚度>719.00m
16. 浅灰色粉砂质黏土岩	10.00m
15. 灰色—灰绿色黑云透辉石角岩	28.00m
14. 灰色白云岩	23.00m
13. 灰白色泥质粉砂岩夹细粒石英岩	74.00m
12. 灰白色斑点状粉砂质黏土岩	57.00m
11. 灰白色细粒石英砂岩与斑点状粉砂质黏土岩互层	56.00m
10. 深灰色含磷硅质岩与灰白色泥岩、钙质白云岩互层	40.00m
9. 灰绿色含石英透闪石透辉石角岩	37.00m
8. 灰白色中细粒石英砂岩	72.00m
7. 浅灰色斑点状泥质粉砂岩	49.00m
6. 浅灰色凝块状粉砂质黏土岩	90.00m
5. 浅灰色斑点状泥质粉砂岩	29.00m
4. 灰黑色含磷透闪石钠长角岩	70.00m
3. 灰白色中细粒石英砂岩	47.00m
2. 灰白色泥质粉砂岩	14.00m
1. 浅灰色角岩	>23.00m

未见底

（2）三亚市白石岭剖面

该剖面为孟月岭组的参考剖面，由宜昌地矿所、海南地质大队 1989～1991 年开展 1∶5 万《南山岭、马岭、崖县》幅区调时测制。剖面位于三亚市白石岭西 397.2m 高地，起点坐标：东经 109°35′14″，北纬 18°11′20″；全长 2195m。自上而下为：

未见顶

孟月岭组

四段

8. 浅灰至深灰色灰岩和硅化灰岩	34.85m

三段

7. 乳白色含绢云变质中粒石英砂岩或石英岩状砂岩	564.44m

二段

6. 上部为灰色堇青石（?）斑点状石英粉砂质板岩，下部为灰色含堇青石（?）斑点状含粉砂绢云板岩	122.32m
5. 灰白色含绢云变质中粗砂细粒石英砂岩	175.55m

一段

4. 白或浅灰色石英粉砂质黑云母水云母黏土岩	585.49m
3. 浅灰至深灰色薄层石英粉砂质黑云母板岩	52.46m
2. 浅灰色薄层变质绢云母石英粉砂岩与灰色薄层石英粉砂质绢云板岩互层	46.29m
1. 灰色含黑云母绢云母变质石英	258.72m

未见底

2. 大茅组剖面

（1）三亚市大茅峒第Ⅵ勘探线剖面

该剖面为大茅组的代表性剖面，由海南地质大队 1962 年开展海南岛崖县大茅磷矿区地质详查时测制，何阳芝等（1980）和宜昌地矿所、海南地质大队 1989～1991 年开展 1∶5 万《南山岭、马岭、崖县》幅区调时先后对剖面进行了修订。剖面位于三亚市大茅矿区，起点坐标：109°37′47″，北纬 18°21′01″；自上而下为（图 3-1）：

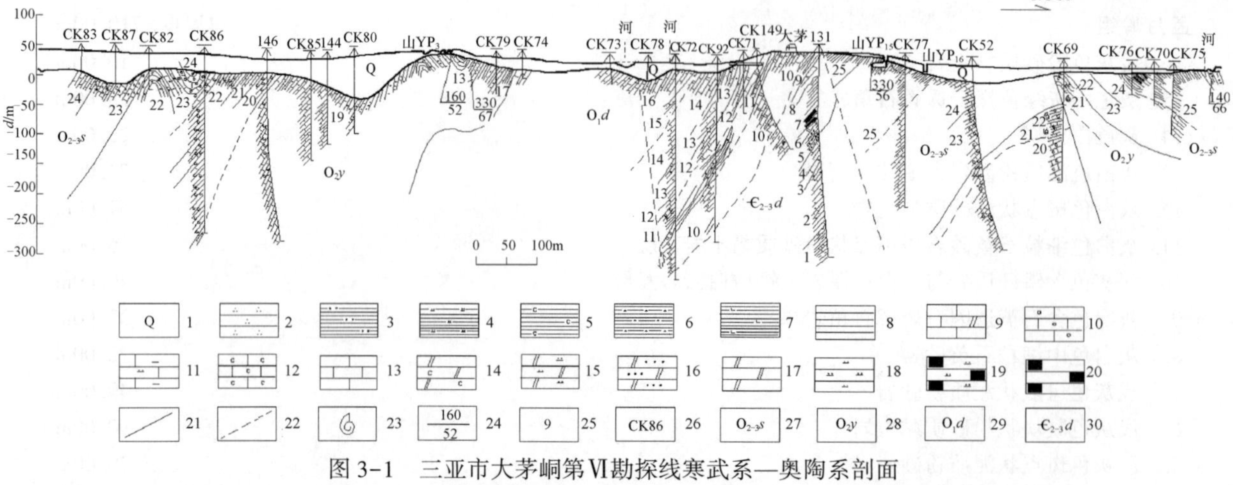

图 3-1　三亚市大茅峒第Ⅵ勘探线寒武系—奥陶系剖面

Fig. 3-1　Cambrian and Ordovician section along Exploratory Line Ⅵ, Damaodong, Sanya City

（据曾庆銮等，1992，略修改）

1—第四系；2—石英砂岩；3—粉砂质页岩；4—硅质页岩；5—炭质页岩；6—绢云母页岩；7—含钙质页岩；8—页岩；9—白云质灰岩；10—砾状灰岩；11—硅质砂质灰岩；12—炭质灰岩；13—石灰岩；14—含炭质白云岩；15—硅质白云岩；16—砂质白云岩；17—白云岩；18—硅质岩；19—硅质磷块岩；20—磷块岩；21—地质界线；22—推测断层；23—动物化石采集位置；24—产状；25—分层号；26—钻孔编号；27—沙塘组；28—牙花组；29—大葵组；30—大茅组

上覆地层 大葵组灰色—青灰色不等粒含砾石英砂岩

------假 整 合------

大茅组 总厚度>309.00m

14. 灰—灰黑色厚层灰岩、白云质灰岩。在灰岩和白云岩中产有丰富的小壳化石：软舌螺类 *Microconus* cf. *elongatus*, *Cupitheca* sp., *Pseudorthotheca* sp., *Bucanotheca* sp.；似软舌螺类 *Hyolithellus tenuis*；开腕骨类 *Eiffella*? *bispanica*, *Archiasterella pentactina*. *Chancelloria altaica*, *Dimidia simplea*, *Zigzagella uietus*, *Miriella hainanensis*；单板类 *Aegides* cf. *places*；刺形壳类 *Cambroclavus fangxianensis*, *Rhombocorniculum*? sp., *Zhijnites claviformis*, *Z. cardiformis*；棱管壳类 *Siphogonuchites* cf. *intermodium*；喙壳纲 *Rostroconcha*；单壁古杯类*Tumuliolynthus tubexteruns*, *T. macrospinosus* 39.00m
13. 灰白色—灰色厚层钙质石英砂岩 23.00m
12. 灰色—黑灰色白云质灰岩 10.00m
11. 黑灰色硅质岩夹硅质页岩、硅质白云岩，以及2~5层磷块岩。在磷块岩顶、底的硅质岩以及勘探线东侧SK5槽的相同层位中产有多门类而又丰富的化石。三叶虫 *Xystridura hainanensis.*, *X. yaxianensis*, *X. orientalis*, *Galahetes hainanensis*, *G. pyrus*, *Kootenia* sp., *Pagetia* sp.；腕足类 *Lingulella atava*, *L. ferruginea.*, *Acrotreta kutorgai.*, *A. schmaleneei.*, "*Botsforida*" *yaxianensis*, *Obolella* sp.；双壳类 *Cycloconcha*? *subovata*；古介形类（高肌虫目）*Ophio sema strumatum*, *O. sicyoideum*, *O. subrotundum*, *Ophiosema* (*Sinophiosema*) *chinense*, *O* (*S.*) *transversum*, *O.* (*S.*) *deltatum*, *O.* (*S.*) *paradoxum*, .*Houlongdongella disulcata.*, *H. inflata.*, *H. sphaerica*, *Yaxianella sulcata*, *Y. punctata*. *Guangdongella obesa*, *Indota acuta*, *Plesidielymella* sp., *Bradoria* sp., *Mannocosmia torquata*；另外还有 *Protoleiosphaeridium* sp., *Microconcentrica* sp., *Brocholaminaria* sp., *Trachysphaeridium* sp., *Leiomarginata* sp. 等微古植物 36.00m
10. 灰白色白云岩夹白云质灰岩、灰色硅质白云岩，部分具鲕状结构 46.00m
9. 黑灰色含粉砂质页岩或含砂钙质页岩夹薄层硅质岩 14.00m
8. 深灰色含硅质粉砂质白云岩或深灰色薄层状泥质灰岩 10.00m
7. 灰白色石英砂岩夹深灰色粉砂质页岩或钙质页岩 18.00m
6. 浅灰色薄层白云岩 6.00m
5. 黑灰色炭质粉砂质页岩 10.00m
4. 灰白色石英砂岩 15.00m
3. 浅灰色白云质灰岩或白云岩 4.00m
2. 灰白色石英砂岩夹薄层粉砂岩 48.00m
1. 浅灰色白云质灰岩与棕灰色含钙绢云母粉砂岩或含炭质粉砂质页岩互层 30.00m

未见底

（2）村子-罗郁下村剖面

该剖面为大茅组的参考剖面，由宜昌地矿所、海南地质大队1989~1991年开展1：5万《南山岭、马岭、崖县》幅区调时测制。剖面位于三亚市孟果岭东北侧山，起点坐标：东经109°32′13″，北纬18°14′42″；总长2496m；自上而下为（图3-2）：

未见顶

大茅组

二段 >408.66m

7. 灰白色绢云母板岩 65.03m
6. 黄灰色泥质粉砂岩或粉砂质水云母黏土岩，偶夹少量粉砂质绢云母板岩 343.63m

一段 486.56m

5. 乳白色中细粒石英砂岩 259.76m

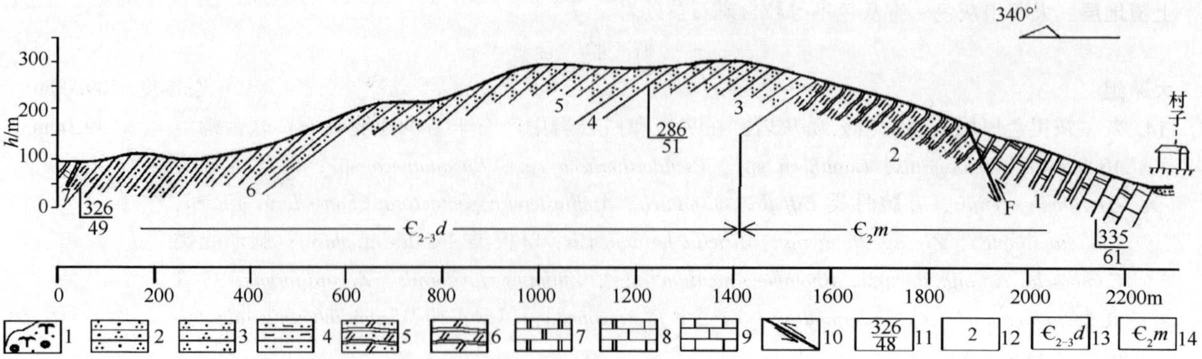

图 3-2　三亚市村子—罗郁下村寒武系实测剖面图

Fig. 3-2　Measured section of Cambrian strata along Cunzi-Luoyuxiacun, Sanya City

1—覆盖层；2—中细粒石英砂岩；3—细粒石英砂岩；4—泥质粉砂岩；5—粉砂质绢云母板岩；6—绢云母板岩；7—硅化白云岩；8—白云质灰岩；9—灰岩；10—逆断层；11—产状；12—分层号；13—大茅组；14—孟月岭组

4. 灰白色条带状细粒石英砂岩　　21.07m

3. 灰白色变质中细粒石英岩状砂岩，偶夹少量粉砂质绢云母板岩　　205.73m

——— 整　合 ———

孟月岭组

五段　　306.4m

2. 灰色粉砂质绢云母板岩　　306.4m

═══ 断　层 ═══

四段

1. 上部为灰白色硅化白云岩；据钻孔资料，下部为浅灰色、深灰色灰岩和硅化灰岩　　>347.28m

3. 大葵组剖面

大葵组仅有三亚市大茅峒第Ⅵ勘探线剖面（剖面情况同上述），为代表性剖面。自上而下为(图 3-1)：

上覆地层　牙花组灰白色页岩与黄灰色细粒石英砂岩互层

——— 整　合 ———

大葵组　　总厚度 214.00m

16. 灰色—深灰色厚层灰岩　　178.00m

15. 灰白色—青灰色不等粒石英砂岩，含少量细砾　　36.00m

- - - - - - 假整合？- - - - -

下伏地层　寒武系大茅组灰色—黑灰色厚层灰岩、白云质灰岩，在灰岩和白云质灰岩中产丰富的小壳动物化石

4. 牙花组剖面

(1) 三亚市大茅峒第Ⅵ勘探线剖面

该剖面为牙花组代表性剖面，剖面情况同上述，自上而下为（图 3-1）：

上覆地层　沙塘组灰色钙质砾灰岩夹薄层—块状灰岩

——— 整　合 ———

牙花组　　总厚度 280.00m

22. 黑灰色炭质钙质页岩，产笔石 *Dicellograptus* sp.，*Climacograptus uniformis*　　19.00m

21. 黑灰色厚层结晶灰岩　　8.00m

20. 黑色钙质炭质页岩，产笔石 *Cryptograptus tricornis*，*Glyptograptus intersitus*，*G.* cf. *dentatus*，*Tylograptus globiformis*，*T. geniculiformis*，*Climacograptus* cf. *uniformis* 15. 00m

19. 深灰色—黑灰色含炭灰岩，间夹白云质灰岩 122. 00m

18. 黄白色块状细粒石英砂岩间夹浅灰色页岩 67. 00m

17. 灰白色页岩与黄灰色细粒石英砂岩互层 49. 00m

———— 整 合 ————

下伏地层 大葵组灰色—深灰色厚层灰岩

(2) 落笔洞剖面

该剖面为牙花组参考剖面，由汪啸风等人 1992 年测制。剖面位于三亚落笔洞一带，自上而下为（图 3-3）：

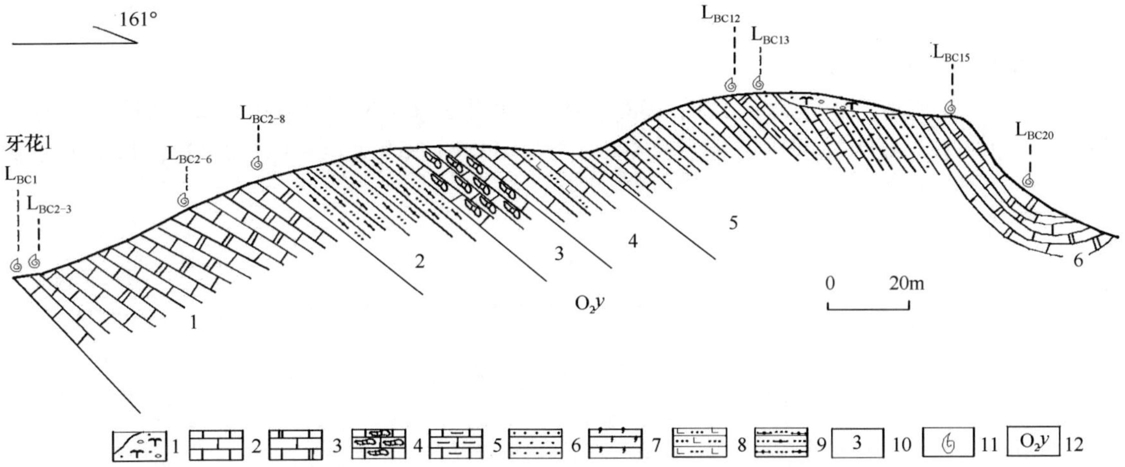

图 3-3 三亚市落笔洞下奥陶统牙花组（二段）实测剖面图

Fig. 3-3 Measured section of the Lower Orodivician Yahua Formation (the Second member), Luobidong, Sanya City

1—覆盖层；2—灰岩；3—白云质灰岩；4—含白云质团块灰岩；5—泥质灰岩；6—石英砂岩；7—粉砂岩；8—钙质砂岩；9—白云质石英细砂岩；10—分层号；11—化石点；12—牙花组

未见顶

牙花组 总厚度 172. 94m

6. 深灰色厚层状白云质灰岩，产牙形石：*Scolopodus* aff. *norami Glyptoconus* cf. *cornuformis*，*Prototanderodus* cf. *leonadi*. *Walliserodus*? sp. 8. 31m

5. 灰色粉砂岩夹 1 层泥灰岩和 2 层灰岩，在灰岩中产牙形石：*Paltodus* sp. 57. 02m

4. 灰色灰岩夹少许灰色粉砂岩 22. 15m

3. 深灰色厚层含白云质团块灰岩 10. 33m

2. 灰色白云质细砂岩，顶部为灰色块状白云质灰岩 23. 08m

1. 下部为灰至深灰色厚层灰岩，中部和上部为深灰色厚层条带状白云质灰岩，夹棕黄色石英细砂岩。灰岩中产牙形石：*Scolopodus euspinus*，*Scolopodus* cf. *rex*，*Scolopodus*? *nogami*. *Glyptoconus* sp.，*Drepanoistodus Iexilis*，*Protopanderodus* cf. *leonadi* 52. 05m

未见底

5. 沙塘组剖面

(1) 三亚市大茅峒第Ⅵ勘探线剖面

该剖面为沙塘组代表性剖面，剖面情况同上述。自上而下为（图 3-1）：

未见顶

沙塘组

上段 总厚度>228.00m

26. 灰色粉砂质页岩 27.00m

25. 上部灰色页岩，下部为黑灰色炭质页岩。下部页岩产笔石：*Dicellograptus* cf. *forceps.*，*D. sextans exilis*，*Glossograptus* cf. *hincksii*，*Glyptograptus dentatus*，*Glyptograptus* sp.，*Climacogruptus uniformls*，*C.* cf. *putillus* 等 56.00m

下段

24. 黑灰色厚层灰岩，上部含炭质 66.00m

23. 灰色钙质砾灰岩夹薄—厚层灰岩 79.00m

———— 整 合 ————

下伏地层 牙花组黑色炭质钙质页岩，产笔石 *Dicellograptus* sp.，*Climacograptus uniformis*

(2) 三亚市鹿回头剖面

该剖面由宜昌地矿所、海南地质大队1989~1991年开展1∶5万《南山岭、马岭、崖县》幅区调时测制。剖面位于三亚市鹿回头，起点坐标：东经109°30′26″，北纬18°12′41″；自上而下为（图3-4）：

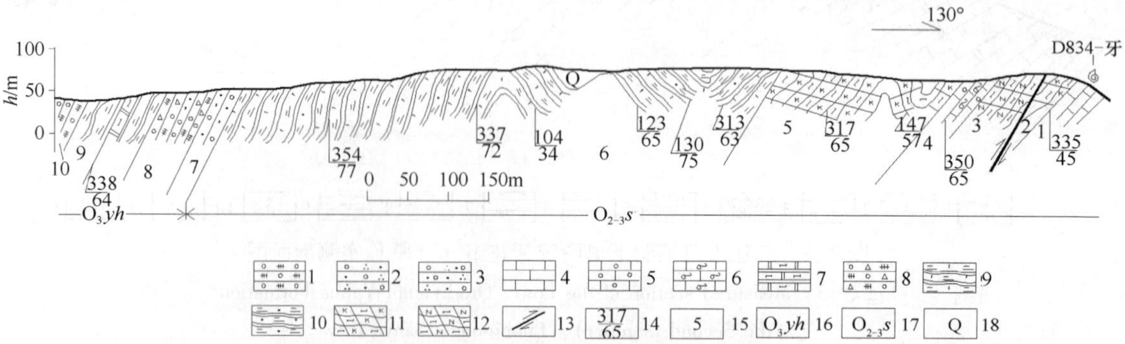

图3-4 鹿回头沙塘组—榆红组实测剖面图

Fig. 3-4 Measured section of the Shatang-Yuhong Formations, Luhuitou

1—复成分砾岩；2—不等粒石英砂岩；3—含砾不等粒石英砂岩；4—灰岩；5—砾状灰岩；6—条带状灰岩；7—透辉石矽卡岩；8—复成分碎裂状砾岩；9—粉砂质绢云板岩；10—细砂质绢云板岩；11—钾长石透辉石角岩；12—斜长石透辉石角岩；13—逆断层；14—产状；15—分层号；16—榆红组；17—沙塘组；18—第四系

未见顶

榆红组

二段 总厚度>254.00m

10. 浅灰色块状角岩化复成分砾岩 >20.0m

9. 灰色或黄灰色粉砂质绢云板岩夹黄绿色滑石透辉石矽卡岩 100.0m

8. 浅灰色、棕黄或砖红色复成分碎裂状砾岩夹浅灰色及灰白色含砾不等粒石英砂岩 120.00m

7. 浅棕黄色含砾中细粒石英砂岩、细粒石英砂岩、粉砂岩组成逆粒序层 14.0m

———— 整 合 ————

沙塘组 总厚度>417.00m

上段

6. 深灰色或黑灰色含炭质粉砂质绢云板岩，间夹细砂质绢云板岩 180.00m

下段

5. 灰色或浅灰色钾长石透辉石角岩，下部夹透镜状或薄层蛇纹岩 121.00m

4. 浅灰色砾状灰岩和条带状灰岩 23.00m

3. 浅灰色或灰色条带状及条纹状斜长石透辉石角岩，上部夹厚层透辉石角岩 39.00m

═══ 断 层 ═══

2. 上部为浅灰色或灰白色厚层钾长石透辉石角岩；中部为青灰色不规则条带状含细粒石英岩状砂岩团块透辉石角岩夹青灰色条纹状钾长石透辉石角岩；下部为灰白色、浅灰色蚀变条带状及条纹状钙质细砂岩 22.00m

1. 上部为浅灰色细晶灰岩，产牙形石 Oistodiform，Drepanodiform；下部为浅灰色竹叶状（钙质砾）细粉晶灰岩 >32.00m

未见底

6. 榆红组剖面

（1）三亚市田独8线剖面

该剖面为榆红组代表性剖面。海南地质大队1979年开展海南岛崖县红花矿区磷矿普查及外围概查时测制，何阳芝等1980年略加修订。剖面位于崖县红沙镇榆红村，起点坐标：109°34′27″，北纬18°18′20″；自上而下为：

上覆地层 尖岭组灰色、浅灰色复成分砾岩

——— 整 合 ———

榆红组	总厚度>452.00m
11. 棕灰色、灰色泥质杂砂质细粒石英砂岩，夹含砾岩屑砂岩	34.00m
10. 浅灰色、灰色中粗粒岩屑砂岩夹含砾岩屑砂岩	43.00m
9. 棕灰色、浅灰色复成分砾岩、砂砾岩、岩屑砂岩、绢云泥质粉砂岩互层，单层厚1.5~2m	47.50m
8. 浅灰色复成分砾岩、含砾岩屑砂岩、页岩互层。韵律厚2~3m	59.00m
7. 掩盖	45.00m
6. 棕灰色、浅灰色泥质细砂岩	24.00m
5. 紫灰色、灰色岩屑砂岩夹粉砂岩，含砾岩屑砂岩	52.50m
4. 掩盖	23.00m
3. 浅灰色、灰色中粗粒石英砂岩，下部含岩屑及砂岩细砾石	47.00m
2. 掩盖	37.00m
1. 浅灰色、灰色复成分砾岩，夹含砾岩屑砂岩	40.00m

未见底

（2）狗岭剖面

该剖面为榆红组参考剖面，由宜昌地矿所、海南地质大队1989~1991年开展1∶5万《南山岭、马岭、崖县》幅区调时测制。剖面起点于红沙欧家园村北侧公路边，起点坐标：东经109°32′52″，北纬18°14′21″；全长3619m；自上而下为（图3-5）：

上覆地层 尖岭组含水云母复成分砾岩或砂砾岩

——— 整 合 ———

榆红组	总厚度>1594.73m
四段	217.63m
17. 浅灰色、紫红色含水云母细粒岩屑杂砂岩	33.54m
16. 黄灰色含褐铁矿水云母黏土岩，靠上部被黄色粉砂质泥岩坡积物掩盖	55.36m
15. 紫灰色含水云母中粒岩屑砂岩	122.68m
14. 灰白色、浅肉红色含水云母复成分砾岩	6.05m
三段	559.08m
13. 青灰色、灰白色含水云母细粒岩屑砂岩	82m

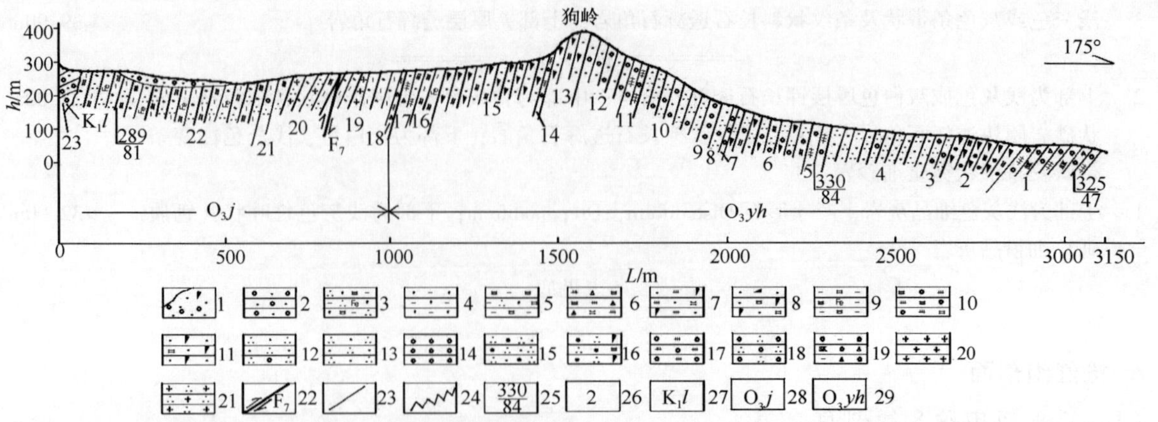

图 3-5　狗岭上奥陶统榆红组—尖岭组实测剖面图

Fig. 3-5　Measured section of the Upper Ordovician Yuhong—Jianling Formations，Gouling

1—浮土；2—砂砾岩；3—含赤铁矿石英粉砂质水云母黏土岩；4—泥质粉砂岩；5—含石英粉砂水云母黏土岩；6—含水云母复成分细角砾岩；7—含水云母细粒岩屑杂砂岩；8—含水云母中粒岩屑砂岩；9—含褐铁矿水云母黏土岩；10—含水云母复成分细砾岩；11—含水云母细粒岩屑砂岩；12—含砾细粒石英砂岩；13—细粒石英砂岩；14—砾岩；15—中细粒石英砂岩；16—含水云母中粗粒岩屑石英砂岩；17—复成分砾岩；18—石英细砾岩；19—硅灰-钙铁辉石矽卡岩化复成分细砾岩；20—花岗斑岩；21—石英斑岩；22—逆断层；23—地层界线；24—不整合界线；25—产状；26—分层号；27—鹿母湾组；28—尖岭组；29—榆红组

12. 灰白色含砾中砂岩（下部被覆盖）	157.36m
11. 灰白色、浅肉红色复成分细砾岩	46.75m
10. 青灰色块状中细粒石英砂岩，偶含砾	178.3m
9. 灰白色砾岩	31.31m
8. 灰白色含砾中细粒石英砂岩	44.01m
7. 灰白色厚层砾岩	19.35m
二段	818.02m
6. 灰白色或浅肉红色含水云母中、粗粒岩屑石英砂岩	232.19m
5. 灰色、浅灰色复成分砾岩	44.16m
4. 灰白色细粒石英砂岩，偶含砾	190m
3. 灰白色石英质砾岩	17.97m
2. 灰白色含水云母中粒岩屑砂岩	136.01m
1. 灰白色硅灰石-钙铁辉石矽卡岩化复成分砾岩	197.69m

未见底

（3）三亚市红花地区晴坡岭剖面

该剖面位于三亚市红花地区晴坡岭，由海南地质大队 1979 年开展海南岛崖县红花矿区磷矿普查及外围概查时测制，宜昌地矿所、海南地质大队 1989～1991 年开展 1∶5 万《南山岭、马岭、崖县》幅区调时重测。剖面起点坐标：东经 109°34′34″，北纬 18°19′12″；自上而下为（图 3-6）：

上覆地层　尖岭组灰白色复岩屑砾岩

———— 整　合 ————

榆红组	>411.04m
四段	234.84m
7. 浅紫红色中粒石英砂岩	71.0m
6. 灰白色、紫红色、黄灰色碎裂状砾岩	5.36m
5. 浅黄色中细粒石英砂岩	57.3m
4. 浅黄色、灰白色复岩屑砾岩	4.81m

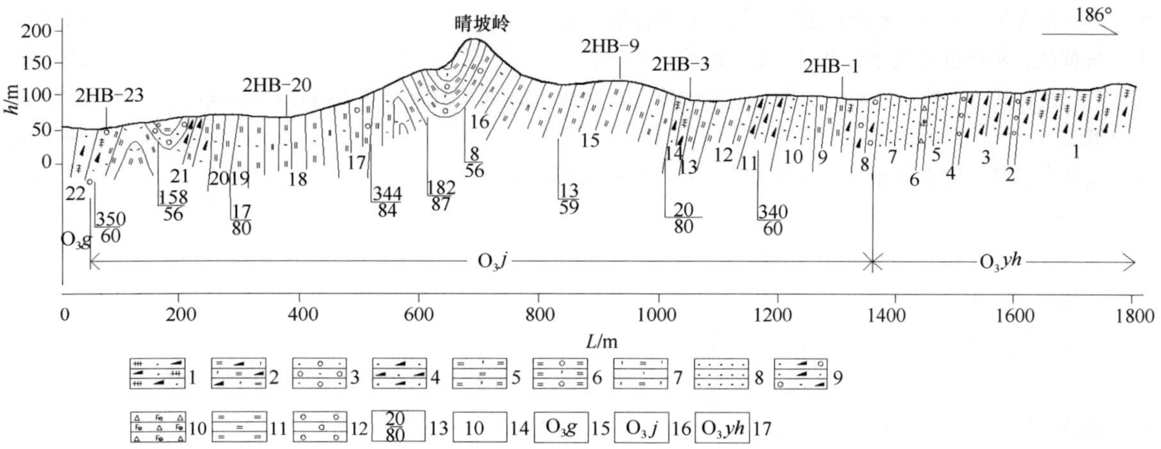

图 3-6　三亚市晴坡岭上奥陶统榆红组至干沟村组实测剖面图

Fig. 3-6　Measured section of the Upper OrdovicianYuhong-Gangoucun Formations, Qingpoling, Sanya City

（据曾庆銮等，1992；略改动）

1—岩屑杂砂岩；2—水云母质岩屑粉砂岩；3—砂砾岩；4—岩屑砂岩；5—粉砂质水云母黏土岩；

6—含砾粉砂质水云母黏土岩；7—水云母质石英粉砂岩；8—石英砂岩；9—复成分不等粒岩屑砂岩；

10—含铁质石英角砾岩；11—水云母黏土岩；12—砾岩；13—产状；14—分层号；15—干沟村组；16—尖岭组；17—榆红组

3. 浅灰色、浅肉红色、浅紫等杂色中粒岩屑砂岩	89. 54m
2. 黄灰色、灰褐色复成分砾岩	6. 83m
三段	>176. 2m
1. 浅灰色、灰白色、浅紫色细粒岩屑杂砂岩	176. 2m

未见底

7. 尖岭组剖面

尖岭组的主要剖面有三亚市红花地区尖岭（Ⅷ线）剖面、三亚市红花地区晴坡岭剖面、狗岭剖面，且以晴坡岭剖面控制程度较好，也比较完整。

（1）三亚市红花地区尖岭（Ⅷ线）剖面

该剖面为尖岭组代表性剖面，由海南地质大队 1976 年开展海南岛崖县大茅磷矿区地质勘探时测制，葛梅钰等（1983）略加修订。剖面位于崖县红花东南部尖岭西埡口，起点坐标：东经 109°35′53″，北纬 18°20′37″；自上而下为：

未见顶

尖岭组	总厚度>935. 9m
15. 浅灰色泥质中粒石英砂岩，局部含砾，部分为浮土掩盖	>119. 5m
14. 浅灰色、微红色绢云母泥质粉砂岩、页岩	56. 20m
13. 掩盖	45. 00m
12. 浅灰色绢云母泥质页岩，局部含粉砂，距底部 50m 处产腕足类 *Zanclorhyncha gentilies*, “*Beitaia*” sp. ；海百合茎 *Cyclocyclicus* sp.	66. 70m
11. 浅黄色泥质硅质石英粉砂岩，局部掩盖	16. 40m
10. 掩盖	45. 0m
9. 褐红、浅灰褐色细、中粒石英砂岩	120. 0m
8. 掩盖	15. 0m
7. 浅灰色黏土岩，距底 15m 处产笔石 *Rectograptus* sp. , *Pseudoclimacograptus* sp. , *Orthograptus* aff. *pauperatus*, *Glyptograptus* cf. *angustifolius*	32. 10m
6. 掩盖	15. 0m

5. 浅红色含绢云母泥质铁质硅质中、细粒石英砂岩，部分掩盖 104.60m

4. 灰黄色、灰白色复成分砾岩，夹中、细粒石英砂岩 20.80m

3. 灰白色泥质铁质中、细粒石英砂岩，局部掩盖。距底 10m 处产海百合茎 *Cyclocyclicus* sp. 135.50m

2. 掩盖 113.00m

1. 浅灰白色复成分石英砾岩，胶结物中产海百合茎化石碎片 >31.10m

未见底

(2) 狗岭剖面

剖面情况见上述，自上而下为（图 3-5）：

上覆地层 鹿母湾组紫红色砾岩

~~~~~~~~ 不整合 ~~~~~~~~

尖岭组

下段 519.76m

22. 黄灰色、棕黄色薄层含赤铁矿石英粉砂质水云母黏土岩 271.15m

21. 黄灰色含砾泥质粉砂岩 2.58 m

20. 黑灰色薄层含石英粉砂水云母黏土岩与黄灰色粉砂岩互层。见花岗斑岩侵入 146.39m

═══════ 断　　层 ═══════

19. 浅灰色泥质粉砂岩。见花岗斑岩侵入 97.14m

18. 浅灰色（局部呈浅紫红色）含水云母复成分砾岩或砂砾岩 2.5m

———— 整　　合 ————

下伏地层 榆红组水云母细粒岩屑杂砂岩

(3) 三亚市红花地区晴坡岭剖面

该剖面为尖岭组参考剖面，剖面情况见上述，自上而下为（图 3-6）：

上覆地层 干沟村组肉红色、黄灰色厚层含细砾中粒岩屑砂岩或砂砾岩

———— 整　　合 ————

尖岭组 总厚度 937.22m

上段

21. 浅灰色水云母岩屑砂岩 53.40m

20. 浅灰色含石英粉砂质水云母黏土岩。产少量化石：腕足类 *Mimella* sp.，*M. yaxianensis*，*Leptellina* sp.，*Sowerbyella* sp.，*S. hainanensis*，*Zanclorhyncha yaxianensis*；腹足类 *Omospira* sp.，*Ophileta* sp.；双壳类 *Eopteria* cf. *richardsoni*，*Similodontasimiles*，*Tancrediopsis dulankarensis*；头足类 *Eurasiaticoceras* sp.，*Michelinoceras* sp.，*Westonoceras* sp.；三叶虫 *Remopleurides* sp.；笔石 *Glyptograptus* sp.，*Prolasiograptus*? sp.；海百合茎化石 *Cyclocyclicus* sp. 46.33m

19. 浅灰色含粉砂水云母黏土岩，产海百合茎化石碎片 25.34m

18. 灰色含石英粉砂水云母黏土岩，产双壳类：*Cymatonota hounghuaensis*；三叶虫 *Ampyxinella* sp. 及海百合茎化石碎片 127.33m

17. 青灰色、浅肉红色含水云母石英粉砂岩或石英粉砂质水云母黏土岩。产较丰富的化石：腕足类 *Zanclorhyncha* sp.，*Atelelasma*? sp.，*Leptellina*? sp.，*Ancistrorhyncha* sp.，*Zanclorhyncha yaxianensis*，*Hallina* sp.，*Protozyga nasuta*，*P. magnicostata*，*Idiospira*? *hainanensis*，*Orbiculoidea*? sp.；双壳类 *Similodonta* sp.，*Pararaphistoma* sp.，*P. cardiformis*，*P.* cf. *imitabilis*；海百合茎 *Trigonotrigonalis* sp.，*Cyclocyclicus* sp.；腹足类 *Lophospira* sp. 93.86m

下段

16. 灰色水云母石英粉砂岩，产海百合茎化石碎片 52.56m
~~~~~~~~

15. 灰黑色含石英粉砂水云母黏土岩。产丰富的化石：三叶虫 *Bulbaspis*（*Parabulbaspis*）sp.，*Dionide* sp.，*Birmanites* cf. *hupeiensis*，*Plumulites* sp.，Raphiophoridae；笔石 *Orthograptus* sp.，*O.* cf. *pauperatus*，*Climacograptus* cf. *brevis*，*Glyptograptus euglyphus siccatus*，*Nemagraptus*? sp.，*Orthograptus* cf. *calcaratus*，*Hallograptus* cf. *bimicronatus*，*Corynoides*? sp.，*Glyptograptus* cf. *euglyphus*，*G.* sp.，*Glyptograptus* sp.，*Nemagraptus* sp.，*Climacograptus* cf. *brevis*；头足类 *Michelinoceras* sp.，*Ancistroceras* sp.，*A.* cf. *clinotum*，*Richardsonoceras* sp.；腕足类 *Protozyga nasuta*，*Zanclorhyncha* sp.；海百合茎 *Cyclocyclicus* sp.；*Trigonotrigonalis* sp. 261.37m

14. 上部为灰色薄层水云母黏土岩，下部为浅肉红色复成分不等粒岩屑砂岩 32.46m

13. 浅灰色薄层水云母黏土岩。产三叶虫 *Dionide* sp.，*Birmanites* sp. 31.11m

12. 掩盖。见粉砂质泥岩残坡积物 51.29m

11. 灰或深灰色薄—中厚层细粒岩屑砂岩 56.11m

10. 黄灰色细砂岩 40.64m

9. 青灰色、黄灰色含石英粉砂水云母黏土岩。产双壳类 *Modiolopsis* sp. 65.42m

8. 灰白色复岩屑砾岩 30.87m

——— 整 合 ———

下伏地层 榆红组浅紫灰色中粒石英砂岩

8. 干沟村组剖面

干沟村组仅有三亚市红花地区晴坡岭剖面（剖面情况见上述），为代表性剖面；自上而下为（图3-6）：

未见顶

干沟村组 总厚度>64.08m

23. 上部为灰白色细粒岩屑石英杂砂岩，中部为青灰色含岩屑石英粉砂岩，下部为灰白色细粒岩屑杂砂岩（未见顶）。产腕足类、三叶虫及双壳类、腕足类：*Lingulella*? sp.，*Mimella yaxianensis*，*Multicostella* sp.，*Zanclorhyncha gentilis*，*Idiospira*? *hainanensis*，*Leptelloidea yaxianensis*，*Sowerbyella* sp.，*S. hainanensis*；双壳类 *Cycloconcha*? *subovta*，*Modiolopsis* sp.；三叶虫 *Prisoceraurus* sp.，*Neseuretus* sp.，*Encrinuroides* sp. >63.64m

22. 肉红色、黄灰色厚层含细砾中粒岩屑砂岩或砂砾岩，产腕足类、双壳类。腕足类 *Sowerbyella yaxianensis*，*S.* cf. *yaxianensis*，*S.* cf. *sericea*，*Zanclorhyncha yaxianensie*，*Z. gentiles*，*Camerella* sp.，*Orthisoid* sp.；双壳类 *Sirnilodonta similes* 0.44m

——— 整 合 ———

下伏地层 尖岭组浅灰色水云母岩屑砂岩

（二）岩石地层特征

1. 孟月岭组（$\epsilon_2 m$）

本书所称的孟月岭组系由中国科学院华南富铁矿科研队地层组（1977）创建的孟月岭组演变而来。创名地点在三亚市红花地区。宜昌地矿所和海南地质大队联合区调队（1991）、曾庆銮等（1992）、汪啸风等（1992）沿用华南富铁矿科研队地层组的孟月岭组并补测剖面，修订完善其岩石地层单位。联合区调队（1991）、曾庆銮等（1992）并以白石岭剖面、孟果村剖面及村子-罗郁下村剖面为代表，同时参考孟月岭24勘探线剖面，将本组自下而上划分为5个岩性段：一段为浅灰色至深灰色石英粉砂质板岩夹浅灰色绢云母石英粉砂岩及灰白色石英粉砂质黑云母、水云母黏土岩；二段为灰白色变质含中粗粒、细粒石英砂岩或石英岩状砂岩及灰白色含堇青石（?）或红柱石斑点状石英粉砂质板岩；三段为一套巨厚的乳白色变质含绢云母中粗至粗粒石英砂岩；四段为浅灰色至深灰色钙质白云岩、灰岩和硅质灰岩；五段为灰色粉砂质绢云母板岩。地层总厚度为2101.48m。陈哲培等（1997）在

《海南岛岩石地层》中采用了联合区调队（1991）厘定的孟月岭组，并指定红花地区孟月岭剖面为代表性剖面，三亚市白石岭剖面和村子-罗郁下村剖面为参考剖面。其下部为粉砂质黏土岩夹泥质粉砂岩或为互层；中部为中细粒石英砂岩与斑点状泥质粉砂岩或粉砂质黏土岩不等厚互层，间夹白云岩及含磷硅质岩；上部为粉砂质黏土岩、泥质粉砂岩，夹白云岩，顶部以中细粒石英砂岩为界与上覆大茅组整合接触。未见底，含磷硅质岩中产微古植物化石。厚719~2101.48m。

孟月岭组分布于三亚市晴坡岭向斜两翼外侧孟月岭、虎头岭、白石岭、藤桥、孟果村等地。孟月岭剖面主要为斑点状泥质粉砂岩、斑点状粉砂质黏土岩与中细粒石英砂岩呈不等厚互层，偶夹含磷硅质岩、白云岩；白石岭剖面中部的灰白色中细粒石英砂岩质地坚硬，常形成高山。区域延伸比较稳定，仅在成层厚度上略有变化，向斜南东翼的安游地区石英砂岩、黏土岩厚度明显增大，其总厚度大于2101.48m。

孟月岭组与岛东万宁东岭地区寒武系美子林组中—下部对比，其沉积类型、岩相等存在明显不同；美子林组一段沉积类型为泥质砂岩型、砂岩型、泥质粉砂岩-粉砂质泥岩型、灰质砂岩型；二段沉积类型为：灰质砂岩-砂岩、灰质泥质粉砂岩-粉砂质泥岩、泥质粉砂岩-粉砂质泥岩，另夹有少量灰岩。总体特征是：粒度较细，一般在细砂级以下，少量可达中砂级，属较低能沉积条件。而三亚地区孟月岭组的沉积类型属于距高隆起区较近且沉降速度较快、碎屑物质来源充足的滨—浅海沉积环境有关。

2. 大茅组（$\epsilon_{2-3}d$）

大茅组系由王树才、金邦权（1962）所创建的大茅群演变而来，中南地区区域地层表编写小组（1974）介绍。创名地点在三亚市大茅峒地区。中国科学院华南富铁矿科研队地层组（1977）改称大茅组，广东区测大队（1964）指定大茅矿区第Ⅵ勘探线剖面为标准剖面。宜昌地矿所和海南地质大队联合区调队（1991）、曾庆銮等（1992）将本组自下而上划分为5个岩性段：一段为乳白色或灰白色中细粒石英岩状砂岩或石英砂岩，偶夹泥质粉砂岩；二段为黄灰色粉砂质水云母黏土岩或泥质粉砂岩，间夹灰白色、乳白色细粒石英砂岩；三段为灰白色白云岩夹白云质灰岩或硅质白云岩，含磷锰矿层；四段为黑灰色硅质岩夹硅质页岩、硅质白云岩，含磷锰矿层；五段为灰色至黑灰色厚层灰岩、白云质灰岩夹白色厚层钙质石英砂岩。陈哲培等（1997）在《海南岛岩石地层》中采用了联合区调队（1991）厘定的大茅组，并指定大茅矿区第Ⅵ勘探线剖面为代表性剖面，参考剖面为三亚市村子-罗郁下村剖面。下部为厚层中细粒石英砂岩，间夹泥质粉砂岩；中部为石英砂岩、粉砂质页岩与白云岩或白云质灰岩不等厚互层；上部为厚层灰岩、白云岩、硅质岩夹硅质页岩、磷块岩及锰矿层，顶部夹一层钙质石英砂岩。上部硅质页岩中产有三叶虫、腕足类、古介形类及小壳类化石。底以厚层中细粒石英砂岩与下伏孟月岭组为界，顶部以厚层灰岩消失与上覆大葵组呈假整合接触。

大区域尺度中，其岩性、岩相类似且所产生物群面貌基本可与澳大利亚昆士兰地区中寒武统Currant Bush组对比。小尺度中，其区域横向岩相变化显著。大茅峒地区中部为石英砂岩、粉砂质页岩与白云岩或白云质灰岩不等厚互层；红花地区中部相变为白云岩、含泥质或含粉砂硅质岩、含磷白云质硅质岩，间夹白云岩；在安游地区中部为厚度较大的泥质粉砂岩或粉砂质页岩。沉积厚度呈现西厚东薄的趋势，在大茅地区为309m，红花地区为752~794m。为浅海沉积环境。

3. 大葵组（O_1d）

广东省区测大队（1964）创名大葵群。葛梅钰等（1983）介绍，改称大葵组。创名地点在三亚市大茅峒地区大葵村。代表剖面为大茅矿区第Ⅵ勘探线剖面。以往资料根据本组下部所产微古植物化石 *Leiminuscula*? sp.，*Archaeodiscina*? sp.，认为其地质时代为寒武纪晚期。汪啸风等（1992）沿用大葵群，在其上部层位（钻孔岩心样品）采集丰富的牙形刺标本，其地质时代为早奥陶世早期。宜昌地矿所和海南地质大队联合区调队（1991）、曾庆銮等（1992）采用了汪啸风等（1992）的地层划分意见，并把大葵群改成大葵组。陈哲培等（1997）在《海南岛岩石地层》中采用了联合区调队（1991）厘定的大葵组，指定代表性剖面为大茅矿区第Ⅵ勘探线剖面，岩石主要由不等粒（含砾）石英砂岩和厚层灰岩组成，厚度>214m。

该组呈隐伏状产出（被第四系掩盖），主要见于三亚市大茅峒一带钻孔中，分布范围小。为滨海-滨外浅海沉积环境。

4. 牙花组（O_2y）

由广东省区测大队（1964）创名牙花群演变而来，中南地区区域地层表编写小组（1974）介绍，并改称牙花组，创名地点在崖县大茅峒地区牙花村。海南地质大队（1976）重新进行厘定，将其顶部含笔石页岩划归沙塘组，该层位以下岩石划归牙花组。何阳芝等（1980）基本上沿用了海南地质大队（1976）的划分方案，仅是把牙花组顶部含笔石页岩与原沙塘组底部的含笔石页岩合并改称为落牙页岩，葛梅钰等（1983）将牙花组顶界上移至原始含义沙塘组底部含笔石页岩上界，以细粒石英砂岩为底界。宜昌地矿所和海南地质大队联合区调队（1991）、曾庆銮等（1992）、陈哲培等（1997）采用葛梅钰等（1983）意见，陈哲培等（1997）并在《海南岛岩石地层》指定大茅矿区第Ⅵ勘探线剖面为代表性剖面。下部为厚层细粒石英砂岩夹页岩，底部为页岩与砂岩互层；中部为厚层含炭灰岩，间夹白云质灰岩、粉砂岩及页岩；上部为钙质炭质页岩夹厚层结晶灰岩，产丰富的笔石化石。顶部以页岩与上覆沙塘组钙质砾屑灰岩呈整合接触；以灰白色页岩与黄灰色细粒石英砂岩互层出现为底界，与下伏大葵组呈整合接触。厚280.0~330.0m。

该组呈隐伏状产出（被第四系掩盖），主要见于三亚市大茅峒一带钻孔中，分布范围小。为滨海-滨外浅海沉积环境。

5. 沙塘组（$O_{2-3}s$）

广东省区测大队（1964）创建沙塘群。中南地区区域地层表编写小组（1974）介绍，并改称沙塘组。创名地点在三亚市大茅峒地区沙塘。海南地质大队（1976）将覆于原始含义的沙塘组之上的山野群并入其中，并把下限移至牙花组上部含笔石页岩之底界。中国科学院华南富铁矿科研队地层组（1977）和葛梅钰等（1983）的沙塘组包括原山野群，但不包括牙花组上部的含笔石页岩。宜昌地矿所和海南地质大队联合区调队（1991）、曾庆銮等（1992）沿用葛梅钰等（1983）沙塘组含义，并划分出上、下两段，地质时代分别为中奥陶世早期和早奥陶世晚期。陈哲培等（1997）在《海南岛岩石地层》中采用联合区调队（1991）意见，指定代表性剖面为大茅矿区第Ⅵ勘探线剖面；参考性剖面为鹿回头Ⅷ号剖面。下段为钙质砾灰岩、厚层灰岩，间夹细砂岩，产牙形石化石；上段为黑色炭质页岩，产笔石 *Dicellograptus* cf. *forceps*。以灰色钙质砾屑灰岩与下伏牙花组呈整合接触，与上覆榆红组底部含砾不等粒石英砂岩呈整合接触。厚228~417m。

沙塘组主要分布于三亚大茅硐、鹿回头、下洋田、抱坡岭等地，岩性稳定，易于对比；为台地边缘斜坡至浅海陆棚沉积环境；沉积厚度呈东北薄西南厚的趋势，在大茅硐厚228m，鹿回头厚417m。

6. 榆红组（O_3yh）

榆红组由何阳芝等（1980）创建，曾庆銮等（1992）介绍。创名地点及代表性剖面在三亚市红沙镇榆红村，参考性剖面为鹿回头剖面和狗岭剖面。该组经宜昌地矿所和海南地质大队联合区调队（1991）、曾庆銮等（1992）重新厘定，并在鹿回头剖面上，将原榆红组底界向下移至首次出现复成分砾岩层。陈哲培等（1997）在《海南岛岩石地层》中采用联合区调队（1991）意见，指定田独8线剖面为代表性剖面，参考剖面为鹿回头剖面和狗岭剖面。由复成分砾岩与中粗粒岩屑砂岩或中细粒石英砂岩互层组成。上部为浅灰色、紫灰色含水云母中、细粒岩屑砂岩夹含褐铁矿水云母黏土岩、水云母复成分砾岩；中部为复成分砾岩与石英砂岩呈不等厚互层；下部为浅灰色粉砂质绢云母板岩，复成分砾岩、不等粒石英砂岩、粉砂岩。底部以粉砂岩、不等粒砂岩、含砾不等粒砂岩与下伏沙塘组呈整合接触，顶部以灰色细粒石英砂岩与上覆尖岭组复成分砾岩呈整合接触。

榆红组在三亚地区分布广泛，主要出露于鹿回头—狗岭—榆红村—大茅峒一带，主要以砾质、砂质岩相为主，属于海岸冲积扇沉积环境。区域上岩性稳定，自东北向西南沉积厚度明显增加，厚452~1831m。

7. 尖岭组（O_3j）

尖岭组由海南地矿局黄香定、陈哲培（1976，手稿）创建，是从原山坡组演变而来的岩石地层单位；许汉奎等（1977）介绍。创名地点为晴坡岭（原山坡组），宜昌地矿所和海南地质大队联合区调队（1991）、曾庆銮等（1992）重新厘定了尖岭组的含义。陈哲培等（1997）在《海南岛岩石地层》中采用联合区调队（1991）意见，并指定代表性剖面为尖岭剖面，参考剖面为三亚市晴坡岭剖面。其底界是以青灰色、黄灰色含石英粉砂水云母黏土岩的出现为划分标志与下伏榆红组呈整合接触。下部为细粒岩屑砂岩，间夹含石英粉砂质黏土岩，或呈不等厚互层状，产笔石、三叶虫、腕足类化石；中、上部为含石英粉砂质水云母黏土岩，间夹含水云母石英粉砂岩，产腕足类、腹足类、双壳类及头足类化石。以灰白色复成分岩屑砾岩出现为底界，与下伏榆红组呈整合接触；顶部以含泥质石英砂岩的消失与上覆干沟村组呈整合接触。厚598~937.22m。

尖岭组主要分布于三亚地层区，出露于晴坡岭向斜轴部死马岭—三公夹，三罗村—晴坡岭—干沟村一带。主体为泥质、砂质岩相，属正常浅海沉积环境；区域上较稳定，沉积厚度自西北向东南逐渐变薄。产十分丰富的古生物化石。

8. 干沟村组（O_3g）

由中国科学院华南富铁矿科研队地层组（1977）创建，许汉奎（1977）、葛梅钰等（1983）介绍。创名地点及代表性剖面均为三亚红花地区晴坡岭。何阳芝等（1980）根据发现的三叶虫化石，把干沟村组和尖岭组上段分别改称为干沟群上亚群和下亚群，时代分别定为早、中志留世。宜昌地矿所和海南地质大队联合区调队（1991）、曾庆銮等（1992）沿用了华南富铁矿科研队地层组（1977）的干沟村组，陈哲培等（1997）在《海南岛岩石地层》中采用联合区调队（1991）意见，指定代表性剖面为晴坡岭剖面，主要由细粒岩屑杂砂岩夹含水云母岩屑石英粉砂岩组成。底部以含砾岩屑砂岩与下伏尖岭组分界，未见顶，厚64.08m。产腕足类、头足类、三叶虫、几丁虫化石。

该组分布于三亚市晴坡岭向斜轴部，小面积出露于晴坡岭北坡。下部浅肉红色、黄灰色厚层含细砾中粒岩屑砂岩，在横向上可相变为砂砾岩层。

二、古生物特征及年代地层

（一）寒武系

1. 孟月岭组

仅发现微古植物化石 *Leiominuscula* sp.（葛梅钰，1983），其时代较长，地层意义不大。主要依据上覆大茅组所产三叶虫、小壳化石的地质时代推测其地质时代为龙王庙阶。

2. 大茅组

产有三叶虫、古介形类、腕足类、双壳类、小壳化石等多门类丰富的化石，主要见于中上部岩石（图版Ⅰ-2、图版Ⅰ-5、图版Ⅰ-8、图版Ⅰ-9）。

三叶虫产于中部岩石中，有 *Xystridura hainanensis*，*X. yaxianensis*，*X. orientalis*，*Galahetes hainanensis*，*G. pyrus*，*Kootenia* sp.，*Pagetia* sp. 等，主要以 *Xystridura hainanensis* 为代表。*Xystridura hainanensis* 延限带以带分子的出现为特征，其他分子有：*Galahetes hainanensis*，*Xystridura yaxianensis*，*X. orientalis* 等。大致相当于汪啸风等（1992）划分的 *Xystridura* 层；可与澳大利亚昆士兰地区 Templetonian 阶下部对比，且含义同华北地区毛庄阶（南京地质古生物研究所，2000）。

古介形虫（高肌虫目）产于上部硅质页岩中，尤其是 *Ophioserna*，*Indota*，*Bradoria* 等三个属均为澳大利亚昆士兰地区寒武纪早期常见分子。

腕足类主要有：*Lingulella*，*Acrotreta*，“*Botsfordia*”，*Acrothele*，*Obolella* 等属。

小壳化石有腹足类 *Pelagiella* cf. *deltoides*；单板类 *Ceratoconus* sp.，*Protowenella flemingi*；软舌螺类 *Paraglobrilus* cf. *condersatus*，*Doliutus zhenbaensis*，奇特骨科类 *Sachites pressus* 等。属于 *Protowenella flemingi* 延限带的常见化石组合。*Protowenella flemingi* 是澳大利亚昆士兰地区 Currant Bush 组（Post-Templetonian）的标准分子，也曾见于丹麦波恩荷尔姆岛的卡尔拜黏土层中（Vivianne Berg-Madsen and John S. peel，1978），地质时代为徐庄期晚期。

在以往地质工作中，宜昌地矿所和海南地质大队联合区调队（1991）、曾庆銮等（1992）将大茅组时代划归武陵世。张喜光（1986）则将澳大利亚 Ordian 阶与我国龙王庙阶（都匀阶）对比，把大茅组下部划归黔东世。在 2000～2002 年开展的 1∶25 万《乐东县、陵水县》幅区调工作中，以 *X. hainanensis* 的出现作为武陵世的底界，把大茅组下部第 5 层灰白色中-薄层粉砂质页岩（产三叶虫：*Xystridura hainanensis*；腕足类：*Lingulella* sp.，*Homotreta* sp.）划归武陵世。*Xystridura* 也见于澳大利亚武陵统 Templetonian 阶（Vivianne Berg—Madsen 等，1978），张文堂、朱兆玲（2000）甚至将毛庄阶（台江阶）与澳大利亚 Templetonian 阶下部直接对比。此外，该类化石也见于新疆霍城，甘肃北山和哈萨克斯坦的武陵统下部。因此大茅组第 5 层以上层位地质时代为武陵世早期无疑，而大茅组第 5 层以下层位地层应为黔东世。因此，大茅组地质时代归为黔东世—武陵世（表 3-1）。

表 3-1　三亚地区与国外寒武纪生物群对比表

Table 3-1　Cambrian biota correlation between Sanya area and some areas in foreign contrios

统	阶	岩石地层	三亚地区（曾庆銮，1992）			澳大利亚 Vivianne B. M.，1978		南极洲 卢衍豪等，1974
			三叶虫	小壳化石	高肌虫			
武陵统	台江阶	大茅组	*Xystridura hainanensis-Galahetes hainanensis* 组合	*Protowenella flemingi-Pelagiella* 组合	*Ophiosema Indota bradoria*	Temp-leouian 阶	*Ptychagnostus gibbus* *Xystridura templetoniensis*	*Amphoton oatesi-Amphoton* *Xystridura glacia-Xystridura* *Xystridura multilinia-Xystridura*
黔东统	都匀阶	孟月岭组				Ordian 阶	*Redlichia*	*Chorbusulina subdita-Chorbusulina*

（二）奥陶系

1. 大葵组

大葵组下部产微古植物 *Leioninuscula* sp.，腕足类 *A-rchaeodiscina*? sp.（葛梅钰等，1983）。因化石较少，未建带，其地质时代相当于早奥陶世早、中期，可与新厂阶大致对比，层位相当于长江三峡地区的南津关组、分乡组和红花园组的下部。

大茅村西侧 0 号勘探线 ZK27 孔和 16 线号勘探线 ZK16 孔大葵组上部含炭灰岩中产牙形虫化石（汪啸风等，1992），以 *Paroistodus parallelus*，*Serratognathus bilobatus*，*Scolopodus* cf. *rex*，*Bergstroemognathus* sp. 为代表。根据该组上部所产牙形石生物群特征（图版Ⅰ-9），其地质时代为早奥陶世中晚期，其层位可与扬子地层区红花园组下部及中部相当。大葵组上部可识别出 *Serratognathus bilobatus* 带，除带化石外，还有 *Bergstroemognathus* sp.，*Oepikodus communis* 等，上述分子常见于华南道保湾阶上部，其层位相当于红花园组上部至大湾组下部。

2. 牙花组

下部灰岩中产的牙形石有（汪啸风等，1992）：*Scolopodus euspinus*，*S.* cf. *rex*，*S.*？*nogamii*，*Glyptoconus* sp.，*Drepanoistodus flexilis*，*Protopanderodus* cf. *leonardi*；中部白云质灰岩中的牙形石有：*Scolopodus*? aff. *nogamii*，*Glyptoconus* cf. *cornuformis*，*Protopanderodus* cf. *leonardi*，*Walliserodus* sp.；上部钙质、炭质页岩中的笔石化石有：*Crytograptus tricornis*，*Glyptograptus intersitus*，*G.* cf. *dentatus*，*Tylograptus globiformis*，

T. genticuliformis。下部共生分子有：*Scolopodus euspinus*，*S.* cf. *rex*，*Drepanoistodus flexilis*，*Drepanodus* cf. *planus*，*Walliserodus* sp. 等（图版Ⅰ-6、图版Ⅰ-8、图版Ⅰ-9），经汪啸风等（1992）研究，虽未见于大湾阶中、上部或者达瑞威尔阶典型分子，但发现了见于Volkhovian—兰维恩期的常见分子 *Glyptoconus* cf. *cornufornis*，其地质时代应为中奥陶世。层位相当于扬子地层区大湾组。牙花组上部有 *Tylograptus globifomis*，*T. genticuliformis*，*Climacograptus* cf. *uniforumis* 等，相当于达瑞威尔阶下部 *Glyptograptus intersitus* 带，是澳大利亚达瑞威尔阶第二化石带，其他分子亦常见于华南中奥陶世中晚期相当地层中。

3. 沙塘组

下段下部灰岩中产少量牙形石和苔藓虫化石碎片（图版Ⅰ-7、图版Ⅰ-8、图版Ⅰ-9）。牙形石有：Oistodiform，Drepanodiform。代表性剖面上段炭质页岩中产笔石化石（葛梅钰等，1983）：*Dicellograptus* cf. *forceps*，*D. sextans* var. *excilis*，*Glassograptus* cf. *hincksii*，*Glyptograptus dentatus*，*G.* sp.，*Climacograptus uniformis*，*C.* cf. *putillus* 等。下段为 *Glossograptus hincksii* 带下部（汪啸风，1982），除带分子外，有 *Climacograptus* cf. *uniformis*，*C.* sp. 等；此外，偶见牙形石碎片。带分子常见于我国江南区是胡乐组下部，其地质时代为中奥陶世晚期。上段为艾家山阶 *Glossograptus* cf. *hincksii* 带，主要分子有 *Dicellograptus* cf. *forceps*，*D. sextans*，*Golssograptus* cf. *hincksii*，*Glyptograptus clentatus*，*climacograptus uniformis* 等（葛梅钰等，1983）。大致可与长江三峡地区 *Nemagraptus gracilis* 带至 *Diplograptus mutidans* 带下部对比，大致相当于艾家山阶下部。

4. 榆红组

组内岩石未见化石产出，其地质时代按上、下地层层位时代确定。

5. 尖岭组

下部产丰富的笔石化石和少量腕足类、双壳类、三叶虫、头足类等门类化石（图版Ⅰ-3至图版Ⅰ-8）。根据动物群共生组合特征，可分为上、下两个组合（曾庆銮等，1992）。

（1）*Climacograptus brevis* 带

产于尖岭组中、下部，除带分子外，笔石有 *Climacograptus wilsoni*，*C.* cf. *brevis*，*Glyptograptus* cf. *siccatus*，*Orthograptus hainanensis*；三叶虫有 *Birmanites* cf. *hupeiensis*；腕足类有 *Zanclorhyncha* sp. 等；*C. brevis* 是胡乐阶至滞江阶常见分子，其层位大致相当于我国庙坡阶或胡乐阶中上部，与卡拉道克期早—中期地层可以对比。

（2）*Mimella yaxianensis* 带

产于尖岭组上部，除带分子外，腕足类有 *Sowerbyella hainanensis*，*Leptelloidea yaxianensis*，*Protozyga magnicostata*，*P. nasuta*；笔石有 *Glyptograptus* sp.，*Prolasiograptus* sp.；可与 *Dicranograptus clingani* 带上部至 *Pleurograptus linearis* 带下部对比，时代为卡拉道克期中—晚期。

6. 干沟村组

干沟村组所产化石主要是介形类化石（图版Ⅰ-3至图版Ⅰ-6），腕足类有 *Sowerbyella* sp.，*S. hannanensis*，*S.* cf. *sericea*，*Zancloorhyncha* sp.，*Z. yaxianensis*，*Z. gentilis*，*Camerella* sp.；双壳类 *Cymatonota hounghuaensis*，*Ctenodonta* sp.，*Similodonta similis*，*Modiolopsis* sp.；头足类 *Michelinoceras* sp.；三叶虫：*Encrinuroides* sp.，*Parisoceraurus* sp.；几丁虫：*Conochitina brevis* 等。组成 *Sowerbyella* 延限带，其地质时代大致与卡拉道克晚期—阿什极尔早期相当。

根据《中国地层指南及中国地层指南说明书》（2001），将中奥陶统底界划在牙形石 *Baltoniodus triangularis* 带之底，中/下奥陶统界线附近牙形石序列由下而上为：*Oepikodus communis*，*Oepikodus evae*，*Baltoniodus triongularis*，*Baltoniodus navis*；将上/中奥陶统界线确定为以笔石 *Nemagraptus gracilis* 首现为标志。从三亚地层区大葵组上部钻孔采获的牙形石属种来看，它们的层位非常接近中/下奥陶统界线层。*Bergstroemognathus* sp. 和 *Oepikodus* cf. *communis* 在长江三峡地区可由红花园组上部延入大湾组底部；*Serratognathodus bilobatus* 是华北地区亮甲山组特有分子；而 *Paroistodus parallelus* 是大湾组下段的典型分子。因此，本区中/下奥陶统界线很可能应划在大葵组与牙花组之间。同时，在尖岭组上部产出笔石 *Glossograptus hincksii*（葛梅钰等，1983），它在我国江南地区胡乐组底部、东南地区湘赣分区龙溪组下部，以及珠江分区下黄坑组均有产出，并且可与扬子地层区庙坡组下部 *Glyptograptus teretiusculus* 带对比（赖才根等，1982）。另在尖岭组下部见及笔石 *Nemagraptus* sp.。据此认为，中/上奥陶统界线附近的层位应相当于尖岭组上部，上/中奥陶统界线划在沙塘组上部是适宜的。

三、岩相古地理特征

（一）沉积环境分析

纵观三亚地区早古生代沉积，总体经历了寒武纪孟月岭期—奥陶纪沙塘期以碳酸盐岩和陆源碎屑岩混合沉积为主和奥陶纪榆红期—尖岭期以陆源碎屑岩沉积为主的两个大的沉积旋回，沉积环境在滨海至浅海间动荡变换。

孟月岭组底部岩石缺失（未见底），就其岩石出露情况而言，孟月岭早期水体较深，位于浅海陆棚滨外环境（图 3-7），沉积了较厚（>883m）的粉砂质泥岩、水云母黏土岩。岩石发育水平层理、小型沙纹层理（层厚 1~2cm），含较多的有机质，且分布不均匀，宏观上呈斑杂状，镜下呈短柱状、斑块状富集，似为生物扰动所致。中期发生海退，滨外环境向近滨环境演化，沉积了一套巨厚层—块状的石英砂岩（厚约 564m）。石英砂岩从下至上颗粒有变粗趋势，上部石英砂岩中见有平行层理，镜下可见石英颗粒呈定向排列，下部过渡为具暗色水平层纹的细粒石英砂岩。岩石结构成熟度和成分成熟度中等，无生物扰动构造等。粒度曲线显示，无明显滚动总体，以跳跃总体为主，曲线斜率陡（图 3-8）。反映该时期陆源补给量丰富，沉积速率较大，盆地性质在该时期并非完全稳定。应为过渡类型。晚期，又发生海侵，进入浅海陆棚滨外环境，沉积了水云母黏土岩、白云岩，岩石发育水平层理、小型沙纹层理。岩石沉积层序与早期沉积层序相似，说明盆地性质并无显著改变。寒武纪孟月岭组时期，从早至晚，沉积环境经历了由滨外→近滨→滨外的演变。

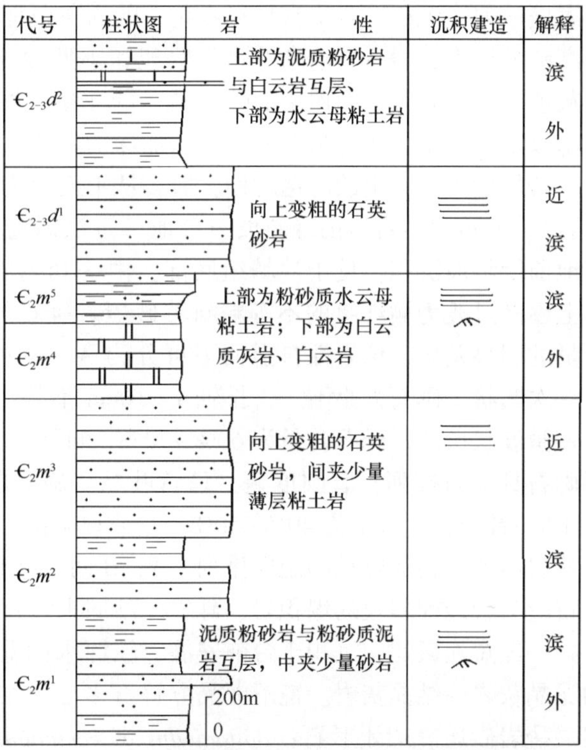

图 3-7 三亚榆林孟月岭组—大茅组沉积序列图

Fig. 3-7 Sedimentary sequence diagram of the Mengyueling-Damao Formations, Yulin, Sanya

大茅期早期，又发生海退，沉积了厚达 486.56m 的近滨相中细粒石英岩状砂岩或石英砂岩。沉积粒度向上逐渐变粗，水平层理发育。岩石磨圆度好，矿物成熟度高。矿物成分几乎都为石英，未见到不稳定矿物组分。中晚期总体上表现为海侵旋回沉积，开始以其底部浅海相水云母黏土岩的形成标志。继之，有几次短暂的海退—海进过程的转换，形成石英砂岩、粉砂岩与水云母黏土岩的互层，盆地演化背景以下陷为主，且具短期动荡的性质。晚期是海侵的高潮，陆源补给量显著减少，形成硅质岩、含磷硅质岩和白云岩的混合沉积。白云岩、硅质岩、硅质页岩中常见有粉砂级的石英颗粒。硅质岩以隐晶结构为主；磷块岩主要为碎屑结构，次为鲕状结构；白云岩类则为微晶结构。沉积韵律性明显，薄层硅质岩、含磷页岩、硅质岩与锰质层呈互层出现。在磷块岩顶、底板的硅质页岩中发现丰富的三叶虫、腕足类、软舌螺等浅海底栖生物化石。在纵向和横向上岩相变化显著，偶夹有碎屑岩层，横向上厚度极不稳定。表明晚期沉积物形成于较局部的浅海环境，沉积盆

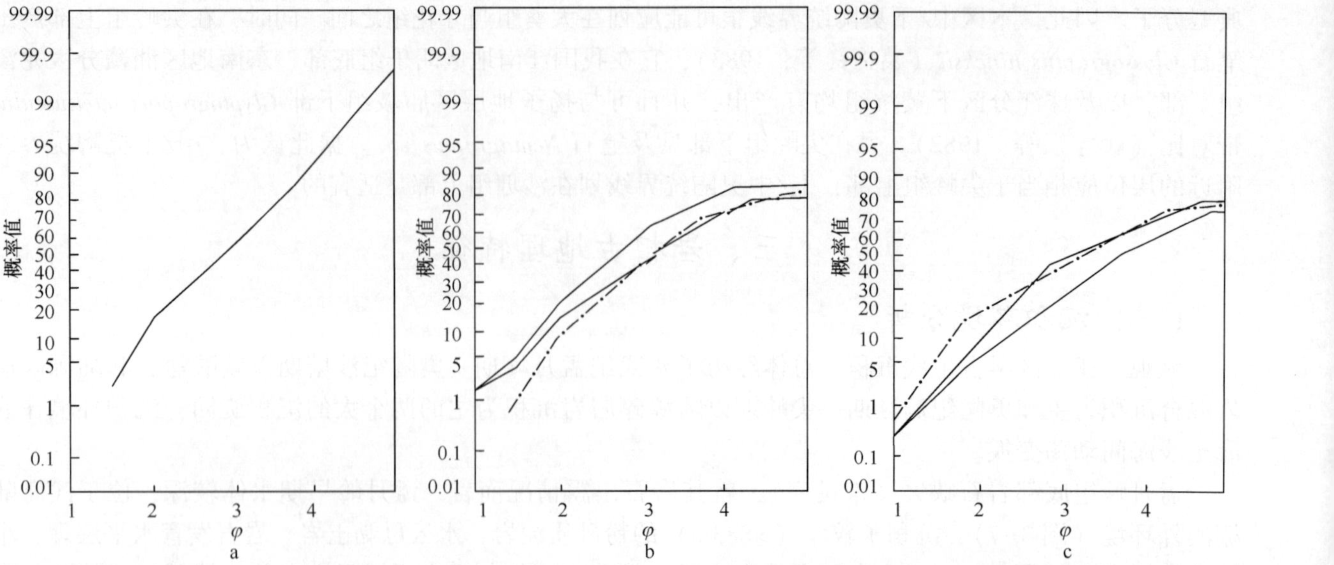

图 3-8　三亚市下古生界岩石粒度分布概率累积曲线图

Fig. 3-8　Cumulative grain-size distribution curve of Lower Palaeozoic rocks, Sanya City

a—孟月岭组底部石英砂岩；b—榆红组岩屑砂岩；c—尖岭组岩屑杂砂岩

地离陆源较近，且具有短期动荡不安的特征。

三亚地区大葵组呈隐伏状产出。大葵期早期，为海侵早期的滨海相石英砂岩，见腕足类碎片及微古植物碎片（葛梅玉等，1983）。大葵期晚期海侵加剧，出现较厚的滨外碳酸盐岩沉积，岩石含少量炭质，产较多的牙形石（汪啸风等，1992）。

继大葵期后，牙花期发生局部海退，出现牙花期早期的近滨相块状石英砂岩沉积。牙花期中晚期为海侵序列，沉积了含丰富牙形石化石的深灰色厚层灰岩等；牙花组顶部含笔石黑色页岩的出现为海侵高潮的标志，岩石具清晰的水平层理和小型沙纹层理。但北东面大茅一带（代表性剖面），中部为深灰色厚层灰岩等，而西南面落笔洞一带（参考性剖面）产牙形石灰岩中常出现薄层砂岩夹层，由此暗示该期沉积水体具有自西南往东北加深的趋势。

沙塘期为较稳定的海退时期，岩石从下而上为钙质砾灰岩→泥晶、结晶灰岩→泥灰岩（已变质为角岩）。显示出由台地前缘斜坡→台地→近滨海盆的海退相序沉积特征（曾庆銮等，1992；图 3-9）。台地前缘斜坡沉积：见于沙塘组底部，由一套钙质砾灰岩组成。沉积环境应为低能的局部还原环境，向上逐渐过渡为榆红组的不等粒石英砂岩。砾石成分为泥晶灰岩—微晶灰岩，由于砾石大小、形状、含量的明显差异，该岩性自上而下可分为 A、B 两小层。A 层砾石具定向排列，分选中等-差，次棱角—次圆状，砾石大小不一，长轴 1～10cm 不等，其中以 1～3cm 者居多。砾石较大处可见基质具水平—微波状层纹，砾石含量为 60%～70%，自下而上砾石具变大趋势，基质为碳酸盐岩砂屑、粉屑。B 层砾石具定向排列，磨圆度差，呈竹叶状—扁豆状，分选性差。砾石长轴为 1～20cm 不等，其中以 5～10cm者居多，含量为 40%～50%，基质以碳酸盐岩砂屑、粉屑为主，具条纹状构造，镜下表现为碳酸盐岩砂屑、粉屑颗粒的定向排列。A、B 两层间具明显的凹凸不平的冲刷界面。上述特征表明，该套岩石应为台地前斜坡的塌积岩、砾岩。砂屑来自台地或上斜坡已胶结的碳酸盐岩层，其受波浪作用而破碎，经短途搬运，沉积于台前安静、较深水的盆地中。台地碳酸盐沉积形成了一套厚层—块状构造的泥晶灰岩—结晶灰岩。镜下泥晶方解石多已重结晶，局部残余可见较均匀的泥—微晶灰岩，含少量碳酸盐岩砂屑。产牙形石：*Baltoniodus* cf. *communis*，*Drepanoistodus* sp.，*Paltodus* sp. 及苔藓虫等。该套灰岩中还偶见清晰的不对称波浪，说明沉积期水体较浅。近滨海盆沉积由下部接触变质形成的透辉石角岩和上部含笔石类化石的水云母黏土岩组成。透辉石是岩浆热液交代钙质沉积物的特征矿物，结合岩石中的残余沉积构造（纹层状构造、“刀砍状”构造等）分析，认为其原岩应为泥灰岩。

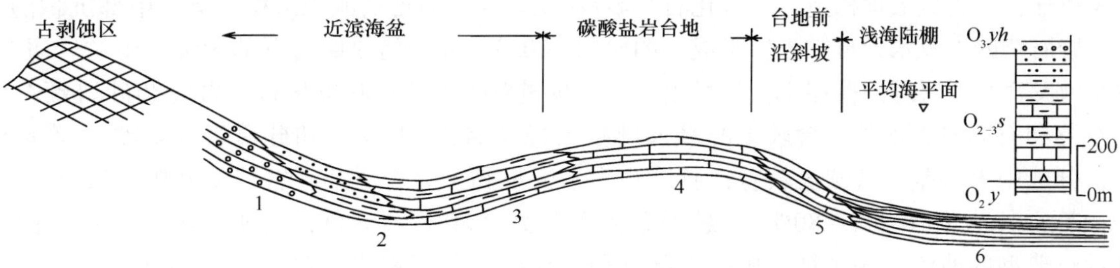

图 3-9　三亚市牙花组顶部—榆红组底部相变示意图

Fig. 3-9　Schematic diagram of facies change of the Yahua Formation Top-Yuhong Formation Bottom, Sanya City

1—复成分砾岩；2—岩屑砂岩、粉砂岩-水云母黏土岩；3—泥灰岩；4—灰岩；5—钙质砾灰岩；6—黑色页岩

榆红期沉积由一套复成分砾岩、岩屑砂岩的多次重复构成，间夹少量水云母黏土岩，厚度>1500m。榆红组下部厚层—块状复成分砾岩的出现是沉积相向陆推进的最高潮。之后，盆地具有总体下陷变深的特征，其标志之一是复成分砾岩层巨厚，砾石的砾径从下至上有明显变小的趋势。此外，自下而上，砾岩层（单层）变薄，砾岩/砂岩比率较低，且砾岩中杂基含量增高。其中，砾岩砾石成分复杂，除稳定的砂岩、粉砂岩砾屑外，还有较多的不稳定组分砾屑，如水云母黏土岩砾屑、灰岩和白云岩砾屑及少量的硅质岩砾屑。砾石大小不一，磨圆度中等—差，无规则排列。颗粒支撑为主，上部层位见杂基支撑。基质为砂级颗粒和水云母黏土。岩石并含透闪石及少量角闪石、橄榄石等不稳定矿物。岩屑砂岩也为厚层—块状构造，砂级颗粒除石英及石英岩屑外（38%~62%），水云母黏土岩屑也占相当比例（9%~38%），此外还见有硅质岩屑（<8%）。其粒度图分析表明，没有明显的滚动总体，以跳跃总体为主（>80%），悬浮总体含量较高（<20%），曲线斜率低，为非正常牵引流的沉积特征（见图 3-8）。上述沉积特征与盆地快速沉降及盆地外侧的快速隆起有关。粗碎屑物质只受水流的短期搬运，快速堆积于安静、低能的近滨海盆，沉积速率快，不宜于生物生存，这也是榆红组不产化石和有机质含量极低的原因。此外，沉积盆地处于良好的补偿环境，因而形成巨厚层由复成分砾岩和岩屑砂岩组成的滨岸冲积扇沉积（曾庆銮等，1992；图 3-10）。

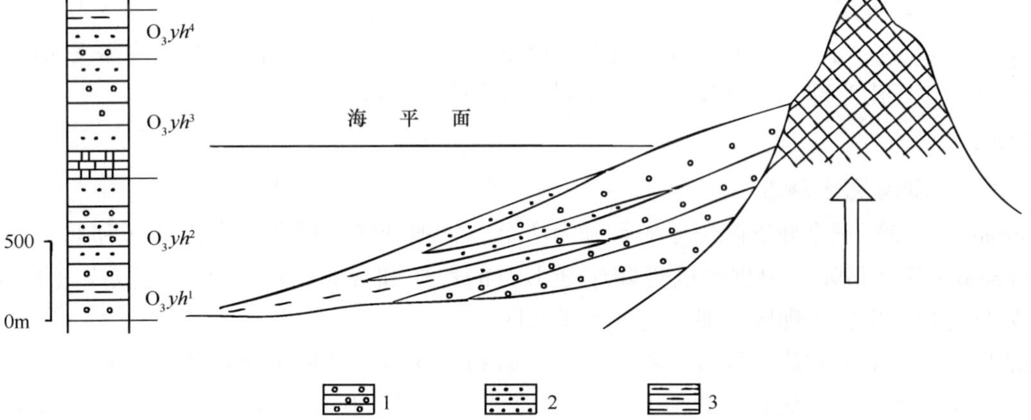

图 3-10　三亚市上奥陶统榆红组海岸冲积扇沉积示意图

Fig. 3-10　Schematic diagram showing sedimentation of the Upper Ordovician Yuhong Formation, Sanya City

1—复成分砂岩；2—岩屑砂岩；3—水云母黏土岩

随着海平面继续相对上升，至尖岭期，沉积环境已由滨岸过渡为浅海环境。沉积了一套韵律构造明显的岩屑砂岩、岩屑杂砂岩、粉砂岩、水云母黏土岩，中、上部产丰富的正常浅海相化石。由于盆地演化的延续性，该时期盆地边缘较陡，泥、砂物质在陡峻的斜坡上一次又一次地受到重力滑动的影响，形成具韵律构造的浊流沉积。底部沉积韵律从下至上由砾岩→砂砾岩→砂岩构成，韵律层厚<1m，

岩石界线清晰，无明显层理构造，不含化石、砂/泥比率高，为近端浊积沉积。下—中部沉积韵律由岩屑砂岩、岩屑杂砂岩及水云母黏土岩组成，韵律层厚0.5~1m。韵律层内下部为细—中粒岩屑杂砂岩，上部主要为发育水平层理的薄层水云母黏土岩，含少量双壳类和三叶虫化石。岩屑杂砂岩颗粒成分复杂，除石英和石英岩岩屑外，含较多的水云母黏土岩（24%~45%）和硅质岩（<4%）等不稳定组分岩屑，分选中等—差，磨圆度较差；胶结物为水云母黏土。岩屑杂砂岩粒度分析结果显示（图3-8），以跳跃总体为主（75%~80%），悬浮总体含量大（20%~25%），无明显滚动总体，曲线斜率低，显示较典型的浊流沉积特征。中—上部沉积韵律由粉砂岩和水云母黏土岩组成，镜下可见清晰的小型韵律，韵律层内粉砂岩见波状水平层理和沙纹层理，粉砂级颗粒中含水云母黏土岩岩屑及硅质岩岩屑等不稳定组分，磨圆度差。水云母黏土岩与粉砂岩呈渐变过渡关系，见水平层理、微粒序层理及微波状层理。含笔石、头足类、三叶虫、腕足类等生物化石，砂/泥比率低，属于远端浊积岩。

从尖岭期至干沟村期，又转换为海退序列。沉积了干沟村组滨—浅海相岩屑砂岩或砂砾岩。岩屑形态以次圆状为主，少量次棱角状，分选性好。产腕足类、头足类、双壳类、三叶虫等化石。之后，由于加里东运动后期的影响，海盆地封闭、隆起，进入剥蚀阶段。

（二）生物古地理区讨论

1. 寒武纪

大茅组产十分丰富的三叶虫、小壳化石、腕足类、双壳类、古介形类等。三叶虫有4属 *Xystridura*，*Galahetes*，*Kootenia* 和 *Pagetia*。仅三叶虫 *Xystriduya* 就有3种。除 *Kootenia* 为广布型分子外，*Xystridura*，*Galahetes* 和 *Pagetia* 均为东方动物群（亚澳生物大区范畴）的重要分子。

小壳化石有27属，*Cupitheca*，*Pseudorthotheca*，*Microconus*，*Hyolithellus*，*Eiffella*，*Bucanotheca*，*Archiasterella*，*Chancelloria*，*Dimidia*，*Zigzagella*，*Miriella*，*Aegides*，*Cambroclavus* 等；主要以 *Cambroclavus-Zhijinites-Rhombocorniculum-Microconus* 组合为特色的 *Cambroclavus* 生物群的分布由哈萨克斯坦-新疆塔里木-湖北西北部，向南与东澳大利亚相连，向北可延入朝鲜半岛；其地理分布与三叶虫东方动物群基本一致。

其他门类化石如腕足类、古介形类、双壳类等，属种单调，但三亚地层区与澳大利亚昆士兰地区古介形类关系密切。至此，本区大茅组与澳大利亚昆士兰地区Bush组同期生物群关系密切。武陵世时期，它们可能均位于南半球（低纬度带）干旱带（陈旭等，2001）。

2. 奥陶系

（1）中、下奥陶统牙形石

Bergström（1973）把全球奥陶纪牙形石分为两个生物地理区，即北美中大陆区和北大西洋区；王志浩和Bergström等（1996）根据中国奥陶纪牙形石的性质和分布，认为在特马豆克（Tremadocian）期之后明显分为两个生物地理区，即华南区和华北区。

三亚地层区中、下奥陶统牙形石主要产于大葵组和牙花组，沙塘组所产牙形石仅为化石碎片。大葵组和牙花组所产牙形石，从属级分类上共有11属，它们是：*Drepanoistodus*，*Serratognathus*，*Walliserodus*，*Paroistodus*，*Scolopodus*，*Bergstroemognathus*，*Drepanodus*，*Glyptoconus*，*Acontiodus*，*Oepikodus*，*Protopanderodus*。除 *Serratognathus bilobatus* 和 *Drepanoistodus flexilis* 为华北生物地理区分子外，其他属常见于华南生物地理区。与澳大利亚中部同期生物群比较（Barry J. Cooper，1981），后者Horn Valley粉砂岩的牙形石共有4种类型：①北美类型；②北大西洋类型；③澳大利亚土著类型；④广布型分子。据此，本区中、下奥陶统牙形刺生物古地理区与华南区相似，与澳大利亚同期生物群有相似性，但并不十分密切，因为在澳大利亚中、下奥陶统有较多土著分子，在海南三亚地层区并未发现。但值得提及的是，三亚地层区大葵组、牙花组和沙塘组所获牙形石资料较少，尚待进一步工作。

(2) 中、上奥陶统

本区中、上奥陶统最大特点是多门类生物的辐射占主导，因此生物地理格局也十分复杂。这里仅对6个门类生物化石属级分类单元的地理分布特征进行初步的分析。

1）笔石

本区笔石动物群见于沙塘组、尖岭组和干沟村组，计有9属，它们是：*Climacograptus*，*Glyptograptus*，*Nemagraptus*，*Orthograptus*，*Hallograptus*，*Corgnoides*，*Prolasiograptus Dicellograptus*，*Glossograptus*；与江南区同期生物群关系密切（汪啸风，1982，1999），应属太平洋动物群，分布在低纬度暖水域（陈旭等，2001）。

2）头足类

主要见于尖岭组和干沟村组，共计5属：*Richardsonoceras*，*Eurasiaticoceras*，*Michelinoceras*，*Westonoceras*，*Ancistroceras*；上述头足类中，除*Westonoceras*为广布型分子外，*Richardsonoceras*和*Michelinoceras*是扬子地层区、塔里木地层区常见分子，前者也偶见于华北地层区。*Ancistroceras*仅见于长江三峡地区。*Eurasiaticoceras*见于华南，在华北地层区见及零星报道；在国外主要见于欧洲中、上奥陶统。从头足类的全球分布来看，早奥陶世该动物群的分区性并不明显，中、晚奥陶世华南与华北分异性明显，但本区缺乏可进行生物地理区对比的特征性分子。

3）三叶虫

主要见于尖岭组和干沟村组，共计9属：*Atractopyge*，*Encrinuroides*，*Parisoceraurus*，*Neseuretus*，*Birmanites*，*Bulbaspis*，*Dionide*，*Anpyxinella*，*Remoplearides*。在上述三叶虫动物群中，除*Encrinuroides*和*Remopleurides*为世界性广布型分子外，*Atractopyge*，*Parisoceraurus*和*Dionide*为华南中、上奥陶统常见分子；*Birmanites*在华南、华北、塔里木及澳大利亚均有分布；*Bulbaspis*和*Ampyxinella*是哈萨克斯坦同期动物群的常见分子，但前者在新疆、宁夏也有分布。最为重要的是，在本区中、上奥陶统发现冈瓦纳周缘生物相带的组合分子*Neseuretus*，该生物相带是由三叶虫*Neseuretus*和笔石*Isograptus*组成，*Neseuretus*的出现，表明本区与冈瓦纳周缘关系密切，古地理位置可能很接近。

4）腕足类

主要见于沙塘组、尖岭组和干沟村组，共计14属：*Mimella*，*Leptellina*，*Lingulella*，*Camerella*，*Sowerbyella*，*Zanclorhyncha*，*Ancistrorhyncha*，*Hallina*，*Protozyga*，*Zdiospira*，*Orbiculoidea*。在腕足类动物群中，世界各地或广布型分子约占50%以上，如*Leptellina*，*Mimella*，*Lingulella*，*Souerbyella*，*Zdiospira*，*Orbicaloidea*；北美类型的分子有*Ancistrohyncha*，*Hallina*和*Protozyga*，*Camerella*主要见于北美和东亚；*Zanclorhyncha*主要分布于扬子地层区、江南过渡带和青海。但总体面貌特征同江南过渡带同期生物群也较相似。

5）双壳类

主要见于尖岭组和干沟村组，共计7属：*Cymatonota*，*Ctenoclonta*，*Similodonta*，*Modiolopsis*，*Eopteria*，*Tancrediopsis*，*Paraphtonia*。在上述双壳类中，*Cymatonota*和*Similodonta*主要分布于北美；*Eopteria*见于北美、法国及华北地区。*Modiolopsis*是广布型分子，不仅在江南过渡带常见，而且奥陶系、志留系均可见分布。*Tancrediopsis*和*Paraphtonia*是哈萨克斯坦Caradoc期常见分子，*Ctenodonta*仅见于江南过渡带。

6）几丁虫

仅见*Conochitina brevis*，但该几丁虫与华南冈瓦纳北缘奥陶纪同期动物群关系密切（汪啸风等，1999；陈旭等，2001）。

从上述中、晚奥陶世6个门类化石动物群的地理分布分析，笔石、几丁虫、头足类、三叶虫的古生物地理指示意义相对重要，总的生物群面貌接近江南过渡区，几丁虫*Conochitina brevis*与华南、冈瓦纳北缘奥陶纪同期生物群关系密切。三叶虫*Neseuretus*是冈瓦纳周缘的生物相带分子。值得提及的是，笔石动物群与江南区同期生物群相比是十分相似的。据此，认为该区中、上奥陶统生物古地理区位于南半球干旱带低纬度区是可能的，与华北、澳大利亚等地在生物古地理区上较为接近，关系密切。

另据冯增昭等人（2001）研究，中国南方寒武纪岩相古地理自西往东主要包括几个主要古地理单元：康滇古陆、扬子碳酸盐岩台地、江南页岩盆地-碳酸盐岩台地、东南碎屑岩台地。奥陶纪岩相古地理与寒武纪基本类似，但从早奥陶世→晚奥陶世，沉积相发生多变，总体呈现出江南页岩盆地-碳酸盐岩台地消亡、而东南碎屑岩台地和康滇古陆的进一步扩宽发展，扬子碳酸盐岩台地也经历了碎屑岩台地→碳酸盐岩台地→页岩盆地的演变。海南岛三亚地层区寒武纪沉积岩为碳酸盐岩-碎屑岩组合，早、中—晚奥陶世为厚层碳酸盐岩-碎屑岩沉积，晚奥陶世则为碎屑岩台地沉积，其演化特点与中国南方奥陶纪岩相古地理变化基本相似。推测寒武纪—奥陶纪时期是连接扬子碳酸盐岩台地和东南碎屑岩台地过渡地带，从而导致了古生物组合特征的多样性。

（三）古地磁资料

古地磁资料显示（汪啸风等，1992），孟月岭组古地理位置为南纬 7°，而大茅组古地理位置为南纬 3. 54°，牙花组古地理位置为南纬 8. 78°，沙塘组古地理位置为南纬 2. 99°，干沟村组为北纬 7. 86°~10. 31°。与石灰顶组最终定位古地理位置（南纬 4°）基本处于同一条纬线上，都处于低纬度的赤道附近。

第二节　五指山地层分区

本地层区由新元古界和古生界一套绿片岩相岩石组成，分布较广，受后期多期次岩浆侵蚀，导致呈残留体状漂浮于花岗岩中。

一、新元古界

本地层区新元古界出露局限，目前仅见分布于海南岛昌江石碌矿区，出露面积约 21km^2。为一套浅变质岩系，自下而上分为石碌群和石灰顶组等两套含铁岩系。石碌群共分为 6 层，为一套具有热变质叠加的高-低绿片岩相浅变质岩；石灰顶组为一套未经区域变质的正常陆源碎屑沉积岩。石碌群未见底，与时代更老的中元古界长城系抱板群为断层接触，与上覆石灰顶组为角度不整合接触，石灰顶组未见顶，与下古生界地层接触关系不清。

自 20 世纪 60 年代初至今，曾有多个生产、科研、教学单位及不少国内知名学者对测区石碌群和石灰顶组地层进行过研究，提出自己的地层划分和时代归属意见（表 3-2）。目前根据石碌群第六层产 *Chuaria-Tawuia* 生物群和第六层铁矿主矿层 Sm-Nd 全岩等时线年龄（841±20）Ma，时代归属为青白口纪。而石灰顶组缺乏年代学、古生物学资料的制约，仅依据其叠覆于石碌群之上而归属为震旦纪。本次根据《〈中国区域地质志〉编制项目“江南古陆”元古宙地层野外地质考察与专题讨论会会议纪要》，新元古代划分为青白口纪（780~1000Ma）、南华纪（635~780Ma）、震旦纪（542~635Ma）。结合海南岛新元古界地质的实际情况，把石灰顶组时代修订为南华纪—震旦纪；而石碌群时代保持不变，为青白口纪。

（一）岩石地层特征

1. 剖面描述

（1）*石碌群（$Pt_3^1\hat{S}$）剖面*

青白口系石碌群赋存有全国闻名的大型富铁矿床，其地质研究程度极高，曾有多家测制过地层剖面。石碌群第 1~5 层采用宜昌地矿所、海南地矿局 1986~1990 年开展《海南岛重要基础地质问题研究》（汪啸风等，1992）时所测金牛岭剖面，第六层采用石碌铁矿区北一矿 344 台阶剖面资料加予补充。陈哲培等（1997）在开展《海南省岩石地层》项目时，把金牛岭剖面和北一矿 344 台阶剖面进行拼接，作为石碌群的代表性剖面。剖面自上而下为：

未见顶

石碌群	总厚度>1215.30m
第六层	
23. 灰色厚层白云岩	>3.30m
22. 黑色薄层炭质板岩，产宏观藻类：*Chuaria circularis*，*Shouhsienia shouhsienensis*，*Tawuia dalensis*，*T. sinensis*	37.80m
21. 灰色厚层白云岩	3.00m
20. 浅灰色透辉透闪岩夹炭质板岩	14.20m
19. 灰色薄至中厚层透闪角砾状灰岩	8.30m
18. 黑色薄层炭质板岩，底部夹2层灰绿色透辉透闪岩	11.90m
17. 灰色、灰绿色绿帘石化条纹状阳起石岩、阳起石硅质板岩，中部夹白色石英岩，顶部夹炭质板岩	22.90m
16. 灰色、灰绿色硅质板岩，微层理发育	30.70m
15. 黄绿色薄层硅化-透闪石化磁铁矿岩，夹泥质绢云灰岩，顶部具条带状构造	28.20m
14. 钢灰色粒状含重晶石石英赤铁矿石，夹黄色砂岩	10.10m
13. 灰色网状重晶石化结晶灰岩，含孔雀石	5.00m
12. 灰色、灰绿色透辉透闪岩，夹白云质灰岩	25.20m
11. 灰色硅质板岩	10.60m
10. 灰绿色含金云母、磁铁矿结晶灰岩	4.70m
9. 灰色透辉透闪结晶灰岩，夹含钙质透闪岩及硅质板岩	21.90m
第五层	
8. 灰色硅质板岩	8.20m
7. 灰色中厚层变质粉砂岩	15.80m
6. 黄色、灰色绢云片岩，夹石英绢云片岩，局部含少量炭质	303.00m
第四层	
5. 灰白色中厚层石英岩、石英片岩，夹灰色斑点状绢云石英片岩	103.60m
第三层	
4. 灰色、深灰色、灰黄色薄—中厚层石英绢云片岩，含绿泥石斑点	293.50m
3. 灰绿色石英绢云片岩夹绢云片岩	25.25m
第二层	
2. 灰色中厚层蛇纹石化白云质灰岩	12.19m
第一层	
1. 灰色含红柱石斑点绢云母片岩与暗灰色绢云母石英片岩互层，局部含砾	>215.00m

未见底

(2) 石灰顶组（$Pt_3^{2-3}\hat{s}$）剖面

1) 石碌铁矿区石灰顶东坡剖面

该剖面引自宜昌地矿所、海南地矿局1986~1990年开展《海南岛重要基础地质问题研究》（汪啸风等，1992）资料，剖面始于石灰顶东坡，止于344隧道口。

未见顶

石灰顶组	总厚度>214.2m
10. 灰黑色含铁质不等粒石英砂岩	>70.00m
9. 下部为灰白色石英岩，上部为棕褐色含赤铁矿不等粒石英岩	28.20m
8. 灰色、浅灰色泥岩夹薄层硅质岩	20.10m
7. 深灰黑色含水云母砂屑赤铁矿粉砂岩	25.60m
6. 黄色薄层粉砂质泥岩	5.70m

5.	含灰质绢云母石英岩	18.80m
4.	灰黑色含赤铁矿水云母不等粒石英砂岩	14.60m
3.	灰绿色含灰质绢云母石英岩	9.00m
2.	棕褐色厚层含赤铁矿细粒石英砂岩	22.20m

══════ 断　层 ══════

石碌群第六层

1. 灰色厚层白云岩

未见底

2）石碌铁矿区红房山

引自宜昌地矿所、海南地矿局1986~1990年开展《海南岛重要基础地质问题研究》（见汪啸风等，1992）资料。剖面位于北一矿南东、石灰顶北。

未见顶

震旦系石灰顶组　　总厚度>214.2m

11.	棕褐色含砾石英细砂岩	>23.80m
10.	砖红色泥岩	3.30m
9.	紫红色厚层含铁质页岩夹白色含砾石英细砂岩	10.70m
8.	白色厚层石英砂岩，岩性坚硬	6.60m
7.	棕褐色铁质石英细砂岩	17.30m
6.	黑色、白色中厚层含黄铁矿石英细砂岩夹紫红色粉砂质泥岩，产宏观藻化石	18.20m
5.	上部为黄色中厚层铁质石英粉砂岩与紫红色砂质泥岩互层，下部为黄色中厚层含砾石英细砂岩（0.60m）	8.70m
4.	棕褐色—紫红色含铁粉砂岩	8.70m
3.	钢灰色赤铁矿岩夹炭质页岩和硅质碎屑磁铁矿，底部1.60m为铁质石英细砂岩	8.90m
2.	黄色、白色泥岩夹重晶石化砂质贫磁铁矿及铁质石英细砂岩，未见底	10.00m

══════ 断　层 ══════

石碌群第六层

1. 灰色微片理化重结晶白云岩

未见底

2. 岩石地层特征

（1）*石碌群*

石碌群为海南地质大队（1964）所创建的岩石地层单位，命名地点为石碌铁矿区，以金牛岭南坡375排土场至北一矿出露较好，原分7层，时代为寒武系—奥陶系，但无任何生物化石依据。之后，不同单位和学者对石碌群提出了各种不同认识和划分方案。其中，中国科学院华南富铁矿科研队地层组（1976）认为石碌群第六层与第七层之间为假整合接触，创建石灰顶组代替原石碌群第七层，修订石碌群含义，将石碌群限定于原剖面的第一层至第六层，修订后的石碌群，内部无重要的沉积间断或不整合面，符合岩石地层单位“群”的含义。张仁杰（1990）将原石碌群第一至第六层划为石碌群，时代为青白口纪，第七层为石灰顶组，时代为震旦纪。陈哲培等（1997）在《海南岛岩石地层》中采用张仁杰（1990）意见，指定以金牛岭剖面拼接北一矿334剖面为代表性剖面，其中第一层至第五层为绢云石英片岩、石英绢云片岩夹结晶灰岩及石英岩，第五层局部地段顶部出现流纹质熔结凝灰岩、流纹质凝灰岩、中酸性沉凝灰岩（图版Ⅰ-10-A、图版Ⅰ-10-B）；第六层为白云岩、结晶灰岩、透辉透闪岩夹富铁矿层，炭质板岩夹层中含宏观藻化石。未见底，顶部以白云岩或透辉透闪岩与上覆石灰顶组呈不整合接触（图版Ⅰ-10-C）。

作为岩石地层单位的划分，石碌群以下直接分层明显不符合岩石地层单位的等级关系，但考虑到石碌群目前在海南及其邻区仅此一地有由石碌群分布，大量的文献资料一直沿用，似乎约定俗成，故暂不改层建组，实际上前人也曾进行过这方面的探索。从上述第一层至第六层岩石类型和层序特征描述，结合各层分布情况，第一层至第三层可划分为一个组，第二层可作为组内非正式地层单位；第四层和第五层为退积型层序或为一个大致对称的他旋回性层序，第四层与第三层之间具有突变性界面或层序转换面，无疑第四层和第五层构成一个组，下分两个段；第六层层序特征虽不清楚，以碳酸盐岩一种岩石为主，在野外具有可填图性，无疑又可分为一个组。

（2）石灰顶组

石灰顶组为中国科学院华南富铁矿科研队地层组（1976）创建的岩石地层单位，杨开庆等（1988）介绍，创建地点为石碌矿区石灰顶隧道北口及三棱山西南坡。后经宜昌地矿所、海南地矿局（1990）修订，代表假整合或不整合于石碌群第六层之上、石炭系南好组第五段（原三棱山组）之下的一套陆源碎屑岩夹铁矿层岩系，即地质大队（1964）所划分的石碌群第七层。张仁杰等（1992）并重测了石灰顶组剖面，厘定其时代为震旦纪。陈哲培等（1997）在《海南省岩石地层》中采用张仁杰等人的划分方案，指定石灰顶东坡剖面为代表性剖面，岩石以石英砂岩、石英岩为主，夹泥岩、硅质岩、赤铁矿粉砂岩。石英砂岩中常含赤铁矿，泥岩中含宏观藻化石，与上覆南好组和下伏石碌群均呈不整合接触。

（二）古生物特征及年代地层

1. 石碌群

1976年以前石碌群中未获任何生物化石，也无同位素测年资料，因而所述石碌群的时代归属意见均属推测。1976年中国科学院华南富铁矿科研队地层组在石碌群第二层及第三层至第五层获得古孢子，古孢子时代从震旦纪至志留纪或更晚，据此将石碌群时代定为寒武纪，底部不排除有震旦纪的可能。1978年广东省冶金地质934队在石碌群第六层中采获大化石石碌虫：*Shiluites hainanensis*，*S. gigantens*，*S. forcipiformis*，*S. rimalis*，*S. sulcifermis*，*S. ropteformis*，*S. nanus*，未提出时代归属意见。1981年欧阳舒、李再平报道石碌矿区CK714孔石碌群第五层中采获孢粉化石，其中无环三缝孢类：*Leitriletes* sp.，*Gulisporltes* sp.，*Punciatisporites*? sp.，*Cyclogranisp rites*?，*Apiculatasporites* sp.；有环三缝孢亚类：*Zonotriletes* sp.；单缝孢类：*Laevigatosporites* sp.，*Punctaosporites* sp.，*Archaeoplerisacus* sp.，*Pericutoporites*? sp.；花粉大类中具囊类：*Cordaitina*? sp.，*Vitreisporites* spp.；具沟类：*Cycadopites* spp.等，欧阳舒、李再平认为该组合中有些种可与欧、美中石炭世的种比较，将石碌群的时代归属于早石炭世晚期—中石炭世早期。张仁杰等（1990）研究了石碌群第六层上部所夹黑色炭质板岩中的宏观藻类化石，计有 *Chuaria circlaris*（圆形查尔藻），*Shouhsienia shouhsienensis*（寿县寿县藻），*Tawuia dalensis*（达尔塔乌藻），*T. sinansis*（中华塔乌藻）等，并命名为 *Chuara-Tawuia* 生物群，认为生物群广泛见于国内外的晚前寒武纪地层中，该生物群的主要分子 *Chuaria circularis* Walcott 最早发现于美国亚利桑那州大峡谷新元古界丘阿尔群，就 *Chuaria* 属而言，我国从长城系团山子组至上震旦统均有发现，时限约15亿~6亿年；这个生物群的另一个主要成员 *Tawuia*，最早发现于加拿大西北部马更些山区小达尔群，它常与 *Chuaria* 共生，地质历史相对较短，约9亿~6亿年。目前在斯瓦巴德群岛（Svalbard Islands）开普罗德组，印度中部凯莫群沙克页岩，我国河北怀来龙凤山青白口系、安徽淮南八公山-凤阳刘老碑组、宿县地区震旦系史家组、辽宁大连复县青白口系钓鱼台组、南芬组，震旦系桥头组、长岭子组等地均有发现；组合中的 *Shouhsienia* Xing 最早发现于安徽寿县刘老碑组，在辽宁复县钓鱼台组、长岭子组，河北怀来青白口系下马岭组及加拿大西北部马更些山区小达尔群均有报道，表明 *Chuaria-Tawuia* 生物群时代属新元古代青白口纪。张仁杰等（1990）还获得石碌群第六层铁矿主矿层Sm-Nd全岩等时线年龄数值（841±20）Ma，与 *Chuaria-Tawuia* 生物群延续时限9亿~6亿年吻合，故张仁杰等将石碌群连同以下连续沉积的第1层至第5层置于新元古界青白口系。

对于中国科学院华南富铁矿科研队地层组所获的古孢子资料，张仁杰等（1990）认为有的属种自中元古代长城纪、蓟县纪和新元古代青白口纪、震旦纪均有分布；欧阳舒、李再平所获孢粉提供的时代不仅与中国科学院华南富铁矿科研队地层组采获的孢粉资料组合面貌出入较大，而且与 *Chuaria-Tawuia* 生物群提供的时代归属及所获同位素年龄资料相差甚远，故不能排除样品采集和处理过程中的污染因素，即所提供的时代归属意见可靠性较差。

近年来，许德如等（2007）在石碌群第六层二透岩中进行碎屑锆石测定。部分锆石内部具均一性振荡环带，属于岩浆来源，其中两个具清晰均一的内部环带锆石点给出约 1670～1890Ma 的 $^{207}Pb/^{206}Pb$ 年龄值，另一颗具清晰内部环带的锆石给出的（1462±29）Ma $^{207}Pb/^{206}Pb$ 年龄，分别与同区中元古界抱板群、中元古代花岗质岩侵位年龄时代接近，认为来源于中元古代地质体。部分锆石内部结构均显示扇形和“幽灵”状振荡环带，反映了变质重结晶特点，普遍出现约 960～1300Ma 的 Pb-Pb 年龄群。推测石碌群沉积上限可能在 960Ma，而沉积下限约在 1300Ma。

此外，Li et al.（2008）获得石碌群变质火山岩 SHRIMP 锆石年龄（1439±9）Ma。同时，采用 SHRIMP 锆石法在石碌群下部获得了大量 1200～2660Ma 的碎屑锆石年龄。似乎暗示石碌群可能包含了中元古代和新元古代的地层，是一个混杂堆积体。但沉积相分析显示，石碌群属于一个沉积有序的地层，不具备混杂堆积成因。后经项目组野外进一步核实，在 Li et al. 采样点未找到变质火山岩，仅见变质碎屑岩分布。

综上所述，本报告采信目前的认识，即石碌群时代为新元古代青白口纪。而碎屑锆石年龄仅作为一个参考年龄，因为它涉及测试对象的代表性（采样层位、样品测试量等）问题。

2. 石灰顶组

1976 年以前，石灰顶组中未获任何生物化石资料，海南地质大队（1964）将其归属于寒武系—奥陶系、奥陶系或寒武系。1976 年，中国科学院华南富铁矿科研队地层组在石灰顶组采获孢子：*Asperatoshaera* sp.，*Leiamarginata maxima*，*Lophodiacrodium* sp.，*Polyedrosphaeridium* sp.，*Microncentrica*? sp. 等，而将它置于下奥陶统。但张仁杰等（1992）认为上述古孢子延续时代较长，从前震旦纪或震旦纪至奥陶纪都有分布。1987 年，南颐在石灰顶组中采获宏观藻类化石，将其时代归属早震旦世。1992 年，张仁杰等记述了石灰顶组下部灰黑色粉砂质泥岩中产 *Shouhsienia shouhsiensis* 及少量 *Chuaria circularis* 化石，认为该宏观藻与石碌群第六层中宏观藻最大区别在于未见第六层中最为丰富的“C”形、“U”形 *Tawuia dalensis*，据此将石灰顶划归为震旦系。

Li et al.（2008）采用 SHRIMP U-Pb 锆石法在石灰顶组中获得了大量 1200～2660Ma 的碎屑锆石年龄，与下伏石碌群碎屑锆石年龄一致。可能显示两者物质来源基本类同，沉积时间基本一致。尽管两岩石地层单位存在沉积间断，但间隔也非常短暂。

综合上述，结合石碌群中碎屑锆石 960～1300Ma 的 Pb-Pb 年龄，认为前人把石灰顶组时代置于震旦纪是合理的。本次仅是根据目前地层年代的划分方案，把其原时代震旦纪修订为南华纪—震旦纪。

（三）原岩恢复和岩相古地理探讨

1. 原岩恢复

由于石灰顶组岩石未发生区域变质作用改造，因此原岩恢复仅涉及石碌群。该岩石地层单位属于一套浅变质岩系，发育有变余层状构造、变余微层理构造、变余粉砂显微鳞片变晶结构、变余粉砂状结构、含变余粉砂片状变晶结构、变余砂砾状结构、变余砂状花岗变晶结构、含变斑晶粉晶化泥微晶结构。副矿物为次棱—半圆状锆石、磷灰石、电气石、金红石、白钛石、赤铁矿等。把碳酸盐岩、二透岩（透闪石透辉石岩）、千枚岩、石英岩等岩石的化学分析结果，作尼格里参数计算与图解分析，其原岩大致有灰岩、泥砂质灰岩、白云岩、泥砂质白云岩、铁质—泥砂质白云岩、石英砂岩、钙质砂岩、泥质砂岩、铁质砂岩、含铁质粉砂岩、含铁质粉砂质泥岩、含铁质泥岩、泥岩、赤铁矿岩及火山岩和火山碎屑岩等（详见第四篇变质岩）。

2. 岩相古地理探讨

（1）*石碌群*

石碌群以富含沉积变质型铁矿而著名，目前在中国及周边地区尚没有找到同时代产出的富铁矿层。尽管与澳大利亚 Hamersley 碧玉赤铁矿建造类似，目前在越南也找到富铁矿层，但它们所形成的时间与石碌群不一致。石碌群和石灰顶组中发现的宏观藻化石分布广，加拿大、印度以及我国的扬子地区、华北地区都有分布（许德如，2003）。因此，很难进行岩石地层单位空间对比分析。古地磁资料显示（汪啸风等，1992），石碌群古地理位置在南纬 11°~16°之间，采样层位自下而上变化无规律，平均为南纬 16°，表明海南古地块（?）青白口纪时古地理位置相对比较稳定。而石灰顶组时，古地理位置变化大，发生大的迁移，由南纬 13°→4°。目前，在江南古陆相继新发现一系列约 820Ma 代表武陵运动时限的年代学资料。青白口系石碌群与南华系—震旦系石灰顶组的不整合接触（图 3-11），以及石灰顶组发生了从南向北的大幅度迁移这些重要事件与 820Ma 的武陵运动在时间上明显具有耦合性，可能表明该时期海南岛也处于武陵运动的作用背景范围。

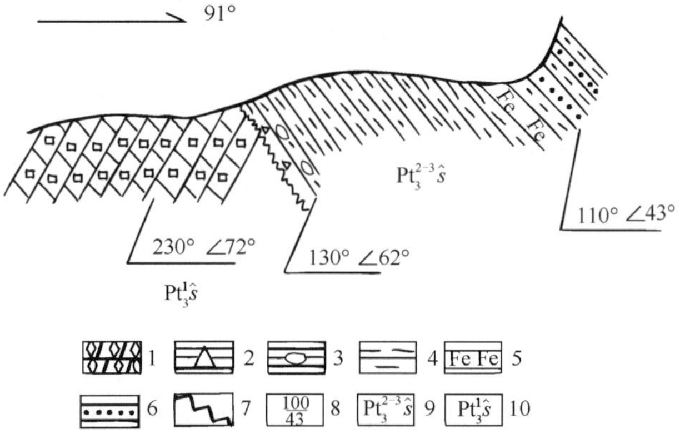

图 3-11　石碌石灰顶西南坡电机车站附近石灰顶组与石碌群第六层角度不整合

Fig. 3-11　Angular unconformity between the Shihuiding Formation and the Sixth Bed, Shilu Group, near the electrical machine station of the southwestern slope of Shihuiding, Shilu

（据徐林，1977）

1—重结晶白云岩；2—含铁锰质屑层；3—含砾泥岩；4—泥岩；5—铁矿层；6—石英砂岩；7—角度不整合界线；8—产状；9—石灰顶组；10—石碌群

广东省冶金 934 地质队（1978）曾对石碌群的沉积相进行过系统的分析，认为石碌群第一层、第四层、第五层和第六层含铁岩系直接底板为浅海相，第二层和第六层下部及上部的碳酸盐岩为浅海—海湾环境，第六层中下部的含铁岩系为浅海—潟湖环境。其他仅有少量文献论及石碌群第六层或第六层中的含铁岩系的沉积相：严国柱等（1979）据含铁岩系中砂岩、粉砂岩粒度分析资料和铁矿层中有铁质内碎屑、粒及团块、矿石具水平—微斜层理构造和含铁岩系上部的碳酸盐岩、铁质泥岩和细碎屑岩中普遍见有纹层构造，认为含铁岩系属海退型具有高盐度混浊浅水中—弱动力条件的海湾潟湖环境，含铁岩系之上以白云岩为主的岩石组合属蒸发潟湖相；陈泽铭等（1979）认为石碌群第六层为浅海—海湾相沉积。陈国达等（1977）指出：含矿沉积无论是碎屑沉积还是“二透岩”，其中均见有斜层理、波状层理及断续层理，同生角砾也有出现，它们是非静水环境产出的，铁矿层是一种浅水相沉积物，成铁前后可能属半封闭性质的海湾环境，原生铁矿形成于海进期之末、海退期之初。

石碌群是一套区域变质作用较低、构造分异作用不强烈、保存较多的变余沉积结构构造的岩石，沉积原岩不难恢复。石碌群第一层至第六层的层序特征和沉积环境描述如下。

1）第一层

由灰色、紫灰色中及薄层状绢云石英千枚岩夹薄层石英绢云千枚岩、含红柱石石英绢云千枚岩、绢云千枚岩，中、上部夹少量电气石化石英片岩。从矿物成分和变余结构恢复原岩应为泥质粉砂岩及少量粉砂夹粉砂质泥岩及泥岩，两者呈厚几厘米频繁互层的韵律层重复出现，在剖面上构成了反映海岸线相对稳定的加积型层序（图 3-12），为浅海相沉积。

2）第二层

为灰色厚层叶蛇纹石化泥粉晶白云岩，从其变余结构恢复原岩应为泥晶碳酸盐岩，属岩性均一的非旋回性基本层序。为浅海相沉积。

3）第三层

底部为棕黄色灰色薄层夹中层石英绢云千枚岩，以上为灰色薄层石英绢云千枚岩、红柱石石英绢云千枚岩与薄层—中层浅灰色绢云石英千枚岩及少量薄层（被顺层剪切变形改造多呈透镜状）石英片岩频繁互层组成韵律结构，在剖面结构上与石碌群第一层基本相似（图 3-12），仅中上部总体岩石粒度略有增粗，略显进积层序的特征。第三层底部从沉积物粒度上表现为渐变关系，但第二层属陆源清水沉积，而第三层为陆源混水沉积，两者性质明显不同；第三层与第四层之间，从沉积物粒度和由岩石结构所反映的沉积介质能量由明显的差异，即存在着突变界面或基本层序转换面。

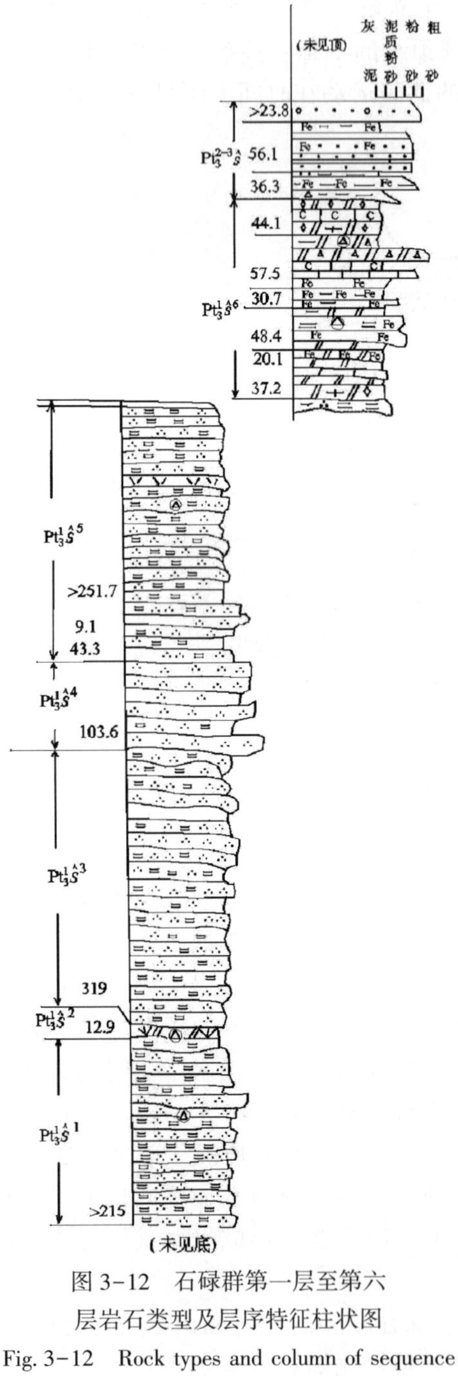

图 3-12　石碌群第一层至第六层岩石类型及层序特征柱状图

Fig. 3-12　Rock types and column of sequence characteristics of the first bed to the sixth bed of the Shilu Group

4）第四层

由灰白色中厚层粗—细粒石英岩夹少量绢云石英千枚岩组成，绢云石英千枚岩夹层随机重复出现。在剖面上底界具有突变界面，自下而上石英岩粒度总体上具有变细的趋势，属退积型层序（图 3-12）。中—厚层石英岩不论在石碌矿区，还是在矿区外围的吉开岭、金牛岭南，单偏光镜下均可见到砂粒形态大小，是一套形成于分选条件良好的矿物成熟度和结构成熟度较高的岩石，其中局部地段还见有异向大型斜层理，这种斜层理一般应为沿岸流形成，个别层系结合面不清、低角度相交者，可能是冲洗层理，其岩石类型及残存的层理类型，岩石部分颗粒出现锯齿状镶嵌、石英普遍具有波状消光和存在微量绢云母，反映岩石遭受过轻变质作用而与云台观组特征不同；石碌群第二层叶蛇纹石化或二透石化碳酸盐岩，其原岩据变余结构判断应为形成于低能环境的泥粉晶白云岩，其沉积或成岩期镁离子浓度较高，应属局限海碳酸盐台地潮下相。

5）第五层

主要为灰色薄层石英绢云千枚岩、含红柱石石英绢云千枚岩和绢云石英千枚岩及少量透镜状石英片岩、千枚岩化含水云母石英细砂岩互层或互为夹层频繁重复组成韵律层，下部夹有一层 9.1m 厚具褶叠层构造片理化细粒石英岩，上部夹有一层 2m 厚浅灰绿色岩屑凝灰岩层（横向上见变厚，如保秀岭北西矿区公路见厚至 6~8m）。在剖面上，第五层底部与第四层为渐变关系，表现在第五层底部向上绢云石英千枚岩含量逐渐降低，石英绢云千枚岩逐渐增多，至下部具褶叠层构造片理化细粒石英岩底部附近，绢云石英千枚岩又逐渐增加，具褶叠层构造片理化细粒石英岩以上，岩石类型的变化与第四层至第五层之间特征相似，但第五层上部薄透镜状石英片岩又有增多的趋势，反映石碌群第五层中、下部和石碌群第四层一样，仍属退积型层序，上部有转换为进积型层序的趋势，第四层至第五层共同构成大致对称的他旋回性层序（图 3-12）。砂质含量较高的岩石还较好地保存原生沉积结构，而且大量水平虫管的存在指示岩石形成于潮下环境。

6）第六层

石碌群第六层主要以碳酸盐白云岩及其接触热变质作用形成的强弱不等的透闪石透辉石化白云岩或二透岩为主。按岩层成层厚度、夹层岩石类型及含矿性的差异，岩性自下而上可分为三部分（图 3-12）：底部为淡紫色、灰色和浅灰色中—厚层细—粗晶白云岩、二透岩化白云岩、二透岩及二透岩化粉晶白云岩，不含铁矿层，层间基本未见非碳酸盐岩及其变化产物，厚 57.3m；中、下部为含铁岩系，由黄色、灰白色、黄绿色及浅灰色薄层粉晶白云岩、透闪石钾长石透辉石岩，透闪石岩、二透石化粉晶白云岩及薄层赤铁矿与中—厚层重晶石化细晶白云岩、石英质细晶赤铁矿岩、二透岩、绿帘石化阳起石矽卡岩及细粒石英岩互层，薄层岩石层面上常见黄绿色黏土薄膜，厚 102m；中上部为浅灰色—深灰色中—厚层泥微晶及细晶白云岩、透闪石或二透石化或二透石化微—细晶白云岩夹少量黑色薄层炭质板岩及深灰色薄层二透石化微晶白云岩，厚>78.5m。炭质板岩及中下部含铁岩石铁矿层的顶底板为薄层状，部分岩层可见水平层理，厚层岩石因白云石重结晶作用或强烈的二透石化多不显层理。含铁岩系除薄层白云岩外尚有薄层铁质板岩、铁质石英砂岩及细粒石英岩，铁矿层中见有变余结构、铁质砂砾屑、石英砂粒及大型板状斜层理。因此，石碌群第六层底部及中上部为局限海碳酸盐台地相，中下部含铁岩系为近滨相，铁矿形成于近滨相下部水下砂坝部位。

（2）*石灰顶组*

石灰顶组为一套含砾石英细砂岩、石英细砂岩、铁质石英细砂岩、含黄铁矿石英细砂岩夹含铁矿粉砂岩、砂质泥岩、含铁泥岩及贫铁矿层（图 3-13），广东省冶金地质 934 队（1978）认为石灰顶组为浅海相沉积。所列剖面均与石碌群第六层为断层接触，底部缺失部分地层。剖面出露岩石为石英细砂岩、含砾石英细砂岩、铁质细砂岩或铁质岩与少量砂质泥岩或铁质泥岩组成韵律层，构成加积型层序。据徐林等（1977）在枫树下北所，见其底部为紫红色含砾泥岩或含角砾泥岩和泥岩。含砾或含角砾泥岩，无层理，分选差、磨圆度差，角砾为原地和微异地来源，应为残坡积相。剖面岩石类型及其组合和沉积层序特征，反映主要形成于浪基面之上，仅少部分粉砂质泥岩和泥岩形成于浪基面之下、氧化还原界面之上，赤铁矿形成于浅水搅动环境，因此所见剖面沉积相为无障壁近滨相。但自残坡积相至近滨相间还应存在有后滨相和前滨相砂岩，在剖面中未见，是否因断层切错而缺失尚需深入研究。就其整体而言，具有磨拉石或类磨拉石建造的特点，属于晚晋宁运动晚期（武陵运动）地壳隆升阶段的沉积产物。

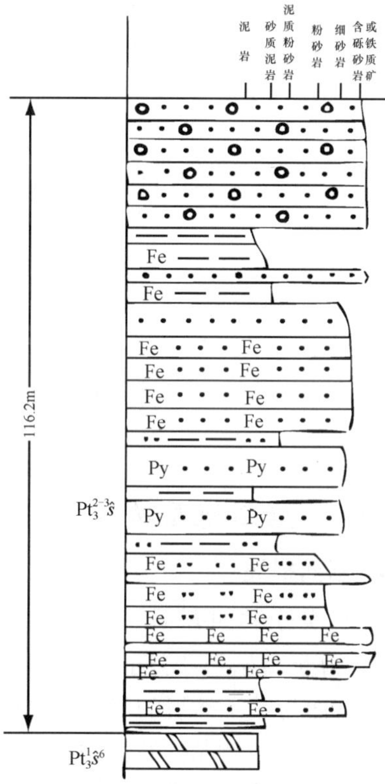

图 3-13　石碌红房石灰顶组柱状剖面图

Fig. 3-13　Columnar section of the Shihuiding Formation, Hongfang, Shilu

二、下古生界

海南岛下古生界有寒武系美子林组、奥陶系南碧沟组和下志留统陀烈组、空列村组、大干村组、靠亲山组、足赛岭组，分布范围较广，东西横穿东方—万宁一线，南北纵贯乐东—儋州，呈近东西向弧状展布。晚期花岗岩浆的侵蚀作用导致局部地段岩石遭到一定程度的改造，使得地层层序出露不全。北面由于东西向王五-文教构造带的下切而被第四系覆盖。岩石普遍经区域变质变形作用的改造，形成一套局部无序而总体有序的岩石组合。其中，陀烈组从海南岛西部的东方市往东至昌江县、白沙县、儋州市、乐东县及岛东的万宁市均有分布；空列村组分布范围从海南岛西部东方市、南部保亭县志仲、东部万宁地区均有出露；大干村组、靠亲山组和足赛岭组主要分布在乐东、三亚、保亭交界的南好地区（图 3-14）。

图 3-14　五指山地层分区下古生界分布及实测剖面位置图

Fig. 3-14　Lower Paleozoic distribution and measured section position of the Wuzhishan stratigraphic subregion

（一）岩石地层特征

1. 剖面描述

（1）美子林组剖面

1）万宁市乐来乡美子林剖面

该剖面为美子林组的代表性剖面，由宜昌地矿所 1991～1995 年开展 1∶5 万《和乐、中原市、博鳌港》幅区调时测制。宜昌地矿所、海南省地质调查院 2000～2002 年开展 1∶25 万《乐东县、陵水县》幅区调时略作修改。剖面位于万宁市乐来乡美子林村南西约 600m 的河沟中，起点坐标：东经 110°24′03″，北纬 18°55′43″。自上而下为（图 3-15）：

美子林组（$\epsilon_1 m$）

三段

10. 掩盖　400m

9. 灰色、深灰色中厚层结晶灰岩，底部 0.06m 为黄灰色中薄层变粉—细石英砂岩。灰岩中产瓶虫化石 *Cambrothyra* sp. 和六射海绵骨针　11.6m

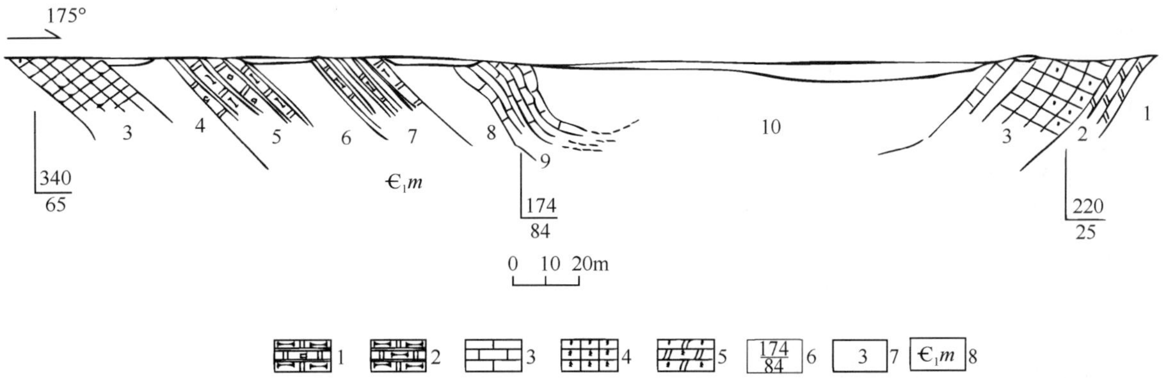

图 3-15 万宁市乐来乡美子林寒武系实测剖面图

Fig. 3-15 Measured section of Cambrian strata, Meizilin, Leilai Town, Wanning City

1—透辉石石英大理岩夹硅灰石大理岩；2—石英透辉石大理岩；3—中厚层结晶灰岩；4—黑云石英角岩；5—变质长石石英砂岩夹白云石英片岩、二云石英片岩；6—产状；7—分层号；8—美子林组

8.	掩盖	12.59m
7.	浅灰色石英透辉石大理岩	29.49m
6.	掩盖	26.31m
5.	黄灰色、灰色透辉石石英大理岩夹硅灰石大理岩	16.26m

二段

4.	掩盖	69.54m
3.	深灰色厚层黑云石英角岩	36.27m

一段

2.	灰黄色变质长石石英砂岩夹白云石英片岩、二云石英片岩	12.08m
1.	掩盖	

2）万宁市乐来乡下田剖面

该剖面为美子林组的参考剖面，由宜昌地矿所 1991~1995 年开展 1∶5 万《和乐、中原市、博鳌港》幅区调时测制。宜昌地矿所、海南省地质调查院 2000~2002 年开展 1∶25 万《乐东县、陵水县》幅区调时略作修改。剖面位于万宁市乐来乡下田一带，起点坐标：东经 110°28′05″，北纬 18°56′11″。自上而下为（图 3-16）：

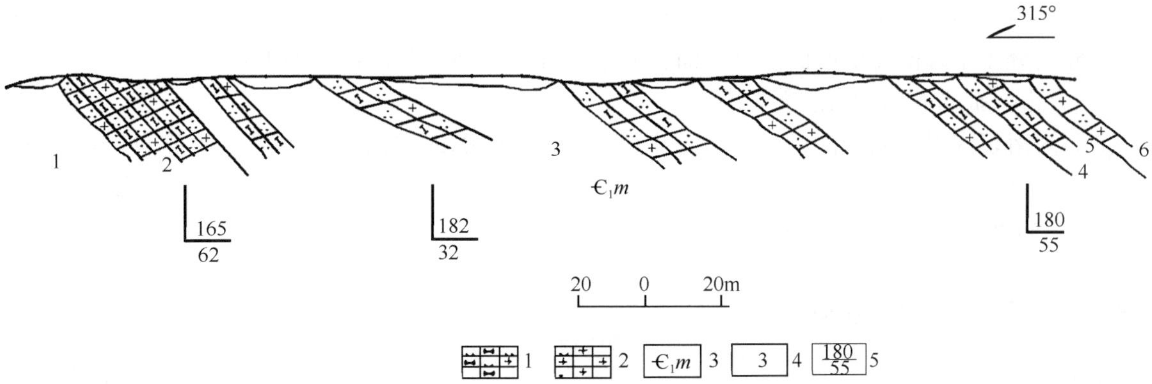

图 3-16 万宁市乐来乡下田寒武系实测剖面图

Fig. 3-16 Measured section of Cambrian strata, Xiatian, Lelai Town, Wanning County

1—黑云透辉石石英角岩；2—黑云石英透闪石岩；3—美子林组；4—分层号；5—产状

美子林组（$\in_1m$）

三段

6. 灰黄色带蓝色黑云石英透闪石岩 3. 34m

二段

5. 掩盖 5. 47m
4. 深灰色黑云透辉石石英角岩 12. 38m
3. 主体掩盖，露头为黑云透辉石英角岩、黑云二透石英角岩 112. 94m
2. 深灰色、灰色带绿色厚层、条带状黑云透辉石英角岩、黑云二透石英角岩 25. 26m
1. 掩盖

3）万宁市大茂乡袁水水库剖面

该剖面为美子林组的参考剖面，由宜昌地矿所 1991～1995 年开展 1∶5 万《和乐、中原市、博鳌港》幅区调时测制。剖面位于万宁市大茂乡袁水水库一带，起点坐标：东经 110°19′52″，北纬 18°51′45″。自上而下为：

美子林组（$\in_1m$）

二段

9. 残坡积物，原岩为灰色、灰绿色透辉石石英角岩、石英角岩、透辉长英角岩 27. 46m

一段

8. 灰黄色变细粒石英砂岩、绢云母化黑云石英片岩夹黑云石英片岩、黑云长英片岩 185. 43m
7. 掩盖 35. 58m
6. 灰色绢云石英片岩 1. 79m
5. 灰黄色变细粒石英砂岩与含绢云母黑云石英片岩不等厚互层，夹变质粉砂岩 216. 74m
4. 掩盖 87. 13m
3. 深灰色角闪石英角岩 4. 79m
2. 掩盖 56. 36m
1. 灰黄色变石英砂岩与绢云母化黑云石英片岩、绢云石英片岩不等厚互层 85. 62m

未见底

（2）南碧沟组剖面

前人测制过的剖面有东方市南捉沟–昌江县南碧沟剖面、白沙县大溪桥剖面、东方市南清河剖面、儋州市洛基大岭剖面、定安县翰林镇蒙花岭剖面和万宁市坑尾–斗牛村剖面。以东方市南捉沟–昌江县南碧沟剖面、白沙县大溪桥剖面、定安县翰林镇蒙花岭剖面和万宁市坑尾–斗牛村剖面岩石出露程度相对较好，分别分布于海南岛南西面、北西面、北东面和南东面，剖面描述如下：

1）东方市南捉沟–昌江县南碧沟剖面

该剖面为南碧沟组代表剖面，位于东方市江边乡东南 10km 的南捉沟至昌江黎族自治县王下乡南碧沟中，由广西区调队 1988～1991 年开展 1∶5 万《猴猕岭、坝王岭》幅区调时测制。陈哲培等（1997）对剖面进行了重新划分，自上而下为（图 3–17）：

上覆地层 陀烈组变质细粒石英砂岩

———— 浮土覆盖 ————

南碧沟组 厚 2032. 40m

41. 浮土覆盖，有变质砂岩零星露头 400. 5m
40. 灰色—深灰色绢云石英千枚岩 32. 3m
39. 灰—深灰色绢云板岩夹变质粉砂岩，或呈互层状产出，具水平纹层 50. 5m
38. 灰色—深灰色绢云板岩夹变质粉砂岩，具韵律性，受构造变形影响有些砂岩形成假砾石 223. 3m

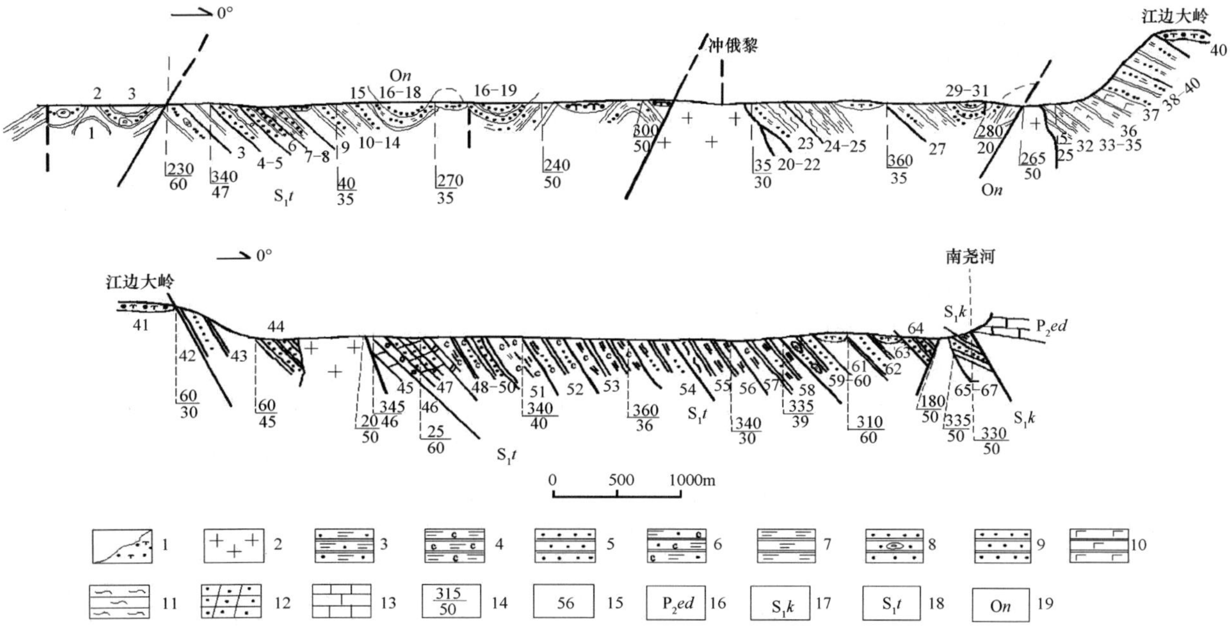

图 3-17　东方市南捉沟-昌江黎族自治县南碧沟奥陶系—志留系实测剖面图

Fig. 3-17　Measured section of Ordovician-Silurian strata, Nanzhuogou (Dongfang City)-Nanbigou (Changjiang Li Autonomous County)

1—浮土；2—花岗岩；3—粉砂质绢云板岩；4—炭质绢云板岩；5—变质粉砂岩；6—粉砂质炭质板岩；7—绢云板岩；8—含炭粉砂质板岩；9—粉砂质板岩；10—钙质板岩；11—阳起石板岩；12—角岩；13—灰岩；14—产状；15—分层号；16—鹅顶组；17—空列村组；18—陀烈组；19—南碧沟组

37. 灰色—灰黑色黑云母化石英绢云千枚岩夹基性火山岩，火山岩具熔岩至凝灰质碎屑岩的韵律，熔岩中具石英质充填的杏仁体，已拉伸变形	39. 3m
36. 灰色—灰黑色透闪石化石英绢云板岩，沿片理面有透闪石、石英及少量黑云母聚集成不规则团块、条带状交代原岩	90. 5m
35. 灰色—灰黑色绢云母板岩与透闪石化基性凝灰质熔岩，具压扁，拉伸的杏仁体构造。熔岩单层厚一般为 20~30cm	75. 6m
34. 灰色—灰黑色黑云母化石英白云母片岩，白云母片状定向聚集呈连续条带，其间分布着石英集合体，具较好的片理构造	41. 1m
33. 灰色—灰黑色绢云板岩夹阳起石化基性火山岩，两者互层状产出，火山岩以凝灰质碎屑岩为主，单层层厚一般为 5~15cm	41. 3m
32. 灰色—深灰色变质黑云母砂岩，粒度 0. 05~0. 1mm，黑云母定向零星分布于长石、石英间，部分黑云母已绿泥石化	31. 4m
31. 上部为灰色—深灰色绿泥石石英岩，下部为灰色—深灰色绿泥石千枚岩	88. 8m
30. 灰色—深灰色绢云石英千枚岩	77. 4m
29. 灰色—灰黑色石英白云母片岩，白云母鳞片沿片理方向聚集呈条带，其间夹杂岩绿泥石鳞片集合体和石英集合体	55. 3m
28. 灰色—深灰色含绿泥石石英板岩，层理发育，以水平层理为主，细层一般厚 3~5mm，石英粒间镶嵌绢云母和绿泥石鳞片	65. 3m
27. 灰色—灰黑色石英白云母片岩，白云母（部分为绢云母）鳞片定向聚集形成片理，其间分布重结晶的石英集合体	75. 7m
26. 浮土覆盖，有零星绢云板岩露头	319. 2m
25. 灰色—灰黑色绢云板岩，水平层理发育，细层一般厚 2~5mm，片理与层理几乎平行，偶夹薄层粉砂质，呈韵律状产出	69. 9m
24. 浮土覆盖	56. 7m

23. 灰色—灰黑色阳起石片岩。原岩为中—基性岩，阳起石纤柱体沿定向聚集，还断续夹杂岩石英及钠长石，使岩石显示片理 71.8m

22. 浮土覆盖 54.9m

21. 深灰色粉砂质绢云板岩，粉砂质含量较多，形成细薄层水平层理，一般纹层厚 3~5mm，片理与层理大体平行 27.6m

20. 掩盖 43.0m

～～～～～～ 侵　　入 ～～～～～～

下伏地层　二叠纪黑云母二长花岗岩

2）白沙县大溪桥剖面

宜昌地质矿产研究所 1991~1994 年开展 1：5 万《昌江县、邦溪》幅区调时测制；2005~2006 年海南省地质调查院开展 1：25 万《东方县》幅区调工作中补做了部分工作。剖面沿白沙县邦溪镇以北海榆西线公路测量，起点坐标：东经 109°07′37″，北纬 19°23′37″。自上而下为（图 3-18）：

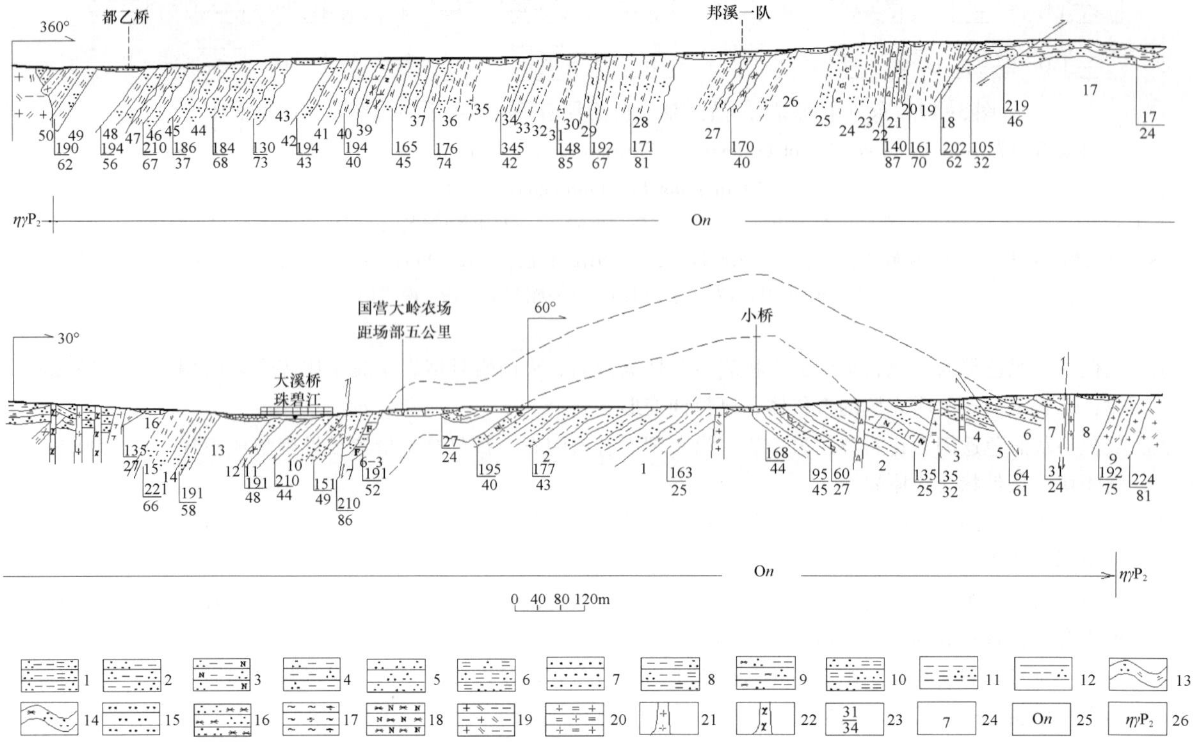

图 3-18　白沙县大溪桥奥陶系南碧沟组剖面图

Fig. 3-18　Section of the Ordovician Nanbigou Formation, Daxiqiao, Baisha County

1—二云母石英片岩；2—黑云石英片岩；3—角闪斜长石英角岩；4—石英黑云片岩；5—石英片岩；6—白云母石英片岩；7—第四系残坡积；8—石英黑云母片岩；9—含阳起石黑云石英片岩；10—石英白云母片岩；11—石墨绢云母千枚岩；12—石墨二云母片岩；13—石英绢云千枚岩；14—含红柱石石英千枚岩；15—变质细砂岩；16—绿泥石化阳起片岩；17—含透闪石绿泥石片岩；18—阳起石斜长片岩；19—斑状二长花岗岩；20—白云母二长花岗岩；21—花岗岩脉；22—煌斑岩脉；23—产状 24—分层号；25—南碧沟组；26—二叠纪花岗岩

未见顶

奥陶系南碧沟组 >1316.32m

50. 深灰色薄层石英白云母片岩夹白云母石英片岩，顶部花岗岩外接触带为深灰色中层含长含矽线石、黑云母石英片岩。本层见有两条含电气石中粗粒花岗岩脉顺层贯入 >73.94m

49. 第四系残坡积层覆盖 62.24m

48. 黄灰色薄层石英白云母片岩夹白云母石英片岩，见一条花岗岩细脉顺层贯入 43.31m

47.	紫红色（风化色）薄片状白云母石英片岩与石英白云母片岩频繁互层	14.34m
46.	第四系残坡积覆盖	25.75m
45.	棕黄灰色石英白云母片岩夹白云母石英片岩	30.44m
44.	灰绿色、黄绿色薄层变质石英粉砂岩、含绢云母白云母石英片岩、二云母石英片岩及石英白云母片岩频繁互层，组成自旋回复理式韵律	88.49m
43.	第四系残坡积层覆盖	57.04m
42.	灰色薄层含白云母石英片岩与白云母石英片岩频繁互层，偶夹薄层石英片岩	19.16m
41.	灰色薄层石英片岩、含白云母石英片岩与石英白云母片岩频繁互层，组成复理式韵律	30.96m
40.	灰色薄—中层石英片岩、白云母石英片岩及石英白云母片岩频繁互层，组成复理式韵律	29.12m
39.	绿灰色阳起石斜长片岩，其中含阳起石 30%～35%，更长石 45%～55%，以及少量金属矿物、黑云母、透辉石、磷灰石等，细粒不等粒他形变晶结构、片状构造	10.88m
38.	灰色薄层石英片岩与白云母石英片岩互层，顶部见一层具底冲刷向上变细的变质细砂岩	17.65m
37.	第四系残坡积层覆盖	8.83m
36.	黄灰色、灰色薄层石英绢云千枚岩、含红柱石石英绢云千枚岩夹薄层变质石英粉砂岩，顶部为含水云母黑云石英千枚岩	54.15m
35.	第四系残坡积层覆盖	48.07m
34.	灰色薄层石英绢云千枚岩夹薄层变质粉砂岩及水云母绢云石英千枚岩	42.56m
33.	灰色薄层石英绢云千枚岩与绢云石英千枚岩互层	12.53m
32.	第四系残坡积层覆盖	21.86m
31.	灰色薄层石英绢云千枚岩	6.26m
30.	第四系残坡积层覆盖	22.85m
29.	灰绿色含透闪石绿泥石片岩	28.93m
28.	灰色薄层石英绢云千枚岩、含黑云母绢云母千枚岩夹绢云石英千枚岩，偶夹薄层变质粉砂岩	101.76m
27.	上部为第四系残坡积层覆盖，在辅助剖面上底部见有灰色薄层状石英绢云千枚岩及绢云石英千枚岩；下部为第四系残坡积层覆盖；在辅助剖面上，见有两层灰绿色绿泥石化阳起石片岩	101.56m 100.00m
26.	第四系残坡积层覆盖为主，仅顶部见 2.79m 厚的薄层石英绢云千枚岩夹薄层变质石英粉砂岩	54.11m
25.	灰色薄层绢云石英千枚岩与石英绢云千枚岩不等厚互层	20.71m
24.	黄绿色含金云母滑石片岩	2.18m
23.	灰色中—薄层变质石英粉砂岩、绢云石英千枚岩与薄层石英绢云千枚岩或含红柱石含水云母石英绢云千枚岩频繁互层，组成复理式韵律	31.82m
22.	灰色薄层绢云母千枚岩	>11.97m
	═══════ 断　层 ═══════	
21.	浅灰色、灰色薄层石英绢云千枚岩与薄—中层变质粉砂岩不等厚互层	>34.43m
20.	第四系残坡积层覆盖	18.34m
19.	灰色、深灰色薄层石英绢云千枚岩、含黑云母千枚岩夹绢云石英千枚岩及变质石英粉砂岩薄层	90.08m
	═══════ 断　层 ═══════	
18.	浅灰色石英白云母片岩、灰色薄层含红柱石石英白云母片岩，含红柱石石英二云母片岩夹灰色中层白云母石英片岩及浅灰色变质细砂岩，见片理切穿成分层现象	>54.34m
	═══════ 断　层 ═══════	
17.	灰色薄层石英绢云千枚岩夹绢云石英千枚岩，下部为白云母石英片岩，顶部为含红柱石石英二云母片岩，本层见有白云母花岗岩、辉绿岩及闪长玢岩岩脉穿切	>65.09m
16.	第四系残坡积层覆盖	11.14m
15.	上部为深灰色条纹状含微晶石墨绢云石英千枚岩；下部为深灰色条纹状微晶石墨二云母石英片岩	73.13m

14. 灰色条纹状白云母石英片岩、石英白云母片岩，上部偶夹灰色薄层石英片岩 28.28m

13. 第四系残坡积层、人工填土及珠碧江河床砂砾层覆盖 101.79m

12. 深灰色中层含阳起石黑云石英片岩，见少量黄铁矿 14.01m

11. 珠碧江现代河床河水掩盖 39.28m

10. 中、上部为深灰色中夹薄层黑云石英片岩；下部为深灰色厚层含硅线石白云母石英片岩夹白云母石英片岩 >53.61m

断　层

9. 棕黄色、灰色中薄层白云母石英片岩、二云母石英片岩、石英二云母片岩夹深灰色中层石英片岩，见少量杆状构造 >18.19m

8. 第四系残坡积层覆盖，仅底部见有3.81m厚灰色薄层石英片岩，其中见白云母花岗岩岩脉顺层侵入 >107.52m

断　层

7. 背斜北翼为灰色薄层白云母石英片岩夹少量石英白云母片岩，见杆状构造；背斜南翼中、上部为棕黄色中—薄层石英白云母片岩、薄层白云母石英片岩夹薄层石英片岩；下部16.3m为灰色中层蓝晶石石英角岩 >56.18m

6. 深灰色黑云母片岩 >12.89m

断　层

5. 灰—深灰色薄层夹层状黑云石英片岩夹石英黑云片岩及薄层石英片岩，见有杆状构造和白云母花岗岩脉穿切 27.72m

4. 第四系残坡积层覆盖 17.06m

3. 深灰色中层黑云石英片岩，上部7.42m为灰色薄层白云母石英片岩，见杆状构造 >26.10m

断　层

2. 背斜北翼为深灰色薄夹中层黑云石英片岩、硅线黑云石英片岩、硅线石红柱石黑云石英片岩，中、下部夹一层1.8m厚绿灰色厚层石英水镁石透辉石角岩，顶部6.18m为深灰色中层角闪斜长石英角岩；背斜南翼为黑云石英片岩，顶部4.50m为深灰色石英黑云斜长角岩。北翼见有杆状构造 135.64m

1. 深灰色中层二云母石英片岩 >31.29m

未见底

3）定安县翰林镇蒙花岭剖面

该剖面由海南省地调院1997~2000年进行1：5万《黎母岭、枫木市、琼中县、白马岭》幅区调时测制。2001~2003年，海南省地质调查院开展1：25万《琼海县》幅区调工作中补做了部分工作。剖面起点位于蒙花岭25°方向1912m处，起点坐标：东经100°13′41″，北纬19°3′24″；剖面长约1306m。剖面自上而下共分25层，自上而下为（图3-19）：

南碧沟组（O*n*） 634.45m

25. 浅灰白色变质细粒绢云母石英砂岩，发育水平层理 12.02m

24. 绢云黑云石英片岩、二云母石英片岩。下部以前者为主，上部以后者为主，绢云黑云石英片岩中含微古植物化石：*Leiominuscula minuta.*，*Leiopsophosphaera* sp.，*Ovulum saccatum*，*O. lanceolatum*，*Triangumorpha* sp.，*Micrhystridium olignum*，*M. brevicornum*，*Aliumella* sp.，*Navifusa* sp. 40.61m

23. 浅白色白云母长石石英片岩，长石呈粒状，大小多<0.5mm 30.37m

22. 浅灰白色千枚状绢云黑云石英片岩，夹浅灰白色石英岩，上部为二者互层 10.76m

21. 浅灰白色含电气石石英二云母片岩、石英二云母片岩 91.97m

20. 浅灰白色（风化色呈浅棕红色）白云母石英片岩，局部夹含红柱石、电气石白云母石英片岩 67.17m

19. 浅灰黑、浅灰绿色厚层—块状阳起石石英（片）岩（原岩为火山岩，下同）与浅灰白色厚层—块状二云母石英片岩互层 39.37m

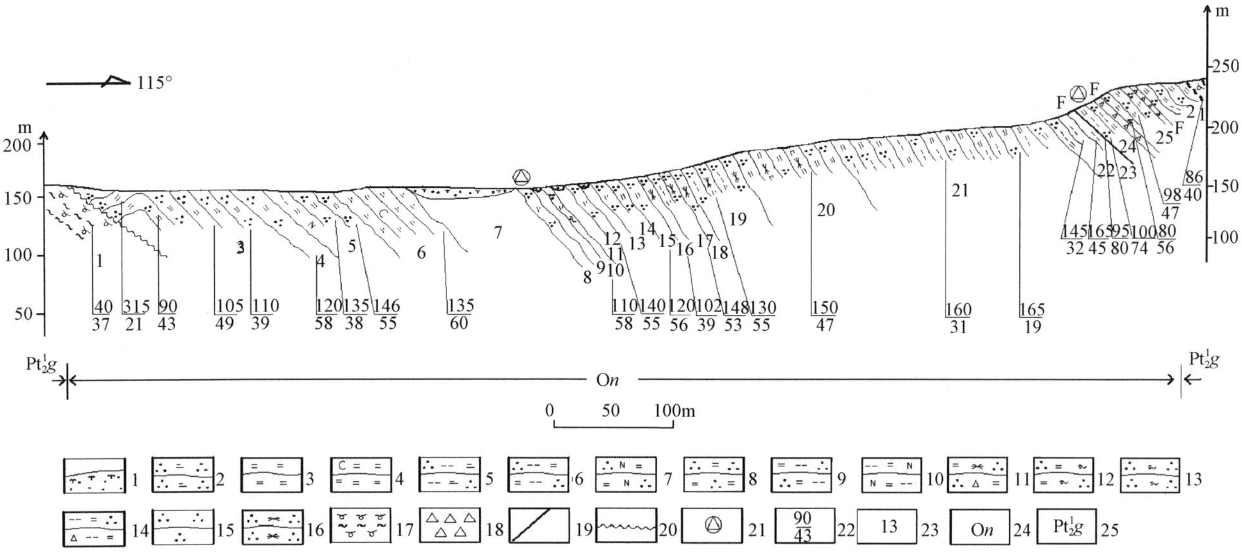

图 3-19　定安县翰林镇蒙花岭奥陶系南碧沟组实测剖面图

Fig. 3-19　Measured section of the Ordovician Nanbigou Formation, Menghualing, Hanlin Town, Ding'an County

1—浮土；2—变质细粒绢云母石英砂岩；3—绢云母千枚岩；4—白云母石英岩；5—含炭绢云母千枚岩；6—二云母石英片岩；7—白云母长石石英片岩；8—白云母石英片岩；9—石英二云母片岩；10—长石二云母片岩；11—含红柱石电气石白云母石英片岩；12—含硬绿泥石白云母石英片岩；13—含硬绿泥石石英片岩；14—含电气石石英二云母片岩；15—石英岩；16—阳起石石英岩；17—眼球状混合岩；18—碎裂岩；19—实测断层；20—不整合；21—孢粉采样位置；22—产状；23—分层号；24—南碧沟组；25—戈枕村组

18.	浅灰绿色阳起石石英岩	7.18m
17.	浅灰绿色厚层—块状阳起石石英岩与浅灰白色厚层—块状二云母石英片岩互层	13.51m
16.	浅灰黑色、浅灰绿色阳起石石英岩，于此层中采人工重砂样一个（锆石 U-Pb 测年）	7.28m
15.	浅灰白色白云母石英片岩	13.86m
14.	浅灰白色厚层—块状白云母石英片岩夹浅灰白色、浅灰绿色细粒阳起石石英岩	25.01m
13.	掩盖	5.36m
12.	浅灰白色厚层—块状白云母石英片英，夹灰黑色中薄层含炭电气石石英岩，局部偶夹浅灰白色中薄层细粒白云母石英岩	16.33m
11.	掩盖	19.56m
10.	浅灰白色、浅灰绿色中厚层含硬绿泥石白云母石英岩与中厚层含硬绿泥石石英片岩互层	5.20m
9.	掩盖	9.74m
8.	浅灰白色中厚层白云母石英片岩与灰黑色中薄层含炭白云母石英片岩互层，底部夹中薄层细粒白云母石英岩，含炭白云母石英片岩中含微古植物化石：*Leiopsophosphaera* sp.，*Micrhystridium spinosum*，*M. lanceolatum*，*M. brevi-cornum*，*Buedingshaeridium* sp.，*Veryhachium* sp.，*Ovulum saccatum*，*Ammonidium* sp.，*Leiofusan* sp.，*Synsphaeridium* sp.，*Legonochitina*	5.20m
7.	掩盖	81.09m
6.	浅灰白色厚层块状绢云母千枚岩夹浅灰黑色中薄层含炭绢云母千枚岩	36.56m
5.	浅灰白色白云母石英片岩，中上部偶夹厚层长石二云母片岩，上部岩层变形强，长石呈细粒状，粒度 0.1～0.4mm	14.45m
4.	浅灰白色白云母石英片岩，夹薄层细粒云母石英岩，偶夹二云母石英片岩	15.51m
3.	浅灰白色白云母石英片岩	54.33m

═══════ 断　　层 ═══════

峨文岭组（Che）

2. 石英二云母片岩，风化后为浅灰褐色、浅褐色　12.27m

———— 推测为整合？————

戈枕村组（Chg）　7.29m

未见底

4）万宁市坑尾-斗牛村剖面

该剖面由宜昌地矿所 1991~1995 年开展 1：5 万《和乐、中原市、博鳌港》幅区调时测制。剖面位于万宁市山根乡坑尾-斗牛村，起点坐标：东经 110°27′03″，北纬 18°56′26″。自上而下共分 23 层，叙述如下。

上覆地层 下志留统陀烈组深灰色—灰黑色板岩

———— 浮土掩盖 ————

南碧沟组（*On*） （厚度>2000m）

23.	掩盖	315.9m
22.	主体为花岗岩，有变泥质粉—细砂岩呈捕虏体产出	36.24m
21.	灰黄色—浅黄绿色厚层变细粒石英砂岩	5.44m
20.	灰色—灰绿色变凝灰质砂岩	85.77m
19.	灰色变泥质粉—细砂岩夹粉砂质板岩	115.95m
18.	掩盖	707.66m
17.	深灰色—绿灰色变泥质粉砂岩	29.39m
16.	掩盖	25.19m
15.	灰色—深灰色变泥质粉—细砂岩	16.79m
14.	掩盖	484.18m
13.	灰色—绿灰色变粉—细砂岩、变泥质粉砂岩不等厚互层，可见水平层理	145.48m
12.	掩盖	190m
11.	灰色—深灰色变泥质粉砂岩、粉砂质板岩、变泥质粉—细砂岩，发育多条顺层断层	76.6m
10.	掩盖	97.49m
9.	灰白色变（含）凝灰质砂岩夹灰-绿灰色变粉砂岩、变泥质粉砂岩	245.21m
8.	掩盖	372.36m
7.	灰色—绿灰色变质粉—细砂岩、粉砂质板岩、变泥质粉砂岩、板岩互层	42m
6.	灰白色变凝灰质砂岩夹灰色、灰黑色、灰绿色、黄绿色变泥质粉砂岩、变粉砂岩	267m
5.	掩盖	846.64m
4.	深灰色变泥质粉砂岩、粉砂质板岩，断层发育	52m
3.	掩盖	150m
2.	灰色—深灰色板岩、泥质粉砂岩，产微古植物化石 *Margominuscula anligua*，*Leipsophosphaera densa*，*L.* aff. *pusilla*，*Trachysphaeridium* cf. *hyalinum*，*T. planum*，*Lophosphaeridium inconfertum*，*L.* cf. *enile*，*Zonosphaeridium* sp.，*Stictosphaeridium* sp.，*Reticulum rugosum*，*R.* sp.，*Polyedrixium minutum*，*Triangumorpha rarum*，*Micrhystridium* sp.1，*M.* sp.2，*Leiofusa bicomula*，*Macroptycha* sp.，*Synsphaeridium conglutinatum*	74.93m
1.	掩盖，偶见花岗岩出露	764m

———— 浮土掩盖 ————

下伏地层 寒武系美子林组变质粉砂岩

（3）陀烈组剖面

陀烈组剖面有娜姆河剖面、南捉沟-南碧沟剖面、尖峰顶-八角岭剖面、南棍园剖面、南坟岭剖面、黄泥沟剖面、财地村剖面、东泰农场加要岭剖面等。娜姆河剖面、南捉沟-南碧沟剖面、尖峰顶-八角岭剖面和东泰农场加要岭剖面岩石出露程度较好，岩石地层单位控制程度较高，剖面叙述如下。

1）娜姆河剖面

该剖面为陀烈组的代表性剖面，位于东方市陀烈水库下游娜姆河段，起点坐标：东经 109°12′28″，北纬 18°57′54″。1986~1989 年由海南地质大队开展 1：5 万《东方、玉道》幅区调时测制；宜昌地矿所、海南地矿局 1986~1990 年开展《海南岛若干重要基础地质问题的研究》（汪啸风等，1992）时重测（图 3-20）。

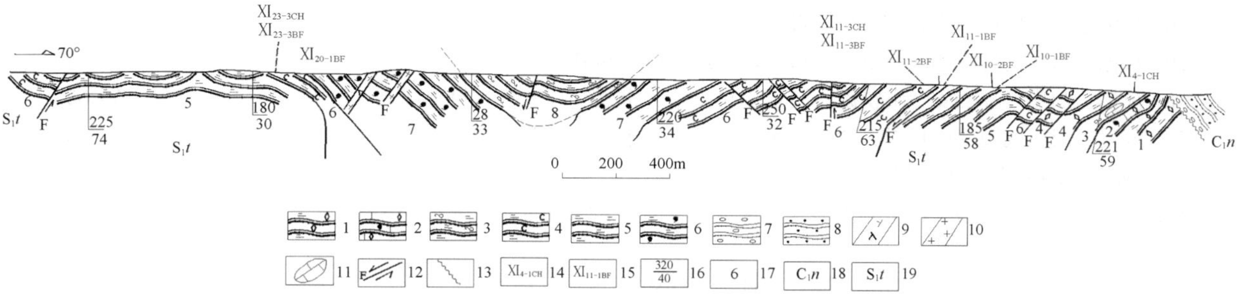

图 3-20 娜姆河下志留统陀烈组实测剖面图

Fig. 3-20 Measured section of the Lower Silurian Tuolie Formation, Namu River

1—含灰岩结核绢云母板岩；2—含灰岩结核碳酸盐化斑点状绢云母板岩；3—条带状或条纹状绢云母板岩；4—炭质绢云母板岩；5—绢云母板岩；6—斑点状绢云母板岩；7—变质砾岩；8—变质砂岩；9—闪长玢岩（脉）；10—花岗岩（脉）；11—灰岩透镜体；12—断裂；13—角度不整合界线；14—几丁虫标本采集点；15—微古化石采集点；16—产状；17—分层号；18—南好组；19—陀烈组

上覆地层 南好组（C_1n）石英质砾岩

~~~~~~~~ 不　整　合 ~~~~~~~~

陀烈组（S_1t） 总厚度>1 879.00m

上段

8. 青灰色中厚层条带状或条纹状绢云板岩 371.00m

中段

7. 暗灰色斑点状炭质绢云板岩 55.00m
6. 黑云炭质绢云母板岩，在距底层面约 54m 处采到几丁虫化石：*Conochitina* cf. *edjelensis*. *C.* aff. *iklaensis*，其下产微古植物化石。主要有 *Leiopsophosphaeridium crassaum*，*Baltisphaeridiumstipiforme*；其次是 *Leiofusa bieornuia*，*Lophosphaeridium crassaum*，*Trachysphaeridium simplex*，*Ynsphaeridium conglutinatum* 787.00m
5. 深灰色绢云母板岩，底部产微古植物化石 168.00m

下段

4. 浅灰绿色绢云母板岩，含少量硅质泥晶灰岩结核及透镜体 274.00m
3. 青灰色钙质或硅质条带与条纹状绢云板岩，含扁豆状、透镜状硅质泥晶灰岩透镜体，下部板岩中产几丁虫：*Conochitina* cf. *edjelensis*，*C. edjelensis armillata*，*C.* sp. 80.00m
2. 灰白色碳酸盐化板岩含黄色泥晶灰岩透镜体，板岩中可见绢云母和石英斑点 86.00m
1. 灰色、灰白色绢云板岩，含少量硅质泥晶灰岩结核、条带或透镜体 76.00m

未见底

2）南捉沟-南碧沟剖面

该剖面为陀烈组参考剖面，位于东方市江边乡东南 10km 处的南捉沟口，起点坐标：东经 109°12′28″，北纬 18°57′54″，广西区调队 1988~1991 年开展 1∶5 万《猴猕岭、坝王岭》幅区调时测制。自上而下为（图 3-17）：

上覆地层 空列村组灰白色变质中粗粒石英砂岩（石英岩）

———— 整　合 ————

陀烈组 总厚度 2843.70m

上段

63. 浮土覆盖 102.20m
~~~~~~~~

62. 灰色、灰黄色轻变质粉砂岩夹变质微层状泥岩，水平层理发育 219.40m

61. 浮土掩盖 191.70m

中段

60. 灰黑色轻变质粉砂岩、含炭粉砂岩，水平层理发育 141.90m

59. 灰色、深灰色具镜体的变质泥质粉砂岩，透镜体以粉砂质为主 67.70m

58. 灰色、灰黑色轻变质含炭粉砂质泥岩、粉砂岩，水平纹理发育，厚3~5mm 60.00m

57. 灰色、灰黑色轻变质页岩，含星点状黄铁矿，水平层理发育，产疑源类 *Leioualia* sp.，*Leiosphaerida* sp. 及几丁虫碎片 47.60m

56. 灰黑、黑色炭质绢云板岩、含炭粉砂质泥岩，产疑源类 *Baltisphaeridium* sp，*Retisphaeridium* sp. 及几丁虫碎片 166.10m

55. 灰色、灰黑色含绿泥石粉砂质泥岩，具火焰状构造 24.50m

54. 灰色、灰黑色砂质板岩、炭质板岩夹变质粉砂岩 185.20m

53. 灰色、灰黑色绢云泥岩夹砂质泥岩 295.70m

52. 灰色、灰黑色炭质板岩、含炭粉砂质板岩 161.20m

51. 灰色、灰黑色含炭绢云千枚岩、炭质粉砂质千枚岩、水平层理发育，单层一般厚1~2cm 186.10m

50. 灰色、灰黑色粉砂质板岩、含炭粉砂质千枚岩，含钙质透镜体 144.70m

49. 灰色、灰黑色微硅化含炭泥岩，下部为厚2.7m白云质结晶灰岩 114.50m

48. 灰色、深灰色角岩化绢云板岩，含炭粉砂质绢云板岩 165.70m

下段

47. 灰色、灰黑色斑点状角岩，斑点成分为黑云母-绿泥石和石英集合体 116.90m

46. 黑色含炭细砾岩，细砾占岩石的60%，成分以石英、石英砂、泥质岩为主 82.00m

45. 灰色、灰黑色黑云石英角岩 33.30m

44. 上部灰黄色石英岩，下部角岩化变质细砂岩 231.90m

43. 浮土掩盖 81.80m

42. 青灰色变质细粒石英砂岩 23.60m

———— 整　合 ————

下伏地层　南碧沟组灰色绢云石英千枚岩

3）儋州市尖峰顶-八角岭剖面

剖面起点位于1∶5万《南丰市》幅尖峰顶南侧约200 m处，起点坐标：东经109°38′57″，北纬19°25′55″；河北区调队、海南地质大队1989~1991年在开展1∶5万《仁兴市、南丰市》幅区调时测制，本次对地层划分做了修改（图3-21）。

上覆地层　青天峡组灰色—深灰色黑云粉砂质板岩

═══ 断　层 ═══

陀烈组（S_1t） 1169.5m

上段（S_1t^3） 8.3m

26. 灰黄色千枚状绢云石英细砂岩 8.3m

中段（S_1t^2） 539.7m

25. 黄灰色铁质粉砂质板岩夹含铁质细砂岩及含炭硅质板岩，夹层厚小于3 m 96.2m

24. 灰黑色含炭硅质板岩 66.7m

23. 灰黄色千枚状粉质细砂岩夹黑色含炭硅质板岩，夹层厚2~4 m 117.4m

22. 黑色含铁质炭质板岩 63.5m

21. 灰黄色千枚状含粉砂细砂岩夹灰色薄层板岩，夹层厚0.1~0.5 m 41.3m

20. 灰黑色含粉砂、含炭质板岩 42.5m

19. 灰黑色含炭硅质板岩 112.1m

下段（S_1t^1） 621.5m

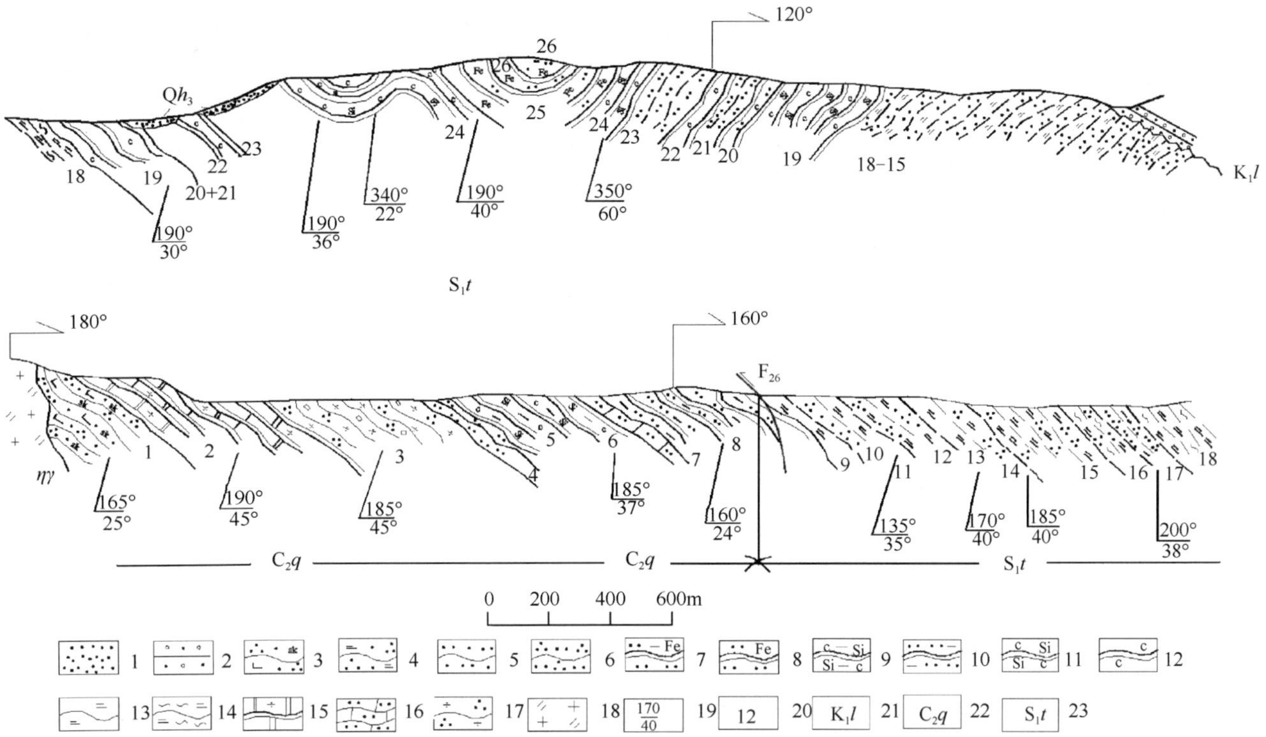

图 3-21 儋州市尖峰顶-八角岭陀烈组、青天峡组实测剖面图

Fig. 3-21 Measured section of the Tuolie and Qingtianxia Formations along Jianfengding-Bajiaoling, Danzhou City

1—第四系坡积物；2—砂砾岩；3—矽卡岩化含钙质石英砂岩；4—千枚状绢云石英细砂岩；5—千枚状粉砂质细砂岩；6—变质石英粉砂岩；7—含铁黑云粉砂质板岩；8—铁质粉砂质板岩；9—含黑云炭质硅板岩；10—黑云粉砂质泥灰岩；11—炭硅质板岩；12—炭质板岩；13—绢云千枚岩 14—绿泥绢云千枚岩；15—大理岩化含透闪石中晶灰岩；16—变质粉砂质泥灰岩；17—含黄铁石英透闪岩；18—二长花岗岩；19—产状；20—分层号；21—鹿母湾组；22—青天峡组；23—陀烈组

18. 灰黑色绿泥绢云千枚岩	146.3m
17. 灰黄色千枚状黑云石英粉砂岩夹绢云千枚岩，夹层厚 0.5~2 m	40.5m
16. 灰黑色绢云绿泥石英千枚岩	36.4m
15. 褐灰色—灰色绢云石英千枚岩夹绢云千枚岩，夹层厚一般 0.5 m	157.6m
14. 灰黄色千枚状绢云石英细砂岩夹绢云千枚岩，夹层厚度小于 2 m	39.4m
13. 浅灰色绢云千枚岩	87.2m
12. 浅灰色千枚状绢云石英细砂岩	32.6m
11. 浅灰色绢云千枚岩	33.0m
10. 黄灰色千枚状绢云石英细砂岩夹灰黑色炭质板岩，夹层厚 0.5~1.5 m	24.0m
9. 灰黄色绢云千枚岩	24.6m

未见底

4）琼海市东泰农场加要岭剖面

该剖面由海南省地调院 1997~2000 年进行 1：5 万《黎母岭、枫木市、琼中县、白马岭》幅区调时测制；2001~2003 年海南省地调院开展 1：25 万《琼海县》幅区调补做了部分工作。剖面起点坐标：东经 110°13′45″，北纬 18°59′47″；剖面长约 9004m（图 3-22）。

上覆地层 下白垩统鹿母湾组（K_1l）

42. 浅灰白色砂砾岩

~~~~~~~~ 不整合 ~~~~~~~~

陀烈组（S_1t）
~~~~~~~~

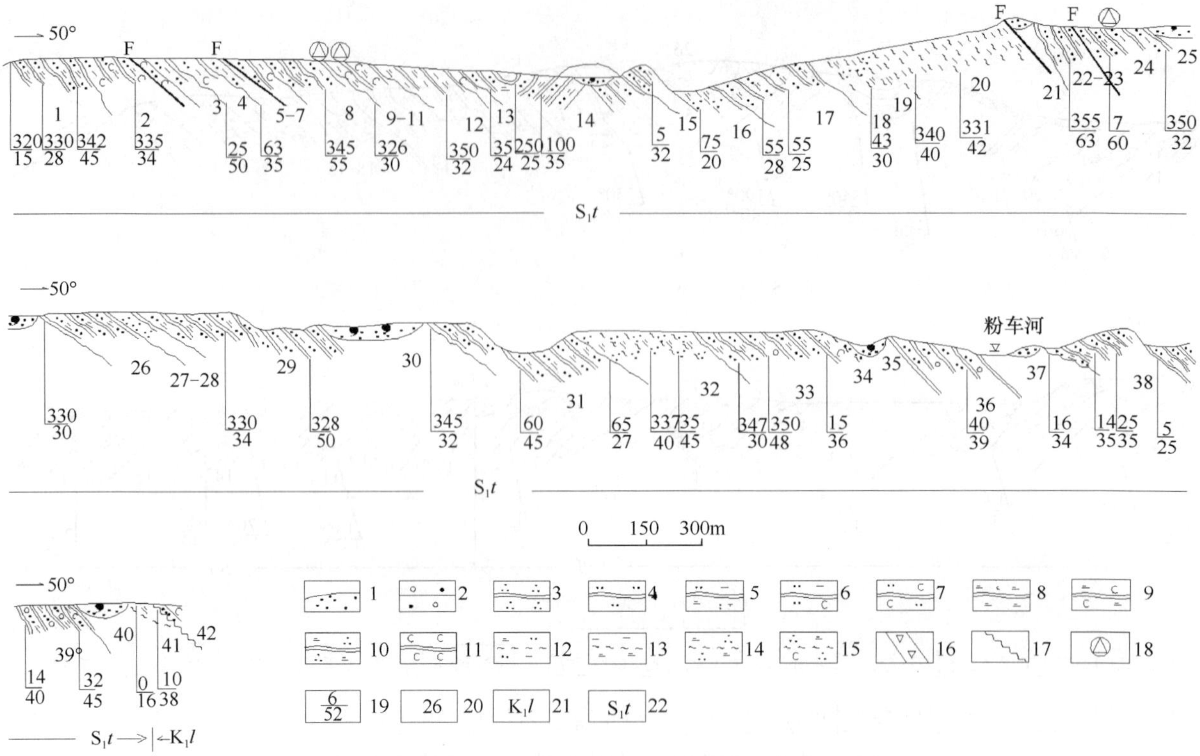

图 3-22 琼海市东泰农场加要岭下志留统陀烈组实测剖面图

Fig. 3-22 Measured section of the Lower Silurian Tuolie Formation, Dongtai Farmland, Qionghai City

1—浮土；2—砂砾岩；3—变质细粒石英砂岩；4—变质粉砂岩；5—变质绢云母粉砂岩；6—变质含炭绢云母粉砂岩；7—变质炭质粉砂岩；8—绢云母板岩；9—含炭绢云母板岩；10—千枚状石英绢云母板岩；11—炭质板岩；12—绿泥绢云千枚岩；13—绢云母千枚岩；14—石英绢云（或绢云母石英）千枚岩；15—炭质绢云石英千枚岩；16—构造角砾岩；17—角度不整合；18—孢粉采样位置；19—产状；20—分层号；21—鹿母湾组；22—陀烈组

上段（S_1t^3）	>85.78m
41. 浅灰色、浅灰白色绿泥绢云千枚岩	18.84m
40. 掩盖	66.94m
中段（S_1t^2）	
39. 浅灰色、浅灰白色块状厚层绢云石英千枚岩夹变质含炭绢云母粉砂岩，近底部见二者互层	188.72m
38. 浅灰白色变质绢云母粉砂岩，上部岩层中见众多小石英脉（脉宽多为 2~10cm）多顺层侵入，中部岩层整体上见弱变形，局部见揉皱现象	190.91m
37. 掩盖	40.52m
36. 浅灰白色变质粉绢云母粉砂岩与黑色炭质绢云石英千枚岩互层	83.64m
35. 掩盖	73.52m
34. 灰黑色、黑色块状炭质绢云母石英千枚岩，中部局部夹变质粉砂岩，下部岩层中见小石英脉（1~2cm 宽）多顺层侵入，小石英脉与岩层一道发生变形（小石英脉受强力挤压呈小透镜状断续相连分布于岩层中）	46.38m
33. 浅灰色中厚层变质绢云母粉砂岩夹黑色中薄层变质炭质绢云母粉砂岩，中、下部岩层变形强烈	103.63m
32. 浅灰色、浅灰白色厚层—块状绢云母千枚岩夹黑色中厚层变质炭质绢云母粉砂岩，岩层变形强，小褶皱发育	174.18m
31. 浅灰白色厚层变质绢云母粉砂岩夹黑色中厚层含炭绢云板岩	177.05m
30. 掩盖	164.41m
29. 浅灰白色变质绢云母粉砂岩。岩石中绢云母含量不均（原岩中泥质含量不均），可见绢云母含量多的岩层与含量少的岩层互层，从而显示出变余层理（水平层理）	230.02m
28. 掩盖	20.46m

27. 变质细粒石英砂岩 5. 70m

26. 浅灰白色变质绢云母粉砂岩，具变余水平层理 73. 37m

25. 掩盖 94. 23m

24. 浅灰白色厚层—块状变质绢云母粉砂岩，夹灰黑色中薄层炭质绢云母千枚岩。前者中具变余水平层理（原岩中泥质含量不同，且见成层显示）。炭质绢云母千枚岩中（XI30-1）含微古植物：*Leiominuscula minuta*，*Lophominuscula* sp.，*Margominuscula* sp.，*Leiopsophosphaera densa*，*Trachysphaeridium dengyingense*，*Micrhystridum* sp.，*Triangumorpha punctulata*，*Conusmorpha brevis*，*Leiofusa Navicулara*，*Nephromorpha regularis*，*Sibiriella prima*，*Bavlinellafaveolata*，*Teophipolia lacerata*，*Ceratophyton* sp.，*Germinosphaera unispinose*，*Taeniatum Crassum*，*Paleamorpha punctulata* 135. 82m

23. 浅灰白色厚层变质绢云母粉砂岩夹灰黑色中厚层变质炭质绢云母粉砂岩 14. 82m

22. 浅灰白色变质绢云母粉砂岩，具变余水平层理 57. 28m

21. 掩盖 61. 43m

20. 浅灰白色绢云母千枚岩、变质绢云母粉砂岩，中下部岩层中见较多小石英脉（宽多为2~6cm）顺层侵入 333. 13m

19. 浅灰白色厚层—块状石英绢云千枚岩夹灰黑色中厚层绢云母石英千枚岩。后者中含炭质，前者中具变余水平层理 93. 12m

18. 掩盖 30. 96m

17. 浅灰白色变质绢云母粉砂岩 36. 66m

16. 浅灰白色绢云母千枚岩、绢云石英千枚岩，上部局部夹薄层变质炭质绢云母粉砂岩 59. 69m

15. 浅灰色变质绢云母粉砂岩，上部夹少量绢云石英千枚岩 65. 70m

14. 浅灰色中厚层变质绢云母粉砂岩与黑色、灰黑色千枚状石英绢云板岩互层。二者均具变余水平层理 65. 38m

13. 灰黑色、黑色绢云母炭质板岩 12. 10m

12. 浅灰白色绢云母板岩，具水平层理，岩层变形强 70. 97m

11. 浅灰色中厚层含炭绢云母板岩与浅黑色中厚层炭质板岩互层 18. 22m

10. 浅灰白色中厚层变质粉砂岩与灰黑色中薄层含炭绢云母板岩互层 9. 93m

9. 黑色块状炭质绢云母板岩，夹灰白色中薄层变质含炭绢云母粉砂岩，炭质绢云母板岩中含微古植物化石：*Leionuscula minuta*，*Margominuscula* sp.，*Leiopsophosphaera dense*，*Trachysphaeridium dengyingense*，*T. Minor*，*T.* sp.，*Granomarginata prima*，*Leiomarginata simplex*，*Triangomorpha punctulata*，*Conusmorphabrevis*，*Nephromorpha regularis*，*Leiofusa Navicулara*，*Turuchaniaovalia*，*Micr hysttridium* sp.，*Synsphaeridium* sp.，*Taeniatum* sp.，*Sibiriella prima*，*Ceratophyton* sp.，*Paleamorpha punctula* 6. 09m

8. 浅灰色变质绢云母粉砂岩夹变质含炭绢云母粉砂岩，前者中具变余水平层理，含炭绢云母粉砂岩中含微古植物：*Leiominuscula minute*，*Trachysphaeridium dengyingense*，*Turuchania ternata*，*T. ovila*，*Leiofusa navicулara*，*Baculimorpha brevis*，*Siphonophycus* sp.，*Melanocyrillium fimbriatum*，*Paleamorpha punctulata*，*Angochitina* 50. 49m

7. 浅灰色铁质绢云母板岩 8. 20m

6. 浅灰色、灰白色变质绢云母粉砂岩，上部岩层中具变余水平层理 20. 94m

5. 浅灰色、浅灰黑色厚层变质绢云母粉砂岩夹浅黑色中薄层变质炭质绢云母粉砂岩 24. 29m

4. 浅灰色、黑色中厚层炭质绢云母板岩与浅灰、浅灰黑色中厚层变质含炭绢云母粉砂岩互层 56. 72m

3. 浅灰色、浅灰黑色变质绢云母粉砂岩 44. 66m

2. 浅灰色、浅灰黑色变质含炭绢云母粉砂岩，局部岩石中绢云母含量变少（<5%）；岩层变形强，具变余水平层理 124. 55m

1. 灰、浅灰白色变质绢云母粉砂岩 100. 09m

未见底

（4）空列村组剖面

空列村组剖面有大干村剖面、昌江县戈枕村剖面、儋州市乐敦岭剖面、万宁斗牛村-黄竹塘高速公路剖面、南捉沟-南碧沟剖面、乐东县木棉头后物村剖面等。其中，戈枕村剖面、乐敦岭剖面、南捉沟-南碧沟剖面、乐东县木棉头后物村剖面等剖面岩石组合特征基本类似，不作一一列举。剖面情况如下。

1）大干村剖面

该剖面为空列村组代表性剖面。剖面位于三亚市雅亮乡空列村南侧，坐标：东经 109°21′00″，北纬 18°35′00″；由海南行政区地质矿产局 1982~1986 年开展 1∶5 万《志仲》幅区调时测制。陈哲培等 1997 年在开展《海南岛岩石地层》时略作修改（图 3-23）。

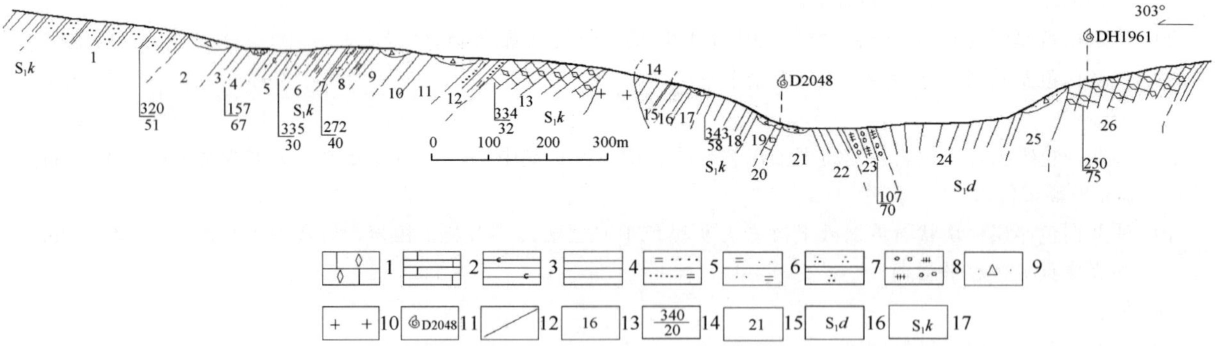

图 3-23　崖县大干村实测剖面图东南段

Fig. 3-23　Southeastern segment of Dagancun measured section，Yaxian County

1—结晶灰岩；2—斑点板岩；3—含炭板岩；4—板岩；5—绢云石英粉砂岩；6—绢云石英细砂岩；7—石英岩；8—变质复成分砾岩；9—掩盖地段；10—斑状花岗岩；11—化石产出相应位置；12—分层界线；13—分层号；14—产状；15—分层号；16—大干村组；17—空列村组

上覆地层　大干村组（S_1d）灰色复成分砾岩

———— 整　合 ————

空列村组（S_1k）　总厚度 855.30m

18. 青灰色薄层板岩，中下部夹绢云石英粉砂岩　62.00m
17. 掩盖　15.10m
16. 深灰色中厚层结晶灰岩，含珊瑚 *Amplexoides*? sp.；海绵? *Receptaculites* sp.　3.80m
15. 掩盖　33.40m
14. 浅灰、灰绿色板岩　78.40m
13. 乳白色厚层石英岩　9.30m
12. 灰黑色薄层板岩，上部夹灰色绢云石英粉砂岩　43.80m
11. 灰色厚层结晶灰岩　63.50m
10. 灰黑色绢云母板岩与灰白色绢云石英粉砂岩互层　24.70m
9. 掩盖　43.00m
8. 灰色绢云母板岩　22.40m
7. 掩盖　27.00m
6. 灰色粉砂质板岩与灰白色绢云石英细砂岩互层　33.70m
5. 浅黄白色石英岩　9.00m
4. 浅灰色粉砂质板岩与灰白色绢云石英细砂岩互层　48.00m
3. 深灰色绢云母板岩　106.20m
2. 掩盖　51.20m
1. 乳白色石英岩、中粗粒石英砂岩，局部含砾，偶夹板岩或含炭板岩　180.80m

———— 整　合 ————

下伏地层　陀烈组（S_1t）灰白色薄层变质粉砂岩

2）斗牛村-黄竹塘高速公路剖面

该剖面由宜昌地矿所1991~1995年开展1：5万《和乐、中原市、博鳌港》幅区调时测制，海南省地质调查院2001~2003年开展《琼海县》幅区调时补做了部分工作。剖面沿万宁肥山根乡斗牛村至黄竹塘的高速公路边测量，起点坐标：东经110°29′02″，北纬18°59′19″。自上而下为：

上覆地层 石炭系南好组变石英砾岩、变砂质砾岩、变砂岩

～～～～～～ 不 整 合 ～～～～～～

空列村组（S_1k）

上段（S_1k^2）	140.87m
28. 紫灰色、灰褐色变质粉砂岩、变质细砂岩夹板岩。砂岩中见小型楔状层理，粉砂岩中见脉状层理	20m
27. 灰黄色—黄白色酸性沉凝灰岩，可见粒序层理和交错层理	30.89m
26. 掩盖	62.44m
25. 灰色变泥质粉—细砂岩与灰色薄层—条带状板岩互层	13.77m
24. 灰色板岩、砂质板岩夹泥质粉—细砂岩	4.47m
23. 灰色厚层泥质砂岩夹灰色板岩	5.30m
22. 灰色板岩、砂质板岩夹泥质粉-细砂岩	4.00m
下段（S_1k^1）	75.83m
21. 掩盖	22.50m
20. 灰黄色—浅灰色厚层—块状细砂岩夹灰色薄—中层板岩	41.78m
19. 浅灰色—灰白色变质细粒石英砂岩、夹少量含泥质细—粉砂岩及少量中粒砂岩	11.02m

———— 整 合 ————

下伏地层 陀烈组

3）南捉沟—南碧沟剖面

剖面情况见上述。自上而下为：

上覆地层 峨顶组燧石生物微晶灰岩

═════ 断 层 ═════

空列村组（S_1k）

上段	
66. 灰黄色—灰绿色变质粉砂岩夹泥质条带，水平层理发育，纹层厚2~3mm	>52.7m
下段	
65. 灰色—灰黄色石英岩。局部见波痕构造	159.6m
64. 灰白色—浅红色变质中粒石英砂岩，具水平层理、微波状层理，低角度交错层理等	221.2m

———— 整 合 ————

下伏地层 陀烈组

（5）大干村组剖面

大干村组剖面有大干村剖面和河叉岭-邦溪八队剖面。

1）大干村剖面

该剖面为大干村组代表性剖面。剖面情况同上述，自上而下为（图3-23）：

上覆地层 靠亲山组（S_1kq）灰色斑点板岩

———— 整 合 ————

大干村组（S_1d） 总厚度380.50m

22. 灰色中厚层结晶灰岩，含珊瑚：*Favosites*，*F.* cf. *gothlandicus*，*Tabularia* sp.，*Catenipora* sp.；海绵 *Receptaculites* sp. 64.00m

21. 掩盖 69.50m

20. 青灰色薄层板岩，上部为浅灰色粉砂质板岩 201.10m

19. 灰色变质复成分砾岩 45.90m

———— 整　合 ————

下伏地层　空列村组（S_1k）青灰色薄层板岩

2）白沙县河叉岭-邦溪八队剖面

该剖面由宜昌地质矿产研究所 1991～1994 年开展 1∶5《昌江县、邦溪》幅区调时测制；2005～2007 年海南省地质调查院开展《东方县》幅区调时补做了部分工作。剖面起点坐标：东经 109°03′57″，北纬 19°24′48″。自上而下为（图 3-24）：

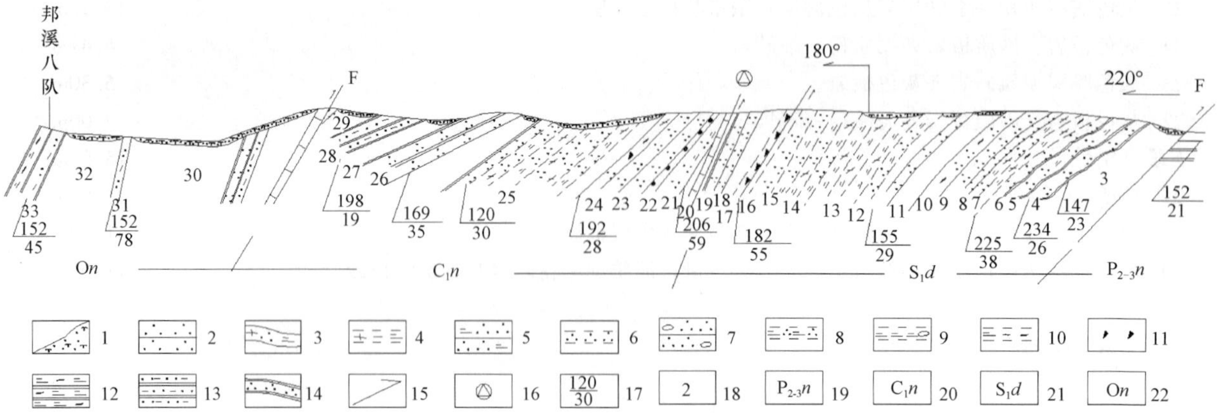

图 3-24　白沙县河叉岭-邦溪八队剖面图

Fig. 3-24　Section of Hechaling-Bangxi No. 8 Brigade, Baisha County

1—第四纪残坡积覆盖物；2—石英砂岩；3—含红柱石绢云石英千枚岩；4—含红柱石微晶石墨绢云千枚岩；5—含绢云石英千枚岩；6—石英片岩；7—含砾石英岩；8—绢云石英千枚岩；9—含砾绢云母片岩；10—含红柱石绿泥石化绢云母片岩；11—断层角砾岩；12—绿泥石化绢云母板岩；13—含泥粉砂质板岩；14—石英岩；15—断裂；16—孢粉采样位置；17—产状；18—分层号；19—南龙组；20—南好组；21—大干村组；22—南碧沟组

上覆地层　石炭系南好组变质石英质砾岩

～～～～～ 不　整　合 ～～～～～

大干村组（S_1d）

上段　总厚度>142.44m

18. 下部为浅灰色薄层石英片岩，中上部为浅灰色薄层含红柱石绢云母石英千枚岩夹绢云母石英千枚岩。产微古植物：*Leiopsophosphaera* cf. *clara*，*L. solida*，*L.* cf. *flava*，*L.* cf. *leguminiformis*，*L.* sp. 1，*L.* sp. 2，*Trachysphaeridium* aff. *angulatum*，*T.* cf. *rugosum*，*T.* sp. 1，*T.* sp. 2，*Lophosphaeraidium* sp.，*Leiofusa* cf. *bicornuta*，*L.* sp. 1，*L.* sp. 2，*Lignum striatum*，*L.* sp.，*Macroptycha* sp.，*Dictyosphaera* sp.，*Dictyosphaera* sp.，*Micrhystridium* sp. 和 *Pseudozonophaeridium* sp. 等 21.89m

17. 第四系残积层覆盖 8.42m

═══ 断　层 ═══

16. 浅灰色薄层绢云石英千枚岩 6.75m

15. 白色石英片岩、含砾石英片岩及浅变质石英质砾岩，自下而上粒度变粗 5.19m

14. 浅灰色薄层含红柱石绢云石英千枚岩夹绢云石英千枚岩 29.54m

13. 剖面上为第四系残积层覆盖，橡胶林中见含红柱石石英绢云千枚岩碎块，剖面向西 200 m 相当层位为浅灰色薄层含红柱石石英绢云千枚岩 70.65m

下段 总厚度>126.0m

12. 灰白色薄层绢云石英千枚岩 3.58m
11. 第四系残积层覆盖，上部见0.69m厚棕黄色厚层石英片岩零星露头 18.48m
10. 白色厚层石英片岩、含砾石英片岩、铁质中粒石英砂岩 13.30m
9. 下部为灰色薄层含绢云母石英千枚岩，中、上部为灰白色中—厚层石英片岩 2.60m
8. 棕黄色中—厚层石英片岩 9.55m
7. 棕红色（风化色）薄层含红柱石微晶石墨绢云石英千枚岩，下部及顶部各夹一层1.80m及1.72m厚—中层石英片岩 20.22m
6. 白色中—厚层石英片岩 13.48m
5. 浅灰色薄层含红柱石绢云石英千枚岩及棕黄色石英片岩，向上粒度变粗 3.15m
4. 灰色浅变质中细粒石英岩 11.74m
3. 第四系残积物覆盖，为韧性剪切带通过部位 >29.90m

═══════ 断 层 ═══════

下伏地层 二叠系南龙组及鹅顶组

1~2. 灰色中层大理岩、深灰色含炭硅质粉晶灰岩，顶部0.98m为深灰色页片状褶劈理化水云母石英千糜岩

（6）靠亲山组剖面

靠亲山组仅有志伦村代表性剖面。剖面位于保亭县毛感乡志伦村，起点坐标：东经109°27′19″，北纬18°39′04″，由海南行政区地质矿产局1982~1986年开展1∶5万《志仲》幅区调时测制。自上而下为（图3-25）：

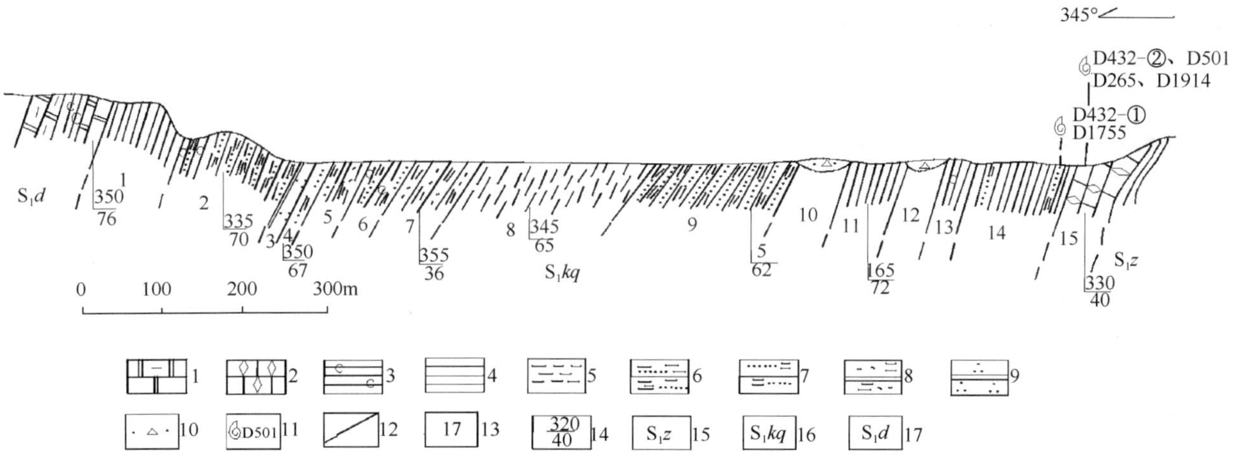

图3-25 保亭县志轮村实测剖面图

Fig. 3-25 Measured section, Shilun Village, Baoting County

1—含泥白云岩、白云岩；2—结晶灰岩；3—含炭板岩；4—绢云母板岩、粉砂质绢云母板岩；5—千枚岩；6—绢云石英粉砂岩与千枚岩互层；7—绢云石英粉砂岩；8—绢云石英细砂岩；9—石英岩；10—掩盖地段；11—化石产出位置或相应位置及化石编号；12—分层界线；13—分层号；14—产状；15—足赛岭组；16—靠亲山组；17—大干村组

上覆地层 足赛岭组（S_1z）灰黑色含炭千枚岩

——————— 整 合 ———————

靠亲山组（S_1kq） 总厚度950.10m

15. 灰黑色薄—厚层含泥结晶灰岩，沿走向变为灰色厚层生物碎屑结晶灰岩。含珊瑚 *Syringopora* cf. *qianbeiensis*, *S.* cf. *yiwaensis*, *S.* cf. *nanshanensis*, *S. fascicualis*, *Syringoporella* sp., *Helilites* sp., *Heliolitella* sp., *H. langgaoensis*, *Squameolites densus*, *Plasmopora* sp.；腕足类 *Atrypa* sp.；三叶虫 *Encrinuridae* 79.00m

14. 灰色、灰黑色粉砂质千枚岩夹灰色绢云石英粉砂岩。含三叶虫 *Coronocephalus* sp.，*C. rex*，*Encrinuridae*；腕足类 *Sowerbyella*? sp.，*Schizophoria*? sp.，*Striispirifer* sp.，*S.* cf. *hsiehi*，*Eospirifer*? sp.，*E.*? *sinensis*，*Atrypa* sp.；腹足类 *Hormotoma* sp.；海百合茎 *Cyclocyclicus* sp. 92.00m

13. 上部灰黑色含炭粉砂质板岩，下部灰色绢云板岩 32.00m

12. 浮土掩盖 39.00m

11. 灰色薄层绢云母板岩 51.00m

10. 浮土掩盖 63.00m

9. 浅灰色薄—厚层绢云石英细砂岩 36.50m

8. 青灰色薄层千枚岩 154.00m

7. 青灰色薄—厚层绢云石英细砂岩 50.00m

6. 灰黑色薄层含炭千枚岩与浅灰色绢云石英粉砂岩互层 48.60m

5. 灰黑色千枚岩与浅灰色绢云石英细砂岩互层 55.00m

4. 灰色薄层绢云石英粉砂岩 15.00m

3. 灰色中厚层石英岩 12.00m

2. 灰黑色薄层千枚岩与灰白色绢云石英粉砂岩互层 135.00m

1. 青灰色粉砂质千枚岩 88.00m

———— 整 合 ————

下伏地层 大干村组（S_1d）白云岩夹千枚岩

（7）足赛岭组剖面

足赛岭组仅有公井岭代表性剖面。剖面位于保亭县南好村东侧公井岭，起点坐标：东经 109°26′00″，北纬 18°37′00″，由海南行政区地质矿产局 1982～1986 年开展 1∶5 万《志仲》幅区调时测制。自上而下为：

上覆地层 南好组含砾不等粒石英砂岩

———— 整 合 ————

足赛岭组（S_1z） 总厚度 819.30m

8. 灰黑色薄层千枚岩 66.70m

7. 浅灰色薄层绢云石英粉砂岩 22.00m

6. 灰黑色含炭粉砂质绢云板岩 15.30m

5. 灰白色结晶灰岩，含珊瑚 *Plasmopora* sp.，*Amplexoides*? sp.；腕足类? *Atrypa* sp. 40.00m

4. 掩盖 161.00m

3. 灰色、青灰色千枚岩 107.10m

2. 灰黑色千枚岩 70.30m

1. 灰黑色含炭千枚岩夹灰黑色千枚岩 336.90m

———— 整 合 ————

下伏地层 靠亲山组（S_1kq）结晶灰岩

2. 岩石地层特征

（1）美子林组（$\epsilon_1 m$）

美子林组是宜昌地矿所 1991～1995 年在开展 1∶5 万《中原市、博鳌港、和乐》幅区调时创建的岩石地层单位，时代为寒武纪；姚华舟等（1999）介绍，并指定万宁县乐来乡美子林剖面为代表性剖面，乐来乡下田剖面及大茂乡袁水水库剖面为参考剖面。剖面上部为结晶灰岩、石英透辉石大理岩，产瓶虫类及小壳化石；下部为变质石英砂岩与云母片岩不等厚互层，夹黑云透辉石英角岩、白云母石英片岩等。顶、底界线因掩盖，接触关系不明，厚度>900m。

（2）南碧沟组（On）

南碧沟组由黎明等（1981）创建的南碧沟群转化而来，命名地点在昌江县南碧沟。1988～1991年广西区调队开展1：5万《猴猕岭、坝王岭》幅区调时，将南碧沟群更名为南碧沟组，其层位相当于原南碧沟群中亚群中下部分。汪啸风等（1992）将南碧沟群划归陀烈群，时代定为早—中志留世。陈哲培等（1997）在《海南省岩石地层》中采用广西区调队命名定义，并指定代表性剖面为岛西南南捉沟-南碧沟剖面。宜昌地矿所、海南省地质调查院2000～2002年开展1：25万《乐东县、陵水县》幅区调时把该剖面第1层至第19层进行了重新厘定，归属为下志留统陀烈组。代表性剖面厚2032.40m，为一套浅变质碎屑岩夹变质基性火山熔岩、基性火山碎屑岩的岩石组合。岩性以千枚岩、绢云板岩为主，夹变质粉砂岩、细砂岩、炭质板岩；中上部夹基性火山熔岩、基性火山碎屑岩。未见底，与上覆志留系陀烈组呈整合接触。

（3）陀烈组（S_1t）

陀烈组系海南地质大队1990年开展1：5万《东方、玉道》幅区调时修订陀烈群而来。陀烈群为广东区测队1964年创建，命名剖面位于东方县陀烈村，岩性为粉砂岩、页岩、硅质岩、夹炭质硅质岩、砂质灰岩，产孢粉化石，时代为寒武纪。夏邦栋等（1979）推测陀烈群为—晚奥陶世—早泥盆世（O_3—D_1），相当于他所建立的旧村岭群中上部。黎明等（1981）通过重测陀烈剖面及昌江县南捉沟-南碧沟剖面，改陀烈群为奥陶系—志留系南碧沟群，认为陀烈群可与南碧沟剖面中南碧沟群中、上部对比。宜昌地矿所、海南地矿局在1986～1990年开展海南岛若干重要基础地质问题研究时（汪啸风等，1992），对其层序、岩石组合特征、时代等进行了详细研究，给予陀烈群以新的含义，并根据早志留世几丁虫化石将地层时代定为早志留世。海南地质大队（1990）在东方幅、玉道幅区调报告中，改陀烈群为陀烈组。陈哲培等（1997）在《海南省岩石地层》中指定东方娜姆河剖面为陀烈组代表性剖面，系指整合于南碧沟组之上、空列村组之下的一套浅变质岩，可分为三段。下段为变质细砂岩、绢云母板岩夹灰岩透镜体；中段以炭质绢云母板岩为主夹变质粉砂岩及绢云母板岩；上段为绢云母板岩夹条带状变质粉砂岩。中下部及中上部产几丁虫及孢粉化石。底部以变质石英砂岩的出现与南碧沟组千枚岩作为划分标志，顶部以变质粉砂岩与上覆空列村组石英岩分界。本组分布广泛，地层厚度为1879～2843m。

（4）空列村组（S_1k）

空列村组系海南行政区地质矿产局1982～1986年开展1：5万《志仲》幅区调时所创建，陈哲培（1987）介绍。空列村组系广东区测大队（1964）创建的逗文庄群进一步划分出来的岩石地层单位（逗文庄群已分解为志留系空列村组、大干村组、靠亲山组、足赛岭组及石炭系南好组）。汪啸风等（1992）对空列村组进一步研究后，推测其时代为早志留世，也可能为奥陶纪沉积。陈哲培等（1997）在《海南省岩石地层》中指定大干村剖面为代表性剖面，底部为石英岩，向上为绢云母板岩与绢云母石英粉砂岩互层，夹结晶灰岩，产珊瑚、海绵化石。顶部以板岩与上覆大干村组呈整合接触，底部以石英岩、石英砂岩与下伏陀烈组呈整合接触。厚140.87～855.3m。

（5）大干村组（S_1d）

大干村组由陈哲培等（1985）创建，陈哲培（1987）介绍。创名地在三亚市雅亮乡大干村。陈哲培等（1997）在《海南省岩石地层》中指定大干村剖面为代表性剖面，以青灰色薄层板岩、粉砂质板岩为主；顶部为灰色中厚层结晶灰岩，产珊瑚、海绵化石。底部为灰色—灰黄色复成分砾岩，具粒序层理。顶部灰岩与上覆靠亲山组灰色斑点板岩呈整合接触，底部以变质复成分砾岩与下伏空列村组板岩呈整合接触。厚142.4～380.5m。

（6）靠亲山组（S_1kq）

靠亲山组由陈哲培等（1985）创建，陈哲培（1987）介绍。创名地在保亭县毛感乡志伦村附近的靠亲山。陈哲培等（1997）在《海南省岩石地层》中指定志伦村剖面为代表性剖面。上部为灰黑色—深灰色薄至厚层泥灰岩、含生屑结晶灰岩，产珊瑚、腕足类、腹足类、三叶虫及牙形石；中部为浅灰色—灰黄色薄层粉砂质板岩、千枚状板岩偶夹粉砂岩，产腕足类、三叶虫及海百合茎化石；下部为粉砂质千枚岩、板岩夹灰色厚层石英岩、绢云石英粉砂岩、细砂岩。该组底界以青灰色粉砂质千枚岩的出现作为划分标志，与下伏大干村组呈整合接触。顶部以灰黑色薄—厚层结晶灰岩结束与上覆足赛岭组呈整合接触。岩性稳定，厚度为950.10m。

(7) 足赛岭组 (S_1z)

足赛岭组由陈哲培（1985）创建，陈哲培（1987）介绍。创名地在保亭县毛感乡南好村附近的足赛岭。指整合于靠亲山组之上，为千枚岩、含炭千枚岩夹结晶灰岩，灰岩产珊瑚、腕足类化石。底界以含炭千枚岩与靠亲山组结晶灰岩分界，顶界以千枚岩与石炭系南好组含砾不等粒石英砂岩、砂砾岩呈不整合接触。厚度>1084m。

在2000~2002年开展1：25万《乐东县、陵水县》幅区调时，在石炭系南好组代表性剖面——南兵—南好公路剖面原下石炭统岩关阶南好组下部发现志留纪化石富集层（张仁杰等，2000；李志宏等，2000；曾庆銮等，2003），有珊瑚、海绵化石、腕足类、珊瑚、三叶虫、腹足类、海百合茎化石等。

(二) 地层划分与对比

1. 美子林组 ($\epsilon_1 m$)

美子林组仅见于万宁地区，是五指山地层分区目前唯一发现寒武纪地层的地区。基岩出露较差，浮土掩盖地段较多，分布面积小，空间延伸差，未见顶、底，与上覆地层及下伏地层接触关系不清。划分为3个岩性段。第一段在美子林及袁水水库一带出露，以袁水水库剖面出露较好，岩性为变质长石石英砂岩、变质细粒石英砂岩、二云石英片岩、黑云石英片岩为主，少量变质石英粉砂岩，局部含酸性凝灰质。未见底，厚度大于673m。第二段在3条剖面中均有出露，以下田剖面出露较好，岩性主要为黑云透辉石角岩、黑云二透石英角岩，厚度大于156m。第三段在美子林和下田等地有出露，以美子林剖面出露较好，岩性为结晶灰岩、石英透辉石大理岩、二透大理岩夹变质粉砂岩、细砂岩。未见顶，厚度大于196m。

2. 南碧沟组 (On)

区域上，南碧沟组岩石组合具有可对比性，以一套细碎屑岩沉积为特点（图3-26），但也具一定的差异性：①洛基大岭剖面和万宁市山根乡坑尾-斗牛村剖面岩石以砂质沉积为主，而其他地区则以泥质沉积占主体，空间相变明显；②南捉沟-南碧沟剖面和大溪桥剖面所夹火山岩属于基性、超基性火山熔岩和火山碎屑岩，翰林镇蒙花岭剖面火山岩为玄武岩、安山岩和英安岩组合，万宁市山根乡坑尾-斗牛村剖面则变为含沉火山碎屑岩，显示了总体由南至北盆地深度逐渐加大，与碎屑岩沉积变化所显示的盆地特点基本一致；③大溪桥剖面、翰林镇蒙花岭剖面岩石多见片岩和石英岩分布，变质程度强于其他地区岩石。

在海南岛西北白沙县大溪桥剖面，南碧沟组厚度大于1316.32m。下部由石英片岩、黑云母片岩、矽线石英片岩等组成；上部为石英千枚岩、绢云母千枚岩、石英片岩、变质粉砂岩、白云母片岩、含红柱石白云母片岩，夹多层变质基性火山岩（图版Ⅰ-10-D、图版Ⅰ-10-E），并偶见具底冲刷和粒序层细砂岩。

在海南岛岛北洛基大岭剖面，南碧沟组厚2148.43m。整体上岩石粒度稍粗。下部岩性为石英片岩、变质细粒石英砂岩、石英岩、石英千枚岩、变质粉砂岩、二云母片岩等；上部岩性为石英千枚岩、石英片岩、石英岩、变质粉细砂岩等，部分千枚岩含炭及片岩中含特征变质矿物红柱石。

在海南岛岛东北翰林镇蒙花岭剖面，南碧沟组厚度大于634m。下部为石英片岩、绢云母千枚岩夹细粒白云母石英岩、长石片岩，部分片岩、千枚岩、石英岩中含炭。中部为浅灰白色石英片岩夹多层中基性火山熔岩、沉火山碎屑岩（图版Ⅰ-10-F）。火山岩夹层厚度不一，薄者5~20cm，厚者4~6m，经区域变质作用，已变质为石英岩、片岩类等，但结构尚有部分保留。上部为变质细粒石英砂岩、石

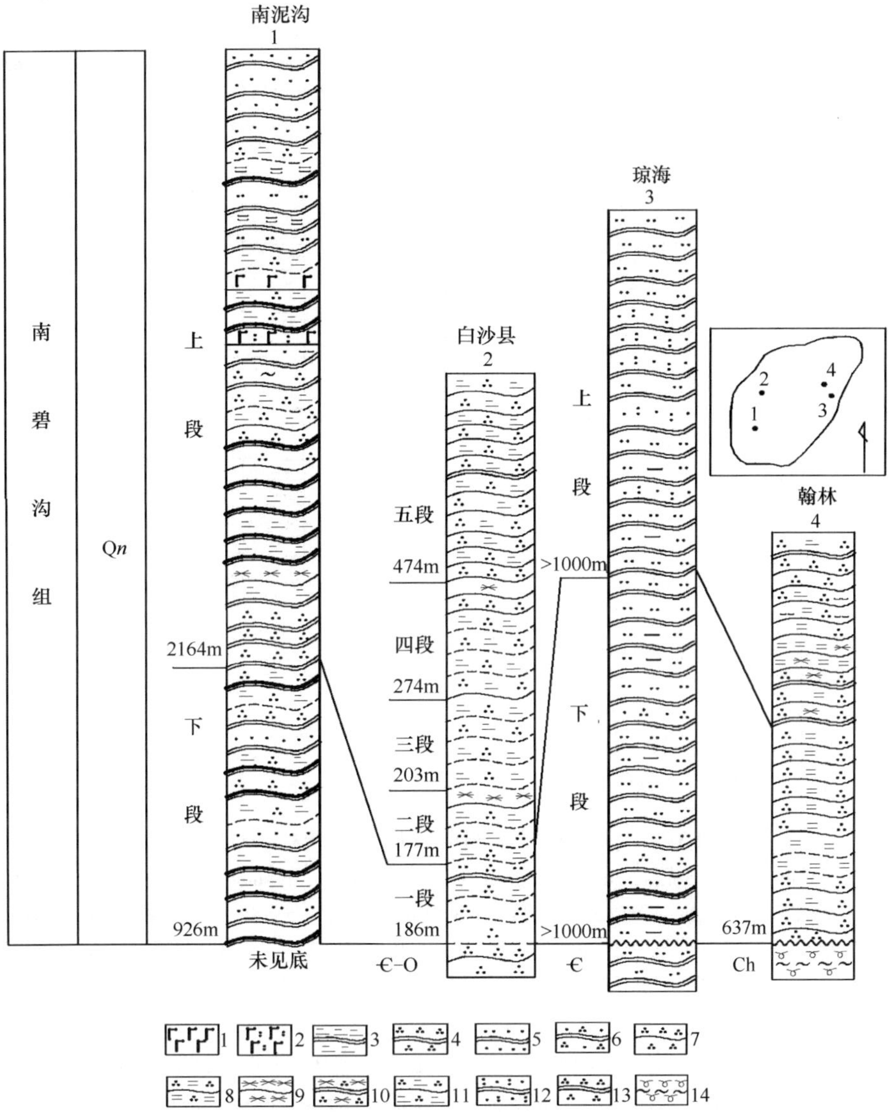

图 3-26 海南岛部分地区奥陶系南碧沟组柱状对比图

Fig. 3-26 Columnar correlation of the Ordovician Nanbigou Formation in some areas of the Hainan Island

1—变质基性火山岩；2—基性凝灰质熔岩；3—绢云母板岩；4—石英板岩；5—变质砂岩；6—变质石英砂岩；7—石英片岩；8—石英白云母片岩；9—阳起石片岩；10—阳起石石英岩；11—绢云石英千枚岩；12—变质含凝灰质砂岩；13—石英岩；14—眼球状混合岩

英片岩、长石石英片岩、二云母片岩、白云母片岩。白云母片岩局部含特征变质矿物红柱石。产似几丁虫与微古植物化石。变余层理发育，泥质岩与砂岩具韵律性互层产出，形成较发育的水平层理，纹层厚 3~5mm，以深浅不同的颜色及粗细颗粒条带显示。

在海南岛岛东南万宁市山根乡坑尾-斗牛村剖面，南碧沟组可分为两段。下段主要为灰色—深灰色粉砂质板岩、板岩、变质泥质粉砂岩，少量变质泥质粉—细砂岩、变质凝灰质砂岩，自下而上砂岩成分增多，厚度大于 1000m。上段下部为灰色—深灰色变质泥质细—粉砂岩、变质泥质粉砂岩、变质粉砂岩夹粉砂质板岩、板岩；上部为灰绿色厚层变质凝灰质砂岩、变质杂砂岩夹质变泥质粉砂岩，厚度大于 1000m。

3. 陀烈组（S_1t）

总体上，陀烈组以沉积粒度较细的泥质粉砂岩、含炭泥质岩相为主体，岩石区域展布比较稳定，基本可以相互对比。尤其是中段含炭（质）岩石往往是良好的标志层（图 3-27）。

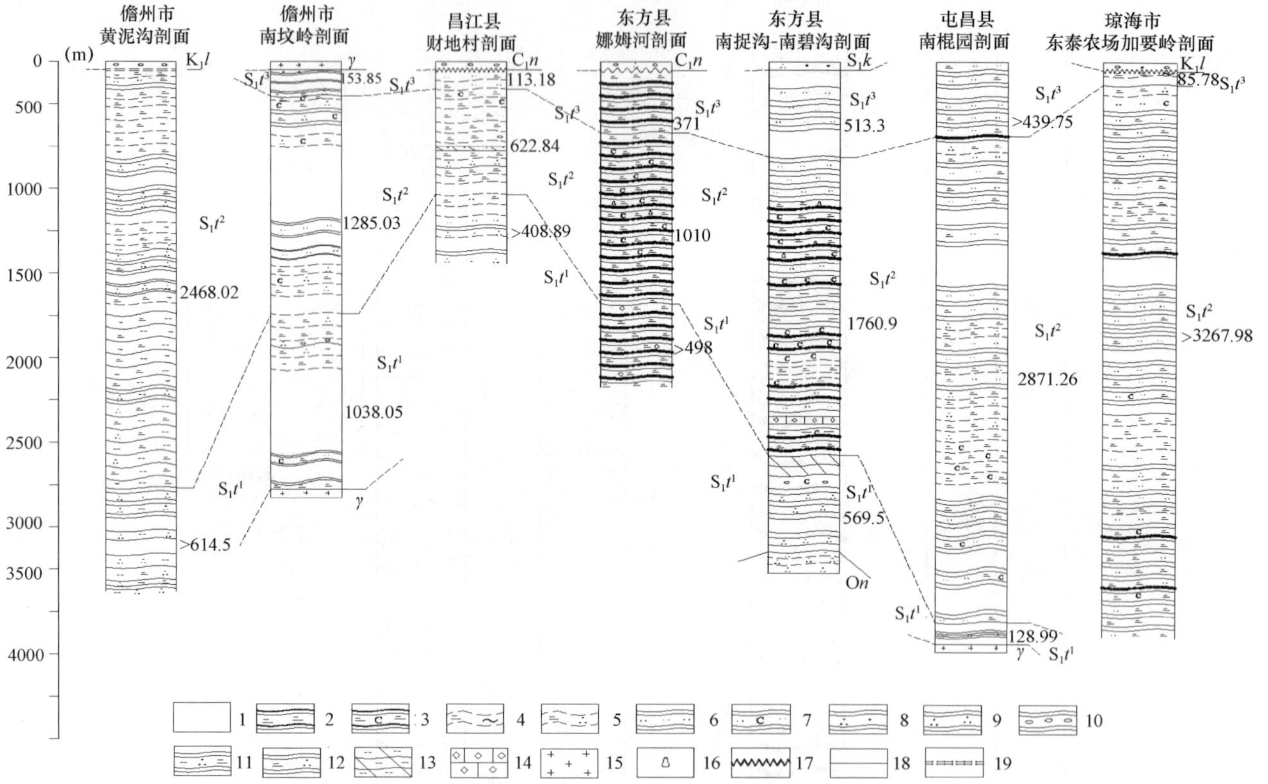

图 3-27　海南岛部分地区志留系陀烈组柱状对比图

Fig. 3-27　Columnar correlation of the Silurian Tuolie Formation in some areas of the Hainan Island

1—掩盖；2—绢云母板岩；3—含炭绢云母板岩；4—绿泥绢云千枚岩；5—绢云石英千枚岩；6—变质粉砂岩；7—变质含炭粉砂岩；8—变质石英砂岩；9—石英岩；10—变质砾岩；11—二云母石英片岩；12—绢云石英片岩；13—角岩；14—结晶灰岩；15—花岗岩；16—几丁虫化石；17—不整合界线；18—整合界线；19—断裂

但该组岩石空间延伸也有一定的差异。①东方江边—儋州西联农场一线地区，上、下部常夹灰岩（或大理岩）透镜体（图版Ⅰ-10-G），而南好和九曲江地区则未见灰岩（或大理岩）透镜体分布；②中段含炭（质）岩石在儋州市以北和南好、九曲江地区非常少；③九曲江地区含多层酸性或中酸性火山碎屑岩（沉凝灰岩或凝灰质砂岩），儋州油文村-屯昌晨星农场一带见基性或超基性火山熔岩夹层，而其余陀烈组分布区没有火山岩出露；④西面陀烈组下部为成熟度较高的变质石英砂岩，层厚大，上部为成熟度低的变质粉砂岩、变质细砂岩层；岛东陀烈组下部为成熟度低的板岩类等泥质岩沉积，沉积粒度相对变细。

4. 空列村组（S_1k）

空列村组分布较广，岩石空间延伸比较稳定，可划分为下、上两段。下段为砂质岩石，由石英岩、变质细粒石英砂岩、变质中粒石英砂岩等组成，岩石质地坚硬，地貌特征明显；上段为砂-泥质沉积，表现为板岩与变质粉砂岩不等厚产出。发育水平层理、微波状层理、低角度交错层理、小型斜层理及透镜状层理等，总体属于水动力较强的滨岸环境退积型沉积（图 3-28），但也存在一定的差异：①下段岩石粒度不一，以东方市南捉沟-南碧沟剖面略粗，为中粒石英砂岩。②上段在三亚大干村代表性剖面中夹结晶灰岩，万宁市斗牛村-黄竹塘高速公路剖面夹酸性沉凝灰岩。③尽管受后期改造，部分剖面未见顶或底，但即其下段沉积厚度而言，变化还是比较明显。其中，三亚大干村厚 232m、万宁斗牛村-黄竹塘厚 75.83m、乐东后物村厚度大于 148.99m、东方市南捉沟-南碧沟厚 380.9m、昌江戈枕村厚>1396.36m、儋州乐敦岭厚度大于 744.69m，呈现南北薄中部厚和东薄西厚的变化趋势。④戈枕村

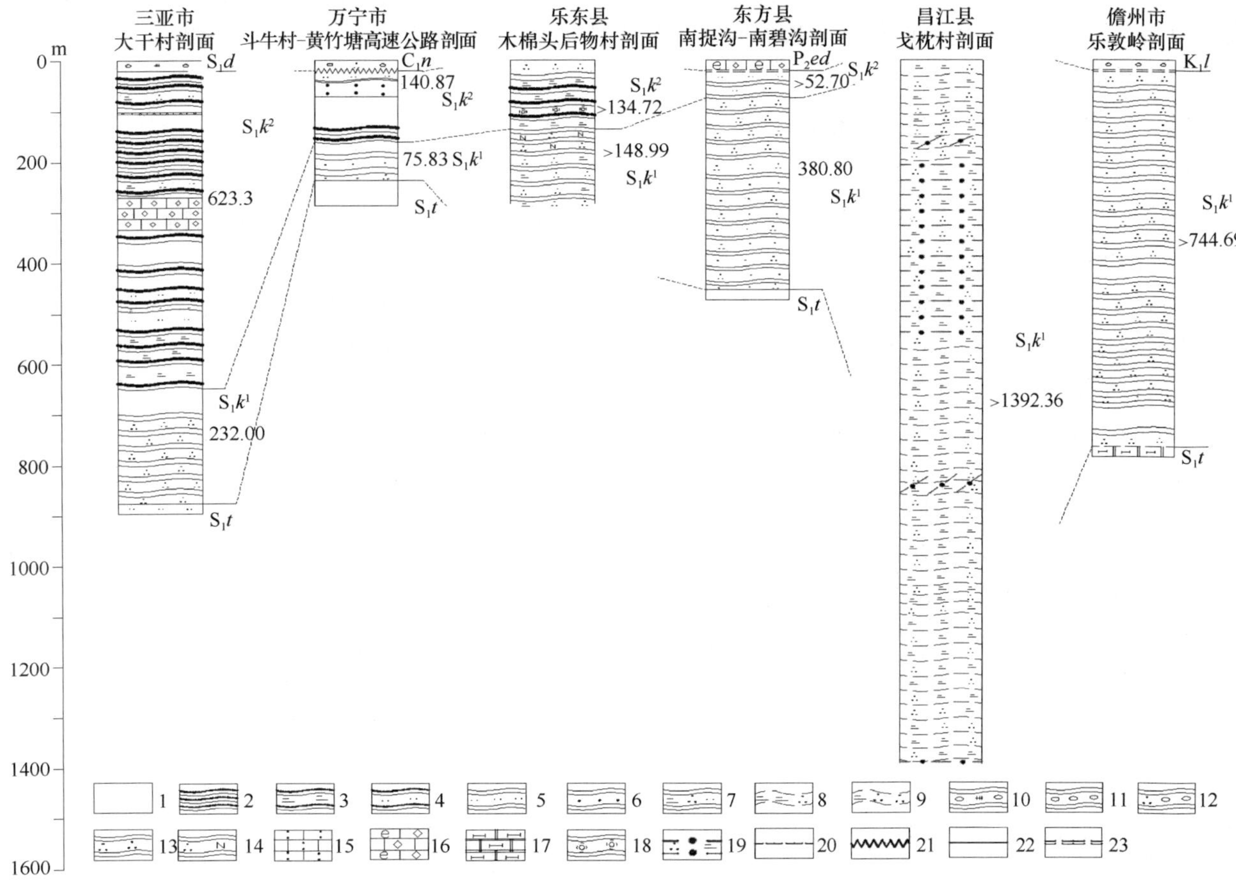

图 3-28　海南岛部分地区志留系空列村组柱状对比图

Fig. 3-28　Columnar correlation of the Silurian Kongliecun Formation in some areas of the Hainan Island

1—掩盖；2—板岩；3—绢云母板岩；4—粉砂质板岩；5—变质粉砂岩；6—变质砂岩；7—绢云母石英粉砂岩；8—石英绢云千枚岩；9—绢云石英千枚岩；10—变质复成分砾岩；11—砾岩；12—石英砾岩；13—石英岩；14—长石石英岩；15—沉凝灰岩；16—燧石生物晶屑灰岩；17—透辉石大理岩；18—硅化岩；19—糜棱岩；20—假整合界线；21—不整合界线；22—整合界线；23—断裂

下段为砂泥质沉积，而其余地区为砂质沉积。结合其沉积厚度，认为空列村组下段沉积时，沉积中心和沉降中心位于戈枕村一带，为一个向西张开的喇叭形盆地。

5. 大干村组（S_1d）

大干村组仅分布于三亚市雅亮乡大干村和白沙县河叉岭-邦溪八队一带。大干村代表性剖面底部为变质复成分砾岩，往上依次为板岩、结晶灰岩。而白沙县河叉岭-邦溪八队剖面岩石总体为石英片岩与石英千枚岩组合，底部为浅变质中细粒石英岩，中上部夹含砾石英片岩、浅变质石英质砾岩。两者岩石大致可以对比，底部为砂质或砾质沉积，往上为泥质沉积，都属于退积型沉积；但也存在明显的差异：白沙县河叉岭-邦溪八队剖面未见碳酸盐岩，且上部夹砂、砾质沉积；而大干村一带未见砂、砾质沉积夹层，并出现厚 64m 的结晶灰岩。

6. 靠亲山组（S_1kq）

靠亲山组分布范围有限，目前仅出露于保亭县毛感乡志伦村一带。上部为灰黑色—深灰色薄—厚层泥灰岩、含生屑结晶灰岩，中部浅灰色—灰黄色薄层粉砂质板岩、千枚状板岩偶夹粉砂岩，下部为粉砂质千枚岩、板岩夹灰色厚层石英岩、绢云石英粉砂岩、细砂岩。

7. 足赛岭组（S_1z）

足赛岭组分布范围有限，目前仅出露于保亭县毛感乡南好村附近。共划分为四段：一段岩性以薄层含炭千枚岩、千枚岩为主，二段为灰白色结晶灰岩，三段为千枚岩夹薄层绢云粉砂岩及粉砂质绢云板岩为主，四段为灰黄色薄层粉砂质板岩间夹薄层至中层中细粒石英砂岩、长石石英砂岩，或呈不等厚互层状产出。

（三）生物特征及年代地层

1. 寒武系

本区寒武系仅有美子林组。其顶部结晶灰岩中产瓶虫 *Cambrothyra* sp.，六射海绵骨针 *Hexactinellid* sp.，小壳化石 *Siphogonuchites triangulatus*，*Lapworthelle* sp.，*Hyolithellus* sp.，*Calcihexactina* sp.。瓶虫 *Cambrothyra* 曾见于湖北房县下寒武统（滇东统）西蒿坪组，丁莲芳、李勇初步鉴定后认为可能是寒武纪早期甚至更晚期生物。六射海绵骨针 *Hexactinellid* 据蒋志文鉴定认为其时代为寒武纪（姚华舟等，1999）。小壳化石中棱管壳类三角棱管壳 *Siphogonuchites* 是我国华南滇东统的特有种类，主要分布于梅树村阶小壳化石第Ⅱ组合带（*Siphogonuchites-Paraglobolirus* 组合）。拉普渥期壳 *Lapworthella* 是梅树村阶小壳化石第Ⅱ组合带的常见分子。似软舌螺 *Hyolithellus* 是我国及世界早寒武世（滇东世）广泛分布的属。张仁杰等（1998）研究认为，美子林组生物群面貌与分布于米仓山-黔北-峡东一带的 *Siphogonuchites-Paragloborilus* 动物群特征相似，应属亚澳生物大区南方古生物地理区。但该组所产 *S. triangulatus* 仅见于扬子地台区，因此，寒武纪滇东世海南岛东部地区与扬子区的关系密切。根据上述化石组合特征，其时代为寒武纪滇东世较为合理，美子林组的年代地层也由原来的寒武系重新厘定为滇东统。鉴于美子林组顶部含寒武纪滇东世梅树村期化石，而化石层位之下尚有厚度大于800m 的地层无地层时代资料，海南岛未见寒武系与震旦系的接触关系，但广东寒武系与震旦系是整合接触。因此不能排除海南岛二者亦为整合接触，也就不能排除美子林组下部包含部分震旦纪地层。

2. 奥陶系

本区奥陶系仅有南碧沟组。迄今未发现大化石，前人在黄竹岭一带南碧沟组中的微古植物 13 属 17 种（包括未定种），主要有：*Leiominuscula minuta*，*Leiopsophosphaera* sp.，*Micrhystridiumspinosum*，*M. lanceolatum*，*M. brevicornum*，*Buedingshaeridium* sp.，*Veryhachium* sp.，*Ovulum saccatum*，*Ammonidium* sp.，*Leiofusa* sp.，*Synsphaeridium* sp.，*Legonochitina*。上述微古植物化石以微刺藻属和卵形藻属的分子出现频率较高，另含少量光球藻、角刺藻、阿莫尼德藻、梭形藻、小型光面小球藻、三角藻、阿利乌穆藻、船形藻、连球藻为特征，还发现了个别的瓶形似几丁虫化石。上述化石属种多数产于寒武纪地层中，少数产于志留纪地层中。南京地矿所阎永奎、汪迎平（化石鉴定者）认为化石样品所代表之地层，其时代应为寒武纪至志留纪，但从化石属种的含量看，更有可能属寒武纪。

此外，1∶5 万《东方县、玉道》幅区调在南碧沟组发现微古植物 *Trematosphaeridium* sp. 和 *Heliconemafunicum* 等，多见于前泥盆纪地层。宜昌地矿所（1995）于邦溪农场一带南碧沟组中采集变质火山岩阳气石片岩、绿泥石化阳气石片岩，阳气石斜长片岩、斜长阳气石片岩和透闪石化橄榄石科马提岩一组 5 个样品，进行 Sm-Nd 法等时线测年，获得（458±36.8）Ma 测年数据。与广西区调队（1993）全岩（变质沉积岩）U-Pb 等时线年龄（447±3）Ma 比较接近。

海南省地质调查院（1999）在翰林镇蒙花岭南碧沟组中部（第 16 层）火山岩夹层中采样，经天津地矿所作锆石 U-Pb 法同位素测年，结果发现样品中存在两种不同成因与不同年龄的锆石。主要锆石呈浅棕黄色、浅棕色，晶形为复正方双锥柱状，以（110）柱面发育为主，晶体熔蚀呈港湾状或凹状，含有气泡及擦沟、擦痕等现象。晶体大小不一，大者 0.08～0.12mm，小者 0.03～0.08mm。较老的一组锆石可能为岩石中的岩浆成因锆石，因此其年龄可解释为岩石的生成年龄（即相应地层的年龄）。较新的锆石可能是岩石中的变质成因锆石，或岩石中的锆石受邻近侵入体的影响而导致其 U-Pb 同位素重设的结果。获得较老锆石的表面年龄为（499±6）Ma（表 3-3，图 3-29），谐和图上交点年龄为（527±48）Ma，综合分析认为火山岩生成年龄为（499±6）Ma 至（527±48）Ma。

表 3-3 变质火山岩锆石 U-Pb 同位素测年分析

Table 3-3 Analysis of zircon U-Pb istopic dating results of meta-volcanic rocks

样品情况			浓度		普通铅含量/ng	同位素原子比率*					表面年龄/Ma		
点号	锆石类型及特征	重量 μg	U μg/g	Pb μg/g		$^{206}Pb/^{204}Pb$	$^{208}Pb/^{206}U$	$^{206}Pb/^{238}U$	$^{207}Pb/^{235}U$	$^{207}Pb/^{206}Pb$	$^{206}Pb/^{238}U$	$^{207}Pb/^{235}U$	$^{207}Pb/^{206}Pb$
1	浅黄色透明粗长柱状自形晶体	15	525	20	0. 001	23760	0. 1796	0. 03646 <94>	0. 2484 <96>	0. 04941 <129>	230. 8± 5. 8	225. 3± 7. 8	167±61
2	无色透明短柱状晶体	5	471	22	0. 019	301	0. 1607	0. 03617 <62>	0. 2518 <60>	0. 05048 <99>	229. 1± 3. 9	228. 0± 4. 9	217±45
3	浅黄色透明长柱状自形晶体	15	851	32	0. 014	2066	0. 1444	0. 03531 <67>	0. 2462 <45>	0. 05057 <59>	223. 7± 4. 2	223. 5± 3. 7	221±27
4	无色半透明短柱状晶体	10	486	38	0. 005	4422	0. 1339	0. 07475 <114>	0. 5895 <121>	0. 05719 <71>	464. 7± 6. 8	470. 5± 7. 7	499±27
备注	* $^{206}Pb/^{204}Pb$ 已对实验空白（Pb=0. 050ng，U=0. 002ng）及稀释剂作了校正，其他比率中的铅同位素均为放射成因铅同位素，括号内的数字为（2σ）绝对误差，例如：0. 03531<67>表示 0. 03531±0. 00067（2σ）												

测试单位：天津地矿所同位素室。

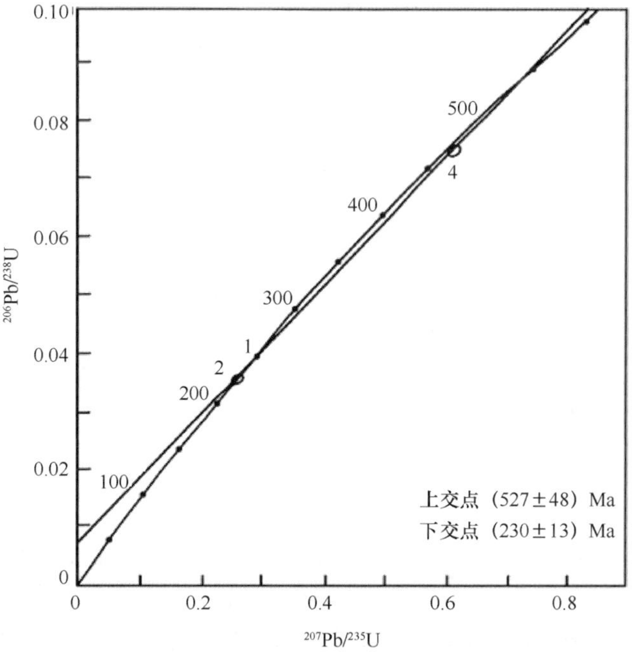

图 3-29 翰林镇蒙花岭奥陶系南碧沟组中火山岩锆石 U-Pb 同位素测量结果谐和图

Fig. 3-29 Concordia plot of zircon U-Pb istopic dating results of volcanic rocks in the Ordovician Nanbigou Formation, Menghualing, Hanlin Town

综合考虑南碧沟组中火山岩夹层同位素年龄、微古植物化石年代以及标准剖面的同位素年龄值，南碧沟组的时代确定为奥陶纪。

3. 志留系

截至目前资料，海南志留系有意义的化石主要集中产出于南好地区，少量分布于儋州红岭农场和东方市娜姆河陀烈组中（几丁虫）化石产出。其余地区则主要是孢粉化石。主要化石反映的时代是：

（1）鲁丹阶

鲁丹阶大致对应于陀烈组下部和中部。东方市娜姆河陀烈组产1个几丁虫 *Conochitina* 生物群。计有 *Conochitina* cf. *edjelensisarmillata*，*C.* cf. *edjelensis*，*C.* aff. *Iklaensis*，*Conochitina* sp.. 其中 *Conochitina edjelensis* 的地质时限可从鲁丹期延至埃隆期晚期，且该生物群在华南地区早志留世鲁丹期—埃隆期地层中有广泛分布，彼此可直接对比（表3-4）。另据曾庆銮等（2004）研究，认为陀烈组的地质时代应为早志留世鲁丹期—埃隆期早期。

（2）埃隆阶

埃隆阶大致对应于陀烈组上部、空列村组和大干村组下部。仅空列村组中上部深灰色中厚层结晶灰岩中产珊瑚：*Amplexoides*? sp.，海绵? *Receptaculites* sp.；大干村组中上部灰色中厚层结晶灰岩中产珊瑚化石 *Tabularia* sp.，*Favosites* cf. *gothlandicus*，*Favocites* sp.. 其中 *Favocites* cf. *gothelandicus* 的地层对比意义重要，它在长江三峡地区主要见于志留系罗惹坪组（表3-4），其地质时代为埃隆期。据此，陀烈组上部、空列村组、大干村组大致可与华南地层区罗惹坪组大致对比，其地质时代为埃隆期。

（3）特列奇阶

特列奇阶大致对应于大干村组上部、靠亲山组和足赛岭组，含1个三叶虫化石带，1个腕足类化石带，1个牙形石组合。

***Coronocephalus rex* 延限带** 见于靠亲山组中上部，以带化石的出现为特征；除带化石外，常见 *Coronocephalus* cf. *rex*，*Encrinuroides* sp.，腕足类化石有 *Eospirifer* cf. *sinensis*；珊瑚 *Syringopora* sp. 等。*Coronocephalus rex* 广泛见于华南地层区志留系纱帽组、宁强组、秀山组等（表3-5），海南岛志留系三叶虫带可直接与上述层位同期生物群对比，其地质时代为特列奇期中期—中晚期。

***Panderodus unicostatus-Ozarkodina* 组合** 见于靠亲山组上部结晶灰岩中，除组合分子外，常见 *Ozarkodina* cf. *parahassi*，*kockelella*? sp.，*Pseudoneotodus bicornis*，*Ozarkodina* cf. *wudangensis* 等。南好矿区锌山志留系足赛岭组二段结晶灰岩也发现 *Panderodus unicostatus*，*Ozarkodedina* sp.，其中 *Panderodus unicostatus* 常见于华南地层区溶溪组—秀山组上段，地质时代为特列奇期中晚期。

***Xinanospirifer* 延限带** 见于保亭南兵-公井岭剖面的足赛岭组四段中上部及上部，产 *Xinanospirifer* cf. *sanduensis*，*Xinanospirifer* sp.；此外，产三叶虫、珊瑚、海百合茎等化石，这些化石常见于我国西南地区志留纪特列奇期晚期地层中，彼此可直接对比。在 *Xinanospirifes* 延限带尚发现中华棘鱼鳍刺化石。

（4）志留系古生物地理分布特征

刘时藩（1997）总结了中国志留纪棘鱼化石的演变趋势和相关生物地层学问题，指出该类化石存在两个组合：第一组合为 *Hunanacanthus-Eosinacanthus* 组合，除组合分子外，也产 *Sinacanthus* 和 *Neosinacanthus*，但 *Sinacanthus* 和 *Neosinacanthus* 在数量上相对组合分子要少些，其层位是溶溪组和纱帽组的下部，相当于笔石带 *Turriculatus-Crispus* 带中上部，其地质时代为兰多弗里世特列奇期早—中期。第二组合为 *Sinacanthus-Neosinacanthus* 组合，以 *Sinacanthus* 为主，*Neosinacanthus* 次之。在该组合中一起保存的有无颌类 *Hanyangaspis*，但不存在 *Hunanacanthus* 和 *Eosinacanthus*，其层位是坟头组和锅顶山组，大致相当于笔石带 *Griestoniensis-Crenulata* 带，其地质时代为特列奇期中晚期。

表 3-4 海南岛五指山地层分区与其他地区志留系对比表

Table 3-4 Correlation of Silurian strata between the Wuzhishan stratigraphic subregion and other regions, Hainan Island

年代地层 系	年代地层 统	年代地层 阶	扬子区(陈旭、戎嘉余,1996) 笔石带	扬子区(陈旭、戎嘉余,1996) 腕足类	扬子区(陈旭、戎嘉余,1996) 三叶虫	川北-陕南	川东南	鄂西	海南岛南好地区	海南岛东方
志留系	普里道利统									
	罗德洛统					车家坝组				
	文洛克统									
	兰多弗里统	特列奇阶	*Crenulata*	*Salopinella*–*Xinanospirifer*	*Cornocephalus* 组合	宁强组	回星哨组			
							秀山组			
			Griestoniensis			王家湾组	溶溪组			
			Crispus	*Naivkinia*–*Nucleospira*	*Ptilillaenus* 组合	崔家沟组	小河坝组		足赛岭组	
			Turriculatus					纱帽组	靠亲山组	
		埃隆阶	*Sedgwicki*			龙马溪组		罗惹坪组	大干村组	
			Convolutus				龙马溪组			
			Triangulatues					龙马溪组	空列村组	
		鲁丹阶	*Cyphus*						陀烈组?	陀烈组?
			Vesiculosus							
			Acuminatus							

足赛岭组四段中上部浅黄色绢云母粉砂质板岩中所产 *Sinacanthus* sp.，虽仅发现一枚较完全的刺体，但它的形态特征与黎作聪（1980）和湖北蒲圻大湖岭坟头组、京山石龙水库纱帽组所发现的 *Sinacanthus* cf. *wuchangensis* 相似，也与刘时藩（1997）在湖南澧县山门水库纱帽组中发现的 *Sinacanthus* sp. 相似。棘鱼鳍刺化石的演化规律随着地质时代的发展，它的刺体向着越来越小的方向发展，而脊沟的数目有所减少（刘时藩，1997）。据此，认为当前的 *Sinacanthus* 的层位应与上述第二组合关系密切，可直接对比，其地质时代为特列奇期晚期。另外，从足赛岭组四段 *Sinacanthus* 相邻层位产出的无脊椎动物化石所提供的地质时代依据也得到充分的印证。足赛岭组四段产腕足类 *Xinanospirifer* cf. *sanduensis*，*X*. sp.，三叶虫 *Astroeproetus* cf. *latilimbtus* 等，*Xinanospirifer* 是我国西南地区（黔东北、黔东南、川北川西南）特列奇期常见标准化石，而且是华南志留系 *Salopinella minutaxinanospiriferfla-bellum* 组合带的特征分子，其地质历程短，形态特征显著，在川陕边境广元一带见于志留系神宣驿段中部；宁强玉石滩扬坡湾段上部至宁强组顶部均有产出（陈旭，戎嘉余等，1991，1996），这表明该区足赛岭组四段所产 *Sinacanthus* 的产出层位的地质时代为特列奇期晚期无疑。特别值得提及的是，我国中华棘鱼化石产出层位的地质时代主要为志留纪兰多弗里世埃隆期晚期至特列奇期，其分布范围不仅包括长江中下游及塔里木盆地西南缘，而且在隶属华南褶皱系的海南岛志留系相同层位也存在中华棘鱼化石的"踪迹"。因此，该类化石的发现，不仅为地层时代的确定提供了可靠化石依据，也对进一步阐明海南岛九所-陵水断裂带以北志留纪生物古地理区系的归属具有重要意义。

此外，岩层中尚产大量的孢粉化石，经鉴定主要有：*Leiopsophosphaera minor*，*L. solida*，*L. flava*，*L. clara*，*L.* cf. *flava*，*Trachysphaeridium* cf. *rugosum*，*T.* cf. *simplex*，*T.* cf. *planum*，*T.* sp.，*Lophosphaeridium* sp.，*Baltisphaeridium*，*Micrhysphaeridium* cf. *spinosum*，*Veryhachium trispinosum*，*V.* sp.，*Leiofusa bicornum*，*L.* sp.，*Macroptycha* sp.，*Favososphaeridium*? sp.，*Nucellosphaeridium*? sp.，*Zonosphaeridium* sp.，*Lignum*? sp.。以上微古植物以球藻类和刺球藻类分子占优势，属种较单调，具少量菱形藻、大褶藻及有核藻。其中 *Leiopsophosphaera*，*Lophosphaeridium*，*Trachysphaeridium* 及 *Baltisphaeridium* 等属为世界各地志留系常见属，显然这些微古化石对地层时代归属的确定意义不是很大。但不同层位、不同地区所分布的微古植物组合在面貌上具有相似性，显示了属于相同盆地的沉积。

使用年代地层单位阶、生物分类属级单元对本区志留纪兰多弗里世特列奇期早期脊椎动物（中华棘鱼鳍刺）、腕足类、珊瑚、三叶虫、牙形石5个化石门类进行了统计，并与华南区相同门类化石的地理分布特征进行了对比。其古生物地理区特征如下。

1）早期脊椎动物（中华棘鱼）

以往资料中华棘鱼鳍刺化石广泛分布于长江中下游。王朴等（1988）、刘时藩（1995）在塔里木盆地西北柯坪、巴楚一带塔塔埃尔塔格组和依木干他乌组也发现志留纪棘鱼鳍刺化石；刘时藩（1997）认为我国现有的4个棘鱼鳍刺属是一个独特的类群，是志留纪中国境内具地方特色的土著鱼群，并总结出志留纪—泥盆纪早期我国早期脊椎动物（包括部分棘鱼类）的地理分区：即扬子区和华南区。其中，扬子区（扬子准地台范围）以无颌类为主导，鱼类中以中华棘鱼为特色；华南区（华南褶皱系的范围）鱼类的生活范围仅局限于滇东一带，以产胴甲类为特色，缺乏相同时期中华棘鱼化石资料。海南岛南好地区首次发现志留纪特列奇期中华棘鱼鳍刺化石，据此，有理由认为特列奇期海南岛南好地区与扬子区和塔里木盆地西北缘彼此接近或关系密切，至少应同属一个海域范围内。此外，华南区在志留纪特列奇期主要发育以地槽型沉积为主的地层，其分布见于钦州、合浦、防城、陆川等地，发育一套厚度巨大的半深海—深海相的浊流沉积为主的地层（即古墓组）；广东与之相当的岩石地层见于粤西区郁南、罗定、廉江等地，是一套以浊流沉积为主的浅海—半深海相类复理石建造（即文头山组）。而海南岛南好地区此时则为滨—浅海相砂质和灰质沉积，与华南区明显不同。而与扬子区沉积特征有密切的联系（陈哲培，1987；汪啸风等，1992；陈孝红、汪啸风，1999；李志宏等，2000；张仁杰等，2002）。

2）无脊椎动物

A. 腕足类

南好地区兰多弗里世特列奇期腕足类化石主要见于靠亲山组和足赛岭组，包括 6 属：*Xinanospirifer*，*Atrpa*，*Sowerbyella*，*Schizophoria*，*Striispirifer*，*Eospirifer*；除 *Xinanospirifer* 外，上述 5 属均为广布型分子，未见华南区土著分子，但足赛岭组四段产丰富的 *Xinanospirifer* cf. *sanduensis*，*Xinanospirifer* sp.，*Xinanospirifer* 的主要特征是中槽、中隆上都发育多条放射褶，是一种特殊的、更为复杂的始石燕类。目前除海南外，主要产地为我国西南地区（戎嘉余、杨学长，1978）。另外，*Striispirifer* 和 *Eospirifer* 亦广泛分布于华南地层区，它们在扬子区特列奇期腕足类序列的下部动物群及上部动物群均可见及（陈旭，戎嘉余，1996）。

B. 珊瑚

兰多弗里世特列奇期珊瑚化石主要见于靠亲山组和足赛岭组，其有 10 属，其中四射珊瑚 2 属，床板珊瑚 8 属。四射珊瑚 2 属为 *Amplexoides* 和 *Tryplasma*；前者见于我国西南地区及湖南、湖北及广东吴川，*Tryplasma* 除见于我国西南地区等地外，也是欧美、澳大利亚相当层位常见分子。李志明（1990）认为 *Tryplasma* 的地理分布为中亚、哈萨克斯坦、新疆和扬子地区，表明华南与西伯利亚关系密切。床板珊瑚计有 8 属：*Favocites*，*Squameolites*，*Syringgopora*，*Heliolites*，*Plasmopora*，*Ningqiangolites*，*Syringoporiella*，*Heliolitella*；其中 *Favocites*，*Squameolites*，*Syringopora*，*Heliolites*，*Plasmopora* 为广布分子，而 *Ningqiangolites*，*Syringoporiella* 和 *Heliolitella* 的地域分布对生物古地理研究十分重要。*Ningqiangolites* 常见于陕西宁强、四川綦江、新疆和布克赛尔；*Syringoporiella* 则常见于贵州、甘肃玉门及哈萨克斯坦；*Heliolitella* 仅见于贵州沿河、贵阳、四川綦江等地。值得提及的是，*Ningqiangolites* 和 *Squameolites* 的地质时限仅限于特列奇期中晚期，广泛分布于扬子地层区。

南好地区志留系靠亲山组、足赛岭组所产珊瑚化石不仅其产出层位大致可与陈旭、戎嘉余（1996）所建立的特列奇期第二珊瑚组合 *Shensiphyllam-Idiophyllum-Erlangbapora-Carnegiea* 组合对比，而且四射珊瑚和床板珊瑚均见于扬子区第二珊瑚组合。这表明兰多弗里世特列奇期海南岛南好地区与扬子区关系较为密切，可能属同一生物地理区，与中华棘鱼鳍刺化石生物地理区指示基本一致。

C. 三叶虫

在靠亲山组、足赛岭组，三叶虫很单调，仅发现 3 属，*Coronocepbalus*，*Astroproetus* 和 *Encrinuroides*。其中 *Coronocephalus* 和 *Astropoetus* 是扬子区地区性属，仅 *Encrinuroides* 是全球广布属。上述三叶虫属中以 *Coronooephalus* 最为重要，它广泛分布于整个华南，甚至于在钦州、防城一带古墓组以浊流沉积为主的半深海沉积相带中也可见该分子。

D. 牙形石

牙形石产于靠亲山组二段及足赛岭组二段结晶灰岩中，由于区域变质和接触变质作用，仅获零星牙形石资料，共计 4 属：*Ozarkodina*，*Pomderodus*，*Pseudoneotodus* 及 *Kockelella*。王成源（1996）指出兰多弗里世鲁丹期、埃隆期和特列奇期最早期，牙形石可分成 3 个生物地理区：①北美-波罗的海-西伯利亚生物地理区；②扬子生物地理区；③澳大利亚生物地理区。而特列奇期中晚期，全球牙形石生物地理区系变得不明显。区内牙形石与扬子生物地理区关系密切。

根据上述南好地区兰多弗里世特列奇期早期脊椎动物、无脊椎动物化石的生物地理统计结果，认为中华棘鱼鳍刺化石、腕足类、珊瑚、三叶虫等可与相同时期扬子区生物群面貌存在着相似的生物发育史，它们很可能同属一个生物地理分区；在特列奇时期，南好地区与扬子区乃至新疆塔里木西北缘均处于低—中纬度带或南半球干旱带有密切关系。

（四）原岩恢复与岩相古地理探讨

1. 原岩恢复

美子林组经历了高绿片岩期的区域变质、热接触变质和接触交代变质作用，因此岩性比较复杂，但大部分岩石尚保留有原岩的结构构造，原岩基本可以恢复。第一岩性段为长石石英砂岩、石英砂岩、粉砂岩、泥质砂岩、少量砂质泥岩，以砂岩、泥质砂岩为主体，局部含酸性凝灰质；第二岩性段为灰质粉砂岩、粉砂岩、灰质砂岩、砂岩；第三岩性段为泥灰岩、灰岩夹变质泥质粉砂岩。

南碧沟组岩石局部发育变余粉砂显微鳞片变晶结构、变余粉砂状结构，变余层状构造、变余沙纹层理、变余微层理构造。副矿物成分为次棱角状—半圆状锆石、磷灰石、电气石、金红石、白钛石等。在 Niggli 四面体图解、K-A 相关图解中，碎屑岩类基本上分别落在残余沉积黏土质沉积区和泥质粉砂岩亚区，火山岩夹层则都落入火成岩区。其原岩主要为泥质粉砂岩、粉砂质泥岩，少量中基性—酸性火山岩、基性—超基性火山岩、沉凝灰岩、含炭泥岩（详见第 4 篇变质岩）。

志留系为一套经历了区域变质作用改造的浅变质岩系岩石，主要由板岩类、千枚岩类、变质砂岩类等岩石组成，夹结晶灰岩或大理岩透镜体及沉凝灰岩。岩石基本保留原岩组构特征，发育变余砂状结构、砾状结构、变余粉砂鳞片变晶结构和微水平层理、微波状、板状交错层理等变余结构构造（图版Ⅰ-10-H）。多含稳定—超稳定副矿物，如锆石、金红石、白钛石、锐钛矿、磷灰石、独居石等，多呈滚圆状、滚粒状，其中锆石的晶棱顶锥均被磨损，呈滚圆状，椭圆状，反映其原岩具正常沉积岩特征。在 Niggli 四面体图解中落入残余沉积黏土质沉积物区。其原岩主要为泥质岩，粉砂质泥岩、石英砂岩、长石石英砂岩、泥炭质粉砂岩、细砂岩、灰岩、沉凝灰岩。

2. 岩相古地理探讨

据冯增昭等（2001）研究，中国南方寒武纪岩相古地理自西往东主要包括以下主要古地理单元：康滇古陆、扬子碳酸盐岩台地、江南页岩盆地-碳酸盐岩台地、东南碎屑岩台地。海南岛五指山地层分区寒武纪沉积岩为碳酸盐岩-碎屑岩组合，奥陶纪为砂质岩-泥质岩组合，志留纪为砂质岩-泥质岩-碳酸盐岩。结合古生物特征分析，认为海南岛五指山地层分区下古生界可能也是处于连接扬子碳酸盐岩台地和东南碎屑岩台地过渡地带。

（1）寒武系

美子林组一段为陆源碎屑岩，二段的碎屑岩中已含较多灰质，三段以碳酸盐岩沉积为主体，自下而上总体上构成一海进沉积序列（图 3-30），属于以近岸陆源碎屑占主导的滨—浅海相碎屑岩-碳酸盐岩混合沉积型。

美子林组一段　包括 4 种沉积类型：泥质砂岩型、砂岩（净砂岩）型、泥质粉砂岩-粉砂质泥岩型、灰质砂岩型。泥质砂岩约占美子林组一段 40% 以上，其粒度一般在细砂级以下，少量达中砂级，泥质含量 5%~25%，分选中等或较差，显示沉积水体能量中等—较低或水流作用持续性差。砂岩（净砂岩）包括长石石英砂岩和石英砂岩两类，一般为细砂岩、细砂—粗粉砂岩，中粒砂岩少见并呈夹层出现，可含较多重矿物（电气石、锆石、磷灰石），并显示一定磨圆度。但作为碎屑颗粒的石英多因变形变质而少见其原始磨圆特点，仅少量可辨识其较好的磨圆形态。重矿物的富集并显示一定磨圆度及碎屑颗粒石英之较好磨圆度说明沉积物曾经历长久的水流簸选并沉积于高能地带，沉积物分选良好，不含泥质杂基，可见变余低角度交错层理等特点则进一步说明它们总体沉积于水体能量高、水流作用持续性较好的开阔前滨—近滨环境。这种砂岩（净砂岩）与上述显示沉积水体能量中等—较低或水流作用持续性较差的泥质砂岩在区域上和层序上是紧密伴生的，按照瓦尔特相律，泥质砂岩无疑是前滨—近滨环境中相对低洼区或浪基面附近的产物。泥质粉砂岩-粉砂质泥岩多风化为灰黄色—黄灰色，新鲜者可见深灰色乃至向灰黑色过渡的色调，显示沉积水体相对较深，但这类岩石仅在泥质砂岩中呈厚度不大的夹层出现，说明不存在长期稳定的较深水环境，只是在滨海沉积环境这一总体背景下曾出现间歇性海平面升降。灰质砂岩仅在乐来乡高速公路边见有 2 处（为修路翻动的巨大转石堆，未见基岩露头），因热变质和热液交代变质而强烈的透闪石、透辉石化，其原岩组构不可辨。从其分布局限、厚度不大、产于上述沉积背景条件下等特征看，显然亦为滨海环境产物，代表陆源物供给减少、碳酸盐岩沉积相对加强的沉积条件。

美子林组二段　主要有 3 种原岩类型：灰质砂岩-砂岩、灰质泥质粉砂岩-粉砂质泥岩、泥质粉砂岩-粉砂质泥岩，另有少量灰岩。灰质砂岩-砂岩约占 50% 以上，灰质多呈条带状富集，经热变质形成

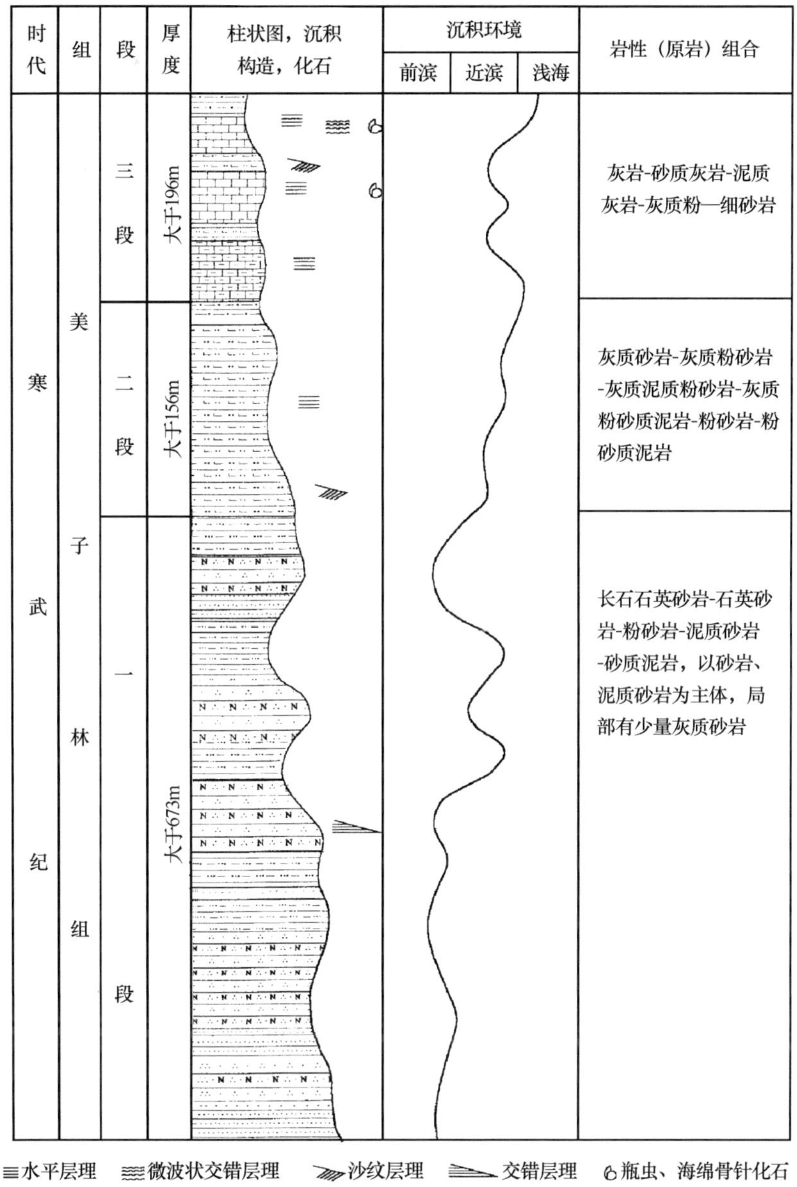

图 3-30　美子林组沉积相柱模式图

Fig. 3-30　Model of the Meizilin Formation sedimentary facies column

条带状透辉石透闪石石英角岩。沉积物分选好，不含泥质杂基，可见变余交错层理，显然代表近滨高能沉积条件。灰质泥质粉砂岩-粉砂质泥岩、泥质粉砂岩-粉砂质泥岩共占美子林组二段50%左右，经热变质分别形成透辉石（透闪石）黑云母石英角岩、黑云母石英角岩，见变余水平层理，均代表较低能沉积条件。结合美子林组总的岩性序列特点，二段的灰质砂岩-砂岩应为近滨高能环境产物，灰质泥质粉砂岩-粉砂质泥岩、泥质粉砂岩-粉砂质泥岩则为滨外浅海低能环境产物。与美子林组一段相比，二段的较深水（浅海）沉积组分（含灰质或不含灰质的粉砂质泥岩、泥质粉砂岩）已显著增多（达50%），与前者仅呈夹层出现明显不同，显示从一段至二段沉积水体总体上加深，沉积场所逐渐远离滨岸，存在明显海侵。二段总体上灰质含量增加、陆源碎屑相对减少并出现灰岩夹层也正是这一环境变迁的反映。

美子林组三段　其原岩主要为灰岩、砂质灰岩、泥质灰岩、灰质粉—细砂岩，美子林组具有时代意义和指相意义的瓶虫化石和海绵骨针即产于灰岩中。灰岩、砂质灰岩等均强重结晶甚至变为大理岩，碳酸盐矿物或颗粒已远非原始形态，不能据此恢复沉积环境。从残留的变余层理特征看，灰岩、砂质灰岩、泥质灰岩中主要为水平层理或为块状，前二者尚偶见微波状层理，显示以浅海沉积为主体、偶有弱水流扰动或浅海沉积为主体、间歇性海平面下降出现滨岸沉积的特点。灰质粉—细砂岩分选较好，见小型沙纹层理，代表近滨中等能量的沉积条件。三段以灰岩、砂质灰岩、泥质灰岩为主体，灰质粉—细砂岩仅呈夹层出现，即该段总体为浅海沉积，滨海沉积仅仅是短暂、间歇的行为。

显然，与一段、二段相比，三段总体上处于海水相对较深、离滨岸较远的环境条件，由一段、二段到三段，代表了海平面多次波动但总体上海水逐步加深、海域不断拓宽的海侵过程。由一段到二段陆源碎屑所占比例逐渐减少，碳酸盐岩沉积相对增多，及至三段碳酸盐岩沉积占主导地位，正是这种环境变化的反映。但始终未脱离滨—浅海这一基本环境条件。

（2）奥陶系

奥陶系沉积盆地相对寒武纪盆地而言，裂解程度加大，沉降盆地拓宽，海水加深，盆地基本成型，呈近东西向展布，沉积作用基本上波及整个海南岛，形成遍布全岛的奥陶系南碧沟组。盆地具有东浅西深的古地理格局。儋州和万宁一带属于陆缘剥蚀区，为盆地沉积提供物源。水流方向可能由西→东（图3-31）。

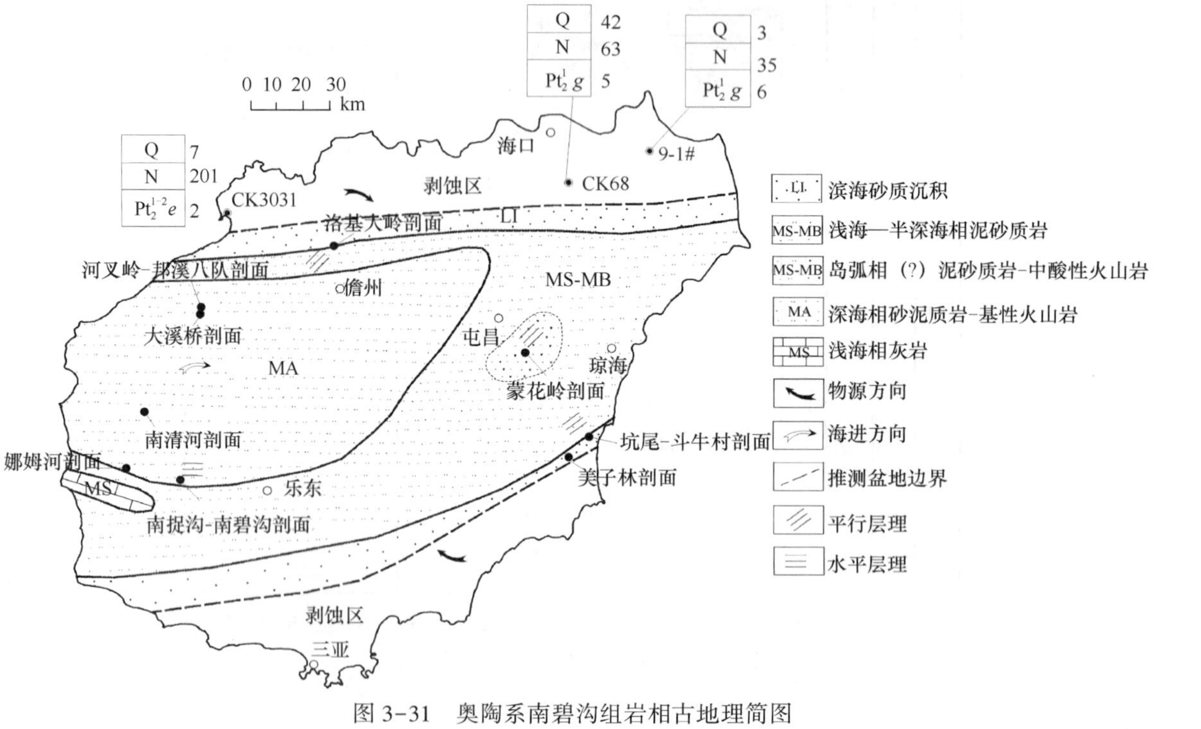

图3-31　奥陶系南碧沟组岩相古地理简图

Fig. 3-31　Sketch map of lithofacies paleogeography of the Ordovician Nanbigou Formation

海南岛西北白沙县大溪桥和岛西南南捉沟—南碧沟一带，下部（大套）地层虽经较强的变质变形，但在其下部个别石英片岩薄层中见有不清晰的变余沙纹层理，其上部白云母石英片岩中偶见变余小型板状斜层理，结合岩石类型及其组合特征，可以认为总体为陆棚环境。邦溪农场—芙蓉田—珠碧江农场一带中下部沉积物的粒度变细，水体能量较低，可能属远陆棚环境；中部据岩石变余结构构造，反映水体能量高、水体变浅，可能是近陆棚环境或内陆棚环境；上部又演化为深陆棚环境，顶部夹有中层变质石英细砂岩，其中见变余小型斜层理，反映水体能量又明显增高、水体相应变浅，变为近陆棚环境。红泉农场一队—东方农场八队—东方红沟村一带的沉积物的粒度整体上较中北部地区粗，显示更近陆地的陆棚环境，自下而上沉积物由粗变细，沉积环境由滨海过渡到滨浅海。上部大套地层虽然经过低绿片岩相区域变质，原始沉积结构构造保留甚少，未见指相化石、指相矿物，但镜下常见变余结构，变余分异作用不强烈，构造面与成分层一致，故此恢复原岩岩石类型及其组合特征来进行沉积环境分析。其中所夹变质火山岩和变质凝灰岩层外，岩石类型及其组合特征比较简单，由一系列自旋回浊积层序反复重复组成（图3-32），每个层序厚度一般为0.1~0.5m，层序中的石英片岩由粉砂岩变质而来，底面平整或微有起伏，粒度自下而上变细，偶见变余沙纹层理，厚度虽然不大但十分稳定、延伸很远，一系列薄层岩石几乎呈平行状态产出，其上为粒度更细并含有白云母的石英片岩，再上为石英白云母片岩，通常由鲍马层序C、D、E段组成，偶见B段，但也常见D、E两段重复现象，即南碧沟组为伴有超基性及中基性火山喷发海槽型浊流沉积，为深水盆地环境，浊积层序中未见A段，B段也十分少，反映该区主要位于深海平原近边部的浊积扇边缘部分。

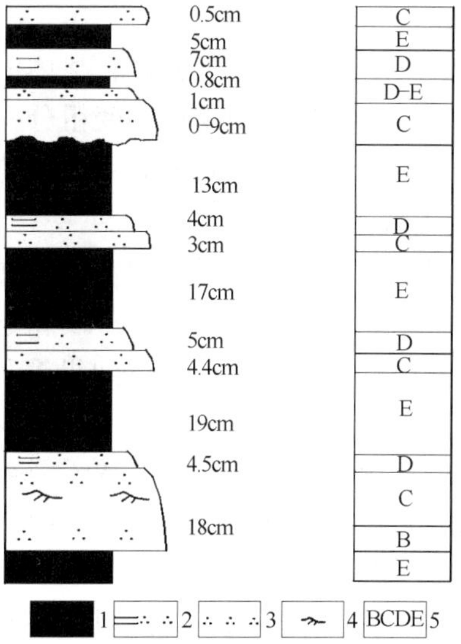

图3-32 南碧沟组上部自旋回浊积层及其解释
（取自大溪剖面44层中一段岩层）
Fig. 3-32 Autocyclicity turbidity and its explanation of the upper Nanbigou Formation
(from the middle-1 part of No. 44 Bed, Daxi Section)
1—石英白云母片岩；2—白云母石英片岩；3—石英片岩；4—沙纹层理；5—鲍马层序B、C、D、E层

海南岛东北翰林镇蒙花岭和岛东南万宁市山根乡坑尾—斗牛村一带，南碧沟组由3种基本层序组成（姚华舟，1999；图3-33）：①层序Ⅰ，泥岩、粉砂质泥岩、泥质粉砂岩无规律互层，可有粉砂岩；此种层序组成了该区南碧沟组约2/3的地层，分布于南碧沟组下部、上部。沉积物细，为泥质-粉砂质正常沉积，无或无明显重力流沉积特征显示；②层序Ⅱ，在相当于上述层序的沉积物中出现泥质粉—细砂岩互层，显示重力沉积特点，即总体上为泥质-粉砂质正常沉积夹重力流沉积；③层序Ⅲ，熔结凝灰岩、沉凝灰岩-泥质粉砂岩、粉砂质泥岩，沉积物细，为泥质-粉砂质沉积与海相火山岩、沉火山碎屑岩沉积频繁而无规律互层，无或无明显重力流沉积特征显示，火山岩层状产出厚度不一，显示火山活动频繁而强度不一的特点。南碧沟组下部主要为泥质-粉砂质沉积。前述基本层序Ⅰ是这种沉积物的基本形式。沉积物粒度细小，为泥岩、粉砂质泥岩、泥质粉砂岩，未见可能代表较强水动力条件的较粗碎屑岩，说明此时还远离波浪作用较强的滨海沉积区，而进入基本不受波浪影响的浅海—半深海环境。中部沉积物粒度与下部一样，较细，发育多层火山岩，说明同时存在较强火山活动。上部基本由前述基本层序Ⅰ和Ⅱ组成，基本层序Ⅱ中出现较粗沉积物，其显示一定粒度变化，具一定浊流沉积特点，但未见典型浊积岩具备的鲍马层序。其沉积特征与沉积环境具一定承上启下的过渡性特点，处于滨—浅海沉积的寒武系美子林组与深海相复理石建造的志留系陀烈组之间，即为浅海—半深海的类复理石-

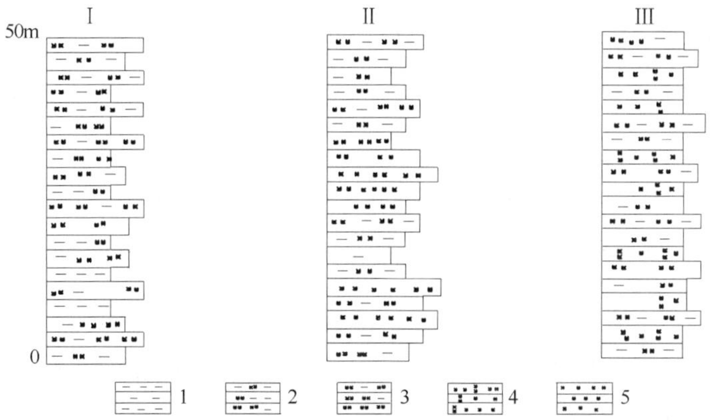

图3-33 翰林镇-山根乡南碧沟组基本层序类型
Fig. 3-33 Basic sequence types of the Nanbigou Formation along Hanlin Town-Shangen Town
1—泥岩；2—粉砂质泥岩；3—泥质粉砂岩；4—沉凝灰岩；5—细砂岩

复理石建造。

由上述可见，奥陶纪沉积时，整体上由南向北、由东至西，总体具有水体变深的特点。结合海南岛西部洛基大岭一带沉积物粒度变粗分析，显示盆地中心大致位于岛西大溪桥和岛西南南捉沟—南碧沟一带，早期由滨海过渡到滨海—浅海陆棚环境，晚期则进入深水盆地环境沉积。

（3）志留系

志留纪是海南下古生界沉积盆地发展成型和闭合的重要地质时期。根据沉积相分析，岛南东面万宁市—保亭和北西面洋浦—海口一线属于剥蚀山地地貌，为沉积盆地提供物质来源（图 3-34）。其沉积环境及其演变可概述如下：陀烈组沉积早期（下段），为砂质岩沉积，相对奥陶纪盆地晚期沉积而言，此时盆地已有所隆升，进入滨海—浅海环境。陀烈组中段，则离大陆斜坡有一定距离的半深海—深海相沉积，处于浊积扇外缘，源于斜坡的浊流对其影响不大，仅偶尔遭受达到浊积扇扇缘的远源浊流的侵扰。之后沉积环境进入浊积扇内，并不断向浊积扇扇根的方向即向着大陆斜坡的方向迁移，浊流作用的频度和强度不断增大。陀烈组沉积晚期（上段沉积期）沉积环境已转入浊积扇根部，浊流活动的强度达到顶峰，形成了陀烈组单层厚度最大、粒度最粗的浊积岩，此时水体已相对变浅，之后进入了浅海环境。空列村组—足赛岭组沉积时本区绝大部分时期为滨岸环境，其间出现过间歇性海平面升降，短期内形成了浅海环境。与此同时，也出现了较强的火山活动。志留纪末期（?），海水全部退出，早古生代海侵结束。

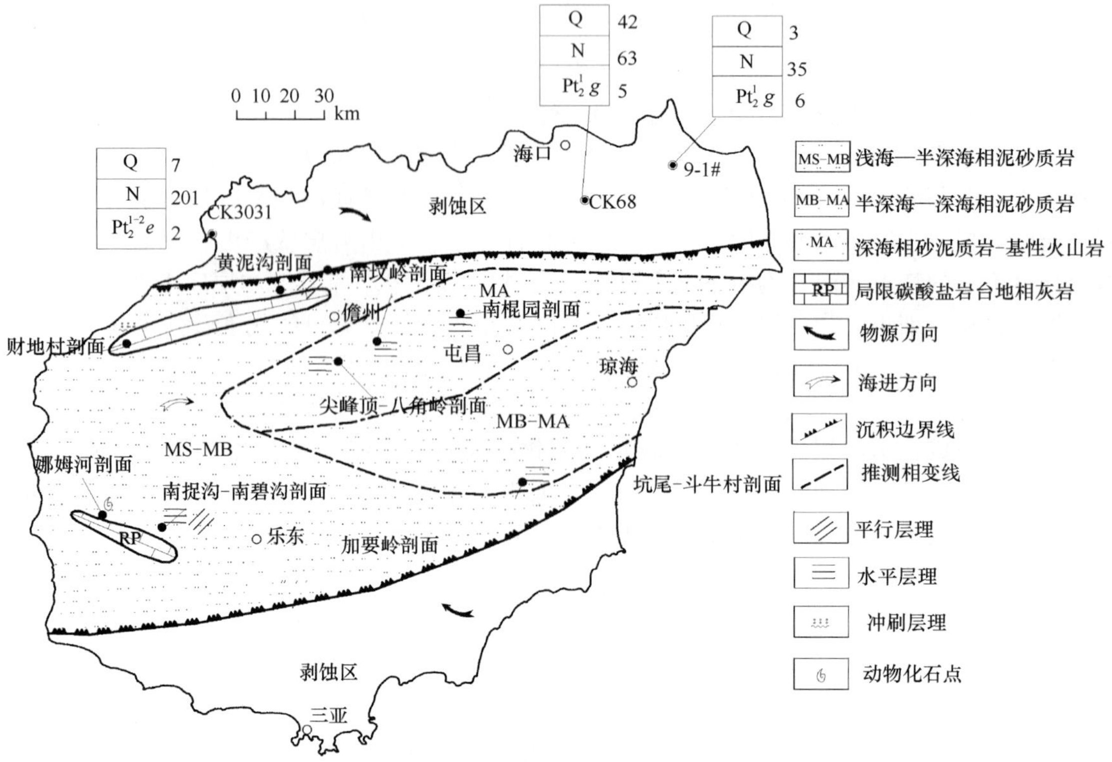

图 3-34　志留系陀烈组岩相古地理简图

Fig. 3-34　Sketch map of lithofacies paleogeography of the Silurian Tuolie Formation

1）陀烈组（S_1t）

根据岩性、单层厚度、沉积物粒度变化等特点，剖面上及填图路线上陀烈组主要分为以下 5 种基本层序（姚华舟，1999；图 3-35）。

a 型　泥岩、粉砂质泥岩为主，偶夹泥质粉砂岩、粉—细砂岩型。这种基本层序构成了陀烈组约 1/3 左右地层，并主要位于陀烈组第一段和第二段下部。其特点是：以泥岩、粉砂质泥岩为主体，泥质粉砂岩、粉砂岩仅偶呈夹层出现，有时厚达数十米的地层全由泥岩或粉砂质泥岩组成，而不见粉砂

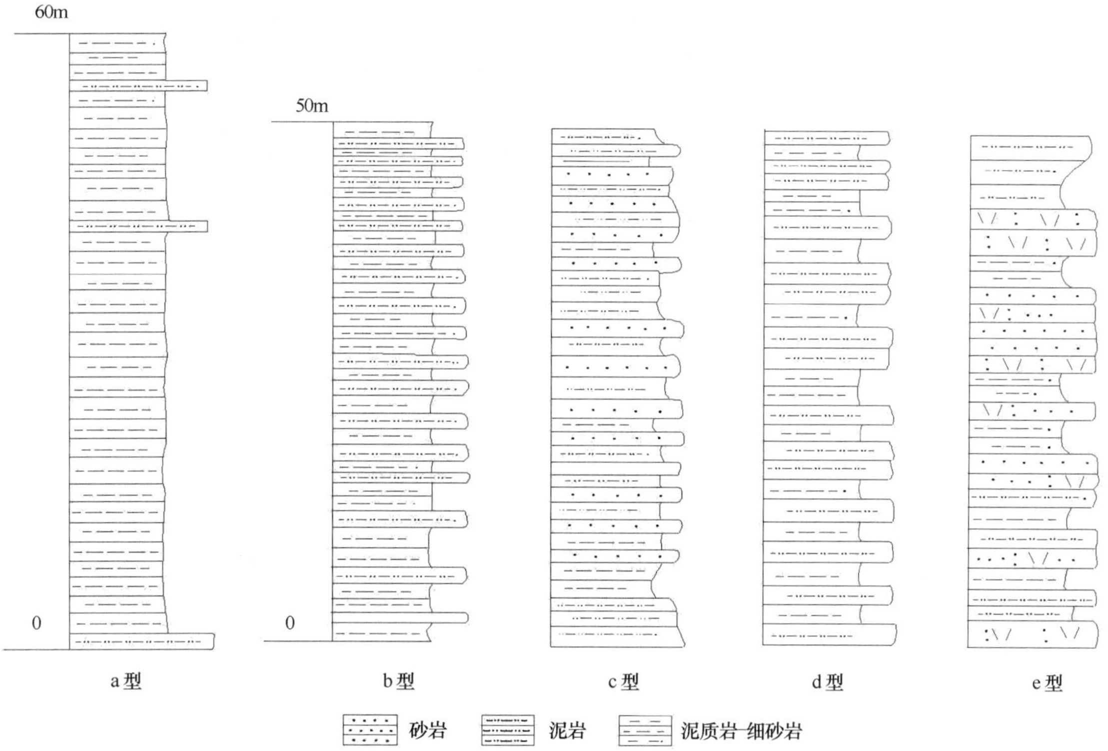

图 3-35　海南岛志留系陀烈组基本层序

Fig. 3-35　Basic sequence of the Silurian Luolie Formation, Hainan Island

岩、泥质粉砂岩。泥质粉砂岩、粉砂岩为这种基本层序中相对较粗的组分，呈薄的夹层出现或构成条带状构造，可见变余粒序层理、变余水平层理。

b 型　泥岩、粉砂质泥岩与泥质粉砂岩、粉砂岩互层型。与 a 型相比，不同之处在于相对较粗物质组分，即泥质粉砂岩、粉砂质泥岩、粉砂岩明显增多，与泥岩、粉砂质泥岩薄层状、条带状频繁互层构成单调的韵律层。这种基本层序组成的地层仅在第二段中、下部发育。

c 型　以泥质粉砂岩、粉砂岩、粉—细砂岩为主，偶夹砂质泥岩、泥岩型。这种基本层序构成了剖面上陀烈组一半以上地层，主要见于陀烈组第二段。其突出的特点是岩性层或沉积物分异不彻底，泥质、粉砂质、粉—细砂质普遍呈分散状出现构成所谓以混积岩为主体的基本层序，不含粉砂质的泥岩层较少。

d 型　粉砂质泥岩与粉砂岩、泥质粉砂岩不等厚互层型。与 b 型相比，粉砂质总体上变多，且沉积成层厚度增大。与 c 型相比，粉砂质所占比例相对较大。

e 型　沉凝灰岩与泥岩、粉—细砂岩不等厚互层型。每一基本层序的下部为沉凝灰岩、凝灰质粉砂岩，上部为泥岩和（或）粉—细砂岩。沉凝灰岩主要为酸性，灰黄色—黄色—黄白色，部分为酸偏中性，绿灰色。由这一基本层序组成的地层总的特点是：沉凝灰岩-凝灰质粉砂岩与泥岩、粉—细砂岩频繁不等厚互层；沉凝灰岩-凝灰质粉砂岩层厚度变化大，薄者不足 1m，厚者达 10 余米，其内部较均一，纵向上看不出明显的成分、结构或粒度变化。宏观上火山岩层与沉积岩层的频繁交替代表了多个火山喷发-沉积韵律，显示其层序变化特点，其分布局限，仅见于九曲江一带陀烈组第三段中。

陀烈组总体上为半深水—深水相复理石-类复理石建造，上述 5 种基本层序已大致反映了其基本沉积特点。

陀烈组下段　整体上出露一套千枚岩、变质泥质粉砂岩、变质粉砂岩，局部夹灰岩透镜体。原岩为富含泥质潮下低能带形成的产物（部分地段覆盖较多），a 型层序大致反映了这种沉积特点。但在岛西面，下部为成熟度较高的变质石英砂岩、石英岩，层厚大，发育变余水平层理，为浅海陆棚环境的产物，向上粉砂质绢云母千枚岩增多，也即砂质泥岩逐渐增多，发育变余粒序层理、变余水平层理，大体上可以达到陆棚边缘靠近斜坡附近，这说明下段开始时水体较浅，随着时间推后，向上水体逐渐加深。而在岛东面，下部为成熟度低的板岩，沉积粒度相对岛西较细，可能暗示当时沉积盆地具有西浅东深的特点。岛西北部儋州西联农场一带陀烈组下段底部沉积了一套成熟度稍高的细粒石英砂岩、粉砂岩，夹少量碳酸盐岩，微水平层理发育，局部砂岩中具微波状、板状等板状交错层理，生物化石罕见。其南西面西庆农场黄泥沟一带见有少量长石石英砂岩分布，表明沉积环境为水体较浅、离岸较近的潮下带至浅海陆棚环境沉积。结合其西北面儋州洋浦 CK3031 钻孔中新近系直接叠覆于长城系—蓟县系峨文岭组之上分析，认为其沉积作用物质来源可能与儋州洋浦一带长城系—蓟县系古陆有关（图 3-34）。

陀烈组中段 突出的特征是颜色暗，主体为灰白色—灰黑色，含较多炭质，显示为静水滞流缺氧沉积条件。常见 a 型—e 型层序。在岛西昌江坡兰沟一带出露比较典型的复理石沉积，这套浊积岩发育鲍马层序的 A、C、D、E 段(图 3-36)，尤其以 D 段和 E 段更为发育，其各段特征如下：①A 段，岩性为砾岩及含砾泥质岩（其变质后为含砾粉砂质绢云母千枚岩），砾石成分有脉石英、变质细砂岩、粉砂岩，为复成分砾。砾石大小一般在 2cm×3cm 左右，最大 4cm×8cm，最小 0.2cm×0.4cm，磨圆度较好，为圆至次圆，砾岩含量 35%～45%，基质为泥质粉砂。含砾泥质岩的砾石含量向上减少。显示一定的分选和具较好的磨圆度，而成分成熟度又较低，说明搬运距离较长而且是快速沉积的。在砾岩底面具槽模，这是鲍马层序 A 段最具特征的构造之一。②D 段，主要为粉砂岩和泥质粉砂岩（变质后为变质粉砂岩、绢云母粉砂质千枚岩），单层厚度小，多为薄层，普遍发育水平层理和微波状层理。③E 段，岩性为粉砂质泥岩、泥岩、含炭泥岩（变质后为粉砂质千枚岩、含炭千枚岩）。泥岩中普遍发育水平层理，单层厚度很小，可以是毫米级的。此段见到遗迹化石，为沿沉积物表层内水平分布的弯曲的觅食迹。上述特征均为确定鲍马层序 E 段的良好标志。另外，这套复理石中还有一些细砂岩，由于还缺乏沉积构造及粒度变化方面的资料，一时还难以确定是 B 段还是 C 段，就其多数单层厚度较小而暂时均归入 C 段。岛中屯昌南棍园发育深水复理石建造，主要表现为：①沉积物中普遍见平行层理、交错层理，发育不完整的鲍马序列，局部岩层沉积底面存在底模构造（槽模），沉积物中既未发现与波浪作用有关的沉积构造，也未发现与风暴作用相关的沉积构造。说明这些沉积构造是深水浊流的沉积作用标志。②沉积物中有一定数量反映深水沉积环境的放射虫化石（中国地质大学冯庆来鉴定）存在。③沉积物中未见生物扰动、潜穴、爬迹等生物活动痕迹，根据生物在同沉积期原地活动的痕迹记录可作为沉积底层充氧状况的一种标志，沉积环境应为一种缺氧不利于底栖生物栖息的环境。④沉积物总体上呈灰白色、灰色、灰黑色、黑色，无论在野外还是薄片中都可见到一定量的

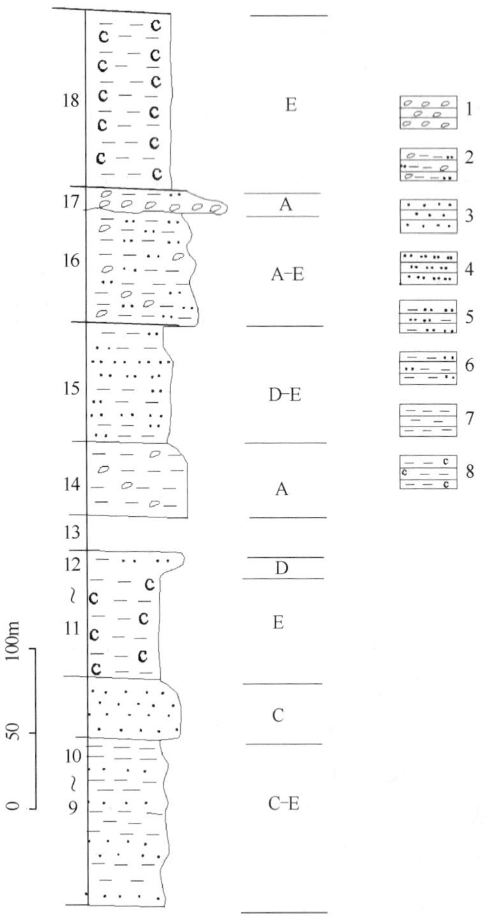

图 3-36　坡兰沟一带财地村剖面陀烈组中段鲍马层序
Fig. 3-36　Bouma sequence of the Middle Member, Tuolie Formation section, Caidi Village along Poland Valley
1—砾岩；2—含砾粉砂质泥岩；3—细砂岩；4—粉砂岩；5—泥质粉砂岩；6—粉砂质泥岩；7—泥岩；8—炭质泥岩

有机质、星点状黄铁矿，沉积物中未见生物扰动构造，根据 C. W. Byers 以黑海为例进行的停滞缺氧生物相分析，剖面沉积物形成于无氧带的深水滞流缺氧环境。岛东琼海东泰农场一带，以基本层序 a 型为代表，是一种沉积速率缓慢的沉积类型，即所谓饥饿沉积或凝缩层，它的出现表明沉积盆地进一步扩展、海域拓宽，形成了相对稳定的深水环境。其中所夹的泥质粉砂岩、粉砂岩薄层含较多泥质杂基，显然为浊流沉积，表明在长时间的静水缓慢沉积条件下偶尔接受小股浊流沉积，具有离大陆斜坡有一定距离的深海平原环境的特点。⑤泥质岩中夹变基性、超基性火山岩和硅质岩。由此显示由西至东，水深具有加大的趋势，与下段沉积古地理格局一致（图 3-34）。

陀烈组上段 主体为浅灰色、灰白色，不含炭质的岩层，从而与中段区分。岛西财地村一带仍具类复理石韵律的泥质岩建造，岩性主体为板岩、千枚岩、变质粉砂岩，但夹有透镜状灰岩（多变质为大理岩类），说明海水已趋于变浅。岛东琼海东泰农场一带，沉积环境已转入浊积扇根部，浊流活动的强度达到顶峰，火山作用也比较活跃，发育 e 型层序。形成了陀烈组单层厚度最大、粒度最粗的浊积岩，此时水体已相对变浅，之后进入了浅海环境。

综上所述，陀烈组下段为浅海陆棚相，中上段为大陆斜坡相，整个陀烈组经历了由浅变深，再由深变浅的环境变迁过程。

2）空列村组（S_1k）

空列村组形成于较开阔的滨岸浅水环境，包括前滨和近滨两种环境。主要由 2 种基本层序组成（图 3-37）。

a 型 厚层砂岩夹泥岩型。空列村组沉积岩的基本层序，有别于上述陀烈组基本层序的特点和主要特征是：沉积物总体上变粗，以砂岩为主体，泥岩总体上呈夹层出现，局部见较厚层的泥岩。砂岩分选好，不含或含少量泥质杂基，主体上为净砂岩；砂岩成分成熟度高，为石英砂岩；砂岩成层较厚，多为厚层乃至块状。砂岩中见小型楔状层理。部分为粉砂岩，可见脉状层理。总体上显示为经历过较长期水流分选的浅水环境产物。

b 型 凝灰岩-沉凝灰岩型。仅见于万宁市山根乡沿斗牛村-黄竹塘高速公路边，见厚达 30 余米凝灰岩、沉凝灰岩出露（浮土掩盖，未见底），并显示一定变化序列，代表了空列村组的一种特殊的基本层序。凝灰岩、沉凝灰岩为灰黄色—黄白色，含较多石英晶屑，部分具凝灰质砂岩、凝灰质泥岩特点。这种基本层序的组成特点是：下部为较粗的凝灰岩、沉凝灰岩，粗—中砂级为主，部分为细砂级，见晶屑和形态各异的岩屑、浆屑，部分发育粒序层理和大型交错层理，显示沉凝灰岩特点。岩屑为泥岩、粉砂岩、流纹岩等；上部为较细的凝灰岩、沉凝灰岩，粉砂—黏土级；自下而上的凝灰质薄层是火山活动间隙期沉积产物。

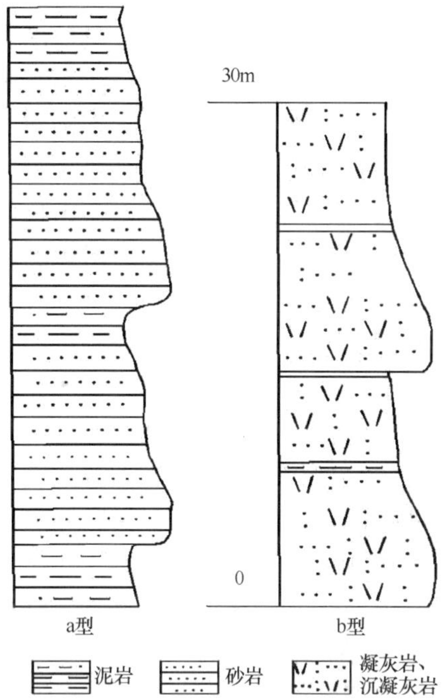

图 3-37 空烈村组基本层序及特征

Fig. 3-37 Basic sequence and characteristics of the Kongliecun Formation

典型前滨沉积的沉积物为浅灰色变质细粒石英砂岩，其特点是：不含泥质杂基，沉积物分选极好，成分成熟度高，石英含量达 95% 以上，呈厚层状。近滨沉积主要为细砂岩，有粉砂岩和少量中—细粒砂岩，可含一定泥质杂基，砂岩中见小型楔状交错层理。另见一定比例的泥岩在砂岩中呈夹层出现或与凝灰岩、沉凝灰岩不等厚互层。这些代表水动力条件较弱的泥岩无疑形成于与滨海毗邻的浅海环境。

空列村组下段为滨海相沉积的石英砂岩、粉砂岩，部分地带底部砂岩中含炭。已变质为石英岩，绿帘云母石英岩、变质石英砂岩，底部夹炭质板岩。沉积物分选性好，几乎不含泥质杂基，属于上述 a 型层序类型，为滨海沉积产物。底部砂岩中含炭可能为一种潟湖相或障壁岛后静水环境沉积产物。但昌江戈枕村一带为砂泥质沉积，且沉积厚度最大（>1396.36m），而其余地区为砂质沉积。从而显示其沉积中心和沉降中心位于戈枕村一带，具有东浅西深的古地理盆地格局，与陀烈组时期东深西浅的盆地格局恰好相反，表明地壳差异性升降运动的存在。空列村组上段沉积时，海南岛各地出露的岩石组合特征基本相似，昌江乌烈林场—南清河一带，岩石以千枚岩或板岩为主，间夹变质砂岩；昌江南界河一带夹碳酸盐岩透镜体，千枚岩常可见水平层理、斜层理及透镜状层理，属于远滨相潮下高能环境形成；岛南面南好地区其上部为板岩与变质砂岩不等厚互层状产出，夹碳酸盐岩透镜体，岛东面山根乡空列村组上部为板岩与变质砂岩不等厚互层状产出，夹沉凝灰岩和凝灰岩，变质砂岩中发育小型楔状层理，沉凝灰岩可见粒序层理和交错层理，都属于远滨相潮下高能环境形成。

3）大干村组（S_1d）

大干村组分布于岛南南好地区和岛西北河叉岭地区。

南好地区以青灰色薄层板岩、粉砂质板岩为主；顶部为灰色中厚层结晶灰岩、产珊瑚化石。底部为灰色—灰黄色复成分砾岩，具粒序层理。由下而上为退积型沉积序列，属滨海—浅海沉积环境。

岛西北河叉岭地区大干村组均以断块形式零星出露，为浅变质陆源碎屑沉积岩，未见碳酸盐岩，岩性上与南好地区大干村组标准剖面的中、下部有一定差异。河叉岭剖面大干村组可划分为下、上两段，其总体以石英片岩和绢云母千枚岩为主，间夹中厚层状石英岩、含砾石英片岩、铁质中粒石英砂岩等。另外，在河叉岭及白石山山顶有厚约20余米变质砾岩夹砂砾岩推覆于二叠系及大干村组上段地层之上，为大干村组底部砾岩滑脱的产物。由此显示该区大干村组在层序自下而上明显为一个旋回退积型层序，总体为一个海进过程。底部的石英质砾岩砾石滚圆度较好、无杂基充填和其中所夹含砾粗砂岩中见板状斜层理，为河流相沉积，其上一般为向上变粗基本层序，故大干村组往上为前滨相—滨外陆棚相。

4）靠亲山组（S_1kq）

靠亲山组仅分布于岛南南好地区。一段上部是一套由板岩、石英粉砂岩及石英岩组成的具水平层理及类复理石韵律结构的浅海陆棚远岸细粒碎屑岩沉积，产三叶虫、腕足类、腹足类和海百合茎化石。二段岩性为深灰色—灰黑色薄—厚层含泥质生屑灰岩、结晶灰岩，产珊瑚、腕足类、三叶虫的腹足类化石，水体变浅，属浅海陆棚近岸沉积。

5）足赛岭组（S_1z）

足赛岭组一段岩性为千枚岩、含炭质千枚岩，属滨海—浅海相。足赛岭组二段岩性为灰白色结晶灰岩，近中部夹薄层或条带状硅质泥岩，偶见珊瑚化石及腕足类化石，属浅海相沉积环境。足赛岭组三段岩性为千枚岩、粉砂岩，偶夹中细粒石英岩，属滨—浅海沉积环境。足赛岭组四段岩性以粉砂岩夹泥岩为主或呈不等厚互层状产出，间夹1~2m厚的薄—中层石英砂岩。实测地层剖面基本层序下部为灰白色中厚层（单层厚度为35~42cm）绢云母石英中细粒砂岩，偶见腕足类化石碎屑，上部由薄层粉砂质泥岩夹泥质粉砂岩或呈不等厚互层状产出，夹3层化石层，共计10~15cm厚不等，整体分析应为退积型沉积序列。发现的20余个化石富集层均产于薄层粉砂质泥岩中，它们共同的特征是：顺层富集，分异度低，相对丰富度较高。这些化石富集层的埋藏特征应为前滨—近滨相，与海南地矿局（1986）所测试的微量元素铁锰比值、锶钡比值反映的沉积环境相吻合，这表明本区兰多弗里世特列奇期所代表的沉积是一个进积型的沉积序列，水体由浅海陆硼远岸→前滨或近滨相。

三、上古生界

上古生界由石炭系南好组、青天峡组和二叠系峨查组、峨顶组、南龙组组成。空间分布常与下古生界毗邻，呈残留状断续分布。主要见于岛南南好地区、岛东东兴农场、岛西北的儋州市，以及岛西东方市和昌江县的江边乡、王下一带（图3-38）。

图 3-38　五指山地层分区上古生界分布及实测剖面位置简图

Fig. 3-38　Sketch map showing the distribution of Upper Paleozoic and the position of the measured section, Wuzhishan stratigraphic subregion

(一) 岩石地层特征

1. 剖面描述

(1) 南好组剖面

南好组剖面较多，基本上涵盖了分布于海南岛各地的南好组。有南兵剖面、青天峡剖面、公井岭（Ⅰ）剖面东南段、南好矿区锌山剖面、如翁-金波剖面、古弄村遗址剖面、坡兰沟剖面、红林农场16队剖面、鸡心村剖面、鸡实村剖面、三棱山剖面、上楼坡-大岭剖面、松坡岭剖面、东乐队-江南队剖面、峰膏岭剖面等。主要剖面描述如下：

1) 南兵剖面

该剖面为南好组代表性剖面。位于保亭县南好地区，起点坐标：东经109°27′48″，北纬18°36′25″。海南地质大队1982~1986年开展1∶5万《志仲》幅区调时测制，汪啸风等（1992）略作修改。剖面沿南好—南兵公路测制，自上而下为（图3-39）：

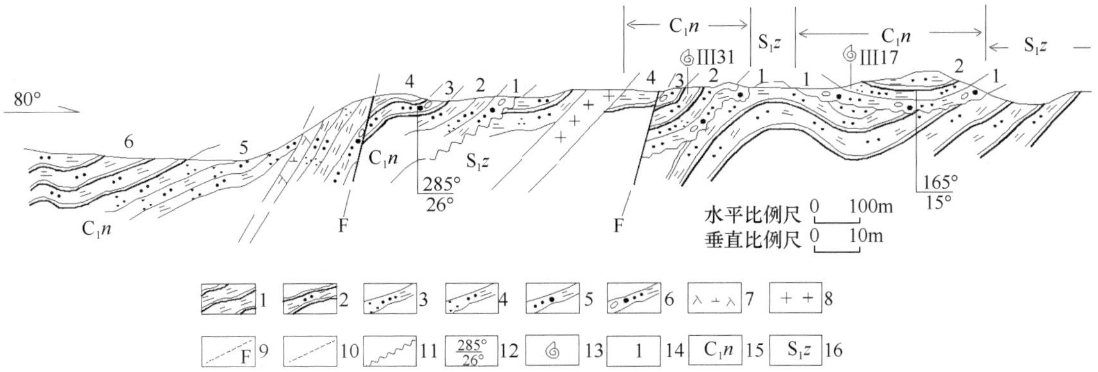

图 3-39　南兵实测剖面图

Fig. 3-39　Nanqiu measured section

1—绢云母板岩；2—绢云母粉砂质板岩；3—绢云石英粉砂岩；4—绢云母石英细砂岩；5—绢云母不等粒石英砂岩；6—绢云母含砾不等粒石英砂岩；7—闪长玢岩；8—花岗斑岩；9—实/推测地质界线；10—实/推测断层；11—不整合界线；12—产状；13—化石位置点；14—分层号；15—南好组；16—足赛岭组

未见顶

南好组 C_1n 总厚度>156.00m

6. 灰色薄层粉砂质板岩 41.90m
5. 浅灰色薄层绢云母石英细砂岩 48.50m
4. 浅灰色绢云母板岩 15.00m
3. 灰色绢云母不等粒石英砂岩 11.70m
2. 中上部青灰色薄层粉砂质绢云母板岩，往下粉砂逐渐增加，过渡到绢云石英粉砂岩，粉砂质板岩中含腕足类化石：（Ⅲ31）*Tenticospirifer* sp.，*Camarotoechia* sp.，*Fusella* sp.，*Ceptago nia* sp.，*Chonetipustula* sp.，下部粉砂岩中产腕足类：（Ⅲ17）*Tenticospirifer* cf. *supervilis*，*T.* sp.，*Ptychomaletoechia* sp. 14.30m
1. 灰色绢云母不等粒石英砂岩、含砾不等粒石英砂岩 24.60m

~~~~~~~~~ 不 整 合 ~~~~~~~~~

下伏地层 足赛岭组灰色薄层粉砂质板岩

2）青天峡剖面

该剖面为参考性剖面。位于东方昌江边青天峡谷（109°00′29″，18°55′16″）。广西区域地质调查大队1991~1993年开展1∶5万《坝王岭、猴猕岭》幅区调时测制，陈哲培等（1997）对剖面进行重新划分。自上而下为（图3-40）：

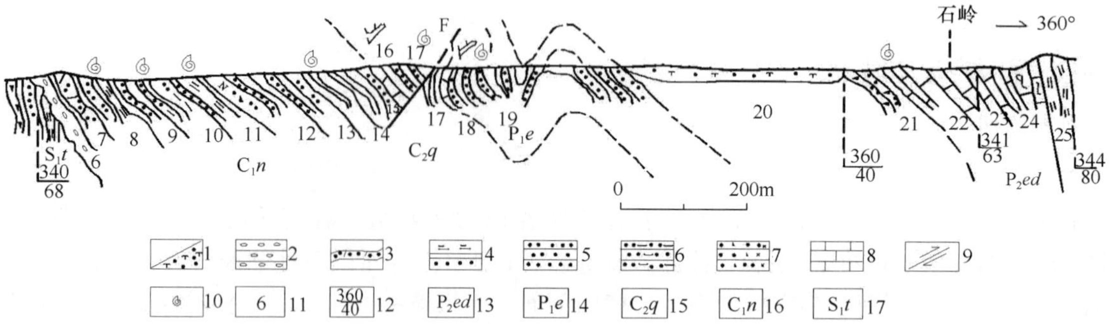

图3-40 东方市昌江边青天峡石炭系—二叠系实测剖面图

Fig. 3-40 Measured section of Carboniferous-Permian strata，Qingtianxia，near Changjiang，Dongfang City

（据广西区域地质调查大队，1991）

1—浮土；2—砾岩；3—沉凝灰砂岩；4—砂质绢云板岩；5—砂岩；6—粉砂质泥岩；7—含岩屑长石砂岩；8—板岩；9—断裂；10—化石点；11—分层号；12—产状；13—峨顶组；14—峨查组；15—青天峡组；16—南好组；17—陀烈组

上覆地层 青天峡组深灰色薄层细砂岩

———— 整 合 ————

南好组 厚404.0m

15. 暗灰色、灰绿色粉砂质板岩与细砂岩呈薄层互层 52.1m
14. 暗灰色板岩，上部夹石英细砂岩，中上部产腕足类 *Pustula scabriculiforms*，苔藓虫、海百合茎化石 46.6m
13. 灰白色、灰黄色石英细砂岩、中粒砂岩，上部夹少量板岩 31.7m
12. 灰色、深灰色、暗灰紫色绢云板岩与石英细砂岩、绿泥石细砂岩、细粒岩屑长石砂岩呈不等厚互层。具水平层理 63.6m
11. 灰绿色、暗灰紫色板岩、硬绿泥石化绢云板岩，富含腕足类 *Megachonetes zimmermanni*，*Eochoristites neipentaiensis*，*Punctospirites* sp.，*Martinia* sp.，*Spirifer* sp.，双壳类 *Schizodus* sp.，苔藓虫 *Fenestella* sp.，海百合茎等化石。底部为薄层细粒岩屑砂岩夹粉砂岩。具水平层理 37.9m
10. 上部为暗灰紫色硬绿泥石化粉砂质泥岩；下部为灰色、暗灰紫色粉砂质板岩。产腕足类 *Punctospirifer* sp.，海百合茎，双壳类等化石 47.4m
~~~~~~~~~

9. 上部浮土掩盖，下部为灰白、灰黄色中厚层细—中粒石英砂岩　30. 0m

8. 上部暗灰紫色硬绿泥石化石英绢云板岩、绢云母板岩、砂质板岩。具水平层理。产腕足类 *Paraphorhynchus*? sp.，头足类 *Michelinocera* sp.，海百合茎等化石；下部为灰白色、灰黄色薄层细砂岩　42. 9m

7. 灰紫色沉凝灰岩、凝灰质砂岩　28. 4m

6. 灰色、灰白色厚层—块状石英质砾岩、岩屑砾岩夹粉砂质泥岩、含砾砂岩　23. 4m

~~~~~~~~ 不　整　合 ~~~~~~~~

下伏地层　陀烈组灰色、深灰色粉砂质绢云板岩夹含炭泥质粉砂岩、灰绿色石英粉砂岩

3）坡兰沟剖面

该剖面由海南地质大队 1991～1993 年开展 1∶5 万《昌化大岭、海尾、昌化港、叉河》幅区调时测制，海南省地质调查院 2005～2007 年开展 1∶25 万《东方县》幅区调时补做了部分工作。剖面起点位于昌江县十月田山村北西 650m 公路边，沿十月田—海尾公路测制。自上而下为（图 3-41）：

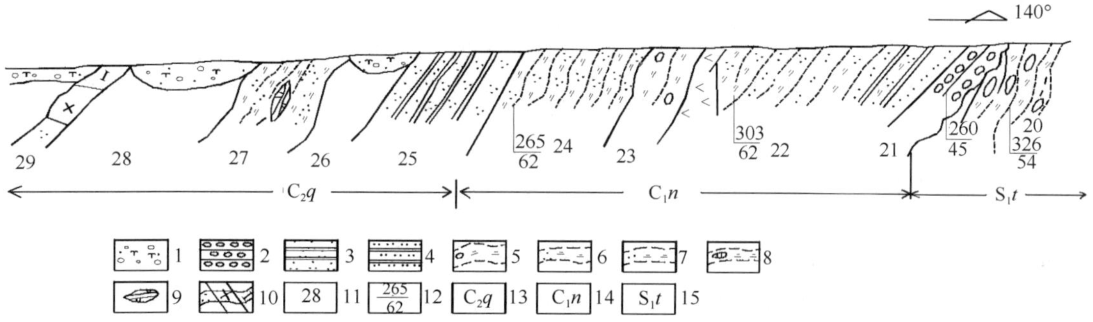

图 3-41　坡兰沟石炭系南好组和青天峡组剖面图

Fig. 3-41　Polangou Section showing the Carboniferous Nanhao and Qingtianxia Formations

1—浮土；2—砾岩；3—变质石英细砂岩；4—变质粉砂岩；5—含砾绢云母千枚岩；6—石英绢云母千枚岩；7—粉砂质绢云母千枚岩；8—含粉砂透镜体绢云母千枚岩；9—大理岩透镜体；10—透辉石透闪石石英角岩；11—分层号；12—产状；13—青天峡组；14—南好组；15—陀烈组

上覆地层　青天峡组灰白色厚层石英岩，变质中细粒石英砂岩

———— 整　合 ————

下石炭统南好组（C_1n）　203. 27m

24. 黄灰色薄层石英绢云母千枚岩　71. 04m

23. 灰绿色薄层含砾绢云母千枚岩　25. 09m

22. 浅黄色薄层石英绢云母千枚岩与变质石英粉砂岩不等厚互层　85. 21m

21. 灰色、灰绿色厚层变质砾岩。顶部变为中薄层含砾绢云母千枚岩　21. 93m

~~~~~~~~ 不　整　合 ~~~~~~~~

下伏地层　下志留统陀烈组

4）昌江县鸡实村剖面

该剖面由宜昌地质矿产研究所 1991～1994 年开展 1∶5 万《昌江县、邦溪幅》区调时测制，主要控制原南好组第四段。海南省地质调查院 2005～2007 年开展 1∶25 万《东方县》幅区调时补做了部分工作。剖面位于鸡实村南约 1km，剖面起点坐标：东经 109°05′28″，北纬 19°13′04″。自上而下为(图 3-42)：

上覆地层　二叠系峨查组粉砂岩、泥质粉砂岩

═════ 断　层 ═════

石炭系南好组　总厚度>208. 63m

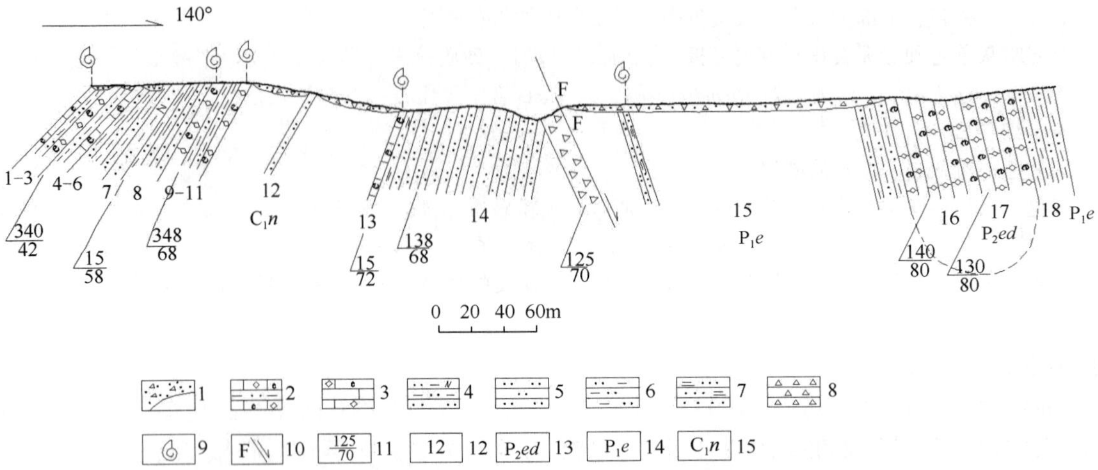

图 3-42　昌江县鸡实村南南好组上部及二叠系峨查组、鹅顶组剖面

Fig. 3-42　Section showing the upper Nanhao Formation and the Permian Echa and Eding Formations, south of Jishi Village, Changjiang County

1—浮土；2—骨屑含泥微晶细粉晶灰岩；3—骨屑结晶灰岩；4—含灰水云母质含长石石英粉砂岩；5—粉砂岩；6—泥质粉砂岩；7—含水云母细粒石英砂岩；8—构造角砾岩；9—双壳类化石点；10—断裂；11—产状；12—分层号；13—鹅顶组；14—峨查组；15—南好组

14. 深灰色中层石英粉砂岩夹少量灰白色细砂岩，顶部为含水云母细粒石英砂岩，见有海百合茎化石　>74. 35m

13. 灰色含骨屑微晶灰岩，含较多自生石英和自生斜长石，产有珊瑚化石 *Cystophrentas kalaohoensis* Yu（革老河泡沫内沟珊瑚），*Kueichowpora baijinensis*（摆金贵州管珊瑚），*Syringopora paralilla*（平行状管珊瑚）；有孔虫化石 *Septabrunsiina kruinica*，（克莱恩隔壁布林斯虫），*S.* sp.（隔壁布林斯虫），*Septatournayella rauserae*（劳棱隔壁杜内虫），*S.* sp.（隔壁杜内虫）等　1. 03m

12. 第四系坡积物覆盖，见零星灰色中薄层粉砂岩露头　64. 6m

11. 灰色中厚层骨屑微晶灰岩，见有珊瑚化石　7. 44m

10. 灰黄色泥质粉砂岩　7. 78m

9. 灰色薄—中厚层骨屑微晶灰岩，含珊瑚 *Neolithostrotion* spp.（新柱石珊瑚），*N.*？sp.，*Amplexus* sp.（包珊瑚），*Pseudouralinia irregularis*（不规则假乌拉珊瑚），*Syringopora gracilis*（细型笛管珊瑚），*S. serenei*（含勒来笛管珊瑚），*Zaphrentites* sp.（内沟珊瑚），*Kueichowpora panxianensis*（盘县贵州珊瑚），*K.* aff. *beijinensis*（摆金贵州管珊瑚近亲种）；含有孔虫 *Septabrunsiina* sp.（隔壁布林斯虫）等化石　6. 23m

8. 灰色石英砂岩积薄层泥质粉砂岩　14. 80m

7. 灰色薄—中厚层含灰水云母质含长石英粉砂岩　3. 90m

6. 第四系坡积物覆盖，可能为粉砂质泥岩　12. 34m

5. 薄层含骨屑泥晶灰岩　0. 67m

4. 第四系坡积物覆盖，可能为粉砂质泥岩　7. 33m

3. 灰色薄—中层骨屑含泥微晶细粉晶灰岩　2. 33m

2. 第四系坡积物覆盖，可能为粉砂质泥岩　5. 87m

1. 灰色中厚层骨屑微晶灰岩，含 *Neolithostrotion* ? sp.（新石柱珊瑚?），*Siphonodella isosticha*（等列始管刺）等化石　0. 50m

5）东乐队-江南队剖面

该剖面由宜昌地质矿产研究所 1992～1995 年开展 1∶5 万《中原市、博鳌港、和乐》幅区调时测制。剖面沿万宁市东岭农场东乐队至江南队的简易公路测制，起点坐标：东经 110°22′10″，北纬 18°59′00″。自上而下为（图 3-43）：

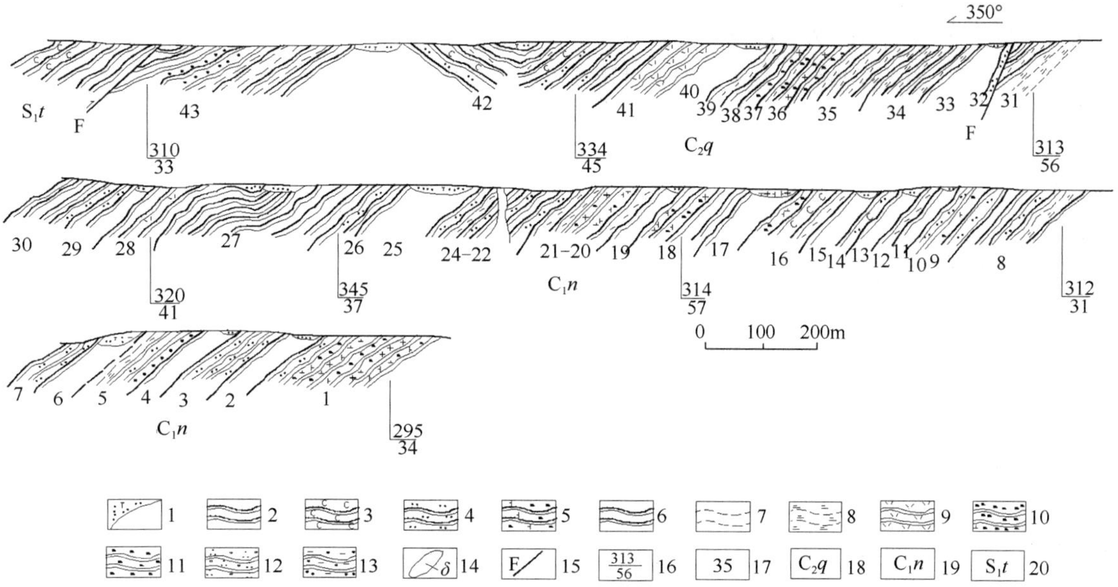

图 3-43 万宁市东乐队-江南队石炭系实测剖面图

Fig. 3-43 Measured section of Carboniferous strata, along Dongle Brigade-Jiangnan Brigade, Wanning City

1—浮土；2—板岩；3—炭质板岩；4—粉砂质板岩；5—硅质板岩；6—千枚状板岩；7—千枚岩；8—绢云母千枚岩；9—变质凝灰岩；10—变质粉砂岩；11—变质细砂岩；12—变质石英砾岩；13—泥质粉细砂岩；14—闪长岩脉；15—断裂；16—产状；17—分层号；18—青天峡组；19—南好组；20—陀烈组

上覆地层 石炭系青天峡组

═══════ 断层接触 ═══════

南好组（C_1n） 厚>1461.24m

31.	灰黄色—灰色片理化砂岩、绢云母千枚岩、千枚状板岩。片理化砂岩中可见顺层展布的石英脉并形成顺层掩卧褶皱	129.92m
30.	紫色、紫灰色—灰色板岩，下部含少量粉砂质板岩、变质粉砂岩	69.41m
29.	主体掩盖，见灰色硅化粉砂岩（2.22m）	34.22m
28.	紫灰色—深灰色板岩夹黄白色变沉凝灰岩	43.67m
27.	主体掩盖，零星露头为灰褐色—深灰色板岩	96.91m
26.	褐色—黄褐色粉砂质板岩、板岩	92.30m
25.	掩盖	57.34m
24.	褐色—黄褐色—深灰色变粉—细砂岩夹灰褐色—深灰色板岩、粉砂质板岩	46.07m
23.	灰色—深灰色千枚状板岩、千枚岩	17.26m
22.	褐色—黄褐色—灰色变粉砂岩、粉砂质板岩	17.59m
21.	褐色—黄灰色绢云千枚岩、千枚状板岩夹褐灰色粉砂质板岩	23.32m
20.	褐色—黄褐色变粉砂岩、粉—细砂岩夹粉砂质板岩、灰黄色千枚状板岩	47.83m
19.	掩盖	24.22m
18.	灰黑色、灰褐色—黄灰色—灰黑色板岩夹黄褐色—黄灰色变质粉砂岩（局部为细砂岩）、粉砂质板岩，向上变粉砂岩（细砂岩）增多	136.61m
17.	掩盖	64.18m
16.	下部为深灰色千枚状板岩、板岩、炭质板岩，上部为灰黄色变粉砂岩	64.38m
15.	主体掩盖，局部露头为灰黑色板岩（1.51m）	61.06m
14.	灰色—灰黑色粉砂质板岩、含炭质板岩、变质细—粉砂岩	32.26m
13.	掩盖	43.8m
12.	浅灰色—深灰色板岩、千枚状板岩，往上见变质粉—细砂岩夹层	6.94m
11.	掩盖	27.46m

10. 灰色—深灰色变质粉—细砂岩、粉砂质板岩 8.16m
9. 灰黑色炭质变质粉—细砂岩、粉砂质板岩 9.56m
8. 灰黄色—灰色—深灰色板岩、千枚状板岩、粉砂质板岩、变质粉砂岩，夹绢云千枚岩 119.96m
7. 灰黄色变质凝灰质粉砂岩 6.82m
6. 掩盖 52.32m
5. 深灰色变粉砂岩、粉砂质绢云千枚岩 13.27m
4. 掩盖 34.23m
3. 灰色薄层粉砂质板岩 23.30m
2. 掩盖 26.4m
1. 灰黄色—浅灰色中—薄层变质细砂岩、粉砂岩 101.01m

未见底

（2）青天峡组剖面

青天峡组剖面有青天峡剖面、南界河剖面、娜姆河剖面、坡兰沟剖面、红林农场16队剖面、华栗山剖面、上番开村剖面、东乐队-江南队剖面等。主要代表性剖面如下：

1）青天峡剖面

该剖面为青天峡组代表性剖面。剖面情况同上述，自上而下为（图3-40）：

上覆地层 下二叠统峨查组（P_1e）

———— 整 合 ————

青天峡组上段 厚92.6m

17. 灰绿色、暗灰紫色、灰色、深灰色绢云板岩、板岩夹灰色石英细砂岩或呈不等厚互层，中上部富含双壳类 *Phestia paundoraeformis*，*Schizodus affinis*，*S. wheeler i*，*S.* cf. *similis*，*S.* cf. *subcircularis*，*S.* cf. *axiniforomis*，*S.* spp.，*Myalina*（*Myalinella*）cf. *exasperata*，*M.*（*Orthomyalina*）*shingensis*，*Septimyalina perattenuala*，*S.* cf. *producta*，*Nuculopsis* sp.，*Palaeoneilo hainanensis*，*Aviculopecten* sp.，*Annuliconcha* sp.，*Selenimyalina dotti*，*S. meliniformis*。腕足类 *Orthotetes*? sp. 及苔藓虫等 47.3m
16. 细—中粒石英砂岩与砂质板岩呈不等厚互层，底部夹一层厚1.4m的灰白色、灰绿色结晶灰岩，具条带状构造。砂质板岩中产腕足类 *Orthotetes* cf. *keokue*。结晶灰岩产牙形石①*Streptognathodus elegantulus*，*S. oppletus*，*S. parvus*，*S.* cf. *symmetricus*，*S. qingtianxiaensis*，*Neognathodus deflexus*，*N. bassleri* 44.3m
15. 深灰色薄层状细砂岩 9.05m

———— 整 合 ————

下伏地层 南好组灰绿色粉砂质板岩与变质细砂岩

2）南界河剖面

南界河剖面为广西区域地质调查大队1988~1991年开展1∶5万《坝王岭、猴狝岭》幅区调时测制，海南省地质调查院2005~2007年开展1∶25万《东方县》幅区调时补做了部分工作。剖面位于海南省昌江黎族自治县七差乡昌化江南界河。自上而下为（图3-44）：

上覆地层 下二叠统峨查组

———— 整 合 ————

青天峡组 厚337.7m

13. 浮土覆盖 94.1m
12. 深灰色板岩 46.3m
11. 灰色中薄层细粒岩屑石英砂岩 16.5m

10. 灰色、深灰色板岩夹石英砂岩、细粒石英杂砂岩，碎屑多为次棱角状 82.6m

9. 灰色、灰黄色中薄层石英砂岩 13.1m

8. 灰色、深灰色板岩夹灰色中薄层石英砂岩 35.5m

7. 灰色、灰绿色石英砂岩夹板岩，底部夹厚4.05m含泥砾砂砾岩、细砾岩，具沉积韵律 49.6m

———— 整 合 ————

下伏地层 下石炭统南好组

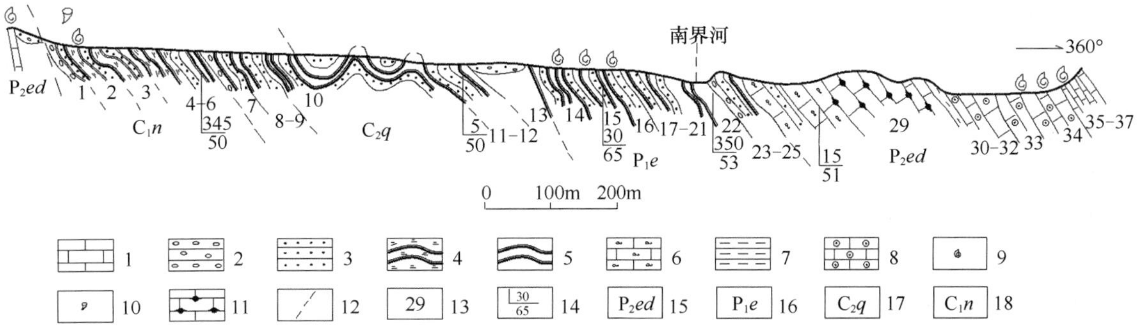

图3-44 昌江黎族自治县南界河石炭系—二叠系剖面图

Fig. 3-44 Section of Carboniferous-Permian strata, Nanjie River, Changjiang Li Autonomous County

1—灰岩；2—砾岩；3—砂岩；4—绢云母板岩；5—板岩；6—条带状灰岩；7—泥岩；8—鲕粒灰岩；9—介壳；10—微管类化石；11—含燧石条带灰岩；12—断层；13—分层号；14—产状；15—鹅顶组；16—峨查组；17—青天峡组；18—南好组

3）昌江县红林农场16队剖面

该剖面由宜昌地质矿产研究所1991~1994年开展1:5万《昌江县、邦溪》幅区调时测制，主要控制原南好组第三段。海南省地质调查院2005~2007年开展1:25万《东方县》幅区调时补做了部分工作。剖面起点位于红林农场16队南南西方向约300m简易公路旁，经由16队向北至鸡实村南石灰岩采石场，剖面起点坐标：东经109°05′18″，北纬19°11′46″。本次根据岩石组合特征对比，把剖面第12~22层原南好组重新划归为上石炭统青天峡组。自上而下为（图3-45）：

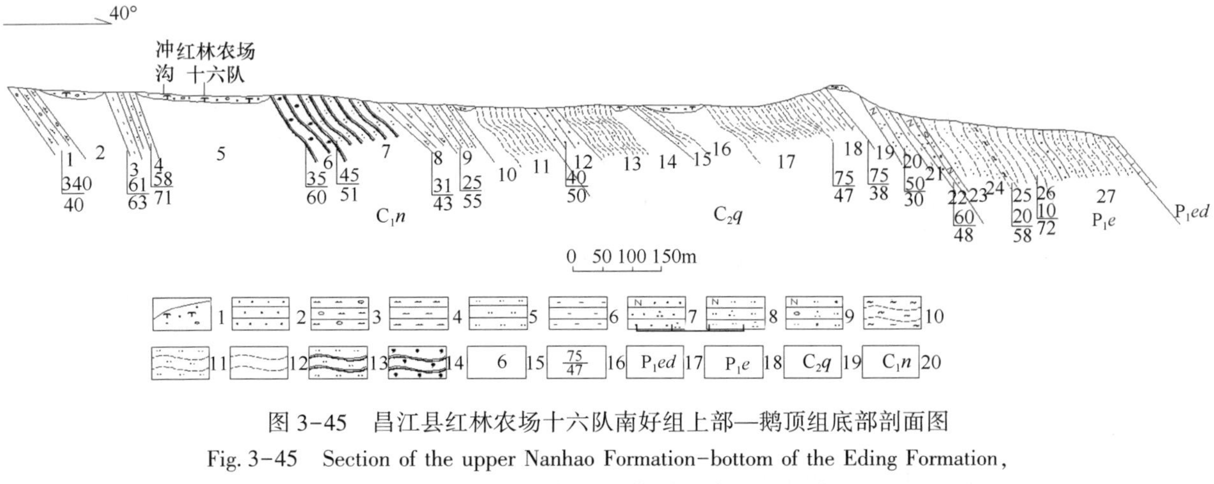

图3-45 昌江县红林农场十六队南好组上部—鹅顶组底部剖面图

Fig. 3-45 Section of the upper Nanhao Formation-bottom of the Eding Formation, No. 16 Brigade, Honglin Farmland, Changjiang County

1—浮土；2—细砂岩；3—含砾硅质岩；4—硅质岩；5—粉砂岩；6—黏土岩；7—长石石英细砂岩；8—长石石英粉砂岩；9—含砾长石石英砂岩；10—绿泥石千枚岩；11—粉砂质千枚岩；12—千枚岩；13—粉砂质板岩；14—斑点状板岩；15—分层号；16—产状；17—鹅顶组；18—峨查组；19—青天峡组；20—南好组

上覆地层 二叠系峨查组

———— 整 合 ————

石炭系青天峡组 总厚度 1145. 53m

22. 灰绿色中—厚层粉晶方解石化长石石英细砂岩，顶部为黄绿色千枚岩 25. 87m
21. 灰绿色薄层千枚岩化石英粉砂质水云母黏土岩夹灰绿色薄层粉砂岩 13. 76m
20. 黄色薄层水云母质含长石石英粉砂岩 58. 37m
19. 第四系残坡积层覆盖 26. 81m
18. 灰色薄层强变形含石英粉砂水云母黏土岩与水云母黏土岩互层，顶部 0. 5m 为灰色中层含内碎屑含石英粉砂微晶灰岩 22. 23m
17. 黄色千枚岩夹粉砂质千枚岩 55. 18m
16. 第四系残坡积覆盖 49. 92m
15. 下部为黄色粉砂岩，上部为黄色千枚岩 19. 60m
14. 第四系残坡积覆盖 9. 05m
13. 紫红色、灰绿色千枚岩夹灰色粉砂质千枚岩 46. 05m
12. 灰色薄层含水云母细粒石英砂岩 56. 57m

———— 整　合 ————

下伏地层　石炭系南好组

4）昌江县华栗山剖面

该剖面由宜昌地质矿产研究所 1991～1994 年开展 1：5 万《昌江县、邦溪》幅区调时测制，海南省地质调查院 2005～2007 年开展 1：25 万《东方县》幅区调时补做了部分工作。剖面位于昌江县石碌铁矿区东缘之华栗山，起点坐标：东经 109°04′02″，北纬 19°13′02″。自上而下为（图 3-46）：

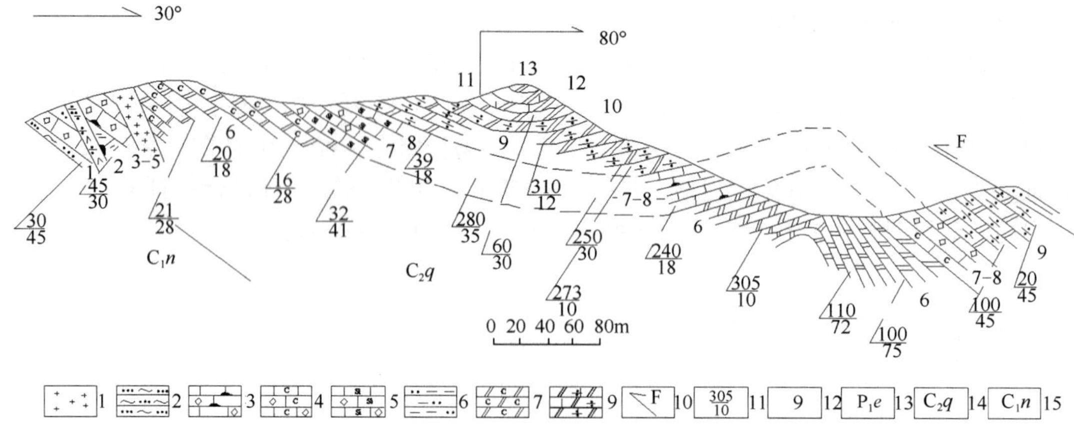

图 3-46　昌江县华栗山青天峡组实测剖面图

Fig. 3-46　Measured section of the Qingtianxia Formation，Huali Mountain，Changjiang County

1—花岗斑岩；2—叶绿泥石化细粒石英岩状砂岩；3—含燧石团块和条带细晶灰岩；4—含炭质和条带细晶灰岩；5—含硅质中细晶灰岩；6—石英水云母黏土岩；7—含炭质粉晶白云岩；8—条带透闪石化泥微晶白云岩；9—含灰微晶白云岩；10—断裂；11—产状；12—分层号；13—峨查组；14—青天峡组；15—南好组

上覆地层　二叠系峨查组

════ 断　层 ════

石炭系青天峡组

13. 灰色厚层—块状碎裂状残余砂屑粉晶含灰白云岩 >5. 80m
12. 灰色厚层—块状含灰微晶白云岩 41. 71m
11. 深灰色厚层—块状弱透闪石化粉晶白云岩，含少量燧石团块，具褶叠层构造 4. 11m
10. 灰色块状含燧石团块及条带透闪石化粉晶白云岩 4. 52m
9. 上部为白色块状含燧石团块及条带透闪石化泥微晶白云岩，下部 8. 63m 为灰白色、白色块状含燧石团块透闪石化粉晶白云岩 23. 22m
8. 灰色含燧石条带残余砂屑粉晶含白云岩，见零星针柱状透闪石 6. 11m

7. 灰色厚层—块状含硅质中细晶灰岩。 32.78m

6. 灰色厚层—块状含炭质粉晶白云岩，见有灰白色花斑岩岩脉侵入 52.25m

5. 灰色厚层具纹层构造含少量燧石团块中细晶灰岩 17.99m

4. 灰色厚层具纹层构造含零星燧石中晶灰岩 9.34m

3. 灰色厚层含燧石团块泥微晶含云灰岩，见有灰黄色钾长花岗斑岩岩脉侵入 19.30m

2. 灰色厚层含燧石团块和条带细晶灰岩 12.43m

———— 整 合 ————

下伏地层 石炭系南好组

5）上番开村剖面

该剖面由海南省地质调查院2005~2007年开展1：25万《琼海县》幅区调时测制。剖面起点位于莲花岭86°方向约1250m处，坐标：东经109°38′58″，北纬19°25′50″；剖面总长约1220m。剖面自上而下共分11层，自上而下为（图3-47）：

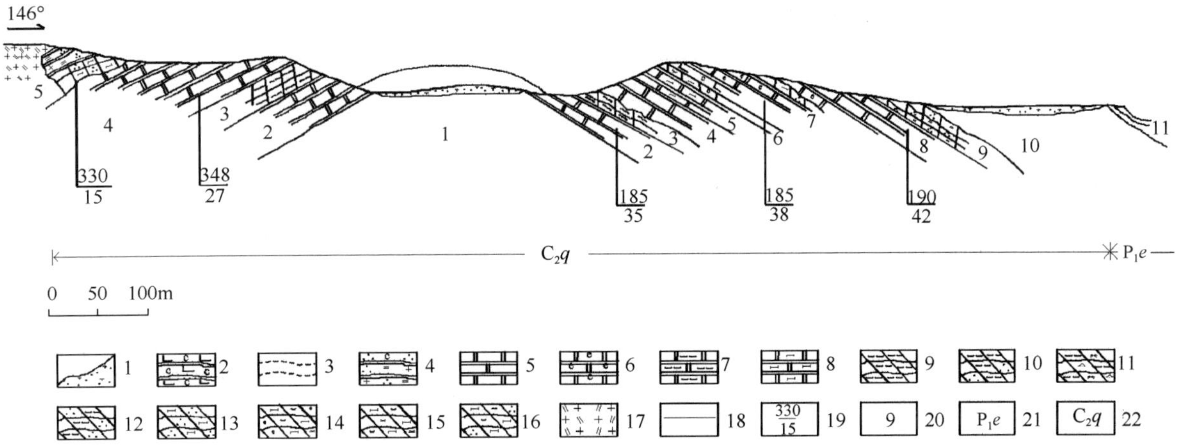

图3-47 儋州市上番开村青天峡组实测剖面图

Fig. 3-47 Measured section of the Qingtianxia Formation, Shangfankai Village, Danzhou City

1—掩盖；2—炭质钙质板岩；3—千枚岩；4—云英岩化绢云母石英岩；5—微晶大理岩；6—含生物微晶大理岩；7—微晶粒状含黑云母大理岩；8—微晶粒状含透辉石大理岩；9—黑云母角岩；10—（石英）黑云母角岩；11—阳起石（堇青石）黑云母角岩；12—含绢云母石英角岩；13—石英透辉石角岩；14—含生物阳起石透辉石角岩；15—阳起石透辉石角岩；16—石英阳起石透辉石角岩；17—中细粒（含斑）黑云二长花岗岩；18—推测整合接触界线；19—产状；20—分层号；21—峨查组；22—青天峡组

下二叠统峨查组（P_1e）

11. 千枚岩，风化色为浅黄色（野外横向追索此层，见此层底部出露复成分砾岩） 7.70m

———— 推测整合 ————

上石炭统青天峡组（C_2q） >304.17m

10. 掩盖 70.20m

9. 浅灰白色块状含绢云母石英角岩 24.64m

8. 浅灰色、浅灰黑色块状微晶大理岩，发育微水平层理 25.31m

7. 浅灰色、浅灰黑色块状含生物微晶大理岩夹厚层块状透辉石角岩，下部局部夹炭质钙质板岩，前者中发育潮汐层理，产珊瑚化石，化石已倒伏与岩层产状一致顺层产出），炭质钙质板岩中含大量海百合茎，少量腕足类化石 49.20m

6. 浅灰白色中薄层含生物（主为海百合茎）阳起透辉石角岩与浅灰色中薄层阳起黑云角岩互层，底部为厚层浅灰色、浅灰白色微晶大理岩，发育潮汐层理。为风暴岩沉积，单层厚多为5~20cm，厚者50cm，薄者5cm，可见底面侵蚀构造、丘状层、生物富集层（海海百合茎）等现象 5.00m

5. 浅灰色块状微晶粒状透辉石大理岩，中部为浅灰色、浅灰白色中薄层黑云角岩与浅灰色中薄层阳起透辉石角岩互层微水平层理发育。透辉石大理岩中发育潮汐层理，见后期液化塑性形变现象 16.65m

4. 浅灰色、浅灰白色块状微晶大理岩，中上部夹浅灰色厚层黑云角岩 26.42m

3.	浅灰色、浅灰白色厚层—块状微晶大理岩与浅灰、浅灰白色中厚层黑云角岩互层	10.14m
2.	浅灰色、浅灰白色块状微晶大理岩	16.22m
1.	掩盖	60.39m

未见底

6）东乐队-江南队剖面

剖面情况见上述。自上而下为（图 3-43）：

上覆地层 志留系陀烈组

═══════ 断　　层 ═══════

青天峡组（C_2q） 厚 505m

43. 灰色—深灰色板岩、千枚状板岩与灰色变泥质粉—细砂岩不等厚互层或后者呈夹层出现 109.33m
42. 灰色—深灰色—灰黑色板岩、炭质板岩、千枚状板岩夹粉砂质板岩、变质泥质粉—细砂岩，见少量硅板岩 141.19m
41. 灰黄色、紫红色—黄白色片理化沉凝灰岩 29.45m
40. 掩盖 40.35m
39. 紫灰色—灰色—深灰色绢云母千枚岩、千枚状板岩 6.12m
38. 掩盖 17.77m
37. 黄绿色—绿灰色变质砂岩，下部为变粉砂岩 10.16m
36. 灰黄色—灰色、灰黄略带蓝色绢云砂质千枚岩、千枚状板岩夹片理化细—粉砂岩 38.94m
35. 紫灰色—灰黄色—灰色、绿灰色绢云千枚岩、千枚状板岩，上部夹片理化粉—细砂岩 159.59m
34. 灰色—灰黄色、灰略带紫色板岩夹片理化粉—细砂岩 14.87m
33. 掩盖 15.53m
32. 变质石英砾岩 2.70m

═══════ 断　　层 ═══════

下伏地层 石炭系南好组（C_1n）

（3）峨查组剖面

峨查组剖面有青天峡剖面、南界河剖面、红林农场三队剖面、红林农场 16 队剖面、大岭农场 14 队剖面等。各剖面对地层单位控制程度较好。

1）青天峡剖面

该剖面为代表性剖面。剖面情况同上述，自上而下为（见图 3-40）：

上覆地层 二叠系鹅顶组中厚层含生物屑粉晶灰岩

———— 整　　合 ————

峨查组 总厚度 384.4m

22. 上部为深灰色、暗灰紫色细砂泥质板岩，下部为灰色中—细粒岩屑石英砂岩夹少量板岩 32.7m
21. 浮土掩盖 269.1m
20. 深灰色砂质板岩 28.5m
19. 灰色石英细砂岩与深灰色板岩呈不等厚状互层。产腕足类 *Rhipidomella* sp.，*Punctospiriter* sp.，以及苔藓虫 *Fenestella* sp.，海百合茎等化石 51.2m
18. 深灰色含生物屑微晶灰岩夹薄层硅质岩。产牙形石 *Sweetognathus whitei*，*S. behnkeni*，*S.* spp.，*Rabeignathus asymmetricus*，*R. hainaensis.*，*Anchignathodus typlcalis*，*Diplognathodus* sp.，*Ozarkodina* sp.，以及腕足类、有孔虫、介形类、骨针、蠕虫管等生物碎屑 2.9m

———— 整　　合 ————

下伏地层 石炭系青天峡组深灰色绢云板岩

2）南界河剖面

剖面情况见上述。自上而下为（图3-44）：

上覆地层 鹅顶组厚层微晶灰岩

———— 整 合 ————

峨查组	厚328.8m
25. 浮土掩盖	20.9m
24. 灰色粉砂质页岩夹中薄层状石英砂岩	16.5m
23. 上部为灰白色厚层细粒石英质砾岩；下部为灰色、灰白色中厚层粗—中粒石英砂岩	8.1m
22. 浮土掩盖	44.2m
21. 深灰色板岩夹钙质板岩	8.1m
20. 浮土掩盖	22.5m
19. 灰色、深灰色板岩夹少量石英砂岩	20.2m
18. 灰白色、灰色含砾中—粗粒岩屑砂岩夹少量板岩	14.2m
17. 灰色、深灰色板岩夹薄层石英砂岩，产支离破碎的海百合茎	26.1m
16. 灰色、灰黄色中薄层石英砂岩夹板岩	19.9m
15. 灰色—浅灰绿色板岩夹中—薄层石英砂岩，产极其丰富 *Lingulella* sp.	79.9m
14. 灰色、暗灰紫色、灰黄色板岩夹石英砂岩，产双壳类、腹足类碎片及细枝状海百合茎等化石	48.2m
13. 浮土掩盖	

———— 整 合 ————

下伏地层 石炭系青天峡组

3）昌江县红林农场三队剖面

该剖面由宜昌地质矿产研究所1991～1994年开展1∶5万《昌江县、邦溪》幅区调时测制，2005～2007年海南省地质调查院开展1∶25万《东方县》幅区调时补做了部分工作。剖面位于红林三队西北约700m的238.9m高地，剖面由CⅣA、CⅣB两部分组成，CⅣA起点坐标：东经109°02′，北纬19°13′。自上而下为：

上覆地层 鹅顶组薄层片理化含炭质泥微晶灰岩

———— 整 合 ————

峨查组	厚度>207.47m
9. 深灰色薄层含斑点状红柱石含炭黄铁矿水云母黏土岩	11.57m
8. 灰色中—厚层含水云母中细粒石英岩状砂岩，含微量透闪石	12.10m
7. 灰色粉砂岩	17.65m
6. 灰白色细—中粒石英砂岩	32.38m
5. 灰色含石英粉砂绢云母黑云母红柱石板岩夹0.34m粗晶灰岩	40.70m
4. 灰白色厚层中粗粒石英岩，底部0.3m为白色含砾石英岩	26.19m
3. 灰色厚层含水云母中细粒石英岩状砂岩	34.71m
2. 深灰色厚层石英岩，未见底	>37.08m

════ 断 层？ ════

下伏地层 上石炭统青天峡组白云石化粉晶白云岩

4）昌江红林农场16队剖面

剖面情况见上述。自上而下为（见图3-45）：

上覆地层 二叠系鹅顶组厚层—块状骨屑泥晶灰岩

——— 整 合 ———

二叠系峨查组 厚 359.53m

27. 青灰色页片状粉砂质千枚岩，下部夹一层 3.22m 灰色厚层石英细砂岩	150.04m
26. 青灰色页片状千枚岩	99.49m
25. 青灰薄层针绿泥石千枚岩	12.47m
24. 灰绿色粉砂质千枚岩，层内揉皱构造发育	93.41m
23. 灰绿色含水云母绿泥石含砾长石石英粗粉砂岩，最大粒径 14cm	4.12m

——— 整 合 ———

下伏地层 石炭系南好组中—厚层粉晶方解石化长石石英细砂岩

5）白沙县大岭农场 14 队剖面

该剖面由宜昌地质矿产研究所 1991～1994 年开展 1∶5 万《昌江县、邦溪》幅区调时测制，2005～2007年海南省地质调查院开展 1∶25 万《东方县》幅区调时补做了部分工作。剖面位于大岭 14 队东约 500m，剖面起点坐标：东经 109°03′23″，北纬 19°26′34″。自上而下为（图 3-48）：

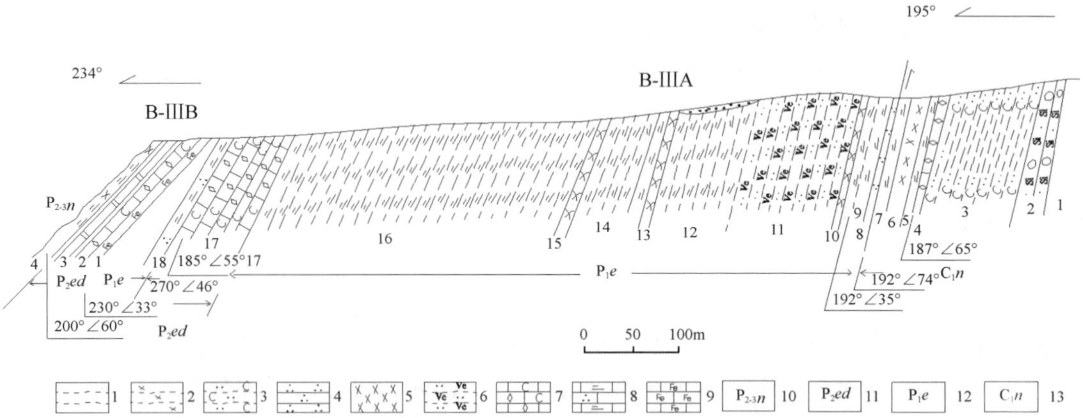

图 3-48 白沙县大岭农场 14 队二叠系峨查组及鹅顶组剖面

Fig. 3-48 Section of the Permian Echa and Eding Formations, No. 14 Brigade, Daling Farmland, Baisha County

1—千枚岩；2—含红柱石千枚岩；3—含炭含细砂质千枚岩；4—石英岩；5—角闪辉绿岩脉；6—变质石英粉砂蛭石千枚岩；7—含炭质细晶灰岩；8—片理化硅化大理岩；9—铁质灰岩；10—南龙组；11—鹅顶组；12—峨查组；13—南好组

上覆地层 二叠系鹅顶组强片理化含炭质细晶灰岩及片理化硅化大理岩

——— 整 合 ———

峨查组 厚度>342.27m

16. 青灰色千枚岩	186.71m
15. 深灰色角闪辉绿岩脉，岩脉宽 6.0m，顺层贯入	
14. 千枚岩	35.19m
13. 灰色角闪辉绿岩脉，岩脉宽 6.0m，顺层贯入	
12. 青灰色薄层千枚岩	49.66m
11. 灰色薄层含变质石英粉砂蛭石千枚岩	52.22m
10. 深灰色角闪辉绿岩脉，岩脉宽 6.0m，顺层贯入	
9. 灰色页片状千枚岩夹 1m 厚灰色薄层石英岩，产 *Agathiceras* cf. *vulgata*，*Alludoceras* sp. 等菊石化石	11.37m
8. 白色块状石英岩，未见底	>7.12m

═══ 断 层 ═══

下伏地层 石炭系南好组深灰色薄层—页片状千枚岩

（4）鹅顶组剖面

鹅顶组剖面有青天峡剖面、南界河剖面、昌江县红林农场三队剖面、白沙县大岭农场 14 队剖面、牛笛岭剖面等。

1）青天峡剖面

该剖面为代表性剖面。剖面情况见上述，自上而下为（图 3-40）：

上覆地层 南龙组灰绿色绢云千枚岩

——————整　合——————

鹅顶组 厚 210.0m

25.	上部为深灰色纹层状灰岩，下部深灰色含生物屑灰岩，具条带状构造	86.1m
24.	灰、深灰色纹层状含燧石灰岩	66.9m
23.	灰、深灰色中厚层状含生物屑粉晶灰岩，局部含燧石团块。产腕足类、苔藓虫及海百合茎等化石碎片	48.0m

——————整　合——————

下伏地层 峨查组泥质板岩

2）南界河剖面

剖面情况见上述，自上而下为（图 3-44）：

鹅顶组 总厚度>543.6m

37.	深灰色厚层含生物屑微晶灰岩夹含炭粉砂质泥岩，产䗴类 *Parafusulina* sp.，海百合茎，未见顶	27.5m
36.	灰色、深灰色页岩（泥岩）夹含生物屑微晶灰岩，产䗴类 *Parafusulina* sp.，海百合茎等化石	5.1m
35.	灰色、深灰色厚层含白云质生物屑微晶灰岩，产䗴类 *Parafusulina* sp.，海百合茎等化石	7.0m
34.	灰色、深灰色厚层含生物屑微晶灰岩，产䗴类 *Parafusulina* sp.，*Neomisellina*? sp.，海百合茎	88.5m
33.	灰色、深灰色厚层—块状生物屑微晶灰岩，产丰富的海百合茎	35.4m
32.	灰色、深灰色厚层含燧石生物屑微晶灰岩，产海百合茎	66.2m
31.	灰色、深灰色厚层—块状微晶颗粒灰岩	31.4m
30.	灰色、深灰色含炭粉砂质泥岩夹深灰色厚层生物屑灰岩	6.4m
29.	灰色、深灰色厚层含燧石微晶灰岩	145.0m
28.	灰色、深灰色厚层微晶灰岩，具纹层状构造	69.2m
27.	灰色页（泥）岩夹中层中—细粒砂岩，碎屑以石英为主，多为次棱角至半滚圆状	13.1m
26.	灰色、深灰色厚层微晶灰岩，具纹层状构造	48.4m

——————整　合——————

下伏地层 峨查组

3）昌江县红林农场三队剖面

剖面情况见上述，自上而下为：

未见顶

鹅顶组

12.	灰色厚层—块状含燧石条带和燧石团块灰岩	>4.23m
11.	灰色厚层—块状水平叠层石灰岩	19.58m
10.	灰色薄层片理化含炭质泥微晶灰岩，*Streptognathodus* sp.（曲颈刺，未定种）	8.01m

——————整　合——————

下伏地层 峨查组水云母黏土岩

4）白沙县大岭农场 14 队剖面

剖面情况见上述，自上而下为（图 3-48）：

上覆地层 南龙组页片状含红柱石千枚岩

——————— 整 合 ———————

鹅顶组 厚 29.29m

18. 深灰色块状片理大理岩 8.23m

17. 灰色薄层状强片理化含炭质细晶灰岩及片理化硅化大理岩 21.06m

——————— 整 合 ———————

下伏地层 峨查组青灰色千枚岩

5）牛笛岭剖面

该剖面由广西区域地质调查大队 1988~1991 年开展 1：5 万《坝王岭、猴猕岭》幅区调时测制。剖面位于东方黎族自治市江边乡青天峡谷至牛笛岭地段，处于宽缓的江边向斜东南部，翼部因受逆断层破坏，地层出露不全或者重复，但地层层序基本清晰。底部与志留系陀烈组呈断层接触，顶部因第四系覆盖而出露不全。自上而下为（图 3-49）：

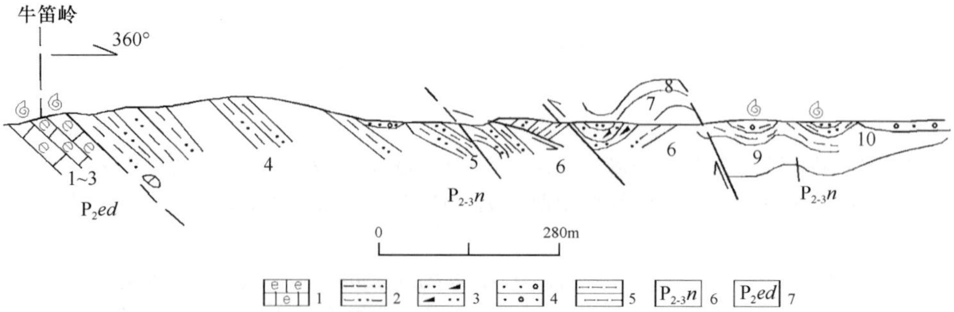

图 3-49 东方市江边至牛笛岭二叠系实测剖面图

Fig. 3-49 Measured section of Permian strata from Jiangbian to Niudiling, Dongfang City

1—生物屑灰岩；2—粉砂质泥岩；3—岩屑粉砂岩；4—含砾砂岩；5—泥岩；6—南龙组；7—鹅顶组

上覆地层 南龙组中厚层粉砂质泥岩

——————— 整 合 ———————

鹅顶组 总厚度>91.4m

3. 浅灰色、灰色薄—厚层含燧石微晶生物屑灰岩，富产蜓类 *Parafusulina* sp.，*Minojapanella* sp.，*Schubertella* sp.，*Chalaroschwagerina* sp.，*Schwagerina* sp. 以及腕足类、海百合茎等化石碎屑 46.5m

2. 深灰色、灰黑色厚层—块状含燧石微晶生物屑灰岩，富产海百合茎、苔藓虫、介形类、有孔虫、珊瑚等化石碎屑，可见蜓类 *Schwagerina* sp.，*Parafusulina* sp. 23.6m

1. 灰白色、浅灰色厚层—块状微晶生物屑灰岩。产蜓类 *Parafusulina* sp.，以及腕足类、介形类、海百合茎等化石碎屑 21.3m

═══════ 断 层 ═══════

下伏地层 志留系陀烈组（S_1t）砂质板岩

（5）南龙组剖面

南龙组剖面有南龙村剖面、石碌水库溢洪道及水库南岸剖面和牛笛岭剖面。

1）南龙村剖面

该剖面为代表性剖面。广东省区测大队（1964）测制，剖面位于东方市南龙村，起点坐标：东经 108°54′44″，北纬 18°53′04″。自上而下为：

未见顶

南龙组 总厚度>448.0m

17.	灰色、深灰色粉砂质泥岩	>48.0m
16.	灰黄色、褐黄色泥岩	18.0m
15.	褐黄色中粒长石石英砂岩	9.0m
14.	灰色、深灰色泥岩，含植物化石 *Pecopteris arborescens*，*P.* sp.，*Annularia crassiuscula*，*A.* sp.，*Pterophyllum* sp.，*Taeniopteris* sp.	15.0m.
13.	掩盖	49.0m
12.	灰褐色中粒长石石英砂岩	10.0m
11.	灰色、褐黄色泥岩，含植物化石碎片	32.0m
10.	褐黄色杂砂岩	17.0m
9.	灰黄色、灰褐色长石石英砂岩	8.0m
8.	灰黄色、灰白色粉砂质泥岩，含植物化石 *Annularia mucronata*，*Pecopteris* sp.，*Taeniopteris* sp.，*Cordaites* sp.	28.0m
7.	灰色、褐灰色中粒杂砂岩，黄褐色粉砂岩与黄红色泥岩互层	175.0m
6.	深灰色薄层泥岩与细砂岩互层	14.0m
5.	灰黑色泥质细砂岩	3.0m
4.	褐灰色泥岩	10.0m
3.	灰褐色泥质粉砂岩，含海百合茎化石	3.0m
2.	灰黄色泥岩与细砂岩互层	3.0m
1.	掩盖	6.0m

———— 整　合 ————

下伏地层　鹅顶组微晶灰岩

2）石碌水库溢洪道及水库南岸剖面

该剖面由宜昌地质矿产研究所1991~1994年开展1：5万《昌江县、邦溪》幅区调时测制，2005~2007年海南省地质调查院开展1：25万《东方县》幅区调时补做了部分工作。剖面起点坐标：东经109°04′25″，北纬19°13′49″。自上而下为（图3-50）：

南龙组 总厚度>1079.89m

30.	第四系残积层覆盖，下部见少量深灰色褶叠层构造石英粉砂岩碎石，顶部以断层与峨查组接触，未见顶	>112.10m
29.	深灰色中层褶叠层构造粉砂岩夹灰色细粒石英砂岩	4.19m
28.	第四系残坡积覆盖，下部及中上部见少量深灰色石英粉砂岩碎石，近顶部见灰色细粒石英砂岩碎石，大碎石见褶叠层构造	147.42m
27.	第四系残坡积覆盖，下部见零星深灰色石英粉砂岩小碎石，上部见有深灰色褶叠层构造含棘屑含石英粉砂含绿泥石透辉石角岩转石	62.86m
26.	深灰色夹浅灰色条带状含棘屑含石英粉砂含绿泥石透辉石角岩，具褶叠层构造	43.94m
25.	第四系残坡积层覆盖，仅中下部见少量深灰色中层石英粉砂岩露头，残坡积层中常见深灰色中层石英粉砂岩及少量石英细砂岩碎石，上部碎石中见褶叠层构造	317.74m
24.	浅灰色中层含水云母细粒石英砂岩，显不清楚的沙纹层理	12.36m
23.	第四系残坡积层覆盖，中下部见灰色薄层泥质粉砂岩及粉砂质泥岩碎块，上部见石英粉砂岩和石英细砂岩碎块，顶部见二透石石英角岩转石	31.31m
22.	第四系残坡积层及淤泥覆盖，下部见深灰色含残余砾屑二透石石英角岩及二透石石英角岩转石，顶部见零星浅灰色石英岩状粉砂岩转石	59.84m
20.	灰色中层含残余砾屑石英透辉石透闪石角岩	47.72m
19.	灰色中厚层石英粉砂岩，略显沙纹层理	5.07m

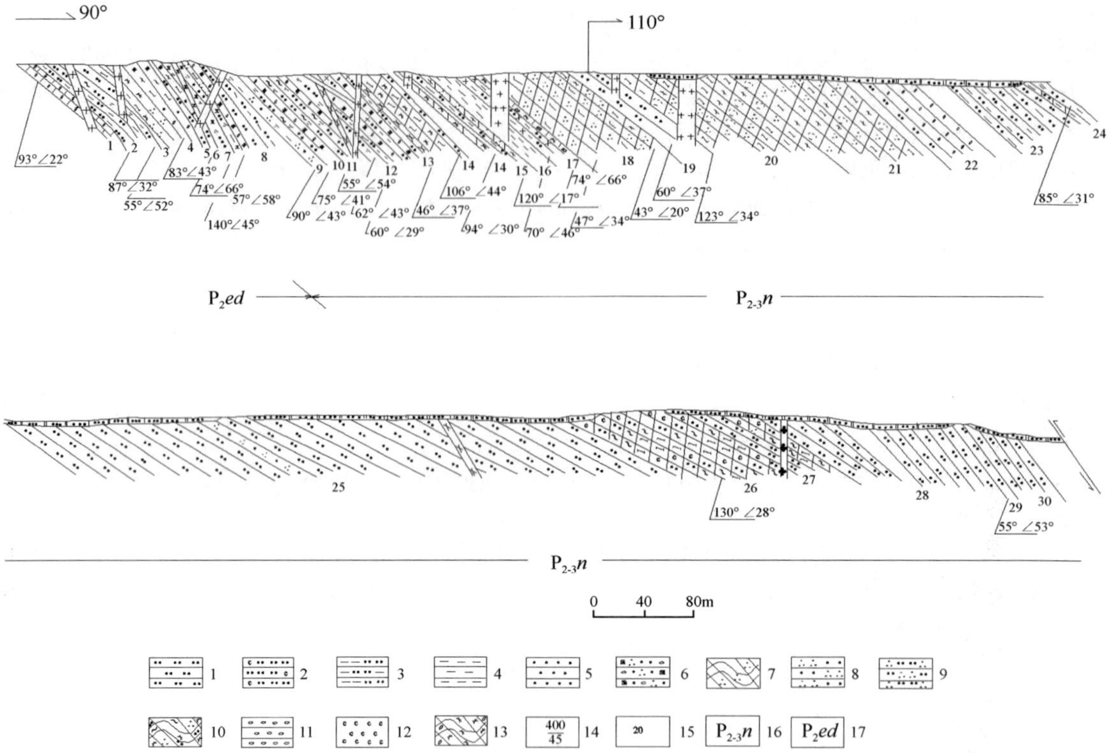

图 3-50 昌江县石碌水库溢洪道及水库南岸南龙组剖面图

Fig. 3-50 Section of the Nanlong Formation in the Shilu reservoir spillway and southern bank of the reservoir, Changjiang County

1—粉砂岩；2—含炭质粉砂岩；3—泥质粉砂岩；4—泥岩；5—细砂岩；6—含石英中粗粒砾屑硅质岩；7—石英角岩；8—细粒石英砂岩；9—石英粉砂岩；10—含石英粉砂绿泥石透辉石角岩；11—细砾岩；12—碳酸盐岩；13—残余细砾屑硅质岩；14—产状；15—分层号；16—南龙组；17—鹅顶组

18. 灰白色、浅灰色中层石英角岩夹浅灰色大理岩条带，顶部 1.12m 为浅灰色中层弱二透石化石英角岩，岩石具褶叠层构造 20.29m

17. 中下部为灰色薄层含红柱石泥质粉砂岩夹薄透镜状石英岩状粗粉砂岩，上部 1.94m 为浅灰色褶叠层构造阳起石角岩 8.79m

16. 黄绿色薄层水云母黏土岩 16.33m

15. 灰色中层透闪石石英角岩 2.02m

14. 灰色中—薄层粉砂岩、泥质粉砂岩夹一层 0.66m 厚灰色中层硅质岩 6.91m

13. 下部 11.65m 为灰色中厚层砾屑硅质岩与透闪石化石英粉砂质水黑云母黏土岩互层；上部为灰色厚层透闪石阳起石角岩 27.13m

12. 深灰色薄层夹中层透闪石化石英粉砂质水黑云母黏土岩 2.90m

11. 浅灰色厚层砾屑硅质岩与深灰色含绿泥石水云母质石英粉砂岩不等厚互层 15.60m

10. 灰色厚层含绿泥石水云母质石英粉砂岩夹透镜状黄白色砾屑硅质岩 21.57m

9. 底部为深灰色厚层弱二透石化含水云母中粒石英岩状砂岩，中上部为深灰色薄层粉砂岩夹 0.23m 厚灰色细粒石英岩状砂岩 4.05m

8. 底部为深灰色薄层含红柱石含石英粉砂水云母黏土岩，下部为灰白色中层含石英砂角砾屑硅质岩，上部为灰色薄层粉砂岩、泥质粉砂岩 11.08m

7. 灰色薄层含红柱石水云母质细粒石英砂岩夹灰色薄层砂岩 13.50m

6. 灰色厚层二透石化含石英中粗砂细砾屑硅质岩，中部夹少量灰色中层细粒石英砂岩及灰色薄层粉砂岩 3.20m

5. 灰色薄层粉砂岩夹 3 层 0.21～0.47m 灰色中层细粒石英岩状砂岩 3.46m

4. 灰色厚层—块状弱二透石化含中粗粒石英砂砾屑硅质岩，部分硅质岩砾屑中见较密集的直径 0.07～0.15mm 的空球虫铸模孔 12.48m

3. 灰色中层细粒石英岩状砂岩、二透石斜黝帘石化细粒石英岩与灰色、浅灰色薄层粉砂岩、黄绿色薄层泥质粉砂岩互层 16.03m

2. 下部为灰色薄层灰质粉砂岩、含炭粉砂岩、泥质粉砂岩及含绢云母斜绿泥石细粒石英砂岩，上部为浅灰色薄层含石英粉砂水云母黏土岩夹数薄层细粒石英岩 20.46m

———— 整 合 ————

下伏地层 二叠系鹅顶组中层细晶灰岩

3）牛笛岭剖面

剖面情况见上述。自上而下为（图3-49）：

未见顶

南龙组 总厚度>495.4m

10. 灰色薄层砂质泥岩夹少量石英细砂岩，含腕足类、苔藓虫、海百合茎等化石 11.8m

9. 灰色、灰黄色中厚层中—细粒长石岩屑砂岩夹含生物屑含砾、钙质中—细粒长石岩屑砂岩、粉砂质泥岩，可见水平层理，富产腕足类 *Leptodus richthofeni*，Spiriferinidae，*Oldhamina* sp.，䗴类 *Parafusulina* sp. 双壳类 *Ensipteria* sp.，*Nuculopsis* sp.，植物 *Tingia* sp.，海百合茎 *Cyclocyclicus* sp. 21.8m

8. 灰色、深灰色薄层粉砂质泥岩夹粉砂岩（4∶1），产海百合茎化石 10.7m

7. 灰色、深灰色中薄层细粒岩屑砂岩夹粉砂质泥岩（4∶1） 27.9m

6. 上部为深灰色含钙质粉砂质泥岩，具条带状构造；下部深灰色中层泥质粉砂岩与粉砂质泥岩互层，并构成条带，夹细粒长石岩屑砂岩，含白云质泥质细粒长石岩屑砂岩 30.4m

5. 浮土掩盖 28.1m

4. 灰黄色、暗灰色中厚层粉砂质泥岩，下部含少量泥质粉砂质结核，条带状构造、水平层理发育；中上部浮土掩盖，露头零星出露。在相对应的层位中产丰富的头足类化石 *Shouchangoceras* sp.，*Shangraoceras* sp.，*Hefengnautilus* sp.，以及海百合茎 364.7m

———— 整 合 ————

下伏地层 鹅顶组薄—厚层含燧石微晶生物屑灰岩

2. 岩石地层特征

（1）南好组（C_1n）

由韩哲元等（1976）创名，陈哲培（1987）介绍。建组剖面在保亭县南好地区，岩性主要为浅灰色、灰褐色和灰白色千枚状绢云母粉砂岩、石英砂岩、粉砂质板岩，底部为含砾石英砂岩或硅质砾岩，厚度>218m，属于岩关期沉积。后经陈哲培等（1985、1987）研究，其最大厚度为1471m或1525m。陈哲培等（1997）在清理岩石地层单位时，赋予南好组以新的含义，将原南好组（广东省地矿局，1988）顶部9.95m厚的深灰色细砂岩划出并入青天峡组底部，并以砂岩的底界线作为南好组与青天峡组的划分界线。南好组代表早石炭世沉积。代表性剖面为南兵剖面，青天峡剖面为参考性剖面，岩性主要为石英砂岩、砂岩与板岩、粉砂质板岩呈不等厚互层并夹少量粉砂岩，顶部为细砂岩与板岩呈薄层互层，底部为砾岩、含砾不等粒石英砂岩。产腕足类、双壳类、头足类、苔藓虫化石。与下伏足赛岭组不整合接触，与上覆青天峡组整合接触。厚62.1~709.9m。

（2）青天峡组（C_2q）

广东省区测大队（1964）在海南东方市江边创名，中南区域地层表编写小组（1974）介绍。标准剖面在东方江边青天峡谷之西公路上，原义系指位于晚泥盆世至早石炭世早期岳岭群之上，下石炭统石岭群之下的一套滨海至浅海碎屑岩夹碳酸盐岩，代表早石炭世晚期沉积。底部为砾岩与岳岭群呈不整合接触，下部为杂色板岩和石英砂岩，上部为灰褐色、深灰色和紫灰色石英砂岩、板岩夹灰岩透镜体。陈哲培等（1997）在清理岩石地层单位时，对青天峡组的定义进行修订，将原青天峡组下部属于大塘阶的部分划归南好组，属上石炭统的部分称为青天峡组，是位于下石炭统南好组之上的一套滨岸至远滨相碎屑岩夹碳酸盐岩，代表晚石炭世沉积。其岩性为板岩、砂质板岩与石英砂岩不等厚互层，底部为细砂岩夹板岩、条带状结晶灰岩。产腕足类、牙形石、双壳类等化石。厚101.55~504m。

（3）峨查组

广东省区测大队（1964）在东方市江边乡峨查村创建，中南区区域地层表编写小组（1974）介绍。建组剖面（代表性剖面）位于海南省东方市江边乡峨查村南光片山，是指上石炭统光片山组燧石灰岩、砂质灰岩之上一套海相碎屑岩地层。广西区调队（1991）认为光片山组应为峨顶组之误，原光片山组之上的峨查组实为南龙组。汪啸风等（1992）把在江边青天峡组剖面由广东区测大队（1964）创立的下石炭统石岭群进行解体，将其下亚群划为下二叠统峨查组，与下伏青天峡组及上覆石岭灰岩（即石岭群上亚群）呈断层接触。广西区调队（1991）进一步把即石岭群上亚群划归峨顶组。陈哲培等（1997）沿用广西区调队（1991）的划分方案，以青天峡剖面为代表性剖面，鹅顶西坡剖面为参考性剖面对峨查组进行厘定，系指覆于上石炭统青天峡组之上、伏于下二叠统鹅顶组之下之间的一个岩石地层单位，岩性以石英砂岩与板岩、砂质板岩呈不等厚互层为主，底部以硅化生屑灰岩的出现作为峨查组的底界，且与下伏青天峡组呈整合接触。顶部以细砂质板岩的消失与上覆峨顶组生屑灰岩分界。产腕足类、牙形石、䗴类及海百合茎化石。

（4）鹅顶组

广东省区测大队（1964）据东方市山猪岭、峨查、鹅顶资料创建。中南区区域地层表编写小组（1974）介绍。陈哲培等（1997）指定青天峡剖面为峨顶组代表性剖面，娜姆河剖面为该组参考性剖面。系指整合覆于峨查组之上、南龙组之下的碳酸盐岩夹少量碎屑岩地层。岩性为生物粉晶灰岩、生物屑灰岩夹含燧石纹层灰岩。富产䗴类、腕足类、有孔虫、海百合茎化石。

（5）南龙组

广东省区测大队（1964）在东方市南龙村创建南龙群；谌建国等（1981）改称南龙组。标准地点在东方黎族自治县江边乡南龙村附近，系指覆于鹅顶组之上的一套碎屑岩地层。广东省地矿局（1988）根据在江边乡昌化江西岸以泥质岩为主的地层中发现菊石动物群，将南龙组下段命名为江边组，代表早二叠世茅口期沉积；上部归为南龙组，代表晚二叠世早期沉积。陈哲培等（1997）在清理岩石地层单位时，对南龙组的定义进行修订，指定南龙村剖面为代表性剖面，岩石上部为泥质岩与中粒长石石英砂岩互层夹少量杂砂岩；下部细砂岩、粉砂岩与泥岩互层夹杂砂岩，产䗴类、腕足类、头足类及植物化石，未见顶。以灰黄色泥岩的出现作为该组底界的划分标志。

（二）地层划分与对比

1. 南好组（C_1n）

控制南好组的剖面除了坡兰沟剖面出露比较完整外，其余剖面或未见底或未见顶。但总体来看，南好组空间延伸上岩石组合特征基本可以对比，下部并常以一套底砾岩分别与下古生界不同层位岩石呈不整合接触（图3-51）。

海南岛南面南好地区（公井岭Ⅰ剖面东南段、锌山剖面）下部为变质石英粉砂岩、粉砂质板岩，偶夹中薄层变质细砂岩；上部为变质长石石英砂岩与粉砂质板岩呈不等厚状。底部以含砾石英砂岩与下伏足赛岭组灰色薄层粉砂质板岩，呈不整合接触关系。

海南岛东面琼海市（东乐队-江南队剖面、峰膏岭剖面）主体为板岩、含粉砂质板岩与变粉砂岩、变细砂岩不等厚互层。变质较强的地段可形成千枚状板岩、千枚岩，有少量弱变质的酸性沉凝灰岩。其岩石组合特征与南好地区基本类似。

海南岛西北面儋州市和庆镇一带（南坟岭剖面）下部为（含炭）绢云母千枚岩、石英千枚岩，底部为变质复成分砾岩，砾石成分复杂，呈次棱-棱角状，且常呈平行状定向排列；上部为（含炭）变质绢云母粉砂岩、绢云母石英岩，夹绢云母千枚岩。往南至兰洋农场（上楼坡-大岭剖面、松坡岭剖面），下部相变为为硅质（砂）砾岩与变质细砂岩不等厚夹弱变质炭泥质石英粉砂岩，上部则为中细粒变质石英砂岩、绢云板岩和透辉大理岩。即由北至南，具有砂质含量增加、沉积粒度变大的趋势。

海南岛西面白沙县金波农场（如翁-金波剖面），岩石整体为一套变质砂岩（细砂岩、粉砂岩）组合，中部见（含长石）粉细杂砂岩分布，偶夹粉砂质泥岩和千枚岩。

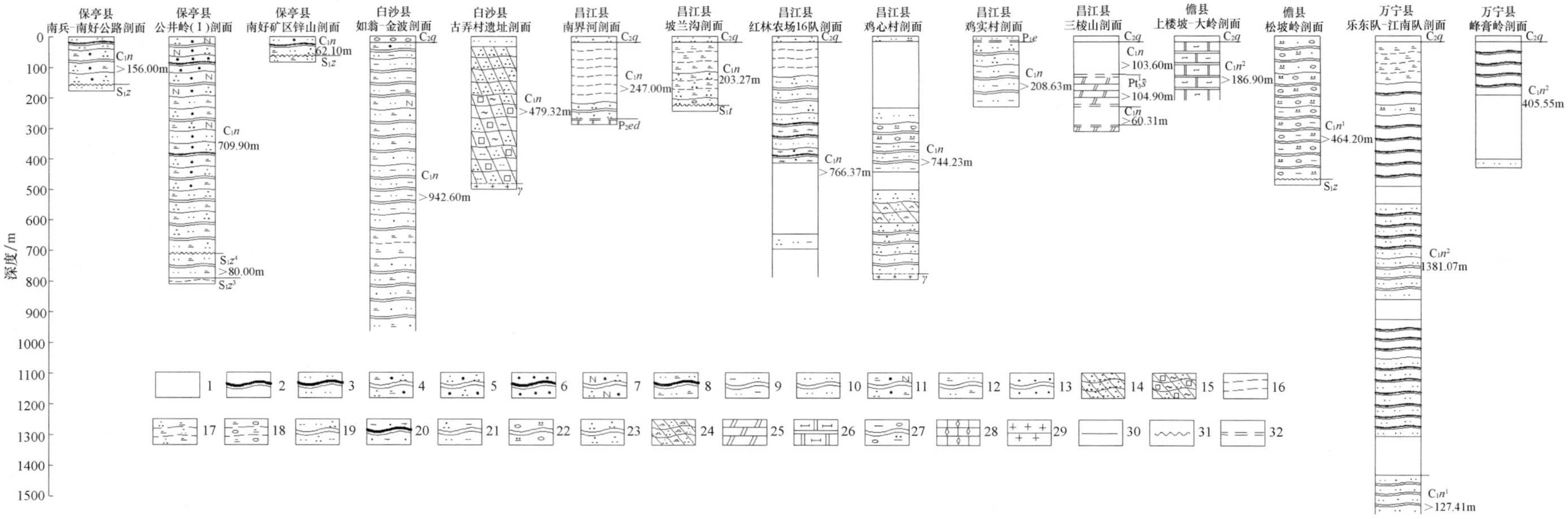

图 3-51　海南岛下石炭统南好组剖面柱状对比图

Fig. 3-51　Columnar correlation of the Lower Carboniferous Nanhao Formation Sections across the Hainan Island

1—掩盖（浮土）；2—板岩；3—粉砂质板岩；4—绢云石英砂岩；5—石英砂岩；6—斑点板岩；7—变质长石石英砂岩；8—粉砂质绢云母板岩；9—变质泥质粉砂岩；10—变质粉砂岩；11—绢云长石石英细砂岩；12—绢云母粉砂岩；13—变质细砂岩；14—石英角岩；15—黄铁矿绿泥石石英角岩；16—千枚岩；17—石英绢云千枚岩；18—含砾绢云母千枚岩；19—黑云石英粉砂岩；20—红柱石斑点板岩；21—石英粉砂水云母粘土岩；22—砾屑硅质岩；23—石英岩；24—堇青石石英角岩；25—白云岩；26—透辉大理岩；27—变质泥硅质砾岩；28—结晶灰岩；29—花岗岩；30—整合接触；31—不整合接触；32—断层接触

海南岛西北面坡兰沟-河叉岭（坡兰沟剖面）一带，下部为厚层变质砾岩；中部为石英绢云母千枚岩与变质石英粉砂岩不等厚互层；上部为薄层（含砾）绢云母千枚岩。往东至鸿沟岭一带，下部则变为厚层含砾石英岩（未见底），向上为厚层石英岩与千枚岩不等厚互层，顶部以石英岩与上覆青天峡组泥晶白云岩整合接触。河叉岭一带，则为变质石英质砾岩、含砾石英片岩、石英片岩、绢云母石英千枚岩等组成曲线河基本层序的沉积。

海南岛西面昌江县石碌镇地区（红林农场16队剖面、古弄村遗址剖面、鸡心村剖面、鸡实村剖面、三棱山剖面），下部由于花岗岩侵蚀，未见底，岩石并发生了接触变质作用，形成角岩-薄层粉砂岩组合。上部岩石为沉积粒度更细的薄层粉砂质泥岩、水云母黏土岩、薄层粉砂岩，夹多层灰岩、砂砾屑硅质岩、粗砂岩。砂砾屑硅质岩多呈透镜状，多具底冲刷现象，砾屑成分为灰岩，为灰岩成岩期硅化的产物。变质程度强部位，出现板岩和千枚岩类岩石。

海南岛南西面（南界河剖面）下部为厚层变质细砾岩；中部为深灰色中厚层变质细砂岩、岩屑石英砂岩、石英砂岩夹硬绿泥石绢云千枚岩、千枚岩；上部为千枚岩夹薄—厚层石英砂岩、粗—中粒石英砂岩、含砾岩屑石英砂岩，具条带状构造，发育变余水平层理。

2. 青天峡组（C_2q）

海南岛西南面（青天峡剖面、娜姆河剖面、南界河剖面），岩石层序基本相似，下部为变质砾岩、岩屑砾岩夹粉砂质泥岩、含砾砂岩、含泥砾砂砾岩、细砾岩，上部总体为砂岩与板岩不等厚产出。青天峡剖面下部夹厚28.4m沉凝灰岩、凝灰质砂岩，中部出现岩屑长石砂岩、岩屑砂岩和特征变质矿物硬绿泥石，水平层理发育；上部夹厚1.4m的灰岩透镜体。南界河剖面中部出现岩屑石英砂岩、细粒石英杂砂岩。产腕足类、头足类、苔藓虫、双壳类、苔藓虫、海百合茎等。

海南岛北西面（坡兰沟剖面）下部为厚层石英岩，变质中细粒石英砂岩，向上为夹灰白色大理岩透镜体的绢云母千枚岩和厚层透辉石、透闪石石英角岩。

海南岛西面华栗山剖面下部为含燧石灰岩，上部为含燧石白云岩。其东面昌江县红林农场16队剖面下部为薄层含水云母细粒石英砂岩，中部为千枚岩夹粉砂岩，上部主体为粉砂质水云母黏土岩与粉砂岩不等厚产出，砂岩局部含长石。

海南岛东北面（上番开村剖面）为厚层—块状微晶大理岩与角岩不等厚产出；发育潮汐层理、微水平层理。

海南岛南东面（东乐队-江南队剖面）主体为板岩、千枚岩夹片理化粉-细砂岩，底部为变质石英砾岩，上部夹厚约29.5m的片理化沉凝灰岩。

青天峡组总体自下而上沉积粒度呈由粗至细的变化，砂岩成分成熟度也趋于增高。底部并常出现石英细砂岩，区域延伸比较稳定，是良好的标志层。沉积厚度变化明显，在江边厚101.55m、王下厚282m，燕窝岭厚337m（图3-52），局部并超覆南好组，直接叠覆于志留系陀烈组之上。但中部昌江华栗山和儋州上番开村一线出露厚200~300m的局限碳酸盐岩台地相灰岩或大理岩类岩石，空间相变明显。

3. 峨查组（P_1e）

主要分布于琼西江边、王下、南界河、河叉岭等地。厚度以王下一带最大，达1001m，往南北两侧变薄，在江边厚56m、青天峡谷厚383m、南界河厚328m、石碌厚328~359m（图3-53）。

岩石整体自下而上，沉积粒度逐渐变细。但岩石横向上变化较大。北面河叉岭—南界河一带，沉积粒度较粗，由砂岩和板岩、千枚岩或页岩组成两个正层序沉积韵律，偶夹灰岩透镜体；底部以含砾砂岩或石英岩叠覆于青天峡组之上。南面王下一带上部夹含生物屑微晶灰岩透镜体，中厚层石英砂岩中双向交错层理极为发育，下部夹1~2m厚的层间砾岩，底部为灰白色、灰绿色结晶灰岩（层位稳定，厚4~7m，灰岩层中化石稀少，仅见少量海百合茎，偶见腹足类化石）。自王下往北至燕窝岭地区，砂岩增多，粒度变粗，上部出现了逆序沉积韵律；中下部以产海百合茎碎屑为主，夹8m厚富含舌形贝类（*Lingulella*）化石层。

峨查组与下伏青天峡组在王下一带及以北地区为连续过渡沉积，分界线上下的岩性变化十分明显。

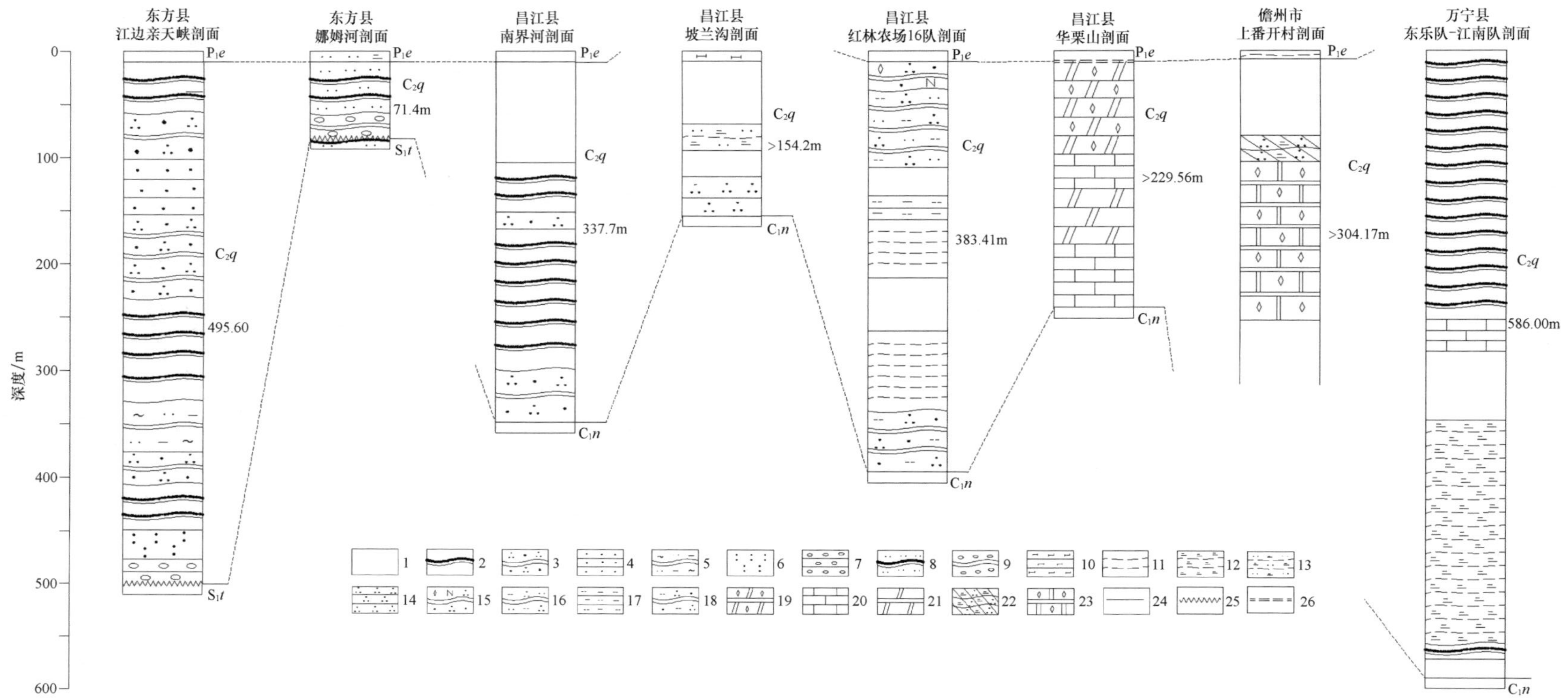

图 3-52 海南岛上石炭统青天峡组剖面柱状对比图

Fig. 3-52 Columnar correlation of the Upper Carboniferous Qingtianying Formation Sections across the Hainan Island

1—掩盖(浮土);2—板岩;3—石英砂岩;4—细砂岩;5—绿泥石化粉砂质泥岩;6—凝灰岩;7—砾岩;8—粉砂质板岩;9—变质砾岩;10—透辉石;11—千枚岩;12—绢云千枚岩;13—粉砂质绢云千枚岩;14—石英岩;15—方解石长石石英细砂岩;16—水云母石英粉砂岩;17—水云母粘土岩;18—水云母石英砂岩;19—微晶白云岩;20—灰岩;21—白云岩;22—绢云母石英角岩;23—微晶大理岩;24—整合接触;25—不整合接触;26—断层接触

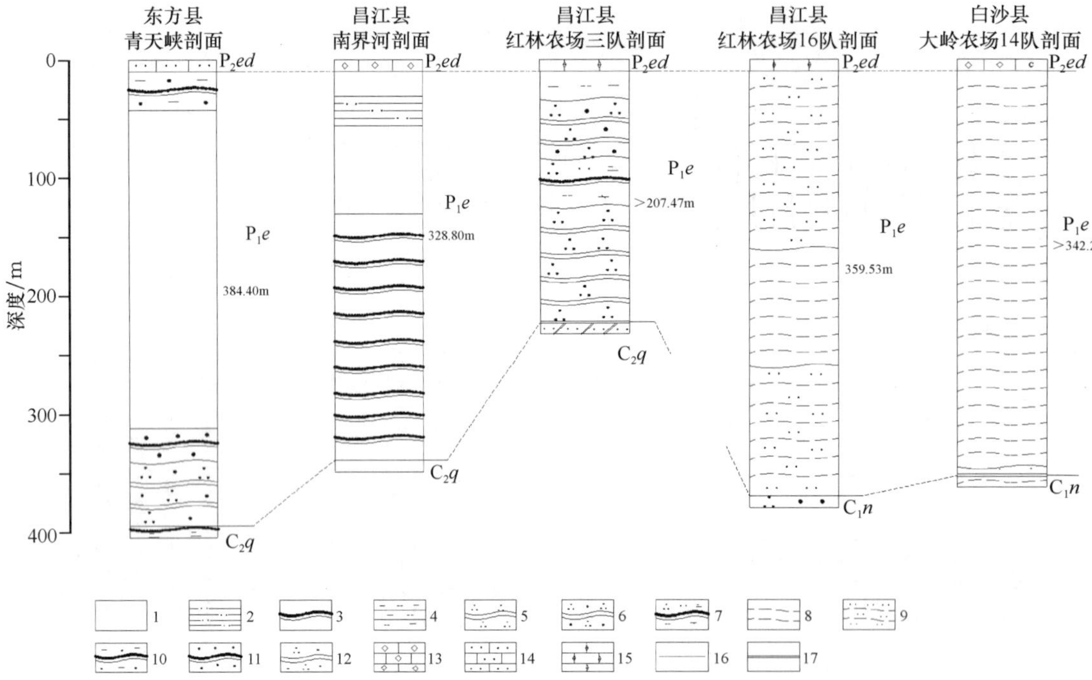

图 3-53 海南岛下二叠统峨查组剖面柱状对比图

Fig. 3-53 Columnar Correlation of the Lower Permian Echa Formation sections across the Hainan Island

1—掩盖（浮土）；2—粉砂质页岩；3—板岩；4—水云母黏土岩；5—石英岩；6—石英砂岩；
7—石英粉砂绢云母黑云母红柱石板岩；8—千枚岩；9—粉砂质千枚岩；10—细砂泥质板岩；11—砂质板岩；12—石英细砂岩；
13—微晶灰岩；14—粉晶灰岩；15—泥晶灰岩；16—整合接触；17—断层接触

下伏青天峡组顶部为灰色板岩夹灰色中厚层石英砂岩或者互层。上覆峨查组底部为灰色、深灰色含生物屑微晶灰岩夹硅质岩（厚 2.9m）；向南西到俄乐村一带为含燧石含生物屑微晶灰岩、生物屑灰岩（厚约 7m）；至王下明望坡、洪水一带为灰白色、灰绿色厚层具条带状含生物屑结晶灰岩（厚4~7m）。灰岩沿走向断续展布，延绵十几千米，成层出现，可连成一线，层位稳定盛产早二叠世牙形石（*Sweetognathus*, *Rabeignathus*）、䗴类（*Parafusulina*），以及海百合茎碎屑等化石。其他地区可见峨查组以较粗的碎屑岩假整合于下伏南好组、青天峡组之上，峨查组底部多为厚度不大的石英质砾岩或含砾石英粉砂岩。

4. 鹅顶组（P_2ed）

主要分布于南部燕窝岭-峨贤岭、王下一带、中部石碌-白沙地区以及北面白沙邦溪地区。鹅顶组岩性标志明显，界线清晰。主要由浅灰色、灰色、深灰色、灰黑色中厚层—块状微晶生物屑灰岩、含燧石条带（或结核、团块）微晶生物屑灰岩、含生物屑微晶灰岩、含燧石微晶灰岩、含白云质生物屑微晶灰岩等组成。碳酸盐岩类经风化剥蚀后形成峰林丛生、秀丽而壮观的岩溶地貌，南界河、峨贤岭、王下一带最为壮观。燕窝岭地区夹 2~3 层粉砂泥岩、石英砂岩及生物颗粒灰岩；白沙元门地区灰岩中水平纹层发育。该组的厚度变化较大，燕窝岭南界河地带为 543.6m，王下地区 408.5m，石碌-白沙邦溪地区厚度变化于 29.29~246.17m。

其生物化石南部地区丰富，中、北部也可见及。在纵、横向上变化大，含量差异显著，局部富集成层。南部燕窝岭南界河地带，䗴类丰富，但属种单调，以 *Parafusulina* sp. 为主，其次为 *Neomisellina* sp.，*Schwagerina* sp.；在王下一带广大地区，生物以海百合茎碎屑为主，仅在底部产丰富的䗴类化石，以 *Robustoschwagerina* sp.，*Nankinella* sp.，*Schwageina* sp. 等为主；其他地区（中、北部地区）仅发现数量不多的珊瑚类、䗴类、牙形石化石。

5. 南龙组（$P_{2-3}n$）

主要分布于江边青天峡谷、娜姆河、石碌、大岭农场等地。

南面江边村北面，上部夹厚约 6m 的粉晶白云岩，岩石水平层理发育。厚度>495.4m。石碌水库一带，由于接触变质作用，形成细砾屑二透石石英角岩、含大理岩条带的石英角岩、透辉石角岩等，厚度>1079.89m（图3-54）。

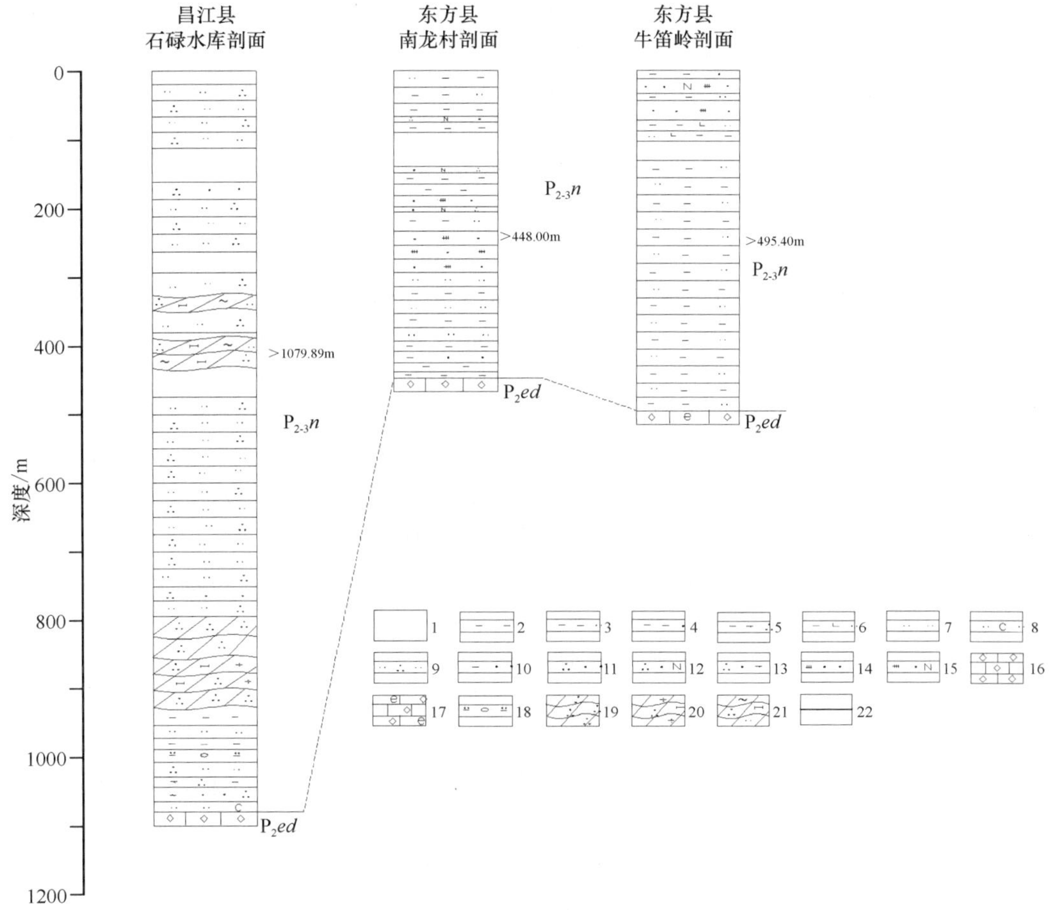

图3-54 海南岛中—上二叠统南龙组剖面柱状对比图

Fig. 3-54 Columnar correlation of the Middle-Upper Permian Nanlong Formation sections across the Hainan Island

1—掩盖（浮土）；2—黏土岩；3—粉砂质黏土岩；4—砂质黏土岩；5—含红柱石石英黏土岩；6—含钙质粉砂质黏土岩；7—粉砂岩；8—含炭粉砂岩；9—石英粉砂岩；10—泥质细砂岩；11—石英砂岩；12—长石石英砂岩；13—含红柱石石英砂岩；14—杂砂岩；15—长石岩屑砂岩；16—结晶灰岩；17—含燧石微晶生物屑灰岩；18—砾屑硅质岩；19—石英角岩；20—石英透辉石透闪石角岩；21—含石英粉砂含绿泥石透辉石角岩；22—整合接触

（三）古生物特征及年代地层

石炭系化石主要分布于东方市江边、昌江县石碌镇、儋州市兰洋镇，保亭县南好地区、万宁市和乐镇一带也见零星分布。化石种类有：腕足类、双壳类、珊瑚、海百合茎、苔藓虫、有孔虫、牙形石、笔石及植物化石等。

1. 南好组（C_1n）

（1）保亭县南好地区

南兵—南好公路剖面2层上部粉砂质板岩中产腕足类化石：（Ⅲ31）*Tenticospirifer* sp.，*Camarotoechia* sp.，*Fusella* sp.，*Leptagonia* sp.，*Chonetipustula* sp.，*Cyrtospirifer* sp.（赵汝旋鉴定），腹足类 *Euomphalus* sp.（南颐鉴定）。其中 *Fusella* 曾见于长江三峡地区下石炭统金陵灰岩（汪啸风等，1992），共生的早石炭世早期牙形石计有 *Siphonodella isosticha*，*S. cooperi*，*Polygnoathus inornatus* 等。桂林南边村剖面边村组上部66层产 *Fusella*，伴生的牙形石（早石炭世杜内期）有 *Siphonodella Praesulcata*，*S. sulcata*，*Polygnathus* cf. *communis* 等（俞昌民等，1988）。以上表明 *Fusella* 的地质时限为早石炭世。*Leptagonia* 见于贵州下石炭统草海组及广东下石炭统大赛组中，赵汝旋（1988）在广东乐昌大赛组中，于 *Eochoristites neipengtaiensis-Martiniella elongata* 组合带（许寿永等，1987）的基础上建立了三个延限带，即

Rhipidomella michelini 延限带，*Leptagonia analoga* 延限带和 *Fusella tornacensis* 延限带，前两个带分布在大赛组上段，后一带主要分布在大赛组顶部（南颐，周国强等，1996）。*Chonetipustula* 目前仅见于下石炭统。*Tenticospirifer*，*Cyrtospirifer* 地质时限为晚泥盆至早石炭世，*Camarotoechia* 地质时限为志留纪—石炭纪。综上所述，Ⅲ31 所产化石时代应为早石炭世。此外，宜昌地质矿产研究所倪世钊等 1977 年考察该剖面时，曾在南好组内采获 *Eochoristites* sp.，*Cyrtospirifer* sp.（汪啸风等，1992）。汪啸风等（1992）在重新研究保存在海南地矿局地质陈列馆腕足类化石标本（D1899-1 化石点，位于南好组代表性剖面南段）后，鉴定有 *Spirifer* cf. *latior*。其中 *Eochoristites* 广泛分布于华南地区下石炭统刘家塘组及金陵灰岩或与之相当的层位中，在湖南湘中地区见于下石炭统马栏边组。*Spirifer* 地质时限为晚泥盆世—早石炭世。由此进一步证实Ⅲ31 化石点及以上地层为下石炭统。

南兵剖面第 2 层底部粉砂岩中产腕足类、腹足类及海百合化石（Ⅲ17）。腕足类有 *Tenticospirifer* cf. *supervilis*，*Ptychomaletoechia* sp.（赵汝纯鉴定），*Fusella* sp.，腹足类：*Euomaphalus*，海百合茎 *Pentagonocyclicus* sp.，*Cyclocyclicus* spp.（南颐鉴定）。腕足类 *Tenticospirifer* cf. *supervilis* 在湖南见于上泥盆统锡矿山组，*Ptychomaletoechia* 则见于泥盆纪；*Fusella* 时限为早石炭世，海百合的地质历程较长，但石炭纪却是最繁盛的时期之一。以上表明Ⅲ17 化石点所产化石，有晚泥盆世化石与早石炭世化石共生的特点。

1966 年海南地质大队在南好矿区进行铁、多金属矿勘查时，在锌山矿段的浅井、地质点及钻孔岩心中发现不少腕足类及海百合茎化石，经中国科学院南京地质古生物研究所腕足类鉴定组鉴定计有腕足类（1511、13A-413-1）*Eochoristites* sp.，（QJ-98、12-13-538、12-13-562）*Fusella* sp.，（12-13-562）*Cyrtiopsis* sp.，（12-13-562）*Spirifer* sp.；海百合茎（ck22-5、6 等）*Cyceocyceins*（穆恩之鉴定）。经宜昌地质矿产研究所和海南省地质调查院（2004）研究，建立 *Eochoristites-Fusella* 组合，该组合以 *Eochoristites* sp.，*Fusella* sp.，*Spirifer* sp. 为主，其层位应大致可与华南地区 *Eochoristites neipentaiensis-Martiniella elongata* 组合带（许寿永等，1987）对比，其共同特点是：*Eochoristites*，*Fusella*，*Spirifer* 等广泛见于华南及西北地区的杜内阶或相当的层位中。此外，*Fusella* 在西欧及前苏联广大地区的杜内阶常见。虽然 *Spirifer* 的地质时限为晚泥盆世—早石炭世，但锌山南好组 *Eochoristites-Fusella* 组合中并无 *Cyrtospirifer*，*Tenticospirifer* 的报道，表明该区南好组第 4 层所产 *Eochoristites-Fusella* 组合，其层位应大致可与华南地区 *E. neipentaiensis-M. elongata* 组合带或 *E. neipentaiensis-M. chinglungensis* 组合带对比，其地质时代为早石炭世早期是适宜的。海百合茎 *Cyceocyceins* 则地质历程较长，不能确定时代。从此化石组合特征看，地质时代应为早石炭世岩关期。

宜昌地质矿产研究所和海南省地质调查院（2004）根据化石出露部位的 12 号勘探线测制了锌山短剖面（图 3-55），剖面中的第 4 层为粉砂岩及粉砂质板岩，厚 62m，前述早石炭世早期的腕足类化石均产在这一层中，第 3 层顶部掩盖中下部为粉砂岩、粉砂质板岩，厚 21m. 在基岩出露的中上部粉砂质板岩中（D3454）发现早志留世特列奇期类标准化石 *Xinanospirifer flabellum*（曾庆銮鉴定）。

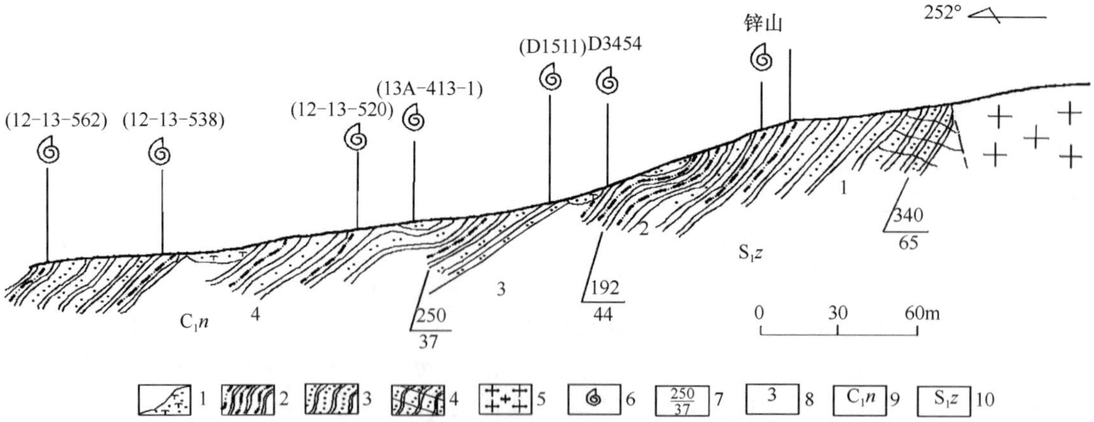

图 3-55　海南岛南好矿区锌山南好组实测剖面图

Fig. 3-55　Measured section of the Nanhao Formation, Xinshan, Nanhao mine, Hainan Island

1—掩盖；2—粉砂质板岩；3—变质砂岩；4—角岩化粉砂岩；5—花岗斑岩；6—化石点；7—产状；8—分层好；9—南好组；10—足赛岭组

综上所述，南兵剖面Ⅲ31化石点及该层位以上所产腕足类及锌山所产腕足类化石的地质时代应为早石炭世岩关期，但考虑南兵剖面未见顶，1：5万志仲幅区调测制的公井岭剖面南好组经宜昌地质矿产研究所和海南省地质调查院（2004）修订，其厚度为709m，也就是在南兵剖面之上至少有500m以上地层无化石资料，因此将南好组时代厘定为早石炭世较合理。关于南兵剖面Ⅲ17化石点所产泥盆纪腕足类化石，化石层位之下尚有24m的含砾不等粒石英砂岩。南兵剖面南好组时代应为晚泥盆世锡矿山期至早石炭世岩关期。但晚泥盆世化石与早石炭世化石共生的特点为泥盆纪的存在提出了质疑。此外，同为南好矿区范围内的锌山剖面并未发现泥盆纪腕足类化石，而且在相邻层位粉砂岩和粉砂质板岩中分别产早石炭世岩关期及早志留世的腕足类化石，该剖面第4层底部粉砂岩几乎是二者的年代地层界线层型，而没有泥盆纪地层存在的空间。南兵剖面北东方向的万宁东岭一带南好组底部所产早石炭世岩关期腕足类化石 *Eochoristites* sp. 可与南好组岩关期腕足类化石对比，也没有发现泥盆纪地层。因此南好组的年代地层暂划归下石炭统。

（2）东方市江边地区

江边剖面中南好组（即1：5万坝王岭、猴猕岭区调报告中的原青天峡组下段），生物化石以富含腕足类为特色，主要有 *Megachonetes zimlnermanni*，*Eochoristites neipentaiensis alatus*，*E.* sp.，*Pustula scabriculiformis*，*Orthotetes keokuk*，*O.* cf. *keokuk*，*O.* cf. *borodencorensis*，*O. hindi*，*O.* sp.，*Dictyoclostus* cf. *pinguis*，*D. semireticulatus*，*D.* cf. *semireticulatus*，*D.* cf. *cranfordsvilensis*，*D.* sp.，*Linoproductus tenuistriatus*，*Antiquatonia* cf. *hindi*，*Balakhonia* cf. *yunnanensis*，*B. kokdschcrensis*，*Paraphorhynchus* cf. *elongatus*，*Neospirifer* cf. *liangchowensis*，*Punctospirifer insculptus*，*P.* sp.，*Marginifera* sp.，*Martinia* sp. 等。双壳类 *Palaeoneilo anthracoenilides*，*Streblopteria* sp.，*Schizodus* sp.，头足类 *Michelinoceras* sp.，苔藓虫 *Fenestella* sp.。

其中腕足类 *Dictyoclostus* cf. *pinguis*，*D. semireticulatus*，*D.* cf. *cranfordsvilensis*，*Linoproduc tustenuistriatus*，*Antiquatonia* cf. *hindi*，*Balakhonia* cf. *yunnanensis*，*B. kokdschcrensis* 等的地史历程为早石炭世晚期。*Orthotetes* cf. *keokuk*，*O.* cf. *borodencorensis*，*O. hindi* 等在南好组中上部均有出现，主要见于杜内阶上部至维宪阶中。*Punctospirifet insculptus*，*Pustulascabriculiformis*，*Paraphorhynchus* cf. *elogatus*，*Neospirifer* cf. *liangchowensis*，*Megachonetes zimmermanni* 等属种均是我国不少地区大塘阶中重要分子和常见化石。*Eochoristites meipentaiensis* 是岩关阶上部 *Eochoristites neipentaiensis*-*Martiniella elongata* 组合中的重要化石。据此广西区调队（1991）认为其地质时代为大塘期至黄龙期底部。

此外，周国强、陈培权（1988）在江边剖面以东的相当层位（剖面的中下部）发现有腕足类 *Temticospirifer* sp.，*Punctospirifer* sp.，*Spirifer* aff. *attenatus*，*Plicochonetes* cf. *elegans*，*Megachonetes* sp.，*Cyrtospirifer* sp.，*Schuchertella* sp.，苔藓虫 *Fenestella* aff. *yabei* 等，认为其化石组合年代为岩关期。

因此，江边剖面所产化石所代表的地质年代为岩关期至德坞期，年代地层划为下石炭统。

（3）昌江县石碌地区

这一地区控制南好组的剖面有三棱山、南界河等8条剖面。

三棱山剖面所含古生物化石较为丰富，有双壳类、苔藓虫、笔石及植物化石。双壳类有 *Palaeoneilo* sp.，*Myalinidae*，*Euchondria*，*Limipecien*。张仁杰等（1990）认为以早石炭世的种为多，仅有少量晚石炭世的种，其时代为早石炭世；笔石有 *Dictyonema* exgr. *frapont*（= *D. shiluense*），*D. hainanense*，*D.* sp. 等，穆恩之等（1981）认为其时代为早奥陶世，但经汪啸风（1992）对其标本进行重新研究，认为穆氏所采标本与比利时早石炭世维宪期的 *Dictyonema fraiponti* 相似，时代属早石炭世晚期似无问题。莫壮观（1981）报道的植物化石有：*Neuropteris* sp.，*Pecopteris* sp.，*Rhodeopteridium* sp.，认为其时代为早石炭世—中石炭世。冯少南（1990）研究了海南地矿局采自三棱山剖面的植物陈列标本，计有 *Mariopterisacuta forma obtuse*，*Neuripteris* sp.，*Cordaites* sp.，认为其时代为早石炭世大塘期。综上所述，三棱山剖面所控制的南好组时代应为早石炭世中期至晚期。

南界河剖面在其底部砾岩之上的千枚岩中产珊瑚、腕足类、双壳类化石。珊瑚据南京地质古生物研究所赵嘉明鉴定，属单带型四射珊瑚，为下石炭统底部的泡沫内沟珊瑚带的常见化石，时代属岩关期。腕足类有 *Dictyoclostus*，*Linoproductus*，*Martinia*，*Rhipidomella*，*Neospirifer*，*Brachythris*，*Brachythyrina* 等，双壳类 *Palaeoneilo*，都是石炭系常见化石。上述化石所反映的地质时代为早石炭世岩关期，将这套地层化石划分为南好组是合理的。

宜昌地质矿产研究所（1995）于昌江鸡实村南剖面上原划归南好组第三段中采获时代意义不大的有孔虫？*Nodosaria* sp.，原划归南好组四段中采获丰富的珊瑚和有孔虫、牙形石化石，珊瑚化石有 *Cystophrentes kalaohoensis*，*Kueichowpora baijinensis*，*Syingopora paralella*，*S. gyacilis*，*S. serenei*，*Kueichowpora panxianensis*，*Kueichowpora* aff. *baijinensis*，*Zophrentites* sp.，*Neolithostrotion*? sp.，*N.* spp.，*N.*？sp. 和 *Pseudouralinia irregularis* 有孔虫有 *Septabrunsiina* sp.，S. *krainica*，*Septatournayella rauserae*；牙形石 *Siphonodella isosticha*；上述所列化石，原第三段中的？*Nodosaria* sp. 延续时间很长（D_3-Q），只能说明产化石的南好组地层晚于中泥盆世；第四段所产的珊瑚等化石，反映含化石层位属石炭系岩关阶无疑，其中下部以 *Cystophrentis kalaohoensis* 为代表的珊瑚化石组合，为我国华南岩关阶中部珊瑚的标准化石带，其层位可与贵州的石炭系革老河组、湘粤的孟公坳组完全对比；上部以 *Pseudouralinia irregularis* 为代表的珊瑚组合为岩关阶上部的重要化石，其层位可与贵州石炭系的汤粑沟组、湘粤的刘家塘组和鄂西的金陵组对比；与珊瑚化石伴生的有孔虫化石，除 *Septabrunsiina* sp. 时限可下延到晚泥盆世外，其他为国内外早石炭世岩关期的常见分子。

2001 年，张仁杰等人对鸡心-鸡实一带地层开展了进一步的研究工作，并于鸡实村一带原划归下石炭统南好组第四段下部灰岩夹层中采获大量牙形石及珊瑚。牙形石有 *Palmastolepis gracilis gracilis*，*P. gracilis sigmoidalis*，*Polygnathus germanus*，*P. communis carinus*，*P.* sp.，*Apatognathus* sp.，*Drepanodus* sp.，*Spathognathodus* sp.，*Ozarkodina* sp.，*Hindeodella* sp. 等 7 属 10 种。珊瑚有 *Cystophrentis kalaohoensis*，*C.* sp.，*Syringopora paralella*，*S. gracilis*，*Kueichowpora baijinensis* 等 3 属 5 种。其中，牙形石 *Palmastolepis gracilis sigmoidalis* 和 *P. gracilis gracilis* 广泛分布在北美洲（美国、加拿大）、欧洲（德国、波兰、奥地利、保加利亚、意大利）、澳大利亚和中国的晚泥盆世法门期（邵东期），它们延续的时限为牙形石标准带 *rhomboidea* 带至 *praesulcata* 带。虽然个别报道后一亚种曾见于下石炭统底部的 *Sulcata* 带，不过一般认为其系再沉积所致，因而不能定论。张仁杰等（2001）认为上述 2 个亚种同时出现可以肯定属于法门期。*Polygnathus* 一属的标本数量最多，其中 *P. germanus* 常见于北美、欧洲、澳大利亚及中国的法门期，时限通常为由 *Rhomboidea* 带至 *Postera* 带，在澳大利亚则由 *Crepida* 带至 *Quadrantinodosa* 带；*P. communis carinus* 在北美、欧洲、澳大利亚及中国的法门期 *Expansa* 带至早石炭世 *Anchoralis* 带均可见及。*Apatognathus* 一属由晚泥盆世延续至早石炭世，*Drepanodus* 和 *Spathognathodus* 的延续时间较长，由奥陶纪至泥盆纪。综观这一牙形石动物群，其时代为晚法门期，很可能属 *Expansa* 带。珊瑚化石以 *Cystophrentis* 为多，包括 *C. kalaohoensis*，*C.* sp.，*S. gracilis*，*Kueichowpora baijinensis* 及 *Syringopora paralella* 3 属 4 种，这些珊瑚化石的时代可能自晚泥盆世—早石炭世（廖卫华论证了 *Cystophrentis* 属晚泥盆世）。

海南省地质调查院（2007）、傅杨荣等（2005）、廖卓庭等（2006）于白沙金波老村原称南好组第四段的泥质粉砂岩中首次发现丰富的腕足类化石，经中国科学院南京地质古生物研究所廖卓庭鉴定为：*Spirifer subgrandis*，*Unispirifer extensus*，*Fusella tornacensis*，*Punctospirifer insculptus*，*Marginatia huaqiaoensis*，*Palaeochoristites* sp.，*Megachonetes* sp.，*Plicochonetes* sp.，*Schuchertella* sp.，*Pustula* sp.，*Sentisia* sp.，*Crurithyris* sp.，*Rhipidomella* spp. 等属种。新发现的腕足动物中 *Spirifer subgrandis*，*Fusellalornacensis*，*Punispirifer* sp.，*Palaeochoristites* sp.，*Sentosia* sp.，*Marginatia* sp. 等种与大量 *Rhipidomella* spp.，*Schuchertella* sp. 等共生，它们具全球性早石炭世早期腕足类组合的典型面貌。该动物群存在大量石燕贝目分子，但没有华南（苏皖湘黔桂等地）晚杜内期 *Eochoristites-Martiniella* 组合中最常见分子，也未见典型的法门期晚期延续至早石炭世的特殊管孔贝 *Syringothyris* 等，因此它既不同于桂林泥盆系—石炭系界线层的南边村组腕足动物群（Yu Chang-ruing，1988），也有别于湘中马栏边组、黔南汤粑沟组、苏皖金陵组的腕足化石组合（张遴信，2000；Liao Zhuo-ling，1995），其层位似应低于后者。虽然岩相的不同也可能导致腕足动物群组合面貌差异。本文仍视之为层位的差别导致腕足动物群组成的差异。与南边村腕足动物群相比，南边村腕足动物群中有许多个体丰度较高、时代意义较重要的属种，是与金波地区腕足动物群共存的，如：*Rhipidomella michenili*，*Aulacella interlineata*，*Fusella tornacensis*，*Unispirifer extensus*，*Spirifer attenuatus*，*Brachythyris chaoi* 等。另有一些属种形态上也很接近，表明两个动物群的地质时代基本同时。总之，此腕足类化石群代表早石炭世早期（杜内期）似无疑问，从而证实金波地区南好组第四段（至少中部）属下石炭统岩关阶。很可能与昌江县鸡实村剖面上的 C－P 间隔带至含珊瑚 *Pseudouralinia irregularis* 牙形刺 *S. cooperi*，*Polygnathus vommunis communis* 等化石的南好组第四段顶部层位大体相当。同时，海南省地质调查院（2007）于白沙金波老村原称南好组第二段、第三段的粉砂岩、砂质板岩中也采获珊瑚、苔藓虫、腕足类等化石（张仁杰等还采获三叶虫），但因保存差，难以准确鉴定。依据其自然下伏于上述含岩关期腕足类化石地层（原称南好组第四段）的关系，此原称南好组第二段、第三段的沉积时代应属早石炭世早期或更早（不排除属晚泥盆世的可能）。值得关注的是，于上述化石层中也首次采获肉鳍鱼（中国科学院古脊椎动物与古人类研究所赵文金鉴定）化石碎片（图版Ⅰ-11-A），但由于所发现的化石碎片太小，未采获关键部位的碎片而不具鉴定意义。有待进一步深入研究。

由上面的讨论可知，昌江鸡实村一带原划归下石炭统南好组第四段岩石中产丰富的大化石及微体化石。其上部含牙形石 *Siphonodella isosticha*，*S. cooperi* 和珊瑚 *Pseudouralin iairregularis*，*Syringoporagracilis*，*S. serenzi*，*Zaphrentis* sp.，*Kueichowpora panxianensis* 等化石，可厘定其属早石炭世早期（岩关期）的沉积；其下部含牙形石 *Palmatolepis gracilis sigmoidales*，*P. gracilis gracilis*，*Polygnathus germanus* 等及珊瑚 *Cystophrentis kalaohensis* 等，可厘定其属晚泥盆世法门期沉积。张仁杰（2001）并据此将下部地层新建上泥盆统昌江组，岩石为碳酸盐岩与陆源碎屑岩的混合沉积，形成于浅海局限台地相至潮坪环境，与上覆下石炭统南好组呈假整合接触。但所划分的昌江组与其上覆的南好组岩石组合特征基本一致，从岩石地层角度出发，本次不采信昌江组的建组方案，主张昌江鸡实村一带南好组为一穿时的岩石地层单位，其沉积时代由晚泥盆世—早石炭世。但考虑到其他地区南好组中目前未采获泥盆纪生物化石（也许工作程度不够或出露太差），因此本次仍将南好组划归早石炭世。

（4）儋州兰洋地区

该区控制南好组的剖面有儋州上楼坡-大岭剖面、松坡岭剖面等 3 条剖面。

上楼坡-大岭剖面在大理岩、结晶灰岩中苔藓虫 *Fistulipora* sp.，孢子 *Foveolatisporites* sp.，*Verrucosisporites microtuberosus* 等。苔藓虫 *Fistulipora* sp. 时代为志留纪—二叠纪。孢粉 *V. microtuberosus* 在我国主要分布在下石炭统，*Foveolatisporites* sp. 产于石炭纪地层中，根据上述化石组合及不整合于下志留统陀烈组之上而将其划为下石炭统。

（5）万宁东岭地区

这一地区控制南好组的剖面有东乐队－江南队剖面、峰膏岭剖面。峰膏岭剖面产四射珊瑚 *Zaphrentites* sp.；腕足类 *Eochoristites* sp.，Spiriferidae；苔藓虫 *Fenestells* sp.；海百合茎 *Cychocyclicus* sp.；植物 *Cordaites* sp.。腕足类 *Eochoristites* 为早石炭世岩关期 *Eochoristites neipentaiensis-Martiniella elogata* 带（许寿永等，1987）或 *Eochoristites chui-Martiniella chinglungensis* 组合（杨式溥，1990）的主要化石，在南好地区南好组选层南兵剖面、锌山剖面及东方市江边剖面均有出现，显示了它与本区的一致性。珊瑚 *Zaphrentites* 在国内外均见于下石炭统。

除上述几个化石产出地外，海南地质综合勘查院（1993）在军营岭南南西 187°约 550m 路线地质点的南好组中找到了较多的腕足类化石及苔藓虫、海百合茎化石。腕足类有（杨德骊鉴定）*Rugosochonetes* cf. *hardrensis*，*R.* sp.，*Schellwinella* sp.，*Punctospirifer* sp.，*Isogramma*? sp.；苔藓虫有（胡兆鉴定）*Polypora* sp.，*Fenestella* sp.；海百合茎有（林彩华鉴定）*Cyclocyclicus* sp.。这些化石保存不够好，鉴定不到种，但基本上可以确定其时代为石炭纪。

最后值得一提的是，2000~2002 年宜昌地质矿产研究所和海南省地质调查院在开展 1∶25 万《乐东县、陵水县》幅区调时，在南兵—南好公路上采获隶属早志留世特列奇期晚期的腕足类、三叶虫、中华棘鱼鳍刺化石等，从而把代表性剖面南兵—南好公路剖面第 1~3 层重新归属为下志留统足赛岭组四段，曾庆銮等（2003）据此认为南好组已具有双重地质时代的岩石地层单位，建议废弃南好组的主张。但本次研究认为，南好组代表性剖面南兵—南好公路剖面产出早石炭世岩关期的 *Eochoristites-Fusella* 化石组合层位为底部第一层（对应于宜昌地质矿产研究所 2000~2002 新测南兵—南好公路剖面的第四层），而采获早志留世特列奇期晚期化石产出层位为新测南兵—南好公路剖面的第二层。上、下地层层位及其相应产出的化石时代是吻合的，两者间并不存在矛盾关系。因此，把代表性剖面南兵—南好公路剖面第1~3层重新归属为下志留统足赛岭组以及废弃南好组的主张并不可取。

综上所述，据已有资料，南好组时代划归早石炭世岩关期证据充足、可靠，昌江鸡实村一带存在晚泥盆世法门期沉积。南好地区代表性剖面产出化石中，也含有泥盆纪的分子，如 *Ptychomaletoechia*，因此南好组应为一穿时明显的岩石地层单位，其沉积时代为晚泥盆世—早石炭世。但考虑到区域上南好组研究程度仍较欠缺，因此本书中仍将南好组划归早石炭世。

2. 青天峡组（C_2q）

前人（海南地质大队，1990；广西区调队，1991；李志宏等，1991）于东方江边青天峡地区青天峡组中采获腕足类、双壳类、牙形石，青天峡组中部—上部，虽产腕足类、双壳类化石，但缺乏具代表性或具重要地层对比意义的典型分子。而青天峡组上部含1个牙形石组合带，主要以 *Streptognathodus elegantulus* 和 *S. oppletus* 为主；*S. parvus*，*Neognathodus bassleri*，*N.* cf. *symmetricus* 等次之，其中 *S. oppletus* 应为达拉阶上部牙形石 *S. oppletus* 带延续而来，因此，该组合带可直接与上石炭统马平阶底部牙形石带直接对比（《中国地层研究二十年》，中国科学院南京地质古生物研究所，2000）。

值得提及的是张仁杰等（2001）于昌江鸡实村原划归称南好组第三段（实为南好组上部原第四段，由于地层变形而使同一处出露不同层位的岩层而易使地层错划）硅化砂屑、砾屑灰岩中采获牙形石 *Mesogondolella clarki* 组合，主要分子包括 *Declinognathodus noduliferus*，*Idiognathoides sulcatus* 和 *Mesogondolella clarki* 等。其中 *Declinognathodus noduliferus* 和 *Idiognathoides sulcatus* 是欧洲、北美和我国华南华北的上石炭统最底部的两个牙形石带的带化石分子。这两个牙形石带自下而上为：*Declinognathodus noduliferus* 和 *Idiognathoides sulcatus* 带。在我国华南如贵州罗甸纳水、广西南丹巴平和甘肃靖远、辽宁本溪等地上石炭统下部（底部）地层中都有报道。*Mesogondolella clarki* 首先见于日本 Kodam 组中部，与蜓 *Fusulinella* 共生。其后在我国贵州罗甸纳水上石炭统下部与 *Profusulinella* 和 *Fusulinella* 共生，由于牙形石带是以最新的分子为准，所以，该层位可能属于 *M. clarki* 带，即上石炭统中下部，可与贵州罗甸纳水剖面的同名带相对比（表3-5）。*Declinognathodus noduliferus* 和 *Idiognathoides sulcatus* 则是 *M. clarki* 带之下层位中的上延分子。

表3-5 海南岛东方市江边青天峡组中上部牙形石动物群与国内外有关地层牙形石带对比

Table 3-5 Correlation of conodont faunas between the central upper Qingtianxia Formation, Jiangbian, Dongfang City, Hainan Island and related stratigraphic conodont zones across the world

海南东方江边			熊剑飞等（1987）贵州罗甸纳水		熊剑飞等（1987）新疆		贵州紫云（康沛泉、王成源、王志浩，1987）		滇西北（董致中、王成源、王志浩，1987）		北美(Scott M. Ritter，1986)		欧洲（Kozur，1977）	牙形刺生物带（中国地层指南，2001，2002）
中二叠统	峨顶组		栖霞组	*Gon.* sp. A And B-*Sw. whitei* A. Z.	棋盘组	*N. sulcaplicatus* Z. *N. pequopensis* Z.	栖霞组	*N. pequopensis-N. toriyamai* A. Z.	砚山组	*N. pequopensis* Z. *Sw. whilei-Neo. bisselli* A. Z.	*Artinskian*	*N. pequopensis* Z.	*Baigendzhinian*	*N. pequop-ensis*
下二叠统	峨查组	*Sw. whitei-R. asymmetricus* 组合	钠水组	*Gon. bisselli -Sw. whitei* A. Z.	克孜里奇曼组	*Sw. whitei* Z.	羊场关组	*Sw. guizhowensis-Sw.* sp. nov. b-Gen. et sp. nov. . A. Z.			*Sakmarian*	*Neo. bisseui-Sw. whi tei* A. Z.	*Aktastinian*	*Sw. whitei* *M. bisselli* *St. barskovi* *St. wabaaunsensis*
上石炭统	青天峡组	*St. elegantulus-St oppletus* 组合	马平组	*St. elongatus-St. gracilis* *St. wabaunsensis* A. Z.	塔哈奇组	*St. elongatus-L. simpler* A. Z.	紫松镇组	*Neo. bisselli* Z. *Neo. bisselli-St. elongatus-St. fucnengen-Sis-St. barskovi* A. Z.	马平组	*St. wabaunsensis St. elongatus* A. Z.			*Sterlita-makian-Tastubian*	*St. wabaun-sensis* *St. elongatus* *St. elegantulus*

此外，儋州兰洋一带，该组上部（二段）产有丰富的珊瑚 *Auloporida*，*Chaetetes* sp.，Gorskyitidae，Syringoporidae，*Syringopora* sp.，Tetraporellidae，*Tetraporinus* sp.，*Gorskyites*? sp.，*Chia*? sp.；层孔虫? *Striatoporinae*；苔藓虫 *Fistulipora* sp.；腕足类；孢粉；虫颚等多门类古生物化石。较为重要的珊瑚见4个属（图版Ⅰ-11-B）：*Gorskyites*，*Chia*，*Tetraporinus*，*Syringopora*。*Gorskyites* 见于世界各地的中泥盆世至早石炭世，在我国和苏联主要见于早石炭世杜内期。该属在我国所见有4种，即 *G. giganteus*，*G. multispinus*，*G. qidongensis* 和 *G. wumingensis*，其中仅前一个种见于华南等地的中泥盆统中，而后3个种均见于华南下石炭统岩关阶。*Chia*，产于我国和苏联的志留纪—早石炭世（林宝玉认为主要见于中泥盆世—早石炭世），迄今在我国共发现15种（含2个亚种），有3种见于华南等地的中泥盆统，另3种见于下石炭统大塘阶寺门组，其余9种皆产于下石炭统岩关阶。所采的 *Chia* 标本虽因重结晶不能进一步详细鉴定，但从其半球状外形，复体中等大小，个体中残留少量床板以及发育联结管等特征来看，最为接近 *Chiacystosa huishuiensis*，该种仅出现于华南下石炭统岩关阶。至于 *Tetraporinus* 在我国虽见于晚志留世—早二叠世，但在苏联主要见于晚志留世—早石炭世。*Syringopora* 虽分布于世界各地的奥陶纪—二叠纪，但我国主要见于志留纪—石炭纪。综上所述，珊瑚组合所提供的时代以归属于早石炭世岩关期（相当于杜内期）为宜，珊瑚专家林宝玉研究员亦倾向于这一意见。同时，河北区调队、海南地质大队（1991）于该段中发现大量孢粉化石，孢粉组合的时代为早石炭世。

3. 峨查组（P_1e）

广西区调队（1991）于南界河、王下地区鹅查组中发现了较为丰富的䗴类（*Parafusulina*）、腕足类（*Lingulella*）等化石（表3-6），但由于化石延续时限较长，不能给出令人信服的时代归属依据。紧邻南侧的江边青天峡地区鹅查组中牙形石很好地解决了此问题。牙形石产于青天峡地区鹅查组底部含生物屑结晶灰岩中，据王志浩、李志宏等人鉴定，主要有 *Sweetognathus whitei*，*S. behnkeni*，*S.* spp.，*Rabeignathus asymmetricus*，*R. hainanensis*，*Anchingnathodus typlcalis*，*Diplognathodus* sp.，*Ozarkodina* sp. 等属种。李志宏（1991）称之为 *Sweetognathus whitei-Rabeignathus asymmetricus* 牙形石动物群（见表3-5）。其中以 *S. whitei* 占主导地位，是北美狼营阶上部 *Gondolella bisselli-S. whitei* 组合带及苏联下阿丁斯克阶 *S. whitei-Gnathodus simplex* 组合带中重要化石。在我国，常见于下二叠统下部，如贵州罗甸纳水组、紫云羊场关组，滇西北砚山组下部，新疆塔里木盆地克孜里奇曼组上部，华北地区太原组下部以及广西“上马平阶”等。*S. behnkeni* 仅见于北美犹他州、得克萨斯西部 Aktastinian 阶中部。拉贝石属 *R. asymmetricus* 在北美得克萨斯等地 Aktastinian 阶 *N. bisselli-S. whitei* 带的顶部至 *Baiqendzhinian* 阶最底部。而在南美洲哥伦比亚早二叠世更为常见。上述牙形石不仅地理分布较广，而且层位稳定。峨查组下部 *S. whitei-Rabeignathus asymmetricus* 牙形石动物群面貌与王成源（1987，1989）所建立的 *N. bisselli-S. whitei* 组合带和北美萨克马尔阶顶部 *N. bisselli-S. whitei* 带以及苏联下阿丁斯克阶 *S. whitei-Gnathodus simplex* 组合带极为相似，层位大致相当，时代应属早二叠世早期（隆林阶）无疑（表3-7）。

中、北部大岭农场、石碌-金波地区仅于峨查组发现少量生物化石（宜昌地质矿产研究所，1995），有头足类 *Agathiceras* cf. *vulgata*，*Altudoceras* sp. 和腕足类 *Rhipidomella* sp.，*Schuckertella*? sp. 及双壳类 *Septimyalina*？sp.，上述生物化石中某些属种，如 *Altudoceras* sp.，仅出现在下二叠统，因此存在着早二叠世沉积是可信的。

4. 峨顶组（P_2ed）

广西区调队（1991）同时在王下地区鹅顶组发现了较为丰富的化石，计有䗴类、牙形石、腕足类、苔藓虫等。以 *Parafusalina-Misellina* 动物群大量富集为主要特征，主要特征分子有 *Parafusulina grape-raensis*，*P. langdaiensis*，*P. sapperi* 等和 *Misellina ovalis*（表3-7）。这些特征分子在华南、华北、中亚、东南亚等地广泛分布，彼此间可直接对比。时代为中二叠世栖霞期。

其中，底部所产䗴类有 *Robustoschwagerina*，*Nankinella*，*Schwagerina* 等，*Robustoschwagerina* 见于广西德保多敬地区下二叠统马平组顶部、栖霞组下部、隆林常么组，贵州普安、晴隆一带包磨山组至栖

表 3-6 海南岛二叠纪生物群对比表

Table 3-6 Correlation of Permian biotas, Hainan Island

地层系统			牙形石	蜓类	腕足类	菊石	珊瑚	古植物
上二叠统	长兴阶	?南龙组						
	吴家坪阶							P. rigida Taeniopteris sp. Annularia mucronata A. crassiuscula Pecopteris arborescens P. hemitelioides Ptcrophyllum sp. Cordaites sp.
中二叠统	茅口阶			*Paratusulina osei-P.* spp. 组合	*Leptodus nobilis-L. richthofeni* 组合	*Shouchangoceras Shouchangense -Altudoceras orientale* 组合		
	栖霞阶	峨顶组		*Parafusulina gruperaensis-P. langdaiensis-Misellina ovalis* 组合			*Wentxellophyllum yalongensies-Polythecalis densicolumis* 组合	
下二叠统	隆林阶	峨查组	*Sweetognathus whitei-Rabeignathus asymmetricus* 组合					

（据宜昌地质矿产研究所、海南省地质矿产局，1992）

表 3-7　海南岛与国内外一些地区二叠系对比表

Table 3-7　Correlation of Permian strata in the Hainan Island and other areas across the world

本书			贵州郎岱 (杨振东,1985)			江苏宁镇地区 (周建平等,1984;詹立培等,1982)			福建西南地区 (福建省地质矿产局,1985)			中亚达尔瓦兹 (Leven,1979~1981)		美国得克萨斯 (Ross & Ross,1979)	
上二叠统	长兴阶	南龙组	上二叠统			上二叠统	大隆组 / 长兴组		上二叠统	大隆组 / 长兴组		上二叠统	?		
													多腊沙姆阶		
	吴家坪阶						龙潭组			翠屏山组			朱尔法阶	Ochonan	
中二叠统	茅口阶		下二叠统	茅口阶	*Neomisellina mulirioluta* 带 *Yabeina gubleri* 带 *Afganella schencki-Neoschwagerina craticulifera* 带 *Cancellina-Neoschwagerina simplex* 带	下二叠统	堰桥组	*Neomisellina-Codonofusiella* 带	下二叠统	童子岩组	*Shouchangoceras-Altudoceras* 带		米底阶	达鲁普阶	Capitanian 亚阶 Wordina 亚阶
							孤峰组	*Waagenoceras* 带 *Kufengoceras* 带					木尔加布阶		
	栖霞阶	鹅顶组		栖霞阶	*Misellina clandiae* 带		栖霞组	*Parafusulina multiseptata* 带 *Schwagerina chihsiaensis* 带 *Nankinella orbicularia* 带 *Misellina clandiae* 带 *Darvasites ordnaitus* 带		文笔山组	*Waagenoceras-Kufengoceras* 带	中二叠统	库别尔干德阶	翁拿得阶	Rood Canyon 组 Cathedral Mt. 组 Skinner Ranch 组
													博罗尔阶		
下二叠统	隆林阶	峨查组		末建阶	*Pseudofusulina kueichowensis* 带					栖霞组	*Cancellina* 带 *Misellina-Nankinella* 带	下二叠统	雅赫塔什阶		

(据宜昌地质矿产研究所,海南省地质矿产局,1992)

霞组之中。据吴望始、张遴信（1979）研究认为，*Robustoschwagerina* 层位一般较 *Sphaero schwagerina* 带高，并建立了 *R. schellwieni* 带作为晚石炭世最顶部的一个蜓化石带。但值得提及的是，在我国 *Robustoschwagerina* 的层位已不仅限在上石炭统上部马平组中发现，而在更高层位中也有发现，如德保多敬、隆林常么，普安晴隆花贡、包磨山等地的 *Robustoschwagerina spatiosa*，*R. longlinensis* 等均和 *Misellina claudiae*，*M. ovalis*，*M. longlinensis*，*M. changmoensis*，*Parafusulina* 等共生，因而 *Robustoschwagerina* 的层位，在我国应是始于晚石炭世晚期，向上可延至中二叠世早期栖霞组 *Misellina claudiae* 带的底部。海南所发现的 *Robustoschwagerina* 属则产于鹅顶组底部，与早二叠世中常见 *Nankinella*，*Schwagerina* 等蜓类共出现在同一层位。与上述 *Robustoschwagerina* 属所产出层位大体相似，时代相当，可资对比。鹅顶组中上部蜓类化石也极为丰富，以富含短筒形—厚纺锤形壳体的 *Parafusulina* 蜓类占主体，还有 *Schwagerina*，*Pseudofusulina*，*Yangchienia*，*Minojapanella*，*Schubertella* 等蜓类。据盛金章（1965）研究认为，鹅顶组的蜓类以 *Parafusulina* 为主，可称为 *Parafusulina* 动物群。这个动物群的时代归为栖霞晚期。近年来，又新发现有 *Chalaroschwagerina*，*Laxifusulina*，*Misellina ovalis* 等蜓类，经张志存（1988）研究，其时代大体属中二叠世栖霞期。以上化石的分布，从主要的蜓类动物群分布来看，鹅顶组大体属中二叠统栖霞阶，即中二叠世栖霞期沉积。

此外，中、北部大岭农场、石碌-金波地区鹅顶组发现少量生物化石（宜昌地质矿产研究所，1995），采获牙形石 *Streptognathodus* sp.；*Neogondolella* ? sp.，珊瑚 *Protomichelinia multitabulata*，*P. xinchengensis*，*P.* cf. *camara* 和蜓 *Parafusulina* spp.。上述化石由于数量较少不能构成组合带，也无生物化石带的特征分子，但上述生物化石某些属种，如 *Parafusulina* spp. indet，*Protomichelinia multitabulata* 等仅出现在下二叠统，显示属于二叠纪沉积。

综上所述，鹅顶组时代应为中二叠世。

5. 南龙组（$P_{2-3}n$）

该组下段中上部以产菊石为特征，主要有 *Shouchangoceras* cf. *shouchangense*，*S. guangd-ongense*，*S.* cf. *xiamaoense*，*Altudoceras* cf. *ziguense*，*A.* cf. *orientale*，*A.* cf. *lijaense*，*A.* cf. *roadense*，*Gaoyaonites* sp. 等（表3-7）。徐光洪（1979）称之为 *Shouchangoceras-Altudoceras* 菊石组合带，这个菊石组合带广泛分布于华南地区下二叠统茅口阶，与苏皖南部的堰桥组、赣中的官山组、浙江的丁家山组，闽西的童子岩组、粤中的广州组夏茅段和鄂西北的孤峰组等完全可以对比，层位大致相当。它是我国华南地区中二叠世茅口期中晚期具有鲜明的地方特色的菊石动物群，据此，南龙组下段的地质时代应为中二叠世茅口期是无疑的。

南龙组上段的化石丰富，门类较多，以底栖型腕足类、蜓类、双壳类为主，苔藓虫、海百合茎、绿藻和植物次之。蜓类主要有 *Parafusulina* sp.，*P.* spp.，*P. bosei* 等属种。其中 *Parafusulina* 是中二叠世中期地区分布广泛而又重要的代表蜓属。*P. bosei* 则是我国中二叠统茅口阶下部 *Parafusulina* 蜓类中的重要分子，在贵州郎岱茅口阶层型剖面上该种见于 *Cancellina-Neoschwagerina simplex* 带中。在北美南西部，此种是瓜德鲁普统底部 *Parafusulina bosei-P. apiculata* 带中的代表化石，其时代为中二叠世茅口期早期至晚二叠世早期。

南龙组上段的腕足类化石较为丰富，主要有 *Leptodus richthofeni*，*Transnnatia*，*gratiosus*，*Asinoproductus gratiosus*，*A. margaritotus*，*Neochonetes* cf. *tenuiirata*，*N. plicatiformis*，*Pygmochonetes* cf. *fingxianensts*，*Meristoygma* cf. *arctica*，*Neospinomarginifera* cf. *huangi* 等属种。杨德骊认为，以前归入 *Dictyoclostus* 和 *Asinoproductus* 的种，现应定为 *Transennatia gratiosus*，主要见于湘中上二叠统吴家坪阶底部。*Neochonetes* cf. *tenuilirata*，*Pygmochonetes* cf. *fingxianensis*，*Neospinomarginifera* cf. *huangi* 等是中二叠世的特征性分子，地史历程从中二叠世早期至晚期。*Meristorygma* cf. *arctica* 在我国尚属首次发现，它是加拿大二叠纪中的重要分子。另外，尚有 *Spiriferidae*，*Neochonetes* sp.，*Squamularia* sp.，*Dictyoclostus* sp. 等属，虽然地史历程较长，却是我国二叠纪层位中普遍见及的属。它们构成该段一个独特的地区性腕足动物群。这个腕足类动物群中的 *Leptodus richthofeni*，*Transnnatia gratiosus*，（或 *Asinaproductus gratiosus*，

A. margaritotus）在许多地区茅口阶上部也有所发现，甚至用 *Leptodus richthofeni* 来划分石炭系与二叠系界线。从腕足动物群的特征也可以说明南龙组的沉积时代为中二叠世茅口期。

在南龙组上段下部相应层位中富含双壳类化石，除有少量的血石蛤类 *Sanguinolites* aff. *costata* 从石炭纪延续而来之外，绝大多数是早二叠世晚期常见的化石，如短嘴蛤类 *Phestia zhejiangensis*，*P.* cf. *hunanensis*，新裂齿蛤类 *Neoschizodus kitakamiensis*，它们均多见于我国上二叠统龙潭煤系，礼贤煤组的底部，近年来在中二叠统茅口阶也多有发现。特别是在双壳类层位之上尚有较丰富的 *Parafusulina* sp. 等早二叠世重要的蜓类化石分布。根据双壳类化石，南龙组上段时代为中二叠世晚期。

在牛笛岭、南宁田、江边村北西部和石岭等地南龙组上段下部，下段的顶部产较丰富的植物化石，一般保存不好，主要为松柏类 *Coniferales* 的茎干、枝和叶，科达类 *Cordaites*? sp. 和丁氏蕨类 *Tingia* 植物群，并与少量的海百合茎、全脐螺科 *Euomphalidae*? 等共生。李星学认为 *Cordaites* 类标本可与昌江王下地区早二叠世的植物对比，属二叠世；姚兆奇则认为该植物群的面貌可与广西扶绥县南上二叠统合山组下部的植物群相类似。从地层层序分析，这一植物群位于中二叠统鹅顶组或 *Shouchnagoceras-Altudoceras* 菊石组合带之上，其分布时代为中二叠世末期至晚二叠世早期。

综上各门类化石分布来看，南龙组分布时代为中二叠世晚期至晚二叠世早期。

（四）岩相古地理探讨

海南岛石炭系所产牙形石、珊瑚、腕足类、双壳类等动物群中大多数属种同华南地层区同期生物群相似，尤其是隶属特提斯大区类型的腕足类 *Eochoristites*，*Martiniella* 等均见于石炭纪地层中。因此，就动物地理分区而言，应属于古特提斯海动物地理分区（汪啸风等，1992）；石炭纪时期，海南岛同华南海域一样位于低纬度地区的热带—亚热带气候（陈旭等，2001）。

根据张志存（1992）关于二叠系蜓类、珊瑚、菊石等门类生物地理区系归属的研究，认为蜓类 *Misellina*，*Parafusulina*，*Schwagerina* 等，皱纹珊瑚 *Wentzellophyllum*，*Polythecalis*，*Szechuanophyllum* 和菊石 *Shouchangoceras*，它们均具有华夏-特提斯动物区的典型特征；而植物化石 *Arnularia mucronata*，*A. shirakii*，*Peeopteris arborescens* 等则为与华夏-特提斯动物区基本吻合的华夏植物区常见分子。因此，海南岛二叠纪生物群、植物群分别是低纬度华夏-特提斯动物区或华夏植物区的组成部分。

此外，海南岛早二叠世早期牙形石 *Sweetognathus whitei* 动物群与北美、日本、华南等地同期生物群相比极具相似性，而且该动物群牙形石色变指标系数与日本琉球群岛早二叠世同名化石带中的标本相比也完全吻合。据此，认为早二叠世 *Sweetognattus whitei* 动物群的生物地理区也应属于华夏—特提斯大区，而且同上述生物群一样在二叠纪时期，均处于低纬度热带—亚热带古气候环境中。

1. 沉积相

（1）*石炭系*

石炭系可分为 4 种沉积相：开阔滨岸相、浅海相、风暴相、半深海相。

1）开阔滨岸相

南好组一段和青天峡组底部的石英质砾岩、砂砾岩、砾岩主要为这种沉积相。其特点是沉积物粒度粗、分选好、总体上磨圆度好，除少量岩屑外，几乎全由石英组成。部分岩屑磨圆度很差，应为近源底砾岩屑。可进一步划分为滨岸相带、近岸相带。

滨岸相带 主要分布于中部兰洋农场马鞍岭—和庆南坟岭一带，东南部龙滚北埇-番根岭一带南好组底部，以及九曲江千秋岭—龙滚龙角岭一带青天峡组底部。下部以厚层—块状石英质砾岩为主，夹透镜状砂砾岩、含砾砂岩、粉砂质泥岩，上部为细砾岩、石英砂岩、泥岩，构成下粗上细的沉积韵律。砾石成分以石英岩、脉石英、石英砂岩、硅质岩为主，其次为细砂岩、粉砂岩、绢云母板岩等。砾石磨圆度较好，以次磨圆者和磨圆者占优势，这是滨岸砾石的特征。这些砾石的磨圆度反映出砾石经较长距离的搬运，经过了岸流作用，是沿海盆边缘滨岸相带沉积。砾岩厚度较大，分选中等，说明沉积速率较大，碎屑物供应充分。上部砂岩夹层中还见有小型变余斜层理。

近岸相带 主要分布于西部石炭系青天峡组下段、南好组下段中上部，沿“儋州岛”南缘展布。

东南部层位为南好组下部，沿“琼中山地”北缘呈北东向展布；其南北两地岩性、生物群落、厚度有些差异，西部青天峡组下段沉积了一套以灰色—灰黑色含炭泥质岩为主夹钙质石英砂岩的沉积组合，水平层理发育，并见有泥质与粉砂质组成的小型脉状面层理，反映了典型的潮坪沉积环境。泥砂岩中见有黄铁矿，不含生物化石，显示出水流不畅，氧气不足的沉积环境。

2）正常浅海相

该相在沉积序列上位于上述滨岸相之上或之下，基本层序②（图3-56）组成的地层，即东南部南好组二段和青天峡组下部之砂板岩、西部青天峡组上部层位。由静水-低能沉积产物泥岩、粉砂质泥岩、泥质粉砂岩和中—低能沉积产物细砂岩、粉砂岩组成。其特点是每一种沉积物在剖面上都没有独立组成厚度很大的地层，即沉积环境水能量变化频繁。代表一定水深、一定能量的沉积环境都不能连续地持续很长时间，海平面波动频繁，但变化不大，沉积环境始终在浪基面上下一定范围内变化波动。砂岩、粉砂岩是浪基面以上近滨环境产物，其特点是分选较好，不含或含少量泥质，偶见变余斜层理。泥岩和并不多见的硅质岩均为浪基面以下浅海低能环境产物，其特点是颜色深，多为深灰色，部分为灰色和灰黑色，可含炭质。粉砂质泥岩、泥质粉砂岩为浪基面附近产物。总体显示海水稍深，西部兰洋农场一带由一套不纯碳酸盐岩及细碎屑岩组成，东部为静水-低能沉积。

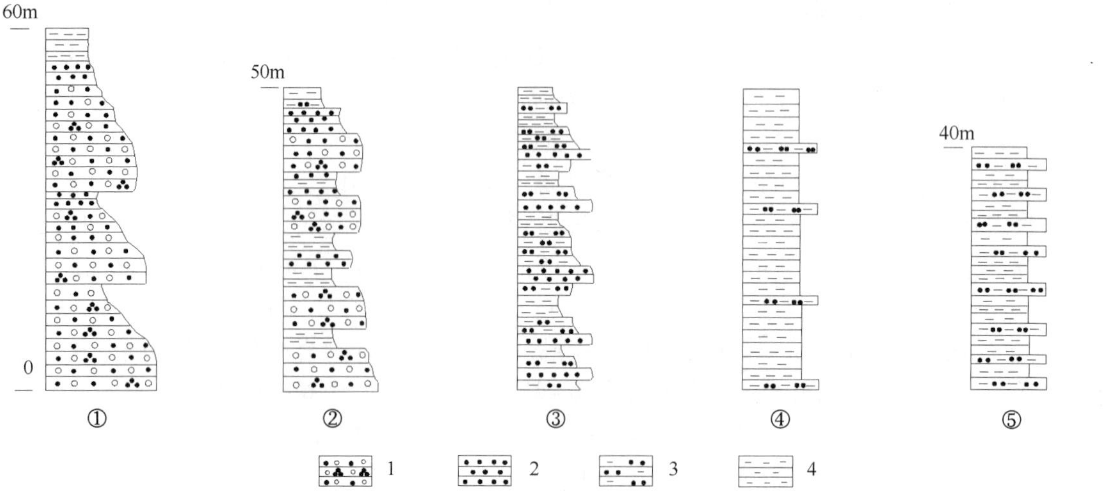

图3-56　石炭系基本层序

Fig. 3-56　Basic sequences of the Carboniferous

1—砾岩；2—砂岩；3—泥质砂岩；4—泥岩

3）风暴相

见于东南部插花岭、汀场岭等地南好组二段中下部、西部兰洋农场青天峡组中上部。基本层序③代表了其在地层中产出的一般特点。沉积物为复成分砾岩、砂砾岩，其特点是：砾屑主要为泥质、粉砂质泥砾，杂基主要为砂质、砂-泥质，砾屑与杂基混杂，无分选性、呈块状，其岩性则与上覆或下伏地层中砂岩、泥岩几乎完全相同。这些特点无疑表明它们是风暴流就近形成的风暴沉积。一个显著的特点是：泥砾、粉砂质泥砾普遍强烈拉伸呈扁平豆状，而不多见的板岩屑则可基本保持其棱角分明的碎屑形态。这无疑进一步说明泥砾、粉砂质泥砾是风暴流搅起的软沉积物再沉积的结果，即泥屑、粉砂质泥屑在未硬化固结时为上覆沉积物所压扁拉长，而板岩岩屑则不因之变形。

该风暴沉积夹于上述近滨—浅海相砂泥质沉积中。与典型风暴沉积相比，它不见典型风暴沉积的一般序列，只相当于一般风暴沉积序列的下部产物。这显然由于它产于近滨地带、波浪对它进一步侵蚀改造的结果。

4）浅海—半深海相

见于东部青天峡组上部地层，相当于前述基本层序④和⑤组成的地层。在相序列上它出现于上述近滨—浅海相沉积之后，代表海侵继续扩大、海水进一步加深。其主要特点是：灰黑色泥岩、炭质泥岩大量出现，厚达140余米，其中不含砂岩，仅夹少量薄层状粉砂质泥岩、泥质粉砂岩（由基本层序④组成），代表了离岸远、延续时间较长的较稳定深水滞流静水环境；之后则出现杂砂岩、泥岩韵律性互层，显示出深水浊积岩序列（由基本层序⑤组成）。杂砂岩为中—薄层，普遍含一定泥质杂基，具粒序层理。

（2）二叠系

二叠系是在继承石炭系基础上随着海侵作用的加强而沉积形成的，由一套碎屑岩夹碳酸盐岩组成（图3-57）。根据岩石组合及沉积构造特征，二叠纪沉积相可划分为滨岸碎屑、浅海陆棚碳酸盐和浅海陆棚边缘盆地3个相（区），可进一步划分为滨岸、近岸、远岸、内陆棚缓坡、外陆棚缓坡和近岸盆地等6个相带。

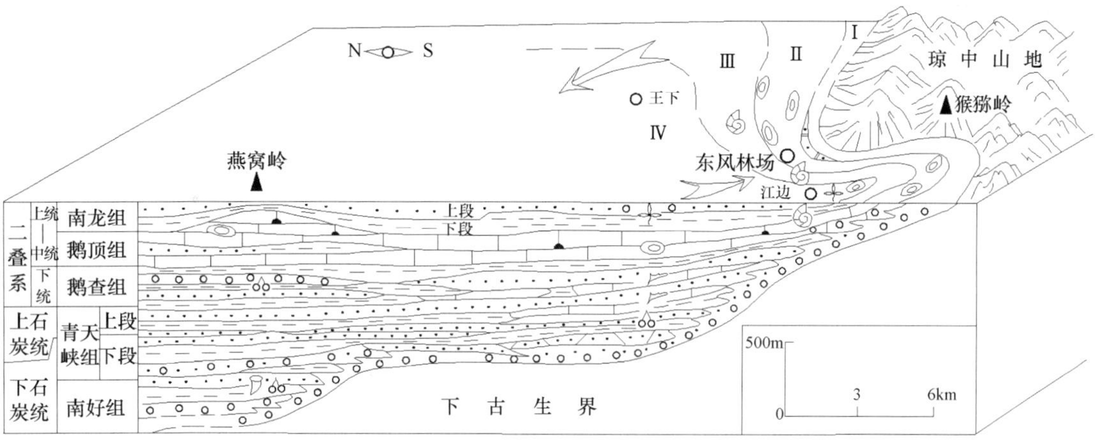

图3-57 二叠纪沉积相、沉积格架图

Fig. 3-57 Permian sedimentary facies and sedimentation framework

1）滨岸碎屑相（区）

滨岸相带 主要分布于中西部的“石碌岛”周边、大岭农场14队及王下一带，属潮间的砂砾质沉积类型，其层位为下二叠统峨查组及中—上二叠统南龙组上段。峨查组底部以厚层—块状石英质砾岩为主，夹透镜状细粒岩屑杂砂岩，厚度一般在30m以下。砾岩成分以石英岩、脉石英、石英砂岩、硅质岩为主，细砂岩、粉砂岩、绢云母板岩等次之。砾径一般为10~33mm，大者达50~60mm。磨圆度较好，砾间多为细砂砾充填。硅质泥质胶结，为滨岸砾岩。峨查组上部及南龙组上段，岩性以灰色中厚层中—细粒长石岩屑砂岩、细粒岩屑砂岩、含砾含生物屑钙质中—细粒长石岩屑砂岩为主，夹粉砂质泥岩、粉砂岩，局部为中—粗粒石英砂岩，含砾砂岩，厚度多为数十米。岩石结构和矿物成熟度中等，水平层理发育，局部可见双向交错层理，逆粒序层理。生物化石以盛产底栖类型的䗴类、腕足类、海百合茎等为特征，植物碎屑异常丰富，两者混合埋藏，一般保存不好，代表潮间中等能量沙滩沉积，局部为潮下带半封闭水体宁静的滨岸潟湖沉积环境。

近岸相带 主要分布于下二叠统峨查组中上部。在南部沿“琼南陆地”北麓呈近东西向展布；中部沿“石碌岛”周缘分布。沉积物主要由砂、泥质组成，呈叠复交替出现。生物不发育，化石甚少，除海百合茎碎屑外，腕足动物仅见丰富的小型舌形贝（*Lingulella* sp.）它生活在潮汐带、潟湖或浅海附近的沙滩上，是典型的指相化石，反映出潮间至潮下带沉积环境。

远岸相带 该相带主要见于下二叠统峨查组，分布于中西部鸡实村-金波农场、西南部王下一带。沉积物以砂、泥质为主，底部为碳酸盐岩，含硅质条带、燧石团块。局部夹层间砾岩，具条带状构造。生物以漂游生活的牙形石为主，底栖类型的䗴类（*Parafusulina*）、腕足类局部富集，海百合茎、苔藓虫等碎屑常见，偶见蠕虫管。这个生物群落面貌与近岸相带生物群完全不同，代表了潮下中低能砂泥质、碳酸盐岩沉积。

2）浅海陆棚碳酸盐相（区）

该相区由一套碳酸盐岩组成，底部为陆源砂砾石与碳酸盐岩混合沉积，上部夹少量砂泥岩。层位为中二叠统鹅顶组。根据岩性特征、生物组合、地形坡度可划分为内陆棚缓坡和外陆棚缓坡两个相带。

内陆棚缓坡相带 主要位于滨岸地带外侧缓坡地带，属浅水中低能带沉积区（潮间带下部至潮下带中等能量的较开阔的沉积环境）。其层位为中二叠统鹅顶组中上部。岩性为灰白色、灰色、深灰色微晶生物屑灰岩、微晶海百合茎灰岩、含生物屑微晶灰岩，北部夹粉砂质泥岩、石英砂岩、生物颗粒灰岩。燧石含量较高，构成燧石条带、团块，多为中厚层—块状岩层，具条带状、纹层状构造，生物骨骼及岩石硅化现象普遍，颗粒组分常多于灰泥组分。产多门类海生底栖生物，有䗴类、海百合茎、腕足类、介形类、红藻、有孔虫、苔藓虫等。其中以䗴类保存较好，含量占生物化石30%~40%，局部呈定向排列。海百合茎常见，多呈粗砾级碎屑状分布，其他均为碎片。

外陆棚缓坡相带 是指缓倾斜坡下部靠向海一侧，处于平均低潮面至浪基面附近。见于西南部峨贤岭-王下、鸡实村-金波农场一带的中二叠统鹅顶组。由深灰色、灰黑色含生物屑微晶灰岩、生物屑粉晶—细晶灰岩、粉晶—细晶灰岩、大理岩等组成。燧石含量较高，构成条带和团块，局部白云石化作用强。沉积构造因受挤压、褶皱、断裂及蚀变等因素影响，面貌皆非。生物多为碎屑，碎屑从南部向中部虽有明显变细、减少，却很难进一步划分出明显的相带界线，这表明了斜坡地带的地势比较平坦。生物比较单调，主要发育于该相带底部，多为圆形厚壳的壮希瓦格䗴、南京䗴和椭圆形希瓦格䗴及海百合茎碎屑，成层分布。这一生物群落和少量陆源石英砂（1%~5%）一起，经搬运、滚动、破碎，然后与碳酸盐灰泥混合沉积于水体较深的缓斜坡上，属潮下带低能的沉积环境。

3）浅海陆棚边缘盆地相区

盆地相区位于“琼南山地”的北西部陆坡、“石碌岛”周边的边缘地带，分布于牛笛岭、江边至江边岭到东风林场一带，石岭以北也见及。整体上大致呈北东东方向展布，其层位为上二叠统南龙组下段。沉积物以泥质为主，含钙质、硅质，下部含粉砂质结核，上部夹碳酸盐岩，白云石化作用强烈，粉砂质条带状构造发育，水平层理常见，纹层状构造屡见不鲜，水下砂质岩脉亦可见及。生物单调，局部富集成层，以壳饰粗强、游泳能力弱的地区性强的 *Shouchangoeras-Altudoceras* 菊石动物群为特色，是局限海近岸盆地相特有的菊石动物群。大体平行和接近滨线，成带状分布。海百合茎、䗴类等生物碎屑颗粒常见，但其硬体磨损、破碎等现象明显，分布杂乱，它们显然是搬运而来的异地组分。

2. 沉积环境分析

加里东运动后海南岛上升为陆，经历了一段时间的风化剥蚀后，地壳复又下沉，海水入侵，开始了石炭纪时的沉积历程。整体上从北西部向南东部发育明显的超覆不整合（当时海陆分界线在岛西南王下—江边一带），地势南东高西北低，海侵自西北向南东扩大。石碌地区、军营-红岭地区西侧出现了一定规模的“陆岛”，陆岛周边沉积物粒度显示出近岸特征。

早石炭世早期（南好组），地壳进入缓慢的沉降时期，海侵自西北向南东推进，南界河地区、昌江石碌—鸡实村—白沙金波、坡兰沟—军营—红岭一带的沉积物多以石英质细砾岩、不等粒石英砂、泥质为主，多交替出现。显示出海水浅、离岸近、坡度缓的滨岸地带，属无障壁型陆源碎屑近岸至远岸相带沉积环境。

西南面南界河地区见适应于软底的、能经受住十分混浊的海水生活的底栖类型，窄盐度单带型四射珊瑚和广海底栖型网格长身贝、线纹长身贝、石燕类、双壳类和海百合茎等生物则较为繁盛。其中珊瑚、海百合茎经搬运后，呈断枝、碎屑状，与砂泥质同时沉积。

西面昌江石碌—鸡实村—白沙金波一带为一套碎屑岩夹多层灰岩，属碳酸盐陆棚环境或远滨相沉积。底部沉积物（白沙金波如翁村一带）原岩为一套灰色—深灰色灰岩，上部夹深灰色薄—中层粉砂岩和泥质粉砂岩，在粉砂岩中见有微波状层理，反映盆地水体能量较低，灰岩中含有炭质和黄铁矿，可能是碳酸盐陆棚环境或远滨相沉积。稍晚（鸡心剖面上）见夹多层（三层）矿物成熟度和结构成熟度很高的灰白色中厚层石英岩或石英岩状砂岩，反映沉积水体变浅、出现高能环境，这种岩石可能是近滨环境潮下高能带。底部和下部两层石英岩之间的灰岩，其 Sr/Ba=23.08（900/39），明显为盐度正常海沉积，应属远滨相或滨外陆棚相；下部和中部主要两层石英岩之间所夹的中—厚层堇青石石英角岩，由于堇青石是富含 Al_2O_3 及 MgO 的泥质岩热变质产物，考虑到夹于两个高能带之间，可能是局限潮下环境。石碌一带南好组下部主要是深灰色薄层细碎屑岩随机组合，明显为碎屑陆棚环境。但顶部出现一层中厚层细粒石英岩状砂岩，反映水体明显变浅、能量突然增高，显示出离石碌“陆岛”距离较近的沉积特点。南好组中部（鸡心—鸡实村一带）为灰色中厚层砾屑硅质岩和残余砂屑硅质岩频繁互层，砾屑硅质岩呈层状、透镜状，具底冲刷。最大砾屑达 25cm，大砾屑一般为灰岩，小砾屑均为硅质岩，砾屑次棱—半圆状，颗粒支撑，其中见有原地生长的苔藓虫化石。砂屑硅质岩中见有大型板状斜层理，应为台地边缘浅滩环境。南好组上部以深灰色含石英粉砂泥晶硅质岩为主的岩石，为低能环境沉积产物。考虑到红林农场 16 队相当层位仍为硅化残余砾屑含灰硅质岩，应为局限潮下环境，而红林农场 16 队一带为台地边缘浅滩环境。尔后，主要为灰色—深灰色薄层泥质粉砂岩、粉砂质泥岩与薄层粉砂岩、含水云母细砂岩随机互层组成加积型层序，产丰富狭盐度珊瑚化石的骨屑微晶灰岩，反映主要为陆棚环境，仅局部出现近滨环境；南好组顶部（最晚期沉积）沉积物在三棱山底部砾岩中扁平砾石呈叠瓦状排列，扁平砾石倾向与岩层倾向接近，倾角大于岩层倾角 12°～24°，其中尚见有赤铁矿砾石，具有滨岸砾岩特征，即属于石质海岸的前滨带（位于石碌“陆岛”周边的沉积产物），其上灰色—灰黑色薄层粉砂岩、泥质粉砂岩、粉砂质泥岩夹具有大型交错层理石英岩状粗砂岩，生物群具有海相动物和陆相植物同层混生现象，显示近滨环境沉积。据零星露头判断，虽然顶部有粒度变粗现象，但均属远滨环境沉积。鸿沟岭一带沉积环境演化与三棱山处相似，从前滨相—近滨相—远滨相，但上部出现厚度达 63m 的白色中厚层石英岩，表明以近滨高能环境结束。

河叉岭一带虽然剖面层序特征为对称性层序，但底部为明显的低曲度河流相，上部一套变质砾岩和石英片岩具有前滨相的特征，顶部可能是近滨相（位于军营地区西侧的“陆岛”附近），再上为断层缺失面貌不清。

军营-红岭地区沉积物粒度整体偏粗，岩层产状及古流向分析表明物源应来自西侧，其西侧存在一定规模的“陆岛”。该区沉积物主要为一套碎屑岩，岩石中的砂岩成分主要为石英，成分成熟度高，磨圆度好，多呈次圆状，石英砂岩中碎屑物多分选性好，杂基少，说明经受了波浪的冲洗、波选和磨圆。其沉积应在波浪带，即正常浪基面以上。南好组底部砾岩较稳定，砾石成分成熟度高，磨圆度好，分选性可达中等，为滨海相（近滨—远滨相）沉积产物。沉积物中所发现的生物化石以广海底栖型腕足类（长身贝、戟贝类、石燕类）、双壳类、海百合茎和牙形石等生物较为繁盛。其中，海百合茎经搬运后，呈断枝、碎屑状，与砂质同时沉积。

南好地区实测剖面附近钻孔资料（张启富等，1991）中，其孔深 2.5～25m 处为变质砾岩、变质粉砂岩。底部为变质砾岩，但横向上不稳定，呈透镜状产出，属曲流河环境滞留砾岩（刘宝珺等，1985）。由下而上为薄层变质粉砂岩夹薄—中层变质细砂岩、薄层粉砂质绢云母板岩则属边滩-堤岸沉积环境。由下而上以薄—中薄层绢云母石英粉砂岩、变质细砂岩夹粉砂质绢云母板岩呈不等厚互层状由下而上叠置，粉砂质板岩中产腕足类化石 *Eochoristites spirifer* 等及海百合茎化石 *Pantagonocyclicus* sp.，*Cyclocyclicus* sp.。此外，变质粉砂岩及细砂岩中发育沙纹层理、板状交错层理、平行层理，应属浅滨环境。据此，南好地区南好组由下而上显示，水动力条件是由浅变深的退积型沉积序列。

晚石炭世沉积时（青天峡组），区域上海侵进一步扩大，海域变得更广，海水普遍变深，原海中“陆岛”亦没入海水之下，由于南临位于测区南东侧的“南部山地”，因此整体上处于“山地”北侧、海水浅的滨海—浅海陆棚环境。紧邻“山地”的王下一带及以北地区，处于滨岸（远滨）—浅海沉积环境，以远岸带的泥质岩、细砂岩夹碳酸盐岩沉积类型最为发育。岩层为中薄层，偶见水平层理。靠滨岸至近岸带一侧底栖类型的双壳类生物极为繁盛。自王下往北至燕窝岭地区逐步过渡到陆源碎屑注入较多的近岸至远岸过渡的沉积类型为主，生物不发育。自燕窝岭向北的广大地区，沉积了一套碎屑岩夹碳酸盐岩，其沉积环境应为浅海环境，浅海之中出现多个碳酸盐台地（石碌一带碳酸盐台地相发育）。东面九曲江千秋岭—龙滚龙角岭一带，因水流作用加强，岸流作用明显，沿“琼中山地”北麓缓坡地带，普遍发育一套近岩带泥质与石英质砾石，石英砂交替沉积旋回。之后，经过频繁的海平面波动、出现频繁的滨海—浅海交替，适应于软底的，并能经受住十分浑浊的海水生活的底栖类型珊瑚很广海底栖型、腕足类、海百合茎等生物则较为繁盛。其中珊瑚、海百合茎经搬运、磨损后，呈断枝、碎屑状，与泥砂质、钙质同时沉积，其分选性差，排列杂乱，这些都反映了在较动荡的水动力条件控制下的沉积。晚期由于伸展幅度加大，形成了一定规模的浅海—半深海环境，沉积了厚达 140 余米的暗色泥岩。

二叠系是在继承石炭纪基础上随着海侵进一步的发展而逐步沉积的，其地势仍南东高西北低，海侵自西北向南东扩大，石炭纪时期南部“琼南山地”部分地域及中部“石碌岛”周边部分地带渐被淹没，开始了二叠纪沉积历程（图 3-58）。

<table>
<tr><td>相区</td><td colspan="6">滨岸碎屑相区</td><td colspan="2">浅海陆棚碳酸盐相区</td><td>浅海陆棚边缘盆地相区</td></tr>
<tr><td>相带</td><td colspan="2">滨岸相带</td><td colspan="2">近岸相带</td><td colspan="2">远岸相带</td><td rowspan="2">内陆棚缓坡相带</td><td rowspan="2">外陆棚缓坡相带</td><td rowspan="2">近岸盆地相带</td></tr>
<tr><td>亚相</td><td colspan="3">滨岸潟湖</td><td colspan="3">滨岸沙滩</td></tr>
<tr><td>模式图</td><td colspan="9">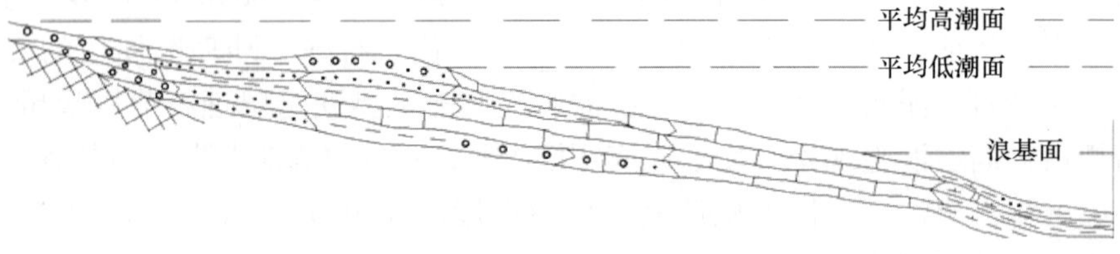
</td></tr>
<tr><td rowspan="2">岩性</td><td colspan="2">石英质砾岩、岩屑砾岩、砂砾岩、砂岩</td><td colspan="2">石英细砂岩、粉砂质泥岩、石英砂岩</td><td colspan="2">石英砂岩、泥岩、含生物屑微晶灰岩夹硅质岩</td><td rowspan="2">生物屑微晶灰岩微晶生物屑灰岩、含燧石灰岩、夹砂泥岩。底部钙质砾岩、含石英质砂砾屑灰岩</td><td rowspan="2">含生物屑微晶灰岩、含燧石灰岩、结晶灰岩</td><td rowspan="2">粉砂质泥岩含菊石泥岩、夹灰质白云岩、砂岩、含钙硅质岩</td></tr>
<tr><td colspan="3">粉砂质泥岩、石英砂岩含钙质粉砂岩</td><td colspan="3">细粒岩、含砾砂岩、石英砂岩、夹粉砂质灰岩</td></tr>
<tr><td>颜色</td><td colspan="2">灰色—深灰色暗灰紫色</td><td colspan="2">灰色、深灰色灰绿色</td><td colspan="2">灰色、深灰色</td><td>灰色、深灰色</td><td>灰色、深灰色、灰黑色</td><td>灰色、深灰色</td></tr>
<tr><td rowspan="2">生物</td><td colspan="2"></td><td colspan="2">腕足类、海百合茎、双壳类</td><td colspan="2">牙形石、腕足类、海百合茎</td><td rowspan="2">蜓类、海百合茎</td><td rowspan="2">蜓类少见、海百合茎为主</td><td rowspan="2">菊石类为主、鹦鹉螺、海百合茎碎片</td></tr>
<tr><td colspan="3">双壳类、腹足类、植物、蜓类、海百合茎、腕足类、角石</td><td colspan="3">腕足类、蜓类、植物碎片、海百合茎</td></tr>
<tr><td>沉积特征</td><td colspan="2">水平层理纹层</td><td colspan="2"></td><td colspan="2">水平层理
交错层理</td><td></td><td>水平纹层示底构造</td><td>水平层理
水下砂质岩脉</td></tr>
<tr><td>所在层位</td><td colspan="6">鹅查组，南龙组上段</td><td colspan="2">峨顶组</td><td>南龙组下段</td></tr>
</table>

图 3-58　海南岛二叠纪沉积相模式图

Fig. 3-58　Model of Permian sedimentary facies, Hainan Island

早二叠世早期（即晚马平期至隆林期），沿滨岸带广泛发育一套以陆源石英质砾石、石英细砂为主的沉积。水动力波动较大，环境不稳定，不利于生物发育生长。峨贤岭至王下一带以北、燕窝岭地区以南之间地域、“石碌岛”以北及以东广大地区，仍处于远岸相带，沉积物厚度大，以砂泥质沉积类型为主，交替叠覆出现。底部往往发育一层碳酸盐岩，分布较为连续，相对稳定，呈近东西向展布，厚 2.9~7m。生物群落的面貌也发生了显著的变化，以代表早二叠世早期的牙形石 *Sweetognathus whitei-*

Rabeignathus asymmetricus 动物群（燕窝岭地区-江边青天峡地区）则开始出现，长纺锤形的厚壳短筒状 *Parafusulina* 蜓属相继出现，大量繁盛，局部地带富集成层。腕足类 *Marctinia*，*Orthotetes*，*Puncto-spirifer*，*Dictyoclostus*，*Rhipidomella*，*Brachythyris*，*Sxhizophoria* 等更为常见，海百合茎、苔藓虫等碎屑亦可见及。靠近岸相带附近植物科达类 *Cordaites* 也有发现（据934队资料）。从天峡谷往北到燕窝岭，然后到石碌地区，逐步过渡到以粗粒的砂泥质为主的近岸相带沉积类型，生物单调、稀少，除海百合茎碎屑外，穴居、钻孔的无铰纲腕足类小型舌形贝（*Lingulella*）极为繁盛，呈原地埋藏状态。晚期，局部地区水体变浅，水流作用加强，双向交错层理发育，逆粒韵律层也见及，形成由石英质细砾岩、石英砂构成的沙滩沉积。

中二叠世早期，大体相当于栖霞期（即鹅顶组）是最大的海侵扩大时期，广大地区被海水淹没，接受沉积。总体上形成近东西向展布的由南向北缓倾斜的较开阔的碳酸盐岩缓坡型浅海陆棚沉积。在这一浅陆棚缓坡上，主要发育内外陆棚碳酸盐岩缓坡相带沉积。

浅海外陆棚碳酸盐岩缓坡相带的主体形成于栖霞早期，分布于南部峨贤岭至王下一带。由于海水相对较深，海水循环受影响，较闭塞，碳酸盐岩沉积作用缓慢，不利于生物繁殖、发育和生长，海百合茎为碎屑状，常与石英砂沉积于灰泥之中。

栖霞期中晚期，随着海侵范围逐步扩大，海域逐渐变得开阔。中、北部地区出现外陆棚碳酸盐岩缓坡相带沉积。南部燕窝岭地区，属浅海内陆棚碳酸盐形缓坡相带，处于潮间下部至潮下带中低能量沉积环境。水体较浅，水质清洁，盐度正常，阳光充足，气候温暖，适应于厚壳体的短筒形、纺锤形、椭球形 *Parafusulina*，*Schubertella*，*Chalaroschwagerina*，*Laxifusulina*，*Misellina* 等蜓类大量的繁殖生长，海百合茎均被打碎，呈粗砾级碎屑状，与蜓类混杂堆积，沿斜坡形成楔状体生物层。除蜓类 *Parafusulina* 和海百合茎等生物较发育之外，还有间歇性强水流将陆源石英砂、粉砂、泥质带入以碳酸盐灰泥为背景的缓斜坡上，形成了砂泥岩与碳酸盐岩交替沉积。

中二叠世茅口期（即南龙组下段），为浅海陆棚边缘相区，属近岸盆地沉积。广大地区发育灰色、深灰色含钙质粉砂质泥岩及灰质白云岩、岩屑质砂岩，水平层理和纹层发育，可见水下砂质岩脉。生物单调（南侧江边地区见含寿昌菊石群 *Shouchangoceras-Altudoceras* 组合为主的动物群），显示这一时期的近岸盆地相沉积的特征。

中二叠世茅口期晚期—晚二叠世吴家坪期（即南龙组上段），是最大的海退高潮时期，水体从南向北逐步变浅，海域面积进一步的缩小，地层厚度变化大，岩性差异也大，反映沉积环境变化迅速。在石碌-鸡实、大岭农场、金波农场等地，由中—细粒石英砂、岩屑、长石以及石英质砾石等构成滨岸沙滩相。局部地段在靠陆一侧或沙滩后侧，因海水运动受到一定阻碍，形成水体宁静的低能区，发育一套粉砂质泥岩夹少量砂岩的沉积，水平层理和纹层发育。晚二叠世中晚期，发生了大规模的海退，到晚二叠世末期可能露出海平面，上升为陆，从而结束了本区沉积历程。

综上所述，二叠纪时，由于受南高北低的古地形的制约，形成了与之相适应的缓斜坡型古地理格局。陆源物质主要来源于“琼南山地”及“海岛”，而海水则由北西向南东入侵。由于海侵、海退周期性的变化，出现不同沉积相类交替叠置和相对并存的局面，总体表现为一个由海侵到海退的旋回。

四、三　叠　系

（一）岩石地层特征

海南岛三叠系仅有岭文组。由于后期多期次岩浆活动的侵蚀，导致三叠纪沉积盆地被分离肢解，岭文组呈残留状分别出露于定安县翰林镇及琼海九曲江镇一带，出露面积约 11.6km^2。

1. 剖面描述

（1）岭文村剖面

该剖面为代表性剖面。广东区测队（1964）测制，汪啸风等（1992）重新测制。剖面位于定安县翰林岭文村，起点坐标：东经 110°14′18″，北纬 19°17′29″；剖面长约 700m。自上而下为：

未见顶

下三叠统岭文组（T_1l） 总厚度>345. 7m

6. 灰略带红色薄层状不等粒长石石英细砂岩、粉砂岩，水平层理发育 >40. 8m
5. 掩盖 74. 8m
4. 灰绿色、灰色、灰黑色、紫色薄层粉砂质泥岩，水平层理发育。产植物化石 *Voltzia* sp.；孢粉 *Alisporites hainanensis*，*Sulcatisporites ovalis*，*Falcisporites zapferi*，*Klausisporipha*，*Cuneatisporites lingwenesis*，*Accinetisporites ligatus*.，*Triangulisaccus hainanensis*，*Fultaesporites jiuquijiangensis*，*Labiisporites songshanensis*.，*Illinites parcus*.，*Jugasporites* sp.，*J. schaubergeroides*，*Lunaisportes novimundi*，*Striatoabieites richteri* 等 21. 0m
3. 浅灰色、棕黄砾岩夹棕红、暗紫色含砾不等粒石英砂岩，层理不清 117. 8m
2. 掩盖 50. 5m
1. 浅灰色、灰色砾岩 40. 8m

~~~~~~ 不　整　合 ~~~~~~

下伏地层　抱板群戈枕村组混合片麻岩

（2）山笈岭剖面

海南省地调院 1997~2000 年进行 1∶5 万《黎母岭、枫木市、琼中县、白马岭》幅区调时测制，海南省地质调查院 2001~2003 年开展 1∶25 万《琼海县》幅区调时补做了部分工作。剖面起点位于翰林镇山笈岭 191°方向约 640m 处，地理坐标：东经 110°13′50″、北纬 19°17′47″，剖面长约 966m。自上而下为（图 3-59）：

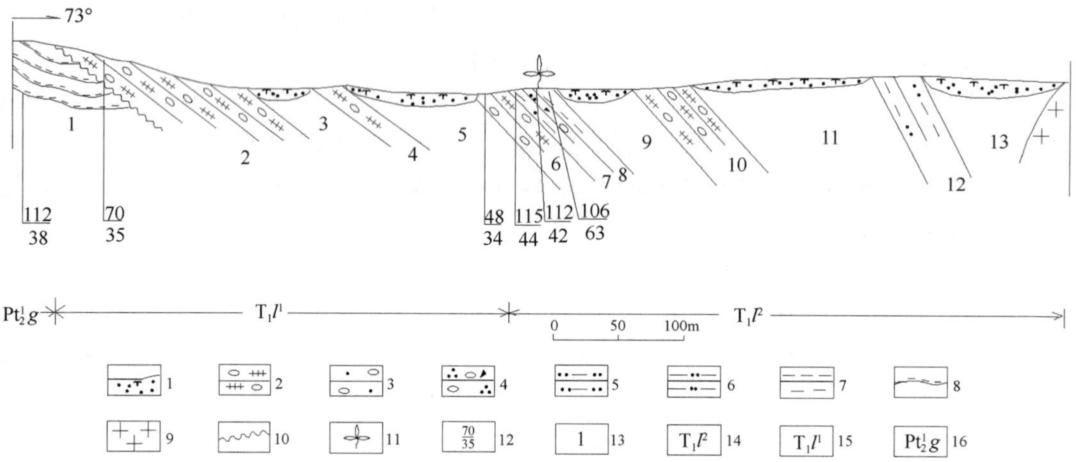

图 3-59　定安县翰林镇山笈岭下三叠统岭文组实测剖面图

Fig. 3-59　Measured section of the lower Triassic Lingwen Formation, Shanjiling, Hanlin Towm, Ding'an County

1—浮土；2—复成分砾岩；3—砂砾岩；4—砾质中细粒含白云母岩屑石英砂岩；5—泥质粉砂岩；6—粉砂质泥岩；7—泥岩；8—片麻岩；9—花岗斑岩；10—不整合接触；11—植物化石；12—产状；13—分层号；14—岭文组上段；15—岭文组下段；16—戈枕村组

未见顶（顶部花岗斑岩侵入）

下三叠统岭文组（T_1l） >463. 62m

13. 掩盖 65. 18m
12. 浅灰白色粉砂质泥岩 26. 10m
11. 掩盖 106. 73m
10. 浅灰白色复成分砾岩 32. 35m
9. 掩盖 48. 52m
8. 浅灰色、浅灰白色泥岩，具水平层理。上部含丰富的植物化石 *Albertica elliptica* Schimper，*A. latifolia* Schimper，*A.* ? sp.，*Brachyphyllum* sp.，*Neuropteridium marginatum* Zhou et Li. 39. 55m
~~~~~~

7. 浅灰、浅灰白色泥质粉砂岩，顶部为浅灰白色砾质中细粒含白云母岩屑石英砂岩，具水平层理　20.58m

6. 浅灰白色、略带浅紫红色复成分砾岩、砂砾岩　6.35m

5. 掩盖　19.78m

4. 灰色复成分砾岩　12.64m

3. 掩盖　27.48m

2. 浅灰、灰白色复成分砾岩　58.36m

～～～～～ 不　整　合 ～～～～～

下伏地层　戈枕村组（Pt_2^1g）浅灰白色片麻岩

(3) 九曲江剖面

海南地质大队（1967）测制，宜昌地质矿产研究所（1995）和海南省地质调查院（2003）先后补做了部分工作。剖面由距海洋村西南 200m 的 TG14、TC8 探槽编录（控制下段岩石）和长边村-尖岭剖面（控制上段岩石）组成。自上而下描述如下（图 3-60）：

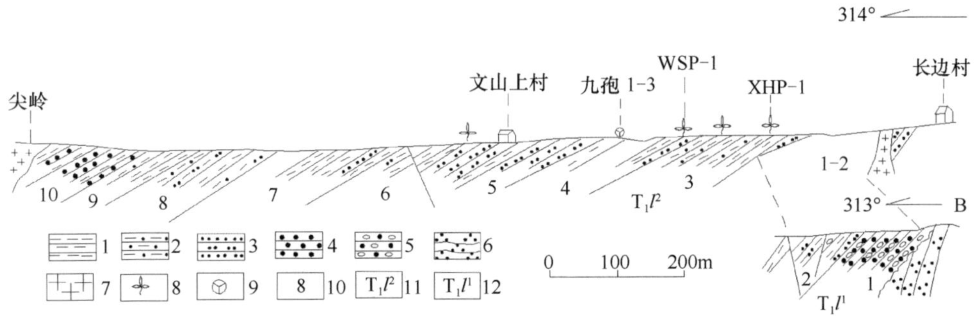

图 3-60　琼海市九曲江岭文组剖面

Fig. 3-60　Section of the Lingwen Formation, Jiuqu River, Qionghai City

（据张振来等，1992，略有修改）

1—泥岩；2—粉砂质泥岩；3—粉砂岩；4—细砂岩；5—砂砾岩；6—石英片岩；7—花岗斑岩；8—植物化石；9—孢粉化石；10—分层号；11—岭文组上段；12—岭文组下段

上覆地质体　被晚印支期迈州岭单元碱长花岗斑岩侵蚀

～～～～～ 侵入接触 ～～～～～

岭文组（T_1l）　厚 586.3m

11. 深灰、灰色薄层泥岩夹粉砂岩，水平层理发育，含植物碎片　28.1m

10. 灰色、灰白色薄层粉砂岩夹泥岩、粉砂质泥岩，水平层理发育，含少量植物化石碎片　31.8m

9. 浅灰色、灰色薄层泥岩、粉砂质泥岩，水平层理及龟裂纹发育，含少量植物化石碎片　15.6m

8. 浅灰色、灰白色薄层粉砂岩夹含砾粉砂岩、泥岩。泥岩见水平层理，局部可见龟裂纹，粉砂岩发育沙纹层理　66.98m

7. 蓝灰色、灰绿色薄层泥岩夹粉砂岩透镜体，水平层理发育，含植物化石（略，详见张振来等，1992）　103.2m

6. 灰绿色、浅灰色薄层泥岩与粉砂质泥岩互层，夹少量粉砂岩，发育水平层理和龟裂纹，含植物化石　111.9m

5. 深灰色、紫灰色薄层粉砂岩夹粉砂质泥岩，发育水平层理及沙纹层理，含植物化石碎片　19.2m

4. 灰色、浅灰色薄层泥岩夹粉砂岩、粉砂质泥岩，发育水平层理及小型斜层理，产丰富的植物和孢粉化石（略），植物化石以 *Voltzia* 和 *Albertia* 两属标本最多，孢粉以裸子植物花粉占绝对优势　54.3m

3. 灰色、深灰色薄层粉砂岩夹泥岩，发育水平层理，下部具龟裂纹，产丰富的植物化石　34.8m

————整　合————

2. 灰白色、浅灰色石英细砂岩、砂砾岩，中部为薄层粉砂质泥岩，局部发育斜层理　53.3m

1. 褐红、灰白色砂砾岩夹粉砂岩　60.8m

～～～～～ 不　整　合 ～～～～～

下伏地层　下志留统陀烈组灰褐色绢云母石英片岩

（4）南强村-教场村剖面

宜昌地质矿产研究所1992~1995年开展1∶5万《中原市、和乐、博鳌港》幅区调时测制，海南省地质调查院2001~2003年开展1∶25万《琼海县》幅区调时补做了部分工作。剖面位于琼海市朝阳乡南强村至教场，起点坐标：东经110°33′30″，北纬19°09′30″。露头连续性好，层序清楚。自上而下为（图3-61）：

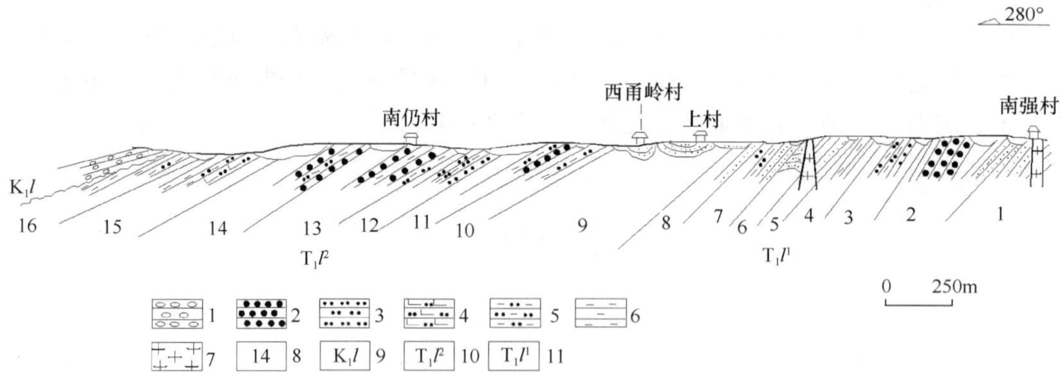

图3-61　琼海市朝阳南强村-教场村岭文组辅助剖面

Fig. 3-61　Auxiliary section of the Lingwen Formation along Nanqiang Village-Jiaochang Village, Chaoyang, Qionghai City

1—砾岩；2—砂岩；3—粉砂岩；4—含钙质粉砂岩；5—粉砂质泥岩；6—泥岩；7—花岗斑岩；8—分层号；9—鹿母湾组；10—岭文组上段；11—岭文组下段

上覆地层　下白垩统鹿母湾组紫红色中细粒凝灰质砾岩

～～～～～ 不　整　合 ～～～～～

岭文组（T_1l）　厚1473.2m

16. 灰绿色薄层含钙质粉砂质泥岩夹极薄层泥岩，局部夹含钙质粉砂岩薄层，具水平层理　67.09m
15. 灰绿色极薄层含钙质粉砂岩、粉砂质泥岩及泥岩互层，单层厚0.5~1.5cm，水平纹层理发育，发育龟裂纹。粉砂岩中偶见小型沙纹层理　89.31m
14. 灰色薄层粉砂质泥岩，偶夹极薄层细砂岩，发育水平及缓波状层理　103.27m
13. 灰色薄层粉砂质泥岩与灰绿色薄层细砂岩互层。细砂岩发育波状层理和小型沙纹层理　36.78m
12. 灰色—灰绿色极薄层粉砂岩、粉砂质泥岩夹极薄层细砂岩，见有波状层理及小型沙纹层理。自下而上，粉砂质泥岩层增加，粉砂岩、细砂岩层减少　95.37m
11. 灰—浅灰色薄—中厚层细粒石英岩屑砂岩夹薄层粉砂质泥岩，砂岩发育冲洗层理、波状层理及沙纹层理　41.30m
10. 浅灰色极薄层—薄层细粒石英砂岩夹灰色极薄层粉砂质泥岩。砂岩发育冲洗层理和波状层理　59.79m
9. 灰白色薄层细粒石英砂岩夹灰色薄层粉砂质泥岩，其上部夹凝灰质细砂岩薄层。砂岩中见小型沙纹层理　74.11m
8. 浅灰色薄层—极薄层细粒石英砂岩，见缓倾角交错层理及沙纹层理，局部发育对称波痕　112.12m
7. 灰色—灰白色中厚层—厚层细粒石英岩屑砂岩，夹深灰色薄层粉砂质泥岩，局部偶夹10~20cm的薄层凝灰质细砂岩。砂岩具平行层理、缓倾角交错层理及沙纹层理，粉砂质泥岩具波状及水平层理。　62.51m
6. 灰色—浅灰色薄—中厚层细粒石英砂岩夹浅灰色中厚层中细粒石英砂岩及粉砂质泥岩，局部偶夹凝灰质细砂岩薄层，具冲洗层理、波状层理、沙纹层　35.37m

5. 灰色—灰白色中厚层—厚层细粒岩屑石英砂岩夹薄层粉砂质泥岩，具平行层理、缓倾角交错层理及波状层理 42.29m

4. 褐黄色薄—中厚层粉砂质泥岩夹细粒岩屑砂岩。粉砂质泥岩具水平-波状层理 154.62m

3. 浅灰色—灰色中薄—中厚层细粒石英岩屑砂岩，夹深灰色薄层粉砂质泥岩，砂岩具平行层理、冲洗层理、波状层理及沙纹层理 147.70m

2. 灰白色中薄层细粒石英砂岩夹灰色—深灰色薄层粉砂质泥岩。砂岩发育缓倾角交错层理，粉砂质泥岩具水平-波状层理 263.86m

1. 浅灰—深灰色中层细粒石英砂岩，局部偶夹黑色薄层粉砂质泥岩，发育冲洗层理及波状层理，局部见有波痕。碎屑颗粒中石英占90%以上，次圆状，分选好 87.71m

~~~~~~~~~ 侵入接触 ~~~~~~~~~

下伏地质体 被印支晚期迈州岭单元花岗斑岩侵蚀

2. 岩石地层特征

岭文组系由广东省地质局区测队（1964）创名岭文群转化而来，命名地点在定安县翰林岭文村，时代定为下侏罗统，海南地质大队（1967）在琼海县九江地区进行煤田地质普查时，获植物化石，新建新昌组，时代改为下三叠统。周志炎等（1979）对九曲江地区所产植物化石进行了系统研究，除沿用岭文群外，命名为下三叠统九曲江组。汪啸风等（1992）认定定安县岭文村与琼海县九曲江地区所产古植物、孢粉化石及岩性均可对比，应属同时代沉积，采用岭文组一名，代表海南岛早三叠世晚期的陆相沉积。陈哲培等（1997）在进行全岛岩石地层单位清理时，指定岭文村剖面为代表性剖面，将岭文组划分为上、下段（表3-8）。下段砂砾岩段，由砾岩、含砾细砂岩组成；上段为泥岩段，岩石为泥岩、泥质粉砂岩，产植物、孢粉化石。

表3-8 海南岛三叠纪地层划分沿革表

Table 3-8 History of division of Triassic strata, Hainan Island

<table>
<tr><th colspan="2">广东区测队（1964）</th><th colspan="2">海南地质大队（1967）</th><th colspan="2">周志炎、厉宝贤（1979）</th><th colspan="2">广东地矿局（1982）</th><th colspan="3">宜昌所、海南地矿局（1992）</th><th colspan="2">陈哲培等（1997）</th><th colspan="3">1：25万《琼海县》幅区调（2001~2003）</th><th>本书</th></tr>
<tr><td rowspan="2">下侏罗统</td><td rowspan="2">岭文群</td><td rowspan="2">下三叠统</td><td rowspan="2">新昌组</td><td rowspan="2">下三叠统</td><td rowspan="2">岭文群或九曲江组</td><td rowspan="2">下三叠统</td><td rowspan="2">九曲江组</td><td rowspan="2">下三叠统</td><td rowspan="2">岭文组</td><td>泥岩段</td><td rowspan="2">下三叠统</td><td>上段</td><td rowspan="2">下三叠统</td><td>上段</td><td rowspan="2">下三叠统</td><td rowspan="2">岭文组</td></tr>
<tr><td>砂砾岩段</td><td>下段</td><td>下段</td></tr>
</table>

（二）地层划分与对比

岭文组下部岩石（下段）在定安翰林地区马革坡、山笈岭、黄竹岭北侧山脊一带，出露的岩性主要为砾岩、砂砾岩，厚度多在100~200m之间，黄竹岭北侧一带厚度有增大趋势。琼海朝阳地区为细粒石英砂岩、石英岩屑及岩屑石英砂岩，厚1081.38m。自下而上岩屑含量增多，粒径变小，中、上部偶夹5~20cm厚的褐黄色凝灰质细砂岩。九曲江地区为含砾中细粒石英砂岩，岩屑石英砂岩，底部有砂砾岩，厚114.1m。砾岩、砂砾岩砾石成分较杂，有脉石英、石英、硅质岩、变质岩、花岗岩，多呈次棱—次圆状，大小多在0.5~3cm之间。岩屑为绢云母板岩、硅质岩、粉砂质板岩、花岗岩等岩屑，次圆—圆状。发育缓倾角交错层理、平行层理、波状层理及沙纹层理，局部发育波痕。

岭文组上部岩石（上段）在翰林地区岭文肚—岭文村一带，厚度为136.6~339.01m，岩性为泥质岩、粉砂质泥岩、泥质粉砂岩，夹复成分砾岩。泥岩、泥质粉砂岩中发育水平层理。泥岩、粉砂质泥岩、泥质粉砂岩中含丰富的植物化石及孢粉化石。琼海九曲江、朝阳地区岩性横向上比较稳定，厚420.07~472.2m。岩石为灰色—灰绿色薄层（含钙质）粉砂质泥岩、泥岩，夹薄层—极薄层粉砂岩、细砂岩，发育水平层理、波状层理和小型沙纹层理和龟裂纹，产丰富的植物和孢粉化石。
~~~~~~~~~

在定安翰林地区本组不整合于长城系抱板群戈枕村组、峨文岭组以及奥陶系南碧沟之上，岩石未遭变质作用；在九曲江地区，本组不整合于下志留统陀烈组之上；在朝阳地区，受花岗斑岩侵入破坏，底部出露不全，石英颗粒受热力及构造挤压作用而具碎裂压溶、次生加大现象，个别颗粒见波状消光。

（三）古生物特征及年代地层

自海南地质大队（1967）在琼海县九江地区首获植物化石以来，周志炎等（1979）对九曲江地区所产植物化石进行了系统研究，汪啸风等（1992）于岭文组上段中发现丰富的植物化石。宜昌地矿所（1995）开展1∶5万中原市等3幅区调联测中，采集到植物化石 *Voltzia* cf. *heterophylla*，*V.* sp.，*Anomopteris*? sp.，*Pagiophyllum* sp.，*Brachyphyllum*? sp.，*Pelleaspermum* sp.，*Alberia latifolia*，*A.* sp. 等。海南地质综合勘察院（2000）及海南省地质调查院（2003）工作中于定安岭文地区山笈岭剖面岭文组中采获的植物化石，经南京地质矿产研究所蓝善先鉴定有：*Neuropteridium marginatum*（图版Ⅰ-11-C），*A. Latifolia*，*A.*？sp.，*Brachyphyllum* sp.（图版Ⅰ-11-D、图版Ⅰ-11-E），*Albertica elliptica*（图版Ⅰ-11-F）等。

截至目前，于该组中发现的植物化石共计28属38种，其中石松类1属1种，有节类5属8种，真蕨类3属3种，种子蕨类3属3种，苏铁类或可能为苏铁类2属2种，银杏类1属1种，科达类2属2种，松柏类6属10种，裸子植物花果及种子4属5种，分类位置不明的形态属1属3种，该植物群基本上反映了当时植物群的自然面貌，并具有以下特征：

1）以松柏类和有节类所占的比例较大。松柏类的化石不仅数量多，分布广，而且属种分异度较高，尤以 *Albertia* 和 *Voltzia* 为特征。

2）真蕨类植物属种虽不多，但化石相当丰富，尤其是 *Neuropteridium marginatum* 在地层中出现多，分布广。

岭文组的植物群可称为 *Pleuromeia*-*Voltzi*a 组合。

从植物组合的主要分子来看，周志炎等（1979）认为其与西欧斑砂岩植物群相近，时代为早三叠世晚期，即奥伦尼克期；周统顺等（1983）认为岭文组的植物群可与我国北方的和尚沟组 *Pleuromeia-Neuropteridium* 组合相比，时代为奥伦尼克期。孟繁松（1990）则将其植物群对比，时代为中三叠世早期（安尼期），其主要依据是岭文组 *Pleuromeia-Voltzia* 组合中出现了华北、华南及西欧的中晚三叠世较重要的 *Todites shensiensis*，*Leuthardtia ovalis*，*Vittaephyllum* sp.，*Voltzia weismanni* 等。

周志炎等（1979）认为琼海九曲江一带岭文组的植物群可与西欧斑砂岩植物群对比，尤其是 *Albertica latifolia*，*A. elliptica*，*Voltzia heterophylla*，*V.* sp.，*Glyptolepis longibracteata* 等具有代表性，时代应为早三叠世晚期，即奥伦尼克期；周统顺等（1983）称岭文组的植物群为 *Neuropteridium*- *Voltzia* 组合，并可与我国北方和尚沟组 *Pleuromeia-Neuropteridium* 组合相比，时代为早三叠世晚期。

汪啸风等（1992）重测岭文组剖面时，于岭文村一带首次获得孢粉化石，孢粉化石共计26属55种，孢粉组合中，裸子植物（包括种子蕨类）花粉占绝对优势，占整个组合的99.2%，其中以本体不具肋纹的双囊粉为主，占51.9%，主要分子有 *Alisporites*（12.2%），*Kausipollenites*（16.4%），*Falcisporites*（4.1%），*Platysaccus*（3.6%），*Sulecatisporites*（0.8%），*Pteruchipollenites*（0.8%），*Vesicaspora*（0.8%），以及 *Pityosporites*（0.8%），*Voltziaceaesporites*（0.8%），*Jausoniuspollenites*，*Vitreisporites* 等。其次为本体具肋纹和具裂缝的双囊粉（40%），主要为 *Triadispora*（26.9%），*Jugasporites*（3.3%），*Protohaploxypinus*（16%），*Striatoabieites*（0.8%），*Illinites*（2.4%），*Limitisporites*、*Labiisporites*（0.8%），*Ovalipollis*（1.6%），*Gardensporites* 等。单脊双囊粉占1.6%，包括 *Chordasporites*（0.8%）和 *Colpectopollis*（0.8%）。此外，还有少量 *Fultaesporites*.（25%）*Triangulisaccas*（1.6%），*Cycadopites*（1.6%）。

蕨类植物孢子在组合中占极次要地位，仅见有少量的 *Endosporites* 等属。

上述组合可称为 *Alisporites*-*Jugasporites*-*Triadrspora* 组合，该组合中的大部分属种为西欧斑砂岩孢粉组合中的重要分子，约一半以上的属种见于国内、外的早三叠世地层中，汪啸风等（1992）认为岭文组所含孢粉组合可与我国华北和尚沟组的孢粉组合对比，与欧洲中上斑岩统孢粉组合相当，因此定其时代为早三叠世晚期（奥伦尼克期）。

综合岭文组所含以早三叠世晚期 *Pleuromeia*，*Voltzia* 等属的植物群及 *Alispor ites-Jugasporites-Triadispora* 孢粉组合为主要分子（50%以上的属种见于国内、外早三叠世）。因此，岭文组的地质时代归于早三叠世晚期，即奥伦尼克期是合理的，并可与国内外同期地层对比（表 3-9）。

表 3-9　不同地区早三叠世地层对比表

Table 3-9　Correlation of early Triassic strata in different areas

本文			华北	新疆北部	西欧	俄罗斯地台	乌拉尔地区	巴基斯坦盐岭	化石组合
下三叠统	奥伦尼克阶	岭文组	和尚沟组	烧房沟组	斑岩统	巴期库查群	列斯坦索尔组	迈瓦里组	植物 *Pleuromeia–Neuropteridium–Voltzia* 孢粉 *Alisporites-Jugasporites-Triadrspora* 组合
	印度阶	?	刘家沟组	韭菜园组		维特曾日群			

（四）岩相古地理探讨

根据汪啸风等（1992）资料，海南岛三叠纪古纬度为北纬 16. 5°，剩余磁偏角为 30. 66°。陈海弘等人（1994）资料显示，三叠纪时，扬子地块古纬度为北纬 9°～17. 5°，华夏板块古纬度为南纬 6. 6°。张用夏（1983）认为，三叠纪时，扬子地块的古地理位置为北纬 15°～30°之间。即三叠纪时，海南岛与扬子地块位置相对比较接近，而与华夏板块相隔较远。从而也导致了海南岛与广东、广西三叠纪沉积作用的巨大差异。

目前所知，海南岛三叠系分布范围有限，零星出露于岛东定安县翰林地区和琼海市九曲江、朝阳地区，给古地理面貌恢复带来了一定的困难。即其沉积特点而言，定安县翰林地区岩石倾向南东，倾角 34°～63°，下部的复成分砾岩、砂砾岩厚 216m，上部的粉砂质泥岩、细砂岩和粉砂岩厚度 137m。琼海九曲江岭文组岩石倾向北西，倾角 20°～45°，下部的细粒石英砂岩、含砾砂岩、砂砾岩厚 114. 1m；上部的粉砂质泥岩、泥岩及含钙质粉砂质泥岩厚 472. 2m。琼海朝阳地区岩石倾向北西，下部的细粒石英砂岩、岩屑砂岩厚 1081. 38m，上部的粉砂质泥岩、泥岩厚 420. 07m。由此初步认为岭文组属于同一盆地沉积，从北西翰林地区往南东至琼海九曲江、朝阳地区，沉积厚度总体加大，沉积粒度变细。说明当时湖盆具有北西浅南东深、北西高南东低的地貌格局。翰林、九曲江一带属于盆地边界位置，而朝阳地区则处于盆地的沉积中心和沉降中心位置，沉积粒度相对细，沉积厚度大。

1. 基本层序

（1）定安县翰林岭文地区

岭文组的地层结构由两种基本层序组成，其中下部由基本层序Ⅰ组成，上段由基本层序Ⅰ、Ⅱ组成（图 3-62）。

Ⅰ. 基本层序为 A、AB 组合，主要见 A 组合，代表了湖滨相近岸地带沉积。

A. 浅灰色、浅灰白色复成分砾岩，砾石大小多 0. 5～3cm，磨圆度较好，分选性一般，成分成熟度、结构成熟度均一般。

B. 浅灰色、浅灰白色砂砾岩，岩石中砾石特征与 A 相同，只是含量比 A 中少而已。

Ⅱ. 基本层序为 AB、AC 组合，代表湖滨远岸地带-浅湖相沉积。

A. 浅灰白色砾质中细粒岩屑石英砂岩，砾石大小多在 0. 5～2cm 间。

B. 浅灰色、灰色、灰白色泥质粉砂岩，发育水平层理。

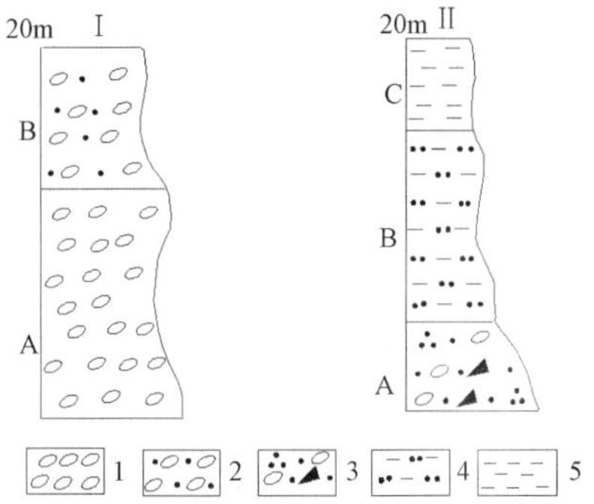

图 3-62　岭文组基本层序类型及特征

Fig. 3-62　Types and characteristics of basic sequences of the Lingwen Formation

1—砾岩；2—砂砾岩；3—砾质岩屑石英砂岩；4—泥质粉砂岩；5—泥岩

C. 浅灰色、浅灰泥岩，发育水平层理。

从上述岩石类型、沉积结构构造及岩石基本层序组成来看，岭文组下部沉积时，其沉积物来源较杂，既有远源碎屑，又有近源碎屑。上段沉积时，以近源碎屑为主，远源碎屑占次要地位。

（2）琼海九曲江、朝阳地区

岭文组下部由向上变细变薄的非旋回性基本层序（图3-63），据岩性组成可分为AB与BC两种，特征如下：

A. 浅灰色—深灰色薄—中厚层细粒石英砂岩及岩屑石英砂岩，局部为含砾砂岩，发育缓倾角交错层理、平行层理、冲洗层理，局部为波状层理和沙纹层理，厚2~8m。

B. 灰色—浅灰色薄—中薄层细粒石英砂岩及岩屑石英砂岩夹灰色—深灰色薄层—极薄层粉砂质泥岩，发育波状层理、沙纹层理及水平层理，厚0.5~2m。

C. 灰黄色中—中厚层粉砂质泥岩夹薄层岩屑砂岩，发育水平—波状层理，厚1.5~3m。

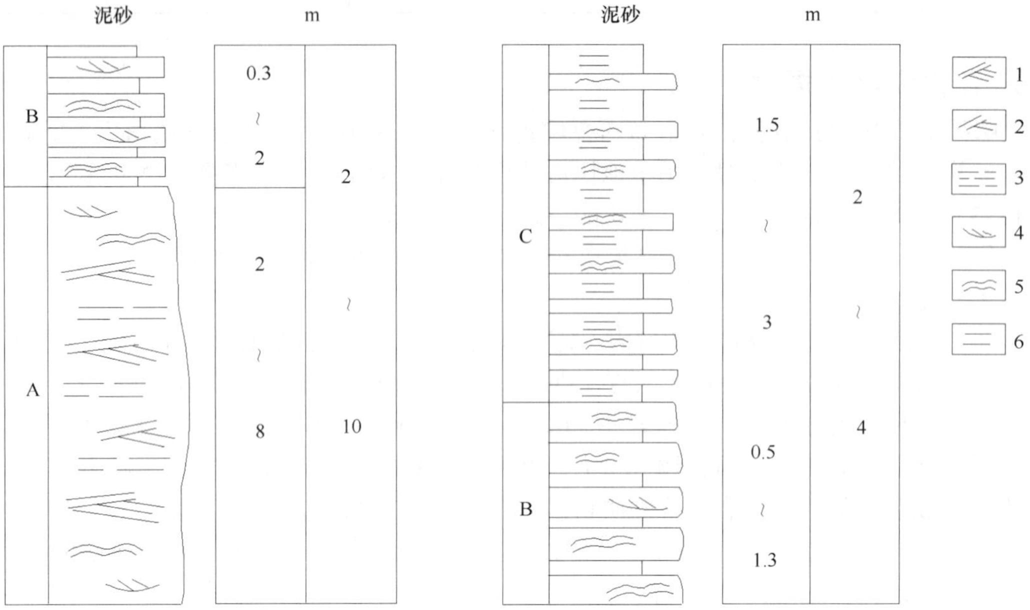

图3-63　岭文组下部基本层序组成及特征

Fig. 3-63　Basic sequence composition and characteristics of the lower part of the Lingwen Formation

1—缓倾角交错层理；2—冲洗层理；3—平行层理；4—沙纹层理；5—波状层理；6—水平层理

岭文组上部的基本层序（图3-64）和特征如下：

Ⅰ. 由浅灰色薄层细砂岩—灰色、灰绿色薄层粉砂岩—粉砂质泥岩组成向上变细的旋回，向上粉砂质含量减少；砂与泥比值在下部近于1∶1，中上部为1∶2~1∶3，发育波状—水平层理及沙纹层理。

Ⅱ. 由灰色薄层粉砂质泥岩偶夹浅灰色极薄层细砂岩组成，向上粉砂质、砂质含量减少；发育波状、水平层理及龟裂纹构造。

2. 沉积环境分析

岭文组下部以异地复成分砾岩、石英碎屑为主，成分成熟度低、分选好，向上含较多的岩屑，表明初期沉积物经历过较强的水流簸选搬运，之后环境能量变弱。根据岩石类型、沉积构造及基本层序特征，结合上部含大量植物化石，下部应形成于山前滨湖—浅湖环境。

岭文组上部以灰色—灰绿色粉砂质泥岩、泥岩、含钙质粉砂质泥岩为主，夹薄层粉砂岩及含钙质粉砂岩。发育波状—水平层理、沙纹层理及龟裂纹，由此显示沉积水动力较弱，应为浅湖—半深湖环境。所夹的深灰色—灰黑色粉砂质泥岩、泥岩中产丰富的植物和孢粉化石，植物化石保存较差，没有完整的宽大的叶部化石和原地生长的根部化石，为异地埋藏植物群落，表明当时地表径流量较大，可能有一较大的河流注入湖区。同时，由植物群及孢粉组合面貌及沉积物颜色反映出当时为半干旱的气候环境。

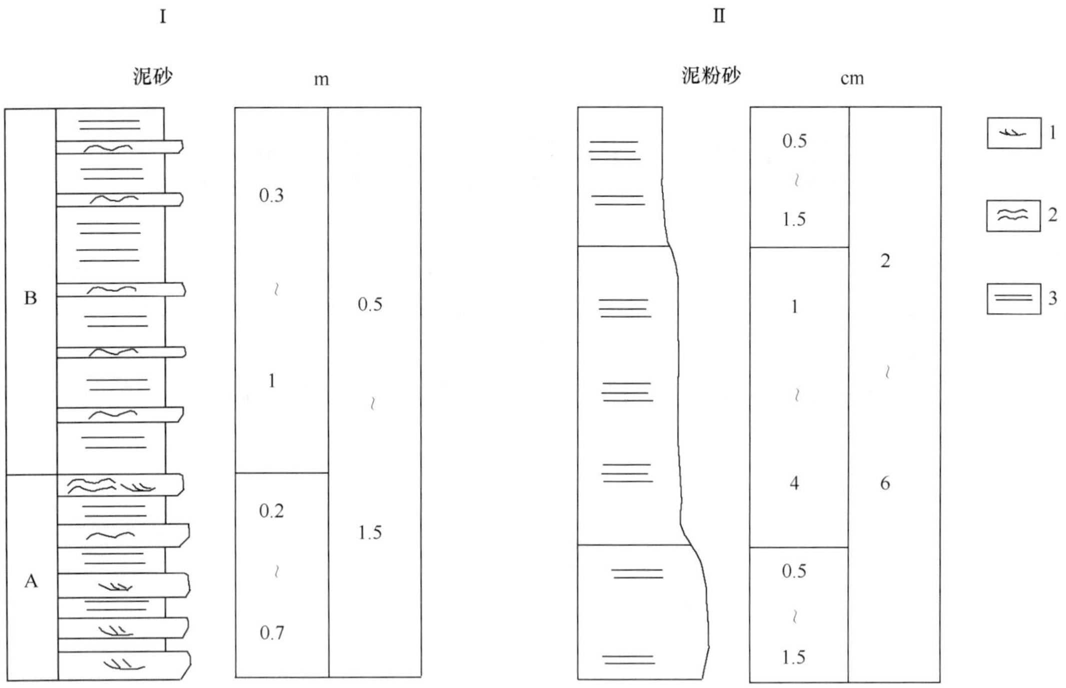

图 3-64　岭文组上部基本层序组成及特征

Fig. 3-64　Basic sequence composition and characteristics of the upper Lingwen Formation

1—沙纹层理；2—波状层理；3—水平层理

第四章 侏罗系—白垩系

海南岛该阶段仅有白垩系，为内陆河、湖相沉积，岩石并不发生区域变质和透入性的变形。沉积地层有白垩系鹿母湾组、报万组，主要分布于中生代陆相盆地（“红盆”）内，为一套河湖相碎屑岩沉积，以白垩系鹿母湾组出露较广。火山岩地层有六罗村组、汤他大岭组、岭壳村组，主要见于南西面同安岭、牛腊岭盆地中（图4-1）。岭文组与鹿母湾组为不整合接触，而鹿母湾组与报万组为整合接触。

图4-1 中生界分布及剖面位置图

Fig. 4-1 Map showing the distribution of Mesozoic strata and the position of sections

此外，《南海地质》（刘昭蜀等，2002）中，认为南沙群岛礼乐滩发育有上三叠统—下侏罗统和下白垩统岩石（未建组），上三叠统—下侏罗统为三角洲相—开阔浅海相砂泥岩及硅质页岩、粉砂岩，下白垩统为滨海—浅海相含煤碎屑岩。显示了中生代时海南岛与南沙群岛一带的沉积环境明显不同，应该属于不同的地层大区或区。但由于本次收集到的地质资料比较零碎而有限，因此未作具体的描述。

海南岛“红盆”地质调查工作由来已久，李承三（1929）曾对白石岭砾岩做过概略地质调查。1960年以来不少单位做过岛内盆地地质调查工作，以汪啸风等（1992）所做工作最为详细。汪啸风等（1992）根据全岛白垩系红盆古生物、同位素年龄等地质资料，将其定为早白垩世。四川攀西地质大队（1991）、陈哲培等（1997）根据岩性、岩相特征及沉积旋回特征、古生物化石组合特征，将海南岛红盆地层划分为下白垩统鹿母湾组和上白垩统报万组，一直沿用至今（表4-1）。

表 4-1 海南岛白垩系划分沿革表

Table 4-1 History of division of Cretaceous strata, Hainan Island

作者（年代）	地区	上部	下部
李承三（1929）	琼北地区	白石岭砾岩	
海南区资源开发委员会（1956）	全岛	第三系—白垩系：红色岩系	
北京矿业学院海南队（1958）	琼北地区	白垩系：南雄层	
中国区域地层表（1958）	全岛	下第三系：报万组	白石岭砾岩
广东省地质局765队（1960）	琼北地区	白垩系：雷鸣岩系	
海南地质队（1960）	阳江盆地	上白垩统：雷鸣岩系	
广东省地质局764队（1964）	全岛	上白垩统：报万群	下白垩统：鹿母湾群
海南地质队（1974）	全岛	上白垩统：报万群	下白垩统：鹿母湾群
吴萍等（1976）	白沙盆地	上白垩统：报万群	下白垩统：鹿母湾群
广东区调队（1978）	福田地区		下白垩统：鹿母湾群
南颐（1978）	全岛	上白垩统：报万群	下白垩统：鹿母湾群
耿良玉等（1979）	东江地区		下白垩统：临高组
关绍曾（1979）	加来盆地	下白垩统：临高组	
	白沙盆地	下白垩统：鹿母湾群	
	雷鸣盆地	下白垩统：报万群	
何阳芝等（1980）	三亚地区	白垩系：上统	白垩系：下统
海南地质队（1983）	全岛	上白垩统：报万群	下白垩统：鹿母湾群；下白垩统—上侏罗统
广东区调队（1986）	全岛	上白垩统：报万群	下白垩统：临高组／鹿母湾群
郝诒纯等（1986）	加来盆地	中白垩统：临高组	
	白沙盆地	中白垩统：鹿母湾群	
	雷鸣盆地	中白垩统：报万群	
宜昌地矿所、海南地矿局（1992）	全岛	下白垩统：鹿母湾群	
陈哲培等（1997）	全岛	上白垩统：报万组	下白垩统：鹿母湾群

第一节　岩石地层特征

白垩系（含火山岩地层）主要分布于阳江盆地、雷鸣盆地、白沙盆地、王五盆地中，文昌、洛基、旺商等小盆地也有出露。可划分为（火山）沉积岩系的鹿母湾组、报万组和陆相火山岩系的六罗村组、汤他大岭组和岭壳村组等5个岩石地层单位。

一、剖面描述

（一）鹿母湾组剖面

本组剖面有鹿母湾-白沙县白水港剖面、南美剖面、琼海市东太农场尖岭剖面、儋州市王五市平地岭剖面、东方市乐安村剖面，其中，前两条剖面分别控制了区域性白沙盆地的北东和南西部，后三条剖面分别控制阳江盆地、王五盆地、乐安盆地。

1. 儋州市鹿母湾-白沙县白水港剖面

该剖面为代表性剖面，由汪啸风等（1992）测制。剖面起自儋州市鹿母湾东北约1500m之鹿母湾河河谷，起点坐标：东经109°30′07″，北纬19°38′11″，沿简易公路经鹿母湾林场、牛场后，沿小径穿越牵裤岭，进入白沙县芭蕉沟，顺小溪而下，至白水港止，全长约13km。自上而下为（图4-2）：

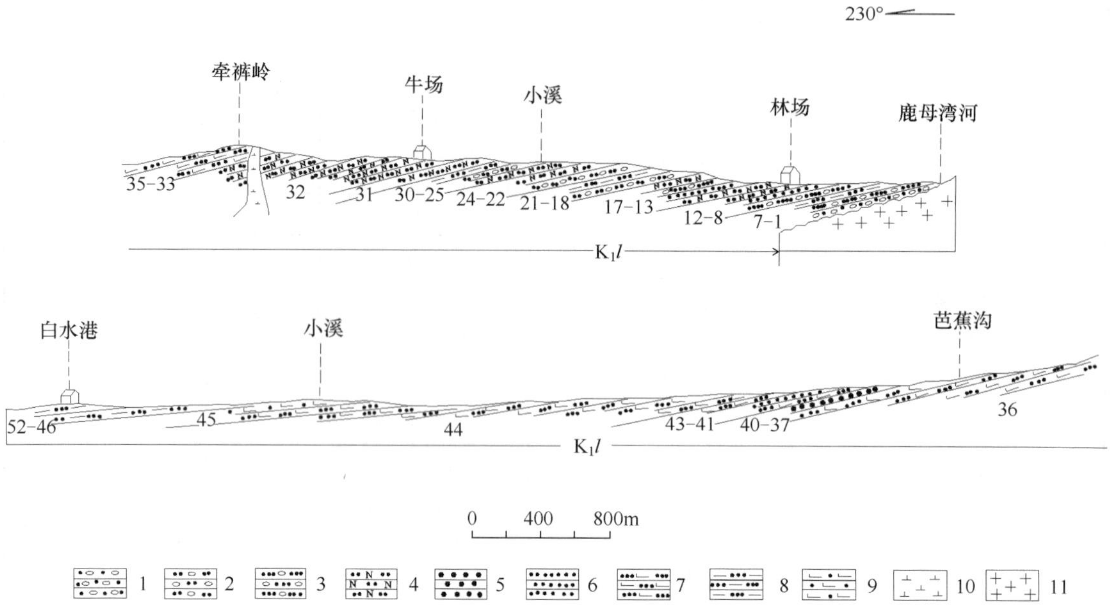

图4-2　儋州市鹿母湾-白沙县白水港下白垩统鹿母湾组剖面

Fig. 4-2　Section of the Lower Cretaceous Lumuwan Formation along Lumuwan-Baisha County, Danzhou City

1—砂砾岩；2—含砾长石砂岩；3—含砾砂岩；4—长石石英砂岩；5—细砂岩；6—粉砂岩；7—钙质粉砂岩；8—泥质粉砂岩；9—含钙粉砂质泥岩；10—闪长岩；11—花岗岩

未见顶

鹿母湾组　　总厚度>1523.72m

上段（K_1l^2）

52. 暗紫色、紫色中—厚层泥质长石粉砂岩　　9.69m
51. 暗紫色中—薄层粗砂屑长石石英粉砂岩，结构松散，水平层理　　7.85m
50. 紫灰色、紫色、暗紫色厚层—块状含钙泥质粉砂岩　　15.18m
49. 灰色、灰绿色、紫灰色厚层—块状含钙泥质长石粉砂岩　　4.45m
48. 紫色、暗紫色、紫灰色厚层—块状含钙泥质石英长石粉砂岩夹石英粗砂岩　　11.00m

47. 黄绿色夹紫灰色中—厚层含钙泥质粉砂岩夹含石英粗粉砂岩 13.59m
46. 暗紫色、紫色中—厚层含钙泥质粉砂岩夹粉砂质泥岩 18.72m
45. 紫灰色中—薄层含钙泥质粉砂岩、粉砂质泥岩互层 69.20m
44. 暗紫色、紫色中厚层含钙质粉砂岩夹粉砂质泥岩 200.00m
43. 黄绿色中—薄层含钙泥质粉砂岩，具水平层理 13.99m
42. 暗紫色、紫色厚层—块状含钙泥质粉砂岩 14.52m
41. 灰绿色、黄灰色中—薄层含钙粉砂质泥岩、泥质粉砂岩互层，具水平层理 36.68m
40. 紫灰色厚层—块状含钙粉砂质泥岩夹泥质粉砂岩 24.77m
39. 灰绿色、黄灰色中—薄层含钙粉砂质泥岩、泥质粉砂岩互层 18.52m
38. 暗紫色、紫灰色中—薄层中粒岩屑砂岩、粉砂岩互层，具水平层理 25.82m
37. 灰绿色厚层—块状含钙粉砂质泥岩，含星点状黄铁矿 4.46m
36. 紫色、暗紫色中—厚层含钙石英长石粉砂岩夹粉砂质泥岩，上部夹一层细砂岩 40.28m
35. 灰绿色厚层含钙石英粉砂岩夹紫灰色、黄绿色泥质粉砂岩 16.42m
34. 紫色、暗紫色厚层—块状含钙长石石英粉砂岩夹灰绿色粉砂质泥岩、泥质粉砂岩 49.26m
33. 掩盖。沿走向局部可见紫灰色、暗紫色粉砂岩，被灰带绿色斑状云辉闪长岩侵入 96.49m

下段（K_1l^1）

32. 紫红色厚层—块状长石石英砂岩夹黄灰色细砂岩 205.29m
31. 紫色中—厚层长石石英细砂岩夹黄灰色含砾中—粗粒砂岩。紫灰色含钙质粉砂质泥岩，砂岩胶结松散 37.16m
30. 紫红色、灰绿色、黄灰色厚层—块状含赤铁矿泥质粉砂岩，上部夹粉砂质泥岩 34.34m
29. 掩盖 14.6m
28. 下部紫色、紫灰色厚层—块状泥质粉砂岩夹黄灰色中—粗粒长石石英砂岩。上部紫色、黄灰色、紫灰色厚层—块状长石石英细砂岩，具椭圆形球状构造 22.53m
27. 紫色、紫灰色厚层—块状长石石英细砂岩夹黄灰色中—粗砂岩，具椭圆形球状构造 12.33m
26. 紫红色厚层—块状泥质粉砂岩，上部夹黄色中—粗粒含砾石英岩 29.62m
25. 紫色、黄灰色厚层—块状长石石英细砂岩夹中—粗粒含砾石英岩 3.23m
24. 紫红色厚层—块状泥质粉砂岩夹紫红色、黄灰色长石石英细砂岩 20.22m
23. 紫色、紫灰色厚层中—粗粒含砾石英砂岩、砂砾岩互层，夹细砂岩，具斜交层理、波痕 37.71m
22. 紫色、紫灰色厚层—块状砂砾岩夹含长石石英粗砂岩，砾石以石英为主，含较多泥砾，滚圆较好，砾径 0.5~10cm，砂岩具大型斜交层理 23.98m
21. 紫红色厚层—块状泥质粉砂岩夹细砂岩、含砾粗砂岩 10.31m
20. 紫灰色厚层—块状含砾长石石英粗砂岩，含紫色、紫红色泥砾，胶结松散 11.19m
19. 紫红色厚层—块状泥质粉砂岩夹粉砂质泥岩、泥岩，下部夹一层含砾粗砂岩 70.27m
18. 紫灰色、紫色厚层—块状不等粒含砾砂岩夹细砂岩、砂砾岩，水平层理发育 20.59m
17. 紫色厚层块状含赤铁矿、含石英粉砂质泥岩，上部夹一层浅紫灰色长石石英细砂岩 34.27m
16. 紫灰色中—厚层水云母质不等粒砂岩夹紫色粉砂质泥岩 17.57m
15. 浅紫灰色厚层—块状含砾水云母不等粒砂岩，夹暗紫色细砂岩、粉砂岩 20.93m
14. 紫色、紫灰色厚层—块状水云母质不等粒砂岩夹粗砂岩，胶结松散 6.65m
13. 浅紫灰色厚层—块状含砾长石石英粗砂岩夹紫灰色细砂岩、灰绿色长石石英细砂岩，紫红色泥质粉砂岩 39.90m
12. 紫红色厚层—块状泥质粉砂岩夹粉砂质泥岩 4.41m
11. 紫灰色厚层—块状长石石英砂岩，胶结松散 5.57m
10. 浅紫灰色厚层—块状不等粒含砾长石石英粗砂岩夹紫红色粉砂质泥岩 15.34m
9. 紫红色、紫灰色厚层—块状长石石英细砂岩夹粗砂岩，胶结松散 27.36m
8. 紫灰色厚层—块状含砾长石石英粗粒砂岩夹紫红色砂质泥岩 27.05m
7. 浅紫色厚层泥质粉砂岩，略显斜交层理 4.74m
6. 浅紫灰色厚层—块状含砾长石石英粗砂岩，胶结松散 6.43m
5. 紫红色、紫灰色厚层—块状含水赤铁矿石英粉砂质泥岩夹含砾石石英砂岩 6.88m

4. 紫灰色厚层—块状石英细砂砾岩、含砾长石石英粗砂岩互层，砂岩具斜交层理 14. 84m

3. 浅紫灰色厚层—块状含砾长石石英粗砂岩，层间夹薄层砂砾岩 15. 56m

2. 紫灰色厚层—块状石英砂砾岩，含砾长石石英粗砂岩互层，砂岩具斜交层理 14. 48m

1. 紫灰色厚层—块状石英砂砾岩，砾石以石英岩为主，其次为板岩、花岗岩等，大小不一，一般为 0. 5~5cm，个别大者可达 20cm，半滚圆状 16. 96m

～～～～～～～～ 不　整　合 ～～～～～～～～

下伏岩体　中三叠纪二长花岗岩

2. 南美剖面

四川省地矿局攀西地质大队 1991 年开展 1 : 5 万《尖峰岭、乐东县》幅区调时测制。剖面位于乐东县山荣乡南美村西北昌化江边。自上而下为：

上覆地层　报万组薄—中厚层细粒泥质长石砂岩

———————— 整　合 ————————

下白垩统鹿母湾组（K_1l） 736. 02m

上段 375. 51m

57. 灰紫色凝灰岩夹同色火山角砾岩及紫红色细粒长石砂岩，三者组成 10 个韵律层。砾石为安山-英安岩及少量石英岩、泥岩等 46. 69m

56. 灰紫色厚层—块状火山角砾岩。风化面呈蜂巢状，砾石为安山-英安岩及少量石英岩、泥岩等，砾径一般为 1~6cm，最大达 30cm。填隙物为凝灰质 16. 27m

55. 紫红色薄层泥质粉砂岩。中下部夹中粗粒安山-英安质凝灰岩，厚度小于 1m，上部含钙质 31. 33m

54. 紫灰色中—厚层安山-英安质粗粒岩屑凝灰岩夹细凝灰岩，中部含砾 24. 40m

53. 紫红色细凝灰岩夹灰紫、灰色中粒凝灰岩。风化面见有针孔大小的空洞，局部见方解石集合体，时有条带构造出现 29m

52. 紫红色玻屑细凝灰岩。风化后显薄层状构造，具水平层理，局部含钙，其风化面呈蜂窝状 5. 62m

51. 灰紫色、紫红色晶屑岩屑凝灰岩夹砾岩。下部见韵律薄层，中部以粗粒为主，上部以中粒为主，具平行层理、楔状层理、水平层理，波痕等沉积构造。底部为凝灰质砾岩，厚 70cm，砾石为石英岩及砂岩 12. 59m

50. 紫红色薄层砂质泥岩夹鲜紫红色含粉砂质泥岩两层，顶部 1.5m 为紫红色空洞状含凝灰质泥岩 7. 41m

49. 紫红色中—厚层安山质岩屑细凝灰岩夹同色粗凝灰岩及鲜红色凝灰质泥岩 15. 03m

48. 灰紫色中层中粗粒英安质岩屑凝灰岩，中上部夹两层鲜红色含凝灰质泥岩和一层细砾岩 22. 39m

47. 鲜紫红色中层安山质细凝灰岩。上部风化后空洞发育 9. 24m

46. 紫灰色中层含砾晶屑岩屑凝灰岩。含红色安山岩、英安岩细砾 2. 72m

45. 鲜紫红色厚层含凝灰质泥岩。中部夹一层厚 1. 6m 之灰紫色晶屑岩屑凝灰岩，空洞发育，局部夹条纹状泥岩，条纹呈平行波状 26. 16m

44. 下部鲜紫红色空洞状粉砂质泥岩；上部浅灰色中层晶屑凝灰岩，含少量粉砂质泥岩团块 5. 16m

43. 紫红色中层晶屑岩屑凝灰岩。底部 30cm 为灰黄色砂砾岩，中部夹一层鲜紫红色厚层状粉砂质泥岩 3. 96m

42. 紫红色含粉砂质泥岩 16. 21m

41. 晶屑凝灰岩。由粗—中细粒组成一个韵律火山沉积层，中下部为紫色，上部为紫红色，含红色安山岩及英安岩岩屑，由下而上增加 5. 54m

40. 鲜紫红色块状粉砂质泥岩。下部有铁质团块，中部夹一层 1m 厚的紫红色中粒长石岩屑砂岩，顶部 30cm 为空洞状粉砂质泥岩 13. 59m

39. 紫灰色含砂质泥岩夹灰绿色细砂岩及砂质泥岩，具绿色泥质斑点 11. 79m

38. 灰绿色中层泥岩，底部 20cm 为灰绿色中粒长石砂岩。产孢粉 *Exesipollenites tumulus*，*E. pseudotriletes*，*Cicatricosisporites* sp. 1. 68m

37. 深灰色中—厚层中粗粒岩屑凝灰岩。底部 80cm 为灰绿色砾质砂岩，顶部 60cm 为紫色钙质结核砂质泥岩 16. 31m

36. 紫红色中层状细粒岩屑长石砂岩夹泥岩，二者无明显界线 12.29m

35. 紫红色中—厚层中—粗粒含砾晶屑岩屑凝灰岩夹中细粒长石砂岩 16.99m

34. 紫色中层细砾质晶屑岩屑凝灰岩。砾石沿层面分布，局部聚集成细砾岩，其厚 1cm 左右，成分为安山岩、英岩 17.63m

33. 灰绿色细砾质沉凝灰岩夹砂砾岩及凝灰岩，底部 30cm 为细砾岩。砂砾岩厚 30~40cm，呈层状中透镜状产出，砾石顺层分布 5.51m

下段 360.51m

32. 灰绿色中层泥岩 0.83m

31. 紫红色厚层泥岩。含粉砂不均匀，局部富集成泥质粉砂岩透镜体，具有不明显的层纹构造 4.65m

30. 上部紫色中层中粒长石岩屑砂岩，层厚 20~30cm，有石英脉贯入；下部为紫色砂砾岩，具有平行层理，砾石为石英、石英岩、安山岩及英安岩 2.70m

29. 紫红色中层泥质岩屑砂质粉砂岩 1.11m

28. 紫灰色细粒长石岩屑砂岩和紫色不等粒长石岩屑砂岩呈不等厚互层 7.54m

27. 灰紫色巨厚层砾质中粗粒长石岩屑砂岩。具正粒序层理，略显大型交错层理 20.80m

26. 灰紫、黄色中厚层中粗粒长石岩屑砂岩，夹细砂岩及细砾岩薄层。层面不清晰，略具正粒序层理 29.20m

25. 灰紫色厚层粗粒岩屑砂岩与同色细砾岩呈韵律不等厚互层。二者厚度比为 4：1，中部夹一层 20cm 厚的紫红色粉砂岩，见交错层理，砾石以石英质岩石为主 22.71m

24. 灰白色巨厚层含细砾粗粒岩屑砂岩，层间夹细砾岩透镜体，中部夹一层 50cm 厚的灰紫色细粒长石砂岩，见大型交错层理 45.48m

23. 紫红色厚层中粒岩屑砂岩，顶部为灰紫色中层长石砂岩，时见泥砾 5.22m

22. 紫灰色巨厚层中粗粒岩屑砂岩。上部夹一层鲜紫红色粉砂质细砾岩，下部夹两层砾岩，沉积构造发育，具大型楔状、槽状层理，冲刷构造等 23.68m

21. 紫灰色厚层含砾中粗粒岩屑砂岩。底部为砾质粗砂岩，砾石沿层面分布，见大型槽状层理 6.45m

20. 紫灰色、浅紫红色含砾粗粒岩屑砂岩与中粗粒岩屑砂岩互层，下部夹三层紫灰色砾质粗砂岩，有石英脉贯入 25.78m

19. 上部鲜紫红色泥质岩屑粉砂岩，下部灰紫色细粒岩屑砂岩 2.24m

18. 紫灰色块状含砾粗粒岩屑砂岩。局部砾石集结成砾质粗砂岩，大型板状层理，槽状层理及平行层理特别发育 16.89m

17. 紫红色厚层细粒岩屑长石砂岩，顶部为紫红色砂质粉砂岩 2.92m

16. 紫灰色中层粗粒岩屑砂岩夹砾岩。中部夹一层 50cm 厚的紫红色细粒长英砂岩，上部夹两层厚 1~2cm 的紫红色泥质岩屑粉砂岩。砾石有 12 层，每层厚 2~100cm，一般厚2~10cm，底部砾石最厚，砾石以细砾为主 38.66m

15. 紫红色中厚层状含砂质泥质岩屑粉砂岩，风化面略显水平层理 7.53m

14. 浅灰色、紫灰色中层岩屑砂岩夹三层砾岩。见韵律层理，夹层厚 2~5cm 7.24m

13. 灰色、紫灰色块状含砾粗粒岩屑砂岩。下部夹 20cm 厚的砾岩，砾石具有成层性。见斜层理、波痕及正粒序层理 8.15m

12. 紫灰色厚层粗粒岩屑砂岩夹砾质粗砂岩及砂砾岩。砾石成层分布，略显正粒序层理，见板状层理、平行层理、冲刷构造，夹层多见于冲刷面上，厚度仅几厘米 18.13m

11. 灰紫红色中层泥质岩屑粉砂岩 10.29m

10. 紫红色细粒岩屑砂岩。底部 1m 为紫色中粒岩屑砂岩，风化面显平行层理构造 6.11m

9. 紫红色细粒岩屑砂岩，中部夹 4cm 厚的泥质条带硅质岩 1.72m

8. 浅紫红色细砾质粗粒岩屑砂岩，略显正粒序层理 1.44m

7. 浅紫红色中厚层中粒岩屑砂岩。局部含泥砾，主要分布于层面上 4.02m

6. 灰紫色含砾粗粒岩屑砂岩。底部为 70cm 厚的砂质细砾岩，砾石分布不均匀，局部构成砂砾岩薄层，见冲刷构造、正粒序层理 5.00m

4~5. 紫红色岩屑砂岩。以细砂岩为主，与中粒砂岩、粉砂岩、泥岩组成两个完整的韵律层 6.54m

3. 紫红色含砾粗粒岩屑砂岩与紫灰色岩屑砂岩不等厚互层。砾石分布不均匀，局部集中成为砾岩薄层 4. 81m

2-2. 鲜紫红色中层泥质岩屑粉—细砂岩 8. 50m

2-1. 暗紫红色砾岩，砾石成分复杂，有石英岩、变砂岩、绢云片岩、千枚岩、石英斑岩、脉石英等，砾径一般为 2~5cm，最大 8cm，圆状至次棱角状。与下伏南碧沟群呈不整合接触 4. 17m

~~~~~~~~ 不　整　合 ~~~~~~~~

下伏地层　南碧沟群浅灰色、紫红色层纹状绢云石英岩

3. 琼海市东太农场尖岭剖面

该剖面由海南省地调院 1997~2000 年进行 1：5 万《黎母岭、枫木市、琼中县、白马岭》幅区调时测制，海南省地质调查院 2001~2003 年开展 1：25 万《琼海县》幅区调工作时补做了部分工作。剖面起点位于河温岭 135. 0m 山包 131°方向约 200m 处，坐标：东经 110°14′11″，北纬 19°3′23″，长约 11666m。自上而下为（图 4-3）：

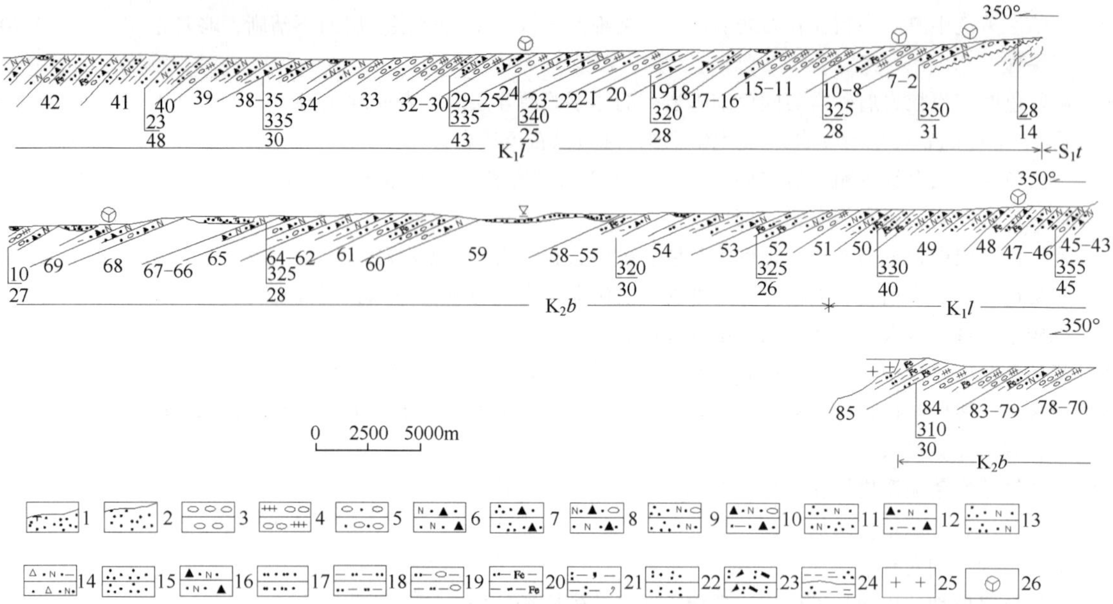

图 4-3　琼海市东太农场尖岭下白垩统鹿母湾组、报万组实测剖面图

Fig. 4-3　Measured section of the Lower Cretaceous Lumuwan and Baowan Formations, Jianling, Dongtai Farmland, Qionghai City

1—浮土；2—河漫滩沉积物；3—砾岩；4—复成分砾岩；5—砂砾岩；6—砾质中粗粒岩屑长石砂岩；7—含砾中粗粒岩屑石英砂岩；8—含砾中粗粒岩屑长石砂岩；9—含砾中粗粒长石石英砂岩；10—含砾中粗粒长石岩屑杂砂岩；11—粗中粒长石石英砂岩；12—粗中粒长石岩屑杂砂岩；13—中细粒长石石英砂岩；14—中细粒长石岩屑杂砂岩；15—中细粒石英砂岩；16—细粒岩屑长石砂岩；17—粉砂质砂岩；18—泥质粉砂岩；19—含砾粉砂质泥岩；20—铁质粉砂质泥岩；21—玻屑凝灰质泥岩；22—沉凝灰岩；23—晶屑玻屑凝灰岩；24—云母石英片岩；25—花岗斑岩；26—孢粉化石

上覆地层　报万组浅灰色、灰白色复成分砾岩

———— 整　合 ————

下白垩统鹿母湾组（K_1l） 2527. 16m

上段

51. 浅灰色块状含凝灰质粗中粒长石岩屑砂岩，杂基中具火山灰，中下部见两条辉绿岩脉侵入 76. 04m

50. 浅紫红色铁泥质粉砂岩 110. 82m

49. 浅灰色、浅灰白色粗中粒长石岩屑杂砂岩，岩石中普遍含少量砾岩石（2%~5%±），中上部夹紫红色厚层—块状泥质粉砂岩（自中部→上部，夹层渐变薄） 193. 49m

48. 浅紫红色块状含铁泥质粉砂岩，上部夹浅紫红色薄层粉砂质中细粒长石岩屑杂砂岩。泥质粉砂岩中含孢粉 *Osmundacidites* sp.，*Cycadopites*. sp.，*Classopollis* sp.，*Laricoidites magnus*
~~~~~~~~

47. 浅灰色、浅灰白色块状粗中粒岩屑长石砂岩，夹浅紫红色块状泥质粉砂岩 65.01m

46. 底部为浅灰白色、浅紫红色含砾中粗粒岩屑长石砂岩，中、上部为浅灰白色含砾中粗粒岩屑长石砂岩夹浅紫红色含砾、长石碎屑泥质粉砂岩。中部泥质粉砂岩中局部具粒序层理，上部泥质粉砂岩中局部含较多凝灰质，上部的含砾砂岩中具同生红层砾石 58.84m

45. 浅灰白色中细粒岩屑长石砂岩，中、上部夹浅灰白色含砾中粗粒长石石英砂岩，局部夹浅紫红色泥质粉砂岩 21.52m

44. 掩盖 5.57m

43. 浅灰色块状细粒长石石英砂岩与浅紫红色块状粉砂质泥岩互层 5.21m

42. 浅灰白色块状含砾中粗粒岩屑长石砂岩与浅紫红色块状铁泥质粉砂岩互层，上部后者层厚渐变薄（变为中厚层） 227.18m

41. 浅紫红色泥质粉砂岩，中上部夹浅灰色厚层—块状中细粒长石石英砂岩，中部局部夹浅灰白色厚层—块状粗中粒长石石英砂岩 233.30m

40. 浅灰白色、浅紫红色中厚层复成分砾岩-浅紫红色中厚层含砾中粗粒岩屑长石砂岩-浅紫红色中厚层含砾中细粒岩屑长石砂岩—浅紫红色中厚层含砾粉砂岩—浅紫红色厚层—块状含长石粉砂质泥岩韵律沉积。每一韵律沉积层见上述岩性中的两三种或以上 20.84m

39. 浅紫红色块状细粒岩屑长石杂砂岩，夹浅灰白色厚层块状含砾粗中粒岩屑长石砂岩、浅紫红色中厚层粉砂泥质岩、含砾砂岩中具同生红层砾石 82.32m

38. 浅紫红色中细粒岩屑长石杂砂岩，上部夹浅紫红色厚层粉砂质泥岩。前者中发育水平沙纹层理 19.55m

37. 下部为浅灰白色粉砂质细粒长石石英杂砂岩，中、上部为浅灰白色、浅紫色含中粗粒岩屑长石砂岩。前者中具水平砂纹层理 12.93m

36. 掩盖 6.85m

35. 浅灰白色粗中粒长石石英砂岩夹浅紫红色厚层块状粉砂质泥岩 8.13m

34. 浅紫红色块状不等粒岩屑长石砂岩与浅灰白色块状中细粒长石岩屑杂砂岩互层 15.38m

33. 浅灰白色复成分砾岩、砂砾岩（由于砾石含量变化大且快，上述两种岩性间界线难以分出。 231.63m

32. 浅紫红色含中细粒长石粉砂质泥岩，砂级长石大小多为 0.1~0.3mm 47.47m

31. 浅紫红色不等粒凝灰质长石岩屑砂岩，上部见夹浅紫红色中厚层—块状含长石粉砂质泥岩，后者中长石碎屑大小不一，一般为 0.1~1mm，少量大者达 2~3mm 21.22m

30. 浅紫红色块状复成分砾岩-浅紫红色块状含长石粉砂质泥岩韵律沉积 40.02m

29. 浅紫红色块状晶屑玻屑凝灰岩，含孢粉：*Monosulcites* sp.，*Araucariaacites* sp.，*Classopollis* sp.，*Abietineaepollenites* sp.，*Ephedripites* sp.，*Magnolipollis* sp.，*Retitricolporites* sp. 77.71m

28. 浅紫红色含砾粗中粒长石岩屑杂砂岩，具同生红层砾石 5.59m

27. 浅紫红色块状复屑凝灰岩夹浅紫红色中厚层含砾复屑凝灰岩 14.0m

26. 浅紫红色、浅灰白色含砾（或砾质）粗中粒长石岩屑杂砂岩夹浅紫红色中层粉砂质泥岩，前者中具同生红层砾石 8.42m

25. 浅紫红色块状复成分砾岩-浅紫红色含砾（或不含砾）玻屑晶屑沉凝灰岩韵律沉积。凝灰岩的层厚见变化，其中砾石成分多为长石，砾岩中具同生红层砾石 20.60m

24. 浅紫红色块状玻屑凝灰质泥岩，下部局部夹浅紫红色厚层中细粒长石石英砂岩，上部局部夹紫红色中厚层含砾中粗粒（或粗中粒）长石石英砂岩 126.98m

23. 浅紫红色复成分砾岩，具同生红层砾石 36.57m

22. 浅紫红色块状粉砂质泥岩夹浅紫红色中厚层复成分砾岩 32.93m

21. 浅紫红色块状复成分砾岩夹浅紫红色中厚层粉砂质泥岩，底部为浅紫红色含凝灰质粗中粒岩屑长石杂砂岩 67.58m

20. 浅紫红色含不等粒碎屑凝灰质泥岩。中下部岩石中，中细粒碎屑（长石、石英、岩屑）多聚集成团块产出；上部岩石中，碎屑（砂级）含量明显减少，粒度也变大 88.97m

19. 浅紫红色块状含不等粒碎屑泥质岩 40.03m

18. 浅紫红色块状粉砂岩 54.10m

17. 掩盖 39.01m

16. 下部为浅紫红色细粒岩屑长石杂砂岩-浅紫红色粉砂质泥岩韵律沉积，中、上部为浅紫红色含砾中粗粒长石英砂岩 16.73m

15. 浅灰白色复成分砾岩 6.42m

14. 下部为浅紫红色与浅灰白色复成分砾岩互层，中、上部为浅紫红色复成分砾岩，局部夹浅紫红色凝灰质泥岩，中、上部的浅紫红色复成分砾岩中发育大型交错层理、平行层理。砾岩中具红层砾石 23.68m

13. 浅紫红色复成分砾岩-浅紫红色含砾粉砂质泥岩-浅紫红色含长石粉砂质泥岩-浅紫红色粉砂质泥岩单韵律沉积（每一韵律中出现上述两种以上岩性）。此层总体上观察，中下部以含砾岩石为主，上部则相反。砾岩中具同生红层砾石 48.60m

12. 浅灰白、浅紫红色块状复成分砾岩，上部夹浅紫红色不等厚层含砾或不含砾晶屑凝灰岩。含砾者与不含砾者构成韵律沉积。中部偏上的浅紫红色复成分砾岩中发育中型交错层理，大型平行层理。砾岩中具同生红层砾石 38.54m

11. 浅紫红色块状沉凝灰岩 30.06m

10. 浅紫红色厚层—块状含细中粒碎屑铁泥质粉砂岩与浅灰白色中厚层细中粒岩屑杂砂岩互层。铁泥质粉砂岩中含孢粉：*Osmundacidites* sp.，*Ephedripites* sp. 5.16m

9. 浅灰白色块状复成分砾岩，中上部岩石中砾石含量见随机变化（15%~35%），具红层砾石 37.29m

8. 浅灰白色复成分砾岩-浅灰白色含砾中粗粒长石石英砂岩韵律沉积，中部夹浅紫红色厚层粉砂质泥岩，砾岩中具同生红层砾石 15.59m

7. 浅灰白色厚层块状中细粒岩屑石英砂岩夹浅紫红色中厚层粉砂质泥岩 10.49m

6. 浅紫红色块状晶屑沉凝灰岩，下部岩石中含孢粉：*Deltoidospora* sp.，*Punctatisporites* sp.，*Toroisporis* sp.，*Schizaeoisporites* sp.，*S. laevigataefomis*，*Monosulcites* sp.，*Inaperturopollenites* sp.，*Classopollis* sp.，*Taxodiaceaepollenites* sp.，*Exesipollenites tumulus Balme* subsp. *triagulus*，*Abietineaepollenites* sp.，*Laricoidites magnus*，*L. araucarites*，*Ephedripites* sp.，*E.*（*Distachyapites*）*fusiforms*，*Quercoidites* sp.，*Tricolpites* sp. 79.95m

下段

5. 浅灰白色砾质中粗粒岩屑杂砂岩，上部为浅灰白色块状复成分砾岩。具同生红层砾石 24.16m

4. 中、下部为浅紫红色块状含砾粗中粒长石岩屑杂砂岩，上部为浅紫色与浅灰白色厚层—块状含砾粗中粒长石岩屑杂砂岩互层 8.84m

3. 浅灰白色块状含砾粗中粒岩屑杂砂岩，上部夹浅紫红色中薄层细中粒岩屑杂砂岩，前者中具同生红层砾岩 5.60m

2. 下部为浅灰白色含砾中粗粒长石岩屑杂砂岩，中上部为浅灰白色复成分砾岩，中部夹浅紫红色薄层粉砂质泥岩 5.22m

~~~~~~ 不　整　合 ~~~~~~

下伏地层　下志留统陀烈组浅灰色、浅灰白色云母石英片岩

4. 儋州市王五市平地岭剖面

该剖面由海南地质大队 1994~1996 年开展 1∶5 万《海头市、排浦市、王五市、白马井镇》幅区调时测制，海南省地质调查院 2005~2007 年开展 1∶25 万《东方县》幅区调时补做了部分工作。剖面起点位于大岭 285°方向 850m 处，坐标：东经 109°21′5″，北纬 19°35′4″，全长 10940m。自上而下为（图 4-4）：

未见顶

鹿母湾组（K_1l）

上段（K_1l^2） 总厚度>440.9m

72. 浅褐色、浅灰色块状砾岩 15.77m

71. 掩盖 16.15m

70. 浅褐色块状砾岩 15.26m
~~~~~~

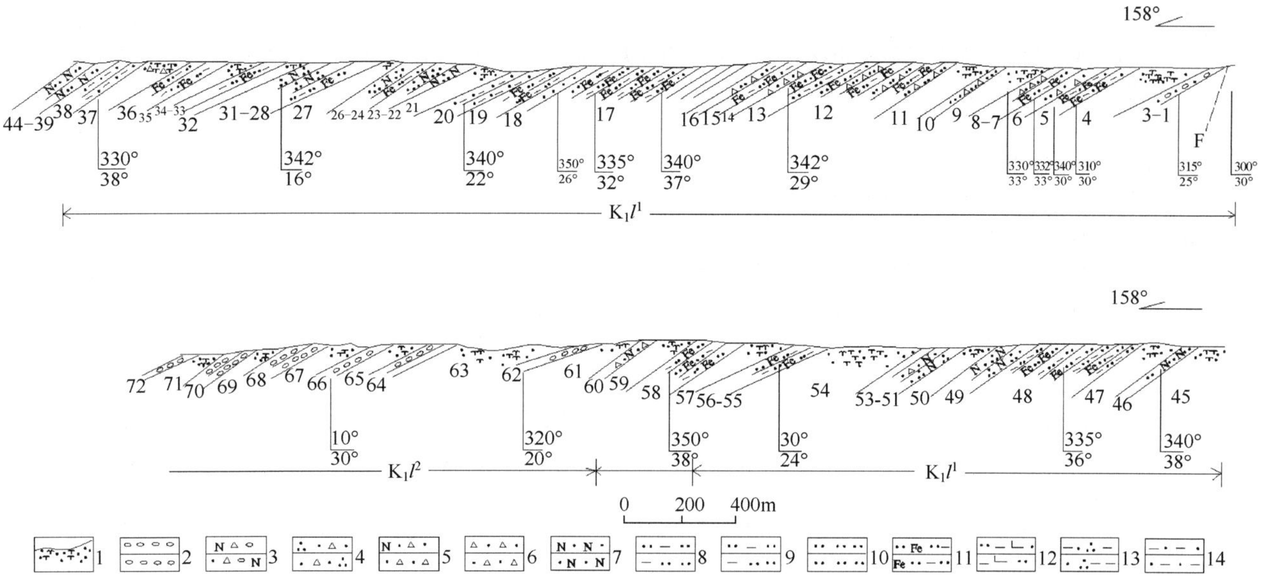

图 4-4　儋州市王五市平地岭下白垩统鹿母湾组剖面

Fig. 4-4　Section of the Lower Cretaceous Lumuwan Formation, Pingdiling, Wangwu District, Danzhou City

1—腐殖土；2—砾岩；3—含砾长石岩屑杂砂岩；4—岩屑石英砂岩；5—长石岩屑砂岩；6—泥质粗粉砂岩；7—长石砂岩；8—泥质粗粉砂岩；9—砂质泥岩；10—石英岩；11—铁泥质粗粉砂岩；12—泥钙质粗粉砂岩；13—泥质石英砂岩；14—泥质砂岩

69. 掩盖　19.8m

68. 浅褐色块状砾岩　2.61m

67. 掩盖　23.48m

66. 浅褐色块状砾岩　16.07m

65. 掩盖　78.99m

64. 浅褐色、浅灰白色块状砾岩夹浅灰白色中粗粒长石岩屑杂砂岩　23.98m

63. 掩盖　204.17m

62. 浅灰白色块状砾岩夹浅灰白色块状砂砾岩，中上部夹层紫红色泥质粗粉砂岩　24.61m

下段（$\mathbf{K_1}\boldsymbol{l}^1$）　总厚度 3129.22m

61. 掩盖　52.32m

60. 浅灰色、浅褐色块状中粗粒长石岩屑砂岩夹浅灰白色中薄层泥质粉砂岩，具有不明显的粒序层理，粒度向上变细　10.13m

59. 掩盖　55.78m

58. 浅紫红色块状铁泥质粗粉砂岩　83.62m

57. 掩盖　74.89m

56. 浅灰色中厚层含砾中粗粒岩屑长石杂砂岩夹浅紫红色中薄层铁泥质粗粉砂岩　16.23m

55. 浅紫红色厚层—块状含铁泥质粗粉砂岩，铁泥质粗粉砂岩夹浅灰白色厚层泥质，砂质粗砂岩，具平行层理　41.57m

54. 掩盖　160.07m

53. 浅紫红色泥质粗粉砂岩　11.15m

52. 浅紫红色、浅灰白色中粗粒岩屑长石杂砂岩，中部夹紫红色薄层泥质粗粉砂岩　28.87m

51. 浅紫红色泥质粗粉砂岩　46.00m

50. 掩盖　103.87m

49. 浅灰白色、浅紫红色砂泥质粗砂岩　37.09m

48. 浅灰绿色、浅灰白色厚层泥质粉砂岩与浅紫红色厚层—块状铁泥质粗粉砂岩互层，具大型水平层理　226.38m

47. 掩盖　85.29m

46. 浅灰白色中厚层粗中粒长石岩屑砂岩与浅紫红色中厚层粗粉砂岩互层，平行层理　30.78m

45. 掩盖 75.62m

44. 上部为浅紫色、浅紫红色砂质粗粉砂岩，下部为浅灰白色中细粒长石石英砂岩，砂质粗粉砂岩中具小型沙纹层理 15.76m

43. 掩盖 19.50m

42. 浅紫红色泥质粗粉砂岩 46.28m

41. 浅灰白色含砾中粗粒岩屑长石砂岩 9.72m

40. 浅紫红色铁泥质粗粉砂岩 14.64m

39. 浅灰白色砾岩。砾岩成分为石英、长石、粉砂岩 8.89m

38. 浅紫红色砂质粗粉砂岩、铁泥质粗粉砂岩，底部岩石中铁质含量少，为浅灰白色泥质粗粉砂岩，砂质粗粉砂岩中具小型沙纹层理 55.83m

37. 掩盖 80.17m

36. 浅紫红色铁泥质粗粉砂岩，上部夹浅灰白色厚层—块状泥质粗粉砂岩 32.88m

35. 掩盖 69.99m

34. 浅灰白色泥质粗粉砂岩 5.80m

33. 浅紫红色含铁、泥质粗粉砂岩，具微平行层理 19.05m

32. 掩盖 19.20m

31. 浅紫红色泥质粗粉砂岩，夹薄层浅灰白色岩屑粗粉砂岩 19.50m

30. 掩盖 14.96m

29. 浅灰白色岩屑粗粉砂岩夹浅紫红色铁泥质粗粉砂岩 52.65m

28. 浅紫红色铁泥质粗粉砂岩夹浅紫红色中薄层细粒长石岩屑砂岩 14.56m

27. 掩盖 30.88m

26. 浅灰白色岩屑粗粉砂岩，下部夹浅灰绿色砂质粗粉砂岩，上部夹薄层浅紫红色铁泥质粗粉砂岩，具不够明显的微水平、平行层理 19.13m

25. 浅紫红色中厚层细粒长石岩屑砂岩与浅紫红色中厚层粗粉砂岩互层 22.07m

24. 浅紫红色铁泥质粗粉砂岩 20.78m

23. 浅紫红色细粒长石岩屑砂岩，夹浅紫红色薄层铁泥质粗粉砂岩 65.37m

22. 浅紫红色中层含泥质粗粉砂岩与浅紫红色中层含砂质、泥质粗粉砂岩 9.98m

21. 掩盖 62.82m

20. 浅紫红色泥质粗粉砂岩夹浅紫红色粉砂质细粒岩屑杂砂岩 56.97m

19. 浅紫红色铁泥质粗粉砂岩，底部砂质含量增多 42.82m

18. 浅灰白色砂质粗粉砂岩，夹浅紫红色薄层铁泥质粗粉砂岩，具微平行层理 11.37m

17. 浅紫红色铁泥质粗粉砂岩，底部铁、泥质含量少，成为含泥质粗粉砂岩 290.07m

16. 浅灰白色中层微粒岩屑杂砂岩与浅紫红色中层铁泥质粗粉砂岩互层 54.61m

15. 浅紫红色铁、泥质粗粉砂岩 52.41m

14. 浅灰白色微粒岩屑杂砂岩 31.95m

13. 浅紫红色中厚层铁泥质粗粉砂岩，下部夹薄层浅灰白色中粒长石岩屑砂岩 95.23m

12. 浅灰白色中厚层—块状中粒、中细粒岩屑杂砂岩与浅紫红色泥质粉砂岩互层 170.59m

11. 掩盖 138.42m

10. 浅灰白色中粒长石岩屑砂岩 7.38m

9. 掩盖 73.87m

8. 浅灰白色、浅紫红色中厚层—块状中细粒长石岩屑砂岩与浅紫红色铁泥质粗粉砂岩互层 23.80m

7. 浅紫红色铁泥质粗粉砂岩 40.51m

6. 浅灰白色块状中细粒长石岩屑砂岩夹浅紫红色厚层—块状铁泥质粗粉砂岩 42.23m

5. 浅紫红色铁泥质粗粉砂岩 30.54m

4. 掩盖 77.02m

3. 浅灰、浅紫灰色复成分砾岩，夹浅紫红色薄层铁泥质粗粉砂岩。砾石成分主要为粉砂岩、次为泥质岩及石英岩 7.94m

2. 掩盖 8.60m

══════ 推测为断层 ══════

下伏地层 陀烈组浅灰色、浅灰黑色长石石英岩

5. 东方市乐安村剖面

该剖面由海南地质大队1990-1993年开展1:5万《昌化大岭、海尾、昌化港、叉河》幅区调时测制。起点位于东方市乐安村170°约750m处，终点止于新村135°约500m处。全长约3225m，总厚度大于1837.11m。自上而下为（图4-5）：

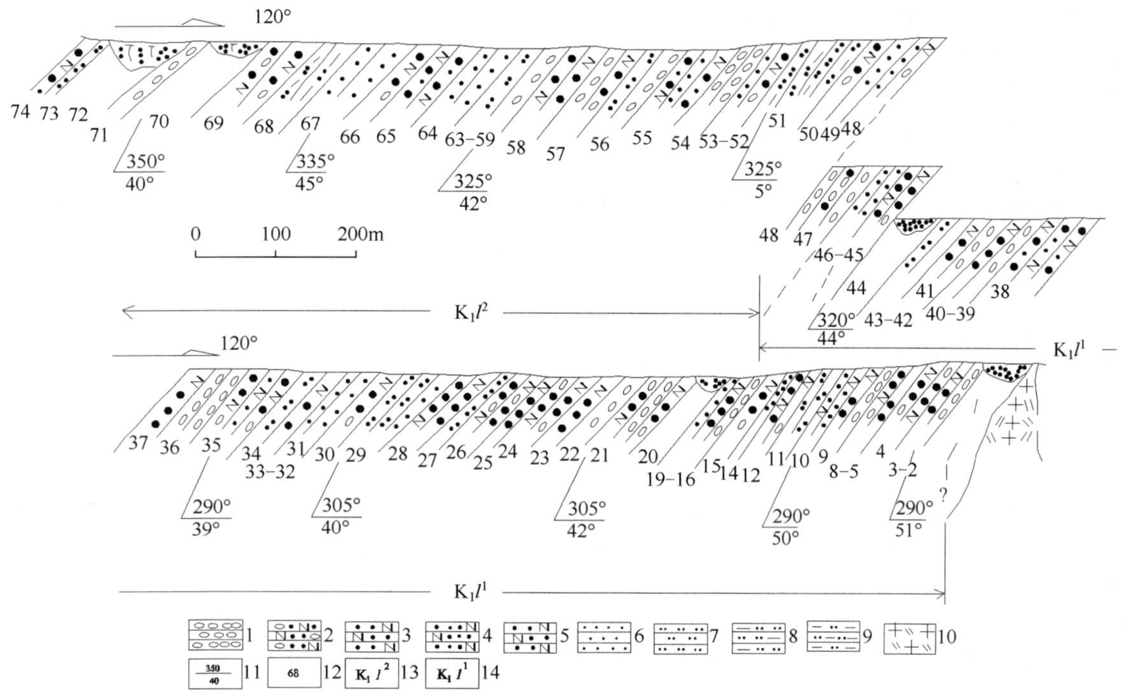

图4-5 东方市乐安村鹿母湾组实测剖面图

Fig. 4-5 Measured section of the Lumuwan Formation, Le'an Village, Dongfang City

1—砾岩；2—含砾粗粒岩屑长石杂砂岩；3—粗粒岩屑长石杂砂岩；4—中粗粒岩屑长石杂砂岩；5—中粗粒岩屑长石杂砂岩；6—细砂岩；7—粉砂岩；8—泥质粉砂岩；9—粉砂质泥岩；10—三叠纪中粒白云母二长花岗岩；11—产状；12—分层号；13—鹿母湾组二段；14—鹿母湾组一段

未见顶

鹿母湾组（K_1l） 总厚度>1837.11m

上段（K_1l^2） 总厚度>846.08m

74. 紫红色中薄层含砾粗粒岩屑石英杂砂岩，局部砾石可富集成砾岩 4.08m
73. 暗红色薄层粉砂岩 5.24m
72. 覆盖 195.83m
71. 紫灰色中薄层砾岩 4.02m
70. 覆盖 60.22m
69. 上部红色薄层岩屑长石细砂岩夹粗砂岩，下部为灰黄色中—厚层含砾粗粒岩屑长石杂砂岩 15.09m
68. 棕红色中厚层泥质粉砂岩 16.14m
67. 紫红色薄层岩屑长石细砂岩夹粗砂岩 35.17m
66. 紫红色中层含砾中粗粒岩屑长石杂砂岩。底部为厚约1m的砾岩 9.36m
65. 紫红色中粗粒岩屑长石杂砂岩夹细砂岩及砾岩 17.94m
64. 紫红色薄层细砂岩夹中粒岩屑长石杂砂岩 25.01m
63. 紫红色厚层含砾粗粒岩屑长石杂砂岩，底部为砾岩 5.13m
62. 紫红色中—厚层砾岩，顶部为粉砂岩夹含砾细砂岩 1863m
61. 紫灰色中厚层粗粒岩屑长石杂砂岩 6.58m
60. 紫红色中—厚层细粉砂岩 7.11m
59. 紫灰色中—厚层砾岩 6.85m

58. 紫灰色中层砾岩，含砾粗粒岩屑长石杂砂岩，向上粒度变小 12. 23m
57. 紫灰色中层砾岩，含砾粗粒岩屑长石杂砂岩，向上粒度变小 29. 95m
56. 灰紫色中—厚层砾岩与粉砂岩不等厚互层 30. 18m
55. 紫红色中层中粗粒岩屑长石杂砂岩夹细砂岩 59. 27m
54. 紫红色厚—巨厚层砾岩，局部夹中层中粗粒长石岩屑杂砂岩及薄层粉、细砂岩 35. 27m
53. 紫灰色中—厚层中—细粒长石岩屑杂砂岩 24. 3m
52. 紫灰色中—厚层含砾粗粒岩屑杂砂岩夹粉砂岩 51m
51. 紫红色薄层粉砂质泥岩 40. 88m
50. 紫红色中层含砾粗粒岩屑长石杂砂岩夹粉砂岩 20. 35m
49. 紫红色厚层含砾粗粒长石岩屑杂砂岩夹薄层粉砂岩。顶部变为粉砂岩 24. 21m
48. 紫灰色中厚层—厚层砾岩、含砾粗粒杂砂岩与粉砂岩不等厚互层。粉砂岩发育水平层理 39. 04m
47. 紫红色中厚层—厚层砾岩、含砾粗砂岩夹细砂岩及粉砂岩 47. 05m

下段（K_1l^1） 991. 03m

46. 紫灰色中厚层细砂岩，上部变为薄层粉砂岩 20. 96m
45. 紫红色含砾粗粒岩屑杂砂岩夹中薄层细砂岩 26. 98m
44. 覆盖 46. 95m
43. 紫红色薄层细砂岩 27. 27m
42. 紫红色粉砂岩 5. 23m
41. 紫灰色中厚层—厚层粗粒与中、细粒砂岩互层，底部夹砾岩 39. 95m
40. 紫灰色中—厚层中粒长石岩屑杂砂岩，向上变细为细砂岩。中粒长石岩屑杂砂岩具水平层理 15. 4m
39. 紫红色砾岩、含砾粗粒长石岩屑杂砂岩夹细砂岩 19. 88m
38. 紫红色中薄层中细粒砂岩夹中粗粒岩屑杂砂岩 11. 71m
37. 紫灰色中厚层中—粗粒长石岩屑杂砂岩，局部含砾。与 36 层之间为霏细岩脉 45. 01m
36. 紫红色砾岩，砾石向上减少过渡为含砾粗粒长石岩屑杂砂岩，具粒序层理 11. 91m
35. 紫红色中厚层粗粒长石岩屑杂砂岩，含砾粗粒长石岩屑杂砂岩与中薄层细砂岩不等厚互层 56. 79m
34. 紫红色厚层中细粒砂岩夹粗粒长石岩屑杂砂岩 20. 1m
33. 紫红色厚层含砾粗粒长石岩屑杂砂岩夹细砂岩，下部含砾 5. 26m
32. 紫红色中厚层粗粒长石岩屑杂砂岩，局部含砾或夹薄层砾岩，底部有两层厚分别为 1. 8m 和 1. 3m 的火山凝灰岩 13. 28m
31. 红色中—薄层细砂岩夹薄层粗粒长石岩屑杂砂岩 25. 02m
30. 紫红色中厚层粗粒长石岩屑杂砂岩夹细砂岩，下部含砾 16. 6m
29. 紫红色薄层粉砂岩，底部为细砂岩 48. 4m
28. 紫灰色、紫红色中粗粒长石岩屑杂砂岩，下部含砾，夹细砂岩 19. 58m
27. 紫灰色薄层粉砂岩夹中粗粒长石岩屑杂砂岩。中粗粒杂砂岩中见平行层理 5. 26m
26. 紫灰色薄—中厚层含砾粗粒长石岩屑杂砂岩夹细砂岩 28. 44m
25. 暗紫色、紫红色中—厚层中粒长石岩屑杂砂岩夹薄层中细粒砂岩 8. 5m
24. 紫红色中—厚层含砾粗粒岩屑杂砂岩。与 23 层之间为霏细斑岩脉 13. 49m
23. 紫红色中薄层粗粒长石岩屑杂砂岩。局部含砾 13. 95m
22. 紫红色中—薄层砾岩 24. 8m
21. 紫灰色中—厚层含砾粗粒长石岩屑杂砂岩，中部夹砾岩，顶部为细砂岩，砾岩具正粒序层 22. 44m
20. 覆盖 67. 42m
19. 暗紫红色薄层粉砂岩夹细砂岩 6. 93m
18. 红色中—薄层含砾中粗粒长石岩屑杂砂岩 5. 39m
17. 下部紫红色厚层砾岩，中部中薄层粗粒长石岩屑杂砂岩，上部薄层粉砂岩 9. 14m
16. 紫红色中厚层中粗粒长石岩屑杂砂岩夹粉砂岩，底部为 0. 8m 砾岩 12. 75m
15. 紫红色中薄层—中厚层含砾粗粒长石岩屑杂砂岩，顶部为中薄层粉砂岩 34. 07m
14. 紫灰色厚层粉砂岩 17. 64m
13. 紫红色厚层粗粒长石岩屑杂砂岩 6. 12m

12.	底部紫红色厚层含砾粗粒长石岩屑杂砂岩。中上部为中厚层粉砂岩	9. 46m
11.	紫红色厚层粉砂质泥岩	11. 24m
10.	底部紫红色薄—厚层中细粒砂岩，中上部粉砂岩	38. 1m
9.	灰白色砾岩与含砾粗、中粗粒长石岩屑杂砂岩不等厚互层	29. 56m
8.	紫红色中粒长石岩屑杂砂岩，顶部夹泥质粉砂岩	8. 96m
7.	紫红色厚层含砾粗粒长石岩屑杂砂岩，顶部为厚层砾岩	11. 93m
6.	下部紫红色粗粒长石岩屑杂砂岩，上部为粉砂质泥岩夹含砾长石岩屑杂砂岩	8. 95m
5.	紫红色厚层粗粒长石岩屑杂砂岩夹细砂岩、粉砂岩	6. 56m
4.	紫红色含砾粗粒长石岩屑杂砂岩夹砾岩	23. 21m
3.	底部紫红色厚层砾岩，中部粗粒长石岩屑杂砂岩，上部为粉砂岩	8. 34m
2.	底部紫红色含砾粗粒长石岩屑杂砂岩，中上部中—粗粒长石岩屑杂砂岩夹粉砂岩	12. 05m
1.	覆盖	34. 69m

下伏地层　中三叠世中粒斑状黑云母二长花岗岩

（二）六罗村组剖面

1. 三亚市北岭村剖面

该剖面由海南地质综合勘察院 1994 开展 1：5 万《九所、高峰、崖城》幅区调时测制，主要控制牛腊岭火山岩盆下白垩统六罗村组火山岩。剖面起点位于三亚市崖城镇北岭村公路旁，坐标：东经 109°10′35″，北纬 18°25′31″，终点止于抱头村南东约 550m 的河沟处。自上而下为（图 4-6）：

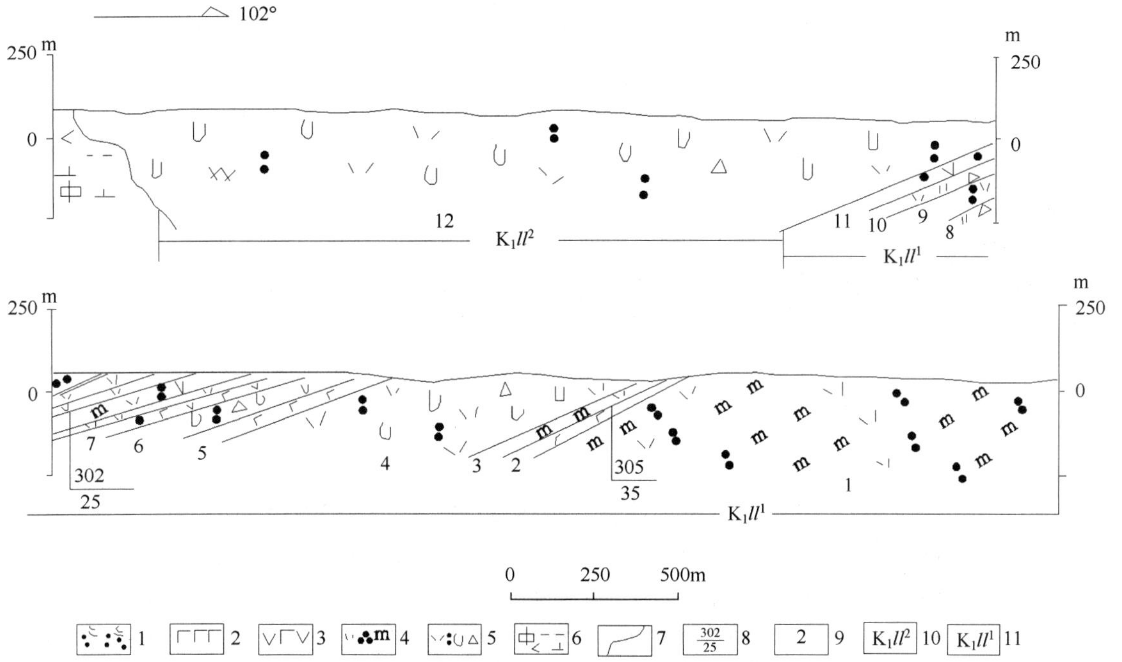

图 4-6　三亚市北岭村下白垩统六罗村组实测剖面图

Fig. 4-6　Measured section of the Lower Cretaceous Liuluocun Formation，Beiling Village，Sanya City

1—浮土；2—玄武岩；3—玄武安山岩；4—流纹质熔结凝灰岩；5—流纹质含砾晶屑凝灰熔岩；
6—晚白垩世中粒角闪黑云花岗闪长岩；7—地质界线 8—产状；9—分层号；10—六罗村组第二段；11—六罗村组第一段

晚白垩世抱浩单元　粗中粒黑云角闪石英二长闪长岩

~~~~~~~~ 侵入接触 ~~~~~~~~

下白垩统六罗村组第二段（K_1ll^2）	总厚度>907. 05m
12. 流纹质含砾晶屑凝灰熔岩	>907. 05m
下白垩统六罗村组第一段（K_1ll^1）	总厚度>818. 52m
~~~~~~~~

11. 玄武安山岩　11.78m

10. 流纹质含砾晶屑凝灰熔岩　6.16m

9. 流纹质熔结凝灰岩　55.67m

8. 流纹质含砾晶屑凝灰熔岩　30.14m

7. 玄武安山岩　4.73m

6. 流纹质含砾晶屑凝灰熔岩　11.97m

5. 玄武岩　34.47m

4. 流纹质含砾晶屑凝灰熔岩　163.18m

3. 流纹质熔结凝灰岩　28.27m

2. 玄武岩　5.13m

1. 流纹质熔结凝灰岩　>467.02m

未见底

2. 三亚市六罗村剖面

该剖面为代表性剖面。由海南地质综合勘察院 1991～1994 年开展 1：5 万《九所、高峰、崖城》幅区调时测制。剖面起点位于三亚市高峰镇六罗村 135°约 400m 处公路旁侧，坐标：东经 109°30′00″，北纬 18°23′42″，终点位于尖岭北东约 2500m 处，全长 17647.98m。控制高峰地区同安岭火山岩盆下白垩统六罗村组、汤他大岭组、岭壳村组。自上而下为（图 4-7）：

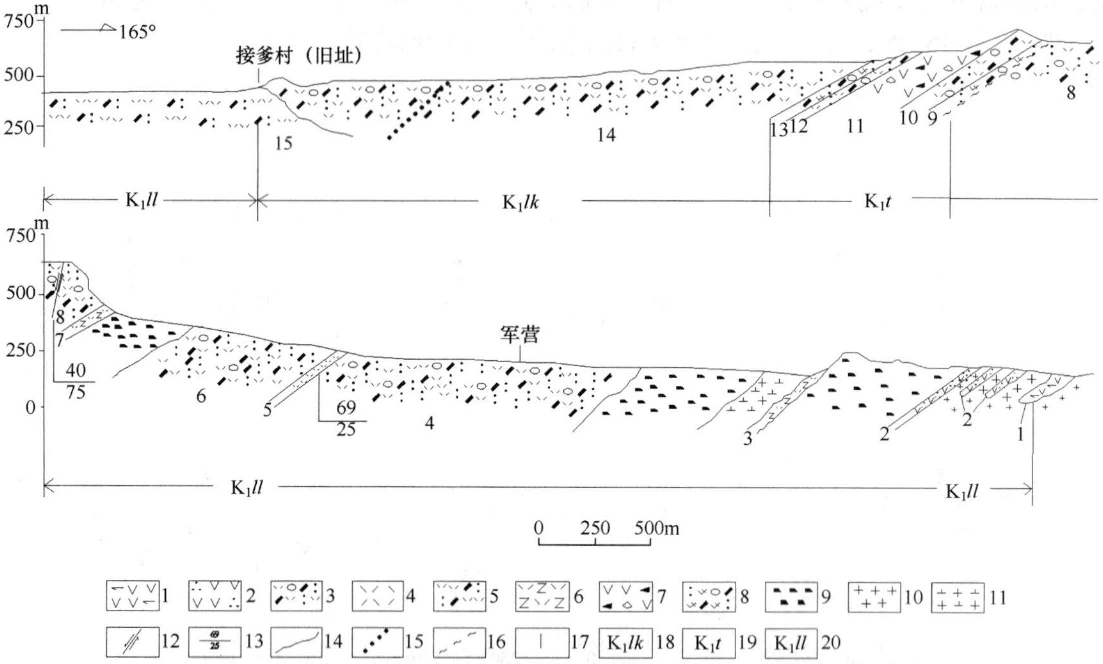

图 4-7　三亚市六罗村下白垩统六罗村组、汤他大岭组、岭壳村组实测剖面图

Fig. 4-7　Measured section of the Lower Cretaceous Liuluocun Formation, Tangtadaling Formation and Qiaokecun Formation, Liuluo Village, Sanya City

1—辉石安山岩；2—石英安山岩；3—流纹质含砾晶屑凝灰熔岩；4—流纹岩；5—流纹质晶屑凝灰熔岩；6—霏细岩；7—安山质岩屑角砾熔岩；8—英安质含砾晶屑凝灰熔岩；9—碎斑熔岩；10—花岗斑岩；11—花岗闪长岩；12—正断层；13—产状；14—地质界线；15—岩相界线；16—推测喷发不整合界线；17—分层号；18—岭壳村组；19—汤他大岭组；20—六罗村组

上覆地层　下白垩汤他大岭组英安质含砾晶屑凝灰熔岩

～～～～～　喷发不整合　～～～～～

六罗村组第二段（K_1ll^2）　总厚度 1681.60m

8. 流纹质含砾晶屑凝灰熔岩　231.30m

7. 霏细岩，底部侵出相的碎斑熔岩　36.10m

6. 流纹质（含砾）晶屑凝灰熔岩 601. 50m

5. 流纹岩 18. 60m

4. 流纹质含砾晶屑凝灰熔岩（U-Pb：107Ma），底部为碎斑熔岩及闪长岩侵入 788. 60m

3. 霏细岩，底部为侵入相碎斑熔岩 5. 50m

六罗村组第一段（K_1ll^1） 总厚度>110. 60m

2. 石英安山岩，潜火山岩相的花岗斑岩侵入多处，地层出露不全 >98. 50m

1. 辉石安山岩 12. 10m

~~~~~~~~ 侵入接触 ~~~~~~~~

下伏地质体 花岗斑岩（侵入）

3. 三亚市周凤园剖面

该剖面由海南地质综合勘察院 1991～1994 年开展 1：5 万《九所、高峰、崖城》幅区调时测制。剖面起点位于三亚市梅西镇周凤园南部约 125 m 的公路旁（坐标：东经 108°59′11″，北纬 18°22′22″），终点位于岭头营村 70°约 950m 处。主要控制牛腊岭火山岩盆下白垩统六罗村组。自上而下为：

上覆地层 第四系含泥质砂

~~~~~~~~ 喷发不整合 ~~~~~~~~

六罗村组第二段 总厚度>606. 14m

14. 流纹质晶屑凝灰熔岩 >606. 14m

六罗村组第一段 厚度>614. 96m

13. 含火山角砾长石石英杂砂岩 76. 6m

12. 硅质岩 7. 96m

11. 含火山角砾长石石英杂砂岩 34. 75m

10. 泥质岩 4. 96m

9. 含火山角砾长石石英杂砂岩 76. 05m

8. 含火山集块角砾复成分砾岩 51. 44m

7. 含火山角砾长石石英杂砂岩 31. 11m

6. 含火山集块角砾复成分砾岩 9. 01m

5. 含火山角砾长石石英杂砂岩 44. 97m

4. 含火山集块角砾复成分砾岩 44. 55m

3. 长石石英砂岩 7. 8m

2. 含砾长石石英砂岩 14. 56m

1. 长石石英砂砾岩 7. 8m

未见底

4. 三亚市阜牯岭剖面

该剖面为海南地质大队 1978 年对《育才 M-Ⅷ-10 航磁异常》验证时所测制的（Ⅳ）剖面。海南地质综合勘察院 1991～1994 年开展 1：5 万《九所、高峰、崖城》幅区调时对其略作修改。剖面位于三亚市立才农场阜牯岭，坐标：东经 109°18′59″，北纬 18°25′57″，自上而下为：

六罗村组第二段 总厚度>170. 17m

12. 流纹质晶屑凝灰熔岩 >170. 17m

═══════ 断　　层 ═══════

六罗村组第一段 总厚度>734. 84m

11. 安山岩 >400m

10. 火山角砾岩 9. 54m

9.	流纹质晶屑凝灰熔岩	31.76m
8.	安山质晶屑凝灰熔岩	30.21m
7.	流纹质晶屑凝灰熔岩	31.76m
6.	火山角砾岩	15.86m
5.	流纹质晶屑凝灰熔岩	57.35m
4.	火山角砾岩	13.93m
3.	流纹质晶屑凝灰熔岩	63.99m
2.	凝灰岩	24.58m
1.	火山角砾岩	55.95m

（三）汤他大岭组剖面

1. 六罗村剖面

剖面情况同上述。自上而下为（图4-7）：

上覆地层　下白垩统岭壳村组流纹质含砾晶屑凝灰熔岩

～～～～～～ 喷发不整合 ～～～～～～

下白垩统汤他大岭组（K_1t）		总厚度 421.90m
13.	英安质含砾晶屑凝灰熔岩	56.20m
12.	流纹质含砾晶屑凝灰熔岩	64.70m
11.	安山质岩屑角砾熔岩	228.20m
10.	流纹质含砾晶屑凝灰熔岩	53.50m
9.	英安质含砾晶屑凝灰熔岩	19.30m

～～～～～～ 喷发不整合 ～～～～～～

下伏地层　下白垩统六罗村组流纹质含砾晶屑凝灰熔岩

2. 汤他大岭剖面

该剖面为汤他大岭组代表性剖面，由海南地质大队一分队1978年开展《海南岛崖县汤他地区地质矿产调查与M-Ⅷ-4航磁异常验证》时测制。剖面位于三亚市高峰乡汤他大岭西南侧，起点坐标：东经109°22′10″，北纬18°21′31″。自上而下为：

上覆地层　岭壳村组流纹质角砾凝灰熔岩

～～～～～～ 喷发不整合 ～～～～～～

汤他大岭组		总厚度 588.75m
20.	英安岩，具流纹结构	38.20m
19.	英安质含砾凝灰熔岩	98.5m
18.	英安岩，流纹结构发育	101.95m
17.	流纹质含砾凝灰熔岩	25.32m
16.	英安质凝灰熔岩	31.58m
15.	英安岩，安山质凝灰熔岩	27.76m
14.	英安质含砾凝灰熔岩，底部为英安质角砾凝灰熔岩	102.38m
13.	英安质粉屑沉凝灰岩	14.32m
12.	英安岩	57.95m
11.	流纹质含砾凝灰熔岩	35.16m
10.	英安岩	40.88m
9.	弱绿帘石化安山质角砾凝灰熔岩	14.75m

～～～～～～ 喷发不整合 ～～～～～～

下伏地层　六罗村组流纹质含砾凝灰熔岩

3. 三亚市抱冲村剖面

该剖面由海南地质综合勘察院1991~1994年开展1：5万《九所、高峰、崖城》幅区调时测制。剖面起点位于三亚市立才农场抱冲村351°约800m的公路边，坐标：东经109°22′44″、北纬18°26′3″，终点在扎浩村南西向约125m处。主要控制高峰地区同安岭火山盆地汤他大岭组火山岩。自上而下为（图4-8）：

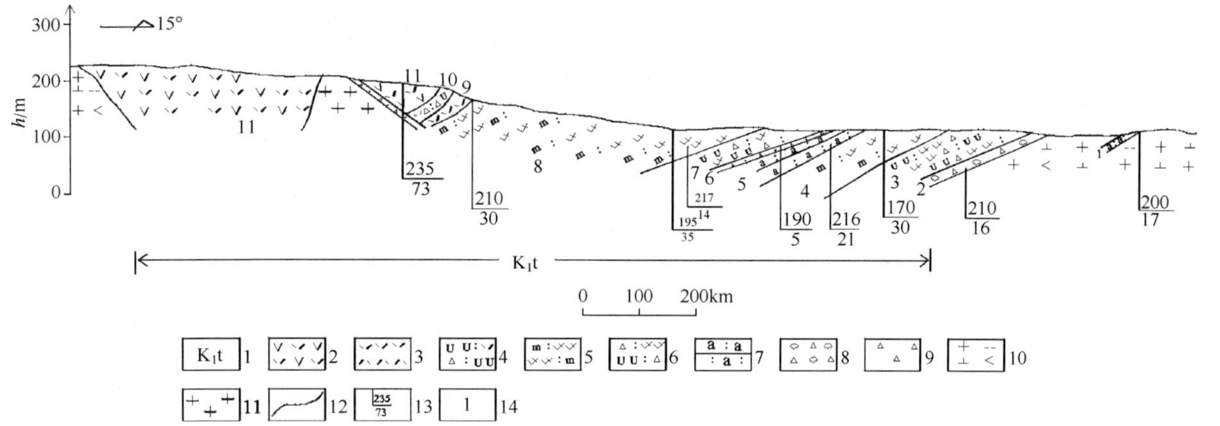

图4-8　三亚市抱冲村下白垩统汤他大岭组实测剖面

Fig. 4-8　Measured section of the Lower Cretaceous Tangtadaling Formation, Baochong Village, Sanya City

1—下白垩统汤他大岭组；2—英安斑岩；3—流纹斑岩；4—流纹质角砾凝灰熔岩；5—英安质熔结凝灰岩；6—英安质角砾凝灰熔岩；7—沉凝灰岩；8—集块角砾岩；9—断层角砾岩；10—角闪黑云花岗闪长岩；11—花岗斑岩；12—地质界线；13—产状；14—分层号

上覆地层　早白垩世细中粒角闪黑云花岗闪长岩

～～～～～　侵入接触　～～～～～

汤他大岭组	总厚度>769.0m
11. 英安岩（Rb-Sr年龄：109.3Ma±7.1Ma），见花岗斑岩侵入	164.15m
10. 流纹质角砾凝灰熔岩	25.32m
9. 流纹斑岩	76.49m
8. 英安质熔结凝灰岩	330.38m
7. 英安质角砾凝灰熔岩	8.63m
6. 角砾（集块）凝灰岩	4.22m
5. 沉凝灰岩	34.59m
4. 英安质熔结凝灰岩	77.26m
3. 英安质角砾凝灰熔岩	39.69m
2. 集块角砾岩。见角闪黑云花岗闪长岩侵入	6.48m
1. 沉凝灰岩。见角闪黑云花岗闪长岩侵入	1.41m

～～～～～　侵入接触　～～～～～

下伏地层　早白垩世细中粒角闪黑云花岗闪长岩

（四）岭壳村组剖面

1. 六罗村剖面

剖面情况同上述。自上而下为（图4-7）：

未见顶

下白垩统岭壳村组（K_1lk）	总厚度1730.10m
15. 流纹质晶屑凝灰熔岩，顶部为侵出相碎斑熔岩	687.20m
14. 流纹质含砾晶屑凝灰熔岩	1042.9m

～～～～～　喷发不整合　～～～～～

下伏地层　下白垩汤他大岭组英安质含砾晶屑凝灰熔岩

2. 岭壳村剖面

该剖面为代表性剖面。由海南地质大队 1982～1986 年开展 1∶5 万《志仲》幅区调时测制。剖面位于三亚市扎南乡岭壳村南面，起点坐标：东经 109°27′49″，北纬 18°30′56″。自上而下为（图 4-9）：

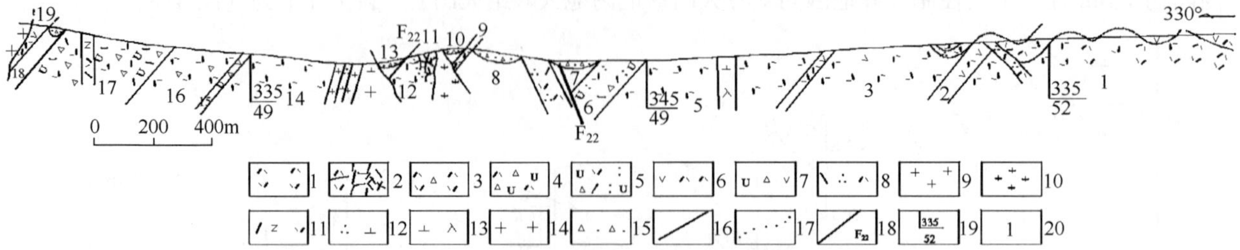

图 4-9 崖县岭壳村组实测剖面图

Fig. 4-9 Measured section of the Lingkecun Formation, Ya County

1—流纹斑岩；2—碎裂流纹斑岩；3—含角砾流纹斑岩；4—流纹质角砾熔岩；5—流纹质角砾凝灰熔岩；6—英安斑岩；7—英安质角砾熔岩；8—石英斑岩；9—花岗细晶岩；10—花岗斑岩；11—霏细斑岩；12—石英闪长岩；13—闪长玢岩；14—花岗岩；15—掩盖地段；16—分层界线；17—岩相界线；18—断层及编号；19—岩层产状；20—分层号

未见顶

岭壳村组

19.	浅灰色英安斑岩，具弱黄铁矿化	>40.19 m
18.	掩盖	47.76 m
17.	灰色、浅灰色含集块、角砾流纹斑岩	315.11m
16.	灰色、浅灰色含角砾流纹斑岩	209.99m
15.	浅灰色英安质角砾熔岩	5.78m
14.	灰色、浅灰色流纹斑岩，顶部夹 20cm 厚的英安质角砾熔岩	235.39m
13.	掩盖	99.30m
12.	流纹斑岩	6.21m
11.	碎裂流纹斑岩，具黄铁矿化	32.69m
10.	掩盖	16.65m
9.	灰、浅灰色流纹斑岩	12.61m
8～7.	掩盖	
6.	灰、深灰色流纹质角砾凝灰熔岩	140.28m
5.	硅化黄铁矿化流纹斑岩，裂隙发育	377.07m
4.	暗灰色英安斑岩	21.49m
3.	暗灰色流纹斑岩，含星点状黄铁矿	356.47m
2.	暗灰色英安斑岩	62.82m
1.	深灰色流纹斑岩	>31.57m

未见底

（五）报万组剖面

1. 山荣-南美剖面

该剖面位于乐东县山荣乡抱湾村东北面，起点坐标：东经 109°4′32″，北纬 18°49′4″，由四川攀西地质大队、海南区调队 1989～1991 年开展 1∶5 万《尖峰岭、乐东县》幅区调时重新测制。自上而下为：

未见顶

报万组（K_2b） 总厚度>816.11m

23. 紫色中厚层细粒长石砂岩 26.18m

22. 紫色中厚层泥质长石粉砂岩夹紫红色薄层状粉砂岩，中下部夹两层含砾粗粒长石砂岩，厚 15～16cm。下部夹紫红色薄层泥岩。见流水波痕，流向 290°～300° 37.22m

21. 灰紫色中—厚层砾质不等粒长石砂岩夹紫色细粒长石砂岩。见大型交错层理 8.17m
20. 紫色中厚层细粒长石砂岩。下部夹泥岩透镜体，平行层理发育 40.47m
19. 灰黄色中厚层细粒长石砂岩。发育三组波痕，指示流向分别为130°、330°、290° 18.20m
18. 紫色中厚层微粒长石砂岩 6.60m
17. 灰、灰绿色中—厚层细粒长石砂岩。具平行层理和斜层理 9.34m
16. 紫色中厚层微粒长石砂岩。平行层理发育 26.03m
15. 灰黄色中厚层细粒长石砂岩 10.67m
14. 紫色中厚层微粒长石砂岩 5.86m
13. 灰黄色中厚层细粒长石砂岩 42.55m
12. 紫色薄层细粒长石砂岩与紫色薄层粉砂岩互层 4.26m
11. 灰黄色中厚层细粒长石砂岩，下部夹两层砂质砾岩，上部夹一层泥砾岩 82.90m
10. 紫色中厚层细粒长石砂岩 55.99m
9. 灰黄色中厚层细粒长石砂岩夹灰绿色、灰黄色泥岩，砂岩底部含泥砾。产孢粉 *Exesipollenites tumulus*，*E. pseudotriletus*，*Cycadopites* sp.，*Schizaeoisporites* sp. 130.05m
8. 灰黄色中厚层细粒长石砂岩与同色薄层粉砂质细砂岩互层，中部夹一层泥岩，具平行层理 9.47m
7. 灰黄色中厚层细粒长石砂岩 49.46m
6. 紫红色、灰紫色中厚层泥质长石粉砂—细砂岩夹紫红色粉砂岩及泥岩。具大型斜层理及平行层理 106.69m
5. 灰紫色中—厚层粉—细粒长石石英砂岩 33.88m
4. 紫红色、灰紫色薄层状粉砂质钙质泥岩夹泥岩，顶部灰绿色粉砂质页岩，厚2m。产孢粉 *Classopollis parvus*，*C. annulatus*，*Exesipollenietes tumulus*，*E. pseudotriletus*，*Schizaeoisporites* sp.，*Interulobits*? sp. 11.30m
3. 灰紫色、黄灰色含泥砾泥质粉-细粒长石砂岩夹紫灰色粗粒长石砂岩，上部具条带构造 14.76m
2. 紫红色薄—中厚层泥质细粒长石砂岩夹粉砂岩。局部见平行层理 46.06m

———— 整　合 ————

下伏地层 鹿母湾组灰紫色凝灰岩夹火山角砾岩

2. 乐东县千家镇长茅岭剖面

该剖面由海南地质综合勘察院1996~1998年开展1：5万《黄流市、千家》幅区调时测制。剖面起点位于豪岗岭43°方向1525 m处小河边，起点坐标：东经109°01′29″，北纬18°39′58″；剖面总长9324.32 m。自上而下为（图4-10）：

未见顶

上白垩统报万组（K_2b） 总厚度>1498.52m

31. 灰黄色中厚层中细粒长石石英砂岩。产孢粉 *Sphagnumsporites* sp.，*Monosulcites* sp.。 13.44m
30. 灰黄色厚层含砾不等粒砂岩夹块状复成分砾岩 46.33m
29. 灰黄浅绿色块状复成分砾岩 28.56m
28. 杂色中厚层含砾长石粗砂岩夹薄层粉砂岩 89 36m
27. 灰黄色中厚层粗—中细粒岩屑长石杂砂岩夹薄层泥岩，产孢粉 *Punctatisporites* sp.，*Densoisporites* sp. 198.00m
26. 灰黄色、褐黄色薄层粉砂质泥岩。微细层理发育部，产孢粉 *Inaperturopollenites* sp.，*Cryptomeriapollenites* sp.，*Stephanocolpites* sp.，*Chenopodipollis* sp. 103.43m
25. 灰黄色中厚层泥质细粉砂岩与青灰色薄层状细粉砂质泥岩呈3：1互层。局部夹厚约10cm水平层理发育的粗砂岩。产孢粉 *Deltoidospora* sp.，*Toroisporis* sp.，*Peripollenites* sp.，*Tricolpopollenites* sp. 31.32m
24. 青灰色、灰黄色中厚层细粒长石砂岩夹青灰色薄层泥岩。水平层理发育 52.22m
23. 紫红色薄层粉砂质泥岩夹长石粉砂岩。微层理发育 39.23m
22. 灰黄色中厚层泥质粗粉砂岩与紫红色中厚层泥岩呈4：1~2：1互层，二者层厚分别为20cm及5~10cm 52.81m

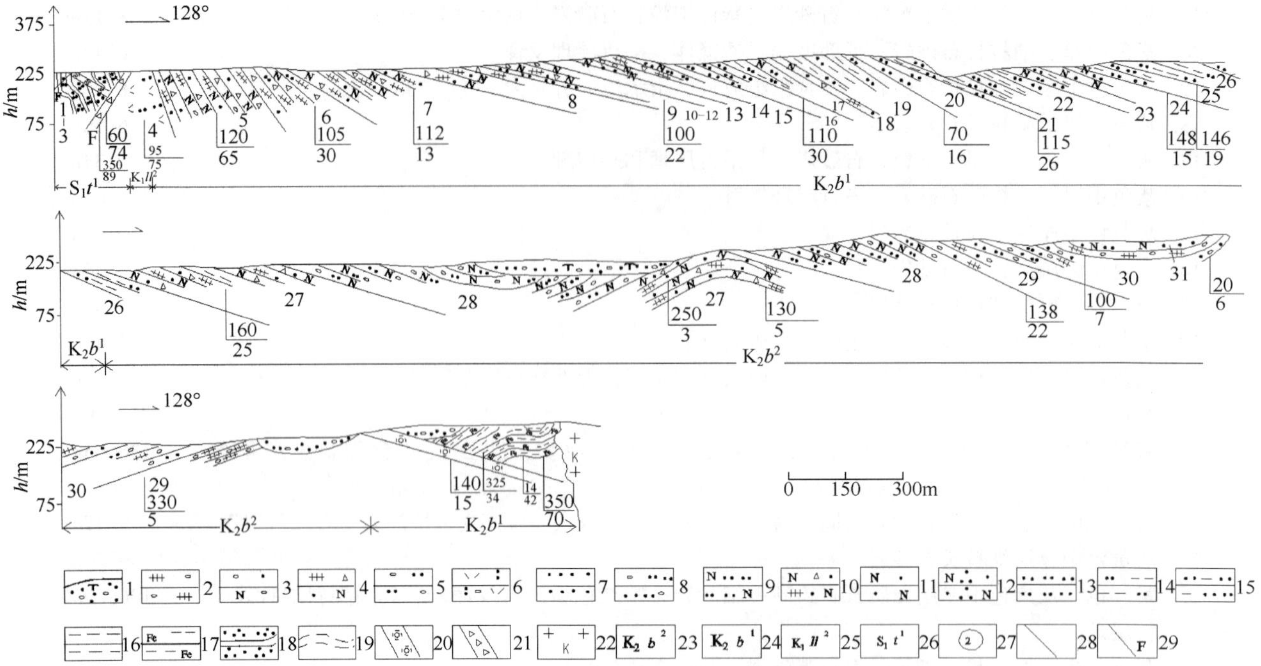

图 4-10 乐东县千家镇长茅岭剖面图

Fig. 4-10 Changmaoling section, Qianjia Town, Ledong County

1—残积物；2—复成分砾岩或巨砾岩；3—含砾长石砂岩；4—含砾长石杂砂岩；5—含砾砂岩；6—含砾凝灰岩；7—砂岩（粗中细）；8—含砾粉砂岩；9—长石粉砂岩；10—岩屑长石杂砂岩；11—长石砂岩；12—长石石英砂岩；13—粉砂岩；14—粉砂质页岩；15—泥质砂岩；16—泥岩；17—铁质泥岩；8—变质石英砂岩；19—千枚岩；20—硅化脉；21—构造角砾岩；22—钾长花岗岩；23—上白垩统报万组上段；24—上白垩统报万组下段；25—下白垩统鹿母湾组上段；26—下志留统陀烈组下段；27—室内分层号；28—岩层界线；29—断层

21. 紫红色薄层粗粉砂质泥岩夹薄层泥质粉砂岩。局部交错层理和微层理发育 0.20m

20. 灰黄色中厚层泥质粗粉砂岩夹多层粉砂质泥岩。下部夹 2 m 厚的细砂岩 49.88m

19. 深紫红色薄层铁质粉砂质泥岩 90.84m

18. 黄绿色、紫红色中厚层中细粒含岩屑长石杂砂岩 7.55m

17. 紫红色薄层粉砂质泥岩夹少量长石粉砂岩。局部见微层理 23.74m

16. 紫红色薄层石英粗粉砂岩。微层理发育。产孢粉 *Alsophilidites* sp.，*Osmundacidites* sp.，*Densoisporites* sp.，*Exesipollenites* sp.，*Podocarpidites* sp.，*Taxodiaceaepollenites* sp.，*Tricolpopollenites* sp. 29.28m

15. 紫红色薄层铁质泥岩夹少量长石粉砂岩。粉砂岩呈团块状、不连续条带状夹于泥岩中 15.06m

14. 浅紫红色中厚层中细粒岩屑杂砂岩。下部夹 1～2 m 厚的棕红色中薄层泥岩 27.25m

13. 掩盖 46.37m

12. 灰黄色中厚层状细粒岩屑长石杂砂岩。产孢粉 *Sphagnumsporites* sp.，*Polypodiaceaesporites* sp. 36.62m

11. 紫红色细粒岩屑长石杂砂岩、米黄色不等粒粉砂岩、紫红色薄层粉砂质泥岩组成两个粒序韵律层 3.81m

10. 青灰色中厚层细粒含岩屑长石杂砂岩 2.59m

9. 紫色中厚层粗粉砂岩夹紫色薄层粉砂质泥岩组成多个韵律层。岩石中含泥砾，呈不续条带状、团块状、角砾状分布 6.16m

8. 青灰色中厚层（局部块状）细粒含长石岩屑杂砂岩夹薄层泥岩。岩石中有砾状泥块状泥砾，层面上波痕发育。产孢粉 *Deltoidospora* sp.，*Tiliaepollenites* sp.，*Tricolpopollenites* sp.。 81.56m

7. 紫灰色厚层细粒长石岩屑杂砂岩夹紫灰色薄层泥岩。产 *Bryophyta* 孢粉，*Sphagnumsporites* sp.，*Deltoidospora* sp.，*Lygodiumsporites* sp.，*Polycingulatisporites* sp.，*Osmundacidites* sp.，*Verrutetraspora* sp.，*Schizaeoisporites* sp.，*Ceratopteris* sp.，*Laricoidites* sp.，*Peripollenites* sp.，*Ilexpollenites* sp. 63.59m

6. 紫灰色厚层细粒含长石岩屑杂砂岩 80.61m

5. 青灰色夹紫色厚层含砾粗粒岩屑长石杂砂岩 209.57m

————— 整　　合 —————

下伏地层　鹿母湾组紫灰色厚层含砾富晶屑凝灰岩夹紫红色凝灰岩

3. 琼海市东太农场尖岭剖面

该剖面由海南省地调院1997-2000年进行1：5万《黎母岭、枫木市、琼中县、白马岭》幅区调时测制。剖面起点河温岭135.0m山包131°方向约200m处，坐标：东经110°14′11″，北纬19°3′23″)，剖面长约11666m。自上而下为（图4-3）：

未见顶（顶部被花岗斑岩侵入）

上白垩统报万组（K_2b）　总厚度>1832.44m

85.	紫红色块状含粉砂褐铁泥岩，局部夹含钙质粉砂质褐铁泥岩（岩石中粉砂含量有变化）	154.00m
84.	浅灰白色（底部同生少量呈浅紫红色）复成分砾岩夹浅紫红色中厚层铁质粉砂质泥岩，砾岩中具同生红层砾石	86.87m
83.	底部为浅灰白色复成分砾岩，中上部为浅灰白色、浅紫色厚层—块状复成分砾岩，夹浅紫红色中厚层含砾中粗粒岩屑长石砂岩，局部夹浅紫红色中厚层铁质粉砂质泥岩。砾岩中砾石呈叠瓦状排列，砾石扁平方向与岩层倾向一致，粉砂质泥岩的顶面可见冲刷现象，砾岩与含砾砂岩中含有红层砾石	33.10m
82.	浅紫红色块状含钙质铁泥质粉砂岩	9.96m
81.	掩盖	5.69m
80.	浅紫红色中厚层含砾中粗粒长石岩屑砂岩	11.97m
79.	浅灰白色复成分砾岩	26.00m
78.	掩盖	40.05m
77.	浅灰白色复成分砾岩，夹青灰色中厚层细粒长石岩屑杂砂岩，细粒长石岩屑杂砂岩底部见不太明显的水平纹层（纹层多厚1~3mm，向上渐无）	3.46m
76.	浅灰白色块状砾质中粗粒岩屑长石砂岩，中间局部掩盖厚	18.60m
75.	掩盖	31.18m
74.	浅灰白色厚层—块状复成分砾岩与浅紫红色厚层—块状粉砂质泥岩互层	5.49m
73.	掩盖	38.32m
72.	浅灰白色厚层中细粒长石石英砂岩与浅紫红色厚层粉砂质泥岩互层	5.33m
71.	掩盖	17.05m
70.	浅灰白色中细粒岩屑长石杂砂岩，上部夹厚层泥质粉砂岩，泥质粉砂岩中含孢粉 *Schizaeoisporites* sp.，*Abietineaepollenites* sp.，*Quercoidites* sp.，*Tricolpopollenites* sp.	36.23m
69.	浅灰色、浅灰白色厚层—块状含砾中粗粒岩屑长石砂岩，下部见与浅灰白色厚层块状中细粒岩屑长石砂岩互层	98.79m
68.	掩盖	169.56m
67.	浅灰色、浅灰白色块状中粗粒岩屑长石砂岩	27.56m
66.	掩盖	33.02m
65.	浅灰白色含砾中粗粒岩屑长石杂砂岩，中部夹浅灰白色、浅灰色厚层复成分砾岩	83.90m
64.	浅黄色（风化色）块状中细粒岩屑长石杂砂岩，中部局部含少量砾石（2%~3%）	79.48m
63.	浅灰色块状中粗粒长石岩屑杂砂岩	12.25m
62.	浅灰色块状泥质粉砂岩，夹浅灰色中厚层复成分砾岩	24.49m
61.	浅灰白色厚层—块状含砾中粗粒岩屑石英砂岩与浅紫红色厚层—块状含铁泥质粉砂岩互层，前者中具同生红层砾石	96.09m
60.	浅灰白色（风化色为浅黄色）块状含砾中粗粒岩屑长石杂砂岩，夹中厚层砂砾岩（前者中砾石含量增多即变为后者岩性）。具同生红层砾石	50.39m
59.	掩盖	226.95m
58.	浅紫红色铁泥质粉砂岩，发育微水平层理	48.65m
57.	浅灰色、灰白色块状中细粒岩屑长石杂砂岩与浅紫红色块状中细粒岩屑长石杂砂岩互层	28.71m

56. 掩盖 52. 33m

55. 浅紫红色泥质粉砂岩夹浅紫红色中细粒岩屑长石砂岩 39. 45m

54. 浅灰色、浅灰白色块状中细粒岩屑长石杂砂岩夹浅紫红色块状泥质粉砂岩。中上部的泥质粉砂岩中普遍含少量长石（3%~5%） 103. 36m

53. 浅紫红色块状铁泥质粉砂岩 69. 49m

52. 底部为浅灰色、灰白色复成分砾岩，中上部为浅灰白色中细粒长石岩屑砂岩，局部夹浅紫红色中厚层泥质粉砂岩 64. 67m

—————— 整　合 ——————

下伏地层　鹿母湾组浅灰色块状含凝灰质粗中粒长石岩屑砂岩

二、岩石地层特征

（一）鹿母湾组（K_1l）

鹿母湾组系广东区测队（1964）创建的鹿母湾群转化而来，命名地点位于白沙盆地的儋州市鹿母湾村。耿良玉等（1979）改称临高组，时代为早白垩世。广东区调大队（1986）将鹿母湾群上部沿用临高组名称，下部改为鹿母湾组，时代为早白垩世。《广东省区域地质志》（1988）修订鹿母湾群为鹿母湾组。河北区域地质调查大队、海南地质大队（1991）和陈哲培等（1997）沿用鹿母湾组名称。陈哲培等（1997）并指定代表性剖面为鹿母湾-白水港剖面，参考性剖面乐东县山荣乡南美剖面。底界以紫灰色厚层石英砂砾岩的出现为划分标志，在不同地区分别与下伏地层奥陶系南碧沟组或志留系陀烈组等呈不整合接触，与上覆上白垩统报万组呈整合接触。岩性下部以砂砾岩、含砾长石粗砂岩为主，夹泥质铁质粉砂岩和泥岩；上部长石石英细、粉砂岩夹钙泥质粉砂岩、粉砂质泥岩。常夹安山-英安质火山岩。

（二）六罗村组（K_1ll）

海南地质综合勘察院（1995）在三亚市高峰乡六罗村所创建。广东省区测队（1964）将同安岭、牛腊岭火山岩系称牛腊岭岩被和同安岭岩被，时代归于中生代末。《广东省区域地质志》（1988）归为上侏罗统高基坪群。海南地质大队（1984）沿用牛腊岭岩被和同安岭岩被名称，时代划归为晚侏罗世。海南地质综合勘察院（1995）将同安岭和牛腊岭火山岩系下部称六罗村组，据同位素年龄值和孢粉化石，定其时代为早白垩世，并将六罗村组分为两段，第一段为沉积相砂砾岩、砂泥岩、玄武岩、安山岩、安山质火山角砾岩；第二段为流纹质火山岩。陈哲培等（1997）沿用海南地质综合勘察院（1995）划分方案，指定三亚市高峰镇六罗村剖面为代表性剖面，北岭村剖面为参考性剖面。岩性主要为流纹质火山岩，下部夹少量玄武岩和安山岩。未见底，顶界以流纹质含砾凝灰熔岩与上覆汤他大岭组呈喷发不整合接触。

（三）汤他大岭组（K_1t）

由海南地质大队（1978）在三亚市高峰镇汤他大岭创名汤他群演变而来。海南地质综合勘察院（1995）将汤他群第 2 段修订为汤他大岭组。以弱绿帘石化安山质角砾凝灰熔岩的出现作为划分标志与下伏六罗村组呈喷发不整合接触，以具流纹构造的英安岩结束与上覆岭壳村组呈喷发不整合接触。陈哲培等（1997）沿用海南地质综合勘察院（1995）划分方案，指定汤他大岭剖面为代表性剖面，岩石以英安质火山岩为主，中部和下部夹少量安山质和流纹质火山岩。顶、底界分别以英安岩和安山质角砾凝灰熔岩与下伏六罗村组和上覆岭壳村呈喷发不整合接触。

（四）岭壳村组（K_1lk）

海南地质大队（1986）在三亚市扎南乡岭壳村创名的岭壳村群演变而来。陈哲培等（1997）将岭壳村群改为岭壳村组，并将上、中、下亚群修订为 1、2、3 段，时代确定为早白垩世。指定岭壳村剖面为代表性剖面，为一套中酸性、酸性火山岩系，自下而上由英安岩—流纹质熔岩及相应的火山碎屑岩组成；至少有 6 个喷发韵律，未见顶，底部以流纹质含砾凝灰熔岩的出现为界线标志与下伏汤他大岭组英安岩呈喷发不整合接触。

（五）报万组（K_2b）

系《中国地质学》编辑委员会、中国科学院地质研究所（1958）在乐东县报万村创建并介绍。广东区测队（1964）对以往资料综合后，改报万组为报万群，时代为晚白垩世。四川地矿局攀西队、海南区调队（1991）恢复报万组，时代为晚白垩世。宜昌地矿所（1992）认为所谓报万群就是鹿母湾群，时代为早白垩世。陈哲培等（1997）进行全岛地层清理时，修订报万群为报万组，时代为晚白垩世。代表性剖面为山荣-南美剖面，参考性剖面为乐东县千家镇长茅岭剖面。以紫红色薄—中厚层泥质细粒长石砂岩夹粉砂岩的出现作为该组底界与下伏鹿母湾组紫灰色凝灰岩夹火山角砾岩呈整合接触，未见顶。厚 385.5~1439m。

第二节 地层划分与对比

一、鹿母湾组（K_1l）

该组区域分布广泛，见于东方市乐安盆地、儋州市王五盆地、琼海市阳江盆地、白沙县—乐东县一线的白沙盆地、文昌市清澜盆地（大部分隐伏）、定安县雷鸣盆地、三亚市三亚盆地，属河流、湖河沉积环境；共划分为上、下两段。尽管不同盆地沉积岩石具有一定的差异，如火山岩层主要分布于白沙盆地、阳江盆地中，乐安盆地也有零星分布，而其余盆地目前没有见到，且白沙盆地火山岩出露于鹿母湾组上部，而阳江盆地则主要见于鹿母湾组下部。盆地早期沉积都以一套近缘碎屑沉积为特点，中期出现粒度相对较细的碎屑沉积，晚期盆地闭合则以发育水平层理、交错层理（图版Ⅰ-11-G）的砾岩、砂砾岩体现（图 4-11）。

白沙盆地南西段乐东县南美剖面下段主体为以含砾岩屑砂岩，上部并见长石分布，底部以复成分砾岩与南碧沟组角度不整合接触，顶部以中—厚层泥岩与上段细砾质沉凝灰岩夹砂砾岩及凝灰岩整合接触。上段以沉火山岩、火山岩为特色，间夹泥岩，岩石主要为沉晶屑凝灰岩、晶屑凝灰岩、火山角砾岩、含砾晶屑岩屑凝灰岩、含凝灰质泥岩等。厚 736.02m。北东段儋州市鹿母湾—白沙县白水港剖面下段以砂砾岩、长石石英细砂岩、含砾长石石英粗砂岩为主，夹泥质粉砂岩、粉砂质泥岩；上段主体为泥质粉砂岩、粉砂质泥岩，偶夹细砂岩、石英粗砂岩，顶部岩石并常见长石分布。厚度>1523.72m。

阳江盆地琼海市东泰农场剖面下段以含砾（质）岩屑杂砂岩为主体。上段主要为复成分砾岩与泥岩不等厚产出，夹沉凝灰岩、晶屑玻屑凝灰岩、凝灰质泥岩。火山物质成分有长石晶屑、玻屑、火山角砾。厚 2527.16m。

王五盆地儋州市王五市平地岭剖面下段以粗粉砂岩为主，间夹长石岩屑砂岩、岩屑杂砂岩，底部以复成分砾岩与陀烈组不整合接触。上段为砂砾岩、砾岩，夹长石岩屑杂砂岩、粗粉砂岩。厚度>3570.1m。

乐安盆地东方市乐安村剖面下段以含砾粗粒长石岩屑杂砂岩、长石岩屑杂砂岩为主，间夹砾岩、粉砂岩、细砂岩，上段以长石岩屑杂砂岩、砾岩为主，夹粉砂岩、细砂岩、粗砂岩。厚度>1837.11m。

二、六罗村组（K_1ll）

该组分布于琼西南同安岭、牛腊岭盆地、琼中五指山盆地、琼西北儋州洛基盆地、琼东北澄迈县旺商等盆地中，以同安岭、牛腊岭盆地研究较系统，属于具三元结构的双峰式火山岩。由下向上为：①沉积相砂砾岩、砂泥岩，②玄武岩、安山岩、安山质火山角砾岩、凝灰岩，③流纹质火山岩类。三层岩性特征明显，相应可划分出三个岩性段。牛腊岭盆地下部为沉火山-沉积碎屑岩，代表了早期边部火山沉积，上部为火山角砾岩、凝灰岩、流纹质晶屑凝灰熔岩、安山质晶屑凝灰熔岩等；同安岭盆地为玄武安山岩、玄武岩、石英安山岩、辉石安山岩、霏细岩、碎斑熔岩、流纹岩、流纹质熔结凝灰岩、流纹质含砾晶屑凝灰熔岩。岩石基本特征总体可以对比，但横向厚度变化大，在周凤园厚 1221m、北

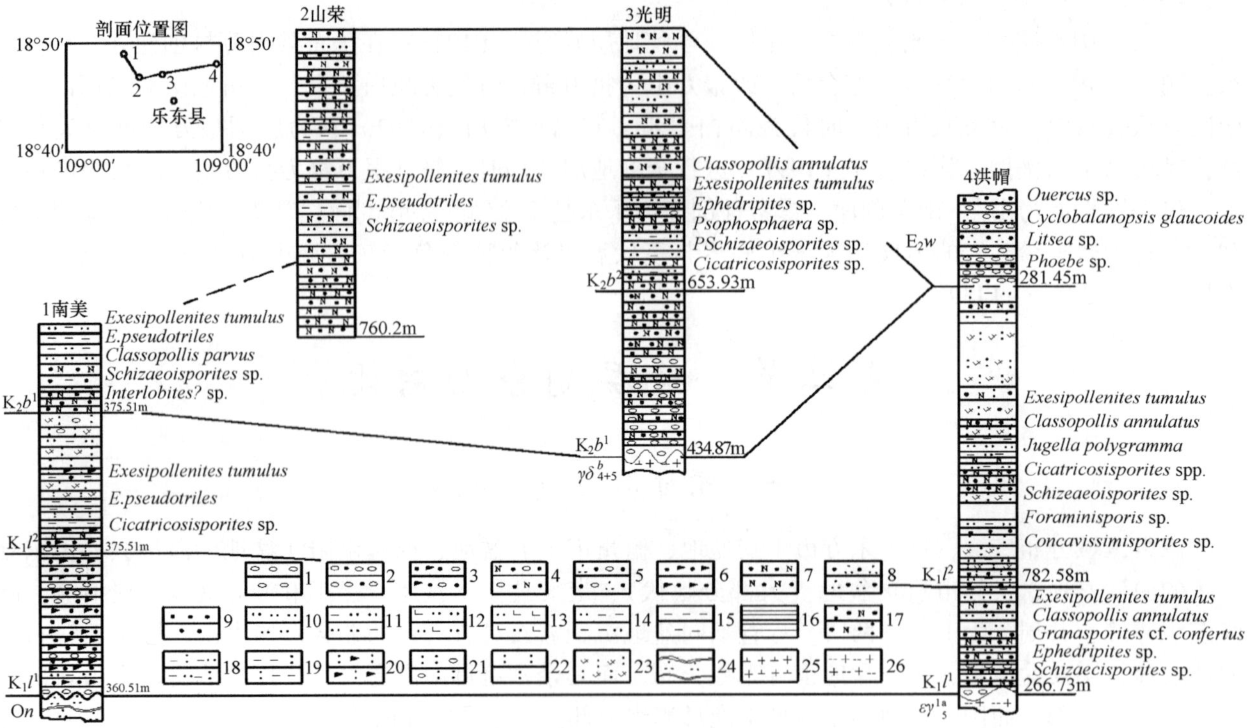

图 4-11　白垩系鹿母湾组-报万组柱状对比图

Fig. 4-11　Columnar correlation of the Cretaceous Lumuwan Formation-Baowan Formation

（据 1∶5 万乐东县幅，1991）

1—砾岩；2—砂质砾岩；3—砾质岩屑砂岩；4—砾质长石砂岩；5—砾质石英砂岩；6—岩屑砂岩；7—长石砂岩；8—石英砂岩；9—砂岩；10—粉砂岩；11—泥质粉砂岩；12—钙质粉砂岩；13—砂质泥岩；14—粉砂质泥岩；15—泥岩；16—页岩；17—凝灰质长石砂岩；18—凝灰质粉砂质泥岩；19—凝灰质泥岩；20—岩屑凝灰岩；21—砾质凝灰岩；22—凝灰岩；23—英安质凝灰岩及英安岩；24—石英岩；25—花岗闪长岩；26—黑云母花岗岩；E_2w—丘糖岭组；K_2b^2—报万组二段；K_2b^1—报万组一段；K_1l^2—鹿母湾组二段；K_1l^1—鹿母湾组一段；On—南碧沟组

村岭厚 1725.57m，在六罗村厚度大于 1792.2m。

三、汤他大岭组（K_1t）

该组分布面积较小，主要分布于崖城、高峰等地，分别属于牛腊岭、同安岭火山盆地。所测制的剖面都分布在同安岭盆地中，以英安质火山熔岩及火山碎屑岩为主，夹数层英安质火山岩，中部和下部夹少量安山质和流纹质火山岩。自下而上为：安山质熔岩-英安质凝灰熔岩-流纹质含砾凝灰熔岩，英安岩、英安质角砾凝灰熔岩夹沉凝灰岩，英安岩夹英安质角砾凝灰熔岩等，组成 5 个喷发韵律。横向上同期喷发韵律岩层厚度变化较大，亦可与东南沿海地区中生代火山岩第Ⅵ喷发旋回对比。

西面汤他大岭-抱冲村一带，岩石为英安岩、集块角砾岩、英安质角砾凝灰熔岩、流纹质角砾凝灰熔岩、流纹斑岩、角砾（集块）凝灰岩、沉凝灰岩。东面六罗村则为英安质含砾晶屑凝灰熔岩、流纹质含砾晶屑凝灰熔岩、安山质岩屑角砾熔岩。

四、岭壳村组（K_1lk）

该组主要分布于同安岭盆地中。南面六罗村岩石为流纹质晶屑凝灰熔岩、流纹质含砾晶屑凝灰熔岩；北面岭壳村岩石为英安斑岩、流纹斑岩、流纹质角砾凝灰熔岩、英安质角砾熔岩、含角砾流纹斑岩、含集块、角砾流纹斑岩。由岭壳村至加跃岭一带，厚度由 63m 增大至 430m。

五、报万组（K_2b）

报万组目前仅分布于白沙盆地南西段、阳江盆地、雷鸣盆地中。属冲积扇、湖河沉积环境。

白沙盆地南西段山荣-南美剖面以细粒长石砂岩为主，夹钙质泥岩、粉砂质细砂岩、长石粉砂岩。长茅剖面，下部主要为含长石岩屑杂砂岩，夹粉砂岩，中部为泥质粗粉砂与粉砂质泥岩不等厚产出，上部为粗中—中细粒岩屑长石杂砂岩、复成分砾岩、含砾长石粗砂岩，夹薄层泥岩、粉砂岩。阳江盆地东太农场尖岭剖面下部为岩屑长石杂砂岩、铁泥质粉砂岩不等厚产出，上部为岩屑长石砂岩、复成分砾岩、粉砂质泥岩、长石石英砂岩。两盆地沉积总体特征相似，岩石基本可以对比（图 4-7）。

第三节　古生物特征及年代地层

一、火山-沉积地层

（一）鹿母湾组（K_1l）

“红盆”的生物地层及年代地层研究以阳江红盆最为详细。宜昌地矿所（1995）、海南省地质调查院（2000）先后在该区采获大量孢粉化石。孢粉组合中以裸子植物花粉为主，其次为蕨类植物孢子，见有个别的被子植物花粉。裸子植物花粉主要有 *Classopollis* sp.，*Exesipollenites tumulus* sub sp.，*triangulus*，*Monosulcites* sp.，*Inaperturopollenites* sp.，*Taxodiacepollenites*. sp.，*Abietineaepollenites* sp.，*Laricoidites magnus*；蕨类植物孢子有 *Deltoidospora* sp.，*S. Toroisporis* sp.，*Schizaeoisporites* sp.，*S. laevigataefomis*；被子植物花粉有 *Magnolipollis*（木兰粉），可称为 *Schizaeoisporites-Classopollis* 组合，组合中的 *Classopollis* 在地史上曾有三个高峰期，即早侏罗世、晚侏罗世和早白垩世，目前许多学者认为 *Classopollis* 一般出现于侏罗系与白垩系界线附近，更多的学者则把它视为早白垩世的起点。*Classopollis* 在昌江、东方南部红水沟一带鹿湾组的 *Klukisporites* - *Classopollis*（Ⅰ）组合（17.2%）和 *Verrucosisporites*-*Classopollis*-*Abietincaepollenites*（Ⅱ）组合（33.3%）均有出现。长江三峡地区下白垩统石门组出现 *Classopollis annulatus*（7.21%），*C. Parvws*（4.51%），五龙组出现 *C. annulatus*，*C. cluaoides*.，*C.* Parvus，*C.* sp 等（余静贤等，1987）。由此可见，*Classopollis* 为鹿母湾组时代归属与地层对比提供了古生物依据。*Exesipollenites* 也是早白垩世的典型代表分子，在琼海中原一带鹿母湾组的孢粉组合中含量丰富（35.93%），属种分异度高，在红水沟鹿母湾组的第Ⅱ孢粉组合中亦有分布，占孢粉组合的 4.2%。长江三峡地区下白垩统石门组也含有 *E. tumulus*（18.02%）分布，且含量丰富。*Schizaeoisporites certus* 在白垩纪地层中也很繁盛，尤其是早白垩世地层中含量最多，在长江三峡地区的石门组和五龙组中都能见到。

汪啸风等（1992）于琼海福田一带鹿母湾组中采获大量植物化石：*Cladophlebis* cf. *exiliformis*，*Cupressionocladus* cf. *gracilis*，*C.* sp. 和 *Podozamites* sp.。海南省地质调查院（2003）在琼海龙江北岸园采获大量植物化石。其中 *Cladophlebis* cf. *exiliformis* 是我国和日本晚侏罗世及早白垩世的常见分子，*Cupressionocladus* cf. *gracilis* 见于我国晚侏罗世及早白垩世地层。此外，耿良玉（1979）于临高加来盆地、汪啸风等（1992）于白沙县元门鹿母湾组采获介形类化石；汪啸风等（1992）还于白沙元门及临高东江采获轮藻化石，介形类和轮藻显示地质时代为早白垩世晚期或中—晚期。

综上所述，鹿母湾组的地质时代属早白垩世的古生物依据是充分的，并可与其他地区同期地层对比（表 4-2）。

表 4-2　华南部分地区早白垩世地层对比表

Table 4-2　Correlation of Early Cretaceous strata，South China area

<table>
<tr><th colspan="3">本书</th><th colspan="3">1∶25 万《琼海县》幅</th><th>1∶5 万《坝王岭、猴弥岭》幅</th><th>粤东</th><th>湘中</th><th>长江三峡地区</th><th>主要化石</th></tr>
<tr><td rowspan="2">下白垩统</td><td>阿尔布阶
|
阿普特阶</td><td rowspan="2">鹿母湾组</td><td rowspan="2">下白垩统</td><td>阿尔布阶
|
阿普特阶</td><td rowspan="2">鹿母湾组</td><td rowspan="2">鹿母湾组</td><td>官草湖群</td><td>神皇山组</td><td>五龙组</td><td>Schizaeoisporites
Classopollis</td></tr>
<tr><td>韦尔登阶</td><td>韦尔登阶</td><td>高基坪群</td><td>东井组</td><td>石门组</td><td>Cladophlebis
Cupressionocladus</td></tr>
</table>

（二）报万组（K_2b）

关于海南岛上白垩统报万组的时代常有争论，海南地质综合勘察院（1996）在长茅水库一带上白垩统报万组中找到孢粉化石，经南京地矿研究所鉴定，共有4类27属。其中蕨类植物孢子共13属，占较大优势（59.09%）；裸子植物花粉8个属，含量次之（19.70%）；被子植物花粉共5属，占15.15%；苔藓类1属，含量最少（6.06%）。这些孢子花粉化石中，由于数量少，无法构成组合特征。它们在白垩纪孢粉分布中均能出现。如*Schizaeoisporites*是白垩纪的繁盛分子，尤其是在早白垩世晚期至晚白垩世早期分布最广。裸子植物花粉*Inaperturopollenites*代表柏科和杉科的分子，它们在古近纪—新近纪获得了极度发展和广泛分布；但在前苏联的阿尔布期，这一类花粉也占有较重要地位（宋之琛等，1981）。*Exesipollenites*最早见于澳大利亚西北部早侏罗世，在北半球广泛分布于白垩纪中，我国南方白垩纪地层中也很发育，在中南地区早白垩世中晚期到晚白垩世早期尤其发育，成为该期孢粉组合中重要代表性分子，到晚白垩世中期开始减少，如湖北下白垩统神王山组（31.8%）、洞下场组（38.2%）、广东晚白垩世三水盆地三水组（2.7%）及海南鹿母湾组（15.4%~88.2%）。*Pojycingulatisporites*亦是白垩纪的特殊分子，如*P. reducus*是前苏联西伯利亚及远东地区早白垩世晚期到晚白垩世早期的关键属种。我国见于湖北玉龙组（2%）及海南鹿母湾组中。综上所述，现已发现的中生代孢粉总貌时代属白垩纪无疑。

值得特别关注的是，在我国早白垩世早期孢粉组合中相当有特征的*Classopollis*、早白垩世中晚期组合中较特征的*Cicatricosisporites*，测区内均未出现。这二属化石在我国早白垩世是相当繁盛的，占有重要位置。海南岛鹿母湾组中这两属化石也很繁盛，在乐东县黎族自治县红水沟一带，*Classopollis*为7.6%~33%，*Cicatricosisporites*为0.4%~1.4%。而乐东黎族自治县南美地区*Classopollis*为5.1%~32.6%，*Cicatricosisporites*最高达11%。*Classopollis*从早白垩世早期发展（17.2%）、中期繁盛（33%）、晚期减弱（7.6%）到晚白垩世开始灭绝的演化过程。而*Cicatricosisporites*由早白垩世早期繁盛（1.4%）、中期减弱（0.8%）、晚期更少（0.4%），到晚白垩世基本灭亡。在早白垩世繁盛的*Exesipollenites*，到晚白垩世明显减少，在南美地区只发现1粒孢粉。在早白垩世晚期到晚白垩世相当繁盛的*Schizaeoisporites*也只发现1粒。从上述孢粉的发展、繁盛、减弱、灭亡的演化历程来分析，排除了孢粉产于早白垩世的可能。孢粉鉴定者认为产这些孢粉的这套地层的时代不会早于早白垩世晚期，当然拉到晚白垩世的证据还不足。

另一方面分析被子植物花粉化石的特征（表4-3），（Ⅱ）（Ⅲ）组合中被子植物花粉仅出现*Claratipollenites*，*Asteropollis*等原始属种，在阿尔布期开始出现的*Tricolpites*仅有个别分子，总量仅有3.2%。而南美地区3个样品中发现*Tricolpites*（代表地层厚580m），表明分布较为广泛，有了较大的发展。而被子植物花粉已达5属，有10粒化石，占66粒化石总量的15.15%，所占比例较大，足以证明晚白垩世已经到来。被子植物花粉在早白垩世早期含量极少或无，早白垩世晚期占10%或更少，晚白垩世含量高。被子植物花粉占15.15%的含量划为晚白垩世理由充分。但鉴定者强调孢粉总量少，统计出百分含量不具代表性，认为含这些孢粉化石的地层时代为早白垩世晚期到晚白垩世早期，可与江苏省浦口组（宋之琛等，1981）对比，可能属于阿尔布期到赛诺曼期。综合考虑海南白垩纪岩石地层时代的具体情况，整合于报万组之下的鹿母湾组已被确定为早白垩世，报万组中含孢粉化石的下段则相应为晚白垩世早期，上段为晚白垩世晚期。

二、火山岩地层

包括六罗村组、汤他大岭组和岭壳村组。海南地质综合勘察院（1995）于六罗村组上部的流纹质含砾凝灰熔岩和流纹质凝灰熔岩中获得锆石U-Pb同位素年龄107Ma和Rb-Sr等时线年龄（121.±14.9）Ma。高岭地区六罗村组底部发育的紫红色碎屑岩中发现地质时代隶属早白垩世的孢粉化石*Classopollis* sp.，*Osmundacidites* sp.，可直接与代表性剖面鹿母湾组下部同期生物群对比，时代为早白垩世。于汤他大岭组上部英安岩中获得Rb-Sr同位素年龄为（109.3±7.1）Ma，其时代大致相当于早白垩世。岭壳村组叠覆于汤他大岭组之上，结合晚白垩世报万组中不含火山（岩）物质，因此，初步认为岭壳村组的时代相当于早白垩世晚期（陈哲培等，1997）。

表 4-3　海南岛白垩纪孢粉对比表

Table 4-3　Correlation of Cretaceoussporopollens, Hainan Island

早白垩世								晚白垩世	
Klukisporites-Classopollis（Ⅰ）组合		Vcrrucosisporites-Classopollis-Abietineacepollenites（Ⅱ）组合		Cycadopites-Cyathidites-Asteropollis（Ⅲ）组合					
孢粉名称	含量/%	孢粉名称	含量/%	孢粉名称	含量/%	孢粉名称	含量/%	孢粉名称	含量/%
Klukisporites sp.	16.0	Klukisporites sp.	6.8	Cyathidites australis	1.6	Psophosphaera grandis	0.8	Spores of Bryophyta	6.1
K. pseudoreticulatus	1.4	K. areolatus	0.8	C. minor	2.4	P. sp.	0.4	Sphagnumsporites sp.	12.1
K. variegatus	1.4	Neoraistrickia sp.	2.6	Cicatricosissporites sp.	0.4	Araucariacites australis	0.4	Deltoidospora sp.	13.6
Neoraistrickia sp.	5.7	Pterisisporites sp.	3.4	C. pacificus	0.4	Inaperturopollenites sp.	3.2	Lygodiumsporites sp.	4.5
N. minor	5.7	Cicatricosisporites sp.	0.8	Pilosisporites baevipapillosus	0.4	Classopollis sp.	4.0	Polycingulatisporites sp.	1.5
Pterisisporites sp.	10.0	Osmundacidites sp.	0.8	Impardecispora sp.	0.8	Classopollis anulatus	3.2	Osmundacidites sp.	7.6
Dictyotriletes sp.	1.4	Multinodisporites sp.	2.6	Concavissmisporites sp.	0.8	C. triangilis	0.4	Verrutetraspora sp.	1.5
Wulongspora sp.	1.4	Lycopodiumsporites sp.	1.7	Baculatisporites comaumensis	0.8	Exesipollenites sp.	0.8	Schizaeoisporites sp.	1.5
Lygodiumsporites sp.	1.4	Verrucosisporites sp.	9.3	Osmudacidites sp.	1.6	Peripollenites sp.	0.8	Ceratopteris sp.	1.5
Lygidioisporites sp.	1.4	Converrucosisporites sp.	3.4	O. wellmanii	0.8	P. elatoide	0.8	Polypodiaceaesporites sp.	3.0
Baculatisporites sp.	1.4	Ischyosporites sp.	1.7	O. micanicus	0.4	P. limbatus	0.4	Alsophilidites sp.	3.0
Leptolepidites major	1.4	Densoisporites sp.	0.8	Verrucosisporites sp.	0.4	Taxodiaceaepollenites hiatus	6.0	Densoisporites sp.	3.0
L. rotundus	1.4	Impardecispora sp.	0.8	Todisporites minor	0.4	T. elongatus	2.0	Toroisporis sp.	1.5
Cicatricosisporites sp.	1.4	Vernutetraspora sp.	6.0	Ceratosporites sp.	0.4	Paleoconiferus sp.	0.8	Punctatisporites sp.	4.5
Osmundacidites sp.	1.4	Stereisporites sp.	3.4	Kubinella sp.	0.4	Protopinus sp.	1.2	Laricoidites sp.	1.5
O. nicanicus	1.4	Deltoidospora regularis	1.7	Aequitriradites cf. echinatus	0.4	Protoconiferus sp.	4.0	Peripollenites sp.	3.0
Multinodisporites sp.	1.4	Tenuicontactosporites sp.	1.7	Densosisporites velatus	1.6	Pseudopinus sp.	0.4	Exesipollenites sp.	1.5
Lycopodiumsporites sp.	1.4	Tripoletes sp.	1.7	D. sp.	0.4	Piceites sp.	1.6	Podocarpidites sp.	3.0
Xinjiangsporites	1.4	Schizosporis sp.	0.8	Stereisporites stereoides	0.4	Pseudopicea sp.	2.4	Taxodiaceaepollenites sp.	1.5
Verrucosisporites sp.	3.0	Biretisporites sp.	0.8	Crybeosporites striatus	0.4	Jiaohepollis verus	0.4	Ineperturopollenites sp.	6.1
Neuesisporites sp.	1.4	Classopollis sp.	19.6	Foraminisporites asymmetricus	0.4	Ephedripite prolongatus	1.2	Cryptomeriapollenites sp.	1.5
Classopollis sp.	3.0	C. anulatus	13.4	Granulatisporites sp.	1.2	E. trinata	0.4	Monosulcites sp.	1.5
C. anulatus	14.2	Exesipollenites sp.	4.2	Laevigatosporites sp.	0.8	Jiaohepollis flxuosus	0.4	Ilexpollenites sp.	3.0
Podocarpidotes sp.	1.4	Abietineaepollenites sp.	2.6	Psilaschizosporites sp.	0.4	Cerebropollenites sp.	0.4	Tiliaepollenites sp.	4.5
Rugubiusiculites sp.	4.5	Pinuspollenites sp.	2.6	Cycadopites sp.	8.0	Jugella sp.	0.4	Chenopodipollis sp.	1.5
Abietineaepollenites sp.	4.5	Monosuleites sp.	0.8	Ginkgoites sp.	2.8	Eucommidites sp.	0.4	Stephanocolpites sp.	1.5
Apinuspollenites sp.	1.4	Callisporites sp.	0.8	Chasmatosporites elegans	0.4	Concentrisporites sp.	1.2	Tricolpopollenites sp.	4.5
Monosuleites sp.	1.4	Quadraeculina sp.	0.8	Abietineaepollenites sp.	4.4	Monosuleites sp.	0.8		
Araucacites australis	3.0	Inaperturopollenites sp.	0.8	Pinuspollenites sp.	3.2	Clavatipollenites hughesii	0.4	苔藓类孢子总含量 6.06%	
Concentrisporites sp.	1.4	Ephedripites trinata.	0.8	Cedripites	2.4	Asteropollis asteroides	0.8	蕨类植物孢子总含量 59.09%	
Palaeoconiferus sp.	1.4	Tricolpopollenites sp.	1.7	Piceaepollenites sp.	12.0	A. microreticulatus	0.4	裸子植物花粉总含量 19.70%	
				Podocarpidites sp.	2.4	Cupuliferoidaepollenites sp.	0.4	被子植物花粉总含量 15.15%	
蕨类植物孢子总含量 64.2%		蕨类植物孢子总含量 51.6%		Pinaceae sp.	2.8	Chenopodipollis sp.	0.4		
裸子植物花粉总含量 35.8%		裸子植物花粉总含量 46.%		Quadraeculina sp.	1.2	Tricolpopllllenites minutus	0.8		
		被子植物花粉总含量 1.7%				蕨类植物孢子总含量 18.8%			
						裸子植物花粉总含量 78.8%			
						被子植物花粉总含量 3.2%			

表 4-4　海南与东南沿海白垩纪火山岩对比表

Table 4-4　Correlation of Cretaceous volcanic rocks in Hainan and China's South-East coastal areas

时代	火山活动旋回	海南				浙东地区				闽东地区					粤东地区				
		组	厚度 m	岩性	年龄 10^6a	组	厚度 m	岩性	年龄 10^6a	群	组	厚度 m	岩性	年龄 10^6a	群	组	厚度 m	岩性	年龄 10^6a
K	第四旋回	岭壳村组	1730 ~ 2294	流纹(斑)岩、流纹质含砾凝灰熔岩,局部夹英安(斑)岩、英安质凝灰熔岩	98.8	赖家组	>884	紫红色砂岩、砂砾岩、粉砂岩、泥岩夹凝灰岩	103.2		石牛山组	330~ 1920	上部钾长流纹岩;下部紫红色熔结凝灰岩夹流纹岩;晚期碎斑熔岩	78 86.7 91.8	南雄群	叶塘组	100~ 400	紫红色粉砂岩、砂岩、砂砾岩	900 ~ 100
																	120~ 1000	流纹岩、火山灰流凝灰岩、英安质碎斑熔岩	
		汤他大岭组	588	英安岩、英安质凝灰熔岩,夹安山质、流纹质含砾(角砾)凝灰熔岩、沉凝灰岩	109.3 120	塘上组	200 ~ 500	浮岩屑、玻璃凝灰岩、熔结凝灰岩、英安岩,夹砂砾岩、流纹岩	105.4 110.9			150~ 720	紫红色凝灰质砂砾岩、砂岩及粉砂岩,凝灰质砂岩夹凝灰岩,顶部时有英安岩				60~ 500	凝灰质砂岩,凝灰岩,砂岩、砂砾岩夹玄武岩	
	第三旋回	六罗村组	1676	第三段:流纹质凝灰熔岩夹流纹岩	107	朝川组	>650	含角砾浮岩屑玻屑凝灰岩、熔结凝灰岩、流纹岩、紫红色砂岩、砂砾岩	118.6	石帽山群	上组	450~ 1200	上部钾长流纹岩、熔结凝灰岩;下部紫红色凝灰质砂砾岩、砂岩、粉砂岩,夹流纹岩、层凝灰岩	101 107.9	官草湖群	上段	200~ 1200	砂岩、粉砂岩、砂砾岩夹凝灰质砂岩	130 ~ 110
			764	第二段:流纹质凝灰熔岩与安山(玄武)岩、火山角砾岩互层。部分地区流纹质凝灰熔岩夹玄武岩、安山岩	121.1	馆头组	200 ~ 600	灰色泥岩、粉砂岩、紫红色砂砾岩、砂岩、凝灰岩、安山岩、玄武岩	110		中组	780~ 1150	上部钾长流纹岩、粗面岩、球粒流纹岩、玄武岩;下部紫红色凝灰质砂砾岩	113.2 116 121		下段	200~ 1000	砂砾岩、砂岩、粉砂岩、凝灰质砂岩夹玄武岩、流纹岩	
			>105	第一段:粉砂质泥岩、砂岩、砂砾岩、砾岩							下组	160~ 1450	上部紫红色熔结凝灰岩、玄武岩;下部凝灰质砂砾岩夹凝灰岩						
J_3	第二旋回					九里坪组		流纹斑岩穹过渡至流纹岩			小溪组		灰、紫灰色熔结凝灰岩、流纹岩、凝灰岩		高基坪群	大安洞组		火山灰流、空落涌流凝灰岩夹流纹岩、安山岩	

海南地质综合勘察院（1995）根据同安岭-牛腊岭火山岩盆地岩石组合、火山喷发旋回及同位素年龄等资料的研究，认为六罗村组、汤他大岭组和岭壳村组与东南沿海地区粤东、闽东、浙东等3个构造-岩浆活动亚区的第Ⅲ、Ⅳ喷发旋回基本可以对比。六罗村组与浙江的馆头组、朝川组、福建的石帽山群，广东的官草湖群相当；汤他大岭组、岭壳村组分别与浙江的塘上组、赖家组、福建的石牛山组、广东的叶塘组相当（表4-4）。

第四节　岩相古地理特征

一、沉　积　相

根据海南省地质调查院（2004）资料，鹿母湾组、报万组可以划分为：冲积扇相、辫状河流相、洪冲积平原相和湖相沉积（表4-5）。从早至晚，冲积扇相从发育到不发育，剖面下部以冲积扇、河流相为主、次为洪冲积平原相，上部以河流相、洪冲积平原相为主，夹滨湖相、冲积扇相沉积。

表4-5　鹿母湾组、报万组沉积相划分表

Table 4-5　Division of sedimentary facies of the Lumuwan and Baowan Formations

相	亚相	主要岩性
冲积扇	河道	复成分砾岩、砂砾岩
	片 汜	含砾中粗砂岩、细砂岩、凝灰质砂岩、含砾砂质泥岩、含砾砂质沉凝灰岩
	泥石流	复成分砾岩、砂砾岩
	泥 流	含砾砂岩、砂岩、含砾砂质泥岩、含砾砂质沉凝灰岩、泥质砂岩、粉砂质泥岩、沉凝灰岩、含砂粉砂质泥岩、凝灰质泥岩
	筛 积	单成分砾岩
瓣状河流相	河 道	复成分砾岩、砂砾岩、含砾中粗砂岩
	心 滩	复成分砾岩、砂砾岩、含砾砂岩、含砾砂质泥岩、泥质粉砂岩、粉砂质泥岩、沉凝灰岩
洪冲积平原相	洪泛决口扇	砂砾岩、含砾砂岩、砂岩、凝灰质砂岩
	洪泛平原	中细砂岩、泥质粉砂岩、粉砂质泥岩、沉凝灰岩、含砾砂质泥岩、含砾砂质沉凝灰岩、凝灰质泥岩
湖相	湖滨相	含砾中粗砂岩、细砂岩、泥质粉砂岩

1. 冲积扇相

以细—粗砾岩、砂砾岩、含砾砂岩、砂岩、泥质沉积为主，可进一步划分为河道、片汜、泥石流、泥流、筛积沉积。

河道沉积：以浅紫红色夹浅灰白色细—粗砾岩为主，多无层理，砾石次棱角状—次圆状为主，棱角状次之，多具定向排列。

片汜沉积：是扇积物的主要沉积类型，为浅紫红色含砾中粗砂岩、细砂岩、凝灰质砂岩、泥质粉砂岩、含砾砂质泥岩、含砾砂质沉凝灰岩。

泥石流沉积：以浅紫红色复成分砾岩、砂砾岩为主。岩石一般呈块状，砾石棱角状—次棱角状，少数次圆状、堆积杂乱，分选极差，砾径多数0.5～10cm，大者可达30～50cm。砾岩、砂砾岩多为泥质支撑，少数颗粒支撑。

泥流沉积：是泥石流的变种，指岩石中泥质含量大大超过砾石者，为浅紫红色含砾砂岩、砂岩、泥质砂岩、粉砂质泥岩、凝灰质泥岩、含砾砂质沉凝灰岩、沉凝灰岩。

筛积：仅在早期扇体中局部发育（鹿母湾组下部），岩性为浅灰白色砾岩，砾石多呈棱角—次棱角状，分选性较好，成分单一，主为变质粉砂岩，砾石大小多为3~10cm，含量高达80%~90%。多为颗粒支撑，少数为泥质支撑。

冲积扇相地层结构，主要由两种基本层序（图4-12）组成。

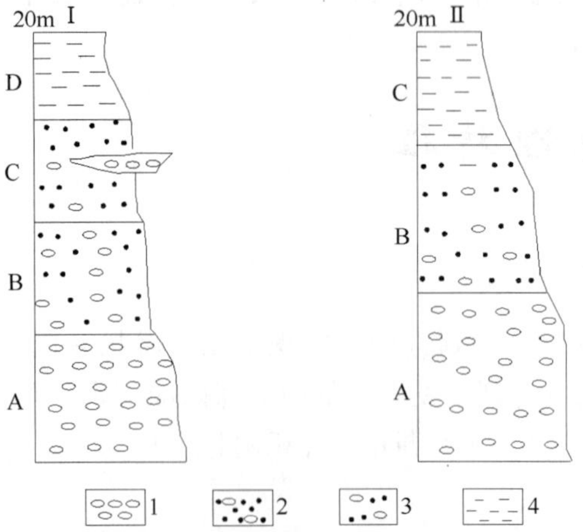

图4-12　冲积扇相基本层序类型及特征

Fig. 4-12　Basic sequence types and characteristics of alluvial fans

1—砾岩；2—砂砾岩；3—含砾砂岩；4—泥岩

Ⅰ. 以AB、ABC、CD组合为主，少数为ACD、ABCD组合，其中A、B代表河道沉积，C代表片汜沉积。

Ⅱ. 以AC、ABC组合为主，少数为BC组合，其中A、B代表泥石流沉积，C代表泥流沉积。

Ⅲ. 以AB、ABC组合为主，少量为BC组合，代表心滩沉积。

2. 辫状河流相

以中—细砾岩为主，可进一步划分为心滩与河道沉积。河道沉积：以浅灰白色夹浅紫红色复成分砾岩、砂砾岩为主，次为含砾中粗砂岩。发育中、大型斜层理、平行层理。砾石呈次棱角状—次圆状，少数次圆—圆状，多具定向排列。

心滩沉积：以浅灰白色浅紫红色复成分砾岩、砂砾岩、含砾砂岩、砂岩为主，次为泥质粉砂岩、粉砂质泥岩。砾石次棱角状—次圆状，棱角状次之，局部富集成层并显示层理。沉积序列下粗上细，底部常为砾岩。

辫状河流相地层结构主要由三种基本层序类型构成（图4-13）。

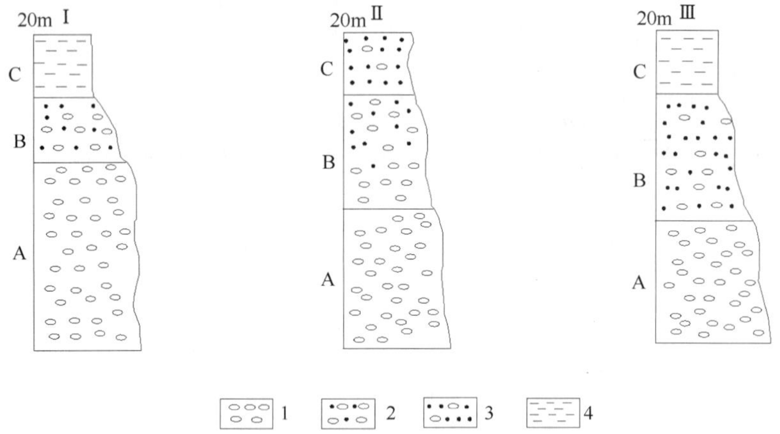

图4-13　辫状河流相基本层序类型及特征

Fig. 4-13　Basic sequencetypes and characteristics of braided river facies

1—砾岩；2—砂砾岩；3—含砾砂岩；4—泥岩

3. 洪冲积平原相

以含砾砂岩、砂岩、凝灰质砂岩、粉砂岩、粉砂质泥岩、凝灰质泥岩、砂质沉凝灰岩、沉凝灰岩为主，可进一步划分为洪泛决口扇、洪泛平原沉积。

洪泛决口扇沉积：以浅紫红色含砾砂岩、砂岩、凝灰质砂岩为主，夹砂砾岩、泥质粉砂岩、粉砂质泥岩、沉凝灰岩。砾石从棱角状至圆状均有，以次棱角状—次圆状为主，分选性差。层理不太发育，局部见平行、水平层理。砾岩、砂岩多为碎屑支撑，少数为泥质支撑，岩性下粗上细。

洪泛平原沉积：以浅紫色泥质粉砂岩、粉砂质泥岩、凝灰泥岩、砂质沉凝灰岩、沉凝灰岩为主，夹中细砂岩。发育水平层理、沙纹层理。

洪冲积平原相地层结构主要由两种基本层序类型组成（图4-14）。

4. 湖相沉积

以浅灰色、浅灰白色含砾中粗砂岩、细砂岩、泥质粉砂岩为主。砾石多呈次棱角状—次圆状，砂级碎屑多呈次棱角状。颗粒支撑为主，少量为泥质支撑。

滨湖相地层结构由一种基本层序类型组成（图4-15）。

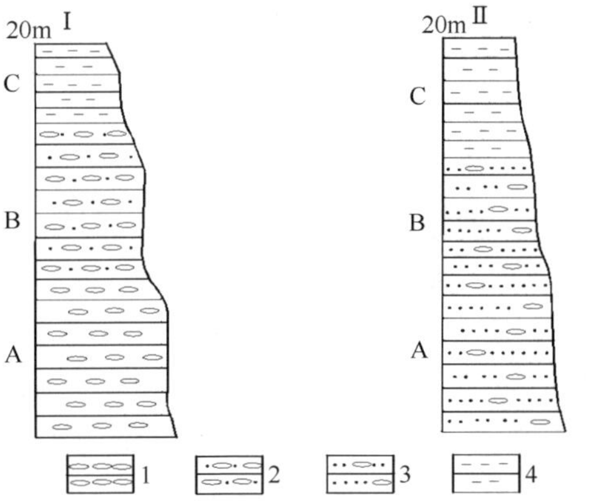

图4-14 洪冲积平原相基本层序类型及特征

Fig. 4-14 Basic sequencetypes and characteristics of pluvial-alluvial plain facies

1—砾岩；2—砂砾岩；3—含砾砂岩；4—泥岩

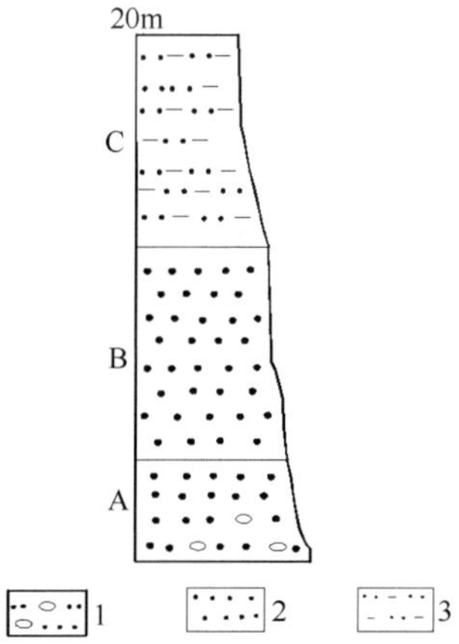

图4-15 滨湖相基本层序类型及特征

Fig. 4-15 Basic sequencetypes and characteristics of shore lake facies

1—含砾砂岩；2—砂岩；3—泥质粉砂岩

鹿母湾组、报万组岩石以浅紫红色—浅灰白色为特征，鹿母湾组中沉凝灰岩、凝灰质等火山物质局部相当发育。而盆地四周剥蚀区岩性为变质岩、花岗岩和花岗斑岩，未见火山岩与火山碎屑岩，说明鹿母湾组沉积时，盆地火山活动频繁，盆内火山碎屑构成了沉积层的部分物源。剧烈频繁的火山活动带来的大量火山物质的加入，使火山碎屑流较为发育，增加了沉积分析和恢复沉积环境的难度，例如造成冲扇相与辫状河流相，冲积扇片汜沉积与冲积平原决口扇沉积难以区分。

二、沉积环境分析

海南岛白垩纪盆地主要有王五、乐安、白沙、雷鸣、清澜（隐伏状）、阳江火山-沉积盆地和同安岭、牛腊岭、五指山火山盆地等，以白沙火山-沉积盆地规模最大。盆地总体呈北东向断续分布，其生成明显与北东向西太平洋构造域作用有关。其中，以王五、乐安、白沙（南西段）、阳江和同安岭、牛腊岭盆地研究比较系统。

从前述可知，鹿母湾组孢粉植物群的组成是较丰富的，其中占主导地位的是松科、掌鳞杉科和麻黄科植物的花粉，也见有水龙骨科、海金砂科等组成的蕨类植物孢子。这些植物主要产于热带、亚热带地区。因此，可以推测在中生代早白垩世时处于热带亚热带气候环境。

值得指出的是，早白垩世植物群与我国南方广大地区中的早白垩世植物群的面貌有着显著不同。张振来（1979）曾系统地总结了华南地区白垩纪孢粉组合特征，他将早白垩世早中期的组合概括为 *Paniophyllum pollenites*，*Psophosphaera-Cicatricosisporites*，*Schizaeoisporites*；没有或仅个别地见到被子植物花粉。其中以 *Paniophyllum pollenites*（同 *Classopillis*）和 *Psophosphaera* 花粉为主，苏铁类花粉常见，带气囊的松柏类花粉较少。湖南汝城文明司组、湖北鄂城灵乡群的孢粉组合可作为代表。早白垩世中晚期的组合被概括为 *Cicatricisisporites*，*Schizaeoisporites-Paniophyllum pollenites*；被子植物 *Exesipollenites* 花粉出现1%~5%。组合中 *Cicatricosisporites* 和 *Schizaeoisporites* 的数量比前一组合明显增加，*Exesipollenites* 大量出现，具气囊的松柏类花粉少量见到，被子植物花粉以三沟型为主，个别见到三孔沟型和三孔型花粉。湖南衡阳神王山组和湖北宜昌五龙组的孢粉组合可作为代表。不难看出，海南孢粉组合与上述两种组合有着较大差异。造成这种差异的原因可能是各自所处的地理位置不同而引起的。根据汪啸风等（1992）资料，海南岛白垩纪古纬度变化于北纬15.11°~21.09°之间（与三叠纪古纬度北纬16.5°基本一致）。其地理纬度与湖北相差近10°，与广东、广西也相差近5°，依据中国科学院编著的《中国自然地理植物地理》（1983）一书提出的中国植物区系分区观点，我国的华中、华南的广大地区均属于泛北植物区中的中国、日本森林植物亚区，而海南岛则属于古热带植物区的马来西亚植物亚区，可见现代植物也分属于不同的植物区系。瓦赫那梅也夫（Baxpameeb，1978）根据古植物资料，将现今的北半球早白垩世划分为4个气候带，自北向南依次为湿温带、湿亚热带、干旱带和潮湿赤道带。他认为孢粉组合中 *Classopollis* 的含量超过50%时，则代表干旱气候。依据这一标准，我国中南、华南的广大地区都基本上属于瓦氏划分的干旱气候带，而海南岛很可能处于干旱气候带与潮湿赤道带之间的过渡地带之中。

火山-沉积盆地早期沉积都以一套近缘碎屑沉积为特点，中期出现粒度相对较细的碎屑沉积，晚期盆地闭合则以发育水平层理、交错层理的砾岩、砂砾岩体现。但是，其也有一定的差异性，如火山岩层主要分布于白沙盆地、阳江盆地中，乐安盆地也有零星分布，而其余盆地目前没有见到，且白沙盆地火山岩出露于鹿母湾组上部，而阳江盆地则主要见于鹿母湾组下部。造成上述差异除了可能与该时期多旋回火山作用有关外，也显示了海南岛白垩纪盆地并不属于一个统一的盆地。由于断陷作用，使得不同盆地边缘成为剥蚀区，为盆地沉积提供了物源。以白沙盆地南西段长茅水库一带南美剖面为例，盆地演化特征如下：

最初在盆地底部形成厚数米的一层陆相复成分底砾岩，不整合在前白垩纪基底之上。向上变为河湖相沉积，随着时间推移，水体不断加深，变为湖相沉积。剖面下部为河流边滩相粗砂岩沉积夹河漫滩相泥质沉积，上部为湖相粉砂质泥岩夹粉砂岩沉积。这些沉积岩中河流相沉积和湖相沉积构造发育，常见到大型交错层理、板状层理、槽状层理、平行层理、冲刷构造、流水波痕等，顶部为湖相灰绿色泥岩夹泥质粉砂岩，岩石中水平层理发育，可见层纹状构造，系微弱水动力条件下形成的产物。另外，沉积厚度较大（360m），这一带可能是断陷盆地水体较深地段。这是早白垩世早期南美地区沉积环境特征。断陷沉积湖盆地雏形基本形成。这段时间，气候炎热、潮湿、多雨，河流发育，湖盆水体丰富，动植物繁盛。这是南美地区鹿母湾组下段所反映的沉积环境和构造盆地特征。

早白垩世晚期，由于太平洋板块向北俯冲到海南古陆之下，激发地幔岩浆上升，使地壳向上拱起，地应力由挤压状态变为拉张变形，在断陷盆地内发生双峰式火山岩浆喷发，地壳运动加剧，沉积盆地向东部迁移，气候继承早白垩世早期特征，持续炎热、多雨，湖盆水体丰富，动植物繁盛，这是该时期沉积环境和盆地演化的总特征。早白垩世晚期开始，南美地区灰绿色层纹状泥岩逐渐消失，变为鲜紫红色安山质（玄武质）、流纹质凝灰岩夹细砂岩、粉砂岩、泥岩，系浅水强氧化条件下形成的产物。颜色鲜艳，以平行层理为主，少见楔状层理和交错层理，表明除湖相沉积外，还有河流相沉积。这阶段，火山喷发异常强烈，到处堆满火山灰，经流水搬运形成水携凝灰岩，成层性好，流水沉积特征明显。凝灰岩与陆源碎屑之比为3：1，表明有河相沉积和丰富的火山物质。东部洪帽一带，鹿母湾组上段，火山岩夹陆源粉砂质、泥质沉积岩中，水平理层极为发育，可见水平层纹状构造。岩石色彩灰暗，以灰绿色为主，厚度较大（398m），表明其是在火山喷发间歇期，水动力微弱，水体较深的环境下形成的。洪帽剖面鹿母湾组上段顶部是流纹质凝灰熔岩，厚63.78m，这是火山喷发减弱，从火山口流出的晚期熔岩，可能指示火山口位置。这一带既是盆地沉积中心，又是构造活跃地带，表明沉降中心由西（南美地区）向东（洪帽）迁移。

晚白垩世早期，火山活动结束，断陷盆地地壳运动继承性继续下降，由震荡下沉变为平稳下降，另一方面气候变得炎热、干燥，雨量不均匀，生物受到恶劣气候制约较难生存，特别是植被明显减少（孢粉化石稀少），致使报万组下段中下部形成紫红色、青灰色、灰黄色细粒岩屑砂岩夹薄层粉砂岩、泥质粉砂岩、粉砂质泥岩，底部形成青灰色夹紫红色含砾粗粒岩屑长石杂砂岩。岩石中有交错层理、冲刷构造，但不发育，而块状构造、粒序层理较发育。砾石约占5%，砾径大小不等，一般1~5cm，成分复杂，次圆状为主；含有较多的经搬运再沉积的火山灰。这是在火山洼地上形成的湖盆底部湖成三角洲沉积（湖河相）。下部为紫灰或青灰色中厚层中细粒长石杂砂岩夹粉砂岩、粉砂质泥岩，岩层波痕发育，含同生深紫色泥砾为特征。这些泥砾沿岩层层面呈不连续条带状、条块状、透镜状、不规则状嵌于岩石中，含量1%~20%间变化。这是在干旱气候条件下，由泥裂破坏后再沉积形成的滨湖相沉积。中部岩性与下部相似，粒度略变细，颜色变浅，平行层理发育；不含泥砾为特征，是一套浅湖相沉积。上部由粉砂岩、粉砂质泥岩、泥岩组成多个韵律层。岩石中微层理、平行层理发育，沉积成因系半深水湖相沉积向深水相沉积过渡（有层纹状层理），但区内未见典型的深湖相沉积。上述几种湖相沉积特征，与刘宝珺等（1985）划分的淡水湖泊亚相基本可对比。因此，长茅岭剖面上由5~26层组成的报万组下段，是火山洼地上生成的湖盆底部湖河相沉积（湖成三角洲沉积）、湖水由浅变深形成的一套湖相岩系，是地壳有规律震荡下降形成的湖相沉积物，含少量孢粉化石。总之，晚白垩世早期地壳运动由强烈运动到震荡运动，即由强变弱，逐渐趋于平稳下降过程。

晚白垩世晚期，由于地壳沉降的不均一性，盆地东部下降较快，继续保持沉降中心地位。另一方面，气候进一步恶化，植被相对减少，风化剥蚀作用增强，地表径流加大，常有山洪暴发，巨大的洪流夹带大量陆源碎屑、砾石、巨砾到盆地中沉积下来，由近而远形成洪流洪积扇砂砾岩、巨砾岩及冲积相粗砂、砂岩、细砾岩沉积，大致平行盆地东岸线呈条带状分布，沿盆地东缘组成一个庞大的洪积扇裙，呈北东向带状分布，宽2~3km。由北东到南西有孔头田、番豆村、昂外、大南海退山、福包岭等多个明显的大扇体（图4-16）。晚白垩世晚期沉积物由报万组上段冲洪积相砂砾岩组成，属近源快速沉积。扇顶由巨砾岩组成，砾石大小悬殊，大者可达1.5m，小者仅0.2cm，杂乱无章堆积，层面不清。扇内由砾岩、细砾岩、含砾粗砂岩组成，砾石一般小于10cm，含砾粗砂岩常见夹于砾岩之间，厚度不大，可见平行层理，可能是洪积间隙期形成的。扇外由含砂粗砂岩、粗砂岩组成，河流相特征明显，常见交错层理和平行层理混生在一起。到晚白垩世晚期末，地壳运动下降速度减慢，冲洪积物把盆地填满。

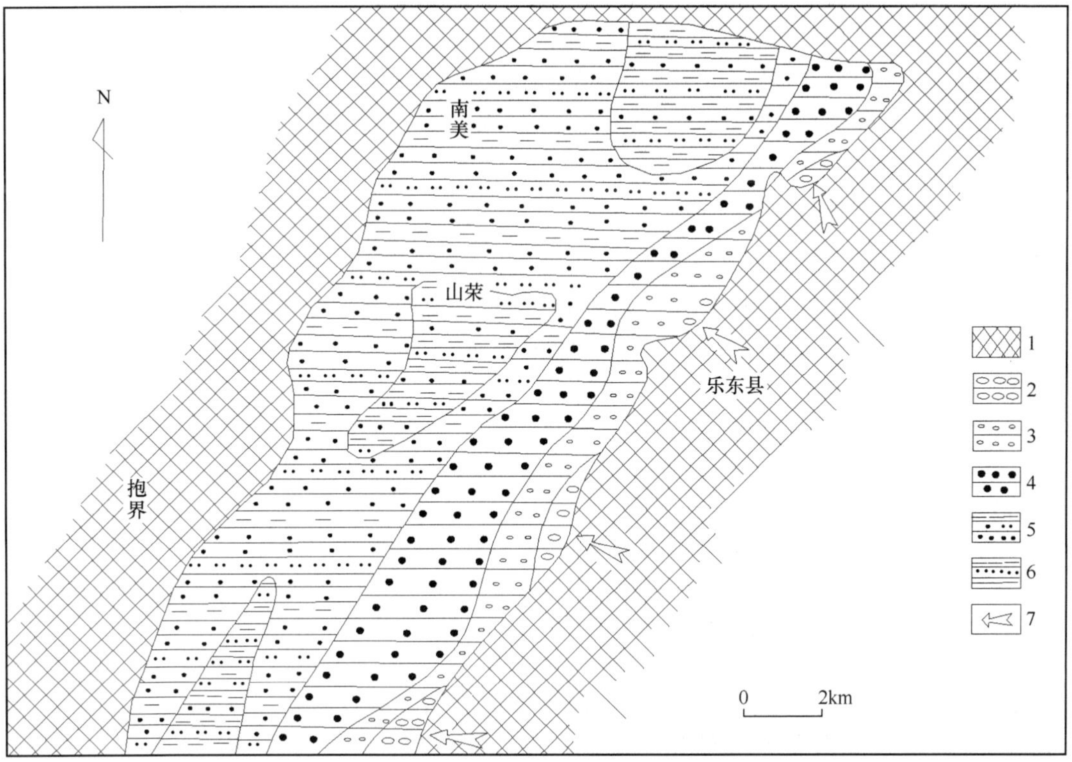

图4-16　报万组岩相古地理简图

Fig. 4-16　Sketch map of lithofacies paleogeography of the Baowan Formation

1—陆缘区；2—巨砾岩；3—粗—细砾岩；4—含砾粗砂岩区；5—砂岩-粉砂岩-泥岩区；6—粉砂岩-泥岩区；7—物源方向

第五章　古近系—新近系

海南省古近系—新近系属于青藏-南海地层大区桂粤-南海地层区、中央海盆地层大区和菲律宾-南沙地层大区南沙地层区。桂粤-南海地层区进一步划分为雷琼地层分区海口地层小区、长昌地层分区、莺歌海地层分区、西沙地层分区；南沙地层区进一步划分为礼乐滩盆地地层分区和曾母盆地地层分区。其中，中央海盆地层大区资料较少，其沉积物未建组，暂不叙述，本次仅保留划分方案。

第一节　雷琼地层分区（海口地层小区）

雷琼地层分区介于遂溪断裂和王五-文教断裂之间，以琼州海峡断裂为界（广东省地质矿产局，1988）划分出南、北两个地层小区，海南省北部命名为海口地层小区。其中，海口地层小区岩石地层单位除海口组和长坡组创建于海口小区外，其余采用雷琼分区岩石地层单位，自下而上划分为：古新统长流组（$E_1\hat{c}$）、始新统流沙港组（E_2l）、渐新统涠洲组（E_3w）和中新统长坡组（$N_1\hat{c}$）、下洋组（N_1x）、角尾组（N_1j）、灯楼角组（N_1d）、上新统海口组（N_2h）。

海口地层小区古近系为陆相碎屑岩夹基性火山岩沉积，是石油、天然气的生油和储油层位。主要隐伏分布于福山盆地、白莲盆地、南宝盆地、加来盆地、长坡盆地一带，其上被新近系—第四系沉积物及第四纪玄武岩所覆盖。新近系主要隐伏在各大盆地（福山盆地、白莲盆地、南宝盆地、加来盆地及长坡盆地等，上中新统—上新统连成一片形成琼北凹陷），岩性为碎屑岩夹基性火山岩，偶夹碳酸盐岩沉积的海相地层，是琼北的主要地下水和海口地区大型矿泉水的含水层位，长坡盆地中并见褐煤产出。

一、剖面描述

（一）长流组（$E_1\hat{c}$）

南宝盆地 ZK9 孔剖面

该孔位于临高县南宝镇西南约 2km 处，由海南地质大队第四分队 1975 年测制；为长流组参考剖面，层位位于 ZK9 孔 18.73~246.09m 处，自上而下为：

上覆地层　上新统海口组灰白色砂砾岩

～～～～～～ 不　整　合 ～～～～～～

古新统长流组	总厚度 217.3m
12. 紫灰色粗砂岩，局部含砾和钙质结核；上部为砖红色、黄褐色泥岩和砂质泥岩	11.2m
11. 暗紫红色砂质泥岩，偶含钙质团块；中部夹紫灰色粉砂岩。产腹足类 *Bithynia* sp.	14.3m
10. 暗紫红色、紫灰色粗砂岩，偶含砾，夹砖红色砂质泥岩和灰白色含砾砂岩	13.5m
9. 暗紫红色砂质泥岩和泥岩，夹暗紫红色、紫灰色细砂岩	29.5m
8. 紫灰色、青灰色细砂岩，夹暗紫红色、砖红杂青灰色含钙质结核砂质泥岩	11.2m
7. 暗紫红色、紫灰色细砂岩与砂质泥互层，含钙质结核。产腹足类 *Mirolaminatus gracilis*, *M. ualidus*	40.1m
6. 暗紫红色含砾粗砂岩与细砂岩互层。产腹足类 *Strobilops* sp.	20.3m
5. 暗紫红色砂质泥岩，含钙质结核	22m
4. 紫灰色含砾粗砂岩	5.2m
3. 暗紫红色杂青灰色砂质泥岩，含钙质结核	15.3m

2. 暗紫红色杂青灰色砂质泥岩夹粗砂岩和砂砾岩，含钙质结核 20.9m

1. 暗紫红色砂质泥岩与紫红色、砖红色砂砾岩互层。砂质泥岩含钙质结核 13.8m

~~~~~~~~~~ 不 整 合 ~~~~~~~~~~

下伏地层 古生界灰色绢云母石英片岩

（二）流沙港组（E_2l）

代表性剖面见于广东省徐闻西连迈参二井2120~3227m。参考剖面位于海南岛福山盆地福8井、福3井。福山盆地福8井位于澄迈县福山镇南西约1.2km处，系茂名石油公司南海石油勘探指挥部所钻，自上而下为（层号略有修改）：

上覆地层 渐新统涠洲组灰白色砂质砾岩、浅灰色砾质砂岩与深棕红色泥岩互层

———— 整 合 ————

始新统流沙港组 总厚度1209.5m

三段 厚212.5m

45. 灰色泥岩，顶为暗棕红色泥岩 29m
44. 灰白色砂质砾岩、浅灰色细砂岩、深灰色泥岩、褐灰色白云质泥岩及黑色炭质泥岩韵律层 25.5m
43. 浅灰色粉砂岩 7.0m
42. 灰白色砂质砾岩与深灰色含粉砂泥岩互层，泥岩具水平层理 40.5m
41. 浅灰色钙质粉砂岩 2.5m
40. 灰色、深灰色泥岩，具微细水平层理 14.5m
39. 浅灰色钙质粉砂岩 2.5m
38. 深灰色泥岩，具微细水平层理，夹炭质泥岩 53.5m
37. 上为灰白色砂质砾岩，下为灰白色砂质泥岩与深灰色泥岩互层，底为浅灰色细砂岩 33m
36. 浅灰色白云质含砾砂岩，含炭屑，顶为深灰色泥岩 4.5m

二段 厚294.5m

35. 深灰色、灰黑色泥岩，夹灰黑色致密块状玄武岩 83.5m
34. 灰白色砾质粗砂岩，含炭屑，夹深灰色泥岩 19m
33. 浅灰色粉砂岩与深灰色泥岩互层 15m
32. 深灰色含粉砂质泥岩，下部夹灰白色含砾砂岩 34m
31. 灰白色白云质砾质砂岩 7m
30. 深灰色、灰黑色泥岩与灰白色含砾砂岩互层 23m
29. 深灰带褐色含粉砂质泥岩 15m
28. 黄灰色含砾砂岩，具油斑 2.5m
27. 深灰色泥岩及浅灰色含砾砂岩 8m
26. 黄灰色含砾砂岩，微含油 1m
25. 黄灰色含砾砂岩，具油斑 10.5m
24. 黄灰色砾质砂岩，具油斑 10.5m
23. 深灰色泥岩及浅灰色砾质砂岩 18m
22. 深灰色泥质砂岩微含油，夹深灰色泥岩 3.5m
21. 灰黄色具油斑的砾质砂岩和砂砾岩，与深灰色泥岩互层 10.5m
20. 上部为浅灰色粉砂岩、砾质砂岩及灰白色白云质砂砾岩，下部为深灰色—灰黑色泥岩，底部为一层黄色白云质泥岩 33.5m

一段 厚702.5m

19. 灰黄色含油砂质砾岩和中砂岩，夹深灰色泥岩 8m
18. 灰黄色微含油砂质砾岩及具油斑的黄灰色砂质砾岩 6m
17. 灰黄色具油斑砂质砾岩及灰白色砂质砾岩 9m
16. 深灰、灰黑色泥岩、灰白色砂质砾岩及浅灰色钙质粉砂岩 32m
15. 灰白色砂质砾岩与灰色泥岩，夹黄色白云质粉砂岩和泥岩 16m
~~~~~~~~~~

14. 灰白色砂质砾岩和砾质砂岩，浅灰色中砂岩和粉砂岩，与深灰色、浅灰色泥岩及黄色白云质泥岩互层 34m
13. 浅灰色粉、细砂岩和黄色白云质粉砂岩与深灰色、灰绿色泥岩互导 23m
12. 灰白色砂质砾岩、砾质砂岩、含砾砂岩、浅灰色粉砂岩与深灰色、绿灰色泥岩互层 28.5m
11. 浅灰色含砾砂岩、粗砂岩和粉砂岩与褐灰色、深灰色、绿灰色泥岩互层 34m
10. 灰白色含砾砂岩、浅灰色砾质砂岩、粉砂岩与绿灰色泥岩互层，下夹2层灰黄色白云质泥岩 21m
9. 灰白色砂质砾岩、浅灰色含砾砂岩、细砂岩、粉砂岩与褐灰色、绿灰色泥岩互层 49.5m
8. 褐灰色、绿灰色泥岩，夹浅灰色中、细砂岩和黄灰色油斑细砂岩 46.5m
7. 灰白色砂质砾岩、浅灰色砾状粗砂岩及灰色、暗棕红色泥岩 30m
6. 灰白色砂质砾岩、浅灰色砾质砂岩和含砾砂岩与绿灰色、灰色和暗棕红色泥岩互层 29.5m
5. 上部为浅灰色、灰白色砂质砾岩，夹绿灰、暗棕红色泥岩；下部为浅灰色、灰白色砂质砾岩与灰色、暗棕红色泥岩互层 115m
4. 浅灰、灰白色砂质砾岩与灰色、绿灰、暗棕红色泥岩互层 57m
3. 灰白色砂质砾岩与棕红色、灰色及杂色泥岩互层 63m
2. 灰白色砂质砾岩与棕红色、灰色泥岩互层 68.5m
1. 灰白色砂质砾岩，含花岗岩块砾 22m

——————— 整　合 ———————

下伏地层　古新统长流组棕红色泥岩与浅紫红色含砾砂岩、含钙砾质砂岩互层 厚>39.4m

未见底

（三）涠洲组（E_3w）

代表性剖面见于广西北部湾涠洲岛涠浅1井。参考剖面位于海南岛福山盆地福8井、福1井。福山盆地福8井系茂名石油公司南海石油勘探指挥部所钻，位于澄迈县福山镇南西约1.2km处，自上而下为（层号略有修改）：

上覆地层　下中新统下洋组

~~~~~~~ 不　整　合 ~~~~~~~

渐新统涠洲组 厚度482.5m

61. 浅灰色砂砾岩与灰绿色、绿灰杂棕红和棕黄色泥岩互层 47.5m
60. 浅灰色含砾砂岩、砂质砾岩与棕红色、浅灰色泥岩互层 20m
59. 浅灰色泥质砂岩与棕红色泥岩互层 29m
58. 浅灰色砂质泥岩与棕红色、浅灰杂棕红色泥岩互层 14.5m
57. 浅灰色砂质砾岩及棕红色泥岩 12m
56. 浅灰色砂质砾岩与棕红色泥岩互层 28.5m
55. 浅灰色砂质砾岩与浅灰杂棕红色泥岩互层 30m
54. 浅灰色砾质砂岩与灰杂棕红色泥岩互层 37.5m
53. 浅灰色砂质砾岩与棕红色泥岩互层 17m
52. 上为浅灰色砂质砾岩与灰杂棕红色泥岩互层，下为浅灰色砂质砾岩夹灰色泥岩 64.5m
51. 浅灰色含砾砂岩、砾质砂岩、砂质砾岩与浅灰色、棕红色泥岩互层 36m
50. 浅灰色含砾砂岩夹棕红色泥岩 23.5m
49. 浅灰色 、灰白色砂质砾岩夹棕红色泥岩 42.5m
48. 灰白色中砂岩与深棕红色泥岩互层 24m
47. 灰白色砂质砾岩与深棕红色泥岩互层 34.5m
46. 灰白色砂质砾岩、浅灰色砾质砂岩与深棕红色泥岩互层 21.5m

——————— 整　合 ———————

下伏地层　始新统流沙港组灰色泥岩，顶为暗棕红色泥岩
~~~~~~~

（四）长坡组（$N_1\hat{c}$）

代表性剖面见于长坡盆地 ZK107 孔。系海南地质大队 1975 年所钻，位于儋州市长坡镇东约 2.5km 处，起点坐标：东经 109°16′30″，北纬 19°42′00″。自上而下为（层号略修改）：

上覆地层 上中新统灯楼角组红黄色—绿灰色含砾黏土质砂

------假 整 合------

中新统长坡组		总厚度 331.40m
14.	灰色泥岩、砂质泥岩与黏土质粉砂岩、细砂岩互层	12.00m
13.	蓝灰色泥岩夹泥质砂岩和含炭泥岩，偶夹褐煤及菱铁矿砂岩，产孢粉化石	29.50m
12.	褐煤夹灰褐色炭质泥岩，常见鱼骨及植物化石碎片 *Pnyuites* sp.	6.20m
11.	黄灰色钙质页岩与褐灰色含炭油页岩互层，夹褐煤、泥灰岩、青灰色泥岩，产介形类 *Cyprinotus* sp.，*C.* cf. *formalis*.，*Ilyocypris* cf. *biplicata*，*Eucypris* sp. 和腹足类	15.90m
10.	青灰色泥岩夹褐灰色油页岩及褐煤，下部油页岩夹层含炭	19.80m
9.	黄灰色钙质页岩与褐煤互层，夹油页岩及泥岩	10.40m
8.	褐灰色油页岩与黄灰色钙质页岩互层，产腹足类	7.60m
7.	褐灰色含炭泥岩夹蓝灰色泥岩和菱铁矿粉砂岩，产鱼骨碎片	12.60m
6.	蓝灰色泥岩、砂质泥岩夹泥灰岩，含黄铁矿结核	5.40m
5.	杂色（以紫红色为主）泥岩夹蓝灰色黏土质粗砂岩及含砾中粗砂岩	34.70m
4.	蓝灰色含砾黏土质中粗砂岩夹砂质黏土岩，局部为黏土质粉细砂岩与杂色泥岩互层，见炭化植物碎片	89.90m
3.	杂色（红杂灰）与灰褐色含砾黏土质中粗砂岩	30.10m
2.	杂色—褐灰色砂砾岩夹浅灰、灰绿色黏土质粉砂岩及褐灰色含砾中粗砂岩；粉砂岩中常见炭化植物化石碎片	21.80m
1.	紫红色、灰绿色砾岩，砾石成分有紫红色砂岩、花岗岩、变质岩，砾径一般 0.5~20cm，最大达 45cm	34.80m

~~~~~~ 不 整 合 ~~~~~~

下伏地层 白垩系鹿母湾组紫红色含砾不等粒砂岩

（五）下洋组（N_1x）

代表性剖面位于广东省雷州半岛，参考剖面见于海南岛福山盆地福 1 井、海口市玉沙村 ZK8 孔。2001 年，为采取地下热水，海南地质地理咨询服务公司在海口黄金海岸花园进行 ZK1 孔钻探，该孔中新近系发育较全。

1. 福山盆地福 1 井剖面

剖面位于澄迈县马村乡大亭村南约 1km 处，为茂名石油公司南海石油勘探指挥部于 1975 年所钻。自上而下为（层号按照原始资料）：

上覆地层 中中新统角尾组浅灰色中、粗砂岩、含砾砂岩及砾状砂岩

———— 整 合 ————

下中新统下洋组		总厚度 241.5m
29.	浅灰色中、粗砂岩不等厚互层，下部夹灰色砂质泥岩、细砂岩及砂砾岩	73m
30.	浅绿灰色粉砂岩	7.5m
31.	浅灰色中、粗砂岩，夹灰色、绿灰色泥质粉砂岩、粉砂质泥岩及白色石膏质粗砂岩	47m
32.	浅灰色粉砂岩和砂岩，上部夹浅灰色泥质粉砂岩和砂质泥岩，下部夹白色石膏质粉砂岩、浅灰色砾状砂岩及砂砾岩	21.5m
33.	灰白色砂岩、夹灰色砂质泥岩	58m
34.	灰白色粗砂岩、砾砂岩和砂状砾岩，下部夹细砾岩，顶部为灰色泥岩	34.5m
~~~~~~

~~~~~~~~~~ 不　整　合 ~~~~~~~~~~

下伏地层　渐新统涠洲组

2. 海口黄金海岸花园ZK1孔剖面

剖面位于海口市秀英区电海村东约800m处，起点坐标：东经110°12′27″，北纬20°03′27″，由海南省地质调查院2001年测制。

上覆地层　中中新统角尾组灰绿色细中粒石英砂岩夹灰色厚层、薄层凝灰岩

——————— 整　合 ———————

下中新统下洋组（N_1x）　总厚度174.65m

15. 灰黑色块状厚层粉砂质泥岩，水平层理发育，上部夹灰白色砂砾岩。含孢粉化石。蕨类植物孢子主要有 *Polypodiisporites*，*Polypodiaceaesporites*，*Lygodiumsporites*；裸子植物花粉主要有 *Dacry diumites*，*Abietineaepollenites*，*Pinuspollenites*，*Tseugaepollenites*，*Ahiespollenites*，*Dacrydiumites*；被子植物花粉主要有 *Caryapollenites*，*Graminiidites*，*Umbellifezaeipites*，*Crucifeiaepites*，*Mazgocosporites*　15.50m
14. 灰白色含砾中粗粒岩屑石英砂岩，局部夹灰黑色薄层泥岩　7.00m
13. 灰色块状泥质粉砂岩与灰黑色块状泥岩互层，下部为灰绿色厚层含砾中粗粒岩屑石英砂岩，局部夹灰黑色薄层泥岩　6.80m
12. 浅紫色含砾中粗粒岩屑石英砂岩，局部含植物化石碎片（树木的枝干）　7.20m
11. 灰白色含砾中粗粒石英砂岩　7.00m
10. 灰色、灰黑色含粉砂泥岩，下部夹灰白色块状中粗粒石英砂岩和灰色中层含砾中细粒石英砂岩　10.00m
9. 浅紫色、灰白色中粗粒岩屑石英砂岩，上部局部夹灰绿色厚层粉砂岩　20.00m
8. 绿色细粒石英杂砂岩，下部局部夹灰色厚层含砾中粗粒石英砂岩　16.00m
7. 灰色、灰绿色含粉砂泥岩。含蕨类植物孢子主要有 *Polypodiisporites*，*Polypodiaceaesporites*，*Polypodiaceoisporites*，*Deltodospora*，*Toroisporis*；裸子植物花粉主要有 *Abietineaepollenites*，*Dacrydiumites*，*Keteleeriaepollenites* 被子植物花粉主要有 *Caryapollenites*，*Cyperaceaepites*，*Potamogetoricites*　6.00m
6. 灰色、灰黑色含砾中粗粒岩屑石英砂岩，上部为绿色块状中细粒岩屑石英砂岩　12.00m
5. 灰色泥岩，中部为灰色层状粉砂质泥岩夹灰黑色厚层含砾砂质泥岩，下部为灰色、灰褐色块状凝灰岩，中部含植物化石碎片。粉砂质泥岩斜层理发育　19.00m
4. 灰色、灰绿色粉砂质泥岩夹绿色巨厚层粉砂岩，上部岩石含大量植物化石碎片。粉砂质泥岩槽状交错层理发育　13.00m
3. 顶部为绿色厚层粉砂岩，中部为灰绿色层理状泥质粉砂岩，中下部为灰绿色块状泥质石英细砂岩，下部为灰色巨厚层含粉砂泥岩，中部局部岩石含植物化石碎片。泥质粉砂岩水平层理发育　13.00m
2. 杂色泥质粉砂岩，顶部为浅灰色、浅紫红色块状粉砂质泥岩。泥质粉砂岩水平层理发育　16.00m
1. 绿色细粒石英砂岩　6.15m

未见底

（六）角尾组（N_1j）

代表性剖面位于广东省雷州半岛，参考剖面见于海南岛福山盆地福1井、海口市长流地区ZK16孔。2001年，为采取地下热水，由海南地质地理咨询服务公司在海口黄金海岸花园进行ZK1孔钻探，该孔中新近系发育较全。

1. 福山盆地福1井

剖面情况同上述。自上而下为（层号按照原始资料）：

上覆地层　上中新统灯楼角组浅灰色砾状砂岩和砂岩及绿灰色砂质泥岩正韵律层
~~~~~~~~~~

——————整　　合——————

中中新统角尾组 总厚度 289m

21. 浅灰色砂岩、绿灰色粉砂质泥岩或泥岩韵律层 36m
22. 浅灰色细砂岩夹浅灰绿色砂质或粉砂质泥岩 46m
23. 浅灰色中、细砂岩和粉砂岩与灰绿色砂质泥岩或浅灰色泥质粉砂岩互层，顶底均夹白色石膏质砂岩和粉砂岩 42m
24. 灰白色砾状砂岩 3m
25. 灰白色中、粗砂岩及灰绿色砾状砂岩、砂砾岩，间夹薄层砂质泥岩 43.5m
26. 浅灰白色中砂岩、含砾砂岩、砾状砂岩，夹灰色砂质泥岩 38m
27. 灰白色细砂岩、含砾砂岩及灰色砾状砂岩，夹灰色砂质泥岩 22m
28. 浅灰色中、粗砂岩、含砾砂岩及砾状砂岩 58.5m

——————整　　合——————

下伏地层　下中新统下洋组灰黑色厚层—块状粉砂质泥岩

2. 海口黄金海岸花园 ZK1 孔剖面

剖面情况同上述。

上覆地层　上中新统灯楼角组灰绿色中细粒岩屑石英砂岩

——————整　　合——————

中中新统角尾组（N_1j） 总厚度 125.50m

22. 灰色粉砂质泥岩，下部夹灰色中薄层沉凝灰岩。含孢粉化石，蕨类植物孢子主要有 *Polypodiisporites*，*Polypodiaceaesporites*；裸子植物花粉主要有 *Dacrydiumites*，*Abietineaepollenites*，*Pinuspollenites*，*Tsugaepollenites*，*Taxodiacepollenites*；被子植物花粉主要有 *Caryapollenites*，*Graminiidites*，*Compositioipllenites*，*Crucpositiopollenites* 35.00m
21. 灰绿色细粉砂岩，局部水平层理发育 20.00m
20. 上部为灰绿色块状泥质粉砂岩与灰绿色块状粉砂质泥岩互层，中部夹灰色块状含砾中粗砂岩屑石英砂岩，下部为灰绿色块状粉砂岩，局部夹中层细粒石英砂岩。含孢粉化石，蕨类植物孢子主要有 *Polypodiisporites*，*Polypodiaceaesporites*，*Polypodiaceoisporites*，*Crassoretitriletes*；裸子植物花粉主要有 *Dacrydiumites*，*Abietineaepollenites*，*Pinuspollenites*；被子植物花粉主要有 *Caryapollenites*，*Graminiidites*，*Potamogetoncites*，*Compositioipllenites*，*Liliacidites*，*Crucifeiaepites* 30.70m
19. 灰绿色厚层细中粒石英砂岩与灰绿色粉砂岩互层，下部为灰色块状中细粒岩屑石英砂岩 9.30m
18. 灰色砂砾岩 5.00m
17. 灰色含砾中粗粒岩屑石英砂岩 16.00m
16. 上部为灰绿色细中粒石英砂岩，夹灰色厚层、薄层凝灰岩，下部为灰绿色中细粒岩屑石英砂岩。前者中含植物化石碎片（树木的枝干）。含孢粉化石，蕨类植物孢子主要有 *Polypodiisporites*，*Polypodiaceaesporites*，*Polypodiaceoisporites*，*Osmundacidites*；裸子植物花粉主要有 *Dacrydiumites*，*Abietineaepollenites*，*Pinuspollenites*；被子植物花粉主要有：*Graminiidites*，*Compositioipllenites*，*Chenopodipollia*，*Hydrocharia* 9.50m

——————整　　合——————

下伏地层　下中新统下洋组灰黑色厚层—块状粉砂质泥岩

（七）灯楼角组（N_1d）

代表性剖面位于广东省徐闻县，参考剖面见于海南岛福山盆地福 1 井、海口市长流地区 ZK16 孔。2001 年，为采取地下热水，海南地质地理咨询服务公司在海口黄金海岸花园进行 ZK1 孔钻探，该孔中新近系发育较全。福 1 井、ZK1 情况如下：

1. 福山盆地福 1 井

剖面情况同上述。

上覆地层 上新统海口组灰色含砾砂岩、砾状砂岩与绿灰色砂砾岩和砂质泥岩互层

———— 整 合 ————

上中新统灯楼角组 总厚度 337m

12. 灰色粗砂岩、砂质泥岩和泥岩互层，含少量细砾和海绿石 20. 5m
13. 灰色砂岩与页状泥质砂岩互层 19m
14. 灰色砂岩及灰色、绿灰色泥质粉砂岩 58m
15. 绿灰色泥质粉砂岩夹细砂岩 51. 5m
16. 绿灰色砂岩及砂质泥岩，顶、底为绿灰色粉砂岩 28. 5m
17. 绿灰色砂岩、泥质粉砂岩韵律层 35. 5m
18. 绿灰色砂岩，顶为泥质粉砂岩夹砂质泥岩和石膏质细砂岩，含少量海绿石 66m
19. 浅灰色含砾砂岩、砾状砂岩和砂砾岩 34m
20. 浅灰色砾状砂岩和砂岩及绿灰色砂质泥岩正韵律层 24m

———— 整 合 ————

下伏地层 中中新统角尾组浅灰色砂岩、绿灰色粉砂质泥岩或泥岩韵律层

2. 海口黄金海岸花园 ZK1 孔剖面

剖面情况同上述。

上覆地层 上新统海口组灰色、灰绿色含贝壳碎屑砂砾岩

———— 整 合 ————

上中新统灯楼角组（N_1d） 总厚度 154. 00m

31. 灰黑色粉砂质泥岩，水平层理发育，上部夹灰黑色巨厚层泥岩 21. 00m
30. 灰绿色细粒石英砂岩 11. 00m
29. 灰色、灰黑色厚层—块状粉砂质泥岩、泥岩，上部夹灰黑色块状含砾砂质泥岩，下部夹灰色块状含砾细粒石英砂岩。粉砂质泥岩，水平层理发育。含孢粉化石，蕨类植物孢子，主要有 *Polypodiisporites*，*Polypodiaceaesporites*，*Polypodiaceoisporites*；裸子植物花粉主要有 *Abietineaepollenites*，*Pinuspollenites*；被子植物花粉主要有 *Caryapollenites* sp.，*Graminiidites laevigatus*，*Potamogetoncites* sp. 28. 00m
28. 灰绿色中细粒岩屑石英砂岩，上部夹灰绿色块状细粒石英砂岩 13. 00m
27. 灰色、灰黑色薄层泥岩与灰色薄层粉砂岩互层，下部为灰色块状含砾中粗粒岩屑石英砂岩 10. 00m
26. 深绿色粉砂岩 25. 00m
25. 深绿色中细粒岩屑石英砂岩 6. 50m
24. 灰绿色中粗粒岩屑石英砂岩 12. 50m
23. 灰绿色中细粒岩屑石英砂岩，上部夹黑色中层泥岩，下部夹灰绿色块状含砾粗中粒岩屑石英砂岩 26. 00m

———— 整 合 ————

下伏地层 中中新统角尾组灰色粉砂质泥岩

（八）海口组（N_2h）

代表性剖面见于长流 ZK18 孔，系海南地质大队五分队 1977 年测制。该孔位于海口市长流镇，坐标：东经 110°07′19″，北纬 20°08′18″。孔中海口组发育较全；自上而下为（层号按原始资料）：

上覆地层 下更新统秀英组灰白色砂砾

-----假 整 合-----

上新统海口组 总厚182.78m

四段

8. 浅灰色薄层状黏土 9.73m
9. 灰色粉砂质黏土，含海绿石，局部含贝壳碎屑，底部含有孔虫 *Amphistegina* sp.，*Pseudorotalis schroeteriana* 25.45m

三段

10. 灰、灰白色贝壳碎屑岩、产双壳类 *Peslten byoritsuensis*，*Barbatis* sp.；有孔虫：*Pseudorotalia schroeteriana*. *Amphistegina* sp. 5.69m

二段

11. 灰色粉砂质黏土 23.01m
12. 灰色含钙质黏土，产双壳类 *Pitar* sp.；有孔虫 *Lenticulina* sp.，*Operculina* sp.，*Pseudorotalia*，*P.* cf. *indopacifica* 49.68m
13. 灰色砂质黏土，含贝壳碎屑 1.75m
14. 灰白色贝壳泥岩，产有孔虫 *Lenticulina*.，*Operculina* sp.，*Pseudorotalia schroeteriana*. *P.* cf. *indopacifica*；双壳类 *Pitar* sp. 4.42m
15. 灰色砂质黏土，含贝壳碎屑 9.41m
16. 灰色凝灰质泥岩夹灰黑色玄武质沉凝灰岩 7.39m

一段

17. 黄褐杂白色、灰绿色含贝壳砂砾与砂砾岩互层，含双壳类 *Limopsis multistriata*，*Scapharca subcrenata*. *Nucula* sp.，*Chione*（*Clausinella*）*isabellina*，*Cadium*（*Cerastoderma*?）*hanzawoi*，*Corbula taiwanensis* 34.40m

———— 整 合 ————

下伏地层 灯楼港组灰绿色中细砂与中细砂岩互层

二、岩石地层单位

（一）长流组（$E_1\hat{c}$）

长流组最早称为红色岩组（南海石油勘探指挥部，1997），由中国石油公司南海分公司（1978）以海康流沙港徐4井1914.0~2270.0m井段剖面创建命名，并作为该组代表性剖面。省内参考剖面为澄迈县福山镇福3井（茂名石油公司南海石油勘探指挥部，1976）、临高县南宝镇ZK9孔（海南地质大队第四分队，1975）。指不整合于前古近纪地层之上，整合于流沙港组之下的一套杂色碎屑岩。

长流组主要隐伏于福山盆地、南宝盆地、加来盆地及白莲盆地一带，岩性较稳定，总体以棕红色、紫红色泥岩与紫红色或灰色砂砾岩、砂岩不等厚互层为特征，仅福山盆地夹数层泥灰岩，加来-南宝盆地含钙质结核。厚度以福山盆地最厚，福3井厚达570m，南宝盆地ZK9孔厚217.3m。

本次以南宝盆地ZK9孔18.73~246.09m井段作为该组代表性剖面。下部岩性为紫灰色、蓝灰色砂岩与含钙质结核的紫红色、砖红色泥岩、砂质泥岩及泥质粉砂岩不等厚互层，间夹数层砂砾岩；上部以砖红色、紫灰色钙质泥岩和泥岩为主，夹灰绿色、紫灰色、灰白色含钙粉砂岩、砂岩及砂砾岩。其底部以砂砾岩不整合于古生代变质岩之上，顶部以砖红色泥岩与流沙港组灰白色砂砾岩整合接触。

（二）流沙港组（E_2l）

流沙港组为茂名石油公司（1975）创建，代表性剖面为徐闻西连迈参二井2120~3227m井段剖面，省内参考剖面为澄迈县福山盆地福8井和福3井（茂名石油公司南海石油勘探指挥部，1971，1976），系指介于长流组与涠洲组之间的一套灰色碎屑岩，以暗色（黑灰色、灰色、褐灰色、深灰色）泥岩为主，夹灰白色、灰黄色砂砾岩的湖泊相沉积，是生油和储油层位。

该组主要隐伏分布于福山盆地、南宝盆地、白莲盆地一带。依岩性特征自上而下划分为三段。一段厚 212.5m，为灰色、深灰色泥岩与灰白色砂砾岩互层，局部夹黑色炭质泥岩，水平层理发育；二段厚 294.5m，为大套深灰色、灰黑色泥岩，含菱铁岩，其间夹薄层玄武岩和灰白色—灰黄色砂砾岩、含砾砂岩，含砾砂岩中微含油或具油斑；三段厚 702.5m，为深灰色、褐灰色、褐红色泥岩与灰白色—黄灰色砂砾岩不等厚互层，上部灰黄色砂质砾岩中普遍含油或具油斑。与下伏地层长流组整合接触，向下以棕红色砂岩和泥岩的同时出现作为进入长流组的标志，顶部以灰色、灰黑色泥岩与涠洲组灰白色砂砾岩整合接触。

流沙港组的灰色碎屑岩相在区域上较稳定，但靠近盆地中心部位泥岩增厚，靠近边部泥岩厚度变小，总厚度变化较大，厚 32.5～1209m。其中福 8 井厚 1209m，福 3 井厚 576m。

（三）涠洲组（E_3w）

涠洲组由地质部第四石油普查大队勘探大队在 1964 年命名，原称涠洲群。1975 年，茂名石油公司地质处将其改称涠洲组，代表性剖面在广西壮族自治区的北部湾涠洲岛涠浅 1 井 695～1084m 井段，省内参考剖面为澄迈福山盆地福 8 井和福 1 井（茂名石油公司南海石油勘探指挥部，1977，1975），代表雷琼地区古近系上部一套杂色碎屑岩，其间的砂岩是储油层位。

该组主要隐伏于福山盆地中，次为白莲盆地，为一套杂色泥岩、粉砂岩与灰白色砂岩、含砾砂岩及砂砾岩不等厚互层。泥岩颜色复杂，有黄褐、棕黄、灰蓝、紫灰、灰绿、棕红等色，但以棕红色为主。灰色、绿色泥岩中常见棕红色斑块，棕红色、紫红色泥岩中含零星灰绿、棕黄色斑块或砂团。砾岩和砂岩的碎屑成分以石英为主，含少量长石，局部含花岗岩和变质岩碎块，碎屑分选程度差—中等，次棱角—次圆状；胶结物以泥质为主，局部为石膏质、钙质或白云质。顶界以大套棕红色泥岩为标志与上覆下洋组粗砂岩、砾岩不整合接触，其底以灰白色砂质砾岩与流沙港组顶部的灰色、灰黑色泥岩整合接触。

在区域上，该组常夹有数层玄武岩，单层厚度 2～24m，中上部局部含有海绿石。厚度变化较大，在福山盆地中一般厚 540～830m，最厚达 1970m，在盆地边缘和次级凸起部位变薄甚至缺失，在福 8 井厚 482m，福 1 井厚 1677m。

（四）长坡组（$N_1\hat{c}$）

长坡组系海南地质大队在琼北儋州市长坡煤矿勘探时命名，陈仁明等（1975）介绍，未指定代表性剖面。代表性剖面为长坡盆地 ZK107 孔 100.49～422.53m 井段，位于儋州市长坡镇，坐标：东经 109°26′30″，北纬 19°42′00″，由海南地质大队 1975 年测制。

该组为陆相含煤沉积，控制着长坡大型油页岩矿床及中型褐煤矿床。根据岩石组合特征，长坡组共划分为 5 段（第四段为褐煤的产出层位）。第一段为灰色或杂色半胶结砾岩、砂砾岩夹少量黏土质砂。砾石成分复杂，以白垩系砂岩及古生界变质岩为主，少量火成岩，胶结物为砂泥质，较松散。第二段下粗上细，下部岩性变化大，为紫红色、蓝灰杂色含砾黏土质中粗砂与杂色块状黏土、砂质黏土或黏土质粉细砂不等厚互层，或以黏土质砂为主夹黏土、砂砾岩及含砾黏土质中粗砂等；上部以杂色块状黏土、砂质黏土为主，夹黏土质中细砂或含砾黏土质砂，偶夹含炭黏土及炭化木，见植物根茎化石。第三段为蓝灰色块状含钙黏土、粉砂质黏土夹泥砂质灰岩或钙质砂岩（单层厚一般小于 1m），局部夹菱铁矿砂岩透镜体，该组边缘夹黏土质粉细砂。第四段中部主要由灰褐色含炭油页岩与含钙油页岩夹炭质油页岩及褐煤组成，向上部与下部逐渐过渡为褐灰色—灰色块状含炭黏土及蓝灰色块状黏土、粉细砂质黏土夹煤线。中部水平层理发育，局部有微波状、透镜状层理，产大量介形类、鱼类、螺类和植物碎片、藻类等化石。第五段为浅灰色、蓝灰色块状黏土、粉细砂质黏土夹黏土质中细—粉细砂，偶夹含砾黏土质中粗砂及紫红色块状黏土。

长坡组顶部以灰色泥岩与上覆灯楼角组绿灰色含砾黏土质砂岩呈假整合接触。底部以紫红色、灰绿色砾岩与下伏鹿母湾组紫红色含砾不等粒砂岩呈不整合接触。

长坡组分布在儋州市长坡盆地及临高县加来-南宝盆地一带。在长坡盆地中部以灰色、蓝灰色黏土岩、粉砂质黏土岩为主，夹油页岩及褐煤，底部为砂岩、砂砾岩，向盆地边缘过渡为以杂色砂砾岩、含砾砂岩为主夹炭质泥岩及煤线。加来-南宝盆地中部岩石组合特征与长坡盆地中部相似，但未见褐煤，仅见炭质泥岩，泥岩中含油也较低。长坡组在两盆地间厚度变化不大，在长坡 ZK107 孔厚 331m，加来 ZK1 孔厚 294m。

（五）下洋组（N_1x）

下洋组系地质部第四石油地质普查勘探大队于 1964 年建立。命名剖面位于广东省徐闻县下洋姑苏村雷琼 3 井 749～892.44m 井段，省内参考剖面为澄迈县福山盆地福 1 井（茂名石油公司地质处，1975）和海口市玉沙村 ZK8 孔（海南地质大队，1988），指位于涠洲组与角尾组之间的一套粗碎屑岩，该组为储油层位。

下洋组主要隐伏分布于福山盆地、白莲盆地一带，为滨海浅滩相沉积。岩石基本为半成岩，局部见松散碎屑物。组合特征以灰白色、浅灰色砂砾层、砂层、粉砂层为主，夹灰色、灰黑色、灰绿色粉砂质泥岩和泥岩，局部呈互层，顶部为灰绿杂棕红色、棕黄色、灰黑色泥岩。碎屑成分以石英为主，含少量长石、暗色矿物及变质岩岩屑，分选中等，次圆状，胶结物主要为泥质，偶见石膏质。普遍含孢粉化石，植物化石碎片（树木枝干）。局部发育平行层理，底部以砂砾岩与下伏涠洲组灰色粉砂质泥岩呈不整合接触，顶部以中砂岩、泥岩与上覆角尾组粗砂岩呈整合接触，厚 50～236m。

在区域上，由福山盆地 1 井至海口市黄金海岸花园 ZK1 孔，该组的岩性与厚度均有变化。泥质成分增加，过渡为砂、粉砂与黏土互层，砂层以灰色、灰白色、灰绿色为主，黏土层大部为蓝灰色、灰黑色，从而与上覆角尾组绿色岩系相区别，厚度也变小。福 1 井剖面厚 236m，黄金海岸花园 ZK1 井厚 175m，玉沙村 ZK18 孔厚 121.4m。

（六）角尾组（N_1j）

角尾组系地质部第四石油地质普查勘探大队于 1964 年创建，代表性剖面为广东省徐闻角尾雷琼 1 井 600～986.8m 井段，原指雷州半岛位于下洋组与灯楼角组两套较粗碎屑岩之间的细碎屑岩，省内参考剖面为澄迈县福山盆地福 1 井（茂名石油公司地质处，1975）、海口市长流地区 ZK16 孔（海南地质大队，1976）。本次以福 1 井与海口市黄金海岸花园 ZK1 孔作为代表性剖面。

角尾组主要隐伏分布于福山盆地、白莲盆地中。岩石特征以半成岩状为主，局部为松散碎屑。在福山盆地岩性远比北部湾海区和雷州半岛要粗，下部岩性为浅灰色砂砾岩、含砾砂岩、中粗砂岩，夹绿灰色泥质粉砂岩、粉砂质泥岩和泥岩，局部含薄层石膏（福 1 井）；上部为绿灰色含砾砂岩、中粗砂岩，夹绿灰色砂质、粉砂质泥岩和泥岩，局部夹多层玄武岩（福 5、8 井）。碎屑成分以石英为主，见有少量的石英片岩岩屑，含少量的海绿石，底部以粗砂岩与下伏下洋组整合接触，顶部以粉砂岩与上覆灯楼角组中细砂岩呈整合接触，厚约 200～500m。在白莲盆地为一套以灰绿色为主的绿色碎屑岩系，与前者相比粉砂泥质成分有所增加，下部灰绿色中细粒岩屑石英砂岩，灰绿色细中粒石英砂岩，灰色砂砾岩，局部夹厚层、薄层凝灰岩；上部为灰绿色细粉砂岩、灰色粉砂质泥岩，夹中层状中细粒石英砂岩及中薄层凝灰岩，局部水平层理较发育，普遍含有孢粉化石、植物化石碎片（树木枝干），底部以灰绿色中细粒岩屑石英砂岩与下伏下洋组灰黑色厚层粉砂质泥岩呈整合接触，顶部灰色粉砂质泥岩与上覆灯楼角组灰绿色含砾粗中粒岩屑石英砂岩整合接触，厚 80～123m。

（七）灯楼角组（N_1d）

灯楼角组系中国石油公司南海分公司（1978）以徐闻竹山徐浅 1 井 304.5～532.5m 井段剖面命名。省内参考剖面为琼北福山盆地福 1 井（茂名石油公司地质处，1975）和海口市长流地区 ZK16 孔（海南地质大队，1976）。指位于角尾组与海口组之间的较粗的碎屑岩，岩石以半成岩为特征，局部为松散状。底部以含砾砂岩为标志与下伏角尾组呈整合接触，顶部与海口组含贝壳砂砾与砂砾岩整合接触。本次研究以福 1 井和海口市黄金海岸花园 ZK1 孔为代表剖面。

该在福山盆地总体岩性较粗，下部岩性为浅灰色、灰绿色含砾砂岩、砂砾岩、砂岩夹灰绿色粉砂质泥岩，上部为灰色、灰绿色砂岩与粉砂岩、粉砂质泥岩互层。普遍含有海绿石，产有孔虫、介形类、苔藓虫、腹足类、双壳类和孢粉等化石，厚337m。底部以浅绿色含砾砂岩与下伏角尾组灰绿色泥岩分界，顶部以灰色砂质泥岩与上覆海口组灰色砂砾岩分界。

在黄金花园ZK1孔，下部以灰绿色、深绿色中细粒岩屑石英砂岩，灰绿色中粗粒岩屑石英砂岩为主，局部为灰色、灰黑色薄—中层泥岩与灰色薄层粉砂岩互层；上部灰色、灰黑色厚层泥岩为主。夹灰色细粒石英砂岩，孢粉化石较发育。底部以灰绿色含砾粗中粒岩屑石英砂岩与下伏角尾组灰色粉砂质泥岩呈整合接触，顶部灰黑色厚层泥岩与上覆海口组灰绿色含贝壳碎屑砂砾岩呈整合接触，厚154m。

在福山盆地福1井以含砾砂岩、砂砾岩、砂岩为主体与北部湾的湾2井相似，但在海口长流地区碎屑变细，泥质增加，表现为以中粗砂、细砂岩与泥岩互层为主，厚度变化也较大，在福1井厚337m，长流ZK16孔厚163m，黄金海岸花园ZK1孔厚154m。

（八）海口组（N_2h）

海口组系海南地质大队在海口市长流地区进行供水水文地质勘查时创名，杨基广等（1977）介绍，未指定层型剖面。代表性剖面为长流ZK18孔19.83~202.61m井段。海口地区上新世海相沉积地层最早划为华场群上部（广东区测大队，1964），1965年地质部第四石油普查勘探大队将海南岛南部创建的望楼港组引入琼北，并广为应用。1997年陈哲培主编的《海南岩石地层》中主张采用海口组，认为二者处于不同的地层区，无论从岩石组合特征、生物面貌、大地构造环境均有很大差异，本文采用此划分方案。

海口组是琼北地区分布最广的地层，主要隐伏于儋州市白马井镇至海口市琼山区三江镇一带，局部出露地表。岩石以半成岩、半松散为主和含贝壳为特征，是琼北最重要的地下水含水层。根据沉积旋回及岩性岩相特征划分为4个岩性段，顶部以硬塑状薄层灰色黏土与上覆秀英组砂砾层呈假整合接触，底部以含贝壳砂砾岩与下伏灯楼角组呈整合接触。

一段为滨海相沉积。岩性为黄褐色、灰色、灰绿色（含）贝壳砂砾岩与砾岩互层，偶夹灰色、灰白色粉砂或粉砂质泥层，局部含海绿石及夹玄武质凝灰岩。岩性及厚度一般较为稳定，分布较广，是重要的区域性标志层，又是重要的含水层，为琼北地区的第二承压含水层。富含化石，长流地区的ZK136、ZK16、ZK18孔，龙桥地区的ZK321、ZK285、ZK289、ZK293、ZK260孔，福山地区福1井，老城地区的ZK220孔，海口市郊ZK175等地区钻孔中均见化石产出。

二段为滨浅海相沉积。岩性以灰色、灰黑色黏土或泥层为主，普遍夹浅黄色贝壳碎屑岩、凝灰岩、沉凝灰岩、橄榄玄武岩等。富含化石，长流地区ZK5-3、ZK136、ZK125、ZK16、ZK119、ZK16、ZK18、ZK224孔、福山地区福1井，龙桥地区ZK287、ZK285、ZK289、ZK293、ZK260，老城地区ZK220，永兴地区ZK293等地区钻孔中均见化石产出。

三段为滨海相沉积。岩性主要为灰色、深灰色贝壳砂岩、贝壳砾岩，局部相变为灰色、灰黄色生物碎屑灰岩，局部顶部为砂层，底部为砾石层，上部偶夹玄武岩。岩性相对较稳定，是区域性标志层，也是重要的地下水含水层，为琼北地区的第一承压含水层。富含化石，长流地区ZK5-3、ZK136、ZK16、ZK119孔，龙桥地区ZK285、ZK289、ZK28，老城地区ZK220，海口市郊ZK175等地区钻孔中均见化石产出。

四段为浅海相沉积。岩性主要为浅灰、灰色粉砂质黏土、黏土，含海绿石岩石，呈硬塑状，具水平层理，上部偶夹灰色、灰黑色凝灰岩、玄武岩，下部偶夹贝壳砂砾岩。富含化石，见于长流地区ZK136、ZK125、ZK18孔和福山地区福1井、龙桥地区ZK285、老城地区ZK220、海口市郊ZK175等钻孔中。

海口组富含海绿石，且其间所产的介形类化石皆为海相介形类的种属，因此，判断该地层为海相沉积。海口组4个岩性段延伸稳定，岩性变化不大，一段和三段较稳定，为含贝壳砂砾岩或贝壳碎屑岩，是海口地层小区海口组划分对比的标志层；二段中的基性火山岩各地发育程度有异，多数发育1~3层沉凝灰岩或拉斑玄武岩，但也有部分钻孔未见；四段局部出现泥灰岩或生物碎屑灰岩（福山ZK7孔），局部赋存基性火山岩（白莲 ZK10 孔、福 1 井）。该组所划分的 4 个岩性段基本可以对比连接（图 5-1）。在海口、白莲一带沉积厚度较大，分别为 182m（ZK18 孔）、146m（ZK10 孔）。往东西两侧厚度变小，西海岸的白马井 ZK10 孔厚度 31m，东部三江一带 ZK19 孔厚 20m（仅发育第一、二段）。

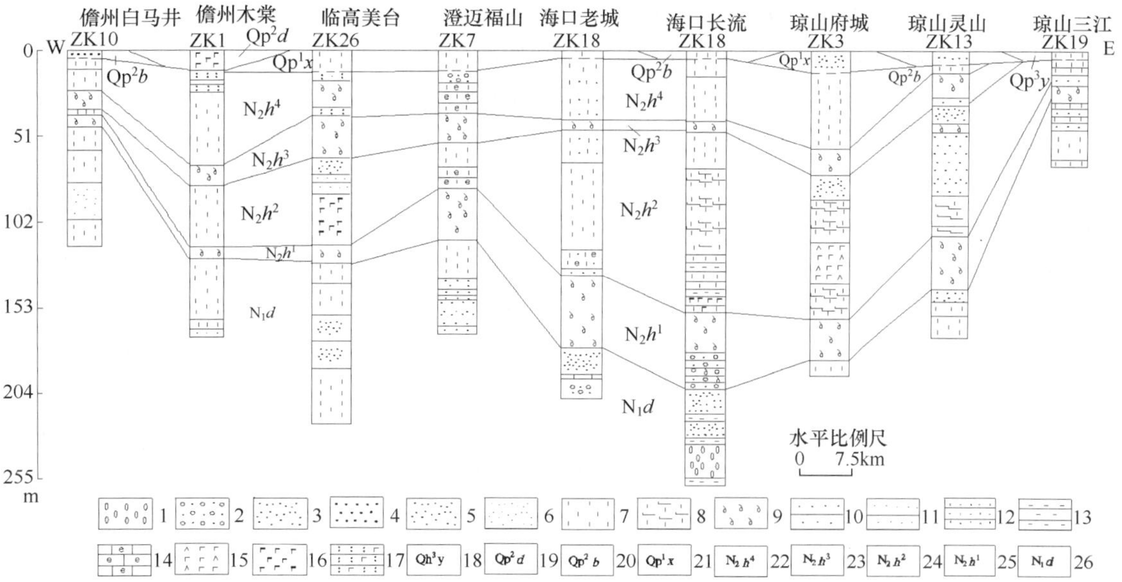

图 5-1 琼北新生代断陷盆地海口组部分钻孔柱状剖面对比图

Fig. 5-1 Correlation of columnar sections of some drillholes of the Haikou Formation, northern Hainan Cenozoic faulted basin

（据海南省岩石地层，1997）

1—砾石；2—砂砾石；3—砂；4—粗砂；5—中砂；6—细砂；7—黏土；8—钙质黏土；9—贝壳碎屑层；10—砂岩；11—细砂岩；12—凝灰质砂岩；13—黏土岩；14—生物碎屑灰岩；15—橄斑玄武岩；16—拉斑玄武岩；17—玄武质沉凝灰岩；18—烟墩组；19—多文组；20—北海组；21—秀英组；22—海口组四段；23—海口组三段；24—海口组二段；25—海口组一段；26—灯楼角组

三、古生物群及时代划分与对比

（一）长流组（$E_1\hat{c}$）

福山盆地的长流组迄今虽未发现化石，对其时代还难以提出可靠依据的意见，不过从区域对比及其上覆流沙港组三段产早始新世孢粉组合（福 8 井 2028~2290m 井段）来看，似应属于古新世沉积。

在加来-南宝盆地中，该组上部在 ZK8 孔 310m、337.5m，ZK9 孔 119m、156.6m 和普 12 孔 153~203m 等深度，产腹足类 *Enteroplax*? *lousigangensis*，*E.*? *pauus*，*Bitynia*? sp.，*Hydrobia* spp.，*Strobilops* sp. 及腹足类口盖 *Minolaminatus gracilis*，*M. validus*，*Assiminea retopercula* 等。

上列种属中，*E.*? *lousigangensis* 最早发现于安徽宣南盆地双塔寺组中段，该组产早始新世早期或晚古新世哺乳类化石。腹足类口盖化石 *Minolaminatlus gracilis*，*M. validus*，*Assiminea retopercula* 是我国南方地区晚古新世地层中常见化石，如广东南雄盆地的浓山组、三水盆地的饰心组，江苏北部地区的阜宁群一、四组，湖南洞庭盆地的沅江组一段，湖北江汉盆地西缘的龚家冲组。

在雷州半岛，长流组中产较丰富的介形类和孢粉化石。介形类发现在雷州半岛西南部的迈 2 井 2893~3214m 井段。所产介形类经勾韵娴等（1981）研究，属于 *Sinocypris* 组合，其时代为晚古新世。孢粉化石产于雷州半岛的迈 2 井 3076~3227m 井段和徐 4 井 1922~2089.5m 井段，据孙湘君等（1981）研究，属于 *Celtidpollenites-Pentapollenites* 组合，常见于华南各地的古新统。

综上所述，长流组的时代为古新世。

（二）流沙港组（E_2l）

流沙港组产丰富的孢粉化石，产于福山盆地流沙港组二段上部及一段的孢粉称为 *Leiospharidia*－*Granodiscus* 组合。该组合中被子植物花粉最为丰富，*Quercoidites micohenrici*，*Q. asper*，*Cupuliferoipollinites pusillus*，*C. oviformis*，*Tricolporopllenites steinensis*，*Corylopsis princeps*，*Liquidambarpollinites minutus*，*L. stigmosus*，*Multipropollenites punctatus* 构成优势分子；其他常见的有 *Alnipollinites verus*，*Juglanspollenites verus*，*Gothanipollis bassensis*，*Operculumpollis operculatus*，*Engelhardtioidites levis*，*Florschuetzia trilobata* 等。裸子植物花粉和蕨类植物孢子含量较少，裸子植物花粉以 *Pinuspollenites* 为主，蕨类植物孢子以 *Leiotriletes adriensis*，*Osmundacidites* 为主。藻类占有一定的含量，平均 10%～30%，主要为 *Leiospharidia hyaline*，*Granodiscus granulatus* 等；*Rugusphaera* 零星出现。其中有较多的始新世及更早时期的分子，又有不少渐新世分子，时代应为早渐新世。在三段中部，称为 *Salixipollenites triletipollenites-Operculumpollis operculatus* 组合，其中被子植物花粉占优势（67%～89%），尤以 *Salixipollenites discoloripites* 含量最为丰富；蕨类植物孢子（0～10%）以 *Polypodiaceaes porites ovatus* 为主；裸子植物花粉零星出现。孙湘君等（1981）认为这一组合时代为中始新世，其中的杂色柳粉曾见于美国西部的中始新统绿河组。三段下部称为 *Monocolpopollenites tranguillus-Crassoretitriletes* 组合，其特征据孙湘君等（1981）研究，以被子植物花粉占绝对优势（89%），其中 *Monocolpopollenites tranguillus* 极为丰富；蕨类植物孢子（10%）中 *Crassoretitriletes nanhaiensis* 含量最高；裸子植物花粉零星见及，与欧洲始新世孢粉组合相似，时代为早始新世。

在北部湾海区的湾 2 井还产介形类 *Chinocythere beibuwanensis*，*Candona* sp. 及 *Eucypris stagnalis*，后者在华南地区普遍出现在始新世地层中。

综合上述各门类化石分布来看，流沙港组的时代为始新世。

（三）涠洲组（E_3w）

涠洲组中孢粉化石分布广泛。分布于福山盆地涠洲组的孢粉称为 *Magnastriatites howardi*－*Trilobapollis*－*Verrutricolporites pachydermus*－*Retitricolpites* cf. *matauraensis* 组合（孙湘君等，1981），本组合以被子植物花粉占优势（47.7%～86.5%），其中 *Cupuliferoipollenites pusillus*，*Quercoidites microhenrici*，*Q. minor*，*Verrutricolporites pachydermus* 最为丰富；蕨类植物孢子含量也较丰富，达 12.9%～44.9%，主要是 *Crassoretitriletes nanhaiensis*，*Polypodiaceaesporites*，*Polypodiaceois porites*，*Polypodiisporites usmensis*，*P.* spp.，*Magnastriatites howardi* 等；裸子植物花粉含量稀少（0.7%～4.3%），其中包括两类分子，一类是欧亚大陆古近纪—新近纪尤其是渐新世孢粉组合中常见的属种，如 *Quercoidites microhenrici*，*Q. asper*，*Caryapollenites simplex*，*Juglanspollenites*，*Cupuliferoipollenitespussillus*，*C. oviformis*，*Ulmipollenites undulosus*，*U. granopollenites*，*Gothanipollis bassensis*，*Verrutricolporites*，*Engelhardtioidites spackmaniana*，*Pinuspollenitespristinipollinia*，*P. strobipites*，*Leiotriletes* 等；另一类是加里曼丹等地渐新世—早中新世孢粉组合中的重要分子，如 *Crassoretitriletes vanraadshooveni*，*Florschuetzia semilobata*，*F. trilobata*，*Magnastriatites howardi*，*Polypodiisporites usmensis*，*Janduflouria seamrogiformis*，*Margodoporites vamwijhei*，*Psilatridolporites operculatus*，*Alnipollenites verus*，*Monosporites annulatus*，*Zonocostites* cf. *ramonae* 等，时代为中、晚渐新世。

吴作基（1975）、杨祖平（1981）、吴进民等（1981）认为涠洲组孢粉属于 *Ceratopteris*－*Polypodiaceaesporites-Polypodiisporites-Pinuspollenites-Quercoidites microhenrici* 组合。此组合含大量 *Ceratopteris* 和草本植物花粉，如 Ranunculaceae，Compositae，Labiatae，Cruciferae，Pedaliaceae 和 Chenopodiaceae 等，还有一定数量的新近纪海岸植物花粉：*Florschuetzia semilobata*，*F. levipoli* 及新近纪热带地区的特征类型 *Crassoretitriletes vanraadshooveni* 等，由此认为涠洲组孢粉组合的时代为中新世或跨晚渐新世。

根据微体古生物组合，本组可与珠江口盆地的珠海组和渤海湾沿岸地区的东营组对比。

综上所述，目前多数地层古生物研究者倾向于将涠洲组归入渐新统，因此，其时代归入渐新世较为合适。

（四）长坡组（$N_1\hat{c}$）

长坡组中古生物较为丰富，产脊椎动物鱼类：*Cyprinidae*，*Siluroidei*，*Hybophtalmichthys*，*Pseudobagrus fulvidraco*；鳖类 *Amyda* sp.；鳄类 *Alligator* sp.；双壳类 *Modiolus*? sp.；腹足类 *Hydrobia* sp.，*Pseudamcicola* sp.，*Bitynia* sp.，*Gyraulus* sp.，*Amnicola* sp.；植物 *Phyllites* sp.，*Betula* sp.，*Alnus* sp.，*Corylus* sp.，*Carya* sp.，等；介形类：*Eucypris* sp.，*Cyprinotus* cf. *formalis*，*C. immanus*，*Ilyocypris* cf. *biplicata*，*I. gibba*，*Erpetocypris* sp.，*Candoniella* cf. *albicans*，*Candonopsis* sp.，*Limnocythere* sp. 等；孢粉为 *Crassoretitriletes-Polypodiisporites usmensis-Quercoidites microhenrici* 组合，分布于该组上部。该组合中蕨类孢子主要分子有 *Polypodiaceaesporites*，*Crassoretitriletes*，*Polypodiisporites usmensis*，常见有 *Magnastriatites*，*Osmundacidites*，*Lygodiumsporites*，*Cyclophorusisporites*，*Ischyosporites* 等；裸子花粉主要有 *Pinuspollenites*，*Abiefineaepollenites*，*Tsugaepollenites*，*Ostryoipollenites*，*Momipites*，*Ulmipollenites*，*Triporopollenites*，*Gothanipollis* 等。组合中以 *Polypodiisporites usmensis* 高含量（13.79%），与北部湾海区和雷州半岛的下洋组组合比较相似，同时以蕨类植物孢子占主导地位的组合面貌可以与福山盆地下洋组对比。

长坡组所含化石大多为新近纪以来生活的分子，其中孢粉组合时代应为早中新世。鉴于长坡组与上覆晚中新世灯楼角组为假整合接触，而孢粉组合又产于长坡组的中部，因此，长坡组划为中—下中新世较适宜。

（五）下洋组（N_1x）

下洋组中产孢粉化石及有孔虫，其中前者较丰富。在福山盆地福1井产出的孢粉称为 *Polypodiisporites usmensis-Crassoretitriletes vanraadshooveni-Graminidites* 组合。其中蕨类孢子以 *Polypodiisporites usmensis*，*Polypodiaceaesporites ovatus*，*Crassoretitriletes vanraadshooveni* 为主；被子花粉主要有 *Quercoidites minor*，*Q. microhenrici*，*Cupuliferoipollenites oviformis*，*Engelhardtioidites*，*Florschutzia trilobata*，*F. levipoli*，*Graminidites* 等；裸子植物花粉中零星见及者有 *Pinuspollenites*，*Abietineaepollenites*，*Taxodiaceaepollenites* 等。这个组合中的优势分子 *Polypodiisporites usminsis* 最初出现在哥伦比亚的 Paz del Rio 剖面的 San Fernando 页岩组和尼日利亚的 Benin 西-1 井的 Akata 组，在其上部地层 Margua 组及 Agbada 组和 Benin 组大量繁盛，特征分子 *Crassoretitriletes vanraadshooveni* 主要分布于加勒比地区和尼日利亚的早中新世晚期，而在婆罗洲 *C. vanraadshooveni* 出现于早中新世早期—晚期。此外，组合还含有一些新近纪分子 *Florschuetzia levipoli* 和 *Graminidites* 等，其时代可能为早中新世（孙湘君等，1981）。

在海口地区（ZK168 孔）产出的孢粉称为 *Polypodiaceaesprites-Tricolpites-Graminidites* 组合。被子花粉中主要有 *Tricolpites*，*Tricolporopollenites*，*Graminidites*；蕨类孢子主要有 *Polypodiaceaesporites*，*Toroisporites*，*Lygodiumsporites*，*Polypodiisporitesusmensis*，*Cyathidites*，*Crassoretitriletes*，*Magnastriatetes* 等；裸子花粉主要有 *Dacrydiumites*，*Pinuspollenites*，*Cedripites*。该组合主要由小型三沟粉、三孔沟粉组成，还含有一些泛热带地区中新世分子 *Polypodiisporites usmensis*，*Crassoretitriletes*，*Magnastriatites*，以及草本植物 *Graminidites*，*Liliacidites* 和 *Potamogetoniaceaepites* 等，其含量超过 6%，其主要成分及组合特征均与北部湾坳陷下洋组可以对比，时代属早中新世。

有孔虫据汪品先等（1981）研究，为 *Globigerina ciperoensis* 组合，因含早中新世标准分子 *Globigerinatella insueta*，并有个别 *Globigerinoides* 出现，时代应为早中新世，其层位与珠江口盆地的珠江组和莺歌海盆地的三亚组可以对比。

综上所述，下洋组时代属早中新世。

（六）角尾组（N_1j）

角尾组中孢粉组合以 *Polypodiisporites usmensis-Florschuetzia* cf. *levipoli-Fonestrites spinosus-Causuarinidites cainozoicus* 为代表，其中被子植物花粉含量为 52%~88%，主要分子与下洋组相似，但出现了 *Florschuetzia levipoli*，*Fonestrites spinosus*，*Causuarinidites cainozoicus* 等新类型。裸子花粉含量为 0~10%，主要是 *Dacrydiumites florinii*，另有少量的 *Podocarpidites*，*Pinuspollenites*。蕨类植物孢子含量为 5.1%~

37%，主要分子为 *Polypodiaceoisporites*，*Polypodiisporites usmensis*，*Polypodiaceaesporites*，偶见 *Distaltriangulisporites hubeiensis*。孙湘君等（1981）认为其时代为中中新世。

有孔虫据汪品先等（1981）研究，以 *Globorotalia siakensis*-*Orbulina suturalis* 组合为特征，主要有 *Globorotalia siakensis*，*G. praemenardii*，*G. peripheroronda*，*Cassigerinella chipolensis*，*Orbulina suturalis*，*Nephrolepidina* sp. 等，其中 *Orbulina* 属出现于中中新世开始，*G. siakensis* 消亡于中中新世末期；*G. praemenardii* 仅限于中中新世。

综上所述，角尾组的时代应为中中新世。

（七）灯楼角组（N_1d）

在福山盆地灯楼角组中化石较丰富，产有孔虫、介形类、苔藓虫、腹足类、双壳类和孢粉等化石。该组的有孔虫以 *Globigerina nepenthes-Globorotalia acostaensis* 为代表，其特征种 *G. nepentehes* 主要分布于中新世晚期；*Globorotalia acostaensis*，*G. multicamerata*，*G. margarite*，*Sphaeroidinellopsis* 等仅生存于晚中新世至上新世，而 *Sphaeroidinellopsis subdehiscens*，*S. seminulina* 等见于中新世。汪品先等（1981）认为这一有孔虫组合的时代为晚中新世。介形类称为 *Spinileberis inflexicostata-Cytheropteron striatituberosa* 组合，数量丰富，属种分异度高，共发现63种，其中20余种常见分子曾发现于日本和印度尼西亚的新近纪至现代沉积物中，如 *Callistocythere subjaponica*，*Cushmanidea* 曾报道于日本仙台上新统 Tatsumokuchi 组；*Krithe sawanensis* 曾见于日本新潟上新统 Sawane 组；*Uroleberis feveolata* 出现印度尼西亚的上新世沉积中；*Cornacoquimba moniwensis*，*Cytheropteron miurense*，*G. uchioi*，*Trachyleberis scabrocuneata* 和 *Hemicytherura cuneata* 曾分别出现于日本仙台中新统 Moniwa 段和 Hatatate 组；上述分子还在相应地区近代沉积物中出现。据此，并考虑到化石的保存状况，勾韵娴等（1981）认为，本组合的时代应归入上新世。腹足类主要是一些营漂浮生活的分子，包括 *Limacina hainanensis*，*Acteocina heraclitica*，*Ringicula obesior*，*Odostomia*（*Odostomia*）*hilgendorfi abmorma*，*Seile*? *trilineata* 等种。其中 *S. trilineata* 和 *A. heraclitica* 曾分别发现于匈牙利皇宫堡和奥地利中新世晚期托尔通期（Tortonian），*R. obesior* 和 *O.*（*O.*）*ancisa* 在新西兰中新统 Lillburnian 层中有过记载，王惠基（1982）认为其时代与欧洲的梅辛（Messinian）相当。双壳类有 *Nucula*（*Leinucula*）*yinggehaiensis*，*Glycymeris cisschuensis*，*Nipponmysella elliptica*，*Cardiocardita granulicosta*，*Carditella hanzawai*，*Pleuromeris minoensis*，*Hybolophus minutissimus*，*Tellina*（*Moerella*）*minima* 等种。该双壳类组合就其性质和化石保存状态而言，都明显不同于长昌地层分区中望楼港组的双壳类组合，兰琇（1981）认为其时代归入中新世晚期较为合适。

灯楼角组孢粉较发育，孢粉组合与北部湾海区和雷州半岛的相应层位的孢粉一起归入 *Cupuliferoipollenites oviformis-Ilexpollenites-Dicolpopollis kockelii-Liquidambarpollenites stigmosus* 组合。该组合被子植物花粉占68.1%~81.4%，含丰富的 *Cupuliferoipollenites* sp.，*Quercoidites microhenrici*，*Ilixpollenites*，*Dicolpopollis kockelii*，*Liquidambarpollenites stigmosus*，常见 *Juglanspollenites verus*，*Alnipollenites*，*Ulmipollenites undulosus*，*Myricipites*，*Florschuetzia* cf. *levipoli*，*Ericipites*，*Umbelliferae* 等。裸子植物花粉含量少，约为0~9%，仅见少量的 *Dacrydiumites florinii*，*Tsugaepollenites igniculus*，*Pinuspollenites*；蕨类植物孢子占12%~13%，主要有 *Polypodiisporites usmensis*，*Polypodiaceaesporites* sp. 等，零星出现 *Leiotriletes*，*Lycopodiumsporites*。该组合未发现加里曼丹弗氏粉第三化石带的代表种 *Florschuetzia meridionalis*，但 *F.* cf. *levipole* 在组合中占一定比例，且个体较大。孙湘君等（1981）认为该组合与加里曼丹的 *F. meridionalis* 带相当，时代为中新世晚期。

在海口地区该组产有孔虫 *Globigerina bulloides*，*G. caliclaplummer*，*Globigerinoides immuturus*，*G. triloba*，*G. sinensis*，*Hanzawaice manlaensis*，*Flurinus costiferus*，*Quinqueloculina* sp.；还有介形类 *Loxoconcha*，*Trachyleberis*，*Aurila*，*Bairdia*，*Callistocythere*，以及鲨鱼牙齿 *Carcharinidae* 及孢粉等。具有地层意义的是有孔虫，时代为中新世。

此外，灯楼角组整合覆盖在中新统角尾组之上，又与上覆上新统海口组呈整合接触，因此，将灯楼角组的地质时代划归晚中新世是合理的。

（八）海口组（N_2h）

海口组富含有孔虫、介形类、双壳类、腹足类及孢粉等化石。

在福山盆地，有孔虫以底栖类型为主，浮游类型少量。底栖有孔虫中以 *Pseudorotalia*, *Asterorotalia*, *Pulleniatina* 占优势，主要有 *Pseudorotalia subtrispinosa*, *P. indopacifica*, *Asterorotalia subtrispinosa*, *Pulleniatina praecusor*, *P. primalis*, *P. obliqueleculata*；浮游有孔虫有 *Globigerinoides extrimus*, *G. obliquus*, *Globigerina picassiana*, *Globoquadrina altispira* 等属种。上列分子中，*Pseudorotalia indopacifica* 曾见于菲律宾上新统，*Asterorotalia subtrispinosa* 在我国台湾发现在上新统与上中新统顶部 N18 带，但其主要分布范围是上新统。其余分子出现在中新世晚期至上新世（汪品先等，1981；吴进民等，1981）。介形类被归入 *Hemikrithefoveata-Neomonoceratina delicata* 组合（勾韵娴等，1981），包括 67 种。常见的分子是 *Aurina uranouchiensis*, *Callistocythere subjaponica*, *Campylocythereis tomokoae* *Cornucoquimba moniwensis*, *Cushmanidea subjaponica*, *Cytheropteron miurense*, *C. uchioi*, *Echinocythereis? bradyformis*, *Hemicytherura cuneata*, *Krithe sawanensis*, *Neomonoceratina delicata*, *Trachyleberis subrocuneata*, *Uroleberis foveolata* 等。上述属种曾发现在日本和印度尼西亚的新近纪—第四纪沉积物中，考虑到已知属种的分布和保存状况，勾韵娴等（1981）认为含该组合的地层应属上新统。腹足类属种繁多，主要有 *Odostomia* （*Odostomia*） *helgendorfi*, *Turritella filioda*, *T.* （*Stiracolpus*） *vigilan*, *Paracirculus novemcarinatus*, *Erato* （*Eratopsis*） *pernana*, *Raphitoma embrylirata*, *Cylichna* （*Cylichna*） *cylindracea* 等，其中 *T. filioda*, *T. pereoa*, *C. kokaiensis* 是我国台湾上新统苗栗群的种，*E.* （*E.*） *pernana* 曾报道于意大利艾米利亚的 Piacentian 阶、阿斯提的 Astian 阶和托斯卡纳的上 Astian 阶；*R.* （*R.*） *embryolirata* 在帝汶岛的上新统有记载；*Odostomia*, *Canaliculata* 在特立尼达岛的上新统下部亦有发现。据此，王惠基（1982）认为该腹足类组合时代属上新世。双壳类组合以 *Bellucinacivica-Cardiocardita granulicosta-Carditella hanzawai* 为代表（兰琇，1981），其中的重要分子如 *Barbatia* （*Acar*） *plicata*, *Carditella hanzawai*, *Cardiocardita granulicosta*, *Hawaiarca uwaensis* 等都是我国台湾苗栗群、日本大桑动物群和宫崎动物群及帝汶、爪哇等地中新世—上新世沉积中的重要成员；此外，*Corbula* （*Caryocorbula*） *howesi* 曾见于新西兰上新统。孙湘君等（1981）将孢粉划为 *Cupliferoipollenites oviformis-Polypodiaceaesporites-Chenopodipollis microporatus* 组合。该组被子植物花粉占 48%~91%，以 *Cupliferoipollenites oviformis* 和草本植物花粉高含量为特征，主要有 *Chenopodipollis microporatus*, *Fonestrites spinosus*, *Artermisia* 等；裸子植物花粉占 0~5%、零星见及 *Pinuspollenites*, *Dacrydiumites florinii*, *Podocarpidites*, *Tsugaepollenites*；蕨类植物孢子含量为 7%~50%，主要为 *Polypodiaceaesporites*，常见 *Polypodiisporites usmensis*, *Polypodiaceoisporites* 等。孙湘君等（1981）认为孢粉组合中木本植物花粉大量消亡，草本植物花粉大量出现，表明当时古植被发生了重大变革，其时代应为上新世。

在海口地区，该组有孔虫主要有 *Amphistegina* sp., *Ienticulina* sp., *Pseudorotalia indopa cifica*, *P. tikutoensis*, *P. schroterina*（底栖），*Globigerinoides ruber*（浮游）等。其中，*Pseudorotalia indopacifica* 在北部湾海区和雷州半岛以及菲律宾等地的上新世地层中均有发现。腹足类有 *Raphitoma embryoliarata*, *Conus kokanensis*, *Phalium* （*P.*） cf. *cancellianum*。其中 *Raphitoma embryolirata* 曾见于帝汶岛的上新统。双壳类 *Cadium* （*Cerastoderma?*） *bauzawaia*, *Vesesicasdia* cf. *csenulicostata*, *V. gsanulicostata*, *Corbula taiwanensis*, *Chlamys satoi*, *Amusium japonicum*, *Chtamys satoi* 等为上新世的标准分子，该分子在我国台湾新竹等地苗栗组中常见。

海口组孢粉称为 *Polypodiaceaesporites-Tricolpites-Graminidites* 组合。其中，被子植物花粉主要有 *Tricolpites*, *Graminidites*, *Tricolporopollenites*, *Momipites*, *Juglanspollenites*, *Caryapollinites*, *Betulaceaepollenites*, *Quercoidites*, *Triporopollenites*, *Nyssapollenites*, *Chenopodipollis*, *Potamogetonacidites*；蕨类植物孢子主要为 *Polypodiaceaesporites*, *Polypodiisporites usmensis*, *Toroisporites*, *Polypodiaceoisporites* 等；裸子植物花粉有 *Pinuspollenites*, *Cedripites*, *Dacrydiumites*。其特征及主要分子与福山盆地一致。

另外，海南省地质调查院龙文国等在海口市黄金海岸花园 ZK1 孔中，在该组一、三段中获得孢粉，该孢粉组合中，以蕨类植物孢子为主（52%~62%），主要分子有 *Polypodiisporites*, *Polypodiaceaes-*

porites, *Polypodiaceoisporites* 等；裸子植物花粉占 17%~39%，主要分子为 *Abietineaepollenites* 和 *Pinuspollenites*；被子植物少量，见有 *Liquidam larpollenites* 和 *Oseryoipollenites* 等。以上组合特征与海口 ZK168 孔 216m 井段的孢粉组合特征近似，时代应为上新统。

根据上述生物组合特征，海口组时代为上新世。

四、岩相古地理特征

新生代以来，由于南海扩张，海南岛形成北面断陷盆地区和南面隆起剥蚀区。北面断陷盆地区向北延入北部湾海区，主要为海口地层小区中的福山盆地、加来-南宝盆地及长坡盆地等。盆地沉降中心大致位于海口-福山一带，呈东西向展布，盆地内接受了厚达 3000 余米的古近纪以来的沉积。

古新世盆地开始接受洪冲积相物源沉积，以棕红色、紫红色泥岩与紫红色或灰色砂砾岩、砂岩不等厚互层为特征，盆地沉降中心大致位于海口-福山一带，在福山地区沉积厚度达 550 余米（福 3 井）；而在加来-南宝地区沉积厚度较薄，只有 220 余米（ZK9 孔），当时气候潮湿炎热，多雨天气，河流发育。进入始新世盆地进一步加深，形成湖泊相，盆地地势非常平坦，气候温暖湿润，湖水深浅交替，此时的动、植物化石大量繁殖，并接受一套以暗色（黑灰色、灰色、褐灰色、深灰色）泥岩为主，夹灰白色、灰黄色砂砾岩的碎屑物质沉积，是形成油气和储油的层位。

渐新世由于地壳差异性升降运动明显，断陷作用加剧，盆地进一步拓宽和加深，至中新世时，波及西面长坡地区。

在福山及白莲盆地，渐新世时，盆地进一步拓宽，湖水加深，局部发生海侵，沉积物为一套杂色泥岩、粉砂岩与灰白色砂岩、含砾砂岩及砂砾岩不等厚互层，中上部局部含有海绿石，属于湖泊相沉积，局部夹海相沉积。进入中新世早—中期，盆地沉降作用加剧，海水完全侵入，湖盆消失，主要接受滨海相沉积的物源，沉积物以碎屑建造为主，碎屑成分主要为石英，含少量海绿石等，局部发育平行层理。普遍含有孢粉化石、植物化石碎片（树木枝干），有孔虫、介形类、腹足类及双壳类等较繁盛，表明了当时气候温暖、阳光充足。

在长坡盆地，从中新世早期至中期，长坡组沉积物呈现出由粗至细的演化，沉积中心大致位于长坡镇-东成镇一带，厚度达 300m 以上。长坡组一段沉积时期，盆地地形高差显著，离蚀源区近，大量陆内碎屑物质快速搬运向盆地注入堆积，形成了灰色或杂色的砂砾岩、砾岩、泥质砂岩等一套碎屑沉积物，此时气候干燥、炎热，不利于生物的生长和繁殖、保存；二段—三段沉积时，盆地进一步拓宽加深，进入滨湖—浅湖环境，水动力作用强烈，不利于植物及造油生物的大量繁殖与保存，沉积物以蓝灰色钙质黏土、含钙砂质黏土为主，并夹有泥灰岩、钙质砂岩；四段时，盆地地势非常平坦，离蚀源区较远，气候温暖湿润，湖盆仍以相对下沉为主。振荡较为频繁，在小幅度的上升中造成了富含植物，（包括草本植物及藻类）、鱼类的泥炭沼泽相环境，动、植物化石大量繁殖，形成了褐煤、炭质油页岩及含炭油页岩。在小幅度的下降中则形成了富含鱼类、介形类及少量藻类的沼泽相环境，沉积了含钙油页岩。岩石发育水平层理、微波状层理、透镜状层理；五段沉积时，沉积范围已明显缩小，接受了含砾黏土质中粗砂、含砂（质）黏土、黏土的沉积，岩石顶部发育水平层理，为滨湖相→浅湖相沉积。

而在加来-南宝盆地，长坡组在加来一带缺失一段沉积，二段沉积直接与古近系呈假整合接触，加来-南宝盆地沉积中心位于加来镇及其东面一带，厚度达 300m 以上，缺乏五段的沉积（图 5-2）。

晚中新世—上新世，由于海侵加剧，盆地沉降等作用，海口地层小区中的福山、加来-南宝及长坡等盆地开始连成一片，形成统一的琼北断陷盆地。晚中新世时期，盆地下部主要为浅灰色、灰绿色含砾砂岩、砂砾岩、砂岩夹灰绿色粉砂质泥岩，上部以灰、灰绿色砂岩与粉砂岩、粉砂质泥岩互层为特征，普遍含有海绿石。有孔虫、介形类、苔藓虫、腹足类、双壳类等生物较为繁盛。进入上新世，盆地继续沉降，海侵加剧，其间发生过火山喷发，并接受其物源沉积。沉积物为灰绿色、灰色、灰白色砂砾岩、砂质黏土、黏土等，局部夹玄武质凝灰岩，具水平层理，有孔虫、介形类、双壳类、腹足类等大量繁殖生长，表明当时沉积环境水能量变化频繁。

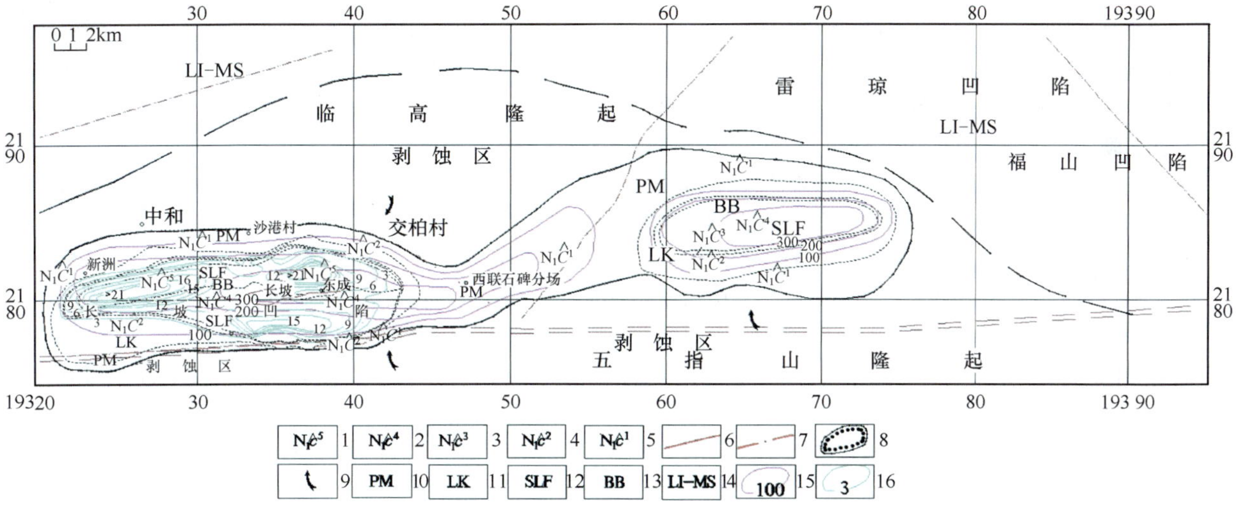

图 5-2 长坡组岩相古地理图

Fig. 5-2 Lithofacies paleogeography map of the Changpo Formation

1—长坡组五段；2—长坡组四段；3—长坡组三段；4—长坡组二段；5—长坡组一段；6—实测正断层；7—物探推测断层；8—坳陷盆地边界；9—物源方向；10—山麓堆积相；11—滨湖相；12—浅湖相；13—泥炭沼泽相；14—滨湖—浅海相；15—沉积厚度等值线；16—黏土/砂砾比值

第二节 长昌地层分区

长昌地层分区为本次工作中新建立的三级区划名称，介于王五-文教断裂和红河断裂、琼东南盆地北缘断裂之间。

根据岩性组合特征、岩相、生物特征、接触关系，在充分利用前人研究成果的基础上，对区内进行了划分与区域对比，自下而上划分为古新统昌头组（$E_1\hat{c}t$）、始新统长昌组（$E_2\hat{c}$）和瓦窑组(E_2w)、渐新统—中新统丘糖岭组（E_3N_1q）、中新统佛罗组（N_1f）和石马村组（$N_1\hat{s}$）、上新统望楼港组（N_2w）和石门沟村组（$N_2\hat{s}m$）。

一、剖面描述

（一）昌头组（$E_1\hat{c}t$）

长昌盆地 ZK0802 孔古新统剖面

该孔位于海口市琼山区甲子镇昌头村南约 500m 处，坐标：东经 110°28′12″，北纬 19°27′34″，系海南地质大队 1980 年测制，为古新统昌头组代表性剖面。

上覆地层 第四系浅棕黄色细砂土

~~~~~~~~ 不 整 合 ~~~~~~~~

古新统昌头组	厚 364.1m
油质页岩段	厚 249m
20. 浅棕色含钙砂质泥岩和泥质粉砂岩，夹含砾黏土质砂岩和砂砾岩	8.6m
19. 棕红色钙质泥岩	25.6m
18. 棕红色钙质泥岩与灰色钙质页岩略等厚互层，灰色钙质页岩含介形类	19.5m
17. 灰色钙质页岩与灰褐色油质页岩互层，产介形类	11.8m
16. 灰色钙质页岩与棕红色钙质泥岩互层	14.1m
~~~~~~~~

15. 灰色钙质页岩，含介形类	10.9m
14. 灰褐色油质页岩夹灰色钙质页岩	20.6m
13. 灰色钙质页岩	10.9m
12. 灰褐色油质页岩与灰色钙质页岩互层，油质页岩中见鱼化石碎片	26.7m
红色岩段	
11. 灰白色长石石英细砂岩与灰褐色含油页岩互层，夹棕灰色钙质泥岩	9.1m
10. 棕红色钙质泥岩与灰褐色油质页岩互层，顶部油质页岩含黄铁矿结核	15.4m
9. 灰棕色含钙泥岩，夹绿灰色薄层含钙泥质粉砂岩，普遍含钙质结核	57.5m
8. 灰白色薄层细粒钙质长石石英砂岩，夹棕灰色钙质泥岩、灰色含油页岩和棕灰色角砾岩	14.5m
7. 灰色、棕灰色含钙泥质粉砂岩和钙质泥岩，局部含钙质结核，底部夹含油页岩	3.8m
红色岩段	厚 115.1m
6. 棕红色钙质泥岩与灰白色薄层钙质细砂岩互层，夹棕红色泥质粉、细砂岩	17.3m
5. 灰白色、棕红色块状—薄层含钙泥质长石石英砂岩，夹棕红色泥岩、粉砂岩和砂砾岩	17.4m
4. 棕红色含砾钙质泥岩	15.9m
3. 灰白色砂砾岩与含钙砂质粉砂岩互层	19.9m
2. 棕红色砂砾岩	24.7m
1. 灰棕色砾岩	19.9m

~~~~~~~~ 不　整　合 ~~~~~~~~

下伏地质体　燕山期花岗闪长岩

（二）长昌组（$E_2\hat{c}$）

1. 长昌盆地 CK48 孔 276.64~430.37m 井段始新统长昌组含煤段剖面

该孔位于海口市琼山区长昌煤矿区，坐标：东经 110°07′09″，北纬 19°38′58″，系海南地质大队 1963 年测制。为始新统长昌组的代表性剖面。

上覆地层　瓦窑组灰白色砂砾岩

———— 整　合 ————

始新统长昌组	总厚度 137.30m
含煤段	
19. 褐色、棕灰色油页岩夹灰白色粉砂岩及褐煤（0.19m）	5.07m
18. 灰白色粉砂岩	5.25m
17. 褐色油页岩、黑色炭质页岩夹褐煤条带	0.97m
16. 灰白色粗砂岩与灰白色、褐灰色细砂岩、粉砂岩不等厚互层	47.46m
15. 灰白色粗砂岩与粉细砂岩不等厚互层，中部夹褐煤（0.15m）	19.86m
14. 黑色炭质页岩夹 0.08m 褐煤层，下部含油质，上部深灰色含油粉砂岩	2.91m
13. 灰色细砂岩、粗砂岩夹深灰色油质页岩	8.28m
12. 深灰色含油粉砂岩夹粗砂岩	3.94m
11. 深灰色油质页岩夹浅灰色粉砂岩及薄层褐煤	2.09m
10. 灰白色细砂岩夹粉砂岩	9.32m
9. 灰褐色油页岩夹高灰分煤及灰白色粗砂岩	6.32m
8. 高灰分煤夹薄层油质页岩及粉细砂岩	3.79m
7. 棕褐色含油粉砂岩夹薄层煤	3.60m
6. 高灰分煤夹棕褐色含油粉砂岩	3.19m
5. 灰白色细砂岩夹粗砂岩	2.04m
4. 高灰分煤夹薄层炭质页岩、油页岩及粉砂岩	6.98m
3. 浅灰色粉砂岩夹黑色炭质页岩	4.41m
2. 深灰色页岩	1.82m
~~~~~~~~

——————— 整　　合 ———————

杂色岩段

1. 浅绿色粉砂岩　6. 33m

未见底

2. 长昌盆地 CK106 孔 39. 43~369. 8m 井段始新统长昌组杂色岩段剖面

该孔位于海口市琼山区长昌煤矿排坡村北西约 400m 处，坐标：东经 110°27′27″，北纬 19°38′09″，系海南地质大队 1963 年测制，为始新统长昌组的参考剖面。

上覆地层　始新统长昌组含煤段暗棕色泥岩

——————— 整　　合 ———————

始新统长昌组杂色岩段　厚度 332. 3m

17. 灰白色、灰绿色、紫红色粉砂岩与灰色、灰白色、灰绿色杂紫红色砂岩不等厚互层　44. 17m
18. 灰绿色、紫红色泥质粉砂岩与灰白色砂砾岩不等厚互层　13. 00m
19. 紫红色泥质粉砂岩与灰绿、灰白色泥质粗砂岩、砂砾岩不等厚互层　16. 51m
20. 灰绿色砂砾岩　8. 08m
21. 绿灰色杂紫红色粗砂岩，底部灰白色砂砾岩　20. 75m
22. 灰绿色、灰白色杂紫红色粗砂岩与薄层泥岩互层，底部夹砂砾岩　16. 39m
23. 灰白色砂砾岩与青灰色杂紫红色细砂岩互层　7. 01m
24. 灰绿杂紫红色粗砂岩夹灰白色砂砾岩及紫红杂灰绿色粉砂岩　13. 66m
25. 灰白色砂砾岩夹粗砂岩、细砂岩　113. 91m
26. 灰白色粗砂岩与青灰杂紫红色细砂岩互层　19. 64m
27. 灰白色砂砾岩与灰黑色、紫红杂灰绿色砂岩互层　59. 18m

——————— 整　　合 ———————

下伏地层　古新统昌头组浅棕色含钙砂质泥岩和泥质粉砂岩

（三）瓦窑组（E_2w）

长昌盆地瓦窑溪古近系剖面（图 5-3）

该剖面位于海口市琼山区长昌煤矿瓦窑井田长昌向斜东翼，起点位于长昌煤矿区的长昌河岸，坐标：东经 110°27′，北纬 19°38′。剖面沿瓦窑溪向西往竹粮山，止于福田村东约 600m 处的长昌向斜核部。系雷奕振等 1992 年测制，海南省地质调查院（2004）对该剖面进行了野外查证。出露的地层为始新统长昌组含煤矿段及瓦窑组。为瓦窑组的代表性剖面。岩性自上而下为：

上始新统瓦窑组　总厚度 276. 20m

59. 黄灰色块状含细砾石粗砂岩　6. 4m
58. 灰白色、浅红色块状含粗砾石细砾岩。砾石成分主要为脉石英，次为花岗闪长岩、硅质岩和紫红色细、中粒砂岩　8m
57. 粉红色、灰绿色、灰白色块状含粉砂质泥岩，具蠕虫状结构，夹 2 层灰黄色含砾泥质粗砂岩　11. 4m
56. 灰白色块状含粗砾石细砾岩，夹青灰色泥岩　5. 9m
55. 灰黄色中层含砾中—细粒砂岩，往上含石英质细砾石、渐变为细砾岩　38. 2m
54. 灰白色中层—块状中、粗砾石砾岩，夹含砾砂岩透镜体和叉状体　12. 6m
53. 浅灰黄色块状细砾岩和砂砾岩，顶部为灰白色泥岩　32. 9m
52. 深黄色块状含砾泥质粗砂岩，夹黑褐色薄层中粒铁泥质砂岩　12. 5m
51. 灰绿色—紫红色块状含粉砂质泥岩，夹黄灰色泥质细砂岩　4. 3m
50. 黄褐色薄层粗粒铁泥质砂岩与灰黄色砂砾岩—细砾岩不等厚互层　5. 5m

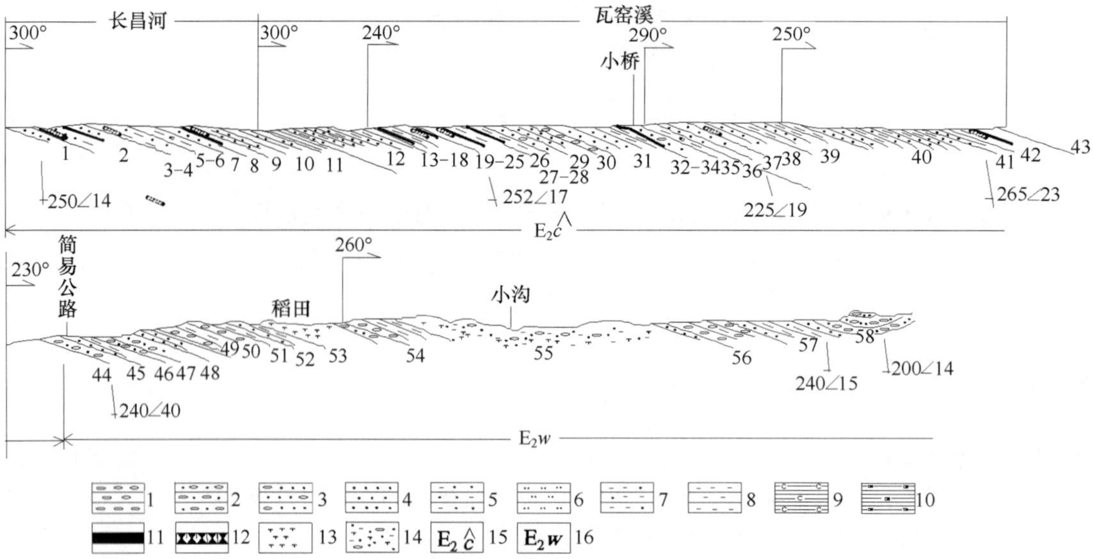

图 5-3　琼山长昌盆地瓦窑溪古近系剖面图

Fig. 5-3　Section of Paleogene strata，Wayaoxi，Changchang basin，Qiongshan

（据 1∶25 万《琼海》幅区调报告，2004）

1—砾岩；2—砂砾岩；3—含砾砂岩；4—砂岩；5—泥质砂岩；6—粉砂岩；7—砂质泥岩；8—泥岩；9—炭质页岩；10—油质页岩；11—煤；12—菱铁矿岩；13—耕作土；14—松散残坡积；15—长昌组；16—瓦窑组

49.	灰黄色、灰绿色厚层状粗砂岩与砂砾岩叉状交互，夹薄层分叉的灰色含粉砂质泥岩，顶部为青灰色泥岩夹褐色泥岩，产植物和孢粉化石	14.9m
48.	灰色杂棕红色蠕虫状含细砾粗砂岩，砾石局部富集成细砾岩或砂砾岩。顶部为厚 0.6m 的灰紫色油质泥岩，含大量植物碎片，产孢粉化石	5.8m
47.	青灰色块状—中层含砾中、粗粒砂岩，夹紫灰色油质泥岩透镜体，产植物化石	8.8m
46.	灰黄色中—厚层铁泥质砂砾岩或细砾岩，底部为鲜黄色薄层铁质细砂岩和粉砂质泥岩	3.6m
45.	青灰色厚层含砾泥质细—中粒砂岩，顶部为黑灰-灰紫色油质泥岩，产植物和孢粉化石	8.1m
44.	灰黄色块状含砾泥质不等粒砂岩，含灰紫色油质泥岩团块	5m
43.	灰白色块状不等粒泥质砂岩，偶含炭化植物碎屑。顶部为青灰色泥岩夹灰紫色油质泥岩透镜体。产孢粉化石	14m
42.	灰黑色薄层含炭泥岩与暗棕色薄层状泥质褐铁矿-赤铁矿化菱铁岩互层，夹灰紫色油质泥岩及 2 层灰黄色中层细—中粒铁质砂岩。产孢粉化石	11.8m
41.	绿灰杂黄褐色、紫红色蠕虫状泥质细砂岩，局部含细砾石，夹砂质泥岩	6.1m
40.	深灰色中层含砾不等粒泥质砂岩，含黄褐色泥质团块，底部见灰黑色泥岩	22.9m
39.	紫红色、淡红杂绿灰色蠕虫状含粉砂质泥岩，含大量钙质结核，下部夹 1 层黄褐色—灰白色中层细—中粒泥质砂岩	16.6m
38.	黑紫色薄层炭质页岩，夹 4 层黄褐色薄层铁质泥岩-粉细砂岩。产孢粉化石	8.8m
37.	黄褐色薄层铁质细砂岩与灰黄色含砾粗砂岩及青灰色泥岩不等厚互层。铁质细砂岩中含大量褐铁矿化菱铁矿结核	6.6m
36.	黄褐色块状泥质不等粒砂岩，局部含细砾石	5.5m

—————— 整　　合 ——————

上—下始新统长昌组含煤段　　总厚度>125m

35.	暗紫褐色油质泥岩，局部变为炭质页岩，夹黄绿色泥质菱铁岩和含菱铁矿结核的铁质粉—细砂岩	4.4m
34.	黑色炭质页岩，产孢粉化石	1.5m
33.	灰白、黄褐色块状细砾岩—含砾粗砂岩，中部夹褐紫色油质泥岩，产鱼类棘刺和鳞片	2.1m
32.	黑褐色油质泥岩，夹灰黑色炭质页岩及褐煤。产植物和孢粉化石	1.4m

31.	黄褐、灰白色薄—中层不等粒泥质砂岩与青灰色、灰紫色泥岩和青灰色泥质细砂岩不等厚互层。产孢粉化石	15.2m
30.	黄褐色块状细砾岩，顶部为一层厚 0.3m 的浅紫色铁质细砂岩	2.3m
29.	青灰色块状中粒泥质砂岩，含少量石英质细砾石和炭质碎屑	4.4m
28.	褐紫色油质泥岩，底部及中部夹黑色炭质页岩薄层，底部为褐煤。产孢粉化石	2m
27.	灰白色块状含砾泥质不等粒砂岩，顶部为青灰色泥质粉砂岩和砂质泥岩	0.4m
26.	青灰色、灰白色泥岩与灰黑色炭质页岩互层，夹薄煤层和煤线及黄褐-紫红色砂质泥岩，底部为褐紫色油质泥岩。产植物和孢粉化石	3m
25.	青灰色粉砂质泥板岩	0.35m
24.	灰白色块状泥质粗砂岩	0.25m
23.	灰紫色油质泥岩与灰黑色炭质页岩互层，中部夹 0.1m 的煤线。底部油质泥岩中产鱼化石碎片及孢粉化石，中、上部炭质页岩产植物叶和种子化石	1.6m
22.	青灰色块状水云母黏土岩	1.1m
21.	暗紫红色硅质菱铁岩	0.9m
20.	青灰色水云母黏土岩夹灰紫色油质泥岩和灰黑色炭质页岩	0.3m
19.	黄褐色薄层状-灰白色中层含砾不等粒砂岩，下部夹硅质菱铁岩	0.3m
18.	浅紫灰色油质泥岩夹灰黑色炭质页岩及青灰色水云母黏土岩，含植物碎片	0.6m
17.	暗紫红色薄层状硅质菱铁岩，含砂质团粒及植物碎片	0.3m
16.	褐紫色油质泥岩夹灰黑色炭质页岩，含植物碎片。产孢粉化石	1.4m
15.	黑色煤层夹炭质泥岩	1.3m
14.	浅灰绿色薄层状菱铁质细砂岩-砂质泥岩	0.6m
13.	灰黑色炭质页岩与紫褐色铁质黏土岩互层，夹铁质粉—细砂岩透镜体。产孢粉化石	1.6m
12.	青灰色—灰紫色含铁泥质细砂岩，含植物碎屑，往上变为黄绿杂紫红色蠕虫状不等粒泥质砂岩	22.3m
11.	灰黑色炭质页岩。产孢粉化石	8.4m
10.	青灰色黏土岩夹泥质粉、细砂岩，含炭屑	9.4m
9.	灰紫色薄层硅泥质菱铁岩，含炭化植物碎片	5.9m
8.	灰黑色炭质页岩夹煤层	3.9m
7.	灰白色中层状含泥粗砂岩，夹灰紫色黏土岩和灰黑色炭质泥岩透镜体。产孢粉化石	4.5m
6.	紫褐色油质泥岩夹煤线，顶部为青灰色黏土岩夹黄褐色—紫红色粉砂质菱铁岩，产孢粉化石	2.4m
5.	青灰色黏土岩夹紫褐色油质泥岩和煤线	2.4m
4.	暗紫红色薄—中层硅质菱铁岩	0.8m
3.	黄褐色块状不等粒泥质砂岩	8.6m
2.	黑色炭质页岩与紫褐色油质泥岩互层，夹薄层粉砂质菱铁岩。产孢粉化石	8.5m
1.	青灰色泥质细砂岩 本层以下地层被掩盖	>1m

（四）丘糖岭组（E_3—N_1q）

琼西南乐东县丘糖岭剖面（图 5-4）

剖面位于乐东县城北东约 8km 的永明乡道汉村丘糖岭，属洪帽剖面的上半段。剖面终点坐标：东经 109°13′18″，北纬 18°49′01″（四川攀西地质大队，1991 年测制，宜昌所、海南省地质调查院略有修改）。

未见顶

渐新统—下中新统丘糖岭组（E_3—N_1q） 总厚度 281m

55.	灰色硅质石英不等粒砂岩。致密坚硬，具油脂光泽，风化后具孔状，显微层纹构造，上部含铁质较多，且夹黄色硅质砾岩	36.78m
54.	黄色硅质中细砾岩。异常坚硬，球形风化，砾石为燧石、硅质岩、石英岩，次圆状，硅质胶结	10.01m

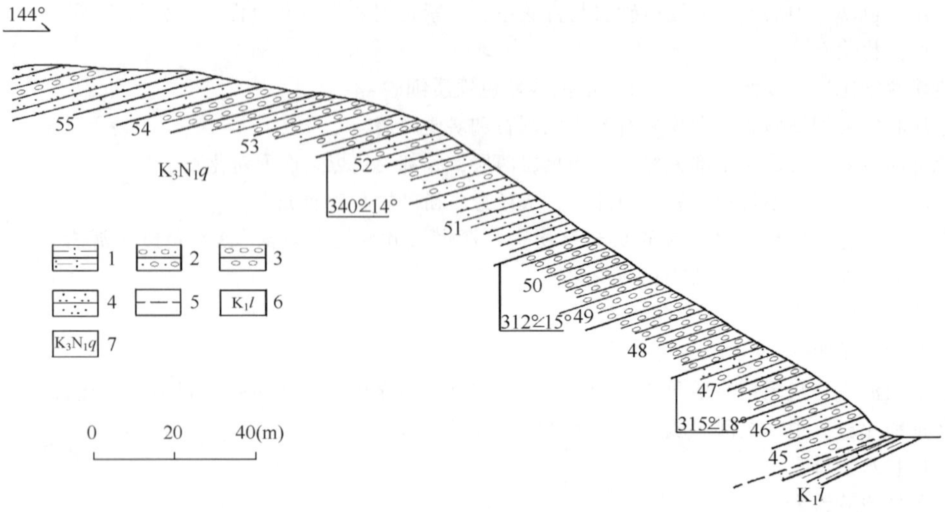

图 5-4　乐东县丘糖岭组剖面图

Fig. 5-4　Section of the Qiutangling Formation, Ledong County

（据四川攀西地质大队，1991，略有改动）

1—凝灰质泥岩；2—砂质砾岩；3—砾岩；4—石英砂岩；5—平行不整合；6—鹿母湾组；7—丘糖岭组

53. 灰色石英质砂砾岩、中砾岩互层夹长石粗砂岩，三者组成韵律沉积，以砂砾岩为主。砾岩呈块状及厚层状。于剖面西南侧相当层位中采到植物化石（宜昌地矿所孟繁松研究员鉴定）：53 层中上部（D1576-3）产 *Magnolia miocenica*, *M.* cf. *miocenica*, *Cinnamomum bodinieri*, *C.* cf. *lanceolatum*, *Litsea* cf. *grabaui*, *Lindera*? sp. 等。53 层下部（D1576-2）产 *Magnolia* cf. *miocenia*, *Cinnamomum* sp., *Marchilus* sp., *M.*? sp. 等　29.31m

52. 灰色块状砂质细砾岩夹含砾长石不等粒砂岩。含铁质结核或团块，砾石为 0.2~0.4cm，之石英细砾，夹层中见斜层理。D1576-1 产 *Quercus simulata*, *Q.* sp., 等　29.71m

51. 灰白色石英不等粒砂岩。风化后很疏松，呈黄色，具平行层理　38.15m

50. 灰白色厚层石英细砾岩。含砾 60%，分选较好。砾径 0.2~0.6cm，砾石以具油脂光泽的石英为主，次有少许石英岩、硅质岩，硅质胶结坚硬　31.94m

49. 浅灰色块状砾岩，硅质胶结。砾石为石英、石英岩，砾径 0.2~0.6cm，大于 1cm 者占 20%~25%，小于 1cm 者占 30%~40%，填隙物为石英及长石砂粒　12.63m

48. 石英质细砾岩。风化后为黄灰色，含砾 70%，硅质胶结，坚硬，砾石以 0.2~0.5cm 直径的石英为主，少许石英岩，偶见千枚岩，具强烈油脂光泽，填隙物为石英砂及玉髓　31.88m

47. 灰黄色含细砾石英粗砂岩与细砾岩互层。砾岩含砾达 60%，成分单一，均为硅质岩类岩屑，其中以石英为主，次有石英岩及燧石，砾径为 0.2~1cm　29.88m

46. 浅灰色、紫灰色块状砂质细砾岩。岩石风化为黄灰色，含砾为 35%~45%，泥铁质胶结。砾石主要是石英、长石，次为石英岩、变质细砂岩，偶见灰绿色凝灰质砂岩，其分选差。砾石直径为 0.2~5cm，但以小于 1cm 为主，偶见 5cm 以上者。细砾的滚圆度较差。岩层具不明显的板状层理　19.48m

45. 灰色、紫色块状砂质细砾岩。岩层单层厚度大于 1m，胶结松散，砾石为石英，砾径小于 1cm，充填物为粗砂，含砾率为 40%，具大型斜层理。偶见植物茎干化石　11.68m

－－－－－假　整　合－－－－－

下伏地层　下白垩统鹿母湾组（K_1l）紫红色含凝灰质粉砂质泥岩夹灰色长石粗砂岩　>20m

（五）佛罗组（N_1f）

乐东县新安第 5 孔剖面

该剖面位于乐东县佛罗镇新安村，坐标 108°51′40″，北纬 18°28′，由广东省地质局 765 队于 1960

年测制，为佛罗组的代表性剖面。岩性自上而下为：

上覆地层 上新统望楼港组灰黑色黏土质粉砂岩

——————整　合——————

上中新统佛罗组	总厚度 60.31m
8. 灰色石灰质	1.30m
7. 褐色泥岩，含介形类 *Pelecyplda* sp.	3.00m
6. 灰绿色含砾砂岩	1.35m
5. 灰绿色、褐色粉砂质泥岩夹深灰色砂砾岩	20.90m
4. 深灰色细砂岩	5.40m
3. 深灰色、灰白色砂砾岩、黏土质砾岩	6.60m
2. 白色高岭土质砂岩	7.00m
1. 白色高岭土质砂砾岩、砾岩夹砂岩	14.76m

~~~~~~不　整　合~~~~~~

下伏地质体 燕山期花岗岩

（六）石马村组（$N_1\hat{s}m$）

1. 文昌市蓬莱镇石马村剖面

该剖面位于文昌市蓬莱镇石马村，起点坐标：东经 110°31′10″，北纬 19°37′10″，系海南地矿局 1994 年测制，为石马村组的代表性剖面，石门沟村组的副层型剖面。

未见顶

上新统石门沟村组	总厚度>95.50m
13. 灰黑色辉斑橄榄玄武岩，含橄榄岩包体	>3.8m
12. 褐灰色玄武质角砾凝灰岩，顶部夹玄武质集块岩	7.50m
11. 灰黑色橄榄玄武岩，含少量二辉橄榄岩包体	15.00m
10. 棕红色玄武质角砾凝灰岩，角砾成分为花岗岩、玄武岩	66.00m
9. 黄灰色凝灰质砂岩，具水平层理	3.20m

－－－－－假　整　合－－－－－

上中新统石马村组	总厚度 52.90m
8. 铁质黏土岩，含铁化木及果实化石：*Dracontomelon dao*	0.30m
7. 灰白色、浅蓝灰色玄武质凝灰岩	2.20m
6. 灰黑色—灰绿色橄榄玄武岩	12.00m
5. 风化呈棕红色玄武质角砾凝灰岩，角砾为花岗岩屑及玄武岩屑等	3.00m
4. 灰黑色橄榄玄武岩，发育柱状节理	12.50m
3. 灰黑色玻基橄辉岩	22.50m

~~~~~~喷发不整合~~~~~~

下伏地质体 海西期—印支期花岗岩

2. 文昌市蓬莱镇后田园村剖面

该剖面位于文昌市蓬莱镇后田园村，起点坐标：110°31′16″，19°33′36″，系海南地矿局 1994 年测制。该剖面作为石马村组的副层型剖面。

未见顶

上中新统石马村组	>39.50m
4. 棕红色（风化）玄武质角砾凝灰岩，角砾成分为花岗岩、玄武岩等	>7.00m

3. 灰黑色玻基橄辉岩 12. 50m

2. 上部为黏土岩，下部为砾岩，产果实化石、植物化石碎片 7. 00m

1. 玻基橄辉岩 13. 00m

~~~~~~~~ 不 整 合 ~~~~~~~~

下伏地质体 海西期—印支期花岗岩

（七）望楼港组（N_2w）

望楼港第4号孔剖面

该剖面位于乐东县九所镇望楼港，起点坐标：108°51′50″，18°26′08″，系广东省地质局765队1960年测制，为望楼港组的代表性剖面。

上覆地层 全新统烟墩组棕黄色砂

------ 假 整 合 ------

上新统望楼港组	总厚度176. 15m
2. 蓝灰色砂质黏土与黏土质砂互层，含贝壳碎屑	6. 00m
3. 棕黄色细砂	7. 00m
4. 蓝灰色砂质黏土、黏土夹黏土质粉砂层，含钙质结核	42. 20m
5. 蓝灰色、黄色砂，含贝壳碎屑	7. 19m
6. 蓝灰色粉砂质黏土，底部含砂质灰岩结核，中部含植物及贝壳碎片	37. 57m
7. 黄色砂	5. 48m
8. 蓝灰色粉砂质黏土与细砂互层，夹厚0. 6m的黑褐色泥炭土，含植物碎片	32. 13m
9. 蓝灰色细砂	6. 08m
10. 蓝灰色粉砂质黏土、黏土，中部含贝壳	25. 06m
11. 蓝灰色细砂	7. 44m

———— 整 合 ————

下伏地层 上中新统佛罗组蓝灰色砂砾与灰白色、淡绿黄色砂砾岩互层

（八）石门沟村组（$N_2\hat{s}$）

文昌市蓬莱镇石门沟村剖面

该剖面位于文昌市蓬莱镇石门沟村东北侧，起点坐标：110°32′26″，19°29′26″，系海南地质大队1986年测制，为石门沟村组的代表性剖面。

未见顶

上新统石门沟村组	总厚度>127. 60m
4. 深绿色、灰黑色粗玄岩，含橄榄辉石岩包体	40. 50m
3. 灰黑色辉斑橄榄玄武岩，含橄榄岩包体	30. 00m
2. 褐灰色玄武质角砾凝灰岩与玄武质集块岩不等厚互层。集块及角砾成分为玄武岩、花岗岩、偶见橄榄岩	37. 10m
1. 灰黑色橄榄玄武岩，含少量二辉橄榄岩包体	

未见底

二、岩石地层单位

（一）昌头组（$E_1\hat{c}t$）

昌头组系海南地质大队1980年在开展琼山长昌煤矿普查时所创建，广东省地矿局（1988）介绍。
~~~~~~~~

标准剖面位于昌头村南约500m的ZK0802孔。最早称为红色碎屑岩组。隐伏分布在海口市长昌盆地及树德盆地中，仅零星出露于盆地东部吴村至潭白村一带的溪沟、水渠和田坎边。根据岩性特征划分为两个岩性段：下部称红色岩段，为洪积相沉积，底部为棕红色、灰白色砾岩、砂砾岩夹粉砂岩，往上棕红色泥岩与灰白、棕红色长石石英砂岩、钙质细砂岩互层，碎屑成分主要为石英，次为花岗岩和变质岩岩屑，分选性差，次棱角状—次圆状。厚115.1~124m，与下伏燕山期花岗闪长岩和古生代绢云母石英片岩呈不整合接触。上部称油页岩段，岩性为灰褐色油页岩与棕红色、灰色泥岩、页岩不等厚互层，夹数层灰白色细砂岩、灰绿色粉砂岩。含介形类及孢粉，属湖泊相沉积，厚96~249m。

该组下段岩性较稳定，上段岩性、厚度变化均较大。盆地南部谭白村一带，为灰色、深灰色、灰褐色油质页岩与浅紫红色、绿灰色页岩、棕红色块状泥岩及粉砂质泥岩不等厚互层；下部夹数层灰白色薄层细砂岩，厚249m（ZK0802孔）。北部高山村一带，相变为棕红色厚层泥岩与长石石英细砂岩互层，夹众多的石膏条带（条带厚度小于2cm），厚96m。

（二）长昌组（$E_2\hat{c}$）

长昌组最早系海南地质大队（1959）划分为长昌煤系和杂色砂岩系，但未指定层型剖面。中南地区区域地层编写小组（1974）介绍。1964年广东区测大队将长昌煤系与下部的杂色砂岩系合并，易名长昌群。1980年海南地质局4队将其改为长昌组。代表性剖面位于海口市长昌煤矿区为CK48孔（276.64~430.37m井段），参考剖面为海口市长昌煤矿区CK106孔（39.43~369.8m井段）。底部以灰白色砾岩与昌头组紫红色砂岩或粉砂岩呈整合接触。顶部以褐色油页岩与瓦窑组灰白色砂砾岩呈整合接触。该组是长昌煤矿的控矿层位。

长昌组主要分布在长昌盆地。根据岩性组合特征划分为两个岩性段。下段称为杂色岩段，属湖相沉积，其下部为灰白色砂砾岩与灰色、灰绿色、紫红色砂岩不等厚互层，夹紫红杂灰绿色粉砂岩，上部为灰白色、灰绿色、紫红色粉砂岩与灰白色、灰绿杂紫红色粉砂岩不等厚互层，夹泥岩，厚250~320m；上段称含煤段，为暗色湖沼相含煤建造。其下部为褐色油页岩、黑色炭质页岩、为主，夹油页岩、炭质页岩及褐煤层，厚190~230m。

区域上，该组厚度变化较大。CK106孔厚350.27m，CK7孔厚226.33m，盆地边部厚25.18m。岩性变化不大，其中下部杂色岩段以砂岩相为主体延伸较稳定，上部含煤段以油页岩、泥岩相为主体变化不大，在瓦窑溪剖面为油质页岩、炭质页岩、泥岩夹砂岩和煤层，在CK48孔砂岩的厚度、层数明显增加。

（三）瓦窑组（E_2w）

瓦窑组最初称为“白色碎屑岩层”，系海南地质大队（1980）所创建，雷奕振等（1992）介绍。建组时未指定层型剖面。现以长昌盆地瓦窑溪剖面作为代表性剖面。底部以含砾不等粒砂岩与下伏长昌组紫褐色油质泥岩整合接触。该组为长昌砂质高岭土矿床的控矿层位。

瓦窑组主要分布在长昌盆地中，属河湖相沉积。下部为砂岩、含砾砂岩及砂砾岩夹青灰色泥岩、紫灰色油质泥岩及泥质菱铁矿层，泥岩中含较多的钙质结核；上部为灰白色含砾粗砂岩、砂砾岩与灰色、灰绿色中细砂岩、粉砂质泥岩互层，具二元结构，中粒砂岩中见板状交错层理，砂岩中含较多的长石，多为长石石英砂岩，风化后形成砂质高岭土矿。产植物和孢粉化石。总厚334~650m。

该组岩性在区域上变化不大，以砂岩、砂砾岩相为主体；厚度变化较大，长昌盆地由中心向边部厚度变小，瓦窑溪剖面厚276.2m，长昌矿区14线剖面厚264.92m，在CK7孔厚10.82m。

（四）丘糖岭组（$E_3—N_1q$）

丘糖岭组原为四川省地矿局攀西地质大队和海南地质大队1991年开展1：5万尖峰岭幅、乐东县幅区调时所新建的岩石地层单位（后根据报告审查意见，改为瓦窑组），李孙雄等（2011）介绍。创名地点位于乐东县西北面丘糖岭，代表性剖面为丘糖岭剖面，为古近纪河流相砂砾岩沉积。2004年，宜昌地矿所、海南省地质调查院在开展1：25万乐东县幅、陵水幅区调时，在丘糖岭剖面及附近发现大量渐新世、中新世植物化石，从而将这套地层划为渐新世—中新世长坡组。本次对这套地层进行了深入研究后（海南省地质调查院，2011），认为这套地层既不能与瓦窑组对比，也不宜与长坡组对比。其一，岩石组合特征有较大差异，丘糖岭剖面所代表的地层为河流相砂岩、砂砾岩沉积，而长坡组为湖相以黏土岩、粉砂质黏土岩、泥质粉砂岩为主的含油页岩的煤系地层；其二，地层时代差异，丘糖岭剖面所代表的地层时代应为渐新世—早中新世，而长坡组时代为早中新世—中中新世；其三，二者不在同一地层分区。因此，本文采用丘糖岭组，但时代应为渐新世—早中新世，并置于佛罗组之下。

丘糖岭组假整合于早白垩世鹿母湾组之上，未见顶，厚度281m。由一套灰色粗碎屑岩组成。岩性为灰色、黄灰岩石英质细砾岩、砂砾岩、砾岩与石英不等粒砂岩、粗砂岩不等厚互层。产植物化石。底部以块状砂质细砾岩与下伏鹿母湾组紫红色含凝灰质泥岩分界，以一套河流相灰色粗碎屑岩与下伏鹿母湾组湖相红色碎屑沉积相区别。岩石中的碎屑呈次圆状、圆状，胶结物以硅质为主，砂岩及砂砾岩中长石碎屑极少。岩石的结构成熟度及成分成熟度均较高，说明是经过长距离搬运、筛选后的沉积物。大型交错层理发育，板状交错层理比较常见。

（五）佛罗组（N_1f）

佛罗组系1960年广东省地质局765队以乐东县佛罗镇新安村第5孔174.52~242.55m井段命名，李云通等（1984）介绍。雷奕振（1992）曾将与佛罗组相当层位的黄流组（曾鼎乾等1981年在莺歌海盆地建立的第三纪地层系统）引入海南岛西南沿海地区。海南省地质矿产勘查开发局（1997）进行岩石地层清理划分时，采用佛罗组命名方案。

该组下部岩性为淡黄绿色砂砾岩与乳白、浅黄色细—粗粒砂岩不等厚互层，上部以蓝灰色砂砾岩与灰白色、浅黄绿色砂砾岩互层，间夹浅黄绿色中—粗粒砂岩为主。隐伏分布于东方、乐东、三亚等地沿海地带。底部以砾岩与下伏燕山期花岗岩呈不整合接触，顶部以砾岩与上覆望楼港组呈整合接触。产介形类化石，厚度小于100m（60.31~84.75m），在区域上较稳定。

（六）石马村组（$N_1\hat{s}m$）

石马村组系海南地质大队根据文昌市蓬莱镇石马村剖面和后田园村剖面创名，陈哲培等（1995）介绍。其中前者为代表性剖面，后者为副层型剖面，系指中新世时期一套陆相基性火山岩及其沉积夹层与下伏海西期—印支期花岗岩呈喷发不整合接触，与上覆石门沟村组呈假整合接触。

该组主要分布在定安县居丁、文昌市蓬莱、新桥及海口市三门坡一带。下部为玻基橄辉岩夹砂砾岩及黏土岩，前者含二辉橄榄岩包体，后者产植物碎片及果实化石；上部橄榄玄武岩与角砾凝灰岩互层，玄武岩柱状节理发育。顶部为厚0.3m的铁质黏土岩（图版Ⅰ-12-A），层面上见铁化木及果实化石印模。

区域上，该组岩性变化不大，但居丁一带石马村组顶部玻基橄辉岩之上未见橄榄玄武岩及凝灰岩，并出现古红土化，上覆石门沟村组橄榄玄武岩喷发不整合于玻基橄辉岩之上。底部的玻基橄辉岩则与下伏早白垩世鹿母湾组呈喷发不整合接触。厚度变化较大，蓬莱石马村一带厚72.5m，居丁厚142.3m。

石马村组的玻基橄辉岩是钴土矿的控矿层位，橄榄玄武岩是钴土矿、铝土矿的控矿层位。居丁钴土矿床即赋存于石马村组的玻基橄辉岩、橄榄玄武岩的风化壳中。橄榄玄武岩、火山碎屑岩与宝石矿床有密切的成因联系。蓬莱宝石矿床的部分矿体赋存于上述岩石的残坡积层中。

（七）望楼港组（N_2w）

该组由广东省地质局765队（1960）根据乐东县九所镇望楼港4号钻孔6.7~183.55m井段命名。李云通等（1984）介绍。雷奕振等（1992）更名为莺歌海组。陈哲培等（1997）在《海南岛岩石地层》中采用望楼港组命名方案。岩性为蓝灰色、黄色薄层细砂与蓝灰色粉砂质黏土、黏土不等厚互层，局部夹含砾砂岩、含砾泥质砂岩。代表性为乐东县九所镇望楼港第4号钻孔6.7~183.5m井段。区域上岩性较稳定。

该组隐伏分布于黄流、九所望楼港、三亚等地沿海地带，由北向南碎屑粒度变粗。黄流一带以深灰色黏土为主，含少量粉细砂，中部含少量贝壳。九所望楼港一带，下部为细砂，上部为粉砂质黏土或砂质黏土，含贝壳碎屑，不等厚互层，由下而上叠置呈韵律性沉积特征。至三亚下抱坡村一带，该组岩性夹有4层含砾砂岩。区域上该组底部与下伏地层佛罗组呈整合接触，顶部与上覆第四纪秀英组含砾亚砂土、含砾杂色黏土质粗砂呈假整合接触。厚6.95~176.15m。

（八）石门沟村组（$N_2\hat{s}$）

石门沟村组由海南地质大队根据文昌市蓬莱镇石门沟村剖面及石马村剖面创名，陈哲培等（1995）介绍。前者为该组的代表性剖面，后者为该组副层型剖面，由海南地质大队1994年测制。原始定义是上部为粗玄岩、橄榄玄武岩，下部火山碎屑岩夹橄榄玄武岩，未见顶。与下伏石马村组呈假整合接触。

该组主要分布在海口市大坡、文昌市蓬莱、南阳及定安县永丰一带，为一套陆相基性火山岩和火山碎屑岩。

三、古生物群及时代划分与对比

（一）昌头组（$E_1\hat{c}t$）

昌头组上部油页岩段含丰富的介形类和孢粉化石。介形类经操晓萍研究，以 *Sinocypris*（37.3%）和 *Eucypris*（35.5%）两属的繁盛为特征，主要有 *Sinocypris excelsa*（11%），*S. elliptica*（11%）*S. reticulata*（8.2%），*Eucypris serrata*（10%），*E. hengyangensis*（8.2%），*E. stagnalis*（6.4%），*Limnocythere hubeiensis*（6.4%）等，属于华南地区晚古新世常见的 *Sinocypris excelsa*-*Eucypris hengyangensis* 组合。值得注意的是，这一组合中还见有极少量的古老类型 *Cypridea* 属的成员，*Cypridea* 属主要存在于中生代，特别是白垩纪，在我国南方可延续到早—中古新世。*Cypridea* 的出现无疑表明，昌头组的介形类组合的时代不会晚至早始新世，或许比雷州半岛的长流组上段、合浦盆地的下洋组、三水盆地的㘵心组、南雄盆地的浓山组、池江盆地的池江组及衡阳盆地的霞流市组等层位的介形类组合稍早，很可能是晚古新世早期。

孢粉组合经张清如、何卫研究，以 *Subtriporopollenites*-*Ulmipollenites minor*-*Pentapollenites* 为代表，其中被子植物花粉占绝对优势（88.86%），主要分子有 *Ulmipollenites minor*，*Nyssapolleninites*，*Triporopollenites*，*Cupliferoipollenites*，*Subtriporopollenites granulatus*，*S. firmus*，*Ilexpollenites* cf. *microloacus*，*Faguspollenites*，*Celtispollenites triporatus*，*C. minor*；裸子植物花粉含量为9.53%，主要是 *Abietineaepollenites microlatus* cf. *major*，*Podocarpidites* spp.，蕨类植物孢子稀少，仅占1.61%。本组合的特征分子及主要类型大多见于华南各地古新世孢粉组合中。其中数量丰富的 *Ulmipollenites minor* 在我国南方含早—中古新世哺乳动物群的上湖组和狮子口组很发育。*Subtriporopollenites granulatus* 较早地出现在云南勐也井组、湖南衡阳盆地东塘组和湖北江汉盆地西部龚家冲组等古新世地层中。其他如 *Plicapollis*，*Ulmoideipites tricostatus*，*Sapotaceoidaepollenites tricolporatus*，*Nanlingpollis aspidoporatus*，*Boehlensipollenites qingjiangensis* 等均为华南古新世孢粉组合中的重要分子。总之，昌头组的孢粉组合以富含小榆粉和粒纹亚三孔粉而与华南地区早—中古新世孢粉组合较为接近，其时代似乎比迈陈凹陷的迈2井长流组孢粉组合稍早，可能为早、中古新世—晚古新世早期。

根据上述生物组合特征，该组时代为古新世。

（二）长昌组（$E_2\hat{c}$）

长昌组产大量的孢粉和植物化石，以前者最为丰富。据瓦窑溪剖面不同层位的样品分析结果，孢粉大致可分为3个组合。

第一组合以 *Porocolpopollenites miikensis*-*Polypodiaceaesporites haardti* 为代表，见于长昌组含煤段下部，即瓦窑溪剖面第2~10层，本组合以蕨类植物孢子为主（52.5%），主要分子有 *Polypodiaceaesporites haardti*，*Osmundacidites crassiprimarus*，*Cyclopherusisporites*，*Polypodiisporites* 及 *Extrapunctatosporites*。其次为被子植物花粉（34.86%），其中 *Porocolpopollenites miikensis*，*Gothanipollis bassensis*，*Caryap ollenites tuiangulus*, *Quercoidites microhenrici*, *Liquidambarpollenites* 较为丰富；裸子植物花粉占 12. 64%，主要由 *Abietineaepollenites microalatus* cf. *minor* 和 *Pinuspollenites banksianaeformis* 组成。本组合中的优势种 *Porocolpopollenites miikensis* 发现在日本九州北部的七浦层和 *Hooshu* 层；在匈牙利 *Bakony* 下始新统也有这种类型报道。*Gothnipollis bassensis* 是早、中始新世的繁盛分子。再者，该组合与北部湾盆地流沙港组三段在许多成分上可以对比。因此，其时代可能为早始新世。

第二组合产于长昌盆地长昌组含煤段中、上部，瓦窑剖面第 12～34 层，称为 *Abietineaepollenites Momipites triletipollenites-Operculumpollis operculatus* 组合。该组合被子植物花粉较多（57. 91%），其中 *Momipites triletipollenites*, *M. coryloides*, *Quercoidites microhenrici*, *Q. minor*, *Cupuliferoidaepollenites quisqualis*, *Ulmipollenites undulosus*, *Liquidambarpollenites mangelsdorfianus*, *Operculumpollis operculatus* 的含量较为突出。与早始新世相比，裸子植物花粉已有明显地增加（23. 20%），主要有 *Abietineaepollenites microalatlus* cf. *minor*, *Pinuspollenites banksianaefornis*, *Laricoidites*, *Cedripites*, *Taxodiaceapollenited* 等。蕨类孢子（18. 89%）以 *Polypsdiaceaesporites haardti*，*Osmunda cidites* 为主。本组合可同福山盆地流沙港组三段中部组合对比，时代为中始新世。植物化石仅见于含煤段，经郭双兴等鉴定，计有 *Sabalites* sp.，*Cinnamomum* spp.，*Nelumbo nucifera*, *Osmunda lignitum* 等 30 余种，认为与茂名盆地尚村组植物群相似，时代为中始新世。1979 年，郭双兴重新研究了这批化石，正式发表的有 *Osmunda lignitum*, *Salvinia* sp.，*Cyclocarya scutelltaa*, *Nelumbo protospeciosa*, *Cinnamomum larteti*, *Ocotea sinensis*, *Citrus niger*, *Sabalites szei*, *S. changchangensis*, *Nordenskioldia borealis* 等 10 种。根据植物群的性质，郭双兴认为长昌组的地质时代可能属晚古新世至早、中始新世。

综上所述，长昌组含煤段包括了早、中始新世沉积。长昌组的杂色岩段迄今虽未发现化石，但该段直接覆于古新统昌头组油质页岩段之上，其上覆的含煤段所产早始新世孢粉化石出现在该段近底部，根据这一层序特点，其时代归入早始新世较为合适。

（三）瓦窑组（E_2w）

瓦窑组中产较丰富的孢粉化石，称为 *Alnipollenites-Quercoidites-Liquidambarpollenites* 组合。该组合中被子植物花粉含量为 73. 99%，其中 *Quercoidites microhemrici*, *Q. asper*, *Q. minutus*, *Alnipollenite sverus*, *A. metaplasmus*, *Liquidambarpollenites orientatisformis*, *L. mangelsdorfianus*, *L. stigmosus* 较为丰富；裸子植物花粉仍较多，达 15. 97%，主要有 *Abietineaepollenites microalatus* cf. *minor*, *Pinuspollenites banksianaeformis*，还有少量的 *Taxodiaceaeplollenites hiatus*；蕨类植物孢子占 10. 04%，主要分子有 *Polypodiaceaesporites*, *Osmundacidites*。本组合的基本面貌与福山盆地流沙港组三段上部—二段下部的孢粉组合接近，时代为晚始新世。

（四）丘糖岭组（E_3—N_1q）

宜昌地矿所、海南省地质调查院开展 1∶25 万区域地质调查时在剖面的第 53 层采集植物化石标本，同时在第 52 层又发现新的植物化石点，这些植物化石经宜昌地矿所孟繁松研究员鉴定计有 4 个种、4 个相似种、9 个未定种。它们分别产于第 52 层、第 53 层下部和第 53 层中上部，同时上述层位也发现大量硅化木化石。上述植物化石可识别出 1 个 *Quercus simulata* 植物群和 1 个植物组合：*Magnoloa miocenica-Cinnamormum bodinieri-Litsea*。上述植物群产在 52 层，属种单调，丰富度较高，同时保存了植物茎干化石。大致可与云南景谷植物群（陶君容等，2000）对比，地质时代为渐新世。上述植物组合大多数属种地质时代为中新世，仅 *C.* cf. *lanceolatus*, *Mindera* ? sp. 是由始新世延续而来，可与云南小龙潭植物群对比，其时代为中新世。

因此，丘糖岭组地质时代定为渐新世—早中新世较适合。

（五）佛罗组（N_1f）

该组区内仅发现有介形类化石 *Pelecyplda* sp. 但区域上与之相当层位的莺歌海分区黄流组产有孔虫、介形类及孢粉组合。根据汪昌先等（1981）对有孔虫的研究，认为上述层位所产特征分子 *Globig-*

erina nepenthes, *Globorotalia acostaensis* 主要分布于中新世晚期的结论，故将佛罗组直接与邻区黄流组对比，其地质时代为中新世晚期。

（六）石马村组（$N_1\hat{s}m$）

该组中没有发现可定年的大化石，拟根据同位素年龄来确定其时代。海南地质大队（1994）在蓬莱镇石马村该组下部玻基橄辉岩获 K-Ar 年龄为（5.43±0.116）Ma（中国科学院地质研究所同位素实验室测定），在三门坡一带，该组的玄武岩中获 K-Ar 年龄为 6.725Ma。海南省地质调查院于临高美夏村海边该组气孔玄武岩中获 K-Ar 年龄为（8.09±0.98）Ma，故将其地质时代定为新近纪中新世晚期。

（七）望楼港组（N_2w）

望楼港组产有丰富的生物化石，其中，介形类有 *Cytheris* sp.，*Cyprideis* sp.，*Paijenborchella* sp.，*Eucythere*? sp.，*Cytherelloidea* sp.，刘宗云（1960）称为 *Ostracoda* 组合，繁盛于古近纪—新近纪海相沉积层中，且与苏北滨海相上新世及上新世后的种属相似；腹足类 *Turritella* ex. gr. *millepunctata* 则是台湾上新世苗粟组常见分子；双壳类有 *Venus* sp.，*Mya* sp.。综合已有资料，依据介形类、腹足类组合特征，望楼港组地质时代为上新世。

（八）石门沟村组（$N_2\hat{s}$）

石门沟村组中未发现可定年的大化石，拟根据同位素年龄来确定其时代。海南地质大队（1994）在蓬莱地区该组上部的粗玄岩中获 K-Ar 年龄为（3.042±0.122）Ma。此外，海南省地质调查院于文昌市蓬莱镇蓬莱该组辉石玄武岩中获 K-Ar 年龄为（3.84±0.30）Ma，于文昌市蓬莱镇高山村橄榄玄武岩中获 K-Ar 年龄为（3.37±0.14）Ma，其地质时代为新近纪上新世。

四、岩相古地理特征

海南省古近系—新近系基本上呈隐伏状产出，尽管有关石油部门曾在海南省周边盆地中开展过钻探地质调查工作。但本次限于资料有限，难以开展全面的岩相古地理面貌恢复。仅对琼北陆相盆地-长昌盆地岩相古地理进行探讨（图 5-5）。

古近纪始新世长昌断陷盆地，位于南华褶皱系五指山带的北缘、东西向的王五-文教断裂带的南侧、南北向的琼山-石合断裂和铺前-博鳌断裂之间。盆地明显受上述东西向和南北向两组断裂构造控制，呈东西向展布，两端向南拐，东西长 21km，南北宽 4~12km，面积 168km^2。上覆有古近纪—新近纪地层和第四纪玄武岩及第四纪沉积层。根据露头和钻孔揭露，盆地底部为一套古近纪陆相地层，其基底为中生代三叠纪花岗岩、闪长岩、志留系陀烈组浅变质岩系、白垩系鹿母湾组和报万组紫红色陆相碎屑岩层。

古新世昌头期，盆地早期接受了一套洪积相红色碎屑岩沉积，其底部主体为砾岩、砂砾岩，向上为泥岩、长石石英砂岩、细砂岩互层，碎屑分选性差，磨圆度中等—良好，总体反映碎屑物经过长距离搬运并快速堆积形成。其沉积厚度变化小，显示基底地势相对比较平坦。之后，进入浅湖环境，接受了含介形类及孢粉的油页岩段沉积。沉积厚度变化大，在盆地南部谭白村一带，厚 249m，沉积粒度细，以油质页岩、页岩、泥岩及粉砂质泥岩为主；在北部高山村一带，厚 96m，沉积粒度有变粗的趋势，为厚层泥岩与长石石英细砂岩互层，夹众多的石膏条带（条带厚度小于 2cm）。表明盆地晚期具有南高北低的古地理格局，沉降中心和沉积中心基本吻合，大致位于谭白村一带。

始新世长昌期早期，长昌组下段（杂色岩段）下部为灰白色砂砾岩与灰色、灰绿色、紫红色砂岩不等厚互层，夹紫红杂灰绿色粉砂岩，上部为灰白色、灰绿色、紫红色粉砂岩与灰白色、灰绿杂紫红色不等厚互层，夹泥岩，厚 250~320m。相对于昌头期晚期，盆地相对上升，水体变浅，沉积物变粗，以砂岩相为主体。由东往西，东面文德一带（ZK7），岩性为粗砂岩、粉砂岩、泥岩，岩相突变大，缺乏规律性，呈“杂乱状”堆积，沉积层序混乱，碎屑物质成分复杂，有时出现钙质和低价的铁（黄铁矿结核），并常见长石矿物分布，颗粒大小不均，属于近源快速堆积的产物，初步认为属于湖口沉积的

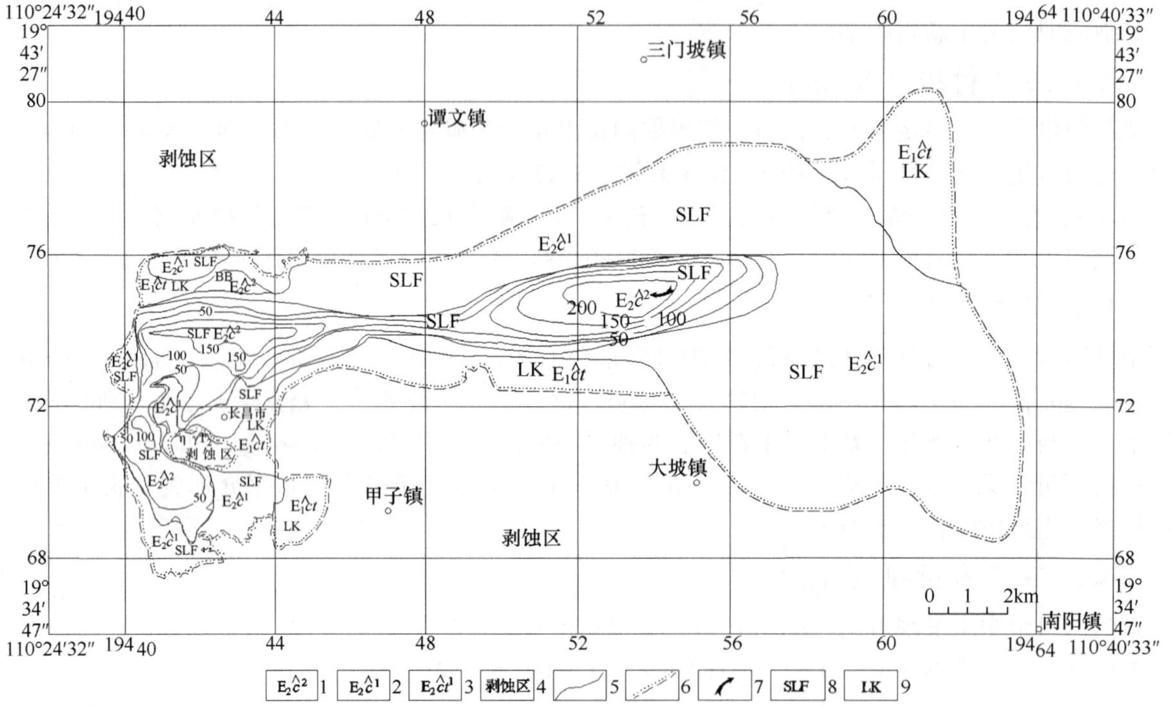

图 5-5 长昌组、昌头组岩相古地理图

Fig. 5-5 Map of lithofacies paleogeography of the Changchang and Changtou Formations

1—始新统长昌组上段：页岩、炭质页岩、油页岩与砂岩、粉砂岩互层夹褐煤；2—始新统长昌组下段：杂色砂岩、粉砂岩为主，夹砂砾岩及泥岩；3—古新统昌头组：砂砾岩、砂岩、泥岩、油页岩；4—中三叠世二长花岗岩；5—实测整合地质界线；6—实测不整合地质界线；7—物源方向；8—浅湖相；9—滨湖相

产物。中部福兴村-文岭一带，CK157 孔下部为细砂岩、粉砂岩，往上为粗砂岩、粉砂岩；CK149 孔下部为细砂岩、粉砂岩，往上总体逐渐递变为粗砂岩、细砂岩、粉砂岩，顶部夹泥岩、粗砂岩。两钻孔岩石组合具有一定的可对比性，下部属于滨湖—浅湖相沉积，上部为滨湖相。西面仁山村周边，CK112 孔为粗砂岩与泥岩不等厚状分布；CK105 孔下部为粗砂岩、细砂岩、泥岩组合，上部为粗砂岩与泥岩不等厚状分布；属于潟湖相或障壁岛后静水环境的沉积产物。CK42 孔为砂砾岩、粗砂岩、细砂岩、粉砂岩“杂乱状”堆积，岩相突变大，沉积规律性差，推断为心滩沉积，即长昌期早期，沉积盆地地势具有东高西低的古地理格局。其沉积物颜色的多样性，显示沉积气候的多变性。古地磁资料研究显示（汪啸风等，1991），古近纪时，位于北纬 12°左右的低纬度区，属于干燥、湿热多雨的气候环境。

始新世长昌期晚期，长昌组上段（含煤段）总体为暗色湖沼相含煤建造沉积。其下部为褐色油页岩、黑色炭质页岩、灰色页岩与灰绿色、灰白色粉砂岩、砂岩不等厚互层，夹褐煤层；上部以砂岩为主，夹油页岩、炭质页岩及褐煤层，厚 190~230m。其中，东面 CK85 孔下部为粗砂岩与油（质）页岩组合，夹煤层，属于潟湖相或障壁岛后静水环境的沉积产物，上部为细砂岩、粉砂岩不等厚状分布，具有滨湖—浅湖相沉积岩石组合特征；西面 CK19 孔岩石比较单一，为炭质页岩与油（质）页岩不等厚状，夹煤层，属于沼泽相沉积，即含煤段沉积时期，由西至东，湖水具有加深的趋势，与长昌期早期盆地格局基本一致。该时期气候潮湿温暖，适宜生物生长，高等植物繁茂，淡水鱼类和贝壳类动物繁殖，为后期成煤作用提供了重要的物质条件。

始新统瓦窑组时期，盆地内继续接受了一套陆相砂砾岩、粗砂岩、细砂岩、泥岩等碎屑岩沉积，直接叠覆于长昌组含煤岩系之上。之后，地壳差异性隆升，并在后期断裂作用下，导致含煤岩系顶面起伏不平。

第三节　莺歌海地层分区

莺歌海地层分区介于西沙海槽北缘断裂和红河断裂、琼东南盆地北缘断裂之间。莺歌海地层分区中的古近系—新近系主要分布在莺歌海盆地，属于海相沉积。

岩石地层单位自下而上划分为渐新统陵水组（$E_3l\hat{s}$）、中新统三亚组（N_1s）、梅山组（N_1m）和黄流组（N_1hl）及上新统莺歌海组（N_2y）。

一、剖面描述

（一）陵水组（$E_3l\hat{s}$）

陵水组代表性剖面见于莺9井2155~2697m井段，由南海石油公司第四海洋地质调查队1981年测制。该孔位于三亚市102°平距83km处海域，坐标：东经110°17′36″，北纬18°05′42″。

上覆地层　下中新统三亚组灰白色砂砾岩

~~~~~~~~~~ 不　整　合 ~~~~~~~~~~

渐新统陵水组	总厚度582.00m
32. 深灰色泥页岩夹浅灰色石英粉砂岩	23.00m
31. 浅灰色含砂砾岩，灰质胶结	7.00m
30. 深灰色泥页岩，含灰质	5.00m
29. 灰白色砂砾岩夹深灰色泥页岩，后者富含黄铁矿和炭质	21.00m
28. 深灰色泥页岩	7.00m
27. 浅灰色含砾粗砂岩，灰质胶结	9.00m
26. 深灰色泥页岩夹浅灰色石英砂砾岩	15.00m
25. 浅灰色含砂砾岩，灰质胶结	6.00m
24. 黑色页岩	7.00m
23. 浅砂色含砾粗砂岩、砂砾岩夹黑灰色页岩，含有孔虫 *Asterigerina* spp.；藻类 *Glaphyocysta yinggehaica*，*G. multiperforata*，*G. combinata*，*Adnatosphaeridium reticulense*，*Emmetrocysta denticulata*，*Cyclopsiella granulata*	10.00m
22. 深灰色泥页岩与浅灰色含砾石英砂岩或砂砾岩不等厚互层。含有孔虫、藻类化石	17.00m
21. 黑灰色泥页岩	19.00m
20. 浅灰色砂砾岩、含砾石英砂岩夹灰黑色泥页岩	6.00m
19. 黑灰色页岩夹灰白色砂砾岩	13.00m
18. 浅砂色细—中粒石英砂岩	14.00m
17. 黑色页岩	7.00m
16. 浅灰色细砂岩，灰质胶结	6.00m
15. 黑色页岩与浅灰色含砾石英砂岩不等厚互层	31.00m
14. 浅灰色含砾砂岩夹黑色页岩	21.0m
13. 黑色页岩与灰白色含砾石英砂岩互层	31.00m
12. 灰白色砂砾岩	10.00m
11. 黑色页岩夹浅灰色粉砂岩、砂岩	16.00m
10. 浅灰色砂砾岩（含少许白垩，夹厚2m的黑色页岩）	49.00m
9. 浅灰色含砾石英砂岩	10.00m
8. 黑色页岩夹浅灰色粉砂岩	33.00m
7. 浅灰色含砾砂岩夹砂砾岩，灰质胶结	42.00m
6. 浅灰色粉砂岩	13.00m
5. 灰色含砾砂岩，含白垩	14.00m
~~~~~~~~~~

4. 浅灰色石英粉砂岩 86. 00m

3. 浅棕红色含长石粗砂岩 7. 00m

2. 浅灰色石英粉砂岩 5. 00m

1. 浅灰色、灰褐色泥晶灰岩夹灰黑色砂质页岩，含双壳类 *Ostrea* sp. 22. 00m

～～～～～ 不 整 合 ～～～～～

下伏地层 燕山期花岗岩

（二）三亚组（N_1s）

三亚组代表性剖面见于莺歌海盆地莺6井1768～1856m井段，系南海石油勘探指挥部第四海洋地质调查队1981年测制。该孔位于三亚市240°平距69km处海域，坐标：东经109°03′05″，北纬17°55′39″。自上而下为（层号按原始编录）：

上覆地层 中中新统梅山组浅灰色白云岩

———— 整 合 ————

下中新统三亚组 总厚度88. 00m

66. 灰白色石英砂砾岩，白云质胶结，夹少许杂色泥岩 6. 00m

67. 杂色泥岩，以棕红色为主，其次为褐黄色、绿黄色 1. 00m

68. 灰白色砂砾岩，含灰白色灰岩块及生物碎屑 8100m

～～～～～ 不 整 合 ～～～～～

下伏地层 白垩系鹿母湾组棕红色砂泥质岩与灰白色砂砾岩互层

（三）梅山组（N_1m）

梅山组代表性剖面见于莺歌海盆地莺6井1532～1768m井段。剖面情况同上述。自上而下为（层号按原始编录）：

上覆地层 下中新统黄流组灰色含灰质泥岩

———— 整 合 ————

中中新统梅山组 总厚度208. 10m

52. 灰白色石英细砂岩，未胶结 3. 10m

53. 灰白色含砾石英粗砂岩夹薄层砂砾岩 29. 00m

54. 灰白色砂砾岩，松散 8. 00m

55. 灰白色含白垩含砾粗砂岩 17. 00m

56. 浅灰色含白垩砂质泥岩 23. 00m

57. 浅灰色白垩质细砂岩，松散 10. 00m

58. 浅灰色粉砂质泥灰岩 16. 00m

59. 浅灰色砂质灰岩，隐晶结构 41. 00m

60. 灰白色砂质白垩 10. 00m

61. 浅灰色砂质灰岩，含有孔虫，底部为厚约1m的灰质细粒长石砂岩 26. 00m

62. 灰白色—浅灰色隐晶质礁灰岩及礁白云岩，具缝合线构造，含岩屑及大量造礁生物残骸，白云岩化自上而下加强 4. 40m

63. 浅灰色、浅褐色含灰质白云岩 9. 60m

64. 灰色砂质白云岩 8. 00m

65. 浅灰色白云岩夹白色砂砾岩 3. 00m

———— 整 合 ————

下伏地层 下中新统三亚组灰白色石英砂岩

（四）黄流组（N_1hl）

黄流组代表性剖面见于莺歌海盆地莺 6 井 1188～1532m 井段。剖面情况同上述。自上而下为（层号按原始编录）：

上覆地层 上新统莺歌海组灰色含生物碎屑岩

——— 整 合 ———

上中新统黄流组	总厚度 344.00m
40. 灰色白垩质泥岩，成岩较好，产孢粉	58.00m
41. 灰白色白垩质泥晶灰岩	4.00m
42. 灰白泥岩，含白垩	30.00m
43. 浅灰色灰质石英粉砂岩	4.00m
44. 灰色泥岩夹砂质泥灰岩（厚 2m），泥岩质纯，成岩好，产介形类化石	86.00m
45. 灰色含砂质泥晶白云岩	6.00m
46. 灰色泥岩	14.00m
47. 灰色泥晶白云岩	4.76m
48. 灰色泥岩夹砂质白垩（厚 2m），泥岩含灰质	32.00m
49. 灰色粉砂质泥岩夹灰白色砂质白垩（厚 3m），含灰质	87.00m
50. 灰色含泥砂质灰岩夹灰质泥	4.54m
51. 灰色灰质泥岩，质纯，成岩好	13.70m

——— 整 合 ———

下伏地层 中中新统梅山组灰白色石英细砂岩

（五）莺歌海组（N_2y）

莺歌海组代表性剖面见于莺歌海盆地莺 6 井 453～188m 井段，剖面情况同上述。自上而下为（层号按原始编录）：

上覆地层 第四系含生物碎屑砂砾层

——— 整 合 ———

上新统莺歌海组	总厚度 735.00m
20. 灰色含生物碎屑黏土，含灰质，产介形类	22.00m
21. 浅灰色含生物碎屑砂砾岩	5.00m
22. 灰色含生物碎屑黏土	45.00m
23. 浅灰色含介壳砂砾岩	5.00m
24. 灰色黏土岩，含灰质及生物碎片，产苔藓虫	50.00m
25. 灰白色含介壳碎片砂砾岩	10.00m
26. 灰色黏土岩	10.00m
27. 灰白色砂砾岩	5.00m
28. 浅灰色含灰质石英砂岩	5.00m
29. 灰色含介壳碎片黏土岩夹含砾石英砂岩（厚 3m），砂岩为灰质胶结，下部黏土岩产介形类、孢粉	122.00m
30. 灰色含介壳碎片石英粉砂岩，泥灰质结构	4.00m
31. 灰色含生物碎屑黏土岩	42.00m
32. 灰色含介壳泥质粉砂岩，夹砂质灰岩条带	8.00m
33. 灰色含介壳碎片泥岩	16.00m
34. 灰色含泥质及灰质粉砂岩	4.00m
35. 灰色含灰质泥岩夹泥质粉砂岩	26.00m
36. 浅灰色含生物碎屑泥质粉砂岩夹灰质泥岩	8.00m

37. 灰色含灰质、介壳碎屑泥岩 56.00m

38. 灰色灰质石英砂岩 6.00m

39. 灰色含生物碎屑泥岩 286.00m

———— 整 合 ————

下伏地层 上中新统黄流组灰色含白垩泥岩

二、岩石地层单位

（一）陵水组（$E_3l\hat{s}$）

陵水组系南海石油分公司（1981）根据莺9井2115~2697m井段剖面创名，主要分布于莺歌海盆地。岩性以灰质砂岩、粉砂岩与灰色泥页岩互层，底部发育一套生物碎屑灰岩，含孢粉、藻类、有孔虫化石，是莺歌海盆地的含石油、天然气层位。

本次研究以莺9井为代表性剖面。底部岩性为浅褐色泥晶灰岩，其上是浅灰色砂砾岩、砂岩与粉砂岩互层；中上部为浅灰色砂砾岩、砂岩与深灰色、灰黑色页岩互层，砂岩、砂砾大都为灰质胶结，局部含白垩；上部灰黑色页岩中含炭、黄铁矿。底部以泥晶灰岩与下伏燕山期花岗岩呈不整合接触，顶部以页岩与上覆地层三亚组砂砾岩呈不整合接触。

（二）三亚组（N_1s）

三亚组系南海石油分公司（1981）根据莺6井1768~1856m井段剖面创名，为滨海浅滩相沉积，岩性较粗，以灰白色砂砾岩为主，上部夹杂色泥岩。与下伏陵水组、下白垩统鹿母湾组呈不整合接触，与上覆梅山组呈整合接触。

本次研究以莺6井和莺9井为代表，西部莺6井为一套粗碎屑岩，厚88m，岩性为灰白色砾岩夹薄层棕红色砂质泥岩，底部砂砾岩松散，少见胶结物，与下伏鹿母湾组呈不整合接触。东部莺9井岩性为灰色砂岩、粉砂岩与灰色泥页岩互层，厚362m。

（三）梅山组（N_1m）

梅山组系南海石油分公司（1981）根据莺6井1532~1768m井段剖面创名，主要分布在莺歌海盆地，为一套陆架浅海相沉积，以砂质灰岩、砂岩夹页岩或大套砂岩与碳酸盐岩为主，含有孔虫、藻类化石。底部以白云岩与下伏三亚组呈整合接触，顶部以细砂岩与上覆黄流组呈整合接触。

本次研究以莺6井和莺9井为代表，莺6井底部岩性为白云岩夹砂质白云岩，其上为薄层礁白云岩、礁灰岩，再上为砂质灰岩夹砂质白垩；上部为细砂岩、含砾粗砂岩夹砂砾岩及砂质泥岩，局部含白垩。上部和底部含有孔虫，下部含造礁钙藻。莺9井岩性以粗砂岩、含砾粗砂岩与泥岩互层为主。该组厚度变化较大，西部莺6井厚208m，东部莺9井厚424m。

（四）黄流组（N_1hl）

黄流组系南海石油分公司（1981）根据莺6井1188~1532m井段剖面创名，分布于莺歌海盆地，为一套浅海相沉积，岩性为粉砂质白垩与白垩质粉砂岩、含白垩质泥岩。该组以富含白垩为特征，含有孔虫、介形类化石。底部以泥岩与下伏梅山组石英细砂岩呈整合接触，顶部以白垩质泥岩与上覆莺歌海组呈整合接触。

本次研究以莺6井孔为代表性剖面。岩性为灰色泥岩与砂质或泥质灰岩，泥晶白云岩、砂质白垩不等厚互层。泥岩中常含灰质及白垩，部分灰岩含白垩，上部夹浅灰色石英粉砂岩。黄流组岩性厚度变化较大，西部莺6井厚344m，东部莺9井厚为230m，岩性以泥岩为主。

（五）莺歌海组（N_2y）

莺歌海组为南海石油分公司（1981）根据莺6井453~1188m井段剖面创名，为一套浅海相沉积，以灰色泥岩、粉砂岩及二者之间的过渡岩性为主，与第四系呈渐变关系。

本次研究以莺6井为代表性剖面，其下部岩性为黏土岩，上部为灰色黏土岩与浅灰色粉砂岩、砂岩、砂砾岩不等厚互层，局部夹砂质灰岩条带，普遍含有孔虫、介形类、腹足类、双壳类等化石，黏土岩、粉砂岩和砂岩常含灰质。底部以含生物碎屑泥岩与下伏黄流组白垩质泥岩呈整合接触，顶部以含生物碎屑黏土被第四系整合覆盖。莺歌海组岩性厚度自西向东有所变化，由西部莺6井向东部莺9井岩性由细变粗，以细砾为主，夹少量粉砂质泥岩、泥质粉砂岩及灰质砂岩。西部莺6井厚735m，东部莺9井厚为531m。

三、古生物群及时代划分与对比

（一）陵水组（$E_3l\hat{s}$）

陵水组产丰富的有孔虫、藻类、双壳类及孢粉等化石。有孔虫仅在中部发现 *Asterigerina* spp.，据汪品先（1981）等研究，为中新世早期。藻类以沟鞭藻占优势，主要有 *Glaphyrocysta yinggehaica*，*G.. multiperforata*，*G.. combinata*，*Adnatosphaeridium retculense*，*Emmetrocysta denticulata*；其次为疑源类 *Cyclopsiella granulata*，偶见盘星属分子；双壳类 *Ostrea* sp.，孢粉称栎粉属（*Quercoidites*）-瘤面水龙骨单缝孢属组合。上述化石组合时代均为晚渐新世—早中新世。陵水组底部以泥晶灰岩与下伏燕山期花岗岩呈不整合接触，顶部以页岩与上覆三亚组砂砾岩呈不整合接触，而上覆三亚组时代为早中新世。因此，将陵水组时代划入渐新世较适合。

（二）三亚组（N_1s）

三亚组古生物特征为上、下部都富含有孔虫。下部以 *Globigerina macrocamerata*，*Cassigerinella chipolensis*，*Asterigerina tentoria* 为主，上部以 *Austrotrillina howchini*，*Miogypsina polymorpha* 为主。以上化石组合特征均属于中新世早期，因此，三亚组地质时代为早中新世。

（三）梅山组（N_1m）

梅山组产丰富的有孔虫、钙藻类、介形类等化石。有孔虫化石在下部为浮游型，以 *Orbulina suturalis* 与 *Cassigerinella chipolensis* 共生为特征，底栖型以 *Heterolepa formosa*，*Martinotiella communis*，*Uuigerina gemmaeformis*，*Boliuina medisulcata* 等为主；上部以 *Nephrolepidina* spp.，*Cycloclypeus* spp.，*Planorbulinella laruata* 等为主。其下部礁灰岩中的钙藻主要有 *Archaeolithothamnium taiwanensis*，*A. myiosporum*，*Lithothamnium saipanense*，*L. toltecensum*，*Dermatolithon saipanense*，*Jania guamensis* 等。介形类上部为 *Loxocorniculum* sp.，*Argilloecia* sp.，*Krithe* sp.。上述生物组合特征均系中新世早期至中期分子，但其底部以白云岩与下伏三亚组、顶部以细砂岩与上覆黄流组均呈整合接触，而三亚组时代为早中新世，黄流组地质时代为晚中新世。依据层序特征，把梅山组地质时代划归中中新世较合适。

（四）黄流组（N_1hl）

黄流组产丰富的有孔虫、介形类及孢粉等化石。孢粉称为 *Quercoidites minor Cupuliferoipollenites* 组合，孙湘君（1981）认为其孢粉组合以新生麻黄粉、刺菊粉、光极弗氏粉（相似种）、伞形花科花粉的出现和繁盛，以及乌斯曼瘤水龙骨单缝孢含量增高为特征；有孔虫浮游型以 *Globorotalia acostaensis*，*Gmulticamenrata*，*Globigerina nepenthes*，*Globigerinoides obliquus*，*S. phaeroidinella dehiscens*，*S phaeroidinellopsis paenedehiscens*，*Pullenialina primalis*，*Ammnonia eqma-toriona*，*Urigerina sehwager* 为主。汪品先等（1981）认为特征分子 *Globigerina nepenthes* 主要分布于中新世晚期；*Globorotalia multicamerata*，*G. margarite*，*S phaeroidinellopsis paenedehiscens* 只出现在中新世晚期至上新世，其时代应属于晚中新世。介形类称为 *Argilloecia-Krithexestoleberis* 组合，主要分子有 *Ambocyhere reticulata*，*Bradleya albatrossia* 等，时代属于中新世晚期。

综合上述古生物化石组合特征，黄流组地质时代应为中新世晚期。

（五）莺歌海组（N_2y）

莺歌海组古生物较丰富，含有孔虫、孢粉、介形类、双壳类及腹足类等。有孔虫有浮游型以 *Globorotalia tosaensis*，*Globigerinoides ruber* 为主，底栖型以含 *Asterorotalia* sp.，*Pseudorotalia* sp. 为特色；

孢粉下部称 *Quercoidites minor-Polypodiaceaesporites* 组合，上部称 *Cupuliferoipollenites ouiformis-Polypodiaceaesporites-ChenoArgilloecia* sp. 组合，还有 *Krithe* sp.，*Paijenborchella malaiensis*，*Atjehellat semiplicata*，*Echinocythereis brabyi*。等；双壳类以 *Anadapodipollis* 组合；介形类称 *Neomonoceratina delicata-Ambocythere elliptica* 组合，主要分子还有 *Rayinggehainsis-Scapharca pilala-Bellucina ciuica* 组合为特征；腹足类有 *Turbo* sp.，*Solarium* sp.，*Limacina* sp.，*Triphora* sp. 等；苔藓虫有 *Entalophora delicatula*，*Cellaria sinuosa*，*C. gracilis*，*Fiabellopora petalosa* 等。

以上生物组合特征，时代均为上新世。故莺歌海组地质时代为上新世。

四、岩相古地理特征

莺歌海盆地是天然气的储藏和生气盆地，朱伟林等（2007）根据其沉积物层序及地震波反射层的分析，对莺歌海盆地岩相古地理面貌进行探讨恢复。

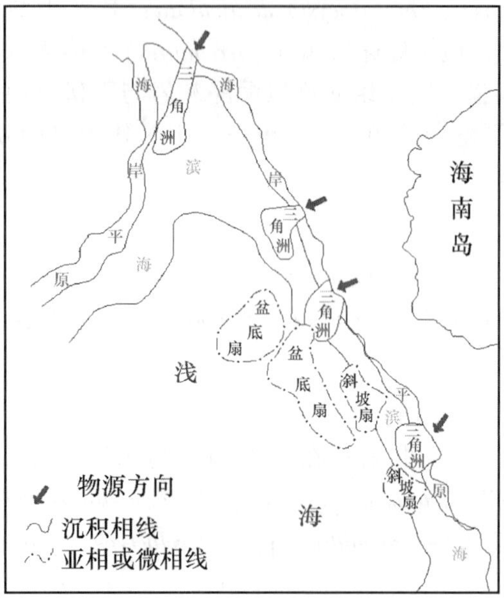

图 5-6　莺歌海盆地早中新世早期沉积岩相古地理图

Fig. 5-6　Map of sedimentary lithofacies paleogeography of the early Early Miocene, Yinggehai Basin

莺歌海盆地在早中新世早期，表现为海退、准平原化，海侵逐渐扩大而成（图 5-6）。盆地西北部均出露地表，周缘发育较窄的滨海相，地震上表现为强反射，向上倾方向尖灭，但局部也可见一定的下超，反映近源三角洲影响，下方可追踪到剖面呈透镜状的低位斜坡扇、盆底扇。红河三角洲沿盆地北部的通道注入，地震表现为杂乱反射。盆地东北和东南部地势平坦，滨海和海岸平原较宽，可有潮坪沉积，泥岩较多；而东部岸坡较陡，滨海较窄，以波浪作用为主，砂岩较多；盆地中部为大片浅海相，底辟活动不多。

早中新世中期，随着海平面上升，沉积范围扩大。海水漫过盆地西北部的坡折，形成大片滨海及海岸平原区，由于地形平坦，可发育障壁岛-潮坪体系。在盆地中部发育了半深海相。盆地西部有两个明显的台阶，大致控制着浅海及半深海的界线；东部则为一个坡度较陡的单斜。受海南岛两个延伸期的影响，中段坡陡而南、北段坡缓。缓坡段也可发育潮坪，陡坡段多为沿岸砂坝或海滩。在深水坡折带附近可有低位滨海砂坝，其下可有斜坡扇或水下滑塌体（图 5-7）。

早中新世晚期，由于在盆地边缘有较大削蚀，原始的沉积范围不详，但水深更大，浅海、半深海相增多，沉降中心继续南移（图 5-8）。盆地东部岸坡较陡，普遍有近源扇三角洲发育，但多沿岸分布，向盆内延伸不远；盆地北部地势相对平缓，三角洲入海后可向南推进较远。推测此期波浪或沿岸流作用较强，可对三角洲沉积物进行搬运改造，形成大致平行岸线的沿岸砂坝。滨海和海岸平原相多沿盆地北部边缘斜坡呈窄条带状分布。浅海多分布在盆地北部区域，半深海主要分布在东方构造附近区域和乐东区中部，其中有低位斜坡扇发育。

进入中中新世早期，沉积范围再次扩大，盆地中地形起伏大，形成隆起、凹陷相间的格局（图 5-9）。盆地东部，南、北段坡缓、水浅，滨海相范围大，可有碳酸盐岩发育；中段坡陡、水深，浊积、滑塌较多。有一系列近源三角洲发育，其中最南边一个规模最大，钻井揭露为砂砾岩。

由于波浪作用较强，三角洲入海后推进不远，沿岸砂坝发育。盆地北部有远源三角洲和水下高地发育，钻井揭露大套细砂岩。半深海相位置在下伏几个层序逐渐南移后又复北移，其间有近东西向的条带状浅海相隔。底辟活动较多并控制沉积，高部位可发育水下高地浅滩砂岩。

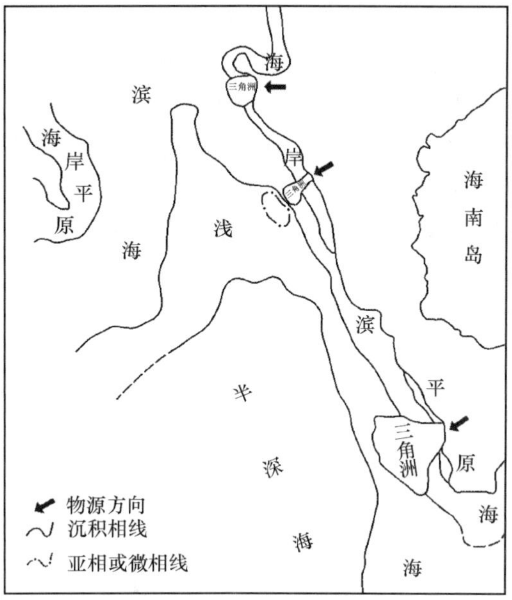

图 5-7　莺歌海盆地早中新世中期沉积岩相古地理图
Fig. 5-7　Map of sedimentary lithofacies paleogeography of the middle Early Miocene, Yinggehai Basin

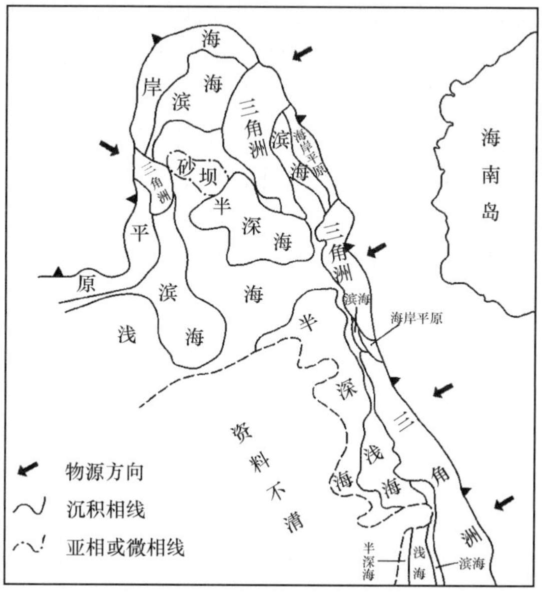

图 5-8　莺歌海盆地早中新世晚期沉积岩相古地理图
Fig. 5-8　Map of sedimentary lithofacies paleogeography of the late Early Miocene, Yinggehai Basin

中中新世中期，浅海相广泛发育，地形起伏更大，在东方和乐东附近区域形成若干局部深水沉积区（图5-10）。盆地东部仍是南、北段滨海相宽，中段较窄，由于晚期剥蚀而缺失海岸平原相。北部红河三角洲退缩且砂质供应量减少，而东部海南近源三角洲则以粗碎屑岩为主，盆地内有低位三角洲发育。

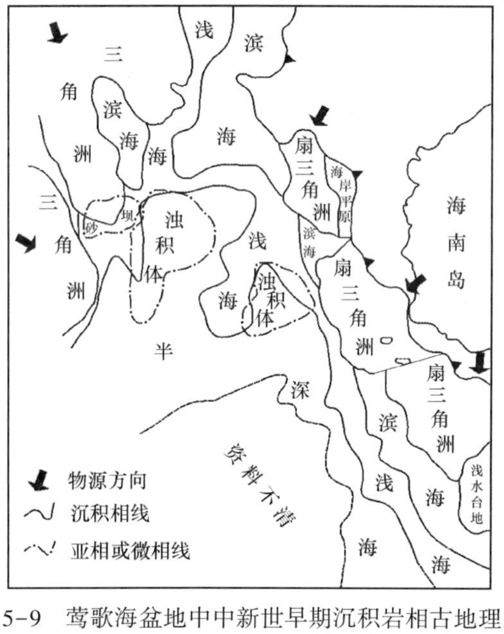

图 5-9　莺歌海盆地中中新世早期沉积岩相古地理图
Fig. 5-9　Map of sedimentary lithofacies paleogeography of the early Middle Miocene, Yinggehai Basin

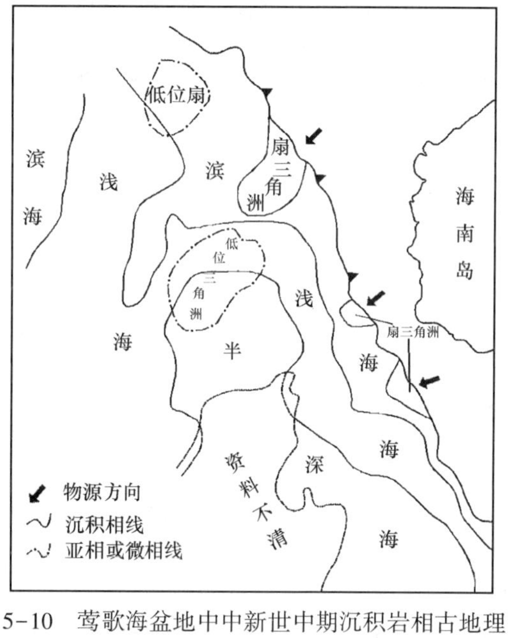

图 5-10　莺歌海盆地中中新世中期沉积岩相古地理图
Fig. 5-10　Map of sedimentary lithofacies paleogeography of the middle Middle Miocene, Yinggehai Basin

中中新世晚期后，浅海相仍广泛发育，滨海相主要分布在盆地边缘、东南缘和中部隆起区。在东方构造附近新出现一近圆形的深水区域，可能与左旋走滑的拉张带有关。在盆底坡折下散布有若干低位盆底扇和斜坡扇（图5-11）。从三角洲的影响范围看，此时西边的物源供应更充足一些。由于晚期剥蚀强烈，地层分布范围小，边缘缺带明显。

莺歌海盆地在晚中新世，早期主要表现为：沉降中心向南移，地势北高南低，北部以海岸平原-滨海为主，南部以浅海为主，半深海位于昌南区附近（图 5-12）。东方构造区成为半封闭的海湾。红河三角洲主要由北西西方向注入，伴之滑塌或重力流沉积。莺东斜坡带此时坡陡，滨海相带窄。中—南段沿一号断层下降盘有若干近源三角洲发育，并可形成一些明显的下切水道。晚期，随海侵扩大，沉积范围明显增大，北高南低的沉积格局未变。北部以滨海较多，南部以浅海—半深海为主。红河三角洲改由盆地西部部注入，但输砂量不足，形成“S”形前积层。盆地东部近源三角洲体系特别发育，但向盆地相变较快（图 5-13）。钻井普遍钻遇砂砾岩，成熟度低、分选差，亦有重力流发育。在盆地中部一般受海南岛和越南双物源影响，有若干低位三角洲发育，但砂岩泥岩含量较高。

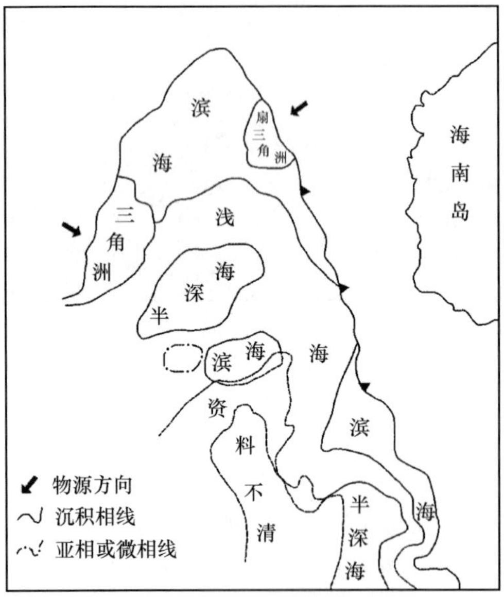

图 5-11　莺歌海盆地中中新世晚期沉积岩相古地理图

Fig. 5-11　Map of sedimentary lithofacies paleogeography of the late Middle Miocene, Yinggehai Basin

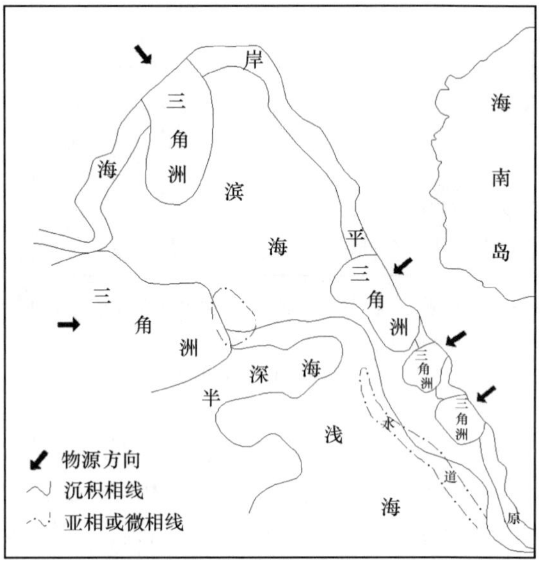

图 5-12　莺歌海盆地晚中新世早期沉积岩相古地理图

Fig. 5-12　Map of sedimentary lithofacies paleogeography of the early Late Middle Miocene, Yinggehai Basin

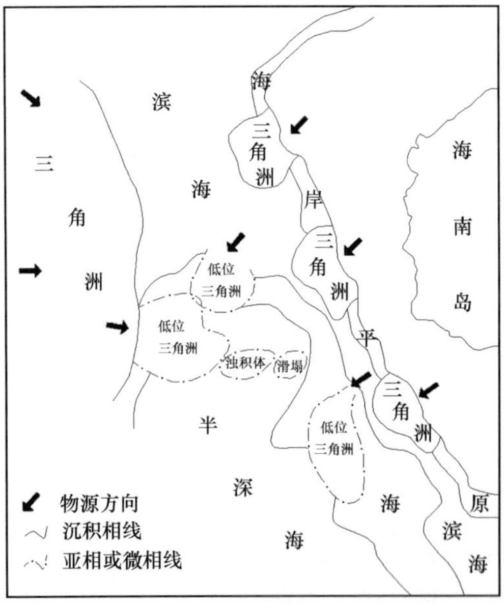

图 5-13　莺歌海盆地晚中新世晚期沉积岩相古地理图

Fig. 5-13　Map of sedimentary lithofacies paleogeography of the late Late Middle Miocene, Yinggehai Basin

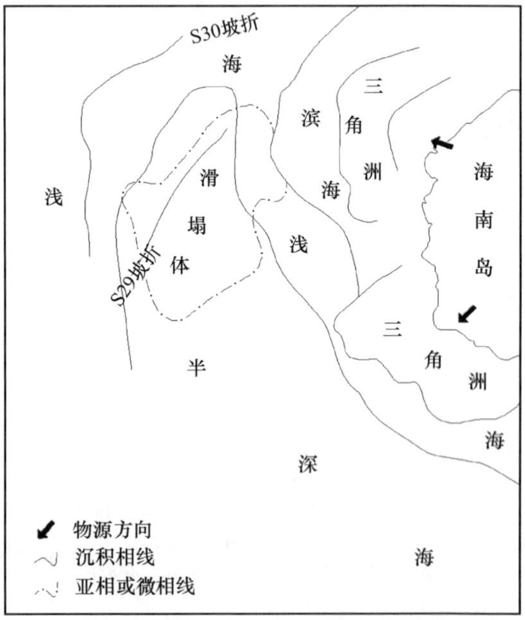

图 5-14　莺歌海盆地上新世早期沉积岩相古地理图

Fig. 5-14　Map of sedimentary lithofacies paleogeography of the early Pliocene, Yinggehai Basin

进入上新世早期，盆地内海平面快速上升，各物源的沉积物供应量相对不足，使全盆地的相对水深变大，普遍发育了“S”形前积的泥质陆架/陆坡体系，具有明显沉积坡折（图 5-14）。北部红河物源供应量大，但以细粒为主；海南岛物源反之，但南段的供应量较大、粗碎屑多，形成较大面积的扇三角洲。因盆底地形仍为北高南低，故盆地西北部以前积为主，陆坡推进较快，而东部则以加积为主，陆坡推进较慢。陆坡区发育滑塌体，盆底半深海中除泥岩外亦有水道体系及盆底扇发育。中期，随着陆架/陆坡前积进展，水深逐渐变浅。浅海、滨海范围扩大，半深海减小（图 5-15）。盆地北部的积坡折继续向南推进，晚些时候盆地东部也开始快速前积，其前缘可见多期砂质滑塌体。因盆底北高南低，东南部陆架坡折的水深及陆坡的高度都明显大于西北部。滨海相在盆地周边分布，但在盆地西北及东南部范围较大。

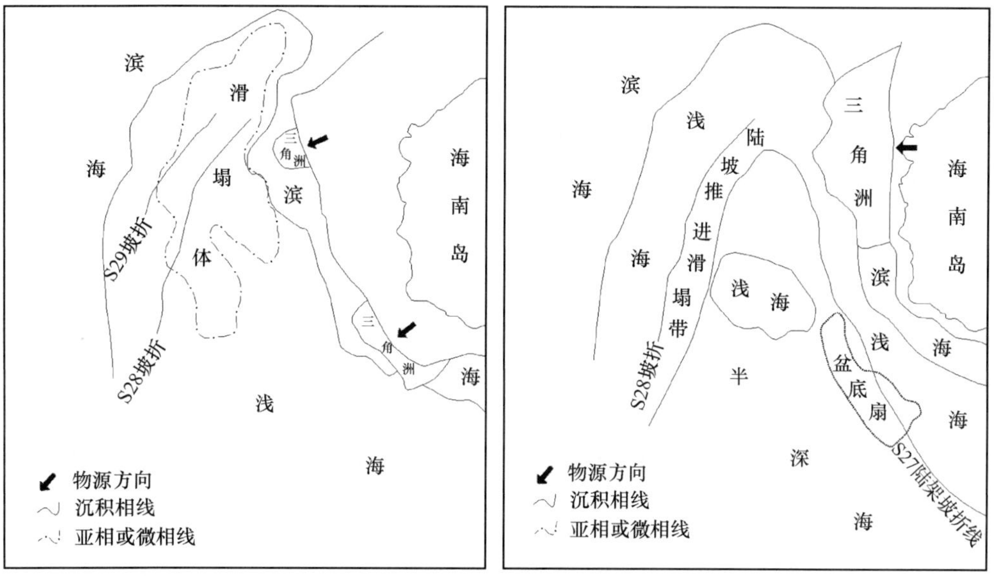

图 5-15　莺歌海盆地上新世中期沉积岩相古地理图

Fig. 5-15　Map of sedimentary lithofacies paleogeography of the middle Pliocene，Yinggehai Basin

海南岛近源三角洲注入点逐渐北移至盆地东北部，规模有所减小。盆底有各类水道和盆底扇。晚期后，盆地东部的陆架坡折开始向盆地中部快速推进，与红河及越南西部来的沉积物汇合后，又一起向南推进（图 5-16）。盆地北部为大片陆架泥岩，南部乐东区多为盆底水深环境，除深水泥岩外，还发育有强制海退滨岸砂、盆底扇、斜坡扇及各种水道体系。盆地周缘三角洲范围缩小，盆底水道沿盆地长轴发育。

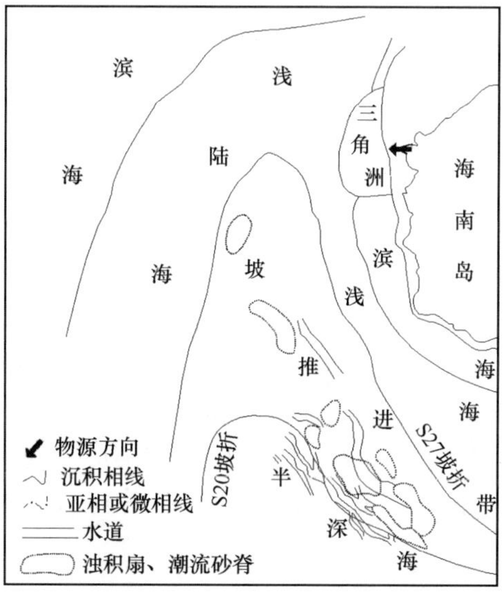

图 5-16　莺歌海盆地上新世晚期沉积岩相古地理图

Fig. 5-16　Map of sedimentary lithofacies paleogeography of the late Pliocene，Yinggehai Basin

第四节　西沙地层分区

据《南海地质》（刘昭蜀等，2002）资料，古近纪在西沙地层分区目前未发现，新近系仅见于永兴岛西永 1 井中，由中新统西沙组（$N_1x\hat{s}$）、宣德组（N_1xd）和上新统永兴组（N_2yx）组成。

一、剖面描述

西沙群岛永兴岛西永 1 井由原石油工业部南海石油勘探筹备处于 1972~1974 年完成，为上新统永兴组、中新统宣德组、西沙组的代表性剖面。

上覆地层　第四系白色造礁灰岩

——————　整　　合　——————

上新统永兴组　　总厚度 150.00m

8. 灰白色造礁（骨架）灰岩与造礁生物碎屑灰岩有少量底栖类有孔虫及介形类化石，底栖类有孔虫有 *Pararotalia calcar*，*Calcarinaspengleri*，*C. hispida*，*Triloculina* spp.，*Quinqueloculina* spp. 等，还有个别浮游有孔虫，如 *Orbulina universa* 等。此外，还见有少许介形类化石，如 *Baidra* spp.，*Macrocypris* sp.，*Aurila* sp.，*Loxoconcha* sp. 等　　50.00m
7. 灰白色含白垩质造礁生物碎屑岩与造礁（骨架）灰岩含有孔虫，底栖类有 *Amphistegina*，*adiata*，*Robulus kimtnensis*，*Calcarina spengleri* 等，浮游类为 *Pulleniatina primails*，*Sphaeroidinella dehiscens* 等。此外，尚见少许介形类，*Bairdra* spp.，*Cytheropteron* sp.，*Xestoleberis* sp. 等　　100.00m

～～～～～　不　　整　　合　～～～～～

中中新统宣德组　　总厚度 460.00

6. 白色超微体软白垩层，含少量浮游类有孔虫，如 *Orbulina universa*，*Globoquadrina altispira* 等　　50.00m
5. 白色有孔虫软白垩层，含有大量底栖类有孔虫化石，底栖类有 *Cycloclypeus* spp.，*Amphistegina madagascariensis*，*Heterostegina depressa*，*Pyrgo* sp.，等。浮游类有 *Orbulina universa*，*Globigerinoides trilobus*，*G. obliquus*，*G. immaturus* 等　　20.00m
4. 造礁碳酸盐岩层上部为白色六射珊瑚礁钙质白云岩与绿藻礁钙质白云岩，下部为浅黄白色钙质白云岩，具生物残余结构。有孔虫化石稀少，属种单调。有 *Cycloclypeus* sp.，*Globoquadrina altispira*，*Globigerinoides trilobus*，*Amphistegina radiata*，*Pyrgo* sp.，*Quinqueloculina* sp.，等　　390.00m

——————　整　　合　——————

下中新统西沙组　　总厚度 591.00

3. 白色中垩虫（屑）灰岩，含大量有孔虫化石，其中，中垩虫（*Miogypsina*）占绝对优势，还有少量 *Lepidocyclina* sp. 及 *Pyrgo* sp，*Amphistegina* sp. 等底栖分子　　190.00m
2. 浅黄白色造礁生物屑灰岩，下部地层可能有含钙质白云岩夹层，有孔虫化石为 *Miogypsina*－*Lepidocyclina* 组合，尚有 *Pyrgo* sp.，*Borelis* cf. *phillippinensis* 等　　255.00m
1. 白色钙藻白云岩层，推测本层下部有含白云质生物礁灰岩夹层 孢粉化石丰富，以热带花粉占优势，如 *Conocephalus* sp.，*Celtis* sp.，*Ficus* sp.，*Palmae*，*Taiwania* sp.，Rhizophoraceae 等　　146.00m

－－－－－假　整　合－－－－－

下伏地层　中元古界灰色、灰绿色花岗片麻岩、黑云二长片麻岩

二、岩石地层特征

（一）西沙组（$N_1x\hat{s}$）

西沙组系中国地质科学院地质所三室（1976）创建的岩石地层单位，创名地为西沙群岛西永一井660~1251m井段，时代为中新世。1984年中国地质科学院王崇友在《中国的第三系》南海诸岛地层分区部分将西沙组上界扩大至孔深300m处（于1979年初成文）。1979年王崇友、何希贤等在《石油实验地质》上发表论文，重新将西沙组限定在西永一井660~1251m井段，即综合柱状图1~3层，但地质时代划为早中新世，本书采用这一划分方案。

西沙组与上覆宣德组（N_1xd）呈整合接触，与陆地风化壳呈假整合接触。上部岩性为中爫虫灰岩层，岩石较坚硬，有小孔洞，局部可见软体动物介壳。中爫虫 *Miogysina* 占绝对优势，成为岩石的主要成分。中爫虫是已灭绝的较大型底栖有孔虫，生活环境推测为正常的浅海陆棚带，代表海侵的高潮。中部为造礁碳酸盐岩层，岩石标本见软体动物介壳印痕。镜下为砂级生物碎屑结构，碎屑以有孔虫（双玦虫、鳞环虫、中爫虫等）、珊瑚藻屑、珊瑚屑为主，尚见石枝藻屑、蟹手藻（*Amphiroa*）？和软体动物壳屑等，其成分为方解石，并见重结晶现象。灰泥发育，似为多种类型的骨屑粉，成分为方解石，其形成环境似为低能的礁缘潮下带。下部为钙藻白云岩层，岩石坚硬、性脆，白云岩化孔洞发育。岩样中含热带植物花粉及陆源矿物（均为沉积变质岩成分，如角闪石、辉石、电气石、石榴子石、长石石英等），岩石形成环境为滨海潮上带半干燥潟湖。

（二）宣德组（N_1xd）

宣德组为中国科学院创建的岩石地层单位，王崇友等（1979）介绍，创名地是西沙群岛西永一井300~660m井段，即综合柱状图中的4~6层，时代为中中新世。本文采用此划分方案。

宣德组上部为超微软体白垩层，中部为有孔虫软白垩层，下部为礁碳酸盐岩。与下伏西沙组（$N_1x\hat{s}$）呈整合接触，与上覆永兴组（N_2yx）呈假整合接触。

超微软体白垩层，由松软而洁白粉尘状方解石颗粒组成，黏附性极强，能书写如粉笔，质轻、孔隙度大；有孔虫软白垩，呈砂粒状，具粉尘颗粒，易染手，胶结疏松，含有大量有孔虫化石及棒状钙质超微体化石及一些软体动物介壳碎屑、珊瑚藻屑、苔藓虫屑、棘皮动物碎片等，岩石成分属含钙质白云岩；两种软白垩分别代表浅海陆棚与深海或远洋环境。礁碳酸盐岩上部为白色六射珊瑚礁钙质白云岩与绿藻礁钙质白云岩，后者岩石较松软，含大量仙掌藻（*Halimeda*）屑、少量海胆刺、软体动物介壳及石枝藻屑等。下部为浅黄白色钙质白云岩，具生物残余结构，其形成环境为滨海礁相带。

（三）永兴组（N_2yx）

永兴组系1979年中国地质科学院地质所三室创建的岩石地层单位，创名地为西沙群岛西永一井150~660m井段，时代为上新世。1984年中国地质科学院王崇友在《中国的第三系》南海诸岛地层分区部分将永兴组范围厘定在西永一井300~150m井段（于1979年初成文）。1979年王崇友等在《石油实验地质》刊物发表论文，永兴组范围仍为西永一井150~300m井段，本书采用这一划分方案。

此外，2002年刘昭蜀等在《南海地质》一书中将西沙群岛第四系下更新统定名为永兴组，与地科院创建的上新统永兴组重名，依据岩石地层单位创名优先权的原则，建议第四系下更新统永兴组停止使用。

永兴组与上覆第四系呈整合接触，与下伏宣德组（N_1xd）呈假整合接触。

上部岩性为造礁灰岩层，主要为白色六射珊瑚礁灰岩，含少量钙藻、棘皮动物、软体动物等碎屑，岩石成分属灰岩，为滨海礁相带；下部造礁生物碎屑灰岩层，上部以白色六射珊瑚礁灰岩为主，下部为白色含白垩质造礁生物碎屑灰岩，含丰富的有孔虫化石及少量超微体化石、软体动物介壳、海胆刺和珊瑚等，有孔虫以底栖类为主，属滨海—浅海相沉积。

三、古生物群及时代划分与对比

（一）西沙组（$N_1x\hat{s}$）

西沙组产中垩虫-鳞环虫组合为代表的有孔虫化石群，为确定地层时代提供了有力的依据。其上部 *Miogysina irregularisra-eeorientalisDouville*，*M. polyznor* 等在日本、菲律宾、印度尼西亚、爪哇、苏门答腊，以及我国台湾、澎湖等地均产在早、中中新世的阿魁阶（Aqtritanian）和布尔迪加阶（Buldigalian），可与澎湖白沙岛钻井相应层位（中垩虫灰岩）进行对比。下部 *Borelis* cf. *phillippinensis* 和 *Lepidocyclina* sp. 共生，汉泽瓦（Hanzawa，1949）曾把它作为菲律宾吕宋岛下中新统的主要分子。这个有孔虫化石组合显然属早中新世。其所含的大量孢粉化石中被子植物花粉占绝对优势，属种繁多，古近纪的古老松柏花粉罕见，杉科花粉含量很少，还出现了新近纪繁盛的禾本科花粉，而且所有孢粉均呈压扁状态，因此，含上述孢粉化石的层位应属新近纪。综上所述，其时代应属早中新世。

（二）宣德组（N_1xd）

根据该组上部层段的含浮游有孔虫 *Orbulina suturazis*，*Globoquadrina altispira*，*Globigerinoides trilobus*，*G. obliquus* 等，其代表了中中新世的组合面貌。其中 *Orbulina suturalis* Bronnimann 从中中新统底部开始出现，如在欧洲产在 Langhian 阶中部、北美产于 Luisian 阶，我国则见于雷琼地区的角尾组及台湾中部的瑞芳群。中、下部化石稀少，其下伏地层（中垩虫灰岩）属早中新世晚期，依照地层顺序，中、下部层段应属中中新世早期。因此，其时代应为中中新世。

（三）永兴组（N_2yx）

该组有孔虫化石除从下部延续上来的浮游类以外，首次出现了 *Sphaeroidinella dehiseens*，*Sphaeroidinella psissubdehiseens*，*Pulleniatina primalis*，*Globorotalia tumida* 等，众所周知 *Sphaeroidinella dehiseens* 在世界各地都从上新统底部开始出现，一般都以它的出现作为上新统的底界，而 *Sphaeroidinella psissubdehiseens*，*Pulleniatina primalis*，*Globorotalia plessiotumida* 从中新世晚期开始出现，消亡于早上新世。因此，含上述化石的层位其时代应为早上新世，永兴组时代应为早上新世。

四、岩相古地理特征

（一）沉积环境

礁相碳酸盐岩夹白垩等是典型的热带海相堆积物，其中所含有孔虫化石群及少量介形类化石属印度洋-西太平洋热带属种，底部所含的孢粉组合也以热带植被为主，有些还具有热带雨林特征，表明当时气候湿热，与现代相近。

（二）岩相古地理讨论

根据南海北部地震波反射层的分析，西沙地区中新世以前与华南古陆、海南古陆连成一片，长期的风化剥蚀，形成了近 30m 厚的风化壳后，中新世早期才开始下沉，并为西沙海槽所隔而与华南古陆、海南古陆分离，西沙群岛的雏形此时才开始出现。

西沙地区与海南、华南二古陆分离后，礁相开始发育，出现了生命力极强的石枝藻等生物，它们逐渐适应了潮波的冲击，形成藻礁坪，而藻礁坪中白云岩的发育，则记录了潮上带半干燥的潟湖环境，属滨海潮上带潟湖白云岩相。其后，海水逐渐加深，底栖大型中垩虫、鳞环虫的出现和白云岩不发育，表明当时的永兴岛已脱离藻坪环境，属滨海潮上带生物屑灰岩相。当海水进一步加深，中垩虫等底栖有孔虫就大量繁殖，形成了中垩虫（屑）灰岩，属浅海陆棚中垩虫灰岩相。早中新世从滨海潮上带潟湖白云岩相发展成为广海陆棚中奎虫（屑）灰岩相，是一次水体深化的海侵过程。

中中新世开始，西沙已再次升起，礁相进入第二个发展阶段，在礁坪台地环境中，先后有含钙藻白云岩、仙掌藻屑钙质白云岩、六射珊瑚骨架礁坪的形成，显示水体有深浅的变化，但总的环境仍属礁坪台地钙质白云岩相。晚期，西沙地区大规模下沉，永兴岛进人广海环境，形成浅海有孔虫软白垩的沉积，其中超微体化石与浮游有孔虫的出现，都是水体较深的标志，苔藓虫、棘皮动物也常见于广海边缘。因此，此时的环境属浅海陆棚有孔虫软白垩相。当海水继续加深，堆积了超微体软白垩，这种几乎全由钙质超微体化石组成的软白垩，其生成环境相当于现代深海或远洋颗石藻软泥，属深海超微体软白垩相。中中新世从礁坪台地钙质白云岩相到深海超微体软白垩相的环境变迁，显示为第二次海侵过程。

此时永兴岛已转为浅海陆棚带，形成了一套含白垩质的生物屑灰岩，其中深海相的指相生物明显减少，出现一些钙藻屑、海胆刺和少数介形类及大量底栖有孔虫，且生物化石的碎屑也大小混杂，这些都是海水变浅的标志。随后，六射珊瑚骨架灰岩的发育表明永兴礁坪再次出现。可见，上新世古地理发展阶段是由海退开始，后又经历了小的海侵波动，因此形成二次海进海退旋回，显示了西沙隆起带的升降运动。

第五节　礼乐滩盆地地层分区

根据《南海地质》（刘昭蜀等，2002）资料，该区古近系由古新统宝滩组（E_1b）、始新统阳明组（E_2y）、始新统—渐新统忠孝组（$E_{2-3}z$）组成；新近系由中新统南湾组（N_1n）、上新统永暑组（$N_2y\hat{s}$）组成。各岩石地层单位岩石组合特征较简单。

中国科学院南海海洋研究所周效中等（1991）按不同时代地层叙述岩石组合特征，并与岩石地层单位一致，因此，本文采用此资料叙述。

1）宝滩组见于礼乐滩东坡礁北缘的桑帕吉塔1井约3053~3353m井段，时代为古新世。该组为断陷期充填型沉积，底部以灰岩与下伏地层白垩系含煤粉砂岩及页岩呈不整合接触，顶部以含砾砂岩与上覆地层阳明组页岩呈假整合接触。宝滩组以三角洲相含砾砂岩为主，底部为浅海陆架灰岩，宝滩组厚约300m。

2）阳明组见于礼乐滩东坡礁北缘的桑帕吉塔1井约2630~3053m井段，时代为早—中始新世。该组为拗陷期超覆披盖型沉积，与上覆地层忠孝组粉砂岩呈不整合接触。该组岩性以页岩为主，少数粉细砂岩，斜层理发育，厚约420m，为外浅海—半深海页岩相。

3）忠孝组见于礼乐滩东坡礁北缘的桑帕吉塔1井约1890~2630m井段，时代为晚始新世—渐新世。与上覆地层生物礁灰岩呈假整合接触，该组岩性主要为粉砂岩、泥岩夹砂岩，水平层理发育，厚约470m，属于滨海相沉积。

4）新近系中新统南湾组和上新统永暑组的创名过程不详，创名地址应为永暑礁南永2井268~369.83m井段及369.83~413.69m井段，前者为永暑组，后者为南湾组。在桑帕吉塔1井忠孝组之上的中新世—第四纪均为生物礁灰岩沉积，称为礼乐组。南湾组和永暑组均为滨海相碳酸盐岩；永暑组总厚度101.83m，产古生物 *Sphaeroidineliopsis*，与上覆更新统南沙组及下伏中新统南湾组均呈假整合接触；南湾组厚度>43m，古生物有：*Nephrolepidina*。

第六节　曾母盆地地层分区

该区古近纪—新近纪地层创名地点及过程不详，主要由古新统—始新统拉让群（$E_{1-2}L$）、渐新统—下中新统曾母组（E_3—N_1z）、中中新统海宁组（N_1h）、上中新统南康组（N_1nk）和上新统—第四系北康组（N_2—Qd）组成。

1）拉让群为浅—深海相沉积，岩性以千枚岩、板岩为主，地层具强烈褶皱和轻微变质，变质程度由南向北减弱，为典型的复理石建造，与菲律宾群岛陆区地槽沉积基本一致，整合于上白垩统丹脑组蛇绿岩之上，与上覆曾母组呈不整合接触。

2）曾母组主要为河流、港湾相沉积，岩性为砂岩夹泥岩和煤层，但在盆地的东北部以海相页岩与灰岩互层为特征，厚度约百米到数千米，与上覆海宁组呈整合接触。

3）海宁组和南康组均为滨海相碳酸盐岩沉积，二者为整合接触。

4）南康组与上覆北康组呈不整合接触。

第六章 第 四 系

海南省第四系属于华南地层大区中南-东南地层区、中央海盆地层大区和菲律宾-南沙地层大区南沙-曾母地层区。省内三级地层区划分为秀英地层分区、莺歌海盆地地层分区、西沙群岛地层分区、南沙群岛地层分区和曾母盆地地层分区。其中，中央海盆地层大区和莺歌海盆地地层分区资料少，其沉积物未建组，暂不叙述。

第一节 秀英地层分区

秀英地层分区的第四系特别发育，是华南地区第四系发育较为典型的地区之一，其分布受新生代晚期形成的新构造格局的控制，形成的沉积物类型多样。主要分布于琼北、琼南和琼西等断陷区内及沿海地带，地层层序发育较全，包括更新世—全新世各期沉积；断隆区内，地层层序发育不全，缺失早更新世沉积。第四系赋存有具重要工业价值的钛铁矿、锆英石、独居石砂矿，以及大型玻璃砂矿床，其储量名列全国前茅，具较为重要的地位。

秀英地层分区内第四系的调查研究始于20世纪50年代，不同时期不同学者对其提出不同的划分方案，本书根据岩性特征、层序关系、年龄数据等，并采用陈哲培等（1997）的划分方案对区内第四系进行划分，自下而上分别为：下更新统秀英组（Qp^1x）、中更新统北海组（Qp^2b）和多文组（Qp^2d）、上更新统八所组（Qp^3bs）和道堂组（Qp^3d），以及全新统石山组（$Qh^1\hat{s}$）、万宁组（Qh^1w）、琼山组（Qh^2q）和烟墩组（Qh^3y）。

一、剖面描述

（一）秀英组（Qp^1x）

1. 秀英医院NW29°500m处冲沟剖面

该剖面位于海口市秀英区医院NW29°500m处冲沟，起点坐标：东经110°15′10″，北纬20°1′40″。系海南地质大队1977年测制，为秀英组代表性剖面。

上覆地层 中更新统北海组黄红色含砾亚砂土夹砾石透镜体

－－－－－假 整 合－－－－－

下更新统秀英组	总厚度11.61m
7. 浅黄、褐黄色含砾中粗砂	2.97m
6. 铁质砾岩	0.27m
5. 灰色页状黏土，间夹褐黄色粉细砂	3.24m
4. 粉红色中细砂夹砾石薄层	2.16m
3. 杂色黏土	2.97m

－－－－－假 整 合－－－－－

下伏地层 上新统海口组杂色页状黏土

2. 黄流市赖园村L346孔剖面

该剖面位于乐东县黄流市赖园村南西南，平距约1300m处（据1∶5万《千家、黄流》幅区调报告）。自上而下为（图6-1）：

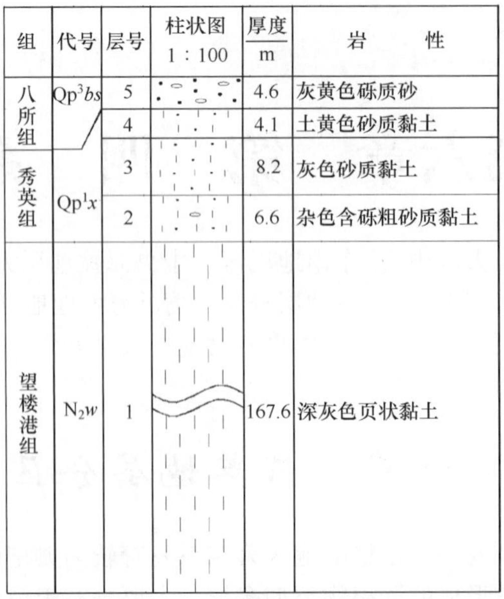

图 6-1 黄流市赖园村 L346 孔 4. 6~167. 6m 井段剖面图

Fig. 6-1 Section of 46~1676m interval, L346 Borehole, Laiyuan Village, Huangliu City

上覆地层 上更新统八所组灰黄色砾质砂 厚度>4. 6m

－－－－－假 整 合－－－－－

下更新统秀英组 总厚度 18. 9m

4. 土黄色砂质黏土 4. 1m

3. 灰色砂质黏土 8. 2m

2. 杂色含砾粗砂质黏土 6. 6m

－－－－－假 整 合－－－－－

下伏地层 上新统望楼港组深灰色页状黏土，中部含少量贝壳 >167. 6m

（二）北海组（Qp^2b）

1. 海口市秀英区秀英砖瓦厂采场剖面

该剖面位于海口市秀英区秀英砖瓦厂，起点坐标：东经 110°16′01″，北纬 19°58′01″。系海南地质大队 1987 年测制，为北海组参考剖面。

中更新统北海组 总厚度>5. 3m

10. 褐红色亚砂土层，含铁质结核 >3. 3m

9. 灰黄色—灰褐色砾石层，含玻璃陨石 0. 9m

8. 褐红色砂砾层，含铁质结核 1. 1m

－－－－－假 整 合－－－－－

下伏地层 下更新统秀英组褐黄色褐铁矿层

2. 王五镇南桥中更新统剖面

该剖面位于儋州市王五镇南桥 274°方向 1100m 处的采坑内，海拔 48m。系海南地质综合勘察院第一勘察处 1996 年测制。自上而下为（图 6-2）：

中更新统北海组 总厚度>10. 15m

5. 棕红色含砾中粗砂夹黏土质粉砂层，产孢粉化石：*Castanopsis* sp.，*Lygodium* sp.，*Pteris* sp.，*Microlepia* sp.，*Cyathea* sp.，*Compositea*，*Concentroystis* sp. 1. 1m

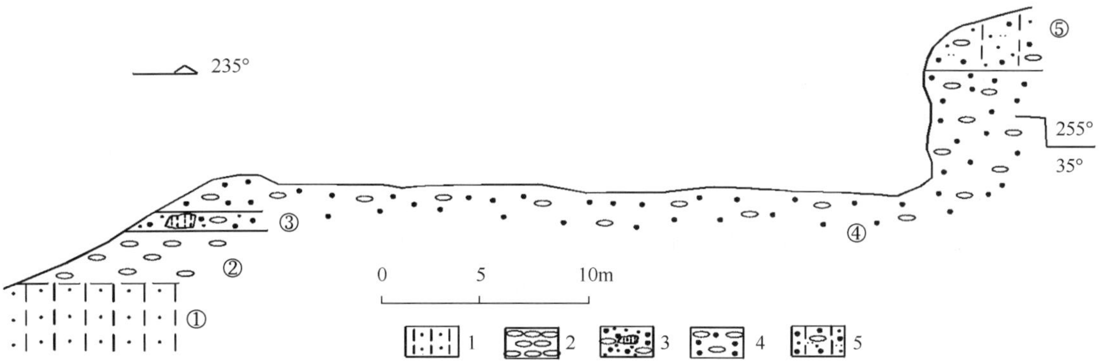

图 6-2　王五镇南桥中更新统北海组剖面图

Fig. 6-2　Section of the middle Pleistocene Beihai Formation, Nanqiao, Wangwu Town

（据 1∶25 万《东方》幅区调报告，2007）

1—灰白色浅黄色亚黏土；2—浅黄色、浅红色、紫红色砾石层；3—浅灰白色、浅红黄色含砾中粗砂夹透镜状灰色黏土层；4—杂色砂砾层；5—棕红色含砾中粗砂夹黏土质粉砂层

4. 杂色砂砾层，具斜层理，上部夹透镜体状灰色黏土　　7m

3. 浅灰白色、浅红黄色含砾中粗砂夹透镜体状灰白色黏土　　1.75m

－－－－－假　整　合－－－－－

下伏地层　下更新统秀英组浅灰白色、浅黄色亚黏土层　　>1.0m

（三）多文组（Qp^2d）

1. 临高县多文镇剖面

剖面起点位于临高县多文镇南兴村北东 50°方向 1300m 处，坐标：东经 109°46′56″，北纬 19°44′18″。为多文组下段代表性剖面，系海南省地质调查院 2003 年实测。共分 4 层，自上而下为（图 6-3）：

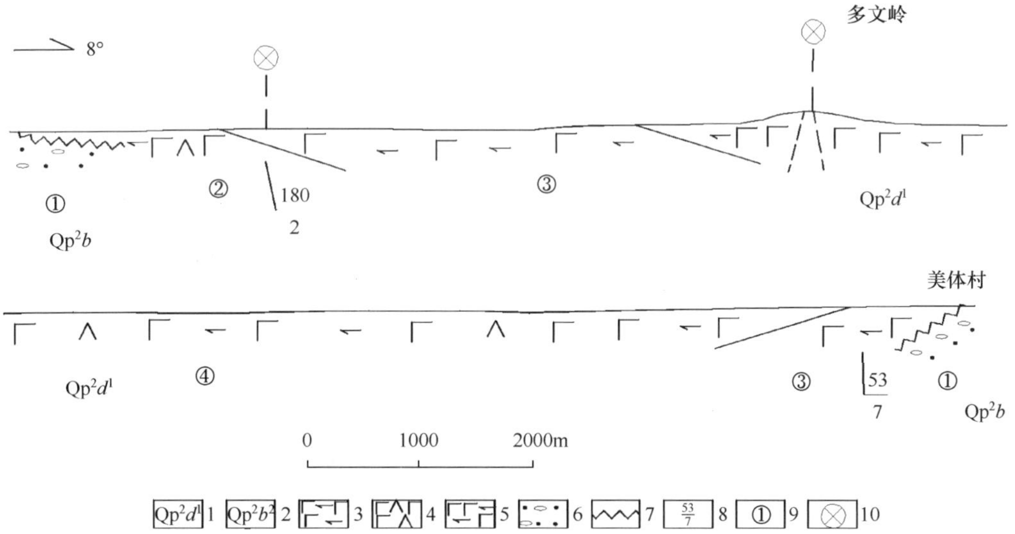

图 6-3　中更新统多文组下段（Qp^2d^1）剖面图

Fig. 6-3　Section of the Lower Member（Qp^2d^1）of the middle Pleistocene Duowen Formation

1—中更新统多文组上段；2—中更新统北海组；3—辉石玄武岩；4—辉石橄榄玄武岩；5—橄榄辉石玄武岩；6—砂砾土；7—火山喷发不整合；8—产状；9—室内分层；10—采样位置

中更新统多文组下段（Qp^2d^1）　　121.7～396.5m

4. 橄榄辉石玄武岩　　28.7～250.3m

3. 粗玄岩（辉石玄武岩）　　54.9～108.1m

2. 辉石橄榄玄武岩 38. 1m

~~~~~~~~~~ 喷发不整合 ~~~~~~~~~~

下伏地层 中更新统北海组

1. 灰白色不等厚砂砾层，含玻璃陨石

2. 临高县东英镇剖面

剖面起点位于临高县波莲镇南朝村南侧，坐标：东经 109°38′27″，北纬 19°52′47″，为多文组上段代表性剖面，系海南省地质调查院 2003 年实测。自上而下为（图 6-4）：

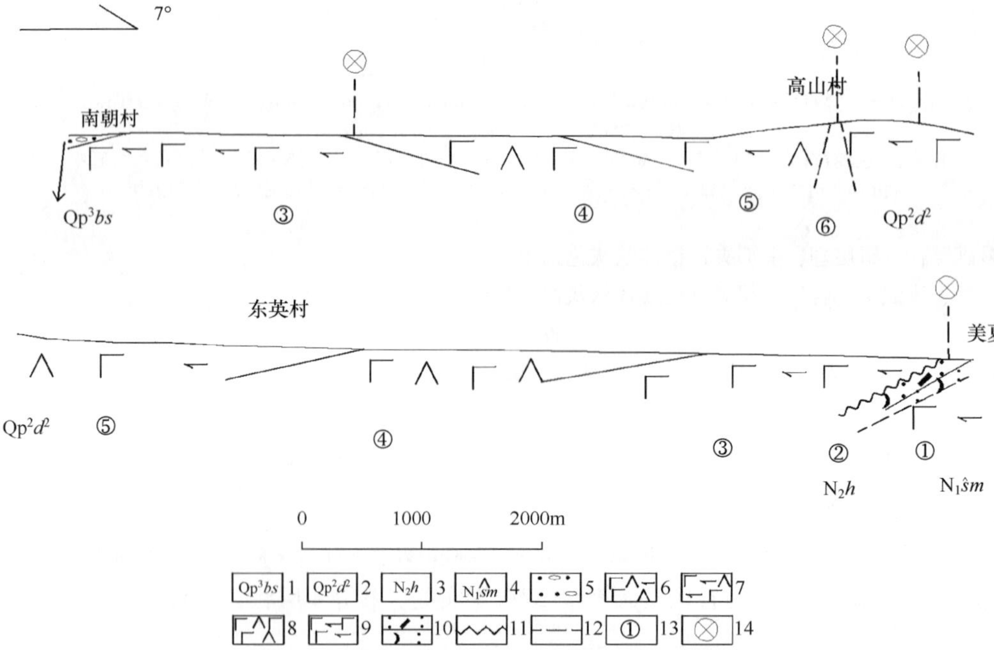

图 6-4 中更新统多文组上段（Qp^2d^2）剖面图

Fig. 6-4 Section of the upper Member（Qp^2d^2）of the middle Pleistocene Duowen Formation

1—八所组；2—多文组上段；3—海口组；4—石马村组；5—含砾砂土；6—辉石橄榄玄武岩；7—橄榄辉石玄武岩；8—橄榄玄武岩；9—辉石玄武岩；10—沉凝灰岩；11—火山喷发不整合；12—假整合；13—室内分层号；14—采样位置

上覆地层 上更新统八所组（Qp^3bs）

6. 含砾泥质粉细砂：浅灰白色，粒径一般小于 5mm，其中粒径为 2～5mm 的砾石含量约 10%左右

------ 假 整 合 ------

中更新统多文组上段（Qp^2d^2） 160. 9～278. 9m

6. 熔渣状辉石橄榄玄武岩 16. 5～27. 5m

5. 橄榄辉石玄武岩 49. 9～118. 6m

4. 橄榄玄武岩 44. 7～67. 2m

3. 辉石玄武岩 49. 8～65. 6m

~~~~~~~~~~ 喷发不整合 ~~~~~~~~~~

下伏地层 上新统海口组（N_2h）

2. 沉凝灰岩

3. 峨蔓镇笔架岭剖面

该剖面位于儋州市峨蔓镇松林岭 192. 8 高地南西约 4700m 处，起点坐标：东经 109°23′37″，北纬 19°45′53″；系海南省地质调查院 2006 年实测。主要是控制多文组与石山组，石山组以橄榄玄武岩与道堂组沉凝灰岩呈喷发不整合接触，但道堂组与多文组的关系不清。自上而下为（图 6-5）：

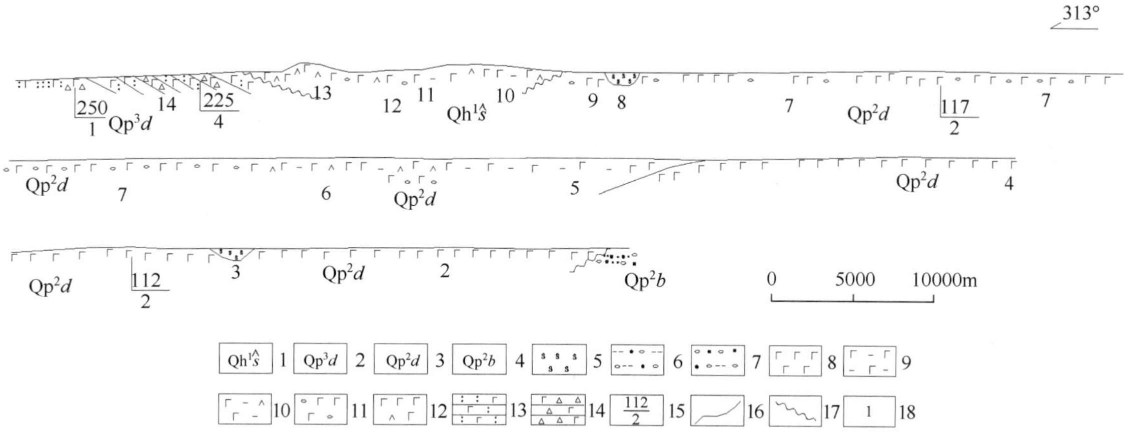

图 6-5 峨蔓镇笔架岭中更新统多文组、全新统石山组剖面图

Fig. 6-5 Section of the middle Pleistocene Wenduo Formation and Holocene Shishan Formation, Bijialing, Eman Town

1—石山组；2—道堂组；3—多文组；4—北海组；5—淤泥；6—含砾砂质黏土；7—砂砾土；8—粗玄岩；9—辉石玄武岩；10—橄榄辉石玄武岩；11—气孔状玄武岩；12—橄榄石玄武岩；13—沉凝灰岩；14—沉火山角砾岩；15—产状（倾向/倾角）；16—岩层界线；17—喷发不整合界线；18—分层号

全新统石山组	总厚度 69. 39m
13. 橄榄石玄武岩	41. 05m
12. 橄榄辉石玄武岩	8. 39m
11. 气孔状玄武岩	8. 72m
10. 橄榄辉石玄武岩	11. 23m

~~~~~~~~ 喷发不整合 ~~~~~~~~

中新统多文组	总厚度 336. 48m
9. 气孔状玄武岩	0. 53m
8. 淤泥	5. 46m
7. 气孔状玄武岩	142. 05m
6. 橄榄玄武岩	18. 16m
5. 橄榄辉石玄武岩	57. 37m
4. 粗玄岩	92. 22m
3. 掩盖	1. 71m
2. 粗玄岩	18. 98m

~~~~~~~~ 喷发不整合 ~~~~~~~~

下伏地层 中更新统北海组砂砾土	总厚度>1m

（四）八所组（Qp^3bs）

1. 八所港北藜村公路旁剖面

该剖面位于东方市八所港北藜村公路旁，起点坐标：东经 108°41′01″，北纬 19°08′37″。系薛万俊等 1981 年测制，为八所组代表性剖面。自上而下为（图 6-6）：

未见顶

上更新统八所组	总厚度>5. 6m
3. 棕黄色、棕红色含粉砂中砂，含少量磨圆度中等的脉石英、花岗岩、伟晶岩质细砾	>3. 10m
2. 黄色含细中砂中粗砂，含铁质结核	0. 9m
1. 灰白色及黄色含砾石和粉砂中粗砂，砾石成分复杂，有脉石英、花岗岩、伟晶岩、砂岩和硅化灰岩，磨圆度中等	>1. 6m

未见底

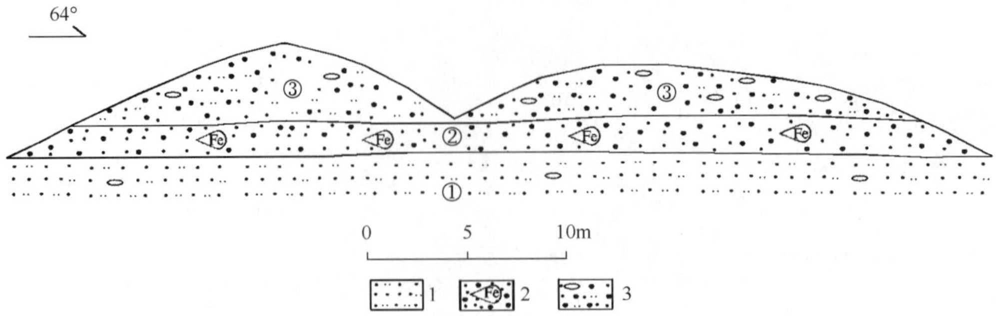

图 6-6　东方市八所港北藜村公路旁晚更新统八所组剖面

Fig. 6-6　Section of the late Pleistocene Basuo Formation, next to the road of Licun, Port of Basuo, Dongfang City

1—含粉砂中砂；2—含细砂中粗砂；3—含砾石粉砂中粗砂

2. 文昌市龙楼镇赤筠村矿区 ZK812 孔上更新统剖面

该剖面位于文昌市龙楼镇赤筠村北东 800m，起点坐标：东经 110°55′13″，北纬 19°42′36″；系海南省地质调查院 2001 年开展 1∶25 万《琼海县》幅区调时测制。自上而下为：

上更新统八所组	总厚度>21.0m
5. 灰白色中细粒石英砂	8.94m
4. 黄色含黏土中细粒石英砂	1.98m
3. 褐黄色—黄色中细粒石英砂	4.15m
2. 褐灰色—褐黑色含炭质中细粒石英砂	5.33m
1. 浅灰色中细粒石英砂	>0.6m
－－－－－ 假　整　合 －－－－－	
下伏地层　下更新统秀英组深青灰色砂质黏土	>1.0m

（五）道堂组（Qp^3*d*）

1. 海口市秀英区道堂–石山剖面

剖面起点位于海口市秀英区博昌村北面 64.5m 高地，坐标：东经 110°10′40″，北纬 19°56′55″；系海南地质大队 1986~1989 年开展 1∶5 万《长流、白莲市、海口市、灵山市》区调时测制，为道堂组和石山组的代表性剖面。海南省地质调查院 2001~2003 年开展 1∶25 万《琼海县》幅区调时做了补充工作，对道堂组重新进行厘定，把原道堂组一段归为多文组，而原二、三、四段依然归入道堂组，即现道堂组含义与陈哲培等（1997）所指的道堂组含义已不同。本次研究沿用 1∶25 万《琼海县》幅区调修订后的道堂组含义，自上而下为（图 6-7）：

上覆地层　全新统石山组第二段含集块、含火山角砾熔渣状橄榄玄武岩

～～～～～ 喷发不整合 ～～～～～

上更新统

道堂组三段（Qp^3*d*^{3}）　厚度 85m

12. 淡黄色薄层玄武质沉岩屑玻屑凝灰岩，含有孔虫 *Globigerina varianta subbtina*（变异抱球虫），*Globigerina* sp.（抱球虫），*Ammonia takanabensis*（高锅卷转虫），*Ammonia annectens*（同现卷转虫），*Ammonia beccarii*（毕克卷转虫），*Ammonia* sp.（卷转虫），*Globigerina triloculinoides*（三室抱球虫），*Globigerina bulloides*（泡状抱球虫），*Globigerinoides trilobus*（三叶似抱球虫），*Ammonia beccarii* var. *lucida*（毕克卷转虫透明变种），*Sphaeroidina chilostomata*（唇口水球形虫），*Lagene* cf. *striata*（线纹瓶虫）*Discorbinella* aff. *montereyensis*（蒙特里小圆盘虫近亲种）（有孔虫化石由宜昌地质矿产研究所张杰存鉴定）　63m

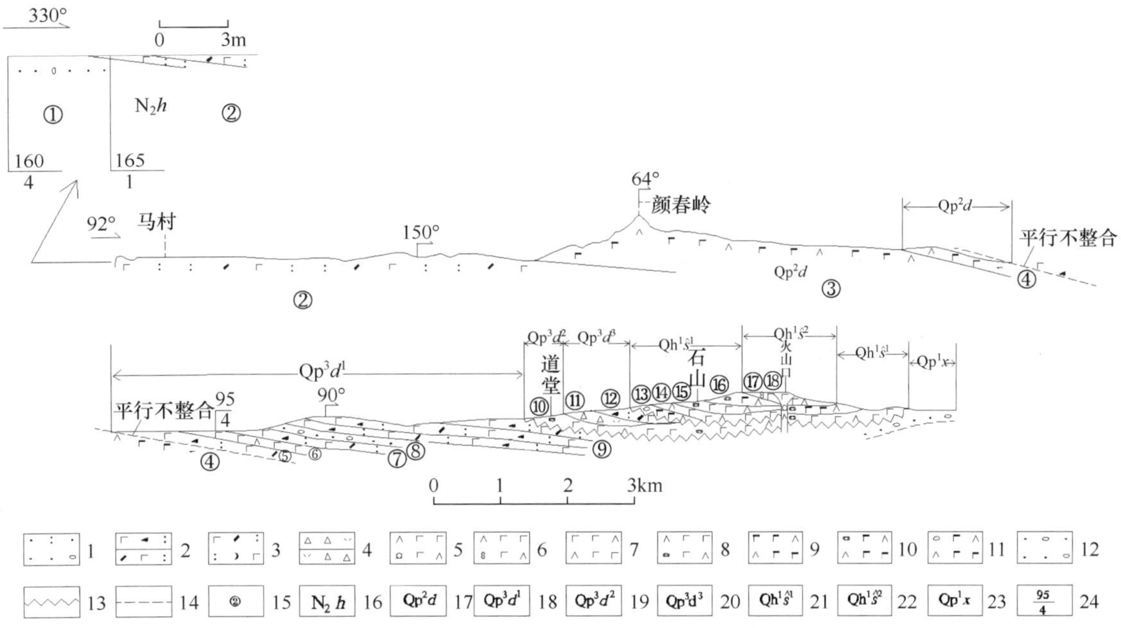

图 6-7　海口市秀英区道堂-石山实测剖面图

Fig. 6-7　Measured section along Daotang-Shishan, Xiuying District, Haikou City

1—凝灰质含砾砂岩；2—玄武质沉岩屑晶屑凝灰岩；3—玄武质沉晶屑玻屑凝灰岩；4—沉火山角砾岩；5—浮岩状橄榄玄武岩；6—熔渣状橄榄玄武岩；7—橄榄玄武岩；8—气孔状橄榄玄武岩；9—橄榄拉斑玄武岩；10—气孔状橄榄拉斑玄武岩；11—含集块、火山角砾熔渣状橄榄玄武岩；12—含砾砂岩；13—喷发不整合界线；14—假整合界线；15—分层编号；16—海口组；17—多文组；18—道堂组一段；19—道堂组二段；20—道堂组三段；21—石山组一段；22—石山组二段；23—秀英组；24—岩层产状

11. 沉火山角砾岩	22m
道堂组二段（Qp^3d^2）	厚度 15m
10. 气孔状橄榄玄武岩	15m

～～～～～～ 喷发不整合 ～～～～～～

道堂组一段（Qp^3d^1）	厚度 129m
9. 玄武质沉岩屑晶屑凝灰岩，中部夹玄武质沉晶屑玻屑凝灰岩、玄武质沉角砾凝灰岩	98m
8. 凝灰质含砾砂岩	8m
7. 灰黄色薄层玄武质沉岩屑晶屑凝灰岩	4m
6. 暗灰色橄榄玄武岩，呈球状风化	4m
5. 灰黄色薄层玄武质沉岩屑晶屑凝灰岩	15m

－－－－－－假　整　合－－－－－－

下伏地层　中更新统 多文组（Qp^2d）橄榄拉斑玄武岩	总厚度>49m

2. 三都镇公堂下村剖面

剖面位于儋州市三都镇莲花山南西 55.7m 高地 203°方向 600m 处；系海南地质综合勘察院第一勘察处 1994~1996 年开展 1：5 万《海头市、排浦市、白马井镇、王五市》幅区调时测制。自上而下为（图 6-8）：

未见顶

上更新统	
道堂组三段	总厚度>4.89m
7. 玻基玄武岩	>2m
6. 黄色、褐黄色玄武质层状火山角砾岩	3.89m

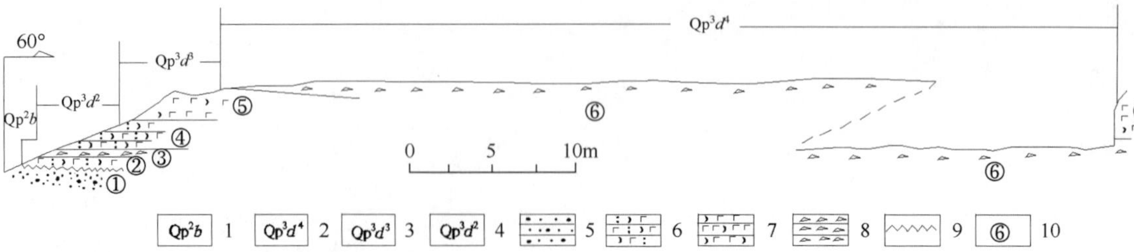

图 6-8 三都镇公堂下村上更新统道堂组剖面

Fig. 6-8 Section of the Upper Pleistocene Daotang Formation, Gongtangxia Village, Sandu Town

（据 1∶25 万《东方幅》区调报告）

1—北海组；2—道堂组四段；3—道堂组三段；4—道堂组二段；5—含砾中细砂；6—层状玻屑凝灰岩；7—玻基玄武岩；8—层状火山角砾岩；9—喷发不整合界线；10—分层号

道堂组二段	总厚度 1.68m
5. 玻基玄武岩	1.68m
道堂组一段	总厚度 1.91m
4. 层状玻屑凝灰岩，上部黄色，下部灰色、灰黄色	1.36m
3. 黄色、褐黄色玄武质层状火山角砾岩	0.23m
2. 浅灰色、灰黄色层状玻屑凝灰岩，含海绿石	0.32m

～～～～～ 喷发不整合 ～～～～～

下伏地层 中更新统北海组、灰黄色、棕黄色含砾中细砂 >0.95m

（六）石山组（$Qh^1\hat{s}$）

石山组仅有海口市秀英区道堂-石山剖面，剖面情况同上述。

全新统石山组

二段（Qh^1s^2）	总厚度>18m
18. 浮岩状橄榄玄武岩	约 8m
17. 熔渣状橄榄玄武岩	约 10m
一段（Qh^1s^1）	总厚度 77m
16. 气孔状橄榄拉斑玄武岩	20m
15. 气孔状橄榄玄武岩	23m
14. 橄榄玄武岩	30m
13. 含集块、含火山角砾熔渣状橄榄玄武岩	4m

～～～～～ 喷发不整合 ～～～～～

下伏地层 上更新统道堂组三段淡黄色薄层玄武质沉岩屑玻屑凝灰岩

（七）万宁组（Qh^1w）

1. 保定 12ZK1 孔剖面

该剖面为代表性剖面，见于万宁保定 12ZK1 孔 7.5～31.5m 井段；系张仲英等 1987 年测制。钻孔位于万宁市北坡乡保定村，坐标：东经 110°27′02″，北纬 18°46′05″（层号按原始资料）。

上覆地层 中全新统琼山组黄灰色中细砂，成分以石英碎屑为主

———— 整　合 ————

下全新统万宁组	总厚度>24.02m
2. 深灰色淤泥质黏土	3.90m
3. 深灰色含粉砂黏土，含贝壳碎片	8.20m

4. 深灰色黏土，局部含粉砂，含少量贝壳碎片和小螺壳　3.82m

5. 深灰色砾质中粗砂　5.20m

6. 深灰色黏土，含贝壳碎片面性　2.90m

未见底

2. 三亚市羊栏南红农场 ZK28 孔全新统剖面

剖面位于羊栏镇 107°约 275m 处水田内，海拔 3.43m；由曾庆銮等 1992 年测制，本文略有改动。剖面中全新统出露较全，其岩性自上而下为：

上覆地层　中全新统琼山组灰白色含砾粗砂层

———— 整　合 ————

下全新统万宁组（Qh^1w）　总厚度 8.88m

2. 浅黄色、浅灰色黏土层，含少量中粗砂，呈硬块状　5.51m

1. 灰白色中粗粒砂层，成分以石英为主，质地松散　3.37m

- - - - - 假　整　合 - - - - -

下伏地层　上新统望楼港组灰色、浅灰色泥岩

3. 海口市美兰区沙头村 7ZK2 孔剖面

剖面位于海口市美兰区东营沙头村，起点坐标：东经 110°24′00″，北纬 20°03′01″；系张仲英等 1987 年测制。自上而下分别为（层号按原始资料）：

上覆地层　中全新统琼山组灰色—深灰色亚黏土，含细砾和砂

———— 整　合 ————

下全新统万宁组　总厚度 7.63m

10. 灰色黏土，顶部为杂色黏土，夹细砂层　2.7m

11. 深灰色亚黏土，含粗砂及细砾　1.25m

12. 上灰白色、下黄色细砾石、砂　3.68m

～～～～ 不　整　合 ～～～～

下伏地层　上新统海口组

（八）琼山组（Qh^2q）

1. 海口市美兰区沙头村 7ZK2 孔全新统剖面

该剖面为琼山组代表性剖面。剖面情况见上述。

上覆地层　上全新统烟墩组浅灰色粗砂

———— 整　合 ————

中全新统琼山组　总厚度 12.02m

9. 灰色—深灰色亚黏土，含细砾和砂　1.61m

8. 深灰色黏土，富含有机质。埋深 15.36m 处的黏土 ^{14}C 年龄为（6750±140）a　1.36m

7. 灰色亚黏土，中部与黏土互层，夹细砂层　5.59m

6. 灰黄色亚黏土　3.16m

———— 整　合 ————

下伏地层　下全新统万宁组灰色黏土

2. 三亚市羊栏南红农场 ZK28 钻孔全新统剖面

剖面情况见上述。

上覆地层 上全新统烟墩组浅灰色含砾粗砂层

——— 整 合 ———

中全新统琼山组（Qh²q） 8.22m

4. 深灰色亚黏土层，主要成分为泥质，含少量石英砂，含贝壳碎屑 5.4m

3. 灰白色含砾粗砂层。砾石成分以石英为主，砾径3~4mm；砂粒成分也以石英为主；砾石和砂粒都呈棱角至次棱角状 2.8m

——— 整 合 ———

下伏地层 下全新统万宁组浅黄色、浅灰色黏土层

3. 三亚市水尾岭中全新统琼山组海滩岩剖面

剖面位于水尾岭北东48°方向约540m处海滩上；曾庆銮等1992年测制。剖面自上而下为（图6-9）：

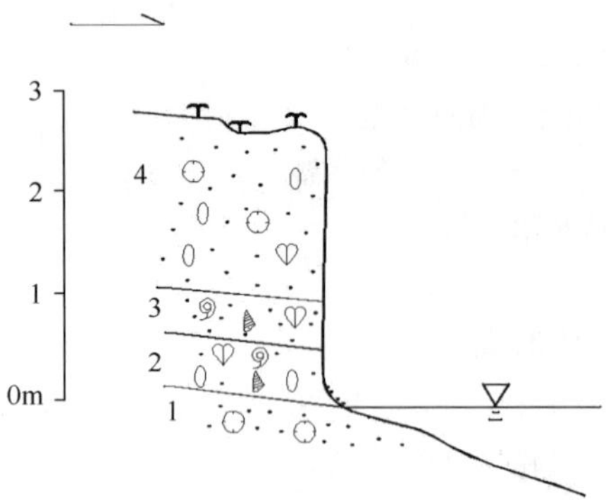

图6-9 水尾岭中全新统琼山组海滩岩剖面图

Fig. 6-9 Section of the Middle Holocene Qiongshan Formation beach rocks, Shuiweiling

1—珊瑚骨屑砂；2—细碎屑角砾海滩岩；3—碎屑砂海滩岩；4—大碎屑砾海滩岩

中全新统琼山组（Qh²q） 总厚度>2.7m

4. 灰白色大碎屑砾海滩岩。碎屑砾石成分主要为珊瑚骨屑、变质岩、花岗岩、石英砂岩等，砾径2~20cm，同时含不少双壳类贝壳，为钙泥质胶结 1.5m

3. 灰白色碎屑砂海滩岩，含生物贝壳 0.4m

2. 灰白色细碎屑砾海滩岩，含生物贝壳，砾石一般为1cm×2cm大 0.6m

1. 灰白色珊瑚骨屑砂 >0.2m

未见底

4. 小东海西采坑珊瑚礁剖面

剖面位于小东海西部约400m处采坑内，海拔约2m，曾庆銮等1992年测制。

上覆地层 上全新统烟墩组灰白色大珊瑚骨屑砾

——— 整 合 ———

中全新统琼山组（Qh²q） 总厚度 >1.1m

8. 灰白色细珊瑚骨屑砾层 0.2m

7. 灰白色中粗珊瑚骨屑砾层 0.1m

6. 灰白色大珊瑚骨屑砾层，珊瑚骨屑^{14}C年龄距今为（2720±100）a 0.1m

5. 灰白色中珊瑚骨屑砾层 0.2m

4. 灰白色细珊瑚骨屑砾层，珊瑚骨屑^{14}C年龄距今为（3775±80）a 0.1m

3. 灰白色大珊瑚骨屑砾层 0.2m

2. 浅紫灰色含灰泥中珊瑚骨屑砾层 0.1m

1. 灰白色大珊瑚骨屑砾层 >0.1m

未见底

（九）烟墩组（Qh^3y）

1. 文昌市烟墩镇人工河剖面

该剖面为代表性剖面，位于文昌市烟墩镇人工河，起点坐标：东经 110°42′27″，北纬 19°25′01″；系赵希涛等 1979 年测制，雷奕振等 1992 年重测。自上而下为（图 6-10）：

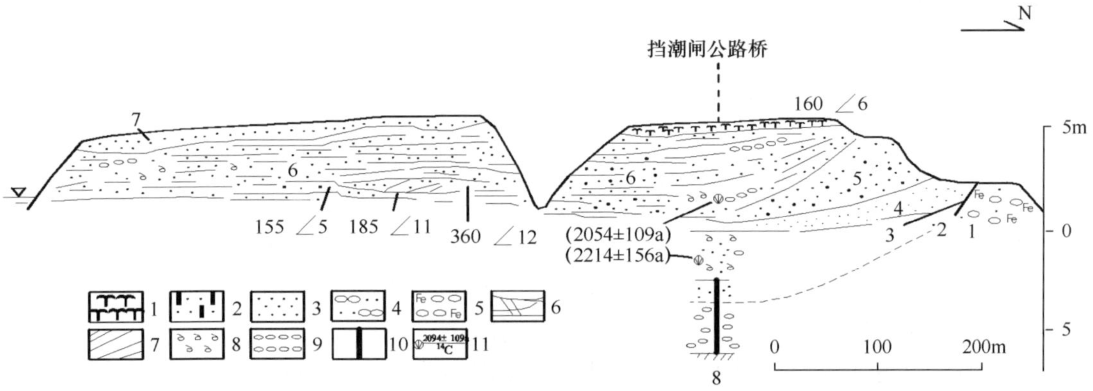

图 6-10　文昌市烟墩第四系剖面图

Fig. 6-10　Section of Quaternary strata，Yandun，Wenchang City

（引自赵希涛等，1979）

1—土壤；2—富含有机质的砂、黏土夹小砾；3—砂；4—砂砾；5—含铁质结核的砾石；6—具斜层理与交错层理的砂岩；7—基岩；8—贝壳与珊瑚砂砾；9—浮石砾石；10—钻孔；11—^{14}C 样品位置及年代

未见顶

上全新统烟墩组	总厚度>14.5m
7. 浅黄色砂，含少量贝壳和珊瑚碎块，组成海拔 4m 的新砂堤	1.0m
6. 褐黄色细砂岩，夹中粗砂层，含贝壳、珊瑚碎片和细砾石，底部为珊瑚贝壳砂砾层。具斜层理、交错层理	5.0m
5. 黄色中、粗砂	3.0m
4. 灰色中、细砂	2.0m
3. 黑色细砂及含有机质黏土，含细砾石和贝壳	>2.5m
2. 灰白色—灰黄色砂砾	>1.0m

------ 假　整　合 ------

下伏地层　上更新统八所组棕黄色砾石层，含铁质结核　>2.0m

2. 海口市美兰区沙头村 7ZK2 孔全新统剖面

剖面情况见上述（层号按原始资料）。

上全新统烟墩组	总厚度 5.93m
1. 浅黄色细砂，含植物根系	1.4m
2. 深色—深灰色中砂，含少量粉细砂	1.35m
3. 深灰色、灰黑色淤泥质亚黏土，含少量细砂。本层所含腐木的^{14}C 年龄为（1565±130）a	1.08m
4. 灰色亚黏土	1.45m
5. 浅灰色粗砂	0.65m

———— 整　合 ————

下伏地层　中全新统琼山组灰黄色亚黏土

3. 小东海西采坑珊瑚礁剖面

剖面情况见上述。

上全新统烟墩组 厚度 0.7m

11. 褐色含砾砂土植被层 0.2m

10. 灰白色含珊瑚碎屑粗砂层。珊瑚骨屑 ^{14}C 年龄距今（2450±75）a 0.3m

9. 灰白色大珊瑚骨屑砾层，上部为含锈褐色浮岩砾。珊瑚骨屑 ^{14}C 年龄距今（2425±85）a 0.2m

——————— 整　合 ———————

下伏地层 中全新统琼山组细珊瑚骨屑砾层

二、岩石地层划分与时代对比

（一）秀英组（Qp^1x）

秀英组系海南地质大队（1977）于海口市秀英一带创建，陈哲培等（1997）介绍。代表性剖面为秀英医院北西 29°900m 处冲沟剖面，岩性以灰色、青灰色页状黏土层或亚黏土层为主，次为浅灰白色砂层、砂砾层、砾石层组成的地层，局部夹一层基性火山岩。顶部往往有褐铁矿层，具古风化壳特征，是制作建筑砖瓦的主要原料之一，也是一个重要的识别标志。

秀英组在琼北、琼东北主要隐伏分布，零星出露在海口秀英、澄迈花场、坡尾岭、临高南宝、儋州新盈等地。由 1~3 个韵律组成，岩性中上部以灰色、青灰色页状黏土层或亚黏土层为主，夹浅灰色砂和砾石层，顶部往往有褐铁矿层，具古风化壳特征，偶夹一层基性火山岩，局部相变为沉凝灰岩。亚黏土具透镜状层理及水平层理。下部为浅灰色砾石层、砂砾层和砂层。局部含双壳类、腹足类及孢粉化石。为滨海潮坪环境，底板标高 -20~30m，地层厚度 5~30m。底部以含砾中粗砂与海口组页状黏土呈假整合接触（图 6-11），顶部以杂色黏土或铁质层与北海组含砾粗砂呈假整合接触，局部地带为多文组火山岩覆盖。

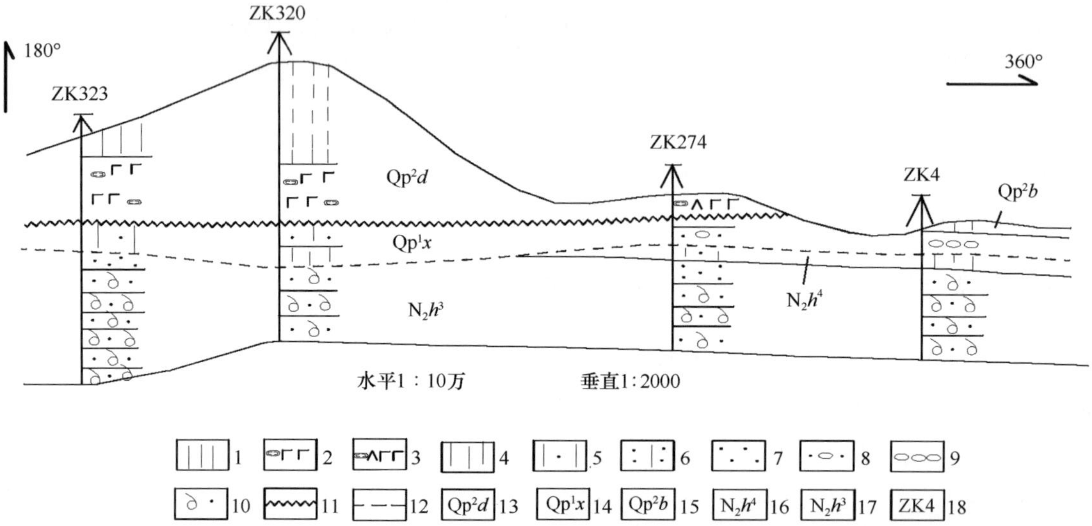

图 6-11　秀英组与海口组呈假整合接触剖面图

Fig. 6-11　Section showing the disconformable contact between the Xiuying Formation and Haikou Formation

（据 1：25 万《琼海县幅》区调报告）

1—红土；2—气孔状玄武岩；3—气孔状橄榄玄武岩；4—页状黏土；5—亚黏土；6—亚砂土；7—砂层；8—砂砾层；9—砾石层；10—贝壳砂岩；11—喷发不整合界线；12—假整合界线；13—多文组；14—秀英组；15—北海组；16—海口组四段；17—海口组三段；18—钻孔编号

在琼西、琼西北主要隐伏分布于沿海地区，局部地表出露。由 1~3 个韵律组成，岩性特征与琼北地区相似，局部含双壳类、螃蟹及孢粉化石，为滨海潮坪及湖泊-河流环境。底板标高 -20~74m，地层厚度 5~30m。

秀英组在琼南、琼西南、琼东南主要隐伏分布于沿海一带，地表零星出露。下部以灰白色粗砂（偶见斜层理）灰白色亚黏土，含少量粉细砂夹灰色黏土；上部以浅灰色、浅黄色亚黏土为主。下部基本层序由粗砂-亚黏土-黏土组成，具有由下向上粒度变细的特征，反映出退积型沉积特征，为湖泊-河流沉积环境。该组在区域上岩性稳定，厚度变化较大，在乐东县佛老村一带厚度为 11. 70m，三亚市红土坎一带为 19. 09m，假整合于望楼港组之上，被北海组假整合覆盖。

秀英组迄今发现少量古生物化石。软体动物化石产于澄迈县金江及桥头附近，包括双壳类 *Cobicula* cf. *fluminea* 和腹足类 *Galba* cf. *elegans*，时代为早更新世（中国科学院南海海洋研究所，1978）。孢粉化石产于海口市秀英砖厂剖面第一层，属于 *Polyopdiaceaesporites*-*Cupuliferoipollenites*-*Graminidites* 组合。主要分子有 *Polypodiaceoisporites*，*Polypodiaceaeaporites*，*Dacrydimites*，*Pinuspollenites*，*Cupuliferoipollenites*，*Tricolpites*，*Graminidites*，*Ilexpollenites*，*Quercoidites*，*Liquidambarpollenites*，*Floschuetzia*，*Palmaepollenites*。上述孢粉组合性质与东海盆地下更新统东海组的孢粉组合一致，时代为早更新世。因此，秀英组地质时代应为早更新世。

（二）北海组（Qp^2b）

北海组原称北海系，系邓植仪 1934 年在广西北海市附近黄屋一带创建，代表广西北海一带的棕黄色砂质土。1978 年中国科学院南海海洋研究所修订为北海组，时代为中更新世。北海组虽然创建于广西北海地区，不在秀英地层分区，但从 60 年代初引于海南岛一直沿用至今，故本文仍保留这一岩石地层单位名称。秀英地层分区的参考剖面为秀英砖瓦厂采场剖面。

该组广泛分布，岩性比较稳定，可分为两个岩性段。上部为橘黄色、棕红色、褐红色亚黏土、亚砂土、含铁质结核；下部为黏质砂、砂砾、砾石层，常含玻璃陨石和铁质结核，偶见不明显的斜层理。砾石成分有石英、脉石英、变质岩、玄武岩（琼北地区）、花岗岩（多见于琼西和琼南）等碎屑，砾石磨圆度差，分选性不好。

海南省地质调查院在文昌地区北海组采集了 6 个沉积物样品作粒度分析，根据曹伯勋（2008）的正态概率曲线（图 6-12），ZK20-1 号样品为多段型，其成因较复杂；其余样品均为粗三段型，以某一组成分占主体，不同程度地含有不同粒级组分，且其余各组分含量往往接近，分选性差，认为属河成。综合所述，北海组沉积环境为冲洪积相沉积。该组底板标高 11~77m，厚 0~18m，底部以砂砾层与秀英组的黏土呈假整合接触（图 6-13），在琼北顶部为多文组或道堂组所覆（图 6-14），在琼西覆于抱板群或中生代花岗岩及白垩系之上。

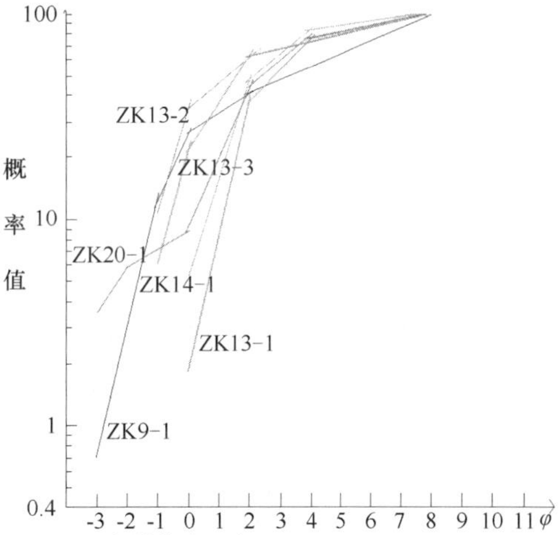

图 6-12　北海组沉积物正态概率曲线图

Fig. 6-12　Normal probability curve of Beihai Formation sediments

北海组迄今尚未发现化石，但常见玻璃陨石（雷公墨）分布。据中国科学院地质研究所及中国科学院贵阳地球化学研究所对东方、儋州、琼山、陵水等 13 处地点的北海组玻璃陨石样品进行裂变径迹法测定，年龄为（527±58）~（765±43）ka 之间。宜昌地质矿产研究所（1995）在北海组上部获 4 个热释光（TL）法测年结果为（227000. 5±11350. 1）~（296212. 2±14810. 5）a。海南地质大队一分队（1993）在昌江县沿海的北海组上部含黏土石英中粗砂中获得热释光年龄为（227001. 5±11350. 1）a。

综合上述分析，北海组的时代为中更新世。

（三）多文组（Qp^2d）

多文组系海南省地质调查院开展 1∶25 万《琼海县》幅区域地质调查时创建，龙文国等（2006）

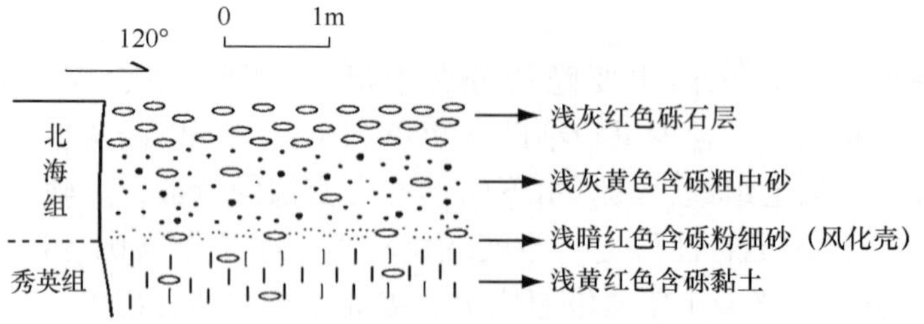

图 6-13　北海组与秀英组假整合接触素描图

Fig. 6-13　Sketch showing the disconformable contact between the Beihai Formation and Xiuying Formation

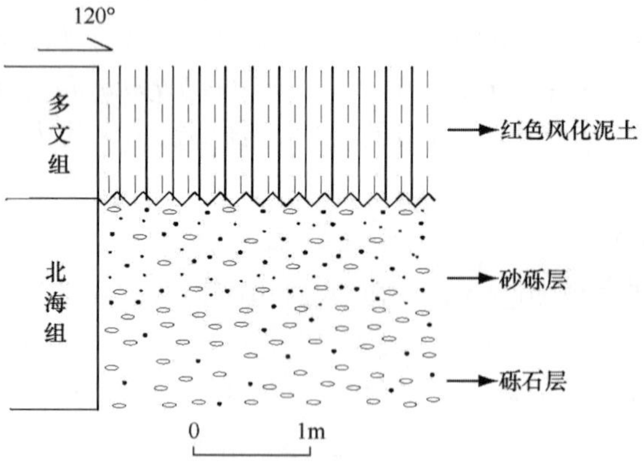

图 6-14　多文组喷发不整合于北海组之上

Fig. 6-14　Duowen Formation Eruptive rocks unconformably lie above the Beihai Formation

介绍，指定临高县多文镇剖面为多文组下段代表性剖面，临高县东英镇剖面为多文组上段代表性剖面，分别代表中更新世多文岭期、东英期形成的一套基性火山熔岩，部分火山口附件的火山碎屑岩。

长期以来，多位地质工作者对琼北第四纪基性火山岩做了不少工作，取得不少科研成果，其中具有代表性的研究者有孙建中、康先济等人。孙建中等（1986）将多文岭、高山岭一带火山岩划为早更新世多文岭期，中更新世—晚更新世东英期。认为多文岭期玄武岩在琼北分布最广，代表性火山口是多文岭火山口，该期玄武岩覆于下更新统湛江组之上，其上被中更新统北海组或后期玄武岩所覆盖。东英期由基性火熔岩及火山碎屑岩组成，分布在德文岭、高山岭一带，该期火山岩覆盖在北海组之上，时代为中更新世中晚期到晚更新世；康先济等（1991）对多文岭期火山岩的认识与前者基本相似，但将文昌市蓬莱地区的玄武岩时代改为第三纪（古、新近纪）。东英期则限于孙建中所划分的东英期下部。时代为中更新世，晚更新世则划为长流期。

龙文国等（2004）在区调工作中发现多文岭期、东英期火山岩均覆盖于中更新世河流相沉积北海组或其他更老地层之上（图 6-14），而均被晚更新世海相沉积八所组所覆盖。两期火山岩均形成于更新世，东英期火山岩形成时代稍晚于多文岭期火山岩。

多文组主要分布在王五-文教深大断裂以北地区，为一套火山地层，岩性划分为上、下段，分别代表中更新世多文岭期、东英期形成的一套基性火山熔岩，部分火山口附近见火山碎屑岩（火山集块岩、火山角砾岩、凝灰岩），代表裂隙-中心式喷发的火山岩系。上、下段之间存在喷发间断，表现为二者风化程度及红土化程度存在明显差异，上段风化程度弱，未见红土化现象，下段风化程度强，普遍见红土化现象，红土化层一般厚 2~10m。

多文组是与北海组层位大致相当的火山地层。多文组下段部分火山熔岩及上段火山熔岩喷发不整合于北海组不同层位沉积层之上，而上部被石山组橄榄辉石玄武岩喷发不整合或被上更新统八所组所覆盖的一套基性火山熔岩。目前对年轻的基性岩类测年多选用 K-Ar 法，但 K-Ar 法测年当前的极限为 0.5~1Ma±（李大明等，1999），那么对年轻于 0.5Ma±的地质体，则 K-Ar 法测年误差极大。因此，多文组上、下年限的确定应主要通过上覆与下伏地质体的形成时代来制约。

以前，部分学者（Ge T. M. et al.，1989；Sun J. S.，1991；Zhu B. et al.，1989；宜昌地质矿产研究所，海南省地矿局，1992）已于研究区新生代更新世（多文岭期、东英期）火山岩中获得大量同位素数据。康先济（1992）综合研究认为：多文岭期火山岩的 K-Ar 年龄集中的时限在 0.77～2.11Ma，喷发时代属早更新世中晚期，东英期火山岩 K-Ar 年龄集中在 0.21～0.67Ma，喷发时代属中更新世中晚期。但早期 K-Ar 法测年技术及仪器设备的精度存在局限，因而所获年龄数据可信度相对较低。

近年来，由于 K-Ar 测年方法的改进及测试仪器的更新，Ho Kungsuan et al.（2000）和 Juang Wenshing（1996）于琼北相当于多文组下段的火山岩中获得了一些 K-Ar 法测年结果（表 6-1）。Ho Kungsuan et al.（2000）还于海口三江、定安龙门多文组下段火山岩中采样（HK28、HK25），分别获得全岩 Ar-Ar 坪年龄为（1.12±0.02）Ma、（0.45±0.01）Ma。龙文国等于琼北多文组下段的火山岩中获得了 K-Ar 法测年结果为（1.16±0.40）Ma（由北京大学地质学系同位素地球化学研究实验室苗国均、黄宝玲测试）。上述测试结果表明 HK28、HK16、IH4-2 等样品年龄明显偏老，只有 HK25 样品年龄较接近。现主要根据其自然叠覆于北海组之上，认为其形成时代应为中更新世中晚期。

表 6-1　琼北更新统玄武岩 K-Ar 法年龄测定结果

Table 6-1　K-Ar age dating results of Pleistocene basalts，Qiongbei

采样地点	样品号	岩性	$w(K_2O)/\%$	$^{40}Ar/(10^{-11}mol\cdot g^{-1})$	年龄/Ma	资料来源
定安龙门	HK25	OT	1.34	0.0830	0.43±0.03	Ho（2000）
临高博厚	Hk16	OT	1.25	0.2176	1.20±0.02	Juang（1996）
临高多文	IH4-2	OB	1.44	0.2808	1.16±0.40	龙文国等（2005）

注：OT 为橄榄拉斑玄武岩；OB 为橄榄玄武岩

（四）八所组（Qp^3bs）

八所组系薛万俊等（1981）于东方县八所港北黎村剖面创建，雷奕振等（1992）介绍。代表性剖面岩性主要为棕红色、棕黄色石英砂，向下递变为灰白色、灰黄色石英砂，部分为含砾亚砂土，未见底。20 世纪 60 年代称为老红砂。该组主要分布于琼东北、琼西、琼南等沿海地带，海拔一般为 15～30m，地貌上表现为Ⅰ级或Ⅱ级海成阶地，是钛铁矿、锆英石砂矿及石英砂矿的主要控矿层位。

八所组在琼东北岩性总体以中细粒为特征，砂粒分选性较好，成熟度高，普遍含有少—微量钛铁矿。下部为灰白色中细砂，中部为褐黄色、黄色中细砂，普遍含黏土或黏土团块，局部出现黏土夹层，上部褐灰色、褐黑色中细砂，局部含有炭质及有机质等，层位较稳定，顶部为一层浅灰色、灰白色中细砂，厚 5～22m。底部以灰白色中细砂与北海组棕黄色细砂土或秀英组灰色、浅蓝绿色黏土-亚黏土呈假整合接触，或不整合于印支期—燕山期花岗岩或志留系之上。

在琼西，八所组岩性主要为棕黄色及白色粉细砂、中砂及含细砾石中粗砂层等。岩性分选性不好，其中粗砂与细砾含量约占 1/2，砾石成分主要为脉石英和花岗岩，次为砂岩和硅化灰岩，上部普遍含有炭质、有机质夹层，含有少—微量钛铁矿。厚 4～25m，底部以灰白色中细砂与北海组深青灰色砂质黏土呈假整合接触。

在琼南，八所组岩性主要为棕黄色、褐红色粉细砂、中砂及含细砾中粗砂偶含铁质结核，岩性稳定，厚度变化不大，厚 3.45～33.04m。底部以灰黄色细砂层与秀英组深灰色含贝壳碎屑泥层呈假整合接触。

在琼西，八所组含有大量的孢粉化石，以草本植物花粉为主，含量达 48%，其中多半是禾本科花粉；木本植物花粉含量为 29%，蕨类植物孢子约占 23%。经张仲英等（1987）研究，认为属于禾本科-蒿属-栎属-枫香属组合，反映了温干气候特点。岩性局部含有机质和炭质，颜色呈灰褐色、褐黑色，层位稳定，且含有黏土，其具有滞水环境沉积的特征，属于潟湖相沉积。

海南省地质调查院在文昌地区八所组不同地点采取 5 个沉积物作粒度分析，根据曹伯勋（2008）

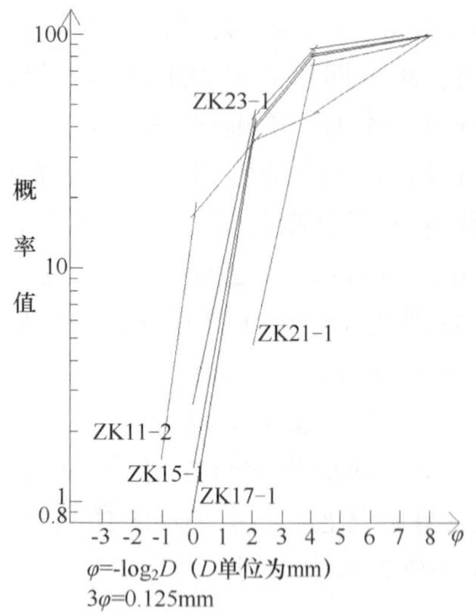

图 6-15　八所组沉积物正态概率曲线图

Fig. 6-15　Normal probability curve of Basuo Formation sediments

的正态概率曲线（图 6-15），ZK21-1 号样品为细粒二段型，以细砂为主，少量粉砂，分选性好，沉积环境一般为片流；ZK11-2 号样品为多段型，以粉砂占主体，其次为中粗砂，少量细砂，分选性差，属风成或河成；其余样品为粗粒三段型，以中粗粒级和细粒级砂为主，含量各分秋色，二者含量占 90%以上，少量粉砂，分选性一般，其沉积环境一般为海浪成因。综上所述，八所组沉积环境应为海相沉积（潟湖-滨海沉积），其间并受到风成或河成的扰动。

前人对八所组不同地点进行过 ^{14}C 和热释光年龄测定，其中文昌市公坡剖面上部的红色砂热释光年龄为（24±0.2）ka（薛万俊等，1981），文昌市翁田 9Zk$_1$ 孔上部红砂（埋深 4.9～6m）热释光年龄为（14.4±0.07）ka（张仲英等，1987），清澜港八所组上部红砂夹层淤泥质砂获 ^{14}C 年龄为（12.130±390）ka（张仲英等，1987），文昌市公坡揽亮村黑色砂 ^{14}C 年龄为（18.73±0.048）ka，东方八所 20ZK2 孔的浅黄色砂（埋深 12.4～14.0m）TL 年龄为（12.4±0.06）×10^4ka（张仲英等，1987），海南地质综合勘察院（1995）在乐东县黄流八所组底部中粗砂层及九所一带该组上部细砂层中分别获 TL 年龄（199332.4±9967）a、（12764.1±638.3）a。

综上年龄数据，八所组年龄与晚更新世上、下界线年龄相近，时代为晚更新世。

（五）道堂组（Qp^3*d*）

道堂组系海南地质大队（1989）以海口市琼山区道堂-石山剖面创建，陈哲培等（1991）介绍。该组为基性火山熔岩和火山碎屑岩沉积，原道堂组分为四段：一段、三段为陆相橄榄拉斑玄武岩、橄榄玄武岩为主，二段、四段为海相玄武质岩屑凝灰岩为主，含海绿石及有孔虫化石，厚 5～278m。与下伏北海组及上覆石山组均呈喷发不整合接触，时代属第四纪晚更新世。

2001～2004 年，海南省地质调查院在开展 1∶25 万《琼海县》幅区域地质调查时重新对道堂组进行了修订，把原道堂组一段归属为多文组。现道堂组相当于原道堂组的二、三、四段。龙文国等（2006）于地质通报月刊发表论文介绍。

本次研究认为将原道堂组一段修订为多文组的理由依据主要有：①白莲以西至福山等地，地表所见大都为玄武岩，而原道堂组一段玄武岩与多文组玄武岩界线难以确定，而以海相火山岩为界则易于识别，这样有利于地质填图及制图。《国际地层指南》中（H. D 赫德伯格主编，张守信译，1979）关于“过去已建地层单位的修订或重新定义”指出：“重新下定义的正当理由在于使单位更有用或便于在它发育的境内去识别、制图和延伸”。将原道堂组一段修订为多文组显然符合国际地层指南的原则要求；②原道堂组一段 TL 年龄为（199609±5988）a，当时晚更新世下界采用 0.2Ma，故划为晚更新世。而国际地层表（1996）将晚更新世下界定为 0.126Ma，因此，原道堂组一段应为中更新世，而将其修订为多文组更合理。

重新修订后的道堂组为一套火山地层，共分为下、中、上段。

下段岩性为薄层玄武质沉岩屑晶屑凝灰岩、沉玻屑晶屑凝灰岩、沉晶屑岩屑凝灰岩及玄武质沉凝灰岩，含海绿石及微体化石。底部偶夹一层微气孔状橄榄拉斑玄武岩和一层凝灰质含砾砂岩。相变为凝灰质含砾砂岩、含砾粉砂岩或亚黏土、黏土层，含炭化木，厚 0～130m，属喷发-沉积相。与下伏多文组呈假整合接触，局部为喷发不整合接触。

中段岩性为暗灰色气孔状橄榄玄武岩夹橄榄玄武岩或橄榄拉斑玄武岩。局部夹集块岩（如十字路的玉里村钻孔），集块为玻基玄武岩屑。在长流的文明村一带和在十字路旧沟分别见气孔状橄榄玄武岩具绳状构造和橄榄拉斑玄武岩具枕状构造。当覆盖在本段上面的道堂组上段缺失时，顶部岩石就发生古红土化，出现 7~8m 厚的古红土层，玄武岩厚度 0~55m，属爆发—喷溢亚相，与下伏下段呈喷发不整合接触（图 6-16）。

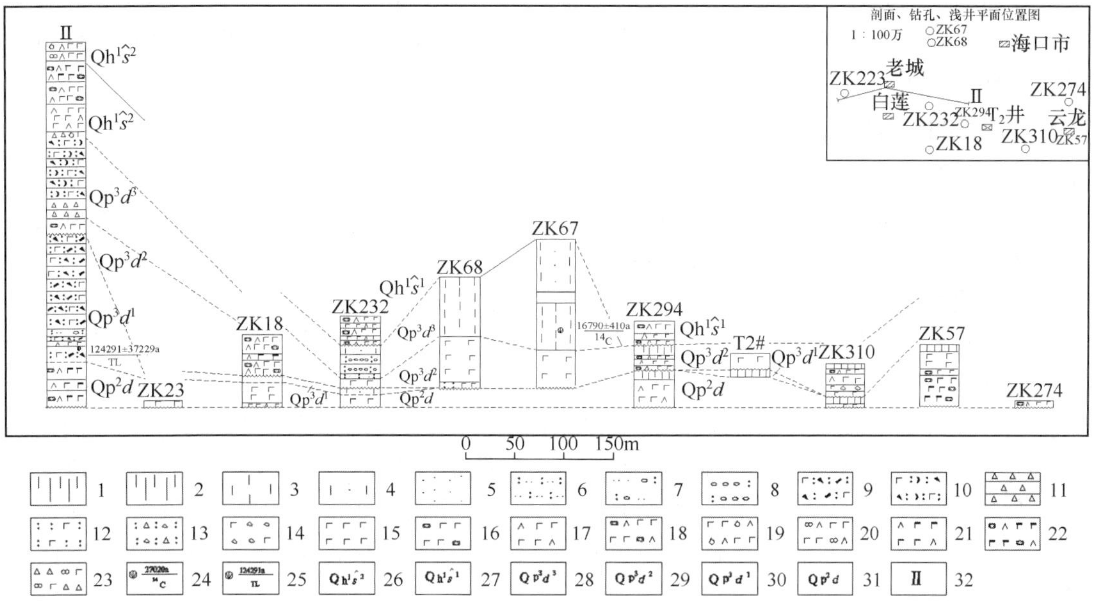

图 6-16 道堂组、石山组柱状对比图

Fig. 6-16 Columnar correlation of the Daotang and Shishan Formations

1—玄武岩地表风化红土；2—古红土化红土层；3—黏土层；4—亚黏土；5—细砂层；6—凝灰质细砂；7—凝灰质含砾砂；8—凝灰质砾岩；9—玄武质岩屑晶屑沉凝灰岩；10—玄武质岩屑玻屑沉凝灰岩；11—沉火山角砾岩；12—玄武凝灰岩；13—含火山角砾、集块凝灰岩；14—玄武质集块岩；15—玄武岩；16—气孔状玄武岩；17—橄榄玄武岩；18—气孔状橄榄玄武岩；19—浮岩状橄榄玄武岩；20—熔渣状橄榄玄武岩；21—橄榄拉斑玄武岩；22—气孔状橄榄拉斑玄武岩；23—含集块\火山角砾熔渣状橄榄玄武岩；24—^{14}C 样品采样位置及年龄值；25—热释光采样位置及年龄值；26—石山组二段；27—石山组 段；28—道堂组三段；29—道堂组二段；30—道堂组一段；31—多文组；32—道堂-石山剖面

上段岩性为玄武质沉岩屑玻屑凝灰岩，含海绿石、有孔虫化石，局部底部沉火山角砾岩。相变为凝灰质砂岩、砂砾岩或黏土、亚黏土层，含炭化木，厚 0~85m，属喷发-沉积相。顶部与石山组呈喷发不整合接触（图 6-16）。

海口地区道堂组含有孔虫化石，主要有：*Globigerina* sp.，*Ammonia takananensis*，*Sphaeroidina chitostomata*，*Lagene* cf. *striata*，*Discorbinella* aff. *montereyensis* 等。1∶5 万海口地区区调（1989）测得下段第②层沉岩屑晶屑凝灰岩热释光年龄为（124.291±3.729）ka。海南地质大队水文队曾测得秀英区新海林场 ZK67 钻孔三段炭化木^{14}C 年龄为（16790±410）a。陈沐龙等（1997）在三都镇剖面下段沉玻屑凝灰岩获 TL 年龄（121790±6080）a，上段沉玻屑凝灰岩获 TL 年龄（41590±2080）a。

综上所述，道堂组时代为晚更新世。

（六）石山组（$Qh^1\hat{s}$）

石山组系海南地质大队 1989 年创建，陈哲培、钟盛中（1991）介绍，代表性剖面为道堂-石山剖面。该组为一套陆相基性火山岩，分布较稳定，其间火山锥体、火山口保存完好，岩石未发生红土化。

该组主要分布在王五-文教深大断裂以北地区，为一套火山地层，按岩性组合特征进一步划分为两段。

第一段岩性上部为气孔状橄榄拉斑玄武岩，下部为气孔状橄榄玄武岩、橄榄玄武岩夹橄榄粗玄岩。局部底部为含集块含火山角砾熔渣状橄榄玄武岩，厚 0~74m，属爆发-喷溢亚相。局部的气孔状橄榄玄武岩发育熔岩洞穴、熔岩隧道。与下伏道堂组呈喷发不整合接触。

第二段岩性为熔渣状、浮岩状橄榄玄武岩，局部玄武质浮岩。厚度约18m。属火山口相。

石山组中未发现生物化石。在下伏道堂组顶部沉凝灰岩获热释光年龄32119a，它反映石山组火山喷发热事件的初始年龄，但年龄数字偏大。由于石山组喷发不整合于道堂组之上，其年龄当晚于道堂组四段年龄16790a。且石山组岩石未发生红土化，火山口及火山锥体保存完好。因此，石山组时代应为全新世，暂划为早全新世。

（七）万宁组（Qh^1w）

万宁组系张仲英等1987年创建，雷奕振等（1992）介绍。标准剖面为万宁保定12ZK1孔19.6~31.5m井段，原始定义为属于河口三角洲相沉积，岩性主要由深灰色黏土夹砾质中粗砂组成。参考剖面为海口市美兰区东营沙头村7ZK2孔17.95~25.58m井段。

该组主要隐伏分布于琼北、琼西及琼南等沿海一带，属河口三角洲-滨海沉积相。在琼北，其岩性主要为下部以细砾石、砂层为主，上部为含粗砂、细砂黏土、黏土层等。在琼南，下部岩性为灰白色中粗粒砂层或含砾细砂，砾或砂成分以石英为主，上部为灰黄色砂质黏土，含少量中粗砂，下部基本层序由中粗粒砂或含砾细砂-砂质黏土组成。同期的生物堆积称西瑁期原生礁，见于三亚西瑁岛和小东海北岸原生礁平台，造礁生物主要由珊瑚组成，次为有孔虫、双壳类和腹足类等。该组地层厚度变化较大，在琼西厚6.58m，在琼南万宁保定厚度大于24.05m，在三亚市羊栏厚8.88m，琼北琼山区东营厚11m。该组与上覆琼山组呈整合接触，在琼北与下伏海口组呈不整合接触；在琼南与下伏莺歌海组呈假整合接触。

前人在该组中获得少量的^{14}C年龄，在该组代表性剖面上部中获^{14}C年龄为（8460±170）a，下部^{14}C年龄为（10230±320）a；在同一时期的沉积西瑁期原生珊瑚礁中（属万宁组同期的沉积产物），其^{14}C年龄值分别为（8420±115）a和（8235±105）a；崖城白坟坡ZK5孔万宁组上部获^{14}C年龄为（9480±290）a（海南地质综合勘察院，1994）。三亚地区万宁组获腐木^{14}C年龄为（10930±370）a。

综合上述年龄分析，万宁组的地质时代应为全新世早期。

（八）琼山组（Qh^2q）

琼山组系张仲英等（1987）创建，原始定义指海口市美兰区东营沙头村7ZK2孔5.93~9.09m井段地层，属滨海沉积，岩性主要为中粗砂，时代为中全新世晚期。1992年雷奕振等对其进行重新划分，赋予新的含义，以海口市美兰区东营沙头村7ZK2孔5.93~17.95m井段作为标准剖面，为滨海-潟湖相沉积，岩性为亚黏土夹少量黏土、砂，富含有机质黏土，含砂、细砾亚黏土，时代为中全新世。

该组主要出露及隐伏分布于海南岛沿海地带，在区域上该组岩性及厚度都有所变化。在琼北该组出露于海口市海甸岛-灵山一带，由含细砾亚黏土、富含有机质黏土、亚黏土、亚黏土与黏土互层夹细砂层等三种基本层序组成，岩性中、上部为灰色、灰黄色亚黏土夹少量黏土及细砂层，下部为深灰色富含有机质黏土，底部为灰—深灰色含砾亚砂土，厚11.72~17.95m，与下伏万宁组及上覆烟墩组均为整合接触。

琼山组在琼西主要出露于沿海一带，岩性为灰色、深灰色、黏土、亚黏土夹少量细砂层，局部富含有机质，厚度>9.57m。

琼山组在琼南主要隐伏分布于西南及三亚等地沿海地带，由含细砾亚黏土、富含有机质黏土；亚黏土与黏土互层夹细砂层等三种基本层序组成，岩性下部为灰白、棕黄色含砾粗砂层，砾石成分以石英为主，上部为深灰色砂质黏土，亚黏土，含贝壳碎屑；同期的沉积还有三亚落笔洞的洞穴堆积及沿海的砂丘岩、海滩岩和珊瑚礁；厚度>8.22m。

琼山组产有孔虫和孢粉化石。有孔虫主要有*Quinqueloculina*，*Triloculina tigonula*，*Ammoniatepida*，*Elphidium*等。孢粉化石分为三带：下部为*Castanopsis*-*Dacrydium*-Proteaceae-*Cyathea*带，以木本植物花粉为主，反映了炎热潮湿的气候；中部为Proteaceae-*Elaeocarpus*-*Castanopsis*-*Castsnes*-*Piper*带，以木本植物花粉占绝对优势，显示气候比下带更温热；上部为*Casuarina equisetifolia*-*Castanopsis*-*Pinus*-*Artemisia*

带，木本植物花粉明显减少，其中木麻黄占重要地位，草本植物花粉和蕨类孢子含量相应增高。整个孢粉带属种较单调，反映气候偏于暖干。

在该组代表性剖面的下部（第 8 层）有机质黏土获^{14}C年龄（6750±140）a；三亚市羊栏南红农场 15ZK3 孔中琼山组^{14}C年龄为（2710±110）~（7280±14）a；大东海、天涯海角等地琼山组^{14}C年龄为（2710±110）~（6345±100）a；在崖城水南一村 ZK6 钻孔中琼山组上获炭化木^{14}C年龄（5680±140）a；在三亚地区琼山组海滩岩获^{14}C年龄（2710±110）a~（6345±100）a，该组上部淤泥获^{14}C年龄（7280±140）a。

综合上述古生物化石及年龄资料，表明琼山组的地质时代应为全新世中期。

（九）烟墩组（Qh^3y）

烟墩组系中国科学院地质研究所赵希涛等（1979）创建并介绍。雷奕振等（1992）重新厘定了原烟墩组，表示海南岛整个晚全新世的滨海砂堤-潟湖系列的沉积。代表性剖面位于文昌市烟墩人工河剖面，原始定义指 2~6 层，属滨海-潟湖相沉积，岩性主要由黄灰白色砂砾、粗砂、细砂及黑色含有机质黏土组成，厚 14.5m，时代为晚全新世早期。现在定义指晚全新世的滨海砂堤-潟湖系列沉积，整合或假整合于琼山组之上，是锆钛矿、金红石及独居石等砂矿的赋矿层位。

该组主要分布于海南岛沿海地带，岩性主要为灰黄色、黄白色粉细砂、细砂、含砾中粗砂、含生物碎屑中粗砂、海滩砂、灰黑色含有机质黏土、淤泥质砂，普遍见贝壳、珊瑚及海滩岩（图版Ⅰ-12-B）和细砂岩等，细砂岩具斜层理和交错层理。地貌上构成滨海平原的上部，并组成沿海砂堤、砂堤封闭的潟湖以及新海滩岩、连岛砂堤等。该组在区域上地层厚度变化较大，在海口沿海地带厚 3m，文昌烟墩沿海一带厚 12.4~18.5m，乐东九所沿海地带厚度大于 4.89m，三亚沿海地区厚 1.8~24.3m。

烟墩组产软体动物和孢粉化石。软体动物化石见于文昌烟墩剖面的第 3 层和第 6 层，据赵希涛等（1979）报道，产腹足类：*Babylonia lutosa*，*B. areolata*，*Polynicespyriformis*，*Strombus canarium*，*Conus betulinus*，*C. textile*，*Bursa rano*，*Olivaispidula* sp.，*Trochus* sp.；双壳类：*Meretrix meretrix*，*Posimia japonica*，*Solen grandis*，*Sanguinoloria chinensis*，*Pophia gallus*，*Circe seripta*，*Tellina* sp.，*Gefrarium* sp.，*Cordium* sp.，*Chama* sp.，*Tecraclita squamosa vividis*，*Ostrea* sp. 等。上述分子分别生存于浅海、滨海及潮间带。孢粉化石发现在琼山东营沙头村 7ZK2 孔和三亚南红 15ZK3 孔，称为 *Casuarina equisetifolia*-*Quercus*-Polypodiaceae-*Pteris*-*Castanopsis* 带，反映湿热气候（张仲英等，1987）。

赵希涛 1978 年在文昌烟墩人工河剖面第 3、6 层分别获得贝壳^{14}C年龄为（2214±156）a，珊瑚碎块的^{14}C年龄为（2054±109）a；张仲英等 1987 年在美兰区沙头村 7ZK2 孔第 3 层获腐木^{14}C年龄为（1565±130）a；在三亚地区相当层位获海滩岩胶结物^{14}C年龄为（1190±70）a，生物骨屑砂砾^{14}C年龄为（2450±75）a（宜昌地质矿产研究所和海南省地质调查院，2002）；雷奕振等（1992）统计烟墩组^{14}C的年龄值在（260±60）~（2360±90）a 之间。

综合以上所述，表明烟墩组地质时代应属于全新世晚期。

第二节　西沙群岛地层分区

根据《南海地质》（刘昭蜀等，2002）资料，西沙群岛地层分区由下更新统永发组（Qp^1y，原为永兴组，但《中国地层（13）中国的第三系》中已建上新统永兴组，造成同名异物，因此本次改为永发组）、中更新统琛航组（$Qp^2\hat{c}$）、上更新统石岛组（$Qp^3\hat{s}$）和全新统西沙洲组（Qhx）组成。

一、剖面描述

剖面见于西沙群岛永兴岛西永 2 井。由原地质部第二海洋地质研究所第十二普查勘探大队 1972~1974 年完成，张明书等 1989 年介绍。

全新统西沙洲组		总厚度>40.00m
6.	礁后岛相生物骨屑砂砾层及礁缘格架相礁格架灰岩、黏结灰岩（未见顶）	10.00m
5.	礁后岛屿相、礁坪潟湖相生物骨屑砂砾层、含砾灰泥层，海岸风成相生物砂屑灰岩与化石土壤层互层	30.00m
	～～～～～～ 不　整　合 ～～～～～～	
上更新统石岛组		厚度 15.00m
4.	礁格架-礁顶相礁格架灰岩、藻黏结灰岩	15.00m
	～～～～～～ 不　整　合 ～～～～～～	
中更新统琛航组		74.00m
3.	礁坪潟湖相粒泥灰岩、泥粒灰岩与斑点礁格架灰岩互层，藻席层发育，局部夹礁石松散沉积物	74.00m
	———— 整　合 ————	
下更新统永发组		总厚度 152.00m
2.	潟湖相藻屑灰岩，夹礁格架灰岩	84.00m
1.	礁坪相泥粒灰岩、藻粒泥灰岩，可见礁石松散沉积物，礁格架灰岩稀少	68.00m
	———— 整　合 ————	
下伏地层	新近系上新统永兴组礁坪及潟湖相藻屑灰岩，白云石化显著	123.00m

二、岩石地层划分与时代对比

（一）永发组（Qp^1y）

本次研究以西永2井孔为代表，岩性分为上、下两段。下段岩性为泥粒灰岩和薄层泥灰岩，有礁石松散沉积物，礁格架灰岩稀少，富含蓝绿藻屑，有时出现核形石，具较强的白云岩化，岩石色调淡，以藻泥屑灰岩为洁白和细腻，犹如白垩灰岩；上段为藻屑灰岩，夹礁格架灰岩，富含红藻藻屑、藻席层和红藻石，岩石色调暗。总厚度变化不大，西永2井厚152.00m，西琛1井厚119.64m。

永发组与下伏永兴组（N_2yx）及上覆琛航组（$Qp^2\hat{c}$）呈整合接触。永兴组地质时代为上新世，琛航组时代为中更新世，依据地层层序特征，把永发组地质时代划归早更新世较适合。

（二）琛航组（$Qp^2\hat{c}$）

琛航组岩性主要为粒泥灰岩、泥屑灰岩与斑点礁格架灰岩互层，藻席层发育，夹有礁石松散沉积物。总厚度变化较小，西永2井厚74.64m，西琛1井厚67.76m。在西石1井获铀系年龄为182000a，西永2井为450000a，西琛1井为600000a，由此可见，琛航组的地质时代为中更新世。

（三）石岛组（$Qp^3\hat{s}$）

石岛组岩性分为两段，下段岩性为礁格架灰岩和薄黏结灰岩；上段为生物骨屑砂砾层和含砾灰泥层，生物砂屑灰岩与化石土壤层互层。总厚度变化不大，在西永2井厚31.36m，西琛1井厚30.60m。在西石1井获铀系年龄为120000a。石岛组与下伏琛航组（$Qp^2\hat{c}$）及上覆西沙洲组（Qhx）呈不整合接触，依层序特征及铀系年龄，石岛组地质时代为晚更新世。

（四）西沙洲组（Qhx）

西沙洲组在西石1井的岩性为沙丘岩，在西永2井和西琛1井的钻孔中均普遍见礁顶现代礁坪和岛洲的松散堆积体，以礁缘格架相、礁坪表层生物繁衍带和灰沙岛及灰砾岛的形式出现，为生物骨屑砂砾层及礁格灰岩、黏结灰岩，厚10m。西沙洲组总厚度大于40m，在西石1井获^{14}C年龄为（8670±250）a、西永1井14m处的样品获^{14}C年龄为（6790±95）a（业治铮等，1985）。由此可见，西沙洲组的地质时代为全新世。

第三节　南沙群岛地层分区

根据《南海地质》（刘昭蜀等，2002）资料，南沙群岛地层分区第四系由更新统南沙组（Qp*n*）和全新统南海组（Qh*n*）组成。主要见于礼乐滩一带。

一、剖面描述

剖面位于南沙群岛于永暑礁南永1井。由中国科学院南沙综合科学考察队1992年测制。

未见顶

全新统南海组	总厚度>17.30m
36. 珊瑚砾块层	4.20m
35. 生物砾砂屑层	11.80m
34. 生物砾屑（砾石滩）层	1.30m

~~~~~~~~~~ 不　整　合 ~~~~~~~~~~

更新统南沙组	总厚度>134.80m
上、中更新统	
33. 珊瑚灰岩	1.10m
32. 珊瑚砾块灰岩	1.70m
31. 生物粗砂屑	1.20m
30. 珊瑚灰岩（块）	0.30m
29. 生物砂屑	0.70m
28. 珊瑚砾块灰岩	1.90m
27. 珊瑚灰岩	0.40m
26. 珊瑚砾块灰岩	0.80m
25. 珊瑚灰岩	5.00m
24. 生物砂砾屑灰岩	4.30m
23. 珊瑚灰岩	3.40m
22. 珊瑚砾块灰岩	4.80m
21. 珊瑚灰岩	1.30m
20. 珊瑚藻石砂砾屑灰岩	15.9m
19. 珊瑚砾块灰岩	1.40m
18. 生物砾砂屑灰岩	18.40m
17. 生物砂砾屑灰岩	5.20m
16. 生物砂屑灰岩	2.70m
15. 生物砂砾屑灰岩	1.80m

~~~~~~~~~~ 不　整　合 ~~~~~~~~~~

下更新统	
14. 珊瑚灰岩	5.4m
13. 珊瑚砾屑灰岩	1.60m
12. 珊瑚灰岩	1.20m
11. 生物砂砾屑灰岩	1.1m
10. 珊瑚灰岩	2.70m
9. 珊瑚（细）砾屑灰岩	0.60m
8. 珊瑚砾块灰岩	9.40m
7. 珊瑚灰岩	2.60m
6. 含（生物）砾砂屑灰岩	7.20m

5. 含生物（细）砂屑灰岩，下段 2m 为溶洞 9.70m
4. 珊瑚灰岩，下段 8.50m 为溶洞 9.70m
3. 珊瑚白云岩 2.60m
2. 生物砾砂屑白云岩（生物砾砂屑灰岩） 2.40m
1. 生物砂砾屑白云岩（生物砂砾屑灰岩） 5.10m

未见底

二、岩石地层划分与时代对比

（一）南沙组（Qp*n*）

南沙组岩性为生物砾砂屑灰岩、珊瑚灰岩、珊瑚白云岩和生物砾砂屑白云岩。其中，以造礁石珊瑚为主，还有多种贝类和藻类。为礁缘-礁坪-潟湖相沉积，南沙组总厚度>250.81m（于永暑礁南永 1 井厚度>134m）。产有仅在第四纪才出现的叶状珊瑚 *Lobophyllia* 和裸肋珊瑚 *Merunlia*。南沙组于南永 2 井中获^{14}C 年龄为（30040±670）a、（33650±770）a、（41780±950）a；南永 1 井中获氨基酸年龄 730ka，时代为更新世。故将南沙组地质时代划为更新世。

（二）南海组（Qh*n*）

南海组下段岩性为未胶结的珊瑚砾块和生物砂屑，以造礁石珊瑚为主，其次为软体动物贝类及珊瑚藻等，系由波浪和潮流作用造成的礁顶潟湖-礁坪相沉积。上段岩性为松散的生物骨壳碎屑砂砾层，以造礁石珊瑚为主，局部夹海滩岩或沙丘岩，为灰沙岛沙坝-干潟湖相沉积，南海组厚度大于 17.30m。于南永 2 井中获^{14}C 年龄为（6420±100）a~（8220±140）a；南永 1 井中获氨基酸年龄为 2900~5740a，时代为全新世。故南海组时代划为全新世。

第四节　曾母盆地地层分区

本分区资料少。据《南海地质》（刘昭蜀等，2002）资料，该区第四系仅有北康组。岩性以生物礁灰岩为主，其沉积环境为河口、海岸平原相沉积，与下伏地层南康组呈不整合接触。

第七章　第四纪地质与地貌

海南省第四系属于华南地层大区中南-东南地层区，第四系主要受到地形、气候、基底性质和新构造运动等因素制约。而地貌的形成主要是受地球内、外营力长期综合作用的结果。内营力是地球内部各种能源所产生的内力（包括地壳运动、岩浆及火山活动等），是形成地壳表面地貌起伏的基本轮廓；外营力是由于受自然环境改变而产生的能量引起地表变化的力（如水流、冰川、太阳能、生物作用及化学作用等），是对基础地貌格架不断进行风化、剥蚀、搬运及堆积，从而形成现代地面的各种地貌形态。海南省地貌成因与新生代以来的地质构造演化密切相关。

第一节　海南省第四纪地质与地貌

一、第四纪地质

海南岛的第四系特别发育，是华南地区第四系发育较为典型的地区之一，其分布受新生代晚期形成的新构造格局的控制，形成的沉积物类型多样。

本次研究根据岩性特征、层序关系、年龄数据等，并采用《海南岛岩石地层》（1997）的划分方案，对区内第四系进行划分，正式命名岩石地层单位16个（详见第六章），非正式命名单位主要为现代沉积物，按其形成的不同机制分为4种成因类型：海陆交互堆积物（Q^{D}）、冲洪积层（Q^{pf}）、坡残积层（Q^{del}）、火山风化土层（Q^{vl}）。

（一）海陆交互堆积物（Q^{D}）

滨海堆积物主要分布于海南岛沿海等地，岩性主要为灰色、灰黄色、黄白色粉细砂、细砂、粗砂、含砾中粗砂、含生物碎屑中粗砂、灰黑色含有机质黏土等，普遍见贝壳、珊瑚等。岩性分选性不好，成熟度一般，普遍含有锆英石及钛铁矿。

（二）冲洪积层（Q^{pf}）

冲洪积层主要为河流一级阶地沉积及现代河流沉积，分布于海南岛主要河流（如南渡江、万泉河、昌化江等）两岸。

岩性特征主要为灰色、灰黄色砂砾、卵石、含砾砂、粉砂、亚黏土、黏土等，含腐殖质。具下粗上细的二元结构。

（三）坡残积层（Q^{del}）

坡残积层主要为前第四系地质体（沉积岩、侵入岩、变质岩）的风化残坡积，以风化土为主，有些地段夹杂有岩石坡残积块分布。成分主要为砂砾、砂、粉砂质黏土等，磨圆度较差，以次棱状—棱角状为主。

（四）火山风化土层（Q^{vl}）

火山风化土层主要为中生代及新近纪火山岩的风化土，岩性以亚黏土、亚砂土为主，局部见火山岩残积碎块。

二、地　　貌

海南岛地貌特点是中间高，四周低，呈穹隆状地形，向外围逐级下降，由山地、丘陵、台地、平原构成层状垂直分布和环状水平分布，梯级结构明显，区内各级冲沟水系发育，呈平行状、树枝状，大小水库密布。

依据地貌成因形态，把海南岛地貌划分为构造侵蚀、剥蚀地貌、火山地貌、河流堆积地貌和滨海堆积地貌等 4 个主要一级地貌单元区。在此基础上，结合地形特征，把不同地貌区进行次级地貌单元划分，共划分出 18 个次一级地貌单元。

（一）构造侵蚀、剥蚀地貌

1. 构造侵蚀中山区

分布于海南岛中部及中南部地区，由五指山、鹦歌岭、黎母岭、吊罗山、尖峰岭、猴猕岭等山峰组成。标高 1000m 以上，其中五指山最高标高为 1867.1m，山脉走向与区域构造线基本一致，山峰陡峻，常形成悬崖峭壁（图版Ⅰ-12-C），岩石剥蚀较强，切割强烈，但覆盖层不厚，水系发育成树枝状，山谷呈 V 形。

2. 构造侵蚀低山区

分布于海南岛中部、中南部地区，标高 500～1000m，山脉走向与区域构造线基本一致，局部地区为山崖，山顶多呈锥形状和锯齿状，侵蚀切割强烈，覆盖层较薄，一般为 1m，水系发育。

3. 构造剥蚀丘陵

分布于海南岛东北部、西部、南部地区，标高 500m 以下，山势和缓，坡度 20°～30°，水系不发育，山顶多呈浑圆状、馒头状，山脉走向北东—南西方向，次为东西向，剥蚀强烈。

4. 溶蚀残丘区

分布于海南岛儋州八一及三亚田独等地区：标高 100～300m，地势和缓开阔，孤丘自立，溶洞发育，具孤峰绝壁地貌景观。

5. 剥蚀堆积平原区

主要分布于海南岛丘陵区前沿山脚至滨海平原地区，标高小于 200m，剥蚀强烈，地势平缓，波状起伏。

（二）火山地貌

火山地貌主要由中生代和新生代火山岩风化土及火山岩残积碎块组成。新生代以火山口、火山群为中心，地形呈阶梯状递降，形成火山台地。

根据其所在区域地形情况不同，将火山地貌进一步划分为 7 个次一级火山地貌。

1. 五指山地区中生代火山中山区

在五指山一带，由中生代火山岩风化土及火山岩残积碎块组成，标高一般大于 1000m，最高标高达 1867m，地形较陡，并遭受不同程度的侵蚀切割，形成冲沟凹谷。

2. 同安岭地区中生代火山低山区

分布于同安岭等地，为中生代火山岩风化土及火山岩残积碎块组成，标高一般为 1000～500m，局部标高达 1144m，地形起伏较大，侵蚀切割强烈，水系发育。

3. 牛腊岭地区中生代火山丘陵区

分布于牛腊岭等地，由中生代火山岩风化土及火山岩残积碎块组成，为火山岩丘陵地形，标高一般小于 500m，局部标高最高达 654m，地形局部较陡，切割较强，水系较发育。

4. 旺商地区中生代火山台地

在澄迈县旺商一带，为中生代火山岩台地地形，台地标高最高 89m，台面比较宽阔平坦。

5. 北部地区新生代火山台地

分布于海南岛北部，由新生代火山岩风化土及火山岩残积碎块组成，台地地形呈阶梯状，可分为四级，各级台地标高因地而异，标高小于 40m 为一级台地，二级台地标高为 40～80m，80～120m 为三级台地及大于 120m 为四级台地，一、二级台地台面平坦，三、四级台地台面较陡，台地呈东西向，中部隆起，向南北两侧变低。

6. 永兴地区新生代火山群

分布于永兴、石山、龙塘等地，火山口形态多样，呈漏斗状、圆锥状，火山口最高标高 220.9m，相对高差 40～100m。正火山口为马鞍山，呈双峰形，两峰间出现一凹地，形如马鞍，北峰标高 189.5m，南峰标高 200.9m，形如烽火台。负火山口为罗经盘，呈圆形，底宽约 500～600m，外围直径约 1000m，深约 20m，自下而上呈阶梯状台阶。永兴火山群三、四级台面较陡，一、二级台面平坦宽广。

7. 北部地区新生代火山锥

分布于海南岛北部，形态上呈似椭圆形、圆锥状、不规则状等（图版Ⅰ-12-D），有的火山锥高而陡，有的低而缓，其顶部为火山口，最高火山口标高为 269.8m。

（三）河流堆积地貌

河流堆积地貌按类型划分为河流侵蚀堆积区和河口三角洲区。

1. 河流侵蚀堆积区

分布于南渡江、昌化江、万泉河等河流中下游平原区。河流中下游阶地发育，河谷宽阔，两级阶地比较明显。一级阶地表面平缓，向下游逐渐变低，标高 2～50m，前缘阶坎高 2～3m，与河漫滩之间呈陡坎或坡度较缓的斜面相连；二级阶地前缘与一级阶地或滨海平原相连，阶面微向河倾，常呈无规则或带状，阶地标高 10～20m；河漫滩在河流的转弯处河床一侧，由砂砾、卵石等组成，高出水面约 1～5m。冲洪积一级阶地位于山间盆地，标高 60m，呈长条状分布。

2. 河口三角洲区

位于昌化江、南渡江、万泉河等河流入海口处，呈三角形状堆积，河流成网状分布，沙滩、沙洲众多，前缘复式砂嘴发育，后缘和两侧为河流冲积一、二级阶地，标高 1.5～9.0m，微向海倾斜。

（四）滨海堆积地貌

滨海堆积地貌根据成因类型划分为海湾堆积区、海积阶地区、海积生物堆积区和海积砂堤区。

1. 海湾堆积区

分布于海南岛沿海，表面平坦，标高一般 2～30m，位于砂堤内侧，呈口袋状、长条状，微向海倾斜，构成岛内海岸最低平原（图版Ⅰ-12-E）。

2. 海积阶地区

位于海南岛沿海地区，多呈条带状平行海岸线，阶地前缘多为砂堤，阶面呈波状起伏，向海倾斜，局部阶面平坦（图版Ⅰ-12-F）。

3. 海积生物堆积区

分布于海南岛文昌昌洒、儋州排浦、三亚鹿回头等沿海地方，其沉积物主要为海滩岩（图版Ⅰ-12-G），呈条带状平行海岸线，发育微斜层理。

4. 海积砂堤区

分布于海南岛沿海各地，砂堤标高 5～60m，多呈条带状平行海岸线（图版Ⅰ-12-H）。

第二节　南海诸岛第四纪地质与地貌

南海诸岛第四纪地质详见第六章第二、三节。根据《南海地质》（刘昭蜀等，2002）资料，南海诸岛的地貌景观以现代珊瑚环礁地貌为主。根据形态和成因的差异及地理位置不同，把南海海底划分为 5 个地貌区（图 7-1）。

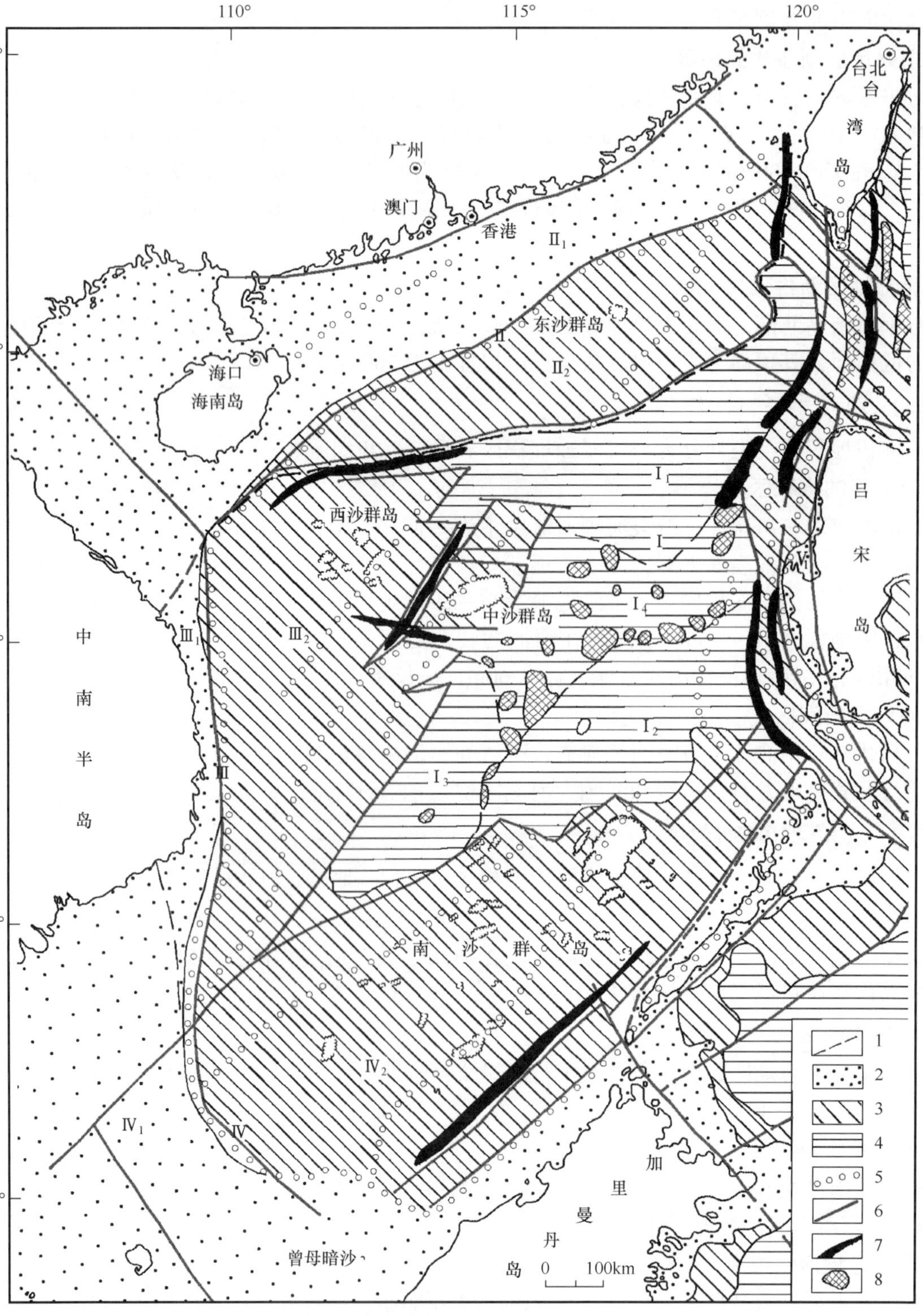

图 7-1 南海海底地貌分区图

Fig. 7-1 Division of submarine geomorphy of the South China Sea

（据刘昭蜀等，2002）

Ⅰ—中央深海盆地貌区：$Ⅰ_1$—北海盆亚区，$Ⅰ_2$—南海盆亚区，$Ⅰ_3$—西南海盆亚区，$Ⅰ_4$—中海盆亚区；
Ⅱ—北部陆缘地貌区：$Ⅱ_1$—北部陆架亚区，$Ⅱ_2$—北部陆坡亚区；Ⅲ—西部陆缘地貌区：$Ⅲ_1$—西部陆架亚区，$Ⅲ_2$—西部陆坡亚区；
Ⅳ—南部陆缘地貌区：$Ⅳ_1$—南部陆架亚区，$Ⅳ_2$—南部陆坡亚区；Ⅴ—东部陆缘地貌区：$Ⅴ_1$—东部陆架亚区；$Ⅴ_2$—东部陆坡亚区
1—分区界线；2—大陆架；3—大陆坡；4—深海盆；5—构造脊；6—大断裂；7—海槽、海沟；8—海山

一、中央深海盆地貌区

南海中央深海盆地貌区的周边均为规模巨大的岩石圈断裂：中央海盆北缘断裂、中央海盆南缘断裂、中央海盆西缘断裂，以及东界的马尼拉海沟断裂。因周缘出现断崖地貌，尤以马尼拉海沟东壁、南沙和中沙海底高原的近海盆一侧边坡，以及盆西脊岭的断崖地貌最为突出。上述断裂把中央深海盆围成一菱形地堑，具有洋壳结构。海盆基底为大洋玄武岩，表面为新生代沉积层覆盖而成深海平原。基底上拱处（如北海盆与南海盆南部）出现丘陵平原地貌。海盆中部为海山链与海山地貌。海盆东缘及西缘则以沟槽地貌或槽谷地貌为特色。

根据地貌组合的差异及地理位置不同，中央深海盆又可分为 4 个亚区：北海盆亚区、南海盆亚区、西南海盆亚区及中海盆亚区。海盆中部的近东西向的黄岩海山链及其附近众多的海山组成的海山群—深海平原地貌组合区应单独划分为中海盆亚区。它将中央深海盆分隔为北海盆亚区及南海盆亚区两部分。如前所述，此两部分地形地貌既有相似性，也有差异。在南海盆亚区中，根据中南海山链一线南延至南沙双子礁北缘陆坡脊的水深 3600~3900m 的海底隆脊这条自然分界线，又可将南海盆分为西南海盆亚区及狭义的南海盆亚区两部分。

二、北部陆缘地貌区

北部陆缘地貌区系指东起澎湖海槽西至西沙海槽之间由北部陆架和北部陆坡共同组成的北东向长条地块。北部陆架是在北东向陆缘大陆壳型基底地堑和地垒系基础上发育形成。陆架外缘有晚更新世低海面时期的滨海残留地貌，如沙垄与小洼地、断续的溺谷以及外缘的海丘链。陆架与陆坡转折处的上陆坡陡坡带地貌，断裂构造上为陆坡北缘断裂。由于受南海北部拉张陆缘控制，故在上陆坡和中陆坡出现三级逐级断落的台阶和张裂式的东沙海底高原，以及在断裂或地堑构造基础上发育的槽谷系统。

三、西部陆缘地貌区

西部陆缘地貌区为一轮廓像直角三角形的地块。两直角边分别为北界的东西向西沙海槽和西界的南北向越东滨外断裂；斜边是中央海盆西缘断裂及其西南端延伸的南沙西缘海谷。西部陆架主体是陡窄的断阶型陆架。陆坡区是以过渡壳地堑带基础上发育的构造地貌为特征，其中有由断裂控制而逐级下降的五级台阶，有以地垒为基础并以台阶面为基座发育成的西沙和中沙海底高原。高原上珊瑚礁地貌发育并形成西沙和中沙群岛。珊瑚礁灰岩共厚 1000~1200m，其底部相当于西沙台阶的深度。此外，区内还有以地堑为基础演化成的槽谷系统，如西沙海槽等。

四、南部陆缘地貌区

南部陆缘地貌区为东起马尼拉海沟，西至南沙西缘海谷之间的南沙陆坡和南海南部陆架区。其特征与北部陆缘地貌区相似，同属北东向长条地块；同样具有受基底地堑带控制的构造地貌类型。陆架发育在外缘的地垒式隆起脊内。断褶声波基底之上堆积发育起广阔的陆架平原和水下三角洲。有保留较完整的水下阶地和水下河网。陆坡地貌相对于深海平原而言，是高达 4000m 以上的南沙海底高原。礁灰岩顶部发育了南海最繁茂的珊瑚礁地貌，形成了约 200 多座珊瑚岛屿礁滩。沿地堑带或断裂发育槽谷网，形成著名的南沙海槽，以及南北向、北西向和近东西向的槽谷系统。

五、东部陆缘地貌区

东部陆缘地貌区为南北向长带状分布的岛架与岛坡地貌区。北起台湾岛南部，往南经吕宋岛至民都洛岛诸岛的西缘，基底构造受南北向的 2 列构造脊和 4 条平行的大断裂控制，造成了陆架狭窄而陡的断阶型地貌类型，陆坡则为脊岭与槽沟或台阶交替地貌，以及上陆坡与下陆坡具断崖地貌的特点。

第二篇 侵 入 岩

第八章　海南省侵入岩概述

海南省辖海南岛、西沙群岛、中沙群岛、南沙群岛及其海域，陆地总面积（主要包括海南岛、西沙群岛、中沙群岛、南沙群岛）3.54 万 km^2。海南岛是海南省最大的岛屿，面积 3.39 万 km^2。根据现有地质调查成果资料，目前在海南岛（包括近岸的七洲列岛、白鞍岛、大洲岛、洲仔岛、蜈支洲、东洲、西洲、西瑁洲等岛屿）见大面积的侵入岩分布，出露总面积 16623.65km^2，约占海南岛总面积的 48.96%。分别归属于早晋宁期、华力西期—印支期和燕山期 3 个构造岩浆旋回。按岩石类型划分，90%以上为二长花岗岩、正长花岗岩和花岗闪长岩类，少量英云闪长岩、闪长岩和辉长岩。最早的中元古代和新元古代侵入岩出露面积分别为 153.47 km^2 和 2.39km^2，仅占海南省侵入岩总面积的 0.94%，绝大部分侵入岩形成于华力西期—印支期和燕山期，其中华力西期—印支期出露面积 12087.24 km^2（二叠系面积 6298.73 km^2，三叠系面积 5788.51km^2），占海南省侵入岩总面积的 73%；燕山期出露面积 4380.55km^2（侏罗系面积 622.57 km^2，白垩系面积 3757.98km^2），占海南省侵入岩总面积的 26%。

广东省区调队（1964）在《海南岛 1：200000 区域地质调查报告》中，首次将海南岛侵入岩划分为 5 期，5 期的侵入时代均为中生代（表 8-1）。

表 8-1　海南岛侵入岩期次划分表

Table 8-1　Division of phases of intrusive rocks in the Hainan Island

时代	T	T	J_1—K_1	K_1—K_2			K_2（?）
期	第一期	第二期	第三期	第四期			第五期
				埠堡笔型	尖峰岭型	花岗斑岩型	
岩性	含辉石石英闪长岩、闪长岩和少量辉长岩	以斑状二长花岗岩、花岗闪长岩和二长花岗岩为主	角闪黑云花岗闪长岩和二长花岗岩，磁铁矿含量高	肉红色钾长花岗岩			以石英正长岩为主，少量钾长花岗岩
				钾长石多有格子状双晶	钾长石罕见格子状双晶	花岗斑岩	
面积比	0.8%	58.5%	11.0%	14.0%	13.1%	1.5%	1.1%
代表性岩体	金竹园、什哺、仙婆岭	儋县、琼中	保城、屯昌、乘坡、税町	埠堡笔、吊罗山、埠石斗	尖峰岭、黑岭、昌化大岭	牙洛、南牛岭	番阳峒、关桥岭

（广东区调队，1964）

此后广东区调队编制的 1：500000（1977 年）和 1：1000000（1984 年）广东省岩浆岩分布图，以及海南地质大队编制的 1：200000 海南岛地质图（1984），在 1964 年划分方案的基础上，根据同位素年龄对少量岩体的侵入时代和先后顺序作了调整。2001～2007 年，由海南省地质调查院、宜昌地质矿产研究所完成的海南 1：25 万区域地质调查，采用“单元—超单元”填图方法对海南岛侵入岩重新进行了划分。在此基础上，根据《中国区域地质志》工作指南要求，本报告采用“岩性+时代”的划分方法，通过综合研究取得的地质资料（主要是近年完成的 1：25 万区调资料），将海南省侵入岩划分为 102 个岩石填图单位，详见海南省侵入岩岩石单位划分表（表 8-2）和海南省侵入岩新、旧岩石单位划分对比表（表 8-3）。必须说明的是，本次划归为同一岩石单位的不同岩体（或侵入体），虽然具有基本相似的岩石成分、结构构造特征和大体一致的侵入时代，但并不一定是同源岩浆演化的产物，其含义不能与前人所划分的单元（或超单元）等同。

表 8-2 海南省侵入岩岩石单位划分表

Table 8-2 Division of intrusive rock units in the Hainan Island

时代	岩石单位		岩性	代号	面积/km^2	主要侵入体	同位素年龄(Ma)及测试方法
	序列	典型命名地					
晚白垩世			花岗斑岩	$\gamma\pi K_2$	364.89	洋淋岭、南牛岭、双客园、大岭	66(K)
		火烧岭	细粒黑云母二长花岗岩	$\eta\gamma K_2$	3.74	火烧岭	67±1(U)
		龙楼	中粒斑状黑云母正长花岗岩	$\xi\gamma m K_2$	24.31	龙楼	65.4(ICPMS)
	吊罗山	大小岭	中细粒黑云母正长花岗岩	$\xi\gamma K_2^c$	41.64	大小岭	82(U)
		吊罗山	粗中粒含斑黑云母正长花岗岩	$\xi\gamma c K_2^b$	726.1	昌化大岭、吊罗山	花岗斑岩
		新村	细中粒斑状黑云正长花岗岩	$\xi\gamma m K_2^a$	149.97	新村	90.93±1(K)
		南牙村	花岗闪长斑岩	$\gamma\delta\pi K_2$		南牙村	
早白垩世	保城	高通岭	细粒黑云母正长花岗岩	$\xi\gamma f K_1^e$	8.59	高通岭	95.6±2(U)
		双顶岭	细中粒斑状黑云母正长花岗岩	$\xi\gamma K_1^d$	32.66	马鞍岭、双顶岭	
		六弓	粗中粒斑状黑云二长花岗岩	$\eta\gamma m K_1^c$	382.36	蒙贡、六弓	107.6±2(ICPMS)
		保亭	细中粒含斑角闪黑云二长花岗岩	$\eta\gamma K_1^b$	644.61	香炉岭、保亭	
		加茂	细中粒角闪黑云花岗闪长岩	$\gamma\delta K_1^a$	879.02	屯昌、加茂、九架岭	102.0±3(ICPMS) 100.9±1(U-Pb)
		乌石	闪长岩	δK_1	69.85	子保村、乌石	100(ICPMS)
		番通	石英正长岩	$\xi o K_1$	40.07	番通、四方岭	106±3(Sh)
	高峰	阜石斗	细中粒黑云母正长花岗岩	$\xi\gamma K_1$	74.6	阜石斗	
		抱跃	细中粒斑状角闪黑云二长花岗岩	$\eta\gamma K_1$	124.96	抱跃	107±1(Rb)
		税町	中细粒角闪黑云花岗闪长岩	$\gamma\delta K_1$	96.87	税町	112±7(Rb)
			花岗斑岩	$\gamma\pi K_1$	60.71		
			花岗闪长玢岩	$\gamma\delta\pi K_1$	20.16		
		拉丁村	细粒辉长岩、辉长闪长岩	νK_1	3.25		97(KR),97.2±2
		半岛	中粗粒角闪黑云正长花岗岩	$\xi\gamma^c K_1$	9.62	半岛	131(K)

续表

时代	岩石单位		岩性	代号	面积/km^2	主要侵入体	同位素年龄(Ma)及测试方法
	序列	典型命名地					
晚侏罗世	三狮岭	马岭	细粒黑云母正长花岗岩	$\xi\gamma^{f}J_3$	25.01		
		马鞍岭	中细粒黑云母正长花岗岩	$\xi\gamma^{m}J_3$	65.17	马鞍岭、朝根坡	
		三狮岭	中粒黑云母正长花岗岩	$\xi\gamma J_3$	407.49	三狮岭、横岭、廖次岭	151(U),150.3±2(U),151(Sh)
		架水	粗中粒斑状角闪黑云二长花岗岩	$\eta\gamma cmJ_3$	30.77	架水	
	南山	冬瓜岭	细粒含斑黑云二长花岗岩	$\eta\gamma fJ_3$	3.48	冬瓜岭	
		椰子园	细粒斑状角闪黑云二长花岗岩	$\eta\gamma J_3$	0.36	椰子园	
		南山岭	中细粒斑状角闪黑云二长花岗岩	$\eta\gamma mfJ_3$	14.9	南山岭	152±6(Rb)
		马翁岭	闪长岩	δJ_3	49.84	老村、托剩岭、马翁岭	157(ICPMS),157±2(U)
中侏罗世	勤赛	八宝坡	巨粒角闪正长岩	ξJ_2	0.64	八宝坡	173±2(Sh)
		黑栋	中细粒石英二长岩	$\eta o J_2$	24.17	黑栋	
		里茂	中粒黑云母辉长岩	υJ_2	0.74	里茂	
晚三叠世		迈洲岭	碱长花岗斑岩	$k\gamma\pi T_3$	3.08	迈洲岭	213.3(K-Ar)
	三架岭	打显村	中细粒少斑黑云二长花岗岩	$\eta\gamma mfT_3$	58.9	打显村	200(U)
		白石岭	细粒含斑黑云二长花岗岩	$\eta\gamma T_3$	31.86	白石岭	
	雅加	黑岭	中粒多斑状黑云二长花岗岩	$\eta\gamma pT_3$	9.1	黑岭	214.7(ICPMS)
		分水岭	中—细粒斑状黑云二长花岗岩	$\eta\gamma fmT_3$	101.6	分水岭	210.8±10(U)
		南宝岭	细粒斑状黑云二长花岗岩	$\eta\gamma mpT_3$	25.12	南宝岭	
	南英岭	五里桥	中细粒黑云二长花岗岩	$\eta\gamma mT_3$	152.76	五里桥、长坡	208(U),219.8±11(U)
		苗圃	细粒角闪黑云二长花岗岩	$\eta\gamma fT_3$	28.96	苗圃	
	黑岭	乌烈岭	中细—中粒斑状黑云母正长花岗岩	$\xi\gamma T_3$	53.32	乌烈岭	240.6±11(U)
		东四	中粗粒角闪黑云正长花岗岩	$\xi\gamma cT_3$	270.88	东四	230.1±0.1(Rb),225.4±11(U)
		七差岭	中细粒黑云母正长花岗岩	$\xi\gamma fT_3$	29.9	七差岭	

续表

时代	岩石单位		岩性	代号	面积/km^2	主要侵入体	同位素年龄(Ma)及测试方法
	序列	典型命名地					
中三叠世		保梅岭	中粗粒黑云母正长花岗岩	$\xi\gamma T_2$	25.27	保梅岭	232.6(ICPMS)
	琼中	洞古岭	花岗斑岩	$\gamma\pi T_2^d$	1.05	洞古岭	232.0±2(U)
		南通岭	中细粒黑云母正长花岗岩	$\xi\gamma T_2^c$	239.08	南通岭	226.5±3(U)
		琼中县	中粒斑状(角闪)黑云二长花岗岩	$\eta\gamma T_2^b$	656.85	琼中县	233.8±2,234.2±2(U)
		黎母岭	粗中粒(巨)斑状(角闪)黑云二长花岗岩	$\eta\gamma p T_2^a$	1256.69	黎母岭	227.2±2,232.5±2,245(U)
	万车	后美	中细粒碱长花岗岩	$X\rho\gamma T_2^b$		后美	
		后龙	中粒碱长花岗岩	$X\rho\gamma T_2^a$		后龙、松美山	233.2±1(U)
	万宁	石墩	中粗粒闪长岩	δT_2	1.17	石墩	
		长安	辉长岩	υT_2	36.79	光头岭、长安	238(Sh),228±14,231.2±3(ICPMS)
	六连岭	石龟岭	细粒黑云二长花岗岩	$\eta\gamma f T_2$	21.52	石龟岭、坡尾坡	
		兵工厂	细粒斑状黑云母(二云母)正长花岗岩	$\xi\gamma f p T_2$	27.25	兵工厂	
		六连岭	中粗粒黑云母正长花岗岩	$\xi\gamma c T_2$	117.74	六连岭、石母屋	239±3(Sh)
	立才	前锋	细粒少斑黑云母正长花岗岩	$\xi\gamma f T_2$	38.86	羊栏、鬼岭	
		结尾	(中)细粒黑云母二长花岗岩	$\eta\gamma m f T_2$	31.99	牛睡岭、虎岭	
		超盆	中粗粒含斑黑云母二长花岗岩	$\eta\gamma c T_2$	316.6	超盆	
		布山村	细中粒含斑-斑状角闪黑云二长花岗岩	$\eta\gamma T_2$	256.65	布山村	239±7(Sh),243.2±2(ICPMS)
早三叠世	袁水	金寮	中细粒黑云正长花岗岩	$\xi\gamma m T_1$	46.02	金寮	
		番阳峒	细中粒(少斑)角闪黑云正长花岗	$\xi\gamma m c T_1$	375.86	番阳峒	241±1(Sh),243.7±3(ICPMS)
		麻山田	中粗粒(少斑)角闪黑云石英正长岩	$\xi o T_1$	109.83	麻山田	243±4(Sh)
		深堀村	中粗粒少斑黑云角闪石英二长岩	$\eta o T_1$	27.38	深堀村	
		深沟	粗中粒少斑霓辉石碱长正长岩	$k\xi T_1$	0.37	深沟	244±7(Sh)
		进岭	花斑岩	$\gamma\pi T_1$	6.18	进岭	247±5(Sh)
	尖峰	瘦岭	中细粒黑云母正长花岗岩	$\xi\gamma f T_1$	169.89	瘦岭	
		黑岭	细粒斑状黑云母正长花岗岩	$\xi\gamma T_1$	180.78	黑岭	
		尖峰岭	中粗粒斑状黑云母正长花岗岩	$\xi\gamma c T_1$	1079.21	尖峰岭	249±5(Sh)

时代	岩石单位		岩性	代号	面积/km^2	主要侵入体	同位素年龄(Ma)及测试方法
	序列	典型命名地					
晚二叠世	西庆	大王岭	中细粒黑云母二长花岗岩	$\eta\gamma mfP_3$	429.21	大王岭	
		高村	中细粒斑状黑云母二长花岗岩	$\eta\gamma pP_3$	192.18	高村、竹山岭	
		两院	中粒斑状黑云母二长花岗岩	$\eta\gamma cpP_3$	445.68	两院、美合	246±4(ICPMS)
		大岭	细中粒少斑黑云母花岗闪长岩	$\gamma\delta pP_3$	50.32	大岭、乾元	252(Rb),251.9(U)
		鹅塘村	中细粒含斑英云闪长岩	$\gamma\delta oP_3$	4.61	鹅塘岭	
		陀类	细粒二云母二长花岗岩	γP_3	34	陀类	250±4(Sh)
		大田	细粒含斑黑云母二长花岗岩	$\eta\gamma P_3$	193.72	大田	236,245,247(U)
		什天	细中粒斑状黑云母二长花岗岩	$\eta\gamma mP_3$	26.72	什天	251±7(Sh)
	禄马	白茶	细粒黑云母二长花岗岩	$\eta\gamma fP_3$	0.53	白　茶	
		东岭	中细粒角闪黑云花岗闪长岩	$\gamma\delta P_3$	192.86	东　岭	254±4(Sh)
	道票岭	那佩	中粗粒斑状黑云母二长花岗岩	$\eta\gamma cP_3$	7.03	那　佩	
		保国	细中粒斑状角闪黑云二长花岗岩	$\eta\gamma mcP_3$	110.98	保　国	259.5±3(Sh)
中二叠世	通什	火岭	中细粒黑云母正长花岗岩	$\xi\gamma P_2$	459.95	火岭、荔枝岭、黄山水库	251.7±18(Rb)
		长塘岭	中粒斑状(角闪)黑云二长花岗岩	$\eta\gamma P_2$	758.43	长塘岭、东崩岭、中林岭	274±5(Sh),265±28(ICPMS)
		大坡	中粒斑状黑云二长花岗岩	$\eta\gamma mP_2$	268.69	大坡	250±15.7,255(Rb)
		乌兰盖	中粗粒含巨斑黑云母二长花岗岩	$\eta\gamma mcP_2$	84.58	乌兰盖	
		通什	中粒(巨)斑状(角闪)黑云二长花岗岩	$\eta\gamma pP_2$	2291.63	通什、邦溪	265±4(Sh)
		山牛塘	细粒闪长岩	δP_2	8.68	山牛塘	
早二叠世		石碌	中粗粒巨斑状黑云母二长花岗岩	$\eta\gamma cP_1$	140.39	石碌	274±6.5(Rb)
	五指山	什败	细中粒含斑石英二长岩	ηoP_1	4.87	什败	
		行干	细中粒石英二长闪长岩	$\delta\eta oP_1$	36.63	行干、什暗	
		方满	片麻状细粒二长岩	ηP_1	0.41	方满	272±7(Sh)
		新风岭	中细粒含电气石黑云正长花岗岩	$\xi\gamma fP_1$	1.90	新风岭	276±4(Sh)
		顺作	弱片麻状细中粒含斑石榴子石黑云(二云)花岗岩	$\xi\gamma mP_1$	3.76		278±7(Sh)

续表

时代	岩石单位		岩性	代号	面积 km^2	主要侵入体	同位素年龄(Ma)及测试方法
	序列	典型命名地					
早二叠世	育才		细粒含斑黑云(二云)二长花岗岩	$\eta\gamma fP_1$	70.69	乘顺岭	
		志仲	巨斑状黑云母二长花岗岩	$\eta\gamma mpP_1$	277.16	志仲	282±11(Pb),287±7(Sh)
		抱郎	中粒斑状角闪黑云二长花岗岩	$\eta\gamma P_1$	43.81	抱郎	
	大乃村	石岭	中细粒花岗闪长岩	$\gamma\delta P_1$	90.96	石岭、羌园、东岭	
		保山村	英云闪长岩	$\gamma\delta o P_1$	26.46	保山村、美马岭	278±1(U),287±10.2(Rb),270±11(ICPMS)
		白石村	闪长岩-石英闪长岩	δP_1	37.64	白石村、坤外、豪老发、什哺	293(K),307±12.5(Rb),298.6±10.3(Rb)
		机告村	辉长岩	υP_1	4.25	机告村、下园、红水岭	
新元古代		公庙	斜长角闪岩	$\Psi o Pt_3$	2.39		955±54,975±8.6
中元古代		亚炮	片麻状二长花岗岩	$\eta\gamma Pt_2$	133.1	亚炮	1430(Sh),1407(Pb)
		戈枕	片麻状花岗闪长岩	$\gamma\delta Pt_2$	20.37	戈枕	1397±46(Pb),1440.87(U),1457(U)

注:1. 各时代界线年龄据中国区域地质年代表(2001):长城纪二分(1800Ma—1600Ma—1400Ma),二叠纪三分(295Ma—277Ma—257Ma—250Ma),三叠纪三分(250Ma—241Ma—227Ma—205Ma)、侏罗纪三分(205Ma—176Ma—159Ma—137Ma)、白垩纪二分(137Ma—96Ma—65Ma)。

2. Sh 为 SHRIMP 锆石 U-Pb 年龄;U 为颗粒锆石 U-Pb 表面年龄;Rb 为全岩或岩石-矿物内部等时线法;K 为 K-Ar 法;Pb 为锆石 Pb-Pb 年龄,ICPMS 为 La-ICPMS 锆石 U-Pb 年龄。数据来源见正文。

3. 代号中 f 为细粒,m 为中粒,c 为粗粒,p 为似斑状。

表 8-3 海南省侵入岩新、旧岩石单位划分对比表

Table 8-3 Correlation of division of new and original rock units, Hainan Island

本书使用岩石单位			岩石类型	1：25 万区调使用岩石单位		
时代	典型命名地	代号		东方县幅	琼海幅	乐东县、陵水县幅
白垩纪		$\gamma\pi K_2$	花岗斑岩	海头	南牛岭、大岭、双客园	$K_2\gamma\pi$
	火烧岭	$\eta\gamma K_2$	二长花岗岩		火烧岭	
	龙楼	$\xi\gamma m K_2$	正长花岗岩		龙楼	
	大小岭	$\xi\gamma K_2^c$	正长花岗岩			大小岭
	吊罗山	$\xi\gamma^c K_2^b$	正长花岗岩	昌化大岭		吊罗山
	新村	$\xi\gamma m K_2^a$	正长花岗岩			新村
	南牙村	$\gamma\delta\pi K_2$	花岗闪长斑岩	南牙村		
	高通岭	$\xi\gamma^f K_1^e$	正长花岗岩		高通岭	
	双顶岭	$\xi\gamma K_1^d$	正长花岗岩	马鞍岭	双顶岭	
	六弓	$\eta\gamma m K_1^c$	二长花岗岩		蒙贡	六弓
	保亭	$\eta\gamma K_b 1$	二长花岗岩		香炉岭	保亭
	加茂	$\gamma\delta K_1^a$	花岗闪长岩		屯昌	加茂
	乌石	δK_1	闪长岩	子保村	乌石	$K_1\delta o$
	番通	$\xi o K_1$	石英正长岩			番通、四方岭
	阜石斗	$\xi\gamma K_1$	正长花岗岩			阜石斗
	抱跃	$\eta\gamma K_1$	二长花岗岩			抱跃
	税町	$\gamma\delta K_1$	花岗闪长岩			税町
		$\gamma\pi K_1$	花岗斑岩			$K_1\gamma\pi$
		$\gamma\delta\pi K_1$	花岗闪长玢岩			$K_1\gamma\delta\pi$
	拉丁村	νK_1	辉长岩、辉长闪长岩			$K_1\nu$、$K_1\delta\mu$
	半岛	$\xi\gamma c K_1$	正长花岗岩	半岛		
侏罗纪	马岭	$\xi\gamma f J_3$	正长花岗岩	马岭		
	马鞍岭	$\xi\gamma m J_3$	正长花岗岩	马鞍岭	朝根坡	
	三狮岭	$\xi\gamma J_3$	正长花岗岩	三狮岭		横岭、廖次岭
	架水	$\eta\gamma cm J_3$	二长花岗岩		架水	
	冬瓜岭	$\eta\ \gamma f J_3$	二长花岗岩			冬瓜岭
	椰子园	$\eta\gamma J_3$	二长花岗岩			椰子园
	南山岭	$\eta\gamma mf J_3$	二长花岗岩			南山岭
	马翁岭	δJ_3	闪长岩	老村	托剩岭	马翁岭
	八宝坡	ξJ_2	正长岩			八宝坡
	黑栋	$\eta o J_2$	石英二长岩			黑栋
	里茂	υJ_2	辉长岩			里茂

续表

本书使用岩石单位			岩石类型	1∶25 万区调使用岩石单位		
时代	典型命名地	代号		东方县幅	琼海幅	乐东县、陵水县幅
三叠纪	迈洲岭	$k\gamma\pi T_3$	碱长花岗斑岩		迈洲岭	
	打显村	$\eta\gamma mfT_3$	二长花岗岩	打显村		
	白石岭	$\eta\gamma T_3$	二长花岗岩	白石岭		
	黑岭	$\eta\gamma pT_3$	二长花岗岩	黑岭		
	分水岭	$\eta\gamma fmT_3$	二长花岗岩	分水岭		
	南宝岭	$\eta\gamma mpT_3$	二长花岗岩	南宝岭		
	五里桥	$\eta\gamma mT_3$	二长花岗岩	五里桥	长坡	
	苗圃	$\eta\gamma fT_3$	二长花岗岩	苗圃		
	乌烈岭	$\xi\gamma T_3$	正长花岗岩	乌烈岭		
	东四	$\xi\gamma cT_3$	正长花岗岩	东四		
	七差岭	$\xi\gamma fT_3$	正长花岗岩	七差岭		
	保梅岭	$\xi\gamma T_2$	长花岗岩	保梅岭		
	洞古岭	$\gamma\pi T_2^d$	花岗斑岩		洞古岭	
	南通岭	$\xi\gamma T_2^c$	正长花岗岩		南通岭	
	琼中县	$\eta\gamma T_2^b$	二长花岗岩		琼中县	
	黎母岭	$\eta\gamma pT_2^a$	二长花岗岩		黎母岭	
	后美	$X\rho\gamma T_2^b$	碱长花岗岩			
	后龙	$X\rho\gamma T_2^a$	碱长花岗岩			
	石墩	δT_2	闪长岩			石墩
	长安	υT_2	辉长岩	光头岭		长安
	石龟岭	$\eta\gamma fT_2$	二长花岗岩		坡尾坡	石龟岭
	兵工厂	$\xi\gamma fpT_2$	正长花岗岩		兵工厂	兵工厂
	六连岭	$\xi\gamma cT_2$	正长花岗岩		石母屋	六连岭
	前锋	$\xi\gamma fT_2$	正长花岗岩			前锋
	结尾	$\eta\gamma mfT_2$	二长花岗岩			结尾
	超盆	$\eta\gamma cT_2$	二长花岗岩			超盆
	布山村	$\eta\gamma T_2$	二长花岗岩			布山村
	金寮	$\xi\gamma mT_1$	正长花岗岩			金寮
	番阳峒	$\xi\gamma mcT_1$	正长花岗			番阳峒
	麻山田	$\xi o T_1$	石英正长岩			麻山田
	深堀村	$\eta o T_1$	石英二长岩			深堀村
	深沟	$\kappa\xi T_1$	碱长正长岩			深沟
	进岭	$\gamma\pi T_1$	花斑岩			进岭
	瘦岭	$\xi\gamma fT_1$	正长花岗岩			瘦岭
	黑岭	$\xi\gamma T_1$	正长花岗岩			黑岭
	尖峰岭	$\xi\gamma cT_1$	正长花岗岩			尖峰岭

续表

本书使用岩石单位			岩石类型	1∶25万区调使用岩石单位		
时代	典型命名地	代号		东方县幅	琼海幅	乐东县、陵水县幅
二叠纪	大王岭	$\eta\gamma mfP_3$	二长花岗岩		大王岭	
	高村	$\eta\gamma pP_3$	二长花岗岩	高村	竹山岭	
	两院	$\eta\gamma cpP_3$	二长花岗岩	两院	美合	
	大岭	$\gamma\delta pP_3$	花岗闪长岩	大岭	乾元	
	鹅塘岭	$\gamma\delta oP_3$	英云闪长岩	鹅塘岭		
	陀类	γP_3	二长花岗岩			陀类
	大田	$\eta\gamma P_3$	二长花岗岩	大田		大田
	什天	$\eta\gamma mP_3$	二长花岗岩			什天
	白茶	$\eta\gamma fP_3$	二长花岗岩			白茶
	东岭	$\gamma\delta P_3$	花岗闪长岩			东岭
	那佩	$\eta\gamma cP_3$	二长花岗岩			那佩
	保国	$\eta\gamma mcP_3$	二长花岗岩			保国
	火岭	$\xi\gamma P_2$	正长花岗岩	火岭、荔枝岭		黄山水库
	长塘岭	$\eta\gamma P_2$	二长花岗岩	长塘岭、东崩岭		中林岭
	大坡	$\eta\gamma mP_2$	二长花岗岩	大坡		
	乌兰盖	$\eta\gamma mcP_2$	二长花岗岩	乌兰盖		
	通什	$\eta\gamma pP_2$	二长花岗岩	邦溪		通什
	山牛塘	δP_2	闪长岩		南坤	山牛塘
	石碌	$\eta\gamma cP_1$	二长花岗岩	石碌		
	什败	ηoP_1	石英二长岩			什败
	行干	$\delta\eta oP_1$	二长闪长岩			行干、什暗
	方满	ηP_1	二长岩			方满
	新风岭	$\xi\gamma fP_1$	正长花岗岩			新风岭
	顺作	$\xi\gamma mP_1$	正长花岗岩			顺作
	乘顺岭	$\eta\gamma fP_1$	二长花岗岩			乘顺岭
	志仲	$\eta\gamma mpP_1$	二长花岗岩			志仲
	抱郎	$\eta\gamma P_1$	二长花岗岩			抱郎
	石岭	$\gamma\delta P_1$	花岗闪长岩	石岭、美园	东岭	
	保山村	$\gamma\delta oP_1$	英云闪长岩	保山村	美马岭	
	白石村	δP_1	闪长岩	白石村	坤外	毫老发、什哺
	机告村	υP_1	辉长岩	机告村	下园	红水岭
元古宙	公庙	ΨoPt_3	斜长角闪岩		公庙	
	亚炮	$\eta\gamma Pt_2$	二长花岗岩	亚炮		$Pt_2\gamma$
	戈枕	$\gamma\delta Pt_2$	花岗闪长岩	戈枕		$Pt_2\gamma$

第九章　海南省侵入岩各论

第一节　元古宙侵入岩

海南省元古宙侵入岩出露面积极少，主要分布于海南岛西部公爱农场至公爱乡（即大蟹岭地区）、乐东县冲坡镇冲卒岭亚南甫沟及琼中县长征地区，分别有中元古代戈枕片麻状花岗闪长岩、亚炮片麻状二长花岗岩和新元古代公庙斜长角闪岩以及一些紫苏花岗岩脉、斜长花岗岩脉等，出露总面积约 155.86km^2（图 9-1）。

一、中元古代戈枕片麻状花岗闪长岩（$\gamma\delta Pt_2$）

1. 地质特征及形成时代

海南岛中元古代戈枕片麻状花岗闪长岩主要分布于海南岛西部抱板镇-戈枕村一带，总面积约 20.37km^2。根据 1∶5 万《叉河》幅区域地质调查报告成果（海南地质大队，1993），戈枕侵入体北侧被亚炮片麻状二长花岗岩侵入（图 9-2），与中元古界抱板群呈侵入接触关系，边界与围岩片理产状一致。前人在戈枕片麻状花岗闪长岩中获得同位素年龄有：1440.87Ma（锆石 U-Pb，俞受鋆，1992）、($1457\pm^{84}_{66}$) Ma（锆石 U-Pb，马大铨、黄香定等，1993）、(1397±46) Ma（锆石 Pb-Pb，丁式江，1995）、1430Ma（锆石 SHRIMP U-Pb，Zheng-Xiang Li，2002），侵入时代厘定为中元古代。

戈枕侵入体普遍遭受不同程度的变质作用和构造改造，变形作用强烈，岩石普遍发育强韧性变形，显片麻状构造。石英拔丝状，云母条纹状，残斑钾长石常呈眼球状，片麻理走向 300°~330°，靠近戈枕断裂带为 30°~60°。

戈枕侵入体发育大量的变形包体，包体岩石类型以细粒黑云母斜长片麻岩为主。包体分布不均匀，局部成群呈带状出现。大小一般 2~3cm，个别大于 40cm，呈扁豆状、透镜状、条带状或椭圆状、不规则状。包体定向排列比较明显，走向与寄主岩片麻理走向基本一致。

2. 岩石学特征

主体岩性为中细粒片麻状黑云母花岗闪长岩。浅灰色，色率 10~15，岩石具中细粒片状、糜棱岩化变晶结构，普遍见有强韧性流态变形，发育片麻理构造，矿物定向排列，片麻理由长英质矿物及暗色矿物呈条带状定向分布组成。矿物变形作用和蚀变作用比较强烈，长石交代结构十分发育；石英粒状、拔丝状；云母条纹状、解理弯曲、扭折；出现大量白云母，长轴方向与片麻理走向大体一致。矿物内有波状消光、变形纹、变形带、变形双晶、亚颗粒化等韧性变形形迹；有白云母、绢云母、绿泥石等新矿物形成及黑云母析出铁质、石英重结晶等低绿片岩相变质变形现象。局部残留有微粒交生体。

上述岩石学特征表明，戈枕侵入体为同构造花岗岩并遭受了后期变质变形改造。

岩石中矿物分布较均匀，粒径多在 0.5~3mm 之间。矿物组成：斜长石（An25~An29）30%~50%、石英 20%~30%、钾长石 15%~20%、黑云母 5%~15%；副矿物主要有石榴子石、磷灰石、锆石、褐帘石、电气石等。锆石主要呈浅黄褐色，以（111）、（100）聚形为主，柱体长宽比 2∶1~3∶1，少数 4∶1，晶面粗糙有裂纹，具熔蚀现象，含黑色包裹体及气泡（大小 0.04~0.25mm）。

3. 岩石地球化学特征

岩石的常量元素含量及特征参数见本章附表 9-1、附表 9-2。SiO_2 含量介于 67.46%~72.536%之间，属酸性岩类（邱家骧、林景仟，1991）；总碱含量中等，K_2O+Na_2O 为 5.04%~5.99%，平均 5.50%，并

图 9-1　海南岛元古宙侵入岩分布图

Fig. 9-1　Distribution of Proterozoic intrusive rocks in the Hainan Island

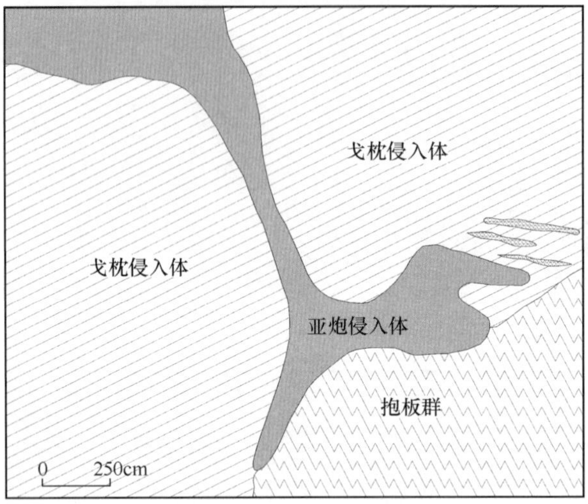

图 9-2 亚炮二长花岗岩呈岩枝状侵入戈枕花岗闪长岩和抱板群

Fig. 9-2 Yapao monzonitic granite intruding the Gezhen granodiorite and Baoban Group as apophysal configuration

（据 1∶5 万《叉河幅》区调报告）

且 K_2O 和 Na_2O 含量相当；Al_2O_3 为 13.29%~14.41%，平均 13.73%，含铝指数 A/CNK 为 1.15~1.41，属强过铝质岩石；Fe_2O_3、FeO、MnO、MgO 含量较低，FeO 含量明显高于 Fe_2O_3；根据里特曼指数（σ=1.04~1.21），该岩石属钙性岩系；巴基耶娃（1976）研究科拉半岛碱性花岗岩时提出碱铝指数 NKA<0.75 为钙碱性岩，地矿部南岭项目花岗岩专题组（1989）则将南岭花岗岩 KAI=0.9~0.4 划分为钙碱性岩，而戈枕花岗闪长岩碱铝指数 NKA=0.48~0.62，属于钙碱性范围内。

岩石的稀土元素含量（本章附表 9-3）在 150×10^{-6}~266×10^{-6}之间，低于维氏世界酸性岩平均值（292×10^{-6}），采用 Sun and McDonough（1989）的 CI 球粒陨石标准化（本书中所有侵入岩均采用该标准），稀土元素球粒陨石标准化配分模式为略向右倾的翼展型（图 9-3）；$(La/Sm)_N$ 为 3.56~5.59，$(Gd/Yb)_N$ 为 1.26~2.06，轻稀土分异较明显，重稀土分异一般；δEu 为 0.46~0.64，铕中等负铕异常。

在原始地幔（Sun and McDonough，1989，下同）微量元素标准化蛛网图上（图 9-4），以强烈富集 Rb、Ba、K、U、Th、Pb，亏损 P、Sr、Ti、Nb、Ta 为特征。

岩石学和岩石地球化学均表现出壳源造山花岗岩特征。

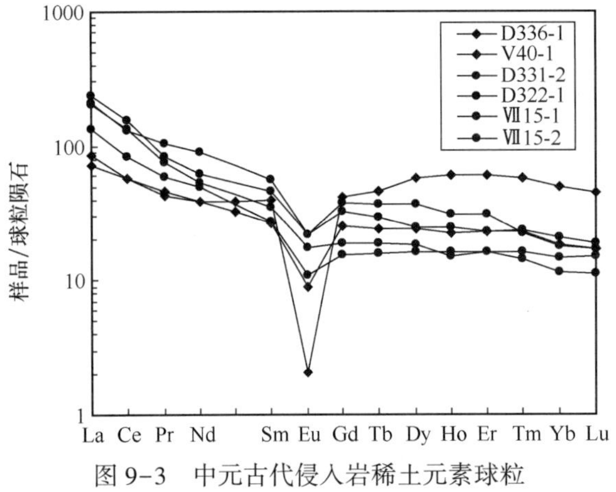

图 9-3 中元古代侵入岩稀土元素球粒陨石标准化分布型式图

Fig. 9-3 Chondrite-normalized REE patterns of Mesoproterozoic intrusive rocks

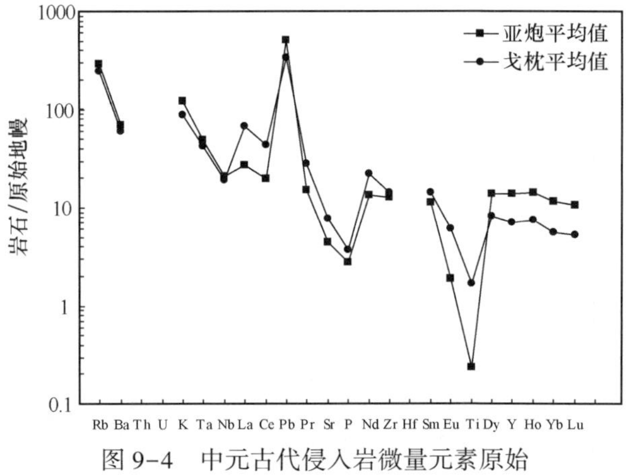

图 9-4 中元古代侵入岩微量元素原始地幔标准化蛛网图

Fig. 9-4 Primitive mantle-normalized spider diagram of trace elements of Mesoproterozoic intrusive rocks

二、中元古代亚炮片麻状二长花岗岩（$\eta\gamma Pt_2$）

（一）地质特征及形成时代

中元古代亚炮片麻状二长花岗岩主要出露于海南岛西南部亚炮村-老烈村及乐东县冲卒岭地区亚南甫沟一带，侵入体形态呈不规则状，出露总面积约133.1km^2。由于受后期构造影响和改造，变形作用强烈，岩石中普遍发育片麻理构造，矿物定向排列，片麻理由长英质矿物及暗色矿物呈条带状定向分布组成。亚炮村-老烈村一带，北侧片麻理走向为300°~330°，南侧靠近戈枕断裂带呈45°~60°，与戈枕断裂带的走向基本一致。

马大铨、黄香定等（1998）在二甲村东约300m的广坝农场4队队部边小沟片麻状细粒二长花岗岩中采样，获得锆石U-Pb同位素年龄（由4个锆石分样拟合测试）（$1457\pm^{84}_{66}$）Ma、锆石Pb-Pb同位素年龄（1407±20）Ma，两个年龄值在误差范围内一致（表9-1）。侵入时代为中元古代（戈枕片麻状花岗闪长岩之后）。

表9-1 亚炮片麻状细粒二长花岗岩锆石分析结果

Table 9-1 Zircon analytical results of Yapao gneissic fine-grained monzonitic granite

分析号	样品重量 mg	粒径 mm	放射成因铅含量 10^{-6}	铀含量 10^{-6}	$^{204}Pb/^{206}Pb$（测定值）	同位素原子比		
						$^{207}Pb/^{206}Pb$	$^{206}Pb/^{238}U$	$^{207}Pb/^{235}U$
B346-1	0.6	0.084~0.105	432.78	2 904.4	0.001 76	0.085 82	0.148 39	1.747 3
B346-2	0.3	0.084~0.105	805.39	5 536.2	0.001 07	0.085 92	0.144 78	1.706 9
B346-3	1.0	0.189~0.21	577.56	3 960.3	0.001 7	0.086 55	0.145 15	1.723 7
B346-4	1.1	0.20~0.232	606.35	3 813.55	0.000 61	0.087 27	0.158 15	1.893 9

注：数据来自《海南岛结晶基底——抱板群层序与时代》（马大铨、黄香定等，1998）。

（二）岩石学特征

主体岩性为片麻状（中）细粒黑云母二长花岗岩，部分地段有正长花岗岩。浅灰色，普遍见有强韧性流态变形（图版Ⅱ-1-A），显片麻状构造，石英拔丝状，云母条纹状，残斑钾长石常呈眼球状。其中大蟹岭地区片麻理走向为南东东—北东向，冲卒岭地区则为北东向。糜棱岩化变晶结构，矿物内有波状消光、变形纹、变形带、变形双晶、亚颗粒化等韧性变形形迹；有白云母、绢云母、绿泥石等新矿物形成及黑云母析出铁质、石英重结晶等低绿片岩相变质变形现象；糜棱碎基含量一般<10%，粒度<0.2mm，局部残留有微粒交生体，表明为同构造花岗岩并遭受后期变质变形改造。残余（中）细粒似斑状或细粒（含斑）花岗结构。前者的斑晶均为微斜长石，含量15%~25%，分布较均匀，大小4mm×6mm~1.5cm×2.5cm；残余基质粒径多为0.5~2mm，少数2~4mm。矿物组成：斜长石（奥长石）25%~50%、石英25%~35%、钾长石（微斜长石）20%~55%、黑云母3%~9%。

副矿物主要有磷灰石、锆石、褐帘石等，正长花岗岩和冲卒岭地区的二长花岗岩中有较多铁铝-锰铝榴石，局部高达2%。

（三）岩石地球化学特征

岩石的常量元素含量及特征参数见本章附表9-1、附表9-2。SiO_2含量平均为75.26%，属酸性岩类；总碱含量$K_2O+Na_2O=6.2\%\sim7.23\%$，平均6.72%，稍高于戈枕花岗闪长岩；$Al_2O_3=12.4\%\sim12.69\%$，平均12.55%，含铝指数A/CNK为1.23~1.26，属强过铝质岩石；Fe_2O_3、FeO、MnO、MgO含量低于戈枕花岗闪长岩，FeO含量明显高于Fe_2O_3；组合指数（里特曼指数）σ为1.18~1.57，碱铝指数NKA为0.66~0.76，属钙碱性序列。

岩石的稀土元素总量为$129\times10^{-6}\sim230\times10^{-6}$，比维氏世界酸性岩平均值低。稀土元素配分曲线为略右倾的 V 形（图 9-3），轻稀土分异较明显，$(La/Sm)_N$ 为 1.83～3.09；重稀土分异不明显甚至反向富集，$(Gd/Yb)_N$ 为 1.35～0.82；δEu 为 0.35～0.05，铕负异常中等至强烈。

微量元素特征与戈枕花岗闪长岩相似，在原始地幔标准化蛛网图上（图 9-4），以强烈富集 Rb、Ba、K、U、Th、Pb 和亏损 P、Sr、Ti、Nb、Ta 为特征，具有造山花岗岩的特点。

研究表明，细粒片麻状二云二长花岗岩初始锶比值为 0.72018，片麻状中细粒似斑状黑云二长花岗岩初始锶比值为 0.70850，显示它们都来源于地壳。其中，二云二长花岗岩来源于成熟度较高的地壳岩石，可能是变沉积岩类；似斑状二长花岗岩可能主要来源于变火成岩类。似斑状黑云二长花岗岩钾长石铅同位素组成为：$^{206}Pb/^{204}Pb$ 17.583±0.002、$^{207}Pb/^{204}Pb$ 15.525±0.002、$^{208}Pb/^{204}Pb$ 38.639±0.001，具有低的放射性成因铅，与其较古老的形成年龄（约 14 亿年）相吻合（宜昌地质矿产研究所、海南省地质调查院，2004）。

三、新元古代公庙斜长角闪岩（$\psi o\ Pt_3$）

（一）地质特征及形成时代

新元古代青白口纪斜长角闪岩主要分布在琼中上安、东方二甲、抱板、乐东峨文岭及屯昌中建农场公庙一带的中元古界中，并见侵入抱板群戈枕村组，出露总面积约 2.39 km^2。

对此类产出于中元古界的斜长角闪岩的成因，目前普遍认为是正变质岩。但对其原岩是火山岩还是侵入岩存在较大分歧。马大铨、黄香定等（1988）认为，产在戈枕村组中的及部分产在峨文岭组中的斜长角闪岩其原岩为浅成侵入体，而产在峨文岭组中的一种结晶较细、具“芝麻点状”的斜长角闪岩，尚不能完全排除其原岩为基性火山岩的可能。为此，马大铨、黄香定等（1998）在峨文岭组中采取斜长角闪岩进行 Sm-Nd 法测年（表 9-2），获得全岩等时线年龄（955±54）Ma。与谭忠福、马国干等（1991）报道的斜长角闪岩 Sm-Nd 等时线年龄（975±8.6）Ma 在误差范围内一致，而比峨文岭组可能形成年龄（1407Ma±20Ma～1637Ma±31Ma）小得多，表明采样测年的斜长角闪岩为较晚侵入的岩脉。可以推断，新元古代早期（青白口纪）在中元古界分布区发生了一次构造岩浆侵入事件，形成了酸性浅成岩和分布更广的基性浅成岩（已变成斜长角闪岩），稍后（841Ma±20Ma 之前）发生的区域变质作用，使这些酸性、基性岩发生角闪岩相变质（马大铨、黄香定等，1998）。

表 9-2　峨文岭组中的斜长角闪岩 Sm-Nd 法测定结果

Table 9-2　Sm-Nd dating results of amphibolites in the Ewenling Formation

序号	样品号	$Sm/10^{-6}$	$Nd/10^{-6}$	$^{147}Sm/^{144}Nd$	$^{143}Nd/^{144}Nd$
1	B238	1.05	3.71	0.171 8	0.512 578±0.000 046
2	B73	4.29	15.51	0.160 4	0.512 496±0.000 021
3	B342	4.05	17.38	0.140 8	0.512 382±0.000 019
4	B342-P1	0.76	4.89	0.094 6	0.512 173±0.000 023
5	B342-（Hb+Ga）	3.25	10.35	0.190 2	0.512 642±0.000 032

注：数据来自《海南岛结晶基底——抱板群层序与时代》（马大铨、黄香定等，1998）。

（二）岩石学特征

岩石类型主要为斜长角闪（片）岩。岩石呈绿色—深绿色，粒状、柱状变晶结构，块状构造为主，部分具片麻状构造或不明显定向构造。主要造岩矿物为普通角闪石 40%～63%、斜长石 25%～45%、石英 3%～7%、钾长石 2%～6%和黑云母 0～12%。普通角闪石大部分呈他形粒状、短柱状，少量呈半自形长柱状，淡黄色—浅绿色，多色性明显；部分角闪石晶体较粗大，晶形发育较好，内部多见长英质矿物包裹体，构成“筛状”结构，常沿轴向近于平行定向排列；斜长石呈他形粒状，粒径

0.1~0.5mm，蚀变较强，部分见聚片双晶；石英呈他形粒状，粒径0.1~0.5mm，具波状消光；钾长石呈他形粒状，多发生黏土化、绢云母化；黑云母呈他形片状，主要分布于浅色矿物颗粒之间。

（三）岩石地球化学特征

斜长角闪岩 SiO_2 含量为49.94%~52.53%，属基性岩类；含铝指数A/CNK为0.61~0.64，里特曼指数 σ 为1.54~1.55，属准铝质钙碱性序列。其 $\varepsilon_{Nd}\approx+6$，暗示原岩来自亏损地幔（马大铨、黄香定等，1998）。

四、紫苏花岗岩脉、斜长花岗岩脉

据马大铨等（1998）和宜昌地矿所、海南省地质调查院2004年完成的1∶25万《乐东县、陵水县》幅区域地质调查成果资料，紫苏花岗岩脉、斜长花岗岩脉产出于琼中长征地区的戈枕村组中。

紫苏花岗岩脉一般厚10余厘米至数十厘米，大致顺片麻理侵入黑云（角闪）斜长片麻岩中，最大出露宽约10m，为无根型脉体。岩脉内有麻粒岩包体，附近往往有紫苏（二辉）斜长片麻岩和麻粒岩的孤岛状退变质残留体，并且可有含麻粒岩小团块的斜长角闪岩。其岩石类型有紫苏二长花岗岩、紫苏花岗闪长岩、紫苏斜长花岗岩，以紫苏二长花岗岩为主。紫苏斜长花岗岩产在变质基性岩中，岩性不均匀，含钾长石斑晶，疑有混染；紫苏花岗岩则与变质基性岩相距稍远，岩性较稳定。不过，这种成分变化也曾见于同一个不大的露头范围内。它们一般为中粗粒花岗结构，或者为似斑状结构（斑晶微斜长石），弱片麻状构造，有时见微粒交生体。矿物含量变化较大，斜长石25%~50%、钾长石5%~45%、石英10%~31%、黑云母5%~20%、紫苏辉石2%~15%、透辉石0~6%、角闪石0~8%、石榴子石0~3%。斜长石半自形板状，发育多种双晶，偶见环带构造，An=38~55；钾长石多为他形粒状，在紫苏斜长花岗岩中可见少量半自形板状斑晶，主要为微斜长石，少量微斜-条纹长石，发育各种交代结构；石英他形粒状，多呈不规则集合体产出，具波状消光；黑云母半自形片状，棕红色或红褐色，多色性明显；紫苏辉石自形—半自形粒状或短柱状，偶见简单双晶，常被（纤维状）角闪石交代；透辉石自形—半自形粒状或短柱状，无色；角闪石柱状，深绿色，多色性明显；石榴子石等轴粒状，部分被黑云母、绿泥石交代；副矿物主要有磷灰石、锆石（常呈浑圆状）、磁铁矿。

斜长花岗岩脉厚10余厘米至数十厘米，基本顺片麻理侵入斜长角闪岩和黑云斜长变粒岩中，并又一起遭受了褶皱。弱片麻状构造，细粒花岗结构，矿物粒径0.1~1mm。主要矿物组成：石英35%、斜长石55%、黑云母8%。副矿物主要有磁铁矿、磷灰石、榍石。

紫苏花岗岩脉和斜长花岗岩脉的常量元素含量及特征参数见本章附表9-1、附表9-2。具有富硅、准铝或弱过铝特点。但两者的 K_2O/Na_2O 比值明显不同，分属钾质和钠质花岗岩。在 SiO_2 - K_2O 图上（图略），紫苏花岗岩脉和斜长花岗岩脉分别落入高钾钙碱性-钾玄岩区和低钾序列中。

第二节　二叠纪侵入岩

海南岛二叠纪侵入岩较发育（图9-5），出露总面积约6298.73km²，占海南省侵入岩总面积的38%。划分为31个岩石单位（表8-2）。岩石类型以正长花岗岩和二长花岗岩为主，少量花岗闪长岩、闪长岩和辉长岩，以岩体内部普遍发育强弱不等的定向构造为主要特征。

一、早二叠世大乃村序列

早二叠世大乃村序列主要分布于海南岛西部昌江县邦溪、坝王岭一带，出露总面积159.31km²。根据侵入体之间的侵入接触关系、同位素年龄及岩石学特征，该序列岩石从早到晚由机告村辉石闪长岩-辉长岩、白石村闪长岩-石英闪长岩、保山村英云闪长岩和石岭花岗闪长岩4个岩石单位组成，成分演化特征明显，具有类TTG花岗岩特征，形成于活动陆缘构造环境，岩石成因属壳幔混合型。

（一）机告村辉石闪长岩-辉长岩（υP_1）

1. 地质特征及形成时代

分布于邦溪、坝王岭、尖峰岭一带，由长塘村、坟山岭、鱼塘村、九工段、俄牙岭、乙劳村和红水岭等一些小侵入体组成，总面积4.25 km²。被稍晚形成的保山村英云闪长岩和石岭花岗闪长岩侵入。坝王岭九工段一带，保山村英云闪长岩呈不规则状岩枝或岩脉插入机告村角闪辉长岩（九工段侵入体）中，两者之间接触界线清楚；长塘村东西向河谷中，见到石岭花岗闪长岩侵入机告村含辉闪长岩（长塘村侵入体），在花岗闪长岩的边部出现4~5cm宽的冷凝边，内部见有含辉石闪长岩的捕虏体。因此，机告村辉长岩形成时间应早于保山村英云闪长岩（约287 Ma）和石岭花岗闪长岩，暂归为早二叠世。

2. 岩石学特征

主体岩性为中细粒角闪辉长岩、石英辉长岩及少量辉石闪长岩，侵入体边缘还有微晶角闪辉长岩。岩石手标本呈黑色或深灰色，辉长结构或含长结构，局部为反应边结构、交代结构。原生矿物组合为单斜辉石（透辉石、普通辉石）0~11%、普通角闪石10%~25%、黑云母8%~30%、斜长石（An12~An78）50%~64%，偶见少量石英（0~5%）和次生钾长石。矿物粒径多在0.4~2mm之间，部分可达3~5mm。斜长石半自形板条状，钠长石双晶发育，常见环带，中心部分An值45~78，边部12~30，属正环带，中心部分常具绢云母化，有时也被绿帘石或碳酸岩交代；角闪石呈自形半自形柱状，N_g深绿色，N_p黄褐色，$C \wedge N_g = 24°$，属普通角闪石，常包裹磷灰石、磁铁矿等副矿物；黑云母呈叠片状，多呈集合体充填在长石颗粒之间，N_g深棕色，N_p黄褐色，常被绿泥石交代并析出他形榍石。见有少量石英、钾长石呈他形粒状充填在斜长石之间。

副矿物有榍石、褐帘石、锆石、磷灰石和磁铁矿等，其中榍石见有他形、楔形或菱形自形晶体。

岩体中偶尔见到颜色更深的闪长质包体（暗色闪长岩）或暗色矿物集中形成的暗色斑块，直径一般为2~5cm。

（二）白石村闪长岩-石英闪长岩（δP_1）

1. 地质特征及形成时代

主要由出露于海南岛西部兰洋-芙蓉田-坝王岭一带的坤外、合罗、北沟、白石村和坡兰沟等一些小侵入体组成，总面积37.64km²。坡兰沟侵入体侵入上石炭统青天峡组，而被早二叠世保山村英云闪长岩、中二叠世邦溪岩体二长花岗岩等后期岩体侵入。已有的同位素年龄为：坡兰沟侵入体K-Ar法年龄293Ma（934地质队，1982），合罗侵入体Rb-Sr等时线年龄（298.6±10.3）Ma（马大铨等，1991）（分析结果见表9-3），上述年龄值与地质侵入接触关系相符，侵入时代厘定为早二叠世。

表9-3 合罗侵入体Rb-Sr分析数据

Table 9-3 Rb-Sr analytical results of Heluo intrusion

样品号	采样位置	$Rb/10^{-6}$	$Sr/10^{-6}$	$^{87}Rb/^{86}Sr$	$^{87}Sr/^{86}Sr$	
HG-298	56.0，52.6	83.345	455.11	0.52805	0.71076	t=（298.6±10.3）Ma I_0=0.70919±20 γ=0.99882
HG-310	56.3，52.2	144.64	191.85	2.17276	0.71835	
HG-574	52.4，50.2	87.774	553.70	0.47424	0.71098	
HG-575	52.4，50.2	129.20	252.35	1.47703	0.71561	
HG-295-1	56.0，53.8	68.505	596.37	0.33122	0.71075	

注：数据来自《海南岛地质（二）岩浆岩》（马大铨等，1991）。

2. 岩石学特征

岩石类型主要为细粒、中细粒闪长岩、石英闪长岩。岩石具他形—半自形晶结构、嵌晶结构，局部交代结构。矿物大小多在1.0~3mm之间，部分可达4mm。矿物组合主要为普通角闪石5%~38%、

黑云母5%、中性斜长石48%~73%、石英8%~15%及少量普通辉石、钾长石。其中斜长石端员组分分别为 Or 0.22%~0.32%、Ab 40.25%~45.58%、An 54.10%~59.53%，属基性斜长石。黑云母以含镁较高为特征。

副矿物有榍石、褐帘石、锆石、磷灰石和磁铁矿等。

岩体中见少量闪长质包体，直径5~15cm，椭圆或透镜状，界线清楚，由暗色闪长岩、含透辉石闪长岩组成。

（三）保山村英云闪长岩（$\gamma\delta oP_1$）

1. 地质特征及形成时代

分布于海南岛西部昌江一带，由大章村、青松等侵入体组成，总面积约26.46km²。保山村英云闪长岩侵入二叠系峨查组、鹅顶组和早二叠世机告村辉石闪长岩-辉长岩和白石村闪长岩-石英闪长岩。与二叠系峨查组、鹅顶组接触处，接触界线呈不规则状，接触面倾向围岩，倾角60°~70°，局部见宽1~2m的细粒边，并可见侵入体呈岩脉贯入于围岩中，内接触带附近见不规则状、棱角状围岩捕虏体。

前人测得保山村英云闪长岩的同位素年龄为（287.2±10.6）Ma（全岩Rb-Sr等时线法，广西区调队，1991）、（278±1）Ma（锆石U-Pb法，广西区调队，1991）、（270±11）Ma（SHRIMP锆石U-Pb法，海南省地质调查院，2007），上述年龄值均与地质侵入接触关系相符，将其侵入时代厘定为早二叠世。

2. 岩石学特征

岩石类型主要为中细粒、细粒英云闪长岩，局部为中细粒花岗闪长岩。岩石具他形—半自形晶结构、细粒花岗结构，矿物大小多在0.5~2mm之间，部分可达2~4mm。矿物定向排列，普遍发育叶理构造，叶理走向近于东西向，南倾为主，倾角50°~80°。矿物组成主要为普通角闪石0~15%、黑云母10%~24%、中酸性斜长石28%~50%、石英18%~42%。零星出现少量钾长石，在靠近后期花岗岩体接触带时钾长石含量明显增多，这种现象可能与花岗岩浆的混染作用有关。其中钾长石的端员组分Or为92.90；黑云母投影在铁质黑云母区；角闪石晶体化学式为：

$$K_{0.180}(Ca_{1.742}Na_{0.216})_{1.958}(Al^{VI}_{0.539}Ti_{0.115}Fe^{3+}_{0.659}Mg_{1.732}Fe^{2+}_{1.847}Mn_{0.061})_{4.953}$$
$$(Si_{6.792}Al^{IV}_{1.208})_8O_{22}(OH.F.Cl)_2$$

根据国际矿物学会矿物分类命名委员会推荐的分类方案（1978），属于钙角闪石类中的低铁-普通角闪石。

保山村英云闪长岩中见有较多的闪长质包体。包体呈灰色—灰黑色，与主岩界线清楚，包体长轴多在2~55cm之间，其中小于10cm的占65%~92%，密度一般2~4个/m²，多者达8~10个/m²。包体含量一般为2%~3%，局部达8%。应变轴率从1.14~2.33不等，并有从边部到中心逐渐变小的趋势。包体形态有长条状、椭圆状、棱角状及不规则状等。长条状及椭圆状包体往往定向排列，与主岩叶理走向一致，可能为同生包体；而棱角状包体则往往成群出现，可能为浅部的捕虏体。部分包体发生弯曲，有时被剪切拉断，发生石香肠化等，上述包体发生的强烈变形现象，可能与区域变形作用有关。

（四）石岭花岗闪长岩（$\gamma\delta P_1$）

1. 地质特征及形成时代

出露于昌江县邦溪至七坊一带，有石岭、金波一队、德立村、白鱼塘等侵入体，总面积约90.96km²。石岭侵入体北部侵入陀烈组变质砂岩、粉砂岩，外接触带具角岩化；西部侵入早二叠世机告村闪长岩；南部被早二叠世石碌二长花岗岩（同位素年龄为274Ma±6.5Ma）侵入。与石碌二长花岗岩接触处，接触界线清楚，在石碌二长花岗岩中，可见有石岭花岗闪长岩捕虏体，而在花岗闪长岩中又见有中粗粒巨斑状黑云母二长花岗岩岩枝穿插（图9-6），因此，石岭花岗闪长岩的形成时间应大于（274±6.5）Ma，将其侵入时代厘定为早二叠世。前人在查苗东高地桥获 Rb-Sr 等时线年龄值为 231Ma（宜昌地质矿产研究所，1995），与地质侵入关系不符，该年龄值应视为石岭花岗闪长岩形成后遭受后期岩浆热事件使 Rb-Sr 同位素局部均一化的年龄。

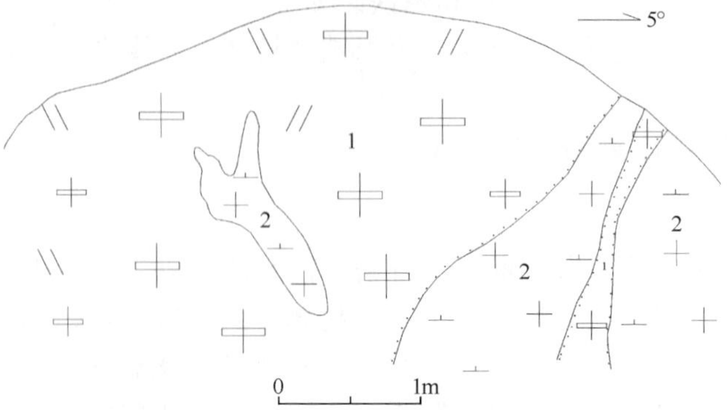

图 9-6　石岭花岗闪长岩被石碌二长花岗岩侵入接触关系剖面素描图

Fig. 9-6　Section sketch showing the contact of Shiling granodiorite intruded by Shilu monzonitic granite

（据 1∶5 万昌江县、邦溪幅区调报告）

1—石碌中粗粒巨斑状黑云母二长花岗岩；2—石岭花岗闪长岩

2. 岩石学特征

主体岩石类型为中细粒花岗闪长岩。手标本呈深灰或灰白色，侵入体与地层围岩接触处为中细粒花岗结构，中心部位为中粒花岗结构。矿物大小多在 0.5～4mm 之间，部分可达 4～5mm。局部出现似斑状结构，斑晶由斜长石组成，大小 3mm×8mm，含量<10%。主要矿物组成：斜长石 38%～56%、石英 20%～31%、钾长石 6%～10%、角闪石 0～6%、黑云母 4%～20%，零星出现少量普通辉石。矿物化学成分为钾长石 An 0.15%、Ab 9.13%、Or 90.72%；斜长石 An 48.30%、Ab 50.51%、Or 1.19%，属中性斜长石；黑云母属铁质黑云母；角闪石属钙质角闪石类。副矿物中普遍出现榍石，呈楔形或菱形自形晶体，粒晶有时达 1mm。

石岭花岗闪长岩中见有暗色包体，包体具典型的岩浆结构，岩石类型以石英闪长质为主，少量二长闪长质和闪长质。包体有圆状、椭圆状及不规则状，直径多为 8～20cm。

（五）岩石地球化学特征

该序列岩石具有比较明显的成分演化特征，从早期的机告村辉石闪长岩-辉长岩、白石村闪长岩-石英闪长岩、保山村英云闪长岩到晚期的石岭花岗闪长岩，随着 SiO_2 含量的增加，Al_2O_3、K_2O 呈递增态势，MgO、Fe_2O_3+FeO、CaO、Na_2O 含量则逐渐降低。含铝指数 A/CNK 在 0.62～1.06 之间，大部分小于 1.0，属准铝质—弱过铝质；碱铝指数 NKA 为 0.36～0.69，组合指数（里特曼指数）σ 最小为 0.96，最大为 5.99，大部分在 1.8～3.3 之间变化，碱铝指数和组合指数均表明该序列岩石属钙碱性序列；K_2O/Na_2O 比值早期小于 1，晚期大于 1，具有从钠质花岗岩向钾质花岗岩演化的特征。在硅-碱图解（图 9-7）和 SiO_2-K_2O 图解（图 9-8）中，该序列岩石亦落在属钙碱性-高钾钙碱性区内。

稀土元素含量平均值在 184.7×10^{-6}～297.31×10^{-6}之间。球粒陨石标准化配分曲线为向右倾斜的曲线（图 9-9），属轻稀土富集型。$(La/Sm)_N$ 平均值为 3.4～3.6，$(Gd/Yb)_N$ 为 1.96～3.4，轻稀土分馏作用略强于重稀土；δEu 为 0.7～0.84，略具负铕异常。微量元素以高 Sr 低 Rb 为主要特征，Rb/Sr 比值均大于 1，具 I 型（壳幔混合型）花岗岩特征。微量元素原始地幔标准化蛛网图中（图 9-10），该序列岩石的标准化曲线总体上向右倾，具有 Rb、Th-U-K、Pb、Nd-Zr-Hf-Sm 的正异常和 Ba、Nb-Ta、P、Ti 的负异常，但晚期石岭花岗闪长岩 Sr、P、Eu、Ti 负异常要比其他早期岩石明显一些，显示存在斜长石、磷灰石和钛铁矿的结晶分离作用。

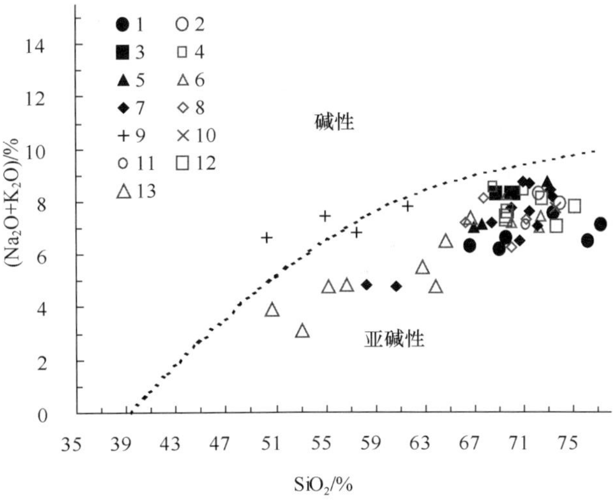

图 9-7　海南岛二叠纪侵入岩硅-碱图解

Fig. 9-7　$SiO_2-Na_2O+K_2O$ diagram of Permian intrusive rocks, Hainan Island

1—西庆序列；2—陀类二长花岗岩；3—大田二长花岗岩；4—什天二长花岗岩；5—禄马序列酸性岩；6—道票岭序列酸性岩；7—通什序列酸性岩；8—石碌二长花岗岩；9—五指山序列中性岩；10—新风岭正长花岗岩；11—顺作花岗岩；12—育才序列酸性岩；13—大乃村序列

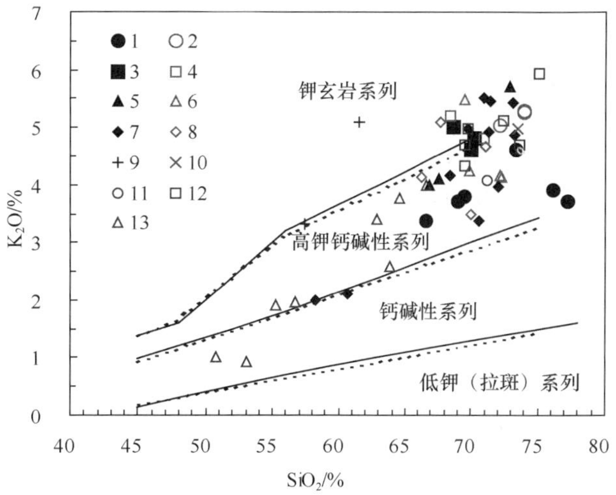

图 9-8　海南岛二叠纪侵入岩 SiO_2-K_2O 图解

Fig. 9-8　$SiO_2-Na_2O+K_2O$ diagram of Permian intrusive rocks, Hainan Island

1—西庆序列；2—陀类二长花岗岩；3—大田二长花岗岩；4—什天二长花岗岩；5—禄马序列酸性岩；6—道票岭序列酸性岩；7—通什序列酸性岩；8—石碌二长花岗岩；9—五指山序列中性岩；10—新风岭正长花岗岩；11—顺作花岗岩；12—育才序列酸性岩；13—大乃村序列

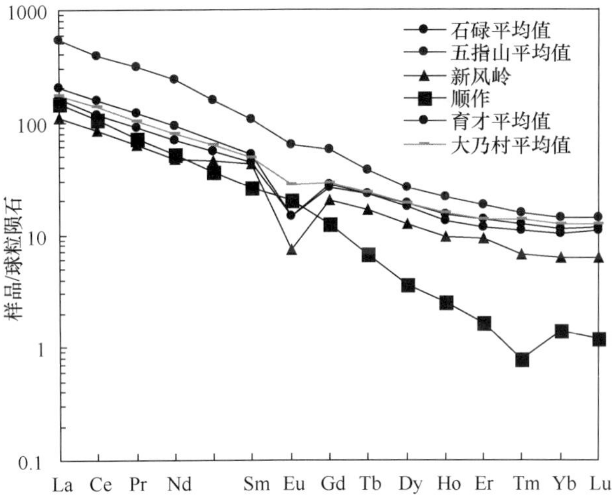

图 9-9　早二叠世侵入岩稀土元素球粒陨石标准化分布型式图

Fig. 9-9　Chondrite-normalized REE patterns of early Permian intrusive rocks

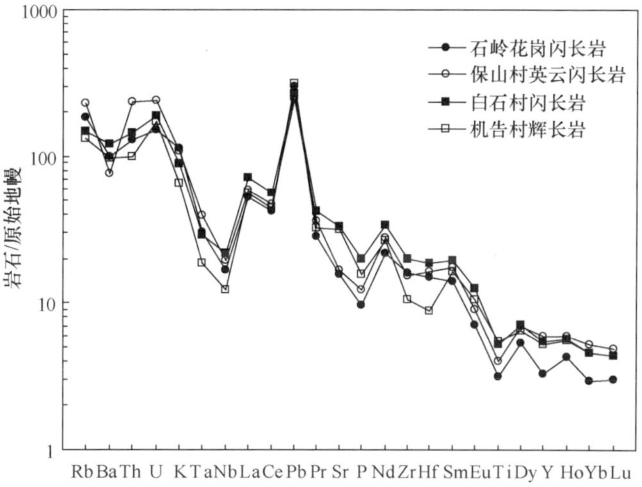

图 9-10　大乃村序列微量元素原始地幔标准化蛛网图

Fig. 9-10　Primitive mantle-normalized spider diagram of trace elements of Danaicun Sequence

二、早二叠世育才序列

该序列岩石主要出露于乐东县大安水库—毛阳—长征农场一带，由抱郎中粒斑状角闪黑云二长花岗岩、志仲巨斑状黑云母二长花岗岩和乘顺岭细粒含斑黑云（二云）二长花岗岩 3 个岩石单位组成，总面积约 391.66km^2。各岩石单位空间上相互依存，志仲二长花岗岩脉动侵入抱郎二长花岗岩，乘顺岭二长花岗岩脉动侵入志仲二长花岗岩、抱郎二长花岗岩（图版Ⅱ-1-B）。该序列岩体侵入的最新地层为下志留统空列村组，并分别被晚二叠世保国二长花岗岩、早三叠世尖峰岭正长花岗岩以及中三叠世后龙碱长花岗岩等侵入。后龙碱长花岗岩侵入志仲二长花岗岩接触界线明显，接触界面处，后龙碱长花岗岩出现2~20cm宽的冷凝边（图版Ⅱ-1-C）；并见有乘顺岭二长花岗岩的捕虏体（图版Ⅱ-1-D）。志仲二长花岗岩中所测同位素年龄有锆石Pb-Pb等时线法年龄（282±11）Ma、U-Pb谐和曲线下交点年龄（253±2）Ma（马大铨等，1991）、SHRIMP锆石U-Pb年龄（287±7）Ma（宜昌地质矿产研究所、海南省地质调查院，2004），形成时代厘定为早二叠世。

该序列岩石普遍具有韧性形变构造，矿物定向排列，方向为25°~85°，与海南岛华力西构造层的主构造线方向基本一致，侵入体边界一般也平行围岩的片理。

（一）抱郎中粒斑状角闪黑云二长花岗岩（$\eta\gamma P_1$）

分布于大安水库—洋淋岭一带，总面积约43.81km²。由斑麦岭和亲亚岭两个较大的侵入体和呆峨岭、保丘村、抗老村等3个面积<1km²的小侵入体构成。主体岩性为片麻状中粒斑状角闪黑云母二长花岗岩，局部为花岗闪长岩。岩石具较强的叶理构造，细中粒似斑状花岗结构。斑晶钾长石10%~20%，粒径一般4mm×9mm~20mm×40mm，长宽比值为1.5~6，常被变形改造成眼球状，长轴方向为北东向或北东东向，内部有时有环带状分布的细粒暗色矿物包裹体。基质粒径多为2~4mm，主要矿物组成：石英20%~23%、斜长石（An29）25%~30%、微斜条纹长石33%~40%、黑云母4%~8%、角闪石1%~4%、单斜辉石0~1%。斜长石环带构造发育，多为韵律振荡环带；黑云母常聚成条纹、条带状或叠片状。副矿物主要有榍石、褐帘石、锆石、磷灰石，榍石有时呈自形的粗大晶体。

岩体中常见微细粒闪长质包体，呈透镜状定向排列，大小为1cm×2cm~50cm×300cm。此外，还常见析离体，即主要由自形程度较差的黑云母、角闪石等暗色矿物组成的矿物团粒或团块，大小为几毫米至几厘米。

（二）志仲巨斑状黑云母二长花岗岩（$\eta\gamma mpP_1$）

主要分布于大安水库—毛阳一带，总面积约277.16km²。主要有志仲、牛龙山、大安、毛阳、永布村、三毛、乙库（只峨岭）、新安等侵入体，岩性为片麻状细中粒—中粗粒巨斑状黑云母二长花岗岩。岩石具片麻状（叶理）构造，细中粒—中粗粒巨斑状花岗结构。斑晶钾长石10%~20%，粒径一般5mm×8mm~20mm×40mm，个别可达3cm×12cm，长宽比值为2~4，粗大的钾长石斑晶和簇状黑云母集合体产出往往具有明显的优选方向（图版Ⅱ-1-E），长轴方向多在20°~80°。基质粒径多为1~7mm，矿物组成：钾长石24%~40%、斜长石（An27~An34）27%~42%、石英24%~30%、黑云母6%~11%。石英常见波状消光和亚颗粒化；黑云母常聚成条纹、条带分布于矿物间隙中。岩石可含少量角闪石，但粒径细小且为他形，肉眼难以识别。副矿物主要有磷灰石、榍石、锆石、褐帘石、磁铁矿、钛铁矿。

岩石中，局部地段见有少量暗色包体（细粒闪长质包体）出现，包体大小3cm×5cm~4cm×12cm，与寄主岩界线清晰。有时可见岩石斑晶跨插于寄主岩与包体的界线之间。由于受构造应力作用，包体常呈长条状或透镜状、扁豆状，产出具有明显的优选方向，其长轴伸展方向与钾长石斑晶排列方向一致。经岩矿鉴定，包体为黑云角闪闪长岩，细粒结构，矿物组成：斜长石70%、角闪石16%、黑云母5%、石英3%、钾长石6%。

（三）乘顺岭细粒（含斑）黑云（二云）二长花岗岩（$\eta\gamma fP_1$）

主要见有南进、以保、毫漫开（内昌）、汉齐岭、毫子门、扎青等侵入体，总面积约70.69km²。岩性为细粒（含斑）黑云（二云）二长花岗岩。岩石具弱叶理构造，含斑结构。斑晶钾长石<5%，一般3%±，大小4mm×6mm~10mm×20mm。基质主要为细粒结构，粒径0.5m×2mm~3mm×2mm，主要矿物组成：石英24%~32%、斜长石（An14~An22）25%~32%、钾长石32%~45%、黑云母2%~6%、白云母1%~3%。副矿物含量较少，主要有褐帘石、锆石、磁铁矿、磷钇矿，少量磷灰石、钛铁矿。

（四）岩石地球化学特征

该序列岩石的SiO_2在69.53%~75.06%之间，均属于酸性岩范围，由早至晚SiO_2含量逐渐增高。

抱郎和志仲二长花岗岩的 Al_2O_3 平均值分别为 14.55%和 13.08%，含铝指数 A/CNK 介于 1.0~1.15 之间，主要为弱过铝；乘顺岭二长花岗岩的 Al_2O_3 平均值为 13.94%，含铝指数 A/CNK1.14~1.39，属强过铝岩石。里特曼指数 σ 为 1.61~2.32，再根据硅-碱图解（见图 9-7）和 SiO_2-K_2O 图解（见图 9-8），该序列岩石属高钾钙碱性系列（晚期岩石落入钾玄岩系列）。

岩石的稀土元素总量多数较低，稀土配分模式为中等右倾曲线（图 9-9）。轻重稀土分异较明显，LREE/HREE 值为 6.2~21；δEu 为 0.77~0.19，具弱至强的负铕异常，且从早到晚负铕异常越来越强烈，表明岩浆结晶过程有斜长石的分异。岩石富集 Rb、Ba、Sr、U、Th、K 等亲石元素，原始地幔蛛网图显示出 Rb、Th-U-K、Pb、Nd-Zr-Hf-Sm 峰和 Ba、Ta-Nb、Sr-P、Eu-Ti 谷，越晚的岩体 Ba、Sr-P、Eu-Ti 的亏损越强，显示存在斜长石、钛铁矿、磷灰石的结晶分异作用（图 9-11）。志仲二长花岗岩的初始锶比值较低，为 0.70852；$\varepsilon_{Nd}(t)$ 较高，为-5.2；T_{2DM} 值 1477Ma，结合闪长质包体的存在，表明它们是壳幔岩浆混合作用的产物。

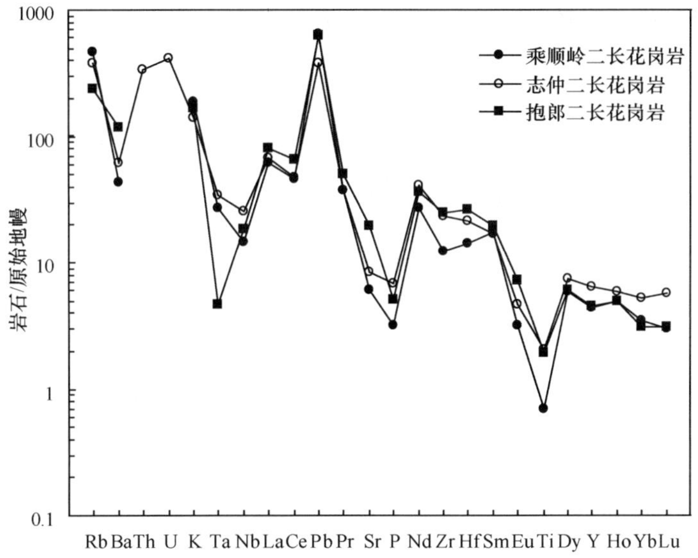

图 9-11　育才序列微量元素原始地幔标准化蛛网图

Fig. 2-2-11　Primitive mantle-normalized spider diagram of trace elements of the Yucai Sequence

三、早二叠世顺作弱片麻状细中粒含斑石榴子石黑云（二云）花岗岩（$\xi\gamma mP_1$）

（一）地质特征、形成时代

见于长征农场东南侧，主要有顺作和长征农场十二队两个小侵入体，总面积约 3.76km^2。顺作花岗岩侵入中元古界戈枕村组，与中二叠世中林岭二长花岗岩相邻，因掩盖而未见接触关系。SHRIMP 法锆石 U-Pb 年龄为（278±7）Ma（宜昌地质矿产研究所、海南省地质调查院，2004），形成时代为早二叠世。

（二）岩石学特性

主体岩性为细中粒含斑石榴子石黑云母（二云母）二长-正长花岗岩，局部为花岗闪长岩或英云闪长岩。弱片麻状-块状构造。细中粒含斑花岗结构，粒径一般 0.5~5mm，约 3%（局部达 10%）的微斜长石粒径为 5mm×7mm~15mm×20mm，构成斑晶。矿物组成变化大，斜长石（An15~An28）20%~63%，微斜微纹长石 0~45%，石英 25%~27%，黑云母 3%~11%（不均匀），白云母 0~1%，石榴子石 1%~5%（粒径 1~10mm）。镜下观察显示，石榴子石被黑云母穿插交代现象（图版Ⅱ-1-F）。

副矿物主要有磷灰石、锆石、钛铁矿，锆石常呈浑圆状。局部见富云包体（石榴子石斜长黑云母片岩）。

（三）岩石地球化学特征

岩石的 SiO_2 含量为 71.16%，Na_2O+K_2O 为 7.15%，CaO 和 Al_2O_3 含量均较高，分别为 2.32% 和 15.53%，A/CNK 为 1.14，里特曼指数 σ 为 1.81。根据硅-碱图解（图 9-7）和 SiO_2-K_2O 图解（图 9-8）判断，岩石属强过铝高钾钙碱性岩系。岩石的稀土元素总量低，平均值为 140.3×10^{-6}，明显低于维氏世界酸性岩平均值。稀土配分模式为右倾曲线（图 9-9），轻重稀土分异明显。$(La/Yb)_N$、$(La/Sm)_N$ 及 $(Gd/Yb)_N$ 分别为 105.1、5.6 和 8.7，重稀土具有较强的分异。其重稀土和 Y 含量低，指示熔融源区有石榴子石、角闪石和单斜辉石的残余。δEu 为 1.06，具弱的正铕异常，结合其 Sr 和 CaO、Al_2O_3 含量较高分析，暗示源区很少有斜长石的残余，其熔融深度较大，也即地壳有过明显加厚。微量元素蛛网图上显示岩石的 Rb、Pb、K 含量较高，P、Ta、Nb、Y 等含量较低，重稀土具有较强的分异作用（图 9-12）。初始锶比值为 0.71098，可能是地壳中以火成岩为主的岩石熔融形成的。

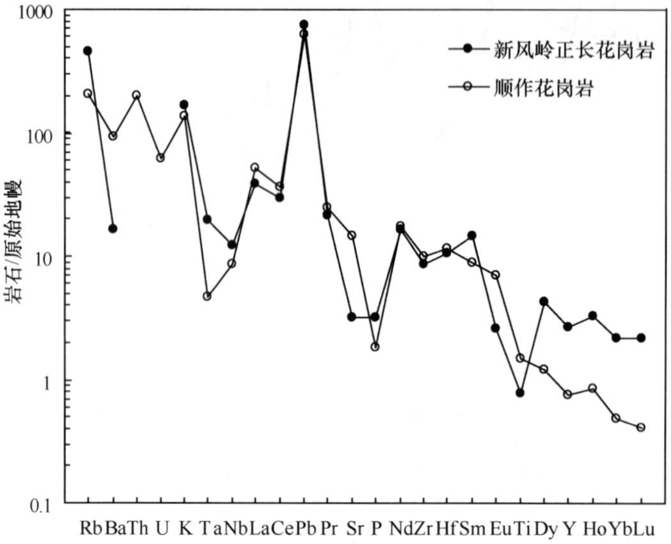

图 9-12 顺作花岗岩、新风岭正长花岗岩微量元素原始地幔标准化蛛网图

Fig. 9-12 Primitive mantle-normalized spider diagram of trace elements of Yucai Shunzuo granite and Xinfengling syenogranite

四、早二叠世新风岭中细粒含电气石黑云正长花岗岩（$\xi\gamma fP_1$）

（一）地质特征及形成时代

分布于海南岛东部万宁水库以北新风岭及衰水水库-岗岭一带，面积 1.9km²，由新风岭和岗岭两个小侵入体构成，新风岭侵入体的 SHRIMP 法锆石 U-Pb 年龄为（276±4）Ma（宜昌地质矿产研究所、海南省地质调查院，2004），侵入时代厘定为早二叠世。

（二）岩石学特性

主体岩性为中细粒含电气石黑云母正长花岗岩。弱叶理构造，中细粒不等粒花岗结构，粒径（0.5~3）mm×4mm。矿物组成：石英 25%~30%、斜长石（An15~An20）15%~22%、微斜条纹长石 48%~49%、黑云母 2%±、电气石 0.5%~4%。副矿物主要为榍石、磁铁矿、钛铁矿、褐帘石、锆石、磷灰石、矽线石、石榴子石等。矽线石呈不规则条带状集合体，可能是同化混染高铝质围岩的产物。

（三）岩石地球化学特征

岩石的 SiO_2、碱质（Na_2O+K_2O）和 Al_2O_3 含量均较高，三者分别为 73.44%、7.74% 和 14.54%，CaO 含量低（0.48%）。岩石化学参数 A/CNK/为 1.35，里特曼指数 σ 为 1.95，DI 为 91.5。根据硅-碱图解（图 9-7）和 SiO_2-K_2O 图解（图 9-8）判断，新风岭正长花岗岩属分异程度高或低熔的强过铝高钾钙碱性岩。岩石的稀土元素总量低（124.3×10^{-6}），其稀土配分模式为中等右倾曲线（图 9-9）。轻重稀土分异较明显，$(La/Sm)_N$ 及 $(Gd/Yb)_N$ 分别为 2.6 和 3.2；δEu 为 0.24，具较强的负铕异常；Sm/Nd 比值较大（0.29）。岩石的 Rb/Sr 值较高（4.24），而 K/Rb 值较低（143）。微量元素蛛网图上显示 Ba、Sr、Zr、Hf、Eu、Ti 含量较低（图 9-12），而 Rb、Th、K 等含量较高，表明它是熔融程度较低或经过了较强分异作用的岩浆所形成。

五、早二叠世五指山序列

该序列岩石分布于五指山东、西两侧山脚，出露总面积约 41.91km²。主要由方满二长岩、行干石英二长闪长岩、什败石英二长岩等 3 个岩石单位构成一个成分演化序列，并且其晚期尚有分异的正长花岗岩脉形成。该序列岩体被中二叠世通什二长花岗岩及燕山期花岗岩体侵入包裹，行干石英二长闪长岩侵入中元古界戈枕村组。方满二长岩中获 SHRIMP 法锆石年龄（272±7）Ma（宜昌地质矿产研究所、海南省地质调查院，2004），形成时代厘定为早二叠世晚期。

该序列岩石普遍发育片麻状构造，具有韧性变形形迹，石英、黑云母、长石遭受韧性变形，角闪石、榍石基本不变形，矿物接合缝和三结点有时含有微粒交生体，表明变形是在岩浆侵位结晶时形成的。片麻理走向为北东—北东东，与海南岛华力西构造层的主构造线基本方向一致。

（一）方满细粒二长岩（ηP_1）

仅见一个侵入体，面积约 0.41km²。在行干石英二长闪长岩中见有方满岩体被侵入切割后残留的小岩块。岩性为弱片麻状细粒黑云角闪二长岩。弱片麻状构造，细粒柱粒状结构，粒径 0.1mm×0.2mm~0.2mm×0.6mm。普通角闪石约 34%、黑云母 9%、斜长石（An27）31%、微斜长石 24%。

副矿物榍石 1%、磷灰石 1%，磁铁矿和钛铁矿常见。

（二）行干细中粒石英二长闪长岩（$\delta\eta o P_1$）

主要分布于通什—五指山乡—行干—长征农场—什暗一带，总面积约 36.63km²。见有什牙口、行干、峨看岭、旧村平、前池、方板、什暗、长征、冲村、罗反等小侵入体。行干石英二长闪长岩侵入戈枕村组，常被通什二长花岗岩侵入。主体岩性为片麻状细粒—细中粒石英二长闪长岩、石英闪长岩。片麻状构造，细粒—中粒结构，粒径 0.2~5mm。矿物组成：斜长石（An20~An44）45%~58%、微斜（微纹）长石 3%~19%、石英 4%~14%、普通角闪石 0~13%、黑云母 11%~28%、普通辉石（或透辉石）0~9%。副矿物主要有榍石、磁铁矿（钛铁矿）、磷灰石、锆石，其中榍石含量可达 1%。

（三）什败细中粒石英二长岩（$\eta o P_1$）

仅见什败一个侵入体，面积为 4.87km²。该侵入体与行干石英二长闪长岩共生，构成复式岩体。但它与毗邻的行干石英二长闪长岩之间的接触关系未见，推测其侵入后者。岩性为片麻状细中粒含斑角闪黑云石英二长岩。片麻状构造，细中粒含斑结构，矿物粒径多为 0.2mm×4mm~2.5mm×4mm，约有 3%的钾长石粒度为 3mm×6mm~7mm×15mm，构成斑晶。矿物组成：斜长石（An27~An35）18%~25%、微斜（微纹）长石 45%、石英 10%~15%、普通角闪石 2%~7%、黑云母 15%~18%。副矿物主要有磷灰石、榍石、磁铁矿（钛铁矿）、锆石。

（四）岩石地球化学特征

五指山序列岩石的 SiO_2 含量在 50.24%~61.54%之间，主要为中性岩，晚期脉岩的 SiO_2 含量为

66.48%～67%；岩石的 Na_2O+K_2O 含量和里特曼指数 σ 值均较高，分别为 6.66%～7.83%和 3.09～5.48，在硅-碱图中落入碱性和亚碱性岩系分界线的两侧；A/CNK 为 0.65～0.87，属准铝质；K_2O 含量相对较高，为 3.32%～5.08%。在硅-碱图解（图 9-7）和 SiO_2-K_2O 图解（图 9-8）上落入钾玄岩系列中。该序列岩石的稀土元素总量高，为 $336.4\times10^{-6}\sim785\times10^{-6}$，稀土配分模式为右倾曲线（图 9-9）。轻重稀土分异较明显，$(La/Yb)_N$、$(La/Sm)_N$ 及 $(Gd/Yb)_N$ 数值都较大；主体岩石 δEu 为 0.95～0.56，具弱—中等的负铕异常。微量元素方面（本章附表 9-4），岩石的 Ni、V、Cr、Co 等亲铁元素含量较高，反映其来源于地幔；Rb、Ba、Sr、K 等大离子亲石元素以及 Zr、Hf、Ta、Nb、Ga、Zn 等高场强元素含量也高，体现了这种钾玄岩系地球化学组成的特殊性。在原始地幔标准化蛛网图上（图 9-13），表现出较强的 Ta-Nb、P、Ti 负异常。该序列岩石（3 个样品）初始锶比值高，为 0.70952～0.71162；$\varepsilon_{Nd}(t)$ 较低，为-7.1～-7.5；T_{2DM} 值相对较大，为 1649～1672Ma，表明它们来源于 EM2 型富集岩石圈地幔。根据其富钾、富大离子亲石元素，还显示它们来源于加厚了的岩石圈地幔的较低程度部分熔融。

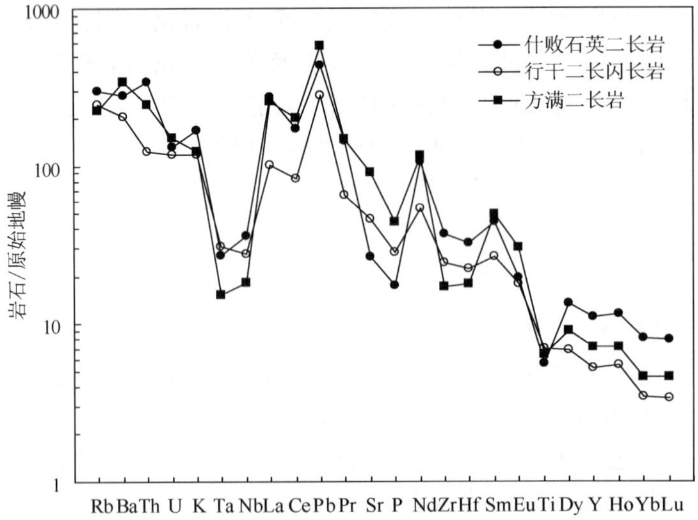

图 9-13　五指山序列微量元素原始地幔标准化蛛网图

Fig. 9-13　Primitive mantle-normalized spider diagram of trace elements of the Wuzhishan Sequence

六、早二叠世石碌中粗粒巨斑状黑云母二长花岗岩（$\eta\gamma cP_1$）

（一）地质特征及形成时代

主要分布于昌江县石碌镇一带，有石碌、查苗等侵入体，总面积约 140.39km²。石碌侵入体南部侵入青白口系石碌群，东部侵入下二叠统地层，北部被晚侏罗世马鞍岭侵入体侵入；查苗侵入体北部侵入早二叠世石岭花岗闪长岩（图版Ⅱ-1-G），南部被晚三叠世东四中粗粒角闪黑云母正长花岗岩侵入。石碌黑云母二长花岗岩虽未与中二叠世邦溪角闪黑云母二长花岗岩直接接触，但在石碌河谷一转石中，见邦溪中粗粒巨斑状角闪黑云二长花岗岩呈岩枝穿插于石碌中粗粒巨斑状黑云母二长花岗岩中（图 9-14）；在鹅毛岭采石场见石碌黑云母二长花岗岩被中二叠世大坡二长花岗岩侵入。前人在石碌二长花岗岩中获得 Rb-Sr 同位素年龄值为（274.0±6.5）Ma，$I_0=0.71119$（马大铨等，1991），与地质侵入接触关系相符，侵入时代厘定为早二叠世晚期。

（二）岩石学特征

岩石类型主要为中粒、中粗粒巨斑状黑云母二长花岗岩。中粒、中粗粒似斑状结构，片麻状构造。岩石均一化程度相对较低，斑晶含量、大小在不同位置变化较大，一般含量在 8%～20%，局部巨斑晶十分密集，斑晶个体既有正常的斑晶，也有巨斑晶（长 3cm 以上），但在每个露头点上均可看到钾长

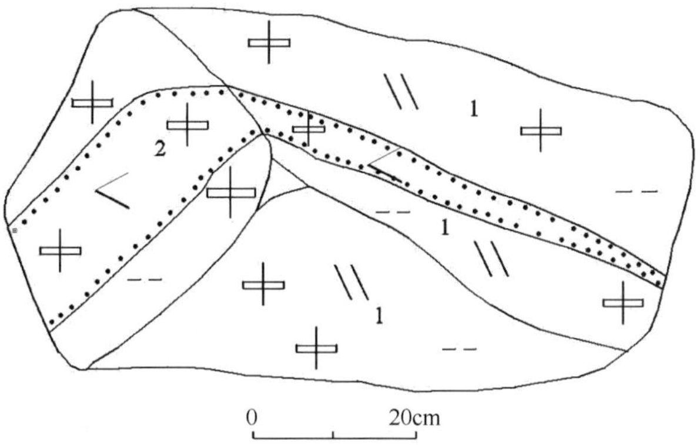

图 9-14 石碌岩体被邦溪岩体侵入素描图（转石）

Fig. 9-14 Sketch showing Shilu intrusion intruded by Bangxi intrusion (according to floats)

1—石碌岩体；2—邦溪岩体

石巨斑晶，粒径 3mm×5mm~15mm×80mm，形成一种“斑杂”状构造特征。斑晶 c/b 之比为3.5~5.5，呈修长的板状晶体。斑晶具有定向分布的特点，走向为 280°~300°。

矿物特征：斑晶为微斜长石，半自形板状，格子双晶比较发育，钠长石条纹极不发育，呈细小的、稀疏的微脉状，镜下见斑晶与基质界线成参差状，斑晶中常包裹石英、斜长石、黑云母等，有时这些包裹体成环带状分布。斜长石为半自形或他形，粒径 1.5mm×2mm~2mm×4mm，有时见简单的环带，中心部分 An 值 30~42，边部 15~25。石英他形粒状或呈集合体充填在长石颗粒之间。黑云母呈叠状集合体，N_g 为黑色，N_p 为棕黄色，常被绿泥石交代。矿物组成：钾长石 24%~44%、斜长石 29%~41%、石英 23%~30%、黑云母 3%~8%。基本不含闪长质暗色包体，有时见有少量富云包体，大小在 1cm×5cm 左右。

石碌二长花岗岩的钾长石端员组分为 An 0.15%、Ab 7.38%、Or 92.47%，斜长石 An 28.07%、Ab 70.65%、Or 1.28%。黑云母属铁质黑云母。

副矿物以褐帘石、锆石、磷灰石最常见，偶尔出现榍石。

（三）岩石地球化学特征

石碌二长花岗岩的 SiO_2 含量和碱含量均较高，分别在 66.23%~71.06%和 6.28%~8.11%之间。但 K_2O 含量并不算太高，在 3.48%~5.08%之间变化，$K_2O>Na_2O$，K_2O/Na_2O 为 1.24~1.77；组合指数（里特曼指数）σ 为 1.45~2.66；A/CNK 介于 1.04~1.06 之间，属弱过铝质。根据硅-碱图解（图 9-7）和 SiO_2-K_2O 图解（图 9-8），石碌二长花岗岩属高钾钙碱性序列。稀土元素总量约为 131×10^{-6}~233×10^{-6}，明显低于维氏世界酸性岩平均值。δEu 为 0.37~0.44，具中等偏强的负铕异常，稀土配分模式为向右倾斜的曲线（图 9-9）。$(La/Yb)_N$ 值为 10.11~26.97，轻稀土富集，轻重稀土分异比较明显。微量元素方面（本章附表 9-14），与维氏酸性岩相比，Pb、Sn、Au、W、Mo、Bi、As、Hf 含量偏高，V、Cu、Be、Sb、Sr、Ba、Nb、Ta、Za 偏低。在原始地幔微量元素标准化蛛网图上表现为高 Rb 低 Sr、Yb，具同碰撞花岗岩的特点（图 9-15）。

七、中二叠世通什序列

该序列岩石主要分布于东方市、昌江县、五指山市及万宁乐来一带，总面积 3871.96km²，整体上呈北东东向展布，从早至晚由山牛塘细粒闪长岩、通什中粒巨斑状角闪黑云母二长花岗岩、乌兰盖中粗粒含斑黑云二长花岗岩、大坡中粒黑云二长花岗岩、长塘岭中细粒斑状（角闪）黑云母二长花岗岩和火岭中细粒黑云母正长花岗岩等 6 个岩石单位组成，以发育强弱不等的原生定向构造为主要特征，岩石成因属壳幔混合型，形成于同碰撞至后碰撞早期阶段。

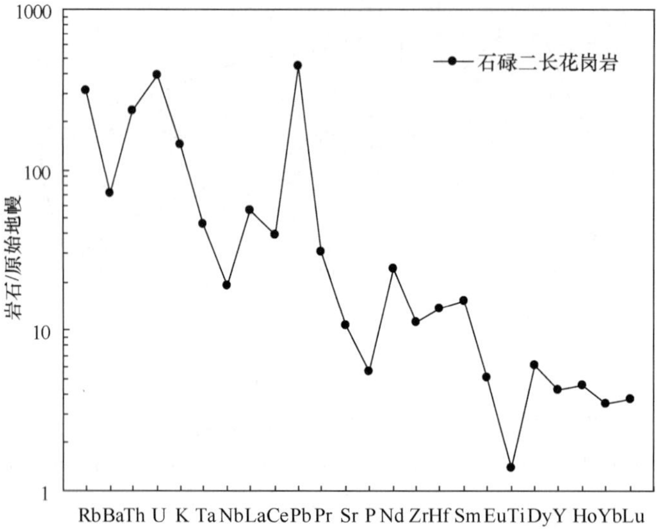

图 9-15　石碌二长花岗岩原始地幔标准化蛛网图

Fig. 9-15　Primitive mantle-normalized spider diagram of Shilu monzonitic granite

（一）山牛塘细粒闪长岩（δP_2）

1. 地质特征及形成时代

呈同深成岩墙、小岩株或包体赋存于通什二长花岗岩中，个别被中林岭二长花岗岩捕掳包裹。主要见有条坑来、加尧队、山牛塘、羌园等侵入体，总面积约 8.68km²。山牛塘细粒闪长岩中既可有通什花岗质岩石的捕虏晶甚至岩石团块，又可被通什花岗质枝脉（反向脉）穿插（图版Ⅱ-1-H），显示其和通什花岗质岩浆是近于同时侵位的，是同深成关系（浆动接触）。昌江一带见其被大坡二长花岗岩侵入，二者呈脉动接触，界线清楚，后者长石斑晶沿接触面定向排列，在大坡二长花岗岩中见羌园侵入体的捕虏体。山牛塘细粒闪长岩已有的锆石 U-Pb 同位素年龄为 264Ma（海南地质综合勘察院第一勘察处，1993），形成时代厘定为中二叠世。

2. 岩石学特征

由于与花岗质岩浆混合及扩散程度的不同，山牛塘细粒闪长岩的成分变化明显，从闪长质至英云闪长质，个别达花岗闪长质，但在较大的岩体或岩墙中一般为闪长岩或石英闪长岩。弱片麻状构造，微细粒—细粒结构，“混染”斑状结构，“斑晶”为钾长石、斜长石、石英等捕虏晶，常有熔蚀和再生长边。矿物组成主要为斜长石（中长石-奥长石）、黑云母、普通角闪石、石英、钾长石。在南礼还见该单元岩浆发生熔离作用，熔离出来的浆泡含量<25%，浆泡大小为 1.5mm×3mm～10mm×24mm，椭球状，均匀地分散分布，少数联通为串珠状或条带状，或聚集为囊包，宽 5～6mm，长几厘米至十几厘米。浆泡矿物组成：黑云母 0～10%、石英 25%±、榍石 5%、斜长石（An 52～An56）60%～70%，属特殊的斜长花岗质（图版Ⅱ-2-A）。浆泡外的基体矿物组成也很特别：由 60%左右斜长石（An 45）及 40%左右的黑云母组成，未见角闪石和辉石。

岩石中局部见闪长质暗色包体，包体长轴以北东向为主，与岩石片麻理方向基本一致。

副矿物主要为磁铁矿和锆石，少量钛铁矿、黄铁矿等。

（二）通什中粒巨斑状角闪黑云二长花岗岩（$\eta\gamma p P_2$）

1. 地质特征及形成时代

分布于通什、昌江一带，由通什、三更罗、什高、大村岭、铜铁岭、中兴、和乐亭、桥头村、老裴、邦溪、大沟、岭尾村、乌烈村、水尾、叉河等侵入体组成，出露面积约 2291.63 km²。通什二长花岗岩分布面积大，岛内多处可见其被后期岩体如中二叠世长塘岭二长花岗岩（图版Ⅱ-2-B）、早三叠世番阳峒正长花岗岩（图版Ⅱ-2-C）等侵入。已有的同位素年龄为锆石 U-Pb 法年龄（248±25）

Ma（广西区调队，1991）、249Ma（海南地质综合勘察院第一勘察处，1993），全岩 Rb-Sr 等时线年龄（272±8）Ma（马大铨等，1991），SHRIMP 锆石 U-Pb 法年龄（265±4）Ma（宜昌地质矿产研究所，海南省地质调查院，2004）。本次研究在吊罗山南部尖石岭侵入体中采集 2 个新鲜样品（D1111-2、D1295-1，分析结果见表 9-4），采用高精度的激光剥蚀电感耦合等离子体质谱（LA-ICP-MS）测得锆石 U-Pb 年龄分别为（259. 1±3. 4）Ma 和（263. 9±2. 9）Ma，将其侵入时代厘定为中二叠世。

表 9-4　通什二长花岗岩 LA-ICP-MA 锆石 U-Pb 测年结果

Table 9-4　LA-ICP-MA zircon U-Pb dating results of Tongshi monzonitic granite

岩石单位	测试点	$^{207}Pb/^{206}Pb$	$^{207}Pb/^{235}U$	$^{206}Pb/^{238}U$	$^{208}Pb/^{232}Th$	$^{238}U/^{232}Th$	$^{207}Pb/^{206}Pb$	$^{207}Pb/^{235}U$	$^{206}Pb/^{238}U$	$^{208}Pb/^{232}Th$
通什	D1111-2-1	0. 0864	0. 5350	0. 0447	0. 0235	0. 404089		435	282	469
通什	D1111-2-3	0. 0515	0. 2986	0. 0419	0. 0133	0. 225769	265	265	264	267
通什	D1111-2-4	0. 0547	0. 3168	0. 0422	0. 0135	0. 388455	467	279	266	272
通什	D1111-2-5	0. 0534	0. 3235	0. 0438	0. 0147	0. 211829	346	285	276	295
通什	D1111-2-6	0. 0526	0. 3068	0. 0422	0. 0135	0. 244069	322	272	266	271
通什	D1111-2-7	0. 0502	0. 2699	0. 0389	0. 0096	0. 256062	206	243	246	193
通什	D1111-2-8	0. 0483	0. 3138	0. 0473	0. 0126	0. 235988	122	277	298	253
通什	D1111-2-9	0. 0565	0. 3826	0. 0495	0. 0148	0. 445295	472	329	312	296
通什	D1111-2-10	0. 0569	0. 2800	0. 0357	0. 0136	0. 431388	487	251	226	273
通什	D1111-2-11	0. 0557	0. 3451	0. 0449	0. 0169	0. 183368	439	301	283	338
通什	D1111-2-12	0. 0531	0. 2903	0. 0397	0. 0129	0. 301277	332	259	251	259
通什	D1111-2-13	0. 0513	0. 2910	0. 0410	0. 0132	0. 20746	257	259	259	265
通什	D1111-2-14	0. 0518	0. 3175	0. 0443	0. 0150	0. 207087	280	280	280	302
通什	D1111-2-15	0. 0518	0. 3004	0. 0420	0. 0132	0. 206851	280	267	265	265
通什	D1111-2-16	0. 0516	0. 3035	0. 0427	0. 0131	0. 284042	265	269	269	263
通什	D1111-2-17	0. 0511	0. 2901	0. 0411	0. 0129	0. 243784	256	259	259	259
通什	D1111-2-18	0. 0504	0. 2943	0. 0423	0. 0130	0. 239286	213	262	267	261
通什	D1295-1-1	0. 0524	0. 3026	0. 0419	0. 0140	0. 175064	302	268	264	282
通什	D1295-1-2	0. 0506	0. 2848	0. 0407	0. 0126	0. 167898	233	254	257	253
通什	D1295-1-3	0. 0494	0. 2680	0. 0393	0. 0122	0. 22749	165	241	248	245
通什	D1295-1-4	0. 0515	0. 2979	0. 0419	0. 0144	0. 290901	261	265	265	288
通什	D1295-1-5	0. 0511	0. 2905	0. 0413	0. 0142	0. 225383	256	259	261	285
通什	D1295-1-6	0. 0508	0. 2866	0. 0408	0. 0130	0. 235627	232	256	258	260
通什	D1295-1-7	0. 0505	0. 2708	0. 0388	0. 0124	0. 221543	220	243	245	250
通什	D1295-1-8	0. 2003	1. 5008	0. 0485	0. 0975	0. 258746	2829	931	306	1881
通什	D1295-1-9	0. 0513	0. 2745	0. 0387	0. 0122	0. 293945	254	246	245	245
通什	D1295-1-10	0. 0517	0. 2726	0. 0381	0. 0116	0. 316776	272	245	241	233
通什	D1295-1-11	0. 0543	0. 2966	0. 0395	0. 0142	0. 153875	383	264	250	284
通什	D1295-1-12	0. 0513	0. 2872	0. 0407	0. 0147	0. 290824	254	256	257	295
通什	D1295-1-13	0. 0589	0. 4257	0. 0523	0. 0174	0. 260688	561	360	329	348
通什	D1295-1-14	0. 0511	0. 2799	0. 0396	0. 0135	0. 265844	243	251	250	271
通什	D1295-1-15	0. 0511	0. 2713	0. 0383	0. 0118	0. 335255	256	244	242	238
通什	D1295-1-16	0. 0532	0. 2747	0. 0373	0. 0120	0. 275292	345	246	236	241
通什	D1295-1-17	0. 0525	0. 2975	0. 0410	0. 0130	0. 261438	306	264	259	260
通什	D1295-1-18	0. 0501	0. 2768	0. 0400	0. 0120	0. 330753	198	248	253	242

2. 岩石学特征

主体岩石类型为中粒巨斑状角闪黑云二长花岗岩，局部见花岗闪长岩或正长花岗岩。似斑状结构，基质以中粒花岗结构为主，局部为中细粒、细中粒或粗中粒结构，普遍具弱—中等的叶理构造。钾长石斑晶及暗色矿物定向排列明显，斑晶为钾长石 10%~20%，局部达 30%。但与闪长质包体及岩墙接触处，或在闪长质包体大量分布的地带，斑晶数量可明显减少，并可使岩性变为英云闪长岩。斑晶粒径一般 7cm×9cm~2cm×4cm，个别可达 5cm×10cm，斑晶长宽比值为 1.5~6，长轴优选方向以北东向为主，斑晶内部有时有环带状分布的细粒暗色矿物包裹体。基质粒径多为 2~7mm，局部为 0.2~2mm。矿物组成一般为钾长石 16%~55%，灰白色，板柱状，卡斯巴双晶、格子双晶和微纹结构发育；斜长石（An20~An46）13%~45%，灰白色，半自形柱状，聚片双晶发育，双晶纹较细密且常见错开，环带构造发育，多为韵律振荡环带；石英 18%~28%，他形粒状，常以集合体分布在长石颗粒间不规则空隙中，石英颗粒间以弧形边、齿形边紧密接触，常见带状消光和波状消光；黑云母 2%~12%，常聚成条纹、条带状或叠片状，黄褐色—黑色等多色性；角闪石 0~12%，柱状，分布不均匀，含量变化较大；有时有少量单斜辉石。

副矿物主要有榍石、褐帘石、磁铁矿、钛铁矿、锆石、磷灰石等，榍石有时呈自形的粗大晶体。

岩体中普遍见有微细粒闪长质包体（图版Ⅱ-2-D），一般 1~2 个/m^2，局部富集达 4 个/m^2，大小 1cm×2cm~50cm×300cm，呈透镜状定向排列，长轴方向与钾长石斑晶长轴方向基本一致。矿物成分主要为角闪石、黑云母、斜长石及少量钾长石、石英。包体内部常见个别钾长石、斜长石、石英呈“斑晶”产出，大小一般 0.5cm×1.0cm，常有熔蚀边，应为来自寄主岩的捕虏晶。此外还常见析离体，即主要由自形程度较差的黑云母、角闪石等暗色矿物组成的矿物团粒或团块，大小为几毫米至几厘米；偶见堆晶包体（由钾长石巨晶和粗大的角闪石构成，少量石英、斜长石、黑云母等矿物填隙）以及斜长角闪岩或黑云母角闪斜长变粒岩深源包体。

副矿物主要为榍石、褐帘石、锆石、磷灰石、磁铁矿等，总量大于 1000g/t。

（三）乌兰盖中粗粒含巨斑黑云母二长花岗岩（$\eta\gamma mcP_2$）

1. 地质特征及形成时代

分布于红岭农场十队至富乃村一带，由乌兰盖、红岭、那历、大发等侵入体组成，出露面积约 84.58km^2。在乌兰盖中粗粒含巨斑黑云母二长花岗岩中见到邦溪二长花岗岩的捕虏体，捕虏体呈浑圆状，两者界线清楚但不截然，表明两者形成时间相近（图 9-16）。空间分布上两者具有紧密共生、相间出现的特征，推测两者具有同源岩浆分异演化的关系，暂将其侵入时代厘定为中二叠世。

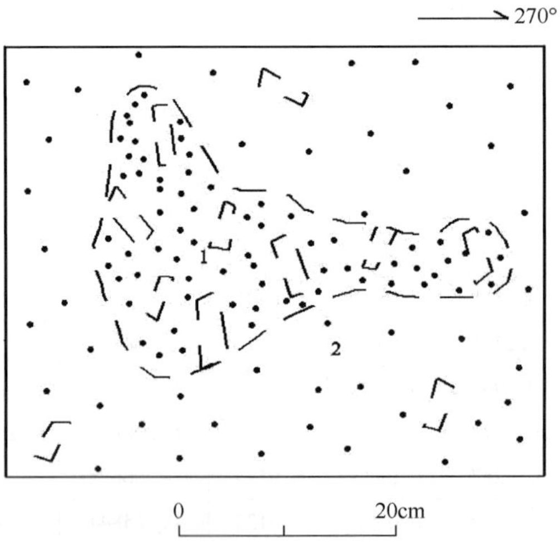

图 9-16 乌兰盖含巨斑二长花岗岩侵入邦溪斑状二长花岗岩接触关系平面素描图

Fig. 9-16 Plan sketch showing the contact that Wulangai megaphenocryst-bearing monzonitic granite intrudes Bangxi porphyritic monzonitic granite

（据 1∶5 万昌江县、邦溪幅区调报告）

1—邦溪斑状二长花岗岩；2—乌兰盖含巨斑二长花岗岩

岩石中偶见石英闪长质包体，长轴一般4~8cm。

2. 岩石学特征

岩石类型主要为中粒或中粗粒含巨斑黑云母二长花岗岩。斑晶定向比较明显（图版Ⅱ-2-E），钾长石巨斑晶个体在1.5cm×4cm~2.5cm×5cm之间，含量2%~3%，最多也不超过5%。矿物组成主要为斜长石22%~37%、钾长石26%~38%、石英20%~33%、黑云母（属铁质黑云母）含量5%~10%，角闪石偶尔出现。钾长石端员组分为An 0.46%、Ab 5.25%、Or 94.29%，斜长石端员组分为An 30.15%、Ab 68.37%、Or 1.48%。

副矿物主要为榍石、磁铁矿、锆石、磷灰石等，总量大于1000g/t。

（四）大坡中粒斑状黑云二长花岗岩（$\eta\gamma mP_2$）

1. 地质特征及形成时代

主要由大坡、青坎、马老山等侵入体组成，分布在昌江县大坡镇及其附近，面积约268.69 km^2。大坡侵入体南部侵入石碌黑云母二长花岗岩，北部侵入邦溪角闪黑云二长花岗岩。大坡二长花岗岩中已有的同位素年龄为（255.3±7.5）Ma（全岩Rb-Sr法，海南地质综合勘察院第一勘察处，1996）、（250.9±15.7）Ma（全岩Rb-Sr法，宜昌地矿所，1995）。根据侵入接触关系和同位素年龄值，暂将其形成时代厘定为中二叠世。

2. 岩石学特征

岩石类型主要为中粒、中细粒似斑状黑云母二长花岗岩，深灰色或灰白色，似斑状结构，定向构造比较明显。斑晶为微斜长石，微带紫色色调的肉红色，自形或半自形厚板状，长轴1.0~1.5cm，很少超过2.0cm，含量在20%~30%之间；格子双晶发育，钠长石条纹不发育，呈稀疏的细纹脉状，斑晶与基质界线呈参差状，晶体内部常见有斜长石、石英、黑云母等。基质为中细粒不等粒结构，主要由斜长石25%~33%、微斜长石32%~46%、石英20%~25%、黑云母3%~12%组成。斜长石呈自形或半自形，个体大小0.4mm×0.6mm~1.5mm×3.5mm，环带构造比较发育，偶见韵律式环带，中心部分An 25~An35，边部15~20；石英为他形粒状或不规则填隙状，粒径0.2~3mm。黑云母多呈叠片状集合体产出，片幅0.4~2.5mm。

岩石中富含闪长质包体，大小一般1cm×4cm~10cm×17cm，个别达数十厘米。统计表明，最密集处8.3个/m^2，少的地方则2.3个/m^2，个体最大为80cm×40cm，最小2cm×0.8cm，绝大部分包体长轴小于10cm。包体岩石类型主要为石英闪长质，少量为富云母石英闪长岩、角闪斜长片岩和石英黑云母斜长片岩。包体一般呈定向分布，长轴方向与斑晶长轴排列一致，走向为280°~300°。

副矿物主要为榍石、磁铁矿、锆石、磷灰石等，以含有大量的榍石为特征。

（五）长塘岭中细粒斑状黑云二长花岗岩（$\eta\gamma P_2$）

1. 地质特征及形成时代

由长塘岭、打金老村、欧弄岭、大叶岭、福来村、可任、中林岭、东坪、罗万村、便文村等侵入体组成，分布在岛西部昌江及通什一带，面积758.43km^2。侵入通什巨斑状角闪石黑云母二长花岗岩。前人在便文村侵入体中获得同位素年龄为（274±5）Ma（SHRIMP锆石U-Pb年龄；宜昌地质矿产研究所，海南省地质调查院，2004）、在可任侵入体中获得同位素年龄为（265±28）Ma（LA-ICPMS锆石U-Pb年龄，海南省地质调查院，2007），形成时代厘定为中二叠世。

2. 岩石学特征

主体岩性为中细粒含斑黑云母二长花岗岩，局部为正长花岗岩。似斑状（含斑—少斑）结构，块状构造，弱叶理构造。斑晶钾长石2%~10%，一般3%±，灰白色，半自形板状，大小大多在6mm×8mm~8mm×15mm之间，个别可达1cm×3.8cm。基质中细粒不等粒结构，粒径0.5mm×5mm~3mm×5mm。矿物组成：石英24%~35%、斜长石（An 20~An45）10%~40%、钾长石37%~64%、黑云母5%~8%。石英他形粒状，常以集合体分布在长石颗粒间不规则空隙中，石英颗粒间以弧形边、齿形边接触，常见波状消光；斜长石灰白色，半自形柱状，聚片双晶和环带构造发育；钾长石柱粒状、半自形板柱状，以灰白色为主，局部略带浅粉红色，卡斯巴双晶和格子双晶发育；黑云母为黑色，片状，具黄色—红褐色—黑多色性，局部含量可高达12%或<5%。有时有微量角闪石。

副矿物主要有磁铁矿、榍石、磷灰石、锆石、钛铁矿等，副矿物组合与通什二长花岗岩类似，但含量有所减少。

偶见微粒闪长质包体，呈椭圆状或不规则状，大小不一，大者直径可达30~50cm，小的仅几厘米。

（六）火岭中细粒黑云母正长花岗岩（$\xi\gamma P_2$）

1. 地质特征及形成时代

由火岭、马夹石、孟果、可保岭、毛头岭、头乐岭、七指岭、罗眉村、上溪村、南嘻村侵入体以及一些面积小于1km^{2}的零星小侵入体或岩脉构成，出露总面积约459.95km^2。空间分布上与通什二长花岗岩、长塘岭二长花岗岩相互依存，并侵入通什二长花岗岩（图版Ⅱ-2-F）和长塘岭二长花岗岩，已有的同位素年龄为（251.7±18.2）Ma（Rb-Sr法；广西区调队，1991），暂将其形成时代厘定为中二叠世晚期。

2. 岩石学特征

主体岩性为中细粒黑云母正长花岗岩，局部为二长花岗岩或碱长花岗岩。弱叶理构造，以细粒花岗结构为主，局部为中细粒或细中粒花岗结构，粒径多为0.5~2mm，少量2~4.5mm。矿物组成：石英20%~30%，他形粒状，透明、半透明，油脂光泽，以粒状或集合体状分布在长石颗粒间不规则空隙中，石英颗粒间以弧形边、齿形边紧密接触，常见波状消光；斜长石（An15~An25）含量一般介于17%~30%之间，局部仅为5%，灰白色，粒状、柱状，聚片双晶发育，双晶纹较细密，为更长石；钾长石37%~69%，灰白色，粒状、柱粒状，卡斯巴双晶和格子双晶发育，边缘常见蠕英结构；黑云母3%~5%±，片状，具黄色—黑色、浅红色—深棕红色、褐色—黑褐色等多色性，大多发生退变且析出黑云母。

副矿物主要有磁铁矿、榍石、磷灰石、锆石、钛铁矿等，副矿物组合与上述的通什二长花岗岩和长塘岭二长花岗岩相似，但含量有所减少。

岩体中偶见闪长质包体，细粒结构，矿物成分主要为角闪石、黑云母、斜长石，少量钾长石、石英。呈椭圆状、不规则状，大小多在2~8mm之间，个别可达15mm，包体长轴与叶理走向基本一致。

（七）岩石地球化学特征

该序列岩石的SiO_2含量为54.94%-74.16%，变化范围大，晚期岩石快速增大。岩石的Na_2O+K_2O含量为4.79%~8.36%，K_2O含量为1.99%~5.33%。σ值为1.27~2.45，根据硅-碱图解（图9-7）和SiO_2-K_2O图解（图9-8），属高钾钙碱性系列；A/CNK在0.83~1.07之间，属准铝质—弱过铝质。稀土元素总量121.7×10^{-6}~270.3×10^{-6}，稀土配分模式为右倾曲线（图9-17）。轻重稀土分异中等—较明显，LREE/HREE值为6.7~16.6，其中重稀土部分的曲线较平缓；δEu为0.73~0.19，整体上具中等负铕异常，但从早到晚负铕异常逐渐增强。在微量元素蛛网图中具Sr-P、Eu-Ti负异常（图9-18）。

山牛塘闪长岩的初始锶比值为0.70766，$\varepsilon_{Nd}(t)$为-2.73，T_{2DM}值为1257Ma；通什二长花岗岩初始锶比值为0.70817，$\varepsilon_{Nd}(t)$为-4.07，T_{2DM}值为1367Ma；长塘岭二长花岗岩初始锶比值为0.70852，反映它们属于壳幔混合来源。

八、晚二叠世道票岭序列

分布于那抗—保国农场—道票岭及大炮—洋老一带。面积约118.01km^2。由保国细中粒—中粗粒斑状角闪黑云母二长花岗岩和那佩中粗粒斑状黑云母二长花岗岩构成，两者的接触关系未见及。该序列岩体侵入下志留统陀烈组和志仲二长花岗岩，被深堀村石英二长岩侵入（马大铨等，1991）。保国二长花岗岩的SHRIMP法锆石U-Pb年龄为（259.5±3）Ma（宜昌地质矿产研究所，海南省地质调查院，2004）。保国二长花岗岩中局部岩石显示定向构造，应属后碰撞阶段产物，形成于挤压-松弛过渡阶段。综合上述地质与同位素年龄资料，将该序列岩石形成时间厘定为晚二叠世。

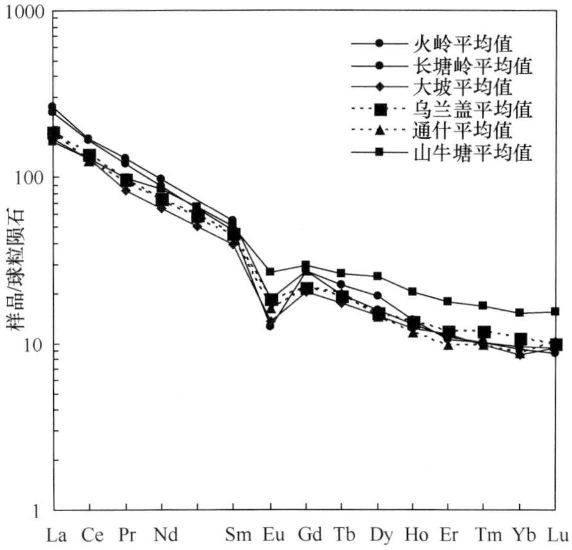

图 9-17 中二叠世通什序列稀土元素球粒陨石标准化分布型式图

Fig. 9-17 Chondrite-normalized REE patterns of middle Permian Tongshi Sequence

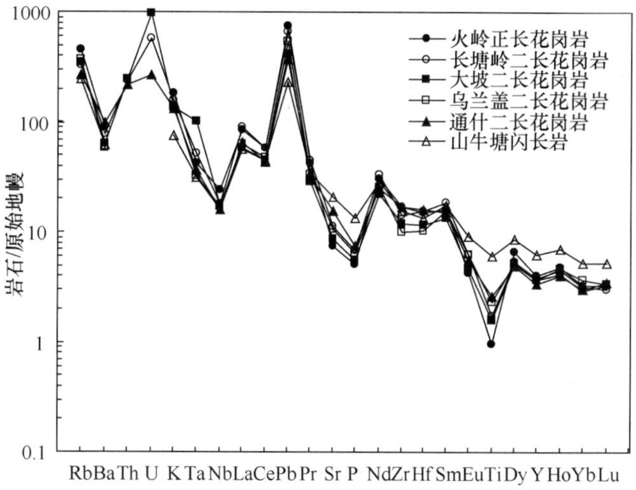

图 9-18 通什序列微量元素原始地幔标准化蛛网图

Fig 9-18 Primitive mantle-normalized spider diagram of trace elements of Tongshi Sequence

（一）保国细中粒斑状角闪黑云二长花岗岩（$\eta\gamma mcP_3$）

见有道票岭、大炮岭、洋老等 3 个侵入体，面积约 110.98km²。主体岩性为细中粒斑状角闪黑云二长花岗岩。一般为块状构造，局部显示定向构造，似斑状结构，斑晶钾长石 10%~15%，局部达 20%~30%，大小 0.6cm×1.2cm~1.5cm×3cm，个别达 9cm×9cm；石英斑晶 3%~5%，大小 6~9mm。基质细中粒花岗结构，粒径以 1~4mm 为主，局部为粗中粒（2~8mm）。矿物组成：石英28%~30%、斜长石（An 25~An30）35%~40%、微斜条纹长石 25%~30%、黑云母 6%~10%、角闪石 2%~7%。副矿物主要有榍石、褐帘石、锆石、磷灰石、磁铁矿，榍石有时呈自形的粗大晶体。岩石中常见微细粒闪长质包体，浑圆状，其中可见钾长石和石英的捕虏晶，包体大小为 2cm×3cm~25cm×50cm。

（二）那佩中粗粒斑状黑云母二长花岗岩（$\eta\gamma cP_3$）

见那佩、那抗 2 个侵入体，面积约 7.03km²。岩性为中粗粒少斑-斑状黑云母二长花岗岩。块状构造，似斑状结构，斑晶钾长石 5%~15%，大小 1cm×2cm~2.3cm×3.5cm；基质中粗粒花岗结构，大小 2~10mm 为主。矿物组成：石英 22%~28%、斜长石（An 23~An 28）24%~36%、微斜条纹长石 32%~42%、黑云母 4%~8%。副矿物主要有褐帘石、磷灰石、锆石。

（三）岩石地球化学特征

该序列岩石的 SiO_2 含量为 65.67%~72.75%；岩石的 Na_2O+K_2O 含量为 7.02%~8.23%，K_2O 含量为 4.14%~5.48%，σ 值为 1.68~2.54。根据硅-碱图解（图 9-7）和 SiO_2-K_2O 图解（图 9-8），属高钾钙碱性系列。保国二长花岗岩 A/CNK 为 0.86~1.02，属准铝质—弱过铝质；那佩二长花岗岩 A/CNK 为 1.04~1.14，属弱铝质—强过铝质。稀土元素总量为 176.6×10^{-6}~432.4×10^{-6}，稀土配分模式为右倾曲线（图 9-19）。轻重稀土分异中等，LREE/HREE 值为 6.7~14.8，其重稀土部分的曲线较平缓。$(Gd/Yb)_N$数值较小，为 1.4~2.9；δEu 为 0.86~0.22（平均 0.47），具中等负铕异常，晚期的那佩二长花岗岩负铕异常较早期的保国二长花岗岩强烈。在微量元素蛛网图上显示较强的 Ba、Ta-Nb、Sr-P、Eu-Ti 负异常和明显的 Rb、Th-U-K、Pb、Nd-Zr-Hf-Sm 正异常（图 9-20）。

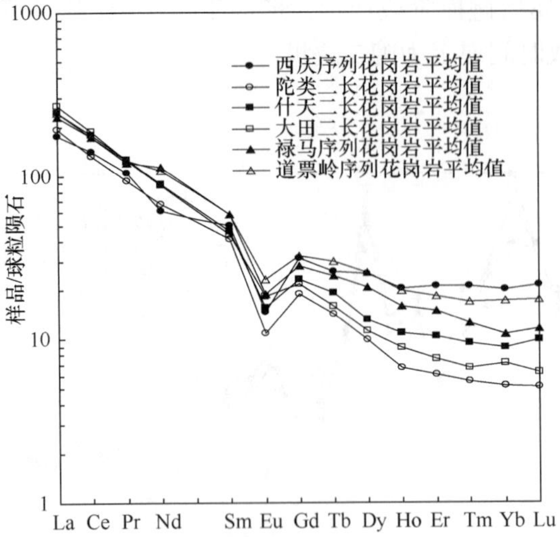

图 9-19　晚二叠世侵入岩稀土元素球粒陨石标准化分布型式图

Fig. 9-19　Chondrite-normalized REE patterna of late Permian intrusive rocks

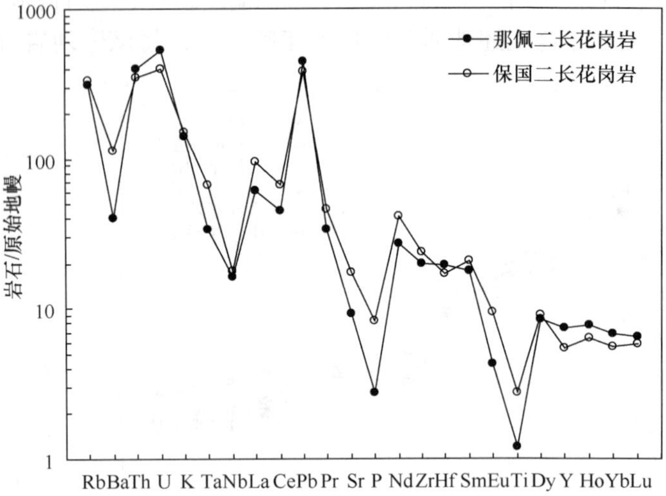

图 9-20　道票岭序列微量元素原始地幔标准化蛛网图

Fig. 9-20　Primitive mantle-normalized spider diagram of trace elements of Daopiaoling Sequence

九、晚二叠世禄马序列

该序列岩石分布于万宁市南桥市—六连岭一带，总面积约 193.39km^2，由东岭中细粒（含斑）角闪黑云母花岗闪长岩及少量白茶细粒黑云母二长花岗岩 2 个岩石单位构成。白茶二长花岗岩脉动侵入东岭花岗闪长岩，被该序列岩石侵入的最新地层为下石炭统南好组，侵入的最新岩体为中林岭二长花岗岩，该序列岩石被进岭花岗斑岩（247 Ma）、六连岭正长花岗岩（239Ma）侵入。东岭花岗闪长岩中所获锆石 U-Pb 谐和曲线法上交点年龄为（242±10）Ma（宜昌地质矿产研究所，1995），SHRIMP 法锆石^{206}Pb/^{238}U 平均年龄（254±4）Ma（宜昌地质矿产研究所，海南省地质调查院，2004），形成时代厘定为晚二叠世。

（一）东岭中细粒角闪黑云花岗闪长岩（$\gamma\delta P_3$）

主要由官谋、加朝、南平、鱼湾、立岭、王宅等侵入体构成，还有一些小于 1km^2 的小岩株或岩枝，总面积约 192.86km^2。根据六连岭东侧六连林场一带见有很宽的角岩带，且常见有东岭花岗闪长岩的小岩枝出露，推测其下部有东岭花岗闪长岩的隐伏岩体。该单元侵入体边界主要受北东向构造（倾向北西，倾角 40°~86°）控制，局部受东西、南北、北西向构造控制。东岭花岗闪长岩侵入地层围岩时，常常切割地层层理或板劈理，外接触带出现 550~1200m 宽的热接触变质晕，有时还伴有弱矽卡岩化。内接触带出现 20~80cm 的冷凝边和宽度不等的细粒边缘相。而当其侵入通什二长花岗岩时，切割后者的叶理构造及脉岩，并出现 80~110cm 的冷凝边和宽度不等的细粒边缘相。

主体岩性为二长花岗岩-花岗闪长岩。灰白色、灰色，块状构造，局部具原生（同侵位）叶理和线理构造，这些部位含微粒交生体，而块状构造部位则不含。中细粒—中粒花岗结构，大小以 0.5~4mm 为主。侵入体边缘以细粒为主，向岩体中心以中粒为主。局部可含少量（<5%）钾长石斑晶，粒径 6mm×8mm~10mm×15mm，长短轴比<2。局部见石英交代穿孔结构、文象结构、交代蠕英结构。矿物组成：角闪石 1%~3%、斜长石（An 20~An 45）30%~49%、钾长石 14%~35%、黑云母 10%~17%。角闪石自形—半自形柱状；黑云母自形—半自形鳞片状，局部团粒状；斜长石半自形—自形板状，环带构造非常发育，常为韵律环带，由多个韵律性正环带构成；钾长石为他形不规则状，可分两种，一是条纹正长石，一是微斜条纹长石，钾长石中钠长石条纹很少，一般<10%，个别可达 15%~20%。

副矿物主要为榍石、磷灰石、锆石、褐帘石、磁铁矿、钛铁矿。榍石为棕褐色，晶形很好，晶粒也较大，肉眼常能见及。

岩体中含有大量微细粒闪长质包体，既可呈透镜状定向排列，也可呈不规则棱角状，大小 1cm×2cm～47cm×230cm。此外，还常见析离体；偶见斜长角闪岩或黑云母角闪斜长变粒岩深源包体。

（二）白茶细粒黑云母二长花岗岩（$\eta\gamma fP_3$）

仅由白茶侵入体及少量岩墙构成。面积约 0.53km^2，岩性为细粒黑云母二长花岗岩。灰白色，块状构造，局部定向构造，细粒花岗结构，粒径一般 0.5～1.5mm。矿物组成：黑云母 2%～3%、石英 25%左右、斜长石（An 8～An 20）29%～30%、微斜条纹长石 40%～44%。黑云母鳞片状；斜长石半自形板柱状，见环带构造；微斜条纹长石他形粒状，少数半自形板柱状。副矿物主要为锆石、磷灰石、磁铁矿、榍石、电气石。岩石中见少量石英电气石囊包体（岩浆分异包体）。

（三）岩石地球化学特征

该序列岩石的 SiO_2 含量为 66.88%～72.84%，晚期岩石快速增大；Na_2O+K_2O 含量为 7.03%～8.75%，K_2O 含量为 4.0%～5.71%，全碱和 K_2O 含量都较高；σ 值为2.05～2.56，根据硅-碱图解（图 9-7）和 SiO_2-K_2O 图解（图 9-8），东岭花岗闪长岩属高钾钙碱性系列，白茶二长花岗岩落入钾玄岩系列。A/CNK 在 0.96～1.08 之间，属准铝质—弱过铝质。稀土元素总量为 209.5×10^{-6}～283.7×10^{-6}，稀土配分模式为右倾曲线（图 9-19）。轻重稀土分异较明显，LREE/HREE 值为 12.1～14.0，$(Gd/Yb)_N$为 1.6～4.0；δEu 为 0.57～0.21，具中等—较强烈的负铕异常。微量元素蛛网图中显示较强的 Ba、Ta-Nb、Sr-P、Eu-Ti 负异常和明显的 Rb、Th-U-K、Pb、Nd-Zr-Hf-Sm 正异常（图 9-21）。东岭花岗闪长岩的初始锶比值为 0.70947，$\varepsilon_{Nd}(t)$ 为～6.77，T_{2DM}值为 1577Ma，属于壳幔混合来源。

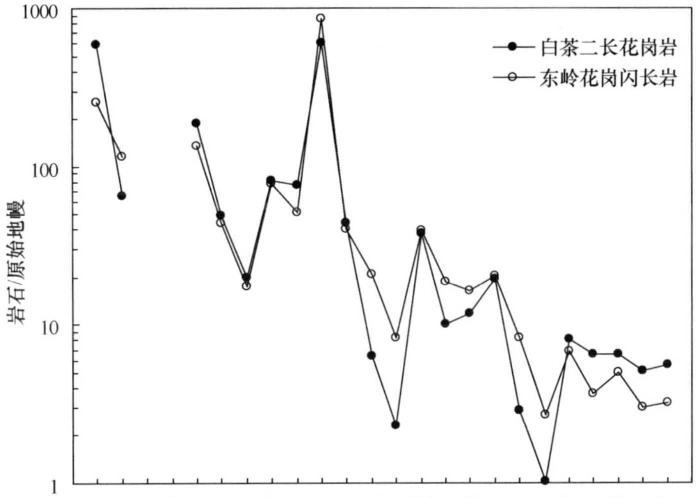

图 9-21　禄马序列微量元素原始地幔标准化蛛网图

Fig. 9-21　Primitive mantle-normalized spider diagram of trace elements of Luma Sequence

十、晚二叠世什天细中粒斑状黑云母二长花岗岩（$\eta\gamma mP_3$）

1. 地质特征与形成时代

主要见有什天、抱隆、红门岭、广坝、半路坡等岩体。总面积约 26.72km^2。侵入峨文岭组和陀烈组，什天和抱隆侵入体被尖峰岭正长花岗岩（249Ma）侵入。什天侵入体 SHRIMP 锆石岩浆结晶年龄为（251±7）Ma（宜昌地质矿产研究所，海南省地质调查院，2004），与地质关系吻合，形成于晚二叠世。

2. 岩石学特征

主体岩性为细中粒斑状黑云母二长花岗岩。块状构造，局部有定向构造，似斑状结构。斑晶钾长石 2%~25%，大小 6mm×12mm~2mm×5cm，有时斑晶聚集成堆晶包体（大小 15cm×40cm~30cm×60cm），有时斑晶定向排列。基质细中粒花岗结构，大小 0.3~2mm 或 1~5mm，矿物组成：石英 20%~30%、斜长石（中长石-奥长石）25%~45%、微斜条纹长石 25%~35%、黑云母 6%~10%、角闪石 0~1%。副矿物主要有磁铁矿、榍石、褐帘石、锆石、磷灰石，次要的有独居石、磷钇矿、萤石。

岩石中含少量闪长质包体，0.2~2 个/m^2，大小 1cm×2cm~4cm×7cm，个别 10~20cm，呈次棱角状—浑圆状。

3. 岩石地球化学特征

什天二长花岗岩的 SiO_2 含量为 68.41%~72.46%；Na_2O+K_2O 含量变化较大，为 7.04%~8.63%；K_2O 含量为 3.88%~5.29%，σ 值为 1.72~2.87，属高钾钙碱性系列。A/CNK 为 0.91~1.09，属准铝质—弱过铝质。根据硅-碱图解（图 9-7）和 SiO_2-K_2O 图解（图 9-8），主要属高钾钙碱性系列。稀土元素总量为 $154.1\times10^{-6}\sim577.5\times10^{-6}$，变化较大，其中抱隆、半路坡侵入体稀土元素含量较高。稀土配分模式为较陡的右倾曲线（图 9-19）。轻重稀土分异较明显，LREE/HREE 值为 12.1~23.3；δEu 为 0.67~0.27，具中等负铕异常。岩石 Rb、Ba、Sr、K 等亲石元素含量较高，在微量元素蛛网图上显示较弱的 Ba 负异常、较强的 Ta-Nb、Sr-P、Eu-Ti 负异常和明显的 Rb、Th-U-K、Pb、Nd-Zr-Hf-Sm 正异常（图 9-22）。初始锶比值为 0.70813~0.70859；$\varepsilon_{Nd}(t)$ 为 -5.21~-5.30，T_{2DM} 值为 1447~1455Ma，为壳幔岩浆混合产物。

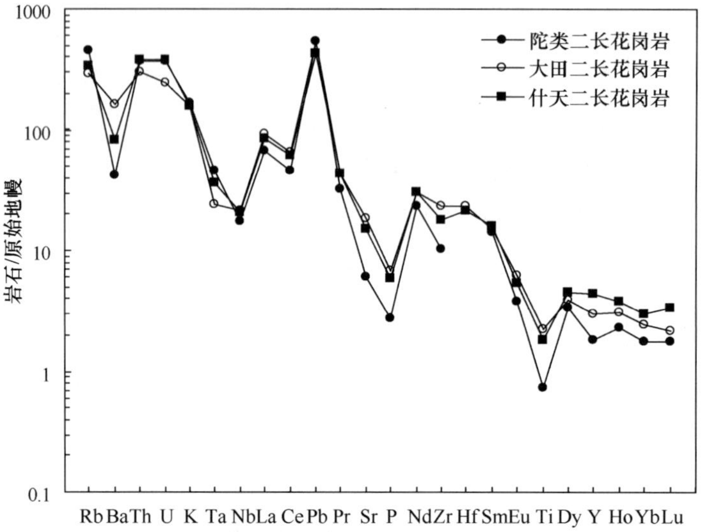

图 9-22 什天、大田、陀类二长花岗岩微量元素原始地幔标准化蛛网图

Fig. 9-22 Primitive mantle-normalized spider diagram of trace elements of Shitian, Datian and Tuolei monzonitic granites

十一、晚二叠世大田细粒含斑黑云母二长花岗岩（$\eta\gamma P_3$）

1. 地质特征与形成时代

分布于海南岛西部陀牙—玉龙印一带，见有大田和玉龙印两个侵入体，出露面积为 193.72km^2。大田侵入体侵入峨文岭组、南碧沟组、陀烈组，玉龙印侵入体侵入什天二长花岗岩的广坝岩体及中—上二叠统南龙组。大田侵入体测得锆石 $^{206}Pb/^{238}U$ 表面年龄为 236Ma，$^{207}Pb/^{235}U$ 表面年龄为 245Ma（海南地质大队，1990）。单颗粒锆石 U-Pb 化学法 $^{206}Pb/^{238}U$ 最大年龄为 247Ma（宜昌地质矿产研究所，

海南省地质调查院，2004），可能接近其形成年龄。结合其地球化学特点与什天二长花岗岩相似，暂将其形成时代厘定为晚二叠世。

2. 岩石学特征

主体岩性为细粒含斑黑云母二长花岗岩。块状构造，局部有线理构造或片麻状构造，细粒或中细粒含斑结构，斑晶钾长石2%～4%，大小4mm×5mm～8mm×12mm；基质粒度以0.3～2mm为主，有时可有部分达3mm。矿物组成：石英24%～30%、斜长石（An 20～An 32）24%～30%、微斜长石（无钠长石条纹）40%～48%、黑云母5%～7%。副矿物主要有榍石、褐帘石、锆石、磷灰石、磁铁矿。

岩石中可含少量闪长质包体，大田侵入体中有时见熔离出来的斜长花岗质浆泡，大小3mm×25mm～8mm×25mm，主要由长石、石英和榍石构成，榍石位于中心，含量占浆泡体积的2%～10%。

3. 岩石地球化学特征

岩石的SiO_2含量为68.70%～70.62%；Na_2O+K_2O含量为8.10%～8.33%；K_2O含量为4.42%～5.00%，σ值为2.36～2.68。根据硅-碱图解（图9-7）和SiO_2-K_2O图解（图9-8），属高钾钙碱性系列；A/CNK为0.99～1.07，多属弱过铝质。稀土元素总量为209.6×10^{-6}～300.1×10^{-6}，稀土元素含量中等；稀土配分模式为较陡的右倾曲线（图9-15）；轻重稀土分异较明显，LREE/HREE值为18.9～38.7；δEu为0.80～0.51，具中等负铕异常。岩石Rb、Ba、Sr、K等亲石元素含量较高，微量元素蛛网图显示弱的Ba负异常和较强的Ta-Nb、Sr-P、Eu-Ti负异常和明显的Rb、Th-U-K、Pb、Nd-Zr-Hf-Sm正异常（图9-22）。

十二、晚二叠世陀类细粒二云母二长花岗岩（γP_3）

1. 地质特征与形成时代

分布于海南岛西部益公村南、陀类村北一带，出露面积34km²。SHRIMP法锆石U-Pb年龄为（250±4）Ma（宜昌地质矿产研究所，海南省地质调查院，2004），形成时代厘定为晚二叠世。

2. 岩石学特征

岩性为细粒二云母二长花岗岩。线理构造很发育，叶理（片麻理）发育程度稍差。细粒花岗结构，粒度以0.3～2mm为主，有时可有部分矿物达3mm。矿物组成：石英23%～26%、斜长石（An 24～An 32）27%～37%、微斜条纹长石和微斜长石34%～42%、黑云母1%～5%、白云母1%～4%。副矿物主要有锆石、磷灰石、独居石，钛铁矿、石榴子石。岩石中含少量薄饼状富云包体。

3. 岩石地球化学特征

岩石的SiO_2含量较高，为72.18%～74.35%；Na_2O+K_2O含量为7.83%～8.28%，K_2O含量为4.83%～5.28%；σ值为1.94～2.34；根据硅-碱图解（图9-7）和SiO_2-K_2O图解（图9-8），属高钾钙碱性系列；A/CNK为1.09～1.33，属强过铝质；DI值较高，为87.67～92.72，分异程度较高。多数样品的稀土元素总量、配分曲线（图9-19）、LREE/HREE值都与大田二长花岗岩相似，只是δEu值较低（0.44～0.33）。稀土含量很低，为85.6×10^{-6}，LREE/HREE值也明显低一些（9.7）。微量元素蛛网图中（图9-22）显示其Rb、Th、U含量比大田和什天二长花岗岩稍高，而Ba、Sr、P、Zr、Hf、Eu、Ti、重稀土等元素含量比什天和大田二长花岗岩低，反映它经过了更强的分异作用。岩石初始锶比值为0.70991，$\varepsilon_{Nd}(t)$为-6.09，T_{2DM}值为1518Ma。

十三、晚二叠世西庆序列

分布于儋州市以西王五镇、西培农场一带，总面积约1122km²。由1∶5万《白马井镇、海头市、排浦市、王五市》幅中的鹅塘村单元、大岭单元、两院单元、高村单元、黑岭单元和陈桶岭单元等组成。根据各侵入体间接触关系、岩石特征及同位素年龄从早到晚划分为鹅塘村中细粒含斑英云闪长岩、大岭细中粒少斑花岗闪长岩、两院中粒斑状黑云二长花岗岩、高村细粒含斑黑云二长花岗岩及大王岭中细粒黑云母二长花岗岩等5个岩石单位。已有的同位素年龄为：大岭花岗闪长岩（252.2±17.2）Ma（Rb-Sr等时线法，海南地质综合勘察院第一勘察处，1993）、两院二长花岗岩（237.2±10.7）Ma（Rb-Sr等时线法，海南地质综合勘察院第一勘察处，1993）和246Ma（锆石ICPMS法，海南省地质调查院，2007）、高村二长花岗岩（218.4±1.9）Ma（Rb-Sr等时线法，海南地质综合勘察院第一勘察处，1996），表明该序列岩石为晚二叠世至早三叠世岩浆活动产物，结合其岩石学特征，本报告暂将其置于晚二叠世。

（一）鹅塘村中细粒含斑英云闪长岩（$\gamma\delta o P_3$）

由新村、鹅塘村、巴黎村等三个小侵入体组成，总面积4.61km^2。

岩石中普遍发育叶理构造和富含暗色闪长质包体。叶理主要由黑云母、长石斑晶、包体长轴的优选方向显示，走向285°~292°左右。包体一般呈灰色—深灰色，椭圆状、次圆状，少量被压扁拉长呈长条状，大小3cm×8cm~10cm×15cm，个别长轴达100~130cm，包体与寄主岩界线清楚。包体分布不均匀，局部富集达12个/m^2。包体内部有时见含有少量长石斑晶。

主体岩石类型为中细粒含斑英云闪长岩。深灰色，似斑状结构，叶理构造发育。矿物组成：石英20%~25%、斜长石（An 27~An 32）54%~63%、钾长石3%~7%、黑云母9%~15%、角闪石1%~2%。斑晶以斜长石为主，粒径6~10mm，含量5%，具弱反环带构造。基质中，钾长石主要为微斜长石，呈他形—半自形粒状或板状，卡氏双晶常见；斜长石半自形柱状，An值为27~32，常见卡钠复合双晶或聚片双晶，与钾长石接触处具蠕英结构，局部发生绢云母化、高岭土化；黑云母呈半自形叠片状，普遍发生绢云母化；角闪石呈半自形柱粒状。副矿物主要为磁铁矿、钛铁矿、榍石、磷灰石等。

（二）大岭细中粒少斑黑云母花岗闪长岩（$\gamma\delta p P_3$）

分布于儋州市桂兰村、尧头山、大成镇一带，由桂兰村、南华村两个侵入体组成，总面积50.32km^2。大岭细中粒斑状花岗闪长岩被高村二长花岗岩脉动侵入。二者界线清楚，高村二长花岗岩一侧见3~5cm的细粒边，黑云母带沿接触面分布，在高村二长花岗岩中见有大岭花岗闪长岩的捕虏体。

大岭花岗闪长岩中叶理构造不甚发育，含有少量闪长质包体。叶理主要由黑云母、长石斑晶、包体长轴的定向方向显示，走向30°~65°左右。包体一般呈灰色—深灰色，椭圆状、透镜状、次圆状，少量被压扁拉长呈长条状，个体小，一般为0.5cm×4cm~4cm×7cm，个别长轴达15~30cm。包体分布不均匀，与寄主岩界线清楚，内部偶见少量长石斑晶。

岩石类型为细中粒少斑花岗闪长岩。深灰色，似斑状结构，弱叶理构造。斑晶以钾长石为主，粒径7~15mm，含量8%，具弱反环带构造。矿物组成：石英20%~23%、斜长石（An 26~An 30）48%~55%、钾长石19%~23%、黑云母5%~8%、角闪石1%~2%。

副矿物主要为磷灰石、磁铁矿、锆石、绿帘石等，总量大于1000g/t。

（三）两院中粒斑状黑云母二长花岗岩（$\eta\gamma cp P_3$）

分布于华南两院、桑田大岭、西培农场、黑岭一带，面积445.68km^2。见高村二长花岗岩呈细脉状侵入于两院二长花岗岩中。

岩石类型主要为中粒斑状黑云母二长花岗岩。似斑状结构，线理构造发育，局部发育叶理构造。斑晶为钾长石，粒径5~15mm，含量10%~15%。基质中粒花岗结构，矿物粒径3~5mm。矿物组成：石英20%~30%、斜长石（An 24~An 26）28%~37%、钾长石28%~40%、黑云母5%~8%、角闪石0~1%。

副矿物主要为榍石、磁铁矿、钛铁矿、磷灰石、锆石、绿帘石等，总量大于1000g/t。

岩石中发育闪长质包体（图版Ⅱ-2-G）。包体呈椭圆状、透镜状，大小一般为1cm×3cm~5cm×10cm，个别长轴达15~30cm，其长轴与长石斑晶均呈大致60°~85°定向排列。包体分布不均匀，局部富集。包体与寄主岩界线清楚。

（四）高村细粒含斑黑云母二长花岗岩（$\eta\gamma p P_3$）

由陈桶岭、桃牛岭、打湾岭、马兰村、黄头村等侵入体组成，面积192.18km^2。

岩石类型主要为细粒含斑黑云母二长花岗岩。似斑状结构，发育弱叶理构造，矿物定向不明显，所含暗色包体小且少。斑晶为钾长石，粒径3~10mm，含量3%~5%，不具环带构造。基质细粒花岗结构，矿物粒径1~2mm。矿物组成：石英20%~25%、斜长石（An 26）37%~40%、钾长石29%~35%、黑云母3%~8%。

副矿物主要为榍石、磁铁矿、钛铁矿、磷灰石、锆石、绿帘石等，总量大于1000g/t。

（五）大王岭中细粒黑云母二长花岗岩（$\eta\gamma mfP_3$）

分布于澄迈县昆仑农场一带，面积约429.21km²。

主体岩性为中细粒黑云母二长花岗岩。岩石呈深灰、灰白色，中细粒花岗结构（矿物粒径1~3mm），交代结构或弱糜棱结构，块状构造。矿物组成：钾长石30%~45%、斜长石35%~40%、石英20%~25%、黑云母10%。偶见闪长质包体。

（六）岩石地球化学特征

西庆序列岩石从早到晚，具有明显的成分演化趋势：SiO_2、K_2O+Na_2O增高，Al_2O_3、CaO、Fe_2O_3+FeO、MgO降低，岩石向富硅、碱，贫铝、铁、钙、镁方向演化。SiO_2含量为66.53%~77.19%，属酸性岩类；K_2O+Na_2O含量在6.23%~7.58%之间；σ值在1.29~1.89之间。根据硅-碱图解（图9-7）和SiO_2-K_2O图解（图9-8），属高钾钙碱性系列。A/CNK大部分在0.86~1.07之间，属准铝质—弱过铝质花岗岩。

稀土总量为117.49×10^{-6}~478.74×10^{-6}，不同岩石单位变化较大。稀土配分模式为向右倾的曲线（图9-19）。轻重稀土分异一般，$(La/Yb)_N$值为1.3~21.91，轻稀土富集；δEu为0.09~0.55（大部分在0.4左右），具中等偏强负铕异常。微量元素方面（本章附表9-4），Ba、Rb、Zr、V等元素含量相对较高，Rb/Sr比值变化较大，早期岩石为小于1，晚期岩石大于1，表明岩浆发生了分离结晶作用。在微量元素原始地幔标准化蛛网图中（图9-23），显示较强的Ba、Ta-Nb、Sr、T i负异常和U、Pb、Nd-Zr-Hf-Sm正异常。

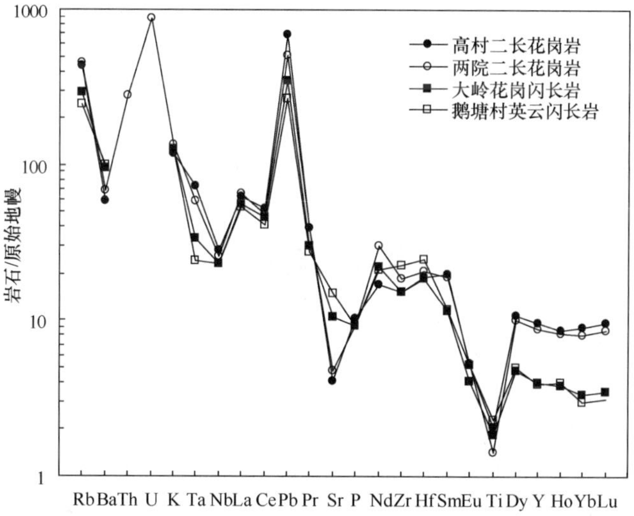

图9-23　西庆序列微量元素原始地幔标准化蛛网图

Fig. 9-23　Primitive mantle-normalized spider diagram of trace elements of Xiqing Sequence

第三节　三叠纪侵入岩

三叠纪侵入岩分布面积较广，出露总面积约5788.51km²，占海南省侵入岩总面积的35%（图9-24）。从早至晚共划分为36个岩石单位（表8-2）。

一、早三叠世尖峰序列

该序列岩石分布于白沙断裂带两侧的尖峰岭、抱由镇-毛岸地区及陵水-龙滚断裂带旁的石栏门-担柴岭一带。该序列岩体在平面上往往呈圆形的杂岩体。由尖峰岭中粗粒斑状黑云母正长花岗岩、黑岭细粒斑状黑云母正长花岗岩及瘦岭中细粒黑云母正长花岗岩共3个岩石单位构成一个结构演化序列。各岩石单位空间上紧密伴生。黑岭正长花岗岩脉动侵入尖峰岭正长花岗岩（图版Ⅱ-2-H），瘦岭正长花岗岩涌动侵入黑岭正长花岗岩。被该序列岩体侵入的最新地层为陀烈组，侵入该序列的最老岩体为早三叠世番阳峒正长花岗岩（马大铨等，1991）。尖峰岭正长花岗岩所测同位素年龄有：锆石U-Pb表面年龄（233±1）Ma（海南地质大队，1998），岩石-内部矿物Rb-Sr等时线法年龄（232±1）Ma及（235±1）Ma（马大铨等，1991）、SHRIMP法锆石U-Pb年龄（249±5）Ma（宜昌地质矿产研究所，海南省地质调查院，2004）；黑岭正长花岗岩所测颗粒锆石蒸发法年龄为244 Ma（海南地质大队，1998），形成时代厘定为早三叠世。

（一）尖峰岭中粗粒斑状黑云母正长花岗岩（$\xi\gamma cT_1$）

主要有尖峰岭、三平、阜堡笔、毛拉、金鸡岭、石栏门、担柴岭等侵入体，总面积1079.21km²。其中三平、阜堡笔侵入体原来可能同属于一个岩体，因后期岩体的侵入而分成两块。

主体岩性主要为中粗粒含斑-斑状黑云母正长花岗岩，局部为二长花岗岩。块状构造，似斑状结构，钾长石斑晶边缘形成蠕英结构（图版Ⅱ-3-A）。斑晶钾长石，粉红色，含量一般3%~25%，粒径1cm×1.5cm~2cm×3cm，分布较均匀，短板柱状，长宽比值小于1.5。基质中粗粒花岗结构，粒径2~10mm。矿物组成：石英30%~35%、斜长石（An12~An29）10%~25%、微斜条纹长石35%~52%、黑云母4%~6%。副矿物主要有磁铁矿、锆石、钛铁矿、磷灰石、独居石、褐帘石，次要的有磷钇矿、萤石、电气石、石榴子石等。

（二）黑岭细粒斑状黑云母正长花岗岩（$\xi\gamma T_1$）

呈不规则状侵入体侵入于尖峰岭正长花岗岩中，主要见有黑岭、狠俊岭、冲俄、头本、文扫等侵入体，总面积180.78km²。

岩性为细粒斑状黑云母正长花岗岩。块状构造，似斑状结构。斑晶钾长石一般5%~10%，局部15%~20%，大小3mm×4mm~15mm×25mm；斑晶石英3%±，局部5%~10%，大小3~9mm；斜长石斑晶1%~5%，大小3mm×4mm~6mm×8mm；黑云母斑晶0~1%，大小3~5mm。基质细粒花岗结构，粒径0.3~1.5mm。矿物组成：石英27%~30%、斜长石（An10~An29）13%~24%、微斜条纹长石45%~52%、黑云母3%~5%。副矿物主要有独居石、锆石、磁铁矿、磷钇矿、萤石。

（三）瘦岭中细粒黑云母正长花岗岩（$\xi\gamma fT_1$）

主要见瘦岭、高园、三独岭、南木、三杨、坡拉村、万好、毛岸、排岭等侵入体，呈不规则状侵入体侵入于尖峰岭、黑岭正长花岗岩或其旁侧围岩中。总面积约117.16km²。

岩性为（中）细粒黑云母正长花岗岩，局部为二长花岗岩。块状构造，局部晶洞构造（晶洞大小不一，一般为0.5~2.5cm，大者达10~30cm，可充填石英和钾长石晶簇）。中细粒花岗结构，粒径绝大多数为0.3~2mm，少数2~4mm。矿物组成：石英30%~35%、斜长石（An8~An27）12%~21%、微斜条纹长石45%~56%、黑云母2%~4%。副矿物主要有独居石、锆石、磁铁矿、磷钇矿、萤石。

（四）岩石地球化学特征

该序列岩石的SiO_2含量高，为70.90%~76.31%，绝大多数>72.09%；Na_2O+K_2O含量也较高，为7.53%~9.40%，K_2O含量介于4.46%~6.21%之间；σ值为1.72-2.93。根据硅-碱图解（图9-25）和SiO_2-K_2O图解（图9-26），属高钾钙碱性-钾玄岩岩系；A/CNK变化范围大，为0.92~1.37，早期岩石属准铝质—弱过铝质，中晚期岩石属准铝质—强过铝质；DI值较高，为86.08~95.87，反映分异程度较高。岩石稀土元素总量变化大，为85.3×10^{-6}~572.6×10^{-6}，其早期岩体的稀土含量高，而晚期岩体的稀土含量相对较低。稀土配分型式为右倾的V字形曲线（图9-27）。轻重稀土分异中等—较

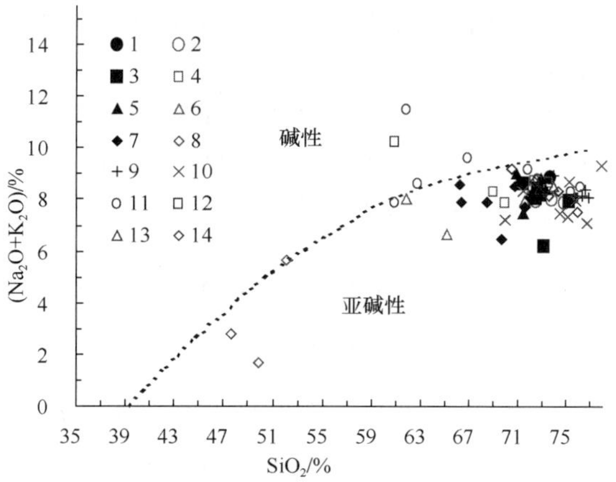

图 9-25　海南岛三叠纪侵入岩硅-碱图解

Fig. 9-25　$SiO_2-Na_2O+K_2O$ diagram of Triassic intrusive rocks from the Hainan Island

1—迈洲岭碱长花岗斑岩；2—三架岭序列酸性岩；3—雅加序列酸性岩；4—南英岭序列酸性岩；5—黑岭序列酸性岩；6—保梅岭正长花岗岩；7—琼中序列酸性岩；8—万宁序列中性岩；9—六连岭序列酸性岩；10—立才序列酸性岩；11—袁水序列酸性岩；12—深沟碱长正长岩；13—进岭花岗斑岩；14—尖峰序列酸性岩

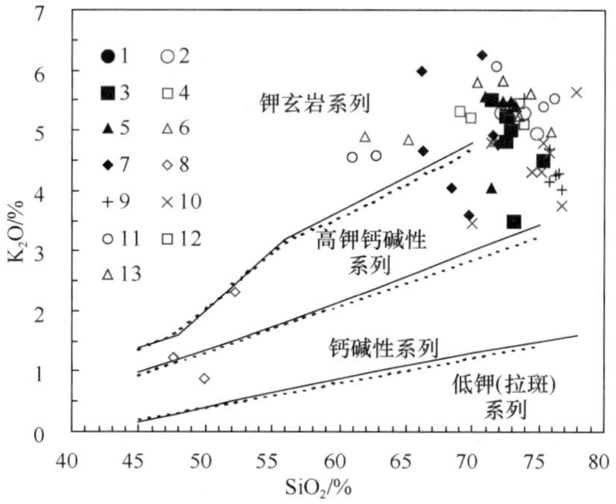

图 9-26　海南岛三叠纪侵入岩 SiO_2-K_2O 图解

Fig. 9-26　SiO_2-K_2O diagram of Triassic intrusive rocks from the Hainan Island

1—迈洲岭碱长花岗斑岩；2—三架岭序列酸性岩；3—雅加序列酸性岩；4—南英岭序列酸性岩；5—黑岭序列酸性岩；6—保梅岭正长花岗岩；7—琼中序列酸性岩；8—万宁序列中性岩；9—六连岭序列酸性岩；10—立才序列酸性岩；11—袁水序列酸性岩；12—进岭花岗斑岩；13—尖峰序列酸性岩

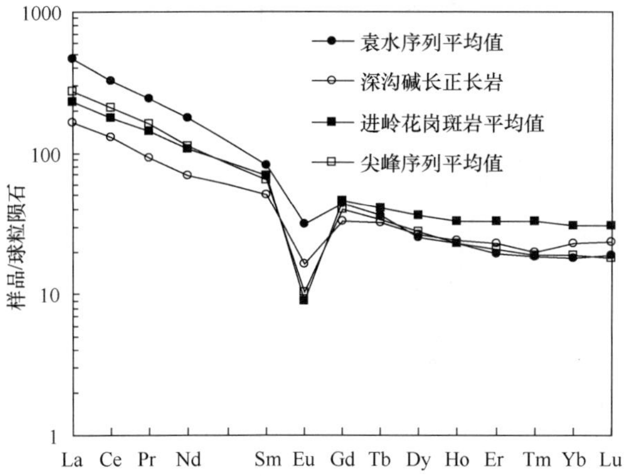

图 9-27　早三叠世侵入岩稀土元素球粒陨石标准化分布型式图

Fig. 9-27　Chondrite-normalized REE patterns of early Triassic intrusive rocks

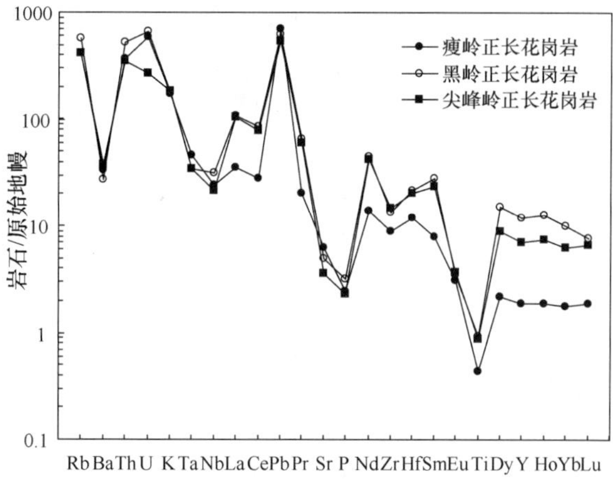

图 9-28　尖峰序列微量元素原始地幔标准化蛛网图

Fig. 9-28　Primitive mantle-normalized spider diagram of trace elements of Jianfeng Sequence

明显，LREE/HREE 值为 5.8~9.5；除极个别样品外，δEu 为 0.45~0.05，具中等—极强烈的负铕异常。微量元素 Rb、Th、U、K 含量较高，Sr、Ba 含量较低，在微量元素蛛网图中显示出很强的 Ba、Sr-P 及 Ti 负异常，而晚期岩石则显示了稀土元素及 Zr、Hf 的含量都低（图 9-28）。尖峰岭正长花岗岩初始锶比值较高，为 0.71658~0.71717；$\varepsilon_{Nd}(t)$ 为-7.59~-8.46，T_{2DM}值为 1639~1709Ma，属于地壳来源，并且可能属于较古老的酸性火成岩或混有沉积岩的地壳岩石熔融所形成。

二、早三叠世进岭花岗斑岩（$\gamma\pi T_1$）

（一）地质特征、形成时代及岩石学特征

分布于海南岛东部坝头-大坡一带，呈南北向的板状岩墙，面积 6.18km^2。进岭花岗斑岩还伴生有一些多斑状花岗斑岩的小岩体及岩脉，万宁附近见其侵入晚二叠世东岭花岗闪长岩（宜昌地质矿产研究所，1995）；六连岭岩体周围见该花岗斑岩的大量岩脉，而六连岭岩体内则未见该种岩脉，因此推测其形成于六连岭岩体（239Ma）之前。进岭花岗斑岩全岩 Rb-Sr 等时线年龄为（236±2）Ma（谢才富，1999），SHRIMP 法锆石 U-Pb 年龄为（247±5）Ma（宜昌地质矿产研究所，海南省地质调查院，2004），后一年龄值与地质侵入接触关系吻合，将其形成时代厘定为早三叠世。

主体岩性为花岗斑岩。在 QAP 图解上投入正长花岗岩区（图略）。岩石呈灰白色—肉红色，块状构造，局部见晶洞构造，充填石英晶芽及暗色矿物。斑状结构，斑晶中几乎没有石英，主要为钾长石和斜长石，含量 1%~30%，一般 4%~7%，少量黑云母或角闪石斑晶，粒径（0.5~2.5）mm×8mm，斑晶可具熔蚀结构。基质多数为球粒结构及显微文象结构，局部显微粒状，由钾长石，石英，斜长石等构成。副矿物主要为萤石、褐帘石、锆石、磁铁矿，也可有少量榍石。

（二）岩石地球化学特征

进岭花岗斑岩的 SiO_2 含量高达 72.69%；Na_2O+K_2O 含量为 8.42%；K_2O 含量为 5.19%；σ 值为 2.38，属高钾钙碱性系列；A/CNK 为 1.02，属弱过铝质；DI 值较高，为 88.26~91.25，分异程度较高；根据硅-碱图解（图 9-25）和 SiO_2-K_2O 图解（图 9-26），属高钾钙碱性系列。稀土元素总量 268.9×10^{-6}，稀土配分模式为较平缓右倾的 V 形曲线（图 9-27）。轻重稀土分异中等，δEu 值低（0.21），具强烈负铕异常。在微量元素蛛网图中显示出很强的 Ba、Sr-P 及 Ti 负异常（图 9-29）。初始锶比值为 0.71020，$\varepsilon_{Nd}(t)$ 为-7.67，T_{2DM}值为 1644Ma，为壳源岩石熔融的产物。

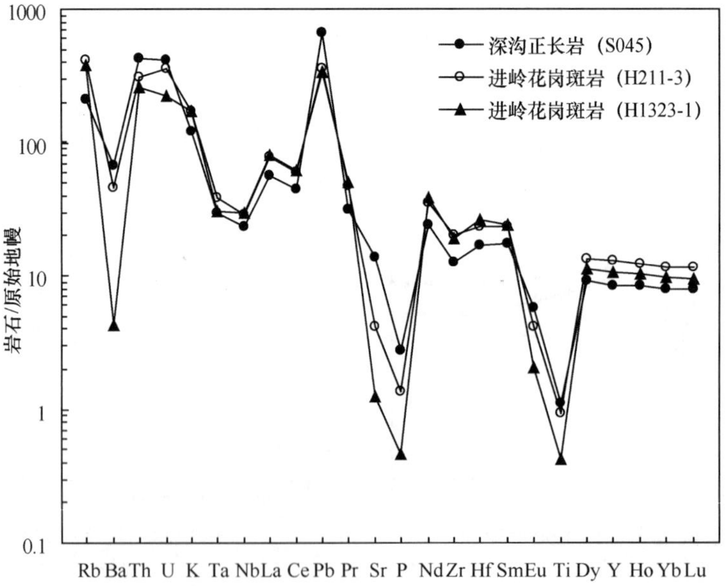

图 9-29　进岭花岗斑岩、深沟正长岩微量元素原始地幔标准化蛛网图

Fig. 9-29　Primitive mantle-normalized spider diagram of trace elements of Jinling granite and Shengou syenite

三、早三叠世深沟粗中粒少斑霓辉石碱长正长岩（$\kappa\xi T_1$）

（一）地质特征、形成时代及岩石学特征

见于深沟和牙笼半岛，呈直径150~300m的巨大捕虏体产于中三叠世布山村二长花岗岩的牙笼村岩体中，总面积约0.37km²。深沟岩体SHRIMP法锆石U-Pb年龄为（244±7）Ma（谢才富等，2005），侵入时代厘定为早三叠世。

岩性为粗中粒少斑霓辉石碱长正长岩。实际矿物在Q-A-P投影图上落在碱长正长岩区。块状构造，似斑状结构。钾长石斑晶6%~9%，大小9mm×15mm~15mm×30mm；基质粗中粒，粒径3~10mm。主要矿物组成：斜长石（奥长石）约5%、微斜条纹长石75%~80%、霓辉石8%~13%、石榴子石7%~11%。斜长石半自形板柱状；微斜条纹长石半自形—他形板状，钠长石条纹极发育；霓辉石矿物中Na_2O含量为5.43%~5.66%，计算的$NaFeSi_2O_6$分子约占40%，属于霓辉石类，并且较富Mn（含MnO 2.82%~3.64%），其准确名称为钠锰辉石；石榴子石富钛，其TiO_2含量2.15%，属黑榴石，镜下观察见霓辉石与黑榴石呈共生关系（图版Ⅱ-3-B）。副矿物主要有磁铁矿、锆石、榍石。

（二）地球化学特征

深沟碱长正长岩的SiO_2含量为60.84%。Na_2O+K_2O含量高，为10.29%；σ值高，为5.79；NKA高，为1.03。根据硅-碱图解（图9-25），属于碱性岩系列。其K_2O含量为3.69%，Na_2O含量为6.60%，K_2O/Na_2O为0.56，虽然仍属钾质碱性岩但接近钠质类型，与福建明溪洋坊霓辉石正长岩富钾有所不同。A/CNK为0.57，属准铝质。稀土元素总量为192.2×10^{-6}。稀土配分模式为较平缓右倾的V形曲线（图9-27）。轻重稀土分异中等，LREE/HREE值6.8；δEu值0.40，有中等负铕异常。微量元素原始地幔蛛网图中显示出Th、U、Pb含量较高，而Ba、Sr-P及Ti负异常比较明显（图9-29）。初始锶比值为0.71037，$\varepsilon_{Nd}(t)$为-6.79，T_{2DM}值为1570Ma，与海南岛相同时代的碱质A型花岗岩类相似。

四、早三叠世袁水序列

该序列岩石分布于白沙断裂南东侧什桥—志仲一带及陵水-龙滚断裂带大花角—九所岭一带，总面积约559.09km²。由深堀村中粗粒少斑黑云角闪石英二长岩、麻山田中粗粒-中粗粒少斑角闪黑云母石英正长岩、番阳峒细中粒（少斑）角闪黑云母正长花岗岩和金寮中细粒黑云母正长花岗岩等4个岩石单位构成。金寮正长花岗岩脉动侵入番阳峒正长花岗岩（图版Ⅱ-3-C）。前人已有的同位素年龄有：麻山田正长花岗岩（牛岭侵入体）所测SHRIMP法锆石U-Pb年龄为（243±4）Ma、番阳峒正长花岗岩（兴隆农场北东三合水侵入体）SHRIMP法锆石U-Pb年龄为（241±1）Ma（宜昌地质矿产研究所、海南省地质调查院，2004）。本次在三合水侵入体中（兴隆南西）采用高精度的La-ICP-MS法测得锆石U-Pb年龄为（243.7±3.1）Ma（分析结果见表9-5）。三个年龄值在误差范围内基本一致，形成时代厘定为早三叠世。

表9-5　番阳峒正长花岗岩LA-ICP-MA锆石U-Pb测年结果

Table 9-5　LA-ICP-MA Zircon U-Pb dating results of Fanyangdong syenogranite

岩石单位	测试点	$^{207}Pb/^{206}Pb$	$^{207}Pb/^{235}U$	$^{206}Pb/^{238}U$	$^{208}Pb/^{232}Th$	$^{238}U/^{232}Th$	$^{207}Pb/^{206}Pb$	$^{207}Pb/^{235}U$	$^{206}Pb/^{238}U$	$^{208}Pb/^{232}Th$
番阳峒	PM1-69-1-1	0.0499	0.2592	0.0376	0.0113	0.37011	187	234	238	226
番阳峒	PM1-69-1-2	0.0497	0.2863	0.0419	0.0134	0.445752	189	256	264	269
番阳峒	PM1-69-1-3	0.0502	0.2678	0.0387	0.0116	0.654065	211	241	244	233
番阳峒	PM1-69-1-4	0.0507	0.2743	0.0393	0.0123	0.446768	228	246	249	247

续表

岩石单位	测试点	$\frac{^{207}Pb}{^{206}Pb}$	$\frac{^{207}Pb}{^{235}U}$	$\frac{^{206}Pb}{^{238}U}$	$\frac{^{208}Pb}{^{232}Th}$	$\frac{^{238}U}{^{232}Th}$	$\frac{^{207}Pb}{^{206}Pb}$	$\frac{^{207}Pb}{^{235}U}$	$\frac{^{206}Pb}{^{238}U}$	$\frac{^{208}Pb}{^{232}Th}$
番阳峒	PM1-69-1-5	0.0557	0.3018	0.0392	0.0125	0.611583	443	268	248	251
番阳峒	PM1-69-1-6	0.0848	0.4510	0.0385	0.0236	0.345805	1311	378	243	471
番阳峒	PM1-69-1-7	0.0523	0.2752	0.0383	0.0115	0.663746	298	247	242	232
番阳峒	PM1-69-1-8	0.0498	0.2655	0.0386	0.0127	0.203799	187	239	244	255
番阳峒	PM1-69-1-9	0.0514	0.2793	0.0394	0.0126	0.478629	257	250	249	253
番阳峒	PM1-69-1-10	0.0564	0.2986	0.0382	0.0157	0.180057	478	265	242	314
番阳峒	PM1-69-1-11	0.0508	0.2656	0.0378	0.0114	0.543537	232	239	239	228
番阳峒	PM1-69-1-12	0.0503	0.2927	0.0420	0.0130	0.445831	209	261	265	261
番阳峒	PM1-69-1-13	0.0517	0.2822	0.0394	0.0111	0.571265	272	252	249	223
番阳峒	PM1-69-1-14	0.0530	0.3198	0.0437	0.0137	0.468647	328	282	276	274
番阳峒	PM1-69-1-15	0.0519	0.2626	0.0363	0.0115	0.362516	283	237	230	231
番阳峒	PM1-69-1-16	0.0501	0.2875	0.0409	0.0126	0.392403	211	257	258	254
番阳峒	PM1-69-1-17	0.0511	0.2766	0.0392	0.0120	0.45108	256	248	248	241
番阳峒	PM1-69-1-18	0.0513	0.2715	0.0384	0.0119	0.379636	254	244	243	239

（一）深堀村中粗粒少斑黑云角闪石英二长岩（$\eta o T_1$）

该岩石单位常呈被侵蚀的岩体残块出现，由深堀村、南郎、从南、五岭等侵入体组成，总面积为 27.38km²。

岩性为中粗粒不等粒—中粗粒少斑角闪黑云石英二长岩。块状构造，中粗粒不等粒结构或似斑状结构。局部斑晶钾长石可达 3%~8%，大小 8mm×10mm~15mm×35mm，长宽比值为 1.2~2.5。基质粒径以 2~7mm 为主。矿物组成：角闪石 4%~8%、黑云母 4%~10%、斜长石（An35~An40）26%~35%、石英 5%~14%、微斜条纹长石 46%~48%。角闪石长柱状，镜下为蓝绿色—淡黄绿色；黑云母鳞片或叠片状，镜下为深棕色—淡黄色；斜长石半自形板柱状，可见环带构造；石英他形粒状；微斜条纹长石半自形—他形，钠长石条纹发育。副矿物主要有榍石、磁铁矿、褐帘石、磷灰石、锆石。局部见细粒闪长质包体。

（二）麻山田中粗粒—中粗粒少斑角闪黑云母石英正长岩（$\xi o T_1$）

主要见有麻山田、牛角岭、牛漏、牛古田、牛岭、关桥岭、志向、空洪、九所岭、英州、港门岭等岩体，总面积约 109.83km²。

岩性为中粗粒—中粗粒少斑角闪黑云母石英正长岩（图版Ⅱ-3-D）。灰白色—肉红色，块状构造，中粗粒不等粒或似斑状结构，矿物粒径一般为 3~10mm。局部含斑晶钾长石 3%~10%，粒径 1cm×1.5cm~1.5cm×4cm，长宽比值为 1.2~2.5。矿物组成：角闪石 2%~5%（个别达 17%）、黑云母 2%~7%、斜长石（An26~An40）13%~35%、石英 5%~15%、微斜条纹长石 53%~70%。角闪石长柱状，镜下为蓝绿色—淡黄绿色；黑云母鳞片或叠片状，镜下为深棕色—淡黄色；斜长石半自形板柱状，可见环带构造；石英他形粒状；微斜条纹长石半自形—他形，钠长石条纹发育。副矿物主要有锆石、磷灰石、磁铁矿、钛铁矿、褐帘石，个别见榍石。见少量闪长质包体。

（三）番阳峒细中粒（少斑）角闪黑云母正长花岗岩（$\xi\gamma mc T_1$）

主要见有大田、大花角、合山村、第一尖岭、三合水、田头、谷六园、下岭、吉村、东高岭、番阳峒、乐中、卡法岭等侵入体，总面积约 375.86km²。

岩石呈灰白色—浅粉红色，块状构造，细中粒不等粒花岗结构或似斑状（少斑）结构，矿物粒径以2~5mm为主，少量0.2~2mm或5~8mm。局部含斑晶钾长石5%~10%，灰白色，粒径6mm×12mm~10mm×20mm，长宽比值多为1.5~2。矿物组成：钾长石45%~67%、斜长石（An5~An37）18%~29%、石英19%~25%、黑云母2%~12%、角闪石1%~13%。钾长石浅灰白色为主（晶体中部常略带暗灰色），部分略显浅粉红色，他形粒状、半自形柱状，钠长石条纹发育，卡斯巴双晶及格子双晶发育；斜长石半自形板柱状，聚片双晶发育，不见环带构造；石英他形粒状；黑云母鳞片状或团粒状，镜下为黑色—淡黄色；角闪石呈半自形—他形粗短柱状、粒状，菱形解理发育，镜下为暗蓝色（N_g）—淡黄绿色（N_p）。岩石中暗色矿物常呈团状集合体产出。副矿物主要为褐帘石、锆石、磁铁矿等。偶见微细粒闪长质包体。

（四）金寮中细粒黑云母正长花岗岩（$\xi\gamma mT_1$）

主要分布于万宁水库至兴隆一带，面积约46.02km^2。

主体岩性为中细粒黑云母正长花岗岩。岩石为灰白色，块状构造，细粒、中细粒不等粒结构、显微文象连晶结构（图版Ⅱ-3-E）。约有20%左右的微斜条纹长石粒径为1.8mm×2.1mm~2.4mm×5.0mm，其余矿物粒径0.2m×2.0m~1mm×2.0mm，两者略有间断。矿物组成：钾长石53%~76%、斜长石2%~10%、石英20%~34%、黑云母1%~7%、角闪石0~1%。钾长石灰白色，半自形—他形板状、粒状，卡斯巴双晶、简单格子双晶和钠长石条纹较发育；斜长石灰白色，他形柱状，偶见环带构造，聚片双晶发育，为更长石；石英他形粒状，分布在长石颗粒间不规则空隙中，波状消光明显，受后期构造作用于碎裂岩中亦可见带状消光；黑云母半自形—他形鳞片状-叠片状，具深褐色—黑、黄色—赤红多色性；角闪石半自形柱状—不规则状，蓝绿色—淡黄绿色，较特别的是他形成较晚，交代黑云母，与A型花岗岩中碱性暗色矿物的结晶顺序相似。

（五）岩石地球化学特征

该序列岩石的SiO_2含量变化大，为58.40%~76.14%；岩石的Na_2O+K_2O含量较高，为7.10%~11.49%；K_2O含量高，为3.78%~7.91%；σ值为2.0~7.93，其中深堀村石英二长岩和麻山田石英正长岩的σ值几乎都>3.3，但番阳峒及金寮正长花岗岩的σ值绝大多数都<3.3；在硅-碱图（图9-25）中，深堀村石英二长岩和麻山田石英正长岩落入碱性区，而番阳峒及金寮正长花岗岩落入亚碱性区。据SiO_2-K_2O图解（图9-26），该序列岩石属钾玄岩系列。A/CNK为0.88~1.12，绝大多数属准铝质；番阳峒及金寮正长花岗岩的NKA值也较大，平均值分别为0.84~0.92。岩石稀土元素含量普遍较高，早期岩石稀土配分曲线型式为较陡的右倾平滑曲线（图9-27），后期岩石为右倾的V字形曲线。轻重稀土分异中等—明显，LREE/HREE值为7.9~36.6；δEu为1.80~0.06，深堀村石英二长岩具弱正铕异常—弱负铕异常，麻山田石英正长岩一般为正铕异常，番阳峒正长花岗岩为中等—极强烈的负铕异常，金寮正长花岗岩为强烈的负铕异常。微量元素Zr、Hf、Ga、Zn含量较高，微量元素蛛网图中深堀村石英二长岩和麻山田石英正长岩具有Ba的正异常，而番阳峒和金寮正长花岗岩具有Ba负异常，且从早至晚Sr-P及Ti负异常呈逐渐增强态势（图9-30）。

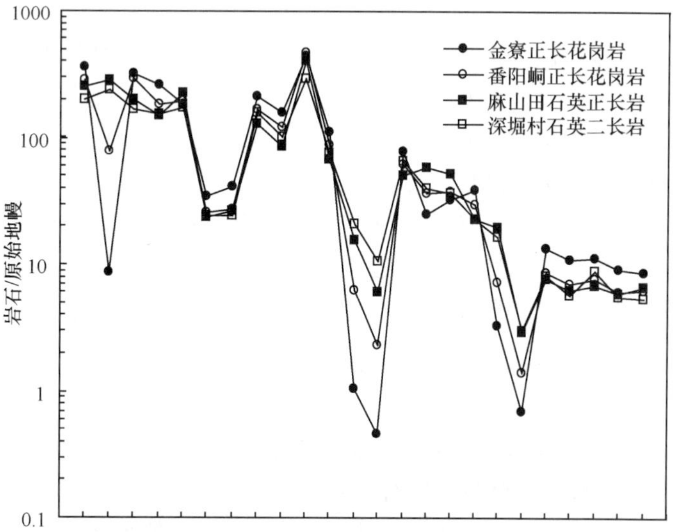

图9-30 袁水序列微量元素原始地幔标准化蛛网图

Fig. 9-30 Primitive mantle-normalized spider diagram of trace elements of Yuanshui Sequence

五、中三叠世立才序列

该序列岩石分布于小洞天至陵水一带，总面积约644.1km^2。由布山村细中粒含斑-斑状角闪黑云母二长花岗岩、超盆中粗粒含斑黑云母二长花岗岩、结尾（中）细粒黑云母二长花岗岩和前锋细粒少斑黑云母正长花岗岩等4个岩石单位构成。超盆二长花岗岩脉动侵入布山村二长花岗岩，结尾二长花岗岩脉动侵入布山村及超盆二长花岗岩，前锋正长花岗岩脉动侵入结尾二长花岗岩及超盆二长花岗岩（海南地质综合勘察院，1994）。布山村二长花岗岩侵入包裹深沟碱长正长岩；侵入该序列的最老岩体为六连岭正长花岗岩（239Ma），被早白垩世火山岩地层不整合覆盖并被花岗斑岩所侵入，在火山爆发相岩石中见该序列岩石的角砾和集块（海南地质综合勘查院，1995）。其已有的同位素年龄有：超盆二长花岗岩锆石U-Pb表面年龄为201.5 Ma及188Ma（海南地质综合勘察院，1994）、布山村二长花岗岩（布山村岩体D3034）SHRIMP法锆石U-Pb年龄为（239±7）Ma（宜昌地质矿产研究所、海南省地质调查院，2004）。本次在陵水以北高坡侵入体中新获得LA-ICPMS法锆石U-Pb年龄（243.2±1.6）Ma，将该序列岩石形成时代厘定为中三叠世早期。

（一）布山村细中粒含斑-斑状角闪黑云母二长花岗岩（$\eta\gamma T_2$）

主要有牙笼村、黑石村、布山村（和尚顶）、墓山村、小洞天、南斗（龙建）、六道岭、田仔、保芬、米市坡等侵入体，总面积约256.65km^2。

岩性为细中粒含斑-斑状角闪黑云母二长花岗岩，局部为花岗闪长岩。多为块状构造，细中粒似斑状结构。斑晶钾长石3%~20%，局部达25%~30%，大小6mm×15mm~20mm×40mm；基质粒径1~9mm。矿物组成：石英19%~33%、斜长石（An22~An40）21%~44%、微斜条纹长石20%~48%、黑云母3%~13%、角闪石2%~6%。斜长石半自形板柱状，环带构造很发育；微斜条纹长石半自形—他形板状；黑云母半自形—他形鳞片状-叠片状；角闪石半自形柱状，深蓝绿色—黄绿色。副矿物主要有磁铁矿、榍石、褐帘石、锆石、磷灰石。

常见闪长质-石英闪长质包体，浑圆状，大小1cm×2cm~15cm×30cm，暗色矿物析离体团块（粒）也较常见。

（二）超盆中粗粒含斑黑云母二长花岗岩（$\eta\gamma cT_2$）

分布于小洞天-立才农场一带，主要有红塘岭、四马、超盆、高岭、北山岭、鹿回头、高土、赤岭、光明等侵入体，出露面积约316.6km^2。

主体岩性为中粗粒含斑黑云母二长花岗岩，局部为正长花岗岩。块状构造，中粗粒含斑-似斑状花岗结构，斑晶钾长石2%~15%，一般3%左右，大小8mm×12mm~20mm×40mm，基质粒径以2~8mm为主。矿物组成：石英24%~33%、斜长石（An16~An34）18%~30%、微斜条纹长石33%~50%、黑云母1%~6%。斜长石半自形板柱状，环带不发育；微斜条纹长石半自形板状—他形；黑云母半自形—他形鳞片状或丛状。副矿物主要有磁铁矿、钛铁矿、锆石、磷灰石、褐帘石，偶见萤石。

可见少量微粒闪长质包体，浑圆状，大小1cm×2cm~8cm×12cm，含有大量长石、石英捕虏晶。

（三）结尾（中）细粒黑云母二长花岗岩（$\eta\gamma mfT_2$）

主要见于牛睡岭、大出水（小熟岭）、牛少坡、虎岭等侵入体，面积约31.99km^2。

岩性以（中）细粒黑云母二长花岗岩为主，少量为正长花岗岩。块状构造，细—中细粒花岗结构，粒径多为0.5~2mm，少数矿物2~4mm。矿物组成：石英30%~34%、斜长石（An12~An24）17%~21%、微斜条纹长石41%~47%、黑云母2%~3%。斜长石半自形板柱状，环带不发育；微斜条纹长石半自形板状—他形；黑云母半自形—他形鳞片状。副矿物主要有磁铁矿、钛铁矿、锆石、磷灰石，此外，还常见有萤石，偶见石榴子石。

（四）前锋细粒少斑黑云母正长花岗岩（$\xi\gamma fT_2$）

主要见有羊栏、床岭、槟榔大队、大烈、田独镇、鬼岭等侵入体。面积约38.86km^2。

岩性为细粒少斑黑云母正长花岗岩。块状构造，细粒斑状花岗结构，斑晶钾长石<10%，大小3~5 mm，还可有少量石英斑晶，基质粒径以 0.3~1.5mm 为主。矿物组成：石英 28%~32%、斜长石（An18~An22）20%~28%、微斜条纹长石40%~46%、黑云母1%~3%、白云母0~1%。斜长石半自形板柱状，环带不发育；微斜条纹长石半自形板状—他形；黑云母半自形—他形鳞片状。副矿物主要有磁铁矿、钛铁矿、锆石，少量磷灰石和萤石。

（五）岩石地球化学特征

该序列岩石的 SiO_2 含量变化大，为65.60%~78.06%，属酸性岩类；岩石的 Na_2O+K_2O 含量变化较大，大多数样品介于7.06%~9.34%之间；K_2O 含量变化较大，为3.16%~5.84%；σ 值为1.49~2.65；根据硅-碱图解（图9-25）和 SiO_2-K_2O 图解（图9-26），属高钾钙碱性岩系；A/CNK 为0.88~1.25（一个样品达1.41），属准铝质—强过铝质。岩石稀土元素含量变化较大，布山村单元稀土总量较高，为 197.6×10^{-6}~489.1×10^{-6}，其他3个单元稀土总量绝大多数较低，为 70.0×10^{-6}~176.0×10^{-6}。稀土配分曲线型式为较陡的右倾曲线（图9-31），LREE/HREE 值为9.3~25.4；δEu 为0.60~0.32；晚期岩石负铕异常比较明显，除个别样品外，其轻重稀土分异中等—不明显，LREE/HREE 值为2.1~8.7，δEu 为0.47~0.08，具中等-强烈的负铕异常。微量元素蛛网图中（图9-32），显示布山村二长花岗岩具有 Ba 的弱负异常，其轻稀土、Sr、Zr、Hf 的含量也较高；而后三个岩石单位具有较强烈的 Ba、Sr-P 及 Ti 负异常，轻稀土和 Zr、Hf 的含量也较低。布山村二长花岗岩的初始锶比值为0.71032、$\varepsilon_{Nd}(t)$ 为-5.83，T_{2DM}值为1488Ma，结合其包体特征，反映其为壳幔混合来源。

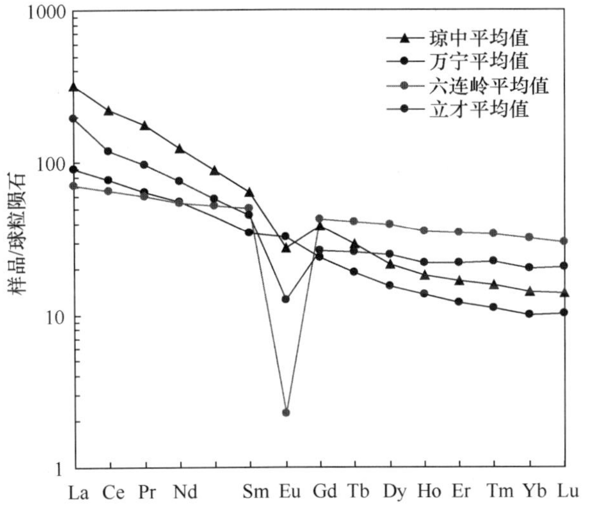

图9-31　中三叠世侵入岩稀土元素球粒陨石标准化分布型式图

Fig. 9-31　Chondrite-normalized REE patterns of middle Triassic intrusive rocks

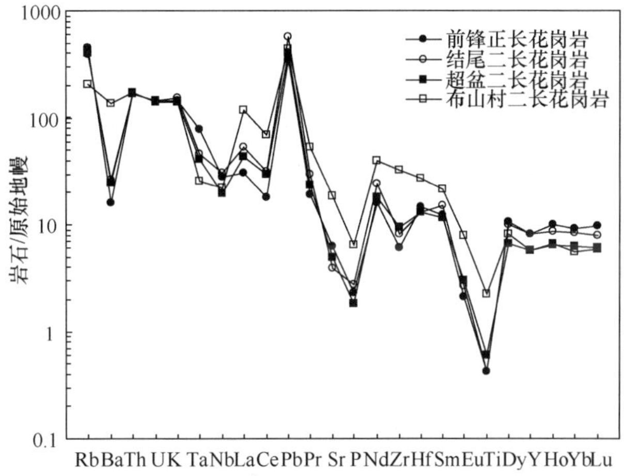

图9-32　立才序列微量元素原始地幔标准化蛛网图

Fig. 9-32　Primitive mantle-normalized spider diagram of trace elements of Licai Sequence

六、中三叠世六连岭序列

该序列岩石分布于陵水-龙滚断裂带两侧，总面积约166.51km²。由六连岭中粗粒黑云母正长花岗岩、兵工厂细粒斑状黑云（二云）正长花岗岩和石龟岭细粒黑云（二云）正长-二长花岗岩三个岩石单位构成。被六连岭中粗粒黑云母正长花岗岩侵入的最新岩体为早三叠世布山村二长花岗岩，陵水县港门村南见六连岭中粗粒黑云母正长花岗岩捕虏中二叠世通什二长花岗岩（图版Ⅱ-3-F）。六连岭岩体测得锆石 SHRIMP 法 U-Pb 年龄为（239±3）Ma（宜昌地质矿产研究所、海南省地质调查院，2004），与侵入接触关系相符，形成时代厘定为中三叠世。

（一）六连岭中粗粒黑云母正长花岗岩（$\xi\gamma cT_2$）

见有六连岭、港门石、东山岭、大溪岭、小沟坡、猴岛、红霞岭、铁炉港等侵入体。总面积约 117.74km^2。

主体岩性为中粗粒黑云母正长花岗岩。新鲜岩石为灰白色，风化后呈肉红色-淡粉红色或黄褐色，块状构造，主要为中粗粒不等粒花岗结构，粒径 0.5~8mm，多为 2~7mm。侵入体边部粒径可变小，为中粒—中细粒；有些侵入体如红霞岭、铁炉港等可为含斑结构。斑晶钾长石约 3%，大小 8mm×12mm~20mm×40mm。矿物组成：黑云母 1%~4%、石英 20%~29%、斜长石 15%~25%、微斜条纹长石 45%~58%。黑云母鳞片状，黑褐色—浅褐黄色，可蚀变为白云母；斜长石分两个世代，早世代为半自形中粗粒更长石-钠长石，晚世代为微细粒梳状钠长石集合体，沿裂隙或粒间交代早期矿物；微斜条纹长石他形，条纹结构极发育，钠长石条纹呈密集的纹脉状产出，可占整个长石颗粒体积的 30%~40%。副矿物见较多萤石（有时达 1%），其次还有少量锆石、磁铁矿、钛铁矿、磷灰石、石榴子石、电气石。红霞岭岩体中的石榴子石有时可超过 1%。岩石结晶晚期富含钠质，出现微粒梳状钠长石或条纹状钠长石的交代作用（图版Ⅱ-3-G），与区内其他花岗岩结晶晚期富含硅质而出现石英穿孔交代明显不同。偶见少量微粒闪长质包体，浑圆状，大小 1cm×2cm~8cm×12cm，含有大量长石、石英捕虏晶。

（二）兵工厂细粒斑状黑云母（二云母）正长花岗岩（$\xi\gamma fpT_2$）

主要有兵工厂、竹落两个侵入体，另有一些同类的岩墙，面积 27.25 km^2。

岩性为细粒斑状黑云母-二云母正长花岗岩。岩石为灰白色，风化后肉红色—灰绿色。块状构造，细粒似斑状结构。斑晶含量 10%~60%，大小（2mm×15mm~7mm×15mm，斑晶矿物种类多，由微斜条纹长石，石英，钠长石（更长石）及少量黑云母组成。斑晶自形程度差，多为他形，少数半自形，边界不平整，常有基质矿物的嵌晶。基质为细粒结构，一般<1mm，多数为 0.1~0.4mm。这是典型的补充期结构，与斑晶为自形—半自形并具熔蚀的斑状结构不同的是，这种似斑状结构的斑晶，其矿物种类、晶形、粒径都与比它稍早的六连岭正长花岗岩中的矿物很相似。因此，它可能是早期岩石的残余岩浆和部分已结晶矿物一起构成的晶粥脉动上侵到较高部位再结晶的结果。矿物组成为：黑云母 2%~4%、白云母 0~2%、石英 25%~37%、斜长石 19%~35%、微斜条纹长石 32%~50%。黑云母棕褐色—淡黄色，常蚀变为白云母，局部边缘变为蓝绿色，可能含锂，有时见填隙状蓝色云母，可能为黑鳞云母（图版Ⅱ-3H）；斜长石 An 牌号变化大，高者可达 28，多数 An 牌号<10，为钠长石；微斜条纹长石格子状双晶常见，钠长石条纹十分发育，有些条纹贯通几个颗粒，所以部分条纹应属交代成因。副矿物常见黄玉，局部达 1%~2%，另有少量锆石、萤石、磷灰石、磁铁矿等，有时见石榴子石。局部见较多伟晶岩囊并可有晶洞构造。偶见微粒闪长质包体。

（三）石龟岭细粒黑云母二长花岗岩（$\eta\gamma fT_2$）

主要见石龟岭一个侵入体，面积约 21.52km^2。

主体岩性为细粒黑云母二长花岗岩，局部为正长花岗岩。块状构造，细粒花岗结构，粒径 0.5~2mm，少数矿物 2~4mm。矿物组成：石英 23%~34%、斜长石（An<28）22%~32%、微斜条纹长石 32%~44%、黑云母 2%~3%，白云母少量。斜长石半自形板柱状，环带不发育，净边结构常见，晚期有少量钠长石沿粒间裂隙交代微斜条纹长石；微斜条纹长石半自形—他形板状；黑云母半自形—他形鳞片状-团粒状。副矿物主要有萤石、石榴子石（有时达 1%）、磁铁矿、锆石、磷钇矿、钍石，偶见褐帘石。

（四）岩石地球化学特征

该序列岩石的 SiO_2 含量高，为 73.88%~77.78%，绝大多数>75.31%。Na_2O+K_2O 含量较高，为 8.08%~8.98%；但 K_2O 含量为 4.02%~5.54%，并不算太高，并且较特别的是从早期向晚期 K_2O 含量总体上降低。σ 值为 1.99~2.61。根据硅-碱图解（图 9-25）和 SiO_2-K_2O 图解（图 9-26），属高钾钙

碱性系列。A/CNK 为 0.93~1.08，从含铝指数看，六连岭正长花岗岩绝大多数样品属准铝质，而兵工厂和石龟岭正长花岗岩属弱过铝质，但从矿物组成来看，该序列岩石普遍出现白云母、石榴子石、电气石、黄玉等富铝矿物，应是强过铝质的。FeO/MgO 值普遍很大，绝大多数为 28.2~284.8；DI 值高，为 90.2~96.4；NKA 值也较大，为 0.89~0.99，按赵子杰等（1989）的划分标准（以 KAI 值>0.90 为界），绝大多数属于碱性系列花岗岩。岩石稀土元素含量为 $197.9\times10^{-6}\sim13.5\times10^{-6}$，从早期往晚期明显降低，晚期岩石可具有很低的稀土含量，可能是遭受了流体交代的结果。该序列岩石还有一个特点是 Sm/Nd 高，为 0.27~0.40。稀土配分曲线为平展的 V 字形或称“海鸥型”曲线（图 9-31）。轻重稀土分异不明显，LREE/HREE 值为 1.0~3.9，晚期岩石重稀土往往反向分异，为向左倾的曲线. $(Gd/Yb)_N$ 常小于 1；δEu 为 0.11~0.02，具有很强烈的负铕异常。微量元素 Rb、Ta、Nb、Ga 含量较高，而 Ba、Sr、P、Eu、Ti 含量很低。Rb/Sr 值大，主要为 24.2~167.3；晚期岩石的 K/Rb 值低，仅为 78~95。微量元素蛛网图中（图 9-33）显示了极强的 Ba、Sr-P、Eu-Ti 负异常，该序列岩石比较特别的是没有 Nb-Ta 负异常，Nb/La 值一般大于 1。

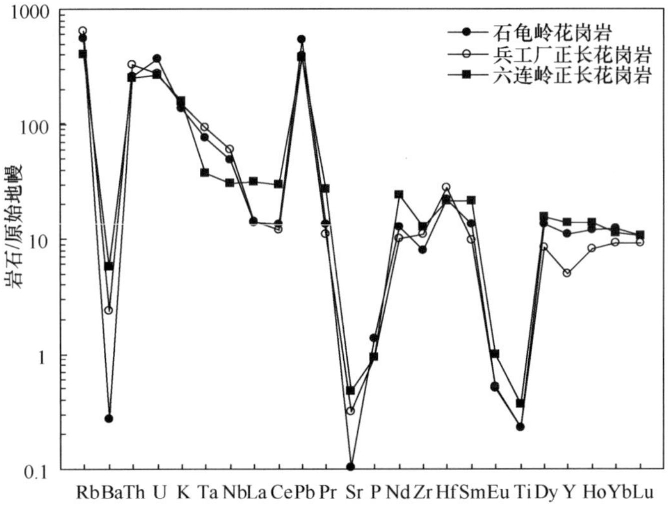

图 9-33　六连岭序列微量元素原始地幔标准化蛛网图

Fig. 9-33　Primitive mantle-normalized spider diagram of trace elements of Liulianling Sequence

七、中三叠世万宁序列

该序列岩石分布于兴隆、长安、长征及儋州市光头岭一带，总面积约 37.96km²。由长安辉长岩和石墩闪长岩两个岩石单位构成，后者脉动侵入前者。被该序列岩石侵入的最新岩体为番阳峒正长花岗岩（241Ma），而晚白垩世吊罗山正长花岗岩又侵入长安辉长岩。前人在长安岩体中采用 SHRIMP 法测得锆石 U-Pb 年龄为（238±3）Ma（宜昌地质矿产研究所、海南省地质调查院，2004），本次采用高精度的 La-ICPMS 法测得长安辉长岩的锆石 U-Pb 年龄分别为（228.1±5.9）Ma（吊罗山场部南侧辉长岩，D2481）、（231.2±2.7）Ma（兴隆农场南西金竹园侵入体，PM1-51）、246±2Ma（兴隆农场北加尧侵入体，D2212）。上述分析结果（分析数据见表 9-6）除了（246±2）Ma（采自加尧队侵入体）明显偏大并与实际侵入接触关系不符外，其余数据均落在中三叠世范围内。

表 9-6　长安辉长岩 LA-ICP-MA 锆石 U-Pb 测年结果

Table 9-6　LA-ICP-MA Zircon U-Pb dating results of Chang'an gabbro

岩石单位	测试点	$^{207}Pb/^{206}Pb$	$^{207}Pb/^{235}U$	$^{206}Pb/^{238}U$	$^{208}Pb/^{232}Th$	$^{238}U/^{232}Th$	$^{207}Pb/^{206}Pb$	$^{207}Pb/^{235}U$	$^{206}Pb/^{238}U$	$^{208}Pb/^{232}Th$
长安	D2481-1-1	0.0527	0.3139	0.0407	0.0143	0.337724	322	277	257	288
长安	D2481-1-2	0.0474	0.2359	0.0340	0.0112	0.40235	78	215	216	226

续表

岩石单位	测试点	$^{207}Pb/^{206}Pb$	$^{207}Pb/^{235}U$	$^{206}Pb/^{238}U$	$^{208}Pb/^{232}Th$	$^{238}U/^{232}Th$	$^{207}Pb/^{206}Pb$	$^{207}Pb/^{235}U$	$^{206}Pb/^{238}U$	$^{208}Pb/^{232}Th$
长安	D2481-1-3	0.2171	2.0862	0.0673	1.2277	0.010918	2961	1144	420	16189
长安	D2481-1-4	0.0642	0.3059	0.0334	0.0121	0.323635	746	271	212	244
长安	D2481-1-5	0.0583	0.2869	0.0353	0.0098	0.693385	539	256	223	196
长安	D2481-1-6	0.0506	0.2373	0.0341	0.0102	0.460617	220	216	216	206
长安	D2481-1-7	0.0720	0.3777	0.0373	0.0189	0.247592	987	325	236	379
长安	D2481-1-8	0.0458	0.2266	0.0357	0.0098	0.415559	error	207	226	198
长安	D2481-1-9	0.0516	0.2570	0.0360	0.0108	0.437694	333	232	228	218
长安	D2481-1-10	0.0525	0.2678	0.0371	0.0109	0.339328	306	241	235	218
长安	D2481-1-11	0.0516	0.2492	0.0351	0.0100	0.585789	333	226	222	202
长安	D2481-1-12	0.0535	0.2699	0.0366	0.0108	0.543006	350	243	232	218
长安	D2481-1-13	0.1020	0.5844	0.0395	0.0220	0.418831	1661	467	250	440
长安	D2481-1-14	0.0552	0.2754	0.0349	0.0138	0.419006	420	247	221	277
长安	D2481-1-15	0.0443	0.2077	0.0343	0.0097	0.39703	error	192	217	196
长安	D2481-1-16	0.0504	0.2328	0.0333	0.0101	0.330284	213	213	211	203
长安	D2481-1-17	0.0502	0.2716	0.0393	0.0124	0.618353	206	244	248	250
长安	D2481-1-18	0.0548	0.2545	0.0336	0.0104	0.575242	406	230	213	209
长安	PM1-51-1-1	0.0522	0.2616	0.0364	0.0115	0.148276	300	236	231	231
长安	PM1-51-1-2	0.0532	0.2726	0.0372	0.0117	0.363069	345	245	236	235
长安	PM1-51-1-3	0.0526	0.2647	0.0367	0.0103	0.57241	309	238	233	208
长安	PM1-51-1-4	0.0835	0.3094	0.0271	0.0160	0.224749	1283	274	172	320
长安	PM1-51-1-5	0.0517	0.2535	0.0355	0.0109	0.239303	272	229	225	219
长安	PM1-51-1-6	0.0568	0.2720	0.0347	0.0114	0.439406	483	244	220	229
长安	PM1-51-1-7	0.0520	0.2664	0.0371	0.0117	0.381665	287	240	235	234
长安	PM1-51-1-8	0.0529	0.2373	0.0325	0.0101	0.356608	324	216	206	202
长安	PM1-51-1-9	0.0433	0.0867	0.0145	0.0042	1.737318	error	84	92.8	85
长安	PM1-51-1-10	0.0535	0.1118	0.0154	0.0053	0.843668	350	108	99	107
长安	PM1-51-1-11	0.0525	0.2729	0.0373	0.0123	0.367647	309	245	236	247
长安	PM1-51-1-12	0.0561	0.1180	0.0153	0.0049	0.711693	457	113	98.0	99.1
长安	PM1-51-1-13	0.0490	0.2448	0.0362	0.0109	0.48714	150	222	229	220
长安	PM1-51-1-14	0.0515	0.2560	0.0361	0.0129	0.296991	261	231	229	260
长安	PM1-51-1-15	0.0508	0.2599	0.0371	0.0108	0.468406	232	235	235	218
长安	PM1-51-1-16	0.0533	0.2560	0.0349	0.0099	1.016157	343	231	221	199
长安	PM1-51-1-17	0.0569	0.2650	0.0337	0.0103	0.753864	487	239	214	207
长安	PM1-51-1-18	0.0535	0.2657	0.0361	0.0136	0.370796	350	239	228	273
长安	D2212-1-2	0.0515	0.2691	0.0379	0.0128	0.328019	261	242	240	257
长安	D2212-1-3	0.0481	0.2589	0.0389	0.0111	0.419586	106	234	246	224
长安	D2212-1-4	0.0516	0.2774	0.0388	0.0121	0.45294	333	249	246	243

续表

岩石单位	测试点	$^{207}Pb/^{206}Pb$	$^{207}Pb/^{235}U$	$^{206}Pb/^{238}U$	$^{208}Pb/^{232}Th$	$^{238}U/^{232}Th$	$^{207}Pb/^{206}Pb$	$^{207}Pb/^{235}U$	$^{206}Pb/^{238}U$	$^{208}Pb/^{232}Th$
长安	D2212-1-5	0.0487	0.2666	0.0396	0.0128	0.688658	132	240	250	258
长安	D2212-1-6	0.0514	0.2771	0.0390	0.0117	0.647291	257	248	247	236
长安	D2212-1-8	0.0503	0.2928	0.0422	0.0136	0.425949	209	261	266	274
长安	D2212-1-9	0.0502	0.2676	0.0384	0.0111	0.710177	206	241	243	224
长安	D2212-1-10	0.0536	0.3025	0.0408	0.0133	0.626684	367	268	258	268
长安	D2212-1-11	0.0531	0.2815	0.0385	0.0129	0.336519	332	252	243	259
长安	D2212-1-12	0.0583	0.3050	0.0381	0.0131	0.330688	543	270	241	263
长安	D2212-1-13	0.0529	0.2724	0.0374	0.0154	0.152307	324	245	237	310
长安	D2212-1-14	0.0557	0.2995	0.0390	0.0136	0.162043	439	266	247	273
长安	D2212-1-15	0.0532	0.2864	0.0391	0.0120	0.186265	345	256	248	241
长安	D2212-1-16	0.0528	0.2851	0.0393	0.0122	0.595557	317	255	248	246
长安	D2212-1-17	0.0521	0.2940	0.0409	0.0138	0.374125	287	262	259	277
长安	D2212-1-18	0.0541	0.3318	0.0445	0.0161	0.403975	376	291	281	322

（一）长安辉长岩（υT_2）

见有长安及岭里角闪辉长岩、赤埠及乘坡农场二十一队苏长辉长岩、金竹园橄榄辉长苏长岩、长征农场中学角闪辉石岩、光头岭辉长岩等侵入体，总面积约36.79km²。

岩石主体为黑色，块状构造，主要为中粒辉长结构。粒径0.4~6mm。主要由斜长石、单斜辉石和（或）角闪石组成，不同的侵入体还可分别有少量紫苏辉石、橄榄石、黑云母、石英。斜长石自形—半自形板柱状，含量一般为40%~55%（长征农场中学侵入体仅有约20%），其牌号一般为An55~An60，但金竹园侵入体An82~An86；单斜辉石半自形短柱状，属次透辉石-普通辉石类，含量1%~25%；角闪石自形—半自形柱状，主要为镁角闪石，含量微量至45%；紫苏辉石见于赤埠、乘坡农场二十一队及金竹园侵入体，含量6%~14%；黑云母红褐色—淡黄色，鳞片状，为镁质-铁质黑云母，含量一般<4%，赤埠岩体中达5%~8%、乘坡农场二十一队岩体达11%；橄榄石主要见于金竹园侵入体（<20%），长安侵入体中局部见及，其Fo67~69，属透铁橄榄石类（马大铨等，1991）；石英含量0~3%，呈他形充填于其他矿物粒间。副矿物见较多磁铁矿、钛铁矿、磷灰石，另有少量榍石、锆石。

（二）石墩中粗粒闪长岩（δT_2）

仅见石墩一个侵入体，面积约1.17km²。

岩性为闪长岩。块状构造，中粗粒结构，侵入体边缘粒度有所减小。矿物组成：斜长石（An40~An55）68%~75%、普通角闪石24%~25%、黑云母≤2%、钾长石0~2%、石英≤2%，局部见微量单斜辉石。斜长石为半自形板状，角闪石为半自形—他形柱状。副矿物主要有榍石、磁铁矿、钛铁矿、磷灰石、锆石。

（三）岩石地球化学特征

该序列岩石的SiO_2含量较低，为45.56%~52.16%，属基性岩。Na_2O+K_2O含量较低，为1.14%~5.66%；K_2O含量也较低，为0.46%~2.32%。σ值为0.19~3.09；据硅-碱图（图9-25）和“An”-Al_2O_3图，属拉斑玄武岩系列；A/CNK为0.21~0.71，属准铝质；K_2O/Na_2O值较低，为0.52~1.06；DI值较低，为9.6~25.92，SI值较高，为59.84~23.8。岩石稀土元素含量为84.5×10^{-6}~303.9×10^{-6}，变化较大。稀土配分曲线多为平滑的右倾曲线（图9-31）。轻重稀土分异中等，LREE/HREE值为

6.4~9.9；δEu 为 0.73~1.97，长安辉长岩为弱的负铕异常，石墩闪长岩为明显的正铕异常。微量元素 Ni、V、Cr、Co、Sc、Zn 等亲铁铜元素的含量较高，Ba、Sr 含量也较高，而 Rb、Th、U、Ta、Nb、Pb、Cs 含量低，Rb/Sr 值小，为 0.10~0.03；K/Rb 值较高，为 208~424（本章附表 9-4）。微量元素蛛网图中（图 9-34）显示了无 Ba 的负异常或具有 Ba 的正异常，Sr 具有正异常或弱负异常，Ti 负异常不明显或具弱的正异常，岩石具有较明显的 Nb-Ta 负异常，和板内玄武岩不同，表明源岩遭受过流体的交代作用或岩浆受到了地壳物质混染。该序列岩石初始锶比值为 0.70846~0.71042（平均为 0.70968），$\varepsilon_{Nd}(t)$ 值为-6.65~-8.28（平均为-7.22），T_{2DM}值为 1554~1686Ma（平均为 1600Ma），反映来源于 EM2 型富集岩石圈地幔。

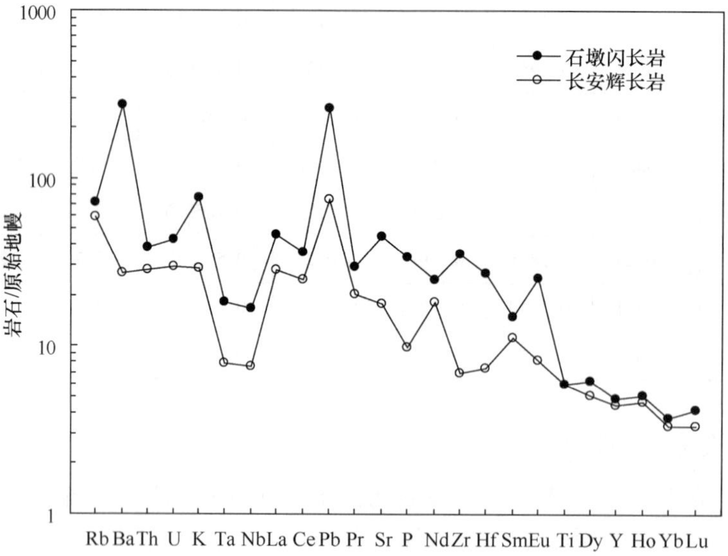

图 9-34　万宁序列微量元素原始地幔标准化蛛网图

Fig. 9-34　Primitive mantle-normalized spider diagram of trace elements of Wanning Sequence

八、中三叠世万车碱长花岗岩序列

万车碱长花岗岩序列由后龙中粒碱长花岗岩（$\chi\rho\gamma T_2^a$）和后美中细粒碱长花岗岩（$\chi\rho\gamma T_2^b$）2 个岩石单位组成。

1. 地质特征及形成时代

该序列岩石属典型的 A 型花岗岩，分布在海南岛西南部乐东县木棉头一带，侵入下志留统空列村组及早二叠世育才序列，并被晚白垩世千家岩体侵入。根据侵入接触关系从早到晚划分为后龙中粒碱长花岗岩和后美中细粒碱长花岗岩 2 个岩石单位。该序列岩石的特点是：结构演化明显，钾长石含量高，暗色矿物含量极少，出现碱性矿物钠闪石和霓石。

前人在后龙碱长花岗岩中获得锆石 U-Pb 同位素地质年龄为（233.2±0.6）Ma（海南地质综合勘察院，1998），与侵入接触关系吻合，形成时代划厘定为中三叠世。

后龙中粒碱长花岗岩由后龙山和松美山 2 个侵入体组成。后龙山侵入体分布在昂浓村-后龙山-万车村一带，呈不规则状。西侧被晚白垩世光头岭侵入体侵蚀，南、东侧侵入下志留统空列村组和早二叠世乘顺岭细粒片麻状黑云（二云）二长花岗岩。在侵入空列村组的地段，侵入体岩石粒度变细，石英含量减少，侵入接触产状 140°∠41°，接触面较为平直。在后龙山，侵入体中见乘顺岭细粒片麻状黑云（二云）二长花岗岩捕虏体，捕虏体大小约 30cm×40cm，发育在其中的节理被后龙山侵入体中粒碱长花岗岩截断，且捕虏体与中粒碱长花岗岩接触处见有 2~3mm 宽的烘烤边（图版Ⅱ-1-D）。

松美山侵入体出露在松美山一带，呈近椭圆状近南北向产出在育才序列岩石中，超动侵入志仲巨斑状二长花岗岩（图版Ⅱ-1-C）。

后美中细粒碱长花岗岩由后美侵入体和红柴侵入体组成，两个侵入体均呈次圆状。后美侵入体与后龙中粒碱长花岗岩呈脉动接触，接触特征表现为接触界线处两侧岩石呈突变关系，且未见冷热差异引起的冷凝边或热液蚀变现象，说明两者的侵位时间很接近，但在靠近界线后美侵入体一侧见有后龙中粒碱长花岗岩的捕虏体，说明后美侵入体形成时间晚于后龙中粒碱长花岗岩；红柴侵入体产出在早二叠世志仲二长花岗岩中，侵入体中见有数厘米到数米大小的志仲二长花岗岩的捕虏体。

2. 岩石学特征

该序列岩石实际矿物在Q-A-P投影图上大多落在碱长花岗岩区，少部分投影点落在正长花岗岩区（靠近碱长花岗岩与正长花岗岩界线附近）。

后龙中粒碱长花岗岩具中粒花岗结构，块状构造，矿物粒径以2~5mm为主，少许钾长石5~6mm。矿物组成：斜长石4%~15%、钾长石62%~70%、石英23%、白云母<1%、钠闪石4%~6%、霓石4%。后美中细粒碱长花岗岩为中细粒花岗结构，块状构造，矿物粒径0.4~2mm为主，部分2~3mm。矿物组成：斜长石10%~13%、钾长石65%~67%、石英20%~24%、白云母<1%、钠闪石4%~5%。矿物组合的最大特点是富含钾长石而极少斜长石和黑云母，出现碱性特征矿物钠闪石及霓石，这与闽东著名的魁歧岩体（A型花岗岩，钾长石65%~75%、斜长石0~2%、石英25%~35%、钠闪石少量至2%、霓石约2%）矿物组合特点十分相似。

钾长石主要以半自形—他形晶出现，少见自形晶，其在后龙中粒碱长花岗岩主要呈厚板状，在后美中细粒碱长花岗岩中呈粒状、板状、柱状等。钾长石普遍见格子双晶、卡氏双晶，成分主要为微斜长石、正长石、微纹长石等，常包裹有斜长石、石英晶体，说明结晶比斜长石及石英要晚，其晶体结合面或边缘常被白云母交代，在与斜长石接触部位偶见蠕英结构。

斜长石为半自形晶，其在后龙中粒碱长花岗岩中多为板状、柱状，在后美中细粒碱长花岗岩中以粒状为主，偶见柱状。钠长石双晶发育，见个别卡钠复合双晶，An11~An12之间，普遍见绢云母化、钠黝帘石化，局部被白云母交代及见包裹有细小的石英晶体。

石英他形粒状，无色透明，不规则充填在长石矿物间隙中，有时聚集成集合体，具弱波状消光。

白云母细叶片状，为后期交代蚀变产物，分布不均匀，多数见一组完全解理，产出于长石间隙中。

钠闪石呈半自形长柱状、纤维状，分布不均匀，大小多在0.05~0.1mm之间，少量为0.2~0.3mm，多色性显著且特殊，呈淡黄绿色—紫蓝色—蓝色，干涉色呈蓝色。晶内嵌有细小石英、长石晶体，局部见与极少量的含钛黑云母共生。

霓石为碱性侵入岩的典型造岩矿物，呈柱状自形晶，后期碎裂现象普遍，多色性强，表现为 N_g = 浅绿色，N_m = 黄绿色，N_p = 深绿色，斜消光，内部常包含细小的碱性长石颗粒或磷灰石。

副矿物总量较低（一般小于1000g/t），出现较普遍、含量相对较高的副矿物主要为独居石、石榴子石、磷钇矿、磷灰石、锆石及磁钛铁矿等，其中出现有滚圆状、粒状的锆石（图版Ⅱ-4-A），表面无玻璃光泽，粒径在0.03~0.2mm，应为地壳重熔过程中残留下来的继承性古老锆石。

3. 岩石地球化学特征

万车碱长花岗岩序列岩石的 SiO_2 含量高，在70.39%~72.43%之间。Na_2O+K_2O 为8.74%~10.46%，K_2O 含量明显高于 Na_2O；σ 值为2.36~3.72，NKA值为0.76~0.93，属碱性（NKA值>0.90）—偏碱性（高钾钙碱性）系列。A/CNK为0.92~1.26，属准铝质—强过铝质花岗岩。与闽东魁歧A型花岗岩体化学成分（SiO_2 76.69%、TiO_2 0.14%、Al_2O_3 11.55%、$FeO+Fe_2O_3$ 1.93%、Na_2O+K_2O 8.53%）相比较，该序列岩石更富碱、铝而贫铁。

岩石稀土元素含量很低，仅为 $84\times10^{-6}\sim147\times10^{-6}$，属稀土强亏损型。稀土配分模式为向右倾斜的曲线（图9-35），且斜率较大；LREE/HREE介于12.50~19.53之间，表明轻、重稀土分异非常明显。$(La/Yb)_N$ 值为30.90~39.79，属轻稀土富集型；$(La/Sm)_N$ 值2.30~4.43，$(Gd/Yb)_N$ 值4.21~6.62，重稀土分馏稍强于轻稀土；δEu 值0.47~0.70，具弱负铕异常。

微量元素Rb、Ta、Nb、Ga含量较高，而Ba、Sr、P、Eu、Ti含量很低。Rb/Sr值大，主要为

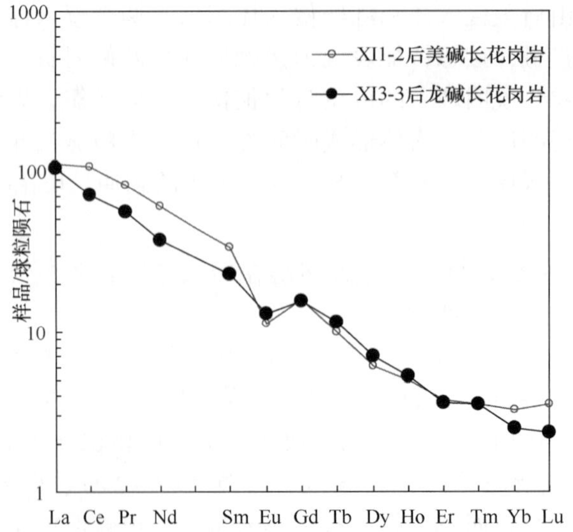

图 9-35 万车序列碱长花岗岩稀土元素球粒陨石标准化分布型式图

Fig. 9-35 Chondrite-normalized REE patterns of alkali-feldspar granite of Wanche Sequence

24.2~167.3；晚期岩石的 K/Rb 值低，仅为 78~95（本章附表 9-4）。在微量元素蛛网图中显示了极强的 Ba、Sr-P、Eu-Ti 负异常，该序列岩石比较特别的是没有 Nb-Ta 负异常，Nb/La 值一般大于 1。

九、中三叠世琼中序列

该序列岩石主要分布在海南岛中部琼中县一带，即由原"琼中岩体"分解而成的一系列岩石单位构成，从早到晚分为黎母岭粗中粒（巨）斑状角闪黑云二长花岗岩、琼中县中粒斑状（角闪）黑云二长花岗岩、南通岭中细粒黑云母正长花岗岩及洞古岭花岗斑岩等 4 个岩石单位，出露总面积约 2153.67km^2。早期岩石原生定向组构发育。已有的锆石 U-Pb 法同位素年龄值分别为：黎母岭二长花岗岩（232.5±1.5）Ma、（227.2±1.8）Ma，琼中县二长花岗岩（234.2±2.3）Ma、（233.8±2.3）Ma，南通岭正长花岗岩（226.5±2.9）Ma，洞古岭花岗斑岩（232±2.2）Ma（海南省地质调查院，2000），侵位时代厘定为中三叠世。

（一）黎母岭粗中粒（巨）斑状（角闪）黑云二长花岗岩（$\eta\gamma pT_2^a$）

由黎母岭、南茂岭、中建等侵入体组成。总面积约 1256.69km^2。被琼中县二长花岗岩、南通岭正长花岗岩及洞古岭花岗斑岩侵入。

主体岩石类型为粗中粒（巨）斑状（角闪）黑云二长花岗岩。岩石呈深灰色—浅黑灰色，似斑状结构，块状、片麻状构造，基质粗中粒花岗结构，矿物粒径以 2~5mm 为主，少量 5~7mm，侵入体边缘矿物粒度稍细，呈中粒或细中粒结构，中心部位则多呈中粗或粗粒结构，从中心往边缘有从粗→中→细中逐渐变小的趋势。主要造岩矿物为：钾长石 20%~53%、斜长石 25%~45%、石英 18%~25%、黑云母 5%~10%、角闪石 0~8%。钾长石以略带浅肉红色为主，部分浅黄色或灰白色。分斑晶和基质两种形式产出，斑晶呈半自形、自形厚板状，发育卡氏双晶，晶形较完整，长轴一般 2~4cm，大的可达 8cm 以上，具明显的原生定向构造，优选方向大多为 30°~50°，局部为北北东向或北西向。斑晶含量一般在 10%~35%之间，局部含量可高达 50%以上而呈多斑或聚斑状（如南茂岭、黎母岭、铜鼓岭等处）。基质中钾长石呈半自形板状、粒状，发育格子双晶或卡氏双晶，晶面普遍泥化，局部（如铜鼓岭侵入体中）钾长石边缘常见"碎粒状"石英分布（微粒交生体）。斜长石含量一般均高于基质中的钾长石，An=26~38，为中长石-奥长石，长石牌号低于东岭花岗闪长岩，呈灰白色，半自形板状、柱状，普遍发育聚片双晶、钠长石双晶，次之卡钠复合双晶，普遍发育环带构造。石英他形粒状，常聚集成集合体，颗粒之间大多呈三点式、弧形接触，波状消光明显。构造作用较强烈部位石英见压扁拉长现象和带状消光现象，在长石与石英接触部位局部见到蠕英结构（如和舍侵入体）。黑云母多呈半自形鳞片状、薄片状或桶状，局部呈宽大叶片状，常呈线状集合体产出，黑绿色—浅黄色，多色性明显。角闪石半自形柱状，呈绿色—浅黄绿色，多色性明显，角闪石式解理发育，为普通角闪石，常聚成集合体或呈线状定向产出，常与黑云母共生，含量变化较大，随机性较强。

副矿物主要为榍石、磁铁矿、锆石、褐帘石、磷灰石等。其中褐帘石呈板状、柱状；磷灰石短柱状或针状，长宽比值在 2：1~1：15 之间；榍石在和舍侵入体中最为明显，手标本上常见到褐色、自形信封状或粒状的榍石分布，局部含量可高达 2%，为岩浆早期结晶产物。

岩石中普遍含有闪长质包体。包体呈深灰色—黑灰色，色率比寄主岩高。细粒至微晶结构，与寄主岩石的中粗粒结构形成鲜明对比，两者之间的界线比较清楚。包体多呈椭圆状、次圆状、长条状或不规则状，大小一般为 3cm×1.5cm~36cm×13cm，在文昌市铜鼓岭东侧海边见有 60cm×250cm 的闪长质包体出现。包体长轴常见定向排列，主体方向与寄主岩石的斑晶定向排列方向大体一致。包体分布不均匀，一般 1~2 个/m^2，局部富集达 4 个/m^2。一般来说，侵入体边部含量多些，中心部位较少。包体岩性主要为闪长（玢）岩，矿物成分为斜长石 55%~75%、黑云母 6%~35%、角闪石 0~35%、石英<18%、钾长石<4%。包体内常见一些钾长石呈斑晶产出，分大、小两种形态，小的一般仅 0.5cm×1.0cm，含量 1%~3%，大的长轴可达 2.5~5cm，呈板状或浑圆状，部分大斑晶外缘见有斜长石包套（环斑结构）。在一些大包体边部常见钾长石巨斑环绕包体分布（图版Ⅱ-4-B）。特别指出的是，大部分包体中普遍含有一些针状的磷灰石（图版Ⅱ-4-C），其长宽比值可高达 20 以上，显示包体岩浆经历了快速冷却的结晶过程。

（二）琼中县中粒斑状（角闪）黑云二长花岗岩（$\eta\gamma T_2^b$）

分布于琼中县一带，总面积约 656.85km^2，区域上整体呈北东向展布。由于遭受后期岩体侵蚀，不同侵入体形状各异、规模大小悬殊。区内多处见琼中县二长花岗岩分别被南通岭正长花岗岩侵入。

主体岩性为中粒—中细粒斑状（角闪）黑云母二长花岗岩，岩石呈深灰、灰白色，似斑状结构，基质以中粒花岗结构为主，基质矿物粒度大多在 2~5mm 之间，少量 1~2mm；以块状构造为主，局部见弱片麻状构造。矿物成分主要有钾长石 20%~60%、斜长石 25%~45%、石英 20%~30%、黑云母 3%~10%及少量角闪石 0~6%等。斑晶成分为钾长石，含量一般在 5%~15%之间，局部<3%，斑晶定向排列明显，其方向主要为 35°~65°，与区域性构造线基本一致。呈灰白色或略带浅肉红色，自形—半自形板状、厚板状，发育卡氏双晶，长轴一般 0.4~1.5cm。基质中，钾长石呈灰白色或略带浅肉红色，半自形板状、他形粒状，卡氏双晶发育，可见发育蠕英结构，含量与斜长石大体相当；斜长石呈灰白色，半自形板状、柱状，An=3~20，长石牌号明显低于黎母岭单元，大部分为碱性程度较高的钠长石-奥长石，普遍发育环带构造和聚片双晶、钠长石双晶，次之卡钠复合双晶，晶面大多由于发生绢云母化、钠黝帘石化而浑浊；石英他形粒状、不规则状充填在长石间隙中，普遍具波状消光；黑云母多呈半自形细薄片状、细叶片状，局部呈宽大叶片状，褐色—浅黄色，多色性明显，{001} 解理极完全，普遍存在于琼中县二长花岗岩中；角闪石分布不均匀，仅局部见到，大部分地区几乎未见到角闪石出现，多呈半自形柱状或柱状集合体，呈深绿色—浅黄绿色等，角闪石式解理发育，常聚成集合体或呈线状定向产出，常与黑云母共生，局部见其包裹黑云母、榍石、磁铁矿、褐帘石等，说明它结晶较晚。副矿物主要为磁铁矿、榍石、锆石、磷灰石等。

岩石中除了少量变沉积岩包体和一些浅色正长花岗岩包体外，更多的是一些富含黑云母的（石英）闪长质包体（图版Ⅱ-4-D）。闪长质包体与寄主岩之间呈突变接触，界线清晰，一般呈次圆状、椭圆状、不规则状、次棱角状、长条状等，形状极不规则，大小悬殊，大者直径可达 30~50cm，小的仅数厘米，大多数包体个体较小；包体分布无规律性，随机性较强，局部可多达 7~8 个/m^2，定向性不明显，仅局部见到一些长条状或椭圆状的包体沿长轴呈现出基本一致的排列方向。闪长质包体呈细粒至微晶结构，块状构造，个别包体中暗色矿物略具定向性或弱片麻状构造，部分黑云母弯曲变形、膝折，斜长石普遍发育环带构造；磷灰石多数呈短柱状，部分呈针状，长宽比值一般介于2~20之间，个别长宽比值可高达30以上，显示包体岩浆经历了快速冷却的结晶过程。个别包体中含有少量2cm×1.5cm~3cm×2mm的石英斑晶，晶面见裂纹，晶体内部见一些磷灰石、黑云母等早期结晶矿物的包裹体，推测这些石英斑晶为来自寄主岩的捕虏晶。上述现象可能暗示琼中县二长花岗岩经历了岩浆混合作用，包体是注入地壳酸性岩浆中的基性岩浆团碎块。

（三）南通岭中细粒黑云母正长花岗岩（$\xi\gamma T_2^c$）

多呈小侵入体产出于各早期岩石单位中，由南通岭、大田仔、南节岭等侵入体组成，总面积约239.08km^2。

主体岩性为中细粒、细粒黑云母正长花岗岩，局部为中细粒、细粒黑云母二长花岗岩。细粒、中细粒花岗结构，矿物粒径以0.2~2mm为主，部分2~3mm，以块状构造为主，局部具微弱片麻状构造。主要造岩矿物有：钾长石25%~58%、斜长石8%~40%、石英20%~30%、黑云母3%~10%、角闪石0~6%。钾长石半自形板柱状、他形粒状，发育卡氏双晶或格子双晶，可见蠕英结构；斜长石An大多在1~19之间，为钠长石-奥长石，个别为An=35的中长石，呈灰白色，半自形板状、柱状，普遍发育环带构造和聚片双晶、钠长石双晶，次之卡钠复合双晶；石英他形粒状，不规则状充填在长石间隙中，普遍具波状消光；黑云母呈细鳞片状，红褐色，多色性显著，常见沿片轴平行定向分布，部分可见磷灰石包晶；角闪石分布极不均匀，大部分侵入体中均未见有角闪石出现。岩石局部可见半自形厚板状的钾长石斑晶出现，呈浅红色或灰白色，大小一般0.4~1cm，含量小于3%。副矿物主要为磁铁矿、钛铁矿、锆石、磷灰石、榍石等。

局部见微晶闪长岩包体，包体矿物成分主要为斜长石、角闪石和黑云母。

（四）洞古岭花岗斑岩（$\gamma\pi T_2^d$）

洞古岭花岗斑岩呈小岩瘤或不规则状产出于早期岩石中，规模较小，总面积约1.05km^2。

岩石呈斑状结构、显微文象结构、熔蚀结构，块状构造，局部可见微弱的片麻状构造。斑晶主要为钾长石，少量斜长石、石英、黑云母等，总含量15%~47%，大小一般2mm×2mm~10mm×10mm，个别钾（斜）长石可达10mm×20mm。其中，钾长石15%~25%，半自形板状、柱状，卡氏双晶发育，发生绢云母化高岭石化或绿帘石化等；斜长石7%~35%，半自形柱状，发育聚片双晶；石英<5%，多呈浑圆状，波状消光明显；黑云母叶片状，片幅0.5~1.5mm，呈黑绿色—浅黄色，多色性明显，多见被绿帘石交代并析出磁铁矿等。基质由钾长石、斜长石、石英和黑云母组成，粒径大部分小于0.2mm。石英和长石形成显微文象结构，文象形成的蠕英石呈蠕虫状、拔丝状或三角形状、不规则状等。副矿物主要为磁铁矿、榍石、磷灰石和锆石等。

局部见闪长质包体出现。包体呈灰色—灰黑色，圆状、椭圆状，大小一般0.7~6cm（直径），个别大者达11cm×20cm；与寄主岩呈渐变过渡关系。成分主要由斜长石、黑云母和少量钾长石、石英组成。

（五）岩石地球化学特征

该序列岩石的SiO_2含量为64.32%~76.52%。K_2O含量一般介于4.04%~6.35%，Na_2O大多为2.51%~3.53%，岩石相对贫钠富钾。σ值为1.33~3.19。根据硅-碱图解（图9-25）和SiO_2-K_2O图解（图9-26），属高钾钙碱性系列。铝饱和指数A/CNK在0.91~1.24之间，大部分属准铝质—弱过铝质。DI值69.29~93.91、SI值3.12~10.31，岩浆分异程度较高。稀土元素总量变化较大，为133×10^{-6}~821×10^{-6}，稀土配分模式为右倾型曲线（图9-31）。轻重稀土分异中等—较明显，δEu为0.25~0.8，具中等—较强的负铕异常，从早到晚负铕异常越来越明显。微量元素蛛网图中具Ta-Nb、Sr-P、Eu-Ti负异常比较明显（图9-36），$\varepsilon_{Nd}(t)$为-4.2~-10.70，初始锶比值为0.70830~0.71104，平均0.71104（云平等，2005）。岩石学和岩石地球化学特征均反映该序列岩石属壳幔岩浆混合来源。

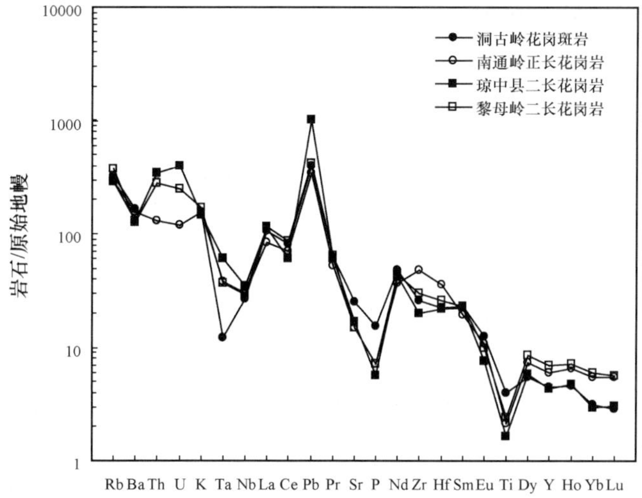

图 9-36 琼中序列微量元素原始地幔标准化蛛网图

Fig. 9-36 Primitive mantle-normalized spider diagram of trace elements of Qiongzhong Sequence

十、中三叠世保梅岭中粗粒黑云母正长花岗岩（$\xi\gamma T_2$）

1. 地质特征及形成时代

仅见保梅岭一个侵入体，分布在海南岛西部昌江县保梅岭北坡、西线公路以南一带，面积 25.27km^2。

关于保梅岭侵入体的形成时代，前人认为其被二叠纪邦溪岩体二长花岗岩侵入，并在保梅岭侵入体中采用单颗粒锆石逐层熔融蒸发法测定^{207}Pb/^{206}Pb 同位素年龄，测定样品 32 个，结果表明，锆石边部和中心^{207}Pb/^{206}Pb 比值差别很大，所对应的年龄值边缘为 245 Ma，中心为 410Ma，而且不是连续变化的，可以分出两个不连续面，边缘一组 7 个数据，年龄在 245Ma～272Ma 之间变化，平均年龄 265Ma；中间一组 14 个数据，年龄在 290Ma～330Ma 之间变化，平均年龄 314 Ma；核心一组 11 个数据，年龄在 343Ma～410Ma 之间变化，平均年龄（369±2.9）Ma，因而认为保梅岭侵入体的形成时代应为晚泥盆世（宜昌地矿所，1995）。本项目调查结果表明，保梅岭侵入体侵入中二叠世邦溪岩体二长花岗岩，证据为：①保梅岭侵入体岩石为块状构造，矿物分布均匀，而邦溪岩体二长花岗岩中斑晶定向明显，保梅岭侵入体并截切邦溪岩体二长花岗岩中的钾长石斑晶（图版Ⅱ-4-E）；②界线附近，保梅岭侵入体一侧常见伟晶岩包裹体，而邦溪岩体二长花岗岩一侧则未见及③本次采用高精度的LA-ICPMS 法测得保梅岭侵入体锆石 U-Pb 年龄为 232.6Ma。因此，笔者认为，保梅岭侵入体应形成于中三叠世、邦溪岩体二长花岗岩之后。

该侵入体除局部可见微弱的片麻状构造外，未经受区域变质作用，属晚造山花岗岩类。

2. 岩石学特征

主体岩性为中粗粒黑云母正长花岗岩。岩石手标本为浅肉红色，风化面为浅黄褐色，中粗粒全他形粒状结构，块状构造，可见微弱的片麻状构造。钾长石 38%～53%，呈肉红色，他形，粒径一般4～8 mm，属微斜条纹长石，常见格子状双晶，钠长石条纹较发育，呈细脉状平行密集分布；电子探针分析结果显示钾长石的端员组分 Or 高达 93.19，Ab 6.59，An 仅为 0.22。斜长石 15%～23%，灰白色，他形或半自形状，粒径 0.4mm×1.2mm～3mm×5mm，只见简单的钠长石聚片双晶，双晶纹细密，An＝10～18，属酸性奥长石，有时见晚世代钠长石（An<5）沿边缘或裂隙交代微斜条纹长石。斜长石常被微斜长石包裹，在斜长石边缘有时出现净边或蠕虫状石英出现。石英 22%～39%，多呈他形粒状集合体，粒径 0.8～5mm。黑云母 1%～3%，呈叠片状，片幅 2～4mm，常见黑云母细鳞片状集合体充填在长石石英颗粒间。黑云母多色性和吸收性均十分明显，N_g、N_m 为黑色，N_p 为棕黄色，常被绿泥石交代。

保梅岭侵入体的黑云母贫镁富铁，在 Mg-($Al^{Ⅵ}$+Fe^{3+}+Ti)-(Fe^{2+}+Mn) 三角图解中投影在铁叶云母和铁黑云母区（图略）。

保梅岭侵入体副矿物总含量高达 3823g/t，其中磁铁矿含量占副矿物总含量的 90%，其他含量相对较高的还有褐帘石、独居石、钍石、锆石等，组合类型为磁铁矿-褐帘石-独居石-钍石型。锆石主要呈棕褐色、黄色或无色，以（110）、（100）、（111）聚形为主，柱体长宽比主要为 2.5∶1，个别4.5∶1。

该侵入体中普遍见有闪长质包体和少量富云包体、黑云角闪斜长片岩包体。与黑云角闪斜长片岩包体接触处，局部见保梅岭中粗粒黑云母正长花岗岩呈细脉贯入并分割黑云角闪斜长片岩，将黑云角闪斜长片岩分割成大小不等的透镜状“包体群”，且在接触处出现边缘混合岩化现象，说明保梅岭中粗粒黑云母正长花岗岩形成时，其部分围岩为黑云角闪斜长片岩。

3. 岩石地球化学特征

保梅岭侵入体的常量元素含量及特征参数见本章附表 9-1、附表 9-2。从表中可看出，其 SiO_2 含量为 71.5%~77.23%，MgO 含量为 0.03%~0.53%、CaO 含量为 0.46%~1.8%，显示出富硅碱、贫镁钙的特征。含铝指数 A/CNK 为 0.98~1.01，碱铝指数 NKA 为 0.77~0.92，组合指数（里特曼指数）σ 为 2.18~2.50，在硅-碱图解（图 9-25）和 SiO_2-K_2O 图解（图 9-26）上，属高钾钙碱性系列。

该侵入体中闪长质包体的 SiO_2 含量为 58.76%~65.16%，MgO 含量为 1.33%~3.07%、CaO 含量为 3.25%~4.54%，显示出与主岩相反的贫硅碱、高镁钙的特征。

岩石的稀土元素总量为 176.58×10^{-6}~291.83×10^{-6}，平均 244.93×10^{-6}，稍低于维氏世界酸性岩平均值（292×10^{-6}）。主岩及闪长质包体的稀土元素球粒陨石标准化配分模式均呈略向右缓倾斜的“海鸥”式曲线，属轻稀土弱富集型（图 9-37）。$(La/Sm)_N$ 为 1.57~4.56，$(Gd/Yb)_N$ 为 1.7~2.16，表明轻稀土分异较明显，但重稀土没有明显的分异作用；δEu 为 0.31~0.41，具中等偏强铕负异常。

微量元素以强烈富集 Rb、Ba、Th 和低 Zr、Yb、Ho、Y 为特征（图 9-38），与同碰撞花岗岩特点相似。

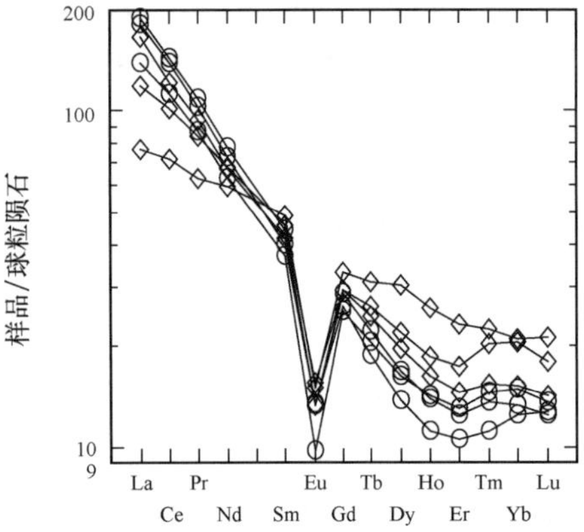

图 9-37 保梅岭侵入体稀土元素球粒陨石标准化图

Fig. 9-37 Chondrite-normalized REE patterns of Baomeiling intrusion

○ 钾长花岗岩；◇ 闪长质包体

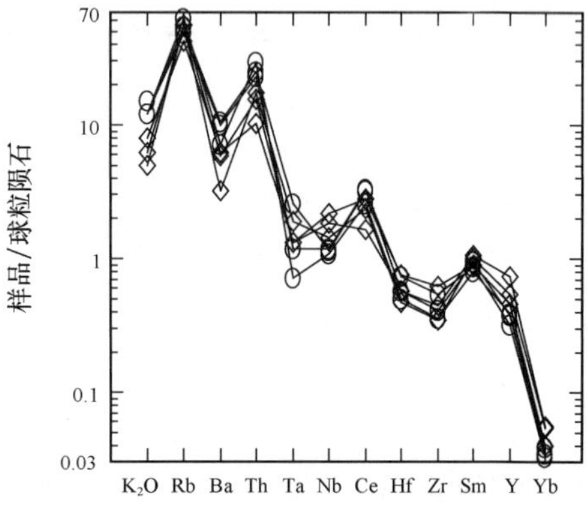

图 9-38 保梅岭侵入体 ORG 标准化蛛网图

Fig. 9-38 ORG normalized spider diagram of Baomeiling intrusion

○ 钾长花岗岩；◇ 闪长质包体

十一、晚三叠世黑岭序列

该序列岩石分布于昌江—坝王岭一带，主要由原 1∶5 万《昌江县、邦溪》幅中的七差岭单元、黑岭单元、乌烈岭单元和 1∶5 万《坝王岭、猴猕岭》幅中的七差岭单元、东四单元和恶涛单元等组成。从早至晚由归并为七差岭中细粒黑云母正长花岗岩、东四中粗粒角闪黑云母正长花岗岩和乌烈岭中（细）粒斑状黑云母正长花岗岩等 3 个岩石单位，总面积 354.10km^2。

（一）七差岭中细粒黑云母正长花岗岩（$\xi\gamma fT_3$）

分布于昌江县以南七差岭一带，仅由七差岭一个侵入体组成，面积约 29.9km^2。区内见七差岭侵入体侵入中二叠世邦溪岩体，并在晚三叠世东四岩体中见有七差岭正长花岗岩的捕虏体，因此，七差岭侵入体形成时间应晚于邦溪岩体而早于东四岩体，暂将其形成时代厘定为晚三叠世。

七差岭侵入体岩石均一性较差，以中细粒不等粒结构为主，局部出现含斑结构，钾长石斑晶长轴一般<1cm，含量少于5%。岩石常具定向构造，这种定向构造由黑云母定向分布、斜长石 C 轴定向排列以及石英集合体长轴排列显示。矿物组成：石英 23%～27%、斜长石 18%～26%、钾长石 45%～51%、黑云母 2%～4%。微斜长石有两个世代，早世代微斜长石中钠长石条纹比较发育，呈细纹脉状平行分布，格子双晶明显，晚世代微斜长石几乎无钠长石条纹。晚世代微斜长石呈镶边状包裹早世代微斜条纹长石，出现环带状构造。有时在斜长石与微斜长石接触处出现蠕虫状石英。斜长石呈自形或半自形板状或长板状，粒度 0.1mm×0.3mm～0.8mm×2.5mm 不等，钠长石律双晶发育，偶见卡钠复合双晶，无环带，An15～28，属奥长石。石英呈他形粒状集合体。黑云母呈细叠片状集合体，N_g 为褐色，N_p 为黄色。副矿物主要为榍石、褐帘石、锆石、磷灰石及磁铁矿，以出现较多的自形楔状榍石为特征。

偶见闪长质暗色包体，个体较小，直径 5～10cm，透镜状。

（二）东四中粗粒角闪黑云母正长花岗岩（$\xi\gamma cT_3$）

分布于昌江县坝王岭林场东四至黑岭一带，由黑岭、雅加林场、东六、七差大岭等侵入体组成，面积 270.88km^2。区内见东四正长花岗岩侵入早二叠世白石村闪长岩、保山村英云闪长岩以及中二叠世乌兰盖中粗粒含巨斑黑云二长花岗岩等，并被晚三叠世苗圃二长花岗岩侵入。东四正长花岗岩与上古生代围岩的侵入接触界线截然清楚，内接触带有时出现中细粒似斑状结构的边缘相，宽度一般 4～5m。与围岩峨查组接触界线十分清楚，围岩为石英角岩，内接触带无边缘相，在 10～20cm 宽范围内冷凝边急速变为中粗粒结构。已有的同位素年龄为（230.1±0.1）Ma（Rb-Sr 等时线法，宜昌地矿所，1995）、（225±11）Ma（锆石 U-Pb 法，广西区调队，1991），形成时代厘定为晚三叠世早期。

岩石中包体比较丰富，包体岩性分有闪长岩、石英闪长岩、石英二长岩、中细粒花岗岩及砂泥岩等，形状有椭圆状、浑圆状、棱角状、不规则状等。包体长轴一般在 1～30cm 之间，个别达 60cm，长宽比一般为（2~3）：1，个别为 4：1。包体与寄主岩界线清楚，部分包体中可见钾长石斑晶及双包体。在南斑河中见有暗色矿物集中的条带体产出，条带中的暗色矿物与主岩相同（以黑云母为主，角闪石次之），粒度也相同，条带体的走向为近南北向，这种现象说明岩浆存在液态分离，它的走向应代表岩浆的流动方向。

主体岩性为中粗粒角闪黑云母正长花岗岩，岩石结构比较单一，以中粗粒结构为主，局部为粗粒结构。钾长石（微斜条纹长石）29%～45%，呈肉红色，以他形为主，个别为半自形，粒径 0.6～1.5cm；斜长石 16%～33%，呈灰白色，自形或半自形板状，呈集合体充填在钾长石粒间，粒度 1.2mm×2mm～4mm×6mm，An=12～15，属奥长石；石英 22%～35%，呈微带紫色的乳白色，多呈粒状集合体产出，粒径 2～8mm；普通角闪石 0～4%，呈自形或半自形板状，粒度 0.5mm×1mm～2mm×3mm；黑云母 3%～8%，呈他形叠片状，片幅在 1～4mm 之间，多色性和吸收性显著，N_g、N_m 为黑色，N_p 为黄棕色，沿节理被绿泥石交代，属铁质黑云母。钾长石端员组分为 An 0～0.63%、Ab 3.84%～35.85%、Or 63.01%～95.56%；斜长石端员组分为 An 4.23%～36.21%、Ab 60.03%～93.79%、Or 1.31%～4.18%。角闪石属钙质角闪石类中的低铁-钙镁闪石质普通角闪石。

副矿物主要为榍石、褐帘石，次为锆石、磷灰石和磁铁矿。

（三）乌烈岭中细—中粒斑状黑云母正长花岗岩（$\xi\gamma T_3$）

分布于乌烈岭、俄牙岭、坝王岭、法叭一带，由乌烈岭、鸡心南、斧丢岭、河坑化、保土岭、二工段等侵入体组成，面积约 53.32km^2。单个侵入体规模不大，出露面积 1～16km^2，呈椭圆状、半圆状或不规则状。

区内见乌烈岭正长花岗岩呈岩枝侵入于东四正长花岗岩内部，说明乌烈岭正长花岗岩形成时间应晚于东四正长花岗岩。已有的同位素年龄值为（240.6±10.6）Ma（锆石 U-Pb 法，广西区调队，1991），暂将其形成时代厘定为晚三叠世。

岩石中包体比较丰富，呈灰白色、灰黑色，大小不一，包体长轴一般在 1~60cm 之间，与寄主岩界线清楚，呈圆状、椭圆状、浑圆状、棱角状、不规则状等。包体岩石类型有闪长岩、英云闪长岩、二长岩、细粒花岗岩等。包体内部常见粗大的钾长石斑晶及更暗色的小包体。

乌烈岭正长花岗岩岩石类型比较稳定，为中细—中粒斑状黑云母正长花岗岩，中细粒—中粒花岗结构，局部出现含斑结构，块状构造。手标本为浅肉红色，钾长石为微斜条纹长石，粒径 0.5~2.5mm，钠长石条纹比较发育，呈细脉状，常见格子状双晶，含量 34%~50%。斜长石 16%~35%，有两个世代，早世代属奥长石（An15~An25），晚世代钠长石（An<5）呈梳状集合体沿微斜条纹长石边缘充填交代。石英 22%~40%，呈他形粒状，粒径 0.2~1.5mm。黑云母 3%~8%，他形片状，片幅在 0.2~1.5mm，N_g、N_m 为黑色，N_p 为黄棕色，属铁质黑云母。角闪石 0~1%。钾长石端员组分为 An（0~0.09）%、Ab 1.25%~17.09%、Or 82.77%~98.75%；斜长石 An 13.65%~29.27%、Ab 66.99%~85.26%、Or 0.95%~3.61%。

副矿物组合为榍石、褐帘石、锆石、磷灰石、磁铁矿等，但与前两个岩石单位相比，榍石含量急剧减少。

（四）岩石地球化学特征

该序列岩石的 SiO_2 含量在 66.78%~74.47%之间，属酸性岩类。碱质含量较高，K_2O+Na_2O 含量 7.45%~10.35%。σ 值为在 1.96~4.5 之间，绝大部分样品 σ 值在 2~2.5 之间。根据硅-碱图解（图 9-25）和 SiO_2-K_2O 图解（图 9-26），属高钾钙碱性岩系；A/CNK 为 0.91~107，整体属准铝质花岗岩。DI 值高达 81.4~90.2，说明岩浆分异程度较高。稀土总量变化较大，在 212.35×10^{-6}~510.98×10^{-6}之间，但绝大部分高于世界酸性岩稀土平均含量。稀土配分模式为铕负异常比较明显的右倾型曲线（图 9-39）；$(La/Yb)_N$ 值为 3.41~73.37，轻重稀土分异程度较高，轻稀土富集；δEu 为 0.17~0.56，具中等至较强铕负异常。微量元素方面（本章附表 9-4），从早至晚，Pb、W、Bi、Sb、Ta 等元素含量和 Rb/Sr、Ba/Sr 比值逐渐增加，V、Ni、Zn、Sn、Sr 等元素含量有逐渐降低趋势。与维氏酸性岩丰度值相比，Cu、Cr、V、Zn、Au、Ag、As、Sr、Ba、Ta 等元素含量偏低，Pb、Sn、Mo、Bi、Rb、Hf 含量偏高。微量元素原始地幔标准化蛛网图中（图 9-40），显示较强的 Ba、Ta-Nb、Sr-P、Eu-Ti 负异常和明显的 Rb、Th-U-K、Pb、Nd-Zr-Hf-Sm 正异常。

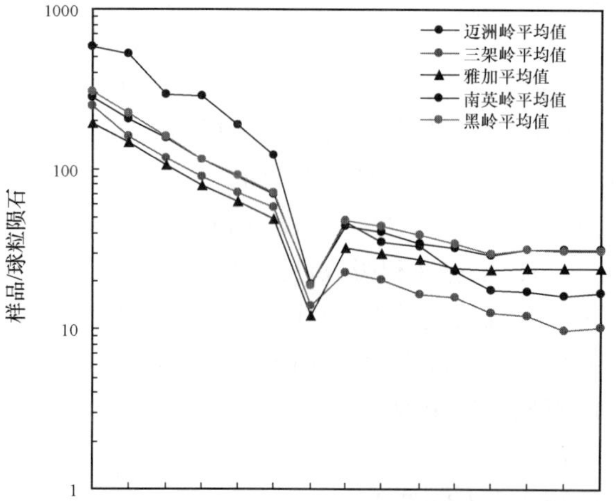

图 9-39　晚三叠世侵入岩稀土元素球粒陨石标准化分布型式图

Fig. 9-39　Chondrite-normalized REE patterns of Late Triassic intrusive rocks

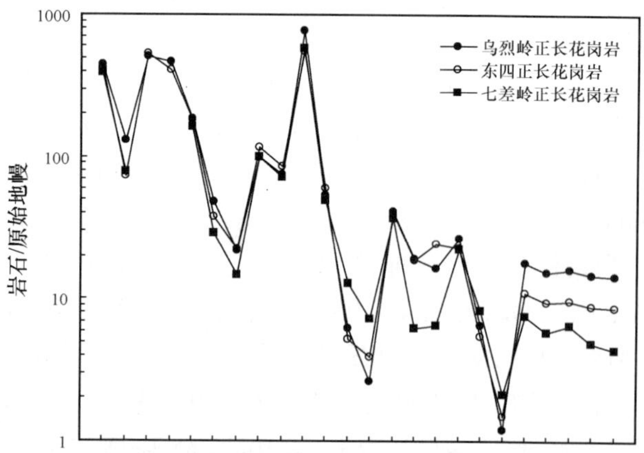

图 9-40　黑岭序列微量元素原始地幔标准化蛛网图

Fig. 9-40　Primitive mantle-normalized spider diagram of trace elements of Heiling Sequence

十二、晚三叠世南英岭序列

该序列岩石分布于坝王岭一带。从早至晚划分为苗圃细粒角闪黑云二长花岗岩和五里桥中细粒黑云二长花岗岩2个岩石单位，总面积181.72km²。

（一）苗圃细粒角闪黑云二长花岗岩（$\eta\gamma fT_3$）

由分布于坝王岭林业公司附近的苗圃侵入体和分布于周爸岭—小岭—扎营村一带的周爸岭侵入体、大讲岭侵入体、小岭侵入体及扎营村侵入体等组成，总面积28.96km²。

苗圃二长花岗岩侵入晚三叠世东四正长花岗岩和乌烈岭正长花岗岩。接触界线截然、清楚，呈弯曲状，接触面倾向围岩，倾角60°~80°，边缘可见3~5cm的冷凝边，局部见其岩枝或岩脉插入围岩中；内接触带常见大小不一、不规则状的围岩捕虏体（图9-41）。因而苗圃二长花岗岩的形成时间应晚于乌烈岭正长花岗岩，将其侵入时代厘定为晚三叠世。

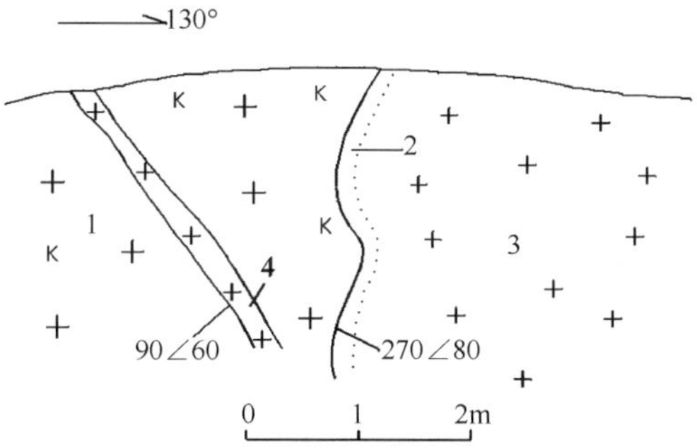

图9-41　苗圃二长花岗岩侵入乌烈岭正长花岗岩接触关系素描图

Fig. 9-41　Sketch showing the contact that Miaopu monzonitic granite intrudes Wulieling syenogranite

（据1∶25万东方县幅区调报告，2007）

1—乌烈岭正长花岗岩；2—冷凝边；3—苗圃二长花岗岩；4—细粒花岗岩脉

岩石中包体较丰富，大小不一，长轴一般在2~30cm之间，少数大于30cm，分布不均匀，往往成群出现。包体与寄主岩界线清楚，呈椭圆状、浑圆状、棱角状、不规则状等。包体岩石类型有闪长岩、英云闪长岩、二长岩、细粒花岗岩等，内部常见粗大的钾长石斑晶及早期的暗色小包体。

主体岩性为细粒角闪黑云二长花岗岩，由于钾长石分布不均匀，局部出现正长花岗岩。岩石具细粒、中细粒花岗结构，块状构造，出现较多钾长石斑晶时具似斑状结构。矿物粒度以细粒（0.2~2mm）为主，少数可达3mm左右。矿物组成：石英22%~32%、斜长石（An10~An38）22%~30%、钾长石33%~44%、黑云母4%~8%、角闪石3%~8%。钾长石端员组分为An 0、Ab 4.8%~9.03%、Or 90.84%~95.07%，斜长石端员组分为An 28.32%、Ab 68.11%、Or 3.44%，黑云母属铁叶云母-铁黑云母。

副矿物常见独居石、褐帘石、钍石、萤石、锡石等。局部有方铅矿、闪锌矿及自然金出现。

（二）五里桥中细粒黑云二长花岗岩（$\eta\gamma mT_3$）

由分布于坝王岭一带的雅加大桥侵入体、保铁侵入体和山鸡岭—春江水库一带的山鸡岭西侵入体、黄蜂岭侵入体及高山东侵入体等组成，总面积152.76km²。

雅加大桥侵入体侵入晚三叠世东四正长花岗岩。接触界面清楚，呈折线状，接触面倾向围岩，倾角60°~70°，边缘可见2~7cm的冷凝边，局部可见3~5cm的伟晶岩外壳，接触带内常见粗粒花岗岩捕虏体，并见侵入体呈岩脉贯入围岩中。已有的同位素年龄值为（219.8±11.0）Ma（锆石U-Pb法；广西区调队，1991），形成时代厘定为晚三叠世。

雅加大桥侵入体中偶见一些闪长质包体和细粒花岗岩包体，多为浑圆状，大小1~18cm，与寄主岩界线清晰。

山脚岭西侵入体侵入苗圃二长花岗岩，两者呈脉动接触。

山脚岭西、黄蜂岭等侵入体中见有闪长岩、花岗闪长岩包体。包体与寄主岩界线清楚，呈椭圆状、长条状、透镜状，大小一般为 2cm×8cm，个别大于 15cm×30cm，包体分布不均匀，局部富集成群出现。

主体岩石类型为中细粒、中粒黑云母二长花岗岩或正长花岗岩。中细粒、中粒花岗结构，块状构造。偶见少量钾长石斑晶，含量<3%。矿物组成：石英 24%～32%、斜长石（An8～An42）20%～28%）、钾长石 35%～48%、黑云母 4%～7%、角闪石 0～3%。钾长石端员组分为 An 1.12%、Ab 56.22%、Or 42.51%，斜长石 An 5.63%～33.75%、Ab 63.14%～92.89%、Or 1.35%～2.97%，黑云母属铁叶云母-铁黑云母。

副矿物方面，五里桥二长花岗岩与苗圃二长花岗岩比较相似，常见副矿物有独居石、褐帘石、钍石、萤石、锡石等。

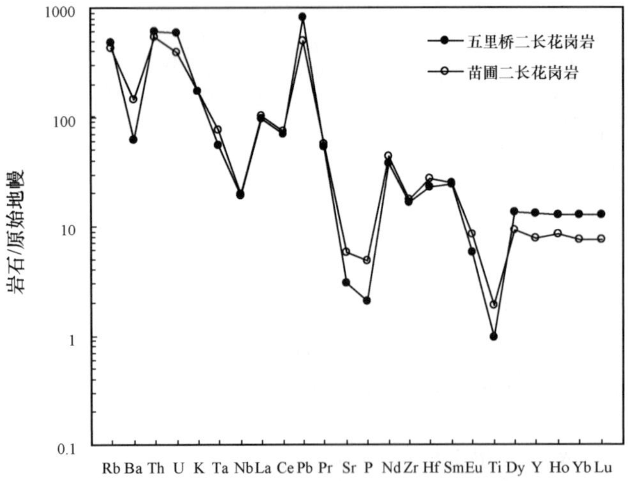

图 9-42　南英岭序列微量元素原始地幔标准化蛛网图

Fig. 9-42　Primitive mantle-normalized spider diagram of trace elements of Nanyingling Sequence

（三）岩石地球化学特征

南英岭序列岩石的 SiO_2 在 66.87%～74.72% 之间，属酸性岩。Na_2O+K_2O 含量较高，为 7.32%～8.92%；σ 值为 1.9～2.98，A/CNK 为 0.91～1.08。根据硅-碱图解（图 9-25）和 SiO_2-K_2O 图解（图 9-26），整体上属准铝质高钾钙碱性岩系。K_2O/Na_2O 值在 1～1.96 之间，属钾质花岗岩。DI 值 74.59～92.26、SI 值 0.37～9.44，岩浆分异程度较高。稀土元素含量为 232.81×10^{-6}～437.59×10^{-6}，其稀土配分曲线为较陡的右倾型曲线（图 9-39）；$(La/Yb)_N$ 值为 4.58～49.71，轻重稀土分异中等，轻稀土分异程度强于重稀土；δEu 为 0.16～0.64，具中等铕负异常。岩石的 Y、Nb、Cu、Pb、Ag 偏高，Sr、Ni、Co、V 偏低；Rb/Sr 平均比值大于 1。原始地幔标准化中显示出较明显的 Ba-Th、Nb、Sr-P 和 Ti 负异常（图 9-42）。

十三、晚三叠世雅加序列

该序列岩石分布于海南岛西部坝王岭一带，由南宝岭细粒斑状黑云二长花岗岩、分水岭中-细粒斑状黑云二长花岗岩和黑岭中粒多斑状黑云二长花岗岩 3 个岩石单位组成。

（一）南宝岭细粒斑状黑云二长花岗岩（$\eta\gamma mpT_3$）

分布于坝王岭林区南宝岭、南宝、艾劈岭、雅加大岭一带，呈一个不完整的椭圆状南北向展布，包括南宝、艾劈岭等侵入体，各个侵入体大小不一，从 0.5～12km^2 不等，总面积约 25.12km^2。

艾劈岭侵入体侵入于晚三叠世东四正长花岗岩中。接触界线清楚，呈弯曲状，接触面倾向围岩，倾角 60°～80°。接触面切割东四正长花岗岩中的钾长石等矿物。接触面附近南宝岭二长花岗岩的斑晶含量减少、基质细小，有 3～12cm 宽的冷凝边。内接触带常见东四正长花岗岩的捕虏体，有时可见南宝岭二长花岗岩岩枝或岩脉贯入围岩中。根据侵入接触关系暂将其侵入时代厘定为晚三叠世。

南宝岭二长花岗岩内部常见包体出现，包体与寄主岩界线清楚，呈浑圆状、圆状、不规则状等，大小 0.5～30cm 不等。包体岩性主要有闪长岩、石英闪长岩、英云闪长岩及细粒花岗岩等。

岩石类型主要为细粒斑状黑云母二长花岗岩。细粒似斑状花岗结构，显微文象结构较常见，局部具潜基结构特点；块状构造为主，局部具定向构造。局部由于钾长石斑晶的不均匀分布，可以出现花岗闪长岩。斑晶含量一般在 10%～30%之间，局部可达 60%以上，而成为多斑状黑云二长花岗岩。矿物组合：石英25%~31%、斜长石（An12~An47）23%~35%、钾长石30%~46%、黑云母4%~8%、角闪石0~3%。钾长石端员组分为An 0~0.79%、Ab 2.17%~21.77%、Or 77.31%~97.7%，黑云母属铁质黑云母。

副矿物主要为磁铁矿、褐帘石、榍石、独居石、黄铁矿、萤石等。局部有方铅矿及自然金出现。

（二）分水岭中—细粒斑状黑云二长花岗岩（$\eta\gamma fmT_3$）

分布于分水岭、雅加大岭、叉河一带，总面积约101.6km^2。

分水岭二长花岗岩1518高地侵入体与南宝岭细粒斑状黑云二长花岗岩呈涌动接触。北侧侵入晚三叠世东四中粗粒角闪黑云母正长花岗岩，接触界线清楚，内接触带斑晶含量减少，可见3~10cm的细粒边，并可见其岩脉穿插入围岩中。已有的锆石U-Pb法同位素年龄为（210.8±10）Ma（广西区调队，1991），形成时代厘定为晚三叠世。

岩石中常见闪长岩包体、英云闪长岩包体及富云母包体，包体与寄主岩界线清楚，多呈浑圆状或椭圆状，大小0.5~20cm。

岩石类型主要为细粒、中粒或不等粒斑状黑云母二长花岗岩（少量正长花岗岩）。斑晶含量10%~20%，局部达40%，斑晶以钾长石为主，个别石英，大小以1~2cm为主，个别达4cm。局部钾长石斑晶呈320°~360°定向排列。矿物组合：石英25%~30%、斜长石（An11~An47）20%~30%、钾长石34%~41%、黑云母4%~8%、角闪石0~2%。钾长石端员组分为An0.29%~0.62%、Ab10.8%~27.48%、Or72.06%~88.78%，斜长石端员组分为An31.44%、Ab 64.62%、Or 3.81%。副矿物常见磁铁矿、锆石、独居石、黄铁矿等。

（三）黑岭中粒多斑状黑云二长花岗岩（$\eta\gamma pT_3$）

分布于黑岭、茅岭一带，由1484高地、1274高地、茅岭等3个小侵入体组成，总面积约9.1km^2。

1518高地和748高地两个侵入体均见涌动侵入分水岭二长花岗岩，两者之间未见截然的分界线，存在5~18cm的过渡带。根据侵入接触关系暂将其侵入时代厘定归为晚三叠世。

主体岩性为细中粒、中粒多斑状黑云二长花岗岩或正长花岗岩。似斑状花岗结构，块状构造。斑晶为钾长石，半自形为主，晶面比较完整，大小多在5~10mm之间，含量多达60%。基质粒度多在2~4mm之间，少数小于2mm。黑云母呈桶状或鳞片状集合体，常见颗粒较粗大的角闪石。矿物组合主要为石英24%~28%、斜长石（An11~An39）24%~30%、钾长石33%~39%、黑云母0~3%及角闪石5%~7%。钾长石端员组分为An 0.1%~0.32%、Ab18.42%~27.48%、Or 72.06%~81.35%，斜长石端员组分为An28.12%、Ab67.5%、Or4.25%。

副矿物常见磁铁矿、榍石、锆石、独居石、黄铁矿、萤石等。局部有辉锑矿出现。

（四）岩石地球化学特征

南宝岭二长花岗岩的SiO_2在70.27%~75.27%之间，属酸性岩。Na_2O+K_2O含量较高，为7.9%~8.1%。σ值为1.97~2.29；A/CNK为0.95~1.04，整体上属准铝质高钾钙碱性岩系。DI值82.62~91.63、SI值2.1~9.58，岩浆分异程度较高。稀土元素含量为252.87×10^{-6}~408.42×10^{-6}，平均318.9×10^{-6}，略高于维氏酸性岩值。稀土配分曲线为较陡的右倾型曲线，轻稀土段较陡，重稀土段相对较平缓。$(La/Yb)_N$值为3.73~12.53，轻重稀土分异中等，轻稀土分异程度强于重稀土。δEu为0.17~0.37，铕负异常较明显。

分水岭二长花岗岩的SiO_2在69.13%~73.12%之间，属酸性岩。Na_2O+K_2O含量6.16%~8.53%。σ值为1.28~2.41，A/CNK为0.96~1.28，整体上属过铝质高钾钙碱性岩系。DI值75.39~88.86、SI值3.01~7.34，岩浆分异程度较高。稀土元素含量为190.95×10^{-6}~346.6×10^{-6}，平均244.78×10^{-6}，略低于维氏酸性岩值，其稀土配分曲线为较陡的右倾型曲线，轻稀土段较陡，重稀土段相对较平缓。$(La/Yb)_N$值为4.73~24.27，轻重稀土分异中等，轻稀土分异程度强于重稀土。δEu为0.04~0.5，具中等铕负异常。

黑岭二长花岗岩的SiO_2在68.62%~72.5%之间，属酸性岩。Na_2O+K_2O含量5.74%~8.65%。σ值为1.29~2.64，A/CNK为0.99~1.22，整体上属过铝质高钾钙碱性岩系。DI值76.03~86.66、SI值3.05~11.77，岩浆分异程度较高。稀土元素含量为210.41×10^{-6}~348.89×10^{-6}，平均294×10^{-6}，略高于维氏酸性岩值。稀土配分曲线为较陡的右倾型曲线，轻稀土段较陡，重稀土段相对较平缓。$(La/Yb)_N$值为8.78~11.86，轻重稀土分异中等，轻稀土分异程度强于重稀土。δEu为0.44~0.52，具中等铕负异常。

该序列岩石从南宝岭二长花岗岩→分水岭二长花岗岩→黑岭二长花岗岩，微量元素Rb、Y、Cu、Pb、Zn、Sn、Cr、Co含量有降低趋势，而Sr、Ba、V则有增高的趋势。微量元素原始地幔标准化蛛网图中，显示出明显的Ba、Nb、Sr-P、Ti负异常和Th-U、Pb、Nd-Zr-Hf-Sm正异常（图9-43）。

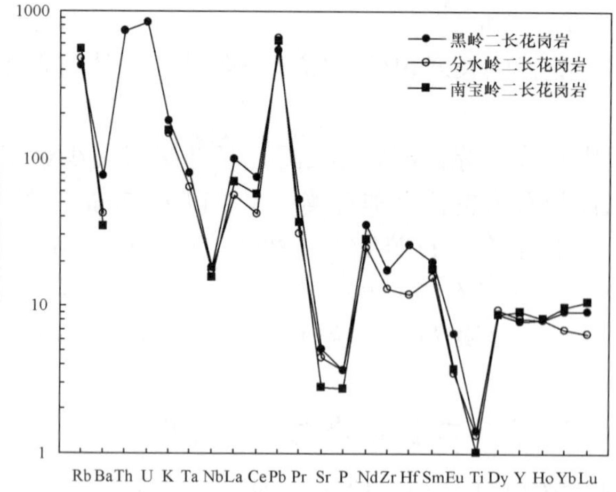

图9-43 雅加序列微量元素原始地幔标准化蛛网图

Fig. 9-43 Primitive mantle-normalized spider diagram of trace elements of Yajia Sequence

十四、晚三叠世三架岭序列

该序列岩石分布于海南岛西部昌江县海尾镇一带，由白石岭细粒含斑黑云二长花岗岩和打显村中细粒少斑黑云二长花岗岩2个岩石单位组成。

（一）白石岭细粒含斑黑云二长花岗岩（$\eta\gamma T_3$）

白石岭细粒含斑黑云二长花岗岩分布于三架岭一带，由白石村、三架岭西、三架岭东三个小侵入体组成，总面积31.86m^2。白石岭二长花岗岩侵入中二叠世大坡中粒斑状黑云二长花岗岩，其接触界面呈锯齿状，且见5~10cm的细粒冷凝边，在内接触带常见大小不一、不规则的大坡二长花岗岩捕虏体，二者呈超动侵入接触。白石岭二长花岗岩内叶理构造不发育，局部见少量的富云母包体、闪长质包体，大小一般为1cm×1.5cm，个别大于5cm×10cm，包体内偶见钾长石斑晶。

岩石类型主要为细粒含斑黑云母二长花岗岩。似斑状结构，块状构造，基质细粒花岗结构，局部见蠕虫结构。钾长石斑晶含量4%，大小3mm×5mm，环带构造明显。基质矿物粒度一般为0.5~2mm，分布均匀，斜长石环带构造发育。矿物组成：石英22%~24%、斜长石（An26~An28）34%~38%、钾长石32%~38%、黑云母3%~4%。

副矿物常见磁铁矿、钛铁矿、锆石、独居石等，总量大于1000g/t。

（二）打显村中细粒少斑黑云二长花岗岩（$\eta\gamma mfT_3$）

分布于白石岭、打显村、小万村一带，由利枝岭、白石岭西等5个侵入体组成，总面积58.9km^2。

打显村二长花岗岩东部与下志留统陀烈组及上石炭统青天峡组接触界线被掩盖。但接触带附近，围岩节理、劈理构造发育，并有1m至几十米厚的角岩化、硅化、绢云母化带，由此推测二者为侵入接触。与白石岭二长花岗岩接触界线清楚，呈岩脉或岩枝状贯入后者中，接触带附近发育伟晶岩脉，在内接触带可见后者的捕虏体，已有的同位素年龄为200Ma（锆石U-Pb法；海南地质综合勘察院第一勘察处，1993），形成时代厘定为晚三叠世末期。

岩石类型主要为中细粒少斑黑云母二长花岗岩。岩石具似斑状结构，块状构造。钾长石斑晶一般2mm×4mm，含量5%。基质中细粒花岗结构，矿物粒径一般为0.7~3mm，个别达4mm。矿物组合成：石英21%~25%、斜长石（An27~An30）34%~39%、钾长石32%~37%、黑云母3%~5%。

副矿物特征与白石岭二长花岗岩相似，常见磁铁矿、钛铁矿、锆石、独居石等，总量大于1000g/t。

（三）岩石地球化学特征

白石岭二长花岗岩的$SiO_2$72.48%~74.75%，属酸性岩。Na_2O+K_2O含量为7.88%~7.94%；σ值为1.96~2.14，A/CNK为1.07~1.11。根据硅-碱图解（图9-25）和SiO_2-K_2O图解（图9-26），属高钾钙碱性岩系。DI值86.3~90.81、SI值4.17~8.82，岩浆分异程度较高。稀土元素含量为230.67×10^{-6}~329.32×10^{-6}，平均280×10^{-6}，略低于维氏酸性岩值。稀土配分曲线为较陡的右倾型曲线（图9-39），轻稀土段较陡，重稀土段相对较平缓。$(La/Yb)_N$值为11.74~27.32，轻重稀土分异中等，轻稀土分异程度强于重稀土。δEu为0.29~0.36，铕负异常比较明显。

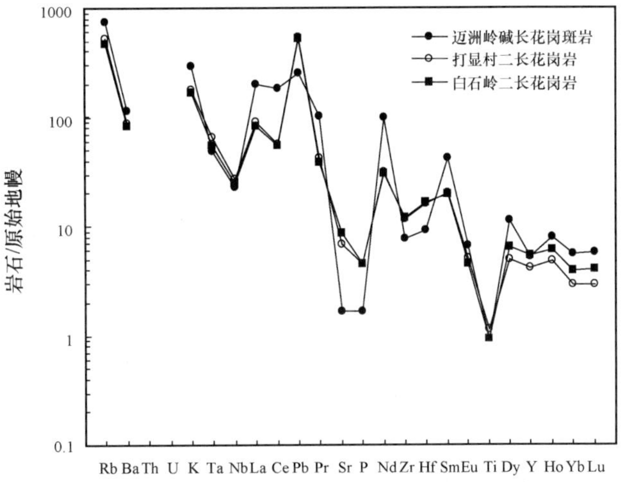

图9-44 三架岭序列、迈洲岭碱长花岗斑岩微量元素原始地幔标准化蛛网图

Fig. 9-44 Primitive mantle-normalized spider diagram of trace elements of Sanjialing Sequence and Maizhou alkali-feldspar granite

打显村二长花岗岩SiO_2含量为72.14%~73.8%，属酸性岩。Na_2O+K_2O含量7.94%~8.04%；σ值为2.1~2.3，A/CNK为1.03~1.07，整体上属弱过铝质高钾钙碱性岩系。DI值86.05~89.19、SI值6~7.95，岩浆分异程度较高。稀土元素含量为242.67×10^{-6}~278.32×10^{-6}，略低于维氏酸性岩值。稀土配分曲线为较陡的右倾型曲线（图9-43），轻稀土段较陡，重稀土段相对较平缓。$(La/Yb)_N$值为24.52~36.52，轻重稀土分异中等，轻稀土分异程度强于重稀土。δEu为0.36~0.38，铕负异常比较明显。微量元素Rb、Sn、Pb偏高，Zr、Be、Ba、Sr、Ta、As、V、Co、Ni偏低。微量元素原始地幔标准化蛛网图中（图9-44），显示出明显的Ba、Nb、Sr-P、Ti负异常和Pb、Nd、Sm正异常。

十五、晚三叠世迈洲岭碱长花岗斑岩（$\kappa\gamma\pi T_3$）

1. 地质特征、形成时代及岩石学特征

分布于琼海市九曲江乡附近的迈洲岭及排岭村一带，总面积约3.08km²。迈洲岭碱长花岗斑岩侵入下三叠统岭文组中，外接触带出现50~100m宽的弱角岩化带。又被下白垩统鹿母湾组沉积覆盖，九曲江一带的鹿母湾组底部见大量迈洲岭碱长花岗斑岩的砾石。

在迈洲岭碱长花岗斑岩中获钾长石K-Ar稀释法年龄213.3Ma（宜昌地矿所，1995），与实际地质侵入关系相符，侵入时代厘定为晚三叠世。

岩性为碱长花岗斑岩。暗紫色、粉紫色、粉红色或肉红色，斑状结构，块状构造。斑晶含量35%~70%。其中，钾长石斑晶20%~45%，粒径2mm×3mm~18mm×25mm，呈单晶或连斑产出，常发生绢云母化或发生高岭土化而呈肉红色或粉红色，有时见这两种蚀变呈环带状出现（图版Ⅱ-4-F）。石英斑晶10%~25%，粒径0.5~0.8mm，自形—半自形六方双锥状，多呈单晶，部分呈集合体，熔蚀

结构发育。偶尔见少量黑云母斑晶。基质矿物主要为正长石和石英，呈显微粒状结构，粒径小于0.1mm。副矿物主要为磁铁矿、锆石、钛铁矿等。

2. 岩石地球化学特征

迈洲岭碱长花岗斑岩高硅富钾、贫钠钙，SiO_2、K_2O 含量分别高达73%和8.7%；而 Na_2O、CaO 含量仅分别为0.19%和0.06%。岩石化学成分特点指示迈洲岭碱长花岗斑岩可能是（变）泥质岩石部分熔融并经过结晶分异形成的产物。稀土元素总量为 702×10^{-6}，是世界酸性岩平均值的两倍多，属稀土强富集型，特别是轻稀土元素总量高达 642×10^{-6}，明显富集，应注意寻找是否有离子吸附型稀土元素矿；LREE/HREE=23，$(La/Yb)_N=34$，轻、重稀土分馏比较明显；$(La/Sm)_N=4.6$，$(Gd/Yb)_N=2.8$，轻稀土元素之间分馏程度强于重稀土；$\delta Eu=0.22$，具强烈负铕异常，稀土元素球粒陨石标化图为略向右倾的翼展型曲线，轻稀土段较陡，重稀土段相对较平缓（图9-39）。与维氏酸性岩丰度值相比，迈洲岭碱长花岗斑岩的Rb、Hf、Mo、Bi较高，而Sr、Li、Be、Ta、Zn、Co、V、Zr较低；Rb/Sr=13.5，远大于维氏酸性岩的Rb/Sr比值；Ba含量高达 800×10^{-6}，与维氏酸性岩值接近，这可能与该岩石中富钾有关；微量元素原始地幔标准化蛛网图中（图9-44），显示出明显的Ba、Nb、Sr-P、Ti负异常和Pb、Nd、Sm正异常。

第四节 侏罗纪侵入岩

海南岛侏罗纪侵入岩出露面积相对较少，出露面积仅622.57km²（图9-45），占海南岛侵入岩总面积的4%，从早至晚共划分为11个岩石单位（表8-1）。

一、中侏罗世勤寨序列

该序列岩石主要分布于海南岛东部长征—文化市一线的勤寨地区，面积约25.55km²。该序列岩石以成分演变跨度较大为特色，由里茂中粒黑云母辉长岩、黑栋中细粒石英二长岩和八宝坡巨粒角闪正长岩3个岩石单位构成一个演化序列。空间分布中，黑栋石英二长岩单独出露，而八宝坡正长岩与里茂辉长岩毗邻，并脉动侵入里茂辉长岩（宜昌地质矿产研究所，1995）。八宝坡正长岩获角闪石K-Ar法年龄151 Ma（谢才富等，1999）、SHRIMP法锆石U-Pb年龄为（173±2）Ma（宜昌地质矿产研究所、海南省地质调查院，2004），将该序列岩石形成时代厘定为中侏罗世。

（一）里茂中粒黑云母辉长岩（υJ_2）

仅见里茂一个侵入体，面积约0.74km²，岩性为黑云母辉长岩。黑色，块状构造，中细粒—中粗粒辉长结构，粒径0.4~8mm。矿物组成：斜长石（An57~An62）33%~47%、单斜辉石33%~43%、黑云母10%~17%、正长石0~2%、角闪石0~3%。斜长石自形—半自形板柱状；单斜辉石半自形短柱状，主要为次透辉石-深绿辉石；黑云母红褐色—淡黄色，粒度变化大，大者常包裹其他矿物，形成筛状包含结构，局部见有辉石巨晶（粒径可达3cm）及辉石巨晶与黑云母填间体构成的堆积体碎块包体，显示有较好的结晶分异作用。与中三叠世长安辉长岩相比，不含斜方辉石，单斜辉石含钙高，常含有少量钾长石，是一种低硅、偏碱性的辉长岩。副矿物磁铁矿、钛铁矿含量达2%~5%，磷灰石含量达1%~2%。

（二）黑栋中细粒石英二长岩（$\eta o J_2$）

主要见黑栋侵入体，勤寨地区见少量同类岩脉，面积约24.17km²，岩性为中细粒石英二长岩。块状构造，中细粒二长结构，粒径0.3~5mm。矿物组成：石英12%~16%、斜长石（An33~An41）41%~47%、微斜条纹长石21%~26%、黑云母8%~17%、角闪石3%~8%、透辉石0~1%。斜长石半自形板柱状；微斜条纹长石半自形板状—他形，钠长石条纹较发育；黑云母半自形—他形鳞片状或叠片状；角闪石半自形柱状，深绿色-淡黄绿色。副矿物主要有褐帘石、磁铁矿、磷灰石、锆石、榍石。

（三）八宝坡巨粒角闪正长岩（ξJ_2）

仅见八宝坡侵入体，出露面积为0.64km²，岩性为巨粒角闪正长岩。青灰色，块状构造，巨粒结

构，粒径主要为 1cm×0.7cm～3cm×4cm。主要由微斜条纹长石 75%～80%、角闪石 10%～12%、斜长石 8%及少量黑云母 1%～2%、石英 3%～4%组成。微斜条纹长石为带蓝色调的白色，巨粒状，一般为 1cm×0.7cm～3cm×4cm，卡氏双晶发育，钠长石条纹含量较高，占 20%～25%，条纹多为不规则细脉、网纹状，少数为补片状；普通角闪石半自形—他形长柱状，粒径 1mm×8mm～2mm×13mm，N_g'—深绿色，N_p'—黄褐色；斜长石自形—半自形板柱状，粒径 0.1mm×0.4mm～2.5mm×4mm，分布不均匀，常分布于微斜条纹长石粒间或被其包裹。副矿物主要有磁铁矿、锆石、磷灰石。

（四）岩石地球化学特征

该序列岩石 SiO_2 含量变化较大，为 40.02%～65.09%。其中，里茂辉长岩 SiO_2 含量仅有 40.02%～43.96%，落入超基性岩范畴；黑栋石英二长岩和八宝坡正长岩的 SiO_2 含量介于 61%～65%之间，属中性岩类。Na_2O+K_2O 含量为 2.39%～11.08%。σ 值变化于 1.05～7.99 之间，多数>3.3；K_2O/Na_2O 值为 0.93～1.32，属钾质碱性岩。在硅-碱图中落入碱性岩系列（图 9-46）。A/CNK 为 0.41～0.96，属准铝质。岩石的稀土元素总量为 135×10^{-6}～248×10^{-6}。稀土配分模式为较平滑的右倾曲线（图 9-47）。轻重稀土分异中等—较明显，LREE/HREE 值为 5.6～14.8；δEu 多为 0.64～1.02，多数具弱的负铕异常。岩石的 Rb、K、Th、U、Zr、Hf 等元素含量低。在微量元素蛛网图中具有较明显的 Sr 正异常，P、Ti 负异常不明显（图 9-48）。该序列岩石具有较明显的 Ta-Nb 负异常，与桂东南的燕山期钾玄岩（李献华等，1999，2000）不同，可能是软流圈岩浆遭受了富集地幔或地壳物质混染的结果。晚期的八宝坡正长岩的 Rb、Th-U-K、Pb、Zr-Hf 及轻稀土元素的含量都明显增加，呈现正异常；而 P、Ti 出现较明显负异常，Ba、Sr、Eu 可出现正异常也可出现负异常，显示存在斜长石、钛铁矿、磷灰石的结晶分异作用。

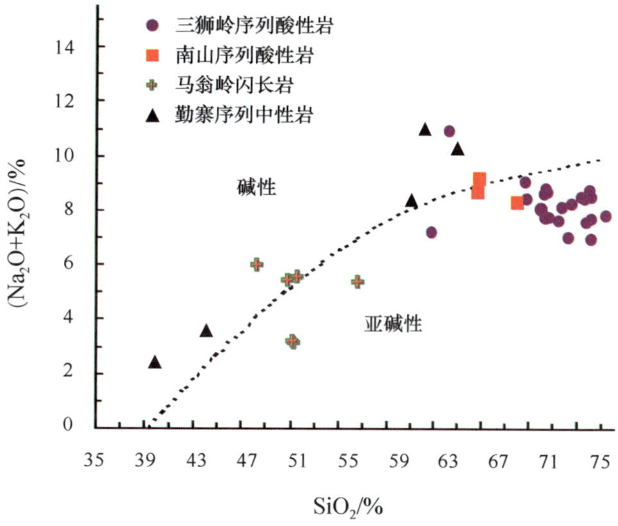

图 9-46 海南岛侏罗纪侵入岩硅-碱图解

Fig. 9-46 SiO_2-Na_2O+K_2O diagram of Jurassic intrusive rocks from the Hainan Island

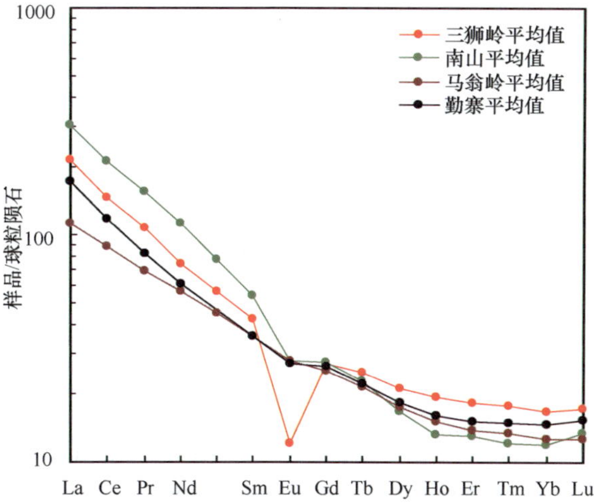

图 9-47 侏罗纪侵入岩稀土元素球粒陨石标准化分布型式图

Fig. 9-47 Chondrite-normalized REE patterns of Jurassic intrusive rocks

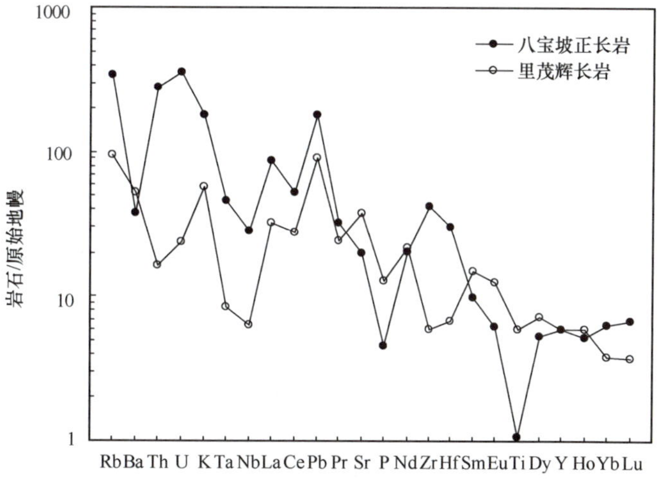

图 9-48 勤寨序列微量元素原始地幔标准化蛛网图

Fig. 9-48 Primitive mantle-normalized spider diagram of trace elements of Qinzhai Sequence

二、晚侏罗世马翁岭闪长岩（δJ_3）

1. 地质特征、形成时代及岩石学特征

分布于琼中县-万宁市及白沙县一带，主要见有马翁岭、托剩岭、牛坡、马龙岭、老村等侵入体，总面积约49.84km²。前人在马翁岭侵入体中获Rb-Sr法年龄为（152±1）Ma（马大铨等，1991）、托剩岭侵入体锆石U-Pb法同位素表面年龄（157.7±1.8）Ma（海南省地质调查院，2000）、白沙县老村侵入体La-ICPMS锆石U-Pb同位素年龄157Ma（海南省地质调查院，2007），将其侵入时代厘定为晚侏罗世。

主体岩性为细粒黑云角闪闪长岩。岩石呈灰绿色，细粒结构，块状构造。矿物组成：斜长石57%、钾长石3%、石英3%、角闪石31%、黑云母5%。斜长石为半自形柱状，粒径大小为0.2~1.75mm，大部分已强烈绢云母化，另一部分蚀变较弱，聚片双晶及环带构造发育，为中长石；钾长石为他形粒状，充填在斜长石颗粒间不规则空隙中，卡斯巴双晶发育；石英为他形粒状，充填在斜长石颗粒间不规则空隙中，波状消光；角闪石为半自形—他形柱粒状，粒径大小为0.2~1.1mm，具黄-褐多色性，柱状解理和菱形解理发育，有的晶粒边缘已退变为黑云母；黑云母为片状，具黄-深红褐多色性，已发生退变且析出磁铁矿。

副矿物主要有榍石、磁铁矿、钛铁矿、锆石、磷灰石。磷灰石长宽比达10∶1。

2. 岩石地球化学特征

主体岩性闪长岩的SiO_2含量变化于50.98%~51.42%，属基性岩类；Na_2O+K_2O含量变化在3.13%~5.39%；σ值为1.08~3.39，A/CNK为0.41~0.57，属准铝质；硅-碱图解（图9-46）中，大多数落入碱性岩系列中，个别样品落入亚碱性岩系列中；SiO_2-K_2O图解（图9-49）中落入低硅高钾钙碱性系列区。岩石的稀土元素总量为79.2×10^{-6}~120.52×10^{-6}；稀土配分模式为较平滑的右倾曲线（图9-48）；轻重稀土分异中等，LREE/HREE值5.1~3.36；$\delta Eu=0.88\sim1.0$，具弱的负铕异常。岩石的Rb、K、Th、U、Zr、Hf等元素含量低，微量元素蛛网图中具有较明显的Sr正异常和Ta-Nb负异常（图9-50），可能是软流圈岩浆遭受了富集地幔或地壳物质混染的结果；初始锶比值较低，为0.70463；$\varepsilon_{Nd}(t)$较高，为1.67，与OIB（洋岛玄武岩）的Sr、Nd同位素组成相似，也与桂东南燕山期钾玄岩（郭新生等，2001）一致，是亏损的软流圈地幔与富集岩石圈地幔混合的产物。

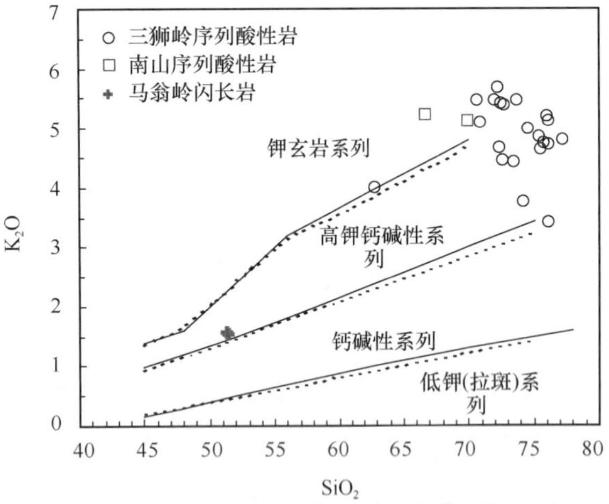

图9-49　海南岛侏罗纪侵入岩SiO_2-K_2O图解

Fig. 9-49　SiO_2-K_2O diagram of Jurassic intrusive rocks from the Hainan Island

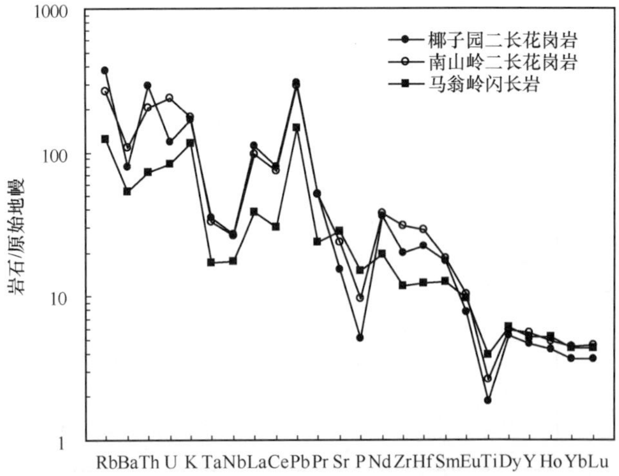

图9-50　马翁岭闪长岩、南山序列微量元素原始地幔标准化蛛网图

Fig. 9-50　Primitive mantle-normalized spider diagram of trace elements of Mawengling diorite and Nanshan Sequence

三、晚侏罗世南山序列

该序列岩石仅出露于海南岛南部南山一带，面积约 18.74km^2。由南山岭中细粒斑状角闪黑云母二长花岗岩、椰子园细粒斑状角闪黑云母二长花岗岩和冬瓜岭细粒含斑黑云母二长花岗岩等 3 个岩石单位构成。各岩石单位紧密伴生。南山岭二长花岗岩被椰子园、冬瓜岭二长花岗岩脉动侵入，冬瓜岭二长花岗岩侵入中三叠世超盆二长花岗岩（红塘岭侵入体）。南山岭二长花岗岩全岩 Rb-Sr 法等时线年龄为（152.2±6）Ma（曾庆銮等，1992），形成时代厘定为晚侏罗世。

（一）南山岭中细粒斑状角闪黑云母二长花岗岩（$\eta\gamma mfJ_3$）

仅见南山岭一个侵入体，面积约 14.9km^2。岩性为中细粒斑状角闪黑云母二长花岗岩，局部为石英二长岩或石英正长岩。块状构造，似斑状结构。钾长石斑晶含量较高，30%~40%，肉红色，大小 5mm×7mm~15mm×35mm，有的包一圈 1~6mm 的白色斜长石边，构成幔状结构（图版Ⅱ-4-G）；斜长石斑晶 1%~4%，大小 4mm×6mm~10mm×15mm。基质大小 0.5~4mm。矿物组成：石英 18%~23%、斜长石（An24~An32）25%~33%、微斜条纹长石 40%~49%、黑云母 5%~8%、角闪石2%~5%。斜长石半自形板柱状，可见环带构造；微斜条纹长石自形—他形板状；黑云母半自形—他形鳞片状-叠片状；角闪石半自形—自形柱状。副矿物主要有磁铁矿、榍石、锆石、磷灰石、褐帘石。常见闪长质包体（3~10 个/m^2），浑圆状，大小 1cm×35cm~20cm×35cm，常有石英、微斜条纹长石、斜长石捕虏晶，其中的钾长石捕虏晶也可有幔状结构。

（二）椰子园细粒斑状角闪黑云母二长花岗岩（$\eta\gamma J_3$）

见椰子园一个侵入体（被冬瓜岭侵入体分割），面积约 0.36km^2，岩性为细粒斑状角闪黑云母二长花岗岩。块状构造，似斑状结构。与南山岭二长花岗岩的区别，一是斑晶数量大为减少，只有 5%~20%；二是斑晶成分复杂，除钾长石、斜长石外，还有少量石英、黑云母、角闪石；三是斑晶粒径变小，为 3mm×4mm~10mm×20mm。基质大小 0.5~2mm。矿物组成：石英 20%~25%、斜长石（An34~An35）35%~37%、微斜条纹长石 31%~33%、黑云母 5%~9%、角闪石一般 0.5%~1%。斜长石半自形板柱状；微斜条纹长石 31%~33%，半自形—他形板状；黑云母半自形—他形鳞片状-叠片状；角闪石半自形—自形柱状。副矿物主要有磁铁矿、榍石、锆石、磷灰石、褐帘石，与南山岭二长花岗岩相似。闪长质包体易见，浑圆状，大小 0.6cm×1.5cm~4cm×7cm。

（三）冬瓜岭细粒含斑黑云母二长花岗岩（$\eta\gamma fJ_3$）

见冬瓜岭和南山角 2 个侵入体，面积约 3.48 km^2，岩性为细粒含斑黑云母二长花岗岩。块状构造，细粒花岗结构。局部含斑晶钾长石，<1%，基质粒径 0.5~2mm。矿物组成：石英 26%~33%、斜长石（An30~An38）34%~45%、微斜条纹长石 20%~25%、黑云母多数为 8%~12%、角闪石 0~2%。斜长石半自形板柱状；微斜条纹长石半自形—他形板状；黑云母多数为半自形—他形鳞片状-叠片状；角闪石半自形柱状。副矿物主要有磁铁矿、榍石、锆石、磷灰石、褐帘石等。见少量闪长质包体，浑圆状，大小与椰子园二长花岗岩中的相似。

（四）岩石地球化学特征

该序列岩石的 SiO_2 含量为 66.76%~70.02%；Na_2O+K_2O、K_2O 含量及 σ 值较高，分别为 9.17%~8.32%、5.42%~5.12%、2.93~3.13；硅-碱图和 SiO_2-K_2O 图解中多数落入亚碱性系列中，少数落入碱性系列中，但靠近分界线，属于钾玄岩系列（图 9-46，图 9-49）；A/CNK 为 0.93~1.01，属准铝质—弱过铝质。该序列岩石的稀土元素总量较高，为 289.1×10^{-6}~303.9×10^{-6}；稀土配分模式为右陡倾曲线（图 9-47）；轻重稀土分异较明显，LREE/HREE 值 16.0~18.9；δEu 为 0.74~0.57，具弱—中等的负铕异常。岩石的 Ba、Sr、Zr、Hf 元素含量较高；微量元素蛛网图中（图 9-50），显示 Ba、P、Ti 中等—较弱负异常，表明岩浆结晶分异不强；南山岭二长花岗岩的初始锶比值较低，为 0.70694；$\varepsilon_{Nd}(t)$ 较高，为-3.58；T_{2DM}值相对较小，为 1234Ma（宜昌地质矿产研究所、海南省地质调查院，2004），结合包体特征，反映其来源于壳幔混合作用。

四、晚侏罗世三狮岭序列

主要分布于文昌市、琼中县及陵水县-保亭县一带，由横岭、雷公岭、什运、三狮岭及福安村等侵入体组成。从早至晚划分为架水粗中粒斑状角闪黑云二长花岗岩、三狮岭中粒黑云母正长花岗岩、马鞍岭中细粒黑云母正长花岗岩和马岭细粒黑云母正长花岗岩等 4 个岩石单位，总面积约 528.44km^2。根据侵入接触关系和同位素年龄，形成时代厘定为晚侏罗世。

（一）架水粗中粒斑状角闪黑云二长花岗岩（$\eta\gamma cmJ_3$）

出露在营根镇以北架水村一带，不规则状，面积约 30.77km^2，被晚期的三狮岭正长花岗岩（横岭侵入体）侵入，两者为脉动侵入接触关系。

主体岩性为粗中粒斑状角闪黑云二长花岗岩。似斑状结构，块状构造。矿物组成：斜长石 25%~45%、钾长石 25%~55%、石英 18%~22%、黑云母 5%~8%、角闪石<8%。斑晶成分为正长石，含量 10%~15%，以浅肉红色为主，部分灰黄色、灰白色，半自形板状，卡氏双晶发育，大小 5mm×10mm~5mm×25mm，分布较均匀，无定向排列，斑晶内部常包含细小的斜长石、石英、黑云母和角闪石等早期结晶矿物。基质呈花岗结构，矿物粒径一般 3~5mm，部分 2~3mm 或 5~7mm。其中，斜长石 An=35，为中长石，灰白色，半自形柱状，普遍发育聚片双晶，少量具格子双晶或卡钠复合联晶，部分发育环带构造；钾长石以浅肉红色为主，部分灰黄色或灰白色，半自形板状、粒状，发育卡氏双晶；石英他形粒状，普遍具波状消光；黑云母半自形—他形鳞片状、叠片状，多色性明显，呈黑绿色—浅黄色，为富镁黑云母；角闪石柱状，呈绿色—浅绿色多色性。副矿物主要为磷灰石、钛铁矿、锆石等。

（二）三狮岭中粒黑云母正长花岗岩（$\xi\gamma J_3$）

分布于昌江县、琼中县及陵水县廖次岭一带，由三狮岭、横岭及廖次岭等侵入体组成，出露面积 407.49km^2。昌江县城附近见三狮岭正长花岗岩侵入石碌二长花岗岩，两者接触界线截然清楚，内接触带见 10cm~15cm 的冷凝边，并出现伟晶岩囊包体（图 9-51）。已有的同位素年龄值有：三狮岭侵入体锆石 U-Pb 年龄 151Ma（海南地质综合勘察院第一勘察处，1993）、横岭侵入体锆石 U-Pb 法表面年龄（150.3±1.8）Ma（海南省地质调查院，2000）、廖次岭侵入体 SHRIMP 法锆石 U-Pb 年龄 151Ma（谢才富，2009），形成时代厘定为晚侏罗世。

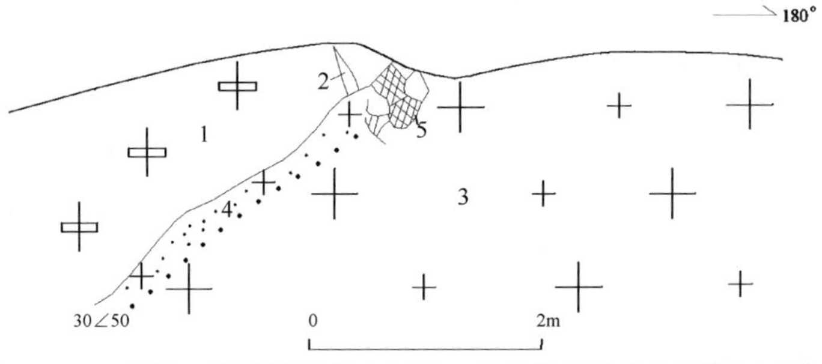

图 9-51　三狮岭正长花岗岩侵入石碌二长花岗岩接触关系剖面素描图

Fig. 9-51　Sketch showing the contact that Sanshiling syenogranite intrudes Shilu monzonitic granite

（据 1∶25 万东方县幅区调报告，2007）

1—石碌二长花岗岩；2—石英脉；3—三狮岭正长花岗岩；4—冷凝边；5—伟晶岩囊包体

岩石类型主要为中粒黑云母正长花岗岩，局部中粗粒黑云母正长花岗岩，少量为二长花岗岩。浅肉红色或灰白色，以中粒花岗结构为主，局部中细粒、中粗粒结构，蠕英结构（图版Ⅱ-4-H），块状构造。矿物粒度一般为 3~4mm，局部 4~6mm。矿物含量分别为石英 20%~30%、钾长石 36%~55%、斜长石 18%~35%、黑云母 2%~5%，个别薄片见有少量角闪石。钾长石浅肉红色，为微斜条纹长石，格子双晶十分发育，钠长石条纹呈脉状或细脉状，个别补片状，常包裹斜长石、石英及黑云母等；斜长石灰白色，多呈板状集合体充填在钾长石粒间，粒径 1.5mm×3mm～2.5mm×5mm，明显有两个世代，早世代为奥长石（An10～An25），半自形板状，晚世代为钠长石（An<10），沿边缘或裂隙交代微斜条纹长石；石英无色，他形粒状或集合体状，大小 1～6mm；黑云母半自形叠片状，片幅 1.5～3mm，N_g、N_m 为黑色，N_p 为棕黄色。副矿物以富磁铁矿、榍石、褐帘石为特征，总量大于 1000g/t。

（三）马鞍岭中细粒黑云母正长花岗岩（$\xi\gamma mJ_3$）

分布于昌江县三狮岭、月大老村、琼中县朝根坡及文昌市福安村一带，总面积约为 65.17km²。马鞍岭侵入体侵入早二叠世石碌二长花岗岩，被马岭正长花岗岩侵入。福安村侵入体中获 La-ICPMS 法锆石 U-Pb 年龄为（142.4±1.0）Ma（海南省地质调查院，2009），形成时代厘定为晚侏罗世。

马鞍岭正长花岗岩岩性不均一，岩石呈中细粒、细中粒花岗结构或似斑状结构，局部见细粒文象结构，块状构造。主要岩石类型为中细粒黑云母正长花岗岩，局部细粒斑状黑云母二长-正长花岗岩，两种岩石中均可见有萤石，但均未见闪长质包体。似斑状结构中斑晶为微斜条纹长石，他形或半自形板状，个体 3mm×6mm，含量 10%～15%。矿物组成：石英 20%～28%、钾长石 38%～55%、斜长石（An15～An49）18%～30%、黑云母 3%～10%、角闪石 0～1%。副矿物以富磁铁矿、榍石、褐帘石为特征。

（四）马岭细粒黑云母正长花岗岩（$\xi\gamma fJ_3$）

分布于长江县马岭、老羊村、昌化大岭一带，包括马岭、老羊岭、从开岭、长山、轰山岭、池头等数个小侵入体，呈椭圆状、长条状或不规则状，单个侵入体出露面积较小，总面积约为 25.01km²。马岭正长花岗岩侵入的最新岩体为晚侏罗世马鞍岭二长花岗岩，接触界面截然清楚，内接触带见 5～8cm 的细粒冷凝边，靠近边界富含伟晶岩囊包体（图 9-52）。因此，马岭正长花岗岩形成于马鞍岭二长花岗岩之后。

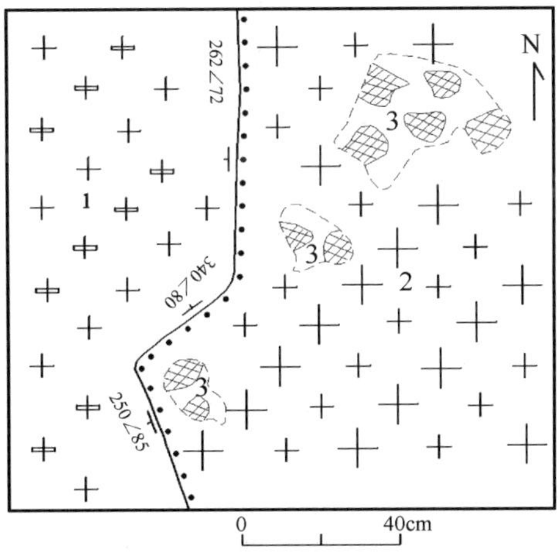

图 9-52　马岭正长花岗岩侵入马鞍岭正长花岗岩接触关系平面素描图

Fig. 9-52　Sketch showing the contact that Maling syenogranite intrudes Ma'anling syenogranite

（据 1∶25 万东方县幅区调报告，2007）

1—马鞍岭正长花岗岩；2—马岭正长花岗岩；3—伟晶岩囊包体

岩石手标本为灰白色，个别带微弱的肉红色或浅黄褐色，岩石类型主要为中细粒黑云（二云）母正长花岗岩，局部为二长花岗岩。细粒—中细粒花岗结构，局部文象结构，块状构造。由微斜条纹长石、斜长石、石英、黑云母及白云母组成。微斜条纹长石粒径 0.5～3.5mm，钠长石条纹十分发育，呈脉状、细脉状或补片状，可见格子双晶，常包裹斜长石、石英及黑云母；斜长石白色，多呈板状集合体充填在钾长石粒间，粒径 0.2mm×0.4mm～0.8mm×1.5mm，明显有两个世代，早世代为酸性奥长石（An10～An20），半自形板状，晚世代为钠长石（An<5），呈梳状集合体沿边缘或裂隙交代微斜条纹长石；石英无色，他形粒状或集合体状，大小 0.1～1.5mm；黑云母部分被白云母交代，也有原生白

云母产出。矿物粒度一般为 0.5～2mm，矿物分布较均匀，含量分别为石英 25%～36%、钾长石 48%～57%、斜长石（An8～An32）12%～23%、黑云母 0～5%及白云母 0～3%。副矿物中，萤石十分普遍，个别岩石薄片中萤石含量达 0.5%～1%；锆石、磁铁矿常见，偶见褐帘石。池头侵入体中见有电气石，马岭侵入体中见有萤石、钨铍矿化等，为区内与稀有、有色金属矿化有关的花岗岩类岩石。

（五）岩石地球化学特征

该序列岩石结构演化特征比较明显，但不同岩石单位之间的岩石化学成分变化不大，演化规律不明显。SiO_2 含量在 62.81%～77.3%之间；Na_2O+K_2O 含量、K_2O 含量以及 σ 值较高，分别为 6.99%～8.81%、3.61%～5.85%和 1.45～5.61；根据硅-碱图解（图 9-46）和 SiO_2-K_2O 图解（图 9-49），属高钾钙碱性岩系；A/CNK 在 0.89～1.07，属准铝质—弱过铝质。稀土元素总量变化较大，为 93.02×10^{-6}～771.6×10^{-6}；稀土元素配分模式为右倾型曲线；轻重稀土分异不明显；δEu 为 0.2～0.85，铕负异常中等至明显（图 9-47）。该序列岩石的 Rb、Ba、Zr 等元素含量相对较高，从早至晚，其 Rb 含量具逐渐增大趋势，Rb/Sr 大部分大于 1；微量元素蛛网图中（图 9-53），显示出明显的 Ba、Nb、Sr-P 及 Ti 负异常，尤其是晚期的马岭正长花岗岩，负异常更是明显，表明岩浆发生了强烈的结晶分异作用。

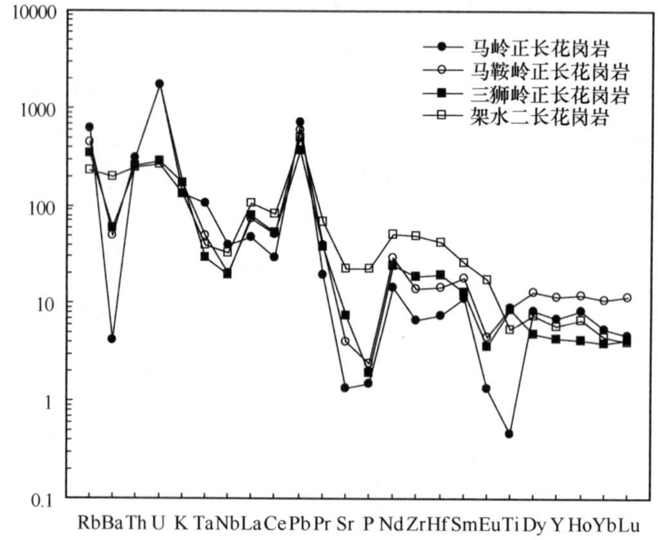

图 9-53 三狮岭序列微量元素原始地幔标准化蛛网图

Fig. 9-53 Primitive mantle-normalized spider diagram of trace elements of Sanshiling Sequence

第五节 白垩纪侵入岩

海南岛白垩纪岩浆活动非常强烈，形成了大量的侵入岩和火山岩。其中最著名的岩体有保城杂岩体、屯昌杂岩体、千家杂岩体和吊罗山杂岩体等（图 9-54），出露总面积约 3757.98km²，占海南岛侵入岩总面积的 23%。从早至晚共划分为 21 个岩石单位（表 8-2）。

一、早白垩世半岛中粗粒角闪黑云正长花岗岩（$\xi\gamma cK_1$）

1. 地质特征、形成时代及岩石学特征

分布于海南岛西部昌江县以北，由半岛、高岭、打金岭 3 个小侵入体组成，总面积 9.62km²。

打金岭侵入体侵入早二叠世石碌二长花岗岩和中二叠世邦溪岩体二长花岗岩。与邦溪岩体接触处，两者接触界线十分清楚，接触界面走向近垂直于邦溪二长花岗岩中的巨斑晶排列方向。半岛侵入体中已有的黑云母 K-Ar 法同位素年龄为 131.7Ma（宜昌地质矿产研究所，1995）。形成时代厘定为早白垩世。

半岛正长花岗岩不同侵入体的岩性十分稳定均一，主体岩石类型为中粗粒角闪黑云正长花岗岩。粗粒或中粗粒结构，块状构造。宏观上的鉴别特征是手标本上常见个体为2mm×6mm～3mm×8mm的自形、半自形角闪石；黑云母呈薄的叠片，片幅1.5～5mm最常见，手标本上也可见到褐帘石。偶尔见有闪长质暗色包体，包体最大仅有5cm×9cm，含量（1～2）个/m^2。镜下观察，钾长石为微斜条纹长石，含量39%～50%，他形，粒径一般5～8mm，最大12mm，钠长石条纹不发育，格子双晶发育；斜长石12%～37%，灰白色，多呈半自形板状集合体充填在钾长石粒间，粒径2mm×4mm～4mm×6mm，常具有净边环带，中心部分An20～An30，边部An10～An15，常被微斜条纹长石包裹，并在边部出现蠕英石；石英18%～38%，无色，他形粒状或集合体状；黑云母1%～6%，叠片状，片幅1.5～5mm，N_g、N_m为黑色，N_p为黄棕色，常与角闪石共生，多被绿泥石交代；普通角闪石1%～2%，具绿色、蓝绿色多色性，N_g为蓝绿色，N_m为绿色，N_p为黄绿色，大小0.5mm×1.5mm～2mm×4mm，常被绿泥石交代。副矿物以富含褐帘石为特征，褐帘石自形楔状，有时见十分发育的环带构造，最大颗粒0.3～1.5mm，锆石、磷灰石、磁铁矿常见，个别薄片见有榍石。

2. 岩石地球化学特征

半岛正长花岗岩的SiO_2含量高，为71.04%～72.5%；Na_2O+K_2O及K_2O含量较高，分别为7.97%～8.6%和4.81%～5.68%；σ值为2.15～2.64；根据硅-碱图解（图9-55）和SiO_2-K_2O图解（图9-56），属高钾钙碱性岩系；A/CNK为0.96～1.02，属准铝质—弱过铝质；DI和NKA都较大，分别为86和0.80，反映岩浆分异程度较高。稀土元素总量变化较大，为292×10^{-6}～423×10^{-6}；稀土配分模式为右倾型曲线，轻重稀土分异较明显，且轻稀土分馏作用强于重稀土；δEu为0.25～0.38，铕负异常比较明显（图9-57）。在原始地幔微量元素上具有较明显的Ba、Ta-Nb、Sr-P、Ti负异常（图9-58）。

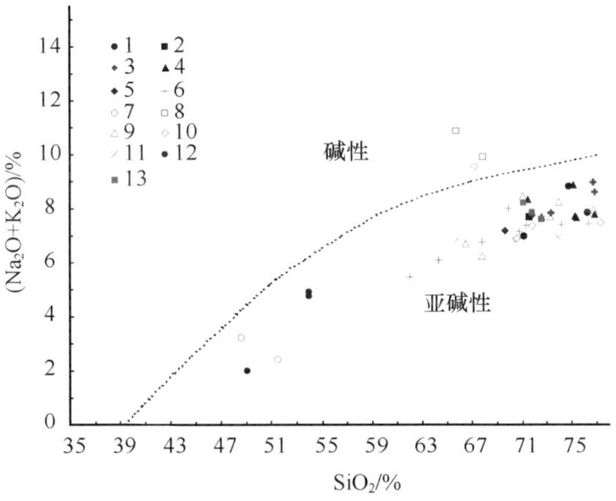

图9-55　海南岛白垩纪侵入岩硅-碱图解

Fig. 9-55　$SiO_2-Na_2O+K_2O$ diagram of Cretaceous intrusive rocks from the Hainan Island

1—$\gamma\pi K_2$花岗斑岩；2—火烧岭二长花岗岩；3—龙楼正长花岗岩；4—吊罗山序列酸性岩；5—南牙村花岗闪长斑岩；6—保城序列酸性岩；7—乌石闪长岩；8—番通石英正长岩；9—高峰序列酸性岩；10—$\gamma\pi K_1$花岗斑岩；11—$\gamma\delta\pi K_1$花岗闪长玢岩；12—拉丁村辉长岩；13—半岛正长花岗岩

二、早白垩世潜火山岩

早白垩世潜火山岩见于早白垩世同安岭及牛腊岭火山盆地中或其旁侧。为潜火山相或火山管道相产物，在火山岩中往往有和其成分相当的对应物。主要见有细粒辉长岩、闪长玢岩、细粒石英闪长岩、花岗斑岩、花岗闪长玢岩等。这些斑岩（玢岩）规模都较小，呈小岩株或岩墙出现。海南地质综合勘察院区调队（1995）认为，中酸性次火山岩形成于第二旋回火山活动早、晚期，早期形成花岗闪长斑岩，与第二旋回早期的英安质火山岩浆活动有关；晚期形成花岗斑岩，与第二旋回晚期的流纹质火山岩浆活动有关。相应地，中基性的细粒辉长岩、闪长玢岩、细粒石英闪长岩即应形成于第一旋回，与玄武质及玄武安山质火山岩浆活动有关。

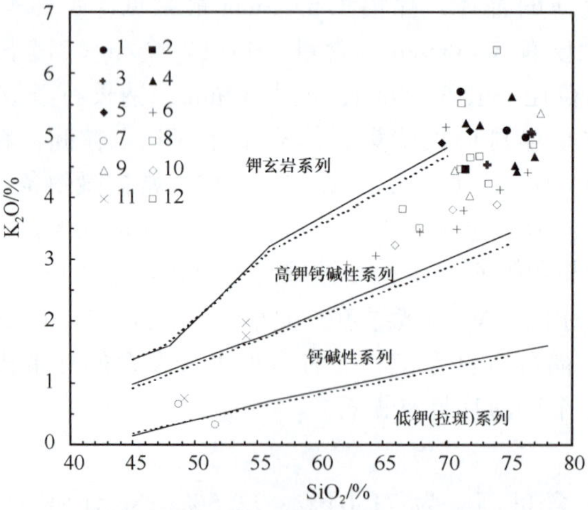

图 9-56 海南岛白垩纪侵入岩 SiO_2-K_2O 图解

Fig. 9-56 SiO_2-K_2O diagram of Cretaceous intrusive rocks from the Hainan Island

1—$\gamma\pi K_2$ 花岗斑岩；2—火烧岭二长花岗岩；3—龙楼正长花岗岩；4—吊罗山序列酸性岩；5—南牙村花岗闪长斑岩；6—保城序列酸性岩；7—乌石闪长岩；8—高峰序列酸性岩；9—$\gamma\pi K_1$ 花岗斑岩；10—$\gamma\delta\pi K_1$ 花岗闪长玢岩；11—拉丁村辉长岩；12—半岛正长花岗岩

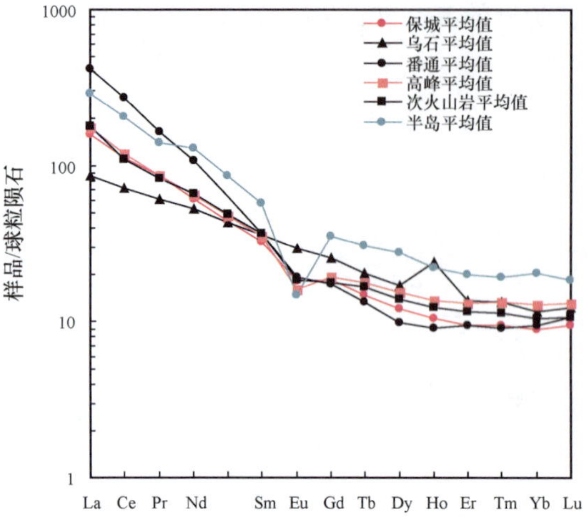

图 9-57 早白垩世侵入岩稀土元素球粒陨石标准化分布型式图

Fig. 9-57 Chondrite-normalized REE patterns of early Cretaceous intrusive rocks

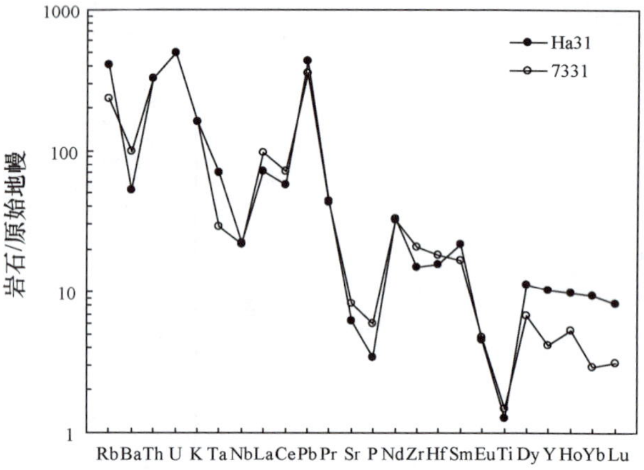

图 9-58 半岛正长花岗岩微量元素原始地幔标准化蛛网图

Fig. 9-58 Primitive mantle-normalized spider diagram of trace elements of Bandao syenogranite

（一）拉丁村细粒辉长岩-辉长闪长岩（νK_1）

拉丁村细粒辉长岩-辉长闪长岩（νK_1）见于同安岭火山岩盆西南侧，见有拉丁村、立才七队北 2 个侵入体，面积约 3.25km²。拉丁村侵入体侵入三叠纪超盆二长花岗岩，被早白垩世抱跃二长花岗岩（文门侵入体，形成年龄 107Ma）侵入，拉丁村侵入体所测 La-ICPMS 锆石 U-Pb 同位素年龄值为（97.2±1.6）Ma（分析数据见表 9-7）。上述年龄值与地质侵入接触先后关系不符，本文暂将其形成时代厘定为早白垩世。

拉丁村侵入体岩性为中细粒辉长闪长岩、二长辉长闪长岩、石英辉长岩。块状构造，细粒结构。由斜长石（An43～An55）55%～64%、普通辉石 4%～21%、普通角闪石 15%～20%、黑云母 8%～12%、微斜长石 2%～7%、石英 2%～8%组成。斜长石半自形板柱状，聚片双晶发育，环带不太发育。普通辉石半自形粒状，自形程度比斜长石稍差，以单个晶体充填在斜长石组成的三角形空隙中，粒径 0.2~2.0mm，发育两组近垂直的解理；有两种辉石，一种具浅绿-浅黄绿弱多色性，发育较多细纹，干涉色达二级紫红，最大消光角为 37°，占辉石的大部分；另一种具无色—浅红多色性，无裂理，干涉色达二级蓝。黑云母自形叠片状，黄-褐-黑多色性，蚀变不强，一般分布在辉石边缘。微斜长石、石英他形填隙。副矿物主要有磁铁矿、钛铁矿、磷灰石，偶见榍石、金红石，有少量斜锆石。

两侵入体的 SiO_2 含量为 49.02%~53.94%；Na_2O+K_2O 及 K_2O 含量分别为 5.25%~2.34% 和 0.88%~2.12%；σ 值为 0.80~2.47，属钙碱性-高钾钙碱性系列；A/CNK 为 0.64~0.81，属准铝质。岩石的稀土元素总量为 65.6×10^{-6}~164.9×10^{-6}；稀土配分模式为右倾的平滑曲线；轻重稀土分异中等，LREE/HREE 值 7.1~9.3；δEu 为 0.99~0.78，基本无铕异常或具弱的负铕异常（图 9-57）。V、Cr、Ni、Co、Sc、Zn 等亲铁铜元素的含量较高，而 Rb、Ba、Th、U、K 等大离子亲石元素的含量较低；微量元素蛛网图中具有弱的 Ba、Nb、P、Ti 负异常，而 Sr 具有正异常。

表 9-7　拉丁村辉长岩 LA-ICP-MA 锆石 U-Pb 测年结果

Table 9-7　LA-ICP-MA Zircon U-Pb dating results of Ladingshu gabbro

岩石单位	测试点	$^{207}Pb/^{206}Pb$	$^{207}Pb/^{235}U$	$^{206}Pb/^{238}U$	$^{208}Pb/^{232}Th$	$^{238}U/^{232}Th$	$^{207}Pb/^{206}Pb$	$^{207}Pb/^{235}U$	$^{206}Pb/^{238}U$	$^{208}Pb/^{232}Th$
拉丁村	DK10-1-1	0.0505	0.1114	0.0161	0.0050	0.896298	217	107	103	99.9
拉丁村	DK10-1-2	0.0482	0.1105	0.0167	0.0051	0.950751	109	106	107	103
拉丁村	DK10-1-3	0.0482	0.1090	0.0165	0.0050	0.76858	106	105	106	100
拉丁村	DK10-1-4	0.0541	0.1237	0.0166	0.0052	0.792267	376	118	106	104
拉丁村	DK10-1-5	0.0519	0.1059	0.0149	0.0045	1.003814	280	102	95.2	91.6
拉丁村	DK10-1-6	0.0497	0.1063	0.0156	0.0049	0.747272	189	103	100.0	99.1
拉丁村	DK10-1-7	0.0516	0.1079	0.0152	0.0047	1.079564	333	104	97.0	94.6
拉丁村	DK10-1-8	0.0477	0.0993	0.0151	0.0045	0.937119	87.1	96.1	96.5	91.1
拉丁村	DK10-1-9	0.0547	0.1193	0.0159	0.0051	0.60361	467	114	101	102
拉丁村	DK10-1-10	0.0476	0.1020	0.0156	0.0047	1.126126	80	98.6	99.7	95.7
拉丁村	DK10-1-11	0.0471	0.0977	0.0150	0.0045	0.990295	57.5	94.6	95.9	90.5
拉丁村	DK10-1-12	0.0480	0.0948	0.0144	0.0043	1.158614	102	92.0	92.5	85.8
拉丁村	DK10-1-13	0.0685	0.1458	0.0153	0.0053	1.130199	885	138	98.1	106
拉丁村	DK10-1-14	0.0540	0.0946	0.0127	0.0043	0.89823	369	91.8	81.4	86.0
拉丁村	DK10-1-15	0.0476	0.0888	0.0136	0.0041	0.929627	76.0	86.4	87.3	83.1
拉丁村	DK10-1-16	0.0694	0.1549	0.0161	0.0063	0.705418	911	146	103	127
拉丁村	DK10-1-17	0.0524	0.1115	0.0155	0.0047	0.938791	302	107	99.4	94.7
拉丁村	DK10-1-18	0.0649	0.1316	0.0147	0.0047	1.386578	770	125	94.4	94.4

（二）花岗闪长玢岩（$\gamma\delta\pi K_1$）、花岗斑岩（$\gamma\pi K_1$）

主要见于同安岭及牛腊岭火山岩盆及其旁侧，花岗闪长玢岩、花岗斑岩分布面积分别为 20.16km² 和 60.71km²。在五指山岩盆中仅有少量花岗斑岩和英安玢岩脉。

岩石呈块状构造，斑状结构。斑晶主要为石英、钾长石、斜长石，有的有黑云母和角闪石斑晶。基质为显微隐晶质或微细粒结构。副矿物主要有磁铁矿，其次有锆石、磷灰石，有的还含有钛铁矿、褐帘石或萤石。

岩石中 SiO_2 含量为 63.48%~78.95%；Na_2O+K_2O 及 K_2O 含量分别为 4.32%~7.86%、3.385%~5.52%；五指山地区的花岗斑岩极其贫钠，Na_2O 含量仅为 0.06%~0.07%，与海南九曲江地区的三叠纪花岗斑岩（宜昌地质矿产研究所，1995）相似；岩石的 σ 值为 0.52~2.54，属高钾钙碱性系列；A/CNK 为

0.97~2.83，绝大多数属弱过铝质—强过铝质。岩石的稀土元素总量为 57.6×10^{-6}~372.7×10^{-6}；稀土配分模式为较陡的右倾曲线；轻重稀土分异较明显，LREE/HREE 值为 10.1~16.6；δEu 为 0.84~0.34，具弱—较明显的负铕异常。除五指山地区的花岗斑岩的 Sr 含量很低外，其他斑岩的 Sr 含量中等或较高。

三、早白垩世高峰序列

该序列岩石分布于同安岭及牛腊岭火山盆地之间的天涯海角-志仲一带，总面积约 296.43km²。由税町中细粒角闪黑云花岗闪长岩、抱跃细中粒斑状角闪黑云二长花岗岩和阜石斗细中粒黑云母正长花岗岩等 3 个岩石单位构成环套状杂岩体。该序列岩石侵入的最晚地层为下白垩统岭壳村组。马大铨等（1991）测得税町花岗闪长岩、抱跃二长花岗岩全岩-内部矿物 Rb-Sr 等时线法年龄为（112±7）Ma，曾庆銮等（1992）获得抱跃二长花岗岩全岩 Rb-Sr 等时线法年龄为（107±1）Ma，两者在误差范围内一致，形成时代厘定为早白垩世。

（一）税町中细粒角闪黑云母花岗闪长岩（$\gamma\delta K_1$）

主要见有高峰、税町、布加那、海鸥、红毛岭、永阳等侵入体，出露面积约 96.87km²，岩性为中细粒角闪黑云母花岗闪长岩，局部为石英二长闪长岩。块状构造，中细粒结构，粒径多为 0.5~4mm，局部含少量（<2%）钾长石斑晶（大小 5~7mm）。岩石中，石英 21%~25%；斜长石（An38~An45）40%~56%，半自形板柱状，环带发育；微斜条纹长石 15%~26%，半自形—他形板状；黑云母 4%~12%，半自形—他形鳞片状-丛状；角闪石 3%~5%；普通辉石 0~0.5%。副矿物主要有榍石、磁铁矿、磷灰石、褐帘石、锆石。见大量微粒闪长质包体，浑圆状，大小 1cm×2cm~40cm×65cm，有时含有长石、石英捕掳晶。

（二）抱跃细中粒斑状角闪黑云母二长花岗岩（$\eta\gamma K_1$）

主要见有文门、三母洞、头隆、三合等侵入体，出露总面积约 124.96km²，岩性为细中粒斑状（或少斑）角闪黑云母二长花岗岩。块状构造，细中粒似斑状花岗结构。斑晶钾长石含量为 5%~15%，一般为 8%~12%左右，大小 6mm×8mm~10mm×25mm。基质粒径 1~5mm，矿物组成：石英 22%~29%、斜长石（An18~An45）20%~40%、微斜条纹长石 24%~43%、黑云母 3%~6%、角闪石 2%~6%。斜长石半自形板柱状，常见环带构造；微斜条纹长石半自形—他形板状；黑云母半自形—他形鳞片状-丛状。副矿物主要有榍石、磁铁矿、磷灰石、褐帘石、锆石。见大量微粒闪长质包体，浑圆状，大小 1cm×2cm~30cm×50cm，其中可含有长石、石英捕虏晶。

（三）阜石斗细中粒黑云母正长花岗岩（$\xi\gamma K_1$）

主要见有阜石斗、布加育、大母岭、岭壳（神人石）、角岭等侵入体，出露面积约 74.6km²，岩性为细中粒黑云母正长花岗岩。块状构造，细或细中粒结构，粒径多为 0.5~4.5mm。矿物组成：石英 23%~35%、斜长石（An15~An29）16%~30%、微斜条纹长石 40%~50%、黑云母 2%~7%。斜长石半自形板柱状，微斜条纹长石半自形—他形板状，黑云母半自形—他形鳞片状-丛状。副矿物种类与抱跃二长花岗岩相似，但含量有所减少。

（四）岩石地球化学特征

该序列岩石 SiO_2 含量变化较大，在 59.62%~77.78%之间；Na_2O+K_2O 含量为 6.07%~8.84%；K_2O 含量多数为 2.76%~5.02%，个别高达 6.55%；σ 值为 1.57~3.81；根据硅-碱图解（图 9-55）和 SiO_2-K_2O 图解（图 9-56），属高钾钙碱性岩系；A/CNK 为 0.78~1.16，属准铝质—强过铝质岩，总体上从早至晚，A/CNK 值略增大（平均值分别为 0.95、1.00、1.06）。岩石的稀土元素总量大部分在 97.8×10^{-6}~260.9×10^{-6}之间变化；轻重稀土分异中等—较明显，LREE/HREE 值 7.2~22.0；δEu 为 1.27~0.24，多数具弱—较强的负铕异常，个别具正铕异常；稀土元素平均值配分模式为具弱负铕异常的右倾型曲线（图 9-57）。微量元素蛛网图上表现出从早到晚 Ba、Sr、P、Ti 等亏损越来越强（图9-59），表明岩浆结晶分异的存在。税町花岗闪长岩初始锶比值较低，为0.70510（马大铨等，1991）；抱跃二长花岗岩（文门侵入体）初始锶比值为0.7077（曾庆銮等，1992）；阜石斗正长花岗岩（角岭侵入体）初始锶比值为0.70859，$\varepsilon_{Nd}(t)$ 较高，为-4.41。T_{2DM} 值相对较小，为1265Ma（宜昌地质矿产研究所，海南省地质调查院，2004），结合包体特征，反映该序列岩石为壳幔岩浆混合来源。

四、早白垩世番通石英正长岩（$\xi o K_1$）

1. 地质特征、形成时代及岩石学特征

分布于六罗村—什龙一带，总面积约40.07km²。番通石英正长岩侵入三叠纪超盆二长花岗岩及下白垩统六罗村组、岭壳村组。已有的SHRIMP法锆石U-Pb年龄为（106±3）Ma（宜昌地质矿产研究所，海南省地质调查院，2004），形成时代厘定为早白垩世。

主体岩性为中粗粒—细粒多斑状（角闪）黑云母石英正长岩，局部为石英二长岩。块状构造，中粗粒或细粒多斑状二长结构，前者粒径主要为1~6.5mm。后者斑晶含量可达70%~75%，粒径1~8.5mm，斑晶主要为钾长石，少量斜长石和石英；基质粒径0.1~0.5mm。岩石中，石英5%~12%，填隙状；斜长石（An20~An30）18%~36%，半自形板柱状，常见环带和净边结构；微斜条纹长石（正长石）51%~67%，半自形—他形板状，钠长石条纹发育；黑云母3%~6%，半自形—他形叠片状-丛状；普通角闪石0~4%。副矿物主要有磁铁矿（钛铁矿）、磷灰石、锆石，褐帘石、榍石。可见少量微粒闪长质包体，浑圆状，大小1cm×2cm~25cm×35cm。

2. 岩石地球化学特征

番通石英正长岩的 SiO_2 含量为65.62%~67.80%；Na_2O+K_2O 含量、K_2O 含量及 σ 值较高，分别为9.94%~11.26%、5.66%~6.05%和4.08~5.56；A/CNK为0.98~1.02，属准铝质—弱过铝质；硅-碱图上落入碱性岩系列（图9-55）。岩石的稀土元素总量在 167.9×10^{-6}~384.2×10^{-6} 之间；其稀土配分模式为向右的曲线（图9-57）；LREE/HREE值15.2~35.9；轻稀土分异明显，$(La/Sm)_N$ 为5.7~11.7；重稀土分异不明显，$(Gd/Yb)_N$ 为1.6~1.9；δEu 为1.17~0.63，具弱的正铕异常-中等的负铕异常。微量元素原始地幔蛛网图中显示出Ba、Nb-Ta、Sr-P、Ti负异常，而K、Sr、Pb、Nd-Zr-Hf为正异常（图9-60）。初始锶比值为0.70787~0.70822，$\varepsilon_{Nd}(t)$ 较高，为-4.86~-4.97；T_{2DM} 值相对较小，为1301~1310Ma，与海南岛白垩纪侵入岩同位素特征相似。

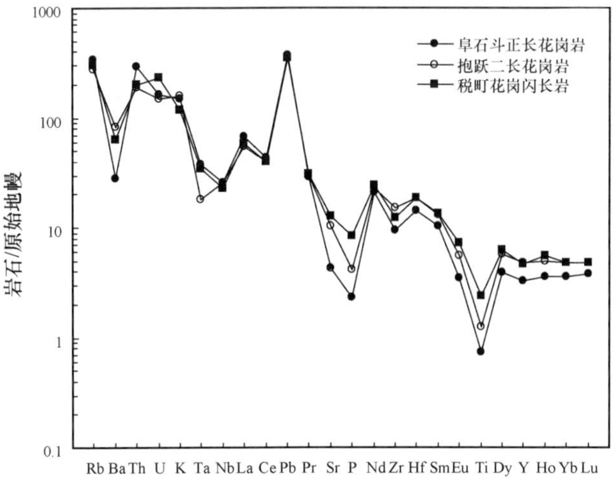

图9-59 高峰序列微量元素原始地幔标准化蛛网图

Fig. 9-59 Primitive mantle-normalized spider diagram of trace elements of Gaofeng Sequence

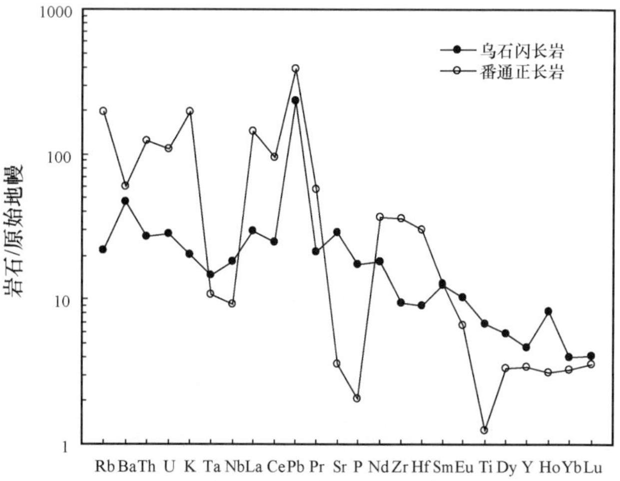

图9-60 番通石英正长岩、乌石闪长岩微量元素原始地幔标准化蛛网图

Fig. 9-60 Primitive mantle-normalized spider diagram of trace elements of Fantong quartz syenite and Wushi diorite

五、早白垩世乌石闪长岩（δK_1）

1. 地质特征及形成时代

由乌石、黎万、祖烈等一些小侵入体组成，总面积约 69.85km^2。空间分布上与屯昌杂岩体、保城杂岩体以及千家杂岩体紧密相伴，矿物组合与上述岩体中的闪长质包体近似，但往往更偏基性，因此乌石闪长岩相当于基性岩浆端元相（也有弱的混合），而稍晚的花岗闪长岩及二长花岗岩等岩石则是基性岩浆和酸性岩浆混合并进一步分异演化的产物。侵入时代厘定为早白垩世。

2. 岩石学特征

不同侵入体（或同一侵入体不同部位）的成分变化明显，从英云闪长岩、石英闪长岩、闪长岩至辉长岩（辉绿玢岩）均有出现，但在较大的侵入体中一般为细粒闪长岩。

闪长岩具半自形粒状结构，块状构造，矿物粒径以 0.2~1.5mm 为主，局部可达 2mm。矿物组成：斜长石一般为57%~74%，局部不足5%；角闪石一般在 15 %左右，局部可高达95%；黑云母 0~10%；辉石在闪长岩中见少量，在辉长辉绿（玢）岩可高达 34%；见少量石英（0~1%）和钾长石。在与早期岩体接触带附近，内接触带常见“混染”现象（常见含有来自围岩的钾长石、斜长石、石英等捕虏晶，含量可高达 20%左右）。斜长石呈灰白色，半自形柱状、板状、他形粒状，钠长石双晶发育，普遍发育不太明显的环带构造，An=35~61，长石牌号变化幅度较大，中性斜长石至基性斜长石均有出现；钾长石大多以不规则填隙状或粒状产出，表明其晶出较晚；石英以他形粒状或不规则填隙状产出，不同部位含量变化较大，常见波状消光；角闪石自形长柱状或六边形，棕色—浅棕色、绿色多色性，柱状解理和角闪石式解理相当发育，可见被绿泥石置换或为纤维状阳起石交代而保留角闪石假象；黑云母叶片状、鳞片状集合体，多呈褐色—浅黄色多色性，常与角闪石共生；辉石自形、半自形短柱状、他形粒状，无色或浅褐色—浅褐绿色多色性。副矿物主要有锆石、磁铁矿、磷灰石、榍石等。其中磷灰石长宽比可达 31∶1 以上，表明岩浆经历了快速冷凝结晶过程。

3. 岩石地球化学特征

主体岩石（细粒闪长岩）的 SiO_2 含量为 48.51%，属基性岩类；K_2O+ Na_2O 及 K_2O 含量分别为 3.56%、0.78%；组合指数（σ 值）为 2.07，属钙碱性系列；A/CNK 为 0.70，属准铝质。稀土元素总量为 134.72×10^{-6}；稀土配分模式为向右倾斜的平滑曲线（图 9-57），Ho 元素凸起明显，暗示源区可能存在基性斜长石和角闪石的堆晶作用；LREE/HREE 值为 5.52，轻重稀土分异中等；δEu 0.93，仅具微弱的负铕异常。岩石的 V、Cr、Ni、Co、Sc、Zn 等亲铁铜元素含量相对较高，而 Rb、Ba、Th、U、K 等大离子亲石元素含量较低，原始地幔微量元素标准化显示出弱的 Th、Ta、Zr-Hf 负异常和 Ba、Pb、Sr、Ho 正异常（图 9-60），而 Sr、Ho 正异常可能与斜长石、角闪石的堆晶作用有关。

六、早白垩世保城序列

保城序列岩石主要由屯昌杂岩体、千家杂岩体和保城杂岩体组成，从早至晚包括加茂细中粒角闪黑云花岗闪长岩、保亭细中粒含斑角闪黑云二长花岗岩、六弓粗中粒斑状黑云母二长花岗岩、双顶岭细中粒斑状黑云母正长花岗岩和高通岭细粒黑云母正长花岗岩等 5 个岩石单位，总面积约 1947.24km^2。同位素年龄值介于 95.6~107.6Ma 之间，形成于早白垩世。

（一）加茂细中粒角闪黑云花岗闪长岩（$\gamma\delta K_1^a$）

1. 地质特征、形成时代及岩石学特征

主要由屯昌杂岩体、千家杂岩体和保城杂岩体中的花岗闪长岩以及九架岭花岗闪长岩组成，空间分布上往往处在上述 3 个杂岩体的外缘，出露总面积约 879.02km^2。已有的同位素年龄为：屯昌杂岩体花岗闪长岩锆石 U-Pb 年龄（100.9±1.3）Ma、（97.1±1.0）Ma（海南省地质调查院，2000）、保城杂岩体花岗闪长岩全岩-黑云母 Rb-Sr 法等时线年龄（96.9±0.2）Ma、（94.4±0.2）Ma（马大铨等，

1991）。本次在保城杂岩体花岗闪长岩（加茂侵入体）中采集新鲜样品采用高精度的LA-ICPMS法测得锆石U-Pb年龄为（102.0±2.5）Ma（图9-61，图9-62，分析数据见表9-8），侵入时代厘定为早白垩世晚期。

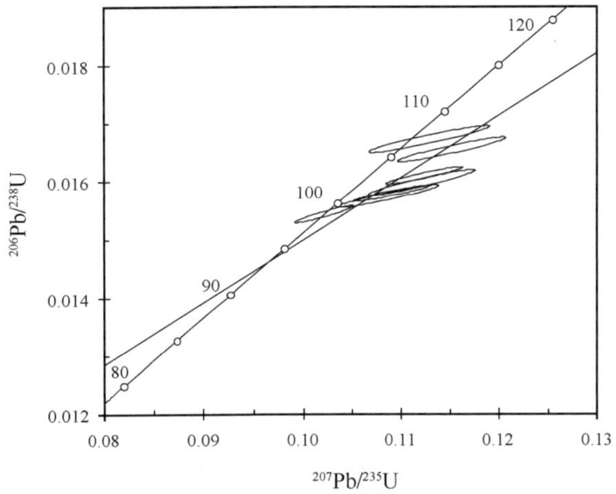

图9-61　加茂花岗闪长岩$^{206}Pb/^{238}U$年龄谐和图（D5009-1）

Fig. 9-61　$^{206}Pb/^{238}U$ age concordia plot of Jiamao granodiorite（D5009-1）

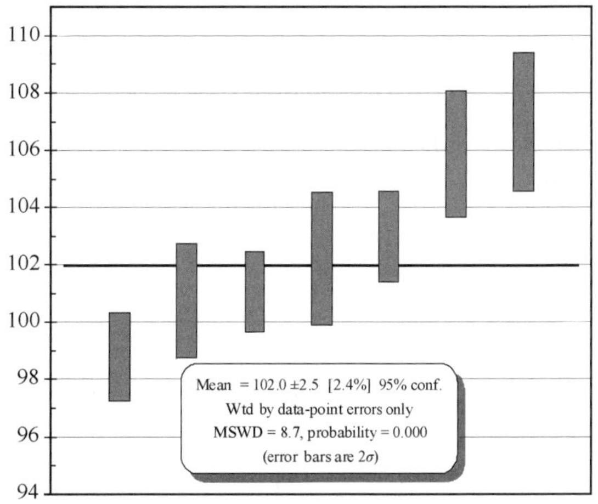

图9-62　加茂花岗闪长岩锆石U-Pb加权图谱（D5009-1）

Fig. 9-62　zircon U-Pb weighted graph of Jiamao granodiorite（D5009-1）

表9-8　加茂花岗闪长岩LA-ICP-MA锆石U-Pb测年结果

Table 9-8　LA-ICP-MA Zircon U-Pb dating results of Jiamao granodiorite

岩石单位	测试点	$^{207}Pb/^{206}Pb$	$^{207}Pb/^{235}U$	$^{206}Pb/^{238}U$	$^{208}Pb/^{232}Th$	$^{238}U/^{232}Th$	$^{207}Pb/^{206}Pb$	$^{207}Pb/^{235}U$	$^{206}Pb/^{238}U$	$^{208}Pb/^{232}Th$
加茂	D5009-1-01	0.0507	0.1124	0.0161	0.0051	1.308387	233	108	103	102
加茂	D5009-1-02	0.0773	0.1704	0.0161	0.0062	0.946074	1128	160	103	125
加茂	D5009-1-03	0.0586	0.1300	0.0163	0.0056	0.761673	554	124	104	113
加茂	D5009-1-04	0.2940	0.9949	0.0248	0.0364	0.518618	3439	701	158	722
加茂	D5009-1-05	0.0516	0.1121	0.0160	0.0053	0.774833	265	108	102	108
加茂	D5009-1-06	0.0592	0.1368	0.0169	0.0056	0.834934	572	130	108	113
加茂	D5009-1-07	0.0548	0.1214	0.0162	0.0049	0.794849	467	116	104	98.8
加茂	D5009-1-08	0.1285	0.2900	0.0160	0.0074	0.744768	2080	259	102	150
加茂	D5009-1-09	0.0569	0.1215	0.0154	0.0049	0.855652	487	116	98.7	99.0
加茂	D5009-1-10	0.0906	0.2039	0.0165	0.0070	0.653467	1439	188	105	140
加茂	D5009-1-11	0.0596	0.1325	0.0160	0.0057	0.769468	591	126	103	115
加茂	D5009-1-12	0.0488	0.1129	0.0167	0.0050	0.763534	139	109	107	101.8
加茂	D5009-1-13	0.0584	0.1366	0.0173	0.0049	0.675493	546	130	111	99
加茂	D5009-1-14	0.0507	0.1152	0.0166	0.0054	0.838364	228	111	106	110
加茂	D5009-1-15	0.0481	0.1021	0.0154	0.0047	0.562398	106	98.7	98.8	94.7
加茂	D5009-1-16	0.0501	0.1087	0.0157	0.0048	0.204712	198	105	101	96.9
加茂	D5009-1-17	0.0501	0.1090	0.0158	0.0049	0.611583	211	105	101	98.4
加茂	D5009-1-18	0.0716	0.1644	0.0167	0.0059	0.833125	974	155	107	119

主体岩性为细中粒角闪黑云母花岗闪长岩，局部为石英二长闪长岩、二长花岗岩。细中粒结构，块状构造，矿物粒径多为0.5~4.5mm，局部含少量钾长石斑晶。岩石中，石英20%~28%；斜长石（An22~An36）35%~48%，半自形板柱状，环带发育；微斜条纹长石13%~22%，半自形板状—他形；黑云母4%~8%，半自形鳞片状-他形丛状；角闪石3%~6%；普通辉石0~0.5%。副矿物主要有榍石、磁铁矿、磷灰石、锆石。

加茂花岗闪长岩的一个显著特征是普遍含有微粒闪长质包体（图版Ⅱ-5-A）。这些包体形态多样，大小悬殊，一般为 3cm×8cm~5cm×10cm，最大者可达 25cm×45cm 以上，小的仅 2cm×3cm。大多数个体较小的包体与寄主岩石呈迷雾状渐变过渡关系，包体分布不均匀，总体上岩体中心含量多些，边缘较少，平均约 2~3 个/m^2。包体岩性主要为石英闪长岩或闪长（玢）岩，矿物组成与寄主岩相同，呈深灰黑色，色率在 34~50 之间，外观上比寄主岩石暗许多。闪长质包体具典型的岩浆结晶结构，块状构造，包体结构（矿物粒径 0.8~1.5mm）总是比寄主岩石细。包体矿物成分为斜长石 50%~70%、角闪石 20%~30%、黑云母 2%~15%、石英 0~5%和极少量钾长石，局部偶见辉石 0~20%。包体中的副矿物主要为磁铁矿、榍石和磷灰石等，其中磷灰石有短柱状和细针状两种结构，后者长宽比多在 1：10~1：30 之间，反映出包体岩浆经历了在较高氧逸度条件和较大温度梯度下快速冷却淬火的结晶环境（云平等，2003）。

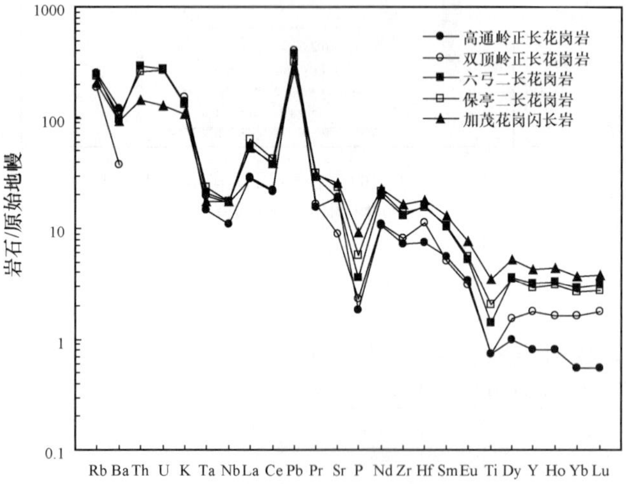

图 9-63　保城序列微量元素原始地幔标准化蛛网图

Fig. 9-63　Primitive mantle-normalized spider diagram of trace elements of Baocheng Sequence

2. 岩石地球化学特征

加茂花岗闪长岩的 SiO_2 含量为 57.40%~62.72%，属中性岩类，与实际矿物含量划分一致；Na_2O+K_2O 含量为 5.73%~7.91%，K_2O 含量为 2.45%~4.93%；σ 值为 1.96~2.43；根据硅-碱图解（图 9-55）和 SiO_2-K_2O 图解（图 9-56），属高钾钙碱性系列；A/CNK 为 0.86~1.10，平均 0.95，属准铝质。岩石的稀土元素总量为 69.24×10^{-6}~243.06×10^{-6}，平均 164.02×10^{-6}；稀土配分模式为向右倾斜的平滑曲线（图 9-57）；轻重稀土分异中等，LREE/HREE 值 7.27~15.43；δEu 为 0.51~0.87，具弱—中等的负铕异常。微量元素蛛网图显示出 Rb、Th-U-K、Pb、Nd-Zr-Hf-Sm 正异常和 Ba、Nb-Ta、P、Ti 负异常（图 9-63）。

（二）保亭细中粒含斑角闪黑云二长花岗岩（$\eta\gamma K_1^b$）

1. 地质特征、形成时代及岩石学特征

分布于屯昌、千家及保亭一带，出露总面积约 644.61km^2。空间分布上位于加茂花岗闪长岩与六弓二长花岗岩之间，呈套环状产出。本次调查发现其后期侵入加茂花岗闪长岩，接触界线内侧见有细粒边（图版Ⅱ-5-B），据此将其侵入时代厘定为早白垩世晚期、加茂花岗闪长岩之后。

主体岩性为细中粒含斑角闪黑云母二长花岗岩。块状构造，细中粒似斑状（含斑）结构。斑晶为钾长石，灰白色—略带浅粉红色，含量 2%~5%，大小 3mm×7mm~5mm×10mm。基质粒径多为0.5~5mm。矿物组成：石英 20%~30%，他形粒状，分布于长石颗粒间不规则空隙中，消光性不明显；斜长石（An20~An36）30%~44%，半自形板柱状，环带构造发育；钾长石 24%~35%，半自形板状—他形粒状，卡斯巴双晶发育；黑云母 3%~7%，半自形鳞片状-他形丛状，具褐-黑多色性；角闪石 1%~6%，柱状，具浅褐—深绿多色性，菱形解理和柱状解理发育，发育简单双晶。副矿物主要有榍石、磁铁矿、磷灰石、锆石等。

岩石中见少量闪长质包体，含量明显要比加茂花岗闪长岩中的包体含量少。包体呈浑圆状，大小 1cm×2cm~20cm×40cm，其中常含有长石、石英捕虏晶。

2. 岩石地球化学特征

岩石的 SiO_2 含量为 66.92%~69.40%，属酸性岩类；Na_2O+K_2O 含量为 7.11%~7.51%，K_2O 含量为 3.59%~4.24%；σ 值为 1.91~2.24；据硅-碱图和 SiO_2-K_2O 图解，属高钾钙碱性系列；A/CNK 为

0.91~98，属准铝质。岩石的稀土元素总量为131.97×10^{-6}~215.73×10^{-6}；轻重稀土分异中等，LREE/HREE值15.04~19.02；δEu为0.57~0.77，具弱负铕异常。微量元素蛛网图显示出Rb、Th-U-K、Pb、Nd-Zr-Hf-Sm正异常和Ba、Nb-Ta、P、Ti负异常（图9-63），地球化学特征与加茂花岗闪长岩相似。

（三）六弓粗中粒斑状黑云母二长花岗岩（$\eta\gamma mK_1^c$）

1. 地质特征、形成时代及岩石学特征

分布于屯昌、千家及保亭一带，出露总面积约382.36km^2。前人所测全岩-黑云母Rb-Sr法年龄为（96±0.4）Ma及（93.3±0.3）Ma（马大铨等，1991），本次在六弓侵入体中采集新鲜样品采用高精度的La-ICPMS法测得锆石U-Pb年龄为（107.6±1.9）Ma（图9-64，图9-65，分析数据见表9-9），将其侵入时代厘定为早白垩世晚期。

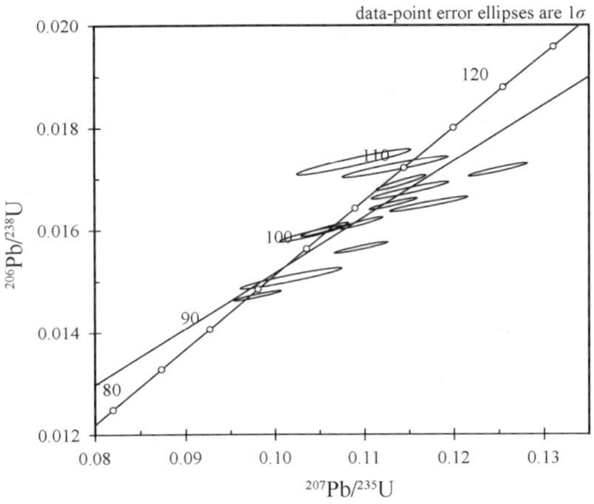

图9-64　六弓二长花岗岩锆石U-Pb谐和图（PM2-101-1）

Fig. 9-64　Zircon U-Pb age concordia plot of Liugong monzonitic granite（PM2-101-1）

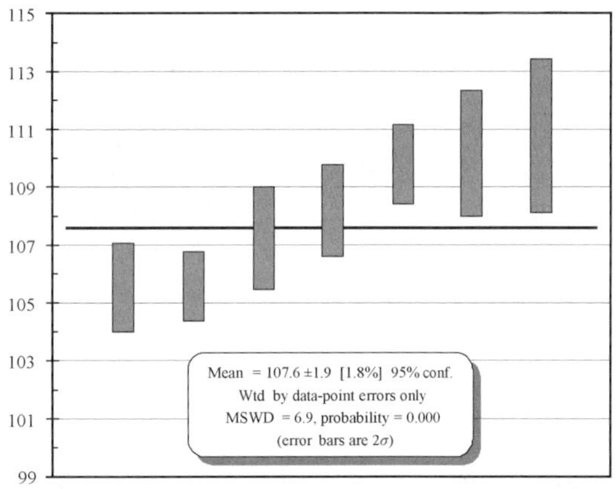

图9-65　六弓二长花岗岩$^{206}Pb/^{238}U$年龄加权图谱（PM2-101-1）

Fig. 9-65　$^{206}Pb/^{238}U$-Pb weighted graph of Liugong monzonitic granite（PM2-101-1）

表9-9　六弓二长花岗岩LA-ICP-MA锆石U-Pb测年结果

Table 9-9　LA-ICP-MA zircon U-Pb dating results of Liugong monzonitic granite

岩石单位	测试点	$^{207}Pb/^{206}Pb$	$^{207}Pb/^{235}U$	$^{206}Pb/^{238}U$	$^{208}Pb/^{232}Th$	$^{238}U/^{232}Th$	$^{207}Pb/^{206}Pb$	$^{207}Pb/^{235}U$	$^{206}Pb/^{238}U$	$^{208}Pb/^{232}Th$
六弓	PM2-101-1-1	0.0482	0.0980	0.0147	0.0044	0.730887	109	94.9	94.2	89.2
六弓	PM2-101-1-2	0.0478	0.1056	0.0160	0.0050	0.596338	87.1	102	102	100.4
六弓	PM2-101-1-3	0.0507	0.1096	0.0157	0.0047	0.736106	228	106	100.2	93.9
六弓	PM2-101-1-4	0.0459	0.1088	0.0173	0.0049	0.86326	error	105	111	98.7
六弓	PM2-101-1-5	0.0481	0.1134	0.0172	0.0058	0.596374	106	109	110	116
六弓	PM2-101-1-6	0.0493	0.1097	0.0161	0.0051	0.442478	161	106	103	102
六弓	PM2-101-1-7	0.2133	0.6450	0.0210	0.0151	0.802697	2931	505.4	134.0	302.0
六弓	PM2-101-1-8	0.0527	0.1249	0.0172	0.0053	0.676819	322	119	110	108
六弓	PM2-101-1-9	0.0494	0.1017	0.0151	0.0048	0.872448	165	98.4	96.4	96.1
六弓	PM2-101-1-10	0.0594	0.7837	0.0950	0.0285	0.536826	581	588	585	567
六弓	PM2-101-1-11	0.0489	0.1141	0.0169	0.0053	0.860882	139	110	108	107
六弓	PM2-101-1-12	0.0514	0.1172	0.0165	0.0055	0.600096	261	112.5	105.5	110.4
六弓	PM2-101-1-13	0.0573	0.1259	0.0160	0.0055	0.689417	502	120	102	111
六弓	PM2-101-1-14	0.0557	0.1238	0.0162	0.0053	0.553342	439	119	103	107
六弓	PM2-101-1-15	0.0472	0.1041	0.0159	0.0054	0.526954	58	101	102	108

续表

岩石单位	测试点	$\frac{^{207}Pb}{^{206}Pb}$	$\frac{^{207}Pb}{^{235}U}$	$\frac{^{206}Pb}{^{238}U}$	$\frac{^{208}Pb}{^{232}Th}$	$\frac{^{238}U}{^{232}Th}$	$\frac{^{207}Pb}{^{206}Pb}$	$\frac{^{207}Pb}{^{235}U}$	$\frac{^{206}Pb}{^{238}U}$	$\frac{^{208}Pb}{^{232}Th}$
六弓	PM2-101-1-16	0.0494	0.1393	0.0205	0.0058	0.64658	169	132	131	116
六弓	PM2-101-1-17	0.0500	0.1151	0.0168	0.0050	0.721449	195	111	107	101
六弓	PM2-101-1-18	0.0497	0.1132	0.0165	0.0050	0.744713	189	109	106	101

主体岩性为粗中粒斑状黑云母二长花岗岩（图版Ⅱ-5-C），局部为正长花岗岩。块状构造，粗中粒—细中粒似斑状结构。斑晶为钾长石，肉红色，卡斯巴双晶发育，含量10%~35%，侵入体边缘斑晶逐渐减少（可在5%以下），大小0.4cm×1cm~3.5cm×4cm。基质粒径一般为1~5mm。矿物组成：石英21%~30%，他形粒状，分布在长石颗粒间不规则空隙中，波状消光；斜长石（An20~An38）22%~40%，灰白色，半自形板柱状，聚片双晶及环带构造发育；钾长石27%~44%，肉红色，半自形板状—他形柱粒状，卡斯巴双晶发育；黑云母3%~7%，半自形鳞片状-丛状，巨褐-黑多色性，有的已退变为绿泥石并析出磁铁矿；角闪石0~1%，柱状，具褐-绿多色性，菱形解理发育。

副矿物主要有磁铁矿、榍石、磷灰石、锆石，磷灰石长宽比可达16∶1以上。见少量微粒闪长质包体，浑圆状，大小1cm×2cm~4cm×5cm。

2. 岩石地球化学特征

六弓二长花岗岩的SiO_2含量为68.99%~72.82%，属酸性岩类；Na_2O+K_2O含量为7.25%~8.57%，K_2O含量为3.37%~4.93%，钾钠含量相当；σ值为1.82~2.46，属高钾钙碱性系列；A/CNK为0.98~1.08，属准铝质—弱过铝质。岩石的稀土元素总量为118.74×10^{-6}~341.96×10^{-6}，平均值约160×10^{-6}；轻重稀土分异中等，LREE/HREE值9.71~19.93；δEu为0.51~0.87，具弱负铕异常。微量元素蛛网图具有Rb、Th-U-K、Pb、Nd-Zr-Hf-Sm正异常和Ba、Nb-Ta、P、Ti负异常（图9-63）。岩石的初始锶比值为0.70463，$\varepsilon_{Nd}(t)$为-5.55，T_{2DM}值较小，为1349Ma，结合其包体特征，反映该岩石为壳幔岩浆混合来源。

（四）双顶岭细中粒斑状黑云母正长花岗岩（$\xi\gamma K_1^d$）

1. 地质特征、形成时代及岩石学特征

主要由屯昌杂岩体中的双顶岭细中粒斑状黑云母二长花岗岩组成，分布于双顶岭、挟牛岭、南吕农场一带，出露总面积约32.66km²。屯昌一带见双顶岭正长花岗岩被高通岭细粒黑云母正长花岗岩（95.6±1.6）Ma侵入两者之间接触界线清楚，两侧未见蚀变或冷凝边等现象出现，高通岭正长花岗岩明显截切了双顶岭中粒斑状黑云母正长花岗岩中的钾长石斑晶，脉动侵入接触关系特征比较明显，根据侵入接触关系将其形成时代厘定为早白垩世晚期。

主体岩性为细中粒斑状黑云母正长花岗岩，中粒似斑状花岗结构，块状构造。矿物成分为：钾长石45%~65%、斜长石10%~18%、石英20%~25%、黑云母2%~8%、角闪石0~2%。斑晶成分为钾长石，含量约5%~12%，呈肉红色，自形厚板状，粒径5mm×10mm~8mm×15mm，卡氏双晶发育，内部包含细小的石英、黑云母等。基质矿物粒径在2~5mm之间。其中，钾长石呈肉红色，半自形板状、他形柱状、粒状，发育卡氏双晶；斜长石灰白色，An=16，半自形柱状、板状、他形粒状，普遍发育不太明显的环带构造；石英他形粒状，波状消光常见；黑云母细鳞片状，呈黑绿色—浅黄色，多色性明显；角闪石柱状，呈绿色—浅绿色多色性，可见简单双晶，分布不均匀，含量较少，与黑云母伴生。副矿物主要为磁铁矿、榍石、锆石和磷灰石等。

2. 岩石地球化学特征

双顶岭正长花岗岩的SiO_2含量为76.29%，属酸性岩类；σ值为1.83，属高钾钙碱性系列；A/CNK为0.97，属准铝质。岩石的稀土元素总量为92.5×10^{-6}；轻重稀土分异中等；δEu为0.79，具弱负铕异常。微量元素蛛网图具有Rb、Th-U-K、Pb、Nd-Zr-Hf-Sm正异常和Ba、Nb-Ta、P、Ti负异常（图9-63）。

（五）高通岭细粒黑云母正长花岗岩（$\xi\gamma fK_1^e$）

1. 地质特征、形成时代及岩石学特征

由分布于南吕农场一带的高通岭、檬正岭等侵入体组成，常呈小岩瘤状产出于杂岩体中，出露总面积约 8.59km²。已有的锆石 U-Pb 年龄为（95.6±1.6）Ma（海南省地质调查院，2000），形成时代厘定为早白垩世末期。

主体岩性为细粒黑云母正长花岗岩，局部见钾长石斑晶呈零星出现，含量<1%。岩石具细粒花岗结构，块状结构，色率<3。矿物粒径 0.2~2mm。主要造岩矿物为：钾长石 45%~65%，肉红色，半自形板状、他形柱状、粒状，发育卡氏双晶；斜长石 10%~30%，灰白色，An=21，半自形柱状、板状、他形粒状，普遍发育不太明显的环带构造；石英 20%~25%，他形粒状，波状消光常见；暗色矿物主要为黑云母，含量 1%~3%，细小叶片状，多色性明显，分布不均匀，含量较少，且多被绿泥石交代。副矿物主要为磁铁矿、锆石和褐帘石等。

2. 岩石地球化学特征

高通岭正长花岗岩的 SiO_2 含量为 74.09%，属酸性岩类；σ 值为 1.95，属高钾钙碱性系列；A/CNK 为 1.02，属弱过铝质，与上述同系列中的其他早期岩石相比较，有从偏铝质向弱过铝质演化的趋势。岩石的稀土元素总量为 87×10^{-6}；轻重稀土分异中等；δEu 为 0.83，具弱负铕异常。微量元素蛛网图具有 Rb、Th-U-K、Pb、Nd-Zr-Hf-Sm 正异常和 Ba、Nb-Ta、P、Ti 负异常（图 9-63），但负异常要比其他早期岩石更明显一些。

七、晚白垩世南牙村花岗闪长斑岩（$\gamma\delta\pi K_2$）

1. 地质特征、形成时代及岩石学特征

分布于昌江县南牙村-马夹石一带的北东—北北东向区域花岗闪长斑岩类岩墙群，南牙村一带的岩墙走向为 45°，陡倾，马夹石一带的岩墙走向为 15°~20°，向北西西陡倾。南牙村一带的岩墙群中，单脉宽 2~65m，最宽处 320m，单脉长 800~1500m，脉带中有尖灭再现现象，每一脉组成为 4~5km 的岩墙。南牙村一带的岩墙侵入邦溪侵入体，马夹石一带的岩墙侵入火岭侵入体。侵入时代暂厘定为晚白垩世。该岩墙岩石类型比较复杂，主要有花岗闪长斑岩、石英二长斑岩，偶尔见石英闪长玢岩。

花岗闪长斑岩手标本为灰色或灰绿色，斑状结构，块状构造。斑晶为斜长石和少量黑云母，偶尔见钾长石和石英，含量约 15%~20%。

石英二长斑岩手标本为灰色或深灰色，斑状结构，块状构造。斑晶为斜长石、钾长石及黑云母，偶见石英斑晶，含量约 25%~30%。基质为全晶质微粒结构，矿物粒度<0.2mm，由钾长石、斜长石、石英及黑云母组成，石英含量小于 15%，钾长石含量大于斜长石。

2. 岩石地球化学特征

岩石以富硅、碱贫镁、铁、钙为特征，其岩石化学成分与区内一些二长花岗岩类似。A/CNK 为 0.93~1.04，属准铝质—弱过铝质。稀土元素总量为 365×10^{-6}~397×10^{-6}，高于维氏世界酸性岩平均值；稀土配分模式为右倾型曲线（图 9-66）；轻重稀土分异较明显，且轻稀土分馏作用强于重稀土；δEu 为 0.29~0.36，铕负异常比较明显。微量元素蛛网图具有较明显的 Ta-Nb、

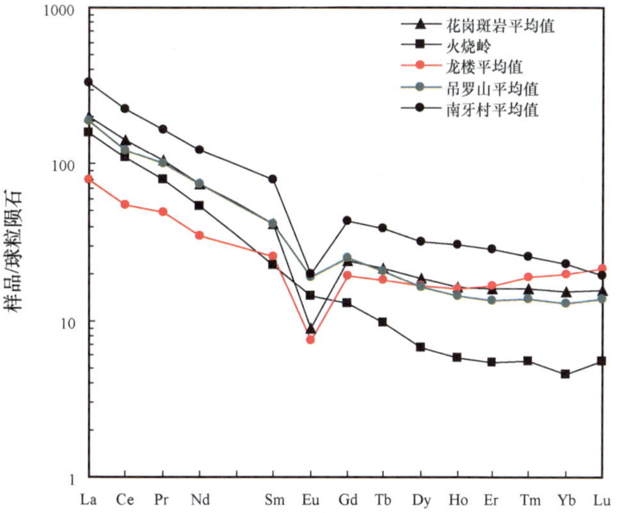

图 9-66 晚白垩世侵入岩稀土元素球粒陨石标准化分布型式图

Fig. 9-66 Chondrite-normalized REE patterns of late Cretaceous intrusive rocks

Hf-Zr 负异常。

八、晚白垩世吊罗山序列

该序列岩石由吊罗山岩体和昌化大龄岩体组成，根据同位素年龄值和岩石学、地球化学特征从早至晚划分为新村细中粒斑状黑云母正长花岗岩、吊罗山粗中粒含斑黑云母正长花岗岩和大小岭中细粒黑云母正长花岗岩等3个岩石单位，出露总面积约917.71km²。

（一）新村细中粒斑状黑云母正长花岗岩（$\xi\gamma mK_2^a$）

1. 地质特征、形成时代及岩石学特征

主要分布于吊罗山地区大里乡新村—兴隆一带，出露面积约149.97km²。空间分布上与吊罗山粗中粒含斑黑云母正长花岗岩、大小岭中细粒黑云母正长花岗岩构成套环状杂岩体。由于覆盖严重，各侵入体之间的侵入接触关系均未找到。前人在新村正长花岗岩中测得全岩K-Ar稀释法年龄为（90±1）Ma及（93±1）Ma（南京地质矿产研究所，1996），侵入时代厘定为晚白垩世早期。

主体岩性为细中粒斑状黑云母正长花岗岩、二长花岗岩，局部为少斑或含斑。块状构造，细中粒似斑状结构。斑晶为钾长石，灰白色—浅粉红色，含量一般在5%~12%之间，侵入体边缘斑晶逐渐变少（<5%），大小0.4cm×1cm~3.5cm×4cm；有时有少量石英斑晶，大小5.5~10mm。基质粒径一般为1mm~5mm。矿物组成：石英20%~25%；斜长石（An25~An27）29%~35%，半自形板柱状，卡钠联合双晶发育；微斜条纹长石38%~41%，半自形板状—他形；黑云母3%~5%，半自形鳞片状-丛状；角闪石分布不均有，含量在0~3%之间。副矿物主要有磁铁矿、榍石、磷灰石、锆石、褐帘石，局部有独居石。微粒闪长质包体较常见，浑圆状，大小1cm×2cm~20cm×40cm，有时含有长石、石英捕虏晶。

2. 岩石地球化学特征

新村黑云母正长花岗岩的SiO_2含量为68.14%~77.34%；属酸性岩类；Na_2O+K_2O含量为7.32%~9.28%，K_2O含量为3.34%~5.80%；σ值为1.82~2.68，属高钾钙碱性系列；A/CNK为0.94~1.10，属准铝质—弱过铝质。岩石的稀土元素总量变化较大，为75.29×10^{-6}~344.03×10^{-6}，但较低值的两个样品可能与后期构造变质作用有关；轻重稀土分异中等，LREE/HREE值5.3~19.09；δEu为0.12~0.78，具中等—较强负铕异常。微量元素蛛网图上具有Rb、Th-U-K、Pb、Nd-Zr-Hf-Sm正异常和Ba、Nb-Ta、P、Ti负异常，与早白垩世加茂花岗闪长岩、保亭二长花岗岩及六弓二长花岗岩的微量元素蛛网图相似，但负异常表现得更强烈一些。

（二）吊罗山粗中粒含斑黑云母正长花岗岩（$\xi\gamma cK_2^b$）

1. 地质特征、形成时代及岩石学特征

主要分布于吊罗山和昌化大岭一带，出露总面积约726.1km²。吊罗山正长花岗岩侵入中三叠世长安辉长岩（图版Ⅱ-5-D）。已有的同位素年龄为（88±1）Ma（全岩-黑云母Rb-Sr法，马大铨等，1991）。本次在吊罗山乡东南约2.5km处吊罗山侵入体中采集新鲜样品采用高精度的LA-ICPMS法测得锆石U-Pb年龄为（95.8±1.5）Ma（图9-67，图9-68，分析数据见表9-10），将其侵入时代厘定为晚白垩世早期。

表9-10 吊罗山正长花岗岩LA-ICP-MA锆石U-Pb测年结果

Table 9-10 LA-ICP-MA Zircon U-Pb dating results of Diaoluoshan syenogranite

岩石单位	测试点	$^{207}Pb/^{206}Pb$	$^{207}Pb/^{235}U$	$^{206}Pb/^{238}U$	$^{208}Pb/^{232}Th$	$^{238}U/^{232}Th$	$^{207}Pb/^{206}Pb$	$^{207}Pb/^{235}U$	$^{206}Pb/^{238}U$	$^{208}Pb/^{232}Th$
吊罗山	D2950-1-1	0.4976	3.2427	0.0423	0.0525	1.140251	4234	1467	267	1034
吊罗山	D2950-1-2	0.0500	0.1030	0.0150	0.0044	0.960523	195	99.6	96.0	88.0
吊罗山	D2950-1-3	0.0653	0.1241	0.0138	0.0044	0.854044	783	119	88.2	89.4

续表

岩石单位	测试点	$\frac{^{207}Pb}{^{206}Pb}$	$\frac{^{207}Pb}{^{235}U}$	$\frac{^{206}Pb}{^{238}U}$	$\frac{^{208}Pb}{^{232}Th}$	$\frac{^{238}U}{^{232}Th}$	$\frac{^{207}Pb}{^{206}Pb}$	$\frac{^{207}Pb}{^{235}U}$	$\frac{^{206}Pb}{^{238}U}$	$\frac{^{208}Pb}{^{232}Th}$
吊罗山	D2950-1-4	0.0524	0.1096	0.0150	0.0046	1.216249	302	106	96.3	91.8
吊罗山	D2950-1-5	0.0610	0.1212	0.0143	0.0046	1.158614	639	116	91.7	93.2
吊罗山	D2950-1-6	0.0474	0.0917	0.0141	0.0044	0.748671	78	89.1	90.3	89.0
吊罗山	D2950-1-7	0.0503	0.0976	0.0141	0.0044	0.727167	209	94.6	90.2	89.0
吊罗山	D2950-1-8	0.0521	0.1130	0.0158	0.0048	0.840619	300	109	101	97.3
吊罗山	D2950-1-9	0.0504	0.1069	0.0154	0.0045	1.032525	217	103	98.7	90.7
吊罗山	D2950-1-10	0.0450	0.0981	0.0145	0.0038	1.054296	error	95.0	92.6	77.4
吊罗山	D2950-1-11	0.0537	0.1188	0.0162	0.0047	1.979414	361	114	103	93.8
吊罗山	D2950-1-12	0.1357	0.2827	0.0152	0.0070	1.305824	2173	253	98	142
吊罗山	D2950-1-13	0.0478	0.0943	0.0143	0.0041	0.458652	100.1	91.5	91.5	83.7
吊罗山	D2950-1-14	0.0465	0.0948	0.0147	0.0043	0.802311	33.4	92.0	94.3	86.3
吊罗山	D2950-1-15	0.0489	0.1021	0.0152	0.0044	0.950661	139.0	98.7	97.3	88.6
吊罗山	D2950-1-16	0.0514	0.1046	0.0151	0.0044	1.160901	257	101	96.5	89.1
吊罗山	D2950-1-17	0.0483	0.1001	0.0150	0.0046	1.164958	122	96.9	96.0	92.5
吊罗山	D2950-1-18	0.0493	0.1149	0.0170	0.0050	0.901388	161	110	109	102

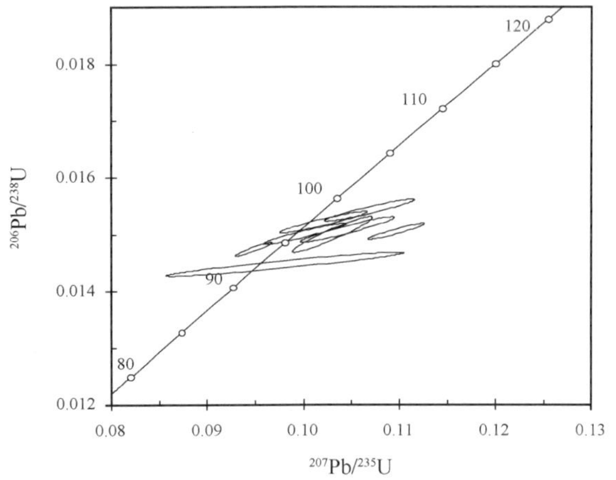

图 9-67　吊罗山正长花岗岩锆石 U-Pb 谐和图（D2950-1）

Fig. 9-67　Zircon U-Pbconcordia plot of Diaoluoshan syenogranite（D2950-1）

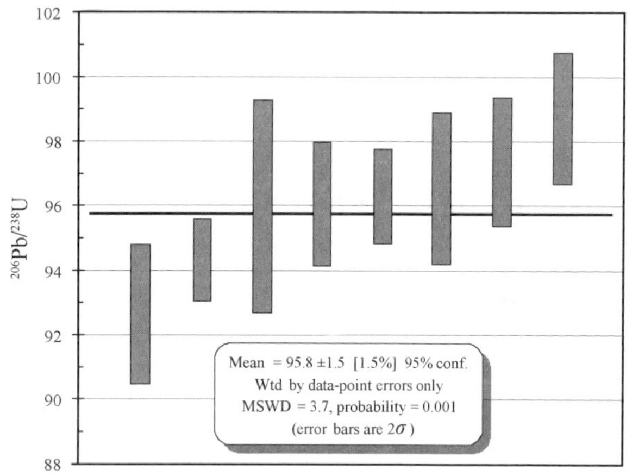

图 9-68　吊罗山正长花岗岩$^{206}Pb/^{238}U$年龄加权图谱（D2950-1）

Fig. 9-28　$^{206}Pb/^{238}U$ age weighted graph of Diaoluoshan syenogranite（D2950-1）

主体岩性为粗中粒含斑黑云母正长花岗岩（图版Ⅱ-5-E），局部为二长花岗岩。块状构造，结构上以粗中粒似斑状为主。斑晶为钾长石，粉红色，卡斯巴双晶发育，粒径 0.5cm×0.7cm～1cm×2.2cm，含量一般<5%，多为 2%～3%，岩体中心部位斑晶含量逐渐增多，可达 10%，岩体自内而外有由粗中粒似斑状结构向中粒结构渐变的趋势。基质矿物粒径在 1.5～8mm 之间，以 2～5mm 为主。矿物组成：石英 20%～36%，他形粒状，以集合体分布在长石颗粒间不规则空隙中，石英颗粒间以弧形边、多边紧密接触，波状消光；钾长石 31%～68%，粉红色，半自形板状—他形粒状，卡斯巴双晶和钠长石条纹发育，内部见包裹细小的斜长石颗粒；斜长石一般在 18%～28%之间，为中长石（An20～An28），半自形板柱状，双晶发育，常见净边结构，粒间常有晚世代的梳状钠长石集合体；黑云母3%～5%，半自形鳞片状-丛状；角闪石少见。副矿物主要有磁铁矿、榍石、磷灰石、锆石、褐帘石。偶见微粒闪长质包体，浑圆状，大小 1cm×2cm～6cm×8cm，常含有长石、石英捕虏晶。

2. 岩石地球化学特征

岩石的 SiO_2 含量为 63.94%～76.90%，平均值 72.1%，属酸性岩类；Na_2O+K_2O 含量为 7.15%～

9.58%，K_2O 含量为 3.69%~6.56%；σ 值为 1.88~4.44，大多数属高钾钙碱性系列；A/CNK 在 0.93~1.45 之间（大部分小于 1.1，个别大于 1.1 可能与蚀变作用有关），平均 1.04，多数属准铝质—弱过铝质。岩石的稀土元素总量变化较大，为 46.90×10^{-6}~410.58×10^{-6}；稀土配分模式为向右倾斜的曲线（图 9-66）；轻重稀土分异中等，LREE/HREE 值 3.2~18.61；δEu 为 0.02~1.03，具中等—较强负铕异常（个别具正铕异常）。微量元素蛛网图具 Rb、Th-U-K、Pb、Nd-Zr-Hf-Sm 正异常和 Ba、Nb-Ta、P、Ti 负异常。岩石的初始锶比值为 0.70724（马大铨等，1991），$\varepsilon_{Nd}(t)$ 为-4.38；T_{2DM}值相对较小，为 1246Ma（宜昌地质矿产研究所、海南省地质调查院，2004），吊罗山正长花岗岩酸性程度高，但具有较低的初始锶比值、较高的 $\varepsilon_{Nd}(t)$ 值及较小的 T_{2DM}值，可能暗示其源岩为较年轻的板底垫托（底侵）形成的下地壳，结合其包体特征，反映了还有少量地幔岩浆的混合。

（三）大小岭中细粒黑云母正长花岗岩（$\xi\gamma K_2^c$）

1. 地质特征、形成时代及岩石学特征

由大小岭、白水坑、黑石姆、常青队、永丰、糖房、垃圾坪等一些小侵入体组成，出露面积约 41.64km²。大小岭正长花岗岩侵入的最新岩体为吊罗山正长花岗岩。前人所测锆石 U-Pb 表面年龄为 82Ma（海南地质综合勘查院，2000），与实际侵入接触关系相符，侵入时代厘定为晚白垩世。

主体岩性为中细粒黑云母正长花岗岩（图版Ⅱ-5-F），局部见二长花岗岩。块状构造，中细粒花岗结构，矿物粒径 0.3~4mm。矿物组成：石英 20%~33%，他形粒状，分布在长石颗粒间不规则空隙中，波状消光；斜长石（An20~An25）11%~25%，半自形板柱状、长柱状，聚片双晶及环带构造发育，可见净边结构和简单环带构造；钾长石 45%~65%，半自形板状—他形，卡斯巴双晶发育，主要是微斜长石，少量属微斜条纹长石（钠长石条纹不太发育）；黑云母 1%~2%，半自形鳞片状-丛状，具黄褐-深棕多色性。副矿物主要有磁铁矿、褐帘石、磷灰石、锆石等。

2. 岩石地球化学特征

大小岭正长花岗岩的 SiO_2 含量为 74.58%~76.81%，属酸性岩类；Na_2O+K_2O 含量为 8.09%~8.58%，K_2O 含量为 4.68%~5.96%，K_2O/Na_2O 大于 1；σ 值为 1.97~2.32；根据硅-碱图解（图 9-55）和 SiO_2-K_2O 图解（图 9-56），属高钾钙碱性系列；A/CNK 为 1.03~1.15，多数属弱过铝质。岩石的稀土元素总量变化较大，为 51.02×10^{-6}~253.46×10^{-6}；轻重稀土分异中等，LREE/HREE 值 6.59~20.47；δEu 为 0.13~0.70，具中等—较强负铕异；$(La/Sm)_N$ 为 3.50~9.58，$(Gd/Yb)_N$ 为 0.77~1.57，轻稀土分馏作用明显强于重稀土。在微量元素蛛网图上具 Rb、Th-U-K、Pb、Nd-Zr-Hf-Sm 正异常和 Ba、Nb-Ta、P、Ti 负异常（图 9-69），和一般花岗岩不同的是，晚期的大小岭正长花岗岩存在较弱的 P 正异常。

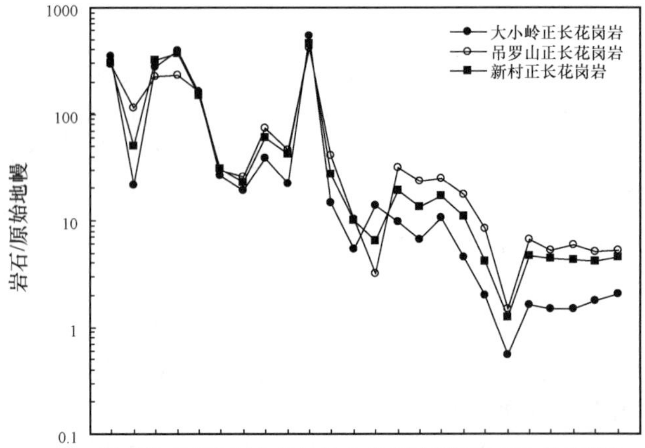

图 9-69 吊罗山序列微量元素原始地幔标准化蛛网图

Fig. 9-69 Primitive mantle-normalized spider diagram of trace elements of Diaoluoshan Sequence

九、龙楼中粒斑状黑云母正长花岗岩（$\xi\gamma mK_2$）

1. 地质特征、形成时代及岩石学特征

分布于铜鼓岭西侧、龙楼镇一带，仅见龙楼一个侵入体，面积约 24.31km²。龙楼侵入体侵入三叠纪铜鼓岭二长花岗岩，接触界线清楚，龙楼侵入体中见大量铜鼓岭二长花岗岩的捕虏体，侵入接触关系清楚。龙楼侵入体 LA-ICP-MS 法锆石 U-Pb 年龄为 65.4Ma（图 9-70，分析结果表 9-11）（海南省地质调查院，2009），侵入时代归属晚白垩世。

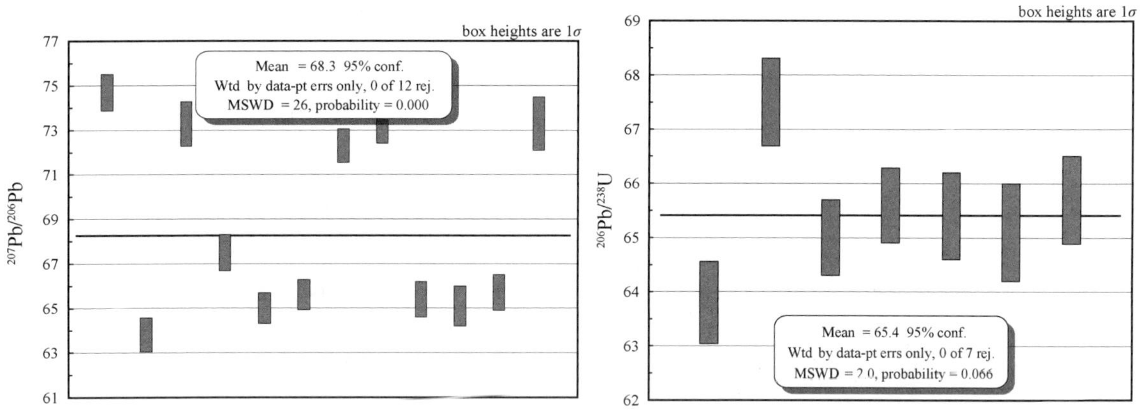

图 9-70　龙楼侵入体 LA-ICP-MS 锆石 U-Pb 加权平均年龄

Fig. 9-70　LA-ICP-MS zircon U-Pb weighted average age of Longlou intrusion

表 9-11　龙楼侵入体 LA-ICP-MS 锆石 U-Pb 年龄数据表

Table 9-11　LA-ICP-MS zircon U-Pb age data of Longlou intrusion

样品号	$^{207}Pb/^{206}Pb$	$^{207}Pb/^{206}Pb$	$^{207}Pb/^{235}U$	$^{207}Pb/^{235}U$	$^{206}Pb/^{238}U$	$^{206}Pb/^{238}U$	$^{208}Pb/^{232}Th$	$^{208}Pb/^{232}Th$
PM03-27-1	152.4	65.89	77.1	1.69	74.7	0.82	77.7	1.25
PM03-27-2	1854.6	39.19	165	2.08	72.3	0.76	203.3	2.25
PM03-27-3	1410.2	43.21	133.6	1.89	73.2	0.77	132.7	1.56
PM03-27-4	483.4	79.85	78.1	2.39	65.4	0.8	73.3	1.48
PM03-27-5	729.4	90.87	86.6	3.3	65.1	0.9	78.4	1.65
PM03-27-6	220.5	89.23	143.5	4.7	138.8	1.75	146	3.28
PM03-27-7	182.9	87.06	68.9	2.25	65.7	0.81	66.2	1.55
PM03-27-8	4091.3	29.37	773.3	5.56	117	1.18	659.3	5.99
PM03-27-9	166.5	141.82	76.2	4.39	73.3	1.21	67.9	1.57
PM03-27-10	169.8	82.84	66.7	2.03	63.8	0.76	64.4	1.37
PM03-27-11	96.9	110.04	74	3.1	73.3	1	66.1	1.33
PM03-27-12	70.8	87.6	67.6	2.14	67.5	0.81	67.9	1.48
PM03-27-13	177	63.18	68.1	1.41	65	0.7	64.2	0.87
PM03-27-14	667.7	51.32	84.9	1.39	65.6	0.69	69.7	0.9
PM03-27-15	261.8	96.77	163.9	5.93	157.2	2.08	81	3.56

主体岩石类型为中粒斑状黑云母正长花岗岩。浅粉红色—灰白色，似斑状结构，块状构造。斑晶成分为钾长石，含量一般在 5%~10%，侵入体边缘斑晶逐渐减少，演变为细中粒少斑或含斑黑云母正长花岗岩。基质矿物成分主要为钾长石 40%~47%、石英 25%~30%、斜长石 20%~25%、黑云母 3%~

6%和角闪石 0~1%。钾长石为半自形柱状，粒径大小以 2.0~5.2mm 为主，0.4~2.0mm 次之，发生弱高岭石化面显浅褐色色调，卡式双晶发育；斜长石为半自形长柱状，粒径大小 2.0mm×1.0mm~4.5mm×2.5mm，一般蚀变较弱，部分从中心发生较强的钠黝帘石化，聚片双晶及环带构造发育，斜消光，用垂直（010）晶带法测得 $N'_p\wedge$（010）≈17°，为更长石-中长石（An33）；石英为他形粒状，一般以集合体分布在长石颗粒间不规则空隙中，少部分颗粒小的石英以穿孔的形式分布在钾长石晶粒中；石英集合体中石英颗粒间以弧形边、多边接触，粒径大小多为 0.2~2.0mm，少量大于 2.0mm，波状消光。黑云母为片状，具褐-黑褐多色性，已向绿泥石退变并析出磁铁矿。

岩石中普遍出现闪长质包体。包体呈黑灰色，透镜状或浑圆状，大小 4~10cm，含量少于 1 个/m^2，色率比寄主岩高，与寄主岩石的中粒结构形成鲜明对比，两者之间的界线清楚。

2. 岩石地球化学特征

龙楼侵入体的 SiO_2 含量为 73.31%~76.73%，属酸性岩；Na_2O+K_2O 含量为 7.81%~8.94%，钾高于钠，K_2O/Na_2O=1.3~1.37；CaO 含量平均为 1.46%，Al_2O_3 含量平均为 13.42%；组合指数（里特曼指数）σ 值为 2.01~2.37；铝饱和指数 A/CNK 为 0.92~1.03；根据硅-碱图解（图 9-55）和 SiO_2-K_2O 图解（图 9-56），属弱过铝质高钾钙碱性系列；DI 值为 82~87.45、SI 值为 6.33~8.15，岩浆分异程度较高。岩石的稀土元素含量为 116.99×10^{-6}~151.02×10^{-6}，低于维氏酸性岩值（292×10^{-6}）；稀土配分曲线为较陡的右倾型曲线（图 9-66）；$(La/Yb)_N$ 值为 1.87~7.32，属于轻稀土富集型；轻稀土段较陡，$(La/Sm)_N$ 值为 1.77~4.32，轻稀土分异明显，重稀土段平缓，$(Gd/Yb)_N$ 值为 0.79~1.22，重稀土弱分异；δEu 0.11~0.67，铕亏损明显。微量元素含量与维氏酸性岩丰度值相比，Rb、Ba、Hf、Pb 较高，而 Sr、Zr、Li、Be、Ta、Sn、Mo、Zn、Co、V 较低；微量元素蛛网图中显示出较明显的 Ba、Ta、Nb、Ti 负异常和 Rb、Th-U-K、Pb、P、Nd-Zr-Hf-Sm 正异常（图 9-71），P 正异常在一般花岗岩中比较少见。

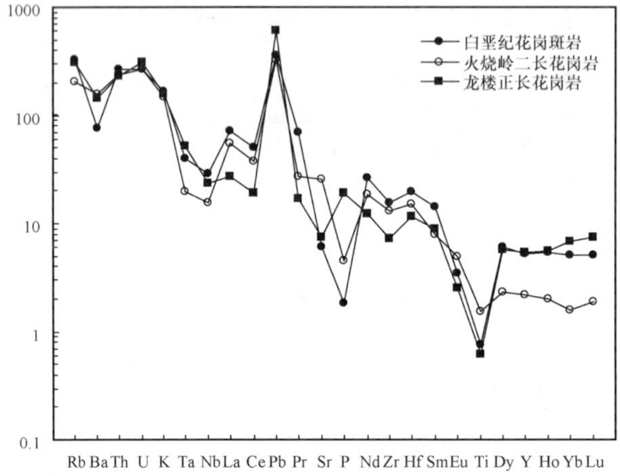

图 9-71 龙楼正长花岗岩、火烧岭二长花岗岩、白垩纪花岗斑岩微量元素原始地幔标准化蛛网图

Fig. 9-71 Chondrite-normalized REE patterns of Longlou syenogranite, Huoshaoling monzonitic granite and Cretaceou granitic porphyry

十、火烧岭细粒黑云母二长花岗岩（$\eta\gamma K_2$）

1. 地质特征、形成时代及岩石学特征

分布在琼中县南方农场南东约 8km 处，出露面积约 3.74km^2。在与下志留统陀烈组的分界线附近，外接触带岩石发生明显硅化。已有的锆石 U-Pb 同位素年龄为（67.0±0.9）Ma（海南省地质调查院，2000），形成于晚白垩世。

主体岩性为细粒黑云母二长花岗岩。岩石呈灰白色，细粒花岗结构，块状构造，矿物粒径 0.3～2mm，与早期岩体接触带附近可变化为微细粒斑状结构。矿物组成：钾长石 30%，呈浅肉红色、灰白色，半自形—他形柱状、板状、粒状，具卡氏双晶，大晶体中常包含有细小的斜长石、黑云母等；斜长石 35%～38%，An=25，为奥长石，灰白色，半自形柱状，个别板状，发育钠长石双晶，次为卡钠复合联晶，部分具简单环带构造，发生绢云母化、钠黝帘石化等；石英 20%～22%，他形粒状，常呈填隙状产出，具弱波状消光；黑云母 10%，呈细叶片状，少量宽大叶片状，片幅 0.05～0.2mm 或 1mm，呈黑褐色色—浅黄色，多色性显著，柱状解理发育完全，局部见被绿泥石置换。此外，人工重砂分析结果显示火烧岭二长花岗岩中有少量辉石存在。副矿物主要为磁铁矿、榍石、独居石和磷灰石等。

2. 岩石地球化学特征

岩石的 SiO_2 含量 71.54%，属酸性岩；Na_2O+K_2O 含量 7.68%，钾高于钠，$K_2O/Na_2O=1.38$，CaO 含量 2.03%；组合指数 σ 值为 2.07；铝饱和指数 A/CNK 为 1.05；根据硅-碱图解（图 9-55）和 SiO_2-K_2O 图解（图 9-56），属弱过铝质高钾钙碱性系列；DI 值 83.9、SI 值 6.9，岩浆分异程度较高。稀土元素总量为 158.5×10^{-6}，低于维氏世界酸性岩平均值；LREE/HREE=20，$(La/Yb)_N$= 32.6，轻、重稀土元素分异程度较高，轻稀土富集；$(La/Sm)_N$=6.2，$(Gd/Yb)_N$=2.7，轻稀土元素之间分馏程度明显强于重稀土；δEu=0.76，具弱负铕异常，稀土元素球粒陨石标准化图为较圆滑的右倾型曲线（图 9-66）。与维氏酸性岩丰度值相比，Sr、Ba、Hf、Pb、Zn、Sn 等元素明显偏高，尤其是 Ba 元素含量高达 2602×10^{-6}，而 Rb、Li、Be、Zr、Ta、Nb、Cu、Ni 等大多数元素则明显偏低，仅 Co 元素与维氏酸性岩丰度值相比大致相当；Rb/Sr=0.2，低于维氏酸性岩值 Rb/Sr 比值；微量元素蛛网图总体上向右倾，并显示出 Ba、Ta、Nb、P、Ti 负异常和 Rb、Th-U-K、Pb、Nd-Zr-Hf-Sm 正异常（图 9-71）。

十一、晚白垩世花岗斑岩（$\gamma\pi K_2$）

1. 地质特征、形成时代及岩石学特征

主要见有牙洛、洋淋林、南下岭、田独、海头、南牛岭、大岭及双客园等侵入体，总面积约 364.89km²。常呈岩墙状或岩瘤状，特别是近南北向岩墙是其常见产状，如牙洛、洋淋林、计架、南脚岭等侵入体都是近南北向的岩墙。南下岭侵入体全岩 K-Ar 年龄为 66Ma（宜昌地质矿产研究所、海南省地质调查院，2004），形成于晚白垩世。

岩性主要为花岗斑岩（图版Ⅱ-5-G）。块状构造，斑状结构。斑晶含量 12%～35%，斑晶主要由正长石（或正长条纹长石）、斜长石、石英构成，有时可有黑云母甚至角闪石斑晶。基质为隐晶质或微细粒结构，局部为玻璃质，常见球粒结构、花斑结构、蠕英结构（图版Ⅱ-5-H），主要由长英质构成。其中规模最大的牙洛岩体（约 72 km²）边部有时斑晶主要为石英而被称为石英斑岩，而岩体中心可为细粒似斑状（角闪）黑云二长花岗岩。

双客园侵入体岩性为钠长白岗岩。灰绿色、灰白色、肉红色，中粗粒—中粒花岗结构，块状构造。矿物粒径 1～10mm，粒度不均匀，以中粒为主，局部见文象结构。矿物成分为：钠长石 60%～80%、石英 20%～40%。钠长石半自形板柱状，常蚀变成高岭土而呈肉红色，或蚀变成绢云母而呈灰绿色；石英他形粒状。

2. 岩石地球化学特征

晚白垩世花岗斑岩的 SiO_2 含量一般介于 68.13%～76.94%之间；σ 值为 1.42～2.53；大多属高钾钙碱性系列。A/CNK 为 0.98～1.11，多数属弱过铝质。该类岩石的稀土元素总量大多在 134.4×10^{-6}～162.1×10^{-6}之间变化（南牛岭侵入体相对较高，为 304.75×10^{-6}）；稀土元素球粒陨石标准化模式为稍向右倾斜的“海鸥式”曲线（图 9-57）；δEu 为 0.68～0.02，具中等至强烈的负铕异常；轻重稀土分异较明显，LREE/HREE 值为 9.7～13.9。微量元素方面（本章附表 9-4），石门山花岗斑岩显示 Ba、V、Cu 含量较高，而 Sr、Pb、Li 含量较低；与维氏酸性岩丰度值相比，双客园侵入体 W、Mo、Sb、Au 等元素丰度略高。南牛岭侵入体 Hf、As、Au、Zr 等元素明显偏高，而 Sr、Ba、Li、Ni、Cu、Zn 等则明显偏低，Rb/Sr=5.6~13，为维氏酸性岩 Rb/Sr 比值的十数倍，显示该侵入体可能为成熟度较高的地壳（变）沉积岩部分熔融的产物。特别指出的是，南牛岭侵入体中有一个样品的 As 含量高达 3842×10^{-6}，为维氏酸性岩值的 2000 多倍，而 As 主要以硫化物形式产于低温热液矿床中，且常与金、银、钼等多金属矿以及稀土元素共生（或伴生），可能预示在侵入体内部或与之相关的围岩、断裂带等有利于成矿的某些部位赋存着极具开发价值的矿产资源。微量元素原始地幔标准化蛛网图中，晚白垩世花岗斑岩微量元素平均值配分曲线与燕山晚期花岗岩大体相似，显示出 Ba、Ta、Nb、P、Ti 负异常和 Rb、Th-U-K、Pb、Nd-Zr-Hf-Sm 正异常。

第六节　脉　　岩

区内脉岩比较发育，在不同的岩体内部和地层中均可见到。脉岩类型繁多，主要有闪长（玢）岩、辉绿（玢）岩、花岗岩、伟晶岩、细晶岩、石英脉、（石英）正长岩、辉长闪长岩、黑云母辉长岩、斜闪煌斑岩，少量二长（斑）岩、碱长花岗岩、钠长石岩、拉辉煌斑岩、白岗岩、碳酸岩脉等。

在上述脉岩中，一般来说，花岗岩、伟晶岩、细晶岩、碱长花岗岩、白岗岩、石英脉等酸性岩脉是花岗质主体侵入岩的派生产物，即为专属性脉岩。辉绿岩、闪长岩、花岗斑岩、煌斑岩、正长岩脉等多属区域性脉岩，区域性脉岩多赋存于区域性的断裂或节理系统中。区域性脉岩具有多期次性。

花岗斑岩　分布最广泛的脉岩，尤其在燕山晚期花岗岩中最多见，宽一般几米至 60m，长几十米至几百米。分布于昌江县一带的俄郎金波花岗斑岩，在马岭以南，岩墙带总体走向为 10°~15°，马岭以北至娥黄岭一带为南东向，自珠碧江至俄郎一带变成 330°走向的平行脉组。岩墙规模很大，南部单脉出露 100~150m，而在娥黄岭一带最宽处为 540m，俄郎一带变小，单脉仅 10~40m。在娥黄岭一带见花岗斑岩岩墙切断南牙村北东向花岗闪长斑岩、石英二长斑岩岩墙。该脉岩侵入马岭岩体，在娥黄岭以南至公路一带，岩墙中见有多个长透镜状邦溪岩体的捕虏体。该岩墙岩石类型十分均一稳定，从南至北几乎无变化。岩石手标本为灰白色，风化面为浅黄褐色，斑状结构，基质中的钾长石和石英成显微文象连晶（花斑结构）非常发育。斑晶主要为自形钾长石、石英，偶尔见斜长石斑晶，粒径一般为 4~5mm，偶尔见最大钾长石斑晶 5mm×8mm，石英斑晶常见被熔蚀成港湾状。基质为全晶质微粒结构，粒径<0.2mm。

闪长（玢）岩　分布最广泛的脉岩之一，大部分侵入岩及地层中都能见到。走向为北西、北东、东西向，有时成群出现，单脉宽 5cm~60m，长几十厘米至上千米。岩脉具多次活动特点，脉壁大多平直，有时见 1~2cm 的冷凝边，脉体中常见围岩捕虏体。脉岩呈灰黑色或暗绿色。块状构造，斑状结构或微细粒结构。斑晶主要为斜长石、角闪石。矿物组成主要为角闪石、斜长石、黑云母、辉石，可有少量石英。

辉绿岩　分布十分广泛，尤其在大乃村、石碌、邦溪岩体及通什岩体中最常见。岩脉往往呈脉群或脉组出现，以北北东走向为主，也有近东西向和北北西向，倾角变化大（倾角 50°~85°），脉宽 5~2m 不等，长度 1m 至数十米。在白沙县南叉河、南斑河中的辉绿岩脉最大脉宽为 7m，追索长度 1.1km。岩石呈灰黑色，块状构造，辉绿结构，暗色矿物 30%~50%，主要为单斜辉石，可有少量角闪石、黑云母，其余主要为斜长石（An 50~An 55），可有少量石英。

斜闪煌斑岩　主要见于华力西期—印支期花岗岩及早古生代地层中，岩脉往往成群产出。以南北走向延伸为主，少量北西、北东向，倾角陡，一般为 60°~80°。脉体规模变化大，脉宽一般 20cm~4m，长 1~1000m 不等。脉体常沿节理或破碎带贯入，呈树枝状、扫帚状、不规则状等。岩脉与围岩界线大多清楚，脉壁平直，边缘常可见 1~2cm 的冷凝边，脉体中常见围岩捕虏体。脉岩为块状构造，煌斑结构，斑晶含量 5%~65%，斑晶主要为自形的角闪石及少量辉石，也可有少量斜长石。基质主要为斜长石、角闪石、黑云母。据葛小月等（2003）研究表明（葛小月、李献华、周汉文，2003），琼南地区（西线高速公路旁）斜闪煌斑岩形成于板内拉张环境，其全岩 K-Ar 同位素年龄为 81Ma，初始锶比值 0.7078~0.7084，$\varepsilon_{Nd}(t)$ 为-3.2~-2.3，来源于富集地幔，为钾玄质系列岩石。

花岗伟晶岩脉 主要见于岩体中，宽几厘米至7cm，长十几厘米至几十米。产状变化不定。块状构造，伟晶结构，主要由微斜条纹长石、斜长石、石英组成，还可含有电气石，电气石含量有时可达15%。

辉绿岩、辉绿玢岩 分布十分广泛，华力西期—印支期、燕山期及地层中均有分布。岩脉往往呈脉群或脉组出现，以北北东走向为主，也有近东西向和北北西向，倾角变化大（倾角50°~85°）。脉宽5cm~7m不等，在南叉河、南斑河中的辉绿岩脉最大脉宽为7m，追索长度达1.1km。该类岩脉侵入围岩的界线大多清楚，脉壁平直，边缘常可见1~3cm的冷凝边，脉体中常见围岩捕虏体。岩脉具分叉复合现象，常切割围岩中的节理或早期岩脉。辉绿岩手标本为黑色或暗绿色，致密块状，镜下见典型的辉绿结构；辉绿玢岩具斑状结构，斑晶为斜长石、辉石、角闪石。

石英脉 石英脉是分布最广、数量最多的岩脉，岩脉大小不一，规模变化大，长数十厘米至几千米，宽几厘米至几千米不等。其中白沙金波一带出露的石英脉沿北东向延伸大于6km。岩脉受构造控制明显，以北东走向为主，次为北西向，也有东西向、南北向。

第七节 花岗岩形成的构造环境、成因类型及侵位机制

花岗岩形成构造环境的划分是花岗岩成因分类研究的基础，本文将碰撞造山岩浆旋回划分为碰撞前（活动及被动陆缘）、同碰撞、后碰撞三个阶段，“同碰撞”的含义是把洋盆消失后的陆-陆碰撞及其后的继续会聚作用均作为同碰撞处理，从某种意义上，其含义相当于与碰撞有关的过程，均看做同碰撞；“后碰撞”则是代表在主碰撞期以后产生的构造带。后碰撞时期通常始于一个陆内环境，主海盆已关闭了，但大陆块体沿巨大剪切带仍然有大量水平方向的运动（肖庆辉等，2002）。后碰撞构造环境产于构造体制转换时期，通常反映的是挤压应力“松弛”并最终趋于稳定的阶段。

按照《中国区域地质志》编制技术要求，本文将海南岛侵入岩划分为I型、S型、A型和M型4个成因大类，其中A型又分铝质A型和碱质A型两个亚类。I型用来表示幔源岩浆与壳源岩浆混合成因的火成岩类，其源区岩石多为变质火成岩，也包括地幔岩部分熔融的贡献；S型是指源区岩石为富铝的变质沉积岩。I型—S型花岗岩的判别标志主要有下列4项：含有钙质闪石类矿物的准铝质序列、普遍含镁铁质微粒包体、Rb/Sr比值大多<1、磁铁矿及榍石含量较高等，原则上具备其中一项的，即为I型花岗岩，完全不具备的则定为S型花岗岩。A型花岗岩系指含有碱性暗色矿物的花岗岩类；M型则泛指幔源原生岩浆及其进化岩浆。

侵位机制是指花岗岩浆从其源地上升、运移直到地壳上部其现今所在位置和空间的全部动力学过程。根据岩浆取得侵位空间的作用，侵位机制通常可分为强力侵位和被动侵位两种类型。强力侵位是指岩浆主动开辟并取得侵位空间，多与区域变形同时发生，在同构造环境下形成一序列“整合”岩体，具体侵位方式包括穹起、底辟和热轻气球膨胀；被动侵位是指岩浆就位空间不是岩浆本身开辟的，属构造期后或构造晚期，多形成“不整合”岩体，具体侵位方式包括顶蚀、破火山口沉陷和断裂扩张。

一、花岗岩形成的构造环境

（一）中元古代花岗岩形成的构造环境

中元古代长城纪花岗岩遭受了强烈的韧性变形作用和绿片岩相变质作用，但局部仍然残留了微粒交生体结构，并且其岩体边界与围岩片理一致；该期花岗岩强烈过铝，属地壳重熔成因，在R_1-R_2多阳离子构造环境判别图上（Batchelor等，1985）（图9-72），花岗闪长岩和二长花岗岩均落入同造山区或附近。而且该套花岗岩的形成年龄约为14亿年，为长城纪末，与扬子板块上的四堡运动或华北板块上的熊耳运动时间一致。综合判断，该套花岗岩形成于碰撞造山环境。

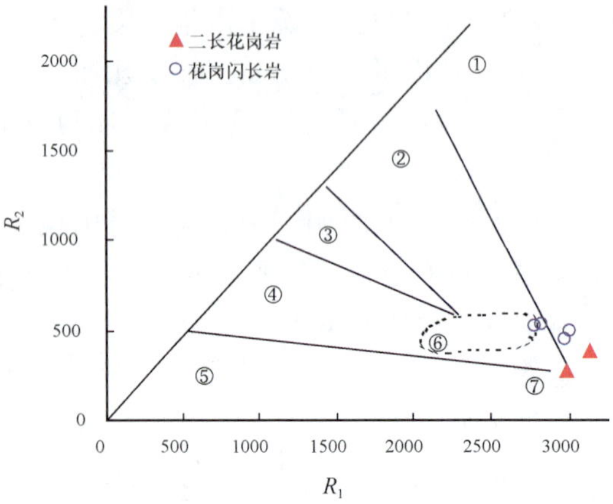

图 9-72 中元古代花岗岩 R_1-R_2 构造环境判别图

Fig. 9-72 R_1-R_2 tectonic setting discriminationdiagram for Mesoproterozoic granites

①地幔斜长花岗岩；②破坏性活动板块边缘（碰撞前）；③碰撞后隆起；④晚造山；⑤非造山；⑥同造山；⑦后造山；R_1=4Si-11（K+Na）-2（Ti+Fe），R_2=6Ca+2Mg+Al

（二）华力西—印支期侵入岩形成的构造环境

海南岛发育大量的华力西期—印支期侵入岩，它们的形成年龄大约在 299~200Ma 之间，不同岩石单位之间没有大的年龄间断，并且从早到晚符合造山岩浆岩的演化规律，为同一次构造岩浆热事件的产物，统一归为华力西期—印支期，形成构造环境可分为活动陆缘、同碰撞和后碰撞三个阶段。

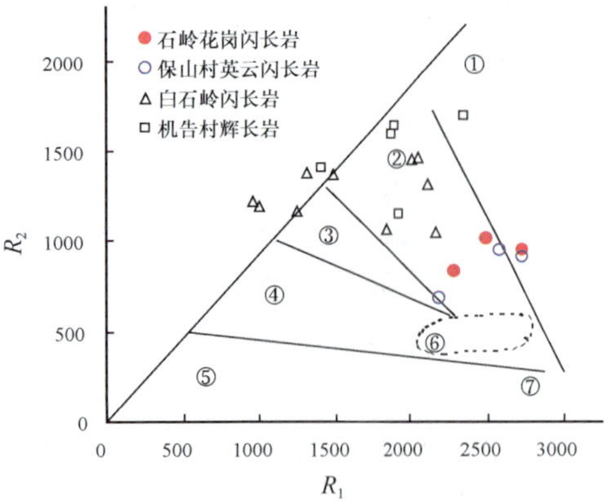

图 9-73 海西期—印支期活动陆缘花岗岩 R_1-R_2 构造环境判别图

Fig. 9-73 R_1-R_2 tectonic setting discrimination diagram for granites of Hercynian-Indosinian active continental margins

其他图例同图 9-72

1. 活动陆缘阶段

区内华力西期—印支期活动陆缘构造岩浆活动时间大致界于 299~287Ma，形成了一系列基性至中酸性的小侵入体（大乃村序列），岩石学和岩石地球化学特征类似于科迪勒拉 I 型花岗岩。在 R_1-R_2 和 Y+Nb-Rb 构造环境判别图上除个别样品投在二氧化硅不饱和区外，大部分投影点落在破坏性板块边缘花岗岩区和岛弧花岗岩区（图 9-73）。

2. 同碰撞阶段

时间大致为早二叠世中晚期至晚二叠世（287~250Ma），该阶段发育了大量壳幔混合型花岗岩和一些壳源重熔型的强过铝花岗岩以及 EM2 型富集岩石圈地幔来源的钾玄岩。它们的共同特点是侵入体接触面与围岩片理和谐，岩石普遍具有同侵位韧性变形构造，走向与区域华力西期变形构造线方向一致，都为北东—北东东向，岩石中普遍见有同构造花岗岩的标型结构-微粒交生体，这些都指示它们形成于碰撞挤压环境（谢才富，2002）；在 Batchelor et al.（1985）的花岗岩类 R_1-R_2 多阳离子构造环境判别图解上，除了通什系列因混入较多的幔源物质而部分落入破坏性板块边缘区外，其余大多数投入同碰撞花岗岩区（图 9-74）。据 Ninkovich et al.（1972）研究，幔源岩浆的来源深度与岩石中的 K_2O 含量成正相关，由于本区同碰撞钾玄岩产出位置与活动陆缘高钾钙碱性辉长岩-闪长岩并无明显变化，所以钾玄岩的出现指示了同碰撞阶段岩石圈的加厚，此外，该阶段产出的强过铝的石榴子石花岗岩具有部分埃达克质岩的地球化学特点，也表明了地壳的加厚。因此，各方面的特点都表明，区内早二叠世中晚期及中二叠世侵入岩形成于同碰撞构造环境，其形成的热动力机制可能是：洋壳的继生性俯冲引起地幔楔的熔融；碰撞挤压使岩石圈加厚引起富集岩石圈地幔的熔融；幔源岩浆底侵（underplating）和地壳挤压增厚导致地壳大规模重熔。

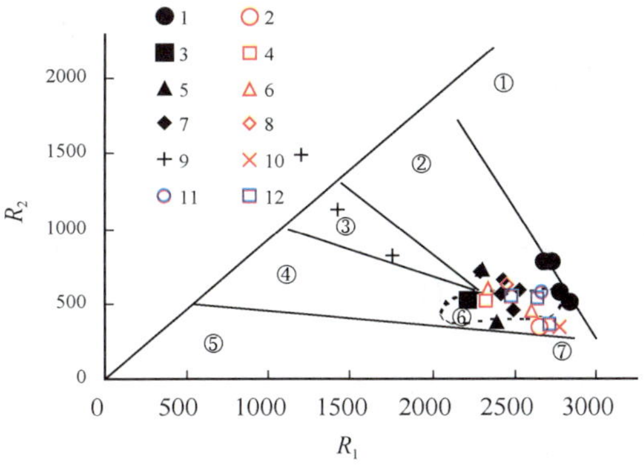

图 9-74 海西期—印支期同碰撞花岗岩类 R_1-R_2 构造环境判别图

Fig. 9-74 R_1-R_2 tectonic setting discrimination diagram for Hercynian-Indosinian syn-collision granitoids

1—西庆序列酸性岩；2—陀类二长花岗岩；3—大田二长花岗岩；4—什天二长花岗岩；5—禄马序列酸性岩；6—道票岭序列酸性岩；7—通什序列酸性岩；8—石碌二长花岗岩；9—五指山序列中性岩；10—新风岭正长花岗岩；11—顺作花岗岩；12—育才序列酸性岩；其他图例同图 9-72

3. 后碰撞阶段

区内经历了早二叠世中晚期至晚二叠世同碰撞阶段后，于三叠纪转入了后碰撞构造岩浆演化阶段，这一时期发育大量壳幔混合型花岗岩（I 型）、壳源花岗岩和 A 型花岗岩、铁镁质岩等。岩体同位素年龄界于 250~200Ma 之间，如尖峰岭序列（S 型）、袁水序列、六连岭序列等 A 型花岗岩以及坝王岭岩体（I 型）等，岩体形态与围岩构造不谐和，多形成“不整合”岩体；但出露于琼中—文昌一带的中三叠世琼中序列岩石中仍然可见到比较明显的原生定向构造。研究表明，海南岛海西-印支期构造活动不均衡，其岩浆作用从西往东呈现出由老变新的演化趋势，反映同期造山活动由西开始，并逐渐往东迁移，具有剪刀式闭合的特征（李孙雄等，2005）。因此，从构造特点来说，海南岛此时总体上虽已处于挤压应力松弛期但仍伴有间歇性的挤压或剪切作用。

在 Batchelor et al.（1985）的花岗岩类 R_1-R_2 多阳离子构造环境判别图解上（图 9-75），个别样品因混入较多的幔源物质而落入破坏性板块边缘区，大部分投影点落在同碰撞至后碰撞（晚造山）花岗岩区域，说明此时地壳熔融环境与同碰撞阶段相近，继承了同碰撞阶段的特点。

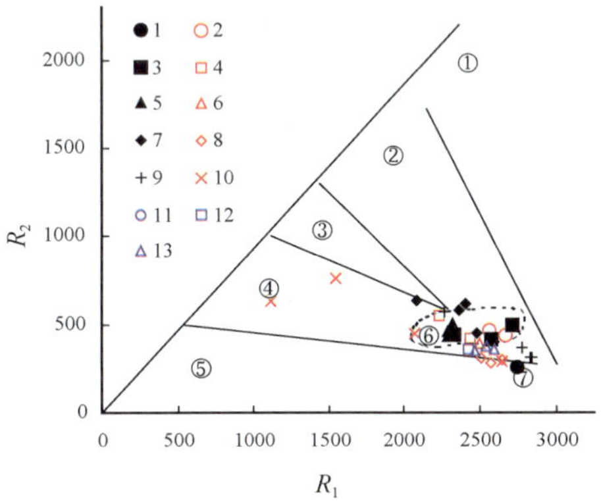

图 9-75 海西期—印支期后碰撞花岗岩类 R_1-R_2 构造环境判别图

Fig. 9-75 R_1-R_2 tectonic setting discrimination diagram for Hercynian-Indosinian post-collision granitoids

1—迈洲岭碱长花岗斑岩；2—三架岭序列酸性岩；3—雅加序列酸性岩；4—南英岭序列酸性岩；5—黑岭序列酸性岩；6—保梅岭正长花岗岩；7—琼中序列酸性岩；8—六连岭序列酸性岩；9—立才序列酸性岩；10—袁水序列酸性岩；11—深沟碱长正长岩；12—进岭花岗斑岩；13—尖峰岭序列酸性岩

其他图例同图 9-72

（三）燕山期侵入岩形成的构造环境

海南岛燕山期岩浆活动非常强烈，形成大面积分布的侵入岩和火山岩，然而，关于这些火成岩形成的构造背景却是地学界长期争论的问题。主要观点有：受太平洋板块俯冲作用控制的活动大陆边缘火山弧系（郭令智等，1983）；板内伸展-裂谷环境（Gilder et al，1991，1996；Li，2000；陈志刚等，2003）、早燕山期（碰撞）挤压晚燕山期裂解伸展（谢窦克等，1989；李献华，1993）、地幔柱（毛景文等，1998）、弧后环境（孙涛等，2003）；大陆边缘后侧（马大铨等，1991），陈培荣等（2007）最近认为，南岭地区晚燕山期以武夷山为界，西区玄武质岩石的源区可能是软流圈物质与岩石圈地幔的混合，形成于大陆裂谷环境；东区玄武质岩石具有明显的板片俯冲印记，与古太平洋板块俯冲作用有成因联系。而张旗（2010）认为，浙闽中生代岩浆岩既不属于岛弧环境，也不是安第斯型活动大陆边缘环境，与板块俯冲作用无关，是大陆内部岩浆活动的产物；笔者倾向于谢才富（2004）提出的本区早燕山期属于被动大陆边缘裂谷环境，而晚燕山期为由板块俯冲引起的弧后拉张环境的观点。理由主要有：

1）海南岛燕山早期岩浆活动很微弱，属于板内构造-岩浆平静期。华南内陆从约180Ma时再次开始的强烈岩浆活动是另一个构造旋回的产物，而且也可能属另一个构造域（太平洋构造域），其构造体制不应是印支期碰撞造山的延续，也不是印支期碰撞造山作用的后造山阶段。

2）华南燕山期岩浆活动有从陆内向洋方向带状整体迁移的现象，不同于东太平洋陆缘（安第斯型）岩浆带向陆内方向迁移的特点。而且华南早燕山期岩浆岩出露宽度深入内陆超过1000km，而安第斯活动陆缘岩浆岩带的宽度<300km，一般俯冲带岩浆弧的宽度则不超过100km。因此，华南早燕山期岩浆岩带的分布特点是活动陆缘或弧后环境所难于解释的。

3）现今中国东南大陆边缘包括广泛的大陆架，不存在或未证实存在180~120Ma间的俯冲带岩石组合（肖庆辉等，2002）。

4）华南内陆早燕山期有较广泛的钾玄岩岩浆的形成，岩石组合显然不同于后造山岩石组合，更与典型岛弧-大陆边缘的玄武岩-安山岩-流纹岩组合不同，也与以发育大规模基性火山岩及A1型花岗岩为主的地幔柱、热点及大陆裂谷岩石组合有所差别。

5）华南早燕山期的花岗岩类在各种构造环境判别图中多落入板内环境（陈培荣等，2002；张佩华，2003），而其玄武岩属碱性玄武岩或拉斑玄武岩，也落入板内环境（陈培荣等，2002），燕山期玄武岩和钾玄岩类一般没有Nb、Ta亏损、富LILE（李献华等，1999；孙涛等，2003），钾玄岩系的Sr、Nd同位素组成类似于OIB（洋岛玄武岩），这些特点也都说明它们形成于板内（裂谷）环境。

6）在南美和北美西海岸的典型大陆边缘区，侵入岩中辉长岩和英云闪长岩的比例很高，如秘鲁海岸岩基中辉长岩占7%~16%，英云闪长岩占46%~55%，花岗闪长岩、二长花岗岩和正长花岗岩三者之和仅占38%~44%（Pitcher，1978）；美国的半岛山脉岩基和内华达山脉岩基中也有许多辉长岩和淡色英云闪长岩（Pitcher，1982）；而海南岛燕山晚期则以大量的花岗岩（包括花岗闪长岩、二长花岗岩和正长花岗岩为主，三者之和占95%以上）与流纹岩普遍地共生，产状上构成火山-次火山-侵入岩组合，仅见少量伴生端元岩石如辉长岩、玄武岩。

7）分布范围上，海南岛燕山晚期岩浆活动整体呈北东向展布，与华南晚中生代岩浆岩带整体上呈北东向、近于平行海岸分布的特征吻合，且由南东向北西，活动强度逐渐减弱、产出层位变高，成分由以流纹岩为主的单峰式变为流纹岩、安山岩平分秋色的双峰式，这种格局显示有另一个板块向海南岛之下消减。

8）岛内广泛发育有早白垩世弧后盆地，由厚度大于4000m的红色砂岩、泥岩、砂砾岩组成，反映了干燥条件的古地理环境。

9）海南岛早燕山期侵入岩不甚发育，主要发育晚燕山期岩浆岩。早燕山期侵入岩为钾玄岩系列的铁镁质岩-正长岩以及壳幔混合型花岗岩；而晚燕山期主要由双峰式火山岩、单峰式高钾钙性中酸性火山岩、壳幔混合型高钾钙碱性花岗岩、石英二长岩-石英正长岩（A型花岗岩）组成，与东南沿海的岩浆岩组合相似。在R_1-R_2图上早燕山期的花岗岩及正长岩、石英二长岩主要落入晚造山区；而晚燕山期的花岗岩类则主要位于破坏性活动板块边缘及其延伸区域（图9-76）。

因此，海南（含测区）早燕山期属于被动大陆边缘裂谷环境，晚燕山期为由板块俯冲引起的弧后

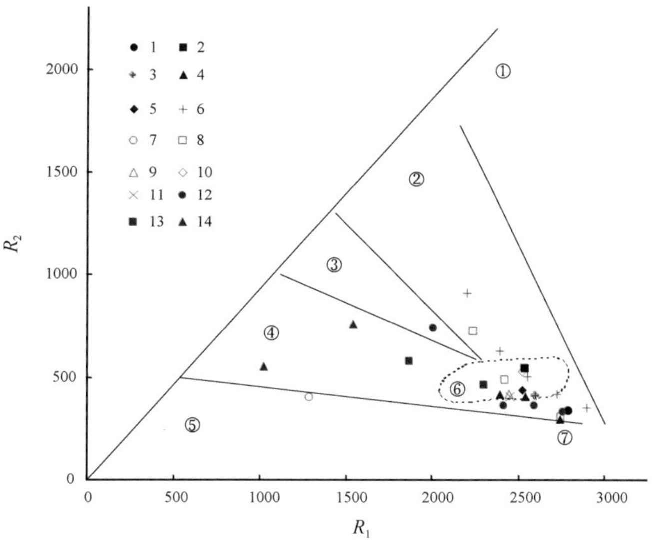

图 9-76　燕山期花岗岩类 R_1-R_2 构造环境判别图

Fig. 9-76　R_1-R_2 tectonic setting discrimination diagram for Yanshanian granitoids

1—$\gamma\pi K_2$ 花岗斑岩；2—火烧岭二长花岗岩；3—龙楼正长花岗岩；4—吊罗山序列酸性岩；5—南牙村花岗闪长斑岩；6—保城序列酸性岩；7—番通石英正长岩；8—高峰序列酸性岩；9—$\gamma\pi K_1$ 花岗斑岩；10—$\gamma\delta\pi K_1$ 花岗闪长玢岩；11—半岛正长花岗岩；12—三狮岭序列酸性岩；13—南山序列酸性岩；14—勤寨序列中性岩

其他图例同图 9-72

拉张环境。推断其岩浆形成的热动力机制可能为：软流圈上涌，岩石圈减薄，引起软流圈或软流圈与岩石圈交界部位的部分熔融，形成碱性玄武岩或钾玄质岩浆，由于软流圈上涌导致的高热流以及软流圈岩浆的底侵，又引起岩石圈不同层圈、特别是地壳的部分熔融作用，形成早燕山期岩浆爆发（主要沿深大断裂分布）。晚燕山期，华南地区岩浆作用主要受古太平洋板块（Izanagi 板块）的俯冲作用控制，岩浆岩带向洋方向整体跃迁迁移。由于板块俯冲引起的弧后拉张，导致大量幔源及壳幔混合源为主的岩浆岩的形成。

二、花岗岩的成因类型

花岗岩的成因分类历来是岩石学家争论不休的问题（战明国，1998；涂绍雄，王桂华，2004；张旗，王焰，李承东等，2006）。马大铨等（1991）曾将海南岛花岗岩类划分为 I 型和 S 型两类。宜昌地质矿产研究所（1995）在 1∶5 万《和乐、中原市、博鳌港》幅区域地质调查报告中将该区内的侵入岩划分为 I 型、S 型和幔熔分异型（富碱侵入岩）三类。而在 1∶25 万《乐东县、陵水县》幅区域地质调查报告（宜昌地质矿产研究所、海南省地质调查院，2004）中，又将测区侵入岩划分为壳源重熔型（C 型）、壳幔混合型（H 型）、A 型及铁镁质岩四个成因大类，其中 A 型又分铝质 A 型和碱质 A 型两个亚类。

在继承前人划分方案的基础上，笔者将海南岛侵入岩划分为壳幔混合型（I 型）、壳源重熔型（S 型）、A 型和 M 型四个成因大类，其中 A 型又分铝质 A 型和碱质 A 型两个亚类。在此基础上再从埃达克质岩石的角度对区内部分岩体的岩石地球化学特征进行阐述。各成因类型的具体含义分别为：I 型用来表示幔源岩浆与壳源岩浆混合成因的火成岩类，其源区岩石多为变质火成岩，也包括地幔岩部分熔融的贡献，对于中酸性岩来说，其岩石学标志主要是含有钙质闪石类矿物、普遍含镁铁质微粒包体、磁铁矿及榍石含量较高等；S 型指源区岩石为富铝的变质沉积岩，其矿物学标志是含有白云母、石榴子石、刚玉等富铝矿物，基本上不含角闪石，多属强过铝质花岗岩；A 型花岗岩系指含有碱性暗色矿物的花岗岩类；M 型则泛指幔源原生岩浆及其进化岩浆。区内侵入岩成因类型划分见表 9-12。

表 9-12　海南岛侵入岩成因类型划分表

Table 9-12　Division of genesis types of intrusive rocks from the Hainan Island

成因大类	I 型		S 型	M 型	A 型	
成因亚类		Adakite			铝质 A 型	碱质 A 型
岩体归类	大仍村系列、育才序列、通什序列、道票岭序列、禄马序列、什天、大田、西庆序列、立才序列、琼中序列、保梅岭、黑岭序列、南英岭序列、雅加序列、三架岭序列、南山序列、三狮岭序列、半岛、高峰序列、南牙村、吊罗山序列、龙楼、火烧岭	保城序列	戈枕、亚炮、顺作、新风岭、石碌、陀类、尖峰序列、进岭、迈洲岭	五指山序列、万宁序列、勤赛序列、马翁岭、拉丁村、乌石	六连岭	深沟、袁水序列、万车序列、番通
岩性组合	闪长岩-花岗闪长岩-二长花岗岩-正长花岗岩		二长花岗岩—正长花岗岩	闪长岩-辉长岩	正长花岗岩	石英正长岩-正长花岗岩、碱长花岗岩

壳幔混合型花岗岩（I 型）最为发育，这类岩体的最主要特点有：早期岩体中含有大量镁铁质微粒包体（MME），暗色矿物析离体也较常见。MME 常为浑圆状外形，包体内常见寄主岩石的长石捕房晶及反映包体岩浆高温快速冷凝结晶的针状磷灰石等。包体和寄主岩中都可见到长石的环带结构、环状包裹体结构、有时见似环斑结构，上述岩相学现象是岩浆混合包体的典型特征，说明有地幔物质混入。副矿物中磁铁矿较多，钛铁矿相对较少，总量一般大于 1000g/t；含铝指数 ACNK 大多数<1，仅个别>1.1，属偏铝质至弱过铝质花岗岩类。

区内壳源重熔型（S 型）花岗岩，岩性组合一般为二长花岗岩-正长花岗岩，一般不含角闪石，暗色矿物主要为黑云母；副矿物中磁铁矿较少，钛铁矿相对较多，总量一般小于 1000g/t；很少见到闪长质包体，主要为变沉积岩的围岩捕虏体（或富云包体）；含铝指数 ACNK 绝大多数>1.0，属弱过铝质—强过铝质花岗岩类。

M 型侵入岩虽然面积不大，但具有比较重要的构造环境和成因指示意义。包括五指山序列、万宁序列、勤寨序列、马翁岭闪长岩、拉丁村辉长岩和乌石闪长岩。其中印支期长安辉长岩属拉斑玄武岩系列，具富集岩石圈地幔（Ⅱ型）来源特征；早燕山期马翁岭闪长岩属高钾钙碱性-钾玄岩系列，属大陆边缘裂谷环境，铁镁质岩浆主要来自上隆的软流圈，幔隆底侵作用和岩石圈伸展减薄是这一时期区内陆壳演化的重要方式（云平，2002）；而晚燕山期乌石闪长岩均属高钾钙碱性系列，铁镁质岩浆可能来自软流圈和富集岩石圈地幔的混合。

区内 A 型花岗岩分为碱质 A 型和铝质 A 型花岗岩两类。碱质 A 型花岗岩具有富碱，贫水，REE、Zr、Hf、Zn、F 含量高及 Ga/Al 高等特点，未发现碱性暗色矿物，但其角闪石较富碱（碱含量一般>3%）、富铁，主要属铁浅闪石和铁浅闪石质角闪石。铝质 A 型花岗岩富硅、碱，贫钙、镁，偏铝-弱过铝，弱亚碱质，富 F、富 HFSE，Ga/Al 高、FeO^*/MgO 高、Rb/Sr 高。与测区其他花岗岩类相比，区内 A 型花岗岩的 Zr、Nb、Y、Ga 等元素明显偏高，V、Cr、Ni、Sr 等元素明显偏低。上述铝质 A 型花岗岩的源岩可能是由变火成岩（长英质麻粒岩）低度部分熔融作用形成的，在此过程中可能有地幔富卤族流体的作用，而碱质 A 型花岗岩可能是由饱满型基性麻粒岩部分熔融的产物。

本次工作的重要进展之一，就是在传统成因分类的基础上，首次厘定了海南岛燕山晚期保城序列（花岗闪长岩-花岗岩组合）属埃达克质侵入岩。岩石地球化学特征上具有高的 SiO_2、Al_2O_3 含量以及高的 Sr 含量和 Sr/Y 比值，低的 Y 和 Yb 含量，Eu 异常不明显等。

埃达克岩（Defant et al.，1990）原指的是新生代与年轻洋壳俯冲有关的、具有独特地球化学特征的一类中—酸性火山岩或侵入岩。埃达克岩没有明确的岩相学标志，不同于传统的岩石学命名原则，但由于其具有重要的成矿与岩浆动力学意义，成为当前地质研究的热点之一。典型的埃达克岩具有以下的地球化学特征：$SiO_2 \geq 56\%$、$Al_2O_3 \geq 15\%$、$MgO<3\%$、$Yb \leq 1.9\times10^{-6}$，$Y<18\times10^{-6}$，$Sr \geq 400\times10^{-6}$，LREE 富集，无 Eu 异常或有弱的负 Eu 异常。埃达克岩的熔融残留相为石榴子石（斜长石少至无）。张旗等（2001）在研究中国东部燕山期岩浆作用时，发现有许多中酸性火山岩和侵入岩类似埃达克岩的地球化学特征，但它们是陆相的，产于板内环境。其成因与板块的消减作用无关，而可能是加厚的陆壳底部基性岩部分熔融形成的，因此，埃达克岩可能分为 O 型和 C 型两类：O 型埃达克岩与板块的消减作用或玄武岩底侵作用有关，C 型埃达克岩则是加厚的地壳底部的中-基性岩部分熔融的产物。

近年来，国内有关埃达克岩的报道已有不少，如北大别山 Adakite 型灰色片麻岩（王强等，2000）、中国东部燕山期埃达克岩（张旗等，2001）以及西天山阿吾格勒 Adakite 型钠质中酸型岩（熊小林等，2001）等，但海南岛目前尚无有关 Adakite 质岩石的公开报道。

从表 9-13 可看出，区内燕山晚期保城序列（包括屯昌杂岩体、千家杂岩体和保亭杂岩体）的大多数指标与中国东部燕山期埃达克质岩（C 型埃达克岩）的特征相似。具体为：$SiO_2>56\%$，属中酸性岩类，Al_2O_3 接近或大于 15%，$MgO<3\%$，$Yb \leq 1.9\times10^{-6}$，Y 大部分小于 18×10^{-6}，$Sr \geq 400\times10^{-6}$，LREE/HREE 为 9.71~19.93，LREE 富集，具中等—弱负铕异常（个别具正铕异常），$(^{87}Sr/^{86}Sr)_i>0.704$，$\varepsilon_{Nd}(t)<0$。$K_2O$ 和 Na_2O 分别在 2.45%~5.31%和 2.65%~3.97%之间，K_2O/Na_2O 比值接近于 1，与环太平洋地区典型 Adakite 明显不同（典型 Adakite 为钠质火成岩，富 Na_2O 贫 K_2O，K_2O/Na_2O 比值仅为 0.4）。在 K_2O-SiO_2 图解中，典型 Adakite 落入拉斑-钙碱性系列岩石区，而本区埃达克质岩石均落入高钾钙碱性系列岩石区（图略）。

表 9-13　埃达克质岩石特征对比

Table 9-13　Comparison of characteristics of adakitic rocks

	典型 Adakite（O 型埃达克岩）	中国东部燕山期 Adakite（C 型埃达克岩）	本区埃达克质岩石
岩石组合	安山岩、英安岩、流纹岩，花岗闪长岩、英云闪长岩	安粗岩、粗安岩、闪长岩、花岗闪长岩、石英二长岩、正长岩	花岗闪长岩、二长花岗岩、正长花岗岩
岩石系列	钙碱性，次为低钾拉斑系列	高钾钙碱性，次为钙碱性和橄榄玄粗岩系列	高钾钙碱性
SiO_2	>56%	>56%	>56%
Al_2O_3	>15%	>15%	>15%
Sr	$\geq400\times10^{-6}$	$\geq400\times10^{-6}$	$\geq400\times10^{-6}$
Y	$<18\times10^{-6}$	$<18\times10^{-6}$	$<18\times10^{-6}$
Yb	$\leq1.9\times10^{-6}$	$\leq2\times10^{-6}$	$\leq1.9\times10^{-6}$
REE	LREE 富集，无铕异常或负铕异常不明显	LREE 富集，无铕异常或负铕异常不明显	LREE 富集，中等至弱负铕异常
$(^{87}Sr/^{86}Sr)_i$	<0.705	>0.704	>0.704
$\varepsilon_{Nd}(t)$	>0	<0，很少>0	<0
部分熔融的源岩	消减的 MORB，低侵的玄武岩	中基性麻粒岩或角闪岩	玄武质岩石
部分熔融残留物	Cpx+Gt	Cpx+Gt	Cpx+Gt
岩浆来源深度	>70~90km	>50km	>50km
混染源	地幔楔	中上地壳	中下地壳
混染的结果	Mg、Cr、Ni、Co 增加	LILE、$(^{87}Sr/^{86}Sr)_i$ 增加，$\varepsilon_{Nd}(t)$ 降低	LILE、$(^{87}Sr/^{86}Sr)_i$ 增加，$\varepsilon_{Nd}(t)$ 降低
成矿作用	Cu、Ag	Fe、Cu、Au、Mo	Fe、Cu、Au、Mo、Pb、Zn

研究表明，埃达克质岩的形成要求原岩必须是基性的、含水、残留相有石榴子石存在。因此，只要有合适的物理化学条件存在，俯冲板片和下地壳熔融都可以形成埃达克岩。目前，有关埃达克质岩石的形成机制有俯冲洋壳熔融、增厚下地壳熔融、拆沉下地壳熔融、玄武质岩浆的结晶分异等几种不同的认识，还有的认为与壳幔岩浆混合与 AFC 模式有关。笔者认为，单一的俯冲洋壳、拆沉下地壳熔融或玄武质岩浆的结晶分异等都不可能形成本区埃达克质岩石。可能的模式是：晚燕山期，华南地区岩浆作用主要受古太平洋板块的俯冲作用控制，岩浆岩带向洋方向整体跃迁迁移，由于俯冲板块后撤，地壳环境处于弧后拉张状态，引发地幔软流圈上涌，源自地幔的镁铁质岩浆底侵于加厚的玄武质下地壳（海南岛二叠纪末—三叠纪时地壳曾强烈增厚，至燕山期虽有减薄，但仍是一个具有山根的加厚陆壳）（邹和平，1997），导致大量幔源、壳幔混合源以及埃达克质岩石的形成。因此，加厚的玄武质下地壳和作为热源的地幔镁铁质岩浆对下地壳的底侵，是产生海南岛燕山晚期埃达克质岩石的两个主要因素。本区埃达克质岩石既不是下地壳玄武质岩单独部分熔融的产物，也不是地幔镁铁质岩原始岩浆分异演化的结果，而是由加厚下地壳玄武质岩部分熔融产物与地幔镁铁质岩浆混合（壳幔岩浆混合）并进一步分异演化的结果，其来源深度类似于中国东部燕山期埃达克质岩（地壳厚度>50km）。

吊罗山杂岩体亦属壳幔岩浆混合成因，部分样品具有正铕异常，可能代表一个加厚陆壳的环境（邓晋福等，1996）。但该岩体的大多数指标与埃达克质岩的岩石地球化学特征不符，而与喜马拉雅型花岗岩特征相似（$Sr<400\times10^{-6}$，$Yb<2\times10^{-6}$），其来源深度类似于喜马拉雅型花岗岩（地壳厚度>40km）。因此，相对于之前早白垩世埃达克质岩形成构造环境，海南岛晚白垩世岩石圈可能发生了更进一步的伸展减薄。

三、岩体构造及侵位机制

根据岩浆取得位空间的作用，侵位机制通常可分成两种类型，强力侵位和被动侵位。强力侵位是指岩浆主动开辟并取得侵位空间，多与区域变形同时发生，在同构造环境下形成一序列“整合”岩体，具体侵位方式包括穹起、底辟和气球膨胀；被动侵位是指岩体侵位空间不是岩浆本身开辟的，属构造期后或构造晚期，多形成“不整合”岩体，具体侵位方式包括顶蚀、破火山口沉陷和断裂扩张。区内侵入岩的侵位机制与其形成的构造环境和区域应力场特征密切相关。按应力场性质可分为挤压、挤压应力松弛-剪切和伸展-拉张 3 种状态。

1. 挤压应力场

主要是同碰撞构造环境阶段侵入的中元古代长城纪花岗岩、早二叠世中晚期至晚二叠世侵入的一系列岩体（如育才序列、五指山序列、通什序列、西庆序列等）。岩体特征主要为：

1）岩体主轴在平面上呈近北东东向，与海南岛华力西构造层主体构造线方向基本一致。

2）岩体外接触带围岩主要形成定向构造明显的片岩，且自远离岩体向接触带方向有板岩、千枚岩→石英绢云母片岩→二云母片岩→红柱石云母片岩等类似于前进区域变质的分带现象。

3）主体岩体中的叶理产状与围岩接触面一般表现为和谐接触关系，叶理不随侵入体的边界发生偏转，具有固态变形面理的特征（马昌前等，1994），说明叶理的形成与区域变形有关。

4）包体测量表明，由侵入体中心到边缘，暗色微粒包体存在明显的递进变形特点，即侵入体边缘包体往往为长条状，向中心逐渐变为椭圆状、圆状，由包体反映出的岩石变形量从侵入体边缘往中心逐渐变小，表明岩体发生过热气球膨胀作用。

5）岩石中普遍见微粒交生结构，即在较大颗粒矿物的结合缝及三结点通道上可见到由微粒石英、钠长石、微斜长石、蠕英石构成的集合体，它们的成分相当于近共结点花岗岩。谢才富（1997，2002）认为，这种微粒交生体是残余熔体在应力作用下的结晶产物，是同构造侵位岩体的结构标志之一。

6）岩体区域上表现为大型的花岗岩岩基。

7）同碰撞早期由于碰撞引起的地壳强烈挤压作用，岩石中同侵位韧性变形形成的叶理构造很发育，表明岩体主要为气球膨胀式强力侵位，同碰撞晚期（晚二叠世），岩石既可为块状构造也可发育同侵位韧性变形形成的线理构造（叶理构造不太发育），说明此时已有部分岩体由剪切应力形成。

2. 挤压应力松弛–剪切应力场

与上述早二叠世中晚期至晚二叠世同碰撞岩体不同，三叠纪侵入的各类岩体，是在区域挤压应力松弛、地壳隆起、抬升阶段定位的，属后碰撞构造环境岩浆活动产物，此类岩体主要具有如下特点：

1）岩体产状大多与围岩的区域性构造走向不整合。

2）见呈岩墙或岩枝侵入围岩，岩体内可见其他早期岩体或地层的岩块等。

3）岩体内部通常以块状构造为主。但也有部分岩体（如琼中序列）矿物定向构造十分发育，因此，从构造特点来说海南岛此时总体上虽已处于挤压应力松弛期但仍伴有间歇性的挤压或剪切作用。

4）其热接触变质围岩地层多为块状构造。

5）出现后碰撞构造环境的各种岩石类型，如壳幔混合型花岗岩、碱性、偏碱性的 A 型花岗岩及基性岩等。

3. 伸展–拉张应力场

海南岛燕山期上地壳一般为脆性状态下的拉张和断块运动，张性断裂作用发育，侵入岩侵位机制属被动侵位。该阶段形成的岩体构造特点主要为：

1）大部分侵入体平面上近圆形，晚期单元的侵入体往往侵入于早期单元侵入体的内部及顶部，构成环套式岩体，岩体剖面上呈现漏斗状形态。

2）侵入体边界与围岩层理、劈理等呈不和谐接触。

3）侵入体边部见大量大小几厘米至几米，甚至可达几十米的棱角状捕虏体，可能是围岩受破火山口塌陷或岩浆顶蚀作用而使碎块落入岩浆中。

4）侵入体内的各类岩脉均表现为被动就位，脉岩矿物杂乱分布，脉壁呈不规则状或阶梯状；

5）侵入体中的暗色包体多呈圆状或浑圆状，无挤压迹象。

6）岩体内岩石为均一构造，矿物随机分布，基本不具定向性，其围岩受较强热变质改造后也呈块状构造。

该阶段岩体的侵位机制主要是受多组张性断裂控制的破火山口沉陷与顶蚀作用，个别主要为岩墙扩展作用。

附表 9-1　海南岛侵入岩岩石化学分析结果一览表

Attached table 9-1　List of petrochemical analysis results of intrusive rocks from the Hainan Island

单位:%

岩石单位	样品号	岩　性	SiO_2	TiO_2	Al_2O_3	Fe_2O_3	FeO	MnO	MgO	CaO	Na_2O	K_2O	P_2O_5	H_2O	CO_2	F	总量	来源
$\gamma\pi K_2$	7329-4	花岗斑岩	73.24	0.13	13.63	0.41	1.38	0.03	0.19	1.02	3.35	5.3	0.11	0.3	0.07	0.08	99.26	1
$\gamma\pi K_2$	Ⅳ12-1	花岗斑岩	76.34	0.075	12.23	0.72	0.44	0.02	0.07	0.48	3.74	4.66	0.04	0.29	0.22	0	99.83	1
$\gamma\pi K_2$	Ⅳ5-1	花岗斑岩	73.76	0.16	13.16	0.92	1.02	0.04	0.26	0.88	3.64	4.84	0.07	0.23	0.36	0	99.93	1
$\gamma\pi K_2$	YD632-1	花岗斑岩	77.64	0.15	11.86	0.74	0.30	0.01	0.19	0.18	2.58	5.26	0.02	0.90	0.01		100.74	2
$\gamma\pi K_2$	D1341-1	花岗斑岩	76.67	0.12	11.56	2.3	0.62	0.07	0.19	0.08	2.03	4.89	0.02	0.04	1.27		101.0	2
$\gamma\pi K_2$	YD445	花岗斑岩	73.77	0.21	13.58	0.82	0.82	0.05	0.39	1.08	3.32	4.56	0.05	0.99	0.07		99.71	2
$\gamma\pi K_2$	YD633-1	花岗斑岩	76.27	0.14	12.81	0.71	0.45	0.04	0.38	0.18	2.89	4.96	0.04	0.96	0.03		100.78	2
$\gamma\pi K_2$	D0377	花岗斑岩	74.75	0.16	13.19	0.62	0.68	0.05	0.18	0.71	3.72	5.08	0.02	0.51	0.10		99.77	本次
$\gamma\pi K_2$	D0514	花岗斑岩	75.26	0.15	14.26	0.67	0.12	0.01	0.07	0.30	1.58	4.77	0.01	2.62	0.04		99.86	本次
$\gamma\pi K_2$	D0523	花岗斑岩	71.12	0.33	15.61	1.39	0.65	0.05	0.55	0.11	1.28	5.70	0.03	2.90	0.06		99.78	本次
$\gamma\pi K_2$	D5028-2	花岗斑岩	77.40	0.09	12.66	0.40	0.22	0.03	0.14	0.11	3.01	4.93	0.01	0.81	0.04		100.66	本次
$\gamma\pi K_2$	D1339	花岗斑岩	70.15	0.36	14.51	1.19	2.02	0.05	0.46	2.07	2.69	4.81	0.08	1.35	0.04		99.78	本次
$\gamma\pi K_2$	D0362-1	花岗斑岩	76.56	0.06	12.47	0.39	0.17	0.02	0.08	0.54	3.38	5.39	0.01	0.50	0.29		99.86	本次
	平均		74.84	0.16	13.19	0.87	0.68	0.04	0.24	0.60	2.86	5.01	0.04	0.95	0.20	0.03	100.07	
火烧岭	XⅪ7-2	二长花岗岩	71.54	0.33	14.53	0.96	1.17	0.05	0.73	2.03	3.23	4.45	0.10	0.59	0.01		100.21	2
龙楼	D6002-3	正长花岗岩	76.87	0.04	12.66	0.09	0.5	0.01	0.09	0.6	3.59	5.01	0.37	0.35	0.06	0.01	99.88	本次
龙楼	D2042	正长花岗岩	76.73	0.09	12.25	0.44	0.27	0.05	0.09	0.77	3.89	5.05	0.26	0.21	0.04	0.01	99.89	本次
龙楼	PM03-11	正长花岗岩	73.31	0.18	14.08	0.45	0.83	0.03	0.46	1.85	3.29	4.52	0.54	0.52	0.12	0.04	99.68	本次
龙楼	PM03-27	正长花岗岩	73.35	0.22	13.94	0.43	0.97	0.05	0.51	1.78	3.35	4.57	0.48	0.45	0.06	0.05	99.73	本次
	平均		75.07	0.13	13.23	0.35	0.64	0.04	0.29	1.25	3.53	4.79	0.41	0.38	0.07	0.03	99.80	
大小岭	D6112-1	正长花岗岩	76.81	0.09	12.61	0.54	0.20	0.03	0.11	0.85	3.34	4.82	0.01	0.45	0.04		100.32	5
大小岭	D3126-1	正长花岗岩	75.24	0.14	13.01	0.09	1.88	0.050	0.24	0.81	3.41	4.68	0.03	0.22	0.1	0.03	99.93	5
大小岭	GXVIII10-1	正长花岗岩	75.88	0.15	12.70	0.44	0.95	0.030	0.26	0.74	3.57	4.84	0.05	0.25	0.05	0.1	100.01	5
	平均		75.98	0.13	12.77	0.36	1.01	0.04	0.20	0.80	3.44	4.78	0.03	0.31	0.06	0.07	100.09	

续表

岩石单位	样品号	岩　性	SiO_2	TiO_2	Al_2O_3	Fe_2O_3	FeO	MnO	MgO	CaO	Na_2O	K_2O	P_2O_5	H_2O	CO_2	F	总量	来源
吊罗山	D0495	正长花岗岩	63.98	0.64	15.41	1.85	4.15	0.10	0.45	2.26	3.42	6.16	0.10	0.91	0.04		99.92	5
吊罗山	D0496	正长花岗岩	63.94	0.62	15.42	1.75	4.18	0.10	0.45	2.27	3.43	6.28	0.09	0.92	0.02		99.91	5
吊罗山	D0503	正长花岗岩	70.10	0.41	15.61	1.33	1.10	0.03	0.89	0.39	2.90	4.90	0.05	2.01	0.06		101.69	5
吊罗山	D0525	石英正长岩	64.38	0.66	15.90	2.38	3.20	0.10	0.52	1.82	3.24	5.88	0.10	1.37	0.04		100.51	5
吊罗山	D1386-1	正长花岗岩	72.91	0.25	13.83	0.70	1.00	0.05	0.68	2.09	3.83	3.69	0.07	0.60	0.08		100.14	5
吊罗山	D1902-1	正长花岗岩	75.60	0.15	12.78	0.48	0.62	0.05	0.32	1.25	3.61	4.52	0.03	0.34	0.06		99.99	5
吊罗山	D1952-1	正长花岗岩	70.64	0.29	14.45	1.04	1.13	0.07	0.79	2.11	3.95	4.50	0.08	0.59	0.08		100.17	5
吊罗山	D2821-1	正长花岗岩	71.56	0.30	13.82	0.59	2.05	0.05	0.45	1.88	3.39	4.84	0.07	0.73	0.08		100.4	5
吊罗山	D2950-1	正长花岗岩	73.35	0.27	13.59	0.73	1.00	0.06	0.48	1.39	3.64	4.74	0.07	0.46	0.04	0.08	100.19	5
吊罗山	D5207-1	二长花岗岩	73.24	0.24	13.76	0.67	0.93	0.06	0.55	1.85	3.68	4.16	0.07	0.50	0.12		99.83	5
吊罗山	PM1-26-1	正长花岗岩	70.92	0.33	14.11	0.46	2.23	0.05	0.47	1.79	3.43	5.23	0.07	0.67	0.06		99.82	5
吊罗山	PM1-30-1	正长花岗岩	71.37	0.32	13.79	0.62	2.12	0.04	0.43	1.76	3.36	5.37	0.07	0.54	0.04		99.83	5
吊罗山	D3124-1	二长花岗岩	75.38	0.12	12.80	0.30	1.48	0.051	0.16	0.94	3.45	4.58	0.01	0.14	0.07	0.04	99.53	5
吊罗山	D3125-1	二长花岗岩	74.90	0.15	12.85	0.13	1.88	0.062	0.22	0.92	3.44	4.42	0.03	0.16	0.06	0.06	99.29	5
吊罗山	S006	二长花岗岩	72.55	0.32	13.51	0.76	2.13	0.060	0.81	1.89	2.99	4.16	0.10	0.27	0.12	0.06	99.73	5
吊罗山	QV13-2	正长花岗岩	74.82	0.15	12.96	1.12	1.00	0.040	0.31	0.11	2.76	5.05	0.04	1.20	0	0.02	99.58	5
吊罗山	QV8-1	正长花岗岩	71.67	0.25	13.52	2.06	0.78	0.060	0.63	1.22	2.62	5.71	0.06	1.18	0.04	0.03	99.83	5
	平均		71.25	0.32	14.01	1.00	1.82	0.06	0.51	1.53	3.36	4.95	0.07	0.74	0.06	0.05	100.02	5
新村	D5269-2	二长花岗岩	75.05	0.07	12.90	0.19	1.05	0.02	0.04	0.74	3.48	5.80	0.01	0.44	0.12		99.91	5
新村	D5325-1	二长花岗岩	77.34	0.11	11.95	0.37	0.38	0.04	0.14	0.65	3.85	4.80	0.01	0.20	0.04		99.88	5
新村	D5392-1	碱长花岗岩	72.52	0.43	13.18	1.23	0.93	0.04	0.32	2.08	2.74	5.34	0.09	0.84	0.09	0.4	101.01	5
新村	D3128-1	二长花岗岩	70.70	0.35	14.34	0.76	2.27	0.063	0.72	2.06	3.43	4.30	0.13	0.16	0.07	0.06	99.417	5
新村	S007	二长花岗岩	72.14	0.29	13.73	0.40	2.13	0.060	0.76	1.95	3.69	3.91	0.08	0.24	0.09	0.05	99.522	5
新村	WH12-7	二长花岗岩	73.52	0.19	13.19	0.93	1.94	0.090	0.46	0.81	4.23	3.34	0.22				98.92	5
新村	WH2-4a	二长花岗岩	68.14	0.42	15.15	0.91	3.05	0.090	1.13	2.55	3.14	4.18	0.42				99.18	5
	平均		72.77	0.27	13.49	0.68	1.68	0.06	0.51	1.55	3.51	4.52	0.14	0.38	0.08	0.17	99.69	5

续表

岩石单位	样品号	岩　性	SiO_2	TiO_2	Al_2O_3	Fe_2O_3	FeO	MnO	MgO	CaO	Na_2O	K_2O	P_2O_5	H_2O	CO_2	F	总量	来源
南牙村	3862-2	石英二长斑岩	71.7	0.45	13.37	0.41	2.08	0.05	0.39	2.13	2.92	5.23	0.03	0.96	0.22	0.12	100.7	1
南牙村	73	花岗闪长斑岩	69.52	0.39	13.16	0.45	3.84	0.05	0.24	1.69	2.52	5.04	0.07	1.34	1.05	0.06	101.6	1
高通岭	YV28-1	正长花岗岩	74.09	0.16	13.65	0.35	0.82	0.01	0.36	1.67	3.51	4.28	0.04	0.54	0.24		100.41	2
双顶岭	YIV7-1	正长花岗岩	76.29	0.16	12.26	0.63	0.52	0.04	0.43	1.30	3.24	4.57	0.05	0.42	0.03		100.3	2
六弓	YIV12-1	正长花岗岩	73.09	0.25	13.38	0.74	1.09	0.05	0.74	1.83	3.41	4.36	0.09	0.49	0.21		100.33	2
六弓	Ⅰ HG3-1	正长花岗岩	73.92	0.27	13.07	0.52	1.55	0.06	0.51	1.37	3.03	4.82	0.09	0.48	0.1		100.14	2
六弓	D0914-1	二长花岗岩	69.93	0.39	14.74	0.92	1.63	0.05	1.16	2.78	3.88	3.37	0.09	0.68	0.16		100.35	本次
六弓	D0972-1	二长花岗岩	72.82	0.24	14.49	0.59	0.92	0.04	0.58	2.21	3.97	3.40	0.06	0.42	0.08		100.16	本次
六弓	D1412-1	二长花岗岩	68.99	0.41	14.95	0.98	1.67	0.04	1.31	3.05	3.63	3.83	0.10	0.64	0.18		100.35	本次
六弓	D1430-1	正长花岗岩	72.76	0.19	14.58	0.58	0.62	0.03	0.35	1.47	3.64	4.93	0.04	0.48	0.08		100.15	本次
六弓	D1823-1	二长花岗岩	71.20	0.32	14.80	0.71	1.18	0.03	0.80	2.47	3.80	3.94	0.07	0.39	0.04		100	本次
六弓	D3119-1	二长花岗岩	70.68	0.32	14.43	0.51	2.27	0.045	0.73	2.32	3.89	3.63	0.08	0.38	0.24	0.06	99.594	本次
六弓	QV10-1	二长花岗岩	70.33	0.35	13.99	1.70	1.57	0.060	0.90	1.97	2.65	4.65	0.09	1.32	0.27	0.05	99.9	5
	平均		71.52	0.30	14.27	0.81	1.39	0.05	0.79	2.16	3.54	4.10	0.08	0.59	0.15	0.06	100.11	
保亭	YIV19-1	二长花岗岩	67.81	0.49	15.07	1.30	2.10	0.06	1.58	3.26	3.58	3.77	0.20	0.41	0.14		100.29	2
保亭	X XI6-1	二长花岗岩	71.16	0.36	14.57	1.05	1.23	0.06	0.93	2.34	3.28	4.08	0.12	0.60	0.01		100.22	2
保亭	Ⅰ HG8-1	二长花岗岩	69.8	0.43	14.19	0.45	2.43	0.05	0.64	1.98	3.08	5.31	0.14	0.77	0.46		100.62	2
保亭	D0919-1	二长花岗岩	67.77	0.48	15.36	1.18	1.83	0.05	1.55	3.43	3.56	3.59	0.12	0.79	0.04		100.3	5
保亭	D1229-1	二长花岗岩	67.97	0.47	15.36	1.04	1.80	0.05	1.45	3.22	3.48	3.89	0.10	0.73	0.19		100.34	5
保亭	D1421-1	二长花岗岩	69.40	0.41	15.06	0.27	2.20	0.04	1.27	3.07	3.51	3.61	0.09	0.67	0.17		100.3	5
保亭	D3117-1	二长花岗岩	69.32	0.36	14.99	0.70	2.40	0.049	1.08	2.78	3.76	3.75	0.09	0.10	0.34	0.07	99.793	5
保亭	D3060-1	二长花岗岩	68.30	0.45	14.47	1.29	2.55	0.063	1.69	3.35	3.13	4.24	0.15	0.62	0.15	0.07	100.527	5
保亭	QV4-1	二长花岗岩	68.21	0.48	14.41	1.11	2.25	0.060	1.88	3.05	2.98	4.13	0.14	1.10	0.08	0.06	99.94	5
保亭	GXVIII9-1	二长花岗岩	66.92	0.55	14.28	2.56	1.75	0.060	1.90	3.23	3.35	4.01	0.10	0.94	0.03	0.07	99.75	5
	平均		68.67	0.45	14.78	1.10	2.05	0.05	1.40	2.97	3.37	4.04	0.13	0.67	0.16	0.07	100.21	

续表

岩石单位	样品号	岩　性	SiO_2	TiO_2	Al_2O_3	Fe_2O_3	FeO	MnO	MgO	CaO	Na_2O	K_2O	P_2O_5	H_2O	CO_2	F	总量	来源
加茂	YD599-1	花岗闪长岩	61.95	0.78	15.87	0.97	4.4	0.1	3.24	5.52	2.77	3.06	0.21	0.89	0.02		100.0	2
加茂	YD612-1	花岗闪长岩	61.26	0.79	16.14	1.15	4.4	0.1	3.33	5.75	2.81	2.89	0.22	0.86	0.06		99.97	2
加茂	D102-1	花岗闪长岩	65.05	0.6	16.44	0.98	2.4	0.05	1.98	4.7	3.09	2.81	0.25	1.33	0.02		100.64	2
加茂	D0355	花岗闪长岩	61.95	0.81	15.73	1.48	3.95	0.09	3.04	5.12	3.20	3.35	0.18	0.79	0.06		100.09	本次
加茂	D0478	英云闪长岩	63.16	0.73	16.19	1.50	3.13	0.08	2.31	4.69	3.41	3.52	0.17	0.81	0.06		100.16	本次
加茂	D0830	英云闪长岩	61.94	0.78	16.33	1.78	3.40	0.09	2.66	5.17	3.33	3.24	0.19	0.83	0.02		100.19	本次
加茂	D0893	花岗闪长岩	64.24	0.62	15.29	1.29	2.97	0.07	2.64	4.28	3.25	3.20	0.13	1.64	0.16		101.14	本次
加茂	D0898	花岗闪长岩	66.09	0.53	15.68	1.24	2.13	0.05	1.87	3.69	3.60	3.70	0.12	0.91	0.14		100.43	本次
加茂	D1248	二长花岗岩	58.89	1.07	16.43	2.54	3.38	0.07	3.24	5.47	3.10	2.99	0.35	1.93	0.25		101.52	本次
加茂	D1914-1	花岗闪长岩	57.40	0.91	17.46	2.11	4.70	0.12	3.46	5.91	3.15	2.45	0.25	1.73	0.04		100.86	本次
加茂	PM2-127-1	花岗闪长岩	67.52	0.47	15.60	0.97	1.97	0.05	1.53	3.44	3.80	3.67	0.12	0.52	0.10		99.76	本次
加茂	D3120-1	花岗闪长岩	65.31	0.58	15.95	1.01	2.98	0.060	2.12	3.92	3.62	3.45	0.14	0.21	0.14	0.08	99.57	本次
加茂	QV13-1	花岗闪长岩	61.31	0.84	16.87	1.65	4.08	0.100	1.94	3.65	3.22	3.20	0.21	2.77	0	0.1	99.94	5
加茂	QV14-1	二长闪长岩	60.52	0.97	16.83	1.05	4.77	0.110	2.05	5.61	2.80	3.16	0.23	1.64	0.1	0.1	99.94	5
加茂	GXVIII3-1	花岗闪长岩	64.32	0.70	14.88	2.36	2.15	0.070	2.41	4.04	3.39	3.81	0.20	1.47	0.03	0.07	100.29	5
	平均		62.73	0.75	16.11	1.47	3.39	0.08	2.52	4.73	3.24	3.23	0.20	1.22	0.08	0.09	100.30	
乌石	YV2-1	石英闪长岩	51.50	1.16	15.76	2.20	7.21	0.16	8.50	9.18	2.29	0.45	0.34	0.62	0.32	0.88	100.57	2
乌石	D1321-1	细粒闪长岩	48.51	1.81	17.32	2.02	7.90	0.20	5.77	10.60	2.78	0.78	0.42	1.44	0.19		100.71	5
番通	D776-1	石英正长岩	65.62	0.31	16.76	0.98	2.53	0.090	0.37	1.05	5.21	6.05	0.06	0.42	0.06	0.032	99.53	3
番通	D3037-1	石英正长岩	67.80	0.23	15.87	0.80	2.29	0.127	0.21	1.05	4.52	5.78	0.03	0.38	0.06	0.029	99.17	3
阜石斗	S017	二长花岗岩	75.17	0.18	12.67	0.65	1.75	0.050	0.36	1.55	3.36	4.32	0.04	0.28	0.16	0.18	100.64	3
阜石斗	HG-167	二长花岗岩	73.46	0.26	13.51	0.19	1.31	0.060	0.54	1.43	3.36	3.99	0.08	0.85	0.22	0.26	99.41	3
阜石斗	GIII5-4	正长花岗岩	77.78	0.10	11.81	1.09	0.12	0.010	0.08	0.30	3.50	4.64	0.01	0.44	0.32	0	100.20	3
阜石斗	GIII4-2	正长花岗岩	73.50	0.20	13.20	0.80	1.53	0.060	0.74	0.90	3.92	4.22	0.08	0.46	0.18	0.03	99.81	3
阜石斗	GIII5-2	二长花岗岩	73.19	0.22	13.29	0.98	1.55	0.060	0.45	1.31	3.68	4.38	0.08	0.42	0.24	0.04	99.87	3
阜石斗	Q155-1	正长花岗岩	77.11	0.08	12.29	0.38	0.65	0.020	0.25	0.20	3.52	4.46	0.02	0.76	0.02	0.02	99.77	3
阜石斗	G1977-1	正长花岗岩	76.72	0.10	12.63	0.80	0.18	0.020	0.10	0.13	3.32	5.02	0.01	0.84	0.02	*		3
	平均		75.28	0.16	12.77	0.70	1.01	0.040	0.36	0.83	3.52	4.43	0.05	0.58	0.17			3

续表

岩石单位	样品号	岩 性	SiO_2	TiO_2	Al_2O_3	Fe_2O_3	FeO	MnO	MgO	CaO	Na_2O	K_2O	P_2O_5	H_2O	CO_2	F	总量	来源
抱跃	D3056-1	二长花岗岩	68. 22	0. 45	14. 93	0. 00	3. 93	0. 065	0. 69	3. 14	3. 06	3. 59	0. 09	0. 76	0. 07	0. 086	99. 04	3
抱跃	QXII1-1	二长花岗岩	73. 86	0. 05	13. 62	0. 89	0. 78	0. 020	0. 22	0. 68	2. 06	6. 55	0. 02	0. 91	0. 04	0. 01	99. 71	3
抱跃	HG-40	二长花岗岩	70. 96	0. 18	14. 83	0. 94	1. 25	0. 110	0. 32	2. 53	4. 22	4. 62	0. 11	0. 26	0. 1	0. 06	100. 46	3
抱跃	N515-1	二长花岗岩	68. 39	0. 48	14. 89	1. 46	1. 80	0. 082	1. 05	2. 58	3. 93	4. 18	0. 14	0. 58	0. 05	0. 056	99. 65	3
抱跃	GIII11-2	二长花岗岩	71. 21	0. 29	13. 51	0. 90	1. 98	0. 060	0. 62	2. 34	3. 28	4. 86	0. 08	0. 48	0. 12	0. 08	99. 78	3
	平均		70. 98	0. 27	14. 32	0. 95	1. 84	0. 062	0. 63	2. 06	3. 34	4. 74	0. 09	0. 53	0. 09	0. 055	99. 94	3
税町	D3055-1	花岗闪长岩	67. 72	0. 54	14. 95	0. 04	4. 16	0. 069	0. 94	3. 35	2. 98	3. 65	0. 10	0. 44	0. 1	0. 093	99. 09	3
税町	GVIII7-2	花岗闪长岩	66. 54	0. 50	15. 53	0. 78	3. 80	0. 050	1. 31	3. 39	2. 98	3. 74	0. 10	0. 66	0. 2	0. 07	99. 62	3
税町	HG-39	花岗闪长岩	59. 62	0. 49	16. 73	3. 18	3. 25	0. 170	1. 84	5. 92	3. 78	4. 20	0. 47	0. 40	0. 09	0. 04	100. 16	3
税町	HG-51	花岗闪长岩	60. 56	0. 53	16. 62	2. 72	3. 35	0. 160	2. 34	6. 09	3. 42	2. 76	0. 42	0. 47	0. 04	0. 03	99. 50	3
税町	QXII4-1	花岗闪长岩	63. 60	0. 78	16. 74	2. 12	3. 17	0. 060	1. 54	3. 88	3. 45	2. 82	0. 21	1. 37	0. 12	0. 11	99. 92	3
税町	GII15-1	花岗闪长岩	69. 04	0. 40	14. 46	1. 57	2. 28	0. 060	0. 87	1. 72	2. 80	4. 80	0. 05	1. 04	0. 24	0. 06	99. 36	3
税町	GIII11-3	花岗闪长岩	66. 38	0. 58	14. 82	1. 17	3. 45	0. 100	2. 13	3. 52	3. 12	3. 96	0. 12	0. 56	0. 18	0. 08	100. 14	3
税町	GXII1-1	花岗闪长岩	66. 06	0. 50	15. 10	2. 06	2. 65	0. 100	0. 89	4. 56	3. 24	2. 83	0. 25	1. 18	0. 06	0. 07	99. 52	3
	平均		65. 49	0. 52	15. 35	1. 69	3. 13	0. 111	1. 59	3. 63	3. 44	3. 56	0. 18	0. 64	0. 15	0. 064	99. 53	3
$\gamma\pi K_1$	GI2-1	花岗斑岩	70. 49	0. 28	12. 96	1. 11	2. 38	0. 060	0. 68	2. 17	2. 68	4. 60	0. 02	0. 87	2. 07	0. 09	100. 42	3
$\gamma\pi K_1$	G21-1	花岗斑岩	71. 69	0. 25	13. 99	1. 45	1. 62	0. 068	0. 61	1. 88	3. 58	4. 18	0. 08	0. 40	0. 01	0. 19	99. 92	3
$\gamma\pi K_1$	GI6-1	花岗斑岩	77. 28	0. 08	12. 09	0. 67	0. 88	0. 020	0. 33	0. 33	2. 34	5. 52	0. 01	0. 40	0. 6	0. 01	100. 54	3
$\gamma\pi K_1$	GI12-2	花岗斑岩	67. 12	0. 16	16. 58	1. 66	1. 52	0. 130	0. 46	1. 82	4. 28	5. 66	0. 04	0. 36	0. 03	0. 03	99. 83	3
$\gamma\delta\pi K_1$	GII8-1	花岗闪长玢岩	73. 84	0. 21	13. 08	1. 37	1. 42	0. 060	0. 54	1. 45	3. 28	4. 04	0. 02	0. 60	0. 03	0. 06	99. 97	3
$\gamma\delta\pi K_1$	GIII3-1	花岗闪长玢岩	65. 76	0. 46	16. 76	1. 64	2. 48	0. 100	1. 04	3. 59	3. 86	3. 38	0. 15	0. 48	0. 18	0. 06	99. 91	3
$\gamma\delta\pi K_1$	G1872	花岗闪长玢岩	70. 34	0. 35	14. 33	1. 59	1. 72	0. 118	0. 87	1. 77	3. 22	3. 95	0. 15	1. 08	0. 09	0. 058	99. 61	3
νK_1	D3045-1	中细粒辉长闪长岩	53. 94	1. 24	16. 76	1. 32	6. 93	0. 127	5. 21	7. 41	3. 21	1. 90	0. 33	0. 58	0. 12	0. 06	99. 11	3
νK_1	N517-1	中细粒辉长闪长岩	53. 90	1. 15	16. 56	8. 10	0. 82	0. 130	5. 31	7. 57	3. 13	2. 12	0. 40	0. 50	0. 07	0. 053	99. 79	3
νK_1	GIX5-1	细粒角闪辉长岩	49. 02	0. 40	16. 45	1. 28	6. 05	0. 150	9. 83	12. 20	1. 46	0. 88	0. 08	1. 38	0. 43	0. 02	99. 62	3
	平均		52. 29	0. 93	16. 59	3. 57	4. 60	0. 136	6. 78	9. 06	2. 60	1. 63	0. 27	0. 82	0. 21	0. 044	99. 51	

续表

岩石单位	样品号	岩　性	SiO_2	TiO_2	Al_2O_3	Fe_2O_3	FeO	MnO	MgO	CaO	Na_2O	K_2O	P_2O_5	H_2O	CO_2	F	总量	来源
半岛	Ha31	正长花岗岩	72.5	0.275	12.85	0.504	2.8	0.07	0.472	1.47	3.15	4.82	0.075	0.589	0.23	0.121	100.1	1
半岛	7331-1	正长花岗岩	71.75	0.32	13.42	0.57	2.34	0.06	0.45	1.73	3.41	4.81	0.13	0.38	0.11	0.09	99.58	1
半岛	HG389	正长花岗岩	71.04	0.22	14.08	0.64	2.22	0.03	0.28	1.6	2.92	5.68	0.06	0.4	0.09	0.07	99.33	1
马岭	177	正长花岗岩	72.38	0.1	14.63	0.25	1.86	0.06	0.08	0.7	4.16	4.65	0.03	0.2	0.19	0.15	99.46	1
马岭	XX1-1	正长花岗岩	74.22	0.18	13.19	1.2	1.08	0.05	0.665	1.14	3.26	3.76	0.05	0.54	0	0	99.98	1
马岭	XⅧ1-1	正长花岗岩	76.11	0.02	12.63	0.44	1.22	0.031	0.15	0.925	3.5	3.42	0.02	0.66	0	0	99.91	1
	平均		74.24	0.10	13.48	0.63	1.39	0.05	0.30	0.92	3.64	3.94	0.03	0.47	0.06	0.05	99.78	
马鞍岭	Ha13-1	正长花岗岩	75.39	0.034	12.91	0.115	1.73	0.047	0.106	0.923	3.65	4.85	0.004	0.24	0.078	0.04	100.14	1
马鞍岭	Ha13-2	正长花岗岩	75.58	0.034	12.84	0.192	1.78	0.067	0.084	0.727	3.8	4.62	0.003	0.32	0.195	0.03	100.31	1
马鞍岭	184	正长花岗岩	72.3	0.35	13.42	0.58	2.3	0.04	0.46	1.42	2.94	5.66	0.08	0.46	0.18	0.14	100.65	1
马鞍岭	187	正长花岗岩	70.88	0.38	13.98	0.4	3.32	0.05	0.43	1.62	3.37	5.08	0.14	0.26	0.13	0.16	100.22	1
马鞍岭	YXV3-1	正长花岗岩	70.67	0.19	15.11	1.02	1.08	0.06	0.35	1.26	3.61	5.46	0.04	0.90	0.03		100.54	2
	平均		72.96	0.20	13.65	0.46	2.04	0.05	0.29	1.19	3.47	5.13	0.05	0.44	0.12	0.09	100.37	
三狮岭	HG400	正长花岗岩	73.72	0.21	12.34	0.59	1.73	0.04	0.22	1.51	2.68	5.46	0.08	0.46	0.14	0.08	99.26	1
三狮岭	Ⅱ28-1	正长花岗岩	76.14	0.09	11.73	0.65	1.44	0.04	0.39	0.22	3	4.7	0.02	0.68	0.48	0	99.58	1
三狮岭	Ⅱ30-1	正长花岗岩	75.77	0.09	12.63	0.81	1.18	0.04	0.41	0.25	2.8	4.75	0.04	0.72	0.34	0	99.83	1
三狮岭	YI1-2	正长花岗岩	71.98	0.31	13.70	0.51	2.20	0.05	0.54	1.71	2.60	5.45	0.08	0.58	0.08		99.79	2
三狮岭	YD438	正长花岗岩	64.35	0.25	18.57	1.64	1.10	0.06	0.24	1.78	5.09	5.85	0.04	0.66	0.07		99.7	2
三狮岭	D1740-2	正长花岗岩	72.58	0.2	13.43	0.55	1.0	0.05	0.72	1.85	4.25	4.45	0.07	0.65	0.04		100.27	2
三狮岭	HN03	正长花岗岩	73.5	0.17	14.17	0.31	0.93	0.05	0.51	1.9	3.23	4.41						2
三狮岭	D958-1	二长花岗岩	72.47	0.4	13.64	0.1	2.2	0.03	0.5	1.74	2.32	5.41	0.08	0.84	0.02	0.44	100.19	2
三狮岭	HN05	二长花岗岩	72.69	0.4	13.33	0.24	2.13	0.04	0.55	1.88	2.4	5.37						2
三狮岭	D0399-1	正长花岗岩	76.18	0.07	12.70	0.48	0.53	0.06	0.13	0.72	3.41	5.11	0.01	0.44	0.06		100.25	本次
三狮岭	D1004-1	正长花岗岩	77.28	0.04	12.40	0.65	0.32	0.06	0.08	0.63	3.03	4.79	0.01	0.57	0.04		100.46	本次
三狮岭	D1213-1	正长花岗岩	74.54	0.13	13.67	0.89	0.45	0.06	0.23	1.02	3.27	4.99	0.01	0.55	0.04		100.39	本次
三狮岭	D1036	二长花岗岩	76.04	0.13	12.62	0.07	2.26	0.040	0.14	0.09	3.54	5.20	0.03	0.36	0.23	0.01	100.76	本次
	平均		73.63	0.19	13.46	0.58	1.34	0.05	0.36	1.18	3.20	5.07	0.04	0.59	0.14	0.11	100.04	

续表

岩石单位	样品号	岩 性	SiO_2	TiO_2	Al_2O_3	Fe_2O_3	FeO	MnO	MgO	CaO	Na_2O	K_2O	P_2O_5	H_2O	CO_2	F	总量	来源
架 水	D3361-1	二长花岗岩	62.81	1.20	15.45	1.20	4.42	0.09	1.59	3.69	3.20	4.00	0.49	1.44	0.07	0	99.58	2
椰子园	D3032-1	二长花岗岩	70.02	0.40	14.30	0.95	2.35	0.056	0.60	1.84	3.20	5.12	0.11	0.40	0.17	0.079	99.57	3
南山岭	S018	二长花岗岩	66.76	0.58	15.15	1.22	3.05	0.080	1.18	2.47	3.43	5.23	0.20	0.48	0.05	0.104	99.94	3
南山岭	N605-1	二长花岗岩	66.89	0.56	15.48	1.64	1.72	0.067	1.11	2.50	3.75	5.42	0.22	0.56	0.11	0.098	100.08	3
	平均		66.83	0.57	15.32	1.43	2.39	0.074	1.15	2.49	3.59	5.33	0.21	0.52	0.08	0.101	100.01	3
马翁岭	Ha37-2	闪长岩	48.36	1.32	19.46	2.24	6.91	0.14	3.94	8.9	3.4	2.54	0.92	0.57	0.22	00	99.13	1
马翁岭	516	闪长岩	56.71	0.98	17.32	8.88		0.13	3.38	6.75	2.84	2.47	0.23	0	0	00	99.82	1
马翁岭	YXVIII6-2	闪长岩	51.66	1.22	16.56	0.89	7.46	0.15	4.94	8.95	3.16	2.33	0.45	1.85	0.1		99.75	2
马翁岭	D1153-1	二长闪长岩	50.98	0.85	15.00	1.94	6.24	0.150	6.58	10.37	2.68	2.71	0.34	1.01	0.45	0.08	99.38	本次
马翁岭	D1670-1	二长闪长岩	51.42	0.84	11.57	1.22	6.07	0.140	9.51	13.09	1.62	1.51	0.23	1.02	0.78	0.05	99.07	本次
马翁岭	D1535-1	角闪辉长岩	51.36	0.74	11.33	1.71	6.10	0.140	10.57	12.94	1.64	1.55	0.18	1.10	0.2	0.05	99.61	本次
	平均		51.75	0.99	15.21	2.81	6.56	0.14	6.49	10.17	2.56	2.19	0.39	0.93	0.29	0.04	99.46	
八宝坡	H1245-2	巨粒角闪正长岩	62.48	0.24	17.75	1.22	3.87	0.206	0.46	2.12	5.40	5.68	0.11	0.37	0.35	0.06	100.29	3
八宝坡	Hlm-2	巨粒角闪正长岩	65.09	0.21	16.49	0.11	3.70	0.190	0.36	2.43	5.08	5.21	0.08	0.38	0.14	0.104	99.53	3
	平均		63.79	0.23	17.12	0.67	3.79	0.198	0.41	2.28	5.24	5.45	0.10	0.38	0.25	0.082	99.91	3
黑栋	D1244	石英二长岩	61.45	0.62	17.54	1.01	4.46	0.090	0.98	3.80	3.70	4.82	0.18	0.43	0.08	0.076	99.20	3
里茂	H1247-2	辉长岩	43.96	0.91	17.08	5.66	7.23	0.213	6.47	12.78	1.56	2.06	0.43	0.22	0.24	0.085	98.86	3
里茂	Hlm-1	辉长岩	40.02	1.67	12.43	5.59	10.34	0.270	9.63	14.60	1.03	1.36	0.13	1.04	0.27	0.132	98.46	3
	平均		41.99	1.29	14.76	5.63	8.79	0.242	8.05	13.69	1.30	1.71	0.28	0.63	0.26	0.109	98.66	
迈洲岭	974-1	碱长花岗斑岩	73.6	0.22	12.28	1.21	2.03	0.013	0.12	0.06	0.19	8.73	0.036	0.92	0.01		99.73	2
打显村	XVI19-1	二长花岗岩	73.8	0.2	13.52	1.08	0.6	0.054	0.62	1.32	2.74	5.3	0.1	0.46	0.21		100.5	1
打显村	XVI2-2	二长花岗岩	72.14	0.3	13.93	1.29	1.18	0.05	0.92	1.68	2.88	5.3	0.15	0.32	0.26		100.8	1
白石岭	XVI16-1	二长花岗岩	74.75	0.12	13.38	0.96	0.58	0.032	0.41	1.04	2.92	4.96	0.07	0.46	0.21		100.4	1
白石岭	XVI2-1	二长花岗岩	72.48	0.3	13.88	1.19	1	0.05	0.98	1.57	2.64	5.3	0.17	0.32	0.12		100.3	1
黑岭	Ha25	二长花岗岩	72.5	0.30	13.08	0.569	2.65	0.05	0.37	1.37	3.2	5.25	0.07	0.38	0.052		100.2	1
黑岭	4086-2	二长花岗岩	71.38	0.32	13.82	0.36	2.45	0.04	0.36	1.56	3.15	5.5	0.09	0.68	0.09		99.8	1
分水岭	XVII10-1	二长花岗岩	73.12	0.22	13.47	0.63	2.22	0.05	0.41	2.21	2.7	3.5	0.075	0.84	0		100.5	1
分水岭	4063-1	二长花岗岩	72.84	0.25	13.57	0.33	2.08	0.1	0.33	1.44	3.24	5	0.07	0.46	0.36		100.9	1

续表

岩石单位	样品号	岩　性	SiO_2	TiO_2	Al_2O_3	Fe_2O_3	FeO	MnO	MgO	CaO	Na_2O	K_2O	P_2O_5	H_2O	CO_2	F	总量	来源
分水岭	4086-7	二长花岗岩	73.19	0.31	12.9	0.6	1.97	0.1	0.51	1.31	3.29	5.24	0.06	0.45	0		100.7	1
分水岭	D5543-1	二长花岗岩	71.09	0.26	14.29	0.43	2.4	0.03	0.61	1.27	2.53	4.37	0.12	1.0	0.07		99.55	1
分水岭	D5545-1	二长花岗岩	69.13	0.40	14.60	0.75	3.06	0.05	0.79	2.99	2.14	4.02	0.15	0.98	0.15		99.36	1
	平均		71.87	0.29	13.77	0.55	2.35	0.07	0.53	1.84	2.78	4.43	0.10	0.75	0.12		100.20	
南宝岭	4602-1	二长花岗岩	75.27	0.11	12.44	0.62	1.65	0.03	0.22	0.78	3.47	4.5	0.03	0.54	0.09		99.75	1
南宝岭	4676-2	二长花岗岩	72.46	0.33	12.89	0.56	2	0.04	1.13	1.66	3.28	4.82	0.09	0.4	0.09		100.5	1
五里桥	Ha22	二长花岗岩	72.35	0.26	12.92	0.543	3	0.06	0.22	1.17	3.4	5.4	0.04	0.38	0.065		100.18	1
五里桥	D887-1	二长花岗岩	70	0.32	13.81	2.38	1.28	0.04	0.51	3.12	2.66	4.66	0.1	0.17	0.24		99.7	1
五里桥	4033-1	二长花岗岩	73.19	0.18	12.98	0.53	2.18	0.1	0.16	1.2	3.24	5.3	0.05	0.4	0.19		100.58	1
五里桥	4077-1	二长花岗岩	74.72	0.17	11.75	0.58	2	0.09	0.04	0.8	3.02	5.3	0.02	0.34	0.4		100.01	1
五里桥	4080-3	二长花岗岩	73.51	0.2	12.66	0.6	1.98	0.1	0.12	1.44	3.24	5.46	0.04	0.36	0.45		101	1
五里桥	4082-1	二长花岗岩	72.45	0.15	13	0.66	2.13	0.1	0.12	1.44	3.14	5.3	0.04	0.43	0.42		100.42	1
五里桥	Ha27	二长花岗岩	73.02	0.19	12.63	0.465	2.57	0.05	0.152	0.99	3.28	5.28	0.02	0.434	0.065		99.63	1
	平均		72.75	0.21	12.82	0.82	2.16	0.08	0.19	1.45	3.14	5.24	0.04	0.36	0.26		100.22	
苗圃	4072-4	二长花岗岩	68.98	0.3	13.54	1.1	3.69	0.14	0.33	2.4	3.01	5.31	0.1	0.54	0.34		100.78	1
苗圃	D887-2	二长花岗岩	69.11	0.4	15.01	0.98	1.82	0.04	0.76	3.17	3.04	4.81	0.12	0.57	0.24		100.87	1
苗圃	D891-2	二长花岗岩	69.84	0.55	14.83	0.92	2.55	0.04	0.63	2.04	2.66	5.22	0.12	0.35	0.24		100.56	1
苗圃	Ha21-1	二长花岗岩	69.68	0.404	13.64	0.492	3.75	0.06	0.39	1.57	3.38	5.54	0.08	0.46	0.091		100.19	1
	平均		69.40	0.41	14.26	0.87	2.95	0.07	0.53	2.30	3.02	5.22	0.11	0.48	0.23		100.60	
乌烈岭	Ha14-2	正长花岗岩	71.45	0.24	12.96	0.342	3.97	0.066	0.096	1.17	3.4	5.77	0.021	0.02	0.143		99.81	1
乌烈岭	Ha15	正长花岗岩	69.53	0.372	13.68	0.5	3.98	0.08	0.196	1.54	3.4	6.08	0.049	0.18	0.065		99.79	1
乌烈岭	4001-1	正长花岗岩	72.74	0.26	13.56	0.48	2.15	0.11	0.3	1.44	3.14	5.3	0.07	0.4	0.28		101.2	1
乌烈岭	4001-7	正长花岗岩	73.19	0.23	12.95	0.52	2.19	0.12	0.2	1.32	3	5.41	0.06	0.47	0.38		101.1	1
乌烈岭	4044-4	正长花岗岩	72.62	0.23	13.85	0.5	1.72	0.12	0.3	1.68	3.31	5.2	0.06	0.49	0.36		101.3	1
乌烈岭	4104-6	正长花岗岩	72.82	0.2	13.6	0.38	1.96	0.11	0.3	1.68	3.26	5.47	0.06	0.18	0.29		101.3	1
乌烈岭	7323	正长花岗岩	73.47	0.29	13.44	0.44	1.98	0.04	0.34	1.27	3.28	5.18	0.08	0.46	0.13		100.9	1
	平均		72.26	0.26	13.43	0.45	2.56	0.09	0.25	1.44	3.26	5.49	0.06	0.31	0.24		100.77	

续表

岩石单位	样品号	岩　性	SiO_2	TiO_2	Al_2O_3	Fe_2O_3	FeO	MnO	MgO	CaO	Na_2O	K_2O	P_2O_5	H_2O	CO_2	F	总量	来源
东四	Ha14-1	正长花岗岩	68.07	0.6	14.29	0.49	4.08	0.065	0.696	2.08	3.42	5.43	0.138	0.25	0.078		99.92	1
东四	Ha23	正长花岗岩	71.66	0.312	13.02	0.39	3.14	0.049	0.348	1.34	3.14	5.3	0.062	0.4	0.143		99.7	1
东四	4091-2	正长花岗岩	72.8	0.2	12.97	0.37	2.25	0.12	0.3	1.93	3.06	5.19	0.07	0.43	0.26		100.9	1
东四	4125-2	正长花岗岩	72.3	0.24	13.58	0.56	2.22	0.12	0.32	1.68	2.96	5.49	0.07	0.53	0.35		101.4	1
东四	7314	正长花岗岩	72.89	0.25	12.92	0.55	2.3	0.04	0.25	1.04	3.37	5.47	0.09	0.54	0.08		99.98	1
	平均		71.54	0.32	13.36	0.47	2.80	0.08	0.38	1.61	3.19	5.38	0.09	0.43	0.18		100.38	
七差岭	6041	正长花岗岩	70.89	0.49	14.47	0.54	2.09	0.04	0.57	1.62	3.46	5.57	0.15	0.8	0.14		101.6	1
七差岭	4964-1	正长花岗岩	71.37	0.42	14.16	0.69	1.85	0.04	0.58	1.76	3.4	4.05	0.16	0.74	0.18		99.4	1
七差岭	4994-1	正长花岗岩	70.68	0.48	14.04	0.59	1.98	0.04	1.13	1.89	3.38	4.96	0.16	0.41	0.1		100.6	1
	平均		70.98	0.46	14.22	0.61	1.97	0.04	0.76	1.76	3.41	4.86	0.16	0.65	0.14		100.53	
保梅岭	Ha30-1	正长花岗岩	71.51	0.29	13.41	0.36	2.92	0.06	0.51	1.8	3.09	4.83	0.08	0.42	0.16	0.09	99.53	1
保梅岭	Ha30-6	正长花岗岩	71.5	0.31	13.38	0.11	3.26	0.06	0.53	1.79	3.15	4.84	0.08	0.47	0.17	0.09	99.74	1
保梅岭	Ha33-1	正长花岗岩	72.84	0.23	13.07	0.29	2.76	0.05	0.36	1.43	3.26	4.91	0.06	0.61	0.12	0.07	100.06	1
	平均		73.76	0.18	12.92	0.21	2.44	0.04	0.30	1.20	3.24	5.12	0.05	0.40	0.17	0.06		
保梅岭	Ha30-2	包体	60.3	0.82	15.49	0.70	6.14	0.13	3.07	4.54	4.56	2.02	0.28	0.45	0.14	0.64		1
保梅岭	Ha30-3	包体	65.16	0.76	15.1	0.73	5	0.09	1.33	3.25	4.16	2.52	0.23	0.60	0.12	0.25		1
保梅岭	Ha30-4	包体	61.84	0.46	16.81	1.02	4.66	0.12	1.56	3.96	4.82	3.2	0.3	0.44	0.15	0.15		1
保梅岭	Ha30-5	包体	58.76	0.55	17.42	0.86	5.87	0.16	1.44	4.14	4.78	3.92	0.28	0.49	0.13	0.23		1
	平均		61.52	0.65	16.21	0.83	5.42	0.13	1.85	3.97	4.58	2.92	0.27	0.50	0.14	0.32		
洞古岭	YⅢ11-1	花岗斑岩	66.28	0.85	14.87	1.96	2.15	0.06	1.44	2.45	3.24	4.67	0.33	1.30	0.02		99.62	2
南通岭	ⅤHG1-7	二长花岗岩	70.7	0.46	13.61	0.79	2.33	0.06	0.63	1.68	2.28	6.26	0.12	0.77	0.04		100.12	2
南通岭	HP8-G-1	二长花岗岩	68.44	0.06	15.36	0.65	2.54	0.04	1.14	2.54	3.84	4.04	0.19				99.52	2
南通岭	HN01-2	二长花岗岩	66.54	0.88	13.08	1.55	4.93	0.18	1.18	3.77	2.28	3.54						2
琼中县	D493-2	二长花岗岩	68.52	0.63	14.5	0.56	2.83	0.06	0.91	2.1	3.12	5.26	0.22	0.9	0.08		100.27	2
琼中县	XXI2-2①	二长花岗岩	75.53	0.19	12.57	1.05	0.73	0.01	0.33	0.3	3.15	4.97	0.04	0.9	0.04		100.67	2
琼中县	YXX1-1	二长花岗岩	73.02	0.45	13.66	1.82	1.17	0.02	0.71	0.57	6.06	1.38	0.16	0.81	0.01		100.56	2

续表

岩石单位	样品号	岩　性	SiO_2	TiO_2	Al_2O_3	Fe_2O_3	FeO	MnO	MgO	CaO	Na_2O	K_2O	P_2O_5	H_2O	CO_2	F	总量	来源
琼中县	YD499	二长花岗岩	72.62	0.33	14.28	0.47	1.06	0.01	0.51	1.3	2.72	5.21	0.09	1.09	0.07		99.76	2
琼中县	YXX6-1	二长花岗岩	71.91	0.28	14.82	0.65	0.86	0.03	0.51	1.26	3.31	4.76	0.1	1.18	0.1		99.77	2
琼中县	D823-1	二长花岗岩	71.5	0.28	14.32	0.21	1.83	0.04	0.9	2.1	2.81	4.92	0.12	0.7	0.08		100.29	2
	平均		72.18	0.36	14.03	0.79	1.41	0.03	0.65	1.27	3.53	4.42	0.12	0.93	0.06		100.22	
黎母岭	D1601-1	二长花岗岩	72.12	0.37	14.21	0.01	1.75	0.04	0.6	2.22	2.74	4.94	0.1	0.66	0.08		100.3	2
黎母岭	D101-3	二长花岗岩	69.43	0.46	14.22	0.65	2.73	0.05	0.56	1.98	2.48	6.26	0.13	0.68	0.1		100.18	2
黎母岭	ⅤHG1-1	二长花岗岩	65.41	0.82	14.52	1.04	4.65	0.09	1.08	2.98	2.58	5.24	0.27	0.95	0.06		100.14	2
黎母岭	D1618-1	二长花岗岩	75.99	0.16	12.2	0.71	1.02	0.02	0.31	1.04	2.52	5.37	0.03	0.43	0.06		100.21	2
黎母岭	ⅢHG2-3	二长花岗岩	66.01	0.67	15.5	0.61	3.1	0.07	1.18	3.47	3.14	4.5	0.25	1.12	0.06		100.43	2
黎母岭	HN01-1	二长花岗岩	64.32	0.79	15.46	0.77	4.67	0.12	1.12	3.32	2.8	5.24						2
黎母岭	HN02	二长花岗岩	67.31	0.58	14.69	0.82	3.7	0.07	0.7	2.18	2.46	6.35						2
黎母岭	HN25	二长花岗岩	69.53	0.43	15.09	0.23	2.52	0.05	0.83	2.81	2.51	4.63						2
黎母岭	HN26	二长花岗岩	69.67	0.46	14.41	0.13	2.73	0.05	1.08	2.4	2.9	3.59						2
黎母岭	HN27	二长花岗岩	66.16	0.75	14.69	2.15	2.43	0.06	1.14	1.96	2.62	5.98						2
黎母岭	HN30	二长花岗岩	78.07	0.21	14.32	0.82	0.67	0.02	0.45	1.5	3.04	4.84						2
	平均	二长花岗岩	69.46	0.52	14.48	0.72	2.72	0.06	0.82	2.35	2.71	5.18	0.16	0.77	0.07		100.25	2
后美	Ⅺ1-5	碱长花岗岩	75.43	0.05	13.32	0.48	0.38	0.02	0.20	0.19	3.30	5.44	0.04	0.96	0.08	0.01	99.90	6
后龙	Ⅺ3-3	碱长花岗岩	70.39	0.15	15.78	0.42	0.90	0.02	0.39	1.11	3.42	5.81	0.11	1.22	0.13	0.03	99.88	6
石墩	H1519-1	中粗粒闪长岩	52.16	1.30	20.48	1.00	7.06	0.123	1.61	7.46	3.34	2.32	0.74	1.66	0.26	0.05	99.6	3
长安	D1653-1	不等粒角闪辉石岩	49.90	0.81	6.61	2.01	8.00	0.170	17.33	11.76	0.98	0.64	0.16	0.44	0.41	0.050	99.3	3
长安	H303-1	中粗粒苏长辉长岩	47.62	1.53	16.06	2.44	9.32	0.160	7.99	10.30	1.56	1.23	0.13	0.62	0.10	0.080	99.1	3
长安	S004-1	中粒角闪辉长岩	45.56	3.19	14.39	5.18	10.10	0.240	5.78	9.14	2.18	1.13	0.69	1.30	0.14	0.144	99.2	3
长安	247-1	辉长玢岩	50.78	0.52	7.82	9.87	0	0.17	13.69	14.71	0.79	0.46	0.07		0	0.091	99.64	1
长安	247-4	辉长玢岩	49.91	0.5	7.79	7.94	0	0.13	11.26	18.92	0.83	0.88	0.03		0		99.52	1
	平均		48.75	1.31	10.53	5.49	5.48	0.17	11.21	12.97	1.27	0.87	0.22	0.79	0.13	0.09	99.35	

续表

岩石单位	样品号	岩性	SiO_2	TiO_2	Al_2O_3	Fe_2O_3	FeO	MnO	MgO	CaO	Na_2O	K_2O	P_2O_5	H_2O	CO_2	F	总量	来源
石龟岭	D3042-1	二长花岗岩	76.74	0.03	12.05	0.00	1.74	0.064	0.04	0.49	4.08	4.02	0.02	0.18	0.1	0.159	99.64	3
石龟岭	N495-1	二长花岗岩	76.51	0.06	12.79	0.37	0.78	0.069	0.03	0.52	4.12	4.28	0.04	0.52	0.07	0.145	100.25	3
	平均		76.63	0.05	12.42	0.19	1.26	0.067	0.04	0.51	4.10	4.15	0.03	0.35	0.09	0.152	99.94	3
兵工厂	H1219-2	碱长花岗岩	76.34	0.05	12.23	0.58	1.97	0.020	0.03	0.12	3.99	4.56	0.01	0.40	0.08	0.153	100.5	3
兵工厂	H1562-1	碱长花岗岩	76.30	0.06	12.18	0.42	1.90	0.015	0.03	0.17	3.89	4.26	0.02	0.48	0.02	0.100	99.9	3
兵工厂	D3044-1	二长花岗岩	75.82	0.03	12.42	0.00	1.89	0.068	0.03	0.48	3.98	4.14	0.05	0.22	0.05	0.711	99.59	3
兵工厂	N673-1	二长花岗岩	76.16	0.04	12.45	0.73	0.45	0.030	0.24	0.50	4.00	4.80	0.01	0.60	0.03	0.067	100.08	3
	平均		76.16	0.05	12.32	0.43	1.55	0.033	0.08	0.32	3.97	4.44	0.02	0.43	0.05	0.258	100.0	3
六连岭	H1207-1	中粗粒黑云母正长花岗岩	75.31	0.07	12.02	0.52	2.38	0.029	0.01	0.53	4.00	4.85	0.01	0.18	0.45	0.206	100.6	3
六连岭	H1207-4	中粗粒黑云母正长花岗岩	75.46	0.07	12.39	0.25	1.99	0.023	0.01	0.64	3.90	4.76	0.02	0.18	0.21	0.320	100.2	3
六连岭	S001	中粗粒黑云母正长花岗岩	75.40	0.07	11.90	0.40	2.36	0.030	0.07	0.68	3.95	4.44	0.01	0.36	0.06	0.278	100.0	3
六连岭	H3357-1	中粗粒黑云母正长花岗岩	75.76	0.12	12.03	0.34	2.09	0.026	0.03	0.64	3.40	4.68	0.01	0.41	0.02	0.140	99.7	3
六连岭	S044	中粗粒黑云母正长花岗岩	73.88	0.15	12.94	0.22	2.88	0.080	0.19	0.84	3.44	5.54	0.04	0.12	0.17	0.166	100.7	3
六连岭	S013-1	中粗粒黑云母正长花岗岩	74.60	0.07	12.66	0.31	2.33	0.050	0.01	0.80	3.74	5.22	0.01	0.32	0.12	0.021	100.3	3
六连岭	D3043-1	二长花岗岩	76.30	0.02	12.35	0.09	1.67	0.071	0.06	0.68	3.91	4.30	0.01	0.20	0.22	0.138	99.96	3
六连岭	N603-1	二长花岗岩	77.78	0.05	11.80	0.31	0.96	0.057	0.04	0.58	4.13	4.48	0.04	0.36	0.02	0.155	100.69	3
六连岭	N7-3	二长花岗岩	76.16	0.08	12.17	0.29	1.21	0.042	0.36	0.58	4.04	4.26	0.04	0.56	0.05	0.225	99.97	3
	平均		75.63	0.08	12.25	0.30	1.99	0.045	0.09	0.66	3.83	4.73	0.02	0.30	0.15	0.183	100.2	3
前锋	GII12-1	正长花岗岩	75.72	0.12	12.40	1.13	1.05	0.050	0.40	0.90	3.32	4.62	0.05	0.34	0.24	0.07	100.38	3
前锋	GII2-3	正长花岗岩	77.63	0.06	12.36	0.62	0.70	0.020	0.24	0.42	3.30	4.12	0.05	0.56	0.16	0.05	100.27	3
前锋	G851	正长花岗岩	76.72	0.10	12.05	0.93	0.93	0.040	0.28	0.53	3.34	3.76	0.05	0.48	0.06	0.09	99.32	3
	平均		76.68	0.09	12.30	0.78	0.98	0.032	0.26	0.51	3.38	4.38	0.05	0.51	0.14	0.058	100.11	3

续表

岩石单位	样品号	岩　性	SiO_2	TiO_2	Al_2O_3	Fe_2O_3	FeO	MnO	MgO	CaO	Na_2O	K_2O	P_2O_5	H_2O	CO_2	F	总量	来源
结尾	G42	正长花岗岩	77.44	0.05	11.53	0.91	0.22	0.010	0.06	0.40	3.44	5.02	0.03	0.40	0.08	*	99.6	3
结尾	GXIV1-3	正长花岗岩	78.06	0.15	12.00	0.89	1.00	0.020	0.35	0.27	2.80	3.16	0.05	0.70	0.06	0.020	99.5	3
结尾	GII1-1	二长花岗岩	77.14	0.08	12.19	0.48	0.78	0.040	0.24	0.61	3.16	4.32	0.25	0.42	0.09	0.02	99.81	3
结尾	GII3-4	二长花岗岩	75.13	0.12	12.76	1.15	1.08	0.040	0.46	1.27	3.04	4.32	0.02	0.50	0.09	0.06	100.01	3
结尾	N604-1	正长花岗岩	77.86	0.07	12.34	0.38	0.52	0.023	0.05	0.31	3.70	5.64	0.04	0.40	0.07	0.037	101.42	3
结尾	Nla-1	正长花岗岩	75.90	0.09	12.74	0.37	0.99	0.017	0.13	0.23	3.63	5.38	0.02	0.62	0.12	0.04	100.26	3
	平均		76.87	0.09	12.22	0.67	0.83	0.027	0.22	0.56	3.29	4.65	0.06	0.50	0.11	0.041	100.1	3
超盆	GIX9-1	正长花岗岩	76.69	0.08	12.59	0.52	0.98	0.000	0.25	0.24	2.92	4.46	0.05	0.42	0.16	0.020	99.4	3
超盆	G1284-2	正长花岗岩	76.18	0.08	12.27	0.50	1.35	0.050	0.28	0.49	3.64	4.14	0.05	0.24	0.08	0.020	99.4	3
超盆	GII6-1	二长花岗岩	74.44	0.18	13.05	0.95	1.62	0.060	0.33	1.39	3.18	4.30	0.05	0.72	0.06	0.06	100.36	3
超盆	G011-2	二长花岗岩	74.24	0.16	12.42	1.00	2.38	0.080	0.45	1.24	3.44	4.32	0.05	0.39	0.18	0.09	100.40	3
超盆	GV1-1	二长花岗岩	72.71	0.20	12.90	0.85	2.38	0.060	0.49	1.81	3.22	3.84	0.05	0.74	0.3	0.1	99.61	3
超盆	GV7-1	二长花岗岩	75.56	0.10	12.35	0.61	1.70	0.060	0.26	1.14	3.08	3.98	0.05	0.46	0.18	0.13	99.61	3
超盆	N7-10	二长花岗岩	75.25	0.06	12.95	0.32	1.87	0.021	0.06	0.24	3.95	4.79	0.02	0.76	0.01	0.066	100.34	3
	平均		75.01	0.13	12.67	0.76	1.66	0.050	0.36	0.93	3.28	4.28	0.04	0.52	0.14	0.063	99.9	3
布山村	N527-1	二长花岗岩	73.17	0.30	13.08	0.73	1.91	0.040	0.35	1.20	3.02	5.84	0.08	0.50	0.02	0.052	100.3	3
布山村	D3034-1	二长花岗岩	69.94	0.40	14.18	0.86	3.13	0.107	0.64	2.25	3.73	3.46	0.12	0.61	0.16	0.125	99.7	3
布山村	N727-1	二长花岗岩	66.09	0.54	15.81	1.31	2.75	0.080	0.90	3.50	3.83	4.03	0.20	0.72	0.08	0.024	99.9	3
布山村	GIX3-2	二长花岗岩	65.60	0.65	15.70	0.94	3.88	0.080	1.48	2.17	3.14	4.24	0.20	1.46	0.24	0.060	99.8	3
布山村	N488-1	二长花岗岩	72.71	0.25	13.63	0.91	1.95	0.065	0.41	1.60	3.92	4.03	0.03	0.10	0.02	0.760	100.4	3
布山村	N743-1	二长花岗岩	71.42	0.29	13.06	0.91	1.60	0.080	0.60	2.08	3.54	4.80	0.09	0.74	0.06	0.024	99.3	3
布山村	N1309-1	二长花岗岩	66.72	0.83	14.93	0.97	3.30	0.070	1.68	3.58	3.35	3.96	0.20	0.96	0.10	0.079	100.7	3
	平均		69.16	0.48	14.51	1.08	2.57	0.073	0.86	2.23	3.51	4.28	0.14	0.84	0.09	0.153	100.0	3
金寮	S039	正长花岗岩	76.08	0.11	11.78	0.01	2.31	0.030	0.03	0.48	2.90	5.46	0.01	0.40	0.14	0.190	99.9	3
金寮	S040-1	正长花岗岩	76.14	0.17	11.72	0.36	2.29	0.050	0.03	0.64	3.01	5.54	0.03	0.24	0.09	0.144	100.5	3
金寮	S040-2	正长花岗岩	75.30	0.16	11.82	0.04	2.75	0.040	0.04	0.52	2.97	5.40	0.01	0.32	0.11	0.130	99.6	3
	平均		75.84	0.15	11.77	0.14	2.45	0.040	0.03	0.55	2.96	5.47	0.01	0.32	0.11		100.0	3

续表

岩石单位	样品号	岩　性	SiO_2	TiO_2	Al_2O_3	Fe_2O_3	FeO	MnO	MgO	CaO	Na_2O	K_2O	P_2O_5	H_2O	CO_2	F	总量	来源
番阳峒	S027	正长花岗岩	69.71	0.33	13.65	0.40	3.82	0.080	0.28	1.43	3.02	6.12	0.08	0.14	0.17	0.086	99.3	3
番阳峒	HG-478	正长花岗岩	67.53	0.54	14.71	1.43	4.02	0.100	0.34	2.07	3.16	5.66	0.10	0.56	0.18	0.120	100.5	3
番阳峒	HG-438	正长花岗岩	75.14	0.18	12.76	1.18	1.72	0.040	0.17	0.82	3.22	4.80	0.05	0.48	0.05	0.080	100.7	3
番阳峒	D467-1	正长花岗岩	70.18	0.34	13.76	0.00	4.19	0.070	0.19	1.35	3.24	5.65	0.05	0.37	0.08	0.158	99.6	3
番阳峒	S029	正长花岗岩	71.78	0.27	13.22	0.29	3.26	0.070	0.19	1.04	3.16	6.06	0.04	0.26	0.27	0.150	100.1	3
番阳峒	D1447-1	正长花岗岩	68.52	0.29	15.55	0.40	3.08	0.058	0.48	2.55	3.28	5.14	0.08	0.40	0.06	0.100	100.0	3
番阳峒	S008-2	正长花岗岩	72.98	0.22	13.03	0.10	2.88	0.040	0.10	0.96	3.36	5.80	0.02	0.16	0.05	0.498	100.2	3
番阳峒	S013-2	正长花岗岩	69.70	0.29	14.49	0.19	3.32	0.061	0.40	1.60	3.57	5.30	0.06	0.52	0.11	0.074	99.7	3
番阳峒	H1527-1	正长花岗岩	70.20	0.32	13.69	0.24	3.90	0.076	0.12	1.48	3.54	5.95	0.03	0.36	0.32	0.015	100.2	3
番阳峒	D33-1	正长花岗岩	71.56	0.34	13.31	0.19	3.22	0.050	0.30	1.33	3.17	5.60	0.07	0.18	0.11	0.154	99.6	3
番阳峒	D35-1	正长花岗岩	73.33	0.22	12.74	0.38	2.78	0.060	0.06	0.81	2.90	6.20	0.01	0.28	0.06	0.054	99.9	3
番阳峒	BD10	正长花岗岩	66.44	0.35	14.98	0.70	4.20	0.094	0.12	2.20	3.61	6.50	0.04	0.66	0.28	0.064	100.2	3
番阳峒	D121-1	正长花岗岩	72.60	0.20	13.48	0.04	2.75	0.040	0.19	1.14	3.40	5.42	0.04	0.22	0.08	0.118	99.7	3
番阳峒	H306	正长花岗岩	74.64	0.17	12.80	0.40	2.34	0.040	0.19	0.94	3.10	5.10	0.04	0.42	0.16	0.049	100.4	3
番阳峒	S043-1	正长花岗岩	69.82	0.32	14.50	0.06	3.33	0.060	0.25	1.62	3.19	6.00	0.06	0.26	0.08	0.055	99.6	3
番阳峒	D0472	正长花岗岩	68.62	0.31	14.66	0.95	2.67	0.07	0.07	1.49	3.68	6.58	0.03	0.61	0.12		100.16	本次
番阳峒	D0476	二长花岗岩	65.94	0.40	15.19	1.62	3.55	0.10	0.07	2.04	3.92	6.30	0.04	0.62	0.04		100.02	本次
番阳峒	D0825	碱长花岗岩	68.38	0.28	15.08	1.02	2.23	0.06	0.11	1.44	3.52	6.98	0.03	0.63	0.08		100.14	本次
番阳峒	PM1-69-1	二长花岗岩	70.24	0.48	13.74	1.47	2.00	0.05	0.48	1.89	2.83	5.86	0.08	0.60	0.10		99.82	本次
番阳峒	PM1-107-2	正长花岗岩	70.56	0.31	15.00	0.76	1.22	0.04	0.75	2.25	3.77	4.48	0.08	0.49	0.06		99.77	本次
番阳峒	PM1-117-1	正长花岗岩	72.92	0.26	13.73	0.63	1.65	0.05	0.08	1.00	3.35	5.61	0.02	0.53	0.06		99.89	本次
	平均		70.51	0.31	14.00	0.59	2.96	0.06	0.24	1.50	3.33	5.77	0.05	0.42	0.12	0.12	99.98	
麻山田	H1498-1	正长花岗岩	63.46	0.52	17.27	0.74	3.92	0.084	0.32	2.04	3.62	7.58	0.06	0.42	0.16	0.046	100.3	3
麻山田	BD2	石英正长岩	60.58	0.63	18.14	0.70	4.28	0.097	0.54	3.09	3.65	6.82	0.16	0.76	0.14	0.036	99.6	3
麻山田	S012-1	石英正长岩	61.84	0.53	16.17	0.42	6.42	0.130	0.28	3.10	5.39	6.10	0.09	0.34	0.04	0.052	100.9	3

续表

岩石单位	样品号	岩 性	SiO_2	TiO_2	Al_2O_3	Fe_2O_3	FeO	MnO	MgO	CaO	Na_2O	K_2O	P_2O_5	H_2O	CO_2	F	总量	来源
麻山田	D182-1	石英正长岩	60.98	0.75	16.42	3.30	4.42	0.130	0.56	2.56	3.38	5.80	0.14	0.76	0.17	0.082	99.5	3
麻山田	HG-828	石英正长岩	59.76	0.83	15.32	1.57	6.46	0.160	0.98	3.51	3.48	5.79	0.20	0.77	*	*	98.8	3
麻山田	IIP10-1	石英正长岩	66.80	0.62	15.70	0.69	3.17	0.070	0.61	2.00	3.10	6.54	0.13	0.20	0.16	0.122	99.9	3
麻山田	D1466-1	石英正长岩	66.92	0.46	16.09	1.02	2.29	0.060	0.51	1.58	3.30	7.53	0.08	0.42	0.13	0.101	100.5	3
麻山田	D759-1	石英正长岩	58.40	0.66	18.65	2.11	4.03	0.110	0.73	2.69	3.26	7.91	0.19	0.44	0.11	0.033	99.3	3
	平均		62.34	0.63	16.72	1.32	4.37	0.11	0.57	2.57	3.65	6.76	0.13	0.51	0.13	0.07	99.85	
深堀村	H1474-2	石英二长岩	58.84	0.82	16.86	1.27	6.09	0.108	1.76	4.80	3.32	3.78	0.18	0.84	0.30	0.090	99.1	3
深堀村	S042	石英二长岩	60.20	0.70	17.31	0.75	6.12	0.120	1.23	4.01	3.25	5.71	0.24	0.32	0.08	0.102	100.1	3
深堀村	HG-18	石英二长岩	64.46	0.75	16.79	1.51	2.27	0.060	0.83	2.89	3.15	5.88	0.26	0.72	0.13	0.190	99.9	3
深堀村	HG-23	石英二长岩	60.88	0.13	17.27	1.79	4.10	0.110	1.82	4.35	3.39	4.54	0.39	0.30	0.13	0.320	99.5	3
深堀村	D430-1	石英二长岩	62.68	0.58	16.38	0.12	6.10	0.120	0.59	2.98	4.06	4.57	0.14	0.32	0.19	0.104	98.9	3
	平均		61.96	0.65	16.61	1.53	4.04	0.090	1.21	3.43	3.40	5.24	0.23	0.42	0.17	0.161		3
深沟	S045	碱长正长岩	60.84	0.24	14.41	3.90	1.19	1.080	0.34	5.76	6.60	3.69	0.06	0.24	1.55	0.025	99.9	3
进岭	H211-3	花斑岩	72.69	0.20	13.17	0.32	2.94	0.022	0.14	1.08	3.23	5.19	0.03	0.81	0.02	0.140	100.0	3
进岭	H1323-1	花岗斑岩	73.84	0.09	12.15	0.35	2.71	0.026	0.06	0.80	3.42	5.12	0.01	0.15	0.42	0.125	99.3	3
瘦岭	D3013-1	正长花岗岩	76.16	0.08	12.31	0.21	1.28	0.067	0.08	0.57	3.65	4.46	0.01	0.24	0.06	0.052	99.2	3
瘦岭	QXI1-1	正长花岗岩	75.43	0.05	13.32	0.48	0.38	0.020	0.20	0.19	3.30	5.44	0.04	0.96	0.08	0.018	99.9	3
瘦岭	QXI3-3	正长花岗岩	70.39	0.15	15.78	0.42	0.90	0.020	0.39	1.11	3.42	5.81	0.11	1.22	0.13	0.030	99.9	3
	平均		73.99	0.09	13.80	0.37	0.85	0.04	0.22	0.62	3.46	5.24	0.05	0.81	0.09	0.03	99.67	
黑岭	QXVI6-2	正长花岗岩	74.98	0.14	12.50	0.46	1.48	0.020	0.22	0.82	2.66	5.55	0.03	0.93	0.10	0.200	100.1	3
黑岭	Y5551-1	正长花岗岩	75.92	0.15	12.46	1.17	0.70	0.040	0.23	0.35	2.55	4.98	0.18	1.06	0.08	*	*	3
黑岭	YXIII35-1	正长花岗岩	75.66	0.16	12.32	0.62	1.31	0.040	0.32	0.71	3.00	5.03	0.02	0.72	0.04	*	*	3
黑岭	YXIII9-2	正长花岗岩	75.21	0.20	12.46	0.70	1.89	0.080	0.37	1.03	2.81	4.74	0.04	0.48	0.14	*	*	3
黑岭	IIP6-1	正长花岗岩	73.46	0.30	13.09	0.00	2.56	0.057	0.24	1.12	2.94	5.24	0.06	0.34	0.10	0.211	99.7	3
	平均		74.54	0.20	12.72	0.70	1.50	0.043	0.33	0.90	2.98	5.36	0.07	0.71	0.09	0.205		3

续表

岩石单位	样品号	岩性	SiO_2	TiO_2	Al_2O_3	Fe_2O_3	FeO	MnO	MgO	CaO	Na_2O	K_2O	P_2O_5	H_2O	CO_2	F	总量	来源
尖峰岭	D3016-1	正长花岗岩	74.26	0.18	12.69	0.09	2.38	0.035	0.13	1.08	2.71	5.62	0.01	0.40	0.08	0.180	99.8	3
尖峰岭	S019	正长花岗岩	72.24	0.24	13.39	0.28	2.91	0.040	0.23	1.11	2.86	5.83	0.04	0.52	0.13	0.040	99.9	3
尖峰岭	QXVI6-1	正长花岗岩	72.19	0.19	13.79	0.40	1.67	0.030	0.27	0.93	2.77	6.21	0.05	1.25	0.09	0.120	100.0	3
尖峰岭	QXVIII1-1	正长花岗岩	70.90	0.35	13.79	0.53	2.34	0.040	0.44	1.67	2.67	6.10	0.08	0.83	0.09	0.120	100.0	3
尖峰岭	Y5550-1	正长花岗岩	73.98	0.15	12.87	0.45	1.46	0.060	0.23	1.12	3.10	5.32	0.12	0.48	0.05	*	*	3
尖峰岭	D3025-1	正长花岗岩	75.71	0.08	12.24	0.08	2.16	0.033	0.06	0.72	3.21	4.70	0.01	0.33	0.14	0.072	99.5	3
尖峰岭	Q1200	正长花岗岩	76.31	0.09	12.17	0.32	1.36	0.020	0.18	0.63	2.92	4.85	0.02	0.92	0.10	0.190	100.1	3
尖峰岭	HG-433	正长花岗岩	72.09	0.32	13.97	1.13	1.73	0.030	0.28	1.05	2.84	5.94	0.08	0.60	0.09	0.230	100.4	3
尖峰岭	HG-467	正长花岗岩	74.70	0.16	12.67	0.55	1.55	0.030	0.33	0.89	2.92	5.06	0.05	0.58	0.09	0.080	99.7	3
尖峰岭	S037	正长花岗岩	73.80	0.15	12.91	0.18	2.00	0.040	0.21	1.15	3.12	5.59	0.03	0.44	0.21	0.072	99.9	3
尖峰岭	D271-1	正长花岗岩	74.78	0.14	12.27	0.29	2.64	0.040	0.06	0.69	3.43	5.00	0.02	0.28	0.07	0.156	99.9	3
	平均		73.83	0.19	12.92	0.42	2.00	0.036	0.25	1.03	2.94	5.48	0.05	0.60	0.10	0.126		3
高村	Ⅷ1-1	二长花岗岩	66.53	0.85	14.27	2.4	2.3	0.053	1.08	3.98	2.94	3.37	0.3	2.24	0.12		102.1	1
高村	D2385-2	二长花岗岩	77.19	0.05	12.51	0.61	0.2	0.001	0.14	1.32	3.44	3.71	0.15	0.96	0.16		101	1
两院	Ⅷ6-1	二长花岗岩	68.91	0.7	14.09	1.85	1.9	0.05	0.64	4.41	3.01	3.97	0.25	1.12	0.2		101.6	1
两院	D2018	二长花岗岩	76.71	0.1	12.04	1.75	0.55	0.03	0.14	1.29	3.35	3.82	0.25	0.84	0.16		101	1
两院	0504	二长花岗岩	73.34	0.24	13.5	2.65	0	0.06	0.4	1.51	2.96	4.62	0.05	0	0		99.64	1
两院	D2385-1	二长花岗岩	76.12	0.2	13.06	1.28	0.22	0.001	0.28	1.29	2.62	3.91	0.25	1.68	0.12		102	1
	平均		73.77	0.31	13.17	1.88	0.67	0.04	0.37	2.13	2.99	4.08	0.20	0.91	0.12		101.06	
大岭	Ⅷ15-2	花岗闪长岩	69.51	0.4	14.23	1.31	1.5	0.03	0.76	4.3	2.85	3.79	0.2	1.24	0.2		101.1	1
鹅塘村	Ⅷ15-1	英云闪长岩	68.98	0.5	14.54	2.01	1.7	0.05	1.06	4.1	2.52	3.71	0.2	1.08	0.27		101.3	1
陀类	D3082-1	片麻状细粒二云二长花岗岩	72.18	0.24	14.26	0.04	2.18	0.036	0.30	1.26	3.26	5.02	0.09	0.14	0.11	0.087	99.20	3
陀类	Y78-1	片麻状细粒二云正长花岗岩	73.95	0.13	14.04	1.33	0.65	0.060	0.32	0.28	2.65	5.28	0.05		0.13		98.87	3

续表

岩石单位	样品号	岩　性	SiO_2	TiO_2	Al_2O_3	Fe_2O_3	FeO	MnO	MgO	CaO	Na_2O	K_2O	P_2O_5	H_2O	CO_2	F	总量	来源
陀类	Y79-1	片麻状细粒二云正长花岗岩	73.91	0.15	13.53	1.81	0.50	0.050	0.30	0.28	2.70	5.26	0.05		0.13		98.67	3
陀类	Y80-1	片麻状细粒二云正长花岗岩	74.35	0.10	13.95	0.84	0.55	0.040	0.25	0.38	3.00	4.83	0.05		0.13		98.47	3
	平均		73.60	0.16	13.95	1.01	0.97	0.05	0.29	0.55	2.90	5.10	0.06	0.14	0.13	0.09	98.97	
大田	D3087-1	中细粒含斑黑云母二长花岗岩	70.62	0.41	14.92	0.00	2.32	0.027	0.50	1.70	3.68	4.42	0.10	0.52	0.28	0.100	99.6	3
大田	D3084-1	细粒含斑黑云母二长花岗岩	68.70	0.53	15.02	0.03	3.46	0.039	0.61	2.01	3.33	5.00	0.19	0.48	0.07	0.100	99.6	3
大田	DT1-1	细粒含斑黑云母二长花岗岩	69.97	0.54	14.72	0.38	2.56	0.042	0.50	2.08	3.66	4.62	0.16	0.21	0.18	0.137	99.8	3
大田	DT2-1	细粒含斑黑云母二长花岗岩	70.16	0.48	14.32	0.20	3.12	0.047	0.40	1.68	3.50	4.80	0.14	0.22	0.26	0.159	99.5	3
	平均		69.86	0.49	14.75	0.15	2.87	0.039	0.50	1.87	3.54	4.71	0.15	0.36	0.20	0.124	99.6	3
什天	YVII9-2	细中粒巨斑状黑云母二长花岗岩	68.41	0.55	14.62	1.11	2.29	0.070	1.02	2.33	3.35	5.21	0.22	0.66	0.04	*	*	3
什天	S031	中粒似斑状黑云母二长花岗岩	70.98	0.34	14.23	0.34	2.44	0.060	0.58	1.71	3.65	4.81	0.10	0.26	0.06	0.066	99.6	3
什天	HG-437	中粒似斑状黑云母二长花岗岩	71.66	0.35	13.76	0.84	2.35	0.030	0.76	1.80	3.16	3.88	0.12	0.68	0.17	0.160	99.7	3
什天	D3020-1	中粒少斑状黑云母二长花岗岩	72.46	0.25	13.59	0.57	2.29	0.060	0.56	1.85	3.14	4.48	0.07	0.20	0.11	0.043	99.7	3
什天	IIP4-1	中细粒似斑状黑云母二长花岗岩	68.48	0.54	14.80	0.39	3.14	0.058	0.74	2.18	3.42	4.80	0.17	0.32	0.25	0.128	99.4	3
什天	S014	细粒多斑状黑云母二长花岗岩	69.46	0.39	13.67	0.83	3.21	0.060	0.51	2.10	3.34	5.29	0.08	0.50	0.10	0.050	99.6	3
	平均		70.41	0.39	14.16	0.67	2.53	0.065	0.68	1.93	3.33	4.71	0.13	0.45	0.16	0.088		3

续表

岩石单位	样品号	岩 性	SiO_2	TiO_2	Al_2O_3	Fe_2O_3	FeO	MnO	MgO	CaO	Na_2O	K_2O	P_2O_5	H_2O	CO_2	F	总量	来源
白茶	H1362-1	细粒黑云母二长花岗岩	72.84	0.22	13.73	0.20	2.11	0.030	0.21	0.87	3.04	5.71	0.05	0.58	0.01	0.030	99.6	3
东岭	H1211-1	中细粒角闪黑云母二长花岗岩	67.52	0.54	14.76	0.83	4.19	0.062	1.43	3.30	3.00	4.10	0.17	0.13	0.05	0.092	100.2	3
东岭	H1267-2	中细粒角闪黑云母花岗闪长岩	66.88	0.63	14.78	0.73	4.25	0.066	1.58	3.33	3.03	4.00	0.19	0.16	0.16	0.095	99.9	3
那佩	D3074-1	中粗粒似斑状黑云母二长花岗岩	72.30	0.29	13.57	0.00	3.14	0.056	0.37	1.62	3.30	4.14	0.06	0.58	0.05	0.081	99.6	3
那佩	G1346	中粗粒似斑状黑云母二长花岗岩	72.12	0.30	13.73	0.39	2.35	0.090	0.73	2.20	2.86	4.16	0.05	0.42	0.16	0.050	99.6	3
	平均		72.39	0.26	13.54	0.46	2.61	0.074	0.60	1.58	3.10	4.27	0.06	0.46	0.11	0.066		3
保国	D3076-1	细中粒似斑状角闪黑云二长花岗岩	69.89	0.41	14.30	0.16	3.36	0.056	0.71	2.47	2.96	4.25	0.10	0.42	0.10	0.125	99.3	3
保国	HG-28	中粒似斑状角闪黑云母二长花岗岩	70.00	0.41	14.44	0.66	2.87	0.060	0.58	2.33	3.18	4.52	0.19	0.29	0.07	0.200	99.8	3
保国	HG-774	中粗粒似斑状角闪黑云二长花岗岩	70.63	0.38	13.84	0.87	2.22	0.050	0.76	2.51	3.43	4.21	0.11	0.45	*	*	*	3
保国	J121-1	中粗粒似斑状角闪黑云二长花岗岩	65.67	1.14	13.98	2.43	3.30	0.050	1.80	3.70	3.06	4.16	0.33	*	*	*	100.2	3
保国	J133-1	中粗粒似斑状角闪黑云二长花岗岩	69.53	0.65	13.70	1.69	2.08	0.080	0.84	2.26	2.75	5.48	0.18	*	*	*	99.9	3
	平均		69.14	0.60	14.05	1.16	2.77	0.059	0.94	2.65	3.08	4.52	0.18	0.39	0.09	0.162		3
火岭	H1311-3	正长花岗岩	74.16	0.14	13.84	0.35	1.52	0.045	0.17	1.20	3.03	5.33	0.03	0.18	0.17	0.049	100.2	3
火岭	VHG1-7	正长花岗岩	70.7	0.46	13.61	0.79	2.33	0.06	0.63	1.68	2.28	6.26	0.12	0.77	0.04	0	100.1	1
火岭	7329-1	二长花岗岩	72.44	0.15	13.8	0.3	2.24	0.04	0.23	1.12	3.13	5.37	0.1	0.32	0.1	0.06	99.42	1
火岭	XI2-1	二长花岗岩	71.29	0.15	14.28	1.18	0.5	0.02	0.28	3.95	3.41	3.99	0.2	0.98	0.14	0	100.87	1
火岭	XII5-1	二长花岗岩	73.07	0.15	14	1.43	0.15	0.004	0.22	1.06	3.03	5.42	0.15	1.12	0.16	0	100.52	1
火岭	4108-1	二长花岗岩	71.38	0.2	14.22	0.27	1.93	0.09	0.33	2.05	3.24	5.46	0.07	0.69	0.46	0.12	101.66	1
	平均		72.58	0.21	13.85	0.65	1.37	0.04	0.31	1.59	2.90	5.43	0.11	0.62	0.15	0.03		

续表

岩石单位	样品号	岩性	SiO_2	TiO_2	Al_2O_3	Fe_2O_3	FeO	MnO	MgO	CaO	Na_2O	K_2O	P_2O_5	H_2O	CO_2	F	总量	来源
长塘岭	H1278-1	二长花岗岩	71.34	0.34	14.01	0.37	3.47	0.057	0.52	1.88	2.70	4.93	0.09	0.15	0.16	0.076	100.1	3
长塘岭	1225	二长花岗岩	70.88	0.38	14.38	0.85	1.74	0.04	0.52	1.35	3.21	5.51	0.1	0.58	0.11	0.15	100.15	1
长塘岭	Ⅸ10-1	二长花岗岩	74.56	0.1	13.35	0.66	0.85	0.05	0.32	0.96	3.16	5.04	0.12	0.6	0.21	0.04	100.44	1
长塘岭	Ⅺ10-2	二长花岗岩	65.7	0.5	14.38	2.18	1.5	0.066	1.41	5.7	2.99	3.23	0.25	1.92	0.37	0	101.8	1
长塘岭	4871-2	二长花岗岩	68.76	0.62	14.89	0.74	2.1	0.04	0.91	2.53	3.18	4.96	0.21	0.8	0.21	0	101.23	1
长塘岭	D823-1	二长花岗岩	71.5	0.28	14.32	0.21	1.83	0.04	0.9	2.1	2.81	4.92	0.12	0.7	0.08	0	100.3	1
长塘岭	YD499	二长花岗岩	72.62	0.33	14.28	0.47	1.06	0.01	0.51	1.3	2.72	5.21	0.09	1.09	0.07	0	99.76	1
	平均		70.52	0.37	14.27	0.81	1.92	0.05	0.74	2.22	3.11	4.75	0.15	0.76	0.16	0.05		
大坡	1389-2	二长花岗岩	69.46	0.36	14.42	0.4	3.24	0.06	0.59	2.16	3.32	4.82	0.09	0.34	0.08	0.07	99.43	
大坡	21	二长花岗岩	71.13	0.19	13.91	1.25	2.25	0.04	0.31	1.78	3.87	4.79	0.05	0.3	0.12	0.04	100.05	1
大坡	HG394	二长花岗岩	69.93	0.45	14.29	1.27	2.48	0.05	1.04	2.46	3.22	3.5	0.15	0.6	0.05	0.1	99.59	1
大坡	Ⅻ7-2	二长花岗岩	69.42	0.4	13.59	2.65	1.05	0.053	0.85	4.13	3.07	3.59	0.25	1.34	0.16	0	101.14	1
大坡	D892-1	二长花岗岩	70.72	0.32	13.77	1.15	1.8	0.065	1	2.21	3.2	3.91	0.1	0.96	0.22	0	100.59	1
大坡	Ⅺ6-1	二长花岗岩	72.61	0.2	14.15	1.28	0.2	0.025	0.31	2.92	2.69	4.48	0.15	1.26	0.12	0	101.26	1
大坡	Ⅻ7-1	二长花岗岩	72.06	0.4	13.26	2.3	0.6	0.03	0.61	1.92	3.09	3.96	0.15	1.58	0.16	0	101.25	1
大坡	Ⅱ8-1	二长花岗岩	69.73	0.2	13.99	0.44	1.8	0.08	0.85	3	3.4	4.98	0.02	0.92	0.24	0	99.65	1
大坡	Ⅱ19-1	二长花岗岩	67.63	0.4	14.95	0.94	2.86	0.09	1.36	2.07	3.3	4.4	0.02	0.88	1.37	0	100.27	1
大坡	XⅣ25-1	二长花岗岩	69.91	0.4	14.33	0.67	2.28	0.061	1.44	2.36	3.48	3.72	0.15	0.7	0.16	0.06	100.33	1
大坡	XⅣ9-1	二长花岗岩	69.64	0.42	14.65	0.92	2.32	0.066	0.83	2.32	3.26	3.84	0.2	0.38	0.16	0.09	99.64	1
大坡	XX12-1	二长花岗岩	72.33	0.28	13.43	1.29	1.22	0.061	1.23	1.85	3.08	3.62	0.075	0.8	0	0	100.37	1
大坡	Ⅸ17-1	二长花岗岩	68.15	0.42	14.04	0.44	3.2	0.08	1.39	3.5	2.41	3.88	0.1	1.52	0.21	0.08	101.2	1
大坡	Ha34-1	二长花岗岩	74.24	0.16	12.63	0.20	2.28	0.04	0.36	1.58	2.65	4.86	0.04	0.4	0.17	0.06	99.69	1
	平均		70.38	0.34	13.99	1.07	2.04	0.06	0.88	2.47	3.19	4.14	0.12	0.83	0.21	0.03		
乌兰盖	3809	二长花岗岩	73.25	0.25	13.89	0.39	1.63	0.04	0.42	1.56	3.35	4.86	0.07	0.34	0.11	0.15	100.56	1
乌兰盖	D45-1	二长花岗岩	68.37	0.50	14.81	1.24	2.38	0.07	1.16	2.87	3.01	4.17	0.14	0.91	0.18	0.08	101.09	1
	平均		68.37	0.50	14.81	1.24	2.38	0.07	1.16	2.87	3.01	4.17	0.14	0.91	0.18	0.08		

续表

岩石单位	样品号	岩 性	SiO_2	TiO_2	Al_2O_3	Fe_2O_3	FeO	MnO	MgO	CaO	Na_2O	K_2O	P_2O_5	H_2O	CO_2	F	总量	来源
通什	H1274-2	二长花岗岩	65.06	0.56	15.70	0.56	4.88	0.081	1.33	3.58	2.75	4.07	0.14	0.17	0.12	0.081	99.1	3
通什	H1287-1	二长花岗岩	62.54	0.61	16.68	0.81	5.89	0.093	1.85	4.28	2.95	3.28	0.17	0.13	0.23	0.089	99.6	3
通什	HG-424	二长花岗岩	65.75	0.74	14.53	1.57	4.16	0.080	1.26	3.48	2.66	4.14	0.21	0.86	0.09	0.160	99.7	3
通什	HG-443	二长花岗岩	66.72	0.64	15.33	0.65	3.58	0.050	1.26	3.15	3.00	4.14	0.10	0.44	0.17	0.100	99.3	3
通什	HG-475	二长花岗岩	65.67	0.92	15.01	0.96	3.98	0.060	1.43	3.35	2.60	4.76	0.13	0.43	0.13	0.180	99.6	3
通什	4220	二长花岗岩	66.82	0.47	14.9	0.94	3.26	0.07	1.22	2.94	3.64	4.28	0.18	0.64	0.08	0.08	99.54	1
通什	4107	二长花岗岩	67.41	0.43	14.56	0.42	3.59	0.07	1.17	3.18	3.73	4.12	0.18	0.38	0.22	0.08	99.56	1
通什	38-1	二长花岗岩	69.92	0.29	14.21	0.43	3.05	0.06	0.5	2.35	3.14	4.63	0.12	0.83	0.15	0.12	99.83	1
通什	Ⅸ8-1	二长花岗岩	68.01	0.4	15.18	0.74	2.48	0.063	1.13	3	2.62	4.72	0.15	0.76	0.21	0.084	100.04	1
通什	XⅣ12-1	二长花岗岩	67.46	0.6	14.91	1.17	2.55	0.068	1.95	3.11	3.44	3.44	0.3	0.64	0.21	0.088	100.53	1
通什	4111-1	二长花岗岩	67.02	0.62	15.14	1.46	2.6	0.07	1.17	2.93	3.75	4.64	0.25	0.6	0.1	0	100.35	1
通什	4577-1	二长花岗岩	72.11	0.4	13.61	0.52	2.2	0.05	0.66	1.96	3.35	3.5	0.12	0.96	0.09	0	99.53	1
通什	Ha35-1	二长花岗岩	70.59	0.43	13.6	0.37	3.77	0.06	0.96	2.62	3.15	3.37	0.13	0.67	0.10	0.09	100.03	1
	平均		66.88	0.57	15.05	0.77	3.70	0.07	1.27	3.24	3.12	4.00	0.16	0.57	0.14	0.09		
山牛塘	H1311-2	细粒石英闪长岩	58.20	1.25	15.54	0.88	8.24	0.180	3.16	6.34	2.85	1.99	0.26	0.80	0.19	0.122	100.0	3
山牛塘	H1397-1	闪长玢岩	54.94	1.22	17.28	1.16	6.74	0.126	3.92	7.26	2.94	2.58	0.23	0.96	0.15	0.100	99.6	3
山牛塘	H209-1	细粒闪长岩	60.64	1.20	16.40	1.07	6.02	0.104	2.07	5.74	2.67	2.12	0.24	1.24	0.08	0.050	99.6	3
山牛塘	H209-3	细粒闪长岩	60.00	1.58	16.10	0.76	6.49	0.098	1.99	5.35	2.87	2.39	0.43	1.20	0.10	0.100	99.5	3
山牛塘	平均		58.45	1.31	16.33	0.97	6.87	0.127	2.79	6.17	2.83	2.27	0.29	1.05	0.13	0.093	99.7	
石碌	137-1	二长花岗岩	67.64	0.39	15.79	0.32	3.62	0.08	0.86	2.41	3.03	5.08	0.13	0.64	0.09	0.14	100.02	1
石碌	Ha29	二长花岗岩	69.92	0.35	14.26	0.28	3.42	0.08	0.76	2.92	2.8	3.48	0.11	0.96	0.17	0.12	100.43	1
石碌	7334-1	二长花岗岩	66.23	0.33	16.46	0.46	3.56	0.07	0.95	3.38	3.06	4.13	0.16	0.89	0.09	0.07	99.86	1
石碌	HG399	二长花岗岩	71.06	0.27	13.87	0.62	2.32	0.04	0.78	2.17	2.64	4.66	0.08	0.62	0.14	0.12	99.39	1
	平均		68.71	0.34	15.10	0.42	3.23	0.07	0.84	2.72	2.88	4.34	0.12	0.78	0.12	0.11	99.93	
什败	D1547-3	石英二长岩	61.54	1.21	15.29	1.40	4.96	0.089	1.42	4.12	2.75	5.08	0.38	0.46	0.26	0.177	99.1	3
行干	D1749-1	石英二长闪长岩	57.46	1.47	16.26	1.12	5.68	0.081	2.85	5.80	3.48	3.32	0.60	0.57	0.42	0.222	99.3	3
行干	D3089-1	石英二长闪长岩	54.90	1.53	17.68	0.78	6.61	0.094	3.14	5.99	3.69	3.74	0.64	0.38	0.07	0.202	99.5	3
	平均		56.18	1.50	16.97	0.95	6.15	0.088	3.00	5.90	3.59	3.53	0.62	0.48	0.25	0.212	99.4	3

续表

岩石单位	样品号	岩 性	SiO_2	TiO_2	Al_2O_3	Fe_2O_3	FeO	MnO	MgO	CaO	Na_2O	K_2O	P_2O_5	H_2O	CO_2	F	总量	来源
方满	D3092-1	片麻状细中粒二长岩	50.24	1.38	15.25	2.72	5.90	0.130	6.37	7.95	2.92	3.74	0.96	0.86	0.29	0.246	99.0	3
新风岭	H5122-1	正长花岗岩	73.44	0.17	14.54	0.16	0.93	0.015	0.16	0.48	2.77	4.97	0.07	0.86	0.10	0.057	98.7	3
顺作	D3136-1	黑云母花岗岩	71.16	0.32	15.53	0.00	1.86	0.017	0.54	2.32	3.06	4.09	0.04	0.32	0.10	0.031	99.4	3
乘顺岭	QIV1-1	二长花岗岩	75.06	0.16	13.58	0.77	0.53	0.010	0.42	0.13	1.88	5.94	0.03	1.19	0.02	0.060	99.8	3
乘顺岭	GXI3-1	二长花岗岩	72.43	0.20	14.43	0.54	1.58	0.040	0.48	1.19	2.98	5.12	0.08	0.48	0.12	0.070	99.7	3
	平均		73.84	0.15	13.94	0.56	1.14	0.040	0.39	0.61	2.41	5.67	0.07	0.76	0.14	0.060	99.8	3
志仲	D3073-1	二长花岗岩	69.54	0.55	14.28	0.00	3.85	0.064	0.78	2.15	3.13	4.34	0.14	0.24	0.10	0.191	99.4	3
志仲	HG-17	二长花岗岩	69.84	0.54	14.10	0.26	3.28	0.060	0.79	2.78	3.10	3.69	0.20	0.64	0.21	0.180	99.7	3
志仲	GVIII18	二长花岗岩	73.63	0.25	13.01	0.48	2.45	0.060	0.70	1.29	2.34	4.70	0.10	0.54	0.16	0.030	99.7	3
	平均		71.00	0.45	13.80	0.25	3.19	0.061	0.76	2.07	2.86	4.24	0.15	0.47	0.16	0.134	99.6	3
抱郎	QIV4-1	二长花岗岩	69.53	0.41	14.71	1.52	1.59	0.060	0.66	2.28	2.59	4.70	0.10	1.58	0.04	0.050	99.8	3
抱郎	QVII2	二长花岗岩	69.87	0.43	14.38	1.24	2.33	0.050	0.81	1.90	2.69	5.23	0.11	0.86	0.04	0.080	100.0	3
抱郎	平均		69.70	0.42	14.55	1.38	1.96	0.055	0.74	2.09	2.64	4.97	0.11	1.22	0.04	0.065	99.9	3
石岭	7315	花岗闪长岩	64.56	0.74	16.34	0.52	4.89	0.08	1.77	3.93	2.74	3.76	0.25	0.44	0.13	0.08	101.24	1
石岭	7316-1	花岗闪长岩	62.78	0.67	16.07	0.79	4.99	0.08	2.08	5.39	2.12	3.4	0.22	1	0.12	0.05	99.78	1
石岭	Ha36	花岗闪长岩	64.13	0.64	15.32	0.57	5.14	0.1	2.2	4.89	1.98	3.09	0.16	1.41	0.12	0.07	100.81	1
	平均		63.82	0.68	15.91	0.63	5.01	0.09	2.02	4.74	2.28	3.42	0.21	0.95	0.12	0.07		
保山村	4758-1	英云闪长岩	61.42	1.1	15.31	2.4	5.4	0.1	2.55	4.68	1.7	3.1	0.37	1.56	0.18	0	99.87	1
保山村	4971-1	英云闪长岩	63.86	0.81	15.64	1.47	4.95	0.08	2.04	4.59	2.15	2.6	0.24	1.34	0.18	0	99.95	1
保山村	Ha32-1	英云闪长岩	66.53	0.67	14.31	0.75	4.21	0.09	1.42	2.95	3.36	3.99	0.21	0.65	0.14	0.13	99.41	1
	平均		63.94	0.86	15.09	1.54	4.85	0.09	2.00	4.07	2.40	3.23	0.27	1.18	0.17	0.04		
白石村	211-3	闪长岩	60.02	0.99	15.95	1.22	5.56	0.11	3.23	5.5	3.76	2.7	0.34	0.48	0.08	0.1	100.04	1
白石村	290	闪长岩	53.37	1.04	14.78	0.43	8.66	0.17	8.82	6.68	2.64	1.38	0.33	1.23	0.13	0.05	99.71	1
白石村	4350-7	闪长岩	52.47	1.48	16.21	1.88	5.77	0.17	5.63	6.91	3.0	3.2	0.63	1.41	0.16	0	98.92	1
白石村	Ha16	闪长岩	51.58	1.39	16.24	3.07	6.06	0.15	5.2	7.28	3.06	3.37	0.64	0.88	0.10	0.15	99.17	1
白石村	Ha17-2	闪长岩	52.68	1.33	19.54	1.96	6.26	0.09	2.96	6.07	4.63	2.67	0.59	0.4	0.09	0.14	99.41	1
白石村	Ha17-1	闪长岩	54.96	1.52	17.81	1.49	5.88	0.12	3	6.14	4.03	3.24	0.74	0.44	0.08	0.15	99.60	1
白石村	Ha18	闪长岩	52.32	1.26	19.46	2.54	5.9	0.14	2.7	6.47	4.38	3.09	0.67	0.58	0.05	0.12	99.68	1

续表

岩石单位	样品号	岩 性	SiO_2	TiO_2	Al_2O_3	Fe_2O_3	FeO	MnO	MgO	CaO	Na_2O	K_2O	P_2O_5	H_2O	CO_2	F	总量	来源
白石村	QIV0-1	石英闪长岩	59.33	1.16	16.95	1.55	5.32	0.160	2.09	5.50	2.82	2.61	0.36	1.82	0.13	0.160	99.96	3
白石村	s028	石英闪长岩	55.22	0.71	16.32	1.00	6.91	0.130	5.86	7.69	2.85	1.92	0.14	0.30	0.23	0.048	99.33	3
白石村	D3019-1	石英闪长玢岩	56.68	1.02	16.47	1.63	5.74	0.122	4.20	7.10	2.86	1.96	0.27	0.82	0.06	0.054	98.99	3
	平均		55.89	1.13	16.81	1.83	5.71	0.14	4.25	6.35	3.33	2.67	0.43	0.84	0.11	0.10		
机告村	1153	辉长岩	57.38	1.12	15.54	2.83	4.98	0.12	4.12	5.81	3.02	2.68	0.45	1.1	0.16	0.06	99.37	1
机告村	4014-1	辉长岩	51.27	0.96	15.61	3.21	6.03	0.15	6.96	8.29	2.92	1.28	0.19	2.3	0.3	0	99.47	1
机告村	Ha19	辉长岩	53.05	1.24	15.79	2.26	5.83	0.16	5.93	7.29	3.2	3.27	0.54	0.44	0.08	0.15	99.23	1
机告村	Ha39-1	辉长岩	50.7	1.52	16.43	0.85	7.81	0.14	6.83	8.69	2.89	1	0.48	1.73	0.10	0.08	99.25	1
机告村	Ha39-2	辉长岩	53.06	0.98	12.52	1.25	7.54	0.15	9.85	8.52	2.16	0.94	0.27	1.41	0.16	0.06	98.87	1
	平均		52.13	1.19	15.32	2.14	7.07	0.16	6.82	8.08	2.48	1.95	0.34	1.40	0.16	0.07		
岩脉	D1650-1	紫苏花岗岩	71.32	0.44	13.77	0.12	2.78	0.04	0.5	2.2	2.76	4.92	0.1	0.45	0.3		99.72	3
岩脉	B326	紫苏花岗岩	69.15	0.33	15.26	0.31	2.62	0.03	1.28	3.36	3.86	3.16	0.06	0.64			100.1	3
岩脉	B200	紫苏花岗岩	71.66	0.29	14.03	0	1.8	0.02	0.62	1.79	2.45	6.54	0.1	0.38			99.68	3
岩脉	S34-7	斜长花岗岩	75.04	0.17	13.61	0	2.03	0.02	0.37	3.78	3.6	0.46	0.035	0.28	0.08		99.49	3
公庙	D4760-1	斜长角闪岩	49.94	1.15	15.04	1.58	7.88	0.16	7.3	10.68	2.68	0.8	0.12	2.26	0.04		99.63	2
公庙	D4761-1	斜长角闪岩	52.53	1.59	13.99	1.84	8.2	0.16	6.03	8.73	2.81	1.2	0.21	2.27	0.02		99.57	2
亚炮	D336-1	二长花岗岩	75.48	0.08	12.69	0.68	1.86	0.03	0.43	1.1	3.08	3.12	0.05	0.82	0	0	100.42	1
亚炮	Ⅴ40-1	二长花岗岩	76.4	0.02	12.4	0.56	1.05	0.05	0.05	0.26	3.08	4.15	0.06	0.55	0	0	98.6	1
	平均值		75.94	0.05	12.55	0.62	1.46	0.04	0.24	0.68	3.08	3.64	0.06	0.69	0.00	0.00	99.51	
戈枕	Ⅴ32-1	花岗闪长岩	67.51	0.5	14.41	0.55	4.98	0.1	1.01	1.88	2.78	2.26	0.12	0.98	0	0	97.1	1
戈枕	Ⅴ37-3	花岗闪长岩	67.46	0.48	13.57	1.22	4.55	0.05	1.45	1.74	2.61	2.68	0.1	0.55	0	0	96.5	1
戈枕	Ⅶ15-1	花岗闪长岩	72.66	0.22	13.29	0.89	2.84	0.06	0.64	1.46	2.97	3.02	0.05	0.98	0	0	99.1	1
戈枕	Ⅶ15-2	花岗闪长岩	72.53	0.24	13.63	0.84	2.43	0.06	0.54	1.88	3.15	2.52	0.06	0.92	0	0	98.8	1
	平均值		70.04	0.36	13.73	0.88	3.70	0.07	0.91	1.74	2.88	2.62	0.08	0.86	0.00	0.00	97.88	

注：样品来源：1—引自 1∶25 万《东方县》幅区调报告；2—引自 1∶25 万《琼海县》幅区调报告；3—引自 1∶25 万《乐东县、陵水县》幅区调报告；4—引自 1∶25 万《昌洒市、文昌县、清澜港、铜鼓嘴》幅区调报告；5—引自 1∶5 万《吊罗山、兴隆、什玲市、陵水县》幅区调报告；6—引自 1∶5 万《黄流市、千家》幅区调报告。

附表 9-2　海南岛侵入岩 CIPW 标准矿物含量（%）及主要岩石化学参数一览表

Attached table 9-2　List of pf mineral contents（CIPW standard minerals）and main petrochemical parameters of intrusive rocks from the Hainan Island

岩石单位	样品号	Q	An	Ab	Or	C	Di	Hy(Ol)	Il	Mt	Ap	K_2O+Na_2O	A/CNK	DI	SI	δ	$\frac{K_2O}{Na_2O}$	KAI	来源
$\gamma\pi K_2$	7329-4	30.75	4.39	28.69	31.7	0.8		2.54	0.25	0.6	0.26	8.65	1.04	91.14	1.79	2.46	1.58	0.82	1
$\gamma\pi K_2$	Ⅳ12-1	35.87	2.15	32.03	27.87	0.26		0.77	0.14	0.79	0.09	8.4	1.01	95.77	0.73	2.11	1.25	0.92	1
$\gamma\pi K_2$	Ⅳ5-1	31.94	3.96	31.19	28.97	0.51		1.63	0.31	1.33	0.16	8.48	1.02	92.1	2.43	2.33	1.33	0.85	1
$\gamma\pi K_2$	YD632-1	42.15	0.77	22.07	31.43	1.66		0.88	0.29	0.68	0.05	7.84	1.15	95.65	2.1	1.77	2.04	0.84	2
$\gamma\pi K_2$	D1341-1	45.59	0.27	17.45	29.36	2.87		2.39	0.23	1.78	0.05	6.92	1.31	92.4	1.92	1.42	2.41	0.75	2
$\gamma\pi K_2$	YD445	34.34	5.1	28.48	27.32	1.36		1.78	0.4	1.08	0.12	7.88	1.09	90.14	3.94	2.01	1.37	0.77	2
$\gamma\pi K_2$	YD633-1	39.83	0.64	24.74	29.65	2.48		1.52	0.27	0.76	0.09	7.85	1.22	94.22	4.06	1.85	1.72	0.79	2
$\gamma\pi K_2$	D0377	31.93	3.42	31.74	30.28	0.33		1.02	0.31	0.91	0.05	8.8	1.02	93.95	1.75	2.43	1.37	0.88	本次
$\gamma\pi K_2$	D0514	48.31	1.46	13.76	29.01	6.15		0.52	0.29	0.46	0.02	6.35	1.71	91.08	0.98	1.24	3.02	0.54	本次
$\gamma\pi K_2$	D0523	41.78	0.36	11.19	34.81	7.45		2.41	0.65	1.26	0.07	6.98	1.84	87.78	5.78	1.71	4.45	0.53	本次
$\gamma\pi K_2$	D5028-2	40.86	0.49	25.73	29.43	2.22		0.64	0.17	0.41	0.02	7.94	1.20	96.02	1.61	1.83	1.64	0.81	本次
$\gamma\pi K_2$	D1339	30.67	9.91	23.13	28.89	1.33		3.43	0.69	1.75	0.19	7.5	1.08	82.69	4.12	2.05	1.79	0.66	本次
$\gamma\pi K_2$	D0362-1	35.24	2.64	28.87	32.16	0.12		0.45	0.12	0.38	0.02	8.77	1.01	96.27	0.85	2.29	1.59	0.91	本次
	平均	37.64	2.74	24.54	30.07	2.12		1.54	0.32	0.94	0.09	7.87	1.21	92.25	2.47	1.96	1.97	0.77	
火烧岭	XXI7-2	30.39	9.5	27.57	26.53	0.96	0	2.8	0.63	1.38	0.23	7.68	1.05	84.49	6.93	2.06	1.38	0.70	2
龙楼	D6002-3	36.11	0.56	30.43	29.66	1.13	0	1.02	0.08	0.13	0.86	8.6	1.02	96.2	0.97	2.18	1.40	0.89	本次
龙楼	D2042	33.84	1.05	32.96	29.88	0	0.9	0.1	0.17	0.49	0.6	8.94	0.92	96.68	0.92	2.37	1.30	0.97	本次
龙楼	PM03-11	33.47	5.68	27.97	26.83	1.71	0	2.07	0.34	0.66	1.26	7.81	1.03	88.27	4.82	2.01	1.37	0.73	本次
龙楼	PM03-27	32.73	5.71	28.45	27.1	1.4	0	2.43	0.42	0.63	1.12	7.92	1.02	88.28	5.19	2.06	1.36	0.75	本次
	平均	34.04	3.25	29.95	28.37	1.06	0.23	1.41	0.25	0.48	0.96	8.32	1.00	92.36	2.98	2.16	1.36	0.84	
大小岭	D6112-1	37.03	4.18	28.44	28.66	0.38	0	0.62	0.17	0.49	0.02	8.16	1.03	94.13	1.22	1.97	1.44	0.85	5
大小岭	D3126-1	34.14	3.84	28.98	27.77	0.94	0	3.85	0.27	0.13	0.07	8.09	1.07	90.89	2.33	2.03	1.37	0.82	5
大小岭	GXVIII10-1	34.33	3.36	30.33	28.71	0.36	0	1.84	0.29	0.64	0.12	8.41	1.02	93.37	2.58	2.15	1.36	0.87	5
	平均	35.17	3.79	29.25	28.38	0.56	0.00	2.10	0.24	0.42	0.07	8.22	1.04	92.80	2.04	2.05	1.39	0.85	

续表

岩石单位	样品号	Q	An	Ab	Or	C	Di	Hy(Ol)	Il	Mt	Ap	K_2O+Na_2O	A/CNK	DI	SI	δ	$\frac{K_2O}{Na_2O}$	KAI	来源
吊罗山	D0495	13.54	8.63	29.37	36.95	0	1.81	5.5	1.23	2.72	0.24	9.58	0.94	79.86	2.81	4.31	1.80	0.80	5
吊罗山	D0496	12.94	8.25	29.46	37.67	0	2.24	5.45	1.2	2.58	0.21	9.71	0.93	80.07	2.8	4.44	1.83	0.81	5
吊罗山	D0503	32.75	1.65	25.12	29.64	5.06	0	3.25	0.8	1.58	0.12	7.8	1.45	87.51	8.02	2.22	1.69	0.65	5
吊罗山	D0525	17.58	8.53	27.92	35.39	1.16	0	4.38	1.28	3.51	0.24	9.12	1.06	80.89	3.42	3.82	1.81	0.74	5
吊罗山	D1386-1	31.13	9.73	32.7	22	0	0.22	2.55	0.48	1.02	0.16	7.52	0.98	85.83	6.87	1.88	0.96	0.74	5
吊罗山	D1902-1	34.32	5.35	30.73	26.87	0	0.57	1.11	0.29	0.7	0.07	8.13	0.98	91.92	3.35	2.02	1.25	0.85	5
吊罗山	D1952-1	25.1	8.49	33.75	26.85	0	1.26	2.36	0.56	1.46	0.19	8.45	0.95	85.7	6.93	2.57	1.14	0.79	5
吊罗山	D2821-1	27.93	8.28	28.97	28.89	0	0.58	3.74	0.58	0.86	0.16	8.23	0.97	85.79	3.98	2.36	1.43	0.78	5
吊罗山	D2950-1	30.34	6.48	31.01	28.2	0.11	0	2.11	0.52	1.07	0.16	8.38	1.00	89.55	4.53	2.31	1.30	0.82	5
吊罗山	D5207-1	31.18	8.79	31.39	24.78	0.01	0	2.26	0.46	0.98	0.16	7.84	0.99	87.35	5.51	2.03	1.13	0.77	5
吊罗山	PM1-26-1	25.53	7.73	29.29	31.19	0	0.66	4.14	0.63	0.67	0.16	8.66	0.97	86.01	3.98	2.67	1.52	0.80	5
吊罗山	PM1-30-1	26.27	6.73	28.65	31.97	0	1.37	3.33	0.61	0.91	0.16	8.73	0.95	86.89	3.61	2.68	1.60	0.82	5
吊罗山	D3124-1	34.74	4.63	29.41	27.26	0.49	0	2.79	0.23	0.44	0.02	8.03	1.04	91.41	1.6	1.99	1.33	0.83	5
吊罗山	D3125-1	34.64	4.41	29.4	26.38	0.81	0	3.8	0.29	0.19	0.07	7.86	1.06	90.42	2.18	1.93	1.28	0.81	5
吊罗山	S006	33.2	8.79	25.48	24.76	0.9	0	4.92	0.61	1.11	0.23	7.15	1.05	83.44	7.47	1.72	1.39	0.70	5
吊罗山	QV13-2	38.97	0.29	23.75	30.35	2.9	0	1.97	0.29	1.4	0.09	7.81	1.27	93.07	3.03	1.91	1.83	0.77	5
吊罗山	QV8-1	31.03	5.75	22.51	34.26	0.97	0	3	0.48	1.87	0.14	8.33	1.06	87.8	5.38	2.4	2.18	0.78	5
	平均	28.31	6.62	28.76	29.61	0.73	0.51	3.33	0.62	1.36	0.15	8.31	1.04	86.68	4.44	2.54	1.50	0.78	
新村	D5269-2	30.66	2.46	29.64	34.5	0	1.03	1.26	0.13	0.28	0.02	9.28	0.97	94.8	0.38	2.68	1.67	0.93	5
新村	D5325-1	35.39	1.15	32.7	28.47	0	1.35	0	0.21	0.52	0.02	8.65	0.94	96.56	1.47	2.18	1.25	0.96	5
新村	D5392-1	31.89	7.98	23.45	31.92	0	1.53	0.76	0.83	1.42	0.21	8.08	0.94	87.26	3.04	2.2	1.95	0.78	5
新村	D3128-1	28	9.45	29.28	25.64	0.61	0	4.92	0.67	1.11	0.3	7.73	1.02	82.92	6.27	2.15	1.25	0.72	5
新村	S007	29.4	9.23	31.49	23.31	0.07	0	5.15	0.56	0.58	0.19	7.6	0.99	84.2	6.98	1.97	1.06	0.75	5
新村	WH12-7	33.47	2.61	36.18	19.95	1.69	0	3.83	0.36	1.36	0.52	7.57	1.10	89.6	4.22	1.87	0.79	0.80	5
新村	WH2-4a	26.15	9.99	26.79	24.91	1.84	0	7.2	0.8	1.33	0.98	7.32	1.06	77.85	9.11	2.12	1.33	0.64	5
	平均	30.71	6.12	29.93	26.96	0.60	0.56	3.30	0.51	0.94	0.32	8.03	1.00	87.60	4.50	2.17	1.33	0.80	

续表

岩石单位	样品号	Q	An	Ab	Or	C	Di	Hy(Ol)	Il	Mt	Ap	K_2O+Na_2O	A/CNK	DI	SI	δ	$\frac{K_2O}{Na_2O}$	KAI	来源
南牙村	3862-2	29.22	8.03	25.02	31.3	0	2.12	2.76	0.87	0.6	0.07	8.15	0.93	85.54	3.54	2.3	1.79	0.78	1
南牙村	73	29.9	8.17	21.99	30.72	0.67	0	6.94	0.76	0.67	0.17	7.56	1.04	82.61	1.99	2.12	2.00	0.73	1
高通岭	YV28-1	33.18	8.11	30.02	25.56	0.31	0	1.89	0.31	0.51	0.09	7.79	1.02	88.76	3.86	1.94	1.22	0.76	2
双顶岭	YIV7-1	36.74	5.44	27.56	27.15	0	0.58	1.33	0.31	0.75	0.12	7.81	0.97	91.45	4.58	1.83	1.41	0.84	2
六弓	YIV12-1	31.64	8.41	29.14	26.02	0	0.14	2.88	0.48	1.08	0.21	7.77	0.98	86.8	7.16	2	1.28	0.77	2
六弓	I HG3-1	33.72	6.26	25.84	28.71	0.6	0	3.38	0.52	0.76	0.21	7.85	1.03	88.27	4.89	1.99	1.59	0.78	2
六弓	D0914-1	26.62	12.99	33.18	20.13	0	0.29	4.48	0.75	1.35	0.21	7.25	0.98	79.93	10.58	1.94	0.87	0.68	本次
六弓	D0972-1	31.1	10.64	33.82	20.23	0.41	0	2.34	0.46	0.86	0.14	7.37	1.02	85.15	6.13	1.82	0.86	0.70	本次
六弓	D1412-1	24.79	13.32	31.04	22.87	0	1.06	4.46	0.79	1.44	0.23	7.46	0.95	78.7	11.47	2.13	1.06	0.68	本次
六弓	D1430-1	29.19	7.09	31.05	29.37	0.68	0	1.33	0.36	0.82	0.09	8.57	1.04	89.61	3.46	2.46	1.35	0.78	本次
六弓	D1823-1	27.44	11.77	32.37	23.44	0	0.09	3.08	0.61	1.04	0.16	7.74	0.99	83.25	7.67	2.12	1.04	0.71	本次
六弓	D3119-1	27.12	11.11	33.28	21.69	0.08	0	5.18	0.61	0.75	0.19	7.52	0.99	82.09	6.62	2.03	0.93	0.72	本次
六弓	QV10-1	31.56	9.35	22.83	27.98	1.25	0	4.09	0.68	2.05	0.21	7.3	1.08	82.37	7.87	1.93	1.75	0.67	5
	平均	29.24	10.10	30.28	24.49	0.34	0.18	3.47	0.58	1.13	0.18	7.65	1.01	84.02	7.32	2.05	1.19	0.72	
保亭	YIV19-1	23.22	14.02	30.53	22.45	0	0.78	5.69	0.94	1.9	0.47	7.35	0.95	76.2	12.81	2.17	1.05	0.66	2
保亭	X XI6-1	30.17	10.92	27.99	24.31	0.8	0	3.42	0.69	1.44	0.28	7.36	1.04	82.47	8.80	1.92	1.24	0.67	2
保亭	I HG8-1	25.56	9.04	26.46	31.86	0.11	0	5.14	0.83	0.66	0.33	8.39	0.98	83.88	5.37	2.6	1.72	0.76	2
保亭	D0919-1	23.68	15.5	30.45	21.45	0	0.74	5.25	0.92	1.73	0.28	7.15	0.96	75.58	13.24	2.05	1.01	0.63	5
保亭	D1229-1	23.65	14.98	29.8	23.26	0	0.43	5.23	0.9	1.53	0.23	7.37	0.97	76.71	12.44	2.16	1.12	0.65	5
保亭	D1421-1	25.76	14.8	30.02	21.56	0.01	0	6.45	0.79	0.4	0.21	7.12	0.99	77.34	11.69	1.91	1.03	0.64	5
保亭	D3117-1	24.5	13.04	32.05	22.32	0	0.21	5.95	0.69	1.02	0.21	7.51	0.98	78.87	9.24	2.13	1.00	0.68	5
保亭	D3060-1	23.9	12.95	26.57	25.14	0	2.24	6.13	0.86	1.88	0.35	7.37	0.91	75.61	13.10	2.14	1.35	0.67	5
保亭	QV4-1	25.41	13.93	25.55	24.73	0	0.39	7.12	0.92	1.63	0.33	7.11	0.97	75.69	15.22	1.99	1.39	0.65	5
保亭	GXVIII9-1	22.64	12.25	28.74	24.03	0	2.69	5.71	1.06	2.65	0.23	7.36	0.91	75.41	14.08	2.24	1.20	0.69	5
	平均	24.85	13.14	28.82	24.11	0.09	0.75	5.61	0.86	1.48	0.29	7.41	0.97	77.78	11.60	2.13	1.21	0.67	

续表

岩石单位	样品号	Q	An	Ab	Or	C	Di	Hy(Ol)	Il	Mt	Ap	K_2O+Na_2O	A/CNK	DI	SI	δ	$\frac{K_2O}{Na_2O}$	KAI	来源
加茂	YD599-1	16.33	22.08	23.71	18.29	0	3.47	12.71	1.5	1.42	0.49	5.83	0.89	58.33	22.44	1.77	1.10	0.50	2
加茂	YD612-1	15.51	23.16	24.06	17.28	0	3.48	12.77	1.52	1.69	0.52	5.7	0.89	56.85	22.84	1.75	1.03	0.48	2
加茂	D102-1	23.15	22.05	26.59	16.88	0.37	0	7.76	1.16	1.44	0.59	5.9	0.99	66.62	17.58	1.56	0.91	0.49	2
加茂	D0355	14.65	18.87	27.38	20.02	0	4.59	10.33	1.56	2.17	0.42	6.55	0.86	62.05	20.24	2.23	1.05	0.57	本次
加茂	D0478	16.23	18.68	29.18	21.04	0	3.03	7.83	1.4	2.2	0.4	6.93	0.90	66.45	16.65	2.35	1.03	0.58	本次
加茂	D0830	15.09	20.26	28.49	19.36	0	3.58	8.65	1.5	2.61	0.45	6.57	0.89	62.94	18.46	2.25	0.97	0.55	本次
加茂	D0893	19.73	18.04	28.07	19.3	0	2.24	9.19	1.2	1.91	0.31	6.45	0.92	67.1	19.78	1.92	0.98	0.58	本次
加茂	D0898	20.32	15.9	30.86	22.15	0	1.49	6.13	1.02	1.82	0.28	7.3	0.94	73.33	14.91	2.28	1.03	0.63	本次
加茂	D1248	13.3	22.65	26.9	18.12	0	2.28	10.5	2.08	3.32	0.83	6.09	0.90	58.32	21.29	2.24	0.96	0.51	本次
加茂	D1914-1	10.54	26.82	27.22	14.79	0	1.18	13.95	1.76	3.12	0.59	5.6	0.94	52.55	21.80	2.09	0.78	0.45	本次
加茂	PM2-127-1	21.59	14.8	32.43	21.88	0	1.32	5.35	0.9	1.42	0.28	7.47	0.95	75.9	12.81	2.26	0.97	0.66	本次
加茂	D3120-1	18.62	17.23	30.9	20.57	0	1.2	8.56	1.11	1.48	0.33	7.07	0.95	70.09	16.08	2.22	0.95	0.61	本次
加茂	QV13-1	18.51	17.24	28.07	19.48	2.04	0	10.05	1.64	2.46	0.5	6.42	1.10	66.06	13.77	2.17	0.99	0.52	5
加茂	QV14-1	15.55	24.49	24.15	19.04	0	1.96	10.84	1.88	1.55	0.54	5.96	0.92	58.74	14.82	1.97	1.13	0.48	5
加茂	GXVIII3-1	18.64	14.38	29.19	22.91	0	3.76	6.55	1.35	2.75	0.47	7.2	0.87	70.74	17.13	2.39	1.12	0.65	5
	平均	17.18	19.78	27.81	19.41	0.16	2.24	9.41	1.44	2.09	0.47	6.47	0.93	64.40	18.04	2.10	1.00	0.55	
乌石	YV2-1	3.23	31.79	19.62	2.69	0	9.76	26.65	2.23	3.23	0.8	2.74	0.75	25.54	41.16	0.84	0.20	0.27	2
乌石	D1321-1	0	33.1	23.98	4.7	0	14.48	11.62	3.5	2.99	0.99	3.56	0.70	28.68	29.97	2.04	0.28	0.31	5
番通	D776-1	8.08	4.52	44.52	36.1	0	0.3	4.3	0.59	1.43	0.14	11.26	0.98	88.7	2.44	5.56	1.16	0.90	3
番通	D3037-1	15.53	5.08	38.75	34.61	0.34	0	3.98	0.44	1.18	0.07	10.3	1.02	88.89	1.54	4.24	1.28	0.86	3
阜石斗	S017	34.33	6.72	28.4	25.5	0	0.6	3.06	0.34	0.94	0.09	7.68	0.97	88.23	3.45	1.83	1.29	0.81	3
阜石斗	HG-167	34.73	6.69	28.96	24.01	1.28	0	3.34	0.5	0.28	0.19	7.35	1.09	87.7	5.75	1.76	1.19	0.73	3
阜石斗	GIII5-4	38.91	1.43	29.8	27.59	0.51	0	0.75	0.19	0.78	0.02	8.14	1.04	96.3	0.85	1.9	1.33	0.91	3
阜石斗	GIII4-2	31.11	3.98	33.45	25.15	0.74	0	3.81	0.38	1.17	0.19	8.14	1.04	89.71	6.60	2.17	1.08	0.83	3
阜石斗	GIII5-2	31.2	6.03	31.39	26.1	0.31	0	2.93	0.42	1.43	0.19	8.06	1.01	88.69	4.08	2.14	1.19	0.81	3
阜石斗	Q155-1	38.84	0.87	30.09	26.63	1.37	0	1.42	0.15	0.56	0.05	7.98	1.12	95.56	2.70	1.86	1.27	0.86	3
阜石斗	G1977-1	37.97	0.59	28.38	29.97	1.54	0	0.69	0.19	0.65	0.02	8.34	1.14	96.32	1.07	2.06	1.51	0.86	3
	平均	35.30	3.76	30.07	26.42	0.82	0.09	2.29	0.31	0.83	0.11	7.96	1.06	91.79	3.50	1.96	1.27	0.83	

续表

岩石单位	样品号	Q	An	Ab	Or	C	Di	Hy(Ol)	Il	Mt	Ap	K_2O+Na_2O	A/CNK	DI	SI	δ	$\frac{K_2O}{Na_2O}$	KAI	来源
抱跃	D3056-1	26.66	15.27	26.38	21.61	0.53	0	8.47	0.87	0	0.21	6.65	1.02	74.65	6.12	1.73	1.17	0.60	3
抱跃	QXII1-1	35.08	3.28	17.65	39.2	1.98	0	1.49	0.1	1.15	0.05	8.61	1.16	91.93	2.10	2.39	3.18	0.77	3
抱跃	HG-40	23.32	7.87	35.68	27.28	0	3.34	0.54	0.34	1.36	0.25	8.84	0.90	86.28	2.82	2.8	1.09	0.81	3
抱跃	N515-1	22.65	10.75	33.6	24.96	0	1.02	3.62	0.92	2.14	0.33	8.11	0.95	81.21	8.45	2.57	1.06	0.74	3
抱跃	GIII11-2	27.61	7.85	28	28.97	0	2.8	2.69	0.56	1.32	0.19	8.14	0.91	84.58	5.33	2.34	1.48	0.79	3
	平均	27.06	9.00	28.26	28.40	0.50	1.43	3.36	0.56	1.19	0.21	8.07	0.99	83.73	4.96	2.37	1.60	0.74	
税町	D3055-1	25.38	16.21	25.6	21.9	0.25	0	9.32	1.04	0.06	0.24	6.63	1.00	72.88	7.99	1.76	1.22	0.59	3
税町	GVIII7-2	23.7	16.37	25.54	22.39	0.66	0	8.98	0.96	1.15	0.23	6.72	1.03	71.63	10.39	1.9	1.26	0.58	3
税町	HG-39	7.85	16.34	32.11	24.92	0	8.28	4.45	0.93	4.01	1.09	7.98	0.78	64.88	11.35	3.81	1.11	0.64	3
税町	HG-51	14.13	22.08	29.25	16.49	0	4.66	7.99	1.02	3.39	0.98	6.18	0.84	59.87	16.08	2.14	0.81	0.52	3
税町	QXII4-1	21.69	18.17	29.68	16.94	1.48	0	6.98	1.51	3.05	0.49	6.27	1.06	68.31	11.76	1.88	0.82	0.52	3
税町	GII15-1	29.02	8.37	24.16	28.93	1.68	0	4.6	0.77	2.32	0.12	7.6	1.12	82.11	7.06	2.19	1.71	0.68	3
税町	GIII11-3	21.14	14.83	26.57	23.56	0	1.61	9.17	1.11	1.71	0.28	7.08	0.94	71.27	15.40	2.13	1.27	0.64	3
税町	GXII1-1	25.58	18.63	27.91	17.03	0	2.29	4.27	0.97	2.7	0.59	6.07	0.91	70.52	7.64	1.57	0.87	0.56	3
	平均	21.06	16.38	27.60	21.52	0.51	2.11	6.97	1.04	2.30	0.50	6.82	0.96	70.18	10.96	2.17	1.13	0.59	
$\gamma\pi K_1$	GI2-1	31.26	10	23.28	27.9	0	0.77	4.53	0.55	1.65	0.05	7.28	0.97	82.44	5.94	1.9	1.72	0.72	3
$\gamma\pi K_1$	G21-1	29.53	8.86	30.48	24.85	0.35	0	3.25	0.48	1.98	0.19	7.76	1.01	84.86	5.34	2.09	1.17	0.74	3
$\gamma\pi K_1$	GI6-1	41.11	1.58	19.89	32.77	1.7	0	1.8	0.15	0.98	0.02	7.86	1.16	93.77	3.39	1.8	2.36	0.81	3
$\gamma\pi K_1$	GI12-2	15.43	8.82	36.43	33.64	0.2	0	2.78	0.31	2.29	0.09	9.94	1.01	85.5	3.39	4.08	1.32	0.79	3
$\gamma\delta\pi K_1$	GII8-1	34.93	7.11	27.95	24.04	0.73	0	3.01	0.4	1.77	0.05	7.32	1.05	86.92	5.08	1.73	1.23	0.75	3
$\gamma\delta\pi K_1$	GIII3-1	20.5	16.96	32.92	20.13	0.59	0	5.26	0.88	2.4	0.35	7.24	1.01	73.55	8.39	2.29	0.88	0.60	3
$\gamma\delta\pi K_1$	G1872	31.43	7.93	27.69	23.73	1.93	0	4.19	0.68	2.07	0.35	7.17	1.12	82.85	7.68	1.86	1.23	0.67	3
νK_1	D3045-1	2.76	26.13	27.61	11.41	0	7.41	19.54	2.39	1.95	0.78	5.11	0.81	41.78	28.06	2.28	0.59	0.44	3
νK_1	N517-1	4.06	25.21	26.84	12.7	0	8.24	15.49	2.21	4.3	0.94	5.25	0.78	43.6	28.00	2.44	0.68	0.45	3
νK_1	GIX5-1	0	36.54	12.63	5.32	0	19.97	15.81	0.78	1.9	0.19	2.34	0.64	17.95	50.41	0.8	0.60	0.20	3

续表

岩石单位	样品号	Q	An	Ab	Or	C	Di	Hy(Ol)	Il	Mt	Ap	K_2O+Na_2O	A/CNK	DI	SI	δ	$\frac{K_2O}{Na_2O}$	KAI	来源
半岛	Ha31	30.41	6.76	26.93	28.78	0	0.1	5.58	0.53	0.74	0.18	7.97	0.98	86.12	4.02	2.14	1.53	0.81	1
半岛	7331-1	28.34	7.18	29.15	28.72	0	0.54	4.3	0.61	0.83	0.3	8.22	0.96	86.21	3.89	2.34	1.41	0.81	1
半岛	HG389	27.49	7.64	25.02	33.98	0.37	0	3.99	0.42	0.94	0.14	8.6	1.02	86.49	2.39	2.62	1.95	0.78	1
	平均	28.75	7.19	27.03	30.49	0.12	0.21	4.62	0.52	0.84	0.21	8.26	0.99	86.27	3.43	2.37	1.63	0.80	
马岭	177	27.7	3.31	35.59	27.79	1.57	0	3.39	0.19	0.37	0.07	8.81	1.11	91.08	0.73	2.63	1.12	0.81	1
马岭	XX1-1	37.39	5.4	27.93	22.5	1.83	0	3.06	0.35	1.42	0.12	7.02	1.15	87.82	6.69	1.57	1.15	0.71	1
马岭	XⅧ1-1	40.25	4.53	30.08	20.53	1.56	0	2.31	0.04	0.65	0.05	6.92	1.13	90.86	1.72	1.44	0.98	0.75	1
	平均	35.11	4.41	31.20	23.61	1.65	0.00	2.92	0.19	0.81	0.08	7.58	1.13	89.92	3.05	1.88	1.08	0.76	
马鞍岭	Ha13-1	32.11	4.53	30.96	28.73	0	0.03	3.37	0.06	0.17	0.01	8.5	1.00	91.8	1.01	2.23	1.33	0.87	1
马鞍岭	Ha13-2	32.75	3.6	32.24	27.38	0.27	0	3.4	0.06	0.28	0.01	8.42	1.02	92.37	0.80	2.17	1.22	0.88	1
马鞍岭	184	28.67	6.55	24.99	33.6	0.07	0	4.41	0.67	0.84	0.19	8.6	0.99	87.26	3.85	2.52	1.93	0.82	1
马鞍岭	187	25.81	7.15	28.62	30.13	0.33	0	6.32	0.72	0.58	0.33	8.45	1.00	84.56	3.41	2.56	1.51	0.79	1
马鞍岭	YXV3-1	25.5	6.06	30.9	32.64	1.08	0	1.87	0.37	1.47	0.09	9.07	1.07	89.04	3.04	2.95	1.51	0.78	2
	平均	28.97	5.58	29.54	30.50	0.35	0.01	3.87	0.38	0.67	0.13	8.61	1.02	89.01	2.42	2.49	1.50	0.83	
三狮岭	HG400	33.57	5.59	23	32.73	0	1.28	2.35	0.4	0.87	0.19	8.14	0.95	89.3	2.06	2.15	2.04	0.84	1
三狮岭	Ⅱ28-1	39.39	0.98	25.79	28.22	1.38	0	3.05	0.17	0.96	0.05	7.7	1.13	93.4	3.83	1.78	1.57	0.85	1
三狮岭	Ⅱ30-1	40.09	0.99	23.99	28.42	2.55	0	2.48	0.17	1.19	0.09	7.55	1.24	92.5	4.12	1.73	1.70	0.77	1
三狮岭	YI1-2	30.53	8.03	22.19	32.49	0.61	0	4.58	0.59	0.75	0.19	8.05	1.03	85.21	4.78	2.23	2.10	0.74	2
三狮岭	YD438	7.94	8.66	43.53	34.94	0.73	0	1.55	0.48	2.06	0.09	10.94	1.04	86.41	1.73	5.55	1.15	0.79	2
三狮岭	D1740-2	26.62	4.46	36.27	26.52	0	3.53	1.24	0.38	0.8	0.16	8.7	0.89	89.41	6.56	2.55	1.05	0.88	2
三狮岭	HN03	32.7	9.5	27.56	26.28	0.63	0	2.56	0.33	0.45	0	7.64	1.05	86.54	5.43	1.91	1.37	0.71	2
三狮岭	D958-1	32.86	8.2	19.85	32.33	1.01	0	4.65	0.77	0.15	0.19	7.73	1.06	85.04	4.75	2.02	2.33	0.71	2
三狮岭	HN05	32.22	9.42	20.51	32.05	0.15	0	4.54	0.77	0.35	0	7.77	1.01	84.78	5.14	2.02	2.24	0.73	2
三狮岭	D0399-1	35.03	3.53	29.03	30.38	0.28	0	0.91	0.13	0.7	0.02	8.52	1.02	94.44	1.35	2.18	1.50	0.88	本次
三狮岭	D1004-1	39.88	3.08	25.83	28.52	1.12	0	0.84	0.08	0.63	0.02	7.82	1.10	94.23	0.90	1.78	1.58	0.82	本次
三狮岭	D1213-1	33.87	5.03	27.88	29.72	1.07	0	1.27	0.25	0.89	0.02	8.26	1.08	91.47	2.35	2.16	1.53	0.79	本次
三狮岭	D1036	33.38	0.25	29.91	30.68	1.07	0	4.29	0.25	0.1	0.07	8.74	1.09	93.97	1.25	2.31	1.47	0.91	本次
	平均	32.16	5.21	27.33	30.25	0.82	0.37	2.64	0.37	0.76	0.08	8.27	1.05	89.75	3.40	2.34	1.66	0.80	

续表

岩石单位	样品号	Q	An	Ab	Or	C	Di	Hy(Ol)	Il	Mt	Ap	K_2O+Na_2O	A/CNK	DI	SI	δ	$\frac{K_2O}{Na_2O}$	KAI	来源
架 水	D3361-1	17.9	15.39	27.59	24.09	0.33	0	9.45	0	2.32	1.77	7.2	0.95	69.58	11.03	2.56	1.25	0.62	2
椰子园	D3032-1	26.18	8.5	27.37	30.58	0.42	0	4.52	0	0.77	1.39	8.32	1.01	84.13	4.91	2.55	1.60	0.76	3
南山岭	S018	18.79	10.56	29.21	31.11	0	0.38	6.58	0	1.11	1.78	8.66	0.96	79.11	8.36	3.14	1.52	0.75	3
南山岭	N605-1	17.87	9.46	31.94	32.24	0	1.27	3.3	0	1.07	2.32	9.17	0.93	82.05	8.14	3.5	1.45	0.78	3
马翁岭	Ha37-2	0	30.91	25.43	15.3	0	6.56	(11.65)	2.55	3.31	2.17	5.94	0.79	42.84	20.70	5.83	0.75	0.43	1
马翁岭	516	10.24	27.46	24.25	14.73	0	3.94	12.65	1.88	4.29	0.54	5.31	0.88	49.22	19.91	2.02	0.87	0.42	1
马翁岭	YXVIII6-2	0	24.67	27.35	14.08	0	14.63	8.02 (6.9)	2.37	1.32	1.07	5.49	0.69	41.43	26.30	3.2	0.74	0.47	2
马翁岭	D1153-1	0	21.36	22.96	16.37	0	23.48	(10.4)	1.65	2.87	0.81	5.39	0.57	39.45	32.66	3.33	1.01	0.49	本次
马翁岭	D1670-1	0	20.41	14.1	9.18	0	36.05	16.03 (0.2)	1.64	1.82	0.55	3.13	0.41	23.28	47.72	1.05	0.93	0.37	本次
马翁岭	D1535-1	0	19.31	14.12	9.32	0	35.92	12.75 (4.2)	1.43	2.52	0.42	3.19	0.41	23.44	49.00	1.14	0.95	0.39	本次
	平均	1.71	24.02	21.37	13.16	0.00	20.10	-3.13	1.92	2.69	0.93	4.74	0.63	36.61	32.72	2.76	0.88	0.43	
八宝坡	H1245-2	2.17	7.45	45.91	33.72	0	2.08	6.18	0.46	1.78	0.26	11.08	0.94	81.8	2.77	6.27	1.05	0.85	3
八宝坡	Hlm-2	8.13	6.88	43.44	31.12	0	4.18	5.5	0.4	0.16	0.19	10.29	0.90	82.69	2.49	4.75	1.03	0.85	3
黑栋	D1244	9.68	17.25	31.74	28.87	0	0.57	8.77	0	1.19	1.48	8.52	0.96	70.29	6.55	3.87	1.30	0.64	3
里茂	H1247-2	0	34.15	4.76	12.4	0	22.63	(13.2)	1.76	5.4	1.02	3.62	0.61	21.87	28.40	7.61	1.32	0.28	3
里茂	Hlm-1	0	26.08	0	0	0	30.27	(19.5)	3.27	5.86	0.31	2.39	0.42	11.37	34.66		1.32	0.25	3
迈洲岭	974-1	38.36	0.06	1.63	52.38	2.53	0	2.73	0.42	1.78	0.08	8.92	1.24	92.37	0.98	2.59	45.95	0.79	2
打显村	X Ⅵ19-1	33.95	5.94	23.35	31.54	1.12	0	2.36	0.38	1.09	0.23	8.04	1.07	88.84	6.02	2.09	1.93	0.76	1
打显村	X Ⅵ2-2	30.09	7.37	24.42	31.38	0.76	0	3.42	0.57	1.63	0.35	8.18	1.03	85.89	7.96	2.29	1.84	0.75	1
白石岭	X Ⅵ16-1	36.04	4.74	24.91	29.55	1.5	0	1.85	0.23	1	0.16	7.88	1.11	90.5	4.18	1.95	1.70	0.76	1
白石岭	X Ⅵ2-1	32.18	6.71	22.44	31.47	1.36	0	3.44	0.57	1.42	0.4	7.94	1.07	86.09	8.84	2.13	2.01	0.73	1
黑岭	Ha25	28.96	5.85	27.24	31.21	0	0.45	4.71	0.57	0.83	0.16	8.45	0.97	87.41	3.07	2.41	1.64	0.84	1
黑岭	4086-2	26.92	7.22	26.92	32.82	0.06	0	4.69	0.61	0.53	0.21	8.65	0.99	86.66	3.05	2.62	1.75	0.81	1

续表

岩石单位	样品号	Q	An	Ab	Or	C	Di	Hy(Ol)	Il	Mt	Ap	K_2O+Na_2O	A/CNK	DI	SI	δ	$\frac{K_2O}{Na_2O}$	KAI	来源
分水岭	XⅦ10-1	37.89	10.62	23.17	20.98	1.42	0	4.37	0.42	0.93	0.18	6.2	1.10	82.04	4.33	1.27	1.30	0.61	1
分水岭	4063-1	30.18	6.74	27.62	29.77	0.38	0	4.17	0.48	0.48	0.16	8.24	1.02	87.57	3.01	2.27	1.54	0.79	1
分水岭	4086-7	29.73	4.98	27.98	31.13	0	0.98	3.59	0.59	0.87	0.14	8.53	0.96	88.84	4.39	2.4	1.59	0.86	1
分水岭	D5543-1	35.61	5.66	21.98	26.51	3.47	0	5.34	0.51	0.64	0.29	6.9	1.28	84.1	5.9	1.67	1.73	0.62	1
分水岭	D5545-1	32.75	14.12	18.46	24.22	1.68	0	6.52	0.77	1.11	0.35	6.16	1.10	75.43	7.34	1.44	1.88	0.54	1
	平均	33.23	8.42	23.84	26.52	1.39	0.20	4.80	0.55	0.81	0.22	7.21	1.09	83.60	4.99	1.81	1.61	0.68	
南宝岭	4602-1	35.16	3.71	29.62	26.83	0.52	0	2.97	0.21	0.91	0.07	7.97	1.04	91.61	2.10	1.96	1.30	0.85	1
南宝岭	4676-2	29.2	6.26	27.96	28.7	0	1.2	5	0.63	0.82	0.21	8.1	0.95	85.86	9.58	2.22	1.47	0.82	1
五里桥	Ha22	27.51	4.07	28.95	32.12	0	1.32	4.63	0.5	0.79	0.09	8.8	0.95	88.58	1.75	2.63	1.59	0.89	1
五里桥	D887-1	29.57	12.12	22.78	27.87	0	2.42	2.11	0.62	2.25	0.23	7.32	0.92	80.22	4.47	1.97	1.75	0.68	1
五里桥	4033-1	30.16	5.27	27.66	31.6	0	0.36	3.7	0.34	0.78	0.12	8.54	0.98	89.42	1.40	2.41	1.64	0.85	1
五里桥	4077-1	34.47	2.89	25.95	31.8	0	0.89	2.76	0.33	0.85	0.05	8.32	0.97	92.22	0.37	2.17	1.75	0.91	1
五里桥	4080-3	30.09	3.9	27.6	32.48	0	2.66	1.92	0.38	0.88	0.09	8.7	0.91	90.17	1.05	2.47	1.69	0.89	1
五里桥	4082-1	29.91	5.81	26.97	31.79	0	1.04	3.11	0.29	0.97	0.09	8.44	0.96	88.67	1.06	2.4	1.69	0.84	1
五里桥	Ha27	30.08	4.2	28.14	31.63	0	0.57	4.26	0.37	0.68	0.05	8.56	0.98	89.85	1.29	2.43	1.61	0.88	1
	平均	30.26	5.47	26.86	31.33	0.00	1.32	3.21	0.40	1.03	0.10	8.38	0.95	88.45	1.63	2.35	1.67	0.85	
苗圃	4072-4	24.24	7.84	25.75	31.73	0	3.09	4.91	0.58	1.61	0.23	8.32	0.90	81.72	2.46	2.65	1.76	0.79	1
苗圃	D887-2	25.13	13.2	25.92	28.64	0	1.53	3.11	0.77	1.43	0.28	7.85	0.94	79.69	6.66	2.35	1.58	0.68	1
苗圃	D891-2	28.18	9.39	22.64	31.03	1.39	0	4.69	1.05	1.34	0.28	7.88	1.08	81.85	5.26	2.31	1.96	0.68	1
苗圃	Ha21-1	22.97	5.74	28.89	33.07	0	1.39	6.24	0.78	0.72	0.19	8.92	0.95	84.93	2.88	2.96	1.64	0.85	1
	平均	25.13	9.04	25.80	31.12	0.35	1.50	4.74	0.80	1.28	0.25	8.24	0.97	82.05	4.32	2.57	1.74	0.75	
乌烈岭	Ha14-2	24.61	3.07	28.92	34.28	0	2.32	5.78	0.46	0.5	0.05	9.17	0.93	87.81	0.71	2.95	1.70	0.91	1
乌烈岭	Ha15	20.89	4.13	28.94	36.14	0	2.85	5.46	0.71	0.73	0.11	9.48	0.91	85.97	1.38	3.38	1.79	0.89	1
乌烈岭	4001-1	29.46	6.72	26.69	31.46	0.21	0	4.09	0.5	0.7	0.16	8.44	1.00	87.61	2.64	2.39	1.69	0.80	1
乌烈岭	4001-7	30.8	5.94	25.59	32.23	0	0.23	3.84	0.44	0.76	0.14	8.41	0.98	88.62	1.77	2.33	1.80	0.83	1
乌烈岭	4044-4	28.59	7.61	28.12	30.86	0	0.32	3.19	0.44	0.73	0.14	8.51	0.98	87.57	2.72	2.44	1.57	0.80	1
乌烈岭	4104-6	27.98	6.33	27.63	32.38	0	1.41	3.19	0.38	0.55	0.14	8.73	0.95	87.99	2.64	2.55	1.68	0.83	1
乌烈岭	7323	30.3	5.79	27.81	30.67	0.32	0	3.72	0.55	0.64	0.19	8.46	1.01	88.78	3.03	2.35	1.58	0.82	1
	平均	27.52	5.66	27.67	32.57	0.08	1.02	4.18	0.50	0.66	0.13	8.74	0.97	87.76	2.13	2.63	1.69	0.84	

续表

岩石单位	样品号	Q	An	Ab	Or	C	Di	Hy(Ol)	Il	Mt	Ap	K_2O+Na_2O	A/CNK	DI	SI	δ	$\frac{K_2O}{Na_2O}$	KAI	来源
东四	Ha14-1	19.97	7.65	29.13	32.3	0	1.57	7.19	1.15	0.72	0.32	8.85	0.94	81.4	4.93	3.11	1.59	0.80	1
东四	Ha23	28.04	5.85	26.9	31.71	0	0.41	5.75	0.6	0.57	0.15	8.44	0.98	86.65	2.83	2.47	1.69	0.84	1
东四	4091-2	29.85	6.37	26.09	30.9	0	2.44	3.23	0.38	0.54	0.16	8.25	0.92	86.84	2.69	2.28	1.70	0.82	1
东四	4125-2	28.82	7.59	25.16	32.59	0	0.28	4.11	0.46	0.82	0.16	8.45	0.98	86.57	2.77	2.43	1.85	0.80	1
东四	7314	28.81	4	28.75	32.6	0	0.53	3.81	0.48	0.8	0.21	8.84	0.97	90.16	2.09	2.61	1.62	0.89	1
	平均	27.10	6.29	27.21	32.02	0.00	1.05	4.82	0.61	0.69	0.20	8.57	0.96	86.32	3.06	2.58	1.69	0.83	
七差岭	6041	24.37	7.06	29.31	32.95	0.16	0	4.08	0.93	0.78	0.35	9.03	0.99	86.63	4.66	2.92	1.61	0.81	1
七差岭	4964-1	31.38	7.8	29.21	24.3	1.39	0	3.71	0.81	1.02	0.38	7.45	1.08	84.89	5.49	1.94	1.19	0.70	1
七差岭	4994-1	25.82	8.39	28.79	29.51	0.06	0	5.28	0.92	0.86	0.37	8.34	0.98	84.12	9.39	2.5	1.47	0.78	1
	平均	27.19	7.75	29.10	28.92	0.54	0.00	4.36	0.89	0.89	0.37	8.27	1.02	85.21	6.51	2.45	1.42	0.76	
保梅岭	Ha30-1	28.86	8.51	26.42	28.90	0	0.05	6.01	0.55	0.53	0.17	7.92	0.99	84.18	4.34	2.20	1.56	0.77	1
保梅岭	Ha30-6	28.11	8.11	26.89	28.91	0	0.32	6.73	0.59	0.16	0.18	7.99	0.98	83.91	4.49	2.24	1.54	0.78	1
保梅岭	Ha33-1	29.86	6.53	27.76	29.26	0	0.25	5.36	0.43	0.42	0.12	8.17	0.98	91.25	3.11	2.24	1.51	0.82	1
	平均	28.94	7.72	27.02	29.02	0.00	0.21	6.03	0.52	0.37	0.16	8.03	0.98	86.45	3.98	2.23	1.54	0.79	
洞古岭	YIII11-1	22.83	10.17	27.89	28.08	0.83	0	5.11	1.64	2.66	0.78	7.91	1.00	78.8	10.71	2.65	1.44	0.70	2
南通岭	VHG1-7	28.21	7.63	19.5	37.4	0.32	0	4.6	0.88	1.16	0.28	8.54	1.00	85.11	5.13	2.62	2.75	0.77	2
南通岭	HP8-G-1	22.41	11.49	32.87	24.16	0.51	0	7.02	0.12	0.95	0.45	7.88	1.00	79.44	9.34	2.42	1.05	0.70	2
南通岭	HN01-2	28.23	15.32	19.7	21.36	0	3.21	8.16	1.71	2.29	0	5.82	0.91	69.29	8.75	1.42	1.55	0.58	2
	平均	26.28	11.48	24.02	27.64	0.28	1.07	6.59	0.90	1.47	0.24	7.41	0.97	77.95	7.74	2.15	1.78	0.68	
琼中县	D493-2	23.56	9.1	26.75	31.49	0.39	0	6.15	1.21	0.82	0.52	8.38	0.99	81.8	7.18	2.73	1.69	0.75	2
琼中县	XXI2-2	37.23	1.24	26.97	29.71	1.58	0	1.62	0.37	1.18	0.09	8.12	1.13	93.91	3.23	2.02	1.58	0.84	2
琼中县	YXX1-1	30.38	1.8	51.81	8.24	1.56	0	3.06	0.86	1.88	0.37	7.44	1.09	90.43	6.40	1.84	0.23	0.84	2
琼中县	YD499	33.54	5.94	23.34	31.23	2.05	0	2.33	0.64	0.69	0.21	7.93	1.14	88.11	5.12	2.11	1.92	0.71	2
琼中县	YXX6-1	31.44	5.68	28.44	28.56	2.2	0	1.94	0.54	0.96	0.24	8.07	1.15	88.44	5.05	2.24	1.44	0.71	2
琼中县	D823-1	29.84	9.73	24.01	29.36	0.85	0	5.09	0.54	0.31	0.28	7.73	1.04	83.21	8.43	2.09	1.75	0.69	2
	平均	31.00	5.58	30.22	26.43	1.44	0.00	3.37	0.69	0.97	0.29	7.95	1.09	87.65	5.90	2.17	1.44	0.76	

续表

岩石单位	样品号	Q	An	Ab	Or	C	Di	Hy(Ol)	Il	Mt	Ap	K_2O+Na_2O	A/CNK	DI	SI	δ	$\frac{K_2O}{Na_2O}$	KAI	来源
黎母岭	D1601-1	30.97	10.45	23.4	29.46	0.56	0	4.2	0.71	0.01	0.23	7.68	1.02	83.83	5.98	2.02	1.80	0.69	2
黎母岭	D101-3	24.86	9.07	21.21	37.39	0.08	0	5.26	0.88	0.95	0.3	8.74	0.98	83.46	4.42	2.87	2.52	0.76	2
黎母岭	ⅤHG1-1	20.53	12.73	22.12	31.38	0	0.4	9.1	1.58	1.53	0.63	7.82	0.95	74.03	7.40	2.7	2.03	0.68	2
黎母岭	D1618-1	37.93	4.99	21.46	31.94	0.43	0	1.84	0.31	1.04	0.07	7.89	1.03	91.33	3.12	1.88	2.13	0.82	2
黎母岭	ⅢHG2-3	20.57	15.13	26.97	27	0	0.57	6.97	1.29	0.9	0.59	7.64	0.95	74.54	9.42	2.51	1.43	0.65	2
黎母岭	HN01-1	16.81	14.34	24.03	31.4	0	2.02	8.76	1.52	1.13		8.04	0.95	72.24	7.67	2.99	1.87	0.66	2
黎母岭	HN02	20.93	10.4	21.06	37.96	0	0.46	6.88	1.11	1.2		8.81	0.99	79.95	4.99	3.17	2.58	0.74	2
黎母岭	HN25	28.6	14.13	21.53	27.74	0.85	0	5.97	0.83	0.34		7.14	1.06	77.87	7.74	1.91	1.84	0.61	2
黎母岭	HN26	31.18	12.22	25.19	21.78	1.43	0	7.11	0.9	0.19		6.49	1.11	78.15	10.35	1.56	1.24	0.60	2
黎母岭	HN27	21.81	9.93	22.64	36.08	0.35	0	4.63	1.45	3.1		8.6	1.02	80.53	7.96	3.14	2.28	0.73	2
黎母岭	HN30	36.28	7.16	24.75	27.52	1.3	0	1.66	0.38	0.94		7.88	1.10	88.55	4.59	1.79	1.59	0.71	2
	平均	26.41	10.96	23.12	30.88	0.45	0.31	5.67	1.00	1.03	0.36	7.88	1.01	80.41	6.69	2.41	1.94	0.70	
后美	Ⅺ1-5	34.61	0.40	27.69	31.87	1.84		0.49			0.09	8.74	1.26	94.1	2.0	2.36	1.65	0.85	6
后龙	Ⅺ3-3	27.90		28.91	34.30	4.17		0.97			0.24	9.23	1.13	91.1	3.5	3.11	170	0.76	6
石墩	H1519-1	2.96	32.97	28.96	14.05	0.7	0	14.58	2.53	1.49	1.76	5.66	0.95	45.97	10.50	3.22	0.69	0.39	3
长安	D1653-1	0	11.94	8.43	3.84	0	36.96	22.37 (11.5)	1.56	2.96	0.38	1.62	0.28	12.27	59.84	0.35	0.65	0.35	3
长安	H303-1	0	33.75	13.42	7.39	0	14.15	19.02 (5.4)	2.95	3.6	0.31	2.79	0.71	20.81	35.45	1.48	0.79	0.24	3
长安	S004-1	0.15	26.81	18.92	6.85	0	12.38	20.63	6.21	6.38	1.64	3.31	0.67	25.92	23.80	3.09	0.52	0.33	3
长安	247-1	0.02	16.75	6.81	2.77	0	45.73	23.57	1.01	3.16	0.17	1.25	0.27	9.6	56.95	0.19	0.58	0.23	1
长安	247-4	0	15.3	6.11	5.33	0	64.45	(4.48)	0.97	2.68	0.07	1.71	0.21	12.03	55.48	0.38	1.06	0.30	1
	平均	0.03	20.91	10.74	5.24	0.00	34.73	13.24	2.54	3.76	0.51	2.14	0.43	16.13	46.30	1.10	0.72	0.29	
石龟岭	D3042-1	35.32	2.32	34.78	23.93	0.14	0	3.39	0.06	0	0.05	8.1	1.01	94.03	0.40	1.94	0.99	0.92	3
石龟岭	N495-1	34.71	2.33	35.01	25.4	0.53	0	1.24	0.11	0.54	0.09	8.4	1.04	95.12	0.31	2.1	1.04	0.89	3

续表

岩石单位	样品号	Q	An	Ab	Or	C	Di	Hy(Ol)	Il	Mt	Ap	K_2O+Na_2O	A/CNK	DI	SI	δ	$\frac{K_2O}{Na_2O}$	KAI	来源
兵工厂	H1219-2	34.03	0.53	33.8	26.97	0.54	0	3.17	0.1	0.84	0.02	8.55	1.04	94.8	0.27	2.19	1.14	0.94	
兵工厂	H1562-1	35.89	0.72	33.17	25.37	0.92	0	3.17	0.11	0.61	0.05	8.15	1.08	94.43	0.29	1.99	1.10	0.90	3
兵工厂	D3044-1	34.66	2.08	34.05	24.74	0.65	0	3.66	0.06	0	0.12	8.12	1.04	93.45	0.30	2	1.04	0.89	3
兵工厂	N673-1	33.16	1.85	34.05	28.54	0	0.48	1.01	0.08	0.81	0.02	8.8	0.98	95.75	2.35	2.33	1.20	0.95	3
	平均	34.44	1.30	33.77	26.41	0.53	0.12	2.75	0.09	0.57	0.05	8.41	1.04	94.61	0.80	2.13	1.12	0.92	
六连岭	H1207-1	31.12	0.52	33.94	28.74	0	1.83	2.94	0.13	0.76	0.02	8.85	0.94	93.8	0.09	2.42	1.21	0.98	3
六连岭	H1207-4	32.02	2.25	33.16	28.27	0	0.72	3.03	0.13	0.36	0.05	8.66	0.97	93.45	0.09	2.31	1.22	0.93	3
六连岭	S001	32.69	1.64	33.66	26.42	0	1.5	3.35	0.13	0.58	0.02	8.39	0.95	92.77	0.62	2.17	1.12	0.95	3
六连岭	H3357-1	35.44	3.14	29.02	27.9	0.23	0	3.51	0.23	0.5	0.02	8.08	1.02	92.36	0.28	1.99	1.38	0.89	3
六连岭	S044	28.45	3.5	29.05	32.67	0	0.35	5.28	0.28	0.32	0.09	8.98	0.98	90.17	1.55	2.61	1.61	0.90	3
六连岭	S013-1	29.74	2.34	31.71	30.91	0	1.4	3.29	0.13	0.45	0.02	8.96	0.96	92.36	0.09	2.54	1.40	0.93	3
六连岭	D3043-1	34.36	3.33	33.26	25.55	0.05	0	3.26	0.04	0.13	0.02	8.21	1.00	93.17	0.60	2.02	1.10	0.90	3
六连岭	N603-1	35.12	0.43	34.87	26.42	0	1.93	0.61	0.09	0.45	0.09	8.61	0.93	96.41	0.40	2.13	1.08	0.99	3
六连岭	N7-3	34.1	2.51	34.45	25.37	0	0.11	2.79	0.15	0.42	0.09	8.3	0.99	93.92	3.54	2.07	1.05	0.92	3
	平均	32.56	2.18	32.57	28.03	0.03	0.87	3.12	0.15	0.44	0.05	8.56	0.97	93.16	0.81	2.25	1.24	0.93	
前锋	GII12-1	35.86	4.15	28.16	27.37	0.42	0	2.25	0.23	1.43	0.12	7.94	1.03	91.39	3.81	1.93	1.39	0.84	3
前锋	GII2-3	41.36	1.77	28.06	24.47	1.84	0	1.4	0.11	0.85	0.12	7.42	1.16	93.89	2.67	1.59	1.25	0.80	3
前锋	G851	41.48	2.33	28.63	22.51	1.66	0	1.89	0.19	1.18	0.12	7.1	1.15	92.62	3.03	1.49	1.13	0.79	3
	平均	39.57	2.75	28.28	24.78	1.31	0.00	1.85	0.18	1.15	0.12	7.49	1.11	92.63	3.17	1.67	1.26	0.81	
结尾	G42	37.56	1.2	29.38	29.94	0	0.52	0.46	0.1	0.75	0.07	8.46	0.98	96.88	0.62	2.07	1.46	0.96	3
结尾	GXIV1-3	48.77	1.03	24	18.91	3.65	0	2.11	0.29	1.12	0.12	5.96	1.41	91.68	4.27	1.01	1.13	0.67	3
结尾	GII1-1	41.1	1.4	26.93	25.71	1.82	0	1.59	0.15	0.7	0.58	7.48	1.11	93.74	2.67	1.64	1.37	0.81	3
结尾	GII3-4	37.17	6.21	25.89	25.69	0.83	0	2.51	0.23	1.42	0.05	7.36	1.07	88.75	4.59	1.68	1.42	0.76	3
结尾	N604-1	33.71	0.4	31.02	33.02	0	0.75	0.3	0.13	0.55	0.09	9.34	0.97	97.75	0.49	2.51	1.52	0.99	3
结尾	Nla-1	33.08	1.02	30.87	31.95	0.58	0	1.73	0.17	0.54	0.05	9.01	1.04	95.9	1.24	2.46	1.48	0.93	3
	平均	38.57	1.88	28.02	27.54	1.15	0.21	1.45	0.18	0.85	0.16	7.94	1.10	94.12	2.31	1.90	1.40	0.85	

续表

岩石单位	样品号	Q	An	Ab	Or	C	Di	Hy(Ol)	Il	Mt	Ap	K_2O+Na_2O	A/CNK	DI	SI	δ	$\frac{K_2O}{Na_2O}$	KAI	来源
超盆	GIX9-1	41.83	0.87	25.01	26.68	2.67	0	1.88	0.15	0.76	0.12	7.38	1.25	93.52	2.74	1.61	1.53	0.76	3
超盆	G1284-2	37.25	2.12	31.1	24.71	1.04	0	2.75	0.15	0.73	0.12	7.78	1.08	93.06	2.83	1.82	1.14	0.85	3
超盆	GII6-1	35.38	6.6	27.03	25.53	0.76	0	2.84	0.34	1.38	0.12	7.48	1.05	87.94	3.18	1.78	1.35	0.76	3
超盆	G011-2	33.04	5.7	29.17	25.59	0	0.12	4.5	0.3	1.45	0.12	7.76	0.99	87.8	3.88	1.93	1.26	0.83	3
超盆	GV1-1	33.73	8.78	27.66	23.04	0.28	0	4.74	0.39	1.25	0.12	7.06	1.01	84.43	4.55	1.67	1.19	0.73	3
超盆	GV7-1	38.98	5.39	26.35	23.78	1.03	0	3.25	0.19	0.89	0.12	7.06	1.08	89.11	2.70	1.53	1.29	0.76	3
超盆	N7-10	32.11	1.06	33.58	28.44	0.88	0	3.27	0.11	0.47	0.05	8.74	1.07	94.13	0.55	2.36	1.21	0.90	3
	平均	36.05	4.36	28.56	25.40	0.95	0.02	3.32	0.23	0.99	0.11	7.61	1.08	90.00	2.92	1.81	1.28	0.80	
布山村	N527-1	29.43	4.9	25.63	34.61	0	0.47	3.12	0.57	1.06	0.19	8.86	0.97	89.67	2.95	2.6	1.93	0.86	3
布山村	D3034-1	27.78	10.5	31.94	20.69	0.5	0	6.24	0.77	1.26	0.28	7.19	1.02	80.41	5.41	1.91	0.93	0.70	3
布山村	N727-1	19.25	14.18	32.72	24.05	0	1.7	4.67	1.04	1.92	0.47	7.86	0.93	76.02	7.02	2.65	1.05	0.67	3
布山村	GIX3-2	22.78	9.64	27.09	25.55	2.53	0	9.29	1.26	1.39	0.47	7.38	1.15	75.42	10.82	2.37	1.35	0.62	3
布山村	N488-1	29.53	7.73	33.33	23.93	0	0.04	3.55	0.48	1.33	0.07	7.95	0.99	86.79	3.65	2.12	1.03	0.79	3
布山村	N743-1	27.84	5.65	30.42	28.81	0	3.52	1.64	0.56	1.34	0.21	8.34	0.88	87.07	5.24	2.43	1.36	0.84	3
布山村	N1309-1	21.25	14.06	28.46	23.5	0	2.03	7.23	1.58	1.41	0.47	7.31	0.92	73.21	12.67	2.25	1.18	0.66	3
	平均	25.41	9.52	29.94	25.88	0.43	1.11	5.11	0.89	1.39	0.31	7.84	0.98	81.23	6.82	2.33	1.26	0.73	
金寮	S039	35.68	2.33	24.74	32.53	0.25	0	4.22	0.21	0.01	0.02	8.36	1.02	92.95	0.28	2.11	1.88	0.91	3
金寮	S040-1	34.57	2.11	25.47	32.75	0	0.78	3.38	0.32	0.52	0.07	8.55	0.97	92.79	0.27	2.21	0.93	1.84	3
金寮	S040-2	34.33	2.54	25.37	32.22	0.17	0	4.97	0.31	0.06	0.02	8.37	1.01	91.92	0.36	2.16	0.91	1.82	3
	平均	34.86	2.33	25.19	32.50	0.14	0.26	4.19	0.28	0.20	0.04	8.43	1.00	92.55	0.30	2.16	1.24	1.52	
番阳峒	S027	23.04	5.67	25.83	36.56	0	0.85	6.62	0.63	0.59	0.19	9.14	0.96	85.43	2.05	3.11	0.85	2.03	3
番阳峒	HG-478	20.46	9.27	26.83	33.56	0	0.33	6.19	1.03	2.08	0.23	8.82	0.98	80.85	2.33	3.16	0.77	1.79	3
番阳峒	HG-438	35.23	3.74	27.22	28.34	0.9	0	2.38	0.34	1.71	0.12	8.02	1.07	90.79	1.53	2	0.82	1.49	3
番阳峒	D467-1	23.59	6.38	27.69	33.72	0	0.05	7.79	0.65	0	0.12	8.89	0.99	85	1.43	2.89	0.83	1.74	3
番阳峒	S029	25.66	4.01	26.91	36.04	0	0.81	5.52	0.52	0.42	0.09	9.22	0.97	88.61	1.47	2.94	0.89	1.92	3

续表

岩石单位	样品号	Q	An	Ab	Or	C	Di	Hy(Ol)	Il	Mt	Ap	K_2O+Na_2O	A/CNK	DI	SI	δ	$\frac{K_2O}{Na_2O}$	KAI	来源
番阳峒	D1447-1	21.67	12.2	27.91	30.55	0.15	0	6.19	0.55	0.58	0.19	8.42	1.00	80.13	3.88	2.77	0.70	1.57	3
番阳峒	S008-2	27.27	3.36	28.58	34.45	0	1.15	4.59	0.42	0.15	0.05	9.16	0.96	90.3	0.82	2.79	0.91	1.73	3
番阳峒	S013-2	22.49	7.62	30.52	31.64	0.12	0	6.64	0.56	0.28	0.14	8.87	1.00	84.65	3.13	2.93	0.80	1.48	3
番阳峒	H1527-1	21.38	3.91	30.09	35.32	0	2.9	5.38	0.61	0.35	0.07	9.49	0.92	86.79	0.87	3.3	0.90	1.68	3
番阳峒	D33-1	26.54	5.6	27.06	33.38	0	0.52	5.81	0.65	0.28	0.16	8.77	0.97	86.98	2.40	2.68	0.85	1.77	3
番阳峒	D35-1	29.13	3.45	24.66	36.83	0	0.46	4.47	0.42	0.55	0.02	9.1	0.98	90.62	0.49	2.72	0.90	2.14	3
番阳峒	BD10	13.95	5.51	30.78	38.71	0	4.62	4.64	0.67	1.02	0.09	10.11	0.88	83.44	0.79	4.33	0.87	1.80	3
番阳峒	D121-1	27.49	5.43	28.97	32.26	0.04	0	5.27	0.38	0.06	0.09	8.82	1.00	88.72	1.61	2.62	0.85	1.59	3
番阳峒	H306	33.27	4.41	26.29	30.21	0.57	0	4.24	0.32	0.58	0.09	8.2	1.04	89.77	1.71	2.12	0.83	1.65	3
番阳峒	S043-1	22.26	7.58	27.21	35.74	0	0.11	6.27	0.61	0.09	0.14	9.19	0.99	85.21	1.95	3.13	0.81	1.88	3
番阳峒	D0472	18.1	4.08	31.37	39.18		2.13	2.81	0.59	1.39	0.07	10.26	0.92	88.7	0.5	4.09	1.79	0.9	本次
番阳峒	D0476	13.68	5.29	33.43	37.53		3.9	2.86	0.77	2.37	0.09	10.22	0.90	84.6	0.5	4.52	1.61	0.87	本次
番阳峒	D0825	17.34	4.77	30.05	41.61		1.97	2.17	0.54	1.49	0.07	10.5	0.95	89	0.79	4.32	1.98	0.88	本次
番阳峒	PM1-69-1	26.54	7.55	24.16	34.94		1.16	2.4	0.92	2.15	0.19	8.69	0.95	85.64	3.8	2.76	2.07	0.80	本次
番阳峒	PM1-107-2	25.44	10.72	32.15	26.68			3.07	0.59	1.11	0.19	8.25	0.99	84.27	6.83	2.46	1.19	0.74	本次
番阳峒	PM1-117-1	28.97	4.86	28.55	33.39			2.39	0.5	0.92	0.05	8.96	1.02	90.91	0.71	2.68	1.67	0.84	本次
	平均	23.98	5.97	28.39	34.32	0.12	1.10	4.65	0.58	0.87	0.12	9.10	0.97	86.69	1.89	3.06	1.09	1.49	
麻山田	H1498-1	6.33	8.52	30.75	44.97	0	1.09	6.14	0.99	1.08	0.14	11.2	0.97	82.05	1.98	6.11	0.82	2.09	3
麻山田	BD2	3.68	13.14	31.3	40.84	0	1.16	7.27	1.21	1.03	0.38	10.47	0.95	75.82	3.38	6.12	0.74	1.87	3
麻山田	S012-1	0	1.9	44.25	35.88	0	11.32	(4.2)	1	0.61	0.21	11.49	0.77	80.75	1.50	7.05	0.96	1.13	3
麻山田	D182-1	11.37	11.97	29.05	34.82	0.27	0	5.88	1.45	4.86	0.33	9.18	1.00	75.24	3.21	4.59	0.72	1.72	3
麻山田	HG-828	6.26	9.26	30.03	34.89	0	6.18	8.98	1.61	2.32	0.47	9.27	0.83	71.18	5.36	4.98	0.78	1.66	3
麻山田	IIP10-1	17.02	9.12	26.38	38.87	0.2	0	5.91	1.18	1.01	0.3	9.64	0.99	82.27	4.32	3.89	0.78	2.11	3
麻山田	D1466-1	13.87	6.86	27.97	44.57	0	0.4	3.79	0.88	1.48	0.19	10.83	0.98	86.41	3.48	4.9	0.84	2.28	3
麻山田	D759-1	0.68	12.26	27.94	47.34	0.29	0	6.67	1.27	3.1	0.45	11.17	0.99	75.96	4.05	7.93	0.75	2.43	3
	平均	7.40	9.13	30.96	40.27	0.10	2.52	5.06	1.20	1.94	0.31	10.41	0.94	78.71	3.41	5.70	0.80	1.91	

续表

岩石单位	样品号	Q	An	Ab	Or	C	Di	Hy(Ol)	Il	Mt	Ap	K_2O+Na_2O	A/CNK	DI	SI	δ	$\frac{K_2O}{Na_2O}$	KAI	来源
深堀村	H1474-2	9.36	20.38	28.72	22.83	0	2.34	12.47	1.59	1.88	0.43	7.1	0.92	60.91	10.85	3.07	0.57	1.14	3
深堀村	S042	5.85	15.84	27.6	33.87	0	2.19	11.67	1.33	1.09	0.56	8.96	0.92	67.32	7.21	4.64	0.67	1.76	3
深堀村	HG-18	16.31	12.79	26.96	35.15	0.62	0	3.91	1.44	2.21	0.61	9.03	1.00	78.42	6.09	3.76	0.69	1.87	3
深堀村	HG-23	10.34	18.73	29.04	27.16	0	0.45	10.48	0.25	2.63	0.91	7.93	0.94	66.54	11.64	3.46	0.61	1.34	3
深堀村	D430-1	10.35	13.19	34.94	27.47	0	0.8	11.62	1.12	0.18	0.33	8.63	0.96	72.76	3.82	3.71	0.71	1.13	3
	平均	10.47	16.94	28.08	29.75	0.16	1.25	9.63	1.15	1.95	0.63	8.26	0.95	68.30	8.95	3.73	0.64	1.53	
深沟	S045	0.09	0	54.69	22.26	0	10.97	0	0.47	2.6	0.14	10.29	0.57	77.04	2.18	5.78	1.03	0.56	3
进岭	H211-3	29.66	5.21	27.6	30.98	0.35	0	5.25	0.38	0.47	0.07	8.42	1.02	88.24	1.18	2.38	0.83	1.61	3
进岭	H1323-1	31.19	2.72	29.36	30.69	0	1.1	4.22	0.17	0.51	0.02	8.54	0.96	91.24	0.51	2.35	0.92	1.50	3
瘦岭	D3013-1	35.95	2.79	31.24	26.66	0.47	0	2.4	0.15	0.31	0.02	8.11	1.04	93.85	0.83	1.98	0.88	1.22	3
瘦岭	QXI1-1	35.02	0.69	28.25	32.52	1.77	0	0.95	0.1	0.6	0.09	8.74	1.14	95.79	2.04	2.35	0.85	1.65	3
瘦岭	QXI3-3	25.48	4.86	29.38	34.86	2.14	0	2.1	0.29	0.62	0.26	9.23	1.13	89.72	3.56	3.09	0.75	1.70	3
	平均	32.15	2.78	29.62	31.35	1.46	0.00	1.82	0.18	0.51	0.12	8.69	1.10	93.12	2.14	2.47	0.83	1.52	
黑岭	QXVI6-2	35.7	3.92	22.77	33.18	0.71	0	2.72	0.27	0.67	0.07	8.21	1.05	91.65	2.12	2.1	0.83	2.09	3
黑岭	Y5551-1	41.52	0.57	21.86	29.82	2.7	0	1.62	0.29	1.2	0.42	7.53	1.22	93.2	2.40	1.72	0.77	1.95	3
黑岭	YXIII35-1	36.54	3.42	25.59	29.97	0.7	0	2.52	0.31	0.91	0.05	8.03	1.06	92.1	3.11	1.97	0.84	1.68	3
黑岭	YXIII9-2	37.02	4.87	23.89	28.14	0.93	0	3.65	0.38	1.02	0.09	7.55	1.07	89.05	3.52	1.77	0.78	1.69	3
黑岭	IIP6-1	32.05	5.21	25.11	31.26	0.7	0	4.96	0.58	0	0.14	8.18	1.04	88.42	2.19	2.19	0.80	1.78	3
	平均	36.57	3.60	23.84	30.47	1.15	0.00	3.09	0.37	0.76	0.15	7.90	1.09	90.88	2.67	1.95	0.80	1.84	
尖峰岭	D3016-1	32.93	5.34	23.12	33.48	0.21	0	4.42	0.34	0.13	0.02	8.33	1.02	89.53	1.19	2.21	0.83	2.07	3
尖峰岭	S019	28.74	5.29	24.4	34.74	0.46	0	5.41	0.46	0.41	0.09	8.69	1.03	87.88	1.90	2.57	0.82	2.04	3
尖峰岭	QXVI6-1	29.37	4.35	23.8	37.26	0.95	0	3.2	0.37	0.59	0.12	8.98	1.06	90.43	2.39	2.74	0.82	2.24	3
尖峰岭	QXVIII1-1	26.81	7.71	22.84	36.45	0	0.12	4.44	0.67	0.78	0.19	8.77	0.98	86.1	3.64	2.74	0.80	2.28	3
尖峰岭	Y5550-1	32.57	4.83	26.53	31.8	0.27	0	2.78	0.29	0.66	0.28	8.42	1.00	90.9	2.18	2.28	0.84	1.72	3
尖峰岭	D3025-1	36.06	3.54	27.44	28.05	0.59	0	4.02	0.15	0.12	0.02	7.91	1.05	91.55	0.59	1.91	0.85	1.46	3
尖峰岭	Q1200	38.67	3.03	24.99	28.99	1.03	0	2.6	0.17	0.47	0.05	7.77	1.09	92.65	1.87	1.81	0.83	1.66	3

续表

岩石单位	样品号	Q	An	Ab	Or	C	Di	Hy(Ol)	Il	Mt	Ap	K_2O+Na_2O	A/CNK	DI	SI	δ	$\frac{K_2O}{Na_2O}$	KAI	来源
尖峰岭	HG-433	29.75	4.71	24.16	35.29	1.16	0	2.48	0.61	1.65	0.19	8.78	1.07	89.2	2.35	2.64	0.79	2.09	3
尖峰岭	HG-467	35.48	4.13	24.98	30.23	0.9	0	3.04	0.31	0.81	0.12	7.98	1.07	90.69	3.17	2	0.81	1.73	3
尖峰岭	S037	30.46	4.75	26.62	33.31	0	0.7	3.54	0.29	0.26	0.07	8.71	0.97	90.39	1.89	2.46	0.87	1.79	3
尖峰岭	D271-1	32.36	3.31	29.21	29.74	0.01	0	4.63	0.27	0.42	0.05	8.43	1.00	91.31	0.53	2.23	0.90	1.46	3
	平均	32.11	4.64	25.28	32.67	0.51	0.07	3.69	0.36	0.57	0.11	8.43	1.03	90.06	1.97	2.33	0.83	1.87	
高村	Ⅷ1-1	27.15	16.11	25.38	20.32	0	1.68	4.28	1.65	2.73	0.71	6.31	0.91	72.85	8.97	1.67	1.15	0.59	1
高村	D2385-2	40.42	5.61	29.32	22.08	0.8	0	0.8	0.1	0.5	0.35	7.15	1.04	91.82	1.73	1.49	1.08	0.77	1
两院	Ⅷ6-1	27.35	13.24	25.53	23.52	0	5.84	0.33	1.33	2.26	0.58	6.98	0.82	76.4	5.64	1.88	1.32	0.66	1
两院	D2018	39.71	4.77	28.36	22.59	0.65	0	1.74	0.19	1.41	0.58	7.17	1.00	90.66	1.47	1.52	1.14	0.80	1
两院	0504	34.4	7.22	25.25	27.53	1.01	0	2.38	0.46	1.61	0.12	7.58	1.07	87.18	3.82	1.89	1.56	0.73	1
两院	D2385-1	43.47	4.81	22.36	23.3	2.79	0	1.42	0.38	0.88	0.58	6.53	1.20	89.13	3.40	1.29	1.49	0.65	1
大岭	Ⅷ15-2	29.41	15.01	24.39	22.65	0	4.3	1.29	0.77	1.68	0.47	6.64	0.86	76.45	7.46	1.65	1.33	0.62	1
鹅塘村	Ⅷ15-1	29.9	17.52	21.47	22.08	0	1.35	4.11	0.96	2.13	0.47	6.23	0.93	73.45	9.69	1.49	1.47	0.56	1
陀类	D3082-1	29.76	5.73	27.9	30.01	1.4	0	4.44	0.46	0.06	0.21	8.28	1.09	87.67	2.78	2.34	1.54	0.76	3
陀类	Y78-1	37.36	1.08	22.72	31.62	3.62	0	1.94	0.25	1.29	0.12	7.93	1.33	91.7	3.14	2.02	1.99	0.72	3
陀类	Y79-1	37.17	1.08	23.2	31.57	3.05	0	2.03	0.29	1.49	0.12	7.96	1.27	91.94	2.86	2.04	1.95	0.75	3
陀类	Y80-1	37.65	1.59	25.82	29.03	3.27	0	1.43	0.19	0.91	0.12	7.83	1.29	92.5	2.65	1.94	1.61	0.73	3
	平均	35.49	2.37	24.91	30.56	2.84	0.00	2.46	0.30	0.94	0.14	8.00	1.25	90.95	2.86	2.09	1.77	0.74	
大田	D3087-1	26.89	7.88	31.55	26.47	1.25	0	4.94	0.79	0	0.23	8.1	1.07	84.91	4.58	2.36	1.20	0.73	3
大田	D3084-1	23.25	8.83	28.49	29.87	0.94	0	7.12	1.02	0.04	0.45	8.33	1.03	81.61	4.91	2.68	1.50	0.72	3
大田	DT1-1	24.81	9.35	31.21	27.51	0.3	0	4.86	1.03	0.56	0.37	8.28	0.99	83.53	4.27	2.53	1.26	0.75	3
大田	DT2-1	25.71	7.51	29.96	28.7	0.65	0	5.92	0.92	0.29	0.33	8.3	1.02	84.37	3.33	2.52	1.37	0.76	3
	平均	25.17	8.39	30.30	28.14	0.79	0.00	5.71	0.94	0.22	0.35	8.25	1.03	83.61	4.27	2.52	1.33	0.74	
什天	YVII9-2	22.27	9.54	28.58	31.04	0	0.55	4.82	1.05	1.62	0.51	8.56	0.95	81.89	7.86	2.87	1.56	0.76	3
什天	S031	25.58	7.89	31.12	28.64	0.15	0	5.23	0.65	0.5	0.23	8.46	0.99	85.34	4.91	2.55	1.32	0.79	3
什天	HG-437	32.79	8.25	27.09	23.23	1.39	0	5.06	0.67	1.23	0.28	7.04	1.09	83.11	6.92	1.72	1.23	0.68	3

续表

岩石单位	样品号	Q	An	Ab	Or	C	Di	Hy(Ol)	Il	Mt	Ap	K_2O+Na_2O	A/CNK	DI	SI	δ	$\frac{K_2O}{Na_2O}$	KAI	来源
什天	D3020-1	31.1	8.78	26.75	26.66	0.38	0	4.86	0.48	0.83	0.16	7.62	1.02	84.51	5.07	1.96	1.43	0.74	3
什天	IIP4-1	23.1	9.83	29.31	28.73	0.43	0	6.59	1.04	0.57	0.4	8.22	1.00	81.14	5.92	2.63	1.40	0.73	3
什天	S014	23.57	6.75	28.56	31.6	0	2.8	4.57	0.75	1.22	0.19	8.63	0.91	83.73	3.87	2.8	1.58	0.82	3
	平均	26.40	8.51	28.57	28.32	0.39	0.56	5.19	0.77	1.00	0.30	8.09	0.99	83.29	5.76	2.42	1.42	0.75	
白茶	H1362-1	30.01	4.03	25.98	34.08	1.1	0	3.96	0.42	0.29	0.12	8.75	1.08	90.07	1.86	2.55	1.88	0.81	3
东岭	H1211-1	22.93	14.71	25.41	24.25	0	0.47	9.56	1.03	1.2	0.39	7.1	0.96	72.59	10.55	2.05	1.37	0.63	3
东岭	H1267-2	22.32	14.99	25.78	23.77	0	0.31	10.12	1.2	1.06	0.44	7.03	0.96	71.87	11.63	2.06	1.32	0.63	3
那佩	D3074-1	31.3	7.73	28.25	24.75	0.87	0	6.39	0.56	0	0.14	7.44	1.06	84.3	3.38	1.88	1.25	0.73	3
那佩	G1346	32.55	10.7	24.45	24.84	0.65	0	5.54	0.58	0.57	0.12	7.02	1.04	81.84	6.96	1.68	1.45	0.67	3
保国	D3076-1	28.2	11.76	25.39	25.46	0.59	0	7.33	0.79	0.24	0.23	7.21	1.02	79.05	6.21	1.92	1.44	0.66	3
保国	HG-28	27.17	10.4	27.11	26.92	0.54	0	5.65	0.78	0.96	0.44	7.7	1.01	81.2	4.91	2.19	1.42	0.70	3
保国	HG-774	27.73	10.03	29.31	25.13	0	1.53	4	0.73	1.27	0.26	7.64	0.94	82.17	6.61	2.1	1.23	0.74	3
保国	J121-1	22.18	12.17	25.99	24.68	0	3.32	5.22	2.17	3.5	0.77	7.22	0.86	72.85	12.21	2.29	1.36	0.68	3
保国	J133-1	26.75	8.92	23.45	32.63	0	0.98	3.14	1.24	2.47	0.42	8.23	0.94	82.83	6.54	2.54	1.99	0.76	3
	平均	26.41	10.66	26.25	26.96	0.23	1.17	5.07	1.14	1.69	0.42	7.60	0.95	79.62	7.30	2.21	1.49	0.71	
火岭	H1311-3	32.38	5.77	25.69	31.56	0.98	0	2.78	0.27	0.51	0.07	8.36	1.07	89.63	1.63	2.24	1.76	0.78	3
火岭	VHG1-7	28.23	7.63	19.5	37.4	0.32	0	4.6	0.88	1.16	0.28	8.54	1.00	85.13	5.13	2.62	2.75	0.77	1
火岭	7329-1	29.86	4.96	26.77	32.08	1.05	0	4.31	0.29	0.44	0.23	8.5	1.06	88.71	2.04	2.44	1.72	0.79	1
火岭	XI2-1	29.07	11.97	29.09	23.77	0	3.23	0	0.29	1.04	0.47	7.4	0.84	81.93	3.01	1.93	1.17	0.70	1
火岭	XII5-1	32.68	4.34	26	32.48	1.6	0	1.23	0.29	1.03	0.35	8.45	1.10	91.16	2.16	2.36	1.79	0.77	1
火岭	4108-1	26.05	8.19	27.63	32.51	0	1.38	3.3	0.38	0.39	0.16	8.7	0.95	86.19	2.94	2.66	1.69	0.79	1
	平均	29.71	7.14	25.78	31.63	0.66	0.77	2.70	0.40	0.76	0.26	8.33	1.00	87.13	2.82	2.38	1.81	0.77	
长塘岭	H1278-1	29.75	8.76	22.91	29.22	1.03	0	6.93	0.65	0.54	0.21	7.63	1.06	81.88	4.34	2.05	1.83	0.70	3
长塘岭	1225	27.13	6.11	27.45	32.9	0.93	0	3.27	0.73	1.25	0.23	8.72	1.05	87.48	4.40	2.71	1.72	0.78	1
长塘岭	IX10-1	34.55	4.01	26.96	30.03	1.25	0	1.76	0.19	0.96	0.28	8.2	1.08	91.54	3.19	2.12	1.59	0.80	1
长塘岭	XI10-2	24.12	16.64	25.86	19.51	0	8.63	1.58	0.97	2.1	0.59	6.22	0.77	69.49	12.55	1.67	1.08	0.59	1

续表

岩石单位	样品号	Q	An	Ab	Or	C	Di	Hy(Ol)	Il	Mt	Ap	K_2O+Na_2O	A/CNK	DI	SI	δ	$\frac{K_2O}{Na_2O}$	KAI	来源
长塘岭	4871-2	24.31	11.3	27.2	29.63	0.19	0	4.61	1.19	1.08	0.49	8.14	0.98	81.14	7.65	2.55	1.56	0.71	1
长塘岭	D823-1	29.84	9.73	24.01	29.36	0.85	0	5.09	0.54	0.31	0.28	7.73	1.04	83.21	8.43	2.09	1.75	0.69	1
长塘岭	YD499	33.57	5.94	23.34	31.23	2.05	0	2.33	0.64	0.69	0.21	7.93	1.14	88.14	5.12	2.11	1.92	0.71	1
	平均	29.04	8.93	25.39	28.84	0.90	1.23	3.65	0.70	0.99	0.33	7.80	1.02	83.27	6.53	2.19	1.64	0.71	
大坡	1389-2	24.36	10.24	28.4	28.8	0.03	0	6.68	0.69	0.59	0.21	8.14	0.99	81.56	4.77	2.49	1.45	0.74	1
大坡	21	25.38	6.46	32.89	28.43	0	1.79	2.72	0.36	1.82	0.12	8.66	0.94	86.7	2.49	2.66	1.24	0.83	1
大坡	HG394	30.45	11.36	27.57	20.93	1.1	0	5.51	0.86	1.86	0.35	6.72	1.05	78.95	9.04	1.67	1.09	0.64	1
大坡	Ⅻ7-2	29.08	12.83	26.26	21.44	0	5.15	1.71	0.77	2.16	0.59	6.66	0.83	76.78	7.66	1.67	1.17	0.66	1
大坡	D892-1	30.84	10.49	27.56	23.52	0.5	0	4.52	0.62	1.7	0.24	7.11	1.02	81.92	9.04	1.81	1.22	0.69	1
大坡	Ⅺ6-1	33.58	13.44	23	26.76	0	0.17	1.4	0.38	0.89	0.35	7.17	0.97	83.34	3.49	1.73	1.67	0.66	1
大坡	Ⅻ7-1	34.28	8.7	26.61	23.81	0.77	0	2.93	0.77	1.75	0.35	7.05	1.03	84.7	5.84	1.7	1.28	0.71	1
大坡	Ⅱ8-1	23.76	8.33	29.21	29.88	0	5.55	2.18	0.39	0.65	0.05	8.38	0.85	82.85	7.41	2.6	1.46	0.78	1
大坡	Ⅱ19-1	23.85	10.34	28.49	26.53	1.06	0	7.52	0.78	1.39	0.05	7.7	1.07	78.87	10.58	2.37	1.33	0.68	1
大坡	XⅣ25-1	27.58	10.86	29.8	22.25	0.65	0	6.75	0.77	0.98	0.35	7.2	1.02	79.63	12.42	1.91	1.07	0.68	1
大坡	XⅣ9-1	29.45	10.36	28.01	23.05	1.41	0	5.07	0.81	1.35	0.47	7.1	1.07	80.51	7.43	1.88	1.18	0.65	1
大坡	XX12-1	34.85	8.83	26.47	21.73	1.28	0	4.59	0.54	1.52	0.18	6.7	1.09	83.05	11.81	1.52	1.18	0.67	1
大坡	Ⅸ17-1	28.56	16.42	20.89	23.49	0	0.58	8.34	0.82	0.65	0.24	6.29	0.97	72.94	12.28	1.55	1.61	0.58	1
大坡	Ha34-1	35.01	7.65	22.64	29	0.24	0	4.77	0.31	0.29	0.09	7.51	1.01	86.65	3.48	1.8	1.83	0.76	1
	平均	29.36	10.45	26.99	24.97	0.50	0.95	4.62	0.63	1.26	0.26	7.31	0.99	81.32	7.70	1.95	1.34	0.70	
乌兰盖	3809	30.42	7.3	28.43	28.8	0.45	0	3.39	0.48	0.57	0.16	8.21	1.02	87.65	3.94	2.23	1.45	0.78	1
乌兰盖	D45-1	26.55	13.5	25.8	24.96	0.47	0	5.61	0.96	1.82	0.33	7.18	1.01	77.31	9.70	2.01	1.39	0.64	1
通什	H1274-2	20.98	17.07	23.57	24.37	0.6	0	11.18	1.08	0.82	0.33	6.82	1.02	68.92	9.79	2.08	1.48	0.57	3
通什	H1287-1	17.28	20.29	25.18	19.55	0.91	0	14.04	1.17	1.18	0.4	6.23	1.03	62.01	12.52	1.97	1.11	0.50	3
通什	HG-424	23.72	15.7	22.83	24.82	0	0.35	8.35	1.43	2.31	0.49	6.8	0.96	71.37	9.14	2.01	1.56	0.61	3
通什	HG-443	23.09	15.18	25.74	24.81	0.43	0	8.33	1.23	0.96	0.23	7.14	1.01	73.64	9.98	2.13	1.38	0.61	3
通什	HG-475	21.42	15.4	22.25	28.45	0	0.46	8.54	1.77	1.41	0.3	7.36	0.97	72.12	10.42	2.37	1.83	0.63	3

续表

岩石单位	样品号	Q	An	Ab	Or	C	Di	Hy(Ol)	Il	Mt	Ap	K_2O+Na_2O	A/CNK	DI	SI	δ	$\frac{K_2O}{Na_2O}$	KAI	来源
通什	4220	20.21	11.83	31.2	25.62	0	1.47	6.96	0.9	1.38	0.42	7.92	0.93	77.03	9.15	2.61	1.18	0.71	1
通什	4107	20.36	10.94	31.93	24.63	0	3.24	7.04	0.83	0.62	0.42	7.85	0.89	76.92	8.98	2.5	1.10	0.73	1
通什	38-1	26.62	11.02	26.92	27.72	0.05	0	6.2	0.56	0.63	0.28	7.77	0.98	81.26	4.26	2.23	1.47	0.72	1
通什	Ⅸ8-1	25.86	14.12	22.51	28.32	0.68	0	6.31	0.77	1.09	0.35	7.34	1.02	76.69	9.67	2.13	1.80	0.62	1
通什	XⅣ12-1	24.51	13.61	29.4	20.53	0.6	0	7.79	1.15	1.71	0.7	6.88	0.99	74.44	15.54	1.92	1.00	0.63	1
通什	4111-1	19.44	10.81	31.84	27.52	0	1.76	4.74	1.18	2.12	0.58	8.39	0.92	78.8	8.59	2.92	1.24	0.74	1
通什	4577-1	33.51	9.08	28.78	21	1.05	0	4.76	0.77	0.77	0.28	6.85	1.06	83.29	6.45	1.6	1.04	0.68	1
通什	Ha35-1	30.24	12.27	26.91	20.11	0.32	0	8.49	0.82	0.54	0.3	6.52	1.00	77.26	8.26	1.53	1.07	0.65	1
	平均	23.63	13.64	26.85	24.42	0.36	0.56	7.90	1.05	1.20	0.39	7.22	0.98	74.90	9.44	2.15	1.33	0.65	
山牛塘	H1311-2	12.1	24	24.39	11.89	0	5.12	18.2	2.4	1.29	0.61	4.84	0.85	48.38	18.46	1.51	0.70	0.44	3
山牛塘	H1397-1	4.64	26.76	25.28	15.5	0	6.88	16.31	2.35	1.71	0.54	5.52	0.83	45.42	22.61	2.45	0.88	0.44	3
山牛塘	H209-1	18.87	26.97	22.99	12.75	0	0.35	13.6	2.32	1.58	0.57	4.79	0.96	54.61	14.84	1.27	0.79	0.41	3
山牛塘	H209-3	17.22	24.2	24.77	14.4	0.1	0	14.09	3.06	1.12	1.02	5.26	0.95	56.39	13.72	1.58	0.83	0.45	3
	平均	13.21	25.48	24.36	13.64	0.03	3.09	15.55	2.53	1.43	0.69	5.10	0.90	51.20	17.41	1.70	0.80	0.44	
石碌	137-1	21.94	11.18	25.81	30.22	1.24	0	8.08	0.75	0.47	0.3	8.11	1.06	77.97	6.66	2.66	1.68	0.66	
石碌	Ha29	31.16	13.99	24.08	20.9	0.86	0	7.64	0.68	0.41	0.26	6.28	1.04	76.14	7.08	1.45	1.24	0.59	1
石碌	7334-1	22.05	15.92	26.21	24.71	1.21	0	8.21	0.63	0.68	0.38	7.19	1.05	72.97	7.81	2.2	1.35	0.58	1
石碌	HG399	31.19	10.4	22.68	27.96	0.74	0	5.4	0.52	0.91	0.19	7.3	1.04	81.83	7.08	1.88	1.77	0.68	1
	平均	26.59	12.87	24.70	25.95	1.01	0.00	7.33	0.65	0.62	0.28	7.22	1.05	77.23	7.16	2.05	1.51	0.63	
什败	D1547-3	14.44	14.63	23.69	30.56	0	3.08	8.28	2.34	2.07	0.9	7.83	0.87	68.69	9.10	3.23	1.85	0.66	3
行干	D1749-1	7.64	19.3	30.01	20	0	4.99	12.13	2.85	1.65	1.42	6.8	0.82	57.65	17.33	3.09	0.95	0.57	3
行干	D3089-1	0.3	20.88	31.6	22.37	0	4.13	15.11	2.94	1.14	1.5	7.43	0.84	54.27	17.48	4.5	1.01	0.57	3
方满	D3092-1	0	17.89	25.33	22.65	0	12.97	0.22 (11.9)	2.69	4.04	2.28	6.66	0.65	47.98	29.42	5.48	1.28	0.58	3
新风岭	H5122-1	37.48	1.97	23.99	30.06	3.99	0	1.76	0.33	0.24	0.17	7.74	1.35	91.53	1.78	1.95	1.79	0.68	3

续表

岩石单位	样品号	Q	An	Ab	Or	C	Di	Hy(Ol)	Il	Mt	Ap	K_2O+Na_2O	A/CNK	DI	SI	δ	$\frac{K_2O}{Na_2O}$	KAI	来源
顺作	D3136-1	31.04	11.37	26.17	24.43	1.97	0	4.31	0.61	0	0.09	7.15	1.14	81.64	5.65	1.81	1.34	0.61	3
乘顺岭	QIV1-1	40.91	0.46	16.15	35.64	3.95	0	1.64	0.31	0.85	0.07	7.82	1.39	92.7	4.41	1.9	3.16	0.70	3
乘顺岭	GXI3-1	31.74	5.43	25.45	30.54	2.03	0	3.43	0.38	0.79	0.19	8.1	1.15	87.73	4.49	2.22	1.72	0.72	3
志仲	D3073-1	26.8	9.87	26.8	25.95	0.87	0	8.32	1.06	0	0.33	7.47	1.04	79.55	6.45	2.09	1.39	0.69	3
志仲	HG-17	29.2	12.66	26.59	22.11	0.44	0	7.09	1.04	0.38	0.47	6.79	1.00	77.9	7.10	1.7	1.19	0.64	3
志仲	GVIII18	37.13	5.8	20	28.05	1.99	0	5.6	0.48	0.7	0.23	7.04	1.15	85.18	6.56	1.61	2.01	0.69	3
	平均	31.04	9.44	24.46	25.37	1.10	0.00	7.00	0.86	0.36	0.34	7.10	1.06	80.88	6.70	1.80	1.53	0.67	
抱郎	QIV4-1	30.72	10.86	22.33	28.3	1.48	0	3.29	0.79	1.95	0.24	7.29	1.09	81.35	5.98	1.98	1.81	0.64	3
抱郎	QVII2	28.29	8.79	22.98	31.21	1.11	0	4.7	0.82	1.82	0.26	7.92	1.06	82.48	6.59	2.32	1.94	0.70	3
石岭	7315	20.54	17.94	23.28	22.31	1.22	0	11.94	1.41	0.76	0.58	6.5	1.04	66.13	12.94	1.95	1.37	0.52	1
石岭	7316-1	20.46	24.64	18.2	20.38	0	0.86	12.48	1.29	1.16	0.52	5.52	0.95	59.04	15.55	1.52	1.60	0.45	1
石岭	Ha36	24.21	23.63	17.06	18.59	0.21	0	13.82	1.24	0.84	0.38	5.07	0.99	59.86	16.95	1.2	1.56	0.43	1
	平均	21.74	22.07	19.51	20.43	0.48	0.29	12.75	1.31	0.92	0.49	5.70	0.99	61.68	15.15	1.56	1.51	0.47	
保山村	4758-1	24.46	21.2	14.66	18.67	1.56	0	12.9	2.13	3.55	0.87	4.8	1.04	57.79	16.83	1.22	1.82	0.40	1
保山村	4971-1	26.56	21.54	18.48	15.61	1.54	0	11.96	1.56	2.17	0.56	4.75	1.06	60.65	15.44	1.06	1.21	0.41	1
保山村	Ha32-1	21.59	12.37	28.87	23.94	0	0.93	9.39	1.29	1.1	0.49	7.35	0.94	74.4	10.34	2.27	1.19	0.69	1
	平均	24.20	18.37	20.67	19.41	1.03	0.31	11.42	1.66	2.27	0.64	5.63	1.01	64.28	14.20	1.52	1.41	0.50	
白石村	211-3	10.07	18.78	32.01	16.06	0	5.32	13.28	1.89	1.78	0.79	6.46	0.84	58.14	19.61	2.43	0.72	0.57	1
白石村	290	1.2	24.82	22.73	8.3	0	5.46	34.07	2.01	0.63	0.78	4.02	0.82	32.23	40.22	1.48	0.52	0.39	1
白石村	4350-7	0.12	21.89	26.08	19.43	0	7.36	17.95	2.89	2.8	1.5	6.2	0.77	45.63	28.90	3.72	1.07	0.52	1
白石村	Ha16	0	21.04	26.41	20.31	0	9.35	10.8 (3.3)	2.69	4.54	1.51	6.43	0.74	46.72	25.05	4.48	1.10	0.53	1
白石村	Ha17-2	0	24.95	39.66	15.97	0	1.35	1.42 (9.8)	2.56	2.88	1.38	7.3	0.91	55.63	16.02	5.29	0.58	0.54	1
白石村	Ha17-1	1.29	21.16	34.47	19.35	0	3.92	12.97	2.92	2.18	1.73	7.27	0.84	55.11	17.01	4.3	0.80	0.57	1
白石村	Ha18	0	24.57	37.08	18.46	0	2.85	(9.1)	2.42	3.72	1.57	7.47	0.87	55.75	14.51	5.77	0.71	0.54	1

续表

岩石单位	样品号	Q	An	Ab	Or	C	Di	Hy(Ol)	Il	Mt	Ap	K_2O+Na_2O	A/CNK	DI	SI	δ	$\frac{K_2O}{Na_2O}$	KAI	来源
白石村	QIV0-1	16.27	25.48	24.39	15.76	0.36	0	12.34	2.25	2.3	0.85	5.43	0.97	56.42	14.52	1.75	0.93	0.44	3
白石村	s028	3.96	26.4	24.42	11.49	0	9.27	21.3	1.37	1.47	0.33	4.77	0.79	39.87	31.61	1.81	0.67	0.41	3
白石村	D3019-1	10.05	26.84	24.68	11.81	0	5.97	15.63	1.98	2.41	0.64	4.82	0.84	46.54	25.63	1.63	0.69	0.41	3
	平均	4.30	23.59	29.19	15.69	0.04	5.09	14.81	2.30	2.47	1.11	6.02	0.84	49.20	23.31	3.27	0.78	0.49	
机告村	1153	11.14	21.35	26.06	16.15		4.09	13.77	2.17	4.18	1.06	5.7	0.84	53.35	23.37	2.18	0.89	0.51	1
机告村	4014-1	1.29	26.54	25.51	7.81		11.82	20.32	1.88	4.35	0.45	4.2	0.73	34.61	34.17	1.89	0.44	0.40	1
机告村	Ha19	0	19.34	27.47	19.61		11.14	11.9 (3.5)	2.39	3.32	1.27	6.47	0.72	47.08	28.94	3.98	1.02	0.56	1
机告村	Ha39-1	0	29.69	25.12	6.07		9.29	23.89 (0.5)	2.97	1.27	1.14	3.89	0.76	31.19	35.24	1.76	0.35	0.36	1
机告村	Ha39-2	3.64	22.31	18.8	5.71		15.64	29.47	1.91	1.86	0.64	3.1	0.62	28.15	45.31	0.88	0.44	0.36	1
	平均	3.21	23.85	24.59	11.07		10.40	21.19	2.26	3.00	0.91	4.67	0.73	38.88	33.41	2.14	0.63	0.44	
岩脉	D1650-1	29.56	10.44	23.6	29.38	0.12		5.66	0.84	0.18	0.22	7.68	0.99	82.54	4.51	2.07	1.78	0.72	3
岩脉	B326	24.23	15.07	32.85	18.78		1.12	6.74	0.63	0.45	0.13	7.02	0.96	75.86	11.4	1.88	0.82	0.64	3
岩脉	B200	26.83	8.03	20.88	38.92		0.27	4.3	0.55		0.22	8.99	0.98	86.63	5.43	2.81	2.67	0.79	3
岩脉	S34-7	42.57	18.71	30.73	2.74	0.4		4.44	0.33	0	0.08	4.06	1.02	76.04	5.73	0.51	0.13	0.47	3
公庙	D4760-1	5.2（ol)	27.38	23.3	4.86		21.41	12.98	2.24	2.35	0.29	3.48	0.61	28.16	36.1	1.54	0.30		2
公庙	D4761-1	3.79	22.63	24.44	7.29		16.78	18.73	3.1	2.74	0.5	4.01	0.64	35.52	30	1.55	0.42		2
亚炮	D336-1	41.51	5.13	26.06	18.44	2.36		3.86	0.14	0.99	0.12	6.20	1.23	87.21	4.69	1.18	1.01	0.66	1
亚炮	V40-1	41.45	0.98	26.06	24.52	2.51		1.65	0.04	0.81	0.14	7.23	1.26	90.83	0.56	1.57	1.35	0.76	1
	平均值	41.48	3.06	26.06	21.48	2.44		2.76	0.09	0.90	0.13	6.72	1.25	89.02	2.63	1.38	1.18	0.71	
戈枕村	V32-1	33.84	8.34	23.52	13.35	4.26		10.56	0.95	0.80	0.28	5.04	1.41	73.55	8.72	1.04	0.81	0.48	1
戈枕村	V37-3	33.39	7.98	22.08	15.84	3.45		10.26	0.91	1.77	0.23	5.29	1.32	74.33	11.59	1.14	1.03	0.53	1
戈枕村	Ⅶ15-1	37.97	6.92	25.13	17.85	2.60		5.82	0.42	1.29	0.12	5.99	1.23	82.50	6.18	1.21	1.02	0.62	1
戈枕村	Ⅶ15-2	38.28	8.93	26.65	14.89	2.44		4.92	0.46	1.22	0.14	5.67	1.15	81.58	5.67	1.09	0.80	0.62	1
	平均值	35.87	8.04	24.35	15.48	3.19		7.89	0.69	1.27	0.19	5.50	1.28	77.99	8.04	1.12	0.92	0.56	

注：样品来源同附表 9-1。

附表 9-3　海南岛侵入岩稀土元素含量（10^{-6}）及主要参数一览表

Attached table 9-3　List of REE contents and main parameters of intrusive rocks from the Hainan Island

岩石单位	样品号	La	Ce	Pr	Nd	Sm	Eu	Gd	Tb	Dy	Ho	Er	Tm	Yb	Lu	Y	ΣREE	δEu	$(La/Yb)_N$	$(La/Sm)_N$	$(Gd/Yb)_N$	来源
$\gamma\pi K_2$	7329-4	33	68	6.7	27	5.8	0.38	4.5	0.73	4.3	0.7	1.7	0.22	1.25	0.2	16	170.48	0.22	17.8	3.58	2.91	1
$\gamma\pi K_2$	Ⅳ12-1	32.5	69.7	5.72	17.2	3.42	0.16	1.79	0.35	2.6	0.65	2.18	0.34	2.06	0.32	18.5	157.49	0.18	10.64	2.98	0.7	1
$\gamma\pi K_2$	Ⅳ5-1	63.4	104	11.1	40.4	6.81	0.88	4.02	0.69	3.37	0.7	1.88	0.29	1.74	0.26	15.1	254.64	0.47	24.57	2.86	1.86	1
$\gamma\pi K_2$	YD632-1	50.7	103.7	11.5	39.4	7.87	0.22	6.82	1.39	9.5	1.86	5.37	0.93	5.75	0.87	55.6	301.5	0.09	5.9	4.0	1.0	2
$\gamma\pi K_2$	D1341-1	43.32	125.6	11.2	35.24	6.96	0.2	6.09	1.28	8.65	1.74	5.56	0.87	5.65	0.8	51.56	304.75	0.09	5.2	3.9	0.9	2
$\gamma\pi K_2$	YD445	73.61	105.6	127	38.6	5.5	1.3	3.8	0.6	2.3	0.49	1.12	0.21	1.21	0.19	12.5	259.53	0.79	41.0	8.4	2.5	2
$\gamma\pi K_2$	YD633-1	9.65	13.2	2.45	8.8	1.48	0.27	1.21	0.19	0.88	0.2	0.57	0.12	0.74	0.14	6.78	46.68	0.6	8.8	4.1	1.3	2
$\gamma\pi K_2$	D0377	44.53	76.91	8.58	26.34	4.19	0.66	3.12	0.46	2.59	0.50	1.43	0.24	1.58	0.25	13.6	171.38	0.53		6.86	1.63	本次
$\gamma\pi K_2$	D0514	102.0	177.1	23.25	81.69	13.66	0.87	9.34	1.27	6.68	1.30	3.34	0.52	3.28	0.50	30.4	424.80	0.22		4.82	2.36	本次
$\gamma\pi K_2$	D0523	38.66	73.07	7.16	23.3	3.56	0.62	2.54	0.35	1.88	0.38	1.04	0.18	1.17	0.19	8.27	154.11	0.60		7.01	1.80	本次
$\gamma\pi K_2$	D5028-2	20.99	27.06	3.87	10.99	1.62	0.26	1.16	0.18	1.01	0.22	0.65	0.12	0.93	0.18	6.43	69.24	0.55		8.36	1.03	本次
$\gamma\pi K_2$	D1339	118.3	175.7	24.6	90.0	14.61	1.57	12.75	1.87	9.80	1.88	4.87	0.69	4.16	0.63	49.7	461.55	0.34		5.23	2.54	本次
$\gamma\pi K_2$	D0362-1	15.05	34.52	4.49	19.7	5.96	0.15	5.84	0.94	5.19	1.00	2.87	0.45	2.82	0.44	29.2	99.49	0.08		1.63	1.71	本次
	平均	49.67	88.78	19.05	35.28	6.26	0.58	4.84	0.79	4.52	0.89	2.51	0.40	2.49	0.38	24.13	214.51	0.37	16.27	4.90	1.71	
火烧岭	XⅪ7-2	37.7	66.7	7.5	24.9	3.48	0.84	2.64	0.36	1.71	0.33	0.90	0.14	0.78	0.14	9.9	158.5	0.76	32.6	6.2	2.7	2
龙楼	D6002-3	14.57	31.94	4.65	17.78	5.17	0.19	5.17	0.96	6.06	1.33	4.19	0.74	5.25	0.85	40.1	139.03	0.11	1.87	1.77	0.79	本次
龙楼	D2042	14.35	27.12	3.77	14.55	3.74	0.25	3.92	0.63	3.93	0.82	2.42	0.41	2.59	0.41	24.65	103.56	0.2	3.74	2.41	1.22	本次
龙楼	PM03-11	17.92	31.49	3.75	12.72	2.61	0.54	2.18	0.35	2.08	0.44	1.35	0.24	1.65	0.28	15.26	92.86	0.67	7.32	4.32	1.07	本次
龙楼	PM03-27	27.83	45.04	6.37	20.6	4.34	0.73	4.62	0.78	4.77	1.06	3.18	0.57	3.9	0.66	19.42	143.87	0.5	4.81	4.03	0.96	本次
	平均	18.67	33.90	4.64	16.41	3.97	0.43	3.97	0.68	4.21	0.91	2.79	0.49	3.35	0.55	24.86	119.83	0.37	4.44	3.13	1.01	
大小岭	D6112-1	16.02	20.45	2.52	7.34	1.08	0.22	0.78	0.12	0.73	0.18	0.47	0.10	0.84	0.17	5.11	51.02	0.70		9.58	0.77	5
大小岭	D3126-1	22.1	39.2	3.9	11.8	1.8	0.34	1.32	0.20	0.98	0.21	0.59	0.12	0.86	0.14	6.9	83.56	0.64		7.93	1.27	5
大小岭	GXVIII10-1	41.7	57.5	5.8	20.1	3.2	0.44	1.69	0.28	1.88	0.34	0.93	0.14	0.89	0.14	8.2	135.03	0.52		8.41	1.57	5
	平均	26.61	39.05	4.07	13.08	2.03	0.33	1.26	0.20	1.20	0.24	0.66	0.12	0.86	0.15	6.74	89.87	0.62		8.64	1.20	

续表

岩石单位	样品号	La	Ce	Pr	Nd	Sm	Eu	Gd	Tb	Dy	Ho	Er	Tm	Yb	Lu	Y	ΣREE	δEu	$(La/Yb)_N$	$(La/Sm)_N$	$(Gd/Yb)_N$	来源
吊罗山	D0495	84.57	145.6	19.65	76.64	14.07	4.14	11.77	1.73	9.64	1.84	5.05	0.80	4.91	0.79	46.43	381.20	0.96		3.88	1.98	5
吊罗山	D0496	50.10	99.20	14.00	58.62	11.91	3.85	10.56	1.61	9.23	1.80	4.93	0.77	4.73	0.78	44.51	272.09	1.03		2.72	1.85	5
吊罗山	D0503	74.61	75.06	13.74	46.67	7.20	1.54	4.73	0.59	2.95	0.55	1.36	0.21	1.30	0.21	11.47	230.72	0.76		6.69	3.01	5
吊罗山	D0525	127.3	152.5	32.28	137.2	28.38	5.55	26.20	3.81	19.50	3.63	9.32	1.37	8.00	1.17	82.94	556.21	0.61		2.90	2.71	5
吊罗山	D1386-1	19.07	33.64	3.96	14.48	2.49	0.61	1.88	0.26	1.42	0.29	0.84	0.14	0.95	0.17	7.76	80.20	0.83		4.94	1.64	5
吊罗山	D1902-1	18.67	33.12	3.56	11.50	1.81	0.33	1.43	0.20	1.11	0.24	0.75	0.14	0.96	0.18	7.13	74.00	0.61		6.66	1.23	5
吊罗山	D1952-1	20.35	38.76	4.99	16.53	2.88	0.69	2.21	0.32	1.70	0.35	0.98	0.17	1.11	0.18	9.46	91.22	0.80		4.56	1.65	5
吊罗山	D2821-1	55.40	103.8	13.18	45.99	9.23	0.99	7.71	1.17	6.56	1.35	3.61	0.56	3.43	0.51	32.13	253.49	0.35		3.87	1.86	5
吊罗山	D2950-1	50.86	78.47	8.66	27.09	4.18	0.76	3.00	0.45	2.30	0.49	1.36	0.23	1.57	0.25	13.57	179.67	0.63		7.86	1.58	5
吊罗山	D5207-1	20.28	38.18	4.53	15.92	2.76	0.63	2.26	0.31	1.63	0.33	0.95	0.16	1.08	0.19	9.41	89.21	0.75		4.74	1.73	5
吊罗山	PM1-26-1	61.53	115.5	13.65	50.91	9.70	1.08	8.33	1.25	6.66	1.39	3.71	0.59	3.53	0.55	35.56	278.38	0.36		4.10	1.95	5
吊罗山	PM1-30-1	66.94	123.9	15.01	56.02	9.78	1.05	7.51	1.03	5.19	1.01	2.35	0.31	1.71	0.24	22.27	292.05	0.36		4.42	3.63	5
吊罗山	D3124-1	21.9	38.5	3.5	11.8	1.6	0.34	1.23	0.19	1.02	0.22	0.66	0.11	0.85	0.16	7.9	82.08	0.71		8.84	1.20	5
吊罗山	D3125-1	23.3	43.1	4.0	12.8	1.9	0.34	1.40	0.17	1.03	0.22	0.62	0.11	0.88	0.16	7.2	90.03	0.61		7.92	1.32	5
吊罗山	S006	31.2	61.6	6.5	22.1	3.5	0.64	2.47	0.36	1.86	0.37	1.06	0.17	1.19	0.19	12.7	133.21	0.63		5.75	1.72	5
吊罗山	QV13-2	23.1	58.1	6.2	19.7	4.1	0.41	3.08	0.47	2.95	0.58	1.63	0.27	1.76	0.26	16.7	122.61	0.34		3.64	1.45	5
吊罗山	QV8-1	101.4	141.2	25.1	90.7	16.2	1.26	11.51	1.78	9.52	1.69	4.37	0.69	4.46	0.70	42.9	410.58	0.27		4.04	2.13	5
	平均	50.03	81.19	11.32	42.04	7.75	1.42	6.31	0.92	4.96	0.96	2.56	0.40	2.50	0.39	24.12	212.76	0.62		5.15	1.92	
新村	D5269-2	18.6	39.79	4.90	19.68	4.90	0.19	4.99	0.76	4.37	0.90	2.54	0.38	2.33	0.36	24.70	104.69	0.12		2.45	1.77	5
新村	D5325-1	18.65	33.73	3.64	10.18	1.67	0.19	1.54	0.26	1.44	0.34	1.16	0.24	1.87	0.38	11.27	75.29	0.36		7.21	0.68	5
新村	D5392-1	77.21	147.2	17.47	61.40	10.88	1.19	9.16	1.43	7.53	1.53	4.07	0.63	3.77	0.56	39.83	344.03	0.35		4.58	2.01	5
新村	D3128-1	51.5	92.6	9.3	32.2	5.1	0.95	3.64	0.44	2.56	0.49	1.25	0.18	1.28	0.20	13.3	201.69	0.64		6.52	2.35	5
新村	S007	20.6	41.6	4.5	15.8	2.5	0.58	1.89	0.29	1.54	0.30	0.89	0.14	1.06	0.18	10.5	91.87	0.78		5.32	1.48	5
新村	WH12-7	33.1	65.6	5.0	15.4	4.0	0.81	3.75	0.76	3.33	0.69	2.10	0.37	1.96	0.35	*	137.22	0.63		5.34	1.58	5
新村	WH2-4a	64.9	105.0	7.2	24.2	5.3	0.95	4.26	0.81	3.11	0.68	2.03	0.38	1.86	0.34	*	221.02	0.59		7.91	1.89	5
	平均	40.65	75.07	7.43	25.55	4.91	0.69	4.18	0.68	3.41	0.70	2.01	0.33	2.02	0.34	19.92	167.97	0.50		5.62	1.68	

续表

岩石单位	样品号	La	Ce	Pr	Nd	Sm	Eu	Gd	Tb	Dy	Ho	Er	Tm	Yb	Lu	Y	ΣREE	δEu	$(La/Yb)_N$	$(La/Sm)_N$	$(Gd/Yb)_N$	来源
南牙村	3862-2	72.6	124	15.9	50.9	11.2	0.97	9.05	1.38	8.1	1.7	4.58	0.66	3.96	0.49	59.6	365.1	0.29	12.36	4.08	1.84	1
南牙村	73	83.7	151	15.8	62.8	13.2	1.35	8.87	1.53	8.16	1.8	4.88	0.64	3.83	0.49	39.3	397.3	0.36	14.73	3.99	1.87	1
高通岭	YV28-1	19.8	38.7	4.3	14.4	2.5	0.57	1.52	0.16	0.73	0.13	0.31	0.05	0.27	0.04	3.6	87.0	0.83	49.3	4.9	4.5	2
双顶岭	YIV7-1	19.0	38.4	4.5	14.7	2.23	0.52	1.68	0.24	1.14	0.27	0.72	0.11	0.81	0.13	8.0	92.5	0.79	15.8	5.4	1.7	2
六弓	YIV12-1	29.7	55.5	6.0	18.0	2.73	0.63	1.83	0.23	1.23	0.26	0.63	0.1	0.61	0.1	7.2	124.7	0.81	32.7	6.8	2.4	2
六弓	I HG3-1	64.46	132.6	15.4	52.59	9.69	0.97	7.4	1.19	6.48	1.4	4.08	0.64	4.45	0.68	39.92	341.96	0.34	9.8	4.2	1.3	2
六弓	D0914-1	37.52	66.01	7.11	23.99	3.69	0.80	2.71	0.36	1.88	0.39	1.01	0.17	1.03	0.17	10.10	146.84	0.74		6.56	2.18	本次
六弓	D0972-1	32.75	49.84	5.58	18.75	2.99	0.71	2.45	0.34	1.98	0.41	1.17	0.20	1.35	0.22	11.82	118.74	0.78		7.07	1.50	本次
六弓	D1412-1	32.09	57.87	6.62	22.40	3.74	0.86	2.92	0.42	2.23	0.47	1.27	0.21	1.26	0.21	12.60	132.57	0.77		5.54	1.92	本次
六弓	D1430-1	35.14	49.33	9.39	33.16	5.79	1.44	4.00	0.53	2.65	0.48	1.20	0.18	1.04	0.16	10.78	144.49	0.87		3.92	3.18	本次
六弓	D1823-1	33.59	59.72	6.88	22.55	3.64	0.80	2.67	0.36	1.83	0.36	0.99	0.15	0.94	0.15	9.84	134.63	0.75		5.96	2.35	本次
六弓	D3119-1	38.4	69.1	6.8	23.1	3.5	0.80	2.60	0.34	1.69	0.34	0.93	0.13	0.92	0.16	10.3	148.81	0.78		7.08	2.34	本次
六弓	QV10-1	30.5	62.6	7.7	28.1	5.7	0.87	4.55	0.68	4.01	0.76	1.89	0.29	1.53	0.24	19.6	149.42	0.51		3.45	2.46	5
	平均	37.13	66.95	7.94	26.96	4.61	0.88	3.46	0.49	2.66	0.54	1.46	0.23	1.46	0.23	14.68	160.24	0.71	21.25	5.62	2.18	
保亭	YIV19-1	42.4	84.3	9.2	31.3	4.68	1.11	3.17	0.33	2.03	0.4	0.97	0.15	0.89	0.14	10.8	191.85	0.83	32.1	5.6	2.9	2
保亭	X XI6-1	35.3	57.2	7.8	27.5	4.64	1.01	3.53	0.51	2.68	0.52	1.51	0.24	1.42	0.23	14.8	158.9	0.73	16.7	4.8	2.0	2
保亭	I HG8-1	76.38	135	14.6	45.08	7.18	1.2	5.26	0.78	3.78	0.71	1.97	0.3	1.92	0.28	19.89	314.37	0.57	26.8	6.7	2.2	2
保亭	D0919-1	35.23	65.59	8.10	27.65	4.65	0.99	3.27	0.48	2.42	0.49	1.16	0.19	1.13	0.18	11.19	151.53	0.74		4.89	2.39	5
保亭	D1229-1	36.50	65.15	8.34	26.71	4.47	0.95	3.20	0.48	2.47	0.49	1.16	0.18	1.12	0.18	11.75	151.40	0.73		5.27	2.36	5
保亭	D1421-1	32.50	57.95	6.87	22.46	3.78	0.84	2.68	0.38	1.87	0.40	0.97	0.15	0.97	0.15	9.43	131.97	0.77		5.55	2.29	5
保亭	D3117-1	33.7	64.8	6.7	23.0	3.5	0.71	2.46	0.34	1.74	0.34	0.91	0.12	0.91	0.14	9.4	139.37	0.70		6.22	2.24	5
保亭	D3060-1	38.3	67.2	7.3	24.5	4.0	0.67	2.89	0.39	2.43	0.47	1.28	0.22	1.47	0.22	13.8	151.34	0.57		6.18	1.63	5
保亭	QV4-1	44.1	74.5	8.6	25.9	4.4	0.84	3.22	0.43	2.58	0.52	1.37	0.23	1.35	0.22	14.2	168.26	0.65		6.47	1.97	5
保亭	GXVIII9-1	66.4	83.1	9.5	35.9	6.3	1.08	3.55	0.64	4.00	0.70	2.01	0.31	1.94	0.30	15.9	215.73	0.64		6.80	1.51	5
	平均	44.08	75.48	8.70	29.00	4.76	0.94	3.32	0.48	2.60	0.50	1.33	0.21	1.31	0.20	13.12	177.47	0.69	25.20	5.85	2.15	

续表

岩石单位	样品号	La	Ce	Pr	Nd	Sm	Eu	Gd	Tb	Dy	Ho	Er	Tm	Yb	Lu	Y	ΣREE	δEu	$(La/Yb)_N$	$(La/Sm)_N$	$(Gd/Yb)_N$	来源
加茂	YD599-1	30.6	60.7	7.6	25.6	5.1	1.2	4.1	0.7	3.3	0.6	1.8	0.3	1.8	0.3	17.7	161.3	0.77	11.8	3.8	1.9	2
加茂	YD612-1	31.0	60.9	7.2	25.9	4.8	1.2	4.2	0.6	3.4	0.6	1.8	0.3	1.8	0.3	18.0	161.9	0.82	11.8	4.0	1.9	2
加茂	D102-1	30.5	50.5	6.5	25.0	4.6	1.5	4.0	0.6	3.1	0.6	1.6	0.2	1.5	0.2	16.3	146.8	1.05	13.9	4.2	2.2	2
加茂	D0355	35.28	66.27	8.22	29.50	5.53	1.28	4.67	0.66	3.72	0.72	2.01	0.31	1.89	0.29	18.79	160.35	0.75		4.12	2.04	本次
加茂	D0478	37.72	67.68	8.84	29.92	5.57	1.29	4.64	0.68	3.58	0.71	1.82	0.30	1.86	0.31	18.38	164.92	0.75		4.37	2.06	本次
加茂	D0830	37.35	69.74	8.40	32.00	5.82	1.39	4.71	0.68	3.76	0.74	1.95	0.30	1.82	0.29	18.20	168.95	0.79		4.14	2.14	本次
加茂	D0893	29.81	59.42	7.51	28.20	5.17	1.03	4.17	0.60	3.35	0.68	1.81	0.29	1.75	0.27	16.09	144.06	0.66		3.72	1.97	本次
加茂	D0898	40.27	72.81	8.69	29.76	4.94	1.05	3.59	0.48	2.62	0.53	1.37	0.21	1.29	0.20	12.67	167.81	0.73		5.26	2.30	本次
加茂	D1248	53.27	97.35	12.79	48.48	8.90	2.37	7.44	0.99	5.35	1.01	2.49	0.35	1.97	0.30	26.38	243.06	0.87		3.86	3.12	本次
加茂	D1914-1	34.28	64.35	8.46	29.28	5.71	1.42	4.80	0.69	3.87	0.77	2.02	0.32	1.92	0.30	20.19	158.19	0.81		3.88	2.07	本次
加茂	PM2-127-1	32.03	59.91	6.73	24.07	3.81	0.92	2.99	0.40	2.02	0.41	1.06	0.17	1.03	0.18	10.34	135.73	0.80		5.43	2.40	本次
加茂	D3120-1	38.6	78.6	8.5	30.9	5.7	1.03	3.88	0.52	2.72	0.52	1.39	0.18	1.27	0.18	14.9	173.99	0.63		4.37	2.53	本次
加茂	QV13-1	35.8	75.9	9.7	34.1	7.2	1.18	6.63	1.02	6.65	1.22	3.30	0.50	2.79	0.42	34.0	186.41	0.51		3.21	1.97	5
加茂	QV14-1	32.7	65.7	8.0	29.8	6.2	1.30	5.32	0.89	5.30	0.99	2.73	0.42	2.52	0.40	28.3	162.27	0.68		3.40	1.75	5
加茂	GXVIII3-1	47.6	80.6	9.5	35.8	7.0	1.34	4.29	0.78	4.74	0.83	2.12	0.34	2.08	0.32	18.0	197.34	0.69		4.39	1.71	5
	平均	36.45	68.70	8.44	30.55	5.74	1.30	4.63	0.69	3.83	0.73	1.95	0.30	1.82	0.28	19.22	168.87	0.75	12.50	4.14	2.14	
乌石	YV2-1	17.6	40.8	4.9	20.8	4.22	1.37	3.79	0.51	3.13	1.6	1.56	0.23	1.35	0.21	16.1	117.3	1.03	8.8	2.6	2.3	2
乌石	D1321-1	22.94	46.96	6.79	28.47	6.82	2.07	6.68	1.02	5.48	1.12	2.94	0.44	2.59	0.40	26.28	134.72	0.93		2.17	2.13	5
番通	D776-1	109.3	185.0	17.6	54.7	6.0	1.27	3.39	0.51	2.46	0.49	1.52	0.22	1.55	0.27	15.1	384.2	0.78	50.6	11.7	1.8	3
番通	D3037-1	89.9	150.0	14.0	45.8	5.3	0.96	3.83	0.48	2.53	0.54	1.59	0.24	1.65	0.26	16.4	317.0	0.63	39.0	11.0	1.9	3
阜石斗	S017	35.0	59.8	5.4	15.9	2.2	0.42	1.55	0.25	1.39	0.29	0.88	0.15	1.10	0.19	10.1	124.5	0.67	22.9	10.4	1.2	3
阜石斗	HG-167	35.0	53.0	7.3	34.3	5.2	0.77	3.40	0.55	3.10	0.53	1.40	0.27	1.60	0.26	13.4	146.7	0.53	15.7	4.3	1.8	3
阜石斗	GIII5-4	89.8	161.0	18.5	57.1	9.5	0.66	6.77	1.19	6.17	1.20	3.35	0.51	3.11	0.44	27.8	359.3	0.24	20.7	6.1	1.8	3
阜石斗	GIII4-2	50.9	88.4	9.2	38.2	6.9	1.12	4.42	0.71	4.49	0.91	2.48	0.39	2.23	0.34	18.3	210.8	0.58	16.4	4.7	1.6	3
阜石斗	GIII5-2	52.2	107.0	8.2	29.8	5.2	0.80	2.82	0.50	2.90	0.63	1.66	0.25	1.50	0.23	13.6	213.7	0.58	25.0	6.5	1.6	3

续表

岩石单位	样品号	La	Ce	Pr	Nd	Sm	Eu	Gd	Tb	Dy	Ho	Er	Tm	Yb	Lu	Y	ΣREE	δEu	$(La/Yb)_N$	$(La/Sm)_N$	$(Gd/Yb)_N$	来源
阜石斗	Q155-1	9.0	20.7	1.8	4.3	0.7	0.06	0.61	0.16	1.22	0.30	1.02	0.20	1.53	0.25	11.9	41.8	0.27	4.2	8.3	0.3	3
阜石斗	G1977-1	52.6	52.2	5.7	15.3	2.6	0.22	1.18	0.20	1.17	0.30	1.13	0.20	1.42	0.24	10.2	134.5	0.34	26.6	13.2	0.7	3
	平均	46.36	77.44	8.01	27.8	4.61	0.58	2.96	0.51	2.92	0.59	1.70	0.28	1.78	0.28	15.0	175.90	0.46	18.79	7.64	1.29	
抱跃	D3056-1	41.4	85.5	9.5	36.1	7.0	1.15	5.74	0.91	5.38	1.07	2.98	0.43	2.94	0.44	31.2	200.6	0.53	10.1	3.8	1.6	3
抱跃	QXII1-1	29.1	72.2	8.4	31.6	6.6	0.75	5.27	0.80	4.64	0.86	2.46	0.40	2.67	0.44	26.1	166.3	0.37	7.8	2.8	1.6	3
抱跃	HG-40	25.6	36.6	5.1	18.0	3.1	0.91	2.30	0.43	2.09	0.48	1.31	0.21	1.49	0.26	9.5	97.8	1.00	12.3	5.3	1.3	3
抱跃	GIII9-1	41.0	64.0	7.8	29.0	4.6	1.00	3.60	0.60	3.80	0.67	2.10	0.32	2.20	0.31	*	161.0	0.72	13.4	5.8	1.4	3
抱跃	GIII11-2	52.5	91.5	9.7	37.7	7.3	0.86	4.75	0.80	5.03	1.00	2.89	0.42	2.52	0.34	20.9	217.3	0.42	14.9	4.6	1.6	3
	平均	37.92	69.96	8.10	30.5	5.72	0.93	4.33	0.71	4.19	0.82	2.35	0.36	2.36	0.36	21.9	168.60	0.61	11.70	4.46	1.50	
税町	D3055-1	40.2	82.9	9.3	35.0	7.1	1.05	5.83	0.97	5.85	1.14	3.09	0.46	3.04	0.45	32.7	196.3	0.48	9.5	3.7	1.6	3
税町	GVIII7-2	51.9	92.4	10.5	41.6	8.2	1.23	6.07	0.98	6.50	1.33	3.48	0.54	3.25	0.48	28.5	228.4	0.51	11.5	4.1	1.5	3
税町	HG-39	30.0	57.0	6.8	26.0	5.0	1.90	4.20	0.71	5.40	0.82	2.20	0.41	2.70	0.38	10.2	143.5	1.23	8.0	3.9	1.3	3
税町	HG-51	21.3	38.3	5.5	24.9	4.4	1.70	3.60	0.53	3.57	0.76	2.01	0.25	1.93	0.30	15.0	109.0	1.27	7.9	3.1	1.5	3
税町	QXII4-1	63.4	112.6	13.5	46.4	7.5	1.18	6.67	0.87	4.19	0.80	1.86	0.28	1.45	0.21	22.0	260.9	0.50	31.4	5.5	3.8	3
税町	GII15-1	44.4	67.7	8.3	31.1	5.7	0.91	3.19	0.55	3.07	0.65	1.82	0.28	1.68	0.26	*	169.6	0.59	19.0	5.0	1.6	3
税町	GIII11-3	29.7	52.5	6.2	24.7	4.9	0.60	3.75	0.68	4.52	1.00	2.94	0.45	2.70	0.42	20.5	135.0	0.41	7.9	3.9	1.1	3
税町	GXII1-1	42.9	60.9	7.3	29.4	5.4	1.23	3.88	0.72	4.25	0.82	2.33	0.37	2.29	0.36	*	162.1	0.78	13.4	5.1	1.4	3
	平均	40.48	70.54	8.43	32.4	6.03	1.23	4.65	0.75	4.67	0.92	2.47	0.38	2.38	0.36	21.5	175.60	0.72	13.58	4.29	1.73	
$\gamma\pi K_1$	GI2-1	55.4	95.2	10.3	40.7	7.1	0.96	4.03	0.62	3.25	0.69	1.88	0.30	1.80	0.28	15.2	222.5	0.50	22.1	5.1	1.9	3
$\gamma\pi K_1$	G21-1	54.3	82.3	9.2	33.6	6.3	0.84	3.67	0.64	3.25	0.59	1.38	0.21	1.28	0.20	18.0	197.7	0.49	30.4	5.6	2.4	3
$\gamma\pi K_1$	GI6-1	32.7	45.8	6.1	22.7	4.1	0.48	2.57	0.46	2.96	0.66	2.03	0.30	1.80	0.28	15.1	122.9	0.42	13.0	5.2	1.2	3
$\gamma\pi K_1$	GI12-2	44.5	70.0	7.5	29.0	5.0	1.58	2.86	0.50	2.81	0.57	1.62	0.25	1.50	0.23	20.1	167.9	1.17	21.3	5.7	1.6	3
$\gamma\delta\pi K_1$	GII8-1	45.9	77.0	9.0	32.8	6.2	0.59	3.96	0.71	4.37	0.95	2.64	0.38	2.28	0.35	*	187.1	0.34	14.4	4.7	1.4	3
$\gamma\delta\pi K_1$	GIII3-1	43.7	73.4	8.3	32.1	6.0	1.39	3.68	0.60	3.47	0.73	2.07	0.33	1.96	0.31	*	178.1	0.84	16.0	4.7	1.6	3
$\gamma\delta\pi K_1$	G1872	68.6	86.8	9.8	37.4	6.5	0.96	3.72	0.68	4.19	0.77	2.16	0.35	2.21	0.34	*	224.4	0.55	22.3	6.9	1.4	3

续表

岩石单位	样品号	La	Ce	Pr	Nd	Sm	Eu	Gd	Tb	Dy	Ho	Er	Tm	Yb	Lu	Y	∑REE	δEu	$(La/Yb)_N$	$(La/Sm)_N$	$(Gd/Yb)_N$	来源
νK_1	D3045-1	29.5	61.2	7.2	29.6	5.6	1.55	4.69	0.75	3.91	0.76	1.99	0.25	1.86	0.26	20.8	149.1	0.90	11.3	3.4	2.1	3
νK_1	N517-1	36.0	59.0	8.2	37.0	6.3	1.50	5.20	0.90	4.90	0.84	2.25	0.34	2.10	0.32	*	164.9	0.78	12.3	3.7	2.0	3
νK_1	GIX5-1	11.5	26.0	3.1	13.2	2.9	0.86	2.26	0.35	2.37	0.44	1.18	0.18	1.12	0.17	9.2	65.6	0.99	7.4	2.6	1.7	3
半岛	Ha31	49	102	12.2	44.9	9.73	0.76	8.93	1.46	8.34	1.62	4.44	0.76	4.62	0.61	47.3	296.67	0.25	7.15	3.17	1.56	1
半岛	7331-1	67	126	12	44	7.5	0.8	5.5	0.86	5	0.87	2.3	0.28	1.45	0.23	19	292.79	0.36	31.15	5.62	3.06	1
半岛	HG389	90	150	16	91	9	1.03	7.3	1.1	7.6	1.3	3.3	0.42	4.3	0.56	41	423.91	0.38	14.11	6.29	1.37	1
	平均	68.67	126.00	13.4	59.9	8.74	0.86	7.24	1.14	6.98	1.26	3.35	0.49	3.46	0.47	35.7	337.79	0.33	17.47	5.03	2.00	
马岭	177	28	53	4.2	12.3	2.2	0.14	1.2	0.2	1.3	0.27	0.74	0.12	0.68	0.11	6.5	110.96	0.24	27.76	8.01	1.42	1
马岭	XX1-1	56	79	7.93	27.9	4.74	0.47	2.27	0.41	2.44	0.47	1.48	0.24	1.5	0.24	12	197.09	0.39	25.17	7.43	1.22	1
马岭	XⅧ1-1	13	23	3.87	19.9	8.1	0.082	11.2	2.13	14.9	3.38	8.75	1.16	5.8	0.7	79	194.97	0.03	1.51	1.01	1.56	1
	平均	32.33	51.67	5.33	20.0	5.01	0.23	4.89	0.91	6.21	1.37	3.66	0.51	2.66	0.35	32.5	167.67	0.22	18.15	5.48	1.40	
马鞍岭	Ha13-1	7.69	19.1	2.86	13.6	4	0.1	7.66	1.73	12.2	2.77	7.88	1.21	7.7	1.26	76	165.76	0.05	0.67	1.21	0.8	1
马鞍岭	Ha13-2	6.13	15.9	2.5	12.2	3.4	0.082	8.07	2.2	15.5	3.28	9.58	1.54	9.6	1.73	96.4	188.11	0.05	0.43	1.13	0.68	1
马鞍岭	184	78.5	144	16.8	58.5	13.4	1.01	10.6	1.46	8.61	1.72	4.82	0.65	3.93	0.49	34.6	379.09	0.25	13.47	3.69	2.18	1
马鞍岭	187	68	153	14	49	9.4	0.8	7	1.1	6	1.05	2.6	0.28	1.65	0.26	21	335.14	0.29	27.78	4.55	3.42	1
马鞍岭	YXV3-1	89.6	125	18.3	63.5	10.3	1.8	8.3	1.3	6.3	1.3	3.5	0.6	3.9	0.6	38.0	372.1	0.56	15.4	5.5	1.7	2
	平均	49.98	91.40	10.89	39.36	8.10	0.76	8.33	1.56	9.72	2.02	5.68	0.86	5.36	0.87	53.20	288.04	0.24	11.55	3.22	1.76	
三狮岭	HG400	32	47	6	21	5.3	0.48	3.4	0.79	3.2	0.62	1.5	0.24	1.6	0.32	19	142.45	0.32	13.48	3.8	1.71	1
三狮岭	Ⅱ28-1	32	82.5	6.95	24.5	5.9	0.137	4.35	0.755	4.05	0.85	2.6	0.325	1.95	0.31	21	188.17	0.08	11.06	3.41	1.8	1
三狮岭	Ⅱ30-1	68	83	15	54	12.5	0.24	10	1.12	7	1.3	3.4	0.45	2.6	0.4	42	301.01	0.06	17.63	3.42	3.1	1
三狮岭	YI1-2	87.5	170.9	18.6	67.2	11.1	1.0	8.7	1.2	6.7	1.2	2.9	0.4	2.3	0.3	32.4	412.6	0.30	25.5	4.9	3.0	2
三狮岭	YD438	244.6	357.5	36.5	98.6	10.6	1.6	5.0	0.7	2.1	0.4	1.2	0.2	1.7	0.3	10.6	771.6	0.57	96.4	14.6	2.4	2
三狮岭	D1740-2	19.6	35.79	4.36	14.6	2.92	0.52	2.75	0.49	2.61	0.51	1.78	0.29	2.21	0.36	18.06	106.87	0.55	6.0	4.2	1.0	2
三狮岭	HN03	17.90	33.68	3.77	13.22	2.45	0.55	2.1	0.34	1.95	0.42	1.32	0.22	1.66	0.27	13.11	93.02	0.72	7.2	4.6	1.0	2
三狮岭	D958-1	49.2	94.2	10.8	37.3	6.8	1.1	5.9	0.9	4.9	0.9	2.9	0.4	2.9	0.4	27.9	246.5	0.52	11.6	4.6	1.7	2

续表

岩石单位	样品号	La	Ce	Pr	Nd	Sm	Eu	Gd	Tb	Dy	Ho	Er	Tm	Yb	Lu	Y	ΣREE	δEu	$(La/Yb)_N$	$(La/Sm)_N$	$(Gd/Yb)_N$	来源
三狮岭	HN05	60.09	113.7	12.26	41.64	7.06	1.18	6.58	0.87	4.88	0.96	2.74	0.39	2.51	0.40	25.76	281.08	0.52	16.1	5.3	2.1	2
三狮岭	D0399-1	21.32	35.11	4.19	13.60	2.87	0.23	2.62	0.43	2.73	0.61	1.79	0.32	2.27	0.40	16.73	88.49	0.25		4.80	0.95	本次
三狮岭	D1004-1	10.98	30.06	3.63	9.92	2.56	0.22	1.97	0.35	2.09	0.44	1.21	0.19	1.29	0.20	10.28	65.11	0.29		2.77	1.26	本次
三狮岭	D1213-1	19.94	41.82	4.09	11.41	2.16	0.44	1.64	0.28	1.46	0.32	0.79	0.14	0.96	0.16	6.70	85.61	0.69		5.96	1.41	本次
三狮岭	D1036	40.7	79.4	8.2	27.9	4.1	0.23	2.91	0.45	2.21	0.38	1.03	0.15	1.06	0.17	11.5	168.89	0.19		6.41	2.27	本次
	平均	54.14	92.67	10.33	33.45	5.87	0.61	4.46	0.67	3.53	0.69	1.94	0.29	1.92	0.31	19.62	227.03	0.39	22.77	5.29	1.82	
架 水	D3361-1	73.8	145.3	18.8	69.0	11.9	3.0	8.7	1.3	5.6	1.1	2.4	0.4	2.2	0.3	26.8	370.6	0.85	22.4	3.9	3.2	2
椰子园	D3032-1	76.4	140.4	14.2	48.6	7.8	1.29	5.66	0.72	3.91	0.70	1.92	0.26	1.80	0.27	21.1	303.9	0.57	30.5	6.3	2.6	3
南山岭	S018	66.5	131.9	14.3	50.9	8.1	1.74	5.79	0.83	4.31	0.80	2.24	0.33	2.19	0.34	25.4	290.2	0.74	21.8	5.3	2.2	3
南山岭	N605-1	74.0	114.0	15.5	58.0	8.8	1.80	5.60	1.00	4.50	0.76	2.30	0.33	2.11	0.40	*	289.1	0.73	25.2	5.4	2.2	3
马翁岭	Ha37-2	38.1	76.6	9.17	36.8	7.38	2.24	6.75	1.03	5.38	1	2.57	0.42	2.7	0.37	27.2	217.71	0.95	9.51	3.25	2.02	1
马翁岭	516	30.79	64.56	7.91	30.14	6.14	1.56	5.61	0.85	4.63	0.9	2.39	0.37	2.24	0.34	24.66	183.09	0.8	9.27	3.15	2.02	1
马翁岭	YXVIII6-2	40.5	74.4	9.9	36.2	6.8	2.0	6.1	0.9	4.9	0.9	2.4	0.4	2.4	0.4	25.8	214.0	0.94	11.4	3.8	2.0	2
马翁岭	D1153-1	21.9	47.2	5.4	22.9	5.1	1.65	4.88	0.76	4.54	0.88	2.46	0.34	2.18	0.33	24.6	120.52	1.00		2.77	1.85	本次
马翁岭	D1670-1	15.2	32.6	3.8	16.6	4.0	1.24	3.78	0.63	3.69	0.69	1.88	0.24	1.70	0.26	19.0	86.31	0.96		2.45	1.84	本次
马翁岭	D1535-1	13.9	29.2	3.3	15.0	3.7	1.09	3.83	0.63	3.54	0.72	1.97	0.29	1.73	0.24	19.6	79.14	0.88		2.43	1.83	本次
	平均	26.73	54.09	6.58	26.27	5.52	1.63	5.16	0.80	4.45	0.85	2.28	0.34	2.16	0.32	23.48	150.13	0.92	10.06	2.98	1.93	
八宝坡	H1245-2	43.6	74.2	7.9	25.9	4.3	1.23	4.13	0.68	4.03	0.87	2.42	0.42	2.86	0.45	25.4	173.1	0.87	10.9	6.5	1.2	3
八宝坡	Hlm-2	75.3	113.2	10.0	29.3	4.4	0.86	3.55	0.63	3.75	0.82	2.59	0.45	3.37	0.56	29.2	248.8	0.64	16.0	11.0	0.9	3
里茂	H1247-2	23.9	50.0	6.6	28.2	6.2	2.11	6.30	0.89	5.09	0.93	2.46	0.32	1.80	0.26	23.9	135.0	1.02	9.5	2.5	2.9	3
里茂	Hlm-1	20.5	48.2	6.8	30.4	7.1	2.10	7.58	1.07	5.70	1.03	2.52	0.34	1.98	0.29	30.6	135.4	0.87	7.4	1.9	3.2	3
迈洲岭	974-1	138	323	27.7	134	18.9	1.11	9.63	1.31	8.36	1.31	2.93	0.44	2.74	0.43	23.8	702.72	0.22	34	4.6	2.8	2
打显村	X VI19-1	60	91	11.2	38	8.1	0.82	4.3	0.68	3.7	0.78	1.9	0.29	1.65	0.25	20	242.67	0.38	24.52	4.66	2.1	1
打显村	X VI2-2	65	112	12.3	48	9.6	0.89	4.7	0.74	3.6	0.78	1.7	0.23	1.2	0.18	17.4	278.32	0.36	36.52	4.26	3.16	1
白石岭	X VI16-1	47	80	9.2	34	7.9	0.64	5.1	0.9	6	1.24	3.1	0.48	2.7	0.41	32	230.67	0.29	11.74	3.74	1.52	1

续表

岩石单位	样品号	La	Ce	Pr	Nd	Sm	Eu	Gd	Tb	Dy	Ho	Er	Tm	Yb	Lu	Y	∑REE	δEu	$(La/Yb)_N$	$(La/Sm)_N$	$(Gd/Yb)_N$	来源
白石岭	XⅥ2-1	65	112	12.3	48	9.6	0.89	4.7	0.74	3.6	0.78	1.7	0.23	1.2	0.18	17.4	278.32	0.36	36.52	4.26	3.16	1
黑岭	Ha25	69.9	138	15	49.7	8.63	0.96	7.87	1.21	6.61	1.35	3.93	0.77	5.37	0.79	38.8	348.89	0.35	8.78	5.09	1.18	1
黑岭	4086-2	65.64	128	13.85	46.51	9.12	1.22	7.24	1.09	6.48	1.28	3.68	0.61	3.73	0.57	33.42	322.62	0.44	11.86	4.53	1.57	1
分水岭	XⅦ10-1	27	47	5.88	26.4	6.42	0.64	6.64	1.13	7.81	1.88	5.78	0.73	3.85	0.51	43	184.67	0.3	4.73	2.65	1.39	1
分水岭	4063-1	41.66	75.7	8.43	30.14	7.16	0.1	7.05	1.4	9.69	1.82	4	0.65	4.39	0.68	46.45	239.28	0.04	6.4	3.66	1.3	1
分水岭	4086-7	62.44	115	12.59	43.11	8.36	0.69	6.59	1.1	6.39	1.4	3.83	0.64	4.39	0.68	44.07	311.62	0.27	9.59	4.7	1.21	1
分水岭	D5543-1	27	61	6.5	32	6.1	0.57	5.8	1	6.7	1	3.1	0.4	3.4	0.38	36	190.95	0.29	5.35	2.78	1.38	1
分水岭	D5545-1	36	76	9.1	36	6.6	0.96	4.9	0.62	3.9	0.52	1.56	0.2	1	0.16	18	195.52	0.5	24.27	3.43	3.95	1
	平均	38.82	74.94	8.50	33.53	6.93	0.59	6.20	1.05	6.90	1.32	3.65	0.52	3.41	0.48	37.50	224.41	0.28	10.07	3.44	1.85	
南宝岭	4602-1	36.77	79.6	9.57	34.31	7.91	0.42	6.65	1.27	7.82	1.7	5.34	0.97	6.64	1.09	52.81	252.87	0.17	3.73	2.92	0.81	1
南宝岭	4676-2	59.1	124	10.95	43.13	7.78	0.84	5.7	0.83	5.14	1.02	2.89	0.47	3.18	0.49	30.24	295.26	0.37	12.53	4.78	1.45	1
五里桥	Ha22	73.3	144	16	56.5	10.8	1.1	9.92	1.62	9.43	1.9	5.31	0.94	6.13	0.84	52.2	389.99	0.32	8.06	4.27	1.31	1
五里桥	D887-1	75.5	131	14.7	57.3	9	1.39	4.81	0.78	3.57	0.7	1.63	0.25	1.49	0.23	14.2	316.55	0.58	34.16	5.28	2.6	1
五里桥	4033-1	54.17	98.25	11.06	41.74	10.95	1.38	11.11	2.16	14.72	2.96	7.24	1.17	7.9	1.22	79.72	345.75	0.38	4.62	3.11	1.13	1
五里桥	4077-1	71.74	128.45	17.22	58.02	12.76	0.86	12.49	2.17	13.07	2.96	8.26	1.33	8.87	1.37	84.68	424.25	0.21	5.45	3.54	1.14	1
五里桥	4080-3	54.93	105.33	14.24	44.65	9.65	0.52	9.46	1.62	9.58	2.09	5.62	0.92	6.32	0.97	64.29	330.19	0.16	5.86	3.58	1.21	1
五里桥	4082-1	55.7	104.52	13.25	44.48	9.57	0.6	9.2	1.59	9.58	2.13	5.88	0.98	6.85	1.06	63.59	328.98	0.19	5.48	3.66	1.08	1
五里桥	Ha27	70.4	144	16.1	57.3	11.2	0.91	10.2	1.66	9.61	1.94	5.48	1.0	6.38	0.91	54.2	391.29	0.26	7.44	3.95	1.29	1
	平均	65.11	122.22	14.65	51.43	10.56	0.97	9.60	1.66	9.94	2.10	5.63	0.94	6.28	0.94	58.98	361.00	0.30	10.15	3.91	1.39	
苗圃	4072-4	41.17	90.11	11.67	40.45	9.05	1.64	8.85	1.52	9.06	2.03	5.62	0.9	6.06	0.93	59.18	288.24	0.55	4.58	2.86	1.18	1
苗圃	D887-2	55.3	99.3	10.5	41.9	6.86	1.15	3.64	0.58	2.21	0.44	0.88	0.15	0.75	0.13	9.02	232.81	0.64	49.71	5.07	3.92	1
苗圃	D891-2	93.1	173	21.4	86.6	15.5	1.31	9.5	1.5	5.96	1.19	2.63	0.38	2.27	0.35	22.9	437.59	0.31	27.65	3.78	3.38	1
苗圃	Ha21-1	85.6	167	18.4	64.9	12.1	1.5	11	1.68	9.46	1.83	4.93	0.89	5.56	0.77	48.3	433.92	0.39	10.38	4.45	1.6	1
	平均	68.79	132.35	15.49	58.46	10.88	1.40	8.25	1.32	6.67	1.37	3.52	0.58	3.66	0.55	34.85	348.14	0.47	23.08	4.04	2.52	

续表

岩石单位	样品号	La	Ce	Pr	Nd	Sm	Eu	Gd	Tb	Dy	Ho	Er	Tm	Yb	Lu	Y	ΣREE	δEu	$(La/Yb)_N$	$(La/Sm)_N$	$(Gd/Yb)_N$	来源
乌烈岭	Ha14-2	82.1	164	18.5	66.8	12.9	1.09	11.7	1.84	10.8	2.12	5.87	1.05	6.85	0.97	55.6	442.19	0.27	8.08	4	1.38	1
乌烈岭	Ha15	69.2	139	15.6	57.3	11.1	1.86	9.9	1.64	9.01	1.72	4.74	0.82	5.37	0.76	46.2	374.22	0.53	8.69	3.92	1.49	1
乌烈岭	4001-1	67.5	122	13.98	53.66	11.42	0.85	13.88	2.6	18.55	3.17	7.1	1.11	7.79	1.08	89.8	414.49	0.21	5.84	3.72	1.44	1
乌烈岭	4001-7	89.25	169.75	17.75	63.75	13.1	0.92	13.17	2.6	18.12	3.55	8.2	1.33	8.53	1.36	99.6	510.98	0.21	7.05	4.29	1.25	1
乌烈岭	4044-4	51.52	103.22	12.46	52.45	10.78	0.69	14.15	2.59	16.64	3.38	8.35	1.43	10.2	1.57	88.3	377.71	0.17	3.41	3.01	1.12	1
乌烈岭	4104-6	56.81	113.72	12.97	50.05	13.53	1.38	13.11	2.13	11.85	2.71	7.67	1.23	8.25	1.27	74.29	370.97	0.31	4.64	2.64	1.28	1
乌烈岭	7323	65.1	110	13	43.7	10.2	0.92	7.81	1.19	7.22	1.54	4.15	0.64	3.84	0.45	32.9	302.66	0.3	11.43	4.01	1.64	1
	平均	68.78	131.67	14.9	55.4	11.86	1.10	11.96	2.08	13.1	2.60	6.58	1.09	7.26	1.07	69.5	399.03	0.29	7.02	3.66	1.37	
东四	Ha14-1	80.1	160	18.4	64.4	12	1.72	10.5	1.6	8.66	1.68	4.47	0.78	4.84	0.66	43.4	413.21	0.46	11.16	4.2	1.75	1
东四	Ha23	86.6	164	17.8	57.7	9.74	1.04	9.05	1.32	7.27	1.44	4.1	0.73	4.79	0.69	39.2	405.47	0.33	12.19	5.59	1.52	1
东四	4091-2	68.72	122.91	13.83	44.14	8.45	0.43	6.59	1.1	6.36	1.53	4.52	0.7	4.48	0.69	47.06	331.51	0.17	10.34	5.12	1.19	1
东四	4125-2	89.14	149.45	16.45	59.65	10.95	0.78	11.54	2.13	13.76	2.55	5.54	0.96	6.85	1.06	74.21	445.02	0.21	8.77	5.12	1.36	1
东四	7314	74	165	15.5	48	9.1	0.57	6.2	0.95	4.2	0.62	1.35	0.16	0.68	0.11	10.3	336.74	0.22	73.37	5.12	7.36	1
	平均	79.71	152.27	16.4	54.7	10.05	0.91	8.78	1.42	8.05	1.56	4.00	0.67	4.33	0.64	42.8	386.39	0.28	23.17	5.03	2.64	
七差岭	6041	75.4	126	15.1	51	10.1	1.52	5.92	0.87	3.91	0.79	1.8	0.28	1.68	0.23	16	310.6	0.55	30.26	4.7	2.84	1
七差岭	4964-1	69.27	127.4	13.97	54.77	12.03	1.4	9.78	1.72	9.53	1.78	4.76	0.73	4.11	0.54	46.79	358.58	0.38	11.36	3.62	1.92	1
七差岭	4994-1	60.39	122.52	11.61	44.36	7.81	1.22	4.86	0.6	3.2	0.6	1.5	0.23	1.44	0.21	16.6	277.15	0.56	28.27	4.86	2.72	1
	平均	68.35	125.31	13.5	50.0	9.98	1.38	6.85	1.06	5.55	1.06	2.69	0.41	2.41	0.33	26.4	315.44	0.50	23.30	4.39	2.49	
保梅岭	Ha30-1	56	112	12.5	43.6	7.76	0.99	6.78	0.89	4.41	0.8	2.22	0.36	2.59	0.41	22.1	273.71	0.41	14.66	4.56	2.11	1
保梅岭	Ha30-6	58	116	13.2	46.7	8.6	0.98	7.43	1.05	5.43	1	2.62	0.44	2.78	0.4	26.6	291.83	0.37	14.21	4.29	2.16	1
保梅岭	Ha33-1	42	90.6	10.6	37.8	7.24	0.72	6.55	0.97	5.25	1.02	2.73	0.47	3.11	0.44	27.4	237.6	0.31	9.26	3.71	1.7	1
	平均	52.00	106.20	12.1	42.7	7.87	0.90	6.92	0.97	5.03	0.94	2.52	0.42	2.83	0.42	25.3	267.71	0.36	12.71	4.19	1.99	
洞古岭	YIII11-1	72.1	146	17.3	64.2	10.1	2.11	6.7	0.78	4.02	0.76	1.75	0.25	1.57	0.21	20.1	347.95	0.74	31.0	4.5	3.4	2
南通岭	VHG1-7	97.51	189.6	22.46	69.38	10.47	1.89	8.12	1.13	5.83	1.15	3.12	0.49	2.98	0.46	30.35	444.93	0.6	22	5.9	2.2	2
南通岭	HP8-G-1	21.4	63	7.1	24.3	4.7	1.1	4.7	0.5	2.3	0.46	0.95	0.14	0.94	0.13	11.1	142.8	0.71	15.4	2.9	4.0	2

续表

岩石单位	样品号	La	Ce	Pr	Nd	Sm	Eu	Gd	Tb	Dy	Ho	Er	Tm	Yb	Lu	Y	ΣREE	δEu	$(La/Yb)_N$	$(La/Sm)_N$	$(Gd/Yb)_N$	来源
南通岭	HN01-2	55.13	119.12	14.32	55.85	10.73	2.5	10.9	1.5	8.2	1.6	4.64	0.63	4.21	0.63	41.03	331.14	0.7	8.8	3.2	2.1	2
琼中县	D493-2	112.4	210.3	22.63	71.99	11.8	1.63	8.29	1.27	6.01	1.1	2.89	0.44	2.52	0.34	29.62	483.24	0.48	30.0	6.0	2.7	2
琼中县	XXI2-2	158.8	84.09	36.32	135.6	23.3	2.42	15.3	2.11	10.1	1.75	4.34	0.66	3.0	0.51	42.39	521.41	0.38	35.7	4.3	4.1	2
琼中县	YXX1-1	48.83	102.7	11.57	39.32	6.08	1.03	4.23	0.56	2.54	0.49	1.26	0.19	1.04	0.18	13.5	233.5	0.62	128.2	5.5	7.3	2
琼中县	YD499	53.86	91.11	10.36	34.47	6.19	0.83	4.32	0.6	2.01	0.36	0.61	0.08	0.42	0.06	8.03	213.31	0.46	86.5	5.5	8.3	2
琼中县	YXX6-1	50.35	78.11	9.83	33.6	6.23	0.97	4.74	0.71	2.88	0.54	1.18	0.2	1.0	0.15	14.15	204.64	0.55	69.8	6.3	3.6	2
琼中县	D823-1	38.83	77.4	8.97	28.53	5.28	0.81	3.6	0.48	2.11	0.39	0.93	0.15	0.88	0.13	10.5	178.98	0.54	30	4.6	3.3	2
黎母岭	D1601-1	48.8	91.91	11.53	37.67	7.28	0.94	5.04	0.77	3.57	0.69	1.89	0.29	1.84	0.27	19.67	232.16	0.45	17.9	4.2	2.2	2
黎母岭	D101-3	89.87	166.4	18.96	60.96	11.65	1.83	8.96	1.45	7.36	1.43	4.0	0.58	3.52	0.51	39.95	417.42	0.53	17.2	4.9	2.1	2
黎母岭	ⅤHG1-1	77.7	150.5	19.16	68.74	12.55	2.64	11.44	1.85	9.36	1.8	5.19	0.76	4.71	0.7	49.85	416.95	0.66	11.1	3.9	1.9	2
黎母岭	D1618-1	79.16	164.6	20.14	64.98	12.15	0.57	9.27	1.47	7.01	1.35	3.69	0.56	3.48	0.5	38.61	407.55	0.16	15.3	4.1	2.2	2
黎母岭	ⅢHG2-3	73.94	131.0	14.59	48.2	7.17	1.72	5.8	0.87	4.03	0.77	2.09	0.33	1.98	0.28	23.02	315.79	0.79	25.2	6.5	2.4	2
黎母岭	HN01-1	64.78	130.17	16.81	62.55	11.88	2.97	12.3	1.67	9.2	1.79	5.01	0.71	4.57	0.66	45.47	370.45	0.75	9.5	3.4	2.1	2
黎母岭	HN02	150.3	287.9	30.6	103.8	16.6	2.17	15.5	1.98	10.88	2.07	5.71	0.77	5.10	0.67	50.09	684.42	0.41	19.8	5.7	2.4	2
黎母岭	HN25	26.76	55.0	6.33	23.8	4.91	1.3	4.68	0.66	3.64	0.69	1.92	0.26	1.72	0.24	18.22	150.16	0.8	10.5	3.4	2.2	2
黎母岭	HN26	39.34	79.96	9.1	32.14	5.75	0.92	5.29	0.67	3.4	0.66	1.86	0.27	1.92	0.28	19.56	201.17	0.5	13.8	4.3	2.2	2
黎母岭	HN27	193.0	367.6	39.03	129.9	18.56	2.28	12.95	1.71	8.33	1.45	3.8	0.5	3.01	0.43	38.56	821.35	0.43	43	6.5	3.4	2
黎母岭	HN30	35.5	63.03	7	23.6	3.7	0.77	2.8	0.3	1.22	0.19	0.45	0.05	0.33	0.04	5.16	144.18	0.7	72	6	6.8	2
后美	Ⅺ1-2	26.44	65.19	7.74	28.04	5.15	0.65	3.27	0.37	1.55	0.28	0.62	0.09	0.55	0.09	6.96	147.01	0.47	32.41	3.22	4.21	6
后龙	Ⅺ3-3	24.79	43.18	5.35	17.39	3.52	0.75	3.22	0.43	1.80	0.3	0.6	0.09	0.42	0.06	8.0	109.9	0.7	39.79	4.43	5.43	6
石墩	H1519-1	31.4	64.4	8.1	33.3	6.7	4.24	6.2	0.85	4.60	0.83	2.27	0.30	1.84	0.31	22.2	165.3	1.97	12.2	3.0	2.8	3
长安	D1653-1	14.3	35.8	4.5	20.1	4.9	1.12	4.3	0.64	3.54	0.67	1.73	0.23	1.36	0.20	17.5	93.4	0.73	7.5	1.9	2.6	3
长安	H303-1	15.2	32.6	4.1	17.2	3.5	1.14	3.4	0.51	2.84	0.57	1.52	0.23	1.45	0.22	16.8	84.5	1.00	7.5	2.8	2.0	3
长安	S004-1	55.5	123.5	15.5	66.4	11.8	3.35	10.3	1.40	7.32	1.39	3.56	0.48	2.91	0.44	41.7	303.9	0.91	13.7	3.0	2.9	3
长安	247-1	6.95	16.7	2.5	10.88	2.8	0.77	2.97	0.49	2.79	0.58	1.7	0.24	1.5	0.22	14.21	65.3	0.81	3.12	1.56	1.6	1

续表

岩石单位	样品号	La	Ce	Pr	Nd	Sm	Eu	Gd	Tb	Dy	Ho	Er	Tm	Yb	Lu	Y	$\sum$REE	δEu	$(La/Yb)_N$	$(La/Sm)_N$	$(Gd/Yb)_N$	来源
长安	247-4	5.53	11.6	1.78	7.88	2.08	0.57	2.17	0.37	2.11	0.58	1.15	0.17	1.1	0.17	10.63	47.9	0.81	3.39	1.67	1.59	1
石龟岭	D3042-1	6.2	18.9	2.7	12.7	5.0	0.03	5.73	1.15	7.96	1.61	4.57	0.71	5.28	0.75	49.3	73.3	0.02	0.8	0.8	0.9	3
石龟岭	N495-1	12.5	29.0	4.6	22.0	6.9	0.14	8.60	1.40	12.00	2.30	7.60	1.10	7.00	0.82	*	116.0	0.06	1.3	1.2	1.0	3
兵工厂	H1219-2	1.6	3.2	0.4	1.2	0.4	0.02	0.6	0.18	1.47	0.42	1.43	0.28	2.07	0.32	8.3	13.5	0.11	0.5	2.8	0.2	3
兵工厂	H1562-1	9.4	12.9	2.6	9.3	2.7	0.04	2.7	0.47	2.83	0.62	2.11	0.40	2.81	0.43	14.9	49.3	0.05	2.4	2.3	0.8	3
兵工厂	D3044-1	10.7	30.8	3.7	16.0	5.6	0.06	5.77	1.07	7.09	1.56	4.59	0.71	5.08	0.75	44.8	93.4	0.03	1.5	1.2	0.9	3
兵工厂	N673-1	17.8	37.4	5.5	28.1	8.5	0.23	9.87	2.02	13.40	2.71	9.13	1.41	8.22	1.18	*	145.4	0.08	1.6	1.4	1.0	3
六连岭	H1207-1	28.4	70.8	9.4	37.7	10.9	0.11	10.7	1.99	11.68	2.36	6.46	0.98	5.61	0.85	65.2	197.9	0.03	3.6	1.7	1.6	3
六连岭	H1207-4	24.5	61.4	8.6	35.7	10.7	0.09	12.0	2.16	13.21	2.68	7.81	1.18	7.53	1.12	75.4	188.8	0.02	2.3	1.5	1.3	3
六连岭	S001	27.0	68.7	9.0	38.0	10.5	0.10	11.7	1.93	11.40	2.35	6.44	0.99	6.26	0.93	78.1	195.3	0.03	3.1	1.7	1.5	3
六连岭	H3357-1	20.3	48.7	6.4	25.6	6.9	0.17	7.6	1.35	8.05	1.61	4.56	0.75	4.56	0.70	47.0	137.3	0.07	3.2	1.9	1.4	3
六连岭	S044	12.4	38.6	6.4	33.0	13.3	0.38	17.8	2.99	17.30	3.33	8.38	1.11	6.12	0.90	93.1	162.0	0.08	1.5	0.6	2.4	3
六连岭	S013-1	19.6	54.2	7.7	34.9	10.9	0.14	12.1	1.82	9.90	1.79	4.23	0.55	3.07	0.47	45.1	161.3	0.04	4.6	1.2	3.3	3
六连岭	D3043-1	14.6	38.7	4.9	20.7	5.6	0.06	5.45	0.99	6.09	1.30	3.68	0.50	3.86	0.58	38.2	107.0	0.03	2.7	1.7	1.2	3
六连岭	N603-1	21.0	44.8	6.7	29.5	8.2	0.14	7.26	1.31	9.31	1.92	5.06	0.75	4.73	0.60	*	141.3	0.05	3.2	1.7	1.3	3
六连岭	N7-3	25.0	41.0	7.7	35.0	9.4	0.29	12.00	2.10	15.00	3.30	9.70	1.32	7.90	1.00	*	170.7	0.08	2.3	1.7	1.3	3
前锋	GII12-1	26.2	42.0	5.8	23.9	5.2	0.49	4.58	0.80	6.44	1.32	4.21	0.62	3.72	0.59	28.8	125.9	0.30	5.1	3.3	1.0	3
前锋	GII2-3	16.9	24.2	4.9	21.5	5.7	0.30	6.28	1.10	8.74	1.79	5.29	0.82	4.92	0.74	41.2	103.2	0.15	2.5	1.9	1.1	3
前锋	G851	19.6	30.1	4.9	20.5	5.3	0.29	5.42	1.04	8.22	1.84	5.37	0.85	5.10	0.83	41.7	109.3	0.16	2.8	2.4	0.9	3
结尾	G42	29.3	53.6	7.8	32.3	8.9	0.25	9.1	1.87	13.10	2.69	7.53	1.19	7.46	1.02	77.3	176.0	0.08	2.8	2.1	1.0	3
结尾	GXIV1-3	88.9	111.0	17.2	60.5	9.3	1.00	4.1	0.74	4.56	0.70	1.68	0.26	1.52	0.24	12.8	301.7	0.43	42.0	6.2	2.2	3
结尾	GII1-1	27.5	40.6	5.7	28.3	6.8	0.20	5.92	1.04	8.05	1.60	4.70	0.73	4.38	0.67	32.0	136.2	0.09	4.5	2.6	1.1	3
结尾	GII3-4	26.1	42.0	5.7	21.9	4.9	0.52	4.34	0.79	6.06	1.32	3.85	0.60	3.60	0.54	28.2	122.2	0.34	5.2	3.4	1.0	3
结尾	N604-1	14.5	26.0	4.2	18.0	3.9	0.33	4.50	0.84	6.50	1.20	4.30	0.70	5.00	0.66	*	90.6	0.24	2.1	2.4	0.7	3
结尾	Nla-1	36.0	61.0	8.0	34.0	6.0	0.40	5.30	0.93	5.70	0.94	3.00	0.44	2.70	0.40	*	164.8	0.21	9.6	3.9	1.6	3

续表

岩石单位	样品号	La	Ce	Pr	Nd	Sm	Eu	Gd	Tb	Dy	Ho	Er	Tm	Yb	Lu	Y	ΣREE	δEu	$(La/Yb)_N$	$(La/Sm)_N$	$(Gd/Yb)_N$	来源
超盆	GIX9-1	10.3	31.2	2.1	7.6	2.4	0.24	2.2	0.55	4.53	1.06	3.35	0.53	3.48	0.47	27.7	70.0	0.31	2.1	2.7	0.5	3
超盆	G1284-2	29.3	49.4	6.0	25.0	5.1	0.34	3.9	0.66	4.03	0.89	2.52	0.38	2.30	0.32	21.0	130.1	0.22	9.1	3.7	1.4	3
超盆	GII6-1	34.6	53.4	6.8	25.6	5.3	0.67	3.89	0.72	4.84	1.06	3.05	0.47	2.82	0.44	22.4	143.6	0.43	8.8	4.2	1.1	3
超盆	G011-2	35.2	64.5	8.2	32.7	6.6	0.61	5.36	0.97	6.62	1.40	3.96	0.61	3.73	0.54	35.6	170.9	0.31	6.8	3.5	1.2	3
超盆	GV1-1	36.9	60.1	7.4	27.4	5.6	0.76	4.03	0.73	4.97	1.07	3.11	0.49	2.94	0.45	22.4	155.9	0.47	9.0	4.3	1.1	3
超盆	GV7-1	28.2	42.0	6.1	21.2	4.7	0.49	3.32	0.65	4.61	1.03	3.12	0.64	3.84	0.58	*	120.5	0.36	5.3	3.8	0.7	3
超盆	N7-10	33.0	62.0	8.3	33.0	6.0	0.38	4.60	0.84	4.60	0.80	2.40	0.38	2.40	0.36	*	159.1	0.21	9.9	3.6	1.6	3
布山村	N527-1	159.0	196.0	23.4	77.7	13.2	1.34	6.1	1.01	4.96	1.02	2.61	0.40	2.00	0.38	*	489.1	0.40	57.0	7.8	2.5	3
布山村	D3034-1	47.5	92.6	9.8	33.5	6.3	1.04	6.2	0.99	5.20	1.14	3.04	0.50	2.95	0.50	35.4	211.3	0.50	11.6	4.9	1.8	3
布山村	N727-1	83.0	137.0	15.3	59.4	8.9	1.73	8.3	1.18	5.72	0.95	2.59	0.46	2.24	0.36	18.9	327.2	0.60	26.6	6.0	3.1	3
布山村	GIX3-2	112.0	156.0	21.2	65.7	15.0	2.17	9.3	1.69	8.41	1.66	4.02	0.56	3.48	0.44	33.7	401.7	0.52	23.1	4.8	2.2	3
布山村	N488-1	58.0	98.0	13.0	48.0	8.6	0.84	6.8	1.14	7.00	1.25	3.80	0.60	3.60	0.59	*	251.2	0.32	11.6	4.4	1.6	3
布山村	N743-1	46.8	81.5	8.8	35.8	5.9	0.83	4.4	0.84	5.08	0.91	2.96	0.49	2.92	0.47	*	197.6	0.48	11.5	5.1	1.2	3
布山村	N1309-1	58.2	105.0	11.9	51.3	7.8	1.27	5.6	0.96	5.17	0.73	2.04	0.37	1.73	0.29	16.8	252.4	0.56	24.1	4.8	2.7	3
金寮	S039	106.7	210.4	22.2	77.9	13.3	0.41	9.9	1.61	9.02	1.70	4.46	0.65	4.71	0.64	47.2	463.7	0.10	16.2	5.2	1.7	3
金寮	S040-1	154.3	305.7	34.3	117.9	18.3	0.69	12.3	1.86	9.78	1.83	5.06	0.73	4.63	0.69	54.4	668.1	0.13	23.9	5.4	2.2	3
金寮	S040-2	176.9	334.7	34.4	121.5	20.4	0.55	15.4	2.16	10.88	1.95	4.86	0.70	4.26	0.60	48.6	729.2	0.09	29.8	5.6	3.0	3
番阳峒	S027	129.3	256.0	28.0	95.9	14.7	1.99	9.9	1.57	8.08	1.51	4.09	0.57	3.55	0.56	43.3	555.7	0.47	26.1	5.7	2.3	3
番阳峒	HG-478	76.0	84.0	19.0	54.0	9.8	1.23	4.6	1.55	4.80	0.90	1.90	0.41	2.80	0.59	21.0	261.6	0.49	19.5	5.0	1.4	3
番阳峒	HG-438	38.0	53.0	7.5	25.0	5.3	0.22	2.7	0.87	3.00	0.54	1.30	0.23	1.30	0.32	13.0	139.3	0.16	21.0	4.6	1.7	3
番阳峒	D467-1	100.1	197.3	21.1	75.3	11.7	1.59	8.5	1.27	6.69	1.34	3.54	0.50	3.42	0.50	36.8	432.9	0.47	21.0	5.5	2.0	3
番阳峒	S029	101.2	200.9	21.7	75.1	11.3	1.40	7.5	1.21	6.27	1.23	3.41	0.50	3.27	0.51	37.7	435.4	0.44	22.2	5.8	1.9	3
番阳峒	D1447-1	67.8	121.9	13.5	47.9	7.5	1.31	6.9	0.99	5.25	1.06	2.95	0.47	3.03	0.42	30.3	281.0	0.54	16.0	5.8	1.9	3
番阳峒	S008-2	178.3	330.2	27.3	130.6	19.4	0.33	12.7	1.64	8.25	1.51	3.58	0.40	2.91	0.39	35.2	717.4	0.06	44.0	5.9	3.6	3
番阳峒	S013-2	103.7	210.6	25.9	82.5	11.6	1.25	9.4	1.14	4.78	0.99	2.45	0.36	1.93	0.34	26.3	456.9	0.35	38.4	5.8	4.0	3

续表

岩石单位	样品号	La	Ce	Pr	Nd	Sm	Eu	Gd	Tb	Dy	Ho	Er	Tm	Yb	Lu	Y	ΣREE	δEu	$(La/Yb)_N$	$(La/Sm)_N$	$(Gd/Yb)_N$	来源
番阳峒	H1527-1	139.0	216.0	28.4	80.8	20.3	1.58	11.7	2.10	11.30	2.40	5.71	0.84	5.10	0.77	57.1	526.0	0.29	19.5	4.4	1.9	3
番阳峒	D33-1	110.8	210.9	25.5	92.5	17.0	0.99	12.9	1.80	9.93	2.02	5.02	0.65	4.69	0.69	50.9	495.3	0.20	17.0	4.2	2.3	3
番阳峒	D35-1	223.2	490.9	49.2	168.5	21.3	0.78	12.6	1.41	5.92	1.01	2.46	0.29	2.07	0.31	23.5	980.0	0.13	77.5	6.8	5.0	3
番阳峒	BD10	77.8	121.1	17.4	62.5	10.1	1.83	8.8	1.22	5.51	1.21	2.91	0.45	2.68	0.46	32.3	314.0	0.58	20.8	4.9	2.7	3
番阳峒	D121-1	54.4	103.9	11.3	43.1	8.3	0.97	7.0	1.12	7.02	1.40	3.92	0.52	3.99	0.59	38.9	247.5	0.38	9.8	4.2	1.5	3
番阳峒	H306	41.4	78.0	8.5	30.3	4.7	0.86	3.6	0.46	2.36	0.44	1.33	0.21	1.27	0.22	11.7	173.8	0.61	23.4	5.6	2.4	3
番阳峒	S043-1	166.2	312.0	33.6	117.7	18.0	1.48	12.7	1.61	7.67	1.49	3.41	0.48	2.88	0.45	36.2	679.6	0.28	41.4	6.0	3.6	3
番阳峒	D0472	123.8	218.8	26.18	89.11	13.22	1.54	9.03	1.12	5.94	1.04	2.84	0.41	2.59	0.44	23.35	496.0	0.41		6.05	2.88	本次
番阳峒	D0476	90.68	161.9	19.91	70.92	12.40	2.25	9.97	1.37	7.35	1.46	3.90	0.62	3.83	0.63	37.16	387.1	0.60		4.72	2.15	本次
番阳峒	D0825	93.53	174.8	20.64	77.80	10.41	0.95	7.05	0.94	4.91	0.92	2.46	0.38	2.39	0.39	20.48	397.5	0.32		5.80	2.44	本次
番阳峒	PM1-69-1	170.8	329.4	38.14	139.2	23.24	1.53	17.87	2.52	12.47	2.51	5.96	0.87	4.94	0.72	56.90	750.1	0.22		4.74	2.99	本次
番阳峒	PM1-107-2	29.95	52.64	5.75	19.7	3.16	0.71	2.27	0.31	1.50	0.30	0.82	0.13	0.83	0.14	8.59	118.2	0.77		6.12	2.26	本次
番阳峒	PM1-117-1	306	535	58.46	187	22.69	0.65	12.84	1.42	5.09	1.13	2.62	0.42	2.51	0.42	22.44	1136.2	0.11		8.71	4.23	本次
麻山田	H1498-1	23.5	49.5	6.2	24.6	4.8	2.67	4.2	0.64	3.75	0.75	2.14	0.32	2.04	0.34	19.0	125.5	1.77	8.3	3.1	1.7	3
麻山田	BD2	80.0	111.0	15.0	52.0	7.0	3.49	6.0	0.84	4.00	0.86	2.28	0.36	2.17	0.37	22.8	285.4	1.60	26.5	7.4	2.3	3
麻山田	S012-1	37.1	80.0	9.9	41.9	8.2	2.89	7.2	0.99	5.25	1.02	2.84	0.42	2.93	0.50	29.2	201.2	1.12	9.1	2.9	2.0	3
麻山田	D182-1	47.6	95.2	10.9	43.1	7.8	4.41	6.8	1.01	6.00	1.18	3.38	0.45	3.53	0.55	32.6	231.9	1.80	9.7	3.9	1.6	3
麻山田	HG-828	55.0	118.0	13.0	50.0	7.3	4.40	9.2	3.00	7.10	1.60	3.30	0.61	3.80	0.77	29.0	277.1	1.64	10.4	4.9	2.0	3
麻山田	IIP10-1	168.5	314.5	34.2	126.0	18.6	2.76	12.8	1.63	7.85	1.40	3.66	0.46	3.09	0.48	37.8	695.9	0.52	39.2	5.9	3.4	3
麻山田	D1466-1	163.8	206.4	35.1	126.1	18.6	2.48	13.9	1.76	8.58	1.62	4.38	0.59	3.76	0.57	44.9	587.6	0.45	31.2	5.7	3.1	3
麻山田	D759-1	132.1	217.2	22.5	79.1	8.3	3.44	5.6	0.59	2.75	0.58	1.63	0.25	1.86	0.34	15.2	476.2	1.45	50.9	10.2	2.5	3
深堀村	H1474-2	44.0	84.7	9.8	36.6	6.8	2.35	6.2	0.90	4.67	0.98	2.48	0.39	2.40	0.33	24.1	202.5	1.09	13.2	4.2	2.1	3
深堀村	S042	190.8	317.0	32.3	105.8	12.8	3.06	7.5	1.06	5.10	0.92	2.70	0.37	2.38	0.39	27.8	682.1	0.88	57.6	9.7	2.6	3
深堀村	HG-18	119.4	119.5	20.7	107.2	9.5	2.85	7.3	1.36	5.49	1.70	2.27	0.31	1.76	0.34	18.1	399.5	1.01	48.7	8.1	3.4	3
深堀村	HG-23	121.9	244.3	27.3	137.2	11.2	3.20	9.5	1.45	7.49	2.20	2.89	0.35	3.42	0.43	22.8	572.9	0.92	25.6	7.0	2.3	3

续表

岩石单位	样品号	La	Ce	Pr	Nd	Sm	Eu	Gd	Tb	Dy	Ho	Er	Tm	Yb	Lu	Y	∑REE	δEu	$(La/Yb)_N$	$(La/Sm)_N$	$(Gd/Yb)_N$	来源
深堀村	D430-1	60.7	123.3	14.3	58.4	10.6	2.38	8.5	1.21	7.07	1.40	3.54	0.50	3.75	0.52	36.4	296.2	0.74	11.6	3.7	1.9	3
深沟	S045	39.0	78.7	8.7	32.5	7.7	0.95	6.7	1.20	6.67	1.36	3.81	0.50	3.91	0.59	38.3	192.2	0.40	7.2	3.3	1.4	3
进岭	H211-3	53.8	107.4	13.2	47.1	10.5	0.70	9.4	1.50	9.87	2.03	5.92	0.86	5.69	0.85	60.0	268.9	0.21	6.8	3.3	1.4	3
进岭	H1323-1	54.4	111.0	13.7	52.2	10.7	0.34	9.6	1.51	8.33	1.69	4.93	0.80	4.80	0.70	47.9	274.6	0.10	8.1	3.3	1.6	3
瘦岭	D3013-1	21.5	41.1	3.6	10.6	1.8	0.15	1.4	0.22	1.41	0.34	1.08	0.18	1.62	0.27	11.0	85.3	0.28	9.6	7.8	0.7	3
瘦岭	QXI1-1	26.4	65.2	7.7	28.0	5.2	0.65	3.3	0.37	1.55	0.28	0.62	0.09	0.55	0.09	7.0	140.0	0.45	34.5	3.3	4.9	3
瘦岭	QXI3-3	24.8	43.2	5.4	17.4	3.5	0.75	3.2	0.43	1.80	0.30	0.60	0.09	0.42	0.06	8.0	101.9	0.67	42.3	4.5	6.3	3
黑岭	QXVI6-2	123.7	235.3	29.7	103.1	18.8	0.67	16.9	2.37	12.80	2.35	6.00	0.92	5.46	0.78	78.4	558.8	0.11	16.3	4.2	2.6	3
黑岭	Y5551-1	80.0	160.0	21.0	65.0	17.0	0.28	17.0	3.00	20.00	3.30	9.20	0.80	5.00	0.78	*	402.4	0.05	11.5	3.0	2.8	3
黑岭	YXIII35-1	40.9	102.8	13.5	36.9	7.3	0.48	8.5	1.28	8.61	1.78	4.33	0.64	7.66	0.37	*	235.1	0.19	3.8	3.6	0.9	3
黑岭	YXIII9-2	73.0	146.0	14.0	57.0	11.0	0.57	12.5	1.50	9.90	1.80	4.40	0.48	3.00	0.46	*	335.6	0.15	17.5	4.3	3.4	3
黑岭	IIP6-1	53.8	104.7	12.1	42.3	7.4	0.95	5.6	0.92	4.61	1.07	2.95	0.45	3.41	0.49	31.7	240.8	0.43	11.3	4.7	1.4	3
尖峰岭	D3016-1	90.3	189.0	21.7	78.6	14.4	0.82	10.9	1.59	9.27	1.70	4.41	0.67	4.08	0.59	47.0	428.0	0.19	15.9	4.1	2.2	3
尖峰岭	S019	121.0	260.2	29.8	105.9	17.8	0.82	12.5	1.97	10.00	1.83	4.94	0.68	4.37	0.65	53.0	572.6	0.16	19.9	4.4	2.4	3
尖峰岭	QXVI6-1	108.6	208.5	24.3	79.2	12.2	0.99	10.6	1.48	8.07	1.59	4.19	0.68	4.16	0.62	51.9	465.2	0.26	18.7	5.7	2.1	3
尖峰岭	QXVIII1-1	119.8	218.8	26.2	85.3	12.2	1.17	10.0	1.33	6.74	1.30	3.19	0.51	2.83	0.43	35.2	489.7	0.32	30.4	6.4	2.9	3
尖峰岭	Y5550-1	95.0	170.0	19.0	65.0	15.0	0.64	13.0	2.00	13.00	2.30	6.50	0.98	6.00	0.90	*	409.3	0.14	11.4	4.1	1.8	3
尖峰岭	D3025-1	32.8	68.4	7.2	25.8	4.2	0.23	2.9	0.39	1.89	0.36	0.90	0.12	0.83	0.12	9.3	146.2	0.19	28.4	5.1	2.9	3
尖峰岭	Q1200	42.4	89.7	12.2	42.0	8.9	0.25	9.0	1.29	6.87	1.34	3.67	0.60	3.67	0.55	40.9	222.4	0.08	8.3	3.1	2.0	3
尖峰岭	HG-433	60.0	83.0	15.0	38.0	9.3	0.70	4.4	1.24	4.10	0.80	1.80	0.35	2.20	0.51	15.0	221.4	0.29	19.6	4.2	1.7	3
尖峰岭	HG-467	32.0	50.0	6.2	24.0	4.8	0.38	2.4	0.79	2.50	0.50	1.10	0.23	1.40	0.30	12.0	126.6	0.30	16.4	4.3	1.4	3
尖峰岭	S037	37.2	76.3	8.7	31.0	5.7	0.57	4.4	0.65	3.57	0.68	1.81	0.28	1.78	0.28	22.3	173.0	0.33	15.0	4.2	2.1	3
尖峰岭	D271-1	52.7	122.2	13.8	51.3	11.2	0.20	8.5	1.19	6.95	1.24	3.05	0.42	2.76	0.39	29.5	275.9	0.06	13.7	3.1	2.6	3
高村	Ⅷ1-1	76.8	161.9	18.75	33.5	13.26	1.62	10.25	1.36	8.41	1.53	4.39	0.65	4.05	0.64	42.66	379.77	0.41	12.78	3.64	2.04	1
高村	D2385-2	9.44	22.62	3.14	13.2	4.39	0.14	5.11	1.02	7.3	1.33	4.48	0.73	4.91	0.79	45.18	123.78	0.09	1.3	1.35	0.84	1

续表

岩石单位	样品号	La	Ce	Pr	Nd	Sm	Eu	Gd	Tb	Dy	Ho	Er	Tm	Yb	Lu	Y	ΣREE	δEu	$(La/Yb)_N$	$(La/Sm)_N$	$(Gd/Yb)_N$	来源
两院	Ⅷ6-1	97.84	200.3	22.18	81.27	13.56	1.52	9.87	1.2	7.31	1.3	3.51	0.51	3.01	0.47	34.89	478.74	0.38	21.91	4.54	2.65	1
两院	D2018	9.77	25.45	3.3	14.13	4.27	0.17	4.59	0.84	5.98	1.15	3.84	0.69	4.93	0.81	37.57	117.49	0.12	1.34	1.44	0.75	1
两院	0504	26.22	53.6	6.44	21.98	4.58	0.61	4.55	0.71	4.08	0.8	2.28	0.35	2.12	0.31	23.68	152.31	0.4	8.34	3.6	1.73	1
两院	D2385-1	45.24	60.45	11.28	47.21	11.12	1.25	11.13	1.9	12.12	2.12	6.3	0.98	6.04	0.95	65.42	283.51	0.34	5.05	2.56	1.49	1
大岭	Ⅷ15-2	38.36	80.9	8.33	30	5.16	0.68	4	0.54	3.53	0.63	1.82	0.28	1.67	0.26	18.3	194.46	0.44	15.49	4.68	1.93	1
鹅塘村	Ⅷ15-1	36.49	72.53	7.69	29.11	5.24	0.87	4.22	0.56	3.67	0.65	1.76	0.26	1.48	0.23	17.94	182.7	0.55	16.62	4.38	2.3	1
陀类	D3082-1	52.8	104.0	11.0	37.9	6.1	0.75	4.0	0.43	1.98	0.32	0.72	0.09	0.61	0.08	8.9	220.6	0.44	62.6	5.6	5.4	3
陀类	Y78-1	57.8	103.0	12.1	42.5	8.7	0.82	5.1	0.69	3.10	0.41	1.12	0.17	1.03	0.16	11.9	236.7	0.35	40.4	4.3	4.1	3
陀类	Y79-1	51.0	80.0	8.8	32.0	7.3	0.66	4.4	0.63	2.70	0.35	0.92	0.13	0.80	0.12	5.2	189.8	0.33	45.7	4.5	4.5	3
陀类	Y80-1	22.0	35.0	3.9	13.5	3.0	0.29	2.2	0.35	2.30	0.43	1.25	0.19	1.10	0.16	7.5	85.6	0.33	14.3	4.7	1.7	3
	平均	45.9	80.5	8.9	31.5	6.3	0.63	3.9	0.53	2.52	0.38	1.00	0.14	0.88	0.13	8.4	183.2	0.36	40.8	4.8	3.9	
大田	D3087-1	56.3	99.1	9.8	32.9	5.1	1.08	2.7	0.31	1.02	0.17	0.46	0.06	0.47	0.06	5.2	209.6	0.80	86.4	7.1	4.9	3
大田	D3084-1	71.2	122.7	12.7	42.5	6.2	1.17	4.6	0.65	3.38	0.60	1.46	0.21	1.46	0.18	17.2	269.0	0.64	34.9	7.4	2.6	3
大田	DT1-1	72.2	133.9	13.7	49.4	8.1	1.18	5.5	0.69	3.43	0.63	1.56	0.20	1.43	0.19	17.8	292.0	0.51	36.2	5.8	3.2	3
大田	DT2-1	73.6	138.8	13.7	49.7	8.0	1.22	5.6	0.75	3.94	0.73	1.77	0.23	1.83	0.25	20.2	300.1	0.53	28.9	5.9	2.5	3
	平均值	63.8	115.0	11.8	41.2	6.7	1.06	4.5	0.59	2.86	0.50	1.25	0.17	1.21	0.16	13.7	250.8	0.57	45.4	6.2	3.4	
什天	YVII9-2	41.1	76.3	10.3	28.3	6.5	0.92	4.0	0.77	1.73	0.39	1.31	0.24	0.59	0.19	*	172.6	0.51	50.0	4.1	5.5	3
什天	S031	43.9	86.9	9.6	33.5	5.9	0.77	4.3	0.62	3.13	0.56	1.56	0.23	1.50	0.23	18.9	192.7	0.45	21.0	4.8	2.3	3
什天	HG-437	33.0	53.0	6.1	25.0	4.4	0.49	3.1	0.56	2.70	0.46	1.30	0.16	1.10	0.22	15.0	131.6	0.39	21.5	4.8	2.3	3
什天	D3020-1	34.8	61.4	6.0	20.6	3.5	0.64	3.0	0.43	2.59	0.54	1.49	0.23	1.96	0.33	16.2	137.5	0.59	12.7	6.3	1.3	3
什天	IIP4-1	79.2	154.5	15.8	56.8	8.2	1.44	5.5	0.64	3.24	0.60	1.65	0.20	1.49	0.21	17.2	329.6	0.61	38.2	6.2	3.1	3
什天	S014	135.1	266.8	28.9	100.3	15.5	1.21	10.8	1.59	8.01	1.45	3.78	0.53	3.10	0.46	42.0	577.5	0.27	31.3	5.6	2.9	3
	平均值	57.8	110.0	12.0	41.4	7.0	0.91	4.8	0.72	3.33	0.62	1.71	0.24	1.50	0.25	19.8	242.2	0.50	30.0	5.3	3.0	
白茶	H1362-1	56.2	136.0	12.2	51.3	8.6	0.48	4.9	0.88	5.95	1.07	2.73	0.42	2.49	0.41	29.4	283.7	0.21	16.2	4.2	1.6	3
东岭	H1211-1	47.0	83.0	10.1	46.0	8.0	1.30	5.5	0.88	4.20	0.64	1.54	0.19	1.04	0.15	12.0	209.5	0.57	32.4	3.8	4.4	3

续表

岩石单位	样品号	La	Ce	Pr	Nd	Sm	Eu	Gd	Tb	Dy	Ho	Er	Tm	Yb	Lu	Y	ΣREE	δEu	$(La/Yb)_N$	$(La/Sm)_N$	$(Gd/Yb)_N$	来源
东岭	H1267-2	61.0	101.0	12.2	61.0	10.1	1.50	7.0	0.97	5.80	1.00	3.10	0.35	1.90	0.32	21.0	267.2	0.52	23.0	3.9	3.0	3
	平均值	54.0	92.0	11.15	53.50	9.05	1.40	6.25	0.93	5.00	0.82	2.32	0.27	1.47	0.24	16.50	238.35	0.55	27.70	3.85	3.70	
那佩	D3074-1	41.7	90.8	10.4	40.2	9.1	0.62	7.7	1.29	7.69	1.62	4.47	0.65	4.55	0.68	45.7	221.5	0.22	6.6	3.0	1.4	3
那佩	G1346	41.9	68.7	8.3	33.1	6.9	0.81	5.3	0.92	4.66	0.92	2.38	0.35	2.10	0.28	21.5	176.6	0.39	14.3	3.9	2.1	3
	平均值	41.80	79.75	9.35	36.65	8.00	0.72	6.50	1.11	6.18	1.27	3.43	0.50	3.33	0.48	33.6	199.05	0.31	10.45	3.45	1.75	
保国	D3076-1	43.7	88.4	9.5	34.7	6.6	0.88	5.7	0.91	5.11	1.11	2.90	0.42	2.90	0.40	32.0	203.3	0.43	10.8	4.3	1.6	3
保国	HG-28	65.7	92.8	9.1	47.0	6.7	0.76	5.2	0.85	5.83	0.73	2.89	0.30	2.28	0.43	26.0	240.5	0.38	20.7	6.4	1.9	3
保国	HG-774	45.0	79.0	10.0	36.0	4.5	0.82	5.4	1.30	5.50	1.00	2.80	0.45	3.20	0.47	24.0	195.4	0.51	10.1	6.5	1.4	3
保国	J121-1	95	188	17.5	86	15	3.6	9.5	1.3	9.1	1.05	2.8	0.4	2.7	0.43	29	432.4	0.86	25.24	4.09	2.91	3
保国	J133-1	74	138	18	76.5	13.8	1.9	8.15	1.26	7.45	1.37	3	0.45	2.62	0.41	14	346.9	0.50	20.26	3.46	2.57	3
	平均	64.7	117.3	12.8	56.0	9.3	1.59	6.8	1.12	6.60	1.05	2.88	0.40	2.74	0.43	25.0	283.7	0.54	17.4	4.9	2.1	
火岭	H1311-3	68.0	104.0	11.0	56.0	10.1	0.56	7.3	0.99	4.90	0.68	1.48	0.22	1.40	0.16	13.0	266.8	0.19	34.8	4.3	4.3	3
火岭	VHG1-7	97.5	189.6	22.46	69.38	10.47	1.89	8.12	1.13	5.83	1.15	3.12	0.49	2.98	0.46	30.35	444.94	0.6	22.06	5.86	2.2	1
火岭	7329-1	49	82	10.5	37	6.7	0.45	6.7	1	5.3	0.75	1.6	0.17	0.85	0.13	14.5	216.65	0.2	38.87	4.6	6.36	1
火岭	Ⅺ2-1	26.5	51.23	5.12	18.53	3.63	0.33	2.59	0.28	1.45	0.22	0.43	0.05	0.26	0.03	5.88	116.53	0.31	68.72	4.59	8.04	1
火岭	Ⅻ5-1	22.47	43.03	4.86	18.32	3.82	0.37	3.24	0.51	3.42	0.58	1.6	0.24	1.39	0.21	16.93	120.99	0.31	10.9	3.7	1.88	1
火岭	4108-1	87.28	141.5	14.8	47.79	8.49	0.71	5.46	1.14	8.22	1.32	2.16	0.38	2.8	0.43	28.27	350.73	0.3	21.02	6.47	1.57	1
	平均值	58.46	101.89	11.46	41.17	7.20	0.72	5.57	0.84	4.85	0.78	1.73	0.26	1.61	0.24	18.16	252.77	0.32	32.73	4.92	4.06	
长塘岭	H1278-1	61.0	106.0	12.2	61.0	11.0	1.15	7.4	0.91	4.80	0.80	2.11	0.25	1.50	0.19	16.0	270.3	0.37	29.2	3.6	4.1	3
长塘岭	1225	101	118	23.4	72.2	13.3	2.1	8.55	1.17	6.22	1.2	3.13	0.45	2.7	0.37	23.6	377.39	0.56	25.22	4.78	2.56	1
长塘岭	Ⅸ10-1	27	46	6.2	23	6.7	0.59	5.2	0.94	6.6	1.38	3.4	0.53	3.2	0.48	36	167.22	0.29	5.69	2.53	1.31	1
长塘岭	Ⅺ10-2	76.59	143.8	12.3	48.37	7.69	0.96	4.85	0.51	3.07	0.54	1.28	0.18	1.08	0.16	14.14	315.46	0.45	47.81	6.26	3.62	1
长塘岭	4871-2	76.59	143.8	12.3	48.37	7.69	0.96	4.85	0.51	3.07	0.54	1.28	0.18	1.08	0.16	14.14	315.46	0.45	47.81	6.26	3.62	1
长塘岭	D823-1	38.8	77.4	8.97	28.53	5.28	0.81	3.6	0.48	2.11	0.39	0.93	0.15	0.88	0.13	10.5	178.99	0.54	29.75	4.63	3.3	1
长塘岭	YD499	53.9	91.11	10.36	34.47	6.19	0.83	4.32	0.6	2.01	0.36	0.61	0.08	0.42	0.06	8.03	213.31	0.47	86.46	5.47	8.3	1

续表

岩石单位	样品号	La	Ce	Pr	Nd	Sm	Eu	Gd	Tb	Dy	Ho	Er	Tm	Yb	Lu	Y	ΣREE	δEu	$(La/Yb)_N$	$(La/Sm)_N$	$(Gd/Yb)_N$	来源
	平均值	62.13	103.73	12.25	45.13	8.26	1.06	5.54	0.73	3.98	0.74	1.82	0.26	1.55	0.22	17.49	262.59	0.45	38.85	4.79	3.83	
大坡	1389-2	39	67	9	31	6.4	0.78	6	0.95	5.4	0.84	2.1	0.22	1.3	0.2	17	187.19	0.38	20.23	3.83	3.72	1
大坡	21	30	62	6.7	28	6	0.64	4.6	0.74	4.9	0.88	2.3	0.26	1.45	0.23	17.5	166.2	0.36	13.95	3.15	2.56	1
大坡	HG394	56	111	11	44	5.9	0.89	4.8	0.89	3.6	0.58	2	0.23	1.7	0.28	20	262.87	0.5	22.21	5.97	2.28	1
大坡	Ⅻ7-2	38.14	78.73	7.89	28.56	5.08	0.71	3.63	0.44	2.73	0.49	1.24	0.19	0.19	0.19	13.29	181.5	0.48	135.34	4.72	15.4	1
大坡	D892-1	32.3	59.9	6.15	23.2	4.87	0.61	3.37	0.61	2.98	0.61	1.79	0.28	1.74	0.26	14.9	153.57	0.44	12.52	4.17	1.56	1
大坡	Ⅺ6-1	35.59	67.6	7.07	25.32	4.66	0.61	3.12	0.28	1.27	0.19	0.41	0.06	0.29	0.04	5.01	151.52	0.46	82.74	4.8	8.68	1
大坡	Ⅻ7-1	40.17	91.74	8.63	30.59	5.12	0.82	4.18	0.51	3.39	0.61	1.66	0.23	1.37	0.22	15.87	205.11	0.53	19.77	4.94	2.46	1
大坡	Ⅱ8-1	42	75	6.6	26	5.3	0.9	3.5	0.62	3.6	0.6	1.8	0.24	1.5	0.24	17	184.9	0.6	18.88	4.98	1.88	1
大坡	Ⅱ19-1	53	83	9	30	6.8	1.05	4.4	0.7	3.4	0.68	1.95	0.23	1.4	0.22	16	211.83	0.55	25.52	4.9	2.54	1
大坡	XⅣ25-1	47	137	8	26	6.4	0.76	2.5	0.41	2	0.41	1.03	0.16	0.88	0.14	7.1	239.79	0.49	36.01	4.62	2.29	1
大坡	XⅣ9-1	53	92	10.4	41	8.9	1.02	5.1	0.82	5	1	2.5	0.35	1.8	0.28	23	246.17	0.42	19.85	3.75	2.29	1
大坡	XX12-1	30	52	6	26.5	5.18	0.52	3.75	0.6	4	0.75	2.22	0.29	1.68	0.24	18	151.73	0.34	12.04	3.64	1.8	1
大坡	Ⅸ17-1	52	92	10.6	42	8.9	1.18	5.2	0.88	5.4	1.15	2.8	0.42	2.5	0.39	26	251.42	0.49	14.02	3.68	1.68	1
大坡	Ha34-1	15.4	32.6	4.07	16	4.22	0.52	3.72	0.65	4.21	0.85	2.36	0.35	2.3	0.43	24.6	112.28	0.39	4.51	2.3	1.31	1
	平均值	40.26	78.68	7.94	29.87	5.98	0.79	4.13	0.65	3.71	0.69	1.87	0.25	1.44	0.24	16.81	193.29	0.46	31.26	4.25	3.60	
乌兰盖	3809	48.3	83.6	9.95	32.9	7.09	0.75	3.95	0.57	3.2	0.61	1.51	0.23	1.38	0.15	13.3	207.49	0.4	23.6	4.29	2.31	1
乌兰盖	D45-1	40.1	82.8	8.33	36.5	6.8	1.36	4.86	0.86	4.54	0.91	2.36	0.37	2.24	0.34	20.4	212.77	0.69	12.07	3.71	1.75	1
	平均值	44.20	83.20	9.14	34.70	6.95	1.06	4.41	0.72	3.87	0.76	1.94	0.30	1.81	0.25	16.85	210.13	0.55	17.84	4.00	2.03	
通什	H1274-2	58.0	98.0	10.5	51.5	8.7	1.30	6.0	0.76	3.80	0.64	1.76	0.22	1.22	0.18	12.0	242.6	0.52	34.1	4.3	4.1	3
通什	H1287-1	61.0	101.0	10.6	50.0	8.6	1.30	6.1	0.80	3.90	0.62	1.50	0.20	1.31	0.19	11.2	247.1	0.52	33.4	4.6	3.9	3
通什	HG-424	48.0	60.0	9.6	29.0	6.8	0.70	4.1	1.00	3.60	0.74	1.60	0.30	1.80	0.40	*	167.6	0.37	19.1	4.6	1.9	3
通什	HG-443	36.0	49.0	6.4	19.0	3.8	0.56	1.9	0.50	2.00	0.36	0.84	0.15	0.92	0.23	*	121.7	0.57	28.1	6.1	1.7	3
通什	HG-475	54.0	69.0	13.0	35.0	8.1	0.74	3.7	1.24	3.80	0.72	1.55	0.32	1.80	0.48	15.0	193.5	0.36	21.5	4.3	1.7	3
通什	4220	39.5	73	7.9	31.5	6.25	0.89	4.45	0.72	3.95	0.65	1.65	0.22	1.3	0.2	13.5	185.68	0.49	20.49	3.98	2.76	1

续表

岩石单位	样品号	La	Ce	Pr	Nd	Sm	Eu	Gd	Tb	Dy	Ho	Er	Tm	Yb	Lu	Y	ΣREE	δEu	$(La/Yb)_N$	$(La/Sm)_N$	$(Gd/Yb)_N$	来源
通什	4107	37	74	7	28	5.8	0.7	3.5	0.5	2.35	0.39	0.8	0.13	0.55	0.09	7.2	168.01	0.44	45.35	4.01	5.14	1
通什	38-1	30	68	6.7	31	7.1	0.63	5.3	0.89	5.4	0.88	2.25	0.31	1.8	0.29	19.5	180.05	0.3	11.24	2.66	2.38	1
通什	Ⅸ8-1	36	62	7.9	31	6.9	1.16	4.4	0.77	4.8	1.04	2.6	0.42	2.6	0.4	25	186.99	0.6	9.33	3.28	1.37	1
通什	XⅣ12-1	48	84	9.9	39	8.7	1.09	5.3	0.86	5.2	1	2.5	0.36	1.9	0.29	23	231.1	0.46	17.03	3.47	2.25	1
通什	4111-1	57.39	107.1	10.7	39.63	6.89	1.36	4.93	0.68	3.68	0.67	1.77	0.27	1.63	0.25	18.86	255.82	0.68	23.74	5.24	2.44	1
通什	4577-1	31.69	59.39	6.75	22.6	4.56	0.77	3.07	0.42	2.1	0.38	0.98	0.15	0.99	0.15	10.86	144.86	0.59	21.58	4.37	2.5	1
通什	Ha35-1	43.3	83.3	8.9	30.7	5.23	1.01	4.49	0.56	2.65	0.47	1.27	0.21	1.31	0.18	13.7	197.28	0.62	22.28	5.21	2.77	1
	平均值	44.61	75.98	8.91	33.69	6.73	0.94	4.40	0.75	3.63	0.66	1.62	0.25	1.47	0.26	15.44	194.02	0.50	23.63	4.32	2.69	
山牛塘	H1311-2	31.2	64.5	8.6	24.5	7.7	1.42	5.1	0.87	5.81	1.26	3.57	0.52	2.86	0.37	30.8	158.3	0.65	7.8	2.6	1.5	3
山牛塘	H1397-1	29.4	62.9	7.7	35.6	6.3	1.36	4.6	0.73	4.88	0.89	1.96	0.28	1.72	0.27	21.5	158.6	0.73	12.3	3.0	2.2	3
山牛塘	H209-1	37.9	79.0	9.0	42.2	6.9	1.55	7.0	1.28	8.77	1.38	3.62	0.54	3.36	0.53	32.6	203.0	0.67	8.1	3.5	1.7	3
山牛塘	H209-3	56.0	111.0	12.4	55.6	9.9	1.80	7.3	1.04	6.12	1.08	2.68	0.39	2.44	0.37	30.3	268.2	0.62	16.5	3.6	2.5	3
	平均值	38.63	79.35	9.43	39.48	7.70	1.53	6.00	0.98	6.40	1.15	2.96	0.43	2.60	0.39	28.80	197.03	0.67	11.18	3.18	1.98	
石碌	137-1	52	90	11.5	39	7.7	0.93	5.5	0.82	4.6	0.7	1.9	0.24	1.3	0.2	15.5	231.9	0.42	26.97	4.25	3.41	1
石碌	Ha29	43.3	87.8	10.4	38.2	7.76	1.06	6.58	0.99	4.98	0.93	2.52	0.42	2.74	0.37	25.5	233.55	0.44	10.65	3.51	1.94	1
石碌	7334-1	33	67	7.6	32	7.1	0.8	5.8	0.94	5.3	0.78	1.75	0.2	1.04	0.17	14.5	177.98	0.37	21.39	2.92	4.5	1
石碌	HG399	27	38	5	22	4.6	0.6	3.5	0.72	3	0.6	1.6	0.25	1.8	0.36	22	131.03	0.44	10.11	3.69	1.57	1
	平均值	38.83	70.70	8.63	32.80	6.79	0.85	5.35	0.87	4.47	0.75	1.94	0.28	1.72	0.28	19.38	193.62	0.42	17.28	3.59	2.86	
什败	D1547-3	186.5	302.8	40.0	144.3	19.4	3.28	15.7	1.84	9.87	1.88	5.08	0.68	3.97	0.58	49.9	736.0	0.56	33.7	6.2	3.3	3
行干	D1749-1	72.7	145.4	17.5	69.7	10.7	2.46	7.8	1.00	4.45	0.78	1.91	0.24	1.52	0.23	20.7	336.4	0.79	34.3	4.4	4.2	3
行干	D3089-1	67.8	150.7	18.4	74.7	12.9	3.58	9.5	1.11	5.52	0.99	2.26	0.28	1.86	0.27	26.9	350.0	0.95	26.1	3.4	4.2	3
方满	D3092-1	176.6	355.1	40.9	155.9	21.8	5.11	14.3	1.58	6.65	1.16	2.87	0.35	2.27	0.34	32.5	785.0	0.83	55.7	5.2	5.2	3
	平均值	125.90	238.50	29.20	111.15	16.20	3.61	11.83	1.38	6.62	1.20	3.03	0.39	2.41	0.36	32.50	551.85	0.78	37.45	4.80	4.23	
新风岭	H5122-1	26.1	51.8	5.9	22.2	6.5	0.43	4.1	0.62	3.13	0.54	1.53	0.17	1.07	0.16	12.3	124.3	0.24	17.5	2.6	3.2	3
顺作	D3136-1	35.0	65.1	6.9	23.8	4.0	1.18	2.5	0.25	0.90	0.14	0.27	0.02	0.24	0.03	3.4	140.3	1.06	105.1	5.6	8.7	3

续表

岩石单位	样品号	La	Ce	Pr	Nd	Sm	Eu	Gd	Tb	Dy	Ho	Er	Tm	Yb	Lu	Y	∑REE	δEu	$(La/Yb)_N$	$(La/Sm)_N$	$(Gd/Yb)_N$	来源
乘顺岭	QIV1-1	37.1	71.0	9.1	28.0	5.9	0.32	4.0	0.49	2.78	0.50	1.20	0.19	0.94	0.13	14.0	161.7	0.19	28.3	4.0	3.5	3
乘顺岭	GXI3-1	47.8	91.5	11.4	45.1	9.7	0.76	7.0	1.14	5.88	1.09	2.86	0.40	2.44	0.31	25.5	227.4	0.27	14.1	3.2	2.4	3
	平均值	42.45	81.25	10.25	36.55	7.80	0.54	5.50	0.82	4.33	0.80	2.03	0.30	1.69	0.22	19.75	194.55	0.23	21.20	3.60	2.95	
志仲	D3073-1	30.7	64.5	7.2	26.6	6.3	0.71	6.1	1.04	6.20	1.30	3.39	0.44	3.03	0.43	36.4	157.9	0.35	7.3	3.2	1.7	3
志仲	HG-17	33.3	59.4	9.1	46.3	6.4	0.79	5.6	1.02	6.88	0.96	3.94	0.38	3.25	0.62	38.6	177.9	0.40	7.3	3.4	1.4	3
志仲	GVIII18	74.5	128.0	14.8	55.5	9.8	0.87	5.1	0.81	3.48	0.66	1.53	0.24	1.47	0.22	13.9	296.9	0.34	36.4	4.9	2.9	3
	平均值	46.17	83.97	10.37	42.80	7.50	0.79	5.60	0.96	5.52	0.97	2.95	0.35	2.58	0.42	29.63	210.90	0.36	17.00	3.83	2.00	
抱郎	QIV4-1	36.1	82.1	10.3	37.8	7.0	1.49	5.3	0.71	4.25	0.77	1.90	0.28	1.50	0.23	20.3	189.7	0.72	17.3	3.4	2.9	3
抱郎	QVII2	73.1	153.2	17.6	60.8	10.3	0.95	7.0	0.97	4.63	0.82	1.90	0.26	1.53	0.23	20.7	333.3	0.32	34.3	4.6	3.8	3
	平均值	54.60	117.65	13.95	49.30	8.65	1.22	6.15	0.84	4.44	0.80	1.90	0.27	1.52	0.23	20.50	261.50	0.52	25.80	4.00	3.35	
石岭	7315	49	99	10	36	7.2	1.1	4.4	0.65	2.9	0.46	0.94	0.14	0.6	0.1	7.6	220.09	0.56	55.06	4.28	5.92	1
石岭	7316-1	30	65	6.8	27	6.2	1.1	4.8	0.75	4.8	0.83	2.1	0.25	1.45	0.23	16	167.31	0.59	13.95	3.04	2.67	1
石岭	Ha36	29.8	59.9	6.86	26.2	5.23	1.35	4.75	0.74	4.16	0.8	2.21	0.36	2.32	0.33	21.6	166.61	0.81	8.66	3.58	1.65	1
	平均值	36.27	74.63	7.89	29.73	6.21	1.18	4.65	0.71	3.95	0.70	1.75	0.25	1.46	0.22	15.07	184.67	0.65	25.89	3.63	3.41	
保山村	4758-1	37.47	76.72	8.88	33.73	6.72	1.55	5.24	0.78	3.8	0.7	1.67	0.25	1.38	0.2	17.36	196.45	0.77	18.31	3.51	3.06	1
保山村	4971-1	37.25	74.82	8.45	31.93	6.17	1.63	4.34	0.61	3.06	0.57	1.45	0.22	1.34	0.21	15.51	187.56	0.92	18.74	3.8	2.61	1
保山村	Ha32-1	46.9	101	12.3	47	10.3	1.34	9.48	1.54	8.6	1.68	4.52	0.78	4.92	0.66	47.4	298.42	0.41	6.43	2.86	1.55	1
	平均值	40.54	84.18	9.88	37.55	7.73	1.51	6.35	0.98	5.15	0.98	2.55	0.42	2.55	0.36	26.76	227.48	0.70	14.49	3.39	2.41	
白石村	211-3	44	80	7.2	41	7.7	1.2	4.8	0.75	3.7	0.56	1.2	0.19	0.9	0.14	10.5	203.84	0.56	32.96	3.59	4.3	1
白石村	290	23	56	5.7	25	5.6	1.2	4.5	0.73	4	0.68	1.75	0.2	1.2	0.2	14	143.76	0.71	12.92	2.58	3.03	1
白石村	4350-7	34.23	78.19	8.8	38.68	7.95	2.11	6.24	0.86	4.54	0.84	2.09	0.33	1.96	0.3	22.51	209.63	0.89	11.77	2.71	2.57	1
白石村	Ha16	48	113	14.7	56	10.7	2.7	8.75	1.22	6.38	1.13	2.98	0.49	3	0.42	30.3	299.77	0.83	10.79	2.82	2.35	1

续表

岩石单位	样品号	La	Ce	Pr	Nd	Sm	Eu	Gd	Tb	Dy	Ho	Er	Tm	Yb	Lu	Y	ΣREE	δEu	$(La/Yb)_N$	$(La/Sm)_N$	$(Gd/Yb)_N$	来源
白石村	Ha17-2	88	167	20.4	78.7	14.6	3.12	11.6	1.57	7.58	1.26	3.07	0.43	2.56	0.29	31.3	431.68	0.71	23.23	3.8	3.66	1
白石村	Ha17-1	58	128	16	59.7	10.6	2.85	8.72	1.16	5.46	0.96	2.52	0.4	2.66	0.35	26.8	324.78	0.88	14.85	3.48	2.65	1
白石村	Ha18	102	190	21.4	77.4	13.2	3.36	10.9	1.42	6.76	1.2	3.2	0.48	3	0.39	33	467.71	0.83	22.92	4.86	2.93	1
白石村	QIV0-1	26.3	53.2	7.6	28.6	5.8	1.52	6.2	0.85	4.64	0.89	2.39	0.37	2.13	0.33	26.7	140.8	0.77	8.9	2.9	2.4	3
白石村	s028	31.7	63.1	8.1	30.8	5.8	1.36	5.1	0.86	4.94	1.00	2.76	0.41	2.65	0.40	29.4	158.9	0.75	8.6	3.5	1.6	3
白石村	D3019-1	30.4	59.5	6.7	26.3	5.3	1.44	4.8	0.73	4.54	0.88	2.37	0.32	2.29	0.35	24.6	145.9	0.86	9.5	3.7	1.7	3
	平均值	48.56	98.80	11.66	46.22	8.73	2.09	7.16	1.02	5.25	0.94	2.43	0.36	2.24	0.32	24.91	252.68	0.78	15.64	3.39	2.72	
机告村	1153	80	155	15.8	62.6	11.6	2.11	6.25	1.06	4.98	0.96	2.66	0.36	1.88	0.28	22.6	368.74	0.69	28.9	4.37	2.68	1
机告村	4014-1	24	39.29	5.17	22.7	4.68	1.23	4.51	0.67	4.32	0.86	2.35	0.34	2.08	0.31	25.32	137.95	0.81	7.82	3.24	1.75	1
机告村	Ha19	49	105	12.1	45.9	8.42	2.36	7.15	0.99	4.86	0.9	2.36	0.37	2.37	0.34	23.5	266.42	0.91	14.17	3.72	2.43	1
机告村	Ha39-1	20	47.1	6.21	26.8	6.18	1.73	5.55	0.96	5.2	0.99	2.59	0.41	2.55	0.35	26.3	153.52	0.89	5.45	2.1	1.76	1
机告村	Ha39-2	20	44.7	5.73	23.3	5.06	1.41	4.74	0.79	4.47	0.86	2.26	0.38	2.39	0.33	22	138.62	0.87	5.7	2.51	1.6	1
	平均值	38.60	78.22	9.00	36.26	7.19	1.77	5.64	0.89	4.77	0.91	2.44	0.37	2.25	0.32	23.94	213.05	0.83	12.41	3.19	2.04	
亚炮	D336-1	20	35	4.02	18.1	4.07	0.52	5.2	0.91	6.11	1.27	3.77	0.59	3.1	0.43	26	129.09	0.35	4.35	3.09	1.35	1
亚炮	V40-1	17	35.5	4.4	18	6	0.12	8.5	1.7	14.5	3.45	10.1	1.45	8.4	1.15	100	230.78	0.05	1.4	1.83	0.82	1
	平均值	18.50	35.25	4.21	18.1	5.04	0.32	6.85	1.31	10.3	2.36	6.94	1.02	5.75	0.79	63.0	179.94	0.20	2.88	2.46	1.09	
戈枕村	D331-2	32	51	5.64	23.3	4.14	0.63	3.19	0.59	4.09	0.91	2.67	0.36	1.92	0.28	20	150.72	0.51	11.24	4.86	1.34	1
戈枕村	D322-1	49	80	10	41.7	8.65	1.25	7.64	1.36	9.34	1.76	5.11	0.57	3	0.43	34	253.81	0.46	11.01	3.56	2.06	1
戈枕村	Ⅶ15-1	56	95	8	29	7.1	1.25	6.7	1.1	6.2	1.38	3.8	0.6	3.5	0.48	46	266.11	0.55	10.79	4.96	1.54	1
戈枕村	Ⅶ15-2	48	82	7.1	25	5.4	1	3.9	0.71	4.7	0.84	2.7	0.41	2.5	0.38	29	213.64	0.64	12.94	5.59	1.26	1
	平均值	46.25	77.00	7.69	29.7	6.32	1.03	5.36	0.94	6.08	1.22	3.57	0.49	2.73	0.39	32.3	221.07	0.54	11.50	4.74	1.55	

注：样品来源：1—引自 1∶25 万《东方县》幅区调报告；2—引自 1∶25 万《琼海县》幅区调报告；3—引自 1∶25 万《乐东县、陵水县》幅区调报告；4—引自 1∶25 万《昌洒市、文昌县、清澜港、铜鼓嘴》幅区调报告；5—引自 1∶5 万《吊罗山、兴隆、什玲市、陵水县》幅区调报告。

附表 9-4 海南岛侵入岩微量元素分析结果一览表

Attached table 9-4 List of trace elements analysis results of intrusive rocks from the Hainan Island

单位：10^{-6}

岩石单位	样品号	Rb	Ba	Sr	Th	U	Ta	Nb	Pb	Zr	Hf	Ga	Zn	Ni	V	Cr	Cs	Co	Li	Cu	Mo	Ag	Au
$\gamma\pi K_2$	Ⅳ12-1	156	537	110		2	2.23	24	21.2	112	3.46		32.3	4.9	8.6			2.9		6.02	0.78	0.6	61
$\gamma\pi K_2$	YD632-1	215	255	38			2.2	32.	22	262	10.		35	3	13			1.7	6.4	8.9	1.3	0.4	7.2
$\gamma\pi K_2$	D1341-1	319	119	23			2.8	39	66	248	7.0		75	9	17			3.9	13	9.8	0.7	0.1	1.1
$\gamma\pi K_2$	YD445	103	1460	246			0.5	11.	32.	135	4.0		44.	6.8	10			4.9	17.	4.9	0.3	0.04	1.8
$\gamma\pi K_2$	YD633-1	222	107	35.			1.1	14.	15.	96	2.9		1.8	4	8.1			1.0	8.7	4.4	0.8	0.07	0.5
$\gamma\pi K_2$	D0377	159	990	155	17.0	2.69	0.94	13.8	34.6	174	6.3	15.9	57.6	4.08	9.09	2.81	3.68	1.67	7.29	8.79	2.17	0.14	0.9
$\gamma\pi K_2$	D0514	162	223	53.1	23.6	5.23	1.79	28.8	31.4	302	9.8	25.7	42.4	4.99	9.77	2.67	1.59	5.98	1.04	11.8	1.18	0.06	0.5
$\gamma\pi K_2$	D0523	224	926	109	21.0	4.45	1.26	17.2	23.5	199	7.3	21.6	63.1	8.46	27.8	7.91	2.65	9.10	12.3	9.07	0.45	0.07	0.4
$\gamma\pi K_2$	D5028-2	232	154	111	26.6	5.56	1.08	14.7	2.2	71.6	3.8	17.0	18.7	4.06	8.03	1.12	3.70	2.10	9.52	2.23	0.06	0.05	0.4
$\gamma\pi K_2$	D1339	222	665	134	23.8	4.12	1.12	16.8	22.1	320	10.3	21.7	55.8	3.76	20.2	2.5	4.36	3.61	27.0	6.04	0.55	0.05	0.6
$\gamma\pi K_2$	D0362-1	230	300	83.6	24.3	4.83	2.42	32.0	7.1	75.3	3.7	18.7	12.5	3.23	6.64	1.53	1.73	1.01	4.44	4.99	0.17	0.23	1.1
$\gamma\pi K_2$	XⅪ7-2	130	1100	537			0.8	11	23	146	4.7		61	4.4	28			5.1	31.	6.8	0.4	0.2	3
$\gamma\pi K_2$	D6002-3	305	64.5	40.3	22.7	16	2.97	14.7	33.5	119	5.4	20.3	12.6	1.16	1.51			0.62	6.18	4.23	1.75	0.5	0.015
	平均	206.1	530.8	128.8	22.7	5.6	1.6	20.7	25.7	173.8	6.1	20.1	39.4	4.8	12.9	3.1	3.0	3.4	12.0	6.8	0.8	0.2	6.0
火烧岭	ⅩⅪ7-2	130	1100	537			0.8	11	23	146	4.7		61	4.4	28			5.1	31	6.8	0.4	0.2	3
龙楼	D2042	173	1640	241	15.5	4.66	1.56	13.9	43.9	89.1	3.8	17.1	39.6	4.26	19.1			2.16	38.9	6.29	0.08	0.5	0.04
龙楼	PM03-11	216	35.7	26.3	25.9	7.66	2.91	19.5	43.5	67	2.9	22	19.6	0.56	2.54			0.43	33.5	1.27	0.25	0.6	0.11
龙楼	PM03-27	208	1328	205	18.1	7.31	1.87	16.1	41	88.9	4.1	17.5	43.4	3.93	20.5			2.48	66.2	2.18	0.27	0.5	0.04
	平均	199	1001	157	20	6.54	2.11	16.50	42.80	81.67	3.60	18.87	34.20	2.92	14.05			1.69	46.20	3.25	0.20	0.53	0.06
大小岭	D6112-1	230	38.3	59.2	28.9	8.40	0.94	12.2	38.0	71.6	2.9	16.6	16.1	0.87	5.05	1.4	2.04	0.59	19.0	7.11	4.56	0.071	0.8
大小岭	D3126-1	223	187	157.0	18.2	7.8	1.10	13.6	40.9	78	3.3	13.3	26.6	18.1	11.8	40.2	2.1	1.4	12.9	15.9			0.4
大小岭	GXⅧ10-1	218	230.0	121.0	*	*	1.20	14.5	37.0	71	3.6	*	23.0	5.0	6.8	*	*	1.4	16.7	6.0			10.8
	平均	223.67	151.77	112.40	23.55	8.10	1.08	13.43	38.63	73.53	3.27	14.95	21.90	7.99	7.88	20.80	2.07	1.13	16.20	9.67	4.56	0.07	4.00
吊罗山	D0495	207	3744	319	18.3	3.99	1.29	25.7	27.2	924	24.6	21.1	98.8	5.36	14.9	2.32	6.35	3.20	23.6	7.21	1.94	0.05	0.4
吊罗山	D0496	198	3800	324	12.8	3.21	1.27	26.1	28.5	903	24.0	21.3	101	5.18	15.2	3.21	6.15	3.23	23.8	7.04	2.13	0.05	0.6

续表

岩石单位	样品号	Rb	Ba	Sr	Th	U	Ta	Nb	Pb	Zr	Hf	Ga	Zn	Ni	V	Cr	Cs	Co	Li	Cu	Mo	Ag	Au
吊罗山	D0503	154	848	384	18.5	4.49	0.98	14.1	23.6	190	6.7	19.4	76.7	6.80	44.4	11.5	2.78	6.93	19.9	4.36	0.45	0.12	0.5
吊罗山	D0525	236	2554	360	13.3	3.71	1.59	29.5	34.2	655	19.3	23.6	115	4.97	22.0	3.09	5.45	4.39	13.1	9.01	0.52	0.05	0.4
吊罗山	D1386-1	115	456	384	10.8	2.95	0.63	8.97	26.5	79	3.2	16.5	39.5	3.01	26.3	2.3	3.10	3.88	36.1	2.89	0.18	0.03	0.5
吊罗山	D1902-1	159	231	165	13.0	6.75	0.74	13.2	35.0	74.8	3.2	16.3	32.7	2.29	14.0	<1	4.54	2.32	37.3	2.21	0.086	0.05	0.4
吊罗山	D1952-1	139	726	394	14.5	9.53	0.74	10.6	28.8	102	4.5	16.7	50.5	3.71	31.5	3.3	2.43	4.50	42.5	7.13	0.18	0.06	0.5
吊罗山	D2821-1	181	482	168	21.5	4.49	0.90	16.9	21.9	200	6.8	19.0	56.7	3.58	19.3	1.4	4.28	3.27	26.4	3.88	0.32	0.03	0.5
吊罗山	D2950-1	175	412	284	25.6	4.06	1.46	19.6	28.7	146	5.3	17.3	41.4	2.21	18.7	2.4	5.53	2.33	34.6	3.32	0.18	0.031	0.41
吊罗山	D5207-1	128	382	310	8.86	2.4	1.18	11	29.7	100	3.1	18.7	40.9	1.16	20.4	5.5	1.9	2.71	32.0	1.68	0.58	0.08	0.5
吊罗山	PM1-26-1	186	564	170	24.1	3.51	1.20	16.7	21.3	221	7	20.9	66.8	0.43	17.8	2.6	3.41	3.22	32.5	1.56	2.45	0.07	0.4
吊罗山	PM1-30-1	144	458	151	17.1	2.64	0.81	14.6	21.3	206	6.4	18.9	56.2	0.53	15	4.5	2.56	2.72	34.0	1.86	0.38	0.11	0.4
吊罗山	D3124-1	199	139.1	145.2	17.3	6.9	1.05	12.4	35.3	68.5	2.9	12.9	24.5	4.4	10.1	10.5	2.1	1.4	199	4.9			3.0
吊罗山	D3125-1	213	204.7	159.6	19.4	7.1	1.12	15.0	39.9	78.3	3.4	14.4	28.0	4.4	12.3	10.2	2.5	1.8	213	9.6			1.2
吊罗山	S006	137	467.2	332.7	15.1	3.1	0.76	9.4	23.3	123	4.0	15.4	48.0	3.8	28.1	7.5	2.5	3.8	137	*			*
吊罗山	QV13-2	198	253.0	71.0	*	*	0.50	12.3	19.5	140	3.8	*	23.7	0.3	13.3	*	*	1.7	198	2.0			1.2
吊罗山	QV8-1	292	452.0	102.0	*	*	0.80	14.3	37.6	195	6.2	*	73.0	5.2	17.0	*	*	5.4	292	23.5			2.0
	平均	187	805.8	215.2	19.1	4.84	1.20	18.4	29.1		7.57	19.31	53.1	2.92	17.1	4.15	3.55	3.26	68.2	12.09			0.76
新村	D5269-2	233	32.2	18.1	35.5	7.59	1.21	15.1	40.2	134	4.9	24.1	55.7	0.46	3.91	2.6	2.05	0.36	10.6	5.99	1.03	0.12	0.5
新村	D5325-1	193	50.9	40.1	30.2	15	1.50	22.3	37.2	72	3	18.7	13.9	0.62	5.96	2.8	1.63	0.59	32.6	3.56	0.32	0.10	0.5
新村	D5392-1	282	525	219	42.2	8.88	1.84	20.6	29.1	295	10.3	21.0	25.7	3.06	17.0	1.3	4.49	1.89	28.8	9.17	1.54	0.030	0.49
新村	D3128-1	136	581.7	406.3	17.8	3.3	0.98	13.1	31.0	141	4.6	17.8	45.2	10.7	33.4	26.1	1.8	4.0	33.6	9.5			0.5
新村	S007	116	445.0	355.0	11.3	4.3	0.77	8.8	25.2	97	3.5	15.2	37.0	3.3	23.9	6.5	2.5	3.5	31.0	*			*
新村	WH12-7	*	310.0	130.0	*	*	*	*	*	83	3.4	*	*	6.6	38.7	279.0	*	*	*	*			*
新村	WH2-4a	*	517.0	326.0	*	*	*	*	*	234	6.9	*	*	6.0	39.1	142.0	*	*	*	*			*
	平均	192	351	213.5	27.4	7.81	1.26	15.98	32.54	150.86	5.23	19.36	35.50	4.39	23.14	65.76	2.49	2.07	27.32	7.06	0.96	0.08	0.50

续表

岩石单位	样品号	Rb	Ba	Sr	Th	U	Ta	Nb	Pb	Zr	Hf	Ga	Zn	Ni	V	Cr	Cs	Co	Li	Cu	Mo	Ag	Au
南牙村	3862-2	275	530	80			1.2	12.6	39	130	3.5		45	6	25	1		4.1	18.4	11	0.86	38	0.2
高通岭	YV28-1	161	841	404			0.6	7.9	27.	82	2.3		19.	5.1	15.			3.6	6.6	27.	2.4	0.1	0.3
双顶岭	YIV7-1	120	265	186			0.8	12.	29.	91	3.5		5.9	2.7	15.			2.7	39.	5.7	1.0	0.02	0.3
六弓	YIV12-1	145	623	395			0.4	11.	22.	137	3.7		53.	5.8	32.			5.5	9.9	1.8	0.3	0.01	0.5
六弓	I HG3-1	205	698	146	55	11	2.1	28	70	219	6.6	14	47	4	29	15		2.6	75	5.8	0.5	0.04	1.9
六弓	D0914-1	139	620	491	17.8	6.02	1.01	11.6	16.9	171	6.3	20.9	53.4	6.89	41.4	26.6	7.87	6.00	41.4	3.93	0.37	0.05	0.3
六弓	D0972-1	162	407	390	20.4	6.18	0.89	10.4	20.2	117	4.8	20.3	42.7	4.79	23.4	7.45	5.03	3.54	44.9	2.95	0.22	0.03	0.4
六弓	D1412-1	143	767	541	16.7	3.87	0.66	9.46	17.8	152	5.6	20.0	53.3	9.47	49.6	33.1	6.51	6.92	27.6	5.34	0.38	0.04	0.4
六弓	D1430-1	209	1113	405	23.3	5.37	0.75	10.2	23.8	107	3.7	20.2	48.7	4.08	18.0	6.49	4.56	2.57	20.7	7.22	0.50	0.04	0.4
六弓	D1823-1	137	780	511	20.9	3.21	0.71	9.60	22.9	137	5.4	19.4	46.6	6.84	32.1	16.2	4.70	4.44	35.5	3.92	0.34	0.03	0.4
六弓	D3119-1	143	782.0	409.4	19.5	5.0	0.86	10.2	21.9	115	3.7	17.8	32.3	8.9	30.5	24.3	3.4	4.0	37.5	8.6			1.1
六弓	QV10-1	131	294.0	211.0	*	*	0.33	11.6	26.1	163	5.3	*	48.5	1.5	34.0	*	*	4.6	28.1	2.7			1.5
	平均	157.11	676.00	388.82	24.80	5.81	0.86	12.45	26.84	146.44	5.01	18.94	47.28	5.81	32.22	18.45	5.35	4.46	35.62	4.70	0.37	0.03	0.77
保亭	YIV19-1	104	921	683			0.2	9.7	19.	149	3.5		54.	10.	45.			8.1	23.	2.7	0.2	0.02	0.9
保亭	X XI6-1	117	704	463			0.3	9.5	20.	103	3.5		59.	4.4	39.			5.2	38.	7.3	0.4	0.18	2.0
保亭	I HG8-1	191	980	392	31	11	3.6	37	43	269	6.0	20	62	4.2	25	8		3.4	16	13	0.8	1.1	0.03
保亭	D0919-1	133	778	616	17.1	3.26	0.76	8.9	16.3	163	5.7	20.2	51.5	10.2	56.6	38.9	4.12	7.51	29.0	3.33	0.40	0.049	0.5
保亭	D1229-1	144	900	601	18.5	3.41	0.74	9.0	15.2	160	5.9	18.7	58.3	9.31	52.6	39.7	5.25	6.56	24.0	5.23	0.45	0.062	0.4
保亭	D1421-1	142	742	561	17.8	4.14	0.58	8.3	19.3	157	5.6	18.4	51.0	7.65	43.6	30.6	5.56	5.69	25.0	2.41	0.38	0.056	0.4
保亭	D3117-1	150	666.0	495.8	20.4	5.4	0.84	9.1	22.9	121	4.0	18.2	43.7	19.2	39.7	59.0	5.3	5.4	38.6	7.6			5.2
保亭	D3060-1	180	488.8	348.9	29.1	6.8	1.07	10.6	16.7	149	4.0	18.4	40.4	139	60.4	236.1	4.2	8.4	16.8	12.5			*
保亭	QV4-1	170.	640.0	370.0	*	*	0.76	10.7	17.5	133	4.8	*	37.4	10.6	67.3	*	*	7.9	34.0	2.8			1.6
保亭	GXVIII9-1	174	470.0	370.0	*	*	0.80	11.5	38.0	122	4.5	*	172	21.0	59.0	*	*	17.0	27.9	9.0			0.8
	平均	150.50	728.98	490.07	22.32	5.67	0.97	12.43	22.79	152.60	4.75	18.98	62.93	23.56	48.82	68.72	4.89	7.52	27.23	6.59	0.44	0.24	1.31

续表

岩石单位	样品号	Rb	Ba	Sr	Th	U	Ta	Nb	Pb	Zr	Hf	Ga	Zn	Ni	V	Cr	Cs	Co	Li	Cu	Mo	Ag	Au
加茂	YD599-1	136	703	435	14	2. 6	1. 1	17	19	194	4. 2	17	89	29	121	53		19	24	26	1. 5	0. 1	1. 3
加茂	YD612-1	124	683	443	12	2. 2	0. 7	18	22	192	4. 3	16	88	28	122	58		18	22	21	0. 9	0. 03	0. 9
加茂	D102-1	93	976	790	4. 5	1. 1	0. 5	11	8. 6	157	3. 5	23	68	11	78	14		13	7. 5	118	0. 3	0. 35	1. 4
加茂	D0355	122	711	523	12. 9	3. 61	0. 82	12. 4	18. 8	187	5. 8	20. 9	80. 1	20. 8	126	44. 6	5. 41	17. 2	22. 4	21. 8	1. 90	0. 13	0. 9
加茂	D0478	122	672	503	10. 7	2. 20	0. 76	11. 7	18. 4	186	6. 3	19. 9	66. 5	12. 6	98. 6	29. 1	4. 61	12. 4	25. 2	19. 1	0. 32	0. 06	0. 6
加茂	D0830	116	580	550	14. 3	3. 78	0. 88	12. 7	17. 1	148	4. 9	21. 4	75. 0	15. 6	118	38. 0	4. 89	15. 7	26. 9	24. 6	2. 22	0. 06	0. 7
加茂	D0893	120	584	500	17. 4	4. 88	0. 57	10. 5	16. 1	187	6. 1	18. 3	59. 5	16. 8	93. 3	72. 5	3. 35	12. 3	14. 7	18. 7	1. 83	0. 06	0. 5
加茂	D0898	146	743	568	21. 9	3. 86	0. 78	10. 7	14. 6	173	6. 2	21. 1	60. 9	13. 2	69. 6	56. 6	4. 26	9. 34	30. 7	7. 52	0. 48	0. 03	0. 4
加茂	D1248	112	841	762	5. 36	0. 93	0. 66	12. 3	21. 0	224	7. 4	20. 0	70. 1	17. 7	125	51. 6	2. 13	15. 0	26. 1	7. 97	0. 55	0. 05	0. 4
加茂	D1914-1	74. 1	616	636	6. 27	1. 24	0. 52	10. 0	11. 1	162	4. 6	20. 7	86. 6	10. 4	128	26. 4	2. 57	17. 1	16. 3	17. 8	0. 22	0. 04	0. 7
加茂	PM2-127-1	135	732	551	14. 3	2. 74	1. 10	8. 5	15. 3	167	5. 8	20. 3	55	8. 62	53. 7	36. 3	4. 33	7. 25	27. 8	4. 42	0. 52	0. 08	0. 5
加茂	D3120-1	145	729. 3	601. 6	15. 0	3. 4	0. 83	10. 5	19. 3	164	5. 0	19. 9	55. 1	20. 6	70. 6	76. 7	4. 2	9. 1	33. 6	9. 3			0. 7
加茂	QV13-1	188	265. 0	300. 0	*	*	0. 37	15. 4	21. 9	227	6. 4	*	74. 3	4. 3	67. 8	*	*	5. 7	29. 8	3. 3			0. 8
加茂	QV14-1	154	424. 0	433. 0	*	*	0. 37	14. 2	34. 9	237	6. 2	*	123	10. 1	53. 9	*	*	14. 8	43. 8	35. 7			2. 5
加茂	GXVIII3-1	154	500. 0	380. 0	*	*	0. 80	12. 0	24. 0	140	5. 1	*	50. 0	12. 0	78. 0	*	*	10. 0	13. 1	5. 0			0. 2
	平均	129. 41	650. 6	531. 71	12. 39	2. 71	0. 72	12. 46	18. 81	183. 00	5. 45	19. 88	73. 41	15. 38	93. 57	46. 40	3. 97	13. 06	24. 26	22. 68	0. 98	0. 09	0. 83
乌石	YV2-1	3	220	529			0. 4	8. 3	13.	52	1. 8		83.	87.	197			35.	12.	20.	0. 1	0. 03	1. 9
乌石	D1321-1	24. 4	436	683	2. 32	0. 59	0. 81	17. 9	20. 2	157	3. 8	21. 9	108	32. 4	296	99. 0	1. 19	41. 0	7. 34	26. 5	0. 31	0. 09	0. 4
番通	D776-1	130. 7	509. 5	74. 3	10. 4	2. 2	0. 42	6. 1	16. 7	481. 7	10. 5	15. 9	44. 0	2. 1	5. 0	4. 1	4. 5	2. 1	8. 0	*			*
番通	D3037-1	118. 2	328. 4	76. 7	10. 6	2. 4	0. 47	7. 1	38. 5	334. 9	8. 3	16. 2	41. 9	4. 1	4. 3	7. 6	1. 6	1. 4	12. 4	10. 4			1. 6
阜石斗	S017	160. 2	255. 1	161. 0	24. 9	3. 4	1. 13	10. 0	17. 7	96. 9	3. 6	13. 7	27. 0	4. 1	11. 8	7. 1	2. 3	1. 2	2. 1	35. 0	*		*
阜石斗	GIII4-2	163. 0	375. 5	120. 0	*	*	1. 28	14. 5	19. 4	140. 0	4. 9	*	25. 7	1. 9	11. 4	*	*	*	2. 3	10. 5	2. 6		0. 5
阜石斗	GIII5-2	173. 0	240. 8	127. 0	*	*	1. 71	17. 8	21. 5	110. 0	4. 9	*	26. 1	1. 0	9. 6	*	*	*	1. 7	16. 2	2. 6		1. 0
阜石斗	Q155-1	233. 0	89. 0	22. 0	*	*	1. 60	29. 1	28. 5	87. 0	4. 2	*	21. 0	2. 5	5. 0	*	*	*	1. 8	7. 6	24. 2		2. 3
阜石斗	G1977-1	340. 0	18. 0	22. 0	*	*	2. 00	19. 6	45. 0	90. 0	4. 3	*	21. 0	4. 3	7. 7	*	*	*	1. 0	6. 8	3. 4		0. 3
	平均	213. 84	195. 6	90. 40	24. 90	3. 40	1. 54	18. 20	26. 42	104. 78	4. 38	13. 70	24. 16	2. 76	9. 10	7. 10	2. 30	1. 20	1. 78	15. 22	8. 20		1. 03

续表

岩石单位	样品号	Rb	Ba	Sr	Th	U	Ta	Nb	Pb	Zr	Hf	Ga	Zn	Ni	V	Cr	Cs	Co	Li	Cu	Mo	Ag	Au
抱跃	D3056-1	131.8	650.6	289.8	16.2	3.1	0.88	11.2	20.4	189.7	5.9	19.1	55.8	7.2	23.5	16.7	3.3	9.3	4.2	26.3	10.0		0.4
抱跃	QXII1-1	161.0	1137	227.0	*	*	0.30	15.5	37.6	223.0	6.5	*	184.0	5.0	31.6	*	*	*	6.2	36.1	30.8		3.1
抱跃	GIII9-1	199.0	243.5	178.0	*	*	0.83	16.0	16.9	96.7	4.9	*	35.3	2.6	29.8	*	*	*	4.6	21.9	3.9		1.3
抱跃	GIII11-2	199.0	290.8	178.0	*	*	0.97	29.3	23.0	158.7	5.9	*	29.6	1.3	23.1	*	*	*	3.7	31.4	5.4		0.8
税町	D3055-1	158.3	613.0	256.5	16.7	4.9	1.00	11.0	27.0	179.9	5.7	19.3	56.1	5.5	29.0	12.1	5.8	10.6	4.7	36.5	7.3		1.1
税町	GVIII7-2	135.0	641.0	368.0	*	*	0.59	10.0	24.2	141.0	8.1	*	11.8	2.6	26.9	*	*	*	6.1	19.4	0.3		0.5
税町	QXII4-1	159.0	580.0	385.0	*	*	2.03	12.8	21.6	159.0	4.9	*	58.8	6.0	44.2	*	*	*	7.5	30.7	38.1		0.8
税町	GII15-1	185.0	384.0	197.0	*	*	1.67	15.4	17.2	124.6	6.6	*	47.0	5.6	45.7	*	*	*	7.0	28.6	3.7		1.1
税町	GIII11-3	211.0	343.6	350.0	*	*	1.46	22.4	25.6	172.0	7.5	*	64.0	4.3	78.6	*	*	*	9.8	34.7	14.5		1.3
税町	GXII1-1	299.0	93.0	36.0	*	*	1.60	25.5	32.0	49.0	1.2	*	26.0	2.0	3.4	*	*	*	3.0	4.5	7.0		0.0
	平均	191.22	442.4	265.42	16.70	4.90	1.39	16.18	24.60	137.58	5.67	19.30	43.95	4.33	37.97	12.10	5.80	10.60	6.35	25.73	11.82		0.80
	GI2-1	229.0	*	134.0	*	*	1.48	15.2	33.0	*	*	15.9	120.4	4.0	*	8.4	*	*	*	6.6			3.3
	G21-1	*	*	182.0	*	*	*	11.3	21.8	*	*	*	34.3	1.0	*	5.0	*	*	*	7.2			1.1
	GI6-1	213.0	*	100.0	*	*	1.99	6.8	25.2	*	*	17.5	21.2	2.1	*	5.0	*	*	*	5.9			1.3
	GI12-2	140.0	*	105.0	*	*	1.42	5.8	21.4	*	*	17.0	33.9	1.0	*	5.0	*	*	*	39.7			4.9
	GII8-1	189.0	*	148.0	*	*	1.83	31.4	16.0	*	*	*	26.7	1.5	*	*	*	*	*	44.0			1.5
	GIII3-1	136.0	*	386.0	*	*	1.30	21.7	21.0	*	*	*	45.0	2.4	*	*	*	*	*	4.8			0.5
	G1872	170.0	*	265.0	*	*	1.00	11.8	34.0	*	*	*	76.0	4.0	*	*	*	*	*	6.0			38.6
半岛	Ha31	257	363	131	27.9	10.3	2.87	15.8	30.8	169	4.84	19	58	3.61	22.8	25.5	8.29	3.31	40.5	11.8			
半岛	7331-1	149	690	173			1.2	15.6	25	235	5.6		38	12	28	1		2.8	7.3	8	2.7	30	
	平均	203.0	526.5	152.00	27.90	10.30	2.04	15.70	27.90	202.00	5.22	19.00	48.00	7.81	25.40	13.25	8.29	3.06	23.90	9.90	2.70	30.00	
马岭	177	413	53	36			2	42	45	82	2.8		12	4	11			0.9	41.9	7	1.7	40	0.3
马岭	X X1-1	365	17	33			6.76	22.95	42.45	78	0		41.55	2.4	3.85			0.85	80.05	24.65	0.55	490	1.3
马岭	X Ⅶ1-1	415	17	18			4.51	19.8	63.4	72	4.01		32	1.66	5.8			1.43	245	3.08	0.15	50	0.5
	平均	397.67	29.00	29.0			4.42	28.25	50.28	77.33	2.27		28.52	2.69	6.88			1.06	122.32	11.58	0.80	193.33	0.70

续表

岩石单位	样品号	Rb	Ba	Sr	Th	U	Ta	Nb	Pb	Zr	Hf	Ga	Zn	Ni	V	Cr	Cs	Co	Li	Cu	Mo	Ag	Au
马鞍岭	Ha13-1	344	22.4	26.5	28.4	40	0.5	8.2	56.3	41.7	2.25	17	35.9	3.57	1	7.18	26.2	1.01	92.6	21.7			
马鞍岭	Ha13-2	325	18.4	25.5	24.7	32.2	4.98	19.5	58.1	49.1	2.31	16.6	26.4	2.22	1	1	18.9	1	48.7	9.07			
马鞍岭	184	280	780	127			1.6	13.7	39	280	7.2		49	3	32			4.2	40.5	4	2.8	30	0
马鞍岭	187	284	490	72			1.6	15.6	37	190	4.3		41	13	21	4		3.6	39.9	10	1.08	20	0.3
马鞍岭	YXV3-1	189	390	183			1.5	16.	23.	219	6.3		45.	3.0	15.			3.1	18.	7.2	1.5	0.02	0.9
	平均	284.40	340.1	86.80	26.55	36.10	2.04	14.60	42.68	155.96	4.47	16.80	39.46	4.96	14.00	4.06	22.55	2.58	47.94	10.39	1.79	16.67	0.40
三狮岭	Ⅱ28-1	207	67	24			1.25	10.5	14.9	123			35	2.85	3.55			0.9	10.05	4.8	0.38	270	0.85
三狮岭	YI1-2	140	391	106			0.5		18.	217			62.	4.7	13.			4.4	38.	4.8	0.6	0.06	3.0
三狮岭	YD438	194	763	405			2.1	12.	33.	726	20.		47.	6.7	16.			4.6	8.3	4.3	1.3	0.04	6.7
三狮岭	D1740-2	257	371	172			1.5	22	32	133	4.2		41	6	21	8.4		4.9	69	4.1	0.9	0.0	0.2
三狮岭	HN03	214	398	225			1.4	12	41	88	3.3				15			1.8					
三狮岭	D958-1	244	1111	196			0.5	19	31	285	5.9		57	3.7	26	5		6.2	32	5.0	3.7	0.0	3.0
三狮岭	HN05	217	773	208			0.8	11	28	326	7.5				16			2.6					
三狮岭	D0399-1	258	66.3	59.6	35	10.5	1.57	16.8	21.7	87	2.6	15.2	22.7	0.54	2.42	2.4	2.72	0.52	18.6	0.67	0.34	0.056	0.3
三狮岭	D1004-1	241	62.4	44.2	17.7	4.41	1.89	14.3	27.5	69	2.4	14.4	26.9	1.08	1.05	2.3	3.69	0.80	10.6	1.38	0.21	0.054	0.3
三狮岭	D1213-1	215	349	248	17.0	3.94	0.87	13.3	16.5	97	3.1	14.2	24.5	0.88	9.30	1.3	1.32	1.42	6.21	2.06	0.34	0.040	0.3
三狮岭	D1036	167	182.9	73.3	15.3	4.9	0.70	6.9	23.6	147	4.7	17.0	36.0	2.2	2.2	4.8	1.8	1.2	6.0	*			*
	平均	214.00	412.2	160.10	21.25	5.94	1.19	13.78	26.11	208.91	5.97	15.20	39.12	3.18	11.41	4.03	2.38	2.67	22.08	3.39	0.97	33.78	1.83
架 水	D3361-1	148	1375	488			1.6	23.	35.	547	13.		123	10.	71.			12.	34.	12.	0.9		8.2
椰子园	D3032-1	238.5	553.8	324.2	24.7	2.5	1.44	19.5	21.5	224.0	6.8	18.9	26.5	15.4	30.0	21.1	3.3	3.6	43.5	9.8			1.3
南山岭	S018	168.1	754.2	502.8	17.3	5.0	1.36	19.0	20.9	346.0	9.0	16.9	58.0	5.6	54.0	9.3	3.7	6.3	18.0				
马翁岭	Ha37-2	68.4	468	1010	4.23	1.49	0.5	18.3	9.19	152	4.52	23.1	87.8	8.68	339	32.2	1.42	34.4	12.6	65.5			
马翁岭	516	102.5	481.9	417.6	13.19	2.62	0.89	11.25	14.31	203.5	5.98	20.65	91.71	14.14	157.6	29.41	4.25	20.4		27.83			
马翁岭	YXVIII6-2	81	413	620			1.1	17	22	178	4.6		100	23	187			27	31	25	1.3		2.3

续表

岩石单位	样品号	Rb	Ba	Sr	Th	U	Ta	Nb	Pb	Zr	Hf	Ga	Zn	Ni	V	Cr	Cs	Co	Li	Cu	Mo	Ag	Au
马翁岭	D1153-1	90.7	368.3	701.6	5.3	1.5	0.55	12.6	7.1	106	3.1	15.2	67.4	39.3	200.	79.7	2.0	28.4	13.6	59.6			0.2
马翁岭	D1670-1	57.7	247.6	448.0	4.4	1.6	0.72	9.4	6.1	79	2.5	11.4	59.6	105.	360.	392.9	2.1	35.0	22.6	61.2			0.4
马翁岭	D1535-1	66.1	248.0	393.4	4.2	1.6	0.46	6.3	4.8	64	2.2	11.2	57.5	122.	253.	484.1	1.4	38.2	27.5	59.6			0.2
	平均	77.73	371.1	598.43	6.26	1.76	0.70	12.48	10.58	130.42	3.82	16.31	77.34	52.02	249.43	203.6	2.23	30.57	21.46	49.79	1.30		
八宝坡	H1245-2	186.8	294.7	506.5	14.1	5.9	1.51	17.5	11.8	543.1	10.3	21.8	71.0	3.6	14.8	34.9	1.4	4.2	6.9	15.2			8.8
八宝坡	Hlm-2	241.1	229.7	339.8	33.4	9.0	2.25	22.9	13.6	397.0	8.3	19.8	72.0	2.4	10.4	4.0	1.6	3.3	11.0	*			*
黑栋	D1244	165.4	2559	606.7	11.2	2.6	0.53	10.6	21.3	395.2	9.4	18.5	71.7	7.4	33.7	16.0	5.1	6.9	43.6	12.1			0.4
里茂	H1247-2	60.7	486.8	960.6	1.8	0.6	0.51	6.1	5.2	71.4	2.0	19.4	89.6	33.4	338.3	86.8	1.2	36.8	10.4	47.4			0.3
里茂	Hlm-1	59.7	246.0	626.8	1.0	0.4	0.19	2.9	7.7	61.5	2.2	15.4	124.0	60.3	497.1	155.2	2.3	52.8	17.0	*			*
迈洲岭	974-1	473	800	35			2.0	16	18	85	2.8		20	7	26			1.1	20	20	6.7	0.3	0.8
打显村	X Ⅵ19-1	335	614	145			2.69	19.5	38.38	131	4.93		51.03	4.1	13.55			2.1	34.5	3.58	0.9	30	0.8
白石岭	X Ⅵ16-1	292	584	184			2.26	17.83	36.73	134	5.03		40.37	4.19	16.5			2.5	31.67	6.89	0.54	30	0.8
黑岭	Ha25	311	473	105	63.1	17.5	3.27	16.6	36	196	8.2	22.3	56.5	4.12	15.8	13.8	13.8	4.19	80	15.6			
黑岭	4086-2	230	595	112				9.7	40				21	5	155	10		5		22		55	2
	平均	270.50	534	108.5	63.1	17.5	3.27	13.15	38.00	196.00	8.20	22.30	38.75	4.56	85.40	11.90	13.80	4.60	80.00	18.80		55.00	2.00
分水岭	X Ⅶ10-1	275	276	122			2.65	15.63	41.68	116	3.76		48	2.04	17.6			3.26		3.83		60	0.83
分水岭	4063-1	330	360	91				12	54	166.4			45	16	16	52		6.5		155		77	1.2
分水岭	4086-7	300	250	70				10	46	161.8			42	4.6	13	25		5		16		36	0.8
	平均	301.6	295.3	94.33			2.65	12.54	47.23	148.07	3.76		45.00	7.55	15.53	38.50		4.92		58.28		57.67	0.94
南宝岭	4602-1	430	114	30				15.3	48				21	5	51	71		13		11		54	2
南宝岭	4676-2	270	366	88				7	41				50	5	20	86		4.5		6		47	2
	平均	350.0	240	59.0				11.1	44.5				35.5	5.0	35.5	78.5		8.75		8.5		50.50	2.00
五里桥	Ha22	332	550	78.2	55.8	11.8	2.27	18.4	37.8	228	8.53	23.6	70.4	2.72	5.52	31.6	18.2	2.49	84.2	9.53			
五里桥	4033-1	290	550	70				13	34	179			63	4.2	7.4	27		4.3		8.8		50	0.3

续表

岩石单位	样品号	Rb	Ba	Sr	Th	U	Ta	Nb	Pb	Zr	Hf	Ga	Zn	Ni	V	Cr	Cs	Co	Li	Cu	Mo	Ag	Au
五里桥	4077-1	260	168	33				11	35	165			59	6.2	2.1	53		3.1		42		59	0.2
五里桥	4080-3	310	545	75				12	45	182			68	4.8	6.2	40		4.1		9.7		52	0.1
五里桥	4082-1	290	395	62				10	159	163			62	3.7	3.8	10		3.6		10		320	0.4
五里桥	Ha27	340	386	60	48.7	12.7	2.16	17.4	40.6	181	6.38	22.3	68.4	2.98	2.4	38.1	15.2	1.73	92.2	10.2			
	平均	303.67	432.3	63.03	52.25	12.25	2.22	13.63	58.57	183.00	7.46	22.95	65.13	4.10	4.57	33.28	16.7	3.22	88.20	15.04		120.25	0.25
苗圃	4072-4	240	1100	135				11	35	161			69	4.1	7.3	41		6.6		10		61	0.4
苗圃	Ha21-1	302	930	110	45.8	8.33	3.09	17.4	35.8	235	8.14	27.8	75.4	3.3	23.9	15.5	13.2	4.9	22.1	28.1			
	平均	271.00	1015	122.50	45.80	8.33	3.09	14.20	35.40	198.00	8.14	27.80	72.20	3.70	15.60	28.25	13.2	5.75	22.10	19.05		61.00	0.40
乌烈岭	Ha14-2	316	287	33	49.7	10.6	2.68	24.8	68.7	304	8.57	22	83.4	3.53	2.64	38.1	17.8	1.58	41.9	12.3			
乌烈岭	Ha15	277	766	70.4	36	8.88	1.62	17.4	34.2	364	7.24	25.9	97.4	2.58	1	28.3	13.8	2.12	62.7	32.7			
乌烈岭	4001-1	280	300	52				11	32	173			53	22	10	51		4		180		74	0.1
乌烈岭	4001-7	290	710	105				11	65	183			56	4.4	10	26		4.6		9.1		60	0.1
乌烈岭	4044-4	250	2050	345				13	36	200			56	3.1	11	24		3.3		6.3		55	0.5
乌烈岭	4104-6	280	1740	245				14	95	180			63	3.1	11	12		3.7		8.9		145	1.5
乌烈岭	7323	279	460	70			1.6	17.1	54	99	3.2		26	4	17	17		7.6	61.2	9	1.02	6	0.2
	平均	281.71	901.8	131.49	42.85	9.74	1.97	15.47	54.99	214.71	6.34	23.95	62.11	6.10	8.95	28.06	15.8	3.84	55.27	36.90	1.02	68.00	0.48
东四	Ha14-1	226	840	163	32.9	6.3	2.04	21	29.5	276	7.53	25.3	77.5	4.3	40	63.9	6.06	6.44	51.6	25.8			
东四	Ha23	309	478	102	57.5	10.6	1.42	15.5	34.9	218	8.13	21.3	60.9	3.84	17.4	14.4	10.2	3.69	50.9	17.7			
东四	4091-2	280	438	70				11	30	151			62	5.4	16	47		5.2		11		33	0.6
东四	4125-2	240	380	97				10	57	151			46	6	18	28		5.4		36		75	0.3
东四	7314	294	438	113			1.2	21.1	46	260	7		30	5	16			2.4	7.4	8	1.8	20	4.2
	平均	269.8	514.8	109.00	45.20	8.45	1.55	15.72	39.48	211.20	7.55	23.30	55.28	4.91	21.48	38.33	8.13	4.63	36.63	19.70	1.80	42.67	1.70
七差岭	6041	252	370	187			1.2	13	43	70	2		32	13	20	7		6.3	98.1	11	1.25	10	0.1
七差岭	4964-1	260	761	340				12.6	46				63	5	55	31		5		7		66	1

续表

岩石单位	样品号	Rb	Ba	Sr	Th	U	Ta	Nb	Pb	Zr	Hf	Ga	Zn	Ni	V	Cr	Cs	Co	Li	Cu	Mo	Ag	Au
七差岭	4994-1	220	510	280				5.6	34				69	8	20	112		3		5		41	2
	平均	244.0	547	269.00			1.20	10.40	41.00	70.00	2.00		54.67	8.67	31.67	50.00		4.77	98.10	7.67	1.25	39.00	1.03
保梅岭	Ha30-1	193	498	159	18.5	9.1	0.82	11.7	25.1	140	5.17	20.2	57.7	3.86	31.3	11.5	9.7	3.84	52.3	10.7			
保梅岭	Ha30-6	250	363	134	23.3	8.2	0.5	10.9	26.8	123	4.45	19.7	50	4.06	18.4	27.5	8.3	3.12	53.6	14.4			
保梅岭	Ha33-1	226	162	217	14	5.86	1.31	14.4	17.2	119	4.29	18.8	118	18.7	162	67.6	25	17.4	129	22.2			
	平均	223	341	170.00	18.60	7.72	0.88	12.33	23.03	127.33	4.64	19.57	75.23	8.87	70.57	35.53	14.33	8.12	78.30	15.77			
洞古岭	YⅢ11-1	205	1140	539			0.5	19	28	294	7.1		258	12	63			7.9	57	273	2.4	4.4	48
南通岭	ⅤHG1-7	259	1251	254			1.9	21	27	479	10	9	69	3.3	14			0.6	19	8.4	0.5	0.04	
南通岭	HN01-2	109	931	373	11	2.5	1.2	22	21	595	12	20			43			6.5					
	平均	184	1091	313.50	11.00	2.50	1.55	21.50	24.00	537.00	11.00	14.50	69.00	3.30	28.50			3.55	19.00	8.40	0.50	0.04	
琼中县	D493-2	215	1378	495	38	4.5	2.7	39	40	398	9.4	20	93	8.1	39	6.7		8.8	52	8.6	16	0.1	0.7
琼中县	D823-1	150	421	200	20	12	2.3	11	105	53	4.2	17	53	5	36	12		3.8	80	4.3	0.3	0.06	1.6
	平均	182.5	899.5	347.50	29.00	8.25	2.50	25.00	72.50	225.50	6.80	18.50	73.00	6.55	37.50	9.35		6.30	66.00	6.45	8.15	0.08	1.15
黎母岭	D1601-1	309	690	231	28	10	2.1	11	28	218	5.3	19	39	5.7	26	6.9		6.1	53	4.5	0.8	0.02	0.3
黎母岭	D101-3	283	1204	237	23	2.2	1.5	15	22	370	8.2	18	77	5.8	22	5.3		7	27	6.2	2.2	0.05	0.3
黎母岭	ⅤHG1-1	219	1447	335	18	3.2	1.7	31	22	617	13	20	117	9.3	42	5.9		11	23	10	1.0	0.02	0.4
黎母岭	D1618-1	448	205	78	17	6	1.7	41	29	261	7.6	19	21	3.6	21	5		5	23	4.9	2	0.07	0.6
黎母岭	ⅢHG2-3	190	1233	639	28	7.6	2.4	30.3	47	253	6.1	20	104	8.1	42	8.1		8.9	52	9	2.2	0.13	0.6
黎母岭	HN01-1	149	1268	418	14	3	1.1	20	26	472	10	20			30			5.8					
黎母岭	HN02	237	986	258	32	2.3	0.8	16	28	379	8	22			20			3.9					
黎母岭	HN25	163	536	289	11	3	0.8	7.9	24	158	3.7	17			18			3.5					
黎母岭	HN26	163	270	273	31	11	1.6	13	26	221	5.6	21			30			4					
黎母岭	HN27	234	1084	364	35	3.7	1.3	23	46	455	12	24			43			4.4					
	平均	239.5	892.3	312.20	23.70	5.20	1.50	20.82	29.80	340.40	7.95	20.00	71.60	6.50	29.40	6.24		5.96	35.60	6.92	1.64	0.06	0.44

续表

岩石单位	样品号	Rb	Ba	Sr	Th	U	Ta	Nb	Pb	Zr	Hf	Ga	Zn	Ni	V	Cr	Cs	Co	Li	Cu	Mo	Ag	Au
后美	XI1-5	252	353	152				11.5	56.6	120	3.9		56	2.9	13.6			2	3.9	7.9	0.83	0.09	1.06
后龙	XI3-1	231	322	220				7.9	47.2	117	3.6		61	3.6	15.9			3.2	48.2	2.4	0.61	0.05	0.5
石墩	H1519-1	45.5	1928	944.8	3.3	0.9	0.74	11.8	18.6	396.2	8.3	23.5	90.7	7.1	77.6	16.6	1.7	11.2	9.3				
长安	D1653-1	25.6	172.4	264.7	1.6	0.3	0.19	3.8	3.9	64.2	2.1	9.3	81.2	166.5	182.2	1365	1.0	53.9	19.6				0.3
长安	H303-1	46.4	356.3	472.9	2.6	0.7	0.39	6.3	5.3	82.4	2.2	16.1	*	65.0	315.1	220.7	2.9	41.1	*				*
长安	S004-1	25.9	249.3	753.1	2.9	0.6	0.77	13.8	7.3	165.1	4.3	18.6	107.0	2.0	245.7	2.1	0.5	34.9	11.0				*
长安	247-1	16.78	77.1	198.7	2.6	0.67	0.17	1.85		40.8	1.37	9.18	48.26	54.7	244.9	499.5	1.35	49.77		65.09			
长安	247-4	73.08	92.6	211.4	2.27	0.85	0.11	1	4.54	37.7	1.49	9.52	37.42	26.1	218.4	243.4	4.43	30.91		30.04			
	平均	37.55	189.5	380.16	2.39	0.62	0.33	5.35	5.26	78.04	2.29	12.54	68.47	62.86	241.26	466.2	2.04	42.12	15.30	47.57			
石龟岭	D3042-1	353.0	1.9	2.2	22.0	7.8	3.08	35.2	38.0	89.3	6.8	16.8	29.7	3.6	1.2	9.2	4.8	0.7	53.1				0.9
兵工厂	H1219-2	482.6	13.2	2.9	29.3	4.2	4.28	43.4	21.5	136.1	9.6	28.9	30.3	4.6	1.1	34.9	7.8	0.8	181.0				0.5
兵工厂	H1562-1	377.6	20.7	4.1	30.1	4.7	3.59	41.6	29.6	134.9	9.3	27.6	28.1	2.7	1.3	4.3	10.3	2.3	74.0				1.2
兵工厂	D3044-1	373.5	15.7	13.0	24.1	8.2	3.70	44.1	32.0	94.7	6.8	15.3	26.0	5.1	1.2	10.4	4.0	0.7	11.2				1.0
	平均	411.23	16.53	6.67	27.83	5.70	3.86	43.03	27.70	121.90	8.57	23.93	28.13	4.13	1.20	16.53	7.37	1.27	88.73				0.90
六连岭	H1207-1	306.7	9.7	3.2	22.1	5.4	1.93	26.3	24.2	171.3	8.1	24.4	41.2	5.0	1.3	51.4	5.5	0.8	30.4				0.9
六连岭	H1207-4	315.5	6.0	3.6	29.9	6.4	2.46	28.5	25.5	156.9	8.6	25.2	27.8	1.4	1.0	5.7	8.4	0.7	51.6				*
六连岭	S001	282.6	2.6	1.7	23.5	5.3	2.30	27.4	25.5	165.8	8.3	21.6	54.0	2.9	0.7	4.7	4.8	0.7	32.0				*
六连岭	H3357-1	256.4	29.7	10.6	20.1	6.5	1.39	15.9	29.5	99.3	4.5	19.5	37.9	4.2	1.0	10.3	6.0	1.9	54.0				1.3
六连岭	S044	241.8	218.0	38.8	16.8	6.5	0.62	15.8	26.1	158.6	6.0	18.0	64.0	1.8	2.8	1.9	4.2	1.4	48.0				*
六连岭	S013-1	118.3	7.2	4.3	10.3	1.8	0.30	10.3	21.1	136.1	5.7	18.4	62.0	2.9	0.0	6.0	1.4	1.0	4.0				*
六连岭	D3043-1	266.7	3.9	7.6	25.9	7.7	1.88	28.6	35.5	93.6	5.4	15.4	33.1	3.6	0.7	8.5	3.5	0.5	15.2				*
	平均	255.43	39.59	9.97	21.23	5.66	1.55	21.83	26.77	140.23	6.66	20.36	45.71	3.11	1.07	12.64	4.83	1.00	33.60				
前锋	GII12-1	256.0	125.0	133.0	*	*	2.42	16.7	25.2	65.6	4.8	*	22.8	2.7	4.4	*	*	1.6	3.3				1.6
前锋	GII2-3	295.0	94.7	<100	*	*	3.66	26.1	32.7	76.9	3.7	*	12.9	1.9	2.5	*	*	2.4	1.0				3.0

续表

岩石单位	样品号	Rb	Ba	Sr	Th	U	Ta	Nb	Pb	Zr	Hf	Ga	Zn	Ni	V	Cr	Cs	Co	Li	Cu	Mo	Ag	Au
前锋	G851	309.0	110.8	<100	*	*	3.55	16.4	17.0	63.5	5.0	*	13.7	1.4	2.9	*	*	1.2	1.0				1.1
	平均	286.67	110.1	133.0			3.21	19.73	24.97	68.67	4.50		16.47	2.00	3.27			1.73	1.77				
结尾	G42	255.0	152.0	64.2	*	*	2.40	34.8	74.0	43.0	2.4	*	22.0	2.2	6.8	*	*	1.5	2.2				1.3
结尾	GXIV1-3	129.0	300.0	86.0	*	*	0.40	6.0	32.0	120.0	4.5	*	25.0	3.0	4.0	*	*	2.0	2.8				10.6
结尾	GII1-1	321.0	81.0	<100	*	*	2.45	24.1	28.0	84.6	3.5	*	12.6	1.0	1.9	*	*	1.1	3.3				0.3
结尾	GII3-4	284.0	199.3	100.0	*	*	2.33	22.8	28.0	118.8	5.6	*	19.4	1.9	5.8	*	*	2.0	5.9				1.0
	平均	247.25	183.0	83.40			1.90	21.93	40.50	91.60	4.00		19.75	2.03	4.63			1.65	3.55				
超盆	GIX9-1	251.0	160.0	<100	*	*	2.19	10.6	35.2	84.4	3.9	*	22.8	2.6	3.2	*	*	2.3	0.2				0.4
超盆	G1284-2	249.0	170.0	<100	*	*	1.29	9.1	37.0	106.0	3.8	*	28.6	1.5	2.3	*	*	1.8	1.3				0.5
超盆	GII6-1	245.0	210.7	107.0	*	*	1.81	28.8	22.2	172.0	4.0	*	23.5	1.3	7.3	*	*	2.3	9.7				0.5
超盆	G011-2	266.0	157.8	100.0	*	*	1.50	11.3	28.2	84.6	4.2	*	24.6	2.7	8.8	*	*	3.0	15.0				1.1
超盆	GV1-1	219.0	238.9	111.0	*	*	1.36	12.7	24.2	114.3	4.5	*	22.0	1.0	8.1	*	*	2.0	11.8				4.0
超盆	GV7-1	295.0	113.3	100.0	*	*	1.84	10.3	26.5	69.4	4.1	*	22.8	1.6	3.3	*	*	1.7	5.3				0.9
	平均	254.17	175.1	104.50			1.67	13.80	28.88	105.12	4.08		24.05	1.78	5.50			2.18	7.22				
布山村	D3034-1	129.9	461.6	252.0	14.7	2.9	0.95	18.0	32.1	213.5	5.3	17.6	53.3	7.1	23.8	19.1	2.2	4.0	15.0				3.1
布山村	GIX3-2	128.0	1444	520.0	*	*	1.12	13.4	30.4	505.0	11.5	*	115.0	3.0	44.2	*	*	7.2	23.2				0.4
	平均	128.95	952.8	386.00	14.70	2.90	1.04	15.70	31.25	359.25	8.40	17.60	84.15	5.05	34.00	19.10	2.20	5.60	19.10				1.75
金寮	S039	226.1	50.3	23.3	24.2	5.9	1.71	30.3	33.0	202.5	8.2	19.3	48.9	3.4	0.6	7.6	2.6	0.9	12.4				0.1
金寮	S040-1	216.1	67.4	20.8	30.6	5.2	1.44	27.7	30.5	329.4	11.4	19.8	79.0	1.6	0.4	3.2	5.8	0.6	30.0				*
金寮	S040-2	234.3	62.5	21.9	27.5	5.3	1.04	28.2	30.4	293.5	10.5	20.9	77.4	2.3	0.6	8.1	4.9	0.9	20.9				0.1
	平均	225.50	60.07	22.00	27.43	5.47	1.40	28.73	31.30	275.13	10.03	20.00	68.43	2.43	0.53	6.30	4.43	0.80	21.10				0.10
番阳峒	S027	208.4	936.4	191.3	39.6	5.2	1.33	19.1	33.6	392.6	10.8	17.3	62.0	1.8	4.6	3.8	4.9	2.2	29.0				
番阳峒	D467-1	278.4	785.4	164.0	41.2	6.7	1.65	22.3	46.4	306.2	9.0	18.0	70.9	3.5	10.9	7.8	6.2	2.0	51.8				0.4
番阳峒	S029	280.5	653.3	126.4	49.3	7.4	1.50	17.8	40.8	334.2	9.5	18.0	62.0	2.0	1.5	1.6	6.7	1.5	45.0				*

续表

岩石单位	样品号	Rb	Ba	Sr	Th	U	Ta	Nb	Pb	Zr	Hf	Ga	Zn	Ni	V	Cr	Cs	Co	Li	Cu	Mo	Ag	Au
番阳峒	D1447-1	193.9	939.1	285.6	21.2	5.7	0.73	12.0	24.5	247.1	6.5	22.7	59.9	3.8	13.2	9.5	5.0	3.2	53.5				*
番阳峒	S008-2	133.2	41.3	26.0	27.1	2.8	0.64	20.9	86.1	408.8	13.1	18.1	82.2	8.5	8.4	15.3	2.4	1.0	14.3				0.2
番阳峒	S013-2	128.6	1461	193.9	21.0	2.5	0.47	16.1	19.3	300.8	7.4	24.0	73.3	6.1	7.7	10.2	2.6	2.1	11.1				*
番阳峒	H1527-1	218.0	340.0	40.0	22.0	*	1.80	22.8	35.0	570.0	14.0	17.0	110.0	4.0	3.0	*	*	2.8	17.4				2.3
番阳峒	D33-1	227.1	322.7	119.6	32.4	4.5	1.39	27.7	38.5	294.5	9.8	22.8	71.5	4.2	16.0	10.1	4.5	2.7	41.7				0.3
番阳峒	D35-1	159.7	65.5	24.6	30.8	3.1	0.75	20.1	26.6	328.2	10.2	24.7	55.9	6.4	3.3	10.0	3.2	0.8	13.3				0.3
番阳峒	BD10	153.2	858.8	113.9	10.4	2.1	0.91	24.8	35.8	704.9	15.5	29.6	135.8	9.2	2.2	22.2	4.2	0.8	27.9				*
番阳峒	D121-1	197.8	424.7	107.8	15.4	5.6	1.24	15.9	28.5	192.7	6.6	18.8	48.9	12.0	9.3	18.5	4.3	1.7	47.1				0.6
番阳峒	H306	125.4	426.7	95.5	9.6	1.6	0.53	8.2	19.9	161.7	5.0	18.7	42.3	2.0	5.3	4.9	1.6	2.4	8.6				*
番阳峒	S043-1	174.1	491.7	151.1	33.3	8.0	0.83	17.2	24.9	311.6	9.4	21.6	59.9	3.1	8.4	8.6	4.7	2.2	26.1				0.8
番阳峒	D0472	139	269	74.6	11.4	1.04	0.76	19.7	33.3	692	18.2	23.3	96.2	2.35	7.29	2.17	2.44	1.09	6.26	7.44	0.66	0.12	0.6
番阳峒	D0476	150	496	89.0	11.6	1.54	1.13	27.1	30.8	886	22.5	25.9	141	3.55	9.02	3.73	6.00	1.22	13.9	8.66	0.49	0.07	0.4
番阳峒	D0825	160	486	45.2	14.1	1.50	0.75	16.3	35.9	583	16.0	24.0	97.1	3.40	7.70	2.76	3.36	1.17	5.41	7.29	0.23	0.07	0.5
番阳峒	PM1-69-1	180	501	134	28.7	3.11	1.34	30.3	24.7	407	11.7	22.2	79.1	0.1	22.5	3.8	2.58	3.44	28.8	2.83	1.23	0.08	0.5
番阳峒	PM1-107-2	178	765	448	16.8	3.58	1.01	9.2	20.3	136	5.8	20.2	45.4	3.21	28.9	10.7	5.24	3.97	39.1	3.39	0.85	0.10	0.4
番阳峒	PM1-117-1	125	69.2	46.6	32.8	3.68	1.42	15.7	22.9	568	15.6	22.5	60.9	0.1	4.91	2.4	4.69	0.77	8.64	9.89	2.06	0.10	1.3
	平均	179.49	543.8	130.37	24.67	3.87	1.06	19.12	33.04	411.86	11.40	21.55	76.54	4.17	9.16	8.23	4.15	1.95	25.73	6.58	0.92	0.09	0.66
麻山田	H1498-1	136.1	731.6	213.5	7.1	3.6	0.67	11.7	20.7	537.0	13.4	20.7	61.8	7.2	4.6	14.7	2.5	2.2	10.9				
麻山田	BD2	162.3	3315	392.8	10.7	1.7	1.14	23.5	30.3	572.1	14.0	28.5	80.4	5.5	11.0	14.0	2.6	3.6	9.6				
麻山田	S012-1	118.0	2349	278.1	7.0	2.5	0.93	21.4	23.0	947.4	20.9	22.0	144.0	2.2	5.6	5.8	2.5	2.4	5.0				
麻山田	D182-1	137.7	4470	372.0	15.8	3.5	1.02	19.2	27.9	741.6	18.9	22.3	98.4	7.4	9.1	15.6	3.8	3.9	20.2				0.2
麻山田	IIP10-1	194.6	767.0	397.5	34.9	4.3	1.25	25.1	38.1	547.2	14.9	22.5	62.9	7.0	28.6	16.2	5.0	3.6	27.5				0.7
麻山田	D1466-1	243.6	537.8	265.5	32.7	4.6	1.45	25.1	35.3	396.5	10.5	21.5	50.3	3.4	13.0	5.8	5.4	2.2	33.4				
麻山田	D759-1	118.7	1829	407.5	11.0	1.5	0.22	6.1	27.6	783.2	18.6	18.9	73.3	18.4	20.5	32.0	1.8	5.8	6.1				0.1

续表

岩石单位	样品号	Rb	Ba	Sr	Th	U	Ta	Nb	Pb	Zr	Hf	Ga	Zn	Ni	V	Cr	Cs	Co	Li	Cu	Mo	Ag	Au
	平均	158.71	1999	332.4	17.03	3.10	0.95	18.87	28.99	646.4	15.89	22.34	81.59	7.30	13.20	14.87	3.37	3.39	16.10				0.33
深堀村	H1474-2	144.9	1026	446.6	14.2	5.2	1.31	14.9	17.9	242.1	6.8	22.7	94.3	5.7	47.1	18.6	6.7	10.0	32.0				
深堀村	S042	121.2	1967	448.6	18.9	2.3	0.83	16.1	18.3	648.8	14.8	20.2	92.0	2.3	40.3	4.0	3.2	6.8	19.0				
深堀村	D430-1	108.4	1955	416.4	9.0	2.2	0.86	21.3	26.7	428.5	12.0	22.9	108.7	5.1	17.2	9.4	1.7	4.0	14.8				0.7
	平均	124.83	1649	437.20	14.03	3.23	1.00	17.43	20.97	439.80	11.20	21.93	98.33	4.37	34.87	10.67	3.87	6.93	21.93				0.70
深沟	S045	135.6	472.9	290.2	35.9	8.8	1.22	16.8	46.5	140.0	5.3	22.0	229.1	3.3	17.3	6.2	1.0	2.8	48.6				0.2
进岭	H211-3	261.0	317.8	88.2	26.6	7.5	1.57	20.3	25.4	224.3	7.3	20.1	24.6	3.3	5.6	3.6	3.4	3.4	15.0				0.7
进岭	H1323-1	243.5	30.0	26.3	22.0	4.7	1.26	21.4	24.2	214.2	8.1	20.8	27.1	6.6	1.7	120.7	2.4	2.1	14.9				0.2
瘦岭	D3013-1	312.5	20.5	23.8	31.1	12.4	3.63	32.5	47.2	59.5	3.5	14.3	21.6	13.3	7.0	31.5	6.5	0.9	45.1				1.2
瘦岭	QXI1-1	252.0	353.0	152.0			0.56	11.5	56.6	120.0	3.9		56.0	2.9	13.6			2.0	3.9				1.6
瘦岭	QXI3-3	231.0	322.0	220.0			1.51	7.9	47.2	117.0	3.6		61.4	3.6	15.9			3.2	48.2				0.5
	平均	265.1	231.8	131.9	31.10	12.4	1.9	17.3	50.3	98.8	3.67	14.30	46.33	6.60	12.1	31.5	6.5	2.0	32.4				1.1
黑岭	QXVI6-2	374.0	182.0	38.0			0.83	18.9	45.0	176.0	7.0		39.4	2.2	10.8			0.9	24.4				
黑岭	Y5551-1		117.0	110.0				25.3	41.8	165.0			30.0	2.3	9.6	5.0							0.8
黑岭	YXIII35-1		103.0	100.0				24.0	46.2	148.0			20.0	2.3	9.6	5.0							16.5
黑岭	YXIII9-2		180.0	100.0				22.8	41.8	84.0			20.0	2.3	9.6	5.0							28.0
黑岭	IIP6-1	360.9	356.3	172.4	45.0	14.1	1.95	21.8	46.9	171.9	6.5	17.4	36.6	9.3	16.4	20.6	9.4	2.3	79.3				0.7
	平均	146.9	187.6	104.0	45	14.1	1.4	22.6	44.3	148.9	6.7	17.4	29.2	3.6	11.2	8.9	9.4	1.6	51.8				11.5
尖峰岭	D3016-1	296.8	251.9	59.1	40.9	6.0	1.20	14.4	40.0	176.1	6.4	20.2	33.9	4.9	7.0	14.2	9.1	1.7	49.0				0.9
尖峰岭	S019	305.3	269.2	69.9	49.0	6.6	1.42	15.5	38.4	213.7	7.5	19.4	46.0	2.8	7.5	5.2	8.1	2.1	38.0				*
尖峰岭	QXVI6-1	366.0	381.0	81.0			4.53	21.5	57.3	219.0	7.6		50.5	3.7	17.5			3.1	66.7				1.1
尖峰岭	QXVIII1-1	227.0	650.0	121.0			0.33	14.0	34.8	227.0	7.0		44.0	4.3	22.8			4.4	20.1				0.5
尖峰岭	Y5550-1		308.0	118.0				14.0	31.0	90.0			20.0	*	9.6	5.0							0.1
尖峰岭	D3025-1	171.1	46.3	29.9	16.5	5.1	0.56	6.2	26.5	90.9	3.4	11.9	22.0	6.5	2.0	11.9	3.3	0.9	10.8				1.2

续表

岩石单位	样品号	Rb	Ba	Sr	Th	U	Ta	Nb	Pb	Zr	Hf	Ga	Zn	Ni	V	Cr	Cs	Co	Li	Cu	Mo	Ag	Au
尖峰岭	Q1200	342.0	53.0	24.0	*	*	0.97	19.1	46.0	125.0	5.5	*	34.4	1.7	7.3	*	*	0.1	18.3				1.6
尖峰岭	S037	228.2	337.8	144.2	22.9	4.3	1.09	9.1	33.1	116.3	4.2	15.9	40.0	2.0	5.4	4.9	6.9	1.6	21.0				*
尖峰岭	D271-1	199.3	47.5	24.2	19.0	5.7	1.03	22.9	38.9	219.9	8.0	13.5	60.4	5.1	2.8	8.8	3.3	1.0	14.1				0.3
	平均	266.9	260.5	74.5	29.6	5.5	1.4	15.19	38.4	164.2	6.2	16.18	39.0	3.8	9.10	8.3	6.14	1.86	29.7				0.81
高村	Ⅷ1-1	202	766	147			1.07	17.7	32.4	251	7.2		67.3	6.7	41.9			8.8	34.8	4.3	4.95	30	0.4
高村	D2385-2	352	55	25			4.88	22.6	65.7	89	4.5		25.5	1.7	7.1			2.1	7.4	61.9	4.96	106	0.7
	平均	277.0	410.5	86.0			2.98	20.15	49.05	170.00	5.85		46.40	4.20	24.50			5.45	21.10	33.10	4.96	68.00	0.55
两院	Ⅷ6-1	308	946	120			2.17	23.7	44.4	366	9.76		91.4	6.8	41.8			10.1	51.2	6.6	4.7	34	0.4
两院	0504	256.5	247.4	109.6	23.59	18.3	1.55	14.45	31.7	147.3	4.76	16.12	41.35	1.93	13.32	10.82	12.73	2.75		48.45			
两院	D2385-1	299	244	75			3.42	16.9	31.8	111	4.56		37.3	2.3	15.5			2.8	32.7	3.7	1.62	64	0.5
	平均	287.83	479.1	101.53	23.59	18.30	2.38	18.35	35.97	208.10	6.36	16.12	56.68	3.68	23.54	10.82	12.73	5.22	41.95	19.58	3.16	49.00	0.45
大岭	Ⅷ15-2	184	663	222			1.37	16.5	24.9	173	5.78		68.5	9	46.4			9.7	35.4	6.6	4.25	32	0.4
鹅塘村	Ⅷ15-1	155	687	319			1	16.7	19	253	7.75		89.9	14.5	86.9			16.3	48.2	9	3.65	35	0.3
陀类	D3082-1	286.5	47.3	187.0	31.7	7.8	1.86	14.5	43.0	138.8	4.6	21.2	48.3	4.3	13.1	10.3	13.0	1.5	77.7				1.0
陀类	Y78-1	*	379.9	116.0	*	*	*	12.3	42.3	109.4	*	*	50.5	*	*	*	*	*	*				*
陀类	Y79-1	*	465.7	116.0	*	*	*	11.8	39.7	135.9	*	*	50.4	*	*	*	*	*	*				*
陀类	Y80-1	*	278.7	100.0	*	*	*	11.1	28.2	73.8	*	*	52.8	*	*	*	*	*	*				*
	平均	286.50	292.9	129.75	31.70	7.80	1.86	12.43	38.30	114.48	4.60	21.20	50.50	4.30	13.10	10.30	13.00	1.50	77.70				1.00
大田	D3087-1	165.3	1207	427.0	20.8	5.4	0.47	5.7	35.8	231.0	6.5	22.5	70.1	56.4	22.7	21.6	4.7	3.2	24.3				1.0
大田	D3084-1	192.9	1340	409.9	23.8	3.9	1.64	24.5	32.1	263.3	7.1	20.7	30.4	5.2	26.4	12.1	5.4	4.5	24.5				0.6
大田	DT1-1	189.1	1062	416.9	28.0	4.8	0.92	16.1	36.9	260.7	7.4	23.9	74.7	5.6	27.1	13.7	5.3	3.1	54.0				1.4
大田	DT2-1	202.4	935.6	313.4	30.9	6.4	0.98	15.7	47.6	298.2	8.3	23.0	69.8	8.4	25.6	23.3	4.6	3.1	41.9				0.7
	平均	187.43	1136	391.80	25.88	5.13	1.00	15.50	38.10	263.30	7.33	22.53	61.25	18.90	25.45	17.68	5.00	3.48	36.18				0.93
什天	YVII9-2	*	590.0	500.0	*	*	*	11.5	24.8	148.0	*	*	*	3.1	36.0	5.0							

续表

岩石单位	样品号	Rb	Ba	Sr	Th	U	Ta	Nb	Pb	Zr	Hf	Ga	Zn	Ni	V	Cr	Cs	Co	Li	Cu	Mo	Ag	Au
什天	S031	258.3	353.0	201.3	31.8	9.2	1.72	13.7	35.5	138.1	4.6	18.6	58.0	2.1	14.6	4.1	10.4	2.9	82.0				
什天	D3020-1	233.8	309.3	217.0	35.5	14.0	2.30	19.3	31.7	103.2	4.0	16.6	28.9	12.6	36.0	31.6	9.8	3.6	26.7				1.1
什天	IIP4-1	225.3	861.8	413.8	31.6	5.5	1.09	14.9	37.4	257.7	7.3	21.2	64.4	9.0	40.2	21.1	5.8	4.9	44.7				0.3
什天	S014	141.0	763.2	267.5	30.5	3.1	0.85	15.4	22.1	371.7	10.5	18.6	38.0	3.5	10.9	5.2	2.0	4.5	8.0			*	
	平均	214.60	575.4	319.92	32.35	7.95	1.49	14.96	30.30	203.74	6.60	18.75	47.33	6.06	27.54	13.40	7.00	3.98	40.35				0.70
白茶	H1362-1	371.0	455.0	134.0	*	*	2.00	14.3	43.0	113.0	3.6	*	28.0	7.0	24.0	*	*	0.4	30.0				0.7
东岭	H1211-1	161.0	750.0	395.0	*	*	1.60	12.4	96.0	220.0	6.0	*	83.8	11.0	46.0	*	*	6.3	33.6				1.3
东岭	H1267-2	167.0	865.0	480.0	*	*	2.00	12.6	27.0	200.0	4.5	*	98.4	15.0	58.0	*	*	9.1	52.7				0.6
那佩	D3074-1	190.4	264.2	110.7	34.1	11.2	1.74	15.4	31.8	173.9	6.4	19.0	43.7	7.1	12.2	19.4	5.2	3.2	46.0				1.9
那佩	G1346	203.0	298.0	278.0	*	*	1.05	8.0	31.6	268.0	5.6	*	82.0	1.2	8.0	*	*	3.9	44.5				0.4
保国	D3076-1	233.8	324.5	245.8	29.8	7.5	1.55	12.8	27.2	156.7	5.3	19.8	70.0	7.4	26.1	17.1	10.2	4.5	67.9				0.8
保国	J121-1	194.0	1400.0	570.0	*	11.5	1.60	*	*	252.0	*	*	*	3.0	*	10.7	8.0						
保国	J133-1	213.0	660.0	292.0	*	6.0	5.00	*	*	390.0	*	*	*	2.0	*	7.0	6.0						
	平均	213.60	794.8	369.27	29.80	8.33	2.72	12.80	27.20	266.23	5.30	19.80	70.00	4.13	26.10	11.60	8.07	4.50	67.90				0.80
火岭	H1311-3	299.0	495.0	187.0	*	*	2.20	19.4	54.0	120.0	3.2	*	74.0	6.0	10.5	*	*	2.0	79.5				
火岭	VHG1-7	259	1251	254	0	0	1.9	21	27	479	10	9	69	3.3	14	0	0	0.6	19	8.4	0.5	40	0
火岭	7329-1	289	320	100			1.2	20.4	53	88	2.6		68	7	11			4.4	97.9	4	1	90	0
火岭	XI2-1	322	273	77			1.69	13	44.5	73	2.56		59.9	2.3	4.5			2.2	50.4	2	9.36	85	0.4
火岭	4108-1	260	790	180				13	85	185			37	5.3	14	37		4.3		7.8		37	1.4
	平均	285.80	625.8	159.60			1.75	17.36	52.7	189.00	4.59	9.0	61.58	4.78	10.80	18.50		2.70	61.70	5.55	3.62	63.00	0.45
长塘岭	H1278-1	190	735	278.0	*	*	1.60	10.7	31.0	202.0	5.2	*	93.1	6.0	16.0	*		2.2	63.8				0.8
长塘岭	1225	265	730	320			2	19.7	45	330	6.5		17	12	34	9		1.2	32.8	14	1.54		
长塘岭	IX10-1	302	50	51			3.63	15.75	47	79	3.64		41.3	4.7	7			1.95	24.45	2.75	0.08	50	1.45
长塘岭	XI10-2	202	469	175			1.19	12.4	31	128	4.41		53.7	6.7	28.7			5.5	47.5	3	9.81	35	0.5

续表

岩石单位	样品号	Rb	Ba	Sr	Th	U	Ta	Nb	Pb	Zr	Hf	Ga	Zn	Ni	V	Cr	Cs	Co	Li	Cu	Mo	Ag	Au
长塘岭	4871-2	150	834	416				7.2	22				35	5	22	73		4.5		5		25	2
长塘岭	D823-1	150	421	200	20	12	2.3	11	105	53	4.2	17	53	5	36	12		3.8	80	4.3	0.3	60	1.6
	平均	209	539	240.00	20.0	12.0	2.1	12.7	46.8	158.4	4.79	17.0	48.85	6.57	23.95	31.33		3.19	49.71	5.81	2.93	42.50	1.27
大坡	1389-2	251	584	193			0.8	13.6	40	162	3.6		67	2	24			7.2	70	7	1.3	70	0
大坡	21	152	465	213			0.8	14.4	23	140	3.4		36	19	22			1.7	29.5	6	1.2	70	1.1
大坡	Ⅺ6-1	194	536	169			15.9	12.6	30.2	122	4.07		53.2	5.9	29			5.3	44.9	3.7	7.52	39	0.4
大坡	Ⅱ19-1	227	446	223			2.27	11.87	12.6	137			47.8	5	46.3			5.63	24.17	6	0.22	40	1.13
大坡	Ha34-1	268	175	117	20.8	20.2	1.08	9.52	41.8	103	3.22	16	40.2	4.13	11.7	31	16.3	2.88	69.4	12			
	平均	218.40	441.2	183.00	20.80	20.20	4.17	12.40	29.52	132.80	3.57	16.00	48.84	7.21	26.60	31.00	16.30	4.54	47.59	6.94	2.56	54.75	0.66
乌兰盖	3809	252	370	187			1.2	13	43	70	2		32	13	20	7		6.3	98.1	11	1.25	10	0.1
乌兰盖	D45-1	216	444	266			1.46	11.5	32.8	157	4.46		56.9	12.8	47			10.5	71	10.1	1.24	190	0.2
	平均	234	407	226.5			1.33	12.2	37.90	113.5	3.23		44.45	12.90	33.5	7.0		8.40	84.55	10.55	1.25	100.00	0.15
通什	H1274-2	144.0	870.0	362.0	*	*	1.60	9.8	18.0	230.0	7.0	*	98.4	11.0	37.0	*	*	10.8	95.1				2.6
通什	H1287-1	161.0	795.0	410.0	*	*	1.60	12.3	18.0	250.0	5.4	*	89.2	8.0	52.0	*	*	10.8	83.8				4.7
通什	4220	165	715	290			1.6	13.1	29	162	3.8		78	7	57			7	63	1	0.8	90	7.3
通什	4107	188	665	278			1.6	13.5	28	155	4		56	7	56			6.6	60.4	15	1	390	3.2
通什	Ⅸ8-1	178	558	272			1.66	12.46	22.93		5.05		61.41	6.73	38.13			6.39	50.14	9.23	0.41	400	0.11
通什	4111-1	170	1000	529				11.3	39				66	6	56	74		4		11		47	4
通什	4577-1	170	484	244				7.1	34				76	7	60	74		15		12		57	2
通什	Ha35-1	188	467	270	18.3	5.62	0.96	12.2	21.5	157	3.82	21	53.8	7.07	44	24.6	11.6	6.89	30.6	22.4			
	平均	170.5	694.2	331.8	18.3	5.6	1.5	11.47	26.3	190.80	4.85	21.0	72.35	7.48	50.02	57.53	11.60	8.44	63.84	11.77	0.74	196.80	3.42
山牛塘	H1311-2	153.0	240.0	220.0	*	*	0.80	10.6	12.0	140.0	2.7	*	142.0	18.0	69.0	41.0	*	11.7	133.5				2.4
山牛塘	H1397-1	166.0	410.0	480.0	*	*	1.20	8.6	13.0	120.0	3.0	*	94.0	15.0	117.0	*	*	17.0	135.0				1.6
山牛塘	H209-1	120.0	570.0	610.0	*	*	1.60	11.5	24.0	230.0	5.8	*	119.0	4.0	60.0	*	*	8.2	94.0				1.0

续表

岩石单位	样品号	Rb	Ba	Sr	Th	U	Ta	Nb	Pb	Zr	Hf	Ga	Zn	Ni	V	Cr	Cs	Co	Li	Cu	Mo	Ag	Au
山牛塘	H209-3	183.0	490.0	470.0	*	*	1.60	14.8	16.0	201.0	4.9	*	135.0	4.0	93.0	*	*	9.7	133.0				1.0
	平均	155.50	427.5	445.00			1.30	11.38	16.25	172.75	4.10		122.50	10.25	84.75	41.00		11.65	123.88				1.50
石碌	137-1	234	740	230			1.6	15.4	34	120	4.4		74	12	38			6.2	52.8	19	1	160	7.4
石碌	Ha29	171	392	249	19.8	8.18	1.68	12.8	27	76.5	3.5	21.5	80.6	3.99	30.2	13.6	8.69	4.54	63.7	12.4			
石碌	7334-1	195	350	196			2.4	12.9	33	180	4.8		54	4	42			5.8	25.7	8	2	30	0
	平均	200	494	225	19.8	8.18	1.89	13.70	31.33	125.50	4.23	21.50	69.53	6.66	36.73	13.60	8.69	5.51	47.40	13.13	1.50	95.00	3.70
什败	D1547-3	191.1	1943	558.8	29.2	2.8	1.11	25.5	31.0	415.0	9.9	21.8	94.7	10.7	80.3	18.4	2.7	8.9	23.5				*
行干	D1749-1	135.0	1177	865.0	9.2	1.5	0.99	18.4	19.0	259.3	6.2	23.7	111.1	10.9	115.7	33.7	2.3	15.3	26.5				*
行干	D3089-1	176.6	1707	1072.9	11.9	3.5	1.54	21.0	20.7	288.6	7.6	25.6	117.4	19.1	120.6	46.6	6.7	15.3	49.6				0.9
方满	D3092-1	142.7	2362	1912.1	20.9	3.2	0.63	13.1	41.2	190.6	5.5	22.5	103.7	41.9	173.6	74.4	4.1	26.1	61.9				1.0
新风岭	H5122-1	288.0	114.0	68.0			0.80	8.7	53.0	96.0	3.3		87.0	6.0	5.0	4.0		2.1	23.0				0.2
顺作	D3136-1	129.6	651.1	304.5	16.8	1.3	0.19	6.2	44.4	110.9	3.6	17.5	38.5	9.6	24.6	26.8	0.9	3.1	26.8				0.7
乘顺岭	QIV1-1	307.0	233.0	65.0	*	*	0.82	13.3	38.5	146.0	4.5	*	59.2	0.9	8.6	*	*	1.0	16.8				0.9
乘顺岭	GXI3-1	284.0	380.0	190.0	*	*	1.43	7.6	53.5	127.0	4.4	*	48.2	1.8	4.4	*	*	2.5	23.6				0.8
	平均	295.50	306.5	127.50			1.13	10.45	46.00	136.50	4.45		53.70	1.35	6.50			1.75	20.20				0.85
志仲	D3073-1	341.1	281.9	141.5	28.5	8.8	2.32	22.6	33.2	212.6	6.7	22.0	70.6	4.6	26.6	14.0	27.2	5.0	158.0				0.6
志仲	GVIII18	143.0	592.0	215.0	*	*	0.49	14.3	21.3	319.0	6.7	*	52.7	4.1	8.7	*	*	3.7	25.0				0.8
	平均	242.05	436.9	178.25	28.50	8.80	1.41	18.45	27.25	265.80	6.70	22.00	61.65	4.35	17.65	14.00	27.20	4.35	91.50				0.70
抱郎	QIV4-1	119.0	906.0	569.0	*	*	0.08	12.2	57.6	277.0	7.8	*	131.1	3.2	41.3	*	*	6.7	36.1				2.0
抱郎	QVII2	183.0	739.0	256.0	*	*	0.30	14.2	30.9	282.0	8.6	*	56.5	4.0	17.5	*	*	4.8	28.4				1.8
	平均	151.00	822.5	412.50			0.19	13.20	44.25	279.50	8.20		93.80	3.60	29.40			5.75	32.25				1.90
石岭	7315	100	885	355			1.2	16.2	22	242	4.9		65	10	73			10.4	21	29	0.9	110	6.6
石岭	7316-1	124	690	335			1	10.6	21	160	4.1		53	3	105			9.3	17.6	18	0.9	60	3.4
石岭	Ha36	124	505	310	10.9	3.16	1.54	8.9	20.6	130	4.78	20.7	74.7	7.61	122	33.1	7.81	11.4	26.3	17.5			
	平均	116.00	693.3	333.33	10.90	3.16	1.25	11.90	21.20	177.33	4.59	20.70	64.23	6.87	100.00	33.10	7.81	10.37	21.63	21.50	0.90	85.00	5.00

续表

岩石单位	样品号	Rb	Ba	Sr	Th	U	Ta	Nb	Pb	Zr	Hf	Ga	Zn	Ni	V	Cr	Cs	Co	Li	Cu	Mo	Ag	Au
保山村	4758-1	140	560	304				13	14				116	8	71	62		13		16		60	1
保山村	4971-1	120	517	305				10.2	12				91	7	63	123		12		13		58	3
保山村	Ha32-1	181	513	459	19.8	5.08	1.6	18.1	26.3	170	5.08	21.6	89.5	7.05	62.6	56.5	9.95	9.52	30.3	22			
	平均	147.00	530.0	356.0	19.80	5.08	1.60	13.77	17.43	170.00	5.08	21.60	98.83	7.35	65.53	80.50	9.95	11.51	30.30	17.00		59.00	2.00
白石村	211-3	92	710	422			1.6	15.4	17	220	4.9		72	17	123			15.7	38.9	14	0.8	30	0
白石村	290	33	515	364			0.6	19.6	10	160	3.2		96	150	150			34.4	33.2	30	0.2	90	1.9
白石村	4350-7	85	1044	1128				10	34				167	48	130	80		24.7		14		50	1
白石村	Ha16	120	1190	873	9.59	6.72	1.08	14	14.2	191	4.5	30.3	139	10.4	200	34.6	4.23	27	31.4	28.9			
白石村	Ha17-2	111	992	993	14	3.01	1.09	13.8	14.2	285	7.3	31.4	159	4.27	142	24.2	5.2	16.7	46	42.6			
白石村	Ha17-1	134	1130	858	13.6	6.88	2.32	22.4	18.6	204	4.86	31.8	132	8.54	149	36.4	6.3	17.8	45	24.7			
白石村	Ha18	116	1940	1250	18.7	2.86	1.89	19.8	19.8	198	6.14	40.8	148	5.2	110	31	3.51	14.7	33.2	30.3			
白石村	QIV0-1	99.2	393.0	513.0	*	*	0.29	17.6	31.8	404.0	11.4	*	120.0	4.6	74.5	*	*	12.2	26.2				1.4
白石村	s028	47.1	224.6	195.7	7.7	1.7	0.82	9.7	12.6	192.8	5.3	16.3	69.0	68.4	126.5	29.6	3.5	30.3	34.0				*
白石村	D3019-1	90.2	401.5	438.4	10.4	2.5	1.05	14.2	15.9	156.9	4.7	19.5	74.0	25.5	170.4	54.7	5.7	21.2	44.4				0.1
	平均	92.75	854.0	703.51	12.33	3.95	1.19	15.65	18.81	223.52	5.81	28.35	117.60	34.19	137.54	41.50	4.74	21.47	36.92	26.36	0.50	56.67	0.88
机告村	1153	110	722	655							0.16			39	134	4		17.2					
机告村	4014-1	37	308	504				5	27				84	80	155	285		40.5		32		43	3
机告村	Ha19	146	1430	992	12.9	6.04	1.02	11.4	14.4	125	2.47	30	132	20	196	58.9	11.4	27.2	65.4	27.3			
机告村	Ha39-1	45	267	527	3.92	1.16	0.5	10	25.2	112	5.64	18.5	90.7	57.8	303	240	4.24	37.4	46.3	41.8			
机告村	Ha39-2	44	280	346	4.32	1.31	0.92	8.35	5.46	111	4.97	15.6	92.3	149	275	512	4.22	46	43.4	47.8			
	平均	76.40	601.4	604.80	7.05	2.84	0.81	8.69	18.02	116.00	3.31	21.37	99.75	69.16	212.60	219.9	6.62	33.66	51.70	37.23		43.00	3.00
亚炮	平均（8）	183	490	94	0	0	2.02	14.81	36.33	142	0	0	57	6.64	41.8	0	0	5.08	30.9	14	0.68	110	127
戈枕村	平均（11）	156	421	160	0	0	1.71	13.75	23.67	157	0	0	61.14	9.39	56.5	0	0	8.12	51.7	23.05	0.54	190	10.15

注：Au 的单位为 10^{-9}；样品来源同附表 9-1。

第三篇

火山岩及火山作用

概　　述

海南省火山岩分布范围广，火山活动频繁，中元古代、新元古代、古生代及中生代、新生代均有火山活动，具有多期性和多次性的喷发历史。一直以来，地质工作者对海南省火山岩地质做了大量的研究工作。1929 年，李承三等人在海南岛北部进行地质调查，对分布于琼北的火山岩进行了粗略划分。边兆祥（1958）、郭丙奎（1959）、黄玉昆（1963）、丁国瑜（1964）等也先后对分布于琼北的火山岩进行了探索性的研究。1964，广东省地质局区域地质测量大队在进行 1 ∶ 20 万海南岛区域地质调查过程中，对出露于海南岛的喷出岩进行了较为系统的划分和对比，按时代及岩石类型分为两套：第一套为晚白垩世—上新世由 4 个喷出体构成的流纹岩类及安山岩类，第二套为上新世—第四纪玄武岩类（大致划分为上新世玄武岩和第四纪玄武岩，前者隐伏于地下，后者出露地表）。20 世纪 70 年代后期，袁奎荣等（1977）发现分布于昌江、儋县一带的古生代浅变质岩系中夹有阳起石岩、细碧岩，认为其原岩为拉斑玄武岩。《广东省区域地质志》（1982）将分布于海南岛的火山岩划分为加里东期、华力西期—印支期、燕山期和喜马拉雅期。进入 20 世纪 80 年代以来，海南岛相继开展了多幅 1 ∶ 5 万区域地质调查和全岛 1 ∶ 25 万区域地质调查以及矿产地质调查与勘查评价、物化探测量、遥感、生态环境地质调查、水文地质调查与地质科研等多项地质工作，大大提高了海南岛地质研究程度，在火山岩研究方面也取得一些新的认识和成果。

海南省火山岩以出露于海南岛北部的新生代火山岩分布最广，其次为分布于陆相火山岩盆地及火山-沉积岩盆地中的白垩纪火山岩，中元古代、新元古代、古生代火山岩仅零星出露，南海新生代火山岩由于淹没于海水之中而分布不明。中元古代火山岩以中酸性火山岩为主，产于长城系戈枕村组中；新元古代火山岩以中酸性火山岩为主，产于青白口系石碌群中；古生代火山岩从超基性至酸性均可见，产于古生代地层中；中生代火山岩的岩性以中性和酸性熔岩为主；琼北和南海新生代火山岩的岩性以基性熔岩为主，夹凝灰岩和沉火山角砾岩。其中，中元古代、新元古代和古生代火山岩均发生了变质作用，其原岩的岩石原貌及火山岩相均已不存在；中生代火山岩虽然没有发生变质作用，但受风化剥蚀影响比较严重，火山机构已残缺不全，火山口相大部分完全剥蚀殆尽，火山颈（火山口）相部分已不存在，主要残存溢流相及喷发-沉积相；新生代火山岩的大部分火山机构保存完好，岩石的矿物成分没有发生明显蚀变。

根据海南省构造单元划分方案，考虑到火山岩的分布特点，可将火山岩划分为以下几个带和亚带：①华夏地块构造岩浆带前寒武纪海南火山岩亚带；②东南造山区构造岩浆带古生代五指山火山岩亚带；③东部中生代构造岩浆带中生代海南火山岩亚带；④印支-南海火山喷发岩浆带。印支-南海火山喷发岩浆带以中央海盆北缘断裂为界，分为雷琼火山喷发岩浆亚带，中央海盆火山喷发岩浆亚带。分别对应海南省 4 个大的火山活动期：前寒武纪火山活动期、古生代火山活动期、中生代火山活动期和新生代火山活动期。

本次研究在系统收集和分析前人成果资料的基础上，从岩石学特征、岩石化学特征、岩石稀土地球化学特征、岩石微量元素地球化学特征及副矿物特征等角度对不同时期火山岩进行研究，分析和探讨其各自形成的构造环境，为重塑海南省不同时期大地构造背景提供依据。

第十章 前寒武纪火山岩

海南省前寒武纪火山岩包括中元古代长城纪和新元古代青白口纪火山岩。长城纪火山岩因喷发时代较早且遭受区域变质作用，后又遭受不同程度的混合岩化，致使岩石的结构、构造及矿物成分等已彻底改变，火山岩原貌难以恢复，仅从岩石学特征、岩石地球化学特征及岩石组合等方面对火山岩某些性质及形成环境进行判别。

长城纪火山岩产于戈枕村组和峨文岭组中，出露于海南岛西部冲卒岭、大蟹岭、抱板和中部上安（长征）及东部黄竹岭等地。戈枕村组中上部主要由混合岩化黑云（角闪）斜长片麻岩（图版Ⅰ-1-B）、黑云（角闪）斜长变粒岩、斜长角闪（片麻）岩组成，下部为紫苏（二辉）斜长片麻岩、麻粒岩；峨文岭组夹有少量变余斑状黑云斜长变粒岩和石英斜长浅粒岩。根据岩石组合、野外产出特征，结合岩石化学方法判别（详见第四篇第一章），片麻岩、变粒岩的原岩主要为中酸性火山岩，斜长角闪（片麻）岩和麻粒岩的原岩主要为中基性火山岩。其中，黑云（角闪）斜长片麻岩和黑云（角闪）斜长变粒岩的主要氧化物含量为（表 10-1）：$SiO_2$58.94%~75.39%、$TiO_2$0.19%~1.30%、Al_2O_3 12.5%~16.8%、MgO 0.36%~3.75%、FeO 0.57%~6.86%、CaO 0.79%~5.13%、Na_2O 2.04%~4.32%、K_2O1.63%~5.03%，成分上大多类似于英安岩，少数相当于流纹岩，个别属于安山岩；麻粒岩的主要氧化物含量为：SiO_2 49.83%~51.34%、TiO_2 0.79%~1.87%、Al_2O_3 12.57%~17.18%、FeO 8.43%~9.39%、MgO 6.20%~9.21%、CaO 7.81%~9.88%、Na_2O 2.06%~3.06%、K_2O 0.68%~2.33%，成分上类似于玄武岩或玄武安山岩。多数样品属钙碱性系列，总体上构成单峰式的高钾钙碱性玄武岩-玄武安山岩-英安岩组合，可以判断形成于岛弧或陆缘弧环境。

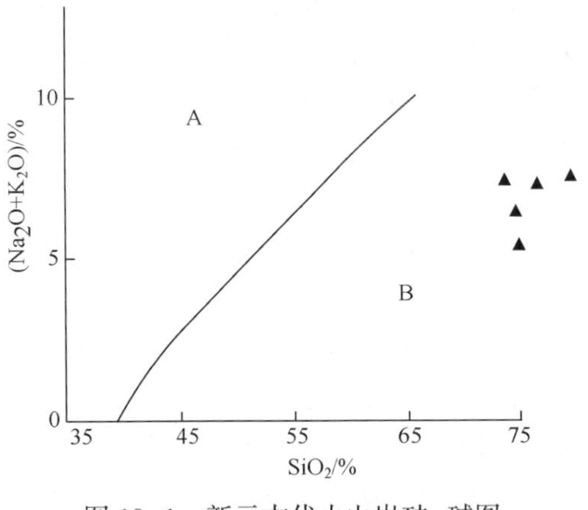

图 10-1 新元古代火山岩硅-碱图

Fig. 10-1 SiO_2-Na_2O+K_2O diagram of Neoproterozoic volcanic rocks

A—碱性系列；B—亚碱性系列

新元古代火山岩出露于海南岛西部昌江县石碌一带青白口系石碌群中，以产于石碌群第五层为主，第六层也有分布。第五层片岩中上部的火山岩层为似层状（图版Ⅰ-10-A，Ⅰ-10-B），岩石以灰绿色层状凝灰岩和含有火山泥球的钾流纹质熔结凝灰岩为主。产于第六层的火山岩岩性有富钾粗安质球粒凝灰岩和富钾流纹质球粒凝灰岩，其规模比产于第五层的火山岩小一些。火山岩主要呈透镜状、薄层状顺层夹在碎屑沉积变质岩中，与上下岩层呈过渡关系但有明显的岩性分界。此外，在靠近火山岩的碎屑沉积变质岩中不同程度地也含有火山凝灰质成分。火山岩主要氧化物含量为（表 10-1）：SiO_2 74.78%~77.53%、TiO_2 0.08%、Al_2O_3 11.33%~13.62%、MgO 0.30%~0.66%、$Fe_2O_3$0.04%~1.05%、FeO 0.46%~0.63%、CaO 0.15%~0.23%、Na_2O 0.31%~0.48%、K_2O 5.58%~7.11%，显示新元古代火山岩为酸性火山岩，全碱含量为 5.89%~7.59%，钠/钾之比值为 0.05~0.08，属富钾并极度贫钠的岩浆岩，SI 值一般在 3.21~8.10 之间，平均值为 5.69，δ 值为 1.08~1.71（均小于 4）。按里特曼岩系划分主要属于钙碱性系列；按 Irvine 等提出的两分法，新元古代火山岩在 Irvine 等人的硅-碱图中投影点均落入亚碱性系列（图 10-1），在 AFM 图解中投影点均落入钙碱性系列（图 10-2）；在里特曼-戈蒂里图解中的投影点均落入岛弧及活动大陆边缘区火山岩（图 10-3）。表明新元古代酸性火山岩系无典型的非造山环境火山岩，而属于造山环境火山岩。

表 10-1　前寒武纪（变）火山岩岩石化学成分表

Table 10-1　List of petrochemical component of Pre-Cambrian（meta-）volcanic strata

单位：%

层位	样品号	序号	岩石化学成分															产地
			SiO_2	TiO_2	Al_2O_3	Fe_2O_3	FeO	MnO	MgO	CaO	Na_2O	K_2O	H_2O	P_2O_5	CO_2	灼失	总量	
戈枕村组	D1650-2	1	49.94	0.79	14.98	1.12	8.43	0.18	8.62	9.88	2.16	1.88	0.67	0.18	0.55		99.38	长征
	B323	2	51.34	1.87	17.58	0.48	9.39	0.16	6.2	7.81	3.06	0.68	0.64	0.69	0.49		100.39	长征
	B317	3	49.83	1.45	12.57	0.64	8.73	0.17	9.21	9.93	2.06	2.33	1.06	0.83	0.62		99.43	长征
	1~3 平均值		50.37	1.37	15.04	0.75	8.85	0.17	8.01	9.21	2.43	1.63	0.79	0.57	0.55			
	Ⅱ12-2	4	75.26	0.03	13.92	0.43	0.57	0.05	0.12	0.43	3.42	5.43					99.66	抱板
	D326-3	5	6103	0.575	15.8	1.45	3.82	1.56	0.087	3.81	3.14	2.62	2.5	0.075		3.16	101.827	抱板
	Ⅱ22-1	6	74.4	0.13	13.66	0.86	1.21	0.05	0.35	0.5	2.72	5.11					98.99	抱板
	0616	7	68.25	0.44	14.64	4.97		0.06	1.18	3.16	2.63	106				0.97	99.56	抱板
	0625	8	67.03	0.49	15.36	4.91		0.07	1.26	4.03	3.02	2.16				1.27	99.6	抱板
	0626-1	9	65.49	0.51	15.09	5.35		0.06	1.44	4.59	2.67	2.58				0.09	97.87	抱板
	D4	10	68.96	0.5	13.1	0.2	4.64	0.24	1.1	4	2.54	2.46	0.98	0.15			98.87	抱板
	XⅣ6-2	11	76.00	0.32	12.47	1.42	0.7	0.01	0.65	0.13	1.29	3.78	2.87	0.1	0.1	2.15	101.99	黄竹岭
	D1662-1	12	73.04	0.5	13.02	0.08	2.08	0.01	1.21	1.45	1.61	5.89	0.67	0.05	0.17	0.6	100.38	黄竹岭
	B192	13	69.14	0.74	13.53	0.04	5.62	0.13	1.5	2.14	2.72	2.51	1	0.11	0.14		99.32	冲卒岭
	B193	14	66.34	0.79	12.75	1.31	6.86	0.18	1.63	114	2.04	118	1.1	0.2	0.16		100.08	冲卒岭
	B194	15	66.03	0.47	14.99	0.18	6.12	0.06	1.79	2.96	3.4	2.67	0.62	0.15	0.16		99.6	冲卒岭
	XIV16-1	16	75.39	0.24	12.77	1.37	0.57	0.03	0.41	0.79	2.21	4.9		0.03		0.96	99.67	冲卒岭
	XIV17-1	17	74.92	0.24	12.83	0.92	1.27	0.02	0.36	0.8	2.49	4.99		0.02		0.83	99.69	冲卒岭
	XIV19-1	18	68.37	0.19	14.65	0.84	3.92	0.07	1.96	2.1	4.32	2.06		0.09		0.55	99.12	冲卒岭
	L932	19	67.47	0.85	13.14	1.83	4.81	0.1	1.3	2.74	2.47	6.57		0.16			101.44	冲卒岭
	DH72	20	69.97	0.71	1119	1.03	3.73	0.06	1.11	1.61	1.98	4.29		0.09		1.21	99.18	冲卒岭
	XIV4-1	21	67.68	0.7	13.59	0.72	4.8	0.06	2.78	2.37	2.38	3.45		0.06		0.73	99.32	冲卒岭
	S022-1	22	67.88	0.07	14.18	0.22	5.36	0.08	1.45	1.91	2.09	5.03	0.72	0.11	0.07		99.17	冲卒岭
	DH127	23	69.3	0.58	13.12	0.91	4.22	0.08	1.51	1.77	2.35	3.89	0.21	0.1		1.19	99.23	冲卒岭
	DH3076	24	67.87	0.77	13.65	1.57	3.93	0.05	1.3	1.94	2.32	4.21		0.12		1.91	99.64	冲卒岭

续表

层位	样品号	序号	岩石化学成分															产地
			SiO_2	TiO_2	Al_2O_3	Fe_2O_3	FeO	MnO	MgO	CaO	Na_2O	K_2O	H_2O	P_2O_5	CO_2	灼失	总量		
戈枕村组	DZK	25	70.17	0.58	12.86	1.64	2.44	0.06	1.43	1.77	2.45	4.18	0.19	0.12		1.57	99.46		冲卒岭
	DH155	26	75.2	0.12	13.14	1.49	0.52	0.02	0.2	1.11	102	4.78		0.02		0.68	100.5		冲卒岭
	B1	27	77.52	0.17	11.37		3.17	0.03	0.33	1.33	3.67	2.32	0.34	0.03			100.28		冲卒岭
	S034-5	28	58.94	1.3	16.8	0.93	6.46	0.09	2.97	5.8	2.64	2.32	0.61	0.32	0.1		99.28		长征
	S034-6	29	66.24	0.59	12.5	0.22	5.05	0.1	3.75	5.62	2.14	1.63	0.34	0.13	0.2		98.51		长征
	GⅡ-9-1	30	67.78	0.64	14.96	0.34	11	0.04	0.91	2.85	3.44	4.13	0.52	0.17	0.28		99.36		长征
	XIV5-1	31	67.08	0.61	14.51	0.82	5.05	0.08	1.77	3.03	3.43	2.2		0.13		0.61	99.32		冲卒岭
	Ⅸ16-1	32	71.13	0.35	13.45	0.65	2.98	0.05	1.03	1.00	2.61	5.15	1.34	0.41	0.22		100.37		抱板
	D43-1	33	68.81	0.38	14.65	0.77	3.1	0.04	1.81	1.86	3.5	107	1.4	0.09	0.13		99.81		抱板
	D92-1	34	74.81	0.05	14.86	0.87	0.48	0.05	0.15	1.58	4.3	1.32	0.2	0.85	1.18		100.7		抱板
	S036-1	35	68.39	0.52	14.5	0.33	3.97	0.07	1.2	3.08	2.81	4.24	0.16	0.18	0.14		99.59		长征
	D95-1	36	62.54	0.1	19.28	1.16	0.85	0.04	0.47	1.1	4.6	7.58	1.54	0.05	0.2		99.51		抱板
	T3	37	61.18	1.04	16.02	0.72	5.68	0.07	1.14	5.13	2.96	2.36	0.46	0.28	0.23		99.47		长征
	S034-3	38	70.10	0.5	14.6	0.13	3.42	0.06	0.91	2.8	3.8	2.36	0.28	0.14	0.07		99.17		长征
	4~38平均值		69.26	0.50	14.09	1.20	3.46	0.11	1.22	2.42	2.84	3.69	0.86	0.15	0.22	1.16			
	1~38平均值		67.77	0.55	14.26	1.16	3.93	0.11	1.71	2.93	2.82	3.50	0.83	0.18	0.27	1.16			
石碌群	样品号	序号	SiO_2	TiO_2	Al_2O_3	Fe_2O_3	FeO	MnO	MgO	CaO	Na_2O	K_2O	H_2O	P_2O_5	总量	N/K	σ	τ	SI
	H-24-3（1）	39	75.18	0.08	13.62	1.02	0.58	0.06	0.66	0.15	0.31	5.58	1.97	0.02	99.23	0.06	1.08	166.38	8.10
	H-24-3（2）	40	77.53	0.08	11.33	0.04	0.57	0.04	0.51	0.22	0.48	7.11	0.70	0.06	98.67	0.07	1.67	135.63	5.86
	H-24-4	41	76.15	0.08	12.98	0.66	0.46	0.05	0.47	0.23	0.31	6.85	1.55	0.02	99.81	0.05	1.55	158.35	5.37
	H-24-5	42	74.78	0.08	13.41	1.05	0.63	0.06	0.30	0.22	0.43	6.95	1.69	0.20	99.80	0.06	1.71	162.25	3.21
	H-24-6	43	75.14	0.08	13.54	0.96	0.6	0.06	0.51	0.18	0.46	6.10	2.22	0.02	99.87	0.08	1.34	163.50	5.91
	39~43平均值		75.76	0.08	12.98	0.75	0.57	0.05	0.49	0.20	0.40	6.51	1.63	0.03					

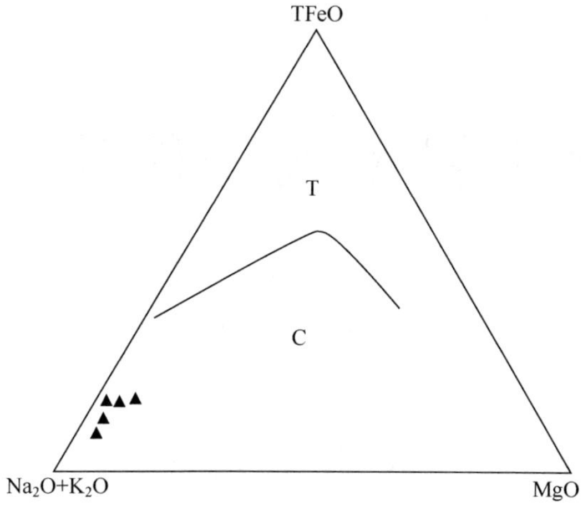

图 10-2　新元古代火山岩 F-A-M 图

Fig. 10-2　F-A-M plot of Neoproterozoic volcanic rocks

T—拉斑玄武岩系列；C—钙碱性系列

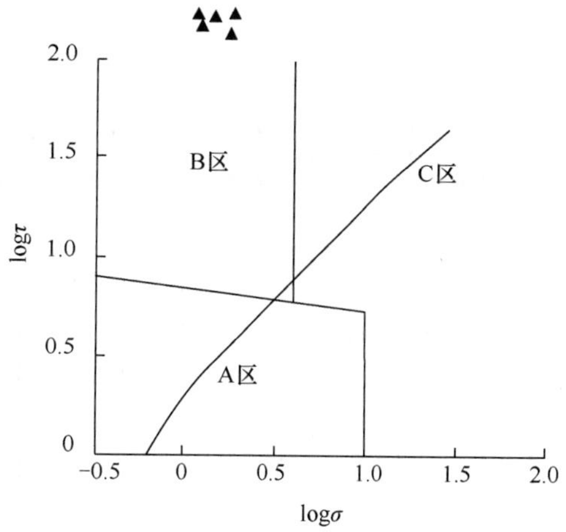

图 10-3　里特曼-戈蒂里图解

Fig. 10-3　Rittmann-Gottini diagram

（据 A. Rittman，1973）

A 区—非构造带（板块内部稳定构造区）火山岩；

B 区—造山带（岛弧及活动大陆边缘区）火山岩；

C 区—A、B 区火山岩派生的碱性岩

第十一章　古生代火山岩及火山作用

海南省古生代火山岩分布于五指山地层分区，多呈薄层状或透镜状赋存于古生界奥陶系南碧沟组、志留系陀烈组和空列村组及石炭系南好组、青天峡组中，多处可见多层火山岩与正常碎屑沉积岩呈互层的形式产出，显示火山活动具有间歇性的特点。

第一节　火山岩主要类型、火山岩建造

一、奥陶纪火山岩分布概况

古生代奥陶纪火山岩主要断续出露于定安县黄竹岭、昌江县邦溪农场、儋州市八一农场及东方市江边乡冲俄村等地，呈夹层状产于奥陶系南碧沟组中。

定安县黄竹岭，火山岩分布于南碧沟组（为一套浅海—半深海沉积的泥质粉砂岩、粉砂岩、粉细砂岩沉积）中部，可见11层火山岩夹层，岩石为酸性至基性火山熔岩（图版Ⅰ-10-F）、沉火山碎屑岩（图11-1）。火山岩夹层厚度不一，薄者5~20cm，厚者4~6m。经区域变质作用，火山岩均已形成阳起石石英片岩、阳起石石英岩、绿帘斜长石片岩、含黑云母石英片岩、含阳起石钠长石绿帘绿岩等，但结构尚有部分保留。原岩恢复显示，层位自下而上，呈现出由变玄武岩→变安山岩→变流纹岩的演化，表明火山作用具有由基性→酸性的喷发作用特点。

昌江县邦溪农场、儋州市八一农场一带火山岩，呈北西西向夹层状产于奥陶系南碧沟组中（图版Ⅰ-10-D，图版Ⅰ-10-F），岩性主要为超基性、中基性变质火山岩及少量变质凝灰岩。火山岩层厚度不一，薄者0.6~0.8m，厚者4~6m，与上、下盘围岩呈整合接触关系。火山岩遭区域变质作用后形成斜长阳起石片岩、阳起石斜长片岩、绿泥石阳起石片岩等变质岩。据夏邦栋（1977）报道，在松涛林场北东石边角河一带发现有流纹斑岩，但河北省地矿局区调队与海南地质大队组成的联合区调队对该处岩石样品进行重新切片及鉴定，鉴定结果认为该处岩石为变质含砾砂岩或含砾千枚岩，不存在流纹斑岩。

东方市江边乡冲俄村，火山岩产于奥陶系南碧沟组上段中部和上部。中部为少量灰黑色变质中—基性火山岩，上部夹三大层变质基性火山熔岩、基性火山碎屑岩。上部三大层火山岩厚度自下而上分别为411m、75.6m、39.3m。每一层火山岩具明显的韵律，往往形成多个由熔岩-凝灰岩-绢云板岩或变质粉砂岩组成的韵律，为一套变质碎屑岩夹变质基性火山熔岩和基性火山碎屑岩的岩石组合(图11-2)。熔岩、凝灰岩单层厚一般为10~30cm，最薄为几厘米，有时可见变基性火山熔岩呈舌状体产于绢云板岩中。熔岩中常见被石英、阳起石等充填的浅色杏仁体，杏仁体大小一般2~5mm，常拉伸变形，拉伸后长可达5~8cm，拉伸方向一般平行于层理（片理）方向。经区域变质作用，火山岩原岩的矿物已全被显微纤柱状透闪石、阳起石及绿泥石、斜黝帘石等取代，但结构尚有部分保留。

此外，宜昌地质矿产研究所（1995）于万宁市山根乡一带奥陶系南碧沟组第二段上部泥质粉砂岩、粉砂岩、粉细砂岩中也见夹有灰带黄绿色厚层变凝灰质砂岩。

二、志留纪火山岩分布概况

古生代志留纪火山岩主要出露于屯昌县牙石园-西昌园、儋州市南丰镇油文村一带。此外，海南万宁市一带志留纪地层中也见凝灰岩、沉凝灰岩夹层。

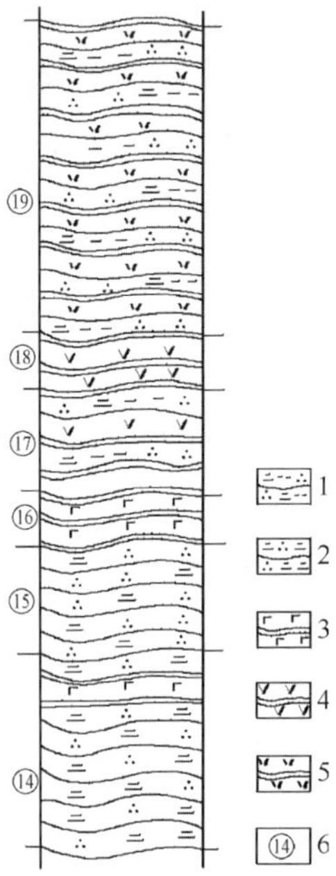

图 11-1　定安县翰林镇奥陶系南碧沟组剖面柱状图
Fig. 11-1　Section column of the Ordovician Nanbigou Formaiton, Hanlin Town, Ding'an County
（据海南省地质调查院，2004）
1—二云母石英片岩；2—白云母石英片岩；3—变玄武岩；4—变安山岩；5—变流纹岩；6—室内分层编号

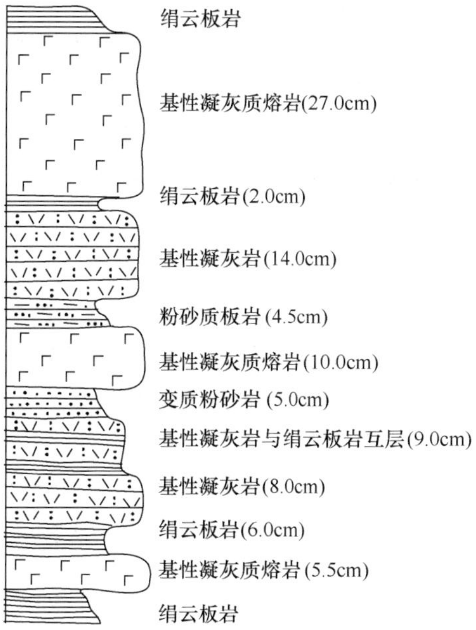

图 11-2　东方市南捉沟南碧沟组火山岩层序
Fig. 11-2　Nanbigou Formation volcanic sequence, south of Dongfang City
（据广西区调队，1991）

屯昌县牙石园-西昌园一带，火山岩呈夹层状产于陀烈组一套半深海—深海相沉积的炭质、泥砂质板岩、炭质板岩、千枚岩中，岩石为基性至超基性火山岩。1991 年陈惠芳、侯威等一度报道在屯昌县发现科马提岩类岩石，具有典型科马提岩矿物成分、矿物晶体形态及鬣刺结构。而许德如、林舸等（2001）的野外和室内鉴定认为岩石大多具绿泥石化、阳起石化，原生矿物遭受了较强烈的变质和变形改造，实际所见到的鬣刺结构已完全不同于低变质的绿岩带中所见的鬣刺结构，不过是一种鬣刺结构的假象。该处火山岩经变质后形成蛇纹石化透闪石岩、绢云绿泥透闪石片岩及透闪石阳起石片岩等。

出露于儋州市南丰镇油文村附近的火山岩，空间上呈脉状、薄层状或透镜状赋存于志留系陀烈组海相沉积岩中。1977~1978 年，原海南地质大队在该地区进行矿产普查工作时，发现在正常碎屑岩中有顺层的中基性—基性火山岩夹层，但未进一步的工作。而 1991 年河北省地矿局区调队与海南地质大队组成的联合区调队在 1∶5 万区调工作中于南丰镇油文村一带发现两条强蛇纹石化透闪石岩及斜长阳起石岩，岩石具球状风化，局部产状与围岩有一定交角，具有呈脉状产出的特征，认为该处变基性岩为潜火山岩。而海南省地质调查院（2004）在 1∶25 万区调工作中根据岩石的变形变质特征与围岩变形变质特征一致的特点，判断该处变基性岩为火山岩类。

海南地质大队（1990）在东方市南清河剖面陀烈组下部绢云母石英粉砂岩、石英岩、绢云母千枚岩中见夹两层变火山岩（角闪石片岩）。

宜昌地质矿产研究所（1995）在万宁市东岭农场江南队-前哨队一带志留纪陀烈组三段见多层沉凝灰岩和凝灰质砂岩。沉凝灰岩主要为酸性，灰黄色—黄色—黄白色，部分为酸偏中性，呈绿灰色。其地层基本层序组成的特点是：沉凝灰岩-凝灰质粉砂岩与泥岩、粉—细砂岩频繁不等厚互层。沉凝灰岩-凝灰质粉砂岩层厚度变化大，薄者不足 1m，厚者达 10 余米，纵向上看不出明显的成分、结构或粒度变化。宏观上火山岩层与沉积岩层的频繁交替代表了多个火山喷发-沉积韵律。

宜昌地质矿产研究所（1995）在万宁县山根乡沿斗牛村-黄竹塘高速公路边空列村组见较多沉凝灰岩、凝灰岩。岩石弱绢云母化，与变砂岩、板岩互层产出或独立大套出现，累积厚度达 30 余米。凝灰岩、沉凝灰岩为灰黄色—黄白色，含较多石英晶屑，部分具凝灰质砂岩、凝灰质泥岩特点。沉火山岩基本层序的组成特点是：下部为较粗的凝灰岩、沉凝灰岩，以粗—中砂级为主，次为细砂级，见晶屑和形态各异的岩屑（泥岩、粉砂岩、流纹岩）、浆屑，部分发育粒序层理和大型交错层理，显示沉凝灰岩特点；上部为粒度较细的凝灰岩、沉凝灰岩，粉砂—黏土级。自下而上沉凝灰岩粒度变细、火山碎屑减少，代表了一种火山活动的韵律性产物。

三、石炭纪火山岩分布概况

与奥陶纪和志留纪火山岩相比，海南岛石炭纪火山岩相对弱，主要为火山碎屑岩和（沉）凝灰岩，分布于海南岛东部万宁市东岭农场东乐队至江南队一带。

石炭系南好组上部见沉凝灰岩呈夹层产出，沉凝灰岩主要由微细粒火山灰组成，含少量石英晶屑，尖棱状，未见岩屑。

石炭系青天峡组上部见厚达39.45m的沉凝灰岩，为含晶屑沉凝灰岩，晶屑为石英，尖棱状、不规则状。可含少量石英砂，呈次圆—次棱角状，显示沉积作用特点。常有强绢云母化并可见少量绿泥石，局部强片理化。

四、古生代火山岩主要类型、火山岩建造

海南岛古生代火山岩类由于经历了海水和海底热液的蚀变、区域变质的改造，以及表生风化作用的影响，原岩的岩石学特征已面目全非，化学成分在较大程度也发生过调整，其中活动性组分（主要是K、Na、Ca、Mg，部分LREE等元素）在发生蚀变或区域变质作用时易发生明显的活动性迁移，而非活动性组分（主要是Al、Ti、Nb、Zr、Y、V、HREE等元素）在岩石发生蚀变或区域变质作用过程中不易发生迁移，它们能反映出岩石的某些原岩性质。因此在对其进行原岩恢复时主要采用以非活动性元素组合为主，多元素综合判别的方法来恢复火山原岩的性质。由图11-3、图11-4综合可知，屯昌县晨星农场、儋州市南丰镇油文村、邦溪农场、八一农场及江边乡冲俄村等地的变火山岩投影点主要落在亚碱性玄武岩区，少量落在碱性玄武岩区；海南岛东部定安县黄竹岭一带阳起石石英片岩主要

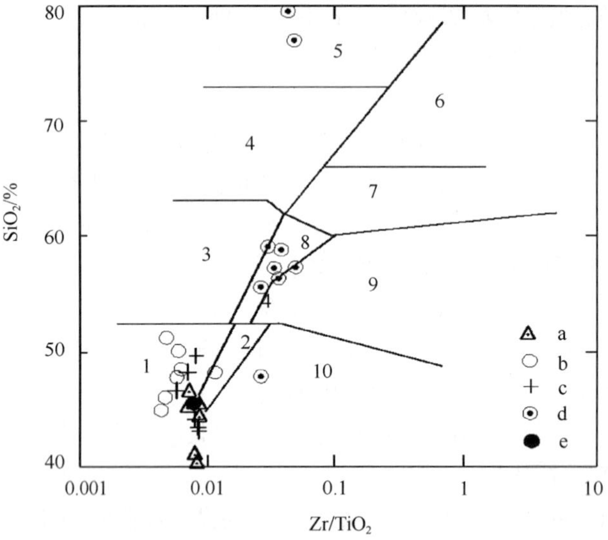

图11-3 SiO_2-Zr/TiO_2图解

Fig. 11-3 SiO_2-Zr /TiO_2 diagram

1—亚碱性玄武岩；2—碱性玄武岩；3—安山岩；4—英安岩；5—流纹岩；6—钠闪碱流岩；7—粗面岩；8—粗面安山岩；9—响岩；10—粗面玄武岩、碧玄岩、霞石岩

a—油文村一带变火山岩；b—屯昌晨星一带变火山岩；c—邦溪-八一一带变火山岩；d—黄竹岭一带变火山岩；e—冲俄村一带变火山岩

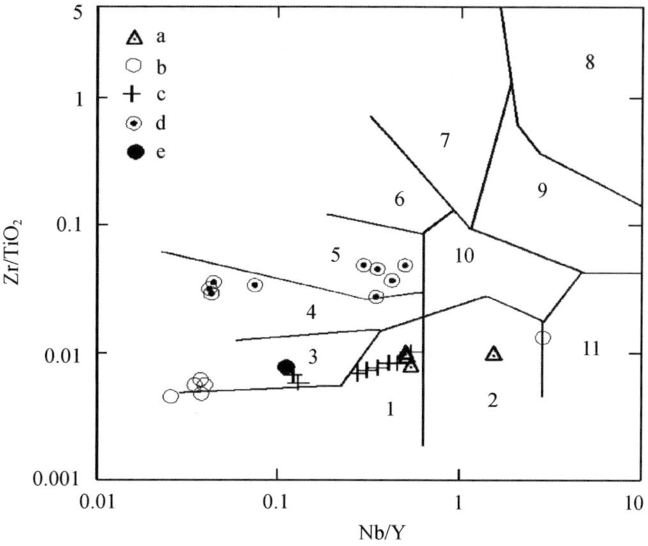

图 11-4　Nb/Y-Zr/TiO_2 图解

Fig. 11-4　Nb /Y-Zr /TiO_2 diagram

1—亚碱性玄武岩；2—碱性玄武岩；3—安山玄武岩；4—安山岩；5—英安岩；6—流纹岩；7—钠闪碱流岩；8—响岩；9—粗面岩；10—粗面安山岩；11—粗面玄武岩、碧玄岩、霞石岩

a—油文村一带变火山岩；b—屯昌晨星一带变火山岩；c—邦溪-八一一带变火山岩；d—黄竹岭一带变火山岩；e—冲俄村一带变火山岩

落在玄武岩区，绿帘斜长石片岩主要落在安山岩区，含黑云母石英片岩主要落在流纹岩区，并且根据对岩石的地质产状、分布以及岩石化学和地球化学特征等方面的综合分析，推测该处火山岩的岩性从早到晚具有由基性→中性→酸性演化特点。

第二节　火山岩的地球化学特征及其形成的构造环境

一、岩石化学特征

海南岛古生代变火山岩岩石化学成分与世界火山岩的平均化学成分值相比（表 11-1），总体上具有 CaO、MgO 偏高，而 Al_2O_3、TiO_2、Na_2O、K_2O 等偏低的特点。不同类岩性之间主要化学成分变化比较大，但同类岩性之间变化范围不是很大。其中，SiO_2 含量在 40.8%~79.6%之间，平均为 48.52%，Al_2O_3 5.16% ~ 16.81%，平均为 12.05%；$TiO_2$0.33% ~ 2.48%，平均为 0.94%；MgO 0.83%~29.40%，平均为 12.43%；CaO 0.93%~22.81%，平均为 9.45%；总铁（FeO+ Fe_2O_3）2.90%~20.67%，平均为 10.92%；Na_2O 含量变化较大，0.015%~6.60%不等；K_2O 含量变化也大，0.01%~4.83%不等。总体上铁镁含量高，钾钠含量低。其中分布于定安县黄竹岭一带的变火山岩显示从基性→中性→酸性的变化特征，而其他地方的均为基性岩。此外，出露于南丰镇油文村及邦溪农场有个别样品（HN-7、D4576）中的 MgO 的含量比较高，可能有铁镁矿物的堆晶。在 SiO_2 对 CaO、Al_2O_3、FeO^*、MgO、TiO_2 和 MnO 的相关图上（图 11-5），SiO_2 与 FeO^*、MgO、TiO_2 呈明显的负相关关系，与 Al_2O_3呈不明显的正相关关系。但在与 CaO 的关系上，定安县黄竹岭一带的呈明显的负相关关系，其他地区的却呈不明显的正相关关系。从 CIPW 标准矿物来看，岩石矿物组合为 Or+Ab+An+Di+Hy+Ol，其中以 An、Di、Dy、Ol 为主，个别样品还出现 Q 或 C，可能由于岩石受变质作用后，NaO_2、K_2O 等迁移所造成的，岩石中均有少量副矿物 Ap+Ilm+Mt 组合。

表 11-1 海南岛古生代火山岩化学成分、CIPW 标准矿物及岩石化学常用参数一览表

Table 11-1 List of chemical composition，CIPW standard minerals and common petrochemical parameters of Paleozoic volcanic rocks from the Hainan Island

地层单位		陀烈组									
岩性		蛇纹石化透闪石岩						透闪阳起石片岩			
样品号		D1891-1	D1891-2	D568-3	D1891-3	D1891-4	D1891-5	D11	CX-1	CX-2	CX-3
序号		1	2	3	4	5	6	7	8	9	10
岩石化学成分/%	SiO_2	47.00	45.66	45.35	43.52	40.80	41.09	48.52	49.97	47.96	47.40
	Al_2O_3	6.87	6.79	7.51	8.86	8.64	10.22	12.98	12.24	14.62	13.43
	Fe_2O_3	6.81	4.72	5.29	8.90	3.65	1.50	3.27	9.80	10.06	9.79
	FeO	7.42	8.65	8.70	6.42	11.00	12.90	7.98			
	MgO	17.36	18.95	18.15	16.86	21.03	18.03	9.92	13.18	12.32	11.64
	CaO	6.36	6.76	6.39	6.10	4.80	6.63	7.05	12.31	12.26	15.31
	Na_2O	0.08	0.10	0.11	0.03	0.06	0.05	3.37	1.74	1.31	1.39
	K_2O	0.06	0.04	0.02	0.03	0.01	0.17	1.82	0.19	0.82	0.63
	MnO	0.24	0.16	0.17	0.19	0.18	0.16	0.15	0.15	0.16	0.17
	TiO_2	1.42	1.41	1.54	1.54	1.83	2.15	2.44	0.56	0.62	0.41
	P_2O_5	0.20	0.22	0.23	0.28	0.29	0.37	0.73	0.05	0.05	0.02
	CO_2	0.04	0.04	0.02	0.04	0.04	0.04	0.08			
	H_2O	5.73	6.12	6.27	6.84	7.30	6.47	1.39			
	总量	99.59	99.60	99.75	99.61	99.63	99.78	99.70	100.18	100.01	99.47
标准矿物成分含量/%	Q	3.41		0.30							
	Or	0.38	0.25	0.13	0.19	0.06	1.08	10.96	1.12	4.84	3.72
	Ab	0.72	0.91	1.00	0.28	0.55	0.45	29.00	14.68	11.05	6.78
	An	19.46	19.22	21.35	25.95	23.97	29.08	15.15	24.95	31.49	28.45
	Di	10.45	12.24	8.95	4.10		3.12	12.89	28.68	23.29	38.07
	Hy	56.94	57.84	59.18	58.72	44.98	36.18	4.09	14.65	7.68	
	Ol		0.91		1.45	19.80	22.51	16.75	11.92	17.46	16.64
	Ap	0.47	0.51	0.54	0.66	0.69	0.87	1.62	0.11	0.11	0.04
	Ilm	2.89	2.87	3.14	3.17	3.77	4.38	4.72	1.06	1.18	0.78
	Mt	5.29	5.28	5.43	5.47	5.73	2.33	4.83	2.84	2.91	2.84
岩石化学常用参数	$Mg^{\#}$	69.55	72.37	70.62	67.57	72.41	69.29	61.82	72.71	70.81	70.20
	AlK	0.14	0.14	0.13	0.06	0.07	0.22	5.19	1.93	2.13	2.02
	Na_2O/K_2O	1.33	2.50	5.50	1.00	6.00	0.29	1.85	9.16	1.60	2.21
	FeO^*/MgO	0.78	0.73	0.74	0.86	0.68	0.79	1.10	0.74	0.82	0.84
	δ							4.88	0.53	0.91	0.93
	τ	4.78	4.74	4.81	5.73	4.69	4.73	3.94	18.75	21.47	29.37
	AR	1.02	1.02	1.02	1.01	1.01	1.03	1.70	1.17	1.17	1.15
	SI	55.30	58.61	56.56	53.19	58.83	55.22	37.63	52.91	50.27	49.64
产地		南丰镇油文村						屯昌县晨星农场			

地层单位		陀烈组									
岩性		透闪阳起石片岩									
样品号		CX-5	CX-6	V-1	K-2	B9-1	B12-1	B14-1	B24-1	B27-1	CX-4
序号		11	12	13	14	15	16	17	18	19	20
岩石化学成分/%	SiO_2	44.78	45.84	46.91	47.62	46.73	44.99	49.71	47.51	48.28	50.92
	Al_2O_3	15.41	15.21	13.95	13.50	8.95	9.13	12.91	11.75	14.03	12.63
	Fe_2O_3	11.00	10.59	1.60	1.28	3.22	3.34	1.27	1.61	1.01	7.97
	FeO			7.36	7.65	8.90	8.55	8.90	6.91	8.28	
	MgO	13.78	11.94	11.01	11.40	17.38	19.34	10.95	15.08	11.37	10.71
	CaO	12.62	14.00	12.31	12.34	9.15	8.48	11.57	12.67	12.35	14.77
	Na_2O	0.98	1.17	1.58	1.13	0.35	0.08	1.52	1.20	1.18	2.12
	K_2O	0.93	0.79	0.28	1.05	0.04	0.02	0.16	0.04	0.22	0.50
	MnO	0.15	0.14	0.20	0.17	0.23	0.19	0.18	0.13	0.15	0.15
	TiO_2	0.50	0.47	0.53	0.55	0.53	0.52	0.72	0.33	0.57	0.39
	P_2O_5	0.03	0.03	0.05	0.04	0.03	0.04	0.05	0.01	0.03	0.02
	CO_2										
	H_2O				2.33	3.97	5.01	1.32	2.07	1.79	
	总量	99.82	100.09	95.78	99.06	99.48	99.69	99.26	99.31	99.26	99.72
标准矿物成分含量/%	Q										
	Or	5.49	4.66	1.73	6.42	0.25	0.13	0.97	0.24	1.34	2.92
	Ab	4.08	5.41	13.94	9.87	3.10	0.71	13.12	10.43	10.23	17.89
	An	34.79	33.81	31.43	29.59	23.78	25.85	28.48	27.28	33.14	23.40
	Di	22.16	28.52	25.81	26.98	18.88	14.66	24.18	29.80	23.90	39.86
	Hy			10.56	10.52	41.56	39.60	27.01	12.20	21.05	0.56
	Ol	27.02	21.15	12.94	13.53	6.99	13.57	2.86	16.98	7.67	12.26
	Ap	0.07	0.07	0.11	0.09	0.07	0.09	0.11	0.02	0.07	0.04
	Ilm	0.95	0.89	1.05	1.08	1.05	1.04	1.40	0.64	1.11	0.74
	Mt	3.18	3.07	2.42	1.92	4.31	4.34	1.88	2.40	1.50	2.30
岩石化学常用参数	$Mg^{\#}$	71.28	69.08	69.05	69.78	72.42	74.90	66.03	76.28	68.81	72.69
	AlK	1.91	1.96	1.86	2.18	0.39	0.10	1.68	1.24	1.40	2.62
	Na_2O/K_2O	1.05	1.48	5.64	1.08	8.75	4.00	9.50	30.00	5.36	4.24
	FeO^*/MgO	0.80	0.89	0.80	0.77	0.68	0.60	0.92	0.55	0.81	0.74
	δ	2.05	1.35	0.88	1.03	0.04		0.42	0.34	0.37	0.87
	τ	28.86	29.87	23.34	22.49	16.23	17.44	15.82	31.97	22.54	26.95
	AR	1.15	1.14	1.15	1.18	1.04	1.01	1.15	1.11	1.11	1.21
	SI	51.63	48.75	50.44	50.64	58.15	61.73	48.03	60.71	51.54	50.28
产地		屯昌县晨星农场									

续表

地层单位		陀烈组		南碧沟组						
岩性		透闪阳起石片岩		绿泥石阳起石片岩					变基性火山岩	
样品号		D91001	D91002	0214-1	0218-7	0219-1	0662-2	0662-3	D4576	BⅣ-5
序号		21	22	23	24	25	26	27	28	29
岩石化学成分/%	SiO_2	46.62	46.85	43.32	42.84	39.70	41.76	44.68	40.85	48.67
	Al_2O_3	15.48	15.50	11.68	8.80	11.71	11.86	10.43	6.16	16.16
	Fe_2O_3	0.34	0.97	12.31	12.78	12.58	12.80	10.87	3.36	2.40
	FeO	8.49	8.80						8.99	9.59
	MgO	10.79	7.78	18.07	20.64	22.05	19.91	19.37	29.40	3.57
	CaO	10.38	14.22	9.23	8.30	6.60	6.40	8.91	3.65	7.71
	Na_2O	1.95	2.25	0.55	0.05	0.05	0.29	0.82	0.06	5.36
	K_2O	0.48	0.48	0.06	0.02	0.01	0.01	0.03	0.02	0.22
	MnO	0.12	0.10	0.18	0.18	0.18	0.17	0.17	0.19	0.20
	TiO_2	0.76	0.72	0.97	1.07	1.02	0.56	0.51	0.58	1.70
	P_2O_5	0.08	0.24	0.10	0.16	0.13	0.06	0.03	0.10	1.03
	CO_2								0.19	0.11
	H_2O	1.74	1.02						6.02	1.09
	灼失量			3.41	4.81	5.71	6.51	4.31	5.26	1.09
	总量	99.21	98.80	99.87	99.64	99.74	100.33	100.13		
标准矿物成分含量/%	Q									
	Or	2.97	2.90	0.37	0.12	0.06	0.06	0.19	0.13	1.35
	Ab	17.26	13.05	4.82	0.45	0.45	2.61	7.23	0.50	43.53
	An	33.54	31.39	30.26	24.99	33.67	33.04	25.74	14.90	20.04
	Di	16.14	32.13	13.3	13.94	0.29	0.36	16.14	3.23	11.16
	Hy	8.16		19.09	29.56	21.10	29.98	19.30	30.98	
	Ol	19.72	13.71	26.33	24.52	38.18	28.71	27.04	43.56	12.84
	Ap	0.18	0.54	0.23	0.37	0.30	0.14	0.07	0.24	2.33
	Ilm	1.51	1.40	1.91	2.14	2.06	1.13	1.01	1.20	3.34
	Mt	0.52	1.44	3.70	3.91	3.89	3.96	3.38	5.28	3.60
岩石化学常用参数	$Mg^{\#}$	68.62	58.92	72.76	74.61	76.13	73.89	76.43	81.35	35.14
	AlK	2.43	2.73	0.61	0.07	0.06	0.30	0.85	0.06	5.58
	Na_2O/K_2O	4.06	4.69	9.17	2.5	5	29	27.3	3	24.36
	FeO^*/MgO	0.82	1.24	0.68	0.62	0.57	0.64	0.56	0.42	3.36
	σ	1.63	1.94	1.16				0.43		5.49
	τ	17.80	18.40	11.47	8.18	11.43	20.66	18.84	8.76	6.35
	AR	1.21	1.20	1.06	1.01	1.01	1.03	1.09	1.02	1.61
	SI	48.93	38.36	58.31	61.63	63.56	60.32	62.30	70.29	16.89
产地		屯昌县晨星农场		八一农场			邦溪农场			

续表

地层单位		南碧沟组									
岩性		变基性火山岩					绿泥石阳起石片岩			含黑云母石英片岩	
样品号		H-1	S-3	B-1	HN-7	D8079-1	1167	HNOn	On-1	H10-1	H10-2
序号		30	31	32	33	34	35	36	37	38	39
岩石化学成分/%	SiO_2	46.28	47.76	49.29	43.79	46.04	46.94	50.52	45.8	79.60	76.60
	Al_2O_3	15.79	16.52	13.59	11.12	14.03	16.12	12.65	13.18	8.24	9.35
	Fe_2O_3	12.22	11.78	11.01	14.44	3.38	1.13	2.96	1.63	0.32	1.00
	FeO					7.48	9.47	12.70	19.04	2.58	3.02
	MgO	13.24	8.63	13.72	21.04	12.73	7.04	5.77	6.98	0.83	0.90
	CaO	9.66	9.60	10.28	7.94	10.78	13.35	9.27	4.22	0.93	0.94
	Na_2O	1.78	3.00	1.34	0.58	0.17	1.74	0.35	0.75	0.88	1.14
	K_2O	0.01	0.45	0.06	0.01	0.10	0.90	1.00	0.45	4.57	4.46
	MnO	0.17	0.18	0.18	0.20	0.20	0.2	0.31	0.25	0.05	0.03
	TiO_2	0.95	2.02	0.63	0.94	0.99	0.65	1.70	2.48	0.46	0.54
	P_2O_5	0.09	0.24	0.07	0.12	0.13	0.06	0.20	0.24	0.16	0.16
	CO_2					0.44			0.18	0.10	0.10
	H_2O					1.64			3.82	1.04	1.49
	灼失量	2.80	1.22	2.29	3.50	1.64					
	总量	100.21	99.87	99.31	100.08	99.75	97.60	97.43	99.07	99.76	99.72
标准矿物成分含量/%	Q					1.85		11.71	7.60	53.16	49.07
	Or	0.06	2.66	0.35	0.06	0.63	5.51	6.07	2.80	27.41	26.88
	Ab	15.02	25.31	11.31	4.89	1.48	13.87	3.04	6.68	7.54	9.82
	An	34.96	30.19	30.80	27.63	38.72	34.70	30.78	20.55	3.73	3.86
	Di	9.98	12.97	15.86	8.81	12.96	27.66	12.70			
	Hy	12.62	4.04	29.21	18.00	37.02		27.54	49.85	5.96	6.24
	Ol	21.85	17.07	7.94	34.38		15.92				
	Ap	0.20	0.52	0.15	0.26	0.29	0.14	0.45	0.55	0.35	0.33
	Ilm	1.80	3.83	1.19	1.78	1.96	1.28	3.31	4.96	0.89	1.05
	Mt	3.53	3.42	3.18	4.18	5.10	0.20	4.41	2.49	0.47	1.48
岩石化学常用参数	$Mg^{\#}$	66.34	57.13	69.39	72.61	68.32	56.70	40.10	37.77	34.01	29.04
	AlK	1.79	3.45	1.40	0.59	0.27	2.64	1.35	1.20	5.45	5.60
	Na_2O/K_2O	178	6.67	22.33	58	1.70	1.92	0.35	1.67	0.19	0.26
	FeO^*/MgO	0.92	1.37	0.80	0.69	0.85	1.36	2.66	2.94	3.46	4.36
	σ	0.98	2.50	0.31	0.44	0.02	1.34	0.22	0.31	0.81	0.92
	τ	14.75	6.69	19.44	11.21	14.02	22.12	7.24	5.01	16.00	15.20
	AR	1.15	1.30	1.12	1.06	1.02	20.12	20.82	17.08	3.93	3.39
	SI	48.59	36.17	52.51	58.33	53.35	36.51	25.33	24.19	9.04	8.56
产地		邦溪农场					冲俄村			定安县黄竹岭	

续表

地层单位		南碧沟组							世界火山岩的平均化学成分及CIPW数字特征（R. W. Le Maitre，1976）		
岩性		阳起石石英片岩	绿帘斜长石片岩								
样品号		XⅥ15-1	H7-1	H7-2	H8-1	H8-2	H9-1	H9-2	流纹岩	安山岩	玄武岩
序号		40	41	42	43	44	45	46			
岩石化学成分/%	SiO_2	47.44	58.61	58.49	56.02	56.85	56.60	55.34	72.82	57.94	49.20
	Al_2O_3	10.58	15.06	14.24	16.81	13.92	11.53	9.94	13.27	17.02	15.74
	Fe_2O_3	2.89	0.56	0.90	0.41	1.56	0.59	0.76	1.48	3.27	3.79
	FeO	8.34	4.66	4.86	5.46	6.94	7.38	9.50	1.11	4.04	7.13
	MgO	2.90	2.63	2.34	2.10	2.29	2.99	3.63	0.39	3.33	6.73
	CaO	22.81	8.71	9.47	8.91	10.07	10.70	12.92	1.14	6.79	9.47
	Na_2O	1.62	6.27	6.60	5.51	5.73	2.53	3.56	3.55	3.48	2.91
	K_2O	0.20	1.01	0.42	1.38	0.58	4.83	1.71	4.30	1.62	1.10
	MnO	0.56	0.24	0.28	0.25	0.35	0.36	0.48	0.06	0.14	0.20
	TiO_2	0.49	0.68	0.69	0.87	0.80	0.57	0.53	0.28	0.87	1.84
	P_2O_5	0.09	0.14	0.15	0.23	0.21	0.13	0.09	0.07	0.21	0.35
	CO_2	0.89	0.14	0.14	0.31	0.24	0.51	0.24	0.08	0.05	0.11
	H_2O	1.05	1.00	1.13	1.37	1.11	0.96	1.01	1.10	0.83	0.95
	总量	99.86	99.71	99.71	99.63	99.65	99.69	99.71	99.65	99.59	99.53
标准矿物成分含量/%	Q	7.82		1.25		0.23	0.58		32.87	12.37	
	Or	1.22	6.06	2.52	8.33	3.45	29.09	10.27	28.44	9.60	6.25
	Ab	14.15	53.48	56.66	46.36	48.77	21.77	30.56	30.07	29.44	24.66
	An	21.68	10.09	8.10	17.38	10.61	5.90	6.17	4.76	26.02	26.62
	Ne		0.15		0.64						
	Di	41.69	27.34	27.25	22.20	32.51	39.91	49.33		4.84	14.02
	Hy					0.16	0.48	1.32	1.34	9.49	15.20
	Ol		0.43		2.28			0.01			1.50
	Ap	0.20	0.31	0.33	0.51	0.46	0.29	0.20	0.17	0.50	0.82
	Ilm	0.96	1.31	1.33	1.69	1.53	1.10	1.02	0.54	1.65	3.49
	Mt	4.33	0.82	1.33	0.61	2.28	0.87	1.12	2.14	4.74	5.49
岩石化学常用参数	$Mg^{\#}$	32.09	47.60	42.40	39.10	32.90	40.30	38.90			
	AlK	1.82	7.28	7.02	6.89	6.31	7.36	5.27			
	Na_2O/K_2O	8.1	6.21	15.71	3.99	9.88	0.52	2.08			
	FeO^*/MgO	3.77	1.96	2.42	0.96	3.64	2.65	2.81			
	δ	0.75	3.40	3.18	3.65	2.87	3.98	2.25			
	τ	18.29	12.93	11.07	12.99	10.24	15.79	12.04			
	AR	1.12	1.88	1.84	1.73	1.71	1.99	1.60			
	SI	18.18	17.38	15.48	14.13	13.39	16.32	18.95			
产地		定安县黄竹岭									

注：1~7、35、36、38~46 据海南省地质调查院（2004）；8~12、20、30~33 据李献华（2000）；13~19、21、22 据许德如（2001）；28、29、34 据宜昌地质矿产研究所（1995）；35、36 据广西区调队（1991）；37 据宜昌地质矿产研究所、海南省地质调查院（2004）；23~27 据海南省地质调查院（2007）；$Mg^{\#}=100\times Mg/(Mg+Fe)$。

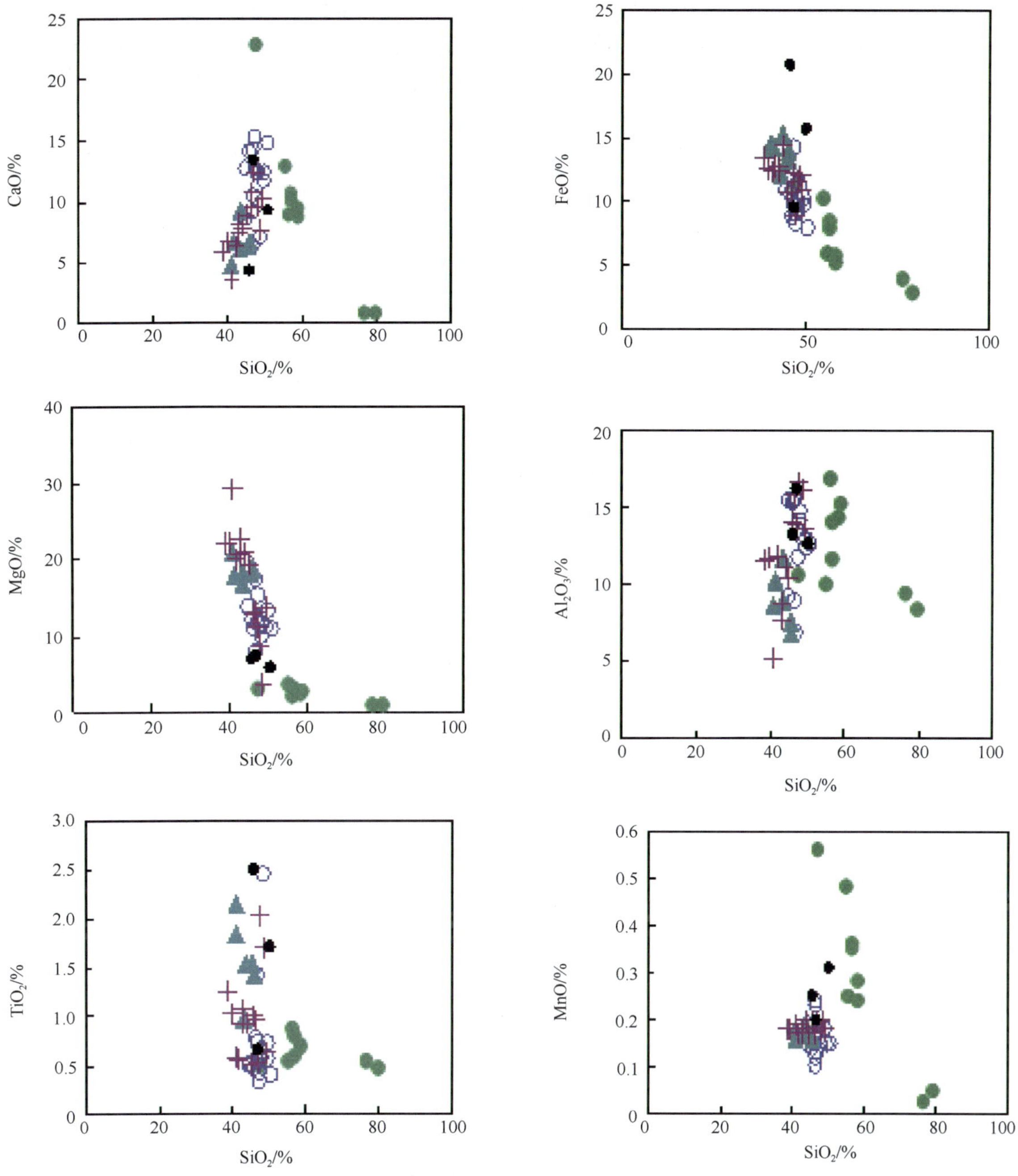

图 11-5 海南岛古生代变火山岩成分变异图

Fig. 11-5 Material composition variation diagram for Paleozoic meta-volcanic rocks from the Hainan Island

二、稀土元素地球化学特征

稀土配分模式显示（表 11-2，图 11-6），相同地区变火山岩岩石具有相似的稀土配分模式，但不同地区的稀土配分模式略有出入。单个岩石样品稀土总量变化较大，由 $12.41\times10^{-6}\sim1424.12\times10^{-6}$。轻、重稀土的比值在 0.48~9.62 之间，除出露于晨星农场部分样品略比 1 小外，大部分样品比 1 大。

油文村一带岩石稀土元素总量 $111.85\times10^{-6}\sim205.24\times10^{-6}$，其配分模式为呈右倾的轻稀土富集型（见图 11-6），与洋岛玄武岩（OIB）的配分模式类似。LREE/HREE 比值为 5.56~7.97，$(La/Yb)_N$值为 15.1~17.72，$(Ce/Yb)_N$值为 5.82~11.78，属轻稀土富集型。δEu 值为 0.81~1.59，大部分均在 1.0 左右，没有明显的 Eu 亏损和富集。Ce 亏损，δCe 值为 0.46~0.95。$(La/Sm)_N$值为 2.57~118，$(Gd/Yb)_N$值为 4.13~4.62，表明轻稀土元素间的分馏作用比较明显，而重稀土元素间的分馏作用比较弱。

表 11-2　海南岛古生代部分变火山岩稀土元素含量及主要特征参数一览表

Table 11-2　List of REE contents and main characteristic parameters of some Paleozoic meta-volcanic rocks from the Hainan Island

地层单位		陀烈组										
岩性		蛇纹石化透闪石岩					透闪阳起石片岩					
样品号		D1891-2	D568-3	D1891-3	D1891-4	D1891-5	D11	CX-1	CX-2	CX-3	CX-4	CX-5
序号		1	2	3	4	5	6	7	8	9	10	11
稀土元素含量/10^{-6}	La	27.14	36.46	41.78	54.56	26.43	44.02	0.72	0.66	0.59	0.47	0.34
	Ce	29.94	35.75	47.87	49.88	53.76	79.40	2.37	2.15	1.39	1.30	0.95
	Pr	5.87	8.18	9.68	11.55	6.68	10.16	0.45	0.41	0.24	0.23	0.20
	Nd	25.57	35.51	41.35	47.79	28.67	41.65	3.09	2.94	1.87	1.71	1.67
	Sm	5.57	7.85	8.93	10.16	6.48	9.16	1.30	1.35	0.89	0.76	0.91
	Eu	1.69	2.23	4.74	2.81	2.08	2.90	0.56	0.32	0.38	0.41	0.41
	Gd	6.24	9.03	9.10	10.89	6.13	7.75	1.99	2.19	1.55	1.37	1.64
	Tb	0.92	1.30	1.30	1.56	0.87	1.14	0.39	0.44	0.32	0.29	0.35
	Dy	4.70	6.49	6.59	8.39	4.54	5.49	2.78	3.15	2.24	2.04	2.61
	Ho	0.85	1.12	1.14	1.49	0.78	0.93	0.63	0.71	0.48	0.47	0.61
	Er	1.84	2.61	2.48	3.31	1.70	2.19	1.88	2.07	1.49	1.41	1.80
	Tm	0.23	0.32	0.33	0.44	0.22	0.27	0.27	0.30	0.23	0.21	0.26
	Yb	1.14	1.59	1.59	2.13	1.18	1.51	1.85	2.03	1.57	1.49	1.76
	Lu	0.15	0.20	0.20	0.28	0.16	0.19	0.30	0.32	0.25	0.25	0.28
	∑REE	111.85	148.64	177.08	205.24	139.86	206.76	18.58	19.04	13.49	12.41	13.79
主要特征参数	LREE/HREE	5.96	5.56	6.79	6.20	7.98	9.62	0.84	0.70	0.66	0.65	0.48
	δEu	0.87	0.81	1.59	0.81	0.99	1.03	1.06	0.57	0.98	1.21	1.01
	δCe	0.55	0.48	0.55	0.46	0.94	0.82	0.98	0.97	0.89	0.95	0.86
	$(La/Yb)_N$	16.05	15.46	17.72	17.27	15.10	18.87	0.26	0.22	0.25	0.21	0.13
	$(Ce/Yb)_N$	6.79	5.82	7.79	6.06	11.78		0.33	0.27	0.23	0.23	0.14
	$(La/Sm)_N$	3.06	2.92	2.94	3.38	2.57	2.76	0.35	0.31	0.42	0.39	0.24
	$(Gd/Yb)_N$	4.42	4.58	4.62	4.13	4.19	4.35	0.87	0.87	0.80	0.74	0.75
	La/Sm	4.87	4.64	4.68	5.37	4.08	4.81	0.55	0.49	0.66	0.62	0.37

地层单位		陀烈组									
岩性		透闪阳起石片岩									
样品号		CX-6	V-1	K-2	B9-1	B12-1	B14-1	B24-1	B27-1	D91001	D91002
序号		12	13	14	15	16	17	18	19	20	21
稀土元素含量/10^{-6}	La	0.45	1.02	1.41	2.54	1.66	1.35	2.28	1.24	3.90	7.70
	Ce	1.29	2.91	3.72	5.08	5.09	3.63	4.56	2.97	8.80	14.00
	Pr	0.24	0.58	0.54	0.89	0.66	0.68	0.50	0.54	1.10	1.90
	Nd	1.86	2.95	2.96	4.08	2.84	3.79	2.58	3.16	4.80	7.30
	Sm	0.89	1.27	1.06	1.31	0.92	1.32	0.71	1.03	1.30	1.80
	Eu	0.42	0.63	0.50	0.48	0.31	0.53	0.30	0.32	0.56	0.78
	Gd	1.58	1.90	1.97	1.80	1.44	1.97	0.98	1.47	2.10	2.30
	Tb	0.33	0.36	0.36	0.33	0.30	0.41	0.23	0.25	0.41	0.44
	Dy	2.45	2.57	2.97	2.50	2.21	2.97	1.48	2.15	2.90	2.90

地层单位		陀烈组									
岩性		透闪阳起石片岩									
样品号		CX-6	V-1	K-2	B9-1	B12-1	B14-1	B24-1	B27-1	D91001	D91002
序号		12	13	14	15	16	17	18	19	20	21
稀土元素含量 10^{-6}	Ho	0.56	0.57	0.62	0.53	0.45	0.60	0.28	0.48	0.62	0.60
	Er	1.68	1.70	1.80	1.51	1.32	1.74	0.89	1.31	1.81	1.70
	Tm	0.25	0.29	0.28	0.23	0.19	0.26	0.14	0.22	0.27	0.26
	Yb	1.74	1.76	1.79	1.38	1.20	1.58	0.78	1.27	1.70	1.60
	Lu	0.27	0.28	0.27	0.20	0.18	0.24	0.12	0.19	0.26	0.25
	ΣREE	14.01	18.79	20.25	22.86	18.77	21.07	15.83	16.57	30.53	43.53
主要特征参数	LREE/HREE	0.58	0.99	1.01	1.70	1.57	1.16	2.23	1.27	2.03	3.33
	δEu	1.07	1.24	1.04	0.96	0.82	1.00	1.10	0.79	1.03	1.17
	δCe	0.93	0.90	1.03	0.81	1.17	0.90	0.99	0.87	1.01	0.86
	$(La/Yb)_N$	0.17	0.39	0.53	1.24	0.93	0.58	1.97	0.66	1.55	3.24
	$(Ce/Yb)_N$	0.19	0.43	0.54	0.95	1.10	0.59	1.51	0.60	1.34	2.26
	$(La/Sm)_N$	0.32	0.51	0.84	1.22	1.13	0.64	2.02	0.76	1.89	2.69
	$(Gd/Yb)_N$	0.74	0.87	0.89	1.05	0.97	1.01	1.01	0.93	1.00	1.16
	La/Sm	0.51	0.80	1.33	1.94	1.80	1.02	3.21	1.20	3.00	4.28

地层单位		南碧沟组											
岩性		绿泥石阳起石片岩					变基性火山岩						
样品号		0218-2	0218-8	0218-7	On-1	1167	D8079-1	HN-7	B-1	H-1	S-3	0219-1	D4576
序号		22	23	24	25	26	27	28	29	30	31	32	33
稀土元素含量 10^{-6}	La	8.217	8.165	2.295	19.3	54.9	6.73	4.27	1.66	2.89	8.96	8.55	7.02
	Ce	18.825	16.77	20.64	47.5	110.1	11.0	9.67	4.58	7.45	23.6	19.71	14.3
	Pr	2.68	2.33	2.832	5.7	12.5	2.28	1.86	0.87	1.32	3.64	2.814	1.90
	Nd	11.77	10.33	12.02	26.1	47.4	8.97	8.67	4.67	6.74	17.0	11.7	7.97
	Sm	2.902	2.501	2.929	7.2	9.3	2.49	2.38	1.80	2.29	4.75	2.837	1.99
	Eu	1.218	1.014	1.02	1.76	1.68	1.19	1.00	0.64	1.08	2.08	0.898	0.98
	Gd	3.15	2.793	3.202	10.23	7.25	2.64	2.84	2.22	3.04	5.38	2.915	2.04
	Tb	0.559	0.477	0.54	1.77	1.33	0.48	0.46	0.40	0.58	0.92	0.522	0.35
	Dy	3.376	2.734	3.341	11.79	7.12	3.31	2.94	2.78	4.15	6.07	3.144	2.86
	Ho	0.693	0.592	0.679	2.61	1.45	0.68	0.60	0.58	0.87	1.21	0.637	0.57
	Er	2.004	1.687	1.936	7.83	4.26	1.90	1.78	1.69	2.53	3.39	1.778	1.67
	Tm	0.293	0.234	0.285	1.14	0.76	0.29	0.26	0.26	0.38	0.48	0.272	0.25
	Yb	1.866	1.534	1.752	7.57	4.33	1.81	1.67	1.69	2.48	3.08	1.645	1.35
	Lu	0.268	0.231	0.271	1.13	0.66	0.26	0.24	0.23	0.37	0.44	0.244	0.14
	ΣREE	57.82	51.39	59.75	151.8	262.9	43.97	38.64	24.07	36.17	81.00	57.67	43.55

地层单位		南碧沟组											
岩性		绿泥石阳起石片岩					变基性火山岩						
样品号		0218-2	0218-8	0218-7	On-1	1167	D8079-1	HN-7	B-1	H-1	S-3	0219-1	D4576
序号		22	23	24	25	26	27	28	29	30	31	32	33
主要特征参数	LREE/HREE	3.74	4.00	3.98	2.44	8.7	2.89	2.58	1.44	1.51	2.86	4.17	3.64
	δEu	1.23	1.17	1.01	0.62	0.60	1.41	1.17	0.98	1.25	1.25	0.95	1.47
	δCe	0.96	0.91	1.02	1.08	0.97	0.67	0.82	0.91	0.92	0.99	0.96	0.93
	$(La/Yb)_N$	2.97	3.59	3.19	1.8	8.55	2.51	1.72	0.66	0.79	1.96	3.50	3.13
	$(Ce/Yb)_N$	2.61	2.83	3.06	1.62	6.58	1.57	1.50	0.70	0.78	1.98	3.10	2.45
	$(La/Sm)_N$	1.78	2.05	1.78	1.7	3.71	1.70	1.13	0.58	0.79	1.19	1.90	2.22
	$(Gd/Yb)_N$	1.36	1.47	1.47	1.09	1.35	1.18	1.37	1.06	0.99	1.41	1.43	1.09
	La/Sm	2.83	3.26	2.83	2.68	5.90	2.70	1.79	0.92	1.26	1.89	3.01	3.53

地层单位		南碧沟组								
岩性		阳起石石英片岩	绿帘斜长石片岩						含黑云母石英片岩	
样品号		XⅥ15-1	H7-1	H7-2	H8-1	H8-2	H9-1	H9-2	H10-1	H10-2
序号		34	35	36	37	38	39	40	41	42
稀土元素含量 10^{-6}	La	32.21	125.0	128.2	65.04	73.07	426.5	595.0	39.06	36.52
	Ce	64.12	71.89	76.93	98.65	88.71	60.58	73.12	56.03	50.41
	Pr	7.76	21.75	20.75	14.22	14.27	77.26	101.0	8.81	8.29
	Nd	27.79	83.37	77.86	48.72	50.32	289.3	385.0	31.33	29.0
	Sm	5.22	15.41	13.58	8.84	8.84	44.25	58.16	5.63	2.19
	Eu	1.08	3.53	3.17	1.51	1.39	9.31	12.55	0.84	0.79
	Gd	4.52	20.46	17.30	8.39	8.96	47.08	64.03	6.05	5.40
	Tb	0.75	3.27	2.62	1.36	1.43	7.23	10.45	0.99	0.90
	Dy	4.35	21.37	16.32	8.11	8.21	39.34	58.82	5.76	4.93
	Ho	0.88	4.46	3.39	1.58	1.61	7.37	10.97	1.13	0.97
	Er	2.60	12.75	9.84	4.20	4.20	18.39	27.25	2.98	2.61
	Tm	0.41	1.70	1.36	0.61	0.59	2.46	3.53	0.39	0.38
	Yb	2.54	9.50	7.79	4.14	3.90	16.44	21.03	2.63	2.39
	Lu	0.37	1.47	1.17	0.64	0.63	2.44	3.21	0.42	0.38
	ΣREE	154.60	395.93	380.28	265.80	266.13	1048.35	1424.12	162.05	148.16
主要特征参数	LREE/HREE	8.42	4.28	5.36	8.16	8.01	6.45	6.15	6.96	7.25
	δEu	0.66	0.61	0.63	0.54	0.47	0.62	0.63	0.44	0.45
	δCe	0.95	0.31	0.33	0.75	0.62	0.07	0.07	0.70	0.67
	$(La/Yb)_N$	8.55	8.87	11.10	10.59	12.63	17.49	19.07	10.01	10.30
	$(Ce/Yb)_N$	6.53	1.96	2.55	6.16	5.88	0.95	0.90	5.51	5.46
	$(La/Sm)_N$	3.88	5.10	5.94	4.74	5.20	6.06	6.44	4.36	4.43
	$(Gd/Yb)_N$	1.44	1.74	1.79	1.64	1.85	2.31	2.46	1.86	1.82
	La/Sm	6.17	8.11	9.44	7.36	8.27	9.64	10.23	6.94	16.68

注：1~6、34~42 据海南省地质调查院（2004）；7~12、28~31 据李献华（2000）；13~21 据许德如（2001）；22~24、32 据海南省地质调查院（2007）；25 据宜昌地质矿产研究所、海南省地质调查院（2004）；26 据广西区调队（1991）；27、33 据宜昌地质矿产研究所（1995）。

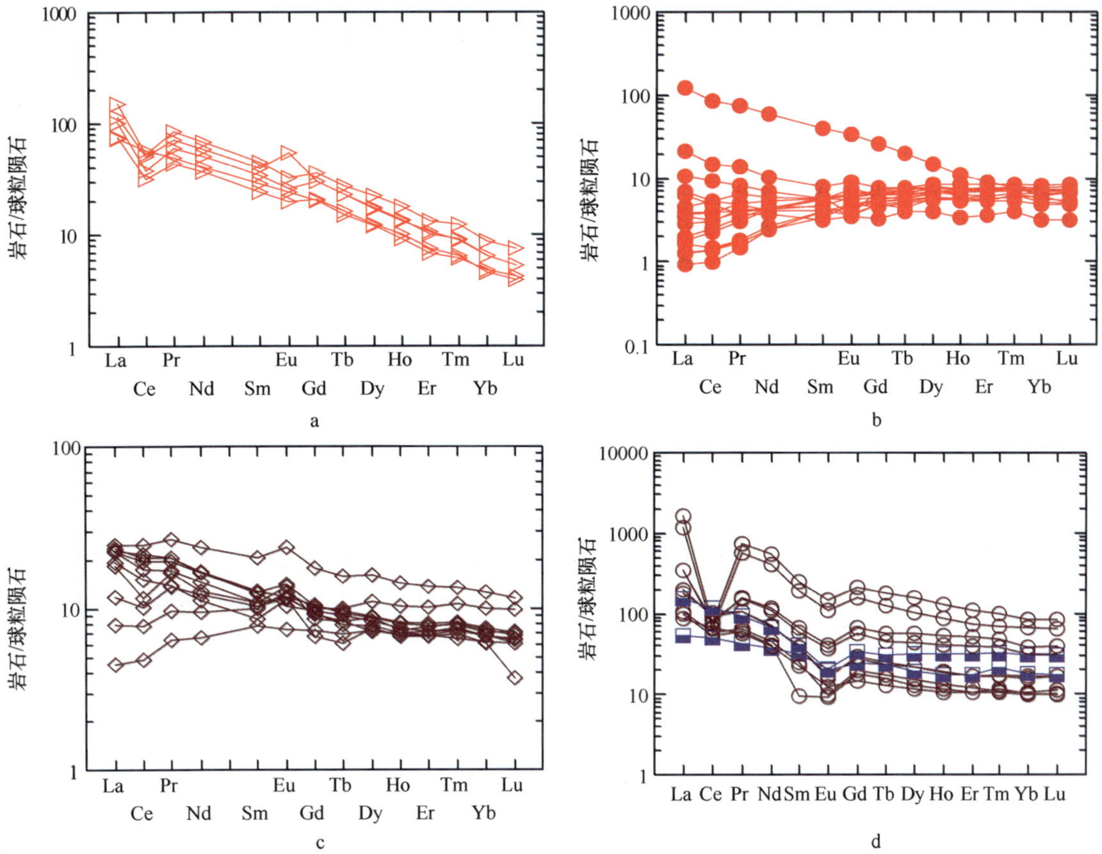

图 11-6　海南岛古生代变火山岩稀土元素球粒陨石标准化图

Fig. 11-6　Chondrite-normalized REE patterns of Paleozoic meta-volcanic rocks from the Hainan Island

a—油文村变火山岩；b—屯昌县晨星农场变火山岩；c—邦溪、八一变火山岩；d—黄竹岭、冲俄村变火山岩

屯昌县晨星农场一带岩石稀土元素总量为 $12.41\times10^{-6}\sim206.76\times10^{-6}$，配分模式分为轻稀土富集型和轻稀土亏损型（图 11-6）。部分样品的配分模式类似于富集型洋中脊玄武岩（E-MORB），另一部分样品的配分模式类似于亏损型洋中脊玄武岩（N-MORB）。轻、重稀土之比值为 0.46~113，平均为 1.24；$(La/Yb)_N=0.13\sim104$，平均为 0.80；$(Ce/Yb)_N=0.14\sim2.26$，平均为 0.69。$\delta Eu=0.57\sim1.24$，平均为 0.98，除有一个样品为 0.57 外，大部分均在 1.0 左右；$\delta Ce=0.81\sim1.17$，平均为 0.93；显示 Eu、Ce 亏损和富集不明显。$(La/Sm)_N=0.24\sim1.02$，平均为 0.87；$(Gd/Yb)_N=0.73\sim1.16$，平均为 0.91，表明轻稀土间分馏作用比较弱，重稀土间分馏作用不明显。

邦溪至八一农场一带，岩石稀土元素总量 $18.79\times10^{-6}\sim81.00\times10^{-6}$，平均为 46.62×10^{-6}，配分模式为轻稀土略亏损到富集型（图 11-6），大部分样品的配分模式类似于富集型洋中脊玄武岩（E-MORB）。轻、重稀土之比值为 0.99~4.17，平均为 2.89；$(La/Yb)_N=0.39\sim3.59$，平均为 2.22；$(Ce/Yb)_N=0.43\sim3.10$，平均为 1.91；表明轻、重稀土分异弱。δEu 在 0.95~1.47，大部分样品均在 1.0 左右；$\delta Ce=0.67\sim1.02$，大部分样品均也在 1.0 左右；没有 Eu、Ce 亏损和富集，表明无明显斜长石的分离结晶作用。$(La/Sm)_N=0.51\sim2.22$，平均约为 1.42；$(Gd/Yb)_N=0.81\sim1.47$，平均约为 1.25；表明轻、重稀土间分馏作用比较弱。

黄竹岭一带，单个岩石样品稀土总量变化较大，由 $148.16\times10^{-6}\sim1428.12\times10^{-6}$，其配分模式为呈右倾的轻稀土富集型（图 11-6），类似于洋岛玄武岩（OIB）的配分模式。轻、重稀土之比值为4.28~8.4，表明轻、重稀土分异明显，δEu 为 0.44~0.65 之间，Eu 略有亏损，但不明显。从阳起石石英片岩（变玄武岩）→绿帘斜长石片岩（变安山岩）→含黑云母石英片岩（变流纹岩），Eu 亏损（δEu）由弱到强。$(La/Yb)_N=8.52\sim19.08$，属轻稀土富集型。$(La/Sm)_N=3.88\sim6.43$ 之间，$(Gd/Yb)_N=1.43\sim2.61$，说明表明轻稀土间分馏作用比较明显，而重稀土间分馏作用不明显。

三、微量元素地球化学特征

海南岛古生代变玄武岩微量元素含量与维氏基性岩（玄武岩）值相对比，Sb、Be、Ni、Co、Hf、Sc、Cr 等元素含量高，尤以 Sb、Co、Sc、Cr 为甚；W、Sn、Li、V、Rb、Zr、Ga、Y 等基本一致；Ag、Au、Cu、Pb、Zn、Bi、Mo、Nb、Sr、Ba、U、Th、Ge、Cs 等元素含量低，特别是 Ag、Au、Cu、Pb、Ba 含量特别低。变安山岩（绿帘斜长石片岩）与维氏中性岩（闪长岩）值相对比，Sb、Bi、Co、Ta、Hf、As、Sc、U、Th 等元素高于维氏值，尤以 Sb、Hf、As、Sc、Th 略显突出；Pb、Zn、Ni、V、Rb、Zr、Ge、Cr 大体一致；Ag、Cu、Nb、Sr、Ba、Ga 等元素含量低，特别是 Cu、Sr、Ba 略显突出。变流纹岩（含黑云母石英片岩）与维氏酸性岩（花岗岩）值相对比，Cu、Sb、Bi、Ni、Co、V、Ta、Hf、As、Sc、Cr 等元素含量高，其中 Sb、Bi、Ni、Co、Hf、As 更突出些；Ag、Pb、Zn、Rb、Ba、Zr、Ge、Cs、Th 等大体相当；Nb、Sr、Ga、U 等元素平均含量低，其中 Sr、Ba 略略突出。从以上对比中可知，大离子亲石元素中 Sr 和 Ba 的含量普遍很低（表 11-3）。

表 11-3　海南岛古生代变火山岩微量元素含量一览表

Table 11-3　List of trace elements of Paleozoic meta-volcanic rocks from the Hainan Island

单位：10^{-6}

地层单位		陀烈组									
岩性		蛇纹石化透闪石岩						透闪阳起石片岩			
样品号		D1891-1	D1891-2	D568-3	D1891-3	D1891-4	D1891-5	D11	CX-1	CX-2	CX-3
序号		1	2	3	4	5	6	7	8	9	10
微量元素含量	Au	0.5	0.4	0.6	0.9	0.3	0.4	1.5			
	Ag	0.027	0.033	0.039	0.022	0.037	0.011	0.078			
	Cu	46.2	23.9	75.	10.8	59.2	12.5	44.0	48.9	29.3	56.8
	Pb	1.0	<1.0	1.0	<1.0	<1.0	<1.0	6.7			
	Zn							155	99.4	109	91.9
	Sb	1.53	1.22	0.71	1.57	2.11	1.21	0.32			
	W	0.3	0.3	0.7	0.5	0.3	0.7	0.43			
	Sn	1.2	1.1	1.0	2.6	0.4	1.4	2.8			
	Bi	0	0.10	1.2	<0.10	<0.10	<0.10	0.05			
	Mo	0.25	0.19	0.9	0.27	0.63	0.58	1.3			
	Be	2.2	1.3	1.1	1.6	1.3	1.2	2.7			
	Li	20	15	12.8	14	10	35	9.8			
	Ni	817.6	907.4	977.9	1206	888.6	334.8	273	382	337	280
	Co	102.4	75.3	78.5	83.8	104.6	67.0	46.8	59.9	59.0	58.0
	V	125	113	122.1	136	159	150	145	197	216	181
	Nb	<2.0	17.5	19.3	15.2	19.9	28.7	62.2	0.48	0.50	0.42
	Ta	1.4	0.9	1.2	2.5	2.4	2.2	5.0	0.028	0.030	0.024
	Rb	4	<3	<3	<3	<3	9	66.0	5.55	24.1	17.5
	Sr	31	28	29	33	27	23	679	141	136	172
	Ba	116	95	102	102	58	58	623	18.9	138	51.7
	Zr	112	127	116	139	176	194	286	34.0	36.5	24.3
	Hf	5.0	10	2.6	4.3	4.1	4.2	6.9	0.91	0.99	0.62
	As	21.6	10	1.7	<0.2	4.3	8.0	1.2			
	Ga	9.5	11.3	11.8	11.4	12.9	18.0	20.0	9.90	12.0	9.42
	Sc	12.2	11.2		16.4	14.7	14.4		32.9	34.7	31.8
	Cr	1556	1339	1480	1093	1299	498	251	887	881	1006
	U	1.4	1.5	1.7	1.9	1.9	2.0	3.3	0.06	0.07	0.11
	Th	0.7	2.3	2.2	2.0	2.9	3.4	6.6	0.02	0.027	0.031
	Y		24.55	35.81	28.85	39.12	17.67	21.12	12.9	14.2	10.5

续表

地层单位		陀烈组			南碧沟组							
岩性		透闪阳起石片岩			绿帘斜长石片岩						含黑云母石英片岩	
样品号		CX-4	CX-5	CX-6	H7-1	H7-2	H8-1	H8-2	H9-1	H9-2	H10-1	H10-2
序号		11	12	13	14	15	16	17	18	19	20	21
微量元素含量	Au											
	Ag				0.037	0.050	0.038	0.044	0.032	0.047	0.073	0.078
	Cu	14.6	80.0	94.8	7.4	7.0	7.7	6.9	8.8	9.4	50.0	44.7
	Pb				16.9	14.4	12.9	13.3	18.2	20.1	16.8	20.9
	Zn	102	130	107	75	61	80	87	123	142	62	66
	Sb				11.7	8.7	9.0	7.8	12.1	12.6	4.7	12.2
	W											
	Sn											
	Bi				0.18	0.14	0.08	0.05	0.09	0.08	2.11	2.79
	Mo											
	Be											
	Li											
	Ni	211	275	322	38.3	32.2	42.2	36.6	38.7	38.9	40.1	39.9
	Co	46.5	60.5	55.5	15.8	15.6	18.9	20.8	18.8	18.6	27.1	25.6
	V	180	261	198	111	100	111	85	87	78	62	77
	Nb	0.35	0.32	0.42	6.7	9.4	19.5	16.5	8.2	12.0	12.2	13.5
	Ta	0.021	0.018	0.025	1.0	1.0	1.7	1.1	1.3	1.1	0.7	1.3
	Rb	13.7	27.7	22.0	73	26	110	51	209	79	169	176
	Sr	189	89.9	154	300	234	475	392	148	156	68	105
	Ba	45.1	80.8	28.7	249	143	326	137	1040	409	629	612
	Zr	20.1	22.6	22.9	193	230	307	369	199	146	202	245
	Hf	0.54	0.59	0.64	7.2	7.2	9.4	10.6	8.3	6.9	6.7	8.5
	As				5.6	4.9	24.5	7.1	11.6	8.9	43.4	57.5
	Ga	8.09	13.0	11.9	5.2	13.7	21.8	11.7	9.7	13.6	11.6	11.8
	Sc	31.5	38.7	38.2	13.0	12.2	14.1	11.9	11.0	9.8	7.2	9.2
	Cr	769	1192	1092	63	60	63	63.9	34.9	42.0	36.7	46.2
	U	0.06	0.34	0.26	2.23	2.66	8.10	3.57	2.02	3.04	1.93	1.68
	Th	0.019	0.029	0.026	22.2	24.0	30.3	29.2	23.8	21.2	18.9	21.2
	Y	9.51	12.7	11.2	173.1	135.4	45.24	50.72	195.6	299.8	32.40	26.78

地层单位		南碧沟组									
岩性		绿泥石阳起石片岩					变基性火山岩				阳起石石英片岩
样品号		0218-2	0218-7	0218-8	0219-1	On-1	B-1	H-1	S-3	HN-1	XⅥ15-1
序号		22	23	24	25	26	27	28	29	30	31
微量元素含量	Au					1.0					2.2
	Ag										0.046
	Cu	43.14	18.13	45.3	28.58	45.5	97.9	86.5	72.0	81.4	7.5
	Pb	0.607	0.901	0.432	0.609	7.0					24.5
	Zn	83.72	79.21	66.93	79.79	211.9	69.1	72.3	86.0	121	99.3
	Sb										13.6
	W										3.81
	Sn										29.17
	Bi										0.18
	Mo					1.2					0.20
	Be										0.20
	Li					85.9					9.2
	Ni	701	615.9	818.5	625.8	12.4	418	354	133	719	39.3
	Co	84.11	62.06	94.77	70.72	43.0	64.8	64.7	49.0	81.9	28.7
	V	160.1	142.9	122	132	375.2	215	259	258	146	89.0
	Nb	7.923	6.633	6.102	7.957	7.2	1.70	2.63	9.00	4.40	9.7
	Ta	0.557	0.49	0.445	0.652	0.72	0.10	0.16	0.58	0.24	0.4
	Rb	0.014	0.319	0.353	0.069	40.5	5.07	3.12	24.3	0.35	10
	Sr	20.78	17.55	59.73	15.3	30.7	143	149	315	97.6	126
	Ba	5.473	8.864	9.399	10.03	48.5	15.5	56.0	198	1.34	61
	Zr	96.64	86.69	78.44	103.9	191.0	48.0	56.8	143	71.1	130
	Hf	2.456	2.094	2.038	2.603	5.7	1.35	1.53	3.89	1.94	4.2
	As										9.39
	Ga	13.16	11.25	9.37	12.68	25.6	13.1	17.8	19.2	13.0	
	Sc	21.07	17.93	16.19	22.11	55.3	35.1	43.6	37.8	22.4	
	Cr	676.6	708.3	852.8	601.1	11.4	859	825	290	1091	
	U	0.169	0.134	0.139	0.174	2.4	0.14	0.07	0.19	0.09	
	Th	0.661	0.634	0.621	0.714	8.4	0.15	0.21	0.75	0.26	
	Y	17.99	17.73	16.96	15.54	75.8	16.4	23.6	33.9	14.2	26.31
	Cs	0.324	0.683	0.611	0.241	11.8	0.82	0.26	1.69	0.26	
	Ge	1.708	2.667	2.215	2.021		1.67	1.91	2.55	1.29	

注：Au 的单位为 10^{-9}。1~7、14~21 据海南省地质调查院（2004）；8~13、27~30 据李献华（2000）；22~25 据海南省地质调查院（2007）；26 据宜昌地质矿产研究所、海南省地质调查院（2004）。

Allégre et al.（1973）在研究地球上主要岩石过渡族元素球粒陨石标准化的分配型式后发现，从超基性的橄榄岩到酸性花岗岩，其分配型式渐近更趋 W 的型式；同时，由地幔岩派生的岩石相对于球粒陨石来说均具不同程度的 Ti 正异常和 Cr、Ni 负异常，而地幔岩及其类似岩石不具这一特征。海南岛古生代变火山岩的过渡族元素球粒陨石标准化型具有相似的配分模式（图 11-7），表现为 Ti 正异常和 Cr、Ni 负异常明显，与 Allégre et al.（1973）提出的玄武岩配分模式相近，暗示了古生代火山岩具地幔来源特点。

在变火山岩的原始地幔标准化的微量元素蛛网图（图 11-8）上，油文村变火山岩大致与洋岛玄武岩（OIB）相似；邦溪、八一农场和黄竹岭地区的变火山岩部分与典型的 E-MORB 相似，部分与典型的 N-MORB 相似；屯昌县晨星农场变火山岩与典型的 N-MORB 相似，只有一个样品（D11）与洋岛玄武岩（OIB）相似，反映出海南岛古生代火山岩的母岩浆不同地区有不同来源。

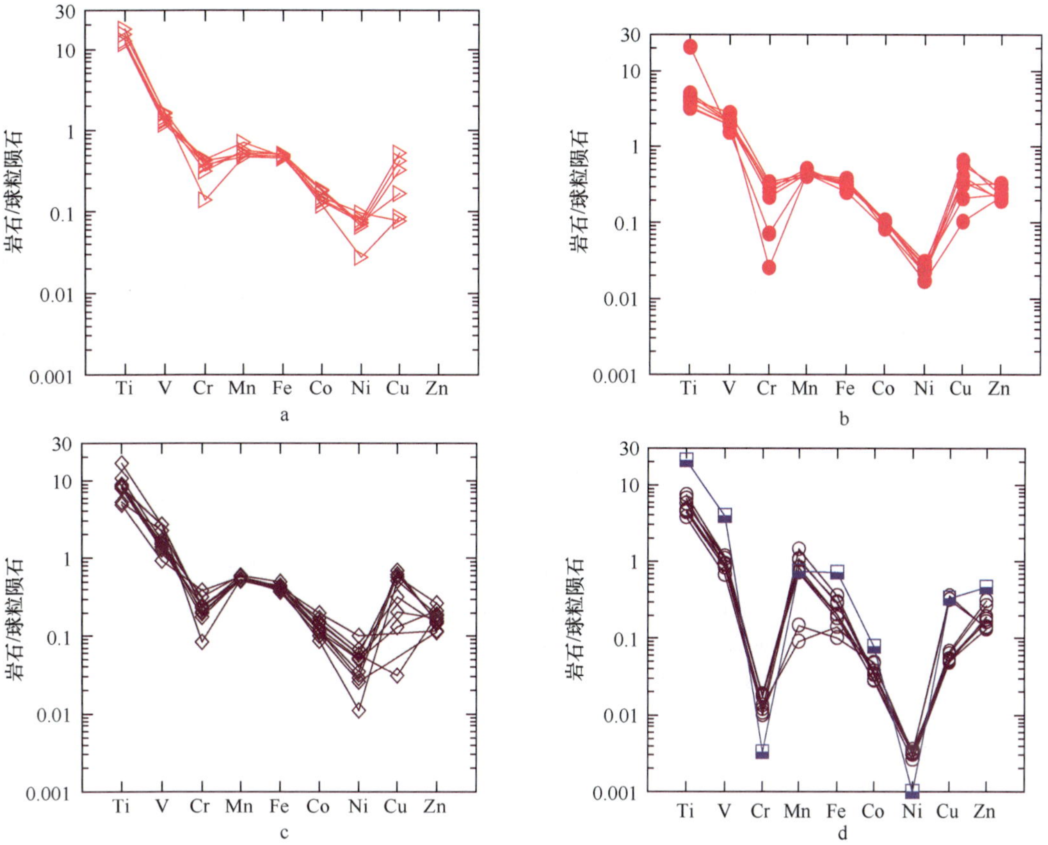

图 11-7　海南岛古生代变火山岩过渡族元素球粒陨石标准化图

Fig. 11-7　Chondrite-normalized patterns of transition metal elements of Paleozoic meta-volcanic rocks from the Hainan Island

a—油文村变火山岩；b—屯昌县晨星农场变火山岩；c—邦溪、八一变火山岩；d—黄竹岭、冲俄村变火山岩

四、火山岩形成的地质构造环境

每个地区火山岩的稀土元素及部分微量元素蛛网配分曲线变化趋势相当一致，说明同一地区火山岩的来源应为同一岩浆源区。此外黄竹岭地区火山岩 Eu 亏损从变玄武岩→变安山岩→变流纹岩由弱到强，也证实这一同源结晶分异的岩浆演化过程。

TiO_2-Zr 图解（图 11-9）中，屯昌县晨星农场和黄竹岭一带变火山岩主要落入火山弧玄武质岩石区，油文村和邦溪、八一变火山岩主要落入火山弧和洋中脊玄武质岩石混合区；在 TiO_2-MnO-P_2O_5 图解（图 11-10）中，古生代火山岩主要落入岛弧玄武质岩石区，屯昌县晨星农场部分样品落入钙碱性玄武质岩石区；在 Nb-Zr-Y 图解（图 11-11）中，变玄武岩主要落入火山弧玄武质岩石区；在 Ti-Zr-Y 判别图中主要落入洋底或洋中脊和洋岛或板内的区（图 11-12）。此外，火山岩微量元素比值蛛网图配分模式与洋岛玄武岩或洋中脊微量元素比值蛛网图配分模式较为相似，综合反映海南岛古生代火山岩应属于岛弧或洋中脊构造背景下形成的产物。

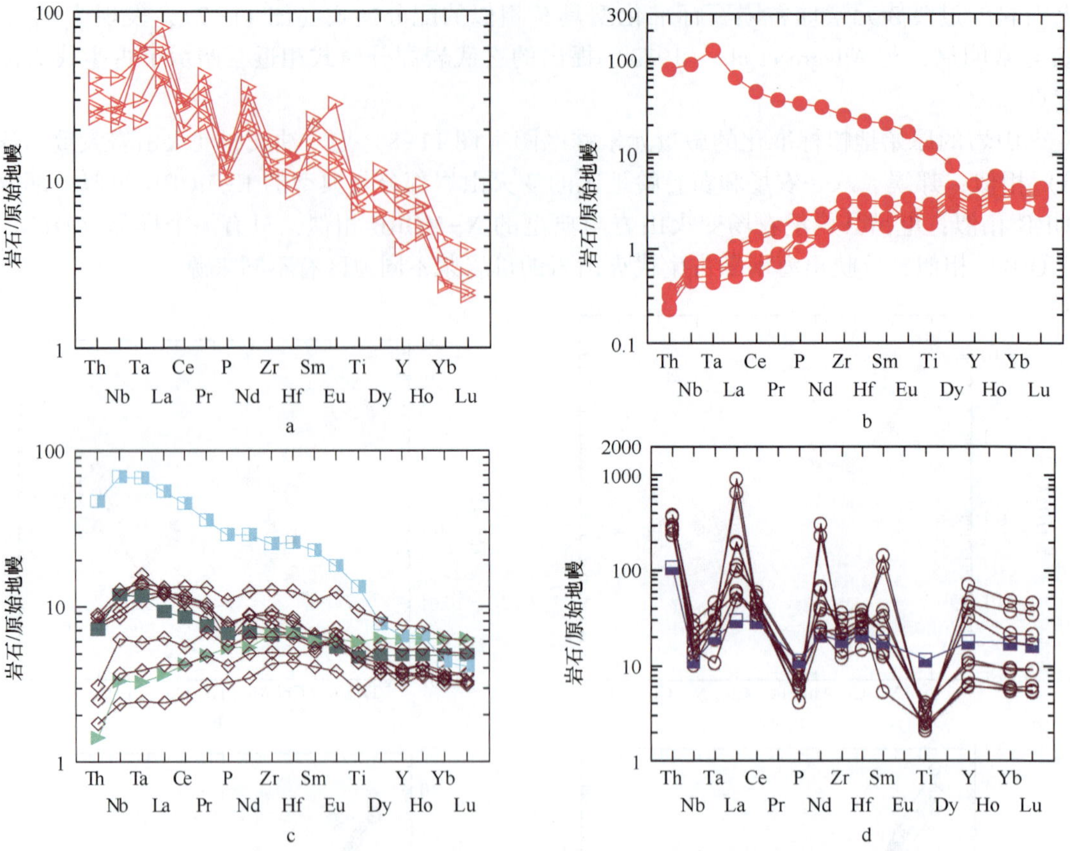

图 11-8 海南岛古生代变火山岩原始地幔微量元素蛛网图

Fig. 11-8 Primitive mantle-normalized spider diagram of trace elements of Paleozoic meta-volcanic rocks from the Hainan Island

a—油文村变火山岩；b—屯昌县晨星农场变火山岩；c—邦溪、八一变火山岩；d—黄竹岭、冲俄村变火山岩

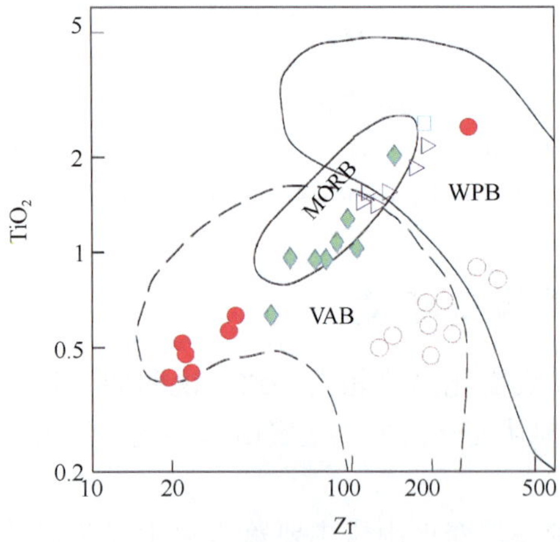

图 11-9 海南岛古生代变火山岩的 TiO_2-Zr 图

Fig. 11-9 TiO_2-Zr diagram of Paleozoic meta-volcanic rocks from the Hainan Island

MORB—洋中脊玄武岩；WPB—板内玄武岩；VAB—火山弧玄武岩

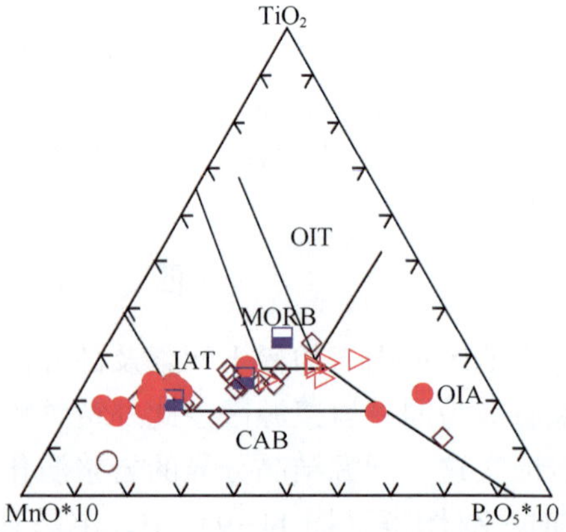

图 11-10 海南岛古生代变火山岩的 TiO-MnO-P_2O_5 图

Fig. 11-10 TiO-MnO-P_2O_5 diagram of Paleozoic meta-volcanic rocks from the Hainan Island

OIT—洋岛拉斑玄武岩；MORB—洋中脊玄武岩；IAT—岛弧拉斑玄武岩；OIA—洋岛碱性玄武岩；CAB—岛弧玄武岩

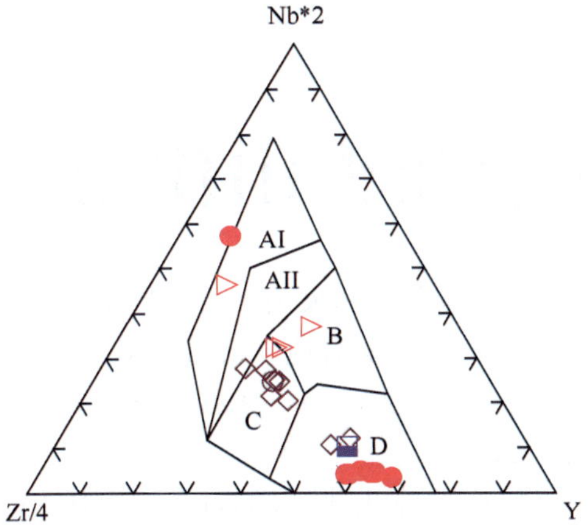

图 11-11　海南岛古生代变火山岩的 Nb-Zr-Y 图

Fig. 11-11　Map of Paleozoic meta-volcanos in the Hainan Island

AⅠ和AⅡ—板内碱性玄武岩；AⅡ和C—板内拉斑玄武岩；B—P-MORB；D—N-MORB；C和D—火山弧玄武岩

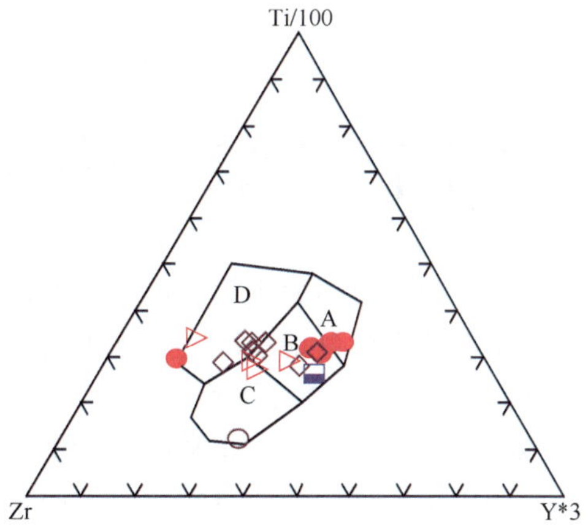

图 11-12　海南岛古生代变火山岩的 Ti-Zr-Y 图

Fig. 11-12　Ti-Zr-Y diagram for Paleozoic meta-volcanic rocks from the Hainan Island

A—火山弧低钾玄武岩；B—洋岛或洋中脊玄武岩；C—火山弧钙碱性玄武岩；D—洋岛和板内玄武岩

第十二章　中生代火山岩及火山作用

海南省中生代火山岩出露广泛，大致可以划分为陆相火山岩盆地和陆相火山-沉积岩盆地。陆相火山盆地有牛腊岭火山盆地、同安岭火山盆地、五指山火山盆地、洛基火山盆地、旺商火山盆地，出露面积约 970km^2，约占全岛陆域面积的 2.85%。陆相火山-沉积盆地有白沙盆地、阳江盆地、王五盆地、雷鸣盆地、乐安盆地等，在正常沉积岩中夹有火山角砾岩、凝灰岩等。

海南岛中生代火山地质有较长研究历史。早期李承三（1929）在琼北地区调查时，将阳江地区红色岩系称为白石岭砾岩；广东省海南地区资源开发委员会（1956）将岛内的红色岩系归入白垩系—第三系；北京矿业学院（1958）将红色岩系称为南雄层，时代为白垩纪；广东省地质局 764 队（1964）在 1∶20 万区调中，开始对海南岛中生代火山地质进行全面调查和研究，在红色岩系中建立了鹿母湾群和报万群，分别归属白垩系下统和上统，同时将分布于五指山、同安岭、牛腊岭、旺商等地火山岩称作岩被，时代定为 K_2—E，并对火山岩的时代和期次进行了划分，为后来的工作奠定了基础；广东省区调队（1977）和耿良玉（1979）在琼海福田、临高东江两地原 764 队归属报万群的地层中，发现早白垩世的植物、介形类和轮藻类化石而改定为鹿母湾群。20 世纪 80 年代至 90 年代以来，研究者日渐增多，工作也更为深入、系统，先后有海南地质大队（1979，1984，1986）、广东区调队（1988）、宜昌地矿所和海南地质大队联合区调队（1991）、四川攀西地质大队和海南地质大队联合区调队（1991）、广西区调队（1991）、宜昌地矿所（1992）、海南地质综合勘察院（1994）、南京地质矿产研究所（1996）、宜昌地矿所（2004）、海南省地质调查院（2004，2007）等单位在完成区调和科研工作中，对中生代火山地质有关问题均作了不同程度的研究工作。

第一节　火山岩主要类型、火山岩建造

一、主要岩石类型

火山盆地以同安岭、牛腊岭、五指山火山盆地研究比较系统。其火山岩可分为 2 个旋回 3 个组。第Ⅰ旋回六罗村组（K_1ll）主要为流纹质火山岩，中下部夹玄武安山岩、玄武质粗面安山岩及少量玄武岩、安山岩，总厚度>1792m，为一套中基性岩-酸性岩双峰式组合。第Ⅱ旋回为（安山岩）英安岩-流纹岩组合，自下往上分为汤他大岭组和岭壳村组。前者以英安质火山岩为主，中部和下部夹少量安山质和流纹质火山岩，总厚度 589m；后者为一套中酸性向酸性演化的火山岩，总厚度>2299m，岩性自下而上由英安质-流纹质熔岩及相应的火山碎屑岩组成，夹一些粗面英安岩。

火山-沉积盆地中的火山岩往往以安山质-英安质-流纹质火山碎屑岩、沉积火山碎屑岩夹层产于鹿母湾组（K_1l）陆源碎屑岩中。

在 TAS 火山岩分类图上（图 12-1），中生代火山岩主要落入流纹岩区，少量落入英安岩、粗面英安岩、玄武安山岩、安山岩、玄武质粗面安山岩区，个别落入玄武岩区。

第Ⅰ旋回火山岩为一套玄武安山岩-流纹岩组合。在 TAS 图上的投影点明显集中于两个区域，酸性部分集中于流纹岩区，中基性部分集中于玄武安山岩和玄武质粗面安山岩区，而两者中间的安山岩、英安岩、粗面安山岩、粗面英安岩区投影点很少，呈双峰式特征。与此相应，在 SiO_2 含量的变化上，表现为 62%~67%间的间断（戴里间断）。第Ⅱ旋回主要为流纹岩类，少量英安岩、粗面英安岩类，局部出现安山岩类，为一套（安山岩）-英安岩（粗面英安岩）-流纹岩组合。

玄武岩　分布于第Ⅰ旋回六罗村组下段。岩石为深灰色，杏仁状构造，斑状结构。斑晶为斜长石

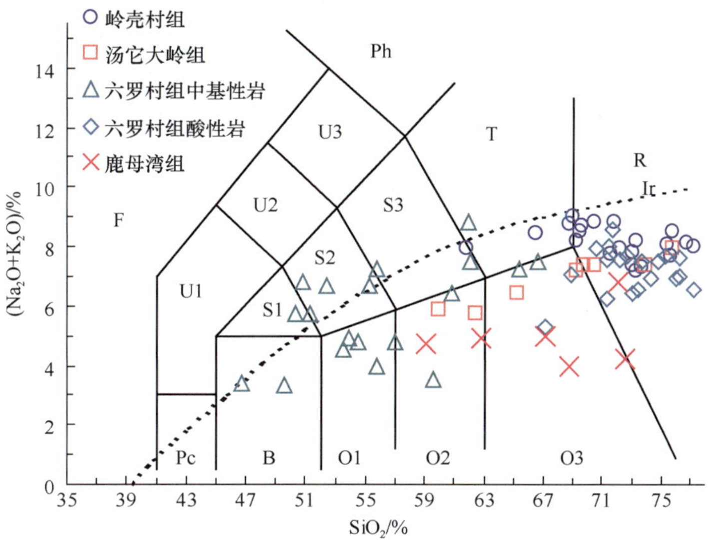

图 12-1　海南岛中生代火山岩硅-碱图

Fig. 12-1　SiO_2-Na_2O+K_2O diagram of Mesozoic volcanic rocks from the Hainan Island

Pc—苦橄玄武岩；B—玄武岩；O1—玄武安山岩；O2—安山岩；O3—英安岩；R—流纹岩；S1—粗面玄武岩；S2—玄武质粗面安山岩；S3—粗面安山岩；T—粗面岩、粗面英安岩；F—副长石岩；U1—碱玄岩、碧玄岩；U2—响岩质碱玄岩；U3—碱玄质响岩；Ph—响岩；

-------- Ir—Irvine 分界线，上方为碱性系列，下方为亚碱性系列

（5%）、辉石（2%）、伊丁石（1%）；基质为间隐结构，矿物有斜长石（36%）、辉石（8%）、玻璃质（18%）及杏仁体（30%，成分为方解石、玉髓、绿泥石、沸石等）。

玄武安山岩　分布于同安岭、牛腊岭、五指山、儋州洛基等地第Ⅰ旋回六罗村组下段。岩石为灰、深灰色，（气孔）杏仁状构造，斑状结构。斑晶为斜长石、辉石、角闪石和石英，含量 10%~15%，局部斑晶及气孔略呈定向排列（图版Ⅲ-1-A）；基质为隐晶结构、玻晶交织结构，成分主要为长英矿物（40%~60%）、角闪石和辉石（10%~20%）。局部见较多玄武安山质角砾或者称捕虏体（25%），副矿物主要为磁铁矿、磷灰石等。

英安岩　分布于汤他大岭、坡岭、儋州洛基等地第Ⅱ旋回汤他大岭组中。岩石为深灰色，块状构造，斑状结构。斑晶为斜长石、石英、钾长石、黑云母等，含量 16%~30%；基质为隐晶结构，由长石和石英（50%~70%）、铁镁矿物（1%~2%）及榍石、金红石、磷灰石、锆石、磁铁矿组成。

流纹岩　广泛分布于火山盆地的第Ⅰ、Ⅱ旋回中。岩石为灰色、灰白色，流纹构造，斑状结构。斑晶细而少，约 1%，有石英、钾长石、斜长石和黑云母等，常见熔蚀状或晶屑状外形；基质为隐晶结构、球粒结构，主要成分为长英矿物，含量一般 60%~70%，最高达 93%；还有磁铁矿、锆石、磷灰石等。

流纹质角砾（凝灰）熔岩　岩石呈紫色、灰绿色，基质具霏细结构的斑状-凝灰结构，块状构造。基质由粒径<0.03mm 霏细状长英质集合体组成，不均匀含有少量玻屑。在基质中分布有斜长石、石英、黑云母等斑晶及晶屑，偶含岩屑及钾长石晶屑。斜长石粒径 0.4~1mm，个别可达 2~3mm，约有 1/2 呈晶屑产出；石英粒径 1~5mm，一般都呈斑晶产出；黑云母多色性，N_g'-暗褐色；N_p'-金黄色；岩屑粒径 1~4mm，呈火山角砾状，岩性为安山岩及霏细岩。

凝灰岩　依据火山碎屑的物态可分为玻屑凝灰岩、岩屑凝灰岩、晶屑凝灰岩、火山尘凝灰岩。岩石为灰绿色，凝灰结构，豆状构造。岩石中碎屑粒径大小一不，从 0.1mm 至 2mm 均有，形态多呈不规则的棱角状，少数呈棱角状—次圆状。岩石含个别粒径 2~5mm（最大可至 15mm）的紫红色安山质角砾。

沉凝灰岩　主要分布于各红盆中，呈浅黄色、浅紫红色，沉凝灰结构，层状构造。岩石由火山碎屑物（60%~70%）和正常沉积物（30%~40%）组成，火山碎屑物主要为火山灰，少量晶屑和岩屑。

火山灰呈浅灰绿色—灰褐色，棉絮状集合体，为均质体，大部分已发生次生变化，分解成方解石、绿泥石、高岭石等，个别甚至脱玻化被细鳞片状黑云母交代，多以胶结物形式存在；晶屑主要为石英，少量钾长石、斜长石和黑云母，均呈尖棱角状—棱角状，粒径一般在0.2~2.0mm之间，个别可达2~6mm，石英晶屑可见裂纹和熔蚀湾；岩屑均为凝灰岩，大部分已重结晶生成绢云母等。

二、火山旋回与岩相

（一）火山旋回

海南地质综合勘察院（1995）、符国祥等（1994）及宜昌地矿所和海南省地质调查院（2004）认为中生代火山岩可划分为两个火山活动旋回，而王中杰等（1996）则划分为一个旋回。根据海南岛中生代两个阶段的火山喷发之后分别有对应的两套次火山岩（浅成岩）及深成侵入岩出现的事实（年龄分别为112~106Ma及96~82Ma），本次研究划分为两个旋回。第Ⅰ旋回由六罗村组组成，从下到上分为泥岩、砂砾岩，玄武质、安山质夹流纹质火山岩类。其中，流纹质火山岩类三层，这种三层式结构，亓润章等（1989）称三元结构，是我国东南沿海双峰式火山岩的主要组合特征。玄武岩、安山岩和流纹岩构成的双峰式火山岩是连续喷发的，玄武岩、安山岩和流纹岩间无沉积层分开，均为直接接触，形成单旋回双峰式火山岩。第Ⅱ旋回由汤他大岭组和岭壳村组组成，主要为中酸性及酸性熔岩、火山碎屑岩夹少量沉积岩和火山-沉积岩。

（二）火山岩相

由于受风化剥蚀的影响及地壳的抬升，大部分火山盆地火山岩相已残缺不全，特别是火山口完全剥蚀，局部尚残存有如下一些岩相。

喷溢相 由玄武岩、安山岩、英安岩、流纹岩组成。产出在第Ⅰ、Ⅱ旋回中。主要见于同安岭、牛腊岭、五指山火山盆地中。

爆发空落相 由集块级、角砾级、凝灰级以及它们的过渡类型岩石组成。在火山盆地、火山-沉积盆地中都有出现。在汤他大岭新村、阜牯岭、尖峰岭等地见到较典型的爆发空落相相序剖面。下部为以粗火山碎屑为主的粗细混杂堆积亚相，上部为细碎屑堆集亚相。从火山口源头到尾端，火山碎屑厚度迅速减薄至尖灭，形成楔状体，粒度由粗变细形成二相式成因模式。

火山碎屑流相 由熔结的和不熔结的集块级、角砾级、凝灰级以及过渡类型的火山碎屑岩组成。分布较广，常与凝灰熔岩共生。抱冲村剖面下段相序清楚，主体岩石为熔结凝灰岩，组成本岩相的核心-熔结带，其上下为未熔结带。自下而上可分为三个堆积亚相，下部涌流堆积亚相（未熔结带）、中部碎屑流堆积亚相（熔结带）、上部灰云堆积亚相（未熔结带），组成三相式成因模式。

爆溢相 组成岩石为凝灰熔岩。属于爆发空落相与喷溢相之间的一种过渡类型。在海南岛中生代火山盆地中分布广、面积大，与成矿关系密切。已知雄岭、富田、六罗山等矿点都产在其中。

喷发沉积相 根据堆积物产出的构造环境，分为破火山口湖盆喷发沉积相和内陆湖盆喷发沉积相两种类型。前者分布在汤他大岭、阜落刀岭、坡岭等地，是破火山口内间歇期间歇性喷发形成。后者主要见于火山-沉积盆地中或产于火山盆地火山爆发前和爆发初期间歇性喷发期，特别是火山盆地的底部，并和沉积相交生，如汤他大岭新村、尖峰岭玉林村等地。

火山颈相 见于火山盆地特别是同安岭、牛腊岭岩盆中。如阜牯岭和尖峰岭破火山口内的安山岩、坡岭破火山口内的花岗闪长斑岩、汤他大岭破火山口内的流纹质凝灰熔岩、抱差岭破火山口内的花岗斑岩等。这些岩石的存在标明了破火山口的位置。

侵出相 由碎斑熔岩组成。分布于火山盆地中。如抱差岭、汤他大岭新村等地。碎斑结构、再生球边结构是碎斑熔岩独有的结构类型，是侵出相的相标志。碎斑熔岩中，既有具侵入岩特征的自形、半自形晶，又有具喷出特征的岩屑、晶屑，是介于侵入岩与喷出岩间的过渡类型岩石。

潜火山相 火山盆地或其旁侧常可有潜火山相，特别是同安岭和牛腊岭岩盆中较发育，而五指山岩盆中规模较小。

三、火山构造

按《〈中国区域地质志〉工作指南》(2009) 的火山构造分类，并参考海南地质队 (1995)、南京地质矿产研究所 (1996)、宜昌地矿所和海南省地质调查院 (2004) 对海南岛中生代火山构造的划分方案，将海南岛中生代火山构造划分为4级。属于东部中生代构造岩浆带 (Ⅰ级) 中的海南火山亚带 (Ⅱ级)。在此基础上，以7个相互独立的火山 (沉积) 盆地为依据分别划分出7个Ⅲ级火山构造。各火山 (沉积) 盆地中的破火山、火山穹隆、火山通道则成为Ⅳ级火山构造 (表12-1)。

表12-1 海南岛中生代主要火山构造类型

Table 12-1 Main rypes of Mesozoic volcanic structures in the Hainan Island

Ⅰ级火山构造	Ⅱ级火山构造	Ⅲ级火山构造	Ⅳ级火山构造		
			破火山构造	火山穹隆构造	火山通道构造
东部中生代构造岩浆带	海南火山岩亚带	五指山火山沉陷	五指山，吊罗山，三更罗	乘坡	五指山，铁钻岭，冲罗排
		同安岭火山沉陷	阜牯岭-尖锋岭，汤他大岭，加跃岭，同安岭	抱差岭南改岭	南败岭抱差岭
		牛腊岭火山沉陷	白石岭-后力岭，笔架山，羊锦岭，白河山	寨阳-马安岭	白石岭 (牛腊岭)，后力岭，白河山，羊锦岭
		琼海-阳江火山构造洼地-火山沉积洼地	南牛岭-大排岭		
		三亚-藤桥火山沉积洼地			
		白沙-乐东火山沉积洼地	红帽地区		
		定安雷鸣火山沉积洼地			

第二节 火山岩的地球化学特征及其形成的构造环境

一、岩石化学特征

中生代火山岩岩石化学分析结果显示见表12-2。

岭壳村组、汤他大岭组火山岩的 SiO_2 含量普遍较高，岭壳村组主要为61.81%~77.26% (平均72.04%)，汤他大岭组为59.95%~75.72% (平均为68.34%)。六罗村组火山岩 SiO_2 含量可分为两个区间：46.68%~66.69% (平均为55.97%) 和67.13%~77.26% (平均为73.30%)，其分异指数DI值也集中于两个区间：35.9~61.5、84~93.1，具有双峰式火山岩的特征。产于白沙盆地中的火山岩夹层岩石 SiO_2 含量为67.10%~72.58%，全为中酸性—酸性岩；而产于阳江盆地中的火山岩夹层岩石 SiO_2 含量为59.05%~68.69%，中性岩至中酸性岩均有。

岩石中 Na_2O+K_2O 和 K_2O 含量普遍较高，里特曼指数 σ 一般为0.59~4.08，少数>11。据硅-碱图 (图12-1)、FAM图 (图12-2) 和 SiO_2-K_2O 图 (图12-3)，除六罗村组中基性岩有少部分属碱性系列外，其他均属亚碱性系列。其中，六罗村组中基性岩大部分属钙碱性-高钾钙碱性系列；而其酸性岩类，以及岭壳村组火山岩大多为高钾钙碱性系列，少数为钾玄岩系列，个别为钙碱性系列；汤他大岭及鹿母湾组中的火山岩夹层属高钾钙碱性系列。除岭壳村组极个别酸性火山岩及六罗村组的部分中基性岩属于钠质岩类外 (Na_2O/K_2O 值>0.5)，其他火山岩属于钾质岩类 (Na_2O/K_2O 值都<0.5)。鹿母湾组中的火山岩夹层及五指山地区六罗村组中的个别酸性火山岩 Na_2O 含量仅为0.05%~

表 12-2 海南岛中生代火山岩化学成分、CIPW 标准矿物及岩石化学常用参数一览表

Table 12-2 List of chemical composition, CIPW standard minerals and common petrochemical parameters of Mesozoic volcanic rocks from the Hainan Island

地层单位		岭壳村组							
岩性		英安岩	粗安岩	粗安岩	英安岩	安粗质角砾熔岩	流纹岩	碎斑熔岩	凝灰熔岩
样品号		D1010-1	D1019-3	D1019-4	D1026-1	D1027-1	D908-1	GI35-12	GI35-4
序号		1	2	3	4	5	6	7	8
岩石化学成分/%	SiO_2	66.56	68.68	69.64	70.48	68.94	75.78	73.35	73.07
	Al_2O_3	16.99	15.29	14.74	14.06	15.02	13.03	13.84	13.84
	Fe_2O_3	1.35	0.88	0.48	0.09	1.2	0.81	0.76	1.51
	FeO	2.4	2.3	2.97	2.89	2.02	1.39	1.48	1.4
	MgO	0.8	0.4	0.42	0.46	0.4	0.08	0.63	0.59
	CaO	1.4	1.95	1.4	1.35	1.49	0.26	1.21	1.03
	Na_2O	6.92	3.92	3.89	3.82	3.92	4.6	2.96	2.84
	K_2O	1.5	4.84	4.8	4.97	5.1	3.91	4.24	4.54
	MnO	0.061	0.083	0.066	0.07	0.07	0.02	0.06	0.12
	TiO_2	0.55	0.34	0.3	0.61	0.32	0.11	0.22	0.22
	P_2O_5	0.15	0.07	0.07	0.07	0.08	0.01	0.2	0.04
	CO_2	0.05	0.17	0.26	0.14	0.19	0.06	0.16	0.22
	H_2O^+	0.56	0.28	0.35	0.58	0.36	0.36	0.68	0.76
	总量	99.29	99.2	99.39	99.59	99.11	100.42	99.79	100.18
标准矿物成分含量/%	Q	15.89	21.65	23.38	24.16	22.33	32.66	36.44	35.95
	C	1.77	0.19	0.74	0.08	0.5	0.77	2.62	2.47
	Or	8.99	28.99	28.74	29.73	30.61	23.13	25.35	27.07
	Ab	59.27	33.55	3108	32.65	33.61	38.88	25.28	24.2
	An	6.15	9.39	6.62	6.36	7.03	1.23	4.88	4.92
	Di								
	Hy	4.55	4.14	5.8	5.56	3.36	1.94	3.45	2.68
	Mt	1.98	1.29	0.7	0.13	1.77	1.17	1.11	2.21
	Hm								
	Il	1.06	0.65	0.58	1.17	0.62	0.21	0.42	0.42
	Ap	0.33	0.15	0.15	0.15	0.18	0.02	0.44	0.09
岩石化学常用参数	AlK	8.42	8.76	8.69	8.79	9.02	8.51	7.2	7.38
	Na_2O/K_2O	4.61	0.81	0.81	0.77	0.77	1.18	0.7	0.63
	FeO^*/MgO	4.52	7.73	8.1	6.46	7.75	26.49	3.43	4.68
	A/CNK	1.09	1.00	1.04	1.00	1.02	1.06	1.19	1.21
	NK/A	0.77	0.76	0.79	0.83	0.80	0.91	0.68	0.69
	DI	84.14	84.18	85.38	86.53	86.53	94.65	87.06	87.21
	SI	6.17	3.24	3.34	3.76	3.16	0.74	6.26	5.42
	AR	2.69	3.07	3.33	3.66	3.41	4.56	2.83	2.97
	ι	18.31	33.44	36.17	16.79	34.69	76.64	49.45	50.00
	δ	3.01	2.99	2.83	2.81	3.14	2.21	1.71	1.81
产地		同安岭							

续表

地层单位		岭壳村组							
岩性		碎斑熔岩	凝灰熔岩	碎斑熔岩	熔结凝灰岩	流纹斑岩		英安斑岩	流纹斑岩
样品号		GI36-1	GI41-3	TH17-3	TH17-4	Z776-1	Z91-1	ZVII1-1	ZVII19-1
序号		9	10	11	12	13	14	15	16
岩石化学成分/%	SiO_2	73.00	71.62	69.52	73.28	72.15	75.65	69.19	76.75
	Al_2O_3	13.19	13.92	14.93	13.09	13.68	13.04	15.16	12.05
	Fe_2O_3	1.78	1.67	0.87	1.17	0.9	0.94	1.73	1.02
	FeO	1.42	1.78	3.07	1.93	1.75	0.44	2	1.02
	MgO	0.94	0.85	0.44	0.36	0.85	0.26	0.99	0.26
	CaO	0.67	1.7	1.48	0.83	1.13	0.23	1.86	0.36
	Na_2O	3.32	3.56	3.62	3.6	3.55	2.85	4.24	3.71
	K_2O	4.48	4.16	4.91	4.57	4.36	4.82	3.97	4.42
	MnO	0.07	0.05	0.1	0.08	0.11	0.05	0.05	0.03
	TiO_2	0.2	0.25	0.33	0.19	0.23	0.1	0.35	0.09
	P_2O_5	0.04	0.05	0.34	0.24	0.11	0.14	0.11	0.04
	CO_2	0.09	0.25	0.31	0.33	0.53	1.5	0	0.43
	H_2O^+	0.66	0.44			0.36	0.39	0.35	0.16
	总量	99.86	100.3	99.92	99.67	99.71	100.41	100	100.34
标准矿物成分含量/%	Q	33.7	29.63	24.75	32.28	30.74	40.72	23.46	36.89
	C	1.75	0.56	1.69	1.22	1.3	3.05	0.73	0.58
	Or	26.74	24.7	29.16	27.21	26.1	28.94	23.56	26.21
	Ab	28.31	30.21	30.71	30.63	30.36	24.45	35.96	31.43
	An	3.12	8.18	5.37	2.73	5.02	0.32	8.62	1.56
	Di								
	Hy	3.31	3.71	5.67	3.33	4.47	0.66	4.24	1.59
	Mt	2.6	2.43	1.27	1.71	1.32	1.31	2.52	1.48
	Hm						0.05		
	Il	0.38	0.48	0.63	0.36	0.44	0.19	0.67	0.17
	Ap	0.09	0.11	0.75	0.53	0.24	0.31	0.24	0.09
岩石化学常用参数	AlK	7.8	7.72	8.53	8.17	7.91	7.67	8.21	8.13
	Na_2O/K_2O	0.74	0.86	0.74	0.79	0.81	0.59	1.07	0.84
	FeO^*/MgO	3.21	3.86	8.76	8.29	3.01	4.95	3.59	7.45
	A/CNK	1.14	1.03	1.07	1.06	1.08	1.26	1.03	1.04
	NK/A	0.78	0.74	0.75	0.83	0.77	0.76	0.74	0.90
	DI	88.75	84.52	84.61	90.11	87.19	94.09	82.98	94.52
	SI	7.87	7.07	3.41	3.1	7.45	2.79	7.66	2.49
	AR	3.57	2.95	3.16	3.84	3.29	3.74	2.86	4.8
	τ	49.35	41.44	34.27	49.95	44.04	101.9	31.2	92.67
	δ	2.03	2.08	2.74	2.2	2.15	1.8	2.57	1.96
产地		同安岭							

续表

地层单位		岭壳村组							汤他大岭组
岩性		流纹斑岩		流纹质角砾熔岩	含角砾流纹斑岩	流纹质角砾熔岩	英安斑岩	平均值	流纹岩
样品号		ZVII19-2	ZVII19-3	ZVII24-2	ZVII39-2	ZVII42-2	ZVII46-2		GXVIII1-1
序号		17	18	19	20	21	22		23
岩石化学成分/%	SiO_2	77.26	77.23	73.64	75.43	71.79	61.81	72.04	75.72
	Al_2O_3	12.05	12.05	13.12	12.57	14.08	15.81	13.89	12.4
	Fe_2O_3	0.92	0.86	0.93	0.61	1.4	1.23	1.05	1
	FeO	1.02	0.98	1.54	0.88	1.21	3.14	1.77	0.9
	MgO	0.13	0.13	1.06	0.65	0.4	2.61	0.62	0.37
	CaO	0.3	0.3	1.09	0.38	0.72	2.95	1.10	0.97
	Na_2O	3.55	3.56	3.4	3.35	3.85	4.6	3.80	3.43
	K_2O	4.45	4.45	3.92	4.7	5	3.33	4.34	4.49
	MnO	0.03	0.03	0.07	0.07	0.07	0.07	0.065	0.07
	TiO_2	0.1	0.1	0.2	0.11	0.21	0.55	0.26	0.15
	P_2O_5	0.03	0.03	0.08	0.04	0.08	0.28	0.10	0.1
	CO_2	0.43	0.36	0.87	0.7	0.78	2.63	0.16	0.03
	H_2O^+	0.18	0.14	0.39	0.23	0.3	0.47	0.40	0.58
	总量	100.45	100.22	100.31	99.72	99.89	99.48	99.96	100.21
标准矿物成分含量/%	Q	38.45	38.42	34.65	36.25	28.3	11.52	29.65	36.05
	C	0.9	0.89	1.47	1.37	1.2	0	1.18	0.34
	Or	26.36	26.4	23.41	28.14	29.93	20.44	25.88	26.66
	Ab	30.05	30.17	29.01	28.66	32.93	40.34	32.48	29.11
	An	1.32	1.32	4.99	1.67	3.14	13.09	4.87	4.25
	Di								
	Hy	1.33	1.31	4.55	2.72	1.87	10.73	3.67	1.64
	Mt	1.34	1.25	1.36	0.9	2.05	1.85	1.54	1.46
	Hm								
	Il	0.19	0.19	0.38	0.21	0.4	1.08	0.50	0.29
	Ap	0.07	0.07	0.18	0.09	0.18	0.63	0.23	0.22
岩石化学常用参数	AlK	8.00	8.01	7.32	8.05	8.85	7.93	8.14	7.92
	Na_2O/K_2O	0.8	0.8	0.87	0.71	0.77	1.38	0.83	0.76
	FeO^*/MgO	14.21	13.49	2.24	2.2	6.17	1.63	6.92	4.86
	A/CNK	1.08	1.07	1.11	1.11	1.08	0.96	1.09	1.01
	NK/A	0.88	0.89	0.75	0.84	0.83	0.71	0.79	0.85
	DI	94.85	94.97	87.06	93.04	91.15	72.29	88.02	91.81
	SI	1.29	1.3	9.77	6.38	3.37	17.51	5.38	3.63
	AR	4.68	4.69	3.12	4.29	3.97	2.46	3.38	3.91
	τ	85.00	84.9	48.60	83.82	48.71	20.38	51.44	59.80
	δ	1.87	1.87	1.75	2	2.72	3.34	2.35	1.92
产地		同安岭							牛腊岭

续表

地层单位		汤他大岭组							
岩性		凝灰岩		英安岩	流纹岩	凝灰岩		凝灰岩	平均值
样品号		GIV12-1	GIV16-1	GIV2-2	GIV5-2	GIV7-1	GIV10-1	GI30-1	
序号		24	25	26	27	28	29	30	
岩石化学成分/%	SiO_2	62.45	69.22	65.29	73.95	70.43	69.73	59.95	68.34
	Al_2O_3	14.73	15.32	16.47	13.78	14.91	14.86	16.3	14.85
	Fe_2O_3	3.03	2.09	1.81	1.34	2.68	2.64	4.26	2.36
	FeO	2.85	1.18	2.78	0.55	0.93	0.83	3.32	1.67
	MgO	3.02	0.74	1.23	0.31	0.54	0.74	2.2	1.14
	CaO	3.99	2.28	3.84	0.85	0.89	1.71	5.03	2.45
	Na_2O	2.44	3.68	3.48	3.92	3.34	3.78	2.94	3.38
	K_2O	3.28	3.48	2.94	3.44	4.02	3.62	2.92	3.52
	MnO	0.117	0.08	0.112	0.024	0.06	0.064	0.14	0.083
	TiO_2	0.6	0.4	0.5	0.25	0.4	0.4	0.86	0.45
	P_2O_5	0.2	0.09	0.15	0.05	0.08	0.1	0.28	0.13
	CO_2	0.92	0.06	0.12	0.18	0.09	0.09	0.16	0.21
	H_2O^+	2.24	0.74	0.62	0.76	0.78	0.96	0.7	0.92
	总量	99.87	99.36	99.34	99.4	99.15	99.52	99.06	99.49
标准矿物成分含量/%	Q	23.02	29.18	23.1	36.41	33.71	29.85	18.17	28.84
	C	0.33	1.55	0.9	2.19	3.67	1.84	0	1.34
	Or	20.06	20.88	17.64	20.66	24.19	21.74	17.59	21.17
	Ab	21.32	31.56	29.83	33.65	28.72	32.44	25.3	29.04
	An	19.27	10.95	18.44	3.99	4.02	8.02	23.02	11.55
	Di							0.58	
	Hy	9.83	1.88	6.15	0.79	1.37	1.88	6.76	3.44
	Mt	4.54	2.95	2.66	1.14	2.07	1.75	6.29	3.47
	Hm		0.09		0.57	1.3	1.47		
	Il	1.18	0.77	0.96	0.48	0.77	0.77	1.66	0.86
	Ap	0.45	0.2	0.33	0.11	0.18	0.22	0.62	0.29
岩石化学常用参数	AlK	5.72	7.16	6.42	7.36	7.36	7.40	5.86	6.90
	Na_2O/K_2O	0.74	1.06	1.18	1.14	0.83	1.04	1.01	0.94
	FeO^*/MgO	1.85	4.14	3.58	5.66	6.19	4.33	3.25	4.23
	A/CNK	0.99	1.10	1.04	1.18	1.30	1.12	0.95	1.07
	NK/A	0.51	0.64	0.54	0.74	0.66	0.68	0.49	0.63
	DI	64.42	81.61	70.56	90.71	86.62	84.04	61.06	79.05
	SI	20.66	6.62	10.05	3.24	4.69	6.37	14.07	9.48
	AR	1.88	2.37	1.92	3.02	2.74	2.61	1.76	2.33
	τ	20.48	29.10	25.98	39.44	28.93	27.7	15.53	30.97
	δ	1.68	1.96	1.85	1.75	1.97	2.05	2.03	1.90
产地		同安岭							

续表

地层单位		六罗村组							
岩性		安山岩	玄武岩	安山岩		粗面玄武岩	粗安岩	玄武安山岩	
样品号		IIIH9-1	G223	XXIV2-1	IIIH7-1	GXIII6-5	GXIII6-4	GXIII7-1	五 1
序号		31	32	33	34	35	36	37	38
岩石化学成分/%	SiO_2	46.68	49.58	50.32	50.83	51.36	52.43	53.56	53.89
	Al_2O_3	14.8	16.49	13.59	15.25	17.39	16.03	18.17	17.48
	Fe_2O_3	8.71	5.78	7.01	7.14	6.39	3.42	8.44	2.77
	FeO	0.25	2.44	0.62	0.23	2.28	4.34	1.4	5.3
	MgO	3.41	4.29	7.87	2.46	3.09	4.08	2.06	4.31
	CaO	9.85	10.6	5.59	7.4	6.09	6.12	7.7	0.76
	Na_2O	1.89	2.42	2.32	3.65	3.98	4.26	3.52	3.13
	K_2O	1.5	0.95	3.42	3.14	1.74	2.39	1.03	1.78
	MnO	0.1	0.16	0.1	0.12	0.14	0.12	0.11	0.14
	TiO_2	1.02	0.88	1	1.05	1.15	1.68	1.3	1.25
	P_2O_5	0.37		0.11	0.6	0.3	0.7	0.25	0.3
	CO_2	4.08		8.09	2.5	1.63	0.36	0.09	0.12
	H_2O^+	7.13	6.22	0	5.32	2.98	3.08	2.19	2.19
	总量	99.79	99.81	100.04	99.69	98.52	99.01	99.82	93.42
标准矿物成分含量/%	Q	9.98	7.85	2.75	3.66	5.9	1.88	11.82	20.86
	C								10.6
	Or	10.02	6.00	22.00	20.22	10.96	14.79	6.25	11.56
	Ab	18.03	21.85	21.32	33.58	35.82	37.67	30.5	29.03
	An	30.96	33.43	17.97	17.32	25.99	18.33	31.47	2.21
	Di	16.93	17.72	8.94	14.46	3.36	7.28	4.83	
	Hy	1.79	3.26	17.26		6.67	9.92	3.04	18.00
	Mt	1.28	6.23	2.53	1.23	4.76	5.19	1.13	4.41
	Hm	8.95	1.88	5.88	6.92	3.52		7.87	
	Il		1.79			2.33	3.34	2.53	2.61
	Ap	0.91		0.26	1.43	0.7	1.6	0.56	0.72
岩石化学常用参数	AlK	3.39	3.37	5.74	6.79	5.72	6.65	4.55	4.91
	Na_2O/K_2O	1.26	2.5474	0.68	1.16	2.29	1.78	3.42	1.76
	FeO^*/MgO	2.37	1.78	0.88	2.71	2.60	1.82	4.37	1.81
	A/CNK	0.65	0.68	0.77	0.67	0.89	0.77	0.87	2.07
	NK/A	0.32	0.30	0.55	0.62	0.48	0.6	0.38	0.4
	DI	38.03	35.71	46.08	57.46	52.68	54.34	48.57	61.45
	SI	21.64	27.02	37.05	14.8	17.68	22.07	12.52	24.93
	AR	1.32	1.28	1.85	1.86	1.64	1.86	1.43	1.74
	τ	12.66	15.99	11.27	11.05	11.66	7.01	11.27	11.48
	δ	3.12	1.73	4.5	5.89	3.91	4.69	1.96	2.21
产地		洛基	同安岭	洛基		牛腊岭			五指山

续表

地层单位		六罗村组							
岩性		安山岩	角闪安山岩	玄武安山岩		石英安山岩	安山岩	粗面英安岩	玄武安山岩
样品号		D1539	IIIH8-1	D3039-1	I5-1	I9-1	XIII3-1	XIII11-1	D3039-4
序号		39	40	41	42	43	44	45	46
岩石化学成分/%	SiO_2	55.28	55.73	55.75	56.98	59.64	60.77	61.95	54.54
	Al_2O_3	17.64	16.73	16.75	18.03	16.83	15.67	15.24	19.34
	Fe_2O_3	6.5	8.2	4.12	3.6	3.7	3.01	1.42	5.02
	FeO	0.52	0.07	4.24	3.72	3.68	3.48	3.85	3.14
	MgO	4.24	1.63	4.08	2.4	3.14	2.4	3.1	3.37
	CaO	5	4.06	8.6	7.76	6.45	1.99	1.61	7.6
	Na_2O	4.17	4.8	2.82	2.9	2.78	3.52	3.58	3.54
	K_2O	2.5	2.47	1.16	1.92	0.72	2.88	5.21	1.22
	MnO	0.14	0.07	0.118	0.1	0.11	0.133	0.1	0.121
	TiO_2	0.82	1.1	1.04	0.84	0.8	0.85	0.6	1.1
	P_2O_5	0.43	0.65	0.25	0.03	0.28	0.22	0	0.34
	CO_2	0.31	1.85	0.06	0.6	0.35	0.95	0.38	0.25
	H_2O^+	1.61	2.45	0.34	0.44	1.2	2.8	2.5	0.44
	总量	99.16	99.81	99.33	99.32	99.68	98.67	99.54	100.02
标准矿物成分含量/%	Q	5.73	9.32	12.36	13.36	22.16	21.86	11.17	9.43
	C		0.17			0.34	3.8	0.79	
	Or	15.21	15.3	6.94	11.56	4.34	17.95	31.88	7.26
	Ab	36.24	42.47	24.09	24.94	23.94	31.34	31.3	30.12
	An	22.61	17.11	29.9	30.99	30.95	9.05	8.27	33.45
	Di	0.25		9.39	6.47				1.96
	Hy	10.79	4.27	8.75	5.68	10.63	9.21	13.28	7.6
	Mt	2.19	0.48	6.04	5.31	5.47	4.6	2.13	7.33
	Hm	5.17	8.26						
	Il			2	1.62	1.55	1.7	1.18	2.1
	Ap	0.97	1.49	0.55	0.07	0.62	0.51	0	0.75
岩石化学常用参数	AlK	6.67	7.27	3.98	4.82	3.5	6.4	8.79	4.76
	Na_2O/K_2O	1.67	1.94	2.43	1.51	3.86	1.22	0.69	2.90
	FeO^*/MgO	1.50	4.57	1.95	2.90	2.23	2.58	1.65	2.27
	A/CNK	0.95	0.93	0.78	0.86	0.99	1.25	1.05	0.92
	NK/A	0.54	0.63	0.35	0.38	0.32	0.57	0.76	0.37
	DI	57.18	67.09	43.38	49.86	50.44	71.15	74.36	46.81
	SI	23.65	9.49	24.85	16.51	22.4	15.7	18.07	20.69
	AR	1.84	2.08	1.37	1.46	1.35	2.14	3.18	1.43
	τ	16.43	10.85	13.39	18.01	17.56	14.29	19.43	14.36
	δ	3.62	4.15	1.24	1.66	0.74	2.31	4.08	1.96
产地		五指山	洛基	同安岭			牛腊岭		同安岭

续表

地层单位		六罗村组							
岩性		英安岩			中、基性岩平均值	火山碎屑岩	熔结凝灰岩	流纹岩	凝灰熔岩
样品号		D401	D405	D1887		H7-4	H12-9	D3036-1	I29-2
序号		47	48	49		50	51	52	53
岩石化学成分/%	SiO_2	66. 69	65. 38	62. 1	55. 97	67. 13	68. 82	70. 6	71. 3
	Al_2O_3	15. 36	16. 73	15. 42	16. 47	17. 3	17. 28	13. 13	13. 7
	Fe_2O_3	0. 91	1. 95	1. 87	4. 735	4. 18	1. 47	0. 34	1. 9
	FeO	2. 18	1. 85	1. 82	2. 39	1. 18	0. 81	2. 05	1. 68
	MgO	1. 84	2. 17	2. 49	3. 286	1	0. 57	0. 47	0. 4
	CaO	2. 95	0. 53	3. 37	5. 475	0. 18	0. 08	1. 79	2. 14
	Na_2O	3. 64	2. 77	3. 3	3. 315	0. 05	1. 58	2. 96	3. 1
	K_2O	3. 86	4. 48	4. 22	2. 452	5. 27	5. 5	4. 98	4. 48
	MnO	0. 06	0. 06	0. 07	0. 109	0. 02	0. 03	0. 06	0. 07
	TiO_2	0. 56	0. 64	0. 64	0. 964	0. 6	0. 28	0. 27	0. 25
	P_2O_5	0. 19	0. 25	0. 23	0. 289	0. 24	0. 07	0. 06	0. 08
	CO_2	0. 07	0. 04	2. 12	1. 255	0	0	1. 54	0. 44
	H_2O^+	1. 38	2. 87	2. 07	2. 485	0. 24	3. 55	0. 64	0. 5
	总量	99. 69	99. 72	99. 72	99. 2	97. 39	100. 04	98. 89	100
标准矿物成分含量/%	Q	21. 44	28. 92	16. 9	12. 49	46. 08	38. 92	29. 66	30. 9
	C	0. 22	7. 11		1. 212	11. 5	8. 88		
	Or	23. 24	27. 37	26. 13	15. 21	32. 09	33. 72	30. 46	26. 8
	Ab	31. 32	24. 18	29. 2	29. 31	0. 43	13. 84	25. 87	26. 5
	An	13. 77	1. 2	15. 44	20. 55	0. 92	0. 41	8. 05	10. 2
	Di			0. 52	4. 848			0. 66	0. 04
	Hy	7. 16	6. 47	7. 18	7. 945	2. 57	1. 48	4. 13	2. 23
	Mt	1. 34	2. 92	2. 84	3. 548	2. 19	1. 96	0. 51	2. 78
	Hm				2. 55	2. 79	0. 17		
	Il	1. 08	1. 26	1. 27	1. 387	1. 17	0. 55	0. 53	0. 48
	Ap	0. 42	0. 56	0. 53	0. 666			0. 14	0. 18
岩石化学常用参数	AlK	7. 5	7. 25	7. 52	5. 767	5. 32	7. 08	7. 94	7. 58
	Na_2O/K_2O	0. 94	0. 62	0. 782	1. 761	0. 01	0. 29	0. 59	0. 69
	FeO^*/MgO	1. 63	1. 66	1. 407	2. 246	4. 941	3. 74	5. 01	8. 47
	A/CNK	0. 99	1. 61	0. 96	0. 982	2. 83	1. 99	0. 97	0. 99
	NK/A	0. 66	0. 56	0. 65	0. 497	0. 33	0. 49	0. 78	0. 73
	DI	76	80. 48	72. 22	57. 02	78. 61	86. 48	85. 99	84. 1
	SI	14. 8	16. 41	18. 18	19. 92	8. 56	5. 74	4. 35	3. 46
	AR	2. 39	2. 45	2. 33	1. 842	1. 88	2. 38	3. 28	2. 84
	τ	20. 93	21. 81	18. 94	14. 65	28. 75	56. 07	37. 67	42. 3
	δ	2. 37	2. 35	2. 96	2. 918	1. 17	1. 94	2. 28	2. 03
产地		旺商				五指山		同安岭	

续表

地层单位		六罗村组							
岩性		火山碎屑岩	凝灰熔岩	火山碎屑岩	凝灰熔岩				熔结角砾岩
样品号		H6-2b	VI4-1	H5-5b	VI3-2	VI7-1	I23-1	XIII8-1	H12-13
序号		54	55	56	57	58	59	60	61
岩石化学成分/%	SiO_2	71.35	71.47	71.74	72.19	72.68	73.09	73.43	73.68
	Al_2O_3	13.79	13.45	14.31	13.19	13.23	13.8	13.05	13.3
	Fe_2O_3	0.29	1.65	1.16	1.99	1.74	1.46	1.08	1.19
	FeO	3.96	1.35	2.25	1.58	1.22	1.68	2.18	1.84
	MgO	0.56	0.37	0.33	0.64	0.32	0.74	0.47	0.5
	CaO	2.64	2.34	0.31	1.59	1.66	1.36	1.18	0.46
	Na_2O	2.79	3.46	2.06	3.54	3.5	3.56	3.56	3.18
	K_2O	3.47	4.52	6.53	4.02	4.2	2.88	3.02	4.32
	MnO	0.08	0.05	0.03	0.05	0.05	0.08	0.07	0.1
	TiO_2	0.24	0.22	0.16	0.22	0.24	0.21	0.3	0.15
	P_2O_5	0.45	0.08	0.39	0.08	0.08	0.05	0.1	0.22
	CO_2		0.18		0.18	0.24	0.13	0.14	
	H_2O^+	0.23	0.9	0.61	0.86	0.5	0.58	1.01	0.81
	总量	99.85	100.04	99.88	100.13	99.66	99.62	99.59	99.75
标准矿物成分含量/%	Q	33.55	29.35	32.5	31.92	32.56	37.13	37.55	36.82
	C	1.6		3.29	0.28	0.06	2.48	2.01	3.05
	Or	20.6	27.02	38.91	24	25.11	17.22	18.15	25.83
	Ab	23.67	29.55	17.54	30.19	29.9	30.42	30.56	27.16
	An	10.5	7.86	1.55	7.49	7.86	6.53	5.35	1
	Di		2.81						
	Hy	8.2	0.4	3.81	2.61	1.31	3.57	3.98	3.62
	Mt	0.42	2.42	1.69	2.91	2.55	2.14	1.59	1.74
	Hm								
	Il	0.46	0.42	0.31	0.42	0.46	0.4	0.58	0.29
	Ap	0.99	0.18		0.18	0.18	0.11	0.22	0.49
岩石化学常用参数	AlK	6.26	7.98	8.59	7.56	7.7	6.44	6.58	7.5
	Na_2O/K_2O	0.80	0.77	0.32	0.88	0.83	1.24	1.19	0.74
	FeO^*/MgO	7.54	7.66	9.98	5.27	8.71	4.05	6.71	5.82
	A/CNK	1.05	0.91	1.3	1.01	0.99	1.21	1.16	1.24
	NK/A	0.61	0.79	0.73	0.77	0.78	0.65	0.7	0.74
	DI	77.83	85.92	88.95	86.11	87.58	84.78	86.26	89.81
	SI	5.06	3.26	2.68	5.44	2.91	7.17	4.56	4.53
	AR	2.23	3.04	3.85	3.09	3.14	2.48	2.72	3.4
	τ	45.83	45.41	76.56	43.86	40.54	48.76	31.63	67.47
	δ	1.38	2.24	2.57	1.96	2.00	1.38	1.42	1.83
产地		五指山	同安岭	五指山	同安岭				五指山

续表

地层单位		六罗村组							
岩性		凝灰熔岩	流纹岩	凝灰熔岩			流纹岩	凝灰熔岩	流纹岩
样品号		XIII15-1	H12-2	五2	XV4-1	XIII5-1	I24-1	XV2-1	D1886
序号		62	63	64	65	66	67	68	69
岩石化学成分/%	SiO_2	74.32	74.77	75.24	76.04	76.22	76.24	77.24	78.4
	Al_2O_3	12.98	13.42	12.57	13.63	12.7	13.1	11.9	13.82
	Fe_2O_3	1.91	0.66	1.91	0.36	0.58	1.04	0.94	0.72
	FeO	0.98	1.54	0.5	1.93	1.5	0.81	0.95	0.22
	MgO	0.4	0.17	0.36	0.34	0.26	0.39	0.38	0.45
	CaO	0.62	0.21	0.27	1.06	0.12	0.21	0.47	0.09
	Na_2O	3.00	3.59	2.9	3.15	3.27	2.7	2.34	0.19
	K_2O	3.95	3.89	4.79	3.75	4.34	4.32	4.24	2.41
	MnO	0.06	0.05	0.04	0.08	0.05	0.03	0.02	0.01
	TiO_2	0.25	0.10	0.10	0.15	0.15	0.17	0.2	0.12
	P_2O_5	0.05	0.11	0.02	0.05	0.05	0.02	0.05	0.03
	CO_2	0.09		0.14	0.11	0.11	0.28	0.11	0.04
	H_2O^+	1.00	1.31	0.81	0.49	0.87	0.9	1.00	3.36
	总量	99.61	99.82	99.65	101.14	100.22	100.21	99.84	99.86
标准矿物成分含量/%	Q	40.57	38.17	39.5	39.09	39.43	43.27	46.25	69.74
	C	2.78	3.19	2.18	2.54	2.52	3.67	2.74	11.19
	Or	23.71	23.36	28.71	22.06	25.87	25.8	25.4	14.78
	Ab	25.74	30.8	24.83	26.48	27.85	23.04	20.03	1.66
	An	2.83	0.4	1.24	4.94	0.3	0.93	2.07	0.28
	Di								
	Hy	1.02	2.67	0.91	3.97	2.79	1.39	1.64	1.17
	Mt	2.67	0.97	1.47	0.52	0.85	1.52	1.38	0.41
	Hm	0.1		0.92					0.47
	Il	0.48	0.19	0.19	0.28	0.29	0.33	0.38	0.24
	Ap	0.11	0.24	0.04	0.11	0.11	0.04	0.11	0.07
岩石化学常用参数	AlK	6.95	7.48	7.69	6.9	7.61	7.02	6.58	2.6
	Na_2O/K_2O	0.76	0.93	0.61	0.84	0.75	0.63	0.55	0.08
	FeO^*/MgO	6.75	12.55	6.16	6.63	7.78	4.48	4.73	1.93
	A/CNK	1.26	1.28	1.2	1.22	1.23	1.38	1.28	4.48
	NK/A	0.71	0.75	0.79	0.68	0.79	0.7	0.71	0.21
	DI	90.02	92.32	93.04	87.64	93.14	92.12	91.68	86.19
	SI	3.91	1.73	3.44	3.57	2.61	4.21	4.29	11.28
	AR	3.09	3.43	3.99	2.77	3.92	3.23	3.27	1.46
	τ	39.92	98.3	96.7	69.87	62.87	61.18	47.8	113.6
	δ	1.54	1.76	1.83	1.44	1.74	1.48	1.26	0.19
产地		牛腊岭	五指山		牛腊岭		同安岭	牛腊岭	旺商

续表

地层单位		六罗村组	鹿母湾组						
岩性		酸性岩平均值	凝灰岩	流纹质凝灰熔岩		沉凝灰岩			平均值
样品号			HMK1	JXb80	JXyQ43	X16-1	X43-1	X8-1	
序号			70	71	72	73	74	75	
岩石化学成分/%	SiO_2	73.30	67.10	72.06	72.58	59.05	62.77	68.69	67.04
	Al_2O_3	13.68	14.17	13.27	14.21	20.08	18.07	15.94	15.96
	Fe_2O_3	1.33	5.44	3.46	2.56	6.75	7.59	5.64	5.24
	FeO	1.51	0.4	0.85	0.28	0.57	0.35	0.32	0.46
	MgO	0.46	0.96	0.69	0.58	2.83	1.09	1.04	1.20
	CaO	0.94	107	0.23	0.76	0.07	0.07	0.06	0.74
	Na_2O	2.72	1.8	3.16	0.12	0.21	0.31	0.46	1.01
	K_2O	4.24	3.16	3.67	4.08	4.5	4.6	3.53	3.92
	MnO	0.05	0.06	0.15	0.075	0.01	0.05	0.01	0.06
	TiO_2	0.22	0.77	0.27	0.31	0.85	0.85	0.69	0.62
	P_2O_5	0.11	1.88	0.07	0.06	0.1	0.1	0.04	0.375
	CO_2	0.17				0.04	0.04	0.07	0.025
	H_2O^+	1.01		2.01	4.13	4.51	3.75	3.16	2.93
	总量	99.76	99.01	99.89	99.75	99.57	99.64	99.65	99.58
标准矿物成分含量/%	Q	38.65	41.27	39.14	56.38	38.1	43.37	52.8	45.18
	C	3.20	5.93	3.9	8.71	15.5	12.98	11.75	9.8
	Or	25.48	18.88	22.18	25.24	28.01	28.39	21.65	24.06
	Ab	23.30	15.37	27.29	1.06	1.87	2.73	4.03	8.73
	An	4.04	5.24	0.75	3.58	0.37	0.36	0.07	1.72
	Di	0.16							
	Hy	2.67	2.42	1.76	1.52	7.45	2.84	2.7	3.12
	Mt	1.63	1.5	2.5	0.26	1.97	1.35	1.1	1.45
	Hm	0.22	4.46	1.81	2.5	5.75	6.99	5.09	4.43
	Il	0.42		0.52	0.62				0.19
	Ap	0.18	4.15	0.16	0.14			0.09	0.76
岩石化学常用参数	AlK	6.97	4.96	6.83	4.2	4.71	4.91	3.99	4.93
	Na_2O/K_2O	0.67	0.57	0.86	0.03	0.05	0.07	0.13	0.29
	FeO^*/MgO	6.44	5.52	5.74	4.45	2.35	6.59	5.19	4.97
	A/CNK	1.45	1.15	1.38	2.38	3.76	3.22	3.40	2.55
	NK/A	0.68	0.45	0.69	0.32	0.26	0.30	0.29	0.39
	DI	87.43	75.52	88.6	82.68	67.98	74.49	78.49	77.96
	SI	4.64	8.16	5.83	7.61	19.04	7.82	9.46	9.65
	AR	2.97	1.79	3.05	1.78	1.61	1.74	1.66	1.94
	τ	57.75	16.06	37.44	45.45	23.38	20.89	22.43	27.61
	δ	1.67	1.02	1.61	0.6	1.38	1.22	0.62	1.08
产地			白沙-乐东			琼海阳江			

注：1~6、39、41、46、52据宜昌地质矿产研究所、海南省地质调查院（2004）；7~10、23~30、32、35~37、42~45、53、55、57~60、62、65~68据海南地质队（1995）；13~22据海南地质队（1986）；11~12、38、50、51、54、56、61、63据南京地质矿产研究所、海南地质大队（1996）；31、33~34、40、47~49、69、73~75据海南省地质调查院（2004）；38、64据符国祥等（1994）；70据广西地矿局；71、72据四川省地矿局。

0.12%，而 K_2O 含量为 5.27%~4.08%，K_2O/Na_2O 值高达 105.4~34，属富钾并极度贫钠的岩浆岩。火山岩的 A/CNK 值变化大。岭壳村组火山岩 A/CNK 值为 0.96~1.26（平均值为 1.09），多属弱过铝质，少数为强过铝质；汤他大岭组火山 A/CNK 值为 0.95~1.30（平均值为 1.07），多属弱过铝质，少数为强过铝质；六罗村组中基性岩的 A/CNK 值多数为 0.65~0.99，属准铝质，个别为1.05~2.07，属过铝质甚至强过铝质；六罗村组酸性岩的 A/CNK 值为 0.91~4.48（平均值为 1.27），多介于 1.16~1.38，少数为准铝质或弱过铝质，多数为强过铝质；鹿母湾组中的火山岩夹层为强过铝质（A/CNK 值为 1.15~3.76）。

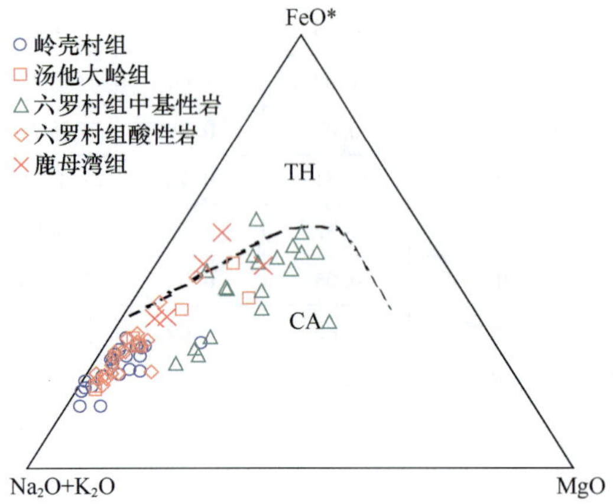

图 12-2 海南岛中生代火山岩 FAM 图

Fig. 12-2 FAM diagram for Mesozoic volcanic rocks from the Hainan Island

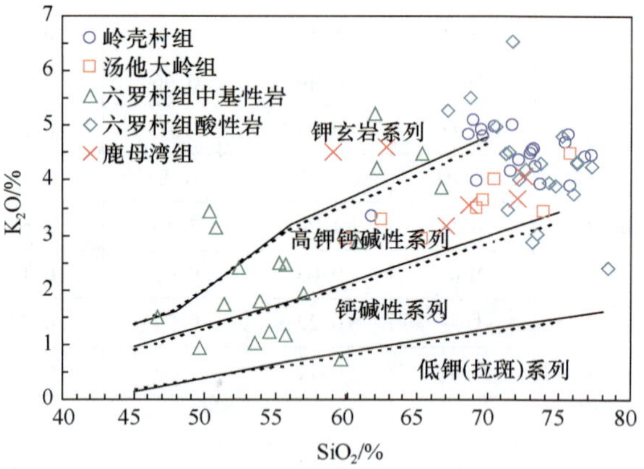

图 12-3 海南岛中生代火山岩 SiO_2-K_2O 图

Fig. 12-3 SiO_2-K_2O diagram of Mesozoic volcanic rocks from the Hainan Island

二、副矿物特征

中生代火山岩副矿物总量高（表 12-3）。其中，磁铁矿、钛铁矿、锆石、黄铁矿、磷灰石、角闪石占 96.23%，而磁铁矿高达 9056.8×10^{-6}。副矿物组合可分为两种类型：①磁铁矿-锆石-磷灰石-褐帘石型，以磁铁矿为主，占总量的 80%以上。该类型比较普遍，分布在六罗村组、汤他大岭组和岭壳村组火山岩中；②磁铁矿-钛铁矿-锆石-褐帘石型，磁铁矿和钛铁矿占 70%以上，主要分布在汤他大岭组和岭壳村组火山岩中。

中生代火山岩的锆石特征基本相似，为透明至半透明，无色或带浅黄色，少数带浅紫色，金刚光泽或玻璃光泽，常有包体，包体成分有锆石、磷灰石及黑色包体。锆石晶形常以柱面（110）、（100）与锥面（111）、柱面（100）与锥面（111）的聚形为主。

三、稀土元素地球化学特征

六罗村组中基性火山岩的 ΣREE 值为 111.18×10^{-6}~327.85×10^{-6}，平均值为 187.86×10^{-6}；六罗村组酸性火山岩的 ΣREE 值为 101.94×10^{-6}~259.55×10^{-6}，平均值为 178.45×10^{-6}；汤他大岭组火山岩的 ΣREE 值为 158.95×10^{-6}~247.15×10^{-6}，平均值为 192.45×10^{-6}；岭壳村组火山岩的 ΣREE 值为 150.94×10^{-6}~211.69×10^{-6}，平均值为 185.87×10^{-6}。绝大多数较低，低于世界酸性岩平均值 258×10^{-6}（Vinigradov，1962），为稀土元素亏损型或强亏损型（表 12-4）。各个组火山岩的配分曲线较一致，为较陡的右倾曲线，轻重稀土分异较明显（图 12-4），LREE/HREE 值为5.7~18.9，多数>10。六罗村组中基性岩的 δEu 值为 1.26~0.61，多数具弱的负铕异常，少数具有弱的正铕异常或中等的负铕异常。其他火山岩的 δEu 值为 0.85~0.18，具弱至强的负铕异常。

表 12-3　海南岛中生代部分火山岩副矿物含量一览表

Table 12-3　List of contents of accessory minerals in some of Mesozoic volcanic rocks from the Hainan Island　（单位：10^{-6}）

层位		岭壳村组										
岩性		流纹岩					英安岩		角砾熔岩			凝灰熔岩
样品号		D91-1	D776-1	Ⅶ19-3	Ⅶ19-2	Ⅶ19-1	Ⅶ1-1	Ⅶ46-2	Ⅶ42-2	Ⅶ39-2	Ⅶ24-2	Ⅰ41-3
序号		1	2	3	4	5	6	7	8	9	10	11
副矿物含量	磁铁矿	微量	77.43	微量	微量	71.32	16.15	1031.8	1398.2	606.0	284.83	3639.45
	钛铁矿	个别			少量		5.78	2.5		112.46		1.86
	锆石	12.12	64.47	11.19	7.13	9.86	78.39	17.81	86.81	30.79	13.97	27.78
	独居石	个别					个别	个别		个别	3.7	
	电气石		28.92									个别
	锐钛矿	个别										
	金红石	微量										个别
	自然铅		微量	20.84	118	少量	微量		微量	微量	微量	3.9
	方铅矿					个别		微量				
	黄铁矿	64.8	21.93	113.58	45.58	9.57	6.27	1.17	25.02	71.7	49.21	0.98
	赤铁矿		7.74	165.8								
	辉银矿											
	赤褐铁矿											
	磷灰石	1.61	32.11				123.42	2.34	6.27	5.07	5.59	15.89
	硬锰矿	63.34	微量							微量	个别	
	绿帘石	个别	6.77				394.04	9.34	微量	10	149.6	少量
	石榴子石	微量		个别	22.62	11.86		30.0	个别	微量		个别
	白钨矿			个别	少量							
	褐钛矿											
	黄玉											66.2
	角闪石											个别
	闪锌矿											微量
	雄黄											个别
	褐帘石											
	萤石											
	自然银											个别
	云母											
	榍石											
	钍石											
	辉钼矿									3.04		
	自然铜											
	尖晶石											个别
	橄榄石											
	刚玉		个别					个别				
	软锰矿					19.96						
	磷氯铅矿											个别
	总量	141.87	239.37	311.41	78.71	122.57	624.05	1094.96	1516.3	832.76	506.9	3756.06

续表

层位		岭壳村组			汤他大岭组							
岩性		凝灰熔岩	碎斑熔岩		流纹岩		英安岩	凝灰岩		熔结凝灰岩		
样品号		Ⅰ35-4	Ⅰ35-12	Ⅰ36-1	XⅧ1-1	Ⅳ5-2	Ⅳ2-2	Ⅳ7-1	Ⅳ12-1	Ⅳ20-1	Ⅳ10-1	Ⅳ16-1
序号		12	13	14	15	16	17	18	19	20	21	22
副矿物含量	磁铁矿	2810	3878	3202	4652	23.4	5882	56.77	868.5	1969	5281	3546
	钛铁矿	8.37	40.37		196.7	329.3	个别	3927	少量	个别	微量	552.2
	锆石	63.54	82.94	78.84	72.1	22.3	57.61	97.97	1.69	10.53	95.0	67.3
	独居石		个别								个别	
	电气石	个别	个别	个别					个别	个别		
	锐钛矿					个别	微量					
	金红石		个别	个别			少量	个别	个别		个别	个别
	自然铅	个别	8.85	13.8	个别	少量	少量	少量	0.11	个别	微量	0.7
	方铅矿											
	黄铁矿	79.82	1.08	微量	微量	个别	3.13	个别	少量	328.5	个别	微量
	赤铁矿											
	辉银矿		个别									
	赤褐铁矿	57.24		54.7	微量	个别		少量		少量		
	磷灰石	26.64	101.7	38.44	26.23		个别	个别	5.08	4.21		
	硬锰矿											
	绿帘石	450.1	少量	个别	微量	个别	微量	个别	1006	12.31	个别	
	石榴子石		63.5				个别			个别		
	白钨矿											
	褐钛矿											
	黄玉		个别	少量			个别					
	角闪石	个别	317.6	6.74	微量		70.8		个别	个别		
	闪锌矿	3.99	25.41	微量						个别		
	雄黄	个别	个别	个别		个别						
	褐帘石				微量						个别	个别
	萤石	个别			微量	微量						
	自然银											
	云母				91.8		个别					22.09
	榍石				141				个别			
	钍石	个别									个别	个别
	辉钼矿		个别									
	自然铜		个别									
	尖晶石											
	橄榄石											
	刚玉											
	软锰矿											
	磷氯铅矿		个别	个别		个别	个别	个别	微量	个别	个别	个别
	总量	3499	4519	3395	5180	375.0	6014	4082	1881	1984	5376	4188

续表

层位		六罗村组									
岩性		凝灰熔岩			流纹岩	凝灰熔岩	英安岩	安山岩	安山岩	安山岩	安山岩
样品号		Ⅵ4-1	Ⅵ7-1	Ⅰ23-1	Ⅰ24-1	Ⅰ29-2	ⅢH7-1	ⅢH8-3	ⅢH9-1	Ⅰ5-1	Ⅰ9-1
序号		23	24	25	26	27	28	29	30	31	32
副矿物含量	磁铁矿	5020.13	4650.9	2595.5	369.29	1902.2	13.79	0.94	120.36	807.46	2470
	钛铁矿			微量	个别		57.93	24.91	2.11		
	锆石	78.99	92.31	82.07	8.57	22.39	9.65	25.81	2.63	1.77	0.91
	独居石							22.15	1.05		
	电气石		个别	微量	个别	个别					个别
	锐钛矿	少量			个别		微量	微量	个别		
	金红石	微量		个别	个别	个别	微量	微量			个别
	自然铅	少量	0.96	7.17	3.0	5.93				29.78	1.17
	方铅矿		个别	个别		个别					
	黄铁矿	9.08	微量	25.0	少量	3.95	1.38	4.53	5.26	3.9	7.7
	赤铁矿						1008	668			
	辉银矿										
	赤褐铁矿			3.83	5.75	微量					个别
	磷灰石	个别	个别	28.04	个别	微量			15.79	12.13	少量
	硬锰矿										
	绿帘石	个别	个别	58.78	0.71	3285.4			微量	2020.5	170.9
	石榴子石	个别	个别		个别		11.59	20.23	1.05	个别	个别
	白钨矿										
	褐钛矿								840.35		
	黄玉		个别	微量	个别		6.9		5.26		
	角闪石	微量	个别	微量	微量	个别	微量	4.98		个别	214.9
	闪锌矿									个别	个别
	雄黄			个别						个别	个别
	褐帘石	个别	17.13	个别							
	萤石	个别			个别						个别
	自然银				个别					个别	
	云母		个别								
	榍石										
	钍石										
	辉钼矿										
	自然铜										
	尖晶石										
	橄榄石									个别	
	刚玉										
	软锰矿										
	磷氯铅矿		个别		个别						个别
	总量	5108	4761	2800	387.3	5220	1109	771.6	993.9	2872.5	2865.6

表 12-4 海南岛中生代部分变火山岩稀土元素含量及主要特征参数一览表

Table 12-4 List of REE contenta and main characteristic parameters of some Mesozoic volcanic rocks from the Hainan Island

地层单位		岭壳村组								
岩性		英安岩	粗安岩	粗安岩	英安岩	安粗质角砾熔岩	流纹岩	碎斑熔岩	凝灰熔岩	碎斑熔岩
样品号		D1010-1	D1019-3	D1019-4	D1026-1	D1027-1	D908-1	I35-12	I36-1	H17-3
序号		1	2	3	4	5	6	7	8	9
稀土元素含量/10^{-6}	La	36.2	41.5	44.4	46.7	42	43.5	47.3	37	48.7
	Ce	71.1	83.7	83.8	91.3	85.5	80.2	97.8	61.1	91.8
	Pr	7.7	9	8.8	9.9	8.9	7.5	8.1	6.5	7.6
	Nd	28.6	33.3	32.7	34.4	32.5	25.2	33.5	27.7	25.6
	Sm	5.2	5.6	6.1	5.9	5.6	3.7	7.9	5	6.2
	Eu	1.28	1.26	1.01	1.06	1.14	0.19	0.67	0.53	1.01
	Gd	4.65	4.45	4.65	4.71	4.5	2.59	4.39	3.33	5.18
	Tb	0.71	0.67	0.73	0.74	0.66	0.35	0.84	0.61	0.96
	Dy	4.81	3.91	4.2	4.11	3.78	2.12	5.01	3.73	4.06
	Ho	0.89	0.8	0.86	0.83	0.78	0.43	0.95	0.74	0.89
	Er	2.72	2.35	2.4	2.41	2.17	1.25	1.88	2.1	2.66
	Tm	0.41	0.36	0.34	0.37	0.34	0.18	0.38	0.3	0.47
	Yb	3	2.67	2.57	2.43	2.19	1.35	2.6	2	2.24
	Lu	0.44	0.4	0.38	0.38	0.33	0.2	0.37	0.3	0.39
	ΣREE	167.7	190	192.9	205.24	190.4	168.8	212	151	198
主要特征参数	LREE/HREE	8.51	11.17	10.96	11.84	11.91	18.92	11.9	10.5	10.7
	δEu	0.78	0.75	0.56	0.6	0.67	0.18	0.32	0.37	0.53
	δCe	0.98	1	0.96	0.97	1.02	0.98	1.11	0.88	1.04
	$(La/Yb)_N$	8.14	10.48	11.65	12.96	12.93	21.72	12.3	12.5	14.7
	$(Ce/Yb)_N$	6.13	8.11	8.43	9.72	10.1	15.37	9.73	7.9	10.6
	$(La/Sm)_N$	4.38	4.66	4.58	4.98	4.72	7.4	3.77	4.65	4.94
	$(Gd/Yb)_N$	1.25	1.34	1.46	1.56	1.66	1.55	1.36	1.34	1.87
	La/Sm	6.96	7.41	7.28	7.92	7.5	11.76	5.99	7.4	7.85

地层单位		岭壳村组		汤他大岭组						
岩性		熔结凝灰岩	平均值	凝灰岩				英安岩		凝灰岩
样品号		H17-4		I30-1	IV10-1	IV12-1	IV16-1	IV2-2	IV5-2	IV7-1
序号		10		11	12	13	14	15	16	17
稀土元素含量/10^{-6}	La	45.8	43.31	37.5	49.2	58.1	55	43.5	59.8	46.7
	Ce	84.4	83.07	61.4	76.8	106	83.1	66.2	75.9	72.9
	Pr	7.13	8.11	7.1	9.2	11.5	9.6	7.4	9.9	7.9
	Nd	23.3	29.68	32.3	36.5	38.8	39.4	30.4	34.9	33.8
	Sm	5.69	5.69	6.3	7.1	10.8	7.4	5.6	6.4	6
	Eu	0.39	0.85	1.5	1.18	1.09	1.38	1.29	1	1.38
	Gd	4.82	4.33	4.49	4.78	5.4	5.19	3.72	4.11	4.23
	Tb	0.92	0.72	0.73	0.79	0.92	0.88	0.62	0.72	0.67
	Dy	4	3.97	3.69	4.7	5.8	5.2	3.29	4.06	3.78
	Ho	0.87	0.80	0.65	0.89	1.34	0.95	0.6	0.8	0.71
	Er	2.69	2.26	1.58	2.33	3.35	2.48	1.58	2.37	1.76
	Tm	0.47	0.36	0.23	0.34	0.5	0.34	0.25	0.37	0.27
	Yb	2.37	2.34	1.28	2.09	3.1	2.14	1.54	2.31	1.67
	Lu	0.41	0.36	0.2	0.31	0.45	0.34	0.24	0.35	0.26
	ΣREE	183	185.87	158.95	196.21	247.15	213.4	166.23	202.99	182.03

续表

地层单位		岭壳村组		汤他大岭组						
岩性		熔结凝灰岩	平均值	凝灰岩				英安岩		凝灰岩
样品号		H17-4		I30-1	IV10-1	IV12-1	IV16-1	IV2-2	IV5-2	IV7-1
序号		10		11	12	13	14	15	16	17
主要特征参数	LREE/HREE	10.1	11.65	11.37	11.09	10.85	11.18	13.04	12.45	12.64
	δEu	0.22	0.50	0.82	0.59	0.39	0.65	0.81	0.56	0.8
	δCe	1.01	1.00	0.85	0.81	0.93	0.8	0.82	0.69	0.84
	$(La/Yb)_N$	13	13.03	19.75	15.87	12.64	17.33	19.04	17.45	18.85
	$(Ce/Yb)_N$	9.21	9.53	12.41	9.5	8.84	10.04	11.12	8.5	11.29
	$(La/Sm)_N$	5.06	4.91	3.74	4.36	3.38	4.68	4.89	5.88	4.9
	$(Gd/Yb)_N$	1.64	1.50	2.83	1.85	1.41	1.96	1.95	1.44	2.04
	La/Sm	8.05	7.81	5.95	6.93	5.38	7.43	7.77	9.34	7.78

地层单位		汤他大岭组		六罗村组						
岩性		流纹岩	平均值	安山岩	玄武岩	安山岩	安山岩	粗面玄武岩	粗安岩	玄武安山岩
样品号		XVIII1-1		IIIH 9-1	D223	XXIV 2-1	IIIH 7-1	XIII 6-5	XIII 6-4	XIII 7-1
序号		18		19	20	21	22	23	24	25
稀土元素含量/10^{-6}	La	43.5	49.16	36.5	21.2	44.6	68.94	29	62.4	31.1
	Ce	70.6	76.61	73	39.1	82.7	139.2	46.7	104	47.1
	Pr	8	8.83	8.93	6.3	10.8	17.43	6.1	13.8	6.5
	Nd	28	34.26	32.9	21.6	40.5	68.05	25.6	48.1	27.9
	Sm	5.6	6.9	5.82	4.9	7.22	11.33	3	5.1	5.5
	Eu	0.49	1.16	1.64	1.52	1.64	3.17	1.28	2.23	1.53
	Gd	4.04	4.50	5.11	5.44	5.27	8.38	3.85	5.75	4.16
	Tb	0.73	0.76	0.81	0.93	0.68	1.12	0.63	0.89	0.75
	Dy	4.78	4.41	4.01	4.37	4.35	4.78	3.65	4.24	4.87
	Ho	1.04	0.87	0.78	0.92	0.76	0.87	0.72	0.82	0.86
	Er	2.88	2.29	2.1	2.38	2.03	2.17	1.89	2.13	2.42
	Tm	0.37	0.33	0.32	0.37	0.3	0.32	0.27	0.31	0.38
	Yb	2.24	2.05	1.94	1.85	1.8	1.83	1.61	1.86	2.39
	Lu	0.36	0.31	0.28	0.3	0.29	0.26	0.24	0.27	0.37
	ΣREE	172.63	192.45	174	111	202.94	327.9	124.54	252	135.8
主要特征参数	LREE/HREE	9.5	11.52	10.3	5.71	12.11	15.62	8.68	14.5	7.38
	δEu	0.3	0.62	0.9	0.9	0.78	0.95	1.15	1.25	0.94
	δCe	0.85	0.82	0.95	0.81	0.88	0.94	0.81	0.82	0.76
	$(La/Yb)_N$	13.09	16.75	12.7	7.73	16.71	25.4	12.14	22.6	8.77
	$(Ce/Yb)_N$	8.15	9.98	9.73	5.47	11.88	19.68	7.5	14.5	5.1
	$(La/Sm)_N$	4.89	4.59	3.94	2.72	3.89	3.83	6.08	7.7	3.56
	$(Gd/Yb)_N$	1.46	1.87	2.13	2.37	2.36	3.7	1.93	2.49	1.0
	La/Sm	7.77	7.29	6.27	4.33	6.18	6.08	9.67	12.2	5.65

地层单位		六罗村组								
岩性		玄武安山岩	安山岩	角闪安山岩	玄武安山岩		石英安山岩	安山岩	粗面英安岩	英安岩
样品号		D3039-4	D1539	IIIH8-1	D3039-1	I5-1	I9-1	XIII 3-1	XIII 1-1	D1887
序号		26	27	28	29	30	31	32	33	34
稀土元素含量/10^{-6}	La	28.8	31.8	65.8	25.1	36.6	31.5	37.7	58.8	49.73
	Ce	59.3	63.1	131.1	55.6	54.6	54.4	57	88.9	91.23
	Pr	6.9	7.4	17.08	6.6	7.8	6.8	7.5	12.1	9.85
	Nd	27.9	27.4	65.66	27.3	34.9	29.5	28.8	39.1	38.3
	Sm	5.3	4.9	11.31	5.3	6.1	5.5	5.3	7.8	6.41
	Eu	1.52	1.2	3	1.53	1.81	1.47	0.99	1.38	1.67
	Gd	4.56	4.25	7.93	4.4	4.11	3.82	3.47	5.56	4.36
	Tb	0.63	0.64	1.06	0.62	0.53	0.6	0.52	0.96	0.57
	Dy	3.37	3.45	4.59	3.5	3.12	3.28	2.59	5.79	2.84
	Ho	0.68	0.7	0.86	0.67	0.66	0.66	0.51	1.2	0.52
	Er	1.63	1.95	2.07	1.61	1.56	1.78	1.2	104	1.23
	Tm	0.24	0.25	0.32	0.2	0.23	0.27	0.17	0.47	0.18
	Yb	1.5	1.82	1.77	1.51	1.38	1.6	1.02	2.77	1.14
	Lu	0.23	0.29	0.26	0.21	0.16	0.26	0.15	0.41	0.17
	ΣREE	143	149.2	312.8	134	154	141.4	146.92	228.48	208.2
主要特征参数	LREE/HREE	10.1	10.17	15.59	9.55	12.1	10.53	14.26	10.2	17.91
	δEu	0.92	0.79	0.92	0.94	1.04	0.93	0.66	0.61	0.91
	δCe	0.98	0.96	0.92	1.02	0.74	0.86	0.77	0.76	0.94
	$(La/Yb)_N$	12.9	11.78	25.06	11.2	17.9	13.27	24.92	14.31	29.41
	$(Ce/Yb)_N$	10.2	8.97	19.16	9.52	10.2	8.79	14.45	8.3	20.7
	$(La/Sm)_N$	3.42	4.08	3.66	2.98	3.77	3.6	4.47	4.74	4.88
	$(Gd/Yb)_N$	2.45	1.88	3.62	2.35	2.4	1.93	2.75	1.62	3.09
	La/Sm	5.43	6.49	5.82	4.74	6	5.73	7.11	7.54	7.76

地层单位		六罗村组								
岩性		英安岩		中基性岩平均值	火山碎屑岩	熔结凝灰岩	流纹岩	熔结凝灰岩	火山碎屑岩	凝灰熔岩
样品号		D405	D401		H7-4	H12-9	D3036-1	I29-2	H6-2	VI4-1
序号		35	36		37	38	39	40	41	42
稀土元素含量/10^{-6}	La	60.34	44.82	42.48	38.8	33.9	53.3	46.6	43.2	58.2
	Ce	109.2	83.2	76.63	73.8	74.3	96.6	78.8	76	80.9
	Pr	12.82	9.44	9.66	6.7	5.7	9.9	8.4	5.9	12
	Nd	46.41	32.41	36.80	23.8	19.2	33.3	32	19.4	38.8
	Sm	7.33	5.2	6.30	6.3	4.9	5.2	5.9	4.5	8.2
	Eu	1.87	1.29	1.71	1.65	0.49	0.93	0.77	0.8	0.89
	Gd	5.02	3.58	4.95	5.25	3.99	3.94	3.75	3.52	6.5
	Tb	0.64	0.49	0.73	0.92	0.79	0.55	0.65	0.7	0.94
	Dy	3.1	2.31	3.79	3.4	3.22	2.91	4	2.49	6.07
	Ho	0.59	0.4	0.73	0.63	0.66	0.62	0.85	0.54	1.3
	Er	1.37	1.03	1.88	1.61	1.87	1.54	2.43	1.52	3.59
	Tm	0.2	0.15	0.26	0.29	0.32	0.23	0.37	0.3	0.4
	Yb	1.3	0.93	1.67	1.25	1.5	1.62	2.23	1.3	2.57
	Lu	0.2	0.14	0.25	0.23	0.25	0.26	0.34	0.24	0.37
	ΣREE	250.39	185.39	187.86	164.63	151.09	210.9	187.09	160.41	220.73

续表

地层单位		六罗村组								
岩性		英安岩		中基性岩平均值	火山碎屑岩	熔结凝灰岩	流纹岩	熔结凝灰岩	火山碎屑岩	凝灰熔岩
样品号		D405	D401		H7-4	H12-9	D3036-1	I29-2	H6-2	VI4-1
序号		35	36		37	38	39	40	41	42
主要特征参数	LREE/HREE	19.16	19.53	12.41	11.12	10.99	17.07	11.8	14.12	9.15
	δEu	0.89	0.87	0.91	0.85	0.33	0.6	0.47	0.59	0.36
	δCe	0.9	0.93	0.88	1.01	1.18	0.94	0.89	1	0.7
	$(La/Yb)_N$	31.29	32.49	18.37	20.93	15.24	22.18	14.09	22.4	15.27
	$(Ce/Yb)_N$	21.73	23.14	12.72	15.27	12.81	15.42	9.14	15.12	8.14
	$(La/Sm)_N$	5.18	5.42	4.33	3.87	4.35	6.45	4.97	6.04	4.46
	$(Gd/Yb)_N$	3.12	3.11	2.48	3.39	2.15	1.96	1.36	2.18	2.04
	La/Sm	8.23	8.62	6.88	6.16	6.92	10.25	7.9	9.6	7.10

地层单位		六罗村组								
岩性		火山碎屑岩	凝灰熔岩				熔结角砾岩	凝灰熔岩	流纹岩	凝灰熔岩
样品号		H5-5b	VI10	VI7-1	I23-1	XIII8-1	H12-13	XIII15-1	H12-2	XV4-1
序号		43	44	45	46	47	48	49	50	51
稀土元素含量/10^{-6}	La	28.2	48.6	49.9	45.9	52.2	25.2	76.3	33.7	54.9
	Ce	52.4	73.1	73.1	75.5	71.4	49.6	91.7	64.8	82.9
	Pr	4.4	8.5	8.14	8.5	10.2	3.8	13.3	4.9	8.9
	Nd	14.2	33.3	32.8	32.6	34	12.7	48.1	16.6	34.5
	Sm	3.5	6.4	6.09	5.9	6.2	3	8.2	3.8	6.9
	Eu	0.52	0.69	0.71	0.77	0.78	0.44	1.08	0.46	0.52
	Gd	2.72	4.28	4.02	3.62	3.49	2.25	5.23	2.68	4.88
	Tb	0.6	0.73	0.71	0.63	0.62	0.5	0.97	0.55	0.88
	Dy	2.07	4.23	4	4.01	4.25	1.64	6.13	1.86	6.02
	Ho	0.46	0.81	0.71	0.8	0.82	0.38	1.2	0.39	1.31
	Er	1.36	2.15	1.98	2.35	2.38	1.08	3.36	1.15	3.83
	Tm	0.27	0.3	0.29	0.37	0.37	0.22	0.5	0.22	0.47
	Yb	1.22	1.86	1.79	2.22	2.29	0.95	2.99	1	2.74
	Lu	0.22	0.31	0.27	0.35	0.36	0.18	0.49	0.17	0.46
	ΣREE	112.14	185.26	184.51	183.52	189.36	101.94	259.55	132.28	209.21

续表

地层单位		六罗村组								
岩性		火山碎屑岩	凝灰熔岩				熔结角砾岩	凝灰熔岩	流纹岩	凝灰熔岩
样品号		H5-5b	VI10	VI7-1	I23-1	XIII8-1	H12-13	XIII15-1	H12-2	XV4-1
序号		43	44	45	46	47	48	49	50	51
主要特征参数	LREE/HREE	11.57	11.63	12.4	11.79	11.99	13.16	11.44	15.49	9.16
	δEu	0.5	0.38	0.41	0.47	0.47	0.5	0.47	0.42	0.26
	δCe	1.02	0.8	0.79	0.86	0.7	1.09	0.64	1.08	0.82
	$(La/Yb)_N$	15.58	17.62	18.79	13.94	15.37	17.88	17.2	22.72	13.51
	$(Ce/Yb)_N$	11.11	10.17	10.56	8.8	8.06	13.5	7.93	16.76	7.83
	$(La/Sm)_N$	5.07	4.78	5.15	4.89	5.3	5.28	5.85	5.58	5
	$(Gd/Yb)_N$	1.8	1.86	1.81	1.32	1.23	1.91	1.41	2.16	1.44
	La/Sm	8.06	7.59	8.19	7.78	8.42	8.4	9.3	8.87	7.96

地层单位		六罗村组					鹿母湾组			
岩性		凝灰熔岩	流纹岩	凝灰熔岩	流纹岩	酸性岩平均值	沉凝灰岩			平均值
样品号		XIII5-1	I24-1	XV2-1	D1886		X16-1	X43-1	X8-1	
序号		52	53	54	55		56	57	58	
稀土元素含量 10^{-6}	La	44.9	46.6	59.6	31.71	45.88	27.2	32.24	33.32	30.92
	Ce	73.2	81.3	91.4	57.5	74.65	52.28	61.9	56.69	56.96
	Pr	9.1	8.1	12.9	6.52	8.20	6.97	8.2	6.99	7.39
	Nd	33.4	28.8	45.2	21.42	29.16	29.75	30.74	25.03	28.51
	Sm	7.4	5.3	8.1	3.41	5.75	7.45	5.76	4	5.74
	Eu	0.52	0.78	0.78	0.71	0.75	2.15	1.41	0.94	1.50
	Gd	5.75	2.67	5.38	2.52	4.02	7.62	4.97	3.04	5.21
	Tb	1.02	0.46	0.93	0.37	0.71	1.25	0.84	0.51	0.87
	Dy	6.7	2.6	5.54	1.87	3.84	6.77	4.67	3.04	4.83
	Ho	1.42	0.96	1.02	0.37	0.80	1.2	0.9	0.65	0.92
	Er	4.23	1.44	2.28	0.95	2.16	3.2	2.58	1.98	2.59
	Tm	0.61	0.22	0.35	0.14	0.33	0.44	0.39	0.34	0.39
	Yb	3.83	1.32	2.19	0.88	1.88	2.56	2.39	2.11	2.35
	Lu	0.6	0.2	0.34	0.13	0.30	0.37	0.37	0.33	0.36
	∑REE	192.68	180.75	236.01	128.5	178.45	149.21	157.36	138.97	148.51
主要特征参数	LREE/HREE	6.98	17.31	12.09	16.77	12.42	5.37	8.2	10.58	8.05
	δEu	0.24	0.57	0.34	0.71	0.47	0.87	0.79	0.79	0.82
	δCe	0.83	0.93	0.76	0.91	0.89	0.89	0.89	0.85	0.88
	$(La/Yb)_N$	7.9	23.8	18.35	24.29	17.74	7.16	9.09	10.65	8.97
	$(Ce/Yb)_N$	4.94	15.93	10.8	16.9	11.54	5.28	6.7	6.95	6.31
	$(La/Sm)_N$	3.82	5.53	4.63	5.85	5.10	2.3	3.52	5.24	3.69
	$(Gd/Yb)_N$	1.21	1.63	1.98	2.31	1.85	2.4	1.68	1.16	1.75
	La/Sm	6.07	8.79	7.36	9.3	8.11	3.65	5.6	8.33	5.86

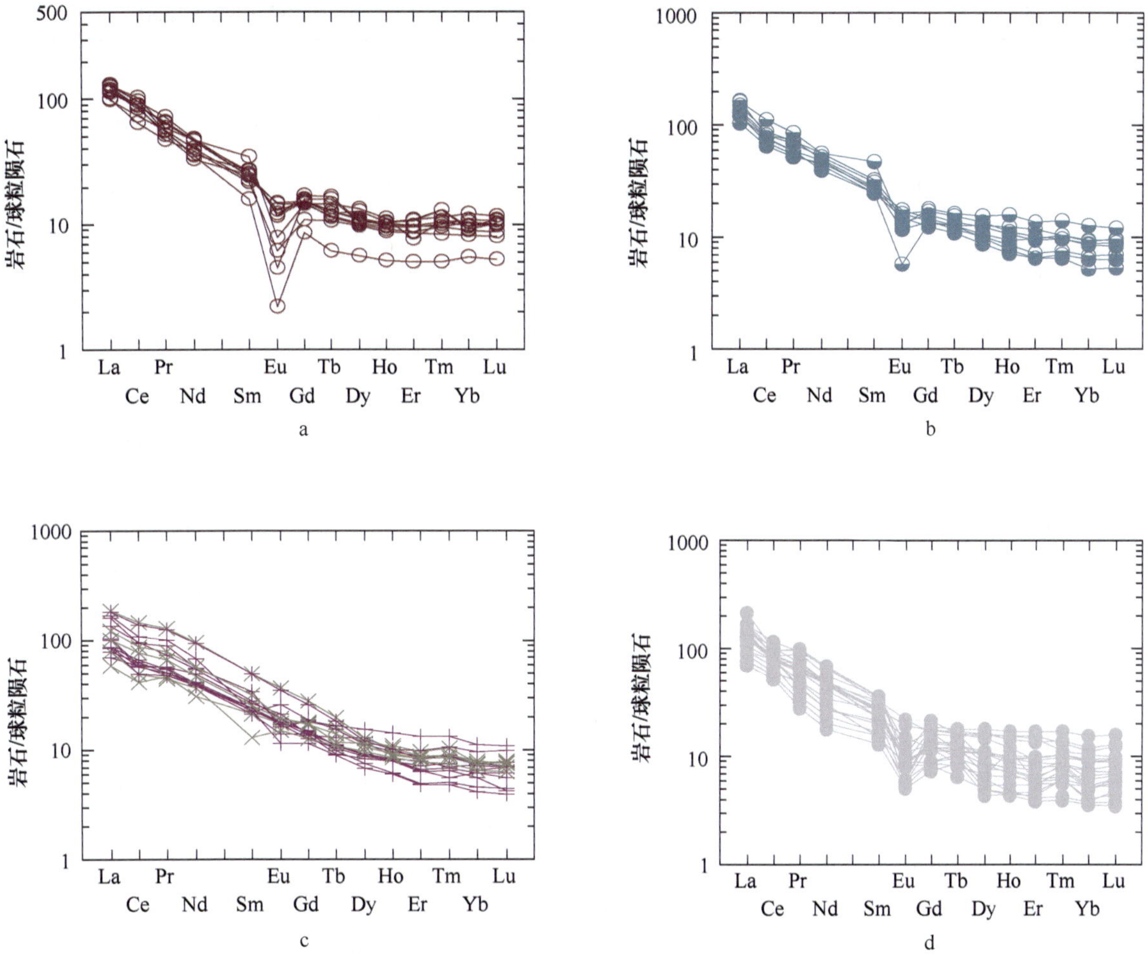

图 12-4　海南岛中生代火山岩稀土元素球粒陨石标准化图

Fig. 12-4　Chondrite-normalized REE patterns of Mesozoic volcanic rocks from the Hainan Island

a—岭壳村组火山岩样品；b—汤他大岭组火山岩样品；c—六罗村组中基性岩样品；d—六罗村组酸性岩样品

四、微量元素地球化学特征

六罗村组中基性岩的 Rb、Ba、Th、U、K 等大离子亲石元素含量较低，而 V、Cr、Co、Ni、Zn 等亲铁铜元素的含量较高，亲钙的 Sr 元素含量也较高，Rb/Sr 值低；而其酸性火山岩的 Rb、Ba、Th、U、K、Y 的含量相对较高，Rb/Sr 值较高，V、Cr、Co、Ni、Zn 及 Sr 的含量多数较低（表 12-5）。汤他大岭组及岭壳村组火山岩相对六罗村组酸性火山岩而言，Rb、Y、Ta 的含量总体上降低，而 Sr、Ba 的含量总体上升高，反映了岩浆熔融深度在不断加大。

火山岩的过渡族元素球粒陨石标准化形式具有相似的配分模式（图 12-5），均表现为 Ti 正异常和 Cr、Ni 负异常明显，与 Allégre et al.（1973）的玄武岩配分模式相似，表明中生代火山岩由地幔岩演化而来。微量元素蛛网图（图 12-6）中，配分曲线基本一致，表现为强不相容元素 Rb、Ba、Th、K 等较中等不相容元素 Sr、Nb、P、Zr 更富集；Rb、Th、K、La、Ce、Nd、Hf 、Zr、Sm、Tb 等元素含量相对高，略显正异常；Ba、Ta、Nb、Sr、P、Ti 等元素含量相对低，显示略的负异常。反映海南岛中生代火山岩具有大陆壳的特点。

表 12-5 海南岛中生代火山岩微量元素含量一览表

Table 12-5 List of REE contents of Mesozoic volcanic rocks from the Hainan Island

地层单位		岭壳村组								
岩性		粗安岩	粗安岩	英安岩	安粗质角砾熔岩	英安岩	英安岩	碎斑熔岩	凝灰熔岩	碎斑熔岩
样品号		D1019-3	D1019-4	D1026-1	D1027-1	D908-1	D1010-1	I35-12	I35-4	I36-1
序号		1	2	3	4	5	6	7	8	9
微量元素含量/10^{-6}	Au	0. 4	0. 9		1. 1	0. 3	1. 5	1. 4	1. 3	1. 5
	Ag							0. 16	0. 23	0. 1
	Cu	12. 6	10. 2		9. 6	12. 2	24. 3	2. 8	5	2. 1
	Pb	32. 3	38. 4	19. 9	21. 9	22. 2	16	27. 8	101	26. 8
	Zn	57. 7	63. 2	44	67. 8	30. 6	112. 5	31. 2	143	30. 8
	Sb									
	W							1. 2	4. 3	1. 4
	Sn							0. 25	1. 15	0. 29
	Bi									
	Mo	4. 7	4. 4	4. 9	1. 7	2. 7	2. 1	1. 8	0. 5	1
	Be									
	Li	27. 8	27. 7	19	24	4. 4	12. 2			
	Ni	5. 8	12. 5	3. 1	5. 5	6. 3	11. 9	1. 5	2. 1	1. 8
	Co	2. 7	2. 5	2	2. 6	0. 8	4. 4			
	V	19. 1	12. 8	10. 1	22. 6	8. 2	42. 8			
	Nb	14. 7	14. 7	13. 52	14. 2	12. 7	14	15. 1	12. 7	10. 6
	Ta	1. 08	1. 15	1. 11	1. 02	1. 03	0. 98			
	Rb	184. 3	191. 2	189. 9	196. 2	126. 2	51. 4			
	Sr	322. 9	241. 9	227. 6	284. 8	46. 7	333. 4	141	173	111
	Ba	823. 6	663. 5	691. 8	805. 9	75. 2	324			
	Zr	250. 2	238. 2	270. 9	249. 8	143. 2	249. 6			
	Hf	7. 1	6. 9	7. 5	7. 2	4. 7	7			
	As							2. 4	2. 6	1. 5
	Ga	19	17. 6	22	17. 4	11. 7	16. 7			
	Sc	7. 2	6. 4	4	6. 9	6. 1	11. 6			
	Cr	13. 4	29. 2	6. 2	15. 5	16. 2	35. 2	5	12. 4	5
	U	5. 5	6. 3	5	5. 2	4. 2	4. 3			
	Th	18. 4	18. 9	17. 4	18. 4	15. 8	14. 3			
	Y	23. 9	24. 4	26. 8	21. 4	12. 8	27. 6	11. 3		13. 3
	Cs	2	2. 4	1. 9	2	2. 4	0. 7			
特征比值	Rb/Sr	0. 57	0. 79	0. 83	0. 69	2. 70	0. 15			
	Ba/Sr	2. 56	2. 78	3. 03	2. 86	1. 61	0. 97			
	K/Rb	218. 01	208. 41	217. 26	215. 79	257. 20	242. 26			
	Nb/Ta	13. 61	12. 78	11. 89	13. 92	12. 33	14. 29			
	Zr/Hf	35. 24	34. 52	36. 12	34. 69	30. 47	35. 66			

续表

地层单位		岭壳村组		汤他大岭组						
岩性		凝灰熔岩	平均值	凝灰岩	凝灰岩	凝灰岩	凝灰岩	英安岩	流纹岩	凝灰岩
样品号		I41-3		I30-1	IV10-1	IV12-1	IV16-1	IV2-2	IV5-2	IV7-1
序号		10		11	12	13	14	15	16	17
微量元素含量 10^{-6}	Au	1. 1	1. 06	1. 7	1. 1	1	1. 9	1. 6	1. 5	0. 8
	Ag	0. 06		0. 13	0. 09	0. 11	0. 13	0. 11	0. 04	0. 09
	Cu	1. 4	8. 91	5. 1	1. 2	1. 2	4. 1	9. 2	1. 6	1. 6
	Pb	15	32. 13	20. 3	25. 7	15. 7	37	23. 5	12. 8	23. 1
	Zn	28. 5	60. 93	85	54. 6	85	74. 7	93. 3	27. 1	40. 3
	Sb									
	W			1. 92	0. 9	1. 45	1. 3	1. 66	2. 02	2. 33
	Sn	1. 2		1. 1	2. 1	4. 0	2. 5	2. 1	1. 0	1. 9
	Bi	0. 03		0. 04	1. 17	0. 93	0. 34	0. 32	0. 12	0. 30
	Mo	1. 7	2. 55	5. 7	0. 5	0. 8	0. 7	1. 9	0. 7	2. 2
	Be	2. 6		3. 0	2. 6	2. 7	2. 6	2. 0	2. 8	2. 6
	Li		19. 18							
	Ni	1. 6	5. 21	7. 9	3. 6	16. 5	2. 5	11	1. 1	1. 7
	Co		2. 5							
	V		19. 27							
	Nb	14. 4	16. 63	9. 9	11. 4	10	11. 8	19. 3	8. 8	7. 3
	Ta		1. 06	1. 11	1. 21	1. 04	1. 67	1. 17	1. 15	1
	Rb		156. 53	108	149	196	176	139	193	177
	Sr	173	205. 53	650	310	375	319	364	112	132
	Ba		564							
	Zr		233. 65							
	Hf		6. 73							
	As	1. 4		7. 4	2. 0	15. 1	1. 2	3. 4	0. 30	0. 9
	Ga		17. 4	19. 6	18	17. 5	16. 4	20. 2	21	18. 5
	Sc		7. 03							
	Cr	5	14. 31	10	9. 9	42. 4	5. 7	6. 6	5	5
	U		6. 1							
	Th		17							
	Y		20. 19	22. 5	35. 8	32. 0	33. 4	20. 0	15. 3	24. 8
	Cs		1. 9							
特征比值	Rb/Sr		0. 73	0. 17	0. 48	0. 52	0. 55	0. 38	1. 72	1. 34
	Ba/Sr		2. 63							
	K/Rb		230. 17	224. 45	201. 69	138. 92	164. 14	175. 59	147. 96	188. 54
	Nb/Ta		12. 88	8. 92	9. 42	9. 62	7. 07	16. 50	7. 65	7. 30
	Zr/Hf		34. 22							

续表

地层单位		汤他大岭组		六罗村组						
岩性		流纹岩	平均值	凝灰熔岩						
样品号		XVIII1-1		XV2-1	XV4-1	XIII5-1	XIII15-1	XIII8-1	VI10	VI4-1
序号		18		19	20	21	22	23	24	25
微量元素含量/10^{-6}	Au	0.2	1.23	5.3	0.8	0.1	1.4	7.4	1	0.8
	Ag	0.04		0.04	0.07	0.10	0.02	0.08	0.05	0.08
	Cu	3	3.38	5	16	6	4	7	1.8	2.8
	Pb	32	23.76	27	40	48	23	33	14.6	23.9
	Zn	72	66.5	34	59	86	46	46	39.8	38.2
	Sb									
	W	2.6		0.92	2.3	3.9	2.2	2.0	1.02	1.42
	Sn	4.2		2.7	4.2	6.6	1.4	2.7	1.4	2.0
	Bi	0.32		0.54	0.64	0.54	0.05	0.14	0.06	0.18
	Mo	1.2	1.71	0.2	2	0.8	0.8	0.8	1.2	1.7
	Be	4.1		2.7	3.8	4.5	2.4	2.1	2.5	2.7
	Li									
	Ni	6	5.33	0	2	4	2	2	2.6	2.8
	Co									
	V									
	Nb	17.2	11.96	11	20.3	25.2	12.8	11.6	8	13.7
	Ta	2	1.29	1.6	1.2	2	0.8	1.6	1.51	1.18
	Rb	274	176.5	225	247	247	158	138	181	182
	Sr	65	290.88	62	60	63	133	148	153	117
	Ba									
	Zr									
	Hf									
	As	4.45		3.86	4.63	4.49	8.73	4.66	2.0	1.4
	Ga	14	18.15	17	15	16	15	15	19.7	17.1
	Sc									
	Cr	106	23.83	96	190	212	1.72	249	8.8	8.8
	U									
	Th									
	Y	24.0	25.98	32.0	28.0	35.0	28.0	21.0	21.7	65.2
	Cs									
特征比值	Rb/Sr	4.22	0.61	3.63	4.12	3.92	1.19	0.93	1.18	1.68
	Ba/Sr									
	K/Rb	136.04	165.56	156.44	126.03	145.86	207.54	181.67	184.38	190.47
	Nb/Ta	8.60	9.27	6.88	25.17	12.60	16.00	7.25	5.30	14.10
	Zr/Hf									

续表

地层单位		六罗村组									
岩性		凝灰熔岩		流纹岩					熔结凝灰岩		碎屑岩
样品号		VI7-1	I23-1	I24-1	I29-2	D3036-1	D1886	H12-2	H12-13	H12-9	H5-5
序号		26	27	28	29	30	31	32	33	34	35
微量元素含量/10^{-6}	Au	0. 9	1. 1	1. 3	1. 7	1. 2	0. 70				
	Ag	0. 07	0. 06	0. 05	0. 09		0. 075				
	Cu	24. 7	1. 5	1. 8	8	17. 7	17. 3				
	Pb	16. 1	25. 1	27. 6	19	31. 8	13. 1				
	Zn	34. 6	33. 9	25. 8	36. 7	39. 3					
	Sb						0. 25				
	W	1. 76	1. 62	1. 64	1. 88		0. 6				
	Sn	2. 1	1. 8	1. 8	2. 2		1. 0				
	Bi	1. 50	0. 36	0. 20	0. 02		<0. 1				
	Mo	1. 6	1. 4	2. 1	0. 7	3. 8	0. 30				
	Be	4. 0	3. 0	2. 5	2. 8		3. 10				
	Li					11	154				
	Ni	11	2. 5	1. 5	3. 5	8. 8	4. 0	3. 8	6. 1	6. 2	7. 3
	Co					2. 2	2. 4				
	V					19. 1	18	22. 4	32. 7	42. 7	35. 2
	Nb	16	13. 2	17. 2	14. 4	15. 3	13. 8				
	Ta	1. 06	1. 19	1. 22	1. 39	1. 12	1. 2				
	Rb	161	177	197	180	216. 3	86. 0				
	Sr	112	134. 5	117	160	189. 7	16. 0	87. 6	98. 8	50. 2	62. 8
	Ba					499. 4	159	422	479	348	334
	Zr					162. 2	65	82. 5	84. 1	121	111
	Hf					5. 2	2. 2	3. 7	3. 5	4. 8	4. 7
	As	0. 8	1. 0	0. 5	1. 2		10. 4				
	Ga	17. 2	17. 3	16. 4	17. 5	16. 4	20. 1				
	Sc					4. 2	1. 9				
	Cr	9. 9	6. 9	5	5	22. 4	5. 0	120	286	107	344
	U					4. 9	2. 0				
	Th					17	9. 7				
	Y		27. 2	13. 9	18. 4	17. 0	9. 64				
	Cs					2. 2					
特征比值	Rb/Sr	1. 44	1. 32	1. 68	1. 13	1. 14	5. 38				
	Ba/Sr					2. 63	10. 00	4. 76	4. 76	7. 14	5. 26
	K/Rb	216. 56	135. 08	182. 04	206. 62	191. 13	232. 64				
	Nb/Ta	15. 09	11. 09			13. 66	11. 50				
	Zr/Hf					31. 19	29. 55	22. 30	24. 03	25. 21	23. 62

续表

地层单位		六罗村组								
岩性		火山碎屑岩		酸性岩平均	英安岩	粗面安山岩			玄武安山岩	
样品号		H6-2	H7-4		XIII11-1	XIII3-1	XIII6-4	XIII6-5	XIII7-1	D3039-1
序号		36	37		38	39	40	41	42	43
微量元素含量/10^{-6}	Au			1.82	0.8	2.1	0.6	0.6	21.2	1.3
	Ag			0.075	0.48	2.2	0.81	0.08	0.03	
	Cu			8.74	15.9	24.2	26.4	1.2	4	13.1
	Pb			26.32	604	2160	267	120	27	11.5
	Zn			43.28	139	440	115	106	93	74.5
	Sb			0.25						
	W			0.6	2.2	4.5	0.6	0.9	1.1	
	Sn			1	3.8	3.7	1.5	1.2	0.76	
	Bi				1.01	1.86	0.31	0.55	0.03	
	Mo			1.34	2.6	0.8	0.4	0.6	0.4	1.5
	Be			3.1	2.1	2.2	2.5	1.8	2.0	
	Li			82.5						6.4
	Ni	10.3	4.1	4.04	20.9	4.5	39.4	6.5	6	11.8
	Co			2.3						15.6
	V	31.6	42.3	30.5						157.3
	Nb			14.81	111	9.1	15	9.4	12.4	10.2
	Ta			1.31	1.2	0.8	1.2	0.8	0.8	0.58
	Rb			184.3	281	170	43	61	34	43.8
	Sr	180	25.2	103.7	332	155	575	682	630	778.6
	Ba	298	326	358.2						343.5
	Zr	130	173	116.1						166.4
	Hf	5.3	5.1	4.31						4.7
	As			10.4	27.0	0.4	3.0	1.8	6.74	
	Ga			16.9				1.8	12	22.1
	Sc			3.05						13.1
	Cr	577	141	126.1					99	18.8
	U			3.45						0.8
	Th			13.35						4
	Y			26.42	27.8	10.2	17.2	16.1	18.0	17.8
	Cs			2.2						4.2
特征比值	Rb/Sr					1.10	0.07			0.06
	Ba/Sr	1.67	12.50							0.44
	K/Rb					140.64	461.41			219.86
	Nb/Ta					11.38	12.50			17.59
	Zr/Hf	24.53	33.92							35.40

地层单位		六罗村组									
岩性		玄武安山岩	玄武岩	安山岩					流纹英安岩		
样品号		ⅢH9-1	D223	I5-1	I9-1	D1539	XXIV2-1	D3039-4	D405	D1887	D401
序号		44	45	46	47	48	49	50	51	52	53
微量元素含量/10^{-6}	Au	0.60	51	1.9	1	1.1	0.76	0.2	0.60	0.60	0.40
	Ag	0.08	0.14	0.09	0.11		0.089		0.018	0.022	0.06
	Cu	18.6	19.2	3.1	1.9	32	18.0	13.9	7.4	7.3	12.9
	Pb	21.9	63.5	7.5	21	24.3	16.5	14.2	5.9	24.5	27.8
	Zn	120	174	120	100	124.9	128	75.8			84
	Sb	0.20					0.29		<0.1	0.23	0.15
	W	0.60		0.92	0.85		0.78		1.0	0.7	0.70
	Sn	1.20		1.2	1.5		1.15		0.9	0.8	<0.1
	Bi	<0.05	9.21	0.17	0.13		0.08		0.1	<0.1	<0.1
	Mo	1.40	0.1	0.6	0.5	0.9	0.68	1.3	0.45	0.51	1.90
	Be	2.30	1.95	2.5	2.8		1.34		2.60	2.30	2.20
	Li	52.7				43.9	510	15.5	33	58	30.5
	Ni	110.2	52.1	9.8	8.4	22.3	144	12.4	14.9	16.4	15.7
	Co	34.8				19.7	35.4	15.9	9.6	9.8	11.0
	V	154				131.3	153	158.1	77.0	70	64.3
	Nb	14.7	5.5	6.8	14.7	16.6	14.0	10.3	11.91	10.8	13.9
	Ta	0.8	1.8	0.51	0.59	0.89	1.25	0.59	0.8	0.4	<0.5
	Rb	38.0	196	92	41	61.6	40.0	45	113.0	117.0	123.8
	Sr	266	459	778	790	820.6	401	796.6	161	439	611
	Ba	263				561.7	716	281.5	1195	859	1243
	Zr	167				141.7	208	167.9	186	168	187
	Hf	4.3				3.9	5.2	4.9	5.3	4.4	3.9
	As	1.3	83.3	2.7	2.9		4.29		0.9	1.4	2.1
	Ga	16.0	22.3	16.2	26.5	23.4		22	25.3	20.8	21.8
	Sc					19.7		111	7.9	8.4	
	Cr	178.9			5	66.8		18.7	37.0	40.0	29.0
	U	1.7				2.5		0.8	2.6	2.0	2.3
	Th	6.2				6.5		4.1	14.8	12.8	13.1
	Y	21.29	25.4	10.5	14.7	17.9	19.5	17.9	14.06	12.34	10.57
	Cs					0.7		4.4			
特征比值	Rb/Sr	0.14	0.43	0.12	0.05	0.08	0.10	0.06	0.70	0.27	0.20
	Ba/Sr	0.99				0.68	1.79	0.35	7.69	1.96	2.04
	K/Rb	327.69	40.24	173.25	145.78	336.91	709.78	225.06	329.12	299.42	258.84
	Nb/Ta	18.38	3.06	13.33	24.92	18.65	11.20	17.46	14.89	28.00	27.80
	Zr/Hf	38.84				36.33	40.00	34.27	35.09	38.18	47.95

续表

地层单位		六罗村组			鹿母湾组				维氏微量元素的平均含量（1962）		
岩性		角闪安山岩		中基性岩平均	沉凝灰岩			凝灰岩	基性岩	中性岩	酸性岩
样品号		ⅢH7-1	ⅢH8-1		Ⅹ8-1	Ⅹ16-1	Ⅹ43-1	Kl-1			
序号		54	55		56	57	58	59			
微量元素含量/10^{-6}	Au	0.60	0.60	4.78	1.10	1.10	1.60		4		4.5
	Ag	0.122	0.068	0.065	0.034	0.023	0.034		0.10	0.07	0.05
	Cu	21.5	15.4	14.22	12.8	4.9	14.9	17	100	35	20
	Pb	26.5	25.5	192.7	35.9	21.2	25.3	69	8	15	20
	Zn	116	86	131.01	52	80	45	60	130	72	60
	Sb	0.10	4.40	0.895	0.5	0.3	0.4		0.1	0.20	0.26
	W	0.50	0.40	0.67	1.1	0.6	1.4		1	1.0	1.5
	Sn	0.80	0.70	0.925	2.8	2.6	11		1.5		3.0
	Bi	<0.05	<0.05	0.09	0.2	0.2	0.4		0.007	0.01	0.01
	Mo	1.10	0.40	0.9	0.3	0.2	0.3		1.4	0.9	1.0
	Be	2.60	2.00	2.19	1.70	1.90	2.60		0.4	1.8	5.5
	Li	35.2	37.0	36.54	20.8	49.8	13.6		15	20	40
	Ni	31.2	28.0	30.81	7.9	11.8	10.6	25	160	55	8
	Co	23.9	18.9	19.46	4.8	12.2	16.0	14.1	45	10	5
	V	179	70	121.4	76.6	111.0	92.7	93	200	100	40
	Nb	13.5	12.9	11.95	12.4	16.9	13.7		20	20	20
	Ta	0.9	0.7	0.86	1.0	1.6	1.0		0.48	0.7	3.5
	Rb	64.0	50.0	89.68	134	180	181		45	100	200
	Sr	447	345	525.93	116	53	97	125	440	800	300
	Ba	1165	335	696.27	400	682	1090	738	300	650	830
	Zr	182	204	177.8	212	227	227		100	260	200
	Hf	5.2	5.2	4.7	6.1	6.6	6.6		1	1	1
	As	1.9	1.9	1.97	0.8	1.1	2.9		2	2.4	1.5
	Ga	17.1	19.1	19.03					18	20	20
	Sc			12.48					2.4	2.5	3
	Cr	27.1	28.4	49.88				29	200	50	25
	U	2.3	1.7	1.86					0.5	1.8	3.5
	Th	9.7	11.3	9.17					3	7	18
	Y	23.02	22.96	17.62	17.90	29.92	24.37		20		34
	Cs			3.1					1		5
特征比值	Rb/Sr	0.14	0.14		1.16	2.40	1.87				
	Ba/Sr	2.63	0.97		3.45	12.50	11.11				
	K/Rb	407.29	410.09		218.69	207.54	210.98				
	Nb/Ta	15.00	18.43		16.90	10.56	13.70				
	Zr/Hf	35.00	39.23		34.75	34.39	34.39				

注：Au 的单位为 10^{-9}。

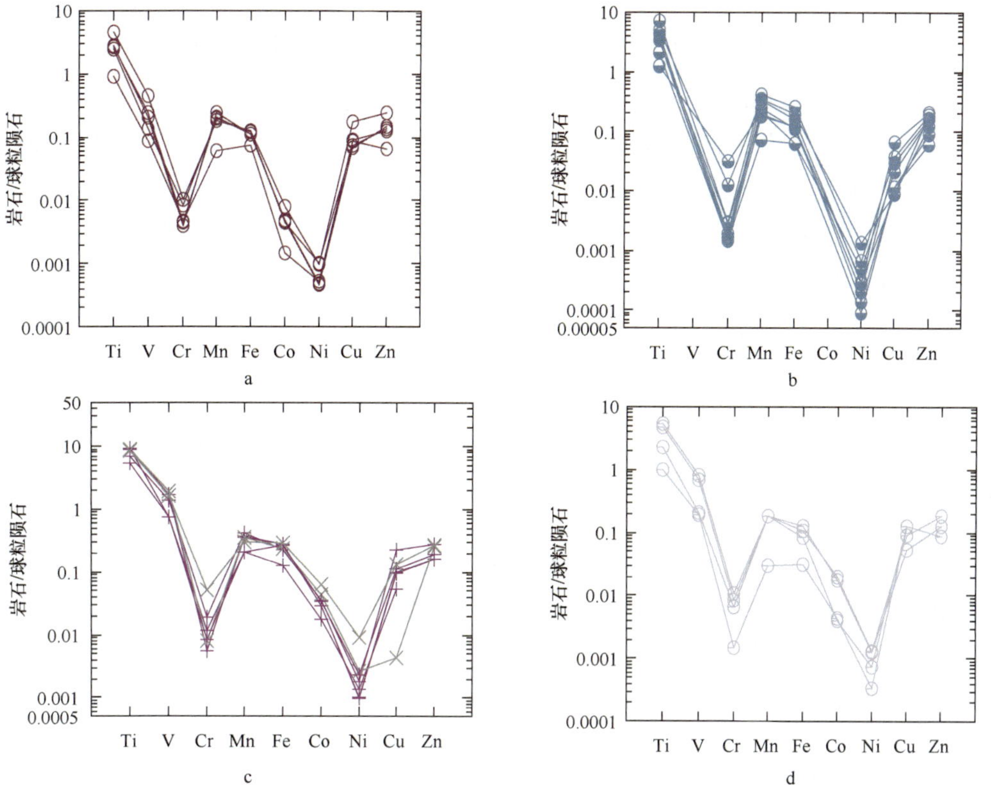

图 12-5 海南岛中生代火山岩过渡族元素球粒陨石标准化图

Fig. 12-5 Chondrite-normalized patterns of transition metal elements of Mesozoic volcanic rocks from the Hainan Island

a—岭壳村组火山岩样品；b—汤他大岭组火山岩样品；c—六罗村组中基性岩样品；d—六罗村组酸性岩样品

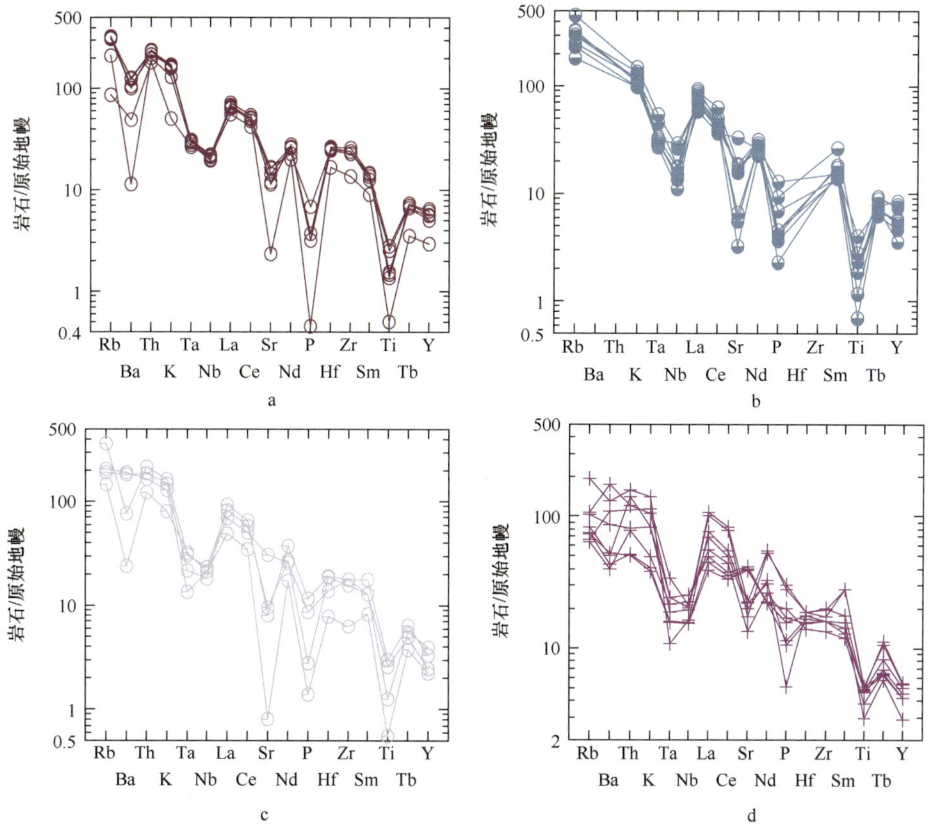

图 12-6 海南岛中生代火山岩微量元素原始地幔标准化蛛网图

Fig. 12-6 Primitive mantle-normalized spider diagram of trace elements of Mesozoic volcanic rocks from the Hainan Island

a—岭壳村组；b—汤他大岭组；c—六罗村组酸性岩；d—六罗村组中基性岩

五、Sr、Nd 同位素地球化学特征

火山岩的 Sr、Nd 同位素样品 8 个（表 12-6）。其中，$^{87}Sr/^{86}Sr$ 的测定值为 0.70683~0.71813，平均值为 0.710766；初始锶比值为 0.70650~0.70869，平均值为 0.70775；ε_{Sr}（t）为 30.4~61.3，平均值为 48.06；$^{143}Nd/^{144}Nd$ 的测定比值为 0.512290 ~ 0.512431，平均值为 0.512339；初始值为 0.512210~0.512346，平均值为 0.512254；ε_{Nd}（t）值为-5.6~-2.7，平均值为-4.58；T_{2DM}值为 1135~1365Ma。从 ε_{Sr}（t）和 ε_{Nd}（t）值可以推测中生代火山岩的源区为中、上地壳。

六、火山岩形成的地质构造环境

中国东南大陆沿海中生代火山带，位于环太平洋构造-岩浆带的西南端，同时又处于欧亚大陆板块的东南缘，自三叠纪印支运动以来，该区受到特提斯构造域和环太平洋构造域双重构造运动的制约和影响，致使中生代构造岩浆活动异常强烈和复杂。海南岛中生代火山岩是中国东南大陆沿海中生代火山岩带的一部分，海南岛中生代，特别是燕山期构造运动以断块运动为主，它引发古老构造带重新活动，呈带状展布特征。海南岛中生代火山岩的稀土配分形式及微量元素蛛网图的配分形式相一致，说明它们形成的地质构造环境相同，75 个火山岩的化学成分在里特曼-戈蒂里图解的投影点主要落入造山带（岛弧及活动大陆边缘区）火山岩及由此派生的碱性火山岩区（图 12-7）；在 R_1-R_2 图（图 12-8）上，六罗村组和鹿母湾组酸性火山岩主要落入造山区，汤他大岭组火山岩主要落入造山区，部分投入破坏性活动板块边缘区（碰撞前），岭壳村组火山岩主要落入造山区，部分投入晚造山区。同时，利用 Pearce et al.（1984）的图解可知，其投影点主要落入火山弧和同碰撞的构造背景区（图 12-9），但其微量元素蛛网图的配分形式与火山弧火山岩的模式略有差别，而与造山带正常大陆弧构造背景下产生的花岗岩相似。综上所述，中生代火山岩是大陆边缘内侧大陆弧构造环境的产物，是增生在大陆边缘的新陆壳。

表 12-6 海南岛中生代火山岩 Sr-Nd 同位素组成

Table 12-6 Composition of Sr-Nd istopes of Mesozoic volcanic rocks from the Hainan Island

地质单位	岭壳村组			六罗村组			
样品号	D908-1	D1019-4	D1010-1	D1539	D3039-1	D3039-4	D3036-4
序号	1	2	3	4	6	7	8
岩性	流纹岩	粗安岩	英安岩	安山岩	玄武安山岩		流纹岩
t/Ma	109	109	109	120	120	120	120
$(^{87}Rb/^{86}Sr)_m$	6.722	2.292	0.4477	0.1934	0.09782	0.1592	1138
$(^{87}Sr/^{86}Sr)_m$	0.71813	0.71176	0.70938	0.70683	0.70819	0.70841	0.71266
ε_{Sr} (0)	193.5	103.1	69.3	33.1	52.4	55.5	115.8
ε_{Sr} (t)	47.5	54.5	61.3	30.4	52.0	53.7	37.0
$f_{Rb/Sr}$	80.28	26.71	4.41	1.34	0.18	0.93	39.36
$(^{87}Sr/^{86}Sr)_i$	0.70772	0.70821	0.70869	0.70650	0.70802	0.70814	0.70697
$(^{147}Sm/^{144}Nd)_m$	0.1302	0.1046	0.1126	0.1081	0.1211	0.1177	0.1005
$(^{143}Nd/^{144}Nd)_m$	0.512322	0.512294	0.51229	0.512431	0.512353	0.512333	0.512353
ε_{Nd} (0)	-6.2	-6.7	-6.8	-4.0	-5.6	-5.9	-5.6
ε_{Nd} (t)	-5.2	-5.4	-5.6	-2.7	-4.4	-4.7	-4.1
$f_{Sm/Nd}$	-0.34	-0.47	-0.43	-0.45	-0.38	-0.40	-0.49
T_{DM}/Ma	1512	1197	1298	1040	1313	1299	1075
T_{2DM}/Ma	1334	1349	1365	1135	1275	1302	1249
$(^{143}Nd/^{144}Nd)_i$	0.512229	0.512219	0.512210	0.512346	0.512258	0.512241	0.512274

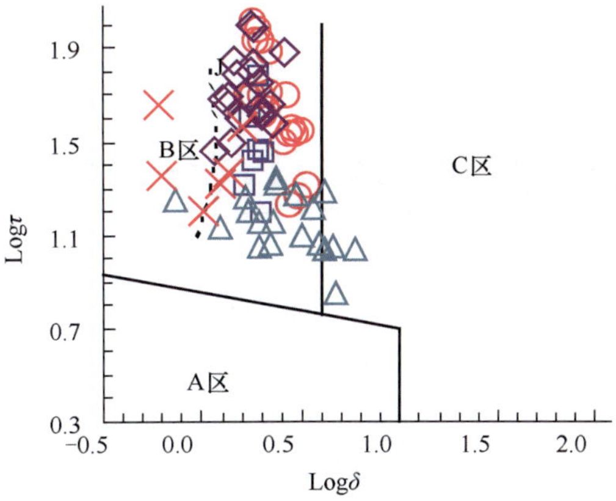

图 12-7　中生代火山岩类里特曼-戈蒂图解

Fig. 12-7　Rittmann-Gottini diagram of Mesozoic volcanic rocks

A 区—非构造带（板块内部稳定构造区）火山岩；

B 区—造山带（岛弧及活动大陆边缘区）火山岩；

C 区—A、B 区火山岩派生的碱性岩

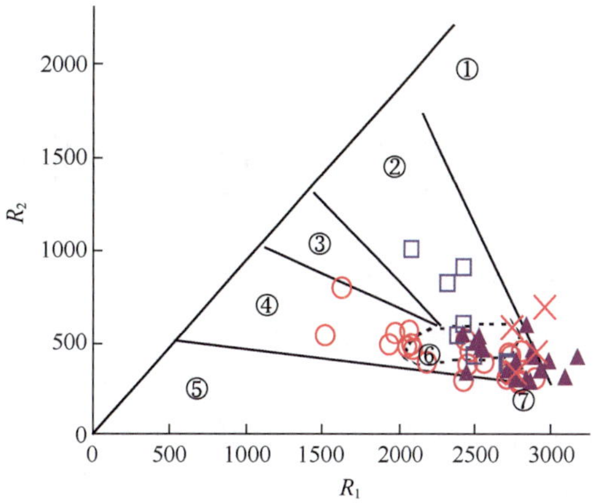

图 12-8　火山岩类 R_1-R_2 多阳离子构造环境判别图

Fig. 12-8　R_1-R_2 polycation tectonic setting discrimination diagram for volcanic rocks

①地幔斜长花岗岩区；②破坏性活动板块边缘（碰撞前）；③碰撞后隆起；④晚造山；⑤非造山；⑥同造山；⑦后造山

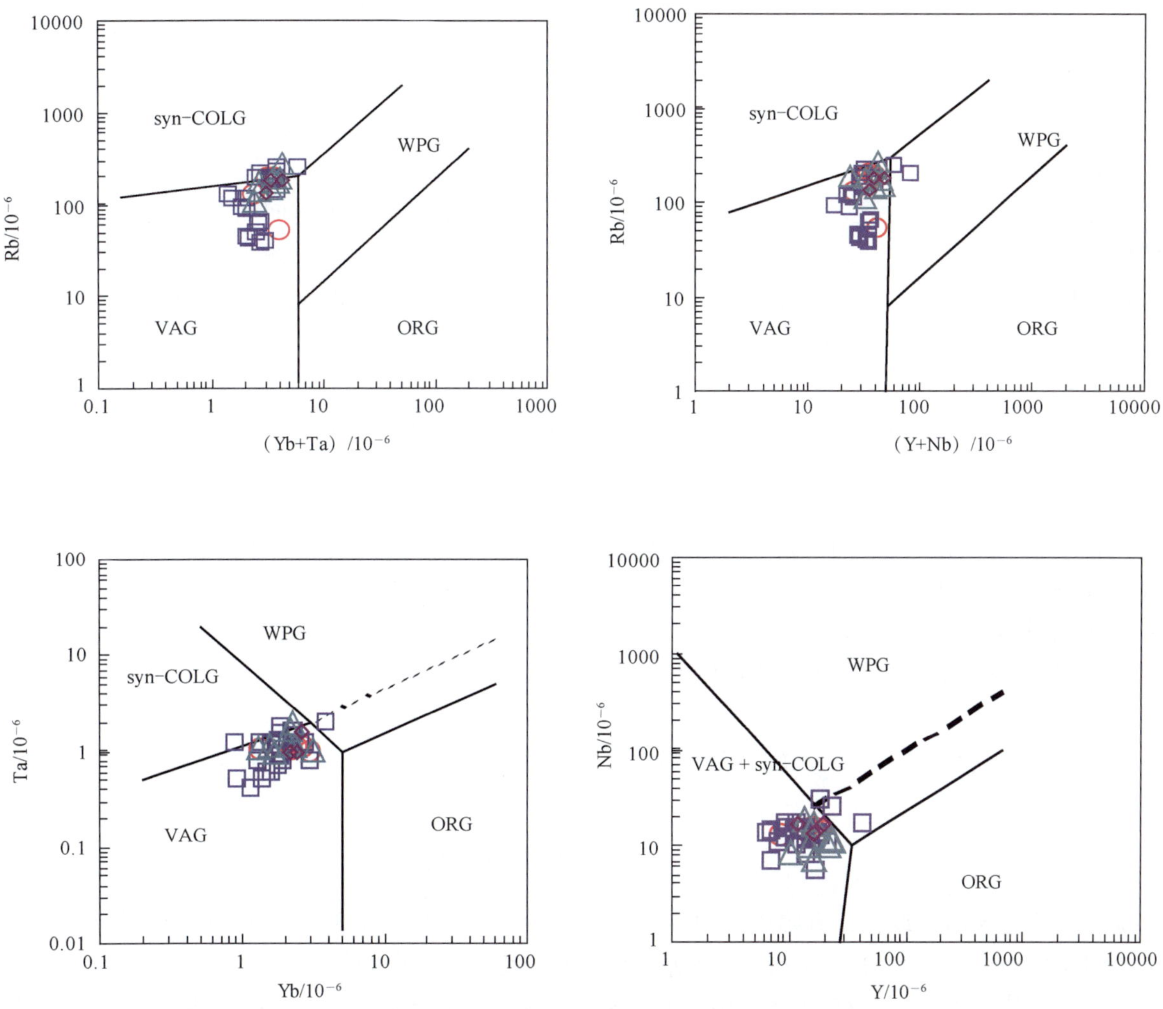

图 12-9　火山岩非活动性元素的环境判别图

Fig. 12-9　Environment discrimination diagram of non-active elements of volcanic rocks

（据 Pearce et al.，1984）

VAG—火山弧花岗岩；WPG—板内花岗岩；Syn-COLG—同碰撞花岗岩；ORG—洋中脊花岗岩

第十三章　新生代火山岩及火山作用

海南省新生代火山岩分布广泛，主要出露于海南岛北部（琼北）和南海两地，大致划分为琼北新生代火山岩和南海（中央海盆）新生代火山岩。琼北新生代火山岩为雷琼火山喷发岩浆亚带的一部分，主要分布于海南岛北部，琼北新生代火山岩在平面上呈“7”字形展布，出露面积约6630km^2，占全岛陆域面积的19.5%，在区域构造上沿王五-文教断裂带分布。新生代火山岩历来备受地质工作者所关注，自1928年以来，许多学者、科研机构及地质单位对琼北和南海等地火山岩分别做了大量工作，特别是琼北新生代火山岩的研究，积累了丰富的资料。但由于对火山岩上、下地层时代认识的不一致，以及缺少合适的同位素年龄数据，导致对火山活动的期次有不同的划分，其中有一、二、三、四、五、八和十一期之分（表13-1）。本次研究通过对已有资料的分析，圈定了琼北不同时代不同喷发期次的火山岩界线，火山活动期次的划分方案大致与海南省地质调查院（2004，2007）划分方案相同，对琼北火山岩的喷发时代、岩相岩性特征等有新的认识。南海盆地的情况比较复杂，在盆地内散布着一些由火山岩组成的海山，火山岩成分以基性岩为主，少量中酸性岩和超基性岩。从现有的资料来看，火山活动主要从始新世末至现代，测得火山岩的K-Ar（或Ar-Ar）法年龄大致为38.72~0.42Ma。王贤觉等（1984）、李兆麟等（1991）认为南海盆地火山岩形成年龄由北向南逐渐变新，由于火山岩呈零星分布，所获得的资料有限，对南海新生代火山活动的期次划分很难确定。

琼北新生代火山岩主要受东西向、南北向、北西向、北东向及北东东向断裂的制约，火山活动有两种喷发类型：裂隙式喷发和中心式喷发。古近纪火山活动以裂隙式喷发为主，形成隐伏于地下的玄武质熔岩和火山碎屑岩；新近纪则以中心式喷发为主，形成了大规模的以玄武岩为主的熔岩被；第四纪以裂隙式喷发为主，兼有少量中心式喷发，形成了大规模的盾形熔岩、熔岩流及熔岩台地上的碎屑锥和复合锥。

琼北新生代是火山活动的强烈时期，火山岩分布极广，出露面积大，自古近纪始新世以来，曾发生多期多次火山活动，火山喷发时断时续，一直延至第四纪全新世。根据火山岩上、下地层的时代（图版Ⅲ-1-B、图版Ⅲ-1-C、图版Ⅲ-1-D）、火山岩所夹地层的时代、火山岩的测年数据、火山岩的风化壳的发育程度及火山机构保存完整程度将海南岛新生代火山岩大致划分为11期（表13-2）。其中，古近纪2期，新近纪4期，第四纪5期。古近纪火山岩均隐伏于地下，新近纪火山岩除蓬莱、居丁和海口金牛岭出露地表外，大部分也隐伏于地下；第四纪除早更新世隐伏于地下外，各时期火山岩地表均见有出露。

第一节　火山岩主要类型、火山岩建造

一、火山活动旋回的划分

琼北新生代火山活动存在有明显的韵律性和旋回性，火山喷发的韵律和旋回反映了火山作用在时间上的演化历史，可以获得有关该区火山活动的机理、发展及演变的信息。

琼北新生代火山作用共有21次火山喷发活动，可划分18个喷发韵律，11个喷发旋回，每个旋回大体与地层单位的一个岩性组相当。

表 13-1 琼北新生代火山岩分期沿革表

Table 13-1 Division History of phases of Cenozoic volcanic rocks of Congbei

地质时代 \ 作者(时间)		李承三 (1929)	李殿臣 (1931)	边兆祥 (1958)	丁国瑜 (1964)	黄金森 (1964)	广东省地质局1:20万说明书 (1964)	钟晋梁 (1964)	海南地质局 (1976)	广东省欧洲队1:50万说明书 (1977)
第四纪 全新世	Q_4	火山活动期与时代未定	中生代末至第三纪初	第四系	晚期玄武岩	第二期	第三期	第三期	βQ_4	Q_4
第四纪 晚更新世晚期	Q_3^2				早期玄武岩	第一期	第二期	第二期	βQ_3^2	Q_3
第四纪 晚更新世早期	Q_3^1							第一期	βQ_3^1	
第四纪 中更新世晚期	Q_2^2								βQ_2^2	
第四纪 中更新世早期	Q_2^1									
第四纪 早更新世晚期	Q_1^2						第一期		βQ_1^2	Q_1
第四纪 早更新世早期	Q_1^1									
新近纪 上新世	N_2									N_2
新近纪 中新世 上	N_1^3									
新近纪 中新世 中	N_1^2									
新近纪 中新世 下	N_1^1									
古近纪 渐新世	E_3									
古近纪 始新世	E_2									

地质时代 \ 作者(时间)		中科院南海海洋所 (1978)	海南地质局 (1981)	第二海洋地质大队 (1981)	姚清尹 (1981)	傅秀银 (1983)	海南地质大队 (1984)	袁宝印 (1984)	张虎男 (1984)	曾广策 (1984)	邹和平 (1984)
全新世	Q_4	近期火山岩	Q_4	雷虎岭火山岩	Q_4	Q_4	β_6^{3D}	第四期	雷虎岭期	第四期玄武岩	第四期玄武岩
晚更新世晚期	Q_3^2	晚期火山岩	Q_3	湖光岩火山岩	Q_3^3 / Q_3^2	Q_3^2	$\beta_6^{3C(2)}$	第三期	湖光岩期	第三期玄武岩	第三期玄武岩
晚更新世早期	Q_3^1				Q_3^1	Q_3^1	$\beta_6^{3(C1)}$			第二期玄武岩	
中更新世晚期	Q_2^2	中期火山岩	Q_2		Q_1	Q_2^2	$\beta_6^{3(B2)}$			第一期玄武岩	第二期玄武岩
中更新世早期	Q_2^1										
早更新世晚期	Q_1^2	早期火山岩	Q_1	石峁岭期		Q_2^1	$\beta_6^{3A(2)}$	第二期	石峁岭期		第一期玄武岩
早更新世早期	Q_1^1			湛江火山岩				第一期	湛江组		
上新世	N_2										
中新世 上	N_1^3										
中新世 中	N_1^2										
中新世 下	N_1^1										
渐新世	E_3										
始新世	E_2										

地质时代 \ 作者(时间)		陈上福 (1985)	陈建中 (1985)	刘公民 (1985)	广东区调队 (1986)	韩中元 (1986)	张仲英 (1989)	宜昌地质矿产研究所、海南省地质矿产局 (1992)	海南省地质调查院 (2004)
全新世	Q_4	第五期 晚期 / 早期	雷虎岭期	第三期	雷虎组火山岩	雷虎岭期	雷虎岭期	雷虎岭期	石山期
晚更新世晚期	Q_3^2	第四期 晚期	东英期 峨蔓亚期	第二期	湖光岩组火山岩	旧州岭期			道堂期
晚更新世早期	Q_3^1	第四期 早期	东英期 高山岭亚期			德义岭期	长流期	长流期	
中更新世晚期	Q_2^2	第三期 早期	东英期 德义岭亚期		石峁岭组火山岩		德义岭期	东英期 木棠亚期 / 德义岭亚期	多文期 东英亚期
中更新世早期	Q_2^1	第三期				多文岭期			多文期 多文岭亚期
早更新世晚期	Q_1^2	第二期 晚期	多文岭期	第一期	胭脂岭组火山岩		多文岭期	多文岭期	
早更新世早期	Q_1^1	第二期 中期				胭脂岭期	光村期		秀英期 下段
上新世	N_2	第二期 早期	金牛岭期		上新世火山岩	金牛岭期	金牛岭期	金牛岭期	石门沟村期 金牛岭期
中新世 上	N_1^3	第一期 晚期			中新世火山岩	蓬莱岭期 / 角尾期		灯楼角期	石马村期 灯楼角期
中新世 中	N_1^2	第一期 中期						角尾期	角尾期
中新世 下	N_1^1	第一期 早期				下洋期		下洋期	下洋期
渐新世	E_3				渐新世火山岩	涠洲期		涠洲期	涠洲期
始新世	E_2				始新世火山岩	丹霞期		流沙港期	流沙港期

表 13-2　琼北新生代火山活动分期表

Table 13-2　Division table of phases of Cenozoic volcanic rocks of Congbei

喷发时代 纪	喷发时代 世	喷发时代 期(组)	喷发序次 旋回	喷发序次 韵律	喷发序次 次序	代号	地层分区	主要岩性	厚度/m	埋深/m	分布情况	代表性火山	接触关系	测年数据	喷发环境
第四纪	全新世	石山期	11	18	21	$\beta Qh_1^1 s^{2+}$	海口地区层小区	熔渣状、浮岩状橄榄玄武岩	>18.0	地表	石山、永兴、道堂、峨蔓	马鞍岭、雷虎岭	未见上覆盖层，下伏为道堂组或北海组		陆相
						$\beta Qh_1^1 s^{1-2}$		橄榄玄武岩、橄榄拉斑玄武岩	7.3						
						$\beta Qh_1^1 s^{1-1}$		含集块、火山角砾熔状橄榄玄武岩	4.0						
	晚更新世	道堂期	10	17	20	$VQp_3^3 d^3$		凝灰岩、层凝灰岩、沉火山角砾岩	0~85	地表	双池岭、雷虎岭、片石市、峨蔓	神岭、永莀岭、双池岭、莲花山、德义岭	上覆为全新世石山期火山岩，下伏为北海组、多文组或更老地层或印支期花岗岩	第二段橄榄玄武岩获K-Ar年龄98300a第三段炭化木^{14}C年龄为(16790±140)a	海相
				16	19	$\beta Qp_3^3 d^2$		橄榄玄武岩	0~55		道堂、罗径盘、十字路、龙桥、莲花山、德义岭、三都				陆相
				15	18	$VQp_3^3 d^{1-3}$		玄武质沉岩屑晶屑凝灰岩	0~120		老城、美选、积涌、社兰、德义岭及莲花山				海相
					17	$\beta Qp_3^3 d^{1-2}$		橄榄拉斑玄武岩	≥4.0						
						$VQp_3^3 d^{1-1}$		玄武质沉岩屑晶屑凝灰岩	8.43						
	中更新世	多文期 东英期	9	14	16	$\beta Qp^2 d^2$		熔渣状橄榄玄武岩		地表	临高县东英镇、定安旧洲岭及定安县龙塘镇-岭口镇	高山岭、旧洲岭、黄岭	上覆为上更新统道堂期或上更新统八所组或全新统烟墩组，下伏为中更新统北海组或更老地层或印支期-燕山期花岗岩	K-Ar年龄0.67Ma、(1.20±0.02)Ma、(1.16±0.4)Ma、(0.43±0.03)Ma、Ar-Ar年龄(0.46±0.01)Ma、(1.12±0.02)Ma	陆相
								辉石玄武岩、橄榄玄武岩、橄榄辉石玄武岩							
		多文期 多文岭期	8	13	15	$\beta Qp^2 d^1$		橄榄辉石玄武岩、辉石玄武岩、辉石橄榄玄武岩			新盈农场、多文镇、福山镇、美兰镇、龙门镇、大路镇等地	多文岭、龙门岭、龙发岭、金鸡岭、岭脚岭			
	早更新世	秀英期	7	12	14	$\beta Qp^1 x$		橄榄玄武岩	4.62~15.87	8.5~91.34	石山、秀英		夹于秀英组地层中		

续表

喷发时代				喷发序次			代号	地层分区	主要岩性	厚度/m	埋深/m	分布情况	代表性火山	接触关系	测年数据	喷发环境
纪	世	期	期（组）	旋回	韵律	次序										
新近纪	上新世	石门沟村期	金牛岭期	6	11	13	βN_2h^{4-2}	海口地层小区	气孔状玄武岩	2.8	26~118	琼山石山、龙桥	金牛岭	海口组第四段夹层		海相
						12	βN_2h^{4-1}		橄榄玄武岩、橄榄拉斑玄武岩	1.86~20.65						
					10	11	βN_2h^{3-1}		橄榄玄武岩	14	135	琼山石山、龙桥		海口组第三段夹层		
					9	10	$\beta\text{-}VN_2h^{2-2}$		石英拉斑玄武岩、橄榄拉斑玄武岩、角砾岩、凝灰岩		0~152.6	海口金牛岭、演丰、澄迈马村		海口组第二段夹层	K-Ar 年龄 3.13~3.82Ma	陆相
						9	$\beta\text{-}VN_2h^{2-1}$		拉斑玄武岩、角砾岩、凝灰岩							海相
					8	8	VN_2h^{1-2}		凝灰岩、火山角砾岩	0.74	189	琼山		海口组第一段夹层		
							βN_2H^{1-1}		玄武岩	0.56~7.17						
					Ⅴ	Ⅴ	$\beta N_2\hat{s}^{2-2}$	五指山地层分区	碱性玄武岩、粗玄岩	>50	地表	文昌市蓬莱镇-石壁镇定安福基田村、南福村	蓬莱岭、庙岭、旧晒岭	上覆多文明火山岩或北海组，下伏为下白垩统鹿母湾组或印支期花岗岩	K-Ar 年龄 3.04Ma、(3.37±0.14) Ma、(3.84±0.3) Ma Ar-Ar 年龄 (3.11±0.12) Ma、(3.14±0.28) Ma、(3.17±0.11) Ma	陆相
							$VN_2\hat{s}^{2-1}$		凝灰岩、角砾凝灰岩、角砾岩	>20						
					Ⅳ	Ⅳ	$\beta N_2\hat{s}^{1-2}$		橄榄玄武岩、碱性橄榄玄武岩	>15						
							$VN_2\hat{s}^{1-1}$		凝灰岩、层凝灰岩	>21						
	中新世	石马村期	灯楼角期	5	Ⅲ	Ⅲ	$VN_1\hat{s}m^{3-3}$		凝灰岩	2.8	地表	三门坡镇-南阳镇-竹村镇一带及定安居丁镇		上覆上新世石门沟村期火山岩，下伏海西期—印支期花岗岩或下白垩统鹿母湾组沉积地层	K-Ar 年龄 (5.43±0.116)Ma (8.09±0.98)Ma Ar-Ar 年龄 (4.46±0.37)Ma、(4.63±0.18)Ma、(5.86±0.35)Ma、(12.89±1.15)Ma	
							$\beta N_1\hat{s}m^{3-2}$		橄榄玄武岩	15.4						
					Ⅱ	Ⅱ	$VN_1\hat{s}m^{3-1}$		凝灰岩、火山角砾岩	1.5						
							$\beta\text{-}\varphi N_1\hat{s}m^{1}$		橄榄玄武岩、玻基辉橄岩	3.1~29.0						
					Ⅰ	Ⅰ	$\varphi N_1\hat{s}m^{1}$		玻基辉橄岩	15.5						

续表

喷发时代 纪	喷发时代 世	喷发时代 期（组）	喷发序次 旋回	喷发序次 韵律	喷发序次 次序	代号	地层分区	主要岩性	厚度/m	埋深/m	分布情况	代表性火山	接触关系	测年数据	喷发环境
新近纪	中新世	石马村期 / 灯楼角期	5	7	7	βN_1d^4	海口层小区	玄武岩	<70	230~485	琼山、澄迈		夹于灯楼角组第四段		海相
新近纪	中新世	石马村期 / 灯楼角期	5	6	6	βN_1d^2	海口层小区	玄武岩	<70	230~485	琼山、澄迈		夹于灯楼角组第二段		海相
新近纪	中新世	石马村期 / 灯楼角期	5	5	5	βN_1d^1	海口层小区	橄榄玄武岩	<70	230~485	琼山、澄迈		灯楼角组第一段		海相
新近纪	中新世	角尾期	4	4	4	βN_1j	海口层小区	玻基玄武岩、橄榄玄武岩，局部见火山碎屑岩	0.7~171	170~495	澄迈、临高		夹于角尾组地层中		海相
新近纪	中新世	下洋期	3	3	3	βN_1x	海口层小区	玻基玄武岩、玄武岩	14.5	469	澄迈福山、海口		夹于下洋组地层中		海相
古近纪	渐新世	涠洲期	2	2	2	βE_3w	海口层小区	玄武岩、基性火山碎屑岩	1.5~9.5		临高美台、澄迈福山、马村		夹于涠洲组地层中		陆相
古近纪	始新世	流沙港期	1	1	1	βE_2l	海口层小区	玄武岩	1.5~10	>900	澄迈福山、老城		夹于流沙港组地层中		陆相

第一喷发旋回 时代为古近纪始新世，为陆相火山喷发。由一个喷发韵律组成，仅出现一层玄武岩，厚 1.5~10mm，主要隐伏于福山盆地、南宝盆地、白莲盆地一带，呈夹层赋存于流沙港组湖泊相杂色砂泥质碎屑岩之中，为海南岛新生代第一次火山喷发。

第二喷发旋回 时代为古近纪渐新世，为陆相火山喷发。见 3 次玄武岩喷发，呈夹层赋存于涠洲组沼泽、河湖相杂色砂泥质碎屑岩中，构成 3 个喷溢-沉积韵律组合。主要隐伏于福山盆地，次为白莲盆地。

第三喷发旋回 时代为新近纪中新世早期，为海相火山喷发。见一次玄武岩喷发，呈夹层赋存于下洋组滨海相灰白色、浅黄色粗碎屑岩中，厚约 15.5m，构成喷溢-碎屑沉积韵律组合。主要隐伏于福山盆地、白莲盆地一带。

第四喷发旋回 时代为新近纪中新世中期，为海相火山喷发。可见 5 次玄武岩喷发（溢），呈夹层赋存于角尾组滨海—浅海相灰绿色碎屑岩中，火山岩主要有玄武岩、火山碎屑岩和凝灰岩，构成 5 个喷发（溢）-碎屑沉积韵律组合。主要隐伏于福山盆地、白莲盆地一带。

第五喷发旋回 时代为新近纪中新世晚期，在东西向王五-文教断裂带以北为海相火山喷发，断裂带以南为陆相火山喷发。断裂带北面见 3 次火山喷发，形成 3 层玄武岩，呈夹层赋存于灯楼角组滨海—浅海相浅灰色—灰绿色碎屑岩中，构成 3 个喷溢（发）-碎屑沉积韵律组合，主要隐伏于福山盆地一带。断裂带南面见 3 次火山喷发（溢），主要分布在定安县居丁、文昌市蓬莱、新桥及海口市三门坡一带，火山岩主要由超基性岩（玻基辉橄岩和橄榄霞石岩）、玄武岩类、火山碎屑岩和凝灰岩构成，形成石马村组一套基性火山岩夹河湖相碎屑岩沉积，构成 3 个喷溢-爆发-碎屑沉积韵律组合。

第六喷发旋回 时代为新近纪上新世，在东西向王五-文教断裂带以北为海相火山喷发，断裂带以南为陆相火山喷发。断裂带北面见 7 次火山喷发，呈夹层赋存于海口组滨浅海红黄色—绿灰色碎屑岩中，主要隐伏于临高县南宝镇至海口市三江镇一带，局部出露地表，构成 4 个喷溢-海相沉积或喷溢-喷发-海相沉积或喷发-喷溢-海相沉积等韵律旋回。断裂带南面见 2 次火山喷发，主要分布在海口市大坡、文昌市蓬莱、南阳及定安县永丰一带，火山岩主要由玄武岩类、火山碎屑岩和凝灰岩构成，形成石门沟村组一套基性火山岩夹河湖相碎屑岩沉积，构成 2 个喷发-喷溢-碎屑沉积韵律组合。

第七喷发旋回 时代为第四纪早更新世，为陆相火山喷发。见一次火山岩喷发，主要分布于临高县东英镇至海口市大坡镇一带，呈夹层赋存于秀英组河湖相杂色泥砂质碎屑岩中，零星分布，构成喷发-喷溢-碎屑沉积韵律组合。

第八喷发旋回 时代为第四纪中更新世。由一个喷发韵律陆相玄武质熔岩构成（多文组），火山岩主要由玄武岩类夹少量火山碎屑岩构成。主要分布于临高县东英镇—澄迈县美亭镇、海口市美蓝镇—旧州镇和定安县黄竹镇—琼海市大路镇，以及定安县金鸡岭和儋州市木棠镇—光村镇一带。

第九喷发旋回 时代为第四纪中更新世，由一个喷发韵律陆相玄武质熔岩构成。主要分布于临高县东英镇、定安县龙塘镇—岭口镇一带。

第十喷发旋回 时代为第四纪晚更新世，由海相玄武质沉凝灰岩-陆相玄武质岩-海相玄武质沉凝灰岩构成，组成 3 个喷发韵律。形成道堂组玄武岩类岩石和火山碎屑岩。主要出露于澄迈县白莲镇—海口市龙塘镇和儋州市三都镇一带。

第十一喷发旋回 时代为第四纪全新世早期，由一个喷发韵律陆相玄武质熔岩构成，形成石山组玄武岩类岩石和火山碎屑岩。主要出露于海口地层小区，分布于海口市石山镇-永兴镇一带和儋州市笔架岭—兵马角一带。

二、岩相特征

琼北新生代火山岩依据喷发时所处的古地理环境，可分为陆相火山岩和海相火山岩，而根据喷发方式又可分为火山通道相、喷发相、喷发-沉积相（表 13-3）。

表 13-3　琼北新生代不同产出层位火山岩岩石类型及其特征一览表

Table 13-3　List of types and characteristics of volcanic rocks of different Cenozoic horizons of Qiongbei

岩相	岩石名称	产出层位	物质成分	结构构造	其他特征
火山口相	熔渣状橄榄玄武岩	Qh^1s^2	斑晶：橄榄石 1%～3%，辉石（主要为普通辉石）<2%，偶见个别斜长石 基质：斜长石（An49）15%～25%，普通辉石 4%～10%，橄榄石 ≤ 1%，玻璃质 <1%～12%，磁铁矿 4%～9%，伊丁石微量；气孔 54%～55%	少斑结构，基质具玻基交织结构，熔渣状构造	橄榄石有两个世代，无辉石反应边
	熔渣状玻基橄榄玄武岩		斑晶：普通辉石<1%，橄榄石 1%，偶见斜长石（An49） 基质：斜长石 5%～8%，辉石 < 1%～2%，橄榄石≤1%，玻璃质 25%～33%，磁铁矿 1%～5%。次生矿物见蛋白石；气孔 60%	玻基斑状结构，熔渣状构造	
	浮岩状橄榄玄武岩		斑晶：橄榄石 1%-4%，透辉石、普通辉石 2%，斜长石偶见 基质：斜长石（An37）4%～8%，玻璃质 15%，辉石 1%，磁铁矿 3%；气孔 70%	少斑结构，基质具玻基交织结构，浮岩状构造	
	玄武质浮岩		矿物成分：橄榄石 1%～2%，普通辉石 ≤ 1%，斜长石 ≤ 1%，基性火山玻璃 27%～56%，次生矿物有伊丁石；气孔40%～70%	斑状结构，基质具玻璃结构，多孔构造，熔渣状构造	橄榄石有两个世代，无辉石反应边。岩石密度小，可浮在水面上
喷溢亚相	熔渣状橄榄玄武岩	Qh^1s^1	斑晶：橄榄石 3%，斜长石个别 基质：玻璃质 20%，透辉石 2%，斜长石（An32）3%，磁铁矿 5%，捕虏体石英+石英岩碎块 1%；气孔 65%；局部见含集块和火山角砾	少斑结构，基质具玻晶交织结构，熔渣状构造	岩石外貌似炉渣。橄榄石不具辉石反应边
	气孔状橄榄玄武岩		斑晶：橄榄石 1%～6%，普通辉石 ≤ 1%，斜长石（An50～An57）≤20%，有时见石英（捕虏晶）<1% 基质：斜长石（An50）35%～48%，易变辉石 10%～28%，橄榄石 < 1%～3%，磁铁矿 1%～15%，钛铁矿 1%，磷灰石微量；气孔 10%～45%	斑状结构，基质具隐晶—微晶结构，微间粒结构，多孔构造	橄榄石无辉石反应边，部分具熔蚀现象

续表

岩相	岩石名称	产出层位	物质成分	结构构造	其他特征
喷溢亚相	橄榄玄武岩	Oh^1s^1	斑晶：斜长石（An44～An46）≤15%，偶见他形钠长石分布于其边缘，橄榄石5%～10%，普通辉石、透辉石没有或少量，偶见石英（捕虏晶） 基质：斜长石（An35～An53）40%～57%，橄榄石1%～12%。普通辉石15%～30%，磁铁矿≤8%，钛铁矿≤2%，磷灰石微量，次生矿物有伊丁石	斑状结构，基质具间粒结构，块状构造	橄榄石有两个世代，无辉石反应边，多被熔蚀成浑圆状。斜长石有两个世代
	气孔状橄榄拉斑玄武岩		斑晶：斜长石（An62）≤15%，橄榄石1%～10%，普通辉石≤2% 基质：斜长石（An50）30%～50%，辉石15%～25%，橄榄石<1%，玻璃质10%～15%，磁铁矿+钛铁矿1%～8%；气孔8%～25%	斑状结构，基质具拉斑玄武结构，气孔状构造	橄榄石有两个世代，无辉石反应边，部分被熔蚀成浑圆状、港湾状。辉石具两个世代
	橄榄粗玄岩		矿物成分：斜长石（An44）58%，普通辉石35%，橄榄石5%，磁铁矿+钛铁矿1%～2%，次生矿物见伊丁石	粗玄结构，块状构造	橄榄石不具辉石反应边
喷发-沉积相	玄武质沉岩屑玻屑凝灰岩	Qp^3d^3	碎屑成分：玻屑30%～36%，玄武质岩岩屑20%～30%，橄榄石1%～6%，辉石<1%～7%，斜长石1%～8%，钾长石≤1%，黑云母、电气石、燧石、微体化石、伊丁石、绿泥石均为微量 胶结物：火山尘+泥质25%～35%	沉岩屑、玻屑砂砾状凝灰结构，块状构造	碎屑分选性差，玻屑呈尖棱角状、鸡骨状，石英多为次棱角状—次圆状
喷溢亚相	气孔状橄榄玄武岩	Qp^3d^2	斑晶：斜长石（An44～An45）10%，橄榄石5%，普通辉石<1%，石英（捕虏晶）<1% 基质：橄榄石1%，辉石+斜长石40%，磁铁矿1%～10%；气孔35%	斑状结构，基质具隐晶质结构，气孔状构造，局部火山口见绳状构造	橄榄石无辉石反应边，有的具熔蚀现象，辉石有两个世代，边缘有反应边，斜长石有两个世代
	微气孔状橄榄拉斑玄武岩		斑晶：橄榄石1%～5%，普通辉石、易变辉石≤1%。 基质：斜长石（An40～An50）46%～50%，辉石20%～25%，橄榄石<1%，玻璃质5%～15%，钛铁矿+磁铁矿1%～2%，次生矿物有伊丁石、绿泥石；微气孔15%	斑状结构，基质具拉玄结构，微气孔状构造	橄榄石辉石常具熔蚀现象，橄榄石有两个世代，无辉石反应边，斜长石正光性，具波状消光

续表

岩相	岩石名称	产出层位	物质成分	结构构造	其他特征
喷发—沉积相	玄武质沉岩屑晶屑凝灰岩	Qp^3d^1	碎屑成分：石英 25%，玻屑 20%，玄武质岩岩屑 15%，长石 5%，橄榄石 3%，绿泥石<1%，海绿石<1%，石英岩岩屑+硅质岩岩屑 1%，伊丁石、锆石、黑云母微量，微体化石<1% 胶结物：火山尘+泥质 40%	沉岩屑、晶屑砂状凝灰结构，层状构造	岩石薄层状，水平层理发育。岩屑次棱角状，长英碎屑次棱角状—次圆状
	玄武质沉玻屑晶屑凝灰岩		碎屑成分：石英 30%，玻屑 20%，长石 5%，橄榄石 2%，玄武质岩屑 8%，绿泥石<1%，黑云母<1%，锆石褐铁矿微量 胶结物：火山尘+泥质 35%	沉玻屑、晶屑砂状凝灰结构，层状构造	岩石薄层状，水平层理发育。长英碎屑次棱角状—次圆状
	玄武质沉晶屑岩屑凝灰岩		斑晶：橄榄石 3%，易变辉石<1%，石英岩（捕虏体）≤1% 基质：斜长石（An45）40%～55%，辉石 29%～30%，橄榄石≤1%，玻璃质 10%，磁铁矿+钛铁矿 1%～2%，次生矿物有绿泥石、伊丁石；气孔 8%～15%	沉岩屑、晶屑砂状凝灰结构，层状构造	岩石薄层状，水平层理发育。碎屑颗粒大小较均匀，分选性较好
	微气孔橄榄拉斑玄武岩		斑晶：橄榄石 3%，易变辉石<1%，石英岩（捕虏体）≤1% 基质：斜长石（An45）40%～55%，辉石 29%～30%，橄榄石≤1%，玻璃质 10%，磁铁矿+钛铁矿 1%～2%，次生矿物有绿泥石、伊丁石；气孔 8%～15%	斑状结构，基质具拉斑玄武结构，微气孔构造	橄榄石无辉石反应边，部分具熔蚀现象
	含砾凝灰质砂岩		碎屑成分：砾石 10%，砂屑 15%，晶屑 5% 胶结物：泥质+火山尘 70%	凝灰砂砾状结构，块状构造	砾石由铁质组成，大小 2～5mm，半圆—滚圆状，砂屑为石英、长石，呈棱角状
	玄武质沉凝灰岩		碎屑成分：晶屑 5%，浮岩岩屑 30%，陆源砂屑 10%，电气石几粒 胶结物：泥质+火山尘 35%，孔隙 20%	沉凝灰结构，层状构造	石英晶屑呈尖棱角状，浮岩岩屑为圆形、偏圆形，陆源砂屑次棱角状—半滚圆状。碎屑分选性较好

续表

岩相	岩石名称	产出层位	物质成分	结构构造	其他特征
喷溢亚相	熔渣状辉石橄武岩	Qp^3d^2	斑晶：橄榄石 10%±，辉石 2%~8%，斜长石≤12% 基质：斜长石 50%~67%，辉石 5%~15%，磁铁矿 3%±，玻璃质 10%~15%	斑状结构，基质具填隙结构，气孔状构造	
	橄榄辉石玄武岩		斑晶：辉石 5%~13%，橄榄石 5%~12%，斜长石 10%~20% 基质：斜长石 43%~58%，辉石 5%~20%，橄榄石 1%~2%，磁铁矿 3%~8%	斑状结构，基质具间粒结构，块状构造	
	橄榄玄武岩		斑晶：橄榄石 10%~12%，偶见辉石和斜长石 基质：斜长石 55%~60%，辉石 20%~28%，磁铁矿 4%，少量橄榄石	斑状结构，基质具间粒结构，块状构造	
	辉石玄武岩		斑晶：辉石 3%~10%，斜长石 3%± 基质：斜长石 50%~67%，辉石 25%~38%，磁铁矿 3%±，隐晶质 3%±	斑状结构，基质具填隙结构，块状构造	
	橄榄辉石玄武岩	Qp^2d^1	斑晶：辉石 10%~20%，橄榄石 5%~8%，斜长石 10%± 基质：斜长石 50%~55%，辉石 10%~12%，橄榄石 0%~2%，磁铁矿 3%~8%	斑状结构，基质具间粒结构，块状构造	
	玄武岩		斑晶：斜长石 5%~6%，辉石 3%~4% 基质：斜长石 43%~60%，辉石 28%~35%，磁铁矿 2%~5%	斑状结构，基质具间粒结构，块状构造	
	辉石橄榄玄武岩		斑晶：橄榄石 10%±，辉石 5%±，斜长石 15%± 基质：斜长石 60%±，辉石 15%，橄榄石 2%±，磁铁矿 3%±	斑状结构，基质具间粒结构，块状构造	
	玄武岩	N_2h^2	斑晶：古铜辉石 3%~5%，斜长石（An50~An62）5%±，橄榄石<1% 基质：斜长石（An42~An50）32%~46%，辉石 25%~60%，橄榄石<1% 副矿物：钛铁矿+磁铁矿 1%±，气孔 30%±	斑状结构，基质具拉斑结构，块状构造	
	辉石玄武岩		斑晶：辉石 5%±，斜长石 3%±，橄榄石 1%± 基质：中斜长石 42%±，辉石 35%±，橄榄石 1%~2%，玻璃质 15%± 副矿物：磁铁矿、钛铁矿、磷灰石	斑状结构，基质具拉斑结构，块状构造	
爆发亚相	玻屑凝灰岩	N_2h^2	碎屑成分：浮岩状玻屑 20%~35%，玻屑 30%~35%，浮岩状玻质玄武岩屑 15%，古铜辉石及石英晶屑 1%~3% 胶结物：火山灰 15%~30%	岩屑玻屑凝灰结构，块状构造	
喷溢亚相	粗玄岩	$N_2\hat{s}$	斑晶：橄榄石 1%±，辉石 5%± 基质：斜长石 50%~57%，辉石 22%~24%，橄榄石 5%~15%	粗玄结构，块状构造	
	橄榄辉石玄武岩		斑晶：橄榄石 3%~7%，辉石 3%~10%，斜长石<1% 基质：斜长石 50%~60%，辉石 10%~28%，橄榄石 1%~7%，玻璃质≤5%	斑状结构，基质具间粒结构，块状构造	柱状节理发育

续表

岩相	岩石名称	产出层位	物质成分	结构构造	其他特征
喷发亚相	含火山角砾晶屑岩屑凝灰岩	$N_2\hat{s}$	火山碎屑：玄武岩屑30%±，橄榄石晶屑10%±，石英晶屑8%±，长石晶屑5% 胶结物：泥质+火山灰+铁质40%±	晶屑岩屑砂砾状结构，块状构造	
喷溢亚相	橄榄玄武岩		斑晶：橄榄石5%，辉石1%~3%，少量斜长石 基质：斜长石50%~60%，辉石20%~30%，橄榄石10%	斑状结构，基质具间粒或间隐结构，块状构造	柱状节理发育
喷发-沉积相	铁质玻屑凝灰质含粉砂泥岩	$N_2\hat{s}$	碎屑成分：铁质玻屑20%，石英粉砂5%，磁铁矿5% 胶结物：泥质+火山岩70%	玻璃粉砂-凝灰结构，块状构造	
	凝灰质粉细砂岩		碎屑成分：石英15%，长石20%，黑云母5%，岩屑+玻屑5% 胶结物：黏土+火山灰55%	凝灰质细砂粉砂结构，块状构造	
	铁质黏土岩		主要为黏土矿物65%，次为褐铁矿30%~35%及少量长石石英等	泥质结构，块状构造	
	凝灰岩	$N_1\hat{s}m$	岩石风化除少量磁铁矿、石英、钛铁矿、晶屑及玄武岩屑外，其他成分被高岭土代替，高岭土含量达80%~90%	泥质结构，块状构造	
喷溢亚相	橄榄玄武岩		斑晶：橄榄石1%~30%，辉石1%~3%，斜长石1%± 基质：斜长石40%~60%，辉石10%~35%，橄榄石1%~3%，局部见二辉橄榄岩包体	斑状结构，基质具间粒或间隐结构，气孔状、杏仁状构造或块状构造	柱状节理发育
	晶屑岩屑凝灰岩		黏土矿物+针铁矿95%，少量橄榄石、辉石晶屑及玄武质岩屑	泥质结构，块状构造	
喷溢亚相	橄榄玄武岩		斜长石55%~65%，辉石20%~30%，橄榄石10%	间粒结构，微气孔构造	柱状节理发育
	玻基橄辉岩		斑晶：橄榄石5%~15%，辉石1%~3%，偶见斜长石 基质：辉石23%~57%，斜长石5%~10%，橄榄石1%~5%，玻璃质15%~56%	斑状结构，基质具似间隐结构，块状构造	
沉积相	黏土岩		岩石由黏土90%、针铁矿5%及少量磁铁矿、石英、长石等组成	泥质结构，块状构造	
	砾岩		岩石由砾石70%、石英砂5%、黏土25%组成	砾状结构，块状构造	
喷溢亚相	玻基橄辉岩		斑晶：橄榄石7%~15%，辉石25%~70%，斜长石≤3% 基质：玻璃质20%~52%	斑状结构，气孔状、杏仁状或块状构造	

（一）火山通道相

火山通道相是指充填于火山通道内的各种岩石的总称，包括火山颈亚相及上部的火山口亚相。

1. 火山颈亚相

多见于形成时代较早、火山机构遭受中等剥蚀的火山岩区。新近纪火山岩分布区的庙岭、日晒岭、蓬莱岭、乌盖岭和金牛岭、第四纪的多文岭、岭脚岭、黄岭、颜春岭、旧州岭（图版Ⅲ-1-E）、峨蔓岭（图版Ⅲ-1-F）、道义岭、莲花山、兵马岭等火山锥的颈筒部分，分布范围不广，组成岩颈的岩石有粗玄岩、橄榄辉石玄武岩、拉斑玄武岩和火山碎屑岩等，火山颈形状似喇叭状或镰刀状，在地貌上多表现为山包。

2. 火山口亚相

火山口亚相主要分布于第四纪全新世形成的火山岩群，中更新世、晚更新世也保存有少量，火山口一般保存较好，如马鞍岭、双池岭、吉安岭、高山岭（图版Ⅲ-1-G）等。按火山口内出露的岩石可分为熔岩型、碎屑岩型和混合型。熔岩型以多孔状熔岩为主，有熔渣状橄榄玄武岩和翻花状玄武岩；碎屑岩常见有火山集块熔岩、熔渣状集块角砾岩等；混合型由熔岩和碎屑岩组成。

（二）火山喷发相

喷发相是火山喷发地表的产物，主要为各种熔岩和正常火山碎屑岩。火山活动在一定时期由于岩浆成分的变化，喷发性质和喷发强度的不同，其形成喷发相的岩石性质也不同，根据这些岩石的形成条件及分布特征，喷发相又可划分为两个亚相：爆发亚相和喷溢亚相。

1. 爆发亚相

爆发亚相是指火山爆发形成的各种粒级火山碎屑物的堆积体。火山碎屑物胶结成岩后形成不同类型的火山碎屑岩，近火山锥为粗粒级的火山角砾岩、集块岩等，远离火山锥渐变为角砾凝灰岩、凝灰岩；在平面上呈不规则环形分布。

2. 溢流亚相

溢流亚相是火山喷溢、泛流的熔岩，在各火山岩相所占比例最大，熔岩流从火山口向外呈面状溢出，形成大面积分布的熔岩被。近火山口熔岩的厚度大，产状陡。远离火山口熔岩逐渐变薄，产状变缓。岩石主要有橄榄玄武岩，橄榄辉石玄武岩、拉斑玄武岩等。

（三）喷发-沉积相

喷发-沉积相是指火山作用叠加沉积作用形成的岩石，由火山岩、沉积火山碎屑岩、火山碎屑沉积岩、沉积岩系组成，局部发育斜层理（图版Ⅲ-1-H）、波状层理（图版Ⅲ-2-A）。根据钻孔揭露，在琼北坳陷区古近系渐新统杂色岩组中部，在沼泽、河湖相的杂色泥质碎屑岩中，局部夹薄层凝灰岩、沉凝灰岩；新近系上新统海口组的海相地层中从海口金牛岭至国社岭北侧的马村一带，隐伏分布沉凝灰岩、凝灰质砂砾岩、凝灰质砂岩等。第四纪喷发-沉积相主要出露于道堂、莲花山、德义岭、老城及双池岭-雷虎岭一带，为第四系晚更新统道堂组，岩性有玄武质沉凝灰岩、含砾凝灰质砂岩等。

三、火山构造

琼北新生代火山喷发带属东南沿海火山岩省（Ⅰ级）南缘海南火山喷发带（Ⅱ级），Ⅲ级构造单元有新近纪、第四纪晚更新世火山盆地和第四纪中更新世、全新世火山洼地。火山盆地和火山洼地中还包含了众多的Ⅳ级火山机构——火山锥体，琼北新生代火山锥有54个（上新世5个、中更新世15个、晚更新世12个、全新世22个），主要形成于晚更新世和全新世。时代愈新，红土化愈弱，锥体保存愈好。火山锥以熔岩锥为主，混合锥和碎屑锥少见（表13-4）。

表 13-4　琼北新生代主要火山锥类型一览表

Table 13-4　List of main types of volcanic cones in Qiongbei

喷发时代			火山锥名称	火山锥分类	海拔标高 m	比高 m	火山锥、火山口的特征及保存情况	缺口方向
纪	世	代号						
第四纪	全新世	$Qh^{1}\hat{s}$	群众岭	混合锥	141. 6	15	锥底座呈椭圆形，长轴北北西向，长 300m，宽 250m，无火山口	
			儒任岭		129	20	锥底座呈椭圆形，长轴北北西向，长 300m，宽 250m，无火山口	
			岭南市	熔岩锥	106. 8	10	锥底座呈椭圆形，长轴北北东向，长 300m，宽 250m，无火山口	
			群仙岭		104. 2	15	锥底座呈椭圆形，长轴北北东向，长 150m，宽 75m，无火山口	
			群墩大岭		140. 4	15	锥底座呈椭圆形，长轴北北西向，长 325m，宽 225m，火山口直径约 10m，深约 7m，呈漏斗状	
			儒群		137. 4	20	锥底座近圆形，直径约 175m，无火山口	
			浩昌		147. 3	15	锥底座近圆形，直径约 175m，火山口直径约 20m，深约 10m，呈漏斗状	
			美玉岭		139. 7	30	锥底座呈椭圆形，长轴北北西向，长 300m，宽 210m，火山口直径约 20m，深约 10m，呈漏斗状	
			卧牛岭		143. 5	20	锥底座近圆形，直径约 400m，火山口直径约 50m，深约 15m，呈盆状	西北
			雷虎岭		160	50	锥底座呈椭圆形，长轴北北西向，长 700m，宽 500m，火山口直径约 220m，深约 52m，呈盆状	30°
			儒洪		135. 6	10	锥底座近圆形，直径约 150m，无火山口	
			阳南岭南		146. 1	10	锥底座近圆形，直径约 250m，火山口长约 50m，宽约 30m，呈漏斗状	东南
			阳南岭西		168. 7	15	锥底座呈椭圆形，长轴北西西向，长 240m，宽 150m，无火山口	
			昌道岭东		160	30	锥底座呈椭圆形，长轴北西西向，长 360m，宽 250m，火山口直径约 200m，深约 10m，呈漏斗状	
			昌道岭西		150	20	锥底座呈椭圆形，长轴北西向，长 300m，宽 250m，无火山口	
			马鞍岭		222. 2	120	锥底座近圆形，直径约 600m，火山口直径约 90m，深约 60m，呈漏斗状	北东
			马鞍岭北		189. 5	25	锥底座近圆形，直径约 300m，火山口直径约 20m，深约 60m	
			马鞍岭南		176. 5	40	锥底座呈椭圆形，长轴北西向，长 450m，宽 300m，火山口直径约 200m，深约 30m，呈漏斗状	东北
			马鞍岭东		143. 5	12	锥底座近圆形，直径约 80m，火山口直径约 38m，深约 10m，呈漏斗状	220°
			博昌		99. 9	30	锥底座呈椭圆形，长轴北北西向，长 590m，宽 350m，火山口直径约 250m，深约 40m，呈漏斗状	240°
			国群		158. 1	40	锥底座呈椭圆形，长轴北西向，长 300m，宽 225m，火山口直径约 100m，深约 15m，呈盆状	西
			春藏		173. 1	25	锥底座呈椭圆形，长轴北西向，长 175m，宽 125m，火山口直径约 70m，深约 10m	

续表

喷发时代			火山锥名称	火山锥分类	海拔标高 m	比高 m	火山锥、火山口的特征及保存情况	缺口方向
纪	世	代号						
第四纪	晚更新世	Op^3d	双池岭	碎屑锥	104.6	20	锥底座近圆形，直径约900m，火山口长约150m，深约30m，呈盆状	
			罗京盘	混合锥			锥底座近圆形，直径约1000m，火山口长约500m，深约300m，呈盆状。火山口中心直径约80m，高约10m的无口火山	
			黑树岭	熔岩锥	100.5	50	锥底座近圆形，直径约200m，火山口长约30m，深约10m，口垣不完整	北
			神岭		141.5	25	锥底座呈椭圆形，长轴北西向，长500m，宽200m，火山口直径约60m，深约5m，口垣不完整	50°
			三雅岭		131.9	10	锥底座呈椭圆形，长轴南北向，长250m，宽170m，火山口直径约4m	
			道堂岭		45.9	15	锥底座呈椭圆形，长轴北西向，长250m，宽200m，无火山口	
			黄里岭西		111.41		锥底座呈椭圆形，长轴北西向，长120m，宽50m，无火山口	
			群香岭		125.2	30	锥底座呈椭圆形，长轴北西向，长310m，宽200m，火山口直径约50m，深约20m，呈漏斗状	
			群香岭东		120	15	锥底座呈椭圆形，长轴北西向，长225m，宽150m，无火山口	
			永茂岭		119.3	30	锥底座呈椭圆形，长轴东西向，长750m，宽600m，无火山口	
			卜亚岭	熔岩锥	95.2	20	锥底座呈椭圆形，长轴北北东向，长1000m，宽850m，火山口面貌不清	
			何儒		70.6	15	锥底座近圆形，直径约100m，呈平顶锥状，无火山口	
	中更新世	Op^2d	黄岭	碎屑锥	224	54	锥底座呈椭圆形，长轴北东东向，长700m，宽300m，火山口完好，深约50m	北
			龙门岭	熔岩锥	228.9	80	锥底座呈椭圆形，长轴北东东向，长900m，宽650m，火山口完好	
			金鸡岭		203.9	60	锥底座呈椭圆形，长轴北东东向，长500m，宽300m，锥顶圆滑，无火山口	
			旧州岭	碎屑锥	199.9	150	锥底座呈椭圆形，长轴北西向，长1000m，宽650m，火山口完好，深约50m	北东
			龙发岭	混合锥	130.5	60	锥底座呈椭圆形，长轴北西向，长500m，宽300m，锥顶圆滑，无火山口	北西
			岭脚岭	熔岩锥	99	40	锥底座近圆形，直径约250m，锥顶圆滑，无火山口	
			夏社岭		73.2	30	锥底座呈椭圆形，长轴北北东向，长350m，宽200m，火山口直径约20m，深约10m，口垣不完整	140°
			颜春岭	熔岩锥	144	70	锥底座近圆形，直径约3000m，火山口直径约25m，深约10m	
			颜春岭东		105	25	锥底座近圆形，直径约250m，锥顶圆滑，无火山口	
			国社岭		57.2	34	锥底座近圆形，直径约400m，锥顶不平，无火山口	

续表

喷发时代			火山锥名称	火山锥分类	海拔标高 m	比高 m	火山锥、火山口的特征及保存情况	缺口方向
纪	世	代号						
第四纪	中更新世	Qp^2d	多文岭	熔岩锥	244.1	80	锥底座近圆形，直径约800m，有浅火山口	北东
			大王岭		187.2	50	锥底座近圆形，直径约450m，锥顶圆滑，无火山口	
			高山岭南		111.4	40	锥底座呈椭圆形，长轴北西向，长750m，宽500m，锥顶圆滑，无火山口	
			高山岭		190.7	80	锥底座呈椭圆形，长轴北西向，长2000m，宽1000m，有浅火山口，口径约42m，深约2m，呈盆状	65°
			高山岭北				锥底座呈椭圆形，长轴北西向，长1500m，宽800m，有火山口，口径约84m，呈盆状（有火山湖）	230°
新近纪	上新世	N_2h	金牛岭	熔岩锥	35	30	锥底座近圆形，直径约400m，锥顶圆滑，无火山口	
		$N_2\hat{s}$	日晒岭	混合锥	171	50	锥底座近圆形，直径约600m，锥顶圆滑，无火山口	
			蓬莱岭		195.2	60	锥底座近圆形，直径约1000m，锥顶圆滑，无火山口	
			乌盖岭		269.8	100	锥底座近圆形，直径约1000m，有火山口，深10~15m	南
			庙岭		210.5	65	锥底座近圆形，直径约800m，锥顶圆滑，无火山口	
			黑岭		156	55	锥底座近圆形，直径约800m，锥顶圆滑，无火山口	

四、岩石学特征

新生代火山岩岩石类型比较简单，未出现次火山岩相岩石，岩浆分异不甚明显。

熔岩部分按 SiO_2 含量大致主要划分为两类。①SiO_2<45%的超基性熔岩类，岩石有橄榄霞石岩和玻基橄辉岩；②SiO_2 在45%~53%之间的基性岩类，岩石中 Na_2O 含量较高，且 $Na_2O>K_2O$，不含实际矿物白榴石，属一般玄武岩类，其命名原则按斑晶成分与结构构造命名，包括橄榄玄武岩、辉石玄武岩、气孔状玄武岩、粗玄岩等。超基性熔岩类在琼北新近纪盆地和南海盆地出现，分布范围不大。新生代火山岩绝大部分属于基性岩类，岩石呈渐变过渡，喷溢与爆发无明显间隔。此外，在南海的尖峰海山和黄岩海山有少量英安岩和安山岩分布。

火山碎屑岩的分类和命名，以《中国区域地质志》工作指南为基础，参考 GB/T17441.1—1998《火成岩岩石分类和命名方案》之《火山碎屑岩类岩石的分类》。

在 TAS 火山岩分类图上（图 13-1，图 13-2），新生代火山熔岩主要落入玄武岩区、粗面玄武岩区和玄武安山岩区，少量落入碱玄岩、碧玄岩区，个别落入副长石岩区和苦橄玄武岩区。

（一）橄榄霞石岩

见于琼山谭文一带和西沙群岛的高尖石岛。岩石呈深灰色，致密块状构造，偶见气孔状、杏仁状构造。貌似玄武岩，肉眼两者极难区分。镜下具斑状结构，斑晶成分为橄榄石（5%~9%），呈自形—半自形柱状粒状，大小0.1~0.5mm，部分晶体见环带结构，常受熔蚀，裂纹发育。基质呈自形—半自形柱粒结构，由钛次透辉石（44%~70%）、霞石（14%~41%）、橄榄石（2%~3%）和火山玻璃（≤1%）组成；副矿物为磁铁矿、钛铁矿、磷灰石。钛次透辉石微晶具浅褐色略带淡紫红色，长柱状，大小0.1mm

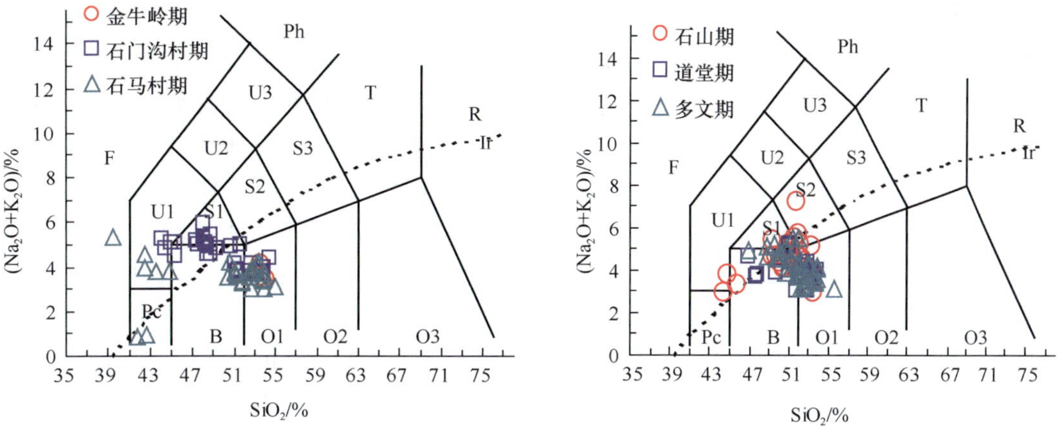

图 13-1 琼北新生代火山岩硅-碱图

Fig. 13-1 $SiO_2-Na_2O+K_2O$ diagram of Cenozoic volcanic rocks from Qiongbei

Pc—苦橄玄武岩；B—玄武岩；O1—玄武安山岩；O2—安山岩；O3—英安岩；R—流纹岩；S1—粗面玄武岩；S2—玄武质粗面安山岩；S3—粗面安山岩；T—粗面岩、粗面英安岩；F—副长石岩；U1—碱玄岩、碧玄岩；U2—响岩质碱玄岩；U3—碱玄质响岩；Ph—响岩；Ir—Irvine 分界线，上方为碱性系列，下方为亚碱性系列

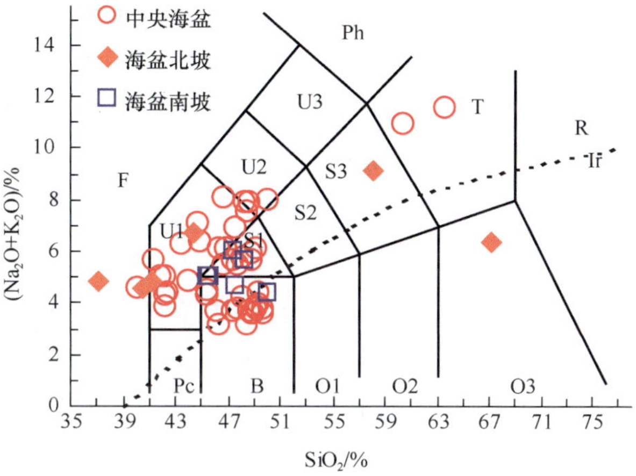

图 13-2 南海新生代火山岩硅-碱图

Fig. 13-2 $SiO_2-Na_2O+K_2O$ diagram of Cenozoic volcanic rocks from the South China Sea

Pc—苦橄玄武岩；B—玄武岩；O1—玄武安山岩；O2—安山岩；O3—英安岩；R—流纹岩；S1—粗面玄武岩；S2—玄武质粗面安山岩；S3—粗面安山岩；T—粗面岩、粗面英安岩；F—副长石岩；U1—碱玄岩、碧玄岩；U2—响岩质碱玄岩；U3—碱玄质响岩；Ph—响岩；Ir—Irvine 分界线，上方为碱性系列，下方为亚碱性系列

左右，常见环带和清晰的砂钟构造；霞石在薄片中呈正方形，有的呈他形粒状，均匀分布在基质中。

（二）玻基橄辉岩

主要出露于文昌蓬莱、定安居丁等地，此外在西沙群岛的高尖石岛也有分布。岩石呈深灰色—灰黑色，斑状结构，致密块状构造。斑晶成分以橄榄石（5%～15%）和单斜辉石（2%～6%）为主，偶见斜长石。橄榄石斑晶为无色或浅黄绿色，板状、等轴粒状或碎屑状，大小 0.1～0.9mm，偶见环带构造，裂纹发育，常见强烈熔蚀现象；单斜辉石斑晶为无色、浅褐色、淡紫红色，自形柱状，大小 0.03mm×0.15mm～0.06mm×0.6mm，普遍见砂钟构造（图版Ⅲ-2-B）和（100）的简单双晶。基质呈玻基柱粒结构，成分主要由单斜辉石（30%～40%）和火山玻璃（20%～45%）组成，少量橄榄石和碱性长石。单斜辉石微晶呈无色、淡褐色，以自形—半自形状为主，部分见他形粒状，大小 0.03～0.18mm，见砂钟构造和（100）简单双晶；橄榄石微晶呈无色或淡黄绿色，他形粒状，粒径小于 0.1mm；火山玻璃呈棕色—棕褐色，不具光性。

（三）橄榄玄武岩

包括橄榄玄武岩、辉石橄榄玄武岩、气孔状橄榄玄武岩等，在各时期熔岩中均见有出露，为新生代火山岩主要熔岩类。中新统石马村组和石门沟村组的橄榄玄武岩柱状节理发育。岩石呈深灰色—灰黑色，斑状结构，块状构造-致密块状构造、气孔或杏仁构造。斑晶成分有橄榄石（5%~15%）、辉石（≤10%）、斜长石（≤15%）。橄榄石斑晶呈厚板状、短柱状、粒状或碎粒状，大小一般在0.1~2.5mm之间，呈正高突起，糙面显著，一般｛010｝解理不完全，干涉色Ⅱ级蓝色—Ⅲ级红色，平行消光，裂纹发育，具熔蚀状，常沿边缘、裂隙、解理缝隙发生伊丁石化或蛇纹石化（图版Ⅲ-2-C），个别晶体完全发生伊丁石化呈棕红色，仅保留橄榄石的假象；辉石斑晶呈短柱状、粒状，大小0.1~3.0mm，正高突起，｛110｝解理发育完全，辉石式解理完全，干涉色为Ⅱ级蓝色，斜消光、对称消光，简单双晶常见，以普通辉石和透辉石为主，个别沿边缘、裂隙发生蛇纹石交代，辉石斑晶常包裹微晶橄榄石、斜长石及磁铁矿等；斜长石斑晶呈半自形板状、柱状，发育聚片双晶、钠长石双晶，双晶单体较宽，大小0.1mm×0.5mm~1mm×2mm，主要为中基性斜长石，个别斜长石斑晶包裹微小的磁铁矿、辉石等。基质呈间粒结构、间隐结构，成分有斜长石（45%~75%）、辉石（15%~30%）、橄榄石（≤5%）、磁铁矿（2%~5%）及玻璃质和隐晶质（≤10%）。斜长石微晶呈长条状、长柱状，大小0.01mm×0.05mm ~0.1mm×1.0mm不等，发育聚片双晶、钠长石双晶，双晶单体粗而宽，排列杂乱无章；辉石微晶呈粒状，少量短柱状，大小0.02~0.20mm，柱面解理完全，有的具辉石式解理，斜消光，以单斜晶系的透辉石，普通辉石为主；橄榄石微量呈近等轴粒状，大小0.05~0.2mm，大部分已完全伊丁石化，磁铁矿呈黑色不透明，呈长柱状、长条状，少量粒状；辉石微晶、橄榄石微晶和磁铁矿充填在斜长石构架的空隙中，构成基质的间粒结构。此外部分岩石尚见有少量玻璃质或隐晶质，玻璃质呈褐色—暗黑色不透明，非晶质体，不显光性；隐晶质呈浅棕褐色，不具光性、全消光，两者均匀充填在斜长石间隙中。

（四）辉石玄武岩

包括辉石玄武岩、橄榄辉石玄武岩、气孔状橄榄辉石玄武岩等，在各时期熔岩中均见有分布。岩石呈灰色—深灰色，斑状结构，块状构造。斑晶成分有辉石（5%~20%）、橄榄石（≤5%）和斜长石（≤10%），辉石斑晶中半自形板状、柱状，大小0.1~3.0mm，为无色—淡的浅绿色，柱状解理，辉石式解理发育完全，斜消光和对称消光，简单双晶常见，偶见聚片双晶，以单斜晶系的透辉石和普通辉石为主，辉石斑晶常包裹微晶橄榄石、磁铁矿等；橄榄石斑晶为半自形短柱状、板状或近等轴粒状，大小0.1~2.0mm，呈无色正高突起，糙面显著，一组解理不完全，平行消光，裂纹发育，具熔蚀状，常沿裂隙、解理隙、边缘发生伊丁石化，个别整个晶体均发生伊丁石化，呈棕红色，仅保留橄榄石假象，此外，个别岩石部分橄榄石斑晶有辉石反应边；斜长石斑晶呈半自形板状、柱状，大小0.1mm×0.3mm~1.0mm×2.0mm不等，发育聚片双晶、钠长石双晶，双晶单体较宽，偶见不发育的环带构造，以中基性斜长石为主。基质呈间粒结构、间粒-间隐结构、交织结构，成分有斜长石（50%~60%）、辉石（20%~35%）、橄榄石（≤8%）及磁铁矿（2%~8%）、玻璃质和隐晶质（≤5%）。斜长石微晶呈长柱状、长条状，大小0.01mm×0.05mm~0.1mm×1.0mm，发育聚片双晶和钠长石双晶，双晶单体粗且宽，主要呈杂乱无章分布，个别岩石可见部分呈平行或近平行排列，构成基质的间粒结构或交织结构；辉石微晶呈半自形—自形柱状、粒状，大小0.01~0.5mm，无色，大部分斜消光，以单斜晶系透辉石和普通辉石为主；橄榄石微晶为等轴粒状，边缘发生伊丁石化；磁铁矿为黑色不规则粒状、长条状、隐晶质和玻璃质为棕褐色—褐色团块，不具光性，辉石微晶、橄榄石微晶、磁铁矿、隐晶质和玻璃质均充填在斜长石微晶空隙之中。

（五）粗玄岩

岩石呈浅灰色—灰黑色，粗玄结构、斑状结构，块状构造或气孔状构造。斑晶成分有辉石

（≤5%）、橄榄石（≤3%）和斜长石（≤8%）。辉石斑晶呈短柱状，大小0.3mm×0.5mm~0.6mm×1.0mm，无色，正高突起，柱状解理发育完全，斜消光，具简单双晶；橄榄石斑晶呈近等轴粒状，大小0.3~0.8mm，无色，正高-正极高突起，糙面明显，干涉色为Ⅱ级绿色，具裂理，裂隙或裂理见伊丁石化；斜长石斑晶呈半自形—自形板状、柱状，大小0.2mm×0.5mm~0.5mm×2.0mm，普遍具钠长石双晶，双晶单体较宽。基质矿物成分为斜长石（50%~65%）、辉石（18%~32%）、橄榄石（≤15%）和磁铁矿（2%~3%）。斜长石呈半自形-自形板条状，长宽之比1：2~1：8，发育钠长石双晶，偶见卡钠复合联晶，双晶单体较宽，呈不规则分布，构成三角形孔隙；辉石呈半自形短柱状、粒状，粒径0.01~0.5mm，呈无色，正高突起，发育柱状解理、辉石式解理，可见简单双晶、聚片双晶，以普通辉石和透辉石为主；橄榄石为近等轴粒状，粒径0.05~0.3mm，无色正高—正极高突起，糙面明显，发育裂理，常沿边缘、裂理纹发生伊丁石化；磁铁矿呈黑色长条状、枝柱状或粒状。辉石、橄榄石、磁铁矿等充填在微晶斜长石构成的三角形孔隙中，构成岩石的粗玄结构。

（六）玄武岩

主要指橄榄石斑晶和辉石斑晶含量均小于1%的玄武岩类，各个时期均可见分布。岩石呈灰色—灰黑色，间粒结构（图版Ⅲ-2-F）、间粒-间隐结构、斑状结构，块状构造、气孔状构造（图版Ⅲ-2-E）。此外分布于南阳农场一带的玄武岩柱状节理发育（图版Ⅲ-2-D）。斑晶成分有斜长石（≤5%）、辉石（≤1%）、橄榄石（≤1%）。斜长石斑晶呈半自形柱状、板状，大小0.1mm×0.8mm~0.3mm×1.0mm，聚片双晶发育，双晶单体较宽；辉石斑晶呈半自形柱状，大小0.2~1.0mm，无色正高突起，柱面解理发育完全，个别见辉石式解理，斜消光，主要为单斜晶系的透辉石和普通辉石；橄榄石斑晶呈短柱状、近等轴粒状，大小0.2~0.6mm，正高突起，粗糙面，个别有一组解理不完全，干涉色呈Ⅱ级蓝色—Ⅲ级红色，裂纹发育，常沿边缘、解理缝隙或裂隙发生伊丁石化。基质成分有斜长石（50%~60%）、辉石（5%~35%）、橄榄石（≤10%）、磁铁矿（2%~4%）及玻璃质和隐晶质(≤40%)。斜长石微晶呈半自形—自形长条状、长柱状，大小0.02mm×0.1mm~0.2mm×1.5mm，普遍发育聚片双晶，个别见卡钠复合双晶，双晶单体较宽，以中基性长石为主，呈杂乱状分布；辉石微晶呈短柱状、粒状，粒径大小0.02~0.5mm，无色正高突起，柱面解理完全，辉石式解理常见，斜消光，对称消光，简单双晶和聚片双晶常见；橄榄石呈无色短柱状，近等轴粒状，大小0.1~0.5mm，糙面显著，一组解理不完全，平行消光，干涉色为Ⅱ级紫红色—Ⅲ级蓝色；磁铁矿为铁黑色，长条状、板状或不规则状粒状；玻璃质或隐晶质呈棕褐色，不具光性非均质体；辉石微晶、橄榄石微晶、磁铁矿及玻璃质和隐晶质均匀充填在斜长石微晶架构的孔隙之中，构成基质的间粒或间隐结构。

（七）粗面安山质浮岩

仅见于南海的尖峰海山。岩石由斜长石（An=35~45）、普通辉石斑晶及火山基质组成。

（八）粗面岩

仅见于南海的黄岩海山。岩石呈浅灰绿色，具少斑、聚斑结构，基质为交织结构。斑晶含量约为2%，主要由透长石组成，偶见透辉石和黑云母。透长石斑晶呈自形板状，正低突起，具细密聚片双晶和简单双晶；透辉石较自形，单偏光下呈淡绿色，无多色性，消光角较大；斑晶黑云母呈棕褐色，正中突起，具有一组细密的解理，多色性较强，平行消光。基质由略呈定向排列的斜长石微晶和玻璃质组成，未见蚀变现象。

（九）英安岩

仅见于南海的尖峰海山。岩石由斜长石微晶（An=30~35）、火山基质及少量聚斑石英组成。

（十）火山集块岩

多分布于火山口附近。岩石呈灰色—灰黑色，集块结构，块状构造。岩石主要由火山集块

（50%~65%）、火山角砾（10%~25%）和胶结物（5%~10%）组成。火山集块成分为玄武岩、火山渣，大小50~1200mm，呈棱角状、尖棱角状；火山角砾成分以玄武岩为主，少量火山渣，大小2~50mm，呈棱角状或次棱角状；胶结物主要由细小的火山碎屑物组成。

（十一）火山角砾岩

分布于火山口及稍远的地方，包括集块火山角砾岩。岩石呈灰色、灰黑色、褐红色等，角砾状结构、集块角砾状结构，块状构造。由火山集块（15%~30%）、火山角砾（50%~80%）和填隙物（5%~20%）组成。火山集块和火山角砾成分以玄武岩为主，次为玄武质火山渣和玄武质火山碎屑，火山集块大小50~300mm，火山角砾大小2~50mm，呈次棱角状—棱角状，少量为浑圆状；填隙物为火山弹、火山灰、岩屑、晶屑和玻屑等。

（十二）角砾凝灰岩

多环绕火山口四周分布，常与凝灰岩呈互层状产出，分布极广。岩石呈灰黄色、灰褐色、灰色等，角砾状凝灰结构，层状构造、块状构造。岩石主要由火山碎屑物组成，成分有岩屑（20%~45%）、晶屑（2%~30%）、玻屑（2%~60%）和胶结物组成。岩屑成分有玻基玄武岩、橄榄玄武岩、辉石玄武岩、火山渣及玄武质凝灰岩，大小一般在2~10mm，以呈棱角状为主，部分呈浑圆状和泪滴状。晶屑成分以石英晶屑为主，次为微斜长石，偶见斜长石、橄榄石、辉石和磁铁矿，石英晶屑呈棱角状，大小0.1~2.0mm，破碎、裂纹发育，边缘不规则；微斜长石呈棱角状，新鲜的可见格子双晶，大小0.4mm左右。玻屑为淡黄色、深褐色，大小一般在0.05~2.0mm，呈多孔状或各种形态的弧面棱角状。胶结物以火山灰为主，由小于0.05mm的火山玻屑和晶屑组成。

（十三）沉凝灰岩

包括沉复屑凝灰岩、沉晶屑玻屑凝灰岩、沉晶屑凝灰岩。岩石呈淡黄色、黄灰色和灰绿色，凝灰结构，层状构造。岩石成分由岩屑（≤40%）、晶屑（≤30%）、玻屑（10%~40%）和凝灰质或胶结物（30%~80%）组成。岩屑以玻璃质玄武岩为主，呈黑色—深褐色，多孔状—浮岩状，外形则呈棱角状—次圆状，大小0.05~2.0mm之间；此外还有少量正常沉积岩屑，成分以硅质岩和生物灰岩为主。晶屑以石英和钾长石为主，次为橄榄石、辉石和斜长石等，呈棱角-次棱角状，大小0.05~1.3mm，石英晶屑边缘不规则，常具不规则裂纹，玻状消光，有的因受熔蚀而呈港湾状；钾长石晶屑普遍具不强的泥化，格子双晶清晰，多见熔蚀边。玻屑形态不规则，大小不一，一般在0.05~0.5mm之间，多呈扁平条带状、撕裂状、扭曲状等。胶结物为火山岩、火山尖及少量细小的晶屑，呈红褐色—褐色，不具光性。

第二节　火山岩的地球化学特征及其形成的构造环境

一、岩石化学特征

岩石（熔岩）化学分析结果、标准矿物成分含量及部分岩石化学特征比值见表13-5~表13-7。其中，在进行标准矿物计算时，对化学成分中过多的Fe_2O_3进行调整，按$Fe_2O_3^* = 0.64TiO_2+2.5$方法控制$Fe_2O_3$的上限。如果岩石中$Fe_2O_3 < 0.64TiO_2+2.5$，可不调整；而$Fe_2O_3 > 0.64TiO_2+2.5$，把多余的$Fe_2O_3$按$0.9\times(Fe_2O_3^* - Fe_2O_3)$，换算成FeO，同时把$CO_2$结合转为$CaCO_3$，在调整CaO、$Fe_2O_3$、FeO及取消$H_2O$、$CO_2$后，再把氧化物之和重新换算成100%，并计算它们的CIPW标准矿物成分。

表 13-5　海南岛新近纪火山岩化学成分、CIPW 标准矿物含量及岩石化学常用参数一览表

Table 13-5　List of chemical composition, CIPW standard minerals and common petrochemical parameters of Neogene volcanic rocks from the Hainan Island

地层单位		海口组			石门沟村组								
岩性		石英拉斑玄武岩			粗玄岩		辉石玄武岩		碱性橄榄玄武岩				
样品号		金 YQ_2	GK-2	1-4	D0046	P2-14	$PM001_5$	$PM001_6$	P2-13	242-1		HNG-7	217-4
序号		1	2	3	4	5	6	7	8	9	10	11	12
岩石化学成分/%	SiO_2	53.76	54.05	53.28	51.34	52.50	51.02	53.22	48.01	50.98	48.46	44.45	47.30
	Al_2O_3	13.91	13.70	14.17	13.53	14.13	14.02	14.01	12.20	13.80	10.85	12.85	14.09
	Fe_2O_3	2.76	2.55	5.11	2.86	2.56	3.28	2.78	3.75	3.43	10.23	1.22	2.77
	FeO	8.18	7.88	5.28	7.70	7.85	7.45	7.65	7.91	7.37	6.03	10.96	8.76
	MgO	7.58	6.63	6.41	8.73	6.40	7.44	6.61	11.13	7.64	9.22	9.54	8.31
	CaO	7.21	8.25	8.17	8.53	8.01	8.31	8.30	6.57	7.09	7.18	9.22	8.41
	Na_2O	3.0	2.85	3.34	2.89	2.83	3.04	2.97	3.25	3.56	3.06	3.05	3.40
	K_2O	0.52	0.60	0.82	0.96	0.92	1.12	0.79	2.03	1.34	1.84	1.80	1.73
	MnO	0.11	0.16	0.14	0.14	0.15	0.19	0.25	0.14	0.14	0.07	0.20	0.30
	TiO_2	2.00	1.84	1.91	1.78	2.12	2.03	1.87	2.52	2.08	2.15	2.68	2.61
	P_2O_5	0.39	0.80	0.25	0.26	0.25	0.34	0.18	0.38	0.32		1.76	0.53
	CO_2			0.10	0.36	0.25	0.10	0.31	0.03		0.08		0.60
	H_2O^+	0.78	0.10	0.50	0.63	0.82	1.31	0.78	0.96	2.16		1.77	1.10
	总量	100.28	99.41	99.68	99.71	98.43	99.65	99.72	98.91	99.92	99.38	99.52	99.91
标准矿物成分含量/%	Q	6.46	8.21	5.64	0.31	5.70	1.26	5.44					
	Or	3.09	3.57	4.90	5.75	5.57	6.74	4.74	12.30	8.11	11.05	10.89	10.42
	Ab	25.50	24.25	28.53	24.74	24.48	26.15	25.45	28.10	30.78	26.27	17.60	24.74
	An	23.06	22.94	21.45	21.35	23.64	21.65	22.84	13.00	18.09	10.57	16.40	18.37
	Ne											4.75	2.45
	Di	8.61	11.02	14.49	16.09	12.62	14.74	14.48	14.40	12.85	20.89	16.17	16.94
	Lc												
	Hy	24.57	21.01	15.32	23.56	19.51	19.93	18.96	0.39	18.45	2.55		
	Ol								20.6	1.88	18.81	23.23	16.77
	Ap	0.86	1.76	0.55	0.57	0.56	0.76	0.40	0.85	0.71		3.93	1.18
	Ilm	3.82	3.52	3.67	3.43	4.12	3.93	3.60	4.91	4.04	4.15	5.21	5.05
	Mt	4.03	3.72	5.45	4.20	3.80	4.84	4.09	5.55	5.09	5.71	1.81	4.09
岩石化学常用参数	$Mg^{\#}$	55.89	53.74	53.64	64.06	52.91	60.01	57.73	63.80	56.57	51.90	58.52	56.83
	AlK	3.52	3.45	4.16	3.85	3.75	4.16	3.76	5.28	4.92	4.90	4.85	5.13
	Na_2O/K_2O	5.77	4.75	4.07	3.01	3.08	2.71	3.76	1.60	2.66	1.66	1.69	1.97
	TFeO/MgO	1.41	1.53	1.54	1.18	1.59	1.40	1.54	1.01	1.37	1.65	1.26	1.35
	δ	1.15	1.08	1.65	1.78	1.48	2.16	1.38	5.56	3.01	4.13	16.34	6.12
	τ	5.46	5.90	5.67	5.98	5.33	5.41	5.90	3.55	4.05	3.50	3.66	4.10
	AR	1.40	1.37	1.46	1.42	1.41	1.46	1.41	1.79	1.61	1.76	1.56	1.59
	SI	34.39	32.33	30.58	37.73	31.13	33.32	31.78	39.65	32.73	30.35	35.91	33.28
	CI	41.03	39.26	37.44	47.28	40.05	43.44	41.13	45.85	42.57	52.01	46.43	42.85
	DI	35.06	36.03	40.60	30.80	35.75	34.16	35.63	40.35	38.71	37.22	33.25	37.61
采样地点		金牛岭			黑山岭	庙岭	蓬莱				日晒岭		蓬莱

地层单位		石门沟村组										
岩性		橄榄玄武岩										
样品号		D1871-1	D1836	D1835-1	P2-11	HNG-3	D0460	蓬-3	福基田-1	81-H-1	81-H-3	HNG-9
序号		13	14	15	15	17	18	19	20	21	22	23
岩石化学成分/%	SiO_2	44.10	54.37	49.36	53.80	48.07	52.64	48.29	48.74	47.49	51.64	45.30
	Al_2O_3	12.48	15.82	13.10	13.85	13.62	14.17	13.61	13.53	15.17	15.49	15.65
	Fe_2O_3	3.77	3.40	2.92	2.35	2.47	3.28	3.57	3.49	2.41	2.34	5.96
	FeO	8.05	5.53	8.25	6.81	8.81	7.50	8.43	8.15	8.78	7.05	7.04
	MgO	11.56	4.73	9.57	6.09	7.62	6.60	8.65	8.36	7.85	5.76	10.30
	CaO	9.06	7.13	7.03	7.67	7.20	8.19	8.10	7.28	7.32	9.25	6.40
	Na_2O	3.18	3.75	2.73	2.68	4.12	3.16	3.43	3.74	124	3.34	3.10
	K_2O	2.11	0.66	2.08	0.98	1.80	0.94	1.62	1.72	1.58	1.64	1.36
	MnO	0.20	0.13	0.16	0.13	0.16	0.15	0.17	0.15	0.13	0.04	0.10
	TiO_2	2.32	1.72	2.24	2.00	3.00	1.91	2.74	2.59	2.47	1.99	2.20
	P_2O_5	1.12	0.28	0.78	0.27	1.98	0.25	0.56	0.66	0.93	0.35	0.70
	CO_2	0.08	0.37	0.23	0.37	0.06	0.20		0.06	0.42	0.01	0.11
	H_2O^+	1.62	0.91	1.26	0.98	1.55	0.77	0.74	1.17	1.42	0.33	1.62
	总量	99.65	98.80	99.91	97.98	100.46	99.76	99.91	99.64	99.41	99.23	99.84
标准矿物成分含量/%	Q		8.49		9.46		3.99					
	Or	12.74	4.00	12.50	6.00	10.77	5.63	9.66	10.34	9.58	9.81	8.22
	Ab	12.13	32.50	23.50	23.44	35.23	27.03	29.23	32.12	29.80	28.55	26.76
	An	13.80	24.97	17.60	23.63	13.48	21.94	17.06	15.26	21.78	22.64	25.26
	Ne	8.30										
	Di	20.26	7.75	10.60	11.42	8.82	14.33	16.19	14.07	7.99	17.60	2.40
	Lc											
	Hy		13.26	16.60	17.98	0.53	18.04	0.06	0.97	3.55	12.01	2.04
	Ol	20.20		8.73		17.41		16.09	15.63	16.83	1.37	23.70
	Ap	2.50	0.63	1.73	0.61	4.37	0.55	1.23	1.46	2.08	0.77	1.56
	Ilm	4.50	3.35	4.33	3.93	5.77	3.67	5.25	5.00	4.81	3.82	4.27
	Mt	5.58	5.06	4.31	3.53	3.62	4.81	5.22	5.14	3.58	123	5.79
岩石化学常用参数	$Mg^{\#}$	64.30	49.54	61.07	54.89	55.18	56.98	56.98	60.83	60.06	56.89	59.69
	AlK	5.29	4.41	4.81	3.66	5.92	4.10	5.05	5.46	5.02	4.98	4.46
	Na_2O/K_2O	1.51	5.68	1.31	2.73	2.29	3.36	2.12	2.17	2.18	2.04	2.28
	TFeO/MgO	0.99	1.82	1.14	1.47	1.45	1.58	1.35	1.35	1.39	1.59	1.20
	δ	0.34	1.71	3.64	1.24	6.91	1.74	4.82	5.19	5.61	2.87	8.65
	τ	4.01	7.02	4.63	5.59	3.17	5.76	3.72	3.78	4.75	6.11	5.70
	AR	1.65	1.48	1.63	1.41	1.79	1.45	1.61	1.71	1.57	1.50	1.51
	SI	40.32	26.18	37.46	32.21	30.70	30.73	33.66	32.84	32.63	28.61	37.10
	CI	50.54	29.41	44.63	36.86	36.15	47.28	42.81	40.82	36.77	38.23	38.33
	DI	33.17	44.99	36.02	38.90	46.00	30.80	38.89	42.46	39.38	38.35	34.90
采样地点		佳永村	庙岭	高塘村	庙岭		蓬莱		福基田		牛姆岭	

续表

地层单位		石门沟村组							石马村组				
岩性		橄榄玄武岩							橄榄玄武岩				
样品号		D0083	九-1	JIUB1	ZK_2-1	$PM001_9$	ZK_3-1	新ZK_1-1	D0752	D1872-1	ZK_{37}-1	P2-4a	
序号		24	25	26	27	28	29	30	31	32	33	34	35
岩石化学成分/%	SiO_2	53.92	48.71	48.38	51.23	48.69	45.39	44.91	52.56	52.23	44.81	51.70	50.52
	Al_2O_3	13.96	12.31	12.68	13.62	13.48	13.19	12.92	14.16	14.10	13.29	14.58	11.97
	Fe_2O_3	1.85	3.30	4.05	3.68	3.90	4.91	4.36	2.00	3.20	6.04	2.93	4.10
	FeO	8.00	8.15	7.23	6.20	8.30	7.20	7.80	8.40	7.08	6.91	7.98	7.18
	MgO	6.59	10.07	10.16	8.44	8.40	7.69	7.46	7.31	7.29	7.12	7.03	7.71
	CaO	8.37	7.24	6.76	8.55	7.93	9.96	10.11	8.72	8.10	10.54	6.31	7.60
	Na_2O	3.09	3.15	2.64	2.77	2.99	3.27	3.22	2.93	3.06	2.30	2.76	2.83
	K_2O	0.88	1.87	1.91	1.03	1.54	1.83	1.85	0.82	1.09	1.47	0.52	1.35
	MnO	0.19	0.156	0.12	0.17	0.18	0.21	0.21	0.15	0.12	0.34	0.14	0.17
	TiO_2	1.89	2.43	2.29	1.93	2.73	2.47	2.62	1.78	1.95	2.63	1.95	2.05
	P_2O_5	0.18	0.71	0.67	0.28	0.42	1.14	1.38	0.24	0.28	0.93	0.26	0.39
	CO_2	0.31	0.13	烧失 2.93	0.31	0.10	0.12	0.02	0.08	0.24	0.89	0.01	0.03
	H_2O^+	0.51	1.19		1.50	1.02	2.30	2.44	0.57	1.04	2.60	2.32	2.52
	总量	99.74	99.42	99.82	99.71	99.68	99.60	99.28	99.72	99.78	99.87	98.49	98.42
标准矿物成分含量/%	Q	4.66			2.63				2.41	3.10		7.20	3.01
	Or	5.26	11.28	11.66	6.22	9.24	11.08	11.30	4.90	6.55	9.03	3.20	8.33
	Ab	26.40	26.14	23.04	23.91	25.64	19.71	20.48	25.00	26.26	20.19	24.26	24.96
	An	21.83	14.07	17.63	22.12	19.05	16.25	15.81	23.25	21.81	22.39	26.86	16.63
	Ne						4.65	4.13					
	Di	15.56	14.59	10.34	15.56	14.75	22.17	22.21	15.47	13.94	21.24	3.35	16.33
	Lc												
	Hy	19.56	7.20	18.27	19.73	12.23			22.11	19.26	3.77	26.28	20.02
	Ol		14.47	7.59		7.15	12.72	11.55			9.79		
	Ap	0.40	1.58	1.51	0.62	0.93	2.55	3.11	0.53	0.62	2.11	0.59	0.89
	Ilm	3.63	4.71	4.49	3.75	5.26	4.80	5.14	3.41	3.76	5.19	3.85	4.06
	Mt	2.71	4.88	5.48	5.45	5.74	6.06	6.26	2.93	4.71	6.30	4.42	5.76
岩石化学常用参数	$Mg^{\#}$	58.85	61.75	66.49	65.05	59.87	54.13	53.15	60.05	56.62	50.70	54.14	55.84
	AlK	3.97	5.02	4.55	3.80	4.53	5.10	5.07	3.75	4.15	3.77	3.28	4.18
	Na_2O/K_2O	3.51	1.68	1.38	2.69	1.94	1.79	1.74	3.57	2.81	1.56	5.31	2.10
	TFeO/MgO	1.47	1.10	1.07	1.13	1.41	1.51	1.57	1.40	1.37	1.73	1.51	1.41
	δ	1.44	4.41	3.85	1.75	3.61	10.88	13.46	1.47	1.87	7.85	1.24	2.32
	τ	5.75	3.77	4.38	5.62	3.84	4.02	3.70	6.31	5.66	4.18	6.06	4.46
	AR	1.43	1.69	1.61	1.41	1.54	1.57	1.56	1.39	1.46	1.38	1.37	1.54
	SI	32.29	37.94	39.11	38.16	33.43	30.88	30.21	34.06	33.56	29.87	33.13	33.28
	CI	44.66	45.84	46.18	44.49	45.14	45.75	45.17	43.92	41.67	46.28	37.90	33.28
	DI	38.28	38.41	34.69	32.77	34.88	35.45	35.92	32.30	35.90	29.22	34.66	36.30
采样地点		水容村	九子陆岭		高塘村	蓬莱岭	新安村		古井塘	古井园	蓬莱		

地层单位		石马村组										
岩性		橄榄玄武岩		石英拉斑玄武岩					玄武岩			
样品号		P2-4b	D1900	531	221-1	194-2	D35-1	D39-1	$PM005_{6-5}$	D2174	D2297-2	D3461
序号		36	37	38	39	40	41	42	43	44	45	46
岩石化学成分/%	SiO_2	50.34	43.61	52.90	53.83	53.29	53.93	54.93	51.06	52.76	51.77	51.01
	Al_2O_3	13.62	12.60	12.95	13.07	14.35	12.63	13.56	13.62	14.05	13.90	14.02
	Fe_2O_3	1.55	5.90	3.10	1.45	1.97	3.23	3.52	4.35	1.35	3.64	3.86
	FeO	10.13	6.05	7.33	8.70	8.80	7.70	6.72	6.35	8.80	7.00	7.05
	MgO	6.45	9.94	6.64	8.01	5.70	7.48	6.55	8.21	7.62	7.46	7.49
	CaO	8.08	10.57	8.73	8.32	8.89	7.75	7.59	8.39	8.27	8.80	8.90
	Na_2O	2.84	3.04	2.98	3.00	3.28	2.49	2.83	2.89	2.92	2.70	2.91
	K_2O	0.64	0.72	0.70	0.70	0.93	0.52	0.30	0.86	1.06	0.66	0.75
	MnO	0.21	0.21	0.18	0.18	0.15	0.17	0.14	0.14	0.14	0.13	0.15
	TiO_2	1.81	2.30	1.98	1.82	1.72	1.60	1.70	1.87	1.84	1.81	1.90
	P_2O_5	0.26	0.86	0.29	0.21	0.28	0.16	0.23	0.33	0.28	0.26	0.29
	CO_2	2.06	0.19	其他：0.07	0.33	0.09			0.12	0.10	0.21	0.10
	H_2O^+	1.02	3.61	0.65	0.59	0.63			1.59	0.55	1.39	1.30
	总量	99.05	99.62	98.40	100.21	100.08	98.67	99.55	99.72	99.74	99.73	99.73
标准矿物成分含量/%	Q	1.81		6.21	3.26	2.88	9.55	11.74	2.64	1.50	5.01	2.64
	Or	4.19	4.45	4.23	4.17	5.54	3.15	1.81	5.21	6.33	3.98	4.51
	Ab	25.01	19.25	25.76	25.54	27.90	21.55	24.39	25.00	24.91	23.25	25.01
	An	23.32	19.43	20.31	20.24	21.79	22.24	23.84	22.13	22.27	24.28	23.34
	Ne		4.13									
	Di	13.92	24.09	17.98	16.35	17.23	13.08	10.62	14.79	14.26	15.00	15.95
	Lc											
	Hy	25.25		16.41	24.38	17.89	22.17	18.59	20.37	24.63	19.03	18.74
	Ol		16.09									
	Ap	0.59	1.96	0.65	0.46	0.62	0.36	0.51	0.74	0.62	0.58	0.64
	Ilm	3.58	4.57	3.85	3.48	3.29	3.11	23.29	3.64	3.53	3.50	3.67
	Mt	2.32	6.02	4.60	2.12	2.87	4.80	5.20	5.49	1.98	5.38	5.49
岩石化学常用参数	$Mg^{\#}$	49.95	60.94	53.91	58.80	49.01	55.70	54.15	58.78	57.56	60.36	55.93
	AlK	3.52	3.76	3.68	3.70	4.21	3.01	3.13	3.75	3.98	3.36	3.66
	Na_2O/K_2O	4.18	4.22	4.26	4.29	3.53	4.79	9.43	3.36	2.75	4.09	3.88
	TFeO/MgO	1.79	1.14	1.52	1.25	1.85	1.42	1.51	1.25	1.31	1.38	1.40
	δ	1.69		1.83	1.26	1.72	0.83	0.82	1.74	1.62	1.29	1.67
	τ	5.96	4.12	5.04	5.53	6.44	6.34	6.31	5.74	6.05	6.19	5.85
	AR	1.39	1.39	1.41	1.42	1.44	1.35	1.35	1.41	1.43	1.35	1.38
	SI	29.79	38.75	32.00	36.64	27.56	34.92	32.88	36.23	35.03	34.76	33.95
	CI	45.09	50.77	42.84	46.33	41.28	43.16	37.71	44.29	44.39	42.90	43.85
	DI	31.01	27.83	36.20	32.96	36.31	34.24	37.94	32.85	32.73	32.24	32.17
采样地点		蓬莱		蓬莱	高明	树德	南阳		大坡	高浪村	南阳农场	禄来村

续表

地层单位		石马村组										平均值	中国东部玄武岩平均值（赵宗溥，1956）
岩性		玄武岩			辉石玄武岩		玻基辉橄岩				橄榄霞石岩		
样品号		D1831-1	D1832-1	D1834-1	PM00119-1	$Ⅱ_{H17-1}$	ZK_2-2	ZK_2-3	HNG-11	212-1	185-1		
序号		47	48	49	50	51	52	53	54	55	56		
岩石化学成分/%	SiO_2	52.65	51.19	53.85	52.53	53.77	41.73	42.65	42.50	42.52	39.40	49.75	47.56
	Al_2O_3	13.99	13.85	14.80	13.76	14.67	13.57	13.22	13.25	12.74	10.74	13.55	14.94
	Fe_2O_3	4.13	3.10	5.70	1.14	2.70	11.92	9.99	3.14	4.55	5.68	3.71	4.75
	FeO	7.20	7.35	4.90	8.60	8.13	3.25	3.98	8.94	7.98	8.82	7.51	6.62
	MgO	7.01	7.60	5.71	6.57	5.95	8.67	9.05	10.50	9.59	11.83	7.89	7.00
	CaO	8.57	8.68	8.64	8.86	8.52	9.13	10.4	10.00	11.16	11.73	8.41	8.50
	Na_2O	2.48	2.71	2.65	2.89	2.88	0.52	0.58	2.80	2.54	3.84	2.92	3.60
	K_2O	0.50	0.87	0.77	0.65	0.45	0.35	0.40	1.16	2.02	1.48	1.16	2.44
	MnO	0.14	0.15	0.14	0.15	0.15	0.21	0.21	0.20	0.22	0.19	0.18	0.14
	TiO_2	1.80	1.84	1.69	1.64	1.59	2.44	2.36	2.63	2.14	2.83	2.12	2.19
	P_2O_5	0.22	0.33	0.24	0.19	0.18	0.76	0.72	1.18	0.90	0.19	0.18	0.14
	CO_2	0.06	0.11	0.07	2.14	0.02	0.31	0.20			其他		
	H_2O^+	1.07	1.30	0.67	0.62	0.81	5.46	5.01	3.05	3.45	1.57		1.82
	总量	99.82	99.80	99.83	99.74	99.81	98.32	98.77	99.35	99.81	99.06		
标准矿物成分含量/%	Q	7.75	3.24	9.27	3.75	7.32	3.12	2.45					
	Or	3.00	5.27	4.61	3.96	2.69	2.26	2.55	7.12	12.41			
	Ab	21.25	23.45	22.35	25.19	24.59	4.79	5.27	14.80	5.52			
	An	25.88	23.57	26.47	23.33	26.00	36.64	34.69	20.90	18.03	7.89		
	Ne								5.30	9.09	18.07		
	Di	12.87	14.95	12.51	17.09	12.79	6.40	13.15	18.82	27.33	37.03		
	Lc										7.05		
	Hy	19.93	20.61	15.47	21.34	19.21	33.52	29.12					
	Ol								20.47	15.53	16.88		
	Ap	0.49	0.74	0.53	0.43	0.40	1.81	1.69	2.68	2.04	2.13		
	Ilm	3.47	3.58	3.25	3.21	3.05	5.05	4.82	5.19	4.22	5.52		
	Mt	5.37	4.60	5.25	1.70	3.95	6.42	6.25	4.73	5.83	6.42		
岩石化学常用参数	$Mg^{\#}$	53.38	57.20	50.37	58.88	50.11	56.55	59.41	61.41	58.61	60.22		
	AlK	2.98	3.58	3.42	3.54	3.33	0.87	0.98	3.96	4.56	5.32		
	Na_2O/K_2O	4.96	3.11	3.44	4.44	6.40	1.49	1.45	2.41	1.56	2.60		
	TFeO/MgO	1.56	1.33	1.76	1.47	1.77	1.61	1.43	1.12	1.26	1.18		
	δ	0.92	1.44	1.08	1.31	1.03	-0.60	-2.74	-31.4	-43.3	-7.86		
	τ	6.39	6.05	7.19	6.63	7.46	5.35	5.36	3.97	4.77	2.44		
	AR	1.30	1.38	1.34	1.37	1.34	1.08	1.09	1.41	1.47	1.62		
	SI	32.88	35.14	28.94	33.10	29.59	36.25	38.68	39.56	35.94	37.38		
	CI	41.63	43.73	26.47	43.35	39.01	51.91	54.39	49.20	52.92	65.85		
	DI	32.00	31.96	36.52	32.90	34.59	8.95	8.76	27.22	27.02	25.12		
采样地点		三门坡	南阳	美文村	墨格村	美夏村	高塘村		蓬莱	黄典园	潭文村		

表 13-6 海南岛第四纪火山岩化学成分、CIPW 标准矿物含量及岩石化学常用参数一览表

Table 13-6 List of chemical composition, CIPW standard minerals and common petrochemical parameters of Quaternary volcanic rocks from the Hainan Island

地层单位		石山组											
岩性		石英拉斑玄武岩				橄榄拉斑玄武岩			橄榄玄武岩				
样品号		长 YQ_1	道 YQ1	文 YQ	412	Ⅱ40	永兴 B	笔架山	HNG-5	雷-1	群-1	马-1	133-1
序号		1	2	3	4	5	6	7	8	9	10	11	12
岩石化学成分 %	SiO_2	52.44	53.23	52.21	49.75	50.17	49.39	51.56	51.00	51.28	51.99	51.69	51.04
	Al_2O_3	13.65	14.48	14.14	14.50	14.06	12.91	16.60	14.85	14.78	14.07	14.47	14.52
	Fe_2O_3	2.34	3.13	3.03	9.55	1.68	1.66	1.56	9.27	4.01	1.47	2.20	3.06
	FeO	9.10	7.94	7.98	1.72	9.10	8.90	7.80	2.43	7.56	8.86	7.90	7.42
	MgO	7.23	6.70	6.56	6.42	8.52	9.20	7.44	6.20	5.56	6.77	6.00	6.45
	CaO	8.47	7.86	7.83	8.67	9.12	8.89	7.44	8.00	9.14	8.94	8.77	9.04
	Na_2O	3.11	3.25	3.31	3.24	2.92	2.90	3.12	3.36	3.38	3.38	3.73	129
	K_2O	1.43	1.92	1.90	1.45	1.48	1.80	1.40	1.36	1.80	1.31	1.74	1.71
	MnO	0.17	0.14	0.16	0.18	0.19	0.20	0.05	0.08		0.15	0.15	0.15
	TiO_2	2.30	2.40	2.40	2.63	2.58	2.45	1.40	2.06	2.25	2.31	2.47	2.23
	P_2O_5	0.45	0.25	0.30	0.45	0.53	0.60	0.30	0.41		0.37	0.46	0.19
	CO_2							其他：					
	H_2O^+	0.04	0.22	0.16	1.14	0.22	0.78	0.69	1.00		0.38	0.34	0.40
	总量	100.73	100.65	99.98	99.70	100.69	99.68	99.36	100.02	99.76	100.11	99.94	100.05
标准矿物成分含量 %	Q	0.32	0.93	0.11	0.60				1.13				
	Or	8.40	11.21	11.26	8.57	8.72	10.77	8.39	8.17	10.67	7.78	10.33	10.15
	Ab	26.10	27.12	28.03	27.94	24.59	24.78	26.72	28.84	28.64	28.68	31.65	29.60
	An	18.90	18.97	18.11	21.22	20.78	17.05	27.48	21.72	19.85	19.39	17.69	18.93
	Ne												
	Di	16.47	14.63	15.43	15.95	17.21	19.41	6.63	13.04	20.72	18.75	18.94	18.74
	Hy	21.12	17.62	17.44	13.42	9.68	4.50	21.27	16.59	8.03	14.50	6.36	7.36
	Ol					10.56	15.04	3.85		1.99	3.55	6.10	5.35
	Ap	0.98	0.54	0.66	1.00	1.15	1.32	0.66	0.91	0	0.81	1.01	1.07
	Ilm	4.34	4.50	4.57	5.10	4.88	4.71	2.70	3.97	4.29	4.41	4.71	4.35
	Mt	3.37	4.48	4.40	6.19	2.43	2.43	2.29	5.63	5.83	2.14	3.20	4.45
岩石化学常用参数	$Mg^{\#}$	53.50	52.62	52.21	52.60	58.87	61.21	59.04	50.64	47.02	54.24	51.99	53.06
	AlK	4.54	5.17	5.21	4.69	4.40	4.70	4.52	4.72	5.18	4.69	5.47	5.20
	Na_2O/K_2O	2.17	1.69	1.74	2.23	1.97	1.61	2.23	2.47	1.88	2.58	2.14	2.04
	TFeO/MgO	1.55	1.61	1.63	1.61	1.25	1.13	1.26	1.74	2.01	1.50	1.65	1.58
	δ	2.18	2.61	2.95	3.22	2.70	3.46	2.39	2.78	3.24	2.45	3.44	3.36
	τ	4.58	4.68	4.51	4.28	4.32	4.09	9.63	5.58	5.07	4.63	4.36	4.84
	AR	1.52	1.60	1.62	1.50	1.47	1.55	1.46	1.52	1.55	1.51	1.61	1.57
	SI	31.15	29.21	28.80	28.72	35.95	37.61	34.90	27.41	24.92	31.07	27.82	29.15
	CI	45.30	41.24	41.84	40.66	44.76	46.08	36.74	39.23	40.84	43.35	39.32	40.24
	DI	34.83	39.26	39.40	37.12	33.32	35.55	35.12	38.14	39.31	36.45	41.98	39.75
采样地点		长流	道堂	文明	石山	乐光	永兴	笔架山	石山	雷虎岭	石山		

续表

地层单位		石山组									
岩性		橄榄玄武岩					碱性橄榄玄武岩				
样品号		HNG-12	D1874	HN9902	106B1	QB011	HN9910	QB013	QB012	128-2	23-3
序号		13	14	15	15	17	18	19	20	21	22
岩石化学成分/%	SiO_2	52.00	50.71	50.51	50.52	50.51	44.77	45.73	44.38	49.37	51.86
	Al_2O_3	16.00	13.80	14.91	14.20	14.33	12.78	13.67	13.58	13.67	14.48
	Fe_2O_3	2.47	2.23	4.96	2.46	2.89	4.82	6.84	5.38	1.99	2.18
	FeO	7.98	8.18	5.31	8.41	7.50	7.65	5.54	6.93	8.60	7.78
	MgO	5.60	8.24	6.71	7.40	6.78	10.01	9.27	9.44	7.55	6.72
	CaO	7.96	8.53	9.11	9.19	9.10	10.63	9.93	10.46	9.52	6.73
	Na_2O	4.04	3.23	2.78	2.68	2.66	2.38	2.03	2.02	3.39	4.56
	K_2O	1.68	1.63	1.44	1.37	1.57	1.41	1.34	0.89	2.04	2.68
	MnO	0.10	0.14	0.25	0.14	0.13	0.31	0.14	0.14	0.16	0.13
	TiO_2	2.12	2.20	2.34	2.25	2.30	2.72	2.54	2.57	2.58	1.98
	P_2O_5	0.37	0.42	0.46	0.47	0.48	0.81	0.65	0.63	0.52	0.56
	CO_2		0.08	烧失量：1.41	烧失量：0.93	烧失量：1.36	烧失量：1.55	烧失量：1.93	烧失量：3.11	0.06	0.02
	H_2O^+	0.22	0.36								0.64
	总量	100.42	99.75	99.73	100.02	99.61	100.11	99.61	99.53	99.97	100.03
标准矿物成分含量/%	Q			2.38		1.36					
	Or	9.91	9.71	8.63	8.18	9.45	8.49	8.14	5.47	12.15	15.91
	Ab	34.04	27.49	23.81	22.90	22.90	17.41	17.61	17.73	24.11	32.07
	An	20.46	18.44	24.23	22.80	22.90	20.35	24.84	26.30	16.14	11.13
	Ne						1.66			2.57	3.58
	Di	13.67	17.54	15.03	16.50	16.30	22.78	17.46	19.00	23.00	15.49
	Hy	5.50	8.82	14.52	20.50	17.30		10.42	7.87		
	Ol	8.04	9.63		0.11		15.99	8.98	10.90	13.18	13.66
	Ap	0.81	0.92	1.02	1.04	1.07	1.80	1.46	1.43	1.14	1.23
	Ilm	4.02	4.21	4.51	4.31	4.45	5.26	4.95	5.07	4.94	3.77
	Mt	3.57	3.26	5.88	3.60	4.27	6.26	6.15	6.23	2.77	3.17
岩石化学常用参数	$Mg^{\#}$	49.46	59.05	55.04	55.40	54.50	59.82	58.56	58.84	56.63	55.15
	AlK	5.72	4.86	4.22	4.05	4.23	3.79	3.37	2.91	5.43	7.24
	Na_2O/K_2O	2.40	1.98	1.93	1.96	1.69	1.69	1.51	2.27	1.66	1.70
	TFeO/MgO	1.82	1.24	1.46	1.44	1.49	1.20	1.26	1.25	1.38	1.48
	δ	3.64	3.06	2.37	2.18	2.38	8.12	4.16	6.14	4.63	5.92
	τ	5.64	4.80	5.18	5.12	5.07	3.82	4.58	4.50	3.98	5.01
	AR	1.63	1.56	1.43	1.42	1.44	1.39	1.33	1.28	1.61	2.04
	SI	25.72	35.05	31.65	33.20	31.70	38.10	37.05	38.28	32.16	28.09
	CI	34.80	43.45	39.93	45.10	42.40	50.29	47.96	49.08	43.89	36.09
	DI	43.94	37.20	34.82	31.10	33.70	27.56	25.75	23.19	38.83	51.55
采样地点		石山	永兴	马鞍岭	博昌村	包子岭	雷虎岭	昌道岭	春藏岭	永兴	笔架山

续表

地层单位		石山组		道堂组								
岩性		玄武岩		碱性橄榄玄武岩		橄榄拉斑玄武岩					橄榄玄武岩	
样品号		EMB4	EMB6	220B3	220B4	305B6	101B1	苍 YQ1	150-1	151-2	HNG-15	10-4
序号		23	24	25	26	27	28	29	30	31	32	33
岩石化学成分/%	SiO_2	53.42	53.37	47.72	47.89	53.03	49.71	50.95	53.12	52.11	46.81	51.09
	Al_2O_3	14.37	13.74	12.90	13.15	14.20	13.92	13.31	14.26	14.71	1124	14.21
	Fe_2O_3	3.53	4.80	3.32	3.17	2.15	4.54	1.93	1.42	2.36	4.33	1.52
	FeO	7.43	6.17	8.00	8.09	8.43	5.63	9.42	9.14	8.60	7.63	9.01
	MgO	6.51	5.91	10.47	9.85	6.89	6.50	8.31	6.90	6.70	9.33	7.46
	CaO	8.35	7.53	9.80	9.76	8.50	8.70	8.36	8.54	8.40	9.13	8.26
	Na_2O	2.40	2.64	2.34	2.32	2.58	2.67	3.28	3.15	3.15	2.92	3.77
	K_2O	0.57	1.31	1.41	1.50	0.87	1.24	1.95	0.92	0.77	1.74	1.36
	MnO	0.12	0.13	0.15	0.14	0.13	0.13	0.18	0.15	0.15	0.18	0.15
	TiO_2	1.53	1.82	2.30	2.37	1.57	2.05	2.40	1.57	1.64	2.74	1.85
	P_2O_5	0.19	0.36	0.47	0.51	0.22	0.41	0.50	0.23	0.90	0.81	0.44
	CO_2	烧失量：1.43	烧失量：2.17	烧失量：0.89	烧失量：0.99	烧失量：0.94	烧失量：4.28		0.06	0.06		0.08
	H_2O^+							0.04	0.58		0.44	0.34
	总量	99.85	99.95	99.77	99.74	99.51	99.78	100.67	100.04	99.81	99.50	99.54
标准矿物成分含量/%	Q	9.39	8.69			5.18	3.89		1.32	1.88		
	Or	3.43	7.93	8.43	8.98	5.22	7.69	11.47	5.47	4.61	10.39	8.12
	Ab	20.61	22.84	20.00	19.86	22.12	23.65	27.56	26.78	26.93	24.16	32.15
	An	27.15	22.26	20.73	21.27	24.92	23.37	15.71	22.15	23.97	18.57	17.96
	Ne										0.41	
	Di	11.20	11.20	20.53	19.84	13.52	15.49	18.38	15.68	13.61	17.83	16.97
	Hy	19.65	17.30	6.81	8.49	22.37	15.12	4.64	23.01	21.82		4.46
	Ol			13.17	11.23			13.85			15.27	13.61
	Ap	0.40	0.80	1.04	1.13	0.49	0.94	1.09	0.51	0.57	1.79	0.97
	Ilm	2.95	3.54	4.42	4.56	3.03	4.08	4.53	3.00	3.15	5.26	3.55
	Mt	5.20	5.43	4.87	4.65	3.16	5.79	2.78	2.07	3.46	6.34	2.22
岩石化学常用参数	$Mg^{\#}$	52.25	50.10	62.95	61.61	54.24	54.40	57.04	54.15	52.69	59.07	56.17
	AlK	2.97	3.95	3.75	3.82	3.45	3.91	5.23	4.07	3.92	4.66	5.13
	Na_2O/K_2O	4.20	2.02	1.66	1.55	2.97	2.15	1.68	3.43	4.09	1.68	2.77
	TFeO/MgO	1.68	1.86	1.05	1.11	1.50	1.49	1.34	1.51	1.60	1.24	1.41
	δ	0.85	1.50	2.98	2.98	1.19	2.28	3.44	1.64	1.69	5.70	3.25
	τ	7.82	6.10	4.59	4.57	7.40	5.49	4.18	7.08	7.05	3.84	5.64
	AR	1.30	1.46	1.40	1.40	1.36	1.42	1.64	1.43	1.41	1.52	1.59
	SI	31.85	28.37	40.99	39.51	32.93	31.58	33.39	32.05	31.05	35.95	32.27
	CI	39.01	37.48	49.80	48.77	42.08	39.66	44.18	43.77	42.04	44.69	40.81
	DI	33.43	39.46	28.44	28.84	32.52	36.05	39.03	33.58	33.42	34.96	40.26
采样地点		峨蔓		永茂岭		永昌村	美造	苍西	龙塘	付伟		公堂岭

续表

地层单位		道堂组									
岩性		橄榄玄武岩		石英拉斑玄武岩							
样品号		D1019	D1894	俯 YQ1	永昌 YQ1	丁 YQ1	Ⅰ166	KH-4	XⅧ5-1	XⅧ7-1	8-1
序号		34	35	36	37	38	39	40	41	42	43
岩石化学成分/%	SiO_2	53. 27	52. 71	53. 3	52. 98	53. 75	53. 55	53. 90	52. 83	51. 78	53. 67
	Al_2O_3	14. 32	14. 04	13. 92	14. 11	13. 70	14. 55	13. 88	14. 28	14. 31	14. 17
	Fe_2O_3	0. 70	1. 30	1. 93	2. 25	2. 20	1. 76	1. 52	2. 70	4. 80	2. 24
	FeO	9. 95	8. 95	9. 42	8. 77	9. 22	9. 14	9. 70	9. 01	6. 95	8. 82
	MgO	6. 94	7. 86	7. 53	8. 17	7. 34	7. 15	5. 96	5. 85	5. 11	6. 50
	CaO	8. 10	7. 87	7. 80	6. 81	7. 95	7. 98	8. 51	8. 20	8. 67	8. 47
	Na_2O	2. 94	3. 00	2. 95	3. 54	2. 86	2. 94	2. 75	2. 56	2. 57	3. 15
	K_2O	0. 92	1. 21	0. 97	0. 86	0. 98	0. 70	1. 20	0. 48	0. 45	0. 49
	MnO	0. 15	0. 14	0. 15	0. 11	0. 17	0. 19	0. 24	0. 15	0. 15	0. 15
	TiO_2	1. 60	1. 85	1. 50	1. 60	1. 60	1. 70	1. 48	1. 68	1. 73	1. 64
	P_2O_5	0. 23	0. 35	0. 25	0. 14	0. 25	0. 25	0. 16	0. 20	0. 20	0. 22
	CO_2	0. 04	0. 04						0. 38	0. 70	0. 04
	H_2O^+	0. 65	0. 45	0. 58	0. 50	0. 60	0. 62	0. 06	1. 44	2. 28	0. 56
	总量	99. 67	100. 40	100. 28	99. 94	100. 62	100. 53	99. 36	99. 76	99. 70	100. 12
标准矿物成分含量/%	Q	1. 95	0. 60	2. 28	0. 08	3. 71	3. 74	4. 17	8. 01	8. 93	4. 60
	Or	5. 49	7. 21	5. 75	5. 12	5. 80	4. 14	7. 15	2. 90	2. 76	2. 91
	Ab	25. 07	25. 54	25. 00	30. 12	24. 17	24. 87	23. 41	22. 09	22. 49	26. 75
	An	23. 33	21. 39	21. 90	20. 17	21. 61	24. 43	22. 10	26. 57	27. 07	23. 16
	Ne										
	Di	13. 07	12. 94	12. 63	10. 58	13. 38	11. 25	16. 10	11. 32	13. 30	14. 56
	Hy	26. 50	26. 11	26. 21	27. 28	24. 57	25. 24	21. 68	21. 41	16. 18	21. 15
	Ol										
	Ap	0. 51	0. 77	0. 55	0. 31	0. 55	0. 55	0. 35	0. 45	0. 45	0. 48
	Ilm	3. 07	3. 54	2. 86	3. 06	3. 04	3. 23	2. 83	3. 26	3. 40	3. 13
	Mt	1. 02	1. 90	2. 81	3. 28	3. 19	2. 55	2. 22	4. 00	5. 42	3. 26
岩石化学常用参数	$Mg^{\#}$	53. 91	58. 07	54. 61	57. 44	53. 88	54. 31	48. 98	47. 69	44. 70	51. 68
	AlK	3. 86	4. 21	3. 92	4. 40	3. 84	3. 64	3. 95	3. 04	3. 02	3. 64
	Na_2O/K_2O	3. 20	2. 48	3. 04	4. 12	2. 92	4. 20	2. 29	5. 33	5. 71	6. 34
	TFeO/MgO	1. 52	1. 29	1. 48	1. 32	1. 53	1. 50	1. 86	2. 00	2. 30	1. 70
	δ	1. 45	1. 83	1. 49	1. 94	1. 37	1. 26	1. 43	0. 94	1. 04	1. 24
	τ	7. 11	5. 97	7. 31	6. 61	6. 77	6. 83	7. 52	6. 98	6. 79	6. 72
	AR	1. 42	1. 48	1. 44	1. 53	1. 43	1. 39	1. 43	1. 31	1. 30	1. 38
	SI	32. 35	35. 22	33. 03	34. 63	32. 48	32. 96	28. 21	28. 40	25. 70	30. 66
	CI	43. 66	44. 50	44. 51	44. 21	44. 17	42. 28	42. 83	39. 99	38. 31	42. 10
	DI	32. 51	33. 35	33. 04	35. 31	33. 67	32. 75	34. 72	33. 00	34. 18	34. 26
采样地点		灵山	永发	府城	永昌	丁村	龙桥				公堂岭

续表

地层单位		道堂组		多文组								
岩性		石英拉斑玄武岩		粗玄岩				石英拉斑玄武岩				
样品号		XⅧ2-1	XⅧ3-2	ⅠH3-3	0510	XⅧ1-1	木 ZK1-1	HNG-6	216	249	32-1	48-1
序号		44	45	46	47	48	49	50	51	52	53	54
岩石化学成分/%	SiO_2	52.76	50.36	53.59	50.67	51.51	51.92	52.90	52.55	52.07	53.86	53.55
	Al_2O_3	14.43	14.87	14.51	14.34	13.92	13.76	15.15	13.55	13.30	14.95	14.67
	Fe_2O_3	3.24	4.25	1.53	Fe_2O_3*:	3.25	3.30	3.20	5.82	4.08	2.96	1.99
	FeO	8.56	7.65	9.50	12.16	7.88	7.60	7.56	6.12	7.78	7.52	8.35
	MgO	6.31	6.62	5.92	7.30	6.85	7.46	4.22	7.25	7.66	5.24	6.40
	CaO	8.07	7.33	8.66	8.16	7.80	8.24	8.35	8.45	8.74	8.35	8.47
	Na_2O	2.76	3.27	2.91	3.37	3.50	3.34	2.91	2.64	2.80	3.53	3.27
	K_2O	0.45	1.29	0.52	1.65	1.50	1.04	0.81	0.58	0.90	0.64	0.54
	MnO	0.15	0.15	0.14	0.14	0.14	0.14	0.12	0.18	0.15	0.18	0.15
	TiO_2	1.70	1.85	1.63	1.76	1.73	1.90	1.90	1.60	1.88	1.59	1.59
	P_2O_5	0.25	0.45	0.19	0.40	0.40	0.40	0.36	0.23		0.20	0.22
	CO_2	0.32	0.32	0.06		0.43	烧失量:				0.05	0.03
	H_2O^+	0.88	1.40	0.46		0.82	0.92	1.09	0.51	0.02	1.20	0.66
	总量	99.88	99.81	99.62	99.95	99.73	100.02	99.62	99.48	99.62	100.30	99.87
标准矿物成分含量/%	Q	7.01	0.31	5.20			1.05	8.95	5.93	3.22	5.33	3.86
	Or	2.70	7.78	3.10	9.85	9.01	6.21	4.91	3.47	5.35	3.82	3.22
	Ab	23.64	28.19	24.82	28.74	30.04	28.48	25.23	22.60	23.76	30.13	27.86
	An	25.97	22.49	25.19	19.27	18.08	19.63	26.52	23.68	21.12	23.25	23.91
	Ne											
	Di	10.72	9.67	14.07	15.52	15.27	15.54	11.35	14.13	16.88	14.23	14.03
	Hy	21.39	21.52	21.83	10.84	16.73	19.73	13.77	21.43	20.12	15.42	20.68
	Ol				6.22	1.86						
	Ap	0.55	1.00	0.42	0.88	0.89	0.88	0.81	0.51	0.57	0.44	0.48
	Ilm	3.27	3.59	3.13	3.37	3.34	3.64	3.70	3.08	3.59	3.05	3.05
	Mt	4.76	5.44	2.24	5.31	4.79	4.83	4.76	5.17	5.39	4.33	2.91
岩石化学常用参数	$Mg^{\#}$	49.50	50.70	49.25	54.32	53.06	55.72	41.88	53.23	54.39	47.85	52.95
	AlK	3.21	4.56	3.43	5.02	5.00	4.38	3.72	3.22	3.70	4.17	3.81
	Na_2O/K_2O	6.13	2.53	5.60	2.04	2.33	3.21	2.59	4.55	3.11	5.52	6.06
	TFeO/MgO	1.82	1.80	1.84	1.50	1.58	1.42	2.47	1.57	1.50	1.94	1.58
	δ	1.06	2.83	1.11	3.29	2.94	2.15	1.40	1.09	1.51	1.60	1.38
	τ	6.86	6.27	7.12	6.23	6.02	5.48	6.44	6.82	5.59	7.18	7.17
	AR	1.33	1.52	1.35	1.57	1.60	1.50	1.38	1.34	1.40	1.44	1.39
	SI	29.60	28.68	29.05	29.82	29.81	32.81	22.57	32.35	32.99	26.34	31.14
	CI	40.14	40.22	41.27	41.27	41.98	43.75	33.58	43.81	45.98	37.03	40.66
	DI	33.34	36.28	33.13	38.59	39.05	35.75	39.10	32.01	32.33	39.28	34.94
采样地点				多文			木棠	云龙	土桥	龙发	乐湾岭	

续表

地层单位		多文组										
岩性		石英拉斑玄武岩							玄武岩			
样品号		60-1	KH-6	205-1	HMG-14	白 YQ1	Ⅱ326-GS	HNG-10	0502	D1828	D1837	ⅡH14-2
序号		55	56	57	58	59	60	61	62	63	64	65
岩石化学成分/%	SiO_2	53.77	55.49	53.85	52.41	53.43	51.89	52.10	52.52	53.72	52.33	52.73
	Al_2O_3	14.43	12.50	14.30	13.85	13.60	13.75	14.95	14.58	14.04	13.87	14.89
	Fe_2O_3	3.26	2.33	0.74	1.60	1.71	2.56	3.08	$Fe_2O_3^*$:	3.86	0.58	2.89
	FeO	7.63	9.15	9.50	8.49	8.83	8.40	7.58	12.22	7.10	10.10	8.23
	MgO	6.34	6.33	6.50	7.29	8.12	7.67	6.70	6.62	6.20	7.54	6.13
	CaO	8.54	8.66	8.55	8.60	6.97	7.26	8.00	8.88	8.07	8.09	8.54
	Na_2O	3.13	2.65	3.34	2.84	2.85	3.15	2.52	2.69	2.63	2.66	2.77
	K_2O	0.52	0.52	0.77	1.16	0.72	0.58	0.60	0.60	0.90	1.25	0.47
	MnO	0.15	0.28	0.15	0.14	0.20	0.19	0.12	0.14	0.14	0.15	0.15
	TiO_2	1.58	1.64	1.77	1.97	2.00	1.95	1.70	1.50	1.83	1.89	1.66
	P_2O_5	0.20		0.24	0.44	0.40	0.50	0.26	0.18	0.25	0.23	0.19
	CO_2	0.05		0.05			0.10	0.08		0.19	0.28	0.02
	H_2O^+	0.64	0.10	0.24	0.55	0.84	1.50	1.74		0.88	0.70	0.85
	总量	100.24	99.65	100.0	99.34	99.67	99.59	99.43	99.93	99.79	99.67	99.52
标准矿物成分含量/%	Q	6.07	9.16	1.82	2.14	4.87	3.25	7.09	5.59	9.13	0.73	6.39
	Or	3.09	3.09	4.57	6.95	4.31	3.50	3.64	3.58	5.39	7.49	2.82
	Ab	26.49	22.50	28.30	24.30	24.37	27.17	21.82	22.95	22.51	22.78	23.73
	An	23.91	20.74	21.78	21.85	22.42	22.08	28.35	26.15	24.12	22.47	27.14
	Di	14.20	18.41	15.79	15.26	8.30	9.48	8.72	14.03	12.07	13.83	11.99
	Hy	18.04	19.57	22.71	22.41	28.50	25.84	21.92	19.36	17.03	27.70	20.06
	Ol											
	Ap	0.44		0.59	0.97	0.88	1.11	0.58	0.40	0.55	0.51	0.42
	Ilm	3.02	3.13	3.37	3.79	3.84	3.78	3.31	2.88	3.52	3.64	3.20
	Mt	4.75	3.39	1.08	2.35	2.51	3.79	4.58	5.07	5.67	0.85	4.25
岩石化学常用参数	$Mg^\#$	51.69	50.09	53.27	56.69	58.27	56.38	53.57	51.77	51.11	55.86	50.23
	AlK	3.65	3.17	4.11	4.00	3.57	3.73	3.12	3.29	3.53	3.91	3.24
	Na_2O/K_2O	6.02	5.10	4.34	2.45	3.96	5.43	4.20	4.48	2.92	2.13	5.89
	TFeO/MgO	1.67	1.78	1.56	1.36	1.28	1.40	1.55	1.66	1.72	1.41	1.77
	δ	1.23	0.80	1.56	1.70	1.22	1.57	1.07	1.14	1.16	1.64	1.08
	τ	7.16	6.01	6.19	5.59	5.38	1.44	7.31	7.93	6.23	5.93	7.30
	AR	1.38	1.35	1.44	1.43	1.42	1.43	1.31	1.33	1.38	1.43	1.32
	SI	30.38	30.17	31.18	30.98	36.53	34.57	32.71	29.91	29.97	34.07	29.92
	CI	40.66	44.51	42.95	43.80	43.15	42.89	38.53	41.33	38.30	46.02	39.50
	DI	34.94	34.75	34.69	33.38	33.55	33.92	32.54	32.12	37.04	31.00	32.94
采样地点		良爱村	美夏	龙州	旧州岭	白莲	国社岭	云龙		云龙	琼海	高山岭

续表

地层单位		多文组											
岩性		橄榄拉斑玄武岩								橄榄辉石玄武岩			
样品号		HNR-1	HK22	木棠	13-1	Ⅰ-GS	100-2	157-1	GK-28	73-1	ⅠH4-2	ⅡH4-3	ⅡH10-2
序号		66	67	68	69	70	71	72	73	74	75	76	77
岩石化学成分/%	SiO_2	51.28	52.02	51.25	51.65	51.24	53.04	51.91	52.98	50.92	48.77	50.11	49.63
	Al_2O_3	14.07	14.09	13.70	14.16	13.88	14.12	13.52	12.47	14.01	13.93	13.30	13.31
	Fe_2O_3	Fe_2O_3*：	Fe_2O_3*：	3.15	2.65	1.71	1.75	1.07	1.59	3.03	2.46	3.13	3.30
	FeO	11.71	9.85	8.32	8.16	9.35	7.43	10.11	9.55	7.72	9.12	8.15	8.17
	MgO	7.54	7.27	7.37	7.61	7.83	7.36	6.75	6.93	7.87	7.34	8.63	8.96
	CaO	8.26	8.27	8.67	7.91	7.11	8.47	8.40	8.83	7.91	8.35	8.40	8.29
	Na_2O	3.57	3.80	3.05	3.76	3.30	3.19	3.16	3.10	3.61	3.24	3.07	3.06
	K_2O	1.47	1.73	1.54	1.33	1.20	1.08	1.03	1.20	1.31	1.91	1.54	1.53
	MnO	0.14	0.135	0.15	0.15	0.30	0.14	0.23	0.19	0.18	0.16	0.15	0.15
	TiO_2	1.76	1.96	1.65	1.73	1.90	1.87	1.77	2.15	1.89	3.08	2.19	2.22
	P_2O_5	0.35			0.35	0.48	0.35	0.27	0.20	0.40	0.72	0.47	0.48
	CO_2			其他：	0.02		0.16	0.71		0.11	0.08	0.02	0.02
	H_2O^+			0.23	0.44	1.10	0.92	0.68	1.22	1.00	0.56	0.59	0.56
	总量	100.15	99.655	99.42	99.92	100.4	99.88	99.63	100.36	99.96	99.72	99.75	99.68
标准矿物成分含量/%	Q						1.95		1.31				
	Or	8.75	10.40	9.21	7.92	7.22	6.47	6.20	7.16	7.84	11.40	9.19	9.13
	Ab	30.37	32.60	26.08	31.21	28.37	27.29	27.19	26.41	30.87	27.64	26.17	26.10
	An	18.11	16.50	19.33	18.32	19.82	21.24	19.99	16.67	18.33	17.95	18.08	18.19
	Di	16.96	20.00	19.63	15.50	10.78	15.56	17.17	21.56	15.33	15.89	17.05	16.46
	Hy	10.71	6.82	14.02	11.76	22.56	20.55	23.02	20.00	11.36	3.37	12.03	10.51
	Ol	5.66	4.43	3.93	7.35	3.99		0.83		7.31	12.65	7.67	9.46
	Ap	0.77			0.77	1.07	0.77	0.60	0.44	0.88	1.59	1.03	1.06
	Ilm	3.37	3.78	3.17	3.31	3.67	3.60	3.42	4.12	3.63	5.91	4.20	4.26
	Mt	5.30	5.52	4.62	3.87	2.52	2.57	1.58	2.32	4.44	3.60	4.58	4.83
岩石化学常用参数	$Mg^{\#}$	56.06	57.23	54.09	56.27	56.18	59.30	52.08	52.94	57.32	53.59	58.39	58.92
	AlK	5.04	5.53	4.59	5.00	4.50	4.27	4.19	4.30	4.92	5.15	4.61	4.59
	Na_2O/K_2O	2.43	2.20	1.98	2.76	2.75	2.95	3.07	2.58	2.76	1.70	1.99	2.00
	TFeO/MgO	1.40	1.22	1.51	1.39	1.39	1.22	1.64	1.58	1.33	1.54	1.27	1.24
	δ	3.07	3.39	2.55	2.89	2.46	1.82	1.97	1.85	3.06	4.60	2.99	3.18
	τ	5.97	5.25	6.45	6.06	5.57	5.84	5.85	4.36	5.50	3.47	4.67	4.62
	AR	1.58	1.66	1.52	1.59	1.55	1.47	1.47	1.51	1.58	1.60	1.54	1.54
	SI	31.04	32.10	31.46	32.49	33.48	35.37	30.52	30.98	33.43	30.49	35.20	35.81
	CI	42.00	40.50	45.38	41.78	43.52	42.28	46.02	48.01	42.08	41.42	45.52	45.52
	DI	39.13	43.00	35.29	39.13	35.59	35.71	33.39	34.88	38.70	39.04	35.36	35.23
采样地点		洋浦		木棠	南滩	龙塘	东水港	美兰	抱舍	雷公岛	多文	高山	东英

续表

地层单位		多文组										平均值	中国东部玄武岩平均值（赵宗溥，1956）
岩性		橄榄玄武岩											
样品号		D0997	D0998	D1204-2	D1802	$ZK_{3-}TW_2$	42-1	237-1	D1830	$Ⅱ_{H3-1}$	YD382		
序号		78	79	80	81	82	83	84	85	86	87		
岩石化学成分/%	SiO_2	50.73	51.96	51.14	51.83	49.58	51.42	49.96	46.87	48.71	50.15	51.46	47.56
	Al_2O_3	13.44	13.72	13.64	13.67	14.20	14.03	15.40	13.62	13.57	13.42	14.04	14.94
	Fe_2O_3	1.32	2.67	3.46	2.60	2.06	2.05	5.40	4.81	2.27	2.96	3.20	4.75
	FeO	10.45	8.73	8.43	8.70	8.64	8.76	5.58	7.45	9.40	8.06	7.84	6.62
	MgO	7.29	7.30	7.53	7.75	7.92	7.41	5.58	7.71	8.80	8.13	7.20	7.00
	CaO	8.26	8.25	8.05	8.17	8.47	7.70	8.47	8.28	8.26	8.51	8.39	8.50
	Na_2O	2.85	2.94	3.01	2.85	3.74	3.70	3.39	3.95	2.83	3.37	3.07	3.60
	K_2O	1.03	0.91	1.10	0.91	1.53	1.85	1.17	4.00	1.85	1.48	1.21	2.44
	MnO	0.17	0.20	0.14	0.16	0.16	0.15	0.15	0.16	0.16	0.16	0.16	0.14
	TiO_2	1.89	1.96	2.02	1.92	2.18	2.07	2.37	3.38	2.62	2.25	2.01	2.19
	P_2O_5	0.26	0.27	0.32	0.26	0.53	0.48	0.44	1.09	0.58	0.46	0.36	0.67
	CO_2	0.62	0.14	0.26	0.06	0.05	0.03	0.02	0.06	0.02	0.07	0.08	
	H_2O^+	0.68	0.67	0.63	0.87	0.34	0.24	1.90	1.30	0.66	0.72	0.79	1.82
	总量	98.99	99.72	99.73	99.75	99.40	99.96	99.80	99.63	99.73	99.74	99.85	
标准矿物成分含量/%	Q		2.37	1.21	2.10			1.38					
	Or	6.24	5.44	6.58	5.45	9.14	10.98	7.08	6.02	11.05	8.85		
	Ab	24.66	25.12	25.74	24.37	29.98	31.47	29.30	33.96	24.15	28.78		
	An	21.30	21.76	20.67	22.05	17.58	16.19	23.84	16.74	19.00	17.27		
	Ne					1.06							
	Di	15.71	14.67	14.45	14.16	17.63	15.73	13.28	14.76	15.34	18.28		
	Hy	24.07	22.37	21.69	23.79		5.45	13.57	0.80	6.65	7.80		
	Ol	1.80				16.25	12.19		11.90	14.19	9.35		
	Ap	0.58	0.60	0.71	0.57	1.17	1.05	0.98	2.42	1.28	1.01		
	Ilm	3.68	3.77	3.88	3.69	4.18	3.95	4.61	6.53	5.03	4.32		
	Mt	1.96	3.91	5.08	3.82	3.02	2.98	5.96	6.87	3.32	4.34		
岩石化学常用参数	$Mg^{\#}$	56.78	57.90	57.77	59.56	57.37	55.47	48.80	53.84	57.83	57.48		
	AlK	3.88	3.85	4.11	3.76	5.27	5.56	4.56	4.95	4.68	4.85		
	Na_2O/K_2O	2.77	3.23	2.74	3.13	2.44	2.01	2.90	3.95	1.53	2.28		
	TFeO/MgO	1.60	1.52	1.53	1.42	1.33	1.43	1.87	1.53	1.30	1.32		
	δ	1.95	1.65	2.08	1.60	4.22	3.67	2.99	6.33	3.84	3.29		
	τ	5.60	5.50	5.26	5.64	4.80	4.99	5.07	2.86	4.10	4.47		
	AR	1.44	1.52	1.47	1.42	1.61	1.69	1.47	1.58	1.55	1.57		
	SI	31.78	32.37	32.00	33.98	33.15	31.16	26.42	30.95	34.99	33.88		
	CI	47.23	44.72	45.10	45.46	41.08	40.30	37.41	40.87	44.52	44.09		
	DI	30.89	32.93	33.53	31.92	40.17	42.46	37.77	39.97	35.19	37.63		
采样地点		福兴村		停竹村	鸭塘村	颜春岭	高山岭	龙塘	龙发	高山岭	龙坡		

表 13-7　南海新生代火山岩化学成分、CIPW 标准矿物含量及岩石化学常用参数一览表

Table 13-7　List of chemical composition, CIPW standard minerals and common petrochemical parameters of Cenozoic volcanic rocks from the South China Sea

岩性		碱性玄武岩		石英拉斑玄武岩				碱性玄武岩	
样品号		10-1	8-1	CB-2	CB-5	CB-7-1	CB-7-2	10-2	10-1
产地		中南	玳瑁	宪北海山				中南海山	
序号		1	2	3	4	5	6	7	8
岩石化学成分/%	SiO_2	41.86	48.08	48.8	42.39	42.34	44.09	41.45	42.28
	Al_2O_3	15.26	14.5	13.66	12.46	13.13	12.4	14.65	15.88
	Fe_2O_3	8.46	9.85	6.75	5.81	5.83	4.82	9.59	7.34
	FeO	4.41	3.5	6.02	5.55	5.88	6.23	3.33	5.5
	MgO	6.8	5.23	6.67	12.29	10.54	12.79	5.4	8.21
	CaO	7.7	8.83	10.02	10	10.33	9.53	6.58	8.83
	Na_2O	2.26	3.3	2.4	2.7	3.03	3.02	3.02	2.41
	K_2O	2.72	0.88	1.19	1.73	2.03	1.85	2.62	1.9
	MnO	0.58	0.31	0.68	0.56	0.89	0.25	0.88	0.28
	TiO_2	3.21	2.64	1.76	2.55	2.77	2.74	3.22	3.2
	P_2O_5	0.65	0.41	0.1	0.65	0.76	0.6	0.65	0.69
	CO_2								
	H_2O^+								
	烧失量	6.30	1.8	1.28	2.61	1.79	1.28	8.44	4.16
	总量	100.21	99.33	99.33	99.3	99.32	99.6	99.83	100.68
标准矿物成分含量/%	Q	0	0	0	0	0	0	0	0
	Or	17.24	5.38	7.21	10.62	12.36	11.16	17.09	11.7
	Ab	13.23	28.82	20.78	4.84	2.69	7.88	17.86	13.3
	An	25.09	22.86	23.52	17.38	16.67	15.08	20.56	27.97
	Ne	3.93	0	0	10.22	12.83	9.84	5.58	4.29
	Lc	0	0	0	0	0	0	0	0
	Di	9.62	16.48	22.11	24.22	25.41	23.63	9.27	10.91
	Hy	0	11.42	13.62	0	0	0	0	0
	Ol	18.98	5.27	5.73	22.95	19.55	22.56	17.39	20.16
	Mt	3.86	3.68	3.39	3.28	3.39	3.21	3.96	3.77
	Hm	0	0	0	0	0	0	0	0
	Il	6.54	5.18	3.43	5.03	5.42	5.31	6.75	6.33
	Ap	1.52	0.92	0.22	1.47	1.71	1.34	1.56	1.57
岩石化学常用参数	$Mg^{\#}$	50.21	42.99	49.58	67.03	62.81	68.33	44.6	54.73
	AlK	4.98	4.18	3.59	4.43	5.06	4.87	5.64	4.31
	Na_2O/K_2O	0.83	3.75	2.02	1.56	1.49	1.63	1.15	1.27
	FeO^*/MgO	1.77	2.36	1.81	0.88	1.06	0.83	2.21	1.47
	FeO^*	12.02	12.36	12.09	10.78	11.13	10.57	11.96	12.1
	δ		3.44	2.22			21.76		
	τ	4.05	4.24	6.4	3.83	3.65	3.42	3.61	4.21
	AR	1.55	1.44	1.36	1.49	1.55	1.57	1.72	1.42
	SI	27.59	22.98	28.96	43.77	38.59	44.55	22.54	32.37
	CI	39	42.03	48.27	55.48	53.76	54.71	37.37	41.18
	DI	34.4	34.19	28	25.68	27.87	28.88	40.53	29.29

续表

岩性		石英拉斑玄武岩								
样品号		8-1	8-2	8-5G	8-6	8-7	8-7G	9-1	9-2	9-3
产地		玳瑁海山						振华海山		
序号		9	10	11	12	13	14	15	16	17
岩石化学成分/%	SiO_2	46.38	47.39	48.58	46.12	45.54	49.12	48.07	47.77	47.39
	Al_2O_3	14.13	14.96	16.57	15	13.67	14.99	15.11	16.08	15.58
	Fe_2O_3	4.68	8.54	1.97	7.22	6.59	2.46	7.62	6.75	7.51
	FeO	7.87	3.52	10.21	3.67	5.71	9.98	4.6	5.11	4.27
	MgO	7.85	6.53	6.04	5.14	5	5.92	3.66	4.02	2.69
	CaO	9.96	10.11	9.67	10.1	6.85	9.69	8.78	8.95	7.91
	Na_2O	0.6	3	2.35	2.95	3	2.9	3.55	3.36	126
	K_2O	2.56	0.64	0.75	0.7	1.28	0.7	2.2	2.11	2.14
	MnO	0.21	0.28	0.17	0.09	0.5	0.17	0.14	0.2	0.11
	TiO_2	1.93	2.09	2.26	2.06	2.04	2.21	3.13	3.45	3.46
	P_2O_5	0.25	0.48	0.46	0.45	0.34	0.34	1.18	1.17	1.30
	CO_2			0.09			0.14			
	H_2O^+			1.04	3.87	6.57	1.23			1.50
	烧失量	3.35	3.11					2.2	1.31	1.29
	总量	99.77	100.65	100.16	97.37	97.09	99.85	100.24	100.28	98.61
标准矿物成分含量/%	Q	0	0	0.44	0	0	0	0	0	0
	Or	15.74	3.91	4.48	4.45	8.41	4.21	13.35	12.67	13.28
	Ab	5.27	26.17	20.05	26.82	28.15	24.89	30.77	28.82	30.69
	An	29.37	26.24	32.72	27.52	22.23	26.19	19.24	22.86	21.64
	Ne	0	0	0	0	0	0	0	0	0
	Lc	0	0	0	0	0	0	0	0	0
	Di	16.75	18.54	10.7	19.35	11.03	17	15.23	12.58	9.64
	Hy	21.26	6.78	22.85	7.82	16.56	17.05	0.3	1.78	10.57
	Ol	3.75	9.93	0	5.66	4.9	2.18	8.84	8.54	0.73
	Mt	3.47	3.26	3.42	3.12	3.61	3.48	3.54	3.5	3.59
	Hm	0	0	0	0	0	0	0	0	0
	Il	3.81	4.1	4.34	4.21	4.3	4.26	6.1	6.65	6.9
	Ap	0.57	1.08	1.01	1.06	0.82	0.75	2.64	2.59	2.98
岩石化学常用参数	$Mg^{\#}$	53.67	50.96	47.33	47.4	43.37	46.4	36.29	39.05	30.31
	AlK	3.16	3.64	3.1	3.65	4.28	3.6	5.75	5.47	5.6
	Na_2O/K_2O	0.23	4.69	3.13	4.21	2.34	4.14	1.61	1.59	1.62
	FeO^*/MgO	1.54	1.72	1.98	1.98	2.33	2.06	3.13	2.78	4.1
	FeO^*	12.08	11.2	11.98	10.17	11.64	12.19	11.46	11.18	11.03
	δ	2.95	3.02	1.72	4.27	7.21	2.12	6.52	6.27	7.14
	τ	7.01	5.72	6.29	5.85	5.23	5.47	3.69	3.69	3.5
	AR	1.3	1.34	1.27	1.34	1.53	1.34	1.63	1.56	1.63
	SI	33.32	29.37	28.33	26.12	23.17	26.96	16.92	18.83	13.40
	CI	49.04	42.61	41.31	40.16	40.41	43.96	34.01	33.06	31.42
	DI	21.02	30.07	24.97	31.27	36.55	29.1	44.11	41.49	43.97

续表

岩性		碱玄岩							
样品号		S04-12-21	S04-12-22	S04-11	S04-12-10	S04-12-11	S04-12-12	S04-12-18	S04-12-20
产地		中央海盆							
序号		18	19	20	21	22	23	24	25
岩石化学成分/%	SiO_2	48.48	42.35	44.80	48.28	47.57	45.43	50.03	46.97
	Al_2O_3	17.5	17.41	16.23	19.04	17.12	18.12	19.15	16.95
	Fe_2O_3	2.61	13	3.29	3.34	4.42	8.72	4.33	5.33
	FeO	6.84	1.47	6.48	4.48	5.2	2.73	2.7	4.85
	MgO	5.61	3.8	7.79	2.6	5.74	3.81	2.13	5.91
	CaO	8.11	8.198	9.00	8.83	8.1	9.87	8.45	7.9
	Na_2O	4.89	2.87	4.33	5.13	4.58	3.21	5.1	4.05
	K_2O	2.61	1.01	2.74	2.74	2.32	1.28	2.83	2.04
	MnO	0.2	0.44	0.19	0.15	0.23	0.23	0.11	0.33
	TiO_2	2.18	2.54	2.97	2.36	2.07	2.59	2.04	2.2
	P_2O_5	0.68	1.01	0.67	0.83	0.85	0.95	0.94	0.7
	CO_2	0.13	0.13	0.09	0.09	0.05	0.13	0.05	0.05
	H_2O^+	0.54	4.13	2.08	1.87	1.93	2.9	2.1	2.77
	烧失量								
	总量	100.38	98.36	100.66	99.74	100.18	99.97	99.96	100.05
标准矿物成分含量/%	Q	0	0	0	0	0	0	0	0
	Or	15.5	6.42	16.48	16.61	14.01	7.86	17.17	12.45
	Ab	19.89	26.07	7.39	21.6	22.13	26.56	27.83	24.57
	An	18.12	33.95	17	21.3	19.67	32.41	21.49	22.71
	Ne	11.7	0	16.16	12.37	9.43	0.88	8.87	5.83
	Lc	0	0	0	0	0	0	0	0
	Di	14.9	2.86	19.63	15.36	13.27	10.51	13.13	11.03
	Hy	0	4.33	0	0	0	0	0	0
	Ol	11.55	14.81	13.15	3.94	12.86	11.28	3.35	14.61
	Mt	2.7	4.01	2.97	2.38	2.72	3.24	2.09	2.9
	Hm	0	0	0	0	0	0	0	0
	Il	4.16	5.19	5.74	4.59	4.02	5.11	3.97	4.31
	Ap	1.49	2.37	1.49	1.86	1.89	2.15	2.1	1.58
岩石化学常用参数	$Mg^{\#}$	52.12	33.97	59.53	38.24	52.72	39.11	36.54	52.21
	AlK	7.5	3.88	7.07	7.87	6.9	4.49	7.93	6.09
	Na_2O/K_2O	1.87	2.84	1.58	1.87	1.97	2.51	1.8	1.99
	FeO^*/MgO	1.64	127	1.21	2.88	1.6	2.78	3.1	1.63
	FeO^*	9.19	13.17	9.44	7.49	9.18	10.58	6.6	9.65
	δ	10.26		27.77	11.73	10.42	8.3	8.95	9.34
	τ	5.78	5.72	4.01	5.89	6.06	5.76	6.89	5.86
	AR	1.83	1.36	1.78	1.79	1.75	1.38	1.81	1.65
	SI	24.87	17.16	31.63	14.22	25.79	19.29	12.46	26.65
	CI	33.3	31.2	41.49	26.27	32.87	30.14	22.54	32.86
	DI	47.09	32.49	40.03	50.58	45.57	35.3	53.87	42.86

续表

岩性		碱玄岩				橄榄玄武岩			
样品号		S04-12-8	S04-14-1B	S04-14-2B	S05-34A	V30B9	V36D-8	V36D10	V36D10
产地		中央海盆							
序号		26	27	28	29	30	31	32	33
岩石化学成分/%	SiO_2	48.75	43.6	44.9	46.8	49.14	49.71	46.73	46.31
	Al_2O_3	19.14	17	17.3	17.51	16.39	16.22	16.74	16.53
	Fe_2O_3	3.21	8.03	6.31	3.85	2.36	1.08	1.01	2.38
	FeO	4.28	1.95	2.87	5.17	8.48	9.72	9.1	8.57
	MgO	2.42	2.7	2.9	4.17	3.13	6.03	6.12	6.32
	CaO	8.52	10.45	10.5	7.44	9.15	10.86	9.52	9.37
	Na_2O	5.13	3.9	3.92	4.75	3.61	3.08	3.28	3.42
	K_2O	2.78	2.35	2.44	3.35	2.46	0.51	2.19	2.67
	MnO	0.14	0.56	0.12	0.28	0.15	0.11	0.19	0.2
	TiO_2	2.33	2.84	2.79	3.14	3.77	2.1	3.4	3.36
	P_2O_5	0.77	2.81	2.77	1.24	1.36	0.3	0.62	0.89
	CO_2	0.09	0.5	0.36	0.05	0	0	0	0
	H_2O^+	2.24	1.85	2.2	1.7	0	0	0	0
	烧失量								
	总量	99.8	98.54	99.38	99.45	100	99.72	98.9	100.02
标准矿物成分含量/%	Q	0	0	0	0	0	0	0	0
	Or	16.9	14.54	14.98	20.32	14.55	3.02	13.08	15.79
	Ab	23.72	21.29	23.21	19.72	30.51	26.08	19.62	15.3
	An	21.52	22.91	23.2	16.94	21.21	28.94	24.69	21.82
	Ne	11.3	7.15	6.05	11.63	0	0	4.54	7.37
	Lc	0	0	0	0	0	0	0	0
	Di	14.27	11.74	11.55	10.96	13.41	19.07	15.66	15.79
	Hy	0	0	0	0	0.95	10.5	0	0
	Ol	3.71	7.35	6.47	8.7	5.86	4.69	11.32	12.28
	Mt	2.3	2.97	2.77	2.84	3.38	3.04	3.19	3.32
	Hm	0	0	0	0	0	0	0	0
	Il	4.55	5.65	5.5	6.12	7.16	4	6.52	6.38
	Ap	1.73	6.42	6.28	2.78	2.97	0.66	1.37	1.94
岩石化学常用参数	$Mg^{\#}$	37.57	34.41	37.69	46.27	34.48	50.14	52.16	51.26
	AlK	7.91	6.25	6.36	8.1	6.07	3.59	5.47	6.09
	Na_2O/K_2O	1.85	1.66	1.61	1.42	1.47	6.04	1.5	1.28
	FeO^*/MgO	2.96	12	2.95	2.07	3.39	1.77	1.64	1.69
	FeO^*	7.17	9.18	8.55	8.63	10.6	10.69	10.01	10.71
	δ	10.88	65.1	21.29	17.27	6	1.92	8.02	11.2
	τ	6.01	4.61	4.8	4.06	3.39	6.26	3.96	3.9
	AR	1.8	1.59	1.59	1.96	1.62	1.31	1.53	1.61
	SI	13.58	14.26	15.73	19.59	15.62	29.53	28.2	27.05
	CI	24.84	27.7	26.29	28.62	30.75	41.3	36.7	37.77
	DI	51.91	42.98	44.24	51.67	45.07	29.1	37.25	38.47

续表

岩性		橄榄玄武岩							
样品号		V36D8-1	V36D8-2	V36D8-2	V36D8-4	V36D9	V36D9-1	V36D9-2	V36D9-3
产地		中央海盆							
序号		34	35	36	37	38	39	40	41
岩石化学成分/%	SiO_2	48.62	49.20	49.44	49.73	48.90	49.13	48.91	49.48
	Al_2O_3	15.71	15.82	15.07	15.97	16.23	16.03	15.77	15.86
	Fe_2O_3	2.76	2.52	2.72	2.42	1.02	2.51	2.58	2.51
	FeO	9.93	9.07	9.79	8.71	9.13	9.05	9.3	9.01
	MgO	5.88	5.65	5.4	6.1	3.06	5.5	3.38	5.69
	CaO	10.51	11.03	9.69	10.77	9.18	11	8.99	10.75
	Na_2O	2.77	3.09	3.46	3.15	3.57	3.02	3.51	3.02
	K_2O	0.89	0.74	0.96	0.59	2.11	0.77	2.42	0.89
	MnO	0.24	0.19	0.22	0.13	0.14	0.14	0.14	0.15
	TiO_2	2.24	2.14	2.8	2.09	3.74	2.19	3.67	2.13
	P_2O_5	0.44	0.54	0.45	0.34	1.11	0.61	1.33	0.48
	CO_2								
	H_2O^+								
	烧失量								
	总量	99.99	99.99	100	100	98.19	99.95	100	99.97
标准矿物成分含量/%	Q	0	0	0	0	0	0	0	0
	Or	5.27	4.38	5.68	3.49	12.7	4.56	14.32	5.27
	Ab	23.42	26.13	29.25	26.63	30.69	25.55	29.67	25.54
	An	27.78	27.09	22.73	27.67	22.37	27.9	20.09	27.07
	Ne	0	0	0	0	0	0	0	0
	Lc	0	0	0	0	0	0	0	0
	Di	18	20.17	18.74	19.5	14.3	19.06	13.88	19.32
	Hy	9.77	6.8	6.4	8.75	4.59	8.62	1.98	8.5
	Ol	7.05	6.99	7.33	6.16	2.36	5.61	6.58	6.02
	Mt	3.49	3.2	3.57	3.08	3.3	3.21	3.61	3.19
	Hm	0	0	0	0	0	0	0	0
	Il	4.26	4.07	5.32	3.97	7.23	4.16	6.97	4.05
	Ap	0.96	1.18	0.98	0.74	2.47	1.33	2.9	1.05
岩石化学常用参数	$Mg^{\#}$	45.78	47.05	44.03	49.97	35.19	46.44	34.15	47.37
	AlK	3.66	3.83	4.42	3.74	5.68	3.79	5.93	3.91
	Na_2O/K_2O	3.11	4.18	3.6	5.34	1.69	3.92	1.45	3.39
	FeO^*/MgO	2.11	2.01	2.27	1.79	3.28	2.06	3.44	1.98
	FeO^*	12.41	11.34	12.24	10.89	10.05	11.31	11.62	11.27
	δ	2.38	2.37	3.03	2.08	5.47	2.34	5.95	2.36
	τ	5.78	5.95	4.15	6.13	3.39	5.94	3.34	6.03
	AR	1.32	1.33	1.43	1.33	1.58	1.33	1.63	1.34
	SI	26.45	26.82	24.18	29.09	16.2	26.38	15.95	26.94
	CI	42.57	41.23	41.36	41.47	31.78	40.67	33.02	41.07
	DI	28.69	30.51	34.93	30.12	43.39	30.11	43.99	30.81

岩性		玄武岩	碱玄岩	粗面玄武岩		碱性橄榄玄武岩		橄榄拉斑玄武岩		平均值
样品号		南-1	S04-16	S08-69-1	S08-69-2	S023-35-6	S023-37-7	S023-37-31	S023-40	
产地		南海	中沙			礼乐滩				
序号		42	43	44	45	46	47	48	49	
岩石化学成分/%	SiO_2	47.79	44.42	45.72	45.46	47.44	48.39	50.11	47.65	46.90
	Al_2O_3	15.34	15.65	15.02	15.02	13.95	14.96	14.46	15.69	15.77
	Fe_2O_3	7.3	4.37	3.6	5.41	2.28	2.21	2.44	2.54	4.78
	FeO	3.96	5.68	6.31	5.83	8.19	7.96	8.77	9.41	6.33
	MgO	5.81	7.89	9.498	8.93	8.46	6.08	7.33	6.82	5.87
	CaO	10.33	9.05	9.79	9.78	10.13	11.17	9.68	10.09	9.39
	Na_2O	2.89	4.18	2.9	2.78	3.86	3.75	3.28	3.6	3.38
	K_2O	0.91	2.55	2.1	2.22	2.15	1.92	1.15	1.05	1.79
	MnO	0.15	0.53	0.16	0.2	0.16	0.16	0.17	0.18	0.27
	TiO_2	1.93	2.88	1.79	1.79	2.59	2.72	2.14	2.61	2.59
	P_2O_5	0.51	0.68	0.47	0.47	0.79	0.67	0.47	0.63	0.77
	CO_2	0.05	0.13	0.09	0.05					0.05
	H_2O^+	2.29	1.65	1.54	1.7					1.768
	烧失量									
	总量	99.26	99.66	98.99	99.64	100	99.99	100	100.27	99.65
标准矿物成分含量/%	Q	0	0	0	0	0	0	0	0	0
	Or	5.58	15.45	12.78	13.46	12.72	11.36	6.8	6.2	10.88
	Ab	25.34	8.11	12.2	11.14	15.03	18.65	27.73	24.7	21.35
	An	27.14	16.77	22.35	22.46	14.36	18.29	21.31	23.46	23.14
	Ne	0	15.22	7.06	7.02	9.54	7.07	0	3.06	4.26
	Lc	0	0	0	0	0	0	0	0	0
	Di	18.82	20.28	19.99	19.84	25.24	27.01	19.54	18.63	16.28
	Hy	8.27	0	0	0	0	0	5.79	0	4.39
	Ol	6.82	14.05	18.29	18.5	13.43	7.99	10.61	14.25	9.70
	Mt	3.07	3.02	2.77	3.05	3.04	3	3.11	3.38	3.21
	Hm	0	0	0	0	0	0	0	0	0
	Il	3.8	5.6	3.5	3.49	4.92	5.17	4.07	4.95	5.04
	Ap	1.16	1.52	1.06	1.05	1.73	1.46	1.03	1.37	1.73
岩石化学常用参数	$Mg^{\#}$	49.59	59.41	63.94	59.81	59.56	52.14	54.37	50.97	47.82
	AlK	3.8	6.73	5	5	6.01	5.67	4.43	4.65	5.17
	Na_2O/K_2O	3.18	1.64	1.38	1.25	1.8	1.95	2.85	3.43	2.34
	FeO^*/MgO	1.81	1.22	1.01	1.2	1.21	1.64	1.5	1.72	2.09
	FeO^*	10.53	9.61	9.55	10.7	10.24	9.95	10.97	11.7	10.64
	δ	3.01	31.9	9.19	10.16	8.14	5.96	2.76	4.65	9.28
	τ	6.45	3.98	6.77	6.84	3.9	4.12	5.22	4.63	5.01
	AR	1.35	1.75	1.5	1.51	1.67	1.55	1.45	1.44	1.53
	SI	27.84	31.98	38.91	35.48	33.92	27.74	31.91	29.12	25.76
	CI	40.78	42.95	44.55	44.88	46.63	43.16	43.13	41.21	38.63
	DI	30.92	38.77	32.04	31.62	37.29	37.09	34.54	33.96	36.50

续表

岩性		粗面岩		副长石岩	英安岩		玻基辉橄岩		霞石岩
样品号		9DG	9DG-2	S04-13-3B	ZF-1	ZF-2	B 高-1	B 高-2	B75-31-1
产地		黄岩岛		中央海盆	尖峰海山		西沙群岛		
序号		50	51	52	53	54	55	56	57
岩石化学成分/%	SiO_2	60.38	63.59	40.08	67.21	58.06	37.09	41.22	40.59
	Al_2O_3	17.56	17.55	16.06	13.22	15.67	9.01	10.74	11.14
	Fe_2O_3	0	0	8.7	2.55	4.23	7.63	5.01	2.78
	FeO	4.84	3.97	2.79	2.43	3.33	3.54	6.98	10.01
	MgO	0.51	0.38	3.58	0.59	1.7	12.73	9.29	13.76
	CaO	1.73	1.16	12.26	2.07	2.85	8.85	12.84	11.37
	Na_2O	5.8	5.52	3.23	4.41	5.46	3.23	3.17	2.76
	K_2O	5.13	6.01	1.38	1.92	3.7	1.65	1.7	1.77
	MnO	0.55	0.22	0.22	0.22	0.65	0.16	0.17	0.2
	TiO_2	0.48	0.31	3.51	0.23	0.82	3.08	3.4	3.86
	P_2O_5	0.15	0.07	3.42	0	0.4	1.37	1.65	1.75
	CO_2	0	0	0.5	0	0	0	0	0
	H_2O^+	1.7	0.66	2.65	4.79	2.22	11.45	3.47	0
	烧失量								
	总量	98.83	99.44	98.38	99.64	99.09	99.79	99.64	99.99
标准矿物成分含量/%	Q	0.04	3.23	0	27.36	1.16	0	0	0
	Or	31.21	35.96	8.63	12	22.66	11.12	10.48	8.06
	Ab	50.42	47.2	18.53	39.37	47.78	1.94	1.84	0
	An	6.87	5.37	26.65	10.85	7.54	5.92	10.46	12.76
	Ne	0	0	5.6	0	0	15.8	14.14	12.65
	Lc	0	0	0	0	0	0	0	1.89
	Di	0.92	0.04	13.58	0	3.97	27.87	36.94	26.68
	Hy	7.96	6.43	0	8.58	12.41	0	0	0
	Ol	0	0	8.55	0	0	23.66	11.98	22.94
	Mt	1.31	1.04	3.52	1.26	1.97	3.63	3.66	3.86
	Hm	0	0	0	0	0	0	0	0
	Il	0.94	0.6	7.05	0.46	1.61	6.67	6.74	7.34
	Ap	0.34	0.15	7.89	0	0.9	3.41	3.76	3.82
岩石化学常用参数	$Mg^{\#}$	15.82	14.58	37.54	18.21	29.81	68.56	59.04	66.22
	AlK	10.93	11.53	4.61	6.33	9.16	4.88	4.87	4.53
	Na_2O/K_2O	1.13	0.92	2.34	2.3	1.48	1.96	1.87	1.56
	FeO^*/MgO	9.49	10.45	2.97	8	4.2	0.82	1.24	0.91
	FeO^*	4.84	3.97	10.62	4.72	7.14	10.41	11.49	12.51
	δ	6.87	6.46		1.66	5.57			
	τ	24.5	38.81	3.66	38.3	12.45	1.88	2.23	2.17
	AR	3.61	4.21	1.39	2.41	2.96	1.75	1.52	1.5
	SI	3.13	2.39	18.19	4.96	9.23	44.23	35.53	44.27
	CI	11.12	8.1	32.7	10.3	19.96	61.82	59.32	60.83
	DI	81.68	86.38	32.76	78.73	71.6	28.86	26.47	22.6

注：1~2、7~17 据王贤觉等（1984）；3~6、53、54 据李兆麟等（1991）；18~29、43~49 据鄢全树等（2008）；50、51 据王叶剑等（2009）；30、33~37、39~41、46~49、57 据涂勘等（1992）；31、32、38、42、55、56 据孙嘉诗（1991）；$Mg^{\#}=100\times Mg/(Mg+Fe)$。

从表 13-5 可知，琼北新近纪火山熔岩的 SiO_2 含量在 39.40×10^{-2}～54.93×10^{-2}之间，属于超基性岩类—基性岩类。超基性岩类有橄榄霞石岩和玻基辉橄岩。橄榄霞石岩与安徽女山霞石岩相对比，Al_2O_3、Na_2O、K_2O、Fe_2O_3略低，而 FeO、MgO、CaO 等略高，但 FeO+Fe_2O_3的含量相当；玻基辉橄岩与内蒙古达尔泊玻基辉橄岩相对比，MgO 略高，Na_2O、K_2O、TiO_2 略低外，其余的基本一致。基性岩类（玄武岩类）SiO_2 43.61%～54.93%（平均 50.58%）、Al_2O_3 10.58%～15.82%（平均 13.63%）、Fe_2O_3 1.14%～10.23%（平均 3.42%）、FeO 4.90%～10.96%（平均 7.62%）、FeO+Fe_2O_3 8.93%～16.26%（平均 11.04%）、MgO 4.73%～11.56%（平均 7.69%）、CaO 6.31%～10.57%（平均 8.19%）、Na_2O 2.30%～4.12%（平均 3.03%）、K_2O 0.3%～2.11%（平均 1.14%）、TiO_2 1.58%～3.0%（平均 2.10%），其平均值与中国东部玄武岩平均值相比，SiO_2、FeO、MgO 等略高，Al_2O_3、Fe_2O_3、CaO、Na_2O、K_2O 等含量低。岩石中 K_2O 含量相差较大，在 SiO_2 对 Al_2O_3、FeO^*、MgO、CaO、Na_2O、K_2O、TiO_2 和 MnO 的相关图（图 13-3）上，SiO_2 与 Al_2O_3呈正相关关系，与 Na_2O 呈不明显的正相关关系，与 FeO^*、MgO、CaO、K_2O、TiO_2 和 MnO 呈明显的负相关关系。

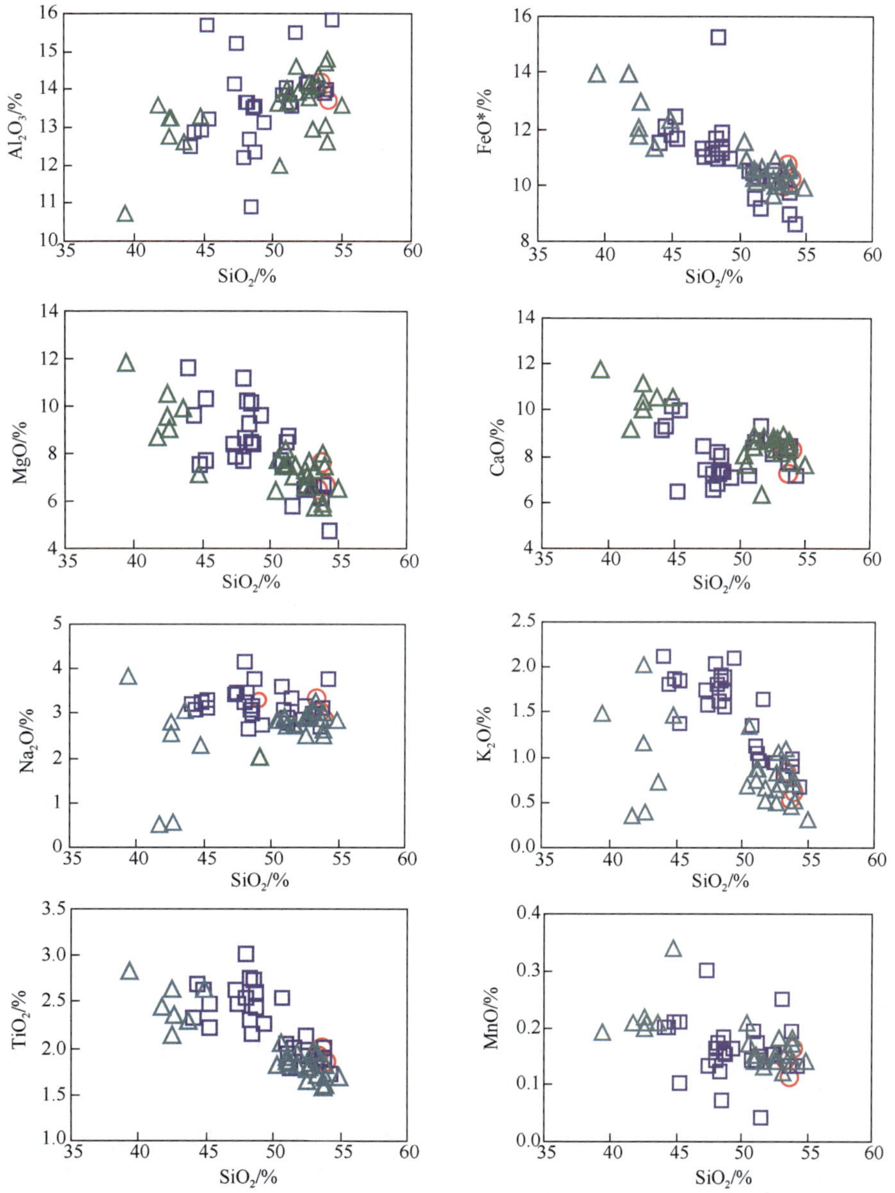

图 13-3 琼北新近纪火山岩成分变异图

Fig. 13-3 Composition variation diagram of Neogene volcanic rocks from Qiongbei

○金牛岭期；□石门沟村期；△石马村期

琼北新近纪火山熔岩的 Mg#值为 49.01~64.30，平均为 55.85。SI 值一般在 26.18~40.32 之间，主要为 30~40，平均值为 33.41，略比原生玄武岩浆的 SI 值低。CI 值在 29.41~52.01 之间，平均值为 42.53。DI 值在 28.92~46.0 之间，平均值为 35.77，说明琼北新近纪火山熔岩的岩浆分离结晶作用程度差。δ 值为 0.82~25.44，除少数样品的 δ 值略高外，大部分的值均小于 4，按里特曼岩系划分主要属于钙碱性系列，石门沟村组有部分样品的 δ 大于 4，属于碱性系列。琼北新近纪火山熔岩在硅-碱图（图 13-1）中投影点一部分落入碱性系列，另一部分落入亚碱性系列。其中，海口组的火山岩均落入亚碱性系列；石门沟村组的火山熔岩大部分落入碱性系列，少部分落入亚碱性系列；石马村组的基性熔岩均落入亚碱性系列，而落入亚碱性系列的岩石在 FAM 图解中投影点主要落入钙碱性系列，但均分布在分界线附近（图 13-4）。从CIPW标准矿物看，琼北新近纪玄武岩主要矿物组合为 Or+Ab+An+Di+Hy，部分样品出现 Q，另一部分出现 Ol，其中有 9 个样品出现 Ne（霞石），岩石中还有少量副矿物 Ap+Ilm+Mt。

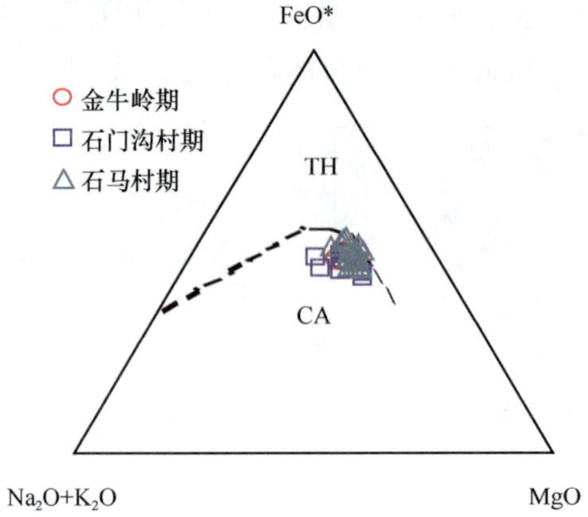

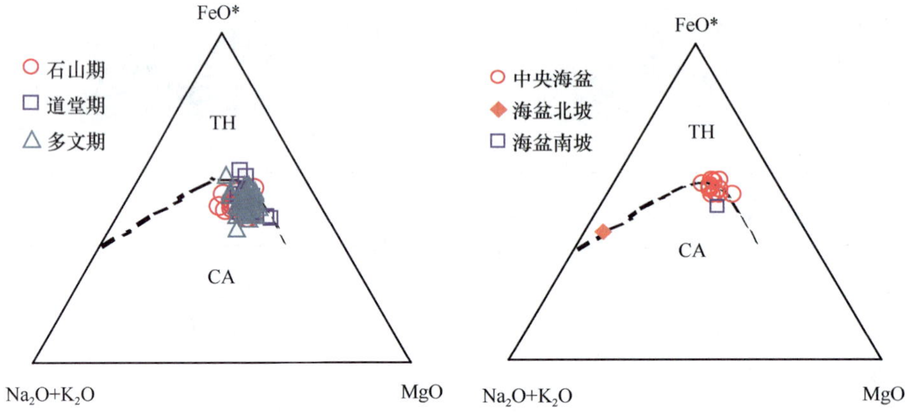

图 13-4　琼北及南海新生代火山岩 FAM 图

Fig. 13-4　FAM diagram of Cenozoic volcanic rocks of Qiongbei and the South China Sea

TH—拉斑玄武岩系列；CA—钙碱性系列

从表 13-6 可知，琼北第四纪火山岩基本属基性岩类。成分含量为 SiO_2 44.38%~55.49%（平均 51.44%）、Al_2O_3 12.47%~16.60%（平均 14.05%）、Fe_2O_3 0.58%~12.22%（平均 3.20%）、FeO 1.72%~10.45%（平均 8.12%）、Fe_2O+FeO 9.18%~12.47%（平均 11.04%）、MgO 4.22%~10.47%（平均 7.21%）、CaO 6.73%~10.63%（平均 8.39%）、Na_2O 2.02%~4.56%（平均 3.07%）、K_2O 0.45%~2.68%（平均 1.21%）、TiO_2 1.40%~3.38%（平均 2.01%）。其平均值与中国东部玄武岩平均值相比，SiO_2、FeO 略高，Al_2O_3、Fe_2O_3、Na_2O、K_2O 略低，但 Fe_2O+FeO 相差不多。而与新近纪玄武岩平均值相比，SiO_2、Al_2O_3、FeO、CaO 略高，Fe_2O_3、MgO 略低，其中 Fe_2O+FeO 相差不大。在 SiO_2 对 Al_2O_3、FeO^*、MgO、CaO、Na_2O、K_2O、TiO_2 和 P_2O_5 的相关图上(图 13-5)，SiO_2 与 Al_2O_3、Na_2O

呈明显的正相关关系，与 FeO^*、MgO、CaO、K_2O、TiO_2 和 MnO 呈明显的负相关关系。琼北第四纪火山岩的 $Mg^\#$值为 41.88~62.95，平均为 54.25。SI 值一般为22.59~40.99，平均为 31.76，略比原生玄武岩浆的 SI 值低。CI 值在 24.17~49.99 之间，平均值为 41.69。DI 值在 23.17~51.55 之间，平均值为 35.67，说明琼北第四纪火山熔岩的岩浆分离结晶作用程度差。δ 值为 0.80~8.12，平均为 2.51，除个别大于 4 外，大部分岩石的 δ 值小于 4，按里特曼岩系划分主要属于钙碱性系列，个别为碱性系列。琼北第四纪火山熔岩在硅-碱图（图 13-1），部分落入碱性一侧，但主要位于碱性与亚碱性的界线附近，而落入亚碱性系列的岩石在 FAM 图解中投影点主要落入钙碱性系列（图 13-4）。从 CIPW 标准矿物看，琼北第四纪火山熔岩主要矿物组合为Or+Ab+An+Di+Hy，部分样品出现 Q，另一部分出现 Ol，其中有 3 个样品出现 Ne（霞石），岩石中还有少量副矿物 Ap+Ilm+Mt 。

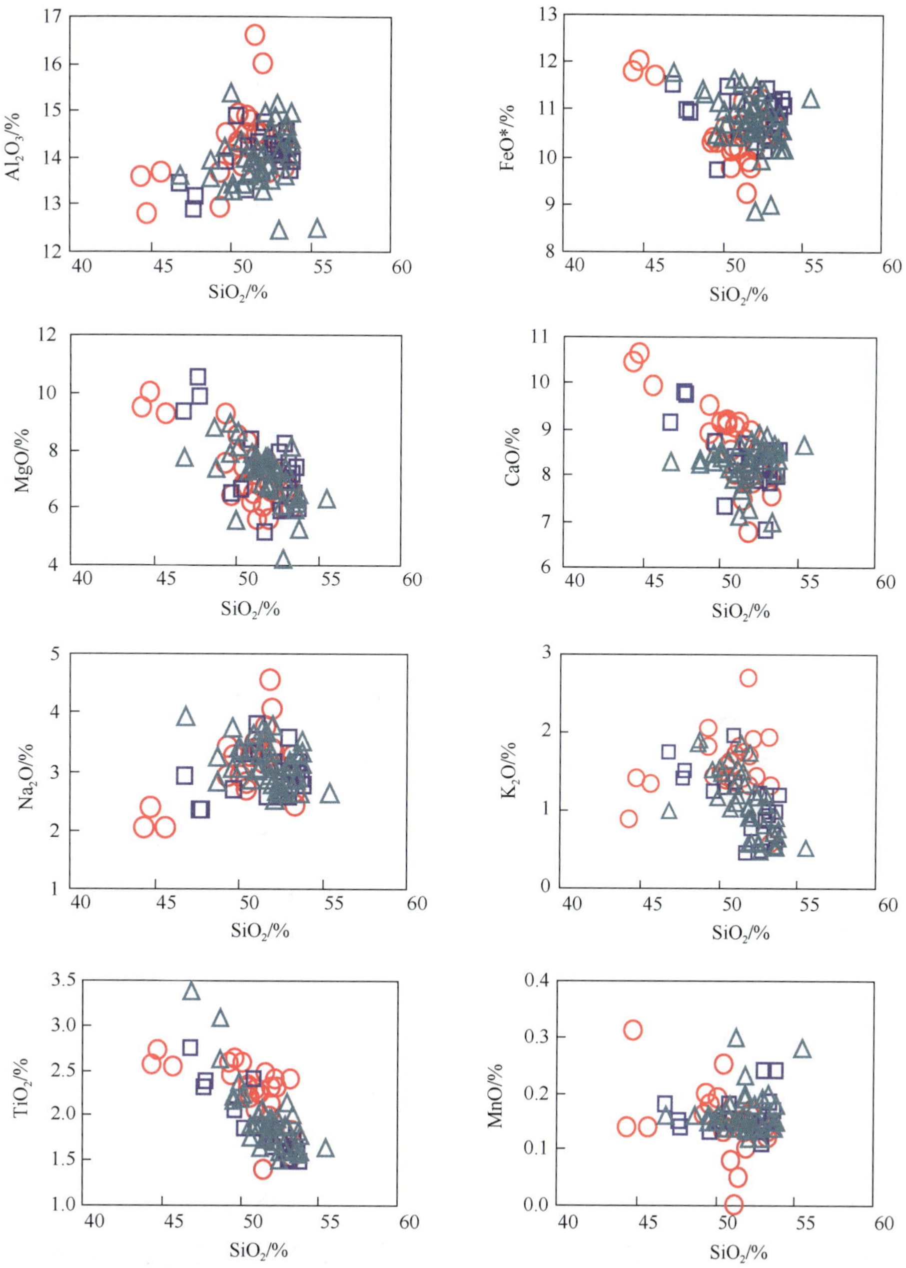

图 13-5 琼北第四纪火山岩成分变异图

Fig. 13-5 Material composition variation diagram of Quaternary volcanic rocks from Qiongbei

○石山期；□道堂期；△多文期

从表 13-7 可知，南海新生代火山岩的 SiO_2 含量在 37.0%~67.21%，除少量中性和超基性岩外，主要属于基性岩类。其中，超基性岩为霞石岩和玻基辉橄岩，中性岩为英安岩和安山岩类。南海新生代基性火山岩其他成分含量为 Al_2O_3 9.01%~16.74%（平均 14.98%）、Fe_2O_3 1.01%~9.59%（平均 4.14%）、FeO 3.33%~10.21%（平均 7.50%）、FeO+Fe_2O_3 10.11%~12.92%（平均 11.64%）、MgO 2.69%~13.76%（平均 6.23%）、CaO 6.58%~12.84%（平均 9.76%）、Na_2O0.60%~3.86%（平均 3.10%）、K_2O 0.51%~2.67%（平均 1.52%）、TiO_2 1.93%~3.86%（平均 2.74%）。其平均值与中国东部玄武岩平均值相对比，FeO、TiO_2、CaO 等略高，MgO、Na_2O、K_2O、Fe_2O_3 等含量低，但总体含量相差不大。在 SiO_2 对 Al_2O_3、FeO^*、MgO、CaO、Na_2O、K_2O、TiO_2 和 MnO 的相关图上（图 13-6），SiO_2 与 Al_2O_3、Na_2O 呈明显的正相关关系，与 FeO^*、MgO、CaO、TiO_2 呈明显的负相关关系，与 K_2O、MnO 呈不明显的负相关关系。

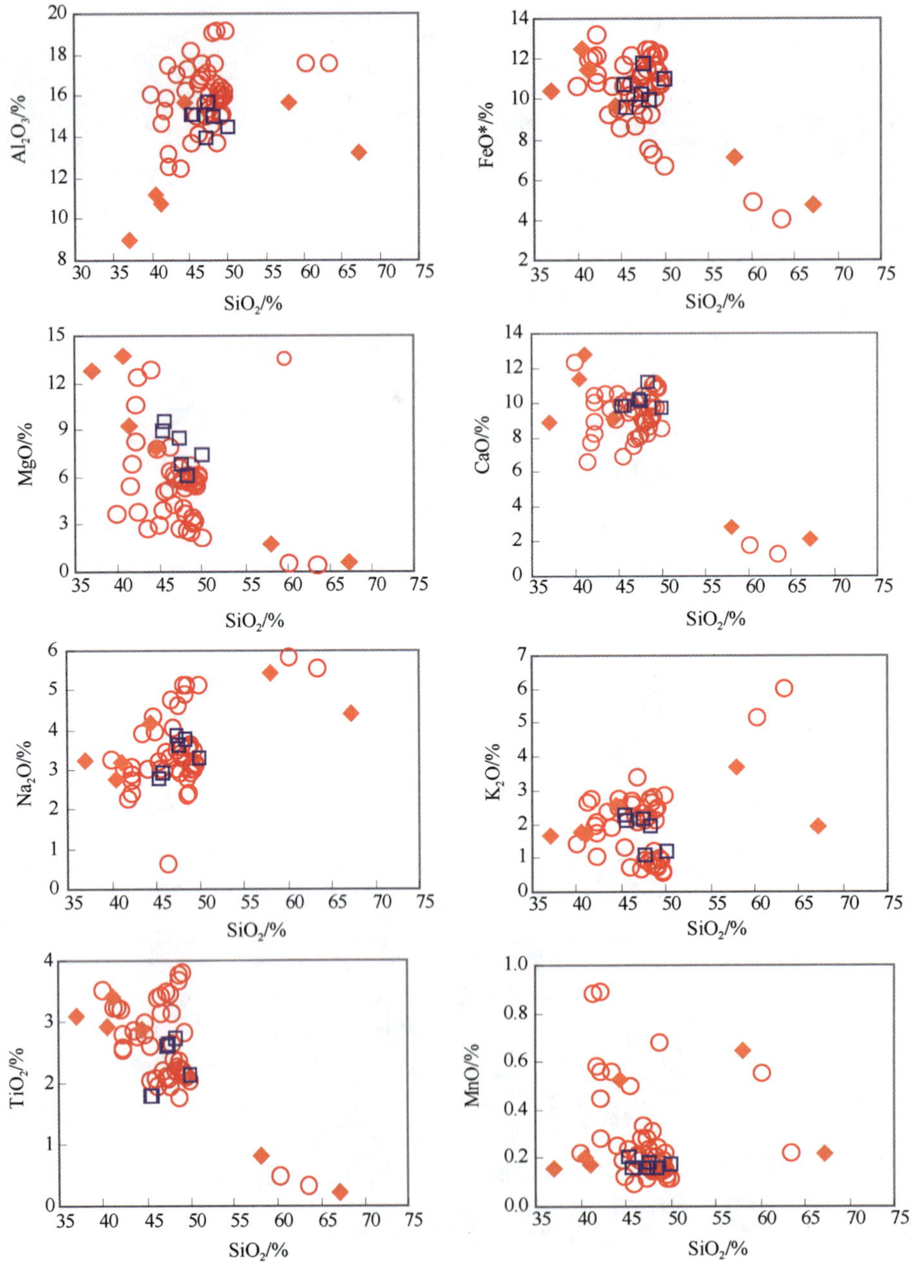

图 13-6 南海新生代火山岩成分变异图

Fig. 13-6 Material composition variation diagram of Cenozoic volcanic rocks from the South China Sea

○中央海盆；◆海盆北坡；□海盆南坡

南海新生代基性超基性火山岩的 $Mg^{\#}$值为 30.31~68.56，平均为 48.15。SI 值一般在 1120~44.27 之间，平均值为 27.04。δ 值为 1.72~11.20，部分样品的 δ 值大于 4，部分样品的 δ 值均小于 4。火山岩在硅-碱图（图 13-1）中投影点大部分落入碱性系列，小部分落入亚碱性系列，且靠近分界线。亚碱性系列的岩石在 Al_2O_3-"An" 图解中投影点主要落入钙碱性系列，且位于分界线旁(图 13-4)。从 CIPW 标准矿物看，南海新生代火山岩主要矿物组合为 Or+Ab+An+Di+Hy，部分样品出现 Q，另一部分样品出现 Ol，其中有 9 个样品出现 Ne（霞石），岩石中还有少量副矿物Ap+Ilm+Mt 。

二、岩石稀土元素地球化学特征

新生代火山熔岩稀土元素分析结果及部分特征值参见表 13-8~表 13-10。

表 13-8　海南岛新近纪火山岩稀土元素含量及主要特征参数一览表

Table 13-8　List of REE contents and main characteristic parameters of Neogene volcanic rocks from the Hainan Island

地层单位		海口组		石门沟村组								
岩性		石英拉斑玄武岩		橄榄玄武岩								
样品号		1-4	金 YQ_2	P2-11	P2-13	D0083	D1835	D1836	D1842-1	D1871-1	$PM001_9$	ZK_2-1
序号		1	2	3	4	5	6	7	8	9	10	11
岩石类型		QT	QT	QT	OB	QT	OT	QT	OT	B	OT	QT
稀土元素含量/10^{-6}	La	12	5.0	20.31	62.79	13.13	40.88	28.67	2.32	94.97	25.25	18.70
	Ce	29	10	37.50	116.2	25.44	76.58	50.18	4.93	164.0	50.32	36.24
	Pr	3.9	1.6	4.72	13.68	3.58	9.77	5.93	0.79	19.77	6.72	5.01
	Nd	19	7.0	21.21	55.43	14.36	39.25	22.42	4.82	74.53	28.70	18.85
	Sm	4.5	2.0	5.89	11.97	4.36	8.14	5.14	1.79	13.64	6.82	4.78
	Eu	1.7	0.75	1.9	129	1.69	2.67	1.85	0.72	4.2	2.40	1.73
	Gd	5.4	2.0	6.86	10.99	4.93	7.78	4.98	2.31	11.11	6.65	5.00
	Tb	0.85	0.24	0.97	1.16	0.75	1.09	0.79	0.42	1.63	0.97	0.77
	Dy	3.8	1.5	5.53	7.28	3.96	5.19	4.13	2.77	7.72	4.81	3.90
	Ho	0.85	0.28	1.08	1.20	0.79	0.86	0.71	0.58	1.40	0.87	0.77
	Er	2.5	0.9	2.55	2.86	1.90	2.07	1.85	1.56	3.15	2.01	1.91
	Tm	0.26	0.09	0.32	0.30	0.27	0.27	0.26	0.24	0.40	0.28	0.26
	Yb	1.6	0.7	2.02	1.92	1.46	1.52	1.51	1.46	2.28	1.39	1.40
	Lu	0.25	0.10	0.37	0.34	0.22	0.19	0.20	0.20	0.30	0.20	0.21
	ΣREE	85.61	32.16	111.26	289.61	77.16	196.26	128.62	24.91	399.10	137.39	99.53
主要特征参数	LREE/HREE	4.52	4.54	4.64	10.12	4.40	9.35	7.91	1.61	13.26	7.00	6.00
	δEu	1.05	1.14	0.91	0.91	1.11	1.01	1.10	1.07	1.01	1.08	1.07
	δCe	1.02	0.85	0.89	0.91	0.88	0.89	0.88	0.87	0.87	0.91	0.88
	$(La/Yb)_N$	5.06	4.82	6.68	22.05	6.06	18.13	12.80	1.07	28.08	12.25	9.01
	$(Ce/Yb)_N$	4.69	3.70	4.73	15.65	4.51	13.03	8.60	0.87	18.61	9.36	6.70
	$(La/Sm)_N$	1.68	1.57	2.17	3.30	1.89	3.16	3.51	0.82	4.38	2.33	2.46
	$(Gd/Yb)_N$	2.72	2.31	2.70	4.62	2.72	4.13	2.66	1.28	3.93	3.86	2.88
	La/Sm	2.67	2.5	3.45	5.25	3.01	5.02	5.58	1.3	6.96	3.7	3.91

续表

地层单位		石门沟村组							石马村组			
岩性		橄榄玄武岩			粗玄岩		辉石玄武岩		橄榄玄武岩			
样品号		D0460	福基田-1	ZK_3-1	D0046	P2-14	$PM001_5$	$PM001_6$	P2-4	P2-6	D1872	D1900
序号		12	13	14	15	16	17	18	19	20	21	22
岩石类型		QT	OB	AOB	QT	QT	QT	QT	QT	OT	QT	AOB
稀土元素含量/10^{-6}	La	17.26	47.14	95.95	19.26	32.89	23.64	12.16	15.83	46.13	18.6	81.58
	Ce	32.35	89.44	170.7	36.27	38.74	43.94	23.67	31.03	87.74	35.13	142.2
	Pr	4.25	11.99	18.70	4.67	5.66	5.82	3.37	3.81	9.88	4.55	15.36
	Nd	17.57	45.17	71.10	18.57	26.22	22.08	14.13	16.75	39.41	18.86	58.22
	Sm	4.76	9.68	13.04	4.72	7.56	5.33	4.20	4.67	8.29	4.78	10.43
	Eu	1.82	3.17	4.19	1.70	2.53	1.91	1.60	1.52	2.46	1.76	3.34
	Gd	5.33	8.27	11.48	4.97	9.43	5.55	4.65	5.34	7.85	5.17	9.27
	Tb	0.81	1.16	1.55	0.77	1.39	0.84	0.71	0.74	0.96	0.77	1.29
	Dy	4.52	5.62	7.19	4.05	8.17	4.31	3.76	4.35	5.65	4.27	6.57
	Ho	0.86	1.05	1.33	0.76	1.67	0.83	0.73	0.86	1.04	0.73	1.20
	Er	2.19	2.33	2.99	1.88	3.89	2.05	1.78	1.98	2.50	1.82	2.82
	Tm	0.30	0.29	0.40	0.26	0.50	0.28	0.25	0.29	0.36	0.26	0.39
	Yb	1.66	1.57	2.20	1.43	3.03	1.59	1.43	1.60	1.91	1.48	2.13
	Lu	0.25	0.22	0.31	0.21	0.50	0.23	0.21	0.24	0.39	0.20	0.29
	ΣREE	93.93	227.10	401.13	99.52	142.18	118.40	72.70	89.01	214.57	98.38	335.09
主要特征参数	LREE/HREE	4.90	10.07	13.61	5.94	3.97	6.55	4.37	4.78	9.39	5.69	12.99
	δEu	1.10	1.06	1.03	1.07	0.92	1.07	1.10	0.93	0.92	1.08	1.02
	δCe	0.88	0.88	0.91	0.89	0.63	0.88	0.88	0.93	0.95	0.89	0.90
	$(La/Yb)_N$	7.01	20.24	29.40	9.08	7.32	10.02	5.73	6.67	16.28	8.47	25.82
	$(Ce/Yb)_N$	5.04	14.74	20.07	6.56	3.31	7.15	4.28	5.02	11.88	6.14	17.27
	$(La/Sm)_N$	2.28	3.06	4.63	2.57	2.74	2.79	1.82	2.13	2.50	2.45	4.92
	$(Gd/Yb)_N$	2.59	4.25	4.21	2.80	2.51	2.82	2.62	2.69	3.32	2.82	3.51
	La/Sm	3.63	4.87	7.36	4.08	4.35	4.43	2.9	3.39	5.56	3.89	7.82

续表

地层单位		石马村组									
岩性		橄榄玄武岩			辉石玄武岩		玄武岩				
样品号		P2-1	D0752	$PM005_{6-5}$	$PM001_{19-2}$	ⅡH17-1	D1831-1	D1832-1	D1834-1	D1016	D2174
序号		23	24	25	26	27	28	29	30	31	32
岩石类型		QT	QT	QT	QT	QT	QT	QT	QT	QT	QT
稀土元素含量/10^{-6}	La	38.44	32.76	22.97	12.10	17.69	8.24	19.53	11.28	12.43	19.41
	Ce	72.21	32.45	40.98	23.56	22.78	16.86	37.13	24.06	24.17	37.32
	Pr	5.74	5.44	5.04	3.17	3.95	2.58	4.58	3.16	3.36	4.83
	Nd	18.89	22.28	20.07	13.61	18.57	11.05	18.57	13.77	13.90	19.27
	Sm	5.34	5.61	5.04	3.97	5.07	3.37	4.62	3.87	3.95	4.95
	Eu	2.46	2.11	1.77	1.54	2.00	1.37	1.74	1.46	1.52	1.81
	Gd	13.98	7.56	4.95	4.65	6.81	4.14	5.09	4.33	4.58	5.27
	Tb	2.82	1.07	0.80	0.68	1.00	0.63	0.76	0.7	0.72	0.80
	Dy	20.84	6.50	3.94	3.85	5.74	3.5	4.13	3.71	3.88	4.23
	Ho	5.71	1.29	0.77	0.75	1.04	0.65	0.78	0.66	0.77	0.82
	Er	15.05	3.32	1.84	1.86	2.83	1.7	2.0	1.76	1.93	2.00
	Tm	2.56	0.42	0.27	0.26	0.37	0.23	0.29	0.25	0.28	0.27
	Yb	12.74	2.39	1.52	1.42	2.11	1.44	1.66	1.43	1.52	1.59
	Lu	2.18	0.33	0.21	0.22	0.30	0.20	0.23	0.20	0.22	0.23
	ΣREE	218.96	123.53	110.17	71.58	90.26	55.96	101.11	70.64	73.23	102.78
主要特征参数	LREE/HREE	1.89	4.40	6.70	4.23	127	128	5.77	4.42	4.27	5.76
	δEu	0.82	0.99	1.07	1.09	1.04	1.12	1.09	1.09	1.09	1.08
	δCe	1.04	0.53	0.88	0.90	0.63	0.87	0.91	0.96	0.88	0.90
	$(La/Yb)_N$	2.03	9.24	10.19	5.74	5.65	3.86	7.93	5.32	5.51	8.23
	$(Ce/Yb)_N$	1.47	3.51	6.97	4.29	2.79	3.03	5.79	4.35	4.11	6.07
	$(La/Sm)_N$	4.53	3.67	2.87	1.92	2.19	1.54	2.66	1.83	1.98	2.47
	$(Gd/Yb)_N$	0.89	2.55	2.63	2.64	2.60	2.32	2.47	2.44	2.43	2.67
	La/Sm	7.2	5.84	4.56	3.05	3.49	2.45	4.23	2.91	3.15	3.92

续表

地层单位		石马村组									
岩性		玄武岩				石英拉斑玄武岩	玻基辉橄岩				橄榄霞石岩
样品号		D2297-2	D3461	D1896-2	ZK_4-1	194-2	ZK_2-2	ZK_2-3	D0744-2	P2-3	185-1
序号		33	34	35	36	37	38	39	40	41	42
岩石类型		QT	QT	OT	QT	QT					NE
稀土元素含量/10^{-6}	La	17.82	17.86	49.59	17.12	15	77.66	76.34	75.30	81.69	94
	Ce	33.15	33.76	89.69	32.35	34	129.8	133.5	136.5	157.7	180
	Pr	4.43	4.68	10.63	3.93	4.3	15.13	15.20	16.29	17.02	18
	Nd	17.57	17.75	40.99	16.67	19	54.57	53.74	58.18	61.78	71
	Sm	4.69	4.82	8.41	4.43	3.8	10.05	9.96	11.01	11.58	14
	Eu	1.72	1.64	2.80	1.67	1.6	3.25	3.20	3.53	3.22	3.2
	Gd	5.04	4.78	7.74	5.18	4.5	8.94	8.74	9.56	10.43	9.0
	Tb	0.77	0.73	1.10	0.78	0.72	1.19	1.17	1.35	1.87	1.3
	Dy	4.18	3.85	5.53	4.13	3.6	5.95	5.83	6.88	7.58	7.3
	Ho	0.82	0.78	1.05	0.78	0.74	1.13	1.12	1.24	1.46	1.23
	Er	1.99	1.85	2.50	1.92	2.2	2.58	2.55	2.90	3.82	2.75
	Tm	0.28	0.25	0.34	0.27	0.29	0.34	0.35	0.40	0.74	0.40
	Yb	1.59	1.41	1.77	1.62	1.80	1.77	1.78	2.05	3.28	1.90
	Lu	0.23	0.22	0.26	0.24	0.28	0.26	0.26	0.30	0.6	0.30
	ΣREE	94.25	94.38	222.40	91.09	91.83	312.62	131.74	325.49	362.77	404.38
主要特征参数	LREE/HREE	5.33	5.80	9.96	5.11	5.5	13.11	13.39	12.19	11.18	15.72
	δEu	1.08	1.03	1.04	1.06	1.18	1.03	1.03	1.03	0.88	0.82
	δCe	0.87	0.87	0.90	0.92	1.01	0.86	0.89	0.90	0.97	0.99
	$(La/Yb)_N$	7.56	8.54	18.89	7.12	5.62	29.58	28.91	24.76	16.79	33.35
	$(Ce/Yb)_N$	5.39	6.19	13.11	5.17	4.89	18.97	19.40	17.22	12.44	24.50
	$(La/Sm)_N$	2.39	2.33	3.71	2.43	2.48	4.86	4.82	4.30	4.44	4.22
	$(Gd/Yb)_N$	2.56	2.74	3.53	2.58	2.02	4.08	3.96	3.76	2.57	3.82
	La/Sm	3.8	3.71	5.9	3.86	3.95	7.73	7.66	6.84	7.05	6.71

表 13-9 海南岛第四纪火山岩稀土元素含量及主要特征参数一览表

Table 13-9 List of REE contents and main characteristic parameters of Quaternary volcanic rocks from the Hainan Island

地层单位		石山组										
岩性		碱性橄榄玄武岩				橄榄玄武岩			橄榄拉斑玄武岩			
样品号		128-2	23-3	QB013	QB012	D1874	乐光 B	永兴 B	HN9902	QB011	HN9910	道 D344
序号		1	2	3	4	5	6	7	8	9	10	11
稀土元素含量/10^{-6}	La	38	37	51.49	49.98	25.34	20	20	25.48	25.52	64.06	11
	Ce	80	75	97.26	97.6	48.58	26	25	50.03	50.56	124.6	15
	Pr	7.6	7.5	12.22	12.2	6.52	4.3	12	6.67	6.692	15.32	1.7
	Nd	38	3.5	48.23	47.99	27.20	16	14	28.87	29.12	59.17	5.5
	Sm	6.8	6.4	8.868	8.862	6.63	3.7	3.1	6.608	6.667	10.28	1.2
	Eu	2.4	2.1	2.66	2.707	2.19	1.0	1.1	2.156	2.197	3.011	0.32
	Gd	6.8	5.3	7.614	7.865	6.14	2.9	2.2	6.496	6.601	8.409	0.65
	Tb	0.98	0.77	1.099	1.12	0.94	0.32	0.23	0.957	0.975	1.213	0.12
	Dy	5.0	3.2	5.77	5.919	4.81	2.5	1.7	5.055	5.098	6.305	0.35
	Ho	0.83	0.55	0.999	1.018	0.79	0.33	0.25	0.866	0.871	1.102	0.004
	Er	2.4	1.80	2.456	2.491	1.99	1.0	0.73	2.05	2.058	2.715	0.1
	Tm	0.31	0.21	0.329	0.329	0.27	0.10	0.07	0.281	0.278	0.372	0.019
	Yb	1.7	1.30	1.956	2.013	1.50	0.85	0.48	1.635	1.626	2.194	0.18
	Lu	0.26	0.21	0.296	0.303	0.20	0.11	0.093	0.241	0.248	0.339	0.027
	ΣREE	191.08	176.34	241	240	133.10	79.10	72.35	137	139	299	36.21
主要特征参数	LREE/HREE	9.45	12.22	10.76	10.42	7.00	8.77	11.58	6.82	6.80	12.20	23.30
	δEu	1.07	1.07	0.97	0.97	1.03	0.90	1.23	0.99	1.00	0.96	1.01
	δCe	1.07	1.03	0.90	0.92	0.89	0.65	0.67	0.90	0.91	0.93	0.75
	$(La/Yb)_N$	15.07	19.19	17.75	16.74	11.39	15.86	28.09	10.51	10.58	19.68	41.20
	$(Ce/Yb)_N$	12.17	14.92	12.86	12.54	8.38	7.91	13.47	7.91	8.04	14.69	21.56
	$(La/Sm)_N$	3.52	3.64	3.65	3.55	2.40	3.40	4.06	2.43	2.41	3.92	5.77
	$(Gd/Yb)_N$	3.23	3.29	3.14	3.15	3.30	2.75	3.70	3.21	3.28	3.09	2.91
	La/Sm	5.59	5.78	5.81	5.64	3.82	5.41	6.45	3.86	3.83	6.23	9.17

地层单位		石山组			道堂组							
岩性		玄武岩			橄榄玄武岩		玄武岩					
样品号		106B1	EMB4	EMB6	220B3	101B1	305B6	XⅧ7-1	10-4	XⅦ2-1	XⅦ3-2	XⅧ5-1
序号		12	13	14	15	16	17	18	19	20	21	22
稀土元素含量/10^{-6}	La	29.52	14.09	18.33	33.76	19.91	12.6	8.96	26	22.74	63.88	9.72
	Ce	57.63	25.986	36.79	66.81	39.48	25	15.32	67	17.86	72.83	16.39
	Pr	7.381	3.308	4.946	8.574	5.322	3.406	2.27	6.2	3.87	10.32	2.42
	Nd	30.78	15.62	21.47	34.7	23.36	15.17	12.51	28	20.74	48.10	13.13
	Sm	6.64	4.079	5.106	6.802	5.501	3.803	3.70	5.5	5.59	10.67	3.73
	Eu	2.111	1.547	1.719	2.109	1.842	1.341	1.42	2.0	1.99	3.40	1.44
	Gd	6.435	5.153	5.273	6.232	5.552	4.315	4.51	5.6	7.21	11.15	4.65
	Tb	0.957	0.771	0.784	0.903	0.834	0.669	0.69	0.85	1.02	1.46	0.73
	Dy	5.223	4.408	4.247	4.839	4.492	3.75	4.01	4.1	5.69	8.18	4.14
	Ho	0.909	0.808	0.728	0.843	0.775	0.688	0.74	0.76	1.06	1.42	0.73
	Er	2.19	1.942	1.739	2.066	1.868	1.679	1.82	2.2	2.60	3.37	1.82
	Tm	0.298	0.274	0.235	0.286	0.257	0.238	0.27	0.29	0.35	0.43	0.28
	Yb	1.782	1.604	1.379	1.724	1.512	1.442	1.43	1.85	1.78	2.20	1.46
	Lu	0.261	0.245	0.212	0.264	0.225	0.217	0.21	0.28	0.28	0.33	0.22
	ΣREE	152	80	103	169	110	74	57.86	150.63	92.78	237.74	60.86
主要特征参数	LREE/HREE	7.42	4.25	6.05	8.90	6.15	4.72	3.23	8.46	3.64	7.33	3.34
	δEu	0.98	1.03	1.00	0.97	1.01	1.01	1.06	1.09	0.96	0.95	1.07
	δCe	0.92	0.89	0.91	0.92	0.91	0.90	0.80	1.23	0.42	0.62	0.79
	$(La/Yb)_N$	11.17	5.92	8.96	13.20	8.88	5.89	4.22	9.48	8.61	19.58	4.49
	$(Ce/Yb)_N$	8.37	4.19	6.90	10.02	6.75	4.48	2.77	9.37	2.60	8.56	2.90
	$(La/Sm)_N$	2.80	2.17	2.26	3.12	2.28	2.08	1.52	2.97	2.56	3.77	1.64
	$(Gd/Yb)_N$	2.91	2.59	3.09	2.92	2.96	2.41	2.54	2.44	3.27	4.09	2.57
	La/Sm	4.45	3.45	3.59	4.96	3.62	3.31	2.42	4.73	4.07	5.99	2.61

续表

地层单位		道堂组					多文组					
岩性		橄榄玄武岩		橄榄拉斑玄武岩			橄榄拉斑玄武岩				玄武岩	
样品号		D1894	D1019	150-1	13-1	安民 B	老城 B	13-1	Ⅱ270-TR1	100-2	D1828	D1837
序号		23	24	25	26	27	28	29	30	31	32	33
稀土元素含量 10^{-6}	La	20. 31	14. 95	16	20	11	11	23	19	16	12. 21	17. 34
	Ce	38. 40	30. 75	34	24	7. 0	17	55	15	34	25. 52	34. 94
	Pr	4. 88	3. 74	4. 0	3. 7	1. 1	2. 7	5. 6	3. 1	4. 0	129	4. 73
	Nd	21. 74	17. 33	19	16	3. 8	11	24	14	19	15. 01	18. 79
	Sm	5. 25	4. 37	3. 7	3. 3	1. 0	2. 7	4. 8	3. 6	4. 3	3. 93	4. 83
	Eu	1. 94	1. 60	1. 5	1. 3	0. 29	1. 1	1. 9	1. 1	1. 8	1. 53	1. 61
	Gd	5. 58	4. 93	4. 3	2. 9	0. 75	2. 5	5. 2	4. 6	5. 4	4. 63	4. 89
	Tb	0. 83	0. 75	0. 7	0. 26	0. 1	0. 27	0. 80	0. 45	0. 8	0. 67	0. 74
	Dy	4. 62	4. 45	3. 3	1. 5	0. 6	1. 6	4. 0	3. 1	4. 4	3. 72	4. 03
	Ho	0. 81	0. 8	0. 66	0. 35	0. 09	0. 32	0. 76	0. 63	0. 8	0. 68	0. 68
	Er	1. 88	2. 03	1. 8	1. 0	0. 23	0. 95	2. 2	1. 9	0. 23	1. 79	1. 77
	Tm	0. 27	0. 29	0. 27	0. 08	0. 026	0. 08	0. 30	0. 15	0. 28	0. 26	0. 25
	Yb	1. 55	1. 65	1. 7	0. 65	0. 20	0. 9	1. 85	1. 4	1. 65	1. 45	1. 44
	Lu	0. 22	0. 24	0. 25	0. 09	0. 047	0. 10	0. 30	0. 15	0. 26	0. 20	0. 19
	ΣREE	108. 28	87. 88	91. 18	75. 13	26. 23	52. 22	129. 71	68. 18	94. 99	75. 09	96. 23
主要特征参数	LREE/HREE	5. 87	4. 80	6. 02	10. 00	11. 86	6. 77	7. 42	4. 51	4. 98	4. 60	5. 88
	δEu	1. 09	1. 05	1. 15	1. 12	0. 98	1. 27	1. 16	0. 83	1. 14	1. 09	1. 00
	δCe	0. 90	0. 96	1. 00	0. 63	0. 39	0. 73	1. 13	0. 43	1. 00	0. 93	0. 91
	$(La/Yb)_N$	8. 83	6. 11	6. 35	20. 74	37. 08	18. 24	8. 38	9. 15	6. 54	5. 68	8. 12
	$(Ce/Yb)_N$	6. 41	4. 82	5. 17	9. 55	9. 05	4. 89	7. 69	2. 77	5. 33	4. 55	6. 28
	$(La/Sm)_N$	2. 43	2. 15	2. 72	3. 81	6. 92	2. 56	3. 01	3. 32	2. 34	1. 95	2. 26
	$(Gd/Yb)_N$	2. 91	2. 41	2. 04	3. 60	3. 03	2. 24	2. 27	2. 65	2. 64	2. 58	2. 74
	La/Sm	3. 86	3. 42	4. 32	6. 06	11	4. 07	4. 79	5. 28	3. 72	3. 11	3. 59

续表

地层单位		多文组										
岩性		玄武岩		粗玄岩			石英拉斑玄武岩			橄榄辉石玄武岩		
样品号		ⅡH14-2	0502	XⅦ1-1	0501	ⅠH3-3	32-1	信文 B	60-1	ⅡH4-3	ⅡH10-2	ⅠH4-2
序号		34	35	36	37	38	39	40	41	42	43	44
稀土元素含量/10^{-6}	La	8.97	9.79	23.44	25.54	7.47	31	5.5	15	26.71	25.88	32.82
	Ce	15.42	19.35	46.16	47.33	15.49	38	3.0	26	52.84	50.72	65.35
	Pr	2.40	2.72	4.16	5.782	2.32	5.7	0.6	3.5	7.03	6.39	8.71
	Nd	12.39	11.85	21.50	22.96	10.44	29	1.9	17	28.98	27.64	26.54
	Sm	124	3.303	5.14	5.396	3.04	6.5	0.5	4.3	6.87	6.43	8.52
	Eu	1.31	1.302	1.75	1.982	1.27	3.2	0.16	1.7	2.57	2.22	2.91
	Gd	4.05	3.699	5.09	5.162	3.68	11.7	0.44	5.5	7.14	6.33	8.08
	Tb	0.63	0.607	0.67	0.713	0.59	1.65	0.09	0.85	1.07	0.94	1.21
	Dy	3.54	3.315	3.64	3.53	3.27	10.5	0.35	4.5	5.36	4.74	2.70
	Ho	0.60	0.618	0.59	0.596	0.57	1.9	0.055	0.77	0.99	0.78	0.96
	Er	1.62	1.615	1.32	1.402	1.59	4.9	0.16	2.2	2.59	1.94	2.33
	Tm	0.23	0.219	0.18	0.179	0.22	0.64	0.025	0.33	0.33	0.26	0.30
	Yb	1.26	1.268	0.93	1.005	1.23	2.90	0.11	2.0	1.77	1.41	1.59
	Lu	0.17	0.191	0.14	0.141	0.17	0.45	0.036	0.28	0.24	0.19	0.20
	ΣREE	56.03	60.29	114.88	121.72	51.35	148.04	12.92	83.93	144.52	135.89	175.22
主要特征参数	LREE/HREE	3.63	4.23	8.15	8.56	3.54	3.27	9.25	4.11	6.40	7.19	7.60
	δEu	1.07	1.13	1.03	1.13	1.16	1.11	1.02	1.07	1.11	1.06	1.06
	δCe	0.79	0.91	1.05	0.90	0.89	0.64	0.33	0.83	0.91	0.92	0.91
	$(La/Yb)_N$	4.80	5.21	16.99	17.13	4.09	7.21	33.7	5.06	10.17	12.37	13.92
	$(Ce/Yb)_N$	3.17	4.04	12.89	12.18	3.26	3.39	7.05	3.36	7.72	9.30	10.63
	$(La/Sm)_N$	1.62	1.86	2.87	2.98	1.55	3.0	6.92	2.19	2.45	2.53	2.42
	$(Gd/Yb)_N$	2.59	2.35	4.42	4.14	2.41	3.26	3.23	2.22	3.26	3.62	4.10
	La/Sm	2.61	2.96	4.56	4.73	2.46	4.77	11.00	3.49	3.89	4.02	3.85

续表

地层单位		多文组										
岩性		橄榄玄武岩									橄榄拉斑玄武岩	
样品号		YD382	D1830	ⅡH3-1	D0997	D0998	D1204-2	D1802	42-1	ZK_3-TW_2	73-1	157-2
序号		45	46	47	48	49	50	51	52	53	54	55
稀土元素含量/10^{-6}	La	29.41	54.28	31.4	17.85	16.74	19.33	16.07	30	24	18	16
	Ce	48.19	111.50	60.23	32.48	33.82	36.91	31.23	62	49	38	36
	Pr	6.48	14.02	7.92	4.42	4.71	5.17	4.42	6.7	5.8	4.6	4.5
	Nd	26.09	57.80	33.36	18.44	18.76	20.59	18.10	34	28	22	20
	Sm	6.08	12.64	7.23	4.76	5.00	5.24	4.98	5.8	5.6	4.3	4.1
	Eu	2.17	4.09	2.60	1.74	1.76	1.84	1.70	2.6	2.1	1.7	1.6
	Gd	6.28	11.05	7.40	5.19	5.30	5.45	5.01	6.4	6.2	4.8	4.5
	Tb	0.96	1.53	1.11	0.78	0.78	0.81	0.78	0.94	0.86	0.74	0.71
	Dy	4.86	7.27	5.39	4.15	4.14	4.27	4.01	4.4	4.6	3.0	3.7
	Ho	0.87	1.21	0.96	0.78	0.79	0.79	0.79	0.75	0.76	0.65	0.64
	Er	2.07	2.75	2.42	1.83	1.97	1.94	1.92	2.0	2.2	2.0	2.0
	Tm	0.30	0.36	0.30	0.25	0.26	0.26	0.26	0.28	0.29	0.23	0.24
	Yb	1.61	1.81	1.61	1.54	1.45	1.47	1.41	1.85	1.70	1.40	1.50
	Lu	0.22	0.23	0.22	0.21	0.22	0.21	0.21	0.26	0.26	0.21	0.24
	ΣREE	135.59	280.54	162.15	94.42	95.70	104.28	90.89	157.98	131.37	101.63	95.73
主要特征参数	LREE/HREE	6.90	9.70	7.35	5.41	5.42	5.86	5.32	8.36	6.79	6.80	6.08
	δEu	1.07	1.04	1.08	1.07	1.04	1.04	1.03	1.30	1.09	1.14	1.13
	δCe	0.81	0.95	0.90	0.86	0.90	0.87	0.88	1.01	0.97	0.98	1.01
	$(La/Yb)_N$	12.32	20.22	13.15	7.81	7.78	8.87	7.68	10.93	9.52	8.67	7.19
	$(Ce/Yb)_N$	7.74	15.93	9.68	5.46	6.03	6.49	5.73	8.67	7.46	7.02	6.21
	$(La/Sm)_N$	3.04	2.70	2.73	2.36	2.11	2.32	2.03	3.25	2.70	2.63	2.45
	$(Gd/Yb)_N$	3.15	4.93	3.71	2.72	2.95	2.99	2.87	2.79	2.94	2.77	2.42
	La/Sm	4.84	4.29	4.34	3.75	3.35	3.69	3.23	5.17	4.29	4.19	3.9

表 13-10　南海火山岩稀土元素含量及主要特征参数一览表

Table 13-10　List of REE contents and main characteristic parameters of volcanic rocks from the South China Sea

岩性		碱性玄武岩	石英拉斑玄武岩				橄榄玄武岩				碱玄岩	
样品号		D10-2	D8-5G	D8-6	D8-7	D9-3	V36D8-2	V36D8-4	V36D9-2	V36D9-3	S04-11	S04-12-10
序号		1	2	3	4	5	6	7	8	9	10	11
稀土元素含量 10^{-6}	La	42.8	9.3	13.92	11.83	33.1	17.47	17.53	51.21	16.2	61.3	64.6
	Ce	187.2	46.25	52.7	51.89	87.18	34.1	40.54	106.44	33.23	103	158
	Pr	11.16	3.2	4.05	3.79	10.32					11.4	12
	Nd	34.04	11.22	14.21	13.87	36.43	15.56	20.06	54.89	23.42	43.4	45.8
	Sm	7.75	3.32	3.82	4.33	9.56	5.12	5.12	11.24	4.8	8.64	8.58
	Eu	2.82	1.27	1.29	1.44	3.31	1.61	1.78	3.76	1.74	2.82	2.81
	Gd	5.33	3.17	3.42	5.39	8.64					7.42	7.58
	Tb	1.19	0.64	0.67	1.04	1.36	0.81	0.75	1.47	0.88	1.08	1.16
	Dy	6.38	3.53	3.35	5.48	7.48					6.06	6.37
	Ho	1.86	1	0.99	1.68	2.66	0.8	0.6	1.5	0.9	1.15	1.24
	Er	2.67	1.65	1.41	2.71	3.27					3.16	128
	Tm	0.31	0.27	0.19	0.28	0.36					0.43	0.48
	Yb	3.02	1.82	1.49	3.25	3.51	2.33	1.91	2.64	1.96	2.67	2.95
	Lu	0.31	0.22	0.16	0.31	0.44	0.38	0.37	0.43	0.36	0.43	0.46
	Y	39.56	18.82	20.59	28.99	41.24	33.6	29.5	45.2	33.8	31.4	35.1
	LREE	285.77	74.56	89.99	87.15	179.9					230.56	291.79
	HREE	21.07	12.3	11.68	20.14	27.72					22.4	23.72
	ΣREE	306.84	86.86	101.67	107.29	207.6					252.96	315.51
主要特征参数	$\frac{LREE}{HREE}$	13.56	6.06	7.7	4.33	6.49					10.29	12.3
	δEu	1.27	1.18	1.07	0.91	1.09					1.05	1.04
	δCe	2.02	2.04	1.67	1.86	1.13					0.88	1.27
	$(La/Yb)_N$	9.55	3.45	6.3	2.45	6.36	5.06	6.19	13.08	5.57	15.48	14.76
	$(Ce/Yb)_N$	16.03	6.57	9.15	4.13	6.42	3.79	5.49	10.43	4.39	9.98	13.85
	$(La/Sm)_N$	3.47	1.76	2.29	1.72	2.18	2.15	2.15	2.87	2.12	4.46	4.74
	$(Gd/Yb)_N$	1.42	1.41	1.85	1.34	1.99					2.24	2.07
	La/Sm	5.52	2.8	3.64	2.73	3.46	3.41	3.42	4.56	3.38	7.09	7.53

续表

岩性		碱玄岩										
样品号		S04-12-11	S04-12-12	S04-12-18	S04-12-20	S04-12-21	S04-12-22	S04-12-8	S04-13-3B	S04-14-1B	S04-14-2B	S05-34A
序号		12	13	14	15	16	17	18	19	20	21	22
稀土元素含量/10^{-6}	La	49.1	48.8	70.2	60.2	57	65.8	66.9	48.1	81.2	57	99.3
	Ce	86.9	87.4	111	96.6	97.1	111	108	96.3	130	104	196
	Pr	9.26	9.64	12.3	10.5	10.6	13	12.4	10.6	15.5	11.4	19
	Nd	34.5	39.4	46	40.1	40.3	52.1	46.6	45.7	61.7	45.4	69.6
	Sm	6.61	7.66	8.4	7.74	7.64	10.2	8.78	9.07	12.4	8.98	12.9
	Eu	2.19	2.52	2.7	2.52	2.45	3.18	2.82	3.02	3.75	2.91	3.81
	Gd	5.68	6.69	7.49	6.71	6.67	9.49	7.48	7.66	12	7.42	10.4
	Tb	0.86	0.99	1.12	1.03	0.95	1.37	1.16	1.13	1.77	1.08	1.45
	Dy	4.85	5.59	6.54	5.9	5.32	7.79	6.32	6.35	11.1	6.12	7.61
	Ho	0.92	1.05	1.21	1.11	1.02	1.46	1.18	1.17	2.22	1.09	1.39
	Er	2.85	3.01	3.54	3.15	2.85	4.16	3.37	3.24	6.58	3.02	3.81
	Tm	0.36	0.39	0.47	0.45	0.38	0.53	0.45	0.43	0.9	0.38	0.49
	Yb	2.24	2.43	3.05	2.65	2.39	3.2	2.88	2.56	5.63	2.3	2.98
	Lu	0.35	0.37	0.48	0.41	0.36	0.48	0.44	0.39	0.92	0.35	0.45
	Y	24.2	30.2	37.1	31	26.5	45.6	34	34.9	83.9	28.6	35.1
	LREE	188.56	195.42	250.6	217.66	215.09	255.28	245.5	212.79	304.55	229.69	400.61
	HREE	18.11	20.52	23.9	21.41	19.94	28.48	23.28	22.93	41.12	21.76	28.58
	ΣREE	206.67	215.94	274.5	239.07	235.03	283.76	268.78	235.72	345.67	251.45	429.19
主要特征参数	LREE/HREE	10.41	9.52	10.49	10.17	10.79	8.96	10.55	9.28	7.41	10.56	14.02
	δEu	1.07	1.05	1.02	1.05	1.03	0.97	1.04	1.08	0.93	1.06	0.98
	δCe	0.92	0.91	0.84	0.85	0.89	0.86	0.84	0.98	0.83	0.93	1.02
	$(La/Yb)_N$	14.78	13.54	15.52	15.32	16.08	13.86	15.66	12.67	9.72	16.71	22.47
	$(Ce/Yb)_N$	10.03	9.3	9.41	9.43	10.51	8.97	9.7	9.73	5.97	11.7	17.01
	$(La/Sm)_N$	4.67	4.01	5.26	4.89	4.69	4.06	4.79	3.34	4.12	3.99	4.84
	$(Gd/Yb)_N$	2.05	2.22	1.98	2.04	2.25	2.39	2.1	2.41	1.72	2.6	2.82
	La/Sm	7.43	6.37	8.36	7.78	7.46	6.45	7.62	5.3	6.55	6.35	7.7

续表

岩性		粗面岩		碱玄岩	霞石岩	粗面玄武岩		橄榄玄武岩				平均值
样品号		9DG	9DG-2	S04-16	75-31-1	S08-69-1	S08-69-2	SO23-35-6	SO23-37-31	SO23-37-7	SO2120	
序号		23	24	25	26	27	28	29	30	31	32	
稀土元素含量/10^{-6}	La	82.246	83.47	67.8	81.13	49.1	50.8	48.04	23.21	36	19.45	49.50
	Ce	154.841	162.3	126	157.26	79.5	79.8	94.41	47.03	73.06	41.35	98.14
	Pr	18.107	16.19	12.1		8.21	8.41					11.01
	Nd	61.193	54.82	47.6	72.17	29.7	31.1	43.32	22.37	35.17	22.82	39.21
	Sm	11.388	8.758	9.89	14.85	5.52	5.68	8.21	5.61	4.33	6	7.90
	Eu	3.02	1.143	3.03	4.84	1.84	1.87	2.82	1.99	2.68	2.16	2.56
	Gd	9.058	6.344	8.54		4.79	4.94					7.06
	Tb	1.318	1.008	1.28	1.2	0.74	0.75	1.33	1.14	1.54	1.09	1.10
	Dy	7.248	5.207	7.41		4.23	4.22					6.11
	Ho	1.432	0.964	1.33	1.3	0.78	0.81	2.4	0.8	1.5	1.2	1.27
	Er	3.756	2.508	3.8		2.15	2.21					3.15
	Tm	0.542	0.389	0.49		0.27	0.3					0.41
	Yb	3.406	2.196	3.04	2.28	1.8	1.8	1.46	1.26	1.77	1.71	2.52
	Lu	0.534	0.345	0.48	0.34	0.26	0.27	0.29	0.24	0.34	0.27	0.38
	Y	31.776	23.11	35.7	48.8	20.8	22.1	30.5	26.7	23.9	29.8	33.19
	LREE	330.8	326.68	266.42		173.87	177.66					227.43
	HREE	27.29	18.96	26.37		15.02	15.3					22.26
	ΣREE	358.09	345.64	292.79		188.89	192.96					249.69
主要特征参数	LREE/HREE	12.12	17.23	10.1		11.58	11.61					1.015
	δEu	0.88	0.45	0.99		1.07	1.06					10.24
	δCe	0.93	1	0.98		0.87	0.85					1.10
	$(La/Yb)_N$	16.28	25.63	15.04	23.99	18.39	19.03	22.18	12.42	13.71	7.67	13.40
	$(Ce/Yb)_N$	11.76	19.12	10.72	17.84	11.42	11.47	16.73	9.65	10.68	6.25	10.24
	$(La/Sm)_N$	4.54	6	4.31	3.44	5.6	5.63	3.68	2.6	5.23	2.04	3.80
	$(Gd/Yb)_N$	2.15	2.33	2.27		2.15	2.21					2.09
	La/Sm	7.22	9.53	6.86	5.46	8.89	8.94	5.85	4.14	8.31	3.24	6.04

琼北新近纪各类岩石具有相似的稀土配分模式，均呈右倾的轻稀土富集型曲线（图 13-7）。单个岩石样品稀土总量变化较大，为 $24.91\times10^{-6}\sim404.38\times10^{-6}$。LREE/HREE = 1.61~15.72，大部分样品小于 10，平均 7.12。$(La/Yb)_N$ = 1.07~33.35，$(Ce/Yb)_N$ = 0.87~24.50，轻、重稀土分异不明显，属轻微—中等型。δEu = 0.82~1.18，铕略亏损—略富集。δCe = 0.63~1.02，平均为 0.91，大部分样品小于 1.0，铈略有亏损，但亏损不明显。$(La/Sm)_N$ = 0.82~4.92，$(Gd/Yb)_N$ = 1.28~4.62，轻、重稀土之间分馏作用不强。

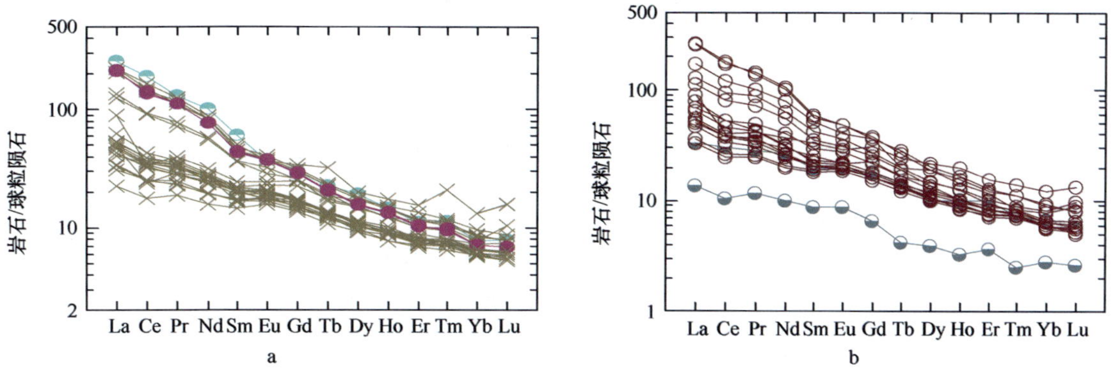

图 13-7　琼北新近纪火山岩稀土元素球粒陨石标准化图

Fig. 13-7　Chondrite-normalized REE patterns of Neogene volcanic rocks from Qiongbei

a—石马村组样品标准化图；b—石门沟村组和海口组样品标准化图

琼北第四纪火山岩各岩石具相似的稀土配分模式，主要呈右倾的轻稀土富集型曲线（图 13-8）。单个岩石的稀土总量变化较大，由 $12.92\sim299.09\times10^{-6}$。LREE/HREE = 3.27~23.30，其中 D344 号样品的比值高达 23.30 外，其余均小于 11.86，平均为 6.86（除 D344 号样品）。$(La/Yb)_N$ = 4.09~41.20，$(Ce/Yb)_N$ = 3.17~21.56，轻、重稀土分异不明显，属轻微—中等型。δEu = 0.83~1.30，平均 1.08，铕略亏损或富集不明显。δCe = 0.33~1.07，除少量样品显示铈亏损外，其余亏损不明显。

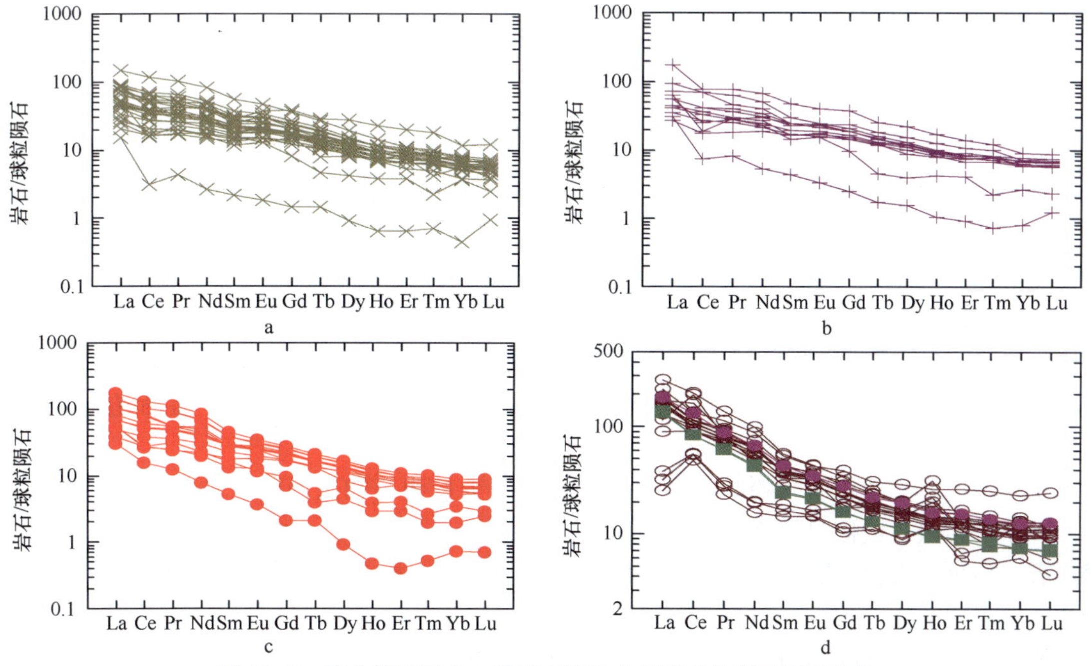

图 13-8　琼北第四纪火山岩及南海稀土元素球粒陨石标准化图

Fig. 13-8　Chondrite-normalized REE patterns of Quaternary volcanic rocks from Qiongbei and the South China Sea

a—多文组样品标准化图；b—道堂组样品标准化图；c—石山组样品标准化图；d—南海样品标准化图

$(La/Sm)_N$ = 1.55~6.92，轻稀土之间分馏作用不强；$(Gd/Yb)_N$ = 1.28~4.62，重稀土分馏作用比较弱。

南海火山岩各岩石具相似的稀土配分模式，主要呈右倾的轻稀土富集型曲线（图 13-8）。单个岩石的稀土总量变化较大，由 86.86×10^{-6}~429.19×10^{-6}，平均为 249.69×10^{-6}。LREE/HREE = 4.33~17.23，平均为 10.24。$(La/Yb)_N$ = 2.45~25.63（平均 1120），$(Ce/Yb)_N$ = 3.79~19.12（平均 10.24），表明轻、重稀土分异明显，属中等型。δEu = 0.45~1.27，平均为 1.02，铕略亏损或富集不明显。δCe = 0.83~2.04，平均 1.10，铈亏损不明显。$(La/Sm)_N$ = 1.72~6.00，平均 3.80，轻稀土之间分馏作用不强；$(Gd/Yb)_N$ = 1.34~2.82，平均 2.09，重稀土分馏作用比较弱。

三、岩石微量元素地球化学特征

新生代火山岩分析结果见表 13-11~表 13-13。

琼北新近纪、第四纪基性火山岩岩石微量元素的含量变化不大，其平均值与维氏基性岩（玄武岩）平均值对比，具有如下特征：①新近纪岩石 Au、Ag、Cu 等元素含量比维氏值略低，而 Ta、Rb、Hf、Sc、U 等元素含量比维氏值略高，其余与维氏值相近；②第四纪岩石 Au、Ag、Cu、As 等元素含量比维氏值略低，而 Be、Ta、Rb、Hf、U 等元素含量比维氏值略高，其余与维氏值相近。南海基性火山岩岩石的微量元素的含量变化不大，其平均值与维氏基性岩（玄武岩）平均值对比，Cu、Co、Li 等元素含量比维氏值略低，而 Be、Nb、Ta、Rb、Hf、U、Th、W 等元素含量比维氏值略高，其余与维氏值相近。

新生代火山熔岩过渡族元素球粒陨石标准化形式具有相似的配分模式（图 13-9），均表现为明显 Ti 正异常和 Cr、Ni 负异常。与 Allégre 等（1973）的玄武岩配分模式相似，表明新生代火山熔岩由地幔岩演化而来。

新近纪火山熔岩的微量元素比值蛛网与第四纪的比较相一致（图 13-10，图 13-11），总体与大陆碱性玄武岩（据 Weaver et al.，1982）具有相似的标准化模式。表现为强不相容元素 Rb、Ba、Th、V、K 等较中等不相容元素 Sr、Nb、P、Zr 更富集。其中，Ba、U、Nb、P、Zr、Sm、Tb 等元素含量相对高，在微量元素比值蛛网图上略显正异常；而 Th、K、Ti、Yb 等元素含量相对低，在元素比值蛛网图显示略的负异常。

四、副矿物特征

海南岛新生代火山岩重砂分析结果见表 13-14。其副矿物组合简单，主要以磁铁矿、钛铁矿、磷灰石、黄铁矿等为主，可称为磁铁矿-钛铁矿-磷灰石-锆石组合。主要副矿物特征概述如下。

锆石　主要锆石为无色或淡红色透明复正方双锥柱状，以柱面（110）发育为主，偏锥面（311）、（131）次之，顶面、柱面发育不对称，颗粒大小、长宽比极不均匀，粒径在 0.02~0.15mm 不等，长宽比在 2∶1~5∶1，玻璃光泽，部分锆石的晶棱及顶锥、晶面有熔蚀或擦痕等现象。

磁铁矿　黑色，多细粒状、粉状、片状或碎屑状，少量为等轴八面体，粒径 0.02~0.2mm 不等，金属光泽，条痕为黑色或棕褐黑色。

磷灰石　白色、无色透明，六方平锥柱状、细柱状、细粒状、圆柱状，粒径 0.03~0.16mm 不等，半油脂光泽或玻璃光泽。

黄铁矿　黄铜色、暗黄铜色，多数为致密块状、粒状、碎块状，少量为立方体、柱状，粒径 0.05~0.30mm不等，金属光泽，条痕灰黑色。

表 13-11　海南岛新近纪火山岩微量元素含量一览表

Table 13-11　List of trace element contents of Neogene volcanic rocks from the Hainan Island

地层单位		海口组		石门沟村组							
岩性		石英拉斑玄武岩		辉石玄武岩		橄榄玄武岩					
样品号		YQ2	1-4	$PM001_5$	$PM001_6$	D1836	D1842	D1871	209-1	D1835	D0083
序号		1	2	3	4	5	6	7	8	9	10
微量元素含量/10^{-6}	Au			0.05	0.4	0.3	1.9	1.4		0.4	0.04
	Ag			0.07	0.05	0.034	0.059	0.055		0.035	0.05
	Cu	34		66.2	70.5	41.1	6.6	57.3		35.0	57.9
	Pb	7		21.96	4.48	3.3	2.9	6.6		2.5	5.08
	Zn	78	104	167	110	135	105	164	141	160	112
	Sb	0.04		0.079	0.084	0.14	0.22	0.27		0.10	0.11
	W			0.33	0.23	1.1	0.68	1.45		0.9	0.55
	Sn			1.4	1.5	1.2	6.9	2.9		1.2	2.1
	Bi			<0.03	<0.03	<0.1	0.26	0.07		<0.1	<0.03
	Mo	0.67		1.23	1.17	1.9	<0.2	3.5		2.5	1.20
	Be			1.00	1.04	1.8	0.8	12		1.5	1.09
	Li			4.00	4.29	8.7	21.7	11.5		9.6	4.21
	Ni	94	145	520	224	78.7	383	303	337	289.7	141
	Co	20	35	131	80.9	31.8	52.0	50.2	40	44.3	53.5
	V	79	103	159	152	129.9	171	168	30	143.9	143
	Nb		16	34.2	23.0	60.4	3.7	118.4	85	66.2	24.2
	Ta		0.6	2.00	1.44	4.2	<0.5	6.8	15	4.1	1.53
	Rb		20	27.9	15.3	42.4	7.1	61.0	55	51	21.1
	Sr		320	354	308	465	138	1031	1000	674	296
	Ba		174	406	302	326	83	814	815	625	373
	Zr		120	149	142	251	44	398	283	271	145
	Hf		6.2	3.3	3.1	5.3	1.6	8.3	19	5.3	12
	As	4.1		< 0.1	< 0.1	1.2	3.2	1.0		1.5	< 0.1
	Ga			20.0	21.6	22.7	12.2	19.0		19.7	21.9
	Sc	4.4	13						3.7		
	Cr	116	176			92	55.9	271	254	266	
	U		1.5	0.71	0.45	2.7	0.5	2.5	1.6	2.7	0.46
	Th		3.0	3.30	1.89	6.2	<1.0	14.3	7.6	5.4	2.23
	Y	8.5	15	19.62	18.37	19.04	13.99	31.75		21.69	19.21
特征比值	Rb/Sr		0.06	0.08	0.05	0.09	0.05	0.06	0.06	0.01	0.07
	Ba/Sr		0.54	1.15	0.98	0.70	0.60	0.78	0.82	0.93	1.26
	K/Rb		340.4	333.3	428.6	129.2	233.8	287.2		338.6	346.2
	Nb/Ta		26.67	17.10	15.97	14.38	7.40	17.41	5.67	16.15	13.79
	Zr/Hf		19.35	45.15	45.81	47.36	27.50	47.95	14.89	51.13	42.65

地层单位		石门沟村组					石马村组				
岩性		橄榄玄武岩					玄武岩				
样品号		ZK_2-1	$PM001_{9-1}$	福基田-1	D0460	ZK_3-1	D1831-1	D1832-1	D1834-1	D1016	D2174
序号		11	12	13	14	15	16	17	18	19	20
微量元素含量/10^{-6}	Au	0.6	0.5	0.6	0.5	0.7	0.1	0.5	1.8	0.5	0.4
	Ag	0.05	0.09	0.03	0.04	0.07	0.028	0.054	0.049	0.04	0.05
	Cu	63.5	68.2	42.6	58.9	46.0	40.6	51.4	48.2	59.4	55.5
	Pb	3.62	4.08	5.82	3.47	7.56	1.0	2.4	3.3	5.04	4.80
	Zn	95.3	115	3.47	120	156	131	123	136	109	114
	Sb	0.073	0.098	0.090	0.057	11.04	0.08	0.09	0.08	0.053	0.063
	W	0.44	0.33	0.39	0.50	0.90	36.9	6.8	6.8	0.55	0.28
	Sn	1.4	3.2	1.5	1.3	1.4	<0.1	1.1	<0.1	1.1	1.0
	Bi	<0.03	<0.03	<0.03	<0.03	<0.03	<0.1	<0.1	<0.1	<0.03	<0.03
	Mo	1.11	1.49	1.94	1.14	3.33	29.6	0.9	1.1	0.99	1.07
	Be	0.93	1.28	2.03	0.98	2.39	0.8	0.9	0.9	0.73	1.07
	Li	3.25	4.41	7.07	4.66	8.41	7.5	8.6	7.1	4.07	4.52
	Ni	203	205	180	145	122	139.5	155.8	106.7	141	148
	Co	49.9	65.3	47.3	52.9	45.3	42.3	39.8	39.8	45.2	46.3
	V	160	179	169	159	147	154.8	150.2	150.8	139	144
	Nb	30.4	43.2	72.1	27.1	116	16.7	29.4	20.1	18.4	25.8
	Ta	1.75	2.65	4.55	1.59	1.58	0.9	1.1	1.0	1.08	1.60
	Rb	27.1	31.6	95.8	22.0	57.5	3.6	14.2	10.6	17.8	26.5
	Sr	341	468	747	330	1032	254	345	306	289	356
	Ba	367	484	689	307	750	171	321	235	405	444
	Zr	131	187	297	139	381	99	122	107	110	143
	Hf	3.2	4.4	8.3	3.0	9.2	3.1	3.0	3.0	2.4	3.1
	As	<0.1	<0.1	0.36	<0.1	7.11	1.1	1.1	1.4	<0.1	<0.1
	Ga	20.2	20.7	23.5	21.8	22.6	18.5	18.3	17.1	20.5	20.5
	Sc										
	Cr					115	179	168	116		
	U	0.59	0.73	1.64	0.56	3.83	1.4	1.5	1.1	0.38	0.63
	Th	2.83	3.15	6.26	2.54	14.6	1.3	2.9	2.1	1.87	2.49
	Y	17.94	19.94	23.08	20.55	32.03	17.28	21.29	17.34	17.77	19.42
特征比值	Rb/Sr	0.08	0.07	0.13	0.07	0.06	0.014	0.04	0.034	0.06	0.07
	Ba/Sr	1.08	1.03	0.92	0.93	0.73	0.67	0.93	0.77	1.40	1.25
	K/Rb	315.5	404.6	149.0	354.7	264.2	1153.0	508.6	603.0	368.4	332.06
	Nb/Ta	17.37	16.30	15.85	17.04	73.42	18.56	26.73	20.10	17.04	16.13
	Zr/Hf	40.94	42.50	35.78	46.33	41.41	31.94	40.67	35.67	45.83	46.13

地层单位		石马村组								
岩性		玄武岩					石英拉斑玄武岩	橄榄霞石岩	玻基辉橄岩	
样品号		$PM005_{6-5}$	D0744-2	D2297-2	D3461	D1896-2	194-2	185-1	ZK_2-2	ZK_2-3
序号		21	22	23	24	25	26	27	28	29
微量元素含量 10^{-6}	Au	0.6	0.7	0.5	0.5	0.7			0.5	0.6
	Ag	0.06	0.04	0.07	0.05	0.04			0.05	0.05
	Cu	51.9	59.3	58.9	57.7	65.7			44.6	48.9
	Pb	3.56	8.02	4.84	3.66	5.23			5.78	6.78
	Zn	111	123	98.3	101	110	114	166	120	125
	Sb	3.65	0.14	0.076	0.080	0.096			0.082	0.17
	W	0.38	1.21	0.28	0.28	0.61			0.55	0.55
	Sn	0.6	1.3	1.2	1.5	1.4			1.2	1.4
	Bi	<0.03	<0.03	<0.03	<0.03	<0.03			<0.03	0.10
	Mo	0.82	2.95	0.92	0.70	1.88			1.76	1.78
	Be	1.11	1.89	0.93	0.96	1.57			1.77	1.90
	Li	4.09	7.61	4.39	5.80	6.83			12.7	12.2
	Ni	189	142	128	124	135	106	226	152	166
	Co	46.2	48.8	46.4	45.6	48.1	34	48	63.8	64.5
	V	143	192	147	159	182	96	150	232	241
	Nb	29.3	70.2	25.8	28.5	69.9	20	110	115	115
	Ta	1.91	1.30	1.43	1.84	4.15	1.8	9.0	6.42	6.57
	Rb	14.5	47.7	18.5	13.5	45.1	20	37	18.1	17.0
	Sr	333	906	331	374	909	380	870	570	696
	Ba	309	914	362	383	636	240	490	1109	1258
	Zr	140	286	126	135	219	125	303	251	243
	Hf	3.0	6.0	3.1	12	5.4	6.8	4.8	5.6	5.4
	As	2.22	0.452	<0.1	<0.1	<0.1			0.76	0.30
	Ga	19.8	21.2	20.5	20.6	21.1			21.6	21.5
	Sc						15	7.2		
	Cr	234					148	224		
	U	0.81	2.08	0.55	0.56	1.44	1.0	1.2	1.65	1.74
	Th	3.65	10.0	2.63	2.97	6.86	4.3	10.9	10.8	11.1
	Y	19.38		19.15	18.28	24.20	16.5	28	25.38	25.23
特征比值	Rb/Sr	0.04	0.05	0.06	0.04	0.05	0.05	0.04	0.03	0.02
	Ba/Sr	0.93	1.01	1.09	1.02	0.70	0.63	0.56	1.95	1.81
	K/Rb	492.4	280.2	296.2	461.2	296.4	386.0	332.1	160.5	195.3
	Nb/Ta	15.34	54.00	18.04	15.49	16.84	11.11	12.22	17.91	17.50
	Zr/Hf	46.67	47.67	40.65	39.71	40.56	18.38	63.12	44.82	45.00

续表

地层单位		石马村组							维氏微量元素的平均含量	
岩性		辉石玄武岩		橄榄玄武岩				粗玄岩	超基性岩	基性岩
样品号		Ⅱ H17-1	$PM001_{19-2}$	D0752	ZK_4-1	D1872-1	D1900	D0046		
序号		30	31	32	33	34	35	36		
微量元素含量 10^{-6}	Au	1.0	0.5	0.4	0.5	1.1	1.0	0.6	5	4
	Ag	0.051	0.05	0.03	0.09	0.059	0.066	0.04	0.05	0.10
	Cu	62.6	65.7	56.9	64.0	62.8	57.4	57.7	20	100
	Pb	4.0	3.83	3.57	3.83	3.4	13.1	3.68	0.1	8
	Zn	105.6	98.7	110	114	127	122	97.5	30	130
	Sb	0.14	0.066	0.053	11.74	0.20	0.25	0.070	0.1	0.1
	W	0.3	0.44	0.61	0.38	0.59	0.8	0.44	0.1	1
	Sn	1.7	1.2	1.4	1.5	2.3	1.6	1.3	0.5	1.5
	Bi	<0.1	<0.03	<0.03	<0.03	0.04	<0.05	<0.03	0.001	0.007
	Mo	1.73	0.87	1.01	1.04	1.0	0.92	1.09	0.2	1.4
	Be	1.2	0.84	1.04	1.08	2.1	3.8	0.85	0.2	0.4
	Li	7.9	3.90	4.39	3.63	8.1	15.2	5.21	0.5	15
	Ni	912	146	144	88.6	152	168	186	2000	160
	Co	34.8	43.4	48.6	38.7	43.3	46.1	47.3	200	45
	V	156.4	134	151	138	154.0	222	150	40	200
	Nb	10.5	17.8	24.9	22.7	25.5	97.6	27.4	1.0	20
	Ta	<0.5	1.06	1.42	1.70	1.5	6.2	1.73	0.018	0.48
	Rb	13	15.2	19.9	22.7	28.5	83.5	25.4	2	45
	Sr	237	292	324	311	240	984	351	10	440
	Ba	152	311	406	254	333	946	364	1	300
	Zr	179	111	124	127	147	256	133	30	100
	Hf	4.5	3.0	2.8	4.0	4.0	5.2	3.4	0.1	1
	As	0.2	<0.1	<0.1	9.47	0.2	0.9	<0.1	0.5	2
	Ga	16.8	19.4	21.4	20.3	20.6	14.4	18.5	1.5	18
	Sc								5	24
	Cr	150			212	158	179		2000	200
	U	1.0	0.44	0.50	0.63	1.0	2.3	0.59	0.003	0.5
	Th	1.1	1.98	2.59	3.03	3.1	9.9	2.70	0.005	3
	Y	36.75	17.82	53.30	19.63	18.50	27.91	17.78		20
特征比值	Rb/Sr	0.05	0.05	0.06	0.07	0.08	0.08	0.07	0.2	0.102
	Ba/Sr	0.64	1.07	1.25	0.82	0.98	0.96	1.04	10	1.333
	K/Rb	287.4	355.0	342.07	307.2	317.5	71.58	313.8	150	184.4
	Nb/Ta	21.00	16.79	17.54	13.35	17.00	15.74	15.84	55.56	41.67
	Zr/Hf	39.78	37.00	44.29	31.75	36.75	49.23	39.12	300	100

注：Au 的单位为 10^{-9}。

表 13-12 海南岛第四纪火山岩微量元素含量一览表

Table 13-12 List of trace element contents of Quaternary volcanic rocks from the Hainan Island

地层单位		石山组									
岩性		碱性橄榄玄武岩				橄榄拉斑玄武岩			玄武岩		
样品号		QB013	QB012	23-3	128-2	HN9902	QB011	HN9910	106B1	EMB4	EMB6
序号		1	2	3	4	5	6	7	8	9	10
微量元素含量/10^{-6}	Au										
	Ag										
	Cu										
	Pb	5.723	5.547			4.93	4.354	6.945	4.059	2.102	4.023
	Zn			166	102						
	Sb										
	W										
	Sn										
	Bi										
	Mo										
	Be										
	Li										
	Ni	187.6	179.7	106	122	89.9	88.38	193.7	121.2	126.7	141.4
	Co	50.67	50.39	32	42	38.93	39.36	50.43	42.03	43.2	41.67
	V	215	219.6	99	158	178.9	184.4	224.3	186.2	125.8	141.6
	Nb	70.76	70.44	61	48	35.6	36.16	87.68	41.06	12.35	27.23
	Ta	4.193	4.174	3.4	9.3	2.088	2.107	5.181	2.44	0.719	1.618
	Rb	25.06	14.15	57	52	24.96	31.79	38.58	28.82	10.52	26.56
	Sr	707.7	716.1	810	705	527.7	541.5	810.6	488.7	281.1	419.6
	Ba	540.1	578.5	580	570	429.2	423.9	584.6	351	154.5	261
	Zr	272.5	271.2	280	210	197.6	202.4	324.8	197.9	95.42	156
	Hf	6.393	6.495	15	11	4.905	4.937	7.568	4.803	2.664	3.921
	As										
	Ga										
	Sc	22.57	22.95	10	20	19.8	19.41	24.71	20.43	19.28	16.65
	Cr	275.6	280.5	152	194	282.4	283.2	296	301	204.6	199.8
	U	1.534	1.283	3.9	406	0.859	0.882	1.93	0.991	0.281	0.691
	Th	7.03	6.992	6.5	6.5	3.956	3.967	8.347	4.293	1.336	3.081
	Y	28.34	29.03	14	19	24.41	24.69	31.07	25.52	25.91	20.69
特征比值	Rb/Sr	0.04	0.02	0.07	0.07	0.05	0.06	0.05	0.06	0.04	0.06
	Ba/Sr	1.31	1.24	1.40	1.24	1.23	1.28	1.39	1.39	1.82	1.61
	K/Rb	443.9	522.1	390.3	325.7	478.9	410.0	3012	394.6	449.8	409.4
	Nb/Ta	16.88	16.88	17.94	5.16	17.05	17.16	16.92	16.83	17.18	16.83
	Zr/Hf	70.83	41.76	18.67	19.09	40.29	41.00	42.92	41.20	35.82	39.79

续表

地层单位		石山组	道堂组								
岩性		橄榄玄武岩							辉石玄武岩		
样品号		D1874	220B3	101B1	D1019	D1894	XⅦ3-2	XⅦ12	XⅦ2-1	XⅧ5-1	XⅧ5-2
序号		11	12	13	14	15	16	17	18	19	20
微量元素含量/10^{-6}	Au	0.9			1.0	0.9	0.6	0.5	0.7	0.5	0.7
	Ag	0.060			0.023	0.041	0.034	0.041	0.027	0.023	0.030
	Cu	56.1			57.8	52.9	47.4	30.2	63.5	38.8	48.0
	Pb	3.2	4.129	2.94	6.6	5.2	1.3	1.8	0.8	0.9	0.7
	Zn	131			115	113	109.2	119.2	106.2	108.3	103.2
	Sb	0.17			0.16	0.14	0.66	1.18	0.55	0.27	1.51
	W	0.68			0.4	0.4	0.53	0.59	0.36	0.36	0.36
	Sn	2.3			1.2	1.7	1.8	1.5	1.0	1.2	1.2
	Bi	0.06			<0.05	<0.05	0.06	0.06	0.06	0.02	0.03
	Mo	1.9			0.96	1.27	1.59	1.76	0.79	0.93	0.77
	Be	2.6			2.2	2.7	1.1	1.1	0.8	0.8	0.9
	Li	9.3			12.8	12.1	8.0	8.0	5.5	6.2	5.1
	Ni	157	235.3	128.1	115	114	140.3	138.4	125.0	127.9	125.1
	Co	47.1	53.39	48.06	39.8	40.5	45.0	44.2	43.2	44.3	43.9
	V	185	211.8	159.7	161	171	135.0	137.4	129.1	129.2	125.7
	Nb	312	50.05	28.54	16.50	24.3	34.0	35.4	11.0	14.1	9.9
	Ta	2.3	2.949		1.0	1.2	2.50	3.06	1.00	0.81	0.97
	Rb	36.3	30.67	23	21.7	21.8	23.0	34	10.0	15	13
	Sr	455	551.8	443.3	280	337	420	417	234.0	262	243
	Ba	407	398.9	317.7	234	301	590	648	121.0	169	320
	Zr	205	206.3	159.4	128	161	127.0	123	97.0	94	83
	Hf	5.1	5.027	3.965	2.6	4.2	3.8	3.6	3.0	3.0	3.2
	As	<0.2			0.3	0.3	0.54	0.64	0.54	0.71	0.71
	Ga	16.9			212	17.6					
	Sc		24.53	18.84							
	Cr	206	343	261.6	158	184					
	U	1.4	1.044	0.658	1.0	1.1					
	Th	4.1	4.755	2.86	1.8	2.6					
	Y	20.78	24.16	22.4	19.51	20.06					
特征比值	Rb/Sr	0.08	0.056	0.05	0.08	0.06	0.05	0.08	0.04	0.06	0.05
	Ba/Sr	1.12	1.38	1.40	1.20	1.25	0.71	0.64	1.93	1.55	0.76
	K/Rb	372.8	381.6	447.6	352.0	460.8	465.6		406.8	265.6	
	Nb/Ta	14.52	16.97	17.27	16.50	20.25	13.60	11.57	11.00	17.41	17.41
	Zr/Hf	40.20	41.04	40.20	49.23	38.33	3122	34.17	32.33	31.33	25.94

地层单位		道堂组			多文组					
岩性		玄武岩		橄榄拉斑玄武岩	玄武岩			粗玄岩		
样品号		XⅧ7-1	305B6	150-1	D1837	ⅡH14-2	D1828	XⅦ1-1	XⅦ1-2	0501
序号		21	22	23	24	25	26	27	28	29
微量元素含量/10^{-6}	Au	0.6			0.3	0.5	0.4	0.6	0.5	
	Ag	0.037			0.077	0.047	0.038	0.013	0.024	
	Cu	66.5			46.6	62.3	45.0	45.0	44.2	5121
	Pb	1.1	3.336		4.7	7.7	2.6	0.9	1.2	1.879
	Zn	106.5		132	136	105.6	137	113.3	116.9	113.8
	Sb	0.34			0.07	0.11	0.07	0.58	10.40	
	W	0.53			1.1	0.39	0.8	0.53	0.45	
	Sn	1.4			1.2	1.5	1.0	1.0	1.5	
	Bi	0.12			<0.1	<0.1	<0.1	0.05	0.06	
	Mo	0.90			1.7	1.77	0.8	0.84	0.99	
	Be	0.8			1.1	1.2	1.0	1.2	1.3	
	Li	5.3			8.7	8.3	6.5	8.3	8.3	
	Ni	116.5	136.1	132	169.3	91.7	121	136.8	156.2	154.3
	Co	41.2	42.5	42	42.8	36.4	39.7	42.9	46.9	43.98
	V	125.1	135.3	103	156	165.3	148.1	127.3	127.8	127.4
	Nb	10.4	15.2	18	29.0	11.5	19.6	39.6	37.7	39.98
	Ta	1.05	0.906	3.4	2.0	<0.5	0.9	2.95	2.71	2.381
	Rb	10	20.08	26	25.7	11	18.8	31.0	34.0	30.83
	Sr	232.0	311.2	340	366	244	292	417	454	530.9
	Ba	139	193.9	235	303	130	243	470	490	423
	Zr	108.0	112.8	110	136	92	129	168.0	162.0	175.8
	Hf	5.5	2.988	5.3	3.1	12	3.2	4.4	5.2	4.314
	As	0.54			0.9	0.2	0.9	0.46	0.54	
	Ga				21.2	14.7	20.4			22
	Sc		19.07	18		20.4				11.07
	Cr		215	198	174	146	142			206.1
	U		0.52	1.6	1.8	0.8	1.5			0.4
	Th		2.261	3.3	3.3	1.8	2.3			4.051
	Y		19.84	15.5	17.78	16.65	18.23			16.27
特征比值	Rb/Sr	0.04	0.06	0.08	0.07	0.045	0.06	0.07		
	Ba/Sr	1.67	1.61	1.45	1.21	1.88	1.20	0.89		
	K/Rb	373.6	359.7	382.1	403.8	354.7	397.4	401.7		
	Nb/Ta	9.90	16.78	5.29	14.50	23.00	21.78	13.42		
	Zr/Hf	19.34	37.75	20.76	43.87	27.06	40.31	38.18		

续表

地层单位		多文组								
岩性		橄榄拉斑玄武岩			橄榄玄武岩				石英拉斑玄武岩	粗玄岩
样品号		73-1		HK22	D0997	D0998	D1204-2	D1802	60-1	Ⅰ H3-3
序号		30	31	32	33	34	35	36	37	38
微量元素含量/10^{-6}	Au				0.5	0.4	0.4	0.4		0.6
	Ag				0.5	0.5	0.07	0.04		0.056
	Cu		57.11	54	51.9	61.8	52.7	54.3		69.1
	Pb		1.412		6.30	4.79	4.21	4.93		14.8
	Zn	104	103.1	121	119	130	121	112	94	107.2
	Sb				0.063	0.11	0.043	0.069		0.18
	W				0.23	0.28	0.28	0.44		0.39
	Sn				1.9	1.8	1.3	1.6		1.5
	Bi				<0.03	0.03	<0.03	<0.03		<0.1
	Mo				1.33	1.25	1.25	1.28		1.08
	Be				0.93	1.04	1.05	0.96		1.2
	Li			8	3.92	4.59	4.84	3.75		8.1
	Ni	232	129.8	150	167	291	175	183	106	96.8
	Co	42	44.05	55	52.3	84.3	512	59.8	36	35.0
	V	105	123.9	126	148	158	163	158	102	156.4
	Nb	32	12.42	49.0	21.8	24.7	28.4	23.7	12	12.1
	Ta	3.4	0.779		1.28	1.44	1.61	1.42	101	<0.5
	Rb	28	10.99	33	22.8	17.8	18.7	20.2	13	14
	Sr	540	276	554	325	240	384	338	297	236
	Ba	300	148.6	457	602	436	402	334	117	109
	Zr	137	103	188	139	142	155	142	92	97.9
	Hf	6.8	2.806	4.22	3.3	3.3	3.4	3.2	409	3.0
	As				<0.1	<0.1	<0.1	<0.1		0.2
	Ga		19.19		21.0	23.1	21.9	21.9		15.0
	Sc	13	15.02	16.0					14	20.5
	Cr	192	174.6	197					154	138
	U	3.3	0.276	0.70	0.51	0.59	0.66	0.60	1.4	0.8
	Th	3.7	1.425	4.41	2.29	2.52	2.94	2.44	2.3	1.2
	Y	14	16.34	20.0	112.7	115.1	123.0	18.23	30	15.74
特征比值	Rb/Sr	0.05	0.04	0.06	0.07	0.07	0.05	0.06	0.04	0.06
	Ba/Sr	1.80	1.86	1.21	0.54	0.55	0.96	1.01	2.54	2.17
	K/Rb	388.4		435.2	375.0	424.4	488.3	374.0	332.1	308.3
	Nb/Ta	9.41	15.94		17.03	17.15	17.64	16.69	10.91	24.20
	Zr/Hf	20.15	36.71	44.55	42.12	43.03	45.59	44.38	18.78	32.63

续表

地层单位		多文组							维氏微量元素的平均含量	
岩性		橄榄玄武岩				橄榄辉石玄武岩			超基性岩	基性岩
样品号		ZK_3-TW_2	YD382	D1830	Ⅱ H3-1	Ⅰ H4-2	Ⅱ H4-3	Ⅱ H10-2		
序号		39	40	41	42	43	44	45		
微量元素含量 10^{-6}	Au		0.9	0.4	0.7	0.5	1.7	0.5	5	4
	Ag		0.031	0.162	0.070	0.069	0.057	0.063	0.05	0.10
	Cu		65.7	44.0	63.0	67.6	63.6	62.3	20	100
	Pb		12.3	7.4	16.0	10.6	6.8	5.8	0.1	8
	Zn	109	132.2	186	119.5	118.1	116.5	108.7	30	130
	Sb		0.14	0.09	0.27	0.22	0.15	0.17	0.1	0.1
	W		0.49	0.9	0.47	0.60	0.52	0.49	0.1	1
	Sn		1.85	2.2	2.7	2.8	2.1	2.1	0.5	1.5
	Bi		0.04	<0.1	<0.1	<0.1	<0.1	<0.1	0.001	0.007
	Mo		1.39	0.9	1.46	1.89	1.22	1.74	0.2	1.4
	Be			2.2	3.0	3.8	2.6	2.9	0.2	0.4
	Li		9.8	10.5	11.9	12.1	11.1	12.0	0.5	15
	Ni	156	183	165.6	109.4	119.3	126.2	132.3	2000	160
	Co	40	50.5	45.6	49.8	51.8	45.8	54.7	200	45
	V	126	202.6	200.7	230.3	274.0	195.4	226.1	40	200
	Nb	38	38.5	83.6	49.1	50.5	36.8	24.1	1.0	20
	Ta	2.5	3.0	4.6	4.2	4.5	2.3	2.0	0.018	0.48
	Rb	36	28	57.4	36	38	33	33	2	45
	Sr	580	486	905	567	569	471	506	10	440
	Ba	390	349	724	593	475	418	982	1	300
	Zr	190	205	426	235	263	198	195	30	100
	Hf	9.5	4.7	8.1	6.3	7.1	5.3	5.5	0.1	1
	As		0.87	1.0	0.6	0.4	0.3	0.4	0.5	2
	Ga			29.3	15.9	15.6	14.2	17.2	1.5	18
	Sc	15			18.3	17.1	20.0	20.1	5	24
	Cr	186		98	163	154	197	186	2000	200
	U	2.3		2.8	1.5	1.5	1.2	1.5	0.003	0.5
	Th	4.5		6.6	4.5	4.9	3.1	3.3	0.005	3
	Y	18	23.76	31.51	29.21	25.32	31.58	20.58		20
特征比值	Rb/Sr	0.06	0.06	0.06	0.06	0.07	0.07	0.07	0.2	0.102
	Ba/Sr	1.49	1.39	1.25	0.96	1.20	1.13	0.52	10	1.333
	K/Rb	352.8	438.8	144.6	426.6	417.3	387.4	384.9	150	184.4
	Nb/Ta	15.20	12.83	18.09	11.69	11.22	16.00	17.05	55.56	41.67
	Zr/Hf	20.00	43.62	52.59	37.30	37.04	37.36	35.46	300	100

表 13-13 海南岛新生代火山岩微量元素含量一览表

Table 13-13 List of trace element contents of Cenozoic volcanic rocks from the Hainan Island

岩性		粗面岩		碱玄岩	石英拉斑玄武岩				碱玄岩		
样品号		9DG	9DG-2	D10-2	D8-5G	D8-6	D8-7	D9-3	S04-11	S04-12-10	S04-12-11
序号		1	2	3	4	5	6	7	8	9	10
微量元素含量/10^{-6}	Rb	87.09	36.38	54	34	38	30	68	97	82.9	66.8
	Sr	235.48	31.62	349	162	170	145	400	773	737	652
	Ba	1633.01	108.3	810	99	*	*	432	973	650	538
	Zr	646.5	419	330	120	135	153	274	250	303	250
	Ta	7		2.5		0.8		1.6	5.79	4.89	4.49
	Hf	16.72							5.72	6.17	5.28
	Nb	99.74		84	158	95	158	79	99.3	81.5	69.3
	Ni	84.25	26.7	237	21	24.5	25	60	135	16.6	100
	Th	14.85		13.6	4	6	5	9.5	11.1	9.78	8.92
	Co	24.8	5.67	79	79	39.5	55	79	54.3	32.4	46.4
	U	1.37		24.8	5.3	5.3	5.3	5.3	2.06	2.27	1.98
	V	17.34	29.15						231	163	148
	Sc	4.63	4.28	23	31	27	25	20	23.9	20.3	18
	Cr	2.01	3.45						110	124	149
	Cu	59.13	25.3	160	64	80	80	24	38.9	41.8	39.8
	Zn	62.35	94.88						98.7	101	102
	Y	31.776	23.11	39.56	18.82	20.59	28.99	41.24	31.4	35.1	24.2
	Cs		28		25	26	21	31	1.21	1.14	0.92
	Li		17		14	56	13	12	7.84	8.69	11.6
	Mo		9.4		1.3	0.3	5.6	0.9	3.94	5.38	4.51
	W								17.8	16.8	20.7
	Be								2.27	3.03	2.56
	Ga								21.1	24	20.9
	Cd								0.12	0.2	0.16
	In								0.09	0.08	0.08
	Tl								0.2	0.26	0.28
	Pb								7.76	7.42	6.82
	Bi								0.15	0.12	0.13
特征比值	Rb/Sr	0.37	1.15	0.15	0.21	0.22	0.21	0.17	0.13	0.11	0.10
	Ba/Sr	6.93	3.43	2.32	0.61			1.08	1.26	0.88	0.83
	K/Rb	489.00	1371.41	402.78	183.12	152.92	354.20	261.25	234.50	274.38	288.32
	Nb/Ta	14.25		33.6		30.62		37.50	17.15	16.67	15.43
	Zr/Hf	38.67							43.71	49.11	47.35

续表

岩性		碱玄岩									
样品号		S04-12-12	S04-12-18	S04-12-20	S04-12-21	S04-12-22	S04-12-8	S04-13-3B	S04-14-1B	S04-14-2B	S05-34A
序号		11	12	13	14	15	16	17	18	19	20
微量元素含量/10^{-6}	Rb	16.9	85.5	54.4	78.8	16.4	83.2	16.9	29	31	95.3
	Sr	597	685	631	661	1929	721	748	754	738	801
	Ba	485	603	492	592	348	625	376	790	783	768
	Zr	232	336	275	279	211	316	205	238	233	436
	Ta	3.57	5.15	4.63	4.33	3.66	5.03	3.07	4.12	3.98	7.36
	Hf	5.17	6.42	5.92	5.61	5.16	6.14	5.12	5.99	5.72	8.06
	Nb	58.8	87.7	71.6	74.5	56.1	82.8	54	69.4	66.3	126
	Ni	67.6	22.4	112	77.2	171	13.6	50.3	107	35.8	37.6
	Th	6.66	10.8	5.95	9.07	8.19	9.85	5	9.91	7.26	12.6
	Co	44.1	28.3	57.2	44.1	76.6	30.1	38.8	120	29.6	43.7
	U	1.27	2.65	1.69	2.02	1.68	2.24	2.33	2.36	2.04	2.65
	V	219	141	146	152	201	150	206	191	181	161
	Sc	23.7	18.5	18.7	18.8	23.7	18.7	20.8	16.4	16.1	12.8
	Cr	195	155	140	121	161	111	58.1	47.5	47.4	5.59
	Cu	42.7	43.4	43.3	42	38.5	42.4	33.3	35.5	26.2	32.8
	Zn	129	84.2	107	91.3	141	96	117	100	103	93.6
	Y	30.2	37.1	31	26.5	45.6	34	34.9	83.9	28.6	35.1
	Cs	0.34	1.26	0.77	1.07	0.53	1.25	0.16	0.12	0.11	1.25
	Li	16.9	9.88	14.4	8.36	19.44	8.34	16.7	13.6	14.6	8.71
	Mo	2.84	5.46	4.21	5.28	5.6	5.12	1.41	8.02	1.23	6.51
	W	9.18	14.1	18.4	17.9	1.4	17	9.43	10.1	26.9	19.3
	Be	1.38	3.37	2.47	2.96	1.6	3.02	1.34	1.34	1.35	2.65
	Ga	21.6	24.4	20.3	23	17.8	23.2	17	19.8	18.5	22.3
	Cd	0.11	0.12	0.14	0.23	0.35	0.09	0.26	0.27	0.14	0.15
	In	0.08	0.07	0.08	0.08	0.09	0.07	0.08	0.08	0.08	0.08
	Tl	0.38	0.57	0.39	0.49	1.1	0.27	0.49	1.82	0.3	0.7
	Pb	7.9	7.16	9.75	7.26	36.5	7.44	9.86	53.6	6.13	20
	Bi	0.14	0.13	0.15	0.14	0.49	0.12	0.21	0.71	0.1	0.21
特征比值	Rb/Sr	0.03	0.12	0.09	0.12	0.01	0.12	0.02	0.04	0.04	0.12
	Ba/Sr	0.81	0.88	0.78	0.90	0.18	0.87	0.50	1.05	1.06	0.96
	K/Rb	628.75	274.77	311.31	274.96	511.25	277.38	677.87	672.71	653.41	291.82
	Nb/Ta	16.47	17.03	15.46	17.21	15.33	16.46	17.59	16.845	16.66	17.12
	Zr/Hf	44.87	52.34	46.45	49.73	40.89	51.47	40.04	39.73	40.73	54.09

续表

岩性		石英拉斑玄武岩									霞石岩
样品号		V30B9	V36 D10	V36 D8-1	V36 D8-2	V36 D8-2B	V36 D8-4	V36 D9-2b	V36 D9-3b	南-1	75-31-1
序号		21	22	23	24	25	26	27	28	29	30
微量元素含量/10^{-6}	Rb	39.8	60.7	26	19.3	20	12.5	53.8	33.8		30
	Sr	783.3	819.2	272	320.6	325.6	313.9	748.8	304.1	363.6	891
	Ba	437.6	853.6		32	147.1	15.1	441.2	18.1	89.5	503.6
	Zr	349.8	348.5	165.8	157.3	204.7	149.8	343.6	154	140	543.3
	Ta		4		1.4		1.2	3.7	1.4		6.3
	Hf		8.1		3.2		3.7	7.4	3.3		9.6
	Nb	55.1	83.5	22.1	23.4	30.3	20.6	52.5	20.6		103.3
	Ni	17.3	76.5	53.3	37.9	31.2	47.7	15.2	43.7	86	312.4
	Th		5.4		1.6		2.2	4.4	2.2		10.7
	Co	51.5	58.3	62.8	54	63.5	59.6	54.2	54.9	120	84.2
	U										
	V	306.3	249.2	317.2	288.1	356.6	288.5	308.1	296.1	330	264.2
	Sc	24.2	23.2	37.4	38.4	31.3	34.5	24.1	41.6	31	21.8
	Cr	50.4	146.4	20.7	32.3	40	174.9	14.7	40.8	50	247.1
	Cu	28.8	47.9	89.1	93.8	74.6	84.8	26.7	93.1	100	56.1
	Zn	154	118.3	147.8	166.6	185.3	128.9	161.3	151.2		140.4
	Y	49.1	43.1	40	33.6	39.4	29.5	45.2	33.8	31	48.8
	Cs										
	Li										
	Mo										
	W										
	Be										
	Ga										
	Cd										
	In										
	Tl										
	Pb										
	Bi										
特征比值	Rb/Sr	0.05	0.07	0.10	0.06	0.06	0.04	0.07	0.11		0.03
	Ba/Sr	0.56	1.04		0.10	0.45	0.05	0.59	0.06	0.25	0.57
	K/Rb	513.11	365.16	284.17	318.30	398.47	391.8	373.41	218.59		489.79
	Nb/Ta		20.88		16.71		17.17	14.19	14.71		16.40
	Zr/Hf		43.02		49.16		40.49	46.43	46.67		56.59

续表

岩性		碱玄岩	玻基辉橄岩		粗面玄武岩		橄榄玄武岩				平均值
样品号		S04-16	高-1	高-2	S08-69-1	S08-69-2	SO23-35-6	SO23-37-31	SO23-37-7	SO23-40	
序号		31	32	33	34	35	36	37	38	39	
微量元素含量 10^{-6}	Rb	91.1			61.7	69.1	49.5	26.9	27.6	13.8	41.43
	Sr	714	397.4	1048.5	539	551	774.3	481.4	180.9	480.9	501.25
	Ba	995	662.8	1012.1	967	985	599.6	284.8	286.5	139.9	543.8
	Zr	229	420	470	87.7	90.7	253.2	158.2	154.9	179.8	154.08
	Ta	5.67			3.62	3.65	4.3	2.2	3.7	1.7	3.20
	Hf	4.95			2.67	2.77	5.8	3.9	5.1	4.7	4.16
	Nb	106			69.8	72.6	66.2	34.3	34.6	28.8	51.05
	Ni	193	320	260	232	224	239.6	151.4	153.2	132.6	188.8
	Th	13.1			9.59	9.64	6.6	3.3	5.3	2.5	6.16
	Co	97.8	140	130	59.8	63.6	74.2	94	99	70.3	76.82
	U	2.24			1.52	1.38					1.45
	V	224	200	310	201	211	252.7	214	216.8	254.1	224.93
	Sc	24.5	23	27	27	28.6	28.6	26.8	29.6	28.8	28.23
	Cr	104	320	290	248	236	404.8	293.7	694.9	199.7	346.18
	Cu	50	62	66	51.8	55.8	72.8	86.7	80	52	66.52
	Zn	107			98.1	107	118.2	122.7	128.4	141.6	119.33
	Y	35.7	38	42	20.8	22.1	30.5	26.7	23.9	29.8	25.63
	Cs	1.15			0.65	0.92					0.79
	Li	7.89			5.02	7.28					6.15
	Mo	10.2			1.69	1.92					1.81
	W	45.6			17.4	34.6					26
	Be	2.14			1.12	1.19					1.16
	Ga	22.2			18.5	19.7					19.1
	Cd	0.22			0.22	0.13					0.18
	In	0.09			0.07	0.07					0.07
	Tl	1.54			0.24	0.26					0.25
	Pb	49			6.7	5.95					6.33
	Bi	0.71			0.16	0.14					0.15
特征比值	Rb/Sr	0.13			0.11	0.13	0.06	0.06	0.15	0.03	0.09
	Ba/Sr	1.39	1.67	0.97	1.79	1.79	0.77	0.59	1.58	0.29	1.14
	K/Rb	232.37			282.55	266.71	360.57	354.90	577.50	631.64	412.31
	Nb/Ta	18.70			19.28	19.89	15.40	15.59	9.35	16.94	16.08
	Zr/Hf	46.26			32.85	32.74	43.66	40.56	30.37	38.26	36.41

注：Au 的单位为 10^{-9}。

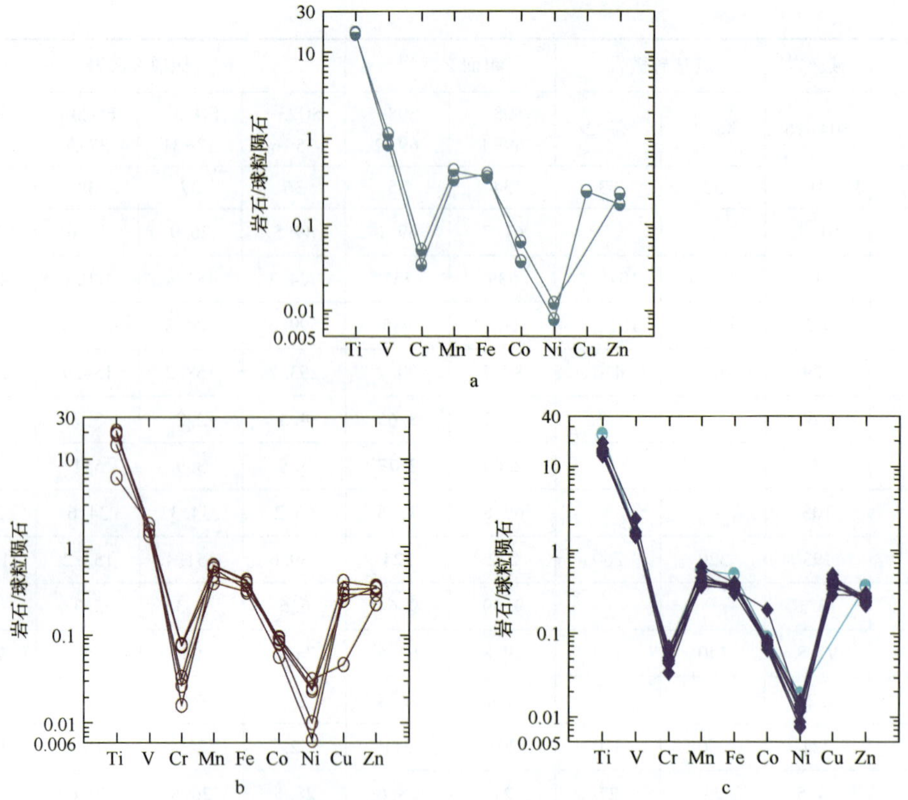

图 13-9　琼北新近纪火山岩过渡族元素球粒陨石标准化图

Fig. 13-9　Chondrite-normalized patterns of transition metal elements of Neogene volcanic rocks from Qiongbei

a—金牛岭期；b—石门沟村期；c—石马村期

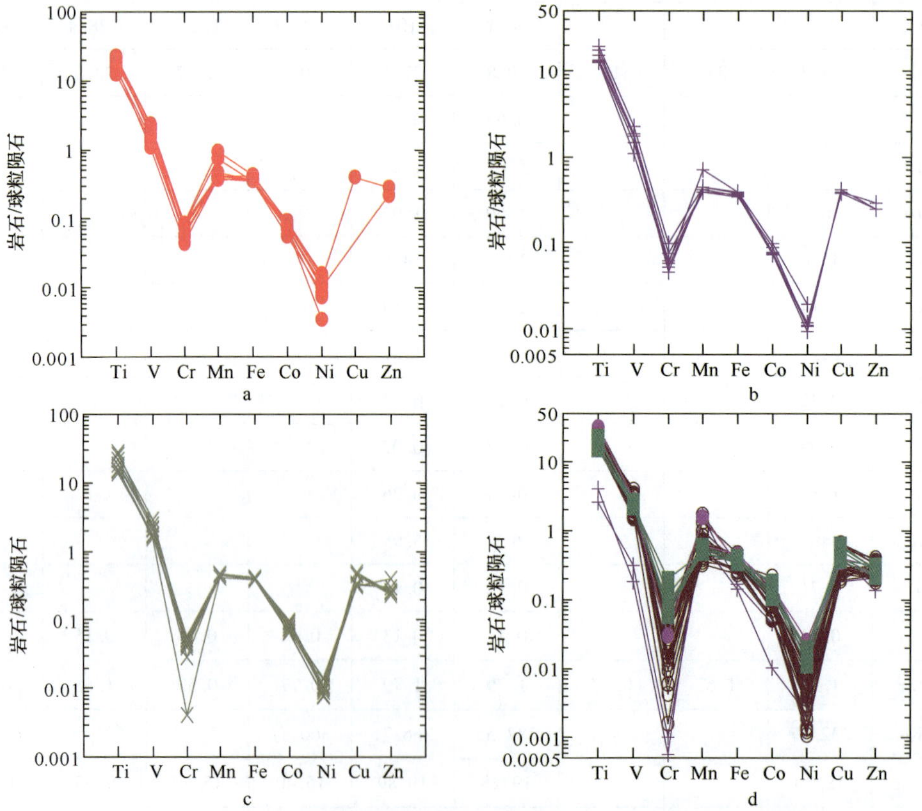

图 13-10　琼北第四纪及南海火山岩过渡族元素球粒陨石标准化图

Fig. 13-10　Chondrite-normalized patterns of transition metal elements of Quaternary volcanic rocks from Qiongbei and the South China Sea

a—石山期；b—道堂期；c—多文岭期；d—南海

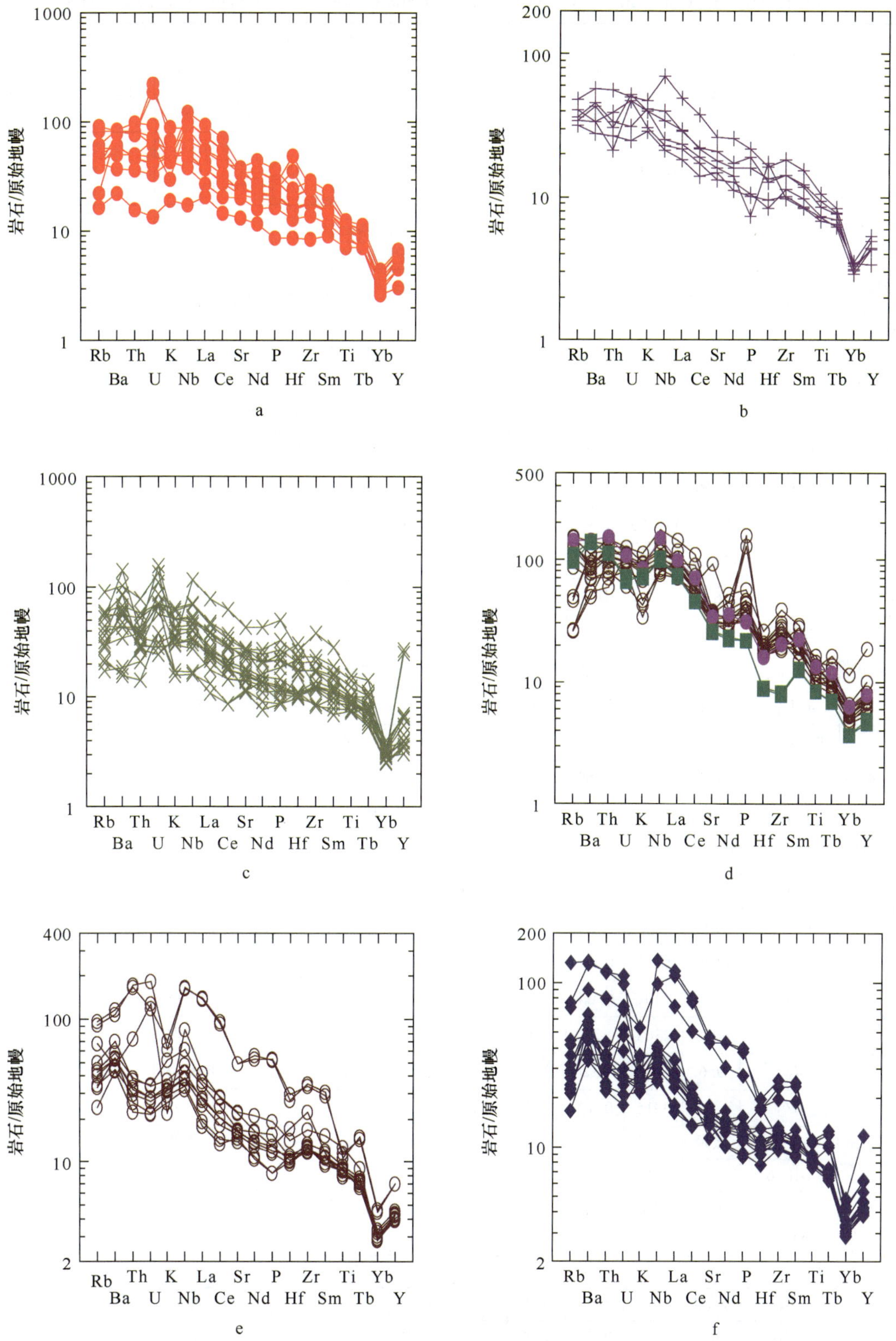

图 13-11　新生代火山岩微量元素比值蛛网图

Fig. 13-11　Spider diagram of trace elements of Cenozoic volcanic rocks

a—石山期；b—道堂期；c—多文岭期；d—南海；e—石门沟村期；f—石马村期

表 13-14 新生代部分火山岩副矿物含量一览表

Table 13-14 List of contents of accessory minerals of some of Cenozoic volcanic rocks

层位	代号	岩性	样品号	副矿物含量/10^{-6}													
				磁铁矿	钛铁矿	锆石	独居石	电气石	板钛矿	金红石	自然铅	方铅矿	黄铁矿	赤铁矿	菱铁矿	钛磁铁矿	磷灰石
石山组	Qh^1s	橄榄玄武岩	Ⅱ40-RZ	151.7	735.6			个别			6.75				个别		个别
道堂组	Qp^3d	玻屑凝灰岩	XⅧ$_{2-1}$	20.7	12.4	4.8	个别	1		3	个别				8		
		橄榄玄武岩	XⅧ$_{3-2}$	1232.7	433.6	微量			个别		微量	微量	微量				微量
		辉石玄武岩	XⅦ$_{2-1}$	1707	817.7	个别				个别			个别				个别
		玻基玄武岩	XⅧ$_{5-1}$	个别	个别	个别				个别	个别	微量	微量				个别
			XⅦ$_{7-1}$	少量	个别	个别				个别	个别	个别	个别				
		拉斑玄武岩	Ⅰ166-RZ	1.26		少量					个别		个别		微量		
			Ⅰ-RZ	0.46	444.73	0.32			个别		个别	个别	个别	个别	微量		
多文组	Qp^2d	橄榄玄武岩	D382	136.29	微量	2.58				微量			7.74			2.58	9.03
			ⅡH_{3-1}	266.67	93.33	微量	26.67			微量			0.69				1.39
		橄榄辉石玄武岩	ⅡH_{4-3}	19.46	23.65	微量	1.89			微量			0.96				1.51
			ⅡH_{10-2}	525.45	30.65	微量	个别			微量			微量			个别	0.65
		玄武岩	ⅡH_{14-2}	226.09	180.21	微量	1087						微量				2.17
			Ⅱ326-RZ	15.68		个别				个别			个别		微量		
		粗玄岩	XⅧ$_{1-1}$	496.0	1180	个别				个别			个别				个别
		辉石玄武岩	ⅠH_{3-1}	1210.96	552.6	微量		少量		微量			0.82				1.64
			ⅠH_{4-2}	226.67	135.67	1.67	微量			微量			3.33				5.0
石门沟村组	N_2s	粗玄岩	I-RZ_1	7670	5277	个别	微量	个别					个别				
			I-RZ_2	1544	804	个别	个别						微量				
		橄榄玄武岩	I-RZ_3	6649	微量	0.32							微量				
			I-RZ_4		40	0.04		2.2					微量				
			RZ_6	4478	2368	7	个别	个别			个别		个别				
			RZ_1	9.6	少量	3.2	个别	微量		微量			0.34				
			RZ_2	4.35	3315	0.12	个别	个别					0.01				
石马村组	$N_2\hat{s}m$	玄武岩	ZK2-RZ	微量	少量	少量	个别	少量	个别				微量				
			D1016	微量	个别	微量							微量				
			D3461	少量	微量	微量							5.1				
			D2174	少量	微量	个别							微量				
			D2297-2	少量	微量	微量				个别							
			I-RZ_7	4600	33	0.6	个别		微量				微量				
			Ⅱ-RZ_6	362	85	0.32	0.08		个别				0.2				
		玻基辉橄岩	RZ_5	329	125	5	2	个别					微量				
			Ⅱ-RZ_3	微量	2.5	微量	个别		个别	微量			个别				
			Ⅱ-RZ_1	28	1.8	0.3	个别		个别	个别			0.06				
		辉石玄武岩	ⅡH_{17-1}	127.5	700.00	0.075	17.5		微量				0.15				0.225

五、岩石 Sr、Nd、Pb 同位素

新生代火山岩的同位素样品共收集 46 个（琼北地区 21 个，南海地区 25 个），分析结果见表 13-15、表 13-16。结果表明，海南省新生代火山岩有着非常相近的 Sr、Nd、Pb 同位素组成，指示了岩浆区同位素组成的均一性。

Sr、Nd 同位素的初始比值可反映火山岩的岩浆源区特征，新生代火山岩的时代较年轻，经过时代校正的与未经校正的结果相差很小，其差值均在分析的误差范围内。一般认为，现在 $^{87}Sr/^{86}Sr$ 比值小于 0.706 的火山物质来源于地幔，比值介于 0.706～0.710 之间的火山物质来源于下地壳，比值大于 0.706 的火山物质来源于上地壳，而来源于地幔未被混染现今大洋中脊和大洋岛屿的火山岩的 $^{87}Sr/^{86}Sr$ 比值分别为 0.7028 和 0.7039。琼北新生代火山岩的 $^{87}Sr/^{86}Sr$ 的比值为 0.703178～0.704523，平均值为 0.703932；南海新生代火山岩的 $^{87}Sr/^{86}Sr$ 的比值为 0.703176～0.7046189，平均值为 0.704119。两者比较接近大洋岛屿的火山岩的 $^{87}Sr/^{86}Sr$ 比值。琼北新生代火山岩的 $^{143}Nd/^{144}Nd$ 的比值变化范围较小，为 0.512787～0.513039，平均值为 0.512852，其 ε_{Nd} 值均大于 0；南海新生代火山岩的 $^{143}Nd/^{144}Nd$ 的比值为 0.512663～0.513035，平均值为 0.512863，其 ε_{Nd} 值均大于 0。显示出两者源于地幔。

表 13-15　琼北新生代火山岩的 Sr-Nd-Pb 同位素成分一览表

Table 13-15　List of Sr-Nd-Pb isotope compositions of Cenozoic volcanic rocks from Qiongbei

层位	样品号	序号	采样地点	岩性	$^{87}Sr/^{86}Sr$	$^{143}Nd/^{144}Nd$	ε_{Nd}	$^{206}Pb/^{204}Pb$	$^{207}Pb/^{204}Pb$	$^{208}Pb/^{204}Pb$
石山组	81-H-22	1	博昌	橄榄拉斑玄武岩	0.704352	0.512934	5.83			
	HN9910	2	雷虎岭		0.704300	0.512819	3.53	18.37492	15.57014	38.59873
	QB012	3	春藏岭		0.704305	0.512871	4.55	18.50276	15.49458	38.45946
	Y-1	4	杨花		0.704512	0.512850	4.14	18.47951	15.50043	38.39894
	HN9902	5	马鞍岭	石英拉斑玄武岩	0.704299	0.512886	4.84	18.57735	15.57239	38.56286
	101B1	6	美造水库		0.704274	0.512877	4.66	18.59185	15.61056	38.76494
道堂组	81-H-18	7	琼山油库		0.704120	0.512797	3.10	18.619	15.596	39.051
多文组	木 ZK_1-1	8	木棠	橄榄玄武岩	0.703178					
	福 ZK_7-1	9	福山		0.703454					
	81-H-16	10	云龙?		0.703780	0.512957	6.26			
	81-H-24	11	福山		0.703775	0.512972	6.53			
	81-H-25	12	福山		0.703645	0.512916	5.42			
石门沟村组	81-H-13	13	金牛岭	橄榄玄武岩	0.703471	0.513039	7.87			
	D1836	14	庙岭			0.512881	4.85	17.653	15.401	37.932
	81-H-1	15	蓬莱绿园村		0.703722	0.512892	5.07	18.488	15.638	38.518
	81-H-3	16	牛姆岭	橄榄拉斑玄武岩	0.703419	0.512909	5.40	18.452	15.556	38.752
	81-H-5	17	蓬莱		0.711237	0.512737	2.04	18.685	15.645	38.782
石马村组	HK06-R1	18	南阳	玄武岩	0.704523	0.512787	3.02			
	D1832	19				0.512804	3.35	17.649	15.482	38.071
	81-H-6	20	蓬莱	碧玄岩	0.703993	0.512851	4.27	18.685	15.466	38.690
	81-H-7	21		霞石岩	0.703763	0.512852	4.29			
平均值					0.703932	0.512852		18.37645	15.54026	38.57083

注：1、7、10～13、15～17、20、21 据朱炳泉等；2～6 据中国地震局火山研究中心（2003）；8、9 据孙嘉诗（1991）；18 据马金龙等；14、18 据海南省地质调查院（2004）。

表 13-16　南海新生代火山岩的 Sr-Nd-Pb 同位素成分一览表

Table 13-16　List of Sr-Nd-Pb isotope composition of Cenozoic volcanic rocks from the South China Sea

样品号	序号	采样地点	岩性	$^{87}Sr/^{86}Sr$	$^{143}Nd/^{144}Nd$	ε_{Nd}	$^{206}Pb/^{204}Pb$	$^{207}Pb/^{204}Pb$	$^{208}Pb/^{204}Pb$
南-1	1	南海中央海盆	玄武岩	0.703368					
V36D8-2	2			0.703182	0.512920	5.7	18.704	15.609	38.325
V36D8-4	3			0.703176	0.512916	5.4	18.600	15.632	38.848
V36D9-3	4			0.703796	0.512922	5.5	18.667	15.535	38.677
V36D9-2	5			0.703897	0.512813	12	18.954	15.588	38.991
V36D10	6		碱性玄武岩	0.703909	0.512805	3.3	18.875	15.593	38.931
S04-11	7		碱玄岩	0.703418	0.512965	6.43	18.692	15.599	38.831
S04-12-12	8			0.704038	0.512863	4.44	18.622	15.596	38.767
S04-12-18	9			0.703961	0.512894	5.05	18.372	15.534	38.443
S04-12-20	10			0.704137	0.512855	4.29	18.501	15.596	38.665
S04-12-8	11			0.703963	0.512901	5.19	18.409	15.558	38.523
S04-13-3B	12			0.70494	0.512847	4.13	18.705	15.634	38.891
S04-14-1B	13			0.705174	0.512675	0.77	18.648	15.718	39.057
S05-34A	14			0.703771	0.51287	4.58	18.703	15.645	38.926
9DG	15		粗面岩	0.704183	0.512827	3.79	18.68668	15.67962	39.00261
S04-16	16	中沙	碱玄岩	0.70366	0.512906	5.28	18.697	15.711	39.085
高-1	17	西沙	玻基辉橄岩	0.706189					
高-2	18			0.705418					
75-31-1	19		霞石岩	0.703218	0.513035	7.7	18.521	15.521	38.402
SO2120	20	礼乐滩	橄榄玄武岩	0.703430	0.512952	6.1	18.601	15.557	38.629
SO23-37-31	21			0.703893	0.512898	5.1	18.543	15.606	38.598
SO23-37-7	22			0.703666	0.512894	5.0	18.481	15.567	38.618
SO23-37-6	23			0.704038	0.512913	5.4	18.411	15.575	38.554
S08-69-1	24	南沙	粗面玄武岩	0.705049	0.512687	1.01	18.783	15.674	39.211
S08-69-2	25			0.705041	0.512663	0.55	18.807	15.667	39.24
平均值				0.704119	0.512863		18.63265	15.60398	38.76808

注：1、17~18 据孙嘉诗（1991）；2~6、20~25 据涂勘等（1992）；7~16 据鄢全树等（2008）。

Haman 和 Graham（1996）对全球的 900 个 MORB 样品的 Pb 同位素进行了统计（太平洋、印度洋和大西洋），发现来自不同地区的样品相交于一个很小的范围，也就是说有一个共同的 Pb 同位素值，他们将其称为 C 组分，并将其解释为地幔柱核部的成分，偏离此值则是由于地幔柱上升时与岩石圈不同程度混合的结果。C 组分 Pb 同位素的变化范围为：$^{206}Pb/^{204}Pb$ 为 19.2~19.8、$^{207}Pb/^{204}Pb$ 为 15.55~15.65、$^{208}Pb/^{204}Pb$ 为 38.8~39.6。琼北和南海新生代火山岩的 $^{206}Pb/^{204}Pb$ 值比 C 组分略偏低，$^{207}Pb/^{204}Pb$ 和 $^{208}Pb/^{204}Pb$ 值与该值相差不大。

六、主要造岩矿物的化学成分

新生代火山熔岩的主要造岩矿物成分以橄榄石、辉石和斜长石为主，在超基性橄榄霞石岩中，还有霞石。本次工作对新生代火山熔岩的主要造岩矿物的化学成分进行电子探针分析，以下分别对每种造岩矿物化学成分特点进行论述。

（一）橄榄石

橄榄石是新生代火山岩的常见矿物，结晶时间较长，即是最早结晶的斑晶产物，也可以延续到晚期于基质斜长石晶出之后才结晶，斑晶含量一般比在基质中的含量高。在薄片下，橄榄石一般呈半自形—他形短柱状、板状、近等轴粒状及碎粒状，大小一般为0.01~3.0mm不等，呈无色或淡的黄绿色，正高突起，糙面显著，一组解理不完全，平行消光，裂纹发育。早期晶出的具熔蚀状，可以是单个晶体，也可见形成联晶状。作为基质部分常与辉石微晶等一起充填在长石微晶构成的格架中。

新生代火山熔岩中橄榄石电子探针分析结果见表13-17、表13-18。琼北新生代火山岩的橄榄石化学成分以 SiO_2、FeO、MgO 为主（含量分别为33.18%~40.60%、7.92%~48.21%、17.17%~50.56%），MnO、CaO等少量。主要化学成分含量变化比较大，形成的种属多。Fo：46~91.9，Fa：8.1~54，从富镁的镁橄榄石到富铁的透铁橄榄石。其中，新近纪火山熔岩的橄榄石不论是斑晶还是基质均以富镁的贵橄榄石为主。第四纪火山熔岩的斑晶以富镁的贵橄榄石为主，而基质以富铁的透铁橄榄石为主。

南海新生代火山岩的橄榄石化学成分以 SiO_2、FeO^*、MgO 为主（含量分别为35.25%~39.67%、13.05%~25.62%、35.22%~47.85%），MnO、CaO、NiO等少量。主要化学成分含量变化不大，Fo：71.0~86.6，Fa：112~29.0，均为贵橄榄石。

（二）辉石

辉石广泛见于新生代火山岩的斑晶和基质，比橄榄石更常见，含量仅次于斜长石。以单斜辉石为主，偶见斜方辉石。在薄片下，辉石一般呈半自形—自形短柱状、微细粒状，粒径0.01~2.5mm，无色正高突起，柱面解理完全，辉石解理完全，双晶十分发育，有简单双晶、聚片双晶和贯穿双晶，绝大部分呈斜消光或对称消光，偶见平行消光。

新生代火山熔岩中辉石电子探针样分析结果见表13-19、表13-20。其中，琼北地区的分析样品50个（单斜辉石44个，斜方辉石6个），南海地区的分析样品20个（均为单斜辉石）。其化学成分以 SiO_2、Al_2O_3、FeO、MgO、CaO 为主，含量分别为47.18%~56.70%、0.95%~8.53%、9.07%~34.07%、0.62%~24.14%，次有 TiO_2、MnO、Na_2O 等。主要成分FeO、MgO和CaO的含量变化范围较大，因此形成的种属多。斜方辉石为古铜辉石和顽火辉石，主要在新近纪火山熔岩的斑晶见及；单斜辉石有普通辉石、透辉石、含钛普通辉石、镁易变辉石、含钛次透辉石等种属。

Thompson（1974）根据实验认为，单斜辉石与玄武质熔体平衡时的 K_d^{cpx}(Fe-Mg)值为0.29，Irving等（1984）和Liotard（1988）根据实验结果认为，当该值在0.2~0.4时，单斜辉石与熔体是平衡的。本区的 K_d^{cpx}(Fe-Mg)值为0.10~1.20，除少量样品符合平衡条件外，大部分样品都不符合平衡条件，琼北地区的分析样品主要表现为单斜辉石与熔体的 K_d^{cpx}(Fe-Mg)值大于0.4的方向偏移，而南海地区的分析样品主要表现为单斜辉石与熔体的 K_d^{cpx}（Fe-Mg）值小于0.2的方向偏移（图13-12）。

根据R. N. 汤普森（R. N. Thompsom，1974）实验推导出的各种玄武岩单斜辉石Al与压力、温度关系公式（周新民，1982）：

$$P\ (\mathrm{GPa}) = (-7.5382+83.1692\mathrm{Al}) \times 10^{-1},\ r=0.95$$

$$t\ (℃) = 1056.8986+902.7978\mathrm{Al},\ r=0.92$$

表 13-17　琼北新生代火山岩橄榄石的化学成分一览表

Table 13-17　List of chemical composition of olivine from Cenozoic volcanic rocks，Qiongbei

层位	样品号	产出岩性	序号	电子探针分析结果/%												端元组分		种属
				SiO_2	TiO_2	Al_2O_3	FeO	MnO	MgO	CaO	Na_2O	K_2O	Cr_2O_3	NiO	总量	Fo	Fa	
石山组	128-2p	碱性橄榄玄武岩	1	40.60	0.07	0.59	15.68	0.18	42.28	0.24	0.02	0.00	0.02	0.24	100.00	52.7	17.3	透铁橄榄石
	128-2m		2	39.51	0.00	0.52	17.32	0.20	41.87	0.31	0.05	0.00	0.03	0.18	100.01	81.0	19.0	贵橄榄石
	23-3p		3	39.28	0.00	0.11	24.21	0.24	35.81	0.22	0.00	0.00	0.06	0.12	99.99	72.5	27.5	
	23-3m		4	37.46	0.04	0.09	31.10	0.40	30.46	0.27	0.04	0.00	0.00	0.14	100.0	612	36.6	透铁橄榄石
	QB014p1		5	39.78	0.02	0.00	13.86	0.11	44.53	0.33	0.03	0.00	0.00	0.36	99.02	85.13	14.87	贵橄榄石
	QB014p2		6	37.38	0.07	0.03	28.76	0.49	33.70	0.38	0.04	0.03	0.00	0.00	100.88	67.62	32.38	透铁橄榄石
	QB014m		7	35.87	0.12	0.03	32.49	0.69	29.43	0.52	0.02	0.06	0.00	0.12	99.35	61.75	38.25	
	219B4p1		8	39.49	0.00	0.01	14.19	0.34	45.28	0.16	0.02	0.00	0.01	0.35	99.85	85.05	14.95	贵橄榄石
	219B4p2		9	35.63	0.08	0.01	37.58	0.62	26.08	0.53	0.01	0.01	0.00	0.01	100.56	55.30	44.70	透铁橄榄石
	219B4m		10	37.99	0.01	0.02	23.84	0.33	37.62	0.45	0.00	0.00	0.01	0.16	100.49	73.77	26.23	贵橄榄石
	106B1p1	拉斑玄武岩	11	37.26	0.00	0.00	24.14	0.13	37.27	0.29	0.00	0.00	0.00	0.13	99.22	73.35	26.65	
	106B1p2		12	36.24	0.01	0.00	33.96	0.47	29.68	0.00	0.05	0.01	0.02	0.13	100.57	60.90	39.10	透铁橄榄石
	106B1m		13	36.13	0.03	0.03	29.64	0.36	32.42	0.33	0.01	0.02	0.00	0.05	99.02	66.10	33.90	
	105B2p1		14	39.90	0.00	0.04	16.31	0.12	41.24	0.28	0.06	0.01	0.00	0.30	100.69	82.60	17.40	贵橄榄石
	105B2p2		15	33.84	0.09	0.00	42.97	0.53	21.59	0.29	0.00	0.05	0.06	0.13	99.55	47.25	52.75	镁铁橄榄石
	105B2m		16	35.59	0.13	0.02	36.61	0.58	25.82	0.37	0.02	0.00	0.00	0.07	99.21	55.70	44.30	透铁橄榄石
	114B4p1		17	38.92	0.00	0.00	18.57	0.34	42.36	0.23	0.03	0.03	0.10	0.16	100.74	80.26	19.74	贵橄榄石
	114B4p2		18	37.52	0.14	0.12	25.76	0.14	35.90	0.34	0.01	0.05	0.09	0.15	100.22	71.31	28.69	
	114B4m		19	33.67	0.13	0.00	43.37	0.64	20.73	0.42	0.02	0.03	0.04	0.00	99.05	46.00	54.00	镁铁橄榄石
道堂组	220B3p1	碱性橄榄玄武岩	20	39.04	0.00	0.05	15.36	0.24	44.66	0.21	0.03	0.00	0.04	0.24	99.90	83.83	16.27	贵橄榄石
	220B3p2		21	36.20	0.00	0.06	27.88	0.44	33.89	0.35	0.04	0.00	0.00	0.05	98.91	68.42	31.58	透铁橄榄石
	220B3m		22	36.51	0.08	0.03	27.28	0.43	34.81	0.35	0.00	0.00	0.00	0.21	99.68	69.46	30.54	
	10-4p	橄榄拉斑玄武岩	23	39.83	0.03	0.06	18.22	0.22	41.39	0.22	0.00	0.00	0.00	0.21	99.97	80.2	19.8	贵橄榄石
	10-4m		24	36.85	0.07	0.00	32.75	0.41	29.41	0.41	0.00	0.00	0.06	0.04	99.93	61.7	38.3	透铁橄榄石
	150-1p		25	39.27	0.05	0.12	19.50	0.24	40.61	0.22	0.00	0.00		0.22	99.95	78.8	21.2	贵橄榄石
	150-1m		26	37.49	0.07	0.10	28.97	0.35	32.78	0.25	0.00	0.00		0.15	99.93	67.3	32.7	透铁橄榄石

续表

层位	样品号	产出岩性	序号	电子探针分析结果/%												端元组分		种属
				SiO_2	TiO_2	Al_2O_3	FeO	MnO	MgO	CaO	Na_2O	K_2O	Cr_2O_3	NiO	总量	Fo	Fa	
多文组	42-1p	橄榄玄武岩	27	37.92	0.03	0.07	27.93	0.35	32.96	0.28	0.05	0.00		0.07	99.59	68.2	31.8	透铁橄榄石
	42-1m		28	33.18	0.10	0.08	48.21	0.71	17.17	0.35	0.00	0.00	0.05	0.02	99.97	60.9	39.1	
	D0997p	橄榄拉斑玄武岩	29	38.18	0.01	0.01	21.75	0.23	39.20	0.22	0.02		0.01	0.36	99.99	76.26	23.74	贵橄榄石
	13-1		30	38.17	0.09	0.07	29.02	0.34	31.94	0.33	0.00	0.00	0.00	0.08		66.3	33.7	透铁橄榄石
	73-1p		31	38.69	0.11	0.11	23.11	0.26	37.43	0.24	0.00	0.00		0.17	99.94	74.5	25.5	贵橄榄石
	73-1m		32	37.51	0.02	0.04	26.91	0.33	34.85	0.32	0.00	0.00		0.13	99.96	69.8	30.2	透铁橄榄石
	ZK_3-TW_3m	碱性橄榄玄武岩	33	36.14	0.07	0.15	33.23	0.34	29.73	0.31	0.00	0.00	0	0.13	99.98	61.8	38.2	
	ZK_3-TW_3p		34	38.29	0.06	0.07	21.14	0.24	39.14	0.22	0.04	0.00	0	0.24	99.20	76.9	23.1	贵橄榄石
	ZK_3-TW_2p		35	36.82	0.04	0.00	30.03	0.38	31.70	0.25	0.00	0.00	0.07	0.13	99.96	65.5	34.5	透铁橄榄石
	ZK_3-TW_2m		36	34.24	0.09	0.00	40.34	0.62	24.30	0.19	0.08	0.00	0.08	0.09	99.94	52.0	48.0	
石门沟村组	209-1p		37	39.79	0.01	0.05	16.25	0.19	43.56	0.05	0.00	0.00	0.05	0.30	99.95	82.9	17.1	贵橄榄石
	209-1m		38	38.70	0.10	0.00	24.58	0.36	35.74	0.34	0.00	0.00	0.09	0.16	99.92	72.3	27.7	
	D1835-1p	橄榄玄武岩	39	40.51	0.01	0.00	8.83	0.08	50.56	0.05	0.00	0.00	0		100.04	91.0	9.0	镁橄榄石
	D1835-1m1		40	38.92	0.01	0.00	23.17	0.36	36.89	0.39	0.00	0.00	0		99.74	73.6	26.4	贵橄榄石
	D1835-1m2		41	37.73	0.04	0.00	23.84	0.36	37.58	0.30	0.00	0.00	0.01		99.86	73.5	26.5	
	D1836-1p		42	35.26	0	0	36.31	0.58	26.34	0.21	0.00	0.00	0		98.70	56.0	44.0	透铁橄榄石
	D1842-1p		43	40.2	0	0	7.92	0.05	50.52	0	0.00	0.00	0.03		98.72	91.8	8.1	镁橄榄石
	D1842-1m		44	37.35	0	0	25.74	0.47	35.64	0.21	0.00	0.00	0		99.41	70.8	29.2	贵橄榄石
	208-9p		45	37.99	0.00	0.00	27.95	0.33	33.56	0.14	0.00	0.00		0.19	99.97	68.8	31.2	透铁橄榄石
	208-9m		46	37.35	0.08	0.00	29.90	0.57	31.87	0.13	0.00	0.00		0.10	99.95	68.6	31.2	
石马村组	D1016m1	玄武岩	47	39.22	0.04	0.11	17.75	0.12	41.80	0.24	0.17	0.02	0.04	0.48	99.99	80.75	19.25	贵橄榄石
	D1016m2		48	39.01	0.03	0.04	17.65	0.10	42.44	0.23	0.05	0.02		0.43	99.96	81.08	18.92	
	D3461m1		49	39.28	0.05	0.05	19.00	0.20	41.72	0.23	0.05		0.02	0.35	100.95	79.65	20.35	
	D3461m2		50	39.20		0.04	18.26	0.18	42.26	0.22	0.04	0.01	0.02	0.30	100.53	80.48	19.52	
	212-1p	玻基辉橄岩	51	40.35	0.01	0.46	14.23	0.22	42.77	0.32	0.04	0.00	0.06	0.24	99.70	84.5	15.5	
	185-1p	橄榄霞石岩	52	40.02	0.00	0.11	18.20	0.49	40.66	0.44	0.00	0.00	0.04	0.14	99.97	79.9	20.1	
	185-1m		53	38.21	0.07	0.00	24.01	0.77	35.81	1.00	0.00	0.00	0.07	0.08	99.94	72.7	27.3	

注：p—斑晶；m—基质；1~4、23~28、30~38、45~46、51~53 据宜昌地质矿产研究所（1992）；5~22 据樊祺诚等（2004）；39~44 据海南省地质调查院（2004）；余者为本院采集。

表 13-18 南海新生代火山岩橄榄石的化学成分一览表

Table 13-18 List of chemical composition of olivine from Cenozoic volcanic rocks, South China Sea

样品号	分析位置	产出岩性	序号	电子探针分析结果/%										$Mg^{\#}$	端元组分		种属
				SiO_2	TiO_2	Al_2O_3	Cr_2O_3	FeO^*	MnO	MgO	NiO	CaO	总量		Fa	Fo	
S04-12-11	pc	碱玄岩	1	38.77	0.00	0.02	0.00	13.15	0.22	47.85	0.04	0.27	100.31	86.6	13.4	86.6	贵橄榄石
	pm		2	38.33	0.00	0.00	0.00	15.77	0.41	43.78	0.15	0.22	99.31	83.2	16.8	83.2	
	pr		3	37.30	0.00	0.74	0.00	21.11	0.69	37.30	0.46	0.48	99.02	75.9	24.1	75.9	
	m		4	35.25	0.23	1.22	0.59	25.62	0.94	35.22	0.07	0.44	99.72	71.0	29.0	71.0	
S04-12-20	pc		5	39.67	0.00	0.00	0.00	13.05	0.00	46.61	0.00	0.17	99.50	86.4	13.6	86.4	
	pm		6	38.12	0.13	0.19	0.13	13.96	0.34	46.91	0.30	0.33	100.40	85.7	14.3	85.7	
	pr		7	37.37	0.22	0.00	0.00	19.09	0.30	43.13	0.24	0.27	100.62	80.1	19.9	80.1	
	m		8	37.89	0.00	0.00	0.06	18.12	0.41	41.93	0.67	0.40	99.49	80.5	19.5	80.5	
S04-12-2B	pc		9	37.26	0.10	0.00	0.12	20.93	0.02	41.10	0.11	0.28	99.93	77.8	22.2	77.8	
	pm		10	36.64	0.35	0.00	0.26	21.02	0.22	40.96	0.00	0.30	99.76	77.6	22.3	77.6	
	pr		11	36.88	0.16	0.00	0.04	21.60	0.00	41.15	0.00	0.35	100.18	77.3	22.7	77.3	
S04-16	pc		12	37.36	0.00	0.00	0.00	18.87	0.26	43.50	0.00	0.19	100.18	80.4	19.6	80.4	
	pm		13	37.46	0.00	0.00	0.00	18.50	0.20	43.68	0.23	0.06	100.12	80.8	19.2	80.8	
	pr		14	37.57	0.11	0.00	0.17	17.09	0.35	44.44	0.00	0.23	99.95	82.2	17.8	82.2	
	m		15	37.45	0.23	0.00	0.00	18.72	0.40	42.56	0.38	0.23	99.97	80.2	19.8	80.2	
S08-69-1	pc	粗面玄武岩	16	37.94	0.24	0.00	0.33	15.51	0.67	44.90	0.18	0.25	100.02	83.8	16.2	83.8	
	pm		17	37.94	0.01	0.00	0.19	14.81	0.41	45.63	0.77	0.21	99.97	84.6	15.4	84.6	
	pr		18	37.96	0.03	0.00	0.00	16.18	0.34	45.23	0.11	0.29	100.14	83.3	16.7	83.3	

注：pc—斑晶核部；pm—斑晶幔部；pr—斑晶边部；m—基质；FeO^*—全铁；$Mg^{\#}=100\times X_{Mg}/X_{Mg+FeO^*}$；测试资料据鄢全树等(2008)。

估算出新生代火山熔岩中单斜辉石斑晶的温度和压力，其中新近世玄武岩单斜辉石斑晶的温度和压力分别为：t=1075.9℃~1318.9℃、P=0.155~1.66GPa，相当于深度约5.11~54.78km；第四纪玄武岩单斜辉石斑晶的温度和压力分别为：t=1086.4℃~1228.0℃、P=0.028~0.823GPa，相当于深度约0.94~27.16km；南海玄武岩单斜辉石斑晶的温度和压力分别为：t=1248.5℃~1525.1℃、P=1.01~3.56GPa，相当于深度约33.33~117.48km。

(三) 斜长石

斜长石是新生代火山熔岩的主要矿物成分，大多产于基质中。在薄片中斜长石多呈半自形-自形板状、柱状、长条状、长柱状，大小0.01mm×0.03mm~1mm×3mm，发育聚片双晶和钠长石双晶，双晶单体较粗，绝大部分呈不规则的分布，个别岩石中偶见近平行排列。

表 13-19 琼北新生代火山岩辉石的化学成分一览表

Table 13-19 List of chemical composition of pyroxene from Cenozoic volcanic rocks, Qiongbei

层位	样品号	产出岩性	序号	电子探针分析结果/%												端元组分			种属
				SiO_2	TiO_2	Al_2O_3	FeO	MnO	MgO	CaO	Na_2O	K_2O	Cr_2O_3	NiO	总量	Wo	En	Fs	
石山组	128-2p	碱性橄榄玄武岩	1	53.47	1.60	8.53	7.63	0.15	9.07	16.90	1.80	0.70	0.00	0.08	99.98	47.8	35.5	16.7	含钛次透辉石
	128-2m		2	54.29	1.00	6.24	9.19	0.17	9.37	16.65	1.86	1.04	0.02		99.99	45.5	35.2	19.3	
	219B4m		3	46.82	3.29	5.41	9.39	0.20	12.94	21.75	0.52	0.01	0.07	0.00	100.4	46.1	38.2	15.7	
	QB014m		4	49.89	1.78	3.59	8.39	0.24	14.33	21.14	0.36	0.03	0.13	0.01	99.89	44.3	41.7	14.0	含钛普通辉石
	106B1m	拉斑玄武岩	5	50.57	1.15	1.91	8.80	0.16	15.64	20.06	0.29	0.01	0.16	0.10	98.85	41.1	44.6	14.3	
	105B2m		6	50.89	1.35	2.41	8.08	0.19	15.45	20.52	0.30	0.00	0.00	0.12	99.31	42.4	44.3	13.3	
	114B4m1		7	49.35	1.97	4.31	7.43	0.29	14.82	20.79	0.42	0.00	0.24	0.08	99.70	43.8	43.5	12.7	
	114B4m2		8	52.18	0.92	2.08	6.95	0.15	16.46	20.84	0.27	0.00	0.00	0.10	100.41	42.3	46.5	11.2	
	114B4m3		9	52.09	1.17	1.74	9.63	0.18	15.43	19.56	0.28	0.00	0.00	0.00	100.08	40.2	44.1	15.7	
道堂组	150-1p	橄榄拉斑玄武岩	10	54.53	0.35	0.48	17.99	0.42	21.96	4.18	0.10	0.00	0.03		100.01	8.8	62.7	28.5	镁易变辉石
	10-4m		11	51.43	1.06	1.10	14.71	0.34	11.83	19.01	0.31	0		0.07	99.79	35.2	24.4	40.4	易变普通辉石
	10-4p		12	52.72	1.04	3.97	6.96	0.08	15.49	19.31	0.45	0		0.08	100.02	46.7	11.5	41.8	低铁次辉石
	8-1m		13	53.48	1.03	1.54	13.39	0.24	15.34	14.77	0	0		0	99.80	46.2	22.1	31.7	
	8-1p		14	55.19	0.36	2.01	12.60	0.19	27.25	2.32	0.00	0.00		0.09	99.92	4.7	76.0	19.3	古铜辉石
	220B3m		15	45.30	3.65	7.81	8.60	0.19	12.26	20.66	0.49	0.00	0.00	0.00	98.96	46.4	38.3	15.3	含钛次透辉石
多文组	ZK_3-TW_2p	碱性橄榄玄武岩	16	51.16	1.35	3.21	7.60	0.00	14.50	21.69	0.28	0.00		0.14	99.93	45.3	42.6	21.1	含钛普通辉石
	ZK_3-TW_2m		17	51.37	1.35	2.17	10.18	0.19	13.84	20.18	0.34	0.00	0.12		99.93	42.4	40.8	16.8	
	ZK_3-TW_3m		18	51.60	1.41	2.54	10.39	0.00	15.71	18.04	0.27	0.00	0.02		99.96	37.5	45.8	16.7	
	42-1p	橄榄玄武岩	19	50.69	1.75	3.18	8.73	0.15	1124	20.65	0.36	0.00			99.15	44.6	40.9	14.5	
	42-1m		20	47.97	1.22	1.95	19.32	0.33	1123	15.43	0.37	0.00		0.02	100.00	31.8	38.8	29.4	
	D0997p		21	50.98	0.89	2.13	8.68	0.15	15.32	20.88	0.28	0.01	0.33		99.65	42.6	43.4	14.0	
	D0997m		22	50.77	1.06	2.73	9.21	0.14	16.80	18.38	0.27	0.01	0.43		99.80	37.5	47.7	14.8	
	13-1p	橄榄拉斑玄武岩	23	51.51	0.91	1.88	8.51	0.16	15.28	21.16	0.30	0	0.21	0.01	99.97	43.6	13.3	43.1	普通辉石
	73-1m		24	52.86	0.93	1.37	9.04	0.23	16.44	19.09	0.16	0.00		0.04	100.12	39.2	46.9	13.9	
	60-1m	石英拉斑玄武岩	25	54.69	0.70	1.25	12.57	0.26	16.32	14.09	0.11	0.00		0.04	99.99	30.3	48.7	21.0	
海口组	1-5p3		26	53.10	0.79	2.53	7.93	0.18	16.67	18.57	0.26	0.00		0.07	100.03	38.4	48.4	13.2	
	1-5p1		27	56.70	0.25	1.03	12.52	0.22	27.28	1.94	0.00	0.00		0.07	99.95	2.7	76.7	20.6	古铜辉石
	1-5p2		28	56.41	0.33	1.03	13.03	0.25	26.79	2.14	0.00	0.00		0.02	99.98	4.2	75.3	20.5	
	1-5m		29	52.26	1.07	1.54	11.66	0.19	14.60	18.45	0.23	0.00		0.02	100.00	38.5	42.7	18.8	含钛普通辉石

层位	样品号	产出岩性	序号	电子探针分析结果/%												端元组分			种属
				SiO_2	TiO_2	Al_2O_3	FeO	MnO	MgO	CaO	Na_2O	K_2O	Cr_2O_3	NiO	总量	Wo	En	Fs	
石门沟村组	D1835-1p1	橄榄玄武岩	30	53. 59	0. 04	4. 52	5. 78	0. 13	33. 33	0. 65	0. 04	0. 00	0. 23		98. 31	1. 3	89. 8	8. 9	顽火辉石
	D1835-1p2		31	50. 42	1. 04	2. 20	7. 05	0. 14	15. 91	21. 73	0. 49	0. 00	0. 32		99. 30	43. 9	44. 7	11. 3	含钛普通辉石
	D1835-1m		32	49. 65	1. 61	2. 78	8. 21	0. 18	14. 99	21. 35	0. 50	0. 00			99. 27	43. 8	42. 8	13. 4	
	208-9p1		33	52. 28	1. 4	2. 87	9. 91	0. 14	14. 82	18. 27	0. 41	0. 00			100. 10	38. 7	44. 1	17. 2	
	209-1p1		34	53. 15	0. 72	6. 81	4. 59	0. 00	13. 53	19. 32	1. 52	0. 00	0. 27	0. 03	99. 90	46. 6	44. 7	8. 7	透辉石
	209-1p2		35	50. 21	2. 18	4. 56	7. 36	0. 10	13. 50	21. 40	0. 40	0. 00	0. 24	0. 02	99. 94	46. 6	41. 0	12. 6	钛次透辉石
	209-1m		36	49. 89	2. 41	4. 22	8. 64	0. 07	13. 62	20. 61	0. 47	0. 00			99. 94	44. 3	41. 1	14. 6	钛普通辉石
	D1836-1p		37	51. 51	0. 93	1. 86	8. 65	0. 16	15. 43	20. 55	0. 41	0. 00			99. 50	42. 0	43. 9	14. 1	普通辉石
	D1836-1m		38	50. 38	0. 92	1. 33	11. 77	0. 20	13. 74	19. 82	0. 45	0. 00			98. 61	41. 1	39. 6	19. 3	
	208-9p2		39	54. 82	0. 53	2. 20	14. 75	0. 25	25. 29	2. 11	0. 00	0. 00		0. 10	99. 95	4. 3	72. 3	23. 4	古铜辉石
	D1842-1p1		40	51. 83	0. 03	2. 37	1. 87	0. 00	16. 79	24. 14	0. 54	0. 00	0. 77		98. 34	49. 3	47. 7	3. 0	透辉石
	D1842-1p2		41	47. 74	1. 19	8. 14	9. 55	0. 06	12. 66	16. 84	1. 85	0. 00			98. 03	40. 1	42. 0	17. 9	普通辉石
	D1842-1m		42	48. 83	2. 18	3. 16	8. 52	0. 08	14. 60	21. 9	0. 71	0. 00			100. 00	44. 8	41. 5	13. 7	
石马村组	D1016m1	石英拉斑玄武岩	43	53. 33	0. 64	1. 00	17. 75	0. 24	20. 66	6. 13	0. 07		0. 01	0. 07	99. 90	12. 7	58. 6	28. 7	镁易变辉石
	D1016m2		44	53. 88	0. 42	0. 59	17. 20	0. 24	23. 19	4. 17	0. 13	0. 01	0. 04	0. 14	100. 01	8. 3	64. 5	27. 2	
	194-2m		45	54. 84	0. 46	0. 95	15. 27	0. 32	22. 61	5. 46	0. 10	0. 00		0. 04	100. 01	11. 1	64. 7	24. 2	
	194-2p		46	55. 88	0. 22	1. 12	12. 02	0. 26	27. 54	2. 01	0. 05	0. 00		0. 01	99. 09	4. 1	77. 1	18. 8	古铜辉石
	D3461m		47	51. 06	0. 75	2. 19	7. 12	0. 13	16. 01	21. 62	0. 29	0. 02	0. 36		99. 55	43. 7	45. 0	11. 3	普通辉石
	212-1p	玻基辉橄岩	48	47. 18	2. 71	7. 74	7. 58	0. 13	11. 71	22. 10	0. 51	0. 00		0. 05	99. 98	50. 0	36. 5	13. 5	钛普通辉石
	212-1m		49	49. 66	2. 07	6. 23	7. 14	0. 14	13. 03	20. 80	0. 59	0. 00			100. 01	46. 9	40. 7	12. 4	
	185-1m	橄榄霞石岩	50	50. 08	2. 16	3. 59	6. 91	0. 17	12. 42	23. 00	0. 5	0. 00		0. 04	99. 03	50. 3	37. 8	11. 9	

注：p—斑晶；m—基质；1~3、10~14、16~20、23~29、33~36、39、45~46、48~50 据宜昌地质矿产研究所（1992）；4~9、15 据樊祺城等（2004）；30~32、37~38、40~42 据海南省地质调查院（2004）；余者为本院采集。

表 13-20 南海新生代火山岩辉石的化学成分

Table 13-20 List of chemical composition of pyroxene from Cenozoic volcanic rocks, South China Sea

样品号	产出岩性	序号	电子探针分析结果/%										端元组分			种属
			SiO_2	TiO_2	Al_2O_3	FeO*	MnO	MgO	CaO	Na_2O	K_2O	总量	Wo	En	Fs	
S04-11-2 p	碱玄岩	1	53. 43	0. 85	5. 54	3. 48	0	19. 42	14. 8	1. 62	0	99. 14	33	61	6	顽透辉石
S04-11-5 p		2	52. 75	1	9. 07	2. 88	0. 17	18. 74	14. 6	1. 51	0. 07	100. 79	34	61	5	
S04-12-1 p		3	49. 19	1. 24	10. 76	5. 35	0. 15	16. 19	14. 12	1. 89	0. 12	99. 01	35	55	10	普通辉石
S04-12-3 p		4	52. 05	1. 03	7. 95	4. 55	0	18	14. 07	1. 75	0. 18	99. 58	33	59	8	顽透辉石
S04-12-4 p		5	51. 21	0. 95	9. 21	4. 97	0. 18	16. 83	14. 7	1. 05	0	99. 1	35	56	9	
S04-12-2-2 p		6	50. 67	1. 29	8. 94	4. 57	0. 12	17. 75	13. 8	1. 37	0. 11	98. 62	33	59	8	
S04-12-5-2p		7	50. 02	1. 37	9. 21	5. 42	0. 05	16. 84	14. 41	1. 62	0	98. 94	34	56	10	
S04-12-11-4 p		8	48. 13	2. 31	12. 15	4. 92	0. 1	15. 69	14. 49	1. 36	0. 01	99. 16	36	54	10	
S04-12-12-1 p		9	52. 07	1. 34	8. 8	5. 15	0	17. 9	14. 33	1. 13	0. 02	100. 74	33	58	9	
S04-12-12-2 m		10	49. 60	1. 74	4. 85	6. 95	0	14. 78	21. 18	0. 69	0	99. 79	45	44	12	透辉石
S04-12-13-1 p		11	50. 88	1. 09	10. 12	4. 52	0	17. 3	14. 27	1. 21	0	99. 39	34	57	8	顽透辉石
S04-12-13-2 p		12	50. 99	1. 19	9. 32	4. 9	0	16. 94	14. 96	1. 62	0	99. 92	35	56	9	
S04-12-13-3 p		13	48. 89	1. 72	10. 45	5. 87	0	15. 65	14. 82	1. 1	0. 12	99. 62	36	53	11	普通辉石
S04-12-18-1 p		14	52. 94	1. 04	6. 85	4. 18	0	19. 84	13. 8	1. 17	0	99. 82	31	62	7	顽透辉石
S04-12-20-4m		15	48. 56	2. 49	7. 29	7. 09	0	12. 07	22. 09	0. 69	0. 07	100. 35	50	38	12	次透辉石
S04-16-1 p		16	49. 43	2. 23	10. 27	3. 42	0	18. 12	14. 58	1. 72	0	99. 77	34	59	6	顽透辉石
S04-16-2 p		17	50. 72	1. 45	9. 41	3. 49	0. 06	19. 15	14. 14	1. 3	0. 07	99. 79	32	61	6	
S04-16-4 p		18	49. 36	1. 67	11. 4	3. 72	0. 04	17. 08	14. 64	1. 37	0. 04	99. 32	35	58	7	
S08-69-2-4 m	粗面玄武岩	19	41. 54	3. 71	11. 38	9. 83	0. 11	9. 45	23. 23	0. 56	0	99. 18	53	30	17	富钙辉石
S08-73A-2-4 m		20	43. 04	4. 22	10. 52	8. 45	0. 23	9. 7	23. 32	0. 51	0	99. 99	54	31	15	

注：FeO*—全铁；p—斑晶；m—基质；测试资料据鄢全树等（2007）。

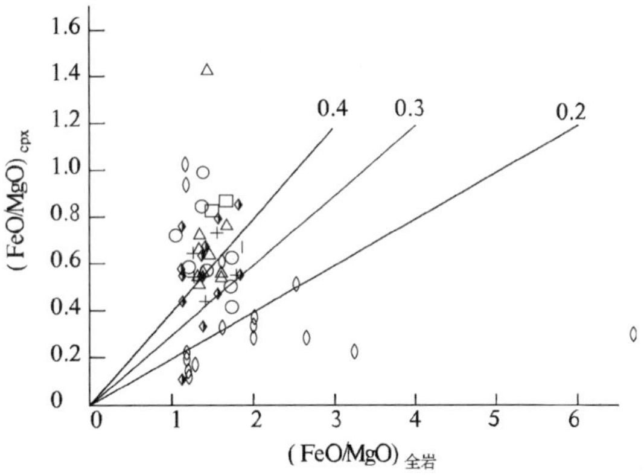

图 13-12 $(FeO/MgO)_{全岩}$-$(FeO/MgO)_{cpx}$关系图

Table 13-12 $(FeO/MgO)_{whole\ rock}$-$(FeO/MgO)_{cpx}$ correlation diagram

新生代火山熔岩斜长石电子探针分析结果见表 13-21、表 13-22。斜长石化学成分以 SiO_2、CaO、Al_2O_3、Na_2O 为主，含量分别为 52. 22%～60. 72%、4. 88%～12. 10%、24. 18%～29. 72%、4. 09%～7. 37%；次要成分为 K_2O、FeO。主要化学成分变化范围不大，其种属以中长石和拉长石为主。

（四）碱性长石

碱性长石在火山熔岩中不常见，仅在个别岩石中有少量，主要在橄榄玄武岩中见及。薄片中呈半自形—自形板状、长条状，大小 0. 03～2. 0mm，表面干净，具卡氏双晶，主要充填于斜长石微晶之中。

琼北新生代火山熔岩中碱性长石电子探针分析结果见表 13-23。碱性长石的化学成分以 SiO_2（63. 09%～67. 11%）、Al_2O_3（18. 31%～21. 84%）、Na_2O（5. 50%～6. 24%）和 K_2O（5. 57%～8. 33%）为主。经计算，将端元组分投影于长石的 Or-Ab-An 三组分类命名图中，其种属为正长石。

表 13-21　琼北新生代斜长石的化学成分一览表

Table 13-21　List of chemical composition of Cenozoic plagioclase from Qiongbei

地层单位	样品号	岩性	序号	电子探针分析结果/%										端元组分			种属
				SiO_2	TiO_2	Al_2O_3	FeO	MnO	MgO	CaO	Na_2O	K_2O	总量	An	Ab	Or	
石山组	128-2m	碱性橄榄玄武岩	1	52.44	0.15	29.72	0.62	0.02	0.09	11.98	4.09	0.34	99.45	60.4	37.5	2.1	拉长石
	23-3p		2	57.79	0.11	26.29	0.39	0.00	0.06	8.59	6.00	0.70	100.00	42.3	53.6	4.1	中长石
	23-3m		3	57.53	0.15	26.41	0.49	0.04	0.09	8.30	5.87	0.50	100.03	42.6	54.3	3.1	
	10-4m		4	62.00	0.12	23.33	0.54	0	0.09	5.02	7.13	1.72	99.94	25.0	64.6	10.4	
	QB014m		5	52.78	0.24	27.83	1.39	0.03	0.08	11.64	4.62	0.27	99.18	56.3	40.4	3.3	高拉长石
	219B4m		6	51.64	0.20	27.85	0.59	0.07	0.09	12.04	4.51	0.38	97.57	56.3	39.5	2.2	
	106B1m	石英拉斑玄武岩	7	63.47	0.20	20.15	0.40	0.00	0.00	2.33	8.00	3.53	98.20	11.1	68.9	20.0	高钠长石
	105B2m		8	57.35	0.15	25.58	0.72	0.00	0.06	8.50	6.63	0.00	99.03	41.5	58.5	0	高中长石
	114B4m		9	55.13	0.08	27.61	0.62	0.00	0.10	10.69	4.83	0.47	99.53	55.2	41.9	2.9	拉长石
	8-1m		10	54.79	0.10	28.69	0.59	0	0	11.20	4.37	0.14	99.88	58.1	40.9	1.0	
道堂组	220B3m	碱性橄榄玄武岩	11	52.84	0.13	27.90	1.05	0.00	0.11	11.49	4.68	0.35	98.60	56.4	41.6	2.0	
	150-1m	橄榄拉斑玄武岩	12	54.17	0.07	29.06	0.60		0.13	11.67	4.06	0.21	99.97	59.4	37.5	3.1	
多文组	13-1m		13	56.13	0.12	26.99	0.58	0.04	0.11	10.41	4.65	0.41	99.43	54.3	43.6	2.1	拉长石
	42-1p		14	53.54	0.13	28.27	0.57	0.00	0.09	11.84	4.36	0.31	99.09	58.6	39.4	2.0	
	42-1m		15	60.43	0.19	24.31	0.73	0.00	0.08	5.98	6.58	1.60	99.94	34.0	59.6	6.4	中长石
	60-1m		16	55.72	0.09	28.53	0.56	0.00	0.00	10.45	4.37	0.20	99.91	56.2	42.7	1.1	拉长石
	73-1p		17	54.64	0.13	28.36	0.44	0.00	0.15	11.55	4.38	0.34	100.00	58.3	39.6	2.1	
	D0997m		18	53.85	0.09	27.97	0.90	0.01	0.18	12.05	4.85	0.31	100.21	56.9	41.4	1.7	
	ZK_3-TW_2m		19	54.24	0.10	27.87	0.78	0.00	0.10	11.17	4.31	0.43	99.00	56.7	40.2	3.1	
	ZK_3-TW_3m		20	55.05	0.20	27.13	0.47	0.09	0.13	11.90	4.72	0.37	100.06	57.4	40.6	2.0	
海口组	1-5p		21	58.25	0.12	26.56	0.63	0.00	0.00	8.81	5.24	0.33	99.93	47.2	50.6	2.2	中长石
	1-5m		22	55.97	0.11	27.54	0.79	0.00	0.04	10.43	4.82	0.27	99.97	53.8	45.2	1.0	拉长石
石门沟村组	D1842-1p		23	60.72	0.00	24.18	0.03	0.00	0.00	4.88	7.34	2.48	99.63	23.1	62.9	14.0	K-更长石
	D1842-1m		24	60.19	0.21	25.42	0.60	0.00	0.10	5.69	7.37	0.82	100.40	28.5	66.7	4.8	高更长石
	208-9p		25	57.52	0.14	27.21	0.61	0.00	0.00	8.68	5.19	0.63	99.96	45.7	50.0	4.3	中长石
	208-9m		26	56.45	0.17	27.49	0.94	0.00	0.05	9.32	5.07	0.49	99.98	48.9	47.8	3.3	拉长石
	209-1p		27	54.8	0.02	28.12	0.34	0.05	0.09	10.67	4.31	0.61	99.01	55.3	40.4	4.3	
	209-1m		28	56.96	0.19	27.03	0.84	0.00	0.31	9.02	4.84	0.81	100.00	47.8	46.7	5.5	
	D1835-1p		29	52.47	0.00	29.32	0.36	0.00	0.00	11.96	4.66	0.44	99.21	57.2	40.3	2.5	高拉长石
	D1835-1m		30	56.76	0.18	26.66	0.80	0.00	0.09	8.75	6.17	0.87	100.28	41.8	53.2	5.0	高中长石
	D1836-1p1-1		31	52.24	0.06	28.93	0.48	0.00	0.00	11.71	4.98	0.27	98.67	55.7	42.8	1.5	高拉长石
	D1836-1p1-2		32	52.22	0.05	29.37	0.47	0.00	0.00	12.10	4.78	0.25	99.24	57.5	41.0	1.5	
	D1836-1p2		33	57.14	0.07	25.13	0.64	0.00	0.00	8.80	5.93	0.43	98.14	44.0	53.5	2.5	高中长石
石马村组	D1016m	石英拉斑玄武岩	34	52.63	0.11	28.39	0.56	0.03	0.14	12.32	4.91	0.19	99.30	57.5	41.4	1.1	高拉长石
	D3461m		35	52.58	0.10	29.06	0.52		0.12	12.73	4.75	0.24	100.15	58.9	39.8	1.3	
	194-2m		36	55.73	0.00	28.53	0.63	0.00	0.09	10.18	4.46	0.30	99.93	54.4	43.3	2.3	拉长石

注：p—斑晶；m—基质；1~4、10、12~17、19~22、25~28、36 据宜昌地质矿产研究所（1992）；5~9、11 据樊祺城等（2004）；23、24、29~33 据海南省地质调查院（2004）；余者为本院采集。

表 13-22　南海新生代斜长石的化学成分一览表

Table 13-22　List of chemical composition of Cenozoic plagioclase from the South China Sea

样品号	产出岩性	序号	电子探针分析结果/%										端元组分			种属
			SiO_2	TiO_2	Al_2O_3	FeO	MnO	MgO	CaO	Na_2O	K_2O	总量	An	Ab	Or	
S04-11 m	橄榄玄武岩	1	51.41	0.00	33.40	0.67	0.00		8.30	6.06	0.20	100.04	42.6	56.2	1.2	高中长石
S04-12-1 m		2	51.00	0.11	32.74	0.26	0.00		8.47	6.62	0.30	99.50	40.7	57.6	1.7	
S04-12-2 m		3	51.12	0.27	31.88	0.45	0.00		8.57	6.74	0.26	99.29	40.7	57.8	1.5	
S04-12-3 m		4	52.43	0.00	32.36	0.00	0.00		7.73	6.34	0.22	99.08	39.7	58.9	1.4	
S04-13-1 p		5	49.55	0.00	35.47	0.47	0.00		10.04	4.68	0.20	100.41	53.5	45.2	1.3	
S04-13-2 p		6	49.75	0.06	34.98	0.44	0.00		9.62	4.52	0.30	99.67	53.0	40.0	2.0	高拉长石
S04-13-3 p		7	49.08	0.00	34.84	0.42	0.04		9.57	5.42	0.15	99.52	48.9	50.2	0.9	
S04-112 m		8	51.75	0.00	32.15	0.74	0.00		8.30	6.35	0.49	99.78	40.7	56.4	2.9	
S04-14-1 p		9	50.12	0.12	34.06	0.31	0.00		9.19	5.47	0.39	99.66	47.0	50.6	2.4	高中长石
S04-14-2 p		10	50.16	0.00	34.35	0.23	0.00		9.17	5.71	0.24	99.86	46.4	52.2	1.4	
S04-14-3 m		11	50.73	0.18	33.31	0.54	0.00		8.39	6.06	0.31	99.52	42.5	55.6	1.9	
S05-34B m	石英拉斑玄武岩	12	50.47	0.38	32.32	0.77	0.00		8.09	6.99	0.28	99.30	38.4	60.0	1.6	
S08-69-1 m		13	50.68	0.14	33.10	0.78	0.20		8.66	5.48	0.19	99.23	46.1	52.7	1.2	
S08-69-2 m		14	50.44	0.32	32.93	0.56	0.00		9.15	5.86	0.46	99.72	45.1	52.2	2.7	
S08-73A m		15	50.72	0.00	33.10	0.57	0.00		8.89	5.55	0.38	99.21	45.9	51.8	2.3	

注：p—斑晶；m—基质；测试资料据鄢全树等（2007）。

表 13-23　琼北碱性长石化学成分一览表

Table 13-23　List of chemical composition of alkali feldspar from Qiongbei

样品号	产出岩石	化学成分/%										端元组分			种属
		SiO_2	TiO_2	Al_2O_3	FeO	MnO	MgO	CaO	Na_2O	K_2O	总量	An	Ab	Or	
永兴 B	橄榄玄武岩	63.09	0.12	21.84	0.44	0	0.05	2.56	6.24	5.57	99.91	12.2	55.5	32.3	正长石
D1836-1		67.11	0	18.31	0.74	0	0	0.02	5.50	8.33	100.01	0.1	50.1	49.8	

（五）霞石

霞石产于橄榄霞石岩的基质中，镜下多呈正方形或粒状颗粒。霞石的电子探针分析结果见表 13-24。矿物单晶体化学式表明，霞石以 Si 略为过剩、其他阳离子缺位为特征。成分中 CaO 含量为 0，Na/K = 2.43，应为钠质霞石。

表 13-24　霞石化学成分一览表

Table 13-24　List of chemical composition of nephelite

样品号	产出岩石	化学成分/%										
		SiO_2	Al_2O_3	TiO_2	FeO	CaO	MgO	MnO	Na_2O	K_2O	Cr_2O_5	NiO	总量
185-1	橄榄霞石岩	42.26	32.75	0.08	1.19	0.00	0.11	0.05	14.32	8.89	0.00	0.00	99.66

样品号	以氧 4 为基准的晶体化学式系数										
	Si	Al	Ti	Fe	Mg	Mn	Ca	Na	K	Cr	Ni	种属
185-1	1.04	0.95	0.002	0.02	0.004	0.001	0.00	0.68	0.28	0.00	0.00	霞石

第三节　新生代玄武质火山岩深源包体

一、深源包体的岩石学特征

深源包体主要分布于琼北文昌市蓬莱岭、庙岭、南福岭、黑山岭、北埇村（九子路岭）、绿塘园村、苦塘村、南阳竹包水库等地，寄主岩石为新近纪金牛岭期（石门沟村组）橄榄玄武岩和粗玄岩。此外，在南海宪北海山北部也含有超镁铁岩包体。包体的种类比较单一，主要以尖晶二辉橄榄岩为主（图版Ⅲ-2-G、图版Ⅲ-2-H），次为尖晶石方辉橄榄岩，少量的尖晶石单辉橄榄岩、尖晶石辉石岩、尖晶石纯橄榄岩等，大小为1~30 cm，一般多为3~5 cm。包体在寄主岩石中的含量一般小于5%，但局部富集，含量达到30%左右。北埇村、绿塘园村两地的包体含量相对丰富，个体也相对大，形态以椭圆状—次圆状为主，棱角状—次棱角状少见，也见有长条状及不规则状。包体与寄主岩石的接触界线清晰，多数无反应边，部分见熔蚀现象，并见玄武质细脉穿切其中。

尖晶二辉橄榄岩呈淡绿色—黄绿色，具有粒状结构、粒状镶嵌变晶结构、塑性变形结构、拉长碎裂结构，块状构造。由橄榄石（50%~90%）、斜方辉石（7%~30%）、单斜辉石（≤20%）和尖晶石（1%~5%）组成，各矿物在岩石中分布不均。橄榄石为淡黄色—黄绿色，镜下无色，呈半自形、等轴粒状，少数呈板状、粒状变晶，并具三连结点，粒径0.2~4mm，最大的可达10mm，正高突起，糙面明显，有的解理不完全，但裂理普遍发育，部分晶体呈波消光，多见扭折带（在一个颗粒上显示不均匀的带状消光），扭折带一般平直清晰，互相平行，滑动方向垂直（100），呈平行消光，干涉色较鲜艳，为Ⅱ级紫红色—Ⅱ级红色；斜方辉石呈灰黄色、黄绿色、墨绿色，多呈他形—半自形粒状，部分呈短柱状，粒径0.1~4.0mm，镜下呈无色或微量淡褐色，正高突起，柱状解理完全，为平行消光，干涉较低为Ⅰ级灰色—Ⅱ级橙黄色；单斜辉石为鲜绿色—翠绿色，短柱状或不规则粒状，常分布在橄榄石和斜方辉石粒间，粒径0.1~1.5mm，在镜下无色—淡绿色，正高突起，(010）解理完全，有的晶体边部见辉石次变边，平行消光，对称消光，干涉色为Ⅰ级灰色—Ⅰ级浅黄色；尖晶石为他形粒状、填隙状，粒径0.08~2.0mm，呈褐黑色—黑色，镜下呈棕褐色、茶褐色，常呈集合体状产出，均质体，全消光，见包裹橄榄石现象，常具钛铁磁铁矿“次变边”（暗化边）。

尖晶石单辉橄榄岩呈淡绿色—黄绿色，以中—中粗粒状结构为主，局部见镶嵌变晶结构、不等粒状结构，块状构造。主要由橄榄石（75%~90%）、透辉石（5%）、普通辉石（20%）和尖晶石（1%~5%）组成。橄榄石为淡黄色-黄绿色，呈等轴粒状，粒径0.5~10mm，正高突起，糙面明显，有的解理不完全，但裂理普遍发育；透辉石呈灰翠绿色，他形不规则状，镜下呈无色，粒径0.16~0.2mm，常分布在橄榄石之间。普通辉石为浅褐色，半自形柱状，边缘有部分熔融现象；尖晶石为他形粒状、填隙状，粒径0.3mm左右，呈红棕色—棕褐色，均质体，全消光，充填在橄榄石、透辉石和普通辉石之中。岩石有时见微晶辉石组成的反应边，并有玄武质微脉渗入橄榄石颗粒之间。

二、深源包体的矿物化学特征

（一）橄榄石

深源包体橄榄石电子探针分析结果见表13-25。橄榄石化学成分以SiO_2、FeO和MgO为主，含量为：SiO_2 38.98%~41.79%（平均40.50%）、MgO 44.60%~53.02%（平均49.21%）、FeO^* 7.54%~13.04%（平均9.54%）、CaO 0.01%~0.45%。$Mg^\#$值较高，其变化范围为86.04~92.29，平均90.16。化学成分变化不大，Fo值较高且成分稳定，值在86.43~92.30之间变化，种属主要属于镁橄榄石或镁质贵橄榄石。

（二）斜方辉石

深源包体斜方辉石电子探针分析结果见表13-26。化学成分以SiO_2、Al_2O_3、FeO、MgO为主，含量分别为：53.42%~56.28%（平均54.93%）、1.15%~5.75%（平均3.60%）、5.12%~7.87%

表 13-25 尖晶石二辉橄榄岩包体橄榄石的化学成分一览表

Table 13-25 List of chemical composition of olivine from spinell lherzolite enclaves

样品号	序号	电子探针分析结果(%)及特征值																端元组分		种属
		SiO_2	TiO_2	Al_2O_3	Cr_2O_3	Fe_2O_3	FeO	MnO	MgO	CaO	Na_2O	K_2O	P_2O_5	NiO	总量	$Mg^{\#}$	$Cr^{\#}$	Fo	Fa	
210-2	1	41.5	0.06	0.03	0.00	∑Fe 9.91		0.16	48.32	0.06	0	0		0.38	99.98	89.68		89.8	10.2	贵橄榄石
庙 1	2	40.93	0.06	0.25	0.10	1.31	8.55	0.00	47.70	0.45	0.40	0.08			99.83	89.73	21.16	90.9	9.1	美橄榄石
绿 1	3	38.98	0.08	0.30	0	3.34	8.24	0.12	47.65	0.32	0.26	0.11			99.40	88.31		88.3	11.7	贵橄榄石
南 1	4	41.79	0.00	0.26	0.00	0.00	10.59	0.12	44.60	0.32	0.70	0.17			98.55	88.25		88.2	11.8	
日 1	5	40.35	0.15	0.13	0.08	0.00	10.62	0.10	48.54	0.01	0.00	0.00		0.37	100.35	89.07	29.22	89.1	10.9	
HND-1	6	40.64	0.00	0.00	0.02	1.40	8.10	0.08	49.70	0.05	0.07	0.00	0.02	0.39	100.47	90.45		91.6	8.4	美橄榄石
8188	7	40.79	0.01	0.31	0.01	0.84	8.83	0.13	47.81	0.08	0.04	0.02	0.02	0.38	99.27	89.89	2.12	90.5	9.4	
吕 1	8	40.35	0.19	0.13	0.08	4.42	6.64	0.10	48.54	0.25				0.47	101.17	89.07	29.22	89.1	10.9	贵橄榄石
石 1	9	40.45	0.01	0.09			8.43	0.12	50.15	0.10	0.02	0.02			99.29	91.38		91.4	8.6	美橄榄石
石 2	10	40.65	0.02	0.05			8.58	0.10	49.92	0.15	0.02	0.00			99.51	91.21		91.1	8.9	
HF98-7	11	40.65	0.01	0.04	0.02		10.04	0.14	49.58	0.08	0.04			0.36	100.93	89.80	25.12	89.8	10.2	贵橄榄石
HF98-5	12	39.29	0.03	0.03	0.01		8.97	0.00	50.98	0.08	0.01				99.40	91.02	18.27	91.0	9.0	美橄榄石
HF98-6	13	39.27	0.03	0.00	0.05		8.21	0.12	51.47	0.08	0.01				99.25	91.79		91.8	8.2	
HF98-3	14	39.25	0.00	0.00	0.07		8.35	0.14	53.02	0.04	0.03				100.90	91.88		91.9	8.1	
HF98-12	15	40.17	0.00	0.03	0.01		9.30	0.04	50.47	0.06	0.00				100.11	90.63	18.27	90.6	9.4	
D1835-2	16	40.51	0.00	0.00	0		9.26	0.05	50.14	0	0				99.96	90.61		90.6	9.4	
D1842-2A	17	41.80	0.00	0.00	0		7.54	0.02	50.63	0	0				99.99	92.99		92.3	7.3	
D1842-2B	18	40.60	0.00	0.00	0		8.24	0.04	51.11	0.01	0				100.00	91.71		91.7	8.3	
D0881-2	19	40.67		0.03			8.99	0.15	49.17	0.05	0.01			0.57	99.62	90.70		90.70	9.30	
D0881-3	20	40.62					9.77	0.11	48.86	0.07	0.02	0.00		0.49	99.94	89.92		89.92	10.08	贵橄榄石
	21	41.29	0.03	0.01	0.07		13.04	0.25	45.08	0.08	0.04	0.00			99.89	86.04	82.24	86.43	13.57	

注:1 据宜昌地质矿产研究所(1992);2~5 据丛柏林(1987);6~7 据曾广策(1986);8 据海南地质队(1977);9~10 据石桂华(1985);11~15 据徐义刚等(2002);16~18 据海南省地质调查院(2004);19~20 据海南省地质调查院(2009);21 据梁德华等(1991);$Mg^{\#}=100*Mg/(Mg+Fe^{2+})$;$Cr^{\#}=100*Cr/(Cr+Al)$。

表 13-26　尖晶石二辉橄榄岩包体斜方辉石的化学成分一览表

Table 13-26　List of chemical composition of orthopyroxene from spinell lherzolite enclaves

样品号	序号	电子探针分析结果(%)及特征值															端元组分			$\frac{100*En}{(En+Fs)}$	种属
		SiO_2	TiO_2	Al_2O_3	Cr_2O_3	Fe_2O_3	FeO	MnO	MgO	CaO	Na_2O	K_2O	NiO	总量	$Mg^{\#}$	$Cr^{\#}$	Wo	En	Fs		
210-2	1	56. 28	0. 12	4. 64	—	—	6. 07	0. 16	32. 05	0. 56	0	0	0. 07	99. 88	90. 40		1. 2	89. 4	9. 4	90. 7	
绿 1	2	55. 10	0. 35	4. 07	0. 29	2. 64	6. 34	0. 17	29. 56	0. 70	0. 30	0. 07	0. 10	99. 69	85. 80	4. 56	1. 4	84. 6	14. 0	85. 8	
南 1	3	56. 26	0. 40	3. 69	0. 35	0. 00	7. 22	0. 17	30. 14	0. 76	0. 39	0. 08	0. 25	99. 71	88. 16	5. 98	1. 5	86. 8	11. 7	88. 1	
日 1	4	53. 52	0. 21	3. 18	0. 50	1. 39	6. 49	0. 14	29. 55	2. 03	0. 50	0. 00	0. 10	97. 61	87. 19	9. 54	4. 2	83. 6	12. 2	87. 3	古铜辉石
HND-1	5	54. 47	0. 08	4. 26	0. 45	1. 04	5. 43	0. 10	33. 41	0. 72	0. 15	0	0. 12	100. 23	90. 34	6. 62	1. 4	88. 9	9. 7	90. 2	
8189	6	54. 75	0. 11	4. 16	0. 26	1. 03	5. 24	0. 16	32. 70	0. 92	0. 16	0. 02	0. 10	99. 52	90. 43	4. 02	1. 8	88. 6	9. 7	90. 2	
HF98-7	7	53. 42	0. 12	5. 75	0. 50	0. 00	6. 88	0. 15	32. 29	1. 03	0. 21	—	0. 07	100. 45	89. 32	5. 51	2. 0	87. 3	10. 7	89. 1	
石 1	8	54. 55	0. 05	3. 38	0. 32	—	5. 40	0. 11	34. 08	0. 74	0. 11	0. 02	—	98. 76	91. 84	5. 97	1. 3	90. 6	8. 1	91. 8	
HF98-5	9	54. 74	0. 09	2. 81	0. 54	—	5. 58	0. 07	35. 16	0. 88	0. 06	—	—	99. 92	91. 83	11. 42	1. 6	90. 2	8. 2	91. 7	
HF98-6	10	54. 49	0. 02	1. 69	0. 56	—	5. 51	0. 04	35. 44	0. 83	0. 05	—	—	98. 63	91. 98	18. 19	1. 5	90. 5	8. 0	91. 9	
HF98-3	11	54. 95	0. 01	1. 15	0. 31	—	5. 70	0. 07	36. 39	0. 46	—	—	—	99. 02	91. 92	15. 31	0. 8	91. 9	8. 1	91. 8	顽火辉石
HF98-12	12	55. 32	0. 01	4. 35	0. 51	—	5. 45	0. 10	34. 18	0. 74	0. 14	—	—	100. 81	91. 79	7. 29	1. 4	90. 4	8. 2	91. 7	
D1842-2A	13	55. 55	0. 18	4. 19	0. 41	—	5. 12	0. 12	32. 11	0. 91	0. 09	0. 00	0. 18	97. 75	90. 5		1. 81	88. 72	9. 47	90. 4	
D0881-3	14	54. 83	0. 06	3. 46			5. 98	0. 12	32. 11	0. 91	0. 09	0. 00	0. 18	97. 75	90. 5		1. 81	88. 72	9. 47	90. 4	古铜辉石
	15	55. 75	0. 07	3. 19	0. 48		7. 87	0. 35	30. 70	1. 03	0. 08	0. 00		99. 52	87. 43	9. 17	2. 06	85. 09	12. 85	86. 9	

注:1 据宜昌地质矿产研究所(1992);2~5 据丛柏林(1987);6~7 据曾广策(1986);8 据石桂华(1985);9~12 据徐义刚等(2002);13 据海南省地质调查院(2004);14 据海南省地质调查院(2009);15 据梁德华等(1991);$Mg^{\#}=100*Mg/(Mg+Fe^{2+})$;$Cr^{\#}=100*Cr/(Cr+Al)$。

表13-27　尖晶石二辉橄榄岩包体单斜辉石的化学成分一览表

Table 13-27　List of chemical composition of monoclinic pyroxene from spinell lherzolite enclaves

样品号	序号	电子探针分析结果(%)及特征值															端元组分			种属
		SiO_2	TiO_2	Al_2O_3	Cr_2O_3	Fe_2O_3	FeO	MnO	MgO	CaO	Na_2O	K_2O	NiO	总量	$Mg^{\#}$	$Cr^{\#}$	Wo	En	Fs	
210-2	1	55.28	0.60	6.54	—	—	2.66	0	15.07	18.26	1.55	0	0	99.96	90.99		44.3	50.6	5.1	顽火透辉石
庙1	2	53.79	0.20	4.01	0.82	1.19	1.82	0.09	15.92	21.37	0.92	0.14	—	100.27	90.76	12.06	46.7	48.4	4.9	透辉石
绿1	3	52.40	0.21	5.18	1.12	2.05	3.25	0.08	16.79	19.07	1.00	0.14	—	100.59	85.44	12.67	42.0	49.3	8.7	
南1	4	53.56	0.30	4.82	0.98	1.27	1.42	0.04	16.30	20.05	1.30	0.14	—	100.18	91.90	12.00	43.6	50.7	5.7	
日1	5	50.77	0.54	5.32	0.83	2.10	3.10	0.11	15.44	19.89	1.00	0.00	0.15	99.10	84.65	9.47	44.0	47.5	8.5	
HND-3	6	51.70	0.25	5.62	0.89	1.44	2.10	0.05	16.13	20.00	1.35	0.00	0.02	99.53	89.43	9.60	44.3	49.8	5.9	顽火透辉石
8190	7	52.25	0.42	5.27	0.75	1.14	1.90	0.10	15.74	19.67	1.52	0.02	0.05	99.83	90.55	8.71	44.8	50.0	5.2	
石1	8	52.20	0.16	4.31	0.65	0.48	2.04	0.07	18.85	20.72	0.95	0.02	—	100	93.15	9.19	42.4	53.7	3.9	
HF98-7	9	50.68	0.59	7.36	0.92	—	3.80	0.09	15.69	18.81	2.10		0.02	100.06	88.04	7.74	43.0	50.0	7.0	
HF98-5	10	50.80	0.08	3.30	1.04	—	2.45	0.04	17.89	22.48	0.56		—	98.66	92.87	17.45	45.6	50.5	3.9	
HF98-6	11	50.99	0.09	3.44	0.82	—	2.58	—	17.84	23.07	0.51		—	99.37	92.50	13.79	46.2	49.8	4.0	
HF98-3	12	52.41	0.07	1.56	0.81	—	1.63	0.04	17.75	24.33	0.84		—	99.45	95.10	25.83	48.3	49.1	2.6	
HF98-12	13	50.01	0.26	5.11	0.84	—	2.84	0.15	15.99	21.54	1.37			99.10	90.94	9.93	46.8	48.2	5.0	
D1835-2A	14	51.75	0.39	6.30	0.63	—	2.49	—	15.33	20.89	1.86			99.64	91.65	6.29	47.3	48.3	4.4	透辉石
D1835-2B	15	51.11	0.36	6.47	0.77	—	2.47	0.04	14.88	20.98	1.79			98.87	91.48	7.39	48.1	47.4	4.5	
D1842-2B	16	51.15	0.18	4.25	0.74	—	2.44	—	17.81	22.72	0.42			99.71	92.86	10.46	46.0	50.1	3.9	
D0881-2	17	51.99	0.04	4.94			2.61	0.04	15.72	21.99	1.27	0.00	0.08	98.68	91.48		47.9	47.6	4.5	
	18	51.79	0.14	4.23	1.10		3.86	0.23	14.52	22.73	1.26	0.00		99.86	87.02	14.85	49.3	43.8	6.9	

注：1据宜昌地质矿产研究所(1992)；2～5据丛柏林(1987)；6～7据曾广策(1986)；8据石桂华(1985)；9～13据徐义刚等(2002)；14～16据海南省地质调查院(2004)；17据海南省地质调查院(2009)；18据梁德华等(1991)；$Mg^{\#}=100*Mg/(Mg+Fe^{2+})$；$Cr^{\#}=100*Cr/(Cr+Al)$。

（平均 6.02%）、29.55%~36.39%（平均 32.79%），次要有 TiO_2、Cr_2O_3、CaO、NaO 等。$Mg^{\#}$值变化范围为 85.80~92.23，平均 90.25；$Cr^{\#}$值为 4.02~18.19，平均 8.38。主要化学成分中，除 Al_2O_3 的变化略大外，其余的变化不大，端元组分的变化范围为 En=84.6~91.1、Fs=7.7~12.2、Wo=0.8~4.2，镁铁比值变化幅度小，在 85.8~91.9 之间，按照 Deer 等（1978）对斜方辉石的分类方案，应属顽火辉石和镁质的古铜辉石。

（三）单斜辉石

深源包体中单斜辉石电子探针分析结果见表 13-27。化学成分以 SiO_2、MgO、CaO、Al_2O_3、FeO 为主，含量分别为：SiO_2 50.01%~55.28%（平均 51.93%）、MgO 14.52%~18.85%（平均 16.31%）、CaO 18.26%~24.33%（平均 21.03%）、Al_2O_3 1.56%~7.36%（平均 4.89%）、FeO 1.42%~3.86%（平均 2.53%），次为 TiO_2、Cr_2O_3、NaO。$Mg^{\#}$值变化范围为 84.65~92.87，平均 90.77；$Cr^{\#}$值为 6.29~25.83，平均 11.50。主要化学成分的变化范围不大，其端元组分的变化范围分别为 Wo=42.4~48.1、En=47.4~53.7、Fs=2.6~8.7，按特吕格（1971）Ca-Mg-Fe 组合分类命名，单斜辉石主要为透辉石和顽火透辉石两类。此外，由于矿物中的 Cr_2O_3 含量相对较高，含量为 0.63%~1.12%（平均为 0.86%），可称为铬透辉石-铬顽火透辉石。但与国外相同产状的单斜辉石（Cr_2O_3 平均 0.94%、TiO_2<0.60%、Al_2O_3 4.31%~6.54%）相比，显示 Cr_2O_3、Al_2O_3 和 TiO_2 含量略偏低特点。

（四）尖晶石

尖晶石呈填隙状分布于橄榄岩包体中，尖晶石电子探针分析结果见表 13-28。化学成分以 Al_2O_3、Cr_2O_3、FeO、MgO 为主，含量分别为：$Al_2O_3$28.92%~61.79%（平均 43.82%）、Cr_2O_3 8.05%~37.85%（平均 22.83%）、FeO 9.40%~20.94%（平均 13.80%）、MgO 14.20%~21.58%（平均 18.77%）。此外含有少量 SiO_2、TiO_2、MnO、CaO 等。$Mg^{\#}$值变化范围为 56.51~79.67，平均 70.72；$Cr^{\#}$值为 8.04~45.80，平均 26.73。Cr_2O_3 和 Al_2O_3 含量变化范围较大，其种属基本属于铝尖晶石（镁

表 13-28　尖晶石二辉橄榄岩包体尖晶石的化学成分一览表

Table 13-28　List of chemical composition of spinell from spinell lherzolite enclaves

样品号	序号	电子探针分析结果（%）及特征值													种属
		SiO_2	TiO_2	Al_2O_3	Cr_2O_3	FeO	MnO	MgO	CaO	Na_2O	NiO	总量	$Mg^{\#}$	$Cr^{\#}$	
210-2	1	0.12	0.15	58.13	9.92	10.71	0.12	20.86	0.00	0.00		100.01	77.64	10.27	低铬镁尖晶石
曾 1	2	0.10	0.11	52.57	13.72	11.98	0.09	21.58	0.01	0.01		100.18	76.26	14.90	
HF98-7	3	0.10	0.25	55.10	11.62	12.71	0.14	20.35	0.41	—		100.64	74.06	12.39	
D1835-2	4	0.00	0.03	61.79	8.05	9.40	0.07	20.66	—	—		100.00	79.67	8.04	低铬镁铁尖晶石
石 1		6.92	0.57	28.92	30.41	18.20		14.20	0.28			99.50	58.18	41.36	铬尖晶石
D1842-2	5	0.00	0.006	48.06	20.37	11.45	0.03	20.02	—	—		99.99	75.71	22.14	低铬镁尖晶石
HF98-12	6	0.10	0.09	45.19	20.76	12.38	0.13	20.37	—	0.01		99.03	74.58	23.56	
HF98-3	7	0.02	—	39.70	29.59	11.91	0.18	18.71	0.02	—		1000.13	73.69	33.33	高铬镁尖晶石
HF98-6	8	0.04	0.10	33.12	33.34	14.77	0.27	18.75	—	0.10		100.49	69.36	40.31	
HF98-5	9	0.00	0.19	33.44	33.30	14.64	0.19	18.14	0.00	0.01		99.92	68.84	40.05	
PM001-9	12	0.10	0.11	30.04	37.85	16.12	0.18	16.56		0.06	0.25	101.27	64.68	45.80	
D0881-2	10	0.05	0.04	49.54	19.04	11.86	0.05	19.53		0.01	0.38	100.50	74.59	20.50	低铬镁尖晶石
D0881-3	11	0.07	0.27	39.52	27.10	16.10	0.11	17.74			0.39	101.30	66.27	31.51	
	13	0.12	0.01	38.37	24.51	20.94	0.14	15.26	0.03			99.38	56.51	30.00	

注：1 据宜昌地质矿产研究所（1992）；2 据曾广策（1986）；3、6~9 据徐义刚等（2002）；4~5 据海南省地质调查院（2004）；10~12 据海南省地质调查院（2009）；13 据梁德华等（1991）；$Mg^{\#}=100*Mg/(Mg+Fe^{2+})$；$Cr^{\#}=100*Cr/(Cr+Al)$。

铁尖晶石）。但部分样的 Cr_2O_3 的含量极高，有向铬尖晶石的过渡趋势。因此大致又可划分为 2 个亚类：第一亚类为低铬镁铁尖晶石，以贫铬、富铝为特征，其中 Cr_2O_3 含量为 8.05%～20.7%、Al_2O_3 为 45.19%～61.79%，Cr/Al 为 0.08～0.30，Cr/（Al+Cr）为 0.081～0.235；第二亚类为高铬镁铁尖晶石，以相对富铬、贫铝为特征，Cr_2O_3 含量为 29.59%～34.84%、Al_2O_3 含量为 33.12%～39.70%，Cr/Al 为 0.616～0.705，Cr/（Al+Cr）为 0.333～0.414。

G. Kurat 等（1980）认为，橄榄岩类包体中的尖晶石富铬，标志着上地幔橄榄岩部分熔融程度较高，因为部分熔融时，铝比铬更易进入熔体。刘若新等（1981）也曾提出，超镁铁岩包体中，铬尖晶石的 Cr/Al 比值，与其赋存包体岩石的基性程度可能有关，即从方辉橄榄岩→二辉橄榄岩→二辉岩，其 Cr/Al 比值逐渐下降。据此推测，研究区二类尖晶石可能来自部分熔融程度不同的上地幔橄榄岩。

三、深源包体的地球化学特征

（一）包体岩石化学特征

包体的化学成分基本在世界橄榄岩包体成分的变化范围之内（表 13-29）。15 个样品的平均值与中国东部尖晶二辉橄榄岩的平均值（张德全等，1982）接近；与 Ringwood（1975）模拟地幔岩比较，Al_2O_3、CaO、Na_2O、K_2O、TiO_2 偏低，MgO、∑FeO 偏高；与世界大陆和大洋的尖晶石二辉橄榄岩包体的平均值（Maaloee，S. and Aoki，K.，1977）比较，包体的 MgO 偏低，∑FeO 偏高，Al_2O_3、CaO、Na_2O 等易熔组分的含量介于世界大陆和大洋尖晶石二辉橄榄岩包体的平均值之间，显示深源包体的化学组成上表现出向大洋地幔过渡的性质。

包体 100 Mg/（$Mg+Fe^{2+}$）值为 89.25～91.29，平均值为 90.30；100 Mg/（Mg+∑Fe）值为 87.38～90.08，平均值为 88.67。位于 Nixon 等（1981）划分的饱满型包体的 100 Mg/（Mg+∑Fe）值（86.5～91）变化方法范围内。镁/铁比值（m/f）为 6.89～9.08，均比 6.5 大，属于镁质超镁铁岩。镁硅比值（m/Si）为 1.41～1.63，平均值为 1.54。在 AlK-CaO 相关图中（图 13-13），深源包体的投影点主要落入根据超镁铁质岩和包体估算的原始地幔范围区及其附近。

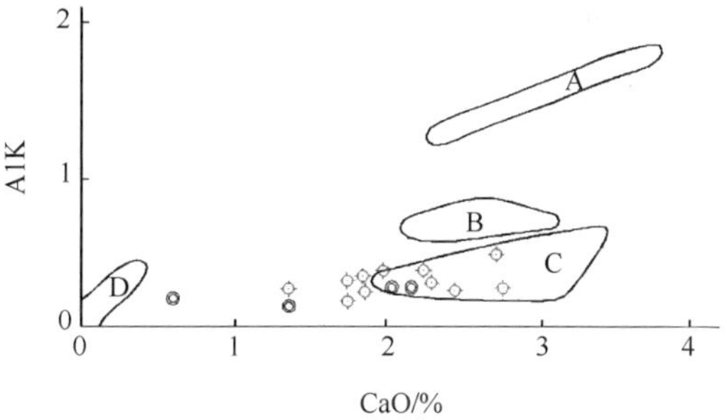

图 13-13　AlK-CaO 相关图

Fig. 13-13　AlK-CaO correlation

（据 Wyllie，1971）

A—据陨石估算的原始地幔范围；B—用 1/3 拉斑玄武岩和 2/3 阿尔卑斯型超镁铁质岩估算的原始地幔范围；C—据超镁铁质岩和包体估算的原始地幔范围；D—亏损（残留）地幔范围

（二）稀土元素地球化学特征

深源包体稀土配分模式均表现呈右倾的轻稀土元素富集型曲线（表 13-30，图 13-14）。稀土总量为 5.02×10^{-6}～20.82×10^{-6}，LREE/HREE 值为 2.44～8.51，表明轻、重稀土略有分异。δEu 值 0.57～1.06，除有一个样铕亏损外，其余 6 个样品的铕亏损或富集不明显；δCe 值为 0.41～0.89，表明铈元素具有略亏损至强亏损；$(La/Yb)_N$=1.93～12.97，$(Ce/Yb)_N$=1.29～8.06，$(La/Sm)_N$=1.57～4.32，轻稀土之间分馏作用不强；$(Gd/Yb)_N$=1.37～3.12，重稀土之间分馏作用也不明显。

（三）微量元素地球化学特征

深源包体微量元素含量大致与维氏超基性岩微量元素的平均含量可以对比（表 13-31），除 Pb、W、Li、V、Ba、Sc、U 等含量略高和 Co、Zr 等含量略低外，其余元素含量非常接近维氏值。

深源包体过渡族元素球粒陨石标准化型式大致具相似的配分模式（图 13-15）。Fe、Co、Ni 的变化范围较小，Ti、V、Mn 的变化范围中等，而 Cr、Cu、Zn 的变化范围较大。7 个样品的大离子亲石元素比值蛛网配分曲线（图 13-16）基本一致呈右倾，主要表现为强不相容元素富集，中等不相容元素弱富集，弱不相容元素亏损，低于原始地幔值。其中，Ba、U、Nb、Nd、Hf、Sm、Tb 等元素的含量相对值高，在大离子亲石元素比值蛛网图中表现为正异常；而 K、Sr、Zr、Ti、Yb 等元素的含量相

表 13-29 尖晶石二辉橄榄岩包体化掌成分一览表

Table 13-29 List of chemical composition of spinell lherzolite enclaves

样品号	序号	岩石化学成分/%																Mg#	mg#	m/f	m/si
		SiO_2	Al_2O_3	Fe_2O_3	FeO	MgO	CaO	Na_2O	K_2O	MnO	TiO_2	P_2O_5	CO_2	H_2O	Cr_2O_3	NiO	总量				
D1835-2	1	45.07	2.09	2.08	6.87	40.40	1.76	0.26	0.03	0.14	0.07	0.04	0.09	0.78			99.68	91.29	89.18	8.24	1.54
D1842-2	2	43.78	3.14	1.83	7.07	39.92	2.76	0.23	0.02	0.14	0.08	0.01	0.08	0.06			99.12	90.96	89.08	8.16	1.59
209-2	3	44.44	3.37	1.58	8.36	39.50	1.35	0.24	0.03	0.092	0.11	0.03		0.19		0.27	99.56	89.39	87.80	7.20	1.54
210-2	4	43.93	2.76	2.33	7.16	40.98	1.85	0.10	0.04	0.088	0.12	0.06		0.24			99.66	91.07	88.76	7.89	1.61
石 1	5	44.07	2.58	1.03	7.49	41.27	1.79	0.18	0.15	0.13	0.14	0.035		0.76	0.29		99.92	90.76	89.74	8.74	1.60
石 2	6	44.20	2.05		8.29	42.21	1.92	0.27	0.06	0.13	0.13	0.03			0.44		99.73	90.08	90.08	9.08	1.63
海 1	7	45.74	0.67	3.25	6.85	38.45	2.22	0.29	0.09	0.16	0.09	0.02		0.51	0.31	0.18	98.83	90.92	87.52	7.01	1.48
海 2	8	46.03	0.79	2.79	7.25	37.71	2.71	0.36	0.12	0.32	0.01				0.31	0.14	98.55	90.27	87.32	6.89	1.41
PM001	9	43.69	2.28	2.13	7.85	38.67	2.28	0.19	0.10	0.54	0.10	0.02	1.12	0.40			99.37	89.78	87.59	7.06	1.57
D0881-2	10	45.30	2.54	2.93	7.07	38.78	1.72	0.13	0.02	0.17	0.07	0.02	0.04	0.59			99.38	90.72	87.69	7.12	1.50
福基田-2	11	45.66	2.84	2.01	6.93	38.04	2.45	0.18	0.04	0.16	0.09	0.01	0.08	0.80			99.29	90.73	88.59	7.76	1.46
N1	12	44.60	2.84	1.45	7.38	40.66	1.95	0.15	0.10	0.15	0.12	0.03			0.16	0.321	100.18	90.76	89.30	8.35	1.57
N2	13	45.20	2.58	1.19	6.75	41.59	2.13	0.14	0.12	0.14	0.11	0.02			0.397	0.314	100.68	91.77	90.46	9.48	1.57
N3	14	43.25	1.58	0.43	10.54	42.25	0.62	0.14	0.07	0.20	0.13	0.00			0.00	0.00	99.21	87.73	87.33	6.89	1.68
CBB-1	15	44.36	1.54	3.75	7.30	40.17	1.40	0.13	0.02	0.16	0.01	0.00	烧失量:0.83				99.67	90.75	87.03	6.71	1.59
中国(29)	16	44.09	2.79	1.63	7.23	40.33	2.07	0.28	0.10	0.18	0.14	0.07			0.35	0.23	99.49				
大陆(301)	17	44.15	2.38	ΣFeO 8.28		42.25	2.08	0.18	0.05	0.12	0.13	0.02			0.44	0.27	99.47				
大洋(83)	18	44.4	1.87	ΣFeO 8.31		42.06	1.34	0.27	0.09	0.17	0.15	0.06			0.44	0.31	99.47				
世界橄榄岩包体成分变化范围	19	42.69～45.59	0.47～4.00	0.38～2.54	5.89～9.27	37.00～45.32	0.57～3.50	0.05～0.38	0.02～0.10	0.05～0.15	0.00～0.28	0.06～0.22									
模拟上地幔岩	20	45.16	3.54	0.46	8.04	37.47	3.08	0.57	0.13	0.14	0.71	0.06			0.43	0.20					

注:1～2据海南省地质调查院(2004);3～4据宜昌地质矿产研究所(1992);5～6据石桂华等(1985);7～8据海南地质队;9～11据海南省地质调查院(2009);12～14据梁德华等(1991);15据李兆麟等(1991);16据张德全等(1982);17～18据 Maaloee and Aoki (1977);19据久野和沧木(1970);20据 Ringwood (1975);$Mg^{\#}=100Mg/(Mg+Fe^{2+})$;$mg^{\#}=100Mg/(Mg+\Sigma Fe)$。

表 13-30　尖晶石二辉橄榄岩包稀土元素含量一览表

Table 13-30　List of REE content of spinell lherzolite enclaves

岩性		尖晶石二辉橄榄岩							地幔岩
样品号		D1835-2	D1842-2	209-2	P2-13	PM001	D0881-2	福基田-2	
序号		1	2	3	4	5	6	7	8
稀土元素含量/10^{-6}	La	1.75	1.88	5.0	4.05	1.48	0.87	0.80	0.63
	Ce	2.35	1.59	8.1	6.07	2.79	1.45	1.61	1.59
	Pr	0.36	0.43	0.83	0.9	0.36	0.20	0.24	0.251
	Nd	1.48	1.98	3.8	4.14	1.52	0.87	0.94	1.21
	Sm	0.32	0.58	0.68	0.83	0.43	0.26	0.32	0.399
	Eu	0.11	0.17	0.22	0.18	0.14	0.09	0.12	0.15
	Gd	0.34	0.64	0.70	1.12	0.44	0.28	0.37	0.533
	Tb	0.06	0.11	0.13	<0.3	0.07	0.05	0.07	0.0974
	Dy	0.33	0.68	0.57	0.59	0.42	0.32	0.43	0.661
	Ho	0.07	0.14	0.10	<0.10	0.10	0.08	0.11	0.148
	Er	0.19	0.37	0.34	0.43	0.28	0.24	0.29	0.432
	Tm	0.03	0.06	0.053	<0.1	0.04	0.04	0.05	0.0676
	Yb	0.02	0.32	0.26	0.29	0.27	0.25	0.28	0.44
	Lu	0.03	0.05	0.04	<0.1	0.05	0.04	0.05	0.066
	ΣREE	7.62	9.00	20.82	<19.20	8.44	5.02	5.67	
主要特征	L/H	5.1	2.80	8.51	5.34	4.02	2.88	2.44	
	δEu	1.01	0.85	0.91	0.57	0.98	1.01	1.06	
	δCe	0.68	0.41	0.87	0.74	0.89	0.81	0.88	
	$(La/Yb)_N$	5.90	3.96	12.97	9.42	3.70	2.35	1.93	
	$(Ce/Yb)_N$	3.04	1.29	8.06	5.41	2.67	1.50	1.49	
	$(La/Sm)_N$	3.44	2.04	4.32	3.07	2.17	2.10	1.57	
	$(Gd/Yb)_N$	1.37	1.61	2.17	3.12	1.32	0.90	1.07	

注：1~2 据海南省地质调查院（2004）；3 据宜昌地质矿产研究所（1992）；4 据石桂华等（1985）；5~7 据本院实测；8 据 Jagoutz et al.（1979）。

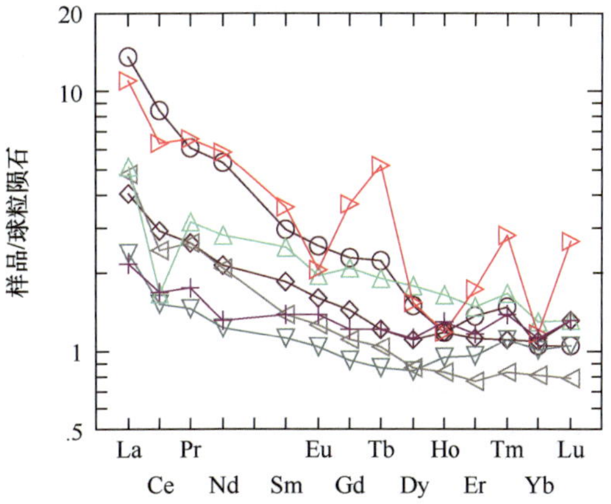

图 13-14　尖晶石二辉橄榄岩稀土元素球粒陨石标准化图

Fig. 13-14　Chondrite-normalized REE patterns of spinell lherzolite

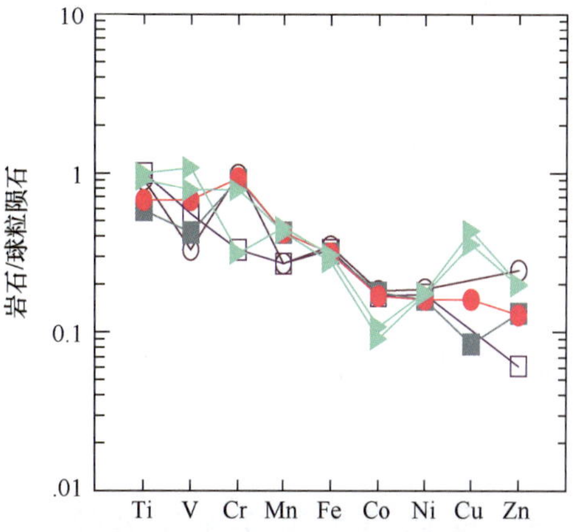

图 13-15　尖晶石二辉橄榄岩过渡元素球粒陨石标准化图

Fig. 13-15　Chondrite-normalized patterns of transition metal elements from spinell lherzolite

表 13-31 尖晶石二辉橄榄岩包体微量元素含量一览表

Table 13-31 List of trace element contents of spinell lherzolite enclaves

岩性		尖晶石二辉橄榄岩							维氏超基性岩微量元素的平均含量
样品号		D1835-2	D1842-2	209-2	210-2	PM001	D0881-2	福基田-2	
序号		1	2	3	4	5	6	7	
微量元素含量/10^{-6}	Au	1.1	1.4			1.4	1.0	1.1	5
	Ag	0.028	0.017			0.02	0.02	0.03	0.05
	Cu	11.7	22.5			35.4	13.1	18.4	20
	Pb	<1.0	<1.0			1.30	0.80	1.18	0.1
	Zn	60	59	112	28	57.6	46.4	44.9	30
	Sb	0.09	0.09			0.061	0.035	0.032	0.1
	W	109.4	0.38			0.22	0.27	0.22	0.1
	Sn	<1.0	0.4			1.1	1.0	1.2	0.5
	Bi	<0.1	0.04			<0.03	<0.03	<0.03	0.001
	Mo	2.9	<0.2			0.095	0.081	0.13	0.2
	Be	0.7	0.3			0.023	0.042	0.006	0.2
	Li	2.8	2.7			4.22	1.62	4.95	0.5
	Ni	1943.0	1956	2260	2097	1535	1549	1531	2000
	Co	98.2	92.0	99	93	115	112	106	200
	V	39.4	64	31	52	48.8	49.3	58.5	40
	Nb	6.7	2.3	0.72	15.4	1.63	1.32	1.45	1.0
	Ta	<0.5	<0.5	0.16		0.099	0.084	0.15	0.018
	Rb	<3.0	<3.0	11	2	4.50	2.80	4.10	2
	Sr	17	10	26	12	23.8	17.0	16.9	10
	Ba	36	60	48	26	93.5	63.2	63.9	1
	Zr	20	9	8.6	16	19.4	12.3	16.1	30
	Hf	0.5	<0.5	0.57		0.5	0.4	0.4	0.1
	As	1.2	<0.2			<1.0	<1.0	0.67	0.5
	Ga	1.3	1.5			2.04	1.99	2.0	1.5
	Sc			12.5	12				5
	Cr	3107	3183	3355	1130				2000
	U	1.4	0.5	0.5	0.4	0.34	0.023	0.038	0.003
	Th	<1.0	<1.0	1.3	0.5	0.31	0.41	0.26	0.005
	Y	1.95	3.79	4.3		2.46	2.18	2.43	

注：Au 的单位为 10^{-9}。1~2 据海南省地质调查院（2004）；3~4 据宜昌地质矿产研究所（1992）；5~7 据本院实测。

对值较低，在大离子亲石元素比值蛛网图中表现为负异常。

四、包体成因探讨

（一）包体形成温度和压力的估算

幔源包体的温度估计对于研究上地幔的物理状态、流变学特征以及上地幔物质结构乃至深部作用具有重要的意义。尽管目前关于幔源包体的温压计有数十种，但对同一个包体样品，采用不同的温压计可以产生 50℃~200℃和 0.2~1.0 Gpa 的误差，不过，对于尖晶石相包体的地质压力计至今没有理想

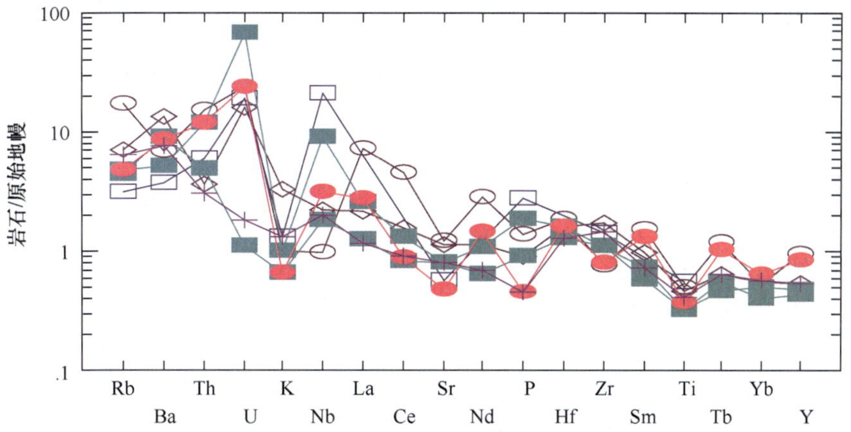

图 13-16 尖晶石二辉橄榄岩微量元素比值蛛网图

Fig. 13-16 Spider diagram of trace element ratios of spinell lherzolite

的方法。本次采用 Mercier（1976）的单斜辉石温度计和压力计估算了包体的平衡结晶温度和压力。对于温度，又采用了 Wood-Banno（1976）二辉石温度计、Mysen（1976）单斜辉石温度计和徐义刚（1993）Ca-单斜辉石温度计加以对比（表 13-32），采用各种方法估算出的温度差值不大，均在热力学计算的允许误差范围内。曾广策（1984）、从柏林（1987）和宜昌地质矿产研究所（1991）对本区尖晶石二辉橄榄岩包体进行过 $P-t$ 条件的估算，其估算值基本一致。结果显示尖晶石二辉橄榄岩包体的温度范围为 954℃~1177℃、压力范围在 1.55~2.50GPa，根据压力所估算的深度为 50~80 km。

表 13-32 尖晶石二辉橄榄岩包体的 $P-t$ 估算一览表

Table 13-32 List of estimated $P-t$ of spinell lherzolite enclaves

序号	方法 / 样品号	Wood 和 Banno（1973） t/℃	Wells（1977） t/℃	Banno（1974） t/℃	Mercier（1980） t/℃	Mercier（1980） P/GPa	徐义刚（1993） t/℃
1	210-2	1026	—	—	990	1.7	1140.32
2	210-2	—	—	—	1113	2.5	—
3	庙 1	—	—	—	964	1.55	954.45
4	绿 1	1177	1119	1139	1094	2.24	1122.71
5	南 1	1108	1042	1059	1055	2.14	1024.64
6	日 1	1121	1050	1062	1063	1.87	1031.22
7	石 1	—	—	—	—	—	1012.50
8	HND-1	—	—	—	—	—	999.05
9	HND-2	1125.7	—	—	1036.1	2.01	—
10	HND-3	—	—	—	1065	2.03	—
11	8189	1127.9	—	—	1064	2.27	998.49
12	8196	—	—	—	1065	2.11	
13	HF98-7	—	—	—	—	—	1009.34
14	HF98-5	—	—	—	—	—	1017.01
15	HF98-6	—	—	—	—	—	748.50
16	D1842-2B	—	—	—	—	—	858.45

注：1、2 引自宜昌地质矿产研究所（1992）；3~7 引自从柏林（1987）；8~12 引自曾广策（1984）；13~15 引自徐义刚（2002）；16 引自海南省地质调查院（2004）。

尖晶石二辉橄榄岩由单斜辉石估算的温度和压力在 Ringwood（1975）的地幔岩矿物相图中呈负相关关系，所构成的地温曲线与大洋地温曲线吻合(图 13-17)。除个别样品外，投影点多数集中分布在尖晶石二辉橄榄岩与石榴子石二辉橄榄岩分界线附近的石榴子石二辉橄榄岩区。这种异常情况的出现，可能是因为尖晶石二辉橄榄岩的稳定区向高压区方向扩展所致。对于这个问题，自然界复杂系统的研究已经得到了证实，但对其解释尚不一致。一些学者（Ringwood，1975）认为，尖晶石橄榄岩区向高压方向扩展是由于尖晶石中 Cr_2O_3 含量增高的结果；而邓晋福等（1980）提出，不仅与尖晶石中 Cr_2O_3 含量有关，而且还与尖晶石橄榄岩中 SiO_2 活度降低有关；还有一些学者认为，全岩中 Cr_2O_3/Al_2O_3 比值增大，将使尖晶石相转变为石榴子石相所要求的压力显著增大（MacGregor，1976）。本区尖晶石二辉橄榄岩中 SiO_2 含量比模拟上地幔岩偏低，尖晶石中 Cr_2O_3 含量高达 8.05%~33.34%，Cr_2O_3/Al_2O_3 =0.13~1.01，高于模拟上地幔的 Cr_2O_3/Al_2O_3 比值（0.12），因此，有利于尖晶石相向高压方向扩展。根据上述，推测本区橄榄岩包体来自 50~80 km 深的尖晶石橄榄岩层。

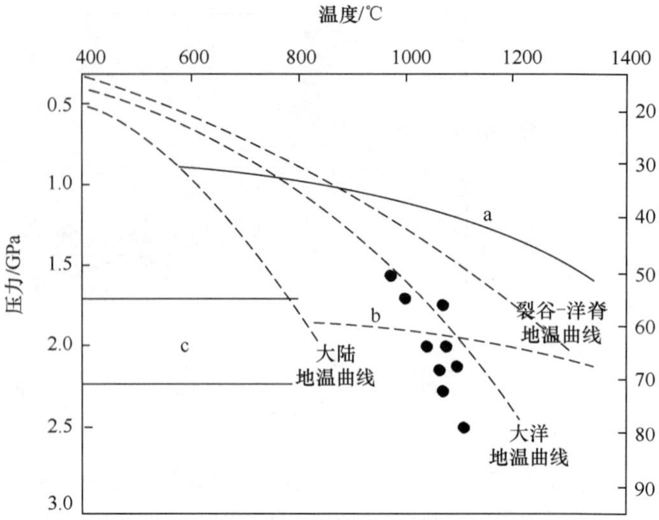

图 13-17　尖晶石二辉橄榄岩包体温度和压力曲线图

Fig. 13-17　Temperature and pressure curve of spinell lherzolite enclaves

a—尖晶石二辉辉石岩与石榴子石二辉辉石岩之间的相转变界线；

b—尖晶石二辉橄榄岩与石榴子石橄榄岩之间的相转变界线；

c—尖晶石相橄榄岩与石榴子石相橄榄岩过渡带

（二）尖晶石二辉橄榄岩包体的成因探讨

尖晶石二辉橄榄岩包体的岩石、矿物、微量元素、同位素及平衡温度和压力估算等研究结果表明，其属于上地幔岩石经低度部分熔融后的残余。依据是：尖晶石二辉橄榄岩包体的矿物组合比较稳定，主要由 Ol-Opx-Cpx-Sp 四矿物组成；岩石化学成分与中国东部及世界其他地区上地幔包体成分基本一致，矿物化学成分变化范围较窄，橄榄石的 Fo 值为 88.2~92.3，略高于原始地幔岩中橄榄石的 Fo 值（86~90）；斜方辉石的 En 值为 83.6~91.1；单斜辉石属透辉石-镁透辉石。南海及琼北尖晶石二辉橄榄岩包体的 $^{87}Sr/^{86}Sr$ 值位于全球平均 Sr 同位素组成 $\varepsilon_{Sr}=0$ 的左边，即 $\varepsilon_{Sr}<0$、$^{87}Sr/^{86}Sr<0.7047$（图 13-18），Nd 同位素比值强烈亏损（比 MORB 源区还亏损）。在 CaO-AlK 图上，尖晶石二辉橄榄岩包体、玄武岩和模拟上地幔岩三者位于一条直线上，玄武岩与包体成分呈现互补关系，说明两者有成因上的联系（曾广策，1986）。

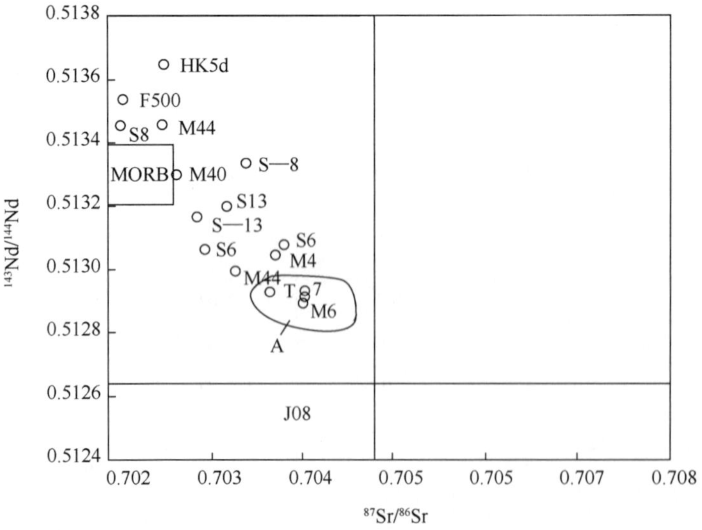

图 13-18　尖晶石二辉橄榄岩 Nd-Sr 同位素图

Fig. 13-18　Nd-Sr isotope diagram of spinell lherzolite

（据赵海玲等，略修改）

A—南海及琼北玄武岩 Nd-Sr 同位素范围

尖晶石二辉橄榄岩包体的结构类型表明，本区地幔岩曾发生过部分熔融作用、高温塑性流动变形作用和固态重结晶作用。此外，对本区包体平衡温度和压力估算表明，包体形成于50~80 km的深度内，大大超过本区莫霍面的深度（31~33 km），显示包体来自上地幔。据二辉橄榄岩熔融实验资料（久城育夫等，1973），在温度为1100℃、压力为1.0~2.0 GPa和H_2O=0.5%时，形成碧玄岩浆，其残留成分为二辉橄榄岩，表明二辉橄榄岩包体是由地幔物质部分熔融后的残留物，这一点，也可由区内包体主要赋存于碱性玄武岩中得到证明。二辉橄榄岩包体的过渡金属元素和稀土元素丰度与球粒陨石值相近，这与估算的包体熔融程度较低的结果一致，稀土元素具明显的LREE富集分离配分型式，而且包体和寄主玄武岩之间的Nd-Sr同位素不平衡，可能与地幔“交代作用”有关。

第四节　新生代火山岩形成的构造环境

南海及海南岛在大地构造上主要位于欧亚板块、印度-澳大利亚板块和太平洋板块的交接部位，因而受不同构造域发展演化的影响，白垩纪—古近纪全球板块运动加速，导致板块边界及板缘地带的活动。中国东南沿海地区在中生代—古近纪受太平洋板块俯冲作用，处于活动的亚洲边缘，新近纪—第四纪俯冲带东移形成了亚洲东部边缘的裂谷盆地。太平洋板块在43 Ma前向北北西方向运动，43 Ma以后，伴随太平洋板块的转向，开创了亚洲东部一个盆地扩张的新纪元，西太平洋边缘盆地相继出现。从始新世中期开始，印度板块和澳大利亚板块一起向北运动，澳大利亚北边的海洋岩石圈板块在印度尼西亚南面的海沟插入，正是在这种弧后俯冲作用，南海板块向南运移，造成南海北部大陆架的引张，南海的大规模扩展，使华南不活动的大陆边缘活化，古近纪的断裂进一步活动，因此有强烈的火山活动。

王五-文教断裂带将海南岛分成海南中部隆起区和琼北新生代断陷盆地两个不同的构造单元。断陷盆地内的断裂主要有北东东向、北西向、北东向和南北向，以北东东向和北西向断裂为主，它们不仅控制断陷盆地的展布方向和盆内次一级隆起和坳陷的分布，同时还控制着新生代基性岩浆的活动。根据区域重力资料，琼北断陷盆地和康拉面较浅，深度为16~17 km，地壳较薄，莫霍面深度为31~33 km；海南中部隆起区，地壳明显增厚，莫霍面深度为34~35 km，从南往北地壳减薄，莫霍面逐渐上隆，反映海南岛上地幔与地壳之间是一个起伏的界面。海南中部隆起属幔坳区，琼北断陷盆地属幔隆区，琼北上地幔的隆起导致岩石圈的拉张，造成区域上地壳张裂下陷，形成大陆裂谷，成为控制琼北新生代火山岩活动的特定构造环境。

南海和琼北新生代火山岩的岩石组合类似，岩石学和岩石地球化学特征相似，显示南海和琼北火山岩的源区特征和岩浆形成环境具有相似性，下面从岩石的主要元素、微量元素及稀土元素等方面对新生代火山岩形成构造环境作详细分析。

一、主要元素

琼北新近纪基性火山岩的平均化学成分分别为SiO_2 50.75%、Al_2O_3 13.61%、Fe_2O_3 3.42%、FeO 7.62%、MgO 7.83%、CaO 8.0%、Na_2O 3.0%、K_2O 1.11%、TiO_2 2.05%、P_2O_5 0.49%；琼北第四纪基性火山岩的平均化学成分分别为SiO_2 50.75%、Al_2O_3 14.06%、Fe_2O_3 2.72%、FeO 8.24%、MgO 7.13%、CaO 8.33%、Na_2O 3.05%、K_2O 1.19%、TiO_2 2.05%、P_2O_5 0.4%；南海基性火山岩的平均化学成分分别为SiO_2 46.90%、Al_2O_3 15.77%、Fe_2O_3 4.78%、FeO 6.33%、MgO 5.87%、CaO9.39%、Na_2O 3.38%、K_2O 1.79%、TiO_2 2.59%、P_2O_5 0.77%。南海及琼北新生代基性火山岩的平均化学成分含量相差不大，总体与世界大陆裂谷型玄武岩的平均值相当，同中国东部裂谷型玄武岩的平均化学成分更加接近，但显示了SiO_2偏高，碱和P_2O_3略低的特点。而同大洋拉斑玄武岩K_2O（0.24%）、P_2O_5（0.21%）相比，南海及琼北的玄武岩K_2O和P_2O_5含量明显偏高，显示出大陆玄武岩化学特征。在TiO_2-P_2O_5-K_2O三角图解中（图13-19），岩石的投影点主要落入大陆拉斑玄武岩区，仅有少数落入大洋拉斑玄武岩区，但比较靠近大陆拉斑玄武岩与大洋拉斑玄武岩的分界线。在里特曼-戈蒂里图解上（图13-20）岩石的投影点主要落入A区（板块内部稳定构造区），有少数分布在B区和C区，但都比

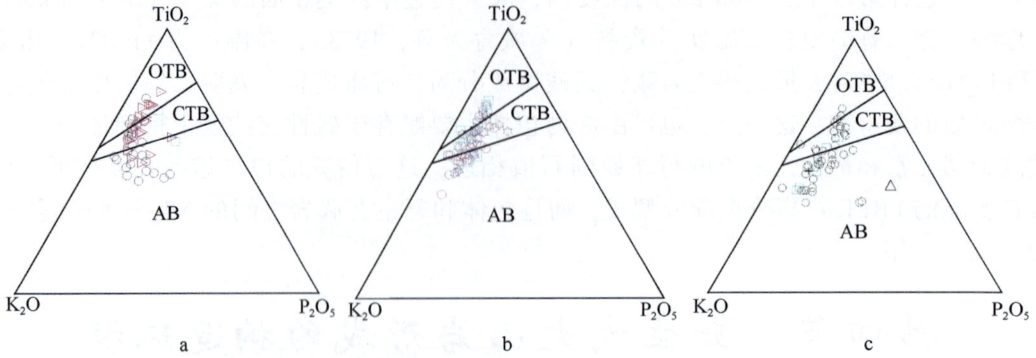

图 13-19 新生代火山岩 TiO_2-K_2O-P_2O_5 图解

Fig. 13-19 TiO_2-K_2O-P_2O_5 plot of Cenozoic volcanic rocks

OTB—大洋拉斑玄武岩；CTB—大陆拉斑玄武岩；AB—碱性玄武岩

a—新近纪；b—第四纪；c—南海

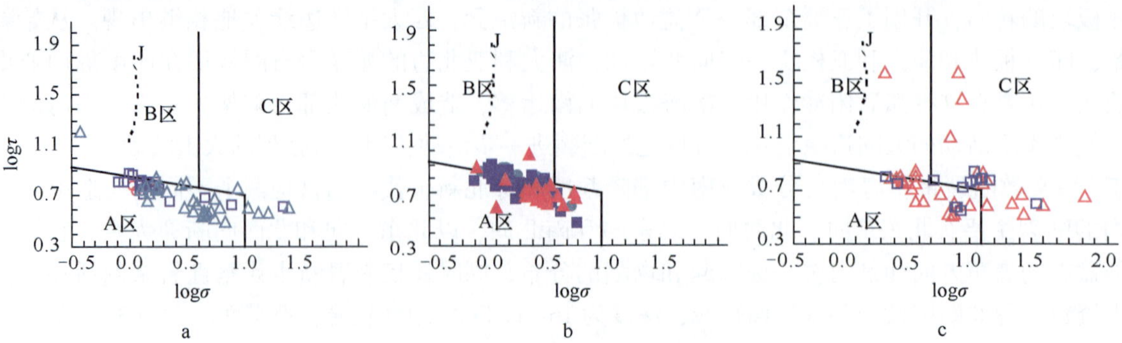

图 13-20 新生代火山岩里特曼-戈蒂里图解

Fig. 13-20 Rittmann-Gottini diagram of Cenozoic volcanic rocks

A 区—非构造带（板块内部稳定构造区）火山岩；B 区—造山带（岛弧及活动大陆边缘区）火山岩；

C 区—A、B 区火山岩派生的碱性岩

a—新近纪；b—第四纪；c—南海

较靠近分界线。此外，利用非碱性玄武岩在 FeO^*-MgO-Al_2O_3 图解中（图 13-21）的投影点主要落入大洋与大陆块内部，这说明新生代火山岩形成于较稳定的大陆构造环境。

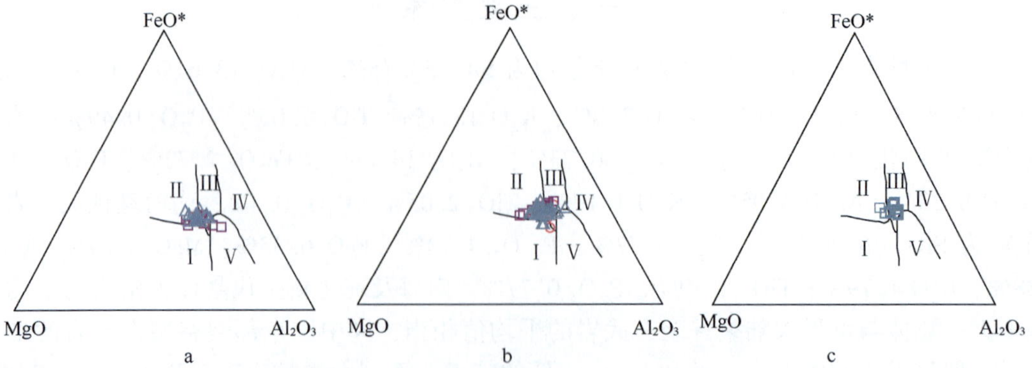

图 13-21 新生代火山岩里特曼-戈蒂里图解

Fig. 13-21 Rittmann-Gottini diagram of Cenozoic volcanic rocks

Ⅰ—洋中脊或洋底；Ⅱ—大洋岛屿；Ⅲ—大陆块内部；Ⅳ—扩张性中央岛屿；Ⅴ—造山带

a—新近纪；b—第四纪；c—南海

二、微量元素及稀土元素

本区玄武岩强烈富集 Ba、Sr、U、Th、Ta、Nb 和 K 等不相容大离子亲石元素，与产于不同构造背景中的玄武岩对比（表 13-33），其丰度和比值与大陆裂谷环境的玄武岩十分相近，而与其他构造背景产出的玄武岩略有差别。此外，利用某些微量元素所构成的构造环境判别图（图 13-22，图 13-23），岩石的投影点主要落入板内环境。

表 13-33　海南岛新生代玄武岩及世界各种构造环境火山岩微量元素平均值一览表

Table 13-33　List of average content of trace elements of Cenozoic basalts from the Hainan Island and volcanic rocks from varioustectonic settings across the world　（单位：10^{-6}）

组分	洋脊	岛弧	大陆裂谷		洋岛拉斑玄武岩	大洋碱性玄武岩	高铝拉斑玄武岩	海南岛		南海
	低钾拉斑玄武岩		拉斑玄武岩	碱性玄武岩				新近纪	第四纪	
Rb	1	5	31	200		33		27.44	26.8	47.5
Sr	135	225	350	1500	350	800	330	434	459	266
Ba	11	50	170	700		500		395	390	621
U	0.1	0.15	0.4	0.5		0.75		1.19	1.41	10.18
Th	0.18	0.5	1.5	4.0		4.5		4.19	3.83	8.53
Zr	85	60	200	800	125	300	100	169	178	223
K	1200	3300	5500	10900		13700		9600	10100	12900
Hf		1.14	124			2.93		4.30	5.05	
Ni	100	25	85	100	150	50	25	183	144	
Co	32	20	30	60	30	25	50	52	45.8	63.13
Cr	300	50	160	400	250	67	40	179	200.7	
Yb	4.0	1.9	4.4	1.7	1.7	1.5	2.7	1.74	1.54	2.82
La	3.9	3.3	33	54	7.2	17	10	34.3	24.8	25.41
Rb/Sr	0.007	0.022	0.089	0.13		0.045		0.06	0.06	0.2
K/Ba	109	66	32	16		28		24.3	25.9	20.8
K/Rb	1200	660	176	55		420		350	376	272
Th/U	1.8	3.3	3.8	8.0		6.0		3.5	2.7	0.8
$(La/Yb)_N$	1.0	1.7	7.5	32	4.2	11	3.7	12.5	11.6	6.17

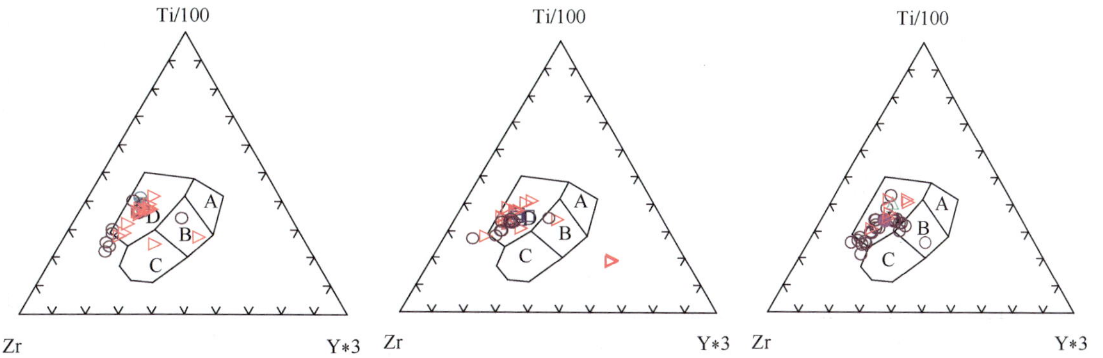

a　b　c

图 13-22　Ti/100-Zr-Y * 3 相关图

Fig. 13-22　Ti /100-Zr-Y * 3 correlation

（据 J. R. Cann，1973）

A—岛弧低钾拉斑玄武岩；B—洋底（洋中脊）玄武岩；C—岛弧钙碱性玄武岩；D—板内玄武岩

a—新近纪；b—第四纪；c—南海

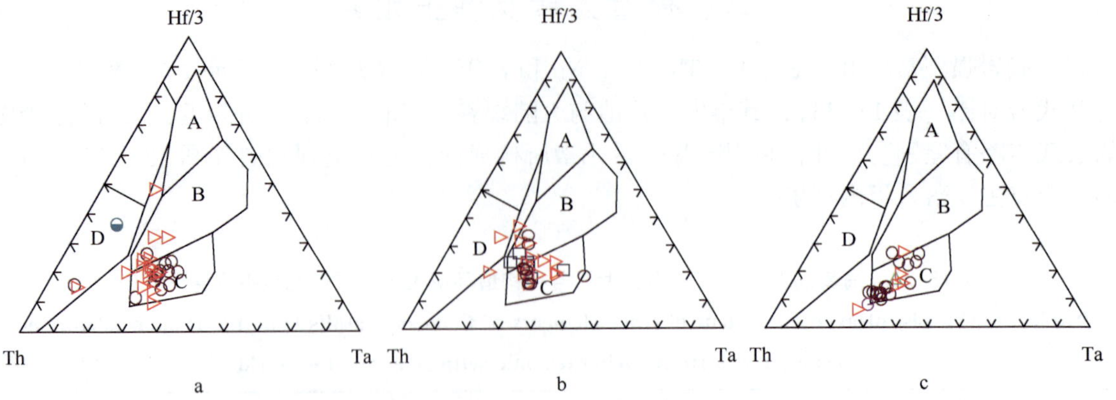

图 13-23　Th-Hf-Ta 相关图

Fig. 13-23　Th-Hf-Ta correlation

（据 D. A. Wood，1980）

A—M-MORB；B—P-MORB；C—板块内部玄武岩；D—岛弧玄武岩

a—新近纪；b—第四纪；c—南海

三、同位素

琼北新生代火山岩的^{87}Sr/^{86}Sr 的比值为 0.703178～0.704523，平均值为 0.703932，比较接近大洋岛屿的火山岩的^{87}Sr/^{86}Sr 比值；南海新生代火山岩的^{87}Sr/^{86}Sr 的比值为 0.703176～0.7046189，平均值为 0.704119，也比较接近大洋岛屿的火山岩的^{87}Sr/^{86}Sr 比值。琼北新生代火山岩的^{143}Nd/^{144}Nd 的比值变化范围较小，为 0.512787～0.513039，其 ε_{Nd} 值均大于 0，岩石源于地幔，南海新生代火山岩的^{143}Nd/^{144}Nd 的比值为 0.512663～0.513035，其 ε_{Nd} 值均大于 0，岩石也源于地幔。火山岩的 Pb 同位素比值与所称 C 组分的变化范围（Haman 和 Graham，1996）相差不大。在^{143}Nd/^{144}Nd-^{87}Sr/^{86}Sr 相关图中（图 13-24）主要位于夏威夷玄武岩区。在^{143}Nd/^{144}Nd-^{206}Pb/^{204}Pb 相关图（图 13-25）、^{208}Pb/^{204}Pb-^{206}Pb/^{204}Pb 相关图（图 13-26）中，主要位于 EM Ⅰ 型地幔。^{87}Sr/^{86}Sr-^{206}Pb/^{204}Pb 相关图中（图 13-27）主要落入洋岛玄武岩。

综上所述，新生代火山岩来源于富集Ⅰ型地幔，且没有明显受到地壳物质的混染，基本保持了原生岩浆特征，其形成构造环境属大陆板内裂谷环境。

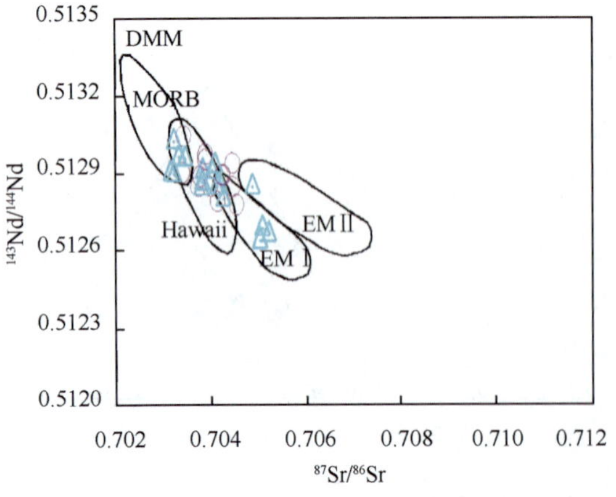

图 13-24　火山岩的 Nd-Sr 同位素相关图

Fig. 13-24　Nd-Sr isotope correlation diagram of volcanic rocks

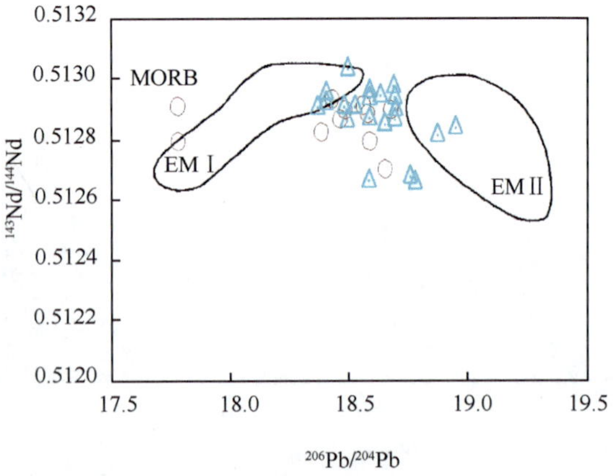

图 13-25　火山岩的 Nd-Pb 同位素相关图

Fig. 13-25　Nd-Pb istope correlation diagram of volcanic rocks

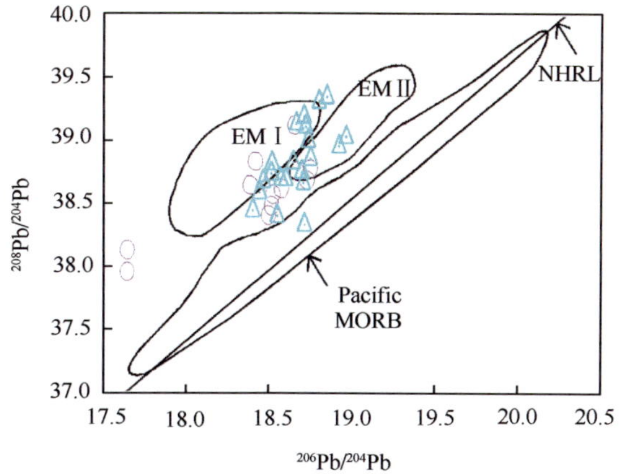

图 13-26　火山岩的 Pb 同位素相关图

Fig. 13-26　Pb isotope correlation diagram of volcanic rocks

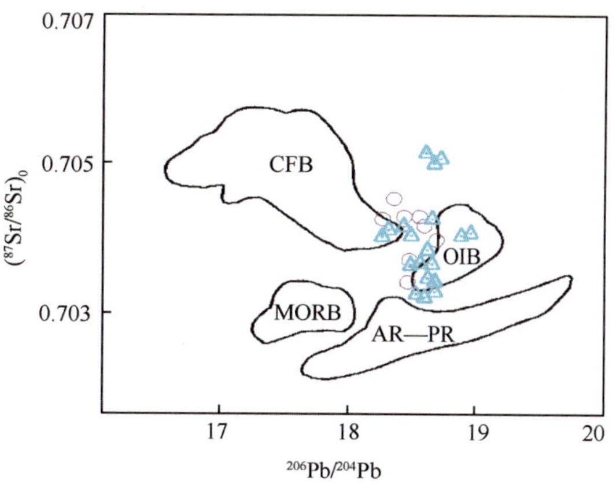

图 13-27　火山岩的 Sr-Pb 同位素相关图

Fig. 13-27　Sr-Pb isotope correlation diagram of volcanicrocks

第四篇
变质岩及变质作用

概　述

海南省变质岩研究历史较长，最早多是伴随大比例尺矿区工作而进行，但中小比例尺的调查不系统。1929 年李承三先生等三人在海南岛北部进行地质调查，对分布于琼北的沉积变质岩进行粗略划分；1952 年后，不同的地质单位和个人对出露于昌江县石碌铁矿的变质岩进行变质和变质作用初步的探讨；1964，广东省地质局区域地质测量大队在进行 1：20 万海南岛区域地质调查过程中，对分布于海南岛抱板、大蟹岭、冲卒岭和黄竹岭地区的混合岩进行了详细的岩性描述，开展了地球化学研究，并对混合岩的成因进行了讨论；1977~1979 年南京大学王赐银等对出露于海南岛东方市（县）抱板地区的变质岩及混合岩进行了岩石学特征、成因、时代的研究，划分了变质岩的相带和混合岩的混合程度。20 世纪 80 年代中期以来，海南岛变质岩的研究进入了一个新的阶段，随着 1：5 万和 1：25 万区调工作的相继开展，系统收集变质岩的岩石学、岩石化学等方面的资料，对区域变质矿物相带进行较详细划分，并对变质岩的原岩恢复，同时结合构造演化史探讨变形变质作用发展史。此外，为深化海南岛基底地层的研究，马大铨等人 1991~1993 年对分布于抱板、大蟹岭、冲卒岭、上安和黄竹岭的抱板群进行了详细调查研究工作，对抱板群的地质演化进行深入的讨论，该研究成果提高了海南省的变质岩研究程度，提供了一系列有价值的实际资料和数据。但至今海南岛变质岩的研究程度总体不高，缺乏变质作用的 $P-T-t$ 轨迹的研究，也从未开展过变质单元划分。

海南省变质岩主要出露于海南岛，西沙群岛、南沙群岛、莺歌海海盆中也有分布。由于西沙群岛、南沙群岛、莺歌海海盆中的变质岩呈隐伏状产出，仅在少量钻孔中见及，资料少，因此，本次工作的重点是海南岛变质岩。以海南岛区域地质调查资料为基础，广泛收集有关单位和个人的研究成果及大比例尺矿区资料，系统总结海南省不同地质时期变质作用类型及其特征，研究海南省不同变质区变质相带的分布规律，探讨变质作用与大地构造环境及地壳演化的关系。

海南岛变质岩相当发育，出露范围广，变质岩类型有区域变质岩（包括混合岩）、接触变质岩、气液变质岩和动力变质岩等，以区域变质岩分布最为广泛。海南岛自中元古代以来，地壳经历了漫长的演变过程，发生了多期次的构造活动，伴随着多期沉积建造及岩浆活动，造就区内多期多样的变形变质作用演化历史。总体上呈现高热流→低热流、深层次→低层次、韧性→脆性的构造变形格局，变质作用程度由强→弱。根据区域构造环境、区域构造变形、岩浆活动及变质矿物组合演变特征，将区内变质作用演化划分为早晋宁期、晚晋宁期、加里东期、华力西期、印支期及燕山期（表Ⅳ-1）。

抱板期变质作用　戈枕村岩组为海南省已知出露最古老的地层，其原岩主要为中酸性、基性火山岩，少量杂砂岩，构成火山-沉积盆地，形成于岛弧或大陆边缘弧环境。在强烈俯冲作用下，地壳明显增厚，一度处于 20~24km 深度（马大铨等，1998）。由于地幔热流和生热影响，地壳温度升高，使上述岩石发生角闪岩相-麻粒岩相变质作用。随着温度的上升，（深部）早期形成的角闪岩相岩石发生部分熔融，形成混合岩，当温度上升到麻粒岩相条件时，大比例熔融形成原地型混合花岗岩。其主导变形机制表现为塑性流变，以深熔作用和流变褶皱为特征，构成下构造层下亚层，构造型式表现为穹隆或短轴背斜，反映了区域中高温变质作用的特点。马大铨等（1998）在混合岩中获得脉体 Rb-Sr 等时线年龄为（1637±31）Ma，本次在混合岩化片麻岩中获得（HN0613）岩浆锆石结晶年龄为（1437±7）Ma，时代为长城纪。

峨文岭岩组的原岩主要为一套陆缘碎屑岩夹火山碎屑岩系，具类复理式的沉积韵律，其形成环境已由陆缘活动带转变为相对较稳定的陆内或陆缘构造盆地。早晋宁末期区内发生大规模造山作用，使已处于较深地壳（15~20km）的峨文岭组发生变质。由于热流上升及热流空间分布上的不均匀性，发生绿片岩相→低角闪岩相变质，最高达矽线石-白云母级。主导变形机制为塑性压扁和塑性变形，以

表Ⅳ-1　海南岛变质变形事件表

TableⅣ-1　Metamorphism and deformation events of the Hainan Island

变形阶段	构造相	主要构造形迹	变质类型	变质相	主要矿物共生组合
喜马拉雅期	表构造相	脆性变形	动力变质		
燕山期	浅构造相	脆-韧性变形 褶皱构造	动力变质 接触变质 气液变质	钠长绿帘角岩相 普通角闪石 角岩相	4. 石英+绿泥石+白云母 3. 石英+红柱石+白云母 2. 石英+黑云母+绢云母+绿帘石±绿泥石 1. 白云母+黑云母+石英+绿泥石
印支期	浅构造相	线理 褶皱构造	动力变质 接触变质 气液变质	钠长绿帘角岩相 普通角闪石角岩 相辉石角岩相	4. 石英+绿帘石±阳起石 3. 石英+黑云母+绢云母+绿帘石±绿泥石 2. 斜长石+黑云母+石英+堇青石 1. 石英+斜长石+紫苏辉石+透辉石+黑云母
华力西期	中浅构造相	脆性变形 褶皱构造	动力变质 接触变质 区域变质	钠长绿帘角岩相 普通角闪石角岩 相辉石角岩相	2. 绢云母-白云母+绿泥石+石英 1. 石英+斜长石+正长石+紫苏辉石+透辉石+黑云母
				低绿片岩相	3. 绢云母+钠长石+黝帘石+石英 2. 绿泥石+绢云母+石英±硬绿泥石 1. 绢云母-白云母+石英
加里东期	中深构造相	固态流变 褶皱构造	区域变质 退化变质	低绿片岩相	3. 绢云母+钠长石+黝帘石+石英 2. 硬绿泥石+绿泥石+绢云母+石英 1. 绢云母-白云母+黑云母+石英
晋宁期-扬子期	深构造相	固态流变 褶皱构造	区域变质	低绿片岩相	3. 透辉石+透闪石+石英 2. 绿泥石+绢云母+石英 1. 绢云母-白云母+石英±黑云母
抱板期	深构造相	固态流变 褶皱构造	区域变质 混合岩化	低-高角闪岩相 麻粒岩相	4. 紫苏辉石+黑云母+斜长石+石英±次透辉石 3. 石榴子石+黑云母+钾长石+斜长石+石英 2. 白云母+黑云母±斜长石+石英 1. 黑云母+斜长石+钾长石+石英±角闪石

发育透入性面理构造和不协调褶皱为特征，构成下构造层上亚层，构造型式表现为线型带状构造。后期热流的变化，引起重熔和重熔岩浆的形成，伴随同构造花岗岩浆活动（片麻状花岗岩）及边缘混合岩化。同时，戈枕村岩组则遭受第一次退变，其中的辉石被镁铁闪石和黑云母交代而成孤岛状残留体，即为后期退变的记录。至此，海南变质基底基本固结和定型。

晋宁期-扬子期变质作用　该时期以青白口系石碌群的变形变质特征为代表。石碌群为一套海相碎屑岩建造夹碳酸盐岩建造，同时，又伴有强烈酸性火山喷发活动。火山岩有钾流纹岩、钾质流纹斑岩、钾粗安质球粒状凝灰岩、钾流纹质凝灰岩。晋宁期构造运动的结果造成石碌群与上覆南华系—震旦系石灰顶组的不整合接触，导致了石碌群发生区域绿片岩相变质和强烈褶皱变形，而石灰顶组变质较浅，基本保持原岩面貌，仅在局部地段出现一些似千枚岩化及结晶灰岩等。

加里东期变质作用　海南岛加里东早期地壳裂解接受古生代巨厚沉积，主要为一套浅海—半深海复理石建造及陆台型石英岩-碳酸盐岩建造。该期变质变形作用与加里东运动有关，主要表现为早古生代地层普遍发生较强烈的区域低绿片岩相变质作用（相当于黑云母带，局部出现高绿片岩相），形成以板岩、千枚岩为主，次为片岩、大理岩、石英岩的岩石组合，体现具有绿片岩相型区域低温动力变质作用的特点。变形作用表现为以新生面理、线理及褶皱构造为主的中深构造相塑性变形。受该期变质作用影响，抱板群叠加了低绿片岩相的区域退化变质作用，表现为黑云母发生绿泥石化、白云母化，斜长石及白云母发生绢云母化等。

华力西期变质作用　加里东运动后，海南岛地壳一度长期为陆，缺失中志留统—泥盆系沉积，上古生界与下古生界呈不整合接触。华力西期地壳活动频繁，早期地壳再度下沉，接受了石炭纪和二叠纪大陆边缘沉积。在华力西运动作用下，晚古生代地层受低绿片岩相型区域低温动力变质作用，发生低绿片岩相的区域变质（相当于绢云母-绿泥石带），形成以变质砂岩、板岩为主，次为千枚岩的岩石组合。变质泥质岩中常见石英+绢云母共生组合，未见白云母及黑云母出现，表明该期与加里东期相比，变质热流已明显降低。岩石中原始沉积层理多为新生面理所置换，并发育弹塑性褶皱和逆冲推覆构造，属中浅构造相变形。在推覆面间的岩石发生动力变质，以韧性变形为主，形成以糜棱岩类系列为主的动力变质岩。晚期，地壳碰撞抬升，结束海相沉积历史。

伴随着大规模的华力西期岩浆活动，晚古生代地层与侵入岩接触部位及在新老花岗岩体之间也形成宽约数厘米至数十厘米的接触变质带，形成了钠长石绿帘石角岩相-辉石角岩相接触变质带，局部出现云英岩化。

印支期变质作用　该时期主要为接触变质和动力变质作用，少量气液变质作用。接触变质见于岩浆岩与古生代地层的接触部位，变质程度为钠长绿帘角岩相至普通角闪石角岩相，分别相当于绢云母-绿泥石带和白云母-黑云母带，局部地段见到少量的褶皱变形。动力变质作用发育于韧性、韧脆性断裂中，主体形成了一套糜棱岩系列岩石组合。其中，北东向戈枕韧性剪切带糜棱岩中绢云母矿物^{40}Ar-^{39}Ar为227Ma（陈新跃等，2006）、228Ma（叶伯丹，1991），北东向大村韧性剪切带糜棱岩中绢云母矿物^{40}Ar-^{39}Ar为230Ma（陈新跃等，2006），北西向公爱韧性剪切带糜棱岩中白云母矿物^{40}Ar-^{39}Ar为245Ma（陈新跃等，2006）。同时，伴随断裂活动，围岩遭受气-水热液作用，形成各种类型的气液变质岩。变形相属中浅构造相。

燕山期变质作用　燕山期变质作用于印支期变质作用相似。在燕山运动作用下，造就了韧脆性断裂变形，形成了以碎裂岩类为主的动力变质岩。同时，发生了大规模的岩浆活动，致使该期岩浆岩与古生代地层及早期岩浆岩的接触部位常发生热接触变质，规模较小，变质程度以钠长绿帘角岩相为主，相当于绢云母-绿泥石带。本次在海南岛北东向、北西向韧性变形岩石中获得变形矿物（黑云母）^{40}Ar-^{39}Ar年龄为156.8~196.5Ma和90.3~104.3Ma。变形相为浅构造相。

喜马拉雅期变质作用　主要见于琼北地区，一是与火山岩活动有关的接触变质，下伏地层因火山岩浆作用被烘烤而出现弱的炭化现象。二是与构造活动有关的动力变质，断裂面常见泥化和羽叶状岩石充填。变形相为表构造相。

第十四章　前寒武纪变质岩及变质作用

海南省前寒武纪区域变质岩分布于海南岛和西沙群岛一带，主要出露于海南岛西部冲卒岭、大蟹岭、抱板、石碌和海南岛中部上安、番阳、毛阳及海南岛东部黄竹岭、琼海烟塘等地区（图 14-1），包括中元古界长城系抱板群戈枕村组、长城系—蓟县系峨文岭组和新元古界青白口系石碌群、南华系—震旦系石灰顶组，面积约 490km²，占海南岛面积的 1.44%。中元古界经受了低角闪岩相—高角闪岩相多期次的区域变质作用，局部达麻粒岩相，其中戈枕村组还遭受区域性混合岩化作用；新元古界为一套低—中压相系的低级变质作用，形成绿片岩相浅变质岩系。

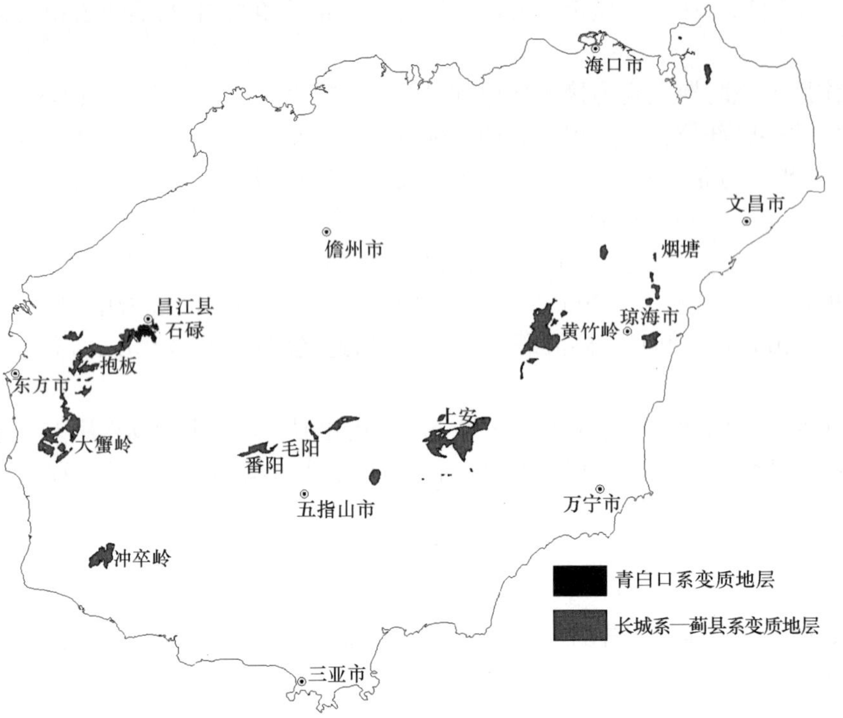

图 14-1　海南岛前寒武纪变质岩分布略图

Fig. 14-1　Sketch map of distribution of Pre-Cambrian metamorphic rocks in the Hainan Island

第一节　岩石类型、岩相特征与原岩建造

一、变质岩石的自然共生组合及原岩建造

（一）中元古界

由长城系戈枕村组和长城系—蓟县系峨文岭组组成。戈枕村组以片麻岩为主体，主要为黑云斜长片麻岩夹黑云角闪斜长片麻岩、二云斜长片麻岩，岩石普遍发生不同程度混合岩化形成混合质片麻岩、混合花岗闪长岩、长英质白云母钾长（条带状）混合岩、长英质黑云钾长眼球状混合岩、长英质二云母钾长条带状混合岩、条带状白云母钾长片麻岩、条带状白云母片麻岩。峨文岭组位于戈枕村组之上，以云母石英片岩、石英云母片岩为主，次为白云母石英片岩、长石云母石英片岩、二云母石英片岩、石英二云母片岩、白云母石英岩、黑云母石英岩、黑云母长石石英岩等，在冲卒岭地区夹数层石墨矿层。

（二）新元古界

由青白口系石碌群和南华系—震旦系石灰顶组组成，出露局限，目前仅见分布于昌江石碌矿区，出露面积约 20.0km^2。

石碌群是一套区域变质作用较低、构造分异作用不强烈、保存较多的变余沉积结构构造的岩石，共划分为 6 层。石碌群第一层为灰色、紫灰色中及薄层绢云母石英千枚岩夹薄层石英绢云母千枚岩、含红柱石石英绢云母千枚岩、绢云母千枚岩，中上部夹少量电气石化石英片岩；第二层为灰色厚层叶蛇纹石化泥粉晶白云岩；第三层主体为石英绢云母千枚岩；第四层为灰白色中厚层粗—细粒石英岩夹少量绢云母石英千枚岩；第五层为灰色薄层石英绢云母千枚岩、含红柱石石英绢云母千枚岩和绢云母石英千枚岩及少量透镜状石英片岩、千枚岩化含水云母石英细砂岩；第六层主要为碳酸盐白云岩及其接触热变质作用形成的强弱不等的透闪石透辉石化白云岩或二透岩。

石灰顶组为一套含砾石英细砂岩、石英细砂岩、铁质石英细砂岩、含黄铁矿石英细砂岩夹含铁矿粉砂岩、砂质泥岩、含铁泥岩及贫铁矿层。变质程度低，除局部见到黏土矿物变为绢云母外，铁矿层状结构保存较好，泥质岩主要由黏土矿物组成，砂质岩砂状结构可辨，岩石未见变质结构构造。

二、岩石类型及岩性特征

根据变质岩的岩石结构、构造特征及其矿物组成、含量，大致将海南省前寒武纪区域变质岩划分为如下 12 种岩石类型。

1. 板岩

出露于新元古界石碌群中，主要为炭质板岩，具变余粉砂状、泥状结构，显微鳞片变晶结构，板状构造、变余微层理构造，变质矿物以绢云母为主，次为石英，局部有少量绿泥石。绢云母多沿片轴平行聚集排列，形成完好的板状劈理，构成岩石的板状构造。

2. 千枚岩

广泛分布于新元古界石碌群中，以绢云母石英千枚岩、石英绢云母千枚岩为主，次为绢云母千枚岩、含红柱石石英绢云母千枚岩。具显微鳞片变晶结构，千枚状构造、变余微层理构造，重结晶作用明显。变质矿物有绢云母、石英，局部见绿泥石等。绢云母呈显微鳞片状，多聚成条带状分布，丝绢光泽明显；石英波状消光显著，常聚成条带与绢云母条带相间排列，构成岩石的变余微层理构造；绿泥石具浅绿—绿色多色性。

3. 片岩

片岩主要见于中元古界抱板群峨文岭组和新元古界石碌群中。

石碌群的片岩呈粒状鳞片变晶结构、显微鳞片粒状变晶结构、变余泥砂状结构，片状构造、变余微层理构造，矿物结晶明显，由石英、白云母、绢云母等组成。

峨文岭组的片岩大致可划分为石英二云母片岩（图版Ⅳ-1-A）和二云母石英片岩两种基本类型，呈粒状鳞片变晶结构、鳞片粒状变晶结构、斑状变晶结构，片状构造。矿物成分有：石英、斜长石、黑云母、白云母等。特征变质矿物有红柱石、石榴子石、十字石、矽线石、蓝晶石和石墨等。石英和长石为他形—不规则粒状变晶，部分被压扁拉长成片状，彼此镶嵌分布，颗粒大小一般在 0.1~2mm 之间。白云母和黑云母共生，呈叶片状，大小一般在 0.1~2mm 之间，最大者可达 3mm，白云母比黑云母稍大，少量云母发生扭曲现象。蓝晶石和石榴子石常呈变斑晶产出，石墨多呈晶质细小片状或不规则粒状，常与云母共生或沿云母解理分布。出露于峨文岭一带的片岩常见有红柱石和矽线石，其中红柱石呈不规则板柱状、粒状，分布不均，有时成聚合状产出，内部常包裹黑云母、石英等小包体；矽线石呈毛发状、针状、成束状、纤维状集合体，个别为细小柱状，长轴沿片理方向平行排列，系黑云母和蓝晶石的蚀变产物，多与云母、蓝晶石共生，推测两者可能为热接触变质的产物。此外，据王赐银等人（1979）报道在东方市尧文地区槟榔村一带的白云母石英片岩含有十字石，十字石为变斑状，常含石英、电气石、锆石等包裹体。

4. 片麻岩

主要见于戈枕村组中，大部分均遭受混合岩化作用，形成混合岩化程度各类的混合岩（图版Ⅳ-1-B至图版Ⅳ-1-E），仅有少部分残存。以黑云母斜长片麻岩为主，次有白云母钾长片麻岩、白云母斜长片麻岩、含石榴子石二云钾长片麻岩、透辉斜长片麻岩、石榴黑云二长片麻岩等。鳞片粒状变晶结构（图版Ⅳ-1-F、图版Ⅳ-1-G），片麻状构造（图版Ⅳ-1-G），矿物成分主要为钾长石、斜长石、石英、黑云母、白云母等，其次有少量次生的绢云母、绿泥石及微量锆石、磁铁矿等副矿物，特征变质矿物有石榴子石和透辉石。钾长石为微斜长石、条纹长石，双晶不发育，多发生绢云母化、高岭土化；斜长石成分主要为中长石（An32~An35），绢云母化也较强烈；石英波状消光强烈，普遍具压扁拉长特征，常与长石聚集定向分布构成浅色条带；黑云母呈细叶片状，多色性显著，多被白云母及绿泥石置换；白云母多为叶片状—宽叶片状，有的发生弯曲变形，有的与黑云母伴生，平行分布于浅色条带中。

5. 石英岩

主要见于中元古界峨文岭组中。主要岩石类型有纯石英岩、黑云母长石石英岩、白云母石英岩、黑云母石英岩、石墨石英岩、钾长石英岩、黑云斜长石英岩及含石墨黑云斜长石英岩等；呈薄层状，等粒（鳞片）粒状变晶结构（图版Ⅳ-1-H），块状或平行构造。主要矿物为石英，次为黑云母、斜长石、白云母等，特征变质矿物有石榴子石。长英矿物为他形—半自形粒状，粒径一般为0.1~0.3mm，少数达0.8~1.0mm，彼此呈紧密镶嵌接触，接触界面较平直，部分具定向排列。其他矿物黑云母、石墨及绢云母、白云母等，总量通常小于20%，一般断续布于长英矿物之中，部分具定向排列。

6. 变粒岩

主要见于戈枕村组和峨文岭组中，呈残留体产出，岩性主要为石榴黑云斜长变粒岩、石榴透辉黑云二长变粒岩、黑云二长变粒岩。岩石呈灰白色、灰色、灰黑色，风化后为暗灰色—肉红色、褐黄色等，镶嵌粒状变晶结构（图版Ⅳ-2-A），块状构造。长英质矿物呈他形粒状变晶镶嵌产出，大小大致相当，粒度一般在0.1~2mm之间。长石矿物总体以斜长石为主，钾长石次之。云母类矿物以黑云母为主，白云母少量。黑云母呈细小鳞片状，不规则状分布，个别略具定向性排列。特征变质矿物有石榴子石和石墨。石榴子石呈不规则细粒状集合体，稀散分布。石墨呈晶质细小片状或不规则粒状，多与黑云母共生或沿黑云母解理分布。

7. 浅粒岩

产出于戈枕村组和峨文岭组中，主要有斜长浅粒岩、钾长浅粒岩、含角闪浅粒岩、含黑云浅粒岩。岩石呈灰白色，风化后多为黄褐色，中细粒变晶结构，块状构造、变余层状构造（峨文岭组见及）。矿物成分以长石和石英为主，两者总和在90%以上，甚至达99%，长石含量在30%~79%之间，长英质矿物彼此间常呈镶嵌状产出，粒度大小一般在0.1~1mm之间，长石矿物呈他形粒状及不规则状，以斜长石为主。石英呈他形粒状，具明显波状消光。岩石中尚见少量黑云母、白云母、石榴子石、石墨及微量锆石和磷灰石等。

8. 白云岩

主要见于新元古界石碌群中，岩石呈浅灰色—灰白色，微晶—细粒结构，块状构造，主要矿物有白云石、方解石、石英、绢云母等。

9. 大理岩类

见于抱板群峨文岭岩组，分布于番阳、毛阳一带，呈厚度数十厘米至数十米的夹层产出。镶嵌粒状变晶结构，块状构造或平行构造。主要矿物成分为方解石68%~82%、石英≤20%、透辉石≤15%、斜长石1%~3%、透闪石≤3%，有时含硅灰石、董青石等。方解石为粒状，粒径0.1~1.0mm；石英他形粒状，粒径0.2~1.5 mm，多呈团块状或条带状分布于岩石中，部分呈脉状交代其他矿物，为后期脉石成分；透辉石他形、半自形粒状，粒径0.2~1.5 mm；斜长石半自形粒状或板状，粒径0.4~1.2 mm；透闪石长柱状，柱长1mm左右，常与透辉石共生，部分为透辉石的退变产物。

10. 钙硅酸盐岩类

见于峨文岭岩组，分布于冲卒岭地区，呈厚度数十厘米至数米的夹层状产出，岩石类型主要有斜长透闪透辉变粒岩和石榴子石英透辉变粒岩。细粒粒状变晶结构，块状或条带状构造。主要矿物组成为透辉石 25%~35%、石英 20%~35%、斜长石 5%~25%、透闪石 3%~10%、石榴子石 3%~25%，常含方解石、微斜长石。透辉石为粒状或短柱状；石英他形粒状，常呈透镜状或条带状集聚；斜长石他形粒状或半自形板状；透闪石长柱状或纤维状，多与透辉石伴生；石榴子石他形粒状，多呈集合体产出，有时集聚成条带。

11. 斜长角闪岩

主要分布在峨文岭组和戈枕村组中。可能为基性脉岩，多沿岩层呈似层状、透镜状或斜交层理呈脉状产出，与围岩界线清楚。以斜长角闪岩为主，此外还有石英斜长角闪岩、含黑云斜长角闪岩、石榴子石角闪岩等。呈绿色—深绿色，具粒状、柱状变晶结构，块状构造，矿物成分以斜长石和角闪石为主，次有少量石英、黑云母、钾长石，局部见特征变质矿物石榴子石，副矿物有磷灰石、钛铁矿、榍石等。角闪石具浅绿—绿色多色性，多聚集成集合体与斜长石、石英镶嵌分布；斜长石双晶不发育，局部见有聚片双晶，绢云母化明显；钾长石只是局部可见，主要为微斜长石，局部见格子双晶，绢云母化、黏土化强烈；黑云母多色性明显，主要分布于浅色矿物中，多发生绿泥石化。

12. 麻粒岩

仅见于长征地区戈枕村岩组，呈孤岛状残留体存在于角闪斜长片麻岩中的紫苏（或二辉）斜长片麻岩夹层或透镜体中，或以包体形式产于紫苏花岗岩中。细粒花岗变晶结构（图版Ⅳ-2-B）、块状或弱片麻状构造。主要矿物成分为斜长石（中、拉长石）50%~70%、紫苏辉石 20%~35%、透辉石≤10%、黑云母 5%~15%、角闪石≤5%、石英 2%~5%。副矿物有磁铁矿、磷灰石等。斜长石一般为他形粒状，少数为半自形板状，粒径 0.2~1.2 mm；紫苏辉石和透辉石多为他形（少数半自形）粒状，粒径 0.3~1.1mm；黑云母他形—半自形叶片状，片长一般为 0.5~0.8mm，具弱定向，呈棕红色，颜色与测区其他岩石类型中的黑云母颜色（褐红色、褐黄绿色）明显不同；角闪石和石英多为填隙状，粒径较小，一般为 0.4mm 左右。岩石常发生退变，此时黑云母和角闪石明显增多，其中紫苏辉石被黑云母、角闪石、绿帘石等矿物取代，有时透辉石也显著增加。

三、区域变质岩的原岩特征

（一）变余结构构造

戈枕村组由深变质的岩系组成，以斜长片麻岩类为主，少量斜长角闪（片麻）岩、变粒岩和麻粒岩。其变形、变质程度高，各类岩石变质结晶较完好，未见变余结构和构造（表 14-1）。片麻岩和变粒岩中少数斜长石呈较粗大的半自形晶，有时还可见复合双晶及（或）成分环带，可能为岩浆结晶的变余残斑。

峨文岭组主要由中深变质的（含石墨）云母片岩夹变粒岩、（含石墨）石英岩、大理岩及钙硅酸盐粒岩等组成，局部可见由不同岩石类型的侧向变化显示的原生层状构造，如石英岩、云母石英片岩和石英云母片岩组成的垂向韵律层（图 14-2），总体显示了沉积建造的特点。云母石英片岩中偶见变余火山玻质结构，黑云斜长变粒岩中则可见变余斑状结构，这表明峨文岭组原岩为砂泥质沉积物夹火山物质及钙质沉积物所组成。

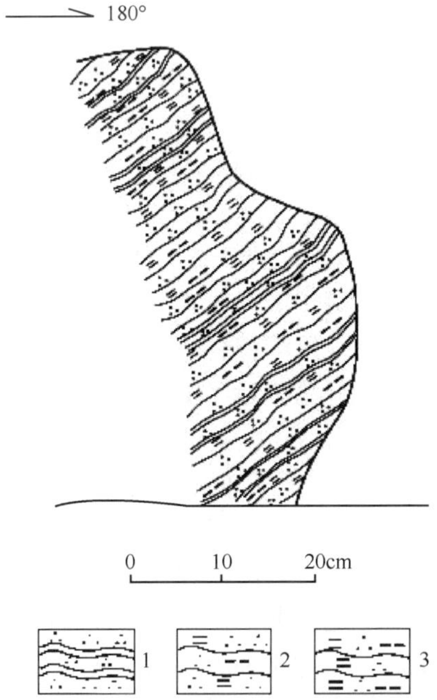

图 14-2　峨文岭剖面中的小型韵律构造素描图
Fig. 14-2　Sketch showing small-sized rhythm structures in the Ewen section
1—石英岩；2—云母石英片岩；3—石英云母片岩

石碌群在野外和镜下均可见变余沉积结构构造，因此可推断峨文岭组和石碌群的原岩为沉积岩类，另从其变质岩类型、变质岩矿物组合，以及变余结构构造等方面分析，其岩类主要应为泥岩、泥质粉砂岩、粉细砂岩、长石石英岩及石英砂岩等。

此外，抱板群各组中都含有斜长角闪岩，它们一般平行围岩片理或片麻理产出，但也有极少数明显切割片理或片麻理，有时还可见内部岩性分带，具变余辉绿结构，故有少量斜长角闪岩是变质的基性岩床或岩墙。

表 14-1 海南岛前寒武纪变质地层变余结构构造及副矿物特征

Table 14-1 Palimpsest textures-structures and accessory mineral characteristics of Pre-Cambrian metamorphic strata from the Hainan Island

地层名称	变余构造	变余结构	副矿物特征
石灰顶组	层状构造	砂状结构	次棱状—半圆状锆石、磷灰石、电气石、金红石、赤铁矿等
石碌群	变余层状构造 变余微层理构造	变余粉砂显微鳞片变晶结构、变余粉砂状结构、含变余粉砂片状变晶结构、变余砂砾状结构、变余砂状花岗变晶结构、含变斑晶粉晶化泥微晶结构	次棱状—半圆状锆石、磷灰石、电气石、金红石、白钛石、赤铁矿等
峨文岭组	变余层状构造 变余微层理构造	未见	磷灰石呈细微浑圆状；锆石呈浑圆状粒状；电气石呈微粒状
戈枕村组	未见	未见	锆石为椭圆状、短柱状；榍石呈不规则粒状、信封状或榍状；磷灰石为短柱状

（二）岩石化学图解

海南岛新元古界华南系—震旦系石灰顶组区域变质作用不明显或未受区域变质作用的影响，除局部见到黏土矿物变为绢云母外，沉积层状结构保存较好，泥质岩主要由黏土矿物组成，砂质岩砂状结构可辨，岩石未见变质结构构造，青白口系石碌群保留有变余原生结构构造，结合岩石组合及野外产出特征，可以直接判断其原岩；而抱板岩群变质岩变形、变质较强，大多不能直接判断其原岩，故采用岩石化学方法进行原岩恢复，主要类型变质岩的常量元素分析结果如表 14-2 所示。

将所收集的化学分析结果（77 个样品）作尼格里参数数值计算（表 14-2）和图解分析，来进行原岩恢复。由图 14-3~图 14-6 中可知戈枕村组的片麻岩和混合岩的投影点主要投入火山岩区及靠近火山岩区的砂质沉积岩区，且在图 14-7 中其投影点主要落入英安质凝灰岩，说明戈枕村组的原岩性质极其复杂，但总体是以中酸性火山岩为主，局部夹有杂砂岩类以及泥岩等。峨文岭组的片岩投影点落入泥质沉积岩区，由此反映出峨文岭组的原岩可能为泥岩、砂质泥岩、砂岩，局部夹有火山岩或钙质岩等。通过对石碌群的碳酸盐岩、二透岩、千枚岩、石英岩等岩石的化学分析结果所作尼格里参数计算与图解分析，其原岩大致有灰岩、泥砂质灰岩、泥砂质白云岩、泥质砂岩、含铁质粉砂质泥岩、含铁质泥岩、泥岩火山岩和火山碎屑岩等。

综上所述，戈枕村组原岩以中酸性火山岩为主，其次为基性火山岩，其中有少量砂质和泥质沉积物的加入，其物质来源与同时代的火山作用密切相关；峨文岭组原岩以砂泥质沉积为主，夹有少量钙质沉积，其物质来源于同一蚀源区，即早先形成的戈枕村组，少量的火山物质与可能间歇性火山喷发作用有关。综合岩石组合关系、野外产状、变余结构和构造，对海南岛前寒武纪区域变质岩的原岩进行了恢复，结果见表 14-3。

表 14-2 海南岛前寒武纪部分变质岩岩石化学成分含量及尼格里参数一览表

Table 14-2 List of petrochemical component contents and Niggli parameters of some of Pre-Cambrian metamorphic strata from the Hainan Island

地层单位		石碌群							
岩性		黑云母绿泥石石英条带二透岩	绢云母石英千枚岩		千枚状二云母石英片岩	白云岩		条纹状二透岩	透辉石石英角岩
样品号		ZK653-26	石碌 6	S-16A	0521	石碌 7	石碌 8	215-8	355-2
序号		1	2	3	4	5	6	7	8
岩石化学成分/%	SiO_2	53. 19	62. 23	75. 57	59. 21	3. 11	3. 68	56. 26	70. 04
	TiO_2	0. 65	0. 85	0. 33	0. 85	0. 02	0. 06	0. 66	0. 38
	Al_2O_3	10. 5	21. 4	14. 89	20. 69	0. 81	0. 50	12. 53	6. 78
	Fe_2O_3	8. 9	4. 18	0. 77	5. 50	0. 17	0. 10	4. 11	0. 81
	FeO	5. 86	0. 29	0. 3	1. 93	0. 23	0. 28	3. 40	2. 73
	MnO	0. 26	0. 06	0. 008	0. 15	0. 038	0. 062	0. 26	0. 23
	MgO	6. 97	0. 87	0. 62	1. 20	21. 25	21. 07	5. 74	5. 68
	CaO	7. 59	0	0. 12	0. 14	29. 69	29. 65	5. 32	9. 03
	Na_2O	0. 25	0. 24	0. 28	0. 15	0. 06	0. 16	0. 39	0. 26
	K_2O	3. 38	5. 48	3. 83	5. 80	0. 18	0. 04	7. 40	2. 10
	H_2O				3. 96				
	P_2O_5	0. 11	0. 068	0. 48	0. 1	0. 01	0. 015	0. 08	0. 24
	CO_2				0. 06				
	灼失								
	总量	97. 66	95. 668	97. 198	99. 74	55. 568	55. 617	96. 15	98. 28
岩石化学常用参数	al	15. 89	59. 84	65. 58	51. 11	0. 74	0. 46	22. 45	14. 9
	alk	6. 16	17. 69	20. 29	16. 12	0. 27	0. 28	15. 5	5. 93
	c	20. 89	0	0. 96	0. 63	49. 33	49. 63	17. 33	36. 08
	fm	57. 05	22. 47	13. 17	32. 15	49. 67	49. 63	44. 73	43. 09
	c/fm	0. 37	0	0. 07	0. 02	0. 99	1	0. 39	0. 84
	si	136. 64	295. 28	564. 85	248. 2	4. 82	5. 75	171. 03	261. 18
	(al+fm) - (c+alk)	45. 89	64. 62	57. 5	66. 51	0. 81	0. 18	34. 35	15. 98
	al-alk	9. 73	42. 15	45. 29	34. 99	0. 47	0. 18	6. 95	8. 97
	k	0. 9	0. 94	0. 9	0. 96	0. 66	0. 14	0. 93	0. 84
	mg	0. 47	0. 27	0. 52	0. 23	0. 99	0. 99	0. 58	0. 73
	K/%	93. 11	95. 80	93. 19	97. 48	75	20	94. 99	88. 98
	A/%	48. 34	78. 91	77. 88	77. 26	2. 64	1. 65	48. 87	37. 31
采样位置		石碌矿区							

续表

地层单位		石碌群		峨文岭组					
岩性		石英片岩		黑云斜长变粒岩		白云母石英片岩	石榴子石白云母片岩	云母片岩	
样品号		0521-3	0521-4	B286	戈 10-11	0612-3	D328-1	D2	D1
序号		9	10	11	12	13	14	15	16
岩石化学成分/%	SiO_2	60.42	64.98	69.73	72.96	74.46	66.29	72.00	70.06
	TiO_2	0.89	0.83	0.38	0.37	0.49	0.625	0.57	0.70
	Al_2O_3	21.16	18.17	15.58	13.18	13.35	16.28	14.07	14.8
	Fe_2O_3	6.7	6.17	0.60	0.84	4.49	5.36	3.65	3.52
	FeO			2.43	2.40		1.34	3.06	2.89
	MnO	0.13	0.12	0.03	0.11	0	1.59	0.04	
	MgO	1.78	1.76	1.33	1.04	0.48	0.076	1.47	0.47
	CaO	0.12	0.11	2.12	1.15	0.06	0.5	0.23	0.08
	Na_2O	0.25	0.23	4.00	3.59	0.03	0.46	0.37	0.4
	K_2O	5.31	4.3	2.78	3.04	3.3	2.84	3.3	3.65
	H_2O			0.70			4.46		
	P_2O_5	0.09	0.07	0.07	0.08	0.05	0.05	0.11	0.04
	CO_2			0.01	1.06				
	灼失	3.55	3.31			2.88	4.68		
	总量	100.4	100.05	99.76	99.82	99.59	104.551	98.3	96.61
岩石化学常用参数	al	51.88	50.6	42.51	41.53	55.56	50.49	44.74	50.44
	alk	15.1	14.02	26.17	28.98	15.07	11.88	13.29	15.71
	c	0.53	0.56	10.52	6.59	0.45	2.82	1.33	0.5
	fm	32.48	34.82	20.80	22.9	28.92	34.81	40.64	33.35
	c/fm	0.02	0.02	0.51	0.29	0.02	0.08	0.03	0.01
	si	251.41	307.1	322.89	390.14	525.86	348.88	388.51	405.22
	(al+fm)−(c+alk)	68.73	70.84	26.62	28.86	68.96	70.6	70.76	67.58
	al−alk	36.78	36.58	16.34	12.55	40.49	38.61	31.45	34.73
	k	0.93	0.92	0.31	0.36	0.99	0.8	0.85	0.86
	mg	0.34	0.36	0.44	0.36	0.17	0.02	0.29	0.12
	K/%	95.50	94.92	41	45.85	99.10	86.06	89.92	90.12
	A/%	78.84	79.66	63.64	62.88	79.75	81.08	78.30	78.18
采样位置		石碌矿区		大蟹岭		红林农场	戈枕村	二甲矿区	

续表

地层单位		峨文岭组							
岩性		二云母石英片岩	石英二云母片岩	二云石英片岩	长石二云石英片岩	黑云斜长变粒岩	石榴透辉变粒岩	斜长透辉变粒岩	浅粒岩
样品号		D410	XⅧ3-1	DH1643	DH71	B12	B132	XⅣ71	戈 10-9
序号		17	18	19	20	21	22	23	24
岩石化学成分/%	SiO_2	83.96	62.73	64.44	69.04	61.1	48.48	61.65	74.24
	TiO_2	0.42	0.81	0.73	0.62	0.8	0.26	0.5	0.32
	Al_2O_3	7.95	19.13	16.63	15.98	16.59	11.17	12.06	12.41
	Fe_2O_3	2.35	6.65	4.40	2.42	0.69	3.83	0.69	1.18
	FeO	0.15	0.38	0.54	2.20	5.47	9.28	4.65	2.08
	MnO	0.14	0.01	0.02	0.03	0.07	1.46	0.25	0.10
	MgO	0.48	0.70	2.30	1.27	3.64	2.27	6.00	0.98
	CaO	0.14	0.10	0.14	0.18	1.78	20.81	7.23	1.66
	Na_2O	0.10	0.29	0.47	0.25	2.86	0.97	0.54	3.18
	K_2O	1.52	4.27	6.37	4.38	4.99	0.14	4.66	3.04
	H_2O	2.46	4.54			0.74	0.52		
	P_2O_5	0.10	0.12	0.05	0.05	0.16	0.31	0.29	0.07
	CO_2	0.07	0.03			0.13	0.43		0.65
	灼失	2.60	4.84	3.81	3.04			1.1	
	总量	99.84	99.76	99.9	99.46	99.02	99.93	99.62	99.91
岩石化学常用参数	al	54.29	54.3	45.21	51.67	34.64	14.57	22.27	39.99
	alk	12.36	14.47	20.85	16.66	21.1	2.28	10.95	27.46
	c	1.74	0.52	0.69	1.06	6.76	49.36	24.27	9.73
	fm	31.61	30.7	33.25	30.61	37.49	33.79	42.5	22.82
	c/fm	0.05	0.02	0.02	0.03	0.18	1.46	0.57	0.43
	si	972.96	302.18	297.27	378.81	216.52	107.32	193.18	405.98
	(al+fm)-(c+alk)	71.8	70.01	56.92	64.56	44.27	-3.28	29.55	25.62
	al-alk	41.93	39.83	24.36	35.01	13.54	12.29	11.32	12.53
	k	0.91	0.91	0.9	0.92	0.53	0.09	0.85	0.39
	mg	0.26	0.16	0.48	0.34	0.51	0.22	0.66	0.35
	K/%	93.83	93.64	93.13	94.6	63.57	12.61	89.62	48.87
	A/%	81.87	80.41	70.44	76.86	63.27	33.76	49.24	61.16
采样位置		中建农场		冲卒岭				大蟹岭	

续表

地层单位		戈枕村组							
岩性		条纹状混合岩		混合片麻岩			石榴黑云斜长片麻岩	条纹状混合片麻岩	黑云母条纹状混合片麻岩
样品号		Ⅸ16-1	D43-1	D92-1	D95-1	Ⅱ12-2	D326-3	Ⅱ22-1	0616
序号		25	26	27	28	29	30	31	32
岩石化学成分/%	SiO_2	71.13	68.81	74.81	62.54	75.26	63.23	74.4	68.25
	TiO_2	0.35	0.38	0.05	0.1	0.03	0.575	0.13	0.44
	Al_2O_3	13.45	14.65	14.86	19.28	13.92	15.8	13.66	14.64
	Fe_2O_3	0.65	0.77	0.87	1.16	0.43	1.45	0.86	4.97
	FeO	2.98	3.1	0.48	0.85	0.57	3.82	1.21	
	MnO	0.05	0.04	0.05	0.04	0.05	1.56	0.05	0.06
	MgO	1.03	1.81	0.15	0.47	0.12	0.087	0.35	1.18
	CaO	1	1.86	1.58	1.1	0.43	3.81	0.5	3.16
	Na_2O	2.61	3.5	4.3	4.6	3.42	3.14	2.72	2.63
	K_2O	5.15	3.27	1.32	7.58	5.43	2.62	5.11	3.26
	H_2O	1.34	1.4	0.2	1.54		2.5		
	P_2O_5	0.41	0.09	0.85	0.05		0.075		
	CO_2	0.22	0.13	1.18	0.2				
	灼失						3.16		0.97
	总量	100.37	99.81	100.7	99.51	99.66	101.827	98.99	99.56
岩石化学常用参数	al	40.91	39.23	52.18	47.04	49.82	39.05	48.19	38.88
	alk	30.02	24.89	29.86	38.48	41.17	19.77	35.3	20.86
	c	5.53	9.06	10.09	4.88	2.8	17.12	3.21	15.26
	fm	23.54	26.83	7.88	9.6	6.2	24.06	13.31	25.01
	c/fm	0.24	0.34	1.28	0.51	0.45	0.71	0.24	0.61
	si	367.19	312.65	445.77	258.94	457.12	265.17	445.38	307.54
	(al+fm)-(c+alk)	28.9	32.11	20.11	13.28	12.05	26.22	22.99	27.77
	al-alk	10.89	14.34	22.32	8.56	8.65	19.28	12.89	18.02
	k	0.56	0.38	0.17	0.52	0.51	0.35	0.55	0.45
	mg	0.34	0.46	0.17	0.3	0.18	0.02	0.23	0.32
	K/%	66.37	48.30	23.49	62.23	61.36	45.47	65.26	55.35
	A/%	60.56	62.93	67.36	59.21	60	62.28	62.12	61.80
采样位置		中建农场							戈枕大坝

地层单位		戈枕村组								
岩性		黑云母条纹状混合片麻岩		黑云变粒岩	白云母钾长片麻岩	黑云母钾长片麻岩	黑云斜长片麻岩	黑云斜长片麻岩	黑云斜长片麻岩	黑云斜长片麻岩
样品号		0625	0626-1	D4	XⅣ6-2	D1662-1	B192	B193	B194	XIV5-1
序号		33	34	35	36	37	38	39	40	41
岩石化学成分/%	SiO_2	67.03	65.49	68.96	76.00	73.04	69.14	66.34	66.03	67.08
	TiO_2	0.49	0.51	0.5	0.32	0.50	0.74	0.79	0.47	0.61
	Al_2O_3	15.36	15.09	13.1	12.47	13.02	13.53	12.75	14.99	14.51
	Fe_2O_3	4.91	5.35	0.2	1.42	0.08	0.04	1.31	0.18	0.82
	FeO			4.64	0.70	2.08	5.62	6.86	6.12	5.05
	MnO	0.07	0.06	0.24	0.01	0.01	0.13	0.18	0.06	0.08
	MgO	1.26	1.44	1.1	0.65	1.21	1.50	1.63	1.79	1.77
	CaO	4.03	4.59	4	0.13	1.45	2.14	3.34	2.96	3.03
	Na_2O	3.02	2.67	2.54	1.29	1.61	2.72	2.04	3.40	3.43
	K_2O	2.16	2.58	2.46	3.78	5.89	2.51	3.38	2.67	2.20
	H_2O			0.98	2.87	0.67	1.00	1.10	0.62	
	P_2O_5			0.15	0.10	0.05	0.11	0.20	0.15	0.13
	CO_2				0.10	0.17	0.14	0.16	0.16	
	灼失	1.27	0.09		2.15	0.60				0.61
	总量	99.60	97.87	98.87	101.99	100.38	99.32	100.08	99.60	99.32
岩石化学常用参数	al	38.84	36.64	35.23	53.32	42.26	36.95	30.63	35.37	35.52
	alk	18.47	17.45	18.4	26.57	29.29	19.64	16.85	20.01	19.64
	c	18.53	20.27	19.56	1.01	8.56	10.63	14.59	12.7	13.49
	fm	24.17	25.65	26.81	19.1	19.89	32.79	37.93	31.92	31.35
	c/fm	0.77	0.79	0.73	0.05	0.43	0.32	0.38	0.4	0.43
	si	287.59	269.86	314.73	551.49	402.29	320.39	270.43	264.36	278.66
	(al+fm)-(c+alk)	26.01	24.57	24.08	44.84	24.3	39.47	37.12	34.58	33.74
	al-alk	20.37	19.19	16.83	26.75	12.97	17.31	13.78	15.36	15.88
	k	0.32	0.39	0.39	0.66	0.71	0.38	0.52	0.34	0.3
	mg	0.33	0.34	0.28	0.37	0.5	0.32	0.26	0.33	0.35
	K/%	41.70	49.14	49.2	74.56	78.53	47.99	62.36	43.99	39.08
	A/%	62.52	60.53	59.28	70.57	59.26	64.74	59.27	62.41	62.62
采样位置		二甲矿区			冲卒岭	中建农场	冲卒岭			

续表

地层单位		戈枕村组								
岩性		黑云斜长片麻岩			角闪黑云斜长片麻岩	含蓝晶黑云斜长片麻岩	黑云斜长片麻岩	眼球状混合岩	眼球状混合岩	眼球状混合岩
样品号		XIV16-1	XIV17-1	XIV19-1	L932	DH72	XIV13	S022-1	DH127	DH3076
序号		42	43	44	45	46	47	48	49	50
岩石化学成分/%	SiO_2	75.39	74.92	68.37	67.47	69.97	67.68	67.88	69.3	67.87
	TiO_2	0.24	0.24	0.19	0.85	0.71	0.7	0.07	0.58	0.77
	Al_2O_3	12.77	12.83	14.65	13.14	13.39	13.59	14.18	13.12	13.65
	Fe_2O_3	1.37	0.92	0.84	1.83	1.03	0.72	0.22	0.91	1.57
	FeO	0.57	1.27	3.92	4.81	3.73	4.8	5.36	4.22	3.93
	MnO	0.03	0.02	0.07	0.1	0.06	0.06	0.08	0.08	0.05
	MgO	0.41	0.36	1.96	1.3	1.11	2.78	1.45	1.51	1.3
	CaO	0.79	0.8	2.1	2.74	1.61	2.37	1.91	1.77	1.94
	Na_2O	2.21	2.49	4.32	2.47	1.98	2.38	2.09	2.35	2.32
	K_2O	4.90	4.99	2.06	6.57	4.29	3.45	5.03	3.89	4.21
	H_2O		.					0.72	0.21	
	P_2O_5	0.03	0.02	0.09	0.16	0.09	0.06	0.11	0.1	0.12
	CO_2							0.07		
	灼失	0.96	0.83	0.55		1.21	0.73		1.19	1.91
	总量	99.67	99.69	99.12	101.44	99.18	99.32	99.17	99.23	99.64
岩石化学常用参数	al	47.68	46.32	37.09	31.37	39.71	33.64	37.11	36.96	37.4
	alk	33.38	34.29	23.64	26.68	23.43	18.94	23.25	22.75	22.95
	c	5.36	5.25	9.67	11.89	8.68	10.67	9.09	9.07	9.67
	fm	13.59	14.14	29.61	30.07	28.18	36.76	30.55	31.23	29.99
	c/fm	0.39	0.37	0.33	0.4	0.31	0.29	0.3	0.29	0.32
	si	477.64	459	293.73	273.31	352.11	284.29	301.49	331.26	315.6
	(al+fm)-(c+alk)	22.53	20.92	33.39	22.87	35.78	40.79	35.32	36.37	34.77
	al-alk	14.3	12.03	13.45	4.69	16.28	14.7	13.86	14.21	14.45
	k	0.59	0.57	0.24	0.64	0.59	0.49	0.61	0.52	0.54
	mg	0.29	0.23	0.42	0.26	0.3	0.47	0.31	0.34	0.3
	K/%	68.92	66.71	32.29	72.68	68.42	59.18	70.65	62.34	64.47
	A/%	61.78	60.78	66.34	52.73	62.95	62.37	61.09	62.09	61.71
采样位置		冲卒岭								

续表

地层单位		戈枕村组								
岩性		条带状混合岩	条痕状混合岩	混合花岗岩	角闪黑云斜长片麻岩	角闪黑云斜长变粒岩	黑云斜长片麻岩	黑云斜长片麻岩	黑云斜长变粒岩	黑云紫苏麻粒岩
样品号		DZK	DH155	B1	S034-5	S034-6	GⅡ-9-1	T3	S034-3	D1650-2
序号		51	52	53	54	55	56	57	58	59
岩石化学成分/%	SiO_2	70.17	75.2	77.52	58.94	66.24	67.78	63.38	70.1	49.94
	TiO_2	0.58	0.12	0.17	1.3	0.59	0.64	1.04	0.5	0.79
	Al_2O_3	12.86	13.14	11.37	16.8	12.5	14.96	16.02	14.6	14.98
	Fe_2O_3	1.64	1.49		0.93	0.22	0.34	0.72	0.13	1.12
	FeO	2.44	0.52	3.17	6.46	5.05	3.3	5.68	3.42	8.43
	MnO	0.06	0.02	0.03	0.09	0.1	0.04	0.07	0.06	0.18
	MgO	1.43	0.2	0.33	2.97	3.75	0.91	1.14	0.91	8.62
	CaO	1.77	1.11	1.33	5.8	5.62	2.85	5.13	2.8	9.88
	Na_2O	2.45	3.22	3.67	2.64	2.14	3.44	2.96	3.8	2.16
	K_2O	4.18	4.78	2.32	2.32	1.63	4.13	2.36	2.36	1.88
	H_2O	0.19		0.34	0.61	0.34	0.52	0.46	0.28	0.67
	P_2O_5	0.12	0.02	0.03	0.32	0.13	0.17	0.28	0.14	0.18
	CO_2				0.10	0.20	0.28	0.23	0.07	0.55
	灼失	1.57	0.68							
	总量	99.46	100.5	100.28	99.28	98.51	99.36	99.47	99.17	99.38
岩石化学常用参数	al	37.94	45.62	41.03	32.18	27.73	39.63	35.81	40.66	20.25
	alk	25.24	36.35	30.85	13.13	11.72	26.83	16.59	24.53	7.55
	c	9.5	7.01	8.73	20.2	22.67	13.73	20.85	14.18	24.28
	fm	27.32	11.02	19.4	34.48	37.88	19.81	26.75	20.63	47.92
	c/fm	0.35	0.64	0.45	0.59	0.6	0.69	0.78	0.69	0.51
	si	351.32	443.02	474.67	191.61	249.34	304.7	240.42	331.31	114.53
	(al+fm)-(c+alk)	30.52	13.28	20.85	33.33	31.22	18.88	25.12	22.58	36.34
	al-alk	12.7	9.27	10.18	19.05	16.01	12.8	19.22	16.13	12.7
	k	0.53	0.49	0.29	0.37	0.33	0.44	0.34	0.29	0.36
	mg	0.39	0.16	0.16	0.42	0.56	0.31	0.24	0.31	0.61
	K/%	63.05	59.75	38.73	46.77	43.24	54.56	44.36	38.31	46.53
	A/%	60.49	59.06	60.83	60.96	57.1	58.94	60.52	61.97	51.83
采样位置		冲卒岭			长征					

续表

地层单位		戈枕村组			抱板群					
岩性		麻粒岩		眼球状混合岩	斜长角闪岩					
样品号		B323	B317	S036-1	B125	YD549-2	D4760-1	D4760-2	D4760-3	Ⅴ213
序号		60	61	62	63	64	65	66	67	68
岩石化学成分/%	SiO_2	51.34	49.83	68.39	52.6	50.40	49.94	52.53	50.20	51.66
	TiO_2	1.87	1.45	0.52	1.19	1.77	1.15	1.59	1.59	1.14
	Al_2O_3	17.58	12.57	14.5	13.57	13.04	15.04	13.99	14.43	12.86
	Fe_2O_3	0.48	0.64	0.33	1.22	1.93	1.58	1.84	1.22	2.6
	FeO	9.39	8.73	3.97	13.03	10.35	7.88	8.20	9.37	11.73
	MnO	0.16	0.17	0.07	0.21	0.22	0.16	0.16	0.15	0.3
	MgO	6.20	9.21	1.2	5.54	7.03	7.30	6.03	7.02	4.54
	CaO	7.81	9.93	3.08	8.65	9.80	10.68	8.73	10.3	9.01
	Na_2O	3.06	2.06	2.81	0.50	2.02	2.68	2.81	2.70	1.86
	K_2O	0.68	2.33	4.24	1.54	0.89	0.80	1.20	0.60	0.08
	H_2O	0.64	1.06	0.16	0.66	0.06	2.26	2.27	1.83	0.95
	P_2O_5	0.69	0.83	0.18	0.14	0.15	0.12	0.21	0.22	0.16
	CO_2	0.49	0.62	0.14	0.64	2.16	0.04	0.02	0.04	
	灼失					1.04	1.35	1.18	0.72	
	总量	100.39	99.43	99.59	99.49	100.86	100.98	100.76	100.39	96.89
岩石化学常用参数	al	26.08	17.15	37.66	20.51	18.52	21	21.44	20.3	20.01
	alk	8.56	8.07	23.93	3.76	6.09	7.36	9.07	7.16	4.9
	c	21.07	24.64	14.54	23.77	25.31	27.11	24.32	26.35	25.49
	fm	44.29	50.15	23.87	51.95	50.08	44.53	45.16	46.19	49.61
	c/fm	0.48	0.49	0.61	0.46	0.51	0.61	0.54	0.57	0.51
	si	129.25	115.38	301.41	134.93	121.49	118.31	136.6	119.85	136.4
	(al+fm)−(c+alk)	40.74	34.59	23.06	44.93	37.2	31.06	33.21	32.98	39.23
	al−alk	17.52	9.08	13.73	16.75	12.43	13.64	12.37	13.14	15.11
	k	0.13	0.43	0.5	0.67	0.22	0.16	0.22	0.13	0.03
	mg	0.53	0.63	0.33	0.41	0.5	0.58	0.52	0.54	0.36
	K/%	18.18	53.08	60.14	75.49	30.58	22.99	29.93	18.18	4.12
	A/%	60.35	46.75	58.87	55.94	50.64	51.51	52.34	51.48	54.01
采样位置		长征			大蟹岭	牛案				红泉农场

续表

地层单位		抱板群								
岩性		斜长角闪岩								
样品号		D262	文 10-13	文 10-21	S034-4	T1	T4	T5	T6	GⅡ-8
序号		69	70	71	72	73	74	75	76	77
岩石化学成分/%	SiO_2	48.29	51.26	51.9	51.34	53.29	55.02	52.46	55.88	54.38
	TiO_2	2.83	0.64	1.59	1.30	1.02	1.13	1.15	1.18	1.18
	Al_2O_3	12.04	13.67	14.06	16.36	16.94	17.18	15.37	16.59	17.05
	Fe_2O_3	3.18	3.04	1.47	0.99	0.84	1.10	1.34	1.08	1.21
	FeO	15.33	7.36	12.86	8.02	6.72	6.5	8.33	6.67	6.63
	MnO	0.43	0.21	0.27	0.16	0.15	0.14	0.17	0.13	0.15
	MgO	5.61	8.46	6.18	5.92	5.86	4.02	5.92	3.91	4.60
	CaO	8.47	10.02	9.17	9.04	8.42	7.06	7.87	7.59	7.34
	Na_2O	1.62	2.5	0.68	2.66	2.74	3.76	2.56	2.81	2.69
	K_2O	0.08	1.24	0.4	1.85	2.22	2.43	2.36	1.88	2.56
	H_2O	0.75	1.14	0.72	0.36	0.58	0.50	0.58	0.70	0.60
	P_2O_5	0.3	0.04	0.13	0.26	0.24	0.32	0.39	0.28	0.35
	CO_2		0.09	0.07	0.44	0.18	0.46	0.42	0.29	0.14
	灼失									
	总量	98.93	99.67	99.5	98.7	99.2	99.62	98.92	98.99	98.88
岩石化学常用参数	al	17	18.63	20.54	24.41	26.14	28.71	23.57	28.62	28.23
	alk	3.89	7.43	2.27	9.52	10.66	14.73	10.38	11.48	11.91
	c	21.75	24.83	24.36	24.52	23.62	21.45	21.95	23.81	22.1
	fm	57.37	49.11	52.82	41.55	39.58	35.1	44.1	36.09	37.76
	c/fm	0.38	0.51	0.46	0.59	0.6	0.61	0.5	0.66	0.59
	si	115.71	118.55	128.7	129.98	139.53	156.05	136.54	163.57	152.79
	(al+fm)-(c+alk)	48.73	35.48	46.73	31.92	31.44	27.63	35.34	29.42	31.98
	al-alk	13.11	11.2	18.27	14.89	15.48	13.98	13.19	17.14	16.32
	k	0.03	0.25	0.28	0.31	0.35	0.3	0.38	0.31	0.39
	mg	0.35	0.59	0.43	0.54	0.58	0.48	0.52	0.47	0.51
	K/%	4.71	33.16	37.04	41.02	44.76	39.26	47.97	40.09	48.76
	A/%	54.21	49.84	57.84	54.7	55.87	56.46	54.58	57.46	57.52
采样位置		二甲	大蟹岭		长征					

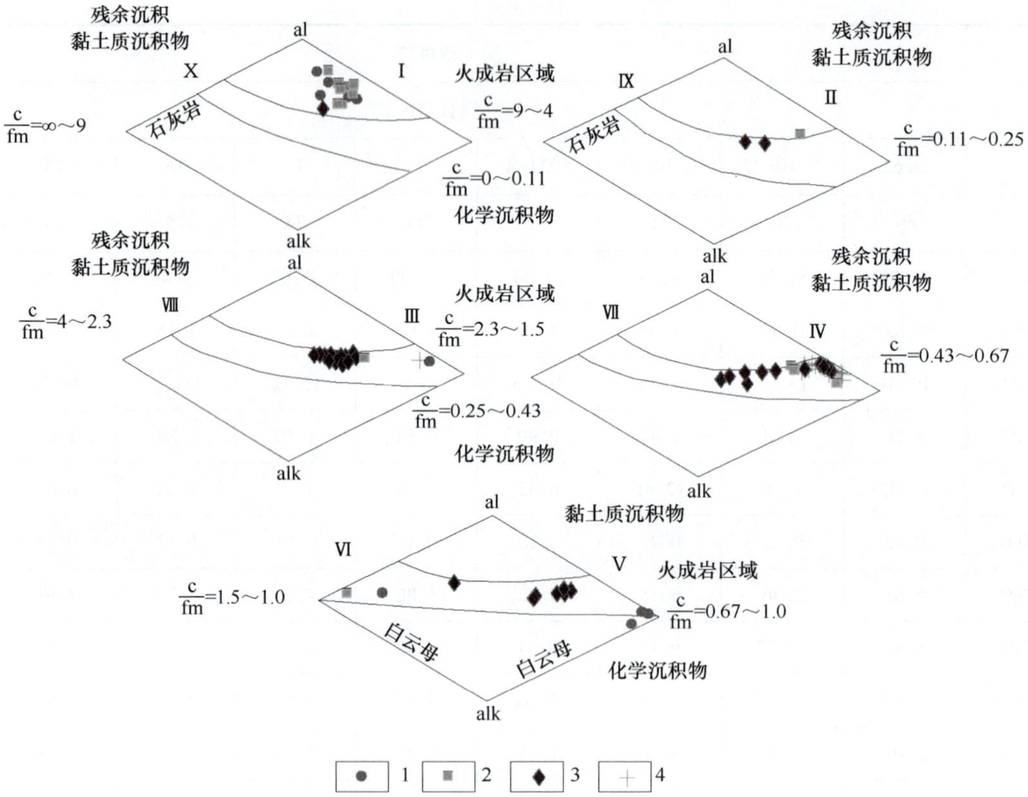

图 14-3 前寒武纪变质岩 Niggli 四面体图解

Fig. 14-3 Niggli tetrahedron diagram of Pre-Cambrian metamorphic rocks

1—石碌群变质岩；2—峨文岭组变质岩；3—戈枕村组变质岩；4—斜长角闪岩

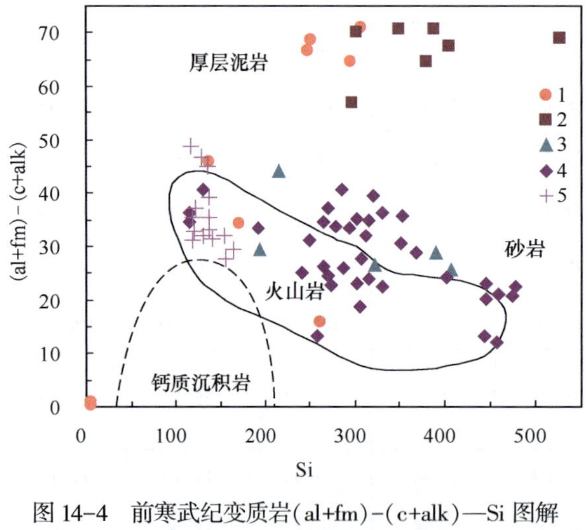

图 14-4 前寒武纪变质岩(al+fm)-(c+alk)—Si 图解

Fig. 14-4 (al+fm) - (c+alk) —Si diagram of Pre-Cambrian metamorphic rocks

(据 A. Simonen，1953)

1—石碌群变质岩；2—峨文岭组片岩类；3—峨文岭组变粒岩；4—戈枕村组变质岩；5—斜长角闪岩

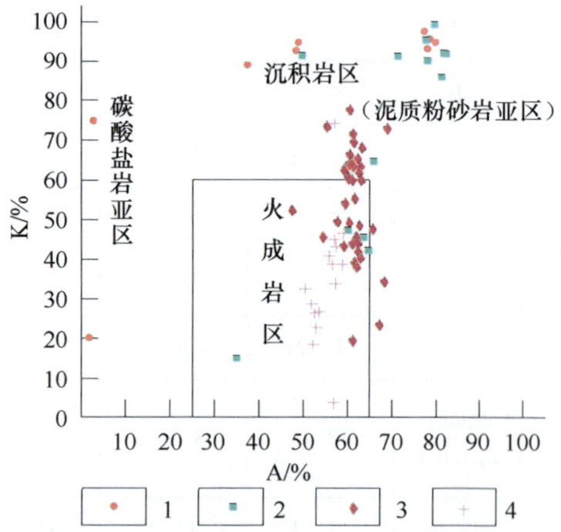

图 14-5 前寒武纪变质岩的 K-A 相关图解

Fig. 14-5 K-A correlation diagram of Pre-Cambrian metamorphic rocks

(据周世泰，1977)

1—石碌群变质岩；2—峨文岭组变质岩；3—戈枕村组变质岩；4—斜长角闪岩

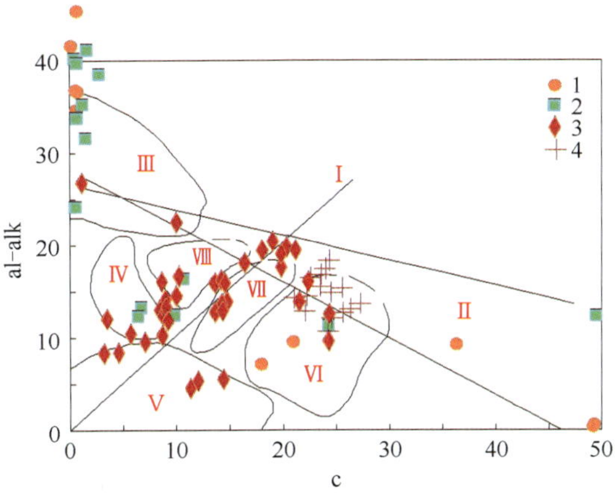

图 14-6 前寒武纪变质岩的（al-alk）-c 相关图解

Fig. 14-6 (al-alk) -c correlation diagram of Pre-Cambrian metamorphic rocks

（据长春地质学院，1975）

1—石碌群变质岩；2—峨文岭组变质岩；3—戈枕村组变质岩；4—斜长角闪岩；Ⅰ—钙质泥灰岩；Ⅱ—白云质泥灰岩；Ⅲ—黏土岩；Ⅳ—中酸性凝灰岩；Ⅴ—角斑岩；Ⅵ—细碧岩-玄武岩；Ⅶ—二长安山质凝灰岩；Ⅷ—英安质凝灰岩

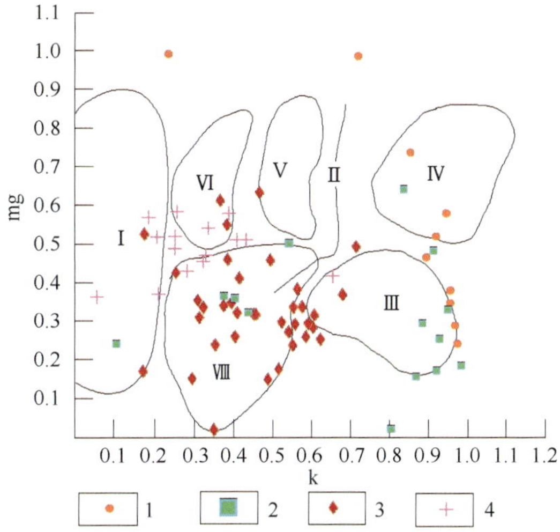

图 14-7 前寒武纪变质岩 mg-k 图解

Fig. 14-7 mg-k diagram of Pre-Cambrian metamorphic rocks

（据长春地质学院，1975）

1—石碌群变质岩；2—峨文岭组变质岩；3—戈枕村组变质岩；4—斜长角闪岩；Ⅰ—细碧岩-角斑岩；Ⅱ—白云质泥灰岩；Ⅲ—黏土岩；Ⅳ—富钾火山岩；Ⅴ—未定区；Ⅵ—玄武岩区；Ⅷ—英安质凝灰岩

表 14-3 海南岛前寒武纪主要区域变质岩类型原岩恢复一览表

Table 14-3 List of protolith reconstruction of main regional metamorphic rock types, Hainan Island

地层单位		变质岩	原 岩
石碌群		板岩类	泥岩、粉砂质泥岩、炭泥质粉砂岩、沉凝灰岩
		千枚岩类	泥质粉砂岩、粉砂质泥岩、含炭泥岩
		片岩类	泥质岩、粉砂质泥岩、泥质粉砂岩
		石英岩类	石英砂岩、长石石英砂岩
		白云岩类	碳酸盐岩
抱板群	峨文岭组	片岩类	泥质砂岩、粉砂质泥岩
		石英岩类	石英砂岩、含炭石英砂岩、长石石英砂岩
		钙硅酸盐粒岩	泥灰岩、混有碳酸盐的基性凝灰岩
		黑云斜长变粒岩	杂砂岩、泥质砂岩
		浅粒岩	中酸性火山岩、杂砂岩
		大理岩	泥灰岩
	戈枕村组	黑云斜长变粒岩	英安质火山岩
		角闪黑云斜长变粒岩	安山质火山岩
		斜长片麻岩	中酸性火山岩、杂砂岩
		角闪（辉石）片麻岩	中基性火山岩
		麻粒岩	基性火山岩
		斜长角闪岩	基性火山岩、基性岩床或岩墙

第二节 区域变质作用

一、区域变质作用类型

根据岩石组合、原岩建造、变质相和变质相系、花岗岩浆活动及构造等特征，可以将海南岛前寒武纪区域变质作用划分为区域中高温变质作用、区域动力热流变质作用和区域低温动力变质作用三种类型。

1. 区域中高温变质作用

中元古代表壳岩（戈枕村组）经历了此种变质作用。其变质相比较简单，以较均匀的中高温角闪岩相为主，仅有少量麻粒岩相存在，属角闪岩相型区域中高温变质作用，变质作用条件为700℃～845℃、5.44×10^8～6.35×10^8Pa（马大铨等，1998）。变质相的界线是过渡的，没有明确的分界，不能划分变质带。

岩石组合以黑云（角闪）斜长片麻岩、黑云（角闪）斜长变粒岩、斜长角闪岩为主，夹少量变粒岩、麻粒岩。麻粒岩相的岩石主要见于琼中长征地区，有黑云紫苏（二辉）麻粒岩、含辉石（紫苏/二辉/透辉）的斜长片麻岩及含辉石的角闪质岩石（斜长角闪片麻岩、角闪斜长片麻岩），后者还见于大蟹岭和冲卒岭地区；原岩建造主要为中酸性、基性火山岩和部分杂砂岩。

区域中高温变质作用是地壳形成和演化早期特有的一种变质作用类型，高的热流值和以热点为中心均匀分布是其主要特点。区域中高温变质作用在构造上表现为穹隆或短轴背斜，穹隆中心（热点）分别位于长征地区的十六队、冲卒岭地区的亚南甫、中建农场、戈枕地区和大蟹岭等地区。与之相伴的混合岩化和原地型花岗岩浆活动强烈，局部出现紫苏花岗岩。

2. 区域动力热流变质作用

中元古代表壳岩（峨文岭组）遭受了此种变质作用，其之前已变质的戈枕村组则发生了相应退变质，变质作用条件为500℃～650℃、3×10^8～5×10^8Pa（马大铨等，1998），属中低压相系。包括低角闪岩相和高绿片岩相，但由于受后期岩浆吞蚀和构造肢解，峨文岭组大多出露不全；以及多期变质作用的叠加改造，使得其变质相带难以完全恢复。

岩石组合以云母片岩类为主，夹变粒岩、石英岩、大理岩等，原岩为泥砂质沉积岩夹钙碱性火山-沉积岩，属扩张弧后环境。它们具有大致呈线型分布的特点，后期伴随同构造花岗岩浆活动（中元古代片麻状花岗岩）和边缘混合岩化，是板块俯冲造山作用的产物。

3. 区域低温动力变质作用

新元古代地层遭受了此种变质作用，为绿片岩相型单相变质，其原岩主要是滨海—浅海相陆源碎屑岩-碳酸盐岩建造，夹有数层中酸性火山熔岩和火山凝灰岩建造。岩石组合以各类千枚岩为主。变质泥质岩中以绢云母+绿泥石+石英为特征组合，最高出现雏晶黑云母。其之前形成的抱板群发生退变或再退变。

区域低温动力变质作用在我国有广泛分布，如晋宁期的板溪群，赣南-粤北的加里东期变质岩系等。它们通常发生于中浅地壳，是元古宙以来构造活动带的一种主要变质作用类型。由于应力较强，常形成各种劈理化、片理化和条带状构造的岩石，褶皱变形十分强烈；热流变化不显著，变质作用较浅，最高可达绿片岩相，一般无混合岩化和混合花岗岩出现，形成于陆缘活动带或陆间、陆内地槽环境。

以上三种类型的区域变质作用具有不同的时空分布特点，反映了海南岛地壳演化的不同阶段和大地构造背景。古元古代洋壳俯冲作用可能引起局部热流的上升，导致了具穹隆或短轴背斜分布特点的区域中高温变质作用的发生；中元古代洋-陆碰撞造山作用形成具线状分布的区域动力热流变质作用带；而新元古代陆弧碰撞造山作用形成具带状分布的区域低温动力变质作用带。

二、变质相及矿物共生组合

1. 绿片岩相

该岩相主要分布新元古代地层中，原岩多属于泥质岩、粉砂岩、砂岩、泥炭岩、碳酸盐岩及中基性火山岩等，经变质作用之后，形成以千枚岩为主，伴有板岩、变质砂岩、石英岩及片岩等岩石组合。镜下常见的变质矿物主要以绢云母（白云母）和石英为主，此外还有少量的绿泥石及偶见雏晶状黑云母。绢云母为显微鳞片状，多沿千枚理定向排列，个别杂乱无章，多为原岩中的泥质物质或长石碎屑物变质而成，变质不彻底之处，仍残留有泥质物质和长石碎屑物。白云母细小鳞片状，常与绢云母伴生。绿泥石为显微鳞片状集合体常与绢云母共生。代表性矿物组合为：绿泥石+绢云母+石英、绢云母-白云母+石英±黑云母、透辉石+透闪石+石英（基性岩）。相当于铁铝榴石带，属于低压相系矿物组合特征。

2. 低角闪岩相

原岩系泥质岩、粉砂质岩、含炭石英砂岩、长石石英砂岩、杂质灰岩、中酸性火山岩等。经变质作用后，形成以云母石英片岩和石英二云母片岩为主，伴有石英岩、变粒岩等岩石组合。在变泥质岩中常见的变质矿物有黑云母、白云母和石英，个别见有斜长石。长英矿物多为他形粒状，大小一般0.1~3mm，以0.5~1mm居多，彼此镶嵌。云母呈叶片状，大小0.1~2mm，白云母比黑云母稍大，两者常一起出现。此外在黑云变粒岩、石英岩和个别片岩中见少量铁铝榴石、蓝晶石、十字石变斑晶，常见石英、斜长石小包裹体。泥质岩中的代表性矿物组合有白云母+黑云母+十字石+铁铝榴石+斜长石+石英（王锡银等，1979）、白云母+黑云母+蓝晶石+斜长石+石英及白云母+黑云母+矽线石+斜长石+石英，分别相当于十字石带、蓝晶石带和矽线石-白云母带，属于中压相系矿物组合特征。由于出露不完整，野外难以实际划分变质带。该变质相见于峨文岭组，零星分布于袁水、乘坡-吊罗山、毛阳-番阳、冲卒岭及大蟹岭等地区。

3. 高角闪岩相

原岩多为中酸性火山岩和杂砂岩，变质后形成以斜长片麻岩和二长片麻岩为主，夹有黑云钾长片麻岩、黑云斜长变粒岩和黑云斜长浅粒岩的岩石组合。高角闪岩相伴有广泛混合岩化，其变质长英质岩类中不含原生白云母，出现透辉石。代表性矿物组合为角闪石+铁铝榴石+斜长石（中长石-拉长石）+石英（基性岩）及黑云母+铁铝榴石+斜长石（中长石）+钾长石+石英（中酸性火山岩），相当于矽线石-钾长石带。见于戈枕村组，分布于冲卒岭、大蟹岭及长征等地区。

4. 麻粒岩相

代表性矿物组合为黑云母+斜长石（中长石-拉长石）+紫苏辉石+透辉石+石英（基性岩）。相当于二辉石带。见于戈枕村组，分布于长征地区。

海南岛前寒武纪麻粒岩中不含石榴子石，可能代表了一种相对低压的形成环境，根据野外观察和镜下鉴定结果，总结出海南岛前寒武纪主要岩石类型的矿物共生组合如表14-4所示，同时利用岩石化学分析数据进行了矿物共生分析（图14-8），二者结果基本一致。值得注意的是，在图14-8c中明显可以看出矿物共生组合依次由云母片岩、变粒岩、钙硅酸盐粒岩到斜长角闪岩的变化，表明低角闪岩相的矿物共生组合主要受原岩总成分的控制；在图14-8d中，矿物共生组合随原岩成分的变化趋于复杂，但斜长角闪岩未出现在含石榴子石的组合中，这与该类岩石中石榴子石少见相一致；在图14-8中，紫苏花岗岩与麻粒岩具有相同的矿物组合，但后者明显偏向暗色矿物组分（F）一侧，仍然保留了受原岩成分控制的印记。

表 14-4 海南岛前寒武纪变质岩的主要矿物组合及变质相

Table 14-4 Main mineral assemblages and metamorphic facies of Pre-Cambrian metamorphic rocks, Hainan Island

岩石类型	绿片岩相	低角闪岩相	高角闪岩相	麻粒岩相
板岩	Ser+Qtz+Chl Ser+Qtz			
千枚岩	Ser+Qtz+Chl Ser（Mu）+Bt+Qtz Ser（Mu）+Qtz			
片岩	Ser（Mus）+Q±Chl Act+Chl+Ep+Q Hb+Ep+Pl+Bt+Q Mu+Bt+Q±Gar±Pl	Mu+Bt+Qtz±Pl Mu+Bt+And+St+ Gar+Pl+Qtz Mu+Bt+Pl+Ky+Qtz Hbl+Bt+Sill+Pl+Qtz		
片麻岩			Bt+Pl+Qtz±Kfs Hbl+Pl+Qtz±Gat Bt+Pl+Qtz+Kfs+Gat±Hb	Di+Pl+Bt+Qtz±Hbl Pl+Opx+Bt+Qtz Pl+Opx+Di+Bt+Qtz
变粒岩		Bt+Pl+Qtz Gar+Bt+Pl+Kfs+Qtz Hb+Bt+Pl+Qtz Pl+Kfs+Qtz		
石英岩	Ser+Mu+Q Ser+Ab+Ep+Q	Mu+Bt+Pl+Qtz±Grp Bi+Pl+Kfs+Qtz±Grp		
斜长角闪岩		Pl+Hb+Bt+Qtz±Epi Pl+Hb+Bt+Qtz+Gat Pl+Hb+Qtz+Epi±Bt		Pl+Hb+Kfs+Qtz Pl+Hb+Kf+Qtz+Opx±Di
大理岩		Cal+Qtz±Mus Cal+Qtz+Di+Tr Cal+Qtz+Pl±Gat		
钙硅酸盐粒岩		Di+Gat+Qtz±Pl Di+Qtz±Tr Di+Wol+Qtz		
麻粒岩				Pl+Opx+Bt+Qtz Pl+Opx+Bt+Qtz±Di Pl+Opx+Bt+Di+Qtz±Hb

注：Act—阳起石；Ab—钠长石；And—红柱石；Bt—黑云母；Grp—石墨；Cal—方解石；Chl—绿泥石；Di—透辉石；Ep—绿帘石；Gat—石榴子石；Hbl—普通角闪石；Kfs—钾长石；Ky—蓝晶石；Mc—微斜长石；Ms—白云母；Opx—斜方辉石；Pl—斜长石；Qtz—石英；Ser—绢云母；St—十字石；Tr—透闪石；Wol—硅灰石。

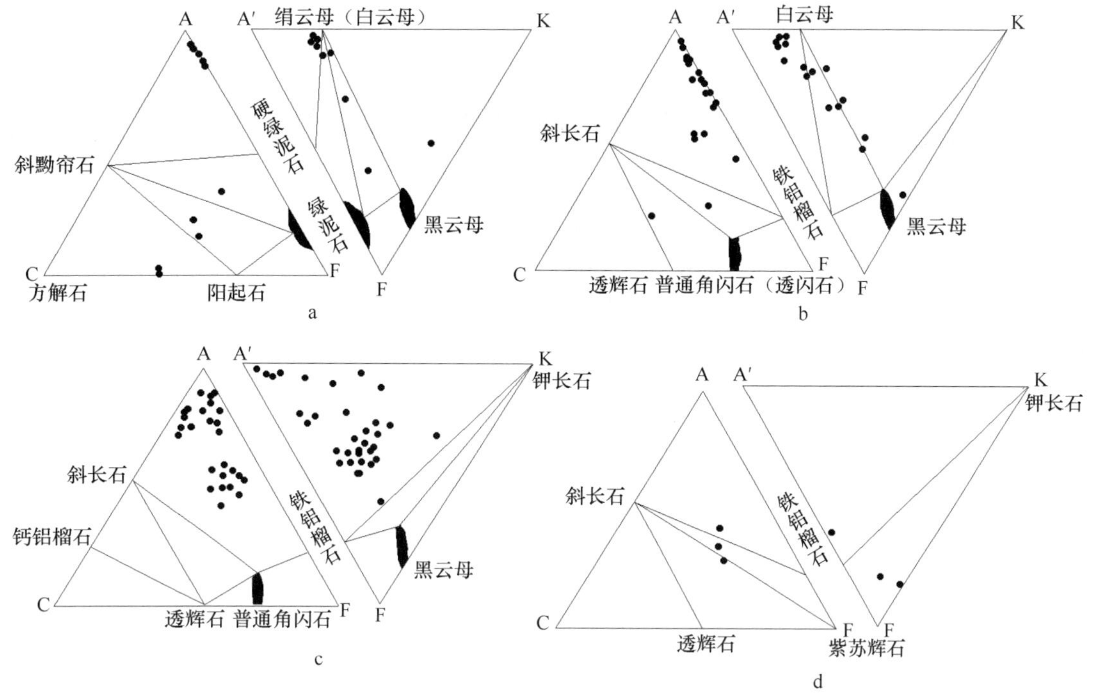

图 14-8　海南岛前寒武纪变质岩变质相 ACF 和 A′KF 图解

Fig. 14-8　ACF and A′KF diagram of metamorphic facies of Pre-Cambrian metamorphic rocks, Hainan Island

a—绿片岩相；b—低角闪岩相；c—高角闪岩相；d—麻粒岩相

第十五章　加里东期变质岩与变质作用

海南岛加里东期变质岩分布广泛，主要出露于海南岛西部儋州市一带（图 15-1）。包括下古生界寒武系美子林组、奥陶系南碧沟组，以及志留系陀烈组、空列村组、大干村组、靠亲山组和足赛岭组，早古生代岩石均遭受不同程度、不同期次的区域变质作用，为一套低—中压相系的低级变质作用，形成浅变质岩系。

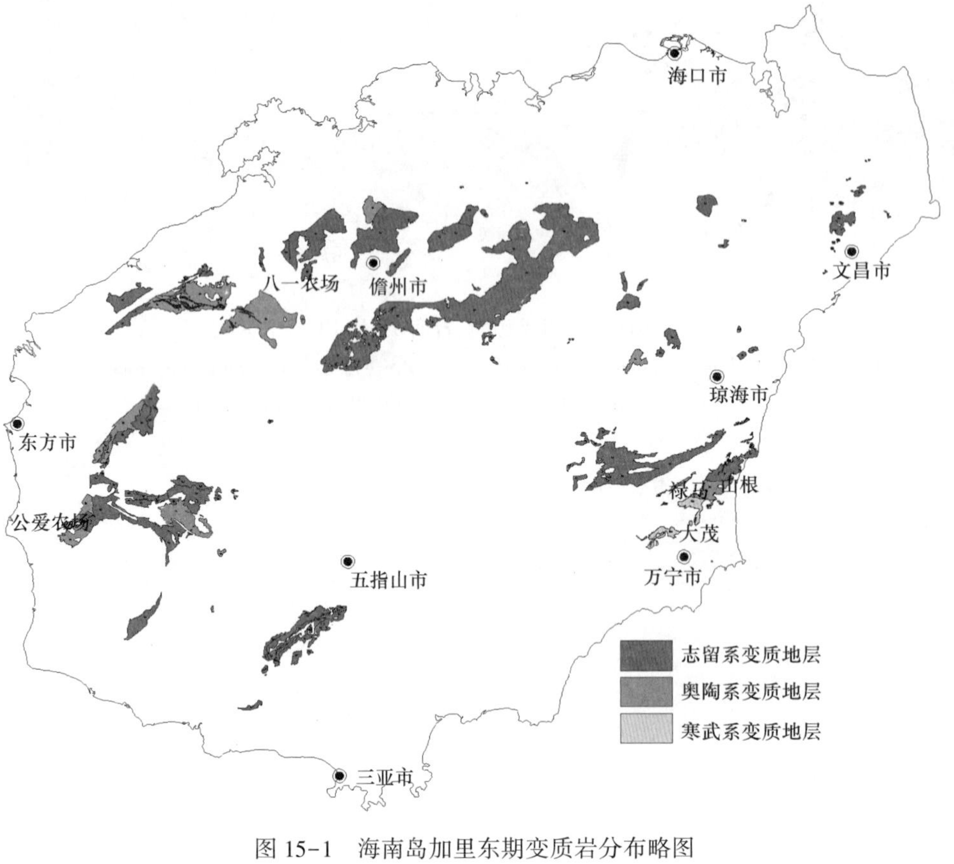

图 15-1　海南岛加里东期变质岩分布略图

Fig. 15-1　Sketch map of distribution of Caledonian metamorphic rocks in the Hainan Island

第一节　岩石类型、岩相特征与原岩建造

一、变质岩石的自然共生组合及原岩建造

1. 寒武系

仅有滇东统美子林组。分布于海南岛东部万宁市大茂-禄马一带，自下而上分为三个岩性段，一段为滨海—浅海相碎屑岩建造，岩石已变质成为黑云母石英片岩，绢云母化黑云石英片岩等。但原岩特征仍清晰可辨的岩石亦大量存在，如变石英砂岩、变长石石英砂岩等。二段为深灰色薄—中厚层状、条带状透辉石石英角岩、长英角岩等，原岩的粉砂岩、砂岩等，沉积物分选好，几乎不含泥质杂基，为滨海沉积产物。三段为结晶灰岩，砂质结晶灰岩、透闪石透辉石大理岩，间夹变质粉砂岩、变质细砂岩。

2. 奥陶系

仅有南碧沟组。分布广泛，主要出露于海南岛西部东方市公爱农场至儋州市八一农场一带，其次为海南岛东部万宁市山根-禄马，主要为一套（滨）浅海—半深海沉积的泥质粉砂岩、粉砂岩、粉细砂岩夹变质基性火山熔岩、变质基性火山碎屑岩，已变质为板岩、千枚岩、片岩、变质粉砂岩、细粒石英岩、阳气石斜长片岩、斜长阳气石片岩，局部岩石中含炭质。变余层理发育，泥质岩与砂岩具韵律性互层产出，形成较发育的水平层理，纹层厚3~5mm，以深浅不同的颜色及粗细颗粒条带显示。变形构造发育，在千枚岩及片岩中层理多被片理置换，常见劈理膝折、小揉皱等构造。变质粉细砂岩、石英岩中石英颗粒虽经重结晶作用，但局部还可保留其磨圆的砂屑外形。

3. 志留系

包括陀烈组、空列村组、大干村组、靠亲山组和足赛岭组。

陀烈组是一套半深海—深海环境沉积的碎屑岩，岩性为粉砂岩、页岩、硅质岩、夹炭质硅质岩，已变质为板岩、千枚岩、变质粉砂岩类，部分板岩、千枚岩、变质粉砂岩含炭。变形构造发育，在千枚岩及片岩中层理多被片理转换，常见劈理膝折、小揉皱、石香肠等，小石英脉受后期构造作用形成肠状弯曲，有些形成假砾石。

空列村组为一套滨浅海沉积环境沉积的碎屑岩，岩性为一套浅变质细碎屑岩夹碳酸盐岩，底部为石英砂岩、含砾中粗粒石英砂岩，已变质为石英岩，绿帘云母石英岩、变质石英砂岩，底部夹炭质板岩。

大干村组主要分布于海南岛乐东县、三亚市和保亭县交接一带和邦溪农场长岭、大岭、白石山和河叉岭等地，属滨海—浅海沉积环境，以青灰色薄层板岩、粉砂质板岩为主；顶部为灰色中厚层结晶灰岩、底部为灰色—灰黄色复成分砾岩，具粒序层理。

靠亲山组主要分布于海南岛乐东县、三亚市和保亭县交接一带，为浅海沉积环境，以粉细砂岩、泥质岩为主体。上部灰黑色—深灰色薄—厚层泥灰岩、含生屑结晶灰岩。中部浅灰色—灰黄色薄层粉砂质板岩、千枚状板岩偶夹粉砂岩。下部粉砂质千枚岩、板岩夹灰色厚层石英岩、绢云石英粉砂岩、细砂岩。

足赛岭组主要分布于海南岛乐东县、三亚市和保亭县交接一带，属浅海沉积环境，以泥质、含炭泥质岩为主体。共分为4段，第四段岩性为灰色—灰黄色薄层粉砂质板岩、粉砂岩偶夹或间夹浅灰黄色薄—中层中细粒石英砂岩，板岩中产腕足类、三叶虫、海百合茎化石。三段岩性为灰色薄层千枚岩、绢云母石英粉砂岩，偶夹浅灰色中细粒石英砂岩；底部为灰黑色含炭粉砂质板岩。二段灰白色厚层结晶灰岩，含白云质灰岩。一段为灰色—灰黄色薄层千枚状板岩、粉砂质板岩，偶夹薄层中细粒石英砂岩，纹层状构造局部发育。

二、岩石类型及岩性特征

根据变质岩的岩石结构、构造特征及其矿物组成、含量，大致将海南岛加里东期区域变质岩划分为如下10种岩石类型。

1. 变质砾岩

岩石具变质粉砂状结构、显微鳞片变晶结构，块状构造，砾石含量为20%~50%，多为次圆状，粒径一般为2~4mm；成分以石英（岩）为主，石英砂岩、粉砂岩次之，少量板岩、泥岩。杂基为泥质、砂及粉砂屑及少量铁质，其中泥质大多已重结晶为绢云母。砂屑成分主要为石英，少量长石和白云母，粒径0.05~0.3mm。砾屑和砂屑多呈飘浮状，少数镶嵌接触，但原碎屑颗粒外形仍可分辨。可见变余粒序层理。

2. 变质砂岩

岩石种类主要有变质绢云母粉砂岩、变质炭质绢云母粉砂岩、变质绢云母粉细砂岩、变质绢云母石英细砂岩等。基本保留原岩的结构、构造特征，呈变余（粉）砂状结构，局部显微鳞片变晶结构，变余微层理构造、块状构造。出现少量新生变质矿物，以绢云母及石英为主，次为少量绿泥石、白云母，局部见雏晶状黑云母。绢云母多呈显微鳞片状集合体均匀随机分布；大部分石英基本保留原岩碎屑特征，部分发生重结晶，具波状消光，颗粒间三点式、齿形紧密接触，局部可见粒径大小不一的石英条带相间平行排列（图版Ⅳ-2-C），反映了原岩的微层理构造。

3. 石英岩

以纯石英岩、绢云母石英岩、阳起石英岩为主，具显微鳞片粒状变晶结构、变余砂状结构，块状构造，主要矿物为石英，次为绢云母、白云母、黝帘石、阳起石、绿泥石等。石英呈他形粒状，大小0.1~0.6mm，呈三点式、多边形接触。

4. 变质泥岩

岩石主要有变质粉砂质泥岩、绢云母泥岩、千枚状含粉砂质泥岩等，岩石具变余泥状结构，显微鳞片变晶结构，千枚状构造、变余微层理构造，岩石的变质作用表现为部分水云母重结晶呈定向排列（图版Ⅳ-2-D），或部分微薄层重结晶变为绢云母，或表现为强变形使岩石的结构构造面貌改变，变质矿物除少量绢云母外，部分岩石可见微量绿泥石、白云母、石英。

5. 板岩

岩石主要有炭质硅质板岩、黏土质板岩、粉砂质板岩、炭质绢云母板岩、炭质钙质板岩等，具变余粉砂状、泥状结构，显微鳞片变晶结构，板状构造、变余微层理构造，变质矿物以绢云母为主，次为石英，局部有少量绿泥石。绢云母多沿片轴平行聚集排列，形成完好的板状劈理，构成岩石的板状构造。

6. 千枚岩

以绢云母石英千枚岩、石英绢云母千枚岩为主，次为绢云母千枚岩、炭质绢云母千枚岩、炭质石英绢云母千枚岩、炭质绢云母石英千枚岩等。具显微鳞片变晶结构，千枚状构造、变余微层理构造，重结晶作用明显。变质矿物有绢云母、石英，局部见绿泥石等。绢云母呈显微鳞片状，多聚成条带状分布，丝绢光泽明显；石英波状消光显著，常聚成条带与绢云母条带相间排列，构成岩石的变余微层理构造；绿泥石具浅绿色—绿色多色性。在陀烈组及南碧沟组中，岩石受后期构造作用，多发生褶皱变形，小褶皱及褶劈理较为发育，其间绢云母多发生弯曲变形，呈皱纹状，顺层石英脉多呈透镜状、香肠状、长条状、无根钩状等。

7. 片岩

主要见于美子林组、南碧沟组和陀烈组中，片岩的种类较多，有绢云母-白云母石英片岩、含长石白云母片岩、石英二云母片岩、石英绢云母-白云母片岩、白云母石英片岩、黑云母石英片岩、透闪石阳起石片岩、角闪石阳起石片岩、绢云母绿泥石透闪石片岩等。呈显微鳞片粒状变晶结构、柱状变晶结构、纤维状变晶结构、柱状鳞片状变晶结构、粒状鳞片变晶结构，片状构造。矿物结晶明显，成分有石英、绢云母、白云母、斜长石、阳起石、角闪石、透闪石、绿泥石、绿帘石、黝帘石、黑云母等，特征变质矿物为硬绿泥石和铁铝榴石。其中，黑云母主要见于南碧沟组中，在陀烈组中零星分布，大部分呈显微鳞片状，晶形不好，呈雏晶状产于绢云母或白云母周边；少部分呈鳞片状—薄片状，具显著的黑褐色—浅红褐色多色性，沿片轴平行定向排列；帘石类矿物往往富集成条带与相对富集的斜长石条带相间排列。特征变质矿物硬绿泥石呈自形柱状，具黄绿色—蓝绿色多色性，铁铝榴石仅见于南碧沟组下部，呈等轴粒状，大小0.05~1.0mm，正高突起，糙面显著。

8. 结晶灰岩

主要见于古生界陀烈组、大干村组、靠亲山组和足赛岭组中，岩石原生结构多已破坏，具花岗变晶结构，局部矿物定向排列，重结晶比较明显，矿物大小0.1~0.5mm，粒间呈镶嵌状，主要矿物为方解石，微量和偶见矿物有石英、白云母、透辉石、透闪石、绢云母和绿泥石。

9. 大理岩

见于古生界陀烈组中，呈透镜状产出，以方解石大理岩为主，次为条带状大理岩、炭质方解石大理岩、含透闪石透辉石大理岩。粒状变晶结构，块状构造、条带状构造，主要矿物为方解石，有时含少量石英、炭质、绢云母、透闪石、白云母等。方解石重结晶明显，颗粒间呈镶嵌状接触。

10. 白云岩

主要见于古生界陀烈组中，岩石呈浅灰色—灰白色，微晶—细粒结构，块状构造，主要矿物有白云石、方解石、石英、绢云母等。

三、区域变质岩的原岩特征

1. 变余结构构造

海南岛加里东期变质岩的沉积结构构造大部分保存完好（表 15-1），甚至局部地段为典型的沉积结构构造，可推断海南岛加里东期变质岩的原岩为沉积岩类，另从其变质岩类型、变质岩矿物组合以及变余结构构造等方面分析，它们的岩类主要应为：泥岩、泥质粉砂岩、粉细砂岩、长石石英岩及石英砂岩等。此外，在奥陶系南碧沟组和志留系陀烈组中有火山岩夹层，其岩性从酸性到基性。

表 15-1 海南岛加里东期变质地层变余结构构造及副矿物特征

Table 15-1 Palimpsest textures-structures and accessory mineral characteristics of Caledonian metamorphic strata, Hainan Island

地层名称	变余构造	变余结构	副矿物特征
足赛岭组	变余叠层状构造变余层状构造	泥微晶结构、粉晶结构、含棘屑微晶及粉晶结构、变余砂状结构	次棱状—半圆状锆石、磷灰石、电气石等
靠亲山组	变余层状构造	变余砂状结构、变余粉砂状结构、含粉砂显微鳞片结构、变余粉砂鳞片变晶结构	棱角状—滚圆状锆石、电气石等
大干村组	层状构造、变余层状构造、交错层理	变余砂状结构、砾状结构、变余粉砂鳞片变晶结构	次棱状—半圆状锆石、磷灰石、电气石等
空列村组	变余层状构造	变余泥状结构、变余粉砂显微鳞片变晶结构、变余粉砂片状变晶结构	磷灰石呈短柱状；锆石呈短柱状；榍石为不规则粒状
陀烈组	变余层状构造 变余微层理构造	变余泥状结构、变余（粉）砂状结构、变余粉砂显微鳞片变晶结构	磷灰石呈浑圆状、短柱状；锆石为浑圆状、短柱状及锥状
南碧沟组	变余层状构造 变余微层理构造	变余粉砂显微鳞片变晶结构、变余粉砂状结构、含变余粉砂片状变晶结构	次棱状—半圆状锆石、磷灰石、电气石、金红石、白钛石等
美子林组	层状构造、变余层状构造、交错层理	变余砂状结构、变余粉砂状结构、含粉砂显微鳞片结构、变余粉砂鳞片变晶结构	次棱状—半圆状锆石、磷灰石、电气石等

2. 岩石化学图解

海南岛加里东期区域变质作用不明显或未受区域变质作用的影响，泥质岩主要由黏土矿物组成，砂质岩砂状结构可辨，保留有变余原生结构构造（图版Ⅳ-2-E、图版Ⅳ-2-F），结合岩石组合及野外产出特征，均可以直接判断其原岩。

本次研究将所收集的化学分析结果（50 个样品）作尼格里参数数值计算（表 15-2）和图解分析，从而进行原岩恢复。其中南碧沟组 29 个、陀烈组分别 20 个、大干村组 1 个，由图 15-2~图 15-4 中可知，绢云母石英千枚岩、绢云母千枚岩、石英绢云母千枚岩、二透石石英角岩、石英角岩、二透岩等主要投入沉积岩区或碳酸岩区以及这两个区的过渡带，而南碧沟组的绿泥石阳起石片岩等主要投入火成岩区，而在图 15-5 中其投影点主要落入细碧岩-角斑岩，说明其原岩有可能是基性火山岩。

综合岩石组合关系、野外产状、变余结构和构造，对海南岛加里东期区域变质岩的原岩进行了恢复，结果见表 15-3。

表 15-2　海南岛加里东期部分变质岩岩石化学成分含量及尼格里参数一览表

Table 15-2　List of petrochemical component contents and Niggli parameters of some of Caledonian metamorphic rocks, Hainan Island

地层单位		大干村组	陀烈组						
岩性		绢云母石英千枚岩	红柱石二云母片岩	含黑云角闪石英片岩	云母石英片岩	云母石英片岩	白云岩		
样品号		TC8	Ⅵ 313	Ⅵ 22-1	243 点	QJ183	重 ZK2-3	重 ZK5-4	重 ZK13
序号		1	2	3	4	5	6	7	8
岩石化学成分/%	SiO_2	63.16	83.66	96.4	69.67	68.09	37.25	9.7	2.9
	TiO_2	0.72	0.5	0.05	0.81	0.95	0.13	0.02	0.02
	Al_2O_3	19.03	6.38	1.64	13.77	16.41	3.27	1.06	0.17
	Fe_2O_3	6.62	3.41	0.93	4.42	5.06	0.93	0.31	0.22
	FeO	0.3	0.15	0.06	1.34	0.58	1	0.2	0.21
	MnO	0.08	0.02		0.108	0.02	0	0.13	0.08
	MgO	0.65	0.64	0.036	2.3	1.83	25.22	17.43	20.68
	CaO	0.33	1.81	0.29	0.36	0.44	19.24	30.98	31.31
	Na_2O	4.16	0.06	0.02	0.43	0.3	0.08	0.22	0.04
	K_2O	0.29	0.84	0.01	3.73	4.01	0.43	3.2	0
	H_2O		2.62				0.13	0.2	0.1
	P_2O_5	0.01	0.2		0.075	0.055	0.01	0.45	0.03
	CO_2		0.16						
	灼失		2.14						
	总量	95.35	102.59	99.436	97.013	97.745	87.69	63.9	55.76
岩石化学常用参数	al	50.85	37.76	45.88	42.12	48.28	3.11	1.03	0.15
	alk	19.12	5.97	1.22	14.52	14.22	0.57	3.6	0.06
	c	1.6	19.48	14.75	2	2.35	33.24	53.05	51.67
	fm	28.43	36.79	38.15	41.36	35.14	63.09	42.35	48.11
	c/fm	0.06	0.53	0.39	0.05	0.07	0.53	1.25	1.07
	si	286.38	840.34	4576.09	361.67	339.97	60.05	15.5	4.47
	(al+fm) − (c+alk)	58.56	49.1	68.06	66.96	66.85	32.39	−13.27	−3.47
	al−alk	31.73	31.79	44.66	27.6	34.06	2.54	−2.57	0.09
	k	0.04	0.9	0.25	0.85	0.9	0.78	0.91	0
	mg	0.15	0.26	0.07	0.43	0.39	0.96	0.98	0.99
	K/%	6.52	93.33	33.339	89.66	93.04	84.31	93.57	0
	A/%	79.92	70.19	83.679	75.29	77.55	14.21	2.99	0.54
采样位置		小尖岭	黄泥沟		红岭				

续表

地层单位		陀烈组								
岩性		白云岩	深灰色白云岩	千枚岩	变质钙质细砂岩	条带状二透岩	纹层状二透岩	石英角岩	透闪石石英角岩	二透长英角岩
样品号		重 ZK9-2	重 ZK12-1	重 ZK2-2	重 ZK2-4	重 ZK2-1	重 ZK1-2	重 ZK8-1	重 ZK1-1	QJ8
序号		9	10	11	12	13	14	15	16	17
岩石化学成分/%	SiO_2	44.51	35.51	69.09	57.52	48.89	46.68	60.7	59.27	56.33
	TiO_2	0.02	0.15	0.76	0.39	0.47	0.38	0.52	0.35	0.52
	Al_2O_3	0	2.08	14.53	7.49	9.01	7	6.34	6.53	4.94
	Fe_2O_3	0.16	2.25	0.19	0.29	0.44	4.2	0.91	0.48	0.54
	FeO	0.26	1.64	6.04	2.96	3.81	4.47	4.31	3.99	3.97
	MnO	0.097	0.044	0.09	0.08	0.11	0.18	0.2	0.12	0.15
	MgO	14.55	35.97	1.84	3.9	13.03	5.4	8.84	8.08	8.57
	CaO	20.16	18.59	0.17	14.46	18.69	17.08	8.11	12.79	12.58
	Na_2O	0.16	0.06	0.35	0.4	0.21	0.16	0.73	0.27	1.37
	K_2O	0.07	0.26	3.04	2.63	1.8	2.2	3.05	2.38	3.04
	H_2O	0.45		0.24	0.04	0.06	0.28		0.22	
	P_2O_5	0	0.09	0.08	0.14	0.11	0.1	0.06	0.07	0.08
	CO_2									
	灼失									
	总量	80.437	96.644	96.42	90.3	96.63	88.13	93.77	94.55	92.09
岩石化学常用参数	al	0	1.57	44.98	14.45	10.68	10.55	11.42	10.94	8.02
	alk	0.45	0.29	11.97	6.76	2.72	3.99	8.11	5.06	9
	c	49.19	25.51	0.96	50.72	40.27	46.82	26.56	38.96	37.15
	fm	50.35	72.64	42.1	28.07	46.33	38.64	53.91	45.04	45.83
	c/fm	0.98	0.35	0.02	1.81	0.87	1.21	0.49	0.86	0.81
	si	101.36	45.47	362.94	188.3	98.33	119.43	185.54	168.49	155.24
	(al+fm) - (c+alk)	0.71	48.41	74.15	-14.96	14.02	-1.62	30.66	11.96	7.7
	al-alk	-0.45	1.28	33.01	7.69	7.96	6.56	3.31	5.88	-0.98
	k	0.22	0.74	0.85	0.81	0.85	0.9	0.73	0.85	0.59
	mg	0.98	0.95	0.34	0.68	0.84	0.53	0.75	0.76	0.77
	K/%	30.43	81.25	89.68	86.80	89.55	93.22	80.69	89.81	68.93
	A/%	0	9.91	80.32	29.98	30.33	26.48	34.78	29.72	22.53
采样位置		红岭								

续表

地层单位		陀烈组				南碧沟组			
岩性		透闪石透辉石岩		千枚状页岩	绢云母石英千枚岩	粉砂质绢云母板岩	变质石英细砂岩		中细粒石英岩
样品号		4	TC10	ZK9	43	Ⅱ139	Ⅱ106	Ⅱ77	Ⅱ151
序号		18	19	20	21	22	23	24	25
岩石化学成分/%	SiO_2	57.07	55.79	60.77	63.33	66.81	69.31	70.71	94.66
	TiO_2	0.31	0.35	1.07	0.72				
	Al_2O_3	2.9	5.19	16.57	18.26	14.96	14.75	13.69	1.59
	Fe_2O_3	0.76	1.85	7.12	7.44	3.48	1.52	2.79	0.38
	FeO	2.33	1.41	1.37	0.55	2.49	4.03	3.33	0.99
	MnO	0.14	0.19	0.11	0.11	0.027	0.073	0.034	0.025
	MgO	12.12	6.55	3.92	1.13	1.22	1.58	1.27	1.03
	CaO	13.47	15.56	0.44	0.33	0.07	0.15	0.3	0.02
	Na_2O	0.83	0.15	0.57	0.15	0.16	0.88	0.4	0.12
	K_2O	2.33	0.69	4.05	3.37	3.88	3.44	3.2	0.1
	H_2O								
	P_2O_5	0.08	0.07	0.118	0.01				
	CO_2								
	灼失								
	总量	92.34	87.8	96.108	95.4	93.097	95.733	95.724	98.915
岩石化学常用参数	al	4.37	9.32	37.83	50.65	48.8	46.16	45.77	24.6
	alk	5.85	1.78	12.15	10.8	14.56	16.18	13.78	4.73
	c	36.87	50.81	1.83	1.66	0.42	0.85	1.82	0.56
	fm	52.91	38.09	48.2	36.89	36.22	36.81	38.62	70.11
	c/fm	0.7	1.33	0.04	0.05	0.01	0.02	0.05	0.01
	si	145.81	170.03	235.43	298.09	369.86	368.06	401.21	2485.11
	(al+fm)-(c+alk)	14.56	-5.18	72.05	75.08	70.04	65.94	68.79	89.42
	al-alk	-1.48	7.54	25.68	39.85	34.24	29.98	31.99	19.87
	k	0.65	0.75	0.82	0.94	0.94	0.72	0.84	0.35
	mg	0.87	0.78	0.47	0.21	0.28	0.34	0.28	0.57
	K/%	73.73	82.14	87.66	95.74	96.04	79.63	88.89	45.45
	A/%	14.85	24.04	76.61	82.59	78.45	76.74	77.83	86.89
采样位置		小尖岭		红岭	小尖岭	东方乡冲金村			

续表

地层单位		南碧沟组							
岩性		细粒石英岩	含炭千枚状石英白云母片岩		千枚状白云母石英片岩	粉砂质绢云母千枚岩		千枚状炭质绢云母板岩	千枚状粉砂岩
样品号		Ⅱ138	H35-2	H35-1	Ⅱ47	Ⅱ72	Ⅱ68	Ⅱ162	0219-6
序号		26	27	28	29	30	31	32	33
岩石化学成分/%	SiO_2	93.45	66.68	71.4	68.67	62.44	70.52	74	65.48
	TiO_2								0.7
	Al_2O_3	3.83	15.15	13.35	15.4	17.18	13.9	11.96	17.75
	Fe_2O_3	0.27	6.85	5.01	3.67	2.79	1.74	2.11	5.96
	FeO	0.17	0.48	0.47	0.59	3.79	3.47	1.77	
	MnO	0.005	0.01	0.01	0.015	0.045	0.039	0.02	0.02
	MgO	0.12	0.91	0.68	0.62	2.74	1.81	1.31	1.3
	CaO	0.05	0.08	0.12	0.11	0.07	0.21	0.11	0
	Na_2O	0.1	0.1	0.12	0.7	0.68	1	0.22	0.04
	K_2O	0.86	4.08	3.9	3.52	4	2.9	3	4.54
	H_2O								
	P_2O_5								0.08
	CO_2								
	灼失								3.94
	总量	98.855	94.34	95.06	93.295	93.735	95.589	94.5	99.81
岩石化学常用参数	al	64.77	47.91	49.84	55.64	44.4	45.06	49.18	52.74
	alk	18.52	14.49	16.5	17.93	14.08	15.51	14.84	14.8
	c	1.54	0.46	0.81	0.72	0.33	1.24	0.82	0
	fm	15.17	37.14	32.85	25.7	41.19	38.19	35.16	32.47
	c/fm	0.1	0.01	0.02	0.03	0.01	0.03	0.02	0
	si	2681.87	357.84	452.33	421.06	273.84	387.94	516.38	330.13
	(al+fm)−(c+alk)	59.88	70.1	65.38	62.69	71.18	66.5	68.68	70.41
	al−alk	46.25	33.42	33.34	37.71	30.32	29.55	34.34	37.94
	k	0.85	0.96	0.96	0.77	0.79	0.66	0.9	0.99
	mg	0.34	0.2	0.2	0.22	0.43	0.39	0.39	0.3
	K/%	89.58	97.61	97.01	83.41	85.47	74.36	93.17	99.13
	A/%	79.13	78.05	76.33	78.05	78.34	77.18	78.22	79.49
采样位置		东方乡冲金村							八一农场

地层单位		南碧沟组							
岩性		绿泥石阳起石片岩					千枚状粉砂岩	绢云母石英千枚岩	
样品号		0662-2	0662-3	02113	0218-7	0219-1	0219-5	0508-1	0532
序号		34	35	36	37	38	39	40	41
岩石化学成分/%	SiO_2	41.76	44.68	43.32	42.84	39.7	62.18	82.42	68.88
	TiO_2	0.56	0.51	0.97	1.07	1.02	0.81	0.39	0.53
	Al_2O_3	11.86	10.43	11.68	8.8	11.71	19.34	7.45	14.37
	Fe_2O_3 / FeO	12.8	10.87	12.31	12.78	12.58	6.73	4.38	7.6
	MnO	0.17	0.17	0.18	0.18	0.18	0.03	0.02	0.03
	MgO	19.91	19.37	18.07	20.64	22.05	1.74	1.09	1.63
	CaO	6.4	8.91	9.23	8.3	6.6	0	0.26	0.08
	Na_2O	0.29	0.82	0.55	0.05	0.05	0.05	0	0
	K_2O	0.01	0.03	0.06	0.02	0.01	4.74	1.63	2.93
	H_2O								
	P_2O_5	0.06	0.03	0.1	0.16	0.13	0.1	0.05	0.03
	CO_2								
	灼失	6.51	4.31	3.41	4.81	5.71	4.23	2.14	3.69
	总量	100.33	100.13	99.88	99.65	99.74	99.95	99.83	99.77
岩石化学常用参数	al	13.04	11.44	12.82	9.48	12.21	51.45	41.24	45.53
	alk	0.54	1.52	1.06	0.11	0.1	13.87	9.77	10.05
	c	12.8	17.78	18.42	16.26	12.51	0	2.62	0.46
	fm	73.63	69.26	67.7	74.14	75.18	34.69	46.38	43.96
	c/fm	0.17	0.26	0.27	0.22	0.17	0	0.06	0.01
	si	77.92	83.19	80.67	78.35	70.24	280.69	774.16	370.38
	(al+fm)-(c+alk)	73.33	61.4	61.04	67.25	74.78	72.27	75.23	78.98
	al-alk	12.5	9.92	11.76	9.37	12.11	37.58	31.47	35.48
	k	0.02	0.02	0.07	0.21	0.12	0.98	1	1
	mg	0.75	0.78	0.74	0.76	0.77	0.34	0.33	0.3
	K/%	3.33	3.53	9.84	28.57	16.67	98.96	100	100
	A/%	63.90	51.66	54.28	51.25	63.75	80.15	79.76	82.68
采样位置		邦溪农场10队		八一农场					东方

续表

地层单位		南碧沟组								
岩性		绿帘石斜长石片岩	含黑云母石英片岩		阳起石石英岩	绿帘石斜长石片岩				
样品号		H9-2	H10-1	H10-2	XⅥ15-1	H7-1	H7-2	H8-1	H8-2	H9-1
序号		42	43	44	45	46	47	48	49	50
岩石化学成分/%	SiO_2	55.34	79.6	76.6	47.44	58.61	58.49	56.02	56.85	56.60
	TiO_2	0.53	0.46	0.54	0.49	0.68	0.69	0.87	0.8	0.57
	Al_2O_3	9.94	8.24	9.35	10.58	15.06	14.24	16.81	13.92	11.53
	Fe_2O_3	0.76	0.32	1.00	2.99	0.56	0.9	0.41	0.56	0.59
	FeO	9.50	2.58	3.02	8.34	4.66	4.86	5.46	6.94	7.38
	MnO	0.48	0.05	0.03	0.56	0.24	0.28	0.25	0.35	0.36
	MgO	3.63	0.83	0.90	2.90	2.63	2.34	2.1	2.29	2.99
	CaO	12.92	0.93	0.94	22.81	8.71	9.47	8.91	10.07	10.7
	Na_2O	3.56	0.88	1.14	1.62	6.27	6.6	5.51	5.73	2.53
	K_2O	1.71	4.57	4.46	0.2	1.01	0.42	1.38	0.58	4.84
	H_2O	1.01	1.04	1.49	1.05	1	1.13	1.37	1.11	0.96
	P_2O_5	0.09	0.16	0.15	0.09	0.14	0.15	0.23	0.21	0.13
	CO_2	0.24	0.1	0.1	0.89	0.14	0.14	0.31	0.24	0.51
	灼失									
	总量	99.71	99.76	99.72	99.96	99.71	99.71	99.63	99.65	99.69
岩石化学常用参数	al	15.18	36.51	36.46	13.44	26.59	24.92	29.23	23.54	19.31
	alk	11.77	28.34	26.14	3.66	20.14	19.8	18.36	17.00	15.75
	c	35.88	7.49	6.66	52.68	27.96	30.13	28.17	30.96	32.59
	fm	37.16	27.66	30.74	30.22	25.30	25.15	24.25	28.51	32.35
	c/fm	0.97	0.27	0.22	1.74	1.11	1.2	1.16	1.09	1.01
	si	143.45	598.57	506.87	102.25	175.63	173.71	165.29	163.11	160.9
	(al+fm) - (c+alk)	4.69	28.34	34.4	-12.68	3.79	0.14	6.95	4.09	3.32
	al-alk	3.41	8.17	10.32	9.78	6.45	5.12	10.87	6.54	3.56
	k	0.24	0.77	0.72	0.08	0.1	0.04	0.14	0.06	0.56
	mg	0.38	0.34	0.29	0.31	0.46	0.41	0.38	0.34	0.39
	K/%	32.45	83.85	79.64	10.99	13.87	5.98	20.03	9.19	65.67
	A/%	35.34	56.36	58.84	30.05	48.5	46.34	51.55	45.94	38.95
采样位置		中瑞农场								

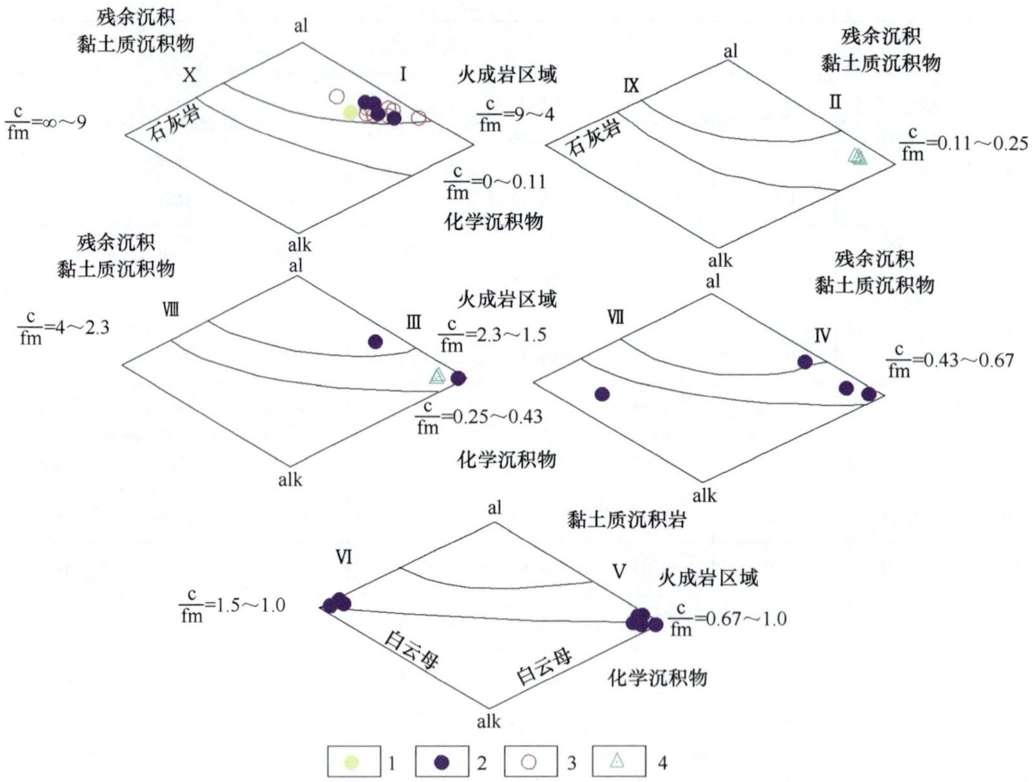

图 15-2 加里东期不同成因类型变质岩 Niggli 四面体图解

Fig. 15-2 Niggli tetrahedron diagram of Caledonian metamorphic rocks of variable origin type

1—大干村组变质岩；2—陀烈组变质岩；3—南碧沟组变质岩；4—南碧沟组绿泥石阳起石片岩

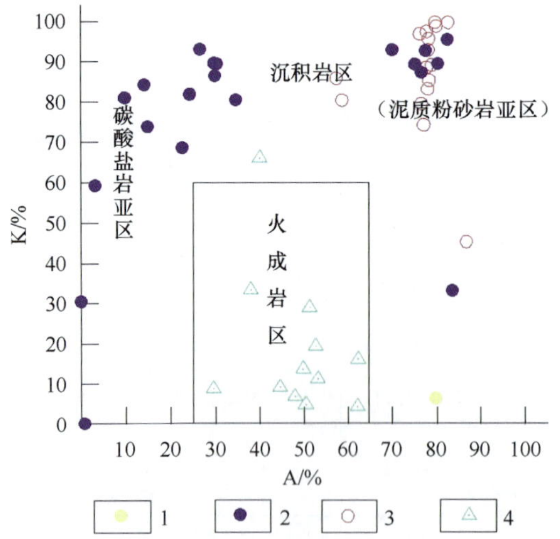

图 15-3 加里东期变质岩的 K-A 相关图解

Fig. 15-3 K-A correlation diagram of Caledonian metamorphic rocks

（据周世泰，1977）

1—大干村组变质岩；2—陀烈组变质岩；3—南碧沟组变质岩；4—南碧沟组绿泥石阳起石片岩

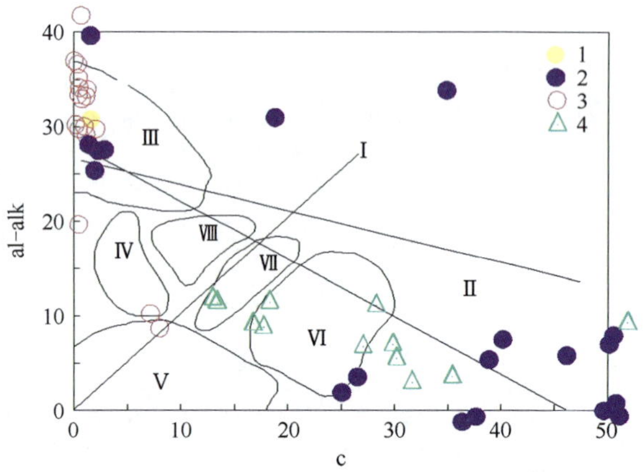

图 15-4 加里东期变质岩的 al-alk 对 c 图解

Fig. 15-4 al-alk vs c diagram of Caledonian metamorphic rocks

（据长春地质学院，1975）

1—大干村组变质岩；2—陀烈组变质岩；3—南碧沟组变质岩；4—南碧沟组绿泥石阳起石片岩

I—钙质泥灰岩区；II—白云质泥灰岩区；III—黏土岩区；IV—中酸性凝灰岩区；V—角斑岩区；斜长角闪岩和麻粒岩；VI—细碧岩-玄武岩区；VII—二长安山质玄武岩区；VIII—英安质凝灰岩区

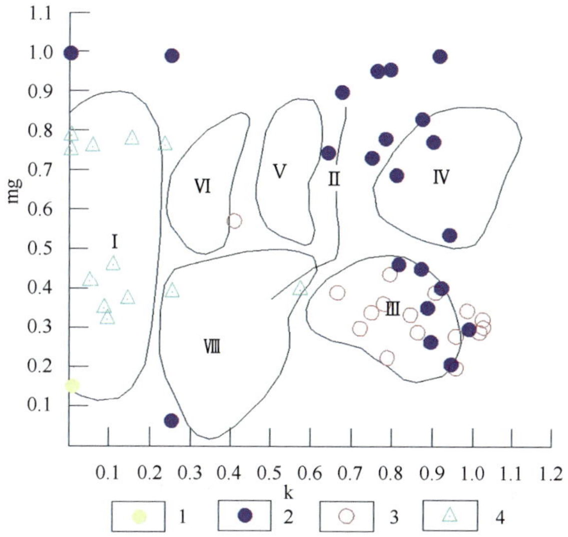

图 15-5 加里东期变质岩的 mg-k 图解

Fig. 15-5 mg-k diagram of Caledonian metamorphic rocks

（据长春地质学院，1975）

1—大干村组变质岩；2—陀烈组变质岩；3—南碧沟组变质岩；4—南碧沟组绿泥石阳起石片岩

Ⅰ—细碧岩-角斑岩；Ⅱ—白云质泥灰岩；Ⅲ—黏土岩；Ⅳ—富钾火山岩；Ⅴ—未定区；Ⅵ—玄武岩区；Ⅷ—英安质凝灰岩

表 15-3 海南岛加里东期区域变质岩主要岩石类型原岩恢复一览表

Table 15-3 List of protolith reconstruction of main Caledonian regional metamorphic rock types, Hainan Island

地层单位	变质岩	原 岩
足赛岭组	变质砂岩类	泥质粉砂岩、细砂岩
	板岩类	含炭泥质岩，粉砂质泥岩
	千枚岩类	泥质粉砂岩、粉砂质泥岩、含炭泥质岩
	结晶灰岩	碳酸盐岩
靠亲山组	石英岩	石英砂岩、长石石英砂岩
	板岩类	含炭泥质岩，粉砂质泥岩
	千枚岩类	泥质粉砂岩、粉砂质泥岩、含炭泥质岩
	变质砂岩类	泥质粉砂岩、细砂岩
	结晶灰岩类	碳酸盐岩
大干村组	变质砂岩类	泥质粉砂岩、细砂岩
	板岩类	泥质岩，粉砂质泥岩
	结晶灰岩类	碳酸盐岩
空烈村组	结晶灰岩类	碳酸盐岩
	变质砂岩	泥质粉砂岩、细砂岩
	板 岩类	泥质岩，粉砂质泥岩
	石英岩	石英砂岩、长石石英砂岩
陀烈组	变质砂岩类	泥质粉砂岩、细砂岩
	板岩类	泥质岩，粉砂质泥岩、泥炭
	千枚岩类	泥质粉砂岩、粉砂质泥岩、泥炭
	大理岩	碳酸盐岩
	石英岩	石英砂岩、长石石英砂岩

续表

地层单位	变质岩	原　岩
南碧沟组	变质砂岩	泥质粉砂岩、细砂岩
	千枚岩类	泥质粉砂岩、粉砂质泥岩、泥质岩、泥炭
	片岩类	泥质粉砂岩、泥质岩、基性-酸性火山岩
	石英岩类	石英细砂岩、酸性火山岩
美子林组	片岩类	泥质岩、泥质粉砂岩
	石英岩类	石英砂岩、长石石英砂岩
	结晶灰岩、大理岩类	碳酸盐岩

第二节　区域变质作用

一、区域变质作用类型

研究不同时期不同地区出现的变质作用，对阐明地壳演化史极为重要。研究一定变质时期形成的变质岩系，包括原岩及原岩建造、变质相及变质相系、混合岩化及构造特征等，是确定变质作用类型的主要依据。根据实际情况，并结合前人的研究成果，海南岛加里东期区域变质作用类型初步划分为区域低温动力变质作用一种类型。

该类型区域变质作用，常形成大面积或带状分布的单一绿片岩相变质地带，是发生于地壳浅部，在强应力和均匀的较低温条件下，变形和重结晶、变质结晶作用的产物。由于遭受强应力作用，岩石中劈理、千枚理及片理较为发育；由于热流变化不显著，变质作用较浅，最高为绿片岩相。根据形成时期的地质构造环境，热流变化程度和形成岩石的矿物共生组合特征，可进一步划分为低绿片岩相型区域低温动力变质作用和绿片岩相型区域低温动力变质作用。

1. 低绿片岩相型区域低温动力变质作用

志留系陀烈组、空列村组、大干村组、靠亲山组和足赛岭组地层遭受的区域变质作用即为此类。该时期地层分布面积较为广泛，其原岩主要为滨海—深海复理石建造。变质后形成的岩石，以变质砂岩、板岩占主要位置，其次为千枚岩。变质泥砂质岩中出现以绢云母+绿泥石+石英为特征的矿物共生组合，未见雏晶黑云母出现。褶皱变形一般，以劈理、千枚理及舒缓褶皱为主，它们往往出现于变质旋回末期，代表该区大陆地壳已趋于稳定，进入结束阶段。

2. 绿片岩相型区域低温动力变质作用

寒武系美子林组、奥陶系南碧沟组地层遭受的区域变质作用即为此类。原岩为浅海—深海复理石夹火山岩建造，火山岩主要分布于南碧沟组中，其所占比例较前者多，具有细碧-角斑岩建造特征，变质后所形成的岩石以千枚岩、片岩及绿片岩为代表。变质泥质岩中出现绢云母、白云母、雏晶黑云母，局部出现鳞片-细小叶片状黑云母及铁铝榴石特征变质矿物；变火山岩中出现阳起石、绿帘石及铁铝榴石等。变质程度相当于绿片岩相黑云母带，局部出现高绿片岩相铁铝榴石带。岩石变形较前者强，以千枚理、矿物拉长线理、石香肠构造、杆状构造及小褶皱为代表。

二、变质相及矿物共生组合

变质岩的矿物组合是在不同温度-压力条件下，岩石组分重组合重结晶的结果，是确定变质相和划分变质相带的依据。海南岛加里东期区域变质岩类型较多，变质矿物共生组合也不尽相同，按矿物组合特征，海南岛加里东期区域变质岩划分为绿片岩相。

该岩相的原岩多属于泥质岩、粉砂岩、砂岩、泥炭岩、碳酸盐岩及中基性火山岩等，经变质作用之后，形成以千枚岩为主，伴有板岩、变质砂岩、石英岩及片岩等岩石组合。镜下常见的变质矿物主要以绢云母（白云母）和石英为主，此外还有少量的绿泥石及偶见雏晶状黑云母和特征变质矿物硬绿泥石。绢云母为显微鳞片状，多沿千枚理定向排列，个别杂乱无章，多为原岩中的泥质物质或长石碎屑物变质而成，变质不彻底之处，仍残留有泥质物质和长石碎屑物。白云母细小鳞片状，常与绢云母伴生。绿泥石为显微鳞片状集合体，常与绢云母共生。硬绿泥石为粒状、细柱状或假六方板状，粒度极小，多色性显著，为黄绿色—浅黄色或近无色，往往具暗色（蓝绿色）中心和浅色（浅黄绿色）边，内常有磁铁矿、金红石等微小包裹体，与片状矿物共生。下面所列的其矿物共生组合与同时利用岩石化学分析数据进行的矿物共生分析（图 15-6），二者结果基本一致。

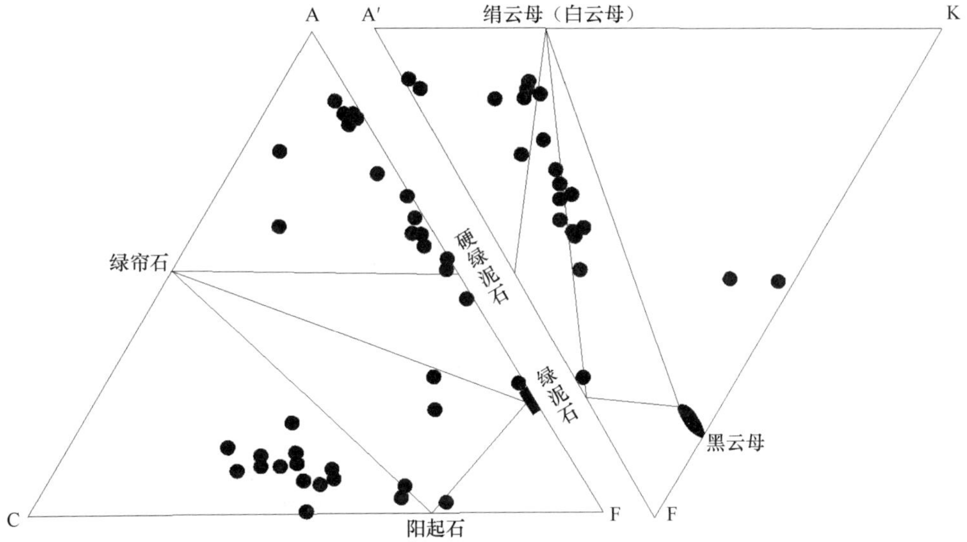

图 15-6　海南岛加里东期变质岩绿片岩相 ACF 和 A′KF 图解

Fig. 15-6　ACF and A′KF diagram of greenschist facies of Caledonian metamorphic rocks, Hainan Island

1. 变泥质岩类

绢云母-白云母+石英±绿泥石±黑云母（雏晶状）

硬绿泥石+绿泥石+绢云母+石英

硬绿泥石+白云母+石英

2. 变粉（细）砂岩类

绢云母-白云母+石英

绢云母+石英

3. 变石英砂岩类

绢云母-白云母+石英±硬绿泥石

绢云母+钠长石+黝帘石+石英

4. 变基性岩

白云母+黑云母+石英+斜长石+石榴子石

斜长石+石英±黑云母

石英+斜长石+石榴子石

第十六章　华力西期变质岩与变质作用

海南岛华力西期变质岩分布广泛，主要出露于海南岛西部（图 16-1）。包括上古生界石炭系南好组、青天峡组和二叠系峨查组、鹅顶组、南龙组，岩石均遭受不同程度的区域变质作用改造为一套低—中压相系的低级变质作用，形成浅变质岩系。

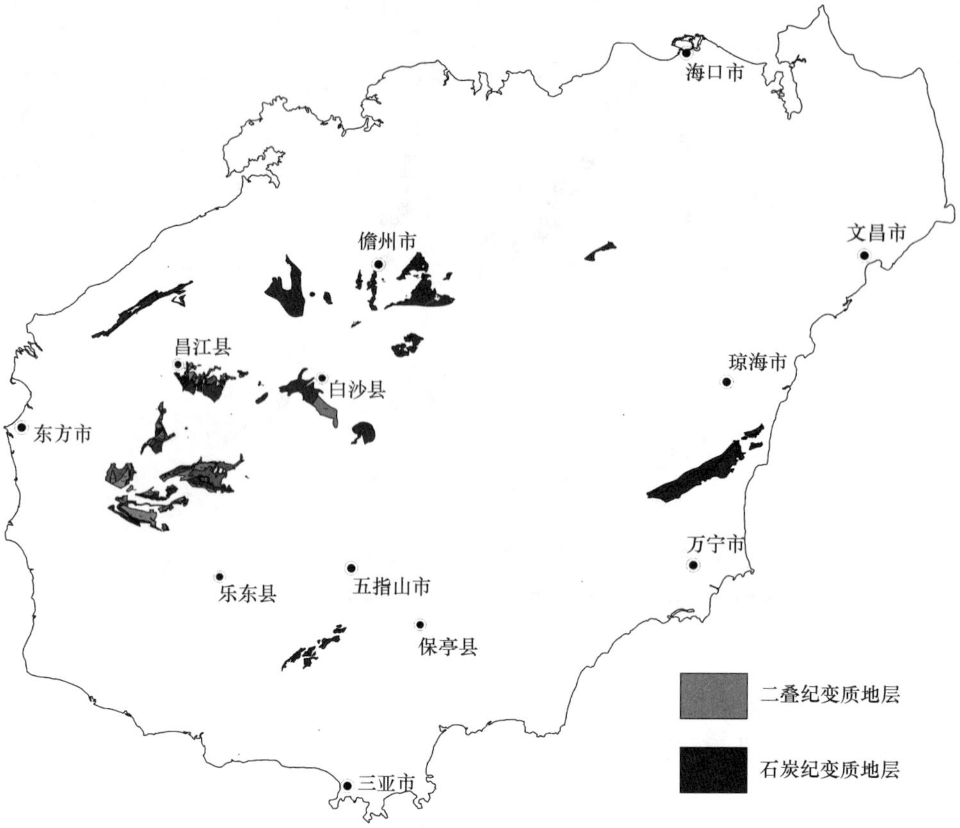

图 16-1　海南岛华力西期变质岩分布略图

Fig. 16-1　Sketch map of distribution of Variscan metamorphic rocks in the Hainan Island

第一节　岩石类型、岩相特征与原岩建造

一、变质岩石的自然共生组合及原岩建造

1. 石炭系

石炭纪变质地层包括南好组和青天峡组。

南好组主要分布于保亭县南好、东方市江边、昌江县石碌、儋州市兰洋、万宁市和乐等地，原岩主要为滨海—浅海砂泥质夹碳酸盐岩建造，底部多为砾岩、含砾砂岩、石英砂岩，下部为灰色—深灰色中厚层细砾岩、岩屑石英砂岩、石英砂岩、细砂岩夹灰泥岩、灰岩透镜体，上部为深灰色薄—中层粉砂岩、泥质粉砂岩和灰岩。岩石已变质成为变质砂质砾岩、变质砂岩、板岩、灰绿色硬绿泥石绢云千枚岩、千枚岩、结晶灰岩、灰白色-深灰色千枚状板岩、千枚岩、变质粉砂岩。

青天峡组为一套浅海相碎屑岩夹碳酸盐岩沉积，以远岸带的泥质岩、细砂岩夹碳酸盐岩沉积类型最为发育，底部多为石英质砾岩、含泥砾砂砾岩、中粒石英砂岩，岩层为中薄层状，偶见水平层理。岩石已变质成为变质粉砂岩、千枚岩、板岩及结晶灰岩透镜体，但原岩特征仍清晰可辨的岩石亦大量存在。

2. 二叠系

二叠纪变质地层包括峨查组、鹅顶组、南龙组。

峨查组主要分布于海南岛西部的东方市、乐东县，此外零星见于白沙县、昌江县及儋州市，原岩主要为滨海—浅海砂泥质夹碳酸盐岩建造，底部多为砾岩、含砾砂岩、石英砂岩，局部夹碳酸盐岩、含硅质条带、燧石团块。下部岩性以灰色中厚层中—细粒长石岩屑砂岩、细粒岩屑砂岩。中上部沉积物主要由砂、泥质组成，呈叠复交替出现，局部夹含生物屑微晶灰岩透镜体。岩石已变质成为变质粉砂岩、板岩、灰白色大理岩透镜体，但原岩特征仍清晰可辨的岩石亦大量存在。

鹅顶组主要分布于海南岛西部的东方市、昌江县一带，原岩主要为滨海—浅海复理石建造，主要由浅灰色、灰色、深灰色、灰黑色中厚层至块状微晶生物屑灰岩、含燧石条带（或结核、团块）微晶生物屑灰岩、含生物屑微晶灰岩、含燧石微晶灰岩、含白云质生物屑微晶灰岩等组成。岩石已变质成为结晶灰岩，但原岩特征仍清晰可辨的岩石亦大量存在。

南龙组主要分布于海南岛西部的东方市、昌江县一带，原岩主要为一套滨海碎屑岩夹碳酸盐岩和硅质岩的碳酸盐岩陆碎建造，下部为粉砂岩、含炭质粉砂岩、泥质粉砂岩、泥岩、石英砂岩、石英中粗砂砾屑硅质岩，中上部主要为石英粉砂岩、石英细砂岩、泥质粉砂岩及碳酸盐岩。部分岩石变质成为变质砂岩、板岩、千枚岩等，但原岩特征仍清晰可辨。

二、岩石类型及岩性特征

根据变质岩的岩石结构、构造特征及其矿物组成、含量，大致将海南岛华力西期区域变质岩划分为如下 9 种岩石类型。

1. 变质砾岩

岩石具变余砾砂状结构、显微鳞片变晶结构，块状构造，砾石多为次圆状，成分以脉石英、石英岩为主，石英砂岩、粉砂岩次之，少量板岩、泥岩。杂基为泥质、砂及粉砂屑及少量铁质，其中泥质大多已重结晶为绢云母。砂屑成分主要为石英，少量长石和白云母。原碎屑颗粒外形仍可分辨。

2. 变质（粉）砂岩

岩石种类主要有变质绢云母粉砂岩、变质绢云母粉细砂岩、变质绢云母石英细砂岩等。呈变余（粉）砂状结构，局部显微鳞片变晶结构，变余微层理构造、块状构造。出现少量新生变质矿物，以绢云母及石英为主，次为少量绿泥石、白云母。绢云母多呈显微鳞片状集合体均匀随机分布；大部分石英基本保留原岩碎屑特征，部分发生重结晶，具波状消光，颗粒间三点式、齿形紧密接触，局部可见粒径大小不一的石英条带相间平行排列，反映了原岩的微层理构造。

3. 变质泥岩

岩石主要有变质粉砂质泥岩、绢云母泥岩、千枚状含粉砂质泥岩等。岩石具变余泥状结构、显微鳞片变晶结构，千枚状构造、变余微层理构造。岩石的变质作用表现为部分水云母重结晶呈定向排列，或部分微薄层重结晶变为绢云母，或表现为强变形使岩石的结构构造面貌改变。变质矿物除少量绢云母外，部分岩石可见微量绿泥石、白云母、微粒石英。

4. 板岩

岩石主要有黏土质板岩、粉砂质板岩、绢云母板岩、硅质板岩、钙质板岩等。具变余粉砂状、泥状结构，显微鳞片变晶结构，板状构造、变余微层理构造。变质矿物以绢云母为主，次为石英，局部有少量绿泥石。绢云母多沿片轴平行聚集排列，形成完好的板状劈理，构成岩石的板状构造。

5. 千枚岩

岩石主要有绢云母石英千枚岩、石英绢云母千枚岩和绢云母千枚岩等。具显微鳞片变晶结构，千枚状构造、变余微层理构造，重结晶作用明显。变质矿物有绢云母、石英，局部见绿泥石等。绢云母呈显微鳞片状，多聚成条带状分布，丝绢光泽明显；石英波状消光显著，常聚成条带与绢云母条带相间排列，构成岩石的变余微层理构造；绿泥石具浅绿—绿色多色性。

6. 结晶灰岩

主要见于青天峡组、南好组、峨查组及鹅顶组中，岩石原生结构多已被破坏。具花岗变晶结构，局部矿物定向排列，重结晶比较明显。矿物大小0.1~0.5mm，粒间呈镶嵌状，主要矿物为方解石，微量和偶见矿物有石英、白云母、透辉石、透闪石、绢云母和绿泥石。

7. 大理岩

见于青天峡组中，呈透镜状产出。以方解石大理岩为主，次为条带状大理岩、炭质方解石大理岩。粒状变晶结构，块状构造、条带状构造，主要矿物为方解石，有时含少量石英、炭质、绢云母、透闪石、白云母等。方解石重结晶明显，颗粒间呈镶嵌状接触。

8. 白云岩

主要见于青天峡组中，岩石呈浅灰色—灰白色，微晶—细粒结构，块状构造。主要矿物有白云石、方解石、石英、绢云母等。

9. 纤闪石岩

主要分布于军营地区青天峡组的上部，呈切层状产出，可能为基性岩脉变质而成。岩石呈黄绿色，聚纤变晶结构，块状构造，主要矿物成分为纤闪石、斜长石，少量绿帘石和石英，纤闪石为雏晶，较粗者呈束状、帚状、放射状生长，不具定向，斜长石多呈单晶体或由两个晶体组成，绿帘石呈粒状聚集体充填状分布，石英呈微粒状。

三、区域变质岩的原岩特征

（一）变余结构构造

大部分上古生界岩石的沉积结构构造保存完好（表16-1），甚至局部地段为典型的沉积结构构造，可推断上古生界的原岩为沉积岩类。另从其变质岩类型、变质岩矿物组合以及变余结构构造等方面分析，它们的岩类主要应为：泥岩、泥质粉砂岩、粉细砂岩、长石石英砂岩、石英砂岩及砾岩等。此外，在青天峡组中有火山岩夹层，其岩性主要为基性火山岩。

表16-1 海南岛华力西期变质地层变余结构构造及副矿物特征

Table 16-1 Palimpsest textures-structures and accessory mineral characteristics of Variscan metamorphic strata, Hainan Island

地层名称	变余构造	变余结构	副矿物特征
南龙组	变余层状构造	角砾屑结构、变余砾屑结构、变余砂状-粉砂状结构、显微鳞片结构、含棘屑变余粉屑结构	次棱状—半圆状锆石、磷灰石、电气石等
鹅顶组	变余叠层状构造、变余层状构造	泥微晶结构、粉晶结构、含棘屑微晶及粉晶结构、变余砂状结构	次棱状—半圆状锆石、磷灰石、电气石等
鹅查组	变余层状构造	变余砂状结构、变余粉砂状结构、含粉砂显微鳞片结构、变余粉砂鳞片变晶结构	棱角状—滚圆状锆石、电气石等
青天峡组	层状构造、变余层状构造	变余砂屑粉晶结构、泥微晶结构、变余砂状结构	次棱状—半圆状锆石、磷灰石、电气石、金红石等
南好组	层状构造、变余层状构造、冲刷构造、砾石叠瓦状构造、交错层理	变余砂屑粉晶结构、泥微晶结构、含微—混结构、变余粉砂显微鳞片变晶结构、变余砂状结构、砾状结构	次棱状—半圆状锆石、磷灰石、电气石、金红石、白钛石、石榴子石、赤铁矿等

（二）岩石化学图解

海南岛华力西期区域变质作用不明显，泥质岩主要由黏土矿物组成，砂质岩砂状结构可辨，保留有变余原生结构构造，结合岩石组合及野外产出特征，均可以直接判断其原岩。

将所收集的化学分析结果（8个样品）作尼格里参数数值计算（表16-2）和图解分析，从而进行原岩恢复。其中鹅查组变质岩1个、青天峡组变质岩2个，南好组变质岩5个，鹅查组绢云母粉砂岩在图16-2～图16-4中沉积岩（黏土岩）区；青天峡组变质岩2个样品为纤闪石岩，在图16-2、图16-3中投入火山岩区，而在图16-4的投影点分布在细碧岩-玄武岩区和二长安山质玄武岩区，说明其原岩有可能是基性火山岩。

表16-2 海南岛华力西期部分变质岩岩石化学成分含量及尼格里参数一览表

Table 16-2 List of petrochemical component contents and Niggli parameters of some of Variscan metamorphic rocks, Hainan Island

地层单位		鹅查组	青天峡组		南好组				
岩性		绢云母粉砂岩	纤闪石岩		变质石英砂岩	黑云母千枚岩	白云质灰岩	结晶灰岩	白云质灰岩
样品号		0606	X31-1	D1292-1	TC7	Q24	ZK3-R2-4	ZK3-R2-6	ZK3-R2-13
序号		1	2	3	4	5	6	7	8
岩石化学成分/%	SiO_2	89.25	46.65	49.35	93.9	53.48	10.9	8.8	38.67
	TiO_2	0.63	0.7	1.08	0.19	0.72	0.03	0.04	0.09
	Al_2O_3	4.18	15.1	12.44	3.16	11.2	0.02	0.05	0.1
	Fe_2O_3	3.9	6.71	2.81	0.76	5.58	0.13	0.14	0.35
	FeO		4.13	7.82	0.35	1.95	0.56	0.26	0.45
	MnO	0	0.17	0.19	0.02	0.1	0.28	0.06	0.11
	MgO	0.04	9.84	9.68	0.37	8.67	17.4	1.05	11.66
	CaO	0.33	7.65	10.57	0.11	3.26	29.25	48.38	25.75
	Na_2O	0.02	2.22	2.42	0	0.1	0	0.1	0.03
	K_2O	0.69	0.08	0.2	0.72	4.5	0.14	0.22	0.11
	H_2O		6.02	1.92					
	P_2O_5	0.04	0.05	0.1	0.02	0.15	0.59	0.03	0.31
	CO_2								
	灼失	1.18	6.96	2.13					
	总量	100.26	99.32	100.71	99.6	89.71	59.3	59.13	77.63
岩石化学常用参数	al	39.28	20.88	16.52	48.09	20.69	0.02	0.05	0.13
	alk	7.33	5.17	5.57	11.86	9.3	0.15	0.44	0.22
	c	5.64	19.23	25.52	3.04	10.95	53.86	95.92	60.16
	fm	47.75	54.72	52.38	37.01	59.06	45.96	3.59	39.5
	c/fm	0.12	0.35	0.49	0.08	0.19	1.17	26.73	1.52
	si	1423.27	109.47	111.21	2424.85	167.65	18.73	16.28	84.32
	(al+fm)-(c+alk)	74.06	51.2	37.81	70.2	59.5	-8.03	-92.72	-20.75
	al-alk	31.95	15.71	10.95	36.23	11.39	-0.13	-0.39	-0.09
	k	0.96	0.02	0.05	1	0.97	1	0.59	0.71
	mg	0.02	0.63	0.62	0.38	0.69	0,97	0.81	0.96
	K/%	97.18	3.48	7.63	100	97.83	100	68.75	78.57
	A/%	80.08	60.28	48.54	79.20	58.76	0.07	0.10	0.38
采样位置		王下乡	波兰沟	军营岭	小尖岭		大岭		

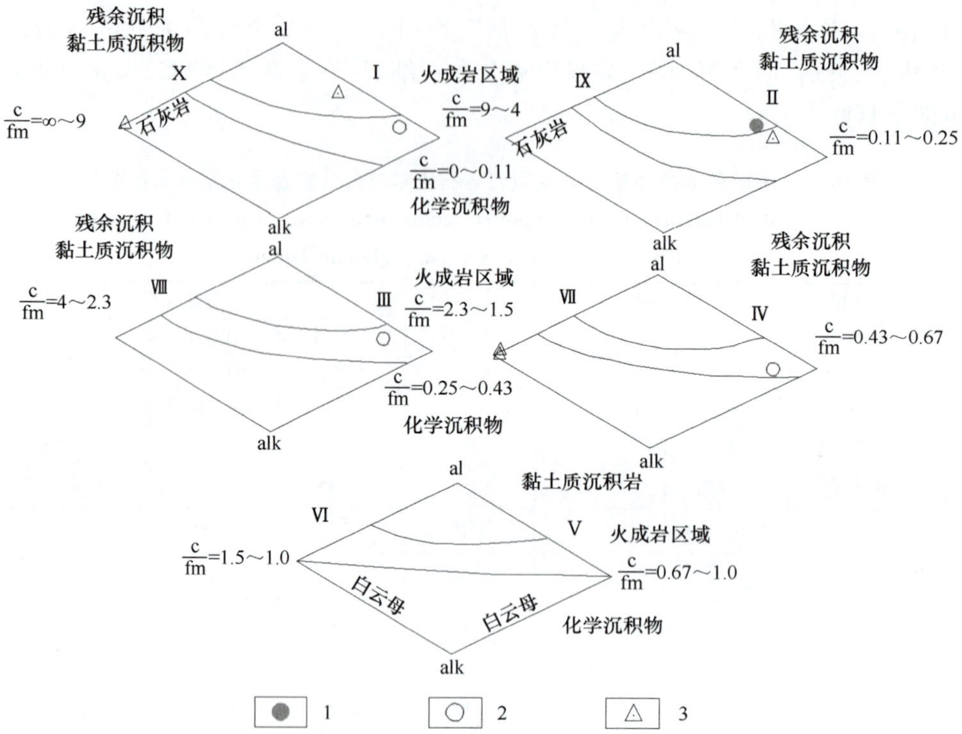

图 16-2 华力西期不同成因类型变质岩 Niggli 四面体图解

Fig. 16-2 Niggli tetrahedron diagram of Variscan metamorphic rocks of variable origin type

1—鹅查组变质岩；2—青天峡变质岩；3—南好组变质岩

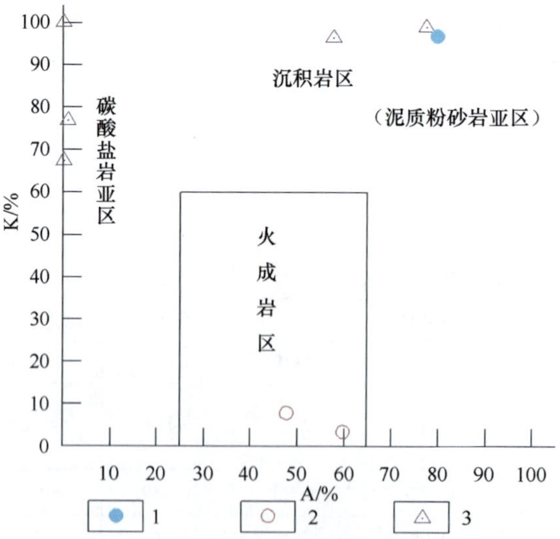

图 16-3 华力西期变质岩的 K-A 相关图解

Fig. 16-3 K-A correlation diagram of Variscan metamorphic rocks

（据周世泰，1977）

1—鹅查组变质岩；2—青天峡变质岩；3—南好组变质岩

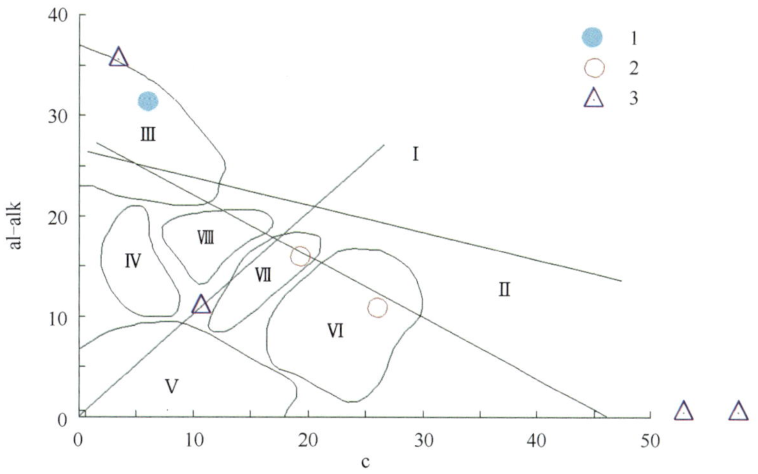

图 16-4　华力西期变质岩的 al-alk 对 c 图解

Fig. 16-4　al-alk vs c diagram of Variscan metamorphic rocks

（据长春地质学院，1975）

1—鹅查组变质岩；2—青天峡变质岩；3—南好组变质岩；

I—钙质泥灰岩区；II—白云质泥灰岩区；III—黏土岩区；IV—中酸性凝灰岩区；V—角斑岩区，斜长角闪岩和麻粒岩；

VI—细碧岩-玄武岩区；VII—二长安山质玄武岩区；VIII—英安质凝灰岩区

第二节　区域变质作用

一、区域变质作用类型

研究不同时期不同地区出现的变质作用，对阐明地壳演化史极为重要。研究一定变质时期形成的变质岩系，包括原岩及原岩建造、变质相及变质相系、混合岩化及构造特征等，是确定变质作用类型的主要依据。根据实际情况，并结合前人的研究成果，海南岛华力西期区域变质作用类型初步划分为区域低温动力变质作用一种类型。

该类型区域变质作用，常形成大面积或带状分布的单一绿片岩相变质地带，是发生于地壳浅部，在强应力和均匀的较低温条件下，变形和重结晶、变质结晶作用的产物。由于热流变化不显著，变质作用较浅，最高为绿片岩相。根据形成时期的地质构造环境、热流变化程度和形成岩石的矿物共生组合特征，确定为低绿片岩相型区域低温动力变质作用。

石炭纪南好组、青天峡组，二叠纪峨查组、鹅顶组、南龙组遭受的区域变质作用即为此类。该时期地层分布面积较为广泛，其原岩主要为滨海—浅海砂泥质夹碳酸盐岩建造。变质后形成的岩石，以变质砂岩、板岩占主要位置，其次为千枚岩、结晶灰岩。变质泥砂质岩中出现以绢云母+绿泥石+石英为特征的矿物共生组合，未见雏晶黑云母出现。褶皱变形一般，以劈理、千枚理及舒缓褶皱为主，它们往往出现于变质旋回末期，代表该区大陆地壳已趋于稳定，进入结束阶段。

二、变质相及矿物共生组合

变质岩的矿物组合是在不同温度-压力条件下，岩石组分重组合重结晶的结果，是确定变质相和划分变质相带的依据。海南岛华力西期区域变质岩类型较多，变质矿物共生组合也不尽相同，按矿物组合特征，海南岛华力西期区域变质岩划分为低绿片岩相。

该岩相的原岩多属于泥质岩、粉砂岩、砂岩、泥炭岩、碳酸盐岩及中基性火山岩等，经变质作用之后，形成以千枚岩为主，伴有板岩、变质砂岩、石英岩等岩石组合。镜下常见的变质矿物主要以绢云母（白云母）和石英为主，此外还有少量的绿泥石及偶见雏晶状黑云母和特征变质矿物硬绿泥石。绢云母为显微鳞片状，多沿千枚理定向排列，个别杂乱无章，多为原岩中的泥质物质或长石碎屑物变质而成，变质不彻底之处，仍残留有泥质物质和长石碎屑物。白云母细小鳞片状，常与绢云母伴生。绿泥石为显微鳞片状集合体常与绢云母共生。硬绿泥石为粒状、细柱状或假六方板状，粒度极小，多色性显著，为黄绿色—浅黄色或近无色，往往具暗色（蓝绿色）中心和浅色（浅黄绿色）边，内常有磁铁矿、金红石等微小包裹体，与片状矿物共生。下面所列的其矿物共生组合与同时利用岩石化学分析数据进行的矿物共生分析（图 16-5），二者结果基本一致。

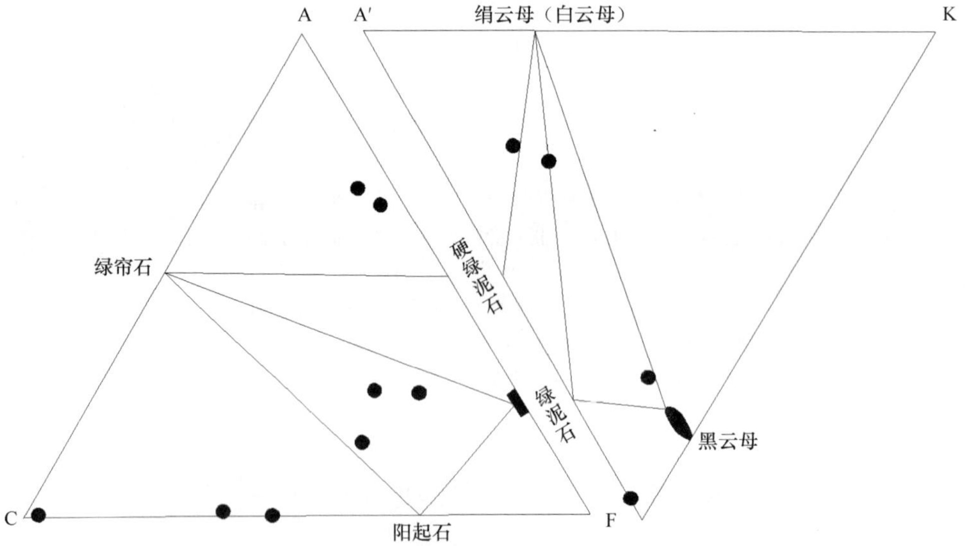

图 16-5　海南岛华力西期变质岩绿片岩相 ACF 和 A′KF 图解

Fig. 16-5　ACF and A′KF diagram of greenschist facies of Variscan metamorphic rocks, Hainan Island

1. 变泥质岩类

绿泥石+绢云母+石英±硬绿泥石

白云母+石英±硬绿泥石

2. 变粉（细）砂岩类

绢云母-白云母+石英

绢云母+石英

3. 变石英砂岩类

绢云母-白云母+石英±硬绿泥石

绢云母+钠长石+黝帘石+石英

4. 变基性岩

阳起石+绿帘石+石英+斜长石

第十七章　混合岩化作用

海南岛混合岩广泛出露于抱板岩群戈枕村组分布区，实际上戈枕村片麻岩大多（约80%）已不同程度混合岩化。在详细研究区内各种变质原岩经受不同程度混合岩化产物的基础上，根据受混合岩化岩石中脉体的含量，将海南岛发生混合岩化的岩石大致划分为弱混合岩化带、中等混合岩化带和强烈混合岩化带。

一、混合岩的岩石类型

1. 混合质片麻岩类

主要有混合质黑云斜长片麻岩、混合质黑云二长片麻岩、混合质含石榴黑云二长片麻岩等。具交代斑状结构（图版Ⅳ-2-G），局部出现交代蠕虫结构、交代碎斑结构，片麻状构造、条带状构造等。脉体含量<15%，成分以眼球状钾长石为主，次为聚集结晶的斜长石和石英，其中石英多呈齿状交生，与其他矿物呈缝合接触。基体为各种黑云斜长质片麻岩，少量的角闪质片麻岩和黑云（斜长）变粒岩。

2. 混合岩类

粒状或鳞片粒状变晶结构，同时可见交代蠕虫结构、交代缝合线结构、交代穿孔结构及交代残留结构等多种交代结构，眼球状、条带（纹）状、片麻状构造。岩性为黑云斜长眼球状混合岩、黑云钾长眼球状混合岩、黑云条带状混合岩、黑云片麻状混合岩等。脉体含量>15%，一般为30%~40%，成分以钾长石为主，次为斜长石和石英，钾长石不再仅以眼球状斑晶为主，也可与斜长石、石英聚集生长，同时发育不同形态（分枝状、补片状、镶边状等）交代条纹，斜长石中常包裹有基体物质，有时可见呈梳状聚集生长的黑云母。基体组分主要为石英、斜长石、钾长石、黑云母，有时含角闪石、白云母及石榴子石等。

3. 混合花岗岩

在空间分布及岩性特征上与其他混合岩呈渐变过渡关系，属原地型混合花岗岩。花岗变晶结构、块状或弱片麻状构造，可见各种交代蠕虫结构、交代碎斑结构（图版Ⅳ-2-H）。主要矿物成分为斜长石40%~54%、钾长石10%~25%、石英20%~35%、黑云母5%~15%、角闪石≤8%。混合花岗岩中仅有少量含云母质岩石的残留体，外观上呈雾迷状（或阴影状）分布，成分以黑云母、斜长石及石英为主，有时可见绿帘石、石榴子石，总量<15%。

二、混合岩化带的划分

由于侵入岩吞蚀、第四系覆盖以及断裂构造的切割剥蚀，海南岛受混合岩化作用形成的岩石出露面积不大，主要产出在戈枕村组中。在详细研究区内各种变质原岩经受不同程度混合岩化产物的基础上，根据受混合岩化岩石中脉体的含量，将海南岛发生混合岩化的岩石大致划分为弱混合岩化带、中等强烈混合岩化带和强混合岩化带。

1. 弱混合岩化带

主要分布于长征南东地区、公爱农场东、新村-冲卒岭及木棉墩-三家村等地。该带岩石类型以眼球状混合质黑云斜长片麻岩和眼球状混合质黑云二长片麻岩为主，岩石以变质基体中含少量长英质脉体（<15%），且基体结构构造没有发生明显变化、二者界线清楚为特征。岩石构造形态以眼球状构造为主，原岩结构基本保留，局部出现交代结构，以钾质交代为主。上述混合质岩石构成海南岛弱混合岩化带。

2. 中等强烈混合岩化带

主要分布于长征南东地区、新安–报告村、屯昌县白银坡等地。该带岩石脉体数量明显增加（>15%），一般为30%~40%，脉体和基体间界线模糊或趋于消失，岩石类型以黑云斜长眼球状混合岩为主，此外还有黑云二长眼球状混合岩、黑云条带状混合岩、黑云条痕状混合岩、黑云钾长眼球状混合岩等。岩石构造形态趋于复杂，以条带状、条痕状、片麻状最为典型，原岩结构遭强烈破坏，各种交代结构发育。该带主体为上述混合岩类，间有少量混合质岩石及混合花岗岩。

3. 强混合岩化带

主要分布于长征北东地区。该带岩石比较均匀，以新生长英质脉体为主，仅有少量的基体物质呈条痕状或阴影状残留。构造形态为片麻状、阴影状、块状，具花岗变晶结构，（石英）缝合线结构及穿孔结构很发育，表明石英交代很强烈，代表混合岩化的晚期阶段。该带以混合花岗岩、混合花岗闪长岩为主，间有少量片麻状混合岩。

三、混合岩成因讨论

对于海南岛混合岩的成因，有两种不同的观点，一是再生成因，如王锡银等（1979）认为，混合岩（混合花岗岩）是片岩类经多阶段交代作用的产物；二是属重熔作用形成，如马大铨等（1993）即认为是片麻岩经深熔作用的结果。我们支持后一种看法，主要证据为：①混合岩中的脉体形态多较规则；②混合岩与片麻岩呈渐变过渡关系，混合岩的条带（条纹）及片麻理与片麻岩的片麻理协调一致；③混合岩的化学成分（包括常量元素、微量元素及稀土元素）和矿物组成有明显继承片麻岩的特点，表明混合岩化在封闭体系下进行，其中较普遍的交代现象并非大量流体的加入而造成，而是熔体与围岩作用的结果；④混合岩具有区域性发育的特点，但又限于高级片麻岩区，意味着其形成与区域高温变质作用有密切联系，具备发生部分重熔（选择性重熔）的温压条件；⑤混合岩中的残留基体或混合花岗岩中的包体，成分相当于片麻岩，但与围岩中的相比，其暗色矿物含量明显增多，说明岩石总体成分虽无大的改变，但有一定程度的分异；⑥混合岩中的锆石多为浑圆状、椭圆状，而其原岩又主要为火山岩，表明其中的原生锆石发生了部分熔蚀。

第十八章　其他变质作用

其他变质作用主要指接触变质作用、动力变质作用、气-液蚀变作用等，这类变质岩一般分布范围不广，除部分接触变质岩略有一定分布规模外，大多呈零星出露。

一、接触变质岩

接触变质作用主要是在岩浆作用的影响下，围岩受岩浆温度的影响而产生的一种变质作用，海南岛花岗岩类岩体分布十分广泛，其围岩均不同程度地遭受接触变质作用形成接触变质带，形成接触变质带的规模与形态取决于侵入体的成分、产状、形态及围岩的岩性、结构构造特征等。接触变质岩按成因分为热接触变质岩和接触交代变质岩。热接触变质岩的出露宽度从几米到数百米，岩石类型主要为角岩和角岩化岩石。此外，早期花岗岩受晚期花岗岩侵入能够观察到的接触变质带，一般仅有数厘米至数十厘米，主要表现为硅化或石英、黑云母等主要造岩矿物的重结晶和蚀变。有时在远离接触带的围岩中也偶见接触变质岩产出，推测其下可能有隐伏岩体的存在。接触交代变质岩主要为矽卡岩，但不发育。

（一）主要岩石类型

1. 热接触变质岩

（1）角岩化岩石类

该类岩石分布广，岩石类型比较多，有角岩化粉砂岩、砂岩、千枚岩、硅质灰岩、流纹岩、英安岩、凝灰岩等。这类岩石基本保留原岩的结构、构造，仅部分出现绿泥石、绢云母-白云母、黑云母、石英、透辉石、透闪石、硅灰石、红柱石等。

（2）斑点状板岩

比较常见，一般分布于热接触变质晕的外带，有红柱石斑点状板岩和绢云母斑点状板岩，岩石呈灰绿色、深灰色，变余粉砂-泥状显微鳞片变晶结构，斑点状构造。矿物成分主要为泥质和粉砂质以及部分重结晶形成的红柱石、绢云母、绿泥石、绿帘石、白云母、黑云母、阳起石、微粒石英等。斑点主要由绢云母组成绢云母斑点状板岩，而主要由红柱石组成红柱石斑点状板岩。

（3）红柱石千枚岩

岩石呈深灰色，显微鳞片变晶结构，斑状变晶结构，千枚状构造。矿物成分主要为绢云母（80%~85%）、红柱石（10%左右）、石英粉砂（3%~6%）及少量白云母、泥质、黑云母、硅质等。变斑晶为红柱石，颗粒大小为3~4mm，岩石经历两期变质作用，前期原岩经过区域变质作用形成千枚岩，后经热变质作用形成红柱石化。

（4）红柱石化石英二云母片岩

岩石具斑状变晶结构，片状构造。主要矿物成分为石英（35%）、白云母（30%）、黑云母（15%）、红柱石（10%）、褐铁矿（10%）、绢云母（3%）。变斑晶为红柱石，颗粒大小为1.2~3.7mm。红柱石变斑晶有推挤先期形成的片理现象，说明岩石经历两期变质作用，前期原岩经过区域变质作用形成石英二云母片岩，后经热变质作用形成红柱石化。

（5）黑云母红柱石角岩

分布于三亚市立才农场、万宁市六连岭，岩石深灰色，斑状变晶结构，基质为角岩结构，显微变晶结构，块状构造。斑晶主要为红柱石（10%±）及少量白云母、黑云母。红柱石粒径1mm左右，个别达4mm，他形，往往包含大量石英、黑云母而呈现包含变晶结构，云母多呈聚合斑晶。基质粒径小于0.4mm，以石英为主，次为黑云母、白云母、钠长石及少量锆石、磷灰石、矽线石。

（6）（云母）长英角岩

分布于保国农场-毛感一带，岩石为灰色—灰绿色，角岩结构，块状构造，部分条带状构造。主要矿物成分为石英（40%~70%）、斜长石（20%~25%）、黑云母（≤15%）、绢云母-白云母（10%~30%），长石石英颗粒大小一般为0.2~0.7mm，呈不规则或等粒状，少量长石具半自形板状，黑云母和白云母呈片状，片幅一般为0.2~0.8mm。

（7）石英角岩

见于琼中县岭头茶场一带，岩石具典型角岩结构，致密块状构造。主要矿物为石英（95%），其次为少量角闪石、绢云母和炭质。石英为细粒状变晶，粒径一般为0.15mm±，少量达0.2~0.3mm，具微弱波状消光，颗粒呈多边形不定向密集分布，构成典型的角岩结构。

（8）（矽线石）黑云母石英角岩

分布于三亚市立才农场、白沙县大岭农场，岩石灰白色，角岩结构，鳞片粒状变晶结构，块状构造。主要矿物成分为石英（45%~70%）、黑云母（30%~40%）、矽线石（≤10%），少量斜长石、红柱石、堇青石，锆石、榍石等微量，石英呈不规则或等粒状，偶有个别具次棱角状，颗粒大小一般为0.05~1.5mm，黑云母为鳞片状聚集，片幅一般为0.1~0.3mm，矽线石为无色毛发状、针状、纤维状集合体，含量多的形成矽线石黑云母石英角岩。

（9）透辉石石英角岩

分布于儋州市西华农场一带、万宁市龙滚，岩石呈灰色—深灰色，细粒状变晶结构（角岩结构），致密块状构造。矿物成分主要为石英（20%~65%）、透辉石（15%~55%）、黑云母（<2%）、斜长石（<1%），局部出现紫苏辉石、透闪石，副矿物有锆石、榍石、磷灰石、电气石等。矿物以粒状、短柱状为主，粒度一般较均匀，颗粒大小一般为0.05~1.0mm，粒状矿物彼此紧密镶嵌，矿物多呈不规则状无定向分布。

（10）透闪石透辉石石英角岩

岩石为灰白色，角岩结构，纤状粒状变晶结构，块状构造。主要矿物成分为石英（45%~65%）、透辉石（20%~40%）、透闪石（10%~30%），少量斜长石、正长石、绿帘石等，副矿物锆石、榍石等微量，石英等轴粒状，颗粒大小一般为0.05~1.0mm，透辉石为无色粒状集合体，颗粒大小一般为0.03~1.0mm，少数为0.2~0.4mm，透闪石呈纤维状集合体，颗粒大小一般为0.3~1.0mm，部分透辉石和透闪石镶嵌分布。

（11）绿帘石石英角岩

分布于儋州市西华农场一带。岩石呈灰色、灰黄色，粒状变晶结构，块状构造，变余微层理构造。主要矿物有石英、绿帘石、黑云母，副矿物有锆石、榍石、磷灰石、黄铁矿、电气石等。石英呈他形粒状，棱角状—次圆状，颗粒大小一般为0.05~0.3mm，呈多边形紧密镶嵌，含量60%~90%，绿帘石呈粒状、柱状，粒径0.1~0.5 mm，含量10%~40%，多呈紧密镶嵌或致密集合产出，黑云母为小叶片状，片幅0.05~0.2mm，呈细条带状及不规则状分布。

（12）黑云母堇青石石英角岩

分布于昌江县乌烈村，岩石呈灰色。矿物成分主要为石英（60%）、堇青石（30%）、黑云母（5%）、白云母（1%），少量绢云母、褐铁矿。具斑状变晶结构，块状构造，变斑晶由堇青石组成，颗粒大小一般为0.3~0.5mm，基质为等粒变晶结构的石英组成，黑云母与堇青石紧密共生并有交代堇青石现象。

（13）含蓝晶石石英角岩

分布于白沙县大岭农场一带，岩石呈灰色。矿物成分主要为石英（81%）、方解石（13%）、蓝晶石（3%）、磁铁矿（2%），及少量或微量黑电气石、锆石、榍石、磷灰石、褐帘石。具粒状变晶结构，块状构造，石英略有拉长定向现象。

(14) 含镁电气石石英角岩

分布于昌江县乌烈村。岩石矿物成分为石英（98%）、镁电气石（1%）、磁铁矿（<1%）、白云母少量。岩石为粒状变晶结构，块状构造，镁电气石无规律分布在石英颗粒间。

(15) 含黑电气石石英角岩

岩石矿物成分为石英（85%）、黑电气石（3%～5%）、黑云母（1%～3%）、炭质（5%～8%）、褐铁矿（4%～5%）；细粒不等粒状变晶结构，块状构造。石英见有明显的不规则状重结晶结构，黑电气石呈自形晶，颗粒大小一般为0.3～0.5mm，零星分布在石英颗粒中，黑云母充填石英粒间。

(16) 含透辉石方柱石石英角岩

见于琼中县岭头茶场一带，具花岗变晶结构，条带状构造。矿物成分为石英（69%）、方柱石（20%）、透辉石（5%）、石榴子石（3%）、绢云母（3%）。石英粒径0.2～1.5mm，呈齿状、弧形紧密接触，常聚成条带与钙硅酸盐矿物条带相间平行排列，构成条带状构造；方柱石呈板状，多聚成条带状集合体；透辉石呈柱状，零星分布在方柱石集合体中。

(17) 含透辉石角闪石石英角岩

分布于乐东县抱伦农场，岩石呈灰黑色，细粒状变晶结构（角岩结构），致密块状构造。矿物成分主要为石英（70%～75%）、角闪石（20%～25%），次为透辉石（3%）、黑云母（≤2%）磁铁矿（2%），斜长石、锆石、磷灰石少量。角闪石呈绿色短柱状，少量针状、纤维状，多色性明显（绿色—棕色），柱体小于0.02mm，大小一致，分布均匀；透辉石为无色细微粒状集合体，为断续条带状，常与针状、纤维状角闪石共生，分布不均匀。

(18) 透辉角闪石英角岩

分布于乐东县抱伦农场，岩石呈灰色，角岩结构，致密块状构造。矿物成分主要为石英（62%）、角闪石（18%）、透辉石（13%），次为斜长石（5%）、磁铁矿（2%），榍石、磷灰石少量。角闪石为短柱状、针状和纤维状，常聚集成集合体；透辉石为细微粒状集合体，分布不均匀，呈断续条带状、团粒状分布，常与针状、纤维状角闪石共生，有时被角闪石包围。

(19) 透闪绿帘长石石英角岩

分布于乐东县抱伦农场，岩石呈深灰色，纤状显微花岗变晶结构，平行构造。主要矿物为石英（49%）、绿帘石（15%）、透闪石（15%）、斜长石（10%）、透辉石（10%），次为绿泥石（1%），磁铁矿少量。

(20) 透辉石长英角岩

分布于乐东县抱伦农场，岩石为灰绿色，显微花岗变晶结构，块状构造。主要矿物成分为石英（54%）、斜长石（20%）、透辉石（25%），次为绿帘石（1%），黑云母、榍石少量。透辉石呈细微粒状集合体，分布极不均匀，呈断续条带状、团粒状分布。

(21) 含白云母绢云母石英角岩

分布于乐东县抱伦农场、万宁市东岭农场，岩石为灰色，鳞片粒状变晶结构，块状构造。主要矿物为石英（50%～80%）、绢云母-白云母（15%～40%）、正长石（≤7%），少量黑云母、绿泥石、红柱石等。石英呈他形近等轴粒状变晶，大致可分为粗、细两部分，细粒石英多自成集合体，而粗粒者常被细粒包围。正长石、绢云母、白云母、黑云母和绿泥石充填在石英颗粒之间。黑云母呈细叶片状，多退变为绿泥石，少量转成白云母。

(22) 含矽线石榴透辉石英角岩

分布于乐东县抱伦农场，岩石为灰色，细粒状变晶结构，致密块状构造。矿物成分主要为石英（71%）、透辉石（20%），次为石榴子石（8%）、矽线石（1%），磷灰石少量。石英呈他形变晶，略具定向排列，总体沿长轴分布；透辉石为不规则粒状集合体，具弱多色性（浅绿色—浅褐绿色），常包含微粒石英；石榴子石为无色致密粒状集合体，集合体连生筛状或网格状，中间常见细小石英、透辉石包体，石榴子石分布不均匀，局部集中生长；矽线石为无色毛发状、针状、纤维状集合体，常与石榴子石伴生，少量与透辉石伴生。

（23）石英透辉石角岩

分布于乐东县抱伦农场、万宁市龙滚，岩石为灰白色，柱粒状变晶结构，致密块状构造。主要矿物为石英（44%）、透辉石（50%），次为方解石（4%），斜长石、正长石、紫苏辉石、磷灰石、榍石、锆石等少量。透辉石为无色粒状集合体；方解石为辉石蚀变产物，常交代细粒的透辉石；紫苏辉石呈淡绿色，多色性明显，夹杂分布于透辉石之间。

（24）黑云母角岩

见于琼中县岭头茶场一带，具显微鳞片变晶结构，致密块状构造。矿物成分主要有黑云母（85%）、石英（15%），局部见少量炭质。黑云母呈显微鳞片状，具红褐色—浅棕红色多色性，呈不规则状分布；石英粒径0.02~0.05mm，粒间呈多边形、齿状紧密接触。岩石中未见变斑晶出现，说明岩石可能由铁镁质较高的黏土受热液作用而形成。

（25）含透辉石大理岩

岩石为灰色—深灰色，均粒状变晶结构，块状构造。主要矿物为方解石（80%~90%）、透辉石（5%~12%）、透闪石（<2%），少量石英、白云母等，方解石呈粒状，颗粒大小为0.3~2.3mm，常紧密镶嵌构成花岗变晶结构，透辉石、透闪石呈粒状，颗粒大小为0.1~1.0mm，局部集中分布于方解石颗粒之间。石英呈细小粒状，白云母呈细小片状，两者常与透辉石连在一起或分布于透辉石附近。

（26）透辉石角岩

分布于昌江县石碌、屯昌县羊角岭。岩石矿物成分主要为透辉石（70%）、透闪石（10%）、斜长石（10%）、斜黝帘石（3%）、石英（3%~5%），少量榍石、电气石及绢云母等。岩石具条带状构造，他形柱粒状变晶结构及斑状变晶结构。

（27）钾长石透辉石角岩

主要分布鹿回头一带，岩石呈灰白微带淡绿色，微粒变晶结构，平行条纹构造。矿物成分为透辉石（50%~85%）、钾长石（10%~45%），少量斜长石、硅灰石、绿帘石、榍石、磷灰石、金属矿物，透辉石和钾长石的粒径一般在0.35 mm左右，少量为0.05~0.12mm，透辉石呈柱状，钾长石多呈细小粒状，常以微细之条纹平行分布。

（28）石英水镁石透辉石角岩

分布于白沙县邦溪，岩石为灰白色，细粒变晶结构，块状构造。矿物成分为透辉石（45%）、水镁石（30%）、石英（20%）、方解石（4%）、榍石（1%）；柱状矿物均匀无序分布构成块状构造。

（29）含透辉石黝帘石绢云角岩

主要分布鹿回头一带，岩石呈灰白色、青灰色，微粒—鳞片变晶结构，条带状构造。矿物成分为绢云母（45%）、黝帘石（30%）、透辉石（30%）、钾长石（5%~10%），含少量阳起石、石英及微量绿帘石、榍石等；矿物颗径细小，一般为0.01~0.025mm，很少达0.05mm；绢云母集合体与黝帘石、透辉石、钾长石集合体互成有规则的平行条带。

（30）含绿泥石透辉石角岩

分布于昌江县石碌。岩石矿物成分为透辉石（60%）、绿泥石（13%）、方解石（7%）、石英（15%）、黄铁矿（4%）、萤石（1%）；强烈透辉石化，但原岩结构仍部分保留，柱粒变晶结构，块状构造。

（31）透辉石阳起石角岩

分布于昌江县石碌。岩石主要矿物成分为透闪石（20%）、阳起石（及透辉石）（30%）、钾长石（30%）、石英（10%）、斜黝帘石（10%）、少量方解石、榍石等，锆石、电气石等微量。岩石具条带状构造和斑杂状构造，条带构造由透辉石-石英-钾长石组合，斜黝帘石-榍石，阳起石-石英-钾长石组合呈条带展布所构成，在透辉石-石英-钾长石条带中以粒状变晶镶嵌结构为主，同时也见筛状变晶结构，透辉石变斑晶可达3mm。

(32) 阳起石方柱石角岩

分布于昌江县石碌。岩石主要矿物成分为方柱石（68%）、阳起石（25%）、石英（4%）、炭质（2%）及少量黄铁矿、褐铁矿。柱粒状变晶结构，方柱石无序分布，镜下见层状残余构造和缝合线构造。

(33) 矽线石角岩

分布东方市乙洞村附近，岩石呈深灰色，显微鳞片花岗变晶结构，局部尚见变余层理构造。主要矿物有石英、斜长石、黑云母、红柱石、矽线石，偶见少量白云母，矽线石往往呈毛发状晶束，与黑云母相伴分布。

(34) 堇青石角岩

分布于昌江县王下村，岩石呈青灰色，显微鳞片花岗变晶结构，定向构造、斑点状构造。矿物组合为黑云母、堇青石、红柱石、绢云母、石英。岩石经多次变形变质作用叠加，堇青石往往被压扁、定向排列。

(35) 钙硅角岩

分布于昌江县王下村，岩石呈灰色—深灰色，显微鳞片花岗变晶结构，变余条带构造。原岩中的条带仍保留，组成各条带的矿物成分也有一定差异，钙质条带中以石英、方解石为主，方解石晶粒往往较石英粗，硅质条带中的矿物成分复杂些；主要矿物有透闪石、方柱石、透辉石，偶尔有符山石、石墨。

(36) 绿泥石长英角岩

三亚市立才农场，灰白色，显微鳞片变晶结构，块状构造。主要矿物成分为石英（90%）、钠长石（8%）、绿泥石（2%）及微量的黑云母、磷灰石、锆石等，石英颗粒大小为0.3~0.6mm，个别达1.5mm，彼此镶嵌，钠长石大小为0.1~0.5mm，多数呈半自形充填于石英粒间，个别自形，少数可见钠长石双晶，绿泥石由含铁镁质胶结物变质而来。

(37) 角闪石角岩

见于琼中县岭头茶场一带，岩石具粒状变晶结构，致密块状构造。矿物成分为角闪石（65%）、石英（30%）、黑云母（5%）及少量磁铁矿、榍石等副矿物。角闪石呈柱状、针状，多色性明显，多为集合体产出；石英为微细粒状变晶，粒径0.07~0.1mm，粒间呈齿状、多边形紧密接触；黑云母呈细鳞片状，具棕红色—浅黄色多色性，常聚成条纹状集合体。

(38) 黑云母石英（片）岩

见于万宁市美子林地区，岩石具鳞片粒状变晶结构，片状构造。主要矿物成分为石英（70%）、黑云母（20%~30%），后期常有绢云母-白云母化。局部可含20%斜长石，形成黑云母斜长石石英岩。此外，在变质强烈处，出现矽线石（3%~10%），矽线石呈针柱状，形成矽线石黑云母石英（片）岩。

(39) 矽线石石英岩

见于万宁市美子林地区，岩石具柱状粒状变晶结构，定向构造。主要矿物成分为石英（55%~80%）、矽线石（20%~45%），矽线石呈针柱状。

2. 接触交代变质岩

(1) 透辉石石榴子石矽卡岩

见于金波农场十八队、汤他村、扎南等地，褐黑色，粒状变晶结构，块状构造、条带状构造。金波农场十八队一带，岩石由石榴子石矽卡岩和透辉石矽卡岩条带相间分布，条带宽5~15cm，石榴子石矽卡岩为全自形变晶结构，石榴子石出现异常消光现象，呈现明显的环带构造，透辉石矽卡岩呈细粒柱粒状变晶结构。分布于汤他、扎南的透辉石石榴子石矽卡岩由自形等轴粒状石榴子石（50%~75%）和粒状透辉石（10%~35%）组成，透辉石充填于石榴子石空隙中，石榴子石淡黄色环带构造，外环带稀而宽，核心环带窄而密；透辉石柱体无色，呈他形充填，有柱状解理和横裂纹出现，沿解理和横裂纹次生分解出碳酸盐（方解石3%）和石英（1%）；赤铁矿（1%）呈红色半透明充填于透辉石裂隙中。

（2）强葡萄石化透辉石矽卡岩

见于石碌水库旁，岩石呈不等粒变晶结构和筛状结构，块状构造。主要矿物成分为葡萄石（45%）、透辉石（30%）、斜绿帘石（10%）、石榴子石（3%）及少量方解石、榍石。葡萄石中见大量透辉石和斜绿帘石交代残余呈筛孔状，局部见石榴子石，石英呈填隙状。

（3）含透辉石硅灰石矽卡岩

见于石碌水库旁。岩石矿物成分主要为硅灰石（95%）、透辉石（3%~5%）及少量方解石、绢云母。岩石具条带状构造，带状构造是由硅灰石粒度不同所显示，不同条带内硅灰石呈纤状变晶、柱粒状变晶或放射状变晶结构，透辉石在硅灰石的放射状集体中呈交代残余结构。

（4）透辉石矽卡岩

见于屯昌县羊角岭水晶矿，岩石呈淡绿色、灰绿色、淡灰色等，不等粒结构，致密块状构造。矿物成分为透辉石、石榴子石、石英、长石、磷灰石、磁铁矿、方解石、绢云母，透辉石与石榴子石常与斜长石伴随在一起。

（5）透闪石矽卡岩

见于屯昌县羊角岭水晶矿、三亚市汤他等地。矿物成分以透闪石为主，含有石榴子石和透辉石，透闪石一般为细长针状集合体，细小的呈纤维状，粗大的呈束状，石榴子石和透辉石均以粒状分布于岩石中。

（6）石榴子石矽卡岩

见于屯昌县羊角岭水晶矿，三亚市田独南下岭、汤他、扎南等地，岩石呈棕色、棕黄色、棕黑色、浅黄褐色，粒状变晶结构，块状—致密块状构造。矿物成分以石榴子石为主，其他矿物有透辉石、透闪石、磷灰石、石英、长石、方解石、黑云母、绢云母、阳起石、绿泥石等，这些矿物以一种或几种与石榴子石伴生在一起。

（7）绿帘石矽卡岩、绿帘石透闪石矽卡岩

见于万宁市美子林、三亚市扎南。绿帘石矽卡岩中绿帘石含量大于90%，另有少量透辉石、石榴子石、磁铁矿、方解石，绿帘石透闪石矽卡岩中透辉石80%±、绿帘石10%±、阳起石5%~10%，石榴子石、磁铁矿、方解石少量。矿物粒径0.5~4mm，阳起石可达4 mm×8mm。绿帘石常呈脉状，阳起石呈囊块或短脉状，方解石、石榴子石呈细脉状。

（8）石榴子石透辉石矽卡岩

见于三亚市田独南下岭，岩石呈深灰绿色，变晶、变嵌结构，块状构造。矿物成分为透辉石（70%）、石榴子石（20%）、绿帘石（5%）、斜长石（3%）、石英（1%），微量的闪锌石、电气石、锆石等，石榴子石呈细粒集合体，多与透辉石共生，有时呈细脉状。

（9）二长透辉石矽卡岩

见于三亚市田独南下岭，岩石呈深灰色，细粒变晶结构，条带状构造、块状构造。矿物成分为透辉石（50%）、斜长石（30%）、钾长石（10%）、石英（4%）、葡萄石（4%），少量黄玉、氟石，微量的斧石、锡石、锆石等。透辉石呈半自形短柱状，多数为细粒集合体，斜长石为半自形至他形，双晶不发育，钾长石多数存在于角岩化结构的细粒变晶集合体中，呈他形，葡萄石呈碟状、放射状或羽状集合体，黄玉呈长柱状。

（10）滑石透辉石矽卡岩

见于三亚市田独南下岭，岩石呈淡黄绿色，鳞片柱状变晶结构，块状构造。矿物成分为滑石（45%）、透辉石（53%）、褐铁矿（2%）。透辉石呈柱状，大小为0.3~3mm，常成不规则团块，滑石细鳞片状集合体，常交代透辉石。

（11）阳起石石榴子石矽卡岩

见于三亚市汤他，纤维状变晶结构，块状构造。矿物成分为阳起石（25%）、石榴子石（60%）及石英（15%）组成。石榴子石等轴自形状，粒径0.5mm，具有环带构造，阳起石呈放射状，长0.5~1mm，多充填在石榴子石晶体间隙中。石英他形，充填于早期晶体空隙中，常包裹纤维状阳起石或充填于石榴子石空隙。

(12) *石榴子石绿帘石矽卡岩*

见于千家镇白脚村、三亚市扎南，出露于白脚村呈细小脉状产出，延伸不远。岩石呈黄绿色—黄褐色，粒状变晶结构，多孔构造，矿物成分主要为石英（5%~25%）、石榴子石（10%~25%）、绿帘石（30%~60%），次为黑云母（3%）、赤铁矿（2%）。绿帘石不规则粒状及半自形粒状，粒径0.1~0.3mm，石榴子石多呈半自形粒状，粒径0.1~2mm，石英他形粒状。

(13) *绿帘石矽卡岩*

见于冲卒岭、扎南，产于冲卒岭的呈小脉状产出。岩石呈灰绿色，粒状变晶结构，块状构造。主要矿物成分为绿帘石-斜黝帘石（40%~90%）、石英（2%~28%），次为石榴子石~绿泥石（3%）、长石（2%）、方解石（2%），蛇纹石、黑云母、磁铁矿、褐铁矿、磷灰石少量。绿帘石-斜黝帘石呈无色或淡黄色柱状晶体或不规则粒状，在其间包含有石榴子石和绿泥石。绿泥石由辉石蚀变而成，具有辉石假象。石英呈聚集体，沿岩石空隙充填。

(14) *绿帘石化透辉石矽卡岩*

见于冲卒岭正西约1000m处，呈团块状产出，出露范围不广。岩石呈灰绿色，包含变晶结构，块状构造。矿物成分主要为透辉石（80%）、绿帘石（10%），次为石英（5%）、钙铝榴石（2%）、榍石（2%）、磷灰石（1%）。透辉石呈半自形—他形粒状、短柱状，粒径一般在1-4mm之间，其中粗大晶体中包含细小石英、钙铝榴石、榍石、绿帘石等矿物，形成包含变晶结构，部分被阳起石、绿泥石等交代；钙铝榴石为无色自形六边形，包含于透辉石之中，大部分已被绿帘石交代。该岩石经历了两期交代作用，早期交代作用形成透辉石、钙铝榴石等矿物；晚期交代作用是早期形成的矿物部分被绿帘石、绿泥石、阳起石等矿物交代。

(15) *绿帘石石榴子石矽卡岩*

见于琼中县百花岭一带，岩石呈残留体产出，形态为扁豆状，分布方向为北东60°。具不等粒状变晶结构，块状构造，矿物成分以石榴子石为主，次为绿帘石、阳起石、符山石及透闪石。

（二）变质带划分

在岩体的外接触带一般应有比较明显和较为完整的分带现象，但研究区（主要是海南岛）接触带的分带现象比较复杂。一方面是区内花岗岩为多期岩浆活动产物，接触变质作用可能存在叠加现象或是在区域变质作用基础上又叠加接触变质作用。另一方面，花岗岩与不同时代、不同岩性的围岩相接触，围岩的矿物成分比较复杂。此外接触变质强度还与侵入体规模、侵入深度、岩浆温度及流体性质等因素有关。因此接触变质带划分困难较大。在局部地区，泥砂质岩类可勉强划分为几个带（图18-1~图18-3），由外向内依次为：角岩化岩石、斑点板岩带，云母角岩带，（矽线石、堇青石、透辉石）角闪石角岩带，分别大致相当Turner（1968）划分的绿泥石-绿帘石角岩相、角闪石角岩相和辉石角岩相。

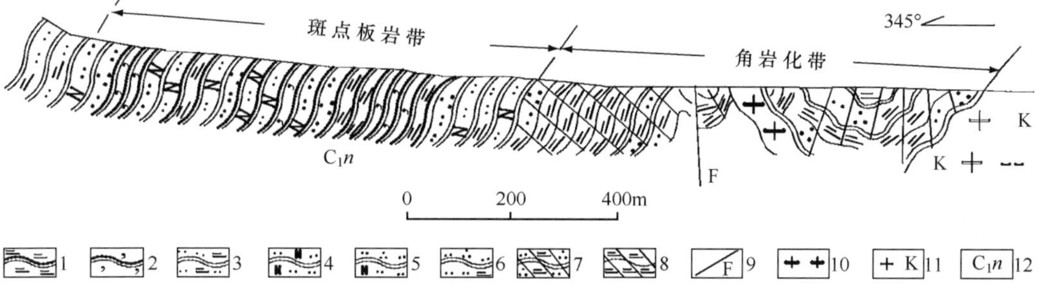

图18-1　保亭县那颂岭接触变质带剖面图

Fig. 18-1　Section of the Nasongling contact metamorphic belt, Baoting County

（据海南地质大队，1986，略修改）

1—绢云母板岩；2—斑点板岩；3—云母或绢云母粉砂岩；4—绢云母含长石石英粉砂岩；5—变质长石石英细砂岩；6—绢云母石英粉砂岩；7—角岩化石英粉砂岩；8—角岩化千枚岩、粉砂质千枚岩；9—断层；10—花岗斑岩脉；11—斑状黑云母正长花岗岩；12—石炭系南好组

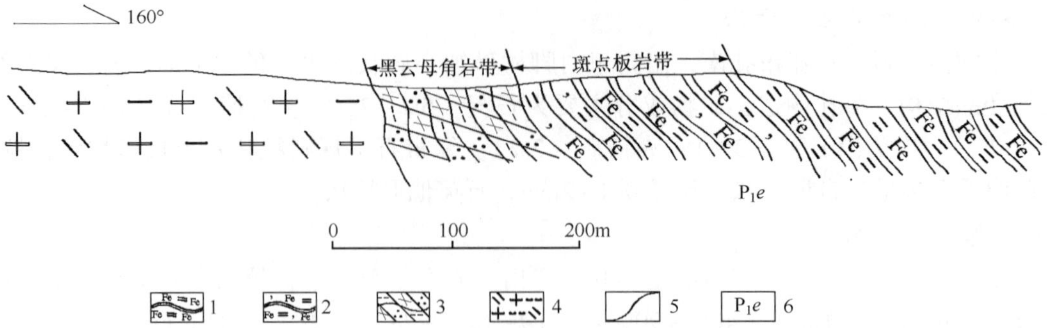

图 18-2　东方市万丁村接触变质带剖面图

Fig. 18-2　Section of the Wandingcun contact metamorphic belt, Dongfang City

（据海南地质大队，1990，略修改）

1—含铁绢云母板岩；2—斑点状含铁绢云母板岩；3—红柱石黑云母石英角岩；4—中细粒黑云母二长花岗岩；5—地质界线；6—峨查组

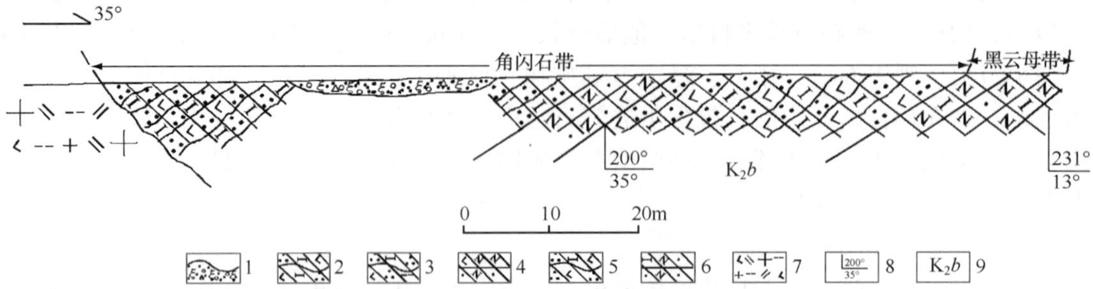

图 18-3　乐东县千家镇接触变质带剖面图

Fig. 18-3　Section of the Qianjiazhen contact metamorphic belt, Ledong County

（据海南地质综合勘察院，1998，略修改）

1—浮土；2—含透辉石角闪石英角岩；3—含黑云透辉石角闪石英角岩；4—角岩化角闪石长石砂岩；5—透辉石角闪石石英角岩；6—角岩化含透辉石粗中粒长石砂岩；7—中细粒角闪黑云二长花岗岩；8—产状；9—报万组

（三）变质相

按特纳尔（F. J. Turner，1968）的变质相分类，海南岛接触变质岩可分为三种变质相：钠长石-绿帘石角岩相、普通角闪石角岩相和辉石角岩相。其矿物共生组合图解如图 18-4 所示。

1. 钠长石-绿帘石角岩相

主要分布于侵入岩与沉积岩的接触变质晕外围，形成的岩石有斑点状板岩、片岩、千枚岩和角岩化岩石等。主要矿物组合有：①石英+绿帘石±阳起石；②石英+绿泥石±绿帘石；③石英+黑云母+绢云母+绿帘石±绿泥石；④石英+绢云母—白云母±黑云母；⑤石英+绢云母+阳起石+绿泥石+绿帘石；⑥石英+黑云母+绿帘石；⑦石英+绢云母—白云母+红柱石±黑云母。

2. 普通角闪石角岩相

分布于侵入岩与沉积岩的接触变质晕内侧，形成的岩石主要包括角岩化岩石、长英质角岩、含透辉石斜长石角岩等。主要矿物组合有：①石英+黑云母+白云母+堇青石；②石英+正长石+黑云母+白云母；③石英+斜长石+角闪石±透辉石；④石英+角闪石+黑云母+透辉石±斜长石；⑤石英+斜长石+透辉石+透闪石；⑥石英+斜长石+黑云母+透辉石；⑦石英+钙铝榴石+透辉石。

3. 辉石角岩相

仅局部见于大型侵入岩与沉积岩的接触带内，以出现透辉石与高级变质矿物紫苏辉石共生及矽线石+红柱石为特征，岩石类型为紫苏辉石石英透辉石角岩、红柱石二云母片岩。其矿物组合为：①石英+斜长石+正长石+紫苏辉石+透辉石+黑云母；②石英+斜长石+正长石+黑云母+白云母+矽线石+红柱石。

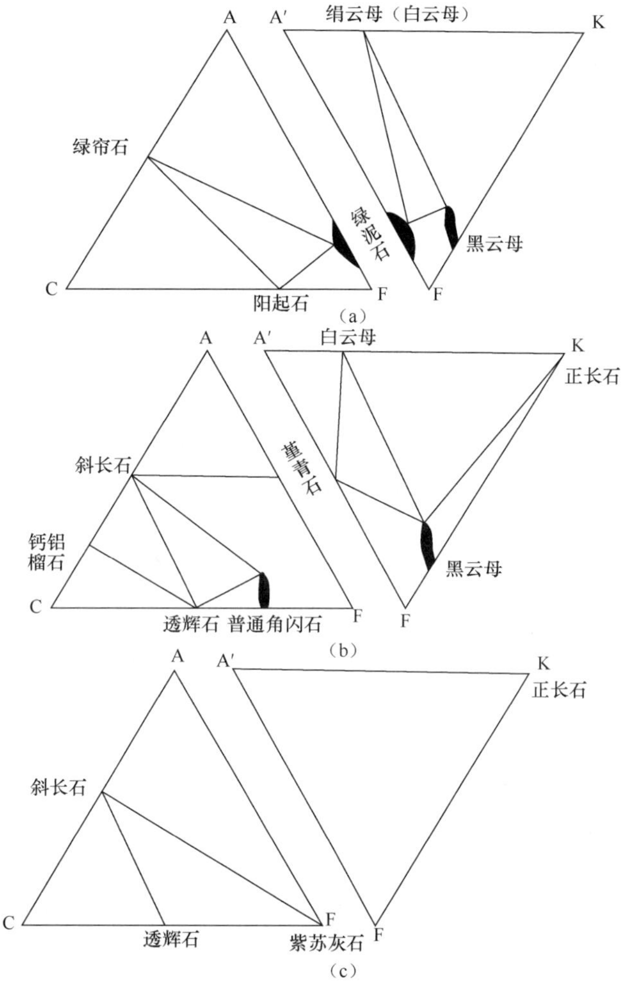

图 18-4　海南岛接触变质岩变质相 ACF 和 A′FK 图解

Fig. 18-4　ACF and A′KF diagram of metamorphic facies of contact metamorphic rocks, Hainan Island

(a) 钠长石绿帘石角岩相 ACF 和 A′FK 图；(b) 普遍角闪石角岩相 ACF 和 A′FK 图；(c) 辉石角岩相 ACF 和 A′FK 图

二、动力变质岩

动力变质岩是各种岩石在不同性质应力影响下，发生破碎、变形和重结晶等作用而成，岩石的性质主要取决于应力的性质和强度，同时也受原岩的性质及物态（脆性或塑性）的影响，其分布严格受构造带的控制。伴随大型构造带偶见各种中低温热液蚀变，主要表现为硅化、黄铁矿化、绢云母化、绿泥石化及碳酸盐化等。根据应力性质、强度及岩石结构特征，将海南岛动力变质岩划分为碎裂岩系列和糜棱岩-构造片岩两大系列。

（一）碎裂岩系列

碎裂岩系列广泛分布于不同规模的断裂带中，以脆性变形为特征。根据岩石的碎裂程度和结构构造等特征，划分为以下两类。

1. 压碎角砾岩类

系岩石受构造应力作用发生破碎所形成的岩石。根据应力性质和角砾形态分为以下两个岩石类型。

（1）构造角砾岩

广泛分布于各断裂带中，平面上常呈带状或透镜状分布。具角砾状结构，砾状构造。角砾为原岩破碎形成，大小不一，粒径一般为 2~15mm，少数达 20~35mm，局部见粒径<1.1mm 的显微角砾，多呈棱角状—次棱角状，极少数呈次圆状，分布杂乱无章，含量 5%~30%；胶结物为原岩的细碎物、硅质、铁质等。

（2）构造圆化角砾岩

具角砾状结构，砾状构造。砾石成分主要有石英、云母石英片岩等，大小以2～5mm为主，个别达15mm，受扭应力作用发生圆化，多呈次圆状—圆状，局部具弱定向性，含量10%～15%；胶结物为碾碎得更细的原岩碎屑及少量次生硅质、绢云母等。

2. 碎裂岩类

以压碎变形为主，岩石不具定向组构。按岩石的破碎程度和形态特征，进一步分为碎裂岩化岩石、碎裂岩、碎斑岩、碎粒岩等4种岩石类型。

（1）碎裂岩化岩石

主要见于脆性断裂带中，在花岗岩区较为发育，岩石类型主要为碎裂岩化花岗岩。岩石受碎裂作用较为轻微，原岩结构构造基本保留，其间裂纹较为发育，裂隙间为少量磨碎长英质及硅质、铁质、绢云母等充填。部分长石双晶发生弯曲变形，并伴有绢云母化、黝帘石化、高岭土化等次生变化；黑云母具弱多色性，多发生绿泥石化，部分为白云母置换。

（2）碎裂岩

海南岛不同方向的断裂中均有分布，以花岗岩地区较为发育，以花岗碎裂岩为主，岩石受碎裂作用较强，具明显碎裂结构。碎块粒径2～15mm，碎块间几乎没有相对位移，裂隙间为磨碎物质及次生绢云母、铁质、硅质等充填；碎基中的石英具强烈波状消光，长石双晶发生弯曲变形乃至断裂，绢云母化、黝帘石化等较为强烈，黑云母叶片弯曲变形，普遍发生白云母化。

（3）碎斑岩

多见于刚性岩石的破碎带中，岩石具碎斑结构，块状构造，由碎斑和碎基组成。碎斑成分主要为钾长石和斜长石，粒径一般为0.5～2mm，多呈棱角状—次圆状，次生变化明显，双晶发生弯曲变形，甚至错断，含量20%～50%；碎基为细碎的长石、石英、黑云母及次生绢云母、绿泥石、硅质、铁质等，常围绕碎斑分布，局部可见微量磁铁矿、磷灰石及锆石等副矿物。根据岩石组合特征，其原岩可能为（斑状）花岗岩。

（4）碎粒岩

主要分布在北东、南北方向的压扭性断裂带中，发育程度较前者弱，但受应力作用强于前者。岩石受强烈改造，原岩结构很少保留，具碎粒结构，块状构造。碎斑成分以石英为主，粒径一般为0.1～0.5mm，形态为次棱角状—次圆状，带状、波状消光明显，常被细碎物质包围，呈“残斑状”，含量5%～10%；碎基为细碎状、粉末状矿物，呈杂乱分布，成分为石英、绢云母、绿泥石、铁质等。受后期热液作用，局部可见岩石发生强烈硅化。

（二）糜棱岩-构造片岩系列

广泛分布于韧性变形带中，尤其在戈枕韧性剪切带、洛基韧性剪切带、石碌水库韧性剪切带、九所-陵水韧性剪切带、什运共轭韧性剪切带、大村韧性剪切带中较为发育。以中-深构造层次塑性变形为特征，具明显的定向性。根据岩石结构构造特征、变形特征以及重结晶程度等情况，进一步划为以下8种岩石类型。

1. 糜棱岩化岩石

主要产于变形带的外带，为糜棱岩化初期的产物，基本保留原岩的一般特征，具微弱糜棱面理。岩石类型主要有糜棱岩化花岗岩、糜棱岩化白云母条带状混合片麻岩、糜棱岩化长英质白云母钾长条带状混合岩等。岩石中，石英波状、带状消光明显，多被压扁拉长成透镜状、扁条状、扁豆状等；长石类矿物则以压碎变形为主，裂隙发育，双晶弯曲变形，蚀变强烈；片状矿物晶体多发生弯曲、膝折等，呈定向断续分布在浅色矿物中。

2. 初糜棱岩

出现于各韧性变形带的边部。以长英质初糜棱岩为主，具弱糜棱面理，矿物变形明显，具初糜棱结构、碎斑结构，片麻状构造、眼球状构造。碎斑成分主要由钾长石、斜长石组成，呈眼球状、透镜状、椭圆状产出，长轴平行糜棱面理，多发生绢云母化、高岭土化，碎斑内部常发生碎裂，裂隙间为次生铁质、绢云母充填，含量50%～70%；碎基有细碎的石英、长石、黑云母、绢云母、绿泥石等，其中石英常被压扁拉长成透镜状、扁豆状等，呈线状断续平行排列，长石常与石英聚成集合体分布，黑云母呈细叶片状集合体或条纹状、线状断续平行排列，常为绿泥石、白云母所置换，重结晶矿物绢云母、绿泥石多聚成线状排列。

3. 糜棱岩

(1) 花岗质糜棱岩

在花岗岩地区的韧性变形带中较为发育，岩石呈现明显的糜棱面理，具糜棱结构、碎斑结构，眼球状构造、条纹状构造。碎斑成分以钾长石为主，斜长石次之，常发生不同程度的旋转变形，呈眼球状、透镜状、椭圆状，长轴平行糜棱面理，含量20%~40%；碎基由磨细及重结晶物质组成，主要为石英、黑云母、绢云母、绿泥石等，其中石英具强烈波状消光，韧性变形明显，常与片状矿物组成不同颜色、不同成分的纹理及细条带。

(2) 石英-绢云母质糜棱岩

主要分布于抱板群出露地区的韧性断裂带中。岩石变形明显，糜棱面理发育，但不平直，具糜棱结构，眼球状构造、流动构造。碎斑成分主要为石英及其集合体，圆化程度较高，多呈眼球状-透镜状、卵圆状，个别眼球发生“S”形旋转，局部呈雁行排列，含量10%~30%；碎基成分为绢云母、绿泥石、白云母及微细粒石英等，片状矿物常聚成集合体呈线状分布，局部与石英混合成条纹状产出，当遇到碎斑时，沿其外形绕行，显示特征流动构造及眼球状构造。

4. 超糜棱岩

岩石受糜棱岩化作用相当强烈，具超糜棱结构，木纹构造。岩石中绝大部分矿物被粉碎成粉末状、霏细状物质，且几乎为绢云母、绿泥石等交代，局部见个别石英残斑，粒径< 0.1mm，多被压扁拉长成伸长状、细条状，与蚀变矿物绢云母、绿泥石等平行相间排列，使岩石具不同成分、不同颜色的条纹，显示强烈的流动构造。

5. 千糜岩

岩石受应力强度与糜棱岩相似，但重结晶明显，具微弱结晶叶理，重结晶程度一般小于50%。具粒状鳞片变晶结构、糜棱结构，千糜构造。矿物成分以绢云母为主，石英次之，还有少量白云母等。绢云母沿长轴呈平行、近平行排列，显示岩石的线状特征，沿其片理面可见强烈丝绢光泽，反映了千枚状构造的特点，新生的绢云母千枚理多斜交或直交残余的云母片理；粒状矿物石英多发生粉碎，波状消光明显，具韧性变形特征，局部可见重结晶现象，部分聚集成条带状与绢云母条带相间平行分布。

6. 糜棱片岩

岩石由糜棱岩受改造而形成，重结晶较为明显，具显著结晶叶理，重结晶程度小于50%，具鳞片粒状变晶结构、糜棱变晶结构，片状构造。矿物成分以石英（50%）、白云母（30%）为主，其次为长石（5%）及绢云母（10%）。石英韧性变形强烈，呈透镜状、长条状、扁豆状等，沿长轴方向平行排列，S形旋转普遍，部分发生重结晶加大；片状矿物白云母晶体多发生弯曲变形，常聚成条带状与石英相间平行分布，使岩石具片状构造。

7. 糜棱片麻岩

岩石受应力作用强度及结晶程度基本与糜棱片岩相同，由于原岩性质不同，形成不同的组构特征，具鳞片粒状变晶结构、糜棱变晶结构，片麻状构造。矿物成分有石英（40%）、长石（30%）、白云母（24%）及少量黑云母、绢云母、绿泥石等。石英具明显的韧性变形，部分发生重结晶；长石以压碎变形为主，多顺石英条带分布；片状矿物晶体多发生弯曲变形及膝折，“云母鱼”较为发育，沿长轴方向断续分布于长英质矿物中，并与浅色条带方向一致，构成片麻状构造。

8. 构造片岩

主要有构造白云母石英片岩、构造绢云母石英片岩，具鳞片粒状变晶结构，片状构造。矿物成分主要为石英、白云母及绢云母等。粒状矿物石英受构造作用，以压扁、拉长变形为主，呈长条状、透镜状、扁豆状沿长轴平行排列，少部分发生破碎、磨细，并有重结晶现象；片状矿物晶体弯曲变形，局部可见白云母呈“云母鱼”产出，多聚成条带平行分布。

三、气-液变质岩

气-液变质作用主要与断裂活动及岩浆活动有关，其次在侵入岩与地层的热接触带的局部地段也有出现。所形成的岩石在规模与产状上受断裂和侵入体规模及产状控制，而其岩石性质则取决于原岩成分、气-液性质及作用方式。

（一）云英岩类

1. 云英岩化花岗岩

主要见于以下地段：①乐东县盗公村、石门山一带，原岩为中粒斑状角闪黑云二长花岗岩（保亭单元千家岩体）；②乐东县五场、脚岭，原岩为（尖峰岭单元）中粗粒黑云母钾长花岗岩；③五指山市牙日村东侧，原岩为（志仲单元）中粒巨斑状黑云母二长花岗岩。此类岩石保留原岩结构构造，矿物成分包括残留矿物和蚀变矿物两部分，蚀变矿物主要为白云母、石英，其含量随云英岩化程度不同而变化，晚期出现绢云母、绿泥石。

2. 云英岩化长石石英砂岩

在乐东县大村东南约3km的田仔岭东侧，见有云英岩化长石石英砂岩，原岩为白垩系报万组长石石英砂岩。岩石为浅灰黄色，鳞片花岗变晶结构、变余砂状结构，平行构造。矿物成分为石英（60%）、白云母（15%）、绢云母（5%）、红柱石（4%）、绿泥石（1%），磁铁矿、白钛矿等微量。白云母成鳞片状，片径0.1~1mm；残余长石表面因黏土化而显污浊，其中有少量鳞片状绢云母。

3. 云英岩化眼球状黑云斜长片麻岩、云英岩化糜棱岩、云英岩化白云母片麻岩

原岩为戈枕村组片麻岩、混合岩等。岩石为浅灰白色略带蓝、褐黄等颜色，聚变晶鳞片花岗变晶结构。矿物成分主要由原岩的石英、长石等残余矿物和蚀变形成的石英、绢云母、白云母、绿泥石等组成，在长石表面因有泥化而显得污浊。大村韧性剪切变形带内的云英岩中的云母（以绢云母为主，白云母次之）常呈条带状集合体分布，一般沿片理定向排列，石英（云英岩化作用的产物）也呈明显条带，颗粒间呈不规则锯齿状紧密镶嵌接触，与云母条带相间分布。

4. 云英岩

岩石为鳞片粒状变晶结构、块状构造。主要矿物成分为石英（30%~70%）、白云母（绢云母）（23%~70%），少量锆石、金红石、独居石、磷灰石、萤石、电气石、磁铁矿等。通常以云英岩为中心向外过渡为云英岩化带、绢云母化带或绢云母绿泥石化带。

（二）绢英岩类

该岩石一般是在中低温气液作用下，发生绢云母化蚀变而形成的，变质程度为低绿片岩相。具显微鳞片变晶结构、交代假象结构，块状构造。矿物成分主要有绢云母（40%~65%）、石英（10%~45%）、绿泥石（≤15%）；其次为少量绿帘石、白云母、黄铁矿、黄铜矿、方铅矿、钠长石、碳酸盐矿物等。其中绢云母呈显微鳞片状，常聚成密集集合体；石英可见次生加大特征；绿泥石常具浅绿色—浅绿蓝色多色性。

（三）硅化岩类

分布广泛，主要出露在挤压强烈的断裂带中，其次在热接触带的局部地段出露。呈乳白色，不等粒状变晶结构，致密块状构造，矿物成分主要为石英（大于95%），次为绢云母、铁质。石英呈不等粒状变晶，粒径0.01~2.0mm，少量达3.0~4.5mm。在高、中温热液作用下，颗粒较为粗大，甚至部分发育成石英脉，脉宽0.1~0.8mm，多呈犬牙状、平行状生长；在低温热液作用时，一般结晶成微细粒石英，呈齿状、三点式接触。

（四）绢云绿帘石岩类

呈深绿色，交代结构，脉状构造、块状构造，矿物成分主要为绿帘石，其次有绢云母、绿泥石、石英及少量黝帘石、方解石。绿帘石呈细小鳞片状集合体，多色性明显，局部呈脉状分布；石英粒径0.02~0.2mm，沿长轴近平行排列，波状消光显著。该岩石呈脉状产出，宽1~7cm，原岩可能为中基性火山岩。

第五篇　区域地质构造

概　　述

海南岛地处欧亚板块、印度板块和滨太平洋板块的交汇地带，这一独特的地质构造位置历来为众多地学工作者所关注。海南岛的基础地质研究工作，最早始于1929年，由李承三先生完成，并编写了《广东海南岛北部地质矿产》。自20世纪50年代以来，先后有原地质部、原地质矿产部、中国科学院、冶金工业部、桂林冶金地质学院、南京大学、中山大学等单位的地质工作者从不同角度对海南岛的地质构造进行了研究。特别是60年代中期以来，1∶20万和1∶5万区域地质调查积累了大量的第一手地质资料，奠定了海南岛地层、构造、岩浆岩、变质岩、矿产和大地构造的研究基础。进入21世纪以来，随着新一轮国土资源大调查的开展，宜昌地质矿产研究所、海南省地质调查院完成了海南岛1∶25万区域地质调查，对海南岛基础地质的研究有了大量新的认识，极大地提高了海南岛基础地质研究水平。

第十九章　构造单元及其主要特征

第一节　构造单元划分

一、以往构造单元划分回顾

关于海南岛地质构造属性，不同学者和单位提出了许多不同的认识。由于不同学者所处时代不同，资料掌握程度不一，所提出的观点也不尽相同，甚至迥然不同。前人对海南岛地质构造性质的认识，归纳起来，主要有以下两类。

1. 槽台论

主要是20世纪80年代末以前。国内多位学者和单位运用槽台理论对海南岛的大地构造与分区问题进行了探讨。中国地质科学院（1962）将海南岛划归南华准地台华夏褶皱带，岛南北分属海南隆起和雷琼新凹陷，发育有加里东准地槽（Є—S），并经历了晚古生代—印支期（D—T）地台前期发展和燕山期—喜马拉雅期（J-Q）地台活化阶段演化。陈国达等（1977）出版的《中国大地构造图》及《中国大地构造概要》中将海南岛划为海南地穹列，属于东南地洼区的雷琼地洼系，认为海南岛早古生代为加里东期地槽褶皱带，晚古生代为地台区，三叠纪末以来为地洼区；同年，陈国达（1977）依据铁矿会战过程所获资料（特别是地层方面的）和对石碌群第七层及其上、下地层的沉积特征、地质时代的认识，以及中国南部地槽回返的地质时代有自西北向东南逐渐推迟的规律性，论述了海南岛存在华力西地槽的可能性，认为海南岛地槽阶段为元古宙—早古生代，或延至晚古生代初，地台阶段为晚古生代初期或中期到中生代初。袁奎荣等（1977）认为岛西存在华力西地槽。陈炳蔚等（1978）、夏邦栋（1979）也先后论述了海南岛华力西地槽的基本特征，把其空间分布扩展至全岛，时间不限于泥盆纪—早石炭世，扩及延续到二叠纪末甚至中生代初。徐安武等（1997）则主张海南岛西部为早古生代裂陷槽，并进一步对其地质特征进行了阐述。王鸿祯（1986）和刘本培（1986）认为海南岛可能属南海-印支地台的北部陆缘带。之后，袁兆亿（1987）从地球物理场角度研究了海南岛和广东的大地构造分区；邹和平等（1987）详细分析了琼北新生代坳陷的构造特征和大地构造类型-裂谷型地洼系。刘以宣（1984）在论述一种新型的陆缘海型地洼区-南海陆缘地洼区时，提及“古生代大部分海区进入地台阶段，海南岛崖县一带构成台坪”。彭格林（1990）从沉积建造角度，把海南岛纳入华夏期地洼区，以东西向九所-陵水断裂带为界，划分出北面的东南地洼区雷琼地洼系和南面的南海地洼区崖县地穹系，前者进一步细分为2个三级构造区（琼北地洼列、琼中地穹列）、10个四级构造分区。汪啸风等（1991）也以东西向九所-陵水断裂带为界，划分出北面的华南褶皱系五指山褶皱带、雷琼断陷和南面的南海地台三亚台褶带，五指山褶皱带进一步划分出3个三级构造单元和6个四级构造单元。

2. 板块论

20世纪80年代以来，板块构造学说和全球构造的观点传入我国以后，为我国学者广泛接受。将海南岛的地质构造研究提高到一个新的高度。陈海泓等（1994）基于石碌群生物面貌及所赋存的富铁矿床，以及该区晚古生代地层中发育的冰水沉积砾岩等，将海南岛纳入冈瓦纳大陆的范畴，属于前白垩纪冈瓦纳大陆裂解碎块北移的地体或复型地体（莫宴情等，1987；虞子治，1989）。乔秀夫（1981）依据海南岛原“陀烈群”中发现的前寒武纪微古植物化石，进而将整个海南岛划入华夏板块范围。郭令智等（1980）并将九所-陵水断裂带以北的琼中地体划归云开地体中。杨树锋（1989）进一步从地层、古地磁角度出发，基于东西向九所-陵水断裂带南北两侧古地磁特征差异性，进而以该构造带为界划分出北面琼中地体和南面崖县地体。雷裕红等（2005）依据Nd同位素示踪结果，以东西向九所-陵水断裂带为界进行构造单元分野，并把该断裂带以北地区全部纳入华南地体中。而陈耀钦等人（1989）通过石炭纪沉积作用的研究，提出广东大陆和海南分属于两个不同的联合地体。汪啸风等（1991）也认为海南岛结晶基底抱板群与云开地体中的云开群具有明显的差异，进而对“海南岛属于华夏古陆”的论断提出了质疑，认为晚古生代地层中的化石组合属于低纬度华夏-特提斯动物区或华夏植物区的一部分，并通过海南与周边地区晚古生代花岗岩特征的对比分析，提出海南岛晚古生代属于特提斯构造域范围内的见解。陈海泓（1994）从古地磁角度主张华南与海南属于不同块体，属于从冈瓦纳大陆裂解形成的华南多岛海的一分子（陈海泓，1998）。夏邦栋（1991）通过对晚古生代变质火山岩的研究，提出海南岛在晚古生代属于裂谷背景。李献华等（2000）认为沿昌江-琼海一带出露的古生代变质基性火山岩具有洋中脊玄武岩特征，从而把昌江-琼海构造带以北划归华夏块体，构造带以南划入印支块体。许德如（2001）在研究元古代变质基性岩体的基础上，主张以北东向白沙构造带（谭爷断陷构造带）为界进行构造单元划分，与Metcalfe I. et al.（1993）以北东向构造进行海南岛大地构造单元划分的认识主张一致。本次工作中，肖庆辉从大地构造相角度建议海南岛以东西向昌江-琼海断裂带和九所-陵水断裂带为界，从北至南划分为琼北俯冲增生杂岩相、琼中岩浆弧相和琼南陆块区大相；葛肖虹则认为海南岛构造形迹主体为北东向，建议采用北东向白沙构造带进行构造单元划分；洪大卫认为海南岛晚古生代岩浆作用与华南内陆不同，而与越南一带岩浆活动相似，认为海南岛晚古生代属于特提斯构造域作用背景。

除上述两大观点外，汪啸风等（1991）按地质力学的观点，认为海南岛属于中国东部新华夏第二隆起带的南部，并划分出纬向、经向、华夏系和新华夏等构造体系。

上述认识可谓众说纷纭，基本上主要涉及海南岛两个重要的基础地质问题，即海南岛与周边地区的大地构造属性关系和海南岛本身的构造分野问题。由于东西向九所-陵水断裂带南北两侧古生代在沉积、变质、变形、古生物面貌等方面具有明显的差异性，以及东西向王五-文教构造带南北两侧新生代火山作用和沉积作用的差异性，长期以来，海南岛以东西向九所-陵水断裂带和王五-文教断裂带为界进行构造分野的主张基本上被地学工作者所认可，成为主流观点，一直沿用至今。值得一提的是，九所-陵水断裂带分布位置经历了多次更改，并总体逐渐北移。广东省地矿局（1986）认为其位置在崖城-陵水南面一带（走向为北东东）；汪啸风等（1992）通过物探资料解译，把九所-陵水断裂带位置更改为九所-陵水一带，走向东西向；2000~2002年，宜昌地矿所和海南省地质调查院在开展1∶25万《乐东县、陵水县》幅区调时，把九所-陵水断裂带改向为北东东向，分布于九所-陵水北面。

但九所-陵水断裂带作为构造单元分野的依据尚不足。汪啸风等（1991）所阐述的九所-陵水断裂带依据是：①重磁异常带；②燕山期岩浆带；③古地磁资料；④发育于燕山期和三亚古生界中的挤压构造形迹。因此许多学者认为属于燕山期地体拼贴的痕迹。遗憾的是宜昌地矿所和海南省地质调查院（2002）根据东西向九所-陵水断裂带南北两侧分布有北东向展布的中三叠世碱性花岗岩带，间接地推断出海南岛南、北两侧地体的拼贴时间为古生代—早三叠世。由此，九所-陵水断裂带构造属性定位受到了质疑：①所依据的重力、航磁资料至多只能显示的是燕山期深部构造轮廓，很难想象经历了侏罗纪以来多期次物质成分调整和重组的地壳，其重力、航磁资料还能反演出古生代—早三叠世的深部构造轮廓；②构造带南面分布有未变质或变质作用非常弱的寒武系—奥陶系，北面寒武系—奥陶系则为一套经历了区域绿片岩相变质作用形成的浅变质岩石，两者间的古地磁资料对比变得毫无地质意义；③作为界线断裂的东西向九所-陵水断裂带，依然未见古生代—早三叠世基性物质组分出露，显然断裂带并不属于深断裂；④南好、三亚古生界中的褶皱、断裂方向总体为北东向，南好地区华力西期—印支期同造山花岗岩中的斑晶和片麻理走向总体也为北东向，这种广泛的北东向构造迹向与东西向构造分野明显极不协调一致。因此，海南岛大地构造单元分野界线成了值得进一步探讨的重大基础地质问题。本次通过工作，尽管重新提出以北东向崖城-万宁断裂带进行构造分野的主张（海南省地质调查院，2011)，但根据海南专家初审意见，认为资料不足而不采信。本次也暂按九所-陵水断裂带作为构造单元分野构造。

二、本次构造单元划分

(一) 划分原则

1) 根据本次工作指南，构造单元划分总体主导思想是以“新全球构造”思想为指导、以板块构造学说为基础、以大陆动力学为线索，参考《〈中国区域地质志〉工作指南》(2009) 中的全国大地构造分区建议方案，从活动论角度出发，按照“三统一”的原则，融大地构造单元分野、地层区划划分为一体，开展海南省大地构造单元划分。

2) 以当前主流观点为主要依据，以海南岛构造演化为主线，以沉积作用、岩浆作用和变质变形作用研究为手段。以最新完成的 1∶25 万区调资料成果和前人地质资料以及各项科研成果为依据，结合海南岛深部构造轮廓特征，进行大地构造单元划分。

(二) 划分方案

海南省经历了多阶段多旋回的地质演化发展过程，地质构造复杂。总体上经历了中元古代结晶基底和新元古代褶皱基底形成、古生代—早中生代（三叠纪）特提斯域演化和晚中生代侏罗纪—白垩纪滨太平洋大陆边缘活动带和新生代南海北缘裂陷等阶段。根据《〈中国区域地质志〉工作指南》(2009) 中的大地构造划分方案，海南以九所-陵水断裂带为界，北面属羌塘-扬子-华南板块华南新元古代—早古生代造山带，南面为菲律宾海板块（?）南海陆块。显然这种划分方案没有涉及至中元古代和晚古生代以来的地质演化。本次结合地层区划划分方案，按中元古代、新元古代—三叠纪、侏罗纪—白垩纪、新生代（古近纪—第四纪）4 个阶段对其进行划分（表 19-1）。

表 19-1　海南岛大地构造单元划分表

Table 19-1　Division of geotectonic units of the Hainan Island

<table>
<tr><th>阶段</th><th>Ⅰ级</th><th>Ⅱ级</th><th>Ⅲ级</th></tr>
<tr><td>中元古代</td><td>华夏古陆</td><td>海南造山带</td><td></td></tr>
<tr><td rowspan="4">新元古代—三叠纪</td><td>羌塘-扬子-华南板块</td><td>华南新元古代—古生代造山带</td><td>五指山造山带</td></tr>
<tr><td rowspan="3">印支-南海块体</td><td>三亚地体</td><td></td></tr>
<tr><td>印支地块</td><td></td></tr>
<tr><td>南海地块</td><td></td></tr>
<tr><td rowspan="4">侏罗纪—白垩纪</td><td>中国东部中生代上叠盆地</td><td>华南上叠盆地带</td><td>海南断陷盆地</td></tr>
<tr><td rowspan="3">菲律宾-南沙群岛裂陷</td><td rowspan="2">南沙裂陷盆地</td><td>礼乐滩-北巴拉望裂陷盆地</td></tr>
<tr><td>曾母-万安裂陷盆地</td></tr>
<tr><td>菲律宾裂陷盆地</td><td></td></tr>
<tr><td rowspan="8">新生代（古近纪—第四纪）</td><td rowspan="4">东亚大陆边缘</td><td rowspan="4">华南大陆边缘</td><td>雷琼裂谷</td></tr>
<tr><td>海南地块</td></tr>
<tr><td>莺歌海裂陷盆地</td></tr>
<tr><td>西沙-中沙地块</td></tr>
<tr><td>中央海盆裂谷</td><td></td><td></td></tr>
<tr><td rowspan="3">菲律宾大陆边缘</td><td rowspan="2">南沙-曾母块体</td><td>南沙地块</td></tr>
<tr><td>曾母地块</td></tr>
<tr><td>加里曼丹块体</td><td></td></tr>
</table>

1. 中元古代构造单元划分

海南岛中元古代主要包括长城纪戈枕村组、长城纪—蓟县纪峨文岭组。该时期岩石发生了区域变质和变形，并见约14亿年同造山期花岗岩分布。但由于受到后期多期次构造作用的改造，岩石呈“残留体”状分布，难以对其进行构造单元划分，本次沿用前人观点，一级构造单元划归华夏古陆（张仁杰等，1990；陈惠芳，1991；谭忠福等，1991；汪啸风等，1991；侯威，1992；张理刚，1994；丁式江等，1995；张业明等，1997；雷裕红等，2005）。但包含在“华夏古陆”中的广东、广西基底地层没有海南岛抱板群的二层结构特点，并缺乏14亿年同造山期花岗岩的记录，表明海南岛与广东、广西中元古代时期的地质作用并不相同（马大铨等，1998），因此，本次在华夏古陆的基础上，再划分出二级构造单元“海南造山带”，以凸显其独特的地质演化过程。任纪舜报道（1984），西沙群岛永兴岛钻孔中见花岗片麻岩和石英云母片岩，片麻岩的矿物Rb-Sr等时线年龄为1465Ma。广东地矿局（1988）认为1465Ma代表了岩石的变质年龄。此外，刘昭蜀等（2002）在《南海地质》中描述：印支断（地）块基底为前寒武纪变质岩，下部为深变质的角闪石、石榴子石、黑云母副片麻岩，夹角闪岩，有时有钙碱性正片麻岩；上部为一套中等变质的二云母硅线石片岩、石榴子石片岩及石墨片岩夹数层石英岩、大理岩及角闪岩。上述两地岩石组合特征与海南岛抱板群相似。另，金庆焕（1989）根据南沙群岛与其北部海域岩石在磁场、地震波速等方面具有相似性，认为在南沙群岛和纳士纳群岛可能有元古宇或新元古界。上述暗示了前南华纪（中—新元古代）时期，海南省、印支地块属于统一的板块范畴。

2. 新元古代—三叠纪构造单元划分

为解决好新元古代—三叠纪阶段构造单元分野断裂问题，本次设立了《海南岛南北地体分野断裂研究》专题（海南省地质调查院，2011），提出了以北东向岩石圈断裂——崖城-万宁构造岩浆带为界进行南北构造分野的新主张（见专题报告），该构造带基本上处于北东向区域性岩石圈断裂——莲花山断裂带（为构造单元分野性断裂；广东省地矿局，1986）的南西延伸部位。但趋于资料有限，本次依然采用前人的观点认识，即九所-陵水断裂带作为构造单元分野的主张。

该时期，海南省在沉积作用、区域变质变形作用、古生物面貌等方面差异性明显。以东西向九所-陵水断裂带为界，划分出北面的羌塘-扬子-华南板块和南面的印支-南海块体。

《〈中国区域地质志〉工作指南》（2009）中，羌塘-扬子-华南板块进一步划分出二级构造单元华南新元古代—古生代造山带。但资料显示，海南岛与广东、广西新元古界—三叠系沉积作用、变质作用等方面具有明显的差异性（陈耀钦，1989；吴浩若，1997，2003）。在此基础上，本次以遂溪断裂为界，划分出南面的五指山造山带。其沉积包括青白口系—三叠系，新元古界—古生界岩石普遍发生了区域变质作用的改造，寒武系与奥陶系为假整合接触，上、下古生界为不整合接触，三叠系为未发生区域变质作用改造的陆相类磨拉石建造沉积。

印支—南海块体沉积为寒武系—三叠系，岩石未发生区域变质和变形。但各地沉积作用差异明显，沉积作用具有南迁的趋势。海南岛三亚地区仅出露中—上寒武统和奥陶系陆表海碎屑岩，寒武系与奥陶系为假整合接触。刘昭蜀等人（2002）资料显示，中南半岛为未变质的含三叶虫化石的中寒武世石英质砂岩和砂质泥板岩（厚400m），与元古宇不整合接触，往上的泥盆系—下石炭统为砂岩和灰岩夹页岩，上石炭统为一套海陆交互相沉积。南沙群岛上古生界二叠系零星分布，与元古宇不整合接触，岩石为灰岩和燧石、硅质碎屑岩、长石杂砂岩，其磁场强度（-70～-100nT）、层速度（5.4～5.7km/s）与北部海区基本一致，上三叠统为变质的深海相沉积岩。因此，本次以北西向红河断裂、南北向越东滨外断裂、北东向陆坡北缘断裂和广雅滩西断裂为界，进一步划分出三个二级构造单元，即印支地块、南海地块和三亚地体（图19-1）。其中，南沙群岛上三叠统变质的深海相沉积岩，是否与西太平洋构造域作用有关，值得进一步关注。

3. 侏罗纪—白垩纪构造单元划分

侏罗纪—白垩纪时，海南岛进入了另一个构造域——西太平洋构造域的发展演化阶段，形成了一系列醒目的北东向线形展布的盆地带和火山岩带，组成了中国东部中生代上叠盆地的一部分。涉及海

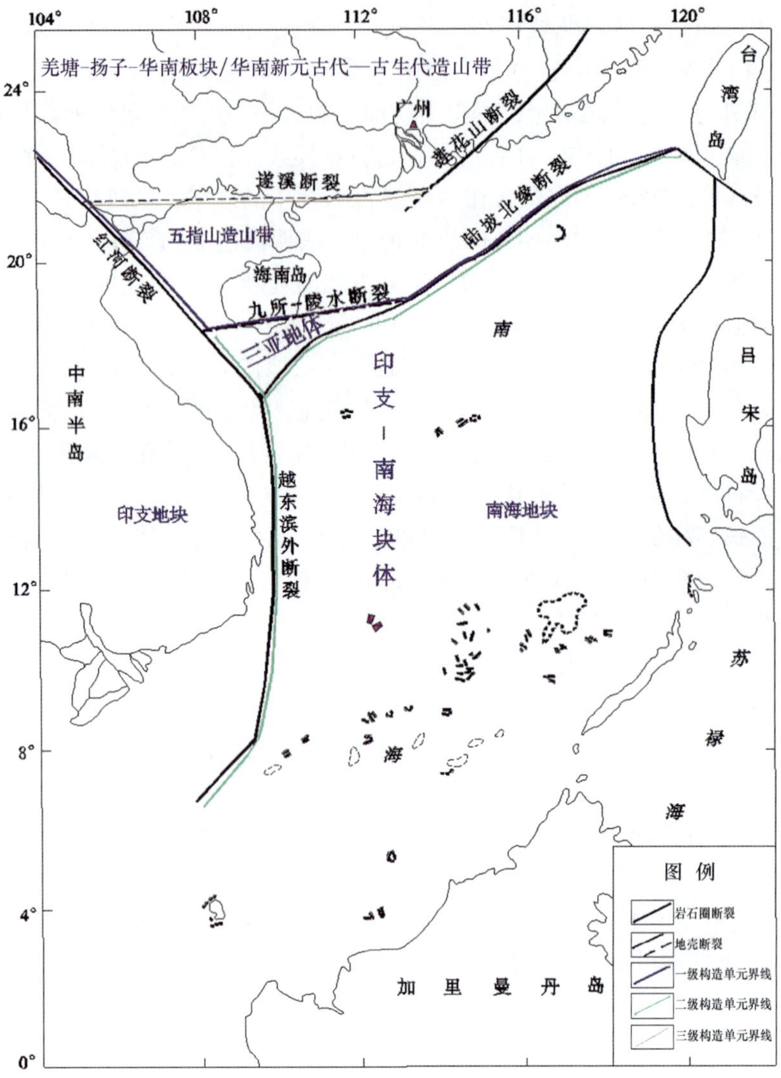

图 19-1　海南省新元古代—三叠纪构造单元划分图

Fig. 19-1　Division of Neoproterozoic-Triassic tectonic units of the Hainan province

南省的盆地主要是海南岛白沙、阳江、雷鸣、王五、乐安等陆内盆地，以及南沙群岛礼乐滩盆地和曾母-万安海相盆地。其中，曾母盆地出露有上白垩统丹脑组蛇绿岩建造；西巴拉望盆地上白垩统—渐新统克罗克组下部为燧石、细碧岩和基性、超基性侵入岩及变质岩、沉积岩，与下伏上侏罗统—下白垩统浅海相沉积为不整合接触；北巴拉望盆地中生代早期为中三叠统利米内康组深海相含放射虫燧石岩，与下伏二叠系和上覆上三叠统—下侏罗统灰岩为不整合接触。礼乐滩上三叠统—下侏罗统为三角洲相-开阔浅海相砂泥岩及硅质页岩、粉砂岩，与上覆下白垩统滨海—浅海相夹蛇绿岩含煤砂页岩和砾岩角度不整合接触。海南岛为陆内红色复陆屑式建造和中酸性火山岩建造的沉积，并发育大量壳源、壳幔混合源和幔源侵入岩（壳幔混合源岩石分布比较普遍）。由此看来，南北应属于不同的构造单元区。本次以北东向南沙海槽北缘断裂和北西向万安滩西南断裂、广雅滩断裂为界，北为中国东部中生代上叠盆地（一级）华南上叠盆地带（二级），南为菲律宾-南沙裂陷带。葛小月等（2003）在通过海南岛燕山期基性侵入岩脉研究，并进行区域对比分析，认为海南岛与粤北、福建一带岩石圈地幔物质成分存在差异性。因此，华南上叠盆地带暂以遂溪断裂为界，划分出海南上叠盆地带（三级构造单元）。菲律宾-南沙裂陷带划分出南沙裂陷和菲律宾裂陷两个二级构造单元，南沙裂陷进一步划分为礼乐滩-北巴拉望裂陷盆地和曾母-万安裂陷盆地两个三级构造单元（图 19-2；菲律宾裂陷跨国界，在此不讨论）。

4. 新生代构造单元划分

新生代古近纪—第四纪，海南岛处于南海陆块作用背景，随着南海扩张，海南岛及其周边裂陷盆地方向具有多方向性。岛内形成了东西向的雷琼裂陷、长昌盆地、长坡盆地，岛周边有北西向莺歌海、北东向琼东南盆地。岛南南海一带并由北至南至少分布有西沙海槽、中沙海槽、南海中央盆地、南沙海槽、北巴拉望盆地、万安盆地、曾母盆地等盆地。形成了一系列断隆区和断陷区。海南省构造单元也依此而 划分出 6 个不同的三级构造单元（图 19–3）。

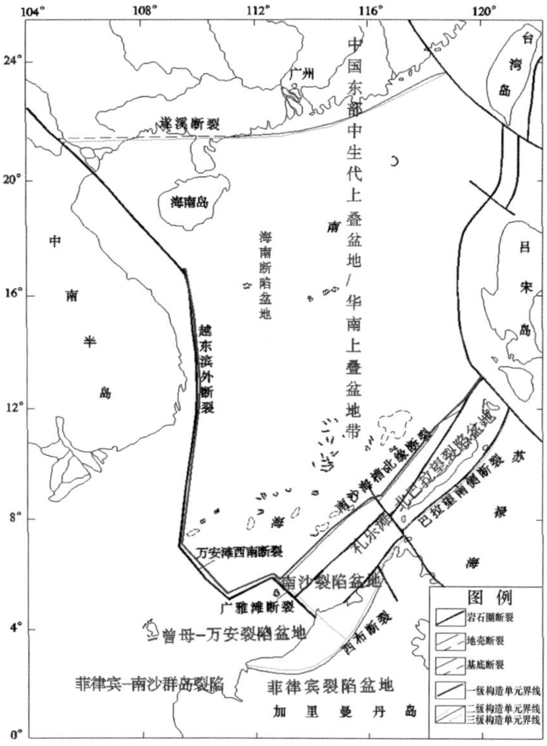

图 19–2 海南省侏罗纪—白垩纪构造单元划分图

Fig. 19–2 Division of Jurassic–Cretaceous tectonic units of the Hainan province

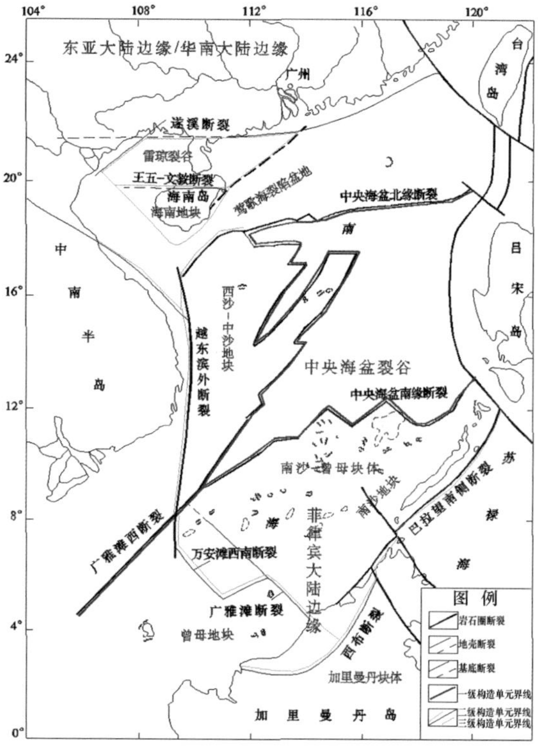

图 19–3 海南省新生代构造单元划分图

Fig. 19–3 Division of Cenozoic tectonic units of the Hainan province

第二节 构造单元特征分述

一、中元古代构造单元特征

海南省中元古代仅有一个二级构造单元，即海南造山带。

地层有长城纪戈枕村组、长城纪—蓟县纪峨文岭组，是海南岛最古老的结晶基底，呈残留体状分布于戈枕、公爱、红五、番阳、什运和黄竹一带，普遍发生了区域绿片岩相、角闪岩相变质作用和透入性的变形。发生于约 1400Ma 的同碰撞型片麻状二长花岗岩和花岗闪长岩的侵位，岩石片麻理与围岩（戈枕村组、峨文岭组）新生片理、片麻理协调一致，暗示了戈枕村组、峨文岭组区域变质和变形与该构造岩浆作用密切相关，即 1400Ma 的造山作用（早晋宁运动），造就了海南省以高温变质和高度塑性变形为特征的中元古代统一结晶基底。

二、新元古代—三叠纪构造单元特征

此时期，构造单元见表 19–1。由于印支地块主体位于中南半岛，因此，海南省的构造单元有五指山造山带（三级）、三亚地体（二级）和南海地块（二级）。

1. 五指山造山带

介于东西向遂溪断裂、近东西向九所-陵水断裂带和北西向红河断裂之间。经历了多旋回的地质演化。沉积记录包括青白口系石碌群、南华系—震旦系石灰顶组、下寒武统美子林组、奥陶系南碧沟组、下志留统陀烈组、空列村组、大干村组、靠亲山组、足赛岭组，以及下石炭统南好组、上石炭统青天峡组和下二叠统峨查组、中二叠统峨顶组、中—上二叠统南龙组等，主体为海相碎屑岩。构造运动有晚晋宁运动、加里东运动和华力西—印支运动、燕山运动和喜马拉雅运动。发育有多个不整合接触面，并伴随大范围华力西—印支期花岗岩侵位和区域绿片岩相变质，最终塑造了区域上醒目的略向南面突出的近东西向弧状构造带。

2. 三亚地体

位于北西向红河断裂、近东西向九所-陵水断裂带和北东向陆坡北缘断裂之间。应相当于任纪舜等所提出的南海地台。地层由黔东统孟月岭组、黔东统—武陵统大茅组、下奥陶统大葵组、中奥陶统牙花组、中—上奥陶统沙塘组、上奥陶统榆红组、尖岭组和干沟村组组成，寒武系与奥陶系为假整合接触。岩石未经历区域变质和透入性变形作用的改造。花岗岩主要是三叠纪富碱性花岗岩、碱性岩，少量钙碱性花岗岩。构造以北东向狗岭逆冲推覆构造带最醒目，席卷整个古生界。晚期并叠加了东西向和北西向断裂形变。构造运动有加里东运动、华力西—印支运动、燕山运动和喜马拉雅运动。

3. 南海地块

位于北东向陆坡北缘断裂和南北向越东滨外断裂之间的广大南海地区。基底为元古宙深变质岩石(片岩和片麻岩类岩石)。盖层为古生界正常沉积岩，与元古宇不整合接触。除了在南沙群岛见二叠纪灰岩和燧石、硅质碎屑岩、长石杂砂岩零星分布外，古生界基本上以隐伏状产出。南、北海区磁场强度（-70~-100nT)、层速度（5.4~5.7km/s）基本一致，显示南、北海区具有相似或相同的沉积构造和物质组成。在西沙群岛永兴岛钻孔中，古近系直接叠覆于元古宇之上。表明南海地块以长期风化剥蚀作用为主导，其间地壳也发生过不均衡升降运动，第一次至少发生于晚古生代二叠纪，第二次发生于白垩纪晚期—古近纪早期。

三、侏罗纪—白垩纪

侏罗纪—白垩纪时海南省进入了另一个构造域——西太平洋构造域的发展演化阶段，形成了一系列醒目的东西向和北东向线形脆性构造带，控制了盆地的形成和多期次的岩浆活动。此时，涉及海南省的构造单元有：海南断陷盆地、礼乐滩-北巴拉望裂陷盆地和曾母-万安裂陷盆地三个三级构造单元(表19-1)。

1. 海南裂陷盆地

北面止于东西向遂溪断裂，南面止于南沙海槽北缘断裂和北西向万安滩西南断裂、广雅滩断裂。西沙群岛以南的广大地区未见该时期的沉积。海南岛由一系列走向北东的白垩纪陆内盆地组成，包括王五断陷盆地、白沙断陷盆地、雷鸣断陷盆地、阳江断陷盆地、牛腊岭火山盆地、同安岭火山盆地等，海南岛西南面莺歌海盆地也有白垩系正常沉积岩分布（肖国林等，2004)。

岛内盆地沉积了白垩系鹿母湾组、报万组湖泊-河流相相杂陆屑式建造和六罗村组、汤他大岭组、岭壳组双峰式火山岩建造。

同期构造岩浆活动强烈，具有多期次性，形成了侏罗纪、白垩纪钙碱性花岗岩，岩石缺乏定向组构，属于伸展构造环境壳幔岩浆混合的产物，其侵位往往受多方向断裂的控制，导致岩体平面中常呈不规则长条状分布。断裂有北东向、北西向、东西向、南北向等，控制了同期钼、铅锌、金、银、铀、萤石等矿产的形成。

2. 礼乐滩-北巴拉望裂陷盆地

位于南沙海槽北缘断裂以南和北西向广雅滩断裂以东地区。发育有礼乐滩盆地、北巴拉望盆地。中国科学院南沙科考队（1991）认为礼乐滩、北巴拉望自中生代以来两者基本上处于相互联系的同一海相沉积区。盆地内沉积了下白垩统滨海—浅海相含煤砂页岩和砾岩及蛇绿岩，底部含中性熔岩、凝灰岩和石英闪长岩砾石，与上覆古近系为不整合接触。

3. 曾母–万安裂陷盆地

位于北西向万安滩西南断裂河广雅滩断裂以西地区，发育有万安盆地、曾母盆地等。据刘昭蜀等（2002）资料，曾母盆地上白垩统沉积了丹脑组蛇绿岩建造，与上覆古近系拉让群为不整合接触。其沉积作用与菲律宾板块作用有关。

四、古近纪—第四纪

该时期，海南省构造单元有中央裂谷盆地（一级构造单元）和雷琼裂谷、海南地块、莺歌海裂陷盆地、西沙–中沙地块、南沙地块、曾母地块6个三级构造单元（表19–1）。

1. 雷琼裂谷

雷琼裂谷分布于王五–文教断裂与琼州海峡断裂之间，接受了一套厚达数千米的古近系—第四系沉积。

白垩纪末—古近纪初，南海北部陆缘由长期的北西—南东向挤压应力场转化为北西—南东向拉张，陆缘解体形成一系列拉张盆地（张训华等，2008），琼州海峡—北部湾一带古近纪时接受了一套海相碎屑岩建造夹可燃性有机岩建造和火山岩建造；新近纪时为滨浅海相碎屑建造和火山建造；第四纪时接受了一套滨浅海相碎屑建造、河流相–三角洲相碎屑建造、火山建造。沉积中心在福山–北部湾一带，钻孔揭示，古近系沉积厚度最大达4000余米，新近系沉积厚度最大达2000余米。

伴随地壳的快速沉降，发生了多期次的火山活动，并在第四纪更新世时火山活动达到高峰，形成琼北地区大面积分布的玄武岩被。

2. 海南地块

介于东西向王五–文教断裂、北西向莺歌海裂陷和北东向琼东南盆地之间。属于海南岛陆地主体部分。主要发育有东西向长昌盆地、长坡盆地等陆内沉积盆地，以及蓬莱火山盆地。沉积盆地内沉积了一套古近系—新近系的河湖相碎屑建造夹可燃性有机岩建造；火山盆地火山岩为基性、超基性火山熔岩和火山碎屑岩。第四纪时接受了一套滨浅海相碎屑建造、河流相–三角洲相碎屑建造、火山建造。

新构造运动主要表现为地壳差异性升降运动，火山喷发活动，并在断裂作用下，导致温泉广布。

3. 莺歌海裂陷盆地

包括北西向莺歌海盆地和北东向琼东南盆地。隐伏于南海之下，盆地内沉积了古新统及其以上地层，地震资料解释新生代沉积层厚度达20000m，其中新近系厚度达10000m以上，以粗碎屑沉积为主，多为冲积扇沉积（张训华等，2008）。经历了由洪冲积相、湖泊相→滨、浅海相→浅海相、半深海相水深逐渐变深的过程（刘昭蜀等，2002）。从老至新，有始新统—下渐新统崖城组、上渐新统陵水组、中新统三亚组、梅山组、黄流组、佛罗组和上新统莺歌海组、望楼港组。

4. 西沙–中沙地块

南海北部在新生代以前与海南陆架同属一体（何起祥等，2006），古近纪晚期由于太平洋板块的俯冲作用，稳定的大陆边缘发生破裂，并解体为许多小陆块，并在地壳的差异性升降运动作用下形成一系列的隆起与裂陷，西沙隆起就是其中之一。由于远离大陆，又有裂陷相隔，陆缘物质缺乏，在前中生代基底之上形成了一套礁相碳酸盐岩建造夹白垩。古近纪以来的沉积有中中新统西沙组、宣德组和上中新统永兴组、下更新统永发组、中更新统琛航组、上更新统石岛组礁相碳酸盐岩建造和全新统西沙洲组砂泥质沉积。除了石岛组与下伏琛航组和上覆西沙洲组为不整合接触外，其余相关联岩石的地层单位间均为整合接触关系，显示其构造运动主要发生于中更新世晚期和晚更新世晚期。

5. 中央海盆

中央海盆是一个具有与正常洋壳相似的地壳结构的洋盆，其沉积基底是海底洋壳（刘昭蜀等，2002）。在洋壳的一些凹陷处可能存在较多的海盆形成初期的火山碎屑、熔岩流成分。火山岩之上，海盆北部可能以陆源碎屑沉积为主；南部则以碳酸盐岩沉积为主；海盆中央区由于平均深度大于碳酸盐岩沉积补偿平衡深度，且离大陆较远，能够到达海盆的主要是黏土物质，故表层沉积物主要是含粉砂质的远洋放射虫黏土或含远洋放射虫黏土的粉砂和砂。

6. 南沙地块

南沙地块新生界以滨浅海相碎屑沉积为主。新生界早期为一套滨海相沉积，以砂砾岩、泥岩为主的古新统宝滩组（原东坡组）不整合于下白垩统之上，其上为阳明组的半深海相页岩，阳明组与东坡组为连续沉积。阳明组之上为滨海相的忠孝组，忠孝组与阳明组之间存在一个仅局部发育的不整合面，被认为是南海开裂作用开始的反映。忠孝组之上接受了一套滨海相台地碳酸盐岩沉积，这套碳酸盐岩沉积以南沙隆起台阶为主体，呈席状向东南越过南沙裂陷延伸至巴拉望-加里曼丹大陆架，表明该时期南沙隆起与南沙断陷的沉积特征趋于一致。

7. 曾母地块

沉积有古近系拉让群、古近系—新近系曾母群和新近系海宁组、南康组（或特隆布组）及新近系—第四系北康组。其中，拉让群为浅海—深海相浊积岩，与下伏白垩系丹脑组和上覆曾母群呈不整合接触；曾母群为河流、港湾汊道相砂岩夹泥岩和煤层沉积，与上覆海宁组整合接触，厚数百米至数千米；海宁组—南康组为台地碳酸盐沉积，与上覆北康组呈不整合接触；北康组为河口、海岸平原相砂泥质沉积。构造运动主要发生于古近纪和新近纪内部，与南沙地块构造运动时间并不一致，反映了构造运动的多次性或构造运动时间具有不均一性。

第二十章 构造阶段、构造层及构造旋回

海南省总体上经历了中元古代块体拼贴、统一基底形成和新元古代—古生代板内裂解、晚古生代—三叠纪碰撞拼贴及侏罗纪以来板内演化等漫长而复杂的地质演化过程，呈现出“合久必分，分久必合”的自然规律，形成了不同时期的构造格局和构造单元特征。不同构造阶段有着不同的构造单元，不同构造单元的构造旋回有明显差异，其构造层也不一。

第一节 中元古代阶段构造层和构造旋回

中元古代是海南省统一结晶基底形成的重要时期，在海南岛见长城系戈枕村组和长城系—蓟县系峨文岭组分布。戈枕村组为混合岩类和片麻岩类岩石组合（图版Ⅴ-1-A），峨文岭组则为石英岩类和片岩类岩石。尽管受后期多期次构造作用的改造，导致中元古界或呈残块状漂浮于花岗岩之中，或与上覆的不同时代地层呈断层接触。但发生于约1400Ma的抱板运动导致了中元古界发生区域角闪岩相变质和高度区域性塑性变形，伴随约1400Ma的中元古代片麻状花岗闪长岩和片麻状二长花岗岩侵位，并使得地壳隆起为陆，缺失待建纪的沉积。因此，中元古界属于一个独立的构造层，即抱板期构造层。

该时期，海南省地壳处于激烈活动时期，火山喷发活动比较频繁，伴随地壳的大幅度急剧下降，接受了一套厚约985m的夹有中酸性浅成岩或熔岩的海相陆缘碎屑沉积，形成了戈枕村组岛弧相中酸性火山岩建造夹碎屑岩建造和峨文岭组海相碎屑岩建造夹火山岩建造，共同构成了抱板旋回的沉积。原岩恢复表明（马大铨等，1998）：戈枕村组属岛弧或陆缘弧钙碱性系列中酸性火山岩夹杂砂岩，火山岩主要是英安质和石英安山质的熔岩、凝灰岩类；峨文岭组原岩为一套夹有少量火山岩的泥砂质岩、砂岩和少量富钙岩石等陆缘碎屑沉积岩系，岩石成熟度低，具类复理石沉积韵律。经区域变质作用，形成下部条带状白云母钾长片麻岩和条带状、眼球状混合岩和上部片岩类、石英岩类，局部发育变余水平层理和韵律性粒序层理。反映了由早至晚，构造环境由强烈活动至逐渐减弱趋于稳定的趋势。中元古代发生的抱板运动，使抱板群戈枕村组和峨文岭组分别发生了不同程度的高角闪岩相和低角闪岩相的区域变质作用及固态流变，戈枕村组并发生混合岩化作用，导致抱板群广泛发育新生的面理-片理、片麻理和线理以及各种类型的褶皱构造。随后褶皱回返隆起为陆，成为海南最古老的结晶基底。

第二节 新元古代—三叠纪阶段构造层和构造旋回

新元古代—三叠纪阶段，海南省地处羌塘-扬子-华南板块和印支-南海块体两个一级构造单元的结合部位。海南省北部属于羌塘-扬子-华南板块华南新元古代—古生代造山带（二级）五指山造山带（三级），南面则为印支-南海块体南海地块（二级）和三亚地体（二级）。

一、五指山造山带

（一）构造层划分

构造层的划分，一般是用构造运动形成的不整合作为划分依据。五指山造山带可识别出的主要不整合面有：青白口系石碌群与南华系—震旦系石灰顶组间的不整合面、上古生界与下古生界间的不整合面、白垩系与古生界间的不整合面等。

1. 不整合面的厘定

海南省经历了多旋回、多阶段的地质演化，导致海南岛北部不整合面由于后期构造作用改造而消失，上、下地层呈现出构造胁迫一致的现象。但在变形弱的部位，该不整合面依稀可见。

（1）青白口系石碌群与南华系—震旦系石灰顶组间的不整合

青白口系石碌群与震旦系石灰顶组之间的不整合面仅在局部地区出露，由于受后期构造作用改造，两者呈断层接触关系，但两者存在沉积建造类型、岩石组合特征、变质变形程度等方面存在差异，仍可确定二者间的不整合关系。石灰顶西北坡见石灰顶组与下伏石碌群呈高角度接触，其间发育了厚0.5~1m的底砾岩，砾石大多棱角清楚，长轴平行接触面排列。石灰顶组具明显的复理式韵律，仅在底部因滑动构造发生轻微变质，主体基本保留了沉积岩的结构特点；而下伏石碌群普遍遭受绿片岩相变质作用，塑性流变变形十分发育。

不整合面见于石碌矿区石灰顶、红房山一带，在石灰顶西北山坡，两者接触面凹凸不平，岩层走向不同，倾角不一（图20-1）。同时，它们间不仅缺失南华系，而且变质变形作用差异性显著。表明两者呈不整合接触，反映了晋宁运动的存在。构造运动的结果直接导致了石碌群发生区域绿片岩相变质和中深层次的变形，形成了海南岛的褶皱基底，构成了海南岛双重基底结构。

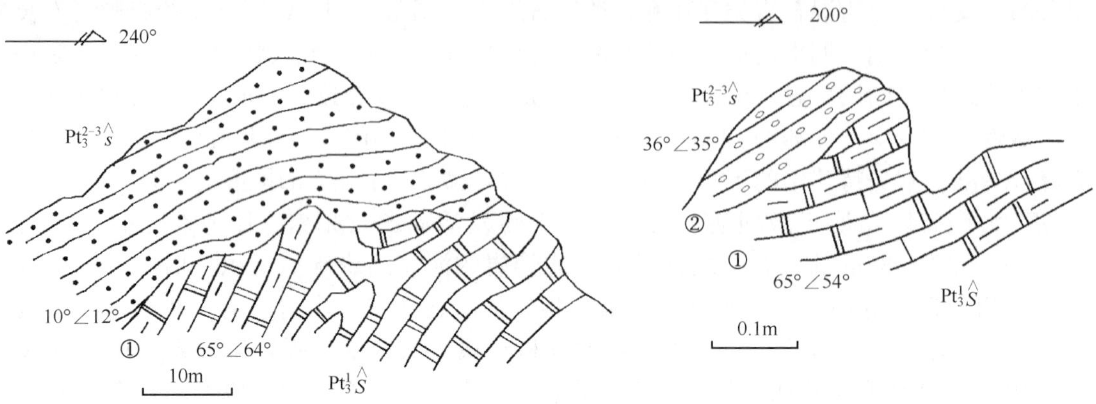

图20-1　南华系—震旦系石灰顶组与青白口系石碌群的不整合接触素描图

Fig. 20-1　Sketch showing the unconformable contact relationship between the Nanhuan-Sinian Shihuiding Formation and the Qingbaikouan Shilu Group

（据宜昌地矿所，1994）

$Pt_3^1\hat{S}$—石碌群；$Pt_3^{2-3}\hat{s}$—石灰顶组；①片理化二透化白云质大理岩；②砾岩

由于沿袭不整合面构造滑动的强烈改造，测区南华系—震旦系石灰顶组和下伏青白口系石碌群之间大都表现为断层接触关系，但依据下列地质事实仍可确定两者间不整合接触。

1）在石灰顶山西北坡，石灰顶组与下伏石碌群呈高角度接触，其间发育了一套厚度0.5~1m不等的底砾岩。砾石含量约60%~70%，成分以脉石英、石英岩、千枚岩为主，局部见铁矿石；砾石少量呈磨圆状，多数棱角清楚，长轴一般平行接触界面排列。

2）石灰顶组岩性、岩相和厚度变化较小，未见明显的复理式韵律，可能代表一次海进旋回开始形成的浅海或海湾相含铁质碎屑建造；而下伏石碌群主要由陆屑泥砂质沉积物、碳酸盐岩和中酸性火山凝灰岩组成，代表了一种浅海海湾-潟湖相的碎屑岩、碳酸盐岩型沉积建造，因此，两者沉积建造差异较大。

3）在变质程度和变形特征上两者差异显著。石灰顶组仅在底部因构造滑动发生轻微变质，主体基本保留了沉积岩的组构特点；而石碌群普遍遭受绿片岩相变质作用。石灰顶组构造相对简单，褶皱多为开阔型，而石碌群塑流变形十分发育，褶皱叠加现象比比皆是。

（2）上古生界与下古生界间的不整合

上古生界与下古生界间的不整合面由于受到后期断裂构造的改造，使得上、下古生界间依然保持宏观尺度中的“整合”一致，两者在变形变质作用方面未具明显的总体差异性，具有比较相似的岩组结构，但在受后期构造改造较弱的部位，如江边青天峡一带，可见到石炭系青天峡组与志留系陀烈组呈不整合接触关系，上覆青天峡组产状为340°∠77°，下伏陀烈组产状为135°∠78°，两者接触面不平整，局部呈明显波状弯曲（图20-2）。

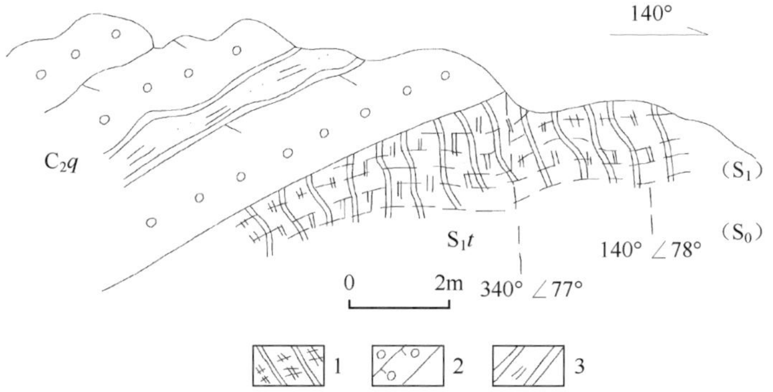

图20-2 青天峡附近石炭系青天峡组（C_2q）与志留系陀烈组（S_1t）的不整合接触素描图

Fig. 20-2 Sketch showing the unconformable contact relationship between the Carboniferous Qingtianxia Formation（C_2q）and the Silurian Tuolie Formation（S_1t），near the Qingtian Gorge

（据广西区调大队，1991）

1—绢云母板岩；2—砾岩；3—粉砂质板岩

（3）三叠系与长城系、古生界间的不整合

该不整合面分布比较广，不整合面之上下两大套岩层变质变形差异明显。

翰林镇马车坡附近，下三叠统岭文组以底砾岩（砾岩或砂砾岩）与抱板群接触，产状分别为48°∠34°、112°∠38°；底砾岩砾石成分较杂，为脉石英、石英、硅质岩、变质岩、花岗岩，呈次棱角状、次圆状，大小一般0.5~3cm。

九曲江海洋村南，下三叠统岭文组与志留系呈不整合接触（图20-3）。接触面之上的岭文组底部为陆相的砂砾岩、含砾中细砂岩，砾石成分以石英为主，少量来自其基底的绢云母板岩、硅质岩、绢云母石英片岩；其下为绿片岩相变质的志留系陀烈组板岩、千枚岩，上、下地层呈交切关系。

（4）白垩系与古生界间的不整合

该不整合面分布比较广，也比较常见，主要沿白垩系红盆边界发育。儋州市红岭一带，白垩系鹿母湾组紫红色砂砾岩呈高角度覆盖于志留系陀烈组三段含铁石英岩和南好组第五段石英岩之上。琼海三州-九曲江一带，表现为下白垩统鹿母湾组底部以陆相沉积的含砾砂岩或砂砾岩分别覆盖在古生界不同的层位之上，上、下两套地层产状不同，呈交切关系（图20-4）。

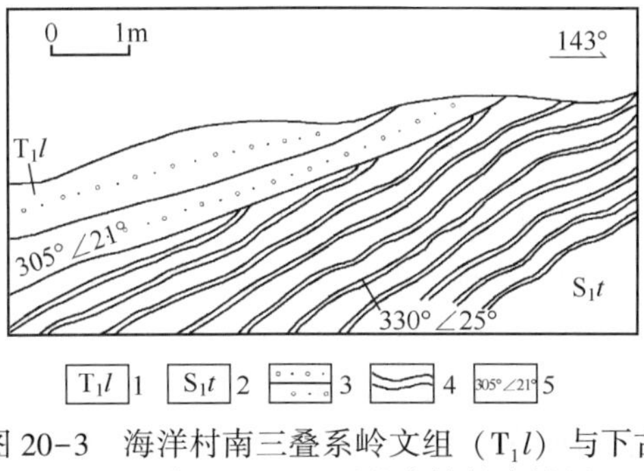

图20-3 海洋村南三叠系岭文组（T_1l）与下古生界陀烈组（S_1t）不整合接触素描图

Fig. 20-3 Sketch showing the unconformable contact relationship betweenthe Triassic Lingwen Formation（T_1l）and the Lower Paleozoic Tuolie Formation（S_1t），south of Haiyang village

（据宜昌地矿所，1995）

1—下三叠统岭文组；2—下志留统陀烈组；3—砂砾岩；4—板岩；5—产状

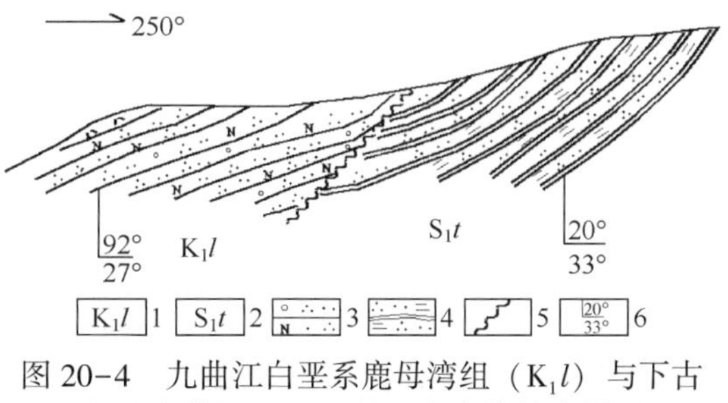

图20-4 九曲江白垩系鹿母湾组（K_1l）与下古生界陀烈组（S_1t）的不整合接触素描图

Fig. 20-4 Sketch showing the unconformable contact relationship between the Cretaceous Lumudu Formation（K_1l）and the Lower Paleozoic Tuolie Formation（S_1t），Jiuqu River

（据宜昌地矿所，1995）

1—下白垩统鹿母湾组；2—下志留统陀烈组；3—含砾长石石英砂岩；4—变质绢云母石英砂岩；5—不整合界线；6—产状

此外，尽管五指山造山带中长城系戈枕村组和长城-蓟县系峨文岭组与青白口系石碌群为断层接触关系。但发生于约1400Ma的中元古代造山运动无疑是两者为不整合接触关系的反证。另外，琼海市中原镇迈州岭白垩系不整合覆盖在三叠系岭文组之上，古近系与白垩系以及白垩纪花岗岩间的不整合接触至今是一个不争的事实（陈哲培等，1997）。

2. 构造层划分

根据上述不整合面划分出6个构造层：晋宁-扬子构造层、加里东构造层、华力西构造层、印支构造层。构造演化划分为晋宁-扬子旋回、加里东旋回、华力西-印支旋回三个构造旋回。

（二）构造旋回

1. 晋宁-扬子旋回

抱板运动后，海南省进入一个相对平静的时期，缺失待建纪（1400～1000Ma）的沉积。晋宁旋回的沉积目前仅在昌江县石碌地区，沉积了巨厚（厚度大于1000m）的青白口系石碌群滨-浅海相陆源碎屑岩-碳酸盐岩建造，夹有数层中酸性火山熔岩和火山凝灰岩建造。晋宁运动的表现是：导致石碌群发生低绿片岩相区域变质作用和固态流变，形成了以横向构造置换为主要形式、顺层构造滑动为主要特色的塑流变形构造形迹组合，成为堆叠于抱板杂岩结晶基底之上的变质基底。根据石碌群上部主矿层中获得Sm-Nd全岩等时线年龄为（841±20）Ma，以及南华系—震旦系与之呈角度不整合接触的事实，晋宁运动发生的时间大致为距今8.4亿年左右，这一构造事件在整个华南地区均有显示，表明晋宁运动波及范围较广。但与晋宁运动相伴随的岩浆活动未见及，反映地壳的克拉通程度不高、稳定性差，并在新元古代南华纪—震旦纪时发生裂解，接受了石灰顶组一套残坡积相-滨海相石英砂岩、铁质石英砂岩、含铁粉砂岩、砂质泥岩、含铁泥岩及贫铁矿等沉积，底部见含砾泥岩、含角砾泥岩。底部的含砾泥岩、含角砾泥岩以其无层理、分选差、圆度差和角砾为原地或微异地等为特征，具有类磨拉石建造特点，其岩石组合及层序特征反映主要形成于浪基面之上，即无障壁近滨相，总体上表现为由陆到海的沉积特征。随后，地壳全体隆升，结束该造山事件记录。

2. 加里东旋回

寒武纪时期的沉积主要见于岛东万宁一带。下寒武统（滇东统）美子林组属于以近岸陆源碎屑沉积占主导的滨海—浅海相碎屑岩-碳酸盐岩混合型沉积，代表了地壳裂解早期的沉积。其下部为长石石英砂岩、石英砂岩、粉砂岩、泥质砂岩，结构成熟度和成分成熟度均较高，局部出现大型低角度的斜层理；上部主要为灰岩、砂质灰岩等，见小型砂纹层理，灰岩中产瓶虫类和海绵骨针类化石，表明裂解早期构造环境相对比较稳定，并呈现出从早至晚（由南至北），岩石粒度逐渐变细，海水逐渐加深，沉积场所逐渐远离滨岸。

经滇东统沉积后，在加里东运动作用下，地壳发生垂直升降运动，导致了寒武系与其上覆的奥陶系呈假整合接触。缺失黔东统—芙蓉统的沉积。

奥陶纪时裂解程度加剧，沉降盆地拓宽，海水加深，沉积作用基本上波及整个海南岛。邦溪、合盛、坝王岭一带地壳持续伸展，裂解程度达到了高峰，沿地壳减薄处发生火山喷发作用，出现生成于板块边缘环境的超镁铁质、镁铁质火山岩夹层，Sm-Nd同位素年龄为（458±36.8）Ma，岩石经区域绿片岩相的变质改造，形成细粒斜长阳起片岩、细粒二长阳起片岩、绿泥石阳起片岩、橄榄质科马提岩等。盆地内并沉积了厚3000余米的深海相粉砂-泥质沉积，自下而上（由南至北），粒度变细，见变余砂纹层理，晚期以频繁地出现复理石韵律为特点，相当于鲍马序列中的C、D、E层，B层少见，A层不发育，相当于浊积扇的边缘部分，具有类复理式-复理式建造夹变质超基性、中基性火山岩建造及少量变质凝灰岩建造的特征。岛东中原镇盆地为厚2000余米的浅海—半深海相陆源碎屑沉积，下部为泥质和粉砂质岩石，上部为泥质粉—细砂岩，局部为细粒石英砂岩、凝灰质砂岩，岩石分选性差，发育粒序层理，可见鲍马序列中的A、B层。儋州市洛基镇洛基大岭一带，下、上部以石英千枚岩、绢云母石英板岩、二云母石英片岩等泥质、粉砂质、砂质岩石为主，夹少量中薄层状的变质粉砂岩、炭质板岩、变质细砂岩、石英岩；中部为中厚层变质粉砂岩、变质细砂岩、石英岩。发育变余水平层理。自下而上，岩石粒度由细→粗→细，沉积盆地呈现出由深→浅→深的变化，总体具有滨海、浅海相碎屑岩建造特征。定安县翰林镇蒙花岭-礼文岭一带，下中部为泥质、粉砂质沉积，岩性为绢云母片岩、石英岩、二云母片岩等，局部夹炭质千枚岩，发育变余水平层理，中部并夹多层基性火山岩、火山碎屑岩，自下而上，火山活动由弱→强，呈现出由安山质沉凝灰岩-酸性流纹岩的演化。上部沉积物粒度变粗，出现浊流沉积。总体为浅海-半深海相碎屑岩建造夹火山岩建造，且从早至晚，其构造环境具逐渐强烈活动的趋势，与之相伴随的火山作用具间歇性的特点。初步认为可能属于同一个盆地的沉积，并呈现出东浅西深的古地理格局（李孙雄等，2007）。

早志留世早期，盆地进一步加深，导致泥质、粉砂质的沉积，且砂质岩石和泥质岩石呈相互交替出现，形成了变质绢云母粉砂岩、变质石英砂岩、绢云母千枚岩、绢云母板岩等。岩石普遍发育变余水平层理、变余粒序层理，显示了半深海—深海相类复理石、复理石建造特征。岛西地区沉积具有对称型层序的特点，从早至晚，岩石粒度呈现出粗—细—粗的演变，沉积环境经历了浅—深—浅的发展，构成了一个完整的沉积旋回，形成浅海陆棚-深海相复理石-类复理石建造。下部为成熟度较高的石英砂岩，层厚大，发育水平层理、微波状层理和板状交错层理；中部以成分成熟度低的粉砂岩、砂岩、泥岩交替出现为特征，发育板状层理、水平层理、粒序层理，发育鲍马序列中的 D、E 层或 C、D、E 层，厚约 1750m；上部为成熟度低的粉砂岩、细砂岩层，见灰岩、粉砂岩透镜体，相当于鲍马序列 A 层或 A 、B 层。而岛东地区则表现为退积型层序，下部为浅海—深海平原相成熟度低的泥质岩-粉砂岩沉积；中部则进入深海平原相，浊流作用频繁，发育粒序层理，属于鲍马序列中的 D、E 层或 C、D、E 层，厚近 3000m；上部为 A 、B 层沉积，常见细砂质层，亦出现中粒砂质层，岩层变厚，多为中厚层状，少量厚层状乃至块状，并普遍见沉凝灰岩或凝灰质砂岩分布，发育粒序层理。显示了西浅东深的古地理格局，与奥陶系盆地西深东浅的格局明显不同，反映了地壳运动强度具有不均一性。早志留世中期，为相对封闭的滞氧还原环境下的沉积，沉积物粒度变细，以泥质为主，并富含炭质。早志留世晚期，盆地逐渐收缩、关闭，进入滨海、浅海环境，沉积物粒度变粗，以砂质为主。

上述分析显示：古生代盆地南面自南至北，沉积物粒度逐渐变细，海水不断变深；北面由北至南，沉积物粒度也逐渐变细。表明该时期海南岛存在一个统一的洋盆，洋盆走向可能为近东西向。盆地中心大致位于石碌镇—八一农场一线，奥陶纪时盆地裂解达到高峰，出现多层洋底基性火山岩，并自东而西，沉积环境由浅海向半深海、深海演化，呈现出东浅西深的古地理格局。而志留纪时，东深西浅的盆地特征与奥陶系明显不同，可能与地壳差异性升降运动有关。自寒武纪至志留纪，盆地经历了多阶段、多旋回的沉积，每个阶段、每个旋回基本上处于早张晚压的背景中，地壳以升降运动为特点，尤其是奥陶纪后，东西两侧差异升降运动更明显，并最终在志留纪末隆起为陆，也使得洋盆消失，缺失泥盆系的沉积。岩石中见洋底变基性火山岩夹层，但缺少或少见内源物质的沉积，洋壳未真正得以生成，反映了该时期地壳具有快速裂解和上升的运动特点，导致了海南岛奥陶纪、志留纪具有不同的沉积盆地格局。据杨坤光研究，华南加里东运动不仅在性质上表现具有多样性，在时间上也具有多阶段性（多幕性），主要分别发生于寒武纪末、奥陶纪末和志留纪末。因此五指山造山带早古生代盆地所表现出来的演化特征具有华南加里东运动的特点。但就其构造运动的强度而言，海南岛表现相对比较微弱，没有华南其他地区强烈。

发生于早志留世晚期的加里东运动使得早古生代岩石发生低绿片岩相的区域变质作用，伴随发生固态流变，形成新生线理、面理和流变褶皱构造。随后，褶皱回返，地壳隆起抬升，遭受剥蚀，缺失中、上志留统—泥盆系的沉积记录。

3. 华力西-印支旋回

华力西-印支旋回的沉积包括石炭系、二叠系、三叠系。

石炭纪的沉积作用以下石炭统南好组、上石炭统青天峡组为代表。早石炭世南好组时期，表现为稳定环境下滨海相碎屑岩相沉积，主要是浅变质作用形成的砂岩类岩石，局部夹弱变质的酸性火山凝灰岩，其底部为变质石英砾岩、砂砾岩、砂岩，沉积物粒度粗，分选性、磨圆度好，结构成熟度和成分成熟度较高，除少量岩屑外，几乎全由石英组成；上部为变质陆源细碎屑岩，总体为板岩、含粉砂质板岩与变质粉砂岩、细砂岩不等厚互层，砂岩、粉砂岩分选性好，不含或含少量泥质，偶见变余斜层理，南好组岩石中普遍富含长石，代表了近源的沉积环境。

晚石炭世早期，沉积环境与南好组相近，为相对稳定的滨、浅海相环境，继承南好组的沉积特点，但底部出现碳酸盐岩，具有滨、浅海相单陆屑式建造夹碳酸盐岩建造特点。晚石炭世晚期，在琼海中原一带，盆地加深，进入浅海—半深海环境，出现厚百余米的板岩与杂砂岩不等厚互层或板岩夹杂砂岩，杂砂岩呈中—薄层状，普遍含一定的泥质杂基，具粒序层理，具有浊流沉积特点。总体显示随地史的演化发展，沉积建造类型由稳定型往次稳定型和不稳定型转化。

二叠纪时的沉积见于测区石碌镇—王下一线，包括下二叠统鹅查组、鹅顶组和上二叠统南龙组。其中，南龙组为海陆交互相沉积。鹅查组、鹅顶组沉积时，为近滨相和碳酸盐岩台地相沉积，相对石炭纪而言，其地壳已有所抬升。原岩恢复表明，该套地层为滨浅海相的泥（碳）、粉砂和细砂等细碎屑沉积，产大量生物化石，包括腕足类、海百合茎及大量孢粉等化石。

三叠系的沉积比较局限，仅出露于定安翰林、琼海九曲江两地，为下三叠统岭文组陆相沉积。呈北东向长条状分布，下部为复成分砾岩、含砾砂岩等，复成分砾岩砾石成分以变质岩、花岗斑岩、脉石英、硅质为主，磨圆度中等—好，而分选性、成分成熟度、结构成熟度差，具快速沉积的特点；上部为泥质粉砂岩、粉砂质泥岩，发育水平层理、砂纹层理，自下而上粒度呈现出由粗→细的沉积变化。产丰富的孢粉、植物化石。

华力西—印支运动是一个具有造山运动性质的构造运动，这次构造运动于海南岛是一个重大的转折点，它结束了海南岛区域变质作用和透入性变形的历史，并进入陆内盆地沉积阶段。华力西运动一方面导致上古生界发生了低绿片岩相区域变质和中—浅层次的变形，使得下古生界叠覆在上古生界之上，形成了区域性崖城-万宁断裂带及其相伴随的弧状构造带，以及北东向戈枕、大村韧性剪切带等主要构造形迹。另一方面拉开了大规模岩浆侵入活动的序幕，导致同期早二叠世俯冲型、早二叠世—中二叠世同碰撞型、晚二叠世后碰撞型和三叠纪隆升型系列花岗岩分布。

值得一提的是，尽管三叠系与石炭系、二叠系分属不同的构造层。但二叠系仅分布于西部，从早至晚，为海相→海陆交互相沉积；而三叠系分布于东部，为陆相类磨拉石、磨拉石沉积。与华力西—印支造山作用时间基本吻合，三叠纪类磨拉石、磨拉石建造的出现代表了华力西—印支造山运动地壳隆升阶段的产物，是造山作用结束的反映。但东部琼中地区中晚三叠世花岗岩普遍发育定向组构，显示了挤压侵位环境；西部晚二叠世—早三叠世花岗岩属于碰撞后-隆升阶段产物分析，初步认为可能属于不同地壳深度的表现。一般认为，隆升阶段应力作用上部为伸展拉张、下部则为挤压环境。由此看来，海南华力西—印支造山运动碰撞后时间应为晚二叠世，而早三叠世则属于隆升阶段，形成于挤压环境的琼中地区中晚三叠世花岗岩属于深侵位的结果。当然并不排除造山作用存在不均衡性和不同步性的可能。

二、印支-南海地块

该地块资料少，也比较零星。即本次所收集到的资料而言，该区分布有下古生界、上古生界和三叠系。下古生界地体出露于中南半岛（原印支地块）和海南岛三亚地区，都为未变质的海相碎屑岩。区域资料显示，中南半岛印支地块古生界、中生界未变质地层与下伏元古宇为不整合接触，上三叠统—下侏罗统与上石炭统—上三叠统也呈不整合接触关系。海南岛三亚地区下古生界由寒武系与奥陶系组成，缺失上古生界，寒武系与奥陶系为假整合接触。南沙群岛上古生界二叠系零星分布，与元古宇不整合接触，岩石为灰岩和燧石、硅质碎屑岩、长石杂砂岩。上三叠统为深海相变质岩。其中，元古宇岩石组合特征与海南岛戈枕村组和峨文岭组基本类似，马大铨等（1998）并认为西沙群岛永兴岛钻孔中所见花岗片麻岩和石英云母片岩甚至就是抱板群，本次研究也持相同的看法。因此，把该地块构造层划分为加里东-印支构造层，经历了加里东-印支构造旋回的演化。

经早晋宁运动，结晶基底形成后，该地块长期为陆，缺失待建系—下寒武统（滇东统）的沉积。黔东世时，地壳开始发生下沉，中南半岛（原印支地块）先后接受了厚400m的含三叶虫的中寒武统（黔东统）海相石英质砂岩、砂质泥板岩沉积，往上为泥盆系—下石炭统为砂岩和灰岩夹页岩、上石炭统—上三叠统为一套海陆交互相沉积，以及上三叠统—下侏罗统一套红色含膏盐的沉积，与下伏上石炭统—上三叠统呈不整合接触。海南岛三亚地区接受了中黔东统孟月岭组、黔东统—武陵统大茅组浅海相陆源碎屑岩夹少量磷块岩和奥陶系大葵组、牙花组、沙塘组、榆红组、尖岭组、干沟村组滨海—浅海相碎屑岩—碳酸盐岩建造等盖层沉积，其沉积总厚度巨大，达7000m以上，远远大于五指山造山带同时代岩石地层单位的厚度，可能显示了该区地壳下沉幅度较大、陆缘物质补给也比较充分的沉积环境。南沙群岛上古生界二叠系零星分布，岩石为灰岩和燧石、硅质碎屑岩、长石杂砂岩；在礼乐滩，上三叠统—下侏罗统为三角洲相-开阔浅海相砂泥岩及硅质页岩、粉砂岩。

该地块早期构造运动以地壳差异性隆升为特点，岩石未发生区域变质和变形，岩浆作用不发育，差异性沉积明显。如海南岛三亚地区寒武系与奥陶系不整合接触，被第四系松散沉积直接上覆；另外，西沙群岛古近系与元古宇直接接触，中南半岛泥盆系—下石炭统直接叠覆于中寒武统（黔东统）之上等。

二叠纪—三叠纪的块体碰撞一方面使得印支-南海地块与五指山造山带拼合为一个整体，另一方面又使得印支-南海块体三亚地体古生界发育走向北东向的褶皱和逆冲推覆构造，导致局部地段寒武系叠覆于奥陶系之上。同期的岩浆作用也比较强烈，育才一带三叠纪花岗岩中普遍发育定向组构，表明其形成于挤压构造环境，结合稍晚期出现碱性花岗岩，说明地体经历了由挤压到伸展的过程。同时，导致了中南半岛上三叠统—下侏罗统与上石炭统—上三叠统呈不整合接触。

第三节　侏罗纪—白垩纪构造层和构造旋回

该阶段，由于菲律宾裂陷盆地已出国境，而南沙裂陷盆地（二级）被海水淹没，资料非常少，因此，本次仅能对涉及海南省范围的海南断陷盆地进行叙述。

海南断陷盆地位于琼州海峡断裂与南沙裂陷之间，其沉积作用目前仅在海南岛见到，为白垩纪陆相红盆沉积，属于燕山构造层，与下三叠统和古近系为不整合接触，为燕山旋回沉积的产物。

燕山阶段是海南岛断块活动非常活跃的时期，并控制了多个白垩纪断陷盆地的形成，同期酸性火山活动强烈且持久，持续整个早白垩世，具有多期次的间歇性喷发活动特点。早白垩世时，在断陷盆地内接受了一套陆相碎屑沉积，形成了复陆屑建造夹酸性火山岩建造，岩性有砾岩、含砾砂质泥岩、粉砂质泥岩夹凝灰岩，部分泥岩含丰富酸性火山碎屑物质；岩石整体上，碎屑成分较杂，分选性往往差—中等，多呈次棱角状—次圆状，结构成熟度和成分成熟度较差，具近地快速堆积的特点。发育中、大型交错层理、大型平行层理、水平砂纹层理、粒序层理、水平层理、韵律层理，可见冲刷痕。同时，火山喷发作用强烈，形成牛腊岭、同安岭、五指山等白垩纪火山岩盆地。总的沉积环境为火山活动频繁、气候干旱的山间盆地冲积扇及河道相。晚白垩世时，火山活动衰竭，断陷盆地内仅接受了白垩系一套浅紫红色—浅灰白色的陆相碎屑沉积，岩性组合为泥岩、粉砂岩、长石岩屑杂砂岩、岩屑长石粗—细砂岩、复成分细—中砾岩等，见微水平层理和冲刷痕。下部以岩屑长石砂岩、泥质粉砂岩为主；中部以杂砂岩为特色，见砾岩、铁泥质粉砂岩夹层；上部则以复成分砾岩为主。总体上，碎屑成分杂，分选性和磨圆度差，结构成熟度和成分成熟度较低，具陆相复陆屑建造沉积特征。自早白垩世到晚白垩世，具有火山活动由强到弱、沉积建造类型由不稳定型到次稳定型、构造环境由动到静转化的趋势。

燕山运动掀起了韧脆性断裂活动的高峰，塑造了北东、北西、南北向等不同尺度、不同规模的断裂构造带，奠定了海南岛的断裂变形格局，控制了同期岩浆侵入活动，同时引起地壳抬升隆起，使得白垩纪形成北东向的单斜构造。

第四节 新生代构造层和构造旋回

新生代，海南省共划分出6个三级构造单元。但由于部分构造单元资料有限，仅进行粗略的概述。

雷琼裂谷位于海南省北部东西向王五-文教断裂带以北地区，省内喜马拉雅旋回沉积有古新统长流组、始新统流沙港组、渐新统涠洲组和中新统下洋组、角尾组、长坡组、灯楼角组及上新统海口组等碎屑岩建造夹可燃性有机岩建造；火山岩有中更新统多文组、晚更新统道堂组和全新世石山组基性火山岩和火山碎屑岩；第四系包括秀英组、北海组、八所组、万宁组、琼山组和烟墩组松散沉积。其中，古新统—中新统呈隐伏状产出，上新统海口组也以隐伏状产出为主，出露比较零星。总沉积厚度大于3000m。根据渐新统涠洲组与中新统下洋组为不整合接触，把其构造层粗略划分出古新世—渐新世构造层、中新世—第四纪构造层两个不同的构造层，属于两个亚旋回（?）的沉积。雷琼裂谷是海南省新构造运动迹象表现丰富、样式多样的地区，有地壳升降运动、断裂活动及与其相伴随的地震、火山喷发活动等。

海南地块古新统昌头组、始新统长昌组和瓦窑组、渐新统—中新统丘糖岭组、中新统佛罗组、上新统望楼港组碎屑岩建造夹可燃性有机岩建造，丘糖岭组与佛罗组为假整合接触；火山岩有中新统石马村组、上新统石门沟组基性火山岩、火山碎屑岩；第四系包括秀英组、北海组、八所组、万宁组、琼山组和烟墩组松散沉积。总体属于一个旋回同一构造层沉积。其新构造运动表现强度相对雷琼裂谷比较弱，迹象也少，主要以地壳总体升降运动占主导，并发生了断裂活动及与其有关的地震、火山喷发活动和温泉出露等。

莺歌海裂陷盆地沉积有渐新统陵水组和中新统三亚组、梅山组、黄流组及上新统莺歌海组等碎屑岩建造夹可燃性有机岩建造。第四系为松散沉积（未建组）。其中，陵水组与上覆的三亚组为不整合接触，据此初步划分出两个构造层，即渐新世构造层和中新世—第四纪构造层，属于两个亚旋回（?）的沉积。发生于渐新世—中新世的构造运动在雷琼裂谷中也有体现，反映了该期构造运动具有一定的波及范围。

西沙-中沙地块沉积有中新统西沙组、宣德组和上新统永兴组礁碳酸盐岩建造沉积，第四系有下更新统永发组、中更新统琛航组、上更新统石岛组和全新世西沙洲组松散沉积。石岛组与下伏琛航组、上覆西沙洲组为不整合接触，据此，初步把其构造层划分为中新世—中更新世构造层、晚更新世构造层和全新世构造层三个构造层。

南沙地块沉积古新统宝滩组、始新统阳明组、始新统—渐新统忠孝组海相碎屑岩建造和中新统南湾组、上新统永暑组、更新统南沙组和全新统南海组礁碳酸盐岩建造。根据阳明组与上覆忠孝组为不整合接触，其构造层有古新世—始新世构造层和始新世—全新世构造层。

曾母地块沉积古新统—始新统拉让群、渐新统—中新统曾母组、中新统海宁组和南康组、上新统—全新统北康组。拉让群以浅海—深海相千枚岩、板岩为主，曾母组主要为河流、港湾相砂岩夹泥岩和煤层，海宁组和南康组礁碳酸盐岩沉积，北康组为河口、海岸平原相沉积。其中，拉让群和曾母组、南康组和北康组为不整合接触，由此把其构造层划分为古新世—始新世构造层、渐新世—中新世构造层、上新世—全新世构造层。

纵观整个南海地质，其演化特点是：由于南海的扩张作用，地壳拉张下陷，形成一系列断陷盆地，盆地内接受了厚达数千米的新生代沉积。古近纪时南海北部断陷盆地内接受了一套陆相类磨拉石、火山碎屑、膏岩和含油页岩沉积；南海南部的水体较深，接受了一套海相碎屑沉积，且在下部见三角洲-河流相碎屑岩，上部见半深海相页岩，海水呈逐渐加深的趋势。新近纪时地壳拉张下陷加剧，南海北部转变为海相，接受了一套滨海相—浅海相的陆源碎屑沉积及三角洲相沉积；南海南部水深进一步加深，接受了一套深海相泥质沉积或浅海生物沉积；中央海盆由于地壳强烈拉张减薄，地幔物质沿岩石圈断裂上升，出现了大洋玄武岩。第四纪南海的拉张作用结束，现代南海格局基本形成，在大陆架沉积了一套滨海相—浅海相的陆源碎屑，在大陆坡为生物礁和生屑碳酸盐岩沉积，中央海盆由于远离大陆，陆源物质难以到达，沉积了一套含粉砂质的远洋放射虫黏土或含远洋放射虫黏土的粉砂和砂。总体来看，南海的新生代沉积序列表现为一个巨大的海侵序列，从古近纪早期的陆相、海陆交互相到新近纪以来的滨浅海相、深海相，表现出从陆到海、海水逐渐变深的演化序列。

第二十一章　构造形变

第一节　结晶基底变形特征

海南岛出露的最古老的岩石是中元古界戈枕村组和峨文岭组，由高角闪岩相、低角闪岩相的区域变质岩组成。其最显著的特点是广泛发育各种新生面理、线理和小褶皱构造。从野外露头观察到镜下鉴定，均能见到上述变形。

一、次生面理构造

一是见于峨文岭组能干性较弱的泥质岩石中广泛发育的透入性新生置换面理-片理构造，反映了岩石受应力作用变形较强，构造置换程度较高；二是见于戈枕村组岩石中，原生构造不复存在，在混合岩化过程中，岩石处于高温高压条件下呈固熔体或半固熔体状态，可塑性较大，受应力作用变形较强，发育透入性的新生面理-片麻理构造。

二、线理构造

主要有矿物拉长线理、石香肠构造、杆状构造。

（一）矿物拉长线理

发育于具片理、片麻理构造的岩石中，微观尺度上，石英矿物多呈拉长状、透镜体状、压扁状等，其长轴平行片理或片麻理定向分布；小型尺度上，见有透入性顺层长英质脉体呈拉长状、眼球状断开，与新生面理平行排列。

（二）石香肠构造

一是发育于峨文岭组软硬相间的岩层中，小范围内，相对能干性较强的岩石，如脉石英、石英岩等，呈细颈化、透镜体化、拉长、拉断现象屡见不鲜，其长轴顺新生面理面平行排列。大范围内，伟晶岩型白云母矿及斜长角闪岩脉呈透镜体状沿片理面分布，总体具藕节形和透镜状石香肠构造的特点(图版Ⅴ-1-B)。二是见于戈枕村组上部片麻岩中，为长英质脉体顺新生面理发生拉长减薄、增厚，甚至拉断变形，形态复杂、很不协调，呈不规则状、肠状、黏滞状，属不规则黏滞型石香肠构造，反映了岩石变形的温压条件都较高的构造环境（图 21-1）。

（三）杆状构造

主要分布于戈枕村组混合岩化片麻岩中，为混合岩化过程中，岩石发生强烈变形变质，并分异出长英质脉体，之后脉体沿垂直于褶皱轴方向发生了滚动变形，呈拉长状、透镜体状、杆状等（图 21-2，图版Ⅴ-1-C）。

三、褶皱构造

包括宏观尺度褶皱与小尺度褶皱两种。

（一）小尺度褶皱

按形态、特点划分见有片内无根型褶皱、强揉皱型褶皱、层内褶皱、圆柱状褶皱、倒转型褶皱等。

1. 片内无根型

其轴面劈理为新生的片理或片麻理，属剪切流动过程中形成。以峨文岭组岩石较常见，也最有特色。

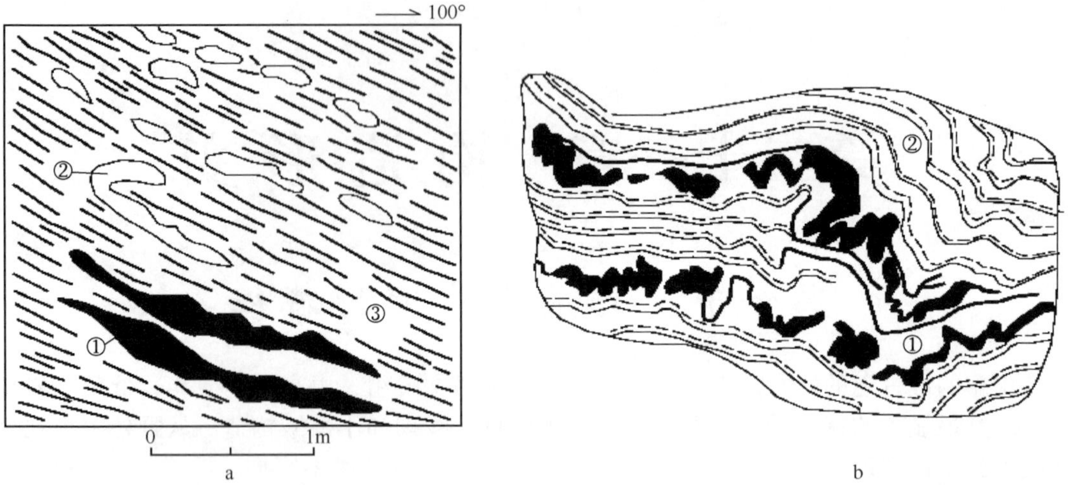

①石英脉；②云母石英片岩；③二云母片岩　　①伟晶岩；②片麻岩

图 21-1　乐东红五地区石香肠构造剖面素描图

Fig. 21-1　Sketch sketch showing the Boudinage structures in the Hongwu area，Ledong

（据海南地质队一分队，1972~1973）

a—藕节状、透镜型 石香肠构造；b—不规则黏滞型石香肠构造

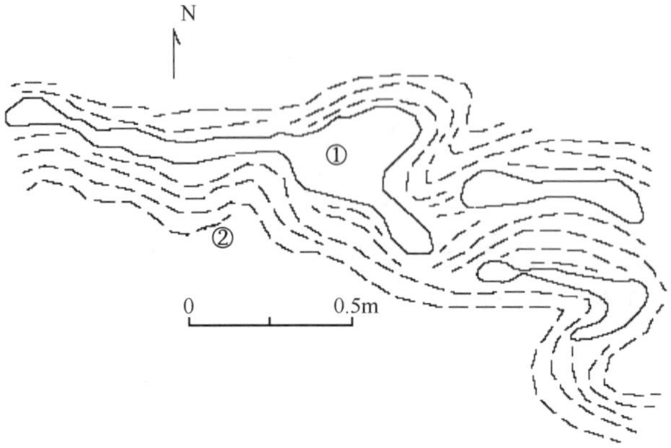

图 21-2　戈枕村组杆状构造平面素描图（乐东县红五）

Fig. 21-2　Plan sketch showing the rodding structures of the Gezhencun Formation（Hongwu，Ledong County）

①长英质脉；②混合岩化片麻岩

褶皱特点是：由多个小的无根钩状褶皱组成，形态较复杂、非常紧闭，极不完整、不连续，已不能恢复总体褶皱形态。两翼长短不一，翼部产状与新生面理近于平行，而核部则与之呈大角度相交。单个褶皱波长、波幅一般为几厘米（图 21-3，图版Ⅴ-1-D）。

2. 强揉皱型

属剪切流动过程中形成。一是由长英质脉体强揉皱变形而成（见于戈枕村组中），其轴面劈理为新生片麻理。褶皱特点是：由多个小的大小不等的褶皱组成，形态较复杂、不规则，呈紧闭尖棱状。单个褶皱形态完整，尚保持一定的连续性，两翼长短不一，翼部产状与新生面理近于平行，而核部则与之呈大角度相交，波长、波幅一般为几十厘米，且波幅大于波长。二是以原生层理为变形面，普遍发育于戈枕村组和峨文岭组中。其特点是：褶皱连续性好，形态完整，但较复杂，翼间角紧闭，两翼长短不一，波长、波幅一般为 1m 至几厘米，且波幅略大于波长。

3. 层内褶皱

发育于粒状矿物与片状矿物组成的岩石中，在应力作用下，粒状矿物呈剪切褶皱变形，而片状矿物则发生剪切活动，褶皱轴与界面有一个交角，其形态紧闭，两翼长短不等，倾角不一，能指示物质运动的方向，表明了岩石具有顺层置换的特点。强揉皱型褶皱普遍发育于戈枕村组和峨文岭组中，其特点是褶皱连续性好，形态完整，但较复杂；翼间角紧闭，两翼长短不一，一翼缓，而另一翼陡；波长、波幅一般为几厘米至 10cm，属剪切流动过程中形成（图 21-4）。

4. 圆柱状褶皱

见于戈枕村组中，褶皱特点是岩层呈圆弧形弯曲，褶皱连续性好，形态完整；翼间角开阔，两翼长短不一，一翼较缓，而另一翼较陡；核部发育许些小的张性节理裂隙，节理面呈锯状，不平整，为纵弯褶皱作用形成（图 21-5，图版Ⅴ-1-E）。

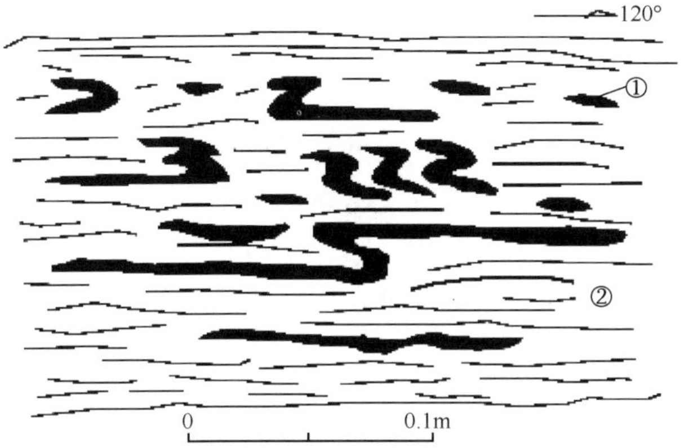

图 21-3　D1644 点片岩内无根型图褶皱剖面素描图

Fig. 21-3　Section sketch showing rootless folds developed inside schist, Point D1644

①长英质脉体；②二云母片岩

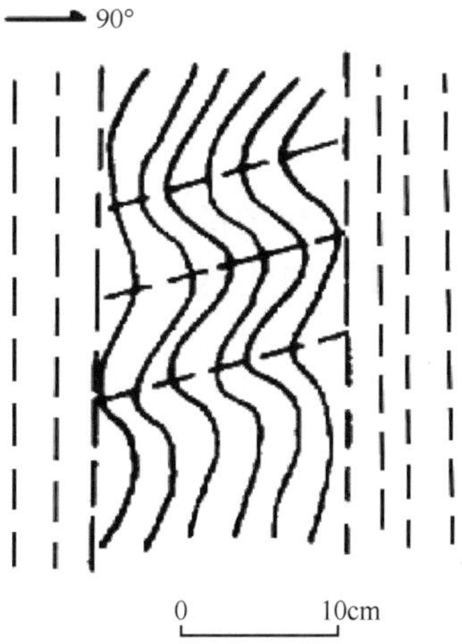

图 21-4　层间褶皱

Fig. 21-4　Interstratified folds

5. 倒转型

褶皱以新生面理为变形面形成。形态较规则、完整，连续性好，轴面倾斜，一般倾伏向北东或北西。两翼基本对称，向同一方向倾斜，翼间角紧闭，波长、波幅一般为 1m 至几米（图 21-6，图版Ⅴ-1-G）。

图 21-5　圆柱状褶皱素描图

Fig. 21-5　Sketch showing cylindrical folds

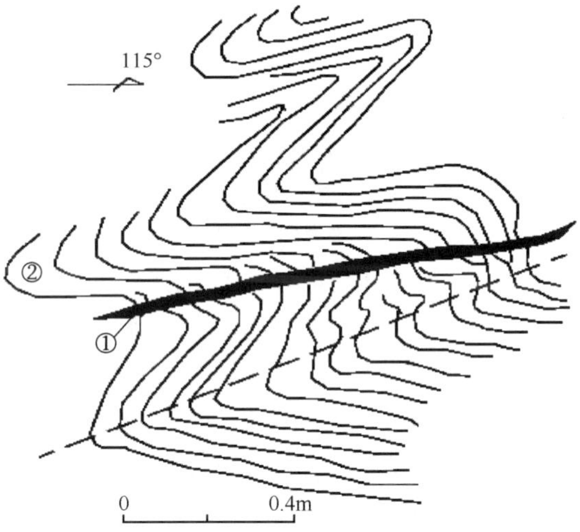

图 21-6　倒转型褶皱剖面素描图

Fig. 21-6　Sketch showing the section of inverted folds

（据海南地质队一分队，1972~1973）

①伟晶岩脉；②云母石英片岩

6. 倾伏型

褶皱以新生面理为变形面形成，分布面积广，基本上赋存于整个抱板群中，但其规模较小，往往呈单个褶皱产出，其特点是：形态简单、较完整，连续性好，轴面近直立，枢纽倾伏，但缺乏方向性，一般倾伏向为北东或北西。两翼基本对称，翼间角紧闭，波长、波幅一般为 1m 至几米，并叠加有最早期褶皱变形（图版Ⅴ-1-F）。

7. 直立型

以新生面理为变形面形成，属亚南甫背斜形成的从属小褶皱。褶皱特点是：形态简单、较完整，连续性好，往往呈单个向斜或背斜产出。其轴面近直立，枢纽近水平，轴迹走向较稳定，具明显的方向性，为近南北向。两翼基本对称，翼间角紧闭，波长、波幅一般为 1m 至几米。

（二）宏观尺度褶皱

主要有屯昌县中建北东向蒙花岭-白银坡褶皱带、乐东县红五近南北向亚南甫背斜和东方市布磨二甲褶皱。

1. 蒙花岭-白银坡褶皱带（f_{11}—f_{18}）

分布于屯昌县中建蒙花岭-白银坡一带，长 3.2～6.5km，宽达 12km，轴迹走向北东，南西端扬起，而北东端倾伏。褶皱带表现为峨文岭组和戈枕村组多次交替重复出现，形成一系列倒转型和对称型小褶皱，倒转型小褶皱位于褶皱群北西面，轴面倾向南东；对称型则处于褶皱群中部和南东端，轴面近直立。受后期花岗岩侵蚀和断裂作用的影响，使得地层产状凌乱，褶皱残缺不全，难以恢复早期褶皱形态，但褶皱带整体具有北东走向的格局。

2. 布磨-二甲褶皱带（f_{24}—f_{31}）

分布于东方市布磨—二甲村一带，发育于中元古界戈枕村组和峨文岭组中，由多个不规则状背斜、向斜组成。由于受到后期花岗岩的侵蚀破坏，导致褶皱出露不完整。长约45km，影响宽度约达20km，总体走向北东，沿走向呈蛇型弯曲，初步认为属于与戈枕韧性剪切带相伴随生成起来的剪切褶皱带。

南西段布磨一带峨文岭中，褶皱走向北东，自北西往南东，由一个向斜和一个背斜组成。褶皱两翼倾向相反。其中，向斜一翼倾角为34°～37°，另一翼为35°～50°。背斜一翼倾角为35°～50°，另一翼为56°。属于不对称向、背斜，即褶皱具有由北西往南东运动的特点，指示戈枕韧性剪切带逆冲活动特点。在公爱一带，褶皱发生扭曲变形，走向转为北西向和南北向，也可能与戈枕韧性剪切带剪切活动有关。

北东段二甲一带，褶皱走向北东或北西，结合中元古代花岗岩片麻理中的向形和背形构造分析，认为属于一个弧状褶皱，从北西往南东，褶皱方向由北西至南东。褶皱两翼倾向相反，倾角基本一致，一般为40°～70°间，在靠近戈枕剪切带时，倾角具有变陡的趋势。

3. 亚南甫背斜（f_{65}）

分布于乐东县红五冲卒岭西南的亚南甫一带，轴向近南北，轴长约 2.5km，宽约 5.1km。核部为长城系戈枕村组条带状、眼球状混合质黑云斜长片麻岩和条带状、眼球状混合岩，两翼依次为峨文岭组云母石英片岩、含石墨长英片岩、二云母石英片岩、白粒岩等。东翼地层走向 310°～14°，倾向北东—南东东，倾角 12°～50°，优选产状 64°∠22°，西翼地层走向 0°～30°，倾向 NWW—NW，倾角 6°～50°，优选产状 300°∠22°。据优选产状求得枢纽倾伏向 360°，倾伏角 10°，轴面走向南北，倾角直立（90°）。为一向南扬起、向北倾伏的倾伏背斜。

背斜中既包含有早期构造变形，又有晚期构造变形的叠加，并发育轴迹走向近南北的从属小褶皱。同时，于其东翼新村-后溪村一带形成席卷整个东翼岩层的次级倒转向斜和次级倒转背斜。

次级倒转向斜 核部岩性为峨文岭组变粒岩、片岩类，两翼为戈枕村组片麻岩类岩石。岩层产状变化较大，东翼倾向 60°～120°，倾角 18°～68°，优选产状 78°∠36°，西翼倾向 55°～120°，倾角 6°～54°，优选产状为 87°∠42°。为一轴迹走向近南北的倒转向斜。由于受晚期花岗岩的侵蚀破坏，导致该向斜出露极不完整，往往缺失相对应的岩层。

次级倒转背斜 核部岩石为戈枕村组眼球状混合岩，两翼为峨文岭组石英岩类、片岩类。东端与陀烈组呈不整合接触。岩层总体走向北北东—近南北，倾向南东东，倾角 30°～70°，轴迹走向与倒转向斜轴迹近平行。局部见三叠纪尖峰超单元侵入。

总体而言，该构造层最早期构造变形，戈枕村组和峨文岭组除了皆形成新生面理外，其各自最大特点是：戈枕村组变形基本上以线理构造最发育为特征，而峨文岭组则以褶皱变形最具特色。综合上述构造形迹的形态特点及其组合，从构造层次观点出发，通过变形相分析（单文琅等，1991），认为该构造层具多期（次）构造变形作用的叠加，经历了固态流变构造群落和弹塑性变形构造群落的发展阶段，岩石具有伸展—挤压—伸展的动力学演化过程。早次变形为伸展环境下的产物，形成了具分层剪切流动特征的固态流变景观；同造山期花岗岩的侵入，意味着变形环境向挤压体制转换，形成了弹塑性变形构造群落。

第二节 褶皱基底变形特征

海南省褶皱基底主要分布于海南岛北面（崖城-万宁断裂带以北地区），岩石发生多阶段多旋回的褶皱变形，包括晋宁期—扬子期、加里东期、华力西期—印支期三个阶段变形。

一、晋宁-扬子旋回变形特征

晋宁-扬子旋回变形特征包括青白口系石碌群和南华系—震旦系石灰顶组。其中，石灰顶组区域变质和透入性变形作用非常弱。因此，该旋回的变形特征以石碌群为代表。其变形特征如下：

（一）次生面状构造

次生面状构造根据形成先后和构造物理意义可分为三种。

透入性顺层剪切面理（S_1） 石碌群地层在早期横向构造置换中，原始层理多为 S_1 所代换，而今呈现的石英纹层和绢云母为主组成的纹层多已不具有原始水流沉积特点。宏观而言，S_1 面理即千枚理或片理与标志层（S_0）呈小角度交切；在露头尺度，到处可见 S_0与 S_1 的交切关系（图 21-7），在微观领域，S_0和 S_1 交角的大小直接与构造置换程度密切相关，置换程度越高，两者越趋向一致的产状。

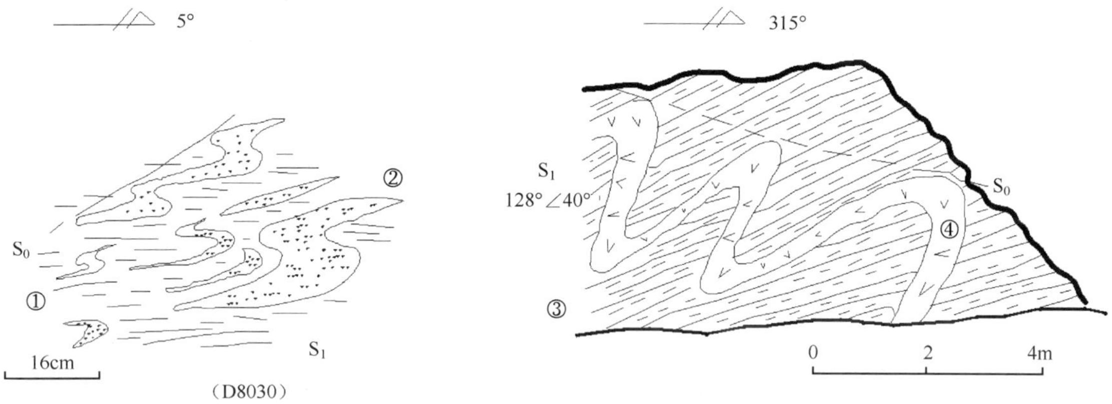

图 21-7 石碌群中的顺层剪切面理素描图

Fig. 21-7 Sketch showing the along-bedding shearing foliations in the Shilu Group

①富绢云母成分层；②富石英质成分层；③石英绢云母片岩；④酸性火山凝灰岩

褶皱轴面劈理（S_2） 在石碌复式向斜转折端处比较发育，以透入性新生剪切面理（S_1）为变形面，形成轴向呈北西—南东向叠加褶皱，伴生轴面劈理（S_2）（图 21-8）。沿 S_2 面理并未出现大量新生矿物，局部仅见鳞片状绢云母沿此定向或为石英细脉充填。

横向褶皱的间隔劈理（S_3） 多出现在北北东向褶皱叠加改造的面状构造发育的泥质岩中，其劈理间隔较大，具非透入性，几何上 S_3与 S_0或 S_1 大角度相交，多表现为压熔劈理或膝折面理，S_3劈理在北一铁矿体中也可见及（图 21-9）。

（二）线状构造

石碌群的线状构造主要形成于早期横向拉伸过程中，表现形式多种多样。宏观规模上，夹于第一层和第三层之间的薄层白云质结晶灰岩、白云岩等强干层空间上呈透镜状展布，构造上具有石香肠特点；露头上由相对能干性的矿物集合体构成的拉伸线理屡见不鲜，而在微观领域内，单矿物或矿物集合体拉伸变形而成的 a 型线理更是层出不穷，几乎见及各个层序之中。

（三）褶皱构造

根据构成褶皱变形性质和叠加形成先后次序，石碌群地层中可划分出三期褶皱。

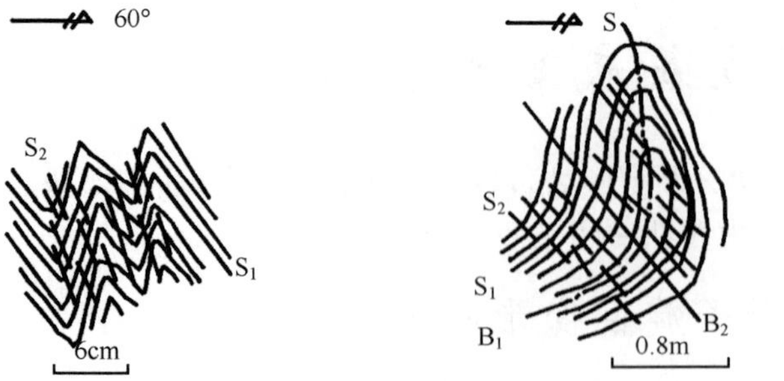

图 21-8　石碌群变质岩系中的褶皱轴面劈理（S_2）素描图

Fig. 21-8　Sketch showing cleavages of the fold axial plane（S_2）in Shilu Group metamorphic rock series

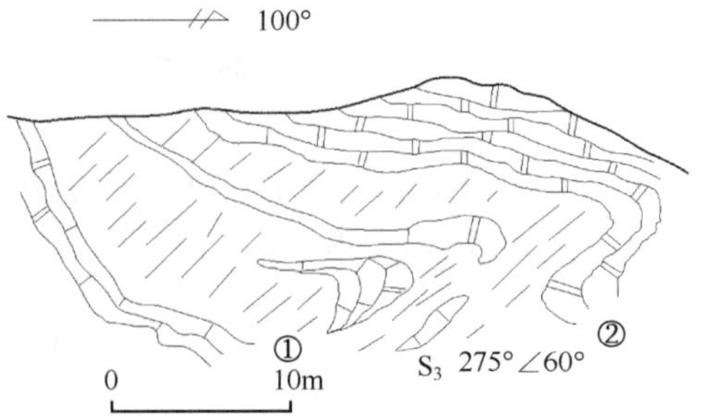

图 21-9　石碌群北一铁矿中的 S_3 劈理素描图

Fig. 21-9　Sketch showing S_3 cleavges in one iron deposit，north of the Shilu Group

①铁矿体；②二透岩

第一期褶皱（B_1）　仅局限于露头尺度和显微尺度，在第五层中尤为发育。B_1 褶皱反映了伸展体制下顺层构造滑动的一般特点。

第二期褶皱（B_2，f_{44}—f_{51}）　石碌群和上覆石灰顶组组成的最主要褶皱变形，前人确定的自北而南的北一向斜、红房山背斜和石灰顶向斜组合而成的轴迹呈北西西—近东西向展布、西端翘起并收敛、东端倾没并撒开的石碌复向斜构造是其代表。在原石碌群第一层中发现的 4 层石英岩或石英片岩，其特点是与第四层石英岩、石英片岩完全一致，结合构造解析，证实原巨厚的第一层实际是由 3 个倒转背斜（形）和 2 个倒转向斜（形）组成的褶叠体，并成为石碌复向斜构造的重要组成部分（图 21-10）。

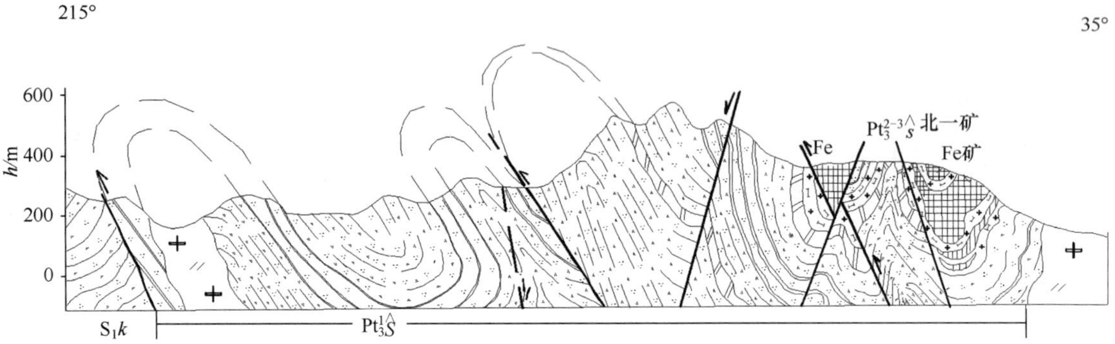

图 21-10　石碌复向斜构造剖面图

Fig. 21-10　Section of Shilu Syclinorium structure

S_1k—志留系空列村组；$Pt_3^1\hat{S}$—青白口系石碌群；$Pt_3^{2-3}\hat{s}$—南华系—震旦系石灰顶组

第三期褶皱（B_3）　褶皱轴迹呈北东—北北东向，褶皱高角度横跨于早期北西西—近东西向褶皱之上，有人称这种叠加褶皱为 W 型褶皱（贾精一，1985）。调查表明，这种北东—北北东向褶皱一般分布在戈枕韧性剪切带和石碌水库韧性剪切带附近，因此，有理由推测其成因可能与这两条主干剪切带的形成有关。

石碌群的变形构造按形成先后划分为三个变形序列（除脆性断裂变形外），每个变形序列都有其特定的构造形迹组合：

第一变形序列的共生形迹组合为顺层剪切片理（S_1）、各种拉伸线理（L_1）和顺层剪切褶皱（B_1），该变形序列的形成环境依据变形期的变质矿物组合而定，根据白云母-绿泥石地质温度计，其形成温度为 450℃±。

第二变形序列形成了以石碌复向斜（B_2）和轴面劈理（S_2）为特征的构造形迹组合。根据卷入地层为南华系—震旦系石灰顶组，因此，其生成为震旦纪之后。结合褶皱被二叠纪花岗岩侵蚀分析，其生成时间应为加里东期或华力西期。

第三变形序列形成了石碌群北东—北北东向褶皱及其轴部零星发育的膝折构造和间隔劈理。

二、加里东旋回变形特征

海南岛的加里东旋回包括奥陶系与志留系，为低绿片岩相区域变质岩，经历了两期变形作用的叠加，早期由新生面理、线理和顺层掩卧褶皱组成的固态流变构造群落，晚期形成对称型和不对称型褶皱，伴随发育皱纹线理、折劈理构造。相对而言，奥陶系变形比志留系稍强，奥陶纪岩石原生面理完全被新生面理彻底改造，原生石英碎屑全部发生重结晶，具强烈的波状消光，呈拉长粒状，变形以线理、面理为主，褶皱少见；而志留系尚可见原生面理构造，石英重结晶程度差，局部保持不规则状、棱角状的沉积碎屑外形，变形过程中，除形成线理、面理外，褶皱也常见。反映了由志留系至奥陶系，岩石具有递进变形作用特征，形成了不同变形样式的组合，相伴随的变质作用也逐渐减弱，表明变形、变质作用呈正相关关系。

（一）小尺度构造形迹

1. 奥陶系

奥陶系的变形特征是在露头尺度表现为广泛发育新生面理外，而在微观尺度中表现为面理、线理、褶皱等。大溪桥地区出露的奥陶系南碧沟组，可识别出两次变形事件。

（1）早期伸展变形构造

剖面上按主要构造样式可自下而上划分为三个带。

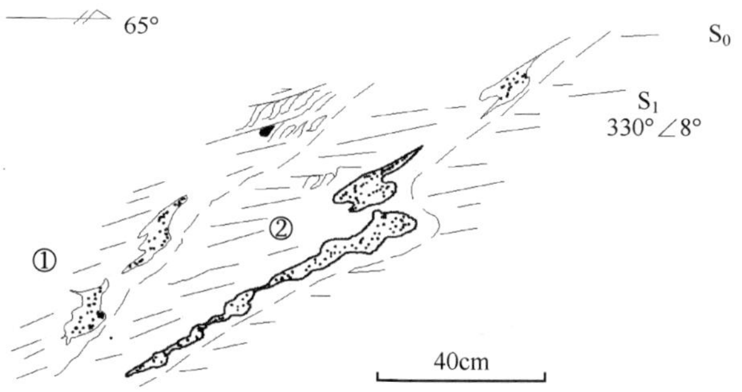

图 21-11　奥陶系中的残存层理构造素描图

Fig. 21-11　Sketch showing residual bedding structures in Ordovician strata

①富砂质成分层；②黑云母石英片岩

拉伸线理带　该带以拉伸线理最为常见，仅发育于南碧沟组下部，夹于云母石英片岩中，由石英角岩、钙质角岩组成的杆状构造十分醒目，杆的长轴基本与片理倾向一致。同能干性强的杆状体相比，云母石英片岩的拉伸变形多由单矿物集合体显示出来。在拉伸线理发育的部位偶见由原始成分层显现的层理构造（图 21-11），而今展示的面理实际上是变质-变形联合作用形成的片理构造。受后期构造影响，拉伸线理产状沿岩层走向有所变化，而在大岭农场一队附近产状稳定，倾角在 5°～15°之间，倾向在 270°～280°之间，这种产状可能代表了同变形期的真实产状。在拉伸线理带内局部发育 a 型褶皱（图 21-12），并在三度空间上清楚看到拉伸线理、褶皱枢纽和杆状体长轴三者之间互为平行的几何关系。由此看来，由相对能干性岩石构成的杆状构造同样具有 a 型线理之特点。

褶皱层构造带　南碧沟组下部广泛发育以褶皱层为主的构造群落（图 21-13）。褶皱形态通过富含砂质的成分层表现出来，枢纽产状近水平，轴面也很平缓。轴面片理在翼部大致平行于成分层，而在转折端部位高角度切割原始层理。在显微镜下，该带中的主要岩石云母石英片岩、千枚岩发育片状构造或千枚状构造。云母石英片岩多具变质分异结构，其中的石英大都动态结晶；在绢云母、石英定向形成的千枚状构造中可见显微褶叠层构造形迹，反映了微观与宏观上的相对一致性。

同斜褶皱带　同斜褶皱带见于南碧沟组下部，同斜褶皱呈不对称状（图 21-14，图版Ⅴ-1-F），一翼长而缓，一翼短而陡，其轴面与区域片理、千枚理近于平行。从原始富砂质细层构成的褶皱的不同部位与片理的关系推测地层具有正常层序特点。南碧沟组以往据阳起石片岩重复出现划为一倒转背斜构造，现经岩石学研究发现目前所见的三层阳起石片岩岩性特征差别较大，认为自下而上呈一单斜。这套地层露头上片理极其发育，局部见有钩状褶皱沿片理产出，而微观域内构造置换十分强烈，现今展示的片理构造为完全顺层构造置换作用形成的透入性置换片理（图 21-15）。

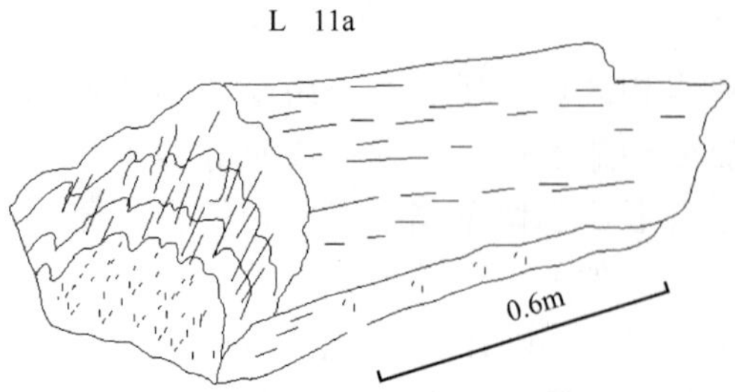

图 21-12 奥陶系二透化石英角岩中的 a 型褶皱素描图

Fig. 21-12 Sketch showing a-type folds in Ordovician diopside-pla-tremolized quartz hornfels

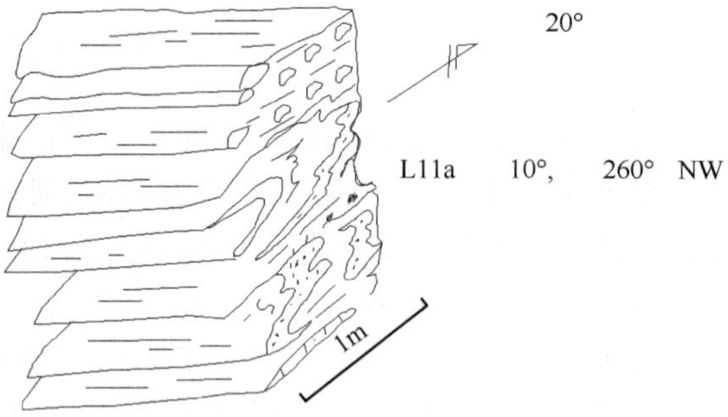

图 21-13 奥陶系下段各变形要素几何关系示意图

Fig. 21-13 Schematic diagram of geometrical relationships of various deformation factors of the Lower Ordovician strata

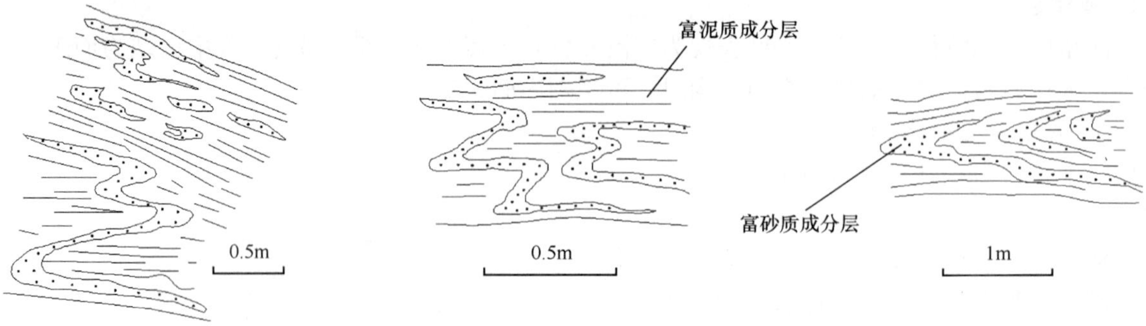

图 21-14 奥陶系下部褶皱叠加构造素描图

Fig. 21-14 Sketch showing superimposed fold structures in the Lower Ordovician strata

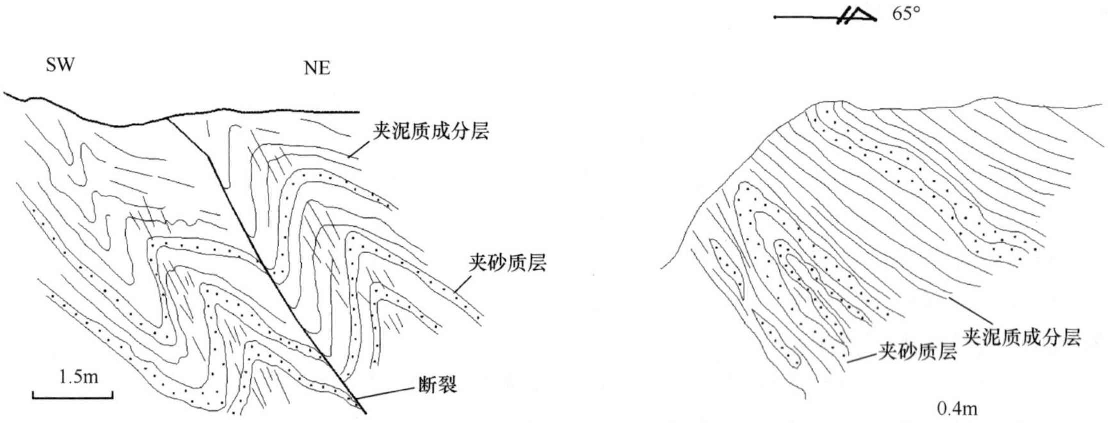

图 21-15 奥陶系上部同斜倒转褶皱素描图

Fig. 21-15 Sketch showing homoclinal inverted folds in the Upper Ordovician strata

上述三个构造带主要构造要素概括如图 21-16。

(2) 晚期挤压变形构造

晚加里东运动使南碧沟组地层强烈褶皱，尖棱状小褶皱、对称状小褶皱发育。区域上形成了芙蓉田背斜构造。

露头范围多处可见该期褶皱共轴叠加于前期褶皱之上，先期褶皱的两翼再次褶皱，而褶皱轴并未发生波状弯曲（图 21-17）。就小褶皱的特点来看，这期变形并未形成透入性区域片理构造，仅在局部发育微弱的矿物定向。

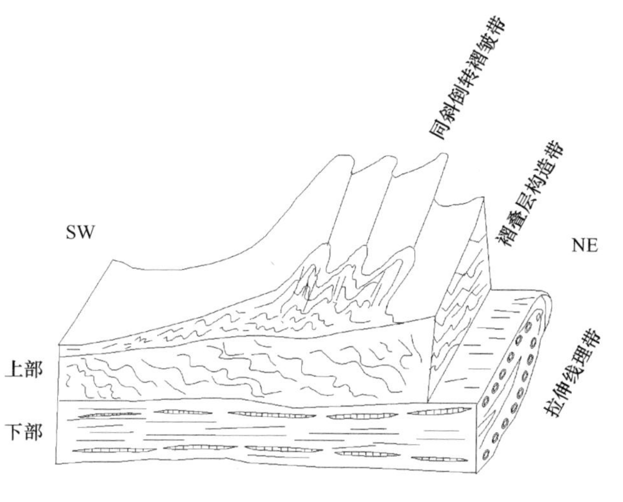

图 21-16 奥陶系主要构造要素空间组合示意图

Fig. 21-16 Schematic diagram of spatial combinations of main structural elements in the Ordovician strata

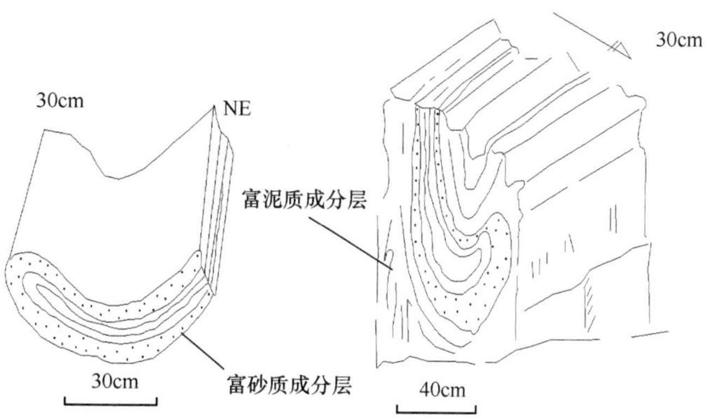

图 21-17 奥陶系中的叠加褶皱素描图

Fig. 21-17 Sketch showing superimposed folds in the Ordovician strata

2. 志留系

志留系变形迹象有新生线理、面理和褶皱。

（1）线理构造

线理构造按线理类型划分为矿物拉长线理、石香肠构造、杆状构造等。

矿物拉长线理常发育在具有千枚理构造的岩石中。微观尺度上，石英矿物呈拉长状、透镜体状、板条状、眼球状，其长轴与千枚理平行定向排列，显示了矿物拉长线理特征。

石香肠构造包括透镜状藕节形石香肠构造、黏滞型石香肠构造和褶皱型石香肠构造三种类型（图21-18，图版Ⅴ-2-A、图版Ⅴ-2-B）。软硬相间分布的岩石具有明显韧性差，在构造作用下，相对能干性较高的岩石，发生变形，呈细颈化，透镜体状拉长断开，其长轴沿千枚理展布，具透镜状藕节形石香肠构造特点；长英质脉体或石英脉顺新生面理发生拉长减薄、增厚，甚至拉断变形，形态复杂，很不协调，呈不规则状、肠状、黏滞状，属不规则黏滞型石香肠构造；赋存于千枚岩中的长英质脉体，在构造作用下，呈拉薄状、增厚状、黏滞状、拉断状，并与围岩一起进一步发生旋转扭动褶皱变形，具褶皱型石香肠构造特征。杆状构造是与千枚理平行的细小长英质脉等，在应力作用下，强烈变形，发生拉长状断开，呈杆状与千枚理平行排列，其断口呈次圆状、方状等。

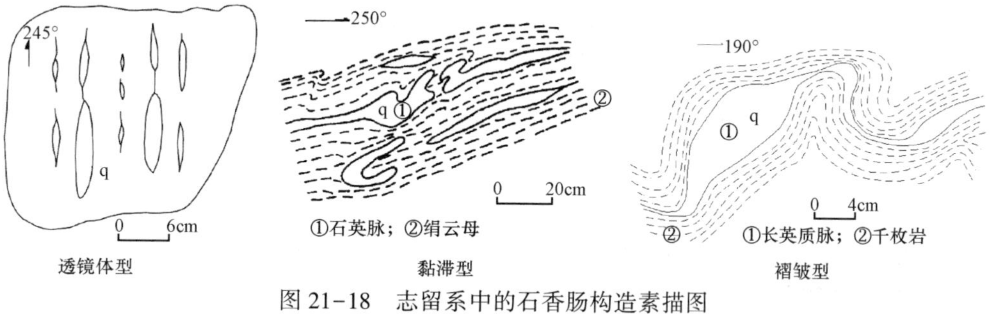

图 21-18 志留系中的石香肠构造素描图

Fig. 21-18 Sketch showing boudinage structures in the Silurian strata

（2）面理构造

志留系的面理构造包括原始层状构造与新生面理构造两类。

原始层状构造 主要见于变质石英粉砂岩、粉砂质千枚岩、绢云母板岩等相对能干性较高的岩石中，由于该类岩石可塑性较差，在构造应力作用下，其变形程度相对较弱，新生面理构造不甚发育，见有变余平行层理、变余粒序层理等；微观上，粒状矿物与泥质矿物相间分布的岩石，相对能干性弱的泥质矿物强烈变形发育顺层新生面理，而可塑性差的粒状矿物变形很弱，整体保留原始层理构造，显示原始沉积的韵律式层理特征。

新生面理构造 包括千枚理与劈理。千枚理构造发育在相对能干性较弱的泥质岩石中，由于其可塑性较大，在应力作用下变形较强，构造置换程度高，广泛发育透入性的顺层新生面理构造——千枚理构造。劈理构造又包括了滑劈理和层间劈理两种类型（图 21-19）。滑劈理多见于相对能干性较弱的岩石中，为一组透入性切过先存千枚理的差异平行滑动面，与千枚理斜交，劈理间距一般较均匀，微劈石发生揉皱式弯曲变形。层间劈理是一种受岩性及层理控制，与层理斜交的劈理，多见于粒状矿物与片状矿物相间分布的岩石中，劈理被限制仅发育于片状矿物中，而未能切割粒状矿物；微观尺度上呈透入性密集发育，劈理面近平行，并为新生片状矿物充填，劈理间距大小一般均匀，微劈石略发生揉皱变形，属分层剪切作用的产物。

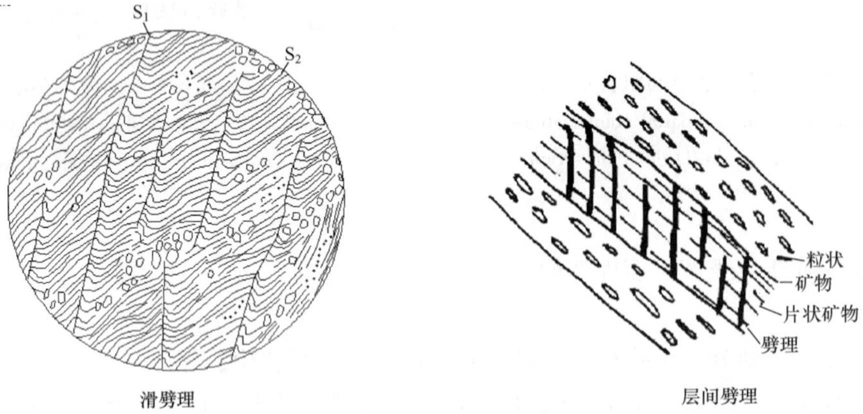

图 21-19 志留系中的劈理示意图

Fig. 21-19 Schematic diagram of cleavages in the Silurian strata

(3) 小尺度褶皱

根据褶皱的形态特征，小尺度褶皱大致可以划分为片内无根型褶皱、层内褶皱、W 型褶皱、强揉皱型褶皱、箱状褶皱、不对称褶皱和对称褶皱等类型。

片内无根型褶皱为早期剪切流动变形的产物，其特点是轴面劈理为新生的千枚理，形态复杂、不完整、连续性差，难以辨认其原来形态；两翼长轴不一，翼部产状与新生面理近于平行，而核部则与之呈大角度相交，翼间角非常紧闭，其波长、波幅均较小，一般仅几厘米（图 21-20，图版Ⅴ-2-C）。

层内褶皱发育于由粒状矿物与片状矿物组成的岩石中，层面控制了褶皱变形。通过层间流动使粒状矿物发生褶皱变形，而片状矿物则呈剪切活动。褶皱连续性较好，形态较紧闭，两翼产状不一，长短不等，轴面与层面有小角度相交（图 21-21），能指示物质运动的方向，反映了岩石具顺层置换的特征。

图 21-20 片内无根型褶皱示意图

Fig. 21-20 Schematic diagram of rootless folds developed inside schist

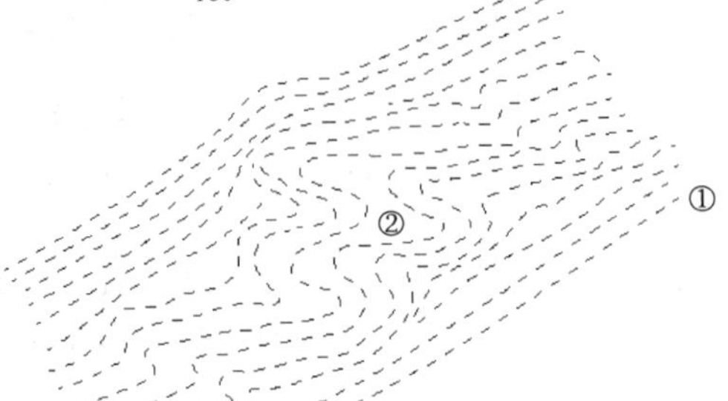

图 21-21 层内褶皱示意图

Fig. 21-21 Schematic diagram of intrafolial folds

W 型褶皱发育于薄的强硬岩层中，经纵弯褶皱作用形成。褶皱形态规则，翼间角较紧闭，呈尖棱状，两翼基本对称，并由许多指示岩石运动的次级不对称褶皱组成（图 21-22，图版Ⅴ-2-D）。

强揉皱型褶皱的特点是以原生层理为变形面，形态完整，但较复杂，不规则；翼间角紧闭，两翼长短不一，往往由多个大小不等的向斜、背斜组成（图 21-23，图版Ⅴ-2-E），属剪切流动过程形成。

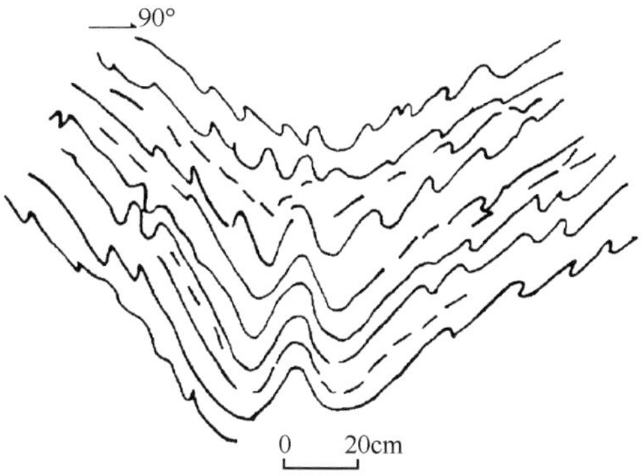

图 21-22　W 型褶皱示意图

Fig. 21-22　Schematic diagram of W-type folds

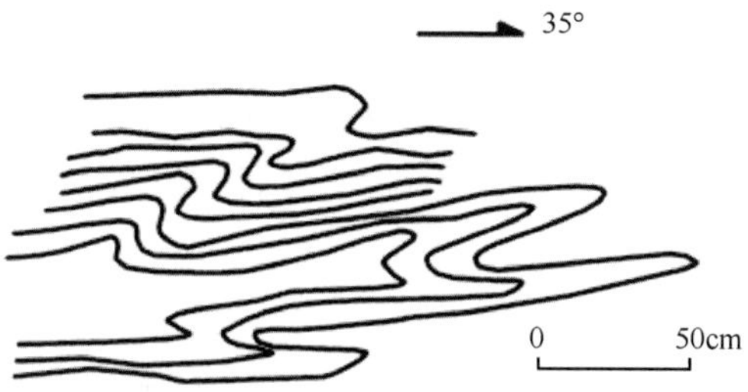

图 21-23　强揉皱型褶皱示意图

Fig. 21-23　Schematic diagram of strong lyrumpled folds

箱状褶皱发育于千枚岩中，褶皱连续性好，形态完整，呈箱状，但较复杂，两翼陡而转折端平直，具有一对共轭轴面，翼间角紧闭（图版Ⅴ-2-F），属早期剪切流动变形的产物。

不对称褶皱分布比较普遍，它以新生面理为变形面形成。褶皱形态规则，一翼较长而平缓，另一翼较短且陡（图 21-24），具有指示岩石剪切活动的特点。

对称褶皱以新生面理为变形面形成，表现为褶皱的轴面与水平面近垂直，而且两翼长度也基本相等，翼间角开阔（图 21-25），褶皱形态规整。

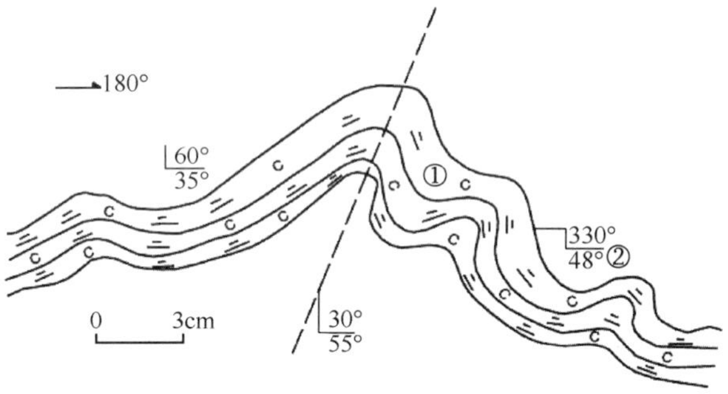

图 21-24　不对称褶皱示意图

Fig. 21-24　Schematic diagram of asymmetric folds

①炭质绢云母千枚岩；②岩层产状

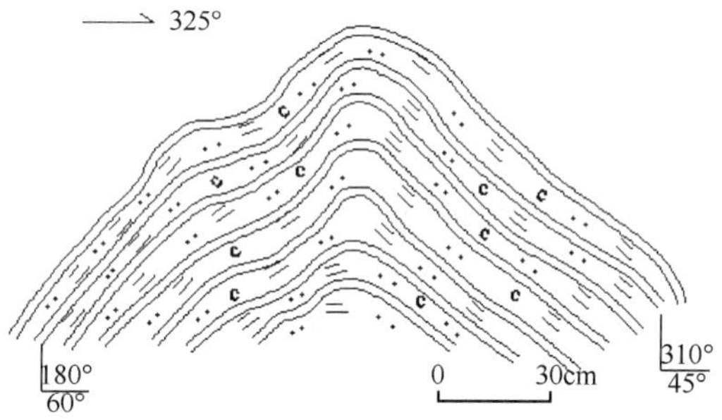

图 21-25　对称褶皱示意图

Fig. 21-25　Schematic diagram of symmetrical folds

（二）大尺度褶皱变形

有北东向旧村岭褶皱带、军营复式背斜、南棍园复式向斜、亳稔松向斜、陀类-布温褶皱带和北西向江边倒转背斜、亳岗岭背斜。

1. 旧村岭褶皱带（f_4—f_7）

发育于戈枕断裂带南东盘下古生界奥陶系南碧沟组和志留系陀烈组中，前人命名为“旧村岭褶皱带”（1∶5 万《东方、玉道》幅区调报告）。北东起自峨理赛，经旧州岭，南西至马岭一带。走向北东，略显 S 形弯曲，影响长度约 22km，宽 4~8km。核部为南碧沟组，两翼为陀烈组。北西翼倾角一般 34°~63°，并发育 Z 型次级褶皱，次级褶皱由 2 个紧闭不协调向斜和 1 个紧闭不协调背斜组成；南东翼倾角略陡，为 40°~75°。属于一个轴面倾向北西的不对称复式背斜。

该复式背斜属于与戈枕韧性剪切带共同生成起来的相互配套的褶皱构造。略显 S 形弯曲的平面走向，指示物质具有右行的运移，与戈枕韧性剪切带的右旋活动性相协调。剖面上，由北西向南东运移的运动学指示，与戈枕韧性剪切带由北西往南东逆冲的运动方向也一致，并在靠近戈枕韧性剪切带时，背斜北西翼发育次级紧闭不协调向斜，也显示戈枕韧性剪切带下盘（南东盘）具有下滑、而上盘（北西盘）逆冲的特点。

2. 军营复式背斜（f_1—f_3）

位于逆冲断裂带的下盘（南东盘），其南西起自红田农场石灰厂，经南田水库、红田农场十八队，北东至红田农场二十队。发育于下志留统陀烈组中。轴迹呈蛇形展布，南西段红田农场附近，轴迹近东西向。北西翼倾向 320°~330°，倾角缓（30°~50°），核部和南东翼倾角较陡（>50°），为一个轴面总体倾向北西的不对称背斜，显示物质整体具有由北西→南东运移的活动特点，与军营-红岭逆冲推覆构造带北西→南东推覆有关。

3. 南棍园复式向斜（f_{10}）

发育于志留系陀烈组中，向斜北东端被第四系掩盖，南西端延伸出图外，北西、南东两翼被花岗岩侵蚀。区内长约 60km，宽 40 余千米，轴迹走向北东，略呈 S 形弯曲。加月村-黑墨岭段，两翼地层倾向相反，倾角近一致，为 40°~60°，为对称向斜。黑墨岭-王帝殿段，被白垩系掩盖和花岗岩侵蚀，核部未出露。王帝殿-细水一带，两翼地层倾角不一，北西翼略陡，为 35°~55°，南东翼较缓，为 20°~35°，属不对称向斜。

复式向斜局部发育从属小褶皱。小褶皱形态规则、简单，褶皱过程中，岩层的铅直厚度基本保持不变，属弯滑作用形成。向斜翼部为不对称的 S 形或 Z 形，而转折端为近对称的 M 形，具有指向性特征。

4. 毫稳松向斜（f_{32}）

分布于毫稳松一带志留系陀烈组中，长约 6.2km，出露宽度 3.0km，由南西→北东向，轴迹走向北东—近南北，略呈弧形弯曲，其凸端指向南东。核部为陀烈组中段炭质千枚岩、含炭千枚岩，两翼为陀烈组下段石英绢云母千枚岩和绢云母石英千枚岩等。北西翼地层走向 16°~28°，倾向南东，倾角 35°~78°，南东翼走向 10°~35°，倾向北西，倾角 41°~83°。为一近对称向斜。

向斜北西翼被三叠纪尖峰超单元侵蚀；南东翼以断裂与白垩系接触，并可能受其影响，导致向斜 SE 端轴迹走向 NE，与断裂走向一致。

5. 芙蓉田复式背斜（f_{37}—f_{43}）

分布于区内石碌、军营等地区奥陶系南碧沟组中，走向北西，由多个平行状分布的褶皱组成。但由于受到岩体的侵蚀，褶皱带出露不完整、不连续，以单个背斜、向斜孤零产出，褶皱带整体面貌难以恢复。

北面军营地区，褶皱发育于奥陶系南碧沟组中，走向北西。北西段被晚期北东向构造带切割，南东段被岩体侵蚀，长约 10km，核部为南碧沟组下段，翼部为南碧沟组上段。两翼倾向相反，北东翼倾角 30°~60°，南西翼倾角略陡，为 45°~75°。从属小褶皱比较发育，主要由多个北西走向、平行状分布的开阔型褶皱组成，单个小褶皱长约 1~1.5km，两翼倾向相反，一翼倾向 20°~35°，倾角 60°~65°；另一翼倾向 200°~210°，倾角 45°~75°；轴迹走向 290°~325°，倾角 55°~70°。

6. 陀类-布温褶皱带（f_{58}—f_{63}）

该褶皱带分布于光益村-娜姆河-国界村一带，长约 20km，宽 km。主要由陀烈复向斜和村头岭倒转背斜所组成，卷入地层为下古生界志留系陀烈组，总体走向近东西，北翼岩层产状较陡，倾角 30°~51°，南翼岩层产状略缓，倾角 25°~31°。向斜紧闭斜歪型，背斜向北倒转，总的形态为一较紧密型褶皱带特征。因受后期戈枕韧性剪切带活动的影响，致使褶皱带西段走向向北偏转，形成向南突出的弧形。

陀烈复向斜　分布于陀类-布温东西向褶皱带的南西侧，西起光益村，沿娜姆河，东至国界村一带。褶皱轴向为北东—北西，呈弧形弯曲，弧顶向西突出，长 18.3km，宽 8.6km。组成核部地层为下志留统陀烈组上段，两翼为陀烈组中段及下段，北东或志东翼的地层倾角较陡，一般为 30°~50°，南西或北西翼的地层较缓，一般为25°~31°，枢纽呈波状延伸向南东倾伏，轴面略倾向南西或北西，为紧密弧型褶皱。

村头岭倒转背斜 位于陀类-布温东西向褶皱带北侧，东起村头岭北侧，西至黑运泽一带、轴迹为近东西向的波状延伸，长约7km，宽3.8km。下志留统陀烈组下段组成核部，两翼为陀烈组中上段及空列村组的下段。南北两翼岩层都向南倾，北翼为倒转翼，岩层倾角较陡，一般为50°~65°，南翼为正常翼，岩层倾角较缓，一般为30°~45°。背斜的轴面倾向南，枢纽呈波状延伸，向西倾伏，为紧密线型褶皱，在倒转背斜中发育有大量的小褶皱及劈理。倒转背斜被后期的志北向断裂切割，轴迹发生移动。

此外，在保亭南好地区，志留系与上古生界石炭系同时发生褶皱变形，形成了北东向的南好褶皱带（见后述）。

三、华力西-印支旋回变形特征

（一）小尺度构造变形

1. 石炭系的变形构造

石炭系南好组和青天峡组按岩石能干性相对差异可以划分为较强能干性、较弱能干性和弱能干性三个基本物理单位，不同能干性岩石变形特征迥然有别。

（1）较强能干性硅化灰岩的构造变形

南好组第三段硅化灰岩厚度近100m，成分以灰岩为主，后期硅化作用使其较一般的灰岩具有较大的刚硬度。岩石变形微弱，常见韵律构造、交错层理等。同碎屑岩相比，硅化灰岩整体表现出主动变形的特点。

（2）较弱能干性岩石的构造变形

此类岩石包括南好组第一段的角岩化粉砂岩（原岩可能含灰岩）和薄层灰岩、青天峡组条带状白云质灰岩等，由于其能干性相对较弱，成层性好且被厚层碎屑岩所夹，变形构造极其复杂。

1）褶皱构造

剪切褶皱构造 露头上可见a型褶皱和鞘型褶皱两种类型（图21-26），褶皱面上发育的拉伸线理平行于褶皱枢纽，并与鞘褶皱的舌端方向一致。

图21-26 石炭系南好组中的剪切褶皱示意

Fig. 21-26 Schematic diagram of shear folds in the Carboniferous Nanhao Formation

a—a型褶皱；b—鞘褶皱

褶叠层构造 表现为限定在层内的多级顺层掩卧褶皱群，其中褶皱发育特点不尽相同，进而细分为协调型顺层掩卧褶皱、半协调型顺层掩卧褶皱和片内无根褶皱（图21-27），褶皱轴面平行于层理或劈理。无根钩状褶皱是横向构造置换发展到一定阶段的产物；褶叠层构造的发育一般揭示了小范围（局部）地层无序而整体有序的特点。

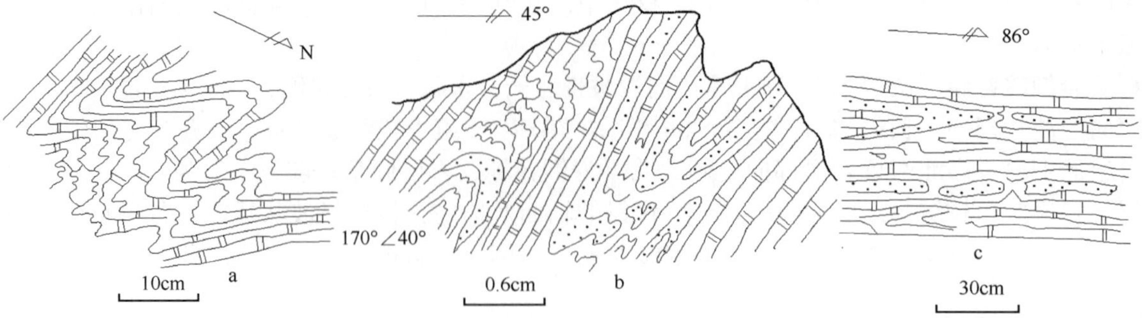

图 21-27　石炭系南好组较弱能干性岩石中的褶叠层构造素描

Fig. 21-27　Sketch showing folded stromatolitic structures in relatively weak dry rocks of the Carboniferous Nanhao Formation

a—协调型；b—半协调型；c—钩状褶皱型

2）石香肠构造

在含燧石条带的南好组地层中广泛发育石香肠构造，按其成因划分为张裂型、黏滞型和褶皱型三类（图 21-28）。

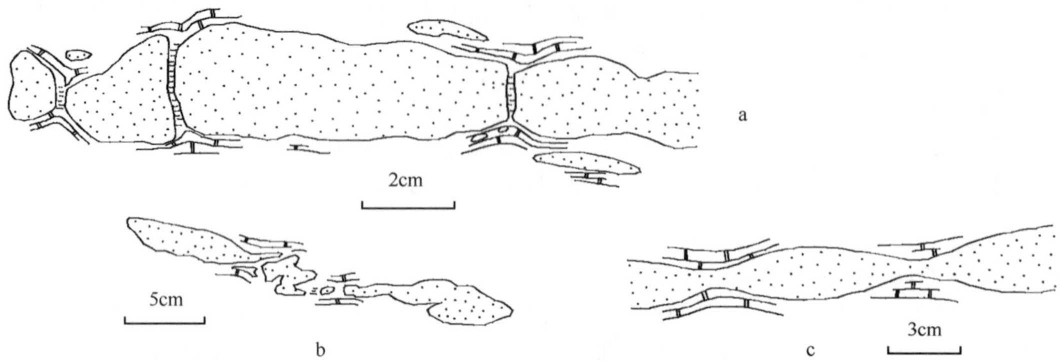

图 21-28　石炭系南好组中的石香肠构造示意图

Fig. 21-28　Schematic diagram of boudinage structures in the Carboniferous Nanhao Formation

a—张裂型；b—黏滞型；c—褶皱型

张裂型石香肠是垂直于层面的脆性张裂造成的，单体呈矩形。在石香肠块段的拉开结合部位造成低压空间，常被其他矿物充填。

黏滞型石香肠断面呈藕节状、透镜状、蝌蚪状等，平行层面分布。

褶皱型石香肠一般在燧石层强烈褶皱的基础上发育而成，这种石香肠与强烈褶皱形迹紧密伴生，其平面形态极不规则。

3）弱能干性碎屑岩的构造变形

南好组和青天峡组碎屑岩的变形多数是以近似平行层面的断层滑动（劈理化）表现出来的，露头上褶皱构造并不多见，但微观域内褶皱构造、剪切滑动构造极其发育。通过横向劈理化作用，地层总体厚度有所减薄。这种劈理化现象在灰岩接触面附近更为突出。

2. 二叠系的变形构造

海南岛出露二叠系有峨查组、鹅顶组和南龙组。鹅顶组为条带状灰岩，局部因热液变质而成为较厚层的大理岩。峨查组和南龙组以碎屑岩为主，在石碌水库坝下南龙组底部见一层厚度变化较大的角砾状硅质灰岩。

二叠系的变形特点与石炭系颇为相似，同样具有顺层构造滑动的构造特点。

峨查组和南龙组碎屑岩除在断裂带附近出露较强的变形现象外，露头上一般难以看到褶皱构造形迹，但小型顺层平行滑动面屡见不鲜，并在微观领域中表现得尤为清楚。值得一提的是，在与鹅顶组接触地带多表现出断层活动迹象，如鹅顶组灰岩之下的峨查组平行于片理的褶皱常呈无根钩状，层理构造消失殆尽（图 21-29）；而在鹅顶组之上的南龙组下部的角砾状硅质灰岩、硅质角砾岩为棱角状，长轴显示较好的方向性，并平行于一组脆性裂隙面，而充填的石英表面洁净，彼此以“三边结构”接触，且强烈波状消光（图 21-30），因此，这套角砾状岩石并不排除早期沉积之可能，但后期顺层构造滑动也是可以肯定的。

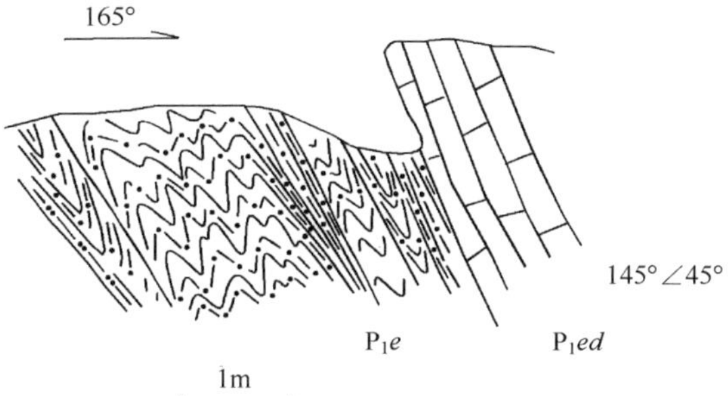

图 21-29　二叠系峨查组与鹅顶组接触面变形素描图

Fig. 21-29　Sketch showing the contact deformation between the Permian Echa Formation and the Eding Formation

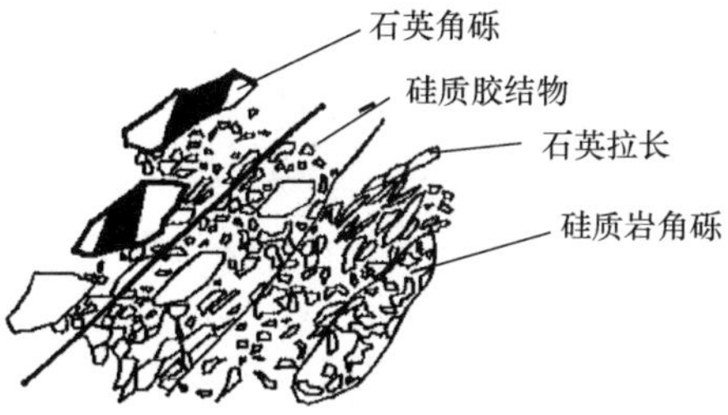

图 21-30　南龙组角砾状硅质灰岩显微组构素描图

Fig. 21-30　Sketch showingthe microscopic fabric of brecciated siliceous limestone of the Nanlong Formation

鹅顶组的变形具有以下特点：

1）薄层条带状灰岩较厚层大理岩化灰岩显示较强的易变性。河叉岭鹅顶组下部为中厚层大理岩化灰岩，其上为薄层灰岩，前者变形简单，后者变形复杂并形成强烈的塑性流动褶皱（图 21-31）。

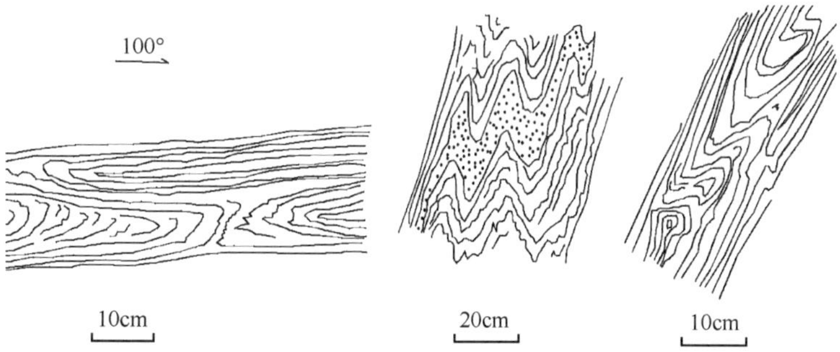

图 21-31　二叠系鹅顶组薄层灰岩塑流褶皱素描图

Fig. 21-31　Sketch showing rheomorphic folds of the Permian Eding Formaiton thin-bedded limestone

2）大型岩体附近的灰岩较远离岩体的灰岩变形更为强烈。如石碌水库坝下石碌河谷出露的鹅顶组条带状灰岩，顺层剪切现象到处可见，形成了以顺层剪切片理带、顺层剪切褶皱、S-C 组构、构造石香肠等反映主体伸展环境的构造形迹组合（图 21-32）；而在鸡实南鹅顶组灰岩的变形比较微弱，地层中尚保存了大量完好的蜓类化石。

3）主干断层带附近的灰岩变形强烈。产于鸿沟岭-红岭断裂带和金波-俄朗断裂带之中及其附近的灰岩，既保留了早期褶皱面貌，又叠加了断裂和褶皱构造，表现出多期变形特点。

4）显微变形现象异彩纷呈，不仅展示了宏观常见的构造形态，而且还清楚地显现了肉眼难以辨别的劈理构造和矿物变形亚结构（图 21-33）。

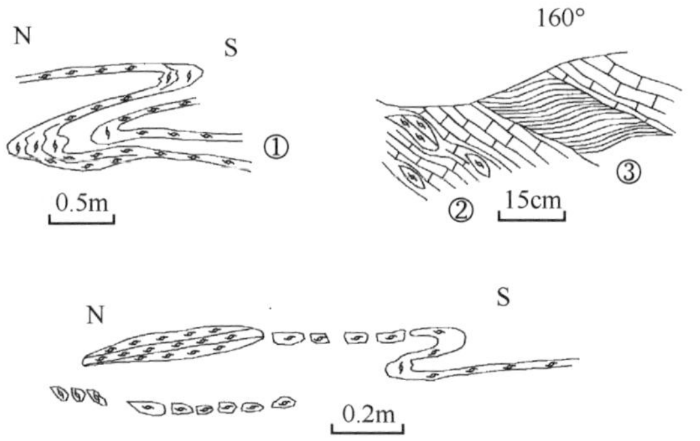

图 21-32　二叠系鹅顶组灰岩变形样式示意图

Fig. 21-32　Schematic diagram of deformation styles of the Permian Eding Formation limestone

①燧石条带；②灰岩；③片理化灰岩

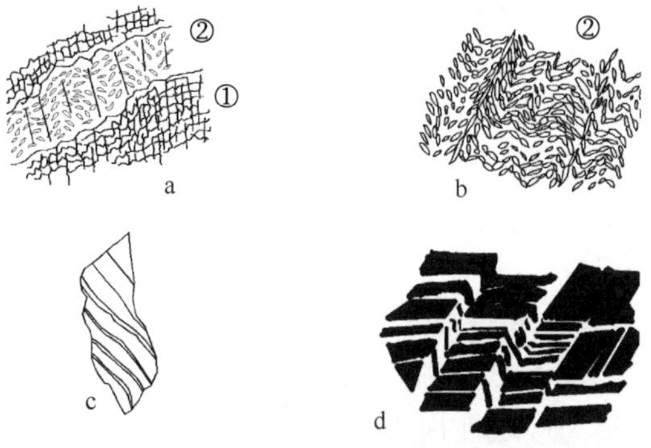

图 21-33　二叠系鹅顶组灰岩显微变形结构示意图

Fig. 21-33　Schematic diagram of microscopic deformation structures of the Permian Eding Formation limestone

3. 三叠系的变形构造

三叠系仅有下三叠统岭文组，为华力西—印支造山作用地壳隆升阶段沉积的产物。岩石未经历区域变质作用改造，其结构构造仍保留完好。岩石中褶皱构造与透入性面理及线理构造不发育，变形特征主要表现为发育少量的韧脆性断裂构造，属于构造层次的韧-脆性变形的范畴。

（二）大尺度褶皱构造

发育于上古生界中，有北东向坡兰沟向斜、南好褶皱带和北西向江边复式向斜，及近东西向王下复式向斜、南北向燕窝岭向斜。

1. 坡兰沟向斜

发育于军营-红岭逆冲推覆构造带中，沿南洞岭—坡木岭—军营岭一带分布，长约15km。发育于上古生界中，核部为上石炭统青天峡组，两翼为下石炭统南好组。轴迹走向北东，北西翼优势产状120°∠57°，南东翼优势产状325°∠50°，β图解显示，其轴面近直立，褶皱轴向北倾伏，倾伏角约21°。属于近对称型倾伏向斜。

2. 南好褶皱带（f_{33}—f_{35}）

分布于保亭县南好一带，长约30km，宽近10km。由大致平行的鹅格岭倒转向斜、那通岭倒转背斜、振海山倒转背斜、什茋-情安岭倒转向斜等组成。总体呈北东向，北东段轴向往东偏呈北50°~60°东，南西段轴向往南偏，呈现出一略向西突出的弧形。褶皱带由志留纪和早石炭世地层组成，地层均倾向北西，正常翼地层倾角较缓，一般为30°~40°，倒转翼倾角较陡为60°~70°，具紧密线状褶皱特点。整个褶皱带被华力西期、印支期的岩体包围，带内又为一系列北北东向断裂和燕山脉岩所穿插，使其变得支离破碎。

鹅格岭倒转向斜 鹅格岭-空猴岭倒转向斜：东段、西段以南好组第一段为核部，两翼为足赛岭组第三、第二段；中段以足赛岭组第三段为核部，足赛岭组第二段构成西翼；在加油南地段，核部为南好组第一段，东南翼为足赛岭组第三段，西北翼因毛拉岩体破坏而缺失。岩层总体走向40°~60°，基本上向北西倾斜。东南翼为正常翼，倾角30°~40°，西北翼为倒转翼，倾角60°~75°。在加油南西北翼倾向南东，东南翼倾向北西，倾角20°~67°，变为正常向斜。向斜轴线约北东50°，往东北部变为70°~80°

那通岭倒转背斜 位于鹅格岭-空猴岭倒转东南侧。以足赛岭组第二段为核部，两翼为足赛岭组第三段。但在加油南地段则以足赛岭第三段和南好组第一段分别组成核部和翼部。岩层走向北东20°~60°，倾向北西。西北翼为正常翼，倾角30°~40°；东南翼为倒转翼，倾角60°~70°。枢纽走向北东50°~70°。

振海山倒转背斜 位于什茂-情安岭倒转向斜南东侧的振海山至狗岭一带。核部为足赛岭组第二段，两翼为足赛岭第三段在狗岭一带由南好组第一段组成核部，两翼遭受断裂破坏而仅零星出现南好组第二段。岩层走向北东35°~74°，倾向北西。西北翼为正常翼，倾角19°~78°，东南翼为倒转翼，倾角34°~66°。背斜轴线北东55°~60°，在狗岭段约为北东35°。

什茋-情安岭倒转向斜 位于那通岭-白土岭倒转背斜的东南侧。核部由南好组第一段、两翼由足赛岭组第三段组成。岩层走向北东65°~75°，基本向北西倾斜。西北翼为倒转翼，倾角66°~71°，东南翼为正常翼，倾角46°~78°。但在南后山和情安岭两地段，组成核部的南好组第一段，在轴线北西侧的岩层走向约北东40°，倾向南东，倾角30°~40°；东南侧岩层走向北东40°~65°，倾向北西，倾角43°~63°。向斜轴线北东段为北东75°，中段为北东65°，狗岭段为北东40°。

3. 江边复式向斜（f_{55}—f_{57}）

向斜为区域性构造，位于陀类-布温东西向褶皱带中部，峨查村-田头村一带。两翼各有次一级的山猪岭向斜、峨文岭-黑运泽向斜。影响长度为15km，宽约8km。组成核部地层为中—上二叠统南龙组，两翼地层为下二叠统峨查组和中二叠统鹅顶组。北翼地层倾向南，倾角较陡，为38°~60°；南翼地层倾向北，倾角较缓，为38°~45°。褶皱枢纽呈近东西向波状延伸，向东倾伏，倾伏角25°左右，轴面向北倾斜，为正常的开阔短轴状斜歪褶皱。

向斜被北西向断裂切割，北翼抬起上升，遭受剥蚀，缺失了部分地层。此外还被近南北向的横向断裂切割，轴迹发生位移。

山猪岭向斜 位于国界村一带，是江边复式向斜南翼的次一级褶皱，轴迹为近东西向延伸，影响长13.5km，宽1km左右，向斜的核部为峨顶组两翼为峨查组，两翼倾角均为40°左右，为轴面直立的对称褶皱、枢纽波状起伏，略向南东方向倾伏，倾角大约8°左右。

峨文岭-黑运泽向斜 为江边复式向斜北翼的次一级褶皱，西起于峨文岭，东至黑运泽一带。向斜总体走向为近东西向，延伸长6.5km左右，宽2.5km。核部为鹅顶组，两翼为峨查组。南翼地层较陡，倾角为35°~50°，北翼地层较缓，倾角为30°左右，轴面倾向南，为开阔的斜歪褶皱。褶皱枢纽呈波状延伸，向西倾伏。该向斜被南北向的一组断裂切割，使其轴面发生位移。

4. 王下复式向斜（f_{52}—f_{54}）

位于王下、俄贤一带，卷入地层主要为二叠系峨查组和鹅顶组。走向近东西向，北翼产状170°~180°∠60°~80°，南翼产状0°~30°∠20°~30°。北翼靠近花岗岩而发生角岩化。属于不对称褶皱。

5. 燕窝岭向斜（f_{64}）

发育于燕窝岭一带二叠系中，挟持于两条近南北向断裂之间。走向近南北，呈蛇形弯曲。核部为二叠系鹅顶组灰岩，两翼为二叠系峨查组碎屑岩。北面西翼倾向南东，倾角40°~54°，东翼倾向北西，倾角53°~60°。为不对称斜歪向斜，属于与南北向断裂相伴随而形成。

第三节 上叠盆地变形特征

上叠盆地包括燕山期—喜马拉雅期的沉积。燕山旋回—喜马拉雅旋回岩石未经历区域变质作用，岩石的原生结构构造仍保留完好，仅白垩纪岩石局部受岩浆活动影响，在与侵入岩接触部位发育小型褶皱变形及接触变质作用。燕山旋回—喜马拉雅旋回岩石中褶皱构造与透入性面理及线理构造不发育，变形特征主要表现为发育少量的韧脆性断裂构造，属于浅表构造层次的韧-脆性变形的范畴。其中，白垩系中的褶皱主要有北东向石头岭向斜、沟门背斜、青龙坡向斜和近南北向松涛向斜等。此外，在白沙盆地中，前人并命名有北东向光明向斜、山荣背斜、南美向斜、长茅向斜等。但本次通过的系统分析，认为其属于向、背形构造，在走向上呈斜列状产出，显示盆地具有拉分盆地成因。

1. 石头岭向斜（f_{19}）

该向斜位于阳江盆地石头岭一带，发育于下白垩统鹿母湾组中。北东端为第四系覆盖，长约8km，波及宽度约5km，轴向50°~60°，略呈波状弯曲。其东段两翼产状较陡，北西翼25°左右，南东翼55°；西段产状较平缓，北西翼10°~17°，南东翼10°~25°，为北西缓南东陡的不对称褶皱，轴面略向南东倾斜。

2. 沟门背斜（f_{20}）

发育于下白垩统鹿母湾组，背斜北东端在山心岭一带消失，南西端在区内下莱一带倾伏。长度6km，波及宽度1.5km，轴向45°，在沟门见其核部出露。南翼产状为21°~40°，北翼产状18°~36°，形态基本对称，轴面直立，为一开阔型褶皱。

3. 青龙坡向斜（f_{21}）

该向斜与沟门背斜平行，仍发育在鹿母湾组第四段中，向斜两端与北东向大岭断层斜接，延伸长度5.5km，宽仅1km，枢纽走向约50°，略呈弯曲状。北西翼产状倾向南东，倾角20°~40°；南东翼倾向北西，倾角35°，轴面产状近直立，为一直立开阔褶皱。

4. 松涛向斜（f_{66}）

分布于松涛水库一带，由于受到北西向断裂左旋活动的影响，白垩系局部发育一轴向近南北向的向斜构造，向斜长约8km，宽约4km，轴向近南北向，轴面近直立，两翼产状均较平缓，倾角约为15°，属于平缓开阔型向斜构造。

第四节　断裂体系

海南省大地构造位置十分独特，在漫长的地质历史发展过程中经历了多期次的构造运动。并形成了东西向、北东向、北西向和南北向等主要构造体系相互交织的复合构造格局。

一、区域性主干构造带

区域性主干构造带是具有区域性展布、影响大的构造带，往往是不同构造单元的分野性构造带。由于南海部分地质资料零星且有限，本次仅针对海南岛区域性主干构造带进行阐述。纵观全岛，其区域性主干构造带主要是东西向王五-文教断裂带、北东向崖城-万宁断裂带，以及分别发育于五指山造山带、印支-南海地体中的石碌-石浮弧状构造带和狗岭逆冲推覆构造带等。

（一）王五-文教断裂带

王五-文教断裂带位于海南岛北部，是新生代海南隆起与雷琼裂陷的分野性断裂带。由东西向平行状排列的 F_1、F_2 和 F_3断裂组成。大致展布在北纬 19°45′左右，横跨儋州、澄迈、定安、文昌等地区，东西两端延伸入海，陆上延伸约 210km。其构造形迹除了在铜鼓岭北面宝陵港见及外，其余地区地表未见出露，呈隐伏状产出，为物探推测构造带。

1. 大尺度构造形迹

1）重力图中，构造带以北，重力场与北部湾一致，其重力高、低反映了基底起伏和新近系沉积，构造带以南是重力低异常区，负异常反映的是莫霍面的起伏，地壳增厚，达 34～35km。重力异常显示构造带由一系列呈东西向分布的断裂组成，其断面倾向北，倾角大于 60°，呈阶梯状下降，具有正断层特征。

2）航磁 ΔT 平面图中，反映为一个剧烈变化的正异常交接带。卫片中表现为一清晰的东西向亮带，为南北两种不同影像的分界线。在 TM 及雷达融合遥感影像上，可以看到该断裂经过区域表现为近东西向延伸的隐晦与清晰相间的条带，线性影像平直。

3）沿构造带发育多个东西向的新生代凹陷盆地，反映了盆地的形成受其制约，自西往东、自南往北，盆地的下拗幅度逐渐增大，长坡盆地、福山-多文盆地和海口盆地下拗幅度分别为 300～400 m、2000 m 和 3000 m，盆地内接受了新生代以来的沉积。

4）控制了琼北新生代多期次的火山喷发或喷溢作用，形成大面积东西向展布的玄武岩被。

5）沿构造带地震活动比较强烈。据 1463～1834 年间的不完全统计，琼山、定安、澄迈、文昌等地发生过 37 次地震，其中破坏性地震 5 次。1970 年临高县附近海域发生的 1. 8 级弱震也与该构造带有关。

2. 小尺度构造形迹

小尺度上的构造形迹主要由节理化带和构造破碎带来显示。

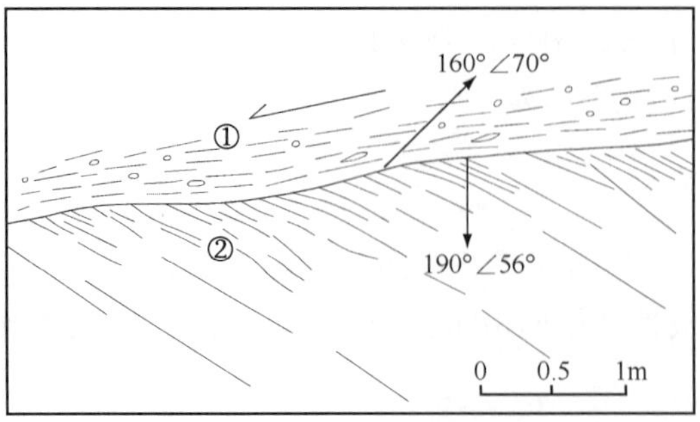

图 21-34　宝陵港东西向构造形迹素描图
Fig. 21-34　Sketch showing EW-trending structural traces in the Baoling Port
①糜棱岩；②劈理

铜鼓岭北面宝陵港三叠纪花岗岩中发育近东西向构造破碎带（图 21-34），破碎带宽约 10m，倾向南，倾角 60°～70°，由构造角砾岩、糜棱岩、糜棱岩化岩石组成。破碎带南侧围岩中发育一组挤压破碎带形成时派生的劈理，根据围岩中发育的劈理与主断面的关系分析，构造带具有左旋压扭性活动特征。铜鼓岭南面尖岭一带三叠纪花岗岩中，断裂以节理化带显示，节理走向近东西向，节理面平直，倾角近直立，节理密度 5～20 条/m，属于一组剪性节理。节理切割北东向构造形迹（花岗斑岩脉），并左行切错小石英脉，错距约 15cm，表明构造带具有左行剪切性质。

铜鼓岭北尖山村西侧海边的三叠纪花岗岩中发育近

东西向的构造破碎带，破碎带宽约 50m，倾向北，倾角 80°～85°，断裂带内岩石碎裂岩化，钾长石斑晶由于构造作用碎裂呈近圆状或眼球状旋转碎斑。破碎带内 S-C 组构发育，显示构造带具右行剪切特征。宝典村北侧花岗岩中，见发育有韧性剪切变形残迹。岩石面理发育，长石斑晶发生剪切破碎，呈眼球状、δ 型旋转碎斑状，指示构造带具有右旋剪切活动的特点。

此外，在儋州林场附近发育一断裂，走向近东西向，倾向西，倾角 70°，属于正断层，并切错第四系，断距达 3.75m（图 21-35）。沿构造带见有华力西期闪长岩脉侵入，反映了构造带至少形成于华力西期。由此表明，王五-文教断裂带具有多期次活动特征，可能形成于前华力西期；华力西期—印支期发生过强烈压扭性活动，导致幔源岩浆侵入；燕山期控制了同期岩体的侵位，导致岩体呈东西向展布，根据岩体未发育定向组构，可能显示该期活动具有张性或张扭性；新生代以来，构造带活动非常强烈，多次活动特征明显，并表现为张性特征，控制了新生代盆地的形成和火山喷发、喷溢活动以及地震活动。

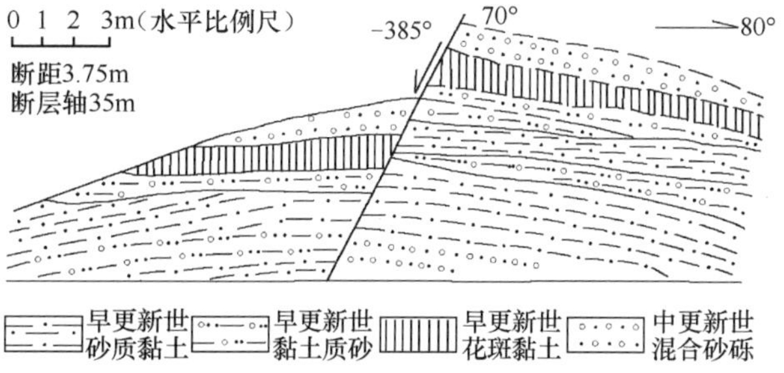

图 21-35　王五-文教断裂切错更新统剖面图

Fig. 21-35　Pleistocene section showing transecting of the Wangwu-Wenjiao fault

（据刘瑞华等，1989）

（二）崖城-万宁断裂带

崖城-万宁断裂带断续展布于海南岛南部崖城-万宁一带，平面上略呈向南东凸出的弧状。长大于上百千米，两端延伸年入海。由 F_{74} 和 F_{75} 断裂组成。走向 30°，倾向 300°，倾角 50°。早期韧性变形形迹发育于二叠纪花岗岩中，因后期燕山期岩浆作用而呈残迹状断续见于陵水县芒三水库、吊罗山林场和万宁市兴隆等一线广大地区。芒三水库一带，由于覆盖严重，剪切带内韧性变形岩石露头零星出露，地表出现的岩石绝大多数为转石。北东面止于白垩纪花岗岩，长约 1.5km，影响宽度约 1km，倾向 120°，倾角 68°。吊罗山林场南东面河沟中，延伸长约 1km，影响宽度约 1km，走向 30°，倾向北西，倾角约 50°，两端止于白垩纪花岗岩。大南泥村一带二叠纪花岗岩中，出露长约 500m，可见宽度约 62m，倾向 130°，倾角约 74°。

1. 岩石学特征

变形带中心为构造片岩、构造片麻岩（?），两侧总体为糜棱岩化花岗岩（图 21-36，图 21-37）。

糜棱岩化花岗岩　糜棱结构，弱定向构造。碎斑和基质的含量分别约占 60% 和 40%，石英和云母类等弱能干性矿物普遍经受韧性变形的改造，其中石英的拉伸比一般在 4∶1 以上，可出现多晶条带构造，黑云母可见“云母鱼”变形组构；斜长石等能干性矿物除发生旋转变形外，主要发育了剪切阶步和晶内裂隙等脆性变形现象。变形环境相当于绿片岩相-低角闪岩相。

构造片岩　糜棱结构，片状构造，由石英（50%）、斜长石（20%）和黑云母（30%）组成。石英为粒状变晶，变晶粒度为 0.1～0.3mm，一般被黑云母围绕而呈右旋 δ 型碎斑，定向分布，波状消光；黑云母为片状变晶，沿片轴方向均匀定向分布，显示岩石的片状构造，并因剪切作用而多呈“云母鱼”状；斜长石为粒状变晶，变晶粒度为 0.1～0.2mm，沿岩石片理方向均匀定向分布，边缘弱泥化，机械双晶发育，局部发生细颈化而呈透镜体状。

构造片麻岩（?）　具糜棱结构，片麻状构造（图版Ⅴ-2-G）。碎斑成分为钾长石（8%）、斜长石（5%），大小为 2.2～12.0mm，蚀变不强，钾长石卡斯巴双晶发育，斜长石聚片双晶和机械双晶发育，聚片双晶纹发生强烈扭曲；岩石基质矿物成分为钾长石（34%）、斜长石（10%）、石英（30%）、黑云母（12%）及少量角闪石。钾长石为他形粒状，粒径大小为 0.2～1.5mm，蚀变不强，双晶不发育，分布具一定方向性；斜长石为半自形柱状，粒径大小为 0.2～0.8mm，未蚀变，聚片双晶发育，双晶纹发生扭曲且见尖灭现象，与钾长石接触定向分布，在它们接触处见蠕英结构；石英为他形粒状，粒径

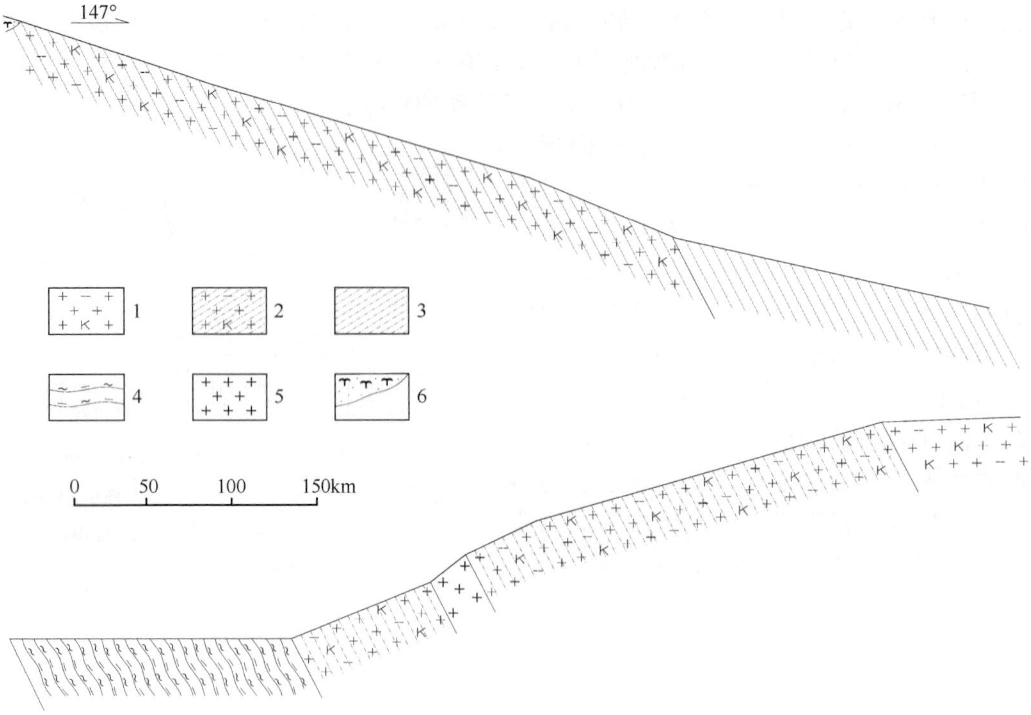

图 21-36　藤桥镇芒三水库实测构造剖面图

Fig. 21-36　Measured structural section of the Mangsan Reservoir，Tengqiao Town

1—中粒斑状黑云母钾长花岗岩；2—糜棱岩化中粒斑状黑云母钾长花岗岩；3—初糜棱岩；4—构造片麻岩；5—长英质岩脉；6—掩盖

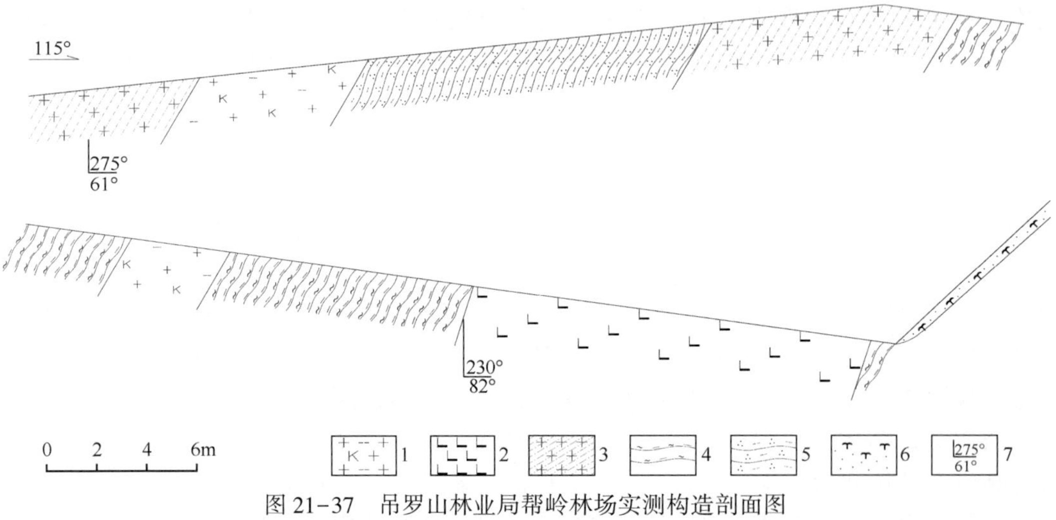

图 21-37　吊罗山林业局帮岭林场实测构造剖面图

Fig. 21-37　Measured structural section of the Bangling Forest Farm，Diaoluoshan Forestry Bureau

1—细粒黑云钾长花岗岩；2—煌斑岩；3—糜棱岩化花岗岩；4—构造片麻岩；5—构造黑云石英片岩；6—掩盖；7—产状

大小为 0.2~2.0mm，以多晶条带状分布在长石颗粒间，与长石一起定向分布；黑云母为片状，其片轴方向与石英的分布方向一致，具褐色—黑色多色性，解理见扭曲，局部见“云母鱼”。

2. 变形组构

野外宏观尺度上的变形组构有不规则褶皱和鞘褶皱、剪切褶皱、肿缩状石香肠构造、长石 δ 型旋转碎斑系及右行书斜构造等（图 21-38）。其中，不规则褶皱和鞘褶皱、肿缩状石香肠构造见于构

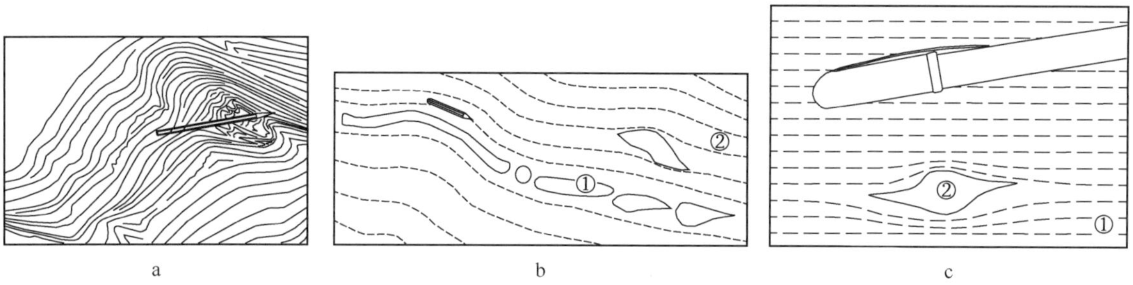

图 21-38 野外尺度变形构造图

Fig. 21-38 Map of deformation structures based on a field scale

a—不规则褶皱；b—石香肠素描图（①石香肠；②构造片岩）；c—旋转碎斑（①构造片麻岩；②旋转碎斑）

造片岩中，而剪切褶皱和长石 δ 型右旋碎斑系则发育于构造片麻岩中。此外，在变形带边部糜棱岩化花岗岩中，并见长石斑晶线理与基质矿物面理（剪切面理）呈锐角相交，指示平面右旋的运动学特点。

微观尺度中，长石斑晶韧性变形强烈，呈眼球状、长条状、拖尾状、δ 型旋转碎斑状。基质中，长石发育变形纹、剪切阶步、扭曲变形；石英基本上以多晶条带产出，晶体呈锯齿状接触，动态重结晶明显；黑云母可见“云母鱼”、沿解理面剪切活动及膝折等变形组构（图 21-39）。

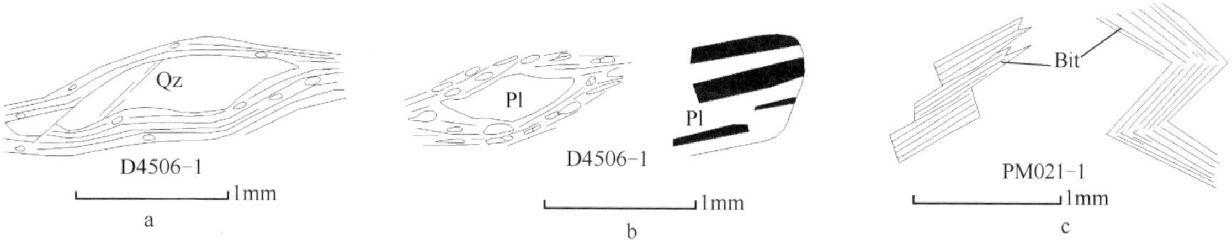

图 21-39 微观变形组构图

Fig. 21-39 Microscopic deformation fabric diagram

a—剪切阶步；b—旋转碎斑（左），机械双晶（右）；c—黑云母沿节理面剪切（左），膝折（右）

3. 岩组学特征

芒三水库岩石从长石的旋转和 S-C 组构判断，变形带具有向北逆冲（南盘）的运动学矢量。采自该剪切带内的 2 个变形岩石的石英岩组图（图 21-40）表明：2 个样品的组构图均为近似单斜对称，并受到了压应力兼扭动作用力的影响，其中，从一个样品岩组图极密中可得出，主压应力在 10°左右的方位，两组剪切面分别为 320°和 50°，运动方向为平面右行，剖面由南东东向北西西逆冲；另两个样品的主压应力方位在 30°的位置，各自发育的两组剪切面走向大约应在 340°和近南北的方位上。从野外调查看，带内发育有北东和北西向两组面理，构成一网脉状。局部并见剪切褶皱，指示逆冲并兼平面右行走滑的运动学特征。

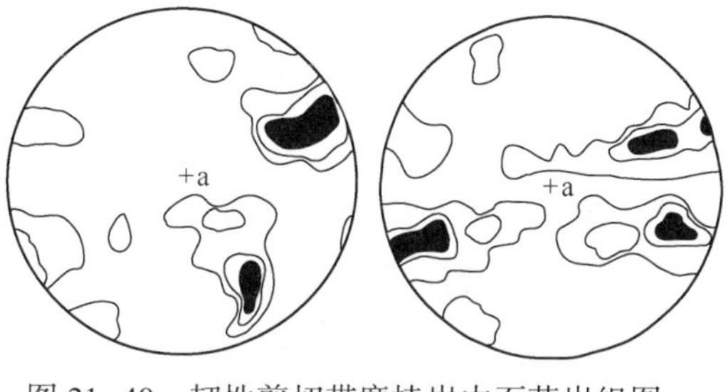

图 21-40 韧性剪切带糜棱岩中石英岩组图

Fig. 21-40 Quartz fabric in mylonite of ductile shear zones

（据 1 : 25 万《乐东县、陵水县区调报告》，2003）

4. 运动学特征

北东段韧性变形岩石中，沿走向（平面）方向，矿物变形比较强烈，常见长石 δ 型旋转碎斑指向性组构，指示右旋活动的运动学特点；而沿剖面（倾向）方向，岩石变形相对较弱，指向性构造不甚发育。反映了变形带以剪切作用为主，倾向面走滑运动分量较小，岩石基本上通过剪切变形而形成。

5. 年代学特征

变形带发育于二叠纪花岗岩中，沿走向止于白垩纪花岗岩，并见早三叠世富碱侵入岩和中三叠世EMⅡ型镁铁质岩充填。镁铁质岩、富碱侵入岩分别代表了伸展、隆升的构造背景。因此，变形带生成时间应为二叠纪—早—中三叠世。

基于Ar-Ar测年的适用性，结合变形带岩石形成的特点，本次参照花岗岩定年测试方法，采用锆石激光ICP-MS法对两处构造片麻岩进行年代学测定，南西段芒三水库构造片麻岩（D0944点）锆石ICP-MS年龄变化于264~294 Ma，年龄加权平均值为278.5Ma；北东段吊罗山构造片麻岩锆石ICP-MS年龄除了一个属于早三叠世外（246 Ma），其余都属于二叠纪（259~312 Ma），其年龄加权平均值为265.9Ma（图21-41）。可能暗示着该构造带经历了265.9~278.5Ma的演化过程，与谢才富（2007）所主张的海南岛造山作用时间（270~290Ma）基本一致。

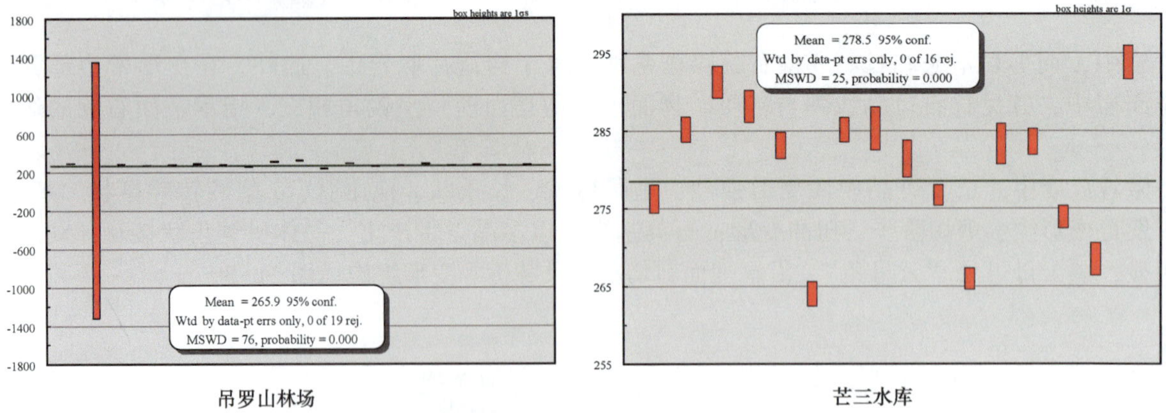

图21-41 变形带锆石ICP-MS年龄

Fig. 21-41 Zircon ICP-MS ages of deformation zones

6. 活动期次

沿变形带走向早白垩世花岗岩中未见相应的韧性变形迹象，而韧性剪切变形带南东面相继发育有北东走向的中三叠纪EMⅡ型基性—超基性的镁铁质岩带（谢才富，2007）、碱性岩带等花岗岩浆带，共同组成了一条呈南南东向凸出的贯穿于崖城-万宁一带的构造-岩浆带，属于一条切割深、影响范围大的重要的超壳构造带，经历了华力西期—印支期早期挤压和印支期晚期伸展过程。

万宁市兴隆一带，其旁侧见大量的流纹质含角砾晶屑凝灰熔岩、流纹质碎斑熔岩坡残积块分布，这些火山岩是海南岛白垩纪火山岩的常见岩石，推断其与牛腊岭、同安岭早白垩世火山盆地为同一火山岩带产物。由也可能暗示了构造带白垩纪再次活动并控制了火山喷发。

（三）石碌-石浮弧状构造带

发育于五指山造山带中，横纵整个五指山造山带。横穿在石碌镇、合盛镇、南丰镇、石浮镇一线的广大地区，自西往东，其走向由北西—近东西—北东向，呈现出向南东向突出的弧形。弧状构造带由弧状褶皱带和断裂带组成（李孙雄等，2006）。根据卷入褶皱的最老地层为青白口系石碌群，最新地层为二叠系，并被二叠纪花岗岩侵蚀的事实，认为褶皱形成时代为二叠纪。此外，陈新跃等（2006）在公爱一带北西向断裂中获得$^{39}Ar-^{40}Ar$年龄为245~250Ma，时代为早三叠世。因此，构造带形成于华力西期—印支期，属于华力西-印支期造山运动生成起来的席卷整个海南岛北部的重要构造迹象。

受后期构造岩浆作用改造，导致弧状构造带残缺不全，连续性差。西段并由于受到晚期北东向军营-红岭构造带和戈枕断裂带的走滑活动影响，使得古生界岩石面理受到强烈改造，走向转向北东。但在影响较弱或未受到影响地段，岩石依然保持面理走向北西的格局。构造带东面，由于东西向王五-文教断裂带切割和第四系覆盖而淹没。小尺度中弧状构造带由北西向河叉岭构造带、石碌构造带、王下构造带和北东向兰洋构造带、东岭构造带、崖城-南好构造带等组成。

1. 河叉岭构造带

位于河叉岭珠碧江西侧，主要由 F_{91}—F_{94}断裂及其相伴随的褶皱（f_{37}—f_{43}）组成。剖面自上而下分成倒转背斜推覆体、推覆型韧性剪切带和原地岩片三部分。

（1）倒转背斜体推覆体

河叉岭倒转背斜推覆体核部为大干村组地层。不整合于大干村组之上的南好组构成了该倒转褶皱的两翼。褶皱轴迹呈325~330°延伸，轴面倾向南西，倾向40°左右。褶皱为不对称状，倒转翼产状较陡，倾角在40°~50°之间。该倒转褶皱与相邻地层均为断层接触，它们可能代表了并非原地的外来系统。

（2）推覆型韧性剪切带

沿大干村组与上古生界接触面发育，平面上呈一向南西凸起的弧形，在不同地段被卷入的岩石类型相差较大。河叉岭一带，大干村组直接推覆于鹅顶组灰岩和峨查组之上，两者之间发育的剪切带宽大于30m，主体是走向325~338°，倾角50°~60°。倾向南东，其中以糜棱岩化石英片岩和糜棱大理岩为主，并到处可见大干村组石英片岩、含砾石英片岩、红柱石绢云母片岩形成强剪切褶皱和a型褶皱、窗棂构造和拉伸线理等（图21-42），并一致反映了由南西向北东推覆的剪切指向。在长岭一带，剪切带主要发育于大干村组下部的石英片岩中，镜下鉴定结果表明，这套片理极其发育的岩石，石英普遍塑性变形，长轴与短轴之比达5∶1，实际上是糜棱岩作用的产物。在河叉岭一带尚见飞来峰和构造窗。由质地坚硬且抗风化的石英质砾岩为主要组成的飞来峰常“飘浮”于古生界地层之上。尽管其根带难以追索，但岩石学和构造学研究证实它不可能是白垩系不整合产物，与大干村组所含的砾岩极其相似，两者可能源于一物，而以断层关系为下古生界环绕的上古生界峨查组、鹅顶组、南龙组岩片可能是残余的构造窗，为推覆面之下的原地或准原地系统。

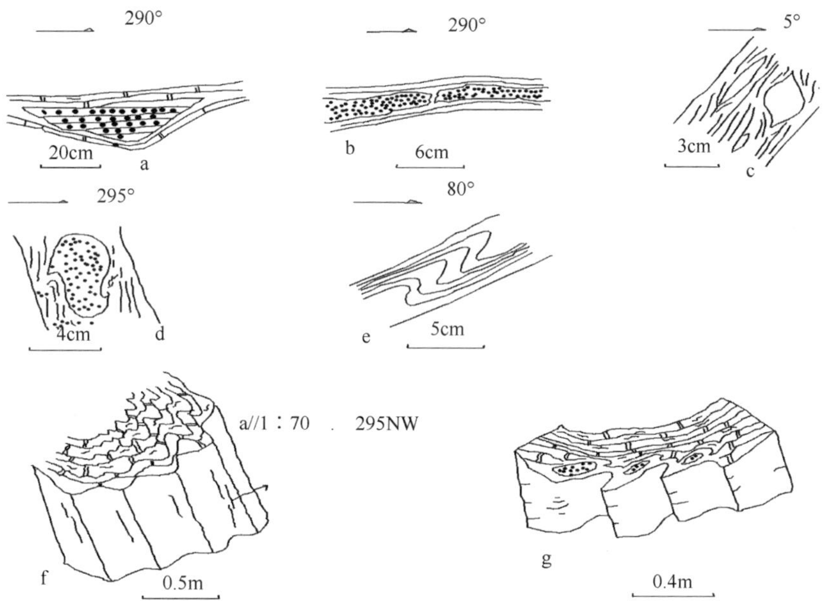

图21-42　河叉岭推覆构造构造变形特征图

Fig. 21-42　Structural deformation characterisitics of Hechaling Nappe Structure

a—石英片岩挤入糜棱大理岩中；b—含砾石英片岩石香肠构造；c、d—含砾石英岩旋转碎斑；
e—剪切褶皱；f—糜棱大理岩中的a型褶皱；g—糜棱大理岩中的窗棂构造

至于这次推覆事件的发生时间目前缺乏可靠的地质资料和同位素测龄数据而难以确定，仅从卷入上古生界地层，并被二叠纪花岗岩侵蚀分析，其形成时间为二叠纪。

2. 石碌构造带

出露于石碌地区，走向北西，主要由小英山断裂（F_{106}）、石碌岭-枫树下断裂（F_{107}）和石碌复向斜（f_{44}—f_{51}）组成。

小英山断裂（F_{106}） 西起石碌铁矿选矿厂南侧，经小英山、大修石东至桦栗山西侧，走向北西向 320°，全长 3. 5km。断带宽 3～5m，产状 50°∠50～80°，局部倾向南西。表现为发育硅化破碎带，构造岩由碎裂岩、角砾岩、透镜体及片理等组成，角砾为所切地层成分，呈扁豆状拉长与断面方向一致，硅化强烈。局部地段可见晚期活动的断层泥及片理未成岩，显示多期次活动特征。早期张扭性，并有花岗斑岩及煌斑岩等脉岩充填；晚期为右旋压扭性逆冲断层。根据断裂切过石碌群、石灰顶组及南好组第五段（原三棱山组），切错南北向断裂，其形成时代应为海西期，燕山晚期至喜马拉雅期右旋压扭活动叠加。

石碌岭-枫树下断裂（F_{107}） 西起石碌岭西侧，经枫树下、枫树顶东至桦栗山南侧，基本沿石碌群第五层和第六层界面发育，延伸长 5km。断层表现为一挤压破碎带，宽 10～25m，总体走向 310°～320°，局部走向 260°～280°。断面舒缓波状，倾向南西，西端转为北东倾向，倾角 29°～80°。破碎带由碎裂岩、角砾岩及挤压透镜体、片理以及断层泥等组成。沿断层两侧岩石强烈变形，表现强烈的挤压和多次活动特征。

石碌复向斜由一系列平行状分布的背、向斜组成，其各自特征见表 21-1。

表 21-1 石碌复向斜各次级褶皱主要特征

Table 21-1 Main characteristics of individual secondary folds of Shilu syclinorium

复向斜位置	褶皱名称	编号	分布位置	规模	翼部产状			轴面产状		组成褶皱地层		褶皱形态
					部位	倾向	倾角	倾向	倾角	核部	翼部	
核部	北一向斜	f_{44}	自北一铁矿，经石灰顶至桦栗山北，东面被北北东向断裂所截	延伸长约 4km	N	200°	60°	轴向 290°～295°	东段近直立，中、西段略向南倒转	石碌群第六层	石碌群第五层	直立开阔型
					S	35°	70°					
	红房山背斜	f_{45}	位于石碌岭西侧，经红房岭至大修厂西，东面被南北向断裂所截	延伸长约 2. 5km	N	30°	30～60°	轴向 295°～300°	近直立	石碌群第五层	石碌群第六层	直立紧闭型
					S	210°	60°					
	石灰顶向斜	f_{46}	自南一铁矿，经石灰顶至枫树坪铁矿，东面被南北向断裂所截	延伸长约 2. 5km	N	200°	50°	轴向 295°～300°	近直立	石灰顶组	石碌群第六层	直立开阔型
					S	20°	62°					
南西翼	金牛岭背斜	f_{47}	西起石碌铁矿仓库北侧，向东经金牛岭，南西至 375 排土场。被中二叠世花岗岩侵蚀	延伸长约 3. 5km	N	40°	35°～70°	40°～45°	40°～60°	石碌群第一层	石碌群第二层	同斜倒转型
					S	45°	60°					
	亲变岭向斜	f_{48}	位于亲变岭东，西为南北向断裂所截，东被中二叠世花岗岩侵蚀	延伸长约 1. 2km	N	50°	75°	50°	70°	石碌群第五层	石碌群第四层	同斜倒转型
					S	50°	60°					
	亲变岭背斜	f_{49}	沿亲变岭产出。西为北东向断裂所截，东被中二叠世花岗岩侵蚀	延伸长约 2. 5km	N	45°	60°	40°	60°	石碌群第三层	石碌群第四层	同斜倒转型
					S	35°	40°～70°					
	昂累岭向斜	f_{50}	沿昂累岭产出。西为北东向断裂所截，东被中二叠世花岗岩侵蚀	延伸长约 2. 0km	为北北东向褶皱叠加，产状变化较大					石碌群第五层	石碌群第四层	同斜倒转型
	昂累岭背斜	f_{51}	沿昂累岭产出。西、东被中二叠世花岗岩侵蚀	延伸长约 1. 2km						石碌群第三层	石碌群第四层	同斜倒转型

3. 王下构造带

发育于海南岛西南部王下-江边一带，由一系列近东西向展布的断裂（F_9—F_{10}、F_{108}—F_{117}）和褶皱（f_{52}—f_{63}）组成。从北至南，主要有王下断裂（F_9）、牙劳断裂（F_{10}）、牙迫村断裂（F_{117}）、王外-马眉断裂（F_{114}）、冲俄黎-佳西断裂（F_{115}）、田头-布温断裂（F_{111}）等组成。北面断裂倾向北或北东，南面断裂倾向南或南西，组成对冲式推覆构造；导致下古生界叠覆于上古生界之上；具有由北→南的运动学特征（丁式江等，1994）。

王下断裂（F_9） 发育于王下一带二叠纪变质碎屑岩和灰岩中，常形成线状沟谷或断隆地貌，被南北向断裂切割和中生界鹿母湾组覆盖；呈近东西向延伸，长约10km，宽1~2m，产状0°~30°∠30°~52°。发育挤压片理和构造透镜体，透镜体定向排列，并见石英脉充填。脉体具铜铅锌矿化及金矿化。

牙劳断裂（F_{10}） 位于王下断裂南面牙劳村一带，发育于二叠纪变质碎屑岩和灰岩中，常形成线状沟谷或断隆地貌，被南北向断裂切割和中生界鹿母湾组覆盖；呈近东西向延伸，长约11km，宽4~60m，产状350°~15°∠30°。发育构造透镜体和片理化岩石，透镜体成分为石英脉，被片理环绕，平行定向排列；见金、银矿化。

田头-布温断裂（F_{111}） 位于田头村-布温一带，长约11.5km，宽5~20 m。东段走向东西，西段走向北西西，产状22°∠49°。断面发育擦痕线理，破碎带由构造角砾岩和碎裂岩组成。角砾岩角砾成分为糜棱岩化石英岩，呈尖棱角状、棱角状、次圆状，胶结物为铁质。糜棱岩化石英岩石英矿物呈透镜体状，略显定向排列；显示断裂具有两次活动，早次为韧性变形，晚次为脆性破碎。

王外-马眉断裂（F_{114}） 位于王外-马眉一带，发育于古生界中。东段被白垩系掩盖，大致沿昌化江呈近东西向延伸；西段左行切错北东向断裂。长大于23.5km，破碎带并见石英脉充填。在公爱一带，断裂出露韧性变形岩石（糜棱岩），由定向的云母、板状石英及长石等组成的糜棱面理以外，还广泛发育由拉长的石英、线状云母和少量长石棒构成的拉伸线理。矿物拉伸线理向南东倾伏，侧伏方向95°~115°，倾伏角15°~35°。糜棱岩中被褶皱的早期石英脉、S-C构造和不对称旋转残斑系等指向构造指示该剪切带具右旋剪切特征。显微镜下，石英颗粒表现为韧性变形特征，波状消光、重结晶作用和塑性拉长强烈；长石颗粒为脆性或韧-脆性变形，呈棒状、碎斑状，具不连续波状消光（图版Ⅴ-2-H）。对该剪切带内三个糜棱岩样品的石英c轴组构的测量表明（陈新跃，2006）：剪切带内石英c轴岩组为具点极密的非对称Ⅰ型交叉环带分布，指示该剪切带的运动性质为右旋剪切，且其点极密靠近c轴，部分样品的石英c轴组构在靠近b轴发育次极密，这表明石英滑移系以｛0001｝<a>底面滑移系为主的同时也存在｛10$\bar{1}$1｝<a>菱面滑移，其变形温度约为300℃~450℃。岩石白云母^{40}Ar-^{39}Ar坪年龄约为245 Ma。由此看来，该断裂具有多次活动性，其特点与田头-布温断裂可能相似，早次为韧性变形，晚次为脆性破碎。

冲俄黎-佳西断裂（F_{115}） 断裂位于测区北部，南东起自乙丙村，往北西经佳西村、可美山至冲峨黎村附近逐渐消失。长18km，走向300°~310°，往南西倾，倾角65°。平面上呈波状延伸，南东段受北东向断裂切割和第四系掩盖，地表断续出露。断裂切割了奥陶系南碧沟组和白垩系鹿母湾组、报万组，航片上线状影像清晰，地貌上形成一系列线状洼地或陡崖。

断层破碎带宽约85m，局部地段剪切变形影响宽度可达数百米，主要为碎裂岩、硅化岩、块状角砾硅化岩、片理化岩石及硅化花岗岩质糜棱岩等组成。岩石中长石、石英集合体定向排列，其中石英颗粒明显定向拉长、糜棱面理发育，局部见石英颗粒具压力影现象。断裂南西盘下古生代变质地层发生硅化，硅化带内见方铅矿化、黄铁矿化。据花岗质粗糜棱岩中的δ型残碎斑晶的剪切指向判断（图21-43），其上盘上升，为压扭性逆断层。断裂带在佳西村一带，发生分枝复合现象，形成宽250~400m的构造透镜状夹块，夹块岩石破碎、挤压强烈，发育有密集片理化带，显示了压性特征，硅化构造岩中宏、微观均可见硅化成分破

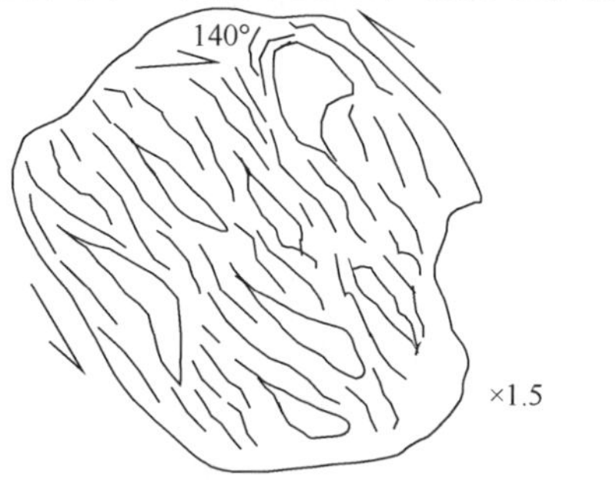

图21-43 冲俄黎-佳西断裂带内花岗质粗糜棱岩中的δ型石英残斑素描图

Fig. 21-43 Sketch showing δ-type residual quartz phenocrysts in coarse-grained granitic mylonite in the Echongli-Jiaxi fault belt

碎后再胶结的二次构造作用，表明其具两期活动，早期活动可能发生于华力西期—印支期，变形以韧性为主，到燕山期又发生了以挤压为特征的脆性活动。

牙迫村断裂（F_{117}） 分布于牙迫村一带，沿袭下、上古生界接触面发育，北盘为二叠系含燧石灰岩，南盘为志留系石英岩、板岩等，常形成线状沟谷或断隆地貌，被南北向断裂切割和中生界鹿母湾组覆盖。呈近东西向延伸，长约12km，宽约15m，倾向北，倾角较缓。岩石为构造角砾岩，呈透镜体状，长轴定向排列，角砾成分为粉砂质页岩、石英砂岩、含燧石条带灰岩等，胶结物为钙质、硅质、铁质。

4. 东岭构造带

由F_{71}—F_{73}和其相伴随的褶皱（f_{22}、f_{23}）组成。

主断面分布在东北角古生界中，东自长岭，向南西经石头岭至燕岭一带，走向北东，表现为下古生界志留系陀烈组直接盖在上古生界石炭系青天峡组之上。北东端鲤鱼岭一带被第四系掩盖，南西向延伸出图外，断裂带在航卫片中有较清晰的线性影像，地貌上显示出分划性特征，即北西盘表现为以北东向山脊为主并总体构成略向南东凸出的弧形，南东盘为不规则的山脊曲线。尽管宏观尺度中未见呈飞来峰存在的志留系或呈构造窗产出的石炭系，这使得该推覆构造更似一个与石炭系以断层接触的逆冲岩片或就是一个逆冲断裂与断盘。但变形分析表明，同一时代同一岩石单位具有不同的变形作用，呈现出北强南弱的变形分区。沉积相分析也显示，志留系直接盖在石炭系一套深水相地层之上，缺失代表深水相之上浅水相的沉积记录。反映了该断裂发生了大规模逆冲运移活动。

（1）外来系统变形特征

外来系统包括主断裂北西面出露的古生代地层。岩石变形强烈，主要以褶皱来体现，而断裂比较少见。褶皱具有多期、次的变形叠加，表现是在早期伸展环境中形成的流变构造基础上叠加发育晚期挤压背景下形成的弹塑性变形。靠近主断裂，志留系岩石以走向北东的不对称小褶皱比较常见，褶皱轴面倾向北西，反映了推覆体具有由北西→南东向逆冲的运动学指示。远离主断裂，小褶皱则以对称型居多。

（2）原地系统变形特征

原地系统自南东→北西主要为志留系陀烈组和石炭系南好组、青天峡组。其宏观尺度中的变形表现为发育一系列走向北东的近对称褶皱。其中，发育在陀烈组中的褶皱出露于加晓岭一带，长7~15km，总宽约6.5km，轴向60°~70°，枢纽近水平（宜昌地矿所，1995）。

（3）逆冲断裂系统

长19.0~21.5km，宽30~35m，倾向330°~340°，倾角40°~70°。构造岩石为构造角砾岩、构造片岩和糜棱岩化岩石等。断裂下盘下石炭统在靠近主断裂时倾向由北西转向南东，具有牵引褶皱的特点，与该断裂逆冲活动有关。

南面东岭农场青年队北西侧，自北西→南东，大致可划分出剪切褶皱带和石英脉带。褶皱层面为区域变形作用形成的新生面理面，轴面走向45°，倾角近直立，枢纽直立，两翼不对称，长短不一，翼间角紧闭，指示断裂平面中具有右行剪切活动的特征（图21-44）。石英脉带由众多密集呈平行状分布

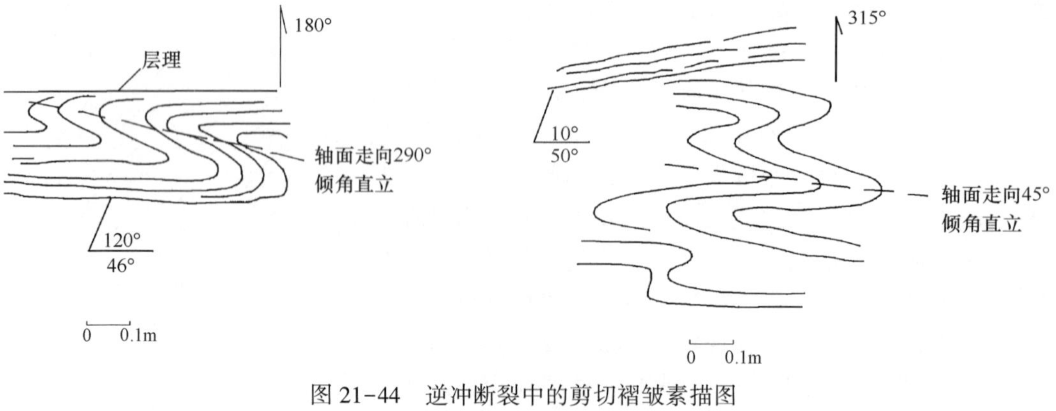

图21-44 逆冲断裂中的剪切褶皱素描图

Fig. 21-44 Sketch showing shear folds in reverse thrusts

的石英脉组成，宽度上百米，脉体变形强烈，呈膨缩状、拉断状顺层理面分布，形成杆状构造。江南队旁，断层由灰黑色千枚状板岩和少量强劈理化粉—细砂岩组成，见劈理面形成的紧闭褶皱。沿破裂面有石英脉大量贯入并进一步形成不对称剪切褶皱、石香肠、构造透镜体。

北面九曲江南部，应变带出露宽达 200 余米，主断面见花岗岩侵入。岩石主要由强劈理化的板岩、千枚状板岩以及强糜棱岩化的条带状白云母片岩组成，局部出现变质分异纹层，有石英脉贯入并形成强揉皱型不对称褶皱。上盘地层靠近断面产状较陡，远离断面产状变缓，离主断面约 20m 出现强揉皱型硅化板岩，褶皱以平卧褶皱为主体。

5. 南棍园构造带

分布于海南岛中北部白沙县-儋州市兰洋镇-屯昌县南棍园一带的古生界中，由 F_{43}、F_{44}断裂和夹持于其中的南棍园向斜（f_{10}）组成。

兰洋断裂（F_{43}） 分布于白沙县-儋州兰洋镇一带，沿上、下古生界接触面发育。长度约 100km，宽约 200m，近东西向沿山脊展布，断面南倾，倾角 70°。构造岩为角砾岩，角砾成分为炭质千枚岩、含铁千枚岩、脉石英等，角砾大小一般为 2~4cm。角砾和碎基均发生硅化作用，硅质为微晶石英，少部分为细晶石英。破碎带内至少有两次石英脉贯入，使构造角砾岩发生硅化，形成沿断裂带分布的硅化带，在地貌上形成一系列的山脊。断裂两侧还见片理化带，表明为压扭性断裂。由于受该断裂逆冲作用，导致下古生界叠覆于上古生界之上。

此外，在上番开村下盘石炭系青天峡组大理岩中，发育近东西向的不对称背斜，南翼倾向 185°~190°，倾角 35°~42°；北翼倾向 330°~348°，倾角 15°~27°。大理岩变形强烈，形成肠状剪切褶皱、钩状褶皱等（图 21-45）。岩石中方解石 e 双晶极发育，一般认为方解石的 e 双晶的发育程度与差异应力成正相关关系，e 双晶的发育组数、比例含量越高，其所受的差异应力就越大（郑亚东等，1985）。只有当差异应力达到最小临界值（10MPa）时，才会出现 e 双晶滑动。通过对大理岩中 e 双晶进行分析，其差异应力达 58. 8~133. 3Mpa（表 21-2），显示岩石受到极为强烈的变形，其变形表明岩石应产于大型断裂带之中。

表 21-2 石炭系青天峡组大理岩差异应力分析表
Table 21-2 Analysis of differential stress of Carboniferous Qingtianxia Formation marble

样品号	总颗粒数	一组颗粒数	二组颗粒数	三组颗粒数	差异应力/MPa	岩性	时代	采样地点
III7-1	75	63	25	15	133. 3	大理岩	石炭纪	兰洋
HN01	110	76	48	17	58. 8	大理岩	石炭纪	兰洋

牙石园断裂（F_{44}） 位于牙石园一带，长 33. 5km，宽 100~200m，走向 40°，倾向 130°，倾角 50°~70°。岩石主要为片理化构造角砾岩，呈扁平状、透镜状，定向排列，片理与主断面呈锐角相交，指示断裂具有右旋活动性质（图 21-46）。断裂下盘，并见有一系列走向 30°的不对称褶皱。

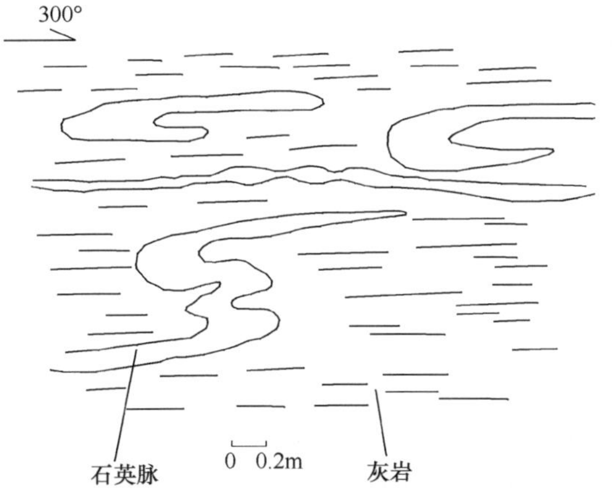

图 21-45 上番开青天峡组大理岩变形构造图
Fig. 21-45 Deformation structures in Qingtianxia Formation marble, Shangfankai

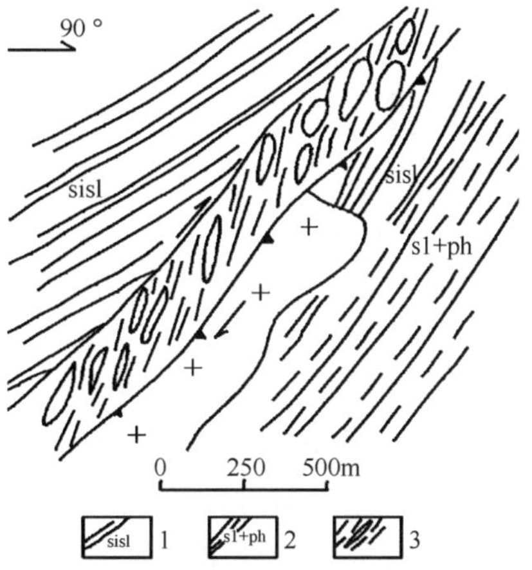

图 21-46 牙石园断裂构造破碎带素描图
Fig. 21-46 Sketch showing the Shiyayuan fault structural fractured belt
1—硅质板岩；2—千枚岩；3—片理化带

6. 九所-南好构造带

分布于九所-南好一带，走向北东，南西段由 F_{79}—F_{81}组成，北东段则以南好褶皱带显示（褶皱特征见上述）。根据构造带切割的最新地质体为白垩纪，局部并控制华力西期—印支期花岗岩边界，初步认为属于华力西期—印支期生成、燕山期进一步发展和活化起来的构造带。主要断裂特征如下：

小立村断裂（F_{79}） 位于孔峨村、南雅三队、小立村至镇海一带，在航片及遥感解译图上显示极为清晰。长约 21km，宽 8～25m，北段产状 125°∠56°、中段产状 315°∠70°、南段产状 303°∠70°～80°。断裂切割早二叠世、早侏罗世花岗岩和早白垩世牛腊岭火山岩盆，并控制了白垩纪花岗岩的边界。断裂带内岩石强烈破碎，北东向挤压劈理、节理极为发育，且透入性强。在断裂通过的南黎水库一带，发育宽约 8m 的断裂破碎带，其两侧岩性截然不同；断层面上见有明显的擦痕，显示该断裂为挤压逆冲断层。另外，断裂在多地段中还见有硅化角砾岩、石英脉、花岗斑岩脉。角砾呈棱角—次棱角状，脉石英多为犬齿状，对称生长。以上特征表明，断裂形成于燕山晚期。早期表现为压性的逆冲断层，晚期又有强烈活动，表现为张性特征。

宁远河断裂（F_{81}） 位于那刀旧址-抱古-崖城一带，在航片及遥感解译图上显示清楚。长约 25km，宽约 10～30m，产状 305°∠60°，切割了志留纪地层（S_1k）、早二叠世和三叠纪花岗岩，被早白垩世牛腊岭火山岩盆掩盖了西南段，被掩盖部分在遥感解译图上仍可见线性构造，并与物化探资料吻合，在地形上宁远河沿该线性构造穿梭。断裂带内岩石破碎，发育透入性密集劈理和挤压片理，普遍见有糜棱岩化现象，糜棱岩化花岗岩后期又被硅化。断面上亦发育硅化岩，并见有擦痕，指示南西盘（上盘）上升，为压扭性断裂。局部地段可见有透镜状的硅化带，产状 175°∠70°，宽 5cm，反映了南北向的挤压。断裂带内岩石变形强烈，岩石由碎基和碎斑两部分组成，碎斑包括钾长石、石英、钠长石，均有不同程度变形，如钠长石的双晶挠曲、位移，石英的位移及沿破裂面重结晶等，部分碎基中有新生石英，呈弱定向排列。岩石具绢云母化和碳酸盐化、绿泥石化等蚀变，石英还具波状消光。另外，断裂带中常见有后期充填的煌斑岩脉、花岗细晶岩脉，发育有棱角状构造角砾岩，反映了断裂后期的张性活动。在那刀旧址一带，该断裂切割了北西向高峰断裂，又说明该断裂燕山晚期有复活现象，且具右旋运动。由上述特征可知，该断裂早期为压扭性，形成糜棱岩化岩石和挤压片理、透镜状硅化岩；晚期则为张性，形成大量的棱角状构造角砾及各种岩脉的充填。断裂形成时代可能为燕山早期。

红卫断裂（F_{82}） 见于红卫大队经济场-那受大队经济场-立才农场二十五队一带，长约 10km，宽约 2m，产状为 340°∠85°。切割了早二叠世花岗岩、早白垩世同安岭火山岩盆，并控制了三叠纪花岗岩边界。断裂带中岩石强烈破碎，见构造角砾岩分布，局部出现糜棱岩。角砾岩角砾呈棱角状，局部有煌斑岩脉充填，在煌斑岩脉两侧发育有梳状石英脉，石英脉宽 0.5～2cm，呈网格状，煌斑岩脉亦被破碎成角砾岩。糜棱岩石英颗粒发生重熔，呈次圆形，具波状消光；长石发生绢云母化。断裂生成于燕山期，具两期活动性。早期为压性，形成糜棱岩，晚期为张性，充填煌斑岩脉和梳状石英脉，并形成角砾岩。平面几何学特征为右旋（顺时针）性质。

（四）狗岭构造带

发育于印支-南海地体北部三亚一带古生界中，因后期三叠纪、白垩纪花岗岩侵蚀和北西向、南北向断裂切割改造，以及第四系掩盖，导致构造带出露不完整。露头中，以北东向的晴坡岭倒转向斜（f_{36}）和狗岭断裂带（F_{84}—F_{86}）最醒目，共同组成了一条北东走向的由北西→南东逆冲的狗岭逆冲推覆构造带。

晴坡岭向斜发育于下古生界中，席卷整个三亚地区古生界，周边多被三叠纪花岗岩侵蚀及第四纪松散沉积覆盖。褶皱总体轴向为北东向，长度大于 20km，宽约 18km，向斜南端没入大海。核部为奥陶系干沟村组，两翼依次为奥陶系尖岭组、榆红组、沙塘组、牙花组、大葵组、大茅组和盂月岭组。根据海南地质大队（1979）红花矿区资料，认为属于轴面倾向北西的倒转向斜。此外，相伴随的次级小褶皱指示了具有北西→南东逆冲运动特征。

狗岭断裂南起鹿回头，往北东经狗岭、田独至三公曲以北逐渐消失，全长约 20km，总体呈 35°～

50°，向北西倾，倾角约 80°，在死马岭、狗岭、大会岭及南边岭等地均可见挤压破碎带。破碎带宽约 20~30m，最宽达 50m，主要由断层泥砾岩带，挤压片理化带，密集节理带组成。在局部露头较好的地段，可见断裂分带现象，从中心往盘由断层破碎带、挤压片理化带及密集节理带组成（图 21-47）。其中，破碎带宽 50~80cm，岩石有断层磨砾岩、构造透镜体、角砾岩及破碎岩块等，透镜体略具定向，与挤压带平行，产状 320°∠80°；挤压片理带宽 30~50cm，片理产状与断裂一致，整个片理化带略呈波状弯曲；密集节理带宽约 80~100cm，由密集节理、劈理及裂隙组成，将岩石切割成沟槽状和棋盘格式状，岩石较破碎。断裂上盘岩层变形、破碎较下盘强烈，反映了挤压逆冲特征。航片上线状影像清晰，地貌上反映出一系列北东向山脊和冲沟。

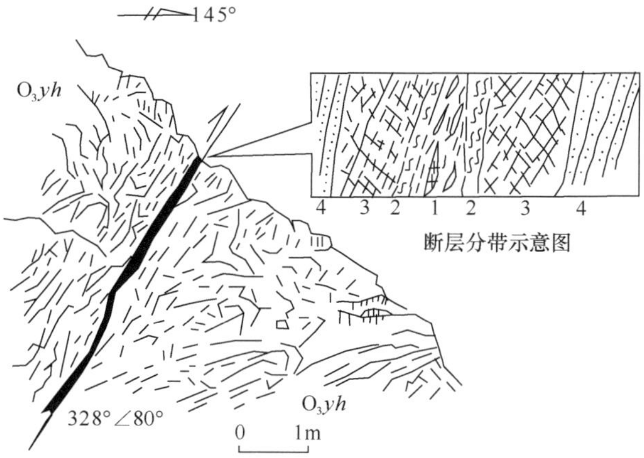

图 21-47　狗岭北东向断裂剖面素描图

Fig. 21-47　Section sketch showing the Gouling NS-trending fault

1—断裂破碎带；2—挤压片理；3—密集裂隙带；4—砂岩

逆冲推覆楔状体由中—上奥陶统沙塘组和中奥陶统牙花组，以及位于狗岭南东盘的上奥陶统牙花组共同组成，属于外来系统。断裂北东段大茅硐一带钻孔资料显示，中—上奥陶统沙塘组和中奥陶统牙花组由于逆冲作用而楔入到中—下寒武统大茅组中。而南面鹿回头，也由于逆冲推覆作用，导致中—上奥陶统沙塘组叠覆于上奥陶统牙花组之上。

二、区域性构造带

本次所指的区域性构造带是具有区域性展布、影响大的构造带，尽管不是构造单元的边界，但往往控制着岩体侵位、沉积盆地形成和各类矿产资源的产出。由于南海部分地质资料零星且有限而不进行阐述外。海南岛区域性构造带主要是北东向谭爷构造带、戈枕构造带、军营-红岭逆冲推覆构造带。谭爷构造带主要控制着白垩纪红盆的形成以及燕山期花岗岩的侵位；戈枕构造带控制了海南岛著名的二甲大型金矿带的产出；军营-红岭逆冲推覆构造带主要控制了华力西期—印支期、燕山期花岗岩侵位。

（一）谭爷构造带

谭爷断裂带位于海南岛中部儋州、白沙、乐东等县市，总体走向 25°~30°，长约 180km，宽 20~30km，由东、西两侧断裂带组成，其间夹有北东向白垩纪红盆。

1. 白沙-和庆断裂带

分布于澄迈县中兴镇-乐东县抱界一带，由多条断续分布的平行状断裂（F_{30}—F_{42}）和褶皱（f_8、f_9）组成。

儋州市和庆-中兴一带，单断裂长 8~34km，宽 0.5~25m，走向北东，倾向南东，倾角 40°~80°，主要表现为糜棱岩带和硅化石英脉带，硅化石英脉常充填于压扭性断裂中，岩石中发育的主次节理关系显示断裂具有右旋压扭性活动特征（图 21-48）。

白沙-南丰一带挤压破碎带发育，断裂倾向南东，倾角约 45°，构造岩石为硅化角砾岩、压碎岩和糜棱岩等。

白沙县打雅儒一带断裂带宽达 200m，由挤压断层角砾岩、硅化角砾岩和糜棱岩等组成，并见大量石英脉充填，断裂倾向南东，倾角达 60°以上。沿断裂带见华力西期—印支期和燕山期花岗岩充填，同时又切过岩体，反映了断裂带的多期次活动。断裂带控制了白垩纪红盆的西界，又在盆地边缘形成强烈的北东向挤压破碎带。

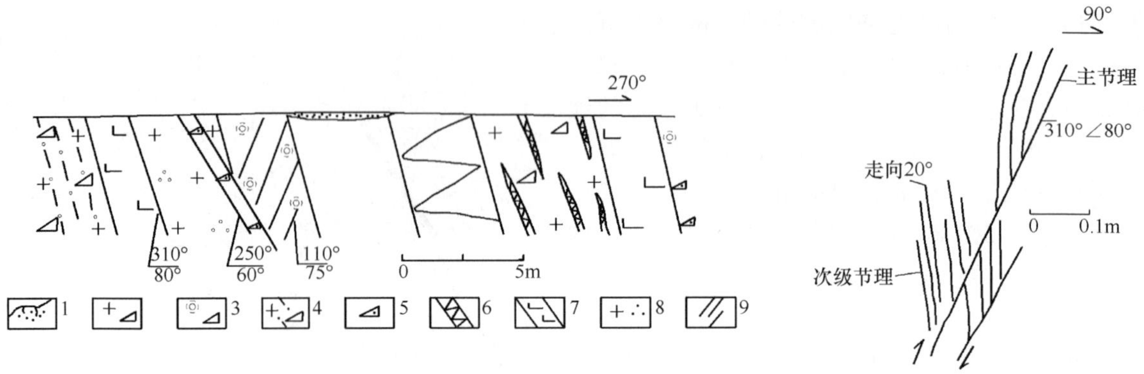

图 21-48　坡后岭北东向构造破碎带素描图

Fig. 21-48　Sketch showing the Pohouling NE-trending structural fractured belt

1—浮土；2—碎裂花岗岩；3—硅化碎斑岩；4—糜棱岩化碎裂花岗岩；
5—碎粒岩；6—石英脉；7—煌斑岩脉；8—石英斑岩脉；9—节理

2. 乐东-西昌断裂带

分布于屯昌县西昌—乐东县千家一带，由多条断续分布的平行状断裂（F_{54}—F_{62}、F_{76}—F_{78}）组成，单断裂长 3~25km 不等，宽 1~20m，局部可达上百米，走向北东，倾向北西，倾角 40°~80°，断裂带主要表现为发育构造破碎带或糜棱岩化带。

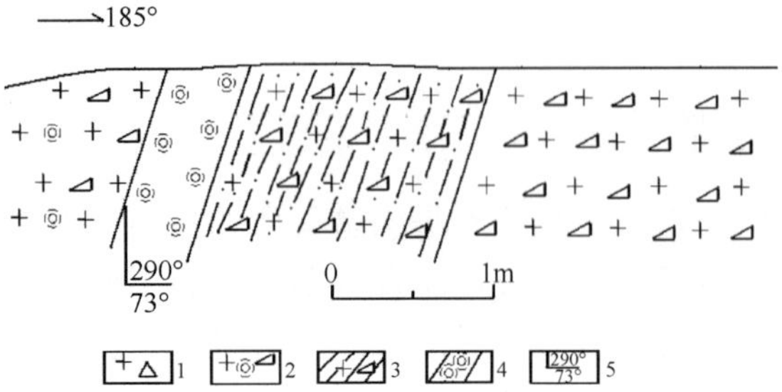

图 21-49　深水北东向构造破碎带素描图

Fig. 21-49　Sketch showing the Shenshui NE-trending structural fracture belt

1—碎裂花岗岩；2—硅化碎裂花岗岩；3—糜棱岩化碎裂花岗岩；4—硅化岩；5—产状

鸡口—横岭一带，构造破碎带倾向南东或北西，断面呈舒缓波状、刀切状，上窄下宽，构造岩石为碎裂花岗岩、构造角砾岩、硅化岩、糜棱岩、绢云母化超糜棱岩、糜棱岩化碎粒岩等，并见后期花岗斑岩脉、石英细脉、闪长岩脉充填。脉体后期破碎明显，并发育指示右行剪切活动的摩擦镜面和擦痕。断裂带内局部岩石呈现出分带现象（图 21-49），中心一般为碎粒岩、糜棱岩，两侧为花岗质碎裂岩、碎裂花岗岩。构造角砾岩角砾成分为花岗岩，形态呈透镜状、椭圆状。糜棱岩中，石英矿物具带状消光，形态为浑圆状、眼球状、透镜状，有的眼球呈 S 型旋转变形和雁列式排列（图版Ⅴ-3-A）；可见 S-C 面理，两者呈 20°锐角相交，C 面理并右行切错与其近直交的早期细小石英脉。碎裂花岗岩的石英矿物具波状消光；长石破碎，发生绢云母化蚀变；黑云母呈弯曲变形，被白云母置换。

鹦哥岭—马定岭一带，断裂带断面倾向北西，表现为岩石发生碎裂化、糜棱岩化，形成宽达几千米的碎裂花岗岩、碎裂岩、碎斑岩、碎粒岩和糜棱岩化碎裂花岗岩等韧脆性变形岩石，并控制了白垩纪红盆的东界。沿断裂带见晚侏罗世和早白垩世花岗岩体呈现出北东走向的展布格局，明显与该断裂带有关，反映了断裂带最早生成于晚侏罗世—早白垩世，并控制了白垩纪红盆的形成，根据发育在中三叠世花岗岩中破碎带的主剪节理（产状 296°∠74°）与其派生的一组张性密集节理（产状 225°∠59°）的关系，认为该期活动具有左旋压扭性。晚期活动导致早期沿断裂充填的脉体发生破碎，并切割晚侏罗世和早白垩世花岗岩体，表明断裂带晚期活动时代为燕山晚期，脉体中的摩擦镜面和擦痕，显示其力学性质为右旋压扭性。

（二）戈枕构造带

戈枕构造带分布于戈枕村—公爱一带，具有多期次的活动性。由 F_{45}—F_{48} 断裂和 f_4—f_7、f_{24}—f_{31} 组成。早期生成于印支期，形成了戈枕韧性剪切带，剪切带由北东向的剪切褶皱带和韧性断裂组成，褶皱带有布磨-二甲褶皱带和旧村岭褶皱带。晚期并叠加发育脆性变形，并发生逆冲推覆活动，导致了中元古界叠覆于新元古界、古生界之上。

韧性剪切带一般倾向北西，局部（红甫门岭）倾向南东，倾角 34°~70°不等。带内岩石侧向分带性明显，从边部往中心依次出现糜棱岩化带、初糜棱岩带、糜棱岩带和超糜棱岩带，各带之间为渐变过渡关系（图 21-50）。构造岩石有超糜棱岩、糜棱岩、初糜棱岩、千糜岩和糜棱片岩、糜棱岩化混合岩，以糜棱岩带出露最宽。叶伯丹（1990）、陈新跃（2006）等人先后在糜棱岩中分别获得绢云母 $^{39}Ar-^{40}Ar$ 年龄为 228Ma、227Ma，时代为印支期。本次在韧性变形岩石中获得黑云母 $^{39}Ar-^{40}Ar$ 年龄为 293Ma。

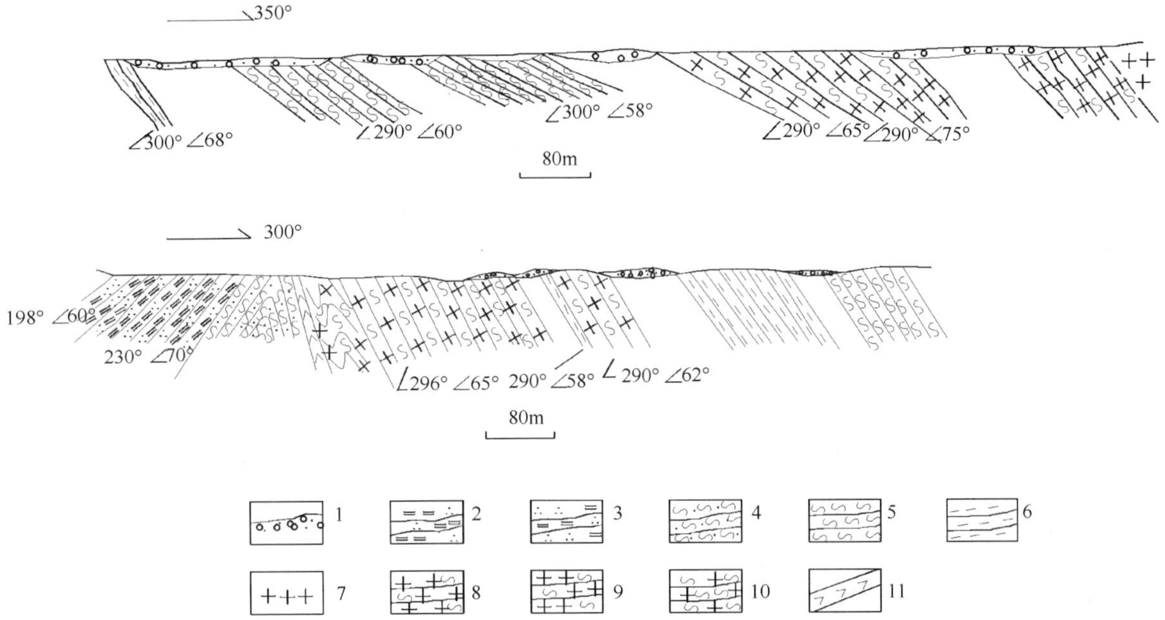

图 21-50　戈枕韧性剪切带实测剖面图

Fig. 21-50　Measured section of the Gezhen ductile shear zone

1—第四系覆盖层；2—石英黑云母片岩；3—白云母石英片岩；4—长英质初糜棱岩；5—长英质糜棱岩；6—长英质超糜棱岩；7—二长花岗岩；8—糜棱岩化花岗岩；9—花岗质初糜棱岩；10—花岗质糜棱岩；11—闪长玢岩

1. 构造岩岩石学特征

超糜棱岩　发育于断裂带中心，包括石英绢云母超糜棱岩和石英超糜棱岩。超糜棱结构，条纹（带）构造。石英碎斑含量 5%~40%，重结晶明显；碎基含量 60%~95%，发生绢云母化、硅化，绢云母含量可达 50%。

糜棱岩　发育在断裂带中部。糜棱结构、碎斑结构，眼球状构造、条纹（带）构造。碎斑含量 10%~50%，碎基含量 50%~90%。碎斑成分为石英、长石，呈眼球状、透镜状、磨圆状定向排列，发育拔丝构造、核幔构造；碎基为石英、绢云母的糜棱物，石英边界呈锯齿状、缝合线状。

初糜棱岩　发育在断裂带边缘。与糜棱岩相似，但变形程度略差。碎斑含量大于 50%，碎基含量小于 50%，石英重结晶现象少。

千糜岩和糜棱片岩　见于断裂带边缘及中心，常与糜棱岩交替出现。鳞片变晶结构，千枚状或片状构造。片状矿物为绢云母、白云母，含量 50%~90%，发育微褶皱；粒状矿物石英多被拉长成透镜体状、条带状，与片状矿物相间排列。

糜棱岩化混合岩　见于断裂带边缘。糜棱岩化花岗变晶结构，条带状、透镜状、眼球状构造。矿物以脆性破碎为主，韧性变形略差，矿物具弱的定向排列，基本保留原岩的结构构造和矿物形态。

2. 变形构造类型

剪切带内岩石普遍发育线理、面理和剪切褶皱构造。

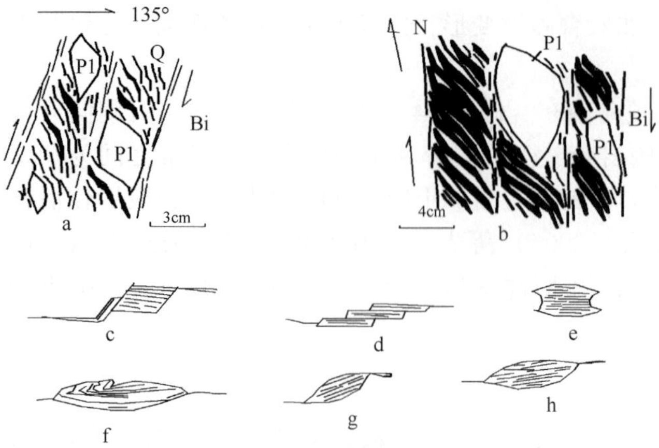

图 21-51　戈枕韧性剪切带 S-C 组构图

Fig. 21-51　S-C fabric of the Gezhen ductile shear zone

a、b 为Ⅰ型；c、d、e、f、g、h 为Ⅱ型

(1) 面理构造

剪切带内广泛发育糜棱面理（Sm）和剪切面理（Sc）（图 21-51），特别是在剪切变形的乌烈岩枝中更是比比皆是。根据 S-C 组构的组成特点，Listert 等（1984）将其划分出Ⅰ型和Ⅱ型两种基本类型。前者广泛出现在受剪切变形改造的乌烈花岗岩枝中，Sm 面理构成岩石的主面理，由石英、黑云母、钾长石、角闪石等矿物定向组成；后者发育于退化变质的抱板群片岩中，通过云母类矿物和石英规律展布显示出来，由云母香肠化和微断层作用形成的“云母鱼”较为发育，Sc 面理构成的岩石主面理，而 Sm 面理不太常见。统计表明糜棱面理产状在剪切带边部与剪切带边界交角较大（≥45°），向中心部位交角渐小甚至近乎一致，总体呈 S 型展布。剪切面理多出现在剪切带中心的高应变带或递进应变域内，既可以是由矿物晶体的形态优选而成，也可以是由矿物晶格优选定向表现出来，微观域中还可以是由“云母鱼”类的微香肠体或微断层作用产生。

(2) 线理构造

剪切带中常见拉伸线理和交面线理两种线理类型。拉伸线理构成云母石英质超糜棱岩中的最主要构造型式，由石英、云母拉伸显示的平行线状体在露头上很容易辨识，线理产状与剪切带总体方向具较大的交角；花岗质糜棱岩面理上的拉伸线理露头表现不太明显，仅由拉伸矿物或矿物集合总体展布趋势而定。交面线理在高应变的具 S-C 组构的花岗质糜棱岩中最为发育，几何上表现为由递进剪切过程中产生的 Sm 面理和 Sc 面理的交线。

(3) 褶皱构造

剪切带中褶皱构造相对不太发育，剪切褶皱形态复杂，两翼极不对称，翼间角紧闭，其规模一般较小，大者仅数十厘米，小者在显微镜下才能见到。按露头褶皱几何特征和运动方式划为 b 型褶皱、a 型褶皱和膝折构造等三种基本类型（图 21-52）。

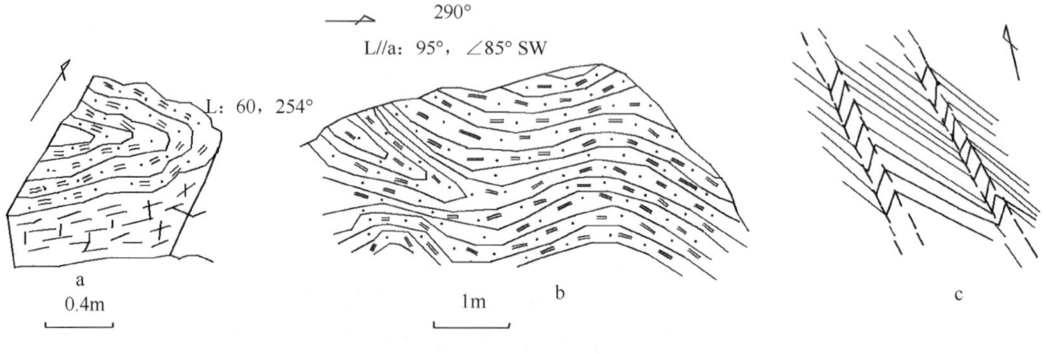

图 21-52　戈枕韧性剪切带的褶皱构造素描图

Fig. 21-52　Sketch showing fold structural elements in the Gezhen ductile shear zone

a—糜棱岩化白云母石英片岩；b—糜棱岩化白云母石英片岩中叠加 a 型褶皱；c—长英质糜棱岩中的膝折构造

b 型褶皱发育于较低应变的剪切带边部，褶皱面上矿物微弱拉伸定向而成的拉伸面线理与褶皱枢纽近于直交；a 型褶皱仅在强应变的抱板群片岩中见及，偶尔还可见到叠加于抱板群早期平卧褶皱之上的 a 型褶皱；膝折构造出现在面状构造发育完好的长英质糜棱岩中，可能代表了剪切带演化过程晚期阶段的变形产物。

(4) 韧性剪切带中刚性体的变形

花岗质糜棱岩中的暗色角闪质包体，同周围花岗岩相比具有较刚性的物理性状，在高应变域中常被改造成石香肠状、透镜状（图版Ⅴ-3-B），局部因塑性和剪切强度等差异而形成的蝌蚪状；而在极高应变带中，卷入超糜棱岩带中的早期斜长角闪岩脉因剪切则形成了透入性的剪切面理，甚至进一步发育露头可见的褶皱构造（图21-53）。

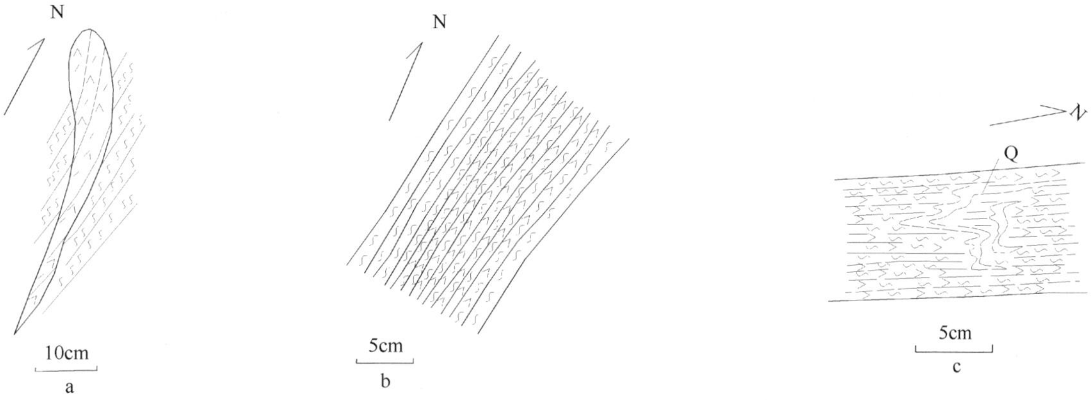

图21-53 戈枕韧性剪切带中的斜长角闪岩变形素描图

Fig. 21-53 Sketch showing amphibolite deformation in the Gezhen ductile shear zone

a—蝌蚪状斜长角闪岩包体；b—强剪切的斜长角闪质糜棱岩；c—斜长角闪质超糜棱岩中的褶皱

3. 变形显微组构

（1）*矿物变形亚结构*

石英普遍以基质形式出现，少量呈碎斑的石英呈压扁拉长状，强烈的波状消光、带状消光，发育变形纹、变形带、边缘粒化和边缘亚颗粒等动态重结晶明显。基质石英随韧性剪切变质作用的加强，由细粉状石英集合体构成的条痕-细粒齿状、蠕状镶嵌的石英组成细条带-细粒粒状镶嵌的石英构成的条带到由板状石英组成的拔丝石英条纹，显示一定的规律性。

呈浑圆状或椭圆状镶嵌于糜棱面理之中的斜长石和微斜长石碎斑常见的变形结构有：变形双晶、剪切阶步、膝折、滑动位移、分享结构和细颈化结构等（图21-54），这些变形结构的存在，足以说明长石矿物经受了一定的塑性改造，并为该剪切带形成于较深的构造层次提供了佐证。

碎斑角闪石和白云母的变形差异较大，前者以剪切阶步结构为主，表现为脆性破碎行为，而白云母矿物变形结构主要取决于解理与主剪切面的关系，当白云母解理平行于主要剪切运动方向时，常常表现出较强的沿解理面的滑移；如果早期白云母解理与主剪切面斜交时，其晶体往往被剪切破碎呈"鱼状体"产出（图21-55）。

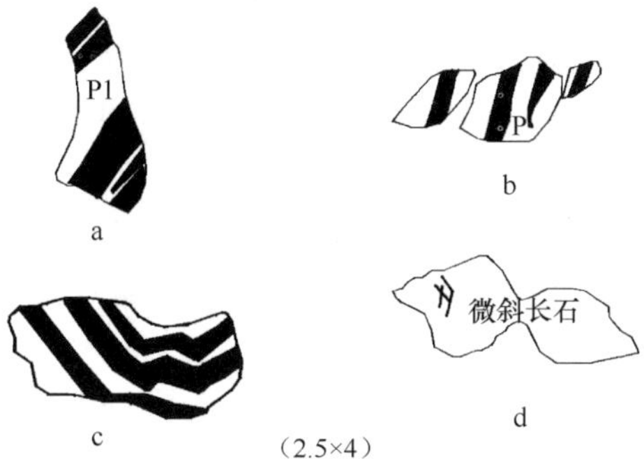

图21-54 长石变形结构素描图

Fig. 21-54 Sketch showing feldspar deformation structures

a—机械双晶；b—剪切阶步；c—双晶膝折；d—细颈化

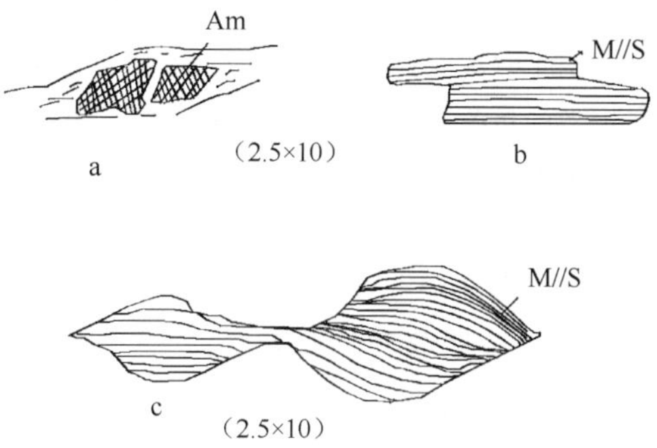

图21-55 角闪石、云母的变形结构素描图

Fig. 21-55 Sketch showing the deformation structures in hornblende and mica

a—角闪石剪切阶步；b—白云母沿节理面滑动；c—"云母鱼"

（2）*变形显微构造*

通过构造岩薄片的详细观察，将该带的显微变形构造归纳为以下几种（图21-56）。

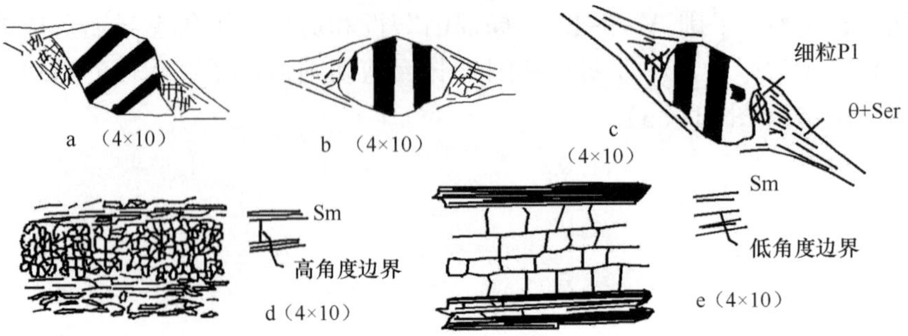

图 21-56　戈枕韧性剪切带内糜棱岩系岩石显微变形组构图

Fig. 21-56　Microscopic deformation fabric of mylonite series in the Gezhen ductile shear zone

a、b—对称和不对称压力影；c—山羊须构造；d、e—高角度和低角度边界

压力影构造　出现在各种类型的糜棱岩中，表现为嵌于糜棱面理中长石碎斑的两端垂直生长了细粒状石英、绿泥石等新生矿物。据两端压力影发育程度的差异进一步分为对称型和不对称型两种。

山羊须构造　与压力影构造表现形式颇为相似，不同之处在于山羊须构造端部新生矿物部分是来自碎斑矿物。

等粒结构　仅见于超糜棱岩中，岩石大都由细小等粒的重结晶石英组成，大颗粒边界基本消失。这种结构是高温-低应变速率条件下的岩石流变的产物，边界滑移是应力释放的主要方式。

拔丝构造　拔丝状石英在超糜棱岩中可以呈单体存在，也可呈集合体形式产出（即多个拔丝单体组合构成一个拔丝集合体），长、短轴之比在 10：1 以上。高角度边界与低角度边界，为云母-石英质糜棱岩和超糜棱岩中最常见的显微构造。高角度边界是石英颗粒中延伸较短并与主面理近于垂直、但局部却可以与之斜交或平行的一组边界，边界一般呈复杂的锯齿状或缝合线状；低角度边界是石英颗粒中与主面理以小角度相交或近于平行的那组边界类型，边界一般平直规整且延伸较好。

褶皱构造　多由石英细脉塑性流动而成，褶皱形态复杂，但轴面大致平行于主面理，转折端较厚而翼部较薄，物质具有自翼部向转折端运移的运动特点，一般见于云母-石英质糜棱岩中。

分异条带状构造　这种构造在遭受韧性变形的抱板群片岩中相当普遍，表现为在粒度上或矿物成分上间带分布的排布特点，其中以层状矿物与石英矿物彼此平行相间排列最为发育。这种构造以边界平直、矿物变化较小且塑性变形与层理构造相区别。

（3）显微变形机制

岩石的变形机制主要包括扩散物质迁移、晶质塑性和颗粒边界滑移及破碎和碎裂作用等。角闪石表现的剪切阶步及晶体分离组构标志着这种较刚性矿物基本以碎裂作用为主；斜长石香肠化的产生为其晶质塑性变形的产物，而石英等亚颗粒和拔丝构造的广泛出现，意味着剪切带具有相当高的晶质塑性变形；扩散物质迁移形成了压力影山羊须等典型组构；通过颗粒边界的滑移，产生了超塑性流动变形极其重要结构类型——等粒结构。因此，该剪切带的显微变形是上述四种主要机制共同作用的结果，同时从边部到中心，从低应变带到高应变带，变形机制显示出由碎裂化→扩散迁移→晶质塑性→边界滑移变化趋势。

4. 有关变形参数的确定

（1）变形温度

通过测定糜棱岩中同剪切变形期形成的绿泥石、白云母的化学成分，并利用绿泥石-白云母共生矿物对地质温度计估算戈枕韧性剪切带形成温度在 450℃±。

（2）古应力值和应变速率

古应力值和应变速率的估算主要基于下面两个公式：

$$\triangle\sigma=40d^{-0.8}\quad \text{(Mercir, 1977)} \tag{1}$$

$$\varepsilon=A\exp\ (-Q/RT)\ (\triangle\sigma)^{n} \tag{2}$$

式中：$\triangle\sigma$ 为古应力值，单位为 MPa；d 为动态重结晶石英颗粒直径，单位为 μm；A 为 Dorn 常数，取值 1.10~7；Q 是活化能（188406J/mol，Goetze）；R 为理想气体常数［7.919J/（mol·K）］；T 为绝对温度；$n=2$。

计算结果显示，自边部往中心，古应力值不断加大，应变速率也呈逐渐升高趋势（表 21-3）。

表 21-3　戈枕韧性剪切带有关参数计算表

Table 21-3　Calculation of parameters related to the Gezhen ductile shear zone

部位	古应力差值 $\Delta\sigma$/MPa	应变速率/Sec^{-1}	包体测温/K	资料来源
边部	41.58	4.2×10^{-18}		1∶5 万《昌江县、邦溪》幅
中部	72.85	3.7×10^{-17}		
中部	128	3.1×10^{-16}		
中心	67	1.21×10^{-13}		
北东段边部	162.92		623	1∶5 万《昌化大岭、昌化港、海尾、叉河》幅
北东段中心	245.60		723	
南西段中心	149.13		623	

（3）富林指数

选择强烈拉伸变形的超糜棱岩切 x/y 和 y/z 切片，镜下系统测定 $x/z=10$；$y/z=4.52$，进而求得 $k=(x/y-1)/(y/z-1)=0.36$，为压扁型（图 21-57）。

5. 运动方向和剪切带类型的确定

剪切带运动方式有拆滑型（丁式江，1991）、逆冲型（何龙清等，1996）、走滑型（涂绍雄等，1994）。本次依据露头和镜下的拉伸线理、a 型褶皱、剖面 S-C 组构、旋转残斑等指向构造综合分析，认为戈枕韧性剪切带具有剖面逆冲兼平面右行走滑的运动学特点。

岩组学研究表明，该带的多期变形组构较清楚。基质岩组图呈三斜对称，由两个大圆环带组成，环带轴极点近于 S 面，以北西—南东向挤压最为强烈；残斑石英的岩组图也主要表现出由北西向南东挤压的特点（图 21-58）。

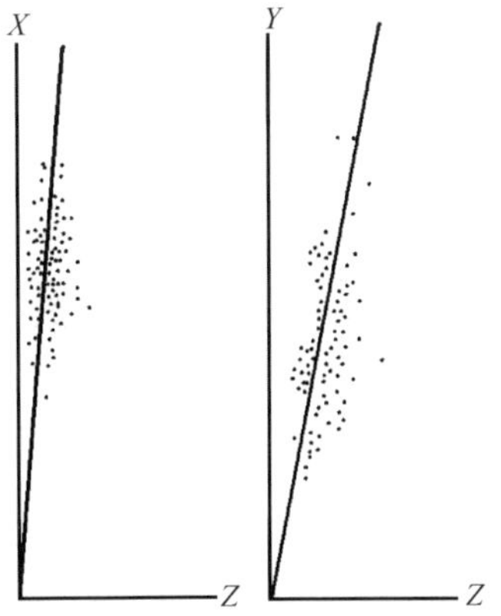

图 21-57　超糜棱岩石英有限应变分析图

Fig. 21-57　Analysis of limited strain of quartz in ultramylonite

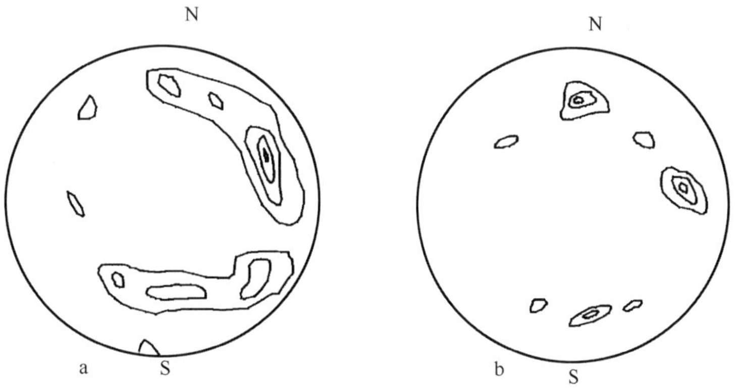

图 21-58　戈枕韧性剪切带石英变形组构图

Fig. 21-58　Quartz deformation fabric of the Gezhen ductile shear zone

a—基质石英（200 个颗粒，等值线 1%~3%~6%~10%）；
b—残斑石英（30 个颗粒，等值线 1%~2%~3%~4%）

综上所述，可以判定戈枕构造带早期活动具有剖面逆冲兼具平面右行走滑的逆冲的。

继印支期戈枕韧性推覆型剪切带形成后，燕山期又在其东部边界部位叠加了一次较强烈的浅层次推覆构造事件，形成了碎裂岩系列岩石，叠加于早期的糜棱岩系列岩石中，构造岩石有碎裂岩、构造角砾岩。构造角砾岩角砾成分为糜棱岩、混合岩、伟晶岩，呈次棱角状、次圆状、透镜体状，局部定向排列。肖志峰（1989）在抱板蚀变碎裂岩中获得 Pb-Pb 等时线年龄为 130~170 Ma，涂绍雄（1993）

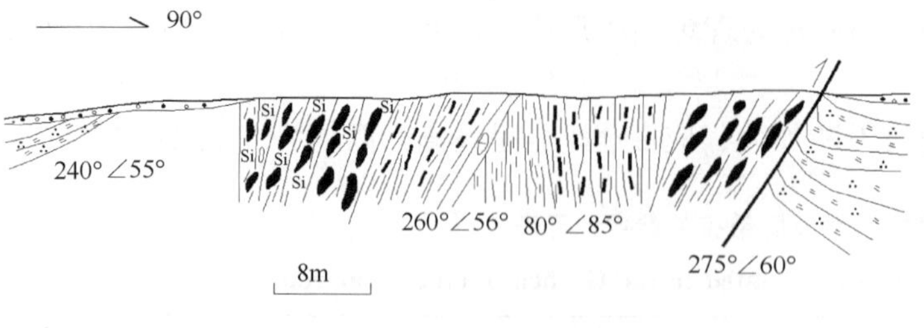

图 21-59　逆冲推覆构造带主断面构造剖面图

Fig. 21-59　Section of the main cross-section structure of a reverse thrust structure zone

1—第四系；2—峨文岭组片岩；3—石碌群千枚岩；4—硅化角砾岩；5—挤压片理化带；6—石英脉；7—断裂

也在含金糜棱岩获得绢云母 K-Ar 年龄为 197Ma，时代都为燕山期。该推覆构造主要表现为中元古代花岗岩和抱板群逆冲于新元古界石碌群、志留系空列村组及奥陶系南碧沟组之上。

（1）*逆冲推覆构造的组成*

1）外来系统和原地系统：外来系统可能由中元古代花岗岩和抱板群组成，抱板群为一套以石英和云母类矿物为主的片岩系岩石，常见膝折构造。原地系统可能包括石碌群、空列村组、南碧沟组等，其构造形迹以前推覆期的构造群落为主，推覆期新生构造不明显（微弱）。

2）主推覆断裂构造：过去人们一直对石碌地区抱板群和石碌群的接触关系争论不休，宜昌地矿所（1993）开展 1∶5 万填图时初步查明了两者间系断层接触。这条较大规模的逆冲断层呈北东 30°~40°展布，倾角 60°~70°，向北西倾斜，断裂带宽 30~100m 不等，其中发育的脆性断层显示较明显的分带性，自边部向中心依次出现硅化角砾岩、碎裂岩、挤压片理化岩石等（图 21-59）。碎裂化成分包括石碌群、抱板群变质岩和糜棱岩等。从断层带内出现大量挤压透镜体、片理化等推测，断层具逆冲性质。钻孔揭露表明，这条脆性断层产状上部较陡，为 275°∠80°，下部变缓，倾角约 15°~30°，为典型逆冲铲状断层（图 21-60）。

（2）*逆冲推覆构造的有关特征*

依据其主推覆面内含印支期戈枕韧性剪切带糜棱岩透镜体和角砾以及又被后期（晚燕山期）同向、右行走滑断层叠加改造等特征分析，推覆时间为燕山期。推覆方向为北西向南东，即垂直于主干推覆断裂的方向。戈枕逆冲推覆构造属前展式，推覆距离在数千米以上（海南地质大队，1993）。

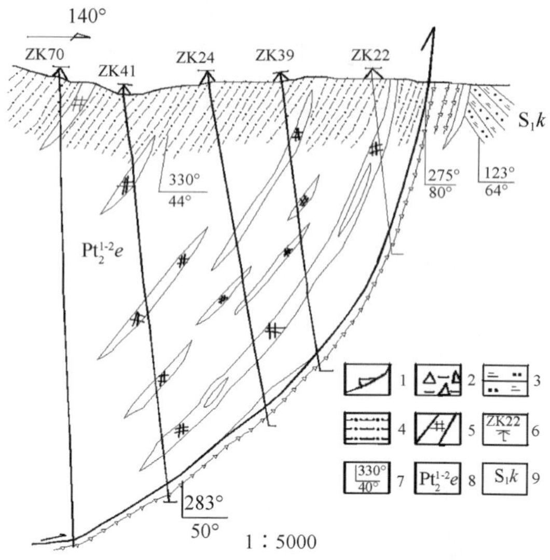

图 21-60　土外山 101 线戈枕逆冲推覆构造逆冲型铲状断层素描图

Fig. 21-60　Sketch showing reverse-type shovel-like faults of Gezhen reverse thrust structure, 101 Line, Tuwaishan

（据 934 队资料）

1—戈枕断裂带；2—碎裂岩带；3—糜棱岩化石英绢云母千枚岩；4—糜棱岩；5—破碎型含金石英脉；6—钻孔及编号；7—产状；8—抱板群；9—志留系空列村组

（三）军营-红岭构造带

军营-红岭构造带南西起自峨沟岭，经南蛇水库、军营老村一带向北东延伸至洛基一带，形成了一条长近百千米、影响宽度在 7km 以上的构造带。由 F_{20}—F_{26} 断裂和 f_1—f_3 褶皱组成。具有逆冲推覆特征。

1. 外来系统

红岭地区经钻孔揭露，外来系统下白垩统鹿母湾组呈楔状体逆冲推覆到原地系统志留系中。十月田椰子村一带槽探揭露结果显示，挟持在两断裂间的外来系统——灰岩透镜体呈楔形体充填于原地系统志留系陀烈组中。楔状体沿北东向分布，长轴与断裂带近平行。楔状体的形成，可能反映了构造带燕山期的大距离逆冲作用（陈沐龙等，2009）。

2. 原地系统

峨沟岭-大岭一带原地系统包括军营-红岭断裂带南侧的地层、岩体等；外来系统因出露有限，除岭屋村、那历、德立村、大发等岩体和志留系外，一些呈叠置岩片和岩锲产出的石炭系、二叠系地层基本局限在宽度不足2km的推覆断层带内，所以难以确定是原地产物抑或外来岩片。

3. 逆冲构造系统

由多条北东向平行展布的断裂带和北东向褶皱带组成，属于与同逆冲期生成起来的构造迹象。在遥感图像上的线性影像特征最为明显，由地形上突出的北东向长条状山体断续排布显示出来，整个断裂带呈现出暗灰色调，并被北西向的线性影像切割。逆冲构造由北东向军营复式背斜、坡兰沟向斜和一系列平行状产出的断裂束组成。断裂特征如下：

鸿沟岭-红岭逆冲断裂 分布于马岭、珠碧江、大岭十四队、红岭九队至红岭一带，走向40°~60°，向北西倾斜，倾角25°~60°。经钻孔揭示，断面总体上陡下缓而呈铲状。断层带宽10~30m，由碎裂岩、角砾岩、挤压透镜体、挤压片理化岩石等组成，断面发育斜向擦痕，旁侧可见指示上盘（北盘）上升的牵引构造。断层上、下盘倒序现象广泛存在，露头也可见及石炭系和二叠系之间逆冲推覆的构造景观（图21-61）。

十四队-七队断裂 位于鸿沟岭-红岭断层以南并与其大致平行展布，间距在0.5~1km之间，这条主断层西起高岭以南，经珠碧江、大岭十四队、南华坡、红岭十三队、红岭七队至富仍水库，长达16km。单断层带宽5~30m，走向10°~60°，波状弯曲，中段在南华坡为北北东—近南北向断层所切，至红岭七队分支多条向北东—北北东撒开。向南西收敛，呈扫帚状展布。断面倾向北西，倾角也显上陡下缓特点。断层带以强烈硅化破碎为特征，发育碎裂岩、角砾岩及挤压透镜体，清楚表现出压扭性力学性质。根据海南地质大队四分队在红岭的钻孔揭露显示该断层带由多条断面构成，切割了陀烈组、白垩系，控制了红岭花岗岩枝的发育，并使陀烈组逆冲推覆在下白垩统鹿母组之上，为一条典型的犁式冲掩断层，由于逆冲推覆作用，在长岭北南好组地层中形成了大型的翻卷褶皱，从而显示出推覆体前峰带的典型构造型式（图21-62）。

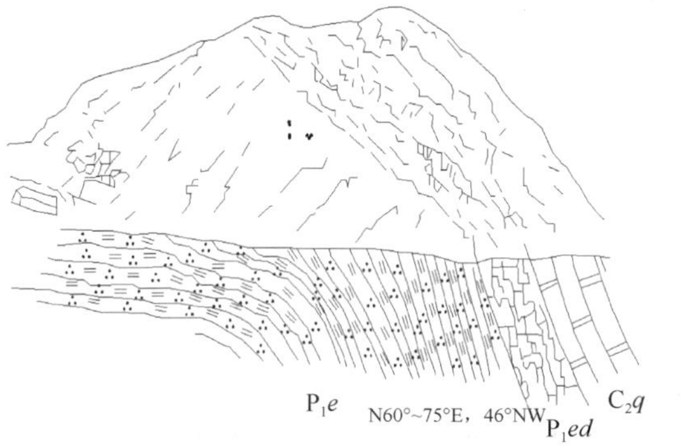

图21-61 鸿沟岭-红岭逆冲断裂破碎带素描图

Fig. 21-61 Sketch showing the Honggouling-Hongling reverse thrust fractured belt

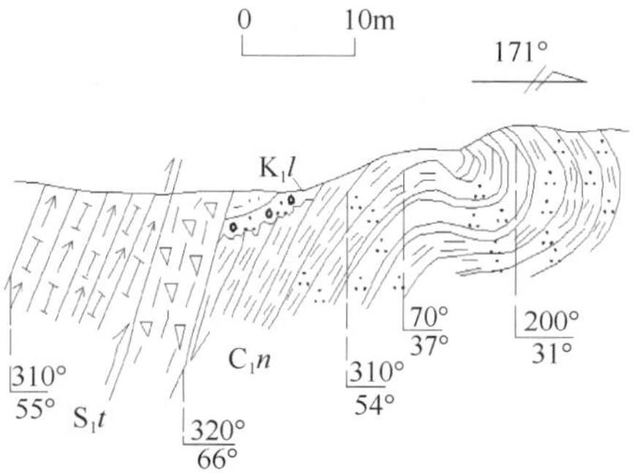

图21-62 前峰带的翻卷褶皱示意图

Fig. 21-62 Schematic diagram of convolute folds in the zone of surging

峨沟岭-军营村断裂 南西端起自峨沟岭，经南蛇水库向北东端延伸至军营村一带，出露长约27km，宽约70~120m，产状310°∠40°。按内部组构可分为三个带：硅化碎裂岩带、糜棱岩化揉皱千枚岩带、揉皱千枚岩带。揉皱千枚岩中广泛发育各种不对称褶皱，小褶皱轴面的优势产状为295°∠21°，反映了该断裂曾经历由北西向南东的逆冲。

军营-红岭断裂 断裂南西端起于峨沟岭，经南蛇水库、军营老村，向北东延伸至红岭一带，出露长度约27km。该断层穿过二叠纪花岗岩、志留系陀烈组及石炭系青天峡组，并切断志留系和石炭系之间的不整合界面，平面上呈舒缓波状展布，总体走向北东60°，倾向北西，倾角50°~60°，断裂带宽约80~100m，在断层带内千枚岩产状零乱，形成揉皱带、破碎带、劈理化带、石英透镜体带，局部可见糜棱岩化。带内广泛发育不协调褶皱、尖棱褶皱和不对称褶皱，石英、硅质条带被拉长呈透镜体顺千枚理分布。断裂带内可见后期构造作用形成的构造角砾及断层泥，角砾成分为早期形成的糜棱岩，表现该断裂具有明显的逆冲性质。十月田椰子村一带经槽探揭露，断裂带走向北东，由两条近平行状分布的断裂组成。

断裂带呈弧状弯曲，凸端指向南东，并向南西发散。断裂带倾向 300°~335°，倾角48°~75°，单断裂宽 1.5~6m。岩石有千糜岩、构造角砾岩、碎裂岩、硅化岩等，局部见花岗斑岩脉、石英脉充填。自南东往北西，依次为构造破碎带、灰岩透镜体带、千枚岩揉皱带、石英脉透镜体带、灰岩透镜体带、碎裂岩带。破碎带岩石由千糜岩、构造角砾岩、碎裂岩组成，具有逆冲活动的运动学特征。千枚岩揉皱带表现为常常发育轴面倾向北西的小同斜褶皱；石英脉透镜体带由多条倾向北西、相互平行的石英脉透镜体组成，带内石英脉与千枚岩发生同步变形，呈香肠状、透镜体分布，上、下盘岩石（千枚岩）并发育牵引褶皱，指示断裂具有逆冲性质；碎裂岩带由千枚岩发生破碎变形及硅化蚀变形成。

峨沟岭断裂 位于峨沟岭一带，沿峨沟岭山脚呈北东向展布。该断裂发育于二叠纪中粒黑云母二长花岗岩中，走向北东 50°，全长约 3km，宽约 50m，产状为 320°∠80°。构造分带明显，从中心向两侧岩性分别为：花岗质碎粉岩、花岗碎斑岩、碎裂花岗岩。花岗质碎粉岩宽约 5m，主要由长英质碎粉组成，碎粉粒度<0.1mm，含少量长英质残斑，粒度 0.5mm 左右，岩石局部硅化，硅化体呈不规则团块状分布。花岗碎斑岩宽约 5m，碎斑结构，碎斑主要为钾长石，石英（含量 20%~30%），呈不规则碎块状，基质主要为长石、石英及少量绢云母，岩石具硅化、绢云母化及黄铁矿化。碎裂花岗岩宽 30m，碎裂结构，可见钾长石斑晶破碎成大小不等的碎块，并发生位移。

4. 岩石组构特征

通过对岩石组构特征分析可以进一步分析逆冲运动的性质及断裂构造形成的热动力条件。军营-红岭断裂带中采取的岩石组构标本分析结果显示，其组构图为小圆环带与大圆环带交叉复合型，单斜对称，反映了早期运动面近水平，以柱面、菱面滑移为主，为北北西向南南东方向的右旋简单剪切；晚期运动面近直立，σ 近南北方向，反映左旋脆性平移剪切。而采于断裂下盘的石炭系青天峡组石英绢云千枚岩组构图呈小圆环带型，反映为北西向南东方向的右旋剪切运动，$T>800℃$、$\varepsilon<10^{-7}/\mathrm{Sec}$ 的低应变速率，并伴有重结晶任用；晚期可能叠加左旋剪切，运动面近直立。采于断裂下盘的下志留统陀烈组云母石英片岩组构图呈小圆环带型与中心点极密综合型，反映逆冲时期为高温低应变速率的动态重结晶作用，以及运动面近水平的简单剪切运动。

5. 运动学特征

逆冲推覆方向主要基于主干逆冲推覆断裂带的几何结构而定，即大致垂直于主干断裂的方向，为北西向南东推覆。平面上，北侧的花岗岩具有中带特点，十四队-七队主断面以北的古生代地层可能构成了逆冲推覆构造的前锋带。

岩石组构特征显示，军营-红岭断裂带在早期运动面为近水平的，由北西向南东逆冲，是一种以平面应变为主的简单剪切运动，并伴有糜棱岩化；后期叠加了运动面近直立的左旋剪切平移运动，形成目前军营-红岭断裂带之产状。

在燕山晚期—喜马拉雅期，军营-红岭断裂带叠加了强烈的平移剪切作用。该时期内，整个海南地块已固化，进入大陆边缘活动带发展阶段；地壳进一步抬升，岩浆活动强烈，完全进入脆性变形阶段，而且变形主要集中于断裂带上。此时期的早阶段，断裂运动主要表现为以逆冲为主的升降运动和以左旋平移剪切为主的水平运动。升降运动使早期形成的深层次至中深层次的混合岩、糜棱岩、碎裂岩及各种类型，金矿床上升与浅层次构造出露于同一水平面上；同时形成透镜状构造砾岩带，走向北东向，角砾长轴近于直立，扁平面也近于直立。左旋平移剪切形成运动面近直立，运动方向近水平的分层剪切带，构造透镜体长轴近水平，且左行雁列。与此同时，由于左旋平移剪切作用，形成诸多脆性 P、D、R、T 裂隙。

晚阶段，南北向挤压高潮已过去，断裂作用带有拉张性质，并存在右旋剪切迹象，这时，P、B、R、T 裂隙被硅质热液、岩浆充填，形成北西西、北西、北东向展布的石英脉及花岗岩脉、花岗闪长岩脉等。

6. 生成时代

燕山阶段是中国东南部强烈的板内造山活动时期，其常见的表现形式为发育逆冲推覆构造。根据逆冲推覆构造的原地系统的最新地层为下白垩统鹿母湾组以及沿主断面有燕山期花岗质岩脉贯入等方面判断，该逆冲推覆构造形成于早白垩世之后，即燕山晚期。

三、主要断裂带

（一）主要韧性剪切带

1. 洛基韧性剪切带

分布于儋州市洛基西侧一带，发育于早三叠世花岗岩中，由 F_{27}和 F_{28}组成。剪切带内岩石有碎裂花岗岩、糜棱岩化花岗岩、初糜棱岩和糜棱岩（图版Ⅴ-3-C、D），在横、纵向中其变形强弱有别，自北西向南东，岩石变形逐渐减弱，总体呈现出由糜棱岩向初糜棱岩、糜棱岩化花岗岩、碎裂花岗岩过渡。糜棱面理倾向北西或南东，自北西向南东，剖面上构成一个背形和一个向形构造。碎裂花岗岩中局部尚保留倾向南西的原生花岗岩面理构造（李孙雄等，2003）。

晚期叠加发育脆性变形，走向北东，倾向北西或南东，以岩石的碎裂化以及岩脉充填而体现，并常导致早期糜棱面理形成牵引褶皱，指示断裂有下滑的运动学特征（图 21-63）。根据共轭关系分析，该时期的应力方向为近垂直的挤压（拉张应力近水平）。可能与后期地体隆升，深部岩浆上涌过程中，对围岩造成的挤压有关，在此应力作用下，根部糜棱面理变陡，倾角大于 55°。

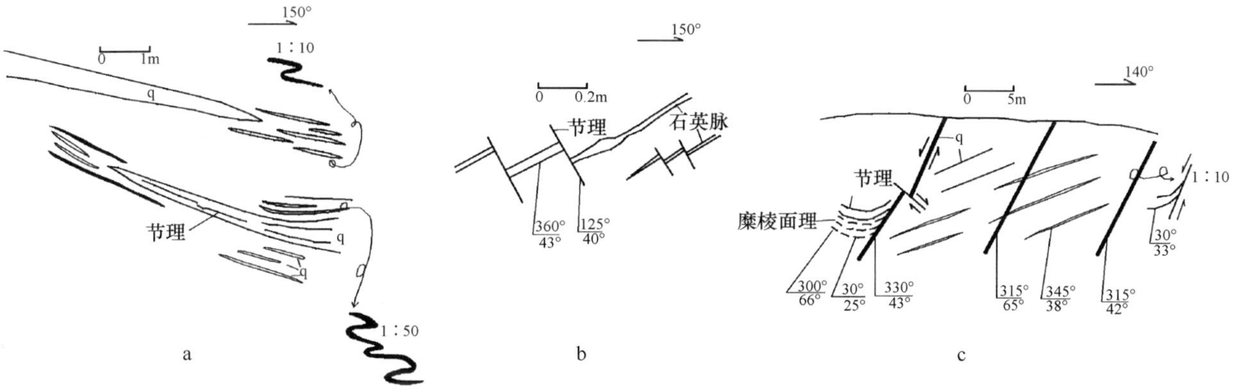

图 21-63 晚期拆滑构造形迹素描图

Fig. 21-63 Sketch showing late-stage detachment gliding structural traces

a—石英脉的肠状褶皱；b—切错东西向石英脉；c—糜棱面理形成的牵引褶皱

（1）构造岩岩石学特征

糜棱岩 糜棱结构、碎斑结构，眼球状构造、条纹（带）构造。碎斑含量 30%~40%，成分为长石，大小 0.2~3.8mm，呈眼球状、透镜体状、磨圆状，沿长轴平行定向排列；碎基含量 60%~70%，由细碎的长英质和大量次生的绢云母组成，围绕碎斑生长。

初糜棱岩 岩石特征与糜棱岩相似，但变形程度略差。碎斑含量 55%，碎基含量 45%。

糜棱岩化花岗岩 与上述两种岩石相比，岩石以脆性变形为主，韧性变形程度较差，矿物具弱的定向排列，局部可见钾长石眼球残斑排列方向与石英拉长定向有小角度相交，具左旋活动性质。

碎裂花岗岩 碎裂结构，不定向构造。岩石变形较弱，主要是发生脆性破碎，相伴随的蚀变作用也弱，新生绢云母、绿泥石等矿物少见。

（2）变形特征

1）面理：面理主要有剪切面理和糜棱面理。剪切面理微观尺度中表现为一组平行状分布的剪节面，面理被新生的细小绢云母矿物充填，原生石英、长石等粒状矿物沿剪切面发生破碎呈细小的碎粒状。糜棱面理为矿物或集合体的长轴优选方位所显示的不连续面状构造，表现为拉长状的石英或其集合体以及长石定向排列。剪切面理与糜棱面理基本上一致，局部呈大角度相交，形成大角度边界。

2）线理：发育在强应变的岩石中，由石英矿物或其集合体以及暗色的铁质矿物呈拉长状沿剪切面定向排列而体现。

3）褶皱：受浮土掩盖影响，变形带小尺度（露头尺度）中的褶皱未见及。宏观尺度中，褶皱走向北东，由一个背斜和一个向斜组成。

背斜位于变形带的北西段，北西翼倾向280°~340°，倾角多变化于50°~65°间；南东翼倾向100°~160°，倾角45°~65°。由翼部向核部，面理倾角逐渐变陡，可能与后期地壳隆升有关，属近对称背斜。

向斜分布于变形带的南东段，南东翼倾向约290°，倾角较缓，为33°，与背斜中的南东翼组成一个不对称向斜。

（3）显微变形构造

矿物变形机制主要通过扩散物质迁移、晶质塑性和颗粒边界滑移及破碎和碎裂作用等，形成了显微破裂结构、晶内纹带状构造、晶内粒状结构以及与晶粒位移有关的结构和与压溶有关的结构。区内变形带岩石主要表现为发育显微破裂结构、晶内纹带状构造（图21-64），反映了其形成于中低温的变形环境，与岩组学和地球化学分析结果一致。

显微破裂结构 表现为长石、石英等发生脆性破碎，裂纹发育，长石并形成剪切阶步。

晶内纹带状构造 表现是石英矿物具波状消光、带状消光，甚至发育变形纹，斜长石发育变形双晶以及膝折变形，绢云母显示膝折或弯曲变形。

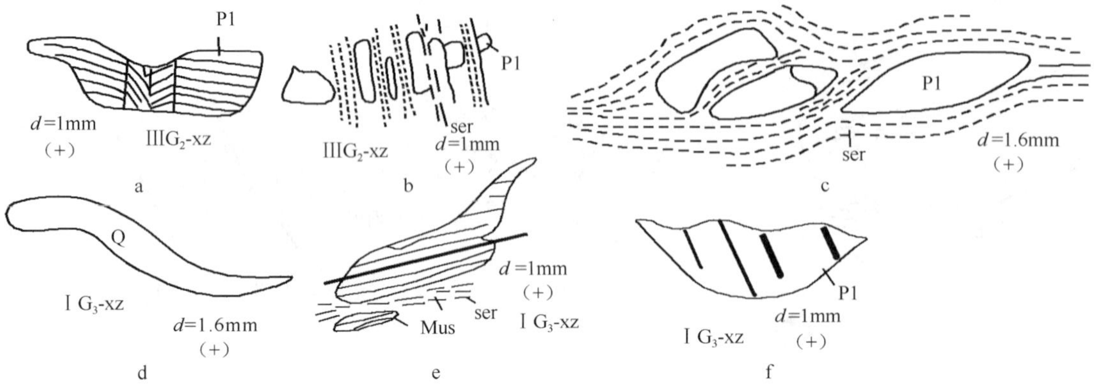

图21-64 显微变形组构图

Fig. 21-64 Fabric of microscopic deformation

a—长石膝折；b—长石剪切阶步；c—旋转眼球残斑；d—石英集合体扭曲变形；e—云母剪切活动；f—长石机械双晶

（4）古应力差值估算

流体力学认为，引起岩石流动（变形）的压力因素为差异应力，而与岩石所处环境的围压大小无直接关系，故差应力大小是决定岩石变形的主要因素。依据动态重结晶石英的粒度与差应力的函数大小关系 $\sigma_1-\sigma_3=6.1D^{-0.68}$（式中 D 单位为mm，σ_1、σ_3 单位为MPa），推算变形带在变形时的差应力变化在19.1~27.9 MPa之间（表21-4），并总体显示出随岩石变形程度的加强，差应力逐渐增高的趋势。

表21-4 洛基韧性剪切带糜棱岩类差应力估算表

Table 21-4 Estimated differential stress of mylonite in the Loki ductile shear zone

采样方向	样品号	岩 性	重结晶石英粒度/mm	差应力（$\sigma_1-\sigma_3$）/MPa
北西→南东	ⅢG1-1	初糜棱岩	0.169	20.4
	ⅢG2-1	糜棱岩	0.136	23.7
	ⅢG2-2	糜棱岩	0.107	27.88
	ⅠG1-1	糜棱岩	0.187	19.1
	ⅠG3-1	糜棱岩化花岗岩	0.138	23.4

（5）变形温压条件分析

韧性变形带一般认为是在大于10~15km地壳深度的温压条件下发生，然而不同的韧性变形带发生变形时的温压条件是有区别的，这些区别在韧性变形岩石的矿物组合、矿物韧性变形表现等方面都有反映（陈柏林等，1996）。

从矿物变形来看，石英矿物表现为韧性变形，具波状消光，发育动态重结晶等，颗粒多呈压扁状、拉长状、透镜体状，晶粒边界多为不规则状、锯齿状。长石矿物则以脆性变形为主，并多发生绢云母化，仅于局部发育剪切阶步以及出现细颈化现象。云母类矿物以脆性变形为特征。表明变形环境已达到了石英发生韧性变形的温压条件（即温度大于350℃，大于10~24km的深度），但未达到长石变形的温压条件（大于600℃和25km的深度）。结合变形带中的新生矿物为石英、绢云母、绿泥石等组合这一特点，共同指示该变形带形成的温压为中低温（400℃~550℃）和中等围压（0.35~0.5GPa）。

（6）三维应变与岩组分析

为了阐明和探讨逆冲推覆运动过程中岩石形成时的条件及其变形历史，前人沿垂直于该构造带走向采集了一些定向标本进行应变和岩组分析。自北西至南东，岩石应变和岩组分析结果如下（图21-65）：

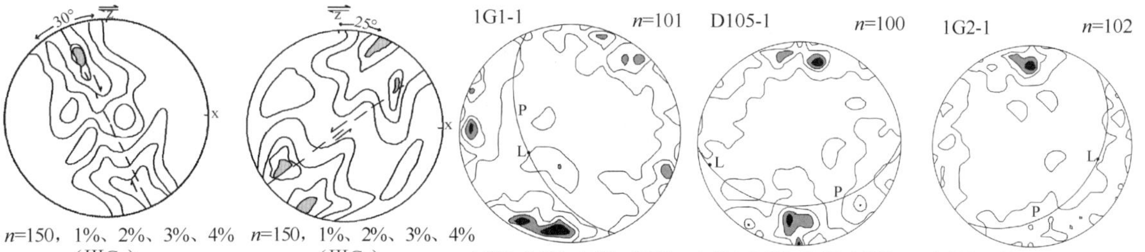

图21-65　洛基韧性剪切带岩组图

Fig. 21-65　Fabric of the Loki ductile-shear zone

1）IIIG1：岩石以挤压变形向平面应变过渡的特点，$K=0.85$，挤压方向北西向南东，略向南东倾伏，拉伸方向北东向南西，近水平。组构具单斜对称，为典型的带点极密的大圆环带型，环带轴在左侧开角30°，石英以菱面滑移为主，反映了浅层次、中低温逆冲。

2）IIIG3：岩石受到强烈拉张，$K=1.89$，拉张方向北西向南东，倾角陡，约为50°，向南东倾伏，挤压方向南东向北西，倾角陡，约为50°，向北西倾伏。发育两期叠加组构，早期为带点极密的大圆环带型，单斜对称，菱面滑移，反映了浅层次中低温左旋挤压逆冲；晚期为带点极密的边缘环带型，单斜对称，菱面滑移，反映中高温、中低应变速率的右旋伸展滑脱。

3）IG1：$K=0.846$，显示岩石以挤压变形向平面应变过渡的特点，挤压方向近东西向，向东倾伏，倾伏角为35°，拉张方向由北东向南西，倾角较陡，为55°。最大变形面的应变达2.024；组构具有不完整斜方对称特点，在北东和南西方向发育近水平次高极密，最高极密位于南西，产状近水平；此外，在近东西方向发育近水平次高极密，最高极密不与任何一个坐标轴一致，沿高极密发育小圆环带，由此显示岩石除了受到北东向南西的强烈挤压外，还受到近东西向的挤压变形，组构中不发育ac大圆环带，沿高极密小圆环带开角很小，由此显示石英以底面滑移为主，变形温度低于500℃。

4）IG2：$K=0.985$，显示岩石以平面应变为特点，挤压方向由北东向南西，倾角达57°，拉张方向由北西向南东，倾角33°。最大变形面的应变达2.25；组构以三斜对称为特点，仅在北部发育近水平高极密，高极密不与任何一个坐标轴一致，沿高极密发育半开角极小的小圆环带；组构中不发育大圆环带，表明石英以底面滑移为主，变形温度低于500℃，挤压应力方向为近南北向。

5）D105：岩石以平面应变为特点，$K=1.07$，挤压方向南东至北西，拉张方向近西向，最大变形面的应变达2.159。组构以单斜对称为特点，对称面为ab面，在北部和南部发育产状约15°的两个高极密，两个次高极密分布在北西和南东方向，近水平。沿高极密与次高极密均发育小圆环带，南部高极密与b坐标轴一致。反映岩石受到多期应力作用，早期是南部与b坐标轴一致的高极密，显示由北至南的挤压，石英以柱状滑动为特点，变形温度大于 550°C。晚期为北西向南东的对称挤压变形，组构中发育不完整 ac 大圆环带，沿高极密小圆环带半开角较小，由此表明石英以底面滑移为主，变形温度不高于 500℃（郑百让等，1989）。

其中，IIIG1、IIIG3、IG1、IG2 位于构造带根部，D105 位于其前峰带。IIIG1、IIIG3 和 D105 都一致表明，岩石受到了北西向南东的挤压，在此方向应力作用下，岩石发育走向北东向的糜棱面理，IIIG3 并叠加了晚期拉张隆升背景下的变形。而 IG1、IG2 基本上反映了晚期叠加变形特征，IG1 显示的北东向南西和近东西向德挤压，露头尺度中，与这两个方向应力作用分别相互配套的构造形迹是走向北西和近南北向的节理、断裂，表明后期变形基本上对早期形变进行了彻底的改造；IG2 显示的近 SN 向挤压，形成了近东西向的节理裂隙。由此表明该构造带叠加了晚期多次不同方向应力形成的构造形迹，导致构造带变得复杂化。

（7）运动学特征分析

沿垂直于变形带走向系统采样切片表明，岩石沿倾向方向变形比较强烈，指向性组构较常见，如石英集合体肠状褶曲，旋转残斑和剪切阶步等，结合岩组分析结果，认为变形带具有由北西向南东的逆冲运动特征。而平行走向的切片中，岩石变形相对较弱，指向性构造不甚发育。反映了变形带以挤压逆冲为主，平行走向的走滑运动分量较小。基本上反映了构造岩石是通过压扁变形而形成的产物。

（8）构造带时代

根据卷入构造带的最新地层为白垩系，初步认为构造带生成于燕山晚期。

2. 石碌水库韧性剪切带

该剪切带（F_{127}）由三条强变形带组成主要产于石碌矿区东北部的华力西期石碌岩体中，南端卷入了部分石碌群地层，并以脆性断层与石碌群、石灰顶组等接触，向北被马鞍岭岩体吞没，并截止于昌江-琼海深大断裂带，其北盘的华力西期花岗岩体中仅以同向脆性断层和岩脉方式出现；东翼发育一条数十米的“杂乱岩”带，与二叠纪地层主体呈断层接触（图 21-66）。

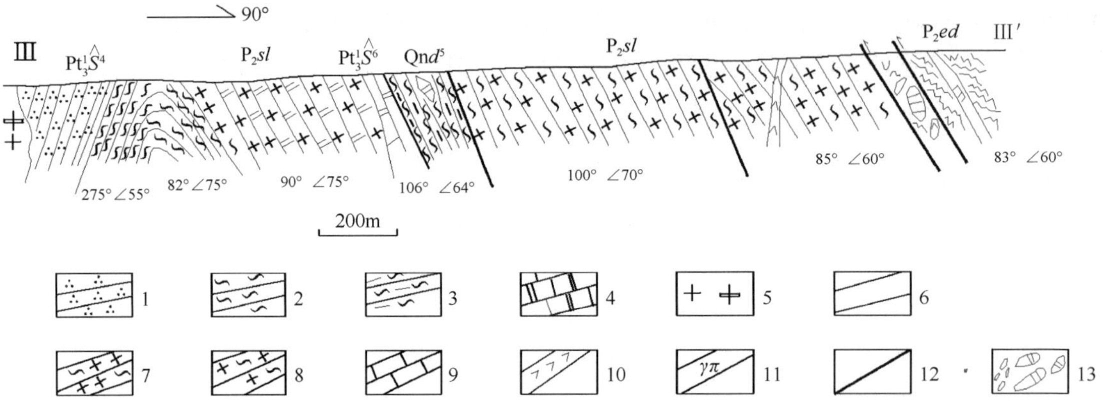

图 21-66　石碌水库韧性剪切带剖面图

Fig. 21-66　Section of the Shilu reservoir ductile-shear zone

1—石英岩；2—长英质糜棱岩；3—构造片岩；4—大理岩；5—二长花岗岩；6—糜棱岩化花岗岩；7—花岗质初糜棱岩；8—花岗质糜棱岩；9—灰岩；10—闪长玢岩；11—花岗斑岩；12—断层；13—杂乱构造带

（1）露头变形构造特征

1）面理构造：该剪切带中面理构造的发育类型决定于原岩类型和剪切变形强度。花岗质糜棱岩化岩石和花岗质糜棱岩中糜棱面理成为主面理构造，并与剪切面理组成了 I 型 S-C 组构；而石碌群片岩（千枚岩）叠加剪切变形的产物——强直片岩，主要发育了剪切面理构造，几何上表现为强烈剪切的条带状构造（图 21-67）。

2）褶皱构造和拉伸线理：卷入剪切带中的石碌群地层，褶皱现象极其丰富，众多褶皱面上的拉伸线理与其枢纽的平行关系，揭示了褶皱变形的剪切成因机理。

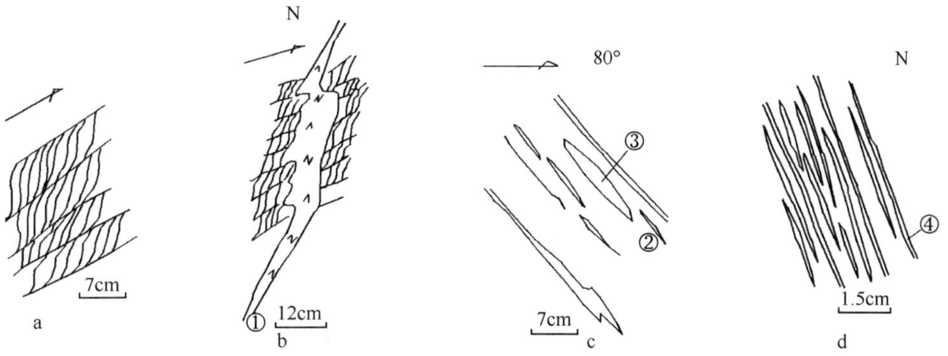

图 21-67　石碌水库剪切带中的面理构造素描图

Fig. 21-67　Sketch showing foliation structures in the Shilu reservoir shear zone

a，b—花岗质糜棱岩中的 S-C 组构；c，d—强直片岩中的条带状构造；

①斜长角闪岩包体；②暗色石英条带；③浅色石英条带；④石英集合体拉长条带

常见的褶皱类型，无论是平面上还是三度空间上都可找到早期褶皱的踪迹，它们或呈钩状包容于 a 型褶皱中，或以重褶形式表现出叠加褶皱形态，还可以多级褶皱组合出现。这种褶皱进一步演变形成了露头上罕见的箭鞘褶皱（图 21-68）。

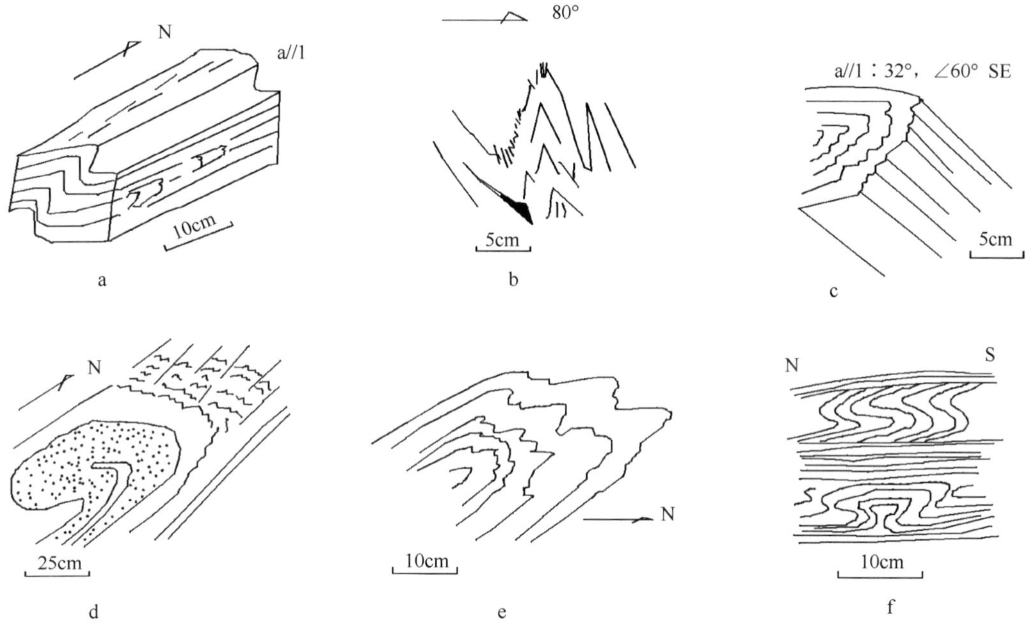

图 21-68　石碌水库剪切带中的褶皱构造素描图

Fig. 21-68　Sketch showing fold structures in the Shilu reservoir shear zone

a—a 型无根褶皱；b—剪切褶皱；c—a 型褶皱；d—叠加剪切褶皱；e—多级褶皱；f—鞘褶皱

强烈褶皱部位一般常伴生由石英等矿物拉长构成的拉伸线理，拉伸线理总是大致平行于褶皱枢纽，并随枢纽产状的变化而变化。

3）较刚性矿物或岩石的变形：在花岗质糜棱岩中，长石相对于石英、云母矿物而言显示了较大的刚性，露头上广泛发育旋转构造，同时这些被韧性物质包裹的斜长石残斑中，普遍出现一组仅限于晶体内的脆性裂隙面，从而说明了长石类矿物变形发生于脆性构造域中（图 21-69）。花岗岩中的斜长角闪岩刚性包体随变形的不断加强，形态上由棱角状、不规则状—椭球状—扁豆状。早期的花岗岩岩墙，卷入剪切带后被强烈拉薄，而弱变形域内仍保持原始产状和位态（图 21-70）。

（2）显微变形构造

依据变形行为、发育形式和形成机制等将该带内丰富多彩的显微变形组构划分为 5 种类型（图 21-71）。

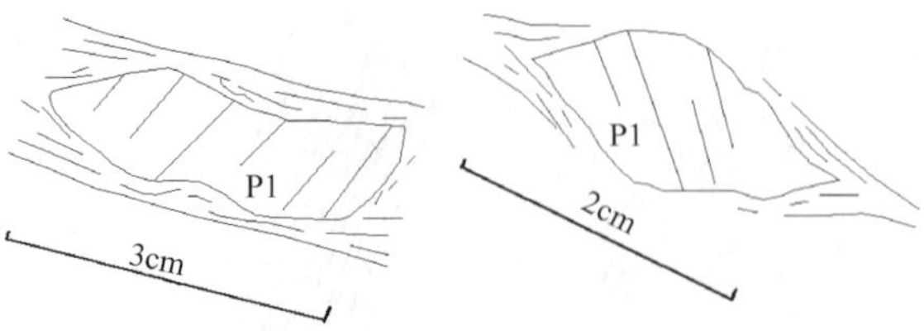

图 21-69　斜长石残斑脆性破碎示意图

Fig. 21-69　Schematic diagram of brittle fracturing of residual plagioclase phenocrysts

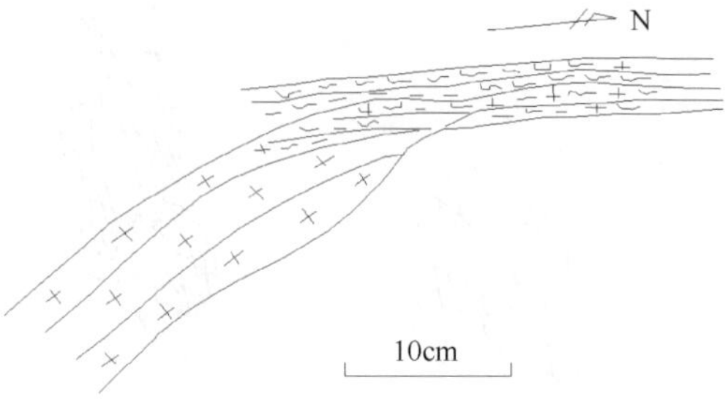

图 21-70　早期花岗岩墙的剪切变形示意图

Fig. 21-70　Schematic diagram of shear deformation of early-stage granite dikes

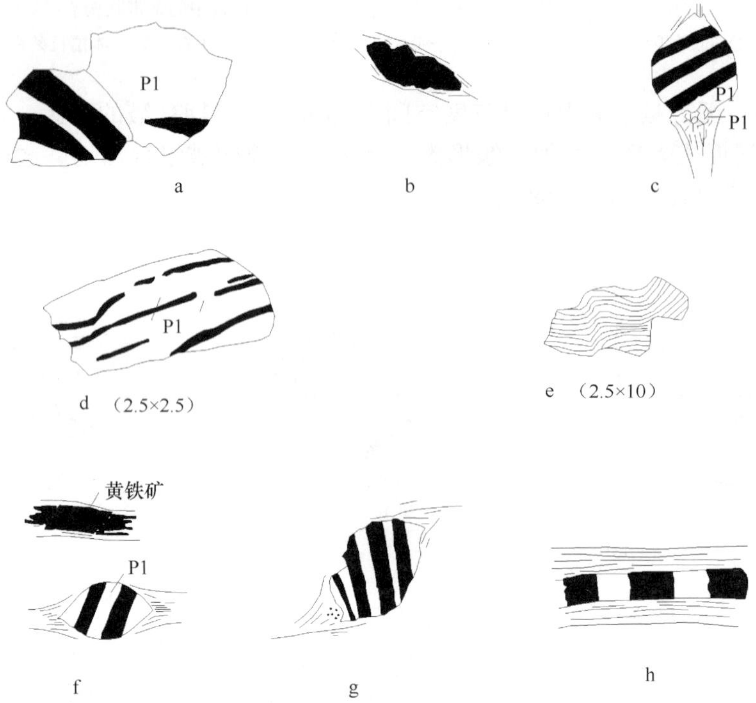

图 21-71　石碌水库剪切带糜棱岩中的显微变形构造图

Fig. 21-71　Microscopic deformation structures in mylonite of the Shilu reservoir ductile-shear zone

a—斜长石显微破碎；b—黄铁矿剪切阶步；c—斜长石“山羊须”构造；d—斜长石机械双晶；
e—黑云母解理弯曲；f—黄铁矿和斜长石压力影；g—斜长石旋转结构；h—石英矩形边结构

1）显微破裂结构：这类结构代表了剪切带边部糜棱岩化岩石的主要变形特点，在石碌水库剪切带中普遍发育，表现为长石、黄铁矿等矿物的脆性破碎，形成了剪切阶步、山羊须等组构类型。

2）晶内纹带状构造：最常见的在石英等矿物的波状消光、变形纹和变形带，斜长石多见变形双晶，层状矿物则主要显示显微膝折和弯曲。

3）晶内粒状结构：仅发育石英矿物中，如亚颗粒、核-幔构造和等粒结构，特别是等粒结构的出现，意味着岩石动态重结晶比较彻底，代表了一种较高温-低应变率条件下的岩石流变。其成因可能与超塑性流变有关，晶粒通过扩散调节，边界滑动而相互转移，最终导致了晶体本身无应变或弱应变，但整体经历了较大应变的变形效果。等粒结构局限于卷入剪切带的原石碌强变形带绢云石英片岩中，而花岗质糜棱岩中未曾显示。

4）与粒间位移有关的结构：表现形式有压力影、旋转结构和矩形边结构等。矩形边结构的成因机理可以从恢复作用角度加以探讨。当恢复速率急剧增加，矿物的生长速度也急剧增加，在温度和压力较高的情况下，一旦应力解除，矿物就有足够的时间和速度进行恢复和生长，使得晶内的变形结构消失，晶粒生长成矩形或多边形。

5）与压溶有关的结构：卷入剪切带的石碌群地层，镜下常见不同矿物聚合而面的条带状构造，呈锯齿状镶嵌的石英晶粒多数可能与压溶作用有关。据 R. Kerrich 等研究，在韧性变形过程中矿物颗粒减少到一定程度时（50~20μm），由于其表面积大大增加，使压深作用为为进一步变形的主要机制。此时在流体相的参与下，组分在应力差异驱动下应变能大的矿物与应变能小的矿物分离，形成条带状构造。这种条带状构造仅发育于强变形的强直片岩中。

（3）有关变形参数的确定

1）温度：依据同变形变质的共生矿物对——白云母-绿泥石地质温度计，测得剪切带的形成温度为 300℃~350℃。

2）古应力差值和应变速率：利用动态重结晶石英的平均粒径，并据前述公式（1），得出了不同部位古应力差值，又将温度 T=600K 代入公式（2），得出了不同部位的应变速率（表 21-5）。

表 21-5　石碌水库韧性剪切带应变参数表

Table 21-5　Strainparameters of the Shilu resevoir dutcile-shear zone

岩性	部位	古应力差值 $\Delta\sigma$/MPa	应变速率/Sec^{-1}	资料来源
花岗质初糜棱岩	边部	93.02	2.5×10−14	1：5 万《昌江县、邦溪》幅
强直片岩	中部	131.3	6.2×10−16	
花岗质糜棱岩	中部	102.9	5.4×10−16	

3）富林指数：强直片岩中石英颗粒有限应变测量结果，以此求得富林指数 $K=(x/y-1)(y/z-1)=0.28$，为压扁型（图 21-72）。

（4）运动方向和剪切带类型的确定

依据露头大量出现的拉伸线理，S-C 组构、a 型褶皱和鞘褶皱以及显微旋转组构等综合分析，推断剪切带具有平面右行剖面逆冲的运动矢量，为自南东向北西逆冲而成的韧性推覆型剪切带。

长英质糜棱岩的岩组学研究结果也证实了上述推论。石英光轴投影呈单斜对称，由一对强极密区和一对不太明显的次极密组成，并构成一对小圆环带。而较强的极密可能反映沿 S 面强烈右旋运动（图 21-73）。

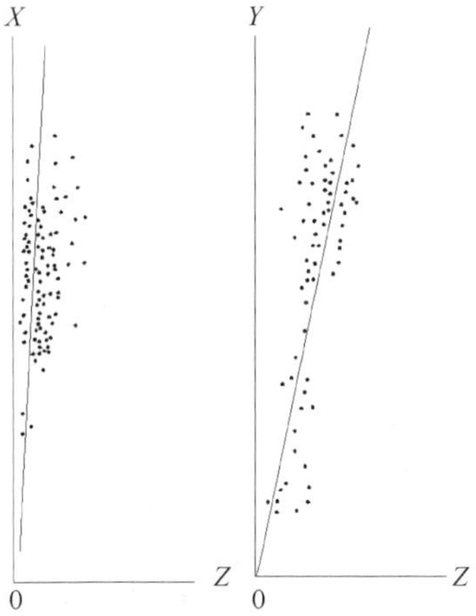

图 21-72　强直片岩中石英颗粒有限应变分析图解

Fig. 21-72　Analysis diagram of limited strain of quartz grains in highly straight schist

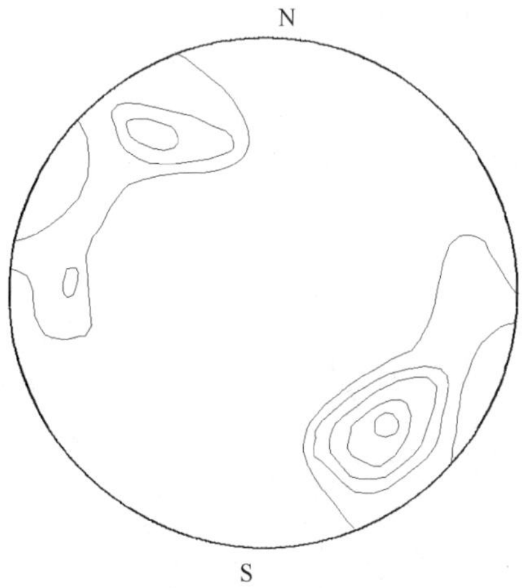

图 21-73　石英光轴岩组图

Fig. 21-73　Quartz optic axis fabric

（200 颗，等值线 1%~3%~15%~25%）

3. 什运共轭韧性剪切带

什运共轭韧性剪切带（F_{53}）分布于海南岛中部什运北西面谭爷断陷构造带的南东缘，并局限于白垩系陆相红色碎屑岩（砂岩、砂砾岩）内，南东面与华力西期—印支期、燕山期花岗岩呈断层接触。

由北东向韧性剪切带和北西向韧性剪切带组成，一组走向 28°~58°，倾向南东，倾角 20°~64°；另一组走向 320°~330°，倾向南西，倾角 22°~46°（李孙雄等，2006）。

（1）构造岩类型及特征

什运共轭剪切带内发育一套糜棱岩系列岩石，由糜棱岩化砂岩、初糜棱岩、糜棱岩组成。岩石具有一定的分带，中心为糜棱岩，两侧依次为初糜棱岩、糜棱岩化砂岩。受两方向韧性剪切带相互作用影响，局部地段，岩石分带不明显。总体表现是由南东向北西，岩石变形逐渐增强，靠近花岗岩接触带，变形相对减弱。

初糜棱岩 糜棱结构、碎斑结构，眼球状构造。碎斑含量 50%~55%，成分为石英、长石、石英岩、石英砂岩、硅质岩等，大小 0.2~3.8mm，呈眼球状、透镜体状，沿长轴平行定向排列；碎基含量 45%~50%，由细碎的长英质和大量次生的绢云母组成，围绕碎斑生长。

糜棱岩 与初糜棱岩相比，岩石碎斑含量减少，并发育较明显的流动构造。碎斑成分为石英、长石、石英岩，含量 30%左右，粒径 0.1~3mm，石英碎斑带状消光常见，有的发育变形纹；碎基含量约 70%，由绢云母和强烈细碎的石英组成。两者相间定向排列形成条带状构造。

糜棱岩化砂岩 与上述两种岩石相比，岩石以脆性变形为主，韧性变形程度较差。碎斑含量>65%，大多保持原生砾石的碎屑形态，少量见韧性变形，呈弱的定向排列。碎基含量<35%，成分为细碎的石英和非常细小的次生绢云母，石英具波状消光，部分呈压扁拉长状，沿长轴定向排列，方向与剪切面理一致，局部与之有小角度相交，构成岩石的 S-C 面理组构。

（2）变形特征

1）面理：剪切带面理主要有糜棱面理（S）和剪切面理（C）。糜棱面理为矿物或集合体的长轴优选方位所显示的不连续面状构造，表现为拉长状的砾石残斑定向排列。剪切面理由细小的重结晶绢云母矿物定向排列而显示。弱变形域两者具小角度相交，而强应变带两者则呈近平行状产出。

2）线理：发育在强应变的岩石中，由石英矿物或其集合体和暗色的铁质矿物呈拉长状定向排列而体现。

3）褶皱：剪切带内褶皱发育，为剪切作用过程形成。露头尺度上的褶皱按形态划分有钩状褶皱和肠状褶皱（图 21-74，图版Ⅴ-3-E），前者见于北东向剪切带，后者分布于北西向剪切带中，可能反映了两者变形程度具有一定的差异。褶皱轴面为新生的面理面，形态复杂，两翼极不对称，翼间角紧闭一翼长而缓，另一翼短而陡，翼部略拉薄、拉断状，核部则呈增厚状，物质具有由翼部向核部运移的运动特点。褶皱指向表明两个方向的剪切带都具有下滑的运动学性质。微观尺度中表现为绢云母矿物沿剪切面发生膝折变形。

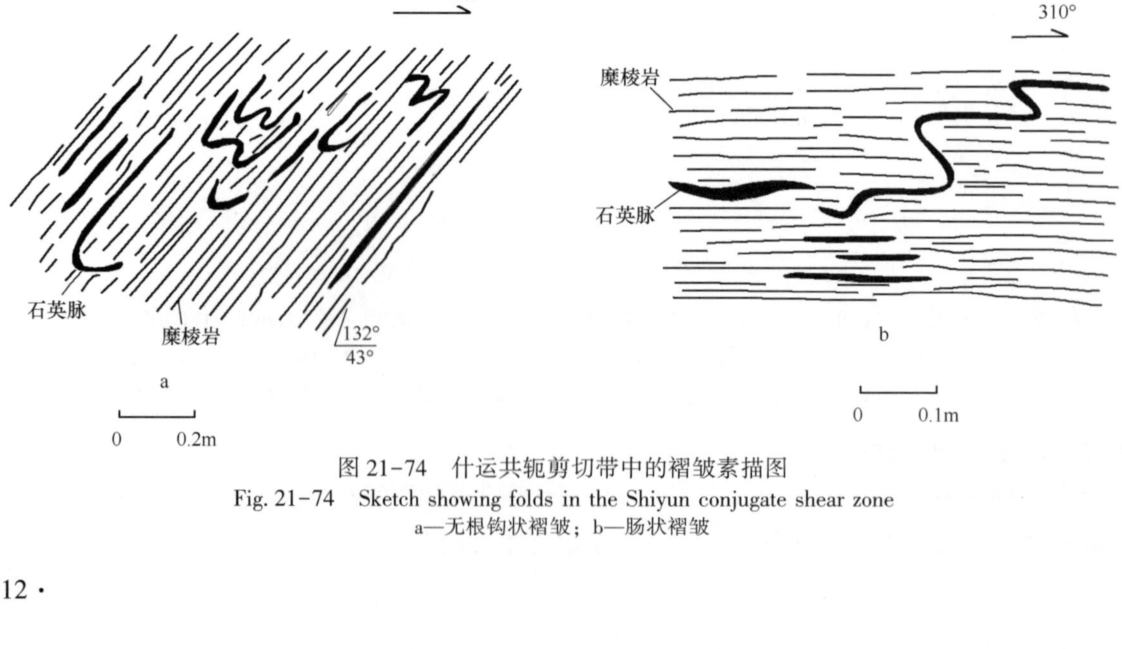

图 21-74 什运共轭剪切带中的褶皱素描图

Fig. 21-74 Sketch showing folds in the Shiyun conjugate shear zone

a—无根钩状褶皱；b—肠状褶皱

4）显微变形构造：变形带内矿物变形机制主要通过扩散物质迁移、晶质塑性和颗粒边界滑移及破碎和碎裂作用等，形成了显微破裂结构、晶内纹带状构造、晶内粒状结构以及与晶粒位移有关的结构（图 21-75，图版Ⅴ-3-F）。其中晶内粒状结构的出现代表了一种较高温、低应变速率条件下的变形，其成因可能与塑性流变有关，晶粒通过扩散调节、边界滑动而相互迁移，最终导致了晶体本身无应变或弱应变，但整体经历了较大应变的变形效果。

a（D_{1672}）　b（D_{1672}）　c（$Ⅱ_{38}$）　d（$Ⅱ_{38}$）

图 21-75　什运共轭剪切带中的显微变形组构图

Fig. 21-75　Microscopic deformation fabric in the Shiyun conjugate shear zone

a—石英矿物形成的不对称压力影；b—石英矿物沿剪切面剪切活动；c—长石机械双晶；d—长石剪切阶步

显微破裂结构：表现为长石、石英砂岩等发生脆性破碎，裂纹发育，长石并形成剪切阶步、山羊须构造。

晶内纹带状构造：表现是石英矿物具波状消光、带状消光，甚至发育变形纹，斜长石发育变形双晶以及膝折变形，绢云母显示膝折或弯曲变形。

晶内粒状结构：仅见于石英矿物中，如亚颗粒和等粒结构，代表了一种较高温、低应变速率条件下的变形。

与晶粒位移有关的结构：表现形式有压力影、旋转结构和矩形边结构。

（3）三维应变与岩组分析

三维应变分析结果：岩石受到强烈挤压，$K=0.11$，挤压方向由南东向北西，呈近垂直状，拉伸方向北东向南西，近水平。最大应变面（XZ 面）的应变达 2.445。岩组结果（图 21-76）：组构具斜方对称特点，发育两组高极密，一组为南东—北西，另一组为北东—南西。最高极密在南西，与 c 轴近垂直。组构中发育不完整大圆环带，半开角近 42°，显示岩石变形温度高于 500℃。

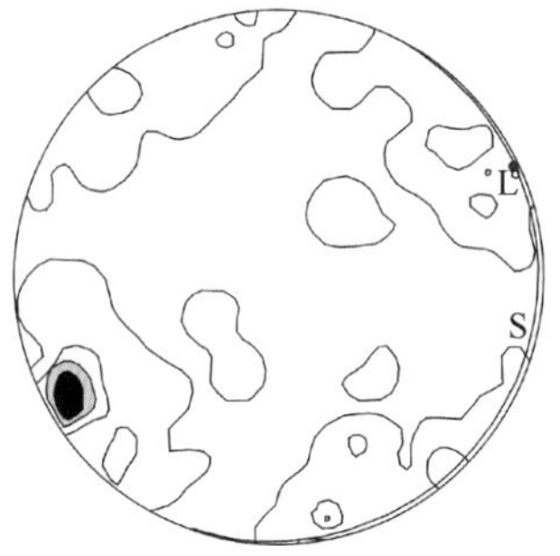

0.8%~4.0%~6.0%~8.0%

图 21-76　什运共轭剪切带岩组图

Fig. 21-76　Fabric of the Shiyun conjugate shear zone

（4）应力差值估算

剪切带岩石重结晶作用不一，北西向变形带石英基本上显示静态重结晶颗粒特征，而北东向变形带石英则多表现为动态重结晶，依据动态重结晶石英的粒度与差应力的函数大小关系 $\sigma_1-\sigma_3=6.1D^{-0.68}$（式中 D 单位为 mm，σ_1、σ_3 单位为 MPa），推算北东向变形带在变形时的差应力变化在 27.2~66.2MPa 之间。

（5）运动学特征

宏观尺度中，岩石发育指示下滑运动的剪切褶皱。微观尺度中，岩石沿走向和倾向两个方向的变形比较强烈，各种指向性小构造形迹普遍可见及，如不对称压力影、剪切阶步等。表明共轭剪切带发生下滑运动的同时，平面上兼有右行走滑的运动特点。

4. 大村韧性剪切带

剪切带（F_{83}）分布于后溪村-大村一带，长约 6.2km，宽 0.75~1.0km，产状 115°~140°∠55°~75°。变形带在平面上呈舒缓波状弯曲。发育于抱板群与陀烈组的接触带中，并局限于戈枕村组内，东面以韧-脆性断裂与陀烈组接触。沿走向终止于燕山期、海西期—印支期花岗岩。

(1) 构造岩分带

该韧性变形带主要发育糜棱岩和糜棱岩化岩石，于野猫隆-抱告小学一带，岩石变形作用相对较强，属主应变带变形，宽约365m，以多次交替出现糜棱岩为特征，其宽度一般约10~15m，两侧为糜棱岩化岩石。沿主应变带水平横、纵向，则过渡为糜棱岩化岩石，进入低应变域的变形，东、西两侧边部并出现脆-韧性断裂。反映了变形带的岩石分带性较明显，且自中心→边部，构造应力减弱，温度、压力逐渐降低，变形由韧性→韧脆性，由高应变→低应变的演化。微观尺度上，糜棱岩系列岩石，主要由长石、石英的破碎蚀变残余和云母、绿泥石等矿物组成，后者多为热动力下重结晶作用的新生矿物。变形作用中，石英矿物表现为韧性变形，具波状消光，发育变形纹、动态重结晶等，其颗粒多呈压扁状、拉长状、透镜体状、豆荚状，晶粒边界多为不规则状、锯齿状；长石矿物则以脆性破碎为主，多发生绢云母化蚀变，糜棱岩内仅于局部发育δ型旋转碎斑、剪切阶步及细颈化等；云母类矿物则以韧脆性变形为特征，既有脆性破碎呈细小鳞片状，又有“云母鱼”、S型扭曲。

(2) 变形特征

沿解理面右旋剪切滑动、消光带、膝折等韧性变形，且后者多见于高应变的糜棱岩中。

韧性变形带内，矿物定向组构发育，矿物拉长线理、面理、褶皱以及小的指向性构造也较常见，呈现出中浅构造层次的变形特征。

1) 矿物拉长线理：发育于强应变的糜棱岩中，石英矿物集合体呈拉长条带状及磁铁矿呈拉长状、“S”型扭曲状等沿片理方向分布，与剪切面理近平行。

2) 面理构造：变形带内发育糜棱面理和剪切面理。糜棱面理（Ss）为矿物或集合体的长轴优选方位所显示的不连续面状构造，由粗粒的片状矿物和拉长的粒状长英质矿物组成；剪切面理（Sc）是一种低级别小尺度的剪切带，平行剪切面，是剪切过程中的产物，主要由细小重结晶的绢云母组成。在弱变形域糜棱岩化岩石中，剪切面理具有继承和迁就了早期面理（片麻理）特点，与糜棱面理呈大角度相交（≥30°），而在强应变带糜棱岩中，原生面理构造不复存在，新生面理对早期面理进行彻底的改造和置换，且Ss与Sc的交角较小，基本上近平行。此外，尚见及“云母鱼”类Ss-Sc组构。

根据Ss与Sc的交角，指示变形带具有右旋活动特征。

3) 褶皱：发育于强应变的糜棱岩中。一为石英矿物集合体发生剪切作用变形，形成各种不协调的褶皱构造，褶皱形态极紧闭，两翼长短不一，翼部呈拉薄、拉断状，而核部则呈增厚状，物质具有翼部向转折端运移的运动特点，轴面与剪切面理平行；另一为云母矿物在剪切变形中发生的膝折变形（图21-77）。

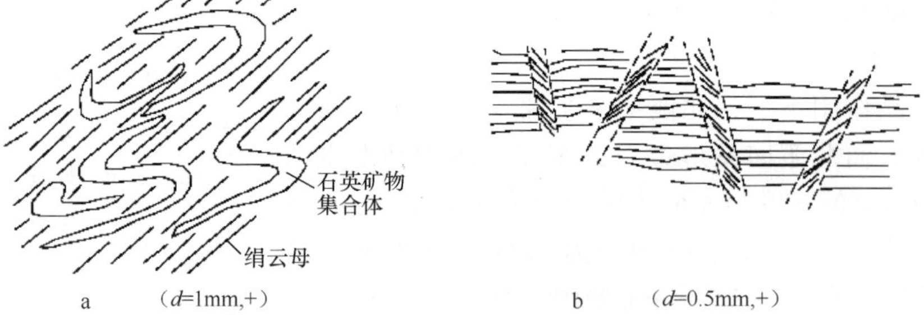

图21-77 褶皱构造素描图

Fig. 21-77 Sketch showing fold structural elements

a—石英矿物集合体的剪切褶皱变形；b—云母矿物的膝折

4) 指向性构造：露头尺度上，见有长英质眼球残斑形成δ型右行旋转碎斑，并右行切错细晶岩脉，错距约3m，脉体与糜棱面理呈70°锐角相交（图21-78）；微观尺度上，主要有：石英、长石矿物形成的δ型右行旋转碎斑系，云母矿物沿解理面发生右行剪切活动，石英矿物S型拖尾，长石矿物的剪切阶步和细颈化，云母矿物S型扭曲和“云母鱼”等等，且多发育于强应变的糜棱岩中。反映了该韧性变形带具右旋剪切活动的特点（图版Ⅴ-3-G）。

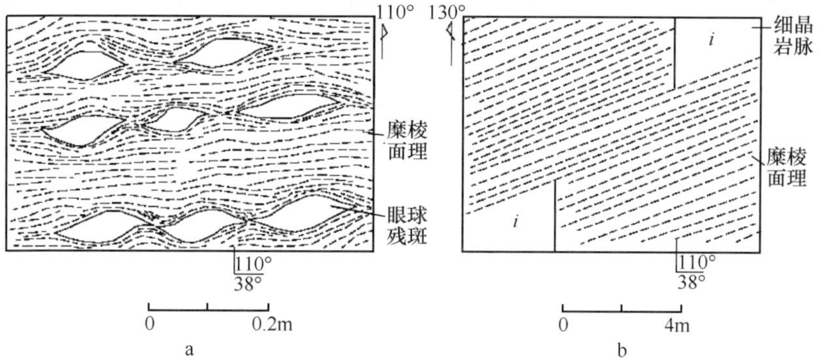

图 21-78　大村韧性变形带指向性构造平面素描图

Fig. 21-78　Plan sketch showing directing structures of the Dacun ductile deformation belt

a—眼球残斑右行旋转变形；b—糜棱面理右行切错细晶岩脉

（3）蚀变特征

微观尺度上，糜棱岩矿物组合为：石英（45%~65%）+绢云母+绿泥石+长石（少量）+白云母（少量），其中绢云母+绿泥石的含量20%~45%；糜棱岩化岩石矿物组合为：石英+钾长石+斜长石+绢云母+白云母+黑云母（少量）；而正常混合岩矿物组合为：石英（10%~15%）+斜长石+钾长石+黑云母（20%）+白云母。上述不同岩石的矿物组合特征，明显反映了随应变作用的加强，长石、黑云母等矿物逐渐消失，而绢云母、绿泥石、石英等矿物则逐渐增多，蚀变作用加深，表明变形作用与变质作用为正相关关系。

（二）其他主要韧脆性构造

其他主要韧脆性构造有北东向、北西向、南北向和东西向等。

1. 北东向

（1）南界断裂带

出露于江边—南界一带，由左雁状展布的F_{49}—F_{52}断裂组成，影响长度约39km。根据其切割的最新地质体为白垩系，生成时间为燕山期。主要断裂特征如下：

大炎新村断裂（F_{49}）　沿潘丁—大炎老村—大炎新村一线展布。走向47°，长度约12km，宽约50~150m，产状130°∠60°。破碎带内具破碎角砾、石英脉及褐铁矿化。角砾以板岩、砂岩及脉石英等，呈棱角状，铁质胶结。石英细脉沿破碎带充填。断裂旁侧的劈理中具铅、锌等多金属矿化。断裂切错了东西向的断层及东西向褶皱，平面位移约200m，为右旋压扭断裂，形成于燕山晚期。

南会断裂（F_{51}）　走向45°，长约12.5km，宽1~10m不等。倾向南东，倾角70°。沿断裂形成明显的断层崖。岩性以断层角砾岩为主。在南会附近断层分为两支，形成长约1.5km左右、由鹿母湾组构成的透镜体。两盘岩石较破碎，并发育一组走向近东西、倾角83°的剪节理。属于压扭性质。

（2）中平断裂带

分布于中平—加定一带，由F_{64}、F_{65}断裂组成，长约18.5km。单断裂宽0.95~13.5m，产状280°~320°∠38°~70°。断面平直，构造岩石有碎裂花岗岩、糜棱岩化碎裂岩、碎粒岩，岩石见硅化、黄铁矿化、高岭土化、绿泥石化现象。具有压扭性质。根据其切割的最新地质体为燕山期花岗岩，认为其生成时代为燕山期。

（3）加盆-龙江构造带

分布于加盆—龙江一带，由多条平行状分布的断裂（F_{66}—F_{70}）和褶皱（f_{19}—f_{21}）组成，走向35°~65°，倾向北西或南东，倾角45°~65°，单断裂长9~16km，宽2~30m不等，各断裂呈平行状展布。

站岭一带表现为南西端收敛、而北东端发散的“帚”状。断裂带两端主要发育一套碎裂岩系列岩石，局部以密集劈理带体现。构造岩石包括构造角砾岩、碎粒岩、碎裂花岗岩、硅化碎裂岩等，构造角砾岩角砾形态呈椭圆状、次棱角状—次圆状。断裂带中段站岭一带则表现为韧性变形，由糜棱岩系列岩石组成。断裂带内岩性分带明显，自南东至北西依次为糜棱片麻岩、初糜棱岩、千糜岩、糜棱岩（图 21-79）。

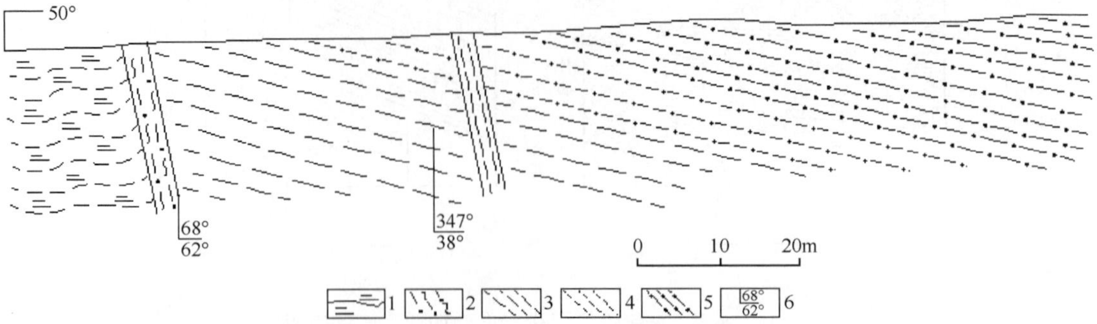

图 21-79　加盆-龙江构造带构造剖面图

Fig. 21-79　Structral section of the Jiapen-Longjiang strucutrual belt

1—变质绢云母粉砂岩；2—糜棱片麻岩；3—千糜岩；4—初糜棱岩；5—糜棱岩；6—产状

构造岩石中石英矿物普遍呈波状消光，带状消光也常见，常常被强烈拉长压扁成透镜状、扁豆状、眼球状；长石矿物发生破碎的同时，也具一定的韧性变形，如双晶弯曲等，多被绢云母交代；黑云母退变为绢云母、白云母、绿泥石。韧性变形岩石中还发育剪切褶皱，褶皱形态极紧密、不协调，轴面与剪切面理平行，两翼长短不一，翼部呈拉薄、拉断状，而核部呈增厚状，物质具有翼部向转折端运移的运动特点。此外，指示性小构造也常见，如石英矿物 S 型拖尾、长石矿物右旋残斑和右行剪切阶步、云母矿物右行“云母鱼”等，反映了断裂具右旋压扭性特点。

2. 北西向

北西向断裂带是海南岛分布范围最广，最醒目的构造形迹，有儋州-万宁断裂带和龙波-翰林断裂带。

（1）儋州-万宁断裂带

分布于儋州市—万宁市一带，由一组断续展布的平行状断裂（F_{95}—F_{97}、F_{104}、F_{105}、F_{123}、F_{124}）组成，走向 292°~345°，倾向 22°~55°，倾角 40°~75°，单断裂长 5~30km，宽一般为 10m，局部可达上百米。断裂带主要发育在华力西-印支期和燕山期岩体中，局部控制了白垩纪红盆的边界。

儋州市兰洋一带，破碎带由构造角砾岩带、石英脉带和花岗斑岩脉带组成，受后期活动影响，角砾岩带再次发生破碎，见一组北西向石英脉充填，并左行切错北东向石英脉，错距近百米。

琼中县腰子镇—加鲜村一带，构造破碎带由构造角砾岩、碎斑岩、碎粒岩、糜棱岩化花岗岩、硅化岩等组成，左行切错近南北向断裂。腰子镇附近，断裂出露较完整，断面较平直，略呈弧弯曲，构造岩石分带明显（图 21-80），中心为断层泥、碎粒岩，两侧依次为碎斑岩、花岗质碎裂岩、碎裂花岗

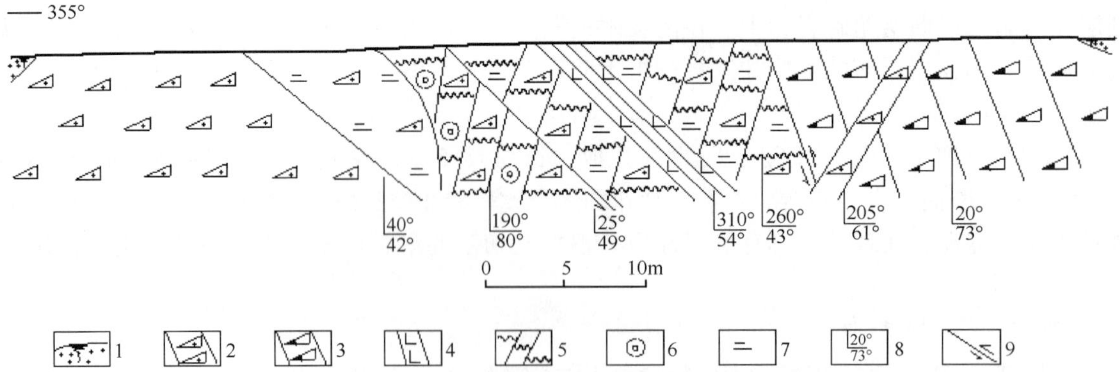

图 21-80　腰子镇南侧构造破碎带素描图

Fig. 21-80　Sketch showing the structural fractured zone, south of Yaozi Town

1—浮土；2—碎粒岩；3—碎斑岩；4—煌斑岩脉；5—节理；6—硅化；7—绢云母化；8—产状（倾向/倾角）；9—断裂活动方向

岩，见硅化脉充填。花岗碎裂岩中发育两组节理，节理面被石英细脉充填，第一组产状225°∠75°，呈右雁状，脉壁平整，第二组产状260°∠43°，呈锯齿状弯曲。硅化脉较破碎，脉体中见一组剪节理，节理产状190°∠80°，与该节理以约70°交角发育一近水平张性硅化脉，脉体形态不完整，脉壁呈锯齿状。碎粒岩碎粒成分为石英，具显著的波状消光和带状消光；碎基发生绢云母、硅化、黄铁矿化等蚀变。碎斑岩碎斑成分为石英、长石，石英具强烈的波状消光和带状消光，发生破碎，长石完全被绢云母交代；碎基成分为石英、长石、次生绢云母等。

牛仔岭-中平镇一带，构造岩石主要为构造角砾岩、碎裂花岗岩、硅化碎裂岩、硅化碎粒岩、糜棱岩化碎裂花岗岩、硅化岩、构造片岩、糜棱岩等，并被花岗斑岩脉、石英脉、闪长岩脉充填。笔架岭旁侧，构造岩石具一定的分带性，自北东至南西依次为构造角砾岩、糜棱岩化碎裂岩、硅化岩。构造角砾岩角砾成分为硅化岩、碎粒岩，呈次棱角状—次圆状，弱定向排列。糜棱岩化碎裂花岗岩中，石英矿物具强烈的波状消光，有的带状消光，被拉长、压扁成长条状，沿长轴方向平行排列；斜长石双晶发生弯曲、断裂；黑云母绝大部分被绿泥石交代而仅保留外形，沿片轴与石英平行定向排列。碎裂花岗岩的石英具强烈波状消光，发生破碎，呈强烈细碎的集合体，有的拉长、压扁成透镜状、扁豆状；长石发生破碎、错断，斜长石双晶见弯曲变形，绢云母化强烈；黑云母几乎被绿泥石置换，受断裂控制，派生一组产状为140°∠70°的压性节理，反映了该断裂在逆冲活动的同时具有右行平移活动特点。

上述特征表明断裂带至少经历了两期活动，早期为左旋压扭性，晚期为右旋压扭性，且近期依然发生过活动，控制了温泉的出露，水温高达90℃，沿断裂延续延伸几千米。

（2）龙波-翰林断裂带

分布于临高县龙波至琼海市翰林一带，呈断续状展布，由一系列北西走向的平行状断裂（F_{98}—F_{103}）组成，走向280°~335°，倾向10°~65°，倾角40°~60°。单断裂长约1~27km，宽度变化较大，一般为2~10m，最宽可达几百米。发育在古生界、白垩系和华力西期—印支期、燕山期花岗岩中，龙波和翰林镇一带被第四系掩盖。

南蛇岭、银瓶岭等地，破碎带宽达130m，岩石以脆性破碎为特色，形成碎裂花岗岩和构造角砾岩。断裂带中石英脉成群成带发育，呈左行尖灭再现排列，组成北西向突出的山脊。沿断裂带见多个温泉点出露，在石英脉群展布地段，还有金、银和铅锌矿化，是一条控制多金属矿化的断裂带。

翰林镇南面表现为发育构造破碎带，由糜棱岩化花岗岩、花岗质糜棱岩、碎粒岩、构造角砾岩、构造片岩、碎裂岩、碎斑岩、石英脉组成。构造岩石分带不太明显，自南西至北东，大致为糜棱岩化碎裂岩、硅化碎裂岩。构造角砾岩角砾成分为硅化脉、糜棱岩，呈次圆状、椭圆状。构造片岩中石英具强烈波状消光，被拉长压扁，沿伸展方向聚成石英条带定向排列；白云母可见右行“云母鱼”和沿解理面活动。

加东田一带，形成叠瓦状逆掩断裂体系，导致抱板群戈枕村组逆冲盖于峨文岭组之上，出现地层多次重复现象。上盘戈枕村组基本上发生糜棱岩化，形成糜棱岩化片麻岩，岩石中，石英矿物发生动态重结晶，被拉长压扁呈长条状，长石发生右行剪切破碎，呈细颈化和不对称右行δ型碎斑，绢云母沿解理面右行拆离而成“云母鱼”。下盘峨文岭组变形则表现为发育倒转型褶皱。同时并使得抱板群面理产状由走向北东、倾向南东往走向北西、倾向北东转变，反映了该构造带发生逆冲推覆的同时，并兼有左行走滑活动的特征。

琼海官塘一带破碎带由北西向的石英脉带组成，脉体后期破碎，形成硅化破碎带或构造角砾岩带，并控制了官塘温泉的出露。

（3）毫岗岭断裂（F_{122}）

分布于毫岗岭一带，总长约1.9km，影响宽度约达1.3km，走向330°~355°，倾向240°~265°，局部倾向60°~75°，倾角55°~82°。切割下志留统陀烈组和三叠纪尖峰超单元。断裂主要表现为由多条相互呈近等距平行状破碎带组成。破碎带间距一般为50~80m，长400~1300m，宽一般为10~30m，最宽80m。破碎带具有良好的分带性，中心一般为千糜岩、糜棱岩，被石英脉充填，两侧为碎裂硅化千枚岩，直至正常岩石——千枚岩。断裂控制了毫岗岭金矿的展布、规模，属一控矿构造。岩石蚀变有硅化、绢云母化、绿泥石化、碳酸盐化、金属硫化物矿化等。

与该断裂相伴随的毫岗岭背斜长约1.5km，宽约0.8km，轴迹北北西向。核部为陀烈组下段绢云母石英千枚岩，两翼为陀烈组中段含炭千枚岩，东翼因断层切割而未出露。东翼地层产状：61°~88°∠58°~88°；西翼地层产状：230°~245°∠60°~80°，向南倾伏，倾伏角48°~60°。为一紧闭倾伏背斜。

岩石三维应变分析结果显示：①弗林参数K值小于1，表明断裂系挤压环境下的产物；②拉伸主轴（X轴）方位较稳定，为北北西向，断裂走向一致；③南西→北东，X轴倾伏角度陡，而压缩轴（Z轴）则恰恰相反，同时，应变强度（r）值逐渐增高，北东端岩石变形更强烈，反映了断裂活动过程中矿物迁移变形有可能从南西→北东向推覆的趋势。

根据该断裂切割三叠纪尖峰超单元，结合于矿石中对伊利石矿物进行K-Ar法测龄结果为（216±3）Ma、205Ma（海南综合勘察院，2000）和Rb-Sr年龄为（236±17）Ma，认为该断裂形成于华力西期—印支期，力学性质表现为压扭性。

（4）千家-三亚断裂带

分布于乐东千家-三亚一带，影响长度约59km。由心银山-后友山断裂（F_{118}）、奋跃断裂（F_{119}）、高峰断裂（F_{120}）、落马咋断裂（F_{121}）断裂组成。根据其切的最新地质体为早白垩世岩体，认为属于燕山期生成起来的左旋压扭性断裂。主要断裂特征如下：

心银山-后友山断裂（F_{118}） 分布于心银山—后友山一带，长约6.2km，宽1~1.5m，局部大于120m，产状220°~230°∠70°~80°。断裂内岩石有构造角砾岩、碎斑岩、硅化岩、糜棱岩化岩石及糜棱岩等。角砾岩角砾成分为硅质岩，呈次棱角状—次圆状。于大安水库库坝边，该断裂出露程度较好，由一系列大致平行的密集小断裂组成，总宽达120m以上，呈左行切错早次活动的北东向断裂，错距达5m。各小断裂构造岩石具明显的分带性，其中心一般为糜棱岩化碎粒岩（或碎裂岩）、糜棱岩，局部尚见断层泥，两侧为碎裂花岗岩。下盘断面发育指示上盘左行平移剪切的擦痕构造。糜棱岩化岩石和糜棱岩中，矿物变形强烈，石英矿物强烈波状消光，呈拉长压扁状、透镜体状、豆荚状、甚至是叶片状韧性变形。断裂局部见石英脉充填。受断裂影响，岩石蚀变较强裂，主要为绢云母化、白云母化、硅化、绿帘石化等，靠近强应变部位，黑云母、长石矿物消失，而白云母、绢云母矿物常见，它们含量有所增高，蚀变作用加深、加强。

高峰断裂（F_{120}） 位于高峰—那受—那刀旧址一带，长约27km，宽约5~30m，产状240°∠20°、210°∠70°。构造岩石为构造角砾岩和糜棱岩化、碎裂岩化岩石，并见硅化石英脉、煌斑岩脉、闪长斑岩脉等充填。角砾岩角砾呈椭圆状，直径为0.5~2cm，成分多为二长花岗岩，由长英质及铁质胶结；糜棱岩化、碎裂岩化岩石发育透入性密集挤压劈理，局部形成透镜体化，透镜体长轴与断面平行。煌斑岩脉、闪长斑岩脉普遍发育绿泥石化、黄铁矿化，有时还见有绢云母化现象。沿裂有多处泉水和温泉出露。

落马咋断裂（F_{121}） 见于羊栏村—落马咋红卫大队经济场一带，长约18km，宽约10~40m，北西段产状为60°∠70°，南东段产状为225°∠70°。断面平整、平直，见有擦痕及阶步，指示断裂为平面左行走滑断层，同时上盘伴随有向下的滑落（?）。断裂带中岩石破碎，发育有大量的构造角砾岩发及碎裂岩，角砾呈棱角状，但变质变形不强，局部地段见有硅化、黄铁矿化现象。破碎带内并见宽0.5~1m呈串珠状分布的煌斑岩脉充填。

3. 南北向

南北向断裂带主要集中分布在昌江、儋州和屯昌等地区，前人曾分别命名琼西南北构造带、琼中南北构造带和琼东南北构造带，构造形迹断续分布，单断裂长度一般较短，且以韧脆性变形为主。

（1）红地岭-天安断裂带

断断续续分布于儋州红地岭—昌江天安一带，走向近南北向，由F_{126}、F_{147}、F_{148}组成，全长约48km，切割奥陶系、志留系及华力西期—印支期花岗岩体，其生成时代可能为燕山期，具有压扭性特征。

红地岭一带断裂总体走向近南北向，局部产状 60°∠54°，破碎带宽约 20~50m，构造岩石为构造角砾岩，角砾岩成分为石英岩，大小 3~5cm，呈次棱角状—磨圆状，定向排列，胶结物为硅质。局部地段岩石见糜棱岩化现象。

抱板一带断裂总体走向南北，呈波状延伸，产状 70°∠65°，破碎带宽 1~6m，断面见擦痕。构造岩石有碎裂岩、构造角砾岩和构造透镜体，构造角砾岩角砾多呈次棱角状、次圆状，少数为棱角状，胶结物为硅质。

天安一带断裂总体走向南北，呈波状延伸，破碎带宽 1~6m，构造带内岩石主要是构造角砾岩，局部有闪长岩脉、石英脉充填。构造角砾岩角砾成分为石英砂岩，呈次圆状，少数为透镜状，具有不同程度的定向排列；胶结物为糜棱质及硅质。

（2）芙蓉田-王下构造带

展布于芙蓉田—王下一带，由一组平行状展布的近南北向断裂（F_{128}—F_{134}、F_{149}—F_{153}）组成，走向 5°~15°，单断裂长 6~19km，宽 5~50m。

石碌一带断裂总体走向 5°~10°，倾向东，倾角 50°~80°，破碎带宽 5~20m，断面波状弯曲，具斜向擦痕，构造岩石有碎裂岩、角砾岩、片理和大小构造透镜体（图 21-81），角砾略具定向排列，具有压扭性质。

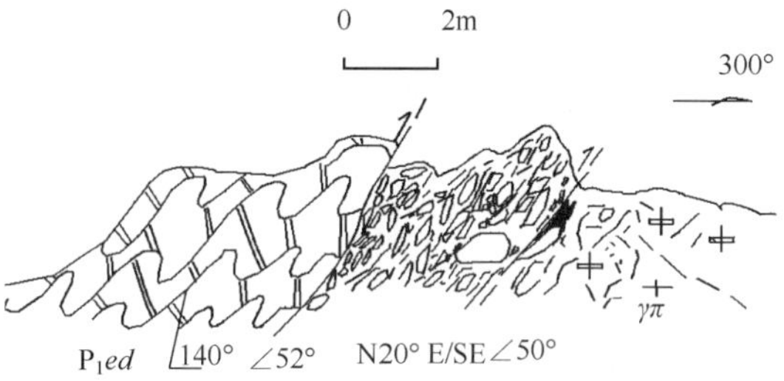

图 21-81 石碌一带构造破碎带素描图
Fig. 21-81 Sketch showing the Shilu structural fractured belt

在石碌水库坝下石碌河谷中，断层带内岩石成分混杂，由石碌花岗岩、石碌水库剪切带内的糜棱花岗岗岩、鹅顶组灰岩以及酸性、基性脉等不同成分岩石破碎混杂而成，在“杂乱岩”带之上的鹅顶灰岩中，常见同斜褶皱，局部可见小型翻卷褶皱（图 21-82）。沿断裂见铅锌矿化现象，并控制了邦溪温泉的出露，显示该断裂不仅是一条控矿断裂，也是一条尚在活动的断裂。

金波农场西部一带断裂带由一组平行状产出的燕山晚期花岗斑岩脉带充填显示，是一条规模巨大的断裂岩浆带，脉体走向近南北，总体呈 S 型弯曲，单脉长几百米至数千米不等，整个脉带延伸长约 32km，脉体在平面中表现为尖灭再现和分枝复合现象，属于燕山晚期生成起来的一条张扭性断裂岩浆带。

用望一带发育断层崖地貌，断层产状变化大，长约 9km，宽约 5m，构造岩石为构造角砾岩，角砾成分杂，有灰岩、砂岩、板岩、石英岩及脉石英，形态为透镜体状、棱角状—次棱角状，岩石局部发生片理化，见铅、锌矿化。

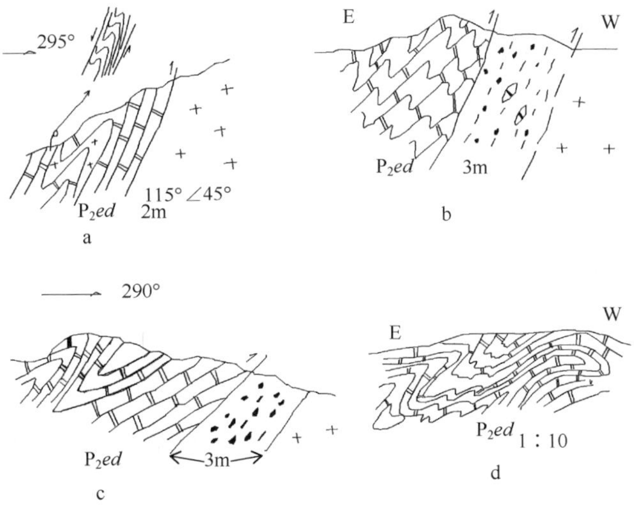

图 21-82 大英山-长岭断裂运动学特征图
Fig. 21-82 Movement characteristics of the Dayingshan-Changling fault

燕窝岭一带断裂带产状 230°~305°∠40°~70°，构造岩石为碎裂岩系列岩石，中心为硅化石英岩、硅化石英角砾岩、强硅化碎裂岩，边缘为碎裂岩、碎裂花岗岩，并见闪长岩脉和石英脉充填。角砾岩角砾形态为棱角状，石英脉常呈雁列状产出，属于张扭性断裂。根据断裂切割二叠系和印支期早期花岗岩体，认为其生成于印支期晚期—燕山期。

坝王岭一带断裂带产状 85°~115°∠75°~85°，长约 15km，宽 40~50m。破碎带岩石由硅化石英岩、硅化碎裂花岗岩、构造角砾岩等组成。构造角砾岩角砾形态为棱角状、次棱角状，碎裂花岗岩见石英脉、煌斑岩脉贯入，石英脉中的石英晶体呈梳状对生。破碎带普遍见绢云母化、绿泥石化、黄铁矿化，伴随铅、锌、铜、银、金等多金属矿化。根据断裂切割印支早期花岗岩体，认为其生成于印支期晚期—燕山期。力学性质可能为张扭性。

（3）洛基-南加六断裂带

分布于儋州市洛基镇—南丰镇南加六村一带，全长30km，走向近南北向，局部走向北北西向，倾向近北北东或南南西，倾角40°~70°，由多条近南北走向的平行状断裂（F_{135}—F_{136}）组成，单断裂长2.5~15km，宽1~11m，局部宽达上百米，甚至上千米。

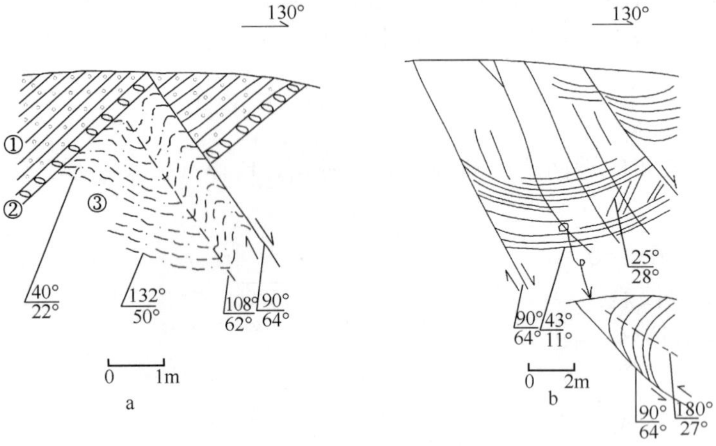

图21-83　洛基东侧南北向断裂形迹素描图

Fig. 21-83　Sketch showing NS-trending fault traces, east of Loki

a—下滑切错NE向断裂；b—上盘形成的次级弧状节理；

① 砾岩；② 构造透镜体；③ 初糜棱岩

洛基镇—两院一带断裂带内岩石变形强烈，以韧脆性破碎为主，形成构造角砾岩、糜棱岩、碎裂岩、碎裂花岗岩等，常见石英脉、花岗斑岩脉充填。洛基镇东侧断裂带表现为密集节理化带，面理倾向近东向。断裂带多期性活动特征明显，早期为左旋压扭性，导致上盘北东向糜棱面理形成轴面产状120°∠27°的牵引褶皱，指示断裂具有逆冲活动性质；晚期活动切错北东向构造（图21-83，图版Ⅴ-3-H），错距约3m，下盘北东向糜棱面理形成牵引褶皱，褶皱轴面与断面近平行，上盘派生一组产状为43°∠11°的密集弧形节理，弧形次级节理并切割早期生成的主断裂，表明断裂沿剖面下滑的同时平面中兼有右旋剪切活动的运动特征。先锋农场北侧，宏观尺度中，断裂带平面上左行切错洛基白垩纪红盆，与断裂带早期活动有关。

在儋州市人民烈士公园入口处，早三叠世花岗岩中发育宽约11m的构造破碎带，破碎带内岩石自西往东由硅化构造角砾岩、构造透镜体、断层泥、构造角砾岩、硅化岩、碎裂花岗岩等岩石组成。构造透镜体成分为硅化岩，长轴与断面呈锐角相交，指示断裂具有逆冲活动特征；硅化构造角砾岩角砾成分为硅化岩、石英脉，呈磨圆状、次圆状，胶结物为钙泥质、硅质，反映了断裂具有两期活动。

两院—南加六一带构造带内岩石普遍发育片理化，形成糜棱岩和糜棱岩化岩石，宽达1~2km，剪切片理产状90°~110°∠70°。由西至东，变形程度总体增强，强变形带和弱变形域交替出现，呈逐渐过渡关系。糜棱岩中，原岩矿物全部发生破碎，碎斑主要为长石、石英，长石碎斑裂纹发育，强波状消光，拉长定向明显；石英呈粒状变晶结构的集合体，有一致的消光位，属粒化重结晶的产物，石英集合体常呈透镜体分布于新生片状矿物、泥质矿物中，除少量碎斑外，长石类矿物大部分变成了绢云母、铁质泥质物。糜棱岩化岩石具有碎斑结构，碎斑成分为斜长石、钾长石，强波状消光，裂纹发育，石英破碎后重结晶形成粒化颗粒，其他被压碎的长石类物质强绢云母化，变形的暗色矿物破碎形成不规则的集合体，呈线状分布，并沿剪切方向形成一些新生绢云母条纹。

根据变形带中全岩K-Ar法同位素年龄测定结果为144~174Ma，且切穿了印支期花岗岩（203 Ma），认为其生成于燕山早期。

（4）石浮-屯昌断裂带

分布于石浮镇—屯昌县一带，走向近南北向，局部走向北北西向，倾向北东东向或南西西向，倾角30°~65°，由多条近南北向的平行状断裂（F_{137}—F_{138}、F_{145}—F_{146}）组成，单断裂长7.5~16km，宽2~15m。

断裂带内岩石比较简单，以碎裂花岗岩、碎裂岩等一套脆性变形岩石为主。北部龙坤村西面山边，于白垩系与上古生界接触处，构造破碎带倾向北东东，自西向东，岩石由绢云千糜岩、构造角砾岩、碎裂岩组成（图 21-84），上盘白垩纪砾岩碎裂化，形成碎裂岩；下盘晚古生代岩石褶皱变形强烈，轴面走向近南北向，由众多平行状分布的小褶皱组成。角砾岩角砾成分为石英，呈次棱角状—次圆状，胶结物为铁质、泥质。

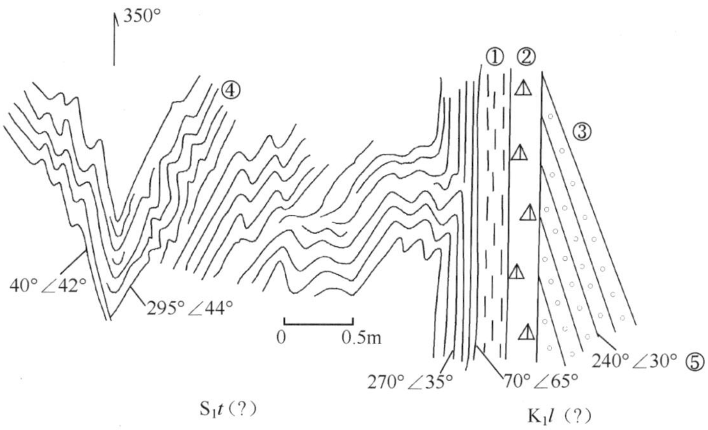

图 21-84 白垩系与上古生界接触处南北向构造破碎带素描图

Fig. 21-84 Sketch showing the NS-trending structural fractured belt at the contact between Cretaceous and Upper Paleozoic

①千糜岩；②构造角砾岩；③碎裂化砾岩；④千枚岩；⑤产状

屯昌县东面，近南北向的断裂非常发育，分布密集，切割长城系抱板群、下三叠统岭文组和中三叠世琼中县单元。断裂倾向 268°~276°，倾角 55°~74°。各断裂宽度变化悬殊，最小 80cm，最大可达 30m。主要以一套脆性岩石的出现和岩脉的充填而反映，包括构造角砾岩、碎斑岩、硅化岩、碎裂花岗岩、碎粒岩等、闪长岩脉、煌斑岩脉、石英脉等。局部呈现韧性变形，形成糜棱岩化花岗岩、糜棱岩化片岩、变晶糜棱岩。构造角砾岩角砾成分为石英，呈棱角状—次棱角状、次圆状，显示挤压作用下角砾岩的形态特点。碎斑岩碎斑成分为长石、石英，呈次棱角状、次圆状；碎基成分为长英质，呈碎粉状，局部被挤压成条带状，发生绿泥石化、硅化等蚀变。碎粒岩中，石英具强烈的波状消光，有的被压扁或拉长呈扁状、透镜体状。糜棱岩化花岗岩中发育眼球残斑右行旋转变形，且见右行剪切揉皱。

黎村旁侧，断裂带内岩石具一定的分带现象，自东向西，依次为碎裂闪长岩、花岗斑岩脉、构造角砾岩、花岗斑岩脉，直至正常岩石。于南斗一带，发育多个轴迹近平行的小褶皱，纵横宽度约 1.7km。各小褶皱轴面近直立，走向近南北向；两翼基本对称，长度也基本相等，翼间角较紧闭。

断裂带控制了早白垩世屯昌岩体的侵位。银水园村西侧，围岩（三叠纪花岗岩）破碎现象明显，形成宽约几十米至上百米的近南北向碎裂岩带。同时，屯昌岩体内部包体呈拉长状沿长轴平行定向排列，长轴走向近南北向，反映了岩体侵位过程中对两侧围岩产生一定的挤压应力作用。

（5）文昌断裂带

分布于文岭—文教一带，由 F_{141}—F_{144}断裂组成。其构造形迹主要见于文昌-南阳一带。走向 0°~2°，倾向 90°~92°，倾角 32°~90°，单断裂长 1.5~18km，宽 0.5~5m。据其切错晚三叠世花岗岩和白垩系鹿母湾组，初步认为其形成于燕山期。

新桥东部—迈众东部，破碎带出露宽约 5m，走向近南北向，产状近直立。构造岩石为构造角砾岩。角砾成分为细粒花岗岩，大小 1~10cm 不等，棱角状；胶结物为铁质。

丰隆村—湖尾村，构造破碎带宽约 50cm，产状 92°∠32°。构造岩石分带明显，自西往东，依次为构造角砾岩、长英质糜棱岩。构造角砾岩角砾成分为石英，呈次棱角状—次圆状，胶结物为砂质、泥质。受该断裂控制派生一组产状为 70°∠34°的压性节理，指示断裂具右旋压扭性特征。

（6）石门山断裂带（F_{154}）

分布于亳角利—石门岭一带，长 8.5km，宽大于 1.0km。走向 330°~350°，倾向北东（或南西），倾角 60°~85°。1：10 万侧视雷达航片中具清楚的线性构造影像，表现为一系列北北西走向近平行展布的低缓带状山脊。切割晚白垩世花岗岩。

断裂平面上由多条密集平行展布的脉岩组成，脉岩类型有石英斑岩、花岗斑岩、闪长玢岩、石英（流纹）斑岩、斜长花岗斑岩等，单脉宽 2~15m，最宽可达 80m。

据前人研究，该断裂至少具两次活动。早期活动较弱，见有石英脉、闪长玢岩和煌斑岩脉充填；于石门岭一带，派生次级近南北向密集平行排列的扭张性小断裂，断面倾向西，倾角约 80°，上盘发育指示上盘以 21°~32°的角度往南西方向斜落的擦痕构造，次级扭张性小断裂中见有石英脉充填，脉带延伸较长，蜿蜒曲折和膨缩变化较大，分枝复合常见，在剖面上呈楔形尖灭。晚次活动引发一系列大致平行展布的石英斑岩、花岗斑岩岩墙侵入充填，岩墙走向较为稳定，但有拐弯转折和膨缩现象，局部，岩墙壁不平整，呈锯齿状，于其两侧，见有构造角砾岩分布，角砾成分为花岗岩、煌斑岩、闪长玢岩（沿早次活动充填的闪长玢岩脉）等，呈次棱角状—次圆状；同时，导致早次派生的次级近南北向扭张性断裂再次发生压扭性活动，其构造岩石具明显的分带：中心为构造角砾岩，并为石英脉充填，两侧则为碎斑岩、碎裂花岗岩，构造角砾岩角砾成分为脉石英，磨圆度好，西盘断面发育指示上盘右行斜冲的擦痕构造。

此外，该断裂南东面邻区芒果头一带的尾端，于其北东盘派生次级右旋张扭性弧形旋扭构造。

岩石蚀变较强烈，见有闪锌矿化、方铅矿化、黄铁矿化、辉钼矿化以及绢云母化、云英岩化、钠黝帘石化、硅化等。蚀变分带明显，自中心往两侧依次为云英岩带→云英岩化带→绢云母化带，表明变形作用与蚀变作用为正相关关系。断裂控制了钼及多金属矿的展布，属一控矿构造。

（7）牙日-南好-狗岭断裂带

断裂带自狗岭向北北东穿过南好，延伸至牙日，长约15km，展布范围宽约6km。主要由才共岭断裂（F_{155}）、南好-足赛岭断裂（F_{156}）、那颂岭断裂（F_{157}）、毛位断裂（F_{158}）组成。切割的最新地质体为燕山期花岗岩，初步认为其生成于燕山期。断裂特征如下：

才共岭断裂（F_{155}）　断裂自狗岭西北侧向北北东延，经才共岭至红水岭，全长约10km多，总体走向北北东15°~25°，呈弯曲状，中段和北段走向近南北。断裂以脉体充填显示，脉体主要为花岗斑岩脉，局部见闪长玢岩脉。中段破碎带岩石为构造角砾岩，见闪长玢岩脉充填。角砾岩成分为板岩和云母石英细砂岩，最大直径为4~5cm的胶结物为糜棱状物质；胶结物及角砾呈显微定向排列。为压扭性断裂。

南好-足赛岭断裂（F_{156}）　分布于南好—足赛岭一带，走向约10°~20°，长约9.5km，破碎带宽约1~5m，产状123°∠64°、93°∠83°、300°∠63°。中段近南北向，沿走向、倾向呈波状弯曲。中段倾向北西，北段和南段倾向南东。主要以花岗斑岩脉、石英斑岩脉充填为特征。破碎带断面光滑平直，见低角度擦痕。岩石为构造角砾岩、断层泥和碎裂石英斑岩脉。角砾岩角砾成分为绢云母石英粉砂岩、板岩、脉石英，多呈次棱角状和次圆状，大小为1~4cm，其排列方向基本上与断层走向一致。断裂左行切错了古生代地层，使得岩层在走向上不连续，错距100m。由此显示断裂早期可能以脉体充填显示，根据脉体空间展布形态，认为属于张或张扭性；晚期发生压或压扭性活动，导致早期脉体发生破碎，形成构造角砾岩、断层泥。

那颂岭断裂（F_{157}）　自草训西北侧向北北东经那颂岭至牙日一带，长约9.5km，宽约1m，产状为295°∠75°。构造岩石为糜棱岩，并见宽约300m石英斑岩脉充填。左行切错燕山期花岗岩与古生界的接触界线，错距200m。为左旋压扭性。

毛位断裂（F_{158}）　北起毛位西侧250m处，南至毛胆北面，长约3.2km，宽大于3m，倾向110°~115°，倾角58°~80°。断面见擦痕，向南西倾状，倾角为44°。岩石为蚀变花岗碎粒岩、构造角砾岩，岩石见密集的挤压片理，其产状与断层平行。断裂切错了花岗岩和变质岩界线，从断裂走向舒缓波状、挤压片理和擦痕看，断裂以压性为主，显示一定的扭动，指示断裂为左旋扭动。

（8）五指山断裂带

分布于五指山东西两侧，总长约58km。由牛奶场断裂（F_{159}）、南流断裂（F_{160}）、什万章断裂（F_{161}）组成。根据其切割的最新地质体为白垩纪岩体，认为其生成时代为燕山期。各断裂特征如下：

牛奶场断裂（F_{159}）　平面呈Z型展布，宽约2m，北部倾向65°，倾角50°。由一组密集的劈理化带和碎裂花岗斑岩脉组成，硅化强烈，石英脉沿劈理充填，局部见褐铁矿化。具有右旋走滑运动特征。

南流断裂（F_{160}）　走向南北，宽2~3m，产状85°∠60°。主要由碎裂岩、碎斑岩及破碎岩块所组成，见硅化、绿泥石化。具有逆冲运动性质。

什万章断裂（F_{161}）　走向南北，宽2~2.5m，产状270°∠80°。主要由碎裂花岗岩、挤压片理化带、碎裂岩和破碎岩块组成，硅化强烈，见褐铁矿化。为压扭性断裂。

4. 东西向

(1) 昌江-琼海断裂带

昌江-琼海断裂带位于北纬19°15′~19°20′左右，横贯昌江、东方、白沙、琼中、屯昌和琼海等县市，东西长约200km，由F_4—F_8等东西向断裂带组成，由于受后期构造作用和岩浆侵入活动的改造，使得构造带呈断续状分布。

1）断裂带在航磁异常上表现为由一系列串珠状局部异常构成一条近东西向的异常带，为一正负异常交接带，其北侧为负异常，南侧为正异常，引起这些局部异常的原因是断裂构造复合改造使得磁性矿物沿应力集中地段富集所致，反映为明显的断裂带特征。在重力异常上，断裂带正好与东西向重力异常相吻合，其北侧为正异常，南侧为负异常，也反映出明显的断裂带特征。在航片、卫片上，兰洋、新兴等地也可见清晰的线性构造。

2）沿断裂带分布有大量近东西向的燕山期花岗岩，组成一条近东西向展布的花岗岩穹隆。此外，沿断裂带还见幔源玄武岩浆分异的华力西期闪长岩体出露。

3）断裂带展布区，尤其是断裂带中段，东西向断裂十分发育，断裂带中挤压破碎现象非常强烈，常见构造透镜体发育，具有压性断裂特征。岛西粉车河一带发育一组平行展布的断裂，断裂带岩石主要为构造角砾岩和硅化岩（图21-85），局部地段见构造片岩、糜棱岩、碎裂岩分布。构造角砾岩角砾形态以次棱角状、次圆状为主，显示了压应力作用下角砾岩的形态特点。构造片岩中，石英具明显的波状消光，长轴与铁质聚集条带呈平行或近平行排列，与构造方向一致。昌江长塘村一带的构造破碎带倾向南，破碎带内岩石主要发生韧性变形，形成糜棱岩化系列岩石，中心为糜棱岩或强片理化花岗岩，往两侧依次为初糜棱岩、碎斑岩（图21-86），显示由中心向边部变形程度逐渐减弱。强片理化花岗岩中的石英矿物发生塑性变形成条带状、碎斑状，有时与片状矿物一起形成δ型旋转碎斑系和S-C构造，具有压扭性的特点。东方二甲一带构造破碎带向北倾，由断层泥、硅化岩、构造角砾岩和碎裂岩组成（图21-87），角砾岩角砾成分为板岩、花岗岩，呈次棱角状、透镜体状，胶结物为粉砂质、泥质、硅质。左行切错北东向戈枕构造带，错距约300m。

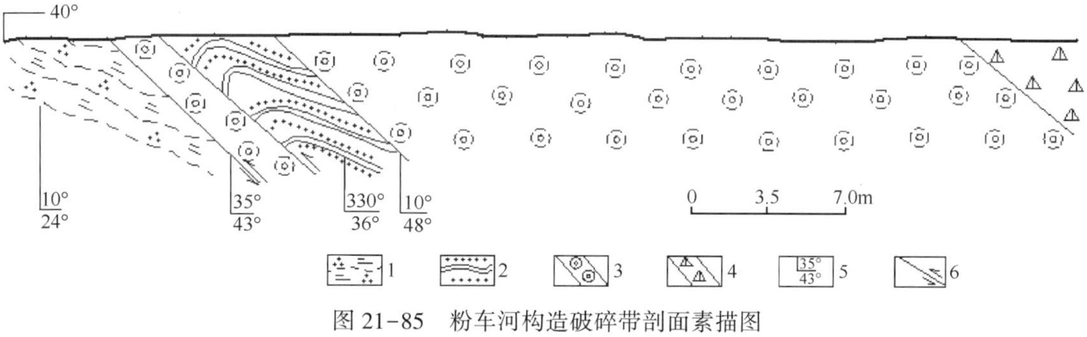

图21-85 粉车河构造破碎带剖面素描图

Fig. 21-85 Sketch showing the section of the Fenchehe structural fractured belt

1—石英绢云母千枚岩石；2—变质粉砂岩；3—硅化岩脉；4—构造角砾岩；5—产状；6—断裂运动方向

昌江-琼海断裂带可能形成于燕山期，表现为压性特征，导致强烈岩浆侵入活动，最终形成巨大的东西向花岗岩穹隆。

(2) 尖峰-吊罗断裂（F_{11}）

尖峰-吊罗断裂带位于北纬18°40′~18°52′，横贯乐东、保亭、陵水河万宁等县市，东西长约190km，两端延伸入海。构造带内各种压性或压扭性构造形迹十分发育。

1）大尺度构造形迹

a. 断裂带展布区航磁异常表现为东西向带状异常，其北侧为负异常，最低峰值可达-400nT，称为万宁-尖峰岭剧烈变化的低异常带，在此异常背景上又叠加了抖动的和不规则的次级异常，被认为是“基底破碎带”或“深部早期发育的构造破碎带”引起的。沿构造带的重力异常则表现为重力低值带。

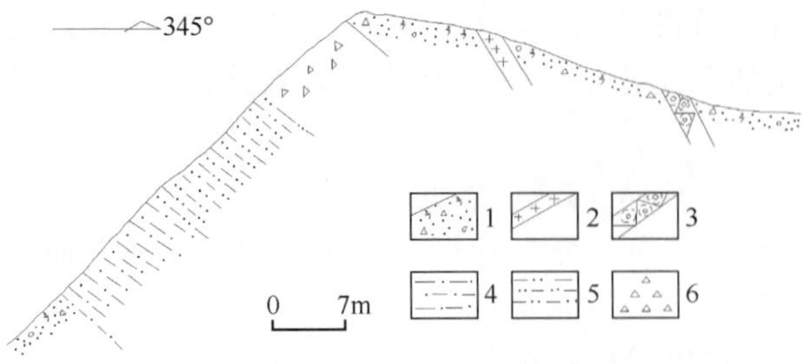

图 21-86 长塘村断裂剖面图

Fig. 21-86 Section of Changtangcun fault

1—第四系；2—细晶岩脉；3—硅化岩脉；4—糜棱岩；
5—初糜棱岩；6—碎斑岩

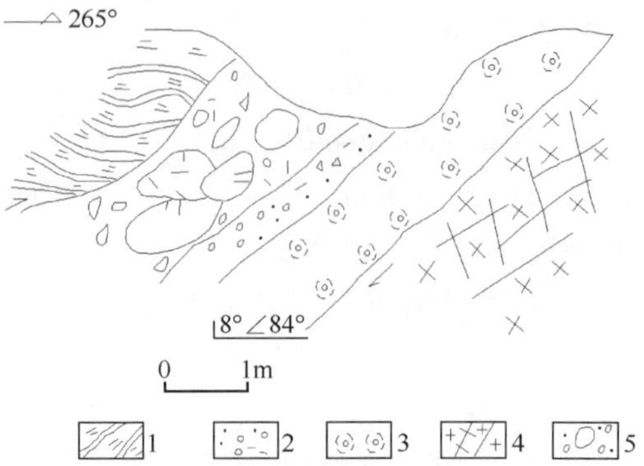

图 21-87 二甲断裂破碎带素描图

Fig. 21-87 Sketh showing the Erjia fault fractured belt

1—板岩；2—断层泥；3—硅化岩
4—碎裂花岗岩；5—构造角砾岩

b. 断裂带中分布有大量燕山期花岗岩，组成一条巨大的东西向花岗岩穹隆构造带，岩体多呈东西向分布于断裂带中。由此可见，尖峰-吊罗断裂带在燕山期都发生过强烈的活动，控制了同期岩体的侵位。

2）小尺度构造形迹

在断裂带展布范围内可见大量压性或压扭性构造形迹发育。邦岭农场公路边燕山期花岗岩中见挤压破碎带发育，破碎带走向东西向，倾向北，倾角约85°；万宁市兴隆农场合口桥下见华力西期花岗岩中斑晶定向排列，定向方向为东西向，并见东西向挤压破碎带和密集片理化带发育；尖峰岭林场坤岭北西侧的陈龙沟中见5条东西向的破裂面构成宽约20m的挤压破碎带；尖峰岭林场六连东北见一走向东西向、倾向南、倾角约73°的断裂，断裂面上擦痕显示断裂具有上盘向上斜冲的压扭性断裂特征；大凯林场-陆岭一带以及南崖林场一带常见东西向挤压破碎带发育，断面上的擦痕均显示上盘向上斜冲。上述现象说明，尖峰-吊罗断裂带在燕山期有强烈活动，并表现为压性或压扭性断裂特征。

（3）九所-陵水断裂带

九所-陵水断裂带位于北纬18°25′~18°35′，前人认为是海南岛地体南北分野断裂，该认识自提出以来一直占据着海南岛大地构造单元划分的主导地位。但对其分布位置有不同的认识，《广东省区域地质志》中该断裂带位于崖城-陵水一带，走向北东东；而《海南岛地质之三构造地质》中依据重、磁资料把断裂带位置置于九所-陵水南面一带，走向东西向。1∶25万《乐东县、陵水县》幅中依据芒三水库一带的韧性变形岩石，把其位置定为九所-陵水一带，走向北东东向。本次通过野外观测，发现芒三水库一带韧性变形岩石走向为北东—北北东向，而不是北东东向，因此把芒三水库一带出露的韧性变形岩石作为九所-陵水断裂带的构造形迹依据不足。经沿该韧性变形岩石走向进行野外追索，在其北东向吊罗山林场南侧、兴隆北面等地区都有韧性变形岩石出露，依此有理由认为芒三水库-兴隆一带存在一条北东向韧性剪切带，属于区域性主干构造带——崖城-万宁断裂带的构造形迹表现。本次以重、磁资料为依据，结合地表构造形迹，把九所-陵水断裂带分布位置重新恢复至《海南岛地质之三构造地质》所认定的位置，即位于九所-陵水南面一带。

断裂带横贯乐东、三亚和陵水等县市，由F_{12}—F_{14}断裂组成。东西总长100余千米，构造带建立的依据主要有重磁异常、古地磁和近东西向的构造形迹。

1）大尺度构造形迹

a. 沿断裂带分布有短促、稳定的带状磁异常，异常轴向呈东西向，其正异常强度远远大于海南岛中部和北部地区。有的异常比较平缓圆滑，有的梯度则变化很大。异常带内中基性岩脉和燕山期侵入岩较发育。带内的侵入岩与喷发岩的磁场混在一起，在空间分布上又与东西向断裂带联系在一起，因此，它们同受九所-陵水断裂带控制。沿构造带呈东西向带状分布的特殊的磁异常应是各种地质体组成的巨大的花岗岩穹隆构造带的综合反映。

b. 沿断裂带布格重力异常也呈东西向分布，梯度较陡。莫霍面的变化形态也和布格重力异常有着明显的对应关系。在莫霍面等深度图上表现为幔坡区与幔凹区的过渡带，北面为幔凹区，南面为幔坡区。重力异常在此带上也表现为梯度变化带与重力低的过渡带，深部构造的这些特点反映了隐伏断裂带的存在。且断裂带北侧莫霍面的埋深较深，而南侧埋深较浅，北侧属于大陆型地壳，南侧属于过渡型地壳，由此推断构造带两侧的基底是不同的。

2）小尺度构造形迹

小尺度上的构造形迹主要表现为构造带展布区内断裂带、挤压破碎带十分发育。并常切割白垩纪地质体和晴安岭，属于燕山期生成起来的压扭性东西向构造形迹。主要有：荔枝沟断裂、马岭断裂等。

荔枝沟断裂（F_{13}） 见于荔枝沟—死马岭一带，长约10km，宽度约3m。构造岩石从中心往两侧大致可分为：①断层泥砾岩带，由构造角砾岩、磨砾岩、碎裂岩、碎斑岩和断层泥砾组成；②挤压片理带，由片理化细粒石英砂岩显示。

马岭断裂（F_{14}） 展布于南山、红塘岭和马岭一带，全长24km，走向80°~90°，总体往南倾，局部反倾，倾角78°，断裂大部分被第四系所覆盖，仅见一些零星露头，破碎带宽约10~20m，局部可达50m，主要由挤压破碎带、糜棱岩带、碎裂岩带组成，沿断裂走向沿线具有压碎岩、碎裂岩、糜棱岩等构造岩滚石分布，地貌特征明显，为一近东西向狭长槽地。在露头点，可见断裂分带现象（图21-88），一般可分为三带：①断层泥砾岩带，宽约60~100cm，由磨砾岩、碎粒等混杂在一起，砾石大小混杂，呈现圆状，大小在2mm×3mm，略呈定向排列，其方向与断裂走向一致，显示压扭性特征；②片理化带，宽约20~30cm，岩石呈糜棱状，挤压明显，矿物具定向排列，整个片理化带呈波状弯曲；③碎裂岩带：靠近内侧，压碎、碎裂现象明显，向外减弱，主要为发育裂隙、节理、劈理、石英脉等次生构造，原岩特征基本能辨认。带内硅化强烈，局部见绿泥石化现象，并有基性岩脉贯入。

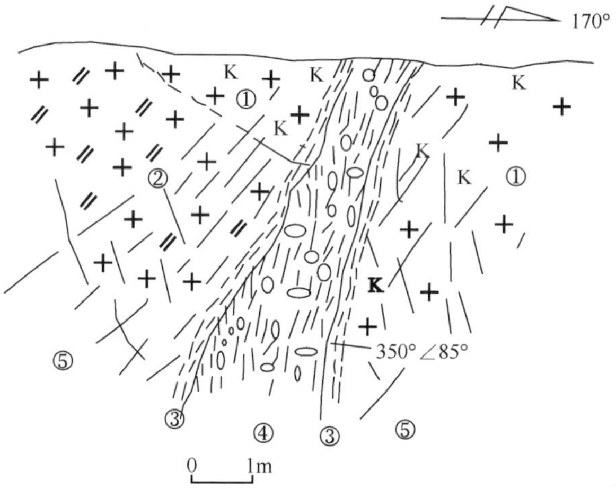

图21-88　马岭断裂剖面素描图

Fig. 21-88　Sketch showing the Maling fault section

①细粒钾长花岗岩；②中粒二长花岗岩；③挤压片理化带；④磨砾岩带；⑤裂隙节理带

第二十二章 新构造运动特征

第一节 新构造运动表现

在喜马拉雅运动的作用下，海南岛新近纪以来的新构造运动活跃且频繁，表现样式多样。在地壳间歇性上升的背景下，发生了强烈的断裂活动，在琼北形成了火山岩台地和火山锥地貌，以及沿海岸带的海积地貌与河流两岸的河积地貌景观。受东西向王五-文教断裂控制，断裂带南北两侧新构造运动的表现形式不同，北面琼北断陷区表现为地震活动和剧烈的火山喷发，形成东西向展布的玄武岩被，南面五指山隆起区则主要表现为地热活动，见多处温泉出露。新构造运动迹象主要有地壳升降运动、断裂活动、地震活动、火山活动、地热活动等。根据近年来的形变测量结果显示，该时期的形变轴有近东西向、北东向和北西向，以近东西向最多，规模大，组成了相间排列的抬升区和沉降区（张虎男等，1984）。

一、地壳升降运动

地壳升降运动总体具有间歇性上升运动的特征，导致新近系与第四系间以及第四系内部存在多个平行不整合接触。新近系与第四系间的平行不整合面是一次重要的地质事件，新近系除了局部见到火山岩出露地表外，大部分被第四系“超覆”而呈隐伏状产出，反映了新近纪—第四纪早期，地壳的沉降幅度和海侵规模最大。

阶地是地壳升降运动的一种表现形式，海南岛可见海积阶地与河积阶地发育。海积阶地分布于沿海一带。地壳差异升降运动导致上新统与更新统北海组、八所组间普遍存在假整合接触，中更新统北海组分布于滨海平原区和山前平原区，成为海成三级阶地，标高大于20m；上更新统八所组则分布海成二级阶地上，标高在5~15m；全新统烟墩组分布于沿海地带，为海成一级阶地，并可分成一级沙堤沉积（标高3~13m）与一级阶地海漫滩沉积（标高0~2m）。

河成阶地见河流两侧。东泰农场场部附近的万泉河边，地貌上为河成一级阶地，相对高差0~5m；南渡江边的一级阶地高出海平面2~15m（海南省地质矿产勘查开发局，1989）。在昌化江、珠碧江、南尧河河谷两侧发育两级河流阶地，一级阶地高出河床5~15m，二级阶地高出河床15~72m，上升总幅度约70m。昌化江流经单一的花岗岩地区，出现二级跌水，第一级跌差25m，第二级跌差15m，在200m范围内总跌差为40m。

在南洞岭等灰岩出露地区发育两层水平溶洞，溶洞距河床高度分别为10~25m、65~85m。峨沟岭等地区分布有海蚀岩，其海拔分别为15~20m和75~95m。峨蔓一带的海滩岩（^{14}C年龄为20070~22540a）现已被抬升到2.5~6.3m的海拔（吴泽龙，1985）。昌洒东部海岸边沙堤上见数层海滩岩分布，单层海滩岩厚度可达数米，出露标高约20~60m，其形成年代为距今1000~3000a之间（吴克刚等，1988），年均抬升约6~60mm。铜鼓岭南部石头公园一带，见第四系烟墩组灰黄色粗中砂被抬升高出海平面达50余米，并在同样高度的花岗岩岩体上见大量海蚀穴、海蚀龛、海蚀槽等海蚀地貌。峨沟岭等地区也分布有海蚀岩，其海拔分别为15~20m和75~95m。

二、断裂活动

海南岛第四纪以来的断裂活动主要发育在琼北地区，其他地区的断裂活动主要表现为老断裂的继承性活动。断裂活动，尤其是老断裂的继承性活动，直接控制了地震活动、火山喷发和地热活动的发生。根据已有区域地质调查资料，琼北地区的活动断裂主要有王五-文教断裂带的再次活动以及北西向、北东向断裂活动。

王五-文教断裂带（F_1—F_3） 走向近东西向，控制了琼北断陷盆地的南界和琼北新生代多期次的火山喷发活动。断裂带以北接受了厚300~3000m的新生代沉积，并形成了东西向展布的玄武岩被，是1605年琼州大地震的主要控震断裂。

公堂断裂（F_{15}） 位于三都公堂—峨蔓一带，走向大致55°，属于物探推测断裂，在重力二次导数图上，断裂北西侧为走向北东的局部正异常，南东侧为走向近东西的局部负异常。断裂总体表现为10~130γ断续延伸的线性正异常带，沿走向，见有火山口分布，控制了更新世火山岩的喷发和喷溢。受该断裂作用，导致公堂岭北面海岸中新近系海口组由于抬升而直接出露。

白坟-黄烟断裂（F_{16}） 位于干冲白坟—黄烟一带，总体走向75°，重力资料显示为一条重力高和重力低的异常密集梯度带，为雷琼凹陷的次一级临高隆起与峨蔓凹陷的分界断裂。电测深资料推断断裂倾向北西，属于正断层。人工地震测深揭示，它由2~3条断层组成，白垩系被错断，错距达400~600m。

颜春岭-道崖断裂（F_{88}） 走向大致320°，为花场隆起和白莲凹陷的分界断裂，在更新世发生过多次活动，沿断裂有一系列火山口呈线状分布。

荣山-岭南断裂（F_{89}） 走向大致324°，据重力资料显示，断裂两侧存在有相当大的东升西降的落差，为白莲隆起和云龙凹陷的分界断裂，控制新生代地层的沉积作用和第四系多次火山喷发，沿断裂见线状火山口分布。

琼华-莲塘村断裂（F_{90}） 走向大致330°，在地貌上，永庄水库走向与断裂一致。控制第四系多次火山喷发，沿断裂见线状火山口分布。

南渡江断裂（F_{125}） 走向近南北向，控制了河流阶地的形成和上更新统道堂组火山岩第三段的分布。

三、地震活动

海南省邻近著名的环太平洋地震带，是地震活动相对活跃的地区。据1463~1834年间不完全统计，琼山、定安、澄迈、文昌等地区发生过37次有感地震，5级以上的地震有7次（吴泽龙，1985），多集中在海口至文昌一带，呈近东西向分布。1605年琼州大地震，是海南岛有记载以来震级最大的地震，震中位于东部塔市附近，发震断裂主要是近东西向的光村-铺前断裂，其次是北西向断裂，两断裂的交汇处即震中（陈惠民等，1987）。1988年的文昌县潭牛镇附近的2.5级有感地震，其震中位于北西向铺前断裂和东西向王五-文教构造带的交汇处。由此表明，区内地震活动与东西向和北西向断裂有关，结合地震等值线呈东西向展布的特点，显示东西向断裂起着发震、控震构造的作用。依据广东省地震局等（1986）所绘制的《海南省地震烈度区划图》（图22-1），海南岛的地震烈度具有北高南低的特点，南部地震烈度为Ⅵ级或Ⅵ级以下地震活动不活跃；北部地震烈度为Ⅶ、Ⅷ级，尚处于地震活动的活跃时期。

从海南岛的历史地震震中分布图上可以看出，其历史地震震中的空间分布表现为北强南弱，琼北凹陷区历史地震和有感地震相对密集，岛内所有历史强震几乎都发生于琼北凹陷内，而现代地震活动则主要集中于琼南及其邻海一带（图22-2）。

南海地区缺少历史地震资料，地震台网的布设也难以捕捉小震和弱震的有关信息，根据《南海地质》（刘昭蜀等，2002）及有关资料，1900年以来，南海地区6级以上地震不多，地震活动相对平静，6级以上地震主要分布于我国台湾—菲律宾一线。

统计资料显示，南海及周缘地震绝大多数属于浅源地震，少数为中源地震，个别为深源地震。平面分布规律是自西向东、由陆向海深度增大。南海及其周缘大于6级的地震大多数沿板块边缘分布，集中分布于欧亚板块与菲律宾板块相互作用地带，属于板缘地震活动。少数分布于海南岛和南海地区，

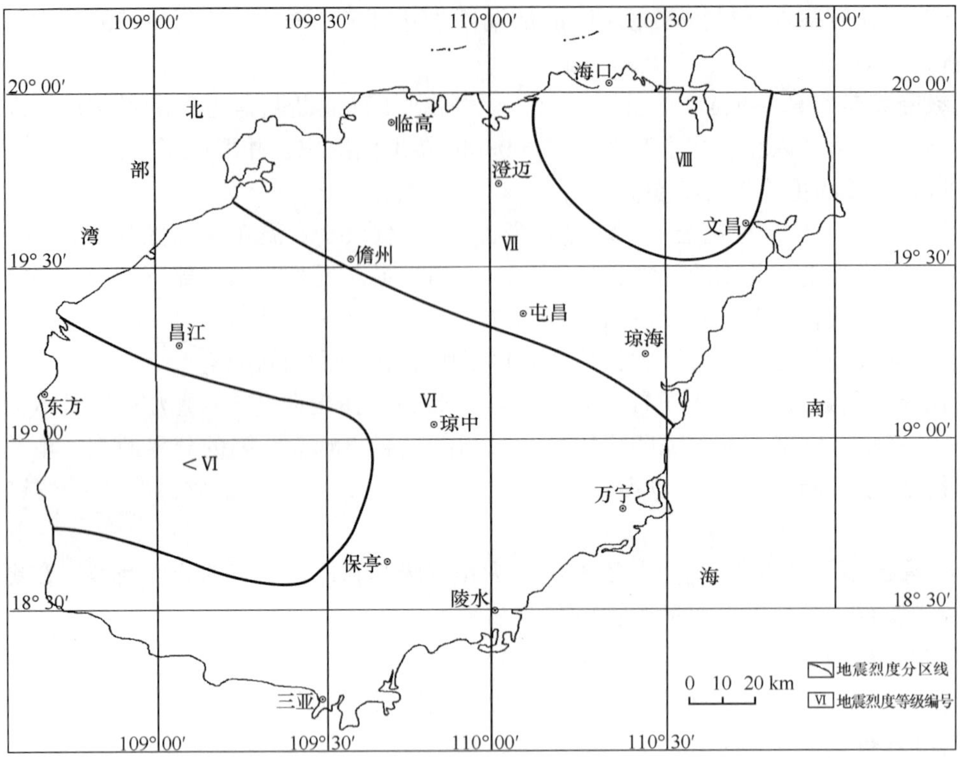

图 22-1 海南岛地震烈度区划图

Fig. 22-1 Map of division of seismic intensity of Hainan Island

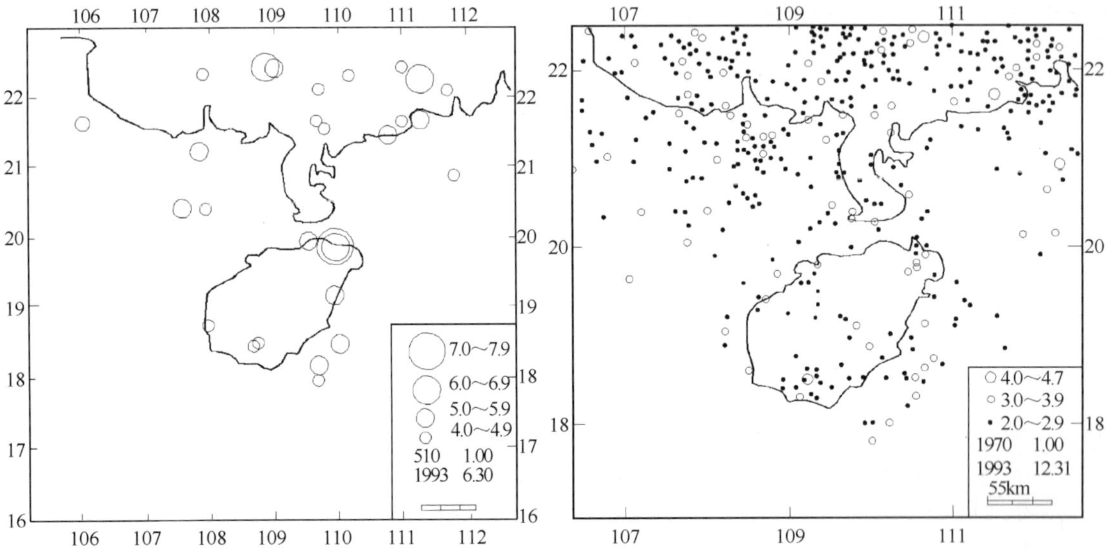

图 22-2 海南岛及邻区历史地震强震震中及区域小震震中分布图

Fig. 22-2 Map of distribution of epicenters of seismic intensity and regional small earthquakes in the Hainan Island and its adjacent area

属于板内地震活动。板内地震活动主要与北西向断裂的活动有关。在震源深度30km以内，由北西向断裂引起的强震（≥6级）与切割较深、规模较大的北西向或北东向断裂有关；中强震（≈5级）则与切割深度、规模均较小，但近期活动明显的北西向断裂有关；小震则具有明显的随机性。从某种意义上说，北西向断裂是地震活动的“能动断层”（刘昭蜀等，2002）。

根据刘昭蜀等（2002）编制的南海及周缘地震区划图，海南省划归板内地震活动区，地震活动的特点是强度小、频度低、震源浅，震中分布随机性大。其中海南岛周边海域属于南海北缘地震活动亚区，地震活动较活跃，有7级以上地震和多次6级以上地震。海南岛以南的广大南海海域属于相对平静亚区，该区虽然地质构造复杂，地震活动却相对平静。即使在超壳断裂上也只有一次6级地震。

四、火山活动

火山活动发育在王五-文教断裂带以北，火山活动频繁，具有多期次喷发的特点。就其分布规模大小而论，从早到晚，火山活动呈现出弱—强—弱、并趋于衰竭的特点。最早的火山喷发活动发生在中新世—上新世，形成了呈半环状分布的石门沟村组、石马村组火山岩。上新世火山活动减弱，以夹层状产于海口组中。中、晚更新世火山喷发活动达到高峰，呈中心式群体喷发，形成了多文组、道堂组大面积东西向展布的玄武岩被，火山口呈北西向排列，反映了在东西向构造控制背景下，沿北西向断裂喷发的活动特征。全新世，火山活动具有局部性，仅发生于峨蔓、石山一带，呈中心式群体喷发，其火山口排列方向亦为北西向，明显受北西向断裂的制约。

五、地热活动

地热活动的直接表现形式是发育温泉。海南岛的温泉众多，分布范围广，儋州兰洋农场、琼海官塘、白沙邦溪-光雅、东方新街-大田、昌江七差、三亚南田农场、陵水南平农场-英州和万宁兴隆农场等地均见有温泉出露，其中的儋州兰洋温泉、琼海官塘温泉、万宁兴隆温泉及三亚南田温泉已等以成为旅游胜地。

兰洋温泉发育在印支期花岗岩与石炭系接触部位，流量大、水温高、延伸远，沿北西向断裂呈线状延续延伸达3km多，水温高达80℃以上，流量3.92L/s，与区域性北西向儋县-万宁断裂带有关。

官塘温泉发育在燕山期花岗岩与白垩系的接触部位，水温48℃~58℃，流量0.3~3.31L/s，前人认为与北东向文塘断层有关（海南地质大队，1981），海南省地质调查院在开展1∶25万区域地质调查时发现该地区北西向断裂非常发育，而北东向构造痕迹未见出露。北西向断裂密集发育，呈带状分布，具多期活动，早期表现为石英脉充填，晚期活动导致石英脉破碎，形成硅化破碎带，由此初步认为官塘温泉可能与北西向断裂有关。

兴隆温泉发育在印支期花岗岩与燕山期花岗岩接触部位，有10多个泉眼，水温常年保持在60℃以上。前人认为与东西向九所-陵水断裂带或北东向断裂带有关。海南省地质调查院在开展1∶5万区域地质调查时发现兴隆农场一带北西向断裂十分发育，其他方向构造形迹极少见，因此，初步认为兴隆温泉与北西向白沙-陵水断裂带有关。

此外，海口地区存在的中、低温热水层，因其地温梯度小，为3.58℃/100m~4.89℃/100m，接近正常的地温梯度，认为是在比较正常的地热增温情况下形成，地热来源推测与活动断裂有关的火山喷发残留的余热有关。

上述特征表明海南岛尚处于断裂活动和地热活动非常活跃时期。

第二节　新构造运动特征

1）在地壳间歇性上升运动的背景下，差异性升降运动明显。据1970~1974年大地水准测量，海南岛东西向王五-文教构造带以北地壳年均上升1.4mm，而其南面地壳年均上升3.2mm，显示了差异运动的存在。另重复水准测量资料表明，1957~1969年海南岛西部沉降的最大幅度为105mm（速率8.8mm/a），东部上升最大幅度为47mm（速率3.7 mm/a）。这种西降东升的活动特点在海南岛布格重力异常图中，体现为由西至东地壳厚度从33km逐渐变小到31km。据钻孔资料，区内北部干冲镇一带，古近系直接盖于长城系之上，而南面海头镇一线地区，缺失古近系、新近系，第四系直接覆盖在华力西期—印支期、燕山期花岗岩之上。差异性升降运动作用的结果，并形成了雷琼断陷、琼东南盆地、莺歌海盆地等多个沉降区，沉降区内接受了古近纪以来的沉积，是海南岛重要的油、气资源的产出层位。

2）新构造运动强度具有不均一性，具有北强南弱的特点。以东西向王五-文教构造带为界，断裂活动、火山喷发和地震活动主要分布在该构造带北面的琼北断陷区，新构造运动强烈且持久，具有多期次性，琼北凹陷区历史地震和有感地震相对密集，岛内所有历史强震几乎都发生于琼北凹陷内。1605 年琼州大地震是海南岛有记载以来震级最大的地震，震中位于东部塔市附近，与东西向王五-文教构造带密切相关。现代地震活动主要集中于海南岛南部及其邻海一带，其中较大的地震活动包括 1969 年 12 月海南岛东南海域的 3.1 级地震群、1982 年崖县的 4.0 级地震、1992 年东方的 3~4 级地震群和 1988 年北部湾的 5 级地震等。研究表明，海南岛地震均属浅源地震，震源深度北深南浅，变化于 12~17km，地震动峰值加速度为 0.05~0.3m/s^2，由南至北，逐渐增大。而该构造带南侧的五指山隆起区则相对较弱，主要是地热活动，见多处温泉出露。

第二十三章　区域地质发展史

海南省地处特提斯构造域、西太平洋构造域和欧亚大陆板块构造域的交合部位，具有多阶段、多旋回的演化。即其现有的地质记录而言，其地质发展历史可追溯至中元古代。在漫长而复杂的地史演化过程中，伴随各时期不同的地质事件形成了一系列不同类型的沉积建造、变质建造和构造相等。总体上可概括为中元古代结晶基底形成、新元古代—古生代板内多阶段裂解与造山作用、中生代陆内盆-山演化和新生代裂解沉陷4个重要阶段（表23-1）。

第一节　中元古代结晶基底形成阶段

该阶段是海南岛基底形成的主要演化阶段。约18亿年，地壳发生裂解，沉积了一套陆缘弧相的火山-碎屑岩，其下部为戈枕村组钙碱性系列的英安质和石英安山质中酸性火山岩；上部峨文岭组原岩为一套成熟度较低的陆缘碎屑岩夹陆缘弧相中酸性火山岩，发育类复理石沉积韵律。经区域变质作用改造，形成下部戈枕村组混合岩类-片麻岩类岩石组合和上部峨文岭组石英岩类-片岩类岩石组合。

另外，据任纪舜报道（1984），西沙群岛永兴岛钻孔中见花岗片麻岩和石英云母片岩，片麻岩的矿物Rb-Sr等时线年龄为1465Ma。广东地矿局认为（1988）1465Ma代表了岩石的变质年龄。该变质岩与抱板群岩石可以对比，变质时代与抱板群变质时代也基本一致，马大铨（1998）认为甚至就是抱板群。由此看来，海南省中元古代可能具有统一的基底。此外，刘昭蜀等人（2002）在《南海地质》中描述：印支断（地）块基底为前寒武纪变质岩，下部为深变质的角闪石、石榴子石、黑云母副片麻岩，夹角闪岩，有时有钙碱性正片麻岩；上部为一套中等变质的二云母硅线石片岩、石榴子石片岩及石墨片岩夹数层石英岩、大理岩及角闪岩。其岩石组合特征与抱板群相似，暗示了中元古代时期，海南岛、西沙群岛、印支地块属于统一的板块范畴。

发生于约14亿年的抱板运动（混合岩化片麻岩和侵入其中的片麻状花岗岩中获得锆石SHRIMP年龄约为14亿年），使得地壳物质局部发生重熔，形成以片麻状二长花岗岩为主体岩性的壳源型钙碱性花岗岩体，这些规模不等的花岗岩体总体与中元古界围岩一同经历了强烈的韧性变形改造。导致早期沉积发生了角闪岩相-麻粒岩相的区域动力热流变质，下部并发生混合岩化。但这期造山作用事件，对海南省不同地区的改造程度不一。如琼西地区、西沙群岛主体为角闪岩相；琼中地区中元古界局部达麻粒岩相，并相伴发育了与其具有成因关联的紫苏花岗岩侵入体。经过中元古代的花岗岩焊接和变质变形再造，塑造了以高温变质和高度塑性变形为特色的海南岛中元古代结晶基底，海南岛基底陆壳宣告形成。

由于中元古界出露面积小，主要呈残块状“漂浮”于花岗岩之中，受后期地质作用改造强烈。因此，很难追寻1400Ma造山事件的动力学背景。目前资料显示，中国境内、尤其是华南地区元古宙造山带形成时间集中于1.0~0.8 Ga（李江海等，1999；Li X H，1999），缺乏1400Ma造山作用的记录（与陆松年老师交谈）。而劳伦大陆南部（北美西部）普遍存在中元古代构造-岩浆活动事件（Nyman et al.，1994），Condie et al.（2005）还认为产于劳伦大陆中南部和巴西西部中元古代花岗质岩（1.50~1.30 Ma）代表了一次地壳增长事件。Li Z.-X et al.（1995，1996，1999，2002）并认为是连接南极洲、澳大利亚东部、劳伦大陆西部的纽带，位于Rodinian超级大陆的核心位置。

表 23-1 海南省大地构造演化阶段特征表

Table 23-1 Characteristics of geotectonic evolution stages of the Hainan Province

构造旋回	构造阶段	构造属性	沉积作用特征	火山作用特征	侵入作用特征	变质作用特征	变形构造特征	成矿特征	造山作用时间
喜马拉雅山期	新生代裂解沉陷阶段	离散	盆地内砂砾岩、砾岩、砂岩、泥岩、页岩和沿海一带松散碎屑沉积	裂谷型基性火山岩				风化型重晶石矿、铝土矿、钴土矿、镍矿轻稀土矿和沉积型褐煤矿、油页岩矿、锆英石砂矿、钛铁砂矿、石英砂矿、独居石砂矿	
燕山期	中生代陆内盆-山演化阶段	离散	盆地内砂砾岩、砾岩、砂岩、泥岩沉积	双峰式火山岩	板内离散环境壳源型、壳幔混合型钙碱性花岗岩，及幔源型辉长岩、闪长岩等	角岩相接触变质	线形韧性、脆性断裂，并发生由北西往南东的逆冲推覆	热液型金、银、钼、钨、铜、铅、锌、萤石矿和接触变质型铁矿等	
华力西期—印支期	晚古生代特提斯构造域演化	汇聚	陆内盆地砂砾岩、砾岩、砂岩、泥岩等类磨拉石、磨拉石建造沉积		壳源型、壳幔混合型钙碱性花岗岩，及幔源型辉长岩、闪长岩等	北部地区发生了低绿片岩相区域动力热流变质	北部地区发育中深-中浅构造相线理、面理和剪切流变褶皱及弧状构造带，南部发育逆冲推覆构造	热液型重晶石矿	290～270Ma
	新元古代—早古生代板内多阶段裂解与造山作用	离散	海相砂岩、泥砂质岩、泥岩夹碳酸岩沉积						
加里东期		汇聚？				发育深构造相线理、面理和剪切流变？	发育深构造相线理、面理和剪切流变褶皱？		下志留统？
		离散	海相砂岩、泥砂质岩、泥岩沉积	北部地区出露板内基性火山岩、岛弧中酸性火山岩		绿片岩相区域动力热流变质		沉积型锰矿、磷矿	
晋宁期—扬子期		汇聚	残坡积相-滨海相含铁质类磨拉石建造沉积			高绿片岩相区域动力热流变质	发育深构造相线理、面理和剪切流变褶皱	沉积变质型铁、铜、钴、镍矿（共生）	大致800Ma
		离散	滨、浅海相泥砂质夹铁质层和碳酸岩层沉积	陆缘弧钾质中酸性火山岩	斜长角闪岩（脉）？				
抱板期	中元古代结晶基底形成阶段	汇聚			壳源型钙碱性片麻状二长花岗岩、紫苏花岗岩	角闪岩相、麻粒岩相区域动力热流变质	以发育深构造相的新生线理、面理为特点	沉积变质型石墨矿、铁矿，伟晶岩型白云母矿	1400Ma
		离散	海相泥砂质岩、砂岩和少量富钙岩石等陆缘碎屑沉积岩	陆缘弧相钙碱性系列英安质和石英安山质的熔岩、凝灰岩类等中酸性火山岩					

第二节　新元古代—古生代板内多阶段裂解与造山作用

一、新元古代板内裂解与造山作用

新元古代海南省基底陆壳发生裂解，海南岛北面石碌地区局部形成一个陆缘盆地，作为裂解事件记录的建造序列仅见于昌江石碌地区，以青白口系石碌群为代表。为具类复理石韵律结构的碎屑岩、碳酸盐岩沉积，其中夹有少量的酸性火山岩。从早至晚，沉积环境由浅海→海湾潟湖相。火山岩有钾流纹岩、钾质流纹斑岩、钾粗安质球粒状凝灰岩、钾流纹质凝灰岩。发生于距今800Ma左右的晋宁运动，导致石碌群发生低绿片岩相区域变质作用和固态流变，成为堆叠于抱板杂岩结晶基底之上的变质基底。这一构造事件在华南地区也有显示，也即晚晋宁运动波及范围较广，此时海南岛可能与华南大陆同属一个构造事件的作用背景中。但与晚晋宁运动相伴随的岩浆活动未见及，反映地壳的克拉通程度不高、稳定性差，并在新元古界南华系—震旦系时发生裂解，形成陆表海环境，接受了石灰顶组一套残坡积相-滨海相含铁质的类磨拉石建造沉积。其底部为含砾泥岩、含角砾泥岩，岩石以其无层理、分选差、磨圆度差和角砾为原地或微异地等为特征。随后，地壳全体隆升，结束该造山事件记录。

二、早古生代板内裂解与造山作用

寒武纪时期，地壳发生裂解，海南岛北面万宁一带裂解方向可能为由南至北，美子林组代表了裂解早期相对稳定的沉积，具滨海—浅海相碎屑岩夹碳酸盐岩建造特点。从早到晚，海水逐渐加深，沉积场所逐渐远离滨岸，海侵加强（宜昌地矿所，1995）。海南岛南面三亚地区沉积作用则始于黔东世，为滨海—浅海环境含磷锰砂泥质岩-碳酸盐岩沉积。

奥陶纪时，海南岛北面裂解程度加剧，沉降盆地扩宽，海水加深，形成了近东西向的海盆，盆地东高西底，南北浅、中部深，接受了奥陶系南碧沟组一套厚达上千米的具有复理石、类复理石特征的海相碎屑岩夹基性、中基性火山岩建造沉积，形成海相碎屑岩建造夹基性火山岩建造，具有活动型建造的特点。呈夹层状产出的火山岩具有多期次的喷发活动特点。早期为陆缘弧环境的玄武岩、变安山岩、变流纹岩，岩石锆石U-Pb等时线年龄为（499±6）~（527±48）Ma（丁式江等，2002；龙文国等，2005）；晚期为板块边缘大洋构造环境的超镁铁质、镁铁质火山岩，岩石Sm-Nd同位素年龄为（458±36.8）Ma。但同期缺少或少见内源物质的沉积，反映了该时期地壳可能具有快速裂解和上升的运动特点。此时，在海南岛南面三亚地区则沉积了滨海—浅海环境砂泥质岩-碳酸盐岩沉积，后期并进入海陆交互相复成分砂砾岩、含砾砂岩、含铁黏土岩混杂沉积。之后，地壳隆升为陆，进入漫长的风化剥蚀时期。

经奥陶系地壳震荡性隆升后，志留纪的沉积仅分布于海南岛北面，包括下志留统陀烈组、空列村组、靠亲山组、大干村组、足赛岭组。早期为滨海相砂质岩沉积，中期进入浅海—半深海相砂泥质、泥砂质沉积，晚期进入滨海、浅海砂质岩、泥砂质岩和碳酸盐岩沉积，构成了一个相对完整的沉积旋回。但西部总体上接受了一套陆棚-大陆斜坡碎屑岩组成的复理石建造，上部并出现碳酸盐台地相的沉积，岩石中上部发育鲍马序列中的A、C、D、E层。而东部地区则为深海平原相泥质、粉砂质沉积，形成了深水相复理石-类复理石建造。显示盆地具有东深西浅的格局，与奥陶纪盆地东浅西深格局有所不同，暗示地壳震荡性隆升作用具有不均衡型。

加里东运动在海南省表现为多次性和多样性。第一幕发生于早寒武世晚期，以垂直隆升作用为特点，使得海南岛北部奥陶系南碧沟组以假整合叠覆于下寒武统美子林组之上。第二幕发生于晚寒武世中—晚期，与第一幕表现形式一样，形成了海南岛南面三地区下奥陶统大葵组与中—上寒武统大茅组间的假整合。第三幕发生于晚奥陶世晚期，地壳继续以垂直隆升作用为特点，导致海南岛地区南面三亚地区隆升为陆，缺失志留纪的沉积。此幕活动在海南岛北面的表现是：奥陶系半深海—深海相沉积后，志留纪早期为滨海沉积，显示地壳曾发生过垂直隆升。第四幕发生于早志留世晚期，从少量零星的资料来看，海南岛北部上古生界与下古生界呈不整合接触，表明第四幕可能具有造山性质，但缺乏同期岩浆作用的记录。伴随该幕构造作用，海南省全面隆升为陆，缺失中、上志留统沉积。

据杨坤光研究（华南加里东运动与印支运动范围北部锋带问题），华南加里东运动不仅在性质、表现上具有多样性，在时间上也具有多阶段性（多幕性），主要分别发生于寒武纪末、奥陶纪末和志留纪末。海南岛加里东运动与华南加里东运动在时间上基本同步，但就其构造运动的强度而言，海南岛表现相对比较微弱，没有华南其他地区强烈，因此海南岛早古生代时不处于加里东运动的峰带位置，而是处于加里东运动影响的边缘区域。

三、晚古生代特提斯构造域演化

经加里东运动后，海南省地壳长期隆升为陆，直至石炭纪才发生沉积作用。

下石炭统南好组沉积时期，整个海南岛主要为稳定环境下滨浅海相碎屑岩相沉积，主要是浅变质作用形成的砂岩类岩石，其底部为变质石英砾岩、砂砾岩、砂岩，沉积物粒度粗，分选性、磨圆度好，结构成熟度和成分成熟度较高，除少量岩屑外，几乎全由石英组成；上部为变质陆源细碎屑岩，总体为板岩、含粉砂质板岩与变质粉砂岩、变质细砂岩不等厚互层，砂岩、粉砂岩分选性好，不含或含少量泥质，偶见变余斜层理，南好乡岩石并普遍富含长石，代表了近源的沉积环境。青天峡组沉积时，沉积环境呈现出一定的差异性，北面石碌镇-琼中石浮镇一线石炭系为海相碎屑岩相和碳酸盐岩台地相沉积，由浅变质作用形成的砂岩、泥质砂岩、灰岩等组成，岩石的分选性、磨圆度好，结构成熟度和成分成熟度较高，具有稳定型滨、浅海相单陆屑式建造夹碳酸盐岩建造的特点。南面中原镇该时期则为海相碎屑岩相，缺乏碳酸盐岩台地相沉积，沉积作用从早至晚（南至北）海水逐渐加深，早期为滨海、浅海相（局部达大陆斜坡相）碎屑岩沉积，中晚期为大陆斜坡相，出现浊流作用形成的杂砂岩，杂砂岩呈中—薄层状，普遍含一定的泥质杂基，具粒序层理；地壳活动性也逐渐加强，上部岩石常见酸性凝灰岩或沉凝灰岩夹层，沉积建造类型由稳定型往次稳定型过渡。

二叠纪时的沉积一是见于海南岛西面石碌镇-东方市江边一线，包括下二叠统鹅查组、鹅顶组和上二叠统南龙组。其中南龙组仅出露于岛西石碌镇地区，为海陆交互相沉积。鹅查组、鹅顶组时，岛西石碌地区为近滨相和碳酸盐台地相沉积，相对石炭纪而言，其地壳已有所抬升。二是见于南沙群岛一带，岩石为灰岩和燧石、硅质碎屑岩、长石杂砂岩。

晚古生代岩石中产海百合茎、珊瑚、层孔虫、苔藓虫、腕足类、疑源类、虫颚以及丰富的孢粉化石。其中，珊瑚 *Gorskyites chia*，*Syringopra* 也常见于与湘中、粤北一带相应时代的地层中；植物和腕足类可以与湘中、赣中、粤北和粤西对比，均以产 *Archaeocalamites dictyoclostus* 为特征。汪啸风等人（1992）根据晚古生代中的化石组合（蜓类 *Parafusulina gruperaensis*-*P. langdaiensis*-*Misellina ovalis*；植物 *Annularia mucronata*，*A. shirakii*，*Pecopteris arborescens*，*P. rigida*，*P. hemitelioides*，*Taeniopteris* sp.，*Cordaites* sp.；珊瑚 *Wentzellophyllum*-*Polythecalis*；菊石类 *Shouchangoceras* 等），认为属于华夏-特提斯动物区或华夏植物区的一部分；并通过海南岛与周边地区华力西期—印支期花岗岩特征的对比分析，认为海南岛华力西期—印支期花岗岩与滇西临沧岩基（或澜沧江岩带）最为相似，提出海南岛晚古生代属于特提斯构造域范围内的见解。

发生于约290~270Ma的华力西-印支运动（谢才富，2007），不仅导致了海南岛北部晚古生代地层发生了区域绿片岩相变质和透入性变形，也使得新元古代、早古生代地层卷入其中，形成了区域尺度上展布的褶皱带和断裂带，塑造出一幅非常醒目的海南岛弧状构造带。同时，伴随的花岗岩浆侵入活动强烈而持久，一直持续至晚三叠世，从早至晚，形成了俯冲型、同碰撞型、后碰撞型、隆升型花岗岩，构成了一个比较完整的造山作用花岗岩浆演化系列；而海南省南面南沙群岛出露的二叠系未见变质变形作用的改造。

尽管海南岛此时处于特提斯构造域范围，但对于其造山运动的动力学背景认识目前依然不明朗。

尤其是广泛的二叠纪花岗岩分布，与滇西临沧岩基（或澜沧江岩带）最为相似，而与同处特提斯构造域的广西、广东缺乏二叠纪花岗岩形成鲜明的对比。且广西、广东一带南盘江盆地发育有三叠纪深水相碎屑岩夹玄武岩，与海南岛早三叠世陆相沉积也完全不一。目前认为，特提斯洋到达广西后出现两个分支，一支东延到达北部湾，另一支南延到达曼谷一带（丘元禧，2000）。因此，本志认为，随着特提斯洋二叠纪的闭合，由于洋盆沿滇西澜沧江—曼谷一线发生俯冲，位于转折部位的南盘江盆地（东支）由于受到俯冲作用而发生进一步的裂解伸展，继续接受三叠纪深水相沉积（吴浩若，2003）。海南岛此时处于东面和东南面挤压应力作用下，一方面使得晚古生代东西向盆地二叠纪时进一步继续发生裂解，广大地区被海水淹没，接受了碳酸盐岩缓坡型浅海陆棚沉积；另一方面导致南面印支-南海地体沿区域性崖城-万宁深大构造带俯冲于五指山造山带之下，海南岛北部五指山造山带（仰冲盘）发育广泛的华力西期—印支期花岗岩，并伴随区域动力低温变质作用的发生。从而形成海南岛二叠纪伸展与挤压共存的环境，并最终以海南岛南北两地体相互挤压作用占主导，且其波及范围可能逐渐西迁，使得特提斯洋东支南盘江盆地闭合和广西、广东三叠纪花岗岩的分布。

此外，作为本次造山作用隆升标志的下三叠统岭文组陆相磨拉石建造，与东部琼中地区中晚三叠世挤压侵位的花岗岩构造环境并不协调，初步认为可能属于不同地壳深度的表现。一般认为，隆升阶段上部为伸展拉张，下部则为挤压环境。

第三节　中生代陆内盆-山演化阶段

海南省经历强烈的华力西期—印支期造山作用后，大约在 175~200Ma 期间进入构造-岩浆平静期：构造活动、岩浆作用微弱，沉积作用也缺失。这一时期也是本区乃至华南从特提斯构造域向滨太平洋构造域转折的时期。早燕山期（约 150~175 Ma），此时在爆发了属另一个构造域（滨太平洋构造域）、另一个构造旋回的强烈构造-岩浆活动。由于软流圈地幔上涌导致岩石圈裂解拉张，形成了一系列幔源和壳幔混合源侵入岩。

晚燕山期，受太平洋板块俯冲引起的弧后扩张带控制，海南省断块活动强烈，此起彼伏，断陷成盆，断隆成山，形成一系列陆内上叠沉积盆地和火山盆地，盆地有白垩纪王五盆地、白沙盆地、雷鸣盆地、阳江盆地、乐安盆地等火山-沉积盆地，及牛腊岭、同安岭火山盆地，沉积了下白垩统鹿母湾组湖、河相杂陆屑式建造夹中酸性火山岩建造和上白垩统报万组湖、河相杂陆屑式建造，以及下白垩统六罗村组双峰式火山岩建造和下白垩统汤他大岭组、岭壳组中酸性火山岩建造。同期构造岩浆活动强烈，形成了北东向、北西向、东西向、南北向韧脆性断裂，奠定了海南岛的大地构造格局。白垩纪晚期并发生了由北西往南东的逆冲推覆作用，塑造了醒目的北东向军营-红岭和戈枕逆冲推覆构造带。花岗岩浆侵入活动具有多期次性，发育有侏罗纪、白垩纪花岗岩，岩石缺乏定向组构，属于伸展构造环境壳幔岩浆混合的产物。其侵位往往受多方向断裂的控制，导致岩体平面中常呈不规则长条状分布。

第四节　新生代裂解沉陷阶段

新生代，由于南海在南北向拉张应力作用下发生持续扩张（中国科学院南海海洋研究所，1990），南海北缘深部岩石圈受软流圈的快速上涌而发生快速减薄，导致了新岩石圈的增长，地幔流的上升导致了浅部岩石圈的张裂下陷，以及南海北缘众多拉张盆地的形成（许义刚等，2002）。盆地内接受了古近纪以来的沉积，形成古近纪河湖相红色碎屑岩建造夹可燃性有机岩建造、膏岩建造、火山岩建造和新近系滨浅海相碎屑岩建造、火山岩建造以及第四系滨浅海相碎屑岩建造、河流相-三角洲相碎屑岩建造、火山岩建造。该时期地壳升降运动频繁，导致多个不同岩石地层单位间普遍存在不整合接触关系，发育海积、河积阶地。受断裂活动影响，发生多次的地震活动和剧烈的玄武岩浆喷发或喷溢，并沿北西向断裂出露多处温泉。

南海地区在中生代断陷盆地的基础上，地壳拉张下陷加剧，形成一系列北东向的沉积盆地。古近纪时南海北部接受了一套陆相-浅海相碎屑岩；南海南部水深较深，以一套厚达数千米的浅海相—半深海相页岩、砂岩为主，东南部靠近岛弧处常发生褶皱、变质作用。新近系下部为一套浅海相—半深海相砂岩与泥岩的互层，海底高原为礁灰岩。新近系上部—第四系下部主要为砂岩、页岩夹火山岩与火山碎屑岩夹层，海底高原为礁灰岩与生物碎屑岩，南海中央海盆则为火山灰与放射虫粉砂岩。全新世以来南海地区的沉积以珊瑚礁与生物碎屑岩为主。

在喜马拉雅运动作用下，海南省新构造运动强烈，表现形式多样，包括地壳升降运动、断裂活动、地震活动、火山活动和地热活动，造就了现今地势格局和各类地貌景观。

第六篇

地球物理地球化学特征

第二十四章 区域地球物理场特征

海南岛开展的区域性地球物理工作较少，涉及本区的工作主要有《华南地区 1：100 万物化遥综合解释成果编图》、《雷琼地区地壳结构研究》、《琼西北地区人工地震测深研究》、《海口市活断层探测与地震危险性评价》等项目，以及由海南省地质综合勘察院完成的全岛 1：20 万区域重力测量工作等。此外，南海海域自 20 个世纪 80 年代起，南海海洋地质队和石油等部门，就在该区开展过许多区域性的海洋地球物理测量工作，但是由于跨部门等其他原因，本次收集到的资料不多，主要收集到的资料有《中国南海海域岩石圈三维结构及演化》、《中国海域构造地质学》、《南海地质》等。以上资料对海南省的地球物理特征进行了详细的研究，为本次工作打下基础。

第一节 海南岛地球物理场特征

一、岩（矿）石地球物理特征

（一）弹性波特征

1984 年广东省地震局联合江苏省地震局在海南岛西北部地区进行了 4 条剖面总长度为 162km 的人工地震测量工作（图 24-1），解释的有效波为 T_2、T_3、T_4、G、G_1。它们与地质界面之间有着如表 24-1 所示的对应关系。此外，1984 年 12 月至 1985 年 2 月国家地震局组织广东省地震局、福建省地震局、江苏省地震局、地壳力学研究所、地球物理研究所等五家单位在雷琼地区进行了地震测深工作，共布设了 7 条测线，其中有Ⅳ、Ⅴ和Ⅶ三条剖面分布在琼北地区（图 24-2）。根据地震波识别标志，对所绘记录剖面图进行了反复对比，共识别出了琼北地区 P_g、t_1、t_2、t_3（P_n^0）、t_4、t_5、和 P_n 等 7 组地震波震相（表 24-2，图 24-3）。

表 24-1 琼西北地区浅部地壳分层概况
Table 24-1 Overview of shallow crustal layering in the Qiongxi area

层序号	地质时代	层速度/(km/s)	界面
1	上新世—第四纪	1.6~2.2	T_2
2	中新世—上新世	2.0~3.1	
3	渐新世中晚期	3.1~4.2	T_3
4	中生代	4.4~5.13	T_4
5	部分晚古生代—中生代	5.15~6.38	G
6		6.5~6.7	G_1

表 24-2 琼北地区地壳速度模型参数表
Table 24-2 List of parameters of the crustal velocity model of the Qiongbei area

白延				树德			
序号	层厚/km	$v_上$/(km/s)	$v_下$/(km/s)	序号	层厚/km	$v_上$/(km/s)	$v_下$/(km/s)
1	0.3		3.10	1	1.2		3.10
2	5.9	5.85	5.91	2	5.8	5.85	5.91
3	1.3	5.91	5.50	3	0.8	5.91	5.50
4	1.3	5.50	5.50	4	1.2	5.50	5.50
5	3.6	6.15	6.15	5	3.4	6.15	6.15
6	8.0	6.15	6.60	6	8.0	6.15	6.60
7	4.6	6.9	6.90	7	5.3	6.9	6.90
8	上地壳	8.0		8	上地壳	8.0	

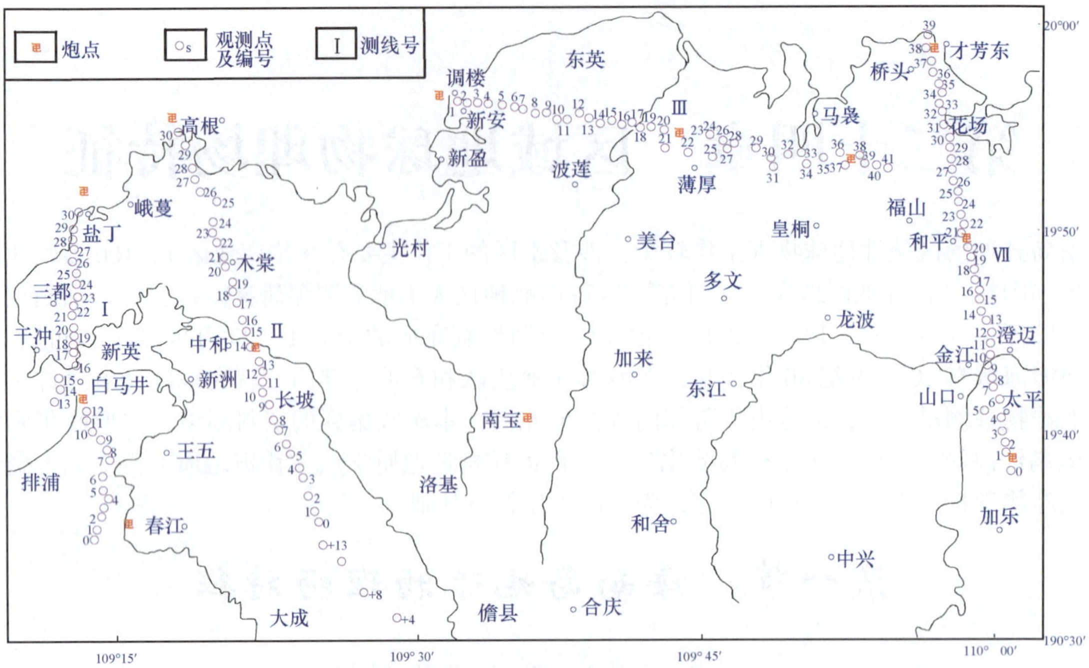

图 24-1　琼西北地区人工地震测深测点、炮点分布图

Fig. 24-1　Map of distribution of artificial seismic depth sounding points and shot points in the Qiongxibei area

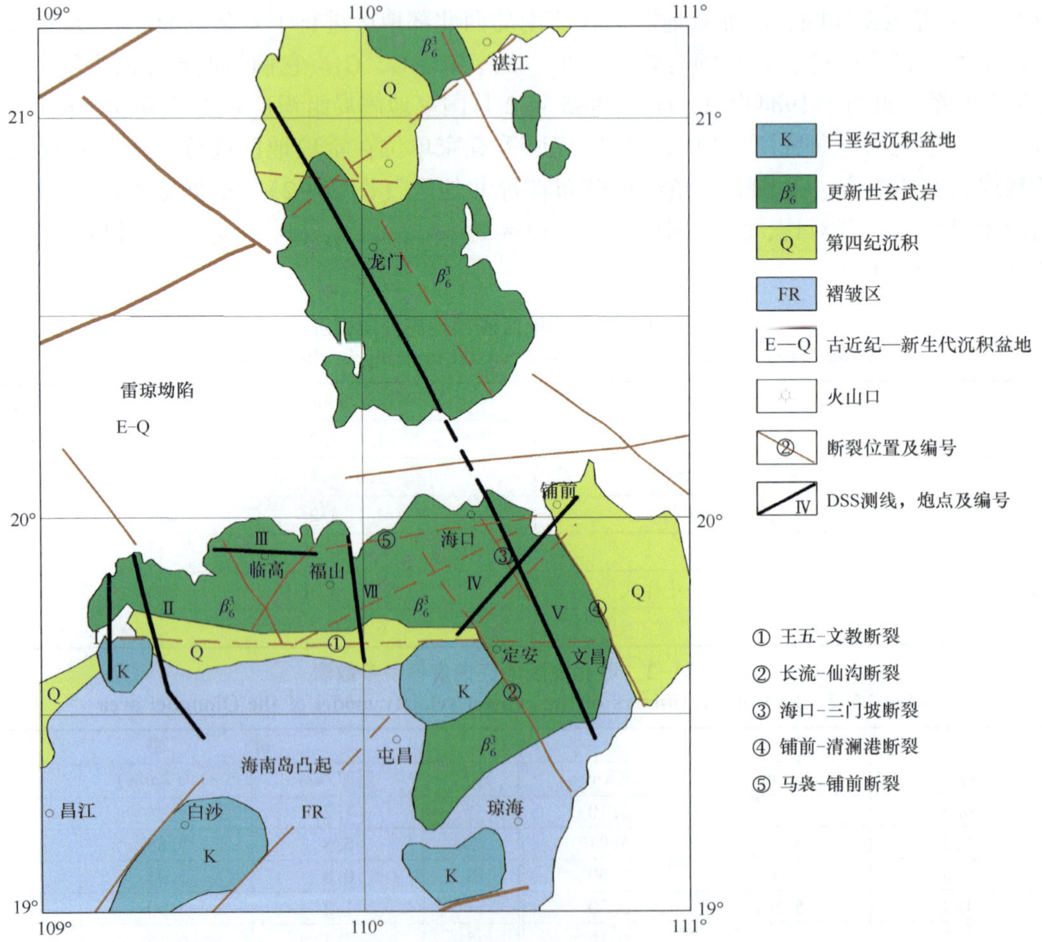

图 24-2　白延-江洪（Ⅴ）、定安-东坡（Ⅳ）、常树-才芳东（Ⅶ）人工地震测深剖面图

Fig. 24-2　Artificial seismic depth soundingsection along Baiyan-Jianghong（Ⅴ），Ding'an-Dongpo（Ⅳ）and Changshu-Caifangdong（Hu）

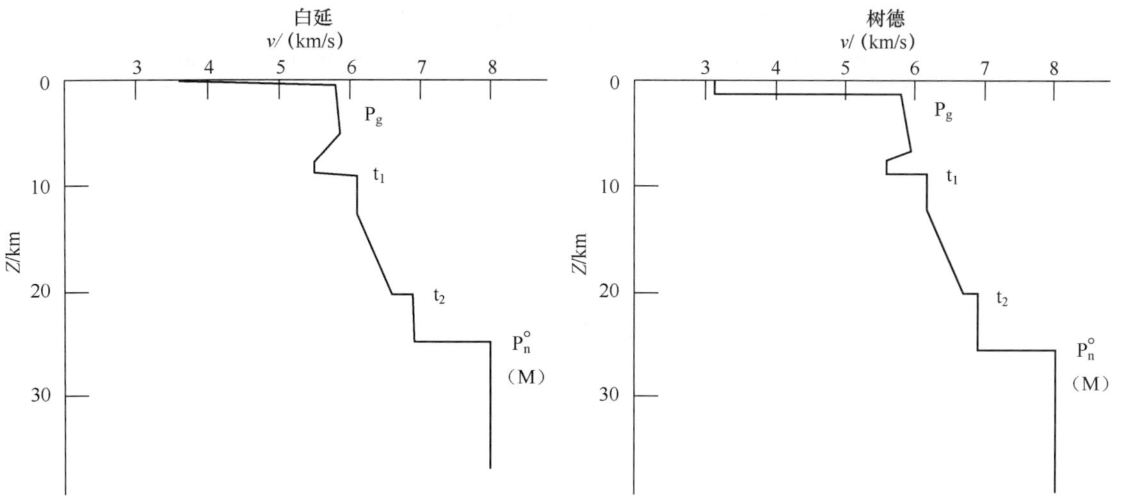

图 24-3　雷琼地区人工地震测深白延、树德炮点的速度模型图

Fig. 24-3　Velocity model of Baiyan and Shuden artificial seismic shotpoints in the Leiqiong area

(二) 磁性特征

海南岛的地层、岩体、磁性矿产（磁铁矿）及其围岩都有较大的磁性差异，磁铁矿石都具有较高的磁化率和剩余磁化强度，但局部分布不均匀，因含矿岩性的不同，磁化率和剩余磁化强度也表现出不均匀性。全岛岩（矿）石磁性特征见表 24-3～表 24-5。

表 24-3　海南岛地区地层磁性参数统计表

Table 24-3　Statistics of magnetism parameters of strata from the Hainan Island

地质时代			岩性	磁化率/10^{-5}SI		
				极大值	极小值	均值
新生代	第四纪		第四系沉积物	136	17	61
	新近纪		长英质砂岩	18	2	5
中生代	白垩纪	晚白垩世	长石石英杂砂岩夹含砾砂岩	186	42	95
			中粗粒杂砂岩	108	0	32
			红色粉砂岩	15	8	12
		早白垩世	砂泥岩	15	5	11
	三叠纪		粉砂岩、泥岩	35	10	22
			砾岩	20	2	10
古生代	二叠纪	早二叠世	变质粉砂岩	15	3	11
	石炭纪	晚石炭世	变质砂岩、砾岩	30	2	13
		早石炭世	变质砂岩	11	2	6
	志留纪	早志留世	变质砂岩	15	4	9
	奥陶纪		石英砂岩、灰岩、千枚岩	64	8	21
			炭质板岩	14630	0	1510
	寒武纪		变质石英砂岩	87	7	45
			页岩	41	3	16
			二云石英片岩	20	0	8
			结晶灰岩	12	0	4
元古宙	南华纪—震旦纪		变质石英砂岩	91317	0	2738
	青白口纪		白云岩	30	0	15
			石英岩	257196	0	8553
	长城纪		石英云母片岩	3993	0	703
			斜长角闪片岩	49638	1809	4725

表 24-4 海南岛地区岩浆岩磁性参数统计表

Table 24-4 Statistics of magnetism parameters of magmatic rocks from the Hainan Island

构造期		岩浆岩分类	岩性	磁化率/10^{-5}SI		
				极大值	极小值	平均值
喜马拉雅期（N—Q）		基性岩	多孔状玄武岩	120952	0	10849
			辉斑橄榄玄武岩	11304	1256	3517
			玄武岩	1507	251	754
			细粒玄武岩	82080	683	11289
			橄榄玄武岩	7724	254	1526
燕山期（J—K）	晚燕山期（K）	酸性岩	花岗岩	19	0	4~11
			角岩化花岗岩	851	41	55~448
			粗粒花岗岩	81	0	4~9
			钾长花岗岩	948	3	11~471
			黑云母钾长花岗岩	781	21	91~327
			角闪钾长花岗岩	291	0	29
		中酸性岩	二长花岗岩	1021	10	109
			花岗闪长岩	2361	10	130~1270
			流纹岩	377	12	31
			流纹斑岩	2575	16	45
			含砾英安斑岩	5778	20	2680
			英安流纹斑岩质凝灰-角砾熔岩	14418	6582	7109
			安山岩	69833	56	13565
			流纹质英安质岩屑晶屑凝灰岩	13183	62	3730
			流纹质晶屑凝灰岩	9420	42	75
			次安山岩	2034	16	56
			花岗闪长岩、花岗岩	1353	22	677
			石英正长岩	47	1	15~26
		中性岩	闪长岩	4789	154	1851
			石英闪长岩	27946	28	1582
		基性岩	辉长岩	33912	1256	6908
	早燕山期（J）	酸性岩	花岗岩	33	2	6~15
			钾长花岗岩	1126	1	4~700
		中性岩	闪长岩	3895	1194	2204
印支期（T）		酸性岩	混合花岗岩	160	0	32
			花岗岩	165	5	15
			粗粒钾长花岗岩	145	58	94
		中性岩	闪长岩	20774	60	4157
			石英闪长岩	7536	502	3479
		基性岩	辉长岩	35168	3140	6235
华力西期（P）		酸性岩	混合花岗岩	607	3	361
		中性岩	闪长岩	12560	9	3454
			石英闪长岩	6569	12	2236
		基性岩	辉长岩	90432	19	5740

表 24-5　海南岛铁矿石磁参数统计表

Table 24-5　Statistics of magnetism parameters of iron ores，Hainan Island

岩（矿）石名称	块数	磁化率/10^{-5}SI		剩余磁化强度 J_r/(10^{-3}A/m)		备注
		平均值	变化范围	平均值	变化范围	
磁铁矿	113	62800	18262~504912	43000	1330~39800	岛南地区（田独铁矿区、红石铁矿区）
铁锰矿	2	49612	47100~51998	44000	25400~62600	
磁黄铁矿	15	41950	197~144440	35900	1090~77300	
褐铁矿	37	12560	0~46220	10900	0~45800	
赤铁矿	6	754	126~2386	1200	300~4300	岛东地区
褐铁矿	34	628	206~2374	440	290~4980	
赤铁矿	528	3303	0~951985	2250	0~701120	石碌矿区

1. 地层的磁性特征

1）地层的磁性特征主要取决于磁性岩石的含量比例，即便是在同一地层中的各种岩石，其磁性相差悬殊，例如西联农场地区志留纪的大理岩、南棍园向斜志留系中黑云母石英片岩，因含有磁黄铁矿，具有较高的磁性，与其他地区志留系的大理岩、黑云母石英片岩的低磁性完全不同。而且地层作为岩石建造组合，因此不能以物性层划分地层。但岩石磁性特征又反映岩石形成和变形的构造环境，磁性火山岩反映地壳比较活跃的大陆边缘环境，非磁性或弱磁性岩石往往是稳定环境的被动大陆边缘沉积。

2）沉积碎屑岩、碳酸盐岩及变质岩等地层，岩石磁性普遍较低，形成大片平稳磁场或负磁场，如白沙盆地、江边等。

3）琼北地区分布的新生代火山岩（玄武岩等）地层、琼南地区分布的中生代火山岩（流纹岩、英安岩等）地层具有变化较大的离散磁性，往往形成高低相间变化的磁场。

2. 火山岩的磁性特征

火山岩磁性变化很大，总体上具有从中酸性到基性逐渐增强的特点。中酸性火山岩具有两种磁性，一种为无—微弱磁性，如流纹岩、火山碎屑岩；另一种具强磁性，如英安质流纹岩、流纹斑岩和安山岩。基性火山岩中的玄武岩、凝灰岩也具有较强磁性，而且变化较大，可引起跳跃变化的磁场或局部异常（表 24-4）。

3. 侵入岩的磁性特征

侵入岩的磁性特征整体上具有从酸性→中性→基性磁性逐渐增强的规律。辉长岩磁化率最大平均值 6908×10^{-5}SI，具有较强的磁性。闪长岩具有两种不同磁性，一种为无—弱磁性，另一种为中等磁性，其磁化率平均值为 2638×10^{-5}SI 左右。花岗岩和花岗闪长岩也具有两种磁性，一种为中等磁性，磁化率平均值为 948×10^{-5}SI 、1758×10^{-5}SI；另一种为无—微弱磁性（表 24-4）。

4. （磁）铁矿的磁性特征

铁矿主要包括磁铁矿、磁赤铁矿、赤铁矿、褐铁矿、菱铁矿等 5 类，一般独立存在，有时共生。其中以磁铁矿磁性最强，其强度与品位关系密切，磁化率变化范围（38465~357583）$\times10^{-5}$SI，当埋藏较浅、具有一定规模时，在地表可引起较强的磁异常；磁赤铁矿磁化率变化范围（0~25120）$\times10^{-5}$SI，其磁化率平均值为 2512×10^{-5}SI，为中等磁性；赤铁矿（126~2386）$\times10^{-5}$SI，其磁化率平均值为 754×10^{-5}SI，为弱磁性。其他的褐铁矿、菱铁矿一般为无磁性—微弱磁性（表 24-5）。

（三）密度特征

1982 年 1 月~1996 年 12 月海南省地质综合勘查院进行了全岛 1：20 万区域重力测量工作，共采集密度标本 5795 块，其中不同时代地层标本 2889 块（钻孔岩心 1690 块，地表 1199 块），岩浆岩类标本 2906 块（钻孔岩心 310 块，地表 2596 块）。

从密度值统计结果看（表 24-6，表 24-7），沉积岩类地表标本比钻孔岩心标本的密度值小 0.2t/m^3，这是岩石易风化，密度值变小之故。而岩浆岩类地表与钻孔岩心标本的密度值差异不明显。因此不同时代地层的密度值以钻孔岩心测定、统计的结果为主。全岛各地层、岩浆岩的密度特征归纳为：

1）地层从新至老，密度值逐渐增大，但古生代各地层的密度值差异不明显（2.70～2.74t/m^3），平均值为 2.72t/m^3。

表 24-6 海南岛地层密度统计表

Table 24-6 Statistics of density of strata from the Hainan Island

地层				代号	标本块数		密度/(10^3kg/m^3)				标准离差 10^3kg/m^3	
							平均值		变化范围			
界	系	统	组		地表	钻孔	地表	钻孔	地表	钻孔	地表	钻孔
新生界	第四系			Q	6	38	1.78	1.91	1.67～1.88	1.66～2.31	0.07	0.15
	新近系	上新统	望楼港组	N_2w		51		1.95		1.65～2.37		0.17
		始新统	瓦窑组	E_2w	12	19	2.41	2.23	2.24～2.53	2.10～2.48	0.09	0.11
		古新统	昌头组、长流组	$E_1\hat{a}$,$E_1\hat{c}$	6		2.43		2.38～2.47		0.03	
中生界	白垩系	下统	鹿母湾组	K_1l	271	29	2.53	2.54	2.20～2.78	2.44～2.70	0.10	0.07
			岭壳村组	K_1lk	19		2.63		2.58～2.81		0.06	
			六罗村组	K_1ll	67		2.62		2.55～2.76		0.03	
古生界	二叠系	中—上统	南龙组	$P_{2-3}n$	5		2.53		2.42～2.71		0.10	
		下统	鹅顶组	P_2ed	55		2.62		2.45～2.76		0.07	
			鹅查组	P_1e								
	石炭系	上统	青天峡组	C_2q	218	44	2.52	2.70	2.07～3.05	2.46～2.82	0.10	0.07
		下统	南好组	C_1n	83	314	2.46	2.70	2.10～2.78	2.30～3.30	0.08	0.05
	志留系	下统	足赛岭组	S_1z	6		2.70		2.68～2.82		0.05	
			靠亲山组	S_1kq	10		2.65		2.61～2.69		0.03	
			空列村组	S_1k	185		2.60		2.29～2.87		0.10	
			陀烈组并层	S_1t								
	奥陶系		南碧沟组	On	60	56	2.50	2.74	2.17～2.76	2.58～3.03	0.06	0.10
		上统	干沟村组	O_3g	28		2.57		2.43～2.72		0.08	
			尖岭组	O_3j	57		2.64		2.30～2.85		0.13	
	寒武系	上统	大茅组	$Є_{2-3}d$	15		2.61		2.58～2.63		0.02	
		中统	孟月岭组	$Є_2m$	21		2.63		2.55～2.74		0.05	
元古宇	南华系—震旦系		石灰顶组	$Pt_3^{2-3}\hat{s}$		56		2.76		2.68～2.93		0.06
	青白口系		石碌群	$Pt_3^1\hat{S}$	75	746	2.58	2.77	2.18～2.93	2.56～3.00	0.08	0.05
	长城系—蓟县系		峨文岭组	$Pt_2^{1-2}e$		299		2.73		2.50～2.95		0.05
			戈枕村组	Pt_2^1g								

表 24-7　海南岛岩浆岩密度统计表

Table 24-7　Statistics of density of magmatic rocks from the Hainan Island

构造期	主要岩性	标本块数		密度/($10^3kg/m^3$)				标准离差 $10^3kg/m^3$	
				平均值		变化范围			
		地表	钻孔	地表	钻孔	地表	钻孔	地表	钻孔
喜马拉雅晚期	火山角砾岩		41		2.05		1.81~2.23		0.10
	玄武质火山角砾岩		5		2.55		2.25~2.70		0.17
	气孔状橄榄玄武岩	146		2.50		2.18~2.71		0.13	
	块状玄武岩	86		2.71		2.53~2.83		0.08	
燕山期	石英正长岩	24		2.55		2.47~2.66		0.05	
	花岗斑岩	41		2.50		2.32~2.57		0.05	
	钾长花岗岩	90		2.56		2.44~2.71		0.05	
	二长花岗岩	67	50	2.60	2.56	2.40~2.71	2.48~2.62	0.08	0.03
	花岗闪长岩	124		2.65		2.48~2.82		0.06	
	黑云母二长花岗岩	87		2.60		2.55~2.65		0.02	
	花岗岩	5		2.55		2.52~2.57		0.02	
	英安岩	47		2.45		2.30~2.57		0.04	
	石英闪长岩	10		2.56		2.46~2.70		0.08	
	辉长岩	7		2.74		2.73~2.76		0.01	
	花岗岩	159		2.58		2.40~2.72		0.05	
华力西期—印支期	钾长花岗岩	45		2.58		2.54~2.63		0.02	
	二长花岗岩	255	52	2.58	2.60	2.32~2.81	2.56~2.64	0.07	0.01
	花岗岩	125		2.56		2.52~2.82		0.05	
	花岗闪长岩	207		2.61		2.31~2.98		0.07	
	黑云母二长花岗岩	390	81	2.59	2.62	2.28~2.85	2.46~2.84	0.09	0.04
	石英闪长岩	27		2.83		2.74~3.01		0.08	
	闪长岩	4	81	2.68	2.84	2.56~2.79	2.79~2.89	0.11	
	辉长岩	650		3.06					

2）元古宙地层的密度平均值为 2.75 t/m^3，与古生代地层密度差异不明显。

3）地层从新至老，划分为 4 个密度层，3 个密度界面。第 1 密度层为第四系和古近系上新统，平均密度为 1.93 t/m^3；第 2 密度层为古近系始新统和古新统，平均密度为 2.42 t/m^3；第 3 密度层为白垩系下统鹿母湾组，平均密度为 2.54 t/m^3；第 4 密度层为古生界和元古宇，平均密度为 2.72~2.75 t/m^3。3 个密度界面分别为：①古近系始新统与上覆地层间的界面，其密度差为 0.49 t/m^3；②白垩系与上覆古新统之间的界面，密度差为 0.12 t/m^3；③古生界或元古宇与上覆白垩系鹿母湾组之间的界面，其密度差为 0.18~0.21 t/m^3。若新生代疏松盖层下伏基岩为古生界或元古宇，则密度差更为明显。地层密度分层及密度界面见表 24-8。

4）各类花岗岩（包括花岗岩、二长花岗岩、钾长花岗岩、黑云母二长花岗岩）的密度值与侵入时代有关，即时代早比时代晚的花岗岩的密度值要大些，但差异不明显。如华力西期—印支期花岗岩的密度为 2.61t/m^3，而燕山期花岗岩为 2.55~2.60t/m^3。

5）不同期次的岩浆岩，其密度值均从酸性→中性→基性增大；花岗岩为 2.55~2.61t/m^3；花岗闪长岩、闪长岩为 2.63~2.68t/m^3；辉长岩为 2.74~3.06t/m^3（表 24-9）。

表 24-8　海南岛地层密度分层和密度界面表

Table 24-8　Density layering and density interfaces of strata from the Hainan Island

界	系	统	组	密度层	平均密度/(t/m^3)	密度界面及密度差
新生界	第四系			第 1 密度层	1.93	第一密度界面，$\triangle\sigma=0.49t/m^3$
	古近系	上新统				
	新近系	始新统		第 2 密度层	2.42	第二密度界面，$\triangle\sigma=0.12t/m^3$
		古新统				
中生界	白垩系	下统	鹿母湾组	第 3 密度层	2.54	第三密度界面，$\triangle\sigma=0.18-0.21t/m^3$
古生界	二叠系—寒武系			第 4 密度层	2.72~2.75	
元古宇	震旦系—长城系					

6）岩浆岩中以华力西期—印支期辉长岩的密度为最大（3.06 t/m^3），燕山期英安岩的密度最小（2.45t/m^3）。

7）华力西期—印支期和燕山期花岗岩的密度平均值分别为 2.61t/m^3和 2.58t/m^3，两者差异不大，但与古生代、元古宙地层有较明显的密度差异（0.17~0.11t/m^3）。

8）岛北部广泛分布的喜马拉雅期喷出岩，块状和气孔状构造玄武岩的密度分别为 2.71t/m^3和 2.50 t/m^3，但呈厚度不大的岩被状产出。

表 24-9　海南岛主要岩浆岩密度值一览表

Table 24-9　List of density values of main magmatic rocks from the Hainan Island

构造期	主 要 岩 性	密度平均值/(t/m^3)
喜马拉雅晚期	气孔状橄榄玄武岩	2.50
	块状玄武岩	2.71
燕山期	花岗岩、钾长花岗岩、二长花岗岩、黑云母二长花岗岩	2.58
	花岗闪长岩	2.65
	辉长岩	2.74
华力西期—印支期	钾长花岗岩	2.60
	花岗闪长岩	2.63
	二长花岗岩	2.60
	花岗岩	2.61
	黑云母二长花岗岩	2.62
	石英闪长岩	2.83
	闪长岩	2.84
	辉长岩	3.06

二、断面地球物理特征

（一）东方-琼海断面

此成果系 1995 年 6 月海南省地质矿产勘查开发局完成的《华南地区 1：100 万物化遥综合解释成果编图》项目的成果，剖面位置见图 24-16。该剖面的布格重力异常特征为：整条剖面以三低三高之异常为特征，具体为 0~13km 范围为一个重力低异常，极小值为$-25\times10^{-5}m/s^2$；13~35km 段为重力高异常，其极大值为$-20\times10^{-5}m/s^2$。昌江东侧的 35~75km 段为一规模相对较大的重力低异常，极小值为$-37\times10^{-5}m/s^2$，43km、70km 处梯度较大。75~110km 之间为重力高异常，其极大值为$-26\times10^{-5}m/s^2$。110~135km 之间为重力低异常，其极小值为$-38\times10^{-5}m/s^2$。135~225km 段为重力高，其极大值为 $2\times10^{-5}m/s^2$。

磁力 ΔT 异常场特征：磁力 ΔT 异常场特征与重力场特征相似，分为 6 个特征段，第一段位于最西端的 0~25km 段，呈现宽度不大的磁力低异常，幅值变化相对不大的异常，异常在 0~60nT 之间变化。第二段位于 25~54km 段，呈现宽缓的磁力高异常，其极大值为 40nT。第三段位于 54~82km 段，为规模稍大，且宽缓的磁力低异常，其极小值为-40nT。第四段自 82km，止于 120km，以规模较大，幅值大，梯度大为特征，其极大值为 80nT。120km 以东至 150km 为第五段，为规模较大的低缓异常，且幅值大，梯度也大，其极小值为-60nT。第六段自 150km，止于 225km，以幅值大，梯度大为特征，其极大值为 120nT（图 24-4）。

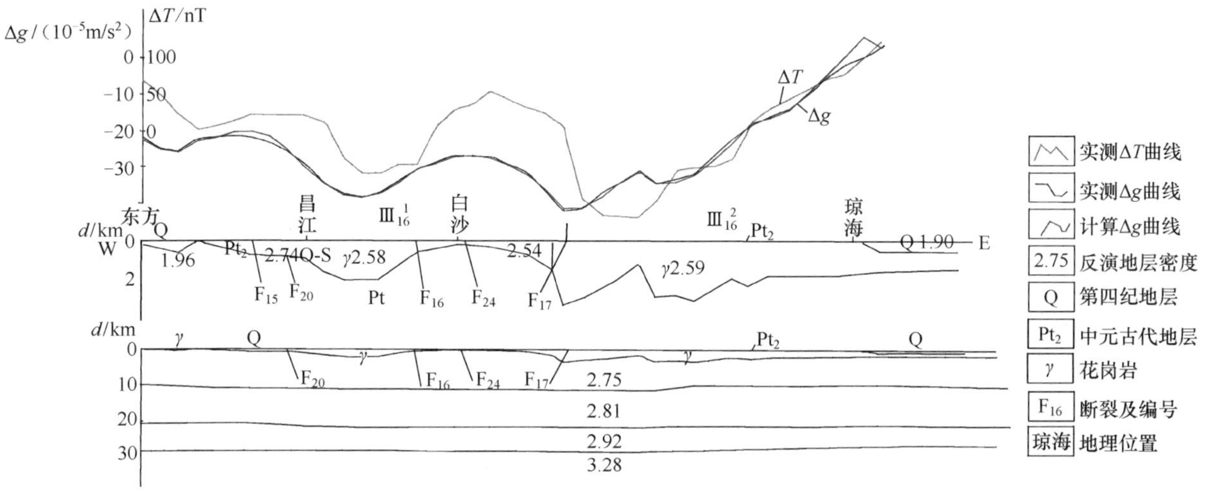

图 24-4　东方-琼海地壳断面图

Fig. 24-4　Crustal cross-section of Dongfang-Qionghai

地质解释：整条剖面的莫霍面深度 27~28. 5km。东方市至黎母山一带莫霍面深度较大，往东逐渐变浅（见图 24-18）。上地壳厚度为 10~11km，下地壳厚度 5~7km。东方市至红泉农场一带为基底出露区，白沙一带为基底隆起区；坝王岭至青松及乌石农场一带为基底凹陷区。隆起区基底埋深0. 3~0. 8km；凹陷区为 2. 0~3. 5km。凹陷区对应浅部花岗岩（图 24-4）。

（二）儋州-保亭断面

此成果为 1995 年 6 月海南省地质矿产勘查开发局完成的《华南地区 1：100 万物化遥综合解释成果编图》项目的成果，剖面位置见图 24-16。该剖面的布格重力异常总体上为两头高中间低之特征，具体为 0~8km 范围为一个重力低异常，极小值为$-20\times10^{-5}m/s^2$。8~30km 段为重力高异常，其极大值为 $8\times10^{-5}m/s^2$。30~165km 段，本段总体上为重力低异常，从儋州北侧至保亭县附近，布格重力异常值从北往南逐渐降低，在保亭附近极小值为$-55\times10^{-5}m/s^2$。165~175km 段，布格重力异常值从北往南逐渐升高，极大值为$-12\times10^{-5}m/s^2$。

磁力 ΔT 异常场特征：整条剖面分为 3 个特征段，第一段位于最西端的 0~25km，为低负变化异常段，呈梯度较大的负值磁异常，场值在-40~10nT 之间变化。第二段位于 25~45km 段，呈幅值变化大，梯度变化大的异常，异常在 0~90nT 之间变化。第三段位于 45~175km 段，为总体变化不大，稍有起伏的磁异常，场值在 20~100nT 之间变化（图 24-5）。

地质解释：该剖面总体上莫霍面深度具有北浅南深之特征（见图 24-18），其莫霍面深度为

26~29.5km。儋州以北为26~27km，儋州以南至三道为28~29.5km，三道以南为28km。上地壳厚度8~10km，中、下地壳厚度17~19km。该断面南北两端及中部细水、五指山一带为基底隆起区；儋州及保亭附近为基底凹陷区。隆起区基底埋深为0.5~0.8km；凹陷区为2~4km。凹陷区对应浅部花岗岩。中生代盆地分布于基底隆起区（图24-5）。

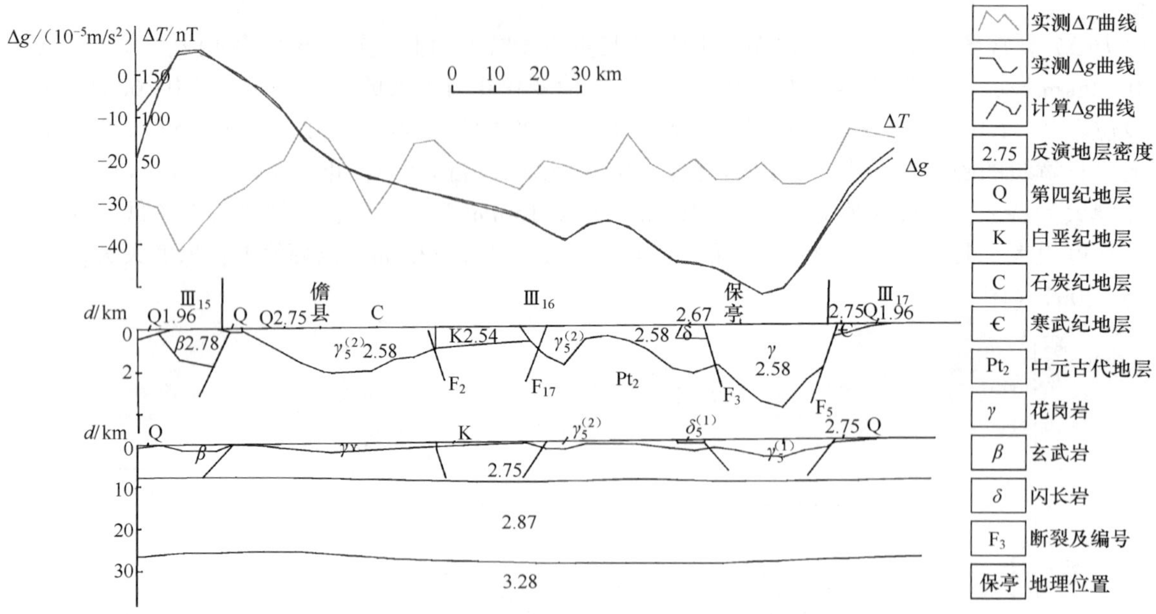

图24-5　儋州-保亭地壳断面图

Fig. 24-5　Crustal cross-section of Danzhou-Baoting

（三）白延-江洪断面

人工弹性波场特征：据《雷琼地区地壳结构研究》项目成果，剖面位置见图24-2。白延-江洪剖面资料显示雷琼区域地震相主要有5组，即结晶基底折射 P_g、Moho界面反射 P_m、上地幔顶部折射 P_n 以及壳内反射 P_{c_1} 和 P_{c_2}，其中 P_g、P_m 为强震相，P_{c_1}、P_{c_2} 和 P_n 为弱震相，树德炮点还显示出清晰的二次波 P_{g_2}、P_{m_2} 和 P_{n_2}。在江洪和龙门炮点首先记录到了可能来自古近纪—新近纪地层的首波，主要出现在纪家凹陷和乌石凹陷中。纪家凹陷最深达2.4km，乌石凹陷为2.5km。结晶基底或古生代变质岩以上为沉积层，其平均速度为3.1km/s，但厚度变化较大。该剖面在海南岛部分，基底在白延附近出露地表，在树德炮点附近最深约2km，至云龙变浅，仅400m左右，至福田又变深。在雷州半岛部分，基底面的埋藏深度在徐闻及龙门附近最浅，不超过1.0km；而在下桥与龙门之间和唐家与江洪之间其埋深则近4.0km。故沿该剖面，基底呈“四隆四凹”之特征，即江洪凹陷、乌石凹陷、琼州海峡凹陷、树德断陷和龙门凸起、下桥-龙塘凸起、云龙凸起、海南隆起。基底沿该剖面被13条断层切割，说明本区基底断裂十分发育（见图24-7）。

布格重力异常场特征：据《华南地区1:100万物化遥综合解释成果编图》项目成果，整条剖面为四低三高之异常场为特征。0~40km范围为一个平缓重力低异常，极小值为 $-2\times10^{-5}m/s^2$；40~70km段为一个北缓南陡的重力高异常，其极大值为 $10\times10^{-5}m/s^2$；海口北侧的70~95km段为一个北陡南缓的重力低异常，极小值为 $-21\times10^{-5}m/s^2$；95~135km之间为平缓的重力高异常，极大值为 $-11\times10^{-5}m/s^2$；135~170km段为一个南北梯度较陡的重力低异常，其极小值为 $-40\times10^{-5}m/s^2$；170~190km段为平缓的重力高异常，极大值为 $-20\times10^{-5}m/s^2$；190~210km段为一个北缓南陡的重力低异常，其极小值为 $-37\times10^{-5}m/s^2$。

磁力 ΔT 异常场特征：据《华南地区1:100万物化遥综合解释成果编图》项目成果，磁力 ΔT 异常特征大致可分为6个特征段，第一段位于最西端的0~15km段，呈现宽度不大，北边梯度陡的负值磁异常，异常值在-100~-10nT之间变化。第二段位于15~50km段，呈现宽缓的磁力高异常，其极大值为25nT。第三段位于50~80km段，为规模稍大，北陡南缓的尖陡负磁异常，其极小值为-95nT。第四段自80km，止于130km，以规模较大，幅值高，南北梯度陡为特征，极大值为100nT。130km以东至155km为第五段，为一个尖陡的负磁异常，极小值为-45nT。第六段自155km，止于210km，为一个规模较大，幅值高，北陡南缓之异常，极大值为75nT（图24-6）。

地质解释：该剖面莫霍面深度25~27.5km。上地壳厚度7~12km，中地壳厚度10~15km，下地壳厚度7~9km（图24-7）。琼州海峡一带，中地壳明显加厚，下地壳明显减薄。该断面白延附近为基底出露区，云龙-徐闻一带为基底隆起区；龙门及江洪附近为基底凹陷区。隆起区基底埋深为0.4~0.6km；凹陷区为1.4~2.2km。新生代沉积盆地对应于基底凹陷区（图24-8）。

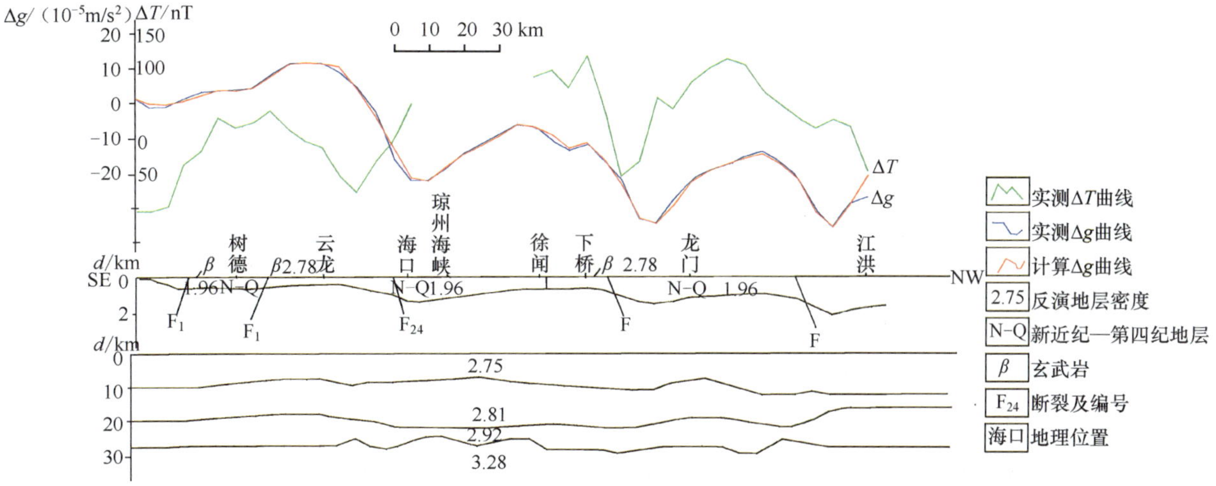

图24-6 白延-江洪地壳断面图

Fig. 24-6 The Baiyan-Jianghong crustal section

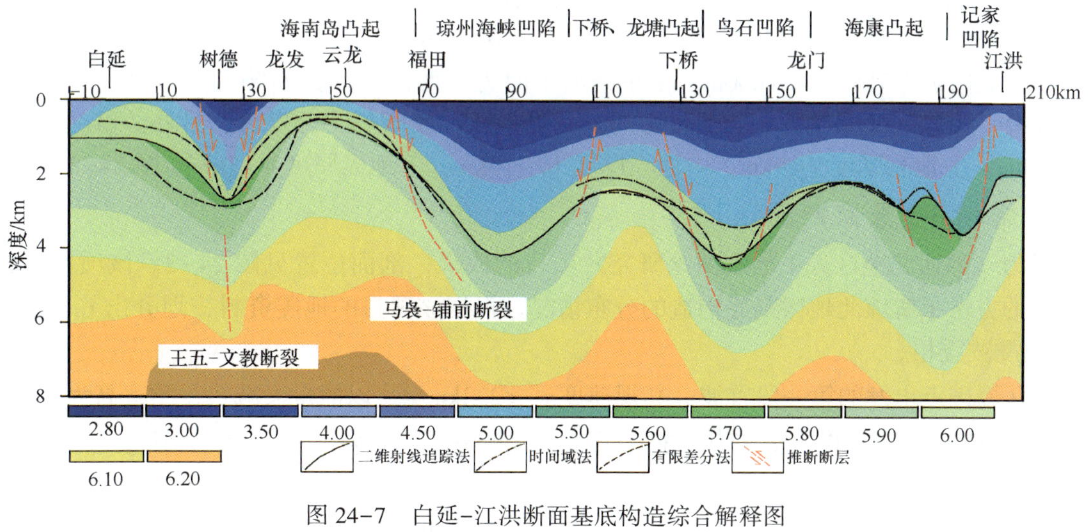

图24-7 白延-江洪断面基底构造综合解释图

Fig. 24-7 Comprehensive interpretation map of basement structures of Baiyan-Hongjiang

（四）定安-东坡断面

此成果为《雷琼地区地壳结构研究》项目的成果，剖面位置见图24-2。根据其地震资料，共划分P_g、P_m、P_{c_1}和P_{c_2}等四组地震波震相。

P_g波是结晶基底或古生代变质岩界面上的首波，在所有炮点记录上均能观测到，视周期为0.1~0.14 s，波形较稳定。其界面速度为6.0~6.05km/s。沉积层的平均速度为3.1km/s。

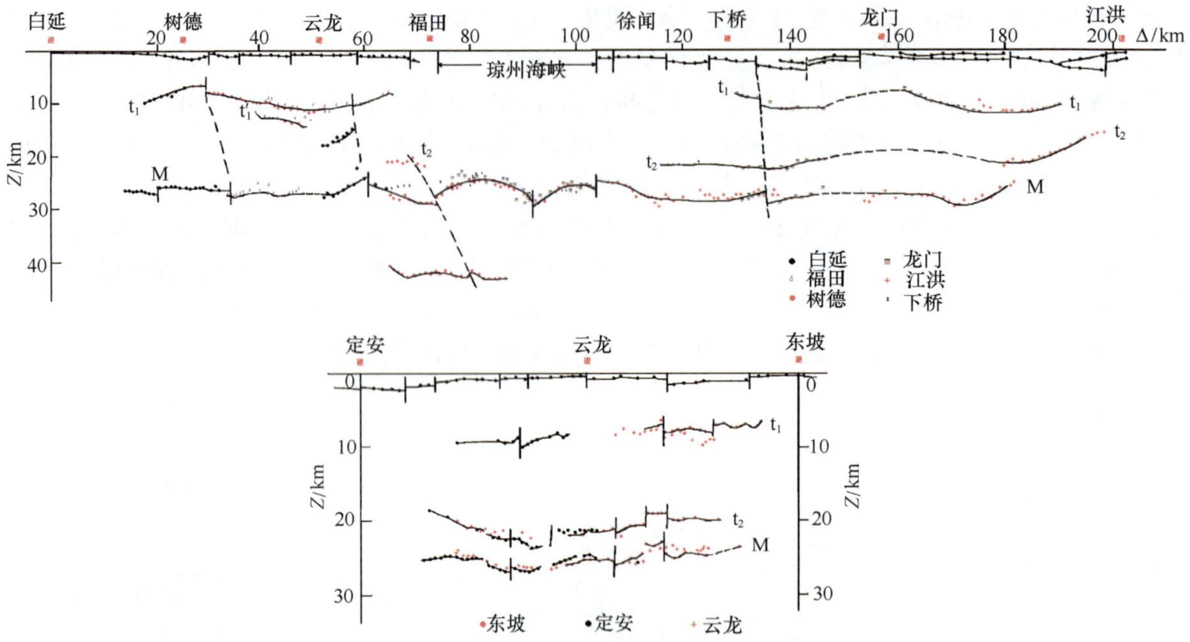

图 24-8　白延-江洪（Ⅴ上图）、定安-东坡（Ⅳ下图）断面地壳内部结构图

Fig. 24-8　Cross-section of the inner structure of crust in Baiyan-Jianghong（Ⅴ upper）and Ding'an-Dongpo（Ⅳ lower）

P_{c_1} 波组几乎与 P_g 波平行，但振幅较大，频率偏低，视周期在 0.12～0.20 s 之间，其平均速度值偏低（5.4km/s），为基底以下低速层下界面的反射。

P_{c_2} 波组整条剖面能记录到，但能量较弱，它出现在临界角内，为地壳内部的反射波。其上覆地层的平均速度为 5.94km/s。

P_m 波组为整条剖面记录到的能量最强的震相，走时曲线呈双曲线形。除个别地段外，波形变化不大，视周期为 0.08～0.16 s。在定安、东坡炮点的记录上则在临界角内出现。其地壳平均速度为 6.13km/s。

沿着定安-东坡剖面，基底在定安附近为一凹陷，最大深度为 1.5km，该凹陷向西南和东北方向逐渐抬升，在云龙附近上升到 300～600m，至东寨港下面为一西深东浅的断陷，过东寨港后基底接近地表（图 24-9）。沿该剖面基底被 8 条断裂切割，说明本区基底断裂十分发育。另外，该剖面莫霍面深度 25～27.0km。上地壳厚度 7～9km，中地壳厚度 10～14km，下地壳厚度 4～7km（图 24-8）。

（五）常树-才芳东断面

此成果为《琼西北地区人工地震测深研究》项目的成果，剖面位置见图 24-2。该项目的目的是：利用折射波的方法了解琼北地区基底构造的分布情况。根据该项目的地震资料，划分出 t_1、t_2、t_3、t_4、P_g 等 5 组地震波震相。

t_1 波：从爆破点出发的第一段首波，其视速度 $v^* < 2.2$km/s，以能量衰减快、频率高为其特点。

t_2 波：2.0km/s $< v^* <$ 3.0km/s 左右，能量衰减快，频率高，优势频率为 25～10Hz。

t_3 波：3.0km/s $< v^* <$ 4.0km/s 左右，初至振幅大，优势频率为 6～8Hz。

t_4 波：4.0km/s $< v^* <$ 5.0km/s 左右，初至振幅大，持续时间长。

P_g 波：5.15km/s $< v^* <$ 6.38km/s，初至振幅明显比沉积层中折射波振幅小，但在远离炮点时随距离衰减较慢，追踪区较长。因其界面速度变化较大，故其界面性质存在明显差异。

该剖面呈近南北走向，结果展示了海南隆起与福山凹陷中段的接触关系及过渡带内东西走向王五-文教断裂构造。福山凹陷为新、中生代沉积两层结构，埋深分别约为 2.5km、4.5km；澄迈至和平段基底迅速下陷 4.0km，为海南隆起与福山凹陷过渡带；在过渡带的顶、底部基底界面出现强烈褶曲变化，推测为王五-文教主断裂控制（图 24-10）。

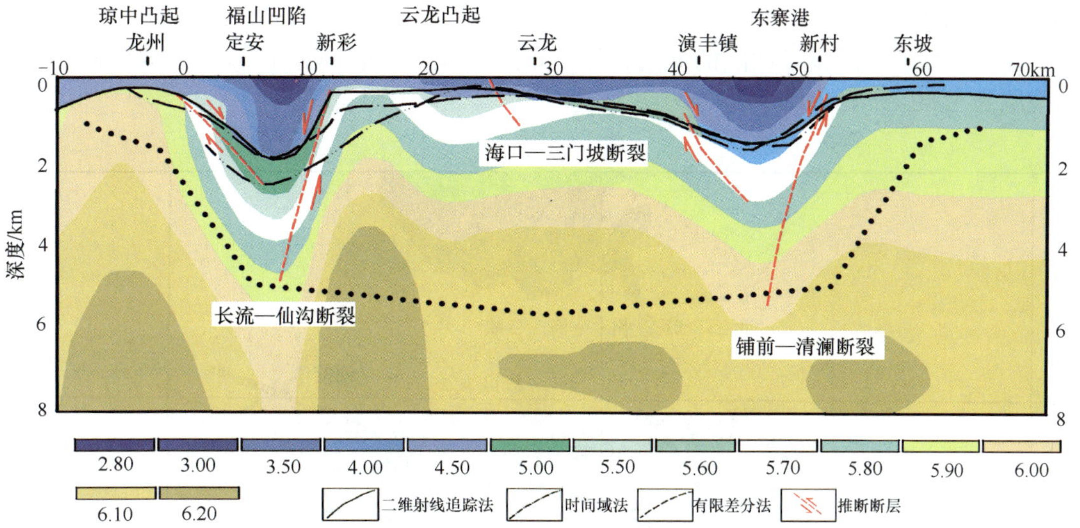

图 24-9 定安-东坡断面基底构造综合解释图

Fig. 24-9 Comprehensive interpretation cross-section showing basement structures of Ding'an-Dongpo

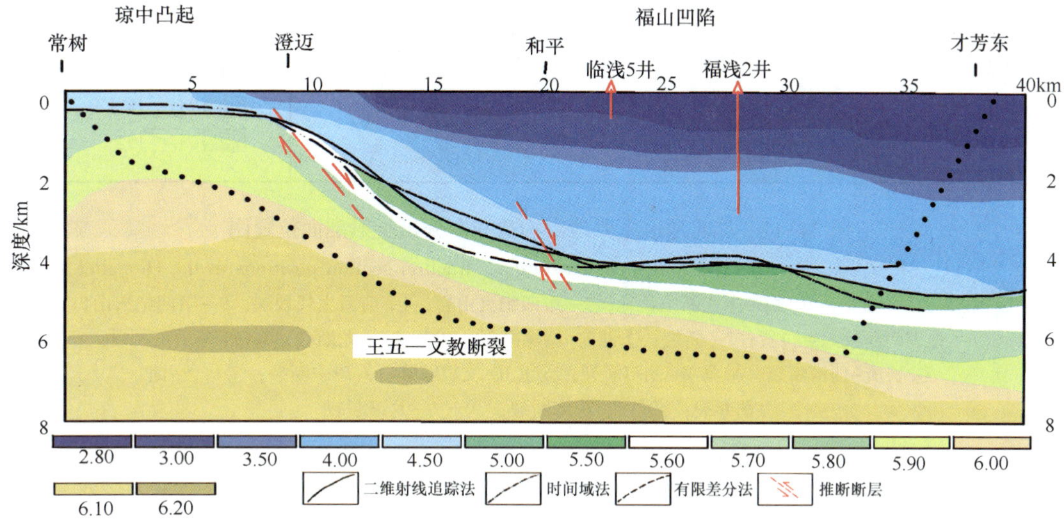

图 24-10 常树-才芳东断面基底构造综合解释图

Fig. 24-10 Comprehensive interpretation cross-section showing basement structures of Changshu-Caifang

（六）儋州-澄迈-冯坡断面

此成果为《海口市活断层探测与地震危险性评价》项目的成果。该剖面位于海口市南部地区，呈北东走向，长度为100~120km（图24-11）。其目的是：通过宽角/折射方法获取莫霍面以上的地壳深部结构和深部界面形态资料。采用高分辨地震折射方法，探查隐伏断裂位置及其随深度的变化特征。

1. 人工弹性波场特征

根据其地震宽角/折射方法数据处理资料，划分出 P_m、P_{c_1} 和 P_{c_2} 等三组地震波震相，其中 P_{c_1}、P_{c_2}、P_m 三个震相与地壳纵向上的13km、19km和25km左右的三个间断面对应。其中25km左右的间断面为莫霍面，其深度从西向东基本上由深逐渐变浅的形态，界面起伏不大，深度范围约为25.5~28.0km。另外两个间断面分别为壳内的 C_1 和 C_2 界面，其界面形态相对更为平坦，起伏不大。由于地壳内速度较低（小于6.0km/s），上地幔速度相对较大（大于8.0km/s），因而，界面上下速度跳动较大，P_m 震相清晰。在19~28.0km之间速度变化为6.30~6.60km/s，其地壳速度相对较高且梯度陡；

13~19.0km 之间速度变化为 6.10~6.20km/s，其速度梯度较小；13.0km 间断面以上的浅层速度结构较为复杂，纵向速度变化除雷琼凹陷的低速区外，5.0km 以上速度一般为 4.10~5.80km/s，5~13.0km 速度变化范围为 5.80~6.15km/s。

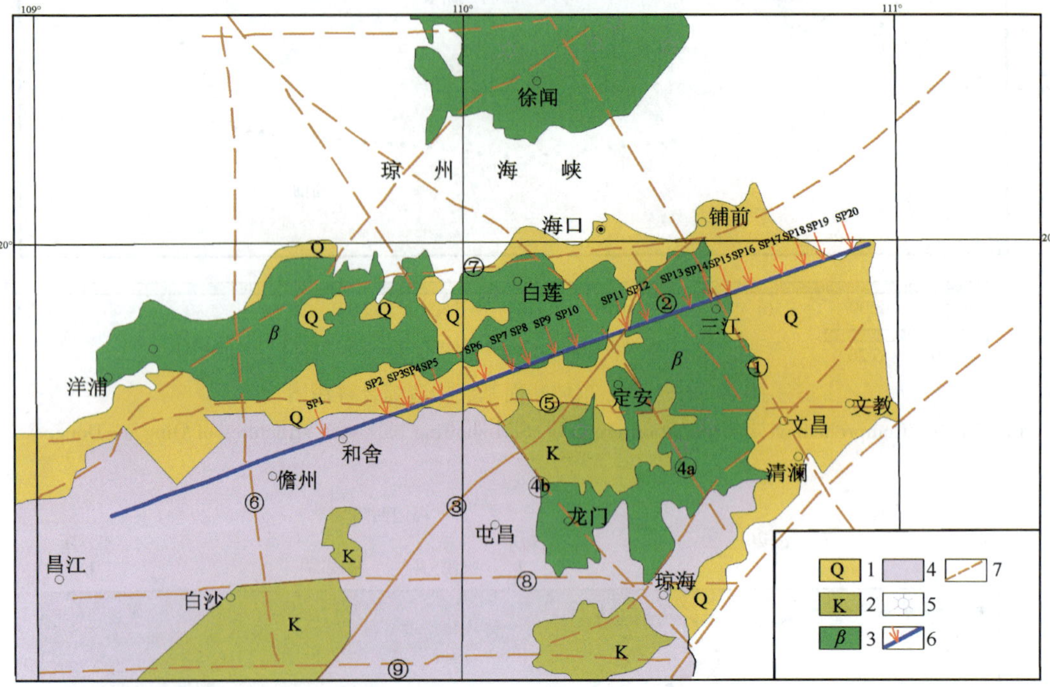

图 24-11 琼北地区主要地震构造和高分辨折射剖面位置图

Fig. 24-11 Main seismic structures and high-resolution refraction section positions in the Qiongbei area

1—第四纪活动盆地；2—晚中生代断陷盆地；3—第四纪火山岩；4—前新生代地层；5—第四纪火山口；6—地震剖面及炮点位置；7—断裂；①铺前-清澜断裂；②海口-云龙断裂；③铺前-乐东断裂；④a长流-仙沟断裂；④b马袅-龙门断裂；⑤王五-文教断裂；⑥琼中断裂；⑦马袅-铺前断裂；⑧昌江-琼海断裂；⑨东方-龙滚断裂

2. 二维速度结构特征

（1）上地壳结构

第一层：该层由基底折射波 P_g 所确定，沿剖面基底界面起伏明显，其深度在 1.8~4.5km 之间变化。基底界面之上的速度结构主要由一个强速度梯度层所组成，由近地表的 2.50~5.00km/s 变化到基底之上的 5.55~5.80km/s，基底面起伏比较明显，主要是受沉积盖层的影响所致。基底面的起伏变化反映了地质条件的复杂程度（图 24-12）。

第二层：基底面以下至 C_1 界面之间，C_1 界面为上地壳底部的一个反射界面，该界面起伏较为明显，其深度为 11.5~14.5km。地震波速度为 5.75~6.15km/s，速度梯度由上到下逐渐变弱，在该界面上速度跳跃为 6.05~6.15km/s 至 6.15~6.35km/s，跳跃差为 0.10~0.20km/s 之间（图 24-13，图 24-14）。

（2）中地壳结构

C_1 与 C_2 界面之间的层位，由壳内反射波 P_{c_1} 和 P_{c_2} 确定，C_2 界面与 C_1 界面相比沿剖面起伏变化较小，而且两个界面几乎相互平行。速度结构上为一弱的速度梯度层，由上到下速度变化为 6.15~6.38km/s，C_2 界面速度从 6.16~6.38km/s 跳跃至 6.30~6.45km/s，跳跃差为 0.11~0.14km/s。该层厚度为 5.6~8.5km，埋深为 18.0~21.7km（图 24-13，图 24-14）。

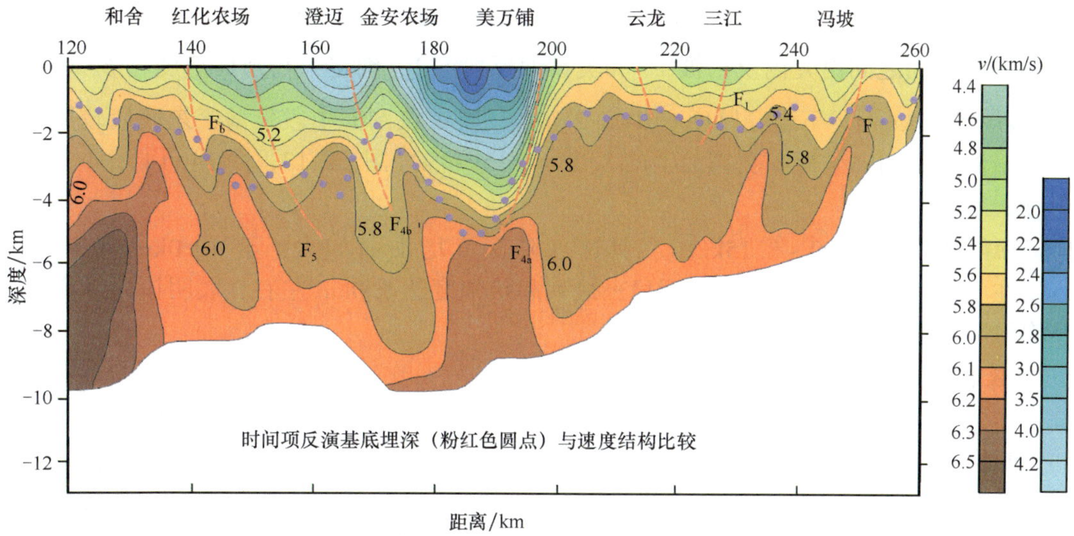

图 24-12 儋州-澄迈-冯坡剖面高分辨折射 P_g 波速结构与推断基底断裂图

Fig. 24-12 High-resolution refraction velocity structure and inferred basement faults of Danzhou-Chengmai-Fengpo section

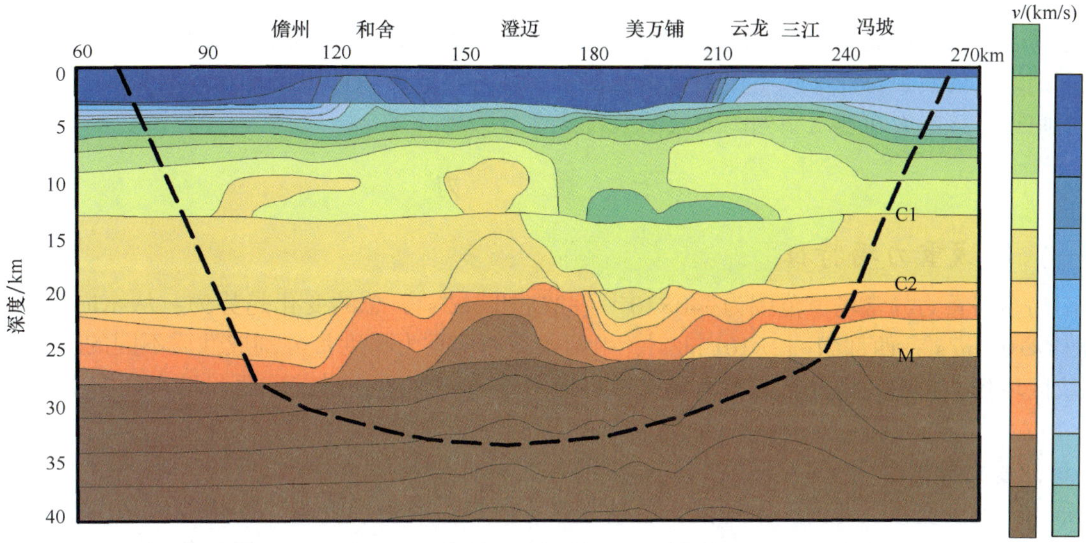

图 24-13 Zelt 方法反演儋州-澄迈-冯坡剖面地壳速度结构图

Fig. 24-13 Crustal velocity structure of Danzhou-Chengmai-Fengpo section obtained using the Zelt method

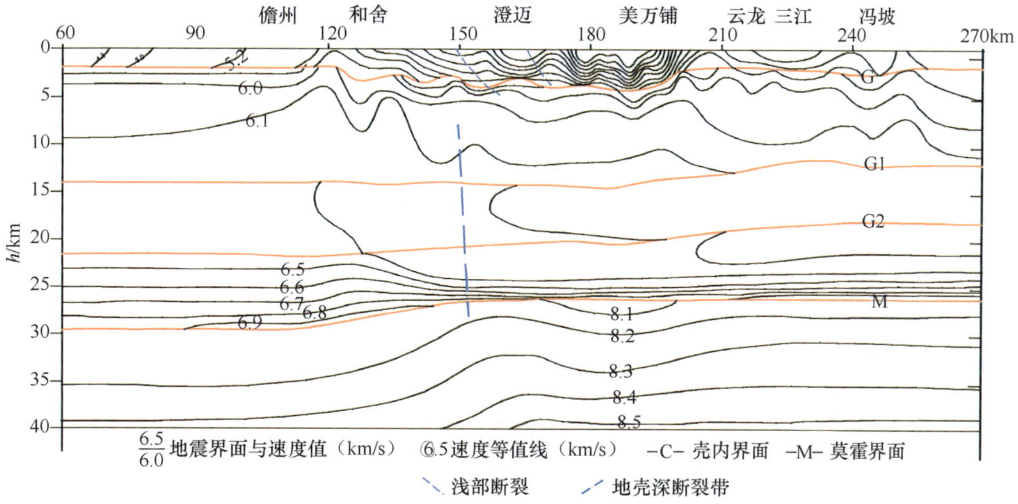

图 24-14 儋州-澄迈-冯坡地震测深剖面地壳上地幔二维速度结构和构造图

Fig. 24-14 Two-dimensional velocity structure and tectonics of crust-upper mantle of Danzhou-Chengmai-Fengpo seismic depth sounding section

(3) 下地壳结构

C_2 界面与莫霍面之间由壳内反射波 P_{c_2} 和莫霍面反射波 P_m 确定，其顶面速度为 6.30~6.45km/s，底面速度 6.80~6.97km/s。下地壳厚度 6.0~8.2km，反映了该剖面中段较薄，东西段较厚。速度由下地壳底板的 6.80~6.97km/s 跃至莫霍面上的 8.00~8.15km/s（图 24-13，图 24-14）。

(4) 上地幔上部

上地幔顶部速度为 8.00~8.15km/s，其速度随深度增加而加大，至 40.0km 深速度为 8.38~8.56km/s。在中段速度等值线起伏变化明显，最大起伏幅度达 8.0km 左右，表明该段纵横向上速度存在强烈非均匀性（图 24-14）。

(5) 断裂特征

该剖面高分辨探测结果，共推断 8 条断裂，这 8 条断裂的特征为：①F 断裂位于冯坡东北侧，倾向西南的正断层，东升西降，为基底断裂；②F_1 断裂与铺前-清澜断裂的中北段相吻合，为穿过基底的正断层，其下端点埋深 1.6km 左右；③F_2 断裂与海口-云龙断裂南段相吻合，为基底断裂，倾向北东，下端点埋深 1.6km 左右；④F_{4a} 断裂与长流-仙沟断裂相吻合，倾向西南，近于直立，基底断距约 3.5km，受断裂的控制，福山凹陷在该区域沉积巨厚，其厚度约 5.0km；⑤F_{4b} 断裂位于澄迈凹陷内，上端点埋深约 1.0km 左右，为一倾向东的隐伏基底断裂；⑥F_5 位于澄迈和红华农场之间，与王五-文教断裂相吻合，为一向北倾的超基底正断层；⑦F_a、F_b 断裂：F_a 断裂位于金安农场东北部附近，该断裂在地震记录上的断裂特征反映不很明显，但在速度结构上具有较明显的断裂结构特征；F_b 断裂位于红华农场附近，该断裂在速度结构图上其断裂特征十分明显（图 24-12）。

三、重　　力

（一）区域重力场特征

海南岛布格重力异常场值范围为 $-40\times10^{-5}\sim+28\times10^{-5}m/s^2$，总的变化趋势为：从东向西逐渐减小，东部为 $18\times10^{-5}m/s^2$，西部为 $-12\times10^{-5}m/s^2$，由两条东西向重力梯度带分割成性质不同的琼北、琼中和琼南三块区域场，再由一条北东向重力梯级带将琼中重力异常区分割成琼西和琼东两个重力异常小区（图 24-15）。

1. 琼北重力异常区

分布于王五—文教以北的广大地域，包括文昌、海口、澄迈、临高、儋州等地区，布格重力异常等值线呈东西、北西或北东走向，以相对变化幅度较大的重力高与重力低相间分布，其间多具明显线性梯级带为特征。其中福山重力低、云龙重力高为岛内最为醒目的重力异常之一。

福山—白莲一带和新盈—三都一带重力低，分别为福山凹陷和邻昌凹陷的反映，福山凹陷具有良好的油气勘探前景。云龙—三门坡和临高地区以重力高为特征，分别属云龙隆起和临高隆起的反映。该地区重力高与重力低之间的线性梯级带为断裂的反映。

2. 琼中重力异常区

位于王五—文教以南，九所—陵水以北地区。为布格重力异常的负值区，以布格重力低为主，并间有布格重力高；布格重力异常等值线走向和重力低长轴多呈北东向。昌江、尖峰、保亭重力低和琼中重力高，是区内规模最大的重力异常。

昌江重力低异常　分布于昌江县石碌—白沙县金波—雅加大岭一带，布格重力异常总体呈北北西向的长椭圆状展布，长大于 30km，宽 10~12km，北和西两侧等值线较密集，异常幅度大，峰值明显。该重力低异常与花岗岩体相对应，是昌江二长花岗岩、雅加大岭正长花岗岩的反映。

尖峰岭重力低异常　分布于佛罗、板桥、尖峰岭一带，布格重力异常总体呈东西向椭圆状展布，长大于 40km，宽大于 20km，北、东和南侧等值线密集，异常幅度大，峰值明显。异常为尖峰岭正长花岗岩的反映。

保亭县重力低异常　分布于五指山-保亭县一带，布格重力异常总体呈北东向似椭圆状展布，长大

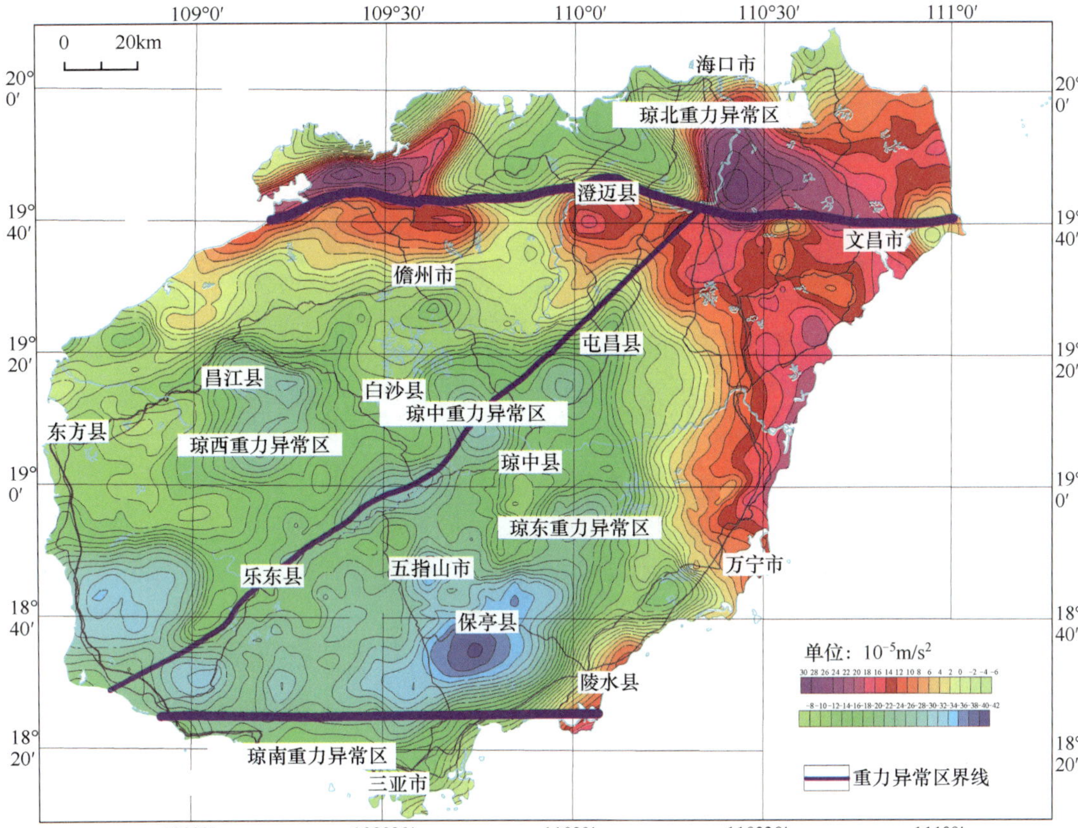

图 24-15 海南岛布格重力异常及分区示意图

Fig. 24-15 Schematic diagram and division of Bouguer gravity anomalies in the Hainan Island

于 40km，宽大于 20km，异常幅度大，东南侧等值线密集，梯度大，异常应是保城二长花岗岩和吊罗山二长-正长花岗岩所引起。

琼中重力高异常 分布于琼中县长征至长安一带，异常总体呈近南北向的椭圆状展布，异常幅度大，南部、东部等值线较密集，分别向南和向东延伸。异常为长城系和中元古代花岗岩的反映。

另外，根据区内的重力场特征，大致以乐东—定安一线为界分为两个亚区，即乐东—定安一线以西为琼西异常小区，乐东—定安一线以东为琼东异常小区。琼西异常小区主要分布抱板重力高、昌江重力低、尖峰重力低等。而琼东异常小区主要分布琼中重力高、保亭县重力低、千家重力低等。

3. 琼南重力异常区

分布于九所-陵水以南地区，布格重力异常等值线多呈东西走向，且重力场变化较平稳。区内分布田独重力低、三亚-罗蓬重力高、榆林重力高等。重力低为花岗岩的反映，重力高与古生代地层有关。

（二）沉积基底面重力场特征与沉积基底面

从海南岛布格重力异常小波分解三阶细节图（图 24-16）可见，重力低均对应了已知的坳陷、凹陷，重力高对应了基底隆起区。重力异常总体上呈北东走向，在北部的雷琼断陷区，从西往东相间分布着三低两高异常，它们分别对应着峨蔓凹陷、临高凸起、福山凹陷、云龙凸起、东寨港凹陷。在中部地区分布着相对变化幅度不大的重力低和重力高，其中昌江、尖峰、保亭重力低为区内最醒目的异常。这些重力低异常与花岗岩对应，重力高为古生代地层的反映。但这一地区的花岗岩和古生代地层，多已出露地表，并受到不同程度的剥蚀。另外，在其中部还分布有雷鸣、阳江-琼海、白沙等中生代盆地。

根据常树-才芳东和儋州-澄迈-冯坡等断面基底构造解释结果，以及《海南岛 1：20 万区域重力

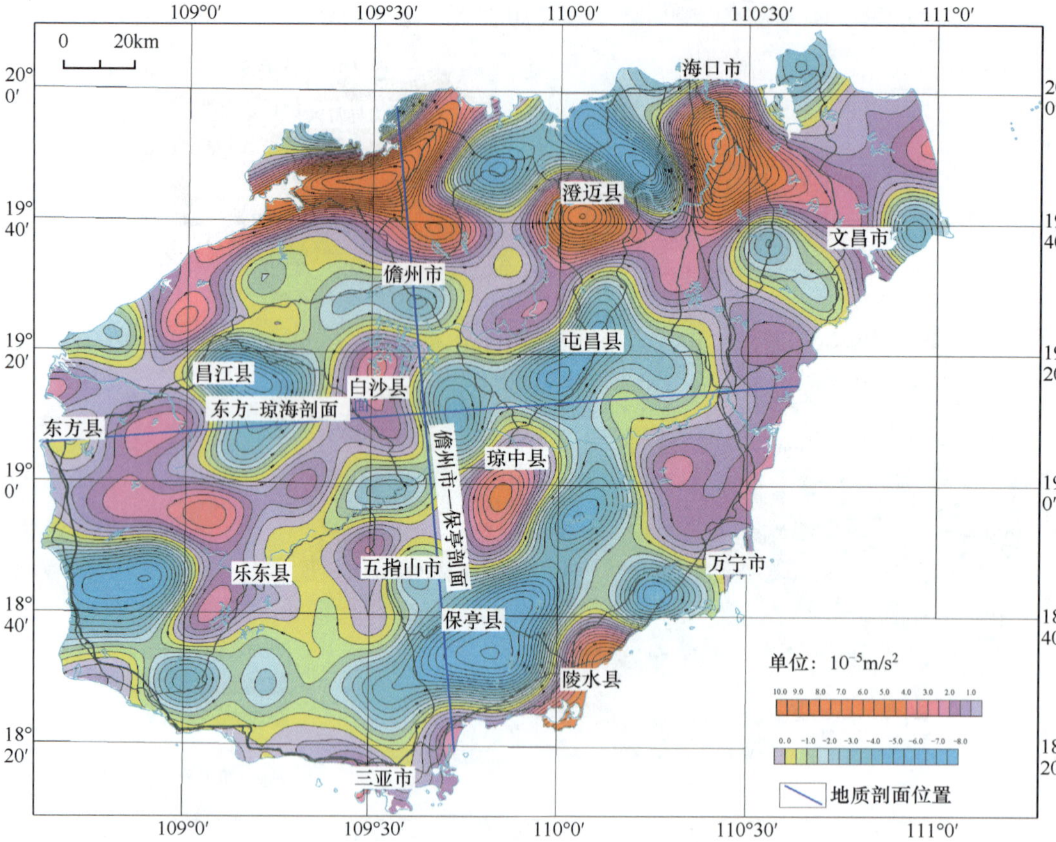

图 24-16　海南岛布格重力异常小波分解三阶细节图

Fig. 24-16　Thrid-order detailed diagram of wavelet decomposition of Bouguer gravity anomalies in the Hainan Island

调查报告》解释结果，海南岛西北部的峨蔓凹陷呈南东浅北西深，陆地上凹陷最大深度为 2100m 左右；长坡凹陷呈南翼缓，北翼稍陡，上宽下窄近于对称的 V 形，最大深度为 1100m 左右；北部的福山凹陷为北深南浅的箕状，其最大深度可达 5900m 左右，该凹陷中还分布有美台断阶带，老城断阶带，南部斜坡带，皇桐凹子，福山-花场拱起和白莲凹子等次级构造；海口凹陷为南浅北深的斜坡状，陆地上最大深度为 1600m 左右；北东部的东寨港凹陷呈西翼缓东翼陡、上部宽下部窄的倒三角形状，最大深度为 700m 左右；文岭凹陷呈上宽下窄近于对称的 V 形，最大深度为 1500m。而海南岛中部的雷鸣中生代盆地，基底埋深在 1100~1800m；白沙中生代盆地基底埋深 800~1300m，其深度自北西往南东增大，最大深度为 2300m 左右。

（三）莫霍面重力场特征与莫霍面

海南岛布格重力异常小波分解四阶逼近图（图 24-17）与断面反映的莫霍面形态，与以往认知的本岛莫霍面形态相似，因而将其作为莫霍面相应的重力场。由图 24-17 可见，重力异常等值线总体呈北东—北东东走向，呈两低两高之特征，即北东部、北西部高，中部和北部低。自西南向北东，重力值由 $-20\times10^{-5}m/s^2$ 左右逐渐增大，在文昌的蓬莱、会文一带为 $12\times10^{-5}m/s^2$；在本岛中部至西南部的五指山-乐东千家一带重力值最低为 $-22\times10^{-5}m/s^2$ 左右，多条重力异常等值线呈北东走向的椭圆状圈闭；重力异常等值线在昌江-屯昌呈近东西走向，在定安-万宁呈近南北走向，在万宁-三亚为北东走向。

据《华南地区 1：100 万物化遥综合解释成果编图》项目成果，认为海南岛莫霍面等深线（图 24-18）在琼西北部的白马井一带呈一未圈闭的半椭圆形，到琼东北部的文昌、定安、澄迈一带莫霍面等深线由北东向往南转为北西走向。整个琼北地区莫霍面总的变化趋势为东部浅西部深，其东部的文昌-

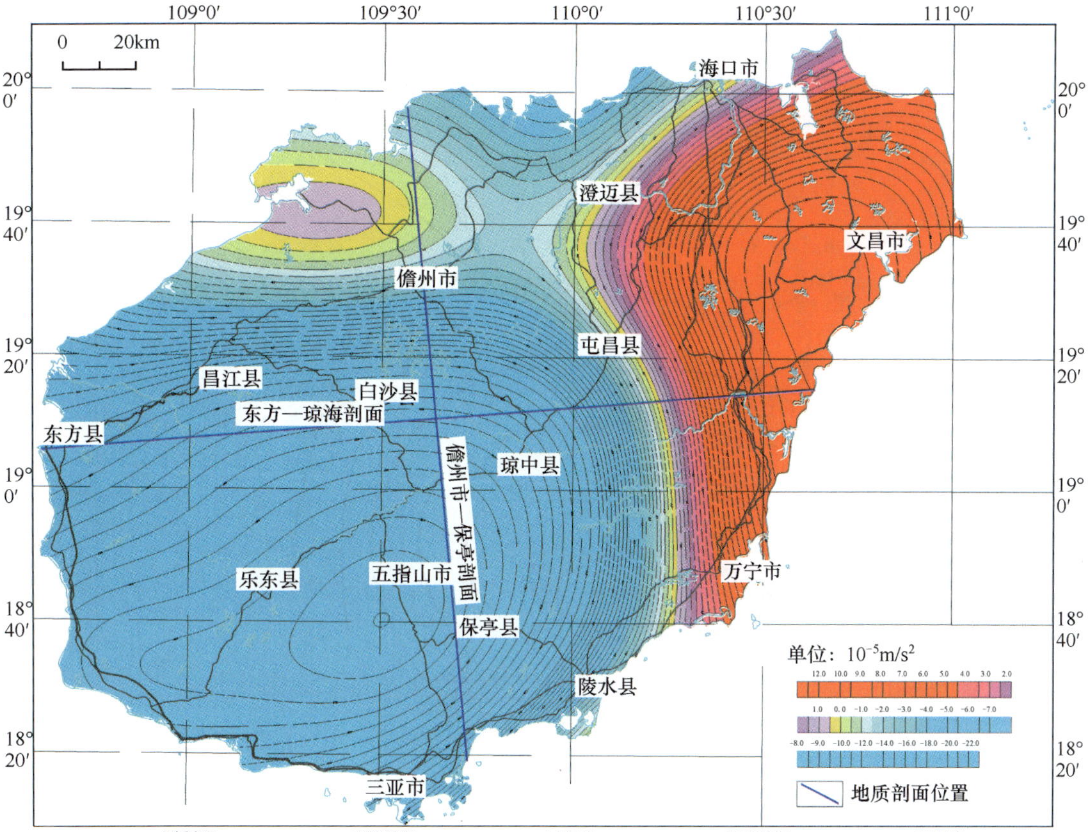

图 24-17 海南岛布格重力异常小波分解四阶逼近图

Fig. 24-17 Fourth-order approximation diagram of Bouguer gravity anomalies in the Hainan Island

琼海潭门一带为 26.5km，到西北部的白马井一带为 28.0km，起伏不大，为莫霍面隆起区。在岛中部地区莫霍面等深线呈北东走向的椭圆状。其莫霍面总的变化趋势为中部深四周浅，在五指山及琼西南为莫霍面坳陷区，莫霍面深度为 27.5~29.5km。

另外，海南岛为地幔隆起背景上的凹陷区，幔凹的中心在琼中至乐东一带。根据莫霍面的幔坡变化带，表现为重力梯级带的特征和规律，将海南岛划分为如下深部构造变化带：昌化-屯昌缓坡带、尖峰-万宁幔凹带、九所-陵水缓倾带、琼海-万宁缓倾带、万宁-陵水陡倾带、昌江-感城缓倾带。

而本次项目组通过收集各种资料研究，认为海南岛莫霍面等深线（图 24-18）存在两个明显特征：一是见于乐东-文昌一带，等深线走向北东，等深线整体为封闭的椭圆状，乐东-琼中一带莫霍面深度最大，为 29~29.5km，往四周逐渐减低至 26.5km，属于幔凹区或地壳增厚区。二是位于海南岛北部儋州—文昌一线，等深线走向近东西向，由西至东等深线值减低，由 28km 减至 26.5km；儋州-澄迈一带等深线梯度变化小，莫霍面深度基本一致，介于 27.5~28km 间，相对北东向等深线，该地区可能属于幔隆区或地壳减薄区。就两个方向的规模而言，以走向北东的等深线为主，具有区域性，东西向等深线具有局部性。由此判断，北东向的等深线显示出来的地壳深度结构属于早期形成，而东西向等深线则属于后期叠加形成。

上述两个不同方向的等深线总体与海南岛地质演化基本吻合。现有地质资料显示，海南岛华力西期—印支期南、北地块发生碰撞拼贴，并由于南面地块的俯冲，直接导致了北面地块地壳的增厚和褶皱变形及区域性变质，也因地块碰撞具有由西向东、由点到线的剪性拼贴的特点，使得位于西部乐东-琼中地区地壳厚度最大。喜马拉雅期，由于南海扩张，软流圈上涌，地壳减薄，海南岛沿东西向王五-文教构造发生裂解，北部发生强烈的基性火山喷溢或喷发，形成了海南岛北部东西向基性火山岩带（被）。

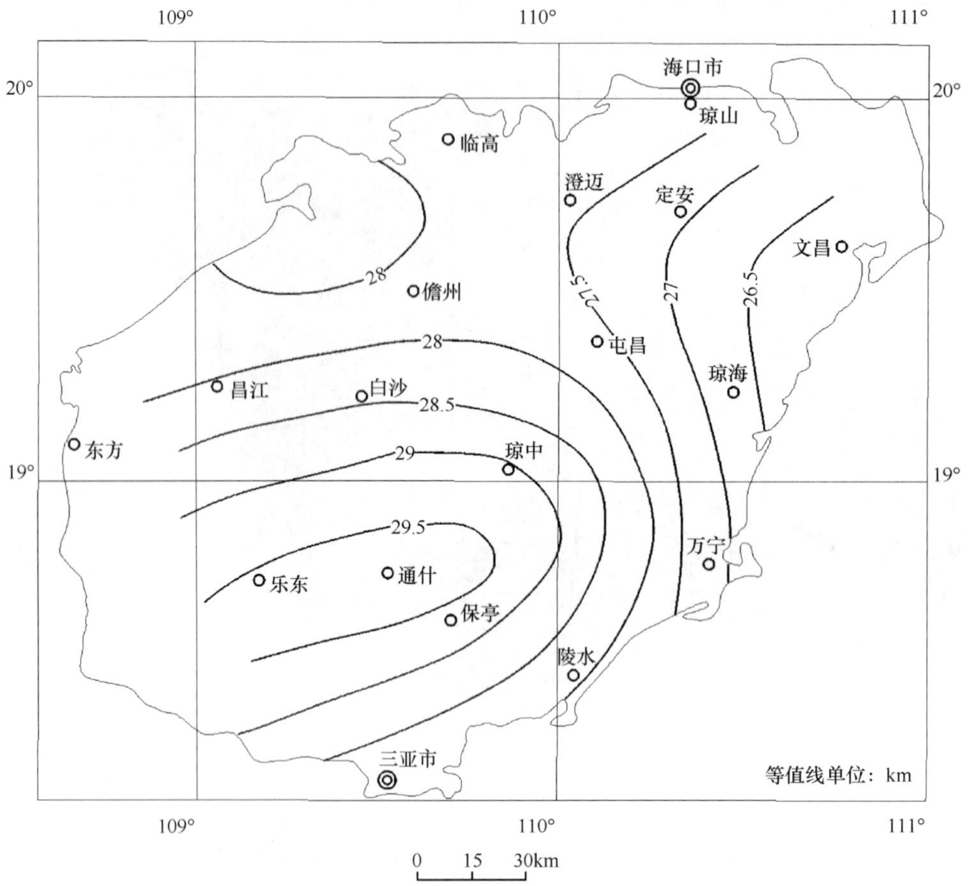

图 24-18　海南岛莫霍面等深线分布图

Fig. 24-18　Distribution of Moho depth contour in the Hainan Island

（四）岩石圈结构特征

1. 沉积层

（1）琼北地区沉积层

琼北地区沉积主要为古近系—新近系、第四系沉积盆地，其沉积厚度变化较大，总体厚度在630~5900m 内变化，自西往东分布 10 个凹陷（图 24-19）。

1）峨蔓凹陷：位于岛西北部，向海域延伸，为迈陈凹陷的一部分，包括三都、峨蔓、新盈、东英等地，岛上面积约 350km^2。经对重力异常正演拟合计算结果，该凹陷为南东浅北西深，陆地上凹陷最大深度为 2100m 左右。

2）福山凹陷：位于岛北部，是区内规模最大的凹陷，包括马袅、多文、福山、美亭、遵潭、白莲等地，面积约 1500km^2。根据重力场特征和前人在该区勘查石油工作结果，为北深南浅的箕状，经对重力异常正演拟合计算结果，凹陷最大深度可达 5900m 左右。该凹陷又分为美台断阶带、老城断阶带、南部斜坡带、皇桐凹子、福山-花场拱起和白莲凹子等。

3）海口凹陷：位于岛北部。包括海口市、叶里村等地，岛上面积约 200km^2。经重力异常正演拟合计算结果，凹陷呈南浅（700m）北深（1600m）斜坡状。

4）东寨港凹陷：位于海南岛东北部，包括演丰、三江、东寨港等地，面积约 170km^2。经重力异常正演拟合计算结果，凹陷呈西翼缓东翼陡、上部宽下部窄的倒三角形状，最大深度为 700m 左右。

5）长坡凹陷：位于儋州市长坡镇一带，面积约 160km^2。经重力异常正演拟合计算结果，凹陷呈

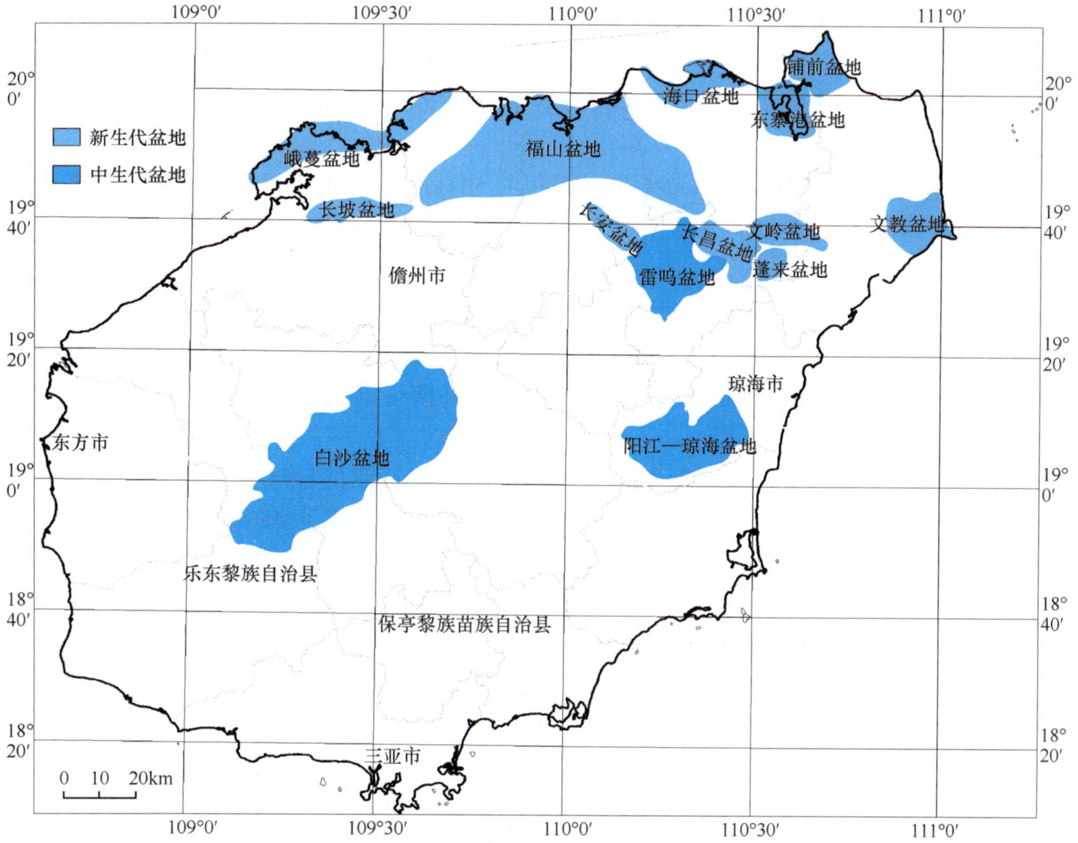

图 24-19 海南岛重力推断沉积盆地分布图

Fig. 24-19 Distribution of sedimentary basins inferred by gravity, Hainan Island

南翼缓，北翼稍陡，上宽下窄近于对称的 V 形，最大深度为 1100m 左右。据长坡煤矿地质勘探资料，该断陷盆地上覆地层为古近系，且含褐煤和油页岩，基底为白垩系上统报万群。

6）文岭凹陷：位于琼山区文岭、树德一带，面积约 100km^2。经重力异常正演拟合计算结果，凹陷呈上宽下窄近于对称的 V 形，最大深度为 1500m。据海南地质大队在该区普查找煤资料，该盆地为新生代沉积盆地，盆地呈东西向不对称向斜构造。凹陷中主要为古近系始新统长昌组、古新统昌头组。基底为花岗岩、闪长岩。

7）长昌、蓬莱凹陷：位于长昌—蓬莱一带，面积约 80km^2。据长昌煤矿地质勘探资料，该凹陷为长昌古近系含煤盆地。

8）长安凹陷：位于澄迈县长安镇一带。凹陷呈北西向，其北东向和南西侧均以断裂为界，面积约 35km^2。地表为第四系覆盖；重力推测凹陷最大深度为 630m。

9）铺前凹陷：位于测区东北部。南侧以博厚-铺前断裂为界，北部、东部向海域延伸。包括铺前镇、赤坡、双龙坡等地，面积约 150km^2。该凹陷有两个沉积中心，其一：位于潭文村附近；其二：位于白石村至滨海一带。经正演拟合计算，白石村沉积中心呈上宽下窄近对称状，最大深度为 850m 左右。

10）文教凹陷：位于文昌市文教镇一带，北侧和西部均以断裂为界，南部向海域延伸，陆地面积约 60km^2。重力推测凹陷最大深度为 1200m 左右。

上述凹陷与临高凸起、云龙凸起组成雷琼裂谷盆地（海南岛部分），其中福山凹陷为主要找油区域，油气勘探已取得了突破。长坡凹陷、长昌凹陷储存着储量较大的褐煤和油页岩。

（2）五指山隆起区沉积层

五指山隆起区主要分布中生代沉积盆地，它们分别为规模较大的雷鸣盆地、阳江-琼海盆地、白沙盆地（图 24-19）等，其沉积层总体厚度在 650~1300m 内变化。

1）雷鸣盆地：位于居丁、雷鸣和新兴镇以东一带，面积约 412km^2。经重力异常正演拟合计算结果，其中生代白垩纪地层沉积厚度在1100~1800m之间。

2）白沙盆地：位于白沙县南溪乡、乐光农场、番阳、鹦哥岭一带，面积约1570km^2。经重力异常正演拟合计算结果，其中生代白垩纪地层沉积厚度在800~1300m之间，其厚度自北西往南东增大，最大厚度为2300m左右。

3）阳江-琼海盆地：位于阳江—琼海一带，面积约504km^2。经重力异常正演拟合计算结果，其中生代白垩纪地层沉积厚度在650~1000m之间。

2. 地壳厚度

琼北地区地壳厚度从西往东逐渐变薄，即由岛西北部的白马井一带28.0km到琼海潭门一带的26.0km，但起伏不大。岛中部的五指山隆起区为四周高中间低，在五指山市—乐东县一带，地壳厚度29.5km，为坳陷区。到岛南部的三亚地区，地壳厚度由北往南逐渐变薄，由29.5km减薄到27.0km。

3. 岩石圈

海南岛缺少研究地壳深部的地球物理资料，尤其是研究岩石圈的地球物理资料。根据朱介寿教授等编制的《中国及相邻陆域海域岩石圈厚度图》（图24-20），资料显示海南岛北部岩石圈厚度为70~75.0km，其南部为65~70.0km，从南往北逐渐增厚，呈北厚南薄之特征。

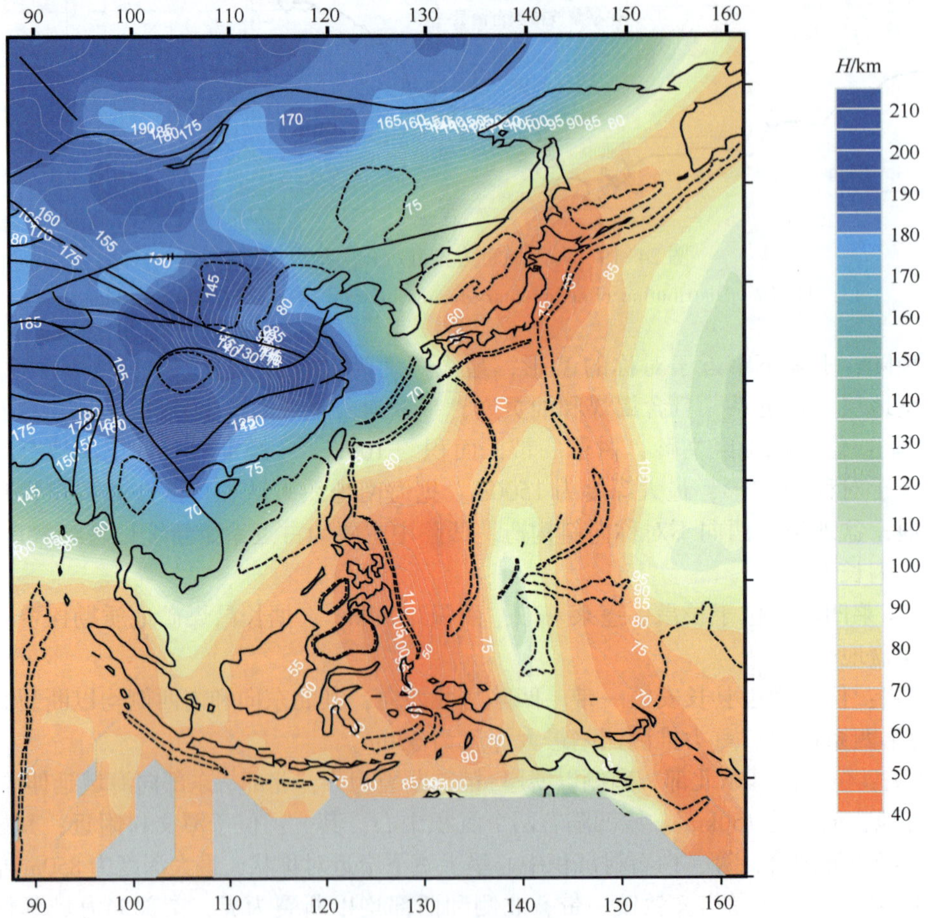

图24-20　中国及相邻陆域海域岩石圈厚度图

Fig. 24-20　Lithosphere thickness of China and its adjacent continental and marine areas

四、航　　磁

从海南岛航磁异常平面等值线图（图24-21）上可以看出，磁场具有明显的南北分块特征。根据九所-陵水、王五-文教两条线性异常带可将测区分为三个磁场区，分别是琼南高磁异常带、琼中升高

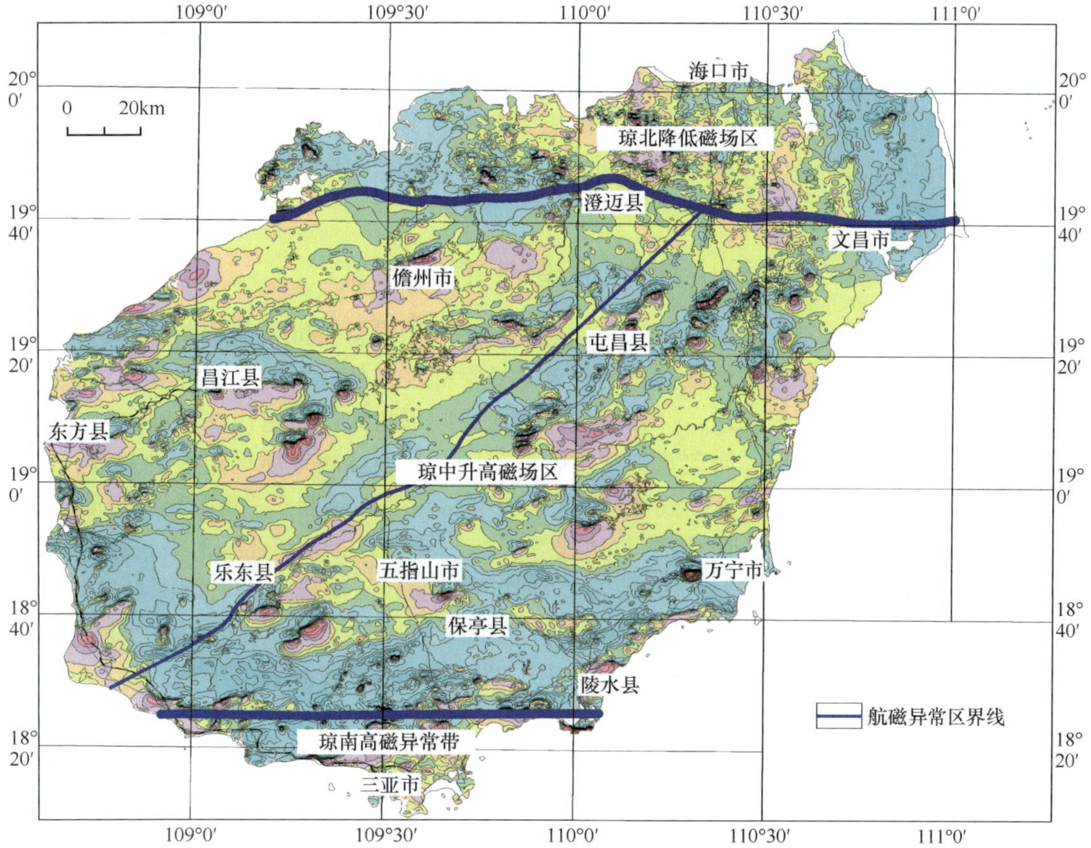

图 24-21 海南岛航磁 ΔT 异常及分区示意图

Fig. 24-21 Schematic diagram and division of ΔT aeromagnetic anomalies in the Hainan Island

磁场区、琼北降低磁场区，再由一条北东向磁异常带将琼中升高磁场区分割成琼西和琼东两个磁异常小区。

(一) 琼南高磁异常带

分布于南部沿海地区，东起陵水，经同安岭、牛腊岭直到九所一带。磁场强度大，强度和梯度变化都很复杂的正负伴生异常带。在伴生负磁场中叠加有大量之局部异常。崖城以西异常呈北西走向，崖城以东异常带呈北东走向，组成琼南复杂的高磁异常带。区内的航磁异常主要与岩体基性成分增高有关。

(二) 琼中升高磁场区

北界王五—文教一线，南至九所—陵水一带，横贯海南岛中部。磁场强度较大，梯度变化也较大，东西两端有变窄趋势，大致在儋州—屯昌一线以南出现一系列无固定走向之平缓起伏航磁异常，即被一系列北东向磁场和局部异常破坏干扰，以北为一条近东西向之正磁场带，但在屯昌地区中断，在琼东地区又复出现。

另外，根据区内的磁场特征，大致以乐东—定安一线为界分为两个亚区，即乐东—定安一线以西为琼西异常小区，区内航磁异常多呈北东和东西走向，规模和强度相对较大，梯度相对较陡；乐东—定安一线以东为琼东异常小区，区内航磁异常多呈北东走向，规模相对较小，强度稍大、梯度相对较缓（图 24-21）。琼西异常小区主要分布儋州市磁力高、军营磁力高、昌江磁力高、尖峰和打金村磁力低等。而琼东异常小区主要分布乐东磁力高、五指山市磁力高、琼中磁力高、屯昌磁力低、万宁磁力低等。该区磁力高多与中酸性、基性侵入岩体关系密切，磁力低与燕山期花岗岩有关。

（三）琼北降低磁场区

分布于新州-加来-定安以北沿海地区及琼东一线东侧。区内锦山-清澜一线以西为变化复杂的以负值为主的磁场带，其表现为负磁场背景上出现剧烈跳动的强度变化特别大、梯度变化特别陡之磁场，反映新生代沉积盆地和火山喷发岩（玄武岩）的磁场特征。以东却出现一片全岛极小值最大的负磁场，其中有轻微的起伏，并有一些局部异常显示。负磁场反映花岗岩地区的磁场特征，局部异常与基性侵入岩体有关。

五、断裂构造与重磁异常

利用重磁异常解释断裂构造是行之有效的方法之一。用重磁异常特征来判别断裂构造的原则是：①线性延伸的重磁异常梯级带或梯级带的线性排列，重磁异常梯级带的存在说明物性界面沿某个水平方向发生了剧变，这是断裂构造的明显标志；②等值线的同向或同形扭曲带，这往往反映物性界面的位移或错断；③不同特征重磁场的分界线，两侧重磁异常特征不同，往往反映地质构造上的差异；④局部重磁异常呈串珠状线性排列；⑤重力高、低的线性过渡带；⑥异常宽度突变带；⑦重磁异常错动带。

根据上述原则，结合海南岛的地质构造和地震等资料，全岛共推测断裂 43 条，其中东西向 5 条，南北向 4 条，北东向 18 条，北西向 16 条。按规模和切割深度分，规模大、切割深度达地幔顶部的壳断裂 7 条，规模较大的基底断裂 27 条，规模较小的一般断裂 9 条。与地质已知（含推测）断裂吻合或基本一致的 15 条，部分基本一致的 12 条（表 24-10，图 24-22，图 24-23）。

表 24-10　海南岛重磁推断断裂构造及其特征一览表

Table 24-10　List of faults inferred by gravity and magnetism, and their characteristics, Hainan Island

序号	断裂编号	断裂名称	断裂长度/km	断层面走向	断裂类别	断裂构造重磁场特征	性质
1	F_1	王五-文教断裂带	210.00	东西	壳断裂	线性延伸重力梯级带；不同特征的重磁场分界线；人工地震资料；航磁不同磁场交变带，北侧为变化复杂的以负值为主的磁场带，南侧为平缓变化的正负伴生的磁场带	压-扭
2	F_2	儋州-新兴断裂	140.00	东西	基底断裂	东西向线性排列的重力梯级带；不同特征重力场分界线；线性延伸的航磁异常带	压-扭
3	F_3	昌江-琼海断裂带	210.00	东西	壳断裂	重力梯级带呈东西向线性排列；不同特征的重磁场分界线；剩余重力负异常和航磁异常呈串珠状东西向排列	压-扭
4	F_4	尖峰-吊罗断裂带	195.00	东西	壳断裂	剩余重力负异常呈串珠状东西向线性排列；通什至吊罗段呈梯级带；航磁资料为不同区域场的分界线和梯度带。南侧以较强的负异常为主；北侧以叠加在区域磁场背景上的局部正异常为主，异常一般为 75nT	压-扭
5	F_5	九所-陵水断裂带	120.00	东西	壳断裂	东西向重力梯级带的线性排列；不同特征重力场的分界线；由一系列近东西向排列的长轴状异常组成，构成一条近东西走向的磁异常带；南北两侧航磁异常特征明显不同，北侧以负异常为主，南侧则为升高正异常，并控制着南部中生代火山岩的分布	压-扭
6	F_6	昌江-抱伦断裂	83.00	南北	基底断裂	重力梯级带呈南北向线性排列；不同特征的重力场分界线；航磁线性延伸的异常带；异常突变带	张-扭
7	F_7	屯昌-兴隆断裂	100.00	南北	基底断裂	重力梯级带呈南北向线性排列；航磁低纬度化极线性延伸异常带；航磁异常梯级带	压-扭

续表

序号	断裂编号	断裂名称	断裂长度 km	断层面走向	断裂类别	断裂构造重磁场特征	性质
8	F_8	十字路-石壁断裂	75.00	南北	基底断裂	南北向线性延伸的重力梯级带；不同特征重力场的分界线；航磁异常串珠状异常带	张-扭
9	F_9	谭文-九曲江断裂	70.00	南北	基底断裂	北段：重力场特征不同；中段：分布有由中基性岩体引起的局部重力异常；南段：南北向重力梯级带；航磁线性异常带	张
10	F_{10}	东英-干冲深断裂	60.00	北东	壳断裂	呈北东向密集梯级带线性延伸；人工地震资料	压-扭
11	F_{11}	王五-昌城断裂	90.00	北东	基底断裂	重力梯级带呈北东向线性排列；剩余重力负异常呈北东向线性排列。航磁异常线性梯级带	张-扭
12	F_{12}	博厚-三更断裂	147.00	北东	基底断裂	北东和南西段重力梯级带呈北东向线性延伸，陈桶岭至乐洒段，为等值线的同向扭曲；乐海至大岭农场段，表现为异常宽度突变带。由叠加在区域磁场背景上的一系列串珠状局部异常组成。这些局部航磁异常构成一系列北东向的航磁异常带	压-扭
13	F_{13}	戈枕断裂	21.00	北东	一般断裂	重力高、低间的线性过渡带。不同磁场的分界线；北西侧为平稳正磁场，南东侧为一群呈北东向展布的航磁异常	张-扭
14	F_{14}	和舍断裂	64.00	北东	基底断裂	重力梯级带呈北东向线性排列	压-扭
15	F_{15}	白沙断裂	92.00	北东	基底断裂	重力梯级带呈北东向线性排列；重力高、低间的线性过渡带。航磁异常呈串珠状异常带；异常突变带	压-扭
16	F_{16}	南棍园断裂	30.00	北东	一般断裂	北东向重力梯级带及等值线同向扭曲	压-扭
17	F_{17}	东营断裂	17.00	北东	基底断裂	北东向线性延伸的重力梯级带	压-扭
18	F_{18}	博厚-铺前断裂	110.00	北东	基底断裂	线性延伸的重力梯级带；航磁化极（$\Delta Z\perp$）异常梯级带；人工地震资料	压-扭
19	F_{19}	锦山-居丁断裂	86.00	北东	基底断裂	重力梯级带的线性排列；等值线同形扭曲带；异常宽度突变。北东和南西段为航磁异常梯级带，中段为航磁异常突变带	压-扭
20	F_{20}	定安-乐东深断裂	220.00	北东	壳断裂	线性延伸的重力梯级带；串珠状负异常呈北东向排列；异常宽度突变带；均衡异常不同特征重力场分界线。串珠状航磁异常带；航磁异常突变带	压-扭
21	F_{21}	琼中-保港断裂	115.00	北东	基底断裂	重力高、低间的线性过渡带；北东向串珠状负异常带。航磁异常（ΔT）低纬度化极水平一阶导数（135°）图上，为线性延伸的航磁异常带	张-扭
22	F_{22}	保亭断裂	90.00	北东	基底断裂	重力梯级带呈北东向线性排列。航磁异常（ΔT）低纬度化极水平一阶导数（135°）图上，为线性延伸的航磁异常带；异常错动带	张-扭
23	F_{23}	虎山水库断裂	13.00	北东	一般断裂	重力高、低间的线性过渡带	压-扭
24	F_{24}	林塘村断裂	11.00	北东	一般断裂	重力高、低间的线性过渡带	压-扭
25	F_{25}	新市-大坡断裂	108.00	北东	基底断裂	重力梯级带呈北东向线性排列	张-扭

续表

序号	断裂编号	断裂名称	断裂长度/km	断层面走向	断裂类别	断裂构造重磁场特征	性质
26	F_{26}	龙滚-东洲断裂	135.00	北东	基底断裂	重力梯级带呈北东向线性排列。航磁异常（ΔT）低纬度化极水平一阶导数（135°）图上，为线性延伸的航磁异常带	张-扭
27	F_{27}	陵水-三亚深断裂	100.00	北东	壳断裂	北东向线性延伸的重力梯级带；均衡异常梯级带。航磁线性延伸异常带；不同磁场的分界线	压-扭
28	F_{28}	椰东林场断裂	10.00	北西	一般断裂	重力高、低间的线性过渡带	压-扭
29	F_{29}	罗豆断裂	24.00	北西	基底断裂	等值线同形扭曲带	张-扭
30	F_{30}	东阁断裂	12.00	北西	一般断裂	北西向线性延伸的重力梯级带	张-扭
31	F_{31}	三江-东郊断裂	68.00	北西	基底断裂	北西向线性延伸的重力梯级带；人工地震资料	张-扭
32	F_{32}	龙华断裂	23.00	北西	基底断裂	重力异常等值线同形扭曲带；航磁异常梯级带	张-扭
33	F_{33}	树德断裂	13.00	北西	一般断裂	北西向线性延伸的重力梯级带；重力高、低间的线性过渡带	张-扭
34	F_{34}	石山-长坡断裂	100.00	北西	基底断裂	北西向线性延伸的重力梯级带；等值线收敛变窄；重力高、低间的线性过渡带；人工地震资料。航磁异常（ΔT）低纬度化极水平一阶导数（45°）图上，为异常突变带；异常梯级带	张-扭
35	F_{35}	长安断裂	26.00	北西	一般断裂	重力高、低间的线性过渡带	张
36	F_{36}	马袅-长安断裂	45.00	北西	基底断裂	重力高、低间的线性过渡带	张
37	F_{37}	山口-岭口断裂	65.00	北西	基底断裂	重力梯级带呈北西向线性排列	张-扭
38	F_{38}	光村断裂	20.00	北西	基底断裂	等值线同向扭曲带	张-扭
39	F_{39}	番加-东澳断裂	110.00	北西	基底断裂	重力梯级带呈北西向线性排列；航磁线性异常带；异常突变带	张-扭
40	F_{40}	雅星断裂	50.00	北西	基底断裂	等值线的同向扭曲；异常宽度突变；重力梯级带呈北西向线性延伸；重力高、低间的线性过渡带	张-扭
41	F_{41}	五指山-本号断裂	60.00	北西	基底断裂	等值线同向扭曲；北西向线性延伸的重力梯级带；异常宽度突变带；航磁线性异常带和异常突变带	张-扭
42	F_{42}	十月田断裂	17.00	北西	一般断裂	重力高、低间的线性过渡带	张-扭
43	F_{43}	高峰断裂	40.00	北西	基底断裂	重力异常等值线同向扭曲带；航磁异常突变带	张-扭

1. 东西向断裂带

这组断裂十分发育和突出，从北往南主要有：王五-文教、昌江-琼海、尖峰-吊罗、九所-陵水深大断裂带，其中九所-陵水构造带是Ⅰ级、Ⅱ级构造单元分界线，王五-文教构造带是Ⅲ级构造单元分界线。这些构造带由东西向压性或压扭性断裂带、构造-岩浆岩带和褶皱带组成，在空间展布上，它们均横贯全岛，长达200km，宽数千米，切割深度达上地幔35km以上，最深达65~75km，大致每带间隔25′左右呈等距分布。在重磁异常图上（图24-22，图24-23）一般表现为线性延伸的重磁异常梯级带、不同特征的重磁场分界线、串珠状的重力负异常带和航磁异常带等。以王五-文教深大断裂带

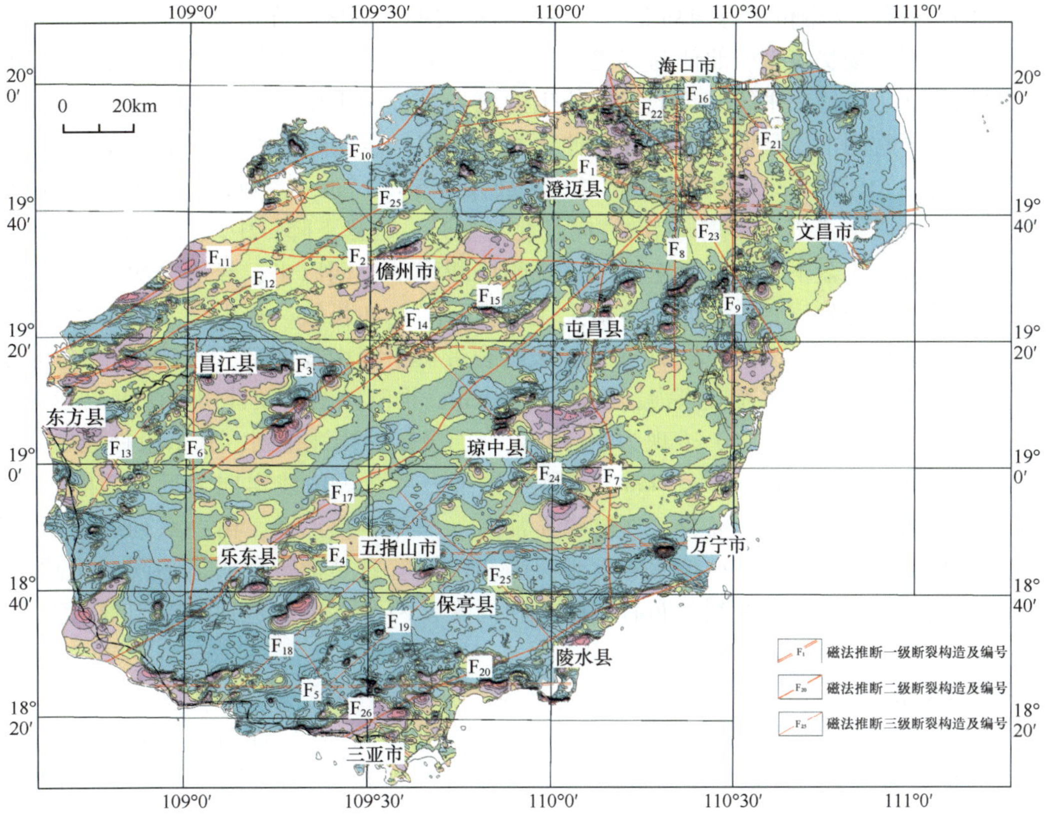

图 24-22 海南岛推断断裂及航磁 ΔT 等值线平面图

Fig. 24-22 Plan of inferred faults and (ΔT) aeromagnetic contour lines of the Hainan Island

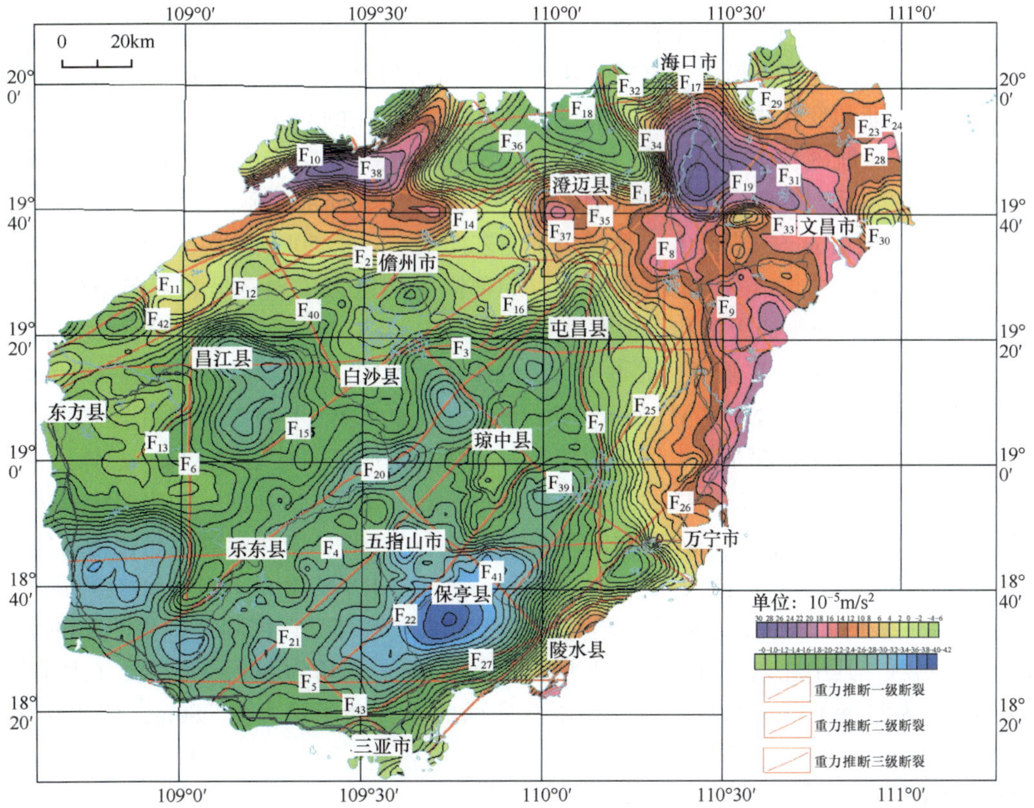

图 24-23 海南岛推断断裂及布格重力异常平面图

Fig. 24-23 Plan of inferref faults and bouguer gravity anomalies in the Hainan Island

为例（表24-10），该断裂带位于海南岛北部，东西向横贯整个测区，在磁场图上反映十分清晰，磁场特征表现为不同类型磁场交变带，北侧为变化复杂的以负值为主的磁场带，反映新生代沉积盆地和火山喷发岩的磁场特征。南侧为平缓变化的正负伴生的磁场带，为中生代、古生代沉积岩和岩浆岩的磁场特征。布格重力异常图上，呈明显的线性延伸重力梯级带，走向近东西，长170km，宽5～8km。重力水平梯度一般为2.5×10^{-5}～$3.5\times10^{-5}m\cdot s^{-2}/km$，最大为$5\times10^{-5}m\cdot s^{-2}/km$。人工地震测深等有关资料表明，该断裂的西段（王五-加来）错动方向是北盘上升，南盘下降；中段（加来-定安）是北盘下降，南盘上升；东段为南盘下降，北盘上升。这说明该断裂带长期以来，经历多次活动。而且是雷琼裂谷与五指山褶冲带的分界线，控制着众多新生代沉积凹陷（如长坡、福山、文岭、文教）的发生和发展。又如九所-陵水深大断裂带（表24-10），该断裂带位于海南岛南部，大致展布在18°25′。西起九所，往东经育才、英州至陵水黎安入海，岛上长120km，推测该断裂及其切割深度已达到地幔顶部。在布格重力异常图（图24-23）上，表现为东西向重力梯级带的线性排列，其东段梯级带长80km，等值线密集，水平梯度为$2.2\times10^{-5}m\cdot s^{-2}/km$。南翼布格重力异常等值线走向为近东西向，北侧为北东向，反映了两侧重力场特征的明显差异。在航磁异常图（图24-22）上，由一系列近东西向排列的长轴状异常组成，构成一条近东西走向的磁异常带；南北两侧航磁异常特征明显不同，北侧以负异常为主，南侧则为升高正异常，沿着断裂分布有牛腊岭和同安岭两个火山岩盆地，控制着南部中生代火山岩的分布。

2. 北东向断裂带

北东向构造形迹在岛内普遍发育，主要以形成北东向的压性-压扭性断裂带为特征，展布方向为50°～60°。北东向断裂带从西往东主要分布有：东英-干冲、王五-昌城、博厚-三更、马袅-铺前、定安-乐东、陵水-三亚等断裂带。这些断裂带在重磁异常图上（图24-22，图24-23）一般表现为线性延伸的重磁异常梯级带、不同特征的重磁场分界线、重磁异常突变带、重磁异常等值线同向扭曲、串珠状的重磁异常带等。以定安-乐东断裂带为例（表24-10），该断裂带位于海南岛中部，北东起定安，往南西经西昌、黎母山、毛阳、乐东至黄流，总体走向45°，长220km。在航磁异常图上（图24-22），该断裂为一条规模较大的北东向航磁异常带，由一系列轴向为北东向的长轴状局部航磁异常组成。这些局部异常的分布特征和变化规模与该区出露的侵入岩有着明显的对应关系，反映为侵入岩引起的航磁异常特征。另外，从航磁异常带的整体特征看，它又明显反映断裂带的航磁异常特征。并且沿这条北东向断裂带侵入的一系列岩体，导致局部地段磁性升高，形成局部航磁异常。在重力异常图上（图24-23），该断裂北东段（定安-黎母山）呈梯级带北东向展布，长50km，宽5～10km，水平梯度为1×10^{-5}～$2\times10^{-5}m\cdot s^{-2}/km$；中段（黎母山-乐东）以局部圈闭的重力负异常呈北东向排列为特征；南西段（乐东-黄流）则显示密集梯级带和异常宽度突变带。在1∶100万均衡重力异常图上显示不同特征重力场的分界线，即东南侧异常幅值比北西侧大5×10^{-5}～$10\times10^{-5}m\cdot s^{-2}/km$。而且认为该断裂为琼西岩浆弧（$Ⅵ_1$级）和琼东陆内盆地（$Ⅵ_2$级）的分界线。又如东英-干冲深断裂带（表24-10），该断裂带位于海南岛西北部。北东起东英，往南西经光村北侧至干冲，两端延伸入海，总体走向为65°，岛内长60km。布格重力异常图上（图24-23），呈北东向密集梯级带线性延伸，北东端跨越琼州海峡延伸至广东省雷州湾以北（据广东物探大队1∶100万布格重力异常图），南西端延至昌化港北侧，是一条长250km以上，规模大的梯级带；岛内长60km，宽6～8km，重力水平梯度为4×10^{-5}～$8\times10^{-5}m\cdot s^{-2}/km$。据人工地震探测等资料表明，该断裂是峨蔓凹陷与临高凸起的分界线，它控制着峨蔓凹陷的生成和发展。而且该断裂倾向北西，其中北东段（抱东村北西侧1km附近）倾角为84°；南西段（大文村南东2km附近）倾角64°，为西盘下降，东盘上升，切割基底面的垂直断距最大为980m，切割深度已达地幔顶部。

3. 南北向断裂

岛内南北向构造也很发育，但规模较小，连续性较差。依次有：昌江-抱伦、屯昌-兴隆、十字路-石壁、谭文-九曲江等断裂。它们控制了燕山期岩体侵入和中新生代断陷盆地的分布和沉积，有的成为新生代玄武岩喷溢的通道。这些断裂在重磁异常图上（图 24-22，图 24-23）一般表现为线性延伸的重磁异常梯级带、不同特征的重磁场分界线、重磁异常突变带、串珠状的重磁异常带等。以昌江-抱伦断裂为例（表 24-10），该断裂位于西南部，北起昌江，往南经广坝、江边至抱伦，走向近南北，长 83km。在航磁异常图上（图 24-22），该断裂表现为线性延伸的异常带，不同磁场特征的分界线，该断裂以东为平静稳定宽缓的正磁场区，以西变化较大，为正负伴生的磁场区。在重力图上（图 24-23），表现为重力梯级带呈南北向线性排列，即北段梯级带长 35km，水平梯度 $1.2\times10^{-5}m\cdot s^{-2}/km$；南段梯级带长 25km，水平梯度 $2.0\times10^{-5}m\cdot s^{-2}/km$。且东、西两侧等值线走向明显不同，即东侧以南北向、北东向为主，而西翼多为东西向。又如屯昌-兴隆深断裂（表 24-10），该断裂北起大同北侧，往南经屯昌、南间、牛路岭水库至兴隆西侧，总体走向近南北，长 100km。在重力异常图上（图 24-23），布格重力异常以全岛唯一、规模较大的南北向密集梯级带为其特征。梯级带长 90km，宽 5~8km，重力水平梯度为 $2\times10^{-5}m\cdot s^{-2}/km$。以梯级带为界，东、西两侧重力场特征明显不同，东侧等值线呈较规则的南北向展布，而西侧为多个呈椭圆状的局部负异常。在莫霍面等深图上，28km 等深线在屯昌附近，由近东西向转为近南北向。另外，在海南岛航磁（ΔT）异常图上（图 24-22），该断裂表现为串珠状航磁异常带。

4. 北西向断裂

岛内北西向断裂十分发育，达 10 余条之多。该组断裂主要分布有：三江-东郊、石山-长坡、马袅-长安、山口-岭口、五指山-本号、高峰等断裂，展布方向 310°~335°，为张扭性，主要发生于燕山期。该组断裂在重磁异常图上（图 24-22，图 24-23）一般表现为线性延伸的重磁异常梯级带、重磁异常突变带、重力高低间的线性过渡带、重磁异常等值线同向扭曲带等。以三江-东郊断裂带为例（表 24-10），该断裂位于海南岛东北部，西北起琼山区中村，往南东经三江至文昌东郊入海，走向 315°，长 68km；在布格重力异常图上（图 24-23），重力梯级带呈北西向展布，长 30km，重力水平梯度最大为 $2.2\times10^{-5}m\cdot s^{-2}/km$。在航磁异常化极（$\Delta Z\perp$）异常图上表现为线性异常带。据人工地震探测结果，在演丰附近证实有断裂存在，错动方向为西盘上升，东盘下降，断距为 450m。该断裂控制着东寨港凹陷的生成和发展。又如石山-长坡断裂（表 24-10），该断裂北西起荣山，往南东经石山、遵潭、仙沟至琼海长坡，走向 327°，长 100km。布格重力异常图上（图 24-23），在荣山至仙沟段，具北西向明显梯级带，长 50km，宽 5~10km，水平梯度为 $5\times10^{-5}\sim10\times10^{-5}m\cdot s^{-2}/km$；而在蓬莱西侧表现为等值线收敛变窄；在长坡附近反映为重力高、低之间的线性过渡带。在南东段为重力推测中基性岩体沿断裂侵入。在航磁异常化极（$\Delta Z\perp$）图上，表现为不同磁场特征的分界线，该断裂以东为杂乱的异常区，以西为大面积的宽缓异常区。另外，据人工地震探测结果，在梯级带的南东端（新村附近）有断裂存在，其断距为 550m，东盘上升，西盘下降。该断裂的北西段是福山凹陷与云龙凸起的分界线。经正演拟合计算结果，在儒云外附近，倾向西，倾角为 65°左右。在遵潭附近，倾角为 75°。

综上所述，海南岛不同级别的断裂带，对本岛的重磁异常排列组合面貌起着重要作用。反之利用重磁异常来分析断裂构造的存在多了一种地球物理的证据。

六、侵入岩的分布及重磁异常特征

海南岛的侵入岩分布广泛，以酸性和中酸性为主，它们多呈北东—南西方向延伸，其间以白沙盆地-南棍园向斜分隔，向北东方向越过东西向王五-文教断裂带后，被新生代玄武岩和沉积岩覆盖。这些酸性、中性、基性侵入岩与重磁异常关系密切，它们主要表现为：酸性侵入岩与重力低和弱磁性相对应，多显重力低和负磁异常；中性侵入岩与中等强度的重磁异常有关，多为剩余重力正异常和正负相伴磁异常；基性侵入岩为强度较高的重磁异常，一般为强度稍大的剩余重力正异常和强度较高正负相伴磁异常。依据剩余重力异常、重力垂向二阶导数和磁力垂直分量 ΔZ（航磁 ΔT 转换）与侵入岩之关系，综合地质、物性等资料圈定其分布范围（图 24-24），其中花岗岩推断 36 处（详见表 24-11），中酸性和基性岩推断 24 处（详见表 24-12）。

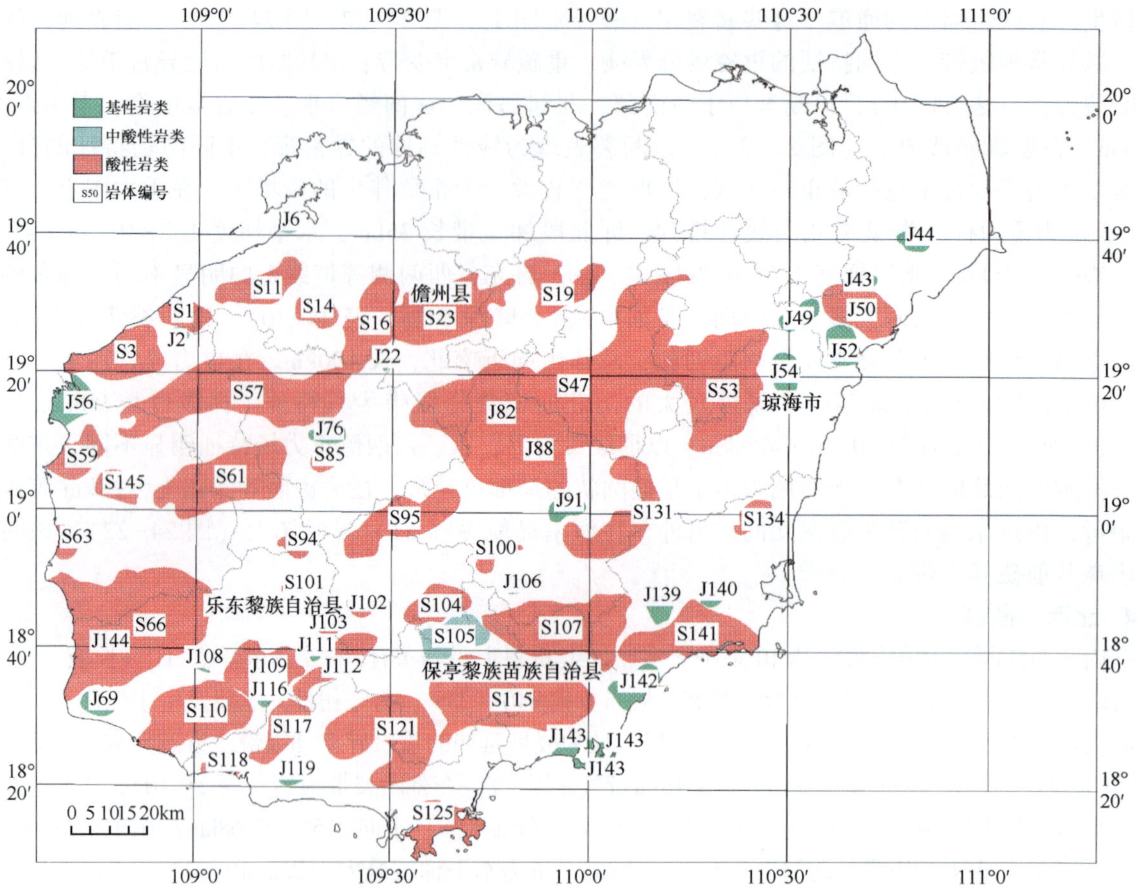

图 24-24　海南岛重磁推断侵入岩分布图

Fig. 24-24　Distribution of intrusive rocks inferred by gravity and magnetism, Hainan Island

从图 24-24 中可以看出，海南岛的侵入岩主要分布于王五-文教断裂带以南的五指山褶冲带。该区琼西部的儋州市蓝洋至东方叉河一带，分布有一条连续的，两头大中间小，规模较大，呈北东走向的花岗岩带，该花岗岩带与地质上推断的儋州花岗岩基相吻合。在中部的新兴—屯昌—黎母山一带，分布有一条规模较大，呈北东走向的花岗岩带。在西部的尖峰、雅加大岭、海尾、海头，以及东部的千家、保城、吊罗山、南桥等地分布着多个规模较大的花岗岩石，这些花岗岩石与铅锌、钼、铜、钨等多金属矿关系密切，是铅锌、钼、铜、钨等多金属矿的成矿有利部位。另外，在该区分布的中基性侵入岩，多呈岩株状侵入，并且规模较小，在数量上具有东部多于西部之特征。

表 24-11　海南岛重力（1∶20 万资料）推测花岗岩石一览表

Table 24-11　List of inferred granites (1∶20 scale data), Hainan Island

序号	岩体编码	名称	走向	顶面面积 km^2	出露情况	重力异常特征	推断依据
1	S 1	海头花岗斑岩石	北西	39.000	半隐伏	布格重力异常显示为等值线向东（重力高）方向突出。剩余负异常呈北西向，且北西部向海域延伸	剩余异常与大岭花岗斑岩引起的负异常基本相当
2	S 3	三架岭二长花岗岩石	东西	38.000	半隐伏	布格重力异常呈东西向椭圆状的重力低。剩余负异常呈东西向不规则状	异常与三架岭二长花岗岩的分布范围基本吻合
3	S 11	丰猛二长花岗岩石	北东	120.00	半隐伏	布格重力异常显示多条等值线向北突出。剩余负异常呈北东走向的长椭圆状	异常分布于丰猛二长花岗岩石中

续表

序号	岩体编码	名称	走向	顶面面积/km^2	出露情况	重力异常特征	推断依据
4	S 14	西华钾长花岗岩石	东西	37.000	半隐伏	布格重力异常为南宝重力高的西南部分，多条等值线向南（重力低）方向弯曲。剩余正异常呈北东向不规则的似球拍状	异常区出露花岗岩
5	S 16	那大二长花岗岩石	北东	239.00	出露	布格重力异常是蓝洋重力低的北部，多条等值线向北（重力高）方向突出。剩余负异常呈北东向的似椭圆状	异常区分布二长花岗岩、黑云母二长花岗岩
6	S19	里万-牛潭门二长花岗岩石	北东	179.00	半隐伏	布格重力异常表现为等值线向北东（重力高）方向突出。剩余负异常呈北东向长椭圆状	异常区分布二长花岗岩，与里万-牛潭门二长花岗岩相吻合
7	S23	蓝洋二长花岗岩石	北东	376.00	出露	布格重力异常表现为等值线呈北东向椭圆状圈闭。剩余负异常呈北东向的瓢状，异常峰值较明显	异常区南侧 ZK1 孔 2.6 ~ 266.6m 见花岗岩
8	S47	屯昌-乌石花岗闪长岩-石英二长闪长岩	北东	1780.0	半隐伏	布格重力异常显示重力低。剩余重力负异常呈南北向条带状展布	异常与屯昌花岗闪长岩-石英二长闪长岩相吻合
9	S50	白延二长花岗岩石	北东	185.00	半隐伏	布格重力异常为正值场背景下的重力低。剩余负异常呈茄子状，总体北西走向	花岗岩与周围之间具有较大的密度差
10	S53	中瑞-岭口花岗斑岩石	北东	573.00	半隐伏	布格重力异常为多条不同等值线向重力高方向突出。剩余负异常呈东西向的不规则状	剩余异常区主要分布低密度的花岗岩
11	S57	七坊-叉河钾长花岗岩石	北东	728.00	出露	布格重力异常显示为等值线向南西西（重力高）方向突出。剩余负异常呈北东向的不规则条带状	异常分布特征与叉河钾长花岗岩石相对应
12	S59	新隆二长花岗岩石	北东	167.00	半隐伏	布格重力异常等值线呈似椭圆状圈闭。剩余负异常西部向海域延伸；往南西方向延伸至大田岩体中	异常范围与大田岩体西延（隐伏）的二长花岗岩石相对应
13	S61	广坝-雅加大岭二长花岗岩石	北东	430.00	出露	布格重力异常两条等值线向南西（重力高）方向突出。剩余负异常总体呈南北向的弧状	异常与广坝二长花岗岩石相吻合
14	S63	新村二长花岗岩石	南北	54.000	隐伏	布格重力异常等值线向北东（重力高）方向突出。剩余负异常西部向海域延伸	异常区地表为第四系覆盖。根据重力异常特征分析
15	S66	尖峰钾长花岗岩石	北东	1053.0	半隐伏	布格重力异常等值线呈西部向海域圈闭的东西向椭圆状。剩余异常呈东西向不规则的长椭圆状，异常幅度大，峰值明显	异常区与尖峰钾长花岗岩石相吻合
16	S85	南针岭石英闪长岩石	北东	76.000	隐伏	布格重力异常等值线向东（重力高）方向突出。剩余负异常呈北东向椭圆状	异常与由二长花岗岩引起的G-84负异常毗邻
17	S94	乱岭二长花岗岩石	北西	113.00	隐伏	布格重力异常等值线向北西（重力高）方向突出。剩余负异常呈北西向雪梨状	从异常特征分析
18	S95	什运二长花岗岩-钾长花岗岩石	北东	399.00	半隐伏	布格重力异常等值线呈北东向条带状圈闭。剩余重力负异常与布格异常形态相同	异常是与什运二长花岗岩-钾长花岗岩石相对应

续表

序号	岩体编码	名称	走向	顶面面积 km^2	出露情况	重力异常特征	推断依据
19	S100	牙代二长花岗岩石	北东	36.000	半隐伏	布格重力异常等值线向南东（重力高）方向突出。剩余负异常呈南北向不规则条带状展布	异常与牙代二长花岗岩石相对应
20	S101	洋老村钾长花岗岩石	东西	249.00	半隐伏	布格重力异常等值线向南西（重力高）方向突出。剩余负异常呈东西向的椭圆状	异常与洋老村钾长花岗岩石对应
21	S102	阜堡笔钾长花岗岩石	北西	64.000	半隐伏	布格重力异常等值线向北突出。剩余负异常呈南北向的椭圆状	异常形态与阜堡笔钾长花岗岩石相对应
22	S103	哥分岭钾长花岗岩石	南北	49.000	半隐伏	布格重力异常等值线向北西（重力高）方向突出。剩余负异常呈北西向的近椭圆状	异常形态与哥分岭钾长花岗岩石相对应
23	S104	青春岭花岗斑岩-二长花岗岩石	北东	184.00	半隐伏	布格重力异常呈南北向的椭圆状。剩余负异常呈南北向的似椭圆状	异常形态与青春岭花岗斑岩-二长花岗岩石相对应
24	S107	吊罗山钾长花岗岩石	东西	487.00	半隐伏	布格重力异常表现为多条等值线向北东方向突出。剩余负异常呈东西向不规则的似长方形状展布	异常与吊罗山钾长花岗岩石相对应
25	S109	志仲二长花岗岩石	北东	221.00	半隐伏	布格重力异常等值线向北东（重力低）方向突出。剩余负异常呈北东向的脚板状	异常特征与志仲二长花岗岩石相当
26	S110	千家二长花岗岩石	北东	373.00	半隐伏	布格重力异常呈东西向的椭圆状。剩余负异常总体呈北东走向的酒瓶状	异常形态与千家二长花岗岩石相对应
27	S112	马咀岭二长花岗岩石	北东	153.00	半隐伏	布格重力异常呈北东向的椭圆状。剩余负异常呈北东向的长条带状	异常形态与马咀岭二长花岗岩石相对应
28	S115	保城二长花岗岩-花岗闪长岩石	东西	827.00	半隐伏	布格重力异常呈近东西向椭圆状。剩余负异常呈近东西向长椭圆状	异常形态与保城二长花岗岩-花岗闪长岩石相对应
29	S117	洋林岭花岗斑岩石	北东	238.00	半隐伏	布格重力异常为北东向展布的椭圆状。剩余负异常呈南窄北宽的似球拍状，走向北东	异常形态特征与洋林岭花岗斑岩石相对应
30	S118	保平花岗闪长岩石	东西	60.000	隐伏	布格重力异常等值线向北东（重力高）方向突出。剩余负异常呈南北向展布的椭圆状	根据异常特征分析
31	S121	新村二长-正长花岗岩	南北	449.00	半隐伏	布格重力异常表现为多条等值线向南（重力高）方向弯曲。剩余负异常呈不规则的弧形	异常一部分与罗蓬二长-正长花岗岩（南部）相对应
32	S125	北山岭二长-正长花岗岩	北西	210.00	半隐伏	布格重力异常等值线向南东（重力高）方向突出。剩余负异常呈北西向的椭圆状	异常形态与北山岭二长-正长花岗岩相吻合
33	S131	白马岭-和平二长花岗岩石	北东	290.00	半隐伏	布格重力异常等值线呈北西向的椭圆状圈闭。剩余负异常呈北西西向的椭圆状	异常区分布花岗岩，从异常特征和地质环境分析
34	S134	东岭二长-正长花岗岩	北东	56.000	半隐伏	布格重力异常等值线向北东（重力高）向突出。剩余重力负异常呈北东向的椭圆状	异常位于六连岭花岗岩中，与东岭二长-正长花岗岩基本吻合

续表

序号	岩体编码	名称	走向	顶面面积 km²	出露情况	重力异常特征	推断依据
35	S141	东奥二长花岗岩石	东西	330.00	半隐伏	布格重力异常等值线呈北东向圈闭和多条等值线向北东（重力高）方向突出。剩余负异常呈北东向不规则状沿滨海地区展布	异常区出露花岗岩，异常形态特征与东奥二长花岗岩石相当
36	S145	观音峰二长花岗岩石	北西	77.000	半隐伏	布格重力异常表现为等值线向北东（重力高）方向凸出，剩余负异常呈北西向的椭圆状	异常区出露花岗岩，异常形态特征与观音峰二长花岗岩石相对应

表 24-12　海南岛重磁（1：20 万资料）推测中基性岩体一览表

Table 24-12　List of inferred medium-basic intrusions（1：200,000 scale data）in the Hainan Island

序号	岩体编码	名称	岩体类型	走向	顶面面积 km²	出露情况	重磁场特征及推断依据
1	J2	波兰沟闪长岩	中性	东西	12.000	半隐伏	布格重力异常等值线向南西（重力低）突出。剩余正异常呈东西向的扁豆状。北东部分布有由青天峡组中磁黄铁矿化炭质千枚岩引的 M-Ⅵ-15 航磁异常，而南西侧有由闪长岩引起的 M-Ⅵ-14 航磁异常
2	J6	白马井角闪岩	中性	东西	93.000	隐伏	布格重力异常表现为多条等值线向东南（重力低）方向突出。剩余正异常呈东西向的椭圆状。伴有 M-Ⅲ-2 航磁异常分布，白马井 ZK1 孔见角闪岩
3	J22	天堂辉长岩	基性	北东	41.000	隐伏	布格重力异常等值线向南西（重力低）方向突出。剩余正异常呈北东向条带状。有由辉长岩引起的 M-Ⅴ-10 航磁异常
4	J43	南阳角闪闪长岩	中基性	北西	50.000	隐伏	布格重力异常等值线向西（重力低）方向突出。剩余正异常呈北西向条带状。伴有 M-Ⅰ-2 航磁异常。异常中 ZK5 孔见角闪闪长岩
5	J44	东阁辉长岩	基性	北西	50.000	半隐伏	布格重力异常等值线向南东（重力低）方向突出。剩余正异常呈北西向规则的椭圆状。东南侧分布有 M-Ⅰ-1 航磁异常。经实地踏勘检查，异常区见辉长岩出露
6	J49	大蒙岩体辉石闪长岩	基性	北东	109.00	半隐伏	布格重力异常等值线向北（重力低）方向突出。剩余正异常呈北东向的“马鞍”状圈闭。伴有 M-Ⅳ-6 航磁异常。钻探证实由辉石闪长岩所引起
7	J52	官回村中基岩	中基性	北北西	145.00	隐伏	布格重力异常呈南北向的椭圆状。剩余正异常呈椭圆状，走向南北。从异常特征及所处地质环境分析
8	J54	大路中基性岩	中基性	南北	110.00	半隐伏	布格重力异常等值线向北（重力低）方向突出。剩余正异常呈南北向不规则的椭圆状，异常幅度较大。伴有 M-Ⅳ-1 航磁异常
9	J69	莺歌海盐场中基性岩	中基性	东西	43.000	隐伏	布格重力异常等值线向北（重力低）方向突出。剩余正异常呈规模不大的椭圆状。根据异常所处地质环境

续表

序号	岩体编码	名称	岩体类型	走向	顶面面积 km²	出露情况	重磁场特征及推断依据
10	J82	握岱闪长岩	中性	北西	33.000	半隐伏	布格重力异常等值线向南（重力低）方向突出。剩余正异常呈北西向展布的椭圆状。异常中心部位有花岗闪长岩出露
11	J88	乌石石英闪长岩	中性	北东	53.000	半隐伏	布格重力异常等值线向北（重力低）方向突出。剩余正异常呈南北向的椭圆状。伴有 M－Ⅴ－4 航磁异常，异常出露华力西期石英闪长岩
12	J91	岭头辉长岩	中基性	北东	42.000	隐伏	布格重力异常等值线向东（重力低）方向突出。剩余正异常呈北东东向长椭圆状。伴有由辉长岩引起的 M-V-3 航磁异常
13	J105	七指岭闪长岩	中性	北东	140.00	隐伏	布格重力异常等值线向西南（重力低）方向弯曲。剩余正异常呈北西向的似“脚板”状。伴有由闪长岩引起的 HC—79-69 航磁异常
14	J106	太平石英二长岩	中性	北东	14.000	隐伏	布格重力异常呈南北向不规则的椭圆状。剩余正异常呈南北向的条带状。伴有 HC-79-2 航磁异常
15	J108	抱尾中基性岩	中基性	北西	4.000	隐伏	布格重力异常等值线向南西（重力低）方向突出。剩余正异常呈北东向似长方形状。南东伴有 M－Ⅸ－10 航磁异常。检查为隐伏中基性岩引起
16	J111	加河中基性岩	中基性	北东	11.000	隐伏	布格重力异常等值线向南东（重力低）方向突出。剩余正异常总体呈南北向不规则的条带状。伴有 M-Ⅷ-20 航磁异常
17	J116	抱头石英闪长岩	中基性	北东	20.000	半隐伏	布格重力异常等值线向北（重力低）呈不闭合的条带状突出。剩余正异常呈南北向的长椭圆状圈闭。南部分布 M－Ⅸ－7 航磁异常，北部有 M-Ⅸ-8 航磁异常
18	J119	立才辉长岩	中基性	北东	46.000	隐伏	布格重力异常等值线向北（重力低）方向突出。剩余正异常呈北西向的椭圆状展布。伴有 M-Ⅸ 2 航磁异常
19	J139	兴隆辉长岩	基性	南北	63.000	半隐伏	布格重力异常等值线向正南（重力低）方向突出。剩余正异常呈南北向椭圆状。南部伴有 C-78-27 航磁异常和化探 As-195 异常，异常区出露辉长岩
20	J140	长安辉长岩	基性	南北	38.000	半隐伏	布格重力异常等值线向西南（重力低）方向突出。剩余正异常呈北西大，南东小的葫芦状。伴有 M－Ⅶ－1 航磁异常。异常与长安辉长岩相对应
21	J142	牛岭中基性岩	中基性	北东	89.000	隐伏	布格重力高表现为多条等值线向北突出。剩余正异常呈北东向的椭圆状。西侧分布 C-76-53 航磁异常。从异常形态特征分析
22	J143	新村中基岩	中基性	东西	98.000	半隐伏	布格重力异常等值线向北东（重力低）方向突出。剩余正异常呈东部、南部向海域延伸的不规则状。伴有 M－Ⅷ－6 航磁异常
23	J56	昌城中基性岩	中基性	北东	115.00	隐伏	布格重力异常等值线向北东弯曲，西南端向海域延伸。剩余正异常呈北西向的似“脚板”状，异常分布多个 1：5 万航磁异常。根据其形态特征分析
24	J76	铂随岭石英闪长岩	中基性	东西	38.000	隐伏	布格重力异常等值线向西（重力低）方向弯曲。剩余正异常呈椭圆状，走向东西；从异常特征及所处地质环境分析

第二节 南海海域地球物理特征

一、南海海底地形特征

南海是东亚陆缘最大的边缘海之一，面积约 $350\times10^4 km^2$。外廓为扁菱形的海盆，北东向长 1600km，宽 700km，水深 3000~4000m。陆架边缘大体沿 200m 水深线延展，陆坡与深海平原的边界大约沿 3000m 水深线延伸（图 24-25）。北部广东陆架宽 150~300km，南部巽它陆架宽 600km，是世界上最宽的陆架之一。东部和西部陆架很窄，约十几千米。陆架地形平坦，在大河岸外分布着水下三角洲，如北部陆架上的珠江水下三角洲，西南部的湄公河三角洲等。陆坡地形复杂，分布着众多的海山、海丘、海槽、海谷和海台。如北部陆坡上，海山、海丘呈线状分布，自神狐暗沙与一统暗沙起，经南卫滩和经北卫滩至台湾浅滩，为一北东向海丘链，海丘高 200~600m。在其南部，自双峰海山经尖峰海山至北坡海山，为一北东向海山链，海山高度大于 1000m。在中沙群岛之南，分布着一系列北东向排列的海山与海谷，山与谷相间排列。海槽有西沙海槽、中沙海槽、礼乐海槽和南沙海槽，海谷有东沙海谷和珠江海谷。南海南部陆坡宽达 400km，地形切割强烈，崎岖不平。这里分布着 230 余座岛屿、沙洲、暗沙、暗礁和暗滩。

南海深海平原的水深在 300~4400m，海底平坦，海山广布。根据水深线及海山的走向，将其分为三个次海盆：西北次海盆、中央海盆和西南次海盆。西北次海盆位于中沙群岛以北，西沙海槽之东，水深为 3000~3800m。海底自西向东倾斜，中间分布着北东向的双峰山。该海盆新生代沉积厚达 1.5~2.0km，因此，层 2 顶部埋深 4.5~5.8km。地壳厚 5.0~5.8km，等厚线沿北东向展布。中央海盆上分布着一系列东西向的海山、海丘，黄岩海山分布于海盆中部。海盆水深为 3000~3900m，个别地方达 4000 m。新生代沉积厚 0.5~2.4km，层 2 顶部埋深 3.5~5.5km，海盆中部高，向南向北下倾，反映海盆中部的洋壳之年龄较小，向南和向北逐渐变老。西南次海盆的水深 4300~4400m，是南海深海平原上最低洼之处。该海盆新生代沉积厚 1.0~2.0km，因此，层 2 顶部埋深 5.3~7.5km。海盆中部分布着北东走向的长龙海山，到 114°E 处与南北走向的中南海山相遇而终止（图 24-25）（姚伯初等，2010）。

二、岩（矿）石地球物理特征

（一）弹性波特征

1. 南海南部岩层地震地层结构

1987~1994 年，中国科学院南海海洋研究所“实验 2”号地球物理调查船在南沙群岛海域先后进行了 6 个航次的综合地球物理调查，其中完成多道反射地震测量 13833km，声纳浮标折射地震测量 23 个站位。根据地震剖面的地震波形态、反射界面的接触关系、层速度变化情况，结合已有的钻井资料和声纳浮标测量资料，以及区域地质构造背景，将该区岩层划分为 3 套构造和 7 个反射层组（图 24-26）。

由图 24-26 可见，T_g 和 T_h 两个反射界面明显地代表两个区域不整合面。以此两个不整合面为界，可划分出下、中、上三套构造层，它们具有不同的层速度（表 24-13）。

表 24-13 各构造层的层速度

Table 24-13 Interval velocity of individual structural layers （单位：km/s）

构造层	层速度	
	地震波速度谱所求得	折射声纳浮标解
上构造层	1.7~3.9	1.6~4.2
中构造层	≥4	4.3~4.8
下构造层	>5	>5

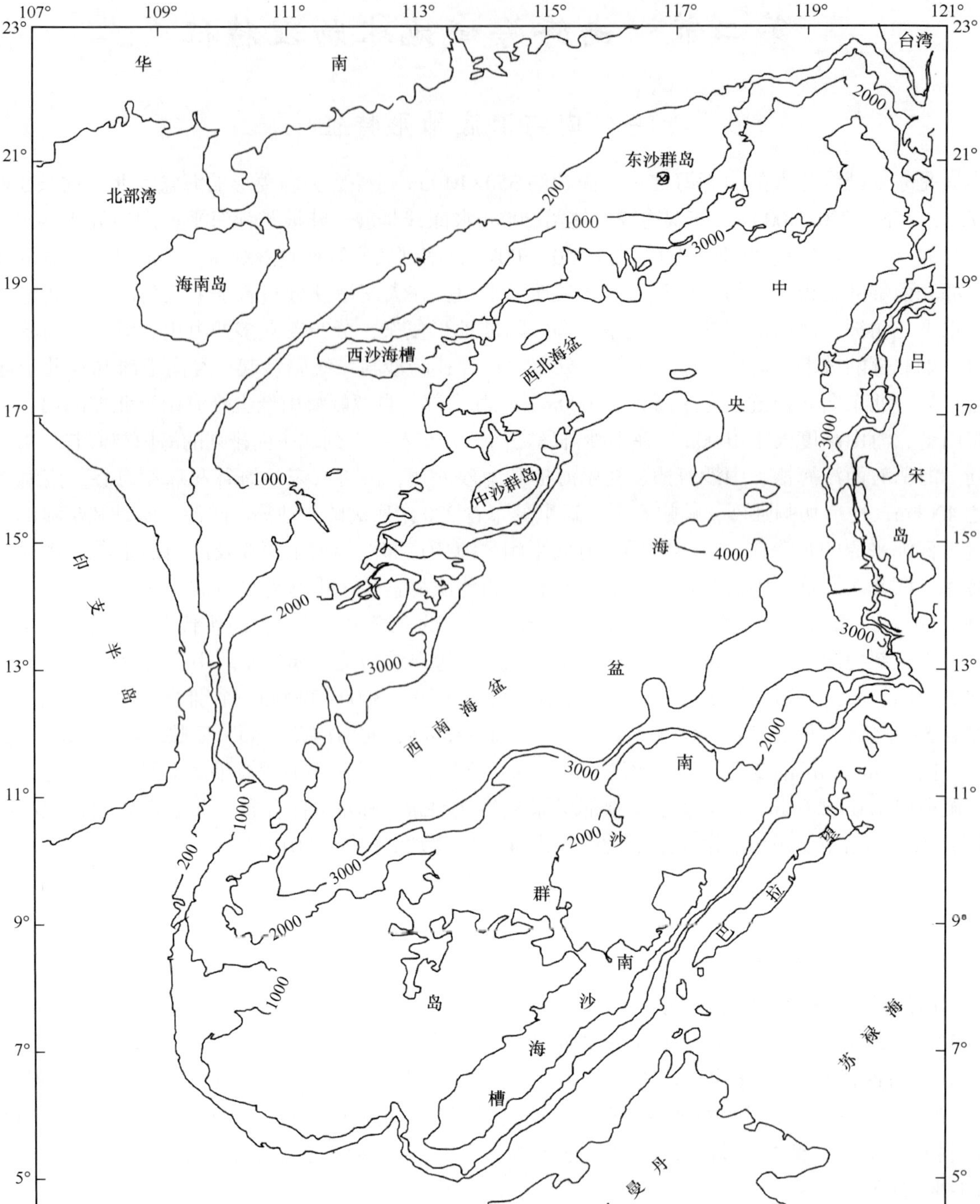

图 24-25　南海地形图

Fig. 24-25　Topography map of the South China Sea

下构造层　T_h 反射界面以下的一套反射层系（图 24-26）。反射波杂乱，层序不清，褶皱强烈，倾角陡，与上覆层组呈不整合接触。一般认为，下构造层是一套中生界岩系，但此岩系分布不连续，部分地区缺失，使更古老的（可能为前中生代的）变质沉积岩或岩浆岩直接与中构造层或上构造层接

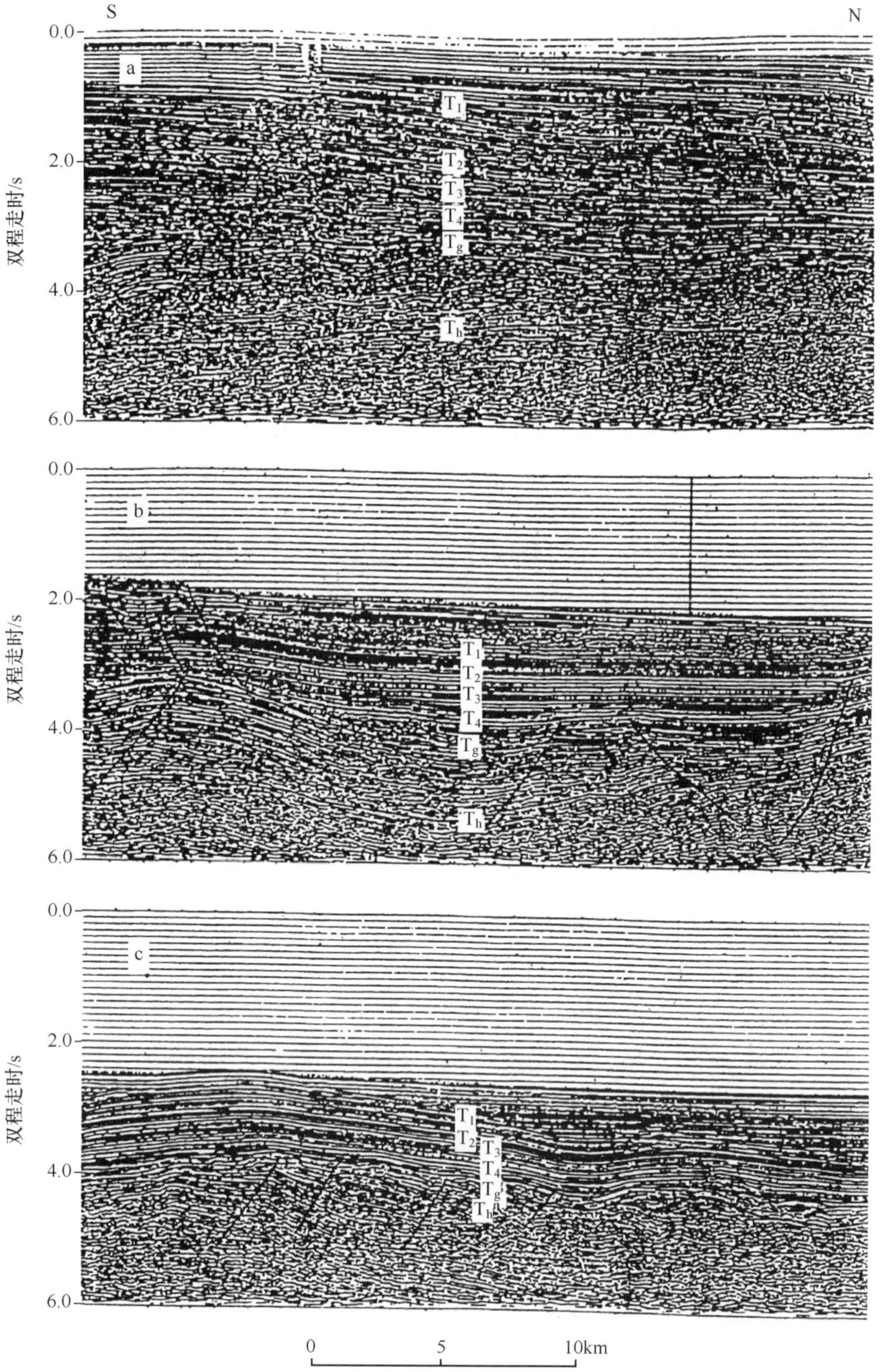

图 24-26　测线 N_4 多道反射地震剖面图

Fig. 24-26　Multi-channel reflection seismic section of Survey Line N_4

触。在礼乐滩，根据 Sampagita-1 和 A-1 两井资料，下构造层相当于早白垩世滨海—浅海相含煤碎屑岩系。但实际上，此两井仅钻达下构造层的上部（孔深小于 1km），而地震剖面反映下构造层厚达 2~4km。因此，下构造层下部可能还有更老的地层。根据在该区采集到的晚三叠世—早侏罗世的三角洲-浅海相砂泥岩，推测该区的下构造层还应包括有侏罗纪和三叠纪岩层。

但在礼乐滩北缘断崖，SO23-36 和 SO23-37 两个站采集到角闪岩、片麻岩和石榴子石云母片岩，在礼乐滩西南五方礁西侧，SO27-21 站采集到片麻岩和石英千枚岩。所有事实，说明礼乐滩及其周围地区，存在更老的由深变质的岩系组成的构造层（姜绍仁等，1993；中国科学院南沙综合科学考察队，1996）。

在南沙群岛西北海区，反射界面 T_h 之下为一套显著褶皱，并在褶皱后发生岩浆岩侵入的层系，据其变形特征，推测可能为三叠纪海相沉积。

中构造层 为 T_g 至 T_h 界面之间的反射层系。反射波振幅弱至中等，频率较低至中等。连续性中等至较好，层次较清晰。中构造层上、下界面与上构造层和下构造层呈不整合接触，其时代早于渐新世。东纳士陆架的 Paus1 井和 2 井钻到古新世和始新世已变质的千枚状沉积岩。然而，曾母暗沙附近的一些钻孔打到井深 2854m 见到未变质的始新世碎屑岩。此外，礼乐滩 Sampagita-1 井揭示该层系为古新世—始新世的三角洲相和外浅海—半深海碎屑沉积。因此，根据地震剖面中构造层的反射特征和上述钻井资料，南沙海区的中构造层主要应为古新世—始新世的未变质的海相沉积岩系，在半地堑底部可能部分为上白垩统。中构造层沉积后发生区域性褶皱隆升及块断掀斜，使部分地区遭受剥蚀夷平。本区的北缘大部分缺失中构造层。自万安-礼乐以南，中构造层基本连片分布，厚度一般为 1~3km；在西南部南康暗沙一带最厚，可达 4.5km；在东北部仁爱礁-礼乐滩一带较薄，大部分小于 1km。

上构造层 T_g 反射界面以上的一套层系。反射波特征是振幅强，频率高，连续性好，层次清晰。大部分产状平缓，呈席状披盖，可追踪对比。厚度一般为 1~3km，最厚超过 9km。根据 Sampagita-1 井资料，上构造层下部为晚始新世至早渐新世的滨海-浅海相碎屑沉积，上部为晚渐新世至现代的浅海厚层灰岩、白云岩化灰岩和致密白云岩。在曾母盆地，根据地震波组特征和反射界面的接触关系，并结合钻井资料，上构造层可细分为 I_1—I_5 共 5 个反射层组（表 24-14）。

表 24-14 地震反射层系划分及时代属性一览表

Table 24-14 List of division and age properties of seismic reflecting horizons

地质时代			地震地层			层深度/km
时代		代号	反射层系	反射层组	反射界面	
第四纪		Q	上构造层Ⅰ	I_1	T_1	1.7~3.9
上新世		N_2		I_2	T_2	
中新世	晚	N_1^3		I_3	T_3	
	中	N_1^2		I_4	T_4	
	早	N_1^1		I_5		
渐新世		E_3			T_g	
始新世		E_2	中构造层Ⅱ			≥4
古新世		E_1			T_h	
白垩纪		K	下构造层Ⅲ			>5
前白垩纪		PreK				

2. 南海北部岩层地震地层结构

南海北部大陆架和大陆坡区，在拗陷部分的地震反射剖面也可划分出 3 套构造层，其地震地层结构特征，与南海南部地区相似。图 24-27 是通过珠江口盆地中部的北西—南东向地震剖面（刘昭蜀，1997）。从图上可见，T_g 反射界面以下的地层其地震波非常杂乱，与上覆岩层呈不整合接触，这些岩层可划分为下构造岩层。根据已有石油勘探资料，珠江口盆地 T_g 以下的岩层可能为中生代火成岩和古生代变质岩系，但从盆地邻近陆区汕头一带出露有上三叠统—下侏罗统海相砂岩和台湾西南海区钻井发现有白垩纪滨海—浅海相含煤碎屑岩系等分析，认为珠江口盆地，尤其是盆地的东部，部分地区下构造层可能有前古近纪（侏罗纪—白垩纪）的沉积层存在。

T_g 至 T_8 两个地震反射界面之间的岩层可划分为中构造层，在地震剖面上表现为大斜层，地震反射波振幅中等，层次较清晰。厚度达 3000 多米，与上覆地层呈不整合接触。中构造层大致与珠江口盆地

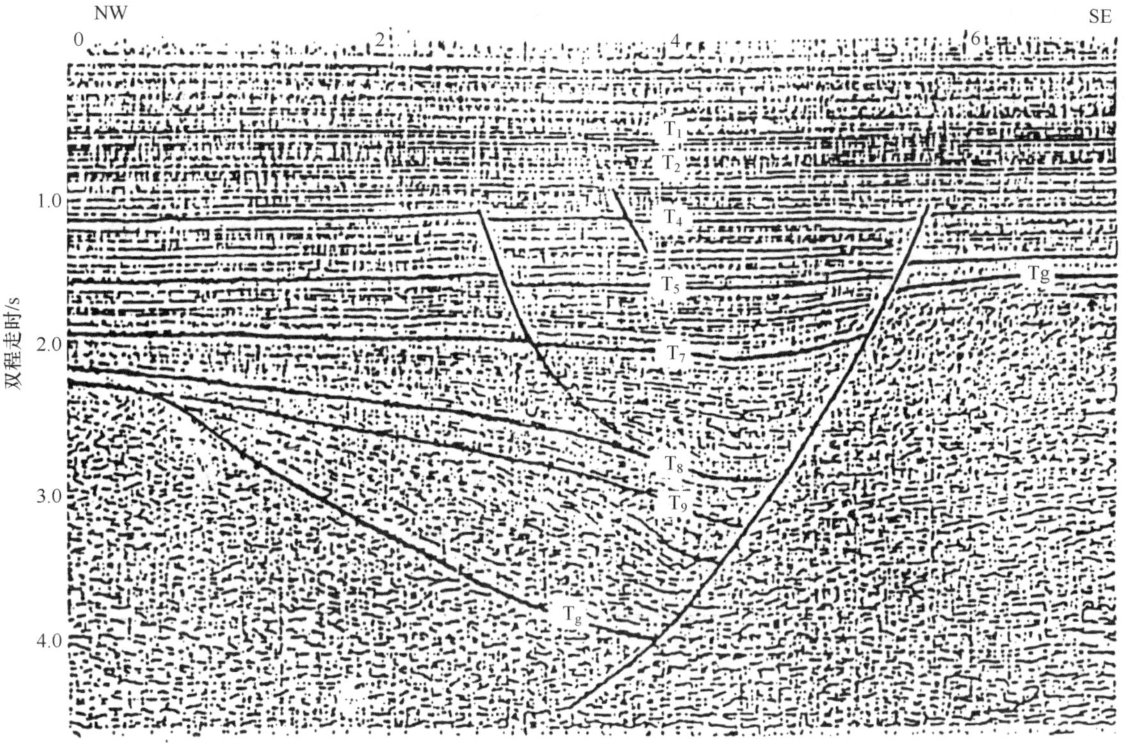

图 24-27 珠江口盆地中部地震剖面图

Fig. 24-27 Seismic section of the central Pearl River Mouth Basin

钻井资料所建立的神狐组（上白垩统—古新统）及文昌组（始新统）相当，为河流-湖泊相碎屑沉积。下部层组（神狐组）呈半地堑充填，上部层组（文昌组）由半地堑向西北的基底高地区超覆。

T_8 反射界面至海底之间的地震层系为上构造层。T_7 以下表现为斜层，属半地堑充填，根据钻井资料，为陆相碎屑沉积，被确定为恩平组（上始新统—下渐新统）。T_7 以上层组，地震波振幅强，层次清晰，呈水平层状，为区域性披覆沉积，沉积环境由近海岸三角洲-滨海相砂泥岩沉积逐渐变为滨海相、浅海相泥砂岩沉积。根据钻孔资料，由地震剖面上的 T_1、T_2、T_4、T_5 和 T_7 等 5 个局部不整合面，将上构造层分为万山组（上新统）、粤海组（上中新统）、韩江组（中中新统）、珠江组（下中新统）、珠海组（上渐新统）和恩平组（上始新统—下渐新统）6 个层组。

3. 南海中央海盆地震地层结构

南海中央海盆基底为晚渐新世—中新世海底扩张所形成的玄武岩层，基底上面主要为中新世至现代的沉积层。沉积层一般厚几百米至 1000 余米。但在中央海盆北缘，其厚度最大，可超过 3000m，且沉积盖层底部还可能存在晚渐新世沉积。图 24-28 是珠江口盆地南缘隆起至中央海盆北缘的地震剖面

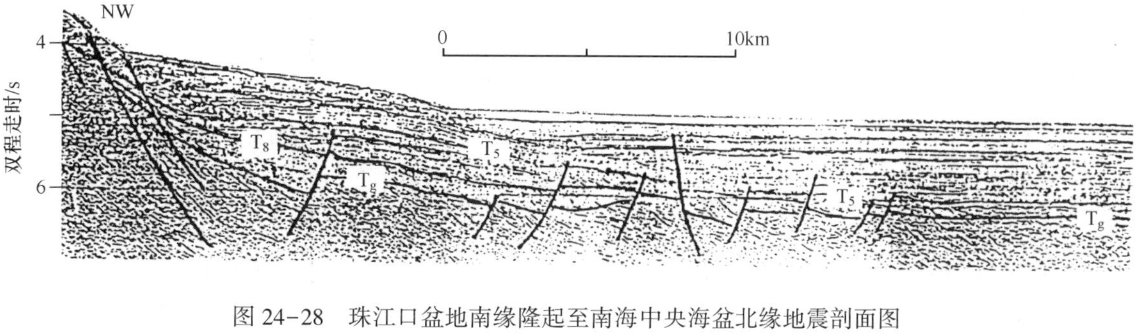

图 24-28 珠江口盆地南缘隆起至南海中央海盆北缘地震剖面图

Fig. 24-28 Seismic section from the southern margin uplift of the Pearl River Mouth Basin to the northern margin of the central sea basin in the South China Sea

图。从图上可见，在剖面的东南段，即中央海盆北缘地区，T_8 之下为玄武岩基底，其上主要覆盖着中新世至现代的深海沉积，而 T_g 与 T_5 之间的层组，即可能为晚渐新世沉积，它代表南海中央海盆海底扩张初期（晚渐新世）的沉积，也是南海洋壳基底之上的最老沉积（刘昭蜀等，2002）。

（二）磁性特征

南海及其邻区的地层和岩浆岩磁性归纳于表 24-15，从表中可以看出，南海周边及中央海盆的岩石磁性特征。能引起局部磁异常的地质体主要为第二、三期侵入岩和具有磁性的玄武岩。由于华南沿海地区构造活动复杂，岩浆活动具有多期性，不同的构造部位存在明显的差异。因此，该地区的磁性基底可能为花岗岩、火山岩、玄武岩、和古老变质岩。南海海盆玄武岩磁性主要由剩磁引起，据分析，中央海盆海山玄武岩为碱性玄武岩，或是介于拉斑玄武岩与碱性玄武岩之间的过渡型玄武岩。

表 24-15　南海海域及围区岩石磁性一览表

Table 24-15　List of magnetism of rocks in the South China Sea area and its peripheries

岩类	时代	岩石名称	块数	磁化率（κ） $4\pi\times10^{-6}$SI		剩余磁化强度 10^{-3}A·m^{-1}		采样地点	备注
				变化范围	平均值	变化范围	平均值		
沉积岩	N—Q	珊瑚礁、砂岩、泥岩、页岩	3524		0		0	南海北部沿海陆地及岛屿	各时代沉积岩一般无磁性
变质岩	Pz_1—J	变质砂岩、板岩、片岩、片麻岩	421	0~2940	0~8920			南海北部沿海陆地及岛屿	各时代浅变质岩一般无磁性，变质岩无磁性或弱磁性，前寒武纪结晶基底一般具有弱磁性
		混合岩化副片麻岩		80~90				西沙群岛西永一井下部钻遇	
火山岩	N	玄武岩	480	50~6870	1033	630~48700	4825	南海北部沿海陆地及岛屿	
	N	石英或橄榄拉斑玄武岩石	1	80~200		2200		玳瑁海山	
		石英或橄榄拉斑玄武岩石	1	130~200		2800		珍贝海山	
		碱性玄武岩	1	30~60		100 以下		中南海山	
	N—Q	玄武岩			1031			越南大叻	
	N	玄武岩		300~500		500~4000		菲律宾	
	K_1	中酸性火山岩	6		500			海南岛	
	K_1	中基性火山岩	12	950~4650				海南岛	
	J—K	安山岩、安山玢岩	186	0~19230		116~68072		南海北部沿海陆地及岛屿	
	J—K	安山岩			407			马来西亚	
	J—K	凝灰岩	780	0~14294		0~160000		南海北部沿海陆地及岛屿	
	Pz_1	凝灰岩	无磁性—弱磁性					越南南部陆地	
	J—K	流纹岩-流纹斑岩	161	0~1240		0~160000		南海北部沿海陆地及岛屿	

岩类	时代	岩石名称	块数	磁化率（κ）$4\pi\times10^{-6}$SI		剩余磁化强度 $10^{-3}A\cdot m^{-1}$		采样地点	备注
				变化范围	平均值	变化范围	平均值		
侵入岩	$\gamma\delta_5$	花岗岩		8~15				菲律宾、马来西亚	
		花岗岩			1138			越南南部陆地及岛屿	各时代侵入岩的磁性极不均匀。印支期和燕山 1、4、5 期侵入岩一般无磁性。燕山 2、3 期花岗侵入岩体具有不均匀磁性
		花岗闪长岩	2527	0~11450	820	0~42269	5408		
	$\lambda\delta_5$	石英闪长岩	367	100~6590	1033	90~1590	461		
		石英闪长岩			92			马来西亚	
		闪长岩			2741			越南南部陆地	
	δ_5	闪长岩	183	250~8080	2457	226~289600	761	南海北部沿海陆地及岛屿	
	σ_5	辉长岩	189	110~32400	3313	850~52400	2060		
		辉长岩			50			马来西亚	
		橄榄岩		38~8900				菲律宾	

（三）密度特征

本次研究收集了华南陆地露头区和盆地钻井岩心实测的各时代岩石密度，以及南海周边地区的岩石密度，归纳于表 24-16。还收集了用南海各海区地震速度换算成的各时代地层的平均速度见表 24-17。从上两个表中可以看出，各时代地层由新到老，岩石的平均密度值增大。各海区新生代的密度平均值也有变化。这主要与地层的物源、物质组成、地层厚度、压实程度等一系列的因素有关。表 24-18 给出了地壳类型和上地幔有关的岩石密度值。各时代的岩浆岩密度相对较稳定，中酸性的花岗岩类一般为 $2.54\times10^3\sim2.64\times10^3kg/m^3$。中基性的玄武岩类密度值较高，一般为 $2.70\times10^3\sim2.75\times10^3kg/m^3$。水层的密度，一般采用 $1.03\times10^3kg/m^3$。

表 24-16 南海海域及围区岩石密度

Table 24-16 Rock densities of the South China Sea area and its peripheries 单位：g/cm³

岩石密度 \ 地区 / 时代	越南南部陆地	北部湾东北部沿岸	中国广西 24°N 以南地区	中国雷琼地区	中国福建省	中国台湾省
Q	2.00~2.20					
N		2.11	2.42	2.18		2.20~2.30
E		2.25		2.37		2.57
K	2.60~2.70	2.60		2.51	2.61	2.50~2.80（前古近纪大南澳变质砂岩密度为 2.62）
J			2.43		2.60	
T			2.50		2.62	
P		2.53~2.70	2.42	2.64~2.91	2.64	
C			2.52		2.64	
D			2.34~2.58		2.62	
S			2.30~2.52			
O			2.52~2.58		2.62	
			2.50~3.09		2.63	
Z			~		2.65	

表 24-17 根据层速度换算的南海各地区平均密度

Table 24-17 Average density of various areas in the South China Sea, converted according to interval velocity

单位：g/cm^3

密度值 / 地区 / 层位	北部陆架	北部陆坡	海盆	礼乐滩	曾母盆地	北康盆地	南沙中部海域	万安盆地
Q—N_2	2.15	2.14	2.12	2.25	2.13	2.09	2.10	1.90
N_1^3	2.23	2.21	2.18		2.28	2.15	2.20	2.05
N_1^{1-2}	2.35	2.35	2.26	2.41	2.42	2.33	2.34	2.20
E_3	2.51	2.51	2.36		2.44	2.46	2.40	2.26
E_2—Mz	2.58	2.58		2.58	2.58	2.48	2.42	
层 2			2.72					
层 3			2.98					
上地幔			3.18					

表 24-18 各密度层参数

Table 24-18 Parameters of individual density layers

单位：g/cm^3

<table>
<tr><th colspan="3">密度层</th><th colspan="3">密度层</th></tr>
<tr><td colspan="2">水层</td><td>1.03</td><td rowspan="2">洋壳</td><td>大洋层 2</td><td>2.75</td></tr>
<tr><td colspan="2">新生代沉积层</td><td>2.2-2.4</td><td>大洋层 3</td><td>3.0</td></tr>
<tr><td rowspan="3">陆壳</td><td>上地壳层</td><td>2.5~2.6</td><td colspan="2">上地幔岩石圈</td><td>3.3</td></tr>
<tr><td>中地壳层</td><td>2.8</td><td colspan="2">上地幔软流圈</td><td>3.2</td></tr>
<tr><td>下地壳层</td><td>2.9~2.95</td><td colspan="2">地壳平均密度</td><td>2.8</td></tr>
<tr><td colspan="2">异常地壳（幔）</td><td>3.05~3.10</td><td colspan="2">岩石圈平均密度</td><td>3.05</td></tr>
</table>

根据上述各表数据综合分析，结合南海及周边地质构造，归纳起来本区有以下几个主要密度界面，这些密度界面与地质构造的不整合面相吻合。

1）水层与海底的密度界面，为几个界面中密度差最大的一个。低密度海水层对海区的空间异常的影响直接，关系密切。

2）新生代沉积盆地基底与前新生代各时代地层之间的界面。该界面随着接触地层时代和岩石性质的不同而有变化，一般其密度差为 $0.2\times10^3\sim0.4\times10^3$kg/m^3。对利用重力资料圈定沉积盆地、分析地质构造性质等有重要作用。在陆架、陆坡区普遍存在。在新生代地层内部也因岩性、岩相、组成物质、地层厚度和压实程度的不同，而存在着次要的密度界面，对分析盆地内次级构造单元起一定的作用。

3）各时代沉积地层与结晶基底之间的密度界面。

4）莫霍面既是地壳与上地幔的分界面，也是一个重要的密度界面。

5）岩石圈底界也应为一密度界面，资料显示其密度差可达 -0.1×10^3kg/m^3。

三、断面地形地貌、地球物理及地质结构构造特征

本次研究收集了《中国南海海域岩石圈三维结构及演化》项目的成果中南海地区的四条地学断面资料（图 24-29）。三条断面呈近南北向，分别起始于南海北部陆缘西部、中部和东部；北部断面为近东西向，西起印支半岛，经南海北部大陆边缘到达台湾恒春半岛东侧，与上述三条南北向断面相交。各断面包括了地质、构造、地球物理场基本特征、岩石圈纵向剖面及时空演化等内容，从不同角度展示了南海海域岩石圈的结构特征。地质构造部分包括断面所经过区域的地形地貌、地层（沉积）和断裂构造，地球物理场包括重力、磁力、地热流和地壳结构。

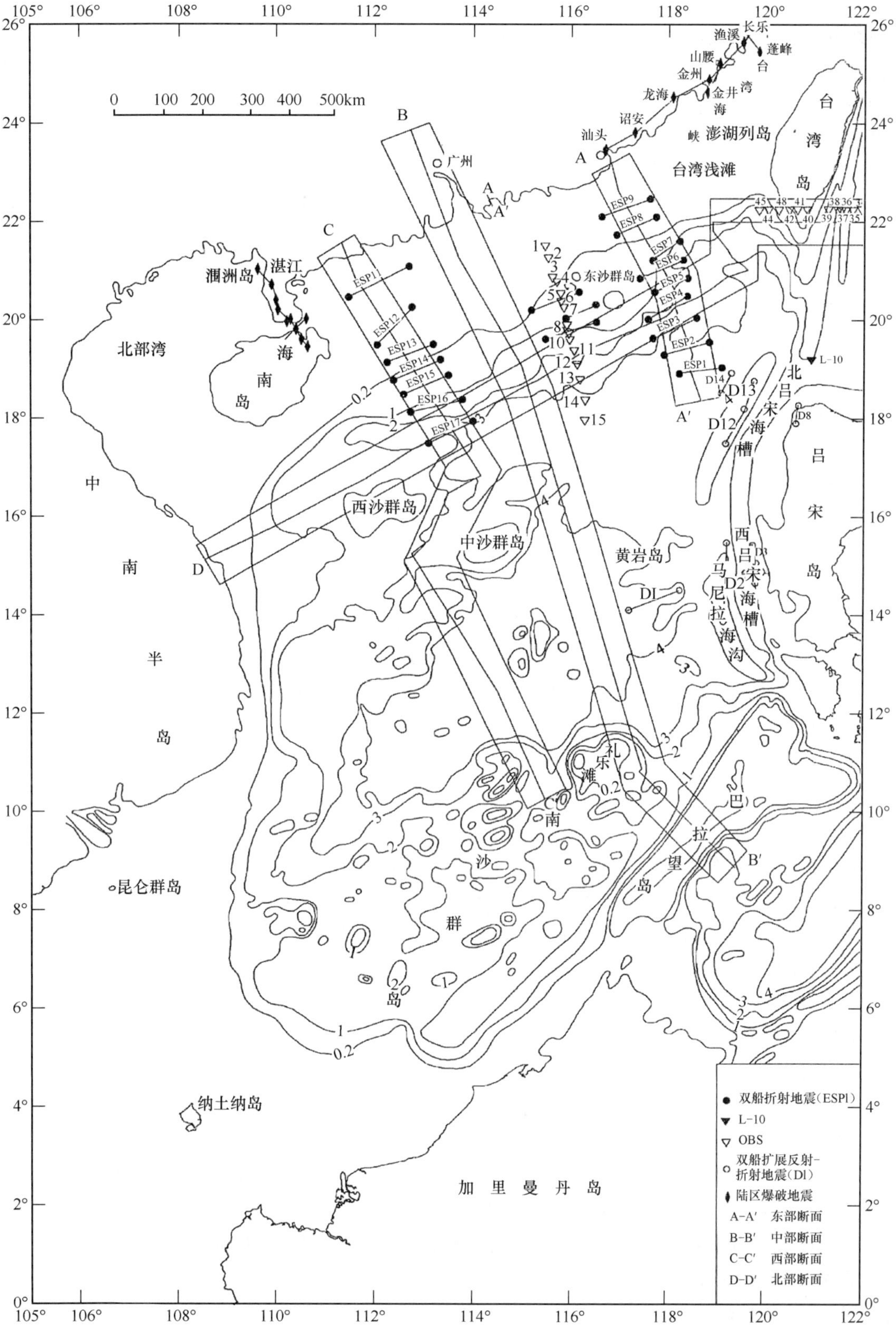

图 24-29 南海地区地学断面位置图

Fig. 24-29 Position of the geosciences cross-section of the South China Sea area

（一）东部地学断面

南海东部地学剖面系收集《中国南海海域岩石圈三维结构及演化》项目的成果。该地学断面位于东沙群岛东侧，北起广东惠来，经过珠江口盆地珠一坳陷东北、东沙隆起、潮汕坳陷笔架南低隆起、笔架南盆地，直至中央海盆，断面总体呈北北东向，全长约600km（图24-29，图24-30）。其剖面的地形地貌特征、地球物理特征及地质构造特征等如下：

1. 地形地貌特征

该剖面东部穿越了南海东部的陆架和陆坡，直达深海洋盆。在断面范围内，陆架平坦，水深小于200m，宽140km左右；而陆坡区地形十分复杂，水深在200~3400 m之间，宽度达270km。断面经过了位于陆坡区的东沙海台、北坡海山、笔架海山西南端。东沙海台为水深200~300m之间的等深线构成的平坦面，其上分布着北卫滩、南卫滩和东沙群岛；北坡海山位于下陆坡，呈北东东方向延伸，与海底间相对高差较大，达1600m；而笔架海山长轴方向为北东向，相对高差较小，仅600m。这些海山皆由火山岩构成，显然它们的形成均与区域构造作用和岩浆活动有关。断面南端进入中央海盆区，水深陡然降到3400~3600m，海底发育现代沉积。

2. 地球物理特征

（1）磁力异常（ΔT）

断面内异常总体特征表现为呈北东—北东东向展布，在东沙隆起以北地区，磁异常呈正负交替变化，异常值较高，异常梯度陡，在剖面上呈波浪状。珠江口盆地以北地区，以负磁异常为主，间夹正磁异常条带，负异常幅值可达-150nT。珠一坳陷北侧是一条0~150nT的高值磁异常带，坳陷内正负异常相间。位于坳陷南侧的东沙隆起为一条高值强磁异常带，向北东和南西方向分别可与台湾浅滩和神狐-统暗沙磁异常带相连。据磁异常计算，隆起基岩内有大块玄武岩存在，推测可能是中生代火山弧，局部可能有新生代基性岩侵入。东沙隆起以南到中央海盆北缘，磁异常变化宽缓，以块状异常为特征，异常值在0至±50nT之间。在中央海盆区，磁异常呈条带状正负相间分布，在剖面上亦呈波浪状，在断面内异常幅值较低。

（2）自由空间重力异常（Δg_F）

东断面自由空间重力异常总体幅值在零值上下波动，广东大陆地区以负值为主，正异常值与山地对应。珠江口盆地以北的陆架区以低幅值正负异常为特征，珠一坳陷以负异常为主，局部分布的重力高反映了坳陷内发育的次级隆起带。位于珠一坳陷南侧的东沙隆起区对应了一条十分显著的北东向高值正异常带，主体部位异常大于$10\times10^{-5}m/s^2$。东沙隆起南侧潮汕坳陷表现为两条北东向负异常带夹一条正异常带，最大负异常值达$-20\times10^{-5}m/s^2$。潮汕坳陷南缘表现为北正南负的重力异常特征，正异常值高达$40\times10^{-5}m/s^2$，负异常值在0至$-30\times10^{-5}m/s^2$之间，这一方面可能反映了该处水深的急剧变化，地形上形成了一个陡坎；另一方面可能是洋陆边界处重力边缘效应的反映。中央深海盆区以正异常值为特征，异常幅值在$0\sim10\times10^{-5}m/s^2$内变化。

（3）地热流

热流值与地壳结构和构造活动性具有明显的相关性，东断面热流由北向南呈起伏变化态势。珠一坳陷以北热流值在$30mW/m^2$左右，在坳陷北缘断阶处，其值开始上升，坳陷内最大值达到约$40mW/m^2$。东沙隆起北侧到隆起，热流值呈逐渐降低趋势。在潮汕坳陷热流值达到断面中的最低值，如此低的热流值，可能反映了10Ma以来该处无大的构造活动，较为稳定之故。在潮汕坳陷南缘又开始上升，直到中央海盆北缘热流值略微降低，在海盆内又逐渐上升至$40\ mW/m^2$。

3. 地质构造及地壳结构特征

（1）中、新生代沉积

根据地震资料反映，东沙隆起以东存在三个清楚的反射界面，可划分出上、下两套构造层。下构造层表现为充填式沉积，厚度变化较大，多发育于断陷内，向隆起方向减薄甚至缺失。钻井揭示这是一套陆相碎屑沉积，时代为古新世—始新世。上构造层呈披覆式覆盖在下构造层之上，沉积厚度较为

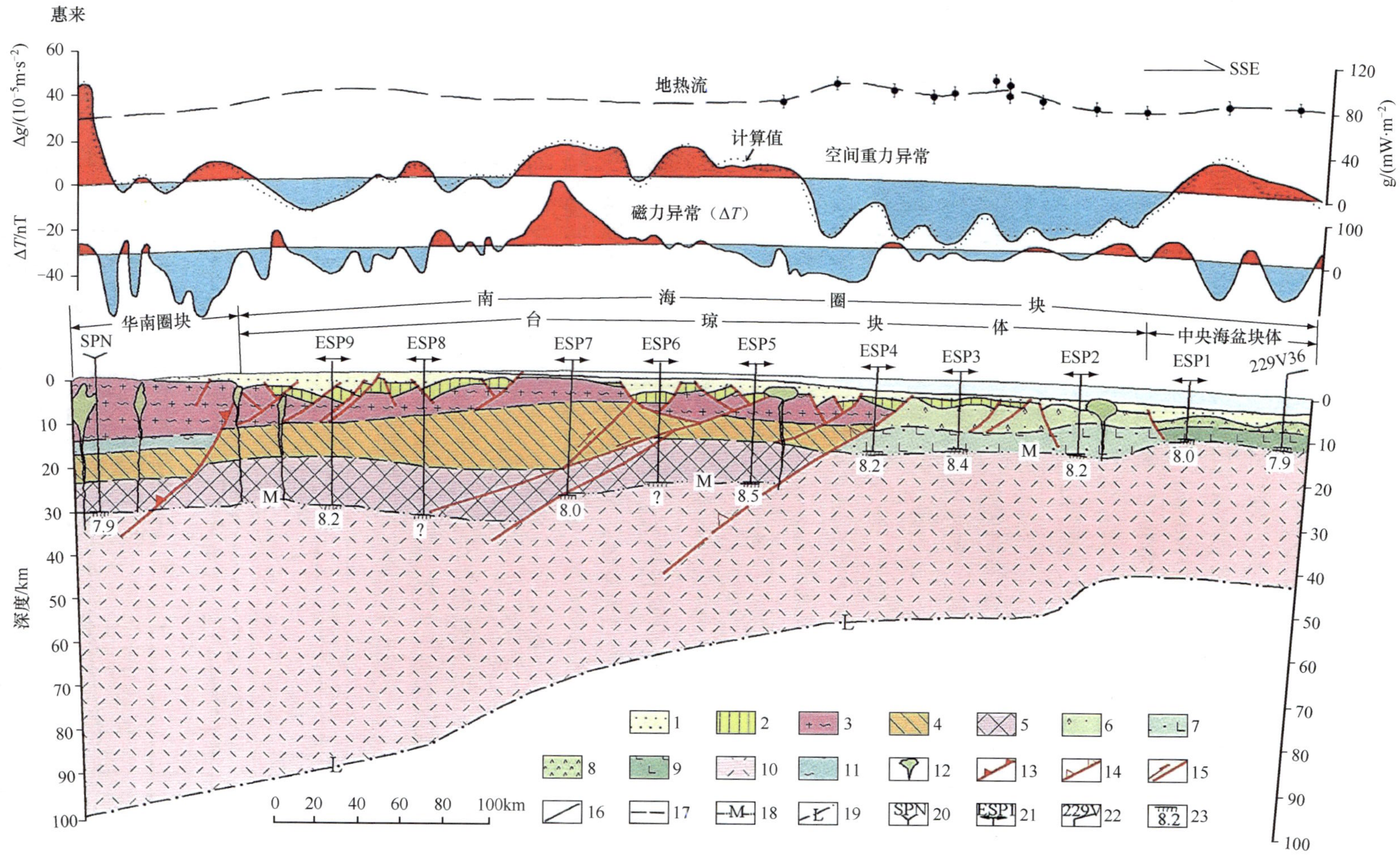

图 24-30　南海东部岩石圈地学断面图

Fig. 24-30　Geosciences cross-section of lithosphere in the eastern South China Sea

1—新生代沉积层；2—中生代沉积层；3—上地壳层；4—下地壳层上部或中地壳层；5—下地壳层下部或下地壳层；6—中生代大洋层 2；7—中生代大洋层 3；8—新生代大洋层 2；9—新生代大洋层 3；10—上地幔固结层；11—壳内低速层；12—新生代超镁铁质层；13—古生代聚敛带；14—中生代聚敛带；15—中生代逆冲断层；16—新生代张性断层；17—壳内分层界线；18—莫霍不连续面；19—岩石圈底界；20—爆破地震站；21—双船扩展排列剖面；22—声纳浮标站；23—上地幔速度值（km/s）

均匀，反映处于相对稳定的沉积构造环境，发育一套海陆过渡相—海相沉积。在珠一坳陷东部，本断面所经过处下构造层厚度最大达3200m，上构造层厚2600m。在东沙隆起区沉积厚度很薄，低于1000m，主要为晚中新世及其以后的沉积。东沙隆起东南的潮汕坳陷，地震反射特征表现为“三层楼”式结构，可划分出三套构造层，在前述上、下构造层之下还发育了一套特征显著的基底构造层。上构造层沉积厚度较大，最大厚度大于5000m。下构造层发育在一系列大小不等的半地堑内，呈楔状充填，厚度一般为2000m左右。基底构造层反射特征与其上覆构造层截然不同，在地震剖面上，这套地层往往发生了强烈的褶皱变形，表现为一大套倾角较大、褶皱较强、连续性较好的反射层系。其沉积厚度分布不均，层速度达5.1km/s，其厚度可达6500~7300m。这套地层以潮汕坳陷最为发育，推测在东沙隆起周缘和珠一坳陷中均有不同程度发育。在邻近的闽粤东南陆区，大片出露下侏罗统—上三叠统地层，这套厚达数千米的海相和海陆交互相沉积已经历不同程度的变形、变质作用，断层发育，局部因岩浆作用而复杂化。在台湾盆地中部隆起区，大量钻井均钻遇白垩系，在台北坳陷钻遇侏罗系—白垩系，二者均属滨浅海相沉积。据此推测基底构造层可能系中生代地层，受中生代末期构造运动影响发生褶皱变形。

(2) 断裂构造

东部断面所在处区域断裂总体呈北东、北东东向展布，其次为北西向，具张性断裂特征。珠一坳陷北缘发育一系列呈雁行状排列的北东东向断裂，构成了坳陷的北界，并控制了沉积发育。东沙隆起被北东向和北东东向两组断裂切割，分别向南北两侧断落。在潮汕坳陷南缘，一条大型犁式断裂以高角度切穿基底后，继续向下以缓倾角度切过上地壳，进入下地壳，并到达壳幔边界附近，推测该断裂可能是一条古缝合带，新生代期间，在张性应力场作用下发生拆离变形。潮汕坳陷与中央海盆之间，断裂多表现为南断北掀，形成众多半地堑。潮汕坳陷及其以南，发育了一组北西向断裂，切割了北东向断裂组，沿该组断裂往往发育了由玄武岩喷溢形成的海山或海丘。中央深海盆内断裂较少，海盆北缘有少量北东东向断裂分布，控制了北缘的沉积发育。

(3) 地壳结构特征

断面穿过了华南圈块、台琼块体、中生代残留洋盆、新生代中央海盆等多个构造单元，自北而南地壳结构存在显著差异。位于断面北端的闽粤沿海地区，人工爆破地震获得的结果显示（图24-31）华南圈块地壳由上、下地壳和上地壳底部的低速层构成。根据速度变化特征，可将上地壳层细分为上、下两层。上层地壳速度为5.46~5.65km/s，横向上变化较为剧烈，层厚度为1.5~4.0km，多为断裂所切割，沿断裂出露燕山期岩浆岩，也可见老的沉积岩（廖其林等，1988）；下层速度为6.1~6.35km/s，厚度比较稳定，为10km左右。在上地壳层底部存在一层厚约3~4km、层速度为5.5~5.9km/s的低速层，在华南大陆内部，鲜见低速层发育。下地壳层也可分出上下两层，上层速度在6.48~6.71km/s之内变化，厚度比较稳定，厚约7km；下层无论层速度还是厚度均存在强烈的纵、横向变化，层速度为

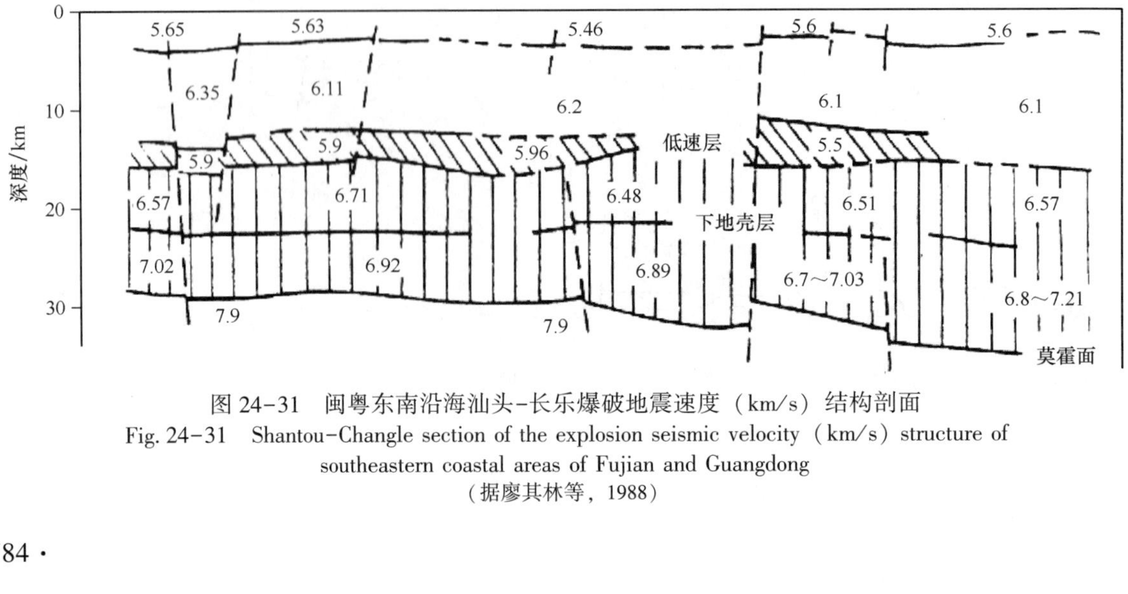

图24-31 闽粤东南沿海汕头-长乐爆破地震速度（km/s）结构剖面

Fig. 24-31 Shantou-Changle section of the explosion seismic velocity (km/s) structure of southeastern coastal areas of Fujian and Guangdong

（据廖其林等，1988）

6.89~7.21km/s，厚度则在6~10km之间。在下地壳底部速度大于7.0km/s的高速层发育处，如泉州、蒲田等地，下地壳层厚度也相应较大。对邻近剖面的普宁地区晚新生代玄武质岩石中所携带的深源捕虏体研究表明，捕虏体主要为麻粒岩相橄榄岩，具有低的$^{87}Sr/^{86}Sr$比值和较高的$^{143}Nd/^{144}Nd$比值，指示其来源于上地幔，可能代表了上地幔部分熔融产生的基性岩浆底侵于地壳的底部，在后期的上地幔岩浆上升过程中被捕获。因此，这层厚数千米的高速致密物质的形成可能反映了壳幔混合作用的结果，同位素资料表明其结晶成岩发生在112Ma前（徐夕生等，1999）。

在地震剖面上所反映的莫霍面速度为7.9km/s，地壳厚度约为30km，且莫霍面多处出现被断裂错断的现象，断距最大可达3km，表明闽粤沿海地带地壳断裂十分发育，前述的壳幔高速层可能是上地幔物质沿这些穿过地壳的断裂上升侵位于下地壳底部所致。

断面经过的南海北部陆缘块体东部地区，中美二阶段合作得到的EPS剖面（图24-32）。珠一坳陷所在的陆架区上地壳层速度为4.7~5.8km/s（ESP9），厚5~7km；东沙隆起上地壳层速度降到4~5.1km/s（ESP8），上地壳厚度仅4km；东沙隆起南侧的上陆坡区速度分别为4.25~5.8km/s（ESP7A）、4.4~6.1km/s（ESP7）、4.8~6.0km/s（ESP6）、4.9~6.0km/s（ESP5），相应厚度分别为5~14km、9km、6km、5~7km；位于下陆坡区的ESP4上地壳层速度6.2~6.3km/s，厚4km。下地壳层表现为强烈的侧向不均一性，在速度结构方面，ESP9-ESP7A之间速度在6.4~6.9km/s之间，厚度在18~26km之内变化，而ESP7与ESP4之间的下地壳明显可分为上、下两部分，下地壳上部速度与ESP9-7A相似，在6.5~6.9km/s之间，但厚度仅数千米，其下部速度为7.0~7.2km/s的异常高速层，该层厚度变化较大，最薄处仅4km（ESP4），最厚处可达12~17km（ESP5）。显然，南海北缘块体的地壳结构特征完全不同于邻近的华南陆区，二者之间可能存在一条重要的分界线（Nissen et a1.，1995；曾维军，2000）。

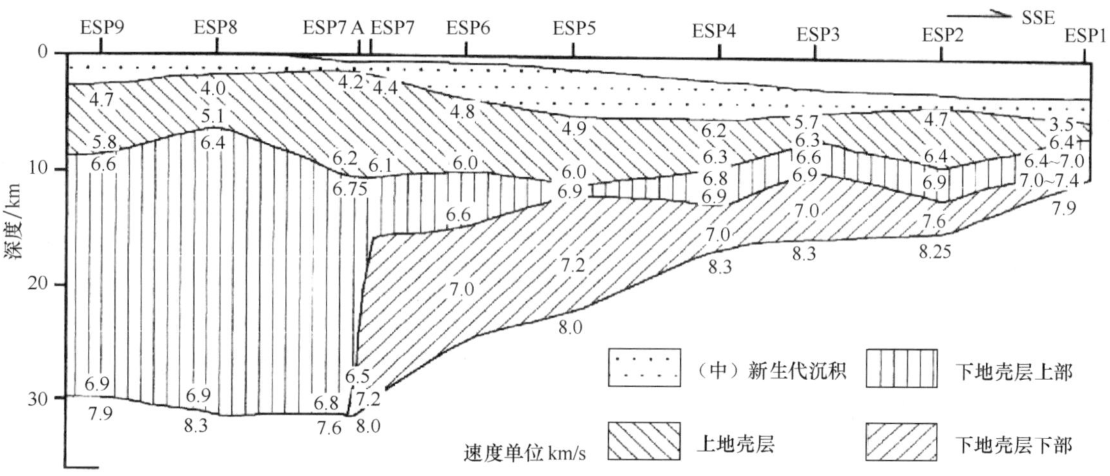

图24-32 南海东部双船地震剖面图

Fig. 24-32 Seismic section of Shuangchuan，eastern South China Sea

（据Nissen et a1.，1995）

关于南海北部陆缘下地壳底部发育的高速层目前存在多种解释，该层主要分布于地壳急剧变薄的拉张边缘，Hayes（1995）认为是南海北部陆缘前张裂期地壳结构变化所引起的，曾维军（2000）则认为该高速层是其南侧中生代古洋壳俯冲增生的结果。

南海北部陆缘上地幔顶部速度在7.6~8.33km/s之间，莫霍面自北而南向上抬升变浅，包括高速层在内的地壳总厚度呈北厚南薄之势，反映了中生代末—新生代期间强烈拉张作用的结果。例外的是东沙隆起处地壳厚度最大，达到30~32km。

中生代残留洋盆和新生代洋盆的速度结构表现为一般洋壳结构特征。ESP3剖面反映其层1速度为1.5~4.65km/s，层2速度为5.7~6.25 km/s，层3速度为7.0~7.65km/s。ESP2位于中央海盆与残留洋盆相交接部位的海丘上，该海丘相当于中央海盆边缘的脊。位于深海平原区的ESP1揭示了中央海盆北缘的洋壳结构特征，层1速度在1.6～3.3km/s之间，沉积层厚约2000m；层2速度为3.5～6.3km/s，由枕状熔岩及少量辉长岩组成，厚约1200m；层3速度为6.8～7.4km/s，由辉长岩为主的超基性岩构成，约5300 m的较大厚度可能是该处洋壳较老的缘故。上地幔顶部速度7.9km/s，与南海北部陆缘区上地幔地震波速度相似。

（二）中部地学断面

南海中部地学剖面系收集《中国南海海域岩石圈三维结构及演化》项目的成果。该地学断面即广州-巴拉望地学断面，北起广州，经南海北部陆缘区、中央海盆区、南部陆缘区，横穿巴拉望岛进入苏禄北缘滨海地带，纵贯整个南海，长约1800km（图24-29，图24-33）。其剖面的地形地貌特征、地球物理特征及地质构造特征如下：

1. 地形地貌特征

本断面所经过的北部陆缘宽约350km，其中陆架、陆坡宽度分别约为220km和130km，200m水深线可大致作为两者的分界位置。陆架平缓，陆坡水深变化在200～3400m之间，缓坡与陡坡相间，平均坡度25×10^{-3}。中央海盆地势平坦，故又称深海平原，局部出现海山、海丘，其中部的海山链把海盆分为南、北两部分。海盆内水深变化在3400～4000m之间，北半部水深普遍小于南半部，代表扩张中心的珍贝海山的比高约为2800m。南部陆缘宽200km，地势复杂，按其地形和地貌类型可分为三部分：巴拉望陆架区、南沙海槽和礼乐海台区。巴拉望陆架宽约50km，水深变化在0～100m之间，陡峻的陆架外缘构成了南沙海槽的南缘；海槽中部水深为1400～1500m，最深处达1700m；礼乐海台周围礁、滩密布，沟、谷纵横，是航海的危险区，礼乐海台面积约9000km^2，海台外缘十分陡峻，平均坡度为170×10^{-3}，陡坡的高差超过4km。

2. 地球物理特征

（1）磁力异常（ΔT）

总观磁异常剖面变化（图24-34），具有明显的三分性：北段（北部陆缘和华南陆区）、中段（中央海盆区）和南段（南部陆缘和外苏禄海）。经高、低通滤波分离后，这三段的特征更一目了然。分窗口求对数功率谱表明，各区的高频和低频段的频率范围和谱值均有所不同。相应频段各自反映的磁源体平均埋藏深度有所差异。对原异常做上延30km和60km处理，发现陆缘区经60km上延后仍可再提取更深一层的磁源异常信息。其结果是：海盆区为二层磁源结构，陆缘区为三层磁源结构。

（2）自由空间重力异常（Δg_F）

自由空间重力异常（图24-35）在北部陆缘和华南陆区以低缓变化的负异常为主，在$0\sim20\times10^{-5}$ m/s^{2}的异常背景上叠加局部异常，其最高值（25×10^{-5}m/s^2）和最低值（-60×10^{-5}m/s^2）都出现在陆区。在海盆区，除中部两个由海山引起的高值异常外，整个异常背景更为宽缓，在海山以北呈低值正、负异常交替出现，以南为低值平缓正异常。剖面南段出现的两个高值正异常，分别对应于礼乐滩和巴拉望岛，异常极值达75×10^{-5}m/s$^2\sim80\times10^{-5}$m/s^2，两者之间的鞍部对应于南沙海槽。苏禄海北缘滨海地带出现一负异常，极小值低于-40×10^{-5}m/s^2。上延30km处理后的空间异常与2～180阶的卫星重力异常相似，只是前者含有较多的局部变化成分，而后者在局部区段的异常值偏大。

布格重力异常与地形变化呈负相关，从陆缘至中央海盆区异常值抬升到320×10^{-5}m/s^{2}左右，陆坡区为异常梯阶带，约200×10^{-5}m/s^{2}的等值线大致反映陆-洋分界线。均衡异常开阔舒缓，变化范围一般为$\pm30\times10^{-5}$m/s^2，局部异常小而少，$\pm10\times10^{-5}$m/s^{2}异常等值线所构成的微弱梯阶带大致反映陆—洋边界。

（3）地热流

按地热流趋势剖面的变化特征（图24-36），大致可分为三段：南段包括礼乐滩及其以南区段，热流值变化较大，其中礼乐滩和巴拉望岛对应低值区（50～60 mW/m^2），南沙海槽为较高值区（80 mW/m^2），苏禄海呈逐渐升高之势；中段即中央海盆区，在总体升高的背景上，残留扩张中心（极值

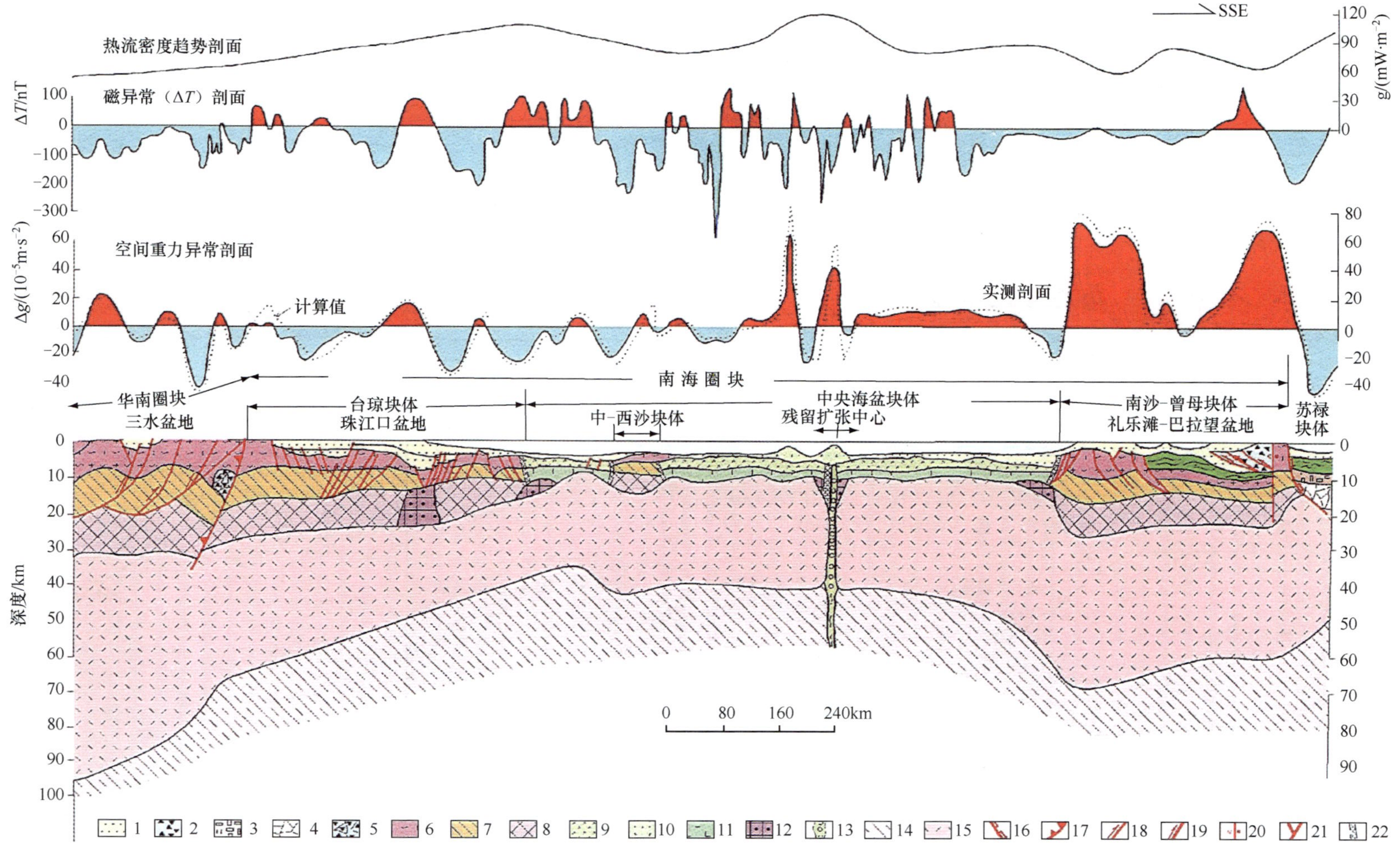

图 24-33　南海中部（广州-巴拉望）岩石圈地学断面图

Fig. 24-33　Geosciences cross-section of lithosphere of the central South China Sea (Guangzhou-Palwan)

1—新生代沉积层；2—中生代混杂岩；3—中、新生代混杂岩；4—中生代增生体；5—古生代蛇绿岩体；6—上地壳层；7—下地壳层上部；8—下地壳层下部；9—大洋层 2；10—扩张期后基性喷发岩；11—大洋层 3；12—异常地壳（幔）；13—岩浆通道；14—上地幔软流层；15—土地幔岩石层；16—新生代聚敛带；17—古生代聚敛带；18—中生代仰冲断层；19—新生代仰冲断层；20—走滑断层；21—张性断层；22—陆—洋边界

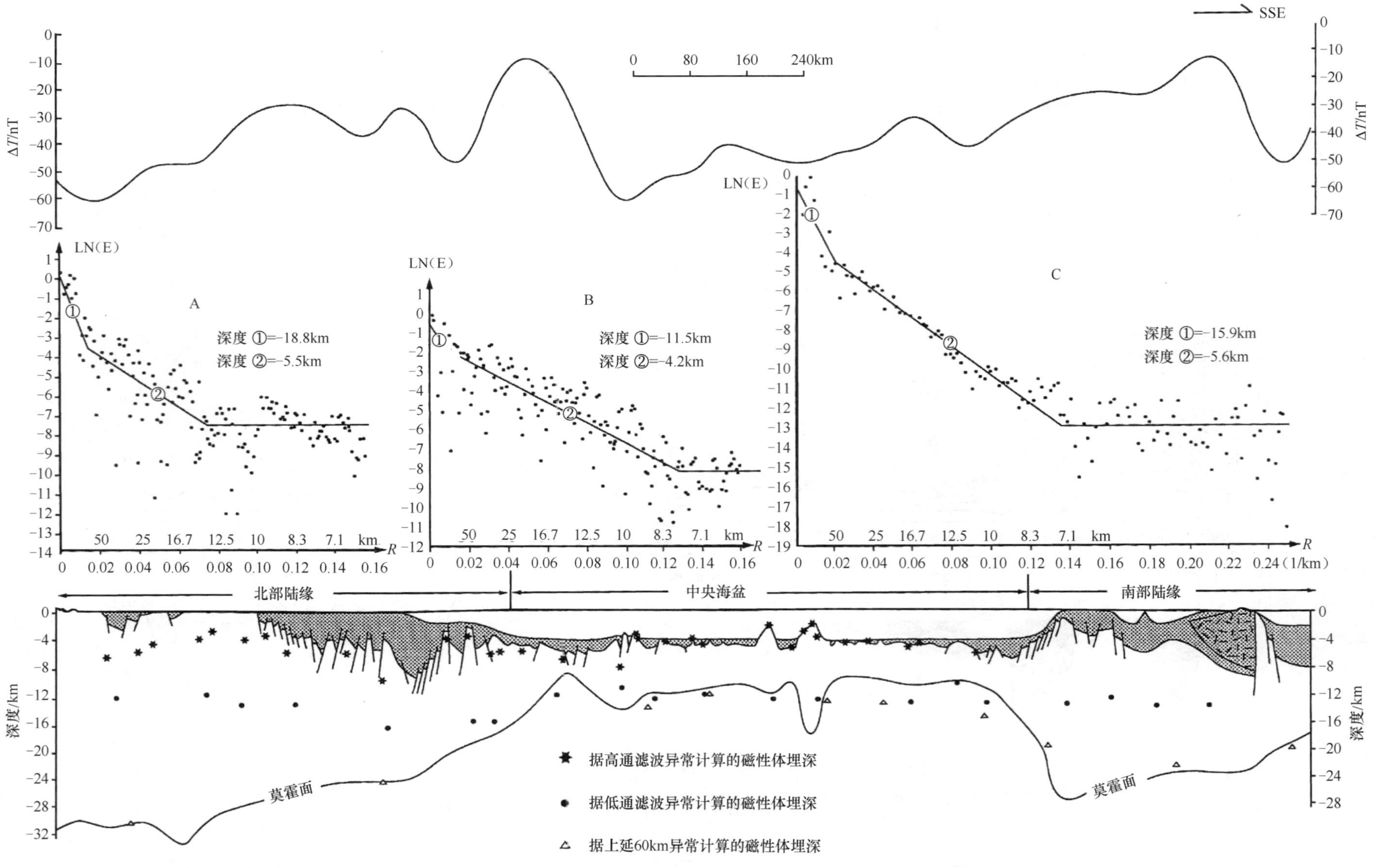

图 24-34 广州-巴拉望磁异常及其分析剖面图

Fig. 24-34 Section of magnetic anomalies and their analysis, Guangzhou-Palwan

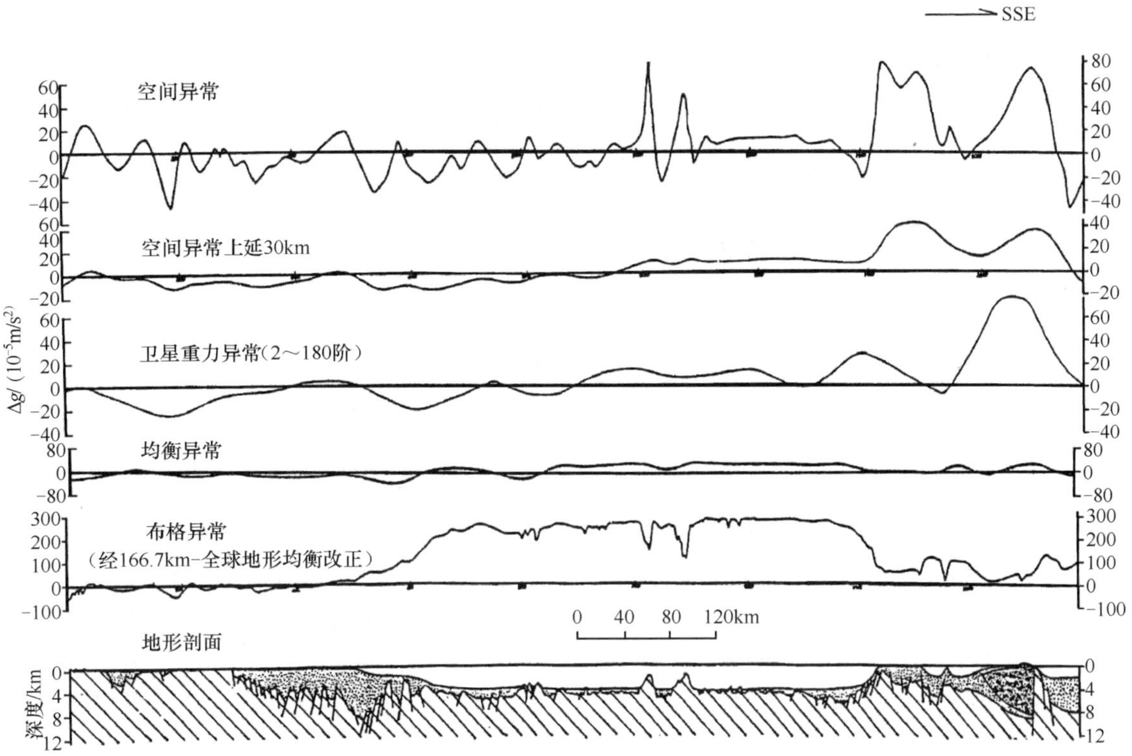

图 24-35　广州-巴拉望重力异常剖面图

Fig. 24-35　Section of gravity anomalies in Guangzhou-Palwan

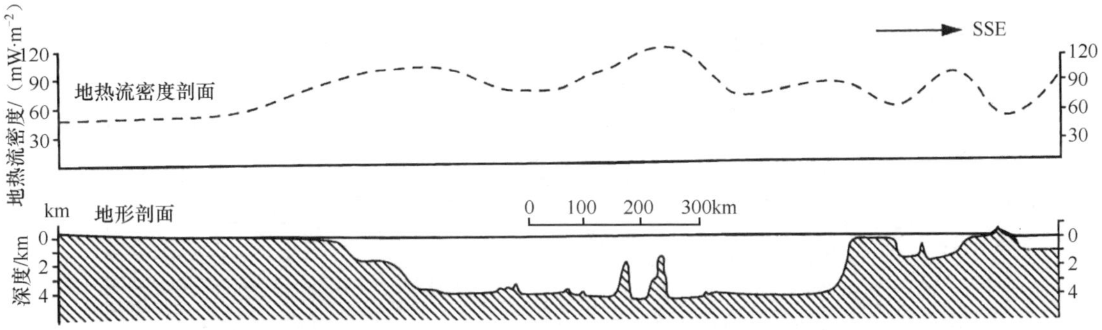

图 24-36　广州-巴拉望地热密度趋势剖面图

Fig. 24-36　Section of the geothermal density tendency of Guangzhou-Palwan

<120mW/m^2）与陆洋交接带处又局部升高；北段包括北部陆缘和华南陆区，总的趋势是由南往北降低，似与地形变化反相关，由陆坡至陆架边缘从 90 mW/m^{2}左右降至 60 mW/m^{2}左右，进入陆区平缓降至 53 mW/m^{2}左右。

3. 地质构造及地壳结构特征

（1）中新生代沉积

沿本断面分布的地震剖面大致可揭示新生界盖层发育特征，并为研究其构造活动提供重要依据。图 24-37 是据地震反射剖面的综合解释结果。从北到南全线新生界厚度变化很大：北缘的珠江口盆地一般厚 2~5km，珠二坳陷中部最厚达 10km 左右；中央海盆从中部 0. 5km 左右向两侧增至 2~3km；南缘礼乐滩-巴拉望一般为 2~4km，呈由北往南增厚之势，至巴拉望陆架边缘最厚达 10km，主要为大套外来混杂岩沉积的结果。

新生界内存在三个可进行区域性对比的不整合反射界面，即北缘的 T_g、T_6 和 T_2 不整合大致相当

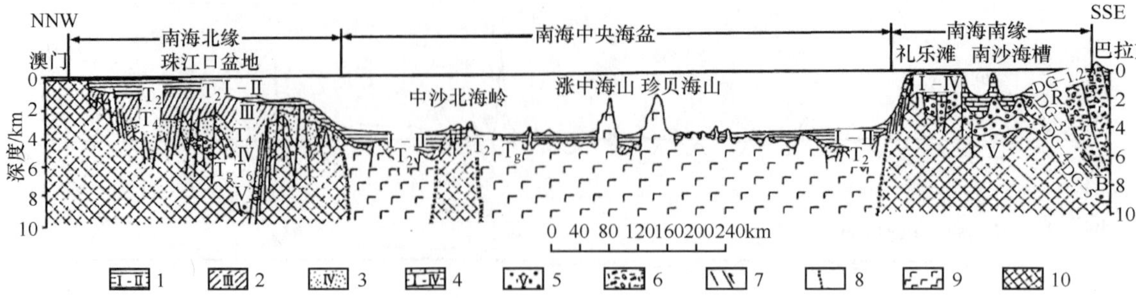

图 24-37 南海北缘-中央盆地-巴拉望新生代沉积剖面

Fig. 24-37 Cenozoic sedimentary section of the northern edge of South China Sea-Central Basin-Palwan

（据反射地震资料解释结果）

1—第四系—上中新统；2—中中新统—下中新统；3—下中新统—上渐新统；4—第四系—上渐新统碳酸盐岩；5—始新统—古新统；6—混杂岩带；7—断层；8—海盆—陆缘边界；9—洋壳基底；10—陆壳基底

于南缘的 V、B* 和 R 不整合。经区域性的追踪对比和钻井资料解释，T_g 或 V 大致相当于声波基底的顶界，在陆缘区主要构成新生界（可能局部包括上白垩统顶部）之底界，反映区域性张裂活动的开始。在海盆区则反映洋壳基底之顶界。T_6 或 B* 出现在渐新统内部，反映南海扩张前的区域性抬升和夷平作用。T_6 或 R 出现在中中新统与上中新统之间，在北缘陆—洋过渡带，T_2 呈一向海盆方向倾斜的上超面（图 24-38），标志这一阶段两者之间曾发生过强烈的差异沉降活动，并伴有玄武质岩浆喷溢活动及热事件（茹克等，1986）。

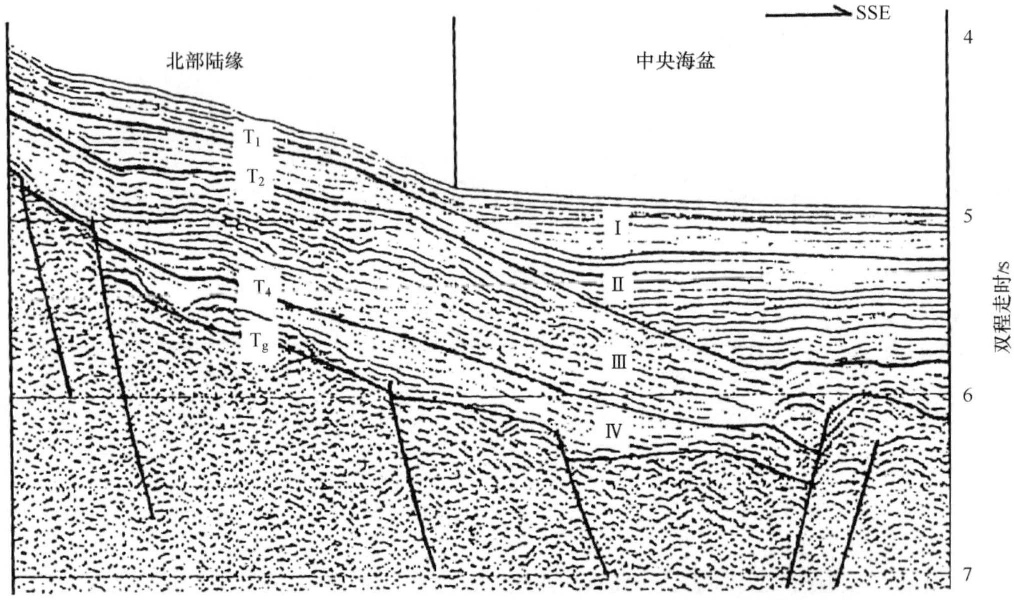

图 24-38 南海北部陆缘—中央海盆过渡带地震反射剖面选段

Fig. 24-38 Selected part of the seismic reflection section of the transition zone from the northern South China Sea continental margin to the central marine basin

（2）地壳结构特征

综观全剖面，地壳结构具有明显的纵向分层、横向分块特征。除去北端陆区和南端岛弧区外，南海可分为三个区段：北部陆缘区、中央海盆区和南部陆缘区。

北部陆缘区地壳大致可以分出上地壳层和下地壳层，后者还可进步分出上部和下部。上地壳层上部沉积层的 v_p 值为 1.6~5.0km/s，结晶层的 v_p 值为 5.0~6.4km/s，平均密度为 2.55g/cm^3；下地壳层上部的 v_p 值为 6.5~7.0km/s，平均密度为 2.7g/cm^3，其下部的 v_p 值为 7.1~7.5km/s，平均密度为 2.9 g/cm^3。上、下地壳层厚度均呈从北向南减薄之势，其莫霍面埋深向海盆方向逐渐抬升，在滨海地带约

30km，至陆—洋交接带变浅到14km左右。

中央海盆区地壳（图24-39）可分出层1（沉积层）、层2（基底）和层3（大洋壳）。层1厚0.5~3km，v_p 值为1.6~3.7km/s；层2厚2~4km，v_p 值为5.0~5.9km/s；层3厚3km左右，v_p 值为6.5~7.4km/s。可进一步分为层3A（v_p 值为6.5~7.0km/s）和层3B（v_p 值为7.2~7.4km/s）。

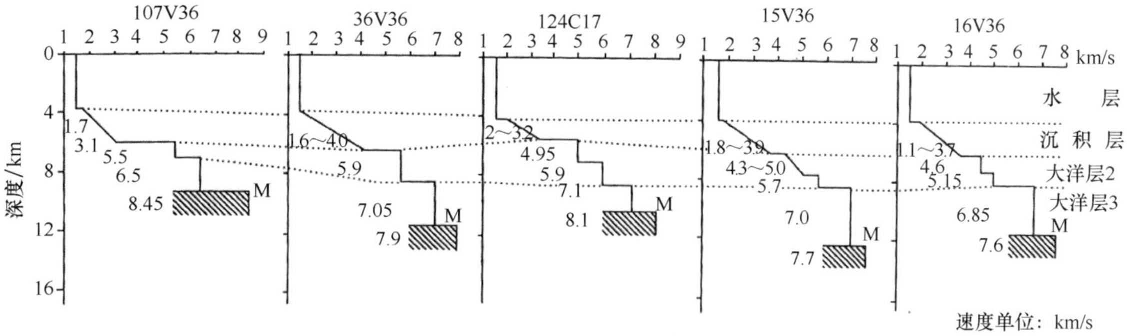

图24-39 南海海盆区地壳速度结构图

Fig. 24-39 Crustal velocity structure of the South China Sea basin region

南部陆缘区尚未开展深部调查。综合解释工作主要参考北部陆缘区。该区也分上、下地壳层，地壳厚度在礼乐滩一带厚约25km，向南减薄至20km左右。

（三）西部地学断面

南海西部地学剖面系收集《中国南海海域岩石圈三维结构及演化》项目的成果。该地学断面位于雷州半岛和海南岛以东，北起广东阳江市沙扒，经珠江口盆地西部珠三坳陷、神狐-一统隆起、西沙海槽、中西沙块体、南海西北海盆、中沙海槽、南海西南海盆，直至南沙-曾母块体，剖面穿越了南海西部多个大地构造单元，总体方向呈北北西—南南东，全长约1300km（图24-29，图24-40）。其剖面的地形地貌特征、地球物理特征及地质构造特征等如下：

1. 地形地貌特征

西部断面穿越了南海北部最宽阔的陆架区，经西沙海槽和西沙群岛东端、西北海盆西南角，沿中沙海槽北侧往西南方向延伸到盆西海岭，之后转为北北东向，进入西南海盆，终止于南沙群岛北部。广东大陆边缘为海岸平原区，广泛分布第四纪沉积，海岸曲折多变，岬湾相间，岛屿密布。陆架区海底平坦开阔，水深等值线平行于海岸线展布，陆架宽逾300km。陆坡区地形地貌复杂多变，发育了西沙海槽、中沙海台、中沙海槽、西沙海台、西南洋盆、南沙海台等地貌单元。西沙海槽呈近东西向展布，水深约3000m，槽底平坦，海槽北坡陡、南坡缓；中沙海槽呈北东向展布，分隔了中、西沙海台，地形复杂，水深2500~3400m，海槽的西北坡缓、东南坡陡；西沙海台位于西沙海槽南侧，由1000 m等深线环围成台面，海台东南侧为西北海盆；中沙海台呈长轴为北东向的椭圆形，为一珊瑚礁，水深200 m以内的海台顶面上峰谷发育、礁滩林立，海台边缘为陡峻的斜坡；西南海盆位于南海海盆的西南部，呈北东向展布，海盆内大部分水深在4300~4400m之间，是南海深海平原上最低洼的部分；南沙海台上广布岛屿、沙洲、浅滩和暗礁，水深2000m左右，是著名的航海危险区。

2. 地球物理特征

（1）磁力异常（ΔT）

磁异常具明显的分区性，北部陆缘为平缓的大片负磁异常，在-50nT的背景上叠加局部伴生异常，海岸带附近局部异常较为发育。神狐隆起至西沙海槽北坡为高值负磁异常带，最大值达-200nT。西沙海槽为高值正异常带，异常峰值为150nT，据磁力资料计算，磁性体顶界埋深仅5.4km，位于基底内部，根据其强磁性特征推测可能为基性岩。西北海盆亦表现为高值正异常特征，峰值为250nT。海盆东南侧出现高值负磁异常，最大异常值为-250nT。中沙海槽及其两侧磁异常正负交替变化，在与盆西海

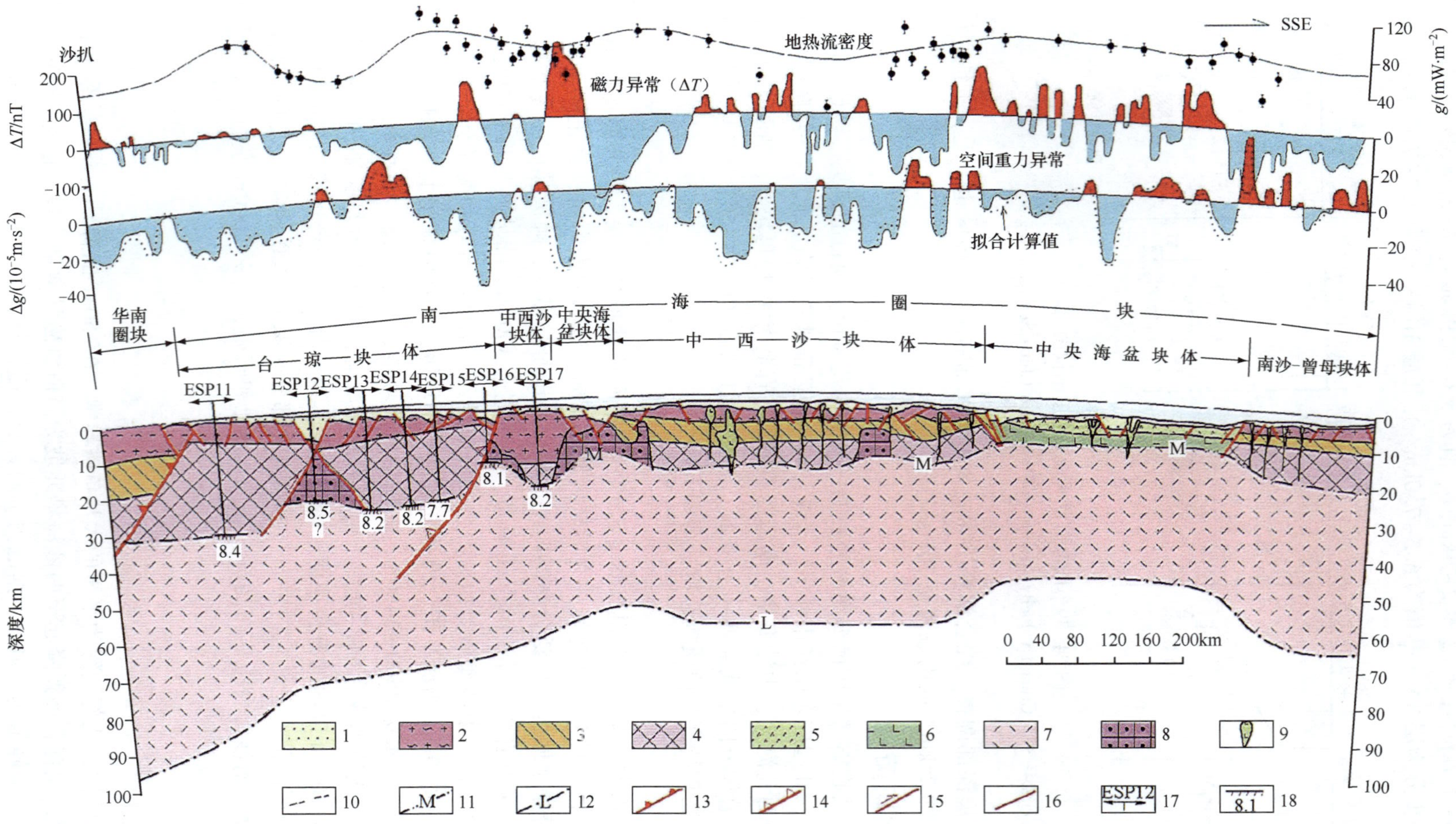

图 24-40　南海西部岩石圈地学断面图

Fig. 24-40　Geosciences cross-section of lithosphere of the western South China Sea

1—新生代沉积岩；2—上地壳层；3—下地壳层上部；4—下地壳层下部；5—大洋层 2；6—大洋层 3；7—上地幔固结层；8—异常地壳；9—新生代超镁铁质岩；10—壳内分层界线；11—莫霍不连续面；12—岩石圈底界；13—古生代聚敛带；14—中生代聚敛带；15—中生代仰冲带；16—新生代张性断层；17—双船扩展排列剖面；18—上地幔速度值（km/s）

岭相交处出现一高值的正异常带，峰值达 250nT。西南海盆磁异常特征表现为正负相间的波浪状异常，在平面上呈北东向线性展布，正负峰值约±150nT。南沙海域为宽缓负异常区，局部叠加伴生异常。

（2）自由空间重力异常（Δg_F）

从陆区至珠三坳陷北侧以低幅值负异常为特征，异常方向以北东东向为主，异常值在（0~20）×10^{-5}m/s^{2}范围变化，重力低对应了断陷的沉积中心。珠三坳陷沉积厚度最大的地方却位于重力高与重力低之间的斜坡地带，异常值在（0~-10）×10^{-5}m/s^{2}之间。坳陷南侧的神狐隆起对应了重力高带，异常值为（0~20）×10^{-5}m/s^2。西沙海槽为重力低带，异常值由北向南降低，至海槽中心降到最低，其峰值达-60×10^{-5}m/s^2，异常方向呈东西向，向东与西北海盆负异常连成一片。西沙海槽南侧和中沙海槽南北两侧陆坡区为重力高值带，异常值在 0~10×10^{-5}m/s^{2}之间。中沙海槽内则由一系列呈斜列状的负异常带组成，异常方向为北东向，异常值最低达 40×10^{-5}m/s^2，在负异常背景上叠加的局部正异常与海山相对应。盆西海岭区表现为高值的正异常带，幅值达 30×10^{-5}m/s^2，但异常方向呈北西西至北西向，与海岭总体北东向迥异，此处正好位于 15°N 附近。西南海盆区重力异常面貌发生了很大的变化，异常总体仍呈北东向展布，但其平静变化的异常背景有别于海盆两侧地区，海盆中轴部位的负异常对应了残留扩张中心，异常峰值达-40×10^{-5}m/s^2，扩张中心两侧为重力高值带。南沙地块重力异常以正值为主，在此正背景场上发育北东向的短轴状局部异常，在与海盆交接处，即水深急剧变化带上，呈现出典型的陆侧正、洋侧负的重力异常边缘效应。

（3）地热流

断面热流变化特征显示具有较大的起伏性。南海北部陆缘，热流值自北而南呈上升趋势，至珠三坳陷北缘断裂附近达 96.06mW/m^2，珠三坳陷热流值降低，平均为 72mW/m^2；到西沙海槽处热流值升高，平均值为 100mW/m^2，显然与新生代期间沿海槽的张裂活动有关；中西沙地区热流值普遍较低，与钻井所揭示的该区前寒武纪基底有关；西南海盆区热流值相对于海盆两侧，为一明显的高值带，其峰值高于 120mW/m^2；南沙海区热流值特征与中西沙地区相似，其值较低，可能也存在前寒武纪的残留陆块。

3. 地质构造及地壳结构特征

（1）中新生代沉积

断面贯穿南海南北陆缘区，跨越范围大，各区沉积发育存在一定差异，根据地震反射层序和区域不整合面划分构造层，并结合石油钻井资料，推断各构造层岩性组成。由于不整合面在各区存在穿时性，因而各构造层顶底界时代在不同地区有所差别。

南海北部陆缘西区，新生代沉积大致可划分为上下两套构造层。下构造层包括古新世—始新统，局部地区可能还包括中生代沉积，其分布具有较大的分割性，大多受晚白垩世—古近纪形成的北东向断陷控制，发育在一些地堑或半地堑型凹陷带内，为一套杂色的陆相沉积，类似于华南陆区同期断陷型盆地内的沉积。上构造层包括渐新统—第四系，从下往上其沉积范围逐渐扩大，该层不整合覆盖于下伏构造层之上，经历了从断陷到断拗、从陆相-滨海湖沼相到海相沉积的发育过程。

断面所经过的珠江口盆地以北内陆架地区，发育一系列小型断陷，呈半地堑样式出现，断陷内上构造层厚约 1000m，下构造层厚 5000m。往南到珠江口盆地北部断阶带，基底隆起，其上沉积较薄，缺失下构造层，晚中新世之后的沉积不整合覆盖在基底之上。在一些小断陷内可能发育有古近系沉积。珠三坳陷为一南断北超的大型断陷（图 24-41），断陷内沉积巨厚，最大厚度达 10km 以上，其中下构造层厚约 5500m，上构造层厚约 4500m。位于珠三坳陷南侧的神狐-一统暗沙隆起，其上仅发育了上构造层，沉积厚度在数百米至 2000m 之间。

在西沙海槽区，沉积厚度不大，最厚约 3000m，也可分出上、下两套构造层，其中下构造层厚度较大，可能为陆相沉积充填。

在西沙群岛，据位于断面西南方的西永一井揭示，在钻穿 1251m 厚的中新世至第四纪珊瑚礁沉积层之后，即为元古宙变质岩基底。该处大套前中新世沉积的缺失反映了该地区经历过长期的抬升和强烈的剥蚀。

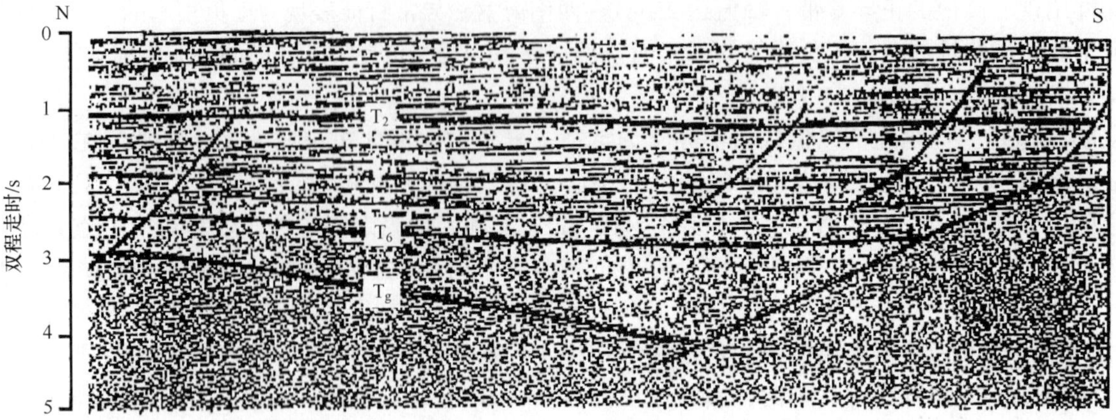

图 24-41　珠三坳陷地震剖面选段

Fig. 24-41　Selected part of the Zhusan depression seismic reflection section

中沙海槽内沉积受北东向隆凹格局的影响，隆起上沉积非常薄，仅数百米厚；但凹陷内沉积厚度也不大，厚度在3000m以内。地震剖面上也可划分出上、下构造层，二者厚度均薄，可能与距离物源较远有关。

西南海盆区新生代沉积也可划分出上、下两套构造层，下构造层为一套在起伏不平的洋壳基底上呈透镜状或楔状充填的大洋沉积，主要分布在海盆两侧及海盆内的残留扩张中心，推测为陆源碎屑及附近的玄武质火山碎屑、熔岩流，时代为晚始新世—早渐新世。上构造层沉积分布范围逐渐扩大，其产状随基底高低呈波状变化，属海相沉积。海盆内沉积厚度一般不超过1000m，多数在200～500m之间，沉积形成于海底扩张期和扩张期后。

南沙中北部海域下构造层仍表现为断陷充填，据断面东面礼乐滩上的桑帕吉塔1井揭示，这套地层为海相沉积，时代为古新世—始新世；中构造层发育在区域隆起背景上的小型局部凹陷内，沉积厚度不大，一般小于1000m，其时代相当于晚始新世—中中新世，属海相沉积，碳酸盐岩发育；上构造层时代为晚中新世—第四纪，呈水平状覆于下伏构造层之上，系南海扩张停止后区域热沉降背景下形成的海相沉积。

（2）断裂构造

西部断面所经区域断裂十分发育，形成复杂的断裂系统。据地震剖面反映，大多数断裂切穿了基底，且以张性断裂为特征，控制了沉积盆地的隆坳格局和新生代沉积的发育。断裂方向多变，以北东向、北东东向、近东西向为主，间或有近南北向和北西向，各组断裂往往呈复杂的交切关系。

在南海北部陆缘西区，断裂以北东、北东东向为主，控制了珠江口盆地西部的隆起与坳陷，珠三坳陷即受其南侧一条大型北倾断裂的控制，断距最大达6500m，深地震揭示该断层已切穿莫霍面，表现为典型的南断北超的半地堑结构。中西沙海域断裂展布仍以北东向为主，近东西向次之。西沙海槽的发育受近东西向阶梯状断裂控制，海槽内发育的新生代沉积亦成东西向展布。中沙海槽的形成与北东向断裂有关，位于海槽南侧的断裂断面北倾，海槽内壁呈南陡北缓之势，海槽内沉积向北西方向超覆于基底不整合面之上。西南海盆内主要发育北东向断裂，在海盆南北边缘断裂将陆壳与洋壳分开，并表现为大的断阶，构成了海盆边界。在海盆北缘与盆西海岭连接处，大约在15°N附近，发育了近东西向断裂，与前述磁力异常梯级带对应，表明该断裂切割深度较大。海盆内还发育了一组北西向断裂，往往将海盆北缘断裂左行错断，因而使西南海盆往北东方向宽度增大。在南沙海域，北东向断裂仍占主导地位，但近东西向断裂活动增强，并且切割了北西和北东向断裂，表现出断裂活动在时间上的先后次序。

（3）地壳结构特征

西断面（图24-42）穿越了南海西部多个构造单元，显示出南海西部地区地壳结构的极其复杂性。揭示地壳结构的深部资料主要集中在断面北部所在的南海北部陆缘区，断面中部和南部的地壳结构则是利用重力资料计算的结果。

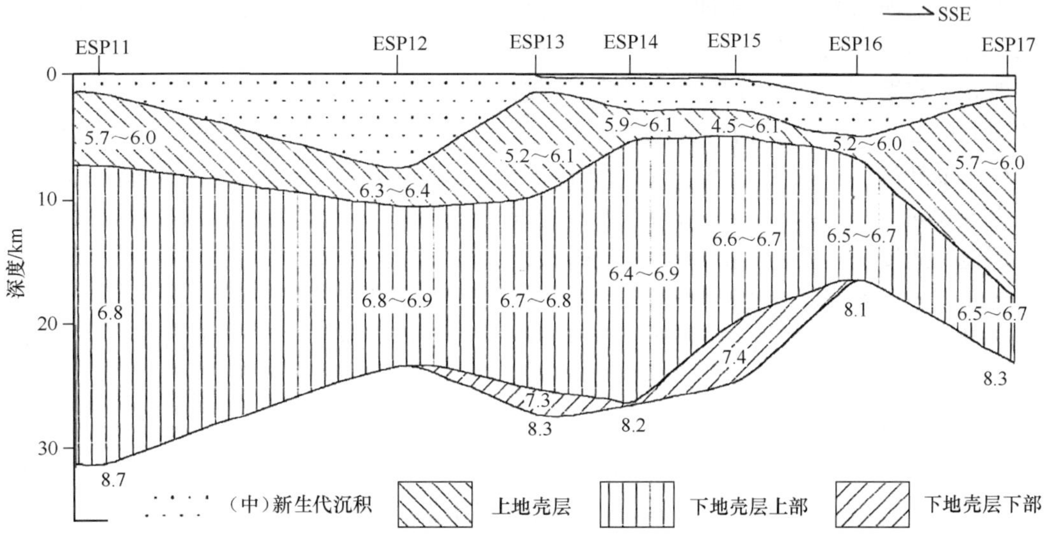

图 24-42　南海北部陆缘西部双船地震剖面图（位置见图 24-1）

Fig. 24-42　Shuangchuan seismic reflection section of the western continental margin of the northern South China Sea

（据 Nissen 等，1995）

地壳结构特征　断面北端中心位置以西约 100km 远的雷琼地区，爆破地震获得的地壳速度结构揭示这一地区地壳为上、下两层结构（图 24-43）。上地壳层可分出三部分，上地壳上部地震速度在 5. 85km/s 与 6. 3km/s 之间变化，包括沉积层在内厚约 6～10km；上地壳下部速度为 6. 15～6. 6km/s，厚约 12～16km；在上地壳内部存在一层约 2～3km 厚的低速层，其速度为 5. 65～6. 05km/s，低速层的发

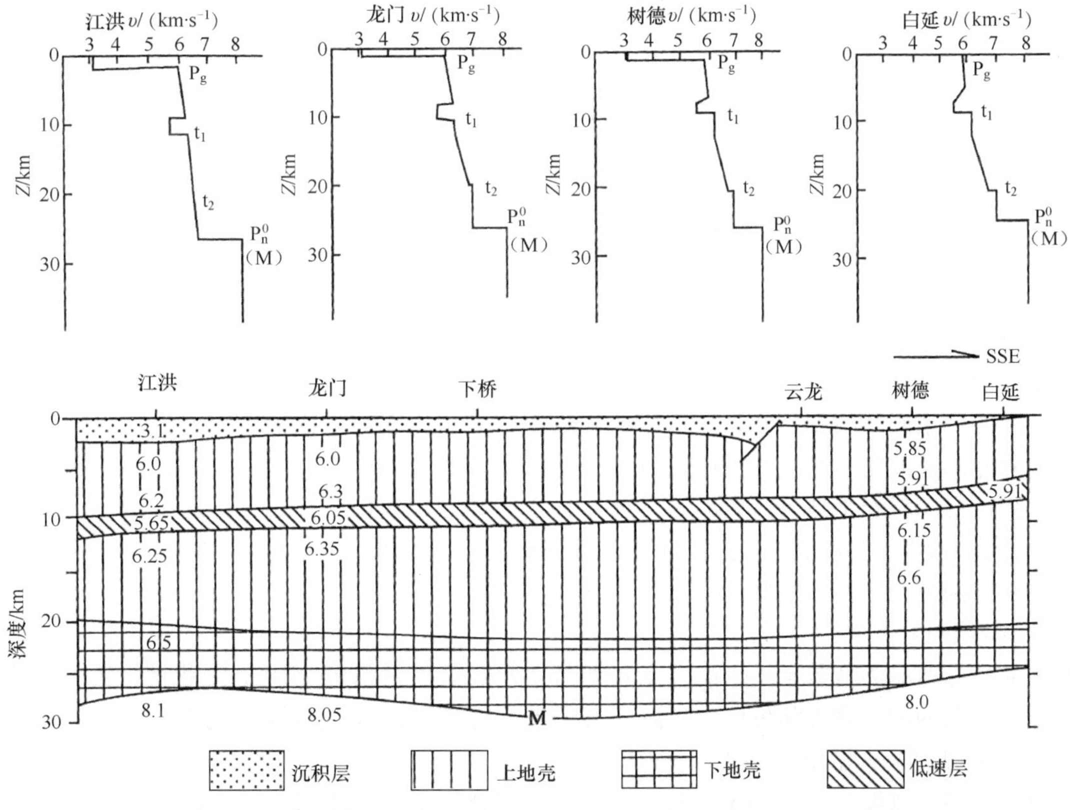

图 24-43　雷琼地区爆破地震地壳结构剖面

Fig. 24-43　Explosion seismic crustal structure section of the Leiqiong area

（据林忠祥，1988）

育类似于闽粤东南沿海的汕头-长乐地区，但其层位上升到上地壳内。下地壳厚度较上地壳薄，多在数千米内变化，与上地壳比较，其横向变化相对较大，莫霍面亦因此存在较大的起伏变化。总体来看，雷琼地区的地壳结构特征与闽粤沿海地区相似，可能代表了华南南部边缘的地壳发育特点。

位于断面北段的7个ESP剖面充分揭示了南海北部陆缘西段陆架、陆坡区的地壳结构特征，上、下地壳层均存在强烈的起伏变化。ESP11位于内陆架，上地壳层速度为5.7~6.0km/s，厚5~6km；下地壳层速度为6.8km/s，厚达22~25km，上下地壳厚度差别甚大。ESP12位于珠江口盆地珠三坳陷沉降中心处，上地壳速度为6.25~6.4km/s，厚度仅2~3km，其上覆厚达8km的新生代沉积；下地壳速度6.8~6.85km/s，厚度为12~13km。显然，这里的地壳发生过极其强烈的拉张减薄，且以壳上减薄为主，在珠三坳陷拉张程度最大，现今地温梯度仍达35C°/km~60C°/km。ESP13位于珠三坳陷南侧的神狐隆起上，新生代沉积层很薄，上地壳层速度为5.2~6.1km/s，厚度增大到8~10km；下地壳厚17~18km，层速度为6.7~6.8km/s。在下地壳底部出现了速度为7.3km/s的薄层高速层，厚2~3km。ESP14位于神狐隆起南侧斜坡上，上地壳厚度依然极薄，仅2~3km，层速度为5.85~6.1km/s；下地壳厚度为20~22km，速度为6.4~6.9km/s，下地壳底部未见高速层，其特征与ESP11相似。ESP15位于西沙海槽北部斜坡，此处上地壳依然很薄，下地壳底部又开始出现了厚4~8km的高速层。ESP16位于西沙海槽中心，上、下地壳均变薄，上地壳厚度仅有1km，下地壳因伸展也减薄至9~10km，此处莫霍面埋深急剧减小，莫霍面呈隆起状态。位于西沙隆起上的ESP17所显示的地壳结构特征截然不同于其北侧，以西沙海槽为中心，莫霍面向南加深，上地壳厚度突然增大到16km，而下地壳却迅速减薄为6km，这一现象暗示在西沙海槽处似乎存在一条重要的地质界线。

断面往南经过西北海盆西南角，该处夹于中沙群岛与南海北部陆缘区之间，64V36、65V36、66V36、67V36、5V36、68V36、107V36、45C14、36V36和108V36等站位揭示了这一地区的地壳结构特征。该区地壳具两个显著特点，一是壳-幔未经充分调整，普遍缺乏层3B速度层；二是可能尚保留部分裂谷期较为复杂的地壳结构特点，短距离内地壳厚度变化大，层2和层3无论是厚度或速度值都很不稳定，而可作为特例的是，处于下陆坡（水深为2970m）部位的65V36却具有典型的洋壳速度结构，其地壳厚度仅4.93km。

莫霍面特征　西断面上莫霍面起伏变化较大，总的说来，一方面与地形起伏（或水深）变化呈负相关，另一方面又反映了新生代隆坳格局的影响，即整体上莫霍面深度自陆区向海盆越来越浅，陆区莫霍面埋深普遍大丁30km；海区由陆架区的28~30km，到陆坡区的16~24km，中沙西沙群岛区的16~23km，南沙群岛区18~20km和西南海盆的10~12km。莫霍面深度反映南海西部地区地壳类型多样，发育有陆壳、过渡壳和洋壳。

（四）北部地学断面

南海北部地学剖面系收集《中国南海海域岩石圈三维结构及演化》项目的成果。该地学断面西起印支半岛东岸的广东群岛附近，经越东陆架进入西沙海槽西南端，沿西沙群岛北缘，从西沙海槽东端穿出，沿南海北部陆坡向北东方向延伸到南海海盆东北部，之后折向北北西，在台湾浅滩南侧斜坡坡脚转向东，穿过台湾恒春半岛及其东侧的台湾纵谷南端和吕宋火山弧，进入菲律宾海盆。断面总体呈北东向，全长约1800km。该断面穿过了印支块体、台琼块体（图24-29，图24-44）。其剖面的地形地貌特征、地球物理特征及地质构造特征等如下：

1. 地形地貌特征

南海北部断面西端为宽数十千米的越东陆架区，水深等值线沿海岸展布，呈北北西向。往东断面进入地形十分复杂的南海北部陆坡区。南海北部陆坡总体特征表现为水深变化大，西宽东窄，珠江口以东陆坡宽142~290km，陆坡下界水深3400~3600m。

珠江口以西陆坡宽度超过300km，在西沙群岛一带宽度最大，中间受西沙海槽切割。断面所经过的西沙海槽西南端，在不到10km距离内，水深从2000m陡然降至500m。西沙海槽在此处由南北向转为北东向，槽内水深向北东方向加深，在大约111°30′E附近，西沙海槽变为近东西向延伸，向东与南

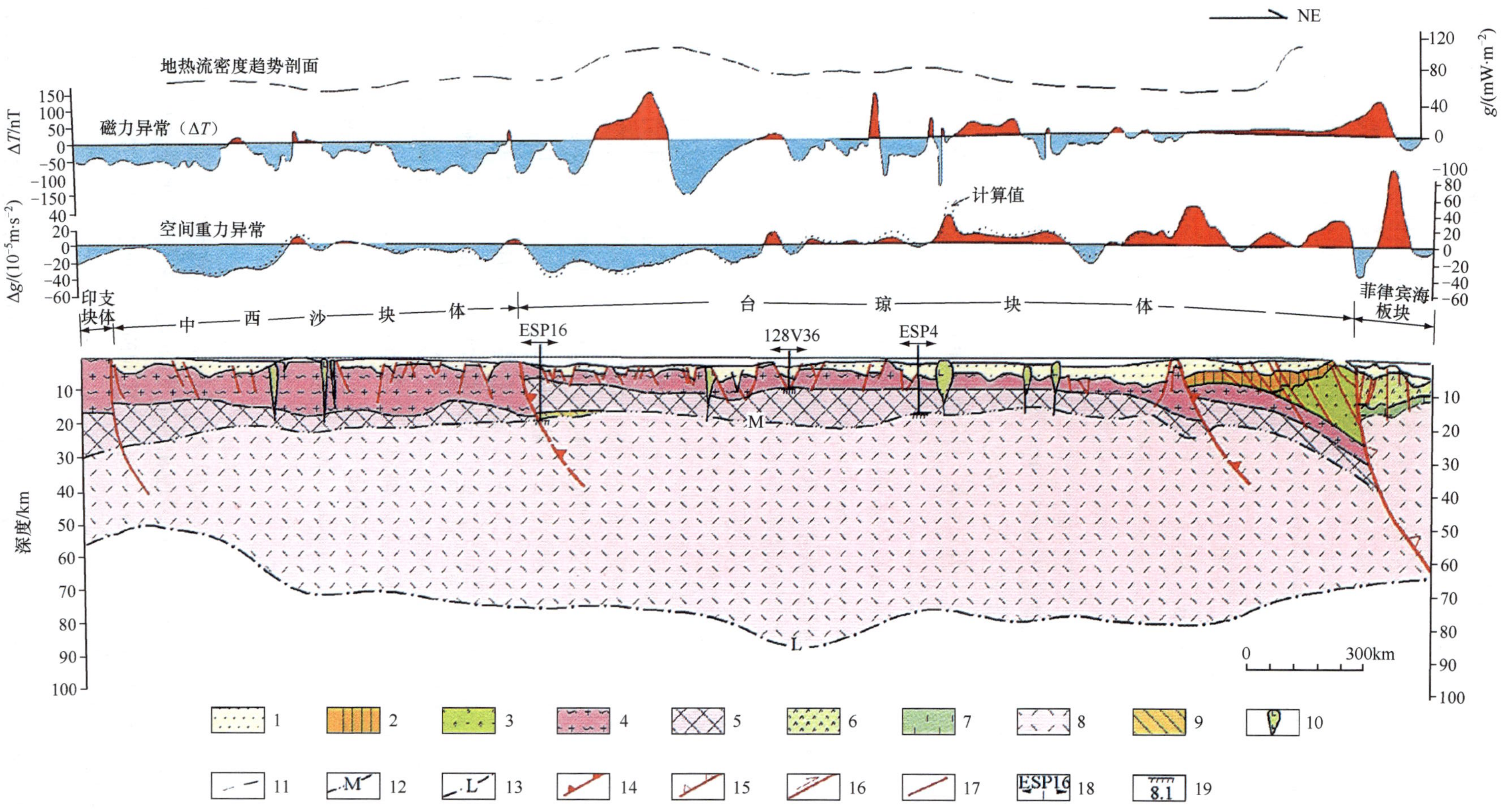

图 24-44 南海北部岩石圈地学断面图

Fig. 24-44 Geosciences cross-section of lithosphere of the northern South China Sea

1—新生代沉积岩；2—前新生代沉积；3—混杂岩；4—上地壳层；5—下地壳层；6—大洋层 2；7—大洋层 3；8—上地幔固结层；9—异常地壳；10—新生代超镁铁质岩；11—壳内分层界线；12—莫霍不连续面；13—岩石圈底界；14—中生代聚敛带；15—新生代聚敛带；16—逆冲断层；17—张性断层；18—双船扩展排列剖面/声纳浮标站；19—上地幔速度层（km/s）

海西北海盆连接。在西北海盆与中央海盆北侧的陆坡区，大约以水深1800m为界，上部地形较平缓，而水深在1800~3400m之间的地形十分陡峭，坡度达50×10^{-3}，构成陆坡带。该陆坡带从东沙群岛南坡折向东沙群岛东至东北部，向北东方向延伸至台湾浅滩南部，向东折向台湾岛西南，陆坡走向变为北西向。断面沿这一陆坡带延伸到东沙群岛南，穿过北坡海山后，进入一个深海平台区，该处水深在3000m左右，地势平坦，其间被北西向的台湾浅滩南峡谷所分割，平台东侧为北北西向的澎湖水道，峡谷和水道共同汇入马尼拉海沟。断面穿过峡谷南段和澎湖水道北段，向东切过台湾西南陆坡，进入台湾南部的恒春半岛。恒春半岛地势呈近南北向，是台湾中央山脉的南延。半岛东侧地势陡峻，在数千米内水深陡然降到1200m。恒春半岛向南延伸到海区，被称作恒春脊，在地形上仍表现为东陡西缓之势。恒春半岛与恒春脊之东水深变化较大，邻近恒春脊是台湾纵谷在海区的南延，这是一条呈近南北向的狭长谷地，水深1200m左右。位于谷地东侧的花东脊隆起东陡西缓，其东侧是北吕宋海槽在台湾东部的北延部分，该处海槽呈狭长带状，紧随吕宋火山岛弧的北段展布，岛弧东西两侧水深变化剧烈，跨入菲律宾海盆，地形渐渐变为平坦，水深增至3000m以上。

2. 地球物理特征

（1）磁力异常（ΔT）

印支和越东陆架陆坡区以大片负磁异常为背景，其上叠加了局部的低值正异常，正磁异常往往是海山或岛礁的反映。西沙海槽往东与西北海盆地区表现为高幅值正磁异常，最大值可达200nT，异常呈近东西向展布，这与此处地壳的强烈拉张引起地幔物质的上涌和洋壳的发育有关。紧邻其北侧是一条亦呈近东西向的负磁异常带，负异常幅值达300nT，具有南正北负正负异常相伴的特征。在南海北部下陆坡与深海洋盆之间的磁异常面貌呈现复杂多变的特点，磁异常类型除南正北负的伴生异常外，还有伴生关系不明显的正负异常，异常方向以北东东向为主。台湾西南海域磁异常极为平静，为大片低值负磁异常区，台湾东南部则以正磁异常为主，沿吕宋火山岛弧是一条南北向的高值磁异常带，幅值在50nT以上。

（2）自由空间重力异常（Δg_F）

在越东陆架区是一条近南北向的重力高带，在广东群岛附近重力值达到$15\times10^{-5}m/s^2$。西沙海槽为重力低值带，最低值为$-40\times10^{-5}m/s^2$。在海槽东面西北海盆北缘出现北正南负的重力异常特征，这一现象在中央海盆北缘亦有反映，被解释为被动大陆边缘重力效应，是海底地形急剧变化和洋陆边界处莫霍面起伏变化所引起的综合反映。位于西沙海槽南侧的西沙隆起区，异常复杂多变，正负异常相间，局部高值正异常对应了一系列岛礁与海山。南海北部下陆坡区以较高幅值异常为特征，东沙隆起及其东南斜坡上异常值最大，异常方向以北东向、北西向为主。台湾浅滩对应了一条重力异常高带，最大值达$50\times10^{-5}\sim60\times10^{-5}m/s^2$，往西南方向与东沙隆起重力高值带相连。位于浅滩南侧的台西南盆地重力特征以正异常为主体，异常值在$\pm10\times10^{-5}m/s^2$之间变化。台湾岛南部的恒春半岛是北部中央山脉的南延，重力上与中央山脉连成一体，表现为异常高幅值，台湾岛东西两侧为两条南北向的异常变化梯级带，尤以东侧变化梯度最大，与该处地形变化和地壳结构的变化有关。台湾纵谷南延端以负异常为特征，吕宋火山弧重力异常为高值正异常，向北与海岸山脉异常带连成一体，形成一条醒目的南北向异常带，是火山弧的反映。

（3）地热流

剖面自西向东具有高的热流背景值，低值区段出现在东沙隆起南侧和台西南盆地，热流值低于$55mW/m^2$，该处异常变化十分平缓。高值热流区段出现在西沙海槽东部西北海盆处，最大值可达$110mW/m^2$，地热流异常呈现大幅度起伏变化，反映了新生代期间所发生的强烈的断裂作用与岩浆活动。

3. 地质构造及地壳结构特征

（1）中新生代沉积

断面西端越东岸外陆架区一条近南北向的坳陷内发育了厚达7000m的新生代沉积。该坳陷属莺歌海盆地的南延部分，沉积发育特征类似于莺歌海盆地，受红河—南海西缘断裂的影响，沉积厚度大，在坳陷北部的盆地主体部位沉积厚度可逾15000m，其中，上中新统—第四系沉积最大厚度达5000~6000m，据钻井揭示，盆地内仅第四纪海相地层厚度约为2000m（汪品先，1991），而邻近的北部湾盆地，沉积特征迥然不同，古近系厚（最大厚度为3000~4000m），其上沉积较薄，整个新近系和第四系沉积总厚度才2000余米。

在南海北部陆坡区因水深较大，且远离物源，沉积层厚度一般较薄。在西沙隆起区、中央海盆北缘的下陆坡地区，沉积层厚度大多不超过3000m，仅在一些断裂附近局部发育更大厚度的沉积层。在一些较大型的坳陷或裂陷中，仍有较大厚度的沉积发育，如西沙海槽内新生代沉积厚度可超过6000m，东沙隆起东南的潮汕坳陷，沉积厚度最大达8000m。位于台湾浅滩南的台西南盆地因邻近台湾岛和台湾浅滩这样的物源区，盆地内中新生代沉积厚度累积最大超过10000m。地震剖面显示台西南盆地基底反射面之下存在一套强烈变形的反射层，在东沙隆起东侧，其厚度可达5000m；而往东厚度减少为1000~3000m，据区域地质和钻井资料反映，该层系属中生代陆相-海陆交互相沉积，可能还包括时代更老的古生代地层（渝何兴，1994）。

恒春半岛南端出露面积为300km^{2}的蛇绿混杂岩，称之为垦丁层，系中晚中新世深海浊积岩与更新世混杂岩组成的增生楔（Huang et al.，2001）。恒春半岛中新统厚4000m，由页岩、粉砂岩和砾岩组成，其上不整合覆盖上新统粉砂岩、页岩，厚度大于100m。台湾岛东侧纵谷南延部分，充填了厚约2000m的火山碎屑岩。

（2）断裂构造

断面西部越东岸外，发育近南北向断裂，系南海西缘断裂的北延，向北可能延伸到莺歌海与琼东南盆地之间的南北向隆起带附近，可能截断了呈北西向的红河断裂，二者均为大型的走滑断裂，近代仍有活动，控制了莺歌海盆地内新生代沉积的发育。西沙隆起区发育北东、北东东向断裂，新生代期间岩浆沿断裂喷溢形成众多的海山与海丘。西沙海槽东端以东西向断裂为主，并切割了北东向断裂。从西沙海槽到东沙隆起南的陆坡区水深变化大，断裂十分发育，断面大多向南倾向海盆，呈近东西向展布，显然其形成与中央海盆的扩张有密切关系。在大断裂下部一般都伴有海丘或海山，系岩浆沿断裂上涌形成的。东沙隆起西南和台湾浅滩西南发育一组北西向断裂，将北东向断裂左行错断，但未见其对沉积的控制，显然其形成时代较晚。

台西南盆地以北东和东西向断裂构造为主，盆地西部地区断裂表现为张性特征，盆地东部台湾滨岸平原地区，受中央山脉隆起的影响，断裂表现为逆冲推覆性质，断裂方向呈北西向。

台湾岛及其东西两侧发育北北东向断裂，与台湾地势走向一致，位于中央山脉与海岸山脉之间的纵谷断裂是菲律宾岛弧与欧亚大陆碰撞的缝合带，纵谷缝合线在纵谷南端大武附近转向南，在恒春半岛东海岸之外与120°E线大体重合（何春荪，1982；林明德等，1987）。

（3）地壳结构特征

南海北部断面地壳厚度总体表现为西部的印支半岛和东部的台湾岛地壳厚度较大，其间的南部和北部陆坡地壳厚度较薄。断面上印支半岛地壳厚约30km，上地壳厚度略大于下地壳，属典型的陆壳。中西沙地壳厚度发生了减薄，约为26km，上下地壳厚度差别较大，上地壳远大于下地壳，表明减薄作用主要发生在下地壳。海盆北部陆坡区地壳减薄剧烈，向海盆方向莫霍面向上抬升，至海盆边缘地壳厚度仅为12~14km，上地壳和下地壳均发生了减薄。特别值得一提的是，北部陆坡区的下地壳底部发育厚薄不等的高速度层，速度值在7.0~7.5km/s之间，高速层的形成可能和上地幔与下地壳之间的相互作用有关。

台湾西南与东南海域近年来作了两条OBS地震测线（Line 29和Line 33）（图24-29），其结果表明台湾恒春半岛东西两侧具有完全不同的地壳结构（图24-45，图24-46）。Line 33位于恒春半岛西侧，6个OBS揭示的地壳速度结构剖面显示：台湾西南海域地壳可分为两层，第一层为沉积层，速度为1.6~3.6km/s，最大厚度为8km，在靠近恒春半岛附近该层厚度变薄；第二层速度为6.0~7.0km/s，厚约11km，该层向东倾，P波速度从6.0km/s逐渐增加到7.0km/s，速度结构与上覆沉积层相比有显著差异，与测线西南侧的双船地震资料获得的地壳厚度相比较，其地壳厚度处在南海北部陆坡区地壳厚度变化范围内。板片发生俯冲的转折部位在 OBS43 附近，俯冲板片的倾角从剖面西段之下的 2°向东变陡，到恒春半岛之下达到 8°，而板片顶部的速度在恒春半岛下由于上覆物质影响而变得有些模糊不清。在 OBS43 以东至恒春半岛区段，在沉积层与俯冲板片之间夹有一个楔形体，结构十分复杂，这种楔形增生体是由于板片由西向东俯冲形成的增生柱。楔形体的平均速度由西向东从 4km/s 增至 5km/s。其顶部有一层速度为 4.5km/s，可能是含有较多洋壳碎片的混杂岩。

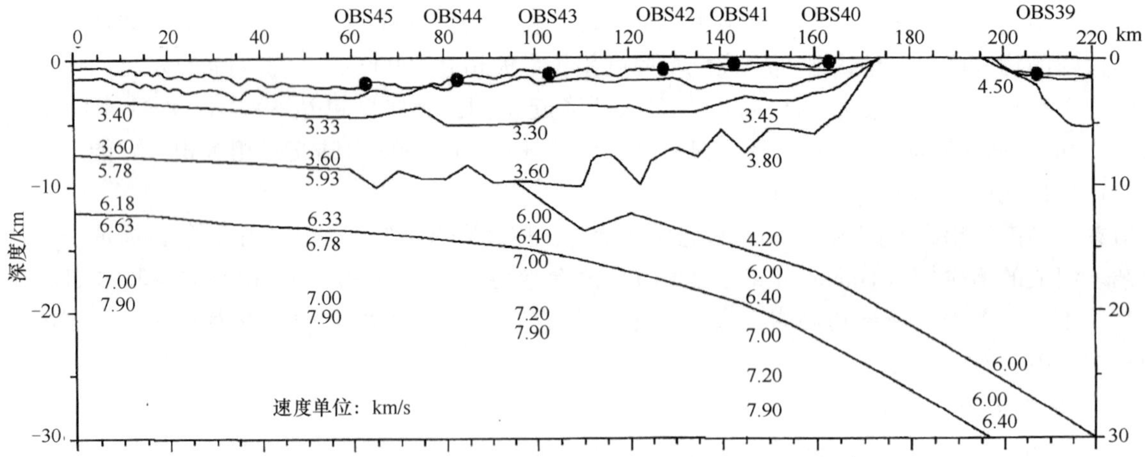

图 24-45　台湾恒春半岛西部海域 OBS 速度（km/s）结构剖面

Fig. 24-45　OBS（km/s）velocity structure section of the western sea area, Hengchun Peninsula, Taiwan

（据 Yosio Nakamura et al., 1998）

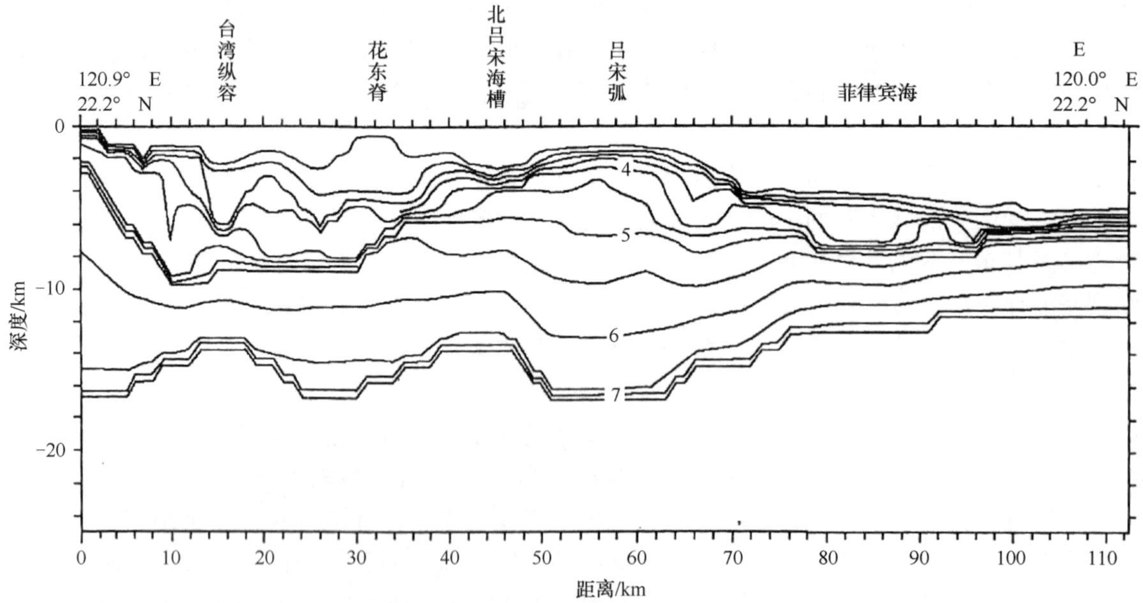

图 24-46　台湾恒春半岛东部海域 OBS 速度（km/s）结构剖面

Fig. 24-46　OBS（km/s）velocity structure section of the eastern sea area, Hengchun Peninsula, Taiwan

（据 Allen et al., 1998）

在剖面西端，上地幔位于 18km 深处，到剖面东端恒春半岛之下则深至 30km。总体特征是，在菲律宾海板块向北西方向移动过程中，中国大陆南缘伸展地壳俯冲于菲律宾海板块之下，并在恒春半岛东侧与菲律宾海板块西缘发生碰撞，俯冲作用早期是南海洋壳的俯冲，在俯冲板片上形成加积增生楔形体，当洋壳俯冲完毕，俯冲板片变成伸展减薄的陆壳时，该俯冲作用受到一定程度的阻碍。而地震震中和震源机制解表明，向东倾的俯冲板片仍然十分活跃，目前仍保持了俯冲板片的向东倾斜。

Line29 位于恒春半岛东侧，6 个 OBS 揭示的地壳结构显示：地壳由东向西增厚，最东端为菲律宾海洋壳，厚 6km；洋壳西侧的吕宋岛弧区莫霍面位于海平面之下 17km 处，弧上沉积非常薄，厚度不足 1km，下地壳速度约为 6.5km/s；吕宋岛弧西侧即为北吕宋海槽（应为吕宋海槽的一支，另一支已与花东脊碰撞），海槽内发育了厚约 2km 的碎屑沉积，该处地壳厚度比吕宋岛弧薄 5km，速度在 4~6.6km/s 之间，上地幔隆起约 3km，莫霍面速度为 7.7~7.8km/s。

吕宋海槽之西的花东脊地壳厚度有显著变化，地形抬升，海底变浅，莫霍面埋深增至 17km，地壳底部速度达 6.85km/s。据钻井资料，花东脊是由强风化沉积岩与晚上新世含放射虫的深水黏土矿物组成的混杂岩（Huang，1992），是南海洋壳向东俯冲于菲律宾海板块时形成的俯冲带杂岩。

台湾纵谷南端莫霍面呈隆起状态，地壳厚度比花东脊薄 3km，纵谷内发育了厚 2.5km 的沉积物。

四、南海海域重力异常特征及其地质解释

本节收集了《中国南海海域岩石圈三维结构及演化》（姚伯初等，2006）项目的成果，该项目详细地对南海地区的重力异常特征、重力异常分区、重力异常的解释及莫霍界面的形态特征等进行较详细的研究，为本次工作提供了较详细的资料。

（一）重力异常特征

根据《南海自由空间重力异常图》（图 24-47）和《南海布格重力异常图》（图 24-48），南海海域重力异常主要与地壳表面的起伏、各大地构造单元和不同地壳结构类型有关，这些因素决定了南海重力异常的宏观特征及局部异常的面貌。自由空间重力异常上延（5km、10km、20km 、30km 和 50km）（图 24-49~图 24-53）显示了不同上延高度异常的特征，大致反映不同深度场源体的重力效应。除自由空间重力异常与海底地形呈正相关、布格重力异常与地形呈负相关的一般特征外，还有以下明显特征。

1）自由空间重力异常的区域背景值南高北低、东高西低。南部礼乐滩-永暑礁海域的区域重力异常值在 10×10^{-5} ~ $30\times10^{-5}m/s^2$，到西侧的巽它陆架区降到（0~20）$\times10^{-5}m/s^2$。中部深海盆一般为 $\pm10\times10^{-5}m/s^2$。北部的台湾海峡-东沙群岛海域的背景值为（0~10）$\times10^{-5}m/s^2$，向西至北部湾降低到 $\pm10\times10^{-5}m/s^2$。在不同上延高度的异常图上尤为明显。

2）自由空间重力异常幅值变化大，具有多种类型。南海自由空间重力异常最高值在南沙群岛的安渡滩，为 $130\times10^{-5}m/s^2$，最低值位于马尼拉海沟的南端，为 $-140\times10\times10^{-5}m/s^2$，变化幅值达 $270\times10^{-5}m/s^2$。

3）几乎可以见到海洋所特有的重力异常类型，如海山、海沟、高地、洋-陆过渡带、扩张脊等异常类型。

4）重力异常的区域走向，无论自由空间重力异常或布格重力异常，中部均以北东—北东东向为主，东西两侧以南北向为主，在西部还有的呈北西走向。

5）除陆架区外，布格重力异常值通常比自由空间重力异常值大，变化幅值也大。最低值在北部湾，为 $-40\times10^{-5}m/s^2$，最高值在海盆区，达 $320\times10^{-5}m/s^2$，幅值达 $360\times10^{-5}m/s^2$。陆架陆坡分界处布格重力异常在 $50\times10^{-5}m/s^2$ 等值线附近，而洋-陆壳分界线则在 $250\times10^{-5}m/s^2$ 附近。

6）岩石圈底界也应为一密度界面，资料显示其密度差可达 $-0.1\times10^3kg/m^3$。

（二）重力异常分区

根据自由空间、布格重力异常特征，异常走向和异常变化梯度，以及与之有相关关系的地形、地貌及地质构造等，将南海海域的重力异常分为 7 个区（带）（图 24-54）：北部陆架、陆坡异常区（Ⅰ）；中沙、西沙岛礁异常区（Ⅱ）；西部陆架、陆坡异常区（Ⅲ）；西南陆架异常区（Ⅳ）；南海海盆异常区（Ⅴ）；南沙岛礁异常区（Ⅵ）和岛弧、海沟（槽）异常带（Ⅶ）。现分别叙述如下：

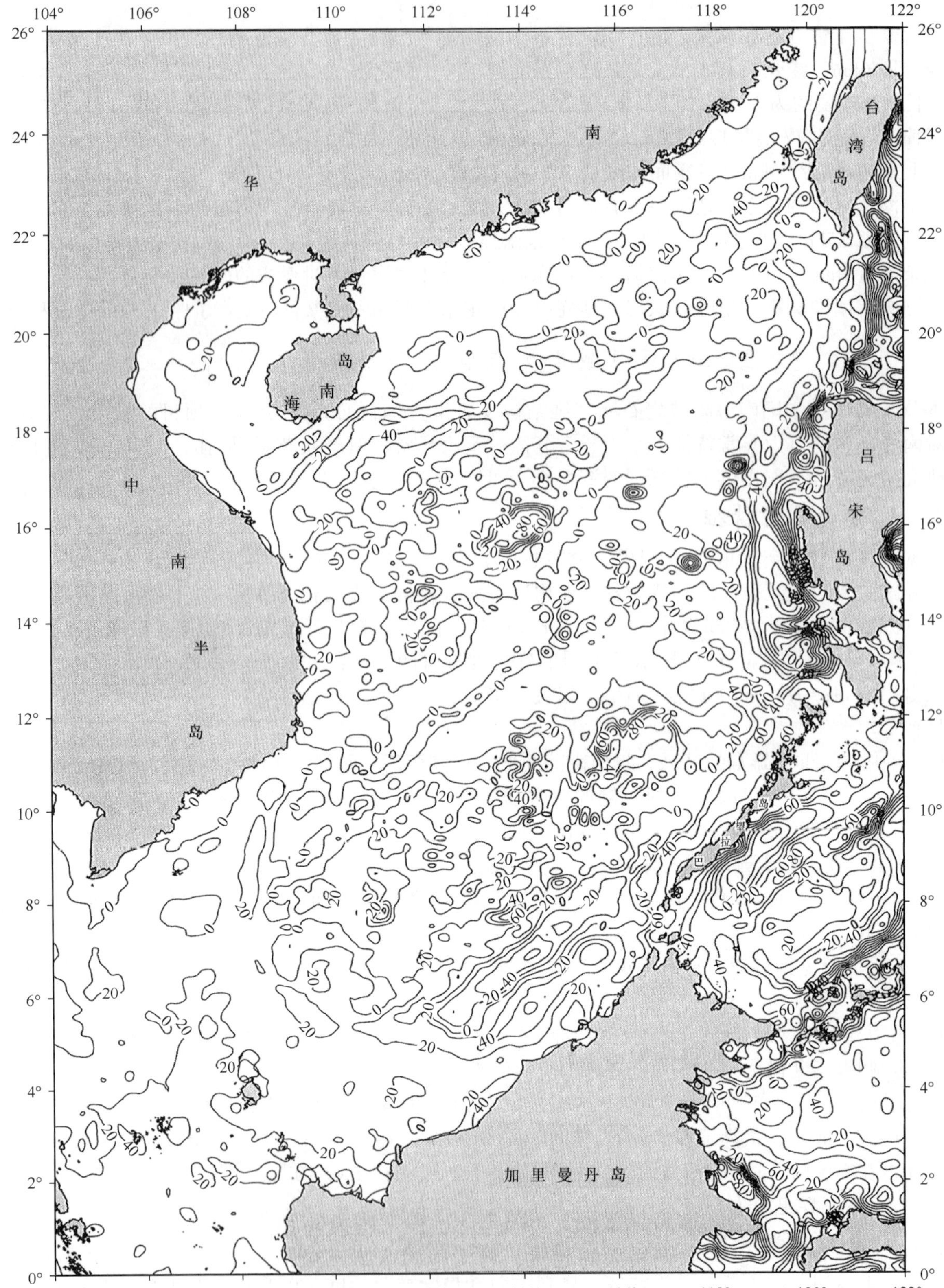

图 24-47　南海自由空间重力异常图

Fig. 24-47　Map of free air gravity anomalies in the South China Sea

（单位：$10^{-5}m/s^2$；底图引自 HY126-03 项目）

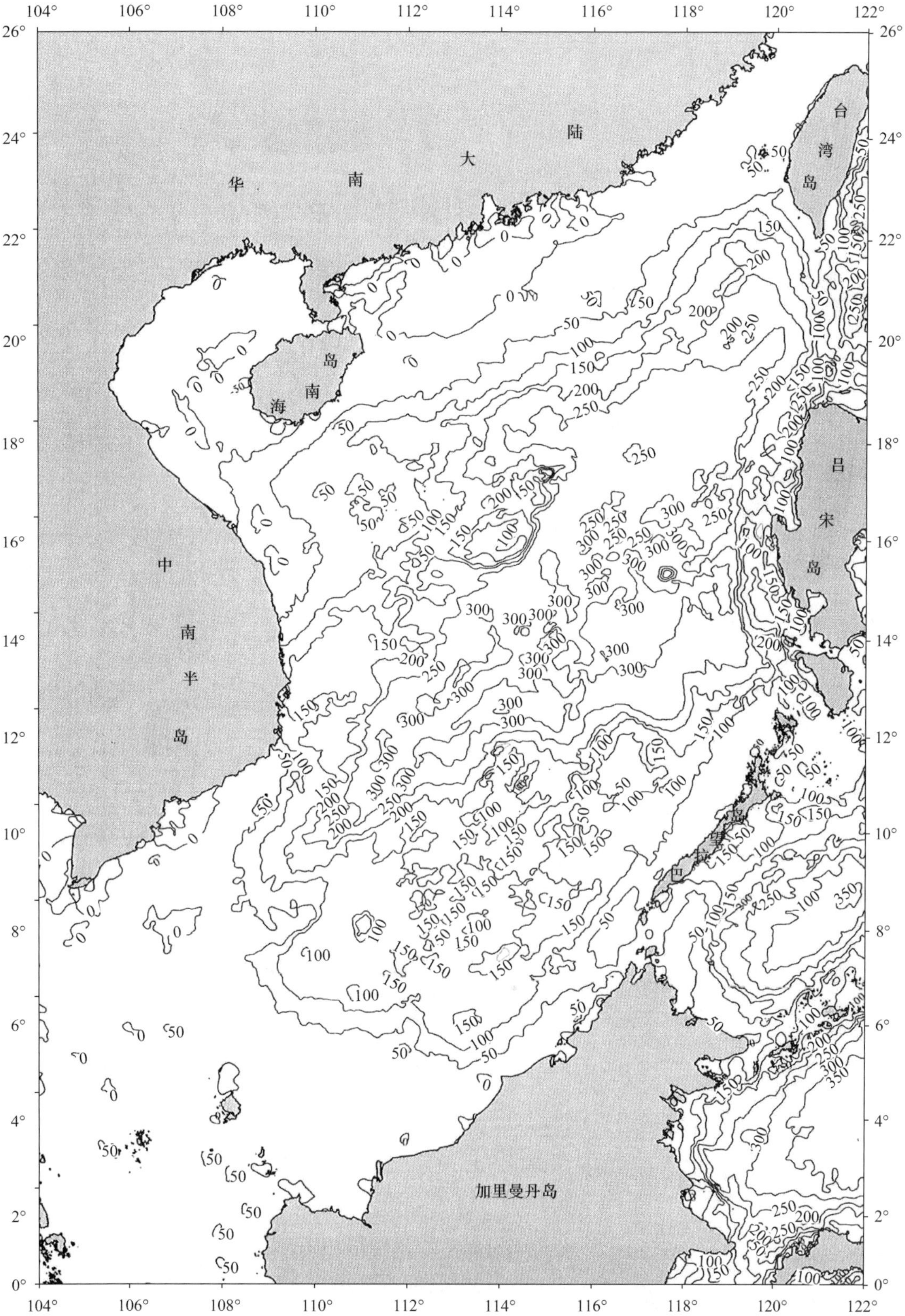

图 24-48 南海布格重力异常图

Fig. 24-48 Map of Bouguer gravity anomalies in the South China Sea

（单位：$10^{-5}m/s^2$；底图引自 HY126-03 项目报告资料，有修改）

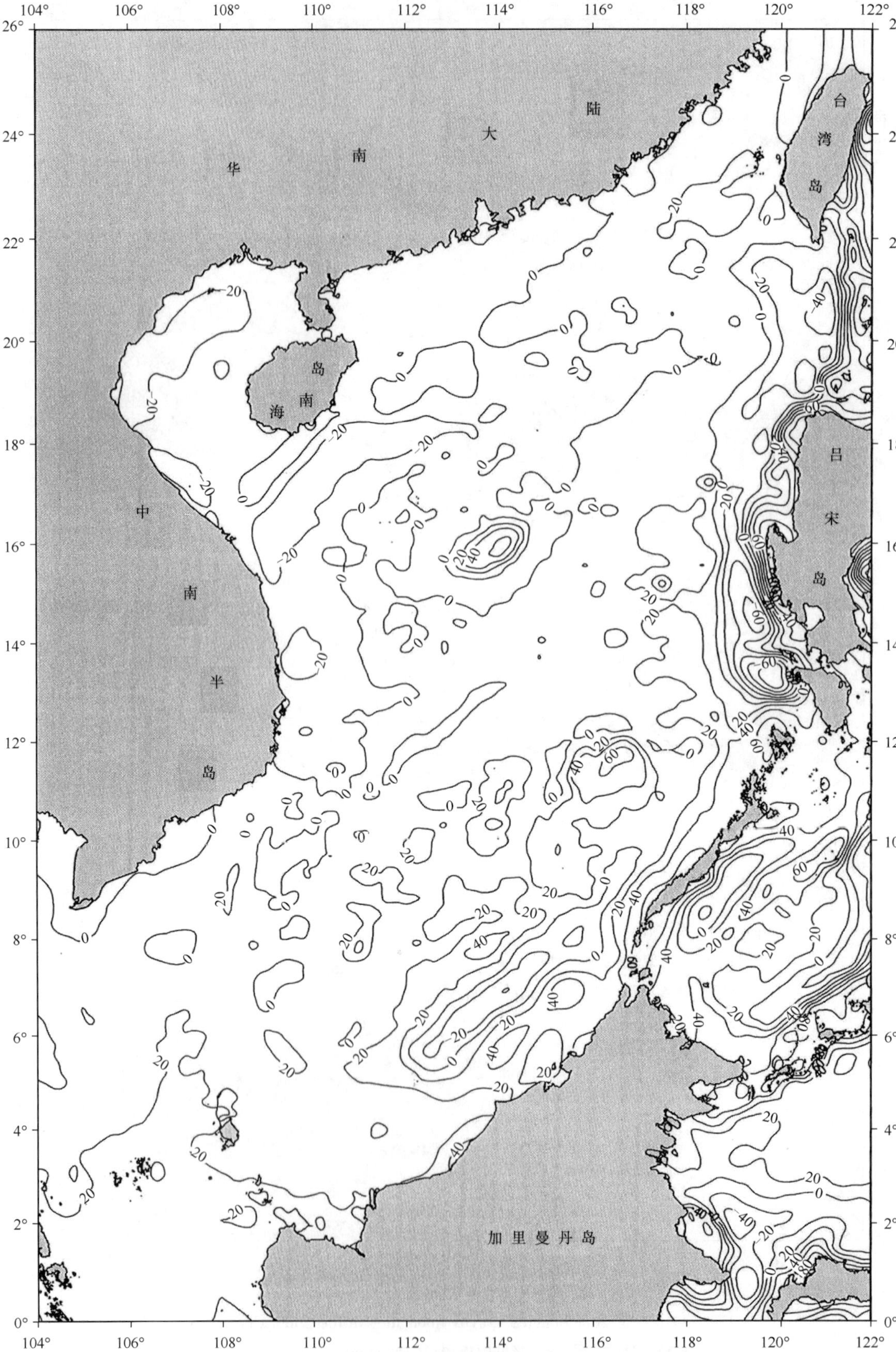

图 24-49　南海自由空间重力异常上延（5km）图

Fig. 24-49　Map of upward continuation (5km) of free air gravity anomalies in the South China Sea

（单位：$10^{-5}m/s^2$；底图引自 HY126-03 项目报告）

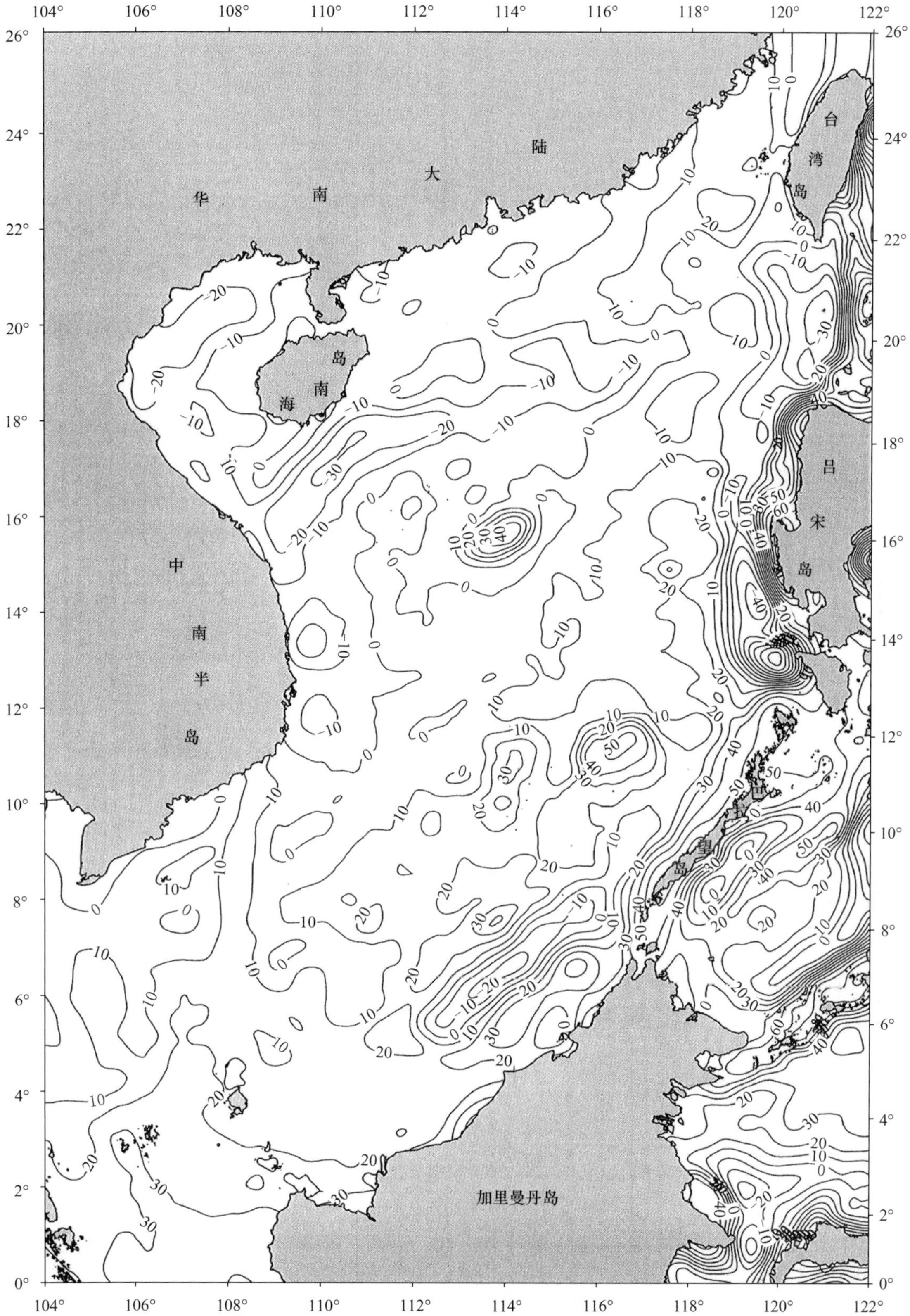

图 24-50　南海自由空间重力异常上延（10km）图

Fig. 24-50　Map of upward continuation (10km) of free air gravity anomalies in the South China Sea

（单位：$10^{-5}m/s^2$；底图引自 HY126-03 项目报告）

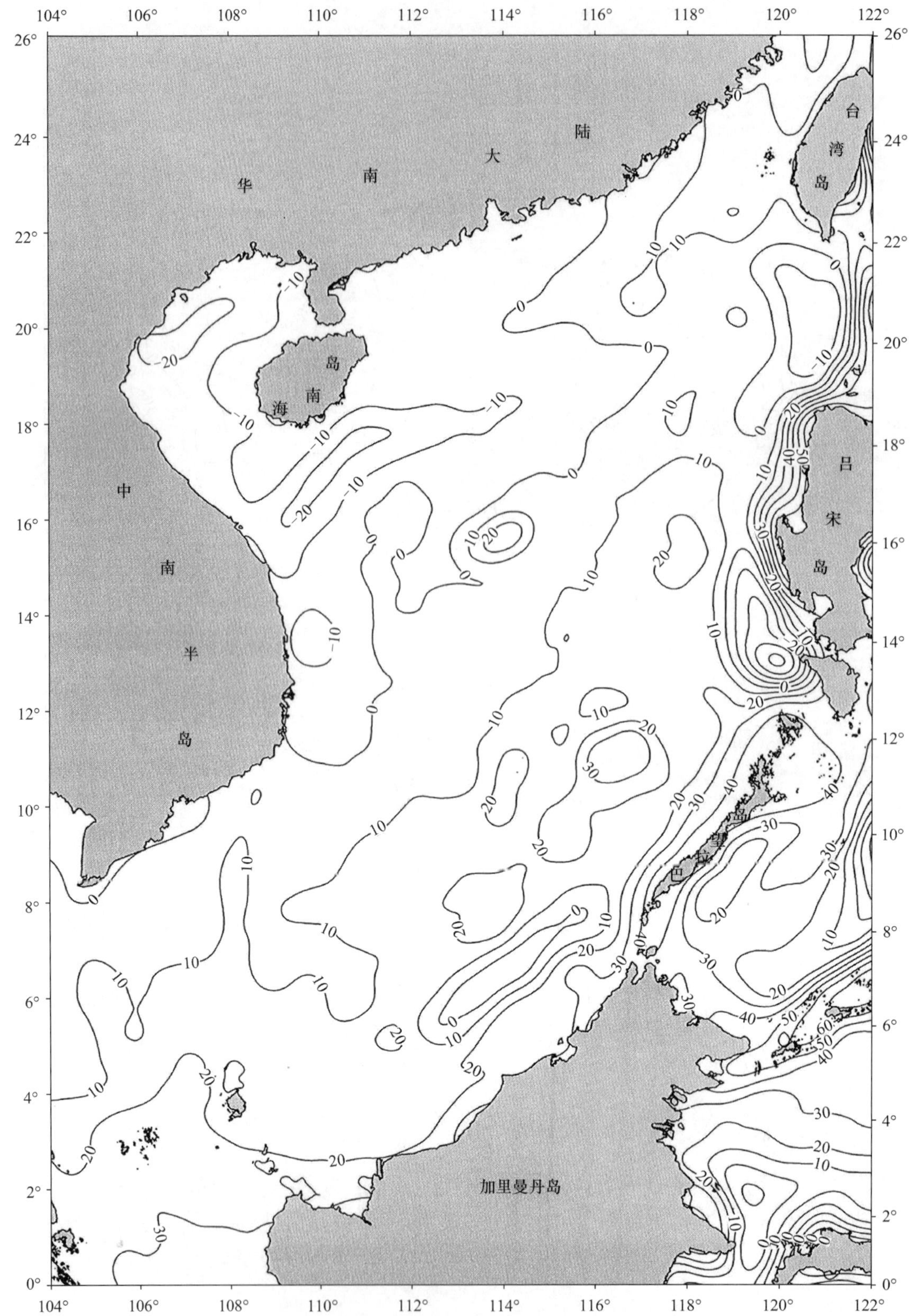

图 24-51　南海自由空间重力异常上延（20km）图

Fig. 24-51　Map of upward continuation（20km）of free air gravity anomalies in the South China Sea

（单位：$10^{-5}m/s^2$；底图引自 HY126-03 项目报告）

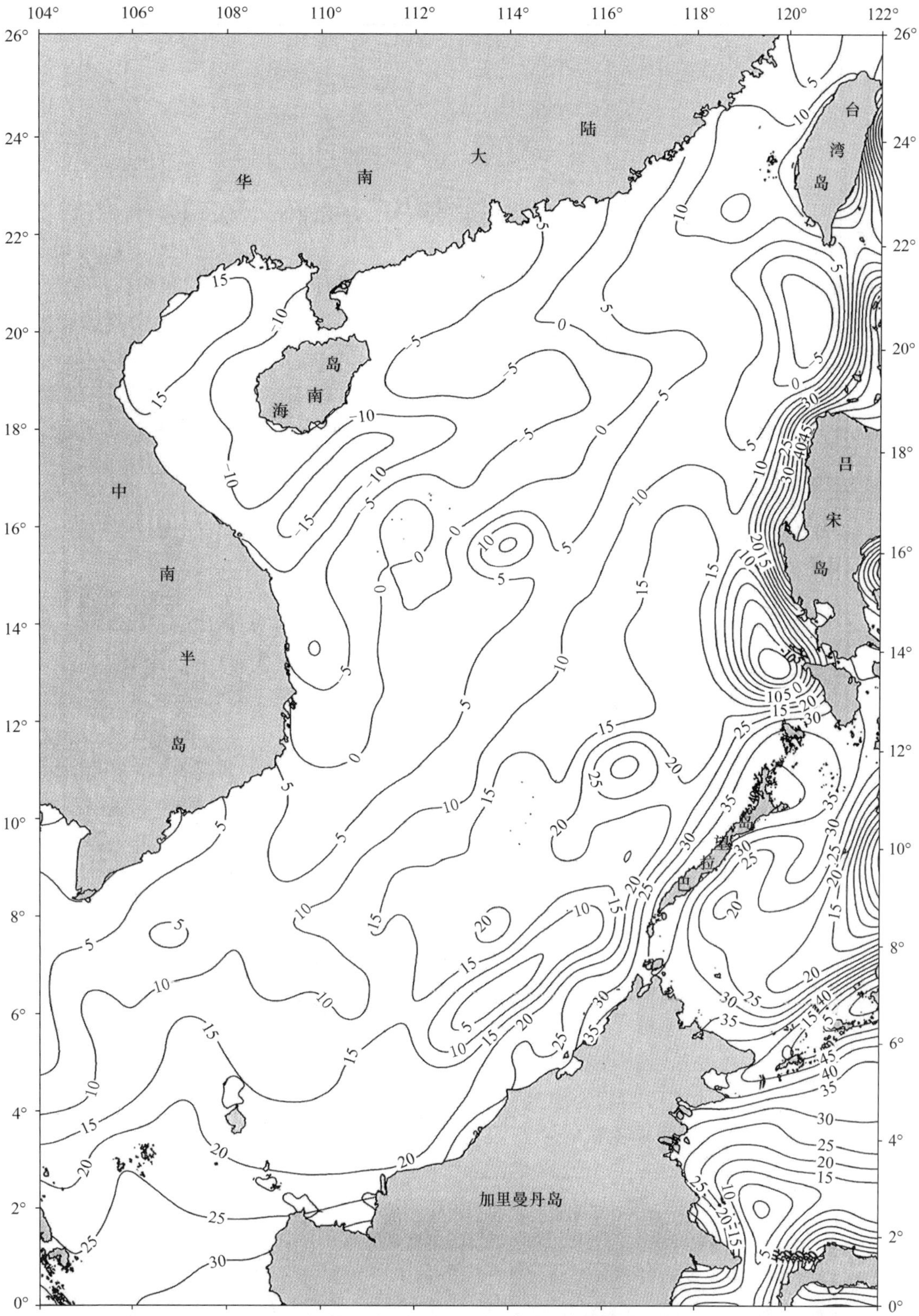

图 24-52 南海自由空间重力异常上延（30km）图

Fig. 24-52 Map of upward continuation (30km) of free air gravity anomalies in the South China Sea

（单位：$10^{-5}m/s^2$；底图引自 HY126-03 项目报告）

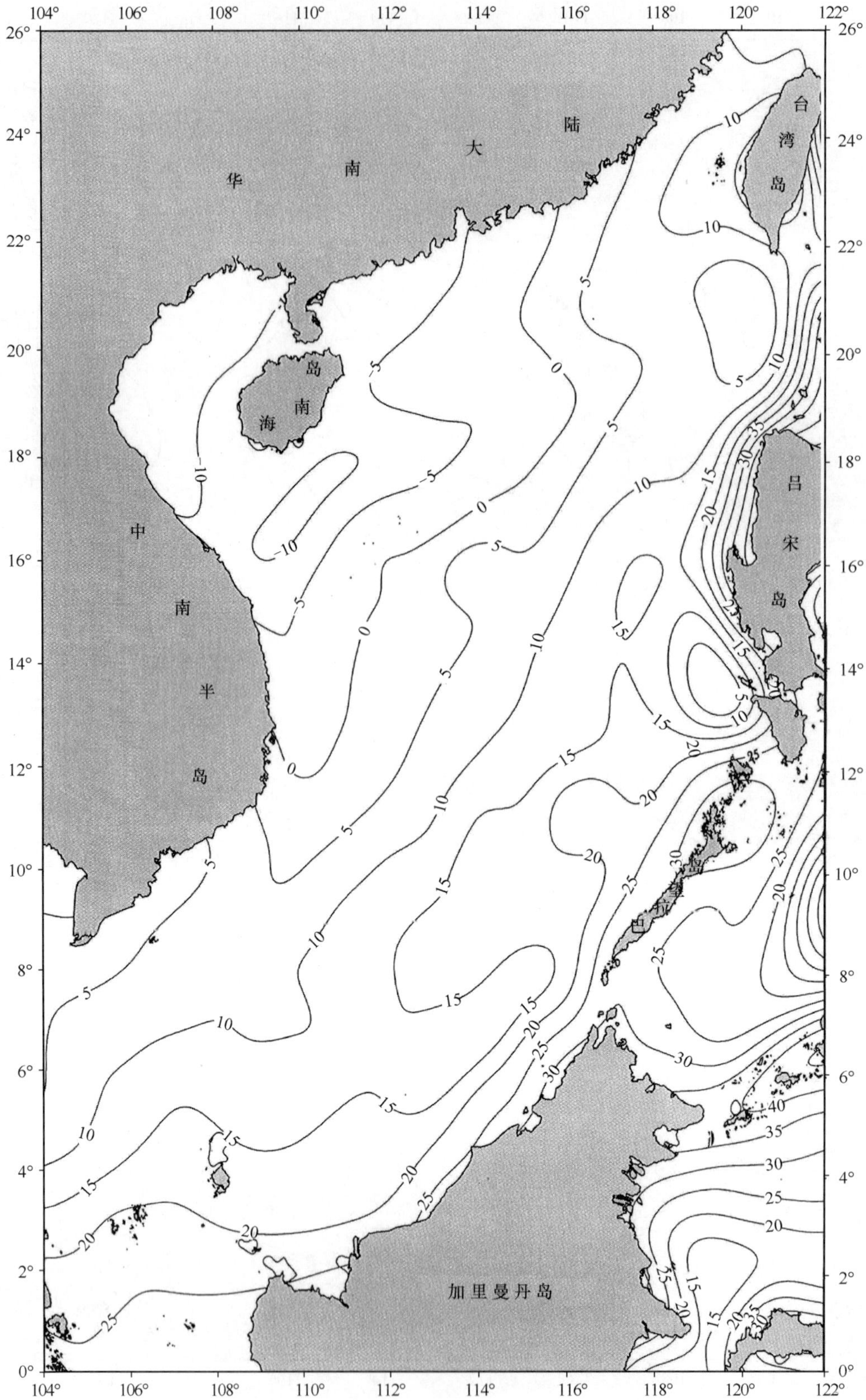

图 24-53　南海自由空间重力异常上延（50km）图

Fig. 24-53　Map of upward continuation（50km）of free air gravity anomalies in the South China Sea

（单位：$10^{-5}m/s^2$；底图引自 HY126-03 项目报告）

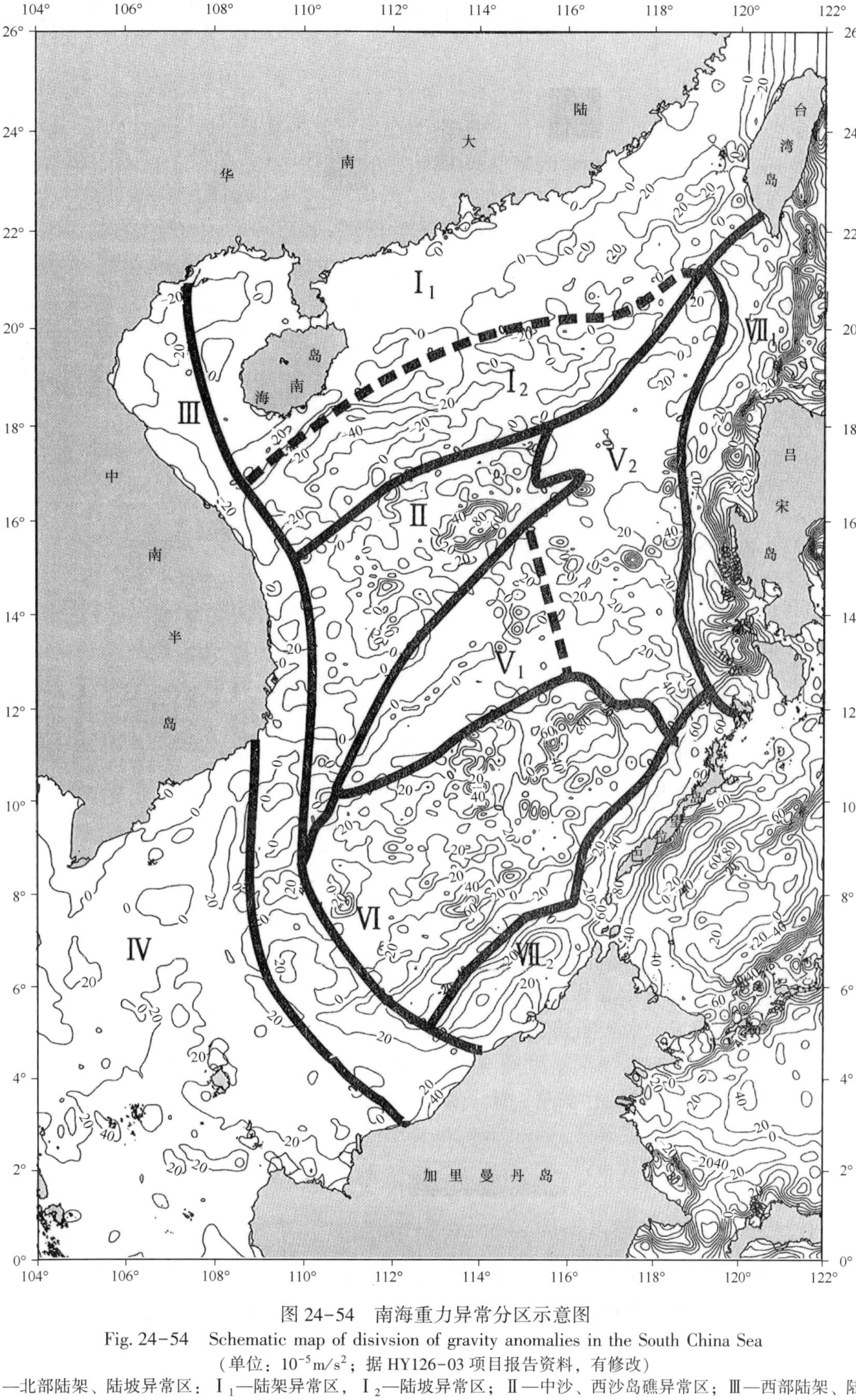

图 24-54 南海重力异常分区示意图

Fig. 24-54 Schematic map of disivsion of gravity anomalies in the South China Sea

（单位：$10^{-5}m/s^2$；据 HY126-03 项目报告资料，有修改）

Ⅰ—北部陆架、陆坡异常区：$Ⅰ_1$—陆架异常区，$Ⅰ_2$—陆坡异常区；Ⅱ—中沙、西沙岛礁异常区；Ⅲ—西部陆架、陆坡异常区；Ⅳ—西南陆架异常区；Ⅴ—南海海盆异常区：$Ⅴ_1$—西南海盆异常区，$Ⅴ_2$—中央海盆异常区；Ⅵ—南沙岛礁异常区；Ⅶ—岛弧、海沟（槽）异常带：$Ⅶ_1$—台湾-吕宋异常带，$Ⅶ_2$—巴拉望-沙巴异常带

1. 北部陆架、陆坡异常区（Ⅰ）

该异常区位于南海北部闽、粤、琼陆架、陆坡和台湾海峡、北部湾海域。区内重力异常可分为两部分。内侧的陆架异常区和外侧的陆架陆坡正负伴生“边缘异常带”。

陆架异常区又可进一步分为内陆架重力高带、外陆架重力高带和它们之间的重力低带。内陆架重力高带，北东自福建南日岛向西至广东南澳岛-担杆列岛至海南岛周围，由一系列的重力高组成，异常值由东段的 10×10^{-5}～$30\times10^{-5}m/s^2$，向西降低，中段为（0～20）$\times10^{-5}m/s^2$，至海南岛周围降至 $\pm10\times10^{-5}m/s^2$。重力高带的两侧为梯级带，南侧的梯级带更为明显。外陆架的重力高带东起台湾的北港、澎湖，向西经台湾浅滩、东沙群岛，再向西与陆架边缘的正异常带相重叠，在莺歌海盆地与南海西部陆架边缘重力高和海南西部重力高带组成一个 Y 形“三连点”。陆架中部重力低带，东起台湾海峡中部西至珠江口外，至琼东南海域，异常值东段为（0～10）$\times10^{-5}m/s^2$，珠江口外为（0～-10）$\times10^{-5}m/s^2$，西段琼东南海域为（0～-20）$\times10^{-5}m/s^2$。另外在北部湾和莺歌海盆地也为负异常构成的重力低，其值为 $-10\times10^{-5}m/s^2$ 至 $-40\times10^{-5}m/s^2$，区内的布格重力异常值与自由空间重力异常值接近，异常形态类似，并向陆架外缘升高。

北部陆架陆坡正负伴生异常带，东起 118°E，向西延伸至西沙海槽。东段受局部海山和沉积盆地干扰，正负伴生特征不明显，异常值一般为 20×10^{-5}～$-20\times10^{-5}m/s^2$。西段与西沙海槽负异常相叠加，异常值为 20×10^{-5}～$-40\times10^{-5}m/s^2$，最低值达 $-50\times10^{-5}m/s^2$。布格重力异常则随着陆坡水深加大，异常值呈阶梯状升高，由 $50\times10^{-5}m/s^2$ 升高至 $250\times10^{-5}m/s^2$。

2. 中沙、西沙岛礁异常区（Ⅱ）

该区由一系列岛礁、海山重力高和海槽、海谷重力低组成。异常总体呈北东向，局部北东东向和近东西向。西沙群岛重力高最为突出，最高可达 $100\times10^{-5}m/s^2$，周围被 -10×10^{-5}～$-30\times10^{-5}m/s^2$ 负值异常围绕。中沙群岛北部和西南部的海山形成 10×10^{-5}～$30\times10^{-5}m/s^2$ 的正异常链。西沙岛礁区则由两列北东向排列的正异常组成，异常值为 10×10^{-5}～$50\times10^{-5}m/s^2$，最大可达 $70\times10^{-5}m/s^2$。

西沙海槽负异常带，由东段近东西走向，向西转为北东至近南北向。异常值由于海底地形和深部因素，边缘效应影响加大，最低达 $-60\times10^{-5}m/s^2$。在海槽的南北两侧为重力梯阶带。中沙北海槽走向东北，异常值为 $-10\times10^{-5}m/s^2$～$-30\times10^{-5}m/s^2$。在中建岛以南海域存在一规模较大的重力低，异常值为 $-10\times10^{-5}m/s^2$～$-40\times10^{-5}m/s^2$。另外在一些海山间的槽谷中，分布规模不人的低异常，异常值为 -10×10^{-5}～$-30\times10^{-5}m/s^2$。

区内的布格重力异常由西北向东南组成逐步升高的台阶，由西北侧的 $50\times10^{-5}m/s^2$ 升高到 $250\times10^{-5}m/s^2$。

3. 西部陆架、陆坡异常区（Ⅲ）

位于中南半岛东部海域，由于陆架狭窄、陆坡陡峻，重力异常呈近南北走向的正负伴生异常带，异常值为 $30\times10^{-5}m/s^2$～$-20\times10^{-5}m/s^2$。北端与西沙海槽边缘异常带相连，南端与巽他陆架边缘异常带相连。在 15°N～13°30′N 的内陆架存在一条重力低异常带，异常值为（0～-20）$\times10^{-5}m/s^2$。6°N～10°N 间正负伴生的重力边缘效应很典型，向北低值带受地形干扰而不完整。布格重力异常向洋侧升高，构成 50×10^{-5}～$150\times10^{-5}m/s^2$ 的梯级带。

4. 西南陆架异常区（Ⅳ）

该区位于南海西南部巽它陆架的东北部。区内以 $10\times10^{-5}m/s^2$～$30\times10^{-5}m/s^2$ 的高背景值和北北西向异常为特征。由陆架边缘的正负伴生异常带和三条重力高异常带及其间的重力低异常带构成了向西南略突出的弧形。正异常带北段南北走向，异常值为 $10\times10^{-5}m/s^2$～$30\times10^{-5}m/s^2$，中段北西走向，异常值为 $10\times10^{-5}m/s^2$～$30\times10^{-5}m/s^2$。东段近东西走向，最东端为北东东走向，异常值有所增大，为 $10\times10^{-5}m/s^2$～$40\times10^{-5}m/s^2$。伴生负异常走向与正异常同步转向，异常值为（0～-10）$\times10^{-5}m/s^2$。东段中部受深部异常干扰而负值升高。

区内的东部重力高异常带，北起昆仑岛，向南为纳土纳岛、苏比岛、塞腊散岛，南端至加里曼丹岛西北的达都角，重力高异常带北低南高。总体走向北北西，异常值在 $10\times10^{-5}m/s^2\sim40\times10^{-5}m/s^2$，最高值在塞腊散岛，为 $50\times10^{-5}m/s^2$。中部重力高异常带北起中南半岛的金瓯角，向南至亚南巴斯群岛、淡美兰群岛，重力高异常带呈北北西走向。中北段异常值为（0~30）$\times10^{-5}m/s^2$，南段抬高到 $10\times10^{-5}m/s^2\sim40\times10^{-5}m/s^2$。西部重力高异常带位于马来半岛东侧近岸到新加坡东部海域，异常值相对较高，一般为 $20\times10^{-5}m/s^2\sim40\times10^{-5}m/s^2$。异常走向北北西至南北向。

三条重力高异常带之间的重力低带，异常总体走向同重力高异常带相一致，局部重力低为东西或南北走向，异常值为（0~-10）$\times10^{-5}m/s^2$。

另外在中南半岛南部海域，分布有昆仑岛-加堆克岛重力高异常带，异常走向与前不同，为北东向展布，由呈串珠状的多个重力高组成。异常值为 $10\times10^{-5}m/s^2$。该带的内侧近岸为一重力低异常带，异常走向也为北东，异常值为（0~-20）$\times10^{-5}m/s^2$。在暹罗湾的东口分布一重力低异常带，走向北西，异常值为（0~10）$\times10^{-5}m/s^2$。布格重力异常值为 $-10\times10^{-5}m/s^2\sim50\times10^{-5}m/s^2$，形态同自由空间异常。

5. 南海海盆异常区（Ⅴ）

该区范围大致以 3200m 水深线圈出。海盆以 115°E 为界，分为东部的中央海盆和西部的西南海盆。重力异常也可分为各不相同的两部分。

中央海盆自由空间重力异常值中部高，向南北两侧降低，东部高，向西降低。中部异常值为 $10\times10^{-5}m/s^2\sim40\times10^{-5}m/s^2$，向南北两侧降低到（0~20）$\times10^{-5}m/s^2$。由东部的 $40\times10^{-5}m/s^2$，向西降低至（0~20）$\times10^{-5}m/s^2$。异常走向海盆北部为北东至北东东向，中、南部为近东西向。海盆内分布有一系列由海山引起的重力高带，15°N 左右的黄岩岛-珍贝海山组成为串珠状重力高带。异常值一般为 $20\times10^{-5}m/s^2\sim80\times10^{-5}m/s^2$，黄岩岛最高值可达 $100\times10^{-5}m/s^2$。在 119°15′E、17°15′N 的一个无名海山最高可达 $130\times10^{-5}m/s^2$。南北走向的中南海山为 $20\times10^{-5}m/s^2\sim100\times10^{-5}m/s^2$。另外，在中央海盆靠近马尼拉海沟，存在一条重力高异常带，其异常值为 $30\times10^{-5}m/s^2\sim40\times10^{-5}m/s^2$，向西降低，比海盆区正常的空间异常值高 $10\times10^{-5}m/s^2\sim20\times10^{-5}m/s^2$。这一重力高带主要是由于海底在向海沟俯冲时，由于洋壳的弯曲上隆而产生的，为俯冲洋壳所特有，被称为“俯冲带洋侧重力高异常带”。

西南海盆自由空间异常总体走向为北东，由海盆中部重力低带、两侧重力高带和南北洋—陆分界处的两条重力低带组成。海盆中部重力低带，异常值为（0~-20）$\times10^{-5}m/s^2$，向西抬升至（0~-10）$\times10^{-5}m/s^2$，东端被中南海山的重力高截断。两侧的两条重力高带，异常值为 $10\times10^{-5}m/s^2\sim20\times10^{-5}m/s^2$，局部重力高可达 $30\times10^{-5}m/s^2$。南北洋—陆分界处的两条重力低，受海盆边缘地形干扰而弯曲。异常值为 $10\times10^{-5}m/s^2\sim-20\times10^{-5}m/s^2$，向西南降低。

海盆区为布格重力高值异常区，异常值为 $250\times10^{-5}m/s^2\sim320\times10^{-5}m/s^2$，向四周降低：在海盆已停止扩张的海底扩张脊仍显示出重力低的特征。

6. 南沙岛礁异常区（Ⅵ）

本区自由空间重力异常由三个重力高带和相间的两个重力低带及南沙海槽重力低带组成，异常走向以北东向为主。高低异常带间有不明显和不连续的重力梯阶带。

北部重力高带，由中业群礁-郑和群礁（太平岛-南薇滩-广雅滩-万安滩）等局部重力高组成，异常走向以北东为主。局部重力高值由北东的 $100\times10^{-5}m/s^2$ 向西南降低到 $30\times10^{-5}m/s^2$。中部重力高带由礼乐滩-费信岛-景宏岛-南薇滩等重力高组成，异常走向北东，局部异常值由东北部礼乐滩的 $100\times10^{-5}m/s^2$ 向西南递减到南薇滩的 $20\times10^{-5}m/s^2$。南部重力高带由海马滩-半月礁-安渡滩-南通礁一南、北康暗沙等局部重力高组成，异常以北东东走向为主，重力值以中部安渡滩为最高，达 $120\times10^{-5}m/s^2$，向东北、西南均降低为 $50\times10^{-5}m/s^2\sim60\times10^{-5}m/s^2$。分布于重力高带之间的重力低带，走向北东，异常值为（0~-20）$\times10^{-5}m/s^2$。在礼乐滩和中业群礁北侧与海盆相邻的洋—陆分界处，分布有正负伴生的边缘效应异常。南沙海槽负异常带，异常值为 $-10\times10^{-5}m/s^2\sim-50\times10^{-5}m/s^2$，最小值可达 $-60\times10^{-5}m/s^2$。异常带南北两侧发育有较明显的梯阶带，局部的海底由形成孤立的重力高，如碎浪暗沙等。

区内布格重力异常值为 $50\times10^{-5}m/s^2\sim200\times10^{-5}m/s^2$，向北部海盆区升高至 $250\times10^{-5}m/s^2$。岛礁分布的部位一般小于 $100\times10^{-5}m/s^2$。岛礁之间纵横交错的槽谷，布格重力异常抬高到 $150\times10^{-5}m/s^2\sim200\times10^{-5}m/s^2$。南沙海槽异常带表现为向南侧降低的梯级带。

7. 岛弧、海沟（槽）异常带（Ⅶ）

该带因重力异常走向和结构的差异可分为两部分，即南海东缘的台湾-吕宋异常带和南海南部的巴拉望-沙巴异常带。南海东缘台湾-吕宋异常带，由南北向海沟（槽）重力低异常带和岛弧重力高异常带组成。海沟（槽）重力低异常带又含马尼拉海沟重力低带和西、北吕宋海槽重力低异常带。马尼拉海沟重力低异常带北起21°40′N，近台湾岛西南，南至12°40′N的民都洛岛西侧。该带又以17°20′N为界，分为南北两段。北段是海沟的北延部分，重力值相对较高，为 $-10\times10^{-5}m/s^2\sim-40\times10^{-5}m/s^2$，异常为南北走向，受地形影响有扭曲。南段是海沟负异常带主体，较狭窄，异常值由北向南降低，一般为 $-10\times10^{-5}m/s^2\sim-80\times10^{-5}m/s^2$。南端止于民都洛岛西侧，是带内异常值最低的部位，极值为 $-140\times10^{-5}m/s^2$。西、北吕宋海槽重力低带北起台湾、台东岸外的绿岛，南至马尼拉湾。可分为三段，北段绿岛至吕宋岛北，是海槽的北延部分，异常为南北走向，异常值北高南低，由一系列的串珠状重力低组成，异常值为 $-10\times10^{-5}m/s^2\sim-80\times10^{-5}m/s^2$，最小值在巴布延岛西侧，达 $-120\times10^{-5}m/s^2$。中段为北吕宋海槽重力低带的主体，异常值为 $-10\times10^{-5}m/s^2\sim-50\times10^{-5}m/s^2$，中部最低可达 $-70\times10^{-5}m/s^2$。南段为西吕宋海槽的主体部分，异常为南北走向，其值为 $-10\times10^{-5}m/s^2\sim-100\times10^{-5}m/s^2$。

岛弧重力高带也可分为南、北两段。北段由台湾东部的绿岛、兰屿和巴士海峡以南的巴坦群岛、巴布延群岛的重力高组成，呈串珠状南北向展布，异常值为 $10\times10^{-5}m/s^2\sim130\times10^{-5}m/s^2$，巴坦岛最高达 $170\times10^{-5}m/s^2$。南段在吕宋岛西侧的近岸岛架，异常值为 $50\times10^{-5}m/s^2\sim120\times10^{-5}m/s^2$，最高值在吕宋岛西北端岛架，可达 $220\times10^{-5}m/s^2$。

带内的布格重力异常呈现由沟槽向岛弧降低的梯级带，异常值从 $250\times10^{-5}m/s^2$ 降低至 $100\times10^{-5}m/s^2\sim70\times10^{-5}m/s^2$。而北段台湾岛-吕宋岛之间，异常由中部岛弧向两侧抬高。

南海南部岛弧带，以巴拉巴克海峡为界分东西两段，东北段为卡拉棉-巴拉望岛架重力高带，异常为北东走向，由一系列正值重力高组成，其值为 $20\times10^{-5}m/s^2\sim80\times10^{-5}m/s^2$，最高可达 $100\times10^{-5}m/s^2$。南段沙巴岛架异常带，由外岛架重力高带和内岛架重力低带组成。外重力高带为北东走向，有多个高点，异常值为 $10\times10^{-5}m/s^2\sim70\times10^{-5}m/s^2$。内重力低带总体走向北东，局部有北西和北东东走向，异常值为 $10\times10^{-5}m/s^2\sim30\times10^{-5}m/s^2$。该带的北侧为南沙海槽重力低带。

该带的布格重力异常为密集的异常梯级带，由海槽的 $120\times10^{-5}m/s^2$ 向岛弧降低到 $20\times10^{-5}m/s^2$，异常走向为北东。

（三）重力异常地质解释

1. 沉积盆地与重力异常

南海分布着一系列的沉积盆地，这些盆地与重力异常之间存在着一定的相关关系。利用重力异常定性解释沉积盆地的特征，在海区一般采用自由空间重力异常。由于沉积盆地所处板块构造位置不同、成盆演化历史不同，各种类型盆地与重力异常的相互关系也各不相同，对新生代盆地而言，相对于盆地基底均为低密度的沉积层，新生代沉积层的密度一般为 $2.10\times10^3\sim2.40\times10^3kg/m^3$，按沉积层厚度的加权平均值为 $2.27\times10^3kg/m^3$，故在重力异常图上一般显示为重力低异常区。但是由于受地形地貌、地壳类型、沉积层的厚度、所处板块构造位置等因素的影响，往往减弱了沉积层的密度效应。

南海新生代沉积盆地与重力异常间有着不同的对应关系，其表现特征如下。

1）板内裂谷、裂陷型沉积盆地与重力异常有较好的相关关系，这类盆地位于大陆型地壳上，在拉张作用下形成，因此，前新生代沉积基底与沉积层之间存在密度差，通常是产生重力低异常的最主要的密度界面。北部湾盆地是典型的例子。高精度的海底重力仪测量，提供的重力异常图所圈定的北部湾盆地与地震圈定的盆地范围极为吻合，甚至盆地内次级构造单元也相同。以后的勘探开发结果进一步证实了这一点。另外，在雷州半岛以东的内陆架，利用重力资料发现了一系列的断陷小盆地（凹陷)；后对重力低异常区部署了地震勘探，发现了 3~5km 厚的沉积凹陷。

2）各种类型的陆缘（板缘）盆地与重力异常有一定的相关关系，但受多种因素的复杂化，使异常幅值和形态发生变化，导致用重力异常圈定盆地的困难。这类盆地处于陆架边缘，有的延伸到陆坡，地壳厚度减薄，深部莫霍面抬高，莫霍（密度）界面的重力作用增强，使异常背景值升高，显示为弱的重力低异常区（带）。分布在南海陆架、陆坡的一些大中型盆地多数属于这种情况。南海北部珠江口盆地中部的一条重力高带，把珠二坳陷分割成一个独立部分。珠一坳陷、珠二坳陷虽然同属一个重力低带，但与地震资料对比，两者也不尽一致。琼东南盆地受西沙重力低叠加影响，盆地面貌发生根本变化。

3）南部海域沉积盆地与重力异常的相关关系不明显。这是因为巽它陆架区重力异常的背景值高，南沙群岛陆坡、岛礁区地形变化剧烈，这些因素均对沉积盆地的重力异常施以影响。盆地中只在局部地区显示重力低异常。万安盆地只在中北部地区显示在 $10\times10^{-5}m/s^2\sim20\times10^{-5}m/s^2$ 背景上的一个（0~10）$\times10^{-5}m/s^2$ 重力低异常。北康、南薇西等盆地与万安盆地类似。

曾母盆地较特殊，盆地南部重力异常较宽缓，显示一个（0~10）$\times10^{-5}m/s^2$ 的重力低，中部为北西向的重力高带，在盆地中沉积厚度达 15km 以上的北部，为一个南北走向的短轴重力高，异常值为 $10\times10^{-5}m/s^2\sim20\times10^{-5}m/s^2$。产生这种特殊的异常，推测与地壳减薄或地壳层缺失、深部上地幔高密度物质上涌直接有关。

4）还有一些沉积盆地与重力异常总体呈负相关。礼乐盆地是一个典型例子。该盆地位于南沙群岛东缘独立的裂离块体之上。据盆地内钻井与地震资料，沉积层厚度 2000~4500m，最大达 6000m 以上，以碎屑岩和生物成因的碳酸盐岩为主。该盆地为重力高，异常值为 $20\times10^{-5}m/s^2\sim100\times10^{-5}m/s^2$，向四周减小，其上分布小的重力低。形成这一特殊重力异常的原因，分析认为，该盆地周围水深较大，构成一个较独立的岛礁高地形区，地形成为构成重力异常的主要因素。其次，新生代沉积层中有厚达 2000 余米密度较高的碳酸盐岩，同时盆地基底为变质程度较高的中基性岩。综合多种因素，礼乐盆地形成独特的重力高异常。

笔架盆地以坐落于洋壳之上为其特征。该盆地位于南海东北部深海盆中，新生代沉积厚度为 2000~4500m。沉积物以浊积层为主。重力异常为 $20\times10^{-5}m/s^2\sim30\times10^{-5}m/s^2$ 高值背景，盆地范围内为 $-10\times10^{-5}m/s^2\sim10\times10^{-5}m/s^2$ 的重力低值区。

综上所述，南海沉积盆地的结构构造与时空演化是十分复杂的，其重力异常呈现出多种样式，二者的关系亦不尽相同。利用中、小比例尺的重力异常资料分析沉积盆地的分布是粗尺度的，它的作用应该在区域油气普查阶段。

2. 区域断裂与重力异常

根据本区重力场特征，结合南海及周围陆地的地质构造和磁力、地震等地球物理资料，划分为北东—北东东、北西、东西和南北向 4 组区域性断裂构造（图 24-55，表 24-19），其中北东向和北西向两组区域性断裂构造全海区均有分布，南北向一组主要分布在东、西两侧和深海盆中，而东西向分布局部。这 4 组断裂构造可以分成两类，I 类为切割岩石圈的深断裂，是不同地壳块体的分界线。Ⅱ类为切割基底、地壳的区域性断裂，是不同地块或大型中新生代盆地的边界断裂。

1）北东—北东东向断裂：这组断裂全海区均有分布，其与重力异常的总体走向一致，因此，在重力图上表现为断裂两侧各不相同的异常面貌，一般一侧为重力高带（区），另一侧为重力低带（区）。密集的重力梯度带与断裂带的走向一致或平行。以南澳-海口深断裂带（Ⅱ）为例（表 24-19），该断裂带分布在北部内陆架，北起福建近岸，经南澳岛南侧向西南延伸，经担杆岛南至海口，穿越琼北到北部湾与红河断裂带相交，全长超过 1660km。断裂带北侧为幅度 $0\sim10\times10^{-5}m/s^2$ 的一系列重力高带，南侧为 $-10\times10^{-5}m/s^2\sim20\times10^{-5}m/s^2$ 重力低区，重力异常梯度带不连续。该深断裂带延伸长度大、切割深，是华南圈块和南海圈块的分界线，具有不同圈块结合带的特征，是华力西期—印支期两个圈块碰撞结合的部位。喜马拉雅期在其南侧发育了珠江口等大型沉积盆地。

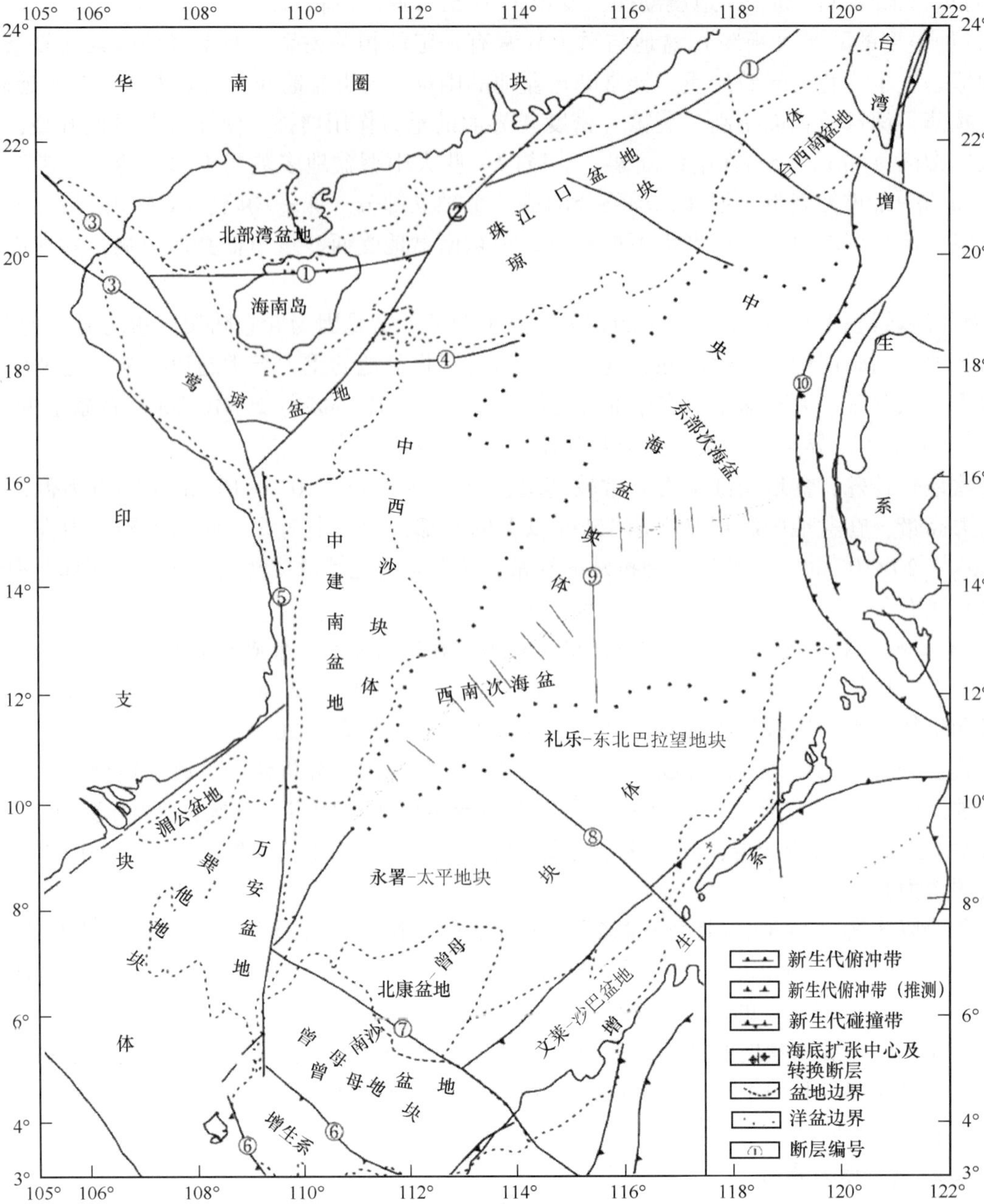

图 24-55　南海主要构造分布及岩石圈构造单元图

Fig. 24-55　Map of distribution of main structures and lithosphere structural units in the South China Sea

（据姚伯初等，2006）

①琼北-珠外-台湾海峡断裂；②莲花山断裂；③黑水河-红河断裂带；④西沙海槽断裂；⑤南海西缘断裂；⑥卢帕尔-武吉米辛断裂带；⑦廷贾断裂；⑧巴拉巴克断裂；⑨中南-礼乐断裂；⑩马尼拉海沟断裂带

2）北西向断裂：该组断裂全海区均有分布，大部分断裂带走向与重力异常的总体走向正交或斜交。因此，往往表现为重力异常被截断、错开或在北东向或其他方向的异常背景上叠置北西向的局部异常，有的北西向局部异常呈串珠状分布。红河深断裂（I_5）向东南延伸为莺歌海盆地中的1号断裂。西南侧为重力低区，东北侧为一条重力高带，有较明显的梯度带。该断裂是印支块体与华南圈块的分界线，是二者在华力西期—印支期的结合带，喜马拉雅期发生剪切拉张，其上叠置了新生代的莺歌海盆地，盆地内沉积了14km以上巨厚的沉积物。该断裂向西北与金沙江断裂相连，向东南的延伸尚有争论，悬而未决。作为一个圈块结合带，由多条断裂组成。南部的马江断裂带也具有类似的性质，该断裂向东南延伸进入莺歌海盆地，成为该盆地的西部边界，在海区也是一条重力梯度带。

表 24-19 南海主要断裂特征一览表

Table 24-19 List of characteristics of main faults in the South China Sea

类型及序号	名称	走向	长度/km	重力场特征	性质	主要活动期
I_1	南澳-海口	北东东	1660	北侧重力高带，南侧重力低带，梯度带明显	早期压扭 晚期张扭	华力西期—印支期、喜马拉雅期
I_2	南海西缘	南北	1350	东侧重力低带，西侧重力高带，梯度带明显	早期压剪 晚期剪切	印支期、喜马拉雅期
I_3	马尼拉海沟	南北	1200	西侧重力低带，东侧重力高带，梯度带明显	挤压	喜马拉雅期
I_4	卢帕尔-沙巴-巴拉望	北西—东西—北东	1540	梯度带、异常区分界	挤压	燕山期、喜马拉雅期
I_5	红河	北西	820	西南重力低带，东北重力高带，梯度明显	早期压剪 晚期剪切	华力西期—印支期、喜马拉雅期
II_1	硇洲	北西	410	截断 NE 向异常，上叠 NW 向异常	张扭	燕山期、喜马拉雅期
II_2	川岛	北西	380	截断 NE 向异常	张扭	燕山期、喜马拉雅期
II_3	韩江口	北西	520	截断 NE 向异常，两侧异常走向改变，NW 向异常	张扭	燕山期、喜马拉雅期
II_4	中建西	北西	630	串珠状重力高，两侧异常差异	张扭	喜马拉雅期
II_5	廷贾	北西	610	两侧异常差异，NW 向异常	张扭	喜马拉雅期
II_6	巴拉巴克	北西	580	改变两侧异常走向	张扭	喜马拉雅期
II_7	北部陆架边缘	北东	1520	改变两侧异常走向	张性	喜马拉雅期
II_8	中沙北	北东	620	改变两侧异常走向	张性	喜马拉雅期
II_9	郑和-万安	北东	7300	改变两侧异常走向	张性	喜马拉雅期
II_{10}	海马-安渡	北东	910	改变两侧异常走向	张性	喜马拉雅期
II_{11}	南沙海槽两侧	北东	8200	改变两侧异常走向	张性	喜马拉雅期

3）南北向断裂：这一组断裂主要分布在南海东、西两侧，是南海圈块的两条边界断裂。在重力异常图上，断裂两侧的异常差异大。南海西缘深断裂带（I_2）位于中南半岛东部陆架边缘，向南延伸成为万安盆地东部边界，南端与卢帕尔深断裂相交。往北与马江、红河深断裂相接，全长达 1350km。重力异常西侧为重力高带，东侧为重力低带，异常梯度较大，连续性好，是南海圈块与印支圈块的分界线。断裂形成于印支期，以挤压性质为主。喜马拉雅期以剪切为主。沿断裂带两侧分布有归仁、中建南、万安等新生代沉积盆地。

马尼拉海沟断裂带（I_3）位于南海东部，构成南海圈块的东部边界。沿海沟的东坡分布，向南延伸到民都洛岛西侧并穿越该海峡，成为苏禄海的东部边界断裂。向北延伸可达台湾南部，在南海的长度可达 1200km。在负值重力低带的背景上，东侧为重力高带，异常梯度大。由多条断裂组成的南北向断裂带，其形成是南海的洋壳向吕宋俯冲消减的结果。此外，在南海中央海盆中分布有一系列南北走向的转换断层，虽然错开了扩张中脊和磁异常条带，但在洋壳内部产生，密度值变化不大，故这些断裂在重力异常上无明显特征。

4）东西向断裂：该组断裂不甚发育，在重力异常上特征不明显。断裂规模相对较大的有西沙海槽东段南、北两侧的断裂带，有较大的重力异常梯度带。另外还有中央海盆分布于15°N左右的扩张脊，由扩张期后的海底玄武岩喷发形成东西向排列的一系列海山，组成串珠状的重力高。

综上所述，南海不同级别的断裂带，对南海的重力异常排列组合面貌起着重要作用。反之利用重力异常来分析断裂构造的存在多了一种地球物理的证据。

（四）南海莫霍界面的基本形态和特征

南海莫霍界面的形态具有明显的三分性：大陆型地壳、洋陆过渡地壳和洋壳。

1. 大陆型地壳

从南海莫霍界面图（图24-56）来看，南海海域的大陆型地壳主要分布于北部陆架地区和南部的巽它陆架。莫霍界面深度基本上在30~26km范围内，南海海域内另一显著特征是：多处存在微陆壳。西沙、中沙和南沙群岛均呈现微陆块性质。西沙群岛地壳厚度在24~26km分布区域较中沙群岛宽阔；中沙群岛地壳厚度可达30km，但分布区域较小；南沙群岛地壳厚度一般在20km以上，几个岛礁的地壳厚度较厚，礼乐滩、安渡滩的地壳厚度可达26~28km。中业-郑和-九章群礁的地壳厚度也有24km。这些岛礁在莫霍界面图上明显表现为微陆块。

2. 洋—陆过渡地壳

南海海域洋—陆过渡地带主要表现为地壳厚度急剧变化的梯度带。在北部陆缘存在一条明显的梯度带。在比较狭窄的区段内，莫霍界面深度由26km急剧抬升到16km，形成明显的梯度带。中央海盆与南沙群岛之间也存在一条较明显的梯度带，地壳厚度由20km急剧减少到12~14km。中央海盆的西侧，中沙、西沙附近海域地壳厚度约20~16km，该较宽阔的海域也是洋-陆壳过渡地带。

3. 洋壳

中央海盆和西南海盆是大洋型地壳区，其范围基本上由12km的莫霍界面等深线所围限。中央海盆内莫霍界面较平缓，深度大部分为12km，也出现一些10km的圈闭。西南海盆莫霍面也较平缓，基本上在10km范围内。西南海盆的地壳厚度比中央海盆薄。

五、南海海域磁力异常特征及其地质解释

本书收集了《中国南海海域岩石圈三维结构及演化》（姚伯初等，2006）的成果，该项目详细地对南海地区的磁力异常特征及解释等进行较详细的研究，为本次工作提供了较详细的资料。

（一）磁力异常特征

根据南海及围区现有岩石磁性资料，引起南海及围区磁异常的主要地质体是火成岩及古老变质岩，南海海盆玄武岩是引起条带状磁异常的主要因素。众多的磁异常类型表现了区内断裂和岩浆活动的复杂性。

纵观南海整个磁场图（图24-55），具有下列5个基本特征。

1）磁异常类型繁多。有南正北负或南负北正的伴生异常，伴生现象不明显的正异常、负异常，叠加异常等。在东沙以东的南海东北部海盆和南沙群岛区域，发育低幅值正负宽缓变化异常，这类异常幅值小（一般小于200nT），且变化平缓，次级异常较少发育。而在东沙、中沙西南、一统暗沙南、巴布延群岛区域等处，均存在大型高值正异常，幅值高，宽度大，延伸长，呈近东西或北东向展布，在图幅中特别醒目。海南岛东北、东南，北部湾、越东、加里曼丹岛北部、南沙海槽等区域，呈现以负异常为主、局部异常较少发育的磁场面貌。此外，还有与火山口、火山锥或玄武岩磁性不均匀有关、在平面等值线图上展现为等值线密集，或形态浑圆，或正负急剧变化的小型异常等。

2）异常显示的磁场强度具有从北部陆坡向南部深海盆、从西部向东部增强的总趋势。由北部陆坡-海盆的宽缓平静变化的磁异常，向南经低幅值、小型正负变化的过渡带，进入南海深海盆的正负剧烈变化异常区。深海盆区异常呈高幅值、高强度、陡梯度变化，局部异常极为发育，磁场面貌显得越来越复杂。由北部陆坡-海盆向东，进入台湾南—吕宋岛之间的海脊区，为剧烈变化异常带，局部异常相当发育。南海西南部异常特征、强度、走向等均有其独特性。

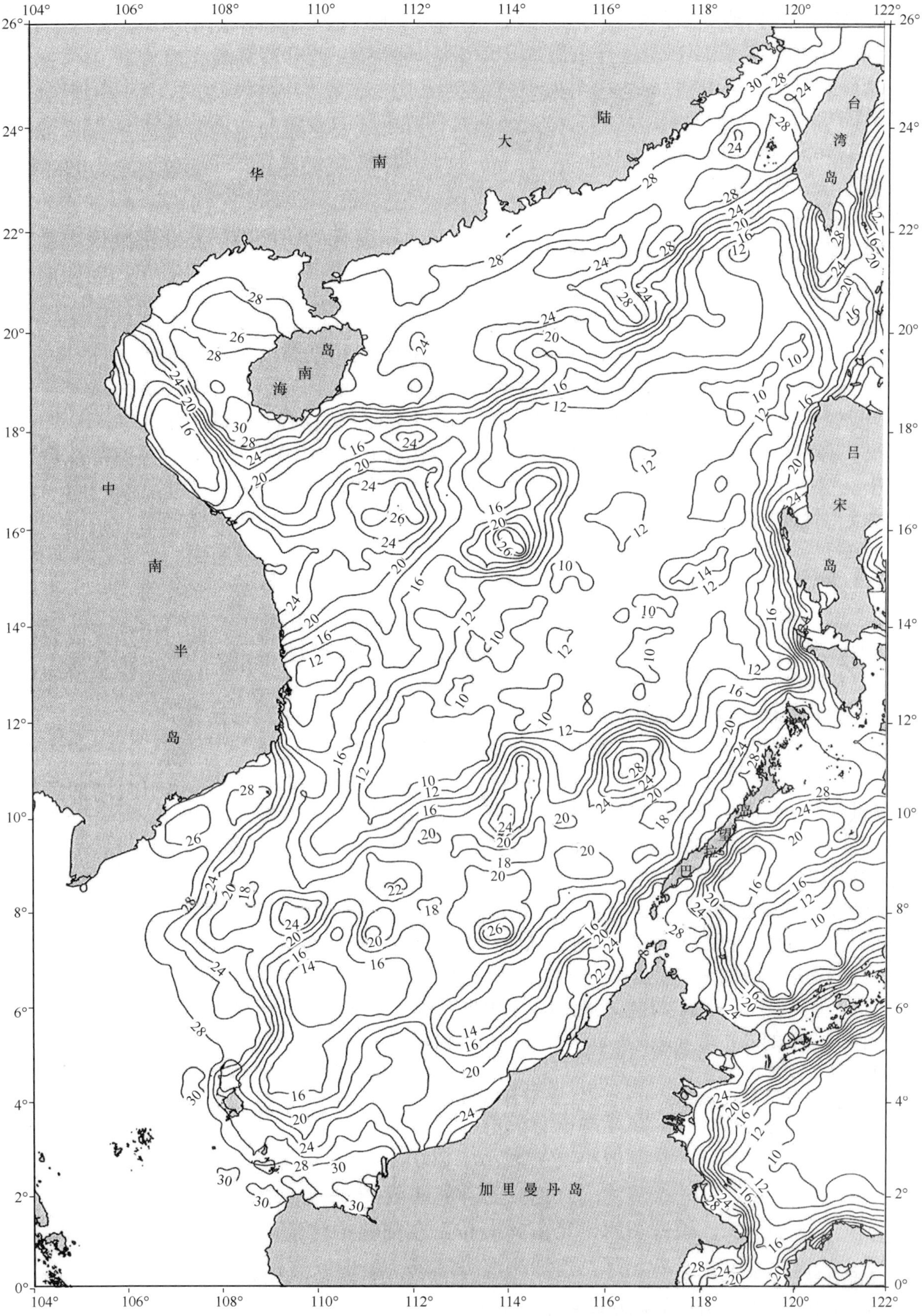

图 24-56　南海莫霍界面深度图

Fig. 24-56　Depth map of Moho Interface in the South China Sea

（深度单位：km；据 HY126-03 项目报告资料，有修改）

3）磁异常走向主要有北东向、北北东向、东西向或近东西向、北西向、北西西向、南北向等。这些异常方向显然是区内地质构造发展的真实记录，客观地反映了该地区基底构造走向、断裂带的展布方向和岩体的长轴方向，基本上揭示了它们与地质构造之间密切的内在联系。

4）南海中央海盆（东区）存在着正负频繁交替变化的、平面等值线图上展示异常长轴以近东西向为主的线性异常带（通称磁条带）。这些异常在平面等值线图上呈正负相间条带排列，异常幅度200~400nT，宽度20~30km，延伸可达数百千米；梯度很陡，最大达40nT/km。

5）南海西南海盆发育一组北东走向的条带状磁异常，其延伸长度较大，可达400~500km；但分布范围较窄，异常变化幅值也较小。

（二）条带状磁异常的地质解释

南海是西北太平洋边缘海中面积最大的海盆，地处欧亚、印澳和太平洋三大板块的汇聚带。近30年来，有关南海，特别是其中的中央海盆的地质构造特征及其形成、演化问题，引起了国内外学者的极大关注。Karig（1971）将南海中央海盆划为具有高于正常热流值的不活动边缘海盆，1983年又提出盆地是晚白垩世—古近纪时向东凸出的菲律宾弧后扩张而形成的弧后盆地。Ben-Avraham等（1973）在南海中央海盆中首次识别出东西向的条带磁异常，而后Tavlor等（1983）的工作进一步证实东西向磁异常为对称分布的磁条带，15°N附近的东西向海山链代表了残留扩张中心。之后，国内众多专家对条带磁异常进行广泛而深入的研究。

1. 中央海盆条带磁异常

在南海中央海盆区，存在着正负频繁交替变化、平面上呈正负相间条带排列、异常长轴以东西向为主的线性磁异常带。一般称之为“条带状”的线性异常。异常呈正负交替起伏变化，短波长、高振幅，变化剧烈，梯度陡。异常变化幅值一般为200~400nT，最大可达700nT。异常长轴或异常组合走向虽以东西向为主，但也有北东东向、北东向；异常展布延伸的轴向与北西向、北东向或南北向的构造线相互错断或扭曲。该区异常特征变化纷繁，十分复杂。

Taylor等（1980）分析了中央海盆的磁异常，并首次对比出5d-11号磁异常条带，由此计算出南海中央海盆是在渐新世—早中新世（距今32~17Ma）期间通过南北向海底扩张形成的。对南海深海平原的形成机制和时代提出了如下的论断：南海深海平原是由于海底南北向扩张形成的，形成的时代大约是渐新世—早中新世（距今约32~17Ma）；扩张是对称性的，其北部边缘是一种被动性的大陆边缘，残留的扩张中心位于北纬15°附近，它与现代形成东西向链状分布海山的位置一致。这一观点就成了南海形成与演化机制的主导观点。

2. 西南海盆条带磁异常

Taylor等（1983）根据有限的单道地震剖面和重力资料，推断西南海盆存在一条北东向的残留扩张中心，随后的调查得以证实。从图24-57可见，该区正负交替变化的条带状异常变化幅值在200~300nT之间。但其强度、规模、梯度变化均小于中央海盆，总体异常特征和走向均与东区有较大的差异。尽管西南海盆的磁异常及其成因机制普遍采用北西—南东向对扩张模式来解释，但此处海盆的宽度较窄且磁异常与典型大洋条带状异常有一定的差异，故造成异常的识别和对比困难，解释结论差别非常大。由此可见，虽然西南海盆磁条带异常是公认的，但基于与中央海盆相似的原因，西南海盆的磁异常解释也有待深化。

六、南海海域基底特征

（一）基底结构与重力异常

根据重力异常和地质构造、断裂等特征，南海基底可分为南海北部陆架陆坡区、中沙-西沙岛礁区，南沙岛礁区，西南（巽它）陆架区和洋壳5个区。

1）南海北部陆架陆坡区：该区基底由三个不同圈块组成，海南岛及其以东为南海圈块的一部分，其上叠置了珠江口盆地、台西盆地、台西南盆地等一系列大型盆地。该区的基底主要由古生代的变质

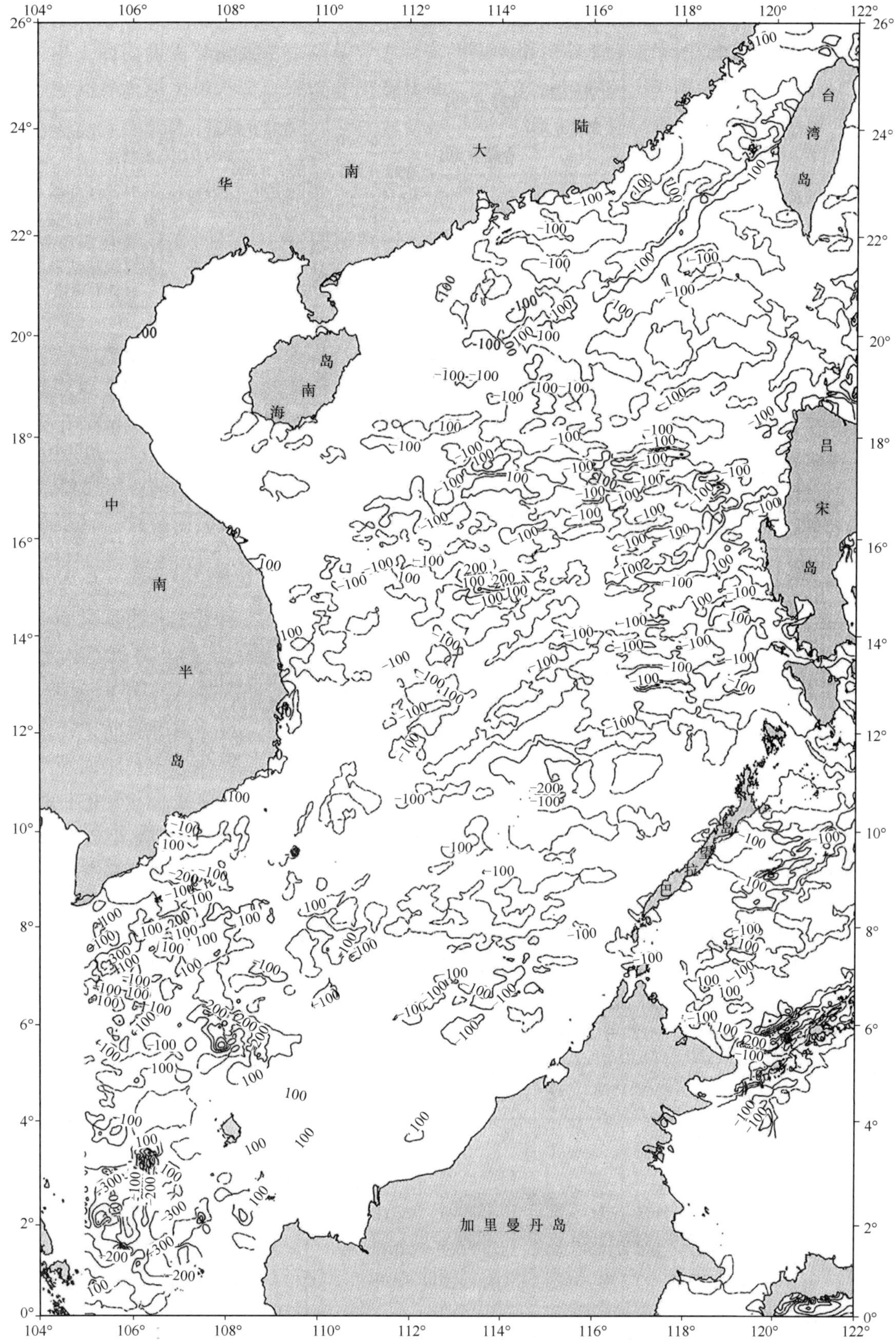

图 24-57 南海磁力异常（ΔT）平面图

Fig. 24-57 Plan of magnetic anomalies（ΔT）in the South China Sea

（单位：nT；据 HY126-03 项目报告资料，有修改，点划线为零线）

岩和中生代的岩浆岩、火山岩等组成，这已得到钻井证实。采用重力资料计算的基底埋藏深度在2~9km之间，在珠一坳陷、珠二坳陷最深达10km和11km。

莺歌海盆地叠置在印支、华南和南海三个不同的圈块上，其基底主要由华力西、印支期褶皱带构成。北部湾盆地是一个华南圈块上的陆内裂谷盆地，其基底主要为晚古生代的碳酸盐岩和早古生代的变质岩，局部为古生代晚期的碎屑岩。

2）中沙-西沙岛礁区。该区的重力异常主要反映了海底的变化，用它来分析基底结构是困难的。据西永1井钻到元古宙片麻岩推断，西沙、中沙的基底主要为元古宙的结晶基底，周围可能有古生代浅变质岩。中沙西南分布一系列重力高和重力低，分析主要为新生代火山岩构成的海底山和山间盆地的反映。

3）南沙岛礁区：中北部与中、西沙岛礁区相类似，但曾母盆地的基底较为复杂。南部为平缓的重力异常，据钻井资料，为古近纪的高度变形、低变质的沉积岩；北部沉积层最厚达15km以上的位置，呈重力高，地壳厚度为16~18km。其可能是一个异常地壳，沉积层或许直接与上地幔相接触。

4）西南（巽它）陆架区：该区重力异常的背景值高，局部异常的变化可能反映了基底的起伏变化。该区的基底是印支块体的南延部分，主要是印支期和燕山期的褶皱带被同期的中酸性花岗闪长岩、花岗岩类侵位而复杂化的基底。

5）洋壳：南海的洋壳是新生代海底扩张而形成的大洋基底，其时代由扩张中心脊向两侧变老，按磁异常条带分析计算，东部海盆形成于32~17Ma，西南次海盆形成的年代尚有争论，这里采用42~3Ma（姚伯初，1994）。重力异常在洋壳区较稳定平缓，一般为（0~±10）$\times10^{-5}m/s^2$。局部因受海底山的作用而形成等轴状重力高。

（二）磁性基底埋深

南海海域磁性体（顶面）埋藏深度图（图24-58）是根据新老磁测资料的磁源体深度计算结果编制而成的，磁性体（层）顶面深度变化反映出该海域主要磁性体（层）上表面的起伏变化，结合岩石类型、磁性资料，可以给出某个地质年代剥蚀面、不同岩相结构间组合等地质构造的时代意义。归纳有以下几点。

1）磁性体（层）顶面深度（从海面算起）起伏变化约15km，最浅的约1km，分布在巽它陆架、暹罗湾、海南岛及其周边；最深处达17km，超过10km的有曾母盆地、万安盆地、莺歌海盆地。其次，中建南、西沙海槽、南薇西、北康、沙巴、礼乐、晋江等盆地凹陷和珠江口盆地珠一坳陷、珠二坳陷的磁性体（层）顶面深度为7~9km。

2）综观南海各具特色的4个边缘、几条区域大断裂及块体划分格架，在磁性体（层）顶面深度图上均有明显的显示：多数呈现出磁性体深度急剧变化的梯度带，或磁性体等深线走向截然不同，如南海深海盆区与周边陆块的界线清晰可辨。

3）中沙-西沙、东沙、南沙三个隆起的磁性体（层）顶面深度为3~5km，虽然内部也有分割，但仍比较稳定。正如图24-58所示，尚有一些细节有待进一步研究，如在东沙隆起东南，下陆坡与东北海盆处的磁性体深度几乎出露海底，而该区磁异常平静宽缓，搞清这个问题有助于评价台西南盆地和潮汕凹陷（姚伯初等，2006）。

七、岩石圈三维结构的初步研究

（一）南海地壳速度结构特征

南海地区由于其特殊的结构位置及复杂的构造发展历史，因此，其地壳和岩石圈结构独具特色，纵向与横向变化极其复杂。

地壳结构的变化特征：在地质发展历史中，由于南海地块是由多个小地块在地质时期经过多次碰撞而缝合起来的，因此，其地壳结构的横向变化很大。如南海北部陆架与陆坡区，穿过西沙海槽，地壳结构发生显著变化。根据资料，该处地壳结构如下：上地壳速度为1.9~6.0km/s，厚2.3km；下地

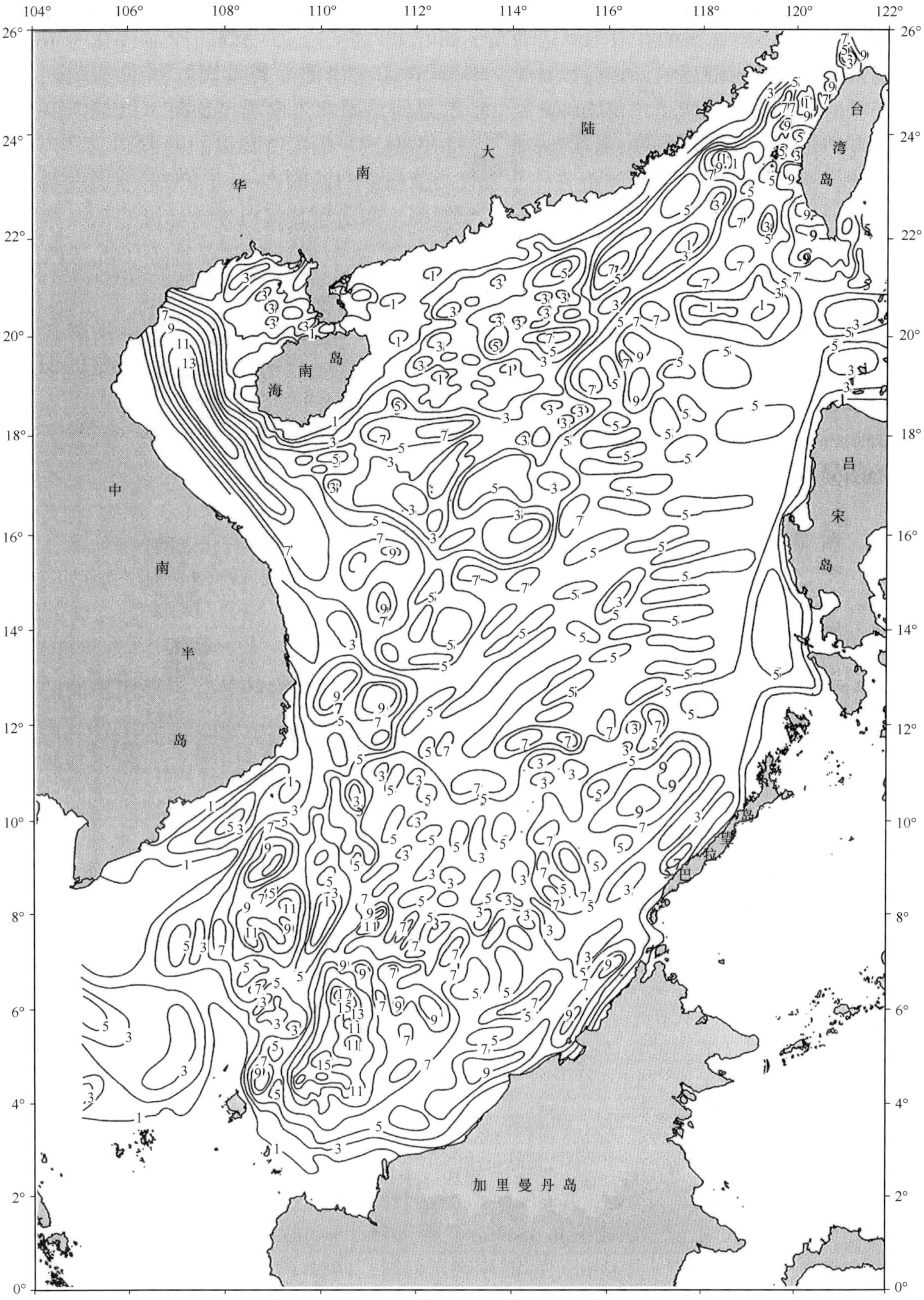

图 24-58　南海海域磁性体（顶面）埋藏深度图

Fig. 24-58　Burial depth map of magnetic bodies (top) in the South China Sea

（单位：km；据 HY126-03 项目报告资料，有修改）

壳速度为6.9km/s，厚21.21km。过了西沙海槽，在中沙-西沙地块上，地壳结构变化很大，上地壳速度为1.6~6.0km/s，厚16.07km；下地壳速度为6.4km/s，厚5.63km。两处的水平距离仅100km左右，但地壳结构却变化如此之大，因此推测西沙海槽可能是印支期的一条缝合线，沿此缝合线，中沙-西沙地块和南海北部陆缘地块发生过碰撞缝合。又如从华南陆地进入南海北部陆架，地壳结构发生明显变化：在汕头地区，地壳30.0km，上地壳速度为5.86km/s，厚16.0km；在16~18km深度处存在一低速层，速度为5.13km/s；下地壳速度为6.61km/s，厚12.0km。在距汕头100km多的陆架区，其地壳结构为：新生代沉积厚5km，速度为1.8~4.8km/s，下伏中生代沉积，速度为5.0~5.5km/s，厚3.5km；向下为中酸性火成岩，厚3.5km，速度为5.6~6.2km/s；再向下是下地壳，速度范围为6.5~7.4km/s，厚14.0km，地壳总厚度为26.0km。由此可见，华南陆区与南海北部陆架的地壳结构之变化很大，反映了它们可能是由两个地块缝合而成的。在中央海盆，沉积层（层1）厚0.5~3.0km，速度为1.8~3.2km/s；层2A的速度为3.5~3.7km/s，层2B的速度为5.1~5.3km/s，层2C的速度为6.0~6.1km/s，层2的厚度为1.99km；层3的速度为6.5~7.4km/s，厚3.22km。由此可见，南海盆地的地壳结构类似于年轻洋壳的速度结构，与地磁异常条带对比出的地壳年龄是一致的。

（二）南海岩石圈速度结构特征

1. 剖面特征

1）图24-59表示沿20°N、深度400km的纵波速度结构剖面。该剖面东起菲律宾海，经吕宋岛弧区、南海东北部海盆、南海东北部陆坡、南海西北部陆架、海南岛、北部湾、红河口至印支半岛。在剖面西部，印支半岛之下的地壳厚38km，岩石圈厚度超过100km。上、中地壳厚30km，下地壳厚8km；在70~80km深度上存在一高速层，速度为8.1~8.2km/s。在107°30′~110°E处，地壳厚30km左右，上、中地壳厚，下地壳薄；在35~48km深度上，存在一低速层，P波速度小于8.0km/s，岩石圈厚度减薄至70km左右。这里对应着红河-莺歌海断裂带，表现为地壳和岩石圈厚度减薄，岩石圈中存在低速层，说明红河-莺歌海断裂带是一条岩石圈断裂。向东至海南岛（109°~111°E），地壳加厚至40km，上、中地壳和下地壳厚度相当；岩石圈厚度80km。从112°E开始，地壳开始减薄，由35km至115°E处的30km，上、中地壳厚20km，下地壳厚10km。此处岩石圈中，50~65km深度上存在一高速层，速度为8.2~8.3km/s。向东到东沙隆起处（116°~118°E），地壳厚度增厚至32km，上、中地壳厚10km左右，下地壳厚度为22km；下地壳底部出现一高速地壳层，速度达7.4km/s。此处岩石圈厚度加厚至95km。在118°~120°10′E处，地壳厚度减薄到20~15km。此段西部为陆坡区，东部南海东北部海盆，是洋壳，正在沿马尼拉海沟向吕宋弧下俯冲。此段岩石圈厚度下部52~80km深度处存在一高速层，厚度30km，速度达8.2km/s。在吕宋岛弧区，地壳厚20km左右，上、中地壳厚15km，下地壳厚5km；岩石圈厚度超过100km。在这条断面上，西部深90~230km上出现速度小于8.0km/s的岩石层，即软流层；而东部深80~220km上出现速度大于8.0km/s的岩石层，表明这里软流层中存在高速岩石层。本剖面在海盆之下，岩石圈结构特征很有构造意义，地壳之下，上地幔速度为7.9~8.1km/s，向下在30~40km，出现7.5~7.8km/s岩石圈层；从40km开始，到70km深度，存在高速岩石圈层，速度8.2~8.5km/s；从70~90km深处，速度又逐步降低，为8.1~8.0km/s。南海洋盆岩石圈的这种层状速度结构，是和海底扩张、新洋壳形成有密切关系的。

2）图24-60表示沿18°N、深度400km的P波速度结构剖面。该剖面从东到西，经吕宋岛、北吕宋海槽、马尼拉海沟、南海中央海盆、西北海盆、中沙-西沙地块，到印支半岛。在剖面西部，印支半岛之下的地壳厚30km左右，上、中地壳厚18km，下地壳厚12km；在岩石圈下部60~80km深处存在一高速层，速度8.1~8.2km/s，它向印支半岛东部外的红河-莺歌海断裂带（108°~109°E）之下减薄至10km。印支半岛之下的岩石圈厚度超过100km。莺歌海断裂带之下的地壳减薄至25km，其上、中地壳厚20km，下地壳厚5km；这里岩石圈厚度减薄至80km。此处在35~45km深度上存在一低速层，速度为7.5km/s，说明莺歌海断裂是岩石圈断裂。在海南岛南部海域（109°~110°E），地壳加厚至35km，其上、中地壳厚30km，下地壳厚5km。向东至西沙海槽，地壳减薄到28km，其上、中地壳厚

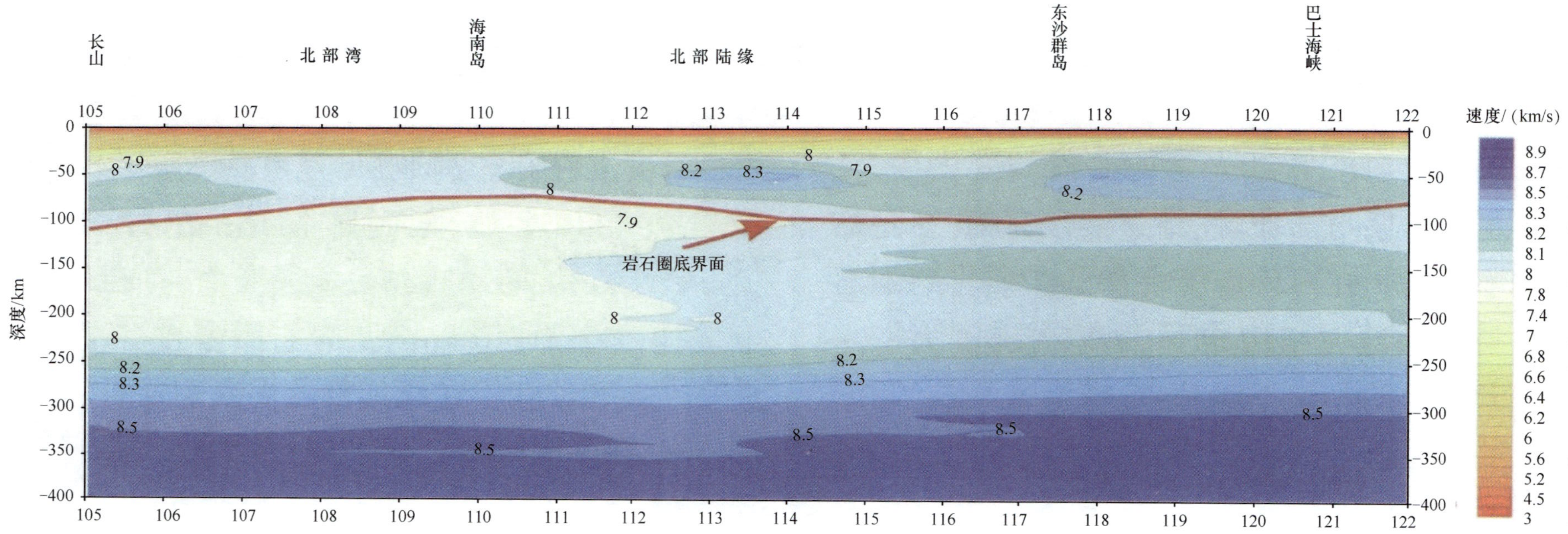

图 24-59 沿北纬 20°穿过南海、深度 400 km 的速度剖面图

Fig. 24-59 Velocity section crossing the South China Sea along N 20°at a depth of 400 km

（据姚伯初等，2010）

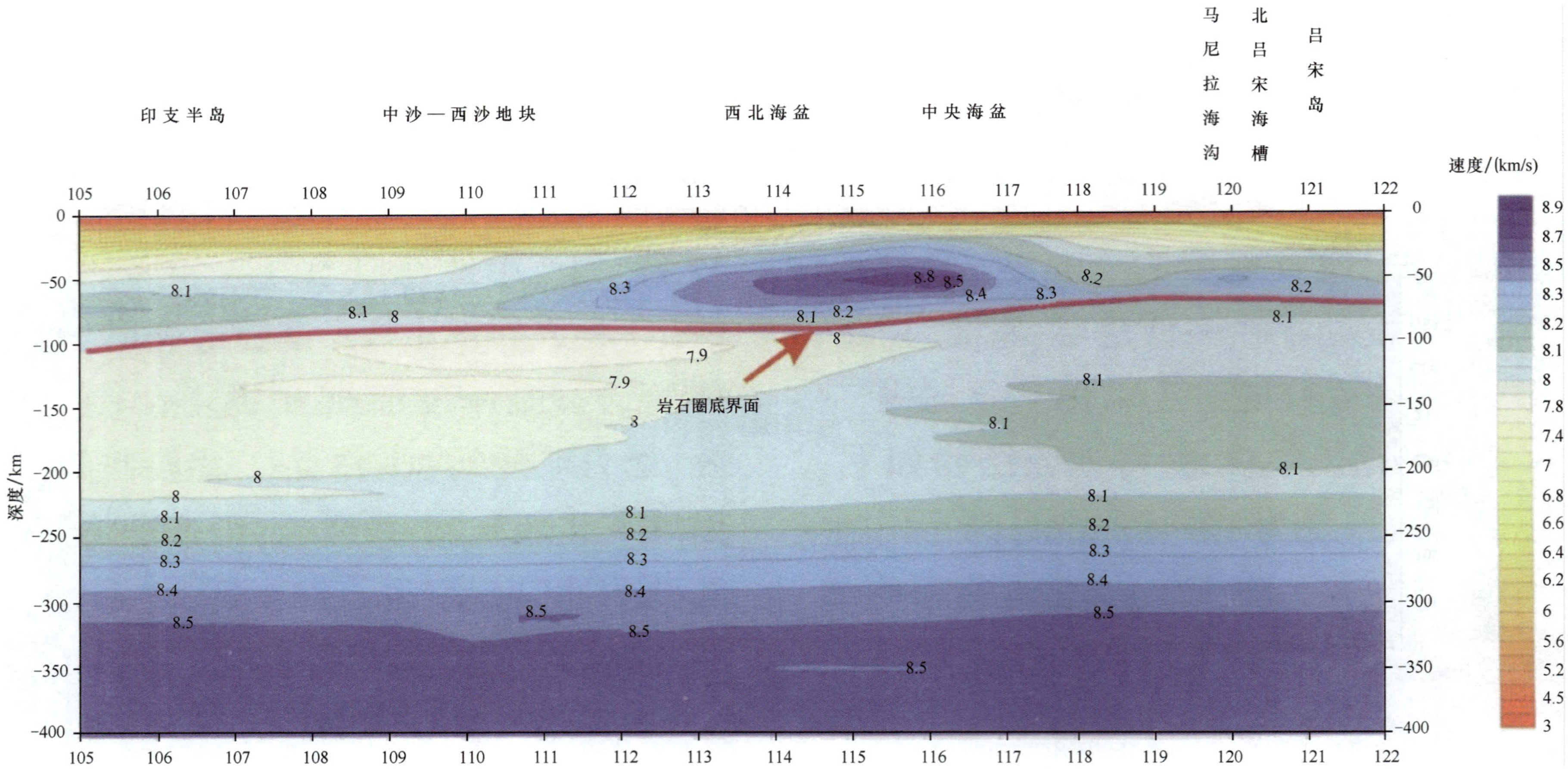

图 24-60　沿北纬 18°穿过南海、深度 400 km 的速度剖面图

Fig. 24-60　Velocity section crossing the South China Sea along N18°, at a depth of 400 km

（据姚伯初等，2010）

20km，下地壳厚 8km。这一段剖面（海南岛之下）的岩石圈厚度在 80～85km 之间，从这里开始，至南海中央海盆东部边缘（118°E），在岩石圈的 60～80km 深度上存在一高速层，其速度为8.2～8.5km/s，岩石圈厚度 90～95km。在南海西北海盆至中央海盆之下（114°～119°E），地壳厚度减薄到 5～6km，莫霍面深度 10km，是典型的洋壳。在马尼拉海沟（119°30′E）之下，莫霍面深度从 10km 向 40km 下降，但岩石圈厚度未见变化。按正常情况，大洋岩石圈向海沟俯冲，可达 700km 深度，从天然地震资料看，这里俯冲岩石圈已到 200km 深处（主要原因是计算层析资料时是用面波层析算法所致）。在剖面上，软流层的变化如下：西部印支半岛至南海海盆之下，80～220km 深度上出现速度小于 8.0km/s 的岩石层，表明这里软流圈厚 140km。而在东部吕宋岛弧之下 110～200km 深度上出现速度大于 8.0km/s 的岩石层，表明这里软流层中存在高速岩石层。

3）图 24-61 表示沿 14°N、深度 400km 的 P 波速度结构剖面。从东到西，剖面穿过吕宋岛南部、北吕宋海槽、马尼拉海沟、南海中央海盆、长龙海山、南海西南部海盆、中沙-西沙地块、南海西缘断裂，至印支半岛的昆嵩高地之下，地壳厚 30km，其上、中地壳厚 25km，下地壳厚 5km；岩石圈厚度 80km。由此向东，从 108°30′E 到 110°30′E，地壳厚度变化不大，但上、中地壳厚 18km，下地壳厚 12km。中沙-西沙地块之下，岩石圈厚度逐渐减薄，在南海西缘断裂带（106°10′E）之下，厚度减至 70km；由此再向东，在中沙-西沙地块之下，岩石圈厚度逐渐加厚，到 113°E 处，进入南海西南海盆，莫霍面深度变至 11km，岩石圈厚度加厚至 95km。但到 115°～116°E 处，地壳结构有变化，层 2 厚度从西南海盆向中央海盆变薄，反映西南海盆的年龄应比中央海盆老；岩石圈厚度由西南海盆的 95km 到中央海盆的超过 100km；在下部岩石圈中的 60～80km 深度上存在一高速层，速度为 8.2～8.5km/s。向东到马尼拉海沟（119°E）处，地壳向下延伸，莫霍面从约 10km 下沉至 20km。在吕宋岛之下，岩石圈中的高速层消失。在整段剖面中，岩石圈之下，软流层底部大约在 240km，底部速度为 8.4km/s。这反映软流层厚度在西部南海西缘断裂之下最厚，约 170km，而在中央海盆之下最薄，厚度 140km。从岩石圈与软流圈厚度变化来看，南海西缘断裂带是一条岩石圈断裂带。

4）图 24-62 表示沿 8°N、深度 400km 而穿过南海的岩石圈速度剖面。剖面西起巽它陆架两兄弟群岛、南沙群岛万安滩、西卫滩（110°E）、南薇滩（112°E）、南海礁与安渡滩（114°E）、都护暗沙（116°E）、南沙海槽、巴拉巴克海峡到苏禄海盆。在剖面西部巽它陆架之下，地壳厚度 26km，其上、中地壳厚 17km，下地壳 9km；岩石圈厚 75～80km。从巽它陆架进入南海南部陆坡，地壳厚度减小，其上、中地壳厚度加厚，下地壳减薄。在巽它陆架之下，岩石圈厚度由西部 95km 向东减薄至 90km；在南沙群岛之下，岩石圈厚度为 85km；由南沙群岛进入南沙海槽（116°E 附近），地壳厚度由 20km 增厚至 26km，主要是下地壳厚度由南沙群岛地区的 4km 增加到南沙海槽之下的 10km。说明南沙海槽曾是地质时期的俯冲带。岩石圈厚度在此处开始增厚，厚度由 90km 增至 95km。从 119°30′E 到剖面东部（苏禄海盆），在岩石圈下部 62～88km 深度上出现一高速层，其速度为 8.2km/s，岩石圈厚度近 100km。在苏禄海盆内（119°～122°E），地壳厚度减薄，莫霍面深度由巴拉巴克海峡的 28km 减至苏禄海盆的 10km，说明东南苏禄海是洋盆，深海钻探已证明了这一推测。

5）图 24-63 表示沿 117°E、深度 400km 的 P 波速度结构剖面。剖面北起广东南澳岛、经南海北部陆架、东沙群岛（20°N）、南海中央海盆、南海南部边缘的礼乐滩（11°N）、南沙海槽（8°N）、巴拉巴克海峡，至婆罗洲上沙巴地区的基纳巴卢山和卡杨河地区。剖面北部陆区地壳厚 28～30km，其上、下地壳厚度相当；陆架区地壳厚度 28km 左右，上、中地壳厚度薄，下地壳厚。陆区岩石圈厚度超过 100km。从东沙群岛经下部陆坡进入中央海盆之后，地壳明显减薄，莫霍面深度从 20km 逐渐减薄至 10km 左右；在岩石圈下部 50～70km 深处存在一高速层，速度为 8.2～8.3km/s；岩石圈厚度超过 100km。在礼乐滩（11°N 附近），地壳厚度约 26km，上、中地壳和下地壳相当；岩石圈厚度 95km 左右。向南进入南沙海槽（8°N），地壳厚度变化不大，地壳层等深线呈水平展布，古俯冲带的痕迹已不明显，反映陆—岛碰撞完成后挤压已停止，或是已变成今天的俯冲带。从巴拉巴克海峡进入沙巴，地壳厚度变化不大，但岩石圈厚度达 100km 左右，说明沙巴地区岩石圈厚度稍有增大，可能它是由增生

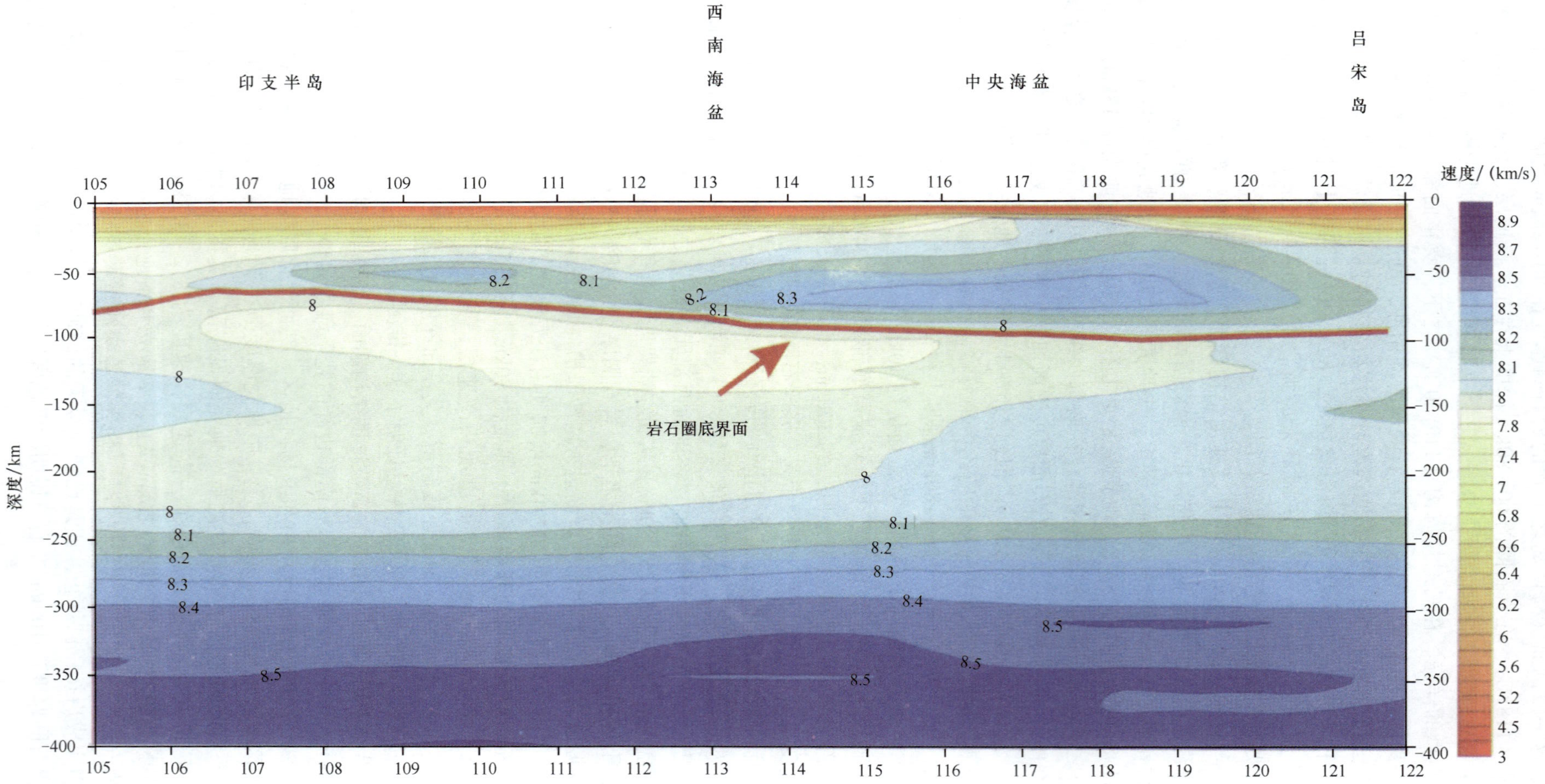

图 24-61　沿北纬 14°、深度 400 km 而穿过南海的岩石圈的速度剖面

Fig. 24-61　Velocity section crossing the South China Sea along N 14°, at a depth of 400 km

（据姚伯初等，2010）

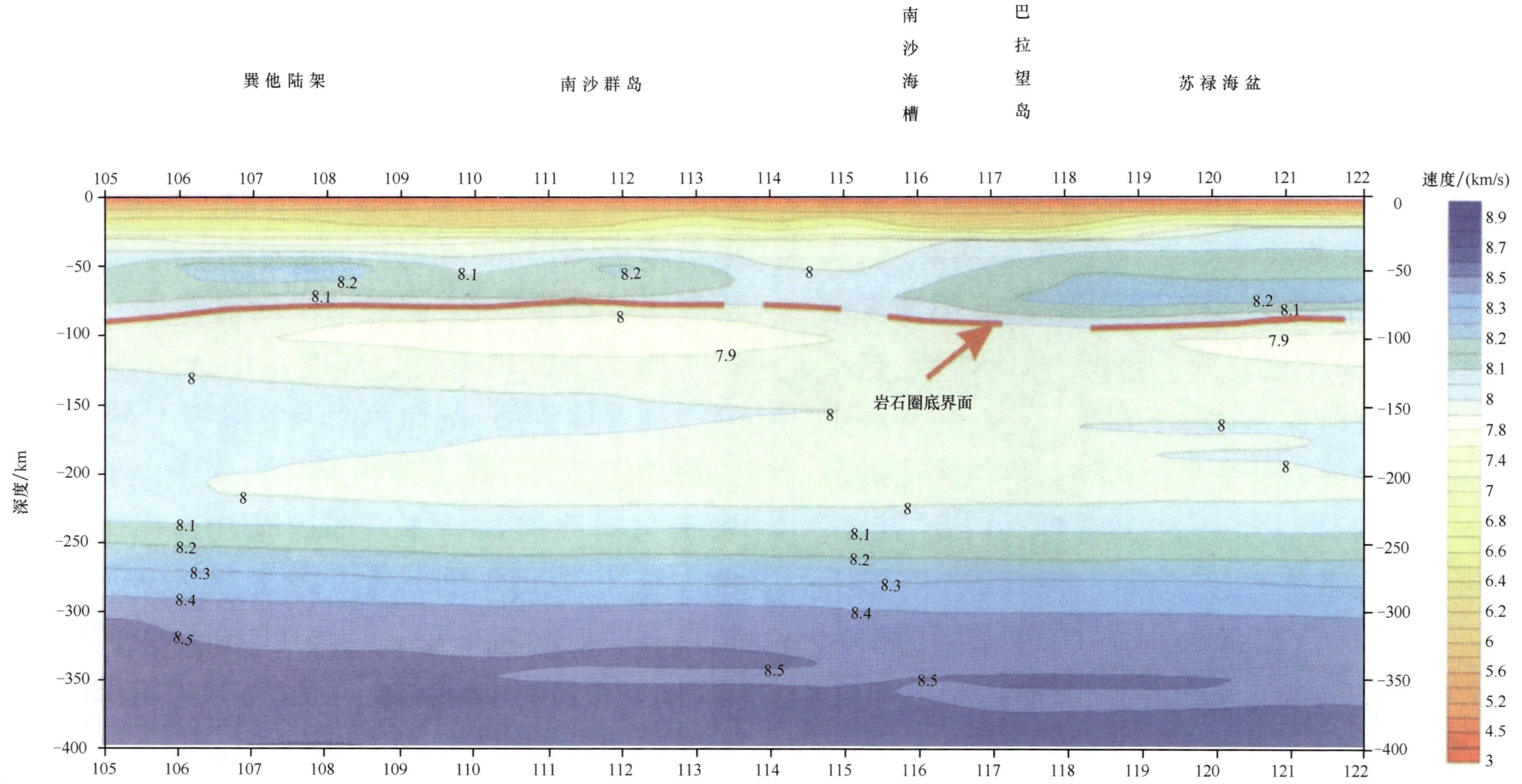

图 24-62　沿北纬 8°穿过南海、深度 400 km 的速度剖面

Fig. 24-62　Velocity section crossing the South China Sea along N 8°, at a depth of 400 km

（据姚伯初等，2010）

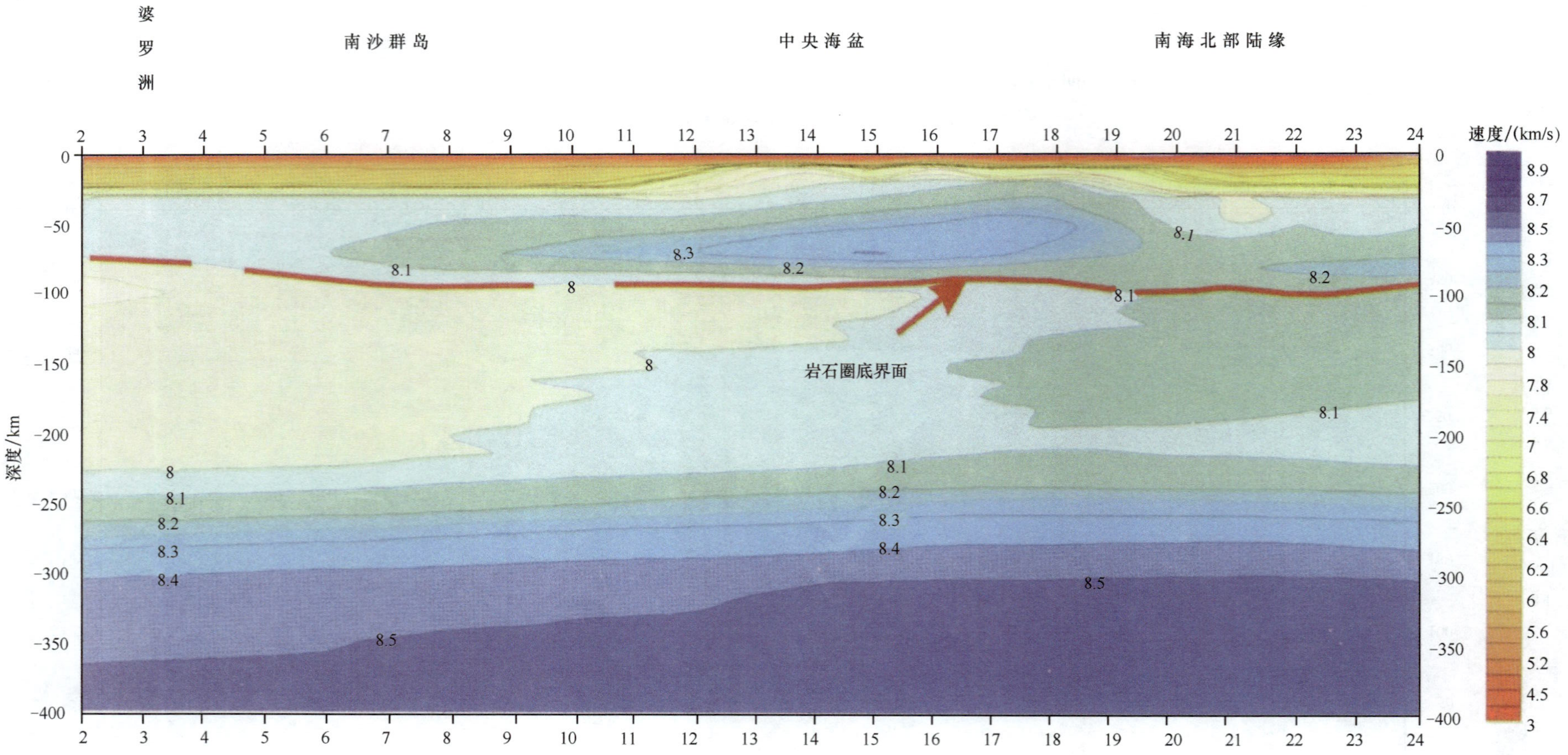

图 24-63　沿北纬 117°、深度 400 km 穿过南海的 P 波的速度结构剖面

Fig. 24-63　Section of the P-wave velocity structure crossing the South China Sea along N117°, at a depth of 400m

（据姚伯初等，2010）

楔发育而成的陆地，南沙海槽的俯冲还未完全结束，南沙地块与沙巴还未碰撞，因此，沙巴地区还未形成造山带，故其岩石圈还比较薄，厚约75km。在这段剖面之下，软流层速度为7.9~8.4km/s，厚度190~205km。

6）图24-64表示沿114°E、深度400km的P波速度结构剖面。剖面北起广东珠江口流域，经南海北部陆架、陆坡上的西沙海槽（17°40′N）与盆西海岭、南海西南海盆（11°40′~14°20′N）、南沙群岛的永暑礁、尹庆群礁、安波沙洲、北康暗沙与南康暗沙、曾母暗沙，至加里曼丹的尹兰山脉。剖面北部广东地区的地壳厚30km，上、中地壳与下地壳的厚度相当；岩石圈厚90~95km。在南海北部陆架上，地壳厚度约28km；在南西北海盆，岩石圈下部47~70km深处存在一高速层，其速度为8.2~8.3km/s，岩石圈厚约90km。在盆西海岭地区，地壳厚度减薄，在15~24km；在西南海盆中，地壳明显减薄，莫霍面深度为10km；这里岩石圈厚80~90km。在南沙群岛地区，地壳厚23km左右，上、中地壳厚，下地壳薄。进入加里曼丹岛，地壳加厚至28km，岩石圈厚70km。岩石圈中，速度随深度增加，没有异常存在。在地质构造上，这里是拉姜群沉积区，为增生楔复理石沉积。这里地壳和岩石圈厚度不太厚，可能是这里碰撞造山带还未达到高潮，山不高，因为地壳与岩石圈厚度还未增厚之故（姚伯初等，2010）。

2. 平面特征

1）图24-65a表示深度为10km切面的P波速度结构平面分布图。由图可见，在华南、中南半岛、加里曼丹、吕宋、巴拉望和台湾等陆区，以及南海广阔陆架和陆坡区，P波速度大都在5.5km/s至6.1km/s之间，说明10km深度仅达上地壳下部；而北部湾及红河地区、加里曼丹岛的东北角、台湾中北部，这里的速度小于或等于6.0km/s，与这些地区的构造结构有关。如沙巴地区是一增生楔，到晚中新世时中生代南海洋壳沿南沙海槽的俯冲活动才停止，这里10km深度以上可能是变形的深海沉积，因此速度较低。台湾岛是年轻的造山带，10km深处的速度较低可能是造山带中深海复理石沉积的速度较低所致。在南海海盆，速度在6.5km/s至7.9km/s之间。因为在南海这样年轻的洋盆中，水深较浅，在3500~4400m之间，地壳厚5~6km，因此，在有些地区，10km深度已达莫霍面；有些地区则达洋壳层3的下部，故有上述速度之分布。

2）图24-65b表示深度为20km切面的P波速度结构平面分布图。由图可见，在华南、中南半岛、加里曼丹、巴拉望和吕宋等陆地，以及广大陆架与陆坡区，P波速度在6.2km/s至6.7km/s之间，说明在这些地区，此深度处于下地壳之中。但在红河流域、北部湾、海南岛西南部、西沙-中沙群岛地区，存在一北西向的低速带，速度为5.7~6.0km/s。向东南，这一低速带被南海西南海盆所切断，主要原因是，由于在南海新生代洋盆中，速度已达7.6~8.0km/s，显然已处于上地幔软流层中。苏禄海盆中也有类似情况。值得注意的是，在112°~116°E，6°~9°N地区，出现一U形地带，速度达6.2~7.6km/s，似乎达到上地幔；在穿过此处的多道地震剖面上，显示这里存在厚度大于15km的新生代沉积。利用重力资料计算该处地壳厚约22km，推测此处曾是一水深4000~5000m的小洋盆，后被沉积充填，洋壳基底由于沉积负载作用而下沉至今日观测的深度。

3）图24-65c表示深度为30km切面的P波速度结构平面分布图。由图看出，在华南、印支昆嵩地块、台湾等处的速度在6.8km/s至7.0km/s之间，处于下地壳的下部；而其他地区，速度在7.8~8.1km/s之间，显然已达岩石圈上地幔之中。从图中速度分布看，从红河流域低速带到北部湾后被一北东向、速度为7.9~8.0km/s的速度带所阻断。在华南出现北东向速度带，而在中南半岛中部（14°N）出现东西向速度带，往南逐渐转为北东向。在台湾岛，速度等值线呈北北东走向，与区域构造走向一致。

4）图24-65d表示深度为40km切面的P波速度结构平面分布图。由图看出，从红河流域、莺歌海、中沙—西沙块体，到巴拉巴克海峡、苏禄海盆，存在一条宽度为400~800km的低速带，速度为7.4~7.6km/s。研究区的其他地区，速度较高，为7.8~8.1km/s，很显然是在上地幔中。值得一提的是，在加里曼丹岛的东北部，出现7.6km/s的岩石圈层，分析可能由于这里是一年轻造山带-沙巴造山带，增生楔沉积还未遭受强烈挤压，变形程度不强，因此速度不高。

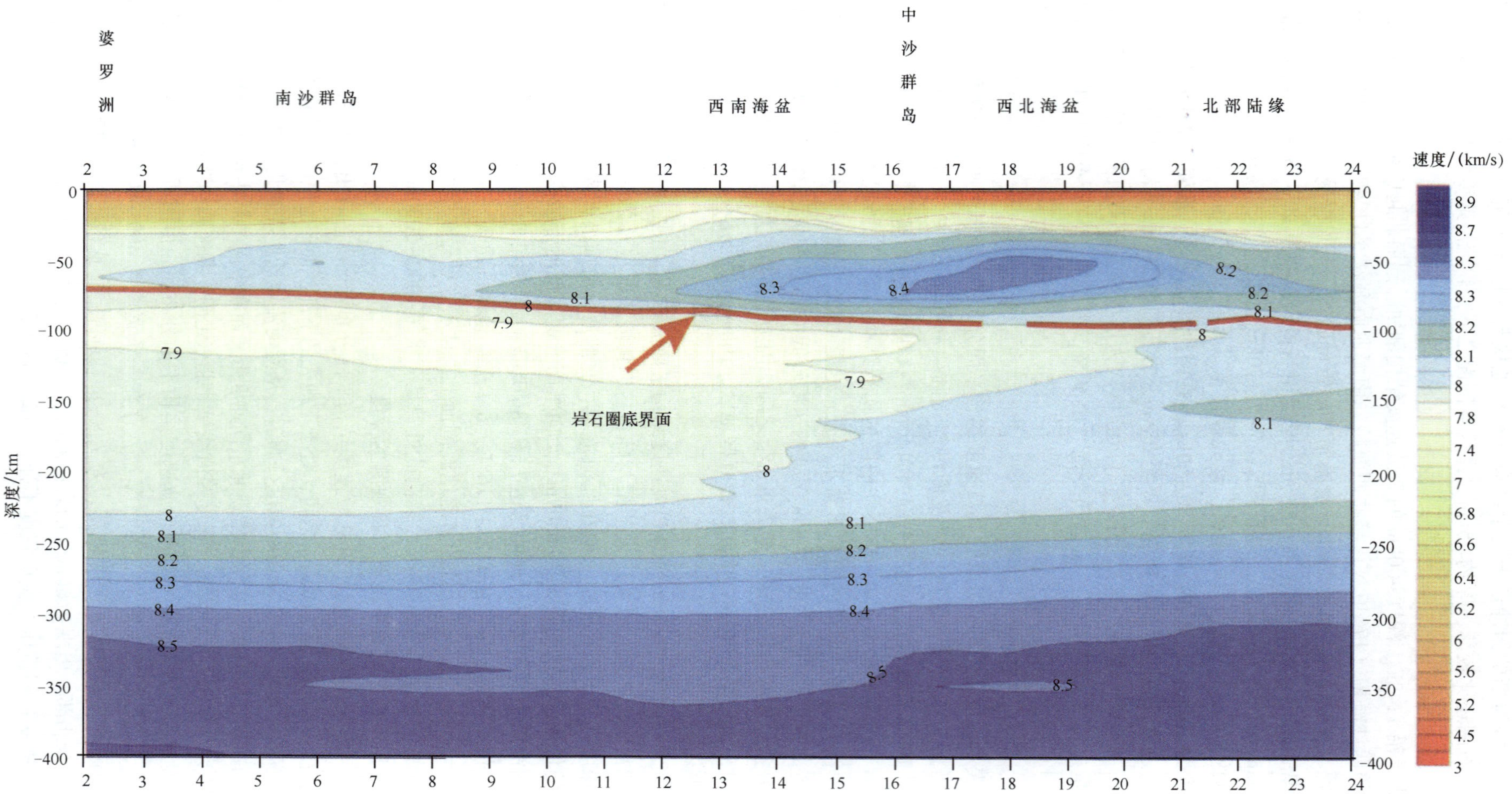

图 24-64　沿北纬 114°E、深度 400 km 穿过南海的 P 波的速度结构剖面

Fig. 24-64　Section of the P-wave velocity structure crossing the South China Sea along 114°E, at a depth of 400m

（据姚伯初等，2010）

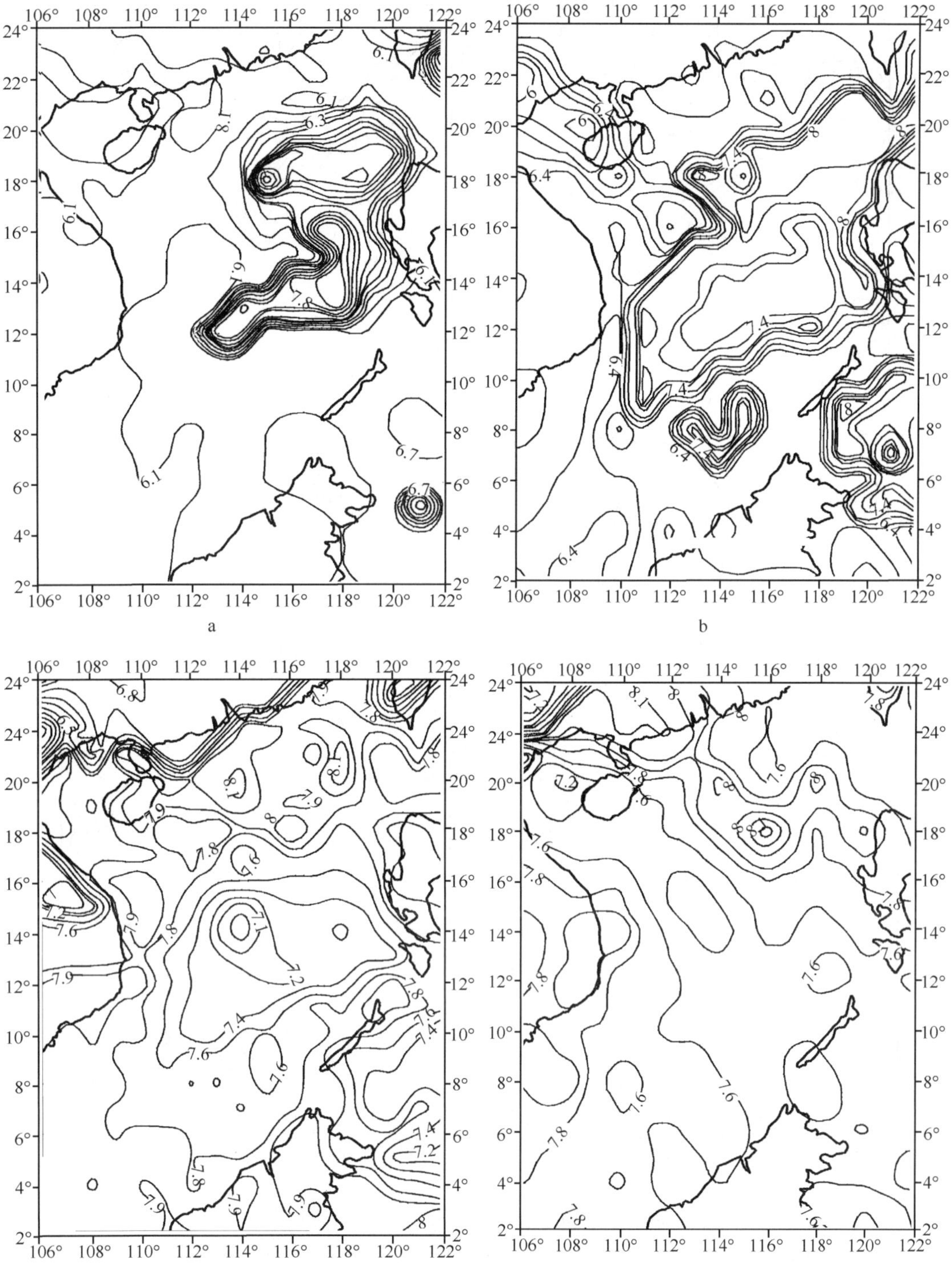

a　　b

c　　d

图 24-65　南海地震纵波速度（km/s）平面图

Fig. 24-65　Plan of seismic p-wave velcity（km/s）of the South China Sea

a—10km；b—20km；c—30km；d—40km

5）图 24-66a 表示深度为 50km 切面的 P 波速度结构平面分布图。由图看出，速度等值线以北西向为主，夹有北东向和南北走向的等值线。从粤东向吕宋岛，存在一断续的 8.1km/s 速度带；往西从粤西经珠江口、南海中央海盆北部、礼乐滩、巴拉望岛到苏禄海，存在一北西向高速带，速度在 8.2km/s 至 8.6km/s 之间；再往西从红河流域经北部湾、莺歌海、中沙西沙块体、南海西南海盆、南

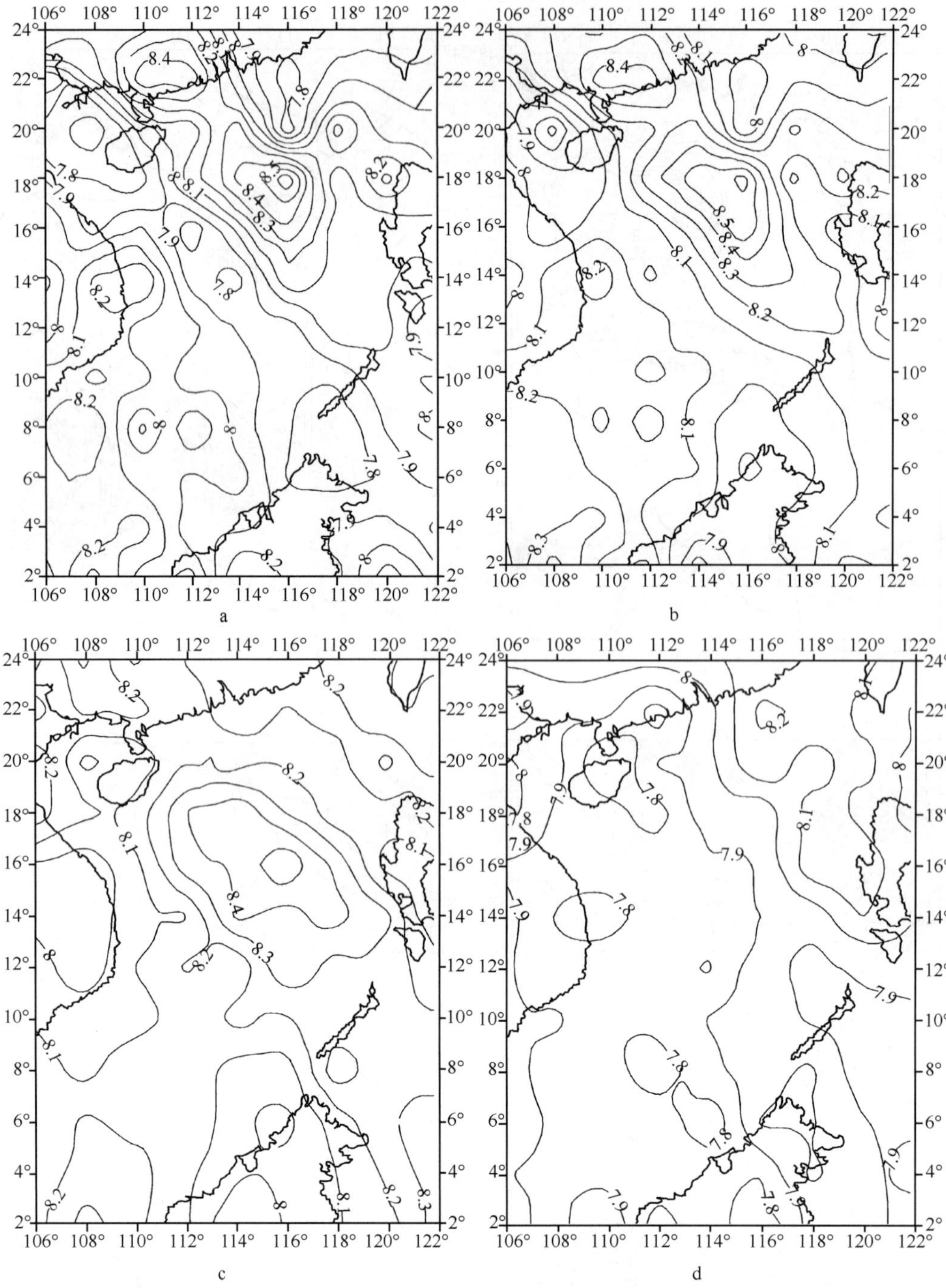

图 24-66　南海地震纵波速度（km/s）平面图

Fig. 24-66　Plan of seismic p-wave velcity（km/s）of the South China Sea

a—50km；b—60km；c—70km；d—100km

沙群岛、巴拉望岛、到苏禄海，存在一北西向低速带，速度为 7.8～8.0km/s。在中南半岛中部存在一北东向低速带，速度为 7.8～8.0km/s；在婆罗洲中部（113°～116°E）出现一南北向低速带，速度小于 8.0km/s。在南海地区，50km 深度还是在岩石圈内，速度等值线走向反映区域构造走向。从整个平面看，北西向速度等值线切割北东或南北向等值线，说明北西向构造活动时间较晚。

6）图 24-66b 表示深度为 60km 切面的 P 波速度结构平面分布图。由图看出，这里速度结构呈北西向分布，与 50km 深度切面一样，中南半岛中部存在一北东向速度带和婆罗洲存在一南北向速度带。从粤西陆地经珠江口、南海中央海盆北部、南沙礼乐滩、巴拉望岛至苏禄海，存在一高速带，速度为

8.2~8.5km/s，此带西部，从红河流域经北部湾、莺歌海、中沙-西沙块体、南海西南海盆、南沙群岛、巴拉巴克海峡至苏拉威西海，存在一低速带，速度为7.9~8.1km/s。在巽他陆架上，存在一南北向高速带，速度为8.2~8.5km/s。由整个平面图看出，60km深度切面速度结构在整体上呈北西向分布，从广东莲花山、广东东部陆架至巴拉巴克海峡、东加里曼丹岛的南北向低速带被粤西陆地、南海海盆至东北巴拉望岛北西向高速带所切断。说明南北向构造的形成时代在先，北西向构造形成时代较晚。

7）图24-66c表示深度为70km切面的P波速度结构平面分布图。由图看出，速度结构总体上呈北西向分布，西部红河地区、中南半岛至加里曼丹岛，有一宽度达400km的低速带，其中包含两个南北向低速带：中南半岛东部和加里曼丹岛，速度在8.0km/s至8.1km/s之间。中部有一高速带，从粤西陆地、南海海盆、苏禄海至苏拉威西海，速度在8.2km/s至8.6km/s之间。再往东，出现3个北西向的速度带：广东东部陆架、东沙群岛海域至吕宋岛低速带（速度为8.0~8.2km/s），粤东陆地、陆架至巴士海峡高速带（速度为8.2~8.3km/s）和福建、台湾海峡至台湾低速带（速度为8.0~8.2km/s），这些带的宽度为100~200km。

8）图24-66d表示深度为100km切面的P波速度结构平面分布图。由图看出，整个图面的速度较均匀，在7.8~8.2km/s之间，切面西部和中部的速度稍低，为7.6~7.9km/s，已处于软流层中；而东北部粤东至南海中央海盆东北部的速度稍高，为8.05~8.2km/s。从整个平面看，北西向构造走向仍然存在，其他方向不太清楚。从整个切面速度分布看，该地区已进入软流层，速度偏低，但北西向构造痕迹仍可分辨出来，说明北西向构造已影响岩石圈之下的软流层。

综上所述，南海岩石圈三维结构上存在两个明显特征：

第一，存在北西向速度结构。在20km深度切面上，从红河流域到海南岛、西沙-中沙群岛地区，到南部巴拉望岛和加里曼丹岛之间。存在一速度7.4~7.6km/s的低速带，走向为北西向.在整个南海海盆地区，速度为7.6~8.0km/s，为高速区。在40km深度切面上，北从广东省雷州湾（110°E，24°N），向东南方向穿过南海和苏禄海至苏拉威西海（121°E，4°N），存在一条北西向低速带，纵波速度为7.4~7.6km/s（横波速度为3.34~3.40km/s）；而在其西南部，在印支半岛经巽它陆架至加里曼丹岛，速度大于7.6km/s（横波速度为3.30km/s）。在东北部，从广东陆地经珠江口外的西南部，在19°N处折向东，在向东南直至吕宋岛，存在一速度为7.8~8.0km/s的高速带。在50km深度上，东部高速带速度稍有增加，纵波速度为8.15~8.6km/s（横波为3.34~3.42km/s）；再其西南部出现一条低速带，北从红河流域，向东南方向穿过南海到婆罗洲的沙巴地区，纵波速度为7.95~8.0km/s。在60km至100km深度上，这种结构特征依然存在。南海岩石圈中存在这一结构特征，首先，从区域构造上分析，这里存在3条主要断裂：红河-莺歌海断裂、南海西缘断裂和西沙海槽断裂。红河-莺歌海断裂是北西走向断裂，南海岩石圈中北西向低速带即沿此断裂展布。该断裂到海南岛南部后转为南北走向的南海西缘断裂和东西走向的西沙海槽断裂，但岩石圈深部低速带则沿北西向穿过整个南海。在南海地区，中生代末至新生代早期发生了神狐构造运动，运动走向为北东—南西，拉张方向为北西-南东。此次运动的西部边界为红河-莺歌海断裂，和南海西缘断裂，在此断裂的东北部，拉张作用强烈，其西南部拉张作用则相对较弱。因此，在神狐运动中，红河-莺歌海断裂为右旋走滑断裂。在此后（30~20Ma前）印支地块发生右旋转动时，它变为左旋走滑断裂。因此，南海岩石圈中的北西向构造可能和神狐运动及以后印支地块的左旋转动密切相关。在神狐运动中，南海地区沿北西—南东向拉张，其西部边界——红河-莺歌海断裂和南海西缘断裂，向东南方向穿过南海至沙巴方向遭受强烈拉伸，岩石圈被拉开，上地幔因减压而发生局部熔融，沿北西向减薄了上覆岩石圈的厚度；熔融的岩浆沿断裂上升到岩石圈中，在岩石圈中形成北西向低速带。在此带的东部，由于岩石圈拉张作用没有沿红河-莺歌海断裂带强烈，因此，岩石圈中表现为相对高速带。当印度板块和欧亚板块碰撞后，亚洲大陆的上地幔中产生了向东南方向的流动，这一流动在上地幔和岩石圈下部产生局部熔融，因此出现北西向低速带。由于上地幔流动在平面上是不均匀的，因此，出现北西向低速和高速带相间的景观。由此可见，南海地区岩石圈及上地幔中存在北西向构造，是新生代南海构造运动的遗迹，至今还未消失。这也说明，在一次构造运动之后，它在岩石圈及上地幔中的影响在相当长的时间内还存在，南海地区发生在中生代末至新生代早期的神狐运动和发生在始新世末的南海运动，至今仍在岩石圈及上地幔中还可找到构造遗迹。

第二，南海岩石圈厚度变化存在明显差异。在南海洋盆的岩石圈底部存在高速岩石层，而且洋盆下的岩石圈之厚度比大陆边缘厚，在海盆岩石圈下部的60~80km深度上存在一高速层，速度为8.2~8.3km/s。无论是中央海盆还是西北海盆与西南海盆，其下部岩石圈中均存在一高速岩石层。而且，南海洋盆岩石圈的厚度大于陆缘及华南与加里曼丹岛的岩石圈厚度。南海边缘海洋盆存在这种岩石圈结构，目前认为是南海新生代海底扩张及其构造演化所造成的。根据Taylor和Hayes（1980）关于南海新生代海底扩张历史的讨论，南海在新生代海底扩张过程中。扩张中心原来位于北纬20°的今日北部陆缘与海盆交界（COB）附近，随着海底扩张的继续进行。南部的礼乐-东北巴拉望地块不断向南运动，扩张中心也不断南移，从而在扩张中心的南北两边不断产生新洋壳，形成了南海的南北对称式扩张。如果扩张中心不移动，那就不能形成对称式扩张。另一方面从地震层析资料看，南海洋盆之下不存在地幔柱，表明在南海海底扩张时不可能从上地幔岩浆囊中得到生成大洋岩石圈的岩浆。那么生成南海大洋岩石圈的岩浆从那里来呢？只能是从上地幔部分熔融的岩浆中得到。可以设想，在南海发生海底扩张时，大陆岩石圈被拉开，上地幔软流圈的部分熔融由于减压而增加，部分熔融的岩浆不断上涌，在海底形成新的大洋岩石圈；由于不是从地幔柱岩浆囊中上升的岩浆，其上升速度较慢，因此形成大洋岩石圈的速率也低。这与南海海底扩张是慢速扩张（扩张速率约为2.5 cm/a）是一致的。由软流层上部的部分熔融岩浆上升而形成了新的大洋岩石圈，余下的超基性岩石之地震波速度较高，这一高速层加入到大洋岩石圈的底部，形成了大洋岩石圈底部的高速岩石圈层，也加厚了大洋岩石圈。这意味着南海洋盆的岩石圈之下的软流层之厚度变薄，与今日观测的结果一致（姚伯初等，2010）。

第二十五章 区域地球化学场特征

第一节 岩石地球化学场特征

根据海南省地质矿产勘查开发局（2001）的统计结果，海南岛岩石中元素背景平均值高于地壳克拉克值的有Bi、Sn、Pb、As、Th、La、Fe_2O_3；B与地壳克拉克值持平，其余均低于地壳克拉克值。水系沉积物中各种元素的含量与相应的岩石含量比较，除CaO、Na_2O、MgO等氧化物，以及F、Sr、Be、Sb等贫化外，其余大多数元素呈现不同程度的富集，其中，B、Ti、W、Zr、Hg富集比达2.1～5.87倍。

根据海南岛地球化学元素的分布规律和表现特点，结合海南岛地质构造特征，将海南岛划分为三亚地球化学区、五指山地球化学区、雷琼地球化学区。

一、三亚地球化学区

位于九所-陵水断裂带以南的海南岛南部陆地地区。该区主要分布华力西期晚期、印支期、燕山早期罗蓬岩体和晚期北山岩体二长花岗岩，燕山期晚期牛腊岭和同安岭英安岩、流纹斑岩，寒武纪石英砂岩、页岩、白云质灰岩、硅质岩、磷块岩、奥陶纪页岩、砂岩、砾岩。地球化学场特征表现为Au、Ag、Hg、Pb、Fe、Cd、Sn、Bi、P、Mn等背景场。主要成矿元素异常面积较小，与之对应分布的矿产，在同安岭火山岩中分布有雄岭、富田、六罗村等与火山岩成矿有关金矿点，中寒武统大茅组分布有化学沉积型磷锰矿床，在田独地区寒武系与燕山期二长花岗岩内接触带分布有矽卡岩型田独磁铁矿床。

二、五指山地球化学区

位于九所-陵水断裂带以北至王五-文教断裂带以南广大地区。根据元素分布规律，结合地质构造特点，以白沙拗陷带为界，进一步划分为五指山隆起地球化学分区、抱板隆起地球化学区和白沙断陷地球化学分区。

（一）五指山隆起地球化学分区

位于白沙拗陷带东侧，王五-文教断裂带东段以南和九所-陵水断裂带东段以北之间的海南岛东部地区。本区广泛分布中元古代片麻状花岗岩，华力西期晚期、印支期、燕山期二长花岗岩，印支期和燕山早期钾长花岗岩，燕山期早期辉长岩、花岗岩和燕山期晚期花岗闪长岩，燕山期晚期五指山中酸性火山岩和同安岭基性-酸性火山岩，喜马拉雅期蓬莱地区基性火山岩；在琼中县长征、定安岭口和琼海市重兴等地区分布长城系抱板群，在保亭南好地区分布志留系、石炭系，在琼海市至万宁市一带分布寒武系、奥陶系、志留系、石炭系、三叠系、白垩系等，在定安县雷鸣地区分布白垩系。区内地球化学场特征表现为Zn、Sr、Na_2O、K_2O、U、Th、La等高背景场；Ba、Sn、SiO_2、Al_2O_3、CaO、Nb、Ta、Pb呈正常背景场；Hg、V、Mn、Bi、Au、Co、Li、Cr、W、Ag、MgO、Zn、F、Mo、Fe_2O_3、Y、Be、P等呈低背景场；Sb、As、B、Ca、Ni呈甚低背景场。在本区内分布的一些大岩体也同样具有明显的地球化学特征，如阜堡笔钾长花岗岩体以K_2O、Zr、Th、Nb、Y、Pb等呈高背景场为特征，千家二长花岗岩体以Na_2O、CaO、Sr等高背景场为特征，吊罗山钾长花岗岩体以Al_2O_3、Th、Pb、Bi、Mo、W等高背景场为特征，保城花岗闪长岩体以Al_2O_3高背景场为特征。

（二）白沙拗陷带地球化学分区

位于儋州市兰洋-白沙县-乐东县一带，区内主要分布白垩系、石炭系和二叠系。地球化学场特征为 Au、Ag、Sb、Cd、Cu、Pb、As、Hg 等高背景场。

（三）抱板隆起地球化学区

位于白沙拗陷带西侧，王五-文教断裂带西段以南至九所-陵水断裂带西段以北之间的海南岛西部地区。区内广泛分布中元古代二甲片麻状花岗岩体，邦溪二长花岗岩、石岭花岗闪长岩，以及一些闪长岩和石英闪长岩，印支期的尖峰岭和黑岭钾长花岗岩、大田二长花岗岩、广坝花岗闪长岩、雅加和大王岭二长花岗岩，华力西期—印支期儋县花岗岩，燕山期早期的马鞍岭、大岭、丰猛岭及晚期的三狮岭等二长花岗岩以及一些钾长花岗岩。还出露长城系抱板群、青白口系石碌群、震旦系石灰顶组、奥陶系、志留系、石炭系、二叠系、白垩系等。区内地球化学场特征表现为 Na_2O、Sr、K_2O、Zr、U、Sn、Ba、Th、SiO_2 等呈高背景场，La、Pb、Ti、Nb、Al_2O_3等呈正常背景场，Sb、Hg、As、Ni、Ca、Au、Cr、Cu、V、Zn、Mn、Mo、Fe_2O_3、Co、MgO、P、Li、B、Ag、F、Bi、Be、W、CaO、Y 等呈低背景场。同时，可看出大部分元素的高背景场往往是某些地质体的表现，例如，雅加二长花岗岩体以 Zr、U、Th、Nb、Sr、Ba 以及 Be、W、Y 等高背景场为特征，尖峰岭钾长花岗岩以 K_2O、La、Pb、W、Be、Sn、Th、U、Y、Nb、Mo、Bi、Au、Ag、Zn 呈高背景场特征，大岭二长花岗岩以 K_2O、Th、La、Nb、Ag 高背景场，丰猛二长花岗岩以 K_2O、U 为高背景场，大王岭二长花岗岩以 Nb 为高背景场，儋县花岗岩体以 Sr、Ba 为高背景场。白垩系中 Zr、K_2O、U、Th、Sn、Bi 均呈低背景场，奥陶系和志留系以 B 为高背景场特征。

三、雷琼地球化学区

位于王五-文教断裂带以北的海南岛北部地区，与琼北玄武岩类火山岩分布地区基本一致，其地球化学场特征表现为 Ti、Ni、Cr、Co、Nb、Fe_2O_3、Ca、V、P、Mo、Mn、Sn、Zr 等呈高背景场，As、K_2O、Sb、Bi、Cd、Li、F、Ba、U、Pb、Na_2O、MgO、Ag、Th、Be、W、Y、Hg、CaO、Au、B、Sr、La、SiO_2 等呈低背景场。主要金属成矿元素 Au、Ag、Cu、Pb、Zn、Sb 等表现为中低背景场，很少出现高背景场或异常。

第二节　水系沉积物区域地球化学特征

根据海南岛 1：20 万水系沉积物测量结果（江西地矿局物化探大队，1988），其地球化学特征如下。

一、区域地球化学背景特征

从海南岛岩石地球化学参数统计及水系沉积物参数统计中可以看出，本区主要金属成矿元素 Au、Ag、Cu、Pb、Zn 及其主要伴生元素 As、Sb，仅 Pb 和 As 的含量高于地壳克拉克值，比值为 1.31 和 1.32，其余元素均明显低于地壳克拉克值，比值小于 0.5，其中 Au 的含量比值仅为 0.09。水系沉积物中，除 Sb 略有贫化、比值为 0.95 外，其余元素均有富集，Ag 和 Au 的富集比分别达 1.77 和 1.83，As、Cu、Pb、Zn 的富集比为 1.21～1.29。As、Cu、Sb 在地层中的背景含量较高，在岩浆岩中的背景含量较低，Pb 和 Zn 则相反，Au 和 Ag 无明显规律。

二、元素在地层中的分布与分配特征

1）中元古界抱板群：岩石中 Au 的平均含量为 1.2×10^{-9}，Pb 的平均含量为 73×10^{-6}；水系沉积物中 Au 的平均含量为 2.37×10^{-6}，Pb 为 35.2×10^{-6}，在全岛各地层中含量最高，Au 一般高出一倍以上。其他元素一般高于全岛平均值，可以认为该地层是金的重要含矿层。

2）新元古界石碌群：该地层中 As、Cu 的含量和水系沉积物中的含量比其他地层高出 1.5 倍至 2

倍以上，Zn也有较高的含量。

3）寒武系：该地层Ag的含量比其他地层略高。水系沉积物中，As、Au、Cu、Sb等元素含量均明显偏低。

4）奥陶系：该地层As和Ag含量略高，而Pb、Cu、Sb等偏低，大多数元素在水系沉积物中富集很显著，Pb的富集比达4以上，Cu、Au、Zn的富集比达2以上。

5）奥陶系—志留系：该地层各种主要金属成矿元素的含量均属正常或偏低。Au、Ag、Zn在水系沉积物中有较明显富集，而Sb呈现贫化现象。

6）志留系：该地层岩石各种金属成矿元素含量在岛内一般正常，仅Sb略高。在水系沉积物中均有较明显的富集。

7）石炭系：该地层多数金属元素背景含量偏低。在水系沉积物中，除Au和Sb外，其余元素有明显富集。

8）二叠系：该地层仅Cu的含量略高，其他元素与岛内其他地层无明显差别。在水系沉积物中，As、Zn、Au和Cu富集比较明显，其余仅略有富集。

9）三叠系：仅在琼海九曲江和翰林地区有小面积出露，岩性为页岩、粉砂岩和砾岩，As、Pb含量较高，Ag、Au含量则偏低。在水系沉积物中，Au、Ag、Zn有明显富集，Cu、Pb略有富集，As、Sb贫化。

10）白垩系：该地层中As、Sb和Au含量偏低，Zn的含量较高。水系沉积物中Au和Pb有较显著富集，富集比为2.7和1.7，Zn和Sb贫化，贫化率为83%和73%。

11）古近系：该地层中Zn、Pb和Ag的含量较高，在水系沉积物中Zn、Pb、As、Ag、Sb均贫化，仅Cu和Au略有富集。

12）第四系：主要分布在沿海地带，主要为松散沙，其水系沉积物中，上述金属成矿元素的含量均明显降低。

三、水系沉积物区域地球化学场特征

根据各元素的分布规律和表现特点，结合地质构造特点，可将海南岛划分为两个地球化学区，即琼北地球化学区和琼中南地球化学区。各区地球化学特征如下：

（一）琼北地球化学区

包括王五-定安以北、东至铺前—文昌一线以西，从定安和文昌往南伸至琼海以北。与琼北玄武岩分布地区基本一致。其地球化学场特征也表现为Ni、Ti、Co、V、Fe、Al、Cu、Mo等元素的明显富集。而对于主要金属成矿元素Au、Ag、Pb以及As、Sb表现为中低背景场，很少出现高背景场或异常。

（二）琼中南地球化学区

本区包括了除琼北地球学区以外的所有地区，该区K、Na、Zr、Sr、Th、U、La、Ba表现为高背景场，F、Sn和Pb为中高背景场，主要反映以酸性岩浆岩为主的地球化学场特征。主要金属成矿元素Pb、Cu、Au、Ag、Zn呈正常或中低背景场，根据元素组合地质特点，大致以乐东-毛阳-黎母岭-屯昌为界，将本区分为两个亚区。

1. 屯昌-三亚地球化学亚区

广泛出露华力西期、华力西期—印支期花岗岩，另有燕山期晚期的中酸性岩体零散分布，局部形成较大岩体，如保城和屯昌花岗闪长岩岩体和钾长花岗岩岩体，以及二长花岗岩岩体，中南部还有面积较大的中酸性火山岩被。在琼海分布一片白垩纪红层和古生代地层，在屯昌东南面尚有小面积的抱板群出露，在保亭县南好和三亚市北部分布古生代地层，区内Pb、Zn背景场正常偏高，As、Sb背景场偏低。

2. 东方-儋州地球化学亚区

主要分布华力西期、华力西期—印支期、印支期、燕山期中酸性岩浆岩和白垩纪、石炭纪地层，尚有面积较大的中元古界抱板群，以及奥陶系、志留系，同时新元古界唯一出露的地区。其特点是主要金属成矿元素中低背景场分布不均匀，异常大小与分布也极不均匀。

四、水系沉积物区域构造地球化学特征

在地球化学图上，表现比较明显的是琼中南两个地球化学区分界线，许多元素如 Ni、Cr、Co、V、Ti、Cu、Fe_2O_3、P 等沿着该线南北呈现高低截然不同的背景场，该线正是公认的王五-文教断裂带，琼中南地球化学区中的两个亚区分界线，也与乐东-定安断裂带基本吻合，在东南侧的屯昌-三亚地球化学亚区上，La 和 Sb 呈现高背景，B 呈现低背景，而在西北侧的东方-儋州地球化学亚区则相反，沿该界线两侧 Th、Pb、U、As、Sb、Co 等也显示不同的背景分布。

第三节　海南岛土壤区域地球化学特征

海南岛大范围系统的土壤样测量由海南省地质调查院在 2004~2010 年开展《海南岛多目标区域地球化学调查》时完成，其地球化学特征如下：

一、区域地球化学背景特征

全岛土壤中元素平均值显著高于全国元素背景值的有 Sn、Br、Zr、N、I、U、Tl；略高于全国背景值的有 S、Cl、Th、Rb、Pb、SiO_2、Mo、Al_2O_3、Ge、Bi；略低于全国背景值的有 Au、Be、Ga、W、Ag、Ce、Hg、K_2O、Nb；其余 24 种元素显著低于全国背景值。在岛内，元素表层含量与深层元素含量比较，P、B、Br、S、N、C、Corg 元素呈现明显的富集，富集比达 2.0~3.8 倍。CaO、Sr、Na_2O 等元素淋失贫化，其中 CaO 的淋失比为 0.35。

二、区域地球化学场特征

根据海南岛表层土壤多目标区域地球化学调查各元素的含量水平、分布规律和表现特点，结合地质构造特点，可将海南岛划分为三个地球化学区，其中儋州-保亭地球化学区又分为两个亚区（图 25-1）。表层土壤中 53 项元素和指标在不同地球化学区和亚区中的原始含量水平及分布特征各有不同，结合地形地貌等因素，不同地球化学区和亚区的矿产资源种类也各不相同。

（一）峨蔓-蓬莱地球化学区

该区铁族元素、Ag、Cu、Hg、Mo、P、Sn、Zn、Ge、Ga、Br、I、N、C、S、Corg 和 Al_2O_3 的含量为全岛最高，碱（土）金属元素、B、Zr、Tl、SiO_2 的含量为全区最低；元素含量均服从正态分布或对称分布。在资源方面，Ag、Cu、Hg、Mo、Zn、S 等元素含量最高，但元素组合特征不明显，因此，形成火山岩型及其次生作用类型的贵金属或有色金属矿产可能性不大。但是，Al_2O_3 的含量为各地球化学区最高，标准差也最大，加上土壤在热带条件下强烈而又持久的脱硅富铝作用，有可能在局部形成铝土矿床。

（二）锦山-羊栏-雷鸣地球化学区

该区除 Cl、CaO 和 SiO_2 外，其他元素均为全岛各地球化学区中最低；元素含量分布大都不均匀。在资源方面，SiO_2 含量最高，可形成滨海石英砂矿；CaO 的最高含量可能来自滨海贝类生物的富集和残留。另外，由于砂矿类矿产资源工业要求品位远低于原生矿产，本区的 Zr 元素含量仅次于屯昌-保亭地球化学亚区，因此可以形成砂矿型锆英石矿。

（三）儋州-乐东地球化学亚区

该亚区元素平均含量 As、B、Bi、Cd、F、Li、Au、Sb、W、Y、Se、MgO 为全岛各区最高；元素

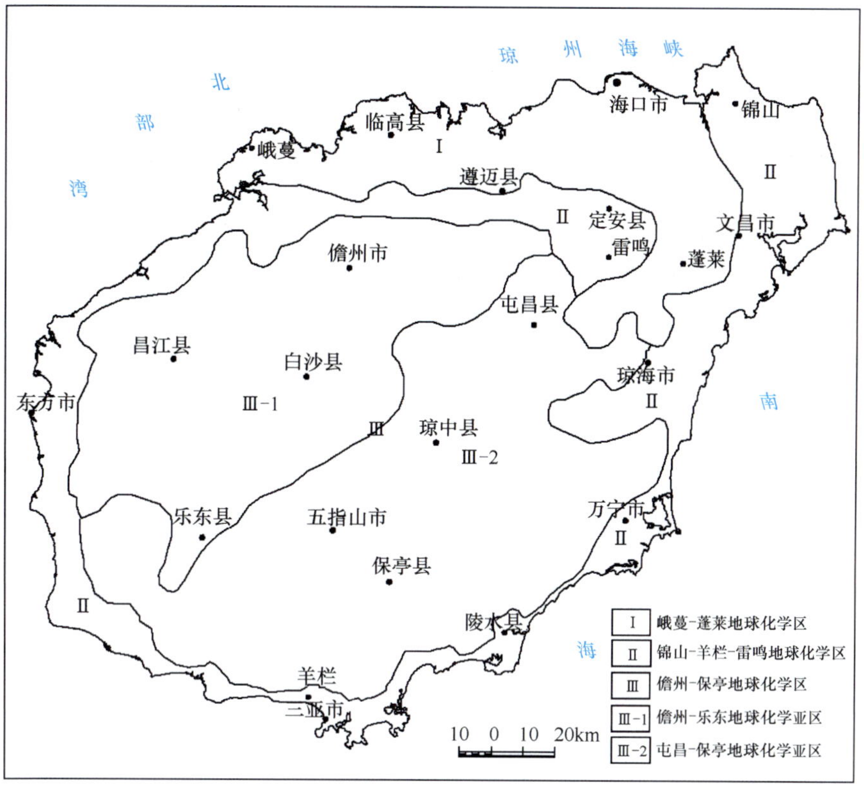

图 25-1 海南岛地球化学分区结果

Fig. 25-1 Results of geochemical division in the Hainan Island

含量分布的均匀性一般较好。在资源方面，Au、As、B、Bi、Cd、Sb、W 含量最高，元素组合特征显著，标准差也为全岛最大，可以在局部形成一定规模的金矿床或有色金属硫化物矿床。

（四）屯昌-保亭地球化学亚区

该亚区元素平均含量 Ba、Be、La、Pb、Rb、Sr、Th、U、Zr、Ce、Tl、Na_2O、K_2O 为全岛各区最高，Cl 为最低；大部分元素含量分布较均匀，较好地服从正态分布或对称分布。在资源方面，稀土元素和放射性元素含量最高，可能形成相关元素的矿产。其他与花岗岩类成矿作用有关的矿产元素（W、Sn、Mo 等）含量较高，变异系数最大，可以形成有规模的矿床。

三、地球化学成果在基础地质研究中的应用

1∶25 万多目标区域地球化学调查所获得的海量数据不仅为农业、环境和资源评价提供了基础资料，同时根据元素的含量差异、组合特征及异常的分布规律等地球化学特征还可以进行某些特定地层的划分、构造形迹的推断解释和大地构造单元的识别等，为基础地质研究提供帮助。

1. 区域地球化学成果为火山岩的进一步精确划分提供帮助

母岩作为风化成土重要的因素之一，在很大程度上制约着风化最终产物——土壤的地球化学特征，因此土壤在一定程度上反映母岩的特征。海南岛火山岩的分布面积约 4627km^2，主要分布于琼北地区。作为海南岛地质历史的重要标志，琼北火山岩在地球化学图上有很好地反映，Ni、Co、Cr、Sc 等在本区呈高背景分布，K_2O、SiO_2、N、Ba、Rb 等在本区呈低背景分布，高低变带分界明显，可作为进一步划分琼北火山岩与其他地质体接触界线的依据。

2. 区域地球化学成果为识别构造形迹提供依据

通常位于较大地质和地球化学单元接触部位的深大断裂除了可以通过地质、地球物理、遥感等方法识别以外，利用区域地球化学成果也可以达到很好地识别效果。如王五-文教断裂带，它是雷琼断陷

与五指山褶皱带的分界线，控制着沿构造带分布的中生代盆地的形成及盆地的沉积作用、岩浆侵入和喷发作用。在地球化学图上，C、S、N、P、Cu、Zn、Mo、Co、Fe_2O_3、Al_2O_3等在断裂带两侧呈东西展布的高、低突变带；SiO_2 在断裂带定安以西地区呈高背景分布。定安以东地区，由于火山岩和滨海砂覆盖，地球化学信息显得不清晰。所反映的信息与水系沉积物地球化学成果一致。

3. 区域地球化学成果为全面认识大地构造单元提供帮助

传统的地质单元绘制基本以路线填图的方式进行，路线之间的内容依靠推测。使用多目标地球化学调查方法获取的数据，覆盖范围大，无遗漏，可以很好地识别构造单元，特别是大地构造单元。在土壤地球化学分区图上，对海南岛的两个主要大地构造单元南海地台和华南褶皱系做了比较清晰的识别，特别是二级构造单元五指山褶皱带、雷琼断陷与三亚台缘拗陷带和三级构造单元五指山隆起区、白沙拗陷带、抱板隆起区反映较好。

第七篇　地质资源

概　　述

地质资源目前尚没有一个统一的概念性定义。但顾名思义，应指与地质作用有关的、并能可供人类利用的物质。随着今后的科技进步，对地质资源的理解也将有所不同。根据人类实际需求，本次初步把地质资源划分为地质矿产资源和地质景观资源两大类。地质矿产资源为人类的实际生产生活服务，满足人的物质需求；而地质景观资源则更多地体现在其观赏性上，主要是满足人的精神需要。

第二十六章 矿产资源

第一节 概 述

海南省矿产资源矿种类型多样，截至2010年底，已发现的各类矿产（包括水资源及海域油气矿产）有煤、石油、天然气、油页岩、铁、锰、钛、铬、铜、铅、锌、铝、镍、钴、钨、锡、钼、锑矿、金、银、铌、钽、锆、铪、镓、镉、稀土元素、铍、铷、钍、铀、石灰岩、白云岩、石英岩、脉石英、耐火黏土、萤石、硫铁矿、重晶石、泥炭、磷、水晶、云母、沸石、宝石（含蓝宝石、红宝石、红锆宝石）、石英砂、硅藻土、水泥配料用页岩、高岭土、膨润土、黏土、饰面花岗岩、大理岩、硅灰石、建筑用玄武岩、地下水、地下热水、饮用天然矿泉水等88种。其中燃料矿产25处，黑色金属矿产63处，有色金属矿产31处，贵金属矿产20处，稀有、稀土、分散元素矿产66处，放射性矿产1处，非金属矿产245处。大型矿床70处，中型矿床126处，小型矿床255处。

按矿床形成的地质作用，海南省矿床可归为四个类型。一是与沉积作用有关的成矿类型，二是与岩浆作用有关的成矿类型，三是与变质作用有关的成矿类型，四是与风化作用有关。沉积型矿有磷、锰、铀、钛、锆、石油、天然气、褐煤、油页岩、石英砂、硅藻土、水泥配料用页岩、泥炭、高岭土、沸石、膨润土等，岩浆型矿有金、银、铌、钽、铁、铜、铅锌、萤石、水晶、钨、锡、钼等，变质型矿有石墨、白云母矿、铁、钴、铜矿等，风化型矿有铪、镓、镉、稀土元素、铍、铷、钍、铝、镍、钴、宝石等（表26-1）。

表26-1 海南岛主要地质矿产资源成矿作用表

Table 26-1 Main mineralization processes of geological mineral resources in the Hainan Island

成因类型	矿床型式	含矿层时代及主要岩性	典型矿床	成矿时代
沉积型	新竹式铀矿	白垩纪砂砾岩、长石石英砂岩、泥岩	定安新竹铀矿	白垩纪
沉积型	长昌式褐煤	古近纪泥岩、粉细砂岩、炭质页岩、褐煤	长昌煤矿	古近纪
沉积型	长坡式褐煤、油页岩	古近纪—新近纪泥岩、粉细砂岩、砂砾岩、油页岩、褐煤	长坡煤矿、油页岩	古近纪—新近纪
沉积型	大茅式磷、锰矿	黔东世—武陵世灰岩、磷块岩、硅质岩、石英砂岩	大茅磷锰矿	黔东世—武陵世
沉积型	金风油气田	古近纪杂色泥岩、粉砂岩、砂岩及砂砾岩	金风油气田、花场油气田	古近纪
火山沉积型	金牛岭式沸石、膨润土	上新世玻屑沉凝灰岩、玄武岩、沉火山碎屑岩	海口市金牛岭沸石膨润土	上新世
沉积砂矿型	铺前锆英石砂矿	全新世烟墩组中细砂层	铺前锆英石砂矿、沙老钛铁矿、锆英石砂矿	全新世
沉积砂矿型	保定钛铁矿、锆英石、独居石砂矿	全新统烟墩组中细砂层	保定钛铁矿、锆英石、独居石砂矿	全新世
沉积砂矿型	长安钛铁矿锆英石砂矿	全新统烟墩组中细砂层	长安钛铁矿、锆英石砂矿	全新世

续表

成因类型	矿床型式	含矿层时代及主要岩性	典型矿床	成矿时代
沉积砂矿型	新隆、八所石英砂矿	晚更新世八所组中细砂层	新隆、八所、龙马石英砂	晚更新世
沉积砂矿型	那大砂锡矿	第四纪冲积砂砾层	那大砂锡矿	第四纪
沉积砂矿型	白沙元门砂金矿	第四纪冲积砂砾层	元门砂金矿	第四纪
沉积砂矿型	羊角岭水晶砂	第四纪冲积砂砾层	羊角岭水晶砂	第四纪
接触交代型	田独式铁矿	黔东世—武陵世白云质灰岩与白垩纪花岗斑岩接触带	田独铁矿	白垩纪
接触交代型	荔枝沟式锌矿	奥陶纪粉砂岩、砂岩、灰岩与白垩纪花岗斑岩接触带	荔枝沟锌矿、南下岭锡铅锌矿	白垩纪
接触交代型	兰洋式钨锡矿，南报式金银矿	石炭纪砂岩、板岩、结晶灰岩和印支期二长花岗岩接触带	兰洋钨锡矿、南报金银矿	印支期
接触交代型	振海山-摩天岭铜矿、硫铁矿	早志留世千枚岩，绢云母粉砂岩，早石炭世板岩，绢云母粉砂岩，燕山期晚期石英斑岩、花岗斑岩、闪长岩、辉长岩	振海山-摩天岭铜矿	白垩纪
接触交代型	红石工铁矿、水晶矿	早志留世千枚岩、白云质大理岩及绢云母粉砂岩，印支期花岗岩、燕山期晚期花岗斑岩、石英斑岩	红石铁矿、水晶矿	白垩纪
接触交代型	南好五〇八铁铜矿	早石炭世板岩、绢云母粉砂岩夹灰岩，燕山期晚期花岗斑岩、石英斑岩	南好五〇八铁铜矿	白垩纪
火山岩型	岭壳式铜矿	早白垩世流纹质火山岩夹英安质火山岩	岭壳铜矿	白垩纪
火山岩型	汤他大岭式铁矿	早白垩世英安质与流纹质火山岩互层	汤他大岭铁矿	白垩纪
火山岩型	雄岭式金矿	早白垩世流纹质含砾晶屑凝灰熔岩、安山质凝灰岩	雄岭金矿	白垩纪
热液型	富文式金矿	早白垩世砂砾岩、砂岩、粉砂岩夹泥岩及中酸性火山岩、白垩纪花岗岩及中酸性火山岩	富文金矿	白垩纪
热液型	黎母岭式铅锌矿	燕山期早期正长花岗岩	黎母岭钨、锡、铅、锌矿、白石岭铅、银矿	侏罗纪—白垩纪
热液型	什统式萤石矿	燕山期早期黑云母二长花岗岩	什统萤石矿	侏罗纪—白垩纪
热液型	石门山式钼、铅、锌、银	晚白垩世二长花岗岩正长花岗岩、花岗闪长岩	石门山钼、铅锌矿、看树岭银矿、后万岭铅锌矿	白垩纪
热液型	烟塘式钼铜矿	晚白垩世花岗闪长岩	烟塘梅岭钼铜矿	白垩纪
热液型	波澎岭式水晶矿	晚白垩世花岗闪长岩	波澎岭水晶矿	白垩纪
热液型	羊角岭式水晶矿	早白垩世花岗闪长岩	羊角岭水晶矿	白垩纪
热液型	铅山-南后山式铅锌矿	燕山期晚期花岗斑岩、石英斑岩	铅山-南后山铅锌矿	白垩纪
热液型	那大式锡、铌、钽矿	华力西期—印支期花岗岩	那大锡、铌、钽矿	二叠纪—三叠纪

续表

成因类型	矿床型式	含矿层时代及主要岩性	典型矿床	成矿时代
热液型	向民铅锌道柴林式萤石矿	二叠纪砂岩、板岩、生物粉晶微晶灰岩，晚白垩世砂岩、粉砂岩、泥岩，燕山期花岗岩	白沙县向民铅锌矿、青松水晶矿、道柴林、孔南坡、抱尾萤石矿	白垩纪
热液型	二甲式金矿	中元古代混合岩	二甲、不磨、土外山、抱板金矿	晚三叠世早期（227~228Ma）
热液型	抱伦式金矿	早志留世千枚岩、印支期花岗岩	抱伦金矿	中三叠世晚期—晚三叠世（205~236Ma）
热液型	王下牙劳式金矿、保由式铅锌矿	石炭纪—二叠纪砂岩、板岩夹结晶灰岩、生物粉晶灰岩夹燧石灰岩，印支期花岗岩	牙劳金矿、保由铅锌矿	三叠纪？
热液型	金波式钨、铍矿（锡）矿、昌化银岭式铅锌矿	燕山期黑云母正长花岗岩、花岗斑岩	金波钨、铍矿，尖峰岭丁司山、红门岭钨，昌化银岭铅、锌矿，锅盖岭水晶矿	侏罗纪—白垩纪
变质型	红五式白云母矿	中元古代混合岩、片麻岩	红五白云母矿	长城纪？
变质型	峨文式石墨矿	中元古代片岩夹石英岩、变粒岩、石墨矿	峨文岭石墨矿、烟塘石墨矿	长城纪？
沉积变质型	石碌式铁、钴、铜矿	新元古代片岩夹石英岩、结晶灰岩，白云岩，透辉透闪石岩夹富铁矿层	昌江县石碌铁、钴、铜矿	青白口纪？
风化型	蓬莱式铝土、钴土、宝石矿	第四纪残坡积砾石层	蓬莱铝土、钴土、宝石矿，居丁钴土矿，南阳钴土矿	第四纪
风化型	那大式铌钽矿	第四纪钠长石化细粒花岗岩风化壳，砂质高岭土	那大钽铌铁矿	第四纪
风化型	坝王岭离子型稀土矿	第四纪印支期二长（钾长）花岗岩风化壳，砂土层	坝王岭离子吸附型稀土矿、尖锋岭稀土矿	第四纪

（据海南省地矿局，2001，略有修改）

第二节　地质作用与成矿作用的关系

一、沉积作用与成矿作用

沉积作用对成矿的控制十分明显，海南省的一些大、中型矿床都与地层有直接或间接的联系。地层对成矿的控制作用主要是：①地层在沉积成岩过程中直接成矿；②地层作为矿源层，在成矿元素初步富集经后期构造形变、热液作用、变质作用，使成矿元素迁移至有利成矿地段进一步富集成矿。

（一）地层在沉积过程中直接成矿

这类成矿作用主要有石油、天然气、褐煤、油页岩、磷锰矿、石英砂矿、锆钛砂矿等。

石油、天然气赋存于古近系—新近系流沙港组、涠洲组、下洋组中，岩石隐伏于琼北盆地内。流沙港组岩性为一套灰色碎屑岩，属于内陆湖相沉积，岩性由深灰色、灰黑色、灰绿色泥岩与灰白色中、细砂岩、含砾砂岩、砂砾岩不等厚互层组成，中部夹薄层玄武岩，含砾砂岩微含油或具油斑。是琼北断陷石油、天然气的生油和储油层位，如花场、金凤、美台等中小型含油区。涠洲组岩性为棕红色、灰绿色、灰黑色、棕黄色、浅灰色泥岩与浅灰色含砾砂岩、砂砾岩不等厚互层，局部夹数层玄武岩，属于湖相，局部为海相沉积；下洋组为一套粗碎屑岩，属于海相沉积，岩性以灰白色、浅灰色砂砾岩、砂岩、粉砂岩为主，夹灰绿色粉砂质泥岩及泥岩。涠洲组和下洋组均为储油层位，

褐煤、油页岩产于古近系长昌组和新近系长坡组中，属于湖盆相沉积形成。长昌组岩性为砂岩、泥岩、油质页岩、炭质页岩和褐煤等。长坡组岩性以灰、灰蓝色黏土岩、粉砂质黏土岩、泥质粉砂岩为主，夹灰色钙质油页岩和褐煤层，底部为杂色砾岩、砂砾岩、泥质中粗砂岩。

磷锰矿赋存于三亚地区黔东统—武陵统大茅组中，其岩性为石英砂岩、粉砂质页岩、白云质灰岩、硅质岩夹磷块岩及碳酸锰，属于滨海环境→较稳定的滨外碳酸盐岩沉积环境。

石英砂矿分布于上更新统八所组海相松散沉积物中，岩性分选性较好，成熟度较高，金属和非金属矿物经过大量的海水冲刷、分选、富集沉积成有用矿床。海南岛东北部该组中石英砂的 SiO_2 含量一般为 96.05%～98.04%，单样最高达到 99.33%，Fe_2O_3 一般为 0.04%～0.19%，Al_2O_3 含量一般为 0.18%～1.96%，矿石平均品位达到一类Ⅲ级品的要求。

锆钛砂矿分布于全新统烟墩组近代海滩、潟湖、砂堤松散沉积物中，已知的大型、特大型锆钛砂矿床主要分布在文昌市及万宁市等沿海一带。

（二）作为矿源层为成矿作用提供物质基础

这类成矿作用以二甲金矿、石碌铁矿、石墨矿等为代表。

二甲金矿赋存于海南岛西部土外山—布磨一线中元古界戈枕村组和峨文岭组变质地层中。研究表明（丁式江等，1993），金矿成矿物质来源与上述两个岩石单位岩石有关。其中，峨文岭组片岩金含量平均值为 5.26×10^{-9}～7.03×10^{-9}，最大含量 34×10^{-9}，离差值 6.96；戈枕村组片麻岩金含量平均值为 5.88×10^{-9}～10.3×10^{-9}，最大含量 98×10^{-9}，离差值 20.65；戈枕村组混合岩金含量平均值为 7.0×10^{-9}～17.88×10^{-9}，最大含量 162×10^{-9}，离差值 31.6。都高于地壳背景值（4.0×10^{-9}）。

石碌铁矿主要分布于青白口系石碌群第六层岩石中，岩性为白云岩、透辉透闪岩夹富铁矿层。对石碌铁矿成因有多种认识，其一为火山-沉积变质型；二是沉积变质型；其三为沉积变质热液改造型。但不管那种认识，都一致主张沉积作用是其主导因素，并在后期构造作用下发生进一步富集。

石墨矿产出于长城系—蓟县系峨文岭组中，矿体多呈似层状、透镜状或豆荚状。含矿岩性有二云母石英片岩、石英片岩、石墨石英岩、石墨绢云母片岩等，石墨矿化普遍，石墨均为鳞片状晶质石墨。石墨矿床成因是含炭质岩层的炭质经区域变质作用下发生重结所形成，原岩中的炭质是形成石墨矿床的物质基础，是成矿作用赖以发生的内在因素。

二、侵入岩浆作用与成矿作用

海南岛具有多旋回的构造岩浆演化，形成了不同时期的侵入岩，伴随发生多期次的地质成矿作用。目前已发现的矿产在空间和成因上与侵入岩浆活动关系密切的矿产资源主要有：能源矿产铀，金属矿产铁、铜、铅、锌、钨、锡、钼，贵金属矿产金、银，稀有金属矿产铌铁矿、铯，稀土金属矿产以及非金属矿产红柱石、白云母、萤石、花岗石饰面石材矿、高岭土矿等。

与花岗岩类直接有关的原生金属矿产，以铅锌铜钼等多金属矿分布最广，成因上主要与壳幔混合型花岗岩有关，尤其是与燕山期晚期壳幔混合型花岗岩关系密切。如与华力西期—印支期坝王岭序列花岗岩体有关的南雅水库、东四、大章村、金炳等铅锌多金属矿点，分布在坝王岭序列内或接触带附近的南北向断裂中，这些铅锌多金属矿的成矿物质主要通过华力西期—印支期花岗岩体的侵入，由深部地壳迁移上来，于岩浆期后热液作用富集成矿。而与燕山晚期壳幔混合型花岗岩有关的矿床（点）主要有高通岭钼矿（与屯昌杂岩体有关）、昌化大岭铅锌矿床和先锋岭铅锌矿点（与燕山期晚期昌化大岭岩体有关）、石门山-看树岭钼多金属（铅锌铜银）矿（与千家杂岩体有关）、昌化铅锌矿（与昌化大岭岩体有关）以及近年来新发现的保亭新村钼矿（主要赋存在燕山期晚期次火山杂岩体中的二长花岗岩、花岗斑岩及其与二长花岗岩、潜流纹斑岩的内外接触带中）、罗葵洞钼矿（成矿时代为燕山晚期，斑岩型浸染-细脉浸染状钼矿床）等。上述已知的燕山期晚期矿床（点），一般具有以下产出特点。

1）一般与复式杂岩体的晚阶段淡色花岗岩有直接的空间或成因关系，岩石分异演化程度较高，矿化往往伴随有黄铁矿化、磁黄铁矿化、云英岩化、硅化、绿泥石化、绢云母、高岭石化等。

2）岩石成因一般属壳幔混合型，并具有埃达克质岩特征。主要是由加厚下地壳玄武质岩石部分熔融岩浆与地幔镁铁质岩浆发生混合（壳幔岩浆混合）并进一步分异演化形成的产物，具有来源深、温度高（大于800°C）、浅侵位（最终定位深度≤3.3km）的特点，岩石化学以高硅、富碱、贫镁、钙为特征，这些特征与东秦岭深源浅成型花岗岩特征相似，而深源浅成型花岗岩是东秦岭钼、钨、金、铜、铁及铅锌的主要成矿母岩，并构成了中国有名的东秦岭有色金属、贵金属成矿带。

3）钾长石多为正长石；斜长石牌号较高，主要为中长石-奥长石，环带构造明显（尤其是早期岩体中）；黑云母多为镁质黑云母，MF值高而稳定（介于49~55之间）。这些矿物特征和南岭地区与铜、钼、铅、锌矿床有关的成矿花岗岩体中的矿物特征相似，而明显不同于南岭地区与钨锡有关的成矿岩体中的矿物特征。

4）成矿微量元素的含量高低，可作为含矿花岗岩重要的评价标志。统计表明，区内燕山期晚期花岗岩中富集系数大于1的微量元素主要有钼、银、铅、锌以及铀、钍等，形成了一个比较好的地球化学背景，为成矿提供了物质基础。

因此，在壳幔混合型花岗岩尤其是燕山期晚期壳幔混合型花岗岩体及其旁侧围岩中具有寻找与岩浆成矿作用有关的钼、金、铁、铅锌及斑岩铜矿等金属矿产的前景，尤其是在分异程度相对较高的晚期岩体中更值得注意。

（一）中元古代岩浆侵入活动与成矿作用

海南岛中元古代发生的火山-沉积作用形成了戈枕村组和峨文岭组，前者以中酸性火山岩为主，后者为陆缘碎屑岩沉积。距今约14亿年发生的抱板运动❶导致了造山花岗岩（戈枕花岗闪长岩、亚炮二长花岗岩两个岩石单位）侵入，伴随戈枕村组和峨文岭组发生区域变质、变形和混合岩化，并产生石墨、白云母矿化作用，生成与中元古代变质作用和岩浆作用有关的石墨、白云母矿床成矿系列。矿床类型有变质型石墨矿床和变质伟晶岩型白云母矿床。同时，也使得海南岛西部抱板—布磨一线戈枕村组、峨文岭组岩石中金元素含量得到初步富集（丁式江等，1993），形成了金矿的矿源层，并在华力西期—印支期韧性剪切带中得到进一步富集成矿。

（二）华力西期—印支期岩浆侵入活动与成矿作用

华力西期—印支期岩浆侵入活动广泛而持久，出露俯冲型（碰撞前）、同碰撞型、碰撞后型花岗岩。与之有关的矿产主要有儋州那大锡矿、临高白虎岭钨矿、儋州蓝洋钨锡矿、南报金银矿、黎母岭钨锡矿、乐东县红门钨矿、乐东县五场锡矿、乐东县脚岭锡矿、抱伦金矿、坝王岭离子吸附型稀土矿、尖锋岭离子吸附型稀土矿等（海南省地质矿产勘查开发局，2001）。

钨、锡、铌、钽矿有热液石英脉型、矽卡岩型、热液充填型，目前仅产于华力西期—印支期花岗岩中，成矿物质均与华力西期—印支期花岗岩浆热液有关。

抱伦金矿产出于下古生界与华力西期—印支期花岗岩的接触部位。研究显示（丁式江等，2007），其成矿年代为（205~216）Ma±3Ma（K-Ar）和（236±17）Ma（Rb-Sr），成矿物质来源与上述两地质体有关。

稀土矿主要分布于华力西期—印支期花岗岩，尤其是碱性组分相对较高的正长花岗岩、石英正长岩类中，如尖峰序列正长花岗岩、袁水序列正长花岗岩-石英正长岩、东四正长花岗岩等，岩石稀土含量明显高于一般花岗岩，如尖峰系列正长花岗岩稀土总量最高值约为558×10^{-6}，东四正长花岗岩的稀土总量

---

❶ 抱板运动：中元古代末期14亿年左右发生的此运动具有三大特点：第一在强烈的构造运动的周期有强烈的岩浆活动；第二，强烈的岩浆活动导致了强烈的混合岩化、形成了眼球状混合岩、混合花岗片麻岩等；第三，在强烈的构造-岩浆活动的后期形成了区域变质晶质石墨矿床和变质伟晶岩型白云母矿床。鉴于南方各省（区）尚未发现可以与其对比的同期构造运动，故暂命名此运动（耿树方，2014年5月26日）。

平均值约为 354×10^{-6}，袁水系列石英正长岩-正长花岗岩最高值达 3187×10^{-6}（海南省地质调查院，2011），而且这些岩体的风化壳均十分发育，为离子吸附型稀土富集成矿提供了极其有利的条件。目前发现的坝王岭离子吸附型稀土矿、尖锋岭离子吸附型稀土矿和其他一些矿（化）点（陵水县兴隆三合水稀土矿、田尾村稀土矿化点以及坝王岭南七河、东二、东四等稀土矿点）均分布在上述花岗岩体中。

此外，部分后碰撞期（三叠纪）岩体基岩露头面积大，节理裂隙发育程度低，矿物分布较均匀，色斑、色线较少，具有致密坚硬，颜色纯正，花纹和谐、稳定的特点，可作为饰面石材矿开发利用，这些岩体主要有尖峰岭正长花岗岩、东四正长花岗岩等。除此之外，部分三叠纪辉长岩体（如儋州市光头岭辉长岩）也可作为饰面石材矿开发。

（三）燕山期岩浆侵入活动与成矿作用

燕山期是中国东部多金属矿成矿作用的重要时期，海南岛此时也发生了钼、银、铜、铅、锌等贵多金属矿的成矿作用。矿体产于岩体内部或其接触带中，形成岩浆热液型和接触交代型两个不同型式的矿体（床）。

1. 岩浆热液型矿

包括乐东千家石门山-看树岭钼多金属矿、文且钼矿、屯昌高通岭钼矿、三亚罗葵洞钼矿床、定安雷鸣地区富文金矿、白沙县金波钨铍矿点、昌江县昌化大岭铅锌矿床、乐东县尖峰锅盖岭水晶矿、琼中县什统萤石矿、道柴林萤石矿、先锋岭铅锌矿、三亚罗葵洞斑岩型钼矿、保亭新村钼矿以及屯昌羊角岭、保亭波澎岭水晶矿、琼海烟塘钼铜矿等，矿体基本上多产出于燕山晚期花岗岩中。千家钼多金属矿、文且钼矿、罗葵洞钼矿辉钼矿 Re-Os 等时线年龄分别为（80.2±0.6）Ma～（88.6±1.0）Ma、（99.7±0.4）Ma 和（103.9±1.0）Ma，廖香俊等（2008）也在屯昌高通岭钼矿中获得了辉钼矿 Re-Os 等时线年龄为（98.4±2.5）Ma。成矿时代都为燕山期晚期。

其中，以产于千家地区晚白垩世花岗岩中的石门山-看树岭钼、银、铅、锌贵多金属矿研究比较系统。其成矿分带性明显，中心为钼，往两侧为银、铅、锌，呈现出由高温→中低温的演化，属于同一成矿作用系列的产物。研究显示（陈沐龙等，2008），该地区岩浆演化晚期花岗斑岩中，主要成矿元素铜、钼、银、钨明显富集，而金、铅、锌则分别在中细粒黑云母正长花岗岩、中粒黑云母正长花岗岩和中粒斑状角闪黑云母二长花岗岩含量居高。锌主要富集于二长花岗岩阶段，铅在正长花岗岩中含量居高，铜、钼、银则强烈富集于晚期斜长花岗斑岩阶段，金也在晚期形成的正长花岗岩、斜长花岗斑岩中含量略高。表明岩浆演化过程中，成矿元素分异明显，为下一步成矿作用提供了重要的基础条件。

2. 接触交代型矿

有白沙青松水晶矿、振海山-摩天岭铜矿、保亭红石铁多金属矿和水晶矿，以及三亚田独铁矿、荔枝沟锌矿等。

红石铁、铅、锌、水晶矿和振海山-摩天岭铜矿产于燕山期晚期花岗斑岩或闪长岩与足赛岭组接触带中；田独铁矿矿体产于燕山期晚期花岗斑岩与中寒武统大茅组白云质灰岩接触带上；荔枝沟锌矿矿体赋存于榆红组、尖岭组的粉砂岩与燕山期晚期花岗斑岩接触交代形成的透辉石石榴子石矽卡岩中。属于接触交代矽卡岩型矿，成矿时代为燕山期晚期。

此外，海南岛燕山期花岗岩类大部分受后期构造破坏作用相对较弱，节理裂隙发育程度低，矿物分布较均匀，色斑、色线较少，可作为饰面石材矿开发利用的岩体相对较多，如目前已勘查评价的金波饰面石材矿、小妹水库饰面石材矿、牛鼻岭饰面石材矿、尖嘴岭饰面石材矿、踢通岭饰面石材矿等，矿床的成矿母岩均为燕山期岩体。因此，在燕山期花岗岩体中具有较好的饰面石材矿找矿前景。

三、火山活动与成矿作用

海南省的火山岩已发现有多种金属与非金属矿产，其中以铁、金、钼、铜、铅、锌等矿产为主。

中元古界长城系戈枕村组（原岩为中酸性火山岩）与金矿关系密切，是海南岛金矿的主要矿源层，已探明大中型金矿床 2 处、小型金矿床 3 处。

新元古界青白口系石碌群为一套海相细粒碎屑岩、泥岩及碳酸盐岩夹火山岩，火山岩主要产于第五层，第六层也有出露，而第六层是富铁矿及铜钴矿的含矿层位，矿体总体呈层状、似层状或透镜状，与围岩整合接触。许多地学工作者（刘裕庆，1981；中国科学院华南富铁矿科学研究队，1988；吕古贤，1988）认为，石碌铁矿成矿物质之基础源于海底火山沉积作用，属火山–沉积变质型矿床。

中生代火山岩主要分布于海南岛南部，岩性主要为中酸性火山岩，自下至上为：六罗村组流纹岩、安山岩、玄武岩；汤他组英安岩、安山岩、流纹斑岩；岭壳村组流纹斑岩、英安斑岩。主要有金(银)、铜、铁、铅、锌等矿点、矿化点10余处。

与新生代火山岩有关的矿产主要分布于文昌市蓬莱镇一带，有大型蓝宝石矿床、大型三水型铝土矿床、中型钴土矿床、小型砂铬矿床。为新近纪火山岩后期风化作用形成。

四、变质作用与成矿作用

除了沉积型矿产外，内生型矿产成矿作用与变质作用关系密切。岩石在发生变质作用过程中，由于温度、压力、溶液等因素的变化，使原岩中的有用元素迁移富集成矿。海南省主要大型、超大型矿床均与面状区域变质作用有关，部分铁、铜、硫、锡等矿点、矿化点及小型矿床的形成与线状接触交代、热液蚀变等变质作用有关。

（一）区域面型变质作用与成矿作用

主要以赋存于中元古界戈枕村组、峨文岭组中的晶质石墨矿、伟晶岩型白云母矿和赋存于青白口系石碌群中的铁矿为代表。

海南省已知的石墨矿属沉积变质晶质石墨矿床，主要分布琼西南乐东县红五地区，石墨矿的赋存层位主要为中元古界长城系抱扳群峨文岭组，矿体多呈似层状、透镜状或豆荚状。含矿岩性有二云母石英片岩、石英片岩、石墨石英岩、石墨绢云母片岩等，石墨矿化普遍，石墨均为鳞片状晶质石墨。石墨矿床的成因是含炭质岩层的炭质经区域变质作用下发生重结所形成，原岩中的炭质是形成石墨矿床的物质基础，是成矿作用赖以发生的内在因素，而变质作用是石墨矿床形成必不可少的条件。

伟晶岩型白云母矿的化学成分与混合岩相近，为混合岩化后期分异出来的富K、Al、SiO_2及挥发份的残余热液在高温高压条件下，沿片理或片麻理“侵入”，并发生同步变形，呈串珠状、透镜体状、不规则肠状沿片理分布。岩石中白云母晶面弯曲、褶皱、皱纹、波纹等变形常见，显示了一种深层次的变形特点。

石碌铁矿产出于青白口系石碌群变质岩系中，矿体呈层状、似层状、透镜状，厚度由数米至数十米，最厚超过400m，与围岩整合接触。富铁矿石（磷、硫等含量低）主要矿物是赤铁矿和少量磁铁矿，脉石矿物为石英和绢云母。以细鳞状结构为主，次为变余粉砂结构；以片状构造为主，次为块状及条带状。具有绿片岩相矿物组合特征。而贫铁矿石（磷、硫等含量高）中并出现透闪石和大量的石榴子石（含量20%~25%）等特征变质矿物组合，属于角岩相变质作用常见的特征矿物组合。根据石碌水库边二叠纪灰岩与二叠纪花岗岩接触处分布有矽卡岩型铁矿体的特征推断，贫铁矿石与晚期接触变质作用有关。

此外，区域变质作用也导致中元古界戈枕村组中金元素发生明显的活化分异，为后期成矿作用提供了重要的基础物质条件。各类混合岩的金元素含量差异明显，混合花岗岩的含金丰度最低，其次是混合岩化斜长角闪片岩，金含量最高的是混合花岗（质）片麻岩。反映区域混合岩化作用愈强，岩石中的含金丰度就越低，金元素在区域混合岩化作用过程中，具有由混合岩化程度高的向混合岩化程度低的方向迁移的趋势。

（二）线型变质作用与成矿作用

此类变质作用成矿比较普遍，以形成线状分布的矿体为特点。部分矿体产在花岗岩接触带，部分矿体产于线型构造带中。

戈枕金矿产在北东向戈枕构造带中，构造岩石有超糜棱岩、糜棱岩、千糜岩等动力变质岩，矿石类型有蚀变糜棱岩型、蚀变碎裂岩型及石英脉型金矿。新生矿物组合为石英、绢云母，属于绿片岩相变质。

千家石门山-看树岭钼多金属矿成矿则热液蚀变有关，矿体产于晚白垩世花岗岩构造破碎带中，断裂形变和蚀变程度愈强，其成矿作用愈明显。热液蚀变在垂直和水平方向上均存在较明显的分带现象。在垂直方向上，深部含矿热液以大范围密集裂隙带渗透交代为主，形成气成高温云英岩化蚀变和辉钼矿化；向上随着温压递降和热液性质变化，过渡为含矿热液沿断裂充填和扩散交代形成的中低温热液绢云母化、硅化、绿泥石化、高岭土化及闪锌矿、方铅矿、黄铜矿、辉银矿等矿化现象。水平方向上，深部云英岩化强烈的中心部位，已蚀变成云英岩，向外过渡为云英岩化带，再往外过渡为绢云母化带；浅部以多金属矿化石英脉为中心，向两侧依次过渡为绢英岩化、硅化、绢云母化碎裂岩和绢云母化绿泥石化二长花岗岩。

三亚罗蓁洞钼矿产出于燕山期花岗岩与白垩纪火山岩接触部位，接触带岩石蚀变有硅化、钾化、黄铁矿化、绢云母化、绢英岩化、云英岩化、黑云母化、绿帘石化、次闪石化、绿泥石化、碳酸盐化等几种，局部尚有叶蜡石化、萤石化。由内向外大致可分为不十分明显的 4 个带：强硅化蚀变带——蚀变表现为石英细脉与石英较大脉并存，岩石中硅化，云英岩化、绢云母化明显；硅钾蚀变带——位于强硅化带的外围，蚀变矿物为石英、钾长石、绢云母等，可见钾长石渗透交代或呈单独脉状或分布在石英脉两侧，是钼矿化主要赋存部位；硅化蚀变带——位于硅钾蚀变带外侧，也可以单独出现，钼矿化较弱；绿帘石、绿泥石化带——常与碳酸盐化、黄铁矿化伴生。

此外，接触变质作用并形成了白沙青松水晶矿、振海山-摩天岭铜矿、保亭红石铁多金属矿和水晶矿及三亚田独铁矿、荔枝沟锌矿等，多为矽卡岩型矿体，其特征变质矿物为透辉石、透闪石、石榴子石等，具有角岩相变质矿物组合特征。

五、构造活动与成矿作用

构造运动是控矿因素中的主导因素。构造运动的结果在大尺度中的表现是控制着沉积盆地形成和岩浆活动。小尺度中发育有断裂、褶皱等构造形迹。从而控制了沉积型、内生型成矿作用的发生。不同规模的构造对成矿的控制作用也不同，区域性构造控制区域性一般成矿特点和矿床分布规律，矿床构造则控制了矿体的形态、产状和分布状况等。各种构造既决定矿体在岩石中的位置，又影响矿体的形态特征。

（一）构造活动与沉积成矿作用

海南岛与沉积作用有关的矿床主要有褐煤、油气、油页岩、石墨、磷、锰、铁等。构造活动对沉积成矿作用的控制表现是：控制了沉积盆地的形成和盆地格局，从而控制沉积成矿作用的发生。其中，磷、锰矿等分布于岛南部三亚一带下古生界黔东统—武陵统大茅组中，该组岩石主体为滨浅海相含磷、锰礁碳酸盐岩和硅质岩、碎屑岩沉积，形成了大茅磷锰矿。铁矿分布于岛西部石碌地区的新元古界石碌群滨浅海相碎屑岩和石灰顶组滨海-海陆交互相碎屑岩中，石墨矿则基于中元古代峨文岭组中沉积的含炭质岩层经重结晶作用形成。褐煤、油气、油页岩等主要赋存在琼北的新生代断陷盆地内，由于太平洋板块的作用，并受东西向王五-文教构造带的进一步控制，海南岛形成了多个东西向展布的新生代断陷盆地，盆地内接受了厚达近千米至几千米的新生代河湖相或海相砂岩、泥岩、页岩等，形成褐煤、油气、油页岩等矿床。

（二）构造活动与区域变质成矿作用

海南岛区域沉积变质型矿产主要有石碌铁矿、红五石墨矿、烟塘石墨矿。

石碌铁矿位于昌江县石碌，含矿层为青白口系石碌群第六层和南华系—震旦系石灰顶组浅变质岩。岩石经历了多期次变形作用的改造，构造迹象复杂，具有多方向、多时代构造形迹的叠加，总体呈现出由褶皱变形往韧脆性断裂变形演化的特征。早期的区域变形变质作用使矿源层中的铁得到进一步的富集，并因变形程度的差异而显示贫矿和富矿之分。无论在手标本上还是显微镜下，都可见到矿石和含矿围岩呈现强烈的塑性流动变形组构和构造分异条带，而且变形强度和构造分异程度往往与矿石品位呈正相关关系，强烈变形的铁矿石品位很高，含磷、硫等组分较少，而弱变形的铁矿石相对富磷、硫，品位也大大降低。铁矿石主要矿物是赤铁矿和少量磁铁矿，脉石矿物为石英和绢云母。以细鳞状结构为主，次为变余粉砂结构；以片状构造为主，次为块状及条带状。具有低绿片岩相矿物组合特征。后期的褶皱变形对矿体的产出形态起到至关重要的控制作用，与主成矿期密切相关的北西—近东西向石碌复向斜，是石碌群中最醒目的向斜构造，造成了已成铁矿物质自翼部向褶折端部位运移，从而构成了矿区矿体进一步加厚的最重要构造。主矿体和富铁矿体多赋存于石碌复式向斜的褶皱紧密、流动构造发育、形变强烈的地段，多位于向斜的谷（槽）部或向斜轴部附近的层间剥离带或褶曲发育处，层间滑动作用和褶皱弯流作用，促成矿体的富集加厚。

石墨矿产于中元古界峨文岭组中深变质岩中，晶质石墨矿的变形主要表现在：矿石矿物中石英多呈压扁拉长状变形岩沿片理走向定向排列，显示矿物拉长线理构造；第三层矿体上下岩层——斜长角闪岩多呈透镜体状顺片理产出，呈现出石香肠构造特征；石墨长石石英岩、石墨石英岩、石墨黑云母石英片岩等含矿母岩在空间分布上已无一定的规律性；各层矿体走向变化大，沿走向、延深方向，其厚度、品位不同，尤以厚度变化较大，矿体膨缩现象比较明显。由此表明成矿过程中不应该排除褶皱变形作用导致了矿体进一步富集的可能。晶质石墨矿的成矿物质基础源于沉积过程中的有机碳岩层，变质变形作用下，有机碳岩层由于具有很大的塑性，并与围岩间的相对能干性存在差异，因而发生强烈的顺层塑性流动变形，呈拉薄、增厚状，导致有机碳得到进一步的富集（李孙雄等，2004）。

（三）构造活动与热液成矿作用

海南岛中生代是内生矿床的鼎盛时期，具有矿种多、矿床类型多样的特点，优势矿种主要有金、银、铜、铅、锌、钼、铁、萤石、水晶等。许多地学工作者或地学单位（冯连顺，1988；涂绍雄等，1993，1994；丁式江等，1994，2001，2007；何龙清等，1996；龙文国等，1999；陈柏林等，2001，2004；李孙雄，2004；陈沐龙等，2008；宜昌地质矿产研究所等，1990，1994；地质力学研究所等，2000；海南省地质矿产勘查开发局，2001；北京大学等，2002）从不同角度对成矿作用进行了不同程度的研究和探讨，提高了构造活动与内生成矿作用认识。其中，以抱伦金矿、二甲金矿、石门山-看树岭钼多金属矿工作程度比较高，研究程度也相对系统。

1. 抱伦金矿

抱伦金矿位于抱伦豪岗岭一带下志留统陀烈组和三叠纪花岗岩的接触部位，矿体赋存于北北西向破碎带中。断裂带具有良好的分带性，中心一般为含金石英脉充填，两侧依次为硅化千糜岩、碎裂岩化硅化千枚岩。矿石以含金石英脉型为主，次为含金蚀变岩型。矿石构造有浸染状构造、网脉状构造、角砾状构造、斑点状构造、似条带状构造等。属于岩浆热液石英脉型矿床。

金矿成矿作用可能分为三个阶段：①加里东期的构造运动及区域变质作用形成了豪岗岭背斜，伴随着背斜的形成，在背斜转折端形成构造裂隙，变质分异热液沿裂隙贯入，形成贫金石英脉；②印支期岩体的侵入进一步改造了豪岗岭背斜，使背斜更加紧闭，其岩浆期后含矿热液充填在构造破碎带内，是本区的主成矿期，一方面岩浆期后含金、硫、氯、氟、H_2O、CO_2 等热液沿构造破碎带上升运移，另一方面岩浆热能又活化萃取了围岩中的金、硫、铁等组分，与岩浆期后热液一起沿构造带向上运移，在断裂带的扩容空间地带，由于氧化—还原势的急剧变化，形成连续的大规模的矿化作用，陀烈组中段的含碳千枚岩又起着隔挡层的作用，因而金矿体主要就位于陀烈组一段北北西向构造中；③燕山期晚期伴随着断裂作用，早期形成的含金石英脉再次破碎，构造分异热液再次充填到破碎的石英脉中，可能为本区金矿的进一步富集起着一定的作用。

2. 二甲金矿

二甲金矿位于岛西部土外山-戈枕-布磨一带北东向戈枕构造带中。构造带发生两次构造活动，早次为韧性剪切变形，形成糜棱岩系列岩石；晚次叠加了脆性破碎变形，形成碎裂岩、角砾岩等碎裂岩系列岩石。金矿床类型有蚀变糜棱岩型、蚀变碎裂岩型、石英脉型。

金矿成矿作用复杂，成矿期次较多，主成矿期可划分为两期。早期为构造分异期，戈枕村组形成后，在华力西期发生韧性剪切动力变质作用引起的绿片岩相退变质，抱板群因韧性剪切产生不同程度的糜棱岩化，形成一套糜棱岩、千糜岩、超糜棱岩，构造分异作用和伴随的热液活动，岩浆活动等使抱板群中的金矿以活化迁移，富集成矿；在此过程中可能有岩浆热液及深部来源的金加入成矿。在韧性剪切后期以脆性构造活动为主，继承性构造热液及岩浆热液在韧性剪切带的边缘和旁侧分别形成蚀变碎裂岩型和石英脉型金矿，并对蚀变糜棱岩型金矿进一步叠加富集。构造分异期形成的蚀变糜棱岩型金矿分布于剪切带中，蚀变碎裂型金矿分布于剪切带边缘，石英脉型金矿分布于剪切带的旁侧。

3. 石门山-看树岭钼多金属矿

石门山-看树岭钼多金属矿发育在千家一带的燕山期晚期花岗岩中，矿体主要赋存在北西向及其派生的南北向—北北西向石英脉绢云母化破碎带内。破碎带经历了先张扭后压扭的活动过程，早期表现为带状分布的石英脉充填，晚期活动导致早期石英脉发生破碎和蚀变，形成绢云母化石英脉破碎带，控制了钼多金属矿的产出。破碎带形变和蚀变程度愈强，其成矿作用愈明显。研究显示，断裂富集砷、锌、钼 、银、铅、铜、金元素，尤以钼、银、铅元素为甚，叠加强度系数高达几十至上百，贫 锑、铋元素，锡元素变化不明显。元素含量波动幅度较大，钼、银、铅等元素最低含量与最高含量相差几十倍至上百倍，反映了这些元素在断裂活动中的分异富集能力极强。砷、锑、钼、铋、银、铜、金等中高温元素含量变化曲线为单峰型，大体上呈同步消长关系，表明了它们间的关系较密切。各元素在断裂中的分带现象较明显，自中心往边部，元素组合呈金、钼→砷、锑、银、铋、铜→铅、锡→锌的变化，高温元素与中低温元素混杂，不具分异性，可能是构造热液多次活动的结果。

六、风化作用与成矿作用

海南岛与风化作用有关的矿产有宝石矿、铝土矿、钴土矿、稀土矿、高岭土矿等。其中，宝石矿、铝土矿、钴土矿分布于文昌蓬莱一带新近纪火山岩风化土层中，稀土矿、高岭土矿则与花岗岩风化作用有关。

花岗岩风化作用一方面导致长石发生高岭土化形成高岭土矿，另一方面使得岩石中的副矿物残积形成稀土矿。稀土矿成矿主要经历以下几个阶段：①母岩风化阶段。海南岛地处热带—亚热带，气候温暖、雨量充沛，有利于物理化学风化作用的发育，风化壳的形成为稀土元素的次生富集奠定了储藏空间。②稀土元素从母岩溶出阶段。在有机物（一元酸、二元酸、非酸有机物）、微生物及其新陈代谢产物、腐殖酸（富里酸）的共同作用下，风化壳处在 $pH=4\sim6$ 的酸性环境中，稀土元素以水溶性配合物 $REE（H_2O）_n^{3+}$ 形式从花岗岩母岩中溶出，并向低势低洼处迁移。③黏土矿物吸附稀土离子阶段。该阶段如风化壳表层的剥蚀速度小于水溶性配合物迁移的速度，则上覆被剥蚀部分的稀土元素继续向下低洼处运移，在稀土元素以离子形式向下运移的过程中，被风化阶段由黑云母和长石风化形成的高岭土等黏土矿物所吸附成矿。稀土矿体主要富集于全风化层的中下部和半风化层的上部。

火山岩（主要是熔岩类）在红土化过程中，一些抗风化能力强的矿物如刚玉、红锆石、铬铁矿、钛铁矿等残留下来富集成残积砂矿，或经短途搬运富集成坡积、洪冲积砂矿，另一些抗风化能力弱的矿物，经过长期、缓慢的化学作用而形成新生矿物，如玄武岩中长石分解和水解形成三水铝石为主的铝土矿、氧化锰从岩石中分解出来，在以胶体状态运移时，吸附从岩石中水解出来的钴、镍、铜等元素，形成钴土矿，在风化壳中富集形成残坡积铝土矿床、钴土矿床。蓝宝石、红锆石主要在金牛岭期（石门沟村组）火山熔岩的风化土中富集成矿，其中橄榄玄武岩、辉斑橄榄玄武岩、粗玄岩和火山碎屑岩是蓝宝石和红锆石成矿的主要母岩。铝土矿和钴土矿在蓬莱期（石马村组）、金牛岭期（石门沟村组）火山熔岩的风化土中富集成矿，蓬莱期的玻基辉橄岩、橄榄玄武岩和金牛岭期的橄榄玄武岩是钴土矿成矿的主要母岩。

第二十七章　地质景观资源

第一节　概　述

海南省地质景观资源丰富。以海南岛为例，有喀斯特地质景观资源（保亭仙安石林、儋州石花水洞）、丹霞地质景观资源（保亭七仙岭、五指山市五指山、琼海白石岭、定安文笔峰等）、水体地质景观资源（保亭七仙岭温泉、万宁兴隆温泉、琼海官塘温泉、儋州兰洋温泉、白沙七差温泉、三亚南田温泉、琼中百花岭瀑布、保亭枫果山瀑布、陵水大里瀑布等）、火山-地震地质景观资源（石山马鞍岭火山口、海口市石山仙人洞、海口市永兴雷虎岭火山口、临高高山岭火山口和多文岭火山口、儋州德义岭火山口、定安旧州火山口等）和海岸带地质景观资源（文昌铜鼓岭、琼海博鳌三江并流汇海和三亚亚龙湾、大东海、大小洞天，以及昌江棋子湾等）。地质景观资源分区明显，喀斯特地质景观资源、丹霞地质景观资源、水体地质景观资源主要分布于中部山地、丘陵剥蚀地貌区，火山-地震地质景观资源见于东西向王五-文教断裂带以北的火山台地地貌区，海岸带地质景观资源则主体分布在东西向王五-文教断裂带以南的堆积平原地貌区（图 27-1，表 27-1）。

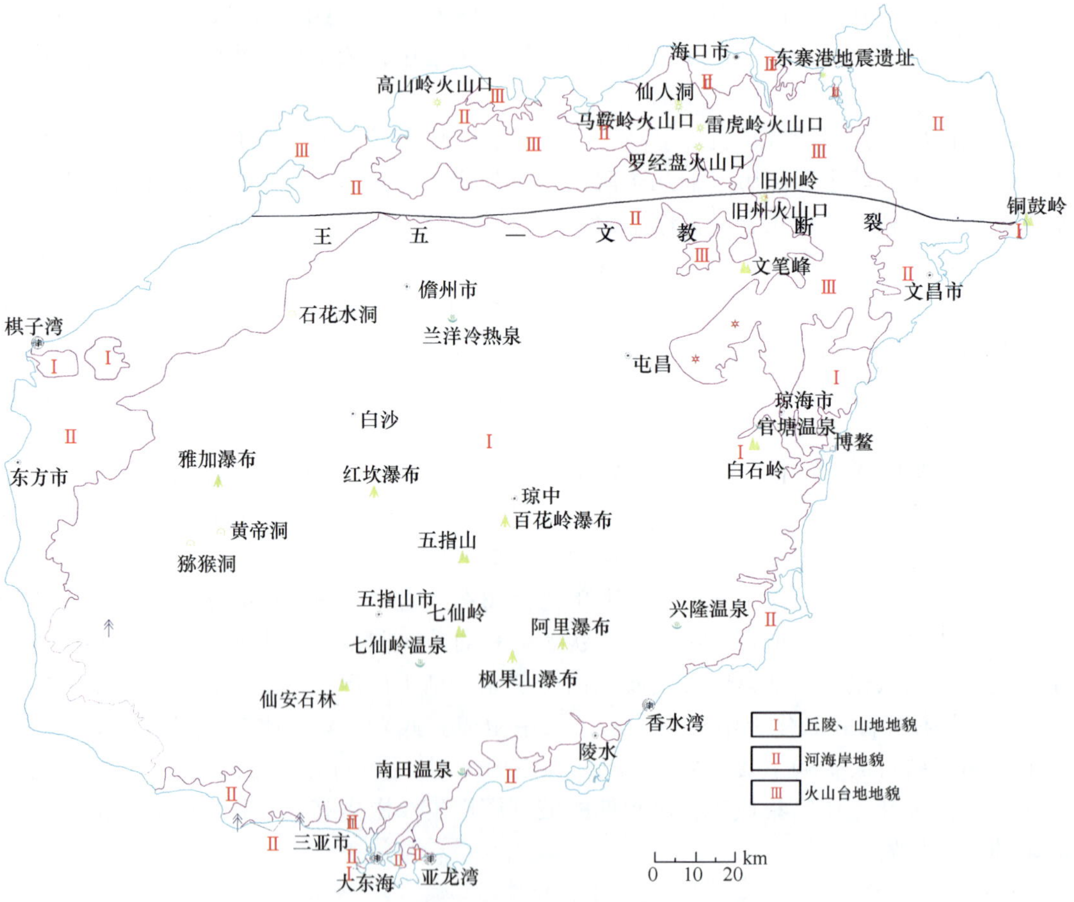

图 27-1　海南岛主要地质景观资源分布简图

Fig. 27-1　Sketch map of distribution of main geological landscape resources in the Hainan Island

表 27-1　海南岛主要地质景观资源特征表

Table 27-1　Characteristics of main geologic landscape resources in the Hainan Island

景观类型	景观资源点	地理位置	地质情况	景观特征
火山-地震地质景观	海口市石山马鞍岭火山口	海口市石山镇	全新统石山组火山岩	位于海口市西南15km处的马鞍岭，由风炉岭和包子岭两座火山锥相连构成，似马鞍状而得名。马鞍岭火山口风炉岭海拔222.2m。火山口底座呈圆形，直径500m，环形火山口直径130m，上宽下窄，呈漏斗状，深69m。在火山口的东北面遗留有一个V形开口通道，这是当时火山喷发火山熔岩浆外流之出口
	海口市石山仙人洞	海口市石山镇	全新统石山组火山岩	是众多熔岩洞中的一处，以其洞穴埋藏浅、规模大，洞中有石花、石床、石椅等多种奇石为特征，熔岩洞高2~5m，宽5~10m，长达千米以上，在中国乃至世界实属罕见，具有极高的科学和旅游观赏价值
	海口市永兴雷虎岭火山口	海口市永兴镇	全新统石山组火山岩	海拔119.3m，状若蹲虎，比马鞍岭火山口大许多倍。岭西北有两个山洞，石景千姿百态
	海口市罗经盘破火山口	海口市永兴镇	全新统石山组火山岩	火山口为负地形，形如体育场，长轴约1000m，宽约800m，深30m，四周为阶梯状，火山岩已风化为红土成为耕地
	临高高山岭火山口	临高县城北高山岭	中更新统多文组火山岩	底座呈椭圆形，长轴北西，长2000m，宽1000m，有浅火山口，口径约42m，深约2m，呈盆状
	定安旧州火山口	定安旧州镇	中更新统多文组火山岩	底座呈椭圆形，长轴北西，长1000m，宽650m，火山口完好，深约50m
	海口东寨港海底地震遗迹	海口市东营港、北创港、东寨港和文昌市铺前港	全新统烟墩组中细砂	1605年琼州大地震导致陆地沉降，72个村庄被海水淹没，成为“海底村庄”。在离东寨港不远的海滩上，能见到一座古戏台，石板上已布满了海蚝等海生动物；从东寨港北上，在铺前湾海岸以北4km左右约10m深水下，有一个叫“仁村”的古村庄，透过海水，村子的庭院及房屋遗迹依稀可辨；在铺前湾与北创港之间的海底有一座“贞节牌坊”屹立于水下
海岸带地质景观	文昌铜鼓岭	文昌龙楼镇	印支期花岗岩、全新统烟墩组中细砂	区内怪石嶙峋，奇洞通幽，丹崖翠壁，傲然矗立在峰顶的硕大的风动石是铜鼓岭的标志物，它是一块花岗岩转石，高约3m，宽约4m，重达20t左右，上圆下尖，当刮起六七级的大风，巨石就会微微摇晃，发出蓬蓬巨响。北面月亮湾登高望远、极目南海、水天一色、浩瀚无垠、湾内波平浪静，海水清澈透明、湛蓝澄碧；湾外一碧万顷，海天无际，涛声不绝，浪花腾雪。南面石头岭公园并见各种海蚀地貌，海蚀石、水臼、波痕石可见。海蚀石形似蜂窝状，水窝形态不规则，以椭圆状、次圆状为主；水臼多以单个产出，形态为椭圆状、次圆状，大小不等，小者约10cm，大者30~40cm；波痕石是水浪作用下在岩石表面刻下的痕迹，波峰、波谷平缓

续表

景观类型	景观资源点	地理位置	地质情况	景观特征
海岸带地质景观	三亚亚龙湾	三亚市亚龙湾	全新统烟墩组中细砂，印支期、燕山期早期及燕山期晚期的二长花岗岩	为一半月形的大型海湾，长度约10km。三面青山环抱，东端为牙龙角，西面为白虎岭，宛如两条游入大海的苍龙，湾内出口处有几个小岛为屏，状若双龙戏珠。沙滩开阔洁白晶莹，细腻柔软，绵延长达8km，沙滩边缘点缀丛丛仙人掌，秀美如画。这里风平浪静，海水清澈湛蓝，能见度达8~9m，是海滨日浴的绝佳场地，被誉为“不是夏威夷，胜似夏威夷”。岸边花岗岩长期受海浪冲击侵蚀被雕塑成“刺猬出洞”、“巨鳄望日”等海蚀岩景物，形态逼真，栩栩如生
	三亚大东海	三亚市区	全新统烟墩组中细砂	三面环山，南面大海，海阔天空。沙滩绵延长约3km，宽20~30m，呈5°~8°向海倾斜。砂层细腻柔软，洁白如雪。冬季海水温度在18℃~22℃，是冬泳避寒休闲胜地
	陵水香水湾	陵水县牛岭	全新统烟墩组中细砂、海西期-印支期花岗岩	因香水岭流来的泉水注入海湾而得名。银白色的沙滩漫无边际，波光粼影景色格外怡人。在铜岭脚下造就出“擎天石”、“龙王椅”、“石亭”等海蚀岩奇特景观
	昌江棋子湾	昌江昌化江入海口	燕山期花岗岩和全新系烟墩组、上更新统八所组松散沉积砂	为一花岗岩质岬角，一条状似帆船的岩体向海中延伸，把湾内湾外分割开来，恰如“一掌划开两重天”的意境。在海湾沿岸的弧形沙滩上和海水中布满了光滑洁净、晶莹明亮、姹紫嫣红的各色卵石。卵石为花岗岩或石英脉遭受风化剥蚀形成小的砾石，再经海水的不断的冲蚀、磨蚀而形成。并因岩石发育两组相互垂直的裂隙而呈“棋盘”状，散落其中的卵石形似棋子而得名
喀斯特地质景观	保亭仙安石林	保亭仙安	晚志留世灰岩	为灰岩经热带雨淋滤溶蚀作用形成，发育有尖状石林，深沟达5~10m，最深25m，石林高10~15m，最高35m，共有90多尊，景区面积500多亩，其周边还有暗洞、陡崖、峰丛等景观
	昌江县皇帝洞	昌江县王下	二叠纪石灰岩	洞口海拔250m，离地面约15m。洞穴由西向东延伸，长122.5m，宽47m，高15m，总面积5647.5m^2，可容万人。洞内石柱、石笋、石钟乳众多，形如仙女、猴子、观音菩萨，千姿百态。在洞内已发现石斧、石网坠，以及汉代印纹硬陶残片等器物
	儋州石花水洞	儋州八一农场	出露晚石炭世深灰色灰岩、大理岩、灰岩	溶洞内因生长着大量的石花而得名，并生长着卷曲石、石芽、石笋、钟乳石等
	东方市猕猴洞	东方市东河	二叠纪石灰岩	洞口海拔380m。洞穴迂回曲折，总长约600m，面积约2000m^2。有两个主要的洞厅。前洞厅里石笋丛生，形态各异，因有的像一群猕猴而得名。后洞厅面积较大，厅内平坦，洞中石钟乳、石幔，晶莹剔透，别具一格。后洞厅还连着两个小洞厅，其内有一石柱，犹如一口古钟，用手重击，可以发出浑厚的响声，余音袅袅

续表

景观类型	景观资源点	地理位置	地质情况	景观特征
丹霞地质景观	保亭七仙岭	保亭七仙岭	下三叠统岭文组砂岩、砂砾岩、砾岩	主峰海拔1126m，是海南著名的名山之一。峰峦起伏成锯齿状，形似七指，故得名
	定安文笔锋	定安龙湖镇文笔锋	早白垩世灰色、暗紫色、紫红色砂砾岩、砂岩	海拔181.4m，长轴呈南北向，长约800m，东西宽约200m。孤立状屹立于琼北定安县龙湖镇。分布有各种风化作用形成的形态各异的岩石，与保亭七仙岭、琼海白石岭景观成因如出一辙
	五指山市五指山	五指山市五指山	下白垩统六罗村组火山岩	是海南第一高山，海拔1867m。峰峦起伏成锯齿状，形似五指，故得名。远眺五指山，群峰如指，直插云间，五峰之间，有深谷相隔，遥相对峙，又有逶迤相连，各自独立
	琼海白石岭	琼海温泉	上白垩统报万组砂岩、砂砾岩、砾岩	矗立在万泉河畔，海拔328m。因岩石裸露风化为白色而得名。白石岭景区由公仔岭、登高岭、衬布岭、三牛岭等群峰组成。群峰凌空，巨石怪立，深涧峡谷，洞壑幽深莫测
水体地质景观	保亭七仙岭温泉	保亭新村	燕山期花岗岩	在东西长700m，宽200m范围内共有热泉47个，泉水从岩石张性节理裂隙中冒出，总流量1269m^3/d，水温70℃左右，高者82℃，钻孔水温最高可达96℃。无色透明，具硫化氢气味，矿化度0.207~0.279g/L，SiO_2含量一般在80mg/L，氟含量7~9mg/L，最高14mg/L，pH值8~9，水化学类型属重碳酸钠水
	万宁兴隆温泉	万宁兴隆	燕山期二长花岗岩，发育有北东走向断裂	有12眼热泉水沿北东及北西向张性节理涌出。热矿水B+C级储量5600m^3/d，自流量2.4L/s，水温45℃~60℃。热水无色，透明，具硫化氢气味，SiO_2含量100mg/L，pH值9.8，属碱性的HCO_3-Na型热水
	琼海官塘温泉	琼海温泉	中侏罗世花岗岩、北西向断裂	热矿水B+C级储量6700m^3/d，自流量3.3L/s，水温34℃~73℃。泉水形成水塘深1~2m，见气泡有节奏的涌出，泉水无色、透明，具硫磺气味，矿化度0.5g/L，SiO_2含量80mg/L，氟含量19.6mg/L，pH值9.6，属HCO_3-Na型热泉。因含氟较高，不能饮用。泉水作为医疗热矿水使用
	儋州兰洋温泉	儋州兰洋	黑云母二长花岗岩、北西向断裂	沿断裂在4km长的范围内分布着三处温泉点。兰洋场部的温泉点中冷热泉共生，冷泉水温30℃，热泉温度达90℃，水质清澈，热雾腾腾。泉自流量3.92L/s，有气泡连续冒出。热水无色透明，具H_2S气味，矿化度0.32g/L，SiO_2含量95mg/L，氟含量12mg/L，pH值7.7，属HCO_3-Na型热水
	三亚南田温泉	三亚市藤桥	华力西期—印支期花岗岩	热矿水B+C级的可开采量8030m^3/d，水温57℃，矿化度1409~2450mg/L，氟含量2.04~7.7mg/L，偏硅酸68.5~117mg/L，氡44.36~59.86贝可/L，水化学类型为氯-钠钙型，具有较高的医疗使用价值

景观类型	景观资源点	地理位置	地质情况	景观特征
水体地质景观	琼中百花岭瀑布	琼中百花岭	华力西期-印支期花岗岩石	主峰海拔1100m，瀑布源头在海拔700m的第二高峰上，由宽20多米、落差达300m的三级瀑布组成。溪水至此，若天河开闸直下三级，蔚为壮观。各级瀑布形态各异，名曰金龙吐珠、神丹妙药、仙女散花，然后泻入深潭，激起一片白雾
	保亭枫果山瀑布	陵水县吊罗山	燕山期花岗岩	瀑布位于陵水河上游的一个溪流中，瀑布有如银练倒悬于叠翠的群山上，排空倾泻100多米，又重新流淌于茫茫的林流峡谷中，四周景观幽静纯朴，春意泱然
	陵水大里瀑布	陵水县吊罗山	燕山期花岗岩	瀑布从两个山体的山坳处川流直下，整体落差约50m，宽约4~6m，气势磅礴。流水从顶倾泻而下约于30m处发生变向，并向东分出一幅宽约1m的分支，两股清流紧贴花岗岩石面以60°倾斜角齐头并进，直至瀑底绿水潭。绿水潭长约10m，宽约8m，站在绿水潭旁仰望两山体，仿佛两个小巨人向绿水潭围拢而来，气势逼人；而从山上往下俯视，犹如一巨人怀抱着一颗碧绿的大珍珠
	白沙县红坎瀑布	白沙县元门乡	白垩纪砂泥砾质岩石	瀑布发源于海拔1101m的红坎岭，落差145m。站在瀑布下仰望，宛如一条巨龙自天而降，化作雪花四溅，璀璨晶莹，分外壮观
	昌江县雅加瀑布	昌江县坝王岭	晚三叠世黑云母二长花岗岩	瀑布直悬雅加大岭西侧，落差110m。四周峰峦雄峙，巨石嶙峋。瀑布犹如飞龙出谷，水雾纷扬，银光耀眼。兼备水、石、潭、瀑、林等自然景观特色，集雄、奇、险、幽和谐于一体

第二节　地质作用与地质景观资源的关系

各类地质景观资源的形成是地球内动力或内外动力共同作用的结果。内动力地质作用是最关键的因素，是地质景观资源形成的根本保障。

一、内动力地质作用与地质景观资源的形成

海南岛经历了多阶段、多旋回复杂的地质演化过程，地壳发生多次的裂解和拼合，凹陷成盆，隆升为陆，海陆变迁，桑海沧田。沉积作用、岩浆活动和构造活动“周而复始”。不同时期的沉积和岩浆活动为地质景观资源提供了物质基础，而新生代以来的新构造活动为地质景观资源的最后定型提供了保障，也最终塑造出火山地质景观资源、海岸带地质景观资源和山地、丘陵地质景观资源等三个不同类型的地质景观资源分区。

研究表明，海南岛新构造活动强烈且持久，表现样式多样，有地壳差异性升降运动、断裂活动及其相伴随的地震活动、火山喷发活动等。

地壳差异性升降运动使得海岸带发生沉积作用，形成沿海岸带分布的全新统烟墩组。因东部海岸主要是属于花岗质海岸，并直面南海，水动力作用相对西部海岸强，海水不断拍击海岸，加快花岗岩风化，长石、黑云母矿物抗风化能力差，发生泥化、高岭土化蚀变，原先致密的花岗结

构变得松散，泥质物被海水带走，而抗风化能力强的砂粒状石英矿物则残留下来，参与到全新统烟墩组松散沉积中，并塑造出海蚀石海岸地貌。烟墩组松散砂经海水强烈的淘洗和筛选，砂质得到净化，砂粒磨圆度和均匀度也大大地得到提高，造就了当前砂粒成分比较均一、磨圆度好、大小均匀的砂质海岸景观。因地壳多次的差异性升降，发育多级夷平面，奠定了海南岛中部高、并向四周逐级降低的现代地貌格局，河流不断下切和冲蚀，流水顺势而下，在陡坎处形成山地、丘陵瀑布景观。

断裂活动控制着海南岛火山地质景观资源、温泉景观资源、地震遗迹景观资源的形成。在琼北表现为地震和火山喷发，琼南则以温泉出露体现。火山喷发活动集中发生于东西向王五-文教断裂带以北地区，具有多期次性，为今天留下了大量的火山地质景观资源。琼北4000多平方千米的火山岩分布区，星罗棋布着数十个火山口、火山熔岩洞、火山隧道等火山机构，其中火山口达50多座，被称为我国南方火山地质博物馆。火山熔岩洞、火山隧道是熔岩流表里冷凝速度不一致所造成的，即熔岩流在流动过程中，表层冷凝成壳，内部的岩流热量不易散失，保持高温而继续流动，当熔岩流来源断绝时，里层岩流“脱壳”而出，留下隧道状的洞穴。

二、内外动力地质作用与地质景观资源的形成

海南岛与内外地质作用密切相关的地质景观资源主要是分布于丘陵、山地中的喀斯特地貌、丹霞地貌。

喀斯特地貌基于古生代沉积的碳酸盐岩经后期风化剥蚀作用而形成。由于海南岛古生代碳酸盐岩主要以夹层状或透镜体状分布于碎屑岩中，岩石出露面积小，因此，形成的喀斯特地貌分布也比较极限，景观面积也小。其中，仙安石林为灰岩经热带雨淋滤、溶蚀作用形成的尖状石芽群。大气降水是仙安石林形成的重要因素。大气降水储存于灰岩溶蚀凹穴内，并沿竖向裂隙下渗，在此过程中溶解空气中的CO_2使水中HCO_3^-离子浓度增加，对灰岩产生溶蚀作用；随着气温升高，凹穴中的水被蒸干，溶蚀残留物留在凹穴表面，又一次的大气降水将残留物从凹穴中冲走，并又一次储存于凹穴内。周而复始的大气降水，不断促使岩溶凹穴加大、裂隙加宽而相连，形成石林雏形，再由大气降水及雨林中凝结水进一步溶蚀、冲蚀，形成“剑状、针状”石林地貌。石花岭水洞形成可划分出三个阶段（海南省地质调查院，2011）：①早期在地下水位以下一定深度处于全充水承压条件下形成早期的潜水带洞穴通道，地下水沿近南北向、近东西向的节理、裂隙发生流动，并通过溶蚀作用拓展空间，形成了最早期的洞穴。②中期随着海南中南部山地的相对抬升，石花水洞的雏形洞体逐渐上升至地下水位附近，风化作用加强，来自大气降水和山地补给区的水流，改造着原先的潜流带通道，并不断扩大，塑造出若干个水平方向延伸的洞穴厅堂，再由地下河通道将它们互相连接。由于石花水洞长期处于地下河水流溶蚀侵蚀的状态，所以在洞壁形成和保留了大量溶蚀形态。③晚期洞穴继续上升而脱离地下水位，洞穴的滴水、洞壁的流水和渗水等开始活动，洞穴之上有数十米厚的碳酸盐岩地层，大量含有过饱和碳酸钙的溶液进入洞穴空间，为了与洞穴CO_2浓度取得平衡而释放出水溶液中的CO_2，于是造成形形色色的洞穴次生化学沉积物的沉积。

丹霞地貌景观资源产出于中生代中酸性火山盆地和陆内沉积盆地中。火山岩主要是火山碎屑岩，少量火山熔岩；陆内沉积盆地岩性主要由巨厚的灰色和紫色砂岩、砂砾岩、砾岩组成。岩石成分成熟度差，分选性差，颗粒大小混杂，结构成熟度低，抗风化能力低。因新构造运动，地壳抬升成山，山体长期遭受风化剥蚀，岩石因结构失衡而容易发生张裂破碎，形成水平和近垂直的裂隙面，大气降水和地表水沿裂隙面不断渗透、侵蚀，风化作用不断得到加强，裂隙逐渐扩大，岩石被分割成各种大小不等、形态各异的块体。随着风化作用的进行，部分块体因重力失衡而发生崩塌或滑落，最终形成孤峰或群峰兀立的险要景观。

结 束 语

本项目最初项目负责人为龙文国，副项目负责人为莫位明。2007 年 3 月，因工作需要，莫位明调离项目组，由李孙雄接替副项目负责人。2010 年 5 月，龙文国也由于工作变动而调离项目组，改由李孙雄接替项目负责人，同时新设云平和林义华两人为副项目负责人。尽管项目负责人发生新的更改和变动，但项目组其他成员基本上未发生变动，保证了项目的顺利进行。

项目实施过程中，海南省地质调查院主要负责人变动情况是：2006~2007 年，院长吴国爱，副院长傅杨荣，总工程师符峰；2008~2009 年，院长吴国爱，副院长傅杨荣，副总工程师杨昌松；2009 年底至今，院长傅杨荣，副院长兼总工程师陈沐龙，副总工程师杨昌松。

本书共 180 余万字。由李孙雄、云平、林义华统稿，陈哲培、王大英提出初步修改意见，最后由廖香俊、陈颖民、余兴邦、傅杨荣、陈沐龙审稿。图件编制资料主要来源于 1∶25 万区调。1∶25 万区调取得的地质成果非常丰富，在地层、岩石、构造等专业领域都有一些新的认识和突破。因此，本项目成果仅能是“精中求精”，围绕 1∶25 万区调存在的地质问题，以及海南省重要基础地质问题开展工作。项目成果是海南省最新基础地质资料的融合和集成，具有资料新、认识新、综合性强、范围广的特点。编图资料截至 2010 年底，把这几年来开展的 1∶5 万区调资料、发表的论文资料和科研资料，以及南海地区地质资料融入其中，是一次多专业、多学科资料的集成；同时较好地解决了海南省一些重要的基础地质问题，得出了一些新的地质认识和见解，提升了海南省基础地质的研究程度。项目成果将对海南省今后的地质工作起到促进作用，也将更好地服务于海南岛国际旅游岛建设和全省经济社会发展。

通过资料收集，结合野外地质调查，项目取得的主要地质成果和存在的地质问题如下。

一、取得的主要地质成果

（一）地层

1）对分布于乐东丘糖岭一带的原长坡组进行了修订，新建了丘糖岭组，时代为渐新世—中新世。把屯昌南棍园一带大面积分布的原二叠系峨查组重新修订为志留系陀烈组，对重新认识海南岛古生代和渐新世—中新世沉积作用和大地构造演化特征具有重要的地质意义。

2）首次系统地引入宝滩组（原东坡组，因与东北东坡组重名而修改）、阳明组、忠孝组、南湾组、永暑组、南沙组、南海组、西沙组、宣德组、永兴组、永发组、琛航组、石岛组、西沙洲组等 14 个岩石地层单位，进一步完善了海南省岩石地层单位，为开展区域地层对比和地层区划提供资料依据，也将促进海南省古近纪以来沉积作用和构造演化的研究。

3）采用高精度锆石 SHRIMP 测年法，在东方一带抱板群混合岩化片麻岩和中元古代花岗岩中分别获得了（1439±7）Ma、（1422±16）Ma，从而显示混合岩化作用与花岗岩侵入事件属于同一地质作用的产物，其动力学背景与距今约 14 亿年的造山运动（抱板运动）密切相关。为重新审视海南省前寒武纪地质构造演化提供了新的同位素测试资料。

4）首次采用断代地层划分法开展海南省地层区划，按中元古代、新元古代—古生代、中生代、古近纪—新近纪、第四纪等 5 个阶段进行了划分。共划分出一级地层区 10 个、二级地层区 9 个、三级地层区 15 个、四级地层区 1 个。

5）在东南面吊罗山林场一带残留状产出的片麻岩类和麻粒岩类岩石中，首次获得了锆石 ICP-MS 年龄为 243.1Ma、251.8Ma、261Ma 等峰值年龄，而缺乏中元古代年龄，表明该类岩石形成于海西期—

印支期。结合西面戈枕和南面红五地区片麻岩类和麻粒岩类岩石形成于14亿年的事实，反映了海南岛存在多期的深变质作用，其变质作用背景可能与同期造山作用伴随的广泛岩浆侵位活动有关。该海西期—印支期变质作用年龄的获得，为深入认识海南岛二叠纪—三叠纪造山运动特征提供了新资料。

（二）岩浆岩

1）首次对海南岛侵入岩进行了统一清理和归并，划分出100个不同的侵入体类型，并探讨其形成的构造环境，为海南省构造-岩浆作用旋回的划分，以及动力学背景研究提供了直观的资料依据。

2）在全面总结海南省岩浆作用的基础上，按抱板期、海西期—印支期、燕山期、喜马拉雅期4个阶段进行了构造-岩浆带的划分。

3）按照不同的物质来源将海南岛侵入岩划分4个大类，即M型（幔源型）、Ⅰ型（壳幔混合型）、S型（壳源重熔型）和相对特殊的A型（碱性、贫水、非造山）；以Ⅰ型成因的岩体居主要地位，次为S型花岗岩。

4）在文昌铜鼓岭南面沿海北东向展布的基性岩（脉）中获得锆石ICP-MS年龄约为244Ma，与万宁一带的基性—超基性镁铁质岩的生成时间（锆石SHRIMP年龄约为241Ma）基本一致，属于一条三叠纪生成起来的北东向岩浆带。为开展岩浆带划分和海西期—印支期构造-岩浆作用研究提供了新的资料，也将对特提斯构造域和滨太平洋构造域构造作用研究带来新的启发。

5）对三亚市天涯海角一带出露的早白垩世辉长岩进行了野外观测，首次获得了其锆石ICP-MS年龄约为100Ma，为岩石时代归属提供了直接的年代学资料。

6）首次对海南岛燕山期花岗岩含矿性及成矿时代进行了研究。采用Re-Os同位素方法，对海南岛三个典型的斑岩型钼矿床（罗葵洞、石门山及文且）中的辉钼矿进行了精确定年，获得其成矿年代为（80.2±0.6）Ma~（103.9±1.0）Ma。

7）从三叠纪至侏罗纪到白垩纪，花岗岩的Nd-Sr同位素组成及T_{2DM}不存在有序的变化，在侏罗纪发生了突变，可能暗示本区在侏罗纪发生了重要的构造环境的转折，显示出幔源物质的明显增多。由侏罗纪至白垩纪Nd-Sr同位素特征又发生了很大的变化，可能是幔源物质的性质发生了改变，即壳幔作用过程中壳源物质增多，壳幔物质可能发生了明显的交代混合作用，这可能与白垩纪具有埃达克质岩石特征的花岗岩的成因有关。

8）对分布于五指山地层分区古生代地层中的基性、超基性火成岩进行了一定的综合研究，为海南岛早古生代大地构造属性及演化特征的研究提供了重要资料。

9）对海南岛第四纪火山岩的喷发期次、分布范围进行了系统的研究。第四纪火山岩是多期次火山活动、多喷发中心形成的一套中基性玄武岩系。形成大陆板块内部、与大陆裂谷活动有关。

10）基本查清了海南岛中生代火山岩盆地中出露的岩石类型、岩性岩相特征、结构构造、产状、厚度等。中生代火山岩是大陆边缘内侧大陆弧构造环境的产物，是增生在大陆边缘的新陆壳。

11）对分布于琼北文昌市蓬莱镇及南海一带的幔源包体进行了系统研究，查清了其产出层位、分布范围，并对其进行了系统的岩石学、岩石化学、地球化学研究。

（三）变质岩

1）根据区域构造环境、区域构造变形、岩浆活动及变质矿物组合演变特征，将海南岛变质作用演化划分为抱板期、加里东期、华力西期、印支期及燕山期。以地质事件和区域构造环境分析为依据，探讨了海南岛地壳演化历史过程中不同时期和阶段变质作用的表现和特征。

2）系统总结研究区区域变质岩、接触变质岩、动力变质岩、气-液变质岩和混合岩的时空分布、岩石类型及特征。系统研究区内变质岩的变质矿物及矿物共生组合，较为详细地划分了区内变质岩的变质相带，将区域变质岩划分为4个变质相，热接触变质岩划分为2个变质相，混合岩划分为3个混合岩化带。通过对变质岩的地质特征、岩石学特征，岩石化学特征的综合研究，恢复了区内变质岩的原岩。

（四）地质构造

1）以板块构造理论为基础，以海南岛构造演化为主线，以沉积作用、岩浆作用和变质变形作用研究为手段，以最新完成的 1：25 万区调资料成果和前人地质资料以及各项科研成果为依据，结合海南省深部构造轮廓特征，从活动论角度出发，按照前南华纪、南华纪—三叠纪、侏罗纪—白垩纪、古近纪—第四纪 4 个阶段进行大地构造单元阶段划分。共划分出一级构造单元 8 个、二级构造单元 11 个、三级构造单元 10 个。

2）重新总结与建立了海南岛构造格架和变质变形序列。把海南岛变质变形期厘定为抱板期、华力西期—印支期、燕山期、喜马拉雅期。抱板期形成了分布于长城系抱板群中的区域角闪岩相变质和深层次的流变褶皱，晚晋宁期由青白口系石碌群中的区域高绿片岩相变质和中深层次的顺层剪切褶皱体现，海西—印支期导致了古生界发生了区域低绿片岩相变质和中浅层次的弹塑性褶皱变形，燕山期则主要是与岩浆侵入活动有关的接触变质和浅表层次的韧脆性断裂变形；喜马拉雅期属于表层次的脆性断裂变形，变质作用未发育。

3）首次开展系统的构造年代学研究。获得北西向构造形迹 $^{40}Ar-^{39}Ar$ 年龄为 211.3～200.6Ma 和 100.2～99.5Ma，北东向韧性变形岩石 $^{40}Ar-^{39}Ar$ 年龄为 293.4Ma、26～244.3Ma、156.8～196.5Ma、90.3～104.3Ma。结合其他地质资料综合分析，认为海南北东向构造和北西向构造的生成时间具有同步性，为研究海南岛该时期构造演化提供了系统的构造年代学资料。

4）在芒三水库-吊罗山林场-兴隆一带原中三叠纪花岗岩中，发现了北东向的韧性剪切变形带迹象。后沿变形带走向追索，在早白垩世花岗岩中未见相应的变形迹象。岩石属于同构造期侵位并发生强烈变形的产物。参照花岗岩测年方法，获得其锆石 U-Pb 年龄为 278.5Ma、265.9Ma，时代为二叠纪。该韧性变形带的发现，无疑将为海南岛大地构造分野及基础地质和成矿作用的研究带来新的启发，也将促进海南岛特提斯构造域作用的研究。

5）通过资料收集和分析，首次把触角延伸到海南省南海广大地区。认为海南省与印支地块具有相似的结晶基底组成，属于统一板块的范畴（可能与劳伦大陆有关）。总体经历了中元古代结晶基底形成、新元古代—古生代板内多阶段裂解与造山作用、中生代陆内盆-山演化和新生代裂解 4 个重要演化阶段。

（五）地球物理

1）在收集前人资料的基础上，本项目对海南岛岩（矿）石地球物理特征（包括弹性波特征、磁性特征、密度特征）做了简单总结。

2）收集了东方-琼海、儋州-保亭、树德-江洪、定安-东坡、常树-才芳东、儋州-澄迈-冯坡等断面的地球物理资料，并绘制相应的地壳断面图 、地壳内部结构图、地壳速度结构图、地壳上地幔二维速度结构和莫霍面等深度图等，为开展岩石圈三维结构研究提供了资料。

3）收集了南海东部、南海中部、南海西部、南海北部等 4 条岩石圈地学断面的地球物理资料，对上述 4 条岩石圈地学断面对南海的地壳内部结构、地壳速度结构、地壳上地幔二维速度结构和莫霍面等深度图等进行了详细的研究，为本项目增添了新的内容。

4）收集了海南岛 1：20 万重磁资料及南海海域的重磁资料，进行了数据处理，划分了重磁异常分区，并通过对重磁资料的研究，指出了海南岛及南海海域沉积基底面和莫霍面的深度，以及侵入岩的分布特征，为今后海南岛及南海海域的深部地质构造和基础地质的研究提供了资料。

5）收集海南岛及南海海域的岩石圈厚度资料，指出了海南岛和南海海域岩石圈厚度及其变化特征，为将来海南岛南海海域岩石圈结构特征的研究提供了依据。

此外，通过对不同图幅间存在的接图问题进行了野外调查及核实，从而使不同图幅间的地质界线不仅在图面上衔接吻合，同时地质体的划分与时代归属也趋于符合实际情况。

二、存在的主要问题

1）几年来，项目组多次对古生界中的变质火山岩进行补充性野外调查，一是调查变火山岩与其围岩间的关系，二是开展变火山岩年代学研究。但遗憾的是，由于岩石风化非常强烈，风化层厚度大，两者接触关系依然未能查明。而多次所选取的锆石样品中，也因所获得的锆石量都太少难以进行锆石SHRIMP 年龄测定（量少，代表性差）而未果。建议下步继续开展专题性的工作。

2）经对琼中长征地区戈枕村组分布区进行系统的观测，发现该地区浆混作用发育，部分地区并不同程度地保留有壳源重融不彻底的地层残留体，有些地段原混合岩（戈枕村组）并见有花岗岩透镜体残留和闪长岩的捕虏体。岩石片麻状构造发育，普遍见石榴子石矿物，但未发育基体和脉体结构，原岩石定名为混合岩值得商榷。考虑到本次所观测的地质资料有限，暂时保留该地区戈枕村组，待以后的地质工作对其进行进一步的厘定。

参 考 文 献

一、专著

地质矿产部南岭项目花岗岩专题组．1989. 南岭花岗岩地质及其成因和成矿作用．北京：地质出版社
广东省地矿局．1988. 广东省区域地质志．北京：地质出版社
广东省地质局．1991. 广东、海南石炭纪沉积相古地理及层控矿产预测．武汉：中国地质大学出版社
海南省地质矿产勘查开发局．1996. 中国矿床发现史（海南卷）. 北京：地质出版社
海南省地方志办公室．2004. 海南省志之地质矿产志．北京：南海出版公司
南京古生物所编著．2000. 中国地层研究二十年．合肥：中国科学技术大学出版社
全国地层委员会．2001. 中国地层指南及中国地层指南说明书（修订版）. 北京：地质出版社
郑亚东，常志忠．1985. 岩石的有限应变测量及韧性剪切带．北京：地质出版社
中国科学院华南富铁矿科学研究队．1986. 海南岛地质与石碌铁矿地球化学．北京：科学出版社
中国科学院南海海洋研究所．1990. 南海陆缘扩张．北京：科学出版社
中国科学院南沙科考队．1991. 南沙群岛及其邻区地质地球物理及岛礁研究论文集（一）. 北京：科学出版社，80~92
鲍才旺．1987. 南海地形图//南海地质地球物理图集．广州：广东科技出版社
曹伯勋主编．2008. 地貌学及第四纪地质学．武汉：中国地质大学出版社
陈圣源．1987. 南海磁力异常图//南海地质地球物理图集．广州：广东科技出版社
陈旭，戎嘉余主编．1996. 中国扬子区兰多弗里统特列奇阶及其与巩英国的对比．北京：科学出版社
陈旭，阮苏萍，A. J. 布科主编．2001. 中国古生代气候演变．北京：科学出版社，35~38，271~290
陈哲培，钟盛中，何圣华等．1997. 全国地层多重划分对比研究（46）海南省岩石地层．武汉：中国地质大学出版社
丁式江等．2007. 海南乐东抱伦金矿地质及矿产预测．北京：地质出版社
丁原章．1988. 海南岛北部地震研究文集．北京：地震出版社
冯增昭，彭勇民，罗璋等．2001. 中国南方寒武纪和奥陶纪岩相古地理．北京：地质出版社
郭令智，施央申，马瑞士．1980. 东南大地构造格架和地壳演化．国际交流地质学术论文集(一).北京：地质出版社
何廉声．1987. 南海地质构造图//南海地质地球物理图集．广州：广东科技出版社
何起祥等．2006. 中国海洋沉积地质学．北京：海洋出版社
黄镇国，蔡福祥，韩中元等．1993. 雷琼第四纪火山．北京：科学出版社
赖才根等．1982. 中国地层 5・中国的奥陶系．北京：地质出版社
李昌年．1992. 火成岩微量元素岩石学．武汉：中国地质大学出版社
李云通等．1984. 中国地层 13・中国的第三系．北京：地质出版社
刘宝珺，曾允孚．1985. 岩相古地理基础和工作方法．北京：地质出版社
刘若新主编．1992. 中国新生代火山岩年代学与地球化学．北京：地震出版社
刘昭蜀，赵焕庭，范时清等．2002. 南海地质．北京：科学出版社
马昌前，杨坤光，唐仲华，等．1994. 花岗岩类岩浆动力学．武汉：中国地质大学出版社
马大铨等．1991. 海南岛花岗岩类．见：汪啸风主编，海南岛地质（二）岩浆岩．北京：地质出版社
马大铨，黄香定，肖志发等．1997. 海南岛结晶基底——抱板群层序与时代．武汉：中国地质大学出版社
莫柱荪，叶伯丹，潘维祖等．1980. 南岭花岗岩地质学．北京：地质出版社
乔秀夫，耿树方．1981. 华南晚寒武纪古板块构造．中国及其邻区大地构造论文集．北京：地质出版社
邱家骧，林景仟主编．1991. 岩石化学．北京：地质出版社
单文琅，宋鸿林，傅昭仁，等．1991. 构造变形分析的理论、方法和实践．武汉：中国地质大学出版社
石桂华，张如玉等．1988. 海南岛北部火山岩宝石及其伴生矿物标型特征．北京：科学技术出版社
宋之琛等．1981. 江苏地区白垩纪—第三纪孢粉组合．北京：地质出版社

汪啸风 . 1982. 江南区的奥陶系 . 见赖才根等 . 北京：地质出版社
汪啸风，陈孝红 . 1999. 中国奥陶纪古生物地理与古气候，地层古生物论文集第二十七辑 . 北京：地质出版社
汪啸风，马大铨，蒋大海等 . 1991. 海南岛地质（二）岩浆岩 . 北京：地质出版社
汪啸风，马大铨，蒋大海等 . 1991. 海南岛地质（三）构造地质 . 北京：地质出版社
汪啸风，马大铨，蒋大海等 . 1992. 海南岛地质（一）地层古生物 . 北京：地质出版社
王鸿祯等 . 1986. 华南地区古大陆边缘构造史 . 武汉：武汉地质学院出版社
肖庆辉，邓晋福，马大铨等 . 2002. 花岗岩研究思维与方法 . 北京：地质出版社
许汉奎等 . 1977. 海南岛崖县大茅中寒武世腕足类 . 中国科学院铁矿地质学术会议论文选集　地层和古生物 . 北京：科学出版社
许汉奎，苏一保 . 1979. 海南岛南部中奥陶世腕足类 . 北京：科学出版社
许寿永，杨德骊 . 1987. 中南地区早石炭世生物地层特征 . 见：全国石炭纪会议论文专集 . 北京：地质出版社，21～46
薛万俊 . 1987. 南海地貌图//南海地质地球物理图集 . 广州：广东科技出版社
姚伯初 . 1991. 南海海盆在新生代的构造演化//南海地质研究 . 武汉：中国地质大学出版社
姚伯初 . 1993. 南海北部陆缘新生代构造运动初探//南海地质研究所 . 武汉：中国地质大学出版社
姚伯初，曾维军，Hayes D E 等 . 1994. 中美合作调研南海地质专报 . 武汉 . 中国地质大学出版杜
姚伯初 . 1996. 南海海槽的构造特征及其构造演化史//南海地质研究所 . 武汉：中国地质大学出版社
姚伯初，万玲等 . 2006. 中国南海海域岩石圈三维结构及演化//中国岩石圈三维结构丛书之八 . 北京：地质出版社
曾庆銮，袁春林，郭真民等 . 1992. 海南岛三亚地区基础地质研究 . 武汉：中国地质大学出版社
张仁杰，马国干，蒋大海等 . 1991. 海南岛前寒武纪地质 . 武汉：中国地质大学出版社
张训华等 . 2008. 中国海域构造地质学 . 北京：海洋出版社
赵子杰，吴良士，郑基俭等 . 1989. 南岭花岗岩地质及其成因和成矿作用 . 北京：地质出版社
郑百让，金淑燕 . 1989. 构造岩组学 . 武汉：中国地质大学出版社
朱介寿等 . 2005. 中国华南及东海地区岩石圈三维结构及演化 . 北京：地质出版社
朱伟林，张功成，杨少坤等 . 2007. 南海北部大陆边缘盆地天然气地质 . 北京：石油工业出版社
Bao Caiwang. 1987. Explanation for map of topography//Atlas of Geology and Geophysics of south China Sea. Guangzhou：Guangdong Science & Technology Press.（in Chinese）
Xue Wanjun. 1987. Explanation for map of geomorphology // Atlas ofGeology and Geophysics of South China Sea. Guangzhou：guangdong Science&Technology Press.（in Chinese）

二、论文

陈炳蔚，艾长兴 . 1978. 海南岛地质构造新情况 . 地质矿产研究，2：98～102
陈柏林，蔡本俊，林仟同 . 1996. 陈建国 . 崇安-石城构造带中段韧性变形特征 . 地质力学学报，2（1）：49～54
陈柏林，丁式江，李中坚等 . 2001. 海南抱伦金矿床成矿时代研究 . 地球化学，30（6）：525～532
陈柏林，李中坚，董诚等 . 2004. 海南抱伦金矿床控矿构造特征及其对金矿化的控制作用 . 中国地质，31（2）：139～146
陈国达，关伊文等 . 1977. 海南岛石碌式铁矿的大地构造成矿条件初探 . 中南矿冶学院学报，（3）：1～12
陈海泓，孙枢，李续亮等 . 1994. 华南早三叠世的古地磁学和大地构造 . 地质科学，29（1）：1～9
陈海泓，肖文交 . 1998. 多岛海型造山作用——以华南印支期造山带为例 . 地学前缘，增刊（5）：95～101
陈惠芳，侯威，梁新权 . 1991. 海南岛两项重要新发现 . 中国地质，（6）：29～30
陈惠民，黄咏茵 . 1987. 1605 年海南岛琼州大地震考察研究（Ⅰ、Ⅱ）. 广东地质快报，（2）
陈沐龙，李孙雄，曾雁玲等 . 2008. 海南岛白垩纪千家岩体岩石地球化学特征及其成矿作用分析 . 矿产与地质，22（1）：36～42
陈沐龙，李孙雄等 . 2009. 海南岛西部燕山期逆冲推覆构造特征 . 矿产与地质
陈沐龙，李孙雄等 . 2011. 海南岛西部青白口系石碌群褶皱变形特征及其成矿分析 . 华南地质与矿产
陈培荣，华仁民，章邦桐等 . 2002. 南岭燕山早期后造山花岗岩类：岩石学制约和地球动力学背景 . 中国科学（D 辑），32（4）：279～289
陈旭，戎嘉余，伍鸿基等 . 1991. 川陕边境广元宁强间的志留系 . 地层学杂志，15（1）：1～25
陈新跃，王岳军，范蔚茗等 . 2006. 琼西南 NW 向韧性剪切带构造特征及其 ^{40}Ar-^{39}Ar 年代学约束 . 地球化学，35（5）：

479~488
陈新跃，王岳军，范蔚茗等 . 2006. 琼西南 NE 向韧性剪切带构造特征及其$^{40}Ar-^{39}Ar$年代学约束．地球化学
陈哲培，黄良钧，钟盛中 . 1985. 南好地区古生代地层划分及时代．广东地质科技，(2)：22~27
陈志刚，李献华，李武显等 . 2003. 赣南全南正长岩的 SHRIMP 锆石 U-Pb 年龄及其对华南燕山早期构造背景的制约．地球化学，32 (3)：223~229
蔡学林，朱介寿，曹学敏等 . 2002. 东亚西太平洋巨型裂谷体系岩石圈与软流圈结构及动力学．中国地质，29 (3)：234~245
陈耀钦，黄宇辉，陈培权等 . 1989. 论海南、广东地体．岩相古地理，42 (4)：13~19
陈泽铭等 . 1979. 海南岛南石铁矿含矿层的划分及其找矿意义．南京大学学报地质学专刊 (一)：105~120
陈哲培 . 1987. 海南岛南好地区发现志留纪地层．地层学杂志，11 (3)：230~233
陈哲培，钟盛中 . 1991. 海南省海口地区第四纪火山岩盆地沉积特征．中国区域地质，(4)：313~322
陈哲培 . 1998. 海南岛上新世地层特征及岩石地层单位．中国区域地质，17 (4)
陈哲培，黄香定，钟盛中 . 1998. 海南岛蓬莱地区新第三纪火山盆地沉积特征与成矿．中国区域地质，17 (1)：94~101
周新民等 . 2007. 南岭地区晚中生代花岗岩成因与岩石圈动力学演化．北京：科学出版社，74~106
丁式江 . 1991. 二甲金矿韧性剪切带及金矿化．矿产与地质，5：357~363
丁式江，彭少梅，董宝林 . 1994. 琼西王下滑覆构造初探．华东地质学院学报，(3)：226~232
丁式江 . 1994. 琼西戈枕断裂带构造演化与金矿化系列．地质找矿论丛，9 (3)：1~7
丁式江 . 1994. 海南岛中西部金矿集中区断裂构造的分形研究．地学前缘，11 (1)：189~194
丁式江 . 1995. 琼西弋枕侵入体的地质特殊性征及其发现意义．大地构造与成矿学，19 (4)：336~343
丁式江，黄香定等 . 2001. 海南抱伦金矿地质特征及其成矿作用，中国地质，5
丁式江，周祖翼，许长海等 . 2002. 海南屯昌变火山岩构造属性及其年代学研究．岩石学报，18 (1)：83~90
付建明 . 1991. 海南岛蓬莱地区超镁铁岩包体及其寄主玄武岩的研究．矿物岩石，11 (4)：22~30
傅杨荣，张仁杰，廖卓庭等 . 2005. 海南岛含杜内期腕足动物群的发现．地层学杂志，29 (1)：34~38
郭令智，施央申，马瑞士．1983. 西太平洋中生代活动大陆边缘和岛弧构造的形成及演化．地质学报，(1)：11~21
葛梅钰等 . 1983. 海南岛崖县地区寒武纪、奥陶纪地层．地层学杂志，7 (1)：41~49
葛小月，李献华，周汉文 . 2003. 琼南晚白垩世基性岩墙群的年代学、元素地球化学和 Sr-Nd 同位素研究．地球化学，32 (1)：11~20
何龙清，余风鸣，刘国庆 . 1996. 海南戈枕剪切带的构造变形与演化．华南地质与矿产，(3)：46~53
侯威，陈惠芳，梁新权等 . 1992. 海南岛前寒武纪地层的确定及其大地构造演化．长春地质学院学报，(2)：133~143
黄香定 . 1998 海南岛山岳地貌景观．海南地质地理，2：1~5
贾精一等 . 1985. 石碌群的"W"型的叠加褶皱及大型组构解析．中国区域地质，12：79~90
雷裕红，丁式江，马昌前等 . 2005. 海南岛地壳生长和基底性质的 Nd 同位素制约．地质科学，40 (3)：439~456
梁德华，李杨 . 1991. 南海宪北海山玄武岩中超镁铁岩包体．南海地质研究 (四)
吕古贤 . 1988. 海南岛石碌铁矿岩系中火山岩类的新发现与研究．中国区域地质，1：52~56
廖其林，王振明，李振江等 . 1986. 福州-泉州-汕头地区结构的爆炸地震研究．地球物理学报，3 (3)：270~280
刘瑞华，张仲英 . 1989. 海南岛的新构造运动特征．热带地理，9 (2)：174~182
刘时藩 . 1983. 对我国早期脊椎动物生物地理特征的一些认识．古脊椎动物与古人类，21 (4)，292~300
刘时藩，1995. 塔里木西北的中华棘鱼化石及地质意义．古脊椎动物学报，33 (2)：85~98
刘时藩，1997. 中国的棘鱼鳍刺化石．古生物学报，36 (4)：473~484
李江海，穆剑．1999. 我国境内格林威尔期造山带的存在及其对中元古代末期超大陆再造的制约．地质科学，(3)：259~272
李孙雄，莫位明，云平等 . 2003. 琼西北洛基韧性剪切带的地球化学特征．广东地质，18 (1)：1~9
李孙雄，莫位明，云平等 . 2004. 海南千家地区主要断裂带特征和控岩控矿作用分析．地质力学学报，10 (2)：137~145
李孙雄，王大英，莫位明 . 2004. 琼西南红五地区结晶基底褶皱变形分析及其找矿地质意义．广东地质，19 (1)：31~42
李孙雄，谢盛周，莫位明等 . 2004. 琼大村韧性变形带地质地球化学特征．华南地质与矿产，(1)：29~34
李孙雄，云平，范渊等 . 2005. 海南岛琼中地区琼中岩体锆石 U-Pb 年龄及其地质意义．大地构造与成矿学，29 (2)：

227~233
李孙雄，范渊，莫位明等．2006. 海南岛古生代弧状构造带的特征及其地质意义．矿产与地质，20（3）：232~236
李孙雄，丁式江，范渊等．2006. 海南岛鹦歌岭白垩系红盆中韧性剪切带的地质特征．地质力学学报，12（3）：387~392
李孙雄，陈沐龙，周进波等．2007. 海南岛中部古生代沉积盆地格局及其演化特征．华南地质与矿产，（2）：41~46
李孙雄，陈哲培，汪焰华等．2011. 海南省乐东县丘糖岭古—新近纪地层特征及岩石地层单位．华南地质与矿产
龙文国．1999. 海南岛中生代地洼期构造与金银多金属成矿特征．大地构造与成矿学，23（4）：308~314
龙文国，黄正壮，王大英．2001. 海南抱板地区片麻状花岗闪长岩锆石 Pb-Pb 同位素年龄及地质意义．广东地质，（2，3）：7~12
龙文国，符策锐，朱耀河．2002. 海南岛东部黄竹岭地区“抱板群”的解体．地层学杂志，21（3）：212~215
龙文国，丁式江，李惠民．2005. 海南岛东部黄竹岭地区变质火山岩锆石 U-Pb 同位素年龄及其地质意义．地质通报，24（1）：92~94
龙文国，丁式江，马大铨等．2005. 海南岛前寒武纪基底组成及演化．地球科学-中国地质大学学报，30（4）：421~428
龙文国，林义华，朱耀河等．2006. 海南岛北部第四纪早中更新世多文组的建立．地质通报，25（3）：408~414
龙文国，林义华，石春等．2006. 海南岛北部更新世道堂组的重新厘定．地质通报，25（4）：469~474
吕文正．1987. 南海中央海盆条带磁异常特征及构造演化．海洋学报，90：69~78
李献华．1993. 华南地壳增长和构造演化的年代学格架与同位素体系制约．矿物岩石地球化学通讯，3：111~115
李献华，周汉文，刘颖等．1999. 桂东南钾玄质岩浆岩带的厘定．科学通报，44（18）：1992~1998
李献华，周汉文，丁式江等．2000. 海南岛洋中脊型变质基性岩：古特提斯洋壳的残片．科学通报，45（1）：84~89
李献华，周汉文，丁式江等．2000. 海南岛“邦溪-晨星蛇绿岩片”的时代及其构造意义-Sm-Nd 同位素制约．岩石学报，16（3）：425~432
梁新权．1995. 海南岛前寒武纪花岗岩-绿岩系 Sm-Nd 同位素年龄及地质意义．岩石学报，11（1）：71~76
梁新权，范蔚茗，许德如．2000. 海南岛屯昌玄武质科马提岩 Sm-Nd 同位素年龄及地质意义．地质科学，35（2）：240~244
廖香俊，王平安，覃海灿等．2008. 海南屯昌地区高通岭钼矿床地质、地球化学特征及成矿时代．地质通报，（4）：560~570
刘月清．1993. 海南岛西部“军营群”发现志留纪微古植物化石．中国区域地质，（1）：94~95
刘以宣．1984. 一种新型地洼区——南海大地构造与演化概述，大地构造与成矿学，8（2）：194~200
李兆麟，丘志力，秦社彩等．1991. 南海海山玄武岩形成条件研究．矿物学报，11（4）
李志宏．1991. 海南岛省青天峡组牙形石研究新进展．广东地质，6（1）：75~81
李志宏，谢才富，付太安等．2000. 海南省南好地区志留纪地层新知．华南地质与矿产，4：23~27
黎作聪，1980. 论湖北含中华棘鱼层位的时代问题．地层学杂志，4（3）：221~225
廖卓庭，张仁杰．2006. 海南岛白沙县金波早石炭世早期腕足动物群．古生物学报，45（2）：153~174
虞子冶．1989. 海南岛早二叠世冰海沉积的确定及其大地构造意义．南京大学学报（自然科学报），25
莫宴情，施央申．1987. 海南岛地体及其毗邻陆缘晚中生代—新生代古地磁研究和构造演化．南京大学学报，23（3）：521~532
马大铨，黄香定，陈哲培等．1997. 海南岛抱板群研究新进展．中国区域地质，16（2）：130~136
毛景文，李红艳，王登红等．1998. 华南地区中生代多金属矿床形成与地幔柱关系．矿物岩石地球化学通报，17（2）：130~132
南颐．1987. 对海南岛石碌铁矿地层时代的再认识．广东地质，2（2）
彭格林．1990. 海南岛沉积建造特征及大地构造问题．大地构造与成矿学，（4）：293~303
冯连顺．1988. 海南岛抱板金矿床同位素地质特征的初步研究．贵金属地质，（3，4）：233~239
孙涛，周新民，陈培荣等．2003. 南岭东段中生代强过铝花岗岩成因及其大地构造意义．中国科学（D 辑），33（12）：1209~1218
樊祺城，孙谦，李霓等．2004. 琼北火山活动分期与全新世岩浆演化．岩石学报，20（3）
丘元禧，张伯友．2000. 华南古特提斯东延问题的探讨．中国区域地质，19（2）：175~180
丘志力，秦社彩，庞学斌．1995. 海南碱性玄武岩中的刚玉巨晶成因探讨．中山大学学报（自然科学版），（3）：95~101
丘志力，秦社彩，陈小明．2001. 海南蓬莱刚玉巨晶中铌（钽）铁矿包裹体及其意义．矿物学报，21（2）：205~208

任纪舜 . 1984. 中国东部构造单元划分的几个问题 . 地质论评，(4)：7~12
孙嘉诗 . 1991. 南海北部及广东沿海新生代火山活动 . 海洋地质与第四纪地质，11（3）
盛金章 . 1965. 海南岛西部的蠖类 . 古生物学报，13（4）
唐红峰 . 1999. 海南岛晨星玄武岩地球化学特征及其构造环境含义 . 大地构造与成矿学，23（4）：323~327
单惠珍 . 1990. 海南抱板地区金矿地层时代归属的研究 . 中山大学学报（自然科学版），29（2）：71~77
谭忠福，蒋大海，马国干 . 1991. 海南抱板群同位素年龄及地质意义 . 中国区域地质，(3)：240~245
涂绍雄 . 1993. 海南抱板群变质原岩建造和地质演化初论 . 广东地质，8（3）：17~22
涂绍雄，战明国 . 1994. 海南戈枕剪切带构造岩石英组构和超微构造研究 . 矿物岩石，14（1）：42~52
涂绍雄，王桂华 . 2004. 花岗质岩石新地球化学分类介绍及有关问题评述 . 华南地质与矿产，(1)
王可伏等 . 1992. 海南岛土外山古元古代玄武岩成因探讨 . 地质地球化学，(6) 72~73
王朴，胡继宗等 . 1988. 新疆柯坪地区 *Sinacanthus* 的发现及其地层意义 . 新疆地质，6（3）：47~50
吴浩若，邝国敦，王忠诚 . 1997. 广西晚古生代构造沉积背景的初步研究 . 地质科学，32（1）：11~18
吴浩若 . 2003. 晚古生代—三叠纪南盘江海的构造古地理问题 . 古地理学报，5（1）：63~75
吴克刚等 . 1988. 海南岛现代风成岩（沙丘岩）的形成及其地质意义 . 地理研究，7（2）：67~72
王强，许继锋，王建新，赵振华等 . 2000. 北大别山 adakite 型灰色片麻岩的确定及其与超高压变质作用的关系 . 科学通报，45（10）
王赐银，季寿元，刘家旺 . 1979. 海南岛东方县抱板群混合岩的岩石学特征及其成岩、时代的探讨 . 南京大学学报-地质学专刊（一），23~42
王贤觉，吴明清，梁德华等 . 1984. 南海玄武岩的某些地球化学特征 . 地球化学
王叶剑，韩喜球，罗照华等 . 2009. 晚中新世南海珍贝-黄岩海山岩浆活动及其演化：岩石地球化学和年代学证据 . 海洋学报，31（4）
王志浩，S. M. Bergstörm 等 . 1996. 中国奥陶纪牙形刺分区和生物地层 . 古生物学报，35（1）：26~59
吴泽龙 . 1985. 海南岛地壳运动和地震活动概况 . 地震地质，(2)
徐安武，张业明，付建明等 . 1997. 海南岛西部早古生代裂陷槽 . 华南地质与矿产，(2)：26~33
夏邦栋 . 1979. 海南岛海西地槽的基本特征 . 南京大学学报-地质专刊（一），57~74
夏邦栋，任震鹏 . 1979. 海南岛石碌及其外围地区的地层及沉积建造 . 南京大学学报-地质学专刊（一），43~55
夏邦栋，施光宇，方中等 . 1991. 海南岛晚古生代裂谷作用 . 地质学报，65（2）：103~115
谢才富，张开明，黄照先等 . 1999. 琼东乐来地区侵入岩岩石谱系单位的地质特征及同位素年代学研究 . 华南地质与矿产，(1)：26~37
谢才富 . 2002. 同构造花岗岩的一种显微构造标记 . 岩石矿物学杂志，21（2）：179~185
谢才富，朱金初，赵子杰等 . 2005. 三亚石榴霓辉石正长岩的锆石 SHRIMP U-Pb 年龄：对海南岛海西-印支期构造演化的制约 . 高校地质学报，11（1）：47~57
谢才富 . 2007. 海南岛海西期同逆冲侵入岩组合的厘定及其构造意义 . 南京大学博士学位论文
许德如，梁新权，唐红峰 . 2000. 琼西抱板群变基性火山岩的地球化学特征及其大地构造意义 . 大地构造与成矿学，24（4）：303~313
许德如，林舸，梁新权等 . 2001. 海南岛前寒武纪岩石圈演化的记录：基性岩类岩石地球化学证据 . 岩石学报，17（4）：599~608
许德如，梁新权，唐红峰 . 2002. 琼西抱板群变质沉积岩地球化学研究 . 地球化学，31（1）：153~160
许德如，陈涛，唐红峰 . 2002. 海南岛存在类 TTG 岩系包体的证据及地质意义 . 矿物岩石地球化学通报，22（3）：209~214
许德如，陈广浩，夏斌等 . 2003. 海南岛几个重要基础地质问题评述 . 地质科技情报，22（4）：37~44
许德如，马驰，李鹏春等 . 2007. 海南岛变碎屑沉积岩锆石 SHRIMP U-Pb 年龄及地质意义 . 地质学报，81（3）：381~393
谢窦克，商玉强 . 1989. 东南大陆岩石圈板块地体构造 . 中国地质科学院南京地质矿产研究所所刊，10（4）：1~12
肖国林，郑浚茂 . 2004. 对南海北部陆坡至东海南部“残留特提斯”的几点认识 . 现代地质，18（1）：103~109
熊小林等 . 2001. 西天山阿吾拉勒 adikite 型钠质中酸性岩及地壳垂向增生 . 科学通报，46（4）
徐光洪 . 1979. 粤中地区早二叠世晚期地层及菊石化石 . 国家地质总局宜昌地质矿产研究所
徐林，沈永欢 . 1977. 海南岛昌江县石碌矿区地层的初步看法和意见 . 中南矿冶学院学报，3：62~71

徐义刚，黄小龙，颜文等 . 2002. 南海北缘新生代构造演化的深部制约（Ⅰ）：幔源包体 . 地球化学，31（3）：230~241
杨开庆，重法先，王建平等 . 1988. 海南石碌铁矿区铁、金、铜、钴构造动力成矿作用的研究 . 中国地质科学院地质力学研究所所刊，11：63~152
姚伯初，曾维军，陈艺中等 . 1993. 南海北部陆缘东部的地壳结构 . 地球物理学报，37（1）：44~53
姚伯初 . 1998. 南海北部陆缘的结构及构意义 . 海洋地质与第四纪地质，18（2）：1~16
姚伯初，万玲等 . 2010. 南海岩石圈厚度变化特征及其构造意义 . 中国地质，（4）
姚华舟，黄照先 . 1999. 海南岛万宁发现奥陶纪地层 . 华南地质与矿产，（1）：13~17
姚华舟，黄照先，张开明等 . 1999. 海南岛万宁地区的志留系 . 华南地质与矿产，（2）：36~42
姚华舟，黄照先，谢才富等 . 1999. 海南岛万宁地区寒武纪地层及其沉积特征 . 地层学杂志，23（4）：270~276
叶伯丹 . 1990. 海南东方二甲抱板群和金矿的时代 . 地质找矿论丛，5（1），12~17
俞昌民，王成源，阮亦萍等 . 1988. 桂林南边村泥盆—石炭系界线层型剖面简介 . 广西地质，1（1）：45~52
云平，莫位任，李孙雄等 . 2002. 海南岛中部侏罗纪基性小岩体岩石地球化学特征及其构造意义 . 广东地质，17（4）：9~16
云平，范渊，莫位任 . 2003. 海南岛晚中生代壳幔岩浆混合作用-来自闪长质淬冷包体的证据 . 华南地质与矿产，（2）：30~35
云平，雷裕红，吕嫦艳 . 2005. 海南岛中北部三叠纪花岗岩源区的锶、钕同位素制约及其意义 . 大地构造与成矿学，29（2）：234~241
阎贫，刘海龄 . 2005. 南海及其周缘中新生代火山活动时空特征与南海的形成模式 . 热带海洋学报，24（2）
鄢全树，石学法，王昆山等 . 2008. 南海新生代碱性玄武岩主量、微量元素及 Sr-Nd-Pb 同位素研究 . 中国科学（D 辑），38（1）
鄢全树，石学法，刘季花等 . 2008. 南海新生代碱性玄武岩中斜长石矿物的化学成分及意义 . 矿物学报，28（2）
鄢全树，石学法 . 2008. 南海新生代碱性玄武岩中橄榄石的矿物化学及南海的地幔潜在温度 . 岩石学报，24（1）
鄢全树，石学法，王昆山等 . 2007. 南海新生代玄武岩中单斜辉石矿物化学及成因意义 . 岩石学报，23（11）
鄢全树，石学法 . 2007. 海南地幔柱与南海形成演化 . 高校地质学报告，13（2）
杨树锋，虞子治，郭令智等 . 1989. 海南岛的地体划分、古地磁研究及其板块构造意义 . 南京大学学报（地球科学），1（1-2）：38~46
俞受鋆，夏萍，邓铁殷等 . 1992. 海南抱板地区中元古代花岗岩副矿物锆石特征及 U-Pb 同位素年龄测定 . 地球化学，3：213~219
袁奎荣等 . 1977. 海南岛西部海西地槽的发现及其对找铁矿的意义 . 中南矿冶学院学报，（3）：13~25
袁奎荣等 . 1980. 再论海南岛及大陆东南沿海的海西地槽与铁矿找矿问题 . 中南矿冶学院学报，（3）：1~10
严正，袁宝印，叶莲芳 . 1979. 海南岛玻璃陨石（雷公墨）裂变径迹年龄的测定 . 地质科学，（1）：37~42
朱炳泉，王慧芳 . 1989. 雷琼地区 MORB-OIB 过渡型地幔源火山作用的 Nd-Sr-Pb 同位素证据 . 地球化学，（3）：193~200
曾广策 . 1986. 海南岛北部碱性玄武质岩石中的深源包体和巨晶矿物 . 矿物岩石学论丛，（2）：98~107
邹和平 . 1993. 试谈南海海盆地壳属性问题 . 大地构造与成矿学，17（4）
邹和平 . 1997. 海南岛后地台造山-造盆模型：火成岩地球化学制约 . 大地构造与成矿学，21（2）：117~123
张虎男，赵希涛 . 1984. 雷琼地区新构造运动的特征 . 地质科学，（3）：276~286
谌建国，徐光洪 . 1981. 广东二叠纪煤系及含煤控制因素 . 中国地质科学院宜昌地质矿产研究所所刊，4：57~79
张理刚，王可法，陈振胜等 . 1994. 论“华夏古大陆”—Pb 同位素研究证据 . 地质论评，40（3）：200~208
曾庆銮，李志宏，谢才富等 . 2003. 论海南岛地区志留系 . 地层学杂志，27（4）：267~275
曾庆銮，李志宏，谢才富等 . 2004. 海南岛早志留世晚期腕足类 *Xinanospirifer* 的发现——兼论南好组 . 古生物学报，43（1）：86~93
战明国 . 1998. 花岗岩类分类与定位机制研究动向和进展 . 中国区域地质，17（2）：182~188
张旗，王焰，钱青等 . 2001. 中国东部燕山期埃达克岩的特征及其构造-成矿意义 . 岩石学报，17（2）
张旗，王焰，李承东等 . 2006. 花岗岩按照压力的分类 . 地质通报，25（11）
张仁杰，马国干，徐光洪等 . 1989. *Chuaria-Tawuia* 生物群在海南岛石碌群的发现及其意义 . 中国科学（D 辑），19（3）：304~311
张仁杰，马国干，冯少南等 . 1992. 海南石碌铁矿的 Sm-Nd 法年龄及其意义 . 地球科学，26（1）：38~43

张仁杰，蒋志文等.1998. 海南岛万宁县早寒武世地层及微体化石. 古生物学报，37（2）：235~244
张仁杰，胡宁，冯少南.2000. 海南岛保亭地区南好组发现志留纪化石. 中国学术期刊文摘（科技快报），6（7）：882~884
张仁杰，王志浩，胡宁. 2001. 海南岛昌江地区石炭纪牙形刺化石. 微体古生物学报，18（1）：35~42
张仁杰，王成源，胡宁等.2001. 海南岛法门期生物地层. 中国科学，21（5）：406~412
张仁杰，胡宁，姚华舟等.2002. 海南岛保亭地区南好组的地质时代. 地层学杂志，26（2）：101~105
张仁杰，胡宁，冯少南等.2000. 海南岛保亭地区南好组发现志留纪化石. 中国学术期刊文摘（科技快报），6（7）：882~883
张业明，张仁杰，姚华舟等.1997. 海南岛前寒武纪地壳构造演化. 地球科学，22（4）：395~400
张业明，谢才富，付太安等.2005. 三亚地体和琼中地体构造边界的直接证据. 科学技术与工程，5（20）：1482~1484
张喜光.1986. 广东省海南岛崖县中寒武世的高肌虫及其地质意义. 古生物学报，25（1）：63~74
张志存.1988. 海南岛二叠纪研究的进展. 科学通报，（9）
张仲英，刘瑞华.1989. 海南岛第四系火山岩的分期. 地质科学，（1）：67~76
Barry J. Cooper, 1981. Early Ordovician conodonts from the Horn Valley siltstone, central Australia. Palaeontology, 24 (1): 147~183, Pls. 26~32
Bergström, S. M. 1973. Ordovician conodonts. In: Hallam, A (ed.), Atlas of Palaeobiogeography, 47~58. Elsevier
Batchelor R A, Bowden P. 1985. Petrogenetic interpretation of granitoid rock series using multi-cationic parameters. *Chem Geol*, 48: 43~55
Cai Xueling Zhu Jieshou, Cao Xueming, et al. 2002. The lithospheric and rheospheric structure of the Asian and the Pacific rifting system and dynamics. Geology in China, 29 (4): 234 ~245 (in Chinese with English abstract)
Defant M J, Drummond W S. 1990. Derivation of some modern arc magmas by young subduction lithospere. Nature, 662~665
Gilder S A, Keller G R, Luo M, et al. 1991. Timing and spatial distribution of rifting in China. Tectonophysics, 197 (2-4): 225~243
Gilder S A, Gill J B, Coe R S, et al. 1996. Isotopic and paleomagnetic constraints on the Mesozoic tectonic evolution of South China. J Geophys Res, 101 (B7): 16137~16154
Huang Xiangding, Chen Zhepei and Zhong Shengzhong (Bureau of Geo. & Min. Resources, Hainan Province, China), Ma Daquan, Xiao Zhifa and Zhang Wangchi (Yichang Institure of Geo. & Min. Resources, Hainan Province, China). 1996. On the Seguence and age of baoban Group in Hainan Province, China. 30th International Geological Congress, Abstracts, 2, Beijing, China
Ludwig W J. 1970. The Manila Trench and West Luzo Trough: Seismic—refraction measurements. Deep Sea Research, 17 (3): 27~45
Ludwig W J. 1979. Profiler-sonobuoy measurement in the South China Sea Basin. J. of Geophysical Research, 84; 3505~3518
Li X H. 2000. Cretaceous magmatism and lithospheric extension in Southeast China. J. Asian Earth Sci, 18 (3): 293~305.
Li X H. 1999. U-Pb zircon ages of granites from the southern margin of the Yangtze Block: timing of Neoproterozoic Jinning: Orogeny in SE China and implications for Rodinia Assembly. Precambrian Research, 97: 43~57
Li Z X, Zhang L, Powell C, Mc A. 1995. South China in Rodinia: Part of the missing link between Australia-EastAntarctica and Laurentia. Geology, 23, 410~707
Li Z X. 1996. Positions of the East Asian cratons in the Neo- proterozoic supercontinent Rodinia. Aust. J. Earth Sci, 43: 593~604
Li Z X, Li X H, Kinny P D, Wang J. 1999. The breakup of Rodinia: Did it start with a mantle plume beneathSouth China? Earth and Planetary Science Letters, 173: 171~181
Li Z X, Li X H, Zhou H W, Kinny P D. 2002. Grenvillian continental collision in south China: New SHRIMP U-Pb zircon results and implications for the configuration of Rodinia. Geology, 30: 163~166
Li Zhengxiang, Li Xianhua, Zhou Hanwen, Peter D. Kinny. 2002. Grenvillian continental collision in south China: New SHRIMP U-Pb zircon results and implications for the configuration of Rodinia. Geology, 30 (2): 163~ 166
Li Z X, Li X H, Li W X, Ding S J. 2008. Was Cathaysia part of Proterozoic Laurentia? - new data fromHainan Island, south China. Terra Nova, 20 (2): 154~164
Metcalfe I, Shergold I H, Li Z X. 1993. Gondvana dispersion and Asian a ccretion: fieldwork on Hainan island. Episodes, 16:

443~447

Nyman M W, Karlstrom K E, Kirby E, Graubard C M. 1994. Mesoproterozoic contractional orogeny in westernNorth America: evidence from ca 1.4 Ga plutons. Geology, 22: 901~904

Pitcher W S. 1978. The anatomy of a batholith. Jour. Geol. Soc., 135: 157~182

Sun, S. S., Mclennan, W. F. 1989. Geochemical and isotopic systematics of oceanic basalts: implications for mantal compositions and processes, In Saunders A. D., and Norry, M. J., eds. Magmatism in ocean basins. Geol. Soc. Lond. Spec. Publ., 42: 313~345

Vivianne Berg-Madsen, John S. Peel. 1978. Middle Cambrian monoplacophorans from Bornholm and Australia, and the Systematic position of the bellerophontiform melluscs. Lethaia, 11 (2): 113~125.

Yao Bochu, Wang guangyu. 1983. Crustalstructure of the South China Sea Basin. Science in Chinese (Series B), 26 (6): 649~661

Yao Bochu. Zeng Weijun, Chen Yizhong, et al. 1993. The crustalstructure in the eastern part of northerm margin of the south China sea. Acta Geophysica Sinica, 37 (1): 44~53 (in Chinese with English abstraca)

三、地质资料

北京 703 航测队第 5 队 . 1978. 海南岛北东地区航空磁测报告

北京大学，海南地质综合勘察院 . 2002. 海南抱伦金矿成矿模式、成矿规律与预测

地质部地球物理探矿局航测大队 905 队 . 1959. 海南岛航空物探工作结果报告

地质力学研究所，海南地质综合勘察院 . 2000. 海南省乐东九所-抱伦地区金、银、钼多金属矿控矿构造及成矿规律研究

广东省区域地质测量大队 . 1964. 海南岛 1：20 万地质图及说明书

广东省区域地质测量大队 . 1964. 海南岛 1：20 万区域地质测量报告

广东省地质局海南地质大队 . 1973. 海南岛基性、超基性喷发岩及其矿产

广东省地质局地球物理探矿队 . 1977. 海南岛昌江-崖县地区航磁航放测量成果报告

广州地理研究所 . 1987. 海南岛北部新生代构造特征及其演化发展

广西区域地质调查大队 . 1991. 1：5 万坝王岭、猴猕岭幅区域地质调查报告

海南地质大队 . 1962. 海南岛崖县大茅磷矿区地质详查总结报告

海南地质大队 . 1964. 海南岛石碌铁矿补充勘探报告

海南地质大队 . 1976. 海南岛崖县大茅磷矿区地质勘探报告

海南地质大队一分队 . 1978. 海南岛崖县汤他地区地质矿产调查与 M-Ⅷ-4 航磁异常验证报告（油印本）

海南地质大队 . 1979. 海南岛崖县红花矿区磷矿普查及外围概查地质报告

海南地质大队 . 1981. 海南岛陀烈群地层剖析

海南地质大队 . 1983. 海南岛铁矿成矿远景区划报告

海南地质大队修编 . 1984. 海南岛 1：20 万地质图

海南地质大队 . 1984. 海南岛铁矿成矿远景区划说明书（1：20 万）

海南地质大队 . 1986. 1：5 万志仲幅区域地质调查报告

海南省地质矿产勘查开发局 . 1989. 1：5 万长流、海口市、白莲市、灵山市幅区域地质调查报告

海南地质大队一分队 . 1990. 1：5 万东方、玉道幅区域地质调查报告

海南地质大队，宜昌地质矿产研究所联合区调队 . 1991. 1：5 万南山岭、马岭市、崖县幅区域地质调查报告

海南省地质矿产勘查开发局海南地质大队 . 1993. 1：5 万昌化大岭、海尾、昌化港、叉河幅区域地质调查报告

海南地质综合勘察院第三勘察处 . 1994. 1：5 万九所、崖城、高峰幅区域地质调查报告

海南省地质矿产勘查开发局 . 1994. 海南岛中生代火山机构及其成矿关系的研究

海南省地质矿产勘查开发局 . 1995. 海南岛 1：100 万物化遥综合解释成果报告

海南省海岛资源综合调查领导小组办公室 . 1996. 海南省海岛资源综合调查研究报告

海南省地质矿产勘查开发局 . 1996. 1：5 万白马井镇、海头市、排浦市、王五市幅区域地质调查报告

海南地质综合勘察院 . 1998. 1：20 万海南岛区域重力调查测量报告

海南地质综合勘察院区调队 . 1998. 1：5 万黄流市、千家幅区域地质调查报告

海南省地质调查院 . 2000. 1：5 万黎母岭、枫木市、琼中县、白马岭幅区域地质调查报告

海南省地质矿产勘查开发局 . 2001. 海南省矿床成矿系列及成矿预测
海南省地质调查院 . 2004. 1：25 万琼海县幅区域地质调查报告
海南省地质调查院 . 2007. 1：25 万东方县幅区域地质调查报告
海南省地震局 . 2007. 海口市活断层探测与地震危险性评价
海南省地质调查院 . 2009. 1：5 万昌洒市、文昌县、清澜港、铜鼓咀幅区域地质调查报告
海南省地质调查院 . 2010. 1：5 万吊罗山、兴隆、什玲、陵水县幅区域地质调查报告
海南省地质调查院 . 2010. 1：25 万海南岛多目标区域地球化学调查报告
海南省地质调查院 . 2010. 海南省铁矿资源潜力评价磁测资料应用研究成果报告
海南省地质调查院 . 2010. 海南省铁矿资源潜力评价重力资料应用研究成果报告
海南省地质调查院 . 2011. 1：5 万吊罗山、兴隆、什玲、陵水县幅区域地质调查报告
海南省地质调查院 . 2011. 海南省乐东县丘糖岭古—新近纪地层特征及岩石地层单位
海南省地质调查院 . 2011. 海南省屯昌县南棍园地区含碳浅变质岩的时代及岩石地层单位探讨
海南省地质调查院 . 2011. 海南岛南北地体分野断裂研究
海南省地质调查院 . 2011. 海南岛白垩纪花岗岩含矿性及钼矿成矿时代研究
河北区域地质调查大队、海南地质大队 . 1991. 1：5 万南丰市、仁兴市幅区域地质调查报告
江西地矿局物化探大队 . 1988. 1：20 万海南岛化探扫面测量报告
南京地质矿产研究所，海南地矿局海南地质大队 . 1995. 海南岛暨五指山地区中生代火山地质与成矿关系
四川省地质矿产局攀西地质大队、海南区调队 . 1991. 1：5 万尖峰岭、乐东县幅区域地质调查报告
冶金部物探公司航测大队三分队 . 1979. 1978 年广东省海南地区航磁报告
宜昌地质矿产研究所 . 1990. 海南金矿类型、形成条件及矿成预测的研究
宜昌地质矿产研究所，海南地质综合勘察院 . 1994. 海南戈枕剪切带构造演化及其金矿控矿关系研究
宜昌地质矿产研究所 . 1994. 1：5 万昌江县、邦溪幅区域地质调查报告
宜昌地质矿产研究所 . 1995. 1：5 万中原市、博鳌港、和乐幅区域地质调查报告
宜昌地质矿产研究所 海南地质调查院 . 2003. 1：25 万乐东县、陵水县幅区域地质调查报告
中南地区区域地层编写组 . 1974. 中南地区区域地层表（内部交流）
中科院华南富铁矿科学研究队地层组 . 1976. 石碌式铁矿地层（石碌群）及古生物研究
中科院华南富铁矿科学研究队地层组 . 1977. 关于石碌铁矿地层的划分、时代及海南岛古生代地层介绍（油印稿）
中国地调局、中国地质科学院地质研究所 . 2009. 中国区域地质志编制工作指南
中国地震局地质研究所 . 2009. 琼北火山探查及喷发危险性研究
丁式江等 . 1993. 二甲布磨成矿带大中比例尺成矿预测
韩哲元 . 1976. 海南岛 1：20 万地质图说明书
何阳芝等 . 1980. 海南岛崖县地区早古生代地层划分与对比
黄香定等 . 1997. 海南蓬莱地区火山岩地质特征及其与蓝宝石矿床形成的关系
黎明等 . 1981. “海南岛陀烈群地层剖析” 专题研究报告
刘公民 . 1985. 琼北第四纪玄武岩
马大铨，黄香定，陈哲培等 . 1993. 海南省东方县二甲-不磨金矿带抱板群时代、层序及含金性研究报告
孙建中 . 1986. 琼北第四纪地层年代学研究报告
吴学益等 . 1991. 琼西金矿带蚀变糜棱岩型金矿成矿机理及其模拟试验（研究报告）
薛万俊等 . 1981. 南海北部沿岸第四纪地质、地貌调查报告
曾鼎乾等 . 1981. 南海北部大陆架第三系
赵鸿 . 1964. 海南岛西部混合片麻杂岩中榴辉岩的发现与初步研究 . 中国地质学会专业学术会议论文摘要汇编（岩石）
邹和平，黄玉昆 . 1987. 海南岛北部新生代构造特征及其演化发展

摘　　要

海南省位于中国最南端，辖管海南岛（含周缘小岛屿）、西沙群岛、中沙群岛、南沙群岛及其海域，陆地（主要包括海南岛和西沙群岛、中沙群岛、南沙群岛）总面积约 35400km^2，海域面积约 2000000km^2。海南岛是海南省最大的岛屿，面积约 33900km^2，是全省政治、经济、文化和交通中心。岛内地形地貌呈中部隆起的穹隆形山体，中部高四周低，以五指山（海拔 1867.1m）为隆起核心，向外围逐级下降，分别由山地、丘陵、台地、阶地和平原等构成环形层状地貌，梯级结构明显。主要河流由昌化江、南渡江、万泉河等。岛内地处热带、亚热带，属海洋气候和季风气候，高温多雨为其主要气候特征。省内各类资源比较丰富，旅游资源极具特色，主要有海岸景观、山岳和热带原始丛林、珍奇异兽、火山、溶洞、温泉、古迹名胜及田园风光等，是著名的国际旅游岛。

海南省地处特提斯构造域、西太平洋构造域和欧亚大陆板块构造域的交合部位，具有多阶段、多旋回的演化。即其现有的地质记录而言，其地质发展历史可追溯至中元古代。在漫长而复杂的地史演化过程中，伴随各时期不同的地质事件形成了一系列不同类型的沉积建造、变质建造和构造相等。总体上可概括为中元古代结晶基底形成、新元古代—古生代板内多阶段裂解与造山作用、中生代陆内盆山演化和新生代裂解 4 个重要阶段。

一、中元古代结晶基底形成阶段

该阶段是海南岛基底形成的主要演化阶段。约 18 亿年，地壳发生裂解，沉积了一套陆缘弧相的火山-碎屑岩，其下部为戈枕村组钙碱性系列的英安质和石英安山质中酸性火山岩；上部峨文岭组原岩为一套成熟度较低的陆缘碎屑岩夹陆缘弧相中酸性火山岩，发育类复理石沉积韵律。经区域变质作用改造，形成下部戈枕村组混合岩类-片麻岩类岩石组合和上部峨文岭组石英岩类-片岩类岩石组合。

发生于约 14 亿年的早晋宁运动（混合岩化片麻岩和侵入其中的片麻状花岗岩中获得锆石 SHRIMP 年龄约为 14 亿年），使得地壳物质局部发生重熔，形成以片麻状二长花岗岩为主体岩性的壳源型钙碱性花岗岩体，这些规模不等的花岗岩体总体与中元古界围岩一同经历了强烈的韧性变形改造。导致早期沉积发生了角闪岩相-麻粒岩相的区域动力热流变质，下部并发生混合岩化。但这期造山作用事件，对海南省不同地区的改造程度不一。如琼西地区、西沙群岛主体为角闪岩相；琼中地区中元古界局部达麻粒岩相，并相伴发育了与其具有成因关联的紫苏花岗岩侵入体。经过中元古代的花岗岩焊接和变质变形再造，塑造了以高温变质和高度塑性变形为特色的海南岛中元古代结晶基底，海南岛基底陆壳宣告形成。由于中元古界出露面积小，主要呈残块状“漂浮”于花岗岩之中，受后期地质作用改造强烈。因此，很难追寻 1400Ma 造山事件的动力学背景。

二、新元古代—古生代板内多阶段裂解与造山作用

1. 新元古代板内裂解与造山作用

新元古代海南省基底陆壳发生裂解，海南岛北面石碌地区局部形成一个陆缘盆地，作为裂解事件记录的建造序列仅见于昌江石碌地区，以青白口系石碌群为代表。为具类复理石韵律结构的碎屑岩、碳酸盐岩沉积，其中夹有少量的酸性火山岩。从早至晚，沉积环境由浅海→海湾潟湖相。火山岩有钾流纹岩、钾质流纹斑岩、钾粗安质球粒状凝灰岩、钾流纹质凝灰岩。发生于距今 8 亿年左右的晋宁运动，导致石碌群发生低绿片岩相区域变质作用和固态流变，成为堆叠于抱板杂岩结晶基底之上的变质基底。这一构造事件在华南地区也有显示，也即晚晋宁运动波及范围较广，此时海南岛可能与华南大陆同属一个构造事件的作用背景中。但与晚晋宁运动相伴随的岩浆活动未见及，反映地壳的克拉通程

度不高、稳定性差，并在新元古界南华系—震旦系时发生裂解，形成陆表海环境，接受了石灰顶一套残坡积相-滨海相含铁质的类磨拉石建造沉积。其底部为含砾泥岩、含角砾泥岩，岩石以其无层理、分选差、磨圆度差和角砾为原地或微异地等为特征。随后，地壳全体隆升，结束该造山事件记录。

2. 早古生代板内裂解与造山作用

寒武纪时期，地壳发生裂解，海南岛北面万宁一带裂解方向可能为由南至北，美子林组代表了裂解早期相对稳定的沉积，具滨海-浅海相碎屑岩夹碳酸盐岩建造特点。从早到晚，海水逐渐加深，沉积场所逐渐远离滨岸，海侵加强。海南岛南面三亚地区沉积作用则始于黔东世，为滨海-浅海环境含磷锰砂泥质岩-碳酸盐岩沉积。

奥陶纪时，海南岛北面裂解程度加剧，沉降盆地扩宽，海水加深，形成了近东西向的海盆，盆地东高西底，南北浅、中部深，接受了奥陶系南碧沟组一套厚达上千米的具有复理石、类复理石特征的海相碎屑岩夹基性、中基性火山岩建造沉积，形成海相碎屑岩建造夹基性火山岩建造，具有活动型建造的特点。呈夹层状产出的火山岩具有多期次的喷发活动特点。早期为陆缘弧环境的玄武岩、变安山岩、变流纹岩，岩石锆石 U-Pb 等时线年龄为（499±6）～（527±48）Ma；晚期为板块边缘大洋构造环境的超镁铁质、镁铁质火山岩，岩石 Sm-Nd 同位素年龄为（458±36.8）Ma。但同期缺少或少见内源物质的沉积，反映了该时期地壳可能具有快速裂解和上升的运动特点。此时，在海南岛南面三亚地区则沉积了滨海-浅海环境砂泥质岩-碳酸盐岩沉积，后期并进入海陆交互相复成分砂砾岩、含砾砂岩、含铁黏土岩混杂沉积。之后，地壳隆升为陆，进入漫长的风化剥蚀时期。

经奥陶纪地壳震荡性隆升后，志留纪的沉积仅分布于海南岛北面，包括下志留统陀烈组、空列村组、靠亲山组、大干村组、足赛岭组。早期为滨海相砂质岩沉积，中期进入浅海-半深海相砂泥质、泥砂质沉积，晚期进入滨海-浅海相砂质岩、泥砂质岩和碳酸盐岩沉积，构成了一个相对完整的沉积旋回。但西部总体上接受了一套陆棚-大陆斜坡碎屑岩组成的复理石建造，上部出现碳酸盐台地相的沉积，岩石中上部发育鲍马序列中的 A、C、D、E 层。而东部地区则为深海平原相泥质、粉砂质沉积，形成了深水相复理石-类复理石建造。显示盆地具有东深西浅的格局，与奥陶纪盆地东浅西深格局有所不同，暗示地壳震荡性隆升作用具有不均衡型。

加里东运动在海南省表现为多次性和多样性。第一幕发生于早寒武世晚期，以垂直隆升作用为特点，使得海南岛北部奥陶系南碧沟组以平行不整合叠覆于下寒武统美子林组之上。第二幕发生于晚寒武世中-晚期，与第一幕表现形式一样，形成了海南岛南面三地区下奥陶统大葵组与中—上寒武统大茅组间的平行不整合。第三幕发生于晚奥陶世晚期，地壳继续以垂直隆升作用为特点，导致海南岛地区南面三亚地区隆升为陆，缺失志留纪的沉积。此幕活动在海南岛北面的表现是：奥陶纪半深海-深海相沉积后，志留纪早期为滨海沉积，显示地壳曾发生过垂直隆升。第四幕发生于早志留世晚期，从少量零星的资料来看，海南岛北部上古生界与下古生界呈角度不整合接触，表明第四幕可能具有造山性质，但缺乏同期岩浆作用的记录。伴随该幕构造作用，海南省全面隆升为陆，缺失中、上志留统沉积。

据杨坤光研究（华南加里东运动与印支运动范围北部峰带问题），华南加里东运动不仅在性质、表现上具有多样性，在时间上也具有多阶段性（多幕性），主要分别发生于寒武纪末、奥陶纪末和志留纪末。海南岛加里东运动与华南加里东运动在时间上基本同步，但就其构造运动的强度而言，海南岛表现相对比较微弱，没有华南其他地方强烈，因此海南岛早古生代时不处于加里东运动的峰带位置，而是处于加里东运动影响的边缘区域。

3. 晚古生代特提斯构造域演化

经加里东运动后，海南省地壳长期隆升为陆，直至石炭纪才发生沉积作用。

下石炭统南好组沉积时期，整个海南岛主要为稳定环境下滨浅海相碎屑岩相沉积，主要是浅变质作用形成的砂岩类岩石，其底部为变质石英砾岩、砂砾岩、砂岩，沉积物粒度粗，分选性、磨圆度好，结构成熟度和成分成熟度较高，除少量岩屑外，几乎全由石英组成；上部为变质陆源细碎屑岩，总体为板岩、含粉砂质板岩与变质粉砂岩、变质细砂岩不等厚互层，砂岩、粉砂岩分选性好，不含或含少量泥质，偶见变余斜层理，南好乡岩石普遍富含长石，代表了近源的沉积环境。青天峡组沉积时，沉积环境呈现出一定的差异性，北面石碌镇—琼中石浮镇一线石炭系为海相碎屑岩相和碳酸盐岩台地相沉积，由浅变质作用形成的砂岩、泥质砂岩、灰岩等组成，岩石的分选性、磨圆度好，结构成熟度和成分成熟度较高，具有稳定型滨海-浅海相单陆屑式建造夹碳酸盐岩建造的特点。南面中原镇该时期则为海相碎屑岩相，缺乏碳酸盐岩台地相沉积，沉积作用从早至晚（南至北）海水逐渐加深，早期为滨海-浅海相（局部达大陆斜坡相）碎屑岩沉积，中晚期为大陆斜坡相，出现浊流作用形成的杂砂岩，杂砂岩呈中-薄层状，普遍含一定的泥质杂基，具粒序层理；地壳活动性也逐渐加强，上部岩石常见酸性凝灰岩或沉凝灰岩夹层，沉积建造类型由稳定型往次稳定型过渡。

二叠纪时的沉积一是见于海南岛西面石碌镇—东方市江边一线，包括下二叠统鹅查组、鹅顶组和上二叠统南龙组。其中南龙组仅出露于岛西石碌镇地区，为海陆交互相沉积。鹅查组、鹅顶组沉积时，岛西石碌地区为近滨相和碳酸盐岩台地相沉积，相对石炭纪而言，其地壳已有所抬升。二是见于南沙群岛一带，岩石为灰岩和燧石、硅质碎屑岩、长石杂砂岩。

晚古生代岩石中产海百合茎、珊瑚、层孔虫、苔藓虫、腕足类、凝源类、虫颚以及丰富的孢粉化石。其中，珊瑚 *Gorskyites. Syrinyopra* 也常见于与湘中、粤北一带相应时代的地层中；植物和腕足可以与湘中、赣中、粤北和粤西对比，均以产 *Archaeocalamites*，*Dictyoclostus* 为特征。汪啸风等（1992）根据晚古生代中的化石组合（蠖类：*Parafusulina gruperaensis*－*P. langdaiensis*－*Misellina ovalis*，植物：*Annularia mucronata*、*A. shirakii*、*Pecopteris arborescens*、*P. rigida*、*P. hemitelioides*、*Taeniopteris* sp.、*Cordaites* sp.，珊瑚类：*Wentzellophyllum*－*Polythecalis* 以及菊石类：*Shouchangoceras* 等)，认为属于华夏-特提斯动物区或华夏植物区的一部分；并通过海南岛与周边地区华力西-印支期花岗岩特征的对比分析，认为海南岛华力西-印支期花岗岩与滇西临沧岩基（或澜沧江岩带）最为相似，提出海南岛晚古生代属于特提斯构造域范围内的见解。

发生于约 290~270Ma 的华力西-印支运动（谢才富，2007），不仅导致了海南岛北部晚古生代地层发生了区域绿片岩相变质和透入性变形，也使得新元古代、早古生代地层卷入其中，形成了区域尺度上展布的褶皱带和断裂带，塑造出一幅非常醒目的海南岛弧状构造带。同时，伴随的花岗岩浆侵入活动强烈而持久，一直持续至晚三叠世，从早至晚，形成了俯冲型、同碰撞型、后碰撞型、隆升型花岗岩，构成了一个比较完整的造山作用花岗岩浆演化系列。而海南省南面南沙群岛出露的二叠系未见变质变形作用的改造。

尽管海南岛此时处于特提斯构造域范围，但对于其造山运动的动力学背景认识目前依然不明朗。尤其是广泛的二叠纪花岗岩分布，与滇西临沧岩基（或澜沧江岩带）最为相似，而与同处特提斯构造域的广西、广东缺乏二叠纪花岗岩形成鲜明的对比。且广西、广东一带南盘江盆地发育有三叠纪深水相碎屑岩夹玄武岩，与海南岛早三叠世陆相沉积也完全不一致。目前认为，特提斯洋到达广西后出现两个分支，一支东延到达北部湾，另一支南延到达曼谷一带（丘元禧，2000）。因此，本次认为，随着特提斯洋二叠纪的闭合，由于洋盆沿滇西澜沧江—曼谷一线发生俯冲，位于转折部位的南盘江盆地（东支）由于受到俯冲作用而发生进一步的裂解伸展，继续接受三叠纪深水相沉积（吴浩若，2003）。海南岛此时处于东面和东南面挤压应力作用下，一方面使得晚古生代东西向盆地二叠纪时进一步继续发生裂解，广大地区被海水淹没，接受了碳酸盐缓坡型浅海陆棚沉积；另一方面导致南面印支-南海地体沿区域性崖城-万宁深大构造带俯冲于五指山造山带之下，海南岛北部五指山造山带（仰冲盘）发育广泛的华力西期—印支期花岗岩，并伴随区域动力低温变质作用的发生。从而形成海南岛二叠纪伸展与挤压共存的环境，并最终以海南岛南北两地体相互挤压作用占主导，且其波及范围可能逐渐西迁，使得特提斯洋东支南盘江盆地闭合和广西、广东三叠纪花岗岩的分布。

此外，作为本次造山作用隆升标志的下三叠统岭文组陆相磨拉石建造，与东部琼中地区中晚三叠世挤压侵位的花岗岩构造环境并不协调，初步认为可能属于不同地壳深度的表现。一般认为，隆升阶段上部为伸展拉张，下部则为挤压环境。

三、中生代陆内盆山演化阶段

海南省经历强烈的华力西期-印支期造山作用后，大约在175~200Ma期间进入构造-岩浆平静期：构造活动、岩浆作用微弱，沉积作用也缺失。这一时期也是本区乃至华南从特提斯构造域向滨太平洋构造域转折的时期。早燕山期（约150~175 Ma），爆发了属另一个构造域（滨太平洋构造域）、另一个构造旋回的强烈构造-岩浆活动。由于软流圈地幔上涌导致岩石圈裂解拉张，形成了一系列幔源和壳幔混合源侵入岩。

晚燕山期，受太平洋板块俯冲引起的弧后扩张带控制，海南省断块活动强烈，此起彼伏，断陷成盆，断隆成山，形成一系列陆内上叠沉积盆地和火山盆地，盆地有白垩纪王五盆地、白沙盆地、雷鸣盆地、阳江盆地、乐安盆地等火山-沉积盆地，及牛腊岭、同安岭火山盆地，沉积了下白垩统鹿母湾组湖、河相杂陆屑式建造夹中酸性火山岩建造和上白垩统报万组湖、河相杂陆屑式建造，以及下白垩统六罗村组双峰式火山岩建造和下白垩统汤他大岭组、岭壳组中酸性火山岩建造。同期构造岩浆活动强烈，形成了北东向、北西向、东西向、南北向韧脆性断裂，奠定了海南岛的大地构造格局。白垩纪晚期发生了由北西往南东的逆冲推覆作用，塑造了醒目的北东向军营-红岭和戈枕逆冲推覆构造带。花岗岩浆侵入活动具有多期次性，发育有侏罗纪、白垩纪花岗岩，岩石缺乏定向组构，属于伸展构造环境壳幔岩浆混合的产物。其侵位往往受多方向断裂的控制，导致岩体平面中常呈不规则长条状分布。

四、新生代裂解沉陷阶段

新生代，由于南海在南北向拉张应力作用下发生持续扩张（中国科学院南海海洋研究所，1990），南海北缘深部岩石圈受软流圈的快速上涌而发生快速减薄，导致了新岩石圈的增长，地幔流的上升导致了浅部岩石圈的张裂下陷，以及南海北缘众多拉张盆地的形成（许义刚等，2002）。盆地内接受了古近纪以来的沉积，形成古近系河湖相红色碎屑岩建造夹可燃性有机岩建造、膏盐建造、火山岩建造和新近系滨浅海相碎屑岩建造、火山岩建造以及第四系滨浅海相碎屑岩建造、河流相-三角洲相碎屑岩建造、火山岩建造。该时期地壳升降运动频繁，导致多个不同岩石地层单位间普遍存在不整合接触关系，发育海积、河积阶地。受断裂活动影响，发生多次的地震活动和剧烈的玄武岩浆喷发或喷溢，并沿北西向断裂出露多处温泉。

南海地区在中生界断陷盆地的基础上，地壳拉张下陷加剧，形成一系列北东向的沉积盆地。古近纪时南海北部接受了一套陆相-浅海相碎屑岩；南海南部水深较深，以一套厚达数千米的浅海相-半深海相页岩、砂岩为主，东南部靠近岛弧处常发生褶皱、变质作用。新近系下部为一套浅海相-半深海相砂岩与泥岩的互层，海底高原为礁灰岩。新近系上部—第四系下部主要为砂岩、页岩夹火山岩与火山碎屑岩夹层，海底高原为礁灰岩与生物碎屑岩，南海中央海盆则为火山灰与放射虫粉砂岩。全新世以来南海地区的沉积以珊瑚礁与生物碎屑岩为主。

在喜马拉雅运动作用下，海南省新构造运动强烈，表现形式多样，包括地壳升降运动、断裂活动、地震活动、火山活动和地热活动，造就了现今地势格局和各类地貌景观。

Abstract

Located in the southernmost of China, Hainan Province controls over the Hainan Island (including its peripheral small islands), Xisha, Zhongsha Islands, Nansha Islands along with their waters and land, covering an area of about 35,400km^2. The sea area is about 2 million km^2. The Hainan Island, being the largest island in Hainan Province and covering an area of about 33,900km^2, is the political, economical, cultural and transportation center of the province. The topography is dome shaped with mountains in the central uplift, high in the middle part and low in the peripheries. The Wuzhi Mountain (with a height of 1,867.1m) is the uplift core and the landform declines towards the peripheries. Mountain land, hills, platforms, terraces and plains form circular stratified landform with clear cascade structure. The main rivers are the Changhua River, Nandu River, Wanquan River, etc. The interior of the island is located in a tropical and subtropical zone, falling into the marine climate and monsoon climate featured by high temperature and rainy weather. The province is rich in all kinds of resources, especially tourism resources, such as coastal landscape, mountains and tropical jungle, exotic animals, volcanos, caves, hot springs, historic sites and idyllic scenery, making it a famous international tourist destination.

Hainan Province is situated in the converge area of the Tethyan tectonic domain, the Western Pacific tectonic domain and the Eurasian continental plate tectonic domain, featuring a multi-stage and multi-cycle evolution. With the existing geological records, its geological development history can be traced back to the Mesoproterozoic. In the long and complex geological evolution process, a series of different types of deposition, metamorphic formation and tectofacies have been formed with geological events in different periods. On the whole, it can be summarized as 4 important stages: crystalline basement formation in the Mesoproterozoic, multi-stage cracking and orogeny within the plate of the Neoproterozoic and Paleozoic, intracontinental basin mountain evolution in the Mesozoic and Cenozoic cracking.

I Crystalline basement formation in the Mesoproterozoic

This stage is the main evolution stage of basement formation of the Hainan Island. About 1.8 billion years ago, the crust cracked, leading to sedimentation of a set of continental marginal arc facies of volcano clastic rock, the lower part of which was intermediate acid volcanic of dacite and quartz rock mountain of Gezhen village calc alkaline; the upper part of which was the Erwenling Formation original rock which was a set of low maturity of the continental margin clastic rock with the continental margin arc phase intermediate acidic volcano rock, development flysch sedimentary rhythm. The regional metamorphism brings the formation of the lower Gezhen village hybrid rocks - gneiss rock assemblage and the upper Erwenling quartzite group - schist rock combination.

The early Jinning movement, occurring in about 1.4 billion years ago (the zircon SHRIMP age is about 1.4 billion years, which is gained from migmatization of gneiss and gneissose granite invaded in) made the local crustal remelt, forming calc-alkaline series rocks dominated by gneissic monzonitic granite. The granite of different scales has gone through intense ductile deformation with the Middle Proterozoic surrounding rocks, leading to early deposition along with regional dynamic heat flow metamorphism of amphibolite facies–granulite facies, and migmatitization in the lower part. But the events of this orogeny have different degrees of transformation in different areas of Hainan Province, for instance, western Hainan and the main body of the Paracel Islands area are of amphibolite facies; Qiongzhong mesoprotozoic is partially granulite facies, and its development is accompanied by causes associated charnockite intrusion. The welding and metamorphism deformation of Mesoproterozoic granites has shaped the Mesoproterozoic crystalline basement of the Hainan Island which is characterized by high temperature metamorphism and highly plastic deformation, for which the basement continental crust of the Hainan Island has been formed. Since the middle Proterozoic exposed area is small, it is mainly a residual mass "floating" in granite, which is strongly transformed by late geological process. Therefore, it is very difficult to trace the dynamic background of 1400Ma orogenic events.

II Multi–stage cracking and orogeny in the Neoproterozoic–Paleozoic plate

1. Cracking and orogeny in Neoproterozoic plate

Due to the cracking of basement continental crust of Hainan Province in the Neoproterozoic, Shilu, the northern part of the Hainan Island, locally forms a continental margin basin. The construction sequence of the cracking event occurs only in Shilu, Changjiang, represented by the Shilu Group of Qingbaikouan. Clastic rocks and carbonate rocks with rhythmic structures in the flysch complex contain a small amount of acid volcanic rocks. From morning to night, the sedimentary environment ranges from shallow water to the Gulf lagoon facies. Volcanic rocks have potassium potassic rhyolite, rhyolite porphyry, potassium trachyandesitic chondrite tuff, rhyolitic tuff potassium. The Jinning movement, occurring about 800 million years ago, has resulted in Shilu Group coming up with low greenschist facies metamorphism and solid flow, becoming metamorphic basement stacked on the Baoban complex rocks in the crystalline basement. This tectonic event is also reflected in the South China area, which means that the late Jinning movement is spread widely, meaning that the Hainan Island and South China continent may belong to the same background of tectonic event. But there is no magmatic activity associated with the late Jinning movement, reflecting that the craton crustal level is not high with poor stability, and that cracking happens in the Neoproterozoic Nanhuan system–Sinian system, forming the epicontinental sea environment, and receiving a set of quasi–molasse sedimentary formation of residual slope facies–littoral facies with iron. Its bottom is composed of pebbly mudstone and breccia mudstone featured by no bedding, poor sorting, poor roundness and breccia in situ or micro – site. Subsequently, the whole crust is uplifted, ending the orogenic record.

2. Intraplate cracking and orogenesis in early Paleozoic era

During the Cambrian period, the crust cracking occurs. The Wanning area in the north of the Hainan Island cracks from south to north, Meizilin Group represents the relatively stable sedimentation in early cracking, featured by littoral neritic facies clastic rock carbonate rock formation. From morning to night, the sea water gradually deepens, the sedimentary site is gradually away from the shore, and marine transgression is intensified. The sedimentation of the Sanya area in the south of the Hainan Island begins in the east of Guizhou Province, featured by phosphorus manganese sand argillaceous–carbonate sedimentary under a littoral neritic environment.

During the Ordovician, the cracking in the north of the Hainan Island was intensified, subsidence basin widening, water deepening, forming a nearly East-West basin which is high in the east, low in the west, shallow in the north and south, deep in the cental, receiving the Ordovician Nanbigou Formation, a set of marine clastic rocks with basic volcanic rock and intermediate-basic volcanic rock formation dedimentary featured by flysch and flyschoid with a thickness up to 1,000 meters. It has characteristics of activity type formation. Volcanic rocks with intercalated strata have the characteristics of multistage eruption. The early environment is of continental margin arc basalt, andesite, rhyolite rocks become variable, zircon U-Pb isochron age is from (499 ±6) to (527±48) Ma; late environment is of ultramafic, and mafic volcano rock of plate edge ocean tectonic environment, rock Sm-Nd isotopic age is (458±36. 8) Ma. However, it has little or lacks endogenous material deposits during the same period, reflecting the rapid cracking and upward movement of the crust during this period. At this time, the Sanya area in the south of the Hainan Island is deposited with littoral neritic sand shale and carbonate rocks sedimentary, and later enters the paralic polymictic conglomerate, pebbly sandstone, clay containing iron rock mixed deposition. Later, the crust is uplifted into land and enters a long period of weathering and denudation.

After the shocking uplifting of Ordovician crustal, the Silurian sedimentary is only distributed in the north of the Hainan Island, including the Lower Silurian Tonie Group, Kongliecun Formation, Kaoqinshan Formation, Dagancun Formation and Zusailing Formation. The early environment is of marine sandy rock sedimentary, the middle is of the neritic bathyal facies sand mud sandy sedimentary, and the late one is into the littoral neritic facies sandy rock, mud sandy rock and carbonate sediments, constituting a relatively complete sedimentary cycle. But the west generally receives a set of flysch formation consisting of shelf - slope clastic rocks, while the upper part has carbonate platform facies sedimentary, the upper and middle parts of rocks develop Bouma sequence in A, C, D, E strata. The eastern area is deep sea plain facies, argillaceous and silty deposits, forming deep-water flysch-flyschoid formation. It shows that the basin has the pattern as deep in the east and shallow in the west, different from the Ordovician basin pattern as shallow in the east and deep in the west, suggesting that shocking uplift movement of the crust function is not balanced.

The Caledonian movement in Hainan Province is characterized by multiplicity and diversity. The first scene is in the late Early Cambrian, featured by the vertical uplift, making the Nanbigou Formation of Ordovician system in the northern Hainan Island with parallel unconformity overlapping on the lower Cambrian Meizilin Formation. The second scene occurs in the middle and late stages of late Cambrian, as with the first act forms, forming parallel unconformity between Lower Ordovician Dakui Formation and the middle-upper Cambrian Damao Formation of three regions in the south of the Hainan Island. The third scene occurs in the late stage of late Ordovician, the crustal is still featured by vertical uplift, resulting in the Sanya area in the south of the Hainan Island is uplift to be continental, lack of Silurian sedimention. This scene activities in the north of the Hainan Island are featured by: after Ordovician bathyal abyssal facies, the early littoral Silurian is coast sedimentary and vertical uplift has occurred to the crust. The fourth act occurs in the late stage of the Early Silurian period, from a small amount of scattered data, the upper Paleozoic and Lower Paleozoic in the northern Hainan Island present angular unconformity contact, showing that the fourth scene may have orogenic properties, but lack of the magmatism record of the same period. Accompanied by the tectonic action, Hainan Province is fully uplifted to be a continen, and the middle and upper Silurian sediments are absent.

According to Yang Kunguang Study (A study on the issues of peak band in the north of the South China Caledonian and Indosinian movement), the Caledonian movement in South China has not only the diversity in nature and performance, but also has multi-stages in time (multi stages), mainly respectively occurred in the end of Cambrian, Ordovician and Silurian. The Hainan Island Caledonian movement and South China Caledonian movement are basicly synchronized in time, but in terms of the intensity of tectonic movement, the Hainan Island performance is relatively weak, not so strong as in other parts of South China, so the Hainan Island in the early Paleozoic ear is not located in peak band position of the Caledonian movement, but in the edge region influenced by the Caledonian movement.

3. Tethys tectonic evolution in the Neopaleozoic

After the Caledonian movement, the crust of Hainan province was uplifted into land for a long time and the occurrence of sedimentation did not come until the carboniferous period.

During the lower Carboniferous Nanhao Formation sedimentary period, the whole Hainan Island was mainly of littoral clastic sedimentary facies under a stable environment, which was mainly sandstone rock developed by shallow metamorphism, the bottom was of metamorphic conglomerate, glutenite and quartz sandstone, the sediments were of coarse granularity, good sorting and roundness, higher texture maturity and compositional maturity, in addition to a small amount of debris, almost entirely composed by quartz; the upper part was of metamorphic terrigenous microclastic rocks, generally containing thick interbedding of slate and silty slate and silty metamorphic sandstone and metamorphic fine sandstone, of which sandstone and siltstone have good sorting, containing no or little mud, occasionally with blasto oblique bedding. The rocks of Nanhao township are generally rich in feldspar, which represents the sedimentary environment of the near source. When the Qingtianxia Group is deposited, sedimentary environment presents a certain difference, the carboniferous system of Shilu town-Qiongzhong Shifu town in the north is Carboniferous marine clastic facies and carbonate platform facies, consisting of sandstone, argillaceous sandstone and limestone formed by shallow metamorphic while the rocks present good sorting and roundness, higher compositional maturity and textural maturity with the characteristics of stable littoral neritic facies continental clast with carbonate rock formation of the type construction. Zhongyuan Town in the south in this period is marine clastic facies, lack of carbonate platform facies sedimentation, from morning to night (south to north), the sea water gradually deepens, the early period is of littoral neritic facies (continental slope facies in local part) clastic rocks, the middle and late period is of the continental slope facies, presenting the greywacke formed by turbidite, greywacke is in moderate thin shape, generally containing certain clay, with graded bedding; the earth crust activity is gradually strengthened. The rocks of upper part are commonly seen acidic tuff or tuffite interbedding, types of sedimentary formation transfer from stable to sub-stable.

One sedimentation in the Permian period is found in Shilu Town of the west of the Hainan Island, the riverside of Dongfang City, including the Echa Group and Eding Group of lower Permian series and Nanlong Group of the upper Permian series, in which the Nanlong Group is only exposed in the Shilu Town area of the west of the island with the sediments of alternative sea and river facies. As the Echa Group and Eding Group are deposited, the Shilu Town area in the west of the island is of offshore facies carbonate platform facies, the crust of which is relatively uplifted to Carboniferous. Another sedimentation is seen in the vicinity of the Nansha Islands, the rocks are limestone and chert, siliceous clastic rock, feldspar, miscellaneous sandstone.

The late Paleozoic rocks are rich in crinoids, corals and stromatoporoids, bryozoans, brachiopods, condensate source, scolecodonts and sporopollen fossils. Among them, *Gorskyites.* Chia and *Syrinyopra* are also common in strata in corresponding era of the central Hunan and Guangdong area; plants and brachiopods can be compared with central Hunan, Jiangxi, and North and West Guagndong, which are all featured by producing *Archaeocalamites Dictyoclostus*. According to the late Paleozoic fossil assemblage (fusulinid: *Parafusulina gruperaensis*–*P. langdaiensis*–*Misellina ovalis*, plants: *Annularia mucronata*, *A. shirakii*, *Pecopteris arborescens*,

P. rigida, *P. hemitelioides*, *Taeniopteris sp.*, *Cordaites sp.*, actinozoan: *Wentzellophyllum–Polythecalis* and ammonoids corals: *Shouchangoceras*, etc.), Wang Xiaofeng et al (1992) hold that it is a part of cathaysian – Tethys animal area or cathaysian flora; through analysis and comparison on Variscan Indosinian granite features between the Hainan Island and the surrounding areas, it is believed that the Hainan Island's Variscan Indosinian granite is most similar to Lincang batholith (or Lancang Jiangyan belt) and proposed that the Hainan Island at the late Paleozoic falls into the scope of the Tethys tectonic domain.

The Variscan Indosinian movement, occurring in about 290~270Ma (Xie Caifu, 2007), not only results in regional greenschist facies metamorphism and penetrative deformation of late Paleozoic strata in the northern Hainan Island, but also leads to the Neoproterozoic, early Paleozoic strata getting involved, forming regional scale distribution of the fold belt and fracture zone, creating a picture of the Hainan Island with a very striking arc structure belt. At the same time, the accompanying granite magma intrusion activities are strong and lasting, continuing to the Late Triassic, from morning to night, forming a subduction type, collision type, post collision type, uplift type granites, constituting a relatively complete orogenic granite magma evolution series. The Permian in the south of the Nansha Islands in Hainan Province has not seen the modification of metamorphism and deformation.

Although the Hainan Island is now in the Tethys tectonic domain, the understanding of its geodynamic background is still uncertain. Especially, the Permian granites are widely distributed, similar to the Lincang batholith (or Lancang Jiangyan band) in the west of Yunnan Province, forming a striking contrast to Permian granites lack in Guangxi and Guangdong provinces in the Tethys tectonic domain. And Triassic deep–water clastic rock with basalt is developed in the Nanpanjiang Basin of Guangxi and Guangdong areas, which is completely inconsistent with early Triassic continental deposits of the Hainan Island. At present, it is held that the two branches of the Tethys ocean appear after arriving in Guangxi, one branch extending to the North Bay along the east trending, another extending to the Bangkok area along the south trending (Qiu Yuanxi, 2000). Therefore, this paper holds that with the closure of the Tethys ocean Permian, due to subduction along the Lancang River in the west Yunnan Province to the Bangkok line, the Nanpanjiang Basin (East Branch) located in the turning parts under the action of subduction occurs further cracking and extension, continuing to receive the Triassic deep water sedimentation (Wu Haoruo, 2003). Under the extrusion stress from the east and the southeast direction, the Hainan Island makes the late Paleozoic east–west basin continue to further crack to the Permian basin, the vast area is inundated, receiving the shallow shelf facies carbonate ramp type; On the other hand, it leads to the Indo–South China Sea body in the south along the cliff city Wanning deep tectonic belt subduction orogenic belt below the Wuzhi mountain orogenic belt, which in the north of the Hainan Island (obduction) develops widely Variscan Indosinian granite, and with regional power for the low temperature metamorphism. Thus the Hainan Island Permian extensional environment and extrusion coexist, and eventually north and south mutually extrusion effect of the Hainan Island dominates, and its scope may gradually move to the west, making the closure of the Nanpanjiang basion (east branch) of Tethys ocean and Triassic granite distribution of Guangxi and Guangdong provinces.

In addition, the continental facies molasse formation of the Lingwen Formation of the Lower Triassic Series as the uplifting sign of this orogenesis, is not coordinated with the granite tectonic environment at extrusion emplacement of Middle–Late Triassic in the Qiongzhong area in the eastern part. The preliminary view is that it may belong to different crustal depth performance. It is generally believed that the upper part at the uplifting stage is stretched and the lower part is an extrusion environment.

III Intracontinental basin-mountain evolution in the Mesozoic

After experiencing strong Variscan Indosinian orogeny, Hainan Province comes into the quiet period of tectonic-magmatic about during 175~200Ma, with weak tectonic activity and magmatic action, and a lack of sedimentation. This period is also a turning point for this region and even South China from the Tethys tectonic domain to the circum Pacific tectonic domain. During the early Yanshanian (about 150~175 Ma), a strong tectonic-magmatic activity occurs in another tectonic domain (the littoral Pacific tectonic domain) and another tectonic cycle. As the asthenosphere mantle upwelling causes the lithospheric cracking to pull up, a series of intrusive rocks of mantle source and crust mantle mixed source are formed.

During the late Yanshanian period, controlled by back-arc extensional belt caused by the subduction of the Pacific plate, fault block activities in Hainan Province are strong, as one falls another rises, fault depression forms into basins, fault uplift into mountains, forming a series of intracontinental superimposed sedimentary basins and volcano basins. There are volcano sedimentary basins such as the Cretaceous Wangwu basin, Baisha basin, Leiming basin, Yangjiang basin, Le'an basin, etc. And volcano basins as Niulaling and Tonganling. The following are deposited: the lower Cretaceous Lumuwan Group lakes and fluvial facies mixed terrigenous clastic type formation with intermediate acidic volcano rock formation, and the upper Cretaceous Baowan Formation lakes, fluvial facies mixed terrigenous type formation, and the lower Cretaceous Liuluo Village Group Shuangfeng volcano rock formation and lower Cretaceous Tangtadaling Formation, Lingke Formation intermediate acidic volcanic formation. During the same period, the tectonic-magmatic activity is strong, forming the NE, NW, EW and NS ductile brittle fracture, and laying the tectonic pattern of the Hainan Island. During the late Cretaceous, thrusting movement occurs from northwest to southeast, shaping the striking northeast barracks-Hongling and Gezhen thrusting structures. The intrusive activity of granitic magma is of multiple stages, developed with Jurassic and Cretaceous granites and rocks lack of oriented fabric, belonging to the mixing of crust, mantle and magma in extensional tectonic environment. The emplacement is often controlled by multi-directional faults, resulting in irregular bars distribution in the plane of rock mass.

IV cracking and subsidence in the Cenozoic

During the Cenozoic, the South China Sea goes through continuous expansion due to the action of tensile stress in the north and south trending (South China Sea Institute of Oceanology of Chinese Academy of Sciences, 1990), the deep lithosphere on the north margin of the South China Sea by asthenosphere rapid upwelling goes through fast thinning, leading to a new lithospheric growth, the rising of mantle flow leading to rifting and subsidence of shallow lithosphere and the formation of numerous extensional basins on the north margin of the South China Sea (Xu Yigang et al, 2002). The interior basin receives deposition since the Paleogene, forming Paleogene lacustrine red clastic rock formation with combustible organic rock formation, gypsum-salt formation, volcano rock formation and Neogene shallow sea facies clastic rock formation, volcano rock formation and quaternary littoral clastic rock formation, fluvial facies-delta clastic facies rock formation, volcano rock formation. During this period, frequent movements of the earth crust result in the existence of unconformity contact between the different lithostratigraphic units, and the development of marine and fluvial deposits terraces. The influence of fault activity brings multiple seismic activities and intense basaltic eruption or outpouring, and exposed many hot springs along the NW faults.

On the basis of the Mesozoic rift basins, the South China Sea has intensified the crustal tension and formed a series of sedimentary basins in thenorth-east trending. During the Paleogene, the northern part of the South China Sea receives a set of continental neritic facies clastic rock; the southern part of the South China Sea is deeper, dominated by a thickness of several kilometers of the neritic facies-semi-deep marine facies shale and sandstone, fold and metamorphism often occur in the southeastern part near the island. The lower part of the Neogene system is a suite of sandstone and mudstone interbedding of neritic facies-semi-deep marine facies, and the submarine plateau is reef limestone. The Neogene upper-lower Quaternary are mainly sandstone and shale interbedded with volcano rock and volcano clastic rock interlayer, submarine plateau is of reef limestone and bioclastic rocks, central basin of the South China Sea is of the volcano ash and radiolarian siltstone. Since the Holocene, the sediments of the South China Sea have been dominated by coral reefs and clastic rocks.

With the movement of the Himalayas, neotectonic movement in Hainan Province is strong with diverse forms, including crustal rising and falling movement, active faults, earthquake activity, volcano activity and geothermal activities, creating present terrain pattern and various geomorphic landscape.

图版 I -1

戈枕村组眼球状混合岩中眼球定向排列（定安县黄竹岭）

戈枕村组眼球状混合岩夹混合岩化片麻岩（乐东红五）

戈枕村组混合岩化片麻岩（乐东红五）

戈枕村组混合岩与片麻岩强烈挤压变形构造（乐东红五）

斜长角闪岩侵入戈枕村组片麻岩中（东方戈枕大坝东）

斜长角闪岩侵入峨文岭组片岩中（居便岭）

图版 I -2

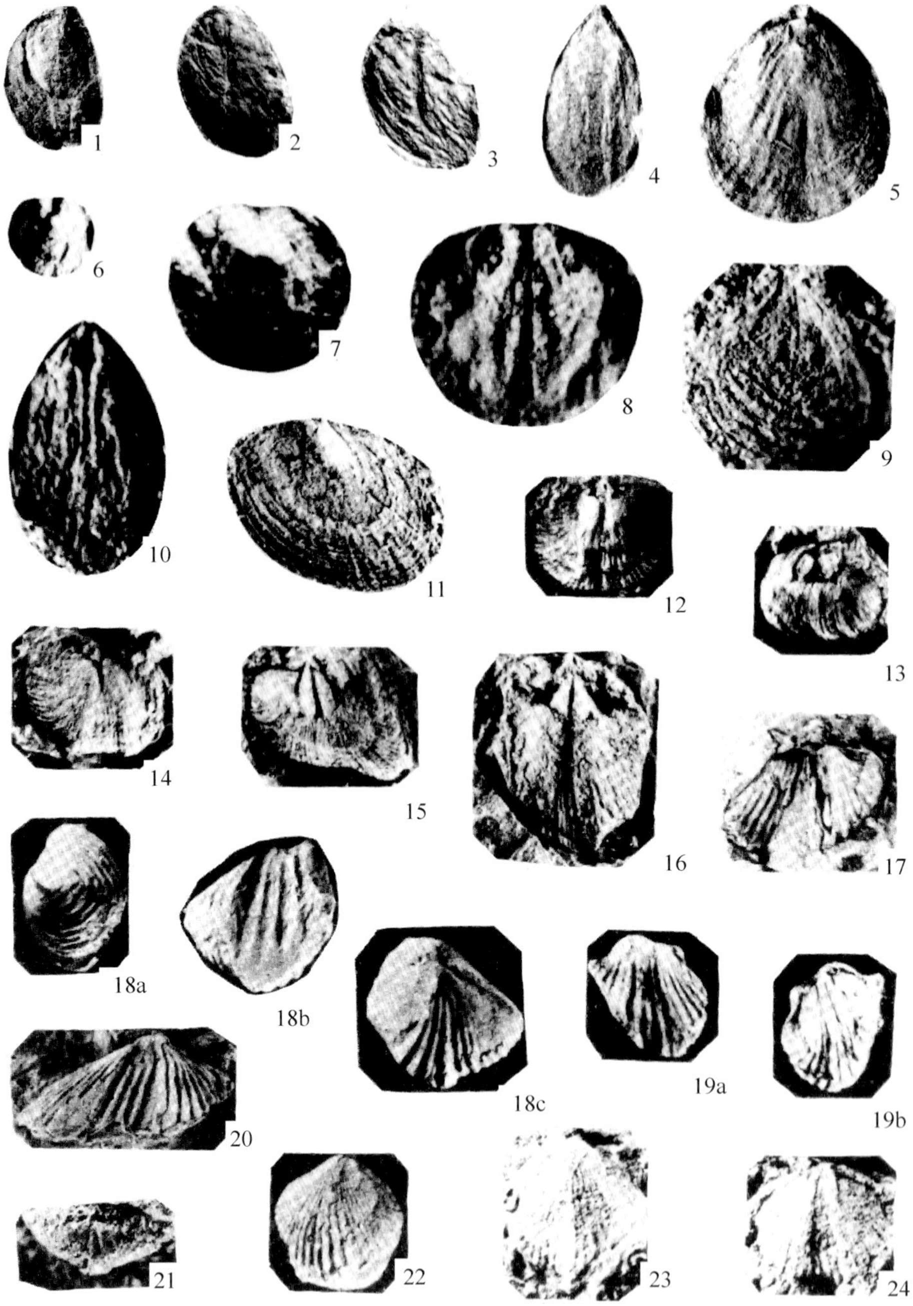

大茅峒一号竖井旁磷矿采坑大茅组第 4 段：1～3. *Lingulella ferruginea* 铁色小舌形贝（1. 腹内模，2. 背内模，3. 背内模）；4. *Lingulella atava* 返祖小舌形贝腹内模；5. *Obolella* sp. 小圆货贝（未定种）腹内模；6～8. *Acrotreta schmalenseei* 施马勒细氏项孔贝（6. 腹视，7. 腹视，8. 背内模）；9、11. "*Botsfordia*" *yaxianensis* 崖县"博茨福迪贝"（9. 北内模，11. 腹内模）；10. *Lingulepis* sp. 鳞舌形贝（未定种）背内模。**晴坡岭、尖岭尖岭组第 2 段**：12～16. *Mimella yaxianensis* 崖县拟态贝（12. 背内模，13. 腹内模，14. 背内模，15. 腹内模，16. 背内模）；17～20. *Zanclor-hyncha gentilis* 同类镰嘴贝（17. 背内模，18a-c. 分别为同一标本的侧视、前视、背视，19a-b. 分别为同一标本的腹视、背视，20. 腹内模）；21～24. *Zanclorhyncha yaxianensis* 崖县镰嘴贝（21. 背内模后视，22. 背视，23. 变形的腹视，24. 腹内模）。（曾庆銮等，1992）

图版 I -3

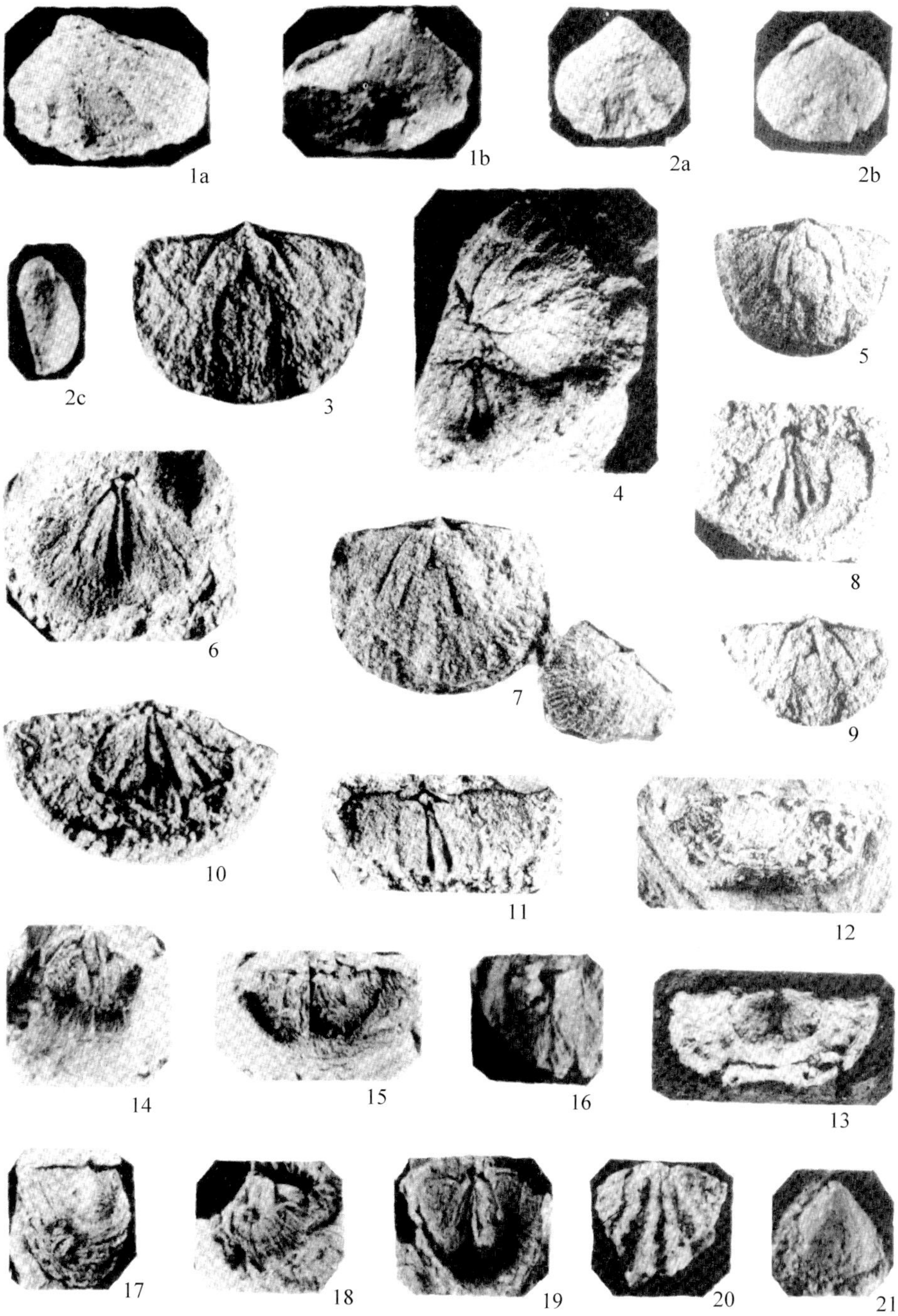

尖岭尖岭组第2段：1~2. *Idiospira*? *hainanensis* 海南自螺贝？（1a-b. 为同一标本的腹视、背视，2a-c. 为同一标本的腹视、背半后视、侧视）。**晴坡岭尖岭组第2段**：12~13. *Leptellina* sp. 准小薄贝（未定种）（12. 腹视，13. 背内模）。**红花干沟村尖岭组第2段**：14~15. 17~19. *Leptelloidea yaxianensis* 崖县拟小薄贝（14. 不完整的腹内模，15. 背内模，17. 背内模，18. 不完整的腹内模，19. 背内模）；20. *Protozyga magnicostata* 巨线初轭贝不完整的背视；21. *Protozyga nasuta* 鼻子初轭贝腹内模。**死马岭尖岭组第1段**：16. *Skenidioides*? sp. 似帐幕贝？（未定种）腹内模。**红花干沟村南侧干沟村组**：3~8. *Sowerbyella hainanensis* 海南小苏维伯贝（3. 腹内模，4. 不完整的背、腹内模，5. 腹内模，6. 不完整的背内模，7. 两个腹内模，8. 不完整的背内模）。9~11. *Sowerbyella* cf. *sericea* 丝小苏维伯贝（相似种）（9. 腹内模，10. 背内模，11. 背内模）。（曾庆銮等，1992）

图版Ⅰ-4

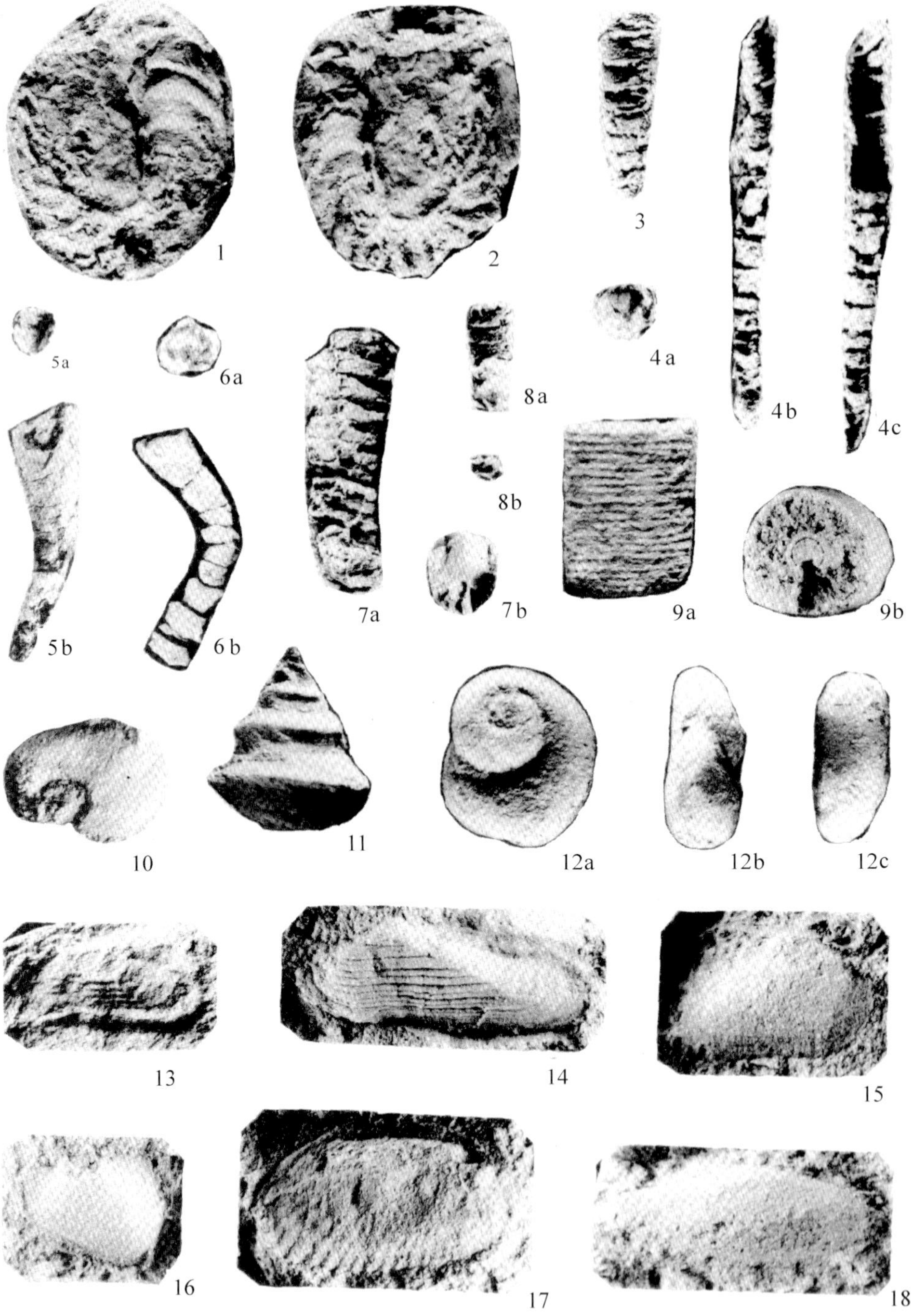

晴坡岭尖岭组第2段：1~2. *Eurasiaticoceras* sp. 欧亚角石（未定种）（1. 侧视，2. 侧视）；9. *Cyclocyclicus* sp. 圆圆茎（未定种）（9a–b. 为茎侧面、茎节面）；10. *Straparollus* sp. 圆脐螺（未定种）顶视；11. *Lophospira* sp. 脊旋螺（未定种）背视；12. *Lesueurilla* sp. 列氏螺（未定种）（12a–c. 分别为顶视、口视、背视）；17. *Paraphtonia* sp. 拟富通蛤（未定种）右壳。**晴坡岭尖岭组第1段：**3，8. *Ancistroceras* sp. 钩角石（未定种）（3-侧视，8a–b. 分别为侧视，横断面）；7. *Ancistroceras* cf. *clinotum* 斜钩角石（相似种）（7a–b. 为侧视、横断面）；15~16. *Modiolopsis* sp. 拟瓢蛤（未定种）（15. 右壳，16. 左壳）。死马岭尖岭组第1段：5. *Richardsonoceras* sp. 雷迦逊角石?（未定种）（5a–b. 为横断面、侧视）；6. *Richardsonoceras*? sp. 雷迦逊角石?（未定种）（6a–b. 为横断面、侧视）；18. *Ctenodonta* sp. 梳齿蛤（未定种）右壳。**红花干沟村南侧干沟村组：**4. *Michelinoceras* sp. 米契林角石（未定种）（4a–b. 分别为横断面、侧视、腹视）；13~14. *Cymatonota hounghuaensis* 红花蕾蛤（13. 右壳，14. 左壳）。（曾庆銮等，1992）

图版 I -5

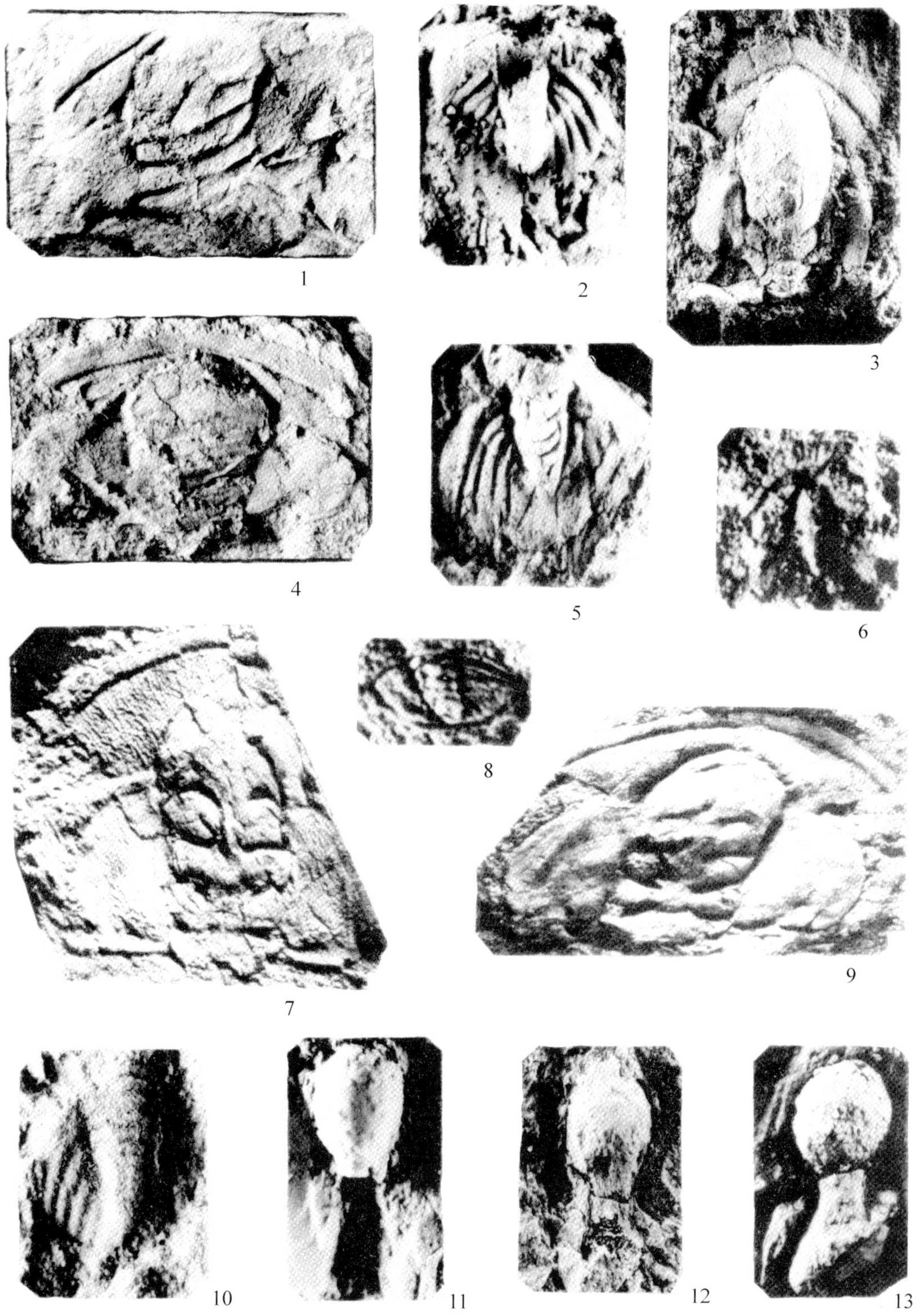

茅峒一号竖井大茅组第4段：1，4. *Xystridura hainanensis* 海南木耙虫头盖；2，5. *Xystridura* sp. 木耙虫（未定种）尾部；3. *Xystridura orientalis* 东方木耙虫头盖；6，8. *Pagetia* sp. 佩奇虫（未定种）（6. 头盖，8. 尾部）；7. *Galahetes hainanensis* 海南加拉赫虫头盖；9. *Galahetes pyrus* 梨状加拉赫虫头盖。**红花干沟村干沟村组**：10. *Encrinuroides*? sp. 似彗星虫？（未定种）尾部。**晴坡岭、死马岭尖岭组第1段**：11～13. *Bulbaspis*（*Parabulbaspis*）sp. 拟球刺虫（未定种）（11-头鞍球刺，12-头鞍球刺，13-头鞍球刺）。（曾庆銮等，1992）

图版 I -6

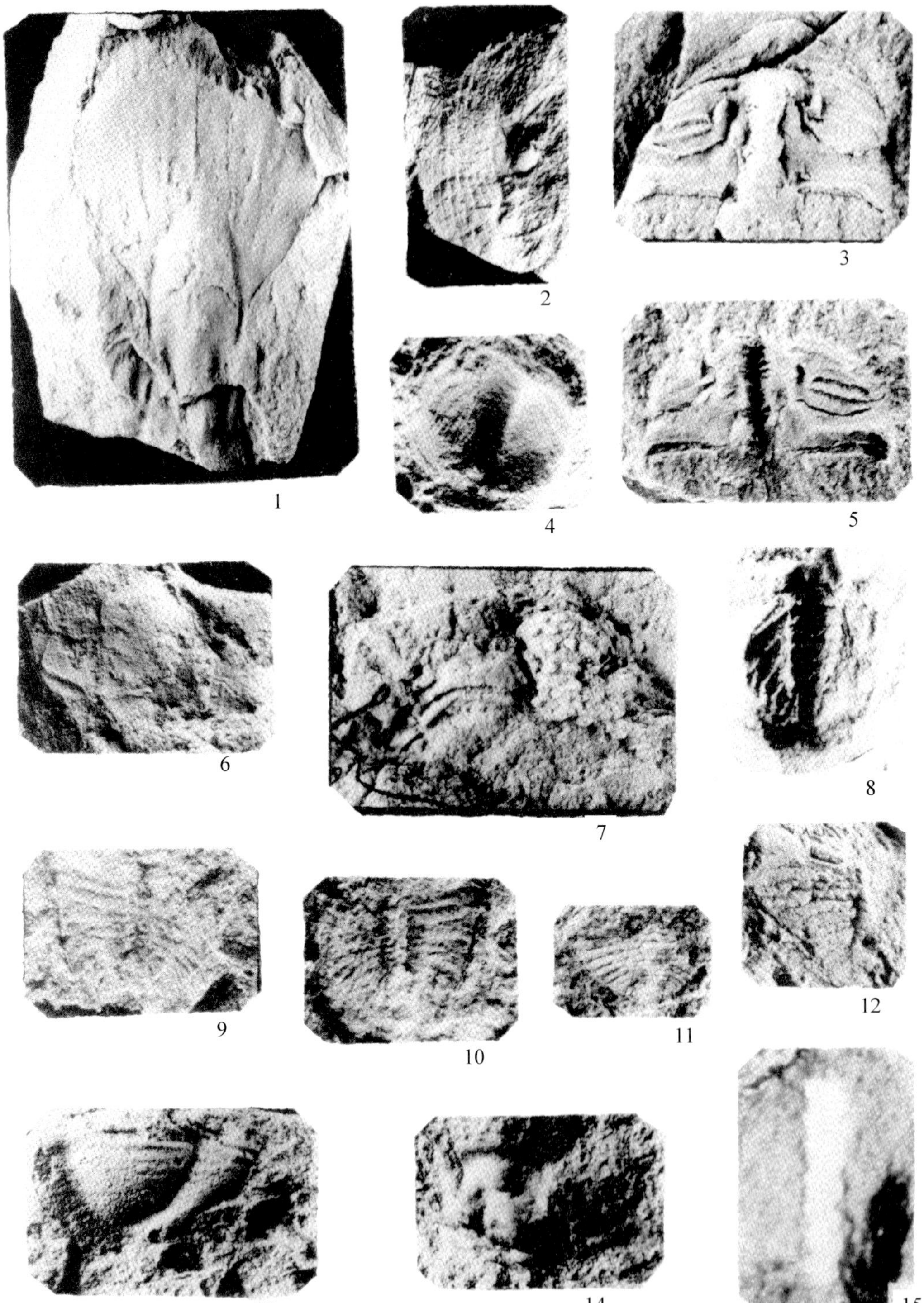

晴坡岭尖岭组第 1 段：1～2. *Birmanites* cf. *hupeiensis* 湖北缅甸虫（相似种）（1. 头盖，2. 活动颊）；8. 属种未定，尾部外模；9～10. 属种未定（9. 带胸节尾部外模，10. 带胸节尾部）；11～12. *Dionide* sp. 美女神母虫（未定种）尾部。**死马岭尖岭组第 1 段**：4. 属种未定，尾部外模；13. 属种未定，尾部。**晴坡岭尖岭组第 2 段**：3，5. *Ampyxinella* sp. 小线头虫（未定种）（3. 头盖，5. 头盖外模）。**红花干沟村干沟村组**：6. *Parisoceraurus* sp. 似角尾虫（未定种）头盖；7. *Encrinuroides* sp. 似彗星虫（未定种）头盖；14. *Neseuretus* sp. 岛头虫（未定种）头盖。**大茅峒牙花组第 3 段**：15. *Glyptograptus* sp. 雕笔石（未定种）。（曾庆銮等，1992）

图版 I -7

死马岭尖岭组第 1 段： 1～2. *Orthograptus* cf. *amplexicaulis* 围笔石式直笔石（相似种）；3～4. *Orthograptus* cf. *pauperatus* 瘦小直笔石（相似种）。**晴坡岭尖岭组第 1 段：** 6～7，10～12. *Corynoides*? sp. 鸡爪直笔石（未定种）；8～9. *Hallograptus* cf. *bimicronatus* 双刺赫氏笔石（相似种）；13. *Climacograptus* cf. brevis 短小栅笔石（相似种）；14. *Glyptograptus euglyphus siccatus* 瘦小精刻雕笔石；15～17. *Orthograptus* cf. *calcaratus* 鸡爪直笔石（相似种）。**大茅峒沙塘组第 2 段：** 5. *Diplograptus* sp. 双笔石（未定种）。（曾庆銮等，1992）

图版 I -8

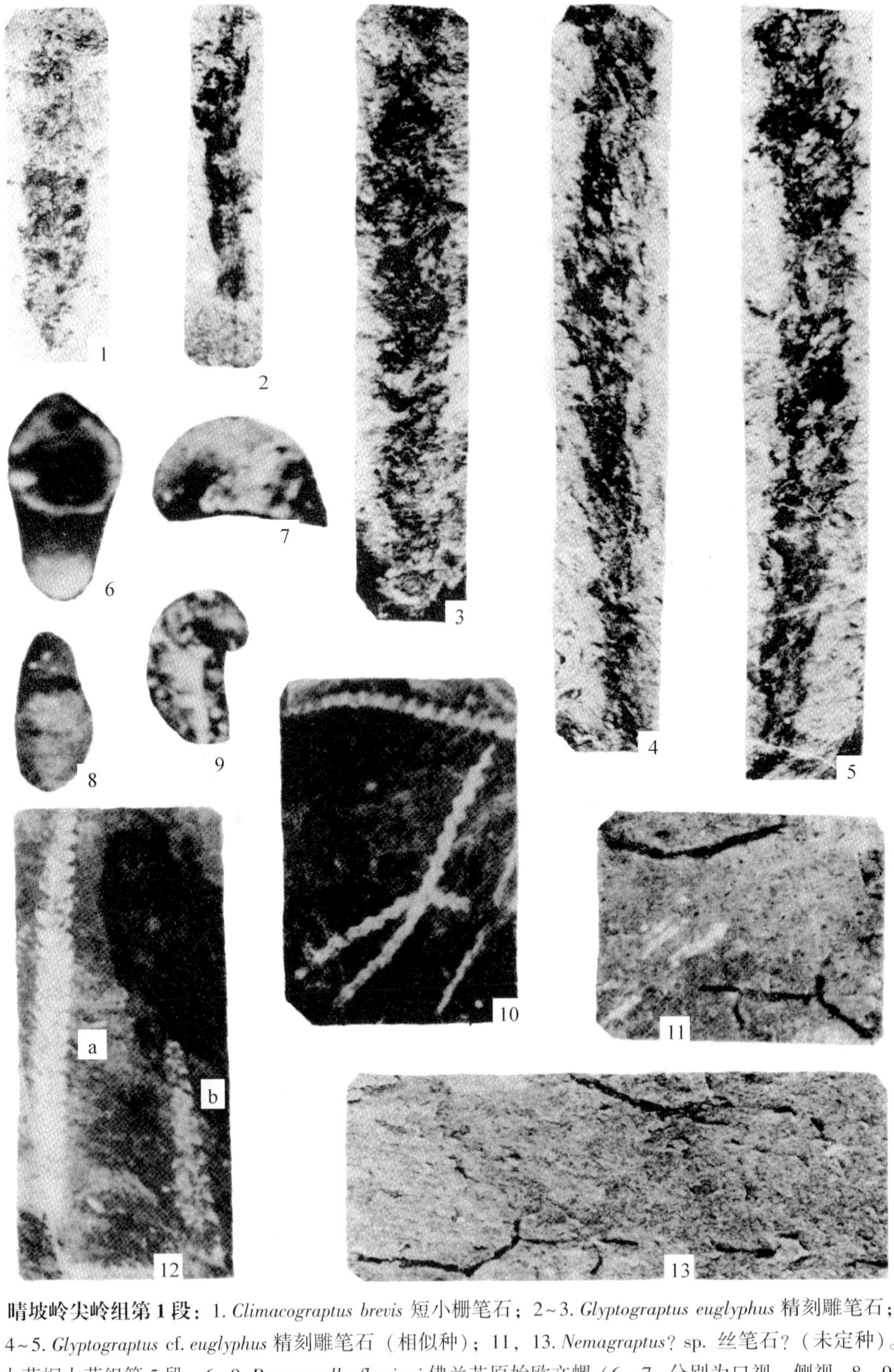

晴坡岭尖岭组第 1 段：1. *Climacograptus brevis* 短小栅笔石；2~3. *Glyptograptus euglyphus* 精刻雕笔石；4~5. *Glyptograptus* cf. *euglyphus* 精刻雕笔石（相似种）；11，13. *Nemagraptus*? sp. 丝笔石?（未定种）。大茅峒大茅组第 5 段：6~9. *Protowenella flemingi* 佛兰芒原始欧文螺（6，7. 分别为口视、侧视；8，9. 分别为侧视、背视）。**大茅峒牙花组第 3 段：**10. *Tylograptus globiformis* 球形瘤笔石。**大茅峒沙塘组第 2 段：**12. a-*Glyptograptus* sp.，雕笔石（未定种），b-*Glossograptus* sp. 舌笔石（未定种）。（曾庆銮等，1992）

图版 I -9

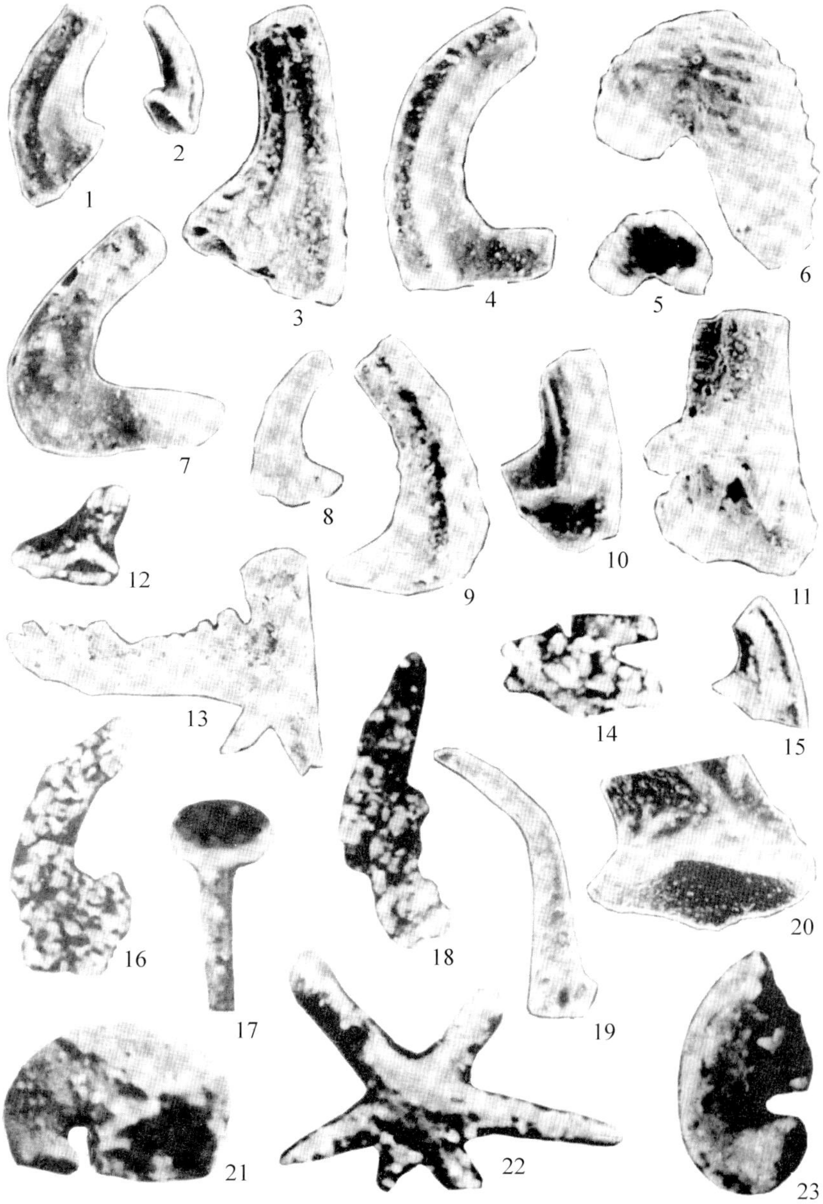

大茅峒大茅组上部：1~2，20. *Scolopodus* cf. *rex* 霸王尖牙形石（1. 近不称型分子，2. 不称型分子，20 为 2 的下部放大）；3，15. *Scolopodus* cf. *rex huolianzhai*ensis 火连寨霸王尖牙形石（相似种）（3. 对称型分子侧视，15. 不对称型分子内侧视）；4. *Glyptoconus* sp. 雕锥牙形石（未定种）近对称型分子内侧视；5 ~ 6. *Serratognathus bilobatus* Lee 双叶锯颚牙形石（不对称型分子的后侧视）；7. *Paroistodus parallelus* 平行副箭牙形石 drepanodiform 分子内侧视；17. *Zhijinites claviformis* 钉状织金钉正视；21. *Pelagiella* cf. *deltoides* 三角似海螺（相似种）口视；22. *Lenastella araniformis*. 蜘蛛形勒拿星骨顶面视；23. *Paragloborilus* cf. *condensatus* 密集拟球螺（相似种）侧视。**落笔洞牙花组第 2 段**：8. *Glyptoconus* aff. *staufferi* 斯托夫雕锥牙形石（亲近种）侧视；10. *Protopanderodus* cf. *leonardii* 列昂纳原潘得尔牙形石（相似种）近对称型分子侧视。**大茅峒大葵组上部**：9. *Protopanderodus* cf. *gradatus* Serpagli 逐渐原潘得尔牙形石（相似种）近对称型分子；11. *Glyptoconus* cf. *acontiodiform* 矢牙形石型雕锥牙形石（相似种）后视；13. *Oepikodus* cf. *communis* 一般奥毕克牙形石（相似种）Prioniodiform 分子内侧视；19. *Scolopodus* cf. *eburnus* 象牙尖牙形石（相似种）侧视。**大东海西岸沙塘组第 1 段**：12. drepanodiform 镰牙形石型，属种未定。**抱坡岭沙塘组第 1 段**：14，18. *Baltoniodus* cf. *communis* 一般波罗的牙形石（相似种）；16. *Drepanoistodus* sp. 镰牙形石（未定种）。（曾庆銮等，1992）

图版Ⅰ-10

A 石碌群第五层中酸性火山岩外貌
（石碌保秀岭北西公路旁）

B 石碌群第五层中酸性火山岩中塑性岩屑(+)

C 石灰顶组不整合于石碌群之上
（石碌主矿区观景平台）

D 南碧沟组中的变基性火成岩外貌
（八一农场309队）

E 南碧沟组中基性火山岩（邦溪十队）

F 南碧沟组中基性火山岩（屯昌黄竹岭）

G 陀烈组下段大理岩（儋州牙拉河）

H 陀烈组中段岩石中的变余平行层理（九曲江）

图版Ⅰ-11

A

南好组中的鱼化石近照（金波老村南）

B

青天峡组上部大理岩中的珊瑚化石（兰洋）

C

岭文组中的*Neuropteridium marginatum* Zhou et Li（翰林）

D

岭文组中的*Brachyphyllum* sp.，3×（翰林）

E

岭文组中的*Brachyphyllum* sp.（翰林）

F

岭文组中的*Albertica elliptica* Schimpler
（示多个叶片特征，翰林）

G

鹿母湾组中发育的交错层理（雷鸣）

H

多文组喷发覆盖于北海组之上（木棠）

图版 I -12

铁质黏土岩

海 滩 岩

中山区地貌

火山锥地貌（旧州岭）

海湾地貌

海积阶地地貌

海积生物地貌

海积砂堤地貌

图版Ⅱ-1

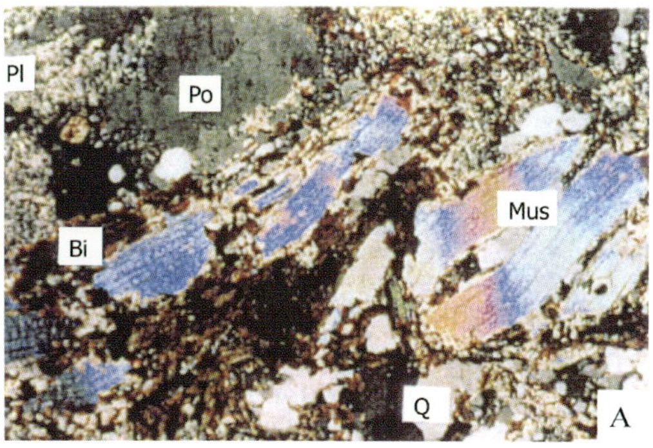

亚炮二长花岗岩镜下的韧性变形（乐东县亚南甫沟）

乘顺岭二长花岗岩中的抱郎二长花岗岩捕虏体

志仲二长花岗岩被后龙碱长花岗岩超动侵入

后龙碱长花岗岩中的乘顺岭二长花岗岩捕虏体

志仲二长花岗岩钾长石斑晶及簇状黑云母
产出具明显的优选方位

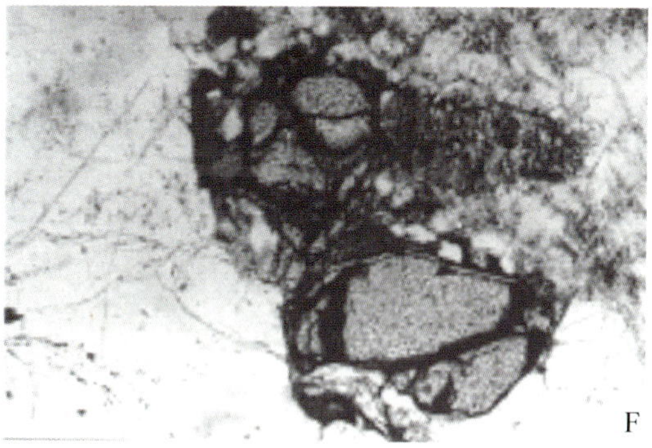

顺作中粒石榴子石正长花岗岩中石榴子石沿裂隙被
黑云母穿插交代

石碌二长花岗岩（浅色部分）捕虏
花岗闪长岩（暗色部分）

石岭通什二长花岗岩侵入山牛塘细粒
闪长岩（暗色部分）

图版Ⅱ-2

山牛塘闪长岩中的花岗质熔离浆泡

通什巨斑状二长花岗岩（上部）被长塘岭二长花岗岩侵入

番阳峒正长花岗岩捕虏通什巨斑状二长花岗岩（中间部分）

通什巨斑状二长花岗岩中的闪长质包体

乌兰盖少斑状角闪黑云二长花岗岩中斑晶定向排列

火岭细粒正长花岗岩捕虏通什巨斑状二长花岗岩

两院中粒斑状黑云母二长花岗岩中暗色包体

黑岭正长花岗岩脉动侵入尖峰岭正长花岗岩

图版Ⅱ-3

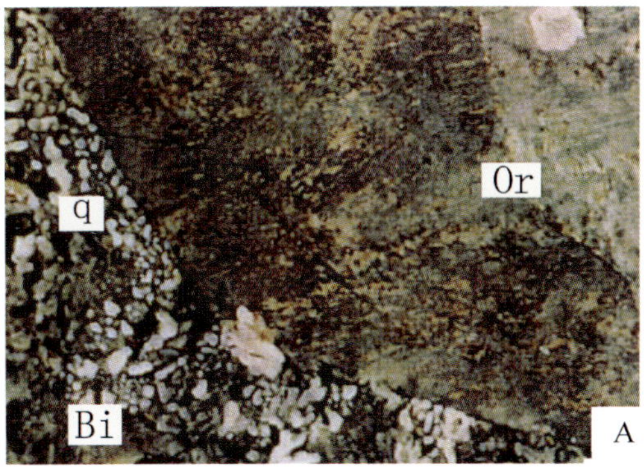

尖峰岭钾长石斑晶边缘形成的蠕英结构

深沟霓辉石黑榴石正长岩中霓辉石（蓝绿色）及黑榴石（红褐色）共生

金寮中细粒正长花岗岩侵入番阳峒正长花岗岩

麻山田石英正长岩外貌

金寮中细粒黑云正长花岗岩中显微文象连晶结构

六连岭正长花岗岩捕虏通什巨斑状二长花岗岩（中间部位）

六连岭中粗粒黑云母正长花岗岩中梳状钠长石沿边缘交代微斜条纹长石

兵工厂细粒斑状二云正长花岗岩中黑鳞云母、萤石

图版Ⅱ-4

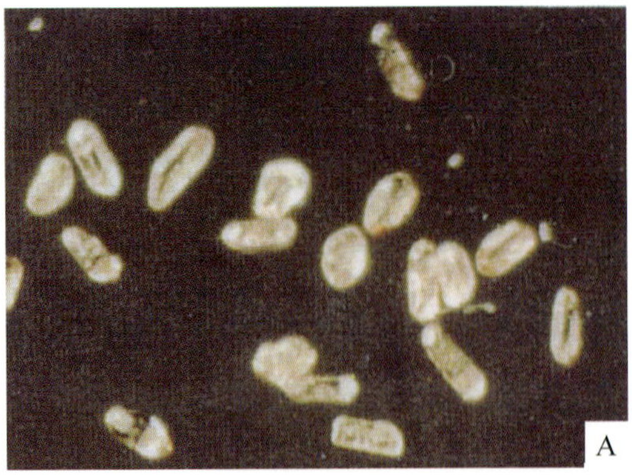

后龙中粒碱长花岗岩中滚圆状、粒状锆石

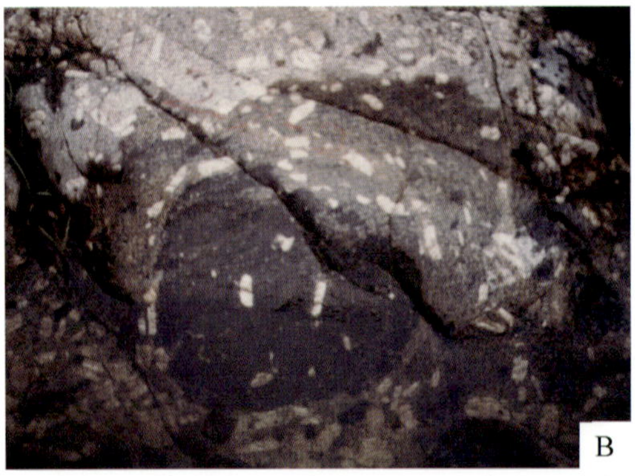

黎母岭二长花岗岩中的钾长石斑晶沿闪长质包体边缘环绕分布

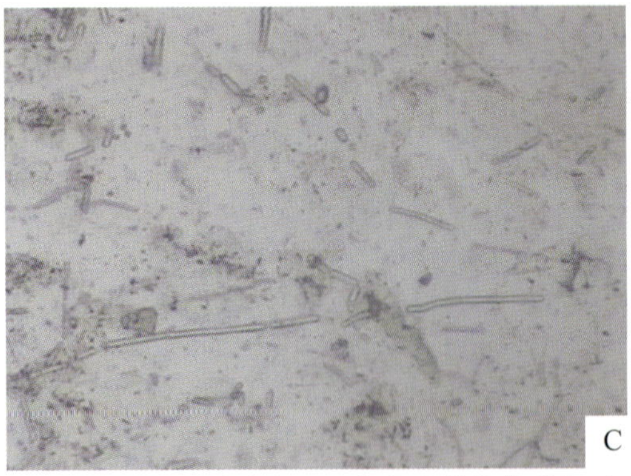

黎母岭二长花岗岩闪长质包体中的针状磷灰石

琼中县二长花岗岩中的闪长质包体

保梅岭正长花岗岩 截切邦溪二长花岗岩的斑晶

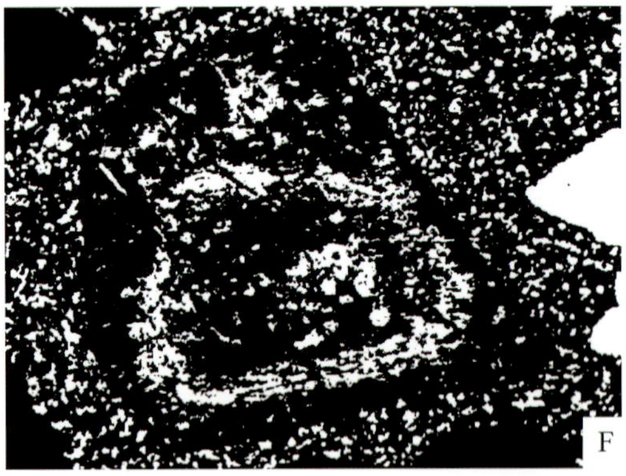

迈州岭碱长花岗斑岩正长石斑晶环带状蚀变

南山岭二长花岗岩中幔状长石

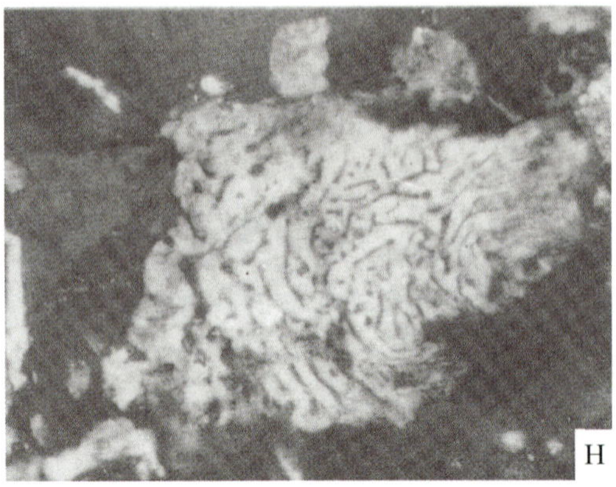

三狮岭正长花岗岩中的交代蠕英结构

图版Ⅱ-5

A
加茂花岗闪长岩中的闪长质包体

B
保亭二长花岗岩（浅色）侵入加茂花岗闪长岩（下部、深色）

C
六弓细中粒斑状（角闪）黑云母二长花岗岩外貌

D
吊罗山正长花岗岩侵入长安辉长岩

E
吊罗山正长花岗岩及覆于其节理面上的辉钼矿（陵水）

F
大小岭细粒正长花岗岩及分布于岩石裂隙面上的辉钼矿（兴隆）

G
花岗斑岩外貌（陵水）

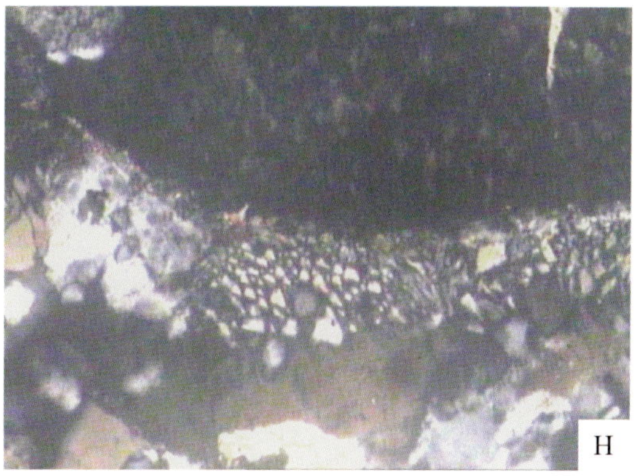
H
花岗斑岩中的蠕英结构（陵水）

图版Ⅲ-1

中生代安山岩斑晶及气孔略具定向排列（儋州洛基）

多文组玄武岩与北海组含砾砂土的接触关系（洋浦）

多文组喷发不整合于海口组之上

石山组与道堂组的接触关系（峨蔓）

旧州岭火山锥远景

峨蔓岭火山锥

高山岭破火山口（火山湖）

道堂组沉凝灰岩斜层理

图版Ⅲ-2

A

道堂组沉凝灰岩波状层理

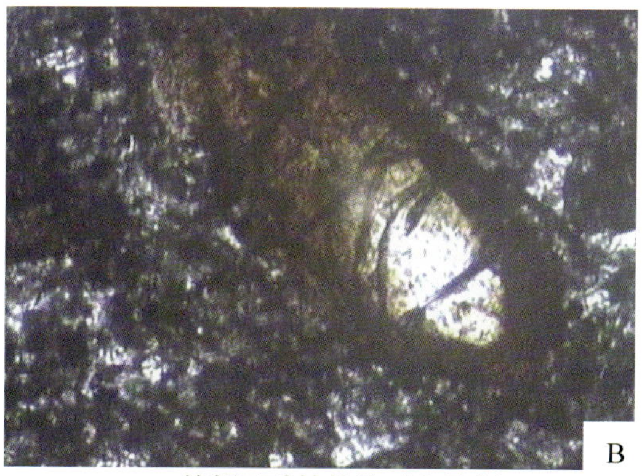
B

钛辉石砂钟构造（文昌市）

C

橄榄石斑晶蛇纹石化（文昌市）

D

石马村组中玄武岩柱状节理（文昌市南阳）

E

多文组玄武岩的气孔

F

玄武岩间粒结构（文昌市）

G

尖晶二辉橄榄岩包体（禄塘园）

H

尖晶二辉橄榄岩包体（北埔村）

图版Ⅳ-1

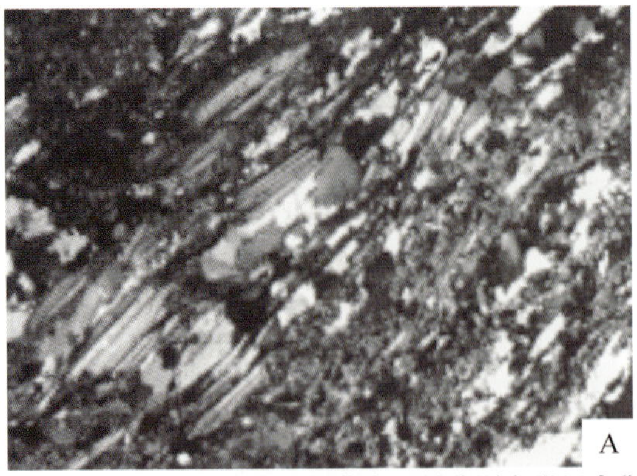

峨文岭组石英二云片岩，石英和云母分别在微层富集

眼球状混合岩（冲卒岭）

戈枕村组混合片麻岩变形强烈，眼球多呈透镜状定向排列（红林农场19队西昌化江中）

眼球状混合岩中眼球密集产出（琼中长征）

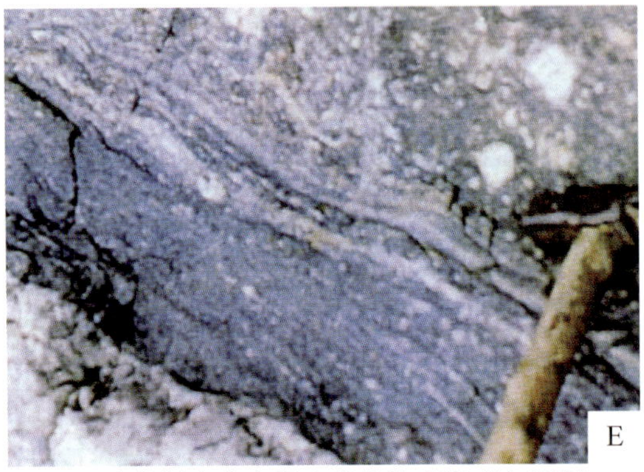

眼球状混合岩夹混合岩化片麻岩（冲卒岭）

含石榴黑云斜长片麻岩，鳞片粒状变晶结构 12.5×2.5（+）

黑云斜长片麻岩，鳞片粒状变晶结构，片麻状构造

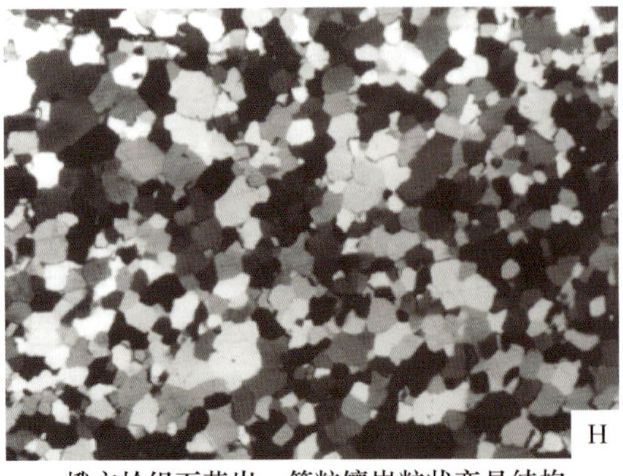

峨文岭组石英岩，等粒镶嵌粒状变晶结构

图版Ⅳ-2

黑云斜长变粒岩，等粒片粒状变晶结构
12.5×2.5（+）

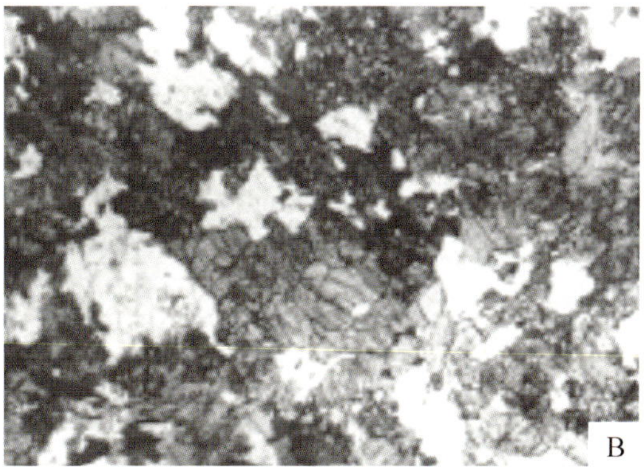

黑云二辉麻粒岩，花岗变晶结构
12.5×2.5（+）

陀烈组变质细粉砂岩中变余石英定向分布

陀烈组变质含粉砂泥岩中绢云母定向分布

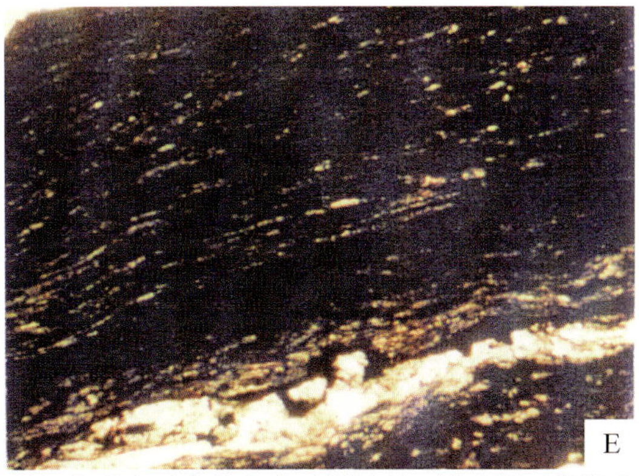

陀烈组中段岩石中的变余平行层理，被晚期石英脉穿插

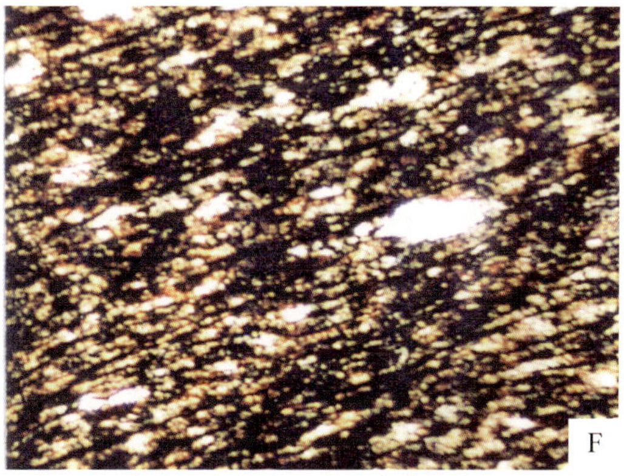

陀烈组中段岩石中的变余平行层理

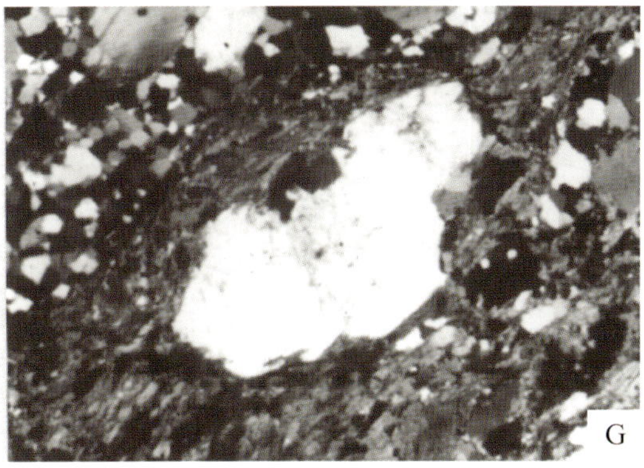

混合质黑云斜长片麻岩，交代斑状结构
12.5×2.5（+）

片麻状混合花岗岩，具交代蠕英、沿缝合线的交代结构（缝合线被交代结构）12.5×2.5（+）

图版Ⅴ-1

A

眼球状混合岩夹混合岩化片麻岩（东方戈枕）

B

峨文岭组中的石香肠构造（乐东红五）

C

戈枕村组中长英质脉的杆状构造（琼中百花岭）

D

峨文岭组的片内无根型褶皱（乐东红五）

E

戈枕村组中的圆柱状褶皱（琼中百花岭）

F

峨文岭组倾伏型小中褶皱，
并见有早期褶皱的叠加（乐东红五）

G

峨文岭组倒转型褶皱（乐东红五）

H

奥陶系南碧沟组中的剪切褶皱（儋州八一）

图版Ⅴ-2

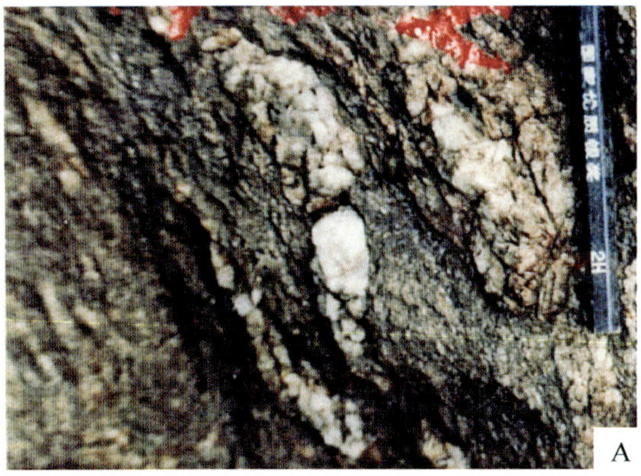
A
陀烈组中的黏滞型石香肠构造（琼海三洲）

B
陀烈组中的褶皱型石香肠构造（琼海三洲）

C
陀烈组中的顺层剪切褶皱（屯昌晨星）

D
陀烈组中的W型褶皱（琼海三洲）

E
陀烈组中的强揉皱型（琼海三洲）

F
陀烈组中的箱状褶皱（琼海三洲）

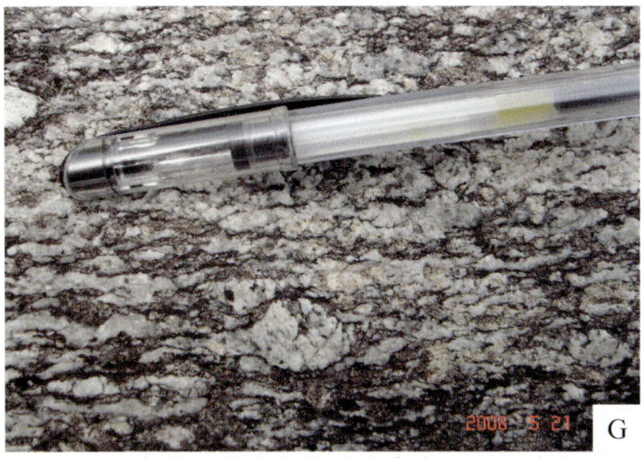
G
崖城-万宁构造带中的构造片麻岩（吊罗山林场）

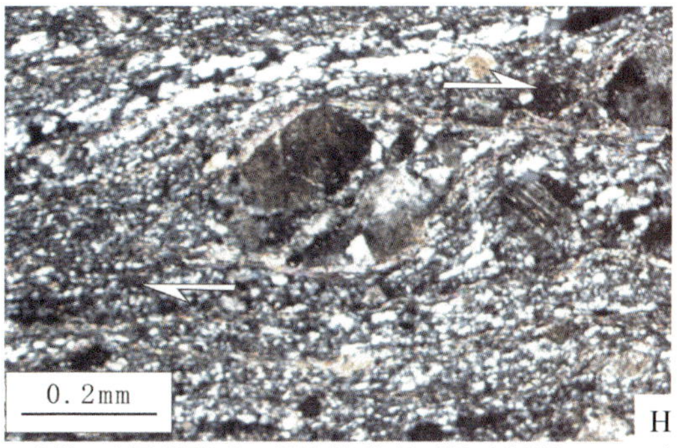

H
王外-马眉断裂中σ型旋转碎斑（据陈新跃，2006）

图版 V-3

A

潭爷构造带东缘破碎带（长茅水库边）

B

戈枕构造带中的“δ”型剪切碎斑（东方戈枕）

C

洛基韧性变形带外貌（儋州洛基）

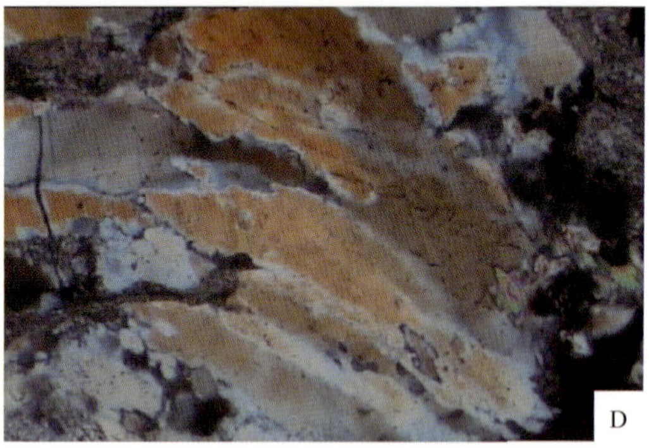

D

洛基韧性变形带中的糜棱岩

E

什运共轭剪切带中的剪切褶皱（什运）

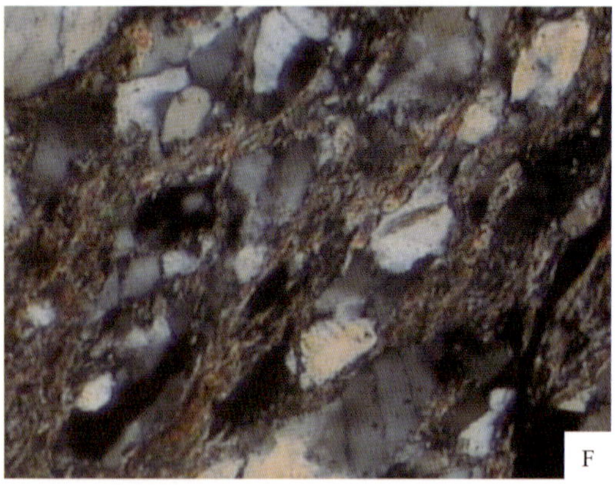

F

什运共轭剪切带中的旋转碎斑结构

G

大村韧性变形带眼球残斑右旋变形

H

洛基-南加六断裂带切错早期北东向断裂